BIOCHEMISTRY

BIOCHEMISTRY

J. David Rawn
TOWSON STATE UNIVERSITY

HARPER & ROW, PUBLISHERS, New York
Cambridge, Philadelphia, San Francisco,
London, Mexico City, São Paulo, Sydney

1817

Sponsoring Editor: Malvina Wasserman
Project Editor: David Nickol
Designer: Helen Iranyi
Production Assistant: Jacqui Brownstein
Compositor: Progressive Typographers
Printer and Binder: The Murray Printing Company
Art Studio: J&R Art Services, Inc.
Stereo Art: Richard J. Feldmann

The cover shows seven bacteriochlorophyll molecules in the configuration in which they are bound to the bacteriochlorophyll protein (see page 710). The cover design is by Richard J. Feldmann of the National Institutes of Health and is based upon x-ray crystallographic studies of B. W. Matthews and his co-workers.

Biochemistry

Library of Congress Cataloging in Publication Data

Rawn, J. David, 1944–
   Biochemistry.

   Includes index.
   1. Biological chemistry. I. Title.
QP514.2.R39 1983   574.19'2   82-15454
ISBN 0-06-045335-4

To Margie and Max

One's kite will rise on the wind as far as ever one has string to let it go.
D. H. Lawrence ("The Rainbow")

Brief contents

Detailed contents

part III MOLECULAR BIOLOGY

Preface

This text has been written to meet the needs of a full-year, comprehensive, introductory course in biochemistry. The application of chemical principles —especially of organic chemistry—to biochemical phenomena forms the foundation upon which the text has been constructed. A full-year course in organic chemistry is therefore assumed as a prerequisite. Physical chemistry has *not*, however, been assumed as part of the background of students taking the course for which this text is intended. A biochemistry text based firmly in chemical principles possesses the twofold advantage of using the chemical background of the students and of reflecting the state of the art in biochemistry, where nearly every process is actively investigated in chemical terms and by chemical methods.

Careful attention to the logical development of topics reflects the structure of the text as a whole as well as the individual chapters. The text is divided into three parts.

PART I: THE CONFORMATION AND FUNCTION OF BIOLOGICAL MACROMOLECULES

The Introduction provides a brief overview of biochemistry, a short survey of cell structure, and some generalizations that apply to all of biochemistry. Chapter 1 provides a review of basic concepts of thermodynamics and some examples of their biochemical applications. Chapters 2—4 present a systematic discussion of protein structure. Chapter 5 treats methods for the isolation, purification, and characterization of biological macromolecules. The following two chapters discuss enzyme catalysis (Chapter 6) and enzyme kinetics (Chapter 7). Chemical principles form the heart of the discussion of enzyme-catalyzed reactions in terms of reaction mechanisms familiar to all students who have taken a full-year course in organic chemistry. The principles developed in these chapters recur throughout the text. A discussion of enzyme kinetics can rapidly escalate in mathematical rigor. The level of this presentation, however, is appropriate for students who have not studied physical chemistry. The chapter on enzyme kinetics further develops the principles introduced in the preceding chapter, and the combination of the two chapters provides a thorough introduction to enzymology. Carbohydrates are discussed in Chapter 8 with the conformations of cyclic sugars and the relationship between conformation and biochemical function receiving particular emphasis. The structures, physical properties, and chemistry of nucleosides, nucleotides, and nucleic acids are discussed in Chapter 9, one of the longest chapters in the text. Among the most current topics in this chapter are methods for determining the primary structures of nucleic acids and methods for chemical synthesis of nucleic acids, both of steadily increasing importance. Having discussed the nucleotides, we turn to coenzymes in Chapter 10. The mechanisms of the major classes of reactions in which coenzymes participate are discussed in detail in this chapter, and stereochemistry is a central topic. The RS system of configurational nomenclature, familiar to all students who have studied introductory organic chemistry, is reviewed, and the concept of prochirality, perhaps new to many students, is introduced. Part I concludes with a discussion of lipids, biological membranes, and the lipid-soluble vitamins in Chapter 11.

PART II: METABOLISM

The pathways for the degradation and biosynthesis of the major classes of biological molecules are discussed in detail in Part II. This part opens with a discussion of bioenergetics in Chapter 12. Coupled reactions, the structural basis of "energy-rich" metabolites, and the enzymology of phosphoryl group transfer reactions are among the topics discussed in this chapter. Chapter 13 considers the design and regulation of metabolic pathways. The many ways by which allosteric enzymes control the rates of metabolic pathways and the covalent modification of proteins in "cascades" of enzymatic activity are discussed. With the fundamental principles in place, the details of glycolysis (Chapter 14), the citric acid cycle (Chapter 15), oxidative phosphorylation (Chapter 16), carbohydrate metabolism (Chapter 17), photosynthesis (Chapter 18), lipid catabolism and biosynthesis (Chapters 19–20), amino acid catabolism and biosynthesis (Chapters 21–22), and nucleotide biosynthesis (Chapter 23) are each treated in turn. These chapters have similar structures: Each pathway is considered as a whole, and then the individual enzymatic reactions of the pathways are discussed. The bioenergetic aspects of the pathways are considered within the context of the individual reactions as well as from the perspective of the entire pathway. The regulation of each pathway is discussed. Metabolism is thus discussed within a unified perspective and is not treated merely as a list of metabolic steps to be memorized.

PART III: MOLECULAR BIOLOGY

The extraordinarily rapidly advancing field of molecular biology occupies center stage in the final chapters of the book as DNA replication (Chapter 24), transcription of DNA (Chapter 25), protein synthesis (Chapter 26), and regulation of gene expression (Chapter 27) are considered. These chapters are as contemporary as possible within the limits of publishing schedules and they have been carefully constructed for a logical development of the subject. Considered are such important topics as DNA repair, split genes and RNA processing, regulation of protein synthesis at the translational level, and the marvelously baroque process by which antibody diversity is generated. Molecular biology has a highly specialized vocabulary, but jargon has been minimized, and the phenomena discussed in Part III are considered in relation to all that has preceded. The chemical basis of molecular biology is explored to the extent that current knowledge makes possible.

To a significant extent, biochemistry consists of processes that involve interactions of macromolecules with one another or with the small molecules to which they bind. Many of the macromolecules whose three-dimensional structures have been determined are drawn in computer-generated stereo pairs provided through the courtesy (and considerable labor) of Richard J. Feldmann of the National Institutes of Health. A deeper understanding of macromolecular structure can be obtained by viewing these images than is attainable in any other way short of a computer data base of one's own. This aspect of the text is absolutely unique and ought to be of great value to both students and instructors. A stereo viewer is provided inside the back cover of the text. To view the stereo images it is necessary only to center the viewer over the stereo pair and focus each eye upon a separate image. (About 13 percent of the population is unable to visualize objects in three dimensions by this method).

Problem solving is a fundamental component of biochemistry courses, and many problems have been provided. A solutions manual with completely worked-out solutions to the problems is available. Dr. Robert N. Lindquist (San Francisco State University) is the co-author of the solutions manual.

Each chapter of the text follows a similar pattern. A brief introduction to the topic at hand is followed by a systematic and logical development in which various topics are separated under their own numbered and lettered headings for ease of cross-reference. A concise summary of the essential points is provided at the end of each chapter. A list of leading references and a set of problems are also included at the end of each chapter.

ACKNOWLEDGMENTS Many persons have labored for many hours to review carefully the contents of the draft manuscript over the years in which it evolved. A critical reviewer, just but unsparing, and intolerant of errors, is the best friend an author ever had. I would like to acknowledge the generous assistance and helpful suggestions of Drs. John N. Aronson (State University of New York at Albany), Nordulf G. W. Debye (Towson State University), Nancy Hamlet (Towson State University), John H. Harrison (University of North Carolina), H. Robert Horton (North Carolina State University), Paul M. Horowitz (The University of Texas, Health Science Center at San Antonio), Robert N. Lindquist (San Francisco State University), and Clarence H. Suelter (Michigan State University). I am also indebted to Floyd A. Blankenship and Alan S. Wingrove who wrote and edited the computer program for preparing the exhaustive index.

I would also like to thank the staff of Harper & Row for their continuous and patient assistance. I am especially indebted to Chemistry Editor Malvina Wasserman, David Nickol, Project Editor, and Helen Iranyi, Designer.

J. David Rawn
Towson, Md.
October, 1982

BIOCHEMISTRY

Introduction:

Surveying the landscape

We are searching for the essence that lies behind the fortuitous.

Paul Klee

From antiquity to the Renaissance, from Aristotle to Leonardo, knowledge of the natural world scarcely changed. This stagnation resulted not from any want of intelligence, but from the profoundly held and unquestioned belief that the world was immutable, governed from outside, and not a suitable object for questioning. Then, quite suddenly toward the end of the fifteenth century, men in widely separated locations began asking questions about the natural world. We recognize ourselves in the men of the sixteenth century; the men of the fourteenth are strangers. We now see the world as governed from inside according to its own laws. This change in perspective marks the principal distinction between the medieval and the modern world. The enor-

Plate 1 *The Unicorn Tapestries.* [The Metropolitan Museum of Art, The Cloisters Collection, Gift of John D. Rockefeller, Jr., 1937.]

mous difference is illustrated by the thirteenth-century *Unicorn Tapestries* (Plate 1) and Albrecht Dürer's *The Young Hare,* pained in 1502 (Plate 2). In the tapestries there is much carefully observed plant life, but at the center of this realistic scene is the fabulous unicorn, whose mythical character disturbed neither the artist nor his audience. Dürer's hare could have been drawn by a naturalist today. The intensity with which Dürer has penetrated the hare is matched by the intensity of today's scientific observation. The significance of this hare lies in the curiosity with which nature is observed, a curiosity that had lain dormant for a millenium.

Plate 2 *The Young Hare,* Albrecht Dürer. [Albertina, Vienna.]

A. The Visible World: Classification by Appearance

There are really only individuals in the world and genera, orders, and classes exist only in our imagination.

Buffon

If we were to take Buffon at his word, the study of living beings would be an impossible task. The diversity of nature would simply overcome us. Every child recognizes the lion in the housecat, the wolf in the dog, the eagle in the sparrow. To divide the world into classes based on appearance is both natural and logical. It is also a very old idea. The classification schemes of present-day biology, one of which (there are several competing ones) is shown in Fig. I.1, are descended from Aristotle by way of medieval scholasticism. In the hierarchy of classification, the *kingdom* is the broadest general class. All living beings are grouped in terms of cellular structure and function, from the unicellular, unnucleated organisms of the Moneran kingdom to the complex, multicellular organisms that are included in the plant and animal kingdoms. Each kingdom in turn contains its own hierarchy of classifications, as illustrated for humans in Table I.1. At the heart of this classification scheme, where what can be seen is *named,* and what can be named has a unique identity, is the *character.* In the eighteenth century, at the dawn of modern biology, each kingdom contained five levels: kingdom, class, order, genus, and species. Any such classification has an arbitrary element, and there has been a continual debate over how classification should proceed. But, the basis of the entire effort is a search for *essences,* for the underlying unity of nature. The species is the fundamental unit of this classification hierarchy: it is what begets its like. In general if two organisms can mate and produce fertile offspring, they belong to the same species.

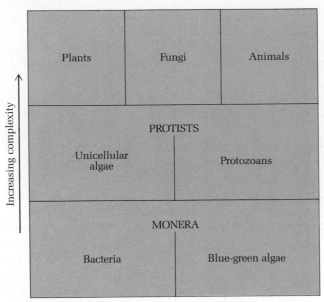

Figure I.1 Five kingdoms of the biological world.

B. Unification of Structure and Organization

Once the living world has been classified, a new question arises. We ask not only "What is it?" (that is, "What is its name?"), but also "How does it work?" Thus the organization of a living being becomes the object of study. An organism is not simply a mosaic of independent structures, but an integrated and self-regulated system. In the nineteenth century the study of organization provided steady occupation for biology. Slowly scientists began to recognize a system of interrelationships buried deep within the organism. If one studies a heart, one may study the substances of which it is composed and its structure in isolation, but to understand how it functions as a living organ, one must study it in relation to the entire organism. The idea of life became synonymous with the idea of organization, and the ancient division of the world into animal, vegetable, and mineral gave way to a broader division into the *organic*, living world and the inorganic, dead one.

Table I.1 Hierarchy of characters within a kingdom illustrated for humans		
Category	**Name**	**Characteristics**
Kingdom	Animalia	Multicellular organisms requiring organic plant and animal substances for food
Phylum	Chordata	Animals with notochord, dorsal hollow nerve cord, gills in pharynx at some stage of life cycle
Subphylum	Vertebrata	Spinal cord enclosed in a vertebral column, body basically segmented, skull enclosing brain
Superclass	Tetrapoda	Land vertebrates, four-limbed
Class	Mammalia	Young nourished by milk glands, breathing by lungs, skin with hair or fur, body cavity divided by diaphragm, red corpuscles without nuclei, constant body temperature
Order	Primata	Tree dwellers or their descendents, usually with fingers and flat nails, sense of smell reduced
Family	Hominidae	Flat face, eyes forward, color vision, upright, bipedal locomotion, with hands and feet differently specialized
Genus	*Homo*	Large brain, speech, long childhood
Species	*Homo sapiens*	Prominent chin, high forehead, sparse body hair

The interests of chemistry and biology converged, and the convergence was marked by the birth of organic chemistry: the chemistry of the living world that we today call biochemistry. With the discovery that chemists could make their own organic molecules (and the effect of "the age of organic synthesis" on the biosphere has yet to be gauged) organic chemistry launched its own career. But it has never strayed very far, even in its remotest theoretical reaches, from its progenitor biochemistry. The enormous proliferation of organic molecules are also classified by their characters, their functional groups. At the close of the last century two new sciences arrived to impart a deep order on organic chemistry. Statistical mechanics found the ordering principles in the chaos of the random disorder of molecules, and physical chemistry provided the key to the structure of molecules. The unification of structure and the organization hidden beneath the surface is the cornerstone of this text, and of all of biochemistry.

C. Time Enters Biology: Evolution

Time was foreign to biology until late in the nineteenth century. If one assumes that an immutable world was created whole in the not too distant past, then living beings have no history, and the study of the living world culminates in the classification of invariant species. Time slowly crept into the biological conception of things, and with the entry of time came the concept of evolution, already the subject of much dispute when Darwin set sail on the *Beagle*. The ultimate result of his observations as the ship's naturalist was *On the Origin of the Species*. Examination of the fossil record had already revealed great changes in flora and fauna. As one looked ever deeper in time, the world became less and less familiar. Darwin's observations unmistakably revealed, and subsequent research has shown beyond any doubt, that a small number of original species diverged into many different descendents. Thus to the original hierarchy of kingdoms (Figure I.1), we add the concept of continuous development (Figures I.2 and I.3).

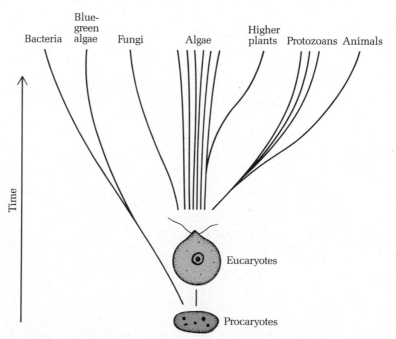

Figure I.2 Evolution of a primordial procaryotic organism to modern bacteria and blue-green algae and of a primordial eucaryotic organism to the modern fungi, algae, higher plants, protozoans, and animals.

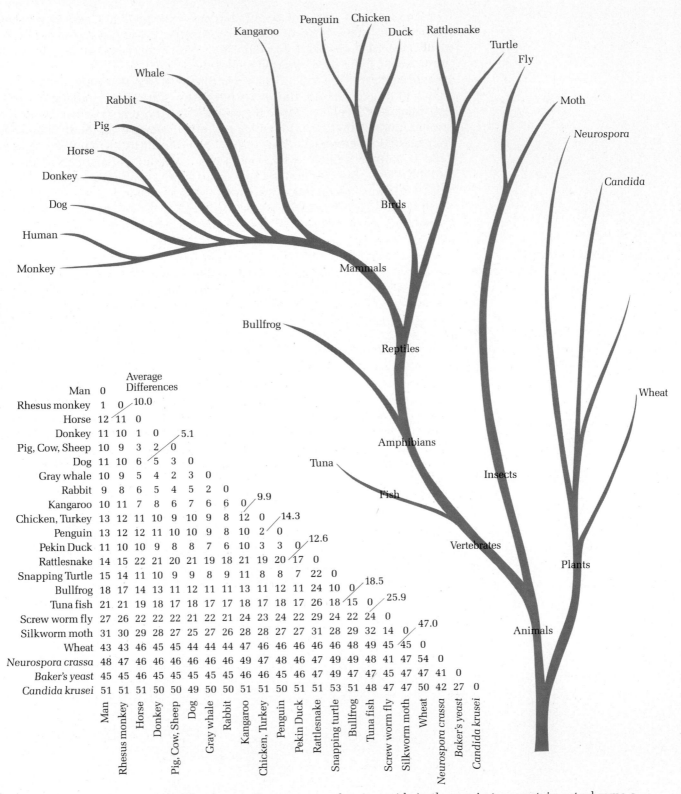

	Man	Rhesus monkey	Horse	Donkey	Pig, Cow, Sheep	Dog	Gray whale	Rabbit	Kangaroo	Chicken, Turkey	Penguin	Pekin Duck	Rattlesnake	Snapping turtle	Bullfrog	Tuna fish	Screw worm fly	Silkworm moth	Wheat	Neurospora crassa	Baker's yeast	Candida krusei
Man	0																					
Rhesus monkey	1	0																				
Horse	12	11	0																			
Donkey	11	10	1	0																		
Pig, Cow, Sheep	10	9	3	2	0																	
Dog	11	10	6	5	3	0																
Gray whale	10	9	5	4	2	3	0															
Rabbit	9	8	6	5	4	5	2	0														
Kangaroo	10	11	7	8	6	7	6	6	0													
Chicken, Turkey	13	12	11	10	9	10	9	8	12	0												
Penguin	13	12	12	11	10	10	9	8	10	2	0											
Pekin Duck	11	10	10	9	8	8	7	6	10	3	3	0										
Rattlesnake	14	15	22	21	20	21	19	18	21	19	20	17	0									
Snapping Turtle	15	14	11	10	9	9	8	9	11	8	8	7	22	0								
Bullfrog	18	17	14	13	11	12	11	11	13	11	12	11	24	10	0							
Tuna fish	21	21	19	18	17	18	17	17	18	17	18	17	26	18	15	0						
Screw worm fly	27	26	22	22	22	21	22	21	24	23	24	22	29	24	22	24	0					
Silkworm moth	31	30	29	28	27	25	27	26	28	28	27	27	31	28	29	32	14	0				
Wheat	43	43	46	45	45	44	44	44	47	46	46	46	46	46	48	49	45	45	0			
Neurospora crassa	48	47	46	46	46	46	46	46	49	47	48	46	47	49	49	48	41	47	54	0		
Baker's yeast	45	45	46	45	45	45	45	45	46	46	45	46	47	49	47	47	45	47	47	41	0	
Candida krusei	51	51	51	50	50	49	50	50	51	51	50	51	51	53	51	48	47	47	50	42	27	0

Figure I.3 An evolutionary tree based upon the sequence of amino acids in the respiratory protein cytochrome *c* (Chapter 16).

D. Cell Theory

In the seventeenth century Robert Hooke, peering through the microscope he had built, observed that plant tissues were divided into small compartments separated by walls. Perhaps reminded of medieval monks, he called the compartments *cells*, meaning "little rooms." Over 150 years passed, however, before Matthias Schlieden proposed that *all* plant tissues were based upon an organization of cells. Shortly thereafter Theodor Schwann proposed that all animal tissues are also organizations of cells and that the fundamental unit of life is the cell. Rudolf Virchov's postulate that only a cell can make another cell overturned the idea of "spontaneous generation" and provided a mechanism for integrating cell theory with evolution. The ever-regenerating cell changes slowly in time, and new species are the result. Modern cell theory is summarized by three statements: (1) all cells arise from other cells; (2) the information required for the production of new cells is passed from one generation to the next; (3) all chemical reactions of an organism, its *metabolism*, take place in cells.

E. Structure of Cells

All cells are divided into two types on the basis of their structure. Cells that do not contain a distinct nucleus are called *procaryotic* (from the Greek *pro*, "before," and *karyon*, "kernel"). Cells that contain a membrane-bound nucleus are *eucaryotic* (Greek *eu*, "true").

a. Structure of Procaryotic Cells

Procaryotic cells arose before the nucleated cells and are descended from the oldest life forms on earth. Not only do they lack a nucleus, they also lack any internal membranes that are not connected to the plasma membrane. A schematic diagram of the structure of the bacterium *Escherichia coli* is shown in Figure I.4. This procaryote, which resides in the human gut, is the best understood of all organisms. Let us begin at the boundary of an *E. coli* cell and work inward, briefly summarizing its principal structure features.

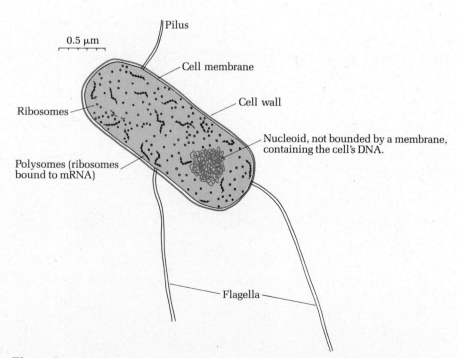

Figure I.4 Structure of an *E. coli* cell. The cell measures approximately $0.8 \times 0.8 \times 2.0$ μm.

i. The Cell Wall The bacterial *cell wall* separates the *E. coli* cell (and all true bacterial cells) from its environment. It is constructed from cross-linked small peptides and complex polysaccharides (Section 8.7A). Bacterial cell walls range from about 40 nm to 80 nm across. The total concentration of metabolites in a bacterial cell is often greater than the concentration of the medium surrounding it, and it would burst were it not for the cell wall. The cell walls of bacteria are encapsulated by an outer membrane composed of complex lipids and carbohydrates.

If the cell wall of a bacterium is removed, for example, by digestion with the enzyme lysozyme (Section 8.7A), a *protoplast* is left behind. The aqueous phase enclosed by the plasma membrane of the protoplast is known as the *cytoplasm*.

ii. Flagella and Pili The surfaces of most strains of *E. coli*, and many other bacteria, are covered with threadlike tentacles, called *flagella* and *pili*, up to 20 μm long and about 12 to 18 nm in diameter (Figure I.5). The flagella (from the Latin *flagellum*, "whip") are composed of the fibrous protein flagellin and are responsible for cell motility. They propel the cell ahead at the prodigious rate of 50 μm s^{-1}. (Since the cell is only about 1 μm long, that is an extraordinary velocity). The *pili* (from the Latin *pilus*, "hair") are shorter and thinner than flagella. The *conjugal pili* of *E. coli* are used for sexual conjugation, and *somatic pili* seem to enable cells to adhere to certain surfaces, but their functions are largely unknown.

iii. Plasma Membrane All cells are enclosed by a plasma membrane (or cell membrane) about 9 nm across. This membrane consists of lipid bilayer in which proteins are embedded (Figure I.6). The types of lipids and proteins of which the membrane is composed vary from cell to cell, but the membranes of all organisms have the same general structure (Chapter 11).

iv. Genome The genetic information of a cell is contained in its complement of *deoxyribonucleic acid* (DNA). This genetic information is known as the genome of the cell. We might regard DNA as the master tape of the cell, the computer program in which all of the information required for the operation of the cell and its reproduction is contained. We have noted that an *E. coli* cell is about 1 μm long. Its DNA is 1 mm long. This DNA encodes for about 4,000 proteins and contains many regions that regulate the expression of these proteins. As we travel through the labyrinth corridors of biochemistry, we shall be led to the manifold processes by which DNA directs the life of the cell.

v. Ribosomes The information contained in the cellular complement of DNA is copied, a process known as *transcription*, as a *ribonucleic acid*

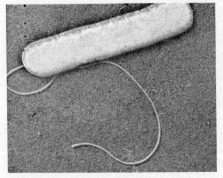 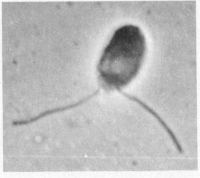

Figure I.5 Bacterial cells having one and two flagella. [Courtesy of J. Pangborn (left) and Drs. Robert Bloodgood and Gregory S. May (right).]

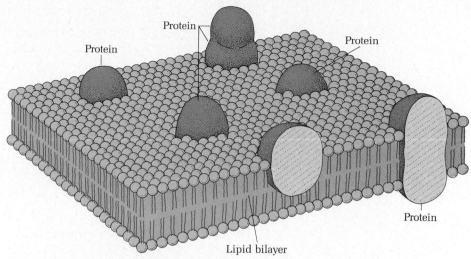

Figure I.6   Schematic diagram of a plasma membrane (Chapter 11). [From S. J. Singer and G. L. Nicolson, *Science*, *175*, 720–731 (1972). Copyright 1972 by the American Association for the Advancement of Science.]

(RNA), which is in turn translated into protein. The subcellular particles upon which protein synthesis takes place are known as *ribosomes*. These complex, rough spherical particles, whose diameter is about 18 nm, are constructed of some 50 proteins and 2 types of ribonucleic acid. An electron micrograph of an *E. coli* that is densely packed with ribosomes is shown in Figure I.7.

b. Structure of Eucaryotic Cells

Eucaryotic cells—protists, fungi, plants, and animals—are much larger and more complex than procaryotic cells. This complexity is reflected in their DNA content. The DNA of eucaryotic cells is enclosed by a double nuclear membrane. While an *E. coli* cell contains a strand of DNA about 1 mm long, a human cell contains 4 m of DNA in 46 pieces: the chromosomes. The relation between the DNA content of an organism and its complexity is shown in Figure I.8. As an organism becomes more complex, more DNA is required to provide a blueprint for the manufacture of all of the parts.

A eucaryotic cell contains many membrane-bound *cell organelles* besides its nucleus. These organelles include the *mitochondria*, the *endo-*

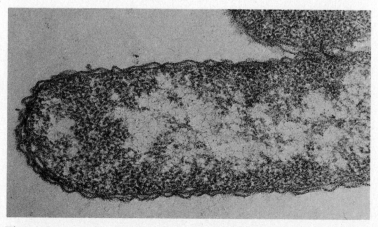

Figure I.7   An *E. coli* densely packed with ribosomes multiplied 100,000 times. [Courtesy of W. Van Iterson.]

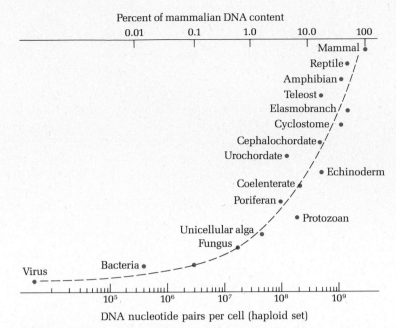

Figure I.8   Relationship between DNA content and the complexity of the organism. [From R. Britten and E. Davidson, *Science*, 165, 349 (1969). Copyright 1969 by the American Association for the Advancement of Science.]

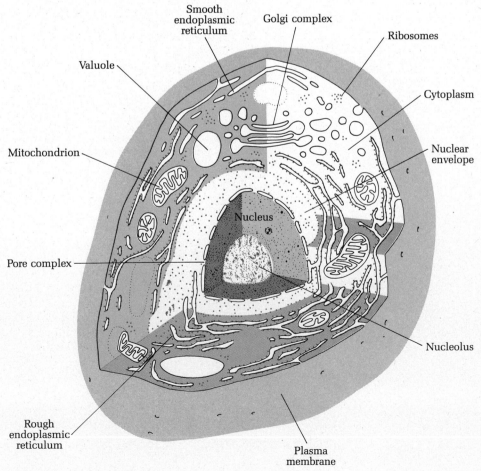

Figure I.9   Schematic diagram of a eucaryotic cell.

plasmic reticulum, and *Golgi bodies.* Plant cells contain still other specialized organelles, the *chloroplasts,* large *vacuoles,* and thick lamellar *cell walls.* The structure of a "typical" eucaryotic cell is shown in Figure I.9, and an electron micrograph of a thin section of a rat pancreas cell is shown in Figure I.10.

i. Nucleus The nucleus is the most conspicuous cell organelle. It is visible through an ordinary light microscope when it is properly stained. The nucleus is separated from the cytoplasm by the *nuclear envelope,* consisting of a double membrane that encloses the *perinuclear space. Pore complexes* in the nuclear envelope permit the passage of large molecules and molecular complexes out of the nucleus.

Chromatin, a complex of DNA and proteins, can be seen as dense fibers inside the nucleus. Also inside the nucleus is the *nucleolus,* a dense granular region containing both fibers and granules, but not surrounded by a membrane. The nucleolus is the site of the assembly of the subunits of ribosomes which are exported to the cytoplasm. The synthesis of proteins is directed by the nucleus.

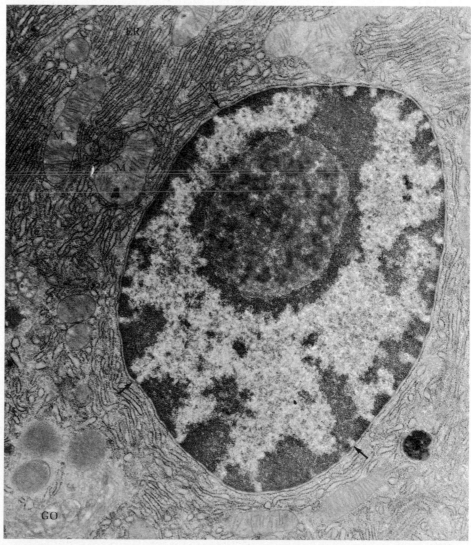

Figure I.10 Electron micrograph of a rat pancreas cell. Cell organelles are abbreviated: N, nucleus; M, mitochondrion; ER, endoplasmic reticulum; Go, Golgi complex; NE, nuclear envelope. Arrows indicate pore complexes. [Courtesy of Dr. Stephen L. Wolfe.]

ii. Endoplasmic Reticulum and Golgi Complexes The cytoplasm of eucaryotic cells is extensively crossed by a network of membranes called the *endoplasmic reticulum.* The *rough endoplasmic reticulum* is lined with ribosomes and the *smooth endoplasmic reticulum* provides channels for the transport of proteins synthesized on the ribosomes of the rough endoplasmic reticulum. Each Golgi complex is a set of flattened discs, or *Golgi bodies,* which look as if they were pinched off from the smooth endoplasmic reticulum. The *Golgi complexes* package proteins for secretion and are the site of complexation of proteins with carbohydrates and some other synthetic reactions.

iii. Mitochondria All aerobic, eucaryotic cells contain football-shaped organelles, about the size of bacteria, roughly $1 \times 1.5 \ \mu$m, called *mitochondria.* The mitochondria are the principal sites of cellular energy production. They have an outer membrane and a complex inner membrane. We shall discuss the structure of the mitochondrion when we discuss the citric acid cycle and respiration in Chapters 15 and 16.

iv. Specialized Organelles of Plant Cells Plant cells (Figure I.11) contain a specialized family of cell organelles known as *plastids.* The most conspicuous members of this family are the *chloroplasts* which carry out photosynthesis (Chapter 18). The membrane-bound structures that develop into chloroplasts if the cell is exposed to light are known as *leucoplasts.* Those that store the starch produced by photosynthesis are called *amuloplasts.*

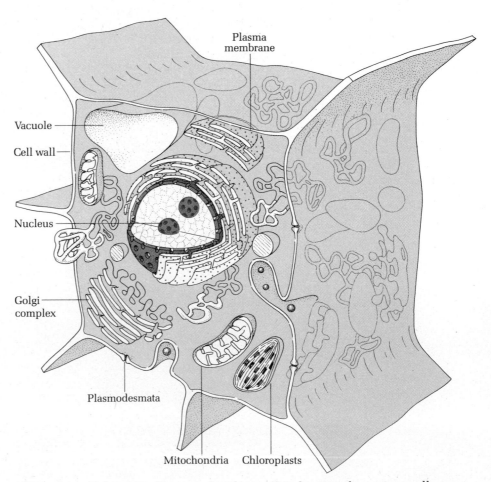

Figure I.11 Schematic diagram of a plant cell indicating the various cell organelles.

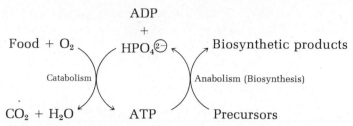

Figure I.12 The energy hypothesis. ATP is the mediator of energy changes which occur in both catabolic, degradative processes and anabolic, biosynthetic processes. All of these reactions are catalyzed by enzymes.

Those that contain pigments that turn color as fruit ripens are known as *chromoplasts.* Plant cells also contain *vacuoles,* small cavities that contain a variety of organic molecules, salts, or pigments. The term is a misnomer that originated because the vacuoles may appear to be empty spaces when viewed through a microscope. The osmotic pressure exerted by the vacuoles against the cell wall helps to provide mechanical support for the cell. The *cell wall* itself is a complex lamellar structure composed chiefly of cellulose. The cytoplasm of adjacent plant cells are in direct contact by means of channels, known as *plasmodesmata,* in the cell wall.

F. Comprehensive Principles

Biochemistry is more of an empirical than a theoretical science. Each organism, indeed each macromolecule, has unique properties of its own. Yet, for all this diversity we can preface our discussion of biochemistry with a handful of generalizations that provide a starting point for our future discussions.

1. *The Genetic Hypothesis.* All of the information required for the reproduction of an organism is contained in its complement of DNA. DNA may either replicate as the cell divides to give a new cell, or its information can be expressed as the protein complement of the cell.

2. *The Enzyme Hypothesis.* All biochemical reactions are carried out by the protein catalysts of cells known as enzymes.

3. *The Energy Hypothesis.* *Adenosine triphosphate* or *ATP* is the biochemical quantum unit of energy. The catabolism (degradation) of all foodstuff is coupled to the generation of ATP, and the biosynthesis (anabolism) of all cell components is coupled to the hydrolysis of ATP (Figure I.12).

4. *The Spontaneous Self-Assembly Hypothesis.* All complexes of macromolecules, such as ribosomes, cell membranes, and cell organelles, assemble spontaneously.

5. *The Carrier Hypothesis.* The transport of most molecules into and out of the cell is mediated by proteins that are integral components of cell membranes.

6. *Organization and Reductionism.* A living organism is the quintessence of a synergistic system; the whole is more than the sum of its parts. As soon as the organization of the cell is destroyed, the most important feature of the system—its life—is lost. Biochemistry has always been a reductionist science, that is, it hs been based on the belief that the secret to the cell's function can be found through an analysis of its macromolecules studied under equilibrium conditions *in vitro.* This approach has made a virtue of necessity, since the complexity of biochemical phenomena precludes analysis of the whole system. As more information accumulates, a movement in the opposite direction is becoming feasible. Eventually, the study of entire systems may supplant the study of individual proteins, but that day is still somewhere in the future.

Part I

The conformation and function of biological macromolecules

Chapter 1

Bioenergetics I: Principles of thermodynamics and their application in biochemistry

Die Energie der Welt ist konstant,
Die Entropie der Welt strebt ein Maximum zu.[1]

Clausius

1–1 INTRODUCTION Clausius's couplet summarizes the vast range of experience that is canonized in thermodynamics, a subject that lies at the heart of biochemistry. The power of thermodynamics lies in its ability to provide a quantitative measure of the energy changes that occur in biological processes. The reactions of metabolism, the transport of material across cell membranes, the assembly of membranes, and the assembly of other types of macromolecular complexes all obey these laws.

[1] The energy of the world is constant,/ The entropy of the world increases to a maximum.

Since every "life process" involves an energy change of some sort, the importance of thermodynamics in biochemistry cannot easily be exaggerated. For example, thermodynamics provides us with a means of determining how much energy is released or absorbed in a process. Thermodynamics also leads us from the concentrations of reactants and products in a biochemical reaction or physical process to the equilibrium constant for the process, and from the equilibrium constant we can determine the magnitude of the "force" that "drives" a process to equilibrium. Thermodynamics also reveals how this force (the free energy) is partitioned between a potential energy change (the enthalpy) and a change in the molecular order (the entropy) of the system.

In this chapter we shall consider the major thermodynamic laws and the equations that relate these laws to each other. Once we have the laws in hand, we shall apply them to various examples as a prelude to their more extensive application in subsequent chapters.

1–2 THERMODYNAMIC SYSTEMS AND THERMODYNAMIC STATE FUNCTIONS

The universe is bigger than our laboratories, and it is convenient to divide the world into *thermodynamic systems* and their *surroundings*. A thermodynamic system is that part of the universe—a test tube, tissue culture, or whatever—in which we are studying a process. Everything else constitutes its surroundings. If matter cannot cross the boundaries between the system and its surroundings, the system is *closed*. If, however, matter can be exchanged between the system and the surroundings, as in all living systems, the system is *open*. If neither matter nor energy can be exchanged with the system, it is *isolated*.

Once we have defined the thermodynamic system, a thermodynamic state must be specified. The state of the system depends upon the thermodynamic variables, sometimes called state variables or *state functions*, such as temperature, pressure, and the number of moles of material in the system. We observe only *changes* in thermodynamic variables. When the state of the system changes, say by increasing the temperature, the change in all other state functions depends only upon the *difference* between the initial and final states and is independent of the pathway by which the process is carried out.

1–3 THE FIRST LAW OF THERMODYNAMICS

The first law of thermodynamics states that *the total energy of the universe is conserved in every physical process*. We shall modestly limit our discussion of this grand statement to chemical reactions. In terms of our arbitrary division of the universe, we can say that the total energy of a system and its surroundings is a constant. Consider the reaction

$$\text{energy} + aA + bB \rightleftharpoons cC + dD \qquad (1.1)$$

The energy absorbed in reaction 1.1 in the forward reaction is exactly equal to the energy released in the reverse reaction. Energy is conserved in the process. The total energy change in reaction 1.1 is

$$\Delta U = Q - W \qquad (1.2)$$

where ΔU is the change in the *internal energy* for the process, Q is the heat absorbed by the reaction, and W is the work done by the system. The type of work done depends on how the reaction occurs. It may involve moving a piston, or it may involve reproducing an organism.

The quantity ΔU is the same for the conversion of A and B to C and D whether the process requires many steps and involves many intermediates, as it well might in a metabolic pathway, or whether it occurs in a single step in a test tube. In essence, the first law of thermodynamics tells us that different forms of energy—chemical, electrical, mechanical, and so forth—can be interconverted. Whenever energy is "lost" in one form, it reappears in exactly the same amount in another form.

Many chemical processes, and almost all biochemical ones, occur at constant pressure. The work done by a system at constant pressure equals the product of the pressure and the change in volume that occurs during the process:

$$W = P\Delta V \tag{1.3}$$

Substituting 1.3 in 1.2 and rearranging we obtain

$$Q_p = \Delta U + P\Delta V \tag{1.4}$$

where Q_p is the heat consumed in the process at constant pressure. Since U, P, and V are state functions, Q_p is also a state function. The heat released or absorbed in the constant pressure process is called the *enthalpy* change for the reaction. Thus, equation 1.4 can be rewritten as

$$\Delta H = \Delta U + P\Delta V \tag{1.5}$$

where ΔH is the symbol for the change in enthalpy.

Heat is measured in either calories or joules. One calorie (cal) or 4.184 joules (J) is the amount of heat required to raise the temperature of 1 gram of water from 14.5 to 15.5°C at a pressure of 1 atmosphere (atm). One kilojoule (kJ) is equal to 1000 J, and 1 kcal is equal to 4.184 kJ. The following conventions are used for enthalpy changes.

$\Delta H < 0$ A reaction is *exothermic* when heat is evolved by the system and enters the surroundings.

$\Delta H > 0$ A reaction is *endothermic* when heat is absorbed by the system.

The specific heat of a substance is the amount of heat required to raise the temperature of 1 gram of the substance by 1°C. The heat required to raise 1 mole of the substance by 1°C is the *molar heat capacity,* equal to the specific heat times the molecular weight of the substance. The units of heat capacity are joules per degree (expressed in kelvins) per mole ($J\ K^{-1}\ mol^{-1}$). The heat capacity at constant pressure (C_p) is defined in terms of the enthalpy (equation 1.6).

$$C_p = \frac{\Delta H}{\Delta T} \tag{1.6}$$

1–4 ENTHALPY CHANGES UNDER STANDARD CONDITIONS

The enthalpy change for a chemical reaction depends upon the number of moles undergoing chemical change, and ΔH is expressed in terms of kJ mol^{-1}. To compare reactions under the same conditions the following conventions are used.

1. The *standard state* of any element or compound is the most stable form of the element or compound at 298 K and 1 atm pressure.

2. In solution the *solute standard state* is 1.0 molar (M).

3. The *standard enthalpy of formation* (ΔH_f°) of any element in its standard state is 0 kJ mol^{-1}. The superscript ° indicates that the reaction occurs under standard conditions.

4. The standard enthalpy of formation of 1.0 M solution of hydronium ions is 0 kJ mol^{-1}.

5. The standard enthalpy of formation of a compound is the enthalpy change when 1 mole of the compound is formed in its standard state from its elements in their standard states.

The enthalpy of formation says nothing about the path or reaction mechanism by which the compound is formed. It depends only on the *difference* in enthalpy between the final state (the compound) and the initial state (the elements). This is true for the enthalpy change of every chemical reaction.

For example, in the reaction

$$aA + bB \longrightarrow cC + dD \qquad (1.7)$$

the standard enthalpy change for the reaction ($\Delta H°$) is given by

$$\Delta H°_{reaction} = \Sigma \Delta H°_{products} - \Sigma \Delta H°_{reactants} \qquad (1.8)$$
$$\Delta \overline{H}°_{reaction} = \{c\Delta \overline{H}°_f[C] + d\Delta \overline{H}°_f[D]\} - \{a\Delta \overline{H}°_f[A] + b\Delta \overline{H}°_f[B]\}$$

The term $\Delta \overline{H}°_f$ in equation 1.8 is the enthalpy change *per mole* of reactant. The units of equation 1.8, therefore, are joules.

1–5 HEATS OF COMBUSTION

The heat content, or in nutritional terms the *caloric content*, of metabolites is obtained by measuring the heat released when the compound is completely burned in oxygen. The *heat of combustion* (ΔH_c) is the heat released for the complete oxidation of the metabolite to its oxidation products. The heats of combustion for carbohydrates, lipids, and proteins are 4.1, 9.3, and 4.1 kcal gm^{-1} respectively.[2] Biochemists seldom refer to thermodynamic data in terms of the number of grams of material, preferring to express such data in kJ mol^{-1} for a given change in state.

The heat of combustion of a given substance can be determined from its standard enthalpy of formation ($\Delta H°_f$) and from the standard enthalpies of formation of the oxidation products, such as carbon dioxide and water. Let us consider the oxidation of glucose (reaction 1.9).

$$\underset{\text{Glucose}}{C_6H_{12}O_6(s)} + 6O_2(g) \longrightarrow 6CO_2(g) + 6H_2O(g) \qquad (1.9)$$

The standard enthalpy change for this reaction is identical to the heat of combustion and is calculated as shown in equations 1.10 and 1.11. Note that the physical state of each reactant is indicated. The heat of combustion depends upon the physical states of all reactants and products.

$$\Delta H_{reaction\ 1.9} = \Delta H_c = [6\Delta H°_f(CO_2)_g + 6\Delta H°_f(H_2O)_g]$$
$$- \Delta H°_f(glucose)_s \qquad (1.10)$$

$$\Delta H_c = [6(-394 \text{ kJ mol}^{-1}) + 6(-242 \text{ kJ mol}^{-1})]$$
$$- (-1268 \text{ kJ mol}^{-1})] \qquad (1.11)$$

$$\Delta H_c = -2548 \text{ kJ mol}^{-1}$$

---

[2] The unit of heat most often used by nutritionists is the "large calorie" (Cal), equal to 1 kilocalorie (kcal). In SI units 1.0 kcal (1.0 Cal) equals 4.184 kJ.

1–6 THE SECOND LAW OF THERMODYNAMICS

The second law of thermodynamics states that *the entropy of the universe is increasing.* The elusive concept of entropy is related to the order, or structure, of the system. In a physical, chemical, or biochemical change, if the final state is more ordered than the initial state, the entropy change is *negative.* On the other hand, if the final state is less ordered than the initial state, the change in state occurs with a positive entropy change. Although positive entropy changes are dictated by the second law of thermodynamics, processes with negative entropy changes are permitted since it is the entropy of the *universe,* including both the thermodynamic system and the surroundings, that must increase for any process (equation 1.12).

$$\Delta S_{process} = \Delta S_{system} + \Delta S_{surroundings} > 0 \tag{1.12}$$

Therefore, ΔS_{system} can be negative if $\Delta S_{surroundings}$ is positive and if $|\Delta S_{surroundings}| > |\Delta S_{system}|$. In a local, microscopic environment, like a cell, order in the system may increase provided that disorder increases in the environment. A cell maintains its low entropy, highly structured state at the expense of increasing the entropy of the environment.

Like enthalpy changes, the change in entropy for a change in state is independent of the path by which the process occurs and depends only on the initial and final states of the system. For the reaction

$$aA + bB \longrightarrow cC + dD \tag{1.13}$$

the standard entropy change is given by

$$\Delta S^{\circ}_{reaction} = [c\overline{S}^{\circ}(C) + d\overline{S}^{\circ}(D)] - [a\overline{S}^{\circ}(A) + b\overline{S}^{\circ}(B)] \tag{1.14}$$

The term $\overline{S}^{\circ}$ is the entropy of formation of 1 mole of a compound under standard conditions.

1–7 ENTROPY, PROBABILITY, AND INFORMATION

"What never?" "No, never." "What never?" "Well, hardly ever!"
W. S. Gilbert and A. S. Sullivan, *H.M.S. Pinafore*

There are several ways of formulating the entropy change for a process. One classical definition relates entropy changes to the heat absorbed in a reversible process and the temperature of the system (equation 1.16).

$$\Delta S = \frac{Q_{rev}}{T} \tag{1.16}$$

where ΔS is the entropy change, Q_{rev} is the heat absorbed in a reversible process, and T is the Kelvin temperature. A second formulation, introduced by Boltzmann, defines the entropy in terms of the *most probable* state of the system. The most probable state is the most random, or least structured state. Entropies are additive (as are enthalpies) because the entropy change for a process depends only upon the initial and final states of the system and is independent of the process by which the change in state is brought about.

Let us now examine the relationship between entropy and probability. Consider a deck of 52 playing cards. The probability of drawing a spade is $\frac{1}{4}$, and the probability of drawing an ace is $\frac{1}{13}$. The probability of drawing an ace of spades is $(\frac{1}{4} \times \frac{1}{13})$. Probabilities are multiplicative. Additive entropies

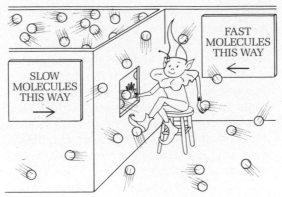

Figure 1.1 Maxwell's demon is able to separate fast and slow moving molecules, but not without expending energy in the form of information, and even then only so long as he himself is not degraded by being continuously bombarded by molecules.

and multiplicative probabilities are related by a logarithmic function (equation 1.17).

$$S = k \ln W \tag{1.17}$$

where S is the entropy, k is the Boltzmann constant (13.8×10^{-24} J K^{-1}, or Avogadro's constant per molecule), and W is the probability that an event will occur. Since entropy in this formulation is a purely statistical law, it can be applied only to large numbers of particles or events.

James Clerk Maxwell (1831–1879), whose monumental achievement in physics was the unification of electricity and magnetism, invented, in a letter to Boltzmann, one of the most significant fantasies in the history of science. Suppose that we appoint a microscopic being, named *Maxwell's demon*, to guard a gate between two flasks containing equal numbers of molecules at the same temperature (Figure 1.1). By letting only fast molecules pass through the gate, the demon (who deserves his appellation) can cause one flask to heat up and the other one to cool down with no expenditure of energy. The entropy of the system would thus *decrease* (in a spontaneous change to a less probable state) in violation of the second law of thermodynamics.

Over fifty years elapsed before the demon was deprived of his paradoxical and magical powers. In 1929 Szilard pointed out that the demon must be endowed with memory to separate "hot" and "cold" molecules, an idea that was altered slightly by Brouillon, who showed that the demon must possess *information* to carry out his duties. The demon has two choices: he must either open or close the gate as a molecule approaches. The *decision* to open or close the gate requires energy equal to $-k \ln 2$. This fundamental quantity is called a *bit* (from binary digit). The loss of energy that accompanies making each decision exactly balances the gain to be had in separating hot and cold molecules[3]. (Alas, poor demon! He, too, is subject to the inexorable second law of thermodynamics. As time passes, the increase in his own entropy will make it impossible for him to tell the difference between hot and cold molecules.)

The demon was vanquished because of an explicit connection between

---

[3] Avogadro's number of bits are required to lower the entropy of 1 mole of an ideal gas by 4.184 J K^{-1} mol^{-1}.

information and entropy. Let us return to our deck of playing cards. When the cards are arranged by suit in ascending order from deuce through ace, the entropy of the system is defined as zero. The second law of thermodynamics says that shuffling the cards will abolish this order, and eventually produce a random distribution of cards corresponding to the state of maximum entropy. Conversely, information is required to restore order to the chaos of the randomly shuffled deck. The information required to restore order is equal in magnitude and opposite in sign to the increase in entropy produced by shuffling the deck originally.

Let us consider a biochemical reaction that is driven by an entropy change in which the statistical concept of entropy comes directly into play. The bacterium *Pseudomonas putida* produces an enzyme, *alanine racemase*, that interconverts L- and D-alanine. From either pure enantiomer the enzyme produces a racemic mixture, that is, a mixture containing equal amounts of the D- and L-stereoisomers (reaction 1.18).

$$
\underset{\text{L-Alanine}}{\overset{\displaystyle \begin{array}{c} CO_2^{\ominus} \\ | \\ H_3\overset{\oplus}{N}-C-H \\ | \\ CH_3 \end{array}}{}} \underset{\text{racemase}}{\overset{\text{Alanine}}{\rightleftharpoons}} \underset{\text{D-Alanine}}{\overset{\displaystyle \begin{array}{c} CO_2^{\ominus} \\ | \\ H-C-\overset{\oplus}{N}H_3 \\ | \\ CH_3 \end{array}}{}} \qquad (1.18)
$$

The enthalpy change for racemization is zero and the equilibrium constant is 1.0, confirming the notion that enantiomers have the same thermodynamic stability. Since the enthalpy change for the reaction is zero, it must be driven by an entropy change. What is the origin of this entropy change? If the reaction begins with the pure L- (or pure D-) enantiomer, converting the starting material into equal amounts of D- and L-alanine produces a more random state than the reactants. The probability that a Maxwell demon can pick out an L-alanine decreases from 1.0 to 0.5; the final state is therefore more disordered than the initial state consisting of pure enantiomer. The entropy change for the reaction is

$$
\Delta S_{\text{racemization}} = S_{\text{products}} - S_{\text{reactants}} \qquad (1.19)
$$

We recall from equation 1.17 that the entropy of a state can be expressed as

$$
S = k_b \ln W \qquad (1.20)
$$

For the products, a racemic mixture, $W = 2$ because there are two possible microscopic states: D-alanine and L-alanine. For the reactants $W = 1$ because there is only a single microscopic state consisting of a pure enantiomer. The Boltzmann constant can be converted to the gas constant R by multiplying it by Avogadro's number, $R = k_b N$. Therefore, the entropy change for racemization of 1 mole of D-alanine at room temperature (298 K) is

$$
\begin{aligned}
\Delta S_{\text{racemization}} &= R \ln 2 - \overset{0}{\cancel{R \ln 1}} \\
&= 5.76 \text{ J K}^{-1} \text{ mol}^{-1}
\end{aligned} \qquad (1.21)
$$

The entropy change for racemization, $\Delta S = R \ln 2$, can be extended to any process in which equal amounts of two pure components are mixed. $\Delta S_{\text{racemization}}$, in fact, is simply a special case of the *entropy of mixing*.

1–8 FREE ENERGY   All physical processes occur with an increase in entropy when changes in both the system and the surroundings are considered. When no further spon-

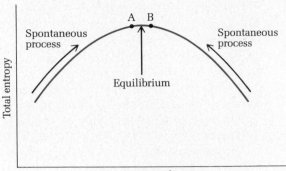

Figure 1.2 Relationship between entropy and composition of a system. When the entropy of the universe is maximum, no spontaneous change is possible, and the system is at equilibrium. [From K. J. Laidler, "Physical Chemistry with Biological Applications," Benjamin/Cummings, Menlo Park, Calif., 1978, p. 206. Reprinted by permission.]

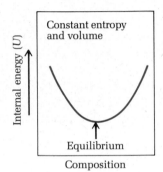

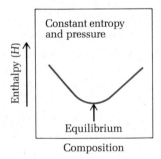

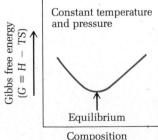

Figure 1.3 Relationship of internal energy, enthalpy, and Gibbs free energy to the equilibrium state. [From K. J. Laidler, "Physical Chemistry with Biological Applications," Benjamin/Cummings, Menlo Park, Calif., 1978, p. 206. Reprinted by permission.]

taneous change is possible, the total entropy has increased to a maximum, and the system is at equilibrium (Figure 1.2). The ability of the system to do work decreases as equilibrium is approached, and at equilibrium there is no *free energy* available to do work. *When an organism is at equilibrium with the surroundings, it is dead.* The *Gibbs free energy* (G) is a thermodynamic state function that defines the equilibrium condition *at constant temperature and pressure* (equation 1.22).

$$\Delta G = \Delta H - T\Delta S \qquad (1.22)$$

The Gibbs free energy determines both the direction and the magnitude of spontaneous change in systems held at constant temperature and pressure. Because biological systems function under these conditions, the Gibbs free energy is of enormous importance in the energetics of biochemical systems. We can regard the free energy as the force that drives a process to equilibrium. The relation of Gibbs free energy as well as internal energy and enthalpy to the equilibrium state is summarized graphically in Figure 1.3. By convention, if ΔG is negative, the process is spontaneous in the direction written and is called *exergonic* (from Greek *ergon*, meaning "work"). When ΔG is positive, the process is not spontaneous in the direction written and is *endergonic*. If the free-energy change for the process is zero, the system is at equilibrium. These conventions are summarized in Table 1.1.

The sign of ΔG is controlled by the balance between ΔH and $T\Delta S$. For processes in which ΔH is negative and $T\Delta S$ is positive, the enthalpy and the entropy act in concert, and both terms favor the spontaneous change. We have now defined two conditions for equilibrium: (1) the entropy of the uni-

Table 1.1	Conventions of the algebraic sign of G and the direction of spontaneous change
$\Delta G < 0$	The change in state is *exergonic* and *spontaneous* in the direction written.
$\Delta G = 0$	The reaction is at *equilibrium*, the system cannot undergo any spontaneous change, and there is no free energy available to do work.
$\Delta G > 0$	The change in state is *endergonic* and is *not* spontaneous in the direction written. (The reverse reaction is spontaneous.)

Table 1.2		**Effect of changes in ΔH and ΔS on ΔG for a reaction**
ΔH	ΔS	ΔG
+	+	The reaction is endothermic, but favored entropically. It may occur if the temperature is high enough.
+	−	The reaction is endothermic and not favored entropically. It will not be spontaneous at any temperature.
−	+	The reaction is spontaneous at all temperatures.
−	−	The reaction is exothermic, but not favored entropically. The process may be spontaneous at low enough temperatures for the $T\Delta S$ term to outweigh the enthalpic contribution to the free energy.

verse is a maximum at equilibrium and (2) the Gibbs free energy of the *system* is a minimum at equilibrium. Since the Gibbs free energy is a system property, it provides us with a measurable criterion of equilibrium in which enthalpy and entropy changes are balanced. Since the enthalpy change is a consequence of the first law of thermodynamics, and the entropy change is described by the second law of thermodynamics, the Gibbs free energy is a tremendous unifying principle. The relationship between the spontaneity of a given change in state and the enthalpy and entropy changes for a given change in state are summarized in Table 1.2.

1–9 STANDARD FREE ENERGY

The standard free energy of a compound is the free-energy change for formation of 1 mole in its standard state (298 K and 1.0 atm) from its elements in their standard states. The standard free energy of formation (ΔG_f°) of any element in its standard state is zero; the standard state for a solute in solution is 1.0 molar; and the standard free energy of formation of a 1.0 M solution of hydronium ions is assigned a value of zero.[4]

The free-energy change for a given process is independent of the pathway by which the change in state is brought about. For the reaction

$$aA + bB \rightleftharpoons cC + dD$$

the standard free-energy change is

$$G_{reaction}^\circ = \{c\Delta \overline{G}_{f_C}^\circ[C] + d\Delta \overline{G}_{f_D}^\circ[D]\} - \{a\Delta \overline{G}_{f_A}^\circ[A] + b\Delta \overline{G}_{f_B}^\circ[B]\} \qquad (1.23)$$

where the term $\Delta \overline{G}^\circ$ is the standard free energy of formation for 1 mole of the compound in question.

1–10 STANDARD FREE-ENERGY CHANGES AND THE EQUILIBRIUM CONSTANT

We have now discussed three related ideas: free-energy changes, standard free-energy changes, and the equilibrium condition. What is the relation among them? The free-energy content of a compound depends upon the number of moles present and is thus an extensive property of the system. The standard free energy of a compound is defined for 1 mole of the compound under specified conditions of constant temperature and pressure. It can be shown that the two are related by equation 1.24.

$$G_A = G_A^\circ + 2.303RT \log [A] \qquad (1.24)$$

[4] Standard free-energy changes in solution at pH 7 are discussed in Section 12.2.

When the concentration of A is 1.0 M, G_A simply equals G_A°, but under all other conditions the standard free energy and the free energy have different values.[5] The second term on the right side of equation 1.24 is a "correction factor" that relates the actual free energy of the compound to the standard free energy. Let us consider reaction 1.1 again.

$$aA + bB \rightleftharpoons cC + dD$$

The free-energy change for the reaction is

$$\Delta G_{\text{reaction}} = (cG_C + dG_D) - (aG_A + bG_B) \tag{1.25}$$

Substituting a term of the form of 1.24 for A, B, C, and D in equation 1.25 we obtain

$$\Delta G_{\text{reaction}} = (cG_C^\circ + dG_D^\circ - aG_A^\circ - bG_B^\circ) + 2.303RT \log \frac{[C]^c[D]^d}{[A]^a[B]^b} \tag{1.26}$$

The term in parentheses in equation 1.26 is simply the standard free-energy change for the reaction ΔG°. Let us call the mantissa of the log term Q, where

$$Q = \frac{[C]^c[D]^d}{[A]^a[B]^b} \tag{1.27}$$

At equilibrium Q is the equilibrium constant for the reaction, K_{eq}, the free-energy change for the reaction is zero, and equation 1.28 becomes

$$0 = \Delta G_{\text{reaction}}^\circ + 2.303RT \log K_{\text{eq}} \tag{1.28}$$

or

$$\Delta G_{\text{reaction}}^\circ = -2.303RT \log K_{\text{eq}} \tag{1.29}$$

Equation 1.29 is one of the most useful in all of thermodynamics. If we take the antilog of equation 1.29, we obtain another useful equation, namely,

$$K_{\text{eq}} = e^{-\Delta G_{\text{reaction}}^\circ / RT} \tag{1.30}$$

Referring to equation 1.26, we see why $\Delta G_{\text{reaction}}^\circ$ is related to the equilibrium constant rather than $\Delta G_{\text{reaction}}$: the log term represents the free-energy change that occurs when the reactants and products are brought from standard state concentrations to equilibrium concentrations. The log term exactly balances the standard free-energy change required to make $\Delta G_{\text{reaction}}$ equal to zero. Thus, $\Delta G_{\text{reaction}}^\circ$ is *not* the criterion of spontaneity for chemical reactions. In many metabolic pathways there are steps whose standard free-energy changes are positive, but which are nevertheless spontaneous. Referring again to equation 1.26, this means that the value of Q determines the spontaneity. If the value of Q is less than 1.0 the log term is negative; if it is sufficiently negative, the process is spontaneous under the prevailing conditions even though $\Delta G_{\text{reaction}}^\circ$ is positive. The numerical relationship between equilibrium constants and standard free-energy changes is shown in Table 1.3.

---

[5] Rigorous thermodynamic treatments use activities, the equivalent of effective concentrations, instead of concentrations. For dilute solutions activities are about equal to concentrations, and we will use concentrations for simplicity.

Table 1.3 Relationship between ΔG°_{298} and the equilibrium constant for the equilibrium $A \overset{K}{\rightleftharpoons} B$, at 298 K

% B	K	$-\Delta G^\circ_{298}$ (J mol^{-1})
50	1.00	0
55	1.22	500
60	1.50	1000
65	1.86	1500
70	2.33	2100
75	3.00	2700
80	4.00	3400
85	5.67	4300
90	9.00	5450
95	19.0	7300
98	49.0	9650
99	99.0	1.1×10^4
99.9	999.0	1.7×10^4
99.99	9999.0	2.2×10^4

1–11 A THERMODYNAMIC ANALYSIS OF THE STRUCTURE OF WATER

An extraordinary range of biological phenomena are influenced by the unique properties of water. The *heat of vaporization* of water (ΔH_{vap}), defined as the enthalpy change for the conversion of liquid water at 373 K to water vapor at 373 K, is +40.88 kJ mol^{-1} (reaction 1.31).

$$H_2O(l) \rightleftharpoons H_2O(g) \qquad \Delta H_{373} = +40.88 \text{ kJ mol}^{-1} \tag{1.31}$$

The endothermic vaporization is driven by a large positive entropy of vaporization, $\Delta S_{vap} = +109.8$ J K^{-1} mol^{-1}. The evaporation of water carries away a considerable amount of heat, as everyone knows who has emerged from bathing on a cool day. This cooling mechanism is important in thermoregulation for both animals and plants. The high boiling point of water and its high heat of vaporization are the result of the ability of each water molecule to form four hydrogen bonds in a tetrahedral array that repeats itself throughout the liquid (Figure 1.4). Each hydrogen bond has a strength of between 12.5 and 21.0 kJ mol^{-1}. It is thought that on the average three hydrogen bonds are broken for each water molecule that enters the vapor phase.

Water is also an excellent solvent for ions and a poor one for nonpolar solutes because of its high *dielectric constant* (D). The dielectric constant of a medium is a measure of the force between two particles that depends upon

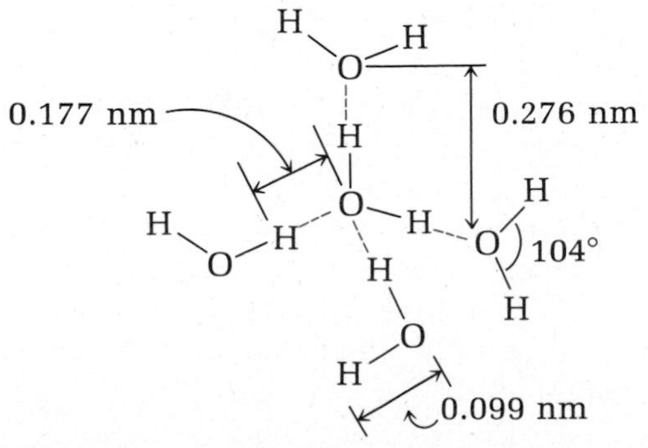

Figure 1.4 Tetrahedral structure of a water molecule hydrogen bonded to four others.

the charges of the interacting particles and the distance separating them. The force between the particles is given by equation 1.32.

$$F = \frac{Q_1 Q_2}{D r^2}$$

(1.32)

where Q_1 and Q_2 are the charges of the ions, D is the dielectric constant of the medium, and r is the distance separating the charges. The dielectric constant of a vacuum is 1.0, for hexane it is 1.9 (at 20°C), and for water it is 78, one of the highest known. This high value means that the attraction between oppositely charged ions, such as $Na^{\oplus}$ and $Cl^{\ominus}$, is only $\frac{1}{78}$ as strong in water as in a vacuum and only about $\frac{1}{40}$ as strong as in hexane. NaCl thus dissolves in water, where the ions interact strongly with water molecules, but is insoluble in hexane.

The tetrahedral array of hydrogen-bonded water molecules is extended throughout the liquid to give an "ice-like" structure to water (Figure 1.5). The lattice is not rigid, but is composed of "flickering clusters" that continually form and disappear (Figure 1.6). Within a cluster a water molecule can be hydrogen bonded to up to four other water molecules or to none. The formation of the clusters is cooperative: the formation of one hydrogen bond encourages formation of others. The disintegration of a cluster is like a zipper running through the hydrogen bonds. Each cluster has a lifetime of 10^{-8} to 10^{-11} s. The mobile network of hydrogen bonds in water is highly *polarizable*, and the introduction of an electric field, in the form of an ion, alters the structure of the liquid. Also note that water has two polar O—H bonds that form an angle of 104° to each other. Water therefore has a large dipole moment. The attraction of the dipoles for each other, weak and highly polarizable interactions, is the source of the high dielectric constant of water.

Solutes are "structure forming" if they increase the internal order in

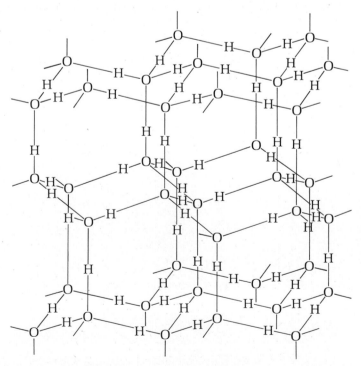

Figure 1.5 Structure of ice I. [From G. Nemethy and H. A. Scheraga, *J. Chem. Phys.*, **36**, 3383 (1962).]

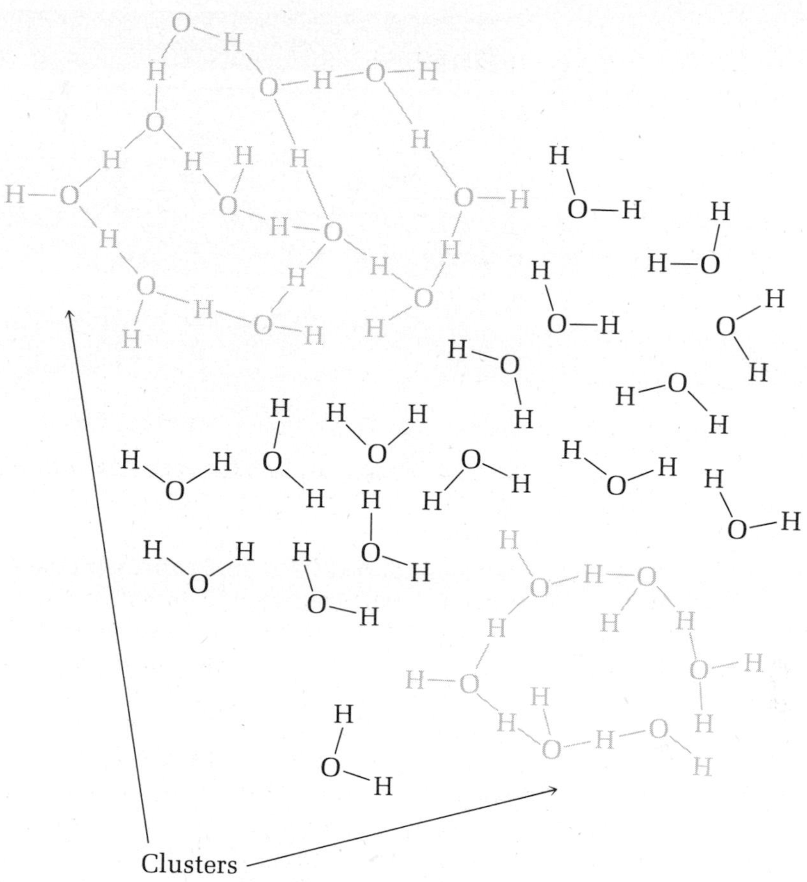

Clusters

Figure 1.6 A "flickering cluster" of water molecules in a temporary arrangement whose lifetime is on the order of 10^{-8} s or less. [From G. Nemethy and H. A. Scheraga, *J. Chem. Phys.*, **36**, 3387 (1962).]

the solvent, thus decreasing the entropy of the system. They are "structure breaking" if they decrease the internal order of the solvent. First, let us consider the effect of a structure-breaking ion. An ion attracts the water molecules that form its *hydration sphere*. These water molecules are more structured than the bulk solvent. Outside the ordered hydration sphere the electric field of the ion disrupts the structure of the bulk solvent increasing its disorder. Thus the structure-breaking ion increases the entropy of the solvent, even though there is considerable ordering of water molecules in the hydration sphere. W. P. Jencks, in a marvelous analogy, compared this effect to introducing the Beatles into the Boston Symphony Orchestra: "The different types of forces compete with each other, and the result is an increase in the disorder and entropy of the system" (Figure 1.7). Structure-forming ions, such as quaternary ammonium ions, $R_4N^{\oplus}$, decrease the entropy of the solvent. The alkyl groups of the ammonium ion decrease the mobility, and thus the entropy, of the water. The effect of hydrophobic solutes on the structure of water and the effect of the hydrophobic side chains on protein structure, both properties of the solvent, shall be taken up again in Chapter 4.

1–12 SUMMARY

In this chapter we have discussed the first and second laws of thermodynamics and the algebraic relations that equate these quite abstract laws to chemical reactions. Table 1.4 summarizes these relationships and conventions regarding the signs of thermodynamic functions. The first law of thermodynamics states that the total energy of the universe is conserved in every

Table 1.4 Thermodynamic laws and the equations that relate them to chemical reactions

Law	Equation	Meaning of algebraic sign
First—the energy of the universe is conserved.	$\Delta H = \Delta U + P\Delta V$ $\Delta H^{\circ}_{\text{reaction}} = \Sigma \Delta H^{\circ}_{\text{products}} - \Sigma \Delta H^{\circ}_{\text{reactants}}$	$H > 0$ endothermic $H < 0$ exothermic
Second—the entropy of the universe is increasing.	$S_{\text{process}} = [S_{\text{surroundings}} + S_{\text{system}}] > 0$ $S = Q_{\text{reversible}}/T$ $S = R \ln W$ $\Delta S_{\text{reaction}} = \Sigma \Delta S_{\text{products}} - \Sigma \Delta S_{\text{reactants}}$	$S_{\text{total}} > 0$ for all real processes
First and second, summarized by Gibbs free energy	$\Delta G = \Delta H - T\Delta S$ $\Delta G^{\circ}_{\text{reaction}} = \Sigma \Delta G^{\circ}_{\text{products}} - \Sigma \Delta G^{\circ}_{\text{reactants}}$ $\Delta G^{\circ} = -2.303\ RT \log K_{\text{eq}}$	$G < 0$ exergonic, spontaneous $G > 0$ endergonic, not spontaneous

physical process. To apply this general principle to chemical reactions we introduce the enthalpy. The enthalpy is the heat released by a change in state at constant pressure. Recall equation 1.5.

$$\Delta H = \Delta U + P\Delta V \tag{1.5}$$

The standard enthalpy of formation ($\Delta H^{\circ}_{\text{f}}$) is the heat released when a compound in its standard state is formed from elements in their standard states at 298 K and 1.0 atm. For any chemical reaction the standard enthalpy change is given by

$$\Delta H^{\circ}_{\text{reaction}} = \Sigma \Delta H^{\circ}_{\text{products}} - \Sigma \Delta H^{\circ}_{\text{reactants}} \tag{1.8}$$

The second law of thermodynamics states that the entropy change for any process is positive when both the thermodynamic system and the thermodynamic surroundings are considered. The entropy change for a process is a measure of the change in the statistical order of the universe before and after the process has occurred.

The enthalpy change and the entropy change are related to changes in the universe, that is, to changes in the system *plus* changes in the surroundings. At constant temperature and pressure the change in Gibbs free energy is the criterion for spontaneous change within a thermodynamic system. It is this system property that confers so great an importance on the Gibbs free energy. The Gibbs free energy combines the first and second laws of thermodynamics.

$$\Delta G = \Delta H - T\Delta S \tag{1.22}$$

The standard free-energy change for a reaction is related to the equilibrium constant for the reaction by equation 1.29.

$$\Delta G^{\circ}_{\text{reaction}} = -2.303 RT \log K_{\text{eq}} \tag{1.29}$$

The actual free-energy change, ΔG, not the standard free-energy change, is the criterion of spontaneity for a chemical reaction.

Water is the cellular solvent. Its structure is important in determining the three-dimensional structures of proteins, nucleic acids, and biological membranes. Bulk water is an extensively hydrogen-bonded network of "flickering clusters" of water molecules. Solutes that increase the internal structure of water, or decrease the entropy of water, are known as

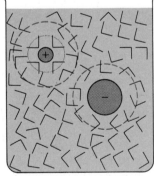

Figure 1.7 Effect of a structure-breaking ion on the structure of water. The water molecules within the dotted circles are well within the hydration sphere and are more ordered than the bulk solvent. Those within the dashed lines are in the region in which disorder has increased. Those in the bulk solvent are in flickering clusters of short lifetimes. [From W. P. Jencks, "Catalysis in Chemistry and Enzymology," McGraw-Hill, New York, 1969, p. 387. Used with the permission of the McGraw-Hill Book Company.]

structure-forming solutes. By contrast, solutes that increase the entropy of the solvent are known as structure-breaking solutes.

REFERENCES

Bray, H. G., and White, K., "Kinetics and Thermodynamics in Biochemistry," 2nd ed., Academic Press, New York, 1966.

Ingraham, L. L., and Pardee, A. B., "Free Energy and Entropy in Metabolism," in D. M. Greenberg, Ed., "Metabolic Pathways," 3rd ed., vol. 1, Academic Press, New York, 1967, pp. 2–46.

Jencks, W. P., "Catalysis in Chemistry and Enzymology," McGraw-Hill, New York, 1969.

Klotz, I. M., "Energy Changes in Biochemical Reactions," Academic Press, New York, 1967.

Morowitz, H. J., "Energy Flow in Biology," Academic Press, New York, 1968.

Van Holde, K. E., "Physical Biochemistry," Prentice-Hall, Englewood Cliffs, N.J., 1971.

Problems

1. Consider a chicken egg in a nest upon which the hen is sitting. (a) Is the egg an open or a closed system? Is the system isolated? (b) The egg spontaneously develops into a chicken. Is the free-energy change for chicken formation exergonic or endergonic? Is the entropy change positive or negative? What "driving force" is responsible for the free-energy change of the spontaneous process?

2. Phosphoglyceromutase catalyzes the reaction

$$\text{3-phosphoglycerate} \underset{}{\overset{K_{eq}}{\rightleftharpoons}} \text{2-phosphoglycerate} \tag{1.33}$$

At pH 7.0 the equilibrium constant is 0.165. In erythrocytes the concentration of 3-phosphoglycerate is 61.2×10^{-6} M (61.2 μM) and the concentration of 2-phosphoglycerate is 4.3 μM. What is the free-energy change under these conditions? How do the actual concentrations of the reactants affect the spontaneity of the reaction?

3. The end product of the anaerobic oxidation of glucose is lactic acid. Using the standard free energies of formation of glucose and lactic acid (-914.5 kJ mol^{-1} and -686.6 kJ mol^{-1}, respectively), calculate the equilibrium constant and free-energy change at 298 K for the reaction

$$\text{glucose(aq)} \longrightarrow \text{2 lactic acid(aq)} \tag{1.34}$$

4. When methane is transferred from benzene to water

$$\text{methane(benzene)} \underset{}{\overset{K}{\rightleftharpoons}} \text{methane(H}_2\text{O)} \tag{1.35}$$

the free-energy change is positive, but the enthalpy change is negative. What does this imply about the entropy change for the reaction? What do the signs of the free-energy change and the entropy change imply about the spontaneity of the reaction? Does your answer make "chemical sense," that is, are hydrocarbons soluble in water?

5. Proteins can be inactivated by heating.

$$\text{protein(active)} \rightleftharpoons \text{protein(inactive)} \tag{1.36}$$

If the inactivation is endothermic and the process is spontaneous, what does this imply about the entropy change? What does the entropy change imply about the structure of the inactive (or denatured) protein vis-à-vis the structure of the active (or native) protein?

6. The intracellular fluid of muscle contains Mg(II) at a concentration of 0.04 M and monohydrogen phosphate at a concentration of 0.14 M. These ions form a 1:1 complex which dissociates according to the following equilibrium:

$$MgHPO_4 \rightleftharpoons Mg(II) + HPO_4^{2-} \qquad K = 1.3 \times 10^{-2} \qquad (1.37)$$

Calculate (a) the concentration of complex at T = 298 K, (b) the standard free-energy change for the reaction at T = 298 K, (c) the fraction of monohydrogen phosphate bound to magnesium (II) at T = 298 K.

7. A lipoprotein monomer called apoA-1 dimerizes with an association constant of 1.3×10^4 M. Calculate the standard free-energy change for dimerization.

8. Studies of the thermal denaturation, or unfolding, of the lipoprotein apoA-1 indicate that the native state is more stable than the denatured state by 10.0 kJ mol^{-1}. (a) Calculate the equilibrium constant for folding of the denatured protein at 298 K.

$$(\text{apoA-1})_{\text{unfolded, denatured}} \rightleftharpoons (\text{apoA-1})_{\text{refolded, native}}$$
$$\Delta G° = -10 \text{ kJ mol}^{-1} \qquad (1.38)$$

(b) The enthalpy of unfolding is endothermic, $\Delta H°_{\text{unfolding}} = 267.8$ kJ mol^{-1}. Using the standard free-energy change above, calculate the entropy of unfolding at 298 K.

9. Thermodynamic data have been obtained for the conversion of glycine(aq) to its dipolar ion.

$$H_2N-CH_2-CO_2H(aq) \rightleftharpoons H_3\overset{\oplus}{N}-CH_2-CO_2^{\ominus}(aq) \qquad (1.39)$$

The standard enthalpy change ($\Delta H°_{\text{aq}}$) is −41.42 kJ mol^{-1}. The standard entropy change ($\Delta S°_{\text{aq}}$) is −31.38 J K^{-1} mol^{-1}. Calculate the equilibrium constant for this reaction.

10. The standard enthalpies of formation of $H_2O(g)$ and $H_2O(l)$ are −242 and −286 kJ mol^{-1}, respectively. Why is the liquid state more stable than the gas state?

Chapter 2

Amino acids, peptides, and the covalent structure of proteins

Philosophy is written in a great book which is always open before our eyes, but we cannot understand it without first applying ourselves to understanding the language and learning the characters used for writing it.

Galileo (*Il Saggiatore*)

2–1 INTRODUCTION

The characters and language of biochemistry consist of the macromolecules of which the biological world is made and the small molecules, the monomers, from which the macromolecules themselves are formed. We shall begin with proteins whose name, derived from the Greek word *protos*, meaning "first," was coined by the German biochemist Mulder in 1838. As this date indicates, biochemistry is a young science, though a most precocious one. Nearly all the proteins in the biological world are made from fewer than two dozen molecules, the *α-amino acids*. These molecules are the "characters" in which the "language" of proteins is written. The "grammar" of the proteins, and the foundation upon which everything else is erected, is their covalent structure.

2–2 THE STRUCTURES OF THE α-AMINO ACIDS

The proteins are made from some 20 amino acids, all but one of which have the general structure 1 and the three-dimensional structure 2 shown below.

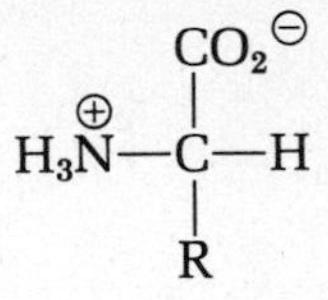

The α-carbon in structure 1 bears four different groups for all amino acids except *glycine*, whose R group is hydrogen. None of these amino acids is superimposable on its mirror image, the α-carbon is *chiral* (Greek *cheir*, "hand"), and the mirror-image isomers are *enantiomers* (Figure 2.1). The three-dimensional representation of structure 1, although accurate, is not very convenient. By convention, molecules having a chiral carbon are drawn with the chiral carbon in the plane of the paper. Groups to the right and left of the chiral carbon project out of the plane of the paper, and their orientation is expressed by heavy lines or wedges. Groups below the plane of the paper are represented by dashed lines. Structure 3 represents the general case.

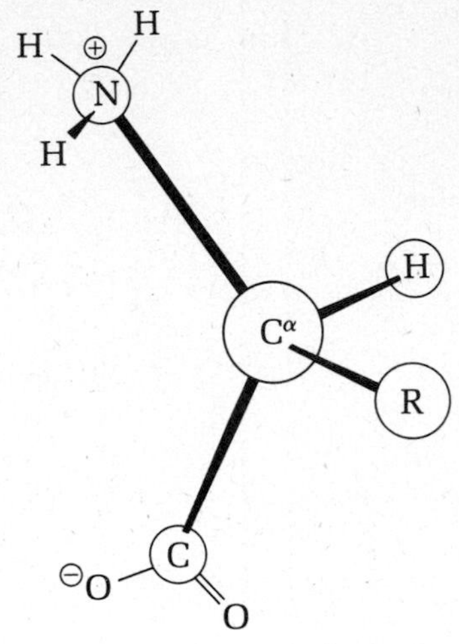

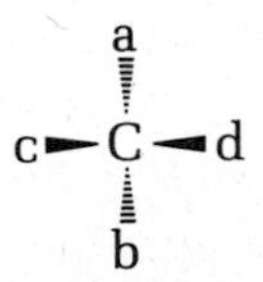

The chirality of the amino acids in proteins is opposite to the configuration of the reference compound D-glyceraldehyde (Figure 2.2). The term D- defines the actual positions in space, or the *absolute configuration*, of the groups surrounding the chiral carbon.[1] The enantiomer of D-glyceraldehyde is L-glyceraldehyde. Again, the L- defines the absolute configuration of the molecule. Amino acids in proteins all have the L- configuration. Since proteins are composed of chiral units, the L-α-amino acids, they, too, are chiral objects. The structures of the amino acids found in proteins, their

[1] An alternate convention of assigning absolute configuration, the R,S system, is used almost exclusively in organic chemistry. We shall discuss this system in Chapter 10. It will not be necessary before then.

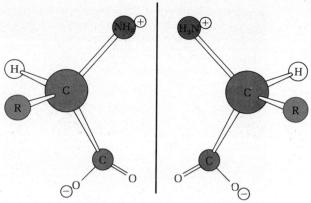

Figure 2.1 A pair of mirror-image amino acids. Amino acids found in proteins have the configuration shown on the right side of the mirror.

names, and their abbreviations are given in Tables 2.1 and 2.2. The carbon chain of amino acids is named by reference to the common system of carboxylic acid nomenclature. As we have seen, the carbon adjacent to the carboxyl group is called α. Moving down the chain the others are called β, γ, δ, and ϵ as indicated below.

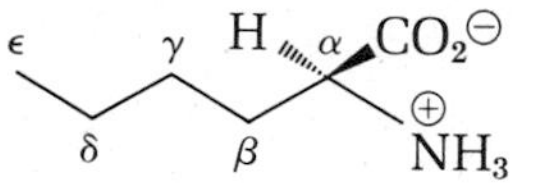

$$\epsilon \qquad \gamma \quad H_{\text{\tiny{///}}} \alpha CO_2^\ominus$$
$$\overset{}{\underset{\delta \qquad \beta}{\diagdown}} \overset{\oplus}{NH_3}$$

2–3 SALIENT PROPERTIES OF INDIVIDUAL AMINO ACIDS

The side chain of an amino acid distinguishes it from its relatives. The side chains of eight of the amino acids are nonpolar and are chemically inert under most experimental conditions. The low affinity for water that is shared by these amino acids is one of their most important physical properties. These nonpolar side chains are *hydrophobic*, (literally, "water hating"). The polar side chains of the remaining amino acids have a high affinity for water. These side chains are *hydrophilic*, (literally, "water loving"). Some of them are acidic, some are basic, and some are neutral. Let us briefly consider the major properties of each of the amino acids.

A. Amino Acids with Nonpolar Side Chains

a. Alanine, Valine, Leucine, and Isoleucine

These four amino acids all are hydrophobic, and all are nearly inert under most experimental conditions. Their side chains tend to order the structure of water in their immediate vicinity, an effect that has profound

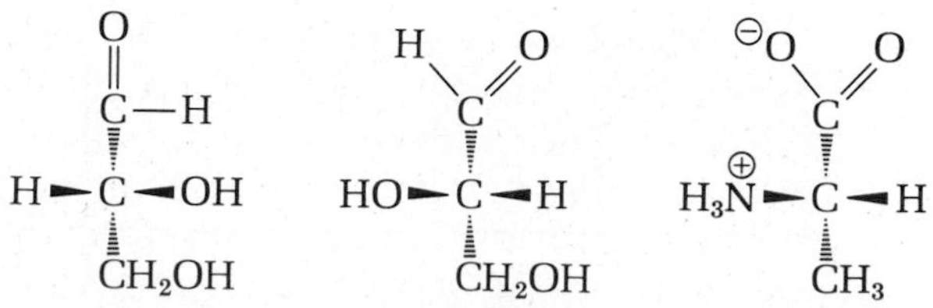

D-Glyceraldehyde   L-Glyceraldehyde   L-Alanine

Figure 2.2 Absolute configuration of groups surrounding a chiral carbon. The amino acids found in proteins have chirality opposite to that of D-glyceraldehyde and the same as that of L-glyceraldehyde.

Table 2.1 Common amino acids found in proteins, showing the ionic forms predominant at pH 7

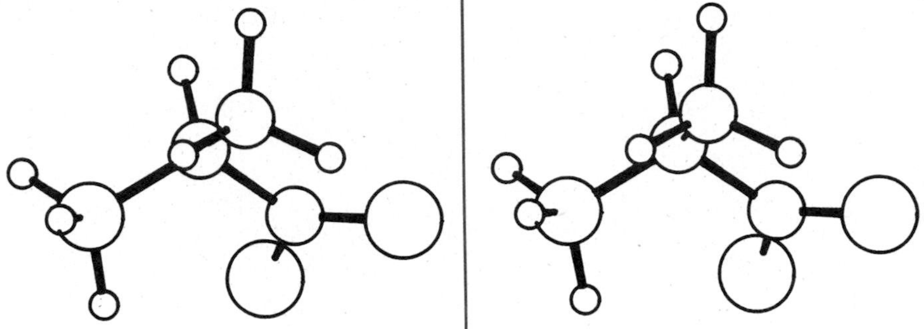

Amino acid	Planar representation	Side chain conformation
Nonpolar side chains		
Alanine Ala A Mol wt 89	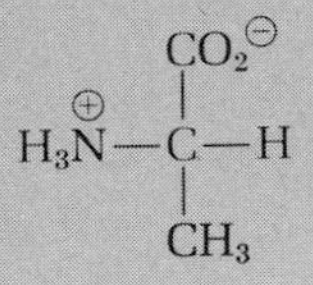 $H_3\overset{\oplus}{N}-\underset{\underset{CH_3}{\mid}}{\overset{\overset{CO_2^{\ominus}}{\mid}}{C}}-H$	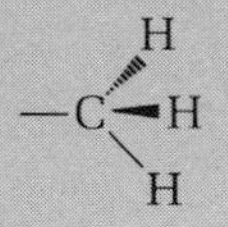 $-\underset{H}{\overset{H}{C}}\cdots H$
Valine Val V Mol wt 117	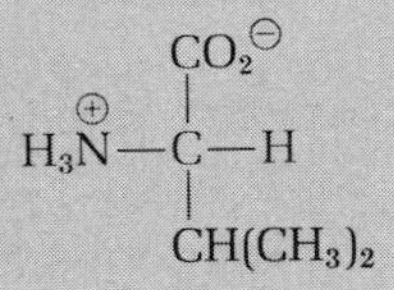 $H_3\overset{\oplus}{N}-\underset{\underset{CH(CH_3)_2}{\mid}}{\overset{\overset{CO_2^{\ominus}}{\mid}}{C}}-H$	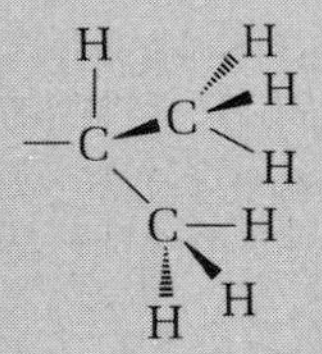 $-\underset{\underset{\underset{H}{\mid}}{\overset{\mid}{C}-H}}{\overset{H}{\overset{\mid}{C}}}\overset{H}{\underset{H}{\overset{\cdots H}{C}}}$

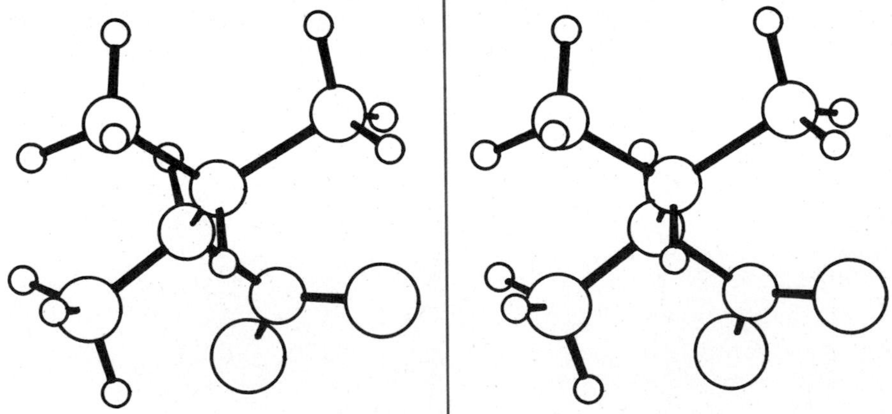

(Continued)

Table 2.1 Common amino acids found in proteins, showing the ionic forms predominant at pH 7 *(Continued)*

Amino acid	Planar representation	Side chain conformation
Leucine Leu L Mol wt 131	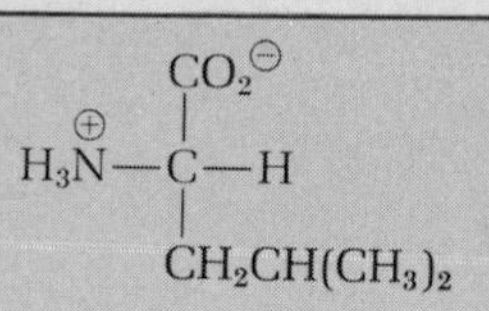 $H_3\overset{\oplus}{N}-\overset{\overset{\displaystyle CO_2^{\ominus}}{\textstyle \vert}}{\underset{\textstyle \vert}{C}}-H$ $CH_2CH(CH_3)_2$	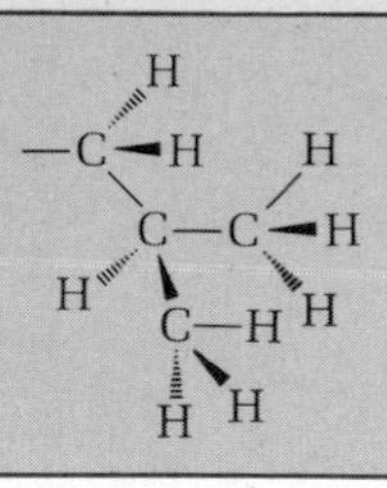
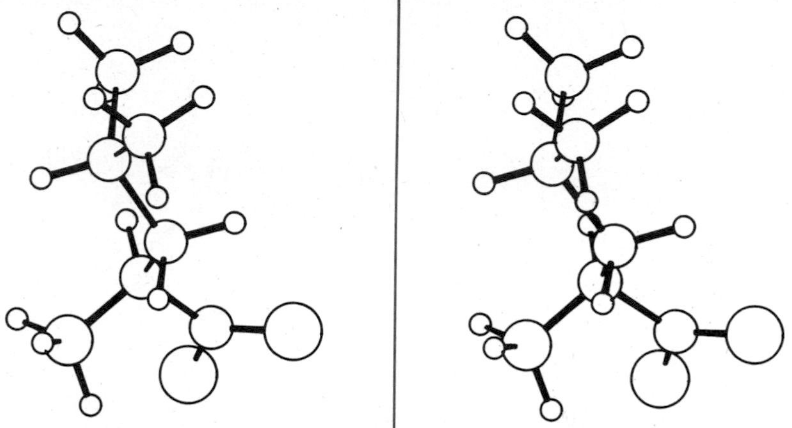		
Isoleucine Ile I Mol wt 131	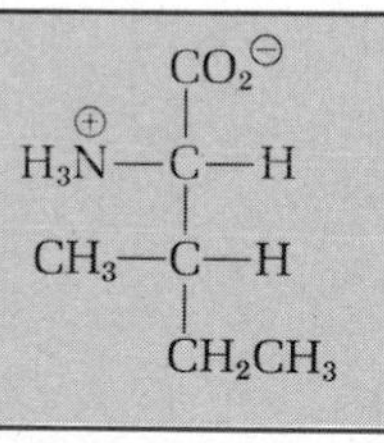 $H_3\overset{\oplus}{N}-\overset{\overset{\displaystyle CO_2^{\ominus}}{\textstyle \vert}}{C}-H$ $CH_3-\overset{\textstyle \vert}{C}-H$ CH_2CH_3	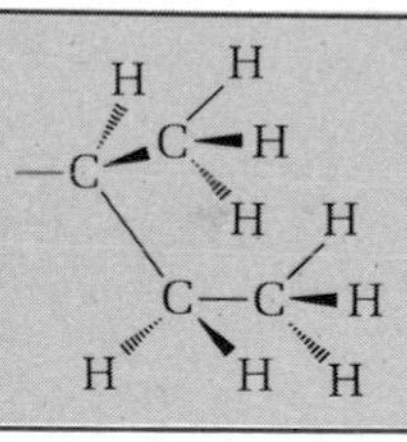
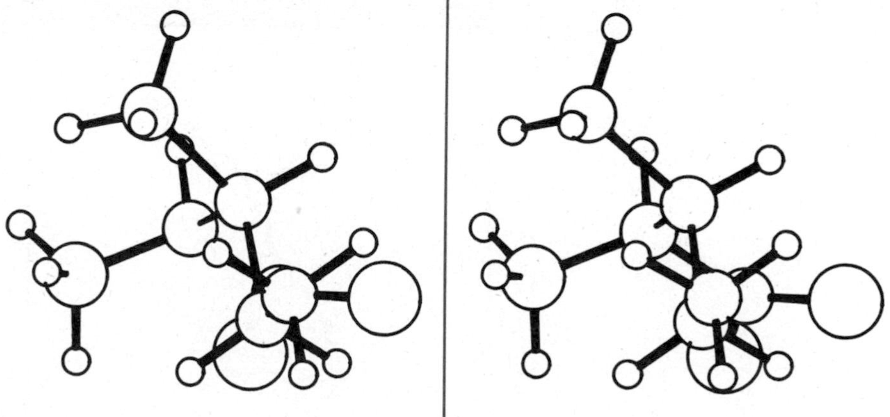		

(Continued)

Table 2.1   Common amino acids found in proteins, showing the ionic forms predominant at pH 7   (Continued)

Amino acid	Planar representation	Side chain conformation
Proline Pro P Mol wt 115	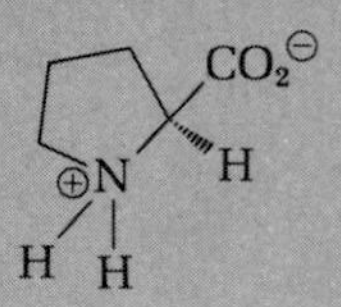	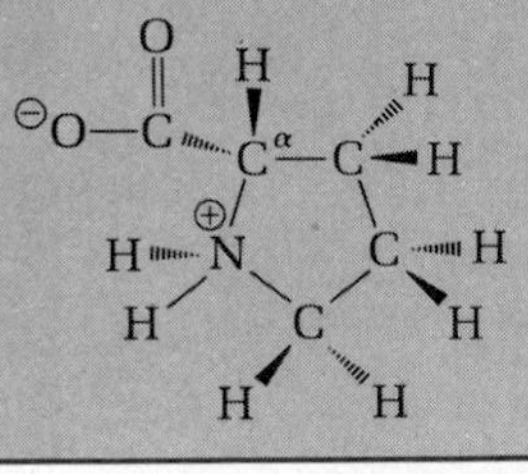
	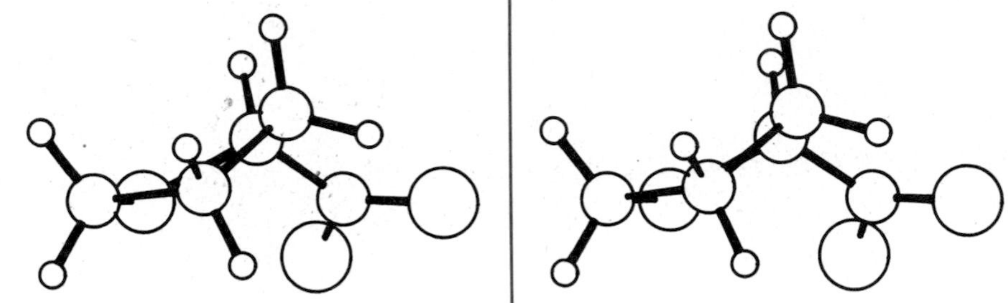	
Phenylalanine Phe F Mol wt 165	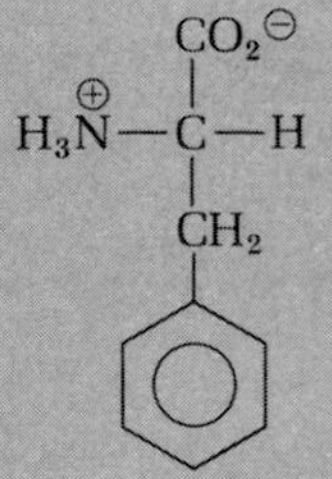	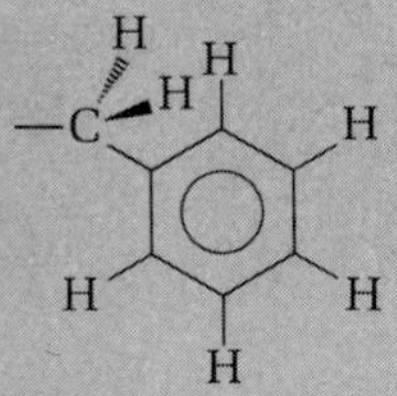
	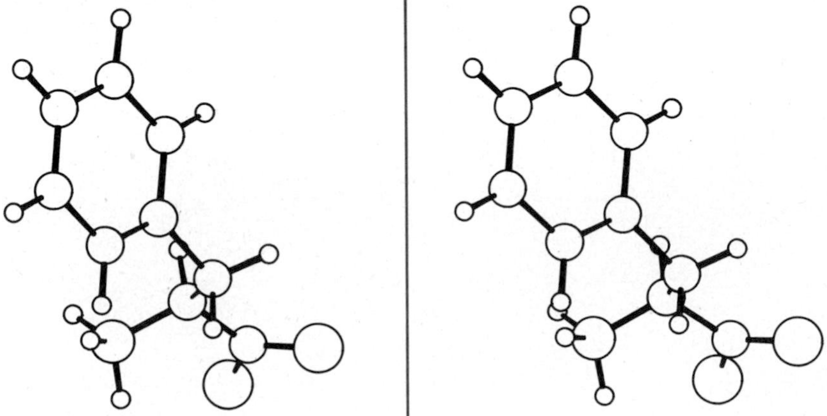	

(Continued)

Table 2.1 Common amino acids found in proteins, showing the ionic forms predominant at pH 7 (Continued)

Amino acid	Planar representation	Side chain conformation
Tryptophan Trp W Mol wt 204	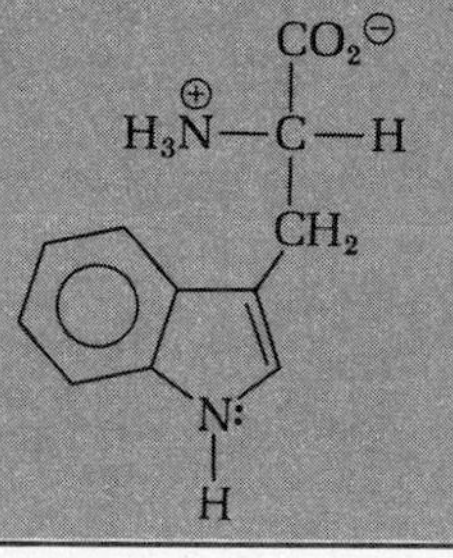	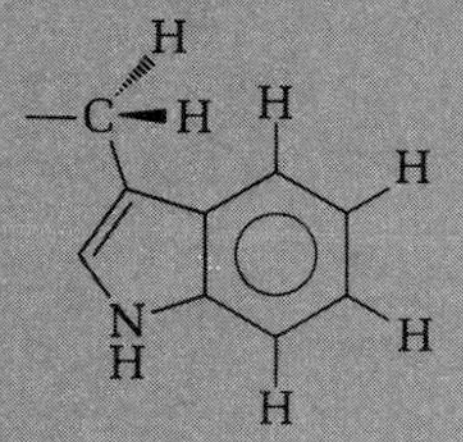
	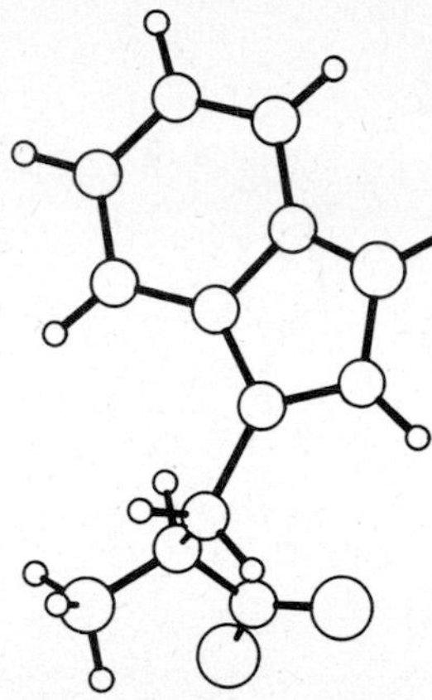	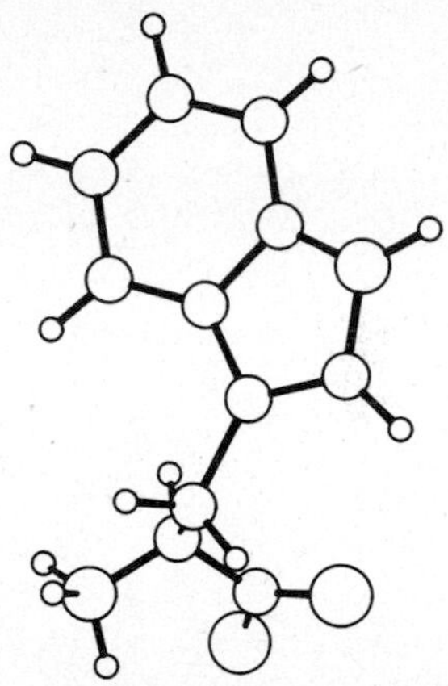
Methionine Met M Mol wt 149	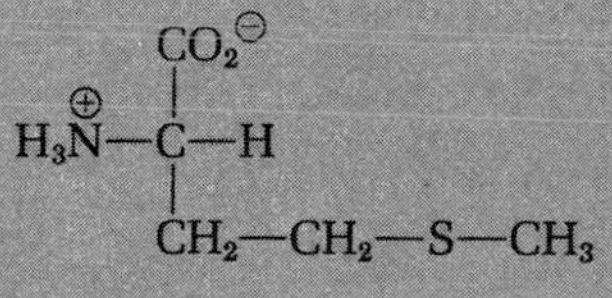	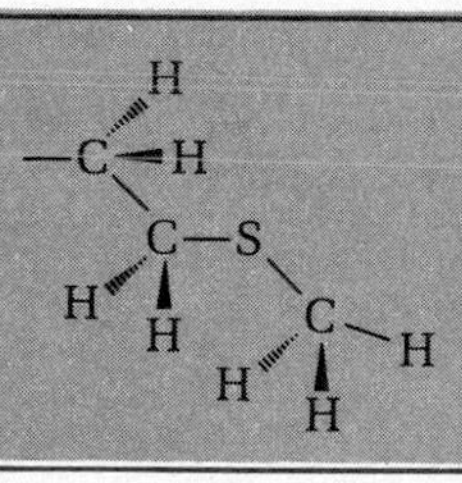
	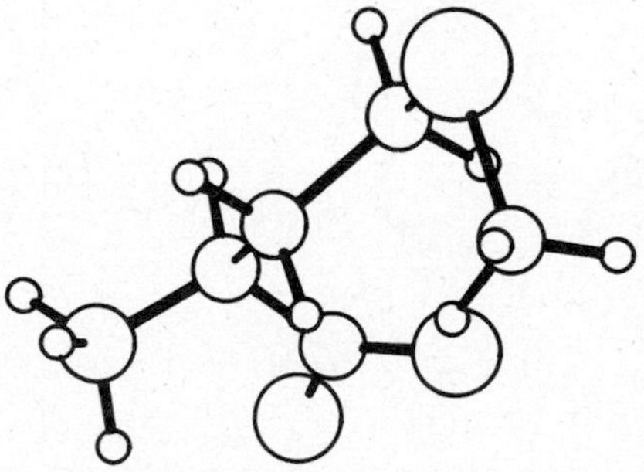	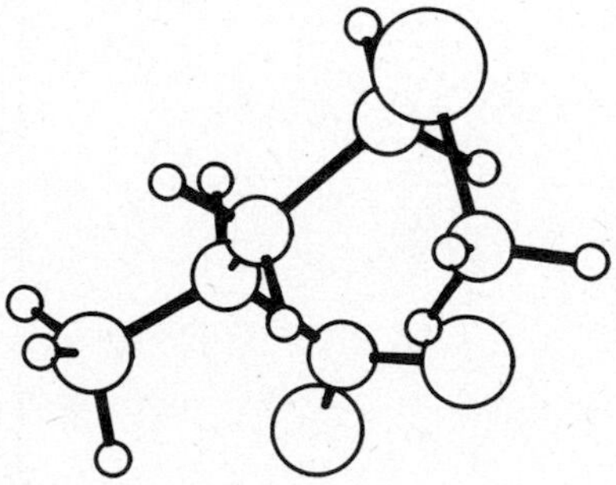

(Continued)

Table 2.1 Common amino acids found in proteins, showing the ionic forms predominant at pH 7 (Continued)

Amino acid	Planar representation	Side chain conformation
Polar side chains		

Glycine
Gly
G

Mol wt 75

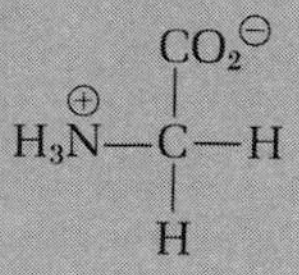

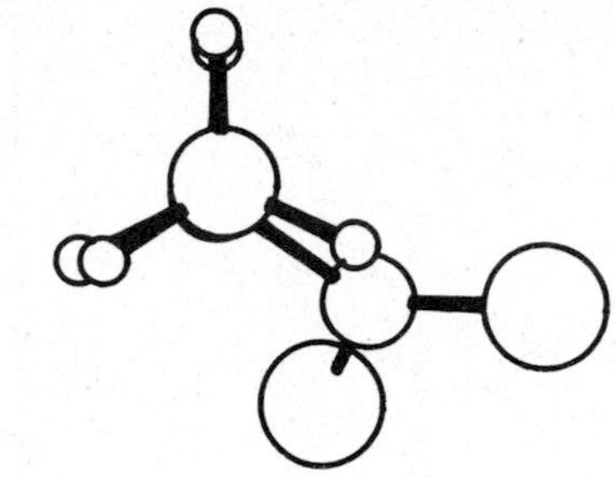

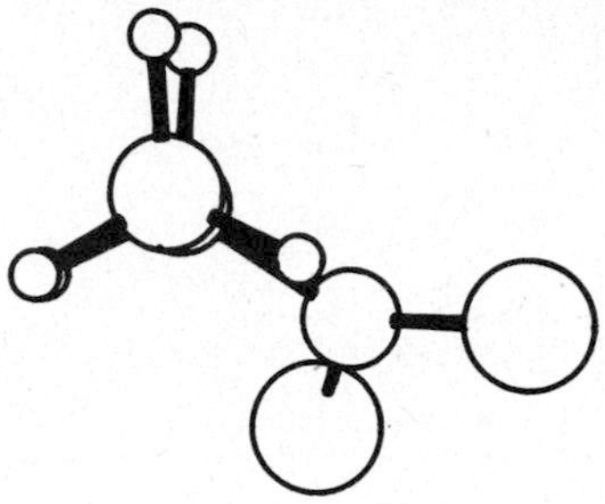

$$\begin{array}{c} CO_2^\ominus \\ | \\ H_3\overset{\oplus}{N}-C-H \\ | \\ H \end{array}$$

Serine
Ser
S

Mol wt 105

$$\begin{array}{c} CO_2^\ominus \\ | \\ H_3\overset{\oplus}{N}-C-H \\ | \\ CH_2 \\ | \\ OH \end{array}$$

$$\begin{array}{c} OH \\ | \\ -C\cdots H \\ | \\ H \end{array}$$

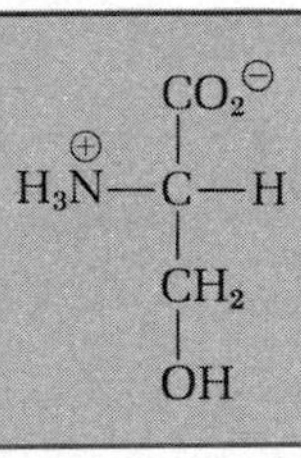

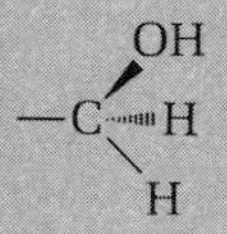

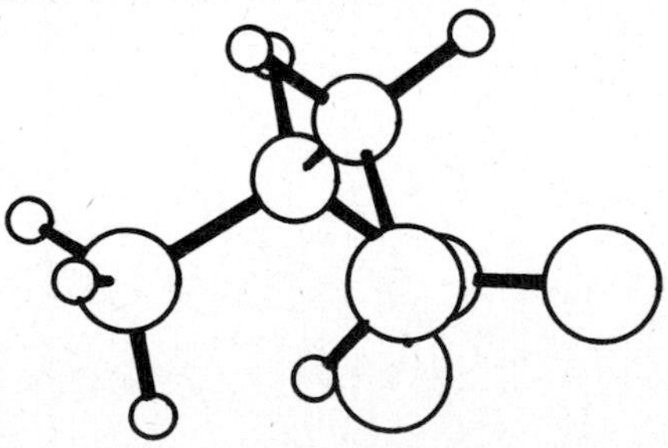

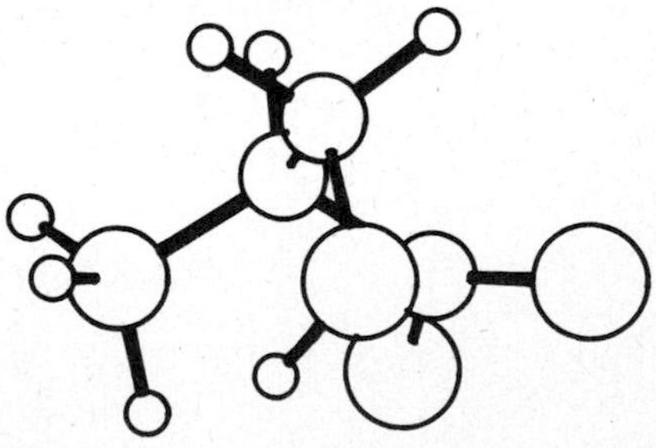

(Continued)

Table 2.1 Common amino acids found in proteins, showing the ionic forms predominant at pH 7 *(Continued)*

Amino acid	Planar representation	Side chain conformation
Threonine Thr T Mol wt 119 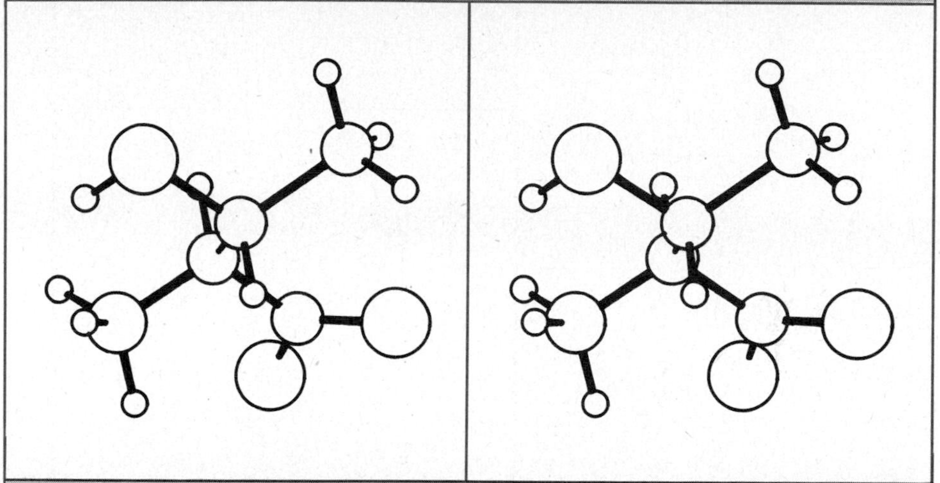	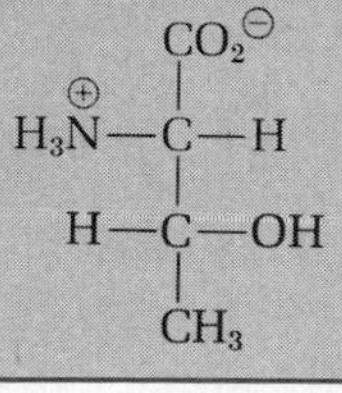 $CO_2^{\ominus}$ $H_3\overset{\oplus}{N}—C—H$ $H—C—OH$ CH_3	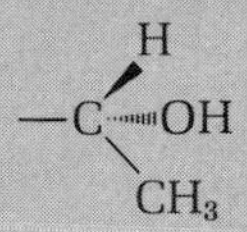 H $—C\cdots OH$ CH_3

| Cysteine
Cys
C

Mol wt 121 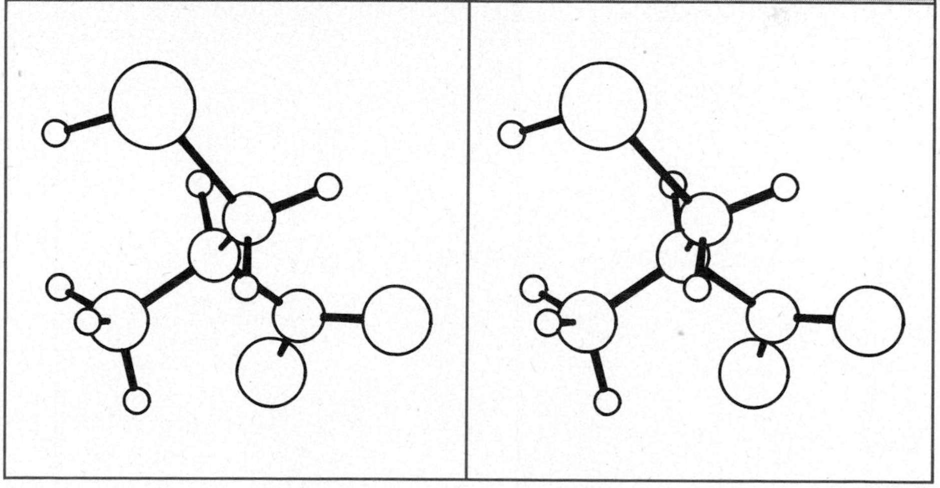 | 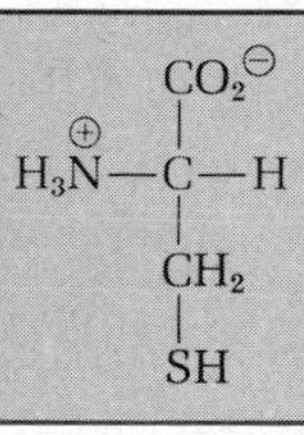 $CO_2^{\ominus}$
$H_3\overset{\oplus}{N}—C—H$
CH_2
SH | 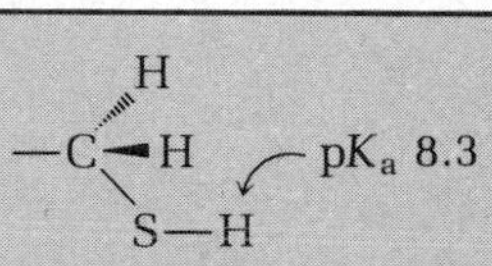 H
$—C—H$ pK_a 8.3
$S—H$ |

(Continued)

Table 2.1 Common amino acids found in proteins, showing the ionic forms predominant at pH 7 *(Continued)*

Amino acid	Planar representation	Side chain conformation
Tyrosine Tyr Y Mol wt 181 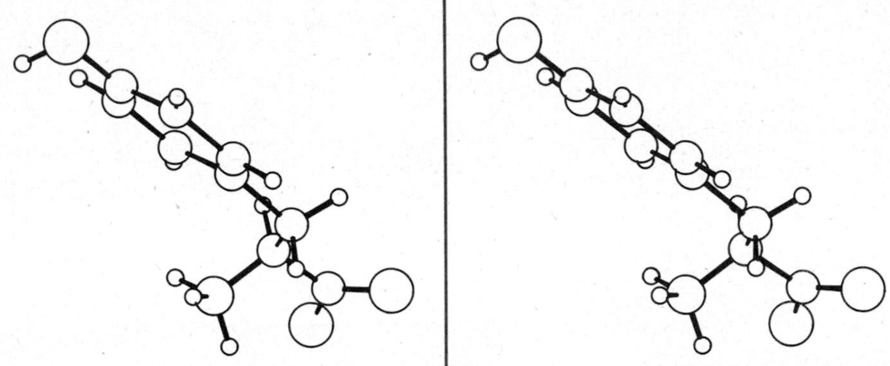	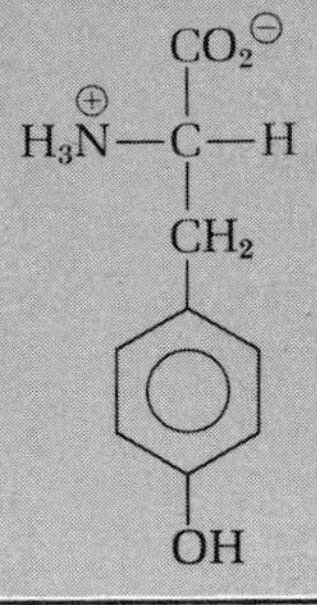 $CO_2^\ominus$ $\overset{\oplus}{H_3N}-C-H$ CH_2 (phenol ring with OH)	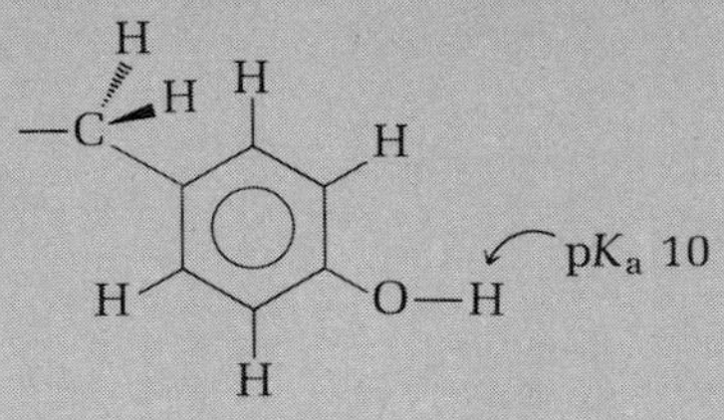 $-C$ (phenol ring) $O-H$ pK_a 10

| Asparagine Asn N Mol wt 132 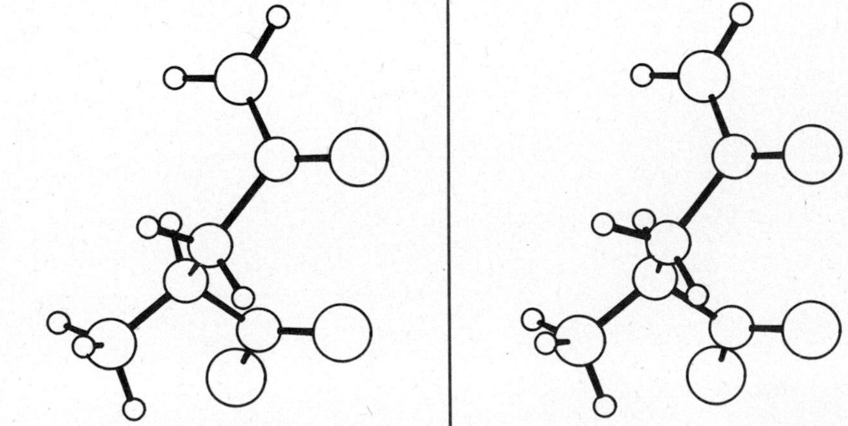 | 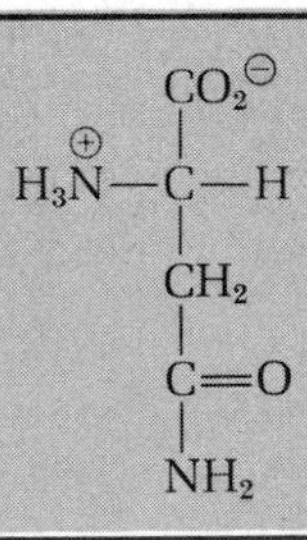 $CO_2^\ominus$ $\overset{\oplus}{H_3N}-C-H$ CH_2 $C=O$ NH_2 | 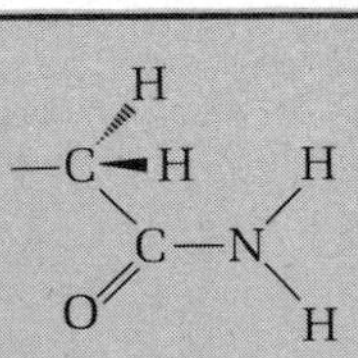 $-C$ $C-N$ O |

(Continued)

Table 2.1	Common amino acids found in proteins, showing the ionic forms predominant at pH 7 *(Continued)*

Amino acid	Planar representation	Side chain conformation
Glutamine Gln Q Mol wt 146 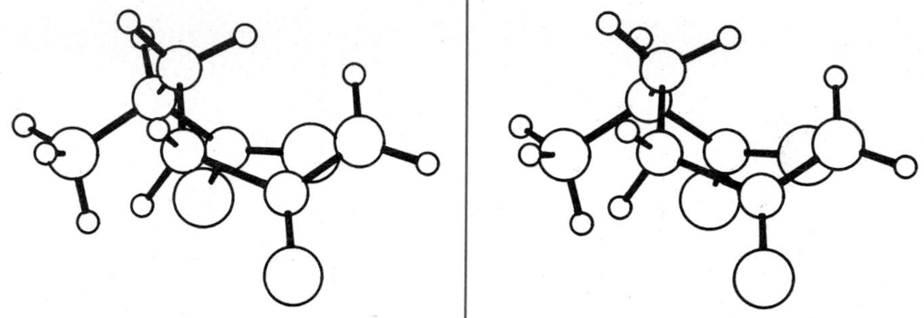	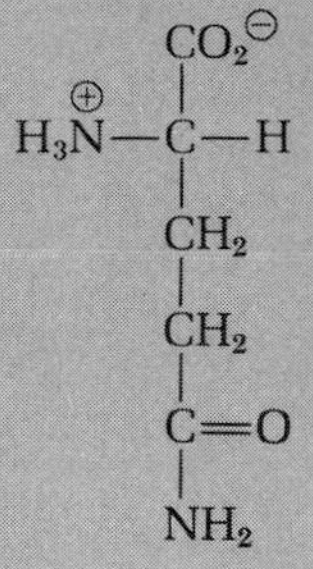 $CO_2^{\ominus}$ $H_3\overset{\oplus}{N}-C-H$ CH_2 CH_2 $C=O$ NH_2	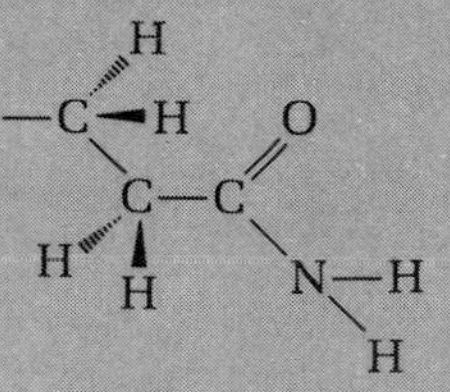

	Acidic side chains	
Aspartic acid Asp D Mol wt 133 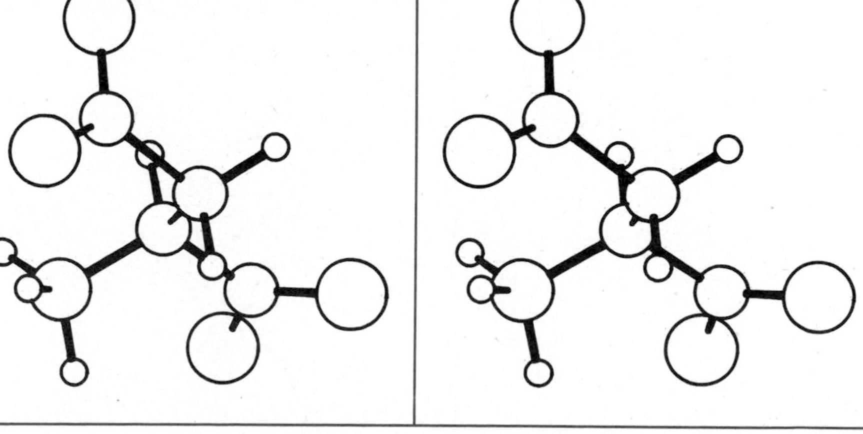	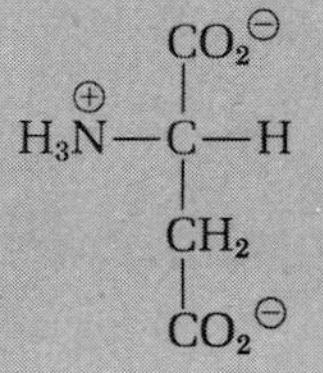 $CO_2^{\ominus}$ $H_3\overset{\oplus}{N}-C-H$ CH_2 $CO_2^{\ominus}$	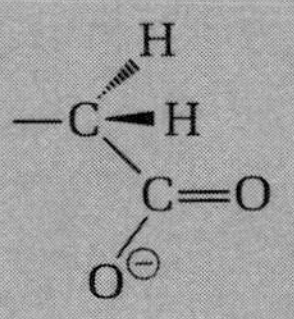

(Continued)

Table 2.1 **Common amino acids found in proteins, showing the ionic forms predominant at pH 7** (Continued)

Amino acid	Planar representation	Side chain conformation
Glutamic acid Glu E Mol wt 147	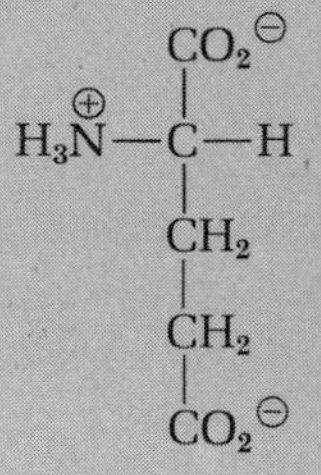 $CO_2^\ominus$ $H_3\overset{\oplus}{N}-C-H$ CH_2 CH_2 $CO_2^\ominus$	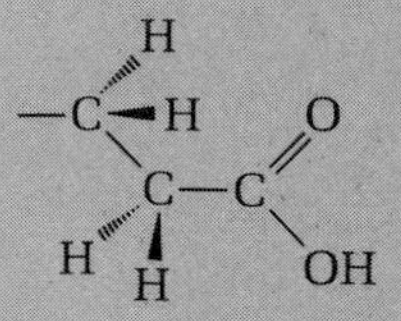
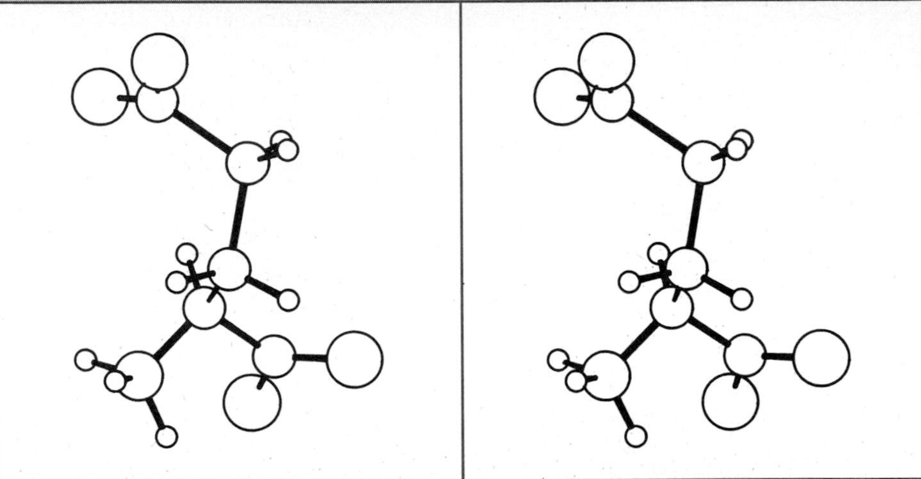		

Basic side chains

| Lysine
Lys
K

Mol wt 146 | 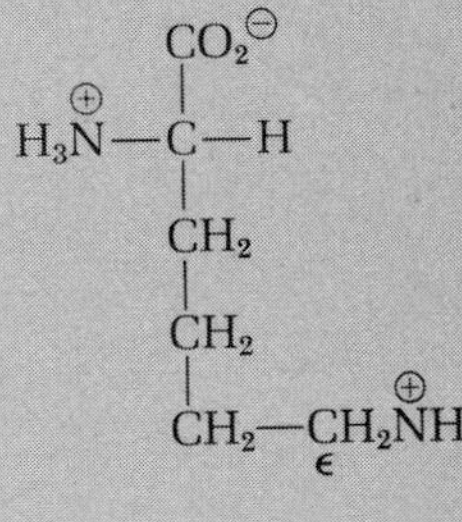 $CO_2^\ominus$
$H_3\overset{\oplus}{N}-C-H$
CH_2
CH_2
$CH_2-CH_2\overset{\oplus}{N}H_3$
ϵ | 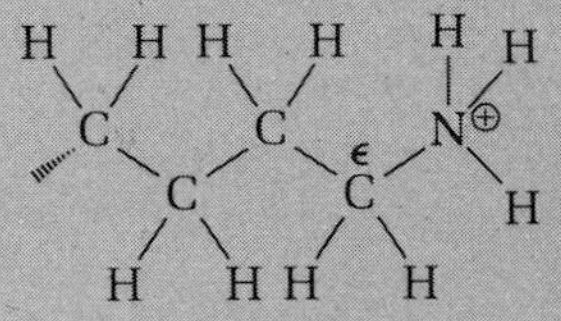 |
| 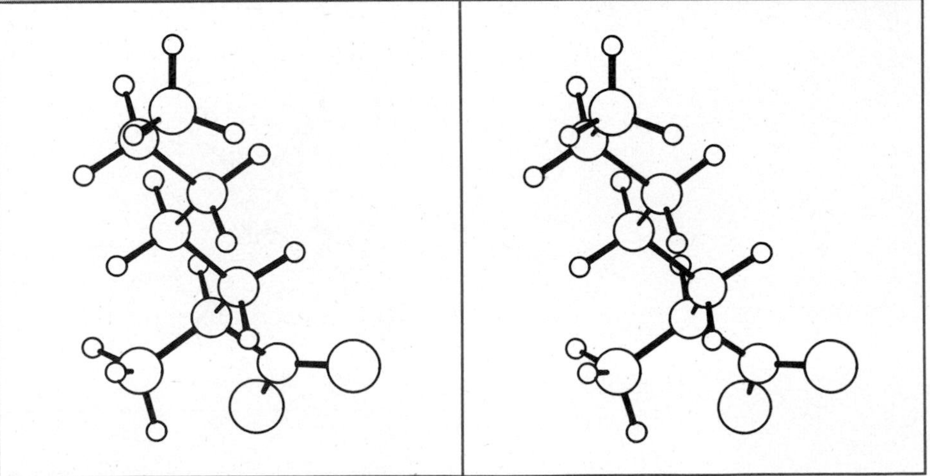 | | |

(Continued)

Table 2.1 Common amino acids found in proteins, showing the ionic forms predominant at pH 7 (Continued)

Amino acid	Planar representation	Side chain conformation
Arginine Arg R Mol wt 174	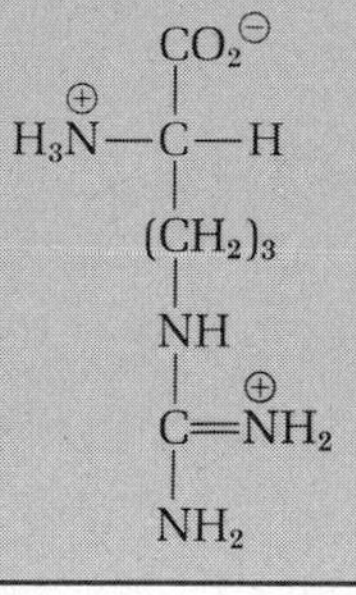	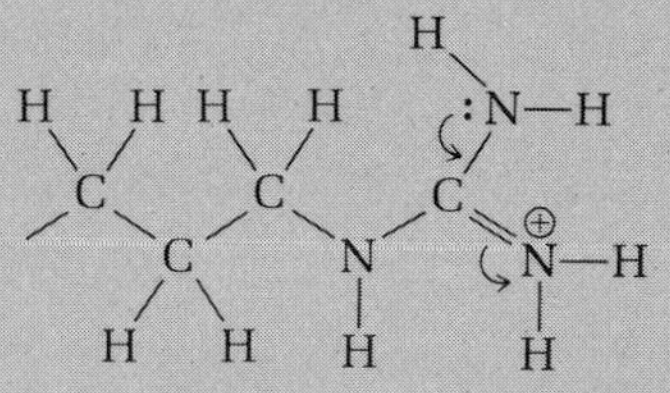
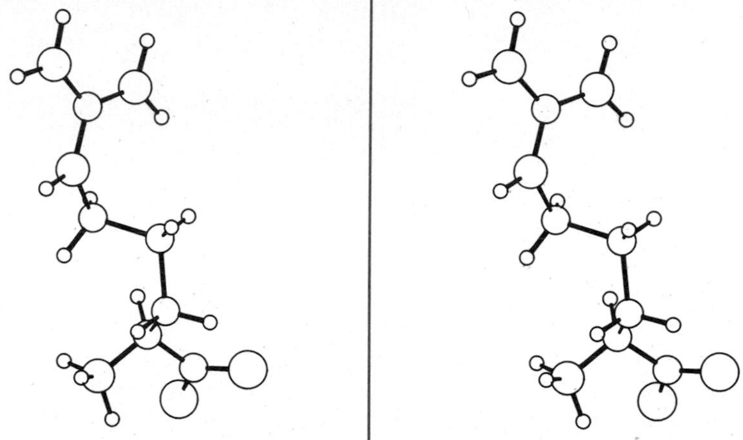		
Histidine His H Mol wt 155	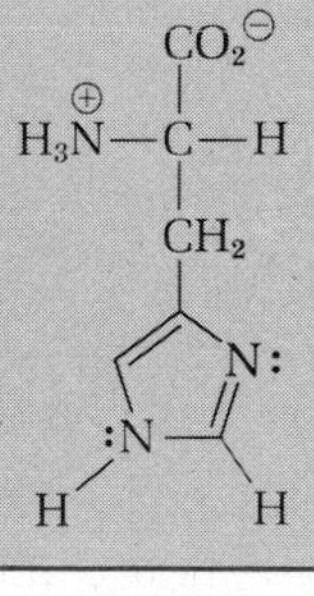	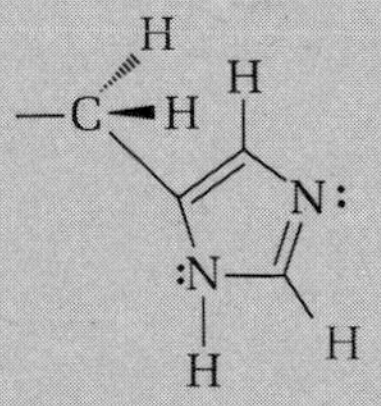
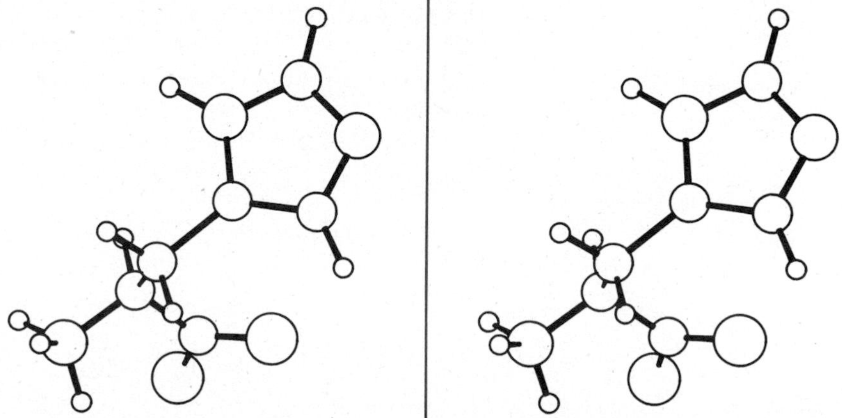		

Table 2.2   Names and abbreviations of the common amino acids

Amino acid	Three-letter abbreviation	One-letter abbreviation
Alanine	Ala	A
Arginine	Arg	R
Asparagine	Asn	N
Aspartic acid	Asp	D
Asparagine or aspartic acid[a]	Asx	B
Cysteine	Cys	C
Glutamine	Gln	Q
Glutamic acid	Glu	E
Glutamine or glutamic acid[a]	Glx	Z
Glycine	Gly	G
Histidine	His	H
Isoleucine	Ile	I
Leucine	Leu	L
Lysine	Lys	K
Methionine	Met	M
Phenylalanine	Phe	F
Proline	Pro	P
Serine	Ser	S
Threonine	Thr	T
Tryptophan	Trp	W
Tyrosine	Tyr	Y
Valine	Val	V

[a] Since amino acid analysis of proteins often involves acid hydrolysis, there is uncertainty about the identity of some aspartic and glutamic acids. Hydrolysis of the amide of glutamine and asparagine gives aspartate and glutamate in the hydrolysate.

implications for the conformation of proteins, as we shall discover in Chapter 4.

Alanine (Ala): R = —CH_3
Valine (Val): R = —$CH(CH_3)_2$
Leucine (Leu): R = —$CH_2CH(CH_3)_2$
Isoleucine (Ile): R = —CH—CH_2CH_3
                                    |
                                    CH_3

b. The Aromatic Amino Acids

Three nonpolar amino acids—phenylalanine, tyrosine, and tryptophan—contain aromatic side chains. All three possess absorption maxima in the ultraviolet region of the electromagnetic spectrum. Since all proteins contain at least one of these amino acid residues, the ultraviolet spectra of proteins provides an easy means of detecting proteins in various purification procedures (Chapter 5).

Phenylalanine (Phe):   R = —CH_2—⟨⟩ λ_{max} = 259 nm

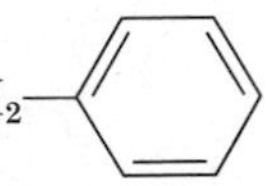

Tyrosine (Tyr):   R = —CH_2—⟨⟩—OH λ_{max} = 288 nm

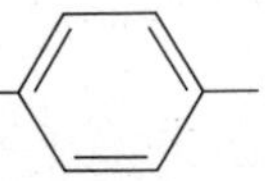

Tyrosine possesses a weakly acidic functional group in its aromatic side chain. The pK_a of the tyrosine hydroxyl group is 10.07. This amino acid is derived metabolically from phenylalanine (Section 22.8D), and it is a precursor of the "fight or flight" hormone, adrenaline (Section 22.8D). The side chain of tryptophan contains an extremely hydrophobic *indole* ring. The indole ring is structurally complex, and the metabolic pathways for the biosynthesis and the degradation of tryptophan involve many steps.

Tryptophan (Trp): $R = -CH_2$ 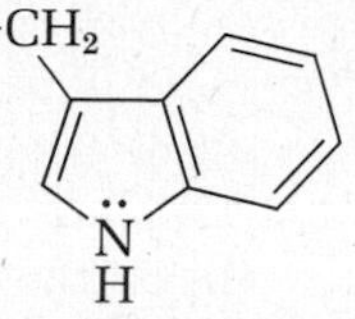 $\lambda_{max} = 279$ nm

c. Proline

Proline is an α-imino acid. The cyclic structure of the secondary amine confers conformational stability upon proline, and this property is reflected in the proteins containing it. Proline is an especially common residue at "hairpin turns" in globular proteins, and it is a major component of the structural protein collagen.

Proline (Pro):

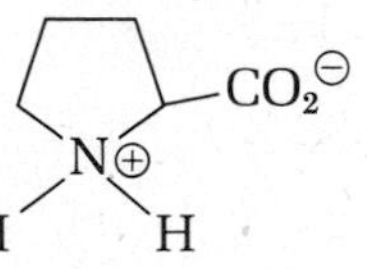

B. Acidic Amino Acids and Their Amides

a. Aspartic Acid and Asparagine

Aspartic acid is one of two acidic amino acids. The β-carboxyl group aspartic acid is ionized at pH 7, and the anion plays an important role in many enzyme-catalyzed reactions. The free acid can act as a proton donor, and the anion can act as a proton acceptor in acid- and base-catalyzed reactions, respectively.

Asparagine is the primary amide derived from aspartic acid. The amide is not very nucleophilic, and it is only weakly basic. Asparagine can, however, form hydrogen bonds to other molecules or functional groups.

Aspartic Acid (Asp): $R = -CH_2CO_2H$ $pK_a = 3.86$

$(\beta-CO_2H)$

Asparagine (Asn): $R = -CH_2\overset{\overset{\textstyle O}{\|}}{C}-\ddot{N}H_2$

b. Glutamic Acid and Glutamine

The γ-carboxyl group of glutamic acid can also act as a base as the carboxylate anion or as a proton donor as the free acid in enzyme-catalyzed reactions. Glutamic acid is also important in the metabolism of amino acids.

Glutamine is the primary amide derived from glutamic acid. The amide nitrogen of glutamine is a biosynthetic source of nitrogen for many metabolites.

Glutamine (Gln): $R = -CH_2-CH_2-\overset{\overset{\textstyle O}{\|}}{C}-\ddot{N}H_2$

Glutamic Acid (Glu):   R = —CH_2CH_2—CO_2H pK_a = 4.25

(γ-CO_2H)

C. Basic Amino Acids

a. Lysine

The basic side chain of lysine is a primary amine. This side chain is a strong base as the free amino group, and its conjugate acid, whose pK_a is 10.53, is a weak acid. The free base is a nucleophile. The "long arm" that bears the ϵ-amino group serves as a flexible attachment by which certain important cofactors are bound to enzymes.

Lysine (Lys):   R = —$(CH_2)_4$—$\overset{\oplus}{N}H_3$ pK_a = 10.53

(ϵ-$\ddot{N}H_2$) group

b. Arginine

Arginine is the most basic amino acid. Its side chain guanidine group is as basic as sodium hydroxide, and under physiological conditions this functional group is always protonated.

Arginine (Arg):   R = —$(CH_2)_3$—$\ddot{N}H$—C—$\ddot{N}H_2$ pK_a = 12.48

$\overset{\oplus}{N}H_2$

Guanidine moiety

c. Histidine

Histidine possesses an *imidazole* ring in its side chain. This unique heterocyclic, aromatic ring can act as a weak acid or as a weak base in enzyme-catalyzed reactions, and it is also a good nucleophile. The imidazole group of histidine is thus quite versatile in enzyme catalysis.

Histidine (His):   R = —CH_2 pK_a = 6.00

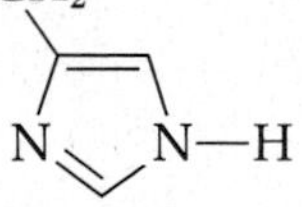

N⟍N—H

Imidazole

D. Neutral Amino Acids

a. Glycine

Glycine is the simplest amino acid, and its side chain hydrogen is involved in very little chemistry. Despite this simplicity, however, glycine is important. There are many places in proteins where the three-dimensional conformation will permit only the smallest of side chains, and glycine admirably fills the bill in this respect.

Glycine (Gly): R = —H

b. Serine

The side chain hydroxyl group of serine is an excellent nucleophile, and it is found at the active sites of many enzymes.

Serine (Ser): R = —CH_2OH

c. Threonine

The secondary hydroxyl group of threonine is polar. It is potentially a nucleophile, but threonine is not known to act as a nucleophile in any enzyme-catalyzed reactions.

Threonine (Thr): R = —CH—CH$_3$
$\qquad\qquad\qquad\quad$ |
$\qquad\qquad\qquad\quad$ OH

d. Cysteine

The sulfhydryl group of cysteine is a potent nucleophile and a weak acid. It and occasionally its conjugate base (—CH$_2$S$^{\ominus}$) participate in a variety of enzyme-catalyzed reactions.

Cysteine (Cys): R = —CH$_2$SH $\qquad$ pK_a = 8.33

The sulfhydryl group is also readily oxidized, and the oxidized dimer of cysteine, known as *cystine*, forms easily (reaction 2.1).

$$2 \; \overset{\oplus}{H_3N}\!-\!\underset{\underset{CH_2SH}{|}}{\overset{\overset{CO_2^{\ominus}}{|}}{C}}\!-\!H \longrightarrow \overset{\oplus}{H_3N}\!-\!\underset{\underset{CH_2-S-S-CH_2}{|}}{\overset{\overset{CO_2^{\ominus}}{|}}{C}}\!-\!H \qquad \overset{\oplus}{H_3N}\!-\!\overset{\overset{CO_2^{\ominus}}{|}}{C}\!-\!H + 2H \qquad (2.1)$$

$\qquad\qquad$ L-Cysteine $\qquad\qquad\qquad\qquad$ L-Cystine

Cystine itself may be oxidized by performic acid to give two moles of *cysteic acid* (reaction 2.2).

$$\overset{\oplus}{H_3N}\!-\!\underset{\underset{CH_2-S-S-CH_2}{|}}{\overset{\overset{CO_2^{\ominus}}{|}}{C}}\!-\!H \quad \overset{\oplus}{H_3N}\!-\!\overset{\overset{CO_2^{\ominus}}{|}}{C}\!-\!H \xrightarrow[\text{Performic acid}]{\text{HCO}_3\text{H}} 2\,\overset{\oplus}{H_3N}\!-\!\underset{\underset{CH_2-SO_3H}{|}}{\overset{\overset{CO_2^{\ominus}}{|}}{C}}\!-\!H \qquad (2.2)$$

$\qquad\qquad\qquad\qquad\qquad\qquad\qquad\qquad\qquad\qquad$ Cysteic acid

This important reaction permanently disrupts the disulfide bonds of proteins and is an important tool in determining the linear sequences of amino acid residues in proteins.

c. Methionine

Methionine provides the methyl group for many biological methylation reactions. To perform in this capacity, methionine first reacts with a molecule of adenosine triphosphate (ATP) to form the important biological methylating agent S-*adenosylmethionine* (Section 20.3B).

Methionine (Met): R = —CH$_2$CH$_2$—S—CH$_3$

2–4 IONIC PROPERTIES OF AMINO ACIDS

Carboxylic acids are weakly acidic, and amines are weakly basic. In water these functional groups are ionized. The ionization of the carboxyl group is written as

$$RCO_2H + H_2O \underset{}{\overset{K_{eq}}{\rightleftharpoons}} RCO_2^{\ominus} + H_3O^{\oplus} \qquad (2.2a)$$

The equilibrium constant for reaction 2.2a is

$$K_{eq} = \frac{[RCO_2^{\ominus}][H_3O^{\oplus}]}{[RCO_2H][H_2O]} \qquad (2.2b)$$

In dilute solutions the concentration of water is not much altered by the ionization of the weak acid, and we incorporate the constant, 55 M, concentration of water in K_{eq} to obtain the *acid ionization constant*, K_a.

$$K_a = K_{eq} [H_2O] = \frac{[RCO_2^{\ominus}][H_3O^{\oplus}]}{[RCO_2H]} \tag{2.3}$$

Equation 2.3 contains a term for the hydrogen ion concentration and a term for the acid ionization constant. The pH is defined as

$$pH = -\log [H_3O^{\oplus}] \tag{2.4}$$

By analogy, the definition of pK_a is

$$pK_a = -\log K_a \tag{2.5}$$

Using these definitions, we take the logarithm of equation 2.5 and rearrange to obtain the *Henderson-Hasselbalch equation* (equation 2.6).

$$pH = pK_a + \log \frac{[RCO_2^{\ominus}]}{[RCO_2H]} \tag{2.6}$$

The pK_a values of the α-carboxyl groups of amino acids range from 1.7 to 2.7 (Table 2.3). Let us take a "typical" pK_a of 2.0, and calculate the fraction of protonated carboxyl groups in an amino acid solution at pH 7.0. From equation 2.6 we find

$$7.0 = 2.0 + \log \frac{[RCO_2^{\ominus}]}{[RCO_2H]} \tag{2.7}$$

Therefore,

$$5 = \log \frac{[RCO_2^{\ominus}]}{[RCO_2H]} \tag{2.8}$$

Table 2.3 pK_a Values of the ionizable groups of amino acids			
		pK_a Values	
Amino acid	**α-CO$_2$H Group**	**α-NH$_3^{\oplus}$ Group**	**Side chain**
Glycine	2.35	9.78	
Alanine	2.35	9.87	
Valine	2.29	9.72	
Leucine	2.33	9.74	
Isoleucine	2.32	9.76	
Methionine	2.17	9.27	
Proline	1.95	10.64	
Phenylalanine	2.58	9.24	
Tryptophan	2.43	9.44	
Serine	2.19	9.44	~13
Threonine	2.09	9.10	~13
Cysteine	1.86	10.78	8.33
Tyrosine	2.20	9.11	10.11
Asparagine	2.02	8.80	
Glutamine	2.17	9.13	
Aspartic acid	1.99	10.00	3.90
Glutamic acid	2.13	9.95	4.32
Lysine	2.16	9.20	10.80
Arginine	1.82	8.99	13.20
Histidine	1.81	9.15	6.05

Taking the antilogarithm we obtain

$$10^5 = \frac{[\text{RCO}_2^{\ominus}]}{[\text{RCO}_2\text{H}]} \tag{2.9}$$

From this calculation we see that the α-carboxyl group of amino acids is virtually completely dissociated at pH 7. In general, if the pH of the solution is greater than the pK_a of the carboxylic acid, the carboxylate anion will be the predominant species in solution, other things being equal.

The fraction of α-amino groups protonated at a given pH can also be determined from the Henderson-Hasselbalch equation. Since we would like to minimize the number of parameters we have to consider, it is convenient to consider the ionization of the conjugate acid of the α-amino group, that is, we will consider this pK_a rather than the pK_b of the free amino group. The ionization is given by

$$\overset{\oplus}{\text{R}\text{N}}\text{H}_3 + \text{H}_2\text{O} \xrightleftharpoons{pK_a} \text{RNH}_2 + \text{H}_3\text{O}^{\oplus} \tag{2.10}$$

The pK_a's of the α-amino groups range from about 9.0 to 10.8. Taking a pK_a of 10, let us calculate the fraction of amino groups protonated at pH 7.0. From equation 2.10 we find that

$$\text{pH} = pK_{a\,\alpha\text{-NH}_3^{\oplus}} + \log \frac{[\text{RNH}_2]}{[\overset{\oplus}{\text{R}\text{N}}\text{H}_3]} \tag{2.11}$$

Therefore, with a pH of 7 and a $pK_{a\,\alpha\text{-}\overset{\oplus}{\text{N}}\text{H}_3}} = 10$ we find that

$$7.0 = 10.0 + \log \frac{[\text{RNH}_2]}{[\overset{\oplus}{\text{R}\text{N}}\text{H}_3]} \tag{2.12}$$

Rearranging and taking the antilogarithm, we obtain

$$10^{-3} = \frac{[\text{RNH}_2]}{[\overset{\oplus}{\text{R}\text{N}}\text{H}_3]} \tag{2.13}$$

At pH 7, therefore, the α-amino group mostly exists as the protonated ammonium ion rather than as the free base. Since the α-carboxyl group is also predominantly ionized at pH 7, the amino acids exist as dipolar ions.

$$\overset{\oplus}{\text{H}_3\text{N}}-\overset{\displaystyle \text{CO}_2^{\ominus}}{\underset{\displaystyle \text{R}}{\overset{|}{\text{C}}}}-\text{H}$$

Dipolar ion

The dipolar ion is the predominant species in solution over a pH range of approximately 4 to 8 for most amino acids that do not have ionizing side chains.

Three amino acids have ionizing side chains that are basic: lysine, histidine, and arginine. The pK_a of the ϵ-amino group of lysine is 10.58, typical of the pK_a values of primary amines.

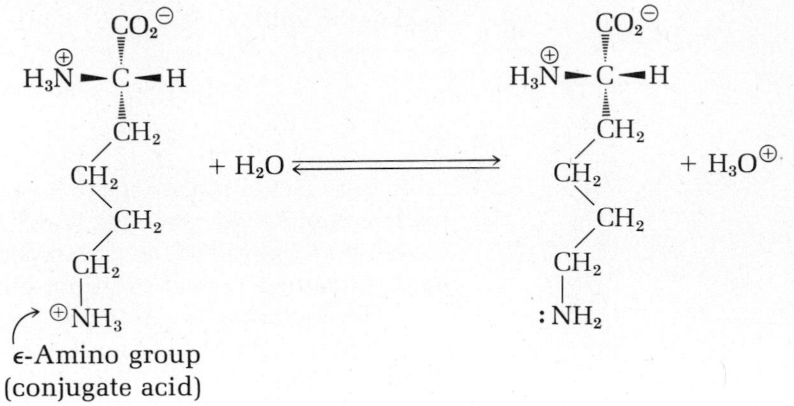

$$\epsilon\text{-Amino group (conjugate acid)} + H_2O \rightleftharpoons \cdots + H_3O^{\oplus} \quad (2.14)$$

The side chain of histidine contains an *imidazole* ring.

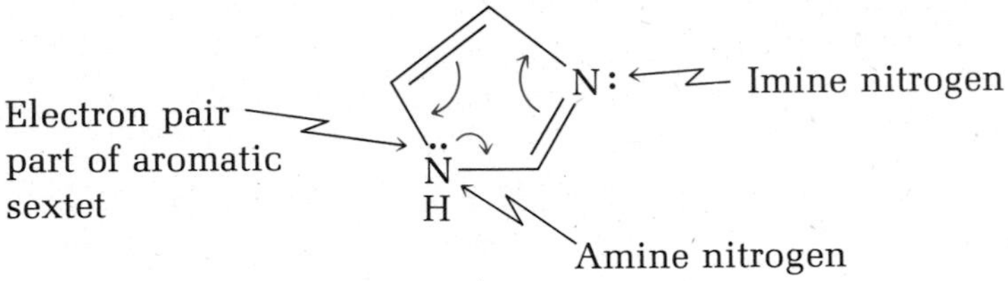

Electron pair part of aromatic sextet

Imine nitrogen

Amine nitrogen

Imidazole

The 4π electrons from the double bonds and the unshared electron pair on the amine nitrogen give a resonance-stabilized, 6π electron, aromatic system. To conserve the aromaticity of the 6π electron system, imidazole is protonated on the *imine* nitrogen. The conjugate acid is called an *imidazolium ion*; in histidine its pK_a is 6.0 (reaction 2.15).

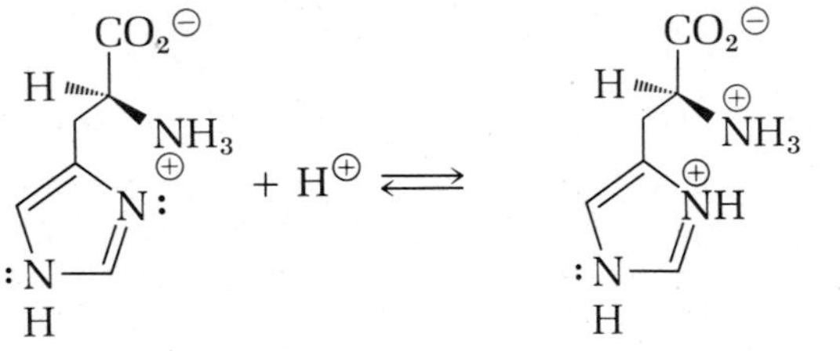

$$+ H^{\oplus} \rightleftharpoons \quad (2.15)$$

Imidazolium ion, pK_a 6.0

The side chain of arginine terminates in a strongly basic *guanidine* group.

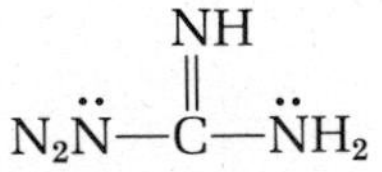

Guanidine 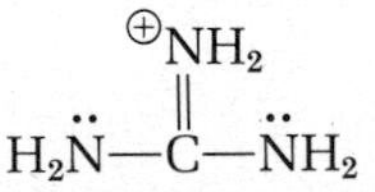   Guanidinium ion

Protonation of the *imine nitrogen* gives a resonance-stabilized *guanidinium ion*. The basic electronic structures of the imidazolium ion and guanidinium ion are identical.

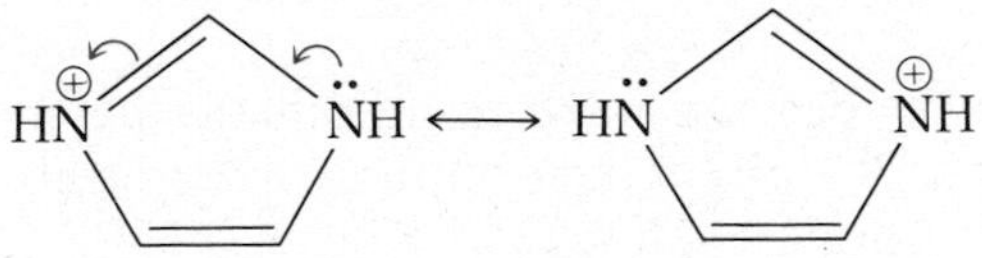

Imidazolium ion 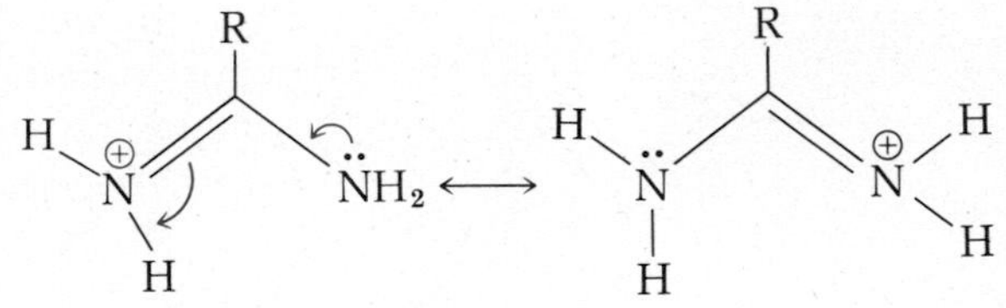   Guanidinium ion

The pK_a of the guanidinium ion of arginine is 12.48 (reaction 2.16).

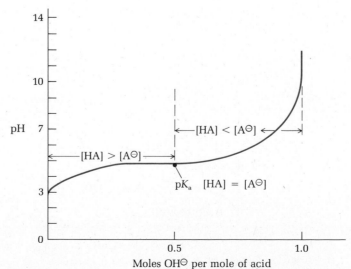

$$+ \; H^{\oplus} \; \rightleftharpoons \qquad\qquad + \; H^{\oplus} \qquad\qquad (2.16)$$

Guanidinium ion, pK_a 12.48

2–5 TITRATION OF AMINO ACIDS

The pK_a values of the ionizing groups of amino acids can be obtained by pH titration. In the titration of a weak acid a sodium hydroxide solution is added in aliquots, and the pH is recorded for each aliquot of base added. The titration curve is generated by plotting the pH of the solution versus the number of moles of titrant added (Figure 2.3). The chemical equation for titration of a weak acid with hydroxide is

$$HA + OH^{\ominus} \longrightarrow A^{\ominus} + H_2O \qquad\qquad (2.17)$$

Weak acid   Conjugate base

The reaction is virtually irreversible since the conjugate base, $A^{\ominus}$, is too weak to remove a proton from water. The *end point* of the titration is reached when 1 equivalent of acid (HA) has been converted to 1 mole of its conjugate base ($A^{\ominus}$) by 1 mole of titrant.

The region of the titration curve over which the pH is nearly constant is called the *buffer* region. The pH of a buffer solution is only slightly altered if small amounts of acid or base are added. Figure 2.3 also shows that a buffer solution is most effective when the pH is equal to the pK_a of the ionizing group in the buffer. The Henderson-Hasselbalch equation demonstrates the buffering action quantitatively.

Figure 2.3   Titration of a weak acid by hydroxide ion.

$$pH = pK_a + \log \frac{[A^{\ominus}]}{[HA]} \tag{2.6}$$

When the acid has been one-half titrated $[A^{\ominus}] = [HA]$ and $pH = pK_a$. If a small amount of base is added, $[A^{\ominus}]$ increases slightly and $[HA]$ decreases slightly. The ratio $[A^{\ominus}]/[HA]$ changes only a little, remains nearly 1.0, and $pH = pK_a$ as before. Therefore, the solution is *buffered* against changes in pH.

Consider the titration of glycine. There are two ionizing groups in glycine, and the titration curve has two distinct segments, one for titration of the carboxyl group and one for titration of the amino group (Figure 2.4). The first segment of the curve is obtained by titration of the carboxyl group.

$$\overset{\oplus}{H_3N}-CH_2-CO_2H \rightleftharpoons \overset{\oplus}{H_3N}-CH_2CO_2{}^{\ominus} + H^{\oplus} \tag{2.18}$$
$$\qquad\qquad 1 \qquad\qquad\qquad\qquad 2$$

Addition of titrant results in the formation of the dipolar ion (2) and water.

$$\overset{\oplus}{H_3N}-CH_2-CO_2H + HO^{\ominus} \longrightarrow \overset{\oplus}{H_3N}-CH_2CO_2{}^{\ominus} + H_2O \tag{2.19}$$
$$\qquad\qquad 1 \qquad\qquad\qquad\qquad\qquad 2$$

The end point of the first segment of the curve is reached upon addition of one equivalent of hydroxide ion. At the inflection point the concentrations of 1 and 2 are equal, and $pH = pK_{a_{\alpha\text{-}CO_2H}}$. Titration of the $\alpha\text{-}\overset{\oplus}{N}H_3$ group gives the next segment of the curve.

$$\overset{\oplus}{H_3N}-CH_2-CO_2{}^{\ominus} + HO^{\ominus} \longrightarrow H_2N-CH_2-CO_2{}^{\ominus} + H_2O \tag{2.20}$$
$$\qquad\qquad 2 \qquad\qquad\qquad\qquad\qquad 3$$

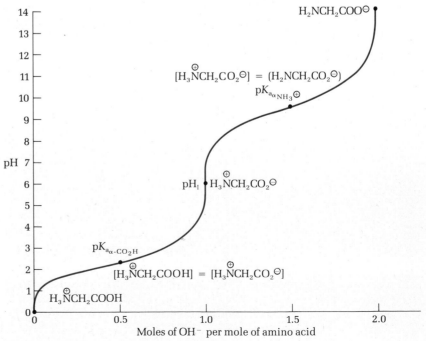

Figure 2.4   Titration curve for glycine.

At the inflection point the concentrations of *2* and *3* are equal, and pH $= \text{p}K_{a_{\alpha\text{-}\overset{\oplus}{N}H_3}}$.

The titration curve of glycine, by yielding the $\text{p}K_a$ values of the ionizing groups, permits us to determine another parameter, the isoelectric pH, designated pH_I, at which the population of amino acids bears no *net* charge. For glycine, and all other amino acids that do not have ionizing side chains, the isoelectric species is the dipolar ion (*2*). At pH_I positively charged glycine molecules (*1*) are balanced by an equal concentration of negatively charged glycine molecules (*3*). We can see from Figure 2.4 that the pH_I falls at a pH midway between the equivalence points for titration of the α-carboxyl and α-amino groups. Therefore,

$$\text{pH}_I = \tfrac{1}{2}(\text{p}K_{a_{\alpha\text{-}CO_2H}} + \text{p}K_{a_{\alpha\text{-}\overset{\oplus}{N}H_3}}) \tag{2.21}$$

Equation 2.21 can be derived by considering the ionization of glycine, or any other amino acid lacking an ionizing side chain, by treating the amino acid as a dibasic acid. The equilibrium constants for ionization of the α-carboxyl and α-amino groups are given by equations 2.22 and 2.23.

$$K_{a_{\alpha\text{-}CO_2H}} = \frac{[H_3\overset{\oplus}{N}CH_2CO_2^{\ominus}][H^{\oplus}]}{[H_3\overset{\oplus}{N}CH_2CO_2H]} \tag{2.22}$$

$$K_{a_{\alpha\text{-}\overset{\oplus}{N}H_3}} = \frac{[H_2NCH_2CO_2^{\ominus}][H^{\oplus}]}{[H_3\overset{\oplus}{N}CH_2CO_2^{\ominus}]} \tag{2.23}$$

Solving equation 2.22 for $[H_3\overset{\oplus}{N}CH_2CO_2^{\ominus}]$ gives

$$[H_3\overset{\oplus}{N}CH_2CO_2^{\ominus}] = \frac{K_{a_{\alpha\text{-}CO_2H}}[H_3\overset{\oplus}{N}CH_2CO_2H]}{[H^+]} \tag{2.24}$$

Substituting equation 2.24 in equation 2.23 and solving for $[H^{\oplus}]$ gives

$$[H^{\oplus}]^2 = K_{a_{\alpha\text{-}CO_2H}} K_{a_{\alpha\text{-}\overset{\oplus}{N}H_3}} \frac{[H_3\overset{\oplus}{N}CH_2CO_2H]}{[H_2NCH_2CO_2^{\ominus}]} \tag{2.25}$$

At pH_1, $[H_3\overset{\oplus}{N}CH_2CO_2H] = [H_2NCH_2CO_2^{\ominus}]$. Therefore,

$$[H^{\oplus}]^2 = K_{a_{\alpha\text{-}CO_2H}} K_{a_{\alpha\text{-}\overset{\oplus}{N}H_3}} \tag{2.26}$$

Taking the negative logarithm of equation 2.26 and setting pH $= \text{pH}_I$, we obtain

$$\text{pH}_I = \tfrac{1}{2}(\text{p}K_{a_{\alpha\text{-}CO_2H}} + \text{p}K_{a_{\alpha\text{-}\overset{\oplus}{N}H_3}}) \tag{2.27}$$

Equation 2.27 is, of course, identical to equation 2.21, the result we obtained by considering the titration curve itself.

In the titration curve of glycine (Figure 2.4), the $\text{p}K_a$'s of the ionizing groups are widely separated and the two segments of the titration curve do not overlap. When the $\text{p}K_a$ values of the ionizing groups are close, the two curves do overlap. This behavior is seen in the titration of aspartic acid and glutamic acid (Figure 2.5). The isoelectric species of aspartic acid contains one protonated and one unprotonated carboxyl group and the protonated α-amino group.

$$
\overset{CO_2^{\ominus}}{\underset{\overset{|}{CO_2H}}{\underset{|}{\underset{|}{CH_2}}{\overset{\oplus}{H_3N}{-}C{-}H}}}
$$

Isoelectric aspartate

The isoelectric point for aspartic acid lies midway between the pK_a's for the two carboxyl groups. In general the isoelectric point for any amino acid can be calculated by taking the average of the pK_a's that bracket the isoelectric species.

Histidine has three ionizing groups whose ionizations are separated enough to yield three distinct, nonoverlapping titration curves (Figure 2.6). The inflection point at pH 1.82 corresponds to the pK_a of the carboxyl group (reaction 2.28).

$$
\underset{\text{pK 1.82}}{\rightleftharpoons} \qquad (2.28)
$$

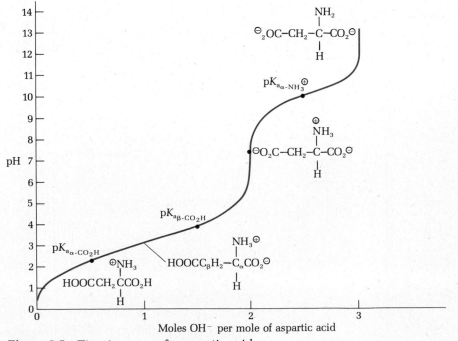

Figure 2.5   Titration curve for aspartic acid.

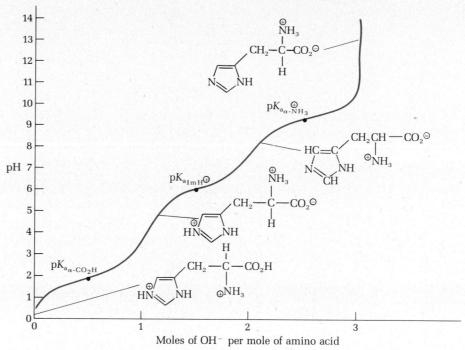

Figure 2.6 Titration curve for histidine.

The inflection point at pH 6.0 corresponds to the pK_a of the imidazolium ion (reaction 2.29).

$$(2.29)$$

The final segment of the titration curve has its inflection point at pH 9.17, corresponding to the pK_a of the amino group (reaction 2.30).

$$(2.30)$$

The isoelectric species of histidine contains a carboxylate anion, a protonated α-amino group, and a neutral imidazole side chain.

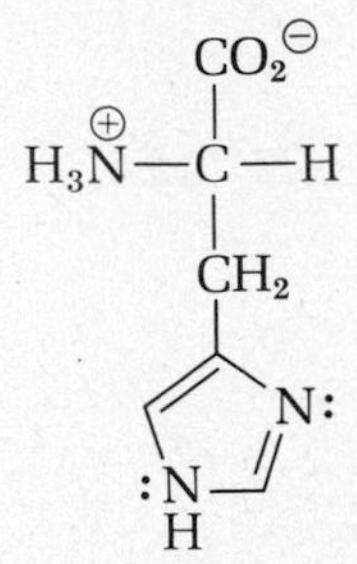

Isoelectric species of histidine

The ionizatons bracketing this species are those of the imidazolium ion and the α-amino group. The isoelectric pH is therefore given by

$$pH_{I_{his}} = \tfrac{1}{2}(pK_a ImH^{\oplus} + pK_{a_{\alpha\text{-}\overset{\oplus}{N}H_3}}) \tag{2.31}$$

2–6 PEPTIDES
A. Nomenclature

Peptides are oligomers of amino acids formed by condensing the α-amino group of one amino acid with the α-carboxyl group of another.

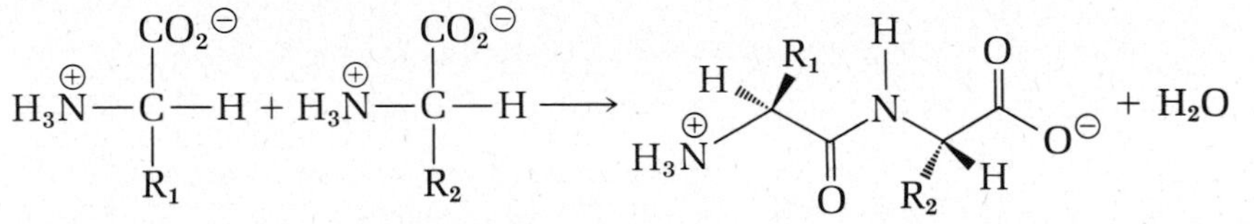

$$\tag{2.33}$$

Dipeptide

The first amino acid residue in the peptide, reading from the left, is the N-terminal residue; the last amino acid in the peptide is the C-terminal residue.

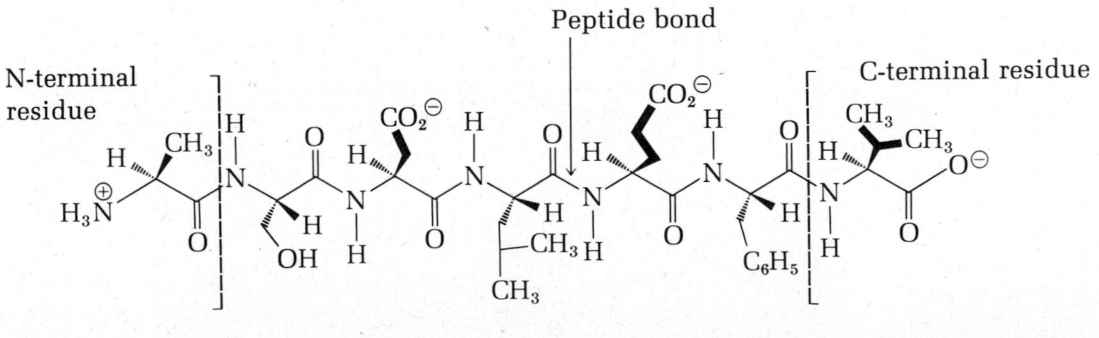

Prefixes such as di-, tri-, and tetra- indicate peptides composed of two, three, and four amino acid residues, respectively. Peptides are named as acyl derivatives of their C-terminal amino acid residue. Some examples are given in Table 2.4.

B. Structure
of the
Peptide Bond

The properties of the peptide bond are largely determined by resonance interaction between the lone pair on the amide nitrogen and the adjacent carbonyl group. In one contributing structure the C—N bond is a single bond with no overlap between the nitrogen's lone pair and the carbonyl carbon (structure 1). In the second contributing structure there is a double bond between the amide nitrogen and the carbonyl carbon. In this contributing structure the nitrogen atom bears a charge of +1 and the carbonyl oxygen bears a charge of −1 (structure 2).

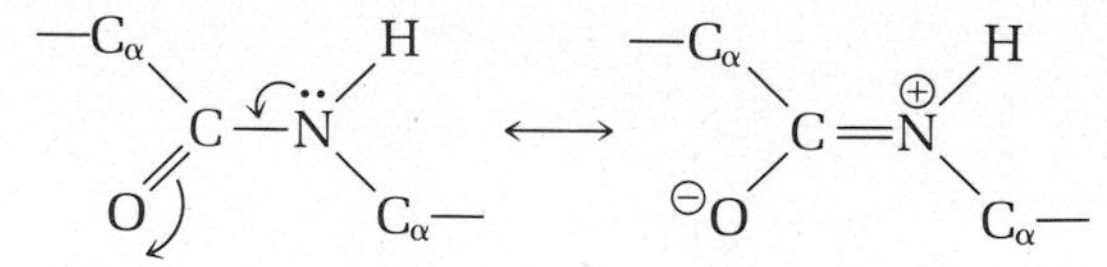

In structure 1 the carbonyl carbon is sp^2 hybridized and is therefore planar, while the amide nitrogen is sp^3 hybridized and is pyramidal. In structure 2 both the carbonyl carbon and the amide nitrogen are sp^2 hybridized, both are planar, and all six atoms lie in the same plane.

The actual structure of the peptide bond is a resonance hybrid (structure 3). As this structure suggests, the peptide bond is a compromise between structures 1 and 2. The bond lengths and bond angle for the common *trans* configuration of the peptide bond are indicated in Figure 2.7a.

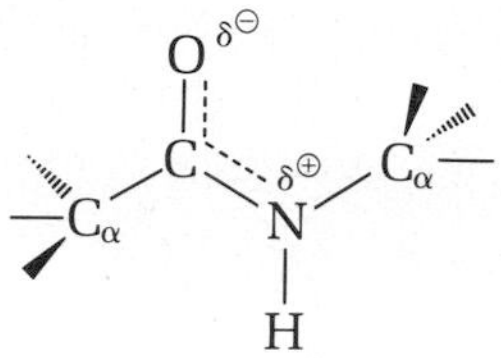

Table 2.4 Some representative peptides and their names

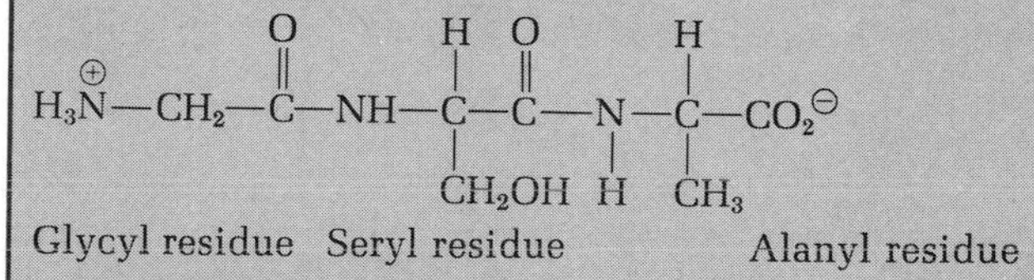

Glycyl residue   Seryl residue   Alanyl residue

Glycylserylalanine

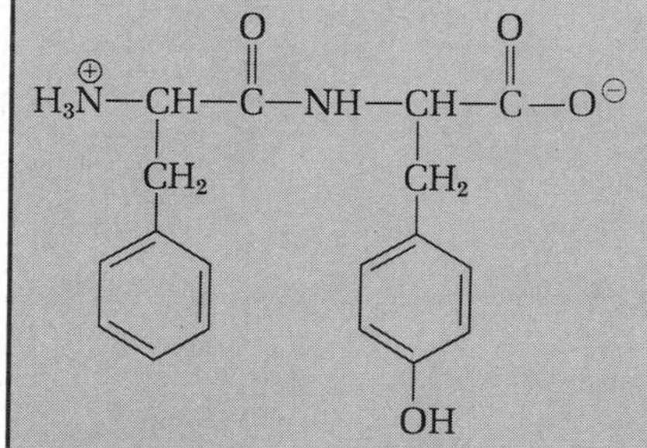

Phenylalanyltyrosine

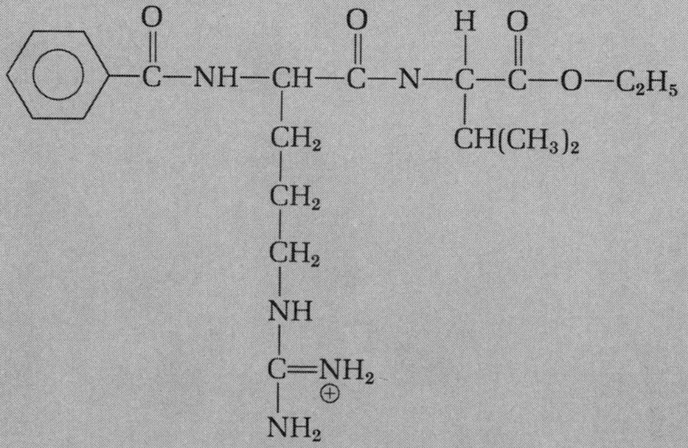

N-Benzoylarginylvaline ethyl ester

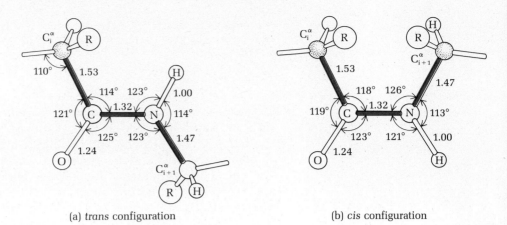

(a) *trans* configuration   (b) *cis* configuration

Trans configuration
of peptide bond

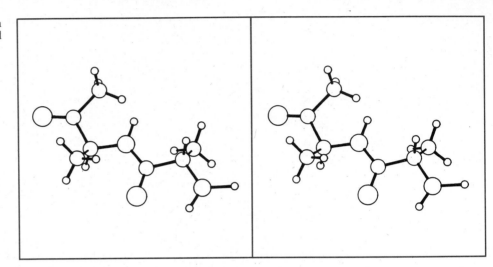

Cis configuration
of peptide bond

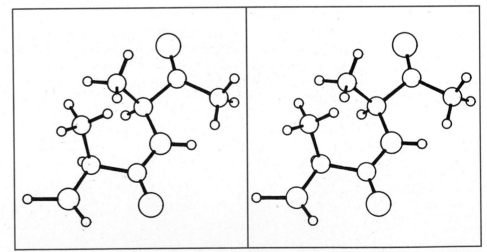

Figure 2.7 (a) Structure and dimensions (in Å) of the peptide bond in its normal *trans* configuration. [From L. Pauling, R. B. Corey, and H. R. Branson, *Proc. Nat. Acad. Sci. U.S.*, *37*, 205–211 (1951).] (b) Structure and dimensions of the peptide bond in the rare *cis* configuration. [From G. N. N. Ramachandran and V. Sasisekharan, *Adv. Prot. Chem.*, *23*, 283–437 (1968).]

As predicted by contributing structure 2 the C—N bond is shortened. In an amine the bond length of the C—N bond is 0.1487 nm versus 0.1325 for the peptide bond and 0.127 nm for a C=N bond. The resonance hybrid 3 has about 50% double bond character. Because of the double bond character of the C—N bond, *the peptide bond is planar*. Another consequence of the partial double bond character of the peptide bond is a relatively high *barrier to rotation* of about 75 kJ mol^{-1}. The energy required for a rotation of θ degrees is given by equation 2.34.

$$E = E_0 \sin^2 \theta \tag{2.34}$$

where E_0 is the barrier to rotation. For peptides this barrier is high enough to effectively prevent rotation at room temperature. The α-carbons can lie either *cis* or *trans* to the peptide bond as shown below. A *trans* configuration places the α-carbons and their substituents far from each other; while the *cis* configuration leads to steric hindrance between R groups on the α-carbons. The *trans* geometry is more stable by about 8 kJ mol^{-1}.

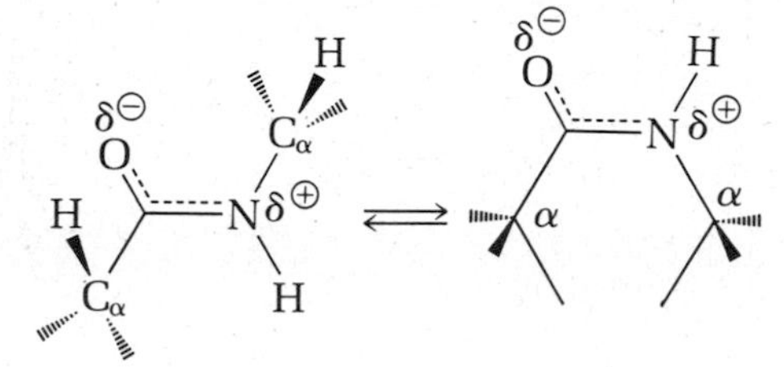

$$\tag{2.35}$$

Trans configuration   *Cis* configuration

The *trans* configuration is therefore favored in peptide chains.

There is one important exception to this generalization. Peptide bonds involving proline may be either *cis* or *trans* because the steric constraints of the tetrahydropyrrole ring eliminate the advantages of a *trans* configuration (equation 2.36, page 62).

C. Acid-Base Properties of Polypeptides

In free amino acids we need concern ourselves only with the ionization of the α-carboxyl group, the α-amino group, and a single ionizing side chain. The acid-base properties of polypeptides, on the other hand, are potentially vastly more complicated. We need to consider the ionization of the N-terminal amino group and the C-terminal carboxyl group and also the ionizations of the several side chains. As the number of ionizing groups increases, the titration curve rapidly becomes too complex to analyze in terms of the ionizations of individual side chains since several may ionize in the same pH range. The Henderson-Hasselbalch equation does, however, permit an easy determination of the predominant ionic state of each side chain at a given pH. We need only apply the rule that when the pH of the solution is greater than the pK_a of the ionizing side chain, the predominant species is the conjugate base of the side chain, and when the pH of the solution is less than the pK_a of the ionizing side chain, the predominant species is the acid, that is,

$$\text{pH} > \text{p}K_a \longrightarrow [\text{HA}] < [\text{A}^{\ominus}]$$
$$\text{pH} < \text{p}K_a \longrightarrow [\text{HA}] > [\text{A}^{\ominus}]$$

Consider, for example, the ionization of glycylglutamyllysylalanine (Figure 2.8). When the pH is less than 3.0, the predominant species in solution has a net charge of $+2$. We obtain this result by consulting Table 2.6 to find the pK_a of each ionizing functional group and then applying the Henderson-Hasselbalch equation to each functional group. Between pH 3.0 and 4.5 the predominant species has a net charge of $+1$ since the C-terminal carboxyl group is mostly ionized. Between pH 4.5 and 7.8 the predominant species has a net charge of zero. Between pH 7.8 and 10.2 the predominant

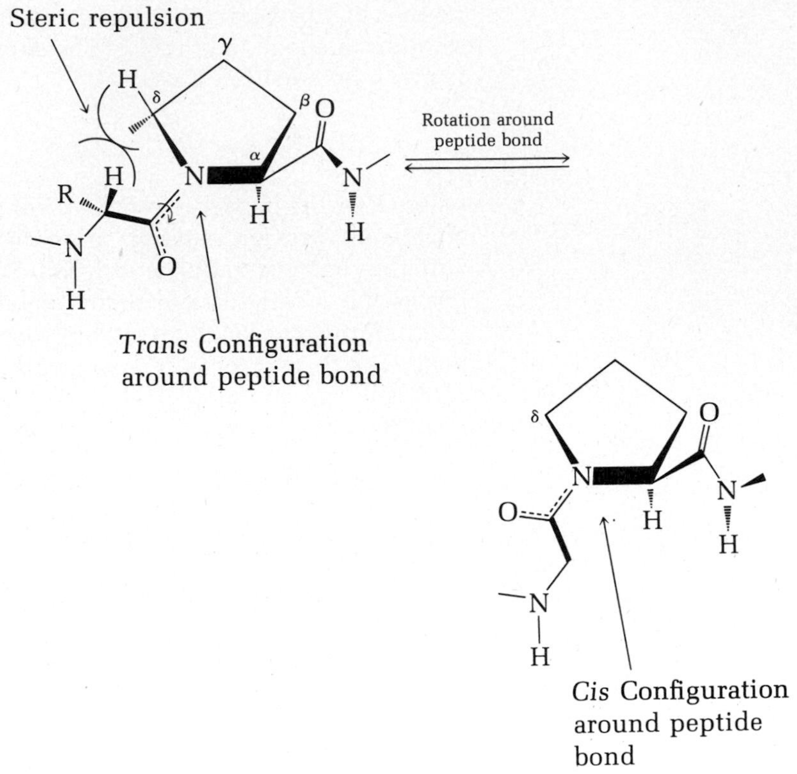

Steric repulsion

Rotation around
peptide bond

Trans Configuration
around peptide bond

(2.36)

Cis Configuration
around peptide
bond

Trans, trans configuration
of peptide bonds involving
proline

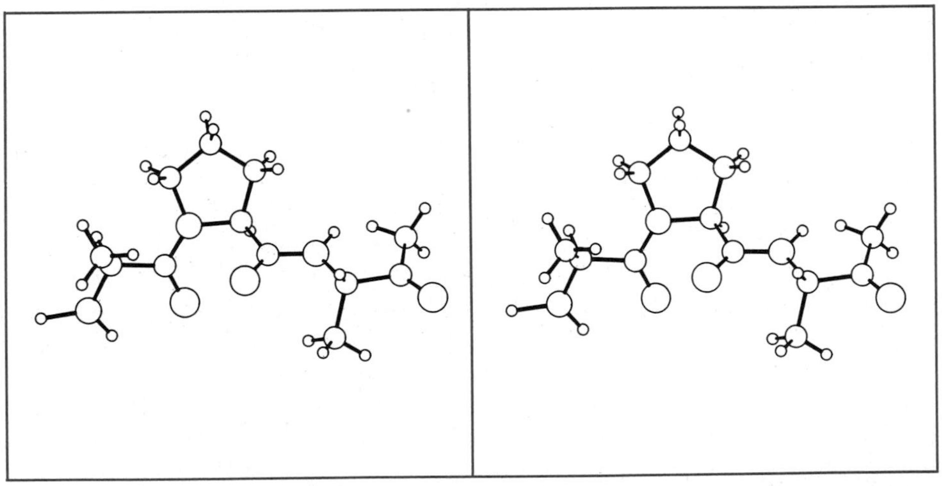

Cis, trans configuration
of peptide bonds involving
proline

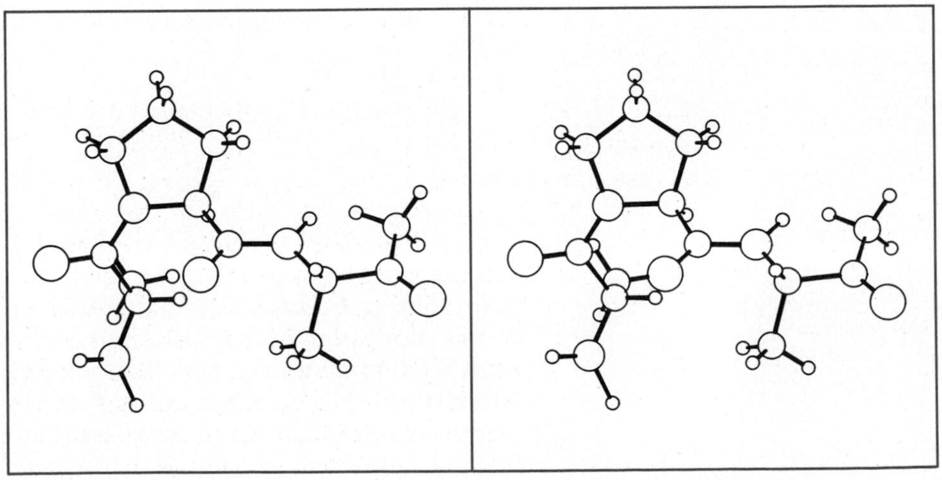

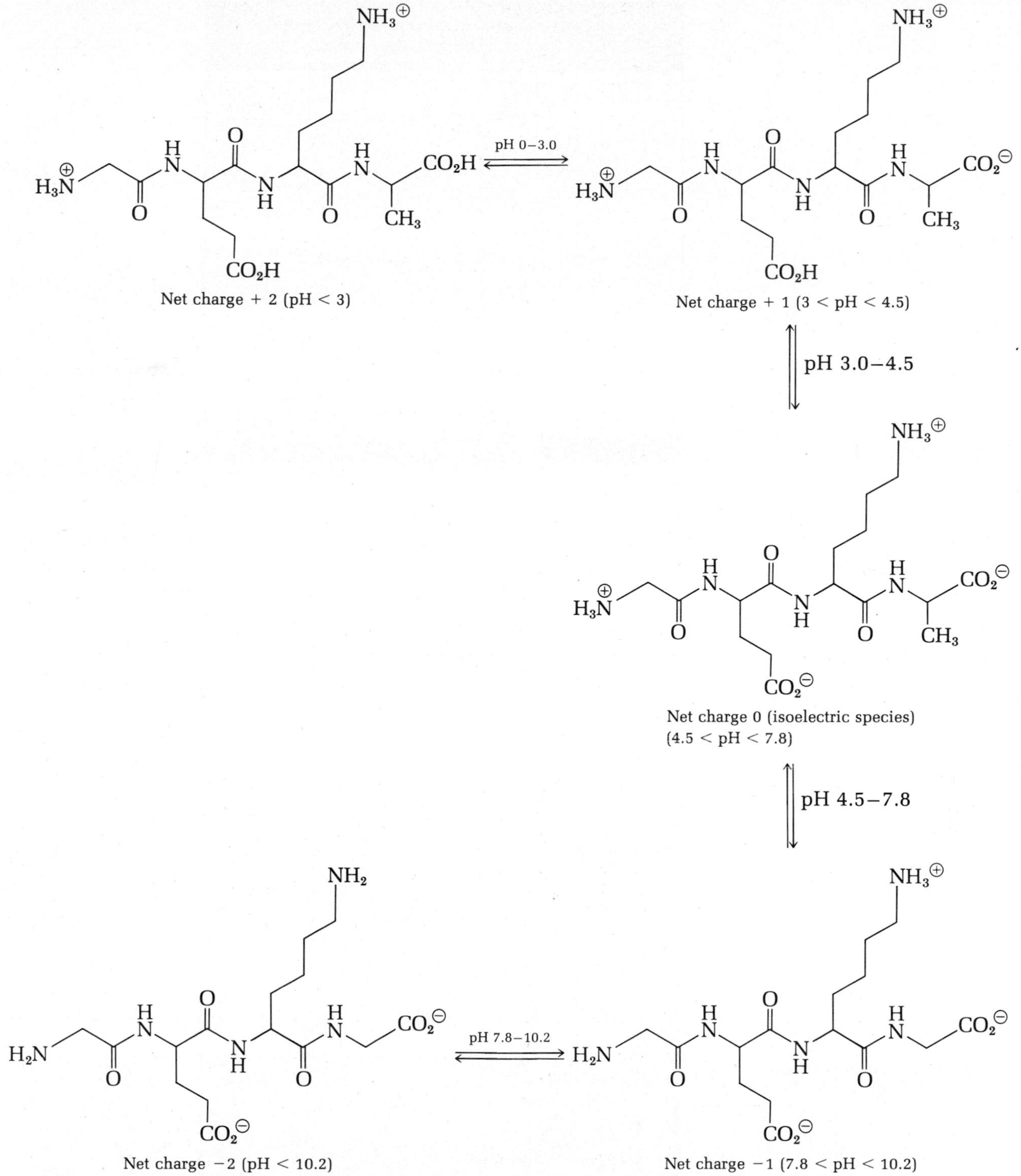

Figure 2.8 Acid-base behavior of the tetrapeptide glycylglutamyllysylalanine.

species has a net charge of -1. At pH values greater than 10.2 the peptide has a net charge of -2.

Table 2.5 summarizes the pK_a values expected for side chains in proteins based upon the pK_a values found in model compounds such as glycylglutamyllysylalanine, and Table 2.6 lists the data obtained for titration of the ionizing side chains in several proteins. It will be seen that the pK_a values of most of the ionizing side chains fall into the range predicted by the model compounds. The more interesting cases are the anomalies. We shall see later

Table 2.5   Expected pk_a values for ionizing side chains in proteins based upon data obtained from model compounds

Ionizing side chain	Expected pK_a
α-CO_2H	3.75
Side chain CO_2H (Glu or Asp)	4.6
Imidazole (His)	7.0
α-NH_2	7.8
—SH (Cys)	8.8
Tyrosyl—OH	9.6
ϵ-NH_2 (Lys)	10.2
Guanidyl (Arg)	>12

Table 2.6   Acidities of functional groups in polypeptides

Type of group	pK
α-Carboxyl	
Insulin	3.6
Side-chain carboxyl	
Serum albumin	4.0
Ovalbumin	4.3
Conalbumin	4.4
Corticotropin	4.6
Insulin	4.7
β-Lactoglobulin (49 of 51 groups)	4.8
β-Lactoglobulin (2 of 51 groups)	7.3
Imidazole	
Ribonuclease	6.5
Myoglobin (6 of 12 groups)	6.6
Chymotrypsinogen	6.7
Ovalbumin	6.7
Conalbumin	6.8
Lysozyme	6.8
Serum albumin	6.9
β-Lactoglobulin	7.4
Phenolic	
Conalbumin (11 of 18 groups)	9.4
Insulin	9.6
Chymotrypsinogen (1 of 4 groups)	9.7
Corticotropin	9.8
Papain (11 of 17 groups)	9.8
Ribonuclease (3 of 6 groups)	9.9
Serum albumin	10.4
Chymotrypsinogen (1 of 4 groups)	10.6
Lysozyme	10.8
Side chain amino	
Serum albumin	9.8
β-Lactoglobulin	9.9
α-Corticotropin	10.0
Ovalbumin	10.1
Ribonuclease	10.2
Lysozyme	10.4
Guanidyl	
Insulin	11.9
α-Corticotropin	~12

Source: Adapted from R. H. Haschenmeyer and A. E. V. Haschenmeyer, "Proteins: A Guide to Study by Physical and Chemical Methods," Wiley, New York, 1973, pp. 266–267.

that the three-dimensional conformation of a protein can fling an ionizing side chain into an environment quite unlike the aqueous solution. In this altered environment the pK_a value of the side chain can shift dramatically. In many cases the ionizing side chains of amino acid residues at the active sites of enzymes have much altered pK_a values, and this can be an important factor in catalysis. These aspects of protein structure and function are discussed in Chapters 4 and 5.

2-7 DETERMINATION OF THE AMINO ACID COMPOSITION OF PROTEINS

The first step in the determination of the structure of a protein is quantitative analysis of its amino acid composition, and the first step in the quantitative analysis is hydrolysis of the protein. This may seem a simple matter. The peptide bond is simply a secondary amide, and amides can be hydrolyzed by refluxing in acidic solutions. However, it is not quite that easy. Because no hydrolytic procedure releases all of the amino acids quantitatively, several samples of the protein must be hydrolyzed under different sets of conditions. Heating the polypeptide in 6 N HCl for 24 to 70 hours at 105 to 110° hydrolyzes all the peptide bonds of the protein. In the process tryptophan is totally destroyed, asparagine and glutamine are converted to their respective carboxylic acids, and cysteine is converted to *cystine* (recall reaction 2.1). Serine, threonine, and tyrosine are also partially destroyed by acid hydrolysis. The degree to which these amino acids are destroyed varies with the incubation time. Therefore, their content in a protein can be reliably estimated by determining their content in samples incubated for varying times and then extrapolating back to zero time (Figure 2.9). Tryptophan can be analyzed by hydrolyzing the protein in 2 to 4 M NaOH. This method is extremely destructive. It causes extensive racemization of the amino acids released by hydrolysis, partially destroys arginine, and wreaks havoc upon serine, threonine, cysteine, and cystine. The amide bonds of glutamine and asparagine are also hydrolyzed in base. Despite this list of drawbacks, base hydrolysis releases tryptophan quantitatively, and it is useful for that reason. Cysteine can be converted to cysteic acid by preliminary oxidation with performic acid followed by acid hydrolysis.

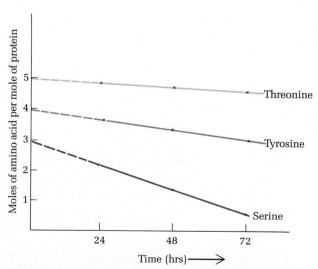

Figure 2.9 Determination of the actual threonine, tyrosine, and serine content of a protein by incubation of samples for varying times and extrapolation to zero time.

Once the protein hydrolysate has been obtained, it is analyzed by *ion exchange chromatography*. Automated instruments, known as amino acid analyzers, combine a sophisticated system for ion exchange chromatography (Section 5.4C) with a system for detecting the amino acids as they emerge from the chromatography column. The amino acids emerging from the column are mixed with *ninhydrin,* which reacts with amino acids to give an adduct that has an intense blue color whose absorption maximum is at 570 nm (reaction 2.37).

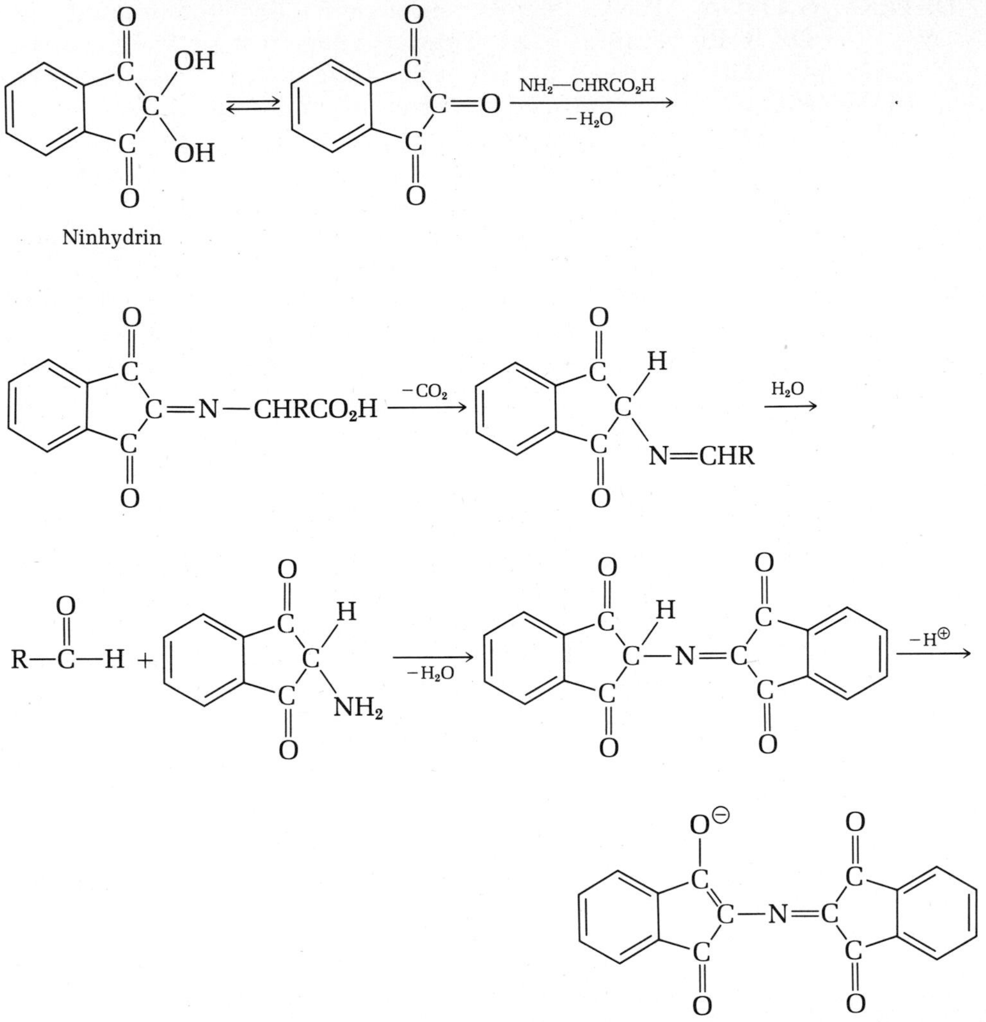

(2.37)

Blue, λ_{max} 570

The color of the adduct is developed by passing the mixture of eluent and ninhydrin through a heated chamber then passing the eluent through a continuously recording spectrophotometer. The chromatogram obtained for a 50 nmol (10^{-9} mol) hydrolysate is shown in Figure 2.10.

The amino acid compositions of many proteins have been determined, and a representative sample is given in Table 2.7. Several conclusions can be reached from these data. (1) Not all amino acids are present in all proteins. Insulin, for example, contains neither tryptophan nor methionine, whereas the calf thymus histone IV protein lacks tryptophan and cysteine. (2) There is not a statistical distribution of amino acids in any given protein. Alanine, for example, is far more abundant than either tryptophan or histidine in nearly all proteins. (3) Somewhere between 30 and 40% of the amino acids in proteins have nonpolar side chains. These side chains might look uninteresting because they are virtually inert in the normal course of events, but

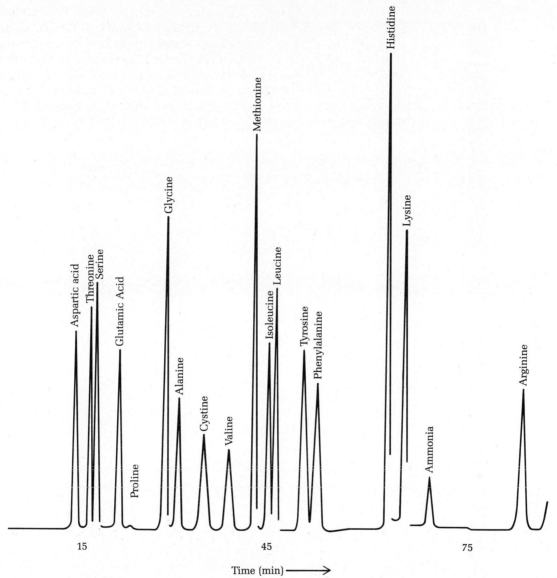

Figure 2.10 Analysis of a peptide hydrolysate by an amino acid analyzer. [Courtesy of Beckman Instrument Company.]

they play a fundamental role in protein structure as we shall see in future chapters. (4) The number of charged side chains varies widely from protein to protein. For example, in calf thymus histone IV, there are a large number of positive residues. By contrast, in spinach ferredoxin, aspartate and glutamate outnumber the combined total of arginines, lysines, and histidines by four to one.

2–8 DETERMINATION OF THE N- AND C-TERMINAL AMINO ACIDS IN PROTEINS

A. Determination of the N-terminal Amino Acid

The N-terminal amino acid of a protein can be identified by any of several techniques, some of which lead into determination of the other residues as well. It is often useful to identify the N-terminal amino acid. For example, may proteins are composed of more than one polypeptide chain, and determination of the N-terminal amino acid can reveal if there is more than one type of polypeptide chain in the protein. If the molecular weight of the protein is known, then the number of moles of reagent required to react with the N-terminus can reveal the number of identical chains that are present in the protein. (There may, of course, be only one chain, but this, too, is revealed by analysis of the N-terminal amino acid.) Three reagents—the Sanger reagent,

Table 2.7 Amino acid composition of a few proteins

Amino acid	Number of residues per molecule of protein								
	Bence-Jones κ (human)	Cytochrome c (human)	Ferredoxin (spinach)	Glucagon (pig)	Insulin (bovine)	Hemoglobin, α (human, gorilla)	Hemoglobin, β (human, gorilla)	Lysozyme (chicken)	Wool (sheep)
Nonpolar									
Ala	12	6	9	1	3	27	15	12	27
Val	13	3	7	1	4	13	18	6	14
Leu	15	6	8	2	6	18	18	8	38
Ile	9	8	4	0	1	0	0	6	12
Pro	11	4	4	0	1	7	7	2	2
Met	1	3	0	1	0	2	1	2	2
Phe	10	3	2	2	3	7	8	3	4
Trp	2	1	1	1	0	1	2	6	0
Cys	5	2	5	0	6	1	2	8	15
Polar, uncharged									
Gly	13	13	6	1	4	7	13	12	14
Ser	28	2	7	4	3	11	5	10	27
Thr	19	7	8	3	1	9	7	7	15
Tyr	9	5	4	2	4	3	3	3	8
Asn	10	5	2	1	0	4	6	13	17
Gln	16	2	4	3	0	1	3	3	25
Polar, negatively charged									
Asp	10	3	11	3	3	8	7	8	17
Glu	10	8	9	0	7	4	8	2	26
Polar, positively charged									
Lys	13	18	4	1	1	11	11	6	18
Arg	6	2	1	2	9	3	3	11	31
His	2	3	1	1	2	10	9	1	2
Total residues	214	104	97	29	54	141	146	129	312

dansyl chloride, and the Edman reagent—are widely used for identification of the N-terminal amino acid, and one of them (the Edman reagent) can be used to obtain long sequences of residues.

a. The Sanger Reagent

The Sanger reagent, *2,4-dinitrofluorobenzene,* reacts with the N-terminal amino acid of peptides to give a 2,4-dinitrophenyl derivative of the peptide (reaction 2.38) in a nucleophilic aromatic substitution reaction.

$$(2.38)$$

2,4-Dinitrophenyl derivative of peptide

Acid hydrolysis of the adduct yields the free amino acids plus the N-terminal amino acid as the 2,4-dinitrophenyl derivative, a bright yellow crystalline solid that is easily isolated and identified by thin layer chromatography (Section 5.4A). Nucleophilic side chains of residues such as lysine also react with the Sanger reagent, but these reactions do not prevent N-terminal analysis since they can easily be separated from the dinitrophenyl derivative.

b. Dansyl Chloride

Treatment of a polypeptide with *dansyl chloride* (1-dimethylamino-naphthalene-5-sulfonyl chloride) gives an N-dansyl peptide derivative. Hydrolysis of the peptide adduct releases the N-dansyl derivative for identification by chromatography (reaction 2.39). N-dansyl amino acids are highly fluorescent, and as little as 1 nmol can be detected. Since it can be used with very small samples, and since side reactions are less of a problem with it than with the Sanger reagent, dansyl chloride is the preferred method for identification of the N-terminal amino acid.

c. The Edman Reagent

When a polypeptide is treated with the Edman reagent, *phenylisothiocyanate,* the free N-terminal amino group reacts to form a *phenylthiocarbamyl* derivative of the polypeptide (reaction 2.40).

The phenylthiocarbamyl adduct can be released *without* hydrolyzing the polypeptide, and the Edman reagent can be used to determine the sequence of many consecutive residues in the polypeptide. Once the adduct has formed, the reaction mixture is neutralized, the product is isolated, and it is heated in an acidic, anhydrous solvent. (If water were present, the entire polypeptide would hydrolyze.) At 100° a cyclization reaction occurs, and the N-terminal amino acid is released as a *phenylthiohydantoin* derivative whose R group is characteristic of the amino acid (reaction 2.41).

We reiterate that the great advantage of the Edman reagent is that the phenylthiohydantoin of the N-terminal residue can be released without hydrolyzing the polypeptide chain, so that the entire procedure can be repeated serially to obtain the sequence of many residues of the protein. Since the

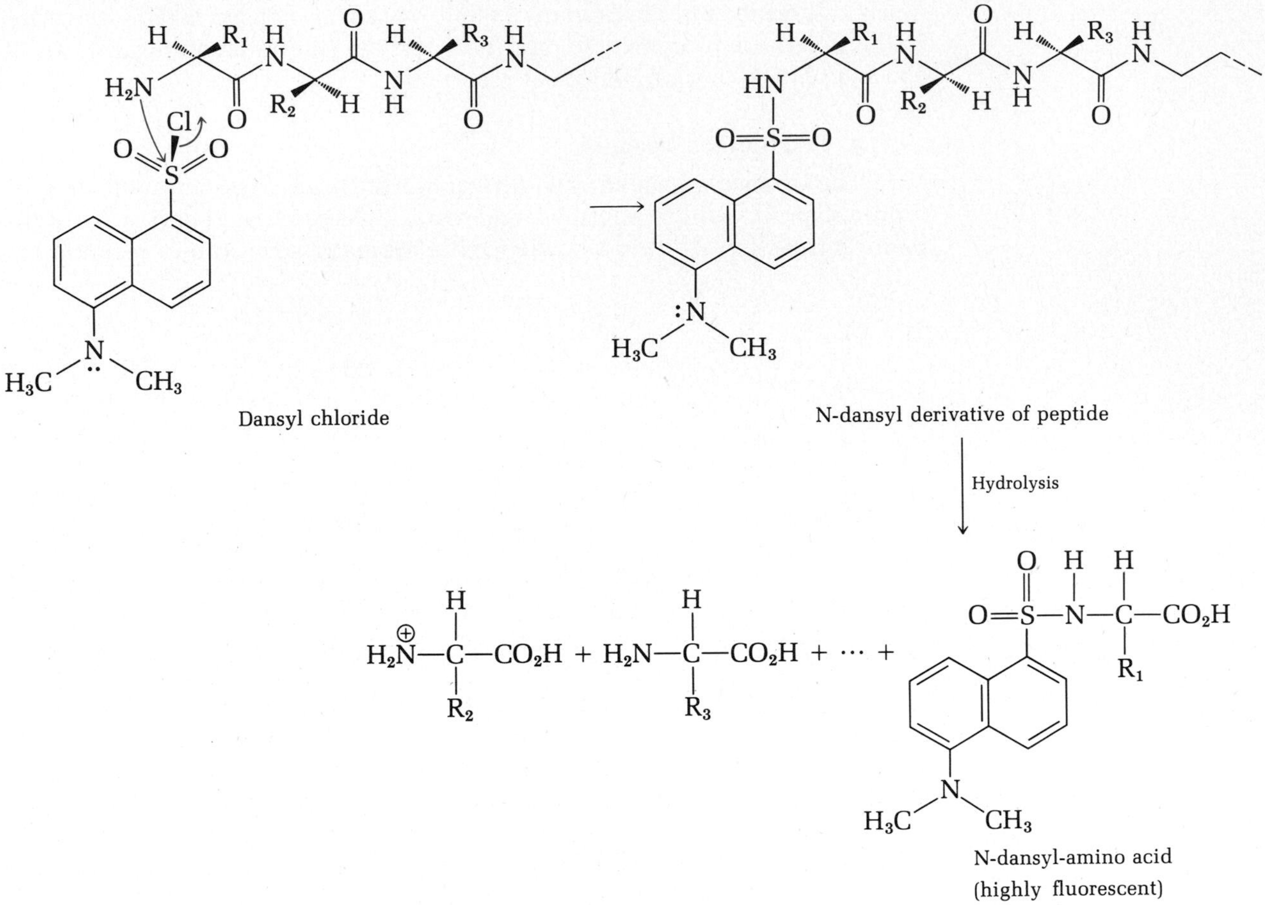

Dansyl chloride

N-dansyl derivative of peptide

Hydrolysis

$$H_2\overset{\oplus}{N}-\underset{R_2}{\overset{H}{\underset{|}{\overset{|}{C}}}}-CO_2H + H_2N-\underset{R_3}{\overset{H}{\underset{|}{\overset{|}{C}}}}-CO_2H + \cdots + \quad\quad\quad (2.39)$$

N-dansyl-amino acid
(highly fluorescent)

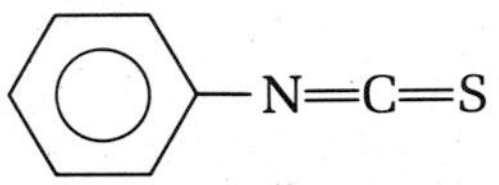

Phenylisothiocyanate

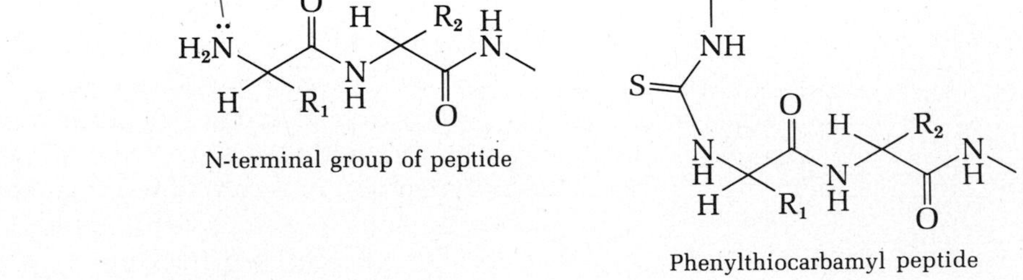

$$\quad\quad\quad (2.40)$$

N-terminal group of peptide

Phenylthiocarbamyl peptide

yield of each step approaches 100%, many residues can be determined by Edman degradation with a relatively small sample. An experienced experimenter can determine a sequence of 20 to 30 residues from a few milligrams of sample. Proteins are much larger than this, however, and other techniques, discussed below, are required to determine the sequence of an entire protein.

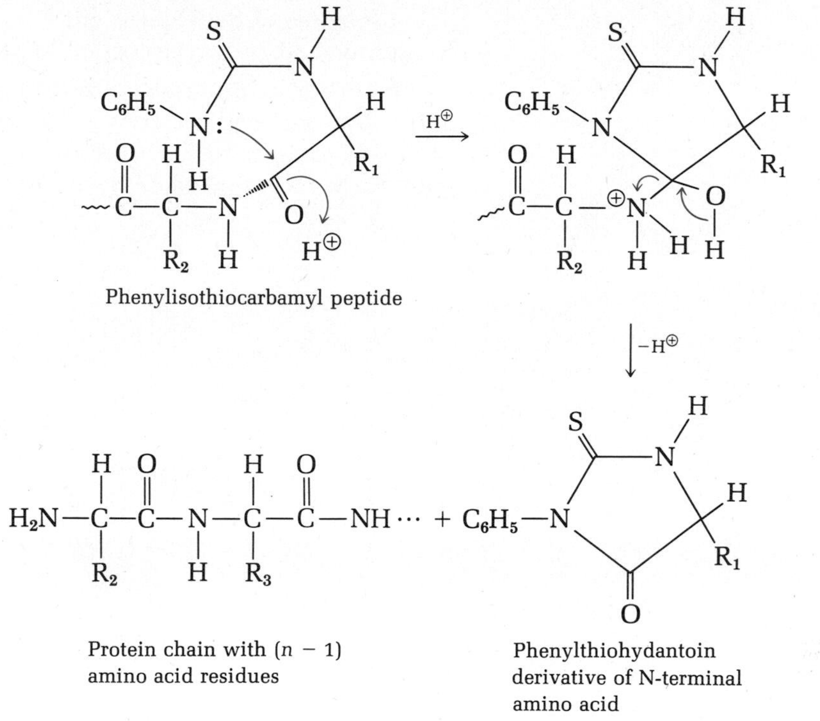

Phenylisothiocarbamyl peptide

$$H^{\oplus} \longrightarrow$$

$$\Big\downarrow -H^{\oplus}$$

(2.41)

Protein chain with $(n-1)$
amino acid residues

Phenylthiohydantoin
derivative of N-terminal
amino acid

B. Determination of the C-terminal Amino Acid

The C-terminal amino acid of a polypeptide is typically determined by one of two methods. The polypeptide is treated with anhydrous hydrazine, or it is hydrolyzed enzymatically by the enzymes carboxypeptidase A or carboxypeptidase B. These methods are neither as reliable nor as facile as methods for determining the N-terminal amino acid. And, although they enable us to determine the identity of the C-terminal amino acid, they are not useful tools for determining the entire sequence of a polypeptide.

a. Hydrazinolysis

The identity of the C-terminal amino acid residue of a polypeptide can be determined by treatment of the polypeptide with *anhydrous hydrazine*, which leads to *hydrazinolysis* of the peptide bonds and release of the free C-terminal amino acid (reaction 2.42). The entire polypeptide is destroyed in this process, and no further sequence determination is possible.

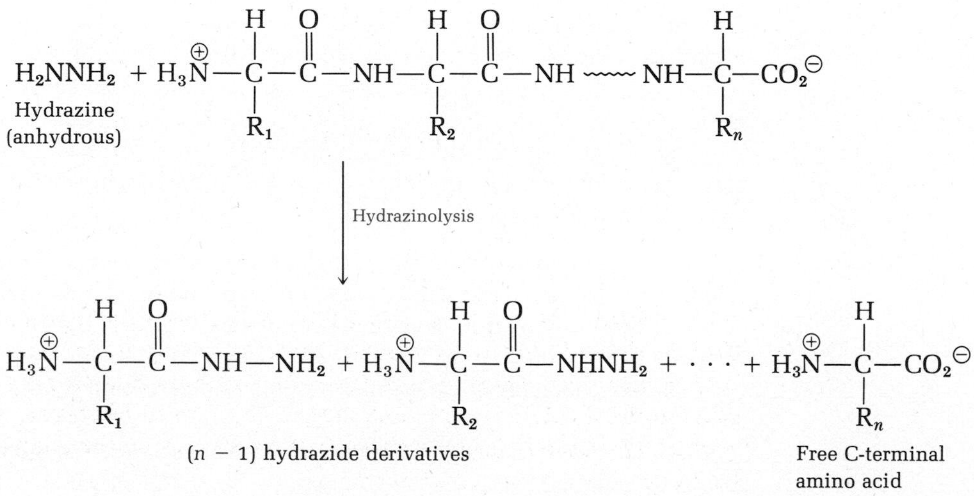

H_2NNH_2 +
Hydrazine
(anhydrous)

$\Big\downarrow$ Hydrazinolysis

(2.42)

$(n-1)$ hydrazide derivatives

Free C-terminal
amino acid

b. Cleavage of the C-terminal
Amino Acid by Carboxypeptidases A and B

The carboxypeptidases hydrolyze polypeptides sequentially from their C-terminal amino acid residues. *Carboxypeptidase A* cleaves all C-terminal amino acids except proline. *Carboxypeptidase B*, on the other hand, is specific for the basic amino acids lysine and arginine. The general reaction is indicated in reaction (2.43).

$$R_1\!-\!\underset{\underset{\oplus}{NH_3}}{\overset{H}{\underset{|}{\overset{|}{C}}}}\!-\!\overset{O}{\overset{\|}{C}}\!-\!NH\!-\!\underset{\overset{|}{R_2}}{CH}\!-\!\overset{O}{\overset{\|}{C}}\cdots\cdots\cdots NH\!-\!\underset{\overset{|}{R_{(n-1)}}}{CH}\!-\!\underset{\overset{\|}{O}}{C}\!-\!NH\!-\!\underset{\overset{|}{R_n}}{CH}\!-\!CO_2^{\ominus} + H_2O$$

Carboxypeptidase (A or B)

$$R_1\!-\!\underset{\underset{\oplus}{NH_3}}{\overset{H}{\underset{|}{\overset{|}{C}}}}\!-\!\overset{O}{\overset{\|}{C}}\!-\!NH\!-\!\underset{\overset{|}{R_2}}{CH}\!-\!\overset{O}{\overset{\|}{C}}\cdots\cdots\cdots NH\!-\!\underset{\overset{|}{R_{(n-1)}}}{CH}\!-\!CO_2^{\ominus} + H_3\overset{\oplus}{N}\!-\!\underset{\overset{|}{R_n}}{CH}\!-\!CO_2^{\ominus} \qquad (2.43)$$

Original C-terminal amino acid

Since the enzyme does not stop to wait for the experimental determination of the first amino acid released, determination of the last C-terminal amino acid hydrolyzed rapidly becomes problematic as more and more residues are released. Quantitative analysis of aliquots withdrawn from the reaction mixture at various times does, however, permit identification of the first few residues.

2–9 DETERMINATION OF THE SEQUENCE OF AMINO ACIDS IN PROTEINS

The *primary structure* of a protein is the linear sequence of amino acids residues that make up its covalent backbone. Insulin was the first protein whose primary structure was determined. Frederick Sanger (using the reagent named for him) required ten years to determine its 51-residue sequence. Advances in experimental techniques, including automated procedures for carrying out the Edman reaction, have reduced the time required to determine the primary structure of a moderate size protein (say, 200 residues in a single chain) from decades to months. Although the task remains formidable, the primary structures of hundreds of proteins have been determined. Since the three-dimensional conformation of a protein is predetermined by its primary structure (Section 4.2F), and since the primary structure of a protein is colinear with the primary structure of deoxyribonucleic acid (DNA), a knowledge of the primary structure is tremendously important.

The techniques employed to determine the primary structure of a protein vary somewhat from case to case, but a broad general strategy might be divided into five steps.

1. The number of chains in the protein is determined, and the protein is separated into separate chains if more than one type of chain is present.

2. The polypeptide is cleaved into small pieces by enzymes or specific reagents.

3. The sequence of each small peptide is determined by the Edman degradation.

4. The order in which the small peptides are arranged in the original polypeptide is determined.

5. The positions of the disulfide bridges (cystines) are determined.

A. Separation of Polypeptide Chains and Disruption of Disulfide Bonds

If analysis of the N-terminal amino acid of a protein reveals that the protein contains more than one type of polypeptide chain, the chains must be separated. Chain separation is most conveniently done by ion exchange chromatography, adsorption chromatography, or gel filtration chromatography (Section 5.4D). Once the chains have been separated (not always an easy task if they are quite similar to each other, as sometimes is the case), any disulfide bonds between cysteine residues must be disrupted. We have already noted that cystines can be oxidized to cysteic acid residues by performic acid (reaction 2.1). This reaction also oxidizes tryptophan, and it is more common to reduce the cystine residues by treating the polypeptide with β-mercaptoethanol (reaction 2.44). Since the reaction is reversible, the free sulfhydryl groups are then covalently modified.

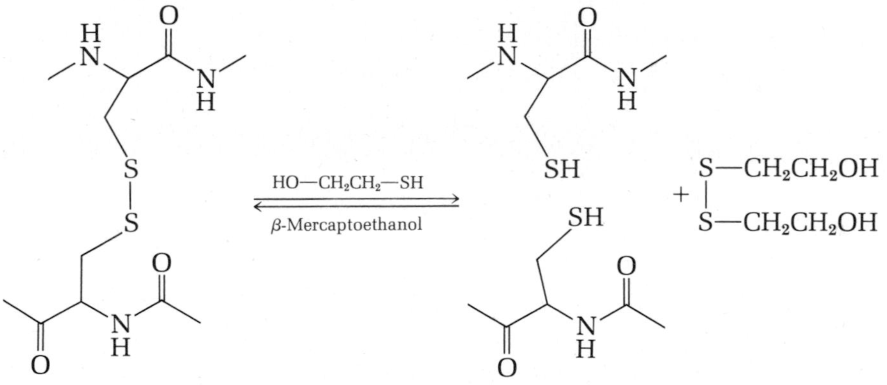

$$(2.44)$$

Three reactions are commonly used to alkylate the sulfhydryl groups. *Iodoacetate* alkylates the sulfhydryl groups to give *S-carboxymethylcysteine* residues (reaction 2.45).

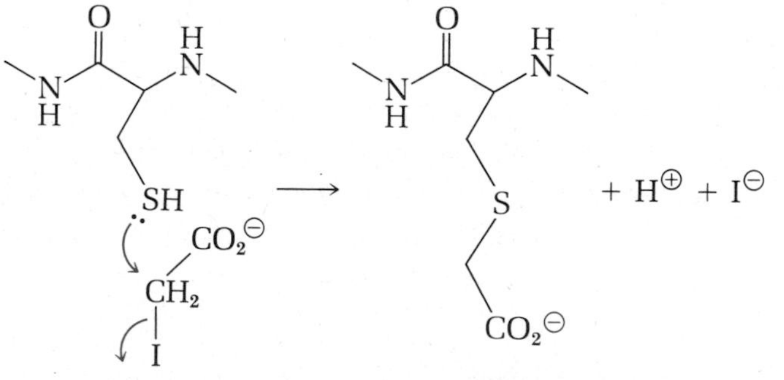

$$+ H^{\oplus} + I^{\ominus} \qquad (2.45)$$

Iodoacetate   S-carboxymethylcysteine residue

Treatment of the polypeptide with *acrylonitrile* converts cysteine sulfhydryl groups to *cyanoethylcysteine* residues (reaction 2.46).

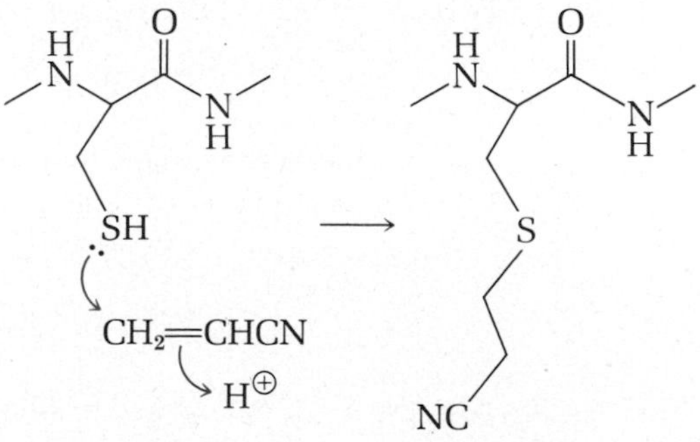

$$(2.46)$$

Acrylonitrile   Cyanoethylcysteine residue

This reaction is quantitative for cysteine residues, but other nucleophilic residues are also modified to some extent. The reduced protein can also be treated with *aziridine* (also known as ethylene amine) to give *S-aminoethylcysteine* residues (reaction 2.47). This reaction has the added advantage of providing new locations for specific enzyme cleavage, since the S-aminoethylcysteine residues look like lysine residues to certain enzymes, such as trypsin (Section 2.9B).

$$\text{(2.47)}$$

S-aminoethylcysteine residue

B. Internal Cleavage of Peptide Bonds

The Edman reagent, we noted earlier, can yield the sequence of 20 to 30 residues in a polypeptide. Proteins are much bigger than this, however, and after the chains have been separated and the disulfide bonds disrupted, the polypeptide is cleaved into smaller peptides whose sequence is determined by the Edman reagent. Both specific, controlled enzymatic hydrolysis and specific chemical reagents are used to chop the polypeptide into smaller fragments.

a. Enzymatic Hydrolysis of Polypeptides

The proteolytic (protein-cutting) enzymes trypsin and chymotrypsin hydrolyze internal peptide bonds in polypeptide chains. *Trypsin* hydrolyzes peptide bonds whose carbonyl group is donated by the basic amino acids lysine and arginine (reaction 2.48).

$$\overset{\oplus}{H_3N}\text{-Gly-Ala-Lys-Ser-Glu-Cys-Arg-Trp-Tyr-CO}_2{}^{\ominus}$$

$$\downarrow \text{Trypsin}$$

$$\text{(2.48)}$$

$$\overset{\oplus}{H_3N}\text{-Gly-Ala-Lys-CO}_2{}^{\ominus} + \overset{\oplus}{H_3N}\text{-Ser-Glu-Cys-Arg-CO}_2{}^{\ominus} + \overset{\oplus}{H_3N}\text{-Trp-Tyr-CO}_2{}^{\ominus}$$

Tryptic peptides

Trypsin is voracious, and controlled conditions are used to limit digestion so that only a few peptides are released. If the hydrolysis is allowed to proceed too long, so many fragments will be obtained that the determination of their original order in the intact polypeptide is quite complicated.

Chymotrypsin hydrolyzes peptide bonds whose carbonyl groups are donated by the hydrophobic residues such as phenylalanine, tyrosine, and tryptophan (reaction 2.48).

$$\overset{\oplus}{H_3N}\text{-Gly-Ala-Trp-Ser-Glu-Cys-Tyr-Arg-Asp-Lys-Gly-CO}_2{}^{\ominus}$$

$$\downarrow \text{Chymotrypsin}$$

$$\text{(2.43)}$$

$$\overset{\oplus}{H_3N}\text{-Gly-Ala-Trp-CO}_2{}^{\ominus} + \overset{\oplus}{H_3N}\text{-Ser-Glu-Cys-Tyr-CO}_2{}^{\ominus} + \overset{\oplus}{H_3N}\text{-Arg-Lys-Gly-CO}_2{}^{\ominus}$$

Chymotryptic peptides

Tryptic peptides: $H_3\overset{\oplus}{N}$-Ser-Ala-Gly-Arg-$CO_2^{\ominus}$

and

$H_3\overset{\oplus}{N}$-Thr-Phe-Ala-Asn-Arg-$CO_2^{\ominus}$

What is the order of these peptides in the original peptide?

To find out, hydrolyze with chymotrypsin, and determine the sequence of each fragment:

Chymotryptic peptides: $H_3\overset{\oplus}{N}$-Ser-Ala-Gly-Arg-Thr-Phe-$CO_2^{\ominus}$

and

$H_3\overset{\oplus}{N}$-Ala-Asn-Arg-$CO_2^{\ominus}$

Line up the fragments:

$H_3\overset{\oplus}{N}$-Ser-Ala-Gly-Arg-$CO_2^{\ominus}$ $H_3\overset{\oplus}{N}$-Thr-Phe-Ala-Asn-Arg-$CO_2^{\ominus}$

$H_3\overset{\oplus}{N}$-Ser-Ala-Gly-Arg-Thr-Phe-$CO_2^{\ominus}$ $H_3\overset{\oplus}{N}$-Ala-Asn-Arg-$CO_2^{\ominus}$

Figure 2.11 Determination of the sequence of a peptide by a combination of specific hydrolysis of the original peptide by proteolytic enzymes and Edman degradation.

If a sample of a polypeptide is hydrolyzed with trypsin, and if a different sample of the same polypeptide is hydrolyzed with chymotrypsin, then overlapping sequences will be obtained by Edman degradation of each of the fragments, and it is relatively simple to deduce the sequence of the original polypeptide. This procedure is illustrated in Figure 2.11.

Other enzymes besides trypsin and chymotrypsin are used to convert a polypeptide into smaller fragments. Two of the more common are *thermolysin* and *pepsin*. Thermolysin cleaves peptide bonds whose carbonyl groups are donated by leucine, isoleucine, and valine residues. Pepsin cleaves peptide bonds whose carbonyl groups are donated by aromatic amino acid residues and those having large hydrocarbon side chains. The specificities of these enzymes and of the chemical reagent cyanogen bromide, considered below, are summarized in Figure 2.12.

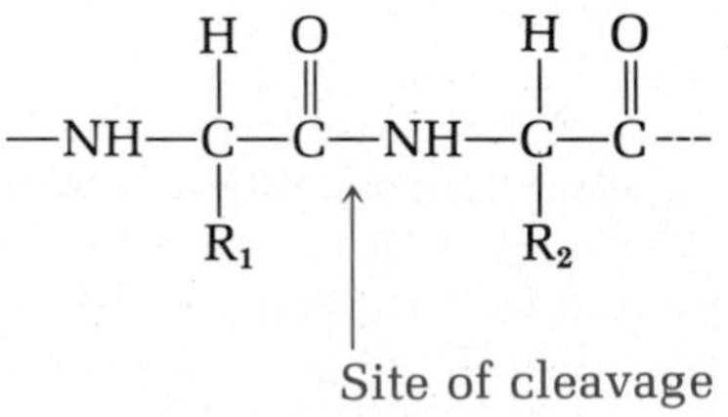

$$-NH-\overset{\overset{\displaystyle H}{|}}{\underset{\underset{\displaystyle R_1}{|}}{C}}-\overset{\overset{\displaystyle O}{||}}{C}-NH-\overset{\overset{\displaystyle H}{|}}{\underset{\underset{\displaystyle R_2}{|}}{C}}-\overset{\overset{\displaystyle O}{||}}{C}---$$

Site of cleavage

Chymotrypsin: R_1 = Tyr, Phe, Trp
Pepsin: R_1 = Tyr, Phe, Trp
Trypsin: R_1 = Lys, Arg
Cyanogen bromide: R_1 = Met
Thermolysin: R_1 = Ile, Leu, Val

Figure 2.12 Specificities of proteolytic enzymes that cleave internal peptide bonds.

b. Cleavage of Peptides by Cyanogen Bromide

Cyanogen bromide (CNBr) is a reagent that reacts quite specifically with peptide bonds in which the carbonyl group is contributed by methionine residues (reaction 2.44).

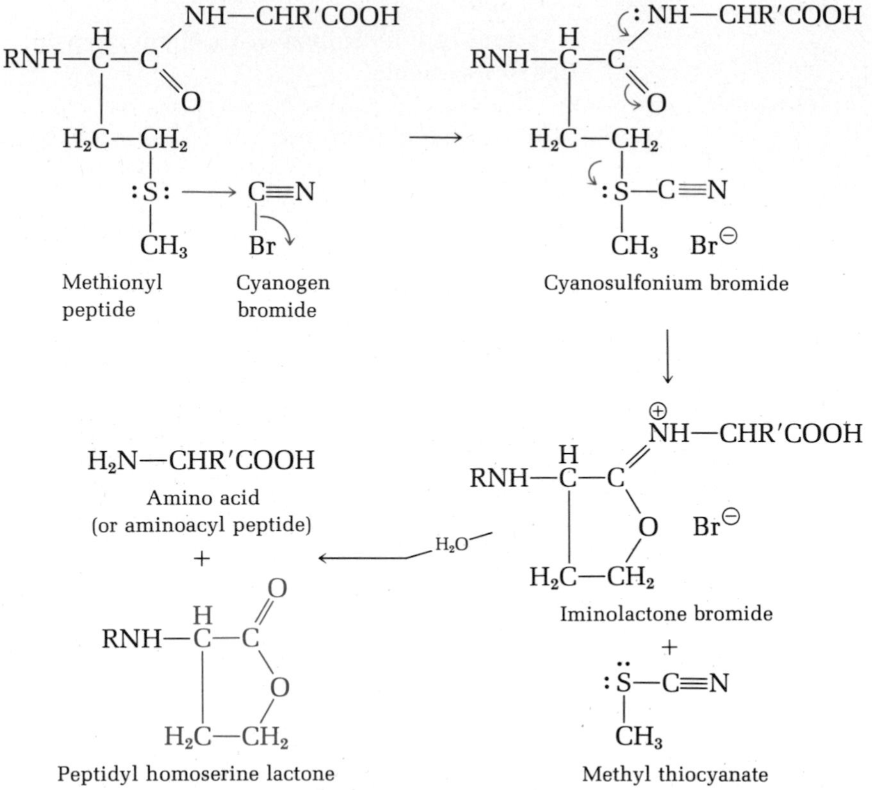

(2.44)

The first step of the reaction is neucleophilic attack of an electron pair of the sulfur atom of methionine upon carbon to give a cyanosulfonium bromide intermediate. Spontaneous cyclization yields an iminolactone bromide that hydrolyzes to yield a peptide whose C-terminal residue is *peptidyl homoserine lactone*. (A lactone is a cyclic ester, and it is called homoserine in this case because the final product is derived from an amino acid whose side chain, —CH_2CH_2OH, is the homolog of serine.) Since the majority of proteins contain few methionine residues, only a few peptides are produced by cyanogen bromide cleavage. These peptides can be further fragmented by treatment with proteolytic enzymes such as trypsin. A flow chart illustrating the use of many of the reagents we have described is shown in Figure 2.13 for the determination of the sequence of the protein pancreatic trypsin inhibitor I.

C. Determination of the Location of Disulfide Bonds

A great many polypeptides contain disulfide bonds, and their location is determined after the sequence of residues in the polypeptide has been ascertained. As the number of cysteine sulfhydryl groups increases, the problem of locating the disulfide bonds becomes increasingly complex since the number of possible disulfide bonds increases as 2^n, where n is the number of cysteine residues in the chain. Of course, not all cysteine residues need be involved in disulfide bonds, and this adds another complication. Once the sequence of residues in the polypeptide has been determined a new sample of the polypeptide is hydrolyzed by trypsin (or another specific reagent) without prior disruption of the disulfide bonds. Peptides that contain disulfide bonds are then isolated, the disulfide bridges are either oxidized or reduced, and the sequence of each of these peptides is determined and

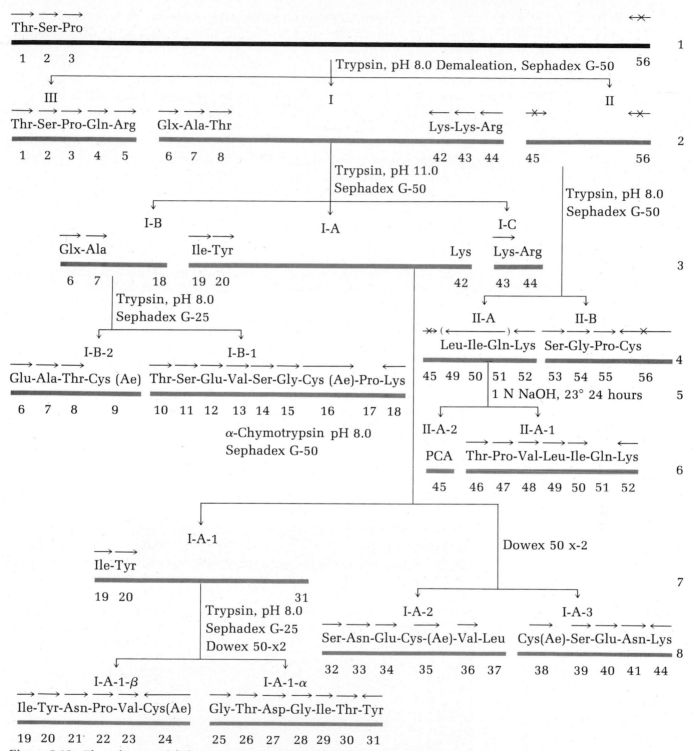

Figure 2.13 Flow diagram for the sequence determination of porcine pancreatic trypsin inhibitor I. Arrows pointing right (→) indicate sequences determined by Edman degradation. Arrows pointing left indicate sequences determined by carboxypeptidase. Arrows with x's indicate that the method failed. Glx indicates either Glu or Gln. [From D. C. Bartelt and L. J. Greene, *J. Biol. Chem.*, 246, 2218 (1971).]

compared to the sequences of the original peptides. If a given peptide contains a cysteine sulfhydryl group that is not in a disulfide bond, then prior alkylation followed by disruption of the disulfide bonds leaves one covalently modified cysteine residue and two free cysteine sulfhydryl groups. These groups are then identified as the ones participating in the disulfide bond.

**2–10 MOLECULAR EVO-
LUTION:
COMPARISONS
OF THE
PRIMARY STRUCTURES
OF CYTOCHROME c**

The availability of methods for rapid determinations of the primary structure of proteins makes it possible to study evolutionary relationships among organisms at the molecular level. The amino acid sequence of a protein is a direct result of the sequence of bases in DNA, so changes in primary structure reflect changes in the genetic material. The availability of rapid methods for nucleic acid sequence determinations (Section 9.9) provides even greater refinement in the analysis of the relationship of primary structure to evolutionary changes in DNA. Proteins, by faithfully representing the genetic heritage of generations, reveal their ancestry in their primary structures. They summon up the past and present themselves as "living fossils."

Cytochrome c, containing 104 residues in vertebrates, is present in the respiratory chain (Chapter 16) of all aerobic organisms. Thus, it is an ideal protein for evolutionary comparisons. Table 2.8 gives the number of sequence variations from human cytochrome c for 18 species. The cytochrome c's of species closely related to man, the chimpanzee and the rhesus monkey, are identical or nearly identical. As species diverge, sequence differences become more pronounced. An evolutionary "tree" for cytochrome c is shown in Figure 2.14. On the evolutionary time scale the greater the time interval since ancestral lines diverged, the greater the number of sequence variations. This generalization seems to apply to all proteins, although the *rate* of sequence divergence can vary widely from protein to protein.

Convergent evolution can also occur. Aquatic mammals, for example, have independently evolved fusiform shapes like the fishes; and it is certainly no coincidence that airplanes and ballistic missiles have similar fusiform shapes; the structure fits the function. The whales, in fact, evolved from land mammals related to the modern ungulates. The cytochrome c sequence of the California gray whale differs from human cytochrome c by 10 residues, but it differs from the identical cytochrome c's of cows, pigs, and sheep by only 2 residues. Convergent evolution can, of course, be seen in

Table 2.8	Number of residues by which the primary structure of a given cytochrome c differs from human cytochrome c
Species	**Number of residues different from human cytochrome c**
Chimpanzee	0
Rhesus monkey	1
Rabbit	9
Kangaroo	10
Whale	10
Cow, pig, sheep	10
Dog	11
Donkey	11
Horse	12
Chicken, turkey	13
Rattlesnake	14
Turtle	15
Tuna	21
Dogfish	23
Screwworm fly	25
Moth (*Samia cynthia*)	31
Wheat	35
Neurospora	43
Yeast	45

SOURCE: From E. L. Smith, in P. D. Boyer, Ed., "The Enzymes," 3rd ed., Vol. 1, Academic Press, New York, 1970.

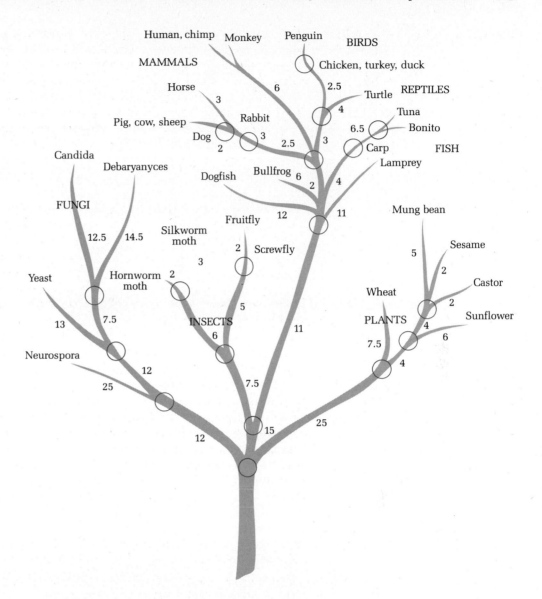

Figure 2.14 A family tree for cytochrome c. Circles indicate branch points for ancestral lines. The numbers beside the branches are the number of sequence differences per 100 residues; for example, dog cytochrome c differs from pig cytochrome c by two residues. [From M. O. Dayhoff, C. M. Park, and P. J. McLaughlin, in M. O. Dayhoff, Ed., "Atlas of Proteins Sequence and Structure," Vol. 5, National Biomedical Research Foundation, Washington, D.C., 1972, p. 8.]

characteristics other than primary structures; Section 6.3 gives the example of three-dimensional structures of enzyme active sites that have converged.

Within the sequence of cytochrome c we find 35 residues that are conserved in all species whose sequences are known (Figure 2.15). These residues are critical to the structure and function of cytochrome c. Residues 70 to 80 are in direct contact with the *heme* group that is reversibly oxidized and reduced during respiration (Section 16.3B). The residues in direct contact with this "active site" heme are essential to the function of cytochrome c. The primary structure of cytochrome c, in revealing the evolutionary development of aerobic organisms, also reveals that the past is not fugitive, it is present.

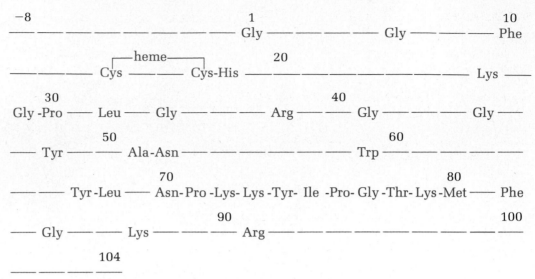

Figure 2.15 Residues known to be conserved in all cytochromes c whose sequences have been determined. [From E. L. Smith, in P. D. Boyer, Ed., "The Enzymes," 3rd ed. Vol. 1, p. 281, Academic Press, New York, 1970.]

2–11 SUMMARY
Proteins are linear polymers of α-amino acids. The free amino acids in aqueous solution exist predominantly as dipolar ions in which the α-amino group and the α-carboxyl group are ionized. The pK_a values of the α-carboxyl groups are in the range of 1.8 to 2.5, while the pK_a values of the protonated amino groups are in the range of 8.8 to 10.0.

The chemical bond that joins the amino acid residues of peptides and proteins is a secondary amide called a peptide bond. The peptide bond is *planar*. The most stable conformation of the α-carbons of the amino acid residues in a peptide bond is *trans* to the C—N bond, and rotation around the peptide bond is slow at room temperature.

The acid-base properties of polypeptides are complex. To a first approximation, the charge of the predominant ionic state of each ionizing side chain can be easily determined. If pH $<$ $pK_{a-\text{side chain}}$, then the acidic form predominates; if pH $>$ $pK_{a-\text{side chain}}$, the basic form predominates. This generalization can however, fail for side chains that are buried in the interior of the protein out of contact with water. The net charge on a polypeptide at a given pH can be estimated by these rules by summing the charge on all ionizing side chains and end groups.

The primary structure of a protein consists of the linear sequence of its amino acid residues. The Sanger reagent, dansyl chloride, and the Edman reagent can be used to determine the N-terminal amino acid. The product obtained by treating the peptide with the Edman reagent can be released without hydrolyzing the peptide, and it can therefore be used to determine the sequences of amino acid residues in peptides. Polypeptides that are too large to be analyzed solely by Edman degradation are specifically hydrolyzed either enzymatically or by cyanogen bromide, which specifically cleaves peptide bonds at methionine residues. The sequences of each of the fragments produced by such specific reagents yields a set of overlapping pieces that reveal the entire primary structure of the protein.

The primary structure of a protein is colinear with the genetic material, DNA, and therefore differences in primary structures of proteins reflect their evolutionary history. As species diverge, the primary structures of their proteins diverge, and by comparing differences in the primary structures of proteins, we are able to study the ancient history of life.

REFERENCES

Barker, R., "Organic Chemistry of Biological Compounds," Prentice-Hall, Englewood Cliffs, N.J., 1971.

Bodanszky, M., Klausner, Y. S., and Ondetti, M. A., "Peptide Synthesis," 2nd ed., Wiley, New York, 1976.

Dayhoff, M., Ed., "Atlas of Protein Sequence and Structure," Vol. 5, National Biomedical Research Foundation, Washington, D.C., 1972.

Dickerson, R. E., and Geis, I., "The Structure and Action of Proteins," 2nd ed., W. A. Benjamin, Menlo Park, Calif., 1981.

Meister, A., "Biochemistry of the Amino Acids," 2nd ed., 2 vols., Academic Press, New York, 1965.

Needleman, S. B., Ed., "Protein Sequence Determination," Springer-Verlag, New York, 1970.

Neurath, H., and Hill, R. L., Eds., "The Proteins," 3rd ed., Vols. 1-3, Academic Press, New York, 1975–1977.

Schroeder, W. A., "The Primary Structure of Proteins," Harper & Row, New York, 1968.

Problems

1. Consider the titration of a 0.05 M solution of alanine. Draw the titration curve obtained when 25 ml of this solution is titrated with 0.1 M NaOH, given that the initial pH of the alanine solution is 0.5 and that pK_{a1} and pK_{a2} are 2.35 and 9.78, respectively.

2. Calculate the isoelectric pH for the following amino acids: (a) alanine, (b) histidine, (c) arginine, (d) aspartic acid, (e) lysine.

3. Calculate the isoelectric pH for the following peptides: (a) glycyl-L-aspartate, (b) glycylaspartate, (c) alanylhistidyltyrosine methyl ester.

4. Draw the titration curves for the peptides in problem 3.

5. Making use of the pK_a's of alanine, show that alanine has a net charge of $+1$ at pH 1.0, a net charge of 0 at pH 7.0, and a net charge of -1 at pH 12.0.

6. Show by writing appropriate "electron dot" structures that the protonated side chain of arginine is resonance stabilized. Should this resonance stabilization increase or decrease the pK_a of the side chain relative to the pK_a of a primary amine?

7. Indicate which species is the predominant one in solutions whose pH's are 1.0, 3.5, 5.0, 7.0, and 9.0: (a) alanine, (b) glutamic acid, (c) histidine, (d) asparagine, (e) serine.

8. At pH 7.0 calculate the fraction of neutral, anionic, and cationic amino acid for (a) alanine, (b) glutamic acid, (c) histidine, (d) asparagine, (e) serine.

9. A peptide has the following composition:

(Ala), (Arg), $(Asp)_2$, $(Glu)_2$, $(Gly)_3$, $(Leu)_1$, $(Val)_3$

The following peptides were isolated after partial acid hydrolysis and their sequence determined.

Peptide	Sequence (NH_2—· ·—CO_2H)
1	Asp-Glu-Val-Gly-Gly-Glu-Ala
2	Val-Asp-Val-Asp-Glu
3	Val-Asp-Val
4	Glu-Ala-Leu-Gly-Arg
5	Val-Gly-Gly-Glu-Ala-Leu-Gly-Arg
6	Leu-Gly-Arg

What is the amino acid sequence of the original peptide?

10. The amino acid sequence of minke whale myoglobin has recently

been determined. A partial sequence is given below. What is the net charge on this peptide at pH 4.0, 7.0, and 10.0?

Phe-Lys-His-Leu-Lys-Thr-Glu-Ala-Glu-Met-Lys-Ala-Ser-Glu-Asp

11. Draw a titration curve for the tripeptide Phe-Gln-Gly. Label each axis and identify all inflection points. Write the equilibria corresponding to each inflection point.

12. (a) The pK_a of the side chain of cysteine is 8.33. Write the equilibrium for ionization of the side chain and calculate the fraction of protonated to unprotonated side chain at pH 7.4. (b) Repeat the analysis of cysteine for the side chain of tyrosine, pK_a 10.0.

13. Draw the predominant electronic structures, including the side chains and charges, for the following tripeptides at the pH values indicated: (a) Phe-Arg-Ser, pH 2 and pH 6; (b) Glu-Met-Tyr, pH 6.0 and pH 12.0; (c) Pro-Asp-Lys, pH 2.0 and pH 8.0.

14. (a) Using three letter abbreviations, show the *products* of the enzymatic cleavages on each of the tripeptides of question 13 when they are treated with the enzymes (i) trypsin (ii) chymotrypsin. (b) Show the *structures* of the products of the CNBr reaction with each of the three tripeptides of question 13.

15. Write the structures for the products of the tripeptide Phe-Arg-Ser when it is treated with 2,4-dinitrofluorobenzene followed by acid hydrolysis.

16. Assuming that N-terminal analysis, total amino acid content, trypsin, chymotrypsin, and CNBr treatments were successful in the sequencing of the tripeptides of questions 1-13, above, which sequence could *not* be fully elucidated?

17. Draw the stepwise mechanism for the reaction of 1 mole of phenylisothiocyanate with 1 mole of Asn-Cys.

18. The side chains of lysine, tyrosine, and histidine react with the Sanger reagent. Write all the products of the reaction of the Sanger reagent with the dipeptides (a) Tyr-Leu, (b) Lys-Phe, (c) His-Ala.

19. (a) Write the ionization reaction for the side chain of histidine. (b) Write the Henderson-Hasselbalch equation for the ionization. (c) Calculate the ratio of protonated to unprotonated imidazole at pH 7.4, assuming a pK_a for imidazole of 6.0.

Chapter 3

The secondary structure of proteins

3–1 INTRODUCTION

The *secondary structures* that are found in proteins are regularly repeating conformations of the polypeptide chain. A secondary structure can involve as few as three residues, or it can involve the majority of residues in the polypeptide chain. Hydrogen bonding provides the major stabilizing force of the secondary structures found in a given protein. Certain combinations of secondary structures form recognizable patterns within a single protein. These closely secondary structures are known as *supersecondary structures*. And, these structures, the elaboration of a simpler structural motif, involve a higher level of structural complexity.

3–2 THE α-HELIX

Many possibilities for intrachain hydrogen bonding are available to a polypeptide chain. If the chain is twisted in just the right way, a structure emerges in which every amide nitrogen and carbonyl carbon is involved in a hydrogen bond (Figure 3.1). The structure is called an *α-helix,* and the regularly repeating, hydrogen-bonded structure of an α-helix is a *secondary structure* of the protein. The chain advances by 0.54 nm per turn of the helix. The rise per amino acyl group is 0.15 nm, and there are 3.6 amino acyl groups per

Figure 3.1 Left-handed and right-handed α-helices. A right-handed α-helix describes a clockwise motion as the chain advances; a left-handed helix describes a counterclockwise motion as the chain advances. [From L. Pauling, R. B. Corey, and H. R. Branson, *Proc. Nat. Acad. Sci. U.S.*, **37**, 205 (1951).

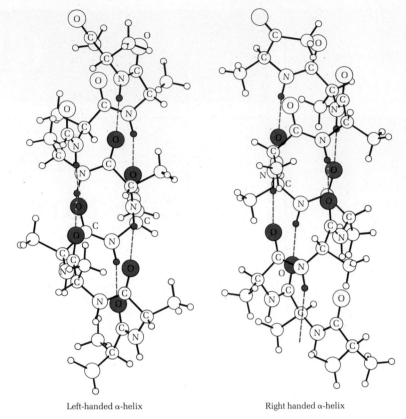

Left-handed α-helix   Right handed α-helix

Left-handed α-helix

Right-handed α-helix

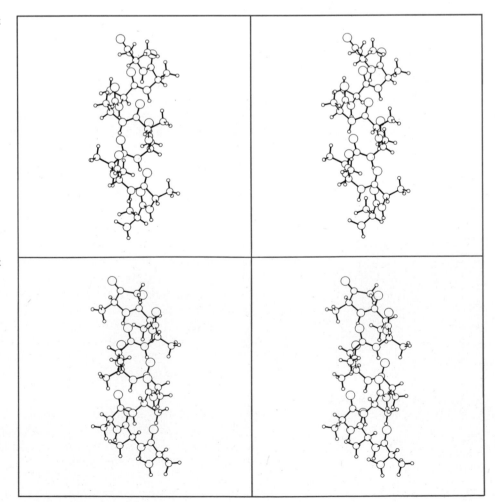

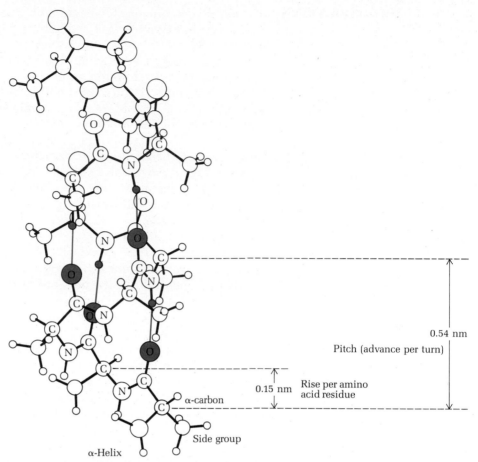

Figure 3.2 Dimensions of the α-helix. The pitch of the helix is 0.54 nm, there are 3.6 amino acid residues per turn of the helix, and each amino acid advances the helix by 0.15 nm.

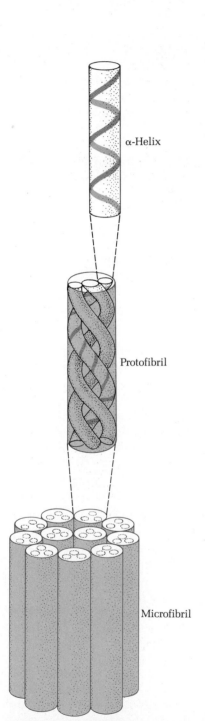

Figure 3.3 Structure of the α-keratins. The α-keratins are assemblies of triple-helical protofibrils arranged in a "9 + 2" array called a microfibril.

turn of the helix (Figure 3.2). Nearly every α-helix found in proteins is right handed. A right-handed α-helix is more stable than a left-handed one because there is more steric hindrance in a left-handed helix.[1] Although rare, the left-handed α-helix does appear occasionally. For example, in the proteolytic enzyme thermolysin (see Figure 2.11), there is a short stretch of left-handed α-helix over the sequence Asp-Asn-Gly-Gly.

A single hydrogen bond is a weak interaction that releases about 20 kJ mol^{-1} when it forms. A covalent bond, by contrast, releases in the neighborhood of 100 kJ mol^{-1} when it forms. An α-helix is stable because it involves many hydrogen bonds whose free energies of formation are approximately additive. Formation of an α-helix is a *cooperative process* because of this additivity, and, as each new hydrogen bond "clicks into place," the formation of the next one is made easier. The α-helix is quite stable, and it is the most abundant secondary structure found in proteins.

---

[1] Perhaps the difference in stability does not seem reasonable at first glance. After all, the stability of an L-amino acid is the same as the stability of a D-amino acid, and in general enantiomers have the same stability. But the left- and right-handed helices are *not* enantiomers. The enantiomer of a right-handed helix of L-amino acids is a left-handed helix of D-amino acids, not a left-handed helix of L-amino acids. The right- and left-handed helices are related as diastereoisomers, not as enantiomers, and diastereoisomers often have significantly different physical, chemical, and biological properties.

3-3 THE α-KERATINS

The major components of hair, skin, fur, beaks, nails, and most of the protective covering of vertebrates are fibrous proteins known collectively as α-keratins. (Many terrestrial invertebrates—snails, insects, worms, and so forth—have hard coverings of other materials.) The α-keratins are mostly composed of α-helical proteins whose polypeptide chains run roughly parallel to their length. Triplets of α-helical proteins are intertwined to form a structure called a *protofibril*, and protofibrils in turn are arrayed in a "9 + 2" structure (Figure 3.3) to form a *microfibril*.[2] Like the coiled cord of a telephone receiver a microfibril is quite elastic; a wool fiber can be stretched to nearly twice its original length. Structural stability is provided by extensive cross-linking of the chains by disulfide bonds; the greater the number of disulfide bonds, the more rigid the fiber. Soft keratins, such as those in skin, have few cysteines and relatively few cross-links, whereas hard keratins, such as those found in nails, have a high cysteine content, are rather inflexible, and do not stretch easily.

The coiled-coil of α-helical polypeptides in the α-keratins is an example of a *supersecondary structure*. The α-keratin superhelix is left-handed, although the monomers are all right-handed α-helices. The helix repeats every 14 nm, and every seventh residue occupies an equivalent position along the axis of the helix. The side chains of the residues in the individual helices mesh with each other. Examination of the primary structures of the α-keratins reveals that nonpolar side chains are turned inward, out of contact with water, and that the polar side chains are on the protein's surface in contact with water. The nonpolar side chains pack tightly with each other, and the stability of the coiled-coil is largely a result of van der Waals interactions between the nonpolar side chains.

3-4 β-PLEATED SHEET

When an α-helix is stretched, the amide nitrogens and carbonyl oxygens are extended at nearly right angles to the long axis of the polypeptide chain. These groups can form hydrogen bonds to another polypeptide chain having the same conformation. The resulting hydrogen-bonded structure is known as a *β-pleated sheet*, or β structure (Figure 3.4). If the C-terminal amino acid of one chain is aligned with the N-terminal amino acid of the second chain, the β-pleated sheet is *antiparallel*; if the C-terminus and the N-terminus coincide, the β structure is *parallel*.

The silkworm, *Bombix mori*, produces *silk fibroin*, a protein whose structure is an extended, antiparallel β structure. Glycine accounts for nearly half the amino acid residues in silk fibroin and alanine and serine for most of the others. There are few residues with charged side chains, and no cysteine or methionine. In silk fibroin a set of residues having the sequence (Gly-Ser-Gly-Ala-Gly)$_n$ repeats many times in the fibroin chain (Figure 3.5). Each carbonyl oxygen and amide nitrogen are involved in a hydrogen bond. The side chains of glycine, which is far more stable in a β-pleated sheet than in an α-helix, lie on one side of the sheet. The methyl groups of alanine residues and the hydroxymethyl groups of the serine residues lie on the opposite side of the sheet from the glycines' hydrogen "side chains." The stacked, β-pleated sheets, therefore, have ideal stereochemistry to accommodate the repeated amino acid sequence of fibroin.

The β-pleated sheet in silk fibroin, and in many other proteins, has a left-handed twist, in contrast to the untwisted plane of a simple parallel or antiparallel β-pleated sheet. In fact, planar β-pleated sheets are rather rare. The most stabled twisted β-pleated sheet forms when the two chains are

---

[2] A 9 + 2 array of fibers is found in other cellular structures such as microtubules and the cilia and flagella of bacterial and mammalian cells.

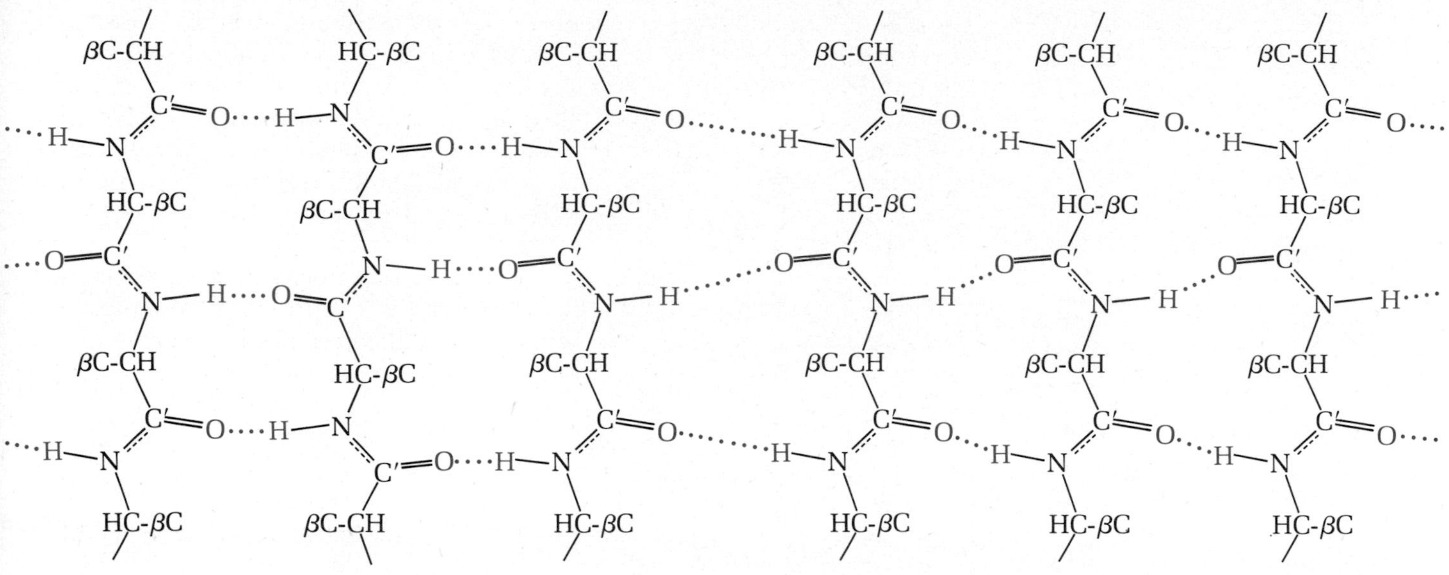

Antiparallel

Parallel

Antiparallel pleated sheet

Parallel pleated sheet

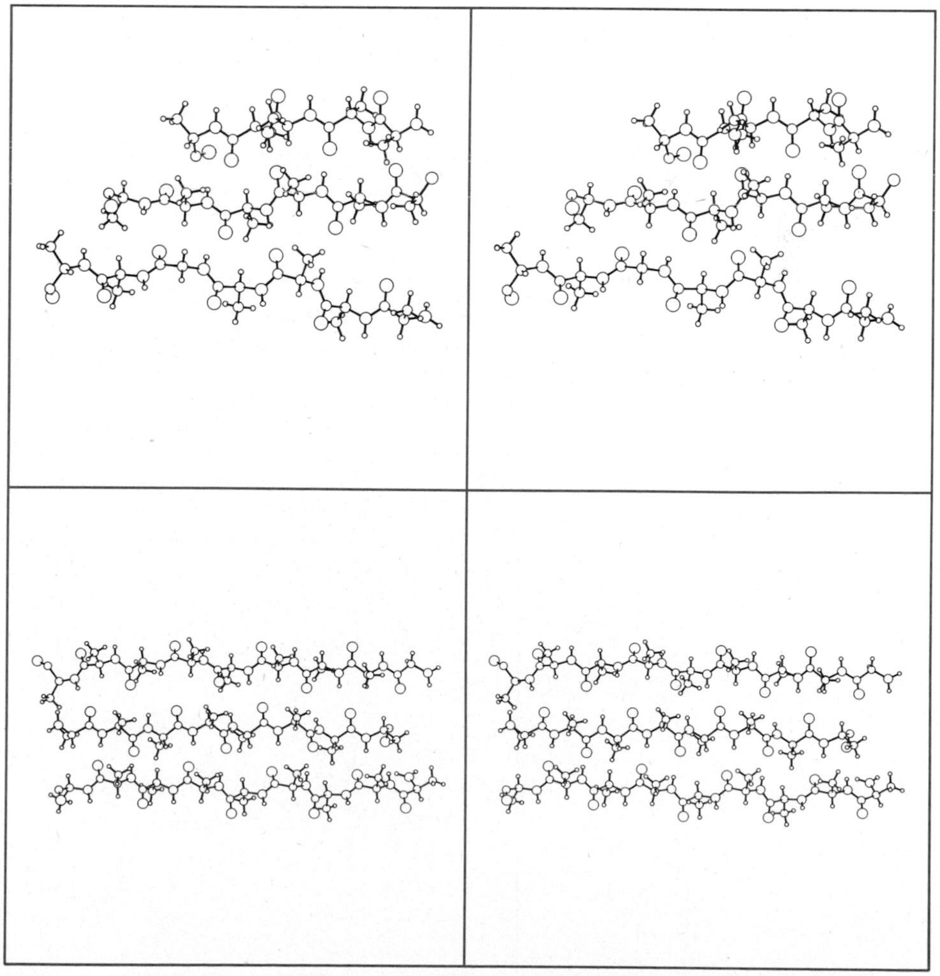

Figure 3.4 β structure. (a) Top view of the parallel and antiparallel β-pleated sheet.

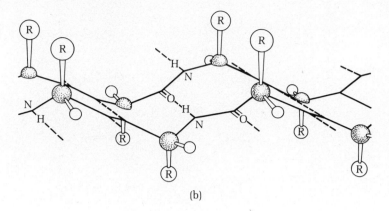

(b)

Side view of pleated sheet

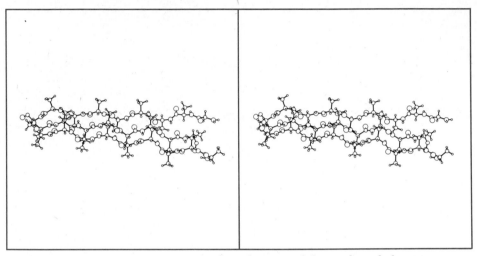

Figure 3.4 *(Continued)* β structure. (b) Side view of the β-pleated sheet.

Figure 3.5 Structure of silk fibroin. The three-dimensional representation shows that (a) alanine residues can interact by van der Waals forces, (b) all carbonyl oxygens and amide hydrogens are involved in hydrogen bonds, and (c) alanine residues from one chain fit nicely into a region above (or below) glycine residues in an adjacent chain.

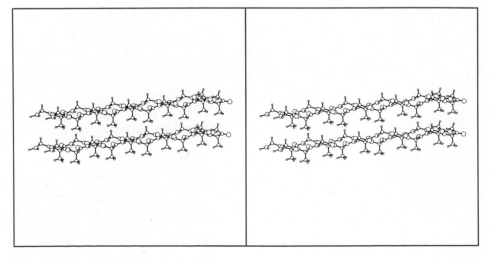

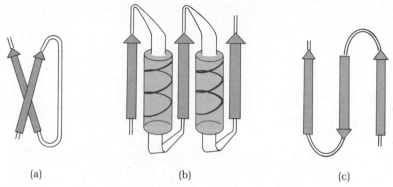

Figure 3.6 Three common supersecondary structures that involve β-pleated sheets: $\beta \times \beta$-unit, Rossman-fold, and β-meander.

twisted by about 25° with respect to the long axis of the sheet. This twisting permits a wide variety of side chains to be accommodated in the β-pleated sheet, whereas a planar β-pleated sheet accommodates bulky side chains rather poorly. A twisted β-pleated sheet nicely fits the large groove of the DNA double helix (Section 9.5C), and such polypeptide conformations may be quite important in protein-DNA interactions.

Higher ordered secondary structures involving β-pleated sheets are common in globular proteins. These structures, which may be designated supersecondary structures, fall into a few general patterns (Figure 3.6). The $\beta \times \beta$-unit consists of two parallel β-pleated sheets connected by an intervening chain. The interval can be a nonregular coil or an α-helix. In nearly every case (there is one known exception) the crossover connection is right-handed. A particularly common elaboration of the $\beta \times \beta$-unit consists of three segments of β-pleated sheet separated by two α-helices. This supersecondary structure is known as the *Rossman-fold*. Another common supersecondary structure is known as the *β-meander*. This structure consists of three strands of β-pleated sheet connected by rather short loops. The connecting pieces, or bends, are discussed in the next section. The β-meander contains nearly as many hydrogen bonds as an α-helix (about two-thirds of all possible intrachain hydrogen bonds are formed), and the great stability of the β-meander no doubt accounts for its high occurrence.

3–5 BENDS IN POLYPETIDE CHAINS

The polypeptide chains of globular proteins are highly folded, and the chain must make many bends to assume its final conformation. The bends, or reverse turns, in the polypetide chain are another aspect of secondary structure. Three types of reverse turns, each of which contains four amino acid residues, have been observed (Figure 3.7). In all three a hydrogen-bonded loop forms between a carbonyl oxygen and an amide hydrogen between the first and third residues in the bend. Type I and type II bends are related by a 180° rotation of the central residue. Glycine is the only allowed residue at position 3 of type II turns because all other side chains are too large to fit into the restricted space. A type III bend is a piece of a helix formed between the first and third residues. This helix contains three residues per turn, and the hydrogen-bonded loop contains ten atoms. This structure is therefore called a 3_{10}-*helix*. Except at reverse turns, the 3_{10}-helix is extremely rare. Type I and type III bends are nearly indistinguishable since their conformations at C_1^{α} are identical, and their differences at C_2^{α} are minute. For the most part, bends are found on the surfaces of globular proteins where there is little resistance to a change in direction of the polypeptide chain.

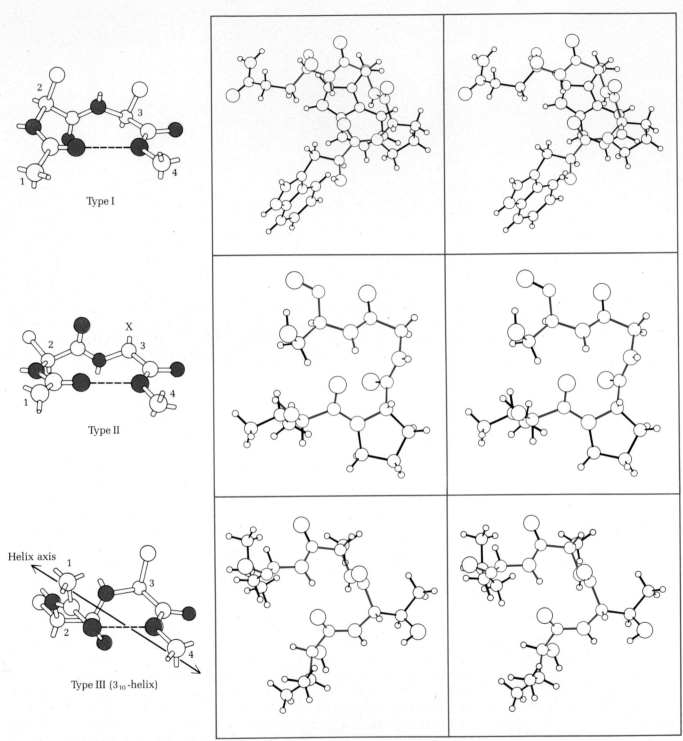

Type I

Type II

Type III (3$_{10}$-helix)

Figure 3.7 Structures of the major types of bends in polypeptide chains [From J. J. Birktoft and D. M. Blow, *J. Mol. Biol., 68,* 205 (1972). Reproduced with permission. Copyright: Academic Press Inc. (London) Ltd.]

Amino acid residues are not uniformly distributed in all types of secondary structures. Glutamic acid residues are found mostly in helices, asparagine and glycine residues are found mostly in turns, and proline, sometimes found at the C-terminal residue of a helix, but never within a helix, is also common in turns. Data from 15 proteins indicating the type of secondary structure in which an amino acid residue is likely to be found are summarized in Table 3.1. We find that 33.2% of the amino acid residues in the sample are involved in turns, 17.1% are found in β-pleated sheets, and 34.3% in α-helices. These three conformations of the polypeptide backbone

Table 3.1 Distribution of amino acids in bends, α-helices, and β structures in 15 proteins

Amino acid	Bends[a]	N-Helix[b]	M-Helix[c]	C-Helix[d]	β-Sheet
Asp	41.3	13.9	13.9	5.6	15.3
Glu	21.4	20.3	24.1	12.7	6.3
Lys	36.6	4.2	16.0	15.1	20.2
Arg	26.3	1.6	11.1	14.3	12.7
His	31.5	10.2	22.4	18.4	6.1
Asn	47.2	8.9	7.9	7.9	12.9
Gln	30.6	9.5	21.6	17.6	14.9
Cys	42.8	8.5	8.5	14.9	17.0
Thr	34.8	11.6	14.3	6.2	19.7
Ser	41.9	10.5	9.9	7.9	11.2
Tyr	44.8	7.2	6.0	3.6	24.1
Trp	40.7	17.1	25.7	2.9	14.3
Gly	45.8	6.4	7.1	6.4	9.6
Ala	20.0	8.5	28.1	9.1	17.1
Val	15.8	5.3	21.4	6.1	27.5
Leu	23.1	6.6	28.7	9.0	26.2
Ile	22.3	8.3	22.2	3.3	30.0
Phe	35.0	5.9	23.6	9.8	21.6
Pro	48.8	15.4	1.5	—	6.1
Met	22.2	4.0	28.0	16.0	24.0
Average	33.2	8.8	16.7	8.8	17.1

[a] Type I, type II, and type III.
[b] N-terminal residue of helix.
[c] Middle of helix.
[d] C-terminal residue of helix.
SOURCE: From B. W. Matthews, X-Ray Structure of Proteins, in H. Neurath and R. L. Hill, Eds., "The Proteins," 3rd ed., Academic Press, New York, 1977, Vol. 3, p. 567.

thus account for 85% of the residues in a mythical "average" protein. The distributions of secondary structures in a few proteins are summarized in Table 3.2. As this table shows, we would be remiss to treat any protein as "average" with respect to its secondary structure. Some proteins, such as hemoglobin and myoglobin, contain a large amount of α-helix; others, such as ferredoxin, contain no α-helix whatsoever. The β-pleated sheet content and the number of bends in a polypeptide chain also vary considerably.

3-6 CONFORMATIONAL ANALYSIS OF PEPTIDE CHAINS

Thus far we have analyzed regularly repeating units of secondary structure, but these by no means exhaust the possibilities for peptide conformations. Consider the conformation of a single amino acid residue in a polypeptide. Since free rotation is not allowed around the peptide bond, the conformation is determined solely by rotation about the C—C^{α} and N—C^{α} bonds (Figure 3.8), and we say that the conformation at a given position has "two degrees of freedom." The α-carbon from one residue and the two amide groups flanking it share one corner of two intersecting planes (Figure 3.8). When both of these planes lie in the plane of the page, the angle between them is defined as 0°. Each plane can rotate independently with respect to the corner shared by the α-carbon. The angle of rotation around the N—C^{α} bond in the counterclockwise direction is called phi (ϕ). The angle of rotation in the clockwise direction around the C—C^{α} bond is called psi (ψ). The values of ϕ and ψ, when specified for every residue, completely define the conformation of the polypeptide chain. In the α-keratins and silk fibroin the values of ϕ and ψ are the same for most of the residues in the peptide chain; this is their most notable conformational property. Some conformations, such as the α-helix and β structure, are highly favored; others are quite unstable. Clockwise rotation about the N—C^{α} bond, ϕ, of 120° places the carbonyl oxygen at

Table 3.2 Distribution of secondary structures in proteins

Protein	Helices %	Helices Number	β-Sheets %	β-Sheets Number of sheets	β-Sheets Number of strands	Bends Number of type I	Bends Number of type II
Carbonic anhydrase	20	7	37	2	13	3	1
Carboxypeptidase A	38	9	17	1	8	10	3
Chymotrypsin (elastase, trypsin)	14	3	45	2	12	11	6
Concanavalin A	2	1	57	3	18	—	—
Cytochrome b_5	52	6	25	1	5	1	1
Cytochrome c	39	5	—	—	—	3	3
Ferredoxin	—	—	4	1	2	—	—
Flavodoxin	30	4	30	1	5	—	—
Insulin	52	3	6	1	1	—	—
Lactate (or malate) dehydrogenase	45	10	20	3	12	6	4
Lysozyme	40	6	12	2	6	4	2
Myogen	57	6	4	1	2	5	—
Myoglobin (hemoglobin)	79	8	—	—	—	6	—
Nuclease	24	3	14	1	3	—	—
Pancreatic trypsin inhibitor	28	2	33	1	3	—	—
Papain	28	5	15	1	7	—	—
Ribonuclease	26	3	35	1	6	1	1
Subtilisin	31	8	16	2	8	16	1
Thermolysin	38	7	22	1	10	—	—

its maximum distance from the side chain of the α-carbon. If, however, we let ψ remain between 0° and 90° the carbonyl oxygens are close to each other, and the conformation $\phi = 120°$, $\psi = 0°$ is unstable because of dipole-dipole and van der Waals repulsion between the carbonyl oxygens (Figure 3.9). If we plot ϕ versus ψ, a graph called a *Ramachandran diagram*, nearly all conformations fall within the areas enclosed in the dotted lines of Figure 3-10. Van der Waals repulsions strongly discourage conformations whose values of ϕ and ψ lie outside the dotted lines. Such conformations are rare, but they are not "forbidden," because the entire protein seeks a free-energy minimum,

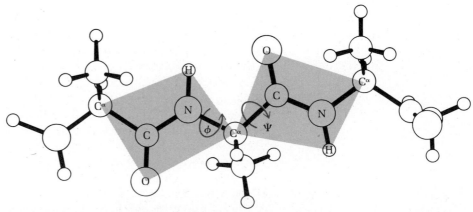

Figure 3.8 Conformation of a residue in a peptide. The α-carbon of a given residue in a peptide is flanked by two amide groups. It shares one corner of two intersecting planes. When ϕ and ψ are both 0°, all atoms except the hydrogen and side chain of the α-carbon lie in the same plane. Rotation of the two planes by the angles ϕ and ψ generates new protein conformations.

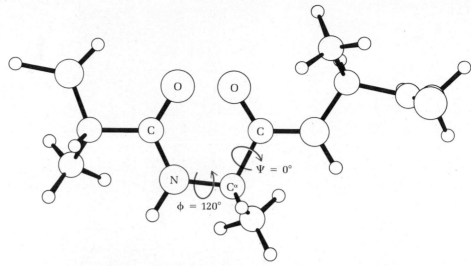

Figure 3.9 Unstable peptide conformation. When $\phi = 120°$ and ψ is between 0 and 90°, the adjacent carbonyl groups are in proximity. This conformation is discouraged by dipole-dipole and van der Waals repulsions.

and a few residues may have conformations that, considered in isolation, are unfavorable. For example, the values of ϕ and ψ for each residue of chymotrypsin are plotted in Figure 3.11 in a Ramachandran diagram. All but 28 of the 245 residues of chymotrypsin have ϕ and ψ values within the "allowed zones" indicated in Figure 3.10. The residues falling outside the zones are mostly glycines that, because of their small side chains, can assume a wide range of ϕ and ψ values without severe steric hindrance.

We have seen that the conformation $\phi = 120°$, $\psi = 0-90°$ is unstable. Continued rotation leads to a conformation with $\phi = 131°$ and $\psi = 154°$ in which the carbonyl oxygen is hydrogen-bonded to the amide nitrogen two

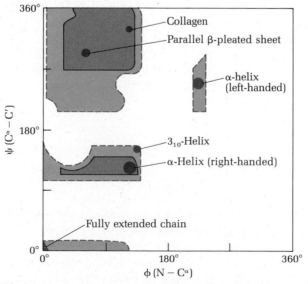

Figure 3.10 Conformational map (Ramachandran diagram) for polypeptides. The solid lines indicate the range of normally observed values of ϕ and ψ in protein structures. The dotted lines give the outer limits observed for most residues in most proteins. The single points are conformations for fibrous proteins in which the conformations of (nearly) all residues are the same.

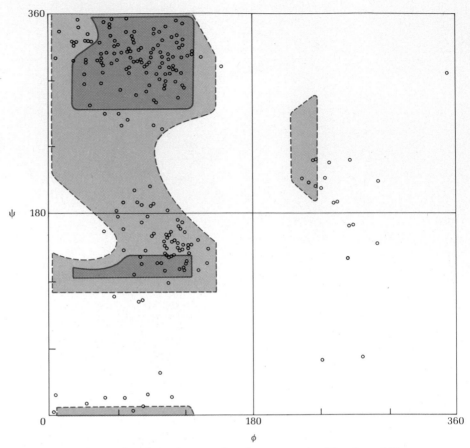

Figure 3.11 Ramachandran diagram for chymotrypsin. The dotted lines represent the outer limits for values of ϕ and ψ. [From J. J. Birktoft and D. M. Blow, *J. Mol. Biol., 68*, 202 (1972). Reproduced with permission. Copyright: Academic Press Inc. (London) Ltd.]

residues away. This is the 3_{10}-helix that we find in type III bends in proteins (Figure 3.12). The α-helix appears when $\phi = 132°$ and $\psi = 123°$. Further rotations of ϕ and ψ produce a π-*helix* in which hydrogen bonds are formed between carbonyl oxygens and amide NH group of every fifth residue (Figure 3.12). Of the three types of helix the α-helix is by far the most common, the 3_{10}-helix is occasionly found in bends, and the π-helix is extremely rare.

3–7 PREDICTION OF SECONDARY STRUCTURE

The secondary structure of a protein is a bridge between its amino acid sequence and its three-dimensional conformation. It would be nice to be able to predict the three-dimensional conformation of a protein given its primary structure, but this goal has not yet been achieved. However, great progress has been made in the preliminary stage of predicting the secondary structure from the primary structure. Perhaps the most successful predictions of secondary structure have involved statistical correlations between the occurrence of a given amino acid residue and the type of secondary structure in which it is likely to be found. Many such attempts have been made. We shall focus upon the results of P. Y. Chou and G. D. Fassman, since their predictions are fairly reliable and easily understood.

Four rules enable us to predict with about 80% accuracy whether a given segment of a polypeptide chain will exist in an α-helix.

1. If four "helical formers," denoted H_α for strong helix formers and h_α for "normal" helix formers, appear over a span of six residues, *helix nucleation* is likely,

3_{10}-Helix

α-Helix

π-Helix

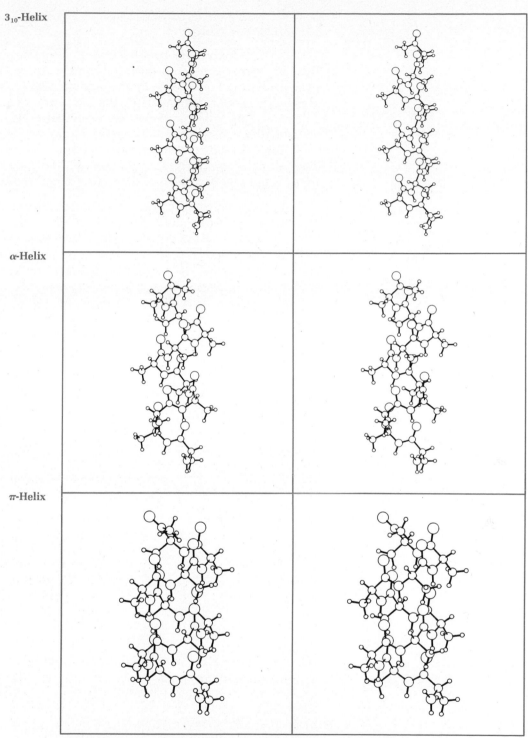

Figure 3.12   Structures of the 3_{10}-helix, α-helix, and π-helix.

and the helix will propagate in both directions. If a sequence of six residues contains two or more "helix breakers," denoted B_α for strong helix breakers and b_α for weak helix breakers, helix formation is unlikely. The I_α residues are weak helix forming residues and the i_α residues are indifferent to helix formation.

   2. An α-helix is terminated by a sequence as short as four residues if there are as many B_α, b_α, I_α, and i_α residues as H_α and h_α residues. Any sequence b_2h_2 in any combination thus breaks an α-helix.

   3. Proline cannot occur within an α-helix or at its C-terminus.

4. Proline and negatively charged aspartate and glutamate all prefer the C-terminus of an α-helix.

Table 3.3 gives for each residue a parameter (P_α) derived from statistical analysis of proteins whose conformations are known. If the sum of P_α's over six residues is equal to or greater than 1.03 (and if $P_\alpha > P_\beta$, a parameter we shall consider momentarily), the formation of an α-helix is predicted. If rules 1 to 4 are followed, however, a calculation is usually superfluous. The statistical parameters should *not* be memorized.

Table 3.4 ranks the stabilities of amino acid residues in β-pleated sheet. The following set of rules, again derived by P. Y. Chou and G. D. Fassman, gives about a 90% chance of predicting whether a β structure will form.

1. If a cluster of three β-forming residues, defined in Table 3.4, exists within a five-residue sequence, β structure is predicted. If the same five-residue sequence contains two or more β-breaking residues (B_β or b_β), β structure formation is unfavorable.

2. β structures are terminated by sequences of four residues containing two or less β-forming residues (H_β or h_β). Thus, a sequence of four residues is required to terminate either an α-helix or a β structure, except proline which always terminates an α-helix and nearly always breaks a β structure.

3. Negatively charged glutamate is rarely found within a β structure.

4. There are few charged residues in β structures. Tryptophan is found, if at all, at the N-terminus of the β structure and almost never at the C-terminus of the β structure.

Over a sequence of five residues a β-pleated sheet is favored if the sum of β-sheet parameters (P_β), like the P_α's derived from a statistical analysis of

Table 3.3	Tendency of amino acids to be found in α-helix and their effect on the stability of the α-helix
Amino acid residues	P_α
Strong-helix forming residues (H_α)	
Glu$^\ominus$	1.53
Ala	1.45
Leu	1.34
Normal helix-forming residues (h_α)	
His$^\oplus$	1.24
Met	1.20
Gln	1.17
Trp	1.14
Val	1.14
Phe	1.12
Weak helix-forming residues (I_α)	
Lys$^\oplus$	1.07
Ile	1.00
Residues indifferent to α-helix formation (i_α)	
Asp$^\ominus$	0.98
Thr	0.82
Ser	0.79
Arg$^\oplus$	0.79
Cys	0.77
Weak helix-breaking residues (b_α)	
Asn	0.73
Tyr	0.61
Strong helix-breaking residues (B_α)	
Pro	0.59
Gly	0.53

SOURCE: Adapted from P. Y. Chou and G. D. Fassman, *Biochemistry, 13,* 223 (1974). Reprinted with permission. Copyright 1974 American Chemical Society.

Table 3.4 Tendency of amino acids to be found in β structure and their effect on the stability of β structure

Amino acid residues	P_β
Strong β-forming residues (H$_\beta$)	
Met	1.67
Val	1.65
Ile	1.60
β-forming residues (h$_\beta$)	
Cys	1.30
Tyr	1.29
Phe	1.28
Gln	1.23
Leu	1.22
Thr	1.20
Trp	1.19
Weak β-forming residues (I$_\beta$)	
Ala	0.97
Residues indifferent to β formation (i$_\beta$)	
Arg	0.90
Gly	0.81
Asp$^\ominus$	0.80
Weak β-breaking residues (b$_\beta$)	
Lys	0.74
Ser	0.72
His	0.71
Asn	0.65
Pro	0.62
Strong β-breaking residues (B$_\beta$)	
Glu$^\ominus$	0.26

SOURCE: Adapted from P. Y. Chou and G. D. Fassman, *Biochemistry, 13*, 223 (1974). Reprinted with permission. Copyright 1974 American Chemical Society.

proteins whose three-dimensional structures are known, is equal to or greater than 1.05, and if the sum of the P_β's is greater than the sum of the P_α's. Again, if rules 1 to 4 are followed, a calculation is usually unnecessary; don't memorize the P_β values.

3–8 COLLAGEN

Collagen, the most abundant protein in many vertebrates and invertebrates, is a structural protein that provides mechanical strength to bone, tendon, cartilage, and skin. The term collagen embraces at least four types of proteins called collagens I, II, III, and IV, whose distribution is summarized in Table 3.5. We will restrict most of our discussion to collagen I. The collagen I of tendons has an incredible tensile strength of 20 to 30 kg/mm², roughly the tensile strength of 12 gauge, hard-drawn copper wire. Collagen fibers in bone provide a matrix about which hydroxyapatite (a calcium phosphate polymer) crystals are arranged. The skin of vertebrates contains rather loosely woven collagen fibers that can expand in all directions. Blood vessels also contain collagen fibers arranged in helical networks. As these few examples show, the structural properties of collagen lead to important physiological functions. Lest we leave the impression that collagen is confined to vertebrates, Table 3.6 lists its distribution in invertebrate phyla.

Collagen I is constructed from a triple-stranded, helical fiber composed of two types of similar monomers called $\alpha1$ and $\alpha2$ in the ratio $(\alpha1)_2(\alpha2)$. The chains are exceedingly long, about a thousand residues each, and determining their primary structures was a difficult task. Table 3.7 gives the sequence of the first 139 residues of the N-terminal end of an $\alpha1$ chain of rat skin collagen. Two hydroxylated amino acids, *hydroxyproline* (Hyp) and

Table 3.5 Structurally and genetically distinct collagens

Type	Distribution	Chain composition	Distinctive features
I	Skin, bone, tendon, ligament, fascia, dentin, blood vessels, interstitial connective tissues	$[\alpha 1(I)]_2 \alpha 2$	Hybrid of two chain types; low (15%) hydroxylation of lysine; low carbohydrate
II	Cartilage, nucleus pulposus	$[\alpha 1(II)]_3$	Intermediate (50%) hydroxylation of lysine; all hydroxylysines glycosylated
III	Same as type I, except bone and tendon; prominent in blood vessels, gastro-intestinal tract, fetal skin	$[\alpha 1(III)]_3$	Contains cysteine; 4-Hyp > Pro; Gly > $\frac{1}{3}$ residues; low (15%) hydroxylation of lysine
IV	Basement membranes	$[\alpha 1(IV)]_3$	High 3-Hyp (1%); contains cysteine; most lysines hydroxylated; all hydroxylysines glycosylated; low alanine; carbohydrate content not limited to glucose and galactose

SOURCE: From P. Bornstein and W. Traub, The Chemistry and Biology of Collagen, in H. Neurath and R. L. Hill, Eds., "The Proteins," 3rd ed., Vol. 4, Academic Press, New York, 1979, p. 430.

Table 3.6 Distribution of collagen in invertebrate phyla

Phylum	Organism
Porifera	*Hippospongia* (sponge)
	Haliclona (sponge)
	Chondrosia (sponge)
	Ircinia (sponge)
Coelenterata	*Physalia*
	Metridium (sea anemone)
	Actinia (sea anemone)
	Balticina (sea pen)
Platyhelminthes	*Fasciola* (liver fluke)
Aschelminthes	*Ascaris* (round worm)
Ectoprocta	*Triphyllozoon*
Acanthocephala	*Macracanthorhynchus*
Annelida	*Lumbricus* (earthworm)
	Octolasium (earthworm)
Echinodermata	*Thyone* (sea cucumber)
	Stichopus (sea cucumber)
	Holothuria (sea cucumber)
	Asterias (starfish)
	Strongylocentrotus (sea urchin)
Arthropoda	*Portunus* (crab)
	Panulirus (lobster)
	Leucophea (cockroach)
Mollusca	*Mytilus* (mussel)
	Octopus (octopus)
	Haliotis (abalone)
	Helix (snail)

SOURCE: From P. Bornstein and W. Traub, The Chemistry and Biology of Collagen, in H. Neurath and R. L. Hill, Eds., "The Proteins," 3rd ed., Vol. 4, Academic Press, New York, 1979, p. 430.

Table 3.7 Amino acid sequence of the first 139 residues from the n-terminal end of the α1 chain of rat skin collagen

pGlu -Met-Ser -Tyr -Gly-Tyr-Asp -Glu-Lys -Ser -Ala-Gly -Val -Ser -Val -15
Pro -Gly -Pro -Met -Gly-Pro -Ser -Gly-Pro -Arg -Gly-Leu -Hyp-Gly-Pro -30
Hyp-Gly -Ala -Hyp-Gly-Pro -Gln -Gly-Phe -Gln -Gly-Pro -Hyp-Gly-Glu -45
Hyp-Gly -Glu -Hyp-Gly-Ala -Ser -Gly-Pro -Met -Gly-Pro -Arg -Gly-Pro -60
Hyp-Gly -Pro -Hyp-Gly-Lys -Asn -Gly-Asp-Asp -Gly-Glu -Ala -Gly-Lys -75
Pro -Gly -Arg -Hyp-Gly-Gln -Arg -Gly-Pro -Hyp-Gly-Pro -Gln -Gly-Ala -90
Arg -Gly -Leu -Hyp-Gly-Thr -Ala -Gly-Leu -Hyp-Gly-Met -Hyl -Gly-His -105
Arg -Gly -Phe -Ser -Gly-Leu -Asp -Gly-Ala -Lys -Gly-Asn -Thr -Gly-Pro -120
Ala -Gly -Pro -Lys -Gly-Glu -Hyp-Gly-Ser -Hyp-Gly-Glx -Asx -Gly-Ala -135
Hyp-Gly -Gln -Met -

Notice that the triplet Gly-X-Y- is repeated many times. Hydroxyproline is found immediately before glycine in the common sequence Gly-X-Hyp-Gly-.

hydroxylysine (Hyl), appear in this sequence. They account for nearly a fourth of the amino acid residues in collagen. Beginning with residue 17, glycine

4-Hyroxyproline   5-Hydroxylysine

is in every third position and hydroxyproline is found immediately before glycine. A typical sequence is Gly-X-Y-Gly-Pro-Y-Gly-X-Hyp-Gly. The sequence Gly-Pro-Hyp-Gly is also common. Collagen is thus the second of three proteins known to have regularly repeating amino acid sequences. The amino acid composition of collagen is given in Table 3.8. There are no

Table 3.8 Amino acid composition of bovine collagens

Amino acid	Number per 1000 residues				
	α1(I)	α2	α1(II)	α1(III)	α1(IV)
3-Hydroxyproline	1	—	2	—	7
4-Hydroxyproline	85	85	91	127	133
Aspartic acid	45	47	43	48	51
Threonine	16	17	22	14	20
Serine	34	24	26	44	37
Glutamic acid	77	71	87	71	79
Proline	135	120	129	106	65
Glycine	327	328	333	366	328
Alanine	120	101	102	82	37
Cysteine	—	—	—	2	1
Valine	18	34	17	12	28
Methionine	7	4	11	7	13
Isoleucine	9	17	9	11	29
Leucine	21	34	26	15	52
Tyrosine	4	3	1	3	2
Phenylalanine	12	16	14	9	29
Hydroxylysine	5	11	23	7	49
Lysine	32	21	15	25	9
Histidine	3	8	2	8	6
Arginine	50	57	51	44	26

SOURCE: From P. Bornstein and W. Traub, The Chemistry and Biology of Collagen, in H. Neurath and R. L. Hill, Eds., "The Proteins," 3rd ed., Vol. 4, Academic Press, New York, 1979, p. 453.

cysteines in collagen, a striking finding when we recall that the polypeptide subunits contain over a thousand amino acid residues.

The enzyme responsible for hydroxylation of proline is *prolylhydroxylase*. Lysine is hydroxylated by *lysylhydroxylase*, an enzyme having many properties in common with prolyhydroxylase. Both enzymes require molecular oxygen, *ascorbic acid* (vitamin C), the dicarboxylic acid α-ketoglutarate,

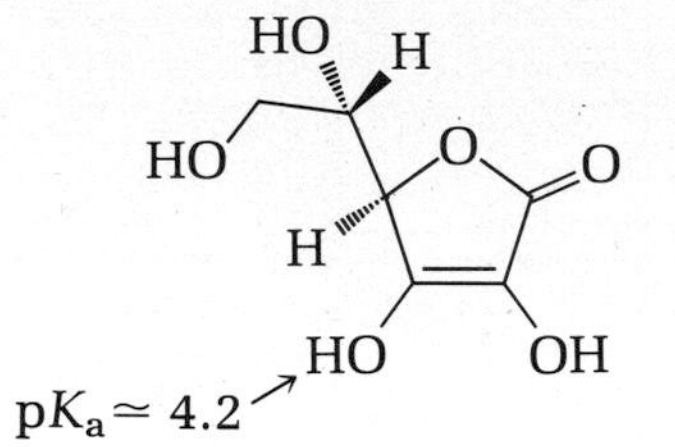

$$^{\ominus}O_2C-CH_2CH_2-\overset{\overset{\displaystyle O}{\|}}{C}-CO_2^{\ominus}$$

α-Ketoglutarate

$pK_a \simeq 4.2$

Ascorbic acid, reduced form

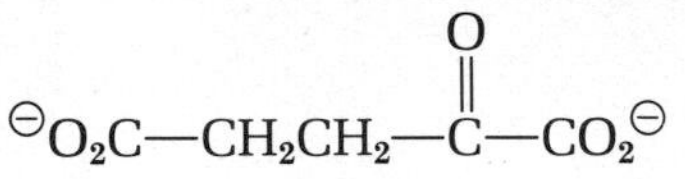

Ascorbic acid

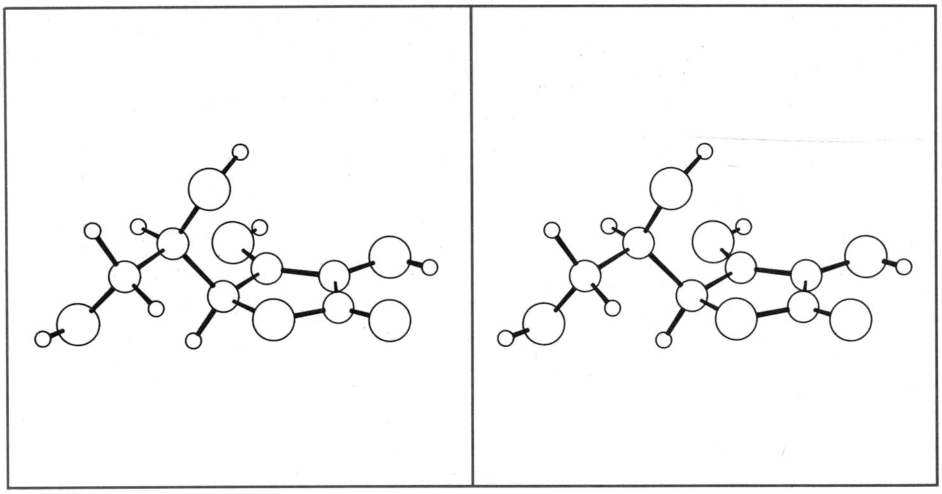

and Fe(II). The hydroxylation reaction for proline is illustrated for the tripeptide Gly-Pro-Pro (reaction 3.1).

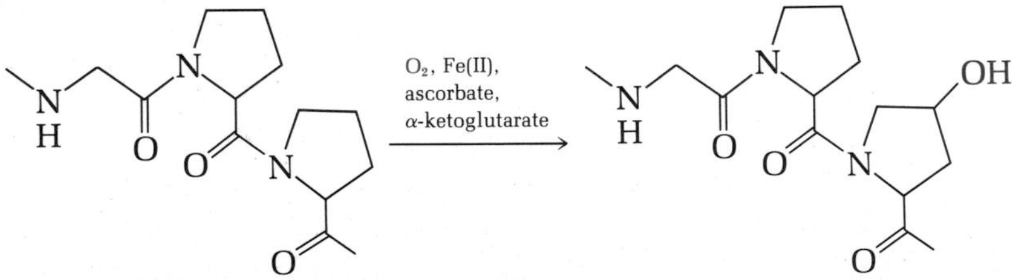

$$\xrightarrow{\begin{array}{c}O_2,\ Fe(II),\\ \text{ascorbate},\\ \alpha\text{-ketoglutarate}\end{array}}$$

(3.1)

The hydroxyl group of hydroxyproline residues in collagen help to maintain the triple helical structure by interchain hydrogen bonding. Procollagen triple helices in which the prolines have not been hydroxylated are far less stable than their hydroxylated counterparts. The hydroxyl groups of hydroxylysine are the sites of attachment of carbohydrates, and they also participate in interchain cross-links to be discussed below.

The role of vitamin C in the hydroxylation of collagen is one of the major physiological effects of this vitamin. Persons deprived of vitamin C, as were sailors on long voyages until the eighteenth century, develop *scurvy*, a disease of collagen metabolism. Persons suffering from scurvy have skin lesions, fragile blood vessels, and bleeding gums. The British Navy solved its scurvy problem by including limes, a fruit rich in vitamin C, in the daily diet of its sailors. The epithet "limey," as slang for British, entered the English language as a result.

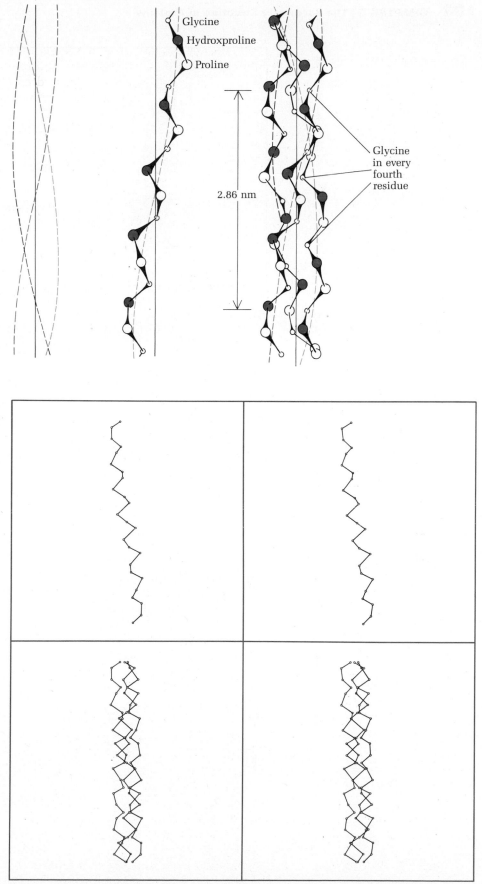

Figure 3.13 Structure of collagen I. The monomers of the tropocollagen triple helix are coiled around each other in a right-handed helix. Glycine lies along the central axis of the helix where its side chain fits nicely. The carbonyl oxygen of X in the triplet Gly-X-Y is hydrogen bonded to the amide hydrogen of glycine. [From A. Rich and F. H. C. Crick, *Nature, 176,* 915 (1955).]

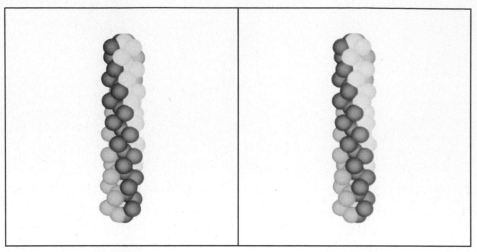

Figure 3.13 (*Continued*)

The structure of collagen I has been tentatively deduced by a combination of X-ray crystallography, electron microscopy, and model building. Each of the α1 and α2 subunits of procollagen has a left-handed helical conformation, and the three left-handed helices coil around each other to produce a right-handed procollagen fibril whose pitch is 0.87 nm (Figure 3.13). The coil advances one turn every 3.3 amino acid residues.

Most of the hydrogen bonding in the procollagen triple helix is *intermolecular*. The amide hydrogen of a given residue is hydrogen bonded to a carbonyl oxygen of a residue in an adjacent chain in which the oxygen occupies the second position of the triplet Gly-X-Y. Glycine lies near the central axis of the triple helix in a position that cannot be occupied by any other residue. Only glycine has a small enough "side chain" to fit into the structure (Figures 3.14 and 3.15).

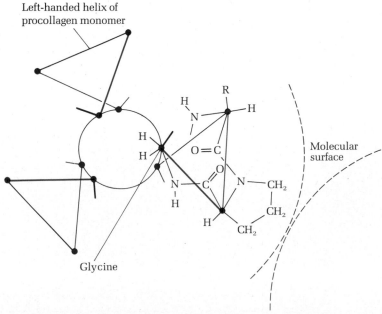

Figure 3.14 Cross section of collagen. Every third residue is glycine, the only amino acid whose side chain is small enough to lie along the central axis of the triple helix. [Courtesy of Dr. Lynn Jelinsky.]

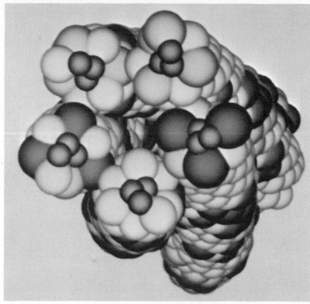

Figure 3.15 A space-filling model of collagen viewed down the long axis in the same perspective as shown in Figure 3.13. [Courtesy of Dr. Lynn Jelinsky.]

Some of the lysine residues of collagen are oxidized by a copper-containing enzyme *lysine oxidase*. The reaction is an *oxidative deamination* and the product is *allysine*.

Lysyl residue   Allysyl residue

An allysyl residue can then react with the ε-amino group of a lysyl residue of an adjacent chain giving a cross-linked fiber. The cross-linked residue is *dehydrolysinonorleucine* (Figure 3.16). The extensive cross-linking of intertwined proteins gives an extremely strong fiber.

Two allysyl residues can react with each other in an *aldol condensation* to produce another type of cross-linked residue (Figure 3.17).

Collagen fibers are built up from cross-linked tropocollagen. An electron micrograph of rat tail collagen reveals regularly spaced bands at right angles to the long axis of the fiber (Figure 3.18). The banding has been interpreted in terms of a "quarter-staggered" array of tropocollagen fibrils. A schematic diagram of the structure of collagen based on this model is shown in Figure 3.19.

3–9 ELASTIN Elastin is another protein component of connective tissue whose most important property—elasticity—gives it its name. Elastin is particularly abundant in blood vessels and ligaments. Elastin is rich in glycine and proline. Unlike collagen, hydroxylysine is not found and hydroxyproline is rare. Two repeating sequences are found: Lys-Ala-Ala-Lys and Lys-Ala-Ala. How these sequences are related to the structure of elastin is not known.

Elastin monomers are cross-linked in two ways. (1) As in collagen, oxidation of lysyl residues is followed by condensation of the aldehyde group

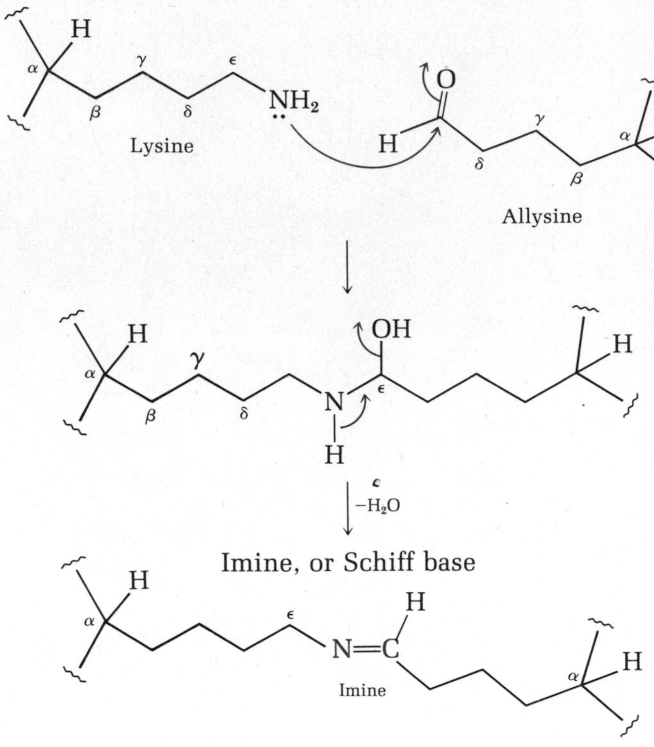

Figure 3.16 Condensation of allysine and lysine to give dehydrolysinonorleucine.

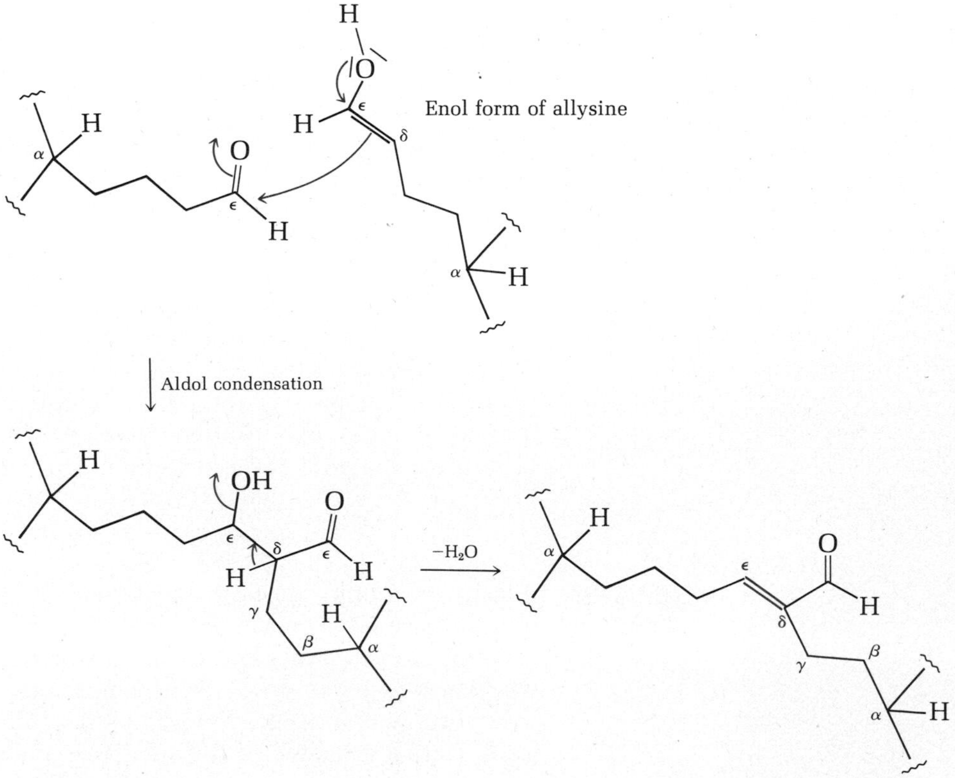

Figure 3.17 Aldol condensation between two allysyl residues producing a cross-link in collagen.

Figure 3.18 Photograph of an electron micrograph of intact calf skin collagen fibrils, deposited from suspension and shadowed with chromium. The magnification is 59,000×. [Courtesy of Dr. Jerome Gross.]

with the ϵ-amino group of another lysine. In elastin the double bond of the imine is reduced to give *lysinonorleucine* (Figure 3.20). (2) A second type of cross-linking involves three allysyl residues which condense with a lysyl residue to give *desmosine* (Figure 3.21).

Desmosine

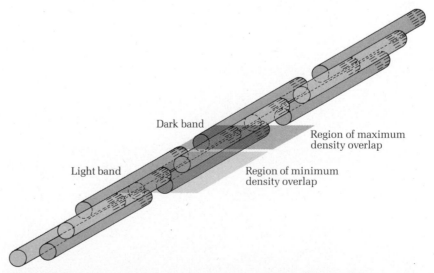

Figure 3.19 "Quarter-staggered" array of tropocollagen fibrils. One possible explanation for the banding observed in collagen fibrils involves the overlap of four tropocollagen units. [From A. Veis, R. S. Bhatnagar, C. A. Shuttleworth, and S. Mussel, *Biochim. Biophys. Acta.*, *200*, 97 (1970).]

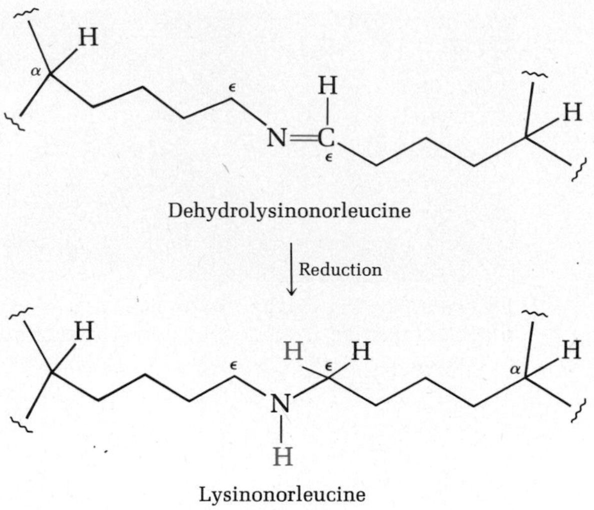

Dehydrolysinonorleucine

Reduction

Lysinonorleucine

Figure 3.20 Dehydrolysinonorleucine residues reduced to lysinonorleucine residues in elastin.

3–10 SUMMARY The secondary structure of a protein consists of regularly repeating conformations of the polypeptide backbone.

Three types of secondary structures are especially important. In the α-helix every carbonyl oxygen and amide nitrogen is involved in a hydrogen bond. These hydrogen bonds are roughly parallel to the long axis of the helix. Each turn of an α-helix spans 3.6 amino acid residues. The α-keratins

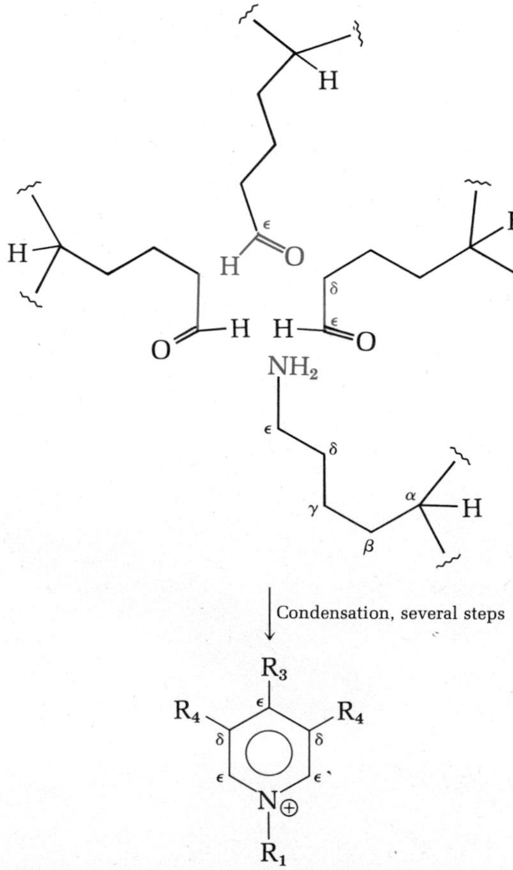

Condensation, several steps

Figure 3.21 Type of cross-link in elastin in which three allysyl residues and one lysyl residue condense to give desmosine.

are composed almost entirely of α-helical polypeptide chains. These chains are coiled round each other in a coiled-coil to give a supersecondary structure. The second major secondary structure is the β-pleated sheet which involves hydrogen bonds between the carbonyl oxygens and amide nitrogens of two separate chains or between two segments of the same chain. The hydrogen bonds are extended at approximately right angles to the long axes of the chains, which may run parallel or antiparallel to each other. Most β-pleated sheets are twisted slightly to alleviate steric strain between the side chains. Silk fibroin is composed almost entirely of a twisted β-pleated sheet. β-Pleated sheets are also a common feature of globular proteins. The third major type of secondary structure involves reverse turns, or bends, in the polypeptide chain. Supersecondary structures are often found in globular proteins in which β-pleated sheets are connected by regions of irregular conformation, by α-helices, or by reverse turns.

The conformation about a given peptide bond is specified by two angles (ϕ,ψ). Specifying these angles for each peptide bond in the protein defines the conformation of the polypeptide backbone. Statistical methods have been used to predict the occurrence of secondary structures with about 80% accuracy.

Collagen, the most abundant structural protein in vertebrates and invertebrates, provides mechanical strength in bone, tendon, cartilege, and skin. Two novel amino acids found in collagen are 4-hydroxyproline and 5-hydroxylysine. The hydroxyl groups are added after the polypeptide chains of collagen have been synthesized. Procollagen I, the immediate precursor of mature collagen, is a triple helical protein that is stabilized by intermolecular hydrogen bonding. Collagen is also stabilized by extensive covalent cross-linking between allysine residues and between allysine and lysine. Collagen fibers are composed of collagen triple helices in a "quarter staggered array" of fibrils generating a structure of enormous mechanical strength.

Elastin as an elastic structural protein that is found in connective tissue, especially in blood vessels and ligaments. Elastin is covalently cross-linked by lysinonorleucine and desmosine. It is not a fibrous protein, but a polymer of globular proteins in a fibrous array.

REFERENCES

Blundell, T., Dodson, G., Hodgkin, D., and Mercola, D., *Adv. Protein Chem.*, 26, 279–402 (1972).

Bornstein, P., The Biosynthesis of Collagen, *Ann. Rev. Biochem.*, 43, 567–604 (1974).

Bornstein, P., and Traub, W., The Chemistry and Biology of Collagen, in H. Neurath and R. L. Hill, Eds., "The Proteins," 3rd ed., Academic Press, 1979, pp. 411–632.

Cantor, C. R., and Schimmel, P. R., "Biophysical Chemistry," Vol. 1, "The Conformation of Biological Molecules," Freeman, San Francisco, 1980.

Dickerson, R. E., and Geis, I., "The Structure and Action of Proteins," 2nd ed., W. A. Benjamin, Menlo Park, Calif., 1981.

Eyre, D. R., Collagen: Molecular Diversity in the Body's Protein Scaffold, *Science*, 207, 1315–1322 (1980).

Gallop, P. M., Blumenfield, O. O., and Seifter, S., Structure and Metabolism of Connective Tissue Proteins, *Ann. Rev. Biochem.*, 41, 617–672 (1972).

Pauling, L., and Corey, R. B., *Proc. Nat. Acad. Sci. U.S.*, 37, 729 (1951).

Shulz, G. E., and Schirmer, R. H., "Principles of Protein Structure," Springer-Verlag, New York, 1979.

Problems

1. Most animals are unable to digest wool because of its extensive disulfide cross-linking and consequent insolubility. The clothes moth, however, has a high concentration of hydrosulfides in its digestive system. Why does this help it to digest wool?

Glu-Pro-Glu-Pro-Asp-Pro-Glu-Ala-Gly-Ile-Gly-Ala-Val-Leu-Lys-Val-Leu-Thr-Thr-Gly-Leu-Pro-Ala-Leu-Ile-Ser-Trp-Ile-Lys-Arg-Lys-Arg-Gln-Gln

Promelitin ──────

Melitin ──────

2. Why does cross-linking lower the solubility of wool and collagen?

3. A partial sequence of minke whale myoglobin is shown below. (a) Where would you expect α-helix to be most stable? (b) Over what sequence of residues would you expect α-helix to be least stable?

Leu-Lys-Lys-Lys-Gly-His-His-Glu-Ala-Glu-Leu-Lys-Pro-Leu-Ala

4. The honey bee synthesizes its venom as an inactive precursor, promelitin, which is enzymatically cleaved to give the active venom, melitin. The amino acid sequences of promelitin and melitin are shown at left. (a) In which regions of promelitin do you expect to find α-helix? (b) In which regions of melitin, if any, do you expect to find β structure?

5. The dimensions of the peptide glutathione are shown in the figures below (recall Section 2.6B). The one on the left gives the bond lengths (in Å) and the one on the right the bond angles of the molecule. (a) Identify the peptide bonds. (b) Decide whether the peptide bonds are *cis* or *trans* in each case. (c) Identify all angles ϕ and ψ. [Structure from W. B. Wright, *Acta Cryst.*, **11**, 632 (1958).]

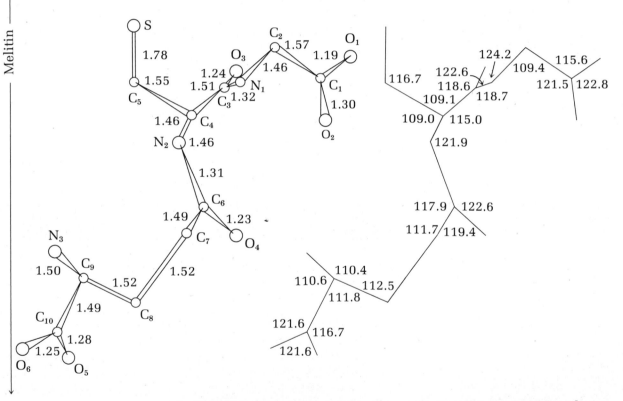

6. The first 15 residues of the plasma lipoprotein apoC-II are shown below. Indicate the region(s) where you expect to find α-helix.

Thr-Glu-Gln-Pro-Gln-Gln-Asp-Glu-Met-Pro-Ser-Pro-Thr-Phe-Leu-

Chapter 4

The conformation and function of globular proteins

4–1 INTRODUCTION

In the preceding two chapters we have examined the three lowest levels of protein structure—the sequence of amino acid residues in the polypeptide chain (the primary structure), the regularly repeating conformations of the polypeptide backbone (the secondary structure), and patterns of secondary structure (the supersecondary structure). We have also glimpsed the concept of structural domains, or assemblies of supersecondary structures, the next higher level of structure in globular proteins (which receive their generic name from their shapes). In this chapter we shall consider the conformations of polypeptide chains, known as the tertiary structure, expand briefly upon structural domains, and then consider the aggregates of polypeptide chains that constitute the quaternary structure of certain proteins.

4–2 FORCES RESPONSIBLE FOR MAINTAINING THE TERTIARY STRUCTURES OF PROTEINS

The forces responsible for maintaining the tertiary structures of proteins are the so-called weak forces consisting of van der Waals forces, hydrogen bonds, and electrostatic interactions. It is nearly impossible to overstate the importance of these forces in the biological world: they are responsible for the folding of polypeptide chains, the conformations of nucleic acids, and the structure of biological membranes. In fact, this section might be entitled "forces responsible for maintaining the structures of biological systems" and not be very wide of the mark.

A. Van der Waals Forces and the Hydrophobic Effect

The folding of a polypeptide chain into its three-dimensional conformation brings most of the nonpolar side chains to the interior where they have minimal contact with water. Simultaneously, the polar side chains tend to move to the surface of the protein where they interact strongly with water. Perhaps surprisingly, it is the interactions of the nonpolar side chains with each other in the interior of the protein that largely determines the final conformation of the protein. The types of interactions available to the nonpolar side chains are the van der Waals forces, consisting of dipole-dipole interactions, dipole-induced dipole interactions, and induced dipole-induced dipole interactions. The van der Waals forces originate in the infinitesimal dipole generated in all atoms by the movement of electrons about the nucleus. Each miniature dipole can exert a small attraction upon a similar dipole in a neighboring atom. (These are the interactions that are responsible for the nonideal behavior of gases.) Individually, the van der Waals forces are in the range of 0.5 kJ mol^{-1}, 200 times weaker than a covalent bond and some 40 times weaker than a typical hydrogen bond. These interactions are, however, both numerous and additive, and their cumulative effect dominates protein folding.

The overall tendency of the side chains of nonpolar residues to collect in the interior of the protein is known as the *hydrophobic effect*. The interactions of the nonpolar side chains with one another are greatly preferred over the interaction of the nonpolar side chains with water. The nonpolar side chains can pack in an enormous number of ways. A few of the possibilities are illustrated in Figure 4.1. (The thermodynamic basis of the hydrophobic effect, a topic that includes the effects of the nonpolar side chains upon the structure of water, is discussed in Section 4.2D.)

B. Hydrogen Bonding

Hydrogen bonds play an enormous role in maintaining the structure of proteins. We have already discussed the hydrogen bonds formed between the carbonyl oxygen and amide hydrogen along the polypeptide that stabilize the secondary structures of proteins. Let us consider the hydrogen bonds that are possible between side chains and water and between different side chains.

All the hydrogen bonding capabilities of a given residue are satisfied. The carbonyl and amide groups of the peptide backbone are either hydrogen bonded in secondary structure, hydrogen bonded to water, or hydrogen bonded to a polar side chain residue. Since the free energy of formation

Figure 4.1 Schematic diagram of some types of hydrophobic interactions observed in proteins between pairs of side chains. [From G. Nemethy and H. A. Scheraga, *J. Phys. Chem.*, 66, 1773 (1962).]

Alanyl-alanyl

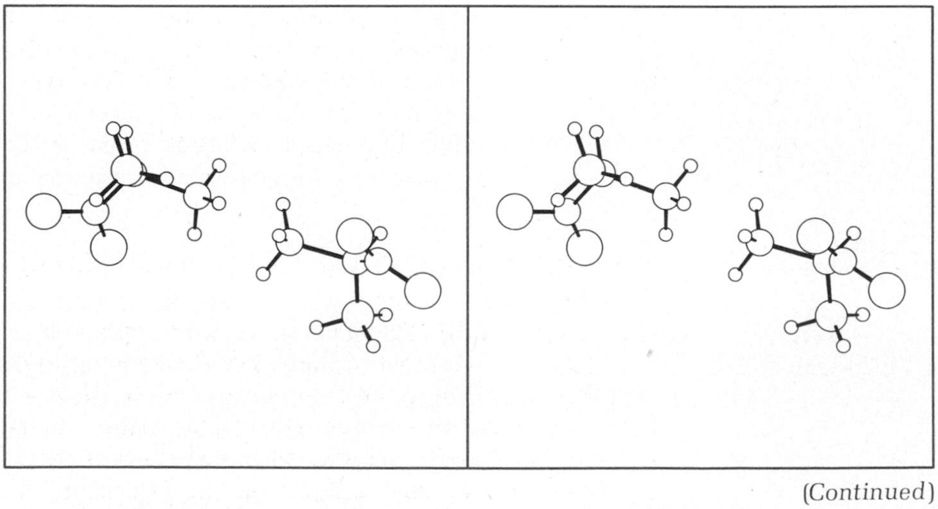

Isoleucyl-isoleucyl

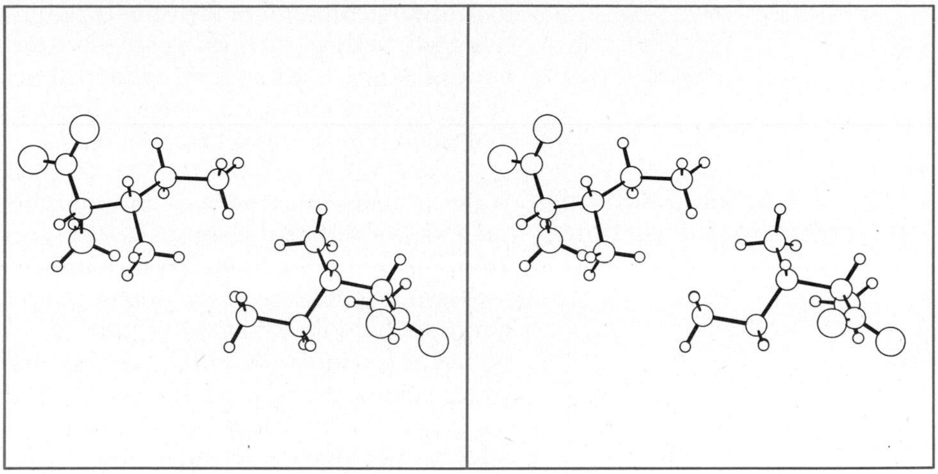

Phenylalanyl-leucyl

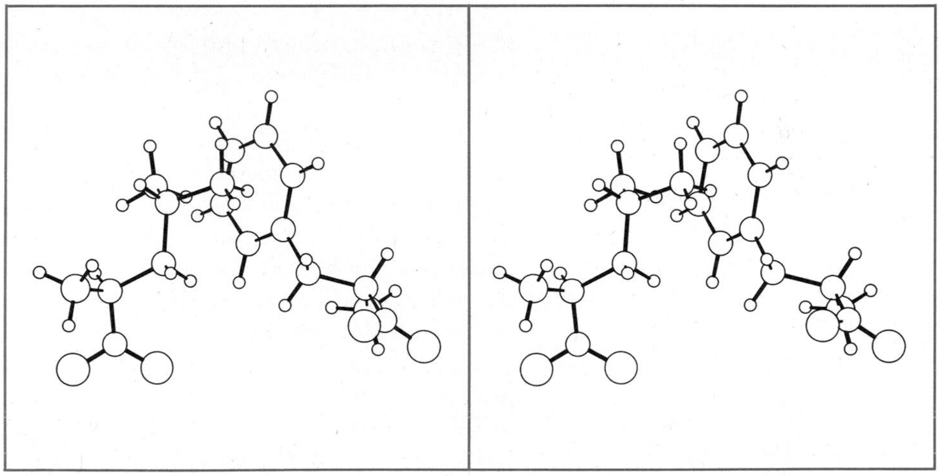

Phenylalanyl-phenylalanyl

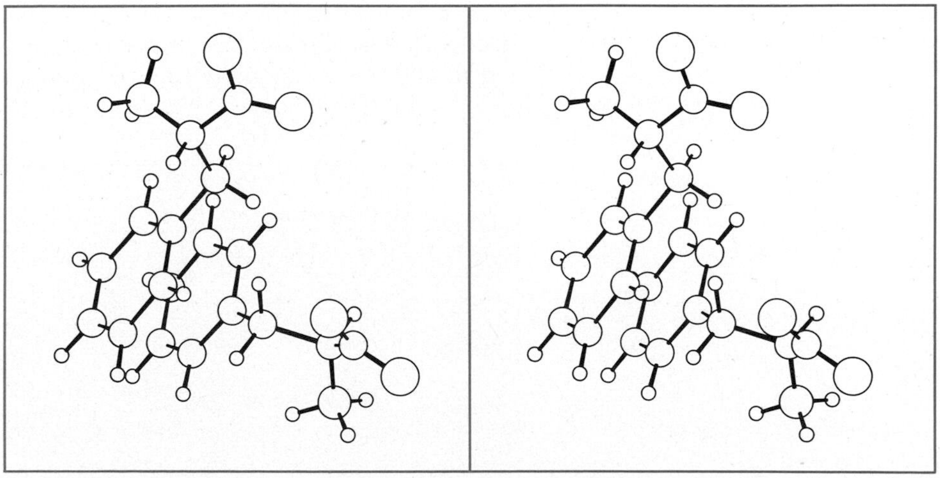

of a hydrogen bond is in the neighborhood of 20 kJ mol⁻¹, and since the contributions of the hydrogen bonds to the stability of the protein are additive, hydrogen bonding is an important factor in the stability of the protein.

We noted in the preceding chapter that α-helices tend to lie on the surfaces of globular proteins. The twist of the helix therefore brings some residues into contact with water, while others point toward the interior. Analysis of the primary structure in such cases reveals a fairly regular pattern in which the nonpolar residues appear in the helix where they can nestle in the hydrophobic interior of the protein, and the polar residues are found at positions where they can form hydrogen bonds with water. Most protein-protein hydrogen bonds are found in the interior of the protein where there is little or no competition with water. Figure 4.2 summarizes the major types of hydrogen bonds found in proteins.

C. Ionic Bonding (Electrostatic Interactions)

Water is an excellent solvent for ions, and in most cases charged side chains, such as the carboxylate anions of glutamate and aspartate and the ϵ-amino group of lysine, are found on the surface of globular proteins. Occasionally, however, the folding of the polypeptide carries an ionic side chain into the hydrophobic interior of the protein where little water is available for solvation. For example, in milk β-lactoglobulin there are two buried carboxyl groups. Since they are in an environment in which they are not well solvated, the pK_a's of these carboxyl groups are shifted higher by two pH units. In other words, the buried carboxyl groups tend to be far more stable as carboxylic acids than as carboxylate anions. The free-energy change for transferring an ionic side chain from the aqueous surface into the nonpolar interior of a protein is unfavorable by about 40 kJ mol^{-1}, and when such residues do appear in the interior, they are usually strongly hydrogen bonded to other groups.

Salt bridges between oppositely charged side chains have been observed in several proteins. The free energy of transfer of two oppositely charged residues from their aqueous environment to the interior is about -4 kJ mol^{-1}. We might expect this value to be much higher, but on the aqueous surface of the protein these charged groups are tightly hydrogen bonded to the solvent, and all of these hydrogen bonds are lost when the side chains move into the interior. In fact, the water is highly structured around these charged residues, and when they move into the interior of the protein, this water of solvation is released to the bulk solvent. The driving force for salt bridge formation is actually entropic and not simply a matter of electrostatic attraction, as we might at first suppose.

In our discussion of protein structure we shall encounter several cases in which charged groups are in the hydrophobic interior and nonpolar groups are on the surface exposed to water. These apparent anomalies are explained if we remember that the stable conformation of a protein depends

Type of hydrogen bond		Distance between donor and acceptor atom (nm)
Hydroxyl-hydroxyl	$-O-H \cdots O-$ $ H$	0.28
Hydroxyl-carbonyl	$-O-H \cdots O=C$	0.28
Amide-carbonyl	$N-H \cdots O=C$	0.29
Amide-hydroxyl	$N-H \cdots O-$ H	0.28
Amide-imidazole nitrogen	$N-H \cdots N \diagup NH$	0.31
Amide-sulfur	$N-H \cdots S$	0.37

Figure 4.2 Hydrogen bonds found in proteins. The strength of the hydrogen bonds decreases with increasing distance between the donor and acceptor atoms. [Adapted from G. E. Shulz and R. H. Schirmer, "Principles of Protein Structure," Springer-Verlag, New York, 1979, p. 35.]

upon a subtle balance of forces in which the energy of the entire system is minimized.

D. Thermodynamics of Protein Folding

We have already noted that nonpolar side chains tend to collect in the interior of globular proteins, a phenomenon known as the hydrophobic effect. Let us consider the transfer of hydrocarbons from an organic solvent to water as a model system for the hydrophobic effect (reaction 4.1).

$$\text{RH(organic solvent)} \xrightleftharpoons{K_{transfer}} \text{RH(aq)} \tag{4.1}$$

$K_{transfer}$ in reaction 4.1 is the equilibrium constant for the process (which is identical to extracting a compound from an organic solvent to water). Data for some representative compounds, models for the side chains of hydrophobic residues, are summarized in Table 4.1. These data are in some respects surprising. The free energies of transfer are all positive, and the equilibrium constants for transferring a nonpolar hydrocarbon from an organic solvent to water are all less than 1.0 as we expect; after all, we know that hydrocarbons are insoluble in water. By analogy, hydrocarbon side chains of amino acid residues in proteins prefer contact with one another to contact with water. But look at the values for ΔH^0, ΔS^0, and $T\Delta S^0$. Every process is *exothermic*; and the unfavorable free-energy change reflects an unfavorable *entropy* change. When the nonpolar hydrocarbon enters the aqueous phase, it causes more hydrogen bonding between water molecules. In an extreme case, the water molecules form a cage, called a *clathrate*, around the hydrocarbon (Figure 4.3). The water molecules are more structured in the clathrate than in the bulk solvent where they can wander about freely, so the entropy of the system decreases when the clathrate forms. If we imagine a protein in an extended random-coil conformation, its folding to a compact globular structure is driven by the *increase* in entropy caused by releasing water molecules from their clathrates to the bulk solvent. (It is likely that the nonpolar side chains have less ordered cages around them than the clathrate shown in Figure 4.3, but the general principle is not altered by that detail.)

In contrast, the ionic side chains interact strongly with water. The charged alkyl ammonium ion of lysine and the carboxylate anion of aspartate have free energies of solvation (defined as the energy released when a molecule is transferred from the gas phase to the solvent) in the range of −290 to −330 kJ mol^{-1}. There are important special cases where ions are found buried in the hydrophobic interior, out of contact with water, but in most cases these groups are exposed to water because of their high solvation energies. Similarly, polar, uncharged residues are also usually exposed to the aqueous phase, not to the hydrophobic interior. A polar uncharged molecule such as ethanol, a model for the side chain of serine, has a solvation energy of about −20 kJ mol^{-1}. The substitution of an amino acid with an ionic or polar side chain for a hydrophobic residue often drastically alters the structure, and thus also the function, of the protein because of the large tendency of the polar side chain to move into the polar solvent.

Table 4.1 Thermodynamic parameters for transfer of hydrocarbons from nonpolar to polar solvents at 25°C

Process	ΔH° (kJ mol^{-1})	ΔS° (J K^{-1} mol^{-1})	$T\Delta S$ (kJ mol^{-1})	ΔG° (kJ mol^{-1})	$K_{transfer}$
CH_4 (in benzene) → CH_4 (in water)	−11.7	−75.3	−22.4	10.9	1.24×10^{-2}
CH_4 (in CCL_4) → CH_4 (in water)	−10.5	−75.3	−22.4	12.1	1.34×10^{-2}
C_2H_6 (in benzene) → C_2H_6 (in water)	−9.2	−83.7	−24.9	15.1	2.28×10^{-3}
C_2H_6 (in CCL_4) → C_2H_6 (in water)	−7.5	−79.5	−23.7	15.9	1.62×10^{-3}
C_2H_6 (pure liquid) → C_2H_6 (in water)	−10.5	−87.9	−26.2	16.3	1.38×10^{-3}

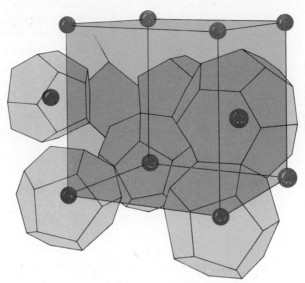

Figure 4.3 Structure of a clathrate. [From "Chemistry" by L. Pauling and P. Pauling, W. H. Freeman and Company, San Francisco, 1975, p. 296. Copyright © 1975.]

The peptide bond itself is polar, having a solvation energy of about -35 kJ mol^{-1}. Its appetite for hydrogen bonds can be satisfied either by the formation of secondary structure or by interaction with water. Intramolecular hydrogen bonds are especially stable in the interior of the protein where water is not a competitor. To illustrate this effect, let us consider the formation of hydrogen-bonded dimers of N-methylacetamide in water and in carbon tetrachloride (reaction 4.2).

$$\text{2 CH}_3\text{—} \quad \rightleftharpoons \quad \text{Hydrogen-bonded dimer} \tag{4.2}$$

N-Methylacetamide   Hydrogen-bonded dimer

In carbon tetrachloride the standard free energy change for dimer formation is -10 kJ mol^{-1}, corresponding to an equilibrium constant of 57.5. In water, however, the formation of a hydrogen-bonded dimer is endergonic, having a standard free-energy change of $+3.1$ kJ mol^{-1} and an equilibrium constant of 0.3 (both equilibria at 298 K). These data imply that intramolecular hydrogen bonds are much stabler in the hydrophobic interior of proteins than on the surface where protein-protein hydrogen bonds must compete with water-protein hydrogen bonds.

E. Denaturation: The Disruption of Noncovalent Forces in Proteins

Certain reagents, changes in pH, heating, and in some cases cooling disrupt the biologically active, or *native*, structure of proteins. Because of the subtle way in which denaturing agents act upon proteins, denaturation can be an excellent probe of the forces responsible for maintaining the tertiary structures of proteins.

The denaturing agents *urea* and *guanidine* compete for hydrogen bonds with the peptide backbone and therefore break up the secondary structure.

One possible type of interaction between urea (or guanidine) and the polypeptide backbone is shown below.

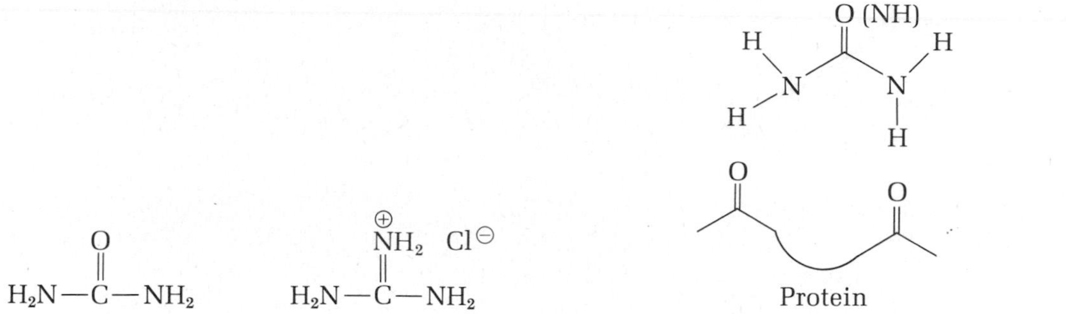

Urea | Guanidine (written as HCl salt) | Protein

H-bonded interaction between urea (guanidine) and carbonyl oxygens of peptide backbone

Perhaps more importantly, urea and guanidine *increase* the solubility of nonpolar side chains in water, thereby decreasing the hydrophobic interactions that are largely responsible for maintaining the tertiary structure.

Concentrated salt solutions also exert a powerful effect on protein structure. Table 4.2 summarizes the effects of various ions on protein solubility and on the denaturation of collagen and ovalbumin. The salt effects can be divided into two phenomena called *salting-in*, a process that increases protein solubility, and *salting-out*, a process that decreases protein solubility. The effects of four salts on the solubility of hemoglobin are shown in Figure 4.4. The increase in solubility caused by sodium chloride and potassium chloride is a salting-in effect, whereas the decrease in solubility

Table 4.2   Effects of ions on protein solubility and denaturation

SOURCE: From W. P. Jencks, "Catalysis in Chemistry and Enzymology," McGraw-Hill, New York, 1969, p. 359. Used with permission of the McGraw-Hill Book Company.

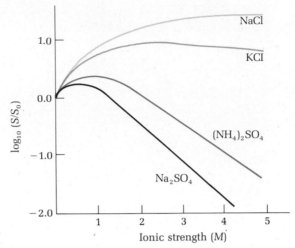

Figure 4.4   Effects of salts on the solubility of hemoglobin. The log term of the y axis is the ratio of the solubility in the absence of salt (s_0) to the solubility at a given salt concentration (s).

caused by ammonium sulfate and sodium sulfate is a salting-out effect. The surface of the protein is exposed to the solution in both cases. The increase in solubility caused by sodium chloride and potassium chloride indicates that the protein surface is better solvated and that the interior residues are largely unaffected by the salt. In ammonium or sodium sulfate solutions, however, the protein surface is less well solvated. Salting-out is roughly equivalent to dehydration of the protein, an effect that causes the interior residues to become exposed to the solvent. Both effects are ultimately caused by changes in the structure of water caused by adding the salt. If the salt increases the interaction of water molecules with one another, it will be more difficult for the protein to "poke a hole" in the water, and the protein is squeezed out of solution. Small, highly charged ions tend to have this effect. Large ions, such as tetraalkylammonium ions, have low charge densities, and they decrease the interactions of the water molecules with one another. Such ions tend to cause salting-in.

Detergents, such as *sodium dodecyl sulfate,* also denature proteins.

$$CH_3(CH_2)_{10}-CH_2-O-\overset{\overset{\displaystyle O}{\|}}{\underset{\underset{\displaystyle O}{\|}}{S}}-O^{\ominus} \quad Na^{\oplus}$$

Sodium dodecyl sulfate

They disrupt the hydrophobic interactions inside the protein, causing the nonpolar groups to become exposed to the solvent. The detergent decreases the free energy of transfer of the nonpolar side chains from the aqueous solution to the hydrophobic interior.

Denaturation is a cooperative process. The change from a native protein to a random coil occurs abruptly over a narrow range of added denaturing reagent or over narrow pH and temperature intervals. The thermodynamic parameters for denaturation of three proteins are summarized in Table 4.3. The values of $\Delta H°$ and $\Delta S°$ are much larger than those for typical chemical reactions in solution, indicating that many interactions within the protein are being more or less simultaneously disrupted. The large increase in heat capacity is another striking feature of denaturation. Since heat capac-

Table 4.3	Thermodynamic parameters for protein denaturation		
Protein	pH	$\Delta H°$ (kJ mol^{-1})	$\Delta S°$ (J K^{-1} mol^{-1})
Trypsin	2.0	282.8	891.2
Soybean trypsin inhibitor	3.0	239.7	753.1
Chymotrypsinogen	2.0	416.7	1322.1

ity is defined in terms of enthalpy ($\Delta H = C_p\Delta T$), the enthalpy changes are quite large, as seen in Table 4.3. Although we might expect large entropy changes to occur upon denaturation, we can be misled, because the formation of ordered water structures around exposed nonpolar side chains can actually cause a decrease in entropy. The entropy changes in Table 4.3 are large and positive, but in principle they can be negative.

F. The Spontaneous Folding of Proteins

How does a protein "know," apparently infallibly, how to fold up into its native conformation? Let us consider the case of *ribonuclease,* an enzyme that hydrolyzes the phosphodiester bonds of ribonucleic acids (Chapter 9). This enzyme contains 124 residues, including 8 cysteines joined by 4 disulfide bonds (Figure 4.5). Urea unfolds the peptide, and β-mercaptoethanol completes the damage by disrupting the disulfide bonds (reaction 2.44). When the urea is removed and a trace of β-mercaptoethanol is present to catalyze disulfide bond formation (reaction 2.44 is reversible), the enzyme *spontaneously* recovers its enzymatically active, native conformation (Figure 4.6). All

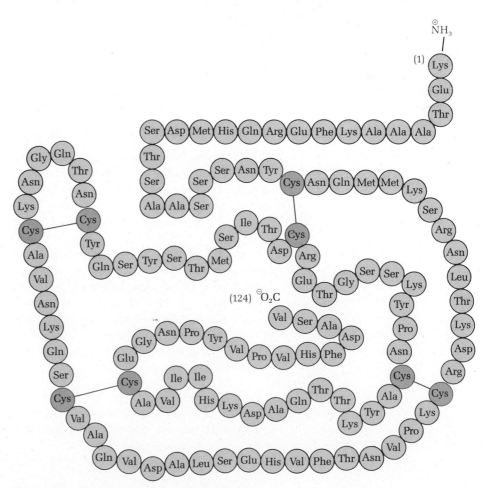

Figure 4.5 Primary structure of ribonuclease.

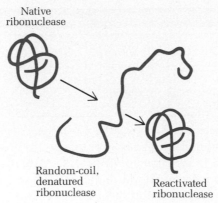

Figure 4.6 Denaturation and reactivation of ribonuclease. Treatment with an 8 M urea solution that contains β-mercaptoethanol unfolds the peptide and disrupts disulfide bonds. When the denaturing agents are removed by dialysis, the protein spontaneously regains its native, active conformation. [From C. B. Anfinson, *Science, 181,* 223 (1973). Copyright 1973 by the American Association for the Advancement of Science.]

four disulfide bonds must form correctly to restore activity. The probability that this will happen randomly is $1/7 \times 1/5 \times 1/3 = 1/105$, since there are 7 possibilities for correct formation of the first disulfide, 5 for the second, and 3 for the third. If we allow 2 rotations (ϕ and ψ) about the polypeptide backbone for each of the 124 residues, there are 2^{124} possible conformations; if we allow 2 or 3 rotations for each side chain, there will be 4^{124} and 9^{124} possible conformations. These numbers are almost too large to imagine. Their reciprocal is the probability that the native conformation will form randomly. Yet the native structure arises spontaneously. Disulfide bonds form only after the molecule has coiled up to bring the correct residues into proximity; disulfide bond formation does not appear to direct the folding of the protein.

Ribonuclease possesses another interesting property. Enzymatic hydrolysis by *subtilisin* (Section 5.3) removes a 21-residue fragment, called the S-peptide, from the N-terminal end of the chain. Neither the S-peptide nor the remaining 103-residue fragment possesses catalytic activity. When the S-peptide is added to the solution, it spontaneously bonds to the 103-residue fragment, and the enzyme regains catalytic activity even though the only interactions between the two peptides are noncovalent.

These results imply that the native structure of the protein is the structure having minimum free energy. Protein structure is thus a *thermodynamic property.* Since the tertiary structure depends upon the nature of the interactions of its side chains, and since these interactions are a function of the side chains of the residues in the primary structure, *the tertiary structure is genetically determined by the primary structure, given the correct initial conditions.*

To illustrate the importance of the correct initial conditions, let us consider the case of pyruvate kinase. The native enzyme is a tetramer (a higher ordered structure discussed in Section 4.5). The tetramer can be dissociated and the monomers denatured by treatment with guanidinium chloride.

$$\overset{\oplus}{N}H_2 \quad Cl^{\ominus}$$
$$\|$$
$$H_2N-C-NH_2$$

Guanidinium chloride

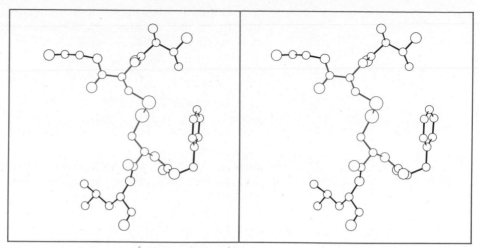

Figure 4.7 Stereo view of a disulfide bond.

When the guanidinium chloride is removed and the denatured protein is restored to an aqueous solution at the correct pH and ionic strength (these, too, being part of the "correct initial conditions"), the monomers reassemble only in the presence of L-valine. Once the monomers have assumed their active conformation, L-valine remains bound (noncovalently) as part of the native structure. The final part of renaturation is spontaneous assembly of the tetramer.

G. Disulfide Bonds in Globular Proteins

Our discussion of the spontaneous folding of ribonuclease, in which the protein assumes its tertiary structure before its disulfide bonds form, leads us into a discussion of the role of disulfide bonds in stabilizing the structures of globular proteins. In this context, let us make a distinction between the thermodynamic stabilization caused by disulfide bonds and the spontaneous folding of the protein. The spontaneous folding of ribonuclease and many other proteins shows rather conclusively that the formation of disulfide bonds does not "direct" the folding of the polypeptide chain. Much other experimental data shows, however, that once a protein has assumed its tertiary structure, that structure is stabilized by the formation of disulfide bridges. The subsequent reduction of *all* of the disulfide bridges in a protein nearly always deprives the protein of its native conformation and biological activity. In many cases disulfide bridges can be selectively reduced, and such experiments have shown that some disulfide bonds are essential for biological activity. Other disulfide bonds, however, can be reduced without loss of activity but with some loss of protein stability. These nonessential disulfide bridges help to stabilize the native structure. In a great many cases disulfide bonds are formed quite near "reverse turns" or β bends in the polypeptide chain (recall Section 3.5), and they generally assume a conformation that is stereochemically favorable within the cramped space of the protein interior. A disulfide bond can assume a variety of conformations, including right and left-handed helical twists. These two conformations are formed about equally. A stereoview of a disulfide bond in a typical left-handed spiral is shown in Figure 4.7. For the most part, half-cysteines are strongly conserved in the primary structure, an indication of the importance of disulfide bonds in stabilizing the tertiary structure. In general, the maximum number of disulfide bonds are formed, another indication of a stabilizing interaction.

H. Structural Domains of Globular Proteins

Hamlet: *Do you see yonder cloud that's almost in the shape of a camel?*
Polonius: *By th' mass, and 'tis like a camel indeed.*

> Hamlet: *Methinks it is like a weasel.*
> Polonius: *It is backed like a weasel.*
> Hamlet: *Or like a whale.*
> Polonius: *Very like a whale.*
>
> Hamlet, III, 2, 346–351

When we examine the convoluted folding of the polypeptide chain of a globular protein, we might feel rather as Polonius must feel as he parries with Hamlet about the shape of a cloud. And, at first glance, globular proteins seem to resemble amorphous clouds, and not crystalline architecture. Analysis of the tertiary structures of many globular proteins, however, has revealed *structural domains* within the conformations of globular proteins. These structural domains consist of patterns of secondary structure: α-helices, β-pleated sheets, β × β-units, Rossman-folds, and so forth. Such structural domains comprise a level of structural complexity between the secondary structure and the tertiary structure of the protein.

Structural domains are the folding units of globular proteins. It appears that the domains fold separately, and that the final step in the folding of a polypeptide chain is the association of domains. In proteins that contain only a single domain, the folding and association are, of course, identical.

Since the primary structure of a protein is colinear with the genetic material, DNA, and since the genes encoding for most of the proteins of eucaryotic organisms are split into pieces that are often widely separated on the DNA, it is intriguing to speculate that the pieces of genes encode for structural domains. And, at least in the case of the immunoglobulins (Section 4.7) and perhaps in other proteins as well, the individual pieces of the split genes do in fact correspond to structural domains. Since structural domains are integral components of globular proteins, we shall not treat each class of structural domains separately, but shall point them out as we find them in our discussions of various globular proteins.

4–3 THE CONFORMATION AND FUNCTION OF MYOGLOBIN

Myoglobin is the oxygen-storage protein of muscle. It is shaped rather like a squashed prism and has dimensions of 4.5 × 3.5 × 2.5 nm (Figure 4.8). Myoglobin contains an extensive amount of α-helix, and these helices define the single structural domain of the protein. For convenient discussion, the eight α-helices of myoglobin are lettered A through H from the N-terminus to the C-terminus (Figure 4.9). A nonpeptidyl moiety, called a *prosthetic*

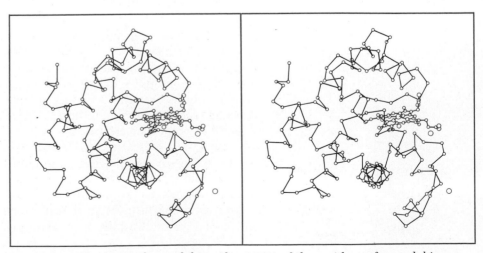

Figure 4.8 Structure of myoglobin. About 77% of the residues of myoglobin are involved in α-helices.

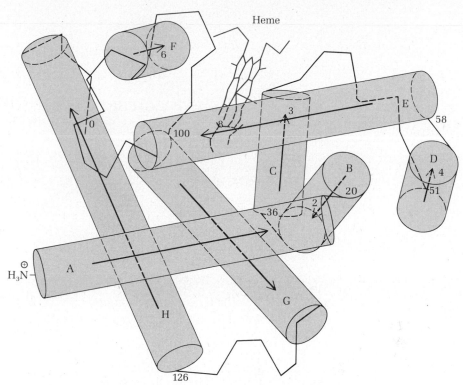

Figure 4.9 Myoglobin α-helices lettered A through H from the N-terminus to the C-terminus. The eight α-helices define the single structural domain of the protein. Residues are numbered by their position in the helices. The seventh residue in helix E is E7, the first residue in a nonhelical segment between helices C and D is CD1, and so forth. [Courtesy of Richard J. Feldmann.]

group, is covalently bound to myoglobin. The prosthetic group is a chelate of Fe(II) and a tetrapyrrole ring system called *protoporphyrin IX*. The chelate is called *heme*.

Protoporphyrin IX

Heme
(Fe-protoporphyrin IX)

The iron atom of heme undergoes reversible *oxygenation*, *not* oxidation, when the heme is bound to the protein (reaction 4.3).

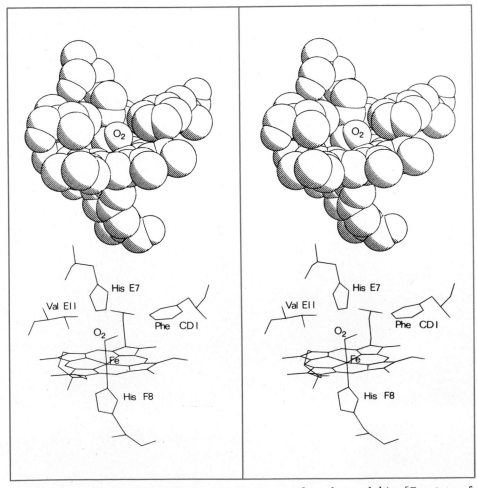

$$\text{(4.3)}$$

The iron atom in heme may be in the Fe(II) or Fe(III) state. In aqueous solution the Fe(II) is rapidly oxidized to Fe(III), and Fe(III)-protoporphyrin IX does not bind to oxygen. The protein is necessary to provide a hydrophobic environment for the heme group in which oxygenation rather than oxidation occurs (Figure 4.10).

Figure 4.10 Stereo view of the heme-oxygen complex of myoglobin. [Courtesy of Dr. S. E. V. Phillips.]

4–4
THE OXYGEN-BINDING
CURVE
OF MYOGLOBIN

Let us consider the quantitative relationship between the *fraction* of myoglobin bound to oxygen and the partial pressure of oxygen. Our derivation of this relationship has the twofold purpose of describing the function of an important protein and of providing us with a general model for analysis of the interactions of proteins with their "substrates."

Myoglobin (Mb) binds oxygen (O_2) with 1:1 stoichiometry (reaction 4.4).

$$MbO_2 \xrightleftharpoons{K_{\text{dissociation}}} Mb + O_2 \tag{4.4}$$

Note that we have written the oxygen-binding equilibrium as a *dissociation constant*, the usual convention in biochemistry. The dissociation constant, K_{diss}, is

$$K_{\text{diss}} = \frac{[Mb][O_2]}{[MbO_2]} \tag{4.5}$$

Since the concentration of oxygen is proportional to the partial pressure of oxygen (P_{O_2}), we can write equation 4.6.

$$K_{\text{diss}} = \frac{[Mb]\, P_{O_2}}{[MbO_2]} \tag{4.6}$$

The concentrations of free myoglobin and the myoglobin-oxygen complex (MbO_2) are difficult to measure. We therefore shall eliminate these parameters from equation 4.6 by introducing a new parameter, θ, for the *fraction* of myoglobin bound to oxygen at a given oxygen pressure (equation 4.7).

$$\theta = \frac{[MbO_2]}{[MbO_2] + [Mb]} \tag{4.7}$$

Substituting equation 4.7 into equation 4.6, we obtain

$$\theta = \frac{P_{O_2}}{P_{O_2} + K} \tag{4.8}$$

Equation 4.8 eliminates all terms containing myoglobin and expresses the fraction of myoglobin bound to oxygen in terms of only the dissociation constant K and the oxygen pressure. A graph of θ versus P_{O_2} is hyperbolic. When θ equals 1.0, all of the myoglobin-binding sites (the hemes) are occupied, and myoglobin is *saturated* with oxygen (Figure 4.11). When $\theta = 0.5$, myoglobin is one-half saturated. Substituting $P_{O_2} = 0.5$ in equation 4.8, we find that $K = P_{O_2}$. This equilibrium constant is often referred to as P_{50}. Our derivation has therefore led us to a fundamental physical constant: the dissociation constant for the myoglobin-oxygen complex, which equals the oxygen pressure at which myoglobin is half-saturated.

Hyperbolic equations are not easy to generate experimentally, since at high oxygen pressure θ approaches the limit of 1.0 asymptotically. We shall therefore rearrange the hyperbolic equation 4.8 into a linear form. The ratio of myoglobin-oxygen complex to free myoglobin is related to the dissociation constant and the oxygen pressure by equation 4.9.

$$\frac{\theta}{1 - \theta} = \frac{P_{O_2}}{K} \tag{4.9}$$

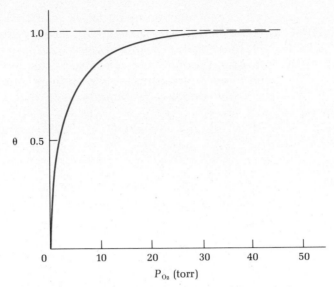

Figure 4.11 Hyperbolic oxygen-binding curve of myoglobin. Since the slope of the curve approaches zero asymptotically, there is considerable uncertaining in finding the P_{O_2} for which θ is zero. Use of equation 4.10 resolves this ambiguity.

Taking the logarithm of equation 4.9 we obtain the *Hill equation* (4.10).

$$\log \frac{\theta}{1 - \theta} = \log P_{O_2} - \log K \qquad\qquad (4.10)$$

A plot of $\log [\theta/(1 - \theta)]$ versus $\log P_{O_2}$ gives a straight line having a slope of 1.0 and an intercept on the $\log P_{O_2}$ axis of $-K$ (Figure 4.12). When $\log [\theta/(1 - \theta)] = 0$, $\log P_{O_2} = \log K$. The slope of Figure 4.12, known as a *Hill plot*, is called the *Hill coefficient*. If the Hill coefficient is 1.0, either (1) the protein has a single binding site or (2) the protein has multiple binding sites that are completely independent of one another and have the same value of K. A Hill coefficient that is not 1.0 implies more than one binding site on a protein and that the sites interact with one another.

4–5 AGGREGATES OF GLOBULAR PROTEINS

Many proteins exist as assemblies of globular proteins that bind one another by noncovalent interactions—van der Waals forces, hydrogen bonds, and electrostatic forces. Such proteins possess a *quaternary structure*. Protein ag-

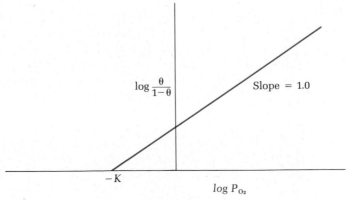

Figure 4.12 Hill plot for binding of oxygen to myoglobin. The Hill coefficient, or the slope of the plot, is 1.0.

gregates assemble spontaneously from their monomers. The assembly of monomers to form *oligomers* (Greek *oligo-*, "few") is similar in principle to the folding of domains to give the conformations of the individual monomers. As in the case of tertiary structure, the assembly of quaternary proteins is an entropically driven process, and the thermodynamics of protein association resembles the thermodynamics of protein folding.

The size and complexity of protein aggregates varies enormously from dimers of identical subunits to ribosomes that contain dozens of proteins (and ribonucleic acid as well). Table 4.4 lists a few of the hundreds of oligomeric proteins that have been discovered. One pattern that leaps out of this table is the preference of oligomeric proteins to form assemblies that contain an even number of subunits. Examples of oligomers composed of an odd number of subunits are known, but they are in the distinct minority. We shall soon see that the subunit composition of an oligomer places a severe restriction on the types of structures that it can have.

The biological properties of oligomeric proteins are often much dif-

Table 4.4 Subunit compositions of some globular proteins

Protein	Source	Molecular weight of protein	Number of subunits	Molecular weight of subunits
Chymotrypsin inhibitor	Potato	39,000	4	9,800
Superoxide dismutase	E. coli	39,500	2	21,600
Hemerythrin	Phascolosoma	40,600	3	12,700
Galactokinase	Human	53,000	2	27,000
Hemoglobin	Mammals	64,500	2	16,000
			2	16,000
Tu·Ts complex	E. coli	65,000	1	41,500
			1	28,500
Malate dehydrogenase	Rat	66,300	2	37,500
Avidin	Chicken	68,300	4	18,000
Alkaline phosphatase	E. coli	86,000	2	43,000
Procarboxypeptidase A	Bovidae	88,000	1	40,000
			2	23,000
Seryl tRNA synthetase	E. coli	100,000	2	50,000
Nucleoside diphosphokinase	Yeast	102,000	6	17,000
Lactate dehydrogenase	Pig	140,000	4	35,000
Tryptophan synthetase	E. coli	148,000	2	45,000
			2	28,700
lac Repressor	E. coli	160,000	4	40,000
Methionine tRNA synthetase	E. coli	170,000	2	85,000
Leucine aminopeptidase	Swine	255,000	4	63,500
Isocitrate dehydrogenase	Yeast	300,000	8	39,000
Aspartate transcarbamoylase	E. coli	310,000	6	33,000
			6	17,000
Nitrogenase	Clostridium	330,000	2	59,500
			4	27,500
			2	50,700
Enolase	T. aquaticus	355,000	8	44,000
Glutamine synthetase	Neurospora	360,000	4	90,000
RNA polymerase core	E. coli	400,000	2	39,000
			1	155,000
			1	165,000
Glutamine synthetase	E. coli	592,000	12	48,500
Isocitrate dehydrogenase	Bovidae	670,000	16	41,000
Hemoglobin	Arenicola	2,850,000	48	54,000
Pyruvate dehydrogenase complex	E. coli	5,000,000	24	91,000
			24	65,000
			24	56,000

SOURCE: Condensed from D. W. Darnall and I. M. Klotz, *Arch. Biochem. Biophys. 166*, 651 (1975).

ferent than the properties possessed by the individual monomers. The functions of such molecules are *synergistic*, and the unique properties of the oligomer could not be deduced from the properties of the monomer. The new properties arise because of specific interactions that exist in the oligomer that are impossible in the monomers. Let us consider, then, some of the theoretical aspects of oligomeric proteins before turning to perhaps the best understood quaternary protein, hemoglobin, in the next section.

A. The Symmetry of Quaternary Proteins

The *molecular symmetry* of a quaternary protein is one of its most important properties. To determine molecular symmetry certain *symmetry operations* are performed. A symmetry operation is any movement of the object such that after the movement (operation), the configuration of the object is indistinguishable from its initial configuration. In other words, if we turned our back on the object while someone else carried out the operation, we wouldn't realize that any operation had been performed. A *symmetry element* is a line, point, or plane with respect to which the symmetry operation is carried out. To possess a given symmetry element, an indistinguishable object must be generated when a symmetry operation is carried out with respect to that element. Four kinds of symmetry elements and their concomitant symmetry operations define the structures of nearly all oligomeric proteins (Figure 4.13).

1. A *plane of symmetry* is a symmetry element whose operation is *mirror reflection*. Since the monomers of oligomeric proteins are chiral, reflection in a mirror plane will generate an enantiomeric protein. A quaternary protein therefore *cannot* have a plane of symmetry.

2. A *center of inversion* is a symmetry element within the object such that changing the coordinates of every atom at every point (x_i, y_i, z_i) into coordinates $(-x_i, -y_i, -z_i)$ generates an equivalent object. This operation equals reflection through a *center of symmetry* along an *inversion axis*. Again, since proteins are chiral, it is impossible for a quaternary protein to have either a center of symmetry or an inversion axis.

3. A *proper axis of symmetry* is a symmetry element corresponding to rotation of the entire object around an axis by $2\pi/n$ degrees, where n is an integer. This is the *only* symmetry element possible for most oligomeric proteins. The symbol for a proper axis of symmetry is C_n, where n is the order of the symmetry axis; it is the value of n in the expression $2\pi/n$. A rotation of $2\pi/n$ generates an equivalent structure. Proper symmetry axes are described as twofold, threefold, and so forth, where the number is the value of n. A C_2 symmetry axis is shown in Figure 4.12.

4. An *improper axis of symmetry* is a symmetry element corresponding to rotation of $2\pi/n$ degrees followed by reflection in a mirror plane. Quaternary proteins cannot possess improper axes.

If all of the symmetry elements of an object intersect at a single point that is not shifted by any symmetry operation, the object possesses *point group symmetry*. Most quaternary proteins possess this type of symmetry. The simplest point group symmetry is *cyclic symmetry* in which rotation around a single C_n axis regenerates the original structure.

If a quaternary protein (or any other molecule) has a C_n axis, and if there is a C_2 axis perpendicular to it, it can be shown that there are n such twofold axes. Molecules that possess such a set of symmetry axes possess *dihedral symmetry*. They are designated D_n, where n is the order of the proper C_n axis. Figure 4.14 shows a schematic diagram of a quaternary protein composed of six subunits. These may be arranged in three different ways to give structures that have D_3 symmetry. A quaternary protein must have $2n$ monomers to have dihedral symmetry. Quaternary proteins composed of an odd number of subunits cannot have dihedral symmetry.

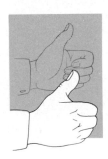

Mirror reflection

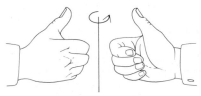

Inversion and reflection at a center of inversion along an inversion axis

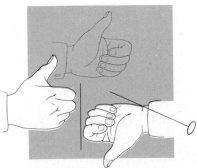

Rotation of 180° around a C_2 axis of symmetry

Figure 4.13 Symmetry operations. (a) Mirror reflection, (b) an inversion and reflection at a center of inversion along an inversion axis, and (c) rotation around a C_2 axis.

The largest group of quaternary proteins are those composed of two identical subunits. These proteins possess C_2 symmetry. Many quaternary proteins are also composed of two nonidentical subunits. These proteins have no symmetry elements. Tetramers constitute the second largest group of quaternary proteins. Only two types of structures are possible for

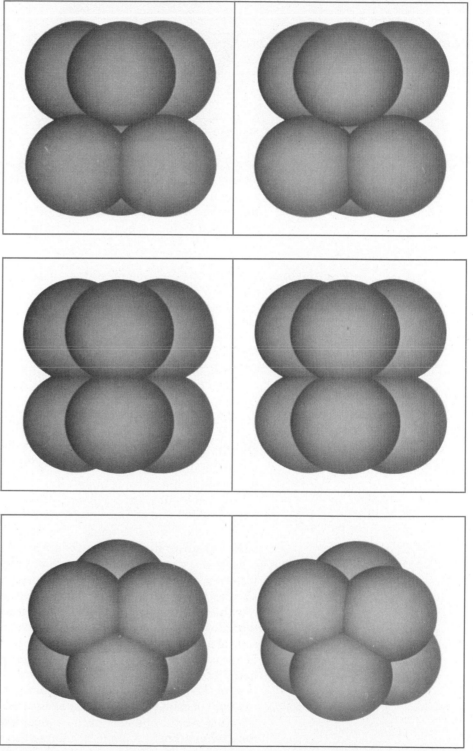

Figure 4.14 Schematic diagrams of hexameric proteins, each of which possesses D_3 symmetry.

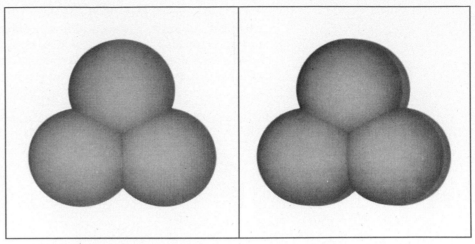

Figure 4.14   (*Continued*)

tetramers having cyclic symmetry: they can exist in a square planar structure or in a tetrahedral structure. A square planar array of monomers has C_4 symmetry, and a tetrahedral array of monomers has D_2 symmetry. Most tetramers have been found to possess D_2 symmetry.

B. Regulation of Biological Function by Quaternary Proteins: Allosteric Proteins

The smooth operation of cells requires a complex regulatory system. The simplest components of cellular regulation are the quaternary proteins that are known as *allosteric proteins*. Allosteric is a word constructed from the Greek words for other, *allos,* and space, *steros.* The term allosteric is applied not only to the protein itself, but also to its binding sites for various metabolites. It is the *interaction* of these sites, mediated by the quaternary structure of the allosteric protein, that accounts for the regulatory properties of the protein.

We can take myoglobin as a paradigm of monomeric proteins that contain a single binding site for a metabolite. We saw that myoglobin has a hyperbolic binding curve for oxygen. In contrast, the interactions between the several binding sites of allosteric proteins lead to S-*shaped,* or *sigmoidal* binding curves for a given metabolite (Figure 4.15). Interactions between identical binding sites are said to be *homotropic.* Besides interactions between the homotropic sites, allosteric proteins are controlled by a second set of metabolites, called *ligands,* that bind to separate sites on the quaternary protein. The stereochemistry of these distinct sites gives the allosteric proteins their generic name. The interactions between allosteric binding sites and substrate binding sites are said to be *heterotropic.* The ligands that bind at the allosteric binding sites are known as *allosteric effectors.* The allosteric effectors alter the binding curve of the substrate (Figure 4.15). If an allosteric effector *increases* the concentration of substrate required to saturate the binding sites for the substrate, it is an *inhibitor.* If, on the other hand, the allosteric effector *decreases* the concentration of substrate required to saturate the binding sites for substrate, it is an *activator.* An allosteric effector is not a switch that turns the activity of the allosteric protein "on" and "off," but is analogous to a light dimmer that regulates brightness. Regulation of this type costs almost no energy. The allosteric effectors are part of the pool of metabolites present in the cell at varying concentrations most of the time. Slight changes in the concentrations of allosteric effectors enable the cell to adjust moment by moment to metabolic demands as the supply of allosteric effectors rises and falls with conditions.

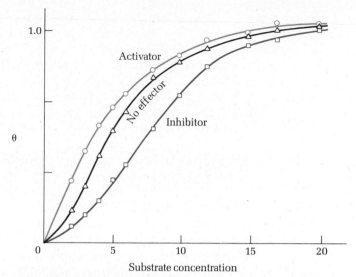

Figure 4.15   Interaction of allosteric sites. An allosteric activator (top curve) increases the affinity of the substrate for the allosteric protein, whereas an allosteric inhibitor (bottom curve) decreases the affinity of the protein for substrate, relative to the affinity of the substrate for protein in the absence of an effector (center curve).

C. The Concerted Model for Allosteric Transitions

The concerted model for the function of allosteric proteins, developed by Monod, Wyman, and Changeux (the MWC model), explains the behavior of allosteric proteins simply and elegantly. The model makes the following assumptions.

1. Allosteric proteins are oligomers whose monomers occupy equivalent positions in the structure. The equivalence of the monomers is a direct result of the symmetry of the oligomer.

2. Each monomer has a stereospecific binding site for each of its substrates. Since the monomers occupy equivalent positions (assumption 1), the substrates occupy equivalent positions. In other words, the symmetry of the binding sites is the same as the symmetry of the protein as a whole.

3. The conformation of each monomer is constrained by its associations with other monomers in the oligomer.

4. The oligomer can exist in either of two states, or conformations. The "tense" state, T, has a lower affinity for substrate than the "relaxed" state, R. The terms tense and relaxed refer to the conformational flexibility of the oligomer. The T and R states are in equilibrium, governed by an equilibrium constant L, called the *allosteric constant* (equation 4.11).

$$T \underset{}{\overset{L}{\rightleftharpoons}} R \tag{4.11}$$

5. When a substrate binds to the T state, a *concerted* change in conformation occurs to produce the R state. The concerted model does not permit any intermediate species in which the monomers have different conformations. Since the transition from the T to the R state is concerted, the molecular symmetry of the oligomer must be conserved in the transition. Figure 4.16 represents a concerted allosteric transition for a tetramer.

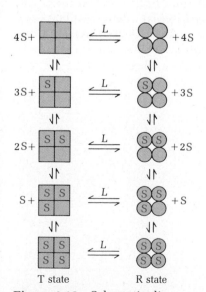

Figure 4.16   Schematic diagram of the concerted (MWC) model of allosteric transitions for a tetramer (S = substrate).

Allosteric inhibitors and activators come into play through their effects on the equilibrium between the R and T states. They involve allosteric ligand dissociation equilibrium constants superimposed on the equilibria involving the substrate. Allosteric effectors that stabilize the R conformation shift the equilibrium toward R, diminish or abolish the interactions between subunits, and are activators. Allosteric effectors that stabilize the T conformation increase the interactions between subunits and are inhibitors.

Perhaps the most trenchant criticism of the concerted model of allosteric transitions is that it oversimplifies allosteric proteins in its insistence that no intermediate conformations containing R and T state are allowed. Although only an approximation of reality, the concerted model correlates the structures of many allosteric proteins to their functions. This structural explanation is a clear advance over preceding theories that provided a mathematical description, but without connection to protein structure.

D. The Sequential Model for Allosteric Transitions

The sequential model for the conformational changes that occur when substrates bind to allosteric proteins was introduced by Koshland, Nèmethy, and Filmer (the KNF model). It replaces the MWC model's assumption of concerted transition with the assumption that the transition from the T to the R state is a *sequential* process. Thus, in place of the single change in quaternary structure proposed by the MWC model, the sequential model substitutes a series of changes in the tertiary structures of the monomers. The sequential model makes only two assumptions.

1. The protein exists in one conformation when no substrate is present.
2. The binding of substrate to one monomer induces a conformational change in an adjacent monomer in the oligomer (Figure 4.17).

We can picture the transition of the oligomer as a row of falling dominoes: in the MWC model the dominoes fall all at once; in the KNF model they fall in rapid succession. If the succession is fast enough, the sequential process merges with the concerted one. The KNF model requires as many parameters as there are binding sites, whereas the MWC model requires only three parameters for all cases. The KNF model is a closer approximation to reality for many proteins than the MWC model. This generality is purchased at the price of conceptual simplicity.

M. Eigen has noted that the two models are special cases of the general binding scheme shown in Figure 4.18. The left- and right-hand columns describe the MWC model, and the diagonal describes the KNF model. Both models predict the same binding curve. How can we distinguish them?

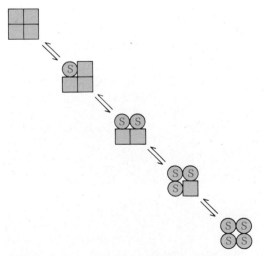

Figure 4.17 Schematic diagram of the sequential (KNF) model of allosteric transitions.

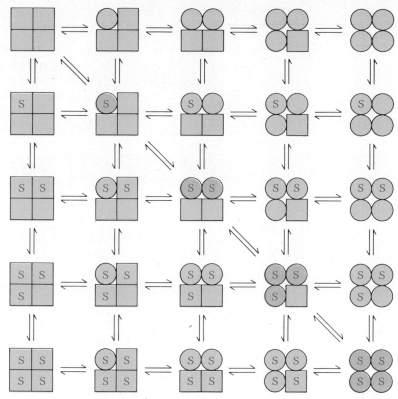

Figure 4.18 General model for the conformational changes of an allosteric protein. The extreme left and right columns represent the simplification of the concerted model; the diagonal represents the simplification of the sequential model. [From G. G. Hammes and C. Wu, *Science*, *172*, 1205-1211 (1971).]

One difference between the two models can differentiate them in some cases. In the MWC model all homotropic interactions must be positive. Positive cooperativity is thus the only type of behavior possible. There is no mechanism for the binding of the first substrate to inhibit the binding of the second one, a phenomenon called *negative cooperativity*. The KNF model permits negative cooperativity. The enzyme *tyrosyl tRNA synthetase* is a symmetric dimer of identical subunits. This enzyme binds to only one mole of tyrosine per mole of dimer even at high substrate concentrations. Thus binding of the first mole of tyrosine inhibits binding of the second: This is one case of negative cooperativity, among several known examples, to which the KNF model applies.

4–6
THE CONFORMATION
AND FUNCTION
OF HEMOGLOBIN

A. Conformational Changes
Caused by
Oxygen Binding

Hemoglobin (Hb) is the oxygen-transport protein of blood, a function that is required because oxygen has extremely low solubility in aqueous solutions. Hemoglobin also transports carbon dioxide and hydrogen ions.

Hemoglobin is an allosteric protein. It is a tetramer composed of two types of subunits, designated α and β, whose stoichiometry is $\alpha_2\beta_2$. The four subunits of hemoglobin sit roughly at the corners of a tetrahedron, facing each other across a cavity at the center of the molecule (Figure 4.19). The subunits of hemoglobin are not identical, and hemoglobin has a C_2 symmetry axis.[1] Each of the subunits contains a heme prosthetic group that can bind oxygen. The stoichiometry of oxygen binding is therefore

$$Hb + 4O_2 \rightleftharpoons Hb(O_2)_4$$

[1] If the subunits were identical, the symmetry would be D_2, and the subunits are so similar that hemoglobin is sometimes said to possess pseudo-D_2 symmetry.

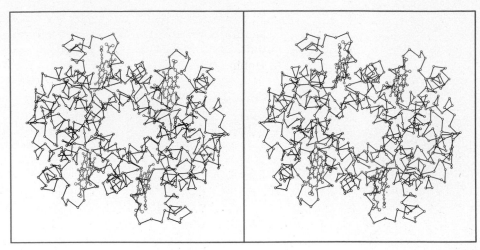

Figure 4.19   Structure of hemoglobin.

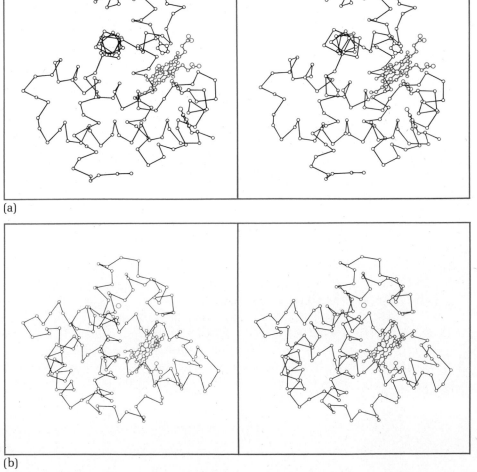

(a)

(b)

Figure 4.20   Tertiary structures of the (a) β chain of hemoglobin and (b) sperm whale myoglobin.

The heme groups are oxygenated, not oxidized. The structures of the α and β subunits are similar, differing primarily by an additional loop in the β subunit. The latter is virtually identical to the structure of sperm whale myoglobin even though only 30 residues of their primary structures are identical (Figure 4.20). The structural similarity of these two proteins reflects a functional similarity: reversible oxygenation. The differences in function between hemoglobin and myoglobin are due to interactions among the α and β subunits of hemoglobin. A tetramer of β subunits acts like myoglobin, not like hemoglobin.

When hemoglobin binds oxygen, its quaternary structure changes dramatically as the conformation of the molecule changes from the T to the R state. In either state the molecule can be regarded as a dimer of $\alpha\beta$ dimers. If one dimer is held fixed, the other rotates by about 15° around an off-center axis and slides along it as the allosteric transition occurs. The rotational movement is shown in Figure 4.21, and the sliding motion is depicted in Figure 4.22.

Why does the conformation of hemoglobin change so dramatically when oxygen binding occurs? This question has been the subject of an enormous volume of research, and the problem is still not completely solved. The change in quaternary structure that occurs upon the binding of oxygen can be divided into electronic and steric effects.

The electronic effects are a consequence of the electronic structure of the Fe(II) atom. The Fe(II) atom in deoxyhemoglobin has four unpaired electrons, whereas the Fe(II) atom in oxyhemoglobin has no unpaired electrons. The Fe(II) in deoxyhemoglobin is said to be "high-spin" iron, and the Fe(II) in oxyhemoglobin is called "low-spin" iron. According to Hund's rule of maximum multiplicity, the lowest energy electronic configuration of an

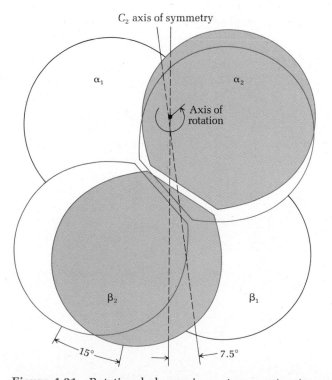

Figure 4.21   Rotational change in quaternary structure of hemoglobin. Oxygenation of hemoglobin causes one dimer to rotate by 15° relative to the other one. The central axis of symmetry itself tilts by 7.5° upon oxygenation. [From M. F. Perutz, *Ann. Rev. Biochem.*, **48**, 327–386 (1979). Reproduced with permission. © 1979 by Annual Reviews Inc.]

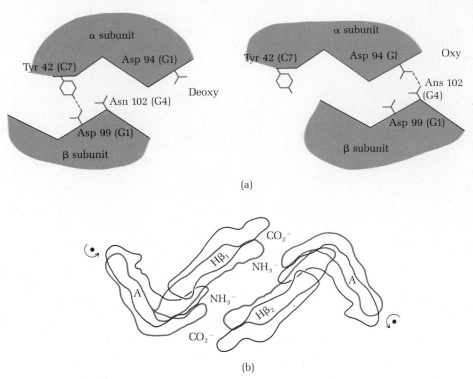

(a)

(b)

Figure 4.22 Sliding change in quaternary structure of hemoglobin. (a) Oxygenation of hemoglobin results in a conformational change as the α_1 and β_2 chains slide past each other, exchanging hydrogen-bonding partners in the process. (b) The shift in subunits that occurs upon oxygenation is shown from the top. The solid lines represent the positions of the deoxy subunits and the dotted lines the positions of the oxy subunits. [Courtesy of Dr. M. F. Perutz.]

atom has the maximum number of unpaired electrons. The change from four unpaired electrons to no unpaired electrons that occurs upon oxygen binding is probably caused by the unfavorable interaction of the unpaired electrons of oxygen with the unpaired electrons of high-spin Fe(II). The energy released upon oxygen binding is therefore great enough to overcome Hund's rule.

The interaction of the iron with the porphyrin ring also changes upon

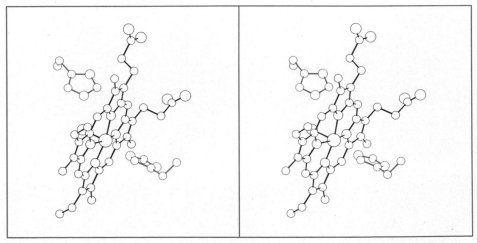

Figure 4.23 Expanded view of the oxygen-binding site in the β subunit of hemoglobin.

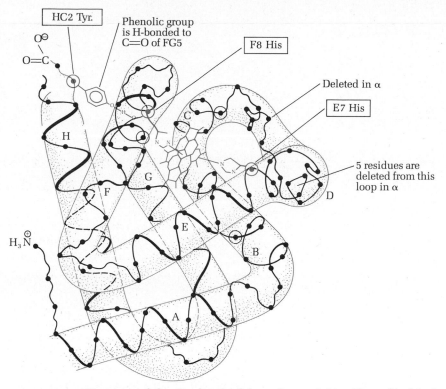

Figure 4.24 Structure of the β subunit of deoxyhemoglobin. The cylindrical regions are α-helical conformations lettered A through H. [From M. F. Perutz and L. F. Ten Eyck, *Cold Spring Harbor Symp. Quant. Biol., 36*, 296 (1971).]

oxygenation. In deoxyhemoglobin the iron atom is bonded to the four nitrogens of the porphyrin ring by ionic bonds, whereas in oxyhemoglobin it is bound to the nitrogens by covalent bonds.

The high-spin iron has a larger atomic volume than the low-spin because the four unpaired electrons are distributed in four orbitals rather than the two orbitals of low-spin Fe(II). At one time it was thought that this larger atomic volume prevented the Fe(II) atom of deoxyhemoglobin from fitting into the hole in the middle of the porphyrin ring, and that the movement of the iron into the plane of the heme triggered the large observed conformational change that occurs upon oxygenation. Extremely exact measurement has shown, however, that both low-spin and high-spin Fe(II) lie in the plane of the heme, and no movement of iron is detected. The trigger that initiates the conformation change probably cannot be explained in terms of a single interaction, but reflects the complex and subtle interaction between the heme, the protein, and the ligand, oxygen.

The oxygen-binding site of hemoglobin is shown in an expanded view in Figure 4.23 and its relation to the subunit of hemoglobin is shown in Figure 4.24. Upon oxygenation helix F moves toward helix H forcing tyrosine HC2 out of a pocket where it is secured by a hydrogen bond between the tyrosyl hydroxyl group and the peptide backbone of helix F at position FG5 (Figure 4.25).[2] The movement of tyrosine HC2 disrupts the ionic bonds that cross-link deoxyhemoglobin. Hemoglobin tends to resist oxygenation because the

---

[2] The helices of the subunits of hemoglobin have the same lettering scheme as the ones in myoglobin. Numbering begins at the N-terminal residue of each helix. Combinations, such as HC, refer to nonhelical segments that connect two helices.

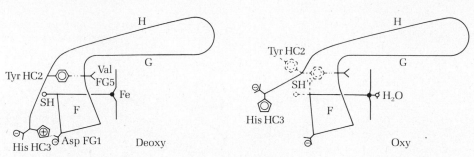

Figure 4.25 Movement of tyrosine upon oxygenation of hemoglobin. Upon oxygenation tryosine HC2 is forced out of its pocket, the ionic bonds are broken, and a large conformation change occurs. [Courtesy of Dr. M. F. Perutz.]

ionic bonds stabilize the deoxy form. Steric hindrance also inhibits oxygen binding, lowering the affinity of deoxyhemoglobin for oxygen. The effects of oxygenation at one heme are "communicated" to the other oxygen-binding sites by the large conformational change that occurs upon oxygen binding.

B. The Oxygen-Binding Curve of Hemoglobin

In Section 4.4 we found that the oxygen-binding curve of myoglobin is described by equation 4.8, and that a plot of θ, the fraction of oxygen bound to myoglobin, versus P_{O_2} gives a rectangular hyperbola. The oxygen-binding curve of hemoglobin is strikingly different. A plot of θ versus P_{O_2} gives a sigmoidal curve (Figure 4.26). The sigmoidal curve signals the appearance of a new set of interactions between the hemes and the oxygens in the quaternary structure of hemoglobin. At low oxygen pressure hemoglobin resists oxygenation because the cross-linked structure is stable. The cross-linked structure corresponds to the tense structure of the allosteric protein. Disruption of the ionic cross-links produces a more relaxed, conformationally mobile state. Hemoglobin binds oxygen much more readily when it is in the R state than when it is in the T state.

The oxygen-binding curves of hemoglobin and myoglobin reflect the functions of these proteins in oxygen transport. To be an effective transporter of oxygen, hemoglobin must have a lower affinity for oxygen than myoglobin. Figure 4.27 shows that it does. Another plot of the ratio of oxy to

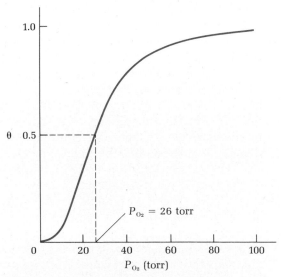

Figure 4.26 Sigmoidal oxygen-binding curve of hemoglobin. In general, a sigmoidal binding curve for a substrate indicates a cooperative interaction among the subunits of an oligomeric protein.

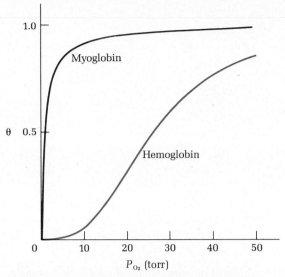

Figure 4.27 Comparison of oxygen-binding curves of myoglobin and hemoglobin. At all oxygen pressures θ is greater for myoglobin than for hemoglobin. Thus, hemoglobin transfers its oxygen to myoglobin. The physiological consequences of these curves can scarcely be overstated: one would die of asphyxiation if hemoglobin and myoglobin oxygen binding both fit the hyperbolic curve of myoglobin.

deoxy hemoglobin (a Hill plot) is shown in Figure 4.28. The tangent to the lower asymptote gives the equilibrium constant for dissociation of the first oxygen, $K_1 = 0.024$. The tangent to the upper asymptote gives the dissociation constant for the last oxygen bound, $K_4 = 7.4$. The affinity of hemoglobin for the fourth oxygen is thus some 300 times greater than its affinity for the first oxygen bound.

C. The Effect of 2,3-Diphosphoglycerate on Hemoglobin

The oxygen-binding curve of hemoglobin is altered by the allosteric effector *2,3-diphosphoglycerate* (DPG).

$$
\begin{array}{c}
CO_2^{\ominus} \\
| \\
H-C-O-PO_3^{\scriptsize{2}\ominus} \\
| \\
CH_2-O-PO_3^{\scriptsize{2}\ominus}
\end{array}
$$

2,3-diphosphoglycerate (DPG)

The central cavity of hemoglobin is lined with positive residues including the N-terminal residues, lysines EF6, and histidines H21 of the two β chains (Figure 4.29). The conformation of the central cavity in deoxyhemoglobin is complementary to the conformation of DPG. The two β chains are cross-linked through DPG by ionic bonds (Figure 4.30). These ionic bonds stabilize the T conformation of deoxyhemoglobin and lower the affinity of deoxyhemoglobin for oxygen. The affinity of DPG for the oxygenated states of hemoglobin is at least an order of magnitude lower than its affinity for deoxyhemoglobin and decreases in the order

$$Hb(O_2) > Hb(O_2)_2 > Hb(O_2)_3$$

DPG does not bind at all to fully oxygenated hemoglobin. Viewed in its simplest terms, the central cavity of fully oxygenated hemoglobin is too small to accommodate DPG.

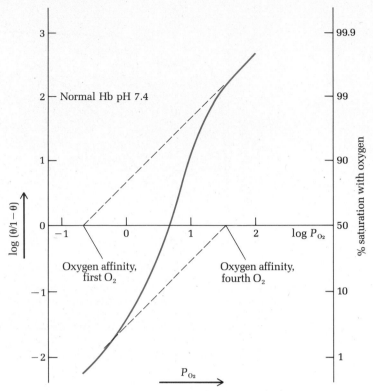

Figure 4.28 Oxygen-binding curve for hemoglobin in a plot of the log of the ratio of oxy- to deoxyhemoglobin versus oxygen pressure. [From J. V. Kilmartin, K. Imai, and R. T. Jones, in "Erythrocyte Structure and Function," Alan R. Liss, New York 1975, p. 21.]

Myoglobin is half-saturated at an oxygen pressure of 1 torr. In the absence of DPG, hemoglobin is also half-saturated at an oxygen pressure of 1 torr, and transfer of oxygen from hemoglobin to myoglobin is not efficient. In the presence of DPG, whose concentration is about 4.5 mM in whole blood, a partial pressure of 26 torr is required to half-saturate hemoglobin. In the lungs the partial pressure of oxygen exceeds 100 torr, and hemoglobin is saturated with oxygen regardless of the presence of DPG, but in the tissues, where the partial pressure is quite low, DPG increases the efficiency of transfer of oxygen to myoglobin.

Figure 4.31 shows the oxygen-binding curves of hemoglobin under varying conditions. In the absence of DPG, hemoglobin is completely saturated by a partial pressure of oxygen of less than 20 torr (curve a). At a DPG concentration of 4.5 mM the oxygen affinity of hemoglobin drops dramatically, and about 30% of the oxygen taken up in the lungs is delivered to myoglobin (curve c). Curve (b) shows the oxygen-binding curve of a unique hemoglobin synthesized by fetuses and designated hemoglobin F (HbF). Hemoglobin F has the subunit composition $\alpha_2\gamma_2$. The γ subunits differ very little in their primary structure from the β subunits of adult hemoglobin. The $\alpha_2\gamma_2$ tetramer, however, has a weaker affinity for DPG and a greater affinity for oxygen than maternal hemoglobin. As a result, oxygen is efficiently transported across the placenta from the mother to the fetus. In early childhood the gene encoding for the γ subunit switches off, and the β chain of adult hemoglobin is synthesized in its place.

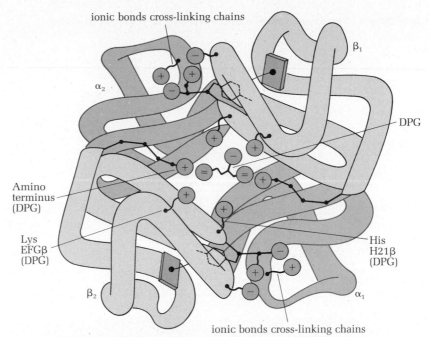

Figure 4.29 Complementary conformation of hemoglobin and 2,3-diphosphoglycerate. The central cavity of hemoglobin is lined with positive residues whose conformation and charge are complementary to the conformation and charge of DPG. [From R. E. Dickerson, *Ann. Rev. Biochem.*, *41*, 815 (1972). Reproduced with permission. © 1972 by Annual Reviews, Inc.]

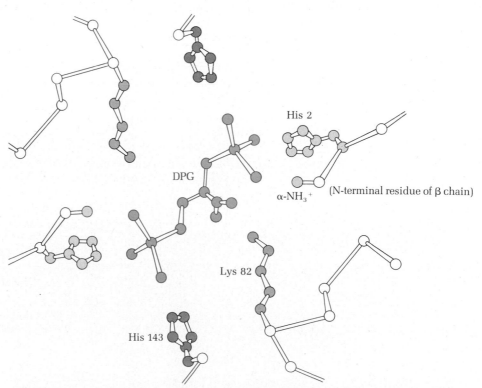

Figure 4.30 Hemoglobin β chain cross-links through DPG. 2,3-Diphosphoglycerate binds hemoglobin along the symmetry axis that passes through the center of hemoglobin. Deoxyhemoglobin binds 2,3-diphsophoglycerate by electrostatic interaction between the positive residues lining the central cavity and the negative charges on 2,3-diphosphoglycerate. As more subunits bind oxygen, the affinity of hemoglobin for DPG decreases, and fully oxygenated hemoglobin does not bind at all to DPG. [From A. Arnone, *Nature*, *237*, 148 (1972). Reprinted by permission. Copyright © 1972 Macmillan Journals Limited.]

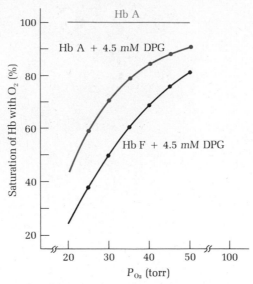

Figure 4.31 Oxygen-binding curves for adult hemoglobin in the presence and absence of DPG and for fetal hemoglobin in the presence of DPG. (Adapted from G. E. Shulz and R. H. Schimmel, "Principles of Protein Structure," Springer-Verlag, New York, 1979, p. 219.]

D. The Bohr Effect

In 1904 Christian Bohr, the father of the great physicist Niels Bohr, found that carbon dioxide alters the oxygen-binding curve of hemoglobin (Figure 4.32). Increasing the pressure of carbon dioxide greatly increases the cooperativity of the hemoglobin subunits, thereby diminishing the affinity of hemoglobin for oxygen. Lowering the pH also increases the cooperativity of the subunits and encourages the loss of oxygen from oxyhemoglobin (Figure 4.33). These linked phenomena, known as the *Bohr effect*, are important physiologically. Rapidly metabolizing tissue releases both $H^{\oplus}$ and CO_2, which act in concert to drive oxygen off oxyhemoglobin. The CO_2 is carried by hemoglobin back to the lungs and released.

The mechanism of the Bohr effect involves a half-dozen or so residues in hemoglobin. We have noted that oxygenation breaks all of the salt bridges

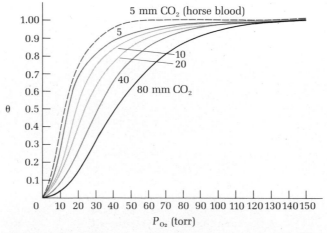

Figure 4.32 Effect of carbon dioxide on oxygen-binding curve of hemoglobin. Carbon dioxide increases the cooperativity of the subunits of hemoglobin and lowers their affinity for oxygen.

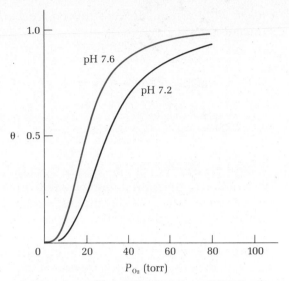

Figure 4.33 Effect of pH on oxygen-binding curve of hemoglobin. Lowering the pH of the solution increases the cooperativity of the subunits of hemoglobin, lowering their affinity for oxygen.

(ionic bonds) that cross-link the subunits of hemoglobin. In the deoxy state the imidazolium ion of the side chain of the C-terminal residue, histidine 146, forms an ionic bond with the carboxylate anion of aspartate 94 (Asp G1) in each β chain (Figure 4.34). By stabilizing the conjugate acid of imidazole (imidazolium ion), the salt bridge decreases its acidity. The pK_a of the imidazolium ion therefore is raised. Upon oxygenation of hemoglobin at pH 7 the salt bridges are broken, the pK_a values of the two imidazolium ions decrease, and the imidazolium ions dissociate. Oxyhemoglobin is thus a stronger acid than deoxyhemoglobin.

Meanwhile, the N-terminal amino groups of all four chains are in salt bridges in deoxyhemoglobin, but are free in oxyhemoglobin. All four can

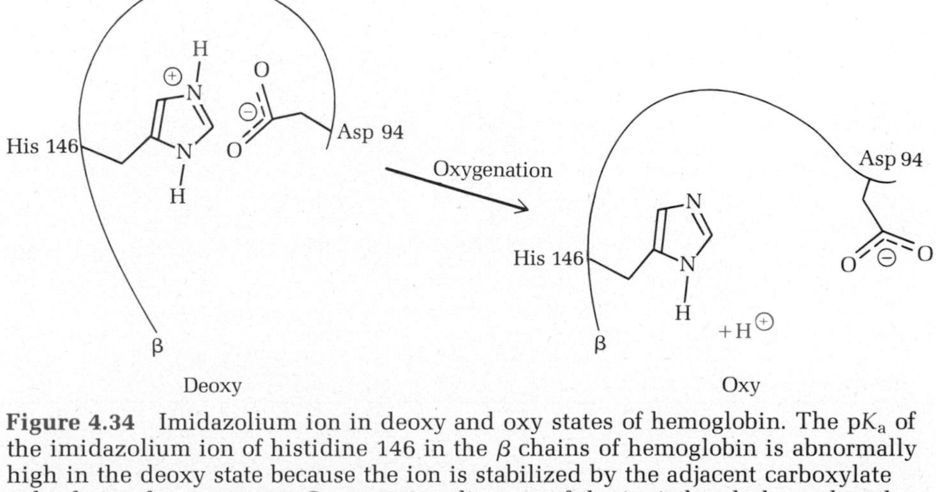

Figure 4.34 Imidazolium ion in deoxy and oxy states of hemoglobin. The pK_a of the imidazolium ion of histidine 146 in the β chains of hemoglobin is abnormally high in the deoxy state because the ion is stabilized by the adjacent carboxylate side chain of aspartate 94. Oxygenation disrupts of the ionic bond, the carboxylate moves away, and a proton is released. The pK_a value of the imidazolium ion is thus lowered in the oxy state.

react with carbon dioxide to give *carbamate* derivatives of hemoglobin (reaction 4.12).

$$R—NH_2 + O{=}C{=}O \rightleftharpoons R—\underset{\underset{H}{|}}{N}—\overset{\overset{O}{\|}}{C}—O^{\ominus} + H^{\oplus} \qquad (4.12)$$

N-terminal amino group of   Carbamate derivative
all four chains (free base)

The formation of carbamates is reversible. In metabolizing tissue the partial pressure of carbon dioxide is relatively high, shifting the equilibrium of reaction to the right. In lungs, where the carbon dioxide partial pressure is much lower, the equilibrium shifts back to the left, carbon dioxide is released, the salt bridges form once again, and we are back where we started.

E. Mutant Hemoglobins

If proof were needed of the subtlety of protein structure, of the fragility of biological systems, of the precision of the molecular architecture of proteins, then hemoglobin might serve as the example. Slight alterations in the quaternary structure of hemoglobin, arising from alterations in the primary structure of its monomers, can change its stability, its solubility, its ability to bind oxygen, or a combination of these properties.

Table 4.5 lists a few mutant hemoglobins whose affinity for oxygen is altered. In every case the decreased affinity for oxygen, and with it impaired ability to transport oxygen and all that such impairment implies for the affected individual, results from a single alteration in the primary structure of either an α or β chain in the oligomer. The hemoglobins in the table have three types of alterations in their quaternary structure. The first is alteration of the contact between α and β subunits. In all but one case the alteration is in the $\alpha_1\beta_2$ contact region that undergoes the most dramatic change upon oxygenation. For example, aspartate 99 in the β chain is hydrogen-bonded to

Table 4.5   Hemoglobins with altered oxygen affinities

Mutant hemoglobin	Residue	Substitution	Region in molecule affected	$P_{(1/2)O_2}$[a]	Bohr effect[b]	n (in Hill equation)[c]	Concentration[d] (g/100 ml)	DPG Interaction
Chesapeake	α92	Arg → Leu (FG4)	$\alpha_1\beta_2$ contact	19	N	1.8	16–18	
Yakima	β99	Asp → His (G1)	$\alpha_1\beta_2$ contact	12	N	1.1	~17	N
Kempsey	β99	Asp → Asn (G1)	$\alpha_1\beta_2$ contact	↑	N	1.1	~20	
Radcliffe	β99	Asp → Ala (G1)	$\alpha_1\beta_2$ contact	12	N	1.1	~18	N
Brigham	β100	Pro → Leu (G2)	$\alpha_1\beta_2$ contact	19.6	N	↓	16–19	N
Denmark Hill	α95	Pro → Ala (G2)	$\alpha_1\beta_2$ contact	↑	—	1.8–2.4	~13	
San Diego	β109	Val → Met (G11)	$\alpha_1\beta_1$ contact	16.4	N	~2	17	N
Kansas	β102	Asn → Thr (G4)	$\alpha_1\beta_2$ contact	~70	N	~1	14	
Ranter	β145	Tyr → Cys (HC2)	Salt bridges	12.9	N	1.1	16–21	
Andrew Minneapolis	β144	Lys → Asn (HCl)	Salt bridges	14	↓	N	~20	N
Syracuse	β143	His → Pro (H21)	DPG site	11	↓	~1	~20	↓
Rahere	β82	Lys → Thr (EF6)	DPG site	18	N	N	19	↓
Providence	β82	Lys → Asp or Asn (EF6)	DPG site	↓	↓	2.5–2.7	—	↓
Heathrow	β103	Phe → Leu (G5)	Heme pocket	↑	N	~1	16–21	

[a] The P_0 required for half saturation of whole blood containing the abnormal Hb. For normal blood this value is 27 ± 2 mmHg at pH 7.4 and 37°C. ↑ = increased. ↓ = decreased.
[b] N = normal.
[c] Normal value for normal whole blood is 2.7 ± 0.2.
[d] Normal adult values are 14 ± 2 for females and 16 ± 2 for males.
SOURCE: From A. White, P. Handler, E. L. Smith, R. L. Hill, and I. R. Lehman, "Principles of Biochemistry," 6th ed., McGraw-Hill, New York, 1978. Used with permission of the McGraw-Hill Book Company.

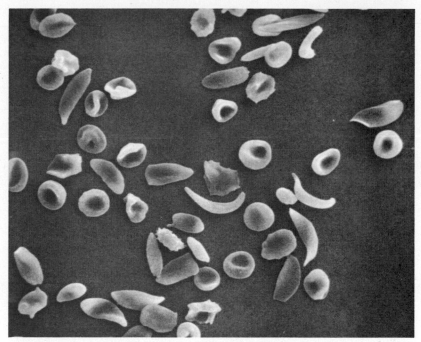

Figure 4.35 Scanning electron micrograph of normal and sickled red blood cells. [Courtesy of Susan E. Shyne.]

tyrosine 42 in the α subunit in deoxyhemoglobin (refer to Figure 4.21). In mutant hemoglobins Yakima, Kemsey, and Radcliffe, the substitution of histidine, asparagine, or alanine upsets this interaction, and the cooperativity of the subunits decreases dramatically. A second type of alteration disrupts salt bridges (ionic bonds) between subunits. A third type disrupts the DPG binding site. In Figure 4.29 we saw that lysine 82 and histidine 143 interact strongly with DPG. Substitution of proline for histidine 143 (hemoglobin Syracuse) destroys the hydrogen bonding and electrostatic interactions that are possible between a phosphate group and an imidazole side chain in the normal hemoglobin, and the oxygen pressure required for half-saturation of hemoglobin drops from 27 torr to 11 torr.

In *sickle cell anemia*, a "molecular disease" affecting about 0.4% of black Americans, the solubility of hemoglobin is dramatically altered. The

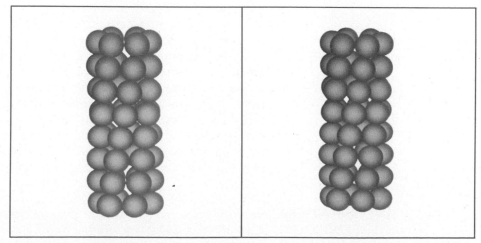

Figure 4.36 Helical structure of the oligomer formed by precipitated sickle-cell deoxyhemoglobin.

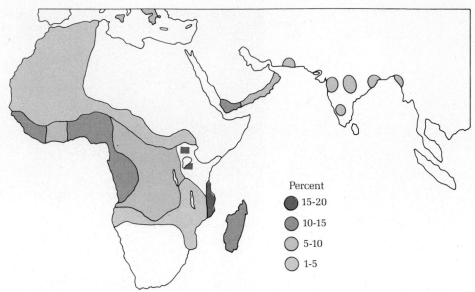

Figure 4.37 Distribution of sickle cell anemia in Africa. High rates of sickle cell anemia correspond to a high incidence of malaria. [From C. F. Herreid, II, "Biology," Macmillan, New York, 1977, p. 63.]

condition, discovered in 1904, derives its name from the sickled shape of red cells in affected persons (Figure 4.35). The sickled cells cannot navigate blood capillaries, impairing circulation and resulting in severe tissue damage, especially in bone and kidney. The sickled cells are unstable, hemolyse (break open) readily, and have short lifetimes. At low oxygen concentrations the solubility of sickle cell hemoglobin (HbS) declines by a factor of 25, and the hemoglobin precipitates forming fibrous oligomers of deoxyhemoglobin S (Figure 4.36). The difference between normal hemoglobin and HbS resides in a single position in the β chain: HbS contains a valine at position B6, whereas normal hemoglobin contains a glutamic acid residue at this position. This mutation in the primary structure of the β chain alters the solubility of deoxyHbS, but not of oxyHbS. The substitution of valine for glutamic acid results in a hydrophobic "tail" that can fit into a hydrophobic pocket in an adjacent sickle cell deoxyhemoglobin, thus causing aggregation.

The most severe form of sickle cell anemia, called *sickle cell disease,* occurs only in persons who are *homozygous* for the altered gene. They produce 90 to 100% defective β chains. Persons who are *heterozygous* for the altered gene produce both normal and altered β chains and are less severely affected. They have *sickle cell trait.*

An intriguing relationship between sickle cell trait and malaria resistance has evolved. In some parts of Africa up to 20% of the population has sickle cell trait. The trait produces an increased resistance to malaria because the malarial parasite cannot feed on the hemoglobin in sickled red blood cells. Persons with sickle cell disease die young; those without sickle cell trait have a high probability of succumbing to malaria. Occupying the middle ground, those with sickle cell trait do not suffer much from the pangs of sickle cell anemia, while avoiding the ravages of malaria (Figure 4.37).

4–7 CONFORMATION AND FUNCTION OF IMMUNOGLOBULINS

Immunoglobulins are serum *glycoproteins* (proteins covalently bound to carbohydrates) that are synthesized by vertebrates as *antibodies* to combat the invasion of various macromolecules called *immunogens* or *antigens.* Antigens may be polysaccharides, nucleic acids, or proteins. Certain low

Table 4.6 Classification of human immunoglobulins

	IgG	IgA	IgM	IgD	IgE
Heavy chains					
Class	γ	α	μ	δ	ϵ
Mol wt	53,000	64,000	70,000	58,000	75,000
Light chains					
Class	κ or λ	κ or λ	κ or λ	κ or λ	κ or λ
Mol wt	22,500	22,500	22,500	22,500	22,500
Formula	$\kappa_2\gamma_2$ or $\lambda_2\gamma_2$	$(\kappa_2\alpha_2)_{n=2,4}$ or $(\lambda_2\alpha_2)_{n=2,4}$	$(\kappa_2\mu_2)_{n=2,4}$ or $(\lambda_2\mu_2)_{n=2,4}$	$\kappa_2\delta_2$ or $\lambda_2\zeta_2$	$\kappa_2\epsilon_2$ or $\lambda_2\epsilon_2$
Total mol wt	150,000	360,000–720,000	950,000	160,000	190,000

molecular weight molecules, while they may bind to antibodies, do not themselves stimulate the production of antibodies, but they may stimulate antibody production if they are tightly bound to macromolecules. Such small molecules are called *haptens*. Nearly all foreign proteins are immunogens, and each protein elicits the production of specific antibodies. Perhaps 10,000 different antibodies exist in a human at any given time.

A. Classification of Immunoglobulins

There are five major types of human immunoglobulins (Table 4.6) spanning a molecular weight range from 150,000 to 720,000. Immunoglobulin M (IgM) is the first antibody that is produced in response to an antigen. Immunoglobulin G (IgG) is the simplest of the immunoglobulins. It contains two identical high molecular weight, or heavy, chains, and two identical low molecular weight or light, chains. These four chains are linked covalently by disulfide bonds in a Y-shaped structure (Figure 4.38). Each immunoglobulin G contains two binding sites for antigens located at the tips of the Y. If an antigen contains more than one *antigenic determinant*, the antibody can form a cross-linked lattice of antibody and antigen molecules that precipitates. The

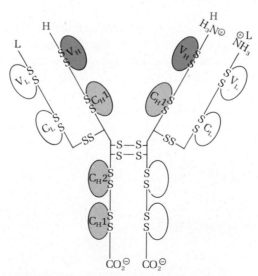

Figure 4.38 General Y-shaped structure of immunoglobulin G. The light (L) and heavy (H) chains are connected by one disulfide bond, and the heavy chains are connected by two disulfide bonds. [From D. R. Davies, E. A. Padlan, and D. M. Segal, *Ann. Rev. Biochem.*, 44, 641 (1975). Reproduced with permission. © 1975 by Annual Reviews Inc.]

largest amount of *immunoprecipitate,* or *precipitin,* forms when equal amounts of antibody and antigen are present (Figure 4.39).

The light and heavy chains of immunoglobulins can be further subdivided. There are two classes of light chains, κ (kappa) and λ (lambda). All five types of immunoglobulins occur in alternate forms containing either κ or λ chains. There are five classes of heavy chains, each found only in one immunoglobulin and designated γ, α, μ, δ, and ϵ for immunoglobulins G, A, M, D, and E, respectively. The much higher molecular weights of immunoglobulins A and M result from the increased number of chains. The functions of the heavier immunoglobulins are not all known with certainty. Immunoglobulin A is the major antibody type found in secretions (tears, mucus, and saliva), and it is the first defense against invading bacteria and viruses. The unique functions of the D and E classes of immunoglobulins are not known.

In the malignant disease multiple myeloma, a cancer of the bone marrow, large amounts of immunoglobulins known as *Bence-Jones* proteins are synthesized. These proteins are present in high concentrations and are excreted in the urine. The Bence-Jones proteins are dimers of κ or λ chains that are joined by a single disulfide bond. The antibodies of a normal person contain thousands of different sequences, but the Bence-Jones proteins produced by a given myeloma are identical.

B. Primary Structure of Immunoglobulins

The primary structures of the light and heavy chains of immunoglobulin G can be divided into structural domains that are classified by their sequence homologies (Figure 4.40). The light chains of each immunoglobulin contain regions whose sequences vary widely from antibody to antibody and regions whose sequences are nearly constant from antibody to antibody. The sequences are called V_L (variable light) and C_L (constant light), respectively. Within the regions of variability there are the so-called hypervariable regions whose sequences are part of the antigen-binding site. Similarly, the heavy chains contain variable and constant sequences denoted V_H, C_H1, C_H2, and C_H3. The V_H sequences of nine human antibodies are shown in Figure 4.41. Nearly all of the changes are found in the hypervariable regions. The relative constancy of the remaining sequence is probably required for formation of the basic structural unit of immunoglobulins (Section 4.7C). The hypervariable regions of the V_H chains combining with hypervariable regions of

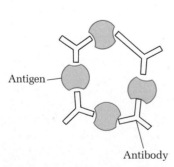

Figure 4.39 Schematic diagram of the cross-linked lattice of an immunoprecipitate between antibodies and antigens with two antigenic determinants. The antigenic determinant is the constellation of atoms that binds to a specific antibody.

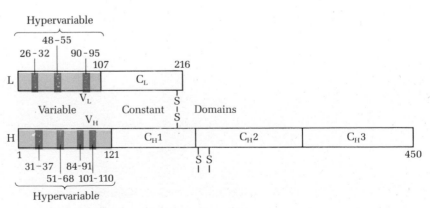

Figure 4.40 Diagram of the light (L) and heavy (H) chains of an immunoglobulin showing the variable (V) and constant (C) regions. The constant regions of the heavy and light chains (C_H and C_L) are labeled. The variable regions of the chains (V_H and V_L) are shaded. Within the variable regions hypervariable sequences are indicated by dark bands. [From A. R. Williamson, *Ann. Rev. Biochem.,* 45, 469 (1976). Reproduced with permission. © 1976 by Annual Reviews Inc.]

```
                                           10                                   20                                        30
Tie  (γ1) Glu  Val  Gln  Leu  Val  Glu  Ser  Gly  Gly  Gly  Leu  Val  Gln  Pro  Gly  Gly  Ser  Leu  Arg  Leu  Ser  Cys  Ala  Ala  Ser  Gly  Phe  Thr  Phe  Ser
Was  (γ1) ——————— Leu ———————————————————————————————————————————————————————— Ser ———
Jon  (γ3) Asp ———————————————————————————— Lys ————————————————————————————————————————
Zap  (α1) ——————————————— Ala ———————————————— Gly ————————————————————————————————
Tur  (α1) ——————— Leu ———————————————————————————————————————————————————————————————
Nie  (γ1) Pca ——————— Gln ——————— Val ——————— Arg ———————————————————————————————————
Gal  (μ)  ——————————————— Asp ——————————— Arg ———————————————————— (Asx  Val  Leu
Lay       Ala ——————— Leu ———————————————————————————————————————————————————————————
Pom       ——————————— Leu ———————————————————————————————————————————————————————————

                                           40                                   50                                        60
Tie       Thr  Ser  Ala  Val  Tyr (   )   Trp  Val  Arg  Gln  Ala  Pro  Gly  Lys  Gly  Leu  Glu  Trp  Val  Gly  Trp  Arg  Tyr  Glu  Gly  Ser  Ser  Leu  Thr
Was       —— Asp —— Met —— (   )   ———————————————————————— Ala —— Lys —— Gln  Glu  Ala —— Asn  Ser
Jon       —— Ala  Trp  Met  Lys (   )   ———————————————————————— Val ——————— Val —— Gln  Val  Val  Glu  Lys
Zap       —— Thr  Ser  Arg  Phe (   )   ———————————————————————— Glu  Phe —— Val  Gln ——————— Ala  Ile  Ser
Tur       Arg  Val  Leu  Ser  Ser (   )   ———————————————————————— Ser  Gly —— Leu  Asn  Ala ——————— Asn  Leu
Nie       Arg  Tyr  Thr  Ill  His (   )   ———————————————————————— Ala  Val  Met  Ser  Tyr  Asx  Gly  Asx  Asx  Lys
Gal       Asx  Asx  Phe  Met  Thr (   )   ———————————————————————— Ala  Asn  Ile  Lys  Glx  Asx  Gly —— Glx  Glx
Lay       Ala ——————— Met  Ser (   )   ———————————————————————— Ala —— Lys ——————— Asn  Gly  Asn  Asp  Lys
Pom       Ser ——————— Met  Ser (   )   ———————————————————————— Ala —— Lys ——————— Asx  Gly  Asn  Asp  Lys

                                           70                                   80                                        90
Tie       His  Tyr  Ala  Val  Ser  Val  Gln  Gly  Arg  Phe  Thr  Ile  Ser  Arg  Asn  Asp  Ser  Lys  Asn  Thr  Leu  Tyr  Leu  Gln  Met  Leu  Ser  Leu  Glu  Pro
Was       —— Phe —— Asp  Thr —— Asn ———————————————————————————————————————— Asn  Arg ——————— Ala
Jon       Ala  Phe —— Asn ——————— Asn ———————————————————————————————————————— Ile —— Val  Thr ——
Zap       ——————— Asp ——————— Ala ————————————————————————————————————————— Asn  Thr  Gly —— Ala
Tur       —— Phe ——————— Ala ————————————————————————————————————————————————————— Gln  Ala
Nie       ——————— Asp ——————— Asn ——————————————————————————————————— Asn —— Asn ——————— Arg ——
Gal       Asx —— Val  Asp ——————— Lys ——————————————— Asp  Asn  Ala ——————— Ser ——————— Asn ——————— Arg  Val
Lay       ——————— Asp ——————— Asn ———————————————————————————————————————— Asn  Gly —— Gln  Ala
Pom       ——————— Asp ——————— Asn ——————————————————————————————————— Leu —— Asn ——————— Gln  Ala

                                   100                                  110                                           120
Tie       Glx  Asx  Thr  Ala  Val  Tyr  Tyr  Cys  Ala  Arg  Val  Thr  Pro  Ala  Ala  Ala  Ser  Leu  Thr  Phe  Ser  Ala  Val  Trp  Gly  Gln  Gly  Thr  Leu  Val
Was       ———————————————————————————————— Phe  Arg  Gln  Pro  Phe  Val  Gln (   )  —— Phe  Asp —— Phe ———————————————
Jon       ——————————————————————————————————— Val  Val  Ser  Thr (        ) Ser  Met  Asp ——————————————— Pro ——
Zap       ———————————————————————————————— Thr  Arg —— Gly  Gly  Tyr (        ) Asp ——————————————————————————
Tur       ———————————————————————————————— Leu  Ser  Val  Thr —— Val (        ) Ala  Phe  Asp ——————————————— Lys ——
Nie       ———————————————————————————————— Ile  Arg  Asp  Thr —— Met (        ) —— Phe —— His ———————————————
Gal       ———————————————————————————————— Gly  Trp  Gly (             ) Gly  Gly  Asp  Tyr ———————————————
Lay       —— Val  Ser —— Ile ——————————————— Asp  Ala  Gly  Pro  Tyr  Val —— Pro ——————— Phe —— His ———————————————
Pom       ——————————— Leu ——————————————— Asp  Ala  Gly  Pro  Tyr  Val —— Pro ——————— Phe —— His  Tyr ————————
```

Figure 4.41 Sequence homologies of the variable regions of the heavy chains of nine human antibodies. Almost all of the variations are in the hypervariable regions. [From J. D. Capra and J. M. Kehoe, *Proc. Nat. Acad. Sci. U.S. 71*, 4032 (1974).]

the C_H chains contribute the amino acid residues that form the antigen-antibody binding site. The specificity of each antibody is determined by the amino acid sequences within the hypervariable regions of the light and heavy chains.

The C_H1 and C_H2 regions of the heavy chains are connected by a "hinge" some 30 residues long that is exposed to solvent. This region can be cleaved by treatment with papain, a proteolytic enzyme (Section 5.2B), to give *Fab fragments*. Cleavage of the hinge by pepsin, another proteolytic enzyme, yields $F(ab')_2$ fragments. The Fab fragments retain their structures and are still able to bind to antigen, showing that the light chains are the sites of antigen binding (Figure 4.42).

There are considerable sequence homologies between the constant regions (C_L, C_H1, C_H2, and C_H3) of antibodies (Figure 4.43). The variable regions of the light and heavy chains are also homologous. Each domain of the primary structure, whether constant or variable, contains a single disulfide bond within the domain. The exceptions are the disulfide bonds that hold the four chains together (Figure 4.44). These similarities suggest that a single primordial gene duplicated several times in succession resulting in modern antibodies. A scheme for gene duplication leading to the variable regions of the κ and λ chains is shown in Figure 4.45.

C. Tertiary Structure of Antibodies

The most prominent feature of the tertiary structure of immunoglobulins is called the *immunoglobulin fold* (Figure 4.46), a structural motif consisting of two stacked β-sheets surrounding an interior packed with hydrophobic residues. The β-sheets, involving about half the amino acid residues of the immunoglobulin fold, are held together by a disulfide bond. The amino acid residues in the β-sheets are largely conserved. Most variations in sequence occur in the loops connecting the β structures. The tertiary structure of a single λ chain is divided into two regions each of which contains an immunoglobulin fold (Figure 4.47).

The immunoglobulin fold recurs, in a structure of ever increasing complexity, in the Fab fragment of a mouse myeloma immunoglobulin G (Figure 4.48). (If a cell of the immune system becomes cancerous, it undergoes runaway replication forming a cancer called a myeloma. The immunoglobulin G produced by the myeloma cell may be present in such large amounts that the blood actually thickens.) The structure of myeloma IgG is divided into a variable region that contains V_H and V_L chains and a constant region that contains C_L and C_H1 domains. The short strand of polypeptide connecting the variable and constant regions is called the *switch region*. The hypervariable sequences of light and heavy chains are on the surface in close contact with each other. Antigen binding occurs in the variable region in a pocket carved out in the hypervariable sequences of the light and heavy chains (Figure 4.49). This binding site is illustrated by the binding of *phosphorylcholine* in a pocket between the hypervariable regions of the light and heavy chains whose dimensions ($1.2 \times 1.5 \times 2.0$ nm) nicely accommodate the antigen (Figure 4.50).

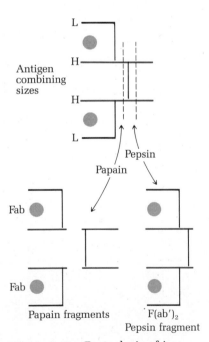

Figure 4.42 Proteolysis of immunoglobulin G. Cleavage by papain gives FAB fragments and by pepsin gives F(ab')₂ fragments, both of which can combine with antigens. [Reprinted by permission from J. D. Watson, "Molecular Biology of the Gene," 3rd ed., Benjamin/Cummings, Menlo Park, Calif., 1976, p. 605.]

$$(CH_3)_3\overset{\oplus}{N}-CH_2-CH_2-O-PO_3^{\circleddash{2}}$$

Phosphorylcholine

The structure of the entire immunoglobulin G is simply an elaboration of the Fab fragment, being built up from repeating immunoglobulin folds (Figure 4.51).

	110									120		

EU C$_L$ (RESIDUES 109–214)   Thr Val Ala Ala Pro Ser Val Phe Ile Phe Pro Pro Ser
EU C$_H$1 (RESIDUES 119–220)   Ser Thr Lys Gly Pro Ser Val Phe Pro Leu Ala Pro Ser
EU C$_H$2 (RESIDUES 234–341)   Leu Leu Gly Gly Pro Ser Val Phe Leu Phe Pro Pro Lys
EU C$_H$3 (RESIDUES 342–446)   Gln Pro Arg Glu Pro Gln Val Tyr Thr Leu Pro Pro Ser

130
Asp Glu Gln — — Leu Lys Ser Gly Thr Ala Ser Val Val Cys Leu Leu Asn Asn Phe
Ser Lys Ser — — Thr Ser Gly Gly Thr Ala Ala Leu Gly Cys Leu Val Lys Asp Tyr
Pro Lys Asp Thr Leu Met Ile Ser Arg Thr Pro Glu Val Thr Cys Val Val Val Asp Val
Arg Glu Glu — — Met Thr Lys Asn Gln Val Ser Leu Thr Cys Leu Val Lys Gly Phe

140   150
Tyr Pro Arg Glu Ala Lys Val — — Gln Trp Lys Val Asp Asn Ala Leu Gln Ser Gly
Phe Pro Glu Pro Val Thr Val — — Ser Trp Asn Ser — Gly Ala Leu Thr Ser Gly
Ser His Glu Asp Pro Gln Val Lys Phe Asn Trp Tyr Val Asp Gly — Val Gln Val His
Tyr Pro Ser Asp Ile Ala Val — — Glu Trp Glu Ser Asn Asp — Gly Glu Pro Glu

160   170
Asn Ser Gln Glu Ser Val Thr Glu Gln Asp Ser Lys Asp Ser Thr Tyr Ser Leu Ser Ser
—Val His Thr Phe Pro Ala Val Leu Gln Ser — Ser Gly Leu Tyr Ser Leu Ser Ser
Asn Ala Lys Thr Lys Pro Arg Glu Gln Gln Tyr — Asp Ser Thr Tyr Arg Val Val Ser
Asn Tyr Lys Thr Thr Pro Pro Val Leu Asp Ser — Asp Gly Ser Phe Phe Leu Tyr Ser

180   190
Thr Leu Thr Leu Ser Lys Ala Asp Tyr Glu Lys His Lys Val Tyr Ala Cys Glu Val Thr
Val Val Thr Val Pro Ser Ser Ser Leu Gly Thr Gln — Thr Tyr Ile Cys Asn Val Asn
Val Leu Thr Val Leu His Gln Asn Trp Leu Asp Gly Lys Glu Tyr Lys Cys Lys Val Ser
Lys Leu Thr Val Asp Lys Ser Arg Trp Gln Glu Gly Asn Val Phe Ser Cys Ser Val Met

200   210
His Gln Gly Leu Ser Ser Pro Val Thr — Lys Ser Phe — — Asn Arg Gly Glu Cys
His Lys Pro Ser Asn Thr Lys Val — Asp Lys Arg Val — — Glu Pro Lys Ser Cys
Asn Lys Ala Leu Pro Ala Pro Ile — Glu Lys Thr Ile Ser Lys Ala Lys Gly
His Glu Ala Leu His Asn His Tyr Thr Gln Lys Ser Leu Ser Leu Ser Pro Gly

Figure 4.43 Sequence homologies in the C$_L$, C$_{H1}$, C$_{H2}$, and C$_{H3}$ regions of an antibody. Deletions indicated by dashes have been added to maximize the homology. Color shading indicates identities in the same positions. [From G. M. Edelman et al., *Proc. Nat. Acad. Sci. U.S.*, **63**, 78 (1969).]

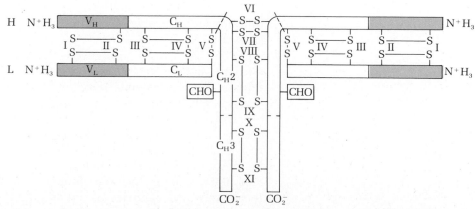

Figure 4.44 Sequence homologies within immunoglobulin G. The variable regions V$_L$ and V$_H$ are homologous to each other. [Courtesy of G. M. Edelman.]

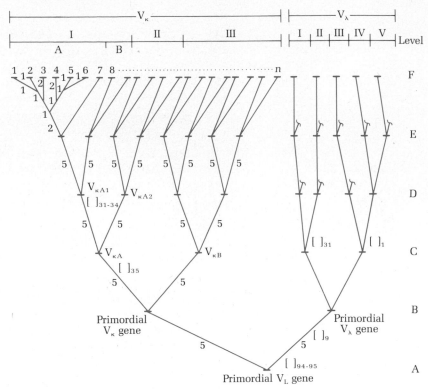

Figure 4.45 A possible genealogic tree for the V_L region of human antibodies. The numbers beside the branches indicate the number of amino acid substitutions between branch points. [From L. Hood, J. H. Campbell, and J. Elgin, *Ann. Rev. Genetics, 9,* 312 (1975). Reproduced by permission. © 1975 by Annual Reviews Inc.]

4–8 VIRUS SELF-ASSEMBLY: A MODEL FOR MORPHOGENESIS

Our discussion of the self-assembly of proteins and the hypothesis that the primary structure contains the information required for "spontaneous generation" of the tertiary structure of proteins leads us to ask how larger scale biological systems develop. This topic is called *morphogenesis,* "the birth of forms." The self-assembly of viruses provides us with phenomena that can be taken as conceptual models, though perhaps not detailed biochemical models, for the morphogenesis of cells and multicellular organisms. We would not like to stretch the analogy too far, since morphogenesis in higher organisms is vastly more intricate, but virus self-assembly does provide a system in which our discussion of protein folding can be extended to more complex objects such as multienzyme complexes (Chapter 14) and ribosomes (Chapter 26).

The Latin word *virus* means poison, a meaning conserved in our word *virulent,* and in fact many viruses are pathogens, responsible for such human diseases as smallpox, yellow fever, rabies, poliomyelitis, measles, and mumps. Despite their extraordinary diversity, all viruses share the property of having a nucleic acid core surrounded by a protein coat. A more recently discovered class of objects, called *viroids,* dispense with coats and consist simply of low molecular weight strands of nucleic acid. Viruses may be viewed as intracellular parasites that operate on the genetic level of cell chemistry. The nucleic acid enters a cell, leaving its coat behind, takes over the metabolic machinery of the cell, and directs its host to synthesize the parts required for the more or less spontaneous self-assembly of new viruses. We enter the drama in the last act, after the synthesis of the macromolecules required to make a new virus has been completed.

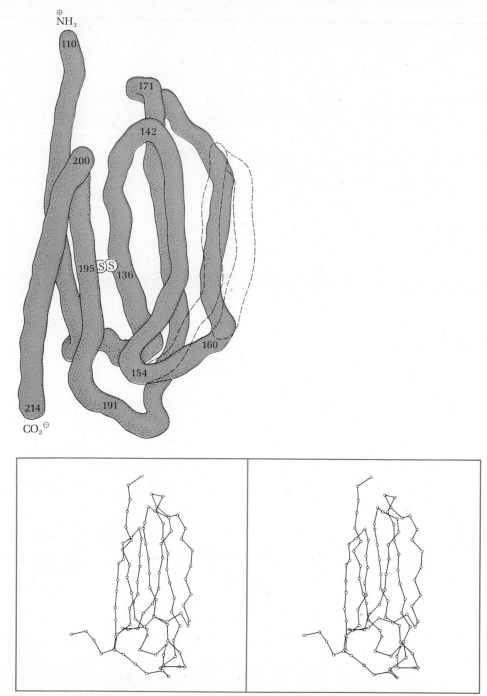

Figure 4.46 Diagram of the immunoglobulin fold. The solid line traces the folding of the protein in the C_L and C_H domains. The numbers designate L chain residues beginning at residue 110. The dotted lines follow the course of the chain in the V_L and V_H subunits. [From R. J. Popjak et al., *Proc. Nat. Acad. Sci. U.S.*, 70, 3305 (1973).]

A. Spontaneous Self-Assembly of the Tobacco Mosaic Virus

The tobacco mosaic virus (TMV) is the paradigm of experimental viruses. It was discovered by M. W. Beijerinck in about 1899 and has been studied continuously ever since. The tobacco mosaic virus is a rod-shaped particle whose dimensions are 30 × 18 nm (Figure 4.52). By weight it is about 95% protein and 5% ribonucleic acid (RNA). Astonishingly, it can be crystallized just like an ordinary chemical compound. The TMV particle is a helical assembly of identical protein subunits. Each particle contains 2,130 protein

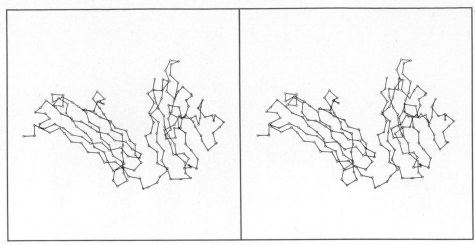

Figure 4.47 Tertiary structure of a single λ chain of a human Bence-Jones protein. The structure consists of two immunoglobulin folds connected by a short region of the polypeptide chain.

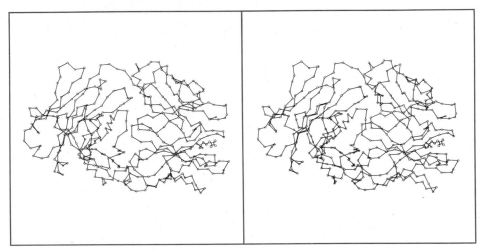

Figure 4.48 Tertiary structure of an Fab fragment of mouse myeloma.

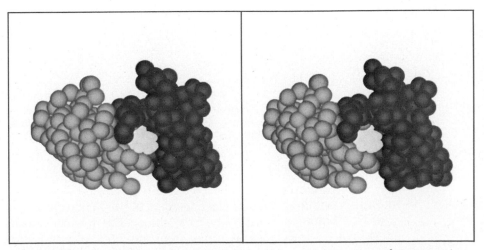

Figure 4.49 Schematic view of the binding region of a human myeloma protein and the hapten vitamin K_1OH. The binding site lies between the light and heavy chains in a shallow cleft.

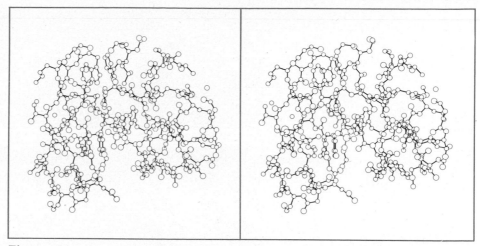

Figure 4.50 Schematic diagram of the specific interactions between phosphoryl-choline and the side chain residues of a mouse myeloma protein. The binding site is located between the L and H chains.

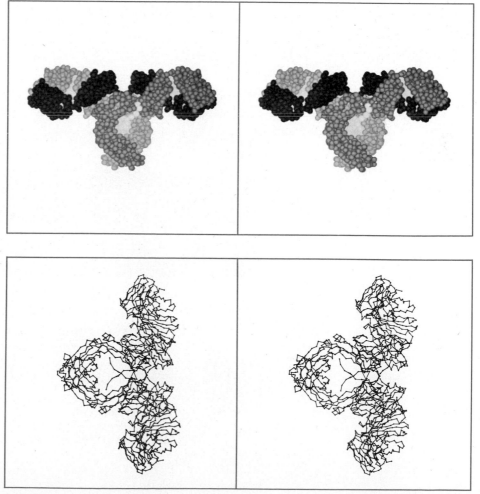

Figure 4.51 The structure of immunoglobulin G Dob. The upper structure is a space-filling model in which one complete heavy chain is white, the othe gray. The large dark section in the center of the upper figure is carbohydrate covalently bound to the antibody. The light chains of the antibody are lightly shaded.

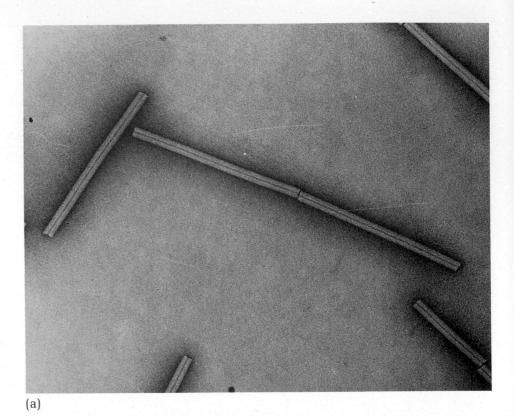

(a)

(b)

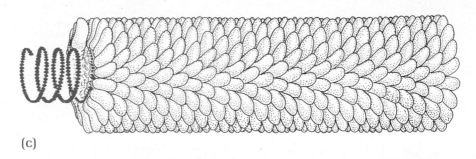

(c)

Figure 4.52   Tobacco mosaic virus. (a) Electron micrograph of TMV particles. [Courtesy of Dr. A. Klug.] (b) Electron micrograph of tabacco mosaic virus. [Courtesy of Robert W. Horne.] (c) Schematic drawing of the quaternary structure of TMV. The protein coat subunits extend radially and surround a single strand of ribonucleic acid.

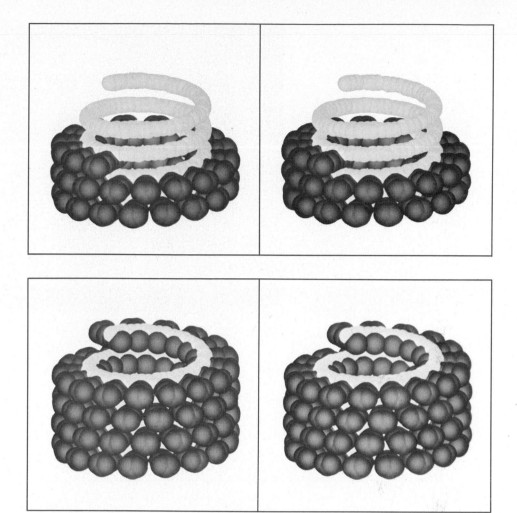

Figure 4.53 Steps in the spontaneous self-assembly of TMV. The protein subunits of the coat first form a 17-subunit disc that is the template for both the ribonucleic acid and other coat subunits.

subunits whose polypeptide chains each contain 158 amino acids. The diameter of the core of the helical coil of subunits is 2 nm. This internal core has a helical groove, like the threads on a metal screw socket, that holds the RNA of the virus.

The assembly of the "viral helix" is endothermic. Nevertheless, the viral sheath, or *capsid,* assembles spontaneously in a process that is driven by a favorable entropy change. The spontaneous self-assembly of the TMV particle further amplifies the hypothesis that the primary structure of a protein determines its tertiary and quaternary structure. The first step in the process, summarized in Figure 4.53, is the formation of a disc of 17 subunits of protein. The disc serves as a template for the assembly of RNA and other subunits.[3] Next, the viral RNA attaches noncovalently to the first disc, and then further subunits form discs and polymerize onto the first. As the discs stack the subunits at one point slide up and over one another so that the disc changes from a flat to a helical, lock-washer shape. Each turn of the helix contains $16\frac{1}{3}$ subunits, rather than 17, because of this conformational change. When all of the RNA has been coiled into its groove, the particle stops growing. The subunits can assemble spontaneously in the absence of RNA, but the helix is less stable than in the RNA-containing virus, and the lengths of the particles that form without RNA vary considerably.

---

[3] This disc has C_{17} symmetry!

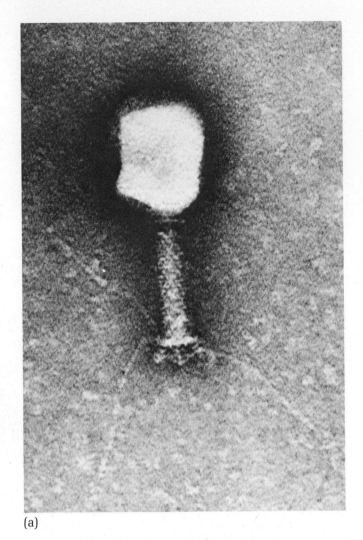

(a)

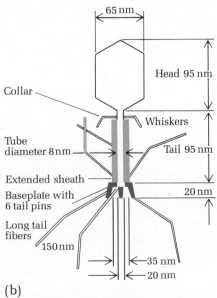

(b)

Figure 4.54 Structure of T-even phage of *E. coli*.
(a) Electron micrograph. (b) Schematic diagram of the
structure of the same phage. [Electron micrograph
courtesy of Dr. Arthur Zachary and Dr. B. Lindsay
Black.]

**B. Assembly
of the
T-even Bacteriophages
of *E. coli***

A virus that infects bacteria is called a *bacteriophage*, often shortened to *phage* by virologists. The name is derived from the Greek *phagein*, "to devour." The general structure of one type of phage, the *T-even phage*, is shown in Figure 4.54. These viruses are far more complicated than the tobacco mosaic virus. *Bacteriophage T4* assembles by the pathway shown in Figure 4.55. The components of T4 (head, tail, and tail fibers) assemble independently, and these preassembled parts are then put together, rather as an automobile is constructed from preassembled parts on an assembly line. The 40 types of proteins in the T4 particle are identified by numbers derived from the *genetic map* of T4. Besides the 40 proteins that constitute T4, there are at least 6 proteins that do not appear in the phage itself, but catalyze phage assembly—they are the workers on the phage assembly line.

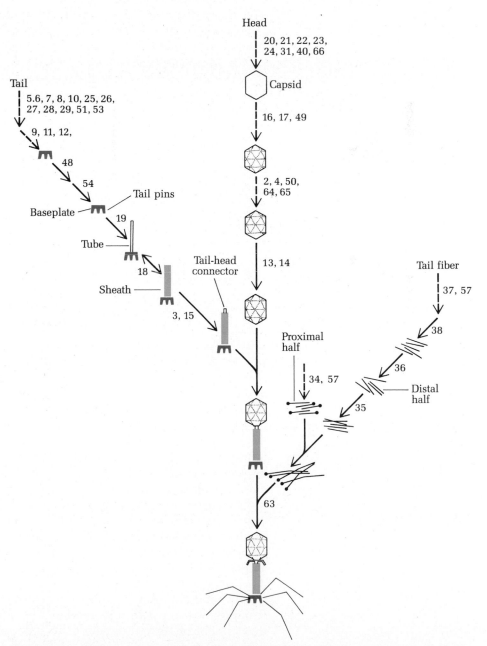

Figure 4.55 Steps in the assembly of bacteriophage T4. The proteins involved are indicated at each step. [From W. B. Wood, in F. H. Ruddle, Ed., "Genetic Mechanisms of Development," Academic Press, New York, 1973.]

Thus, in contrast to TMV, the assembly of T4 is not a completely spontaneous process, although it has spontaneous segments. There are several major sub-assembly processes.

1. The baseplate of the tail contains 15 different proteins that assemble with the aid of at least 2 other phage proteins that are not present in the final tail structure. Some enzymes from the host cell also participate in assembly.

2. The tail tube is a kind of "hypodermic needle" though which the viral nucleic acid is injected into a host cell. The tail-tube protein subunits stack on the completed baseplate to form a structure of 24 stacked discs containing 6 subunits each. Why the tail stops at just this size is not known. Perhaps the tail tube is simply the thermodynamically most stable unit. If that is correct, the information contained in the primary structure must dictate the tail length. This hypothesis seems almost too extraordinary to be true, but if perchance it is, the thermodynamic hypothesis has profound implications for the entire field of morphogenesis. Certain viruses that have undergone a mutation in the sequence of the tail-tube protein grow tails of random lengths instead of the precise tail length of normal viruses.

3. The phage head contains about 1,000 proteins of 10 different types, and its assembly requires some enzymatic steps. One protein that initially occupies the internal space of the head, the future location of the phage DNA, digests itself to give a group of small peptides that are eliminated later in phage assembly.

4. The head and tail units assemble spontaneously both in vivo and in vitro.

5. The tail fibers, constructed from four different proteins, assemble spontaneously, but they can attach to the baseplate only after the head and tail have combined.

6. The phage DNA, compressed into a highly compact ball of unknown topography, enters the empty head after it has been assembled. Perhaps the internal proteins provide a matrix about which the phage DNA coils, although this is not certain.

4–9 SUMMARY The three-dimensional conformation of a polypeptide chain constitutes the tertiary structure of the protein. This conformation is maintained by interactions that include hydrogen bonds, electrostatic forces, and van der Waals forces. Of the weak forces responsible for the tertiary structure, the ones that contribute the most to the stable, biologically active conformation are the van der Waals forces. The thermodynamic parameter that governs the hydrophobic effect is the positive entropy change that accompanies the movement of the nonpolar side chains of amino acid residues from the aqueous solvent to the relatively nonaqueous interior of the protein. The final conformation of the polypeptide reflects a subtle balance of forces: it is the free energy of the entire molecule that is minimized, and a given residue can sometimes be found in an "unfavorable" environment. The tertiary structures of proteins appear to be determined by the "genetically stipulated" primary structures and arise spontaneously given the proper initial conditions.

Myoglobin is the oxygen-storage protein of skeletal muscle. The binding site for oxygen is provided by the heme prosthetic group. The oxygen-binding curve of myoglobin is hyperbolic. A Hill plot of myoglobin has a slope of 1.0, a property shared by all other monomeric proteins that contain a single binding site for a given ligand.

Hemoglobin is the oxygen-transport protein of blood. It is a tetramer whose subunits each contain a heme group that binds to oxygen. The four subunits interact with one another to confer properties upon the tetramer that are not possessed by any of its subunits alone nor by myoglobin. The "action at a distance" in hemoglobin, and in many other quaternary proteins, is a consequence of allosteric interactions.

Allosteric proteins often bind to their various ligands with sigmoidal binding curves. The negative, heterotropic, allosteric effector 2,3-diphosphoglycerate decreases the affinity of hemoglobin for oxygen by cross-linking its subunits by ionic bonds.

The concerted model and the sequential model of allosteric transitions have been proposed to account for the properties of allosteric proteins. The fundamental difference between the two theories is that in the concerted theory the symmetry of the oligomer is conserved throughout the allosteric transition, whereas in the sequential theory there is no symmetry restriction.

Immunoglobulins defend a vertebrate against invading pathogens by forming stereospecific complexes with an antigenic site such as a carbohydrate moiety or some part of the surface of a protein. The tertiary structures of immunoglobulins are constructed from repeating structural units, called domains, whose iteration provides a mechanism for building complicated structures from relatively simple components.

The viruses are parasites that operate at the molecular level of cellular life by expropriating the biosynthetic machinery of the host to produce the parts required for the assembly of new virus particles. The spontaneous self-assembly of viruses may be a simple model for morphogenesis in higher organisms. Extending the thermodynamic principle of protein assembly, we hypothesize that virus self-assembly produces the thermodynamically most stable object in an entropically driven process.

REFERENCES

Anfinsen, C. B., Principles That Govern the Folding of Polypeptide Chains, *Science*, *181*, 223–230 (1973).

Arnone, A., X-ray Diffraction Study of Binding of 2,3-diphosphoglycerate to Human Deoxyhemoglobin," *Nature*, *237*, 146–149 (1972).

Cantor, C. R., and Schimmel, P. R., "Biophysical Chemistry," Freeman, San Francisco, 1980.

Dickerson, R. E., and Geis, I., "The Structure and Action of Proteins," 2nd ed., Benjamin/Cummings, Menlo Park, Calif., 1981.

Haschenmeyer, R. H., and Haschenmeyer, E. V., "Proteins: A Guide to Study by Physical and Chemical Methods," Wiley-Interscience, New York, 1973.

Kendrew, J. C., Myoglobin and the Structure of Proteins, *Science*, *139*, 1259–1266 (1963).

Moffat, K., Deatherage, J. F., and Seybert, D. W., A Structural Model for the Kinetic Behavior of Hemoglobin, *Science*, *206*, 1035–1042 (1979).

Monod, J., Changeux, J. P., and Jacob, F., Allosteric Proteins and Cellular Control Systems, *J. Mol. Biol.*, *6*, 306–329 (1963).

Monod, J., Wyman, J., and Changeux, J. P., On the Nature of Allosteric Transitions: A Plausible Model," *J. Mol. Biol.*, *12*, 88–118 (1965).

Neurath, H., and Hill, R. L., Eds., "The Proteins," 3rd ed., Vols. 1–3, Academic Press, New York, 1975–1977.

Perutz, M. F., Regulation of Oxygen Affinity of Hemoglobin," *Ann. Rev. Biochem.*, *48*, 327–386 (1979).

Perutz, M. F., The Hemoglobin Molecule and Respiratory Transport, *Sci. Amer.*, December 1978.

Perutz, M. F., "Stereochemistry of Cooperative Effects of Haemoglobin," *Nature*, *228*, 726–739 (1970).

Shulz, G. E., and Schirmer, R. H., "Principles of Protein Structure," Springer-Verlag, New York, 1979.

Tanford, C., "The Hydrophobic Effect," Wiley-Interscience, New York, 1973.

Watson, H. C., The Stereochemistry of the Protein Myoglobin, *Progr. Stereochem.*, *4*, 299–333 (1969).

Problems

1. Nozaki and Tanford have studied the following reaction

$$R-\underset{\underset{\oplus}{NH_3}}{\overset{\overset{H}{|}}{C}}-CO_2^{\ominus} \text{ (water)} \underset{}{\overset{K_{transfer}}{\rightleftarrows}} R-\underset{\underset{\oplus}{NH_3}}{\overset{\overset{H}{|}}{C}}-CO_2^{\ominus} \text{ (ethanol)}$$

and obtained the following data.

Side chain	$-\Delta G°_{transfer}$ $(j/mol)^a$
Glycine	0
Tryptophan	1.42×10^4
Phenylalanine	1.04×10^4
Tyrosine	9.62×10^3
Leucine	7.53×10^3
Valine	6.28×10^3
Methionine	5.44×10^3
Alanine	2.09×10^3

^{a} The data are for the additional free energy of transfer relative to glycine.
SOURCE: Y. Nozaki and C. Tanford, *J. Biol. Chem.*, *246*, 2211 (1971).

(a) Which of the amino acids is the most hydrophobic? (b) Tyrosine is slightly less hydrophobic than phenylalanine. Why? (c) Where do you expect to find these amino acids in the native conformation of proteins? Why? (d) Hydrophobic amino acids are sometimes found on the surface of proteins, exposed to the aqueous phase. Given these data, how is that possible?

2. The effect of the triphosphate SLTP on the ability of hemoglobin to bind oxygen has recently been studied. Some of the experimental data is contained in the accompanying graph of oxygen equilibrium binding curves. Interpret this data.

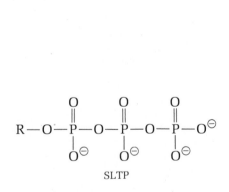

SLTP

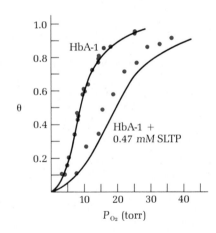

3. An aspartic acid residue is buried in the hydrophobic interior of chymotrypsin. (a) Explain what changes, if any, you expect to find in the pK_a value of this aspartic acid residue. (b) A histidine residue is buried quite near the aspartate residue of part (a) above. What will be the effect of the aspartic acid on the pK_a of the imidazolium side chain of the nearby histidine?

4. In the activation of chymotrypsinogen (Chapter 5) trypsin cleaves a peptide bond between arginine 15 and isoleucine 16. This cleavage activates the enzyme. The α-amino group of isoleucine 16 must be in an ionic bond with the carboxyl side chain of aspartate 194 for the enzyme to be active. How do you expect the pK_a of the isoleucine α-amino group and the pK_a of the carboxyl group of aspartate 194 to be altered by this ionic bond?

5. Enolase (Chapter 14) contains 14 histidine residues. However, titration of enolase reveals only four groups that titrate between pH 7 and pH 8.5. How do you account for this observation?

6. The statement that an allosteric transition is concerted is equivalent to the statement that the symmetry of the oligomer is conserved in an allosteric transition. Why are the statements equivalent?

7. When collagen is heated, a physical change in its structure occurs in which its helical structure is disrupted. At 20°C collagen is almost 100% triple helix; at 40° it is nearly 0% triple helix. A plot of percent helicity versus temperature is sigmoidal. What does this shape of curve and the relatively low amount of energy required to disrupt collagen structure imply about the nature of the interactions that stabilize collagen?

8. The rate vs. substrate (PEP) curves for pyruvate kinase without fructose 1,6-diphosphate and pyruvate kinase with fructose 1,6-diphosphate are shown below. What can you say about the allosteric interaction of fructose 1,6-diphosphate with this enzyme?

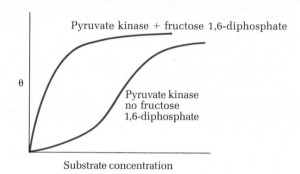

9. (a) Briefly describe the Bohr effect in relation to the carbon dioxide equilibrium:

$$CO_2 + H_2O \rightleftharpoons HCO_3^{\ominus} + H^{\oplus}$$

(b) Bicarbonate is often used intravenously for shock victims. What effect would this treatment have on oxygen transport for these patients?

10. How does the substitution of an aspartate residue for a lysine residue alter the interaction of 2,3-diphosphoglycerate with hemoglobin? (See Figure 4.23 to answer the question and Table 4.5, hemoglobin Providence, to see if your answer is right.)

11. A certain fish has a protein that binds to oxygen with 1:1 stoichiometry. The dissociation constant for the fishglobin-oxygen complex is 0.5 torr. (a) At what concentration of oxygen will fishglobin be one-half saturated with oxygen? (b) Calculate the fraction of fishglobin bound to oxygen at an oxygen pressure of 0.25 torr. What slope would be expected for a Hill plot of oxygen binding to fishglobin?

12. Explain why chloride ion is a negative allosteric effector of oxygen binding in hemoglobin.

13. A typical histidine residue in a protein has a pK_a of about 6.4 and an enthalpy of ionization of about -32 kJ mol^{-1}. Calculate the standard free energy of ionization ($\Delta G°$) and the standard entropy of ionization ($\Delta S°$) for the reaction at 298 K

$$\text{protein-ImH}^{\oplus} \rightleftharpoons \text{protein-Im} + H^{\oplus}$$

14. In the mutant hemoglobin known as hemoglobin Providence lysine EF6, which projects its side chain into the central cavity of hemoglobin, is replaced by an asparagine residue. Predict (a) the effect of this mutation on the affinity of hemoglobin for 2,3-diphosphoglycerate, and (b) whether you expect the affinity of hemoglobin Providence for oxygen to increase, decrease, or remain the same relative to normal adult hemoglobin.

15. The N-terminal residues of certain fish hemoglobins are acetylated. This covalent modification greatly diminishes the Bohr effect, and makes the

binding of oxygen independent of pH. Explain this observation. [Reference: R. G. Gillen and A. Riggs, *J. Biol. Chem.*, *248*, 1961–1965 (1973).]

16. Rats adapt better to high altitudes when sodium cyanate is administered orally. The cyanate ion reacts with the N-terminal residues of hemoglobin to give carbamoyl adducts:

$$R-NH_2 + O{=}C{=}N^{\ominus} \longrightarrow R-NH-\underset{\underset{O}{\|}}{C}-NH_2$$

Carbamoyl adduct

Why does this covalent modification alter the oxygen transporting system of the rats? [Reference: J. W. Eaton, T. D. Skelton, and E. Berger, *Science, 183,* 784 (1974)]

17. The study of amino acid substitutions (mutations) in proteins has revealed that serine is the amino acid that is most often replaced without destruction of protein function, that tryptophan is the amino acid that is least often replaced without the loss of biological activity, and that in a great many cases glycine residues are critical to protein structure. Explain each of these observations.

18. Explain why amino acid substitutions such as Lys → Arg and Ile → Leu are often possible in proteins without the loss of biological activity, but that the change Glu → Val in the sixth position of the β chain of hemoglobin (for example) leads to a defective molecule.

Chapter 5

The isolation, purification, and characterization of biological macromolecules

5–1 INTRODUCTION Many biochemical studies are devoted to the isolation and purification of proteins. This is often no mean task. To obtain a pure protein we first must be able to identify it and to distinguish it from any other protein. In short we require an *assay* based on distinctive properties of the protein. An assay procedure, for example, one based on enzyme activity, must be absolutely specific. If the assay is not specific, we are nearly certain to obtain a mixture of proteins. The assay must also be sensitive, since in many cases the protein is present only in minute concentrations in the source material. The assay must also be reproducible with high precision. If it is not, the work will be difficult to repeat in one's own laboratory, to say nothing of other laboratories attempting to repeat the purification. It is also helpful if the assay technique is simple and convenient, although this is not always possible. Perhaps we need not add that these criteria are not always easily met.

Once the assay and the source material have been determined, the fractionation procedure begins. A wide variety of techniques are employed in cell fractionation and subsequent protein purification. We shall describe only a broad general scheme that is fundamental to the countless variations. Regardless of the detailed procedure, however, the assay is performed at each step to be sure that the preceding step has not destroyed biological activity and to test the effectiveness of the step. At each stage of the purification the specific activity of the enzyme should increase as the concentration of the protein increases.

The first step in fractionation is rupture of the cells. If the protein to be purified is located within an organelle, the rupture of the cell must be done as gently as possible to avoid destruction of the organelles. Many methods are used to break open cells, including grinding with a tissue homogenizer or a mortar and pestle, lysis in a blender, and disruption by ultrasonic vibration, a technique known as *sonication.*

During these procedures the protein can lose biological activity immediately (denature) if the correct pH of the solution is not carefully maintained, if the temperature rises too high, if heavy metals are present, or if the protein oxidizes. Heavy metals can be removed by adding a small amount of *ethylenediaminetetraacetic acid* (EDTA) to the buffer solution.

$$\begin{array}{c} \quad\quad\quad CH_2CO_2^{\ominus} \\ \quad\quad\quad / \\ CH_2-N-CH_2CO_2^{\ominus} \\ | \\ CH_2 \\ | \\ N-CH_2CO_2^{\ominus} \\ \quad\backslash \\ \quad CH_2CO_2^{\ominus} \end{array}$$

Ethylenediaminetetraacetic acid (EDTA)

Oxidation can be minimized by addition of small amounts of thiols, such as β-mercaptoethanol, to the solution. The sulfhydryl group of the added thiol helps to maintain proteins in their reduced states.

$$HO-\overset{\alpha}{CH_2}-\overset{\beta}{CH_2}-SH$$

β-Mercaptoethanol

The next stages of purification usually involve some combination of centrifugation, various types of chromatography, and perhaps fractionation by precipitation or solvent extraction. The protein can be analyzed for purity in many ways, electrophoresis being one of the most important. The molecular weight of the purified protein can be determined using such techniques as sedimentation velocity centrifugation, zonal centrifugation, molecular exclusion chromatography, or SDS-gel electrophoresis. The protein can also be characterized by its optical spectrum, an especially important feature of molecules such as cytochromes that have complex UV-visible spectra. We shall see as we proceed that some of these techniques are as useful for protein characterization as they are for protein purification.

5–2 CENTRIFUGATION

The history of an epoch is the history of its instruments.

Albert Einstein

Perhaps more has been learned about proteins and nucleic acids (Chapter 9) through the use of high-speed centrifugation than by any other technique except x-ray crystallography. High-speed centrifugation can reveal the molecular weight, density, size, and shape of a macromolecule. The centrifuge operates on the principle that molecules of different masses and shapes move through a solution at different rates in the presence of an applied gravitational field, obtained by spinning the sample in a rotor at high speeds. Since molecules of different mass sediment at different rates, the centrifuge is extremely valuable in bulk protein purifications as well. In fact, it is hardly possible to walk into a biochemistry laboratory without discovering a preparative centrifuge.

Svedberg built the first ultracentrifuge in 1926. His instrument was capable of generating a centrifugal force 5,000 times the force of gravity. This is expressed as 5,000 g, where g is the gravitational force constant. Modern ultracentrifuges are typically capable of speeds of 65,000 rpm. At this speed, a force of 250,000 g is generated at a distance of 6.5 cm from the center of rotation. Reduced to its essence, a centrifuge consists of a spinning rotor with holes in it to hold sample tubes of various sizes. Its modern incarnation comes to something more than that, but the underlying principle remains unchanged.

A. Preparative Ultracentrifugation

The preparative ultracentrifuge permits rapid separation of fractions of a tissue homogenate. The principle upon which the separation is based, *velocity sedimentation,* relies upon different rates of sedimentation of various cell parts and subcellular particles, such as cell organelles, cell walls, and ribosomes.

Sample volumes in preparative centrifugation vary from 10 ml to 2 liters (l). Typically a *fixed angle rotor* is used (Figure 5.1). The centrifuge is refrigerated to prevent denaturation of sensitive proteins during centrifugation. The most massive particles, such as intact eucaryotic cells, are first removed by preliminary low-speed centrifugation. The supernatant fraction is then further fractionated by centrifuging at progressively higher speeds, a technique known as *differential centrifugation* (Figure 5.2). Table 5.1 gives the cellular fractions that sediment at various speeds. Those molecules that remain in solution following centrifugation at a force of 100,000 g are the soluble cytosolic proteins. These soluble proteins may be separated from one another by techniques that are described later in this chapter.

B. Analytical Ultracentrifugation

An analytical ultracentrifuge consists of a finely machined mechanical component to generate high speeds and an optical component to monitor events during ultracentrifugation. A block diagram of a Beckmann analytical ultracentrifuge is shown in Figure 5.3. The centrifuge rotor (Figure 5.4) is suspended from the drive shaft of the instrument in an evacuated cooled chamber. Since the rotor would be deadly if it exploded or broke loose, the chamber is armor plated. Both the temperature and the rate of rotation can be precisely controlled. A light source illuminates the sample parallel to the axis of rotation and perpendicular to the flow of material through the cell. A variety of optical systems have been designed to monitor the boundary of the solution as the sample flows through the cell. With this brief introduction to the instrument, let us turn to some of the experimental methods of analytical ultracentrifugation.

Table 5.1	Cellular fractions that sediment at various centrifugal forces	
Centrifugal force	**Time**	**Fraction sedimented**
1,000 g	5 min	Eucaryotic cells
4,000 g	10 min	Chloroplasts, cell debris, cell nuclei
15,000 g	20 min	Mitochondria, bacteria
30,000 g	30 min	Lysozomes, bacterial cell debris
100,000 g	3–10 hr	Ribosomes

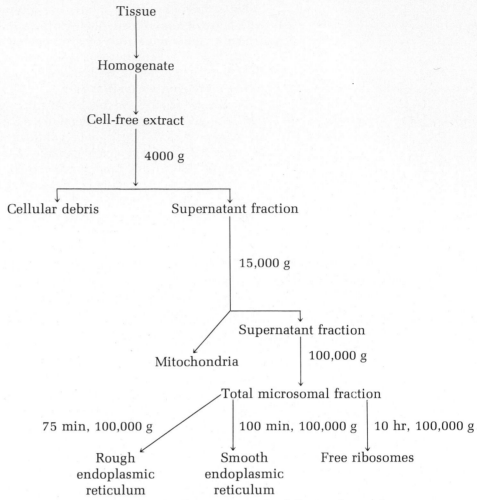

Tissue

Homogenate

Cell-free extract

4000 g

Cellular debris   Supernatant fraction

15,000 g

Supernatant fraction

Mitochondria   100,000 g

Total microsomal fraction

75 min, 100,000 g   100 min, 100,000 g | 10 hr, 100,000 g

Rough   Smooth   Free ribosomes
endoplasmic   endoplasmic
reticulum   reticulum

Figure 5.2 Fractionation of cell components by differential centrifugation.

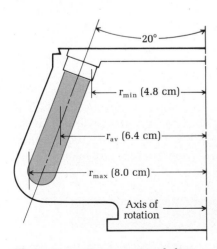

Figure 5.1 Cross-sectional diagram of a fixed angle rotor giving r, the distance from the axis of rotation, to the top, center, and bottom of the centrifuge tube. [Courtesy of Spinco Division, Beckmann Instruments Inc., Palo Alto, Calif.]

First, we shall examine simple diffusion of a protein sample. We shall see that diffusion is hindered by friction and that the coefficient of friction is related to the shape of the molecule. Then we shall "turn on the ultracentrifuge" and add gravitational force to the picture by considering sedimentation velocity centrifugation and sedimentation equilibrium centrifugation.

a. Diffusion

If we put a drop of India ink on the surface of water in a beaker, the color soon spreads throughout the solution as the ink is propelled by the random Brownian motion of the solvent molecules. The thermodynamic driving force for diffusion is the increase in entropy that occurs as the initially concentrated ink droplet becomes more and more dilute, and the system becomes more random. This process, and many others like it with which we are equally familiar, is *translational diffusion*. The rate at which a particle of mass m moves through a cross-sectional area A is proportional both to A and to the *concentration gradient* at x, or dc/dx. These quantities are related by *Fick's first law of diffusion* (equation 5.1),

$$\frac{dm}{dt} = -AD\frac{dc}{dx} \tag{5.1}$$

where t is time and D is the *diffusion coefficient*, whose units are area divided by time ($cm^2\ s^{-1}$). D is the proportionality constant that produces the

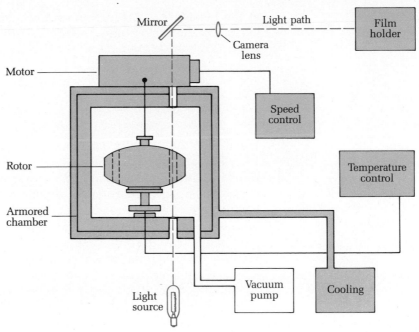

Figure 5.3 Block diagram of a Beckmann analytical ultracentrifuge.

equality. Measurement of simple diffusion is not easy, but a fairly simple "thought experiment," described below, gives the general idea.

Suppose we carefully layer water on a 12 M sucrose solution. At first a fairly sharp boundary exists, but it is slowly blurred as the sucrose solution diffuses into the water layer. We want to know how the concentration varies

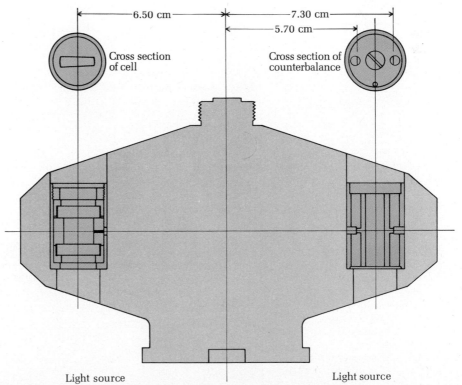

Figure 5.4 The ultracentrifuge rotor. Light from a single source passes through the cell and is reflected onto a photographic plate (as shown in Figure 5.3). Light also passes through the aperature in the counterbalance, producing a superimposed image on the photographic plate from which distances are calculated.

at the boundary as a function of time, that is, we want to know about the change in the concentration with time, dc/dt, as a function of the change in the concentration gradient, d^2c/dx^2. This relationship is given by *Fick's second law of diffusion* (equation 5.2),

$$\frac{dc}{dt} = D \frac{d^2c}{dx^2}$$

(5.2)

where D is the *diffusion coefficient*.

The concentration change can be measured by following the change in refractive index at the boundary. A variety of experimental systems have been developed to make such measurements. Table 5.2 lists the diffusion coefficients of a few molecules. The diffusion coefficient, $D_{20,w}$, is defined with reference to the standard conditions in which water at 20°C is the solvent, rather than the buffer solutions that are usually used in protein studies. We note that small molecules such as glycine have far higher diffusion coefficients than high molecular weight molecules. We also find that globular proteins such as ribonuclease have much larger diffusion coefficients than rodlike fibrous proteins or tobacco mosaic virus.

A diffusing molecule experiences a certain friction as it moves. It can be shown that the *coefficient of friction* (f) is related to the diffusion coefficient by equation 5.3,

$$D = \frac{kT}{f}$$

(5.3)

where k is the Boltzmann constant and T is the absolute temperature. The product kT is the thermal energy responsible for Brownian motion. Gravity also acts on a diffusing molecule, but for a protein whose molecular weight is 50,000, the gravitational force is only 1/200 the thermal energy.

The frictional coefficient brings us closer to molecular properties since it depends upon the size, shape, and solvation of the molecule. For a rigid sphere the frictional coefficient is related to the radius of the sphere by *Stokes' Law* (equation 5.4), where r is the *Stokes' radius* of the particle, and η is the viscosity of the solution.

$$f = 6\pi\eta r$$

(5.4)

We shall see how the Stokes' radius is used when we discuss gel filtration chromatography in Section 5.4D. The frictional coefficient is one of the parameters in sedimentation velocity centrifugation (one that we shall be

Table 5.2 Diffusion coefficients for selected molecules

Molecule	$D_{20,w} \times 10^7\ \mathrm{s}^{-1}$	Molecular weight
Glycine	93.30	75
Sucrose	45.90	342
Ribonuclease (pancreatic)	11.90	13,683
Lysozyme (egg white)	10.40	14,100
Serum albumin (bovine)	6.10	66,500
Hemoglobin	6.90	68,000
Tropomyosin	2.20	93,000
Fibrinogen	1.98	330,000
Myosin	1.10	493,000
Bushy stunt virus	1.15	10,700,000
Tobacco mosaic virus	0.44	40,000,000

glad to be able to eliminate from our equations), and we turn to this topic in the next section.

b. Sedimentation Velocity Centrifugation

A particle in a spinning centrifuge tube is subjected to a *centrifugal force* (F) that depends upon the angular velocity of the rotor (ω), the distance of the particle from the center of rotation (r), and the mass of the particle. The effective mass (m) of the particle is given by equation 5.5,

$$\text{effective mass} = m(1 - \bar{v}\rho) \tag{5.5}$$

where ρ is the density of the solvent and $\bar{v}$ is the *partial specific volume* of the particle. The partial specific volume is the volume change produced when a particle of mass m is added to a large excess of solution. The product $\bar{v}\rho$ represents the total mass of the solvent displaced by the particle. Equation 5.5 is, in fact, simply a quantitative statement of Archimedes' principle. The total centrifugal force experienced by the particle is given by equation 5.6.

$$F_c = (1 - \bar{v}\rho)\omega^2 r \tag{5.6}$$

The particle also encounters frictional resistance as it moves through the solution. The total frictional force depends upon the *sedimentation velocity, v,* where

$$v = \frac{dr}{dt} \tag{5.7}$$

The frictional force therefore is

$$F_r = f\frac{dr}{dt} \tag{5.8}$$

The frictional coefficient (f) depends upon the mass and shape of the particle. At constant sedimentation velocity, attained nearly instantly in the ultracentrifuge, the *net force* on the particle is *zero*, and $F_c = F_r$. Therefore,

$$\frac{dr}{dt} = \frac{m(1 - \bar{v}\rho)\omega^2 r}{f} \tag{5.9}$$

It is common to replace the sedimentation velocity (dr/dt) with the *sedimentation coefficient, s,* defined as

$$s = \left(\frac{dr}{dt}\right)\left(\frac{1}{\omega^2 r}\right) \tag{5.10}$$

The sedimentation coefficient has units of seconds. Biological macromolecules and complexes of macromolecules such as ribosomes have sedimentation coefficients in the range of 10^{-13} seconds. The quantity 10^{-13} s is the *Svedberg unit,* S. Substituting the definition of s (equation 5.10) in equation 5.9, we obtain

$$s = m\frac{1 - \bar{v}\rho}{f} \tag{5.11}$$

Since the frictional coefficient and the partial specific volume are not always easily determined, particles are often classified in terms of their sedimenta-

Table 5.3	Classification of ribosomal RNA molecules by their sedimentation coefficients	
Source	**Sedimentation coefficient (Svedberg units, S)**	**Molecular weight**
E. coli	16.0	5.50×10^5
	22.5	1.07×10^6
Paramecium	18.0	6.90×10^5
	25.5	1.25×10^6
Mammalian cytoplasm	18.0	7.00×10^5
	30.0	1.80×10^6
Yeast cytoplasm	17.0	7.20×10^5
	26.0	1.50×10^6
Yeast mitochondria	16.0	6.60×10^5
	23.0	1.30×10^6
Higher plant cytoplasm	17.0	7.00×10^5
	26.0	1.40×10^6
Higher plant chloroplasts	16.0	5.50×10^5
	22.0	1.00×10^6

tion coefficients. It is especially common to classify nucleic acids (Chapter 9) in terms of their sedimentation coefficients, as shown for ribosomal RNA's in Table 5.3. Ribosomes and large protein aggregates are also often classified in terms of their sedimentation coefficients.

The frictional coefficient in equation 5.11 is the same parameter that we found in equation 5.3. Equating equations 5.3 and 5.11, and recalling that a molecule's molecular weight (M) equals Nm, and that the gas constant (R) equals Nk, we obtain the *Svedberg equation*

$$M = \frac{RTs}{(1 - \bar{v}\rho)D} \tag{5.12}$$

By eliminating f from the equations for sedimentation and diffusion we have arrived at equation 5.12, which gives a true value of the molecular weight of the particle independent of its size, shape, or degree of solvation.

Let us consider the parameters of the Svedberg equation. The gas constant R has cgs units: $R = 8.31 \times 10^7$ erg K^{-1} mol^{-1}. The values of the sedimentation coefficient span a wide range, from about 1S to 8S for proteins and from 5S to 26S for ribosomal RNA's (Table 5.3). Viruses have very high sedimentation coefficients, as we would expect given their large masses (Table 5.4).

Table 5.4	Values of sedimentation coefficients and partial specific volumes for selected proteins		
Protein	**Molecular weight**	**Sedimentation coefficient, $s_{20,w}$ (S)**	**Partial specific volume, $\bar{v}$ (cm³ g⁻¹)**
Ribonuclease A (bovine)	12,400	1.85	0.728
Lysozyme (chicken)	14,100	1.91	0.688
Serum albumin (bovine)	66,500	4.31	0.734
Hemoglobin	68,000	4.31	0.749
Tropomyosin	93,000	2.60	0.710
Fibrinogen (human)	330,000	7.60	0.706
Myosin (rod)	570,000	6.43	0.728
Bushy stunt virus	10,700,000	132.00	0.740
Tobacco mosaic virus	40,000,000	192.00	0.730

Sedimentation coefficients are usually corrected to the values they would have in water at 20°C ($s_{20,w}$). The density of water to be used in the Svedberg equation under these conditions is 0.998 g cm^{-3}. The partial specific volumes of proteins nearly all fall within the range of 0.69 to 0.75 cm^3 g^{-1}, and a typical protein has a partial specific volume of about 0.73 cm^3 g^{-1} (Table 5.4). Nucleic acids are more dense, and their partial specific volumes are about 0.50 cm^3 g^{-1} (assuming that Na$^{\oplus}$ is the counterion).

c. Sedimentation Equilibrium

If we apply a sufficiently high gravitational force to a protein solution, the proteins will steadily increase their concentration at the bottom of the centrifuge tube. But if a slow speed is used (5,000 to 25,000 rpm), so that diffusion and the gravitational force are about equal, an equilibrium will be established in which a smooth concentration gradient exists in the centrifuge cell (Figure 5.5). Between 16 and 24 hours are required to achieve equilibrium in most cases. It can be shown that at equilibrium the molecular weight is given by equation 5.13,

$$M = \frac{RT}{(1 - \bar{v}\rho)\omega^2} \frac{l}{rc} \frac{dc}{dr} \tag{5.13}$$

where v, ρ, ω, all have the meaning we have given them before, and dc/dr is the change in concentration as a function of distance from the center of rotation (r). Rearranging equation 5.13 and integrating, we obtain

$$M = \frac{2RT}{(1 - \bar{v}\rho)\omega^2 r} \frac{\ln c(r)}{c(a)} \frac{1}{r^2 - a^2} \tag{5.14}$$

where $c(r)$ and $c(a)$ are the concentrations of protein at the radius (r) and the meniscus (a). Plotting ln $c(r)$ versus r^2, we obtain a straight line whose slope is M $(1 - \bar{v}\rho)\omega^2/2RT$ (Figure 5.6). The molecular weight that is obtained by equilibrium centrifugation is also independent of frictional forces and can be applied to solutes having molecular weights ranging from 320 (sucrose) to 40 $\times$ 10^6 (tobacco mosaic virus).

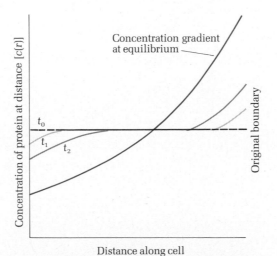

Figure 5.5 Sedimentation of a solute to an equilibrium distribution. [Adapted from C. Tanford, "Physical Chemistry of Macromolecules," Wiley, New York, 1961, p. 385.]

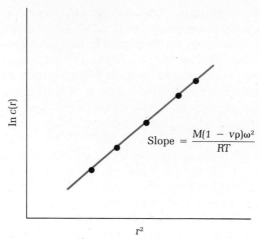

Figure 5.6 Plot of ln $c(r)$ versus r^2. The plot gives a straight line whose slope can be used to calculate molecular weights from sedimentation equilibrium data.

5–3 PURIFICATION BY SELECTIVE PRECIPITATION

Proteins are highly charged molecules that are soluble because their charged and polar side chains are tightly bound to water molecules. Agents that change the degree of protein solvation alter protein conformation, the state of aggregation of proteins in solution, and protein solubility. These solubility differences can be exploited in protein purification work. They also provide information about protein structure.

A. Titration and Precipitation at the Isoionic pH

The acidic and basic side chains of amino acid residues can be titrated. The titration curve of a typical protein, β-lactoglobulin, is shown in Figure 5.7. This curve can be divided into three broad regions in which many residues of similar pK_a are ionizing. The pH range from 1.5 to 6.0 corresponds to ionization of C-terminal and side chain carboxyl groups; the pH range from 6.0 to 8.5 corresponds to ionization of side chain imidazoles of histidine residues; and the pH range from 8.5 to 14.0 corresponds to titration of the N-

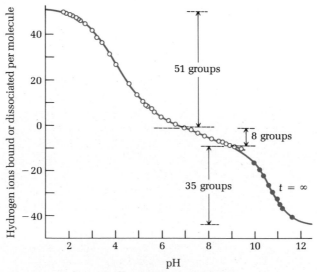

Figure 5.7 Titration curve of β-lactoglobulin. [From R. H. Haschenmeyer and A. E. V. Haschenmeyer, "Proteins: A Guide to Study by Physical and Chemical Methods," Wiley, New York, 1973, p. 264.]

terminal amino group and basic side chains of lysine and arginine residues and the hydroxyl groups of tyrosine residues.

Titration data reveal certain cases in which groups have anomalous pK_a values. For example, β-lactoglobulin contains 51 asparate and glutamate carboxyl groups, 49 of which titrate with a pK_a of 4.8. (This is an average pK_a. The curve is too broad to permit identification of individual residues.) Two of the 51 have anomalously high pK_a values of 7.3, indicating that these residues are buried in the hydrophobic interior of the protein and are inaccessible to the solvent at low pH. They move to the surface of the protein during a conformational change that occurs in β-lactalbumin at about pH 7.5.

The point at which the protein contains an equal number of positive and negative charges, and therefore has a net charge of zero, is called the *isoionic pH* of the protein.[1] When the pH is greater than the isoionic pH, the protein bears a net negative charge; when the pH is lower than the isoionic pH, the protein bears a net positive charge. In either case the protein is more soluble than at the isoionic pH where electrostatic repulsion between protein molecules in solution is minimal. Thus, protein molecules tend to aggregate at the isoionic pH. Adjusting the pH to a given protein's isoionic point therefore often causes its precipitation.

B. Salting-Out: Precipitation in Concentrated Salt Solutions

Protein precipitation upon the addition of a concentrated salt solution, a phenomenon called *salting-out*, results from a complex physical process. The added salt alters the structure of the solvent, which can lead to large changes in protein conformation by altering the electrostatic interaction of charged groups on the protein surface, the solvation of polar, uncharged residues exposed to solvent, and the van der Waals interactions among hydrophobic side chains. The salt also competes with the protein for solvent molecules and thereby lowers its solvation. A sufficient amount of salt essentially dehydrates the protein molecules, which then precipitate when, no longer solvated, they clump together.

One of the most common methods of protein precipitation involves the addition of ammonium sulfate. Slow addition of the salt leads to a succession of precipitates, each of which must be assayed to determine the fraction that contains the desired protein. About a fourfold purification is typically obtained by salt precipitation. An added benefit of precipitation is a sharp decrease in the volume of the sample to be handled. A dilute protein solution can be quickly concentrated by salt precipitation.

C. Precipitation by Organic Solvents

Organic solvents that are miscible with water sharply decrease protein solubility. At room temperature solvents such as acetone, ethanol, and methanol cause not only precipitation, but concomitant denaturation. This problem can be largely avoided by maintaining the temperature of the protein solution at 0°C, cooling the organic solvent to temperatures of $-40°$ to $-60°$, and adding the organic solvent drop by drop with continuous stirring to avoid high local concentrations of organic solvent. Salts increase the solubility of proteins in ethanol. This added variable permits a wide range of experimental conditions for precipitation. Protein solubility in organic solvents also varies with pH, proteins being least soluble at their isoionic point.

The miscible organic solvent brings about precipitation in much the same way as concentrated salt solutions—by competing with the protein for

---

[1] We ought not confuse the terms isoionic and isoelectric pH. At its isoelectric pH a protein undergoes no migration in an applied electric field (see Section 5.5). It might be supposed that the isoionic and isoelectric pH's are identical, but since proteins bind many small ions, both cations and anions, a protein can have no net charge due to ionized side chains, but still possess a net charge due to bound ions. The values of the isoionic and isoelectric pH are, however, usually about the same, and the two terms are often used synonymously, even though it is not perfectly correct to do so.

water of solvation and by depriving the protein of its solvation sphere. The organic solvent also lowers the dielectric constant of the aqueous solution. At low dielectric constant the charged residues on the protein surface are not as well shielded from one another by the solvent. The poorly solvated, charged surfaces of protein molecules interact strongly with one another leading to formation of large aggregates that precipitate from solution.

Water-soluble, nonionic polymers such as polyethylene glycol also cause protein precipitation. The major effect of polyethylene glycol seems to be removal of the solvent sphere of the protein. Protein solubility in polyethylene glycol is nearly independent of the salt concentration of the solution, the pH, and even the absolute solubility of the protein. These observations suggest that the polymer interacts with the hydrophilic groups of the protein and sterically excludes access of the protein to the aqueous phase. This hypothesis is supported by the observation that the solubility of a protein in polyethylene glycol is strongly dependent upon its molecular size.

5–4 CHROMATOGRAPHY

The Russian botanist Michael Tswett invented chromatography in the nineteenth century. He filled columns with such materials as crushed limestone and table sugar and passed solutions containing the pigments of flowers through the columns. The result was a separation of the solutions into bands of colors, hence the term chromatography, which means "graph of colors." Today, the term chromatography embraces a wide range of experimental techniques that still share the goal of Tswett's columns: the separation of complex mixtures. Differences in solubility, polarity, charge, size, and biological specificity are exploited to carry out such separations.

A. Partition Chromatography

Partition chromatography depends upon differences in the *phase distributions* of the components of a mixture. Each component of the mixture establishes an equilibrium between two insoluble phases. Separation occurs by selective removal of a component from one solvent when it is in contact with the second solvent. The procedure, in fact, is simply an extraction. The distribution of a solute, A, between two solvents, S and S′, is expressed as the *partition coefficient* K_A (equation 5.15).

$$K_A = \frac{\text{concentration of A in S}}{\text{concentration of A in S}'} \tag{5.15}$$

Suppose that components A and B are partitioned between solvents S and S′. If the partition coefficient for B (K_B) is not equal to K_A, then A and B can be separated, since A will tend to be extracted into one solvent and B into the other. In partition chromatography a mobile phase flows past a stationary phase that is bound to an insoluble support such as silica gel or paper. The solvent bound to the solid support is water. As the mobile phase, which is not miscible with the stationary phase, passes over the stationary phase, a series of equilibria are established. After many such equilibria have been attained, a complete separation is obtained.

Different techniques of partition chromatography use different materials and arrangement of the stationary phase. *Paper chromatography* and *thin-layer chromatography* are the two most widely used types of partition chromatography.

a. Paper Chromatography

In paper chromatography the solid support is cellulose, the stationary phase is the water of hydration of cellulose, and the mobile phase is a solvent that has a different polarity than water. The sample is applied along the baseline of the paper, and air dried. The paper is then placed in a closed system, and the moving solvent carries the sample with it. Separation occurs by partition between the mobile and stationary phases.

Two types of experimental conditions are used for paper chromatography. In *ascending paper chromatography* the solvent climbs the paper by capillary action; in *descending chromatography* the solvent moves from an elevated tray down the paper by gravity and capillary action (Figure 5.8). Depending upon the distance the solvent moves, between 1 and 24 hours are typically required to develop the chromatogram.

After separation the components of the mixture can be detected by a variety of techniques. For example, the paper can be sprayed with a specific reagent, such as ninhydrin, which reacts with the components of the mixture (amino acids or peptides or both) to give a set of colored spots. If the components of the mixture fluoresce or if they absorb ultraviolet light, they can be visualized simply be exposing the chromatogram to an ultraviolet light in a darkened room or viewing box.

If inadequate separation is obtained by one solvent, the paper can be rotated 90° and the procedure repeated with a second solvent of different po-

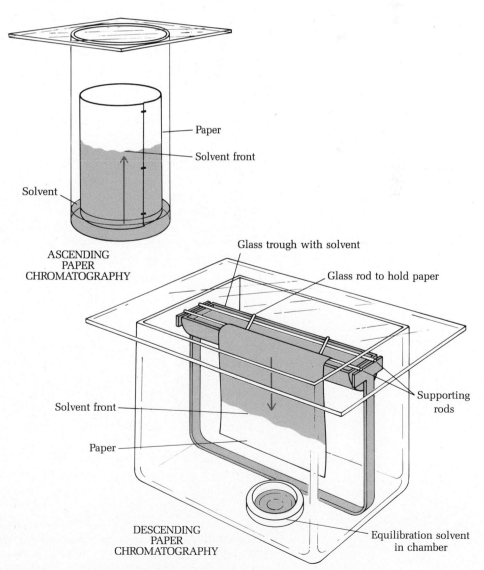

Figure 5.8   Ascending and descending paper chromatography.

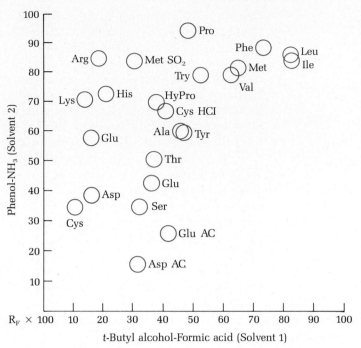

Figure 5.9   Two-dimensional paper chromatography of amino acids obtained from the hydrolysis of lysozyme. The spots were detected by ninhydrin. [From J. L. Bailey, "Techniques of Protein Chemistry," 2nd ed., Elsevier, New York, 1967.]

larity. This technique, known as *two-dimensional paper chromatography*, is an efficient way of separating amino acids in protein hydrolysates (Figure 5.9). It is also used to "fingerprint" a peptide mixture that is produced by limited proteolytic digestion with an enzyme such as trypsin.

   The classic example of fingerprinting is identification of the lesion in sickle cell hemoglobin. In one of the peptides produced by digestion of hemoglobin S with trypsin a valyl residue replaces the glutamate residue of normal hemoglobin A (Figure 5.10). Identification of this unique peptide reveals the genetic defect. With the additional knowledge of the genetic code (Section 26.10), which was not available when the work summarized in Figure 5.10 was done, the genetic lesion can be even more narrowly defined.

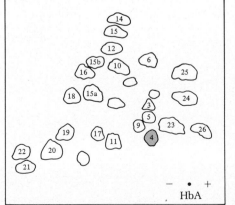
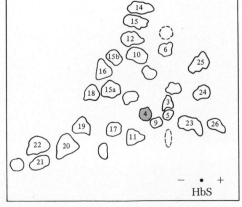

Figure 5.10   Identification of sickle cell hemoglobin by fingerprinting. Schematic diagram of fingerprint of normal (HbA) and sickle cell hemoglobin (Hb S) obtained by hydrolysis with trypsin. The fingerprint reveals that all spots except one (colored) are identical. Valine has preplaced glutamic acid in this peptide. [From C. Baglioni, *Biochim. Biophys. Acta*, 48, 392 (1961).]

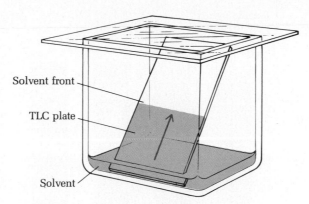

Figure 5.11 Ascending thin-layer chromatography. In two-dimensional thin-layer chromatography the plate is air dried and rotated by 90° and chromatography is repeated in a second solvent.

Modern methods of DNA analysis (Chapter 9) have to some extent eclipsed fingerprinting, but it remains a valuable technique.

b. Thin-Layer Chromatography

In thin-layer chromatography, the second major type of partition chromatography, a slurry of a solid support, such as silica gel, aluminum oxide, or diatomaceous earth, is spread evenly over a glass plate. The chromatogram is obtained in much the same way as the chromatogram in ascending paper chromatography (Figure 5.11). A two-dimensional thin-layer chromatogram is shown in Figure 5.12.

Thin-layer chromatography is faster and gives better resolution than paper chromatography. The use of smaller samples in thin-layer chromatography is an aid to high resolution and a hindrance to large-scale preparations. It is especially valuable for the separation of lipids (Chapter 11), nucleotides (Chapter 9), and amino acids (Chapter 2).

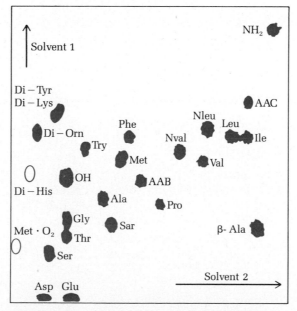

Figure 5.12 Two-dimensional thin-layer chromatogram of DNP-amino acids. [From R. H. Haschenmeyer and A. E. V. Haschenmeyer, "Proteins: A Guide to Study by Physical and Chemical Methods," Wiley, New York, 1973, p. 43.]

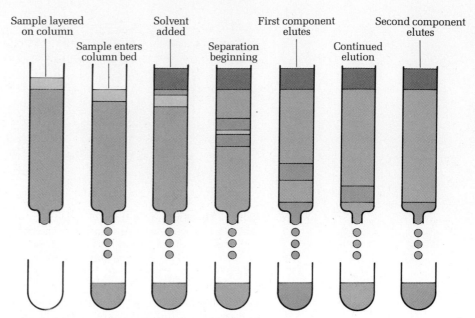

Figure 5.13 Operation of a chromatography column. The same general principles apply all types of column chromatography.

B. Adsorption Chromatography

Finely divided solids, such as activated charcoal, adsorb many types of molecules. The separation of mixtures by selective adsorption is a far less exact science than partition chromatography, but this has not prevented its wide use for certain types of separations. The interactions between the solutes dissolved in the moving phase and the solid involve differences in van der Waals forces, hydrogen bonding, and electrostatic forces between the solid and the various components of a mixture.

The most widely used column materials for adsorption chromatography are hydroxyapatite [$Ca_{10}(PO_4)_6(OH)_2$], charcoal, silica gel, alumina

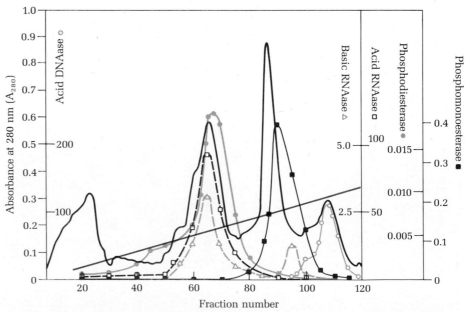

Figure 5.14 Separation of a mixture of spleen enzymes by adsorption chromatography on hydroxyapatite. The proteins were eluted by a gradient of a phosphate buffer at pH 6.8 whose concentration increased from 0.05 M to 0.5 M (dotted line in color). The fractions were monitored by UV absorbance at 280 nm. [From G. Bernardi, A. Bernardi, and A. Chersi, *Biochim. Biophys. Acta, 129,* 1 (1966).]

(aluminum oxide), and calcium phosphate gel. The operation of a typical adsorption column is summarized in Figure 5.13. The sample is loaded in a thin band on the top of the column and *eluted* by passing a solvent through the column. Differential adsorption of the components of the sample separates them into bands. The effluent solution is continuously collected in small fractions, each of which is assayed for activity.

The results obtained by chromatography of a mixture of spleen enzymes on hydroxyapatite are summarized in Figure 5.14. Hydroxyapatite is also used to separate mixtures of nucleic acids and to purify viruses.

C. Ion Exchange Chromatography

Mixtures are separated through ion exchange chromatography by differences in the net charge of each species at a given pH. The resins commonly employed for ion exchange chromatography are substituted

Table 5.5 Major types of ion exchange resins

Type and exchange group	Bio-rad resin	Dowex resin
Anion exchange resins		
Strong base, polystyrene type		
$C_6H_5CH_2N^{\oplus}(CH_3)_3Cl^{\ominus}$	AG 1-X1	1-X1
	AG 1-X2	1-X2
	AG 1-X4	1-X4
	AG 1-X8	1-X8
	AG 1-X10	1-X10
	AG 21K	21K
$C_6H_5CH_2N^{\oplus}(CH_3)_2(C_2H_4OH)Cl^{\ominus}$	AG 2-X4	2-4
$NH^{\oplus}Cl^{\ominus}$	AG 2-X8	2-X8
	AG 2-X10	
	Bio-Rex 9	
Intermediate base, epoxypolyamine type		
$RN^{\oplus}(CH_3)_3Cl^{\ominus}$ and $RN^{\oplus}(CH_3)_2(C_2H_4OH)Cl^{\ominus}$	Bio-Rex 5	
Weak base, polystyrene, or phenolic polyamine type		
$RN^{\oplus}HR_2Cl^{\ominus}$	AG 3-X4	3-X4
Mixed bed resins		
$C_6H_5SO_3^{\ominus}H^{\oplus}$ and $C_6H_5CH_2N^{\oplus}(CH_3)_3OH^{\ominus}$	AG 501-X8	
$C_6H_5SO_3^{\ominus}H^{\oplus}$ and $C_6H_5CH_2N^{\oplus}(CH_3)_3OH^{\ominus}$ plus indicator dye	AG 501-X8 (D)	
$C_6H_5SO_3^{\ominus}H^{\oplus}$ and $C_6H_5CH_2N^{\oplus}(CH_3)_2(C_2H_4OH)OH^{\ominus}$		
Strong acid, phenolic type		
$RCH_2SO_3^{\ominus}H^{\oplus}$	Bio-Rex 40	
Strong acid, polystyrene type		
$C_6H_5SO_3^{\ominus}H^{\oplus}$	AG 50W-X1	50-X1
	AG 50W-X2	50-X2
	AG 50W-X4	50-X4
	AG 50W-X5	50-X5
	AG 50W-X8	50-X8
	AG 50W-X10	50-X10
	AG 50W-X12	50-X12
	AG 50W-X16	60-16
Intermediate acid, polystyrene type		
$C_6H_5PO_3^{2-}(Na^{\oplus})_2$	Bio-Rex 63	
Weak acid, acrylic type		
$RCOO^{\ominus}Na^{\oplus}$	Bio-Rex 70	
Weak acid, chelating, polystyrene type		
$C_6H_5CH_2N(CH_2COO^{\ominus}H^{\oplus})_2$	Chelex 100	A-1

Adapted from T. G. Cooper, "The Tools of Biochemistry," Wiley, New York, 1977, p. 140.

polystyrene beads that are cross-linked by divinyl benzene (Table 5.5). Cellulosic ion exchangers are also widely used (Table 5.6).

Various functional groups (strong or weak acids or bases) are covalently bound to the resin. Resins that contain negatively charged functional groups are *cation exchangers,* and resins that contain positively charged functional groups are *anion exchangers.* Ion exchange resins are supplied in several different ionic forms. The counterions that are bound to the resin can be modified simply by washing the resin with the desired buffer. For ex-

Table 5.6 Cellulosic ion exchangers

Ion exchanger	Ionizable group	Structure
Anion exchangers		
Intermediate base		
AE	Aminoethyl	$-OCH_2CH_2NH_2$
Strong base		
DEAE	Diethylaminoethyl	$-OCH_2CH_2N(C_2H_5)_2$
TEAE	Triethylaminoethyl	$-OCH_2CH_2N(C_2H_5)_3$
GE	Guanidoethyl	$-OCH_2CH_2NHCNH_2$ (with $=NH$)
Weak base		
PAB	p-Aminobenzyl	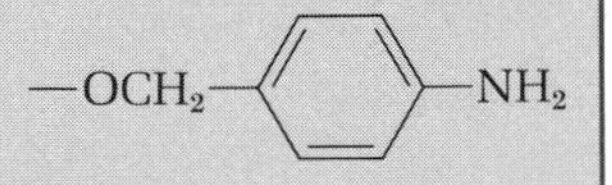 $-OCH_2-\bigcirc-NH_2$
Intermediate base		
ECTEOLA	Triethanolamine coupled to cellulose through glyceryl and polyglyceryl chains mixed groups (mixed amines)	
DBD	Benzylated DEAE cellulose	
BND	Benzylated naphthoylated DEAE cellulose	
PEI	Polyethyleneimine adsorbed to cellulose or weakly phosphorylated cellulose	
Cation exchangers		
Weak acid		
CM	Carboxymethyl	$-OCH_2COOH$
Intermediate acid		
P	Phosphate	$-O\overset{O}{\underset{OH}{\overset{\|}{P}}}OH$
Strong acid		
SE	Sulfoethyl	$-OCH_2CH_2\overset{O}{\underset{O}{\overset{\|}{S}}}OH$
SP-Sephadex	Sulfopropyl	$-C_3H_6\overset{O}{\underset{O}{\overset{\|}{S}}}OH$
Strong base		
QAE-Sephadex	Diethyl(2-hydroxypropyl) quaternary amino	$-C_2H_4\overset{\oplus}{N}(C_2H_5)_2$, CH_2CHCH_3 , OH

Adapted from T. G. Cooper, "The Tools of Biochemistry," Wiley, New York, 1977, p. 143.

ample, if an acidic resin is supplied as the sodium salt, washing with HCl will give the hydrogen ion form.

The tenacity with which the anions in the sample bind to the resin depends upon their net charge. For example, the most negatively charged ion in a mixture will bind most tightly to an anion exchange resin and least tightly to a cation exchange resin. If an anion exchange resin is being used, the anions bound to the resin are eluted by steadily increasing the salt concentration of the eluent. The gradient that is applied to the column is the most important part of the experiment. Many commercial gradient mixers are available to control the gradient precisely. The separation of proteins known as histones (see Chapter 24) by ion exchange chromatography is shown in Figure 5.15.

D. Gel Filtration Chromatography

Macromolecules, such as proteins, are separated by gel filtration chromatography by differences in their molecular sizes and shapes. This type of chromatography is also known as *molecular exclusion chromatography* and *molecular sieve chromatography*.

A gel filtration column is made of porous beads of fairly uniform particle size; the pores also have a uniform size within a given fractionation range. Table 5.7 gives the fractionation range of a widely used gel filtration medium sold as Sephadex®. This material is a cross-linked dextran produced by certain strains of bacteria. Other gel filtration media are also available, but all operate on the same principle.

As the mixture passes through the column, molecules whose dimensions are larger than the largest pore size of the beads that form the gel bed are excluded from the gel and are eluted first. Smaller molecules enter the pores of the gel beads. The extent to which a molecule enters the pores of the gel depends upon its size and shape. Molecules therefore exit from the column in order of decreasing size, and the *elution volume* is logarithmically proportional to molecular size (Figure 5.16).

Gel filtration chromatography can also be used to determine apparent molecular weights of globular proteins. Since we have just stated that separation is based on size, we shall require a slight detour to connect molecular size with molecular weight. This detour leads us back to the equations for

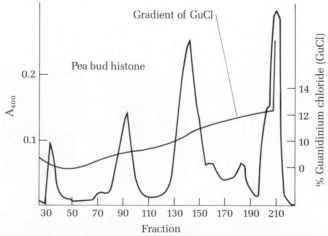

Figure 5.15 Fractionation of histones by cation exchange chromatography. Proteins of pea bud histone were eluted with a gradient of guanidinium chloride and detected by their absorbance at 400 nm (A_{400}). [From D. M. Fambrough, F. Fugimura, and J. Bonner, *Biochemistry*, 7, 575 (1968). Reprinted with permission. Copyright 1968 American Chemical Society.]

Table 5.7 Fractionation range of Sephadex® gel filtration medium (cross-linked dextran polymer beads)	
Sephadex® type	**Fractionation range**
G-10	0–700
G-15	0–1,500
G-25	100–5,000
G-50	1,500–30,000
G-75	3,000–80,000
G-100	4,000–150,000
G-150	5,000–300,000
G-200	5,000–600,000

diffusion that we discussed in Section 5.2A. The flow of protein molecules through a gel filtration column closely resembles diffusion. We recall that the diffusion coefficient is directly proportional to the thermal energy and inversely proportional to the frictional coefficient (equation 5.3). We also recall that the frictional coefficient is proportional to the Stokes' radius of the diffusing particle, assuming a rigid, nonhydrated, spherical particle (equation 5.4). Substituting equation 5.4 for f in equation 5.3, we obtain

$$D = \frac{RT}{6\pi\eta r} \tag{5.16}$$

The Stokes' radius is the radius of a perfect sphere that would elute from a gel filtration column in the same volume of eluent as the protein of unknown molecular weight. If, therefore, a gel filtration column is calibrated with a set of proteins of known Stokes' radii and known molecular weights, then the elution volume of the unknown protein can be converted to an apparent molecular weight if the shape of the unknown is the same as the shape of the standards. The total volume of the gel column, V_t, is the sum of the volume retained by the gel, or internal volume, V_i, and the volume excluded by the gel, or void volume, V_0 (equation 5.17; Figure 5.17).

$$V_t = V_0 + V_i \tag{5.17}$$

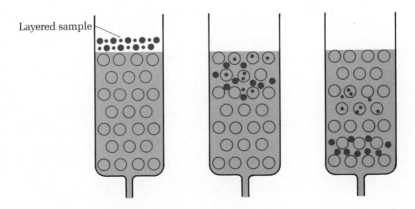

Layered sample

◯ = Gel particle

• = Sample molecule smaller than pores of gel

● = Sample molecule larger than pores of gel

Figure 5.16   Gel filtration chromatography. Molecules larger than the largest pore size of the gel cannot diffuse into the gel pores, and they elute first. Smaller molecules emerge in order of decreasing molecular size.

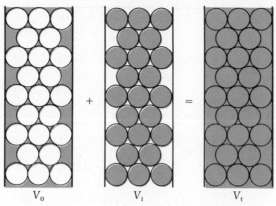

Figure 5.17 Variables used to characterize gel filtration chromatography columns. The void volume, V_0, is the volume that surrounds the gel bed; the internal volume, V_i, is the volume occupied by the gel; and the total volume, V_t, is simply $V_i + V_0$, neglecting the volume occupied by the matrix itself.

The *elution volume* of the column, V_e, is the volume of eluent required to elute a given solute. The distribution coefficient, K_D, of a solute between the pores of the gel, in the internal volume (V_i) and the exclusion volume (V_0) is given by equation 5.18.

$$K_D = \frac{V_e - V_0}{V_i}$$

$$(5.18)$$

Molecules that are totally excluded from the gel have a K_D value of zero, and molecules that completely enter the pores of the gel have K_D values of nearly 1.0. A K_D value greater than 1.0 indicates that the protein binds to the gel. The distribution coefficient depends upon the size of the gel pores and upon the Stokes' radii of the protein molecules in question. A plot of the logarithm of the molecular weight of the set of standard proteins versus their elution volumes is nearly linear over a restricted range (Figure 5.18).

Gel filtration therefore leads from the elution volume of the protein to the apparent molecular weight of the protein, based upon the assumption, by no means perfect, of spherical macromolecules. To obtain apparent molecular weights from gel filtration experiments, only the linear portion of the curve should be used, and the protein of unknown apparent molecular weight should be bracketed by proteins of known molecular weight.

E. Affinity Chromatography

In affinity chromatography proteins are separated by differences in their biological specificity rather than by differences in their physical and chemical properties. Virtually any protein can be purified by affinity chromatography, and a fortnight would be required to recite the list of those that have been. Among the types of proteins that have been separated by this method are enzymes, antibodies, transport proteins (Chapter 11), as well as entire cell organelles and distinct cell types such as virus-induced tumor cells.

A small molecule, to which only one of the components of a mixture of proteins bind with high affinity, is covalently bound to the solid support of the column, and the mixture is passed through the column. Three requirements must be met by the affinity column. First, the binding of the ligand to the protein must be quite tight so that the protein will be held on the column long enough to permit separation. Second, the binding must not be so tight that irreversible denaturation or destruction of the protein is re-

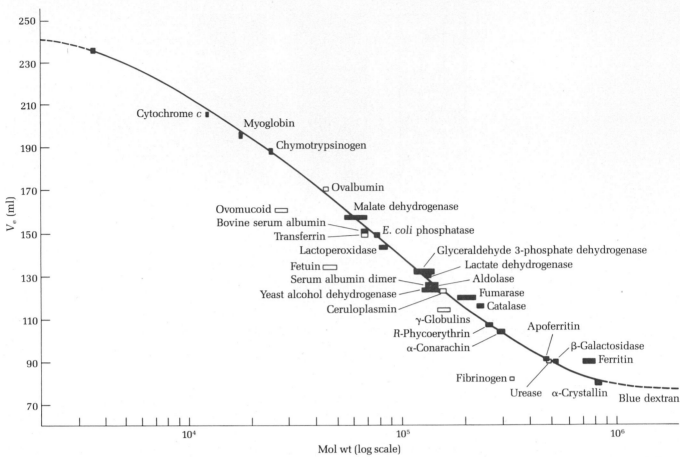

Figure 5.18 Plot of the elution volume of a protein (V_e) versus the logarithm of the apparent molecular weight of the protein. The plot is approximately linear over a wide range of molecular weights. The data of the figure were obtained on the cross-linked dextran Sephadex G-200 at pH 7.5. [From P. Andrews, *Biochem. J.*, *96*, 595 (1965).]

quired to remove it from the column. Third, the matrix support itself must not bind proteins. The many commercial gels available for affinity chromatography contain various functional groups attached to "spacer arms" to hold the ligand away from the gel so that steric hindrance will be unlikely to prevent binding of a specific protein to the column (Figure 5.19).

From hundreds of possible examples, let us consider the purification of chymotrypsin by affinity chromatography. A long spacer arm attached to the gel terminates in an N-hydroxysuccinimide ester. This functional group is treated with D-tryptophan methyl ester (Figure 5.20). The D-tryptophan moiety that is bound to the column support specifically binds to the active site of chymotrypsin. Since, however, the configuration of the D-amino acid is not complementary to the catalytic site of the enzyme, chymotrypsin cannot hydrolyze the amide bond of the substrate analog to which it is bound. The enzyme does, however, bind tightly to the D-amino acid residue, and it is held on the column. Most other proteins bind weakly or not at all to the D-tryptophan residue. The chymotrypsin is eluted by a solution that contains a ligand to which chymotrypsin binds more tightly than it binds to D-tryptophan.

5–5 ELECTROPHORESIS Electrophoresis is an experimental technique in which molecules are separated by differences in their net charge in the presence of an externally applied electric field. Among the molecules that can be separated by electrophoresis are amino acids, proteins, nucleotides, and nucleic acids. The

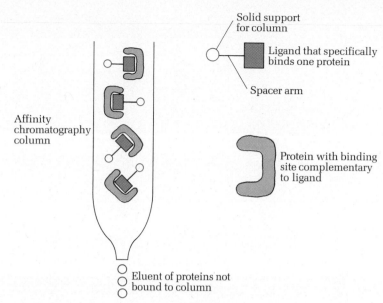

Figure 5.19 Affinity chromatography. A protein that is to be purified by affinity chromatography binds specifically to a ligand covalently attached to the solid support for the column. Proteins that have no affinity for the ligand are not retained on the column. The protein bound to the column is eluted with a solution containing an excess of ligand.

theoretical aspects of electrophoresis are as complex as the basic idea is simple, and we shall keep our theoretical description to a minimum. If an external electric field is applied to a solution of a given molecule, the molecule will move under the influence of the electric field if it bears a net electric charge. The distance that the molecule migrates in a given time interval depends upon its mass, its charge, and its interaction with the solid support to which it is applied.

In *zone electrophoresis* the sample is applied in a spot or band on an inert solid support such as paper, cellulose acetate, or various gels such as

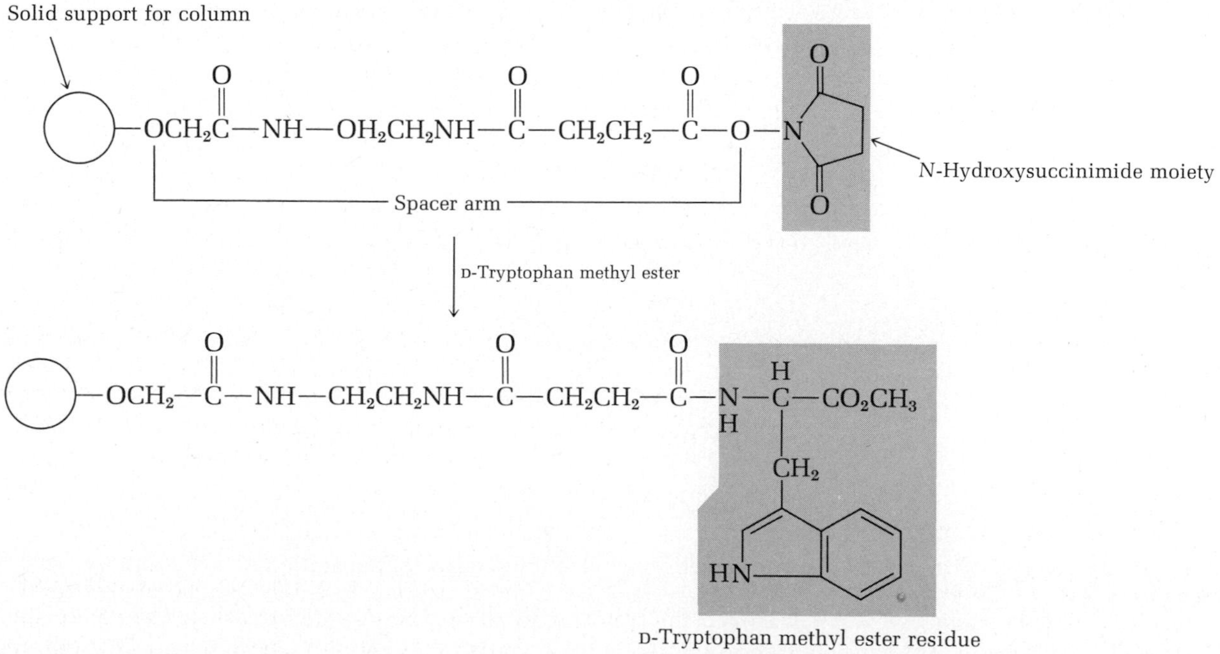

Figure 5.20 Attachment of a ligand to an affinity column solid support that specifically binds chymotrypsin.

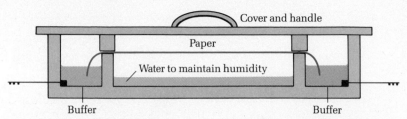

Figure 5.21 Schematic diagram of an apparatus for paper electrophoresis.

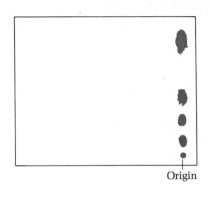

(a)

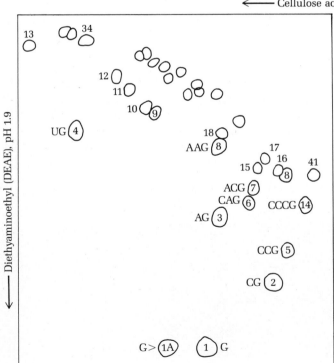
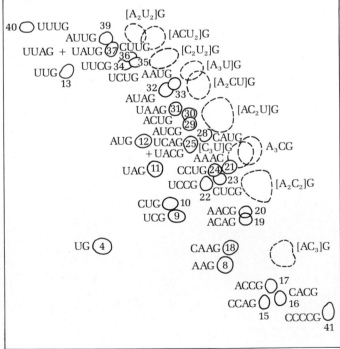

(b)

Figure 5.22 Results of a single- and two-dimensional electrophoresis. (a) Separation of a mixture by electrophoresis in one dimension. (b) Two-dimensional electrophoresis of a mixture of oligonucleotides obtained by hydrolysis of a ribonucleic acid. The row of spots is transferred from the cellulose acetate support to diethylaminoethyl (DEAE) cellulose, the support is rotated 90° and electrophoresis in a different buffer is repeated. [From F. Sanger, G. G. Brownlee, and B. G. Barrell, *J. Mol. Biol.*, *13*, 381 (1965). Reproduced with permission. Copyright: Academic Press Inc. (London) Ltd.]

polyacrylamide or agarose. An experimental apparatus for electrophoresis on a paper or cellulose acetate strip is shown in Figure 5.21. Electrophoresis in one dimension leads to separation of the sample into a row of spots (Figure 5.22a).

Mixtures of amino acids, and especially of oligonucleotides, are often incompletely separated by a single electrophoresis experiment. The row of spots may be transferred to a second support medium, turned at right angles, and subjected to electrophoresis again. The transfer may be accomplished by laying the first electrophoresis sheet on top of the support medium to be used for the second electrophoresis experiment; transfer occurs by blotting.

If cellulose acetate were used for the first separation, a support such as diethylaminoethyl (DEAE) cellulose might be chosen for the second separation. This method, called *two-dimensional electrophoresis*, was invented by Sanger and his coworkers to separate oligonucleotide mixtures (Figure 5.22b). Oligonucleotides differing only by their terminal nucleotide residues can be separated by this method. Individual oligonucleotides are identified by their electrophoretic mobilities, that is, by the distance they move in the electrophoresis experiment.

High resolution is also obtained by electrophoresis on polyacrylamide or agarose gels. The former are favored for proteins and small oligonucleotides and the latter for nucleic acids. The gel may be either in the form of a column or a slab. Since slabs have greater capacity than columns, and since many single-dimensional separations can be performed simultaneously on a slab gel, slabs are often preferred. Schematic diagrams of *gel electrophoresis* systems for columns and slabs are shown in Figure 5.23.

If proteins are subjected to electrophoresis in the presence of sodium dodecyl sulfate (SDS), they migrate through the gel as if they have identical shapes and charge-to-mass ratios.

$$CH_3(CH_2)_{10}CH_2OSO_3^{\ominus}Na^{\oplus}$$

Sodium dodecyl sulfate (SDS)

In *SDS-polyacrylamide gel electrophoresis* the separation is caused by the sieving action of the gel. SDS electrophoresis therefore separates proteins by size and can be used to estimate the molecular weights of individual polypeptide chains (Figure 5.24). Nucleic acids and oligonucleotides have nearly identical charge-to-mass ratios; thus they are separated by molecular weight when they are subjected to gel electrophoresis, even in the absence of

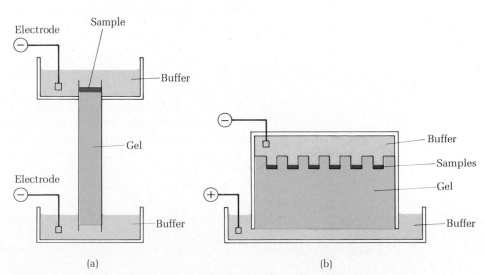

Figure 5.23 Schematic diagram for gel electrophoresis on columns and slabs.

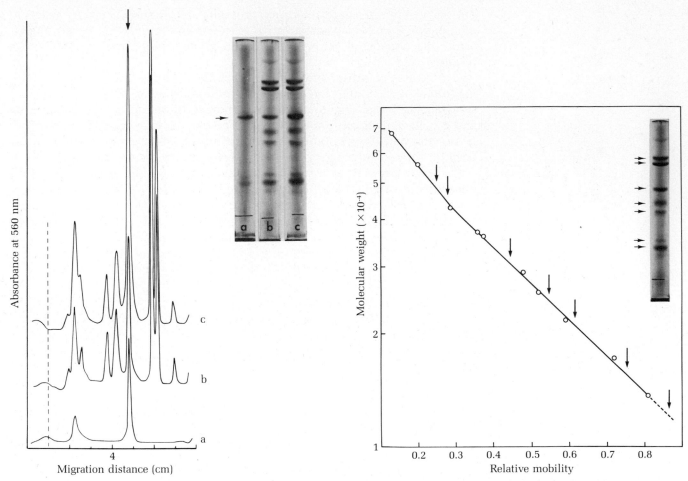

Figure 5.24 Separation and molecular weight determination of proteins in respiratory chain complex III (Section 16.5C) by SDS gel electrophoresis. The standard curve is obtained by electrophoresis of proteins of known molecular weights (open circles). The distances that the complex III proteins migrate in the gel under the same conditions are related to their apparent molecular weights. [Courtesy of Dr. Bernard Trumpower.]

sodium dodecyl sulfate. The type of separation obtainable may be seen in Figure 5.25.

We shall see applications of gel and paper electrophoresis in Section 9.9 when we discuss methods for determining the primary structures of DNA and RNA.

5–6 SUMMARY

Methods for the isolation and purification of proteins and other biological macromolecules are closely related to methods for characterizing them. The first step in protein purification is often preparative ultracentrifugation. Differential ultracentrifugation separates the various components of a crude homogenate as a result of differences in their sedimentation velocities.

Each fraction of the centrifuged cell homogenate is assayed for the protein of interest, and further purification steps are then taken. Separations based on differences in molecular dimensions are carried out by gel filtration chromatography. Separations based on differences in electric charge are carried out by ion exchange chromatography, by isoelectric precipitation, or by electrophoresis. In most cases electrophoresis is used as analytical, rather than a preparative technique. Separations based upon differences in biological specificity can be effected by affinity chromatography. Differences in the

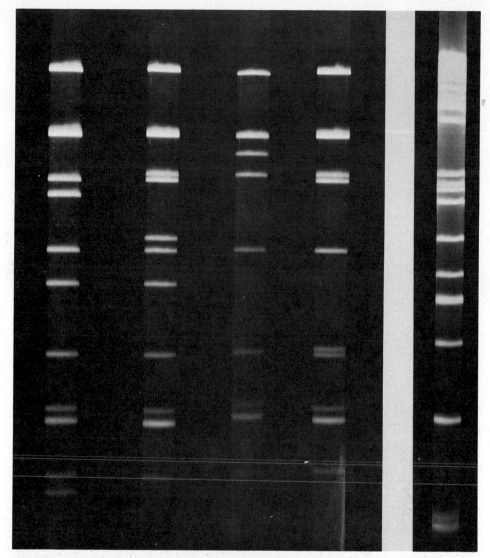

Figure 5.25 Separation on an agarose gel of DNA fragments obtained by enzymatic digestion. [Courtesy of Dr. Nancy Hamlett.]

absorption of the components of a mixture to a solid support such as hydroxyapatite permit separations to be carried out by adsorption chromatography.

The purity of a sample is determined by a variety of techniques. The homogeneity of the preparation may be determined by analytical electrophoresis, since a single band (or spot) will be observed for electrophoresis of the pure protein. Often, several methods should be used to ascertain homogeneity. Various types of chromatography will also reveal the presence of only a single component in the solution when the protein is pure.

The molecular weight of a purified protein can be determined by a variety of techniques including sedimentation velocity centrifugation, sedimentation equilibrium, gel filtration chromatography, or SDS gel electrophoresis.

REFERENCES

Cantor, C. R., and Schimmel, P. R., "Biophysical Chemistry," Vol. 2, Freeman, San Francisco, 1980.

Cooper, T. R., "The Tools of Biochemistry," Wiley, New York, 1977.

Freifelder, D., "Physical Biochemistry," Freeman, San Francisco, 1976.

Haschenmeyer, R. H., and Haschenmeyer, A. E. V., "Proteins: A Guide to Study by Physical and Chemical Methods," Wiley, New York, 1973.

Jakobi, W., Ed., "Enzyme Purification and Related Techniques," Vol. 22, "Methods in Enzymology," Academic Press, New York, 1971.

Van Holde, K. E., "Physical Biochemistry," Prentice-Hall, Englewood Cliffs, N.J., 1971.

Problems

1. Explain why all proteins migrate toward the anode in SDS gel electrophoresis.

2. Suppose that you are studying homologous proteins from different species and comparing their differences by their electrophoretic mobilities. Which of the following substitutions would be detected more readily: alanine for valine, valine for aspartic acid, glutamic acid for aspartic acid, histidine for serine? Why?

3. Given the following data, predict the order of elution of myoglobin, catalase, cytochrome c, myosin, chymotrypsinogen, and serum albumin from a gel filtration column.

Protein	Molecular weight
Myoglobin	16,900
Catalase	221,600
Cytochrome c	13,370
Myosin	524,000
Chymotrypsinogen	23,240
Serum albumin	68,500

4. Given the following facts:

Protein	Isoionic pH	Molecular weight
Chymotrypsinogen	9.4	23,200
Hemoglobin	7.0	64,500
Lysozyme	11.1	14,100
Ovalbumin	4.9	45,000

  a. Draw the elution profile for a mixture of these four proteins that would be obtained by passing them through Sephadex® G-100.

  b. If chymotrypsinogen, hemoglobin, and ovalbumin are mixed and spotted on a strip of cellulose acetate at a centrally located origin and an electric field is applied across the block, draw the final distribution pattern of proteins if the pH of the buffer soaking the cellulose acetate is (i) 7.0, (ii) 9.4.

  c. What would be the order of elution of hemoglobin, lysozyme, and ovalbumin from an ion exchange column of diethylaminoethyl cellulose [cellulose—CH_2—CH_2—$\overset{\oplus}{N}H(C_2H_5)_2$] eluted with a gradient of increasing ionic strength at (i) pH 8.0, (ii) pH 6.0, (iii) pH 4.0?

  d. What would be the order of elution of the same three proteins under similar conditions (i, ii, iii) from an ion exchange column of the cation exchange resin carboxymethyl cellulose [cellulose—CH_2—$CO_2^{\ominus}$]?

5. Ovotransferrin is an iron-containing protein that is found in hen egg whites. There are two binding sites for iron in the protein. There are also two aspartic acids that are covalently bound to identical sugar molecules. The

molecular weight of ovotransferrin is 76,000. The N-terminal amino acid is alanine. Amino acid analysis of the protein reveals 8 methionine residues per mole of protein. Cyanogen bromide cleavage followed by gel filtration chromatography gives the following elution pattern.

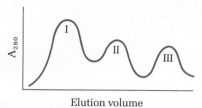

Each peak was concentrated and run on a gel electrophoresis column. The following results were obtained.

Analysis of the components of the mixture yielded the following additional facts.

	Molecular weight	N-terminal amino acid
I	21,000	Alanine
II	9,400	Glycine
IIIa	3,000	Leucine
IIIb	4,100	Phenylalanine

a. How many fragments do you expect for cyanogen bromide cleavage?

b. What product is missing? Why might it be missing?

c. Ovotransferrin contains 56 lysines and 28 arginines per mole. How many spots would you expect from digestion of ovotransferrin with trypsin?

d. Why is cleavage with cyanogen bromide more reliable in this case than cleavage with trypsin?

6. A curve for the solubility of a certain protein versus ammonium sulfate concentration is shown below. Explain why the solubility of the protein increases in going from point A to point B, and why the solubility decreases in going from point B to point C.

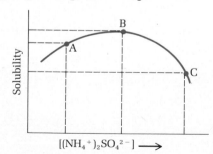

7. Bovine serum albumin has a diffusion coefficient of 6.1×10^{-7} cm^2 s^{-1} at 20°C. The viscosity of water is 0.01002 g cm^{-1} s^{-1}. Calculate the Stokes' radius of serum albumin ($k = 1.38 \times 10^{-16}$ g cm^2 s^{-2} K^{-1}).

8. Human hemoglobin has a sedimentation coefficient of 4.46 S, a partial specific volume of 0.749 cm^3 g^{-1}, and a diffusion coefficient of 6.9×10^{-7} cm^2 s^{-1}. At 20°C the density of water is 0.998 g cm^{-3}. Calculate the molecular weight of hemoglobin from these data. (Pay close attention to the units of the sedimentation coefficient.)

9. Concanavalin A (Section 8.9), a protein that binds carbohydrates, can be purified by affinity chromatography on Sephadex®. The column distribution coefficient for concanavalin A is greater than 1.0. How will this affect a molecular weight determination of concanavalin A by gel filtration chromatography?

Chapter 6

Enzyme catalysis

In every object there is inexhaustible meaning; the eye sees in it what the eye brings means of seeing.

Thomas Carlyle

6–1 INTRODUCTION Like many faceted gems, enzymes sparkle in the light of a thousand perspectives. Two of their properties in particular excite our imagination. First, rates of enzyme-catalyzed reactions are 10^3 to 10^6 times greater than rates of corresponding uncatalyzed reactions. Second, enzymes are so highly stereospecific that most catalyze a single chemical reaction of a single metabolite and are able to distinguish not only between enantiomers, but between apparently identical atoms or groups.

In the late 1940s the mathematician and philosopher Norbert Wiener wrote a little book in which the topic of enzymes appeared. Wiener hypothesized that enzymes were examples of Maxwell demons, since they are able to select molecules having the proper stereochemistry and charge as substrates from a sea of thousands of molecular species, some of them closely resembling each other. Because enzyme catalysis in some respects resembles this demonlike activity, and is partly the consequence of an entropy effect, enzymes have been characterized as "entropy traps." The term *entropy trap* emphasizes the stereospecific nature of the "capture" of a substrate by an enzyme's active site; it is another name for enzyme specificity.

6–2 FACTORS RESPONSIBLE FOR ENZYME CATALYSIS

Any theory of enzymatic catalysis must account for both rapid rates of reaction and extreme stereospecificity. And, since an enzyme-catalyzed reaction obeys the same laws of chemistry as a corresponding uncatalyzed reaction, the theory must also encompass the basic factors responsible for any chemical reaction. Three criteria must be met for any chemical reaction to occur. (1) The reactants, called *substrates* in enzymology, must collide. (2) The molecular collision must occur with the correct orientation for reaction to take place. (3) Given a collision with the correct orientation, the reactants must have sufficient energy, called the *activation energy,* for a reaction to occur. The change in concentration of reactants or products with time, that is, the *reaction rate,* is given by equation 6.1,

$$\text{rate} = Z_{AB}e^{-E_a/RT} \tag{6.1}$$

where Z_{AB} is the collision frequency, E_a is the activation energy for the reaction, R is the gas constant, and T is the absolute temperature. The *Arrhenius equation* expresses the relation between the *specific rate constant,* k, and the activation energy as

$$k = A\ e^{-E_a/RT} \tag{6.2}$$

where A is a statistical term related to the probability that the colliding reactants have the correct orientation for reaction.

In an enzyme-catalyzed reaction the rate constant k is many orders of magnitude greater than the rate constant for the corresponding uncatalyzed reaction. Such a reaction has a combination of a higher probability of correct orientation for reaction and a lower activation energy which, acting in concert, lead to faster rates. *Since enzymes are true catalysts, they do not alter the equilibrium constants for the reactions they catalyze.*

Enzymes also display a remarkable stereospecificity. Enzyme-catalyzed reactions occur at an asymmetric "pocket" of the protein called the *active site.* The conformation of the active site is responsible for the specificity of enzymatic catalysis. The active site can be subdivided into a *binding site,* which includes all amino acid residues in contact with the substrate, and a *catalytic site,* which includes residues directly responsible for catalysis.

The conversion of reactants to products in any chemical reaction is accompanied by a continuous change in energy. Each constellation of reactants has a definite energy which increases as the reactants approach each other and begin to undergo chemical reaction. In the ground state the bond lengths and bond angles and electron distributions of the reactants are at their equilibrium values. At some point the energy of the system reaches a maximum that corresponds to a definite configuration of reactants in the *transition state* for the reaction. In the transition state bond lengths, bond angles, and the electron distribution of the reactants are distorted to some higher energy configuration. As the reaction continues, the energy of the system decreases until it reaches a new minimum in the products. A plot of the energy of the system versus the *reaction coordinate,* or the extent of reaction, gives a *reaction profile* for the reaction. The reaction profile for an *elementary reaction,* that is, a reaction having only one step, is shown in Figure 6.1. The reaction profile for ester hydrolysis, consisting of three elementary reactions, is shown in Figure 6.2. The difference in energy between reactants and products determines the equilibrium position for the reaction. The difference in energy between the reactants and the transition state is the *activation energy* for the reaction. The height of this barrier determines the rate of the reaction at a given temperature. *An enzyme lowers the energy of the transition state for the reaction it catalyzes,* and enzyme-catalyzed reactions are many orders of magnitude faster than the corresponding uncatalyzed reactions.

The reaction profile for an enzyme-catalyzed reaction that occurs in one step is shown in Figure 6.3. The first step in any enzyme-catalyzed reaction is formation of an *enzyme-substrate complex* (ES) which is more stable than the isolated enzyme and its substrate (E + S). The stability of the enzyme-substrate complex is a result of favorable van der Waals interactions, hydrogen bonding, and ionic interactions between the enzyme and substrate. Formation of the enzyme-substrate complex sets the stage for catalysis. Since the enzyme-catalyzed reaction is faster than the uncatalyzed reaction, the energy of the transition state decreases by an amount greater than the lowering of energy in the enzyme-substrate complex. *This means that the interaction of the enzyme with the substrate is stronger in the transition state than in the ground state.* We assume that the structure of the substrate portion of the activated enzyme-substrate complex (ES‡) has the same conformation as the substrate transition state (S‡) in the uncatalyzed reaction. This assumption is not perfect, but it is a reasonable approximation in many cases. Late in the nineteenth century the German chemist Emil Fischer described the interaction of an enzyme and its substrate as analogous to the fitting of a key into a lock. Since the conformation of the enzyme in the transition state is complementary to the conformation of the substrate transition state, the lock and key are both mobile. Enzymes are not rigid molecules, but have considerable conformational flexibility.

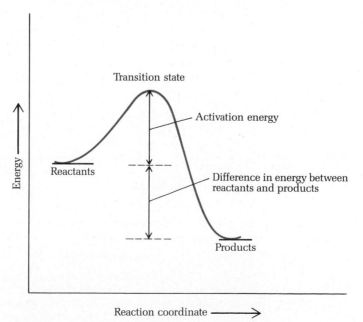

Figure 6.1 Reaction profile for a one-step reaction.

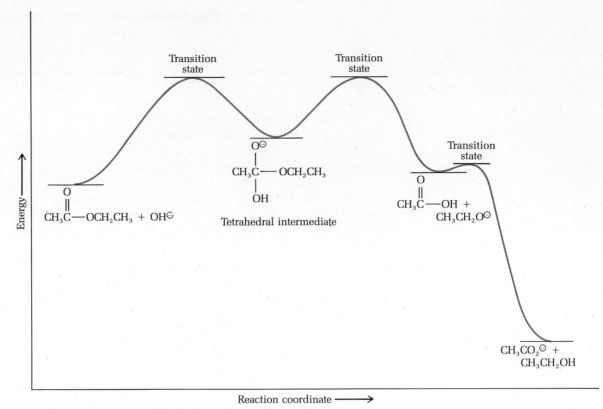

Figure 6.2 Reaction profile for hydrolysis of an ester.

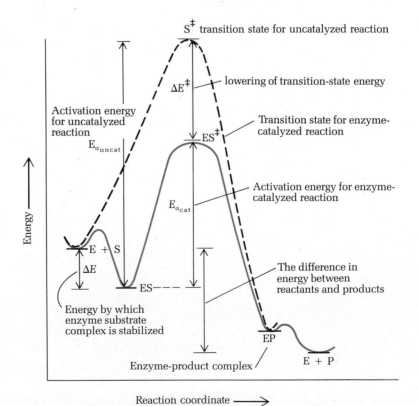

Figure 6.3 Reaction profile for an elementary enzyme-catalyzed reaction compared to an uncatalyzed reaction. Note that the difference in energy between reactants and products is not altered by the enzyme and formation of enzyme complexes. The enzyme therefore does not alter the equilibrium constant for the reaction.

A. Entropic Effects in Enzyme-Catalyzed Reaction

The first step of any enzyme-catalyzed reaction is the formation of a stereo-specific *enzyme-substrate complex*. The formation of this complex has several consequences. The "effective concentration" of the substrate in the enzyme-substrate complex is large and this is reflected in an increased rate. The enzyme-substrate complex has the correct orientation for reaction, since it places the substrate in a position that is accessible to the catalytic apparatus of the enzyme. The effect of the correct orientation is also reflected in the rate and is sometimes called a *proximity effect*.

The importance of the correct orientation of reactants in enzyme-catalyzed reactions can be gauged by results obtained in a model reaction: the reaction of phenyl esters with acetate anion (reaction 6.3).

$$(6.3)$$

Data for a series of these reactions are shown in Table 6.1. Reaction 6 in the table occurs 30 million times faster than reaction 1, a rate enhancement of the magnitude observed in enzyme-catalyzed reactions. Proceeding from the slowest reaction to the fastest one, each new increase in rate is caused by restricting the rotational and translational mobility of the reactants; the more rigidly they are held in place, the faster they react. Enzymatic reactions are similarly facilitated by the favorable effect of stereospecific binding of the substrate to the active site of the enzyme.

The tremendous rate enhancements of the model system we discussed above, and an important aspect of enzyme catalysis, can be ascribed to entropy effects. In an intermolecular reaction two molecules must collide with the proper orientation for reaction to occur, a process that involves a decrease in entropy. Each of the colliding molecules possesses translational, rotational, and internal entropy, the latter largely due to internal rotations and vibrations. A molecule whose molecular weight is between 20 and 200 at a concentration of 1 M possesses about 120 JK^{-1} mol^{-1} of translational entropy and about the same amount of rotational entropy. The internal entropy contributes another 20 to 30 JK^{-1} mol^{-1}. When a molecule binds to an enzyme's active site, most of this entropy is lost. A loss of 190 JK^{-1} mol^{-1} is equivalent to an *effective concentration* in the enzyme-substrate complex of about 6×10^9 M. The rate enhancement caused by enzymes is caused, in part, by this entropy effect.

The combination of an enzyme and a substrate to form an enzyme-substrate complex converts a bimolecular reaction into one that is essentially unimolecular. The model reaction discussed above shows the rate

Table 6.1 Rate changes in reactions of phenyl esters with acetate ion

Reaction	Relative rate

1. $CH_3CO_2C_6H_5 + CH_3CO_2^{\ominus} \longrightarrow$ ... $+ C_6H_5O^{\ominus}$ 1

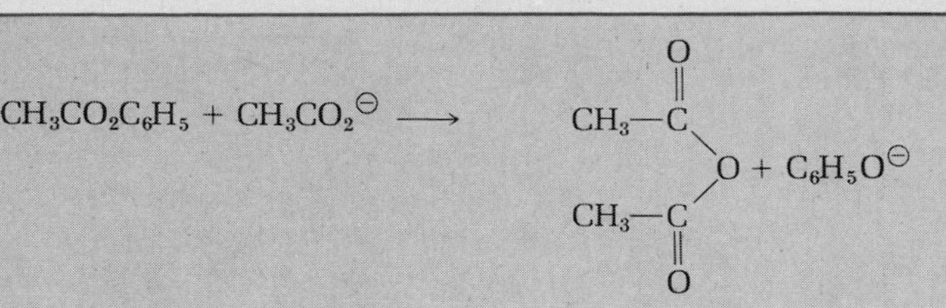

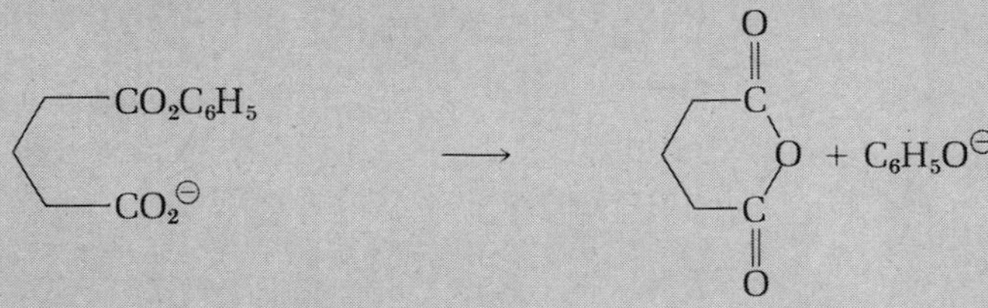

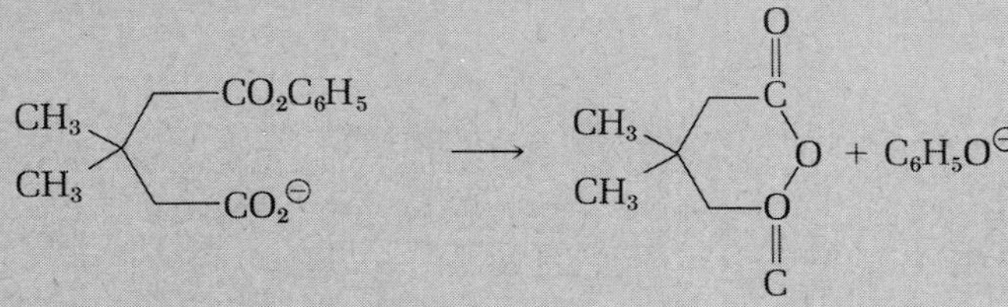

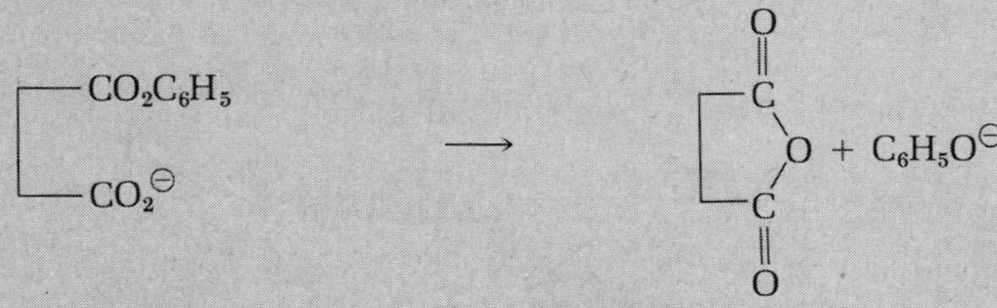

2. ... $\longrightarrow$ ... $+ C_6H_5O^{\ominus}$ 6.0×10^2

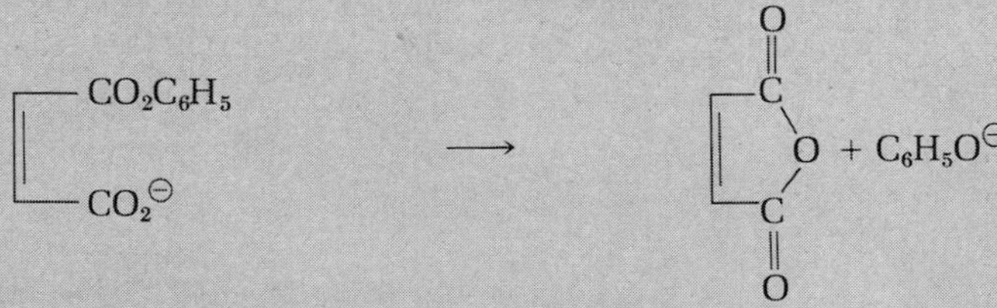

3. ... $\longrightarrow$ ... $+ C_6H_5O^{\ominus}$ 1.2×10^3

4. ... $\longrightarrow$ ... $+ C_6H_5O^{\ominus}$ 1.4×10^4

5. ... $\longrightarrow$ ... $+ C_6H_5O^{\ominus}$ 6.0×10^6

6. ... $\longrightarrow$ ... $+ C_6H_5O^{\ominus}$ 3.0×10^7

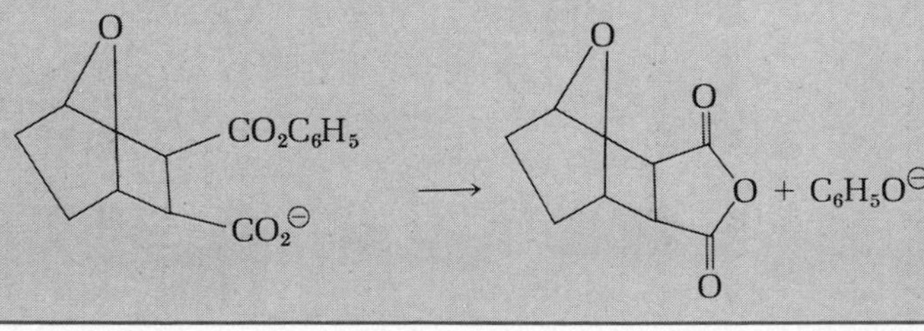

From T. C. Bruice, *Ann. Rev. Biochem.*, 37, 63 (1970). Reproduced with permission. © 1970 by Annual Reviews Inc.

enhancements that can result from such changes. If formation of the enzyme-substrate complex, in which the substrate has lost so much entropy, is entropically unfavorable, how can it occur? The formation of the enzyme-substrate complex is an equilibrium process, therefore ΔG, not ΔS is the criterion of spontaneity. Since $\Delta G = \Delta H - T\Delta S$, an unfavorable entropy change can be "paid for" by a favorable enthalpy change. The van der Waals attractions, hydrogen bonds, and electrostatic interactions between the enzyme and substrate provide the "binding energy" for the enzyme-substrate complex. The net binding energy, ΔG_{ES}, is favorable, but it is accompanied by some entropy loss since two randomly oriented molecules, the enzyme and substrate, combine to form a single, highly stereospecific complex. The loss of entropy that occurs upon formation of the enzyme-substrate complex weakens the binding between enzyme and substrate. This may seem to be a disadvantage for catalysis, but as we shall see in Section 6.3, it is just the opposite: weak binding in the enzyme-substrate complex is a boon to catalysis.

B. Acid-Base Catalysis

Once the substrate is correctly positioned at an enzyme's active site, acidic or basic residues may come into play. Those residues with ionizing side chains—histidine, tyrosine, glutamate, aspartate, the ϵ-amino groups of lysine, and so forth—can either donate protons to or accept protons from the substrate. Many enzyme active sites contain both proton donors and proton acceptors. The *acid-base* catalysis that results is a significant factor in the majority of enzyme-catalyzed reactions.

There are two broad types of acid-base catalysis. *Specific acid-catalyzed* reactions have rates that depend on the concentration of hydronium ion and no other acid; *specific base-catalyzed* reactions depend only on the concentration of hydroxide ion. The rates of *general acid-catalyzed* reactions are increased by *all* acids in solution, and the rates of *general base-catalyzed* reaction are increased by all bases in solution. A *general acid-base–catalyzed* reaction is facilitated by both acids and bases. Most enzyme-catalyzed reactions fall into the category of general acid-base catalysis.

Specific acid (or specific base) catalysis can be detected by studying a reaction under several different conditions. First, it is studied at varying pH and constant buffer concentration. If the rate changes as a function of pH at constant buffer concentration, the reaction is specific acid catalyzed (if pH $<$ 7) or specific base catalyzed (if pH $>$ 7). Next, the reaction is studied at constant pH and varying buffer concentrations. If the rate increases at constant pH as the buffer concentration increases, the reaction is general acid catalyzed (or general base catalyzed, depending on the pH). If the rate is not affected at constant pH by changes in the buffer concentration, then it is specific acid catalyzed (or specific base catalyzed) as determined in the first set of experiments (Figure 6.4).

Most acid-base–catalyzed reactions occur in two or more steps. Let us consider the conversion of a substrate (S) to a product (P) in a specific

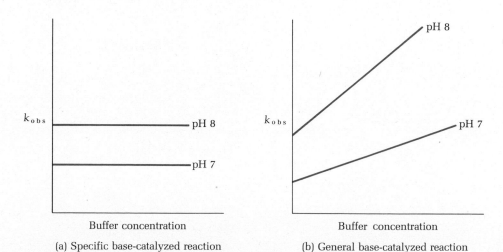

(a) Specific base-catalyzed reaction　　　(b) General base-catalyzed reaction

Figure 6.4　Detection of specific acid and base catalysis. In a specific base-catalyzed reaction the observed rate constant, k_{obs}, increases as pH increases, but is independent of the buffer concentration. In a general base-catalyzed reaction the rate increases as the buffer concentration increases at constant pH.

acid-catalyzed reaction. First, the substrate is protonated in a *rapid, reversible* step.

$$S + H_3O^{\oplus} \underset{K_{eq}}{\overset{\text{Fast}}{\rightleftharpoons}} SH^{\oplus} + H_2O \qquad (6.4)$$

The protonated substrate, $SH^{\oplus}$, then rearranges to product in the *slow, rate-determining step* of the reaction.

$$SH^{\oplus} + H_2O \xrightarrow{\text{Slow}} P + H_3O^{\oplus} \qquad (6.5)$$

The reaction is specified acid catalyzed because the proton-transfer step (reaction 6.4) is much faster than the product-forming step (6.5). Increasing the hydronium ion concentration increases the rate by increasing the concentration of the *conjugate acid* of the substrate ($SH^{\oplus}$). Notice that the proton is released when product forms; the catalyst is not consumed in the reaction.

The preceding statements can be placed on a quantitative footing by considering the rate law (the algebraic expression relating the change in product concentration with time to the concentrations of the reactants). Although we will not discuss kinetics in detail until the next chapter, this introduction will clarify the general statements we have made about acid-base catalysis.

The rate of formation of product depends upon the rate of the slow step (equation 6.6). Therefore,

$$\text{rate} = \frac{d[P]}{dt} = k\,[SH^{\oplus}] \qquad (6.6)$$

where P is the product, k is the specific rate constant, and $[SH^{\oplus}]$ is the concentration of the conjugate acid of the substrate S. The concentration of the conjugate acid is related to the concentration of S by equation 6.7.

$$K_{eq} = \frac{[SH^{\oplus}][H_2O]}{[S][H_3O^{\oplus}]} \qquad (6.7)$$

Hence,

$$[SH^{\oplus}] = \frac{K_{eq}\,[S][H_3O^{\oplus}]}{[H_2O]} \qquad (6.8)$$

Substituting for $[SH^{\oplus}]$ in equation 6.8, we obtain

$$\frac{d[P]}{dt} = \frac{k\,K_{eq}\,[S][H_3O^{\oplus}]}{[H_2O]} \qquad (6.9)$$

Setting

$$\frac{kK_{eq}}{[H_2O]} = k' \qquad (6.10)$$

we obtain the rate law for specific acid-catalyzed reactions (equation 6.11).

$$\frac{d[P]}{dt} = k'\,[S][H_3O^{\oplus}] \qquad (6.11)$$

The final rate law (equation 6.11) contains only terms for the substrate S and hydronium ion $[H_3O^{\oplus}]$. This is the requirement of specific acid catalysis.

Now consider the same reaction at constant pH in an imidazolium ion–imidazole buffer. The specific acid-catalyzed reaction proceeds as above. But suppose that there is also an imidazolium-ion–catalyzed reaction. The first step is protonation of substrate followed by conversion of $SH^{\oplus}$ to product.

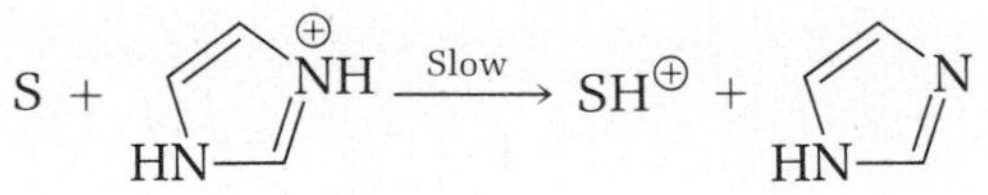

$$S + \underset{HN}{\overset{\oplus}{\bigsqcup}}NH \xrightarrow{\text{Slow}} SH^{\oplus} + \underset{HN}{\bigsqcup}N \tag{6.12}$$

$$SH^{\oplus} \underset{}{\overset{\text{Fast}}{\rightleftharpoons}} P + H^{\oplus} \tag{6.13}$$

Note that the fast and slow steps are reversed when the mechanism changes from specific to general acid catalysis. Imidazolium ion is a weak acid, $pK_a \simeq 7$, and imidazole has a rather tight hold on the proton being transferred. The rate-determining step is proton transfer from the general acid (imidazolium ion) to the substrate. The rate law for reaction 6.13 is therefore

$$\frac{d[P]}{dt} = k\,[S][ImH^{\oplus}] \tag{6.14}$$

The rate laws for general acid-catalyzed reactions can be quite complex. The observed rate constant, k_{obs}, is a composite constant that depends upon all of the acidic species, HA, in solution (equation 6.15).

$$k_{obs} = k_1[HA_1] + k_2[HA_2] + k_3[HA_3] + \ldots\, k_i[HA_i] \tag{6.15}$$

In an enzyme-catalyzed reaction an imidazolium ion might be involved in proton transfer to the substrate in the rate-determining step of the reaction. In such a case the enzyme acts as a general acid catalyst. In a subsequent fast step the protonated substrate, probably still bound to the enzyme, rearranges to product (reaction 6.16).

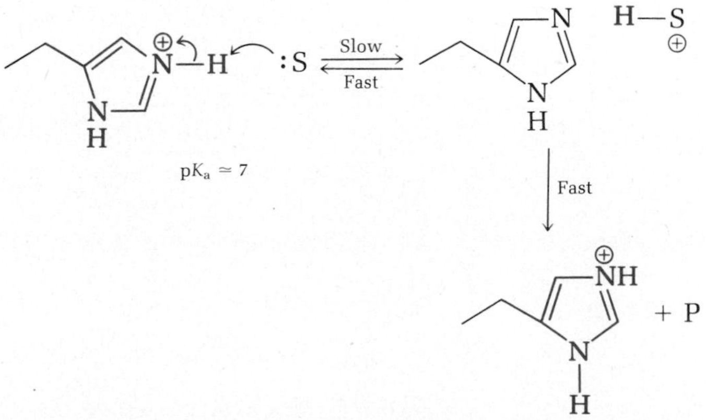

$$\tag{6.16}$$

We next turn to an example of a reaction that is general acid catalyzed—the hydrolysis of esters (reaction 6.17).

$$R-\overset{\overset{\displaystyle O}{\parallel}}{C}-O-R' + H_2O \rightleftharpoons R-CO_2^{\ominus} + R'-OH \tag{6.17}$$

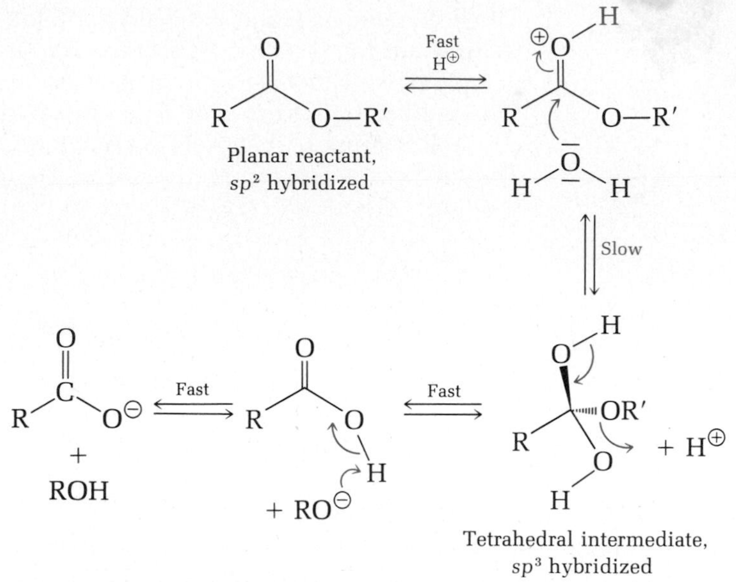

Figure 6.5   Specific acid-catalyzed hydrolysis of an ester. The reaction involves rapid protonation followed by slow attack by water upon the carbonyl carbon. (The terms slow and fast are, of course, relative and refer only to this specific reaction.)

Hydronium ion accelerates the rate by protonating the carbonyl oxygen, making the carbonyl carbon more susceptible to nucleophilic attack by water. Figure 6.5 shows the specific acid-catalyzed hydrolysis reaction.

A general acid (HA) also accelerates the rate by transferring its proton to the carbonyl group. In general acid catalysis this is the rate-determining step, which is followed by rapid attack of water. The *tetrahedral intermediate* is metastable and rapidly rearranges to product (Figure 6.6).

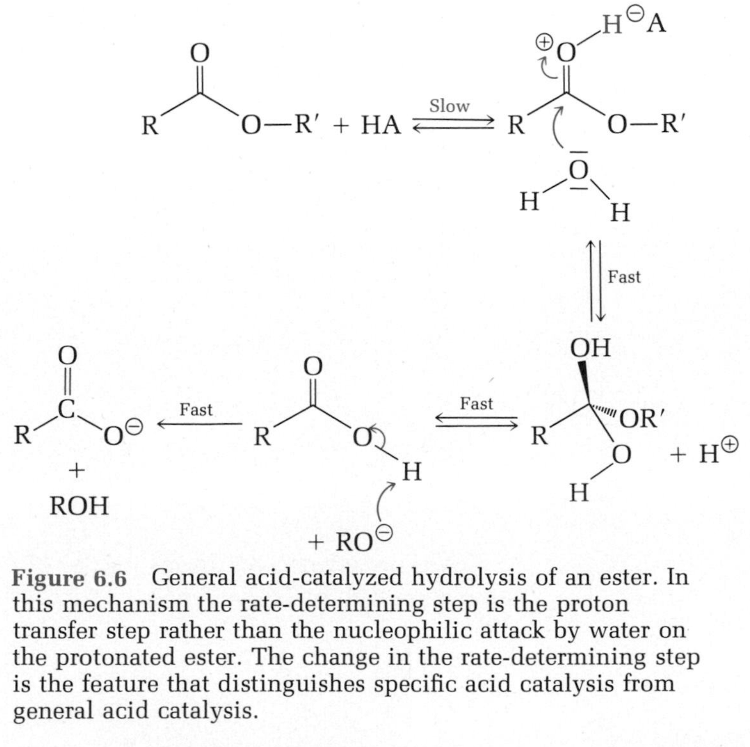

Figure 6.6   General acid-catalyzed hydrolysis of an ester. In this mechanism the rate-determining step is the proton transfer step rather than the nucleophilic attack by water on the protonated ester. The change in the rate-determining step is the feature that distinguishes specific acid catalysis from general acid catalysis.

Papain, an enzyme found in the fruit of the papaya tree, catalyzes the hydrolysis of esters and amides. A plot of the effect of pH on the rate of papain-catalyzed hydrolysis of esters, known as a *pH-rate profile,* is bell-shaped with inflection points at pH 4.2 and 8.2 (Figure 6.7). The inflection points of the pH-rate profile can be interpreted (with caution) as pK_a values of ionizing residues at the enzyme active site. The pK_a value of 4.2 is assigned to cysteine 25, and the pK_a of 8.2 is assigned to histidine 159. At low pH both residues are protonated. At high pH the cysteine exists predominantly as the thiolate anion, $R—S^{\ominus}$, and the imidazole side chain of histidine is predominantly the free base. Neither of these ionic forms is catalytically active. The ionic form of the active enzyme is either an ion pair, $R—S^{\ominus}$ . . . $ImH^{\oplus}$ or the tautomeric form R—SH . . . Im. These two tautomers cannot be distinguished by considering the titration data of Figure 6.7.

$$R—S^{\ominus} \quad H—Im^{\oplus} \Longleftrightarrow R—SH \quad Im \qquad (6.18)$$
$$\text{Cys 25} \quad \text{His 159} \qquad\quad \text{Cys 25} \quad \text{His 159}$$

The inability to distinguish between the two tautomers is known as the *principle of kinetic equivalence.* If the rate law for a reaction contains a term HA, as in the rate law for general acid catalysis (equation 6.14), the transition state of the reaction may involve either $H^{\oplus}$ and $A^{\ominus}$, or it may involve undissociated HA. *Kinetically,* there is no way to tell the difference. Similarly if the rate law involves a term $A^{\ominus}$ (the thiolate anion in the case of papain), the transition state involves either $A^{\ominus}$ or HA and hydroxide ion. Again, there is no way to tell the difference by a kinetic analysis. Other experiments, however, have revealed that the catalytically active ionic form of papain is the imidazolium-mercaptide ion pair ($ImH^{\oplus}$ . . . $^{\ominus}S—R$). The mercaptide anion acts as a nucleophile that attacks the carbonyl carbon of amide and ester substrates. The pK_a of the imidazolium ion at the active site is abnormally high, 4.2 versus the pK_a of 6.0 observed for the free amino acid and for most histidine residues in proteins. The imidazolium ion of histidine 159 in papain is stabilized by forming a *charge-transfer complex* with tryptophan 177, which hovers above it at the active site (Figure 6.8). The pK_a of cysteine 25 is shifted lower by 4 to 5 units at the active site of papain. The thiolate anion is electrostatically stabilized by the imidazolium ion of histidine 159, and this accounts for the low pK_a of cysteine 25.

C. Metal Ion Catalysis

A *metal ion* can catalyze ester hydrolysis in a variation of general acid catalysis in which the metal ion functions as a Lewis acid. The metal forms a complex with the carbonyl group; the formation of the complex polarizes the carbonyl bond and once more the carbonyl carbon is attacked by water (Fig-

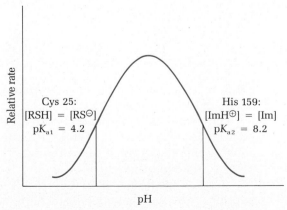

Figure 6.7 pH-rate profile for papain-catalyzed hydrolysis of an ester.

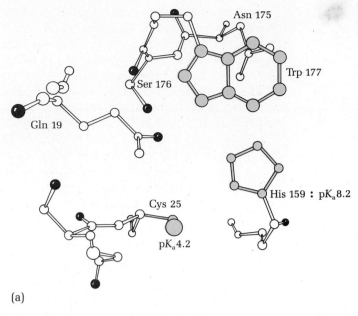

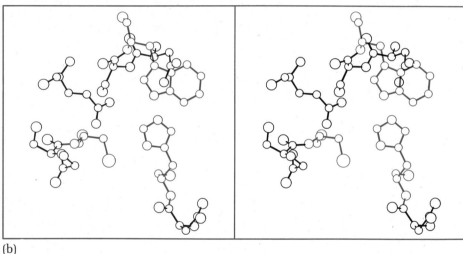

Figure 6.8 (a) Active-site residues of papain. The imidazolium ion of histidine 159 forms a charge transfer complex with tryptophan 177. The stable cation, in turn, stabilizes the thiolate anion of cysteine 25. [From J. Drenth, J. Jansonius, R. Keokoek, and B. Wolthers, in P.D. Boyer, Ed., "The Enzymes," 3rd. ed., Vol. 3, Academic Press, New York, 1971, p. 497.] (b) Stereo view of active site.

ure 6.9). The metal ion is an *electrophile,* and metal ion catalysis is an example of *electrophilic catalysis.* Many enzymes require metal ions for activity. Their catalytic function (they can also have a structural function) is often polarization of a functional group by complexation with lone pair electrons such as the electrons in the carbonyl bond, the lone pair electrons on the oxygen in alcohols, and so forth.

Carboxypeptidase A hydrolyzes the C-terminal peptide bonds of polypeptides. The enzyme contains a Zn(II) ion that forms a complex with the carbonyl oxygen of the peptide bond to be hydrolyzed. The Zn(II) ion, surrounded by four ligands in a distorted tetrahedral array, is held in place by

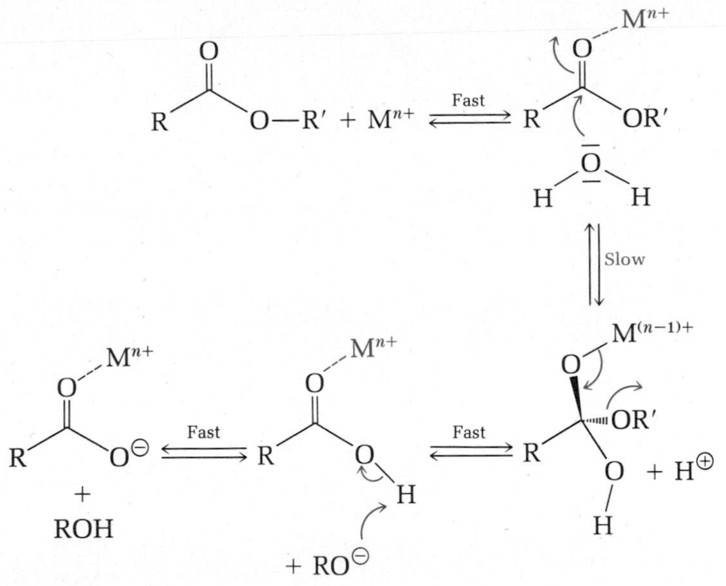

Figure 6.9 Metal ion–catalyzed hydrolysis of an ester. The reaction is facilitated by formation of a complex between the metal ion and the carbonyl group, which polarizes the carbonyl bond, and makes it more susceptible to attack by water.

the imidazole side chains of histidines 69 and 196 and by the carboxylate group of glutamate 72. The fourth ligand is either water or the substrate (Figure 6.10).

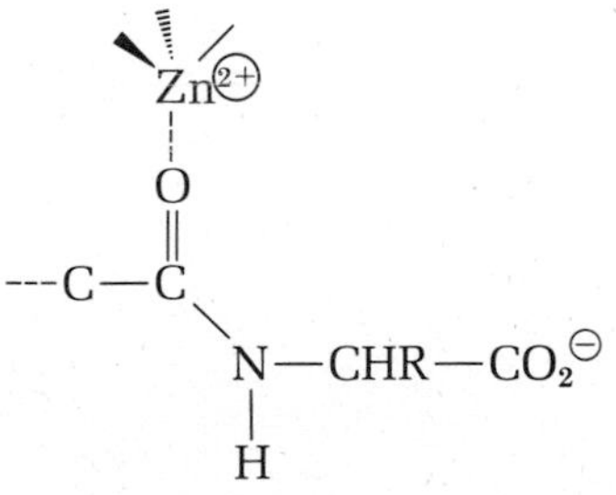

Complex between Zn^{2+} and a carbonyl group

The polarized carbonyl carbon of the peptide substrate is attacked by water in a reaction assisted by the carboxylate anion of glutamate 270 which acts as a general base catalyst (step 1 of Figure 6.11).[1] The tetrahedral intermediate formed in step 1 is thought to be a dianion which collapses to give a free amino acid, which was formerly the C-terminal amino acid residue of the peptide substrate. The leaving group of the C-terminal amino acid picks up a proton from the hydroxyl group of tyrosine 248, an example of general acid catalysis (step 2 of Figure 6.11).

Metal ions can also provide hydroxide ions to enzymes for use in catalysis. Certain of the metal ions that are bound to enzymes, such as iron and zinc, can hydrolyze water. The metal ion is a Lewis acid and water is a Lewis base. When the metal ion interacts with water in its hydration sphere, it can

---

[1] For years the accepted mechanism postulated a covalent intermediate in which the carboxylate anion of glutamate is covalently bound to the substrate. Recent experiments have thrown doubt on such a covalent intermediate. It is possible that the mechanisms of ester and peptide bond hydrolysis for carboxypeptidase A are different. There is strong evidence that a covalent intermediate is formed between glutamate 270 and the substrate when the substrate is an ester, but not when the substrate is a peptide.

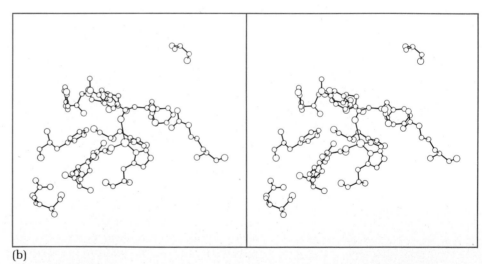

(a)

(b)

Figure 6.10 (a) Schematic representation of the enzyme-substrate complex for carboxypeptidase A and a peptide substrate. [From W. N. Lipscomb et al., *Proc. Nat. Acad. Sci. U.S.*, *28* (1979).] (b) Stereo view of an enzyme-substrate analog complex.

not only polarize the water, but actually split it into a hydroxide ion and a proton (reaction 6.19).

$$M(H_2O)_n^{z+} \rightleftharpoons [M(H_2O)_{n-1}(OH^{\ominus})]^{(z-1)^{\oplus}} + H^{\oplus}(aq) \qquad (6.19)$$

The metal-bound hydroxide ion is a potent nucleophile. *Carbonic anhydrase*, which catalyzes the hydration of carbon dioxide, has a Zn(II) ion at its active site. This Zn(II) ion is coordinated with the nitrogens of three imida-

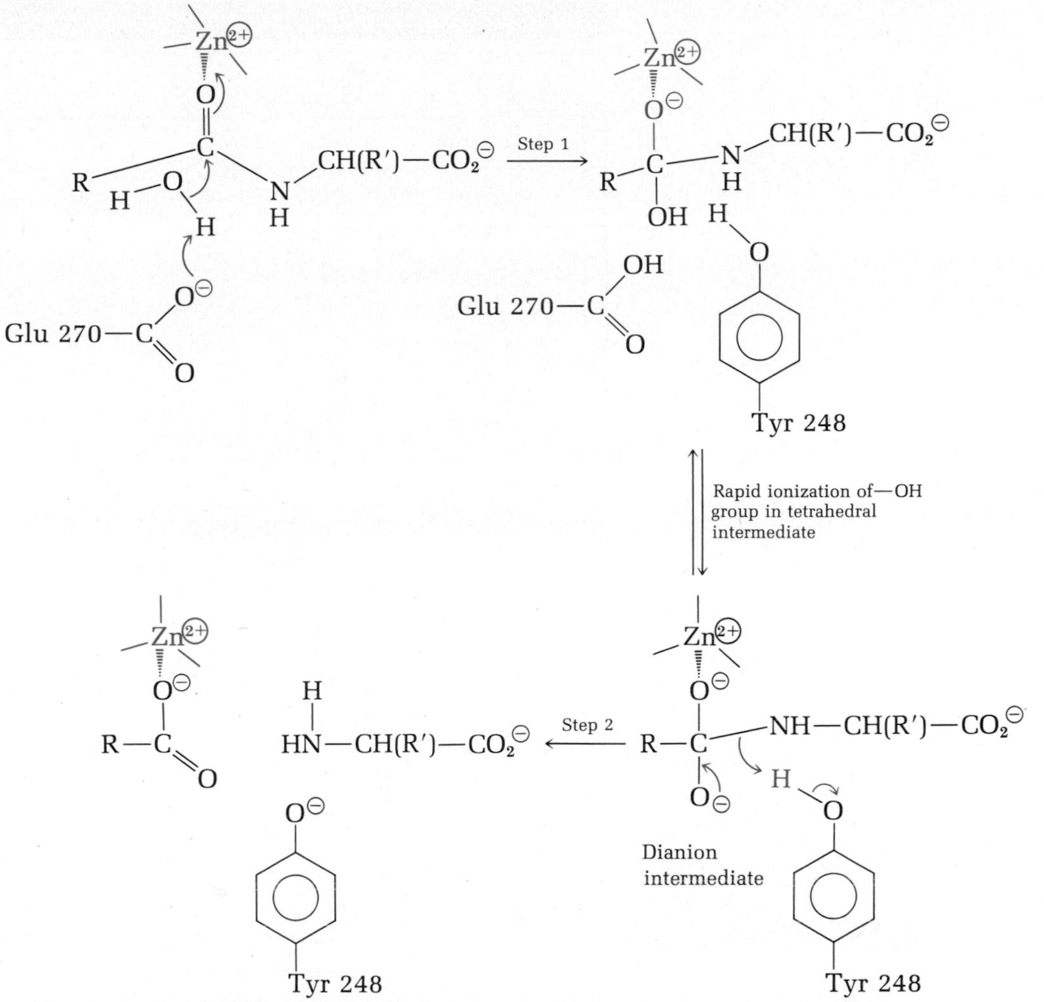

Figure 6.11 Mechanism of action of carboxypeptidase A with a peptide substrate.

zole groups and an hydroxide ion. The hydroxide bound to zinc adds to the carbon dioxide (reaction 6.20).[2]

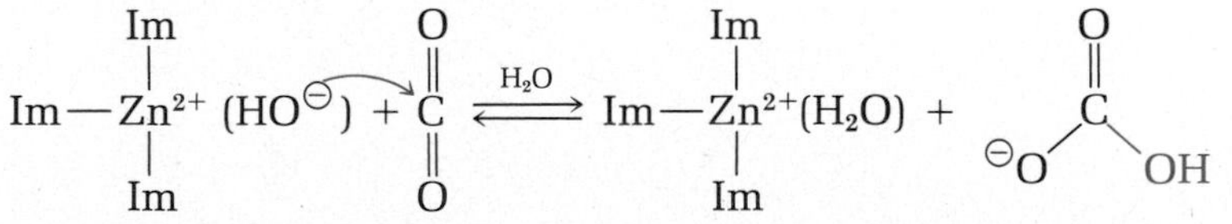

$$ \text{Im—Zn}^{2+}(\text{HO}^{\ominus}) + \overset{O}{\underset{O}{\overset{\|}{C}}} \overset{H_2O}{\rightleftharpoons} \text{Im—Zn}^{2+}(\text{H}_2\text{O}) + \overset{O}{\underset{^\ominus O}{\overset{\|}{C}}}\text{OH} \qquad (6.20) $$

The water molecule bound to the zinc in the product of reaction 6.20 diffuses to the active site from the solvent. This water is hydrolyzed (reaction 6.19) to provide the hydroxide for the next round of catalysis.

D. Nucleophilic Catalysis

The hydroxyl groups of serine and tyrosine, hydroxide ions supplied by zinc, the sulfhydryl group of cysteine, the carboxyl group of aspartate, the imidazole side chain of histidine, and the amino group of lysine can all act as nucleophiles in enzyme-catalyzed reactions (Table 6.2). During catalysis these side chains often become covalently bound to the substrate. Nucleo-

---

[2] Eight or nine different mechanisms have been proposed for carbonic anhydrase. The mechanism of reaction 6.20 is chemically reasonable and shows the role played by Zn(II) in catalysis. It is possible that hydroxide ion is not the nucleophile, but that a basic side chain at the active site (Glu 106) participates with the Zn(II) in generating hydroxide at the active site.

Table 6.2 Nucleophilic side chains in enzymes

Nucleophile	Enzyme	Intermediate	Text reference
—OH (serine)	Serine proteases	Acyl enzyme	Section 6.3
	Phosphatases, phosphoglucomutase	Phosphoryl enzyme	
—SH (cysteine)	Papain, glyceraldehyde 3-phosphate dehydrogenase	Acyl enzyme	Section 6.2
—CO$_2^{\ominus}$ (aspartate)	Pepsin	Acyl enzyme	Sections
—NH$_2$ (lysine)	Acetoacetate decarboxylase, aldolase, pyridoxal enzymes	Schiff base	14.3D and 10.6
Imidazole (histidine)	Phosphoglycerate mutase, succinyl-CoA synthetase, nucleoside diphosphokinase, histone phosphokinase	Phosphoryl enzyme	Section 14.4C

philic catalysis, then, is one type of *covalent catalysis.* The other type is electrophilic catalysis, discussed in Chapter 10. For covalent catalysis to be effective, the enzyme-substrate intermediate must convert rapidly to products. A stable enzyme-substrate intermediate would be fatal to the enzyme. In fact, one way to inhibit enzymes irreversibly is to inactivate them through the formation of just such intermediates.

Covalent catalysis is observed in many enzyme-catalyzed reactions, and various examples of nucleophilic catalysis are distributed throughout the text.

6–3 CHYMOTRYPSIN AND THE SERINE PROTEASES

A rather extensive group of enzymes, known as *proteases,* specialize in the hydrolysis of peptide bonds. The specificity of these enzymes varies markedly. Some proteases, including those discussed in this section, have a fairly broad specificity; others, such as the enzymes of the blood clotting system (Section 11.7B), have an extremely narrow specificity.

Chymotrypsin is a member of a family of enzymes known as *serine proteases,* so named because they have an unusually reactive serine residue at their active sites (Table 6.3). These enzymes have similar amino acid sequences at their active sites (Table 6.4) and share a common catalytic mechanism. They differ primarily in their specificities: *chymotrypsin* hydrolyzes aromatic esters and amides, *trypsin* hydrolyzes basic esters and amides, and

$$O_2N-\!\!\!\!\bigcirc\!\!\!\!-O-\underset{\underset{O}{\|}}{C}-CH_3$$

p-Nitrophenyl acetate

Table 6.3 Some members of the serine protease family

Enzyme	Reactive amino acid on enzyme	Type of covalent intermediate
Chymotrypsin	Serine (OH)	Acylserine
Trypsin	Serine (OH)	Acylserine
Elastase	Serine (OH)	Acylserine
Subtilisin	Serine (OH)	Acylserine
Thrombin	Serine (OH)	Acylserine

Table 6.4 Amino acid sequences at the active site of serine proteases and esterases near the serine residue involved in catalysis

Sequence	Enzyme
Gly-Val- Ser-Ser-Cys-Met-Gly-Asp-Ser-Gly- Gly-Pro-Leu-Val-Cys-Lys	Chymotrypsin
$\quad$ NH$_2$	
$\quad\quad$ Asp-Ser-Cys-Glu-Gly- Gly-Asp-Ser-Gly- Pro-Val-Cys-Ser-Gly-Lys	Trypsin
$\quad\quad\quad\quad\quad\quad\quad\quad$ Asp-Ser-Gly	Thrombin
$\quad\quad\quad\quad\quad\quad\quad\quad$ Asp-Ser-Gly	Elastase
$\quad\quad\quad\quad\quad$ NH$_2$	
$\quad\quad\quad\quad\quad$ Asp-Gly-Thr-Ser-Met-Ala-Ser-Pro- His	Subtilisin (*B. subtilis*)

elastase hydrolyzes proteins at glycyl residues. The catalytic sites of the enzymes are nearly identical, but their binding sites are markedly different. We shall begin our discussion of the serine proteases by considering the catalytic action of chymotrypsin.

$\quad$ Chymotrypsin hydrolyzes *p*-nitrophenyl acetate, a convenient, though nonbiological substrate. The reaction occurs in two distinct steps. First, *p*-nitrophenol is rapidly formed in a "burst phase" that produces an *acylenzyme intermediate*. Second, the covalent intermediate hydrolyzes in a slow "steady state" phase of the reaction (Figures 6.12 and 6.13). The identity of the amino acid side chain bound to acetate was determined by treatment of chymotrypsin with diisopropylphosphofluoridate (DIFP).

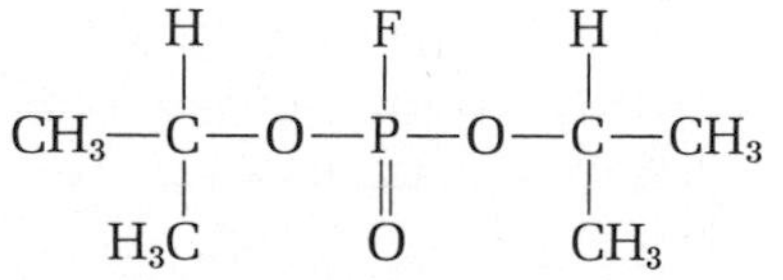

Diisopropylphosphofluoridate (DIFP)
[or diisopropylfluorophosphate (DFP)]

The treated enzyme, diisopropylphosphoryl-chymotrypsin, is inactive. Chemical analysis of the treated enzyme showed that a single residue, serine

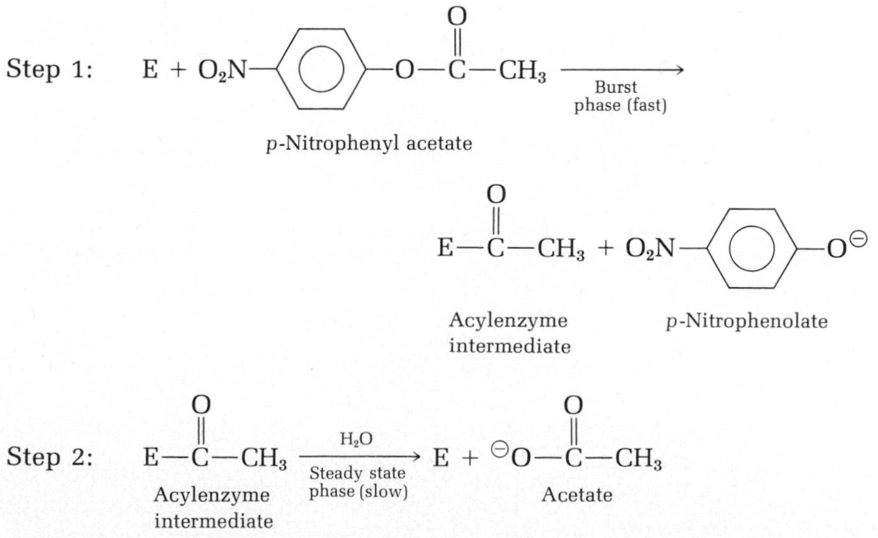

Step 1: *p*-Nitrophenyl acetate

Acylenzyme intermediate + *p*-Nitrophenolate

Step 2: Acylenzyme intermediate → E + Acetate

Figure 6.12 Chymotrypsin-catalyzed hydrolysis of *p*-nitrophenylate esters. The first step is rapid formation of an acylenzyme intermediate which hydrolyzes slowly in the second step.

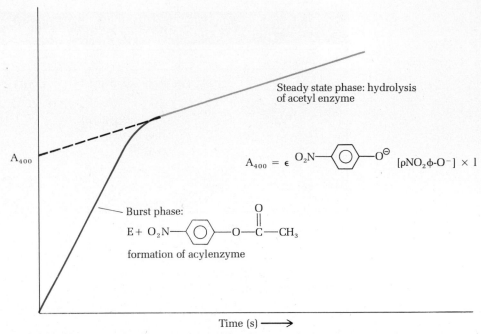

Figure 6.13 Hydrolysis of *p*-nitrophenyl acetate by chymotrypsin. The reaction occurs in two steps: a rapid burst phase in which 1 mole of *p*-nitrophenylate anion is released per mole of chymotrypsin and a steady state phase in which the acetylenzyme is hydrolyzed.

195, had been covalently modified by DIFP. The chemical reaction of chymotrypsin with DIFP is a nucleophilic displacement of fluoride by the serine hydroxyl group (reaction 6.21).

(6.21)

Chymotrypsin is also inactivated by treatment with *N-tosylamido-*L*-phenylethyl chloromethyl ketone*. Analysis of the adduct that forms when chymotrypsin is treated with this reagent reveals that histidine 57 has been covalently modified (Figure 6.14).

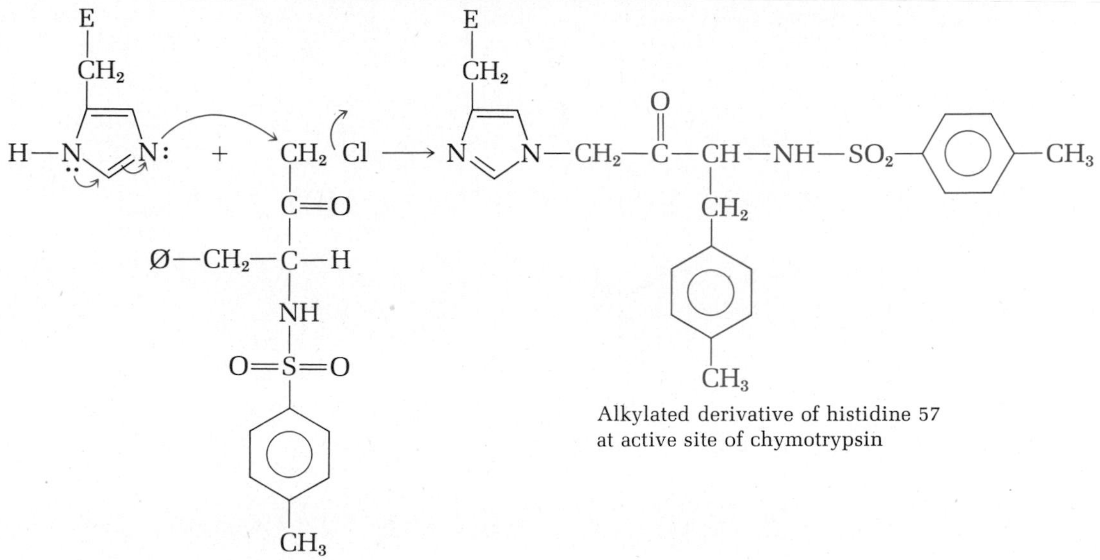

N-Tosylamido-L-phenylethyl
chloromethyl ketone (Tos-Phe-CH₂Cl)

Alkylated derivative of histidine 57
at active site of chymotrypsin

Figure 6.14 Alkylation of histidine 57 and inactivation of bovine chymotrypsin by Tos-Phe-CH₂Cl. The phenyl side chain of the alkylating agent binds in a hydrophobic pocket that helps provide the correct orientation for catalysis or, as in this case, inactivation.

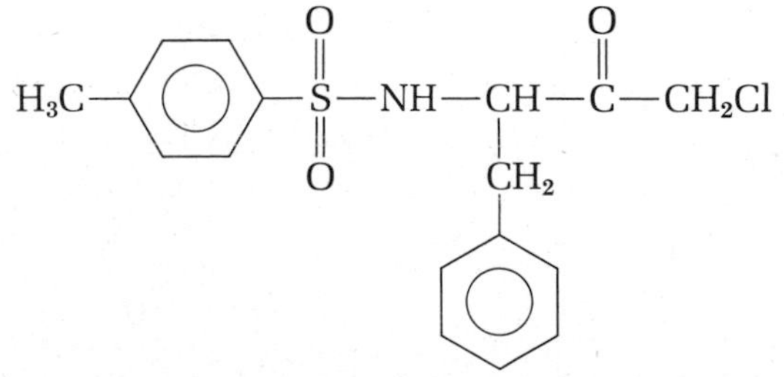

N-Tosylamido-L-phenylethyl chloromethyl ketone (Tos-Phe-CH₂Cl)

The conformation of chymotrypsin is shown in Figure 6.15. Nearly all the charged amino acids lie on the surface but there are some important exceptions. A salt bridge between the ammonium ion of isoleucine 16 and the carboxyl side chain of aspartate 194 is necessary for activity (see Section 6.4B). Aspartate 102 is also buried, and it too is required for catalysis. In the hydrophobic interior of the protein a carboxylate anion requires a counterion to stabilize its negative charge. In chymotrypsin this service is provided by histidine 57, whose imidazole side chain can donate a proton to the aspartate. The imidazole in turn retrieves a proton from the hydroxyl side chain of serine 195. The so-called *charge relay network* makes the serine hydroxyl group a potent nucleophile (Figures 6.16 and 6.17).

The specificity of chymotrypsin is due, in large part, to a hydrophobic pocket whose conformation is complementary to aromatic side chains, and which stabilizes the enzyme-substrate complex by van der Waals interactions.

The binding of a substrate analog, formyl-L-tryptophan, to chymotrypsin is shown in Figure 6.18. The formyl group is hydrogen bonded to main chain residue 193, and the amide nitrogen is hydrogen bonded to main chain residue 214. The hydrophobic pocket holds the indole ring of tryptophan. The enzyme-substrate complex, or in this case the enzyme-substrate

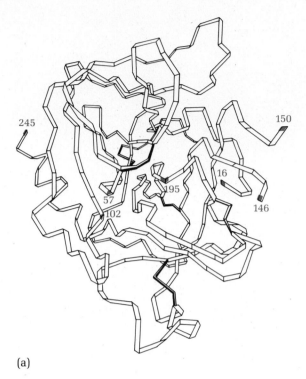

(a)

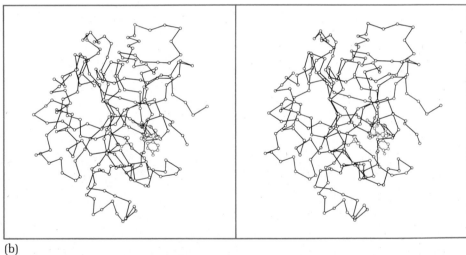

(b)

Figure 6.15 (a) Schematic diagram of the polypeptide backbone of chymotrypsin. [From B. S. Hartley and D. M. Shotten, in P. D. Boyer, Ed., "The Enzymes," 3rd ed., Vol. 3, Academic Press, New York, 1971.] (b) Stereo view of chymotrypsin.

Formyl-L-tryptophan

analog complex, is therefore stabilized not only by van der Waals interactions, but by hydrogen bonds with the polypeptide backbone. The mechanism of action of chymotrypsin can be divided into five steps.

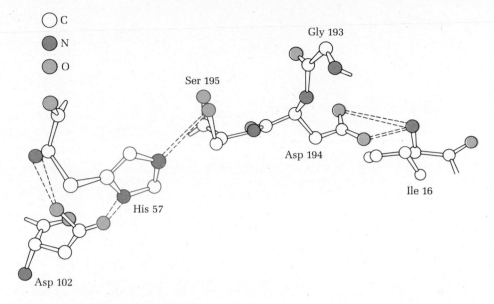

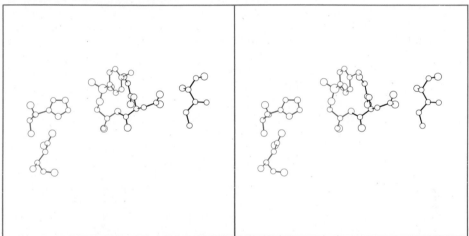

Figure 6.16   Charge relay network formed by active-site residues of chymotrypsin. Dotted lines indicate hydrogen bonds. The network enhances the nucleophilicity of serine 195 toward ester and amide substrates. [From D. M. Blow and T. A. Steitz, *Ann. Rev. Biochem.*, *39*, 63 (1970). Reproduced with permission. © 1970 by Annual Reviews Inc.]

1. The formation of the enzyme-substrate complex positions the substrate with the correct conformation for reaction. Correct binding for reaction leads to a large rate enhancement (Figure 6.19).

2. Formation of a covalent enzyme-substrate intermediate is aided by the charge relay network (Figure 6.20). The first intermediate has a tetrahedral structure. The enzyme stabilizes the intermediate by means of two hydrogen bonds from the amide nitrogens of residues 193 and 195 in the peptide backbone. The oxygen fits neatly into the *oxyanion* pocket of the enzyme. By stabilizing the tetrahedral intermediate the enzyme accelerates the rate of reaction.

3. The peptide bond now breaks as the acylenzyme intermediate is formed. The leaving group picks up a proton from the imidazolium ion of histidine 57, so this step of the reaction is general acid catalyzed (Figure 6.21).

4,5. The acylenzyme intermediate hydrolyzes in a repeat of steps 2 and 3 leaving the enzyme ready for another round of catalysis (Figure 6.22).

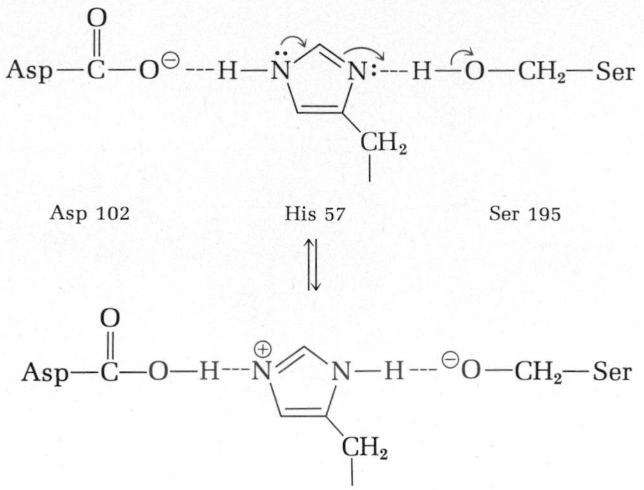

$$\text{Asp}-\overset{\overset{O}{\|}}{C}-O^{\ominus}---H-N\overset{\cdot\cdot}{}=N:---H-O-CH_2-\text{Ser}$$

Asp 102   His 57   Ser 195

$$\text{Asp}-\overset{\overset{O}{\|}}{C}-O-H---\overset{\oplus}{N}=N-H---\overset{\ominus}{O}-CH_2-\text{Ser}$$

Figure 6.17 Charge relay network of chymotrypsin. The network increases the nucleophilicity of the hydroxyl group of serine 195 toward the carbonyl carbon of the substrate. The carboxyl group of aspartate 102 is not basic enough to remove the proton from the imidazolium ion, but it stabilizes the imidazolium ion by hydrogen bonding and electrostatic attraction.

6–4 ACTIVATION OF PANCREATIC PROTEASES

The pancreas, lying just beneath the stomach, produces many proteolytic enzymes. These enzymes are not synthesized in their catalytically active states, but are synthesized as inactive polypeptides, called *zymogens*, that are activated by specific proteolysis. The zymogens *trypsinogen, chymotrypsinogen, proelastase,* and *procarboxypeptidase* leave the pancreas through the pancreatic duct, which empties into the small intestine. The release of pancreatic zymogens is triggered by hormone action. At least five hormones are released when partially digested food from the stomach passes through the small intestine.

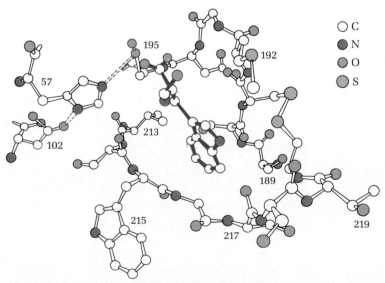

Figure 6.18 Binding of pseudosubstrate formyl-L-tryptophan at the active site of chymotrypsin. The indole side chain of tryptophan fits in the hydrophobic pocket. [From D. M. Blow and T. A. Steitz, *Ann. Rev. Biochem.,* **39,** 86 (1970). Reproduced with permission. © 1970 by Annual Reviews Inc.]

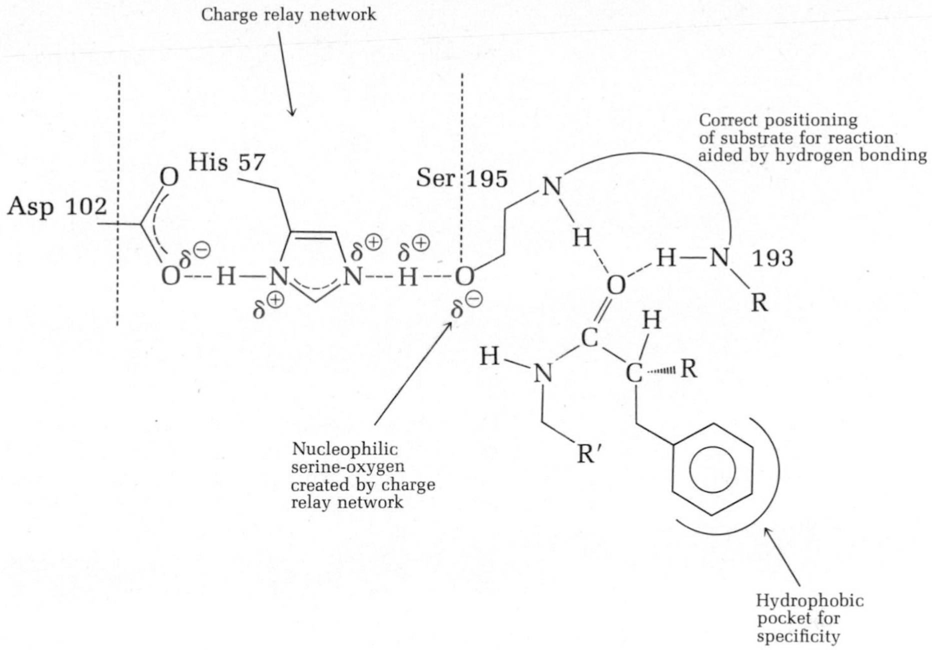

Charge relay network

His 57

Ser 195

Asp 102

Correct positioning of substrate for reaction aided by hydrogen bonding

N 193

R

Nucleophilic serine-oxygen created by charge relay network

Hydrophobic pocket for specificity

Figure 6.19 Mechanism of chymotrypsin action, step 1. The enzyme-substrate complex in chymotrypsin brings the reacting groups of the active site into proximity with the substrate.

A. Activation of Chymotrypsinogen

When chymotrypsinogen, the inactive precursor of chymotrypsin, is transported to the small intestine, it is cleaved by *trypsin* between arginine 15 and isoleucine 16 to give active π-*chymotrypsin* (Figure 6.23). The conversion from inactive zymogen to active chymotrypsin occurs as a result of this single cleavage by trypsin. The N-terminal peptide produced by trypsin cleavage remains bound to the main chain of chymotrypsin by a disulfide bond between residues 1 and 122.

Although it is completely active, π-chymotrypsin is not the final form of chymotrypsin. Proteolytic excision of the dipeptides serine 14−arginine 15 and threonine 147−asparagine 148, catalyzed by chymotrypsin itself, produces α-chymotrypsin.

When the bond between arginine 15 and isoleucine 16 is cleaved by

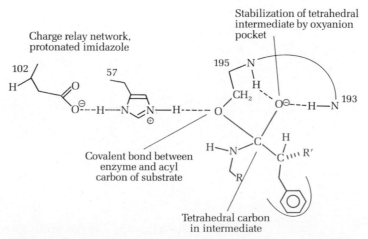

Figure 6.20 Mechanism of chymotrypsin action, step 2. The tetrahedral intermediate is formed, and covalent catalysis begins.

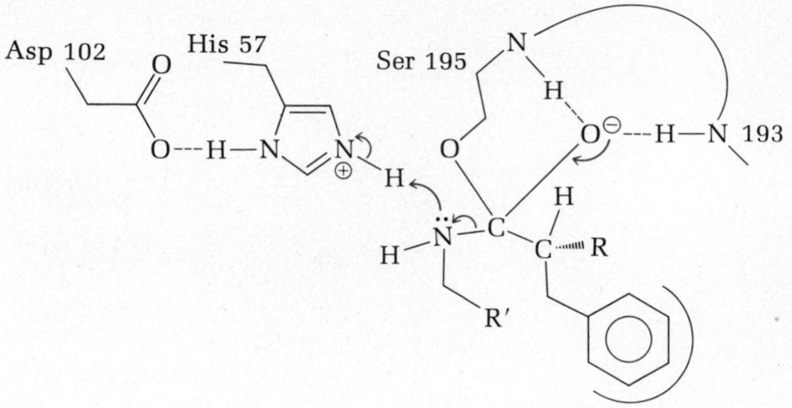

Figure 6.21 Mechanism of chymotrypsin action, step 3. The formation of an acyl-enzyme intermediate is assisted by the imidazoline ion of histidine 57, which acts as a general acid catalyst.

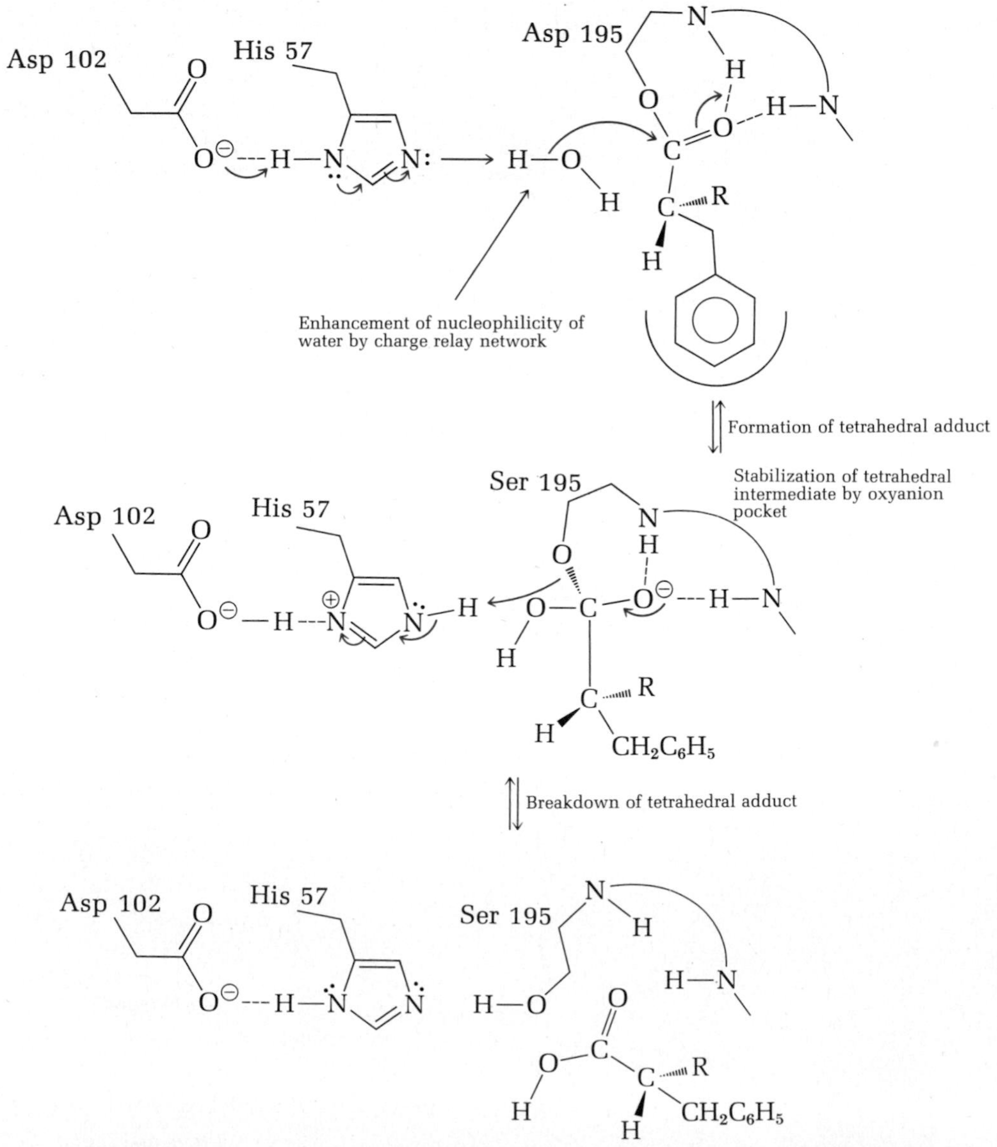

Figure 6.22 Mechanism of chymotrypsin action, steps 4 and 5. The final steps of chymotrypsin-catalyzed hydrolysis of an ester are a "replay" of steps 2 and 3, and result in hydrolysis of the acylenzyme intermediate, regenerating the free enzyme.

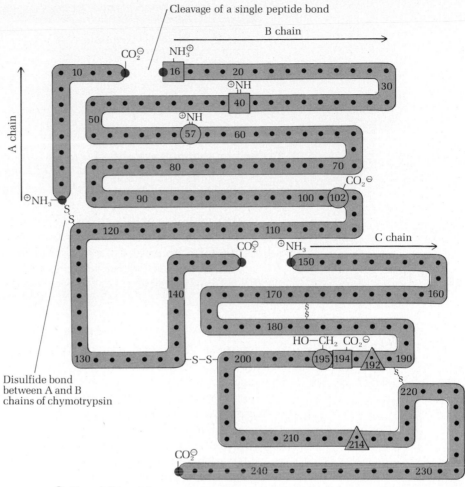

Figure 6.23 Activation of chymotrypsinogen. Chymotrypsinogen in converted to active chymotrypsin by cleavage of a single peptide bond between arginine 15 and isoleucine 16. [Courtesy of Dr. B. S. Hartley.]

trypsin, an ionic bond forms between the free ammonium group of isoleucine 16 and the carboxylate side chain of aspartate 194. The ionic bond is required for catalytic activity. The structure of chymotrypsinogen precludes substrate binding. The active site is not accessible to substrate until the critical cleavage between arginine 15 and isoleucine 16 occurs.

B. Activation of Other Pancreatic Zymogens

Trypsinogen, the *procarboxypeptidases*, and *proelastase* are also synthesized in the pancreas (Table 6.5). Trypsin is the common activator of the pancreatic zymogens, but to perform this function trypsinogen must itself be activated. The enzyme responsible for activation of trypsinogen is *enteropeptidase*, a proteolytic enzyme secreted by the duodenum. This enzyme acts in the small intestine, so the regulation of protease activity occurs at a site remote from the site of pancreatic zymogen synthesis. Enteropeptidase excises a hexapeptide whose net charge is −3 from trypsinogen, producing trypsin. Trypsin activates itself and the other pancreatic zymogens.

The effect of a small amount of enteropeptidase is dramatically ampli-

Table 6.5 Pancreatic zymogens		
Site of synthesis	**Zymogen**	**Active enzyme**
Pancreas	Chymotrypsinogen	Chymotrypsin
Pancreas	Trypsinogen	Trypsin
Pancreas	Procarboxypeptidase	Carboxypeptidase
Pancreas	Proelastase	Elastase

fied by the later activity of trypsin (Figure 6.24). The activation of the pancreatic zymogens is a *cascade* of protease activity. Each newly activated enzyme has the enormous reactivity associated with enzymes, and the ability of the system to hydrolyze proteins rapidly increases as each new zymogen is activated. The active proteases have short lifetimes. They self-destruct by hydrolyzing themselves and one another. The combination of enteropeptidase activation of trypsinogen, autocatalytic activation of trypsinogen, and autocatalytic self-destruction provides a highly self-regulating system of proteolytic activity.

6–5 CONFORMATIONS OF CHYMOTRYPSIN, TRYPSIN, AND ELASTASE

The mechanisms of action of chymotrypsin, trypsin, and elastase are nearly identical. X-ray crystallography reveals that their conformations are also nearly identical, differing only by extraneous loops or deletions on their surfaces (Figure 6.25). The differences in specificity among the three enzymes result from differences in their binding pockets for the side chains of the amino acid residues of the substrates for the three enzymes. Chymotrypsin has a hydrophobic pocket that binds well to nonpolar side chains in its substrates. Replacement of serine 189 in the bottom of this pocket with an aspartate residue in trypsin produces the binding site for the positively charged lysine and arginine side chains of trypsin's substrates. Substitution of valine and threonine residues for two glycines at the entrance to the hydrophobic pocket of chymotrypsin produces a shallow pocket that accommodates the side chains of alanine and glycine in the substrates for elastase. These serine proteases are the result of *divergent evolution* from a primordial protease whose catalytic mechanism has been conserved, but whose specificity has changed considerably.

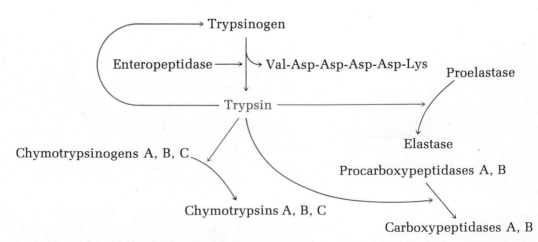

Figure 6.24 Activation of the inactive pancreatic zymogens chymotrypsinogen, procarboxypeptidases, and proelastase by trypsin.

Chymotrypsin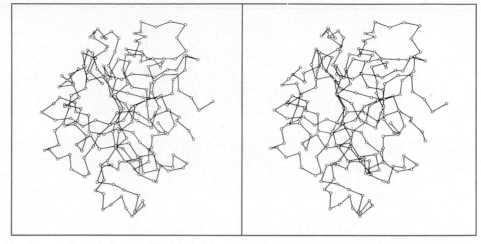

Elastase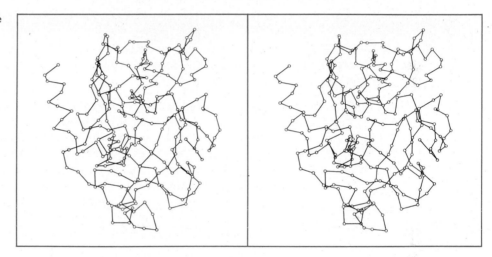

Trypsin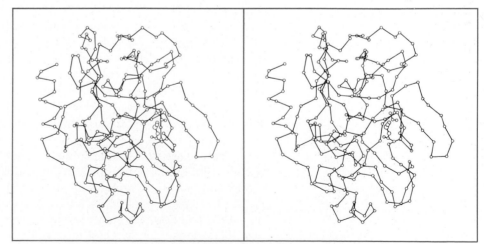

Figure 6.25 Stereo views of chymotrypsin, elastase, trypsin, and (on page 220) superimposed structures of chymotrypsin and elastase (color).

Chymotrypsin superimposed on elastase

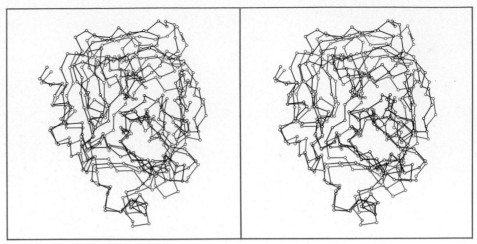

Figure 6.25 *(Continued)*

6–6 CLASSIFICATION OF ENZYME REACTIONS

An international commission on enzymes has devised a systematic nomenclature for enzymes in which all enzyme-catalyzed reactions are divided into six classes.

1. Oxidoreductases: catalyze oxidation-reduction reactions.
2. Transferases: catalyze transfer of a chemical group from one substrate to another or from one part of a substrate to another.
3. Hydrolases: catalyze hydrolysis reactions.
4. Lyases: catalyze addition of groups to double bonds or catalyze cleavage by electronic rearrangements.
5. Isomerases: catalyze isomerizations.
6. Ligases: catalyze bond formation and are accompanied by the hydrolysis of adenosine triphosphate (ATP) or a similar triphosphate.

Alongside this systematic nomenclature is a "common" nomenclature which is in wide use. In this nomenclature the suffix *-ase* is generally added to a term which indicates the substrate of the reaction. For example, carboxypeptidase A, officially α-carboxypeptide aminoacidohydrolase, catalyzes the cleavage of a peptide bond at the C-terminus of a peptide (as discussed in Section 6.2C). However, many of the names in the common nomenclature bear no obvious relationship to the reaction catalyzed. *Catalase*, for example, catalyzes the disproportionation of hydrogen peroxide giving water and molecular oxygen—a function that we would not be likely to infer from the name alone.

Enzymes can also be classified by their mechanisms of action. Consider the following topics: nucleophilic substitution, nucleophilic addition to carbonyl compounds, hydration of alkenes, elimination of water to give alkenes, reactions of enolate anions, aldol and Claisen condensations (Table 6.6). These reactions are catalyzed by enzymes using residues at their active sites, metal ions, certain prosthetic groups, such as heme, and coenzymes. The biochemistry of the additional, nonprotein molecules that share responsibility for catalysis with active-site amino acid residues will be introduced in Chapter 10.

6–7 SUMMARY

Enzymes are the protein catalysts of cells. They are extremely stereospecific, give no side products, and accelerate reaction rates enormously. Since enzymes are true catalysts, they do not alter the equilibrium constants of the reactions that they catalyze. Enzymes lower the activation energies for

**Table 6.6 Classification of
enzyme-catalyzed reactions by reaction mechanism**

Reaction	Example	Text reference
1. Nucleophilic displacement (including addition-elimination)		
A. 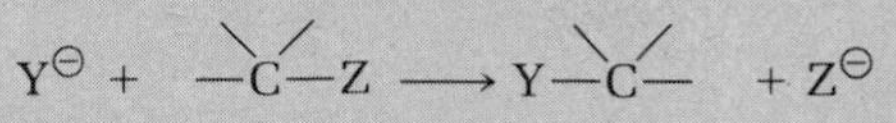 $Y^{\ominus}$ + —C—Z ⟶ Y—C— + $Z^{\ominus}$	Glycosidases	Section 8.4
B. 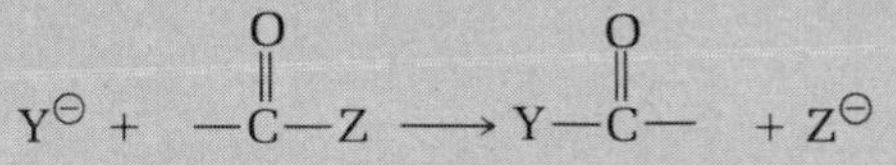 $Y^{\ominus}$ + —C—Z ⟶ Y—C— + $Z^{\ominus}$ (C=O)	Esterases	Chapter 6
C. 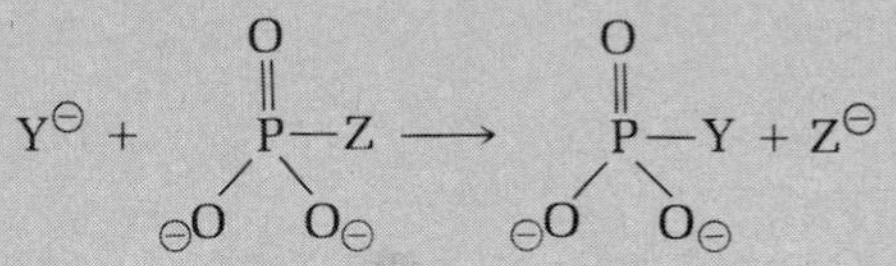 $Y^{\ominus}$ + P—Z ⟶ P—Y + $Z^{\ominus}$	Phosphatases, phosphomutases, phosphokinases	Chapter 17
2. Addition to C=C, C=O, and C=N		
A. 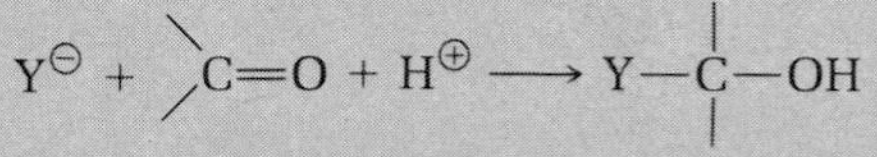 $Y^{\ominus}$ + C=O + $H^{\oplus}$ ⟶ Y—C—OH	Mutarotase	Section 8.3
B. 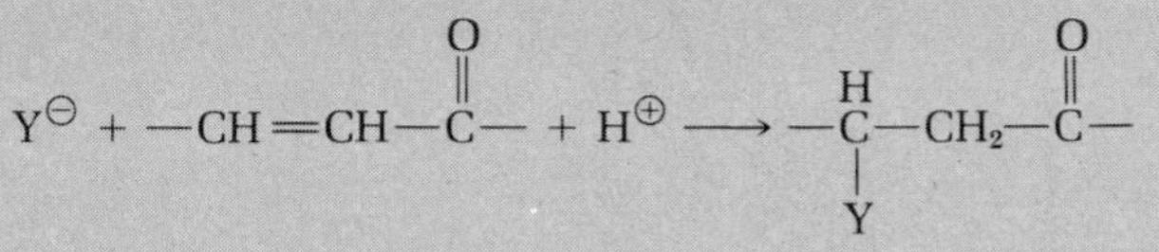 $Y^{\ominus}$ + —CH=CH—C— + $H^{\oplus}$ ⟶ —C—CH$_2$—C—	Fumarase	Section 15.3G
3. Elimination		
—C—C— ⟶ C=C + H_2O (H, OH)	D-β-Hydroxyacyl-ACP-dehydratase	Section 19.7C
4. Formation of enolate anions and enamines and their reactions		
A. Enolate anion formation (one step of complex mechanism)		
—C—C— + $B^{\ominus}$ ⟶ —C—C— ⟷ C=C— + BH	Enolase	Section 14.4E
B. Enamine formation (one step of complex mechanism)		
—C—C— + H_2N—R ⇌ —C—C— + B: ⇌ —C—C— ⟷	Aldolases	Section 14.3D
C=C + :NH$_2$ → C=C— + $BH^{\oplus}$		
C. Isomerization via enediol		
—C—C— ⟶ —C=C— ⇌ —C—C—OH	Sugar isomerases	Section 14.3B
D. Nucleophilic displacement of enolate anion nucleophiles		
—C=C + —C—Y + $H^{\oplus}$ ⇌ —C—C—C— + $YH^{\oplus}$	Thiolase	Section 19.3B

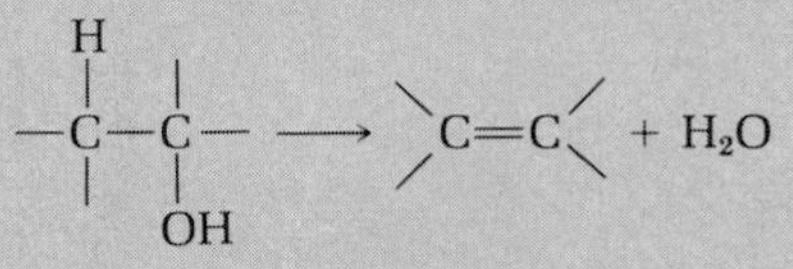

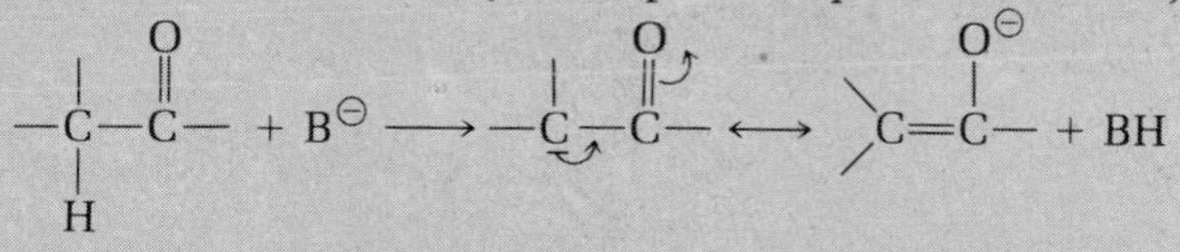

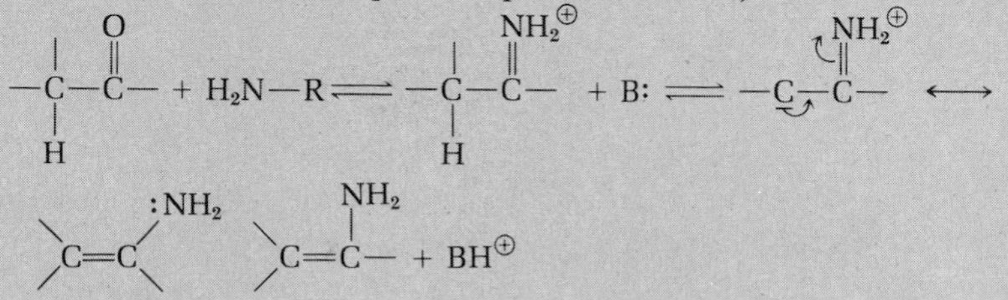

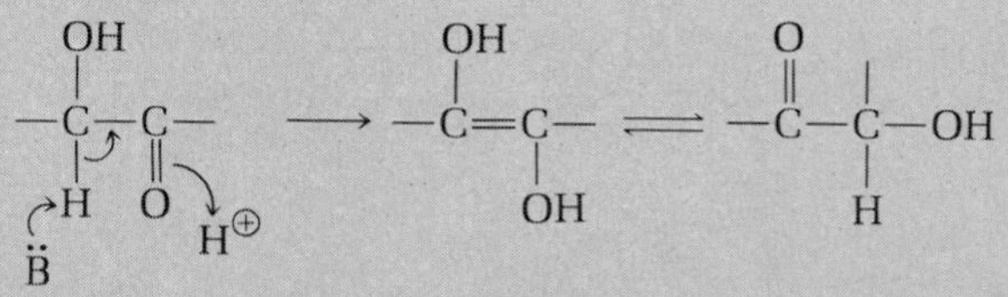

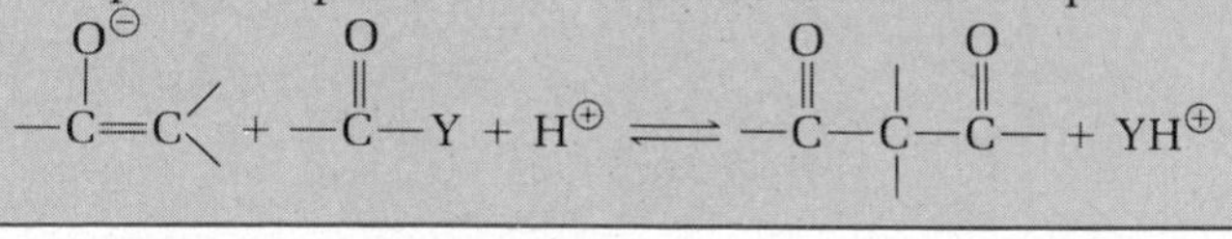

(continued)

Table 6.6 Classification of enzyme-catalyzed reactions by reaction mechanism (continued)

Reaction	Example	Text reference
E. Aldol condensations $-\overset{\overset{O^{\ominus}}{|}}{C}=C\diagdown \ +\ \overset{\overset{O}{\|}}{C}\ +\ H^{\oplus}\ \longrightarrow\ -\overset{\overset{O}{\|}}{C}-\overset{|}{C}-\overset{|}{C}-OH$	Aldolases	Sections 14.3D and 17.3A
F. β-Carboxylation $-\overset{\overset{O^{\ominus}}{|}}{C}=C\diagdown \ +\ CO_2\ \longrightarrow\ -\overset{\overset{O}{\|}}{C}-\overset{|}{C}-CO_2^{\ominus}$	Phosphoenolpyruvate carboxylase	Section 17.3A

chemical reactions by forming stereospecific enzyme-substrate complexes. The interaction of the enzyme with the substrate in the transition state of the reaction is greater than the interaction of the enzyme with the substrate in the ground state.

Bringing the enzyme and substrate together at the active site with the correct orientation for reaction is a major factor in enzymic catalysis. This contribution to catalysis is primarily an entropy effect. The unfavorable entropy change that accompanies substrate binding to an enzyme's active site is offset by a favorable enthalpy change. Enzymes often possess weakly acidic or weakly basic amino acid residues at their active sites. These residues contribute to catalysis by donating protons to the substrate, accepting protons from the substrate, or both, in the transition state of the reaction. This phenomenon is known as general acid-base catalysis.

Electrophiles are provided at some enzymes' active sites by metal ions. These ions act as "electron sinks" and function as Lewis acids, or electron pair acceptors, in many enzyme-catalyzed reactions. Carboxypeptidease A and carbonic anhydrase are two examples of enzymes that contain metal ions at their active sites. Metal ion catalysis is one type of covalent catalysis.

The side chain residues of several polar amino acids are nucleophiles. Nucleophilic catalysis is a type of covalent catalysis in which the enzyme becomes covalently bound to the substrate at some point in the catalytic process. The serine proteases, including chymotrypsin, trypsin, and elastase, provide examples of covalent catalysis. In these examples and in the bacterial enzyme subtilisin a charge relay network enhances the nucleophilicity of a seryl residue at the enzymes' active sites. The seryl residue becomes covalently bound to the carbonyl carbon of ester and amide substrates during the course of the reaction.

The mammalian serine proteases are secreted as inactive zymogens that are activated by a cascade of enzyme activity after their delivery to the small intestine.

REFERENCES

Boyer, P. D., Ed., "The Enzymes," 3rd ed., Vols. 1–13, Academic Press, New York, 1970–1976.

Cold Spring Harbor Symp. Quant. Biol., 36 (1972). (A discussion of the structure and function of enzymes.)

Fersht, A., "Enzyme Structure and Mechanism," Freeman, San Francisco, 1977.

Gutfreund, H., "Enzymes: Physical Principles," Wiley, New York, 1972.

Jencks, W. P., "Catalysis in Chemistry and Enzymology," McGraw-Hill, New York, 1969.

Kassell, B., and Kay, J., Zymogens of Proteolytic Enzymes, *Science*, 180, 1022–1027 (1973).

Kirsch, J. F., Mechanism of Enzyme Action, *Ann. Rev. Biochem.*, 42, 205–234 (1973).

Kraut, J., Serine Proteases: Structure and Mechanism of Catalysis, *Ann. Rev. Biochem.*, 46, 331–358 (1977).

Walsh, C., "Enzymatic Reaction Mechanisms," Freeman, San Francisco, 1979.

Problems

1. Acetylcholinesterase catalyzes the reaction

$$(CH_3)_3\overset{\oplus}{N}-CH_2-CH_2-O-\overset{\overset{\displaystyle O}{\|}}{C}-CH_3 + H_2O \longrightarrow$$

Acetyl choline

$$(CH_3)\overset{\oplus}{N}-CH_2-CH_2-OH + {}^{\ominus}O-\overset{\overset{\displaystyle O}{\|}}{C}-CH_3$$

Choline

The enzyme is inactivated by DIFP. What amino acid side chain is likely to be blocked by DIFP? Suggest a role for this side chain in catalysis. How would you proceed to identify the residue inactivated by DIFP?

2. Carboxypeptidase is a proteolytic enzyme that is synthesized as an inactive zymogen. Dogfish procarboxypeptidase A is converted to active carboxypeptidase A by trypsin. Procarboxypeptidase A is 2 to 25% as active as carboxypeptidase itself. At high concentrations of substrate the active enzyme is inhibited by substrate, but the zymogen is not. Thus at high substrate concentration the zymogen is more active than the so-called active enzyme! Explain the advantage that results from zymogen activity.

3. Fluoride ion is an inhibitor of many serine esterases and proteases. What aspect of the structure of these enzymes is likely to be responsible for this inhibition? (Hint: see Figure 6.21.)

4. Tos-Phe-CH₂Cl reacts at the active site of chymotrypsin. How would you alter the structure of this reagent to design an affinity labeling reagent for trypsin?

5. Ficin is a proteolytic enzyme that is isolated from plants. When the ficin-catalyzed hydrolysis of O-methylthiohippurate was studied it was found that a new absorbance appeared at 313 nm. This absorbance then slowly disappeared. The data are shown below. What is the implication of these data in terms of covalent catalysis?

Time (min)	1	4	17	26	29	32	35
A_{313}	0.38	0.40	0.41	0.31	0.28	0.24	0.23

6. The amino acid sequences at the active sites of several proteases are shown below. Using these data, what catalytic mechanisms would you expect for these enzymes vis à vis chymotrypsin?

Trypsin (bovine):	NSCQGDSGGPVVCSGK
Chymotrypsin (bovine):	SSCMGDSGGPLVCK
Trypsin (pig):	NSCQGDSGGPVVCGQ

7. Since the active sites of enzymes consist of only a small fraction of the total number of residues in a given enzyme, why are enzymes so large?

8. The following data have been obtained for the effects of metal ions

on the activity of carboxypeptidase. What is the principal conclusion to be reached from these data?

Ionic radii of some of the metal ions that can combine with carboxypeptidase A		
Ion(II)	**% Peptidase activity**	**Ionic radius (Å)**
Zn	100	0.74
Co	160	0.72
Ni	106	0.68
Mn	8	0.80
Cu	0	0.96
Hg	0	1.10
Cd	0	0.97
Pb	0	1.21

From B. L. Vallee, R. J. P. Williams, and J. E. Coleman, *Nature,* 190, 633 (1961). Reprinted by permission. Copyright © 1961 Macmillan Journals.

9. The rate of base-catalyzed hydrolysis of glycine ethyl ester increases by a factor of 2×10^6 in the presence of ethylenediamine $_2(Co^{3+})$.

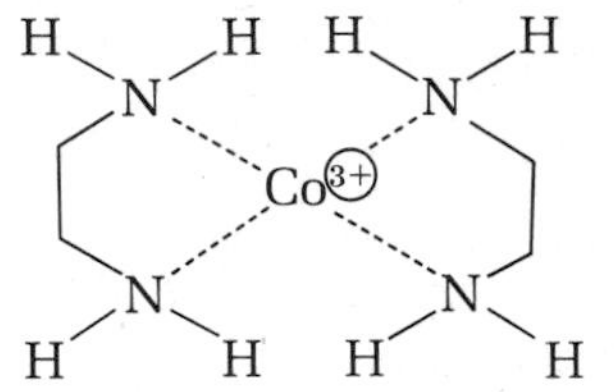

How does Co^{3+} facilitate the reaction? In what respect is this similar to carboxypeptidase?

10. Leucine aminopeptidase catalyzes the hydrolysis of peptide bonds. Relative rate data are given below for three substrates. (a) What accounts for the difference in rate between L-Leu-L-Leu and D-Leu-Gly? (b) What accounts for the difference in rate between L-Leu-L-Leu and Gly-L-Leu?

Substrate	**Relative rate**
L-Leu-L-Leu	100
D-Leu-Gly	0
Gly-L-Leu	10

11. When the papain-catalyzed hydrolysis of *O*-methylthiohippurate was studied the graph shown below was obtained. Interpret this graph in terms of the mechanism of action of papain.

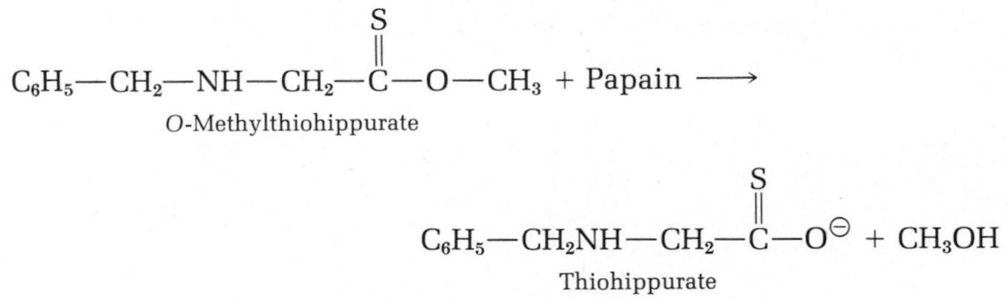

$$C_6H_5-CH_2-NH-CH_2-\overset{\overset{\displaystyle S}{\|}}{C}-O-CH_3 + Papain \longrightarrow$$
O-Methylthiohippurate

$$C_6H_5-CH_2NH-CH_2-\overset{\overset{\displaystyle S}{\|}}{C}-O^{\ominus} + CH_3OH$$
Thiohippurate

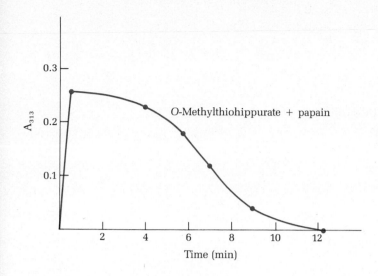

12. Further study of papain showed that there is a cysteine sulfhydryl group at the active site. Write a chemical reaction using this sulfhydryl group that is in agreement with your postulate about the mechanism in problem 11.

13. The rate of reaction of papain with O-methylthiohippurate esters is nearly independent of the leaving group. The same esters are also hydrolyzed by hydroxide ion, where the rate differences are considerable. Interpret the following rate data in light of your postulated mechanisms from problems 11 and 12.

	Relative rate	
O-**Methylthiohippurate esters**	**Enzyme**	**OH**$^\ominus$
p-Nitrophenyl	1.4	170
Phenyl	1.25	18
Methyl	1.0	1

14. Suggest an experimental method that will permit you to distinguish between the tautomers at the active site given in text equation 6.18?

15. An enzyme called cyclic-AMP-dependent protein kinase contains two types of subunits, one of which, called the catalytic subunit, is inactivated by N-tosyl-L-lysyl-chloromethyl ketone. A cysteine sulfhydryl group of the enzyme is involved in this inactivation. Write the structure of the adduct formed by reaction of a cysteine sulfhydryl group with the above reagent.

16. The hydrolysis of acetylsalicylic acid (aspirin) is 100 times faster than the hydrolysis of phenylacetate. Provide a mechanistic explanation and explain how your answer is related to enzyme catalysis.

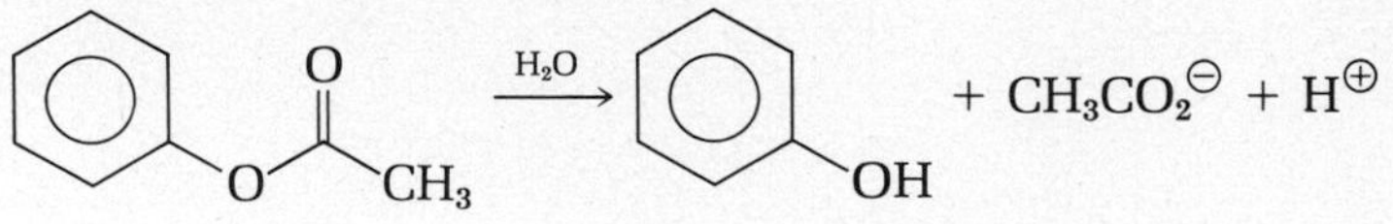

Phenylacetate

17. The compound shown below is an irreversible inhibitor of chymotrypsin. Propose a mechanism for inhibition.

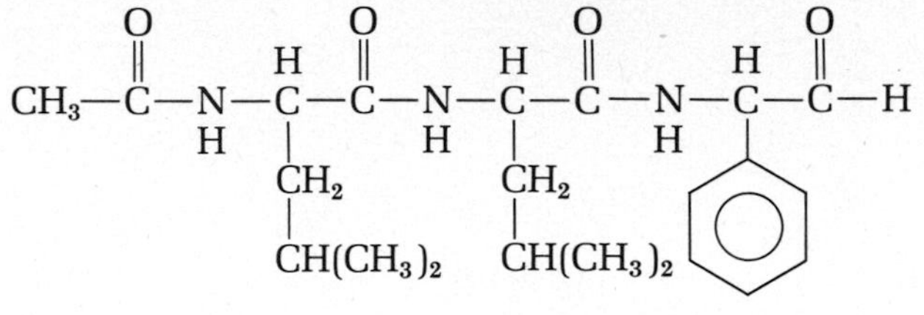

Chapter 7

Enzyme kinetics

7–1 INTRODUCTION

Enzyme kinetics deals with the rates of enzyme-catalyzed reactions. Kinetic studies of enzymatic reactions provide information about the mechanism of the catalytic reaction, insight into enzyme specificity, and several parameters that characterize the physical properties of the enzyme.

The practical applications of enzyme kinetics are widespread. Kinetic assays *in vitro* are widely used in clinical laboratories. An abnormal rate can provide a clue to a pathological condition. Kinetic assays *in vivo* (see, for example, Section 14.6) provide information about the rates not only of individual enzymes, but of entire metabolic pathways, thus enabling us to extend our knowledge of enzymes in isolated systems to living organisms.

Enzyme inhibition, a topic that we shall examine in some detail, is also enormously important in enzymology. The study of enzyme inhibitors helps us to understand how enzymes work. By systematic modification of the structure of the substrate, the active site of an enzyme or an allosteric binding site can be "mapped." Once the detailed chemistry of the active site has been deciphered, it becomes possible to carry out the rational design of inhibitors with therapeutic value.

7-2 ELEMENTARY KINETICS

In a chemical reaction that occurs with no change in the volume of the system, the *reaction rate* or velocity, v, equals the change in concentration of product per unit time. Consider the reaction

$$A + 2B \longrightarrow 3C \tag{7.1}$$

For equation 7.1 the rate is defined as

$$v \equiv -\frac{d[A]}{dt} = -\frac{1}{2}\frac{d[B]}{dt} = +\frac{1}{3}\frac{d[C]}{dt} \tag{7.2}$$

The expression $d[\ ]/dt$ is the infinitesimal change in reactant or product concentration occurring over an infinitesimal time period. The minus signs in front of $d[A]/dt$ and $d[B]/dt$ indicate that the concentrations of these species are decreasing with time. Similarly, the plus sign in front of $d[C]/dt$ indicates that the concentration of C is increasing. The change in concentration of any species with time is divided by the coefficient of that species in the balanced chemical equation.

For reactions occurring in one step, called *elementary reactions*, the rate is a function of the concentrations of one or more reagents. The relationship between reaction rate and reagent concentration defines the *order* of the reaction. Consider the reaction

$$A \longrightarrow P \tag{7.3}$$

If the rate is directly proportional to the concentration of A, then the reaction is *first order* in A.

$$v \propto [A]^1 \tag{7.4}$$

The constant of proportionality is the *specific rate constant*, k_1.

$$v = -k_1[A] \tag{7.5}$$

Equation 7.5, which relates the rate of the reaction to the concentration of reactant, is the *rate law* for the reaction. The time required for one-half of A to react is called the *half-life* for the reaction, τ. For a first order reaction

$$\tau = \frac{0.693}{k_1} \tag{7.6}$$

where k_1 is the *first order rate constant* for the reaction. The half-life of a first order reaction is independent of the concentration of reactant.

For the reaction

$$A + B \longrightarrow P \tag{7.7}$$

the rate depends upon the concentrations of A and B. The rate law is

$$v = -k_2[A]^1[B]^1 \tag{7.8}$$

where k_2 is the *second order rate constant*. The reaction is first order in A, first order in B, and second order overall. Doubling the concentration of A doubles the rate, doubling the concentration of B doubles the rate, and doubling the concentrations of A and B increases the rate fourfold.

Some reactions are independent of the concentration of reactants and they are *zero order*. The rate law for a zero order reaction is

$$v = \text{constant} = k_0 \tag{7.9}$$

7–3 MICHAELIS-MENTEN RATE LAWS

Systematic studies of the effect of substrate concentration upon enzyme activity were begun late in the nineteenth century. The concept of the enzyme-substrate complex was introduced in 1882. Although subject to debate at the time—no enzymes had been purified, and it was not even known that enzymes were proteins—the idea of the enzyme-substrate complex proved critical to the development of enzyme kinetics.

A. Initial Rate of Reaction

If we follow the appearance of product (or the disappearance of substrate in a reaction with 1:1 stoichiometry) as a function of time in an experiment, we obtain the *progress curve* for the reaction (Figure 7.1). *The initial rate of the reaction is equal to the slope of the progress curve at time zero.* The initial rate, or initial velocity, is designated by the symbol v_0. As the reaction continues, the product continues to accumulate, but at ever slower rates as the supply of substrate diminishes. Initial rate studies are therefore carried out only to 1 or 2% completion so that the concentration of substrate remains essentially constant during the experiment. By using initial rates the reverse reaction need not be considered, the enzyme is not inhibited to a significant extent by products, and a much simpler rate law can be used.

The effects of varying initial concentrations of substrate on the initial rate of reaction are determined by carrying out a series of kinetic experiments. A set of plots such as the one in Figure 7.1 are obtained, and the data are replotted in a graph of v_0 versus S_1, S_2, S_3, . . . S_n (Figure 7.2). For experiments in which the initial substrate concentration is low, the rate is directly proportional to substrate concentration, and the reaction is first order in substrate. As the substrate concentration increases, the order gradually decreases until the rate is independent of substrate concentration. The rate of reaction, then, depends upon the fraction of enzyme bound to substrate. At high initial substrate concentrations the enzyme is *saturated* with substrate, that is, every active site is occupied. Increasing the substrate concentration cannot increase the rate. Hence the reaction is zero order, and the rate is at a maximum.

B. Derivation I: Rapid Equilibrium Assumption

In 1913 Michaelis and Menten proposed a kinetic scheme that explains the results of Figure 7.2. (Henri had made much the same proposal in 1902, but, unaccountably, it went mostly unnoticed.) They proposed that the reaction

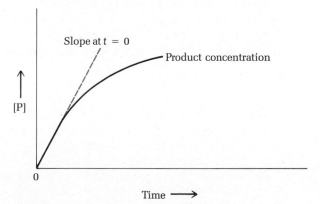

Figure 7.1 Progress curve for the enzymatic conversion of substrate S to product P.

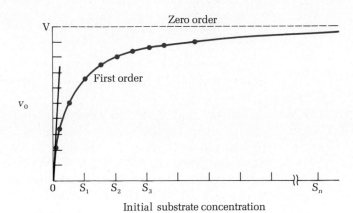

Figure 7.2 Effects of varying initial substrate concentrations on initial rate of reaction. An enzyme-catalyzed reaction with a single substrate approaches first order in substrate at low concentrations of substrate and zero order at high substrate concentrations. The rate curve has the shape of a rectangular hyperbola overall. Each point in the plot is obtained from the slope of a plot such as the one shown in Figure 7.1.

proceeds through an enzyme-substrate complex that forms rapidly and is then slowly converted to product in the *rate-determining* step of the reaction. The kinetic scheme is

$$E + S \underset{}{\overset{K_s}{\rightleftharpoons}} ES \overset{k_{cat}}{\longrightarrow} E + P$$

Scheme 1

The major assumption of scheme 1 is that the enzyme and the substrate remain in thermodynamic equilibrium with the enzyme-substrate complex at all times. K_s is the *dissociation constant* for the enzyme-substrate complex, sometimes called the *Michaelis complex*. K_s is defined as

$$K_s = \frac{[E][S]}{[ES]} \tag{7.10}$$

where [E] is the concentration of free enzyme, [S] is the concentration of the substrate, and [ES] is the concentration of the enzyme-substrate complex.

In most kinetic experiments the concentration of enzyme is much less than the concentration of substrate, and the rate of reaction is directly proportional to the total enzyme concentration, $[E]_t$. We shall make this assumption in the rate laws that we derive. The rate of the reaction depends upon the magnitude of the *catalytic rate constant*, k_{cat}, and upon the concentration of the enzyme-substrate complex. Thus,

$$v = k_{cat}[ES] \tag{7.11}$$

Michaelis-Menten kinetics are formulated for *initial rates*. Since only a small amount of product accumulates initially, we can ignore the reverse reaction of the second step in scheme 1.

The total enzyme concentration is related to the concentration of free enzyme and the concentration of the enzyme-substrate complex by the *conservation equation* (7.12).

$$[E]_t = [E] + [ES] \tag{7.12}$$

Thus,

$$[E] = [E]_t - [ES] \tag{7.13}$$

Substituting for [E] and solving for [ES] in equation 7.10, we find that

$$[ES] = \frac{[E]_t[S]}{K_s + [S]} \tag{7.14}$$

Substituting equation 7.14 in 7.11, we obtain

$$v = \frac{k_{cat}[E]_t[S]}{K_s + [S]} \tag{7.15}$$

The general form of equation 7.15 applies to many enzyme-catalyzed reactions, but the *assumption* of thermodynamic equilibrium is not applicable in all cases, and it must be verified experimentally for each specific enzyme before it is accepted as true.

C. Derivation II: Steady State Assumption

Let us rewrite scheme 1 in terms of the *microscopic rate constants* for the formation and breakdown of the enzyme-substrate complex (scheme 1a).

$$E + S \underset{k_{-1}}{\overset{k_1}{\rightleftharpoons}} ES \overset{k_2}{\longrightarrow} E + P$$

Scheme 1a

In the rapid equilibrium assumption of derivation I, $k_{-1} \gg k_2$. If, however, the rate of conversion of the enzyme-substrate complex to product is comparable to the rate of its dissociation, the rapid equilibrium assumption is not valid. We now assert that the concentration of enzyme-substrate complex will be small and *constant*, provided that $[E]_t \ll [S]$. This assertion is known as the *steady state assumption*. Figure 7.3 shows the change in concentration of enzyme substrate and enzyme-substrate complex versus time. Within milliseconds of mixing enzyme and substrate, the concentration of enzyme-substrate complex reaches a steady state. Since the concentration of enzyme-substrate complex is constant, *its rate of change is zero*. That is,

$$\frac{d[ES]}{dt} = 0 \tag{7.16}$$

The enzyme-substrate complex is formed by one pathway, the combination of E and S, and it disappears by two pathways, either the reversion to reactants or the conversion to product. Thus

$$\frac{d[ES]}{dt} = 0 = k_1[E][S] - k_2[ES] - k_{-1}[ES] \tag{7.17}$$

Substituting equation 7.13 in equation 7.17 and rearranging, we obtain

$$[ES] = \frac{[E]_t[S]}{\left(\dfrac{k_{-1} + k_2}{k_1}\right) + [S]} \tag{7.18}$$

Substituting equation 7.18 in equation 7.11 yields the rate law for the reaction.

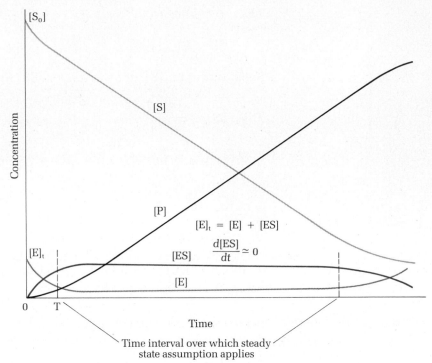

Figure 7.3 Concentration of substrate, free enzyme, product, and enzyme-substrate complex over time. After an initial burst, the concentration of enzyme-substrate complex is almost constant, and the steady state assumption is applicable.

$$v = \frac{k_2 [E]_t [S]}{\left(\dfrac{k_{-1} + k_2}{k_1}\right) + [S]} \tag{7.19}$$

The composite term $(k_{-1} + k_2)/k_{-1}$ in the denominator of equation 7.19 is referred to as the *Michaelis constant*, K_m.

$$K_m = \frac{k_{-1} + k_2}{k_1} \tag{7.20}$$

Comparing equations 7.19 and 7.15, we see that they have the same form. K_m is related to K_s by equation 7.21

$$K_m = K_s + \frac{k_2}{k_1} \tag{7.21}$$

where $K_s = k_{-1}/k_1$. $K_m \cong K_s$ only when $k_{-1} \gg k_2$, and K_m is *not* generally equivalent to a dissociation constant for the enzyme-substrate complex.

Let us rewrite equation 7.19 in terms of K_m.

$$v = \frac{k_2 [E]_t [S]}{K_m + [S]} \tag{7.22}$$

For any given enzyme reaction $[E]_t$, k_2, and K_m are constants. Let us consider two limiting cases.

First, for low substrate concentrations, $[S] \ll K_m$, equation 7.22 reduces to

$$v \cong \left(\frac{k_2 [E]_t}{K_m}\right) [S] \tag{7.23}$$

Since the terms in parentheses in equation 7.23 are all constants the rate law is

$$v \cong k_{obs}[S] \tag{7.24}$$

and the reaction approaches first order as we noted in Figure 7.2.

Second, at high substrate concentrations, $[S] \gg K_m$, equation 7.22 reduces to

$$v \cong k_2[E]_t \tag{7.25}$$

Both k_2 and $[E]_t$ are constant. We define the maximum rate of reaction, V,[1] as

$$V = k_2[E]_t \tag{7.26}$$

Since the maximum rate of reaction is independent of substrate concentration, the reaction is zero order, conforming to the results in Figure 7.1. The specific rate constant (k_2) is the catalytic rate constant (k_{cat}). Using our definition of V, we can rewrite equation 7.22 as

$$v = \frac{V[S]}{K_m + [S]} \tag{7.27}$$

Equation 7.27 is the *Michaelis-Menten rate law*. When the substrate concentration equals K_m, equation 7.27 becomes

$$v = \tfrac{1}{2} V \tag{7.28}$$

The value of K_m, then, is that substrate concentration which gives half-maximal reaction velocity. The enzyme is half-saturated when $[S] = K_m$.

K_m is an important kinetic parameter for several reasons. (1) The value of K_m characterizes the interaction of an enzyme with a given substrate. We might perhaps be tempted to equate enzyme specificity with the value of K_m, but we shall see in Section 7.8 that the value of k_{cat}/K_m is actually the criterion for enzyme specificity. The values of K_m for some glycolytic enzymes (Chapter 14) are listed in Table 7.1. These data show that K_m has different values from tissue to tissue. (2) Many of the enzymes of Table 7.1 have K_m values that are near the physiological substrate concentrations. The rate of reaction is extremely sensitive to substrate concentration near K_m, and slight changes in the concentrations of a few substrates can change the rate of an entire metabolic pathway. (3) The allosteric effectors of regulatory enzymes can alter the rate of reaction by changing the value of K_m for a substrate. Most allosteric regulatory enzymes have altered K_m values in the presence of allosteric effectors.

D. Derivation III: Reactions in Which a Covalent Enzyme-Substrate Intermediate Forms

In Section 6.3 we saw that in the chymotrypsin-catalyzed hydrolysis of esters a covalent enzyme substrate intermediate forms, and covalent intermediates are observed for many other enzymes. The general *form* of the Michaelis-Menten rate law is obeyed in such reactions, but the kinetic constants become more complex. Let us consider the formation of a single covalent enzyme-substrate intermediate, ES′ (scheme 1b).

$$E + S \underset{k_{-1}}{\overset{k_1}{\rightleftharpoons}} ES \overset{k_2}{\underset{X}{\longrightarrow}} E\text{–}S' \overset{k_3}{\longrightarrow} E + P$$

Scheme 1b

[1] The abbreviation V_{max} is often encountered in the literature of enzyme kinetics. The Commission on Biochemical Nomenclature (1973) has recommended that V_{max} be replaced by V.

Table 7.1 Values of K_m and physiological concentrations of metabolites for some glycolytic enzymes

Enzyme	Source	Substrate	Concentration (μM)	K_m (μM)
Glucose phosphate isomerase	Brain	G6P	130	210
	Muscle	G6P	450	700
		F6P	110	120
Aldelase	Brain	FDP	200	12
	Muscle	FDP	32	100
		G3P	3	1000
		DHAP	50	2000
Triosephosphate isomerase	Erythrocyte	G3P	18	350
	Muscle	G3P	3	460
		DHAP	50	870
Glyceraldehydephosphate dehydrogenase	Brain	G3P	3	44
	Muscle	G3P	3	70
		NAD	600	46
		P_i	2000	
Phosphoglycerate kinase	Brain	1,3DPG	<1	9
		ADP	1500	70
	Erythrocyte	3PG	118	1100
	Muscle	3PG	60	1200
		ADP	600	350
Phosphoglyceromutase	Brain	3PG	40	240
	Muscle	3PG	60	5000
Enolase	Brain	2PG	4.5	33
	Muscle	2PG	7	70
Pyruvate kinase	Erythrocyte	PEP	23	200
		ADP	138	600
Lactate dehydrogenase	Brain	Pyr	116	140
	Erythrocyte	Pyr	51	59
		Lac	2900	8400
		NADH	0.01	10
		NAD	33	150
Glycerophosphate dehydrogenase	Mouse	GlyP	170	37
	Muscle	GlyP'	220	190
		DHAP	50	190

SOURCE: From A. Fersht, "Enzyme Structure and Mechanism," Freeman, San Francisco, 1977, p. 256.

The steady state equation for ES is

$$\frac{d[ES]}{dt} = k_1[E][S] - (k_{-1} + k_2)[ES] = 0 \tag{7.29}$$

and the steady state equation for E$-$S$'$ is

$$\frac{d[ES']}{dt} = k_2[ES] - k_3[E-S'] = 0 \tag{7.30}$$

The total enzyme concentration is

$$[E]_t = [E] + [ES] + [E-S'] \tag{7.31}$$

Substituting for [ES] and [E$-$S$'$] in equation 7.30 gives

$$[E]_t = [ES]\left[\frac{k_{-1} + k_2}{k_1[S]} + 1 + \frac{k_2}{k_3}\right] \tag{7.32}$$

The rate of formation of product is

$$v = k_2[ES] \tag{7.33}$$

Solving equation 7.32 for [ES] and rearranging, we obtain the steady state rate law

$$v = [E]_t[S] \left[\frac{(k_2 k_3)/(k_2 + k_3)}{(K_s k_3)/(k_2 + k_3) + [S]} \right] \tag{7.34}$$

Equation 7.34 has the same form as the Michaelis-Menten equation we derived above where

$$K_m = K_s \frac{k_3}{k_2 + k_3} \tag{7.35}$$

and

$$k_{cat} = \frac{k_2 k_3}{k_2 + k_3} \tag{7.36}$$

A consideration of the three widely different forms of the Michaelis-Menten rate law that we have derived in this section illustrates a general principle of great importance: the form of the rate law implies nothing about the mechanism of action of the enzyme and, conversely, quite different enzyme mechanisms can be kinetically indistinguishable.

7–4 TURNOVER NUMBERS AND UNITS OF ENZYME ACTIVITY

In the preceding section we saw that when an enzyme is saturated with substrate, $K_m \ll [S]$,

$$V = k_{cat}[E]_t$$

When the enzyme concentration is expressed in terms of moles of active sites per liter (the molar concentration of enzyme times the number of catalytic sites per enzyme molecule), the first order rate constant k_{cat} is the *turnover number* for the enzyme. The units of k_{cat} are time^{-1}, or the time required for the enzyme to "turn over" a molecule of substrate. The range of values and the magnitudes of k_{cat} for a few enzymes are summarized in Table 7.2.

7–5 EFFECT OF pH ON RATES OF ENZYMATIC REACTIONS

The activities of most enzymes vary with pH because both the active sites of enzymes and their substrates often contain acidic or basic functional groups whose state of ionization is a function of pH. A few examples of the effects of pH on enzyme activity are given in Figure 7.4. The kinetic parameters k_{cat}, K_m, and k_{cat}/K_m can all be affected by pH. As may be seen from Figure 7.4 the types and magnitudes of pH effects differ from enzyme to enzyme.

Table 7.2 Values of k_{cat} for some enzymes	
Enzyme	**k_{cat} (s^{-1})**
Acetylcholinesterase	1.4×10^4
Carbonic anhydrase	1×10^6
Catalase	4×10^7
Chymotrypsin	1×10^2
DNA polymerase I	1.5×10^1
Lactate dehydrogenase	1×10^3
Lysozyme	5×10^{-1}
Penicillinase	2×10^3

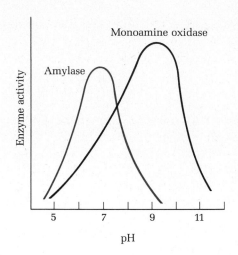

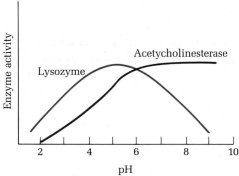

Figure 7.4 Effect of pH on the activity of four enzymes: amylase, monoamine oxidase, lysozyme, and acetylcholinesterase.

Let us consider the effects of pH on kinetic scheme 2.

$$E + S \underset{}{\overset{K_s}{\rightleftharpoons}} ES \xrightarrow{k_{cat}} E + P$$

$$K_{EH} \updownarrow \pm H^{\oplus} \qquad \pm H^{\oplus} \updownarrow K_{ESH}$$

$$EH + S \underset{}{\overset{K'_s}{\rightleftharpoons}} HES$$

Scheme 2

The major types of kinetic observations resulting from analysis of scheme 2 are:

1. If the pH-dependent process is the rate of conversion of enzyme-substrate complex to product, k_{cat} will be affected. The value of k_{cat} in such cases depends upon the pK_a of the enzyme-substrate complex, pK_{aESH}.

2. K_m changes as a function of pH when the dissociation of the enzyme-substrate complex is pH-dependent.

3. When k_{cat}/K_m changes as a function of pH, the rate depends upon the pK_a's of both free enzyme and free substrate. If the substrate does not ionize, the value of k_{cat}/K_m reflects the pK_a of the active site of the *free enzyme* in most cases.

7–6 EFFECT OF TEMPERATURE ON RATES OF ENZYMATIC REACTIONS

The rates of most chemical reactions increase with increasing temperature, and enzyme-catalyzed reactions are no exception. For most elementary reactions the activation energy lies in the range of 0 to 300 kJ mol⁻¹. The rates of very fast reactions having activation energies near zero are virtually independent of temperature, a situation that changes rapidly as the activation energy increases. In more "typical" reactions, if the activation energy is about 50 kJ mol⁻¹, an increase in temperature from 25 to 35°C approximately doubles the rate. If the activation energy is about 85 kJ mol⁻¹, a similar 10° increase about triples the rate. The ratio of the rate constants of two reactions 10 K apart is called the Q_{10} of a reaction. The Q_{10} values for enzyme-catalyzed reactions vary over a range of approximately 1.7 to 2.5. Temperature must therefore be controlled to $\pm 0.1°C$ to obtain reproducible kinetic results.

Figure 7.5 shows the effect of temperature on typical catalyzed and uncatalyzed reactions. The rate of the enzymatic reaction rises to a maximum and then rapidly declines because heat denatures enzymes, and the rate of denaturation increases with increasing temperature.

7–7 TRANSITION STATE THEORY AND ENZYMATIC CATALYSIS

Our qualitative description of transition state theory in the previous chapter will suffice for most of the discussion throughout this text. Yet, there are cases in which a more thorough understanding of transition state theory is helpful. Thus, in this section we shall extend our consideration of transition theory and then apply our analysis to enzyme catalysis.

Let us first consider the uncatalyzed conversion of S to P in a single step. In transition state theory S is assumed to be in equilibrium with the transition state for the reaction (S‡) which then converts to product with rate constant k_r.

$$S \underset{}{\overset{K^{\ddagger}}{\rightleftharpoons}} S^{\ddagger} \overset{k_r}{\longrightarrow} P \tag{7.37}$$

$K^{\ddagger}$ is the equilibrium constant for the interconversion of S and S‡, and is defined as

$$K^{\ddagger} = \frac{[S^{\ddagger}]}{[S]} \tag{7.38}$$

The rate constant k_r of reaction 7.37 is called the *absolute rate constant*. Transition state theory assumes that all transition states are converted to product with the same absolute rate constant. The theory further assumes that reactants are converted to products by the conversion of a vibration in the transition state to a translation. It can be shown that

$$k_r = \frac{k_b T}{h} \tag{7.39}$$

where k_r is the absolute rate constant, k_b is the Boltzmann constant, and h is Planck's constant. The absolute rate constant is 6×10^{-12} s⁻¹, approximately the rate constant for a bond vibration. The rate of the reaction is given by equation 7.40.

$$v = k_r[S^{\ddagger}] \tag{7.40}$$

Substituting for k_r using equation 7.39 and for [S‡] using equation 6.38, we obtain

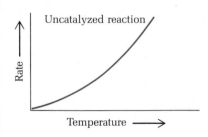

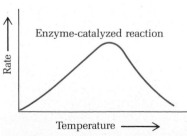

Figure 7.5 Effect of temperature on the reaction rate for an uncatalyzed reaction and an enzyme-catalyzed reaction.

$$v = \frac{k_b T}{h} K^{\ddagger}[S] \tag{7.41}$$

We recall from Chapter 1 that the equilibrium constant for a reaction is related to the standard free-energy change by the relationship

$$\Delta G^0 = -RT \ln K_{eq} \tag{7.42}$$

and that

$$K_{eq} = e^{-\Delta G^0/RT} \tag{7.43}$$

We now make an explicit analogy between the equilibrium constant $K^{\ddagger}$ and the free energy of activation for the reaction $\Delta G^{\ddagger}$. Hence,

$$\Delta G^{\ddagger} = -RT \ln K^{\ddagger} \tag{7.44}$$

and

$$K^{\ddagger} = e^{-\Delta G^{\ddagger}/RT} \tag{7.45}$$

Therefore, the rate of the reaction is

$$v = \frac{k_b T}{h} e^{-\Delta G^{\ddagger}/RT}[S] \tag{7.46}$$

The experimentally determined rate is

$$v = k_1[S] \tag{7.47}$$

where k_1 is the first order rate constant for the reaction. Equating the theoretical rates, we obtain the specific rate constant k_1 in terms of the parameters of transition state theory.

$$k_1 = \frac{k_b T}{h} e^{-\Delta G^{\ddagger}/RT} \tag{7.48}$$

These ideas are directly applicable to enzyme catalysis. Consider the following scheme.

$$E + S \underset{}{\overset{K_s}{\rightleftharpoons}} ES \underset{}{\overset{K^{\ddagger}}{\rightleftharpoons}} ES^{\ddagger} \overset{k_{cat}}{\longrightarrow} E + P$$

Scheme 1c

By applying transition state theory we conclude that

$$k_{cat} = \frac{k_b T}{h} e^{-\Delta G^{\ddagger}/RT} \tag{7.49}$$

The catalytic rate constants for enzyme-catalyzed reactions are many orders of magnitude greater than the rate constants for the corresponding uncatalyzed reactions because the enzyme lowers the free energy of activation for the reaction. We shall see how the catalytic rate constant is related to the free energy of activation and the Michaelis-Menten constant in the next section.

7–8 THE IMPORTANCE OF k_{cat}/K_m

The ratio k_{cat}/K_m is of fundamental importance in discussions of enzyme kinetics and in theories of enzyme catalysis. In our analysis of the Michaelis-Menten rate law in the preceding section we saw that when $K_m \gg [S]$,

$$v = \frac{k_{cat}}{K_m}[E]_t[S] \tag{7.50}$$

In equation 7.50 k_2 of kinetic scheme 1a is equated with the catalytic rate constant k_{cat}. The ratio k_{cat}/K_m is an *apparent* second order rate constant, not a true microscopic rate constant. At sufficiently low substrate concentrations only a small fraction of the total enzyme is bound to substrate. Thus, at low substrate concentrations

$$v = \frac{k_{cat}}{K_m}[E][S] \tag{7.51}$$

and the reaction rate is proportional to *free enzyme*.

Enzyme specificity is also related to the ratio k_{cat}/K_m. Suppose that two substrates are competing for the same enzyme, a common situation in cells. Let us consider the general case (reactions 7.52 and 7.53).

$$E + A \rightleftharpoons EA \longrightarrow E + P \tag{7.52}$$

$$E + B \rightleftharpoons EB \longrightarrow E + Q \tag{7.53}$$

The rates of these two reactions are given by equations 7.54 and 7.55.

$$v_A \cong \frac{k_{cat}^A[E][A]}{K_m^A} \qquad \text{when } [A] \ll K_m^A \tag{7.54}$$

$$v_B \cong \frac{k_{cat}^B[E][B]}{K_m^B} \qquad \text{when } [B] \ll K_m^B \tag{7.55}$$

The ratio of the two rates, v_A/v_B, expresses the relative ability of the enzyme to catalyze the conversion of A and B to products P and Q.

In other words v_A/v_B is a measure of enzyme specificity. When the concentrations of A and B are equal, the ratio of rates is

$$\frac{v_A}{v_B} = \frac{k_{cat}^A/K_m^A}{k_{cat}^B/K_m^B} \tag{7.56}$$

Equation 7.56 tells us that enzyme specificity depends upon both k_{cat} and K_m, not upon K_m alone.

The value of k_{cat}/K_m cannot be greater than any true microscopic second order rate constant on the forward reaction pathway in the catalytic mechanism. The fastest possible reaction is one that is controlled by the diffusion of the substrate to the enzyme's active site and not by any subsequent chemical step. The rates of a few enzymes seem to be diffusion controlled. Acetylcholine esterase, for example, has a k_{cat}/K_m value of $1.6 \times 10^8 \, s^{-1} \, M^{-1}$, close to the diffusion-controlled limit of a second order rate constant.

Let us consider the relationship between the ratio k_{cat}/K_m and theories of enzymatic catalysis. We have stated that the substrate and enzyme are complementary to each other in the transition state of the reaction. The goal of efficient catalysis is to lower $\Delta G^{\ddagger}$. The maximization of k_{cat} is achieved at

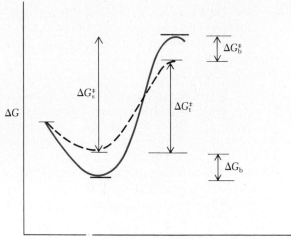

Reaction coordinate

Figure 7.6 Free-energy changes for tight and weak binding of a substrate. Tight binding places the enzyme-substrate complex in an energy well that increases the free energy of activation, $\Delta G^{\ddagger}$, for the reaction. $\Delta G_s^{\ddagger}$ is the activation energy if the substrate is complementary to the active site in its ground state conformation; the smaller activation energy, $\Delta G_t^{\ddagger}$ is the activation energy if the transition state conformation of the substrate is complementary to the transition state conformation of the enzyme. The increment $\Delta G_b^{\ddagger}$ is the energy difference between weak and tight binding. This is also the amount by which the energy of the transition state is lowered if the substrate binds weakly to the enzyme.

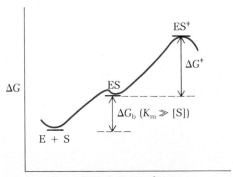

Reaction coordinate

Figure 7.7 When the substrate binds the enzyme in the transition state of the reaction, k_{cat} will be maximized if part of the binding energy in the enzyme-substrate complex is used to distort the substrate toward its transition state conformation. If the value of the Michaelis constant is greater than the substrate concentration ($K_m \gg [S]$), the enzyme-substrate complex is less stable than the isolated reactants, fulfilling the requirement for a low free energy of activation and a large value of k_{cat}.

the expense of weak binding. Tight binding ($K_m \ll [S]$) decreases the catalytic efficiency of the enzyme, lowering k_{cat}, because the free energy of binding (ΔG_b) pulls the enzyme-substrate complex into so deep a potential energy well that it can climb out only by surmounting a high free energy of activation (Figure 7.6). If, however, the substrate binds weakly ($K_m \gg [S]$), then formation of the enzyme-substrate complex pulls the substrate part way along the reaction pathway, lowering $\Delta G^{\ddagger}$ and increasing the catalytic rate constant (Figure 7.7). Part of the energy of binding in the latter case increases k_{cat}/K_m without lowering K_m.

For example, elastase (Section 6.5) catalyzes the hydrolysis of polypeptides. Increasing the chain length of the substrate from Ac-Ala-Pro-Ala-NH$_2$ to Ac-Pro-Ala-Pro-Ala-NH$_2$ lowers K_m from 4.2 to 3.9 mM, but it increases k_{cat} from 0.09 to 8.5 s^{-1}. The value of k_{cat}/K_m increases from 21 s^{-1} M^{-1} to 2,200 s^{-1} M^{-1}. The increased binding energy for the larger substrate is used to increase k_{cat}, not to lower K_m. Those enzymes that have values of k_{cat}/K_m that have reached the diffusion-controlled limit for a second order rate constant are said to be *perfectly evolved enzymes*. Acetylcholinesterase falls into this category, as do carbonic anhydrase (Section 6.2C), triosephosphate isomerase (Section 14.3E), and several others.

7–9 KINETIC ISOTOPE EFFECTS

Enzymes catalyze many reactions in which C—H bonds are broken. Nicotinamide adenine dinucleotide dehydrogenases and flavin-dependent dehydrogenases (Section 10.3) catalyze such reactions. If the slow, rate-determining step of these reactions is cleavage of the C—H bond, substitution of deuterium (^{2}H or D) for hydrogen typically decreases the reaction by factors between 2 and 15. This decrease in rate is called a *deuterium isotope effect*. It is expressed as the ratio of rate constants for the slow step for the hydrogen and deuterium isotopes, k_H/k_D. The k's, then, are the catalytic rate constants of our kinetic schemes. *Tritium isotope effects*, k_H/k_T, are roughly twice as large as deuterium isotope effects.

Kinetic isotope effects reflect differences in the zero-point vibration en-

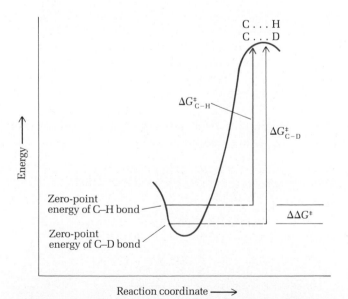

Figure 7.8 Kinetic isotope effect. The isotope effect reflects the differences in zero-point energies of the C—H and C—D bonds in the ground state which vanish in the transition state as a vibration is converted to a translation.

ergies of the C—H, C—D, and C—T bonds.[2] The zero-point energy of the C—D bond is about 4.8 kJ mol^{-1} lower than the zero-point energy of the C—H bond. In the transition state of a reaction a vibration is converted to a translation as the bond breaks, and the zero-point energy disappears for that particular bond. Thus the energies of the transition states for C—H and C—D cleavage are the same, but the free energy of activation is less for the C—H bond than for the C—D bond, and the rate is faster for the C—H compound (Figure 7.8). A difference in zero-point energies of 4.8 kJ mol^{-1} gives a *kinetic isotope effect* (k_H/kD) of about 7.

Horse liver alcohol dehydrogenase stereospecifically catalyzes the transfer of an hydride anion from ethanol to NAD (reaction 7.52).

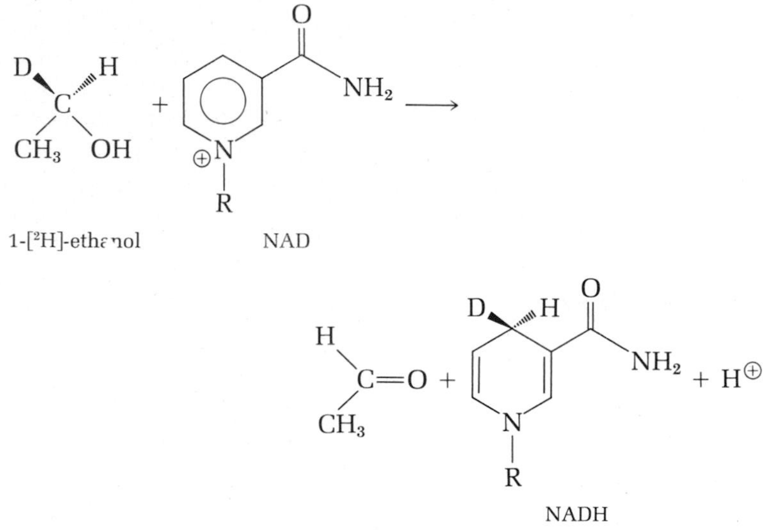

$$1\text{-}[^2H]\text{-ethanol} \qquad NAD$$

$$\tag{7.52}$$

$$NADH$$

The value of k_H/k_D for this reaction is 3.11, indicating that C—H bond cleavage is the rate-determining step of the reaction. (Further kinetic details of horse liver alcohol dehydrogenase are presented in Section 7.12.)

7–10 GRAPHICAL REPRESENTATION OF KINETIC DATA

Since 1913 numerous attempts have been made to represent kinetic data graphically for the Michaelis-Menten equation. The rate approaches V asymptotically, and all graphical procedures must grapple with the problem of finding V accurately. If V is not known accurately, K_m cannot be determined accurately, since $K_m = [S]$ at $v = \frac{1}{2}V$.

Two linear transformations of the Michaelis-Menten equation are widely used. The *Lineweaver-Burk* or double reciprocal plot is obtained by taking the reciprocal of the Michaelis-Menten equation (equation 7.57).

$$\frac{1}{v} = \frac{K_m + [S]}{V[S]} \tag{7.57}$$

Rearranging equation 7.57, we obtain

$$\frac{1}{v} = \frac{1}{[S]}\frac{K_m}{V} + \frac{1}{V} \tag{7.58}$$

Equation 7.58 has the form $y = mx + b$. A graph of $1/v$ versus $1/[S]$ has a slope of K_m/V, a $1/v$ intercept of $1/V$, and a $1/[S]$ intercept of $-1/K_m$

---

[2] At 25°C only the lowest, or zero-point, energies are appreciably populated; otherwise the problem would be much more complicated.

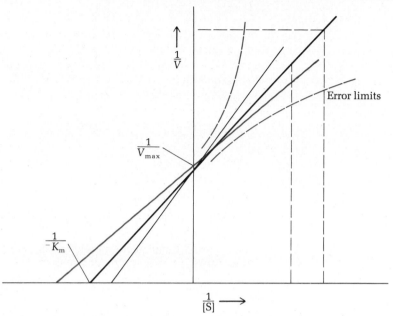

Figure 7.9 Double reciprocal or Lineweaver-Burk plot. Kinetic data may be treated by making a plot of $1/v$ versus $1/[S]$. The intercept on the $1/[S]$ axis equals $-1/K_m$, and the intercept on the $1/v$ axis equals $1/V$. The disadvantage of this graphical procedure is that small experimental errors in the determination of $[S]$ and v result in large errors in K_m.

(Figure 7.9). The Lineweaver-Burk equation suffers from the disadvantage that a small experiment error can lead to a large error in the graphically determined values of K_m and V. Its principal advantage is its "legibility"; the relation between velocity and substrate concentration is clear at a glance. Weighed against this advantage is another defect: the values of data obtained at high substrate concentrations are squeezed into a narrow region near the $1/v$ axis. Because reciprocals of the data are calculated, errors are difficult to analyze by Lineweaver-Burk plots.

A second widely used linear form is the *Eadie-Hofstee* plot which is fitted by equation 7.59.

$$\frac{v}{[S]} = -\frac{v}{K_m} + \frac{V}{K_m} \tag{7.59}$$

A plot of $v/[S]$ versus v has a $v/[S]$ intercept of V/K_m, a v intercept of V, and a slope of $-1/K_m$ (Figure 7.10). Eadie-Hofstee plots do not compress the data at high substrate concentrations, but the relation between velocity and substrate concentration is less obvious than in the Lineweaver-Burk plot. Since the Eadie-Hofstee plot involves only one reciprocal, it is more accurate than a Lineweaver-Burk plot. Both of these plots are useful, but both require plotting reciprocals of experimental data, which inevitably reduces the accuracy of the K_m and V values obtained.

In 1974 Eisenthal and Cornish-Bowden introduced a simple graphical treatment of kinetic data known as the *direct linear plot*. The Michaelis-Menten equation is rearranged to the form

$$V = v + \frac{v}{[S]} K_m \tag{7.60}$$

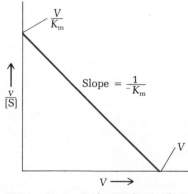

Figure 7.10 Eadie-Hofstee plot. A plot of $v/[S]$ versus v has a slope of $-1/K_m$ and intercepts of V and V/K_m on the v and $v/[S]$ axes, respectively. This plot involves only one reciprocal, but care must be taken in the determination of $[S]$, since a small error is magnified when any reciprocal is taken.

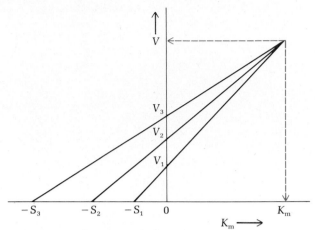

Figure 7.11 Direct linear plot. This graphical procedure provides an accurate and rapid means of determining K_m and V from experimental data. A given kinetic experiment is characterized by an observed reaction rate, or velocity, and an initial substrate concentration.

The rearranged Michaelis-Menten rate law is then treated graphically as illustrated in Figure 7.11 and described in the original paper of Eisenthal and Cornish-Bowden:

> Set up axes, K_m and v, corresponding to the familiar x and y axes, respectively. For each observation (s,v) [the measured substrate concentration and the measured velocity] mark off the points $K_m = -s$ on the K_m axis [we recall that K_m has units of concentration], and $V = v$ on the v axis, and draw a line through the two points, extending it into the first quadrant. When this is done for all observations, the lines intersect at a common point, whose coordinates (K_m, V) provide the values of K_m and V that satisfy the Michaelis-Menten equation exactly for every observation.[3]

The direct linear plot provides us with another convenient and useful feature: the analysis of experimental error.[4] In most experiments the intersection of lines in the first quadrant in Figure 7.11 does not occur at a single point, but over a small range. K_m is the median value along the K_m axis, and V the median value along the V axis. The range of values along the two axes gives the error in K_m and V (Figure 7.12).

7–11 ENZYME INHIBITION

Those metabolites, drugs, and probes of enzyme specificity which reversibly inhibit enzymes are of fundamental importance in biochemistry. The three main types of inhibition—competitive, uncompetitive, and noncompetitive—can all be analyzed in terms of the Michaelis-Menten rate law.

A. Competitive Inhibition

A competitive inhibitor binds reversibly to an enzyme and denies the substrate access to the active site. The kinetic scheme for competitive inhibition is shown in Scheme 3.

---

[3] R. Eisenthal and A. Cornish-Bowden, *Biochem. J.*, *139*, 116 (1974).

[4] Errors can be analyzed in Lineweaver-Burk and Eadie-Hofstee plots by use of a statistical technique called "weighted least squares."

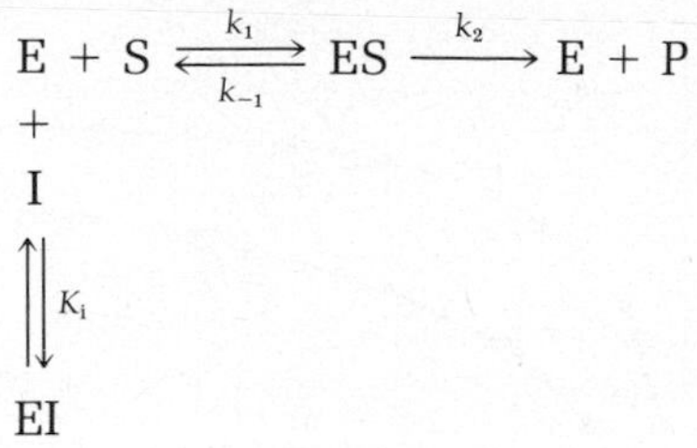

E + S
k_1
k_{-1}
ES
k_2
E + P
+
I
K_i
EI

Scheme 3

In scheme 3 k_1, k_{-1}, and k_2 are microscopic rate constants; the equilibrium constant K_i is the dissociation constant for the enzyme-inhibitor complex, defined in equation 7.61.

$$K_i = \frac{[E][I]}{[EI]} \tag{7.61}$$

The rate of the enzyme-catalyzed reaction leading to product P is directly dependent upon the concentration of the enzyme-substrate complex.

$$v = k_2[ES] \tag{7.62}$$

The total enzyme concentration is now partitioned into three species: free enzyme, enzyme-substrate complex, and enzyme-inhibitor complex.

$$[E]_t = [E] + [ES] + [EI] \tag{7.63}$$

Using equations 7.61, 7.62, and 7.63, the steady state rate law for competitive inhibition is obtained.

$$v = \frac{V[S]}{[S] + K_m\left(1 + \frac{[I]}{K_i}\right)} \tag{7.64}$$

V
True V
True K_m
$-S_3$
$-S_2$
$-S_1$
0
K_m

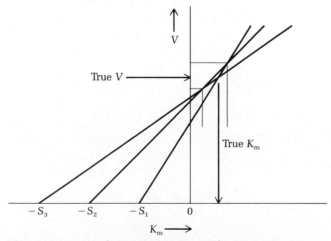

Figure 7.12 Analysis of experimental error with direct linear plot. The error in K_m and V is determined by taking the median value of the range over which the lines intersect. The extremes in either direction give the error.

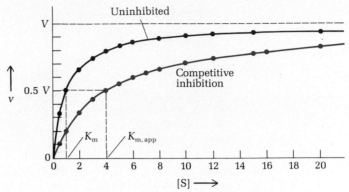

Figure 7.13 Plot of v versus [S] in the presence and absence of a competitive inhibitor. Competitive inhibition can be overcome at high concentrations of substrate, and the velocity (v) approaches V. $K_{m,app}$, obtained graphically, equals $K_m(1 + [I]/K_i)$.

Equation 7.64 differs from the Michaelis-Menten rate law by the term $(1 + [I]/K_i)$ in the denominator. Since this term is always greater than 1.0, the *apparent* K_m, $K_{m,app}$ *increases* in the presence of inhibitor. $K_{m,app}$ is defined as

$$K_{m,app} = K_m \left(1 + \frac{[I]}{K_i}\right) \tag{7.65}$$

The rate law for competitive inhibition in terms of $K_{m,app}$ is

$$v = \frac{V[S]}{K_{m,app} + [S]} \tag{7.66}$$

Equation 7.66 has the same form as the Michaelis-Menten equation. A plot of v versus [S] for a reaction carried out in the presence and absence of a competitive inhibitor is shown in Figure 7.13, and a direct linear plot of the same data is shown in Figure 7.14. It is clear that the apparent K_m increases and that V remains constant. The values of K_m and $K_{m,app}$ are read directly off the direct linear plot, and the value of K_i can be calculated from these values using equation 7.65.

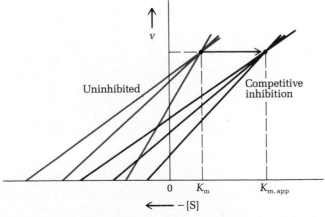

Figure 7.14 Direct linear plot for reaction subject to competitive inhibition. V remains constant, but the apparent value of the Michaelis constant increases.

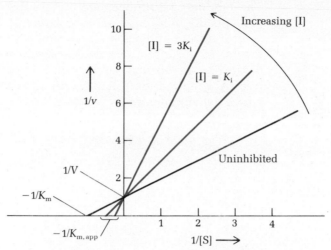

Figure 7.15 Lineweaver-Burk plot for reaction subject to competitive inhibition. The concentration of inhibitor in each plot is constant for all experiments.

A Lineweaver-Burk plot can also be used to determine the dissociation constant for the enzyme-inhibitor complex. Taking the reciprocal of equation 7.64 we obtain

$$\frac{1}{v} = \frac{K_m}{V}\left(1 + \frac{[I]}{K_i}\right)\frac{1}{[S]} + \frac{1}{V} \tag{7.67}$$

Kinetic experiments are carried out in the absence of inhibitor to provide values of V and K_m. Then, experiments are carried out at various fixed concentrations of inhibitor. When the data are plotted as $1/v$ versus $1/[S]$, a series of lines are obtained that intersect at a common point on the $1/v$ axis (Figure 7.15). The common point of intersection indicates that the inhibition is competitive. A competitive inhibitor increases the slope of the Lineweaver-Burk plot. The slope is

$$\text{slope} = \frac{K_m}{V}\left(1 + \frac{[I]}{K_i}\right) \tag{7.68}$$

where K_m and V are the values obtained in the absence of inhibitor. The $1/[S]$ intercept equals $-1/K_{m,app}$ where

$$-\frac{1}{K_{m,app}} = -\frac{1}{K_m\left(1 + \frac{[I]}{K_i}\right)} \tag{7.69}$$

We have already discussed a metabolite, 2,3-diphosphoglycerate, that exerts powerful physiological effects by binding to hemoglobin. It acts as a competitive inhibitor of *phosphoglycerate mutase* (Section 14.4C), the enzyme that interconverts 1,3-diphosphoglycerate, an intermediate in glycolysis, and 2,3-diphosphoglycerate as shown below.

$$
\begin{array}{c}
\overset{\displaystyle O}{\overset{\|}{C}}\!-\!O\!-\!PO_3^{2-} \\
|\\
H\!-\!C\!-\!OH \\
|\\
CH_2\!-\!OPO_3^{2-}
\end{array}
\quad\xrightarrow[\text{mutase}]{\text{Phosphoglycerate}}\quad
\begin{array}{c}
CO_2^{\ominus} \\
|\\
H\!-\!C\!-\!O\!-\!PO_3^{2-} \\
|\\
CH_2\!-\!OPO_3^{\ominus}
\end{array}
\tag{7.70}
$$

1,3-Diphosphoglycerate   2,3-Diphosphoglycerate

The product, 2,3-diphosphoglycerate, binds tightly to the active site, and prevents the binding of another molecule of 1,3-diphosphoglycerate.

Substrate analogs are also competitive inhibitors in many cases. Malonate, for example, is a potent competitive inhibitor of *succinate dehydrogenase*, an enzyme that functions in both the citric acid cycle and oxidative phosphorylation (Chapters 15 and 16). The reaction catalyzed by succinate dehydrogenase is

$$^{\ominus}O_2C-CH_2-CH_2-CO_2^{\ominus} + A \xrightarrow{\text{Succinate dehydrogenase}}$$
Succinate

$$\begin{array}{c} ^{\ominus}O_2C \quad\quad H \\ C=C \\ H \quad\quad CO_2^{\ominus} \end{array} + AH_2 \quad (7.71)$$
Fumarate

The hydrogen acceptor A is flavin adenine dinucleotide, a coenzyme discussed in Chapter 10. The structures of malonate and succinate differ by a single methylene group. Because of its close resemblance to succinate, malonate binds tightly to the active site of succinate dehydrogenase and acts as a competitive inhibitor.

$$^{\ominus}O_2C-CH_2CH_2-CO_2^{\ominus} \quad\quad ^{\ominus}O_2C-CH_2-CO_2^{\ominus}$$
Succinate Malonate

B. Transition State Theory, Transition State Analogs, and the Role of Strain in Catalysis

In Section 6.2 we introduced the idea that in catalysis enzymes are able to induce "distortion" or *strain* in the substrate toward the geometry it will assume in the transition state. If the transition state configuration of the substrate is complementary to the active site of the enzyme, then inhibitors whose configuration resembles the transition state configuration of the substrate should bind much more tightly to the enzyme than does the substrate itself. Such inhibitors are called *transition state analogs*. They are competitive inhibitors whose dissociation constants (K_i in scheme 3) reveal the role of strain in catalysis.

We recall from our discussion of chymotrypsin in Section 6.3 that the hydroxyl group of serine 195 becomes covalently bonded to the substrate in the first step of the reaction. The configuration of the transition state leading to the tetrahedral intermediate is presumably similar to the intermediate itself.

$$\begin{array}{c} O \\ \parallel \\ R-C-O-R' \end{array} + \text{enz-OH} \rightleftharpoons \left[\begin{array}{c} R \quad\quad \overset{\delta}{O}^{\ominus} \\ C \\ R'O \quad\quad O\text{-enz} \end{array} \right]^{\ddagger} \longrightarrow$$
Nearly tetrahedral
transition state

$$\begin{array}{c} R \quad\quad O^{\ominus} \\ C \\ R'O \quad\quad O\text{-enz} \end{array} \quad (7.72)$$
Tetrahedral intermediate

Stable molecules whose structure resembles the structure of the transition state should bind more tightly to the enzyme than planar molecules resembling the substrate, if strain is a significant factor in catalysis. This is, in fact, observed.

For example, phenylethane boronic acid can form stable tetrahedral adducts. Its boronic acid functional group is planar.

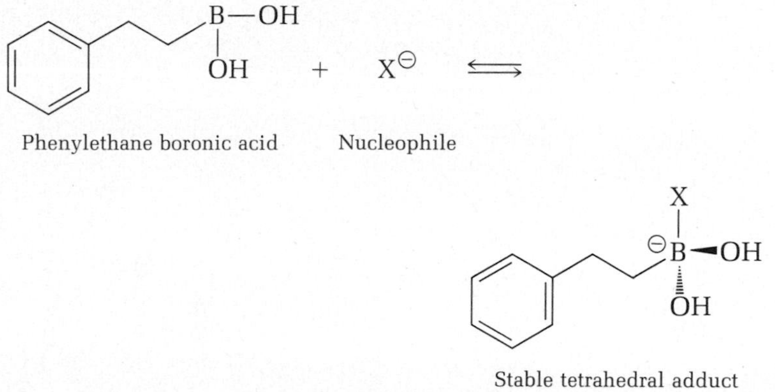

Phenylethane boronic acid   Nucleophile

$$(7.73)$$

Stable tetrahedral adduct

The serine hydroxyl group adds to the boron of phenylethane boronic acid to give a stable tetrahedral adduct that is bound very tightly to the active site (Figure 7.16). Thus the boronic acid is a transition state analog. The x-ray crystallographic picture partially explains why the boronic acid binds to the enzyme orders of magnitude more tightly than ester substrates. The negatively charged oxygen of the tetrahedral intermediate forms strong hydrogen bonds with the amide NH bonds of residues 193 and 195. This hydrogen-bonding interaction is much weaker in the planar ground state than in the tetrahedral adduct. By stabilizing metastable intermediates and the transition states leading to them, enzymes promote catalysis. The inhibitory ability of such transition state analogs is strong evidence for the role of strain in catalysis.

C. Uncompetitive Inhibition

An inhibitor that binds only to the enzyme-substrate complex, and not to the free enzyme, is an *uncompetitive inhibitor*. The binding order is *obligatory* because the required sequence of events is "first bind S, then bind I." The kinetic scheme for uncompetitive inhibition is scheme 4, in which k_1, k_{-1}, and k_2 are rate constants and K_{esi} is the dissociation constant for the enzyme-substrate-inhibitor complex (ESI).

$$E + S \underset{k_{-1}}{\overset{k_1}{\rightleftharpoons}} ES \xrightarrow{k_2} E + P$$
$$+$$
$$I$$
$$\updownarrow K_{esi}$$
$$ESI$$

Scheme 4

The kinetic scheme shows that V will be decreased by the inhibitor since a certain fraction of the enzyme-substrate complex will always be diverted by the inhibitor to the inactive ESI complex.

Since the inhibitor decreases the concentration of the enzyme-substrate complex, it also decreases the apparent K_m.

In deriving the rate law for uncompetitive inhibition we proceed as before. The rate of formation of product is

$$\text{rate} = k_2[ES] \tag{7.74}$$

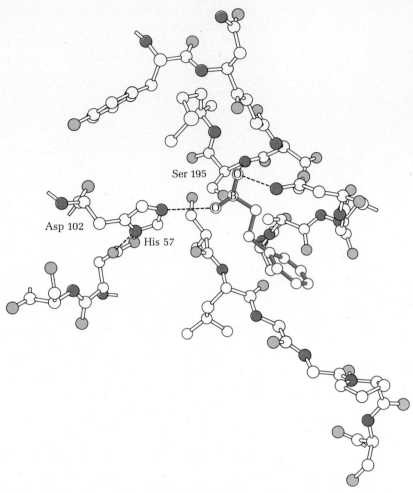

Figure 7.16 Transition state analog binding to chymotrypsin. Phenylethane boronic acid forms a tetrahedral adduct with chymotrypsin that
is similar in structure to the tetrahedral intermediate that forms as a
transient intermediate in the catalyzed reaction. The tetrahedral intermediate is stabilized by the oxyanion pocket. [Courtesy of Dr. G. E. Lienhard.]

The dissociation constant for the enzyme-substrate-inhibitor complex is

$$K_i = \frac{[\text{ES}][\text{I}]}{[\text{ESI}]} \tag{7.75}$$

The analytical enzyme concentration is

$$[\text{E}]_t = [\text{E}] + [\text{ES}] + [\text{ESI}] \tag{7.76}$$

Using equations 7.74, 7.75, 7.76 and the steady state approximation, we obtain the rate law for uncompetitive inhibition (equation 7.77).

$$v = \frac{V[\text{S}]}{K_m + [\text{S}]\left(1 + \frac{[\text{I}]}{K_i}\right)} \tag{7.77}$$

$K_{m,app}$ decreases in the presence of an uncompetitive inhibitor. V also
decreases because of the term $(1 + [\text{I}]/K_i)$ in the denominator of equation
7.77 (Figure 7.17). A direct linear plot for a reaction subject to uncompetitive

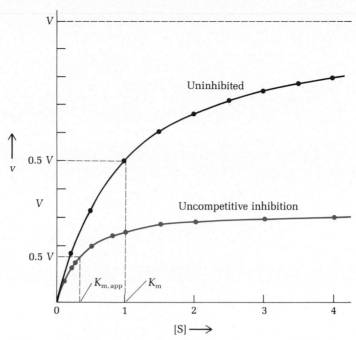

Figure 7.17 Plot of v versus [S] in the absence of inhibitor and in the presence of an uncompetitive inhibitor. Even at very high substrates concentrations the velocity of the inhibited reaction never approaches V.

inhibition is shown in Figure 7.18. The point of intersection in the first quadrant for the reaction in the presence of inhibitor lies on a straight line connecting the origin with the point of intersection for the reaction in the absence of inhibitor.

To obtain kinetic parameters for uncompetitive inhibition from a Lineweaver-Burk plot we first take the reciprocal of equation 7.77 which yields

$$\frac{1}{v} = \left(\frac{K_m}{V}\right)\frac{1}{[S]} + \frac{1}{V}\left(1 + \frac{[I]}{K_i}\right) \tag{7.78}$$

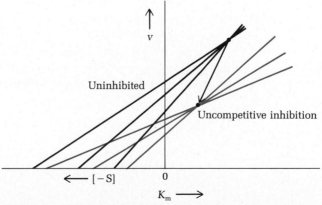

Figure 7.18 Direct linear plot for reaction subject to uncompetitive inhibition. Kinetic data reveal that both $K_{m,app}$ and V_{app} have decreased. The point of intersection in the first quadrant shifts directly toward the origin.

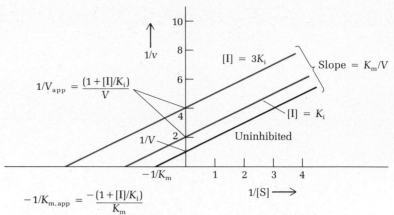

Figure 7.19 Lineweaver-Burk plot for reaction subject to uncompetitive inhibition. Repeated experiments at varying concentrations of substrate at various fixed inhibitor concentrations yield a set of parallel lines.

A plot of $1/v$ versus $1/[S]$ gives a straight line whose slope is K_m/V. An uncompetitive inhibitor, then, does not change the slope of a Lineweaver-Burk plot (Figure 7.19). The $1/v$ intercept is related to V by the relation

$$\frac{1}{V_{app}} = \frac{1 + \dfrac{[I]}{K_i}}{V} \tag{7.79}$$

and the $1/[S]$ intercept is given by equation 7.80.

$$\frac{1}{K_{m,app}} = \frac{1 + \dfrac{[I]}{K_i}}{K_m} \tag{7.80}$$

As in the case of competitive inhibition, the kinetic experiments are first carried out in the absence of inhibitor and then repeated for varying fixed concentrations of inhibitor. Uncompetitive inhibition is extremely rare for reactions having a single substrate, but is common for bisubstrate reactions with ping-pong mechanisms (to be discussed in Section 7.12D).

D. Noncompetitive Inhibition

Noncompetitive inhibition is more complex than either competitive or uncompetitive inhibition. A noncompetitive inhibitor can bind either to the free enzyme or to the enzyme-substrate complex, and it provides an example of a *random* binding order.

If the binding order is truly random, the dissociation constants for the EI complex and for the ESI complex will be the same. Scheme 5 shows the kinetic pattern observed for noncompetitive inhibition.

$$
\begin{array}{ccccc}
E + S & \underset{k_{-1}}{\overset{k_1}{\rightleftarrows}} & ES & \overset{k_2}{\longrightarrow} & E + P \\
+ & & + & & \\
I & & I & & \\
\Big\updownarrow K_i & & \Big\updownarrow K_i' & & \\
EI + S & \underset{k_{-1}}{\overset{k_1}{\rightleftarrows}} & ESI & &
\end{array}
$$

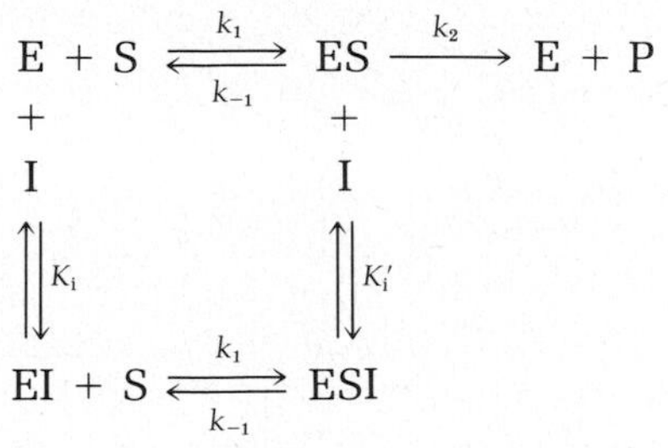

Scheme 5

The rate law for noncompetitive inhibition, $K_i = K_i'$, is

$$v = \frac{V[S]}{K_m \left(1 + \dfrac{[I]}{K_i} \right) + [S] \left(1 + \dfrac{[I]}{K_i} \right)} \tag{7.81}$$

Dividing equation 7.81 by the constant term in the denominator, $(1 + I/K_i)$,

$$v = \frac{V_{app}[S]}{K_m + [S]} \tag{7.82}$$

where

$$V_{app} = \frac{V}{1 + \dfrac{[I]}{K_i}} \tag{7.83}$$

Inspection of equations 7.82 and 7.83 reveals the fundamental kinetic behavior expected for noncompetitive inhibition: the value of $K_{m,app}$ is not altered by a noncompetitive inhibitor, and the apparent maximum rate of the reaction (V_{app}) decreases (Figure 7.20). A direct linear plot of data for a reaction subject to noncompetitive inhibition reveals that the K_m, V coordinate moves toward the K_m axis, parallel to the v axis (Figure 7.21).

The Lineweaver-Burk equation for noncompetitive inhibition is obtained by taking the reciprocal of equation 7.81.

$$\frac{1}{v} = \frac{K_m}{V} \left(1 + \frac{[I]}{K_i} \right) \frac{1}{[S]} + \frac{1}{V} \left(1 + \frac{[I]}{K_i} \right) \tag{7.84}$$

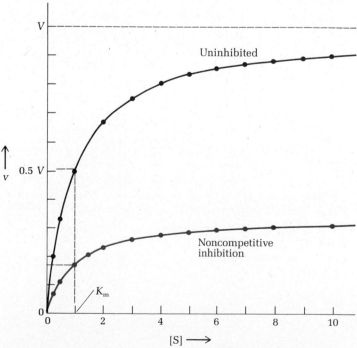

Figure 7.20 Plot of v versus [S] for reaction subject to noncompetitive inhibition. K_m is unaltered by the inhibitor, but V_{app} is lowered.

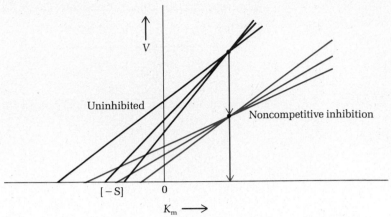

Figure 7.21 Direct linear plot for reaction subject to pure noncompetitive inhibition. V_{app} decreases, but $K_{m,app}$ is unaltered in the presence of inhibitor.

The reciprocal equation shows that both the slope and the $1/v$ intercept are increased by the factor $(1 + [I]/K_i)$ in the presence of inhibitor. Since, however, they both increase by the same factor, the $1/[S]$ intercept is not altered, and K_m is not altered by a noncompetitive inhibitor (Figure 7.22).

If $K_i = K_i'$ in scheme 5 then the intercept of a Lineweaver-Burk plot will be on the horizontal axis as shown in Figure 7.22. However, if K_i and K_i' differ the curves intersect at a different point to the left of the vertical axis and either above or below the horizontal axis. This more complex case is known as *mixed* inhibition. If $K_i > K_i'$, the intercept falls above the horizontal axis; if $K_i < K_i'$ the intercept falls below.

E. Summary of Inhibition for Unisubstrate Reactions

We have now seen three kinds of enzyme inhibition—competitive, uncompetitive, and noncompetitive—which all can be viewed as variations of the Michaelis-Menten rate law.

$$v = \frac{V[S]}{K_m + [S]} \tag{7.85}$$

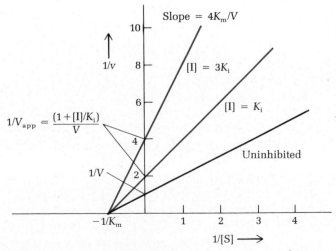

Figure 7.22 Lineweaver-Burk plot for reaction subject to noncompetitive inhibition.

An inhibitor can alter K_m, V, or both. The Michaelis-Menten rate law can be rewritten to reflect these possibilities.

$$v = \frac{V_{app}[S]}{K_{m,app} + [S]} \tag{7.86}$$

For competitive inhibition the rate law is

$$v = \frac{V[S]}{[S] + K_m \left(1 + \frac{[I]}{K_i}\right)} \tag{7.87}$$

The maximum velocity and the apparent maximum velocity are equal, and

$$K_{m,app} = K_m \left(1 + \frac{[I]}{K_i}\right) \tag{7.88}$$

The rate law for uncompetitive inhibition is

$$v = \frac{V[S]}{K_m + [S] \left(1 + \frac{[I]}{K_i}\right)} \tag{7.89}$$

where

$$V_{app} = \frac{V}{\left(1 + \frac{[I]}{K_i}\right)} \quad \text{and} \quad K_{m,app} = \frac{K_m}{\left(1 + \frac{[I]}{K_i}\right)} \tag{7.90}$$

The rate law for simple noncompetitive inhibition is

$$v = \frac{V[S]}{K_m \left(1 + \frac{[I]}{K_i}\right) + [S] \left(1 + \frac{[I]}{K_i}\right)} \tag{7.91}$$

where

$$V_{app} = \frac{V}{\left(1 + \frac{[I]}{K_i}\right)} \quad \text{and} \quad K_{m,app} = K_m \tag{7.92}$$

The various types of inhibitors can be readily detected by using direct linear plots. In Figure 7.23 we consider the effects of inhibitors on the coordinates V, K_m. If the V coordinate is constant and the K_m coordinate moves away from the v axis parallel to the K_m axis, the inhibitor is competitive. If the V coordinate moves toward the K_m axis parallel to the v axis, the inhibitor is noncompetitive. And if the coordinates K_m and V move directly toward the origin in the presence of inhibitor, the inhibitor is uncompetitive. Lineweaver-Burk plots for competitive inhibitors give lines that intersect at a common point on the $1/v$ axis; noncompetitive inhibitors give plots whose lines intersect on the $1/[S]$ axis; and uncompetitive inhibitors yield sets of parallel lines.

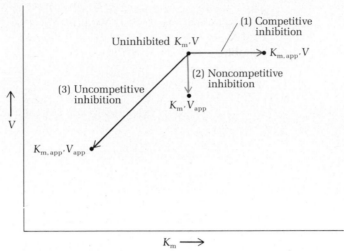

Figure 7.23 Direct linear plot summarizing effects of various types of inhibitors. (1) If K_m increases and V is constant, the inhibition is competitive, and the $K_{m,app},V$ coordinate moves directly away from the v axis, parallel to the K_m axis. (2) If K_m is constant and V decreases, the inhibition is noncompetitive, and the K_m,V_{app} coordinate moves directly toward the K_m coordinate, parallel to the v axis. (3) If $K_{m,app}$ and V_{app} both decrease, the inhibition is uncompetitive, and the coordinate $K_{m,app},V_{app}$ moves straight toward the origin.

7–12 KINETICS OF BISUBSTRATE REACTIONS

Many enzymes require two substrates. A nomenclature devised by W. W. Cleland provides a simple way of diagramming the sequence of events in bisubstrate reactions. In "Cleland diagrams" the following conventions are used:

1. Substrates are designated A, B, C . . . , and products are designated P, Q, R . . . .

2. The enzyme is labeled E. Stable forms of the enzyme that are produced and then disappear during the reaction are labeled F, G . . . .

3. Enzyme-substrate complexes in which all substrates are bound to the enzyme are called *central complexes* and are enclosed in parentheses. Those central complexes in which two substrates are bound to the enzyme are called *ternary complexes*.

4. Reactions in which both substrates must bind to the enzyme before catalysis of any type can occur are *sequential*. If the order of binding is *obligatory*, the mechanism is *ordered sequential*. If there is no specified order of binding, the mechanism is *random sequential*. If one product is released from the enzyme before all of the substrates bind, the mechanism is called *ping-pong*.

A. Ordered Sequential Mechanism

The ordered sequential mechanism is diagrammed below for the case where A must bind before B.

Scheme 6

The steady state initial rate law derived from scheme 6 is

$$v = \frac{V[A][B]}{K_{sA}K_{mB} + K_{mB}[A] + K_{mA}[B] + [A][B]} \tag{7.93}$$

where K_{sA} is the dissociation constant of A from EA. The K_m's must be treated carefully. Only when the enzyme is saturated with A will the apparent K_{mB} be the true K_m for B; likewise with A. Similarly, the "true" value of V is obtained only when the enzyme is saturated with *both* substrates. An example of an ordered sequential mechanism is provided by phosphofructokinase, a critical enzyme in glycolysis (Section 14.3C).

B. Random Sequential Mechanism

In the random sequential mechanism there is no necessary order of addition of the substrates to the enzyme. The Cleland diagram for this mechanism is shown in scheme 7.

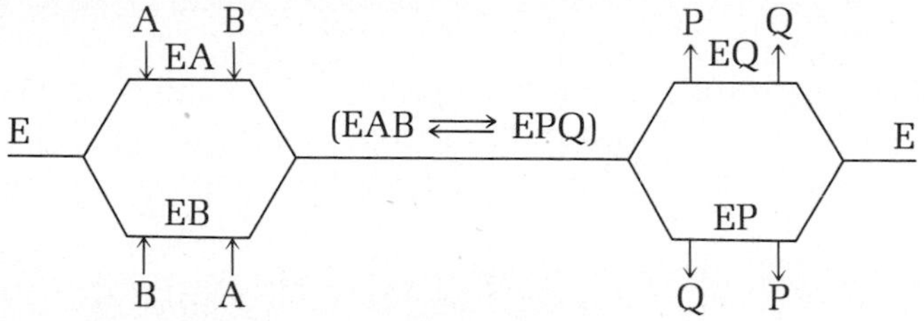

$$
E \underset{\underset{B \quad A}{\underset{\uparrow \quad \uparrow}{EB}}}{\overset{\overset{A \quad B}{\overset{\downarrow \quad \downarrow}{EA}}}{}} (EAB \rightleftharpoons EPQ) \underset{\underset{Q \quad P}{\underset{\downarrow \quad \downarrow}{EP}}}{\overset{\overset{P \quad Q}{\overset{\uparrow \quad \uparrow}{EQ}}}{}} E
$$

Scheme 7

The steady state rate law for initial rates for the random sequential mechanism is the same as for the ordered mechanism. These two mechanisms cannot, therefore, be distinguished by initial rate studies. A wide range of ancillary techniques can, however, yield the information required to identify the binding patterns and thus the kinetic mechanism. Product inhibition patterns and isotope exchange will be discussed below.

The double reciprocal form of equation 7.93 is

$$
\frac{1}{v} = \frac{1}{V}\left(K_{mA} + \frac{K_{sA}K_{mB}}{[B]}\right)\frac{1}{[A]} + \frac{1}{V}\left(1 + \frac{K_{mB}}{[B]}\right) \tag{7.94}
$$

A plot of $1/v$ versus $1/[A]$ at varying fixed concentrations of B gives the results shown in Figure 7.24.

C. Theorell-Chance Mechanism

The Theorell-Chance mechanism is a variation of an ordered sequential mechanism in which the central complex (EAB) does not exist (scheme 8).

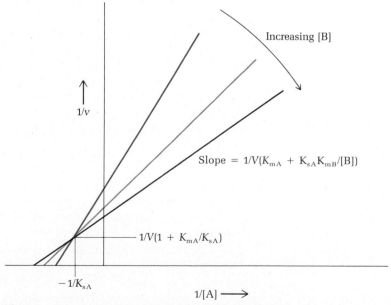

Figure 7.24 Lineweaver-Burk plot for a bisubstrate reaction.

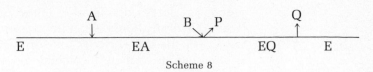

Scheme 8

Horse liver alcohol dehydrogenase is an enzyme that obeys the Theorell-Chance mechanism. NAD is the first substrate to bind. Ethanol then binds to form a ternary complex that does not accumulate, and the binding of ethanol is followed immediately by release of acetaldehyde. In scheme 8, then, A is NAD, B is ethanol, and P is acetaldehyde. In the final step of the reaction the second product, NADH (Q) is released.

D. The Ping-Pong Mechanism

In the ping-pong mechanism one substrate binds and one product is released before the second substrate can bind to the enzyme (scheme 9).

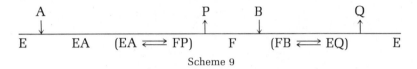

Scheme 9

Aspartate transaminase is an example of an enzyme that obeys ping-pong kinetics (Section 21.2A). The rate law for ping-pong kinetics is

$$v = \frac{V[A][B]}{K_B[A] + K_A[B] + [A][B]} \tag{7.95}$$

The Lineweaver-Burk double reciprocal form of equation 7.95 is

$$\frac{1}{v} = \frac{K_{mA}}{V}\left(\frac{1}{[A]}\right) + \frac{1}{V}\left(1 + \frac{K_{mB}}{[B]}\right) \tag{7.96}$$

Plots of $1/v$ versus $1/[A]$ at varying fixed concentrations of B give sets of parallel lines (Figure 7.25). Graphical analysis thus provides strong evidence for a ping-pong mechanism, although, as we noted above, it cannot distinguish random and ordered sequential mechanisms.

E. Product Inhibition in Bisubstrate Reactions

A few simple rules, a condensation of a much more elaborate treatment by Cleland, can be used to distinguish kinetic mechanisms by patterns of product inhibition. These rules are successful because the binding of a product, P or Q, to the enzyme alters the binding of substrates A and B in different ways for different kinetic mechanisms. In product inhibition experiments, for a given set of kinetic runs, first, the product and the concentration of one substrate (the fixed substrate) are held constant, while the concentration of the other substrate (the variable substrate) is varied. Then, in a second set of experiments, the concentration of the fixed substrate is altered to a new fixed value. (It is then called the variable fixed substrate.) The rules follow.

1. If a substrate and product combine with the same enzyme form, the product is a competitive inhibitor of the substrate. For example, in scheme 6 both Q and A bind to the same form of E. A Lineweaver-Burk plot of $1/v$ versus $1/[A]$ at constant B and at various concentrations of Q gives the results shown in Figure 7.26a, the same inhibition pattern we observed for a single substrate reaction subject to competitive inhibition.

2. If a product and a substrate combine with different forms of the enzyme, the product is a noncompetitive inhibitor if all forms of the enzyme are present at finite concentrations. All reactants and products must also be present for rule 2 to apply. Let us again consider scheme 6 under conditions in which A is the variable sub-

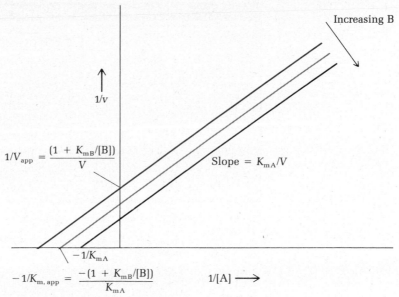

Figure 7.25 Lineweaver-Burk plot for a ping-pong mechanism.

strate, B the variable fixed substrate, and P is the added inhibitor. P combines with EQ, whereas A combines with E. At nonsaturating concentrations of B rule 2 predicts noncompetitive inhibition (Figure 7.26b).

   3. If a substrate and product combine with different forms of the enzyme when (a) one form of the enzyme is not present in a finite concentration or (b) one of the substrates or products is absent, the product is an uncompetitive inhibitor of that substrate. Referring again to scheme 6, rule 3 predicts that when the enzyme is saturated with B, P is an uncompetitive inhibitor of the binding of the variable substrate A. There is a single exception to rule 3. When B is the variable substrate and the enzyme is saturated with A, no free enzyme remains, none is left to bind to Q, and thus the reaction is not inhibited by Q.

   Table 7.3 summarizes the effects of product inhibition on the bisubstrate reactions that we have considered.

F. Isotope Exchange in Bisubstrate Reactions

The mechanisms we have discussed can be divided into two types. In *single displacement reactions* the two substrates are both bound to the enzyme before any reaction occurs. The random sequential, ordered sequential, and

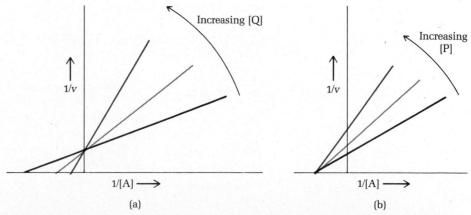

Figure 7.26 Product inhibition patterns. (a) Pattern for product inhibition of a bisubstrate reaction when product Q and substrate A bind to the same form of the enzyme. (b) Inhibition pattern when product P and substrate A bind to different forms of the enzyme.

Table 7.3 Product inhibition patterns for bisubstrate reactions

Mechanism	Inhibitory product	Variable substrate A		Variable substrate B	
		Unsaturated	Saturated with B	Unsaturated	Saturated with A
Ordered	P	NC	UC	NC	NC
sequential	Q	Comp	Comp	NC	—
Theorell-	P	NC	—	Comp	Comp
Chance	Q	Comp	Comp	Comp	Comp
Random	P	Comp	—	Comp	—
sequential	Q	Comp	—	Comp	—
Ping-pong	P	NC	—	Comp	Comp
	Q	Comp	Comp	NC	—

The terms Comp, UC, and NC designate competitive, uncompetitive, and noncompetitive inhibition, respectively.

SOURCE: From K. Plowman, "Enzyme Kinetics," McGraw-Hill, 1972, p. 154. Used with the permission of the McGraw-Hill Book Company.

Theorell-Chance mechanisms are of this type. If, on the other hand, the reaction occurs in two discrete steps in which one substrate binds and a product is released before the second substrate binds (a ping-pong mechanism), the reaction is a *double displacement*. A product can form in the absence of the second substrate in a double displacement, but not in a single displacement reaction.[5] Isotope exchange techniques exploit this mechanistic difference.

Let us consider the general reaction

$$A—X + B \rightleftharpoons A—B + X \tag{7.97}$$

In a single displacement reaction a ternary complex of the enzyme, A—X, and B must be formed for reaction to take place. If A—X and radioisotopically labeled X* are added to the enzyme in the absence of B, no isotope exchange can occur because the A—B bond forms as the A—X bond breaks. Isotope exchange can, however, occur for a double displacement mechanism because the enzyme can react with A—X in the absence of B.

$$A—X + E \rightleftharpoons E—A + X \tag{7.98}$$

If A—X and X* are incubated with the enzyme, X* can exchange with X by the above equilibrium. Thus, the single and double displacement mechanisms can be distinguished by isotopic exchange.

This method was initially applied to the study of enzyme mechanisms in 1947 by Douderoff and his coworkers in their study of *sucrose phosphorylase*, which catalyzes the reaction

$$\text{glucose 1-phosphate} + \text{fructose} \rightleftharpoons \text{sucrose} + HPO_4^{2-} \tag{7.99}$$

When the enzyme is incubated with glucose 1-phosphate and isotopically labeled $H^{32}PO_4^-$ in the absence of fructose, there is an exchange of phosphate, and glucose 1-^{32}phosphate may be isolated from the mixture. Since this exchange does not occur in the absence of enzyme, sucrose phosphorylase acts by a double displacement mechanism. The phosphate exchange reaction is

$$\text{glucose 1-phosphate} + \text{enzyme} \rightleftharpoons \text{enzyme-glucose} + HPO_4^{2-}$$
$$\tag{7.100}$$

[5] Oxidation-reduction reactions can also be ping-pong or sequential, although they do not occur by "displacement" mechanisms.

Addition of $H^{32}PO_4^{2-}$ shifts equilibrium to the left leading to incorporation of radiolabeled phosphate in the substrate. Isolation of the glucosylenzyme is conclusive proof of this part of the mechanism. Similarly, if radioactive fructose is added to sucrose phosphorylase in the presence of sucrose, unlabeled, but in the absence of glucose 1-phosphate, radiolabeled sucrose can be isolated from the reaction mixture (reaction 7.101).

$$^{14}\text{C-fructose} + \text{glucose-fructose} \rightleftharpoons \text{fructose} + {}^{14}\text{C-fructose-glucose}$$
$$\text{(sucrose)} \qquad (7.101)$$

In most cases a combination of initial rate studies, product inhibition patterns, and isotope exchange data is required to establish a kinetic mechanism for bisubstrate reactions.

7–13 SUMMARY

Kinetic studies of enzyme-catalyzed reactions provide important information about enzyme catalysis and characterize the enzyme. The Michaelis-Menten rate law (equation 7.27) describes the kinetic behavior of an enzyme when the initial rate of the reaction is measured. The Michaelis constant, K_m, equals the substrate concentration at which the enzyme is half-saturated with enzyme. At this substrate concentration the initial velocity is one-half maximal. The Michaelis-Menten rate law has the same algebraic form for three conditions: if it is assumed that the enzyme-substrate complex forms in a rapid equilibrium, if the enzyme-substrate complex exists in a steady state, and if a covalent enzyme-substrate intermediate is formed during the course of the reaction.

The reciprocal of the catalytic rate constant, $1/k_{cat}$, has units of time and represents the time required for the enzyme to turn over one molecule of substrate. One international enzyme unit is the amount of enzyme that produces one micromole of product per minute. The specific activity of an enzyme is the number of units per milligram of enzyme.

The pH of the medium has a large effect on the rates of enzymic reactions. If the pH-dependent process is the rate of conversion of enzyme-substrate complex to product, the value of k_{cat} is pH dependent. If the dissociation of the enzyme-substrate complex is pH dependent, K_m will be altered by changes in pH. In most cases, if the ratio of k_{cat}/K_m is altered, it is the pK_a of an active site residue in the free enzyme that is responsible for the effect of pH.

The rates of most enzyme-catalyzed reactions increase with increasing temperature until a point is reached at which the enzyme becomes thermally denatured. If a reaction has an activation energy of about 50 kJ mol^{-1}, increasing the temperature from 25 to 35°C approximately doubles the rate.

The ratio k_{cat}/K_m is a measure of the specificity of an enzyme for two competing substrates. The magnitude of k_{cat}/K_m cannot be greater than the value of the largest microscopic rate constant in the catalytic reaction.

In perfectly evolved enzymes the value of k_{cat}/K_m has reached the diffusion-controlled limit for a second order rate constant. This evolution implies that the enzyme binds much more tightly to the substrate in the transition state of the reaction than in the ground state of the enzyme-substrate complex.

In catalytic reactions in which a C—H bond is broken the substitution of hydrogen for deuterium (^{2}H or D) or of hydrogen for tritium (^{3}H or T) will result in a slower rate of reaction if the C—H bond is broken in the rate-determining step of the reaction.

Several methods are widely used to represent kinetic data graphically. The reciprocal of the Michaelis-Menten rate law (equation 7.54) gives a relationship between $1/v$ and $1/[S]$ that is plotted in a Lineweaver-Burk plot. The major advantage of such a double reciprocal plot is its legibility, but its major disadvantage is that a small experimental error can lead to a large graphical error in the determination of K_m and V. The Eadie-Hofstee plot is a graph of $v/[S]$ versus v. Eadie-Hofstee plots do not compress the data at high substrate concentrations, but they are less legible than Lineweaver-Burk plots. Direct linear plots avoid the pitfalls of distortion of the kinetic data in its graphical presentation. A plot of the varying initial velocities (v_i) on the vertical V axis versus the varying initial substrate concentrations (s_i) on the horizontal K_m axis yields a set of straight lines that intersect. The coordinates (K_m, V) of the intersection point give the Michaelis constant and the maximum velocity for the enzyme.

Enzyme inhibitors can be divided into three major types. Competitive inhibitors increase the apparent K_m for the enzyme, but do not alter V. A transition state analog is a competitive inhibitor whose structure resembles the structure of the substrate of the enzymic reaction in the transition state of the reaction. Uncompetitive inhibitors bind to the enzyme-substrate complex and decrease the values of both K_m and V. A simple noncompetitive inhibitor binds equally well to the free enzyme and to the enzyme-substrate complex. The value of K_m is not altered by a noncompetitive inhibitor, but the value of V decreases.

Bisubstrate reactions can be divided into several basic types. In ordered sequential mechanisms the binding order is obligatory. In a random sequential mechanism the two substrates can bind to the enzyme in any order. The Theorell-Chance mechanism is a variation of an ordered sequential mechanism in which binding of the first substrate is followed immediately by release of a product when the second substrate binds to the enzyme. The second product is then released. In a ping-pong mechanism the first substrate binds to the enzyme and a product is released with formation of an intermediate enzyme species. The second substrate then binds to the enzyme in its altered form releasing the second product and converting the enzyme to its original form.

Bisubstrate kinetic mechanisms can be distinguished by the effects of inhibitory products on the type of inhibition observed (Table 7.3). Isotope exchange experiments can also be employed to distinguish the single displacement mechanisms—random sequential, ordered sequential, and Theorell-Chance—from the double displacement, or ping-pong mechanism.

REFERENCES

Cornish-Bowden, A., "Principles of Enzyme Kinetics," Butterworths, London, 1975.

Gutfreund, H., "Enzymes: Physical Principles," Wiley-Interscience, New York, 1972.

Piskiewicz, D., "Kinetics of Chemical and Enzyme-Catalyzed Reactions," Oxford University Press, New York, 1977.

Plowman, K., "Enzyme Kinetics," McGraw-Hill, New York, 1972.

Problems

1. Fumarase catalyzes the hydration of fumarate to L-malate:

Fumarate   L-Malate

The enzyme is composed of four identical subunits and has a molecular weight of 194,000.

a. The following data were obtained when fumarate was used as the substrate and the initial velocity of the hydration reaction was measured at pH 5.7, 25°C, enzyme concentration 2×10^{-6} M.

Fumarate (mM)	Rate of product formation (mMoles l.$^{-1}$ min^{-1})
2.0	2.5
3.3	3.1
5.0	3.6
10.0	4.2

When the same reaction was studied in a buffer containing 0.06 M phosphate, the following data were obtained with an enzyme concentration of 2×10^{-6} M.

Fumarate (mM)	Rate of product formation (mMoles l.$^{-1}$ min^{-1})
2.0	4.0
3.3	5.0
5.0	5.8
10.0	6.6

Calculate the value of V and K_m for the above conditions. Analyze and discuss briefly any significant differences.

b. The Haldane equation relates the kinetic constants V and K_m for a reaction to the equilibrium constant K_{eq}.

$$K_{eq} = \frac{(V)_f (K_m)_r}{(V)_r (K_m)_f}$$

where r indicates that the constants are for the reverse reaction, f for the forward reaction.

The initial rates of the hydration reaction were measured as a function of pH and for various substrate concentrations. The values obtained, expressed as mMoles per liter of fumarate hydrated per minute, were as follows.

| pH | Fumarate concentration | | | |
	10 mM	5 mM	2.5 mM	1.25 mM
5	0.92	0.81	0.66	0.48
6	4.00	3.28	2.42	1.60
7	5.40	4.30	3.08	1.96
8	2.67	2.40	1.99	1.53
9	0.21	0.17	0.16	0.12

The values of the maximum initial velocity of the dehydration of L-malate catalyzed by fumarase had the following dependence on pH.

pH	5	6	7	8	9
V (mMol $1.^{-1}$ min^{-1})	0.068	0.44	3.00	5.68	2.00

Calculate values for the Michaelis constant and maximum velocity of the hydration reaction for each pH value.

c. Does the maximum velocity in the forward reaction occur at the same pH as the maximum velocity of the reverse reaction?

d. Calculate the values of the Michaelis constant for the dehydration reaction for each pH value.

2. In many enzyme-catalyzed hydrolysis reactions, a proton is released each time a molecule of substrate reacts. An instrument known as a *pH stat* can be used to follow the rate of hydrolysis. A pH stat assay is one in which the pH is held constant by the instrument by continuous neutralization of the acid released in the reaction. The observed rate, then, is the rate of consumption of base.

a. The reaction of chymotrypsin with N-acetyltryptophan ethyl ester was followed at pH 6.0 by a pH stat assay. Determine the order of the reaction with respect to chymotrypsin.

Chymotrypsin (mM)	N-AcTrpOEt (mM)	Initial rate (ml 1 mM NaOH min^{-1})
1.5×10^{-7}	0.4	3.0×10^{-2}
2.0×10^{-7}	0.4	4.0×10^{-2}
2.5×10^{-7}	0.4	5.6×10^{-2}
3.0×10^{-7}	0.4	6.0×10^{-2}

b. The same reaction was also studied to determine the dependence of the rate upon substrate concentration. The pH for all experiments is 6.0, the enzyme concentration is 2×10^{-7} mM. From these data calculate K_m for N-acetyltryptophan ethyl ester and V for the reaction.

N-AcTrpOEt (mM)	Initial rate (mM min^{-1})
0.6	1.41×10^{-7}
0.4	1.33×10^{-7}
0.2	1.14×10^{-7}
0.1	0.88×10^{-7}
0.05	0.62×10^{-7}

3. J. O. Westerik and R. Wolfenden have studied the inhibition of the papain-catalyzed hydrolysis of p-nitrophenylhippurate (N-benzoylglycine-p-nitrophenyl ester). They found that this reaction was inhibited by N-benzoylamino acetaldehyde. The data they obtained are shown below. (a) Calculate K_m for the substrate. (b) Calculate K_i for the inhibitor. (c) Calculate V for the reaction, given an enzyme concentration of 1.2×10^{-7} M for all reactions. (d) When the substrate concentration is 6.6×10^{-5} M, what fraction of the enzyme is bound to substrate in the presence of the inhibitor? (e) If

no inhibitor is present, what fraction of the enzyme is bound to substrate at a substrate concentration of 6.6×10^{-5} M? (f) What type of inhibition is being observed?

p-Nitrophenylhippurate (mM)	Initial rate (M min^{-1})
No inhibitor	
0.71	0.2
0.4	0.18
0.31	0.16
0.098	0.12
0.066	0.10
0.040	0.07
3.3×10^{-5} M inhibitor + substrate	
0.71	0.18
0.4	0.15
0.31	0.11
0.098	0.07
0.066	0.05
0.040	0.04

4. J. S. Fruton and his coworkers have studied the hydrolysis of a series of synthetic peptides by the proteolytic enzyme pepsin. The structure of one set of synthetic peptides is shown below. Cleavage between adjacent phenyl-alanyl residues was followed kinetically. (The arrow indicates the point of bond cleavage.) The effect of chain length for glycine "tails" from one to four units long was investigated. The data are summarized in the table below. What do the data imply about the affinity of the enzyme for the series of peptides? Which one is cleaved most efficiently?

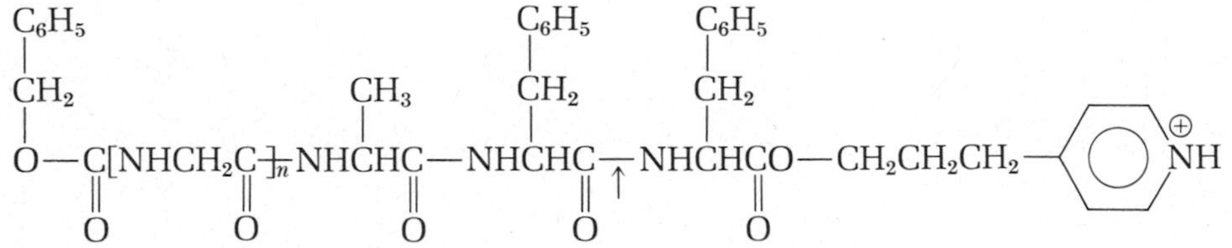

Number of glycine residues n in molecule above	k_{cat} (s^{-1})	K_m (mM)
1	3.1	0.4
2	71.8	0.4
3	4.5	0.4
4	2.1	0.7

5. Subtilisin (mol wt 27,600) is a bacterial protease which specifically hydrolyzes aromatic esters and amides. K_m and k_{cat} for the hydrolysis of N-acetyltryosine ethyl ester are 0.146 M and 548 s^{-1}, respectively. (a) If the enzyme concentration is 0.4 mg/ml and the substrate concentration is 0.25 M, what is V for this reaction? (b) Indole is a competitive inhibitor of subtilisin. K_i for indole is 0.05 M. Calculate the maximum velocity of the hydrolysis of the above substrate in the presence of 6.25 mM indole. (c) What is the reaction velocity in the presence of 1.0 M indole?

6. Show by algebraic manipulation that the Michaelis-Menten equation can be transformed into the linear equation 7.59.

7. Using the data from problem 3, plot equation 7.59 to determine K_m and V. Compare this graphical procedure with the direct linear plot.

8. Many enzyme reactions are inhibited by the product of the reaction. Using the data given below for the inhibition of alcohol dehydrogenase oxidation of ethanol by acetaldehyde, determine (a) the type of inhibition observed and (b) K_m, V, and K_i.

v (μM min^{-1})	Ethanol (mM)	Acetaldehyde (mM)
24.9	100	0.0
23.3	50	0.0
21.3	30	0.0
19.5	20	0.0
19.2	100	4.0
17.2	50	4.0
14.7	30	4.0
11.9	20	4.0

9. One way to study changes in protein conformation is to study the exchange of hydrogen-bonded hydrogens with solvent enriched in tritiated water. The rates of exchange of hydrogen are different for different conformations. When the exchange of hydrogens in deoxyhemoglobin was studied the following data were obtained. (a) Determine the order of the exchange process. (b) Determine the half-life of the process. (c) Determine the rate constant for the process. (d) It has been found that hydrogen exchange is much faster in oxyhemoglobin than in deoxyhemoglobin. Interpret this finding.

Time (min)	Number of hydrogens exchanged per hemoglobin subunit
0	4.70
60	3.15
104	2.35
120	2.10
180	1.42
208	1.17
240	1.00

10. Renin acts as a specific protease that cleaves a decapeptide unit from a serum α_2-globulin. Renin is inhibited by the microbial pentapeptide pepstatin. Kinetic data are shown below. From these data determine K_i for pepstatin and the type of inhibition being observed.

v (μM min^{-1} $\times 10^{-2}$)	Substrate (μM)	Pepstatin (M)
0.48	0.1	0
0.81	0.2	0
1.23	0.4	0
1.45	0.6	0
1.60	0.75	0
0.24	0.1	2.36×10^{-10}
0.45	0.2	2.36×10^{-10}
0.73	0.4	2.36×10^{-10}
0.95	0.6	2.36×10^{-10}
1.08	0.75	2.36×10^{-10}

11. Recent experimental measurements of the rates of binding of certain substrate analogs and transition state analogs to the enzyme adenylate deaminase have shown that the substrate analogs bind rapidly to the enzyme and that the transition state analogs bind slowly. Suggest an interpretation for this observation. [Reference: C. Frieden, L. C. Kurz, and H. R. Gilbert, *Biochemistry*, 19, 5303–5309 (1980).]

12. The reaction velocity is half-maximal when the enzyme is one-half saturated with substrate. Show that this is true algebraically.

13. Predict the inhibitory effects of products P and Q on the bisubstrate kinetic mechanisms we discussed. (The answers are in Table 7.3.)

14. *Proline racemase* catalyzes the reaction shown below. *Pyrrole-2-carboxylate* is a powerful inhibitor of proline racemase. The inhibitor is believed to be a transition state analog. Write a mechanism for the racemization and explain why pyrrole-2-carboxylate might be a transition state analog.

L-Proline   D-Proline   Pyrrole-2-carboxylate

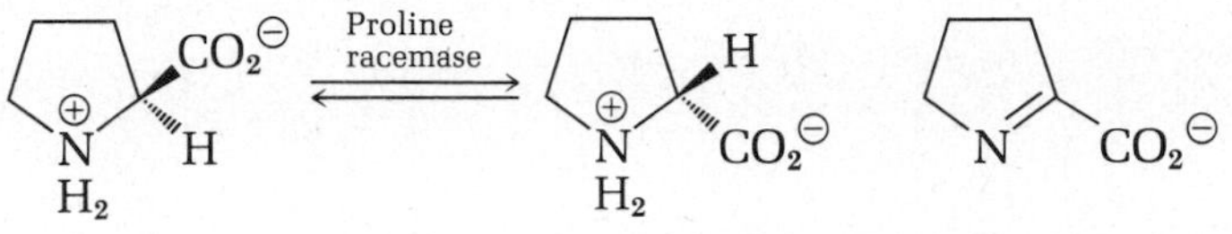

15. Fumarase catalyzes *anti* addition of water to fumarate and to fluorofumarate giving malate and α-fluoromalate, respectively.

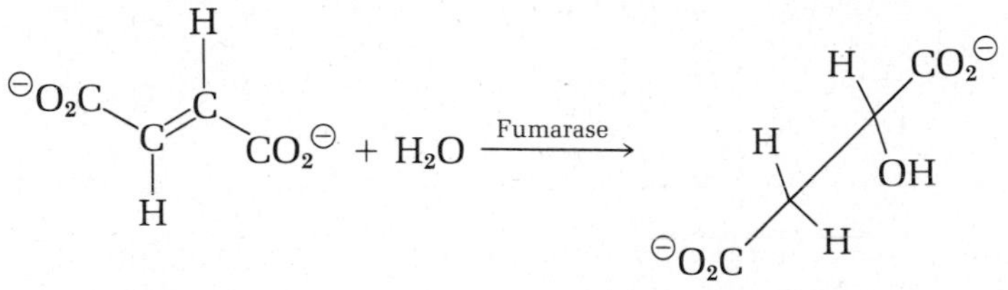

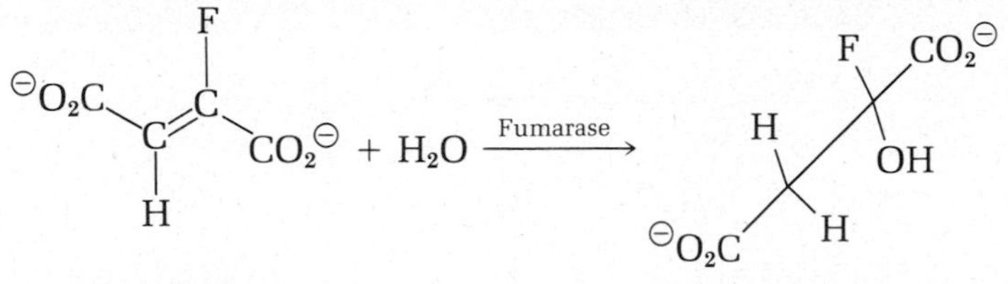

Fumarate   Malate

Fluorofumarate α-Fluoromalate

The values of k_{cat} and K_m for fumarate are 2.7×10^3 s^{-1} and 2.7×10^{-5} M, respectively. For fluoroflumarate k_{cat} is 8×10^2 s^{-1} and K_m is 5×10^{-6} M. For which substrate is fumarase more specific?

16. In problem 7.1b we saw that the Haldane equation relates the equilibrium constant for an enzyme reaction to V and K_m.

$$K_{eq} = \frac{(V)_f (K_m)_r}{(V)_r (K_m)_f}$$

Show that for the reaction

$$S \underset{k_r}{\overset{k_f}{\rightleftharpoons}} P$$

the Haldane equation can be transformed to

$$K_{eq} = \frac{(k_{cat}/K_m)_S}{(k_{cat}/K_m)_P}$$

Chapter 8

Carbohydrates

8–1 INTRODUCTION

Carbohydrates are the most abundant organic molecules in the biosphere, and their functions span an enormous range of biochemistry. Carbohydrate metabolism provides a significant fraction of the energy available to most organisms. The interactions of cells with one another are often mediated by their carbohydrate coats. For example, differences in blood groups are a function of certain blood cell carbohydrates.

Carbohydrates also have structural roles. They are a major component of bacterial cell walls. The exoskeletons of arthropods contain large amounts of the polysaccharide chitin. The structural polysaccharide cellulose is the most abundant organic molecule in plants; indeed, it is the second most abundant molecule (after water) in the biosphere.

Many complex polysaccharides are found in animal tissues as well. As this brief catalog suggests, carbohydrate biochemistry is wide and deep.

The first reference to wine sugar, glucose, appeared in Moorish writings of the twelfth century, a time when Europe languished in the Dark Ages, but we can date modern carbohydrate chemistry to the late nineteenth century, when the German chemist Emil Fischer carried out his researches, including a proof of the structure of glucose, one of the great achievements in chemistry.

The generic formula for the most common carbohydrates is $(CH_2O)_n$, hence the name, "hydrate of carbon." The generic formula, however, does not hint at the structural complexity of carbohydrates nor at the wide range of their chemical and biochemical reactivity. Glucose, for example, contains six functional groups and five asymmetric carbons, each different from the others. In this chapter we shall discuss the structure, chemistry, and some aspects of the biochemistry of some of the most important carbohydrates. Carbohydrate metabolism is treated as a separate subject in Chapters 14 and 17.

8–2 STRUCTURES OF MONOSACCHARIDES

The *saccharose unit* is a common structural feature of low molecular weight carbohydrates. The chemical possibilities of a carbonyl group flanked by an hydroxyl group, with an acidic α-hydrogen added for good measure, are

$$R-C=O$$
$$H-C-OH$$

Saccharose unit

large and varied. Let us begin by considering the case in which the R-group of the saccharose unit is a hydrogen atom. The highest oxidized carbon in this case is an aldehyde, and the carbohydrate is an *aldose*. Tautomerization of the aldehyde group produces a keto group, and a carbohydrate known as a *ketose*. The suffix *-ose* indicates the presence of a saccharose unit (either an aldose or a ketose). The chain length of a carbohydrate is indicated by the prefixes tri-, tetra-, penta-, hexa-, and so forth.

The *aldotriose* D-*glyceraldehyde* is the simplest carbohydrate, or sugar,

$$CHO$$
$$H-C-OH$$
$$CH_2OH$$

D-Glyceraldehyde

that has a chiral carbon. The enatiomers D- and L-glyceraldehyde are the only possible stereoisomers. In general, for a molecule having n chiral centers, 2^n stereoisomers are possible. Any pair of stereoisomers that are not related as enantiomers are *diastereomers*.

Monosaccharides can be drawn as planar *Fischer projections* in which groups to the right and left of the chiral carbon in the planar structure extend out of the plane in the three-dimensional structure, and groups above and below the chiral carbon extend into the plane in the three-dimensional structure (Figure 8.1). Fischer projections are always written with the most highly oxidized carbon, C-1, at the "top." When the highest numbered chiral carbon has an hydroxyl group on the right in a Fischer projection, the prefix D- is added. The stereoisomer with an hydroxyl group on the left at the highest numbered carbon is given the prefix L-. By a stroke of good fortune the Fischer projections of carbohydrates give the actual configuration at all of the chiral carbons, since the choice Fischer made of placing the hydroxyl group on the right in D-glyceraldehyde turned out to be the correct one. The prefixes D- and L-, then, give the *absolute configuration* at the highest numbered asymmetric carbon atom of carbohydrates.

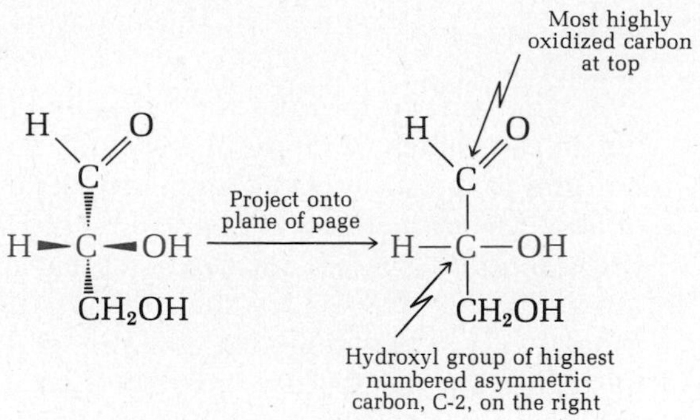

Figure 8.1 Fischer projection of D-glyceraldehyde.

Since hydrogens bound to carbon atoms that are α- to a carbonyl group are slightly acidic, aldoses can *tautomerize* to give *ketoses*. Tautomerization of D-glyceraldehyde gives the ketotriose 1,3-dihydroxy-2-propanone, or dihydroxyacetone (reaction 8.1).

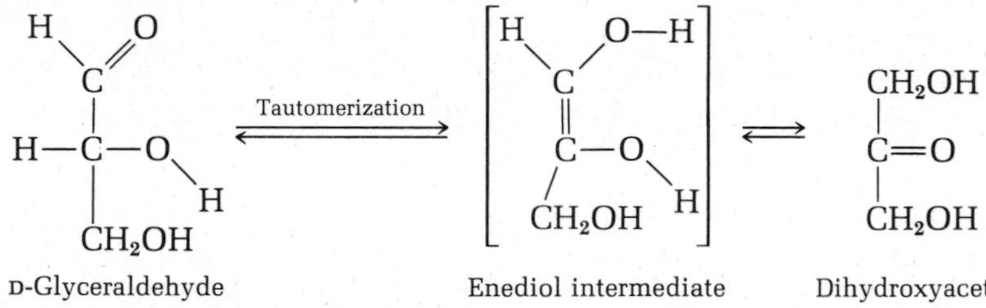

$$(8.1)$$

D-Glyceraldehyde   Enediol intermediate   Dihydroxyacetone

A. Aldo- and Ketotetroses

When the triose D-glyceraldehyde is lengthened by another —CH₂OH group, an *aldotetrose* results which has two chiral carbons and a total of 4 stereoisomers comprising two pairs of enantiomers. The diastereomers of the D family are D-*erythrose* and D-*threose* (Figure 8.2). The names D-*threo-* and D-*erythro-* completely define the stereochemistry of the molecule. D-erythrose and D-

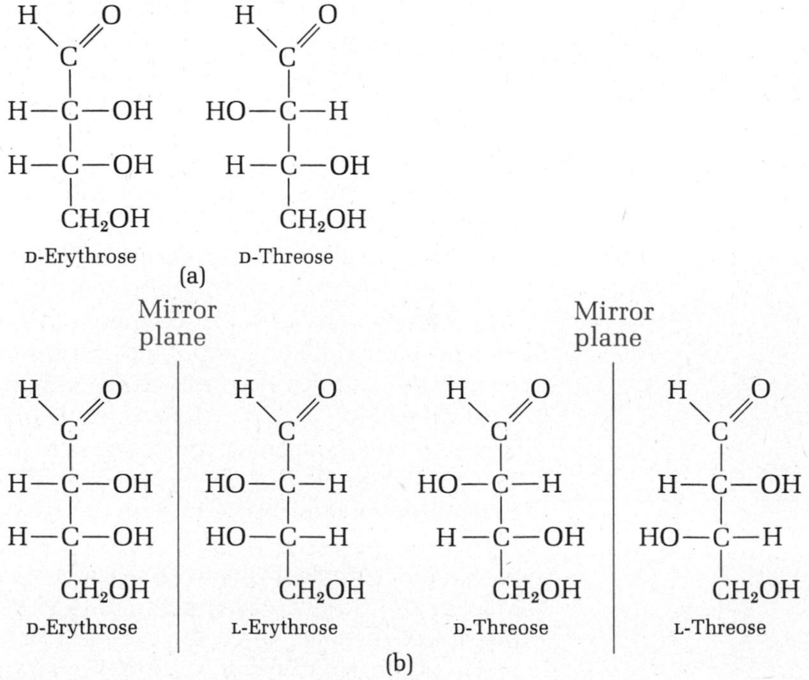

Figure 8.2 Stereoisomers of an aldotetrose (Fischer projections). (a) Diastereoisomers D-erythrose and D-threose. (b) Enantiomers D- and L-erythrose and D- and L-threose.

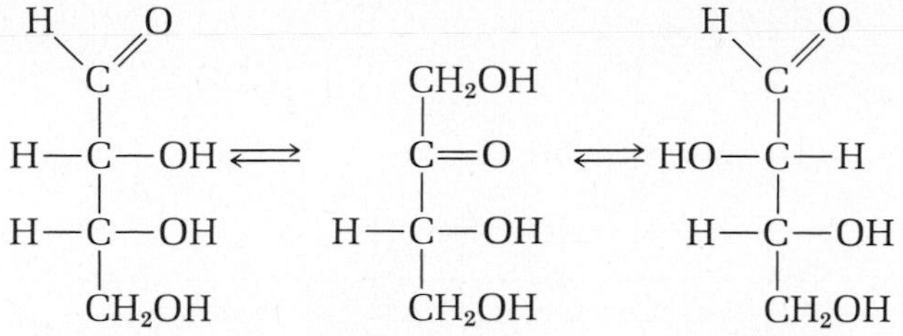

D-Erythrose   D-Glycerotetrulose   D-Threose
               (D-erythulose)

Figure 8.3 D-glycerotetrulose, more commonly known as D-erythrulose. It is the ketose derived from the D-aldotetroses.

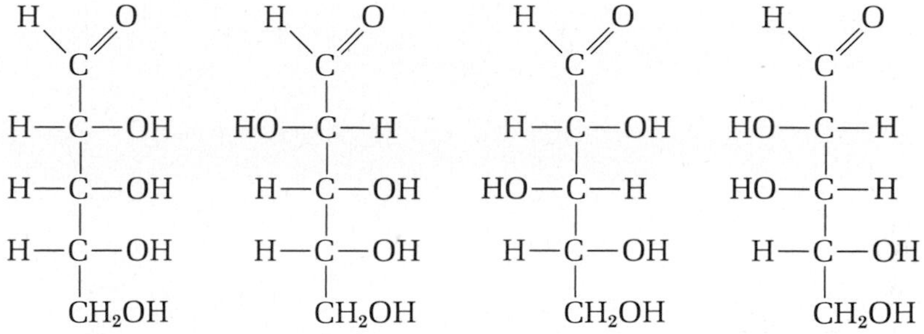

D-Ribose   D-Arabinose   D-Xylose   D-Lyxose

Figure 8.4 Structures of the D-aldopentoses.

threose, each having two chiral carbons, differ from each other in only one asymmetric center. Such diastereomers are called *epimers* (*epi-* means "center").

Tautomerization of either D-erythrose or D-threose gives the same keto sugar, which is identified by the suffix -*ulose*. Hence, the ketotetrulose is D-glycerotetrulose, more commonly (and incorrectly) known as D-erythrulose (Figure 8.3).[1]

B. Aldo- and Ketopentoses

The aldopentose family has three chiral carbons and eight diastereomers. The four D-aldopentoses are D-*ribose*, D-*arabinose*, D-*lyxose*, and D-*xylose* (Figure 8.4). The prefix D- again indicates the chirality of the highest numbered asymmetric center, C-4, and the names D-*ribo-*, D-*arabino-*, D-*lyxo-*, and D-*xylo-* define the stereochemistry of every position. There are two D-ketopentoses: D-*erythropentulose* (D-*ribulose*) and D-*arabinopentulose* (D-*xylulose*) (Figure 8.5).

C. Aldo- and Ketohexoses

The aldohexoses have 4 chiral centers, and there are 16 diastereomers for the family. Again, the D-isomers are by far the most important biochemically. The aldohexoses are *allose*, *altrose*, *glucose*, *mannose*, *gulose*, *idose*, *galactose*, and *talose* (Figure 8.6). Glucose is the most abundant hexose, followed by galactose and mannose. The latter two differ from glucose by their configurations at a single chiral center. Galactose is the C-4 epimer of glucose, and mannose is the C-2 epimer of glucose.

There are four D-hexuloses (Figure 8.7). Of these fructose is the most abundant. The systematic name for fructose—D-*arabinohexulose*—is seldom used. The stereochemical relationships of the trioses, tetroses, pentoses, and hexoses are shown in Figure 8.8.

Having considered the structures of the most common monosaccha-

---

[1] The prefix D-*glycero* implies the stereochemical relation of D-glyceraldehyde. It does not, however, imply that there are three carbons in the molecule.

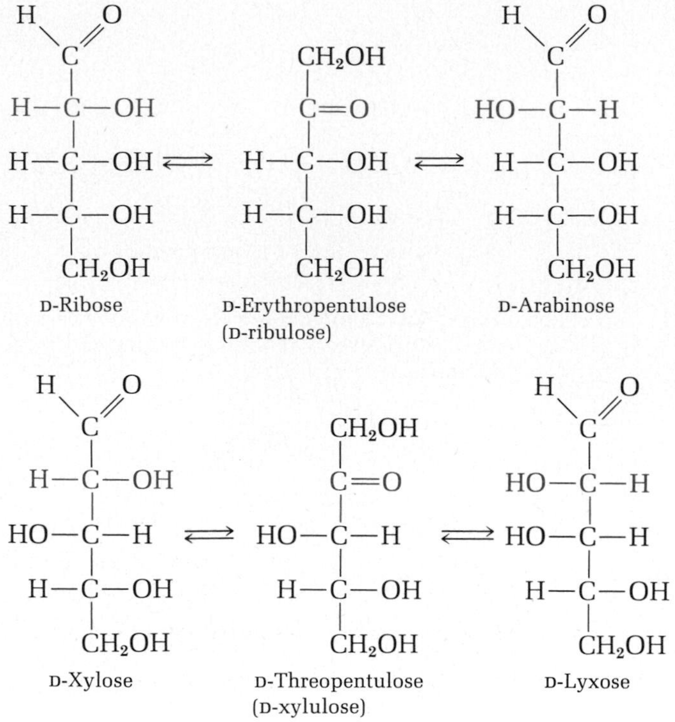

D-Ribose

D-Erythropentulose
(D-ribulose)

D-Arabinose

D-Xylose

D-Threopentulose
(D-xylulose)

D-Lyxose

Figure 8.5 Interconversions of the D-aldopentoses and the D-ketopentoses.

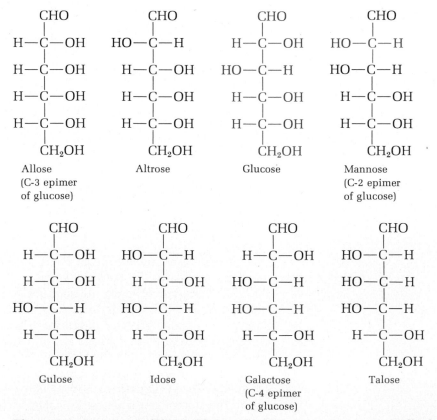

Allose
(C-3 epimer
of glucose)

Altrose

Glucose

Mannose
(C-2 epimer
of glucose)

Gulose

Idose

Galactose
(C-4 epimer
of glucose)

Talose

Figure 8.6 Structures of the D-aldohexoses. The mnemonic device "All Altruists Gladly Make Gum In Gallon Tanks" provides a nice way of learning and remembering the structures. Note that mannose is the C-2 epimer of glucose and that galactose is the C-4 epimer of glucose. These three are the most important aldohexoses.

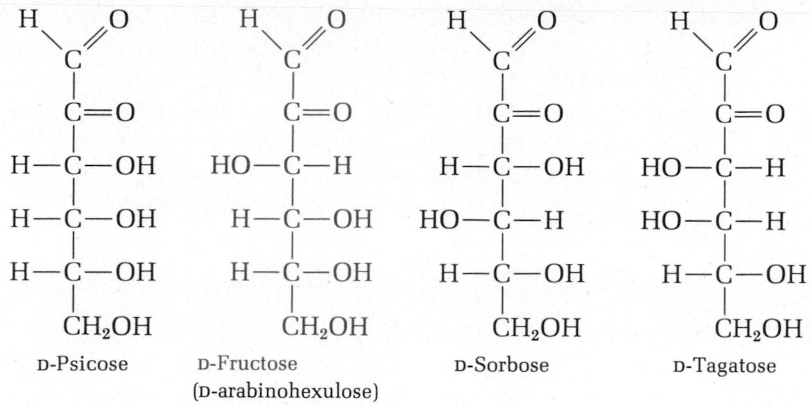

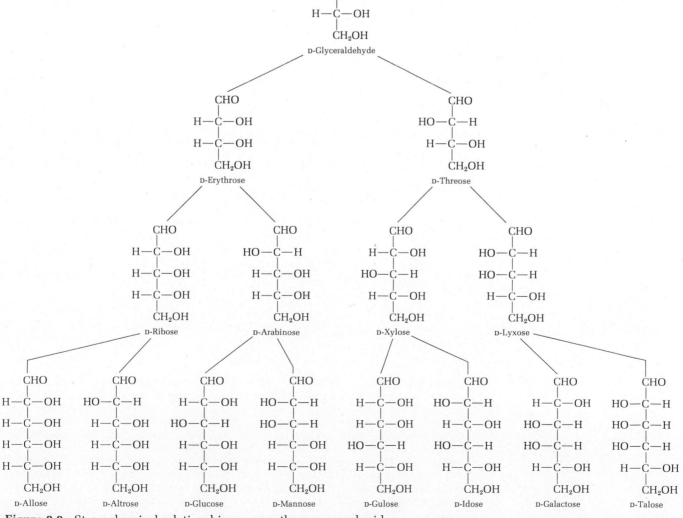

Figure 8.7 Structures of the D-ketohexoses. D-Fructose is by far the most important and most abundant.

rides, we shall now turn our attention to their chemistry. Since the carbohydrate structures in the preceding pages recur often in carbohydrate metabolism they should be learned now. Figure 8.8 is a convenient guide for learning the open chain structures of the monosaccharides.

Figure 8.8 Stereochemical relationships among the monosaccharides.

8–3 CYCLIC SUGARS

A. Pyranoses

D-Glucose can be isolated in two crystalline forms, called α and β, which have different physical properties. Among the differences is their ability to rotate plane polarized light—one of the most easily measured properties of chiral molecules. The *optical rotation* of a solution, or pure liquid, is measured in a *polarimeter* (Figure 8.9). The *specific rotation* of a molecule is defined as

$$[\alpha]_{\text{D}}^{25} = \frac{\alpha_{\text{obs}}}{1 \times c} \tag{8.2}$$

where $[\alpha]_{\text{D}}^{25}$ is the specific rotation at the D line of sodium, 589 nm, at 25°C, α_{obs} is the observed rotation, l is the length of the polarimeter cell in decimeters, and c is the concentration of the sample in grams per liter. The specific rotations for the α and β isomers of glucose are $+112°$ and $+18.7°$, respectively. Samples that rotate plane polarized light to the right (clockwise rotation) are called *dextrorotatory*, symbolized by (+). Samples that rotate plane polarized light to the left (counterclockwise rotation) are called *levorotatory*, or (−). The experimentally observed optical rotation is *not* correlated with absolute configuration in a simple way: a D-isomer of a chiral molecule can be either dextrorotatory or levorotatory.

The α and β crystalline forms are six-membered cyclic ring forms

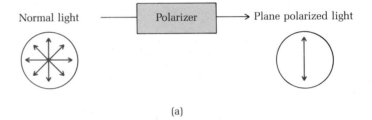

(a)

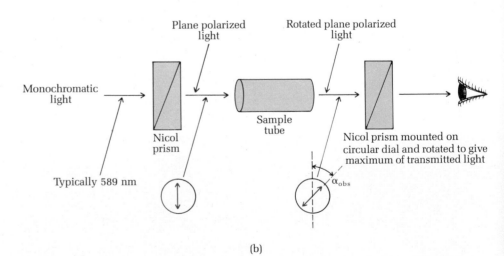

(b)

Figure 8.9 Rotation of plane polarized light. (a) Schematic diagram of the conversion of unpolarized light to polarized light by a polarizing element such as a Nicol prism. (b) Schematic diagram of a polarimeter. Monochromatic light passes through a Nicol prism and emerges polarized in one plane. The light then traverses the sample tube, which rotates the plane of polarized light in the clockwise (+) or counterclockwise (−) direction by some amount. The rotated plane polarized light then passes through a second Nicol prism which is rotated to allow the light beam to pass through it with maximum intensity; this is the observed optical rotation. Modern instruments perform these operations electronically with high precision.

(pyranose rings) of glucose. An alcohol can react with an aldehyde, in 1:1 stoichiometry, to give a *hemiacetal* (reaction 8.3).

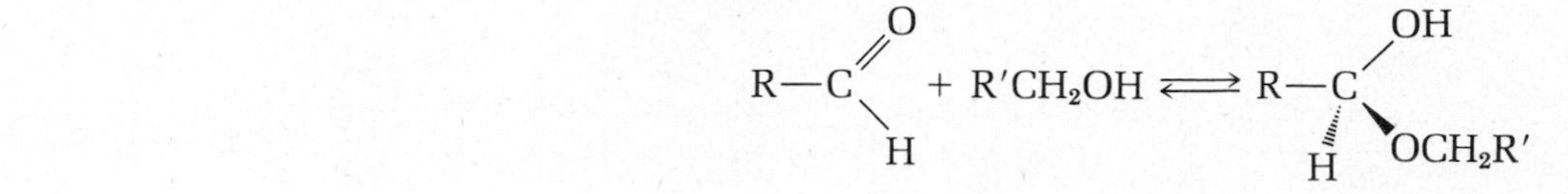

$$R-C\overset{O}{\underset{H}{\diagup}} + R'CH_2OH \rightleftharpoons R-\underset{H}{\overset{OH}{\underset{|}{C}}}OCH_2R' \tag{8.3}$$

Hemiacetal

Consider formation of an intramolecular acetal of glucose. Rotation of the carbon-carbon bonds of glucose results in a conformation in which the C-5 hydroxyl group is close to the carbonyl carbon. Reaction of these two groups gives a *cyclic hemiacetal* (Figure 8.10). The equilibrium constant for hemiacetal formation is greater than 200, so there is little open chain glucose in solution. The reaction is, however, freely reversible.

Hemiacetal formation creates a new chiral center at C-1, the *anomeric* carbon, and the C-1 epimers are called *anomers*.[2] The nomenclature of the cyclic glucoses is based on analogy with the heterocyclic *pyran* ring. The anomers of D-glucose are α- and β-D-*glucopyranose*. These are the two crystalline forms of glucose.

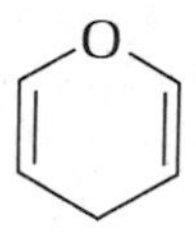

Pyran

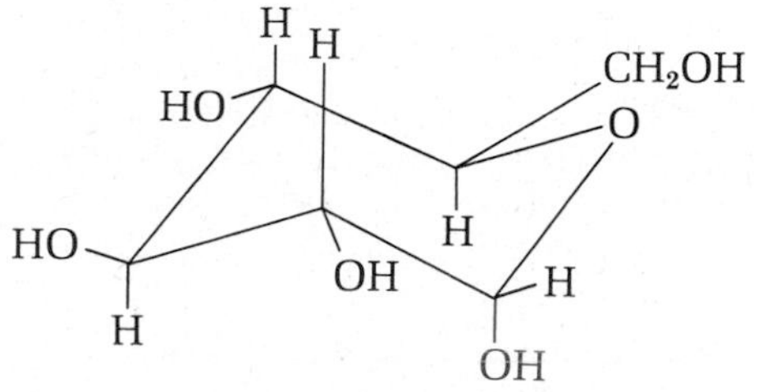

α-D-Glucopyranose

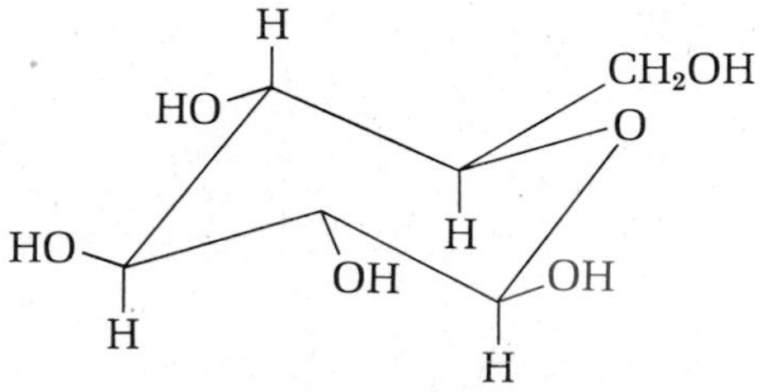

β-D-Glucopyranose

Pure β-D-glucose (as the pyranose is often abbreviated) spontaneously equilibrates in water to give an equilibrium mixture of 64% β- and 36% α-D-glucose, a process called *mutarotation*. Many organisms contain an enzyme, *mutarotase*, that catalyzes this reaction.

We have drawn the structure of glucopyranose in its most stable "chair" conformation. In this conformation the two lone pairs on the ring oxygen bisect the bond to the equatorial hydroxyl group on C-1. The conformation is drawn below both as a chair and as a Newman projection.

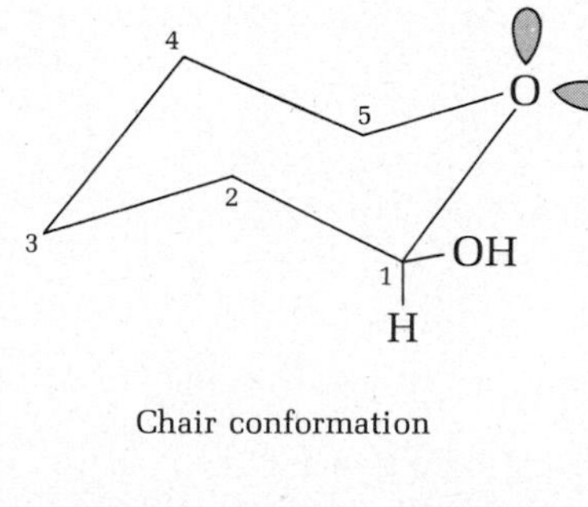

Chair conformation

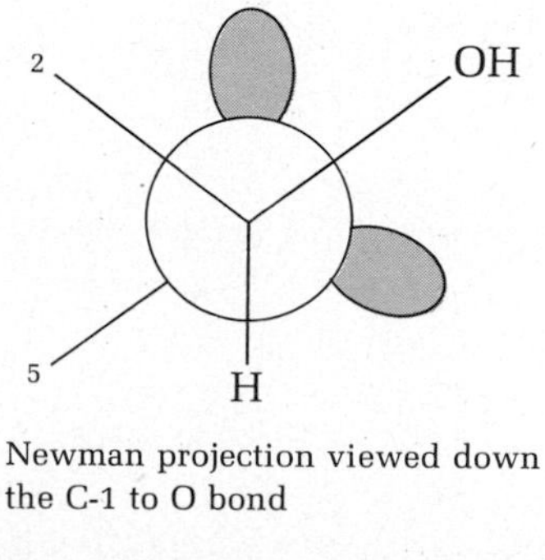

Newman projection viewed down the C-1 to O bond

In glucose the substituents at C-2, C-3, C-4, and C-5 are equatorial in the most stable chair conformation. The hydroxyl group at C-1, the anomeric carbon, can be either axial, as in the α anomer, or equatorial, as in the β anomer. The

---

[2] In ketoses, C-2 is the anomeric carbon (Section 8.3B).

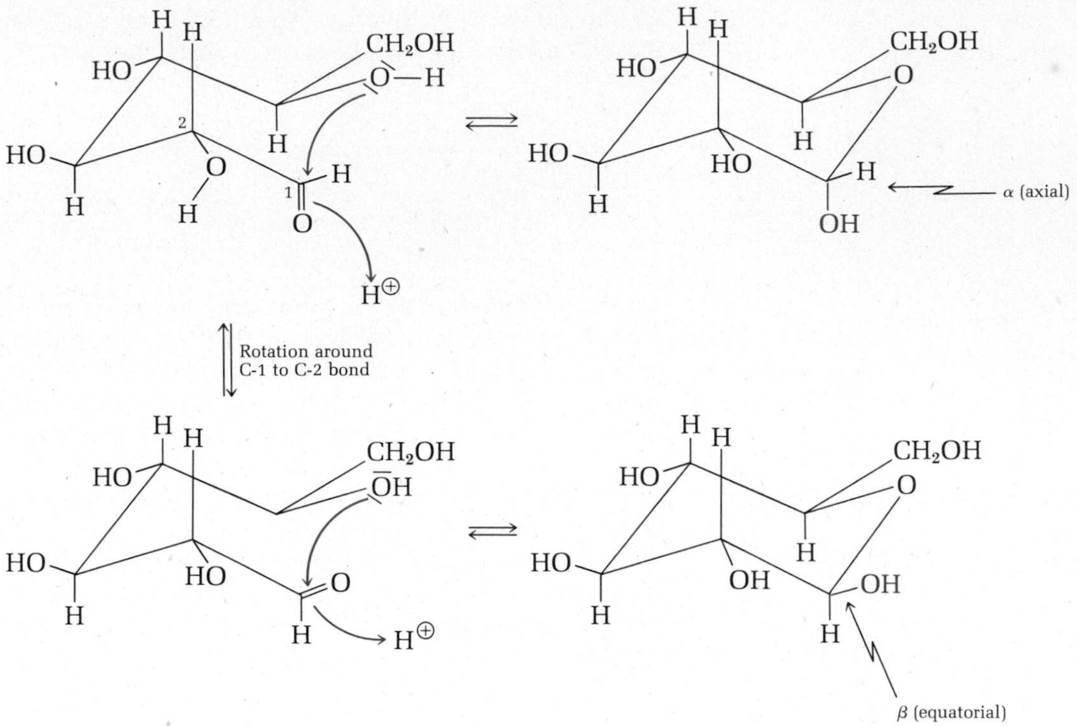

Figure 8.10 Cyclization of glucose. Glucose can form two cyclic hemiacetals, designated α and β, which have different crystalline forms and different optical rotations. The spontaneous inter-conversion of the two forms is known as mutarotation. The α anomer has an axial C-1 hydroxyl group and the β anomer has an equatorial C-1 hydroxyl group.

β anomer of glucose is more stable than the α anomer largely because it is better solvated. There is, however, an unfavorable interaction between the dipoles of the C-1 hydroxyl group and the ring oxygen in the β anomer (Figure 8.11). The dipoles would prefer to be antiparallel, but with an equatorial substituent at C-1 they are nearly parallel. For all aldohexoses except glu-

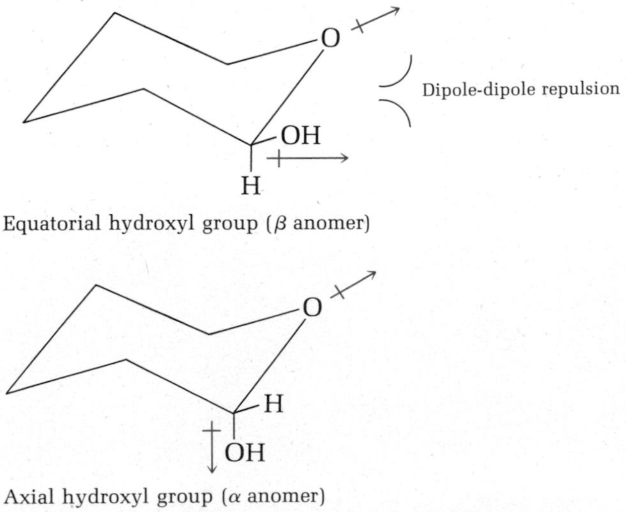

Figure 8.11 Parallel and antiparallel dipoles of anomers. The dipoles between the ring oxygen and an equatorial hydroxyl group are nearly parallel, an unfavorable interaction. In an axial hydroxyl group the dipoles are antiparallel, a more stable configuration. For all hexoses except glucose the α anomer is more stable than the β anomer because of the dipole effects.

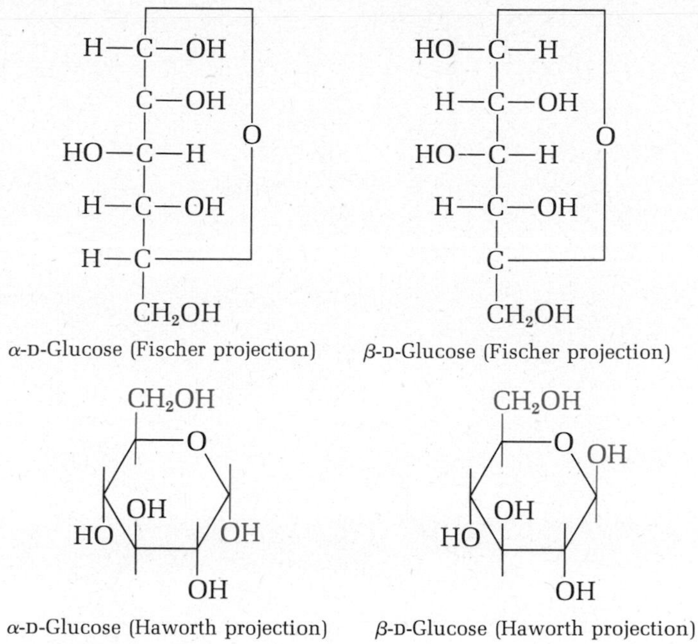

α-D-Glucose (Fischer projection) β-D-Glucose (Fischer projection)

α-D-Glucose (Haworth projection) β-D-Glucose (Haworth projection)

Figure 8.12   Fischer and Haworth projections for α-D-glucose and β-D-glucose.

cose the dipole effect predominates and the α anomer is more stable than the β for all other hexoses. This phenomenon is known as the *anomeric effect*.

Two planar representations of glucose are widely used (Figure 8.12). In a *Fischer projection* of α-D-glucose the hydroxyl group at C-1, the anomeric carbon, is *cis* to the C-5 oxygen; in the β anomer the C-1 hydroxyl group is *trans* to the C-5 hydroxyl group. The actual structure of D-glucose is more accurately represented in a *Haworth projection*, which employs a planar pyran ring with substituents extending above and below the plane of the ring. Note that the —CH$_2$OH group attached to C-5 is above the plane in all D-pyranoses. Any hydroxyl group to the right in a Fischer projection is down in a Haworth projection.

B. Furanoses

a. Fructose

Cyclization of fructose by formation of a *hemiketal* between the carbonyl group at C-2 and the C-5 hydroxyl group gives a five-membered *furanose* ring in which two anomers are possible, α- and β-D-fructofuranose (Figure 8.13).

Fructose can also exist as a pyranose. The possible cyclic fructoses are shown in Figure 8.14.

b. Cyclic Pentoses

Like the hexoses, the pentoses exist as cyclic hemiacetals in aqueous solution. Cyclization between the aldehyde carbon and the C-4 hydroxyl group generates a furanose ring system. For D-ribose two furanoses are possible, α- and β-D-ribofuranose (Figure 8.15). Little free ribose is present in cells, but D-ribosyl moieties are important constituents of ribonucleic acids, nucleosides, and nucleotides (Chapter 9).

Two types of conformations are possible for furanoses. The five-membered ring can be planar without introducing any angle strain in the ring. In a planar conformation, however, the substituents are all eclipsed. If the ring is twisted to relieve this eclipsing, a conformation is produced in which three ring atoms lie in the plane of the page, one ring atom is above the plane, and one ring atom is below the plane. If, on the other hand, one

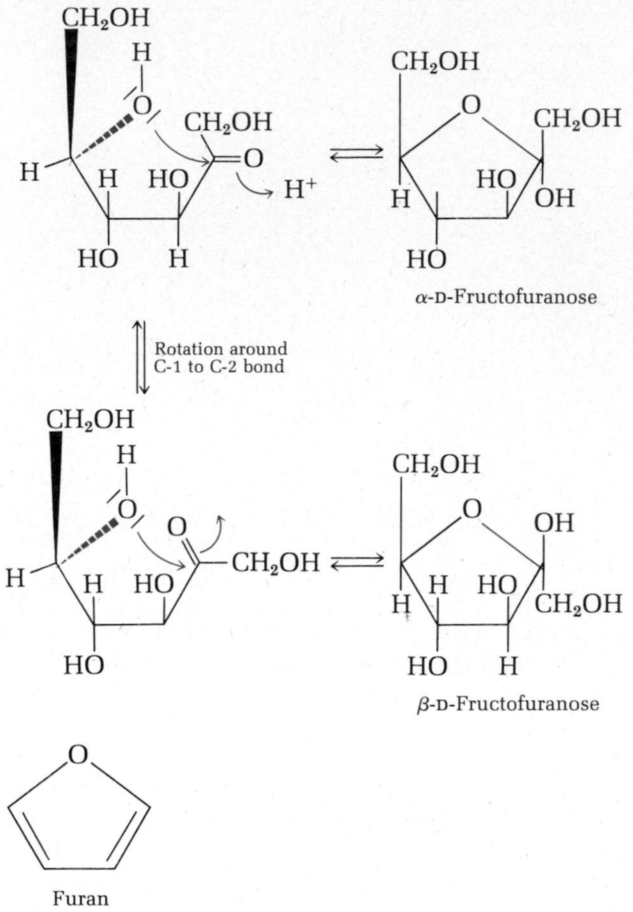

Figure 8.13   Cyclization of fructose to give α- or β-D-fructofuranose, and the furan ring system by which furanoses are named.

atom is placed above (or below) the plane, an envelope conformation results. The envelope conformation in which C-2 is below the plane of the other four ring atoms is observed in nucleic acids (Figure 8.16).

8–4 CHEMISTRY OF MONOSACCHARIDES

A. Oxidation

Aldoses, and aldehydes in general, can be oxidized by an aqueous solution of silver ammonia complex, $Ag(NH_3)_2^{\oplus}$, known as *Tollens' reagent*. The aldehyde form of carbohydrates is oxidized to an *aldonic* acid, while cyclic hemiacetals are oxidized to *lactones* (Figure 8.17). The balanced chemical equation for the reaction is

$$RCHO + 2Ag(NH_3)_2^{\oplus} + 2OH^{\ominus} \longrightarrow RCO_2^{\ominus} + 2Ag^0 + 3NH_3 + NH_4^{\oplus}$$

(8.4)

The deposition of a silver mirror provides a qualitative test for the presence of reducing sugars.

Benedict's solution and Fehling's solution are other common oxidizing reagents for detection of reducing sugars. The two reagents employ cupric sulfate and sodium carbonate with a citrate buffer for Benedict's solution and a tartrate buffer for Fehling's solution. In either case an aldonic acid and a brick-red Cu_2O precipitate are the products of the reaction.

$$\underset{\text{Aldose}}{RCHO} + 2\ Cu(II) + 5\ OH^{\ominus} \xrightarrow[\text{or tartrate}]{\text{Citrate}} \underset{\text{Aldonic acid}}{RCO_2^{\ominus}} + Cu_2O + 3\ H_2O$$

(8.5)

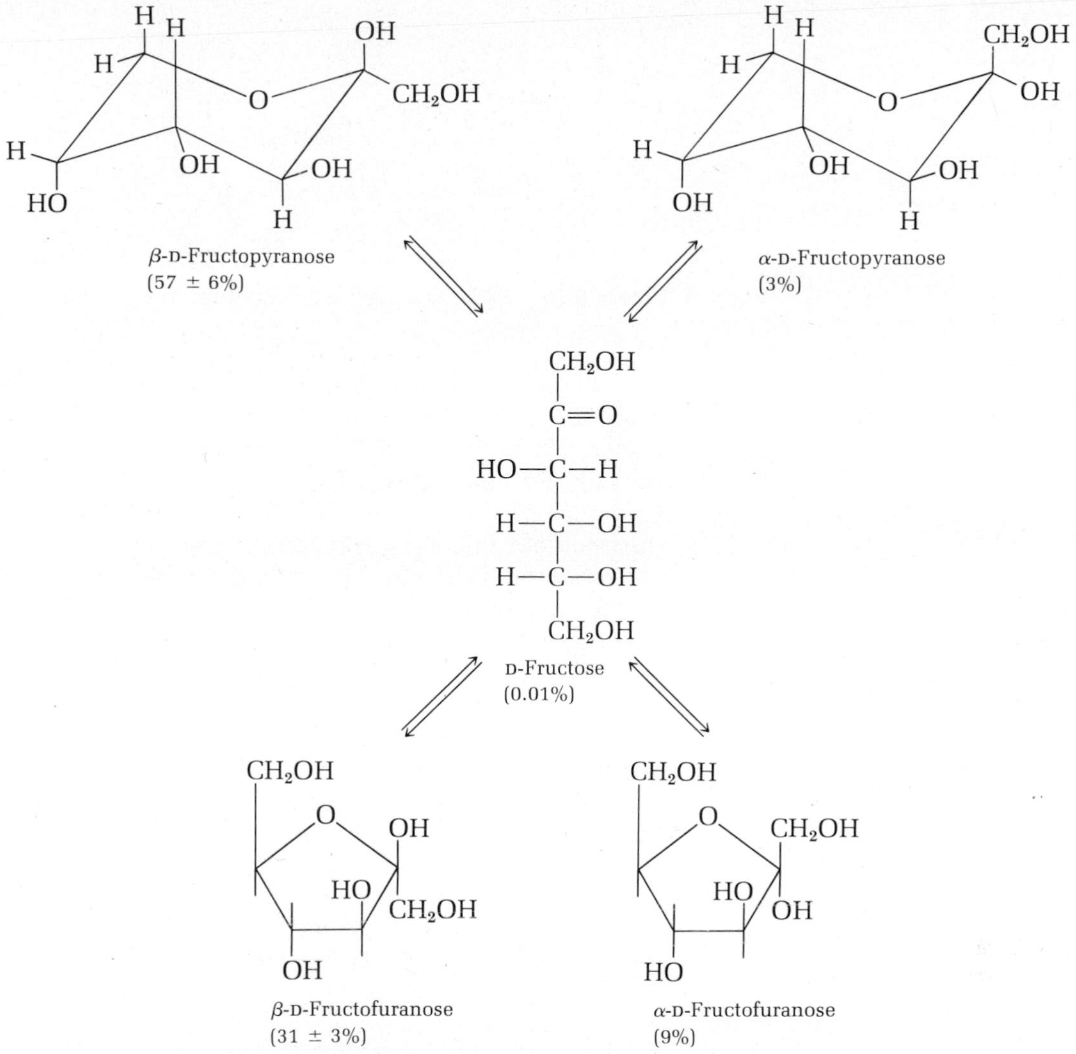

Figure 8.14 Major forms of fructose in aqueous solution and the approximate percent of each at equilibrium. [Data from A. Allerhand and D. Doddrell, *J. Amer. Chem. Soc.*, 93, 2777, 2779 (1971).]

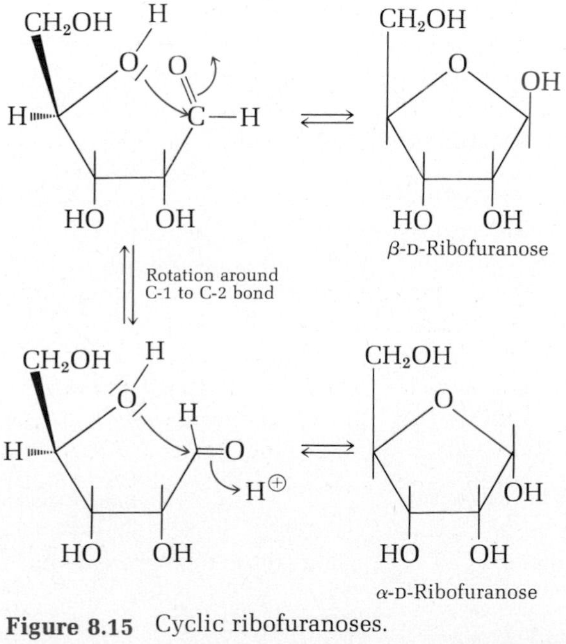

Figure 8.15 Cyclic ribofuranoses.

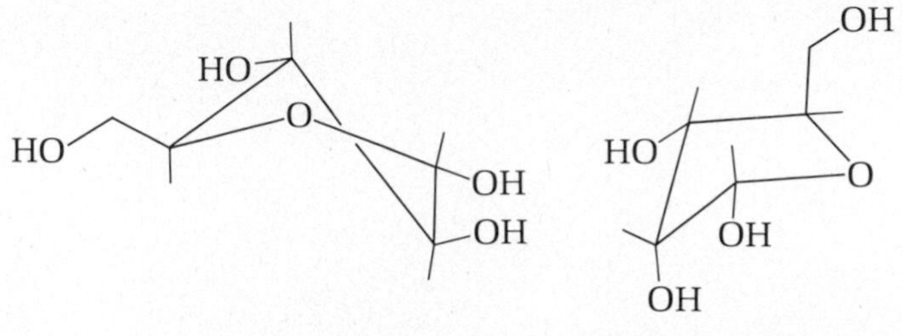

Twist conformation   Envelope conformation

Figure 8.16 Twist and envelope conformations of ribose.

Ketoses also react with Tollens', Benedict's, and Fehling's reagents. They are not directly oxidized, but undergo a base-catalyzed tautomerization known as the *Lobry de Brun−Alberda van Eckenstein* rearrangement of the aldose to the ketose (Figure 8.18). The equilibrium mixture of fructose contains 2.5% D-mannose, 63.5% D-glucose, and 31% fructose. The *enediol intermediate* in the rearrangement is of a type frequently encountered in enzyme-catalyzed isomerizations in carbohydrate metabolism (see, for example, Section 14.3E). The aldoses in the reaction mixture are oxidized by Tollens' and the other reagents, as expected.

Aldoses are also oxidized by bromine water. The reaction yields δ-*lactones* without opening the ring. The reactivities of the α and β anomers differ considerably, and oxidation of the β anomers is faster. In fact, the rate-determining step for oxidation of the α anomer is mutarotation to the β anomer. The reaction mechanism is shown in Figure 8.19. The enzyme glucose 6-phosphate dehydrogenase (Section 17.2) also oxidizes glucose to a δ-lactone. The relative rates of oxidation for the α and β anomers are the same as those observed for bromine oxidation. The more rapid oxidation of the β anomer plays a role in metabolic regulation since the anomeric specificities of enzymes in glucose metabolism differ.

B. Glycosides   A hemiacetal, such as α-D-glucopyranose, can react with an alcohol, such as methanol, to give a *full acetal* or *glycoside* (reaction 8.6). The glycoside is named as an alkyl pyranoside (or furanoside). The glycoside is not a reducing sugar and will not react with Tollens', Benedict's, or Fehling's reagent.

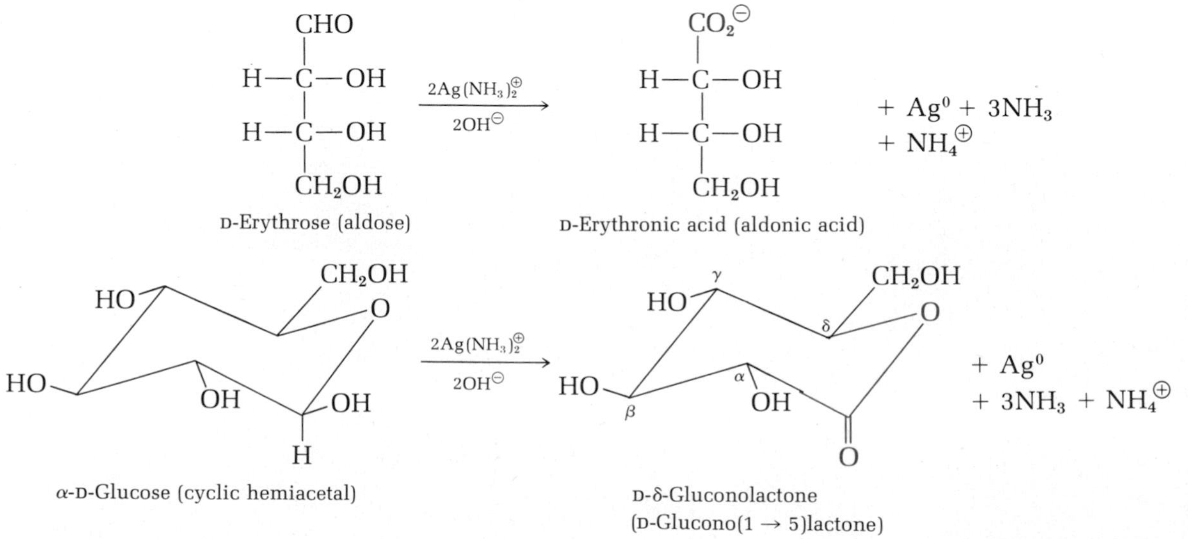

D-Erythrose (aldose)

D-Erythronic acid (aldonic acid)

α-D-Glucose (cyclic hemiacetal)

D-δ-Gluconolactone
(D-Glucono(1 → 5)lactone)

Figure 8.17 Oxidation of an open chain aldose to give an aldonic acid and oxidation of a cyclic hemiacetal to give a lactone.

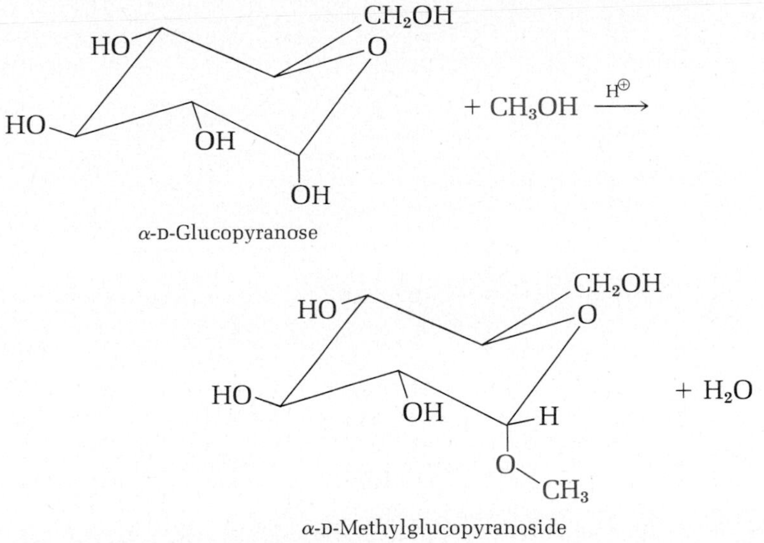

α-D-Glucopyranose

$+ CH_3OH \xrightarrow{H^{\oplus}}$

$+ H_2O$ (8.6)

α-D-Methylglucopyranoside

A hemiketal, such as β-D-fructofuranose, can also react with an alcohol to give a *full ketal* (reaction 8.7). Neither full acetals nor full ketals are reducing sugars.

Glycosides are quite stable in basic solutions (hydroxide is not a powerful enough nucleophile to readily displace alkoxide anion), but they hydrolyze readily in dilute aqueous acid solution. The anomeric α- and β-glycosides can be distinguished by their reactivity with α- and β-*glucosidases*, stereospecific hydrolases of glycosidic bonds. The former are

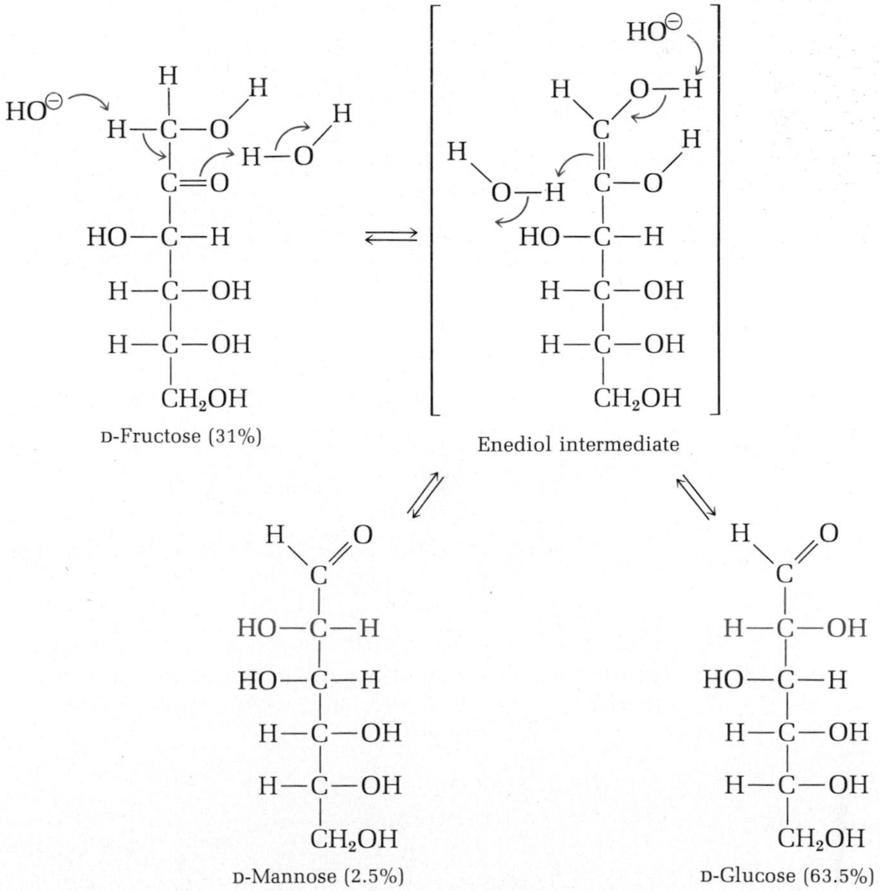

D-Fructose (31%)

Enediol intermediate

D-Mannose (2.5%)

D-Glucose (63.5%)

Figure 8.18   Lobry de Brun–Alberda van Eckenstein rearrangement.

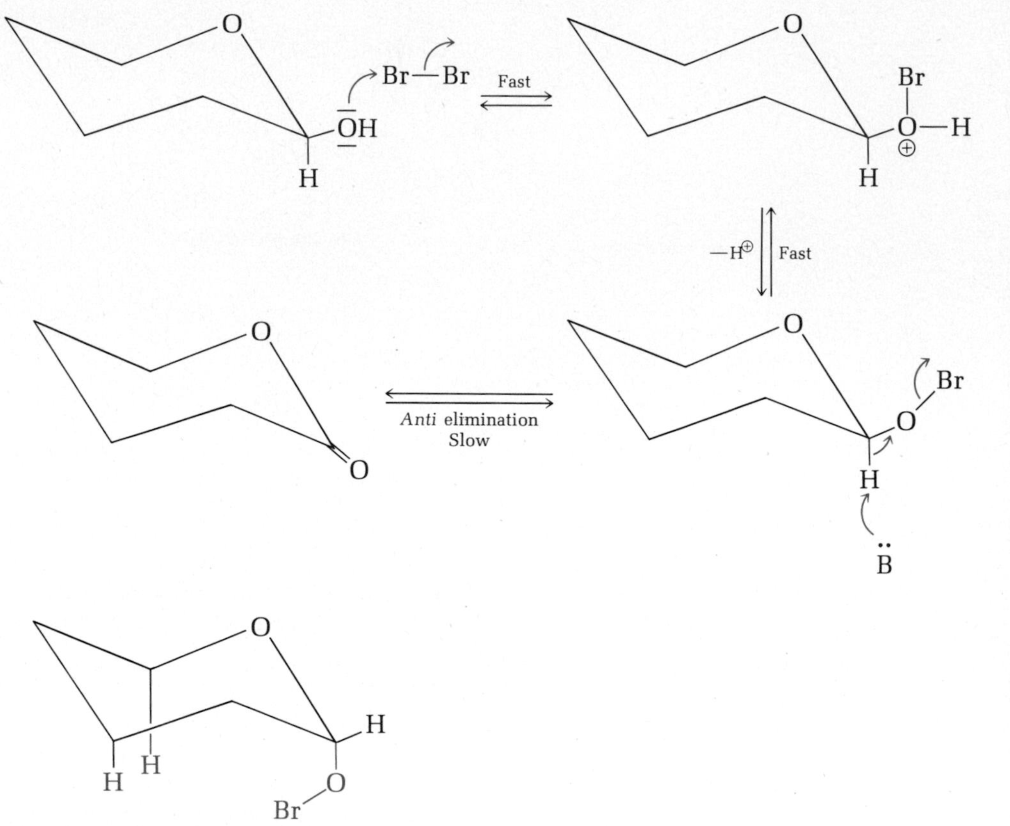

Steric hindrance of conformation
required for anti elimination
of HBr

Figure 8.19 Mechanism for bromine oxidation of a pyranose. The β anomer reacts much faster than the α anomer because the conformation required for *anti* elimination of HBr is sterically hindered in the α anomer. The same effect probably governs the reaction catalyzed by glucose 6-phosphate dehydrogenase (see Section 17.2A).

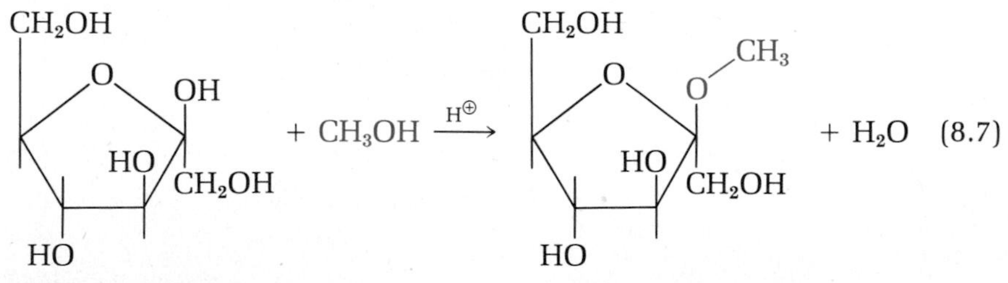

β-D-Fructofuranose $\qquad$ β-D-Methylfructofuranoside

specific for α-glucosides and the latter are specific for β-glucosides (Figure 8.20).

C. Methylation with Dimethyl Sulfate

Reaction of hemiacetals with methanol affects only the anomeric carbon. The remaining hydroxyl groups—either primary or secondary alcohols—can be methylated with *dimethyl sulfate*.

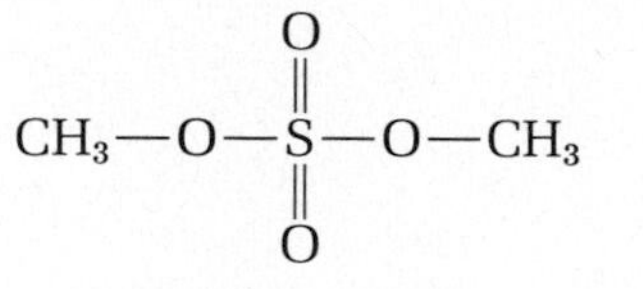

Dimethyl sulfate

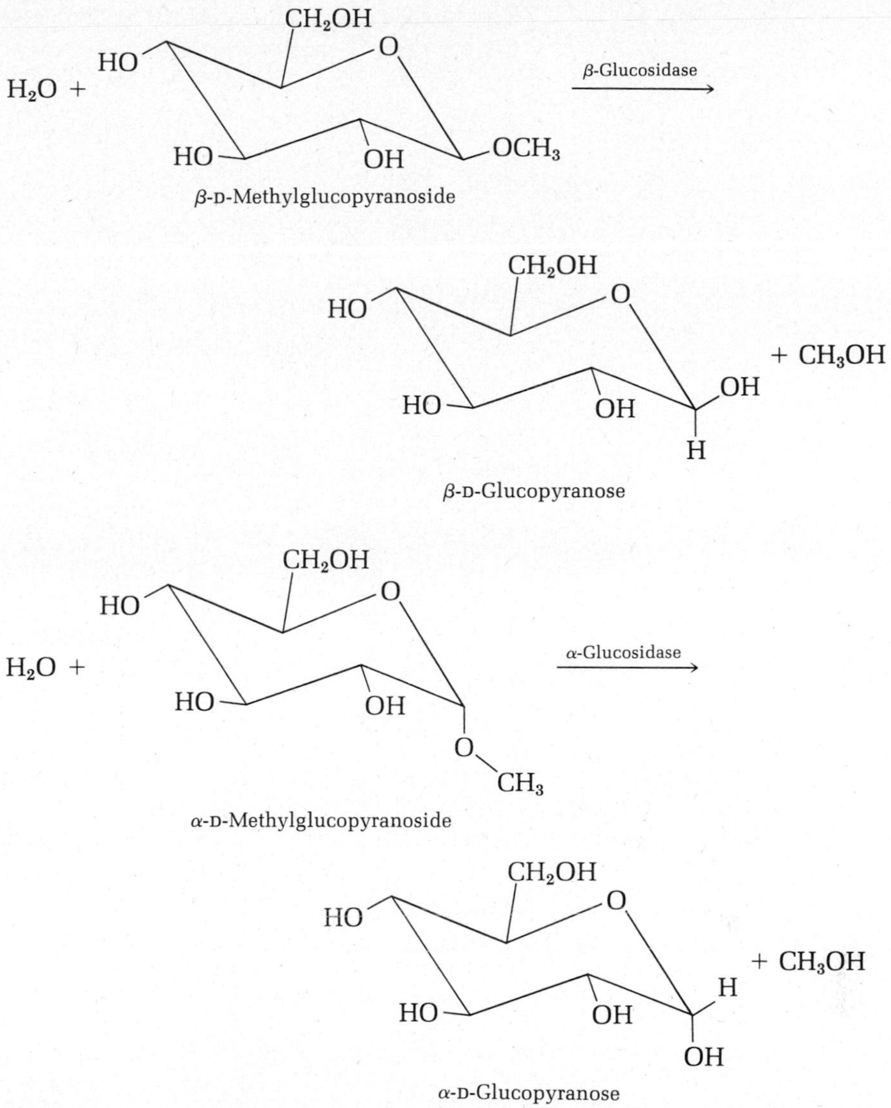

β-D-Methylglucopyranoside

β-D-Glucopyranose

α-D-Methylglucopyranoside

α-D-Glucopyranose

Figure 8.20 Specific reactivity of α- and β-glucosidase for α- and β-glucoside, respectively.

Treatment of glucose or of methylglucosides with dimethyl sulfate results in methylation of every hydroxyl group (*permethylation*) (reaction 8.8).

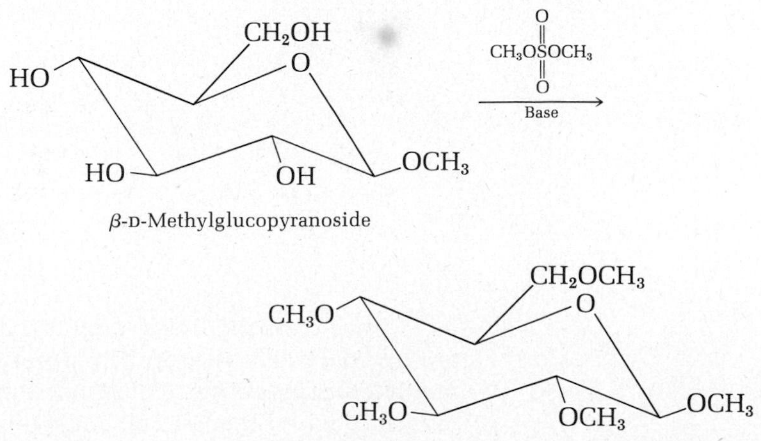

β-D-Methylglucopyranoside

(8.8)

2,3,4,6-Tetra-*O*-methyl-β-D-methylglucopyranoside

The methyl glucoside is easily hydrolyzed in dilute acid and the product is 2,3,4,6-tetra-O-methyl-β-D-glucopyranose (reaction 8.9).

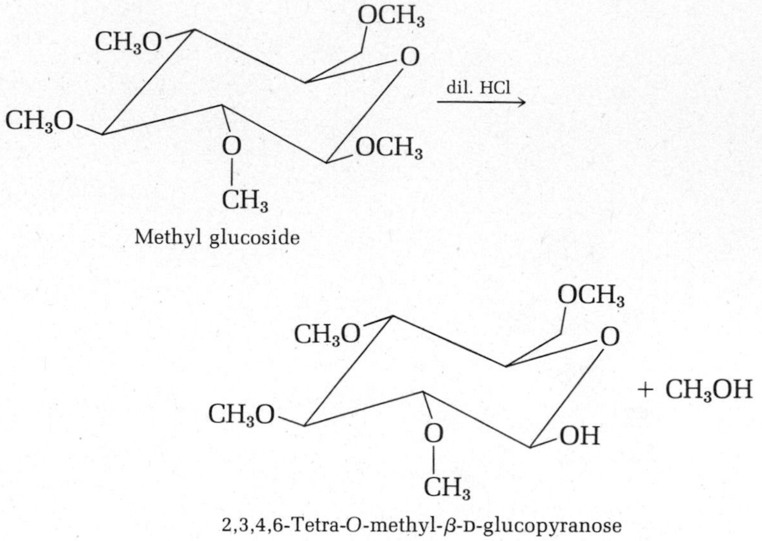

Methyl glucoside

$$\xrightarrow{\text{dil. HCl}}$$

$+ CH_3OH$ (8.9)

2,3,4,6-Tetra-O-methyl-β-D-glucopyranose

D. Oxidation of Carbohydrates with Periodate

Periodate, $IO_4^{\ominus}$, specifically cleaves carbon-carbon bonds of vicinal glycols, α-hydroxy aldehydes and ketones, and other compounds containing adjacent oxidizable functional groups (Figure 8.21). In most cases the reaction proceeds via a cyclic periodate ester, as shown below for a vicinal glycol (reaction 8.10).

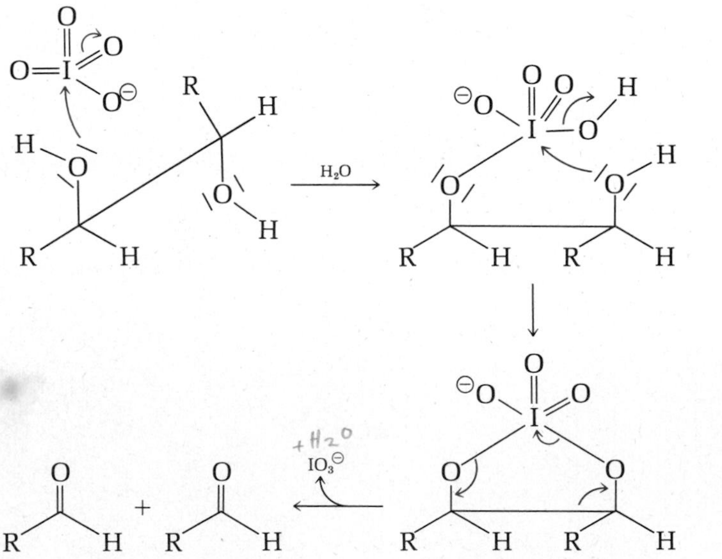

Cyclic periodate ester intermediate

(8.10)

Since carbohydrates contain many adjacent oxidizable functional groups, periodate is useful in structure determinations. For example, β-D-methylglucopyranoside consumes 2 moles of periodate and produces 1 mole of formic acid (Figure 8.22), but β-D-methylfructofuranoside consumes only 1 mole of periodate and produces no formic acid (Figure 8.23). The observations show that fructose is indeed a furanose and that glucose is a pyranose.

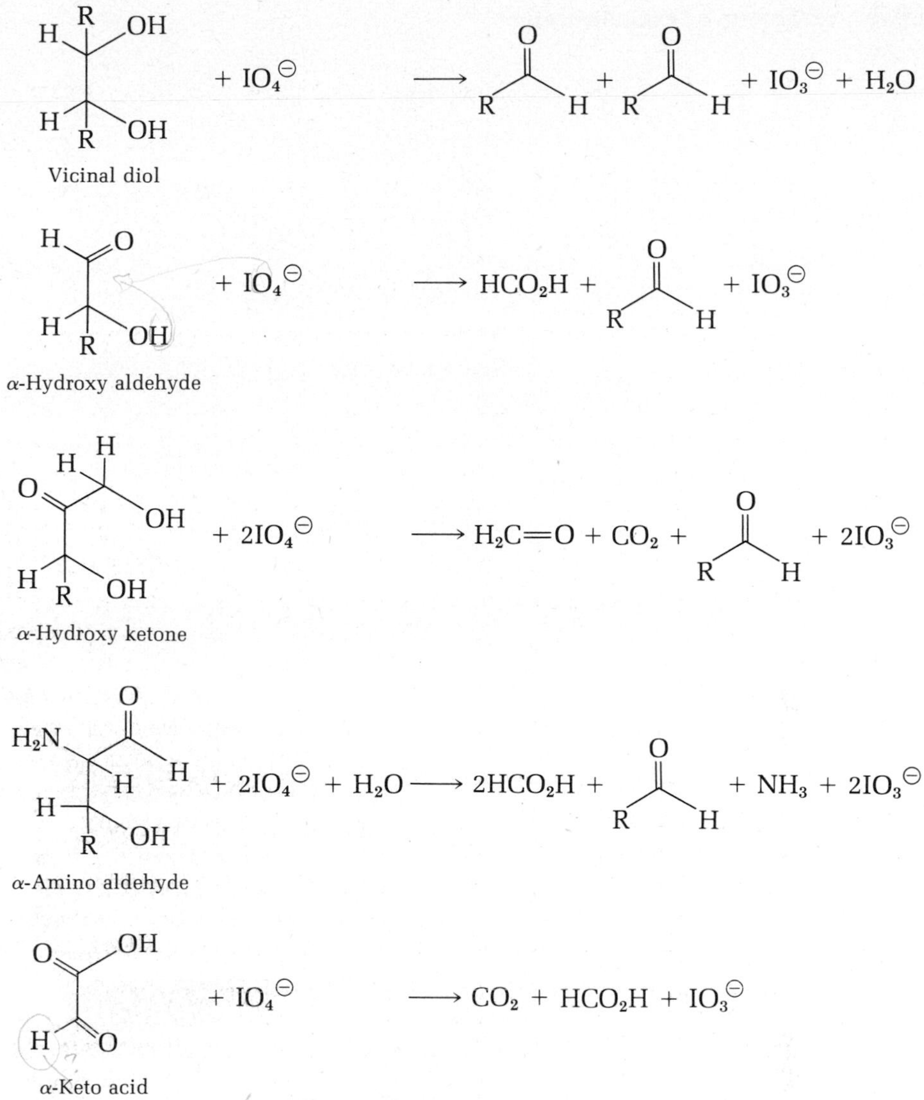

Figure 8.21 Representative functional groups oxidized by periodate.

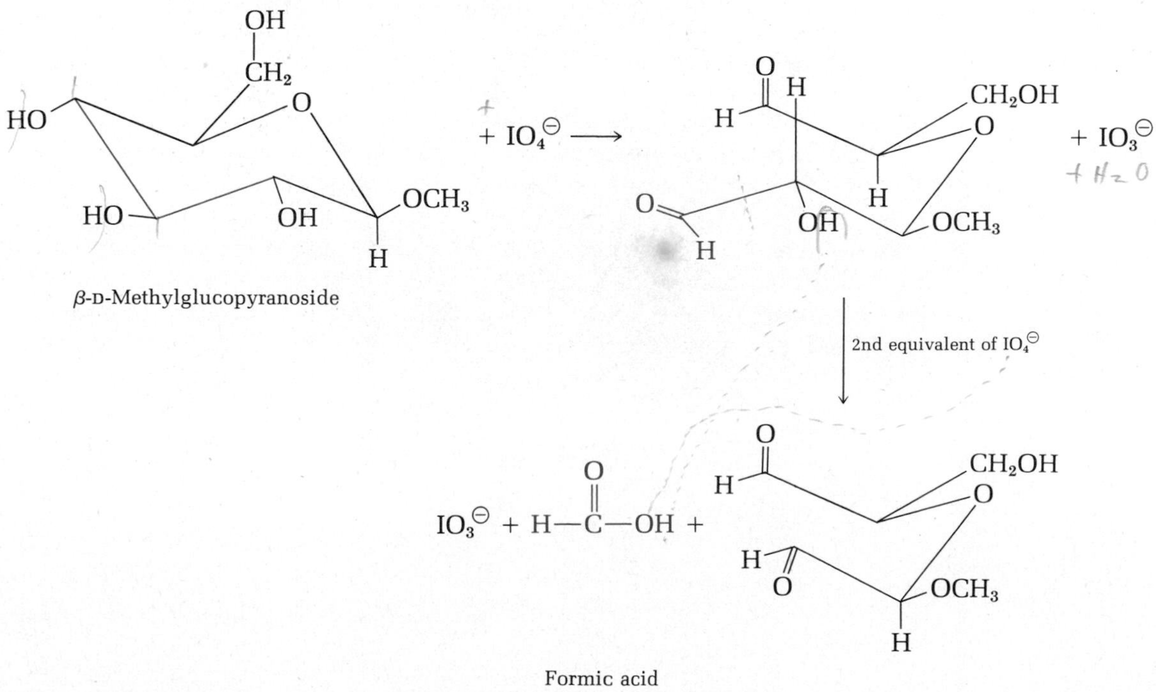

Figure 8.22 Oxidation of β-D-methylglucopyranoside with periodate.

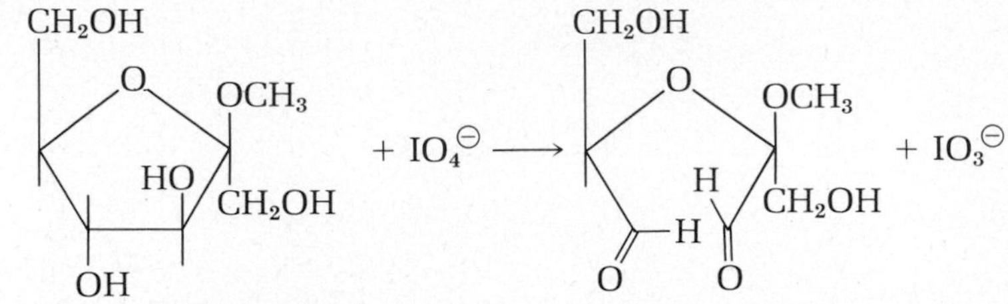

β-D-Methylfructofuranoside

Figure 8.23 Oxidation of β-D-methylfructofuranoside with periodate.

E. Reduction of Carbohydrates

Sodium borohydride, $NaBH_4$, reduces aldoses to *alditols* (reaction 8.11).

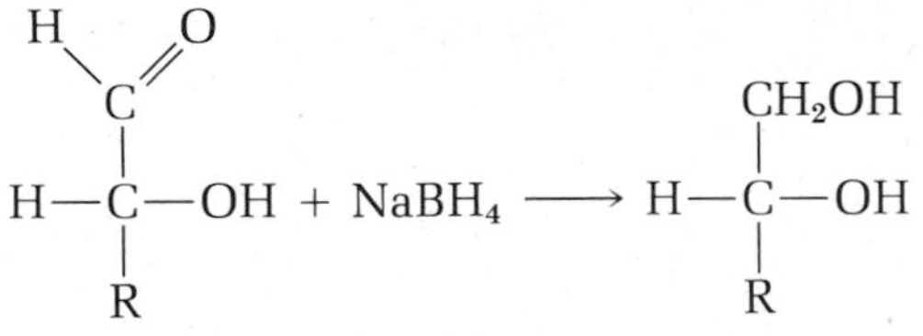

(8.11)

Aldose   Alditol

Alditols are widely distributed in nature. For example, *erythritol* is found in algae, fungi, and lichens, and the hexitols D-*mannitol* and D-*sorbitol* are found throughout the plant kingdom.

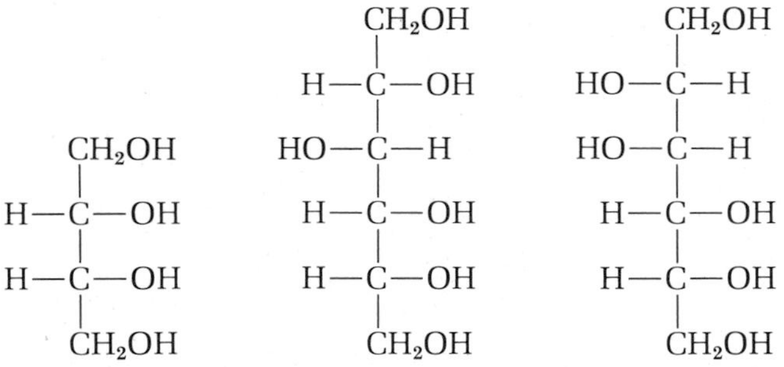

Erythritol   D-Sorbitol   D-Mannitol

8–5 PHOSPHATE ESTERS OF CARBOHYDRATES

Among the large number of carbohydrate derivatives found in nature, phosphate esters assume a singular importance because they are the major class of derivatives formed in carbohydrate metabolism (Chapters 14 and 17).

A. Structures and Acidities of Common Phosphate Esters

The structures and pK_a values for some important phosphate esters are shown in Table 8.1. Sugar phosphates are stronger acids than orthophosphoric acid. The ionizations for these acids are shown in reactions 8.12 and 8.13.

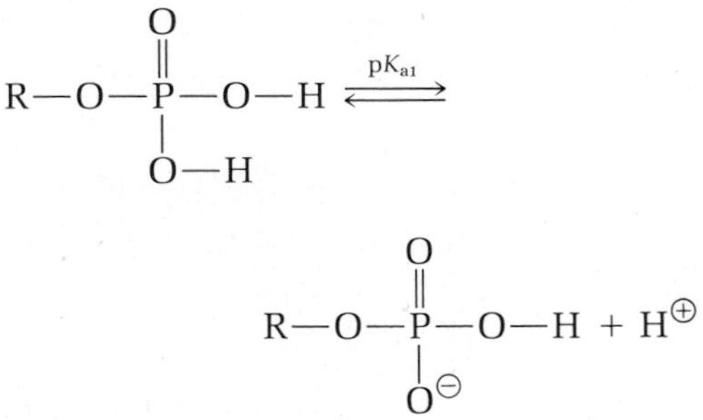

pK_{a1} 0.94 to 2.10   (8.12)

and

$$
R-O-\overset{\overset{\displaystyle O}{\|}}{\underset{\underset{\displaystyle O-H}{|}}{P}}-O^{\ominus} \underset{\longleftarrow}{\overset{pK_{a2}}{\rightleftharpoons}}
$$

$$
R-O-\overset{\overset{\displaystyle O}{\|}}{\underset{\underset{\displaystyle O^{\ominus}}{|}}{P}}-O^{\ominus} + H^{\oplus} \qquad pK_{a2} \text{ 6.1 to 6.8} \qquad (8.13)
$$

From the data in Table 8.1 we can calculate (using the Henderson-Hasselbalch equation) that the ratio of monoanion to dianion, $ROPO_3H^{\ominus}:ROPO_3^{2-}$, varies from 8:1 for glucose 6-phosphate to 1.77:1 for D-glyceraldehyde 3-phosphate at pH 7.0.

B. Hydrolysis of Phosphate Esters

In principle hydrolysis of phosphate esters can occur with either C—O or P—O bond cleavage. The latter is analogous to acyl bond cleavage in carboxylate esters and the former is analogous to alkyl bond cleavage in carboxylate esters.

C—O cleavage   Alkyl bond cleavage

$$
R{\overset{\curvearrowleft}{-}}O{-}\overset{\overset{\displaystyle O}{\|}}{\underset{\underset{\displaystyle OH}{|}}{P}}{-}OH \qquad\qquad R{\overset{\curvearrowleft}{-}}O{-}\overset{\overset{\displaystyle O}{\|}}{C}{-}R'
$$

P—O cleavage   Acyl bond cleavage

Hydrolysis of methyl phosphate in $H_2^{18}O$ shows that the P—O bond is cleaved since all ^{18}O is incorporated in phosphate (Figure 8.24). The configuration of phosphorus changes from tetrahedral to trigonal bipyramidal to tetrahedral during the course of the reaction (Figure 8.25). This configurational change is analogous to the change from trigonal to tetrahedral to trigonal that occurs during hydrolysis of carboxylate esters (Figure 8.26). The hydrolysis of methyl phosphate is strongly dependent on pH and exhibits a maximum at pH 4.0 (Figure 8.27). The predominant species in solution at pH 4.0 is the monoanion, and the maximum rate at pH 4.0 indicates that the monoanion is the species undergoing hydrolysis.

The hydrolysis of glucose 1-phosphate, on the other hand, has the pH-rate profile shown in Figure 8.28. The rate increases steadily as the pH decreases. This behavior indicates that the fully protonated molecule is more reactive than the anionic species. The hydrolysis of glucose 1-phosphate is assisted by the adjacent ring oxygen. The intermediate is stabilized by the neighboring group ring oxygen (Figure 8.29). The intermediate in the reaction is a resonance-stabilized *acylium ion* which has a *half-chair* conformation.

Half-chair conformation of acylium ion

Phosphate esters hydrolyze quite slowly in base since the dianion, which is the predominant species in solution above pH 7.0, repels the approaching hydroxide anion. Also, in basic solutions the leaving group is alkoxide anion, a poor leaving group.

Table 8.1   Structures and pK_a values of some important phosphate esters of carbohydrates

Structure	Name	pK_{a1}	pK_{a2}	pK_{a3}
	β-D-Glucose 1-phosphate	1.10	6.13	
	Glucose 6-phosphate	0.94	6.11	
	Fructose 6-phosphate	0.97	6.11	
	Fructose 1,6-diphosphate	1.48	6.32	
	β-D-Galactose 1-phosphate	1.00	6.17	
H_3PO_4	Orthophosphoric acid	1.97	6.82	12.4
	D-Glyceraldehyde 3-phosphate	2.10	6.75	

continued

Table 8.1 Structures and pK_a values of some important phosphate esters of carbohydrates (*continued*)

Structure	Name	pK_{a1}	pK_{a2}	pK_{a3}
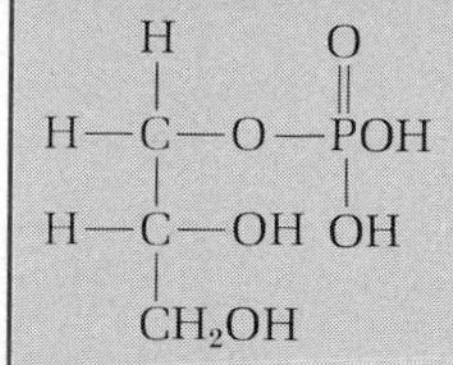 (structure of sn-glycerol 3-phosphate)	sn-glycerol 3-phosphate (glycerol 1-phosphate; α-Glycerol phosphate)	1.44	6.44	
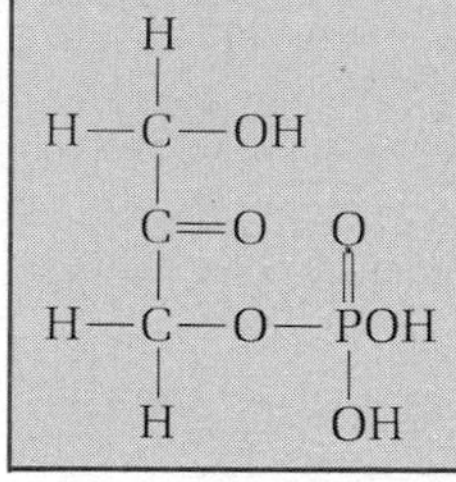 (structure of dihydroxyacetone phosphate)	Dihydroxyacetone phosphate	1.77	6.45	

8–6 OLIGOSACCHARIDES

The oligosaccharides are important sources of energy throughout the biological world. They contain only a few monosaccharide residues (the Greek prefix *oligo* means "a few"), and complete hydrolysis gives either monosaccharides or simple derivatives of monosaccharides.

There is no clear demarcation between oligo- and polysaccharides, but the most abundant oligosaccharides have fewer than 10 residues. The monomers are bonded to one another by glycosidic bonds. If both hemiacetal

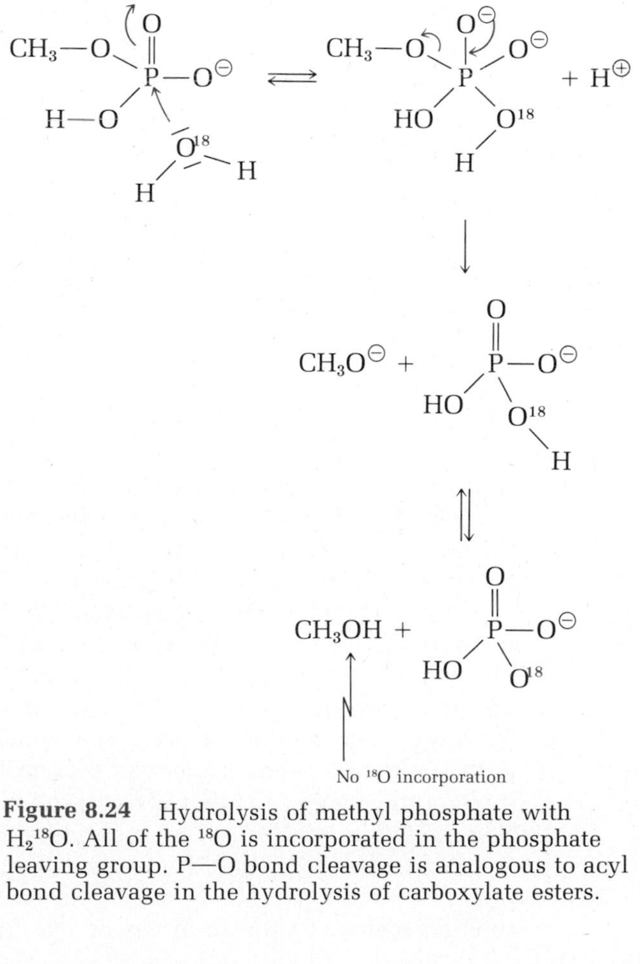

Figure 8.24 Hydrolysis of methyl phosphate with $H_2{}^{18}O$. All of the ^{18}O is incorporated in the phosphate leaving group. P—O bond cleavage is analogous to acyl bond cleavage in the hydrolysis of carboxylate esters.

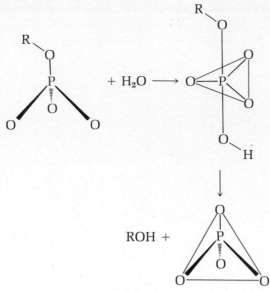

Figure 8.25 Configurational changes in the structure of phosphate in the hydrolysis of phosphate esters. (See Section 12.6 for further details, especially stereochemical considerations.)

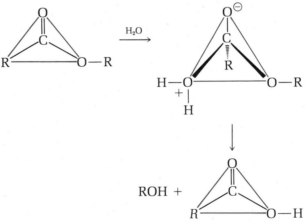

Figure 8.26 Configurational changes in the structure of the carbonyl group in the hydrolysis of carboxylate esters.

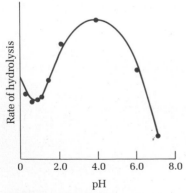

Figure 8.27 pH-rate profile for the hydrolysis of methyl phosphate. [From C. A. Bunton, D. R. Llewellyn, K. G. Oldman, and C. Vernon, *J. Amer. Chem. Soc.*, 80, 3574 (1958).]

carbon atoms are involved in the bond, the sugar is nonreducing. If one of the hemiacetal bonds is free, the oligomer is a *reducing sugar*. The nonreducing end is written to the left in oligo- and polysaccharide structures. Some representative disaccharides are listed in Table 8.2.

Humans, especially Americans, consume vast amounts of *sucrose*, a nonreducing sugar of α-D-glucose bonded (1 → 2) to β-D-fructose. The nonreducing sugar *trehalose* is a dimer of glucose linked (1 → 1) which occurs in three forms: α,α, α,β, and β,β trehalose. These sugars are produced by fungi. Milk sugar, or *lactose*, is a dimer of β-D-galactose bonded (1 → 4) with D-glucose. *Galactosemia* is a condition in humans in which one or more of the enzymes responsible for the metabolism of galactose is missing. The condition leads to severe mental retardation, cataracts, and early death if left untreated. We shall discuss the details of galactose metabolism in Section 17.5B. The reduced form of galactose, called *dulcitol*, is the toxic by-product that accumulates in persons suffering from galactosemia.

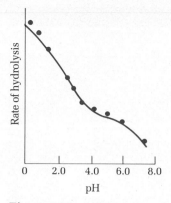

CH₂OH
H—C—OH
HO—C—H
HO—C—H
H—C—OH
CH₂OH
Dulcitol

Figure 8.28 pH-rate profile for the hydrolysis of glucose 1-phosphate. [From C. A. Bunton, D. R. Llewellyn, K. G. Oldman, and C. Vernon, *J. Amer. Chem. Soc.*, *80*, 3574 (1958).]

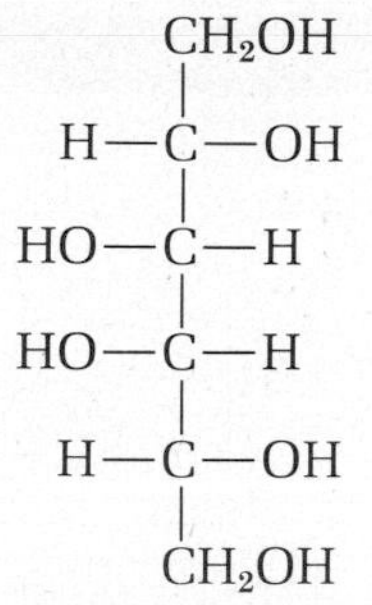

The effects of galactosemia can be mitigated by providing susceptible infants with a diet that does not contain galactose.

Since oligosaccharides are glycosides, they hydrolyze readily in dilute aqueous acid. The enzymes *maltase* and *emulsin* specifically hydrolyze α- and β-glucosides, respectively. The disaccharide maltose (Table 8.2) is produced by the action of maltase on polysaccharides in which the glucosyl residues are bonded α(1 → 4).

8–7 POLYSACCHARIDES

The carbohydrates found in nature consist mostly of high molecular weight polymers of glucose and its derivatives. There are five positions for linking the monosaccharides, and the number of possible structures is enormous. Because all biochemical reactions are enzyme catalyzed, it would take an extraordinary number of enzymes to synthesize and degrade all of the possible structures. Perhaps in the interest of simplicity, nature has limited her-

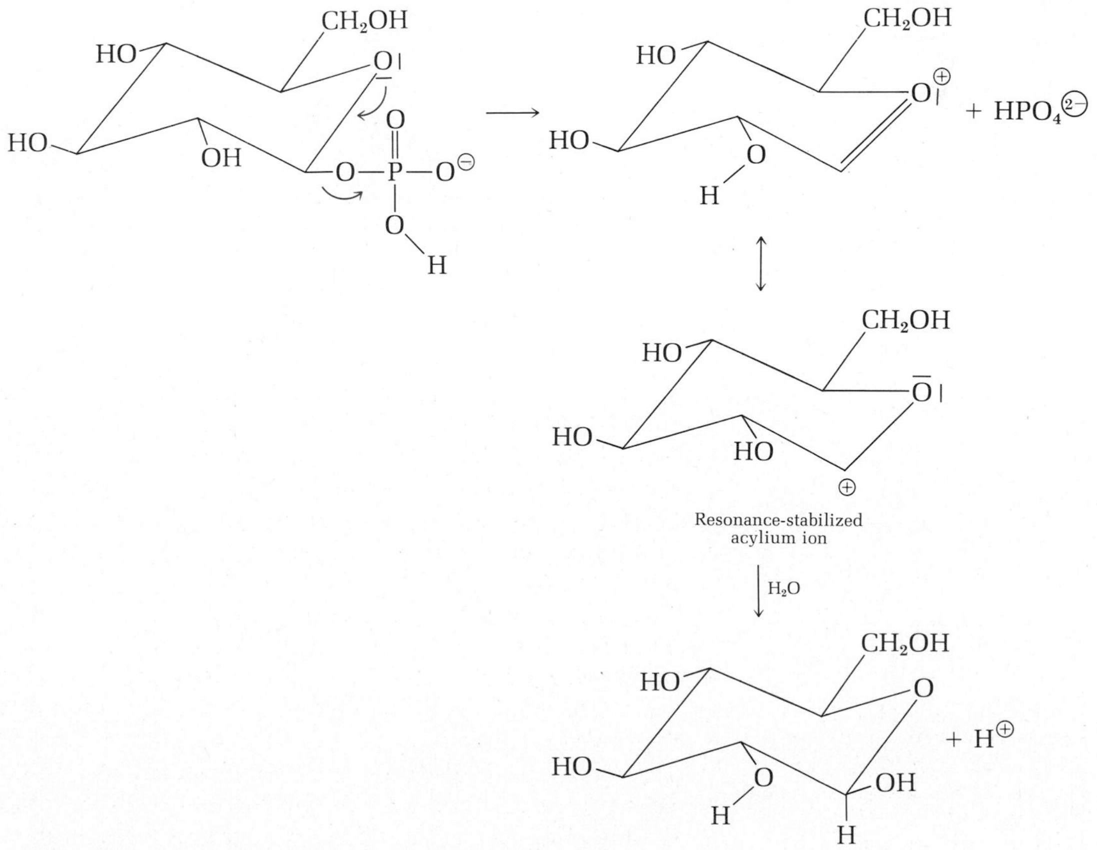

Resonance-stabilized acylium ion

Figure 8.29 Hydrolysis of glucose 1-phosphate at pH 4.0.

Table 8.2 Structures and bonding of some common disaccharides

Structure	Name	Bonding
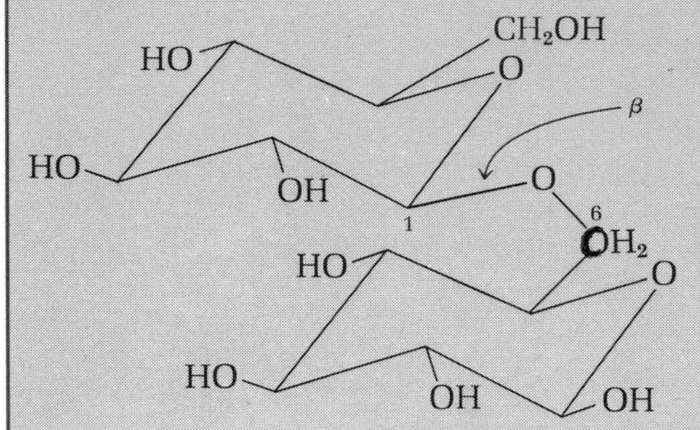	Gentiobiose	$\beta(1 \rightarrow 6)$
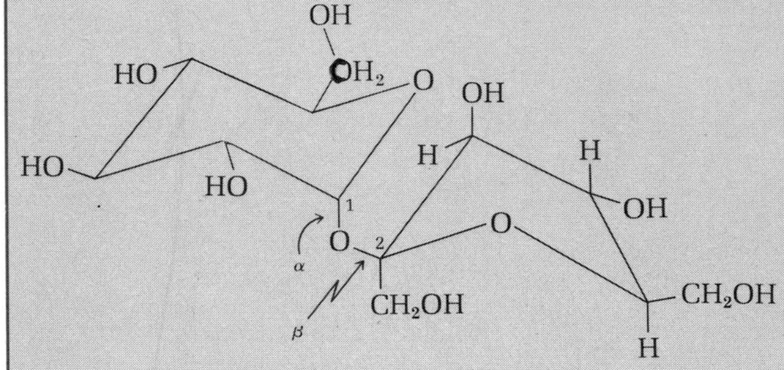	Sucrose	$\alpha\text{-}1 \rightarrow \beta\text{-}2$
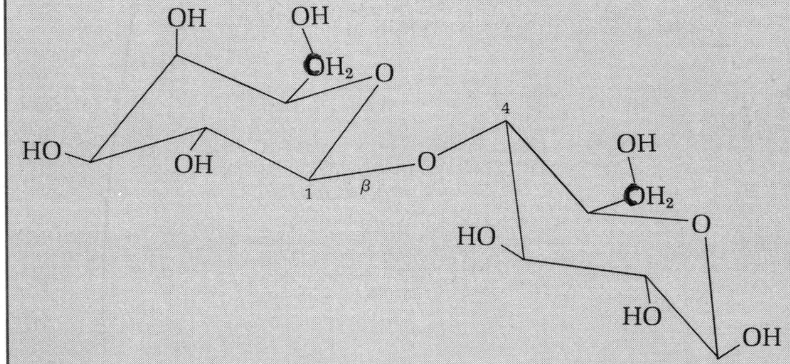	Lactose 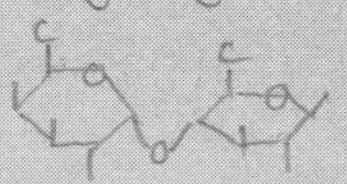	$\beta(1 \rightarrow 4)$
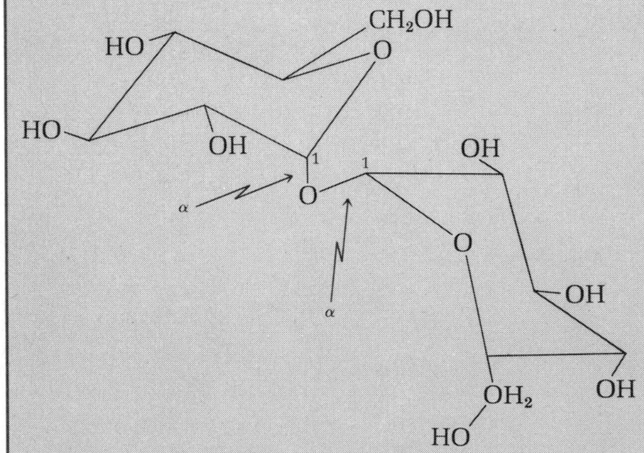	α,α-Trehalose	α,α
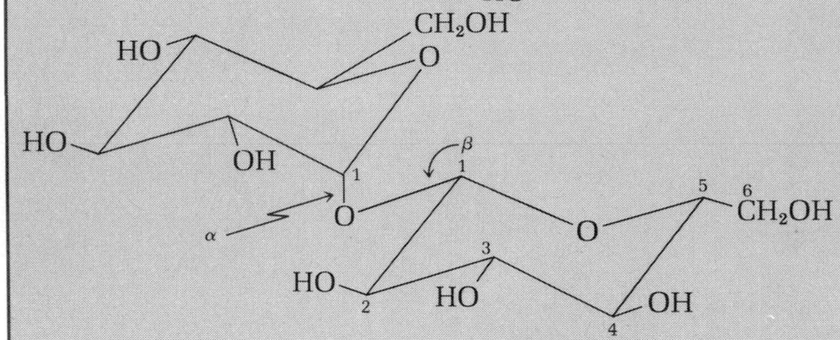	α,β-Trehalose	α,β

continued

Table 8.2 Structures and bonding of some common disaccharides (continued)

Structure	Name	Bonding
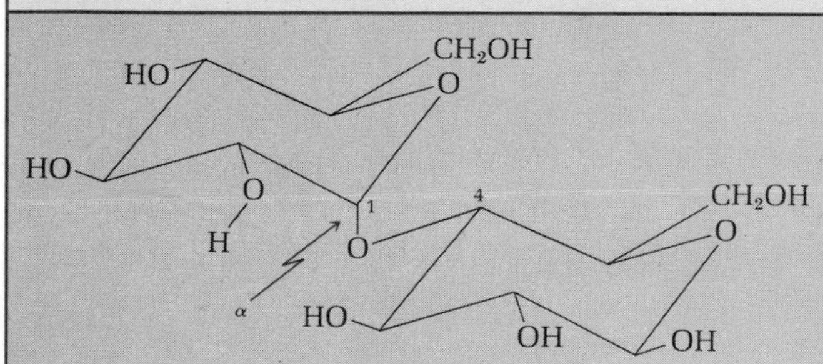	Maltose	$\alpha(1 \rightarrow 4)$
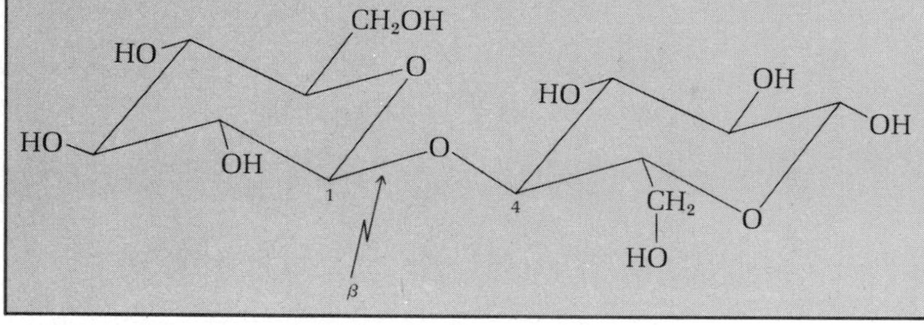	Cellobiose	$\beta(1 \rightarrow 4)$

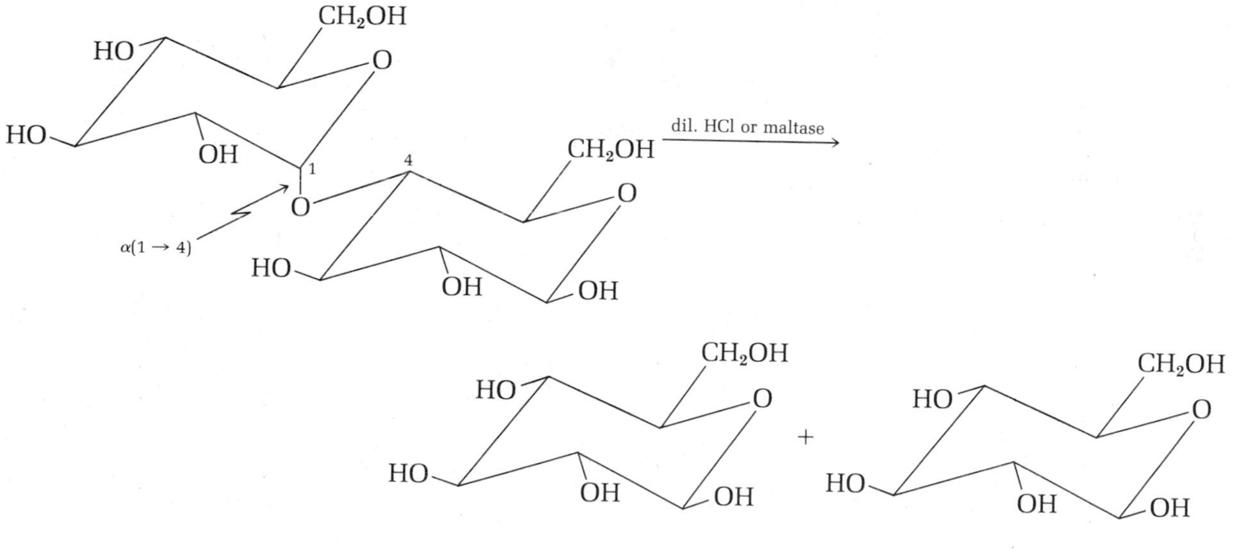

$$\xrightarrow{\text{dil. HCl or maltase}}$$

$$\text{(8.14)}$$

$$\xrightarrow{\text{dil. HCl or emulsion}}$$

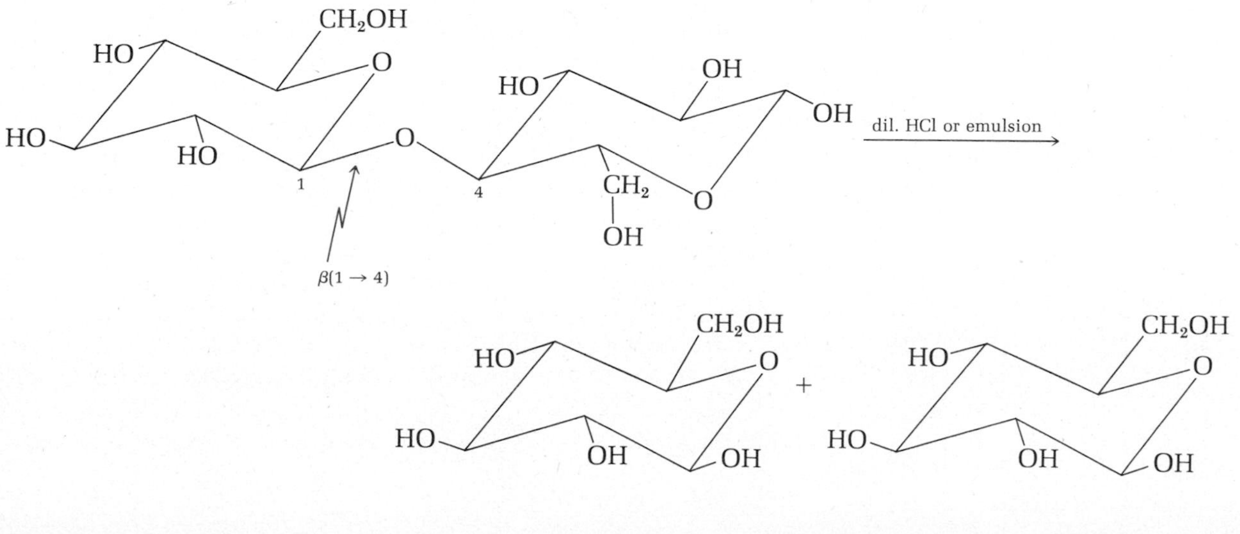

$$\text{(8.15)}$$

Table 8.3 Structures of some common amino sugars and their derivatives

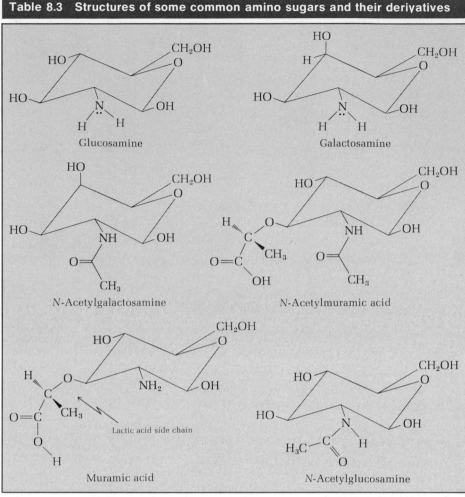

Glucosamine

Galactosamine

N-Acetylgalactosamine

N-Acetylmuramic acid

Muramic acid

Lactic acid side chain

N-Acetylglucosamine

self to a modest number of polysaccharides, of which we will consider only a few.

Complete hydrolysis of polysaccharides gives their monosaccharide components. Those that contain only one type of monomer are *homopolysaccharides,* and those that contain more than one type are *heteropolysaccharides.* The most common monomer is glucose. Polymers containing D-fructose, D-arabinose, D-xylose, D-mannose, and even L-galactose are known. Many hexose derivatives are found in polysaccharides. Among the most common are the *amino sugars* D-glucosamine, N-acetylglucosamine, galactosamine, and muramic acid (Table 8.3). The basic structural features of a number of polysaccharides are summarized in Table 8.4.

A. Structural Polysaccharides

a. Cellulose

Cellulose is the major polysaccharide in wood and fibrous plants. It is a linear polymer of glucose whose monomers are linked by $\beta(1 \to 4)$ glycosidic bonds. Carnivorous animals lack the enzymes specific for $\beta(1 \to 4)$ bonds and are therefore unable to use cellulose as an energy source, whereas ungulates possess celluloytic bacteria with the requisite enzymes.

The molecular weight of cellulose ranges from 3 to 5×10^5, or up to 3,000 glucosyl units.[3] In water cellulose forms *micelles.* A micelle is a *supramolecular (supra,* "more than") assembly in which nonpolar hydrophobic

[3] The largest known cellulose is produced by the alga *Valonia.* It contains 26,000 glucosyl residues.

Table 8.4   Structures of some homopolysaccharides and heteropolysaccharides

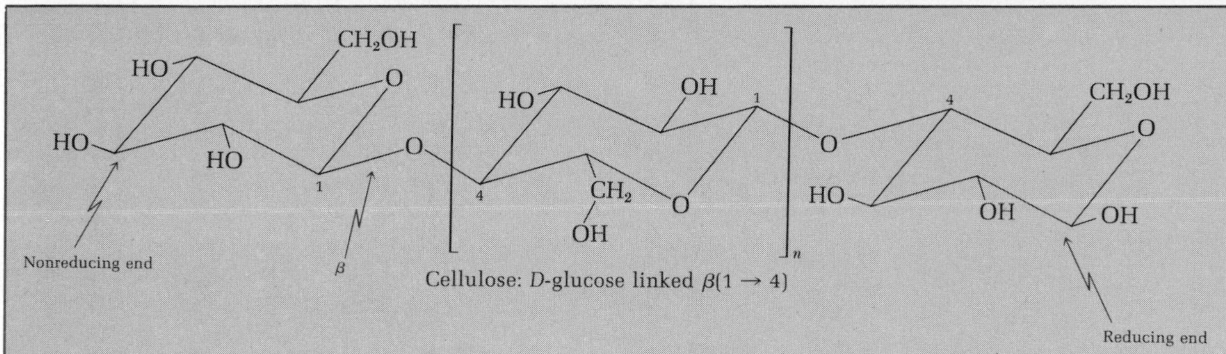

Cellulose: D-glucose linked β(1 → 4)

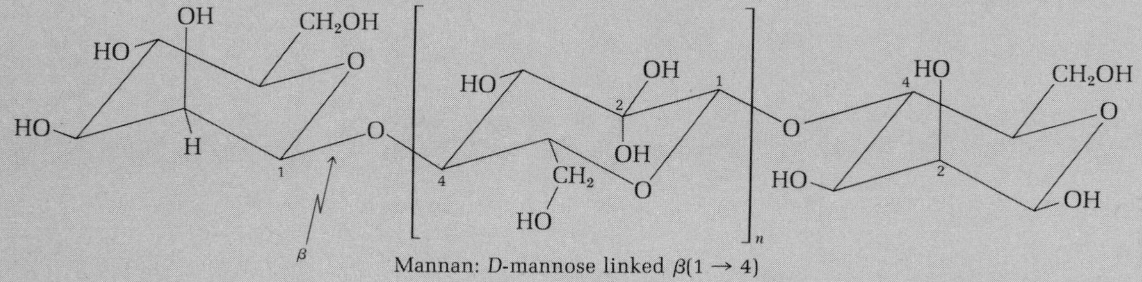

Mannan: D-mannose linked β(1 → 4)

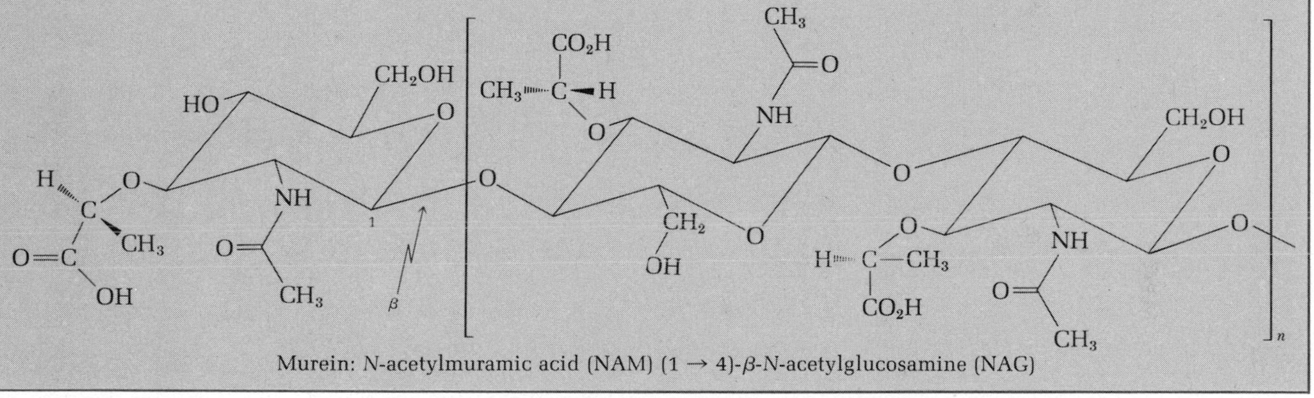

Murein: N-acetylmuramic acid (NAM) (1 → 4)-β-N-acetylglucosamine (NAG)

groups are buried in a region where water is excluded and the hydrophilic groups are exposed to water on the surface. In cellulose micelles this structure is varied slightly. Stacking cellulose chains vertically keeps the hydrophobic interior of the glucosyl units in contact. Extending the chains horizontally enables all of the hydroxy groups to be hydrogen bonded. In glucose all hydroxy groups are equatorial so a regular repeating structure is possible (Figure 8.30). The micelles in cellulose contain 100 to 200 chains of 200 or so glucosyl residues each. The resulting structure has tremendous mechanical strength.

b. Mucopolysaccharides

The mucopolysaccharides are heteropolysaccharides that have a variety of cellular functions. Three of the most abundant mucopolysaccharides are hyaluronic acid, chondroitin sulfate, and dermatan sulfate. Their structures are summarized in Figure 8.31. These polymers, all having acidic functional groups, act as cement in the "ground structure" of connective tissue.

Hyaluronic acid is an alternating polymer of D-glucuronic acid and N-acetylglucosamine that is found in connective tissue, in the vitreous humor of the eye, and in the synovial (lubricating) fluid of joints. *Chondroitin* differs from hyaluronic acid by replacement of N-acetylglucosamine with N-acetylgalactosamine, so the two differ only in their configuration at

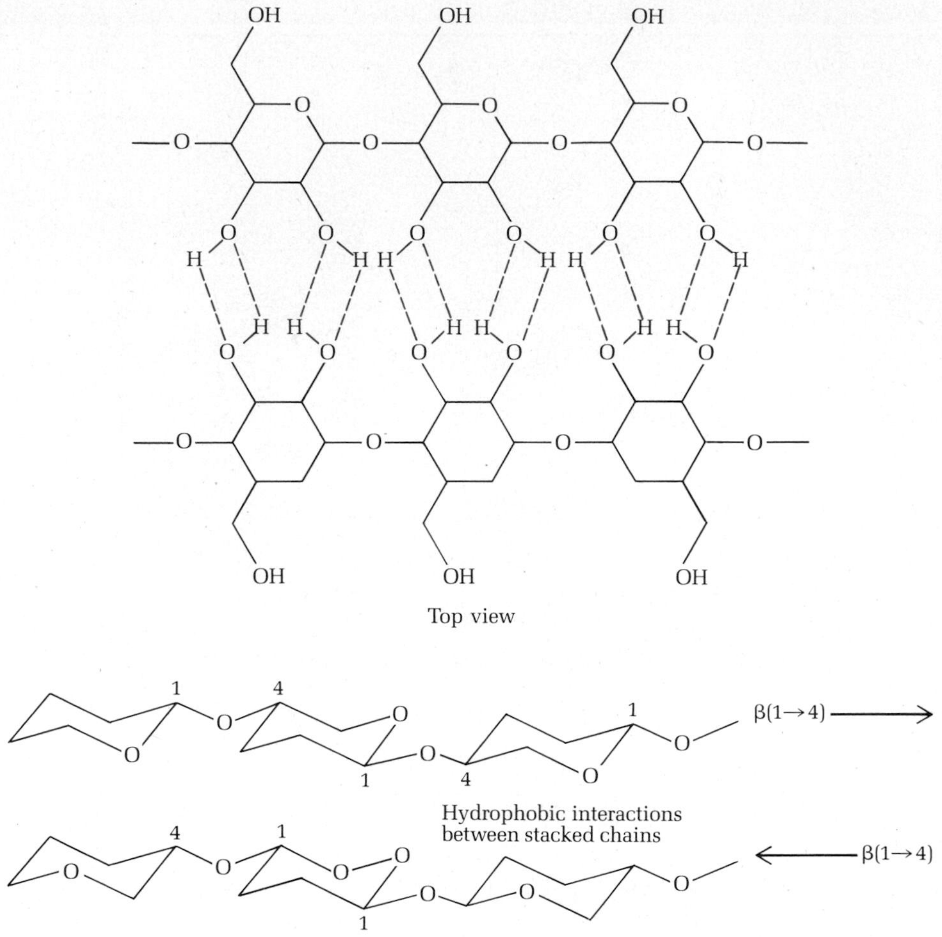

Top view

Side view

Figure 8.30 Structure of cellulose micelle. The cellulose micelle is stabilized by hydrogen bonds extending radially from the main axis of the individual carbohydrate chains (top view). The chains, which run in opposite directions, are stacked with the hydrophobic interiors of the glucosyl rings in van der Waals contact with one another (side view, with side chains omitted for clarity).

C-4 of their amino sugar moieties. Chondroitin itself is a minor component of connective tissue, but its 6-sulfate ester, *chondroitin sulfate,* is widespread in vertebrate connective tissue. *Dermatan,* another alternating polymer, contains L-*iduronic* acid and N-acetylgalactosamine. In *dermatan sulfate,* abundant in skin, the 4 position of N-acetylgalactosamine is present as the sulfate ester. Skin also contains mucopolysaccharides called *keratans* (not to be confused with the proteins, the α-keratins we discussed in Chapter 3) that are alternating polymers, linked $\beta(1 \rightarrow 4)$, of galactose and N-acetylglucosamine-6-sulfate. Connective tissue also contains trace amounts of silicon that bridges adjacent polysaccharide chains through ether bonds.

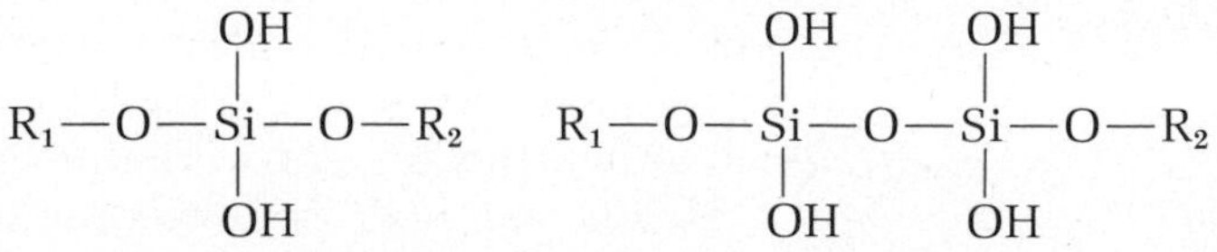

Silicon bridges between adjacent carbohydrate chains

Dermatan sulfate, chondroitin sulfate, and heparin sulfate (discussed below) of connective tissue contain about 1 silicon atom per 128 to 130 carbohydrate monomers.

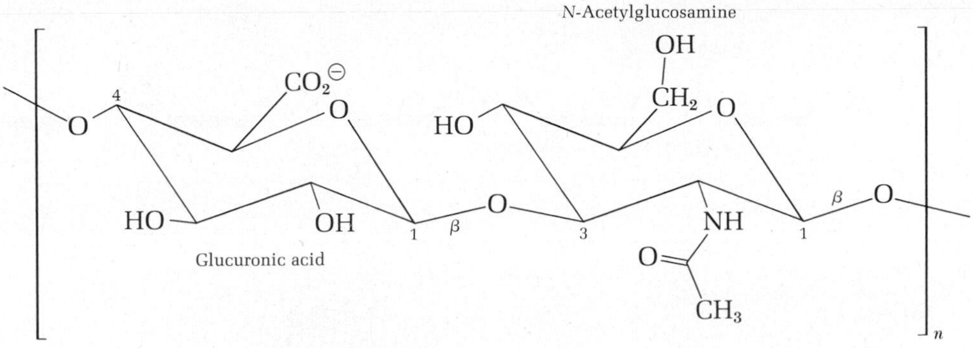

Hyaluronic acid: $[\beta\text{-GlcUA-}(1\rightarrow3)\text{-}\beta\text{-GlcNAc-}(1\rightarrow4)\text{-}]_n$

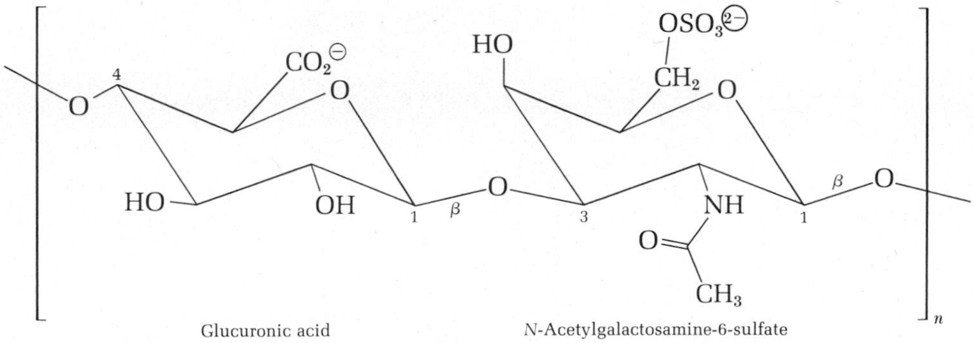

Chondroitin sulfate: $[\beta\text{-GlcUA-}(1\rightarrow3)\text{-}\beta\text{-GalNAc-6-sulfate-}(1\rightarrow4)\text{-}]_n$

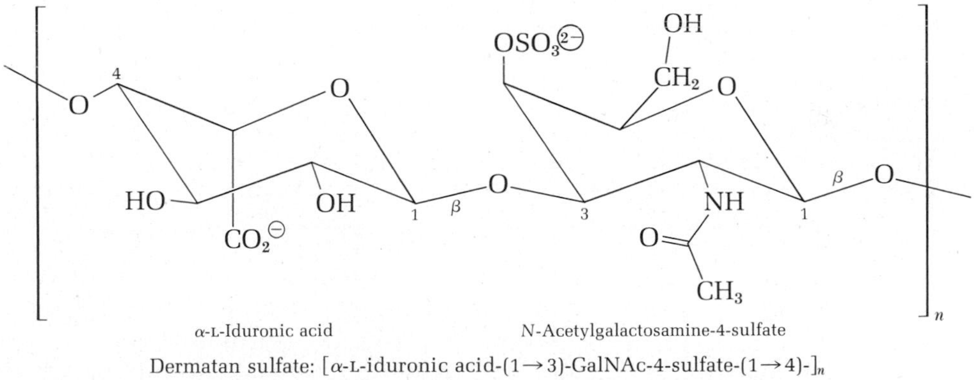

Dermatan sulfate: $[\alpha\text{-L-iduronic acid-}(1\rightarrow3)\text{-GalNAc-4-sulfate-}(1\rightarrow4)\text{-}]_n$

Figure 8.31 Structural formulas and abbreviations for the mucopolysaccharides hyaluronic acid, chondroitin sulfate, and dermatan sulfate.

Heparin is a mixture of polysaccharides with anticoagulant properties that is found in connective tissue. It is prepared commercially from beef lung and pork intestinal mucosa used widely to treat diseases of the heart and blood vessels. The molecular weight of heparin ranges from 7,600 to 19,700. Its structure is shown in Figure 8.32. L-Iduronic acid, also a component of dermatan sulfate, alternates with glucosamine 2,6-disulfate. Heparin acts as an anticoagulant by binding to a plasma protein called *antithrombin III*. The heparin–antithrombin III complex is an inhibitor of the serine proteases of the blood clotting system. (Blood coagulation is discussed in conjunction with vitamin K in Section 11.7B).

c. Peptidoglycans

The major component of bacterial cell walls is a heteropolysaccharide conjugated with short peptide chains called a *peptidoglycan*. The polysac-

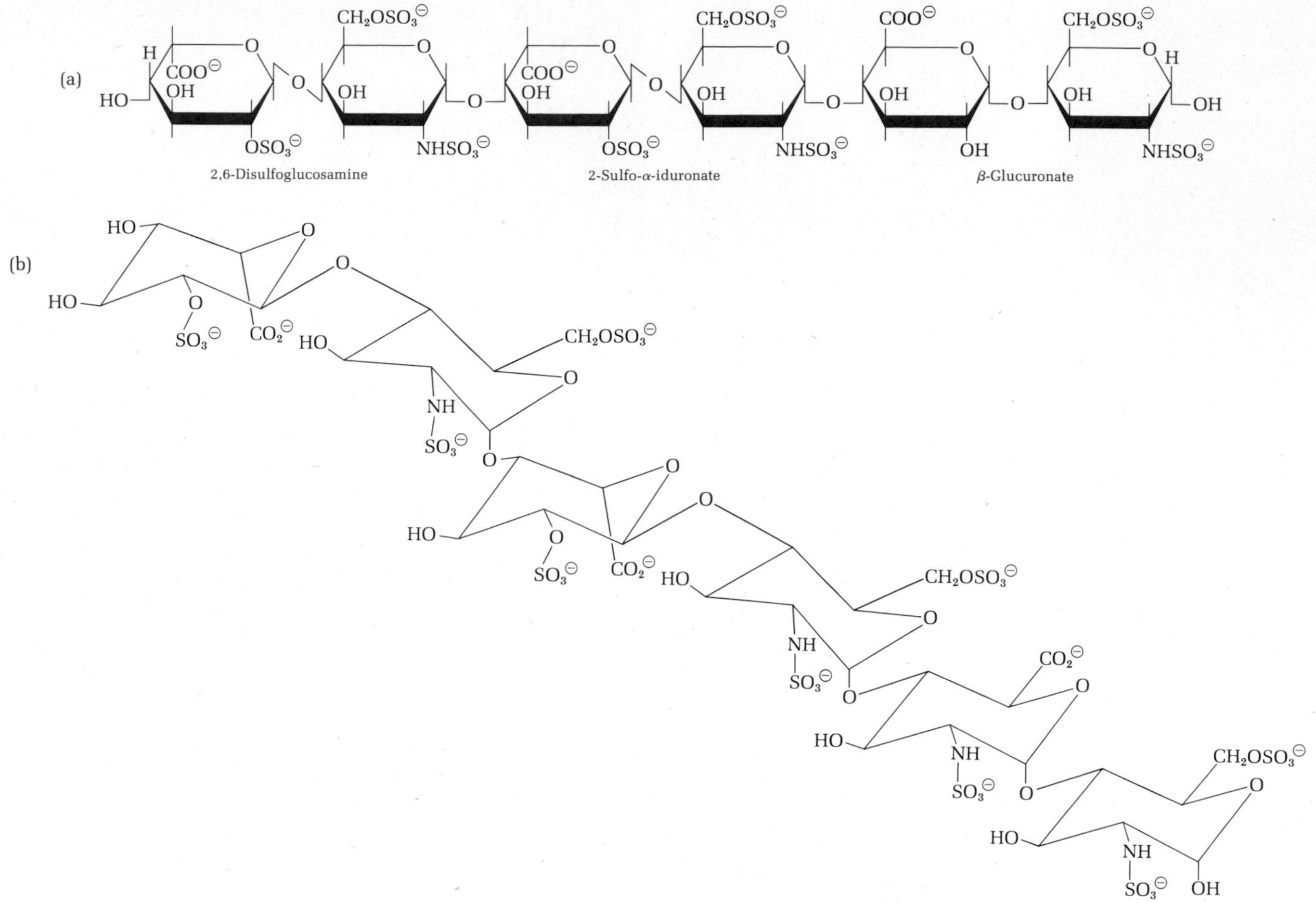

Figure 8.32   Structure of heparin hexasaccharide. Three molecules of 2,6-disulfoglucosamine alternate with three-uronic acid molecules (two of 2-sulfoiduronic acid and one of unsulfated glucuronic acid). The major constituents of heparin as represented by this hexasaccharide: 2 × [2,6-disulfoglucosamine-2-sulfoiduronic acid]-[2,6-disulfoglucosamine-glucuronic acid]. (a) Haworth projection of the hexasaccharide and (b) conformations of the residues in the hexasaccharide. [From L. B. Jaques, *Science*, *206*, 529 (1979). Copyright 1977 by the American Association for the Advancement of Science.]

charide, called *murein* (from the Latin word for wall), is an alternating polymer of N-acetylglucosamine and N-acetylmuramic acid (see Table 8.3) linked by $\beta(1 \rightarrow 4)$ glycosidic bonds (see Table 8.4). The peptides are condensed with the carboxyl group of the lactic acid side chain of N-acetylmuramic acid. They provide bridges between polysaccharide strands in a highly cross-linked structure (Figure 8.33). The cross-linking peptides usually contain four amino acids including D-alanine and D-glutamic acid. The bacterial cell wall is one of the few places where a large number of D-amino acids are found. In the bacterium *Staphylococcus aureus*, the tetrapeptide has the sequence L-Ala-D-Glu-L-Lys-D-Ala.[4] A fifth amino acid, D-alanine, is eliminated in the final step of cell wall synthesis in which the cross-links between polysaccharide chains are formed.

The cross-linking reaction is catalyzed by *peptidoglycan transpeptidase* (Figure 8.34). The first step of the reaction involves formation of an *acylenzyme intermediate* (analogous to the acylenzyme intermediate formed in the reaction of chymotrypsin, Section 6.3) and release of D-alanine (Figure 8.35). The acylenzyme intermediate is then attacked by the N-terminal amino group of the pentaglycine bridge (Figure 8.36).

---

[4] The type of cross-link varies from species to species.

β-Lactam ring

Penicillin

The antibiotic *penicillin* inhibits formation of the acylenzyme intermediate in the cross-linking reaction. The β-lactam ring of penicillin is highly susceptible to nucleophilic attack by the enzyme. Relief of steric strain in the four-membered ring provides the driving force for the formation of a *penicillinoylenzyme* intermediate between the transpeptidase and penicillin (Figure 8.37). The conformation of the D-alanyl-D-alanine portion of the peptidoglycan is similar to the conformation of penicillin, and the enzyme mistakes the antibiotic for the peptidoglycan (Figure 8.38). Since the structure of penicillin resembles the tetrahedral intermediate in the cross-linking reaction, penicillin can perhaps be regarded as a transition state analog (recall Section 7.11B).

d. Lysozyme

Lysozyme is an enzyme found in animal secretions such as tears and saliva and in egg white. It forms part of the antibiotic defense by hydrolyzing

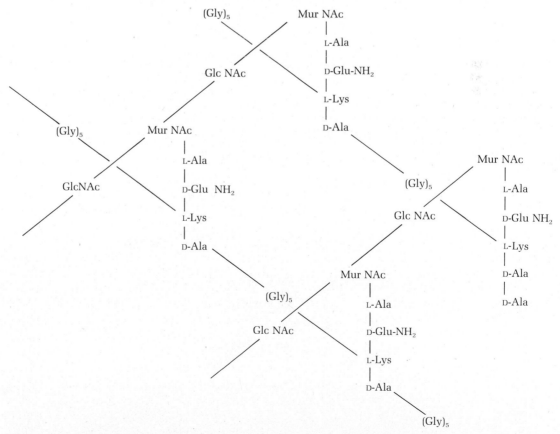

Figure 8.33 Schematic drawing of the peptidolycan of the bacterial cell wall of *S. aureus*. [From M. J. Osborn, *Ann. Rev. Biochem.*, *38*, 501–538 (1969). Reproduced with permission. © 1969 by Annual Reviews Inc.]

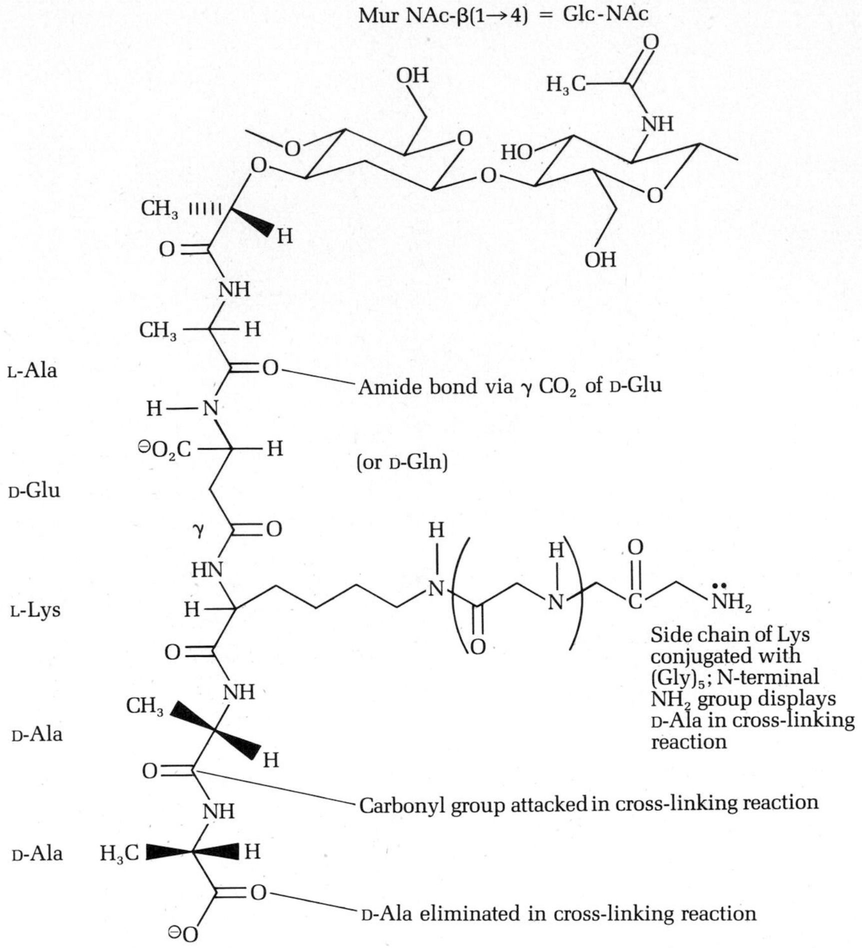

Mur NAc-β(1→4) = Glc-NAc

L-Ala

Amide bond via γ CO₂ of D-Glu

D-Glu

(or D-Gln)

L-Lys

Side chain of Lys conjugated with (Gly)₅; N-terminal NH₂ group displays D-Ala in cross-linking reaction

D-Ala

Carbonyl group attacked in cross-linking reaction

D-Ala

D-Ala eliminated in cross-linking reaction

Figure 8.34 Details of the peptide side chain of a peptidoglycan showing the groups involved in the cross-linking reaction.

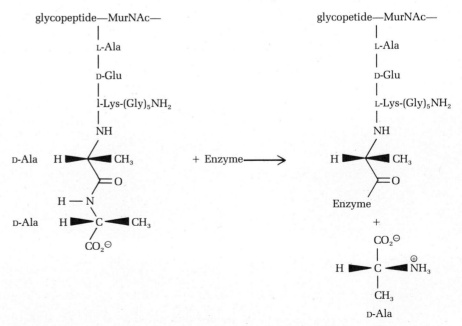

glycopeptide—MurNAc—
|
L-Ala
|
D-Glu
|
l-Lys-(Gly)₅NH₂
|
NH

D-Ala   H — C — CH₃
        ‖
        O

H — N

D-Ala   H — C — CH₃
        |
        CO₂⁻

+ Enzyme ——→

glycopetide—MurNAc—
|
L-Ala
|
D-Glu
|
L-Lys-(Gly)₅NH₂
|
NH

H — C — CH₃
    ‖
    O

Enzyme

+

CO₂⁻
|
H — C — NH₃⁺
|
CH₃

D-Ala

Figure 8.35 Formation of acylenzyme intermediate and release of D-alanine in the cross-linking reaction catalyzed by peptidoglycan transpeptidase.

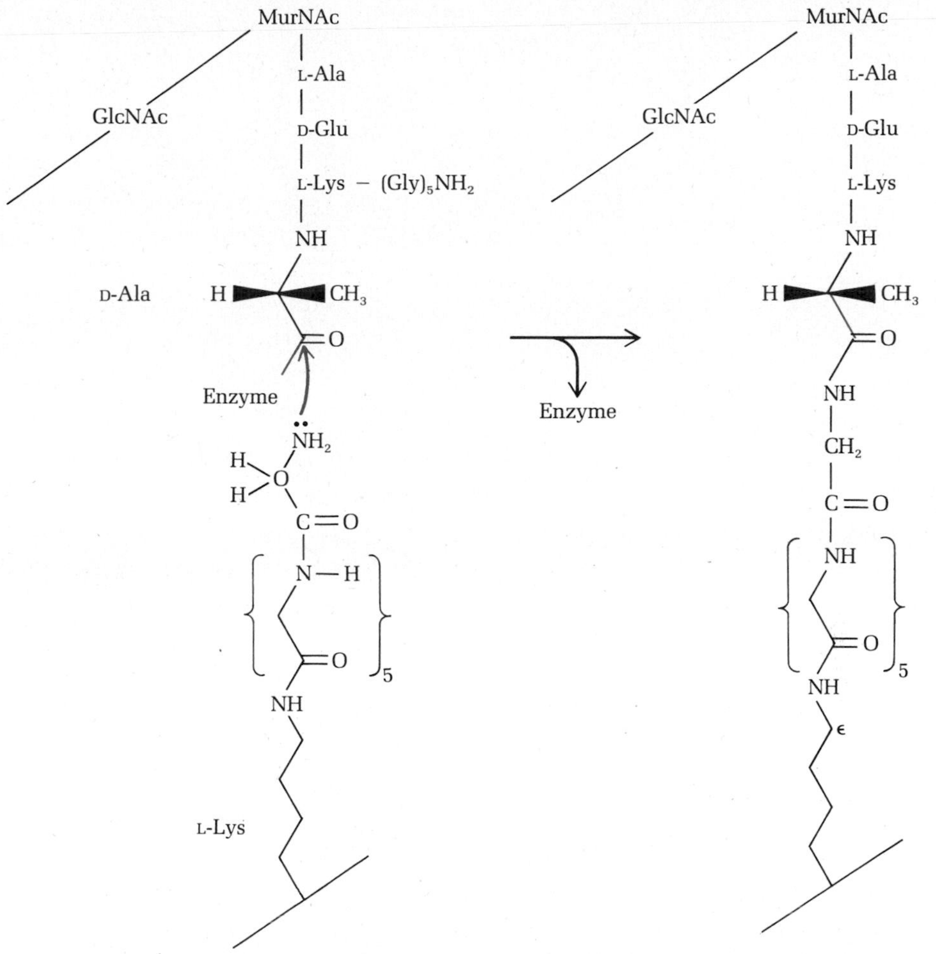

Figure 8.36 Displacement of the enzyme by the N-terminal amino group of a pentaglycine moiety in the final cross-linking step in peptidoglycan biosynthesis.

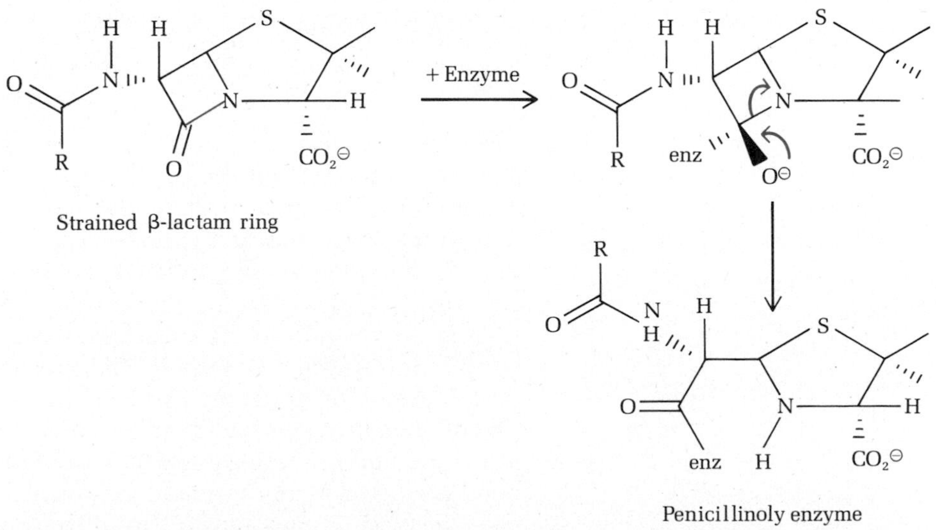

Figure 8.37 Formation of a penicillinoyl-enzyme intermedrate. The formation of the covalent enzyme-substrate bond between penicillin and peptidoglycan transpeptidase irreversibly inhibits the enzyme. A closely related cell wall enzyme that is inactivated by penicillin, D-alanine carboxypeptidase, forms a covalent intermediate with a seryl hydroxyl group. The enzyme has been isolated from *Bacillus stearothermophilus*.

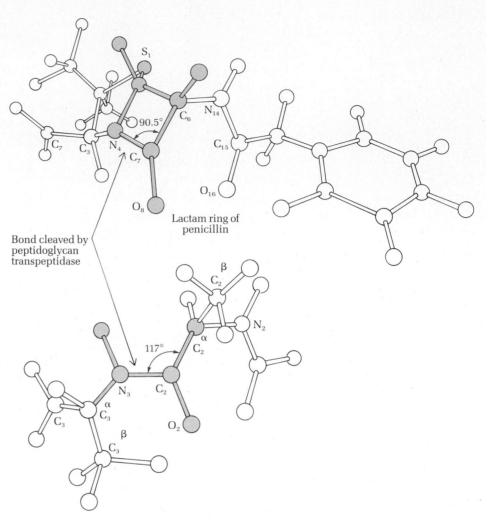

Figure 8.38   Conformations of the lactam ring in penicillin and the D-alanyl-D-alanine portion of the peptidoglycan of bacterial cells walls. The conformations are similar and peptidoglycan transpeptidase confuses the cell wall substrate and penicillin. Arrows indicate the site of cleavage in each case. [From R. R. Rando, *Biochem. Pharmacol., 24,* 1153–1160 (1975).]

the $\beta(1 \rightarrow 4)$ glycosidic bond between *N*-acetylglucosamine and *N*-acetylmuramic acid in the peptidoglycan of the bacterial cell wall. The enzyme binds a hexasaccharide segment of the peptidoglycan (Figure 8.39). One of the monomers in the hexasaccharide, an *N*-acetylmuramic acid residue, is distorted by lysozyme from its usual chair conformation into a *strained half-chair conformation*. The enzyme promotes catalysis by inducing steric strain in the substrate. Bond cleavage occurs at the glycosidic bond of the strained half-chair NAM residue.

The rate of lysozyme catalysis depends upon the pH of the medium. The pH-rate profile is bell-shaped with a maximum at pH 5.0 and inflections at pH 3.8 and 6.7 (Figure 8.40). X-ray data show that the labile glycosidic bond is flanked by the carboxyl groups of aspartate 52 and glutamate 35. The pK_a's of these groups match the inflection points of the pH-rate profile. At pH 5.0, the optimum for lysozyme, aspartate 52 is unprotonated and glutamate 35 is protonated.

The mechanism of action of lysozyme has been deduced from a combination of x-ray crystallographic and chemical studies. The main features of the mechanism are summarized in Figure 8.41. Bond breaking between the two sugars gives a planar, resonance-stabilized acylium ion (recall the hy-

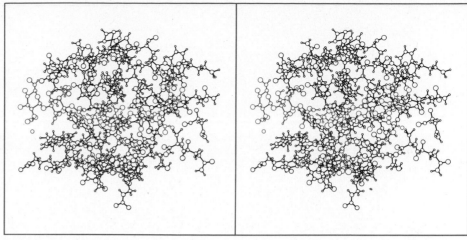

Figure 8.39 Enzyme-substrate complex between lysozyme and a hexasaccharide moiety of murein.

drolysis of phosphate esters, Figure 8.29). The ion is further stabilized by electrostatic interaction with the negatively charged carboxylate anion of aspartate 52. Theoretical calculations suggest that the increase in rate of the lysozyme-catalyzed reaction relative to the noncatalyzed reaction is almost entirely due to stabilization of the acylium ion by electrostatic interactions with the negatively charged carboxylate anion of aspartate 52.

Lactone analogs of the peptidoglycan substrate are potent competitive inhibitors of lysozyme and of other "glycosidases" such as amylase (Section

Lactone analog of peptidoglycan substrate

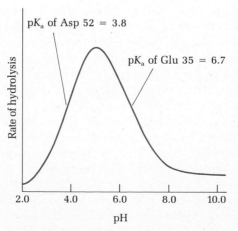

Figure 8.40 pH-rate profile for lysozyme-catalyzed hydrolysis of poly-N-acetyl-glucosamine.

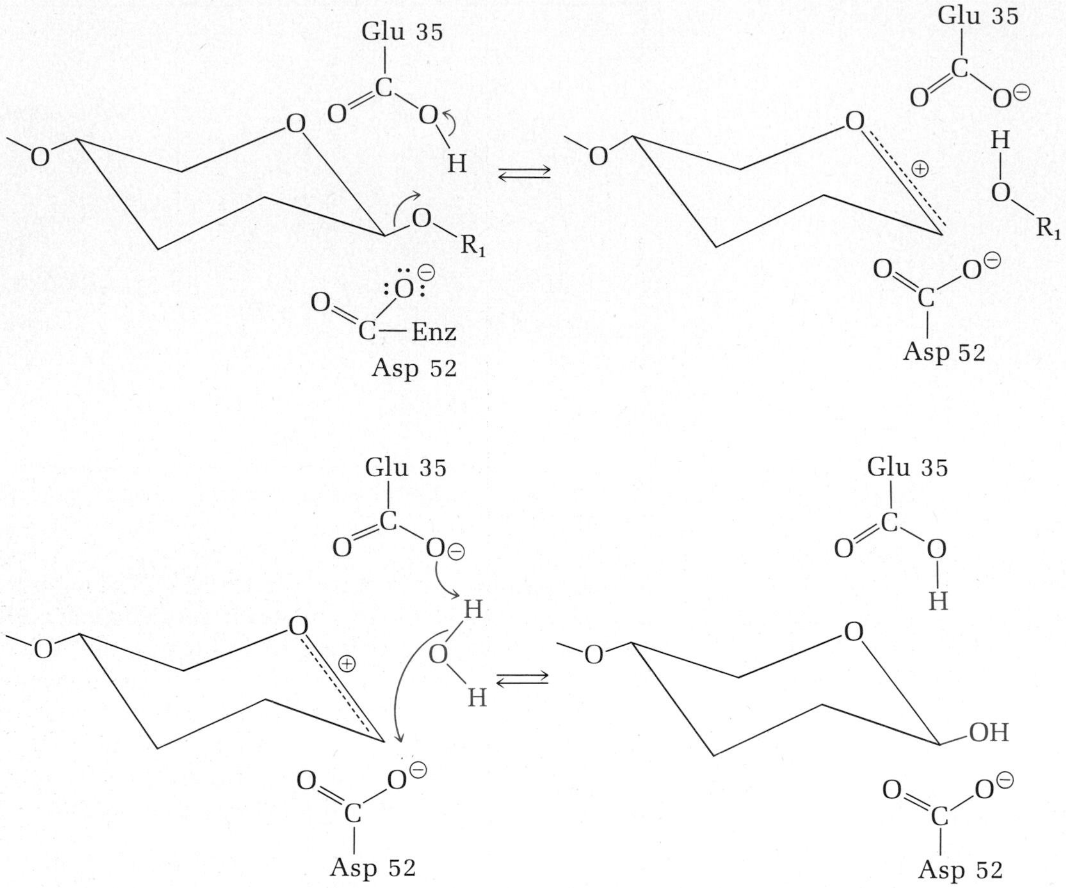

Figure 8.41 Mechanism of action of lysozyme. The mechanism involves formation of a resonance-stabilized acylium ion that is stabilized by the carboxylate side chains of glutamate 35 and aspartate 52. Water then reacts with the acylium ion, and the enzyme is ready for another round of catalysis.

8.7B) and phosphorylase (Section 17.4A). By mimicking the structure of the planar acylium ion intermediate, the lactone fits exactly into the active site, but without having to be distorted into a high energy half-chair conformation.

e. Structure of the Bacterial Cell Wall

The peptidoglycan is the major, but not the only component of the cell walls of all bacteria. *Gram-positive bacteria* also contain *teichoic acids*. These molecules provide the antigenic determinants of gram positive bacteria. The teichoic acids fall into two general types (Figure 8.42). The *ribitolteichoic acids* are polymers of D-ribitol 1-phosphate. The hydroxyl groups of ribitol in *Bacillus subtilis* are conjugated with both β-D-glucose and D-alanine. One hydroxyl group is always conjugated with β-D-glucose. D-Alanine is attached to another one in at least half the units. The phosphate group of the terminal teichoic acid is bound to the peptidoglycan by a phosphodiester bond to the carboxyl group of a muramic acid residue. The *glycerolteichoic acids* are polymers of *sn*-glycerol 3-phosphate. In *Lactobacillus arabinosis* the 2-hydroxy group is most often conjugated with D-alanine, but β-D-glucose replaces it about once every nine residues. As in the ribitolteichoic acids, glycerolteichoic acids are bound to the peptidoglycan by a phosphodiester bond to the carboxyl group of a muramic acid residue. Teichoic acids are also found as components of glycolipids in the plasma membranes of some organisms (Figure 8.43).

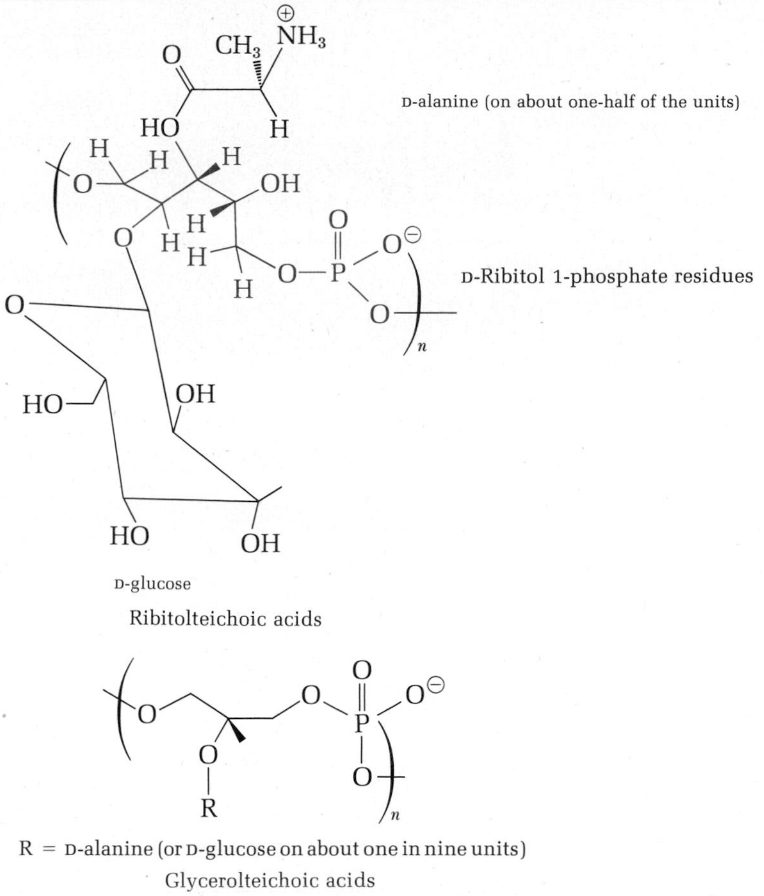

D-alanine (on about one-half of the units)

D-Ribitol 1-phosphate residues

D-glucose

Ribitolteichoic acids

R = D-alanine (or D-glucose on about one in nine units)

Glycerolteichoic acids

Figure 8.42 Structures of ribitolteichoic and glycerolteichoic acids of gram-positive bacteria.

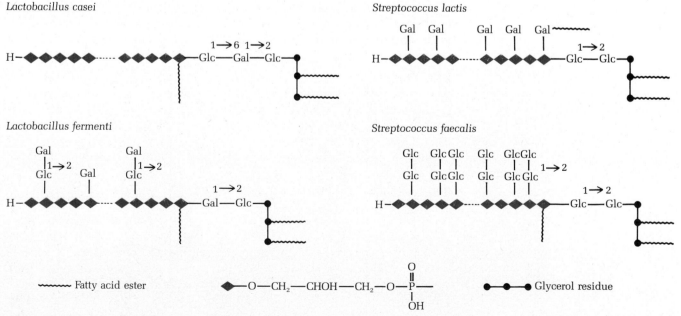

Lactobacillus casei

Streptococcus lactis

Lactobacillus fermenti

Streptococcus faecalis

∿∿∿ Fatty acid ester

◆—O—CH₂—CHOH—CH₂—O—P—OH (Glycerol residue)

●—● Glycerol residue

Figure 8.43 Proposed structures of some lipoteichoic acids. [From A. J. Wicken and K. W. Knox, *Science*, *187*, 1162 (1975). Copyright 1975 by the American Association for the Advancement of Science.]

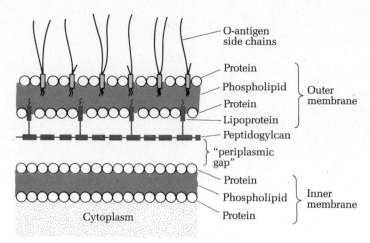

Figure 8.44 Structure of the cell envelope of gram-negative bacteria. [From C. A. Schnaitman, *J. Bacteriol.*, *108*, 553 (1971).]

The cell walls of *gram-negative bacteria* do not contain teichoic acids, but in other respects they are far more complex than their counterparts in the gram-positive bacteria. Only about 10 to 20% of the dry weight of the cell wall of the gram-negative bacteria is peptidoglycan. The peptidoglycan is coated with a second membrane that contains glycoproteins (Section 7.8) and complex lipopolysaccharides (Figure 8.44). The lipopolysaccharides are highly toxic substances, called *endotoxins* because they are integral parts of the cell wall. The endotoxins are responsible for the high fevers caused by gram-negative bacteria. The polysaccharide portion of the lipopolysaccharide is called the *O-antigen*. We will discuss the biosynthesis of peptidoglycans and the O-antigen in Section 17.5.

B. Storage Polysaccharides

a. Starch: Amylose and Amylopectin

In plants glucose is stored in *starch granules* varying in diameter from 3 to 100 μm. Virtually all plant cells contain some starch granules, but in seeds (such as corn) as much as 80% of the cell's dry weight is starch. The molecular weight of "starch" has little meaning since starch is a heterogeneous material. Its principal components are two types of polysaccharides: *amylose* and *amylopectin*. The starch from each plant is unique, but most

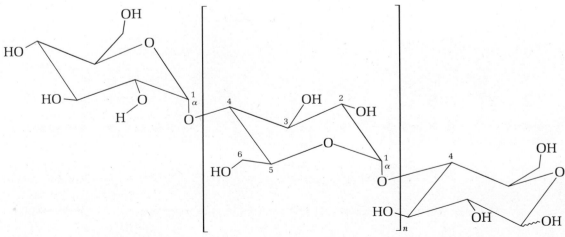

Figure 8.45 Structure of amylose. Amylose is a linear polymer of glucose residues linked $\alpha(1 \rightarrow 4)$. A single chain can have up to 4,000 glucosyl residues.

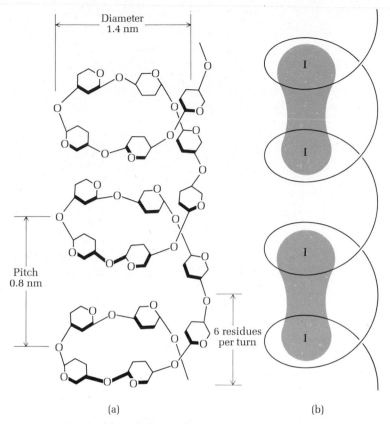

Figure 8.46 Amylose helix and starch-iodine complex. (a) Amylose forms a left-handed helix that contains 6 glucosyl residues per turn. (b) Iodine molecules fit inside the helix parallel to its long axis. Six turns of the helix (36 residues) are required to give the characteristic blue color of the starch-iodine complex.

starches have about the same composition of 20 to 25% amylose and 75 to 80% amylopectin.

Complete hydrolysis of either amylose or amylopectin gives only glucose. These homopolysaccharides are called *glucans*. The glucosyl residues of amylose are linked by $\alpha(1 \rightarrow 4)$ glycosidic bonds in single chains that contain up to 4,000 glucosyl residues (Figure 8.45). Because of the $\alpha(1 \rightarrow 4)$ bonds of the glucosyl residues, amylose cannot exist as a "straight chain." Instead it coils into a left-handed helix having 6 residues per turn, a pitch of 0.8 nm, and a diameter of about 1.4 nm (Figure 8.46a). The structure of the amylose helix nicely accommodates iodine molecules, giving rise to the well-known blue starch-iodine complex (Figure 8.46b). About 36 glucosyl residues, or 6 turns of the helix, are required to achieve the characteristic blue color.

Amylose is degraded by α-*amylase* and β-*amylase*, both of which are found in the pancreatic juice and saliva of animals. The two enzymes have quite different modes of action. β-Amylase is an *exoglycosidase* which sequentially cleaves maltose from the nonreducing end of the chain. The product is β-maltose, and the reaction proceeds with inversion of configuration at the anomeric carbon of the glycosidic bond being cleaved (Figure 8.47). α-Amylase is an *endoglycosidase* which attacks amylose randomly along the chain. The lactone of maltobionic acid, 4-0-(α-D-glucopyranosyl)-5-gluconolactone, is a potent competitive inhibitor of α-amylase suggesting that catalysis occurs via an acylium ion intermediate as shown in Figure 8.41 for lysozyme (recall that lysozyme is also inhibited by lactones).

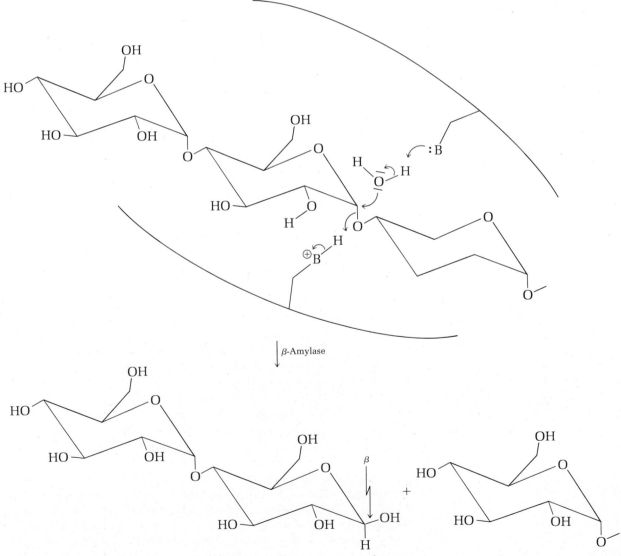

Maltobionic acid [4-O-(α-D-glucopyranosyl)-5-gluconolactone]

Amylopectin is a highly branched amylose having two types of glycosidic bonds. The "main chain" is linked $\alpha(1 \rightarrow 4)$, and the branches are $\alpha(1 \rightarrow 4)$ glucans containing 20 to 25 residues linked to the main chain by $\alpha(1 \rightarrow 6)$ glycosidic bonds (Figure 8.48). The term "main chain" is a slight misnomer since an amylopectin molecule contains so many branches that a main chain can scarcely be distinguished.

β-Amylase

Figure 8.47 Degradation of amylose by β-amylase. β-Amylase cleaves α-glycosidic bonds in amylose from the nonreducing end of the chain. The product, β-maltose, results from inversion of configuration at the anomeric carbon of the glycosidic bond that is cleaved.

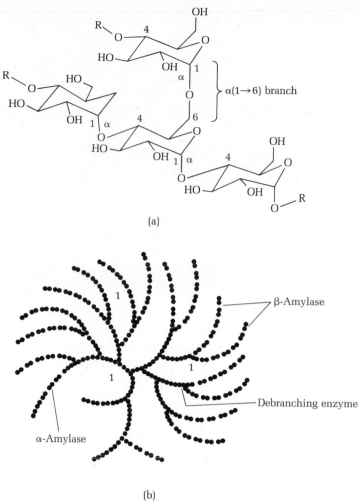

(a)

(b)

Figure 8.48 Structure of amylopectin. (a) Amylopectin is an $\alpha(1 \rightarrow 4)$ glucan that has $\alpha(1 \rightarrow 6)$ branches. The chains that comprise the branches are themselves $\alpha(1 \rightarrow 4)$ glucans. (b) A schematic diagram of amylopectin indicates the points of glycosidic bond cleavage by specific glycosidases.

Amylopectin can also be degraded by α-amylase and β-amylase, but neither can hydrolyze the $\alpha(1 \rightarrow 6)$ branch points. Amylopectin may be attacked on either side of a branch point by α- or β-amylase. The product of their action is a *limit dextrin* (Figure 8.49). Plants and animals possess *debranching enzymes* which specifically hydrolyze $\beta(1 \rightarrow 6)$ bonds. The combination of α- and β-amylase plus debranching enzymes permits the total conversion of starch to glucose. *Phosphorylase* also cleaves the $\alpha(1 \rightarrow 4)$ glycosidic bonds of amylose and amylopection (Section 17.4A).

b. Glycogen

Glycogen, the glucose storage molecule of animals, is found in granules in most cells. Glycogen granules also contain many of the enzymes required for glycogen degradation. It is a highly branched $\alpha(1 \rightarrow 4)$ glucan whose structure is similar to that of amylopectin (refer to Figure 8.4). The principal differences between amylopectin and glycogen are in the degree of branching and in the length of the branches. Glycogen has more, shorter branches (in the range of 20 to 25 residues in amylopectin and 12 to 18 in glycogen), but for all practical purposes the metabolism of the two are identical. The metabolism of glycogen will be discussed in Chapter 17.

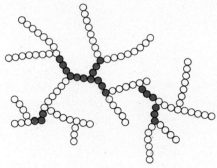

Figure 8.49 Schematic diagram of the limit dextrin produced by the action of α- and β-amylases. Degradation of the limit dextrin requires the action of an $\alpha(1 \rightarrow 6)$ glucosidase to remove the branches.

8–8 GLYCOPROTEINS

Many cellular proteins, perhaps most, are covalently bound to carbohydrates of varying complexity. They are called glycoproteins, from *glykos* the Greek word for "sweet." The range of phenomena in which glycoproteins participate is extraordinary. We saw in Chapter 3 that collagen is a glycoprotein whose carbohydrates are bound through the hydroxyl groups of hydroxylysine in glycosidic bonds. The structure of β-D-glucosylhydroxylysine, one of the major carbohydrate linkages in collagen, is shown below. Galactose is

β-D-Glucosylhydroxylysine

bound to hydroxylysine in the same way. The disaccharide 2-*O*-α-D-glucosyl-*O*-β-D-galactosylhydroxylysine is another major component of col-

2-*O*-α-D-glucosyl-*O*-β-D-galactosylhydroxylysine

lagen. The function of this carbohydrate in collagen is not known, though there has been speculation that it aids in collagen assembly.

Certain proteins found in Antarctic fish act as antifreeze. The antifreeze proteins lower the melting point of aqueous solutions more effectively than equimolar amounts of sodium chloride. The antifreeze proteins contain the repeating sequence Ala-Ala-Thr. A galactosyl-N-acetylglucosamine disaccharide is attached to every threonine residue.

Immunoglobulins are also glycoproteins. The carbohydrate is wedged between the C_H1 and C_H2 domains of the immunoglobulin, preventing close contact between them (Figure 8.50).

The complexity of the carbohydrate bound to glycoproteins varies as much as the proteins themselves. Hen ovalbumin, the major protein in egg white, contains a single oligosaccharide chain. The lubricating glycoprotein produced by sheep submaxillary (salivary) glands, called a *mucin*, contains

$$\alpha Man\text{-}(1\rightarrow4)\text{-}\beta GlcNAc\text{-}(1\rightarrow4)\text{-}\beta GlcNAc\text{-}N\text{-}Asp \text{ (protein)}$$
$$\overset{|}{H}$$
$$(GlcNAc)_{\overline{0,1\,or\,2}}\ (Man)_{4\,or\,5}$$
$$|$$
$$(GlcNAc)_{0\,or\,1}$$

Hen ovalbumin oligosaccharide

as many as 800 chains per protein molecule of α-D-N-acetylneuraminic acid-$(2 \rightarrow 6)$-α-D-N-acetylglucosamine.

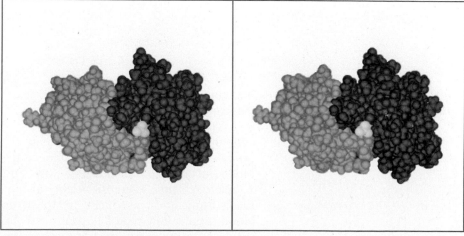

α-D-N-Acetylneuraminic acid (2-6)-α-D-N-acetylglucosamine

Figure 8.50 Space-filling model of the constant region of immunoglobulin G. The carbohydrate chains provide most of the contacts between the C_H2 subunits.

Human blood groups are divided into four antigenic types designated A, B, AB, and O. This classification is based upon differences in oligosaccharides bound to the surface proteins of erythrocytes. These oligosaccharides all contain the unusual sugar L-fucose.

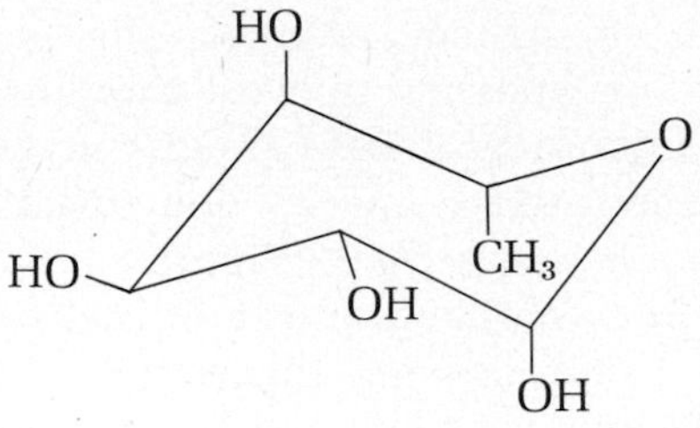

α-L-Fucose
(6-Deoxy-α-L-galactose)

Each blood group type is further subdivided into two types of chains that differ in their glycosidic linkages. The structure of the type 1 chain of the A blood group antigen is shown below.

$$\beta\text{GalNAc}(1\rightarrow3)\beta\text{Gal}(1\rightarrow3)\beta\text{GlcNAc}$$

$$\uparrow 2$$
$$| \; 1$$

α-L-Fucose

A blood group antigen, type 1 chain

If the linkage between β-D-galactose and β-D-N-acetylglucoseamine is (1 → 4) rather than (1 → 3), the blood group is still type A, but the chain is type 2, as shown below. If α-D-galactose replaces N-acetylgalactosamine at the nonreducing end of the oligosaccharide, a B group antigen is the result. This is shown below for a type 1 chain.

$$\beta\text{GalNAc}(1\rightarrow3)\beta\text{Gal}(1\rightarrow4)\beta\text{GlcNAc}$$

$$\uparrow 2$$
$$| \; 1$$

α-L-Fucose

A blood group antigen, type 2 chain

$$\alpha\text{Gal}(1\rightarrow3)\beta\text{Gal}(1\rightarrow3)\beta\text{GlcNAc}$$

$$\uparrow 2$$
$$| \; 1$$

α-L-Fucose

B blood group antigen, type 1 chain

In type O persons, the terminal residue, either galactose or N-acetylgalactosamine, is absent.

The blood type of a given person depends upon the expression of a gene that codes or the biosynthesis of a glycosyltransferase. If a person has type A blood the enzyme transfers N-acetylgalactosamine, if he has type B blood the enzyme transfers galactose, and if he has type O blood the enzyme is inactive. The α-L-fucose moiety of the oligosaccharide is transferred by a fucosyltransferase produced by the H gene. If this gene is inactive a rare blood type called type I is produced. In type Le[a] blood a fucose is linked α(1 → 4) to the β-D-N-acetylglucosamine moiety of the oligosaccharide. In type Le^b blood a fucose residue is bound both to β-galactose and to β-N-acetylglucosamine, as shown below.

$$\alpha\text{-L-Fucose}$$
$$\downarrow \begin{matrix}1\\4\end{matrix}$$

$$\alpha\text{GalNAc}(1\rightarrow3)\beta\text{Gal}(1\rightarrow3)\beta\text{GlcNAc}$$
$$\uparrow \begin{matrix}2\\1\end{matrix}$$

$$\alpha\text{-L-Fucose}$$

Le^b blood group antigen

These carbohydrates are the *antigenic determinants* of their blood groups. Antibodies from type A persons "attack" type B blood, clumping the cells; type B antibodies similarly clump type A blood. Persons who are type A or type B lack antibodies against type O blood, so type O persons are called "universal donors." They are not, however, universal acceptors, since they produce antibodies against both type A and type B blood.

Cell surfaces are coated with glycoproteins. Cell surface recognition depends upon carbohydrates present, blood groups being an example. Normal cells stop growing when they touch each other, a phenomenon known as *contact inhibition*. Cancer cells lack contact inhibition. Some of the unusual properties of cancer cells are related to changes in cell surface glycoproteins.

8-9 LECTINS The *lectins*, a group of proteins first discovered in plants and now known to have counterparts in mammalian cells, have stereospecific binding sites for carbohydrates. In their agglutinating properties they resemble immunoglobulins. *Wheat germ agglutinin* binds N-acetylglucosamine, a lectin produced by red kidney beans binds N-acetylglucosamine, and *concanavalin A,* isolated from jack beans, binds α-D-mannopyranose and α-D-glucopyranose. Concanavalin A also binds cells whose coatings have oligosaccharides

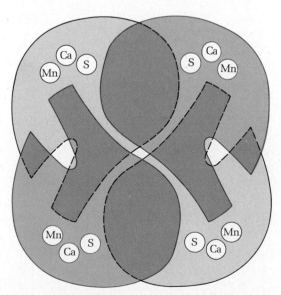

Figure 8.51 Schematic drawing of the concanavalin A tetramer. The Mn(II), Ca(II), and saccharide binding loci are circled. [From J. W. Becker, G. N. Reeke, Jr., B. A. Cunningham, and G. M. Edelman, *Nature*, 259, 407 (1976).]

whose nonreducing ends are glycosides of α-D-mannopyranose and α-D-glucopyranose. Similarly, wheat germ agglutinin binds oligosaccharides whose nonreducing end is an N-acetylglucosamine glycoside, and the red kidney bean lectin binds oligosaccharides whose nonreducing end is N-acetylglucosamine.

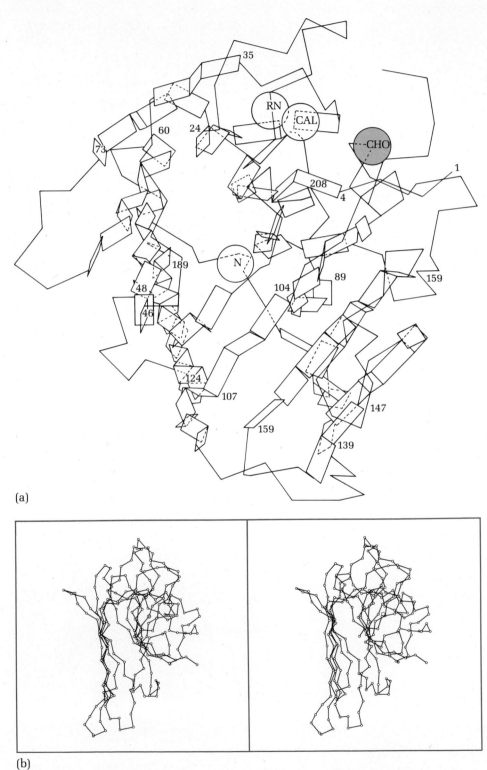

(a)

(b)

Figure 8.52 (a) Backbone structure of concanavalin A. The binding sites for carbohydrate (CHO) and metal ions are circled. β structure is indicated by ribbons. (Concanavalin A contains no α-helix.) (b) Stereo view of concanavalin A.

Concanavalin A is a tetramer whose general structure is shown schematically in Figure 8.51. Concanavalin A has no α-helix, although 57% of its residues are involved in β structure. Concanavalin A also has binding sites for Ca(II) and for transition metals such as Mn(II). Carbohydrates will not bind to concanavalin A in the absence of bound metal ions. The tertiary structure of one subunit of concaravalin A is shown in Figure 8.52.

The biological function of lectins is not known. They might act as "plant immunoglobulins" by binding to bacterial, viral, and fungal invaders. Concanavalin A causes agglutination of human cancer cells, a property that has been much studied. The function of animal agglutinating proteins, sometime called "animal lectins," is also unknown, but has been the object of considerable speculation. Since cells are coated with glycoproteins, the animal agglutinating proteins might be important in cellular recognition and embryonic development.

8-10 SUMMARY

Carbohydrates have the generic formula $(CH_2O)_n$, and many simple, open-chain carbohydrates contain a saccharose unit. Carbohydrates that have an aldehyde as their most oxidized functional group are called aldoses, and those having a keto group as their most oxidized functional group are called ketoses. The simplest carbohydrate having a chiral center is D-glyceraldehyde. The stereochemical designation D- indicates that the hydroxyl group is on the right and the hydrogen on the left in a Fischer projection, in which, by convention, the most highly oxidized carbon is placed at the top. Most biologically important carbohydrates have the absolute configuration D- at their highest numbered asymmetric carbon, numbering from the most highly oxidized end of the molecule. Monosaccharides containing five or six carbon atoms can exist as five-membered rings known as furanoses or as six-membered rings known as pyranoses. The aldopentoses are most commonly furanoses in complex biological molecules such as nucleic acids and certain coenzymes. The aldohexoses commonly exist as pyranoses. Formation of the cyclic hemiacetal produces a new asymmetric carbon at C-1, called the anomeric carbon. The β anomer has an equatorial hydroxyl group and the α anomer has an axial hydroxyl group. The α- and β-anomers interconvert spontaneously, a process called mutarotation. In glucose the stable configuration is β, but in all other pyranoses the α anomer is more stable.

Reducing sugars are oxidized to cyclic lactones by Tollens' reagent (aqueous ammoniacal silver ion), by Fehling's solution (a Cu(II) solution in a tartaric acid buffer), or Benedict's solution (a Cu(II) solution in a citrate buffer). In each of these reactions, the pH is basic, so that ketoses tautomerize to aldoses and thus are also oxidized by either Cu(II) or Ag(I). Bromine also oxidizes pyranoses to δ-lactones. Cyclic hemiacetals and cyclic hemiketals react with alcohols, such as methanol, in the presence of an acid catalyst to give full acetals or full ketals called glycosides. Anomeric glycosides can be distinguished by their reactivities as substrates for various specific glycosidases. Dimethyl sulfate permethylates carbohydrates; periodate oxidizes vicinal diols to dialdehydes; and aldoses and ketoses are reduced to alditols by sodium borohydride.

Phosphate esters of carbohydrates are important in carbohydrate metabolism. The phosphate esters are strong acids with pK_a values of about 1 and 6. At pH 7, therefore, the phosphate esters exist predominantly as dianions. In most cases hydrolysis of phosphate esters occurs with P—O bond cleavage, analogous to acyl bond cleavage in carboxylate esters. The phosphorus of phosphate esters has a tetrahedral configuration. A trigonal bipyramidal intermediate forms during the course of phosphate ester hydrolysis.

Oligosaccharides are compounded of a few monosaccharides. Among the most important oligosaccharides are the nonreducing sugar sucrose, a disaccharide of α-D-glucose bonded $(1 \rightarrow 2)$ with β-D-fructose; lactose, a dimer of β-D-galactose bonded $(1 \rightarrow 4)$ with D-glucose; and maltose, a dimer of α-D-glucose bonded $(1 \rightarrow 4)$ with D-glucose.

Cellulose is the major structural molecule of plants. It is a $\beta(1 \rightarrow 4)$ polymer of D-glucose that can contain thousands of glucose monomers. Three of the most abundant mucopolysaccharides are hyaluronic acid, chondroitin sulfate, and dermatan sulfate. Hyaluronic acid is an alternating polymer of D-glucuronic acid and N-acetylglucosamine that is found in connective tissue, synovial fluid of joints, and the vitreous humor of the eye. Chondroitin sulfate, an alternating polymer of glucuronic acid bonded $\beta(1 \rightarrow 3)$ to N-acetylglucosamine 6-sulfate, is a component of vertebrate connective tissue. Dermatan sulfate, an alternating polymer of α-L-iduronic acid bonded $(1 \rightarrow 3)$ with N-acetylglucosamine 4-sulfate, is abundant in skin. Heparin is a mixture of polysaccharides that acts as an anticoagulant and is widely used to treat diseases of the heart and blood vessels.

The peptidoglycans of bacterial cell walls are complex polysaccharides cross-linked by short peptides that contain D-amino acid residues. Murein is an alternating polymer of N-acetylglucosamine and N-acetylmuramic acid linked by $\beta(1 \rightarrow 4)$ glycosidic bonds. The antibiotic penicillin inhibits formation of the acylenzyme intermediate in the reaction that cross-links the peptide units of the polysaccharide chains of the peptidoglycan. Lysozyme forms part of the defense of animals against bacteria. This enzyme cleaves the $\beta(1 \rightarrow 4)$ glycosidic bond between N-acetylglucosamine and N-acetylmuramic acid in the peptidoglycan of the bacterial cell wall. The peptidoglycan is the major component of the cell walls of gram-positive bacteria. Other important components include ribitolteichoic acids and glyceroteichoic acids. The cell walls of gram-negative bacteria are much more complex than those of gram-positive bacteria. They are coated with a second membrane that contains glycoproteins and complex lipopolysaccharides.

The storage polysaccharides of plants are amylose and amylopectin. Amylose is a homopolysaccharide composed of glucosyl residues bonded $\alpha(1 \rightarrow 4)$. Amylose has a helical conformation whose central pore has just the right dimensions to bind iodine, giving the characteristic starch-iodine blue complex. Amylopectin differs from amylose by having many branches formed by $\alpha(1 \rightarrow 6)$ glycosidic bonds between glucosyl residues. Glycogen is the storage polysaccharide of animals. It differs from amylopectin by having more and shorter branches.

Many cellular proteins have complex oligosaccharides bonded to side chains of amino acid residues in the protein. The immunoglobulins and the blood group substances are among the many important proteins that are covalently bound to oligosaccharides.

The lectins are a group of plant proteins that bind stereospecifically to carbohydrates. Wheat germ agglutinin binds to N-acetylglucosamine; concanavalin A binds stereospecifically to α-D-mannopyranose and to α-D-glucopyranose. Concanavalin A causes agglutination of human cancer cells.

REFERENCES

Bailey, R. W., "Oligosaccharides," Macmillan, New York, 1965.

Barker, R., "The Organic Chemistry of Organic Compounds," Prentice-Hall, Englewood Cliffs, N.J., 1971.

Florkin, M., and Stoltz, E. H., Eds., "Comprehensive Biochemistry," Vol. 5, Elsevier, New York, 1963.

Jeanloz, R. W., and Balasz, E. A., Eds., "The Amino Sugars," 2 vols., Academic Press, New York, 1965.

Pigman, W., and Horton, D., Eds., "The Carbohydrates," 2nd ed., Academic Press, New York, 1970.

Problems

1. Glucan is a β(1 → 3) polyglucose isolated from the cell wall of *Saccharomyces cerevisiae*. Partial hydrolysis of glucan yields the dimer 3-*O*-(β-D-glucopyranosyl)-β-D-glucopyranoside (**1**).

1

a. Write the product(s) for reaction of **1** with (i) Br$_2$, (ii) dimethyl sulfate, (iii) methanol and HCl, (iv) periodate.
b. Show how you would use the reagents of part a and the enzymes α- and β-glucosidase to deduce the structure of **1**.

2. Explain why the α anomer of mannose is more stable than the β anomer.

3. Draw the structures of β-D-arabinofuranose and α-D-xylofuranose.

4. Ribose exists predominantly as a six-membered ring when it is free in solution. Draw the structures of α- and β-D-ribopyranose.

5. Write the reaction for the periodate oxidation of α-D-methylribopyranoside and α-D-methylglucopyranoside. How would the products of these oxidation reactions enable you to tell whether the reactants are pyranoses or furanoses?

6. Why isn't erythritol (see Section 8.4E) written as "D-erythritol"?

7. Write the products for the reactions of α-D-galactopyranose with each of the following reagents: (a) Tollen's reagent, (b) NaIO$_4$, (c) NaBH$_4$, (d) CH$_3$OH/HCl.

8. One of the products from question 7 will not rotate plane polarized light. Which one lacks this ability?

9. An unknown sugar with the formula C$_5$H$_{10}$O$_5$ was treated with NaBH$_4$ and the product was optically inactive. Write the two possible products, assuming that the unknown sugar was D. Are the products enantiomers or diastereoisomers? Are they epimers?

10. L-Threonine has two chiral centers that have the L-threo configuration. Draw the correct stereochemical structure of L-threonine.

11. A student accidentally mixed the labels on two sugars, but he knew he had fructose and glucose. He performed a periodate digestion of one and obtained a periodate to formic acid ratio of 5:3. Which sugar was this one? Show how it is broken down to form this ratio.

12. The structure of the capsular polysaccharide of a strain of *E. coli* has recently been determined. This polysaccharide is composed of the repeating trisaccharide shown on page 318.

a. Write the products obtained for mild acid hydrolysis of the trisaccharide.
b. How many moles of periodate are consumed by the trisaccharide?

c. What information about the bonding of the two ribosyl moieties is obtained from the observations of parts a and b? Write the product obtained for permethylation of the trisaccharide with dimethyl sulfate.

d. Write the product obtained when the permethylated product is hydrolyzed and explain what information about the bonding of the monosaccharide residues to each other is obtained from the hydrolysis products.

13. Chitin [poly(1 → 4)NAG] is virtually insoluble in most solvents, but it can be dissolved in a solvent system consisting of dimethylacetamide that contains 5% LiCl. In this solvent the optical rotation, $[\alpha]_D$, changes from $+33°$ to $-52°$ for chitin isolated from the blue crab. No glycosidic bonds are hydrolyzed during this time period. What possible structural feature of chitin might account for the change in optical rotation? [Reference, P. R. Austin, *Science*, 212, 749–755 (1981).]

14. The following abstract recently appeared in the *Journal of Molecular Biology*.

Phage lysozyme has catalytic activity similar to that of hen egg white lysozyme, but the amino acid sequences of the two enzymes are completely different.

The binding to phage lysozyme of several saccharides including N-acetylglucosamine (GlcNAc), N-acetylmuramic acid (MurNAc) and (GlcNAc)₃ have been determined crystallographically and shown to occupy the pronounced active site cleft. GlcNAc binds at a single location analogous to the C site of hen egg white lysozyme. MurNAc binds at the same site. (GlcNAc)₃ clearly occupies sites B and C, but the binding in site A is ill-defined.

Model building suggests that, with the enzyme in the conformation seen in the crystal structure, a saccharide in the normal chair configuration cannot be placed in site D without incurring unacceptable steric interference between sugar and protein. However, as with hen egg white lysozyme, the bad contacts can be avoided by assuming the saccharide to be in the chair conformation. Also Asp20 in T4 lysozyme is located 3 Å from carbon $C_{(1)}$ of saccharide D, and is in a position to stabilize the developing positive charge on a carbonium ion intermediate. Prior genetic evidence had indicated that Asp20 is critically important for catalysis. This suggests that in phage lysozyme catalysis is promoted by a combination of steric and electronic effects, acting in concert. The enzyme shape favors the binding in site D of a saccharide with the geometry of the transition state, while Asp20 stabilizes the positive charge on the oxocarbonium ion of this intermediate. In phage lysozyme, the identity of the proton donor is uncertain. In contrast to hen egg white lysozyme, where Glu35 is 3 Å from the glycosidic D—O—E bond, and is in a non-polar environment, phage lysozyme has an ion pair, Glu . . . Arg145, 5Å away from the glycosidic oxygen. Possibly Glu undergoes a conformational adjustment in the presence of bound substrate, and acts as the proton donor. Alternatively, the proton might come from a bound water molecule. [W. F. Anderson, *J. Mol. Biol.*, *147*, 523–543 (1981).]

From your knowledge of hen egg white lysozyme, propose a plausible mechanism for T4 bacteriophage lysozyme.

Chapter 9

Nucleosides, nucleotides, and nucleic acids

9–1 INTRODUCTION

The goddess of learning is fabled to have sprung full grown from the brain of Zeus, but it is seldom that a scientific conception is born in its final form, or owns a single parent. More often it is the product of a series of minds, each in turn modifying the ideas of those that came before, and providing material for those that come after.

George P. Thomson

The discovery of the nucleic acids late in the nineteenth century was inauspicious. Nucleic acids are now known to be extremely large and complicated molecules, but they initially gave the appearance of being too simple to be biochemically important. In contrast to proteins, which are formed from a set of 20 monomers, the nucleic acids contain only four monomers. (A substantial number of minor monomers are found in certain nucleic acids, but these were not known at the time the nucleic acids were discovered nor for a long time after.) As late as the 1930s a tetrameric structure for DNA was proposed, a hypothesis that survived in some circles until the 1950s. The explosion of knowledge that followed the discovery of the structure of DNA and its function as the genetic material has led to an understanding of genetics that was unthought of 20 years earlier.

The discovery of the structure of DNA launched the field of "molecular biology," a discipline scarcely 30 years old that has revolutionized the entire framework of biology. The so-called central dogma of molecular biology asserts that deoxyribonucleic acid (DNA) self-replicates, that the genetic information contained in DNA is copied onto a ribonucleic acid (RNA) by a process called *transcription,* and that the information then contained in RNA is *translated* into protein (Figure 9.1).

9-2 PYRIMIDINES AND PURINES

A. Pyrimidines

Nucleic acids, nucleotides, and nucleosides contain heterocyclic bases of two classes: the *pyrimidines* and the *purines.* First, let us consider the pyrimidine ring system. Pyrimidine is an aromatic, six-membered ring that contains two nitrogens at positions 1 and 3. Unfortunately, there are two numbering systems for pyrimidines. The older numbering, still used in Europe, is the Beilstein system. We shall use the IUPAC numbering system in our discussions.

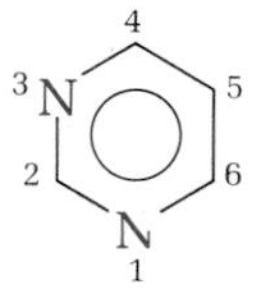

Pyrimidine
(IUPAC numbering)

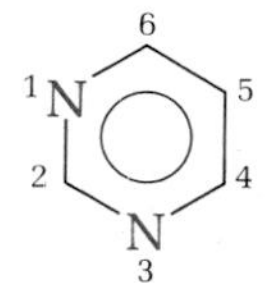

Pyrimidine
(Beilstein numbering)

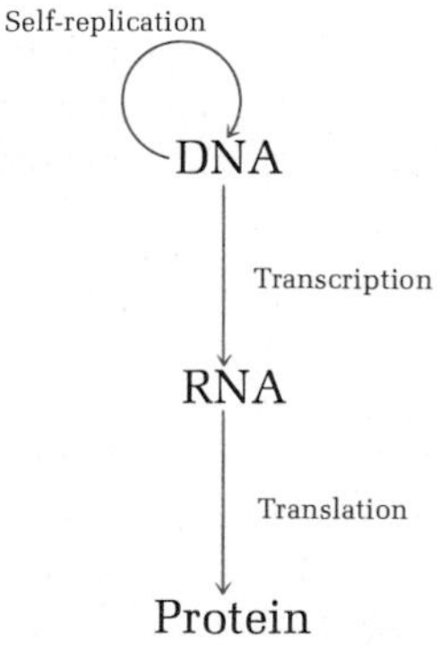

The most abundant pyrimidines are *uracil* (U), found in ribonucleic acids; *cytosine* (C) found in both ribonucleic acids and deoxyribonucleic acids; and *thymine* (T), found in deoxyribonucleic acids. The UV absorption spectra of uracil, thymine, and cytosine are shown in Figure 9.2.

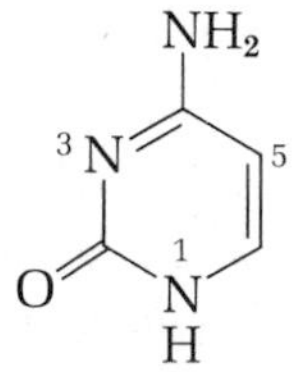

Cytosine

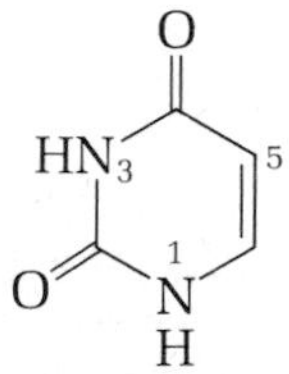

Uracil

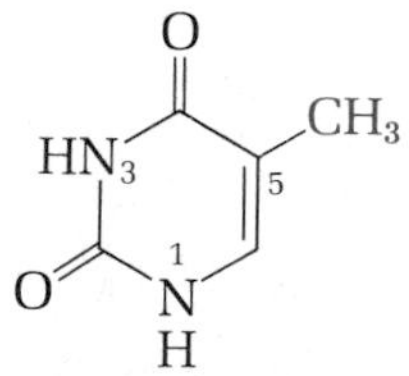

Thymine
(5-methyluracil)

Besides these "major" bases, 5-methylcytosine is sometimes found in DNA, and certain viruses contain 5-hydroxymethylcytosine. Other methylated pyrimidines are present in small quantities in transfer ribonucleic acids (Sections 9.6C and 9.6D).

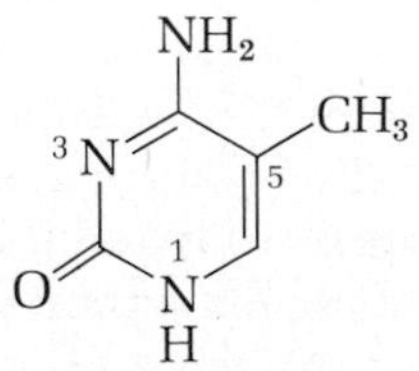

5-Methylcytosine

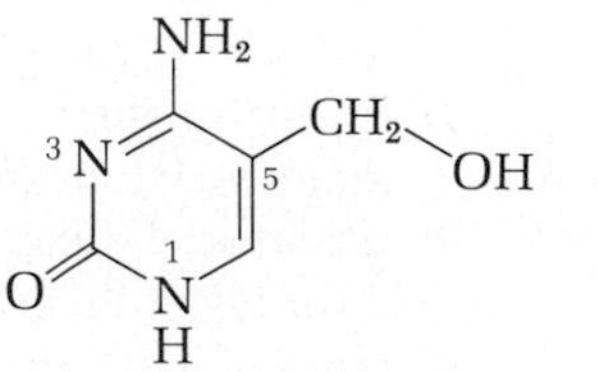

5-Hydroxymethylcytosine

Self-replication

DNA

Transcription

RNA

Translation

Protein

Figure 9.1 "Central dogma" of molecular biology. DNA directs its own replication (Chapter 24), and the flow of cellular information is unidirectional from DNA to RNA to protein (Chapters 25 and 26). These hypotheses have proven correct, but the biochemical processes involved in each step are enormously complicated.

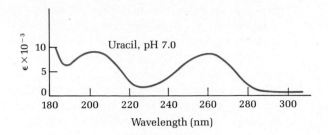

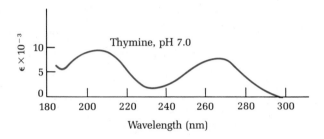

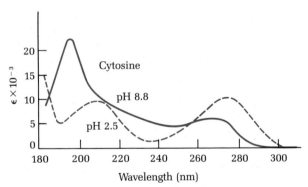

Figure 9.2 The UV spectra of the most common pyrimidine bases. The region of the spectrum at wavelengths less than about 220 nm is inaccessible to most UV instruments, so only absorption near 260 is useful for routine experimental measurements.

We have written the structures of pyrimidines as *lactams*, or cyclic amides. Each can also exist in a tautomeric form called a *lactim* (reaction 9.1). The equilibrium constant for tautomerization lies overwhelmingly on the side of the lactam form. The concentration of the lactim is so small that it is difficult to measure accurately, but the equilibrium constant for tautomerization is in the range of 10^{-4}. In equation 9.1 the lactim forms *3* and *4* are possible for thymine itself, but in nucleic acids N-1 is bonded to C-1 of a ribose (or 2′-deoxyribose), so that *3* and *4* are not possible.

Of the many chemical reactions of pyrimidines one is of particular biochemical importance: exposure of solutions of thymine to ultraviolet light results in *photodimerization* (reaction 9.2).

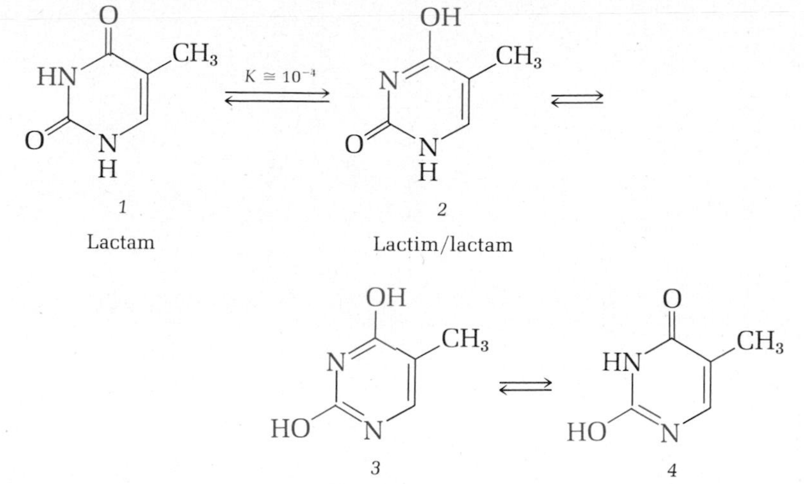

Lactam — Lactim/lactam

$$(9.1)$$

3 — Lactim 4 — Lactim/lactam

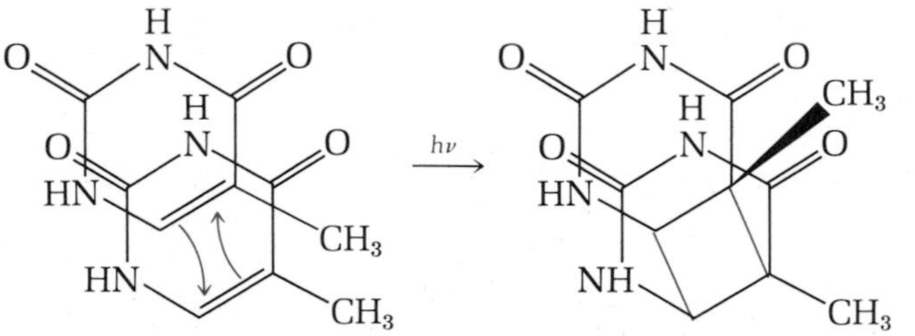

Thymine — Photodimer of thymine

$$(9.2)$$

Adjacent thymines in a strand of DNA also can photodimerize disrupting DNA replication. Other pyrimidines can also dimerize, although thymine dimers are the most important biologically. Inactivation of DNA by ultraviolet light is the basis for germicidal lights used in hospitals.

B. Purines

The purine ring system can be viewed as a pyrimidine fused to an imidazole. Note that the numbering system of purines follows the Beilstein convention with respect to pyrimidine ring. The π electrons of the purine ring are extensively delocalized. The principal contributing structures are shown below. As these structures suggest, positions 2, 6, and 8 are susceptible to attack by nucleophiles, and positions 3 and 7, being electron rich, are susceptible to attack by electrophiles.

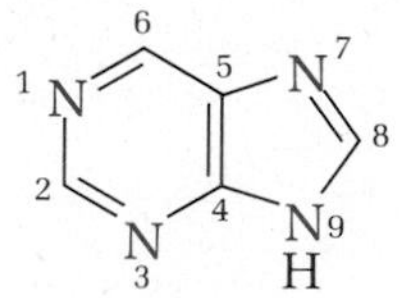

Purine

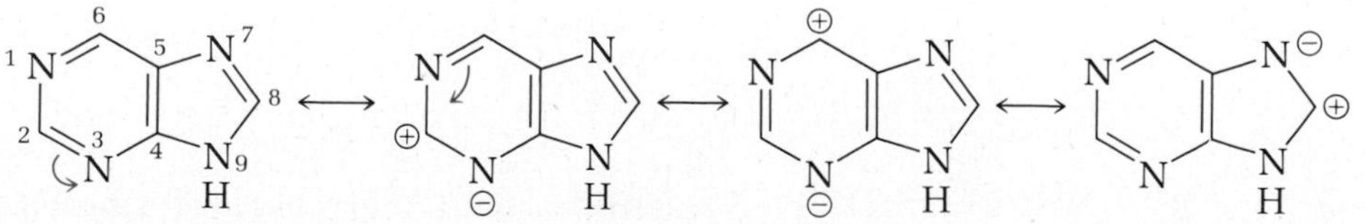

The two most common purines are *adenine* and *guanine*, whose ultraviolet absorption spectra are shown in Figure 9.3. The purines *xanthine*, *hy-*

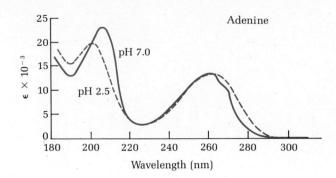

Figure 9.3   UV absorption spectra of adenine and guanine. [From D. Voet, W. B. Gratzer, R. A. Cox, and P. Doty, *Biopolymers*, *1*, 193 (1963).]

poxanthine, and *uric acid* are also important metabolites, although they are far less abundant than adenine and guanine.

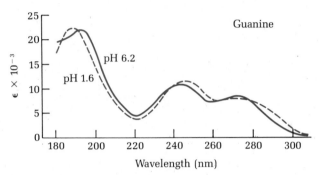

9–3 NUCLEOSIDES   Nucleosides are β-N-glycosides formed between a purine or pyrimidine and a carbohydrate, almost always either β-D-ribofuranose or 2-deoxy-β-D-ribofuranose. The structures of some common nucleosides are shown in Figure 9.4. The numbering systems for the purine and pyrimidines of nucleosides follow the conventions of the free bases. The carbon atoms of the carbohydrate are numbered with primes to distinguish them from the ring atoms. In pyrimidine nucleosides the glycosidic bond is to N-1, and in purine nucleosides it is to N-9. The plane of the purine or pyrimidine ring bisects the plane of the sugar ring.

First, consider the conformation of the furanose, whose conformation can be either a half-chair or an envelope. Carbon atom C(2′) and C(3′) are displaced above or below a plane through atoms C(4′)-O(1′)-C(1′). If the displaced atom lies above the ring (toward the base or on the same side of the plane as the C-5′ atom), the ring conformation is known as *endo*; if it lies below the ring (on the other side), the ring conformation is known as *exo* (Figure 9.5). The C(2′)-*endo* and C(3′)-*endo* conformations are about equally abundant in free nucleosides. The C(3′)-*exo* conformation is found in deoxy-

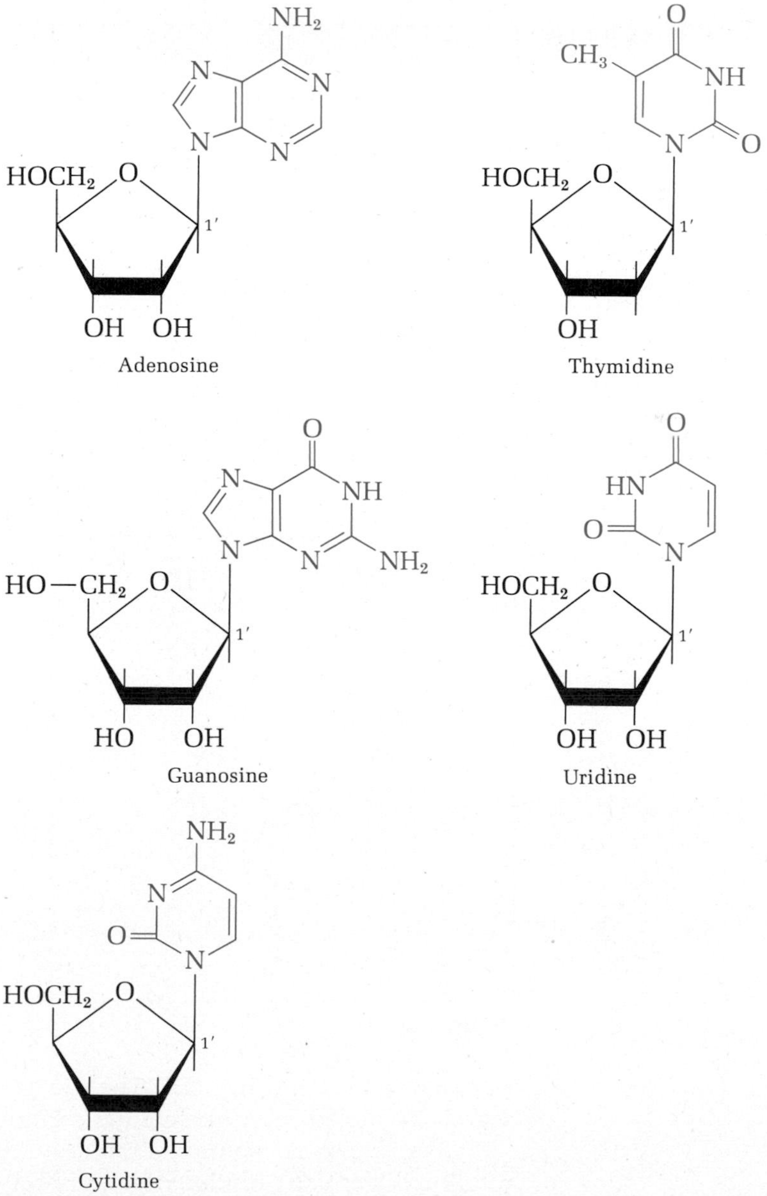

Figure 9.4 Structures of some common nucleosides. The 2′ hydroxyl group is replaced by a hydrogen atom in the 2′-deoxyribonucleosides.

Adenosine

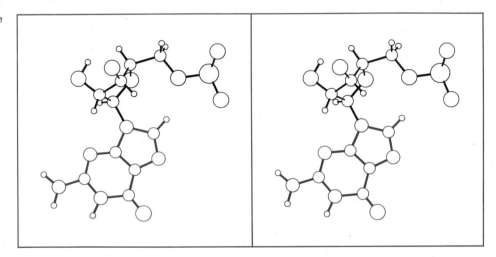

Guanosine

Cytidine

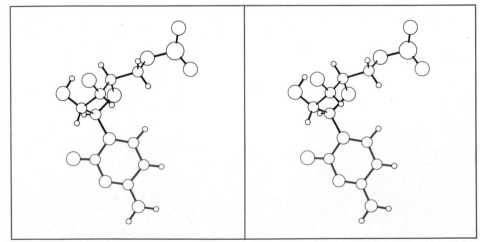

Figure 9.4 *(continued)*

ribonucleosides and in DNA. The C(2′)-*exo* conformation has not been observed in nucleosides, nucleotides, or nucleic acids.

Turning to the conformation around the glycosidic bond, we find that rotation is sterically hindered by a proton at the C(2′)-*endo* position. The nucleosides and nucleoside phosphates (Section 9.4) can therefore exist in either of two conformations designated *syn* and *anti* (Figure 9.6). In the *syn* conformation the O(2) in pyrimidine nucleosides, or the N(3) in the purine

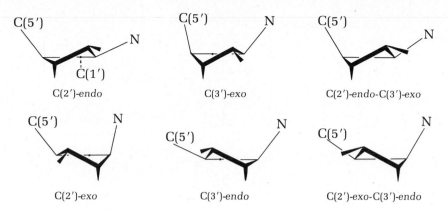

Figure 9.5 Possible conformations of the sugar moiety of nucleosides. [From W. Saenger, *Angew. Chem., Internat. Ed., 12,* 591–601 (1973).]

Syn conformation

Rotation around glycosidic bond

Anti conformation

Sys conformation

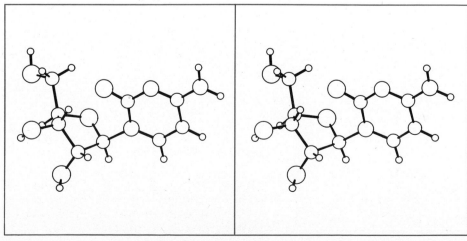

Figure 9.6 *Sys* conformation and *anti* conformation (on page 328) of nucleosides. The rotation around the glycosidic bond relieves van der Waals repulsions between the C-2 substituents and the ribose.

Anti **conformation**

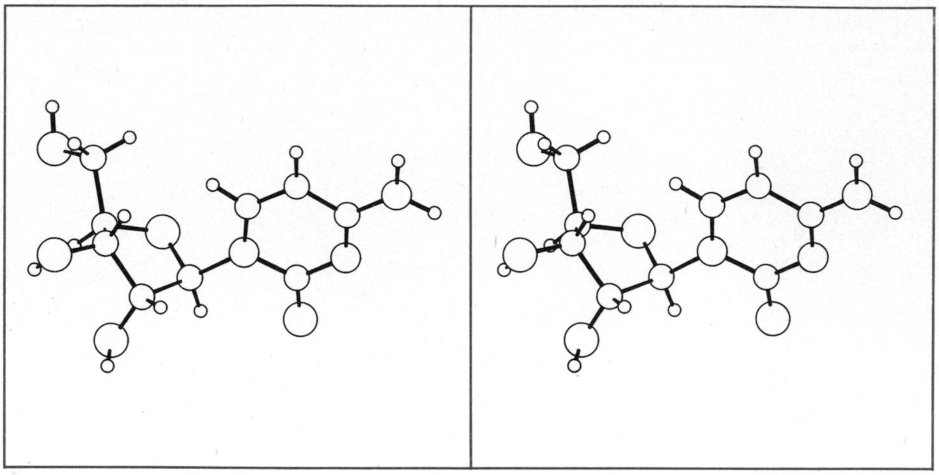

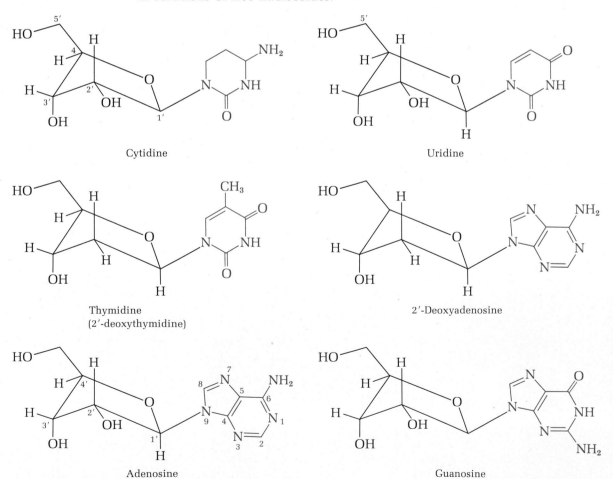

Figure 9.6 (*continued*)

nucleosides, lies above the plane of the sugar. In either case the nitrogenous base is at roughly right angles to the plane of the sugar. The *anti* conformation is more stable for both purine nucleosides and pyrimidine nucleosides (Figure 9.7). At the other end of the sugar rotation about the C(4′)-C(5′) bond can produce three different conformations that have about equal stabilities in solutions of free nucleosides.

Cytidine

Uridine

Thymidine
(2′-deoxythymidine)

2′-Deoxyadenosine

Adenosine

Guanosine

Figure 9.7 The most stable conformations of some common nucleosides. The furanose ring of ribose is in the C(2′)-*endo* conformation, the glycosidic bonds are β, and the most stable conformation around the glycosidic bond is *anti*.

2'-deoxyadenosine

2'-deoxyguanosine

2'-deoxycytidine

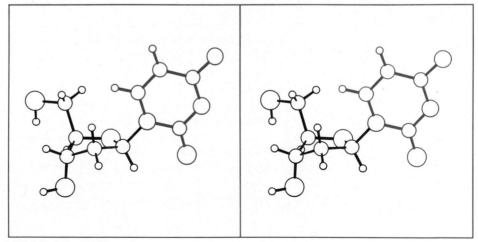

Figure 9.7 (continued)

2'-deoxythymidine

2'-deoxyuridine

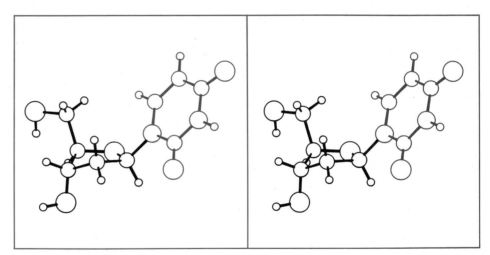

Figure 9.7 *(continued)*

Nucleosides are hydrolyzed in acid and are stable in dilute alkali. Hydrolysis in acid is much slower than the corresponding hydrolysis of *O*-glycosides. The hydrolysis of purine nucleosides is faster than the hydrolysis of pyrimidine nucleosides. In the hydrolysis of guanosine the first step is protonation of N-7, followed by breaking of the glycosidic bond and formation of an acylium ion. Rapid reaction of the acylium ion with water produces ribose (Figure 9.8).

Nucelosides other than those involved in the structures of nucleic acids are also produced by many organisms. Many antibiotics are nucleosides that act by inhibiting protein synthesis (Chapter 26). Such nucleosides include *cordycepin, puromycin,* and *cytosine arabinoside* (Figure 9.9). The latter is an effective drug against myeloblastic leukemia.

9−4 NUCLEOTIDES Nucleotides are sugar-phosphate esters of nucleosides. Those nucleotides whose sugar is ribose are called *ribonucleotides,* and those derived from 2'-deoxyribose are known as *deoxyribonucleotides.* Each of the free hydroxyl groups of ribose can bear a phosphate group, and 2'-, 3'-, and 5'-phosphate esters of ribonucleosides are all known. When the sugar is 2'-deoxyribose, esterification with phosphate is, of course, possible only at the 3' and 5' positions (Figure 9.10). Cyclic 2'-3' and cyclic 3'-5' nucleotides are also found in trace amounts in cells (Figure 9.11). They are important in the

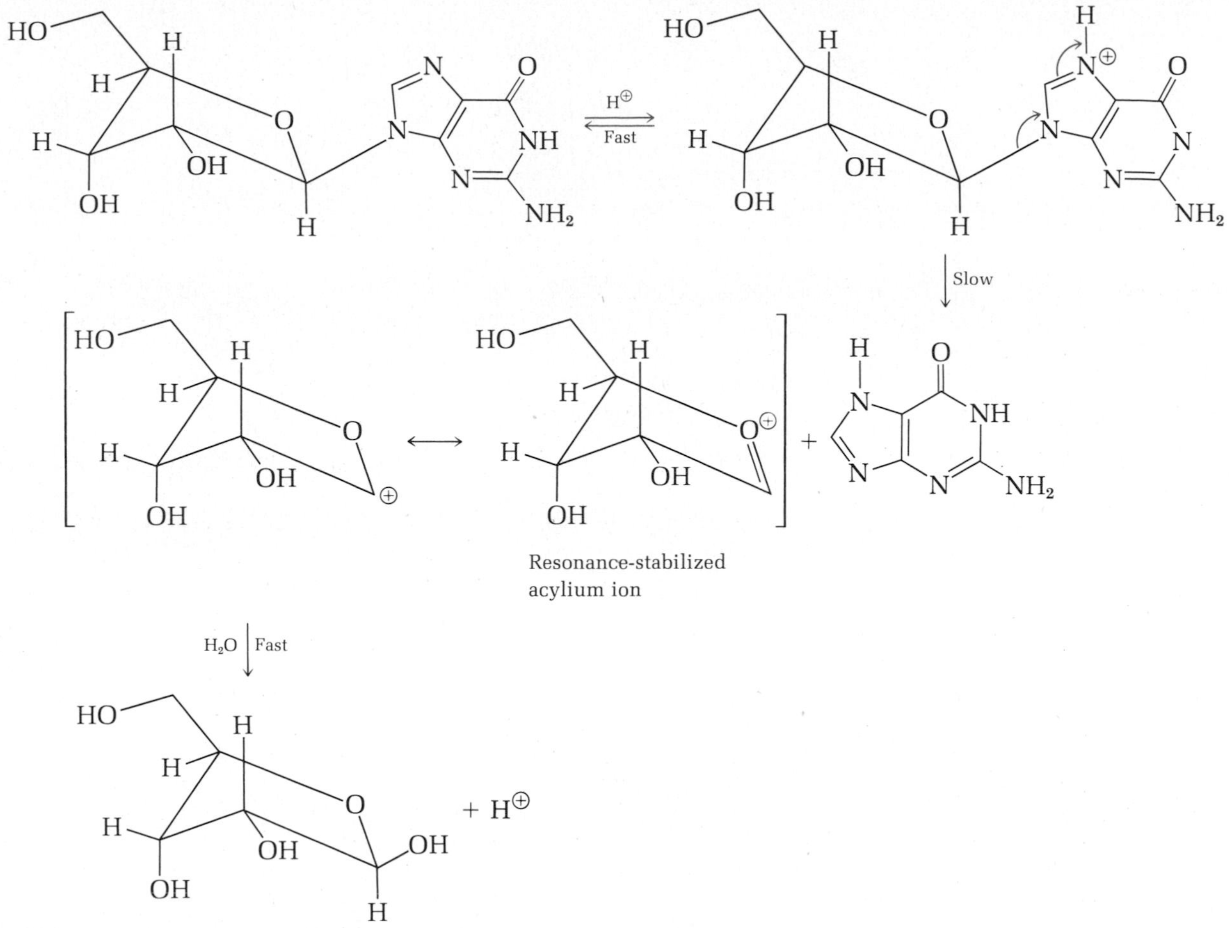

Figure 9.8 Acid-catalyzed hydrolysis of guanosine.

regulation of many biochemical phenomena (see, for example, Section 13.6). The names and abbreviations of the common nucleotides are given in Table 9.1. The phosphate monoesters are strong acids with pK_a values of approximately 1.0 and 6.0. At pH 7.0, therefore, the nucleotide monophosphates exist predominantly as the dianions.

Nucleoside 5′-diphosphates and nucleoside 5′-triphosphates are phosphoanhydride derivatives of nucleotides (Figure 9.12). They are also strong acids and are extensively ionized at physiological pH. ATP has four ionizable hydrogens. The pK_a values of three of them are less than 5 and the pK_a of the fourth, for the equilibrium between ATP^{3-} and ATP^{4-} is 6.95 (reaction 9.3).

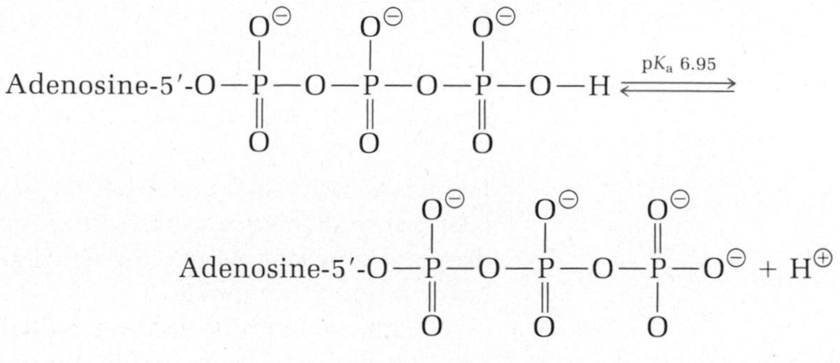

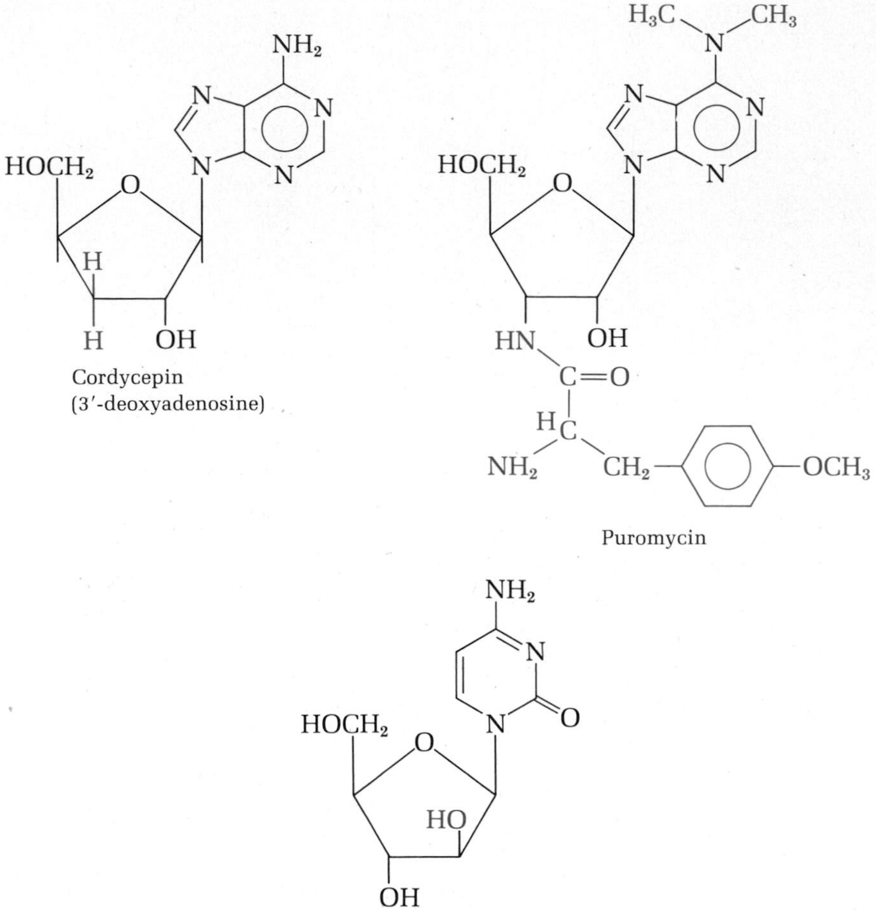

Cordycepin
(3'-deoxyadenosine)

Puromycin

Cytosine arabinoside (*ara*-C)

Figure 9.9 Structures of some nucleoside antibiotics.

Hence, at pH 7.0 ATP exists as 53% of the ATP^{4-} ion and 47% of the ATP^{3-} ion. ADP has three ionizing protons. The pK_a values of the first two are less than 5; the pK_a of the third is 6.7 (reaction 9.4).

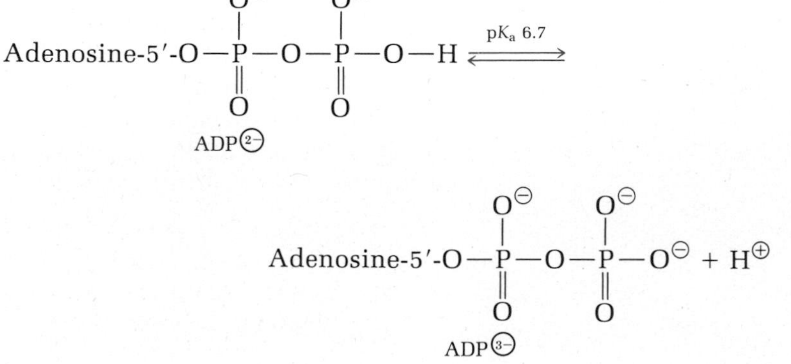

$$\text{Adenosine-5'-O}-\underset{O}{\overset{O^{\ominus}}{\underset{\|}{P}}}-O-\underset{O}{\overset{O^{\ominus}}{\underset{\|}{P}}}-O-H \underset{\longleftarrow}{\overset{pK_a\ 6.7}{\longrightarrow}}$$

$ADP^{\text{②-}}$

$$\text{Adenosine-5'-O}-\underset{O}{\overset{O^{\ominus}}{\underset{\|}{P}}}-O-\underset{O}{\overset{O^{\ominus}}{\underset{\|}{P}}}-O^{\ominus} + H^{\oplus} \qquad (9.4)$$

$ADP^{\text{③-}}$

Hence, at pH 7, 67% of ADP is present as ADP^{3-} and 33% as ADP^{2-}. The bases of nucleosides, nucleotides, and nucleic acids also have ionizing hydrogens. The pK_a values for the bases are shown in Table 9.2. At pH 7 the bases are neutral species.

The nucleotides are enormously important in biochemistry. They are the components from which the nucleic acids are constructed, and they

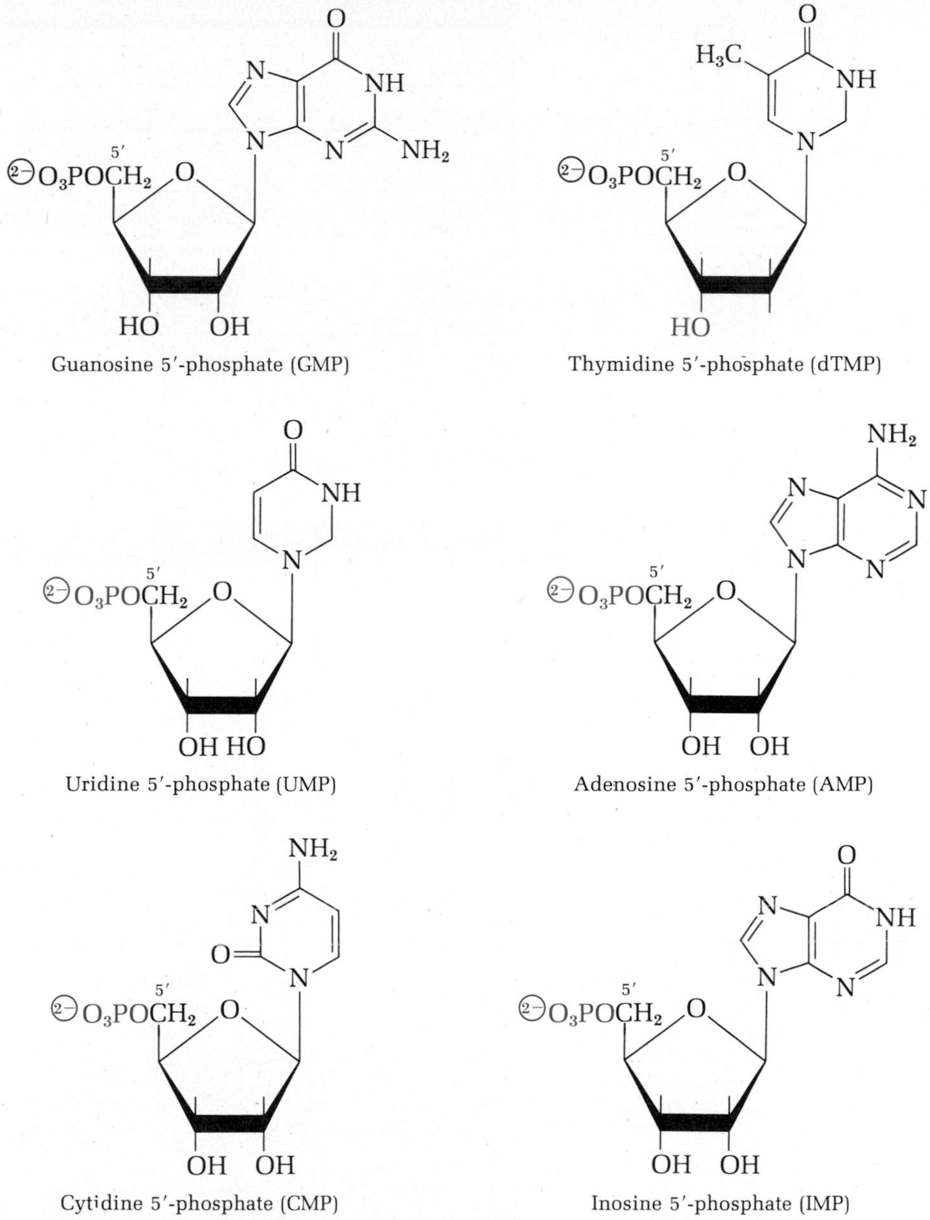

Guanosine 5'-phosphate (GMP)

Thymidine 5'-phosphate (dTMP)

Uridine 5'-phosphate (UMP)

Adenosine 5'-phosphate (AMP)

Cytidine 5'-phosphate (CMP)

Inosine 5'-phosphate (IMP)

Figure 9.10 Structures of some ribonucleotides.

participate in many metabolic reactions. Nucleotides are also components of certain coenzymes: adenosine 3',5'-diphosphate is a component of coenzyme A (Section 10.3A), adenosine 2',5'-diphosphate is a component of the coenzyme NADP, and ADP is a component of the coenzyme NAD (Section 10.3B). Nucleotides are also intermediates in the biosynthesis of complex carbohydrates (Section 17.5).

9-5 STRUCTURE OF DEOXYRIBONUCLEIC ACID (DNA)

A. Primary Structure of DNA

DNA is a linear polymer (double stranded in its native state) of 2'-deoxynucleotide residues linked by phosphodiester bonds between the 3' and 5' positions of the 2'-deoxyribosyl moieties (Figure 9.13). The most common bases in DNA are adenine, thymine, guanine, and cytosine. The *primary structure* of DNA is the linear sequence of nucleotide residues comprising the polydeoxyribonucleotide chain.

Table 9.1 Common nucleosides and nucleotides

Name	Abbreviation
Bases	
Uracil	U (Ura)
Cytosine	C (Cyt)
Thymine	T (Thy)
Adenine	A (Ade)
Guanine	G (Gua)
Nucleosides	
Uridine	Urd
Cytidine	Cyd
Thymidine[a]	dThd
Adenosine	Ado
Guanosine	Guo
Inosine	Ino
Ribonucleoside 5′-phosphates	
Uridine 5′-phosphate or 5′-uridylic acid	UMP (Urd-5′-P)
Cytidine 5′-phosphate or 5′-cytidylic acid	CMP (Cyd-5′-P)
Thymine ribonucleoside-5′-monophosphate or 5′-ribothymi- dylic acid	rTMP[b]
Adenosine 5′-phosphate or 5′-adenylic acid	AMP (Ado-5′-P)
Guanosine 5′-phosphate or 5′-guanylic acid	GMP (Guo-5′-P)
2′-Deoxyribonucleoside 5′-phosphates	dNMP
Deoxyadenosine 5′-phosphate or deoxyadenylic acid	dAMP
Deoxyguanosine 5′-phosphate or deoxyguanylic acid	dGMP
Thymidine 5′-phosphate, deoxythymidine 5′-phosphate thymidylic acid, or deoxythymidylate	dTMP (or TMP)
Deoxycytidine 5′-phosphate or deoxycytidylic acid	dCMP
Ribonucleoside 5′-diphosphates and triphosphates	NDP, NTP
Adenosine diphosphate	ADP
Adenosine triphosphate	ATP
Guanosine diphosphate	GDP
Guanosine triphosphate	GTP
Cytidine diphosphate	CDP
Cytidine triphosphate	CTP
Uridine diphosphate	UDP
Uridine triphosphate	UTP

[a] Thymidine is a deoxyribose derivative (dThd).
[b] "TMP" is used for the deoxyribonucleoside as in "TMP synthetase."

The covalent structure of Figure 9.13 is cumbersome, and abbreviations are often employed. In one shorthand structure the 2′-deoxyribose moieties are represented as vertical lines (Figure 9.14). The top of the line represents position 1 of the sugar, and the base is perched on top. Diagonal lines from roughly the center of the vertical line of one sugar to the bottom of the adjacent sugar represent the 3′–5′ phosphodiester bonds. The sugar residue having a free (or phosphorylated) 5′-hydroxyl group is the 5′ terminus, and the sugar having a free (or phosphorylated) 3′-hydroxyl group is the 3′ terminus of the nucleic acid. The 5′ terminus is conventionally written to the left. This structure can be abbreviated further by omitting the vertical lines.

The primary structures of DNA molecules from various sources range from a few thousand residues in some viruses to 10^6 residues in bacteria to 10^9 residues in a human chromosomal DNA. A single molecule of bacterial DNA is about 1 mm long, and the length of a DNA molecule in a human

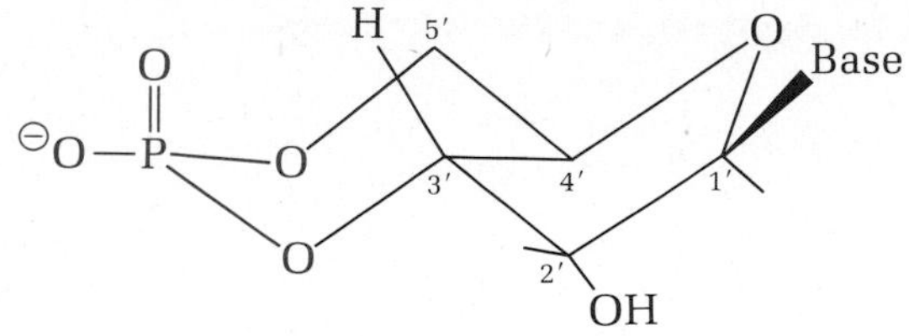

3′-5′ Cyclic ribonucleotide

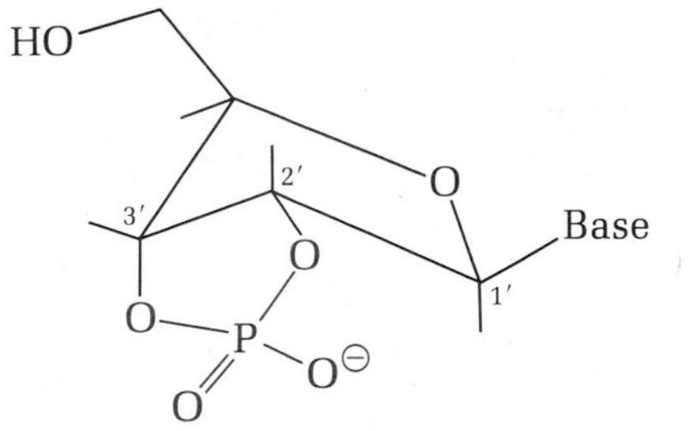

2′-3′ Cyclic ribonucleotide

Figure 9.11 Structures of the cyclic ribonucleotides.

Table 9.2 Ionization constants for the purine and pyrimidine bases of nucleosides, nucleotides, and nucleic acids

Adenine in adenosine, adenylic acids, AMP, etc. and RNA and DNA

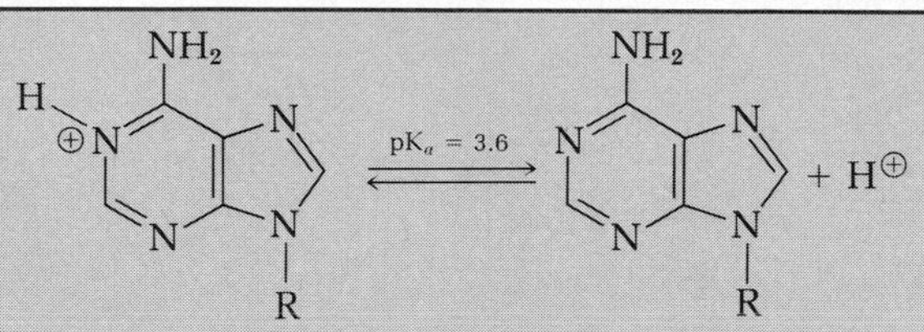

$pK_a = 3.6$

Guanine in guanosine guamylic acids, GMP, etc., and RNA and DNA

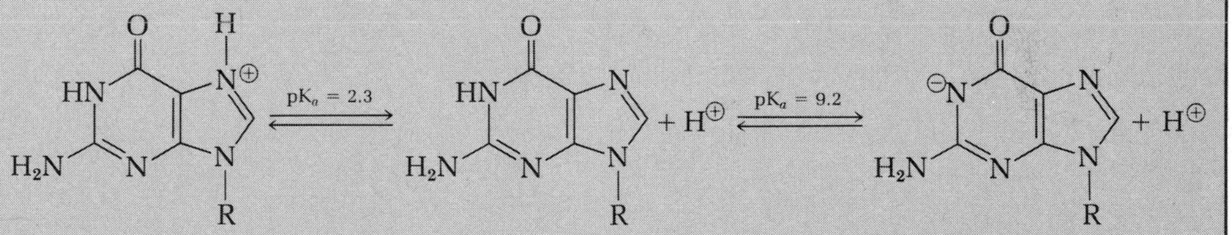

$pK_a = 2.3$

$pK_a = 9.2$

Cytosine in cytidine, cytidylic acids, CMP, etc. and RNA and DNA

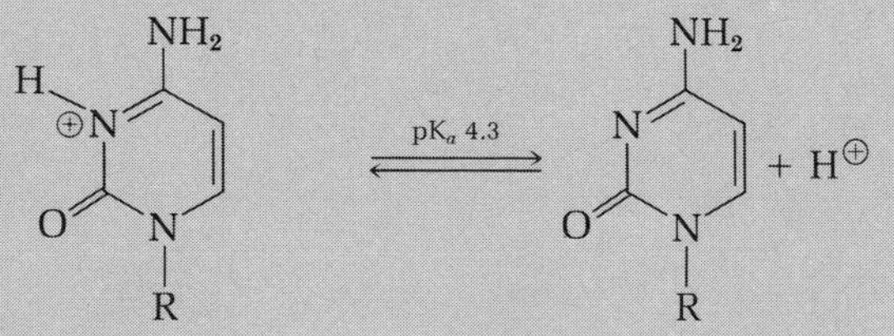

$pK_a\ 4.3$

Uracil in uridine, uridylic acid, UMP, etc., and RNA

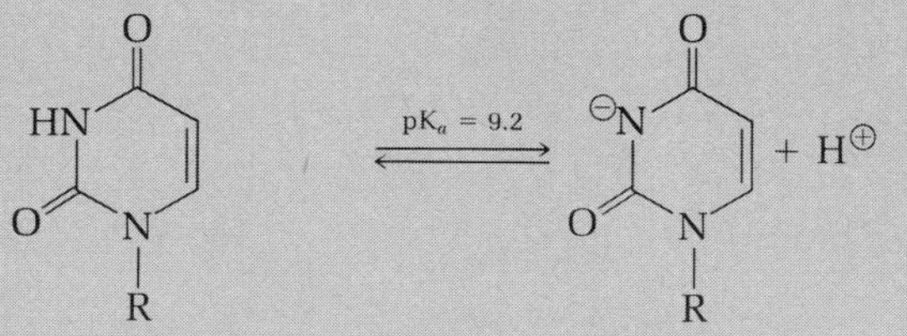

$pK_a = 9.2$

Thymine in thymidine, thymidylic acid, TMP, etc., and DNA

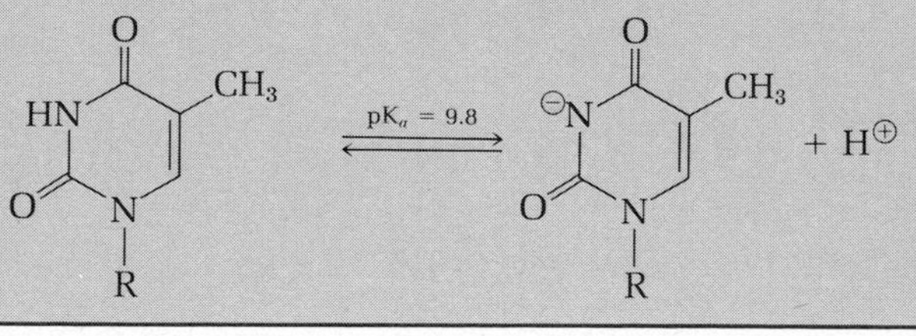

$pK_a = 9.8$

Ribonucleoside 5′-diphosphate

Ribonucleoside 5′-triphosphate

Adenosine triphosphate

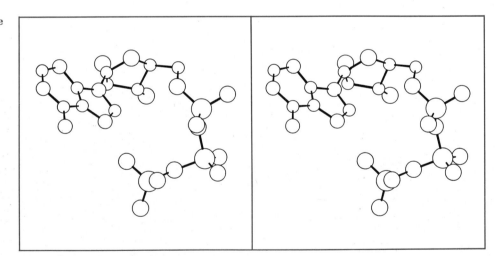

**Adenosine triphosphate
(space-filling)**

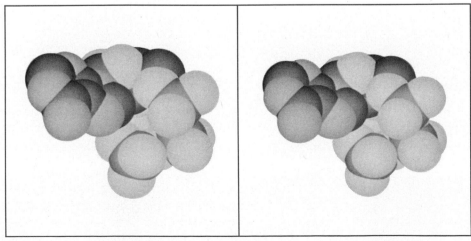

Figure 9.12   Structures of ribonucleoside diphosphates and triphosphates.

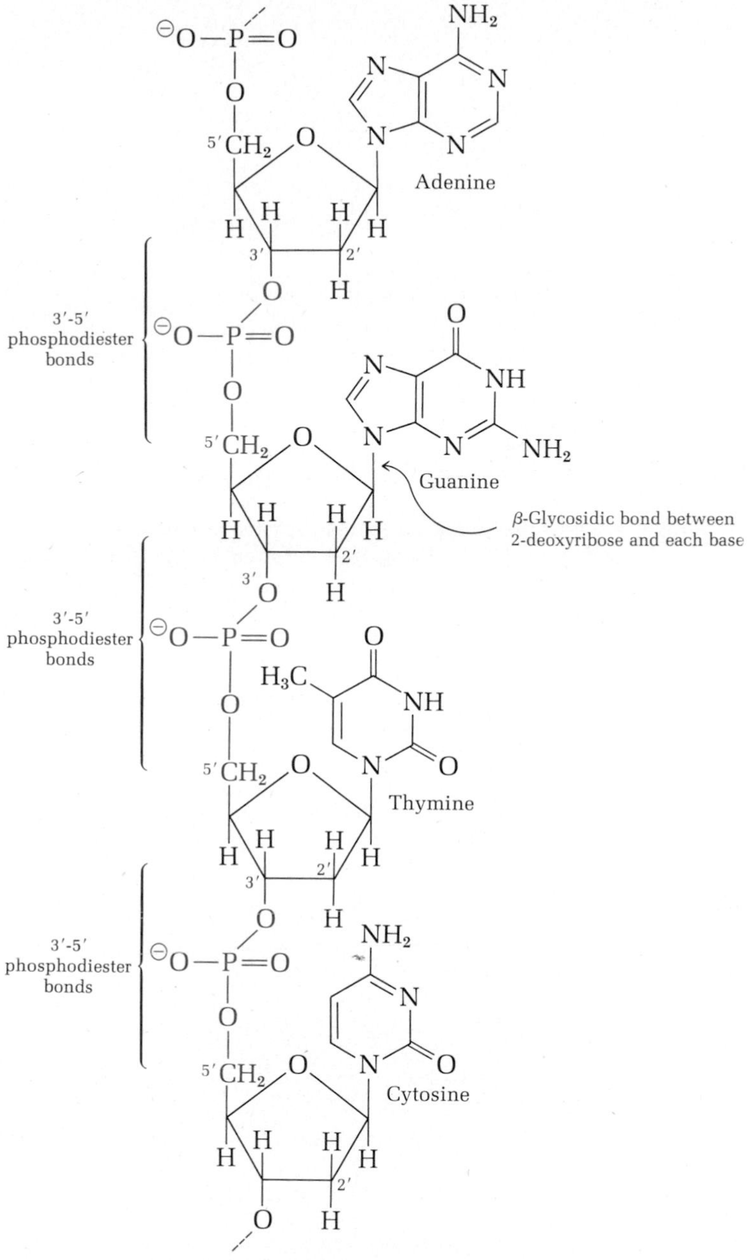

Figure 9.13 Covalent backbone of DNA.

chromosome is estimated to be 8.2 cm. DNA molecules shear easily because of this great length, and the isolation of intact DNA molecules is difficult. The molecular weights and lengths of DNA molecules from several sources are summarized in Table 9.3.

B. Base Composition of DNA

The base composition of DNA remained unknown for nearly a century after its discovery because of the difficulty of separating the products of DNA hydrolysis from one another. The invention of paper chromatography provided a means to overcome this difficulty, and E. Chargaff was able to determine the base composition of many species in 1950. The base composition varies widely from one species to the next (Table 9.4). Each species has a characteristic base composition, and, of course, every organ of higher organisms has the same base composition. Strikingly, for every organism the ratio of pyrim-

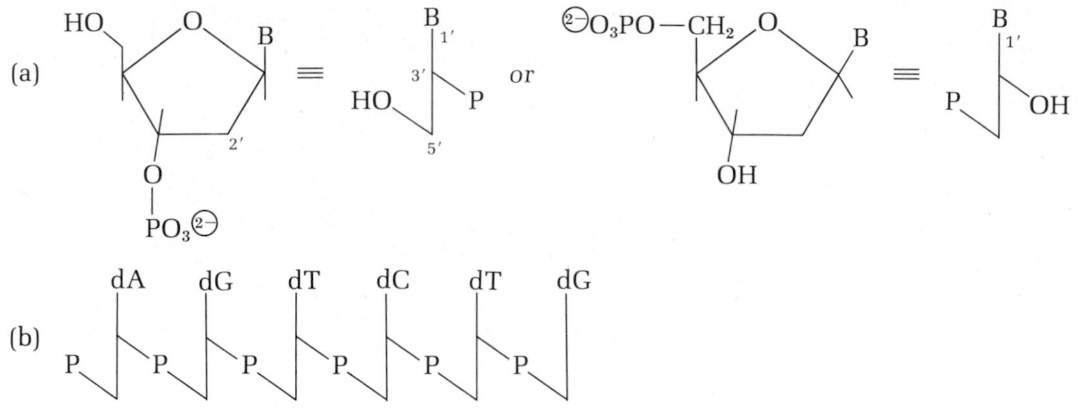

Figure 9.14 Shorthand abbreviations of DNA structure. (a) The sugar is represented as a vertical line with the base on top. (b) A series of vertical lines joined by diagonals representing the 3′-5′ phosphodiester bonds delineates the primary structure of the oligonucleotide. The *d*'s indicate 2′-deoxyribose. (c) A further abbreviation deletes the vertical lines and merely indicates the deoxysugar residues as dA, dG, dC, or dT. Phosphate groups at either terminus are indicated by a p. Thus, deoxyguanosine 5-phosphate is pdG, and deoxyguanosine 3-phosphate is dGp.

Table 9.3 Molecular weights and lengths of DNA's from several sources

Source	Molecular weight	Length	Number of nucleotide pairs	Conformation
Escherichia coli chromosome	2×10^9	1 mm	3×10^6	Circular, double stranded
Haemophilus influenzae chromosome	8×10^8	300 μm	12×10^5	—
Mycoplasma PPLO strain H-39	4×10^8	150 μm	6×10^5	—
Bacteriophage T2 or T4	1.3×10^8	50 μm	2×10^5	Linear, double stranded
Bacteriophage λ	33×10^6	13 μm	0.5×10^5	Linear, double stranded
Bacteriophage $\phi\chi$174	1.6×10^6	0.6 μm	—	Circular, single stranded
Polyoma virus	3×10^6	1.1 μm	4.6×10^3	Circular, double stranded
Mouse mitochondria	9.5×10^6	5 μm	14×10^3	Circular, double stranded
Drosophila melanogaster chromosome	43×10^9	20 mm	65×10^6	—

SOURCE: From J. N. Davidson, "Biochemistry of the Nucleic Acids," 8th ed., Academic Press, New York, 1976, p. 87.

Table 9.4 Base composition of DNA (mol %) and ratios of various components

Source	A	G	C	T	(A + T)/(G + C)	A/T	G/C	(A + G)/(C + T)
Man	30.4	19.9	19.9	30.1	1.53	1.009	1.000	1.006
Ox	29.0	21.2	21.2	28.7	1.36	1.010	1.000	1.006
Pig	29.8	20.7	20.7	29.1	1.42	1.024	1.006	1.014
Mycobacterium tuberculosis	15.1	34.9	35.4	14.6	0.42	1.034	0.985	1.006
E. coli	26.0	24.9	25.2	23.9	1.00	1.034	0.985	1.006
Yeast	31.7	18.3	17.4	32.6	1.80	0.972	1.051	1.000

idines to purines, $(A + G)/(C + T)$, is nearly unity. And, the mole percent ratio of guanine to cytosine, G/C, and adenine to thymine, A/T, are also nearly unity in each case. The $(A + T)/(G + C)$ ratios, however, vary widely. The $(G + C)$ content in mammalian DNA molecules ranges from 40 to 45 mol %; in bacterial DNA molecules it ranges from 30 to 75 mol % (Table 9.5). We shall discuss the relationship between the base composition of DNA and its physical properties in Section 9.7.

C. Secondary Structure of DNA

In 1953 Watson and Crick deduced the structure of crystal fibers of DNA. This discovery resulted from a combination of analysis of x-ray diffraction data, model building, intuition, and inspiration. One key to the structure of DNA was Chargaff's observation that purines and pyrimidines are present in DNA in equimolar quantities (recall Table 9.5). This observation is accommodated in a structure in which each purine in one strand of the double-stranded, or *duplex*, DNA is hydrogen bonded to a pyrimidine in an adjacent strand. The sugar-phosphate backbone of the DNA is exposed to the aqueous solution, and the hydrophobic bases are stacked in the interior, out of contact with the solvent. DNA is a double *helix* in which the two strands are coiled about a central axis. Each chain is a right-handed helix. One chain runs with its phosphodiester bonds advancing in the 3′-5′ direction, while the other is antiparallel, with its phosphodiester bonds running in the 5′-3′ direction (Figure 9.15). Adjacent residues in each chain are rotated by 36° relative to each other. The double helix completes 1 turn every 10 residues, and one complete turn advances the helix by 3.4 nm. Each base pair therefore

Table 9.5   (G + C) Content in DNA from various sources

Source	(G + C) Content (mol %)
Dictyostelium (slime mold)	22
M. pyogenes	34
Vaccinia virus	36
Bacillus cereus	37
Bacillus megaterium	38
Haemophilus influenzae	39
Saccharomyces cerevisiae	39
Calf thymus	40
Rat liver	40
Bull sperm	41
Diplococcus pneumoniae	42
Wheat germ	43
Chicken liver	43
Mouse spleen	44
Salmon sperm	44
Bacillus subtilis	44
Bacteriophage T1	46
Escherichia coli	51
Bacteriophage T7	51
Bacteriophage T3	53
Neurospora crassa	54
Pseudomonas aeruginosa	68
Sarcina lutea	72
Micrococcus lysodeikticus	72
Herpes simplex virus	72
Mycobacterium phlei	73

SOURCE: From J. N. Davidson, "Biochemistry of the Nucleic Acids," 8th ed., Academic Press, New York, 1976, p. 86.

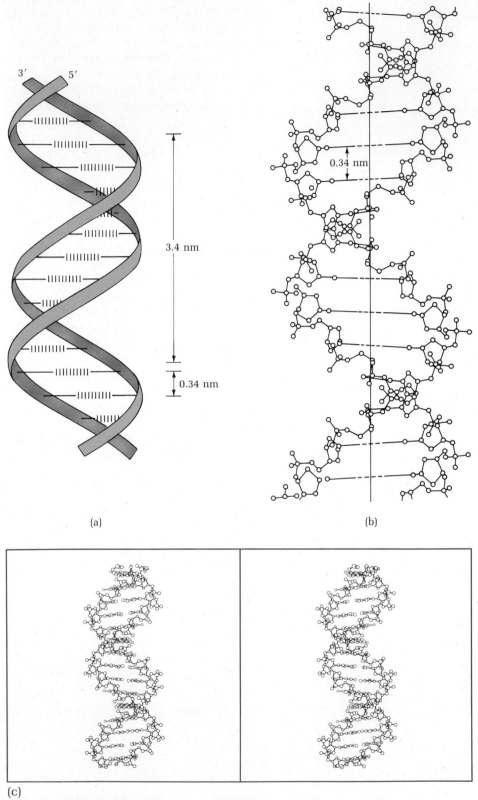

Figure 9.15 Two representations of the structure of DNA. (a) A ribbon structure in which the sugar-phosphodiester backbone advances as a "spiral staircase" with the base-paired purines and pyrimidines as the "treads." Notice the opposite polarity of the chains. (b) A structure in which the conformations of the nucleotidyl residues are indicated. (c) Stereo view of DNA.

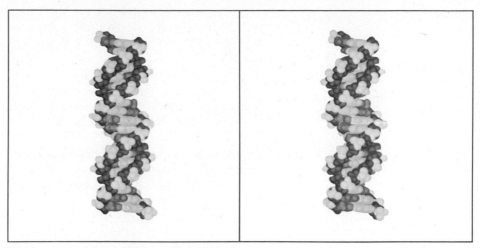

Figure 9.16   Space-filling model of DNA.

advances the double helix by 0.34 nm along the axis. The phosphorous atoms are 1 nm from the central axis, and the entire molecule has a diameter of 2.0 nm. Figure 9.16 shows a space-filling model of DNA.

The secondary structure of DNA is stabilized by hydrogen bonds between the complementary base pairs. The hydrogen-bonding scheme for Watson-Crick base pairing is shown in Figure 9.17. (Watson-Crick base pairing is not the only type found in nucleic acids, although it is the type found in DNA.) Adenine is always paired with thymine, and guanine is always paired with cytosine. The standard free energy of formation for the addition of an adenine-thymine base pair to an existing double helix is -5.0 kJ mol^{-1}; for the addition of a guanine-cytosine base pair it is -21 kJ mol^{-1}.[1] The guanine-cytosine base pair is more stable because it contains three hydrogen bonds as opposed to two hydrogen bonds in the adenine-thymine base pair. The double helix is also more stable when it has a high guanine-cytosine base content.

The double helix is stabilized by the same weak forces that stabilize proteins. Both hydrogen bonds and van der Waals forces play an important role in stabilizing DNA. The bases in the interior of duplex DNA are stacked in van der Waals contact with one another. The rise of 0.34 nm per base pair is, in fact, the thickness of the aromatic ring. Thus, hydrophobic interactions, so important in protein structure, also stabilize DNA. It has been estimated that stacking of two purines is favored by about -4.1 kJ mol^{-1}. This is a small interaction, but in a polynucleotide containing millions of base pairs, the sum of the base-stacking interactions makes a significant contribution to the stability of the double helix.

Since adenine must be paired with thymine and guanine with cytosine, the two antiparallel chains are exactly complementary, and the sequence of one chain automatically specifies the sequence of the other. In a masterful understatement Watson and Crick remarked in their original paper that "the specific pairing [we have] postulated immediately suggests a possible copying mechanism for the genetic material." We shall return to this topic in Chapter 24.

D. Tertiary Structure of DNA

DNA need not exist as a straight rod. In bacteria and some viruses the ends of the DNA are covalently joined to give a closed, circular, duplex molecule (Figure 9.18). The DNA molecules found in the chloroplasts of plant cells and in mitochondria are also closed, circular duplexes.

[1] These data were obtained for an RNA double helix, but are not likely to be much different for a DNA double helix.

Closed, circular DNA can be imagined as having been opened, un-wound, and reclosed. The unwinding and reclosing, however, produces a *supercoiled* DNA (Figure 9.18). Supercoiling produces a tertiary structure in the DNA. The DNA double helix is right-handed, and if the twisting of the double helix around its own axis is also right-handed, the DNA is *positively supercoiled*. If the DNA is twisted around its own axis in a left-handed sense, it becomes *negatively supercoiled*. Negative supercoiling is equivalent to underwinding of the double helix.

Supercoiling is observed for the linear DNA molecules of plants and animals and in the circular DNA molecules of bacteria. In eucaryotic cells the supercoiling arises when the DNA coils around basic proteins known as *histones*. All DNA isolated from natural sources is negatively supercoiled (Figure 9.19). Typically, there is about 1 negative twist in the double helix per 15 turns of the double helix.

The supercoiled DNA is characterized by a *winding number*, α, that is the sum of the number of turns in a straight rod, β, plus the number of super-helical turns, τ (equation 9.5).

$$\alpha = \beta + \tau \tag{9.5}$$

By definition β must be positive, but τ is positive if the turns are right-

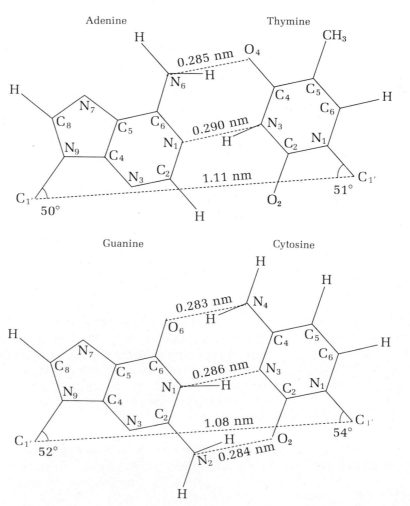

Figure 9.17 Watson-Crick base pairing for adenine and thymine and for guanine and cytosine.

Adenine thymine

Guanine-thymine

Figure 9.17 (*continued*)

handed and negative if the turns are left-handed (as they are in naturally oc-curring duplex DNA). Thus α can be either positive or negative.

Another parameter that describes the amount of supercoiling is the *su-perhelix density*, σ, which equals the number of superhelical turns per 10 base pairs. Therefore,

$$\sigma = \frac{\tau}{\beta} \tag{9.6}$$

The DNA isolated from cells and viruses typically has a superhelix density of about -0.05 (or about 5 superhelical turns per 1,000 base pairs).

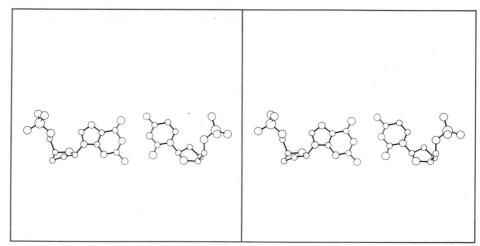

Ethidium   Proflavin   Acridine yellow

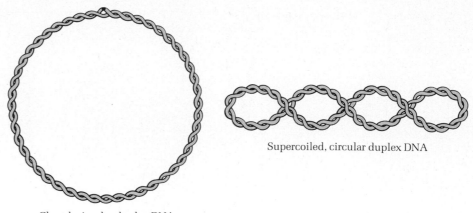

Supercoiled, circular duplex DNA

Closed, circular duplex DNA

Figure 9.18 Schematic diagram circular and supercoiled DNA.

Supercoiling is an endergonic process, and thus any process that causes unwinding of supercoiled DNA is thermodynamically favorable. Dyes such as *ethidium, proflavin,* and *acridine yellow* have been used to study the properties of supercoiled DNA. Ethidium is especially widely used. These planar dye molecules *intercalate* in the double helix, sliding between base pairs of the double helix. Intercalation of a single molecule of ethidium in a superhelical, closed duplex DNA is estimated to cause positive supercoiling of 26°, approximately 0.1 turn of the double helix. If an intercalating agent is added to negatively supercoiled DNA, the underwound double helix begins to rewind. This is a thermodynamically favored process, and it continues until the superhelical DNA is completely relaxed. As more of the agent is added, however, the intercalation of dye molecules produces less stable, positively supercoiled DNA, and the energy of the system increases (Figure 9.20).

Two classes of enzymes have been discovered that are able to alter the supercoiling of DNA. The *topoisomerases,* isolated from bacteria, viruses, plants, and animal cells, nick the duplex DNA by cleaving a phosphodiester bond. This enables the supercoiled DNA to relax to a less strained, less underwound structure that is lower in energy. The relaxed DNA is then resealed by the same enzyme. No outside source of energy is required to drive the reaction. The first topoisomerase to be discovered, and the prototype of these enzymes, is the so-called ω-protein of *E. coli.* The *E. coli* enzyme relaxes only negatively supercoiled DNA. All naturally occurring DNA is negatively supercoiled, but during DNA replication the double helix becomes

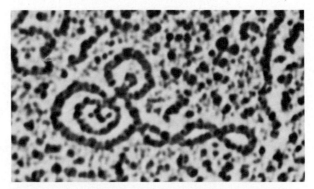

Figure 9.19 Electron micrograph of a negatively supercoiled bacterial DNA. (Courtesy of Dr. Nancy Hamlett.)

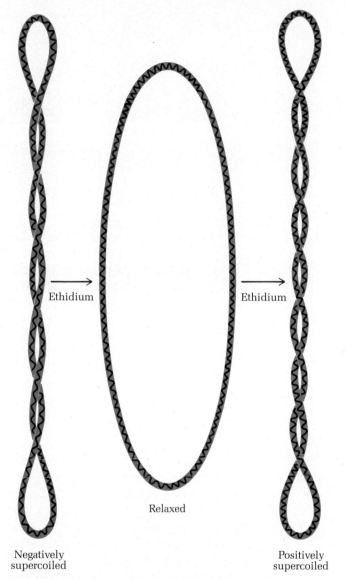

Ethidium

Ethidium

Relaxed

Negatively
supercoiled

Positively
supercoiled

Figure 9.20 Effect of intercalation of dye molecules on DNA coiling. Addition of ethidium to negatively supercoiled DNA causes positive supercoiling to proceed until the circular molecule is completely relaxed, a minimum energy conformation. Further addition of ethidium induces further positive supercoil formation, and the energy of the circular DNA increases. [From W. R. Bauer, F. H. C. Crick, and J. H. White, *Sci. Amer.,* July 1980, p. 132. Copyright 1980 by Scientific American, Inc. All rights reserved.]

positively supercoiled, and the ability to relax positively supercoiled DNA becomes important. Many eucaryotic organisms produce topoisomerases that are able to relax either positively or negatively supercoiled DNA and that are thought to be important in DNA replication. The function of the bacterial topoisomerases is not known.

DNA gyrase performs the opposite operation by introducing supercoils in closed, circular DNA. One molecule of DNA gyrase introduces about 100 supertwists per minute in a closed circular DNA molecule. To date, DNA gyrases have been found only in bacteria. DNA replication in *E. coli* is completely inhibited by inhibition of DNA gyrase. The antibiotics *novobiocin* and *coumermycin A* act by inhibiting DNA gyrase.

Novobiocin

The reaction catalyzed by DNA gyrase is complicated. To drive the endergonic supercoiling process, a source of energy is required. As we shall see in later chapters, the phosphoanhydride bonds of ATP are the primary source of energy used to drive endergonic processes. Hydrolysis of ATP provides the driving force for the introduction of supercoils in DNA.

The introduction of negative supercoils involves three steps, whose net result is termed *sign inversion*. First the enzyme binds to the DNA molecule so that the two double helices cross to form a right-handed node. The formation of a positive twist is counterbalanced by the introduction of a negative twist. In Figure 9.21 the positive twist is represented by (+) and the negative twist is represented by (−). Gyrase then breaks both strands at the back of the right-handed node and passes the front segment through the break. This operation converts the right-handed, positive twist to a left-handed, negative one. Sealing of the double-stranded break completes the task.

9–6 STRUCTURE OF RIBONUCLEIC ACID (RNA)

The covalent backbone of RNA consists of ribonucleotidyl residues linked by 3'-5' phosphodiester bonds (Figure 9.22). The primary structure of RNA differs from that of DNA in two major ways. (1) The sugar residues are ribose rather than 2'-deoxyribose, and (2) uracil replaces thymine as one of the four common bases. For RNA, then, the common bases are adenine, uracil, guanine, and cytosine. Many bases are found in trace amounts in RNA. The most common of the rare nucleotidyl residues found in tRNA are pseudouridylate, inosinate, thymidylate, ^{2}N-methylguanylate, 1-methylinosinate, ^{2}N-dimethylguanylate, and 5,6-dihydrouridylate (Figure 9.23). RNA is single stranded, except in certain viruses in which it is double stranded. However, complementary base pairing can occur between adenine and uracil and between guanine and cytosine if the single strand folds back upon itself.

Ribonucleic acids are found in multiple copies and in multiple forms in a given cell. An RNA molecule is classified by its cellular location and by its function. *Messenger RNA, mRNA*, is an often short-lived molecule that

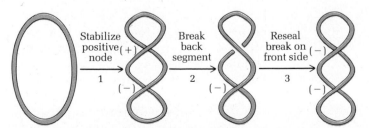

Figure 9.21 Schematic representation of the sign inversion mechanism of DNA gyrase. The enzyme introduces a positive twist, breaks both strands of the duplex DNA at the "back," and reseals the break on the "front." [From N. R. Cozzarelli, *Science, 207,* 953–960 (1980). Copyright 1980 by the American Association for the Advancement of Science.]

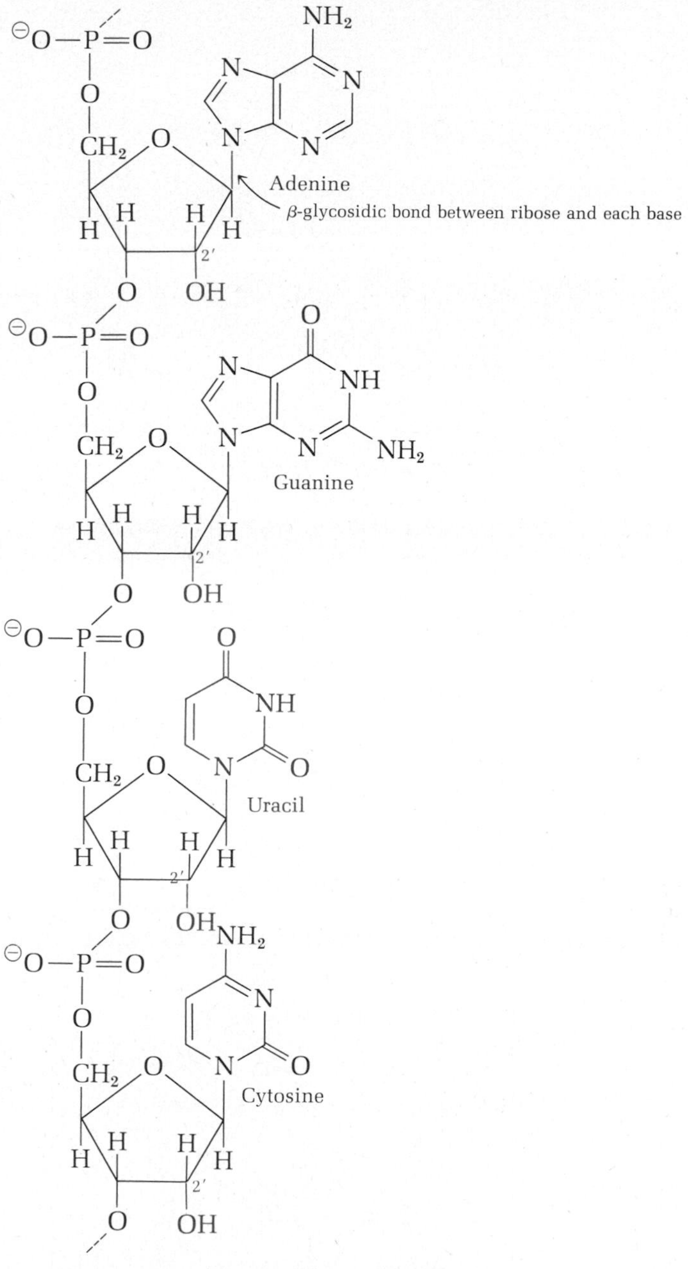

Figure 9.22 Covalent backbone of RNA.

carries genetic information from DNA to the ribosomes where protein synthesis occurs. *Ribosomal RNA, rRNA,* is an integral part of ribosomes that also participates in protein synthesis. About 75% of the cellular complement of RNA is rRNA. Ribosomal RNA molecules span a wide range of molecular weights. The lowest mass rRNA has a molecular weight of about 40,000, while two high mass rRNA molecules have molecular weights between 500,000 and 1.5 million. (These RNA molecules are classified by their sedimentation coefficients, Section 9.7B.)

Transfer RNA, tRNA, serves as a carrier of amino acid residues. An *aminoacyl-tRNA,* in which the RNA is bound to an amino acid residue at its 3′ terminus (Figure 9.24), transfers its aminoacyl residue to a growing polypeptide chain during protein synthesis. Transfer RNA molecules contain 75 to 78 nucleotidyl residues, and their molecular weights range from 23,000

Inosine

1-Methylinosine

Pseudouridine

5,6-Dihydrouridine

Thymine ribonucleoside (ribothymidine)

²N-Methylguanosine

²,²N,N-Dimethylguanosine

Pseudouridine

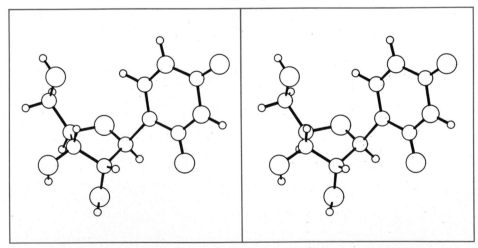

Figure 9.23 Rare nucleosides in RNA. In pseudouridylate uracil is bound to ribose by a bond at C-5 rather than by a glycosidic bond at N-3. Nucleosides of the other rare bases involve β-N-glycosidic bonds at N-9 of the purines and at N-3 of the pyrimidines, as usual.

to 25,000. The primary structure of yeast phenylalanyl tRNA, tRNA[phe], is shown in Figure 9.25 in a "cloverleaf" structure that suggests base pairing. In all functional tRNA molecules the triplet pCpCpA is found at the 3′ terminus. Roughly in the middle of the tRNA molecule is a sequence of three bases called the *anticodon*. These three bases are hydrogen bonded to a complementary sequence in mRNA during protein synthesis. It appears that all tRNA molecules have the same basic L-shaped tertiary structures (Figure 9.26). The hydrogen-bonded base pairs indicated in the cloverleaf structure of

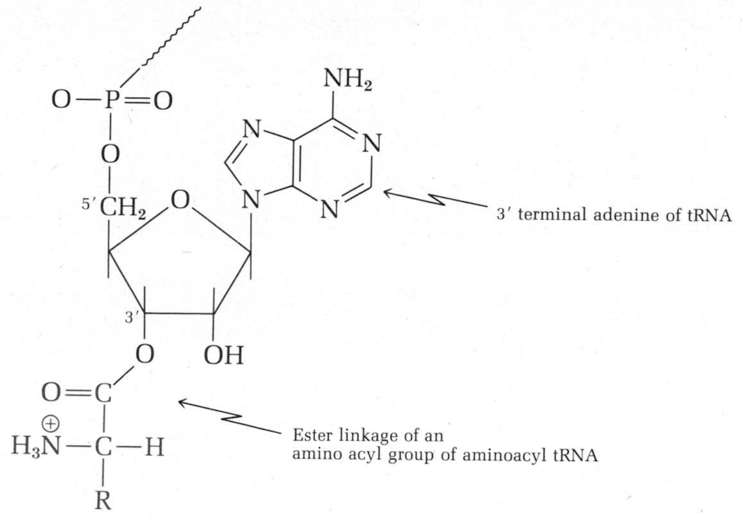

3' terminal adenine of tRNA

Ester linkage of an amino acyl group of aminoacyl tRNA

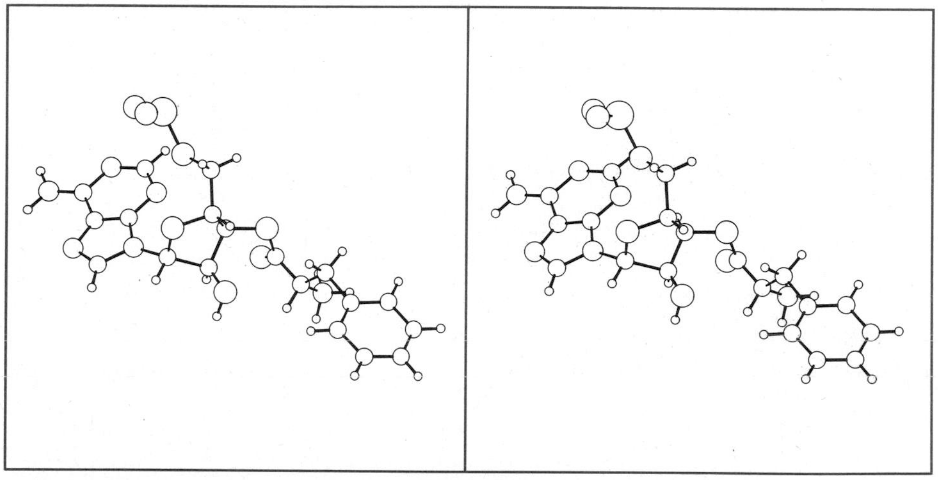

Figure 9.24 Aminoacyl derivative of the 3' terminus of transfer RNA.

tRNA are indicated by the connecting rods in Figure 9.26a. In phenylalanyl-tRNA 71 of the 76 nucleotidyl residues are also involved in base-stacking interactions. The hydrophobic interactions are a major stabilizing force in the tertiary structure of phenylalanyl-tRNA. The 2'-hydroxyl group of the ribose residues of tRNA are involved in intramolecular hydrogen bonds and also contribute to the stability of the tertiary structure.

Cations stabilize the three-dimensional conformation of yeast phenylalanyl-tRNA by interacting with the phosphate groups of the backbone. *Spermine*, a polyamine whose net charge is +4 at pH 7, binds in a groove of tRNA between the D stem and the variable loop. Since the negative charges of four phosphate groups are neutralized by spermine, the strands of the tRNA move closer together. Removal of the spermine would move the anticodon loop 1 nm by altering the conformation of the anticodon stem. The distance between adjacent codons on messenger RNA is also 1 nm, and spermine may play an important part in the regulation of protein synthesis.

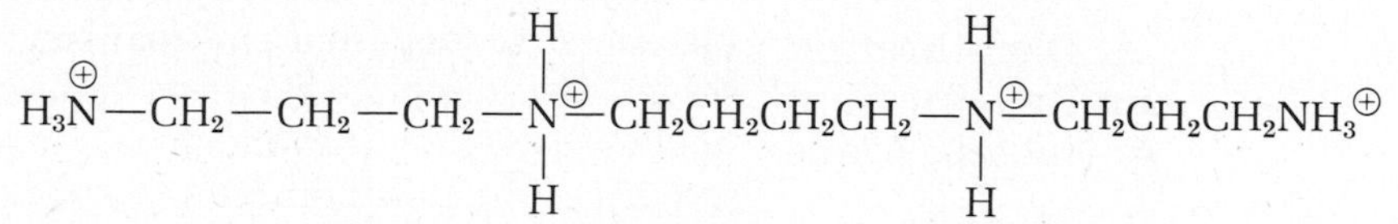

Spermine

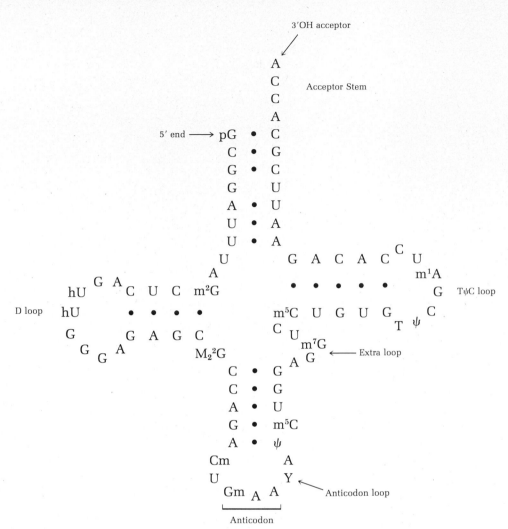

Figure 9.25 "Cloverleaf" structure for ᵖʰᵉtRNA. Base pairing between purines and pyrimidines is indicated by a "dot." The abbreviations for rare nucleosidy residues are: m₂²G, ²N-dimethylguanosine; hU, 5,6-dihydrouridine; m²G, 2-methylguanosine; ψ, pseudouridine; m⁵C, 5-methylcytidine; m¹A, 1-methyladenosine; m⁷G, 7-methyladenosine. (From G. J. Quigley and A. Rich, *Science, 194,* 796–806 (1976).]

Magnesium ions are also tightly bound to tRNA. Each magnesium ion is coordinated with eight oxygens contributed by water molecules and phosphate groups. The magnesium ions bind to tRNA at points where two strands come close together, and magnesium ions therefore stabilize the tertiary structure. If magnesium ions and spermine were removed from tRNA, it is likely that the molecule would unfold because of electrostatic repulsion among the phosphate groups.

9–7 PHYSICAL PROPERTIES OF NUCLEIC ACIDS

A. Denaturation and Renaturation of Nucleic Acids

In their initial paper describing the double helical structure of DNA, Watson and Crick noted that since adenine is always paired with thymine and cytosine is always paired with guanine, specifying the base sequence in one strand defines the base sequence of the complementary strand. This complementarity suggested a mechanism for DNA replication in which one strand serves as a template for the synthesis of the complementary strand. This mode of DNA replication, since shown to be correct, is said to be *semiconservative*. A minimum requirement of semiconservative replication is unwinding of the double helix. Studies of the unwinding, or denaturation, of

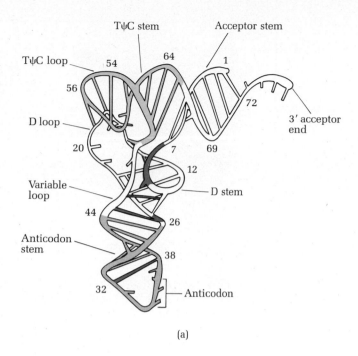

(a)

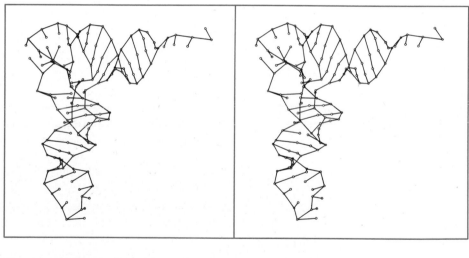

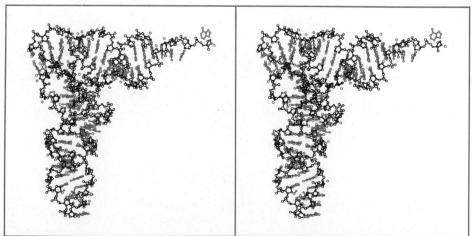

Figure 9.26 Tertiary structure of transfer RNA. (a) Tertiary structure
Rods represent hydrogen bonds between purines and pyrimidines. [From G. J.
Quigley and A. Rich, *Science*, *194*, 796–806 (1976). Copyright 1976 by the
American Association for the Advancement of Science.] (b) Stereoscopic view
of tRNA[phe].

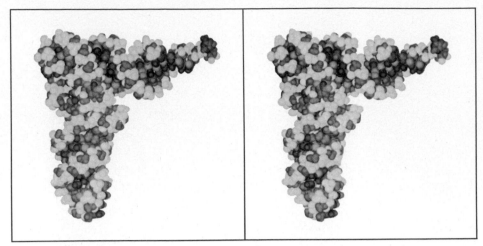

Figure 9.26   (*continued*)

the double helix and its renaturation provide a model for one important aspect of DNA replication. Such studies also provide information about the forces responsible for maintaining the double helix. Studies of the denaturation of nucleic acids parallel studies of protein denaturation (Section 4.2E), as indeed they ought, since the physical principles involved are similar.

The denaturation of DNA involves a transition from a double helix to a random coil conformation (Figure 9.27). The *helix to coil transition* can be easily monitored by following the change in absorbance of DNA in solution as the sample is denatured. DNA absorbs ultraviolet light with an absorbance maximum at 260 nm (Figure 9.28). The total absorbance of DNA increases by about 37% when DNA is denatured. This hyperchromic effect is a result of unstacking the bases in native, duplex DNA. The decrease in the absorbance that occurs when random coil DNA renatures is a *hypochromic* effect. When the bases are stacked in native DNA, their π electrons interact strongly in a way that suppresses the absorption of incident photons.

If a solution of DNA is heated, its absorbance increases sigmoidally as a function of temperature (Figure 9.29). The temperature at which the DNA is half-denatured is called the *melting temperature*, T_m. When $T = T_m$, the absorbance of the sample has increased by 18.5%. The sigmoidal shape of the melting curve indicates that denaturation is a cooperative process. Denaturation begins in regions of the double helix that are rich in adenine-thymine

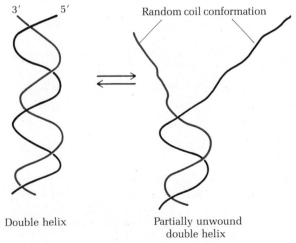

Figure 9.27   Partial unwinding of DNA in a helix to coil transition.

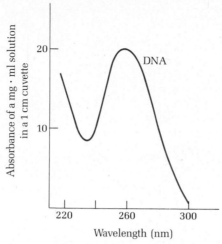

Figure 9.28 Absorbance spectrum of DNA at pH 7.0.

base pairs. The adenine-thymine base pair, having two hydrogen bonds, is less stable than the guanine-cytosine base pair. This decreased stability is reflected in the melting point of DNA, since increasing the guanine-cytosine content of the DNA increases its melting point. The (G + C) content of the DNA from any species can be deduced from its T_m and a standard curve such as the one shown in Figure 9.30. A comparison of the (G + C) content from thermal denaturation and from direct quantitative analysis of the base content is given in Table 9.6.

Returning again to Figure 9.30, we see that T_m increases with increasing ionic strength. The phosphodiester backbone of DNA is extremely acidic, and at pH 7.0 it is highly negative. The electrostatic repulsion of the

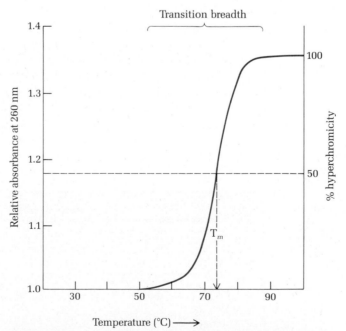

Figure 9.29 Typical melting curve for DNA. The temperature at which the double helix is half unwound is the inflection point in the sigmoidal curve. This point is the melting point of the DNA. At T_m the increased absorbance of the sample is one-half the value of completely melted DNA.

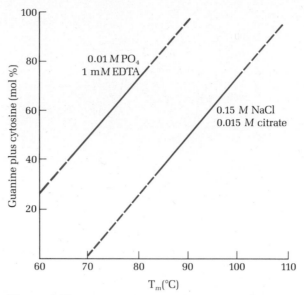

Figure 9.30   Plot of T_m versus guanosine plus cytosine content gives a straight line. This line can be used as a standard curve to determine the base content of an unknown sample of DNA. [From J. Marmur and P. Doty, *J. Mol. Biol.*, **5**, 109–118 (1962). Reproduced with permission. Copyright: Academic Press Inc. (London) Ltd.]

negative phosphodiester groups is neutralized as the concentration of the counterion in the buffer increases. As the concentration of the buffer increases beyond the minimum required to neutralize the charges on the phosphodiester backbone, further salt should not affect the stability of the DNA. In fact, however, increasing the salt concentration decreases the stability of the DNA, and its melting point decreases (Figure 9.31). The salts that have this effect do not disrupt hydrogen bonds. What, then, is the salt doing? Let us recall our discussion of the hydrophobic effect in Section 4.2. The salts that denature DNA do so by breaking the structure of water, and they act by disrupting hydrophobic interactions between stacked bases. Base stacking is a major contributor to the stability of DNA. Those reagents that decrease the solubility of the phosphodiester backbone or increase the solubility of the bases disrupt the hydrophobic interactions between the stacked bases and cause denaturation of DNA. The situation is reminiscent of the factors responsible for protein structure: hydrogen bonds, at first thought to be the primary force responsible for maintaining protein structure, were found to

Table 9.6   (G + C) Content in DNA from various sources determined from T_m and from other methods of analysis

| | | (G + C) Content (mol %) | |
Source	T_m	From T_m	From other analysis
Tobacco leaf	85.5	39	—
Wheat germ	88.5	46	46.6
Human spleen	86.5	41	41.4
Mouse spleen	86.5	41	41.9
Calf thymus	87.0	42	41.9
Chicken liver	87.5	43	41.7
Salmon sperm	87.5	43	41.2
Proteus vulgaris	87.0	37	36.5
Streptococcus pneumoniae	85.5	39	38.5
Clostridium perfringens	80.5	26.5	31.0
Branhamella catarrhalis	86.5	41	40.7

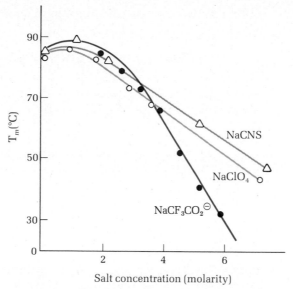

Figure 9.31 Effect of increasing salt concentration of the T_m of DNA. [From K. Hamaguchi and E. P. Geiduschek, *J. Amer. Chem. Soc.,* **84,** 1329–1338 (1962).]

be less important than hydrophobic effects. A parallel set of developments in nucleic acid chemistry leads us to the same conclusion.

If a heated sample of DNA is slowly cooled, the absorbance of the solution decreases, indicating that the double helix is zipping back up again. As long as a single hydrogen bond remains between the base pairs, denaturation is rapidly reversible. Even if the two strands become totally separated, renaturation can occur, but the process is then much slower. The optimum temperature range for renaturation is somewhere between T_m and the temperature of complete denaturation, in the range of $T_m - 15°$ to $T_m - 20°$.

B. Physical Methods in Nucleic Acid Biochemistry

In Chapter 5 we discussed some of the most common methods employed for the isolation and characterization of biological macromolecules. In the study of nucleic acids centrifugation and electrophoresis are among the most important experimental methods. For example, ribosomal RNA molecules are commonly classified by their sedimentation coefficients (Table 9.7). Let us

Table 9.7	Classification of ribosomal RNA's by their sedimentation coefficients	
Source	**Sedimentation Coefficient (S)**	**Molecular weight**
E. coli	16.0	5.5×10^5
	22.5	1.07×10^6
Paramecium	18.0	6.9×10^5
	25.5	1.25×10^6
Mammalian cytoplasm	18.0	7.0×10^5
	30.0	1.8×10^6
Yeast cytoplasm	17.0	7.2×10^5
	26.0	1.5×10^6
Yeast mitochondria	16.0	6.6×10^5
	23.0	1.3×10^6
Higher plant cytoplasm	17.0	7.0×10^5
	26.0	1.4×10^6
Higher plant chloroplasts	16.0	5.5×10^5
	22.0	1.0×10^6

briefly consider some aspects of these methods as they are applied to nucleic acids.

a. Equilibrium Density Gradient Centrifugation

When a concentrated solution of cesium chloride, CsCl, is centrifuged, a concentration gradient is established from the top to the bottom of the centrifuge tube. The centrifugal force pulling the CsCl to the bottom of the tube is exactly balanced by the tendency of the salt to diffuse toward the region of lower concentration at the top of the tube when the system is at equilibrium. The effects of centrifugal force and diffusion differ from point to point in the centrifuge tube, and the result is a steady increase in the density from the top to the bottom of the tube. If a DNA sample is present when the gradient of CsCl is established, the DNA molecules will move to that point in the gradient at which the CsCl density equals the density of the DNA. This technique for analyzing DNA is called *density gradient centrifugation*. We have used DNA as our example because the technique is most often used to separate and analyze DNA. The technique was invented to demonstrate that DNA replication is semiconservative, but for our purposes, the results of

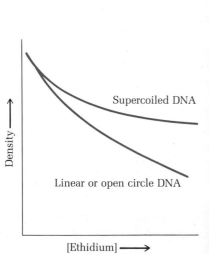

Figure 9.32 Separation of linear and open circular DNA from supercoiled DNA by equilibrium density gradient ultracentrifugation of DNA in CsCl. Negatively supercoiled DNA binds to fewer moles of ethidium than linear or open circular DNA since the topological constraints of the closed, circular molecule permit only a certain number of moles of ethidium to bind before a maximally positively supercoiled DNA is attained. Lacking this constraint, the linear and open circular forms of DNA bind to more moles of ethidium.

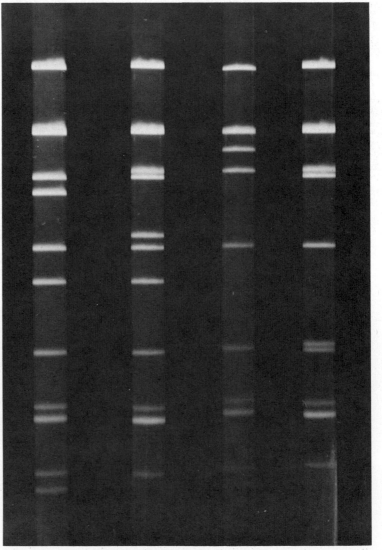

Figure 9.33 Separation of nucleic acid fragments obtained by enzymatic digestion of DNA on an agarose gel. (Courtesy of Dr. Nancy Hamlett.)

density gradient centrifugation are important for the information they provide about DNA structure.

Supercoiled DNA, linear DNA, and open circle DNA all have the same density in CsCl. If, however, ethidium is present in the CsCl, it will bind to DNA in amounts that depend upon the concentration of ethidium. Ethidium binding decreases the DNA density. At a fairly high concentration of ethidium the number of moles of ethidium bound is restricted by the topological constraints of closed circular DNA. More ethidium binds to linear or open circular DNA than binds to closed circular DNA, and their densities therefore decrease more than the density of closed circular DNA (Figure 9.32).

b. Electrophoresis

We noted in Section 5.5 that electrophoresis is a powerful tool for the separation of peptides and proteins. It is also an extremely valuable tool for separating mixtures of oligonucleotides and nucleic acids.

Mixtures of oligonucleotides are often incompletely separated by a single electrophoresis experiment. Oligonucleotides differing only by their terminal nucleotide are separated by two-dimensional electrophoresis (refer to Figure 5.22). Thus A-T-G-C will be separated from A-T-G-C-T. Individual oligonucleotides are identified by their electrophoretic mobilities, that is, by the distance they move in the electrophoresis experiment. Electrophoresis on polyacrylamide is usually favored for proteins and small oligonucleotides and on agarose gels for nucleic acids. Because nucleic acids have nearly identical charge-to-mass ratios, they are separated by molecular weight when they are subjected to gel electrophoresis. The type of separation possible is shown in Figure 9.33. We shall see immediate application of gel and paper electrophoresis in Section 9.9 when we discuss methods for determining the primary structures of DNA and RNA.

9–8 HYDROLYSIS OF NUCLEIC ACIDS

A. Acid- and Base-Catalyzed Hydrolysis of Nucleic Acids

The covalent backbones of both DNA and RNA consist of 5′-3′ phosphodiester bonds. The bases are bound to the deoxyribosyl or ribosyl residues by β-N-glycosidic bonds. Both types of bonds are labile in acid solution. Treatment of DNA with a strongly acidic solution, such as 12 M perchloric acid, for an hour at 100°C cleaves DNA to its constituent bases, phosphoric acid, and 2′-deoxyribose (reaction 9.7). RNA yields its constituent bases, phosphoric acid, and ribose.

$$\xrightarrow{\text{12 M H}^{\oplus},100°C,\ 1\ hr} \quad \text{bases} + H_3PO_4 + 2\text{′-deoxyribose} + \text{degradation products of 2′-deoxyribose} \tag{9.7}$$

The β-N-glycosidic bonds of the purines are more reactive than those of the pyrimidines. Partial acid hydrolysis at pH 1.6 at 37°C liberates the purines, leaving *apurinic acid* (reaction 9.8).

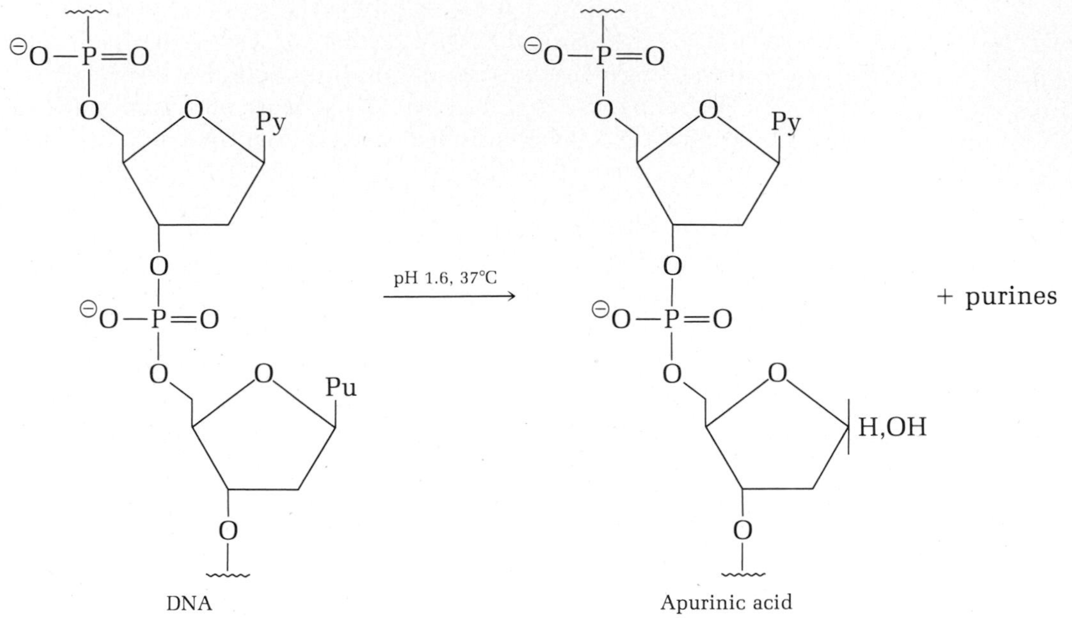

DNA

Apurinic acid

$$\text{(9.8)} \qquad + \text{ purines}$$

The increased rate of hydrolysis of the purine N-glycosidic bond is a result of intramolecular catalysis by the N-3 proton of purines (reaction 9.9).

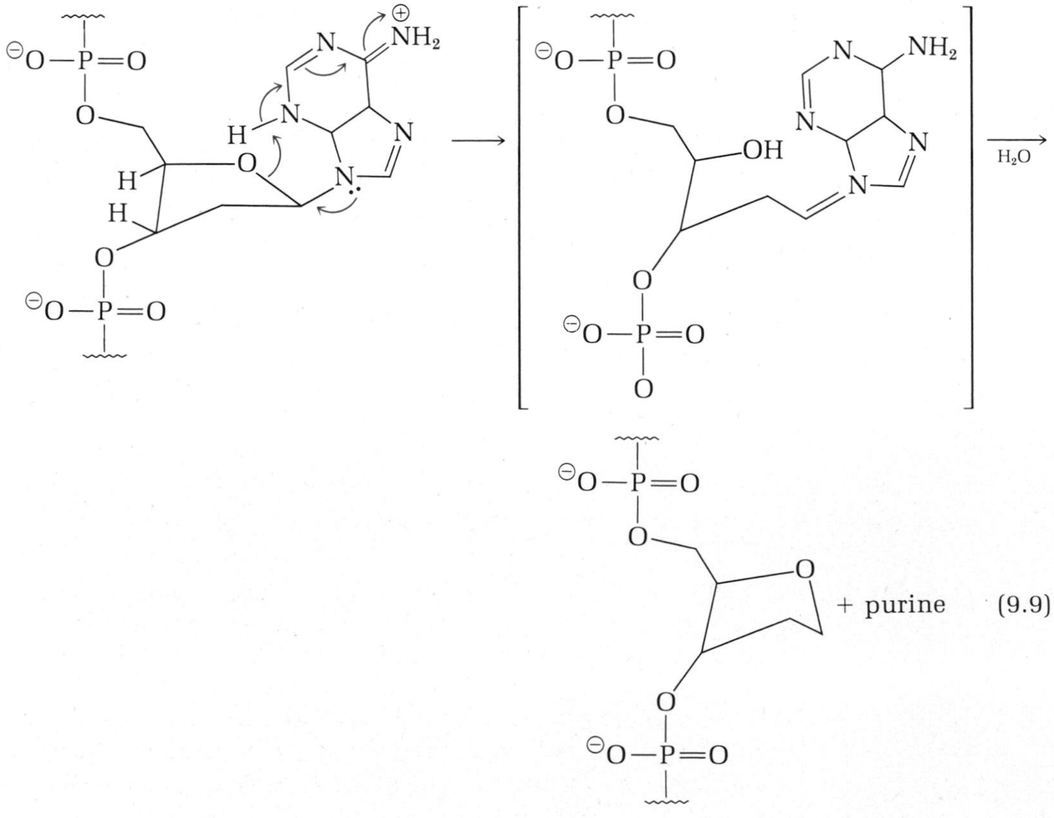

$$\text{(9.9)} \qquad + \text{ purine}$$

RNA does not give an apurinic acid under these conditions.

The behavior of RNA in alkaline solution differs markedly from that of DNA. Treatment of RNA with 0.1 M alkali at 25°C degrades RNA to a mixture of 2′- and 3′-nucleoside phosphates. In the presence of hydroxide ion the 2′-hydroxyl group of the ribosyl moiety of RNA is converted to its conjugate base, alkoxide ion. Intramolecular attack by the 2′-alkoxide, a potent nucleophile, on the phosphodiester gives a 2′-3′ cyclic nucleotide, cleaving the phosphodiester bond in the process. Further attack by hydroxide on the 2′-3′ cyclic nucleotide produces a mixture of 2′- and 3′-nucleoside phosphates (reaction 9.10).

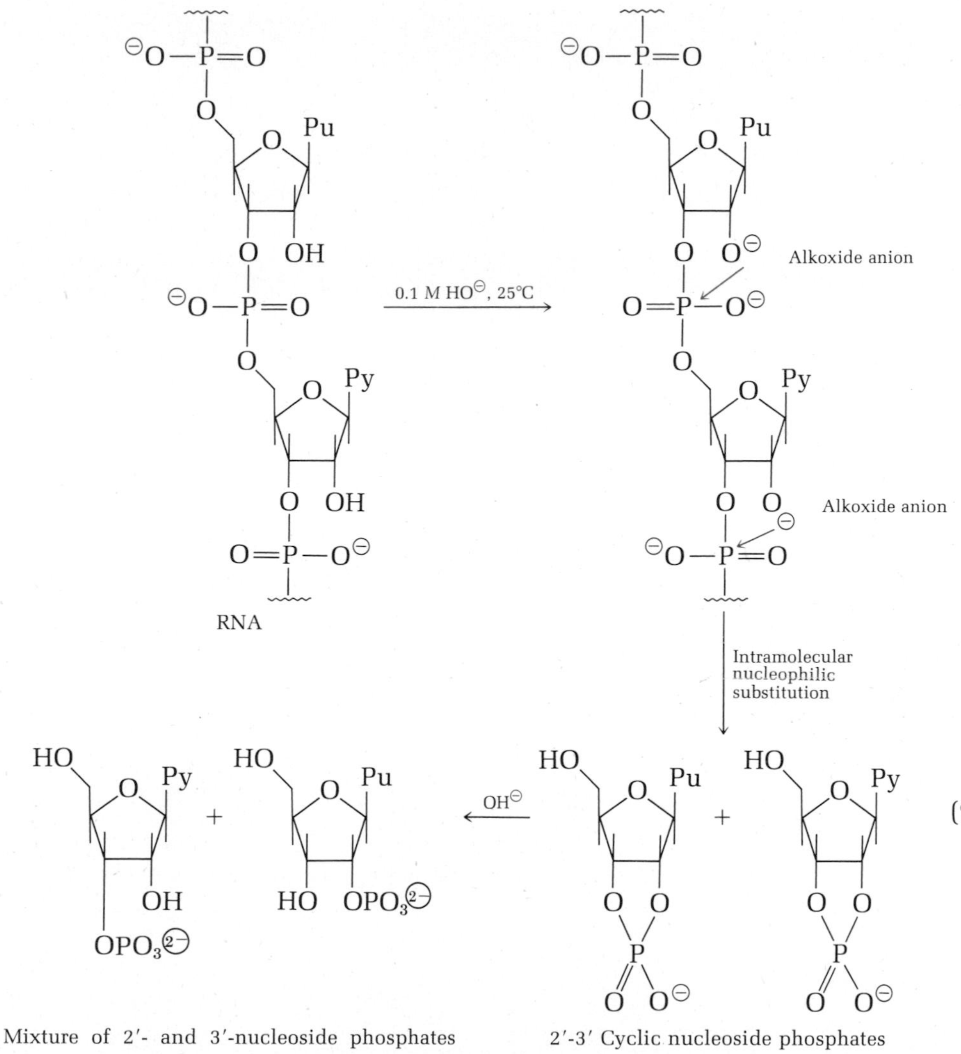

Mixture of 2′- and 3′-nucleoside phosphates   2′-3′ Cyclic nucleoside phosphates

(9.10)

DNA is stable in basic solution, under conditions that hydrolyze RNA, because DNA lacks a 2′-hydroxyl group to carry out intramolecular catalysis.

B. Enzymatic Hydrolysis of Nucleic Acids: Classification of Nucleases

The enzymes that hydrolyze the phosphodiester bonds of nucleic acids are collectively known as *nucleases*. These enzymes, ubiquitous in nature, may be classified in many ways. The most elementary distinction is between those that are specific for ribonucleic acids and those that are specific for deoxyribonucleic acids. The former are called *ribonucleases*, RNases, and the latter are called *deoxyribonucleases, DNases*. A few enzymes lack even this degree of specificity and hydrolyze the phosphodiester bonds of either DNA or RNA.

The nucleases can also be classified by their point of attack upon the polymer chain. Those that attack the polymer at either its 3′ or 5′ terminus and sequentially remove nucleotidyl residues one at at time are known as *exonucleases;* those that attack the polymer within the chain are called *endonucleases*. The 3′-5′ phosphodiester bonds of nucleic acids can be cleaved to yield either a 3′-nucleotidyl product or a 5′-nucleotidyl product. Nucleases that catalyze both types of cleavages are known. Those nucleases that yield 5′-nucleotide monophosphates as products catalyze a-*type cleavage;* those that yield 3′-nucleotide monophosphates catalyze b-*type cleavage.* The specificities of nucleases vary widely from nearly indiscriminate hydrolysis of phosphodiester bonds, to marked preference for attack at a purine or pyrimidine nucleotidyl residue, to the requirement for a given residue, to absolute specificity of sequence.

C. Ribonucleases

a. Pancreatic Ribonuclease A

Pancreatic *ribonuclease A* hydrolyzes RNA between the phosphate residue of a pyrimidine and the adjacent nucleotide to give a 3′-phosphate and the 5′-hydroxyl termini (*b* cleavage). The products may be either mononucleotides or oligonucleotides (reaction 9.11).

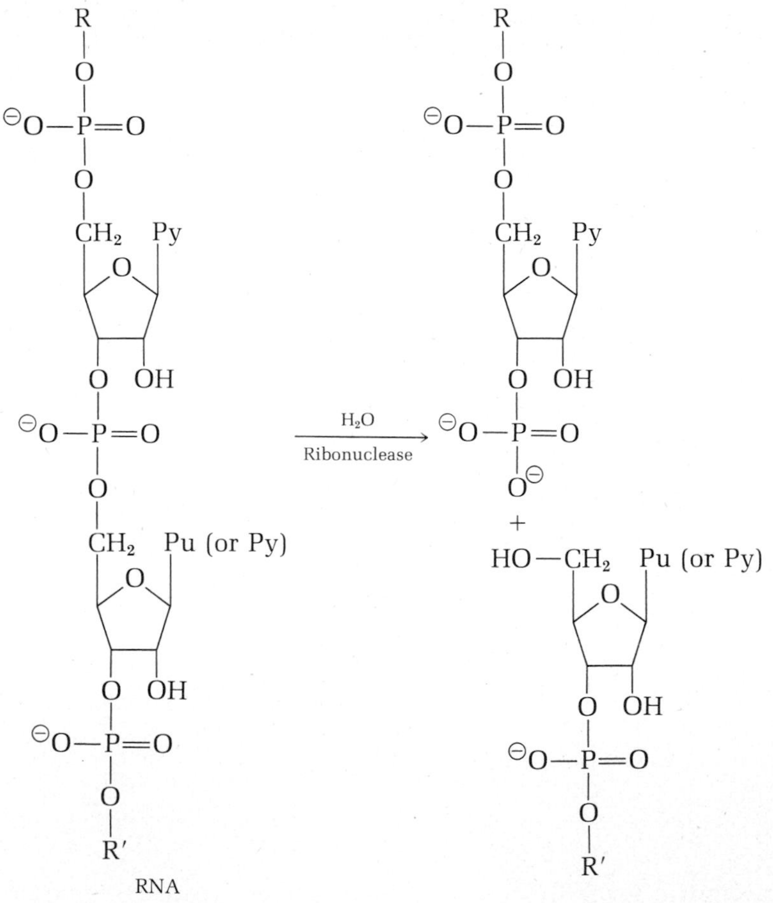

$$(9.11)$$

A schematic diagram of the complex between the substrate analog *uridyl 3′,5′-adenosine phosphonate* and pancreatic ribonuclease is shown in Figure 9.34. The interactions between active site residues and the substrate analog are shown in Figure 9.35. The enzyme-substrate complex is stabilized by stereospecific hydrogen bonds that are possible only when the base occupying the active site is a pyrimidine. The catalytic reaction occurs by intermediate formation of a cyclic 2′-3′ phosphate ester. (We recall that this inter-

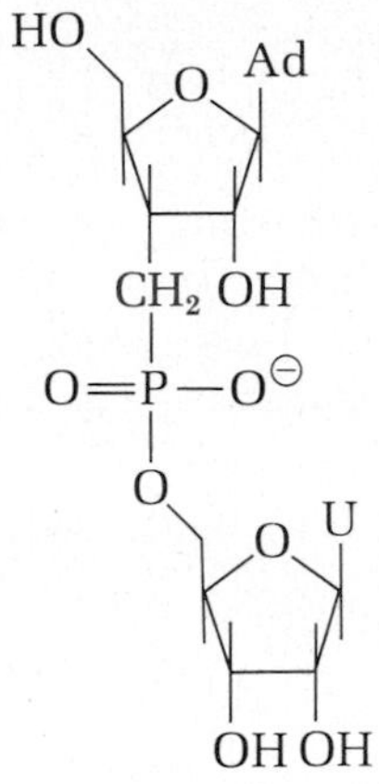

Uridyl 3′,5′-adenosine phosphonate (in the phosphonate a methylene group replaces the ester oxygen)

mediate also forms in the base-catalyzed hydrolysis of RNA.) Two histidine residues (His 12 and His 119) flank the substrate. One of them (His 12) acts as a general base to generate an alkoxide anion at the 2′ position of the ribosyl moiety; the other histidyl residue acts as a general acid catalyst to protonate the leaving 5′-alkoxide anion as the 2′-3′ cyclic phosphate ester forms (reaction 9.12).

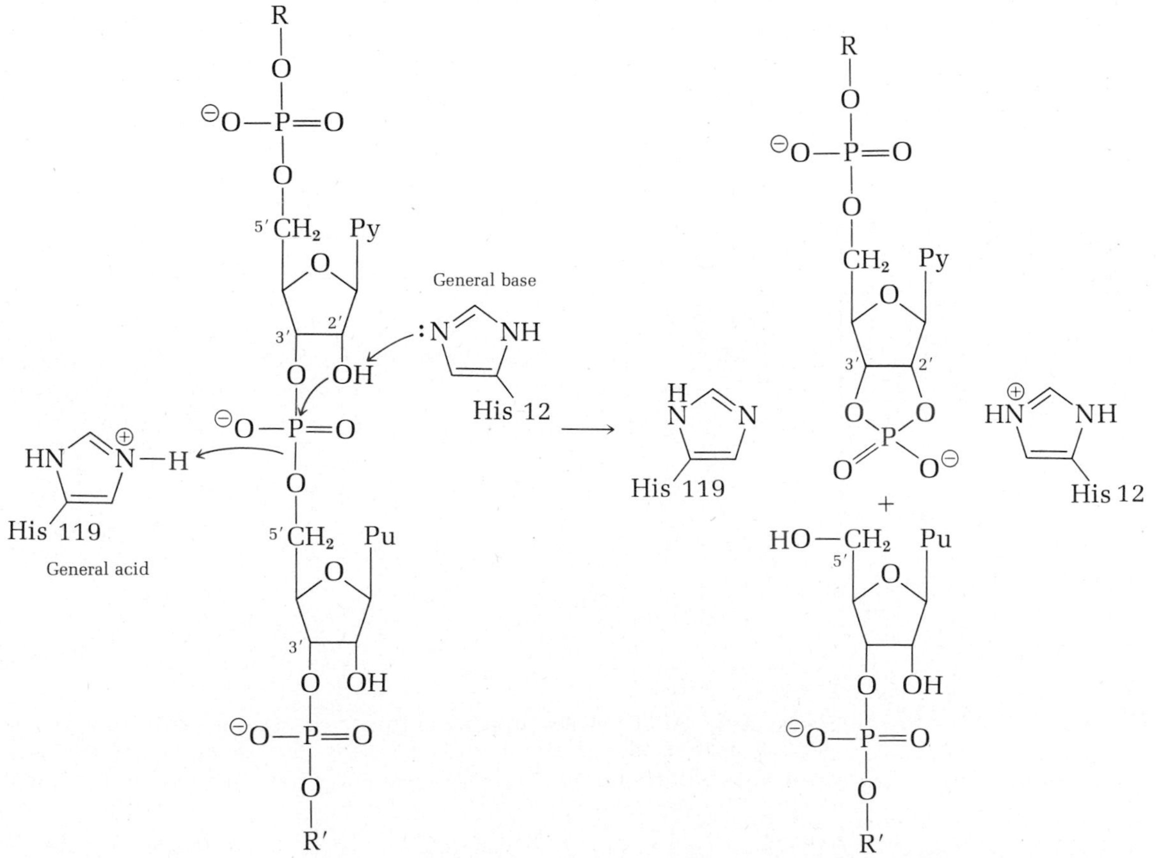

$$(9.12)$$

The opening of the 2′-3′ cyclic phosphate ester reverses the roles of the two histidyl residues. In the first half of the reaction residue 119 was converted to its conjugate base. In the second half, the ring-opening reaction, it extracts a proton from a water molecule, generating a hydroxide anion that attacks

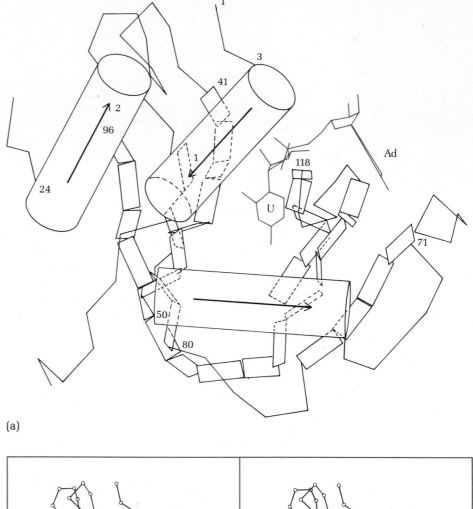

(a)

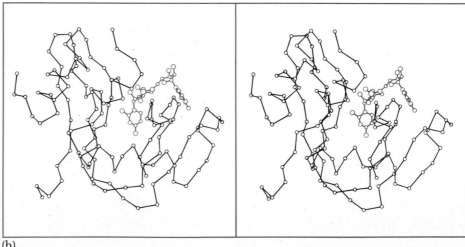

(b)

Figure 9.34 (a) Schematic drawing of the complex between bovine ribonuclease S and the substrate analog uridyl 3′,5′-adenosine phosphonate. The cylinders indicate region of the enzyme that are in α-helical conformations, and the ribbon indicates regions of β structure. (b) Stereo view.

the phosphorus.[2] The leaving 2′-alkoxide anion retrieves a proton from histidyl residue 12, freeing a nucleoside 3′-phosphate and converting ribonuclease back to its original ionic state in which histidyl residue 12 is protonated and histidyl residue 119 is present as the free base (reaction 9.13).

[2] The ring-opening reaction occurs by a so-called "in-line" mechanism, described in Section 12.6.

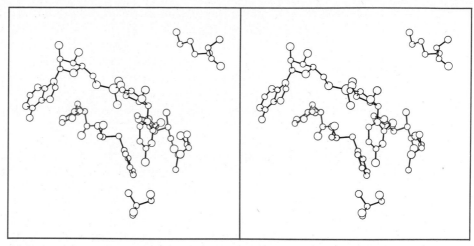

Figure 9.35 Stereo drawing of the complex between the substrate analog uridyl-(3′,5′)-adenosine phosphonate and ribonuclease at the enzyme's active site.

(9.13)

b. Microbial Ribonucleases

The pancreatic ribonuclease discussed in the preceding section is but one of a large family of related enzymes. The formation of 2′-3′ cyclic phosphate esters as intermediates in the hydrolysis of the phosphodiester bond is a general feature of the endolytic RNases that produce nucleoside 3′-phosphate products in *b* cleavage. We single out fungal ribonucleases T1, T2, N1, U1, and U2 because they have been widely used in sequence deter-

Table 9.8   Properties of fungal ribonucleases

RNase	T1	T2	U1	U2	N1
Source	*Aspergillus oryzae*	*Aspergillus oryzae*	*Ustilago sphaerogena*	*Ustilago sphaerogena*	*Neurospora crassa*
Base specificity	Guanine	Nonspecific	Guanine	Purine	Guanine
Molecular weight	11,085	36,200	11,000	10,000	11,000

SOURCE: Adapted from T. Uchida and F. Egami, in P. D. Boyer, Ed., "The Enzymes," Vol. 4, 3d ed., Academic Press, New York, 1971, p. 208.

minations of ribonucleic acids (Section 9.9A). Some of the properties of these enzymes are summarized in Table 9.8. Ribonuclease T1 is isolated from Takadiastase, a commercial product of *Aspergillus oryzae* and an important reagent for sequence analysis of RNA. RNase T1 cleaves phosphodiester bonds between 3′-guanylic acid residues and the 5′-hydroxyl groups of adjacent nucleotidyl residues. RNase T2 is a nonspecific enzyme that cleaves all phosphodiester bonds in RNA. RNase U2 specifically cleaves the phosphodiester bonds between 3′-purine nucleotidyl residues and the 5′-hydroxyl groups of adjacent nucleotides. It, too, is used for sequence analysis of RNA.

Bacteria produce many ribonucleases of varying specificities. For example, *E. coli* produce endonucleases that yield nucleoside 5′-phosphates as products (catalyzing *a* cleavage). *E. coli* also produce exonucleases that cleave single-stranded RNA's in the 3′-5′ direction to liberate nucleoside 5′-phosphates. *E. coli* RNase II attacks ribonucleic acids that contain a free 3′-hydroxyl group to give nucleoside 5-phosphates and can therefore be used in sequence determinations at the 3′ terminus of RNA.

c. Nonspecific Nucleases That Hydrolyze Both RNA and DNA

A few nucleases are able to hydrolyze both RNA and DNA. Two of the better known examples of such enzymes are *snake venom phosphodiesterase* and *spleen phosphodiesterase*. Both of these enzymes are exonucleases. The snake venom enzyme hydrolyzes either DNA or RNA from the 3′-hydroxyl end of oligonucleotides to produce nucleoside 5′-phosphates (reaction 9.14). The enzyme is inactive if the substrate has a 3′-phosphate group.

$$\text{P} \diagdown \text{P} \diagdown \text{P} \diagdown \text{P} \diagdown \text{OH} \xrightarrow[\text{phosphodiesterase}]{\text{Snake venom}} \text{P} \diagdown \text{OH} + \text{P} \diagdown \text{OH} + \text{P} \diagdown \text{OH} + \text{P} \diagdown \text{OH} \quad (9.14)$$

The spleen enzyme hydrolyzes nucleic acids from their 5′-hydroxyl termini to produce nucleoside 3′-phosphates (reaction 9.15). The spleen enzyme is inactive if the substrate has a 5′-phosphate terminus.

$$_{5'}\text{HO} \diagdown \text{P} \diagdown \text{P} \diagdown \text{P} \diagdown \text{P} \,_{3'} \xrightarrow[\text{phosphodiesterase}]{\text{Spleen}} \text{HO} \diagdown \text{P} + \text{HO} \diagdown \text{P} + \text{HO} \diagdown \text{P} + \text{HO} \diagdown \text{P} \quad (9.15)$$

Nonspecific endonucleases have also been isolated from many organisms. For example, *micrococcal nuclease* cleaves nucleic acids to give a mixture of nucleoside 3′-phosphates and oligonucleotides having 3′-phosphate termini. Endonucleases that cleave nucleic acids to give nucleoside 5′-phosphates and oligonucleotides having 5′-phosphate termini have been isolated from bacteria, plants, and mammalian cells.

D. Deoxyribonucleases

Deoxyribonucleases are phosphodiesterases that act on DNA. They can be subdivided into *exonucleases* (with 3′-5′ and 5′-3′ activity), *endonucleases*, and an important subclass of endonucleases known as *restriction endonucleases*. Restriction endonucleases are useful laboratory reagents in the study of DNA. Endonucleases have biochemical functions in DNA replication, repair of defective DNA, and genetic recombination, topics we shall consider in Chapter 24.

The hydrolytic action of nucleases can create several types of products.

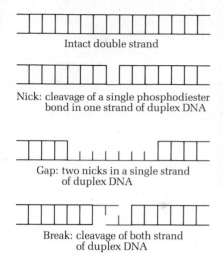

Intact double strand

Nick: cleavage of a single phosphodiester bond in one strand of duplex DNA

Gap: two nicks in a single strand of duplex DNA

Break: cleavage of both strand of duplex DNA

Figure 9.36 Types of scissions in double-stranded DNA.

Table 9.9 Some representative exonucleases

Exonuclease	Substrate	Product
Exonucleases (3'-5')		
Exonuclease I (*E. coli*)	DNA, single strand	Nucleoside 5'-phosphates
DNA polymerase I (*E. coli*)	DNA, single strand[a]	Nucleoside 5'-phosphates
Exonuclease III (*E. coli*)	DNA, double strand[a]	Nucleoside 5'-phosphates
B. subtilis exonuclease	DNA, double strand RNA	Nucleoside 3'-phosphates, oligonucleotides
Exonucleases (5'-3')		
B. subtilis exonuclease	DNA, single strand	Nucleoside 3'-phosphates (also works from 3' end)
Exonuclease VII (*E. coli*)	DNA, single strand	Oligonucleotides (also works from 3' end)
N. crassa exonuclease	DNA, single strand RNA	Nucleoside 5'-phosphates
Exonuclease VI or DNA polymerase I (*E. coli*)	DNA	5'-Nucleotides (also works from 3' end)

[a] Also works on nicked and gapped double-stranded DNA.

A single break in one strand of double-stranded DNA is called a *nick*. If several bases are missing in one strand, the single-stranded region is called a *gap*. If the phosphodiester bonds in both strands are cleaved, the scission is called a *break* (Figure 9.36).

a. Exonucleases

Exonucleases can be divided into those that require a 3' terminus for activity and those that require a 5' terminus for activity. Table 9.9 lists a few examples of 3'-5' and 5'-3' exonucleases.

b. Endonucleases

The endonucleases do not require a 3' or 5' terminus for catalytic activity. These enzymes, a few of which are listed in Table 9.10, vary in their specificity for single-stranded and double-stranded DNA. Most endonucleases possess at least some base sequence specificity, the least specific of them recognizing sequences of three or four bases, some only a single base, and some perhaps none.

c. Restriction Endonucleases

The importance of restriction endonucleases in molecular biology cannot easily be exaggerated because of their involvement in the creation of artificially recombinant DNA, that is, of new DNA molecules that contain spliced-in sequences of DNA from another organism (Chapter 24). The term

Table 9.10 Some representative endonucleases

Endonuclease	Substrate	Product
Neurospora crassa endonuclease	DNA, single strand RNA	Oligonucleotides with 5'-phosphate termini
Endonuclease I (*E. coli*)	DNA, single or double strand	Oligonucleotides with 5'-phosphate termini
DNase I (bovine pancreas)	DNA, single or double strand	Oligonucleotides with 5'-phosphate termini
DNase II (calf thymus)	DNA, single or double strand	Oligonucleotides with 3'-phosphate termini

restriction enzyme derived originally from the observation that certain bacteria destroy the DNA of particular viruses (or other, incompatible bacterial species), that is, these *restricting hosts* allow only certain strains of viruses to replicate within them. It appears that the viruses that are able to replicate have a modified DNA with methyl groups at specific sites. The restriction endonucleases cleave any DNA that is unmethylated at these specific sites, but the methylated DNA is not degraded. Each species of bacteria and even an individual strain has its own highly specific restriction enzymes.

There are two general classes of restriction endonucleases. *Type I restriction endonucleases* possess methylation activity and nuclease activity. The methylation of specific bases requires ATP and the methylating agent S-*adenosylmethionine* (Section 20.3B). These enzymes also possess a third catalytic activity, the rapid hydrolysis of large amounts of ATP. The ratio of moles of ATP hydrolyzed to moles of phosphodiester bonds cleaved in DNA is over 1,000:1, an apparently wasteful expenditure of the cell's major energy supply. The most important and unique property of type I endonucleases is recognition of specific base sequences in a DNA molecule. The recognition sequences for the *EcoK* and *EcoB* enzymes are shown below.[3]

$$5' \ldots \text{TGA*NNNNNNNNTGCT} \ldots$$
$$3' \ldots \text{ACT NNNNNNNNA*CGA} \ldots$$

The asterisks indicate that an adenine residue is methylated by the methylase activity of the restriction enzyme. All of the host DNA is methylated at these positions. The invading virus is not methylated and is cleaved by the nuclease activity of the restriction enzyme.

Type II restriction enzymes are simpler since they contain only a nuclease activity. The type II restriction endonucleases, of which over 200 are known, recognize specific base sequences in DNA. In most cases the base sequences recognized have a *two-fold axis of symmetry*. The nucleases cleave both strands of the DNA. Cleavage by type II restriction endonucleases is symmetrical with respect to the two-fold axis. Table 9.11 summarizes a few of the sequences recognized by restriction endonucleases from various sources. The table also indicates one of the reasons that the type II restriction endonucleases are so valuable in the study of DNA. The DNA molecules of even the simplest organisms are enormous. Attempts to determine DNA sequences (Section 9.10) must begin by hydrolyzing the DNA to pieces of manageable size. Since restriction endonucleases recognize specific sequences and since these sequences are not highly probable, a given DNA molecule may possess only a handful of recognition sequences for a given restriction endonuclease. Nucleolytic cleavage therefore produces only a few fragments. The number of fragments produced by treatment of DNA from various sources with various restriction endonucleases is also included in Table 9.11.

9–9 DETERMINATION OF THE PRIMARY STRUCTURES OF NUCLEIC ACIDS

Since DNA is the genetic material for all organisms (except certain viruses), a knowledge of its primary structure is a prerequisite for an understanding of gene organization and regulation. RNA molecules, too, are critical in biological information transfer, and sequence studies of RNA molecules provide

---

[3] The names of the enzymes consist of an abbreviation for the genus and species of the host organism, a strain designation where necessary, and a letter and Roman numeral to identify one of several enzymes from the same source. For example, *EcoK* for *E. coli* strain K.

Table 9.11   Type II restriction endonucleases

Enzyme and source	Recognition sequences	Number of cleavage sites on viral DNA			
		$\phi\chi$174	λ	Ad2	SV40
	Axis of Symmetry ↓				
EcoRI (E. coli RTFI)	5′pGpApǍpTpTpC CpTpTpApApGp ↑ *	0	5	5	1
EcoRII (E. coli RTFII)	↓CČTGG	2	>35	>35	16
HindII (Haemophilus influenzae Rd)	GTPy ↓ PuǍC	13	34	>20	7
HindIII (Haemophilus influenzae Rd)	ǍAG ↓ CTT	0	6	11	6
HaeIII (Haemophilus aegyptius)	GG ↓ CC	11	>50	>50	19
HpaII (Haemophilus parainfluenzae)	CC ↓ GG	5	>50	>50	1
PstI (Providencia stuartii 164)	CTG ↓ CAG	1	18	25	2
SMaI (Serratia marcescens Sb_b)	CCC ↓ GGG	0	3	12	0
BamI (Bacillus amyloliquefaciens H)	GGA ↓ TCC	0	5	3	1
BglII (Bacillus globiggi)	AGA ↓ TCT	0	5	12	0

The asterisks represent position where bases are methylated. The arrows represent cleavage sites.

information about mRNA, rRNA, and tRNA that allows a correlation of structure to function that parallels the information available from the primary structure of proteins.

A. Determination of the Primary Structure of DNA

Within the past few years rapid progress has been made in techniques of determining DNA sequences. The primary structures of several entire DNA molecules from viruses have been determined. They include E. coli bacteriophage $\phi\chi$174 and the monkey virus SV40, whose sequences each contain about 5,000 residues (Figure 9.37). About 2,000 pages of type of the size in Figure 9.37 would be required to express the genetic content of an E. coli cell, and about a million pages would be required to express the DNA content of a mammalian cell.

Two techniques have been developed for the determination of DNA sequences, a method that relies upon specific chemical reactions and an enzymological method that relies upon DNA polymerase. Restriction enzymes are used in both methods to cleave DNA to pieces of manageable size. Restriction enzymes with different specificities produce overlapping fragments, analogous to proteolytic cleavage of proteins in the determination of the primary structures of proteins. In the chemical method the sequences of the pieces are determined directly; in the enzymatic method the restriction fragments are used as primers.

GGTCTAGGAG	CTAAAGAATG	GAACAACTCA	CTAAAAACCA	AGCTGTCGCT	ACTTCCCAAG
5067	5077	5087	5097	5107	5117
AAGCTGTTCA	GAATCAGAAT	GAGCCGCAAC	TTCGGGATGA	AAATGCTCAC	AATGACAAAT
5127	5137	5147	5157	5167	5177
CTGTCCACGG	AGTGCTTAAT	CCAACTTACC	AAGCTGGGTT	ACGACGCGAC	GCCGTTCAAC
5187	5197	5207	5217	5227	5237
CAGATATTGA	AGCAGAACGC	AAAAAGAGAG	ATGAGATTGA	GGCTGGGAAA	AGTTACTGTA
5247	5257	5267	5277	5287	5297
GCCGACGTTT	TGGCGGCGCA	ACCTGTGACG	ACAAATCTGC	TCAAATTTAT	GCGCGCTTCG
5307	5317	5327	5337	5347	5357
ATAAAAATGA	TTGGCGTATC	CAACCTGCAG	AGTTTTATCG	CTTCCATGAC	GCAGAAGTTA
5367	5377	1	11	21	31
ACACTTTCGG	ATATTTCTGA	TGAGTCGAAA	AATTATCTTG	ATAAAGCAGG	AATTACTACT
41	51	61	71	81	91
GCTTGTTTAC	GAATTAAATC	GAAGTGGACT	GCTGGCGGAA	AATGAGAAAA	TTCGACCTAT
101	111	121	131	141	151
CCTTGCGCAG	CTCGAGAAGC	TCTTACTTTG	CGACCTTTCG	CCATCAACTA	ACGATTCTGT
161	171	181	191	201	211
CAAAAACTGA	CGCGTTGGAT	GAGGAGAAGT	GGCTTAATAT	GCTTGGCACG	TTCGTCAAGG
221	231	241	251	261	271
ACTGGTTTAG	ATATGAGTCA	CATTTTGTTC	ATGGTAGAGA	TTCTCTTGTT	GACATTTTAA
281	291	301	311	321	331
AAGAGCGTGG	ATTACTATCT	GAGTCCGATG	CTGTTCAACC	ACTAATAGGT	AAGAAATCAT
341	351	361	371	381	391
GAGTCAAGTT	ACTGAACAAT	CCGTACGTTT	CCAGACCGCT	TTGGCCTCTA	TTAAGCTCAT
401	411	421	431	441	451
TCAGGCTTCT	GCCGTTTTGG	ATTTAACCGA	AGATGATTTC	GATTTTCTGA	CGAGTAACAA
461	471	481	491	501	511
AGTTTGGATT	GCTACTGACC	GCTCTCGTGC	TCGTCGCTGC	GTTGAGGCTT	GCGTTTATGG
521	531	541	551	561	571
TACGCTGGAC	TTTGTGGGAT	ACCCTCGCTT	TCCTGCTCCT	GTTGAGTTTA	TTGCTGCCGT
581	591	601	611	621	631
CATTGCTTAT	TATGTTCATC	CCGTCAACAT	TCAAACGGCC	TGTCTCATCA	TGGAAGGCGC
641	651	661	671	681	691
TGAATTTACG	GAAAACATTA	TTAATGGCGT	CGAGCGTCCG	GTTAAAGCCG	CTGAATTGTT
701	711	721	731	741	751
CGCGTTTACC	TTGCGTGTAC	GCGCAGGAAA	CACTGACGTT	CTTACTGACG	CAGAAGAAAA
761	771	781	791	801	811
CGTGCGTCAA	AAATTACGTG	CGGAAGGAGT	GATGTAATGT	CTAAAGGTAA	AAAACGTTCT
821	831	841	851	861	871
GGCGCTCGCC	CTGGTCGTCC	GCAGCCGTTG	CGAGGTACTA	AAGGCAAGCG	TAAAGGCGCT
881	891	901	911	921	931
CGTCTTTGGT	ATGTAGGTGG	TCAACAATTT	TAATTGCAGG	GGCTTCGGCC	CCTTACTTGA
941	951	961	971	981	991
GGATAAATTA	TGTCTAATAT	TCAAACTGGC	GCCGAGCGTA	TGCCGCATGA	CCTTTCCCAT
1001	1011	1021	1031	1041	1051

Figure 9.37   Nucleotide sequence of the DNA of *E. coli* phage φX174. [Courtesy of Dr. F. Sanger.]

CTTGGCTTCC 1061	TTGCTGGTCA 1071	GATTGGTCGT 1081	CTTATTACCA 1091	TTTCAACTAC 1101	TCCGGTTATC 1111
GCTGGCGACT 1121	CCTTCGAGAT 1131	GGACGCCGTT 1141	GGCGCTCTCC 1151	GTCTTTCTCC 1161	ATTGCGTCGT 1171
GGCCTTGCTA 1181	TTGACTCTAC 1191	TGTAGACATT 1201	TTTACTTTTT 1211	ATGTCCCTCA 1221	TCGTCACGTT 1231
TATGGTGAAC 1241	AGTGGATTAA 1251	GTTCATGAAG 1261	GATGGTGTTA 1271	ATGCCACTCC 1281	TCTCCCGACT 1291
GTTAACACTA 1301	CTGGTTATAT 1311	TGACCATGCC 1321	GCTTTTCTTG 1331	GCACGATTAA 1341	CCCTGATACC 1351
AATAAAATCC 1361	CTAAGCATTT 1371	GTTTCAGGGT 1381	TATTTGAATA 1391	TCTATAACAA 1401	CTATTTTAAA 1411
GCGCCGTGGA 1421	TGCCTGACCG 1431	TACCGAGGCT 1441	AACCCTAATG 1451	AGCTTAATCA 1461	AGATGATGCT 1471
CGTTATGGTT 1481	TCCGTTGCTG 1491	CCATCTCAAA 1501	AACATTTGGA 1511	CTGCTCCGCT 1521	TCCTCCTGAG 1531
ACTGAGCTTT 1541	CTCGCCAAAT 1551	GACGACTTCT 1561	ACCACATCTA 1571	TTGACATTAT 1581	GGGTCTGCAA 1591
GCTGCTTATG 1601	CTAATTTGCA 1611	TACTGACCAA 1621	GAACGTGATT 1631	ACTTCATGCA 1641	GCGTTACCAT 1651
GATGTTATTT 1661	CTTCATTTGG 1671	AGGTAAAACC 1681	TCTTATGACG 1691	CTGACAACCG 1701	TCCTTTACTT 1711
GTCATGCGCT 1721	CTAATCTCTG 1731	GGCATCTGGC 1741	TATGATGTTG 1751	ATGGAACTGA 1761	CCAAACGTCG 1771
TTAGGCCAGT 1781	TTTCTGGTCG 1791	TGTTCAACAG 1801	ACCTATAAAC 1811	ATTCTGTGCC 1821	GCGTTTCTTT 1831
GTTCCTGAGC 1841	ATGGCACTAT 1851	GTTTACTCTT 1861	GCGCTTGTTC 1871	GTTTTCCGCC 1881	TACTGCGACT 1891
AAAGAGATTC 1901	AGTACCTTAA 1911	CGCTAAAGGT 1921	GCTTTGACTT 1931	ATACCGATAT 1941	TGCTGGCGAC 1951
CCTGTTTTGT 1961	ATGGCAACTT 1971	GCCGCCGCGT 1981	GAAATTTCTA 1991	TGAAGGATGT 2001	TTTCCGTTCT 2011
GGTGATTCGT 2021	CTAAGAAGTT 2031	TAAGATTGCT 2041	GAGGGTCAGT 2051	GGTATCGTTA 2061	TGCGCCTTCG 2071
TATGTTTCTC 2081	CTGCTTATCA 2091	CCTTCTTGAA 2101	GGCTTCCCAT 2111	TCATTCAGGA 2121	ACCGCCTTCT 2131
GGTGATTTGC 2141	AAGAACGCGT 2151	ACTTATTCGC 2161	CACCATGATT 2171	ATGACCAGTG 2181	TTTCCAGTCC 2191
GTTCAGTTGT 2201	TGCAGTGGAA 2211	TAGTCAGGTT 2221	AAATTTAATG 2231	TGACCGTTTA 2241	TCGCAATCTG 2251
CCGACCACTC 2261	GCGATTCAAT 2271	CATGACTTCG 2281	TGATAAAAGA 2291	TTGAGTGTGA 2301	GGTTATAACG 2311
CCGAAGCGGT 2321	AAAAATTTTA 2331	ATTTTTGCCG 2341	CTGAGGGGTT 2351	GACCAAGCGA 2361	AGCGCGGTAG 2371
GTTTTCTGCT 2381	TAGGAGTTTA 2391	ATCATGTTTC 2401	AGACTTTTAT 2411	TTCTCGCCAT 2421	AATTCAAACT 2431

(continued)

TTTTTTCTGA 2441	TAAGCTGGTT 2451	CTCACTTCTG 2461	TTACTCCAGC 2471	TTCTTCGGCA 2481	CCTGTTTTAC 2491
AGACACCTAA 2501	AGCTACATCG 2511	TCAACGTTAT 2521	ATTTTGATAG 2531	TTTGACGGTT 2541	AATGCTGGTA 2551
ATGGTGGTTT 2561	TCTTCATTGC 2571	ATTCAGATGG 2581	ATACATCTGT 2591	CAACGCCGCT 2601	AATCAGGTTG 2611
TTTCTGTTGG 2621	TGCTGATATT 2631	GCTTTTGATG 2641	CCGACCCTAA 2651	ATTTTTTGCC 2661	TGTTTGGTTC 2671
GCTTTGAGTC 2681	TTCTTCGGTT 2691	CCGACTACCC 2701	TCCCGACTGC 2711	CTATGATGTT 2721	TATCCTTTGA 2731
ATGGTCGCCA 2741	TGATGGTGGT 2751	TATTATACCG 2761	TCAAGGACTG 2771	TGTGACTATT 2781	GACGTCCTTC 2791
CCCGTACGCC 2801	GGGCAATAAC 2811	GTTTATGTTG 2821	GTTTCATGGT 2831	TTGGTCTAAC 2841	TTTACCGCTA 2851
CTAAATGCCG 2861	CGGATTGGTT 2871	TCGCTGAATC 2881	AGGTTATTAA 2891	AGAGATTATT 2901	TGTCTCCAGC 2911
CACTTAAGTG 2921	AGGTGATTTA 2931	TGTTTGGTGC 2941	TATTGCTGGC 2951	GGTATTGCTT 2961	CTGCTCTTGC 2971
TGGTGGCGCC 2981	ATGTCTAAAT 2991	TGTTTGGAGG 3001	CGGTCAAAAA 3011	GCCGCCTCCG 3021	GTGGCATTCA 3031
AGGTGATGTG 3041	CTTGCTACCG 3051	ATAACAATAC 3061	TGTAGGCATG 3071	GGTGATGCTG 3081	GTATTAAATC 3091
TGCCATTCAA 3101	GGCTCTAATG 3111	TTCCTAACCC 3121	TGATGAGGCC 3131	GCCCCTAGTT 3141	TTGTTTCTGG 3151
TGCTATGGCT 3161	AAAGCTGGTA 3171	AAGGACTTCT 3181	TGAAGGTACG 3191	TTGCAGGCTG 3201	GCACTTCTGC 3211
CGTTTCTGAT 3221	AAGTTGCTTG 3231	ATTTGGTTGG 3241	ACTTGGTGGC 3251	AAGTCTGCCG 3261	CTGATAAAGG 3271
AAAGGATACT 3281	CGTGATTATC 3291	TTGCTGCTGC 3301	ATTTCCTGAG 3311	CTTAATGCTT 3321	GGGAGCGTGC 3331
TGGTGCTGAT 3341	GCTTCCTCTG 3351	CTGGTATGGT 3361	TGACGCCGGA 3371	TTTGAGAATC 3381	AAAAAGAGCT 3391
TACTAAAATG 3401	CAACTGGACA 3411	ATCAGAAAGA 3421	GATTGCCGAG 3431	ATGCAAAATG 3441	AGACTCAAAA 3451
AGAGATTGCT 3461	GGCATTCAGT 3471	CGGCGACTTC 3481	ACGCCAGAAT 3491	ACGAAAGACC 3501	AGGTATATGC 3511
ACAAAATGAG 3521	ATGCTTGCTT 3531	ATCAACAGAA 3541	GGAGTCTACT 3551	GCTCGCGTTG 3561	CGTCTATTAT 3571
GGAAAACACC 3581	AATCTTTCCA 3591	AGCAACAGCA 3601	GGTTTCCGAG 3611	ATTATGCGCC 3621	AAATGCTTAC 3631
TCAAGCTCAA 3641	ACGGCTGGTC 3651	AGTATTTTAC 3661	CAATGACCAA 3671	ATCAAAGAAA 3681	TGACTCGCAA 3691
GGTTAGTGCT 3701	GAGGTTGACT 3711	TAGTTCATCA 3721	GCAAACGCAG 3731	AATCAGCGGT 3741	ATGGCTCTTC 3751

(continued)

TCATATTGGC 3761	GCTACTGCAA 3771	AGGATATTTC 3781	TAATGTCGTC 3791	ACTGATGCTG 3801	CTTCTGGTGT 3811
GGTTGATATT 3821	TTTCATGGTA 3831	TTGATAAAGC 3841	TGTTGCCGAT 3851	ACTTGGAACA 3861	ATTTCTGGAA 3871
AGACGGTAAA 3881	GCTGATGGTA 3891	TTGGCTCTAA 3901	TTTGTCTAGG 3911	AAATAA	
CCGTCAGGAT 3927	TGACACCCTC 3937	CCAATTGTAT 3947	GTTTTCATGC 3957	CTCCAAATCT 3967	TGGAGGCTTT 3977
TTTATGGTTC 3987	GTTCTTATTA 3997	CCCTTCTGAA 4007	TGTCACGCTG 4017	ATTATTTTGA 4027	CTTTGAGCGT 4037
ATCGAGGCTC 4047	TTAAACCTGC 4057	TATTGAGGCT 4067	TGTGGCATTT 4077	CTACTCTTTC 4087	TCAATCCCCA 4097
ATGCTTGGCT 4107	TCCATAAGCA 4117	GATGGATAAC 4127	CGCATCAAGC 4137	TCTTGGAAGA 4147	GATTCTGTCT 4157
TTTCGTATGC 4167	AGGGCGTTGA 4177	GTTCGATAAT 4187	GGTGATATGT 4197	ATGTTGACGG 4207	CCATAAGGCT 4217
GCTTCTGACG 4227	TTCGTGATGA 4237	GTTTGTATCT 4247	GTTACTGAGA 4257	AGTTAATGGA 4267	TGAATTGGCA 4277
CAATGCTACA 4287	ATGTGCTCCC 4297	CCAACTTGAT 4307	ATTAATAACA 4317	CTATAGACCA 4327	CCGCCCCGAA 4337
GGGGACGAAA 4347	AATGGTTTTT 4357	AGAGAACGAG 4367	AAGACGGTTA 4377	CGCAGTTTTG 4387	CCGCAAGCTG 4397
GCTGCTGAAC 4407	GCCCTCTTAA 4417	GGATATTCGC 4427	GATGAGTATA 4437	ATTACCCCAA 4447	AAAGAAAGGT 4457
ATTAAGGATG 4467	AGTGTTCAAG 4477	ATTGCTGGAG 4487	GCCTCCACTA 4497	TGAAATCGCG 4507	TAGAGGCTTT 4517
GCTATTCAGC 4527	GTTTGATGAA 4537	TGCAATGCGA 4547	CAGGCTCATG 4557	CTGATGGTTG 4567	GTTTATCGTT 4577
TTTGACACTC 4587	TCACGTTGGC 4597	TGACGACCGA 4607	TTAGAGGCGT 4617	TTTATGATAA 4627	TCCCAATGCT 4637
TTGCGTGACT 4647	ATTTTCGTGA 4657	TATTGGTCGT 4667	ATGGTTCTTG 4677	CTGCCGAGGG 4687	TCGCAAGGCT 4697
AATGATTCAC 4707	ACGCCGACTG 4717	CTATCAGTAT 4727	TTTTGTGTGC 4737	CTGAGTATGG 4747	TACAGCTAAT 4757
GGCCGTCTTC 4767	ATTTCCATGC 4777	GGTGCACTTT 4787	ATGCGGACAC 4797	TTCCTACAGG 4807	TAGCGTTGAC 4817
CCTAATTTTG 4827	GTCGTCGGGT 4837	ACGCAATCGC 4847	CGCCAGTTAA 4857	ATAGCTTGCA 4867	AAATACGTGG 4877
CCTTATGGTT 4887	ACAGTATGCC 4897	CATCGCAGTT 4907	CGCTACACGC 4917	AGGACGCTTT 4927	TTCACGTTCT 4937
GGTTGGTTGT 4947	GGCCTGTTGA 4957	TGCTAAAGGT 4967	GAGCCGCTTA 4977	AAGCTACCAG 4987	TTATATGGCT 4997
GTTGGTTTCT 5007	ATGTGGCTAA 5017	ATACGTTAAC 5027	AAAAAGTCAG 5037	ATATGGACCT 5047	TGCTGCTAAA 5057

a. Chemical Method for Determining the Primary Structure of DNA

In 1977 A. M. Maxam and M. Gilbert introduced a chemical method for determining the sequence of DNA that is the simplest and fastest of the available methods. Chemical reactions that cleave a 5'-^{32}P-labeled DNA at a specific nucleotide residue are combined with gel electrophoresis to yield the sequence.

Four separate reactions are used. The first cleaves DNA at guanine in preference to adenine (G > A); the second cleaves at adenine in preference to guanine (A > G); the third cleaves at cytosine and thymine equally (C = T); and the fourth cleaves at cytosine alone (C). Each reaction is carried out on a separate aliquot under conditions that yield an average of only 1 modified base per 100 residues. The products of each reaction are separated by gel electrophoresis, and the DNA sequence is read from the pattern of radioactive bands in the gel.

Reaction 1: Guanine-Enhanced Cleavage (G > A) Treatment of DNA with dimethyl sulfate results in methylation of guanine at N-7 and of adenine at N-3, the most electron-rich nitrogens of each base (Figure 9.38). The methylated purines are removed by adjusting the pH to 7.0. Since a given strand is methylated only once, on the average, there is now one depurinated nucleotide per strand.[4] Hydrolysis of the sugar phosphates at the apurinic sites is carried out in 0.1 M alkali at 90°C. Cleavage of the DNA gives all of the fragments ending just before guanine and just before adenine. Guanine is five times more reactive than adenine in the methylation reaction, so there are more fragments terminating before guanine than before adenine. Separation of the fragments by gel electrophoresis and detection of the 5'-^{32}P-labeled end of each fragment by autoradiography (a technique in which a photographic plate is exposed to the disintegration of radioisotopes)

[4] We assume that there are about a hundred residues per strand of DNA.

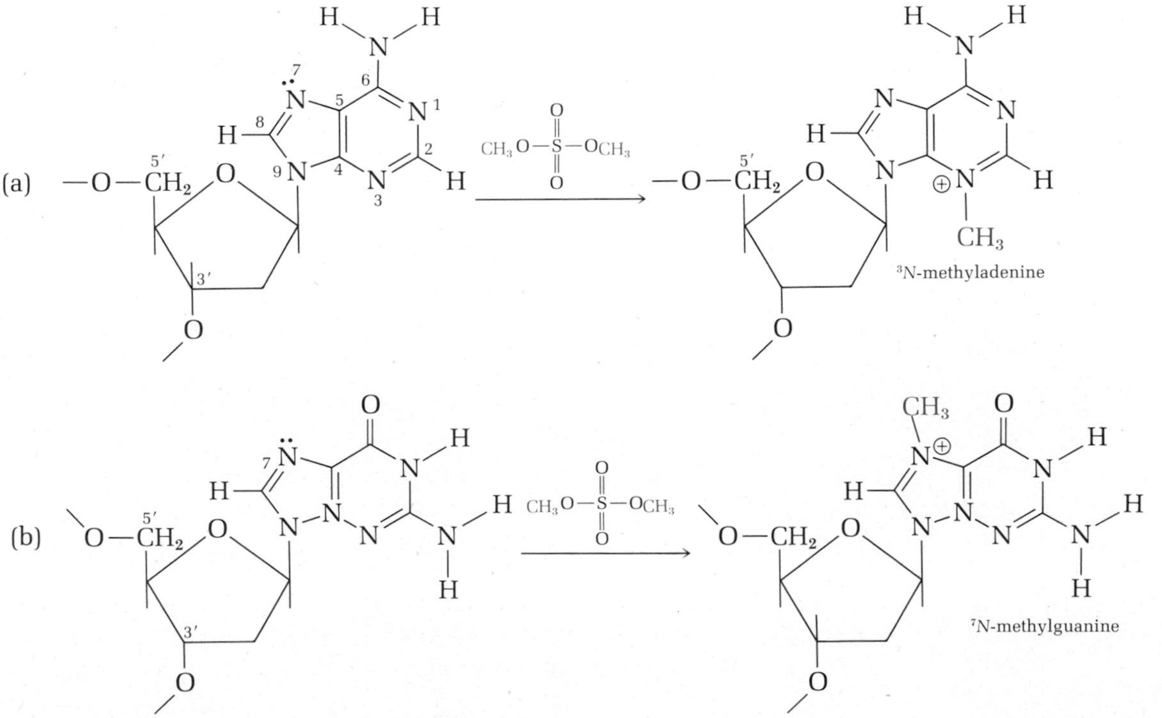

Figure 9-38 Treatment of DNA with dimethyl sulfate resulting in methylation at N-3 in adenine (a) and at N-7 in guanine (b). Guanine is about five times more reactive than adenine. Adjusting the pH to 7.0 results in hydrolysis of the glycosidic bonds of the methylated bases.

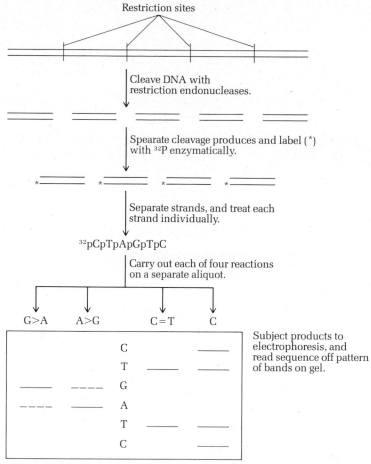

Figure 9.39   Steps in the analysis of the primary structure of DNA by the method of Maxam and Gilbert.

gives a pattern of light and dark bands corresponding to fragments cleaved just before adenine and to cleavage just before guanine (Figure 9.39).

Reaction 2: Adenine-Enhanced Cleavage (A > G) The β-glycosidic bond of ^{3}N-methyladenine is more labile than the glycosidic bond of ^{7}N-methylguanine. Treatment of methylated DNA with dilute acid (as opposed to treatment at pH 7.0 in reaction 1) results in preferential hydrolysis of ^{3}N-methyladenine glycosidic bonds. Subsequent treatment with alkali, electrophoresis, and detection of the 5'-^{32}P-labeled end give a pattern of light and dark bands. This time the dark bands correspond to fragments ending just before adenine and the light bands to fragments ending just before guanine (Figure 9.39). The combination of reactions 1 and 2 provides half of the information needed to obtain the primary structure.

Reaction 3: Cleavage at Cytosine and Thymine (C = T) Treatment of DNA with hydrazine results in cleavage of the polymer chain at cytosine and thymine. Hydrazinolysis at thymine produces a 2'-deoxyribosyl hydrazone, 4-methyl-5-pyrazolone, and urea (Figure 9.40). Hydrazinolysis at cytosine yields a 2'-deoxyribosyl hydrazone, 3-aminopyrazole, and urea (Figure 9.41). Treatment of the hydrazinolysis product with *piperidine* results in cleavage of the 3'-5' phosphodiester bonds. The fragments obtained correspond to cleavage at all sequences terminating just before cytosine or thymine (Figure 9.42). As in the purine reactions the reaction conditions are chosen so that only one base per piece of DNA reacts and a random sample of fragments terminating just before cytosine and thymine is obtained. The fragments are separated by electrophoresis (Figure 9.43).

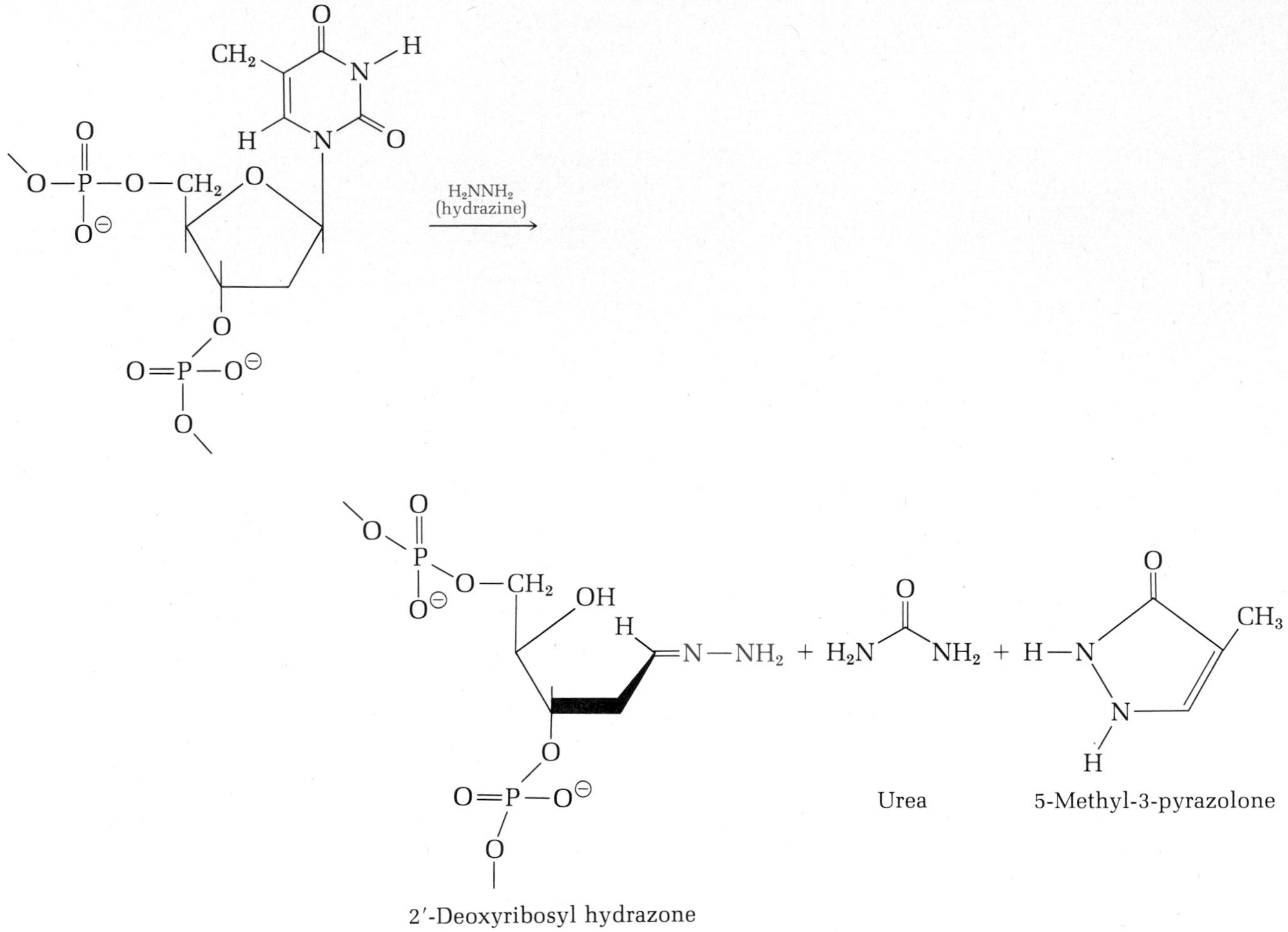

Figure 9.40 Reaction of a thymine base in DNA with hydrazine.

Reaction 4: Cleavage at Cytosine (C) Reaction of DNA with hydrazine in 2 M NaCl leads to hydrazinolysis of cytosines, but not of thymines. Treatment of the cytosine-free DNA with piperidine to cleave the 3′-5′ phosphodiester bonds gives fragments terminating before cytosine (Figure 9.44).

When the fragments from all four reactions are run simultaneously on the same gel, the sequence of the polynucleotide can be read directly from the pattern of light and dark bands (Figure 9.45).

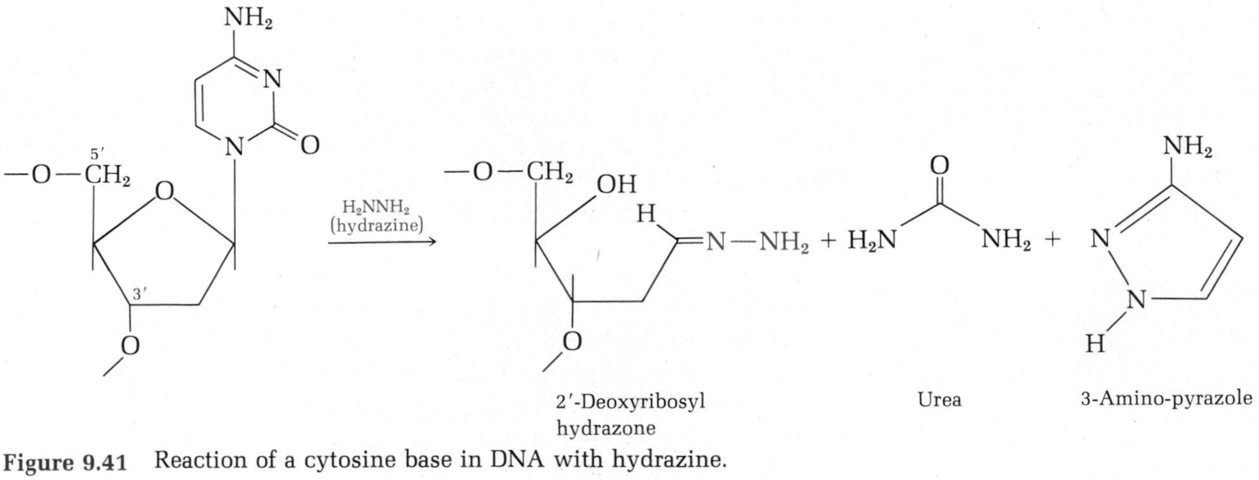

Figure 9.41 Reaction of a cytosine base in DNA with hydrazine.

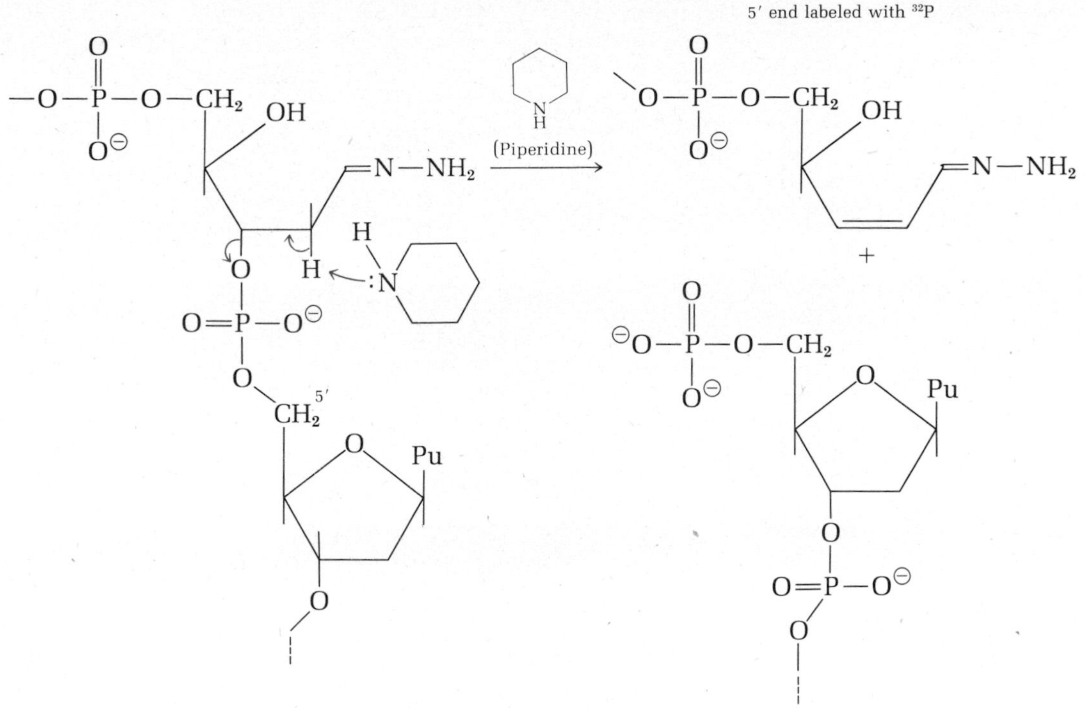

Figure 9.42 Cleavage of the 3′-5′ phosphodiester bond catalyzed by piperidine.

$$^{32}\text{pCpTpApGpTpC} \ldots \xrightarrow{\text{Hydrazine}} \begin{array}{l} ^{32}\text{p-pTpApGpTpC} \quad + \\ ^{32}\text{pCppApGpTpC} \quad + \\ ^{32}\text{pCpTpApGppC} \quad + \\ ^{32}\text{pCpTpApGpTp-p} \end{array}$$

$$\downarrow \text{Piperidine}$$

$$\begin{array}{l} ^{32}\text{pC} \quad + \\ ^{32}\text{pCpTpApG} \quad + \\ ^{32}\text{pCpTpApGpT} \end{array}$$

$$\downarrow \text{Electrophoresis}$$

$$\begin{array}{l} ^{32}\text{pCpTpApGpT} \quad - \\ ^{32}\text{pCpTpApG} \quad - \\ ^{32}\text{pC} \quad - \end{array}$$

Figure 9.43 Cleavage pattern obtained in the cytosine and thymine reaction with hydrazine and piperidine.

$$^{32}\text{pCpTpApGpTpC} \xrightarrow[\text{2 M NaCl}]{\text{Hydrazine}} \begin{array}{c} ^{32}\text{ppTpApGpTpC} \\ + \\ ^{32}\text{pCpTpApGpTpp} \end{array}$$

$$\downarrow \text{Piperidine}$$

$$^{32}\text{pCpTpApGpT}$$

Figure 9.44 Reaction of DNA with hydrazine in the presence of 2 M NaCl resulting in hydrazinolysis of cytosines, but not of thymines.

(a)　(A > G)　　(G > A)　　　(C = T)　　　　(C)　　　Conclusion

　　　　　　　　　　　　^{32}pCpTpApGpT　^{32}pCpTpApGpT　　T

　　　　　　　　　　　　^{32}pCpTpApG　　　　　　　　G

　　　^{32}pCpTpA　^{32}pCpTpA　　　　　　　　　　　A

　　　^{32}pCpT　　^{32}pCpT　　　　　　　　　　　　T

　　　　　　　　　　　　^{32}pC　　　　　　　　　　C

　　　　　　　　　　　　　　　　　　　　　　pCpTpApGpT

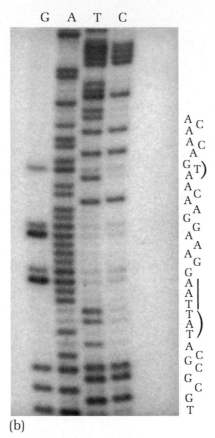

(b)

Figure 9.45　(a) Nucleotide sequence of our hypothetical DNA molecule. (b) Sequence of a segment of DNA from the *lac* repressor. [Courtesy of Dr. T. Platt.]

b. Plus and Minus Method for Determining the Primary Structure of DNA

The "plus and minus" method for determining DNA sequences was developed by F. Sanger and A. R. Coulson. This technique employs *E. coli* DNA polymerase I and a DNA polymerase from bacteriophage T4 infected *E. coli*. DNA polymerase I requires a primer oligonucleotide for activity. In the plus-minus method either a synthetic oligonucleotide or fragments produced by the action of restriction endonucleases can serve as primers. The principle of this technique is shown in Figure 9.46 for a hypothetical DNA sequence. In the presence of a primer, DNA polymerase I copies the template DNA if all four nucleoside triphosphates are provided. In the plus-minus method one of these nucleotides is labeled with ^{32}P. Synthesis is carried out for varying lengths of time to allow production of oligonucleotides of varying lengths. After polymerization the nucleotides that remain are removed and the double helical DNA is treated in one of two ways.

i. The Minus System In the minus system the mixture of random-length oligonucleotides, still hybridized to the template DNA, is again treated with DNA polymerase I, but this time in the presence of only three nucleoside tri-

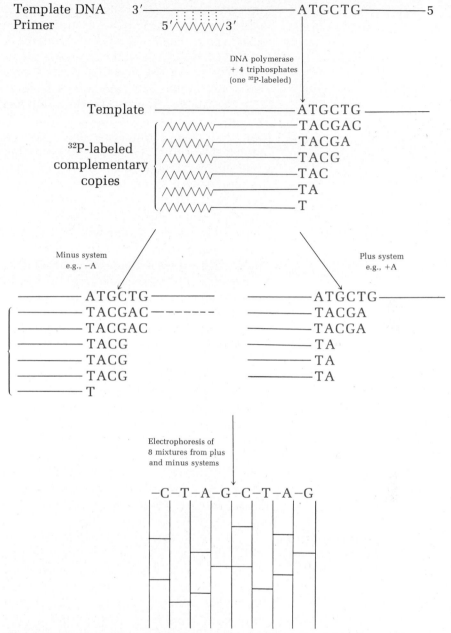

Figure 9.46 Flow diagram showing the major features of the plus and minus method for determining oligonucleotide sequences. [From F. Sanger and C. A. Coulson, *J. Mol. Biol.*, *94*, 442 (1975). Reproduced with permission. Copyright: Academic Press Inc. (London) Ltd.]

phosphates. One is missing, hence the term "minus system." If dATP is missing from the incubation mixture, synthesis proceeds until dA is required, and then it stops. Separate samples, each with a different nucleoside triphosphate missing, are treated in this way. The four reaction mixtures are then individually denatured, and the resulting single-stranded molecules are separated by gel electrophoresis. The movement of the oligonucleotides through the gel depends upon their relative sizes, the smallest moving the most rapidly. In the ideal case each oligonucleotide species is separated from its nearest neighbors, one of them having one more residue and the other one less residue. Since one of the nucleotides is labeled with the radioisotope ^{32}P, exposure of the gel to a photographic plate gives a radioautograph with a band at every position where a radioisotopically labeled nucleotide is present. In the −dA system, for example, the radioautograph will contain bands corresponding to positions before a dA residue in each newly synthesized chain. The positions of each of the four labeled nucleotides are determined in this way, and the sequence can be read directly from the autoradiograph of the gel. To minimize errors a second system is employed to confirm the results obtained by the minus system.

ii. The Plus System The DNA polymerase from bacteriophage T4 has a remarkable property: in the presence of a single nucleoside triphosphate it functions as an exonuclease that hydrolyzes DNA from the 3′ terminus and stops at residues corresponding to the single nucleoside triphosphate present in the mixture. If dATP, for example, is added to the mixture of random-length oligonucleotides described above, in the presence of T4 DNA polymerase all degraded oligonucleotides will have dA as their 3′ terminal residue. Denaturation of the DNA fragments obtained in this +dA system, separation by electrophoresis, and production of the autoradiograph gives the positions of the dA residues in each of the oligonucleotides. These products will be at least one residue larger than the product of the corresponding −dA system. If there are several dA residues in a row, the distance between the bands in the +dA and −dA systems gives the number of consecutive residues.

Referring again to the hypothetical oligonucleotide in Figure 9.46, analysis of the radioautograph from the minus system reveals a band in the −dT position for the smallest oligonucleotide. Therefore, dT is the next nucleotide, reading in the 3′ direction. This is confirmed by the presence of a +dT band in the next largest oligonucleotide one notch up in the gel. The conjunction of a +dT band and a −dA band in nucleotides of the same length shows that the 3′ terminus is dT and that the nucleotide following is dA. Similar analysis shows that the next residue is dC, and so forth. In effect, by reading the bands in the gel we are marching from the 5′ to the 3′ end of the oligonucleotide. The plus and minus technique was used to determine the nucleotide sequence of the entire genome of bacteriophage $\phi\chi$174, a sequence of 5,375 residues (recall Figure 9.37).

B. Determination of the Primary Structure of RNA

The methods employed for determination of the sequences of ribonucleic acids parallel those used for determination of the primary structures of proteins. The basic strategies are the same. Cleavage of the ribonucleic acid by ribonucleases generates fragments of varying lengths and sequences. The sequence of each oligonucleotide is determined, and the entire primary structure is obtained by comparing the sequences of the overlapping fragments produced by cleavage with ribonucleases of different specificities. Progress in determination of ribonucleic acid sequences has been aided by the development of techniques for separating radioisotopically labeled oligonucleotides. The first complete RNA sequence was determined by R. W. Holley and his coworkers for alanyl-tRNA in 1965. Since that time the sequences of about a hundred tRNA molecules as well as the sequences of various rRNA and mRNA molecules have been determined. More recently, a method that parallels the Maxam and Gilbert method for sequencing DNA has been developed for RNA. We shall also discuss this method.

a. Cleavage with Specific Ribonucleases

The first step in determining the sequence of an RNA molecule is cleavage of the ribonucleic acid with specific ribonucleases. Among the most commonly used ribonucleases are pancreatic ribonuclease, which cleaves adjacent to pyrimidine residues, and takadiastase ribonuclease T1, which cleaves RNA adjacent to guanine (and inosine) residues. The large oligonucleotide fragments produced by digestion of alanyl-tRNA with these two enzymes are shown in Figure 9.47.

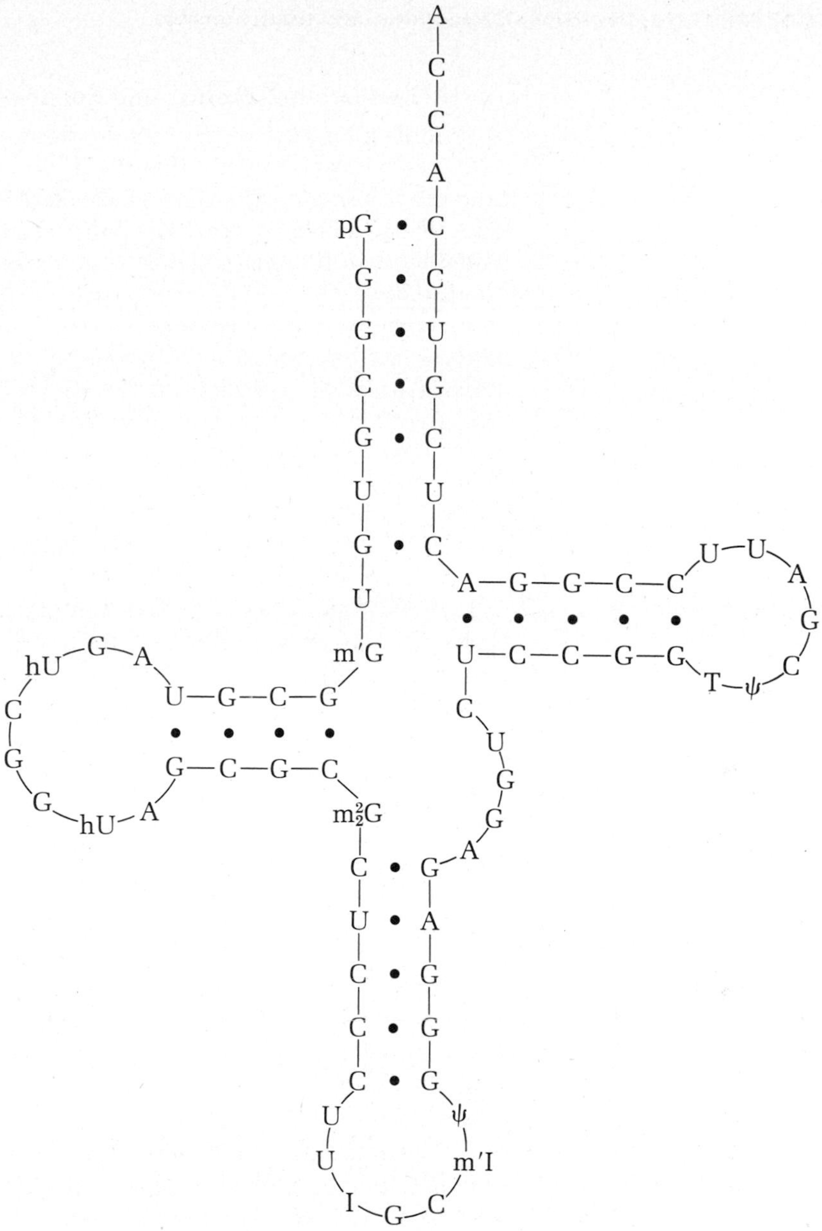

Structure of an Alanyl-tRNA molecule

Large Oligonucleotide Fragments

pG—G—G—C—G—U—G—U—m'G—G—   C—m$_2^3$G—C—U—C—C—C—U—U—Ip!   C—m'I—ψ—G—G—G—A—G—A—G—
  (a)   (b)   (c)

A—C—U—C—G—U—C—C—A—C—C$_{OH}$ —hU—C—G—G—hU—A—G—C—G—C—m$_2^3$G—C—U—C—C—C—U—U—Ip!
  (d)   (e)

C—m'I—ψ—G—G—G—A—G—A—G—U*—C—U—C—C—G—G—T—ψ—C—G—
                                    (f)

A—U—U—C—C—G—G—A—C—U—C—G—U—C—C—A—C—C$_{OH}$
                    (g)

U—A—G—hU—C—G—G—hU—A—G—C—G—C—m$_2^3$G—C—U—C—C—C—U—U—Ip!
                          (h)

C—m'I—ψ—G—G—G—A—G—A—G—U*—C—U—C—C—G—G—T—ψ—C—G—A—U—U—C—C—G—
                                      (i)

pG—G—G—C—G—U—G—U—m'G—G—C—G—C—G—U—A—G—hU—C—G—G—hU—(h)—A—G—C—G—C—m$_2^3$G—C—U—C—C—C—U—U—I—Gp!
                                              (j)

C—m'I—ψ—G—G—G—A—G—A—G—U*—C—U—C—C—G—G—T—ψ—C—G—A—U—U—C—C—G—G—A—C—U—C—G—U—C—C—A—C—C$_{OH}$
                                            (k)

Figure 9.47   Structure of ^{ala}tRNA from yeast and the sequences of the major oligonucleotides crucial to the proof of the primary structure. [From R. W. Holley, J. Apgar, G. A. Everett, J. T. Madison, M. Marquisse, S. H. Merrill, J. R. Panswick, A. Zamir, *Science*, 147, 1463 (1965). Copyright 1965 by the American Association for the Advancement of Science.]

b. End-Group Labeling and Separation of Oligonucleotides

Once a mixture of oligonucleotides has been obtained, the next step is radioisotopic labeling of either the 3′ or 5′ termini. The labeled oligonucleotides are digested with exonucleases from their unlabeled ends giving radioisotopically labeled fragments of different lengths. These fragments are separated by two-dimensional electrophoresis, or other suitable techniques, and identified.

i. Labeling the 3′ Termini The 3′ termini of oligonucleotides in a reaction mixture can be labeled with tritium in two steps. First, the vicinal 2′-3′ diol is oxidized with periodate to the dialdehyde. Second, the aldehyde is reduced with tritiated sodium borohydride, NaB^3H_4 (reaction 9.16).

Vicinal dialdehyde

Tritiated 3′ terminus

(9.16)

The radioisotopically labeled oligonucleotides are separated by two-dimensional paper electrophoresis, or other suitable technique, producing a set of spots that are visualized by their radioautograph. The oligonucleotides are identified by their mobility in the electrophoresis experiment. This task is greatly simplified by the existence of a large "library" of oligonucleotide sequences whose migration in the separation system is known. For large oligonucleotides further digestion with ribonucleases, isotopic labeling, and separation are required.

An alternative method of labeling the 3′ termini of oligonucleotide mixtures produced by nucleolytic digestion uses ^{32}P-labeled phosphate and *polynucleotide phosphorylase,* an enzyme that adds a phosphoryl group to the 3′-OH group of ribonucleotides. The labeled oligonucleotides are digested by spleen phosphodiesterase, an exonuclease that cleaves nucleotide residues from the 5′ terminus of the oligonucleotide. The products are all still labeled in their 3′ termini. Two-dimensional electrophoresis permits separation and identification of each oligonucleotide.

ii. Labeling the 5′ termini *Polynucleotide kinase* phosphorylates the free 5′-OH groups of oligonucleotides in the presence of γ-^{32}P-labeled ATP. The labeled oligonucleotides are separated by electrophoresis and partially hydrolyzed with snake venom phosphodiesterase. The snake venom enzyme removes residues sequentially from the 3′ end leaving fragments differing in length by one nucleotide residue. Separation of these products by electrophoresis permits identification of the oligonucleotides by differences in their electrophoretic mobility.

5′ A-C-U-G-G-OH 3′

γ-^{32}P-ATP +
Polynucleotide
kinase

^{32}PO$_4$H $\ominus$ +
Polynucleotide
phosphorylase

^{32}p-A-C-U-G-G-OH

A-C-U-G-G-^{32}p

Snake venom
phosphodiesterase

Spleen
phosphodiesterase

^{32}p-A-C-U-G-OH +

^{32}p-A-C-U-OH +

^{32}p-A-C-OH +

^{32}p-A-OH

C-U-G-G-^{32}p +

U-G-G-^{32}p +

G-G-^{32}p +

G-^{32}p

Separation and identification
by electrophoresis

Separation and identification
by electrophoresis

Figure 9.48 Schematic diagram for analysis of an oligonucleotide produced by nucleolytic cleavage of an RNA molecule. One end of the oligonucleotide is radioisotopically labeled. The labeled oligonucleotide is digested from the unlabeled end by an exonuclease producing a mixture of all possible oligonucleotide products. These fragments are then identified by their mobility in one- or two-dimensional electrophoresis.

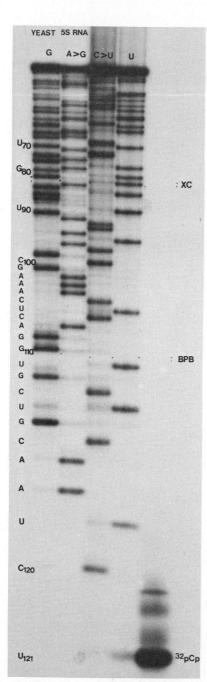

Figure 9.49 Autoradiograph of a polyacrylamide gel in which a 3′-end–labeled 5S RNA has had its sequence determined chemically. The sequence of the RNA is read directly from the gel. [Courtesy of Dr. Debra A. Peattie.]

The steps in end-grap labeling (either 3′ or 5′ terminus) and separation for a hypothetical oligonucleotide fragment are diagramed in Figure 9.48.

c. Direct Chemical Method for Sequencing RNA

A chemical method for determining RNA sequences has been developed that permits the determination of RNA sequences that contain up to 200 residues. The technique is rapid and requires only picomole amounts (10^{-12} moles) of RNA. This experimental technique parallels the Maxam and Gilbert method, but employs a different set of reagents to exploit the unique properties of RNA.

First, the RNA molecule is specifically labeled at its 3′ terminus with [5′-^{32}P]pCp. Second, separate aliquots of the labeled RNA are treated to modify the bases of RNA. These reactions yield specifically labeled uridine residues (U), specifically labeled guanosine residues (G), residues in which adenosine is labeled in preference to guanosine (A > G), and residues in which cytidine is labeled in preference to uridine (C > U). Each of the specifically modified samples is then cleaved by treatment with aniline. Strand scission occurs only at a chemically modified base. The products of the cleavage reactions are then simultaneously run in separate lanes of an electrophoresis gel. Electrophoresis resolves the fragments by length. An autoradiograph of the four lanes of the gel shows a series of dark bands, each of which corresponds to an oligonucleotide produced by cleavage at a specific base. The sequence can therefore be read directly from the autoradiograph of the gel (Figure 9.49).

9–10 CHEMICAL SYNTHESIS OF DNA

The availability of nucleic acid sequences was not long in stimulating organic chemists to attempt their synthesis. At first this was largely an intellectual and technical exercise. The rapid, and totally unexpected, development of techniques for producing artificially recombinant DNA, that is, of splicing a gene from one species into the DNA of another, suddenly brought techniques of nucleic acid synthesis into the mainstream of biological research. There are many applications of the chemical synthesis of DNA. They include shortening, lengthening, or changing naturally occurring genes, as well as synthesizing certain segments of DNA that seem for one reason or another to have special biochemical importance. The techniques of DNA synthesis include an enzymatic method in which polynucleotide phosphorylase is used with a short primer to build oligonucleotides and two chemical approaches known as the phosphodiester approach and the phosphotriester approach (Figure 9.50). Using chemical methods, H. G. Khorana and his coworkers were able to synthesize the structural gene for a tyrosine transfer RNA. This gene contains 77 base pairs. A later synthesis of this gene

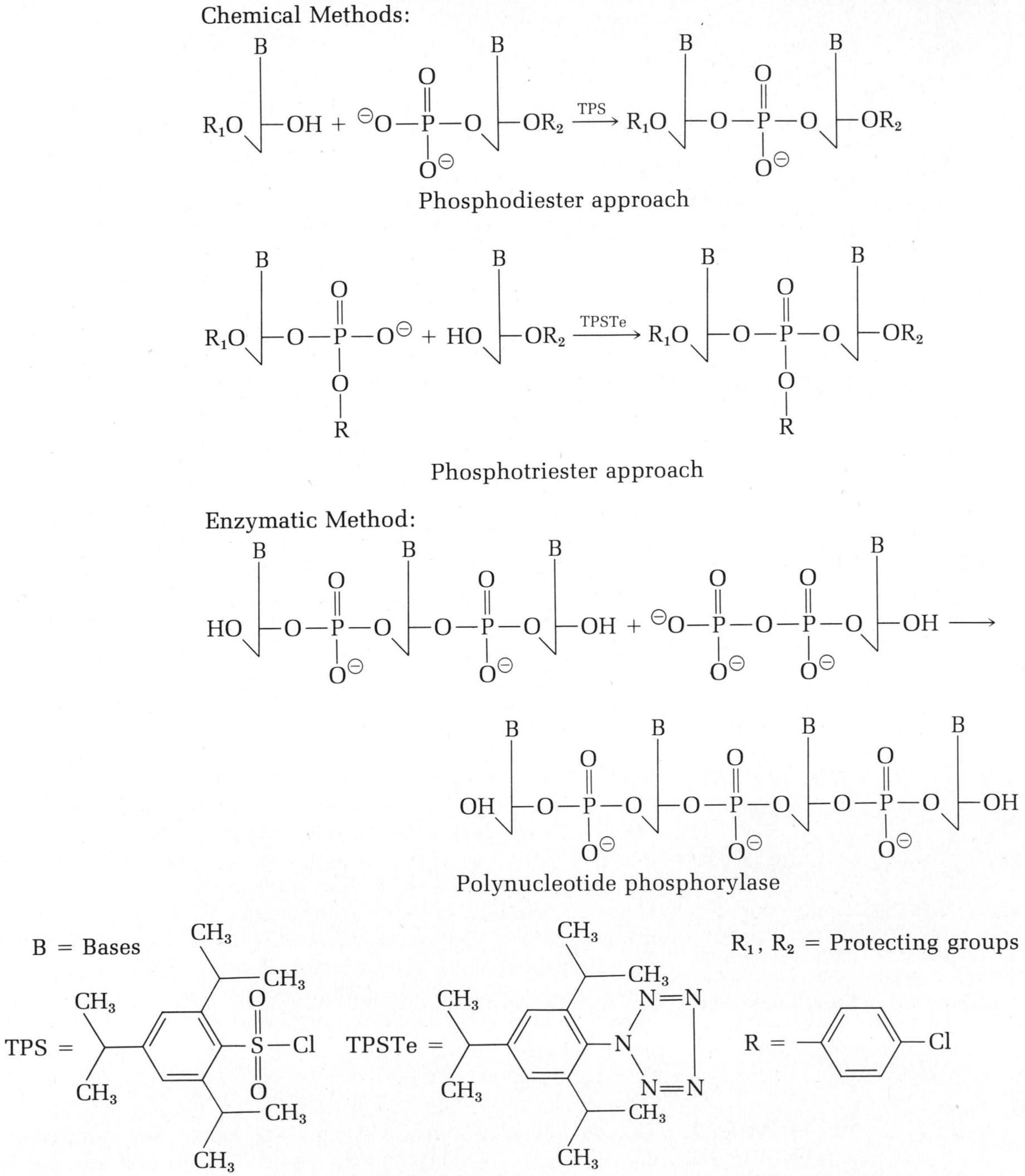

Chemical Methods:

Phosphodiester approach

Phosphotriester approach

Enzymatic Method:

Polynucleotide phosphorylase

B = Bases   TPS =   TPSTe = R_1, R_2 = Protecting groups   R =

Figure 9.50 Three methods for the synthesis of DNA. [From K. Itakura and D. A. Riggs, *Science, 209,* 1402 (1980). Copyright 1980 by the American Association for the Advancement of Science.]

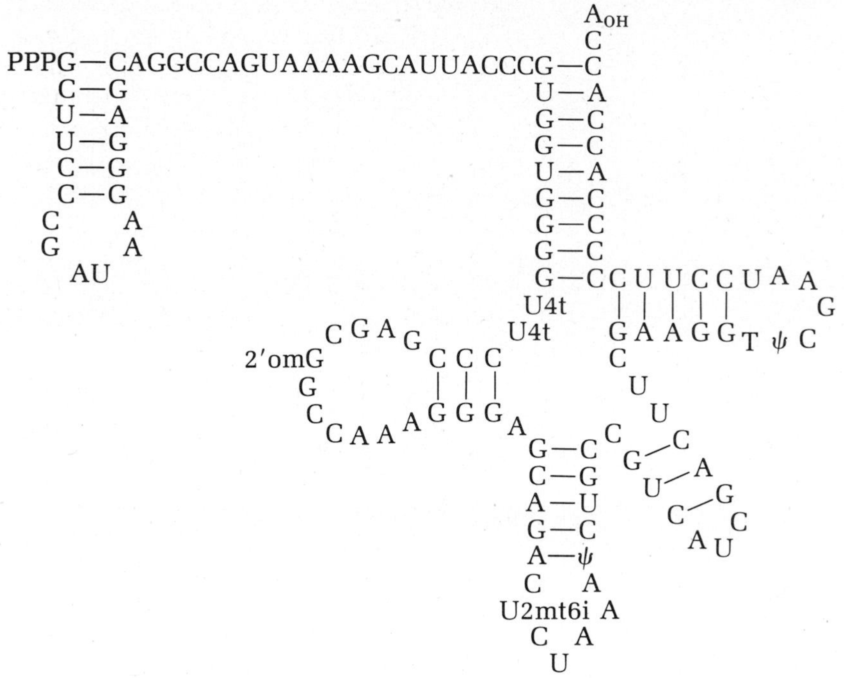

```
                                                A_OH
                                                C
PPPG—CAGGCCAGUAAAAGCAUUACCCG—C
      C—G                               U—A
      U—A                               G—C
      U—G                               G—C
      C—G                               U—A
      C—G                               G—C
    C     A                             G—C
    G     A                             G—C
      AU                                G—CCUUCCUA A
                                        U4t  | | | | |    G
                  C G A G               U4t      G A A G G T ψ C
        2'omG        C C C              C
           G         | | |              U
           C       G G G                U
            C A A A      A         G—C  C  C
                                  C—G G—  —C
                                  A—U   U—A
                                  G—C   C—  G
                                  A—ψ      A U
                                  C     A
                              U2mt6i A
                                  C     A
                                  U
```

Figure 9.51 Primary structure of the precursor of *E. coli* tyrosine tRNA. The modified bases of the completely processed tRNA are indicated. These modifications are not present in the precursor tRNA. Abbreviations: 2mt6iA, N^6-isopentyl-2-methylthioadenosine; 2'omG, 2'-methoxyguanosine; U4t, 4-thiouridine; ψ, pseudouridine. [From H. G. Khorana, *Science, 203,* 617 (1979). Copyright 1979 by the American Association for the Advancement of Science.]

includes adjacent regulatory regions of the DNA and spans 207 base pairs (Figure 9.51).

The chemical synthesis of DNA has three major stages. First, all potentially reactive functional groups are blocked. These include the amino groups of cytosine, adenine, and guanine and the 3'- and 5'-hydroxyl groups. Second, the condensing reaction is carried out (in high yield). Third, the blocking groups are removed under mild conditions that do not cleave the phosphodiester bonds of the oligonucleotide.

A. Protecting Groups

In the chemical synthesis of DNA protecting groups are used to block the 3'- and 5'-hydroxyl groups and the amino groups of adenine, cytosine, and guanine. Some of the most commonly employed blocking agents for amino groups are benzoyl chloride, anisoyl chloride, and isobutyryl chloride (reactions 9.17–9.19).

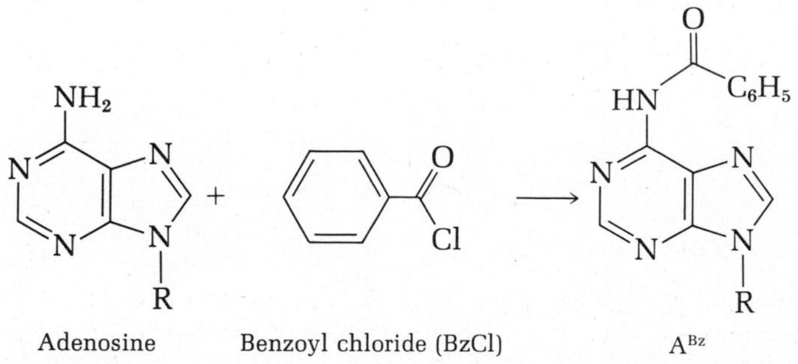

Adenosine   Benzoyl chloride (BzCl)   A^Bz

(9.17)

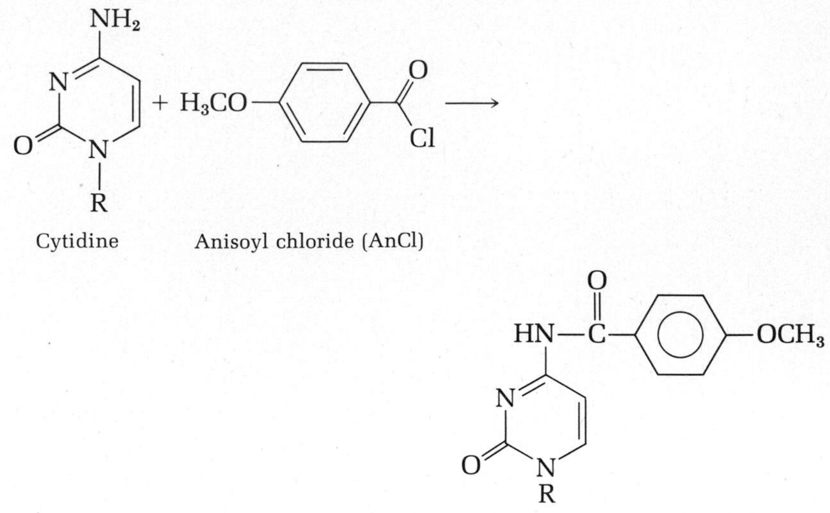

C^{An} (9.18)

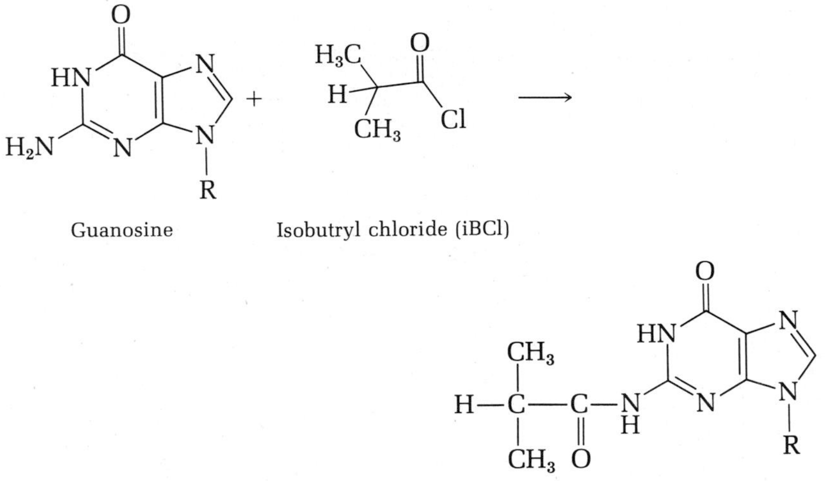

G^{iB} (9.19)

These acyl-protecting groups are removed under mild conditions by treatment with ammonia.

The 3′-hydroxy group can be protected by a variety of blocking agents, including *tert*-butyldiphenylsilyl, acetyl, and benzoyl groups added as their acid chlorides (reactions 9.20–9.22).

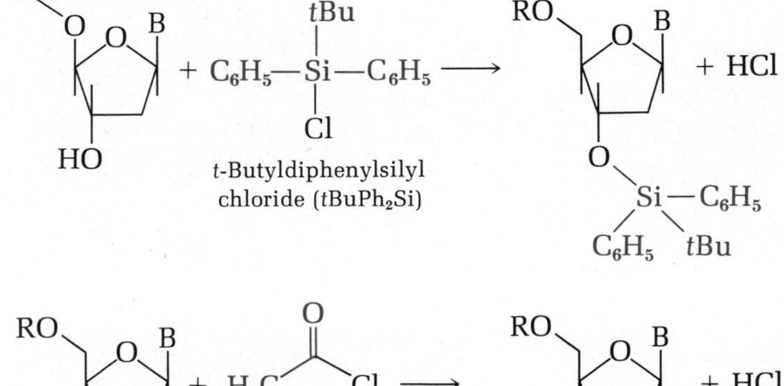

(9.20)

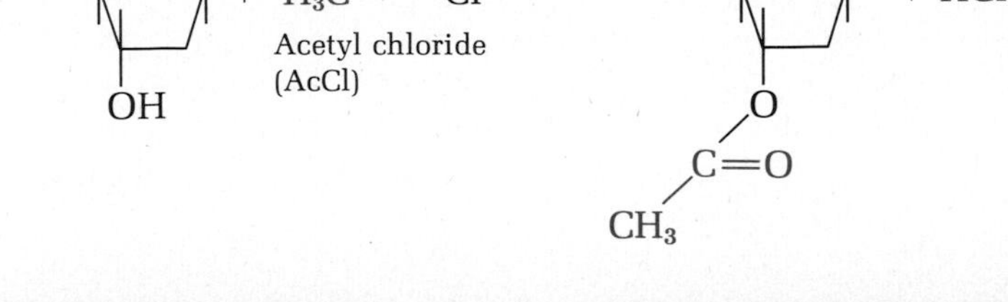

(9.21)

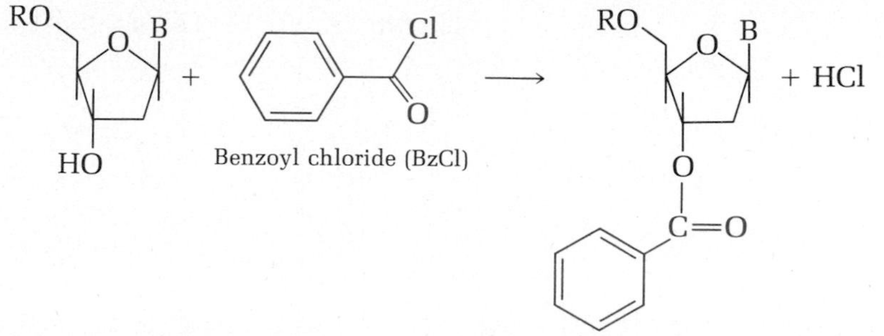

$$\text{(9.22)}$$

The silyl group can be selectively removed by treatment with fluoride, the benzoyl groups are removed by catalytic hydrogenation, and the acetyl groups are removed by hydrolysis in dilute acid.

Among the protecting groups for the 5′-hydroxyl group are the closely related monomethoxytrityl and dimethoxytrityl groups (reactions 9.23 and 9.24).

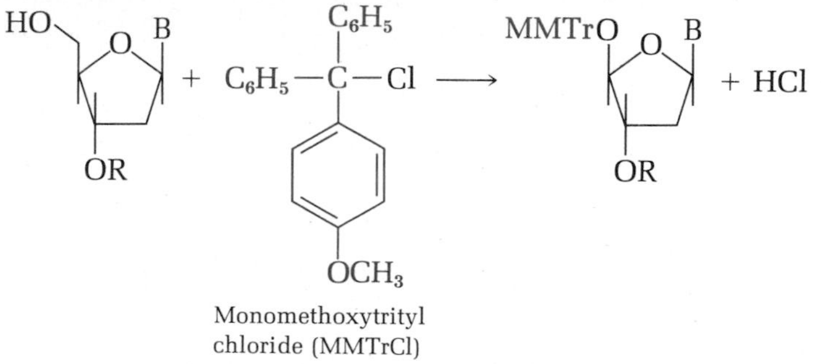

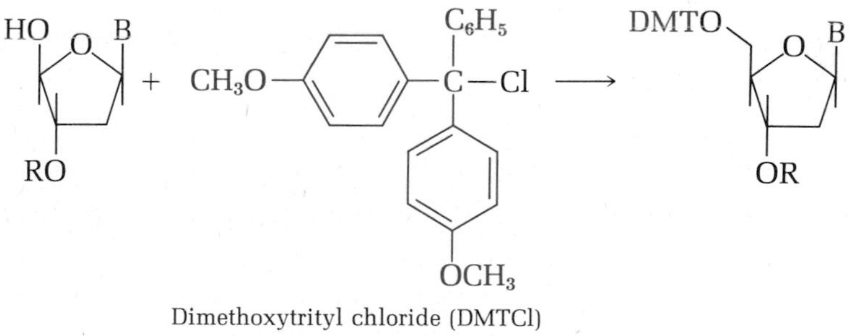

$$\text{(9.23)}$$

Monomethoxytrityl
chloride (MMTrCl)

$$\text{(9.24)}$$

Dimethoxytrityl chloride (DMTCl)

The trityl blocking groups are removed by treatment with dilute acid.

B. Condensing Reactions in Phosphodiester and Phosphotriester Approaches

In the *phosphodiester approach* to synthesis of polydeoxynucleotides the free 3′-hydroxyl group of one mononucleotide condenses with the free 5′-phosphate group of a second nucleotide, all reactive functional groups having been previously blocked. The condensing agents for this reaction are typically either *dicyclohexylcarbodiimide* (DCC) or *triisopropylbenzenesul-*

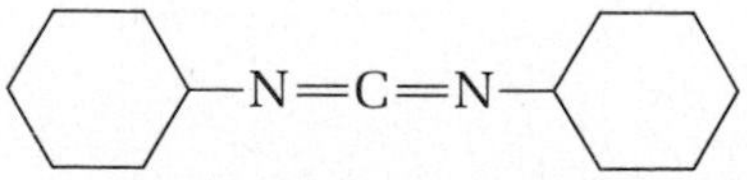

Dicyclohexylcarbodiimide

(DCC)

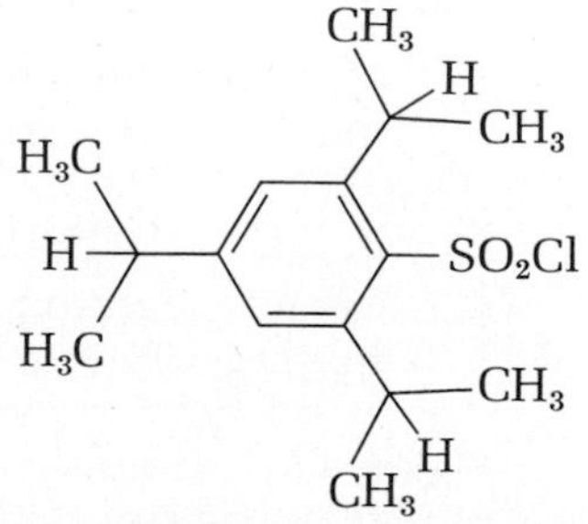

Triisopropylbenzenesulfonyl chloride (TPS)

fonyl chloride (TPS). The synthesis of a dinucleotide with TPS as the condensing agent is shown in reaction 9.25.

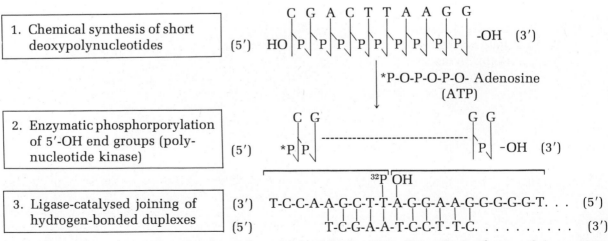

$$(9.25)$$

Continuation of the synthesis involves repeated condensation reactions with TPS and mononucleotides having a free 5′-phosphate group.

The synthesis of polydeoxynucleotides involves three steps. (1) A series of oligodeoxynucleotides is synthesized. (2) The deoxynucleotides are phosphorylated at their 5′ termini by polynucleotide kinase in the presence of γ-³²P-labeled ATP. The radioisotope makes it possible to follow the course of events in the third step. (3) *DNA ligase* catalyzes the formation of the phosphodiester bonds of duplex DNA. The size of the double-stranded product is readily determined by gel electrophoresis. The polydeoxynucleotides are detected by autoradiography (Figure 9.52).

Figure 9.52 Three-step strategy for the total synthesis of duplex DNA. [From H. G. Khorana, *Science*, 203, 617 (1979). Copyright 1979 by the American Association for the Advancement of Science.]

The *phosphotriester approach* is distinguished from the phosphodiester approach by three differences in the first step of the process, namely, in the synthesis of the polydeoxynucleotides. First, *triisopropylbenzenesul-*

Triisopropylbenzenesulfonyl tetrazolide (TPSTe)

fonyl tetrazolide (TPSTe) is used as the condensing agent. Second, the phosphate group is blocked by either a cyanoethyl group or by a *parachlorophenyl* group. Third, the genetic information in DNA that is translated to protein is encoded in triplets of three nucleotides, each triplet corresponding to one amino acid in the protein to be synthesized. In the phosphotriester approach these triplet *codons*, a total of 64 in all, are individually synthesized. All have been synthesized in yields of over 50% by the triester method. The synthesis of one such codon is indicated in reaction 9.26.

Once these codons have been synthesized, they can be assembled to suit the purpose of the experimenter. The general procedure is summarized in Figure 9.53. The gene coding for the mammalian hormone *somatostatin* was synthesized by the triester method (Figure 9.54). It was inserted into an *E. coli* DNA molecule, and then the *E. coli* proceed to manufacture the mammalian polypeptide.

C. Automated Chemical Synthesis

The methods described above for the chemical synthesis of DNA are employed in a modified form in an automated technique of solid phase synthesis. The principle advantage of this technique is that the isolation and purification of synthetic intermediates are eliminated, so that the product is purified only once after the entire oligomer has been assembled.

The technique is outlined in Figure 9.55. The first monomer (A) is attached to an insoluble solid support. Silica beads, which are easy to modify chemically and whose volume does not change when solvents are changed, are currently the preferred solid support. The second monomer (B) is added along with all necessary coupling agents and mixed in a solvent with the solid support. Once A has been attached to B, the unreacted reagents are washed away. A large excess of solvent can therefore be used, and the critical condensation reaction can be driven to nearly quantitative yields. If the yield can be increased to 99%, then a 90% yield of product will be obtained in the synthesis of a decamer. (Should the yield of the condensation reaction drop to 90%, which in many synthetic procedures would be very good indeed, the yield of product after 10 steps would be only 35%.) After the oligomeric DNA has been synthesized, the protecting groups are removed, and the oligomer is released from the solid support. The product is then purified by some combination of chromatographic techniques. A schematic diagram of a DNA (or RNA) synthesizer designed by K. K. Ogilvie is shown in Figure 9.56. In essence, this machine consists of a set of pneumatic valves connected to reagent bottles that are controlled by a set of solenoids. A microprocessor directs the opening or closing of the valves to add reagents according to a preprogrammed set of synthesis instructions. Using this device

(9.26)

(69%)

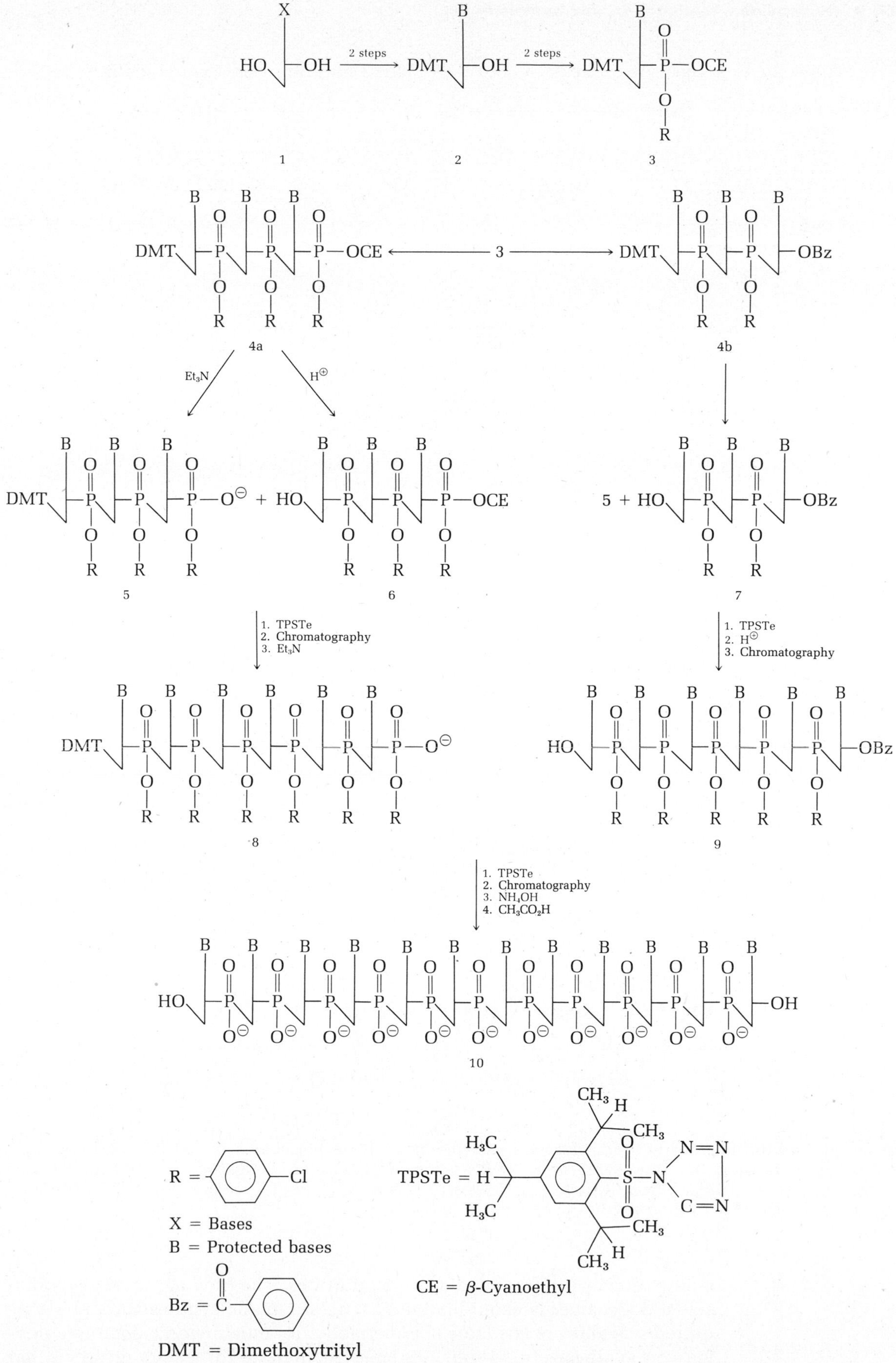

Figure 9.53 General procedure for synthesis of polydeoxynucleotides by the triester method. Triplet codons are successively combined to yield the desired product. [From K. Itakura and A. D. Riggs, *Science*, 209, 1403 (1980). Copyright 1980 by the American Association for the Advancement of Science.]

(a)

Ala Gly Cys Lys   Asn Phe Phe Trp Lys Thr Phe Thr   Ser Cys Stop Stop

EcoRI (A)———→ ←———(B)————————→ ←———(C)————→ ←———(D)———→ Bam I

5′AATTC ATG GCT GGT TGT AAG AAC TTC TTT TGG AAG ACT TTC ACT TCG TGT TGA TAG
　 GTAC CGA CCA ACA TTC TTG AAG AAA ACC TTC TGA AAG TGA AGC ACA ACT ATC CTAG 5′

←————(E)————————→ ←————(F)————→ ←————(G)————→ ←————(H)———→

(b)

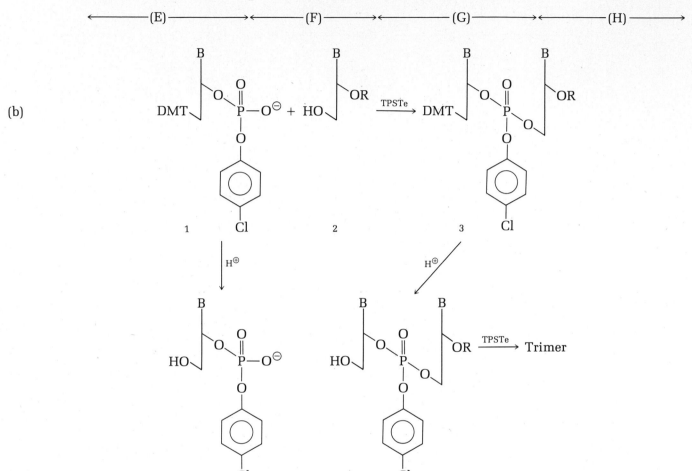

B = Protected Base

DMT = 4,4′-dimethoxytrityl

Figure 9.54 The structure of the chemically synthesized gene for the mammalian hormone somatostatin. (a) Eight oligonucleotides, labeled (A) through (H), were synthesized by the triester method and condensed. (b) The triester method applied to one nucleotide triplet. [From K. Itakura, T. Hirose, R. Crea, A. D. Riggs, *Science*, *198*, 1057 (1977). Copyright 1977 by the American Association for the Advancement of Science.]

a new residue can be added every 30 minutes. The time required to synthesize a dodecamer is about 5.5 hours. The automatic synthesizer and all required reagents are available commercially. Astonishingly, a complete novice can synthesize a 14-unit oligomer in high yield, a tour de force that would have amazed synthetic chemists in the not distant past.

Figure 9.55 Automated chemical synthesis of DNA. (a) Schematic diagram of the major steps of solid phase synthesis of gene fragments. (b) Chemical details of the preparation of the solid support and the attachment of the first and second residues. (c) Time required for each step in the automated process. [Courtesy of Dr. K. K. Ogilvie.]

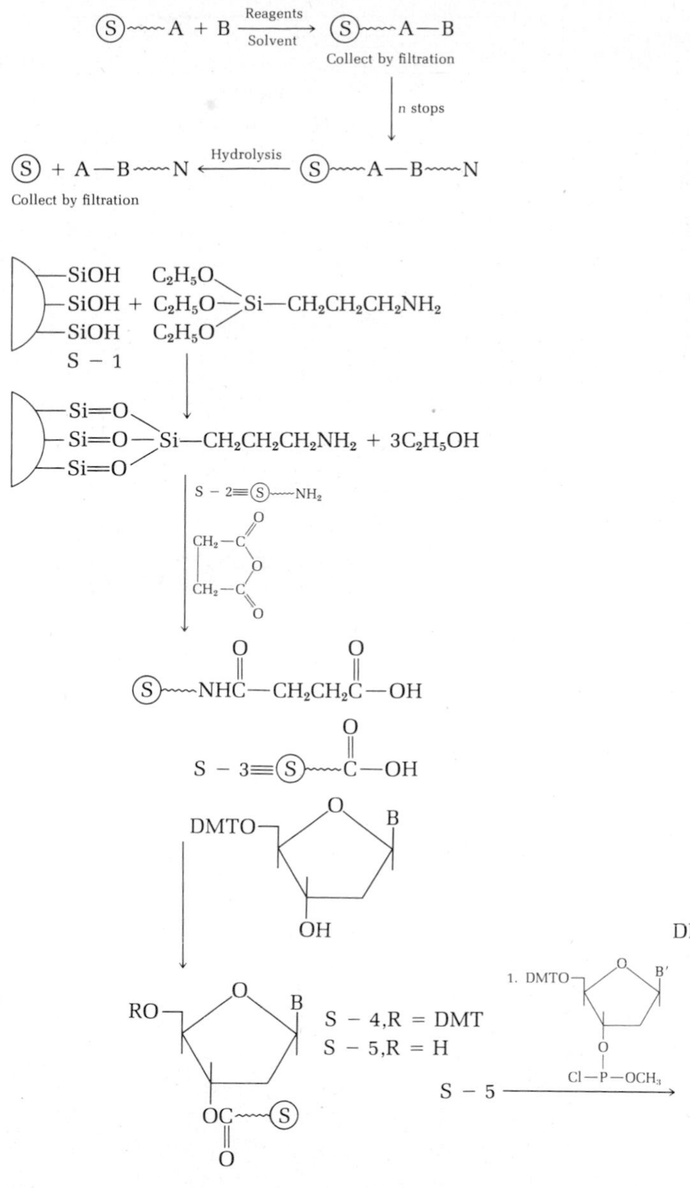

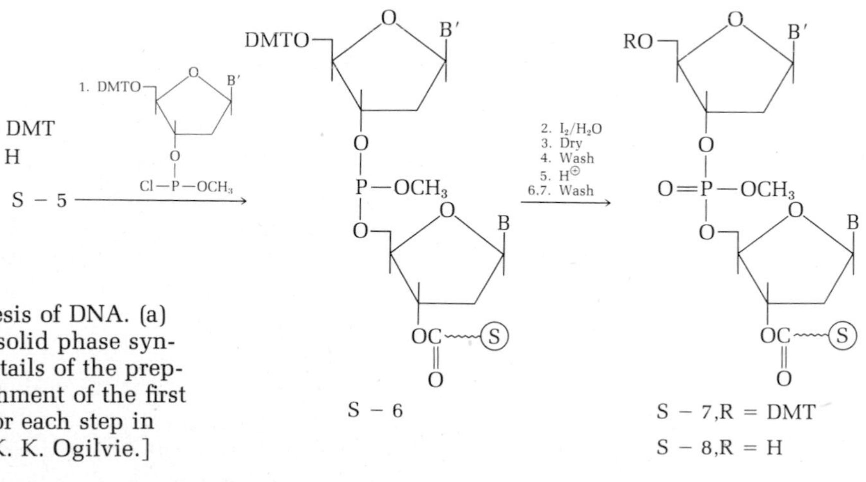

Automated chain extension cycle		
Step	**Process**	**Time (min)**
1	Condensation	7
2	Oxidation	3
3	Drying	4
4	Washing (CHCl$_3$)	3
5	Detritylation	3
6	Washing (CGCl$_3$)	3
7	Washing (pyridine)	7
		Total 30 min
Time to prepare dodecamer, 5.5 hr		

9-11 SUMMARY Nucleosides, nucleotides, and nucleic acids contain two classes of heterocyclic bases: the pyrimidines and the purines. The most abundant pyrimidines are uracil, found in RNA, thymine, found in DNA, and cytosine, found in both DNA and RNA. The most abundant purines are adenine and guanine. Nucleosides are β-N-glycosides formed between either a purine or a pyrimidine and a carbohydrate. In RNA the carbohydrate is ribose, and in DNA the carbohydrate is 2'-deoxyribose. The plane of the base is at approximately right angles to the plane of the sugar, and the orientation of the base to the sugar is *anti*. Nucleotides are sugar-phosphate esters of nucleosides. A phosphate group is most commonly found at either the 3' or the 5' position of the sugar. Nucleoside 5'-diphosphates and nucleoside 5'-triphosphates are

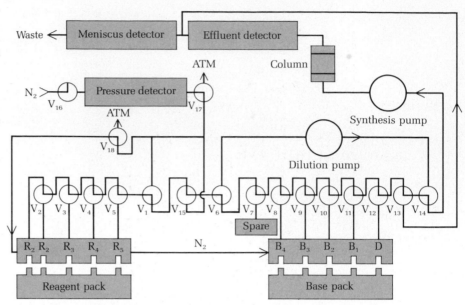

Figure 9.56   Schematic diagram of the DNA (RNA) synthesizer that can be used in the automated synthesis of gene fragments. Abbreviations: ATM, vent to atmosphere; B_1 to B_4, solutions of the four protected nucleoside phosphates; D, denitrifying solution; R_2, pyridine; R_3, chloroform; R_4, drying reagent; R_5, oxidizing agent; V_1 to V_{18}, valves. [Courtesy of Dr. K. K. Ogilvie.]

phosphoanhydride derivatives of nucleotides. Nucleotides are the components from which nucleic acids are constructed, they participate in many metabolic reactions, and adenine nucleotides are components of certain coenzymes.

DNA is a double-stranded linear polymer of 2′-deoxynucleotide residues linked by phosphodiester bonds between the 3′ and 5′ positions of the 2′-deoxyribosyl moieties. The most common base in DNA are adenine, thymine, guanine, and thymine. The adenine/thymine and guanine/cytosine ratios are equal to 1.0 in DNA molecules isolated from all sources. In the double stranded, native structure of DNA adenine is always paired with thymine and guanine is always paired with cytosine. The primary structure of one strand of the double helix automatically determines the primary structure of the complementary strand. The two strands of the DNA are antiparallel. Each turn of the double helix of DNA advances by 10 nucleotide residues. Van der Waals forces, which are responsible for the hydrophobic effect, are the major forces that stabilize the double helix. Hydrogen bonds between base pairs also stabilize the double helix.

Coiling of a DNA molecule around its own axis produces a supercoiled structure, the tertiary structure of the nucleic acid. All DNA isolated from natural sources is underwound, that is, it is negatively supercoiled. Dyes such as ethidium, proflavin, and acridine intercalate in the double helix and are important reagents in the study of supercoiled DNA. DNA topoisomerases relax the supercoiled structure of DNA. DNA gyrase performs the opposite operation by introducing supercoils in DNA.

RNA is constructed from ribonucleotides linked by 3′-5′ phosphodiester bonds. The three major classes of RNA are ribosomal RNA, messenger RNA, and transfer RNA. RNA molecules are single stranded, but they can fold back upon themselves to give base-paired regions. Transfer RNA molecules fold into an L-shaped conformation.

DNA can be denatured by heating or by the addition of various chemical reagents. The total absorbance of a sample of DNA increases by about 37% when the double helix uncoils. The temperature at which the double helix is one-half unwound is the melting temperature, T_m. The melting temperature parallels the base composition of the nucleic acid: the higher the melting temperature, the greater the mole percent of (G + C) in the sample.

Nucleic acids can be hydrolyzed in dilute acid to give various degradation products. At pH 1.6 and 37% hydrolysis preferentially liberates purines, and produces an apurinic acid. DNA resists hydrolysis in base, but RNA is readily hydrolyzed in base by intramolecular base catalysis.

Nucleases are enzymes that hydrolyze DNA molecules. Endonucleases cleave the chain internally, and exonucleases cleave nucleic acids sequentially from either the 3′ or 5′ termini. Pancreatic ribonuclease cleaves the phosphodiester backbone of RNA between purine and pyrimidine residues. Highly specific restriction endonucleases cleave DNA only at those points where a specified base sequence is found. The sites recognized by restriction endonucleases typically are palindromes. Cleavage is symmetrical to a two-fold axis of symmetry.

The primary structure of DNA molecules can be determined either by the chemical methods introduced by Maxam and Gilbert or by the enzymatic method introduced by Sanger. Sequences in excess of 5,000 residues have been determined by these methods. A chemical method for the determination of the primary structure of RNA has also been introduced. In its basic strategy it parallels the method of Maxam and Gilbert.

Oligonucleotides can be chemically synthesized by automated chemical methods. These methods have led to the chemical synthesis of complete genes and transfer RNA molecules.

REFERENCES

Altman, S., Ed., "Transfer RNA," MIT Press, Cambridge, Mass., 1978.

Arber, W., Restriction Endonucleases, *Angew. Chem. Internat. Edit., 17,* 73–140 (1978).

Davidson, J. N., "The Biochemistry of the Nucleic Acids," Academic Press, New York, 1977.

Freifelder, D., Ed., "The DNA Molecule: Structure and Properties," Freeman, San Francisco, 1978.

Itakura, I., and Riggs, A. D., Chemical DNA Synthesis and Recombinant DNA Studies, *Science, 209,* 1401–1405 (1980).

Khorana, H. G., Total Synthesis of a Gene, *Science, 203,* 614 (1979).

Kim, S. H., Three-Dimensional Structure of Transfer RNA, *Progr. Nucl. Acid Res. Mol. Biol., 17,* 181–216 (1976).

Kim, S. H., Quigley, G. J., Suddath, F. L., McPherson, A., Snedon, D., Kim, J. J. P., Weinzierl, J., and Rich, A., Three-Dimensional Structure of Yeast Phenylalanine Transfer RNA: Folding of the Polynucleotide Chain, *Science, 179,* 285–288 (1973).

Peattie, D. A., Direct Chemical Method for Sequencing RNA, *Proc. Nat. Acad. Sci. U.S., 76,* 1760 (1979).

Rich, A., and RajBhandary, U. L., Transfer RNA: Molecular Structure, Sequence, and Properties, *Ann. Rev. Biochem., 45,* 805–860 (1976).

Saenger, W., Structure and Function of Nucleosides and Nucleotides, *Angew. Chem. Internat. Edit., 12,* 591–682 (1973).

Wang, J. C., The Degree of Unwinding of the DNA Helix by Ethidium, *J. Mol. Biol., 89,* 783–801 (1974).

Watson, J. D., and Crick, F. H. C., Genetic Implications of the Structure of Deoxyribonucleic Acid, *Nature, 171,* 964–967 (1953).

Watson, J. D., and Crick, F. H. C., Molecular Structure of Nucleic Acid. A Structure for Deoxyribose Nucleic Acid, *Nature, 171,* 737–738 (1953).

Problems

1. Polyadenylic acid forms a single-stranded helix. What forces might be expected to stabilize such a helix?

2. DNA that is rich in G/C base pairs is denser than DNA that is rich in A/T base pairs. Explain this observation on the basis of DNA structure.

3. How might the lactim form of thymine pair with (a) adenine and (b) guanine?

4. What features of the structure of DNA cause a helix-to-coil transition when the pH is changed (a) from 7.0 to 1.0 and (b) from 7.0 to 13.0?

5. Why is the base sequence of one chain in double-stranded helical

DNA sufficient to completely determine the base sequence in the other chain?

6. Why does the melting point of DNA increase with increasing G/C content?

7. Should a DNA that is rich in G/C base pairs have a larger or smaller partial specific volume than one that is rich in A/T base pairs? How will the difference in partial specific volume affect the sedimentation coefficient for the DNA under consideration?

8. Which do you expect to have the larger frictional coefficient: a globular protein or a DNA fragment having the same molecular weight?

9. Using either the phosphodiester or the phosphotriester synthetic methods, illustrate how you would synthesize the amino acid triplet codon pCpApT.

10. When two completely separated, complementary DNA strands are mixed, a solution having high ionic strength is required for renaturation. Explain this observation.

11. Consider the pentadeoxynucleotide A-G-G-C-T. Write the products for $A > G, G > A, C = T,$ and C reactions of the Maxam and Gilbert method of determining the primary structure of DNA. Show how the results of these reactions yield the sequence.

12. Consider the same nucleotide as in problem 11. Show how you would determine the sequence by the plus and minus method.

13. Consider the oligoribonucleotide UUCUCGAGCCA. What oligonucleotide products would be obtained by digestion with each of the following: (a) pancreatic ribonuclease, (b) ribonuclease T1, (c) ribonuclease T2, (d) ribonuclease U1.

14. Propose mechanisms for the TPS- and TPSTe-catalyzed condensations of an alcohol and a phosphate to give a dinucleotide.

Chapter 10

Coenzymes

10–1 INTRODUCTION

In Chapter 6 we initiated our discussion of enzyme catalysis by focusing on the nature of enzyme active sites and the chemistry of the amino acid residues that participate in catalysis. In this chapter we pick up the tale again, now focusing on the biochemical properties of a small group of low molecular weight molecules that are required for the function of all cells. These molecules, the *coenzymes*, are nonprotein reactants that lend their services to certain enzymes in catalysis. Coenzymes require an enzyme to carry out their reactions at an appreciable rate, although some can catalyze reactions by themselves at slow rates. It is possible that a primordial sea, rich in potential life, contained both coenzymes and polypeptides and that coenzyme binding to polypeptides marked one of the earliest steps in the evolution of enzymes.

We remarked in Section 6.2E that certain polar side chains of amino acid residues are nucleophiles. No amino acid residues bear *electrophilic* side chains, unless we count the protons donated by acidic residues. One of the primary functions of certain coenzymes—pyridoxal phosphate, thiamine pyrophosphate, and tetrahydrofolate—is to provide electrophilic centers in enzyme-catalyzed reactions. Each of these coenzymes contains a quaternary nitrogen whose positive charge is an electrophilic "electron sink." The ability of these coenzymes to stabilize carbanions is central to their chemical reactivity, and hence their biological functions.

The other coenzymes that are discussed in this chapter participate in group transfer reactions, in molecular rearrangements, and in oxidation-reduction reactions.

Three related terms are often met in biochemistry: coenzyme, cofactor, and prosthetic group. The differences between them can be razor thin, but the following conventions can be applied to distinguish them in many cases. A *coenzyme* is a true substrate for an enzyme-catalyzed reaction, but it is recycled in a later step of a metabolic pathway by another enzyme and, therefore, can be used repeatedly. The term coenzyme is sometimes restricted to those molecules that contain a nucleotide residue as part of their structure. A *cofactor* is often taken to mean any nonprotein molecule or ion that is essential for an enzyme-catalyzed reaction. Metal ions, for example, are cofactors by this definition. A *prosthetic group* is similar to a cofactor, but is more tightly bound to the enzyme than either a cofactor or a coenzyme. A prosthetic group is regenerated on the enzyme to which it is bound as part of the enzyme's function. For example, the heme groups of the cytochromes are reversibly oxidized as part of the action of the proteins to which they are bound.

10–2 COENZYMES AND VITAMINS

Vitamins are trace nutrients required in the diet. Not all vitamins are required by all organisms. Table 10.1 lists those required in the human diet, their major nutritional sources and the clinical conditions that result from their deficiency. The vitamins are classified by their solubility into two groups: fat (lipid) soluble and water soluble. The water-soluble vitamins are components of many coenzymes. Table 10.2 lists the coenzymes derived from them and their chemical function. (The fat-soluble vitamins are considered in the next chapter.)

10–3 COENZYMES THAT CONTAIN ADENINE NUCLEOTIDES

Coenzyme A, the *pyridine* dinucleotides, and the *flavin* dinucleotides all contain an adenine nucleotide as one of their components (Figure 10.1). Although it does not participate directly in the reactions of these coenzymes,

Table 10.1 Water-soluble vitamins and coenzymes required by humans		
Vitamin	**Coenzyme**	**Chemical reaction catalyzed**
B_1, thiamine	Thiamin pyrophosphate	α-Cleavage and α-condensation reactions (making and breaking bonds to carbonyl carbon)
B_2, riboflavin	Flavin adenine dinucleotide Flavin mononucleotide	Oxidation-reduction reactions, hydride anion transfer and one-electron transfer reactions
B_3, nicotinic acid	Nicotinamide adenine dinucleotide	Oxidation-reduction reactions, hydride transfer
Pantothenic acid	Coenzyme A	Acyl transfer in Claisen condensations and related reactions
Biotin	Biocytin	Transcarboxylation (carboxyl transfer)
B_6, pyridoxine	Pyridoxal phosphate	Amino group transfer
Folic acid	Tetrahydrofolic acid	One-carbon metabolism
B_{12}	Coenzyme B_{12}	Molecular rearrangements
C, ascorbic acid	—	Hydroxylation

Table 10.2   Major nutritional sources of vitamins required by humans and some physiological effects of deficiencies

Vitamin	Source	Effects of deficiency
Water soluble		
B_1, thiamine	Brain, liver, kidney, heart, whole grains	Beri-beri, neuritis, heart failure, mental disturbance
B_2, riboflavin	Milk, eggs, liver, whole grains	Photophobia, fissuring of skin
B_3, nicotinic acid	Whole grains, liver	Pellagra, skin lesions, digestive problems
Pantothenic acid	Most foods	Neuromotor problems, cardiovascular disorders
Biotin	Egg white, intestinal bacteria	Scaly dermititis, muscle pains, weakness
B_6, pyridoxine	Whole grains, liver, kidney, fish	Dermatitis, nervous disorders
Folic acid	Liver, leafy vegetables	Anemia
B_{12}, cyanocobalamin	Liver, kidney, brain	Pernicious anemia
C, ascorbic acid	Citrus fruits, green leafy vegetables, tomatoes	Scurvy, failure to form connective tissue fibers
Fat soluble		
A, carotene	Egg yolk, green or yellow vegetables, fruits, liver	Night blindness
D_3, calciferol	Dairy products, action of sunlight on skin	Rickets (poor bone formation)
E, tocopherol	Green leafy vegetables	Fragile red blood cells
K, naphthoquinone	Leafy vegetables, intestinal bacteria	Failure of blood clotting

SOURCE: H. Curtis, "Biology," 2nd ed., Worth, New York, 1975, p. 646.

the adenine nucleotide is important because it helps the appropriate enzyme to "recognize" its coenzyme. Since the adenine dinucleotide group has many sites for hydrogen bonding, ionic bonds, and hydrophobic interactions with the enzyme, it contributes to the stability of the enzyme-coenzyme complex.

A. Coenzyme A   Coenzyme A (CoA) may be divided into three parts: adenosine 3'-phosphate-5'-diphosphate, the vitamin pantothenic acid, and β-mercaptoethanolamine (Figure 10.2). Coenzyme A serves as an *acyl carrier* and promotes the condensation of acylthioesters. In its acyl carrier function the thiol group of CoA is bonded to the acyl group in a thioester. The chemistry of acylCoA and of other thioesters often involves abstraction of the hy-

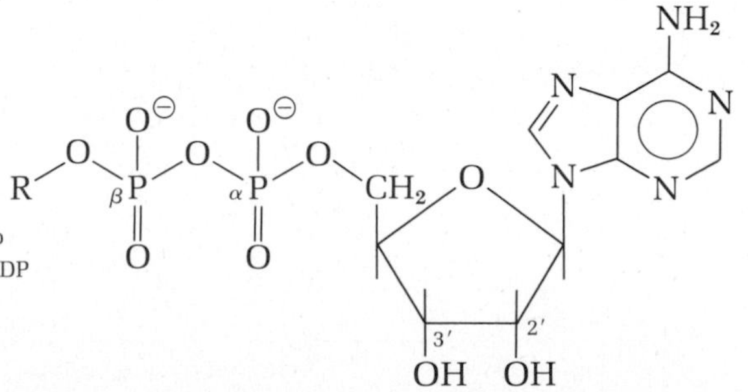

Figure 10.1   Adenosine diphosphate moiety of coenzyme A, the pyridine dinucleotides, and the flavin nucleotides. In coenzyme A there is another phosphate group at the 3' position. In nicotinamide adenine dinucleotide phosphate (NADP) there is another *phosphate group at the 2' position.*

drogen α to the carbonyl group to give the α-carbanion of the thioester (Figure 10.3). A basic group at the active site of the enzyme is capable of removing this proton. The α-hydrogens of thioesters are more acidic than those of corresponding carboxylate esters. The thioester is *less* resonance stabilized than the carboxylate ester. Compare the *relative* stabilities of contributing structures 1 and 2 with the relative stabilities of contributing structures 3 and 4 (Figure 10.4). Sulfur is a row three element and overlap in 2 is much poorer than overlap in 4. The α-carbanion of the thioester is *more* resonance stabilized than the α-carbanion of the carboxylate ester because there is little *opposing resonance* in 5, and considerable opposing resonance in 6 (Figure 10.5). The enhanced acidity of thioesters is illustrated by S-acetoacetyl-N-acetylthioethanolamine and ethylacetoacetate which have pK_a values of 8.5 and 10.5, respectively (Figure 10.6).

Nucleophilic displacement at the carbonyl carbon in thioesters is favored over nucleophilic displacement in carboxylate esters because the thioalkoxide anion, $RS^{\ominus}$, is a better leaving group than alkoxide anion, $RO^{\ominus}$ (Figure 10.7). The biological importance of this chemistry will be discussed in greater detail in our discussions of the critic acid cycle (Chapter 15) and fatty acid metabolism (Chapter 19).

B. Nicotinamide Coenzymes

The two nicotinamide coenzymes—nicotinamide adenine dinucleotide, NAD, and nicotinamide adenine dinucleotide phosphate, NADP (Figure 10.8)—participate in a wide variety of biological oxidation-reduction reac-

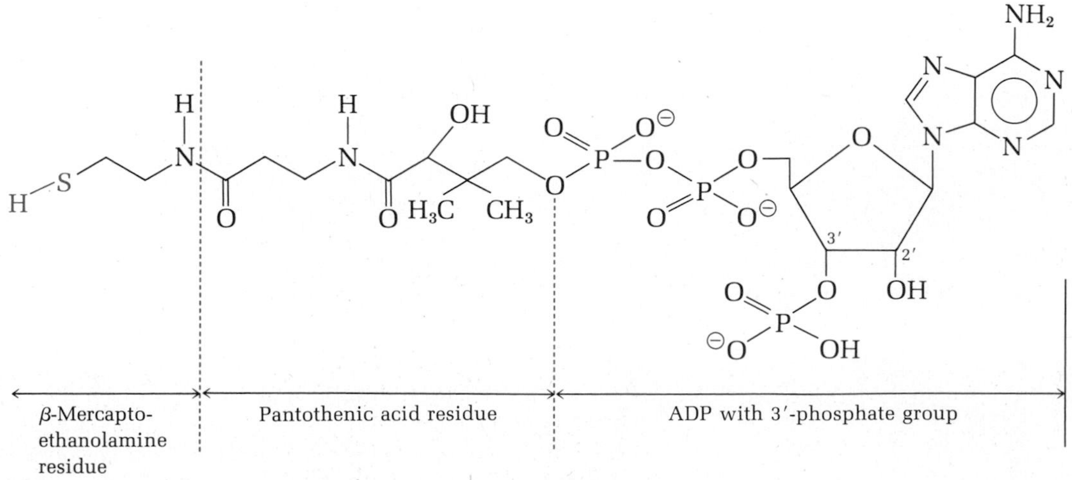

Figure 10.2 Structures of coenzyme A and its components, including the vitamin pantothenic acid.

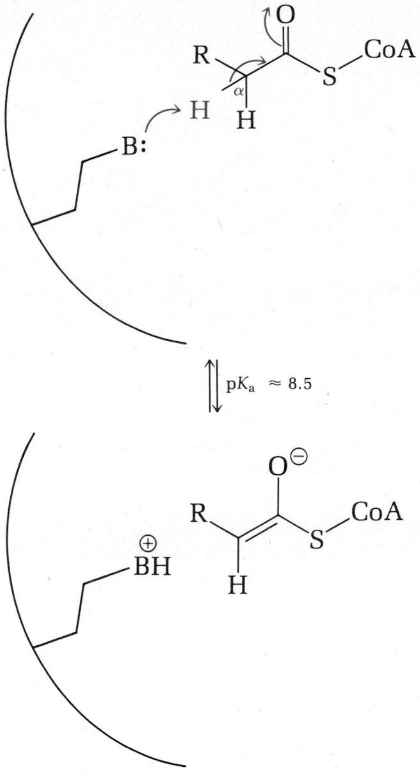

Figure 10.3 Abstraction of the α-proton of acyl coenzyme A by a basic group of a coenzyme A–dependent enzyme.

tions. The reactive portion of both NAD and NADP is the 4 position of the pyridine ring which accepts (or donates) a hydride anion. The positively charged nitrogen of the pyridine ring acts as an electron sink that aids in the transfer of a hydrogen with its bonding electron pair (hydride anion) from the substrate (AH) to NAD or NADP (reaction 10.1).

Reactions involving oxidation of NADH (or the reduction of NAD) are easily followed spectrophotometrically since the UV spectra of the two molecules are distinct (Figure 10.9).

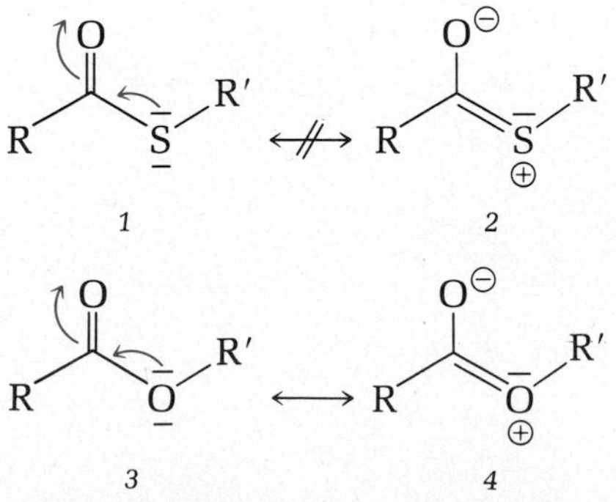

Figure 10.4 Relative stabilities of contributing structures. Thiolesters are less resonance stabilized than esters because structure *2* contributes much less to the resonance hybrid of thiolesters than *4* contributes to the resonance hybrid of esters.

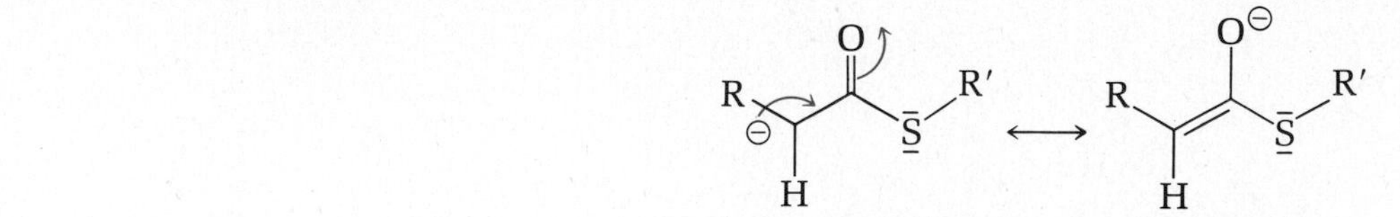

Figure 10.5 Relative stabilities of contributing structures. Structure 5a contributes more to the resonance hybrid of an enolate anion of a thioester than structure 6a contributes to the resonance hybrid of a carboxylate ester since there is no competing resonance in 5, while there is considerable competing resonance in 6.

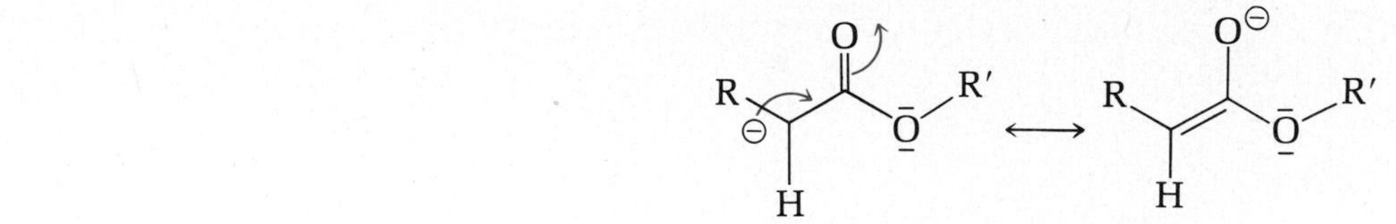

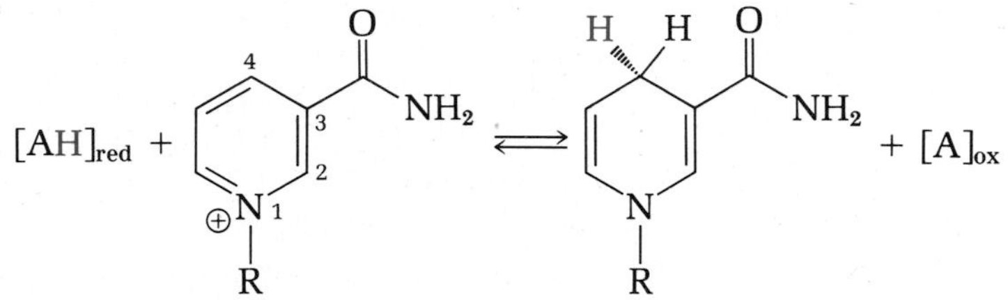

$$[AH]_{red} + \text{(Oxidized coenzyme)} \rightleftharpoons \text{(Reduced coenzyme)} + [A]_{ox} \qquad (10.1)$$

Oxidized coenzyme (NAD or NADP)

Reduced coenzyme (NADH OR NADPH)

Although NAD and NADP differ only by a 2'-phosphate group in the adenine nucleotide, far from the reactive center of either molecule, the metabolic pathways in which they participate are markedly different. NAD is predominantly involved in catabolic, or degradative processes, whereas NADP is predominantly involved in anabolic, or biosynthetic reactions. The presence or absence of the 2'-phosphate group alters the ability of the two coen-

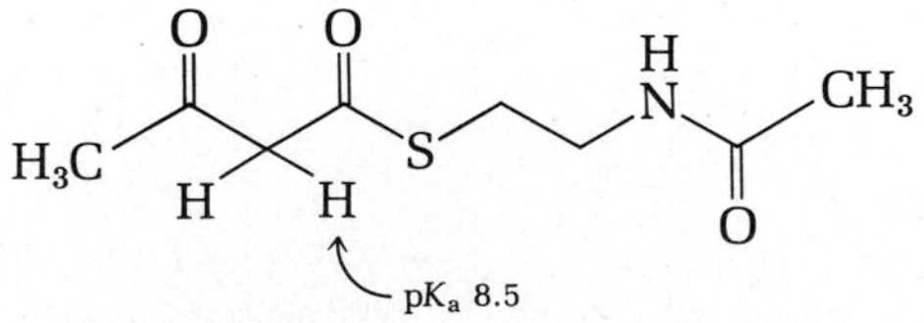

S-Acetoacetyl-N-acetylthioethanolamine

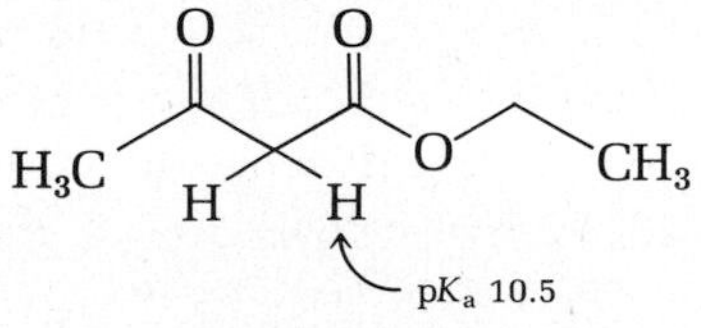

Ethylacetoacetate

Figure 10.6 pK_a values of the α-protons of a carboxylate ester and a thioester.

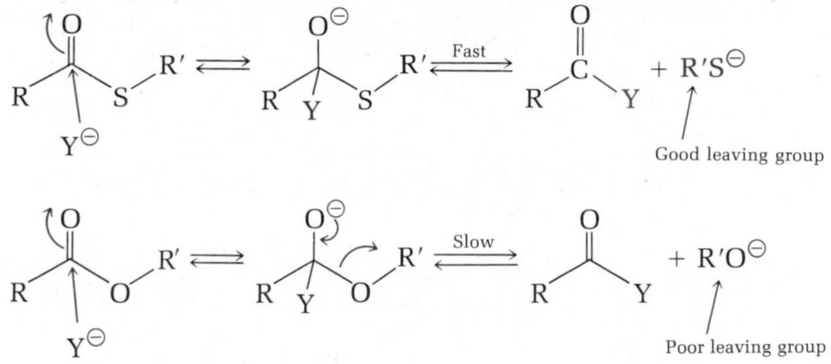

Figure 10.7 Nucleophilic displacement in thioesters and carboxylate esters. Nucleophilic substitution at acyl carbon is faster in thioesters than in carboxylate esters because the thioalkoxide anion, $R'S^{\ominus}$, is a much better leaving group than the alkoxide anion, $R'O^{\ominus}$.

zymes to bind the enzymes that require them. Most enzymes that require NAD have a low affinity for NADP and vice versa.

C. NAD-Dependent Dehydrogenases: Structural and Mechanistic Properties

Nearly 250 known nicotinamide nucleotide–dependent enzymes, called *dehydrogenases*, catalyze the transfer of two reducing equivalents from the substrate to the coenzyme. The fundamental mechanistic property of nearly all NAD-dependent dehydrogenase reactions is the direct transfer of a hydride anion (a hydrogen atom with its bonding electron pair) from the substrate to the C-4 position of the nicotinamide ring, as illustrated in the preceding section.

Let us consider the transfer of hydride anion from ethanol to NAD catalyzed by *alcohol dehydrogenase* (reaction 10.2).

$$CH_3CH_2OH + NAD \xrightarrow{\text{Alcohol dehydrogenase}} CH_3CHO + NADH + H^{\oplus} \quad (10.2)$$

The reaction is *stereospecific* with respect to *both* substrates, even though neither the C-1 of ethanol nor the C-4 position of the nicotinamide ring is chiral. To appreciate more fully the stereochemistry of this reaction, and that of many other enzyme-catalyzed reactions, a brief interlude on stereochemistry will be helpful.

a. Chirality and the *RS* Configurational Nomenclature

In our previous discussions we have assigned stereochemistry to a chiral center by comparing it to D-glyceraldehyde. This convention works quite well for amino acids and carbohydrates, but breaks down for many other metabolites because there is no suitable reference compound. The *RS* nomenclature of configurational isomers overcomes this difficulty. Any carbon atom with four different groups bonded to it is asymmetric, or *chiral*. For amino acids, such as alanine, the absolute configuration is described by reference to D-glyceraldehyde. We can also describe it by assigning each group bound to the chiral carbon a certain priority. The groups are arranged in a *priority sequence* in which the lowest priority is assigned the number 1 and the highest priority the number 4 (the rules for assigning priorities follow shortly). Now, imagine that you are looking at the steering wheel of your au-

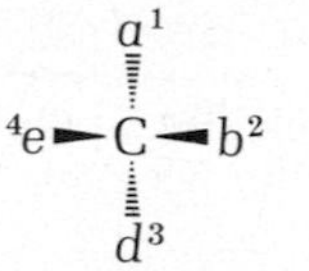

$$\begin{array}{c} a^1 \\ | \\ {}^4e \blacktriangleright C \blacktriangleleft b^2 \\ | \\ d^3 \end{array}$$

Reactive position of pyridine ring

β-Glycosidic bond

NH₂

(PO₃ ²⁻)

Phosphate group in NADP replaces H

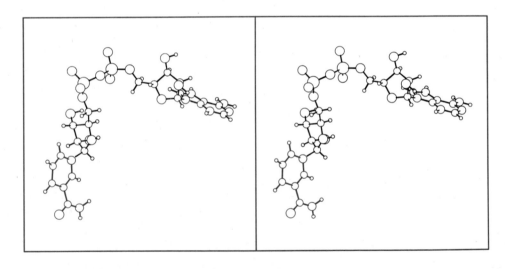

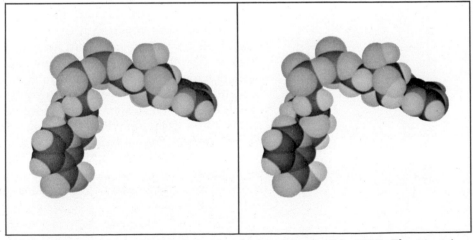

Figure 10.8 Structure of nicotinamide adenine dinucleotide, NAD. The 2′ position is phosphorylated in the coenzyme nicotinamide adenine dinucleotide phosphate, NADP.

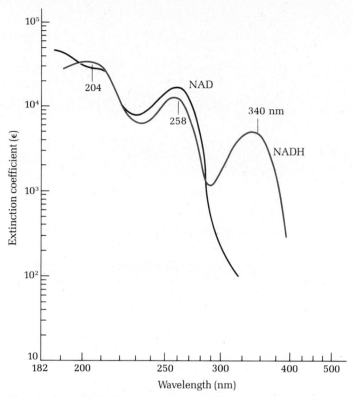

Figure 10.9 UV absorption spectra for NAD and NADH. [From W. B. Gratzer, in G. D. Fasman, Ed., "Handbook of Biochemistry and Molecular Biology," 3rd ed., Vol. 2, CRC Press, Cleveland, Ohio, 1976, p. 167.]

tomobile and that the group with lowest priority, group 1 (in this case a) is away from you at the bottom of the steering column and the other three groups are on the rim of the steering wheel itself. Trace around the rim of the wheel from group 4 to group 3 to group 2. If you have described a clockwise movement the molecule has absolute configuration R (from the Latin *rectus*, "to the right"). If a counterclockwise movement is required, the compound has an absolute configuration S (from the Latin *sinistra*, "to the left"). Thus, the configuration of the molecule shown above is S.

$$^4e{-}C{-}b^2 \;=\; \left(\begin{array}{c}^4e \quad b^2 \\ a^1 \\ d^3\end{array}\right) \;\equiv\; S$$

PRIORITY RULES

1. Consider the atoms that are directly attached to the chiral carbon: the lower the atomic number, the lower the priority. If two isotopes are present, such as hydrogen and deuterium, the lower priority is assigned to the lower mass isotope.

2. If the first atoms of two groups are the same, move to the next set of atoms and again compare atomic numbers. The lower the atomic weight, the lower the priority. Thus, a CH_3 group has a lower priority than a CH_2Br group.

3. If a group is bonded by a double or triple bond, the bonded atoms are considered to be duplicated or triplicated. Thus,

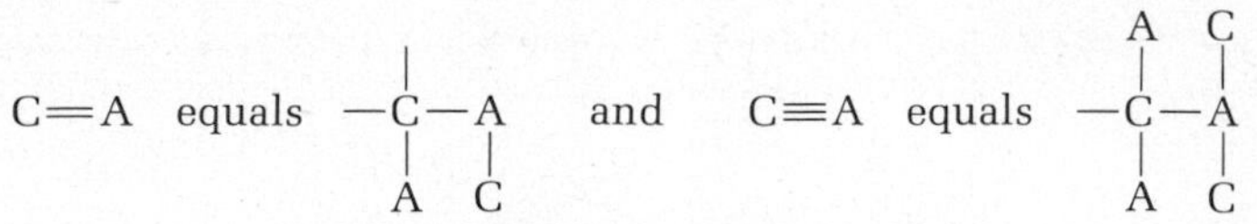

$$C{=}A \quad \text{equals} \quad \overset{|}{\underset{A}{\overset{|}{C}}}{-}A \quad \text{and} \quad C{\equiv}A \quad \text{equals} \quad \overset{A \quad C}{\underset{A \quad C}{\overset{|}{\underset{|}{C}}}}{-}A$$

In glyceraldehyde, for example, the —CHO group equals $-\overset{H}{\underset{O}{\overset{|}{\underset{|}{C}}}}{-}O$. The priority

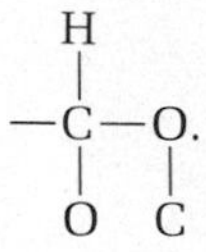

ranking is therefore —OH > —CHO > —CH$_2$OH > H—, since the O,O, H of —CHO takes precedence over the —CH$_2$OH.

4. In a molecule having two or more chiral carbons, each one is assigned an R or S configuration.

Using the general rules outlined above, augmented by the sequence of priorities in Table 10.3, we conclude that D-glyceraldehyde has absolute configuration R and that L-alanine has absolute configuration S.[1]

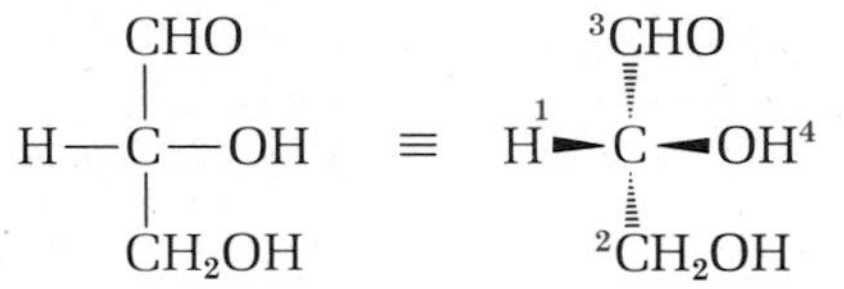

CHO
|
H—C—OH   ≡   H—C—OH
|
CH$_2$OH

D-Glyceraldehyde   R-Glyceraldehyde
(Fischer projection)

CO$_2^\ominus$
|
NH$_3^\oplus$—C—H   ≡   H$_3$N—C—H
|
CH$_3$

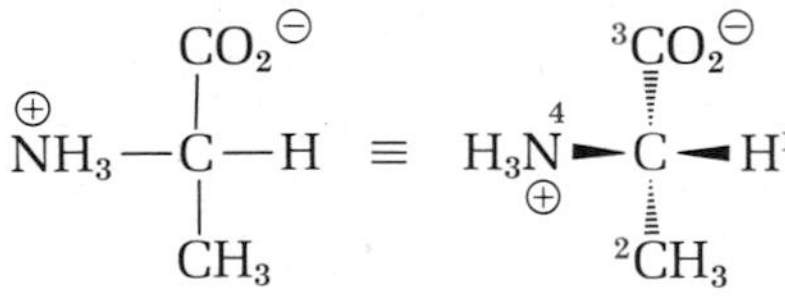

L-Alanine   S-Alanine

b. Prochirality

Consider a molecule C with groups a, a, b, d. Since the carbon is not bound to four different groups, it is not chiral. Suppose that the groups are arranged as below and that the priority is $d > b > a$. If a_1 (the subscript merely enables us to distinguish the a's) is replaced by e, and if the priority of e is greater than the priority of d, then the molecule becomes S. Similarly, replacement of a_2 with e gives the R enantiomer.

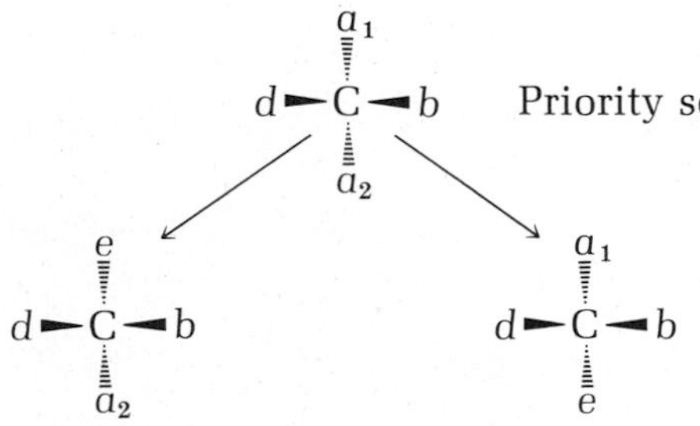

Priority sequence: $d > b > a$ $(e > d)$

Since replacement of a_1 with e gives the S isomer, and replacement of a_2 with e gives the R isomer, we designate the a_1 group as *pro-S* and the a_2 group as *pro-R. In other words, a_1 (now written as a_S) and a_2 (now written as*

[1] The α carbon of all L-amino acids, except cysteine, is S. Cysteine has absolute configuration R because the sulfur atom takes precedence over oxygen (Rule 1).

Table 10.3   Condensed priority sequence from lowest to highest for assigning absolute configuration

Priority number	Group	Priority number	Group	Priority number	Group	Priority number	Group
1	Hydrogen	11	Benzyl	21	Methylamino	31	Phenoxy
2	Methyl	12	Isopropyl	22	Ethylamino	32	Acetoxy
3	Ethyl	13	Vinyl	23	Phenylamino	33	Benzoyloxy
4	n-Propyl	14	sec-Butyl	24	Acetylamino	34	Fluoro
5	n-Butyl	15	t-Butyl	25	Dimethylamino	35	Sulfhydryl (HS—)
6	n-Pentyl	16	Phenyl	26	Nitro	36	Sulfo (HO₃S—)
7	Isopentyl	17	p-Tolyl	27	Hydroxy	37	Chloro
8	Isobutyl	18	Acetyl	28	Methoxy	38	Bromo
9	Allyl	19	Benzoyl	29	Ethoxy	39	Iodo
10	Neopentyl	20	Amino	30	Benzyloxy		

a_R) are in mirror image environments. The carbon is *prochiral*, and the groups a_R and a_S are *enantiotopic*. To distinguish between enantiotopic groups a chiral reagent, such as an enzyme, is required (Figure 10.10).

Our first example has involved sp^3 hybridized carbon. Trigonal or sp^2 hybridized carbon can also be prochiral since addition of a group from above the plane of the carbon gives one stereoisomer and addition from the opposite side of the plane gives the enantiomer (Figure 10.11).

PRIORITY RULES FOR TRIGONAL CARBON

1. For a carbon with groups a, b, d assign each group a priority.
2. Place the central carbon and its three groups in one plane.

Priority sequence: $d > b > a$

If the sequence of groups is clockwise, you are closest to the *re* face. If the sequence of groups is counterclockwise (as it is in the example above), you are looking at the *si* face.

Enzymes can distinguish between pro-R and pro-S groups and between *re* and *si* faces of trigonal carbon because enzymes themselves are chiral. In a chiral environment pro-R/pro-S and *re*/*si* pairs are *not* stereochemically

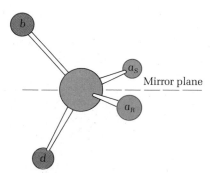

Figure 10.10   Groups a_R and a_S are in mirror image environments. The carbon is prochiral, and the groups a_R and a_S are enantiotopic. Groups a_R and a_S are identical in a symmetric environment, but not in a chiral environment, such as the active site of an an enzyme, and enzymes can distinguish between them.

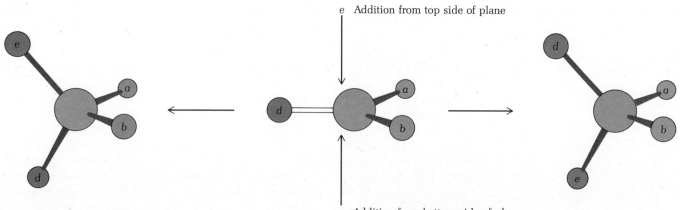

Figure 10.11   Prochiral trigonal carbon. A trigonal carbon bound to three different groups is prochiral since addition of a fourth group from the "top" face yields the enantiomer of addition from the "bottom" face.

equivalent. Our first illustration of this point is alcohol dehydrogenase to which we now return.

The methylene hydrogens of ethanol are prochiral and yeast alcohol dehydrogenase (and most other dehydrogenases) can distinguish between them. Alcohol dehydrogenase stereospecifically transfers the pro-R hydrogen of ethanol to the re face of the nicotinamide ring of NAD at C-4 (reaction 10.3).

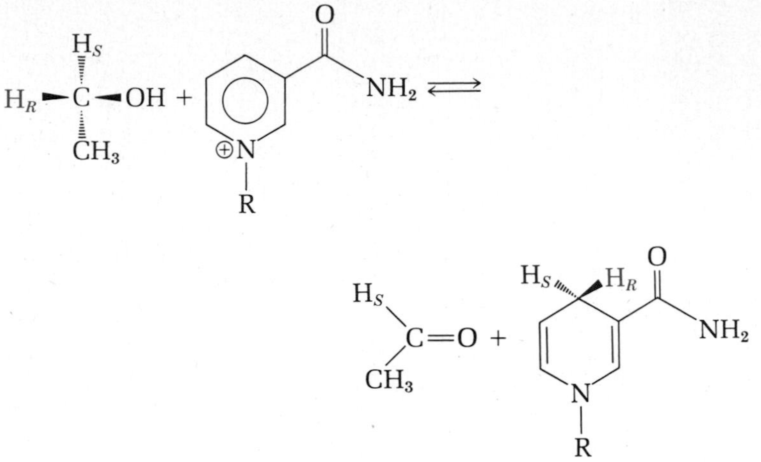

$$(10.3)$$

This stereospecificity is shown by using R-[^{2}H]-ethanol as the substrate. All of the deuterium in the reactant is transferred to C-4 of the nicotinamide ring. In the reverse reaction the pro-R hydrogen (deuterium in the isotopic labeling experiment) is transfered to the re face of acetaldehyde (reaction 10.4).

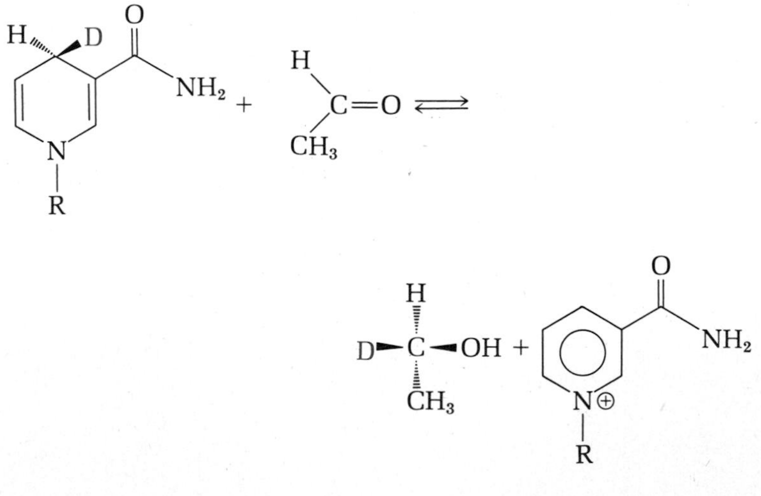

$$(10.4)$$

The enzyme can thus distinguish between the re and si faces of NAD and acetaldehyde and between the pro-R and pro-S position of ethanol and NADH. Table 10.4 gives the stereospecificity at C-4 for several nicotinamide-dependent dehydrogenases. The stereospecificity of these reactions is a result of the interaction between the enzyme and its substrates that allows for hydride anion transfer from only one enantiotopic position of the reactant to only one face of the nicotinamide ring. This is illustrated for lactate dehydrogenase in Figure 10.12. The structure of the enzyme-substrate complex for lactate dehydrogenase is further amplified in Figure 10.13.

Hydride transfer is not the only type of reaction catalyzed by NAD-dependent dehydrogenases. Table 10.5 lists some of the other reactions cata-

Table 10.4 Stereospecificity at C-4 in NAD(P)-dependent dehydrogenases

Enzyme	Nucleotide	C-4 prochiral hydrogen transferred	Product
Alcohol dehydrogenase	NAD	H_R	Acetaldehyde
UDP-glucose dehydrogenase	NAD	H_S	UDP-glucuronate
L-Lactate dehydrogenase	NAD	H_R	Pyruvate
L-Malate dehydrogenase	NAD	H_R	Oxalacetate
Isocitrate dehydrogenase			
Cytosolic	NADP	H_R	α-Ketoglutarate + CO_2
Mitochondrial	NAD	H_R	α-Ketoglutarate + CO_2
D-Glyceraldehyde-3-P-dehydrogenase	NAD	H_S	1,3-Phosphoglycerate
Glutamate dehydrogenase	NADP, NAD	H_S	α-Ketoglutarate + $NH_4^{\oplus}$

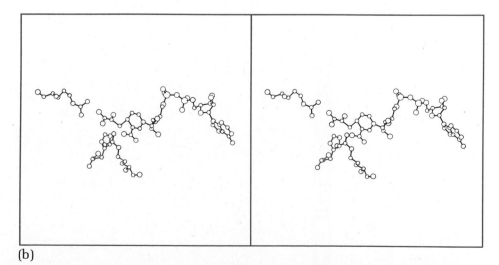

Figure 10.12 (a) Partial view of the active site of lactate dehydrogenase. The hydride anion can be transferred to only one face of the nicotinamide ring system because of the stereochemical restrictions of the active site. [From J. J. Holbrook, A. Liljas, S. J. Steindel, and M. G. Rossman, in P. D. Boyer, Ed., "The Enzymes," 3rd ed., Vol 2, Academic Press, New York, 1957, p. 243.] (b) Stereo view of the active site of lactate dehydrogenase.

Table 10.5 Types of reactions catalyzed by NAD-dependent dehydrogenases

Reaction	Enzyme	Example
$\begin{array}{c} \text{H} \\ \mid \\ -\text{C}- \\ \mid \\ \text{OH} \end{array} \rightleftarrows \begin{array}{c} \\ -\text{C}- \\ \parallel \\ \text{O} \end{array}$	L-Lactate dehydrogenase L-Malate dehydrogenase	$\begin{array}{c} \text{OH} \\ \mid \\ \text{CH}_3\text{CHCO}_2^{\ominus} \end{array} \rightleftarrows \begin{array}{c} \text{O} \\ \parallel \\ \text{CH}_3\text{CCO}_2^{\ominus} \end{array}$ $\begin{array}{c} \text{OH} \\ \mid \\ {}^{\ominus}\text{O}_2\text{CCH}_2\text{CH}-\text{CO}_2^{\ominus} \end{array} \rightleftarrows \begin{array}{c} \text{O} \\ \parallel \\ {}^{\ominus}\text{O}_2\text{CCH}_2\text{CCO}_2^{\ominus} \end{array}$ (OAA) oxaloacetate
$\begin{array}{c} \text{H} \\ \mid \\ -\text{C}- \\ \mid \\ \overset{\oplus}{\text{NH}_3} \end{array} \rightleftarrows \begin{array}{c} \\ -\text{C}- \\ \parallel \\ \text{O} \end{array}$	L-Glutamate dehydrogenase	${}^{\ominus}\text{O}_2\text{CCH}_2\text{CH}_2\underset{\overset{\mid}{\oplus\text{NH}_3}}{\text{CHCO}_2^{\ominus}} \rightleftarrows \text{NH}_4^{\oplus} + {}^{\ominus}\text{O}_2\text{CCH}_2\text{CH}_2\overset{\overset{\text{O}}{\parallel}}{\text{C}}\text{CO}_2^{\ominus}$ (αKG) α-ketoglutarate
$\begin{array}{c} -\text{C}-\text{H} \\ \parallel \\ \text{O} \end{array} \rightleftarrows \begin{array}{c} -\text{C}-\text{O}^{\ominus} \\ \parallel \\ \text{O} \end{array}$	Aldehyde dehydrogenase	$\text{CH}_3\text{CHO} \rightleftarrows \text{CH}_3\text{CO}_2^{\ominus}$
$\begin{array}{c} -\text{C}-\text{N}- \\ \mid\ \ \mid \\ \text{H}\ \ \text{H} \end{array} \rightleftarrows \begin{array}{c} -\text{C}=\text{N}- \end{array}$	Dihydrofolate reductase (NADP-dependent)	(ATP also required)

lyzed by these enzymes that we will encounter in our discussion of metabolism.

D. Flavin Coenzymes

Flavin adenine dinucleotide, FAD, and flavin mononucleotide, FMN (also known as riboflavin 5′-phosphate) are the two flavin coenzymes (Figure 10.14).[2] Vitamin B$_2$, riboflavin, is common to both coenzymes. The flavins

[2] The C—N bond at N-10 of the isoalloxazine ring is not truly a glycosidic bond, but this has not deterred people from calling FAD a dinucleotide and FMN a mononucleotide, a usage we retain.

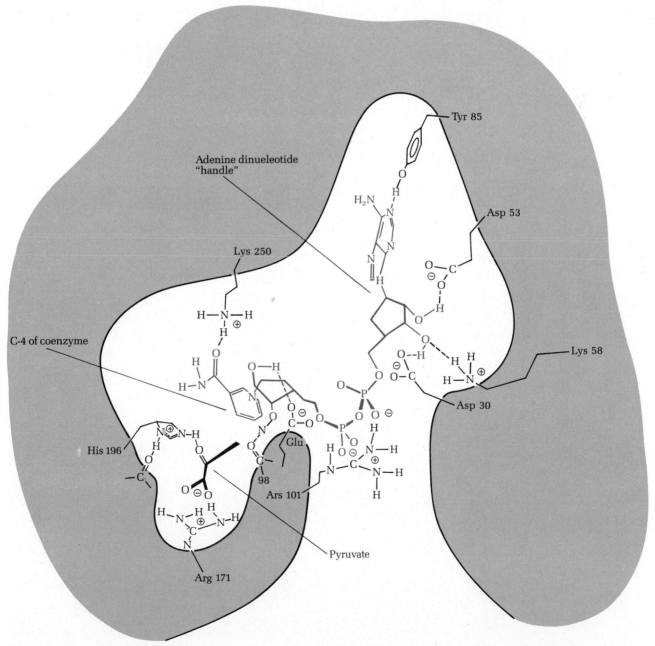

Figure 10.13 (a) Enzyme–substrate–coenzyme complex of lactate dehydrogenase. The interactions of the adenine dinucleotide moiety with the enzyme to hold the coenzyme in place for stereospecific hydride anion transfer from NADH to pyruvate. (b) Stereo view of the enzyme substrate complex (on page 410).

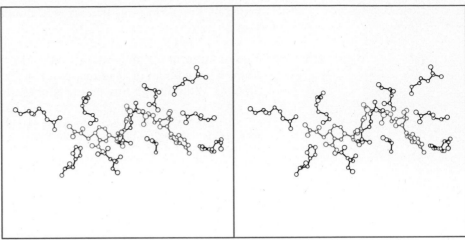

Fig. 10.13 (*continued*)

take their names from their bright yellow color. (*Flavius* is the Latin word for yellow.) The extended π system of the oxidized flavins is lost when the *isoalloxazine* ring of the coenzyme is reduced, and $FADH_2$ and $FMNH_2$ are colorless. The spectra of these coenzymes are shown in Figure 10.15.

Enzymes that require FAD or FMN are called *flavoenzymes*. If the cofactor is absent the protein is called an *apoenzyme* (*apo-* is a Greek prefix that means "away from"). The dissociation constants of flavins from flavoproteins are in the range of 10^{-8} M to 10^{-11} M, and the cellular concentration of flavin coenzymes is extremely low (reactions 10.5 and 10.6).

$$\text{enzyme-flavin} \underset{}{\overset{K_{diss} = 10^{-8} \text{ M to } 10^{-11} \text{ M}}{\rightleftharpoons}} \text{apoenzyme + flavin} \tag{10.5}$$

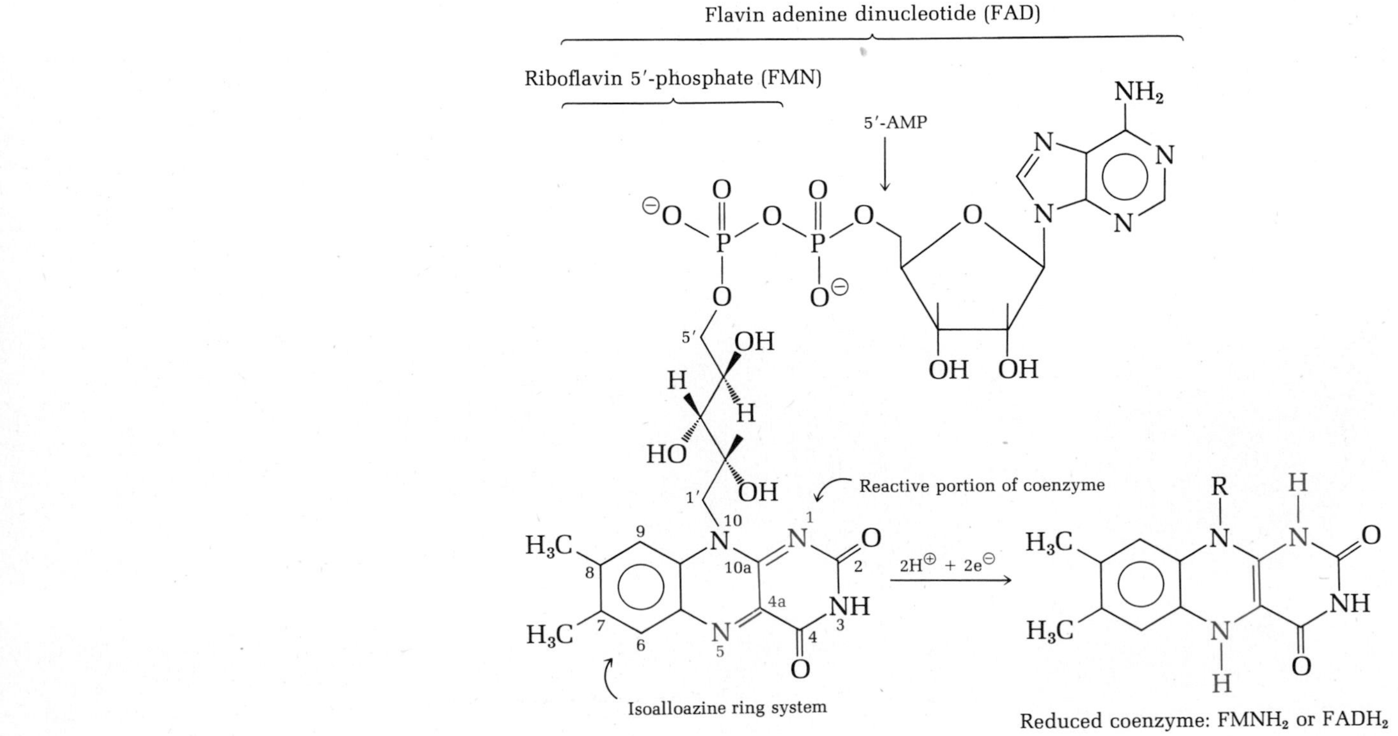

Flavin adenine dinucleotide (FAD)

Riboflavin 5'-phosphate (FMN)

5'-AMP

NH_2

$2H^{\oplus} + 2e^{\ominus}$

Reactive portion of coenzyme

Isoalloazine ring system

Reduced coenzyme: $FMNH_2$ or $FADH_2$

(a)

Figure 10.14 (a) Structures of the flavin coenzymes. (b) Stereo view of riboflavin (on page 411).

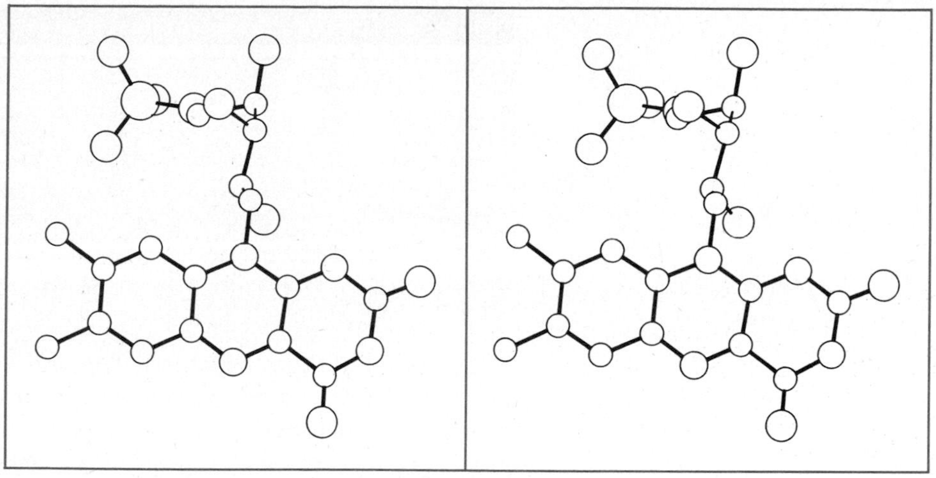

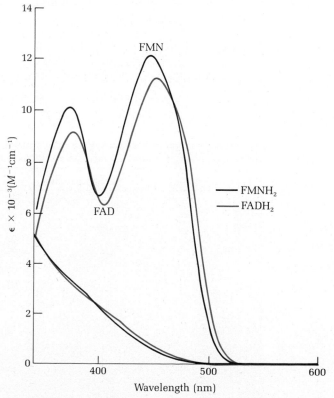

(b)

Figure 10.14 (*continued*)

$$K_{diss} = \frac{[\text{apoenzyme}][\text{flavin}]}{[\text{enzyme-flavin complex}]}$$ (10.6)

Flavoenzymes catalyze a wide variety of oxidation-reduction reactions. The major classes of flavoenzymes are listed in Table 10.6 and the types of substrates oxidized (or reduced) are listed in Table 10.7.

E. Mechanisms of Action of Some Flavoenzyme Dehydrogenases

In contrast to the similar mechanisms of action of most NAD-dependent dehydrogenases, flavoenzymes present us with a wide range of mechanistic phenomena, many of which are not very well understood. No overall gener-

Figure 10.15 Visible absorption spectra of the flavin coenzymes.

Table 10.6 Major classes of flavoenzymes

Dehydrogenases	Use one electron acceptor, other than O_2; often functionally linked to cellular membrane respiratory chains
Oxidases	O_2 electron acceptor, reduced to H_2O_2
Oxidase-decarboxylases	4 $e^\ominus$ oxidations; $O_2 \rightarrow H_2O$
Hydroxylases	$S + O_2 \rightarrow S{-}OH + H_2O$
Metalloflavoenzymes	Require a bound transition metal for catalysis

SOURCE: From C. Walsh, *Ann. Rev. Biochem.*, 47, 886 (1978). Reproduced with permission. © 1978 by Annual Reviews Inc.

alization about the mechanistic behavior of flavoenzymes is possible. Some flavoenzymes catalyze hydride anion transfer, and others carry out successive one-electron transfers with free radicals as intermediates. Yet another large class of enzymes requires metal ions for activity. These *metalloflavoenzymes* catalyze both one- and two-electron transfer reactions. In this chapter we will consider flavoenzyme dehydrogenases and some aspects of one-electron reductions involving flavins. We will encounter many other flavoenzymes in our detailed discussions of metabolism.

a. Direct Hydride Transfer

Dihydroorotate dehydrogenase, which participates in the biosynthesis of uridine monophosphate (Chapter 23), provides an example of direct hydride anion transfer involving a flavin coenzyme. The enzyme also requires NAD. The catalytic reaction occurs by transfer of hydride anion from NADH to FAD, followed by transfer of hydride anion from $FADH_2$ to orotate (Figure 10.16). Dihydroorotate dehydrogenase is a tetramer (mol wt 115,000) that contains FMN, FAD, and two iron atoms whose oxidation states oscillate between Fe(II) and Fe(III) during the reaction. The mechanism of Figure 10.16 accounts for two experimental observations: direct hydride transfer from NADH to FAD and direct hydride transfer from $FADH_2$ to orotate. The intervening steps involving the second flavin and the two iron atoms known to participate in the reaction have not been elucidated.

b. One-Electron Transfer: Semiquinones

Flavins are reduced by successive one-electron transfers in many flavoenzymes. These reactions are far more complex than direct hydride

Table 10.7 Substrate classes for flavoenzyme catalysis

Reduced form	Oxidized form	Examples
H │ —C— │ OH	—C— ‖ O	*E. coli* membrane D-lactate dehydrogenase, glucose oxidase
H │ —C— │ NH₂	C + NH₄⊕ ‖ O	D- and L-amino acid oxidases; amine oxidases

—CH—CH—C—
          ‖
          O
	C=C—C—
	        ‖
	        O
		Succinate dehydrogenase, acyl CoA dehydrogenases, α-glycerophosphate dehydrogenases

NADH	NAD	NADH dehydrogenases, transhydrogenases, dihydroorotate dehydrogenases
SH SH	S—S	Lipoamide dehydrogenase, glutathione reductase (some organisms)

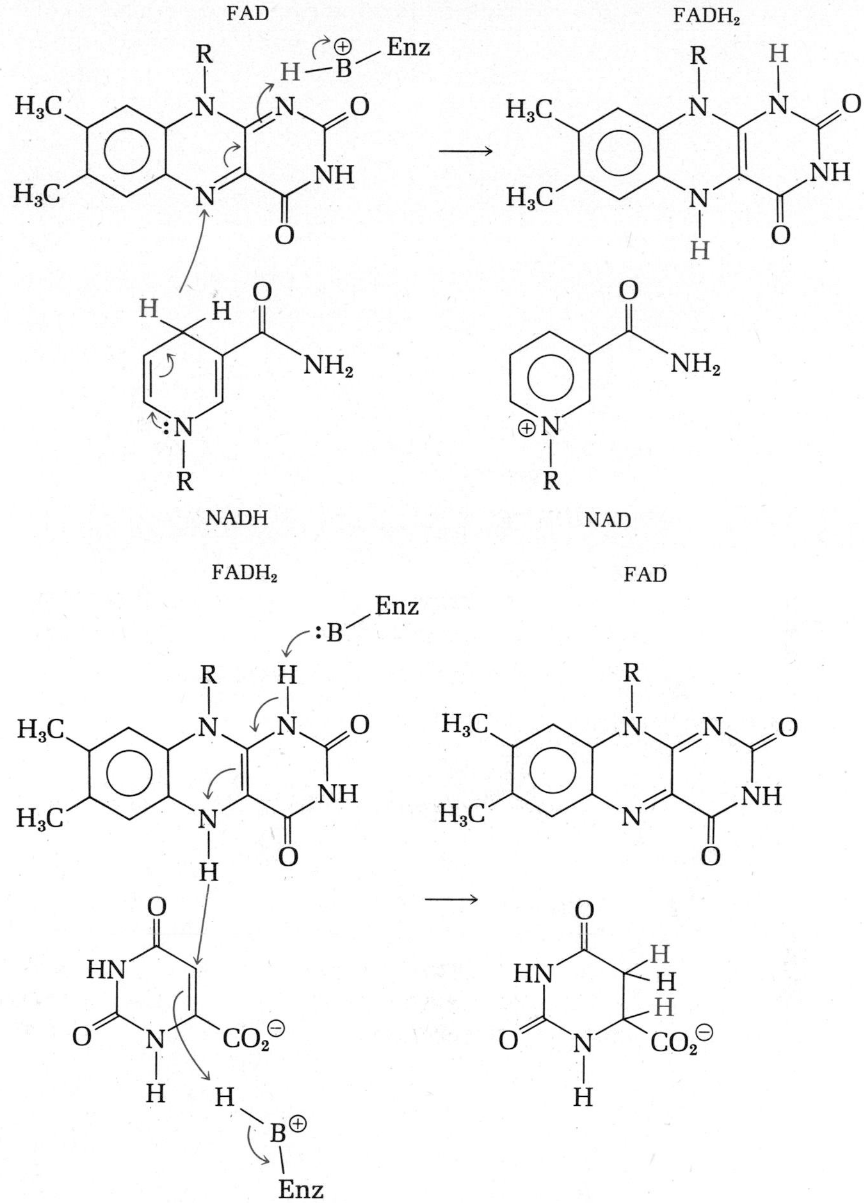

Orotate Dihydroorotate

Figure 10.16 Mechanism of action of dihydroorotate dehydrogenase. The reaction involves successive hydride transfers. The intermediate steps involving oxidation and reduction of an iron atom in the flavoenzyme are not understood.

transfers. A one-electron transfer from a substrate to a flavin produces a *semiquinone free radical* (reaction 10.7).

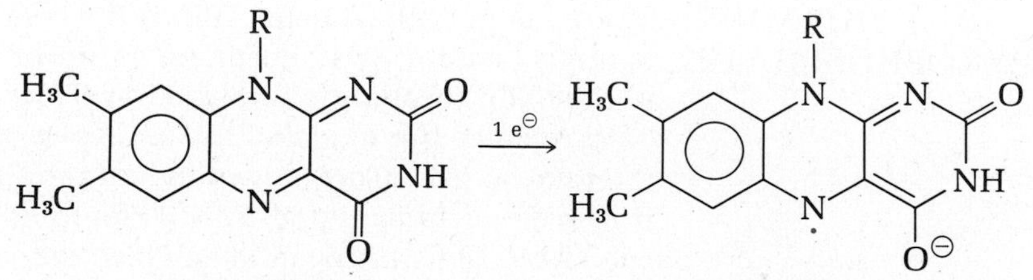

(10.7)

Oxidized flavin

(1) Semiquinone free radical (λ_{max} 480, red)

The isoalloxzine ring has a net charge of -1 in the semiquinone radical (1). The *radical anion* has a λ_{max} of 480 nm and is bright red. Protonation of the radical anion (1) gives the neutral species with a net charge of zero (2) (reaction 10.8.)

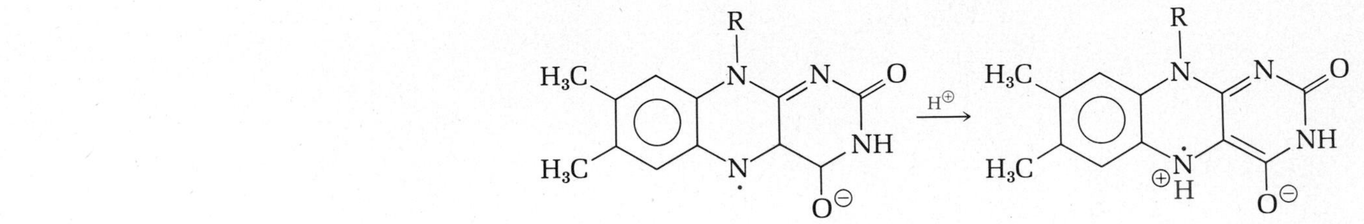

(1) Radical anion

(2) Neutral radical
(λ_{max} 570 nm, blue)

(10.8)

The pK_a of the neutral free radical (2) is about 8.4, and this species is blue. The transfer of a second electron to the alloxazine ring gives the anion 3 (reaction 10.9).

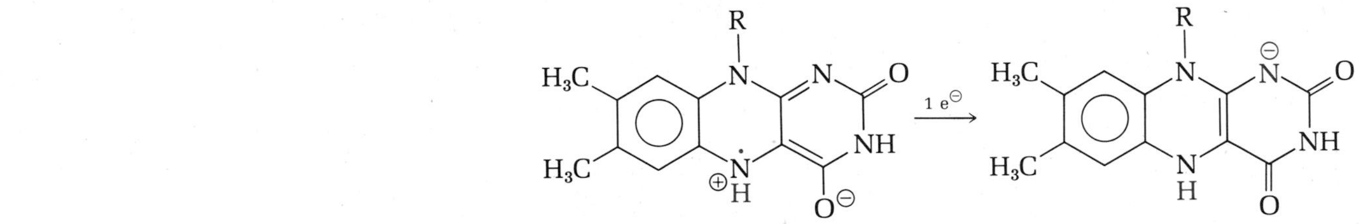

(2) Semiquinone radical

(3) Anion

(10.9)

The anion (3) now picks up a proton to give fully reduced dihydroflavin (reaction 10.10).

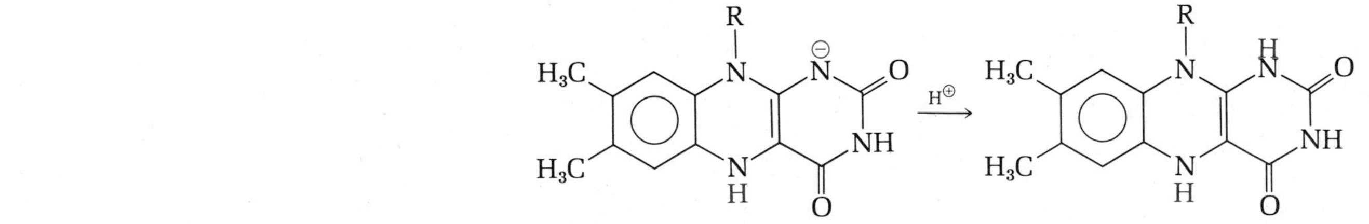

(3) Anion

Fully reduced dihydroflavin

(10.10)

The semiquinone radicals (2 and 3) are stable because their unpaired electron can be extensively delocalized through the extended π system of the isoalloxazine ring. The stability of the semiquinone radicals enables them to act as intermediaries between two-electron and one-electron transfer processes such as those that occur in respiration (Chapter 16).

10–4 THIAMINE PYROPHOSPHATE

Thiamine pyrophosphate, TPP, is the coenzyme derived from thiamine (vitamin B₁) and a pyrophosphate group originating in ATP. The pyrophosphorylation reaction is catalyzed by TPP synthetase (reaction 10.11).

A molecular model of thiamine pyrophosphate and the UV spectrum of thiamine at pH 5.0 are shown in Figures 10.17 and 10.18. The quaternary nitrogen of the *thiazolium ring* of TPP provides the clue to the properties of the coenzyme. This nitrogen is an "electron sink" that enhances the acidity of the hydrogen at C-2, whose pK_a is 12.7. Removal of this proton (reaction

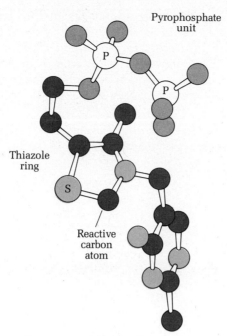

Pyrophosphate
unit

Thiazole
ring

Reactive
carbon
atom

Figure 10.17 Thiamine pyrophosphate.

Thiamine (vitamin B₁)

$+ (MgATP)^{2\ominus} \xrightarrow{\text{TPP-synthetase}}$

Thiamine pyrophosphate (TPP)

$$(10.11)$$

10.12) produces a carbanion that is stabilized by the quaternary nitrogen and perhaps by delocalization of the carbanion electron pair into the vacant *d*-orbitals of the adjacent sulfur.[3] There are two resonance structures for the

$$+ BH^{\oplus} \qquad (10.12)$$

(1) Ylid   (2) Nucleophilic carbene

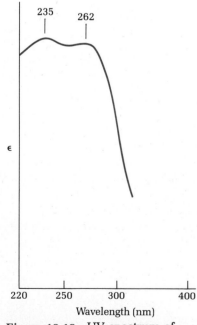

Figure 10.18 UV spectrum of thiamine at pH 5.0. [From W. B. Gratzer, in G. D. Fasman, Ed., "Handbook of Biochemistry and Molecular Biology," 3rd Ed., Vol 2, CRC Press, Cleveland, Ohio, 1976, p. 146.]

ε

Wavelength (nm)

235   262

220   250   300   400

[3] There has been considerable debate about the existence of "d-p π overlap" between an electron pair on carbon and the vacant *d*-orbitals of sulfur. Whether this is a "real" phenomenon or simply an attractive possibility, sulfur ylids exist and are relatively stable.

carbanion. The dipolar ion (*1*) is an *ylid*, and the neutral divalent carbon (*2*) is a *nucleophilic carbene*. The reactions of TPP involve the synthesis or cleavage of a bond to a carbonyl carbon. The bond-making reactions are called α-condensations, and the bond-breaking reactions are called α-cleavages. These processes can be either oxidative or nonoxidative. The oxidative reactions require FAD, NAD, or both.

A. Nonoxidative Decarboxylation: Yeast Pyruvate Decarboxylase

Yeast *pyruvate decarboxylase* converts pyruvate to acetaldehyde and carbon dioxide (reaction 10.13).

$$CH_3-\overset{\overset{\displaystyle O}{\|}}{C}-CO_2^{\ominus} \xrightarrow[\text{decarboxylase}]{\text{Pyruvate}} CH_3CHO + CO_2 \qquad (10.13)$$

Pyruvate

The first step in the reaction is attack of the C-2 carbanion of TPP on the keto group of pyruvate (reaction 10.14).

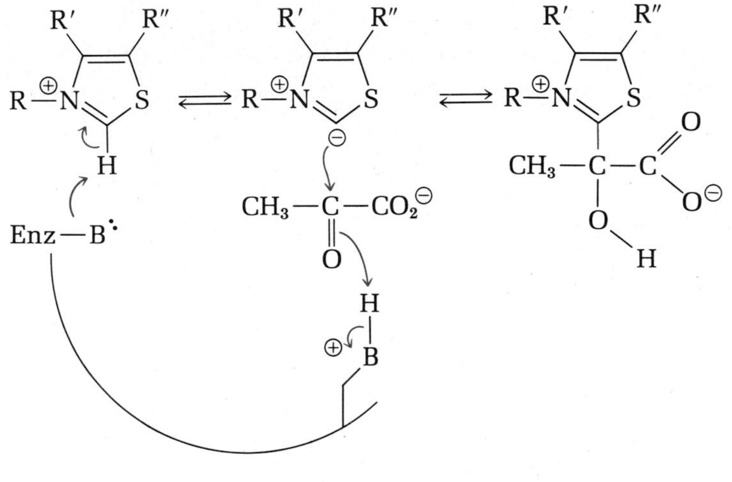

(10.14)

The initial adduct then decarboxylates with electrons flowing toward the quaternary nitrogen. The loss of carbon dioxide leaves *hydroxyethylthiamine pyrophosphate*, HETPP (reaction 10.15).

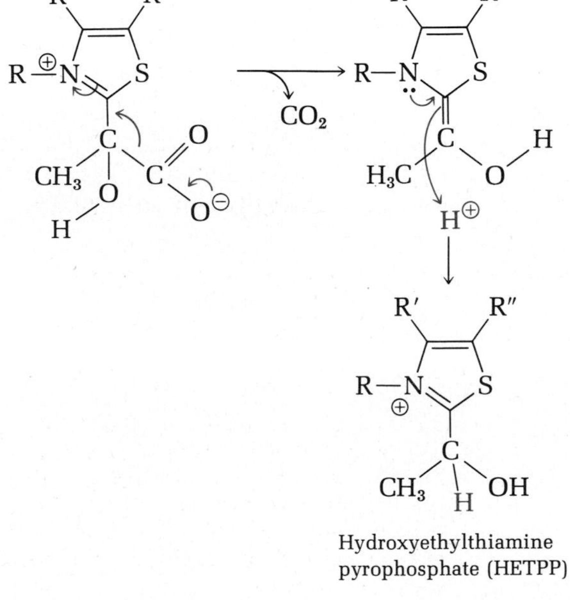

(10.15)

Hydroxyethylthiamine pyrophosphate (HETPP)

This reaction can occur without the aid of an enzyme, but the rate is much slower than that of the enzyme-catalyzed reaction. In the final step of the reaction, the enzyme abstracts the hydroxyl hydrogen of HETPP, electrons flow toward the quaternary nitrogen, and acetaldehyde is produced. The carbanion of TPP is ready for another round of catalysis (reaction 10.16).

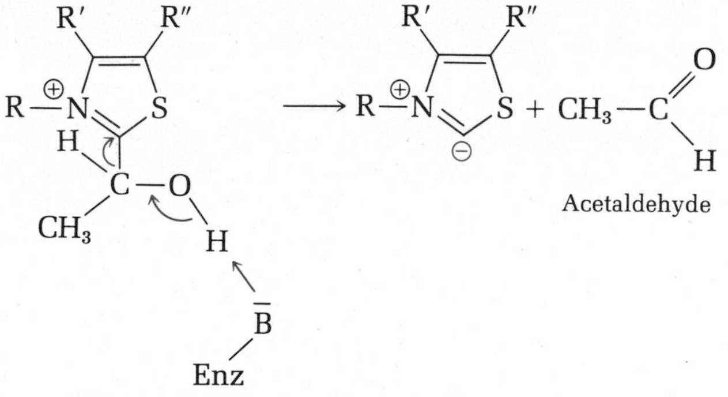

(10.16)

B. Oxidative Decarboxylation: *E. coli* Pyruvate Oxidase

E. coli pyruvate oxidase decarboxylates pyruvate to produce acetate and carbon dioxide instead of acetaldehyde and carbon dioxide. FAD is also required for pyruvate oxidase activity. The first step of the reaction produces HETPP just as pyruvate decarboxylase does.

In the next step the methine hydrogen of HETPP is transferred to FAD producing *acetylthiamine pyrophosphate* and FADH$_2$. Hydrolysis of acetyl-TPP gives acetate and the carbanion of TPP ready for another round of catalysis (Figure 10.19).

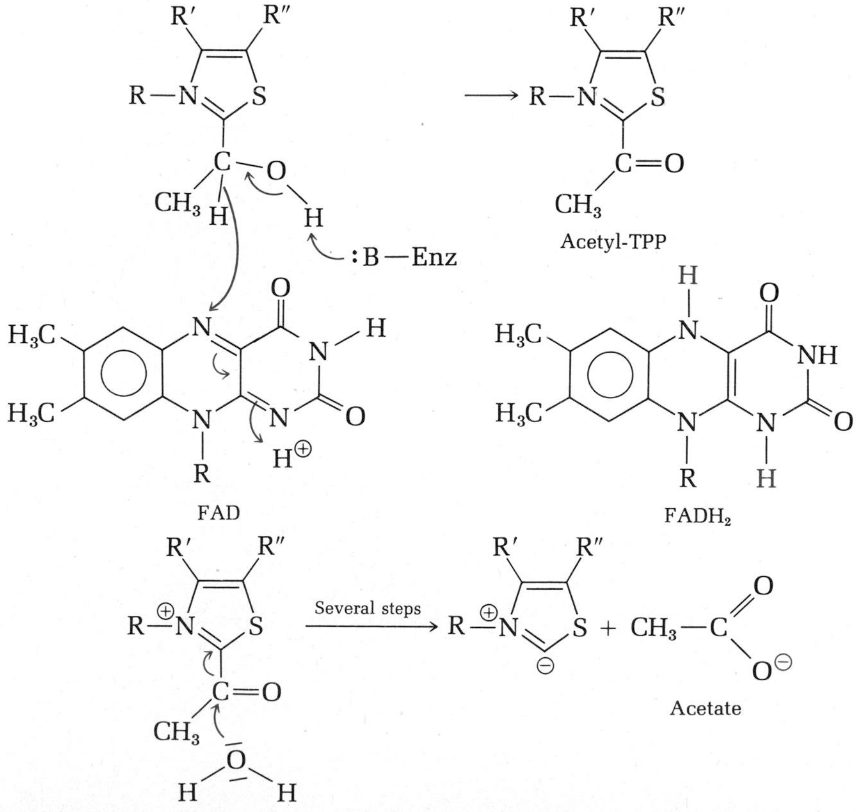

Figure 10.19 Transfer of a methine hydrogen from hydroxyethyl thiamine pyrophosphate (HETPP) to FAD. The acetyl-TPP produced in step (a) is hydrolyzed to give acetate and the TPP carbanion, ready for another round of catalysis. The transfer of the hydrogen from HETPP to FAD is a two-electron reduction. We have shown it as a direct hydride transfer (b), but the exact nature of this step is not known.

10−5 LIPOIC ACID Lipoic acid was first isolated in 1951 by L. Reed and his coworkers. They obtained 30 mg of coenzyme from 10 *tons* of beef liver—a truly Herculean achievement.

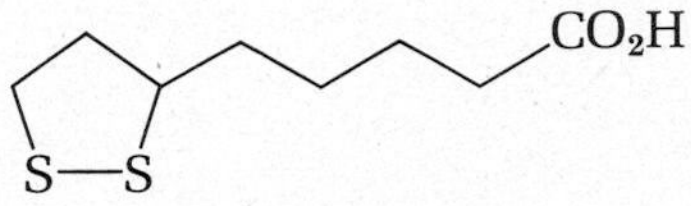

$$CO_2H$$

Lipoic acid

Lipoic acid is found in two multienzyme complexes—*pyruvate dehydrogenase* and *α-ketoglutarate dehydrogenase*—where it serves as an *acyl carrier* and as a two-electron carrier. It is covalently bound to a lysyl residue of *dihydrolipoyl transacetylase*. The disulfide bridge of lipoic acid is 1.5 nm

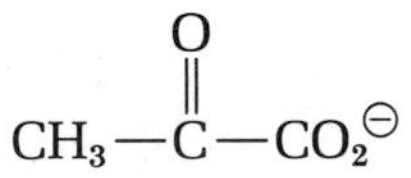

$$CH_3-\overset{\overset{\displaystyle O}{\|}}{C}-CO_2^{\ominus} \qquad \qquad ^{\ominus}O_2C-CH_2-CH_2-\overset{\overset{\displaystyle O}{\|}}{C}-CO_2^{\ominus}$$

Pyruvate

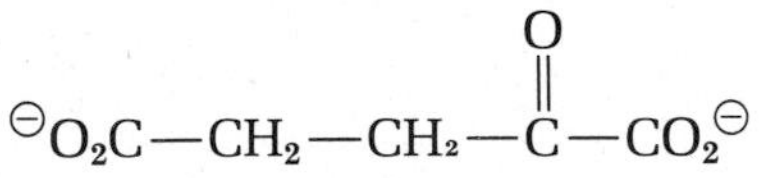

α-Ketoglutarate
(also called 2-oxoglutarate)

from the peptide backbone, and this miniature "arm" serves to swing an acyl group from one part of the enzyme complex to another. In pyruvate dehydrogenase (Section 14.8), a hydroxyethyl group is transferred from HETPP to

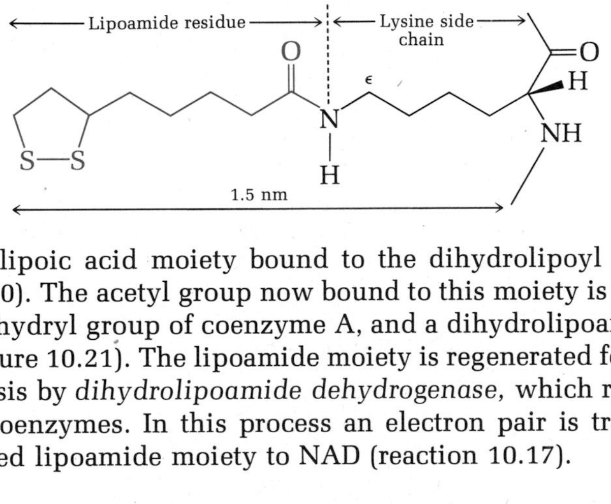

←——— Lipoamide residue ———→ | ←— Lysine side chain —→

1.5 nm

the lipoic acid moiety bound to the dihydrolipoyl transacetylase (Figure 10.20). The acetyl group now bound to this moiety is then transferred to the sulfhydryl group of coenzyme A, and a dihydrolipoamide moiety is formed (Figure 10.21). The lipoamide moiety is regenerated for another round of catalysis by *dihydrolipoamide dehydrogenase*, which requires FAD and NAD as coenzymes. In this process an electron pair is transferred from the reduced lipoamide moiety to NAD (reaction 10.17).

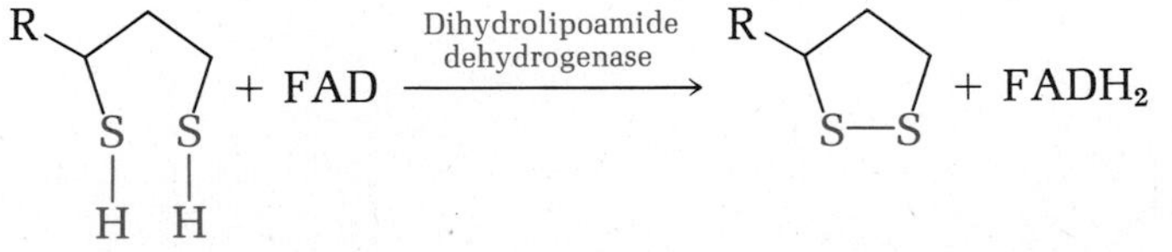

$$R \quad + \text{ FAD} \xrightarrow{\substack{\text{Dihydrolipoamide} \\ \text{dehydrogenase}}} R \quad + \text{ FADH}_2 \qquad (10.17a)$$

$$\text{FADH}_2 + \text{NAD} \longrightarrow \text{FAD} + \text{NADH} + \text{H}^{\oplus} \qquad (10.17b)$$

10−6 PYRIDOXAL PHOSPHATE *Pyridoxine*, *pyridoxal*, and *pyridoxamine* are the three members of the vitamin B_6 family. Pyridoxine, the usual commercial form of vitamin B_6, is

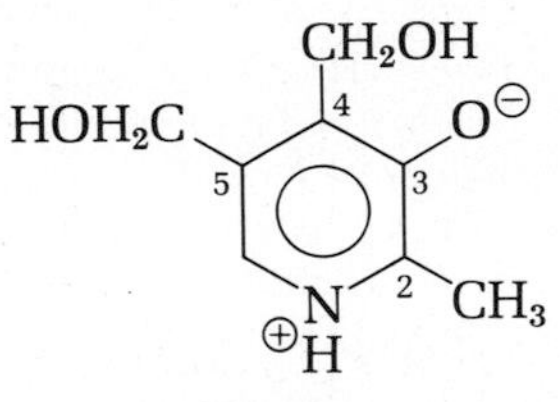

Pyridoxine 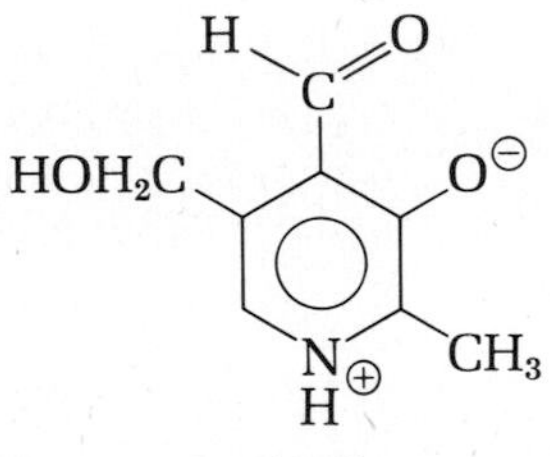   Pyridoxal 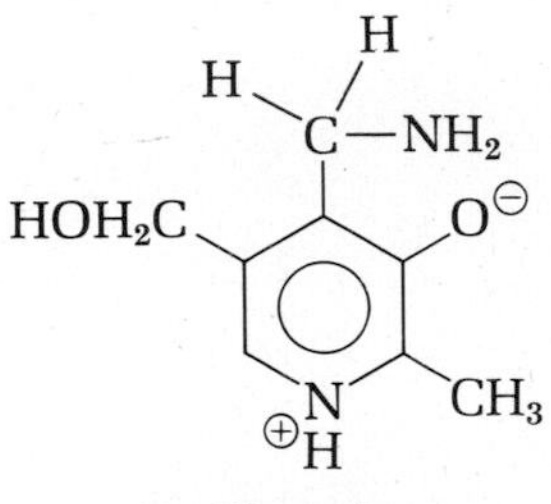   Pyridoxamine

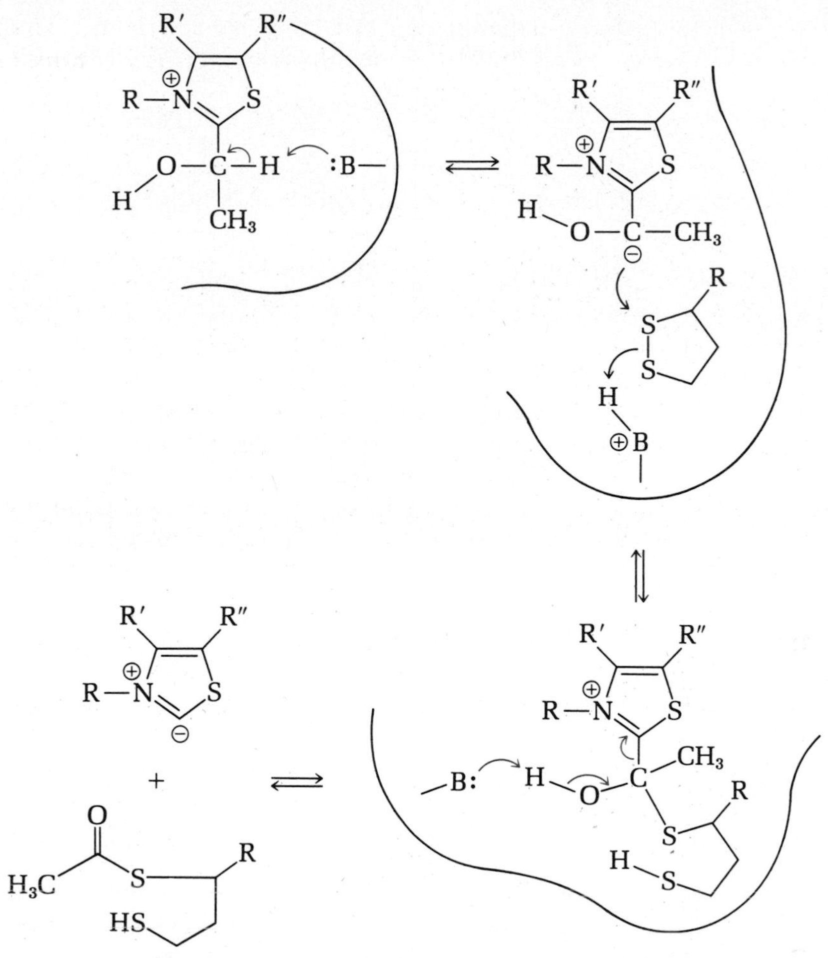

Figure 10.20 Transfer of a hydroxyethyl group from thiamine pyrophosphate (TPP) to lipoic acid.

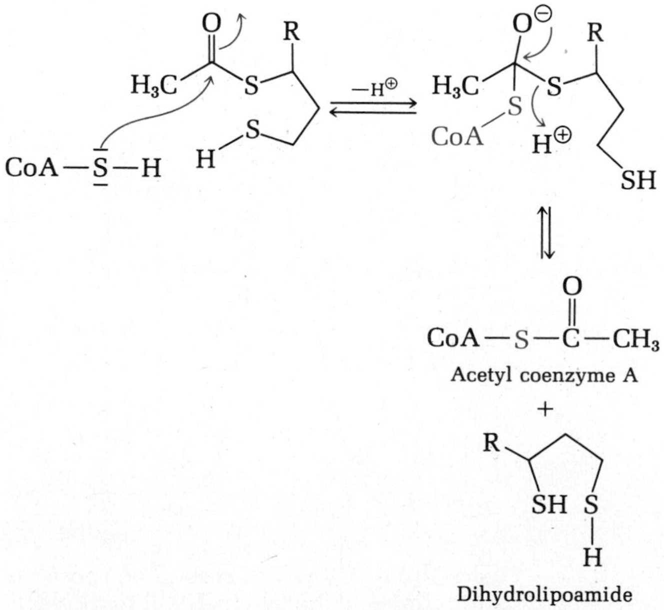

Figure 10.21 Transfer of an acetyl group from lipoic acid to coenzyme A. The reaction is a transthioesterification.

readily converted to pyridoxal *in vivo*, since benzylic alcohols are easily oxidized. *Pyridoxal phosphate* (pyridoxal-P or PLP), the active form of the

Pyridoxal phosphate

coenzyme, results from phosphorylation of the hydroxyl group of the C-5 substituent. The UV absorption spectra of pyridoxal phosphate and pyridoxamine phosphate (pyridoxamine-P or PMP) are shown in Figure 10.22.

At pH 7.0 pyridoxal-P is a dipolar ion in which the C-3 hydroxyl group is ionized and the pyridine nitrogen is protonated (reaction 10.18).

$$\tag{10.18}$$

In most pyridoxal-P–dependent enzymes the electron-deficient carbonyl carbon of pyridoxal-P reacts with the ε-amino group of an active site lysine to give an *aldimine*, or *Schiff base*, between the enzyme and pyridoxal-P (reaction 10.19).

The imine nitrogen is basic, and at pH 7.0 an equilibrium is established between the aldimine and the protonated *aldiminium ion*, whose positive charge is stabilized by the adjacent negatively charged oxygen atom at C-3 of the pyridine ring (reaction 10.20).

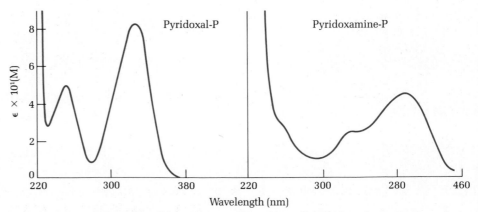

Figure 10.22   UV-visible absorption spectra of pyridoxal phosphate (PLP) and pyridoxamine phosphate (PMP). [From C. Walsh, "Enzymatic Reaction Mechanisms," Freeman, San Francisco, 1979, p. 783.]

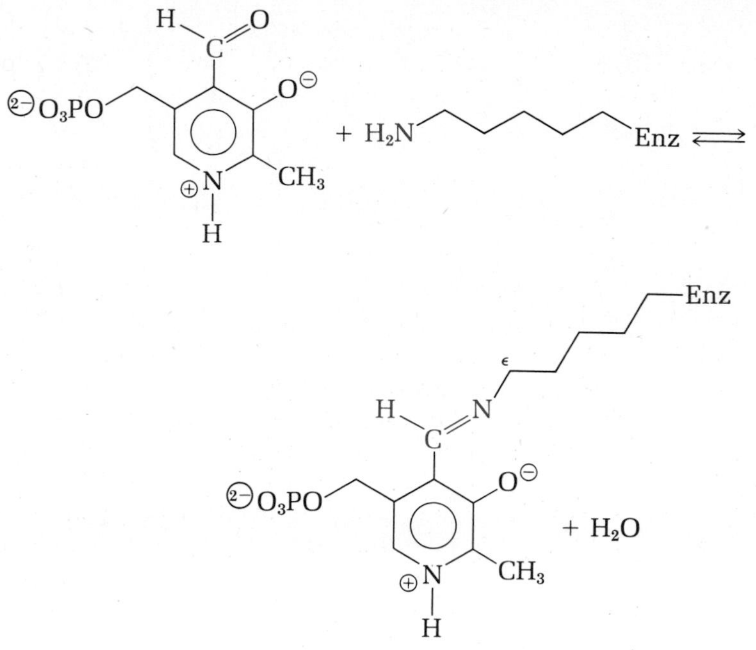

$$+ \; H_2N-(CH_2)_4-Enz \;\rightleftharpoons$$

(structure: pyridoxal phosphate aldehyde, $^{2-}O_3PO$ substituent, $O^{\ominus}$, CH_3, pyridinium $\overset{\oplus}{N}-H$)

$$+ \; H_2O \qquad (10.19)$$

Aldimine (Schiff base)

The positively charged pyridinium ion, in conjunction with the positively charged imine nitrogen, acts as an electron sink in pyridoxal phosphate–catalyzed reactions. The electrophilic aromatic ring of pyridoxal phosphate stabilizes carbanions through an electron withdrawing, delocalized π system.

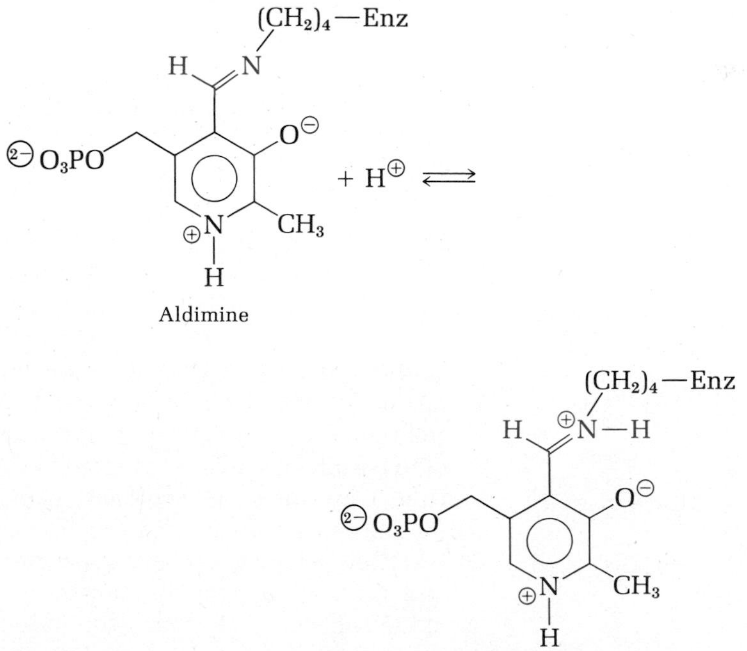

Aldimine

$$+ \; H^{\oplus} \;\rightleftharpoons$$

$$(10.20)$$

Aldiminium ion

A. Transamination In a *transamination* reaction the amino group of one reactant exchanges with the carbonyl oxygen of a second reactant (reaction 10.20).

In all known transaminases an ϵ-amino group of a lysyl side chain of the enzyme forms an aldimine with pyridoxal-P as described above. The aldimine then exchanges with the α-amino group of a substrate amino acid in a *transaldimination* (reaction 10.21).

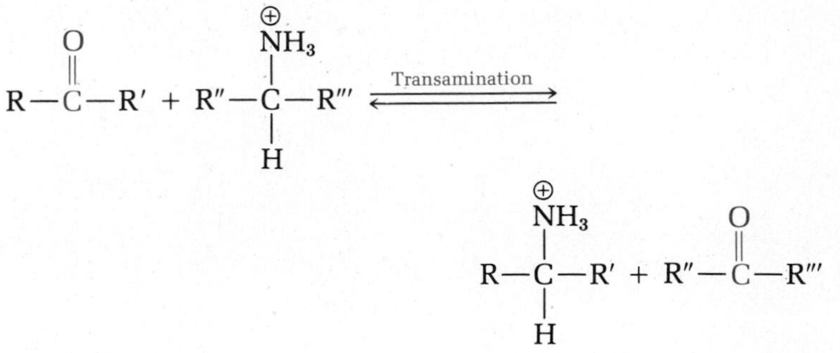

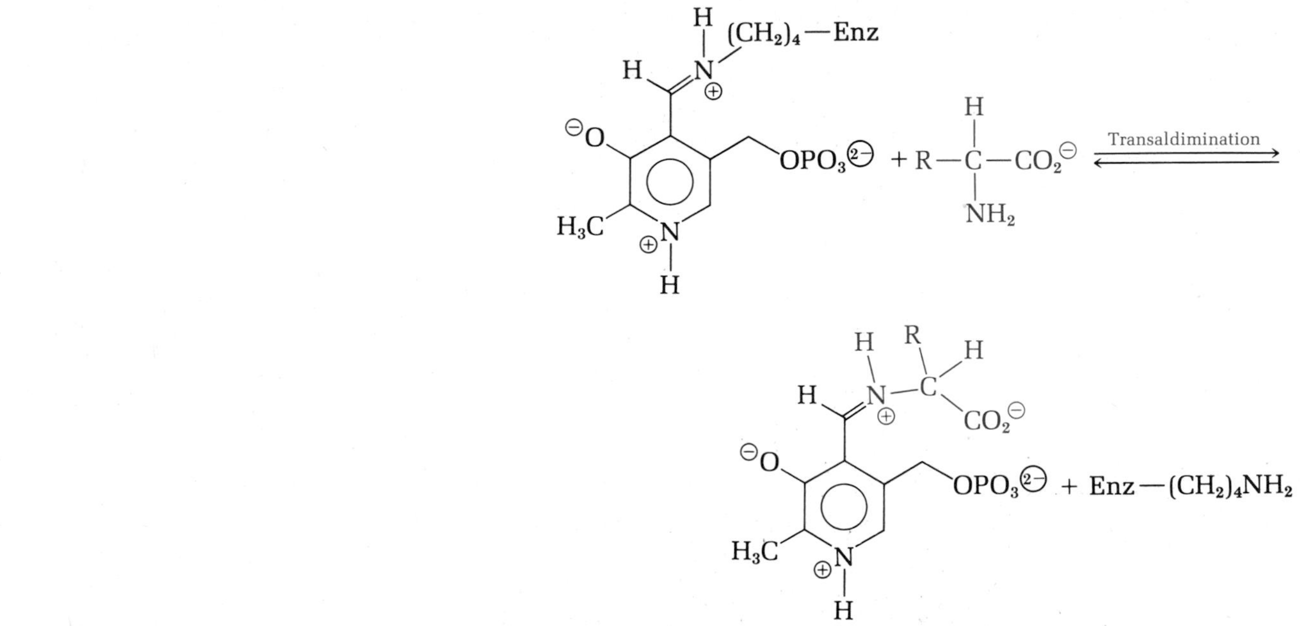

The new aldimine undergoes base-catalyzed tautomerization to a *ketimine*. This tautomerization is aided by the positive charge on the imine nitrogen which joins forces with the positively charged pyridinium ion to withdraw electrons from the C_a—H bond of the substrate amino acid enhancing its acidity (reaction 10.22). The ion formed by abstraction of the α-proton is resonance stabilized. Catalysis thus combines two effects: inductive destabilization of the reactant and resonance stabilization of the product. Both of these effects can be attributed to the quaternary nitrogen of the pyridinium ring of pyridoxal-P. Protonation of the resonance-stabilized ion occurs at the imine carbon of pyridoxal-P to give the ketimine (reaction 10.23). In this reaction the original aldehyde carbon of pyridoxal-P has been reduced, and the α-carbon of the substrate amino acid has been oxidized. No net oxidation occurs because the aldehyde carbon gained two electrons and the α-carbon of the substrate lost two electrons. Hydrolysis of the ketimine, produces an α-keto acid and pyridoxamine-P (reaction 10.24).

Transaminases also use the reduced (pyridoxamine) form of the coenzyme to convert α-keto acids to α-amino acids. We could write out this series of reactions, but that is not necessary. Simply follow the sequence of reactions backwards (reactions 10.24 to 10.27) for the conversion of an α-keto acid to an α-amino acid.

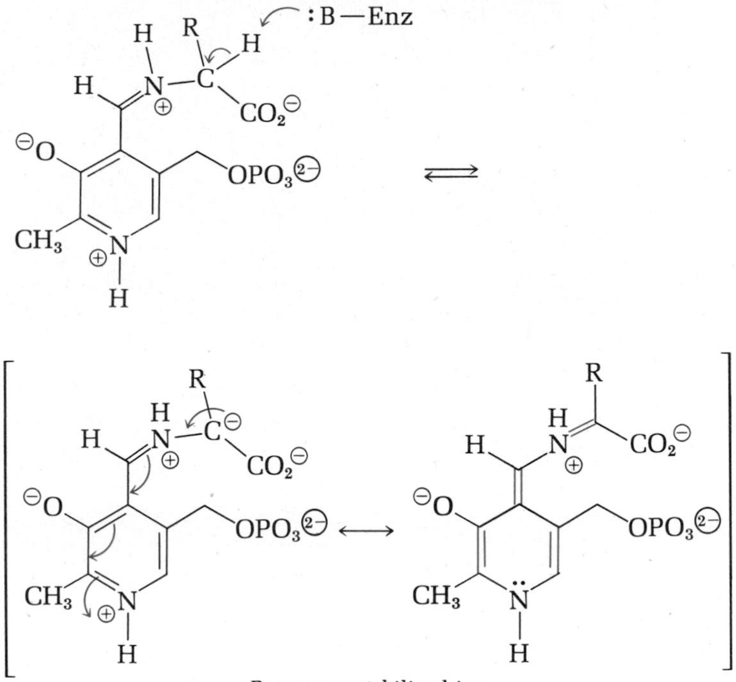

$$(10.22)$$

Resonance stabilized ion

B. Racemization   The bacterium *Pseudomonas putida* synthesizes a pyridoxal-P–dependent enzyme that catalyzes the racemization of L- or D-alanine (reaction 10.25). Alanine racemase is probably the source of the D-alanine required in the bacterial cell wall (Section 8.7). Abstraction of the α-proton of alanine from the aldimine adduct between pyridoxal-P and alanine gives a carbanion that is reprotonated to give a 1:1 mixture of D- and L-alanine from either pure enantiomer (reaction 10.26).

The bacterium *Pseudomonas graveolans* produces another pyridoxal-P–dependent racemase, *arginine racemase*, that interconverts L- and D-arginine.

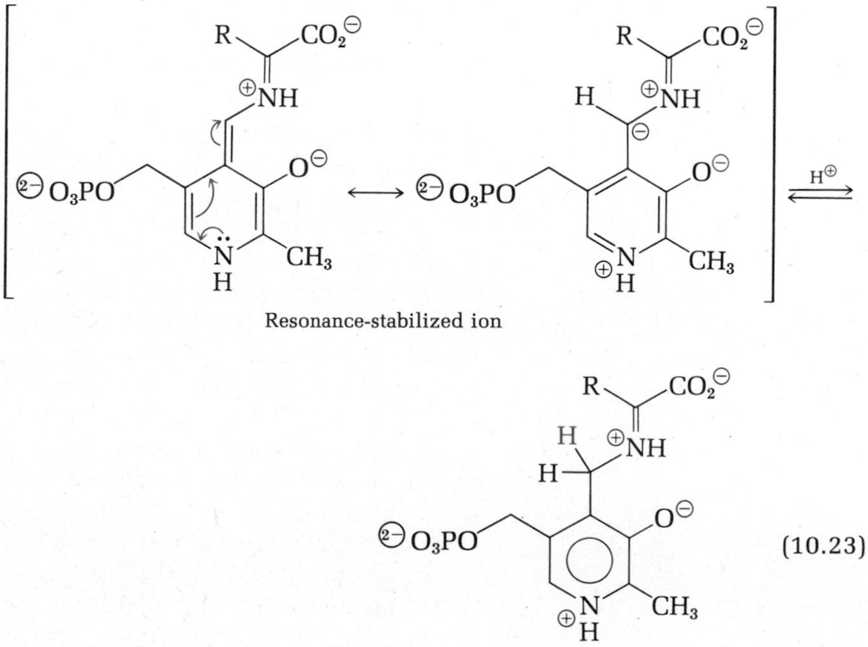

Resonance-stabilized ion

$$(10.23)$$

Ketimine

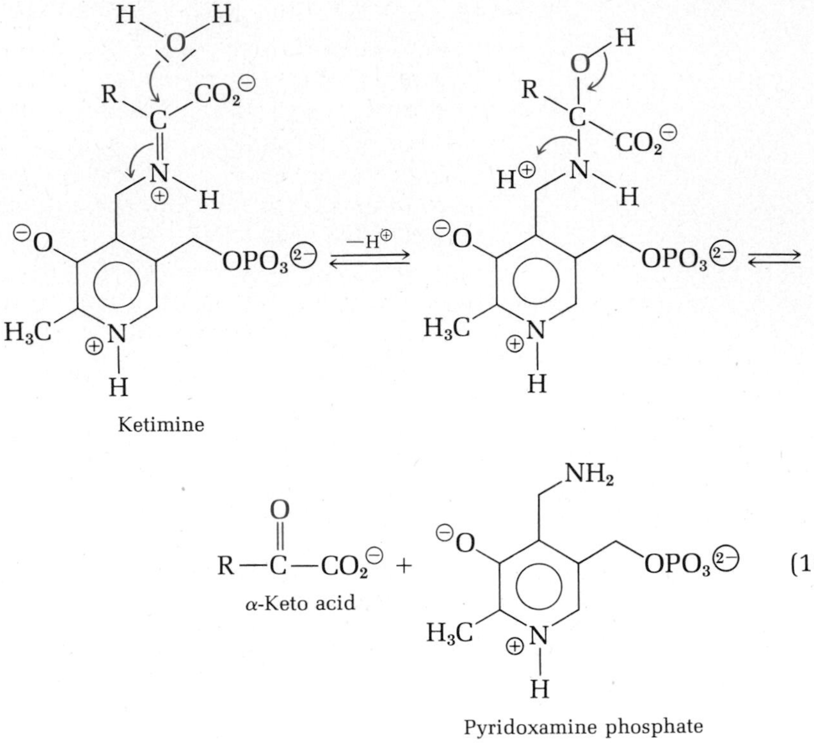

Ketimine

$$R-\overset{\overset{\displaystyle O}{\|}}{C}-CO_2^{\ominus} \; + \qquad\qquad\qquad \qquad (10.24)$$

α-Keto acid

Pyridoxamine phosphate

L-Alanine   D-Alanine   (10.25)

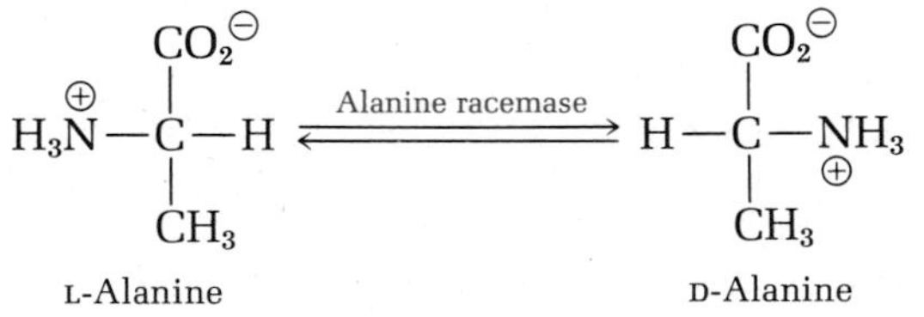

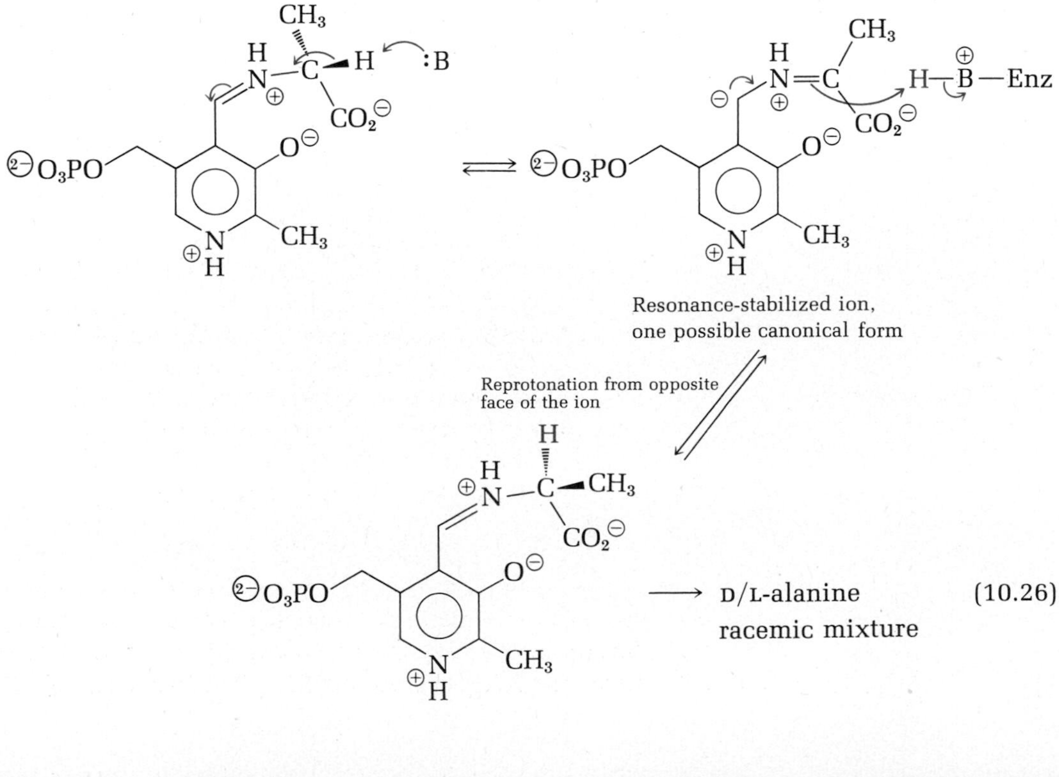

Resonance-stabilized ion,
one possible canonical form

Reprotonation from opposite
face of the ion

$\longrightarrow$ D/L-alanine   (10.26)
racemic mixture

C. Decarboxylation

a. Decarboxylation of the α-Carboxyl Group of α-Amino Acids

Perhaps a dozen decarboxylases have been discovered. All but two require pyridoxal phosphate as a coenzyme. Some of these decarboxylases have important physiological functions. For example, *histidine decarboxylase* produces *histamine,* a molecule that is important in allergic reactions, stimulates gastric secretions, and participates in a variety of hypersensitive reactions. L-*Dopa decarboxylase* participates in the metabolic pathway leading from tyrosine to *dopamine,* an intermediate in the biosynthesis of the neurotransmitters *epinephrine* and *norepinephrine* (Section 22.8C). The same enzyme decarboxylates 5-hydroxytryptophan to produce *serotonin,* another neurotransmitter (reactions 10.27 and 10.28).

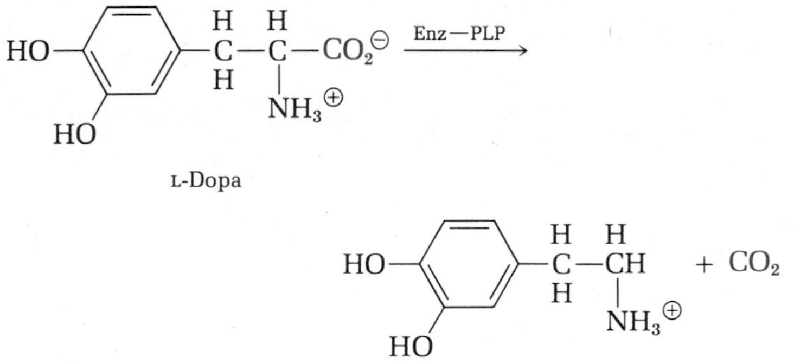

L-Dopa

$$\text{(10.27)}$$

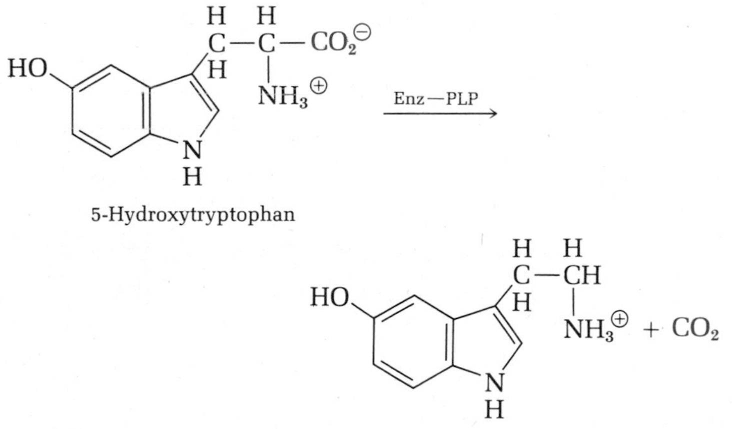

Dopamine

5-Hydroxytryptophan

$$\text{(10.28)}$$

Serotonin

The mechanism of decarboxylation is shown in reaction 10.29. The α-proton is not abstracted in this process. Nevertheless, a resonance-stabilized ion is again produced when carbon dioxide is removed from the α-carbon of the aldimine. Thus, a second central feature of the mechanism, a common feature of pyridoxal-P—catalyzed reactions, is the ability of the pyridinium ring, acting as an electron sink, to stabilize carbanions. Protonation of the resonance-stabilized carbanion at the α-carbon (reaction 10.30a) followed by transaldimination by an ϵ-amino group of an active site lysyl residue (reaction 10.30b) gives the decarboxylated product.

b. Decarboxylation at the β-carbon: Aspartate-β-Decarboxylase

In contrast to the α-decarboxylases, which do not abstract the C_α-proton, but catalyze cleavage of the α-carboxyl group, aspartate-β-decarboxylase begins its catalytic activities by removing the α-proton to produce an extensively delocalized ion (reaction 10.31).

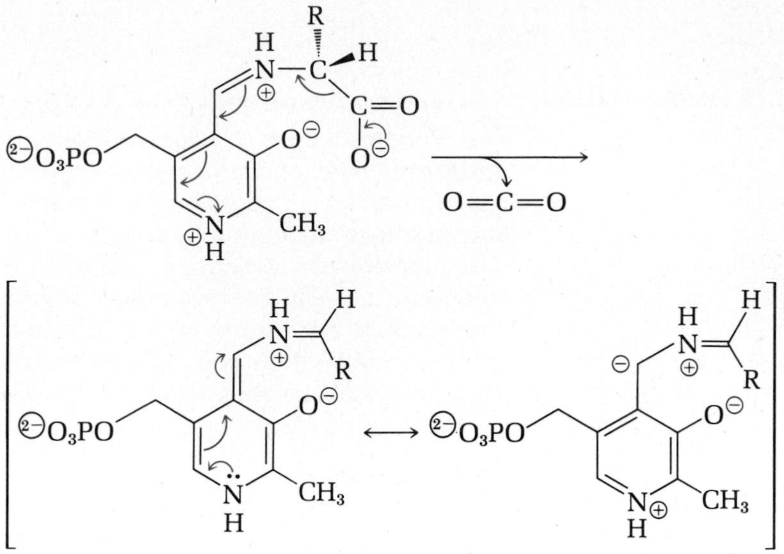

$$\text{(10.29)}$$

Resonance-stabilized ion, two of many possible canonical forms

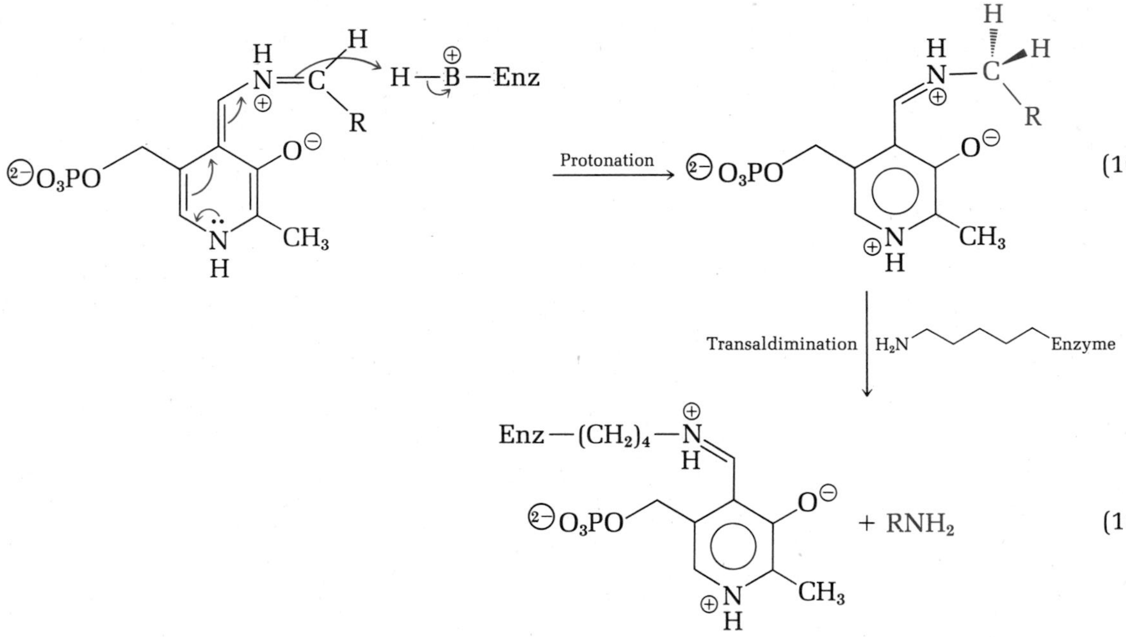

$$\text{(10.30a)}$$

Transaldimination

$$\text{Enz}-(CH_2)_4-N$$

$$\text{(10.30b)}$$

$$+ RNH_2$$

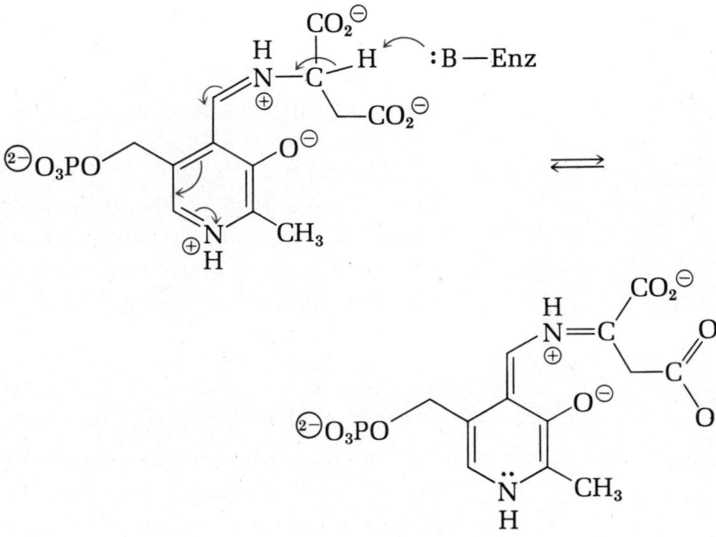

$$\text{(10.31)}$$

(1) Resonance stabilized ion

The decarboxylation step is a 1-4 elimination (reaction 10.32). Once more the major feature of the mechanism is the flow of electrons toward the electropositive quaternary nitrogen.

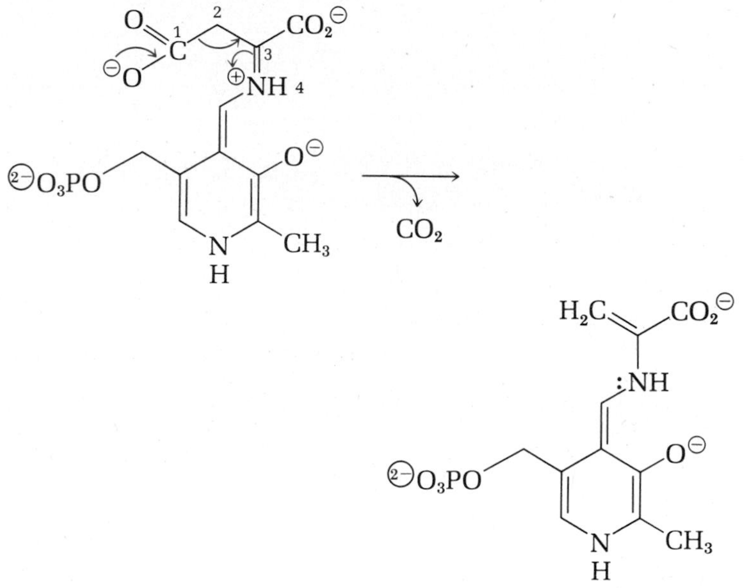

(10.32)

Let us now isolate the amine nitrogen and the remnant of aspartate in structure 2.

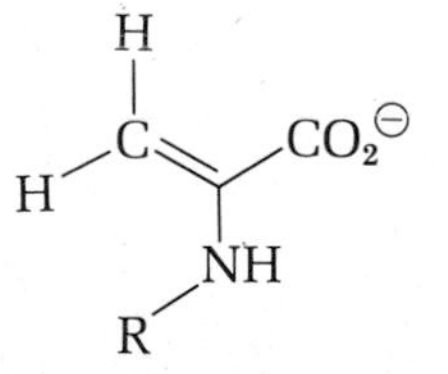

(3) Enamine

Structure 3 is an *enamine*. Tautomerization of the enamine, followed by protonation of the imine nitrogen converts the methylene group to a methyl group (reaction 10.33).

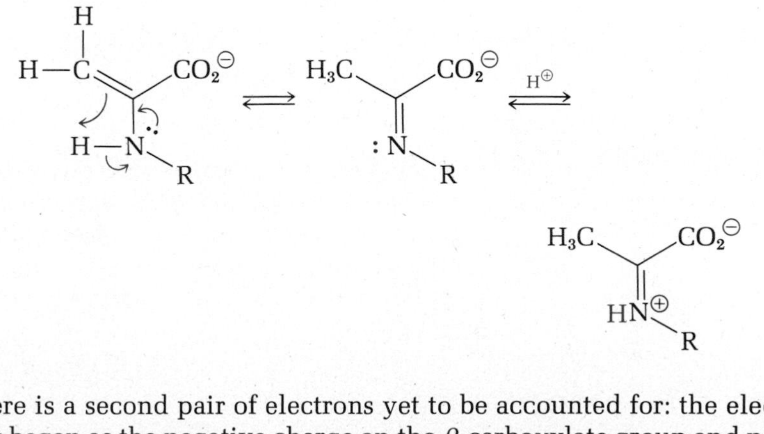

(10.33)

There is a second pair of electrons yet to be accounted for: the electron pair that began as the negative charge on the β-carboxylate group and now stored on the pyridine ring nitrogen. This electron pair flows from the pyridine nitrogen to the original α-carbon of aspartate. The α-carbon is protonated on

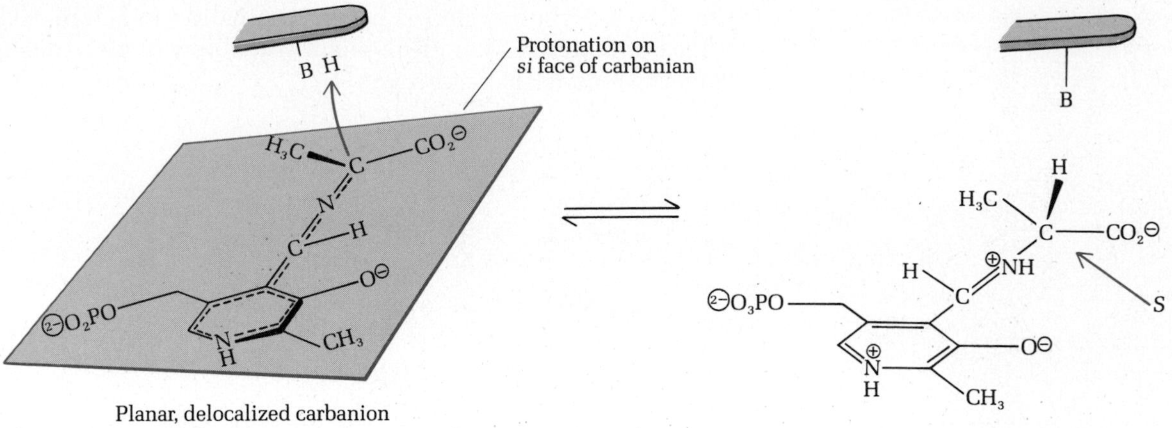

Figure 10.23 Protonation of α-carbon on the *si* face. The delocalized carbanion is produced by decarboxylation of the β-carboxyl group of asparte. Protonation on the *si* face produces the S isomer at the α-carbon of the original aspartate molecule. This carbon will emerge as the α-carbon of L-alanine.

the *si* face to give the S configuration at the α-carbon as shown in Figure 10.23. Transaldimination by an ϵ-amino group of an active site lysine releases L-alanine (reaction 10.34).

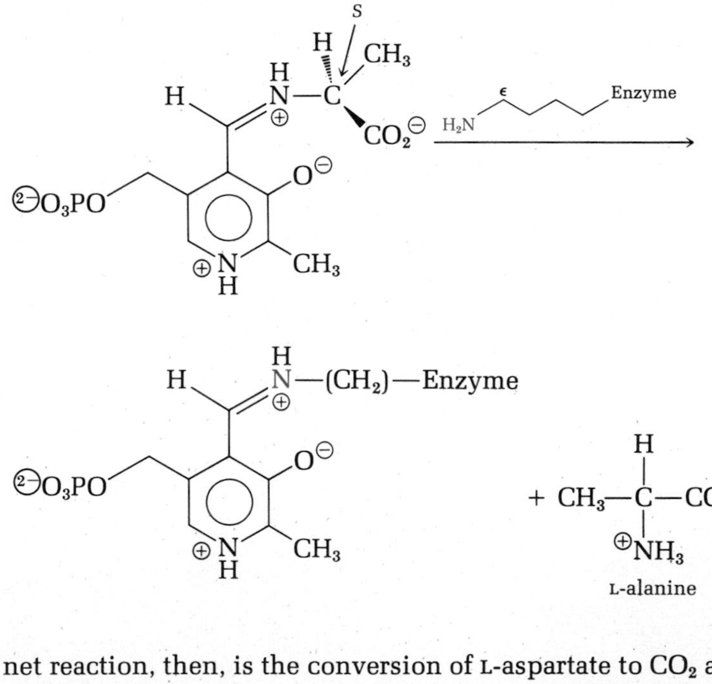

$$\tag{10.34}$$

The net reaction, then, is the conversion of L-aspartate to CO_2 and L-alanine.

Decarboxylation does not exhaust the possibilities for pyridoxal-P–catalyzed reactions at the β-carbon of amino acids. We shall see examples of other types of reactions at the β-carbon for serine, cysteine, tyrosine, and tryptophan when we discuss amino acid metabolism in Chapters 21 and 22.

10–7 BIOTIN   The structure of biotin, also called vitamin H, is shown below. As may be seen from the portion of biotin that is colored, biotin can be regarded as a substituted urea, and the biochemical reactions of biotin revolve around the

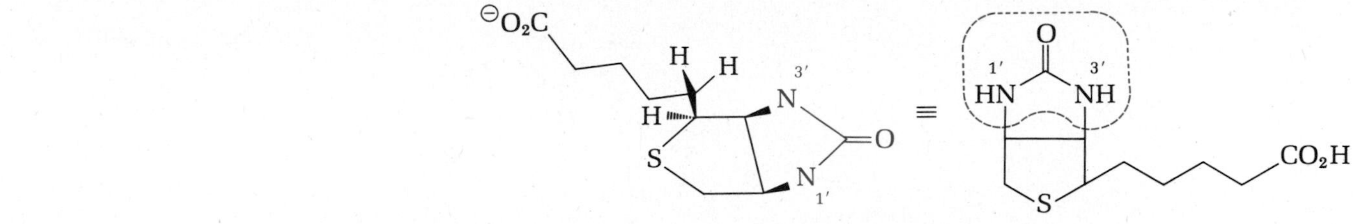

Biotin (vitamin H)

N-1′ nitrogen. The carboxyl group of the biotin side chain is covalently bound to an ϵ-amino group of a specific lysyl residue in biotin-containing enzymes. The ϵ-*N*-biotinyl lysyl derivative of biotin is called *biocytin.*

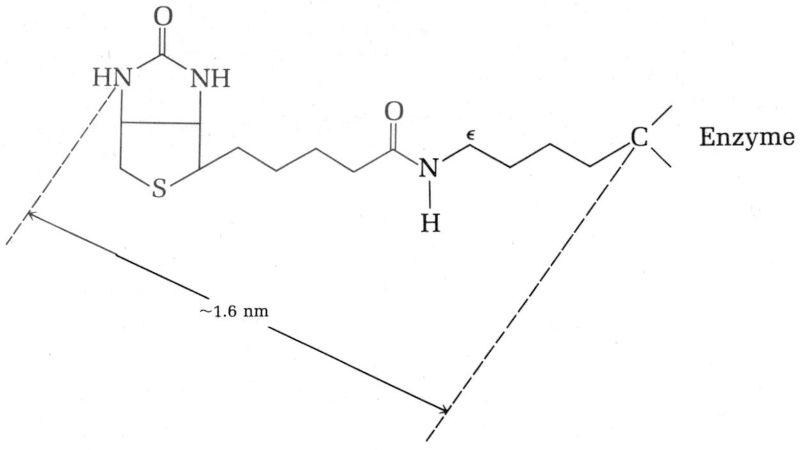

Biocytin

Biotin is a coenzyme in two types of enzyme-catalyzed reactions: (1) transfer of a carboxyl group from one substrate to another, as exemplified by transcarboxylase, and (2) ATP-dependent carboxylations.

Transcarboxylase catalyzes the transfer of a carboxylate group from *methylmalonyl coenzyme A* to pyruvate in two steps (reactions 10.35 and 10.36).

$$CH_3-\underset{\underset{H}{|}}{\overset{\overset{CO_2^{\ominus}}{|}}{C}}-CO-SCoA + \text{Enz-biotin} \rightleftharpoons$$

Methylmalonyl coenzyme A

$$CH_3CH_2\overset{\overset{O}{\|}}{C}-S-CoA + \text{Enz-biotin-}CO_2^{\ominus} \qquad (10.35)$$

Propionyl coenzyme A

$$\text{Enz-biotin-}CO_2^{\ominus} + CH_3\overset{\overset{O}{\|}}{C}-CO_2^{\ominus} \rightleftharpoons {}^{\ominus}O_2C-CH_2\overset{\overset{O}{\|}}{C}-CO_2^{\ominus} \qquad (10.36)$$

Oxaloacetate

Transcarboxylase has been found only in propionic acid bacteria. A proposal for a concerted reaction mechanism for transfer of CO_2 to pyruvate is shown

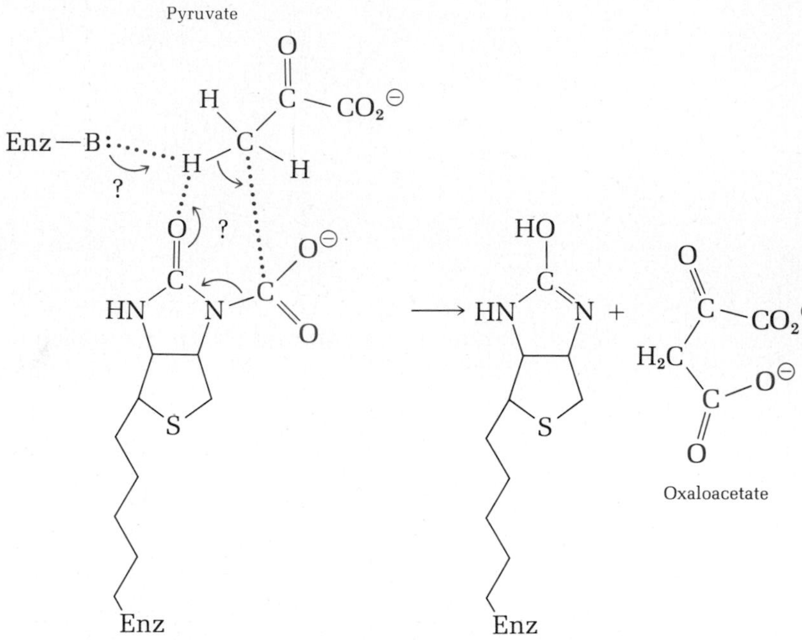

Figure 10.24 Proposed concerted mechanism for the second step of transcarboxylase action (reaction 11.36). In this mechanism proton abstraction from the methyl group of pyruvate is concomitant with the carboxyl transfer to the methyl group that generates oxaloacetate. It is not known whether the proton-abstracting group is the ureido oxygen of carboxybiotin or a basic residue of the enzyme. [From W. C. Stallings, C. T. Monti, M. D. Lane, and D. G. DeTitta, *Proc. Nat. Acad. Sci. U.S., 77*, 1263 (1980).]

in Figure 10.24. We shall see later that a concerted reaction mechanism is not the only possibility.

The *ATP-dependent biotin carboxylases* are much more common. Their reactions can be summarized by reaction 10.37, where the generalized substrate is either an α-keto acid or an acyl-CoA thioester.

$$\text{biotin-CO}_2^{\ominus} + \text{H} - \overset{|}{\underset{|}{\text{C}}} - \overset{O}{\overset{||}{\text{C}}} - \;\rightleftharpoons\; \text{biotin} + {}^{\ominus}\text{O}_2\text{C} - \overset{|}{\underset{|}{\text{C}}} - \overset{O}{\overset{||}{\text{C}}} - \qquad (10.37)$$

"Biotin-$CO_2^{\ominus}$," to which we have now twice referred, is N^1-*carboxybiotin*. N^1-carboxybiotin is formed by biotin carboxylase in a reaction that requires

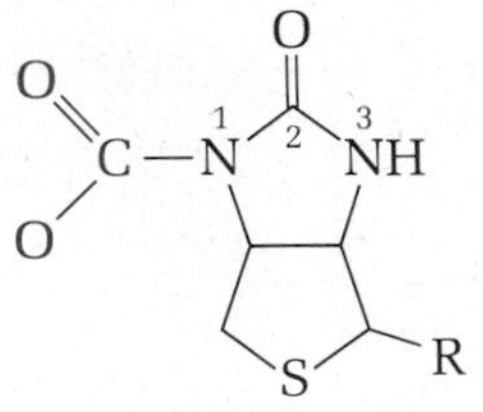

N^1-Carboxybiotin

ATP. A mixed anhydride, *carbonyl phosphate*, is an intermediate (reaction 10.38). Amides are poor nucleophiles. In biotin, however, there is competing resonance between N-1 and N-3. As a result the nitrogens of biotin are about as nucleophilic as the nitrogens of secondary amines. N-1 is probably the reac-

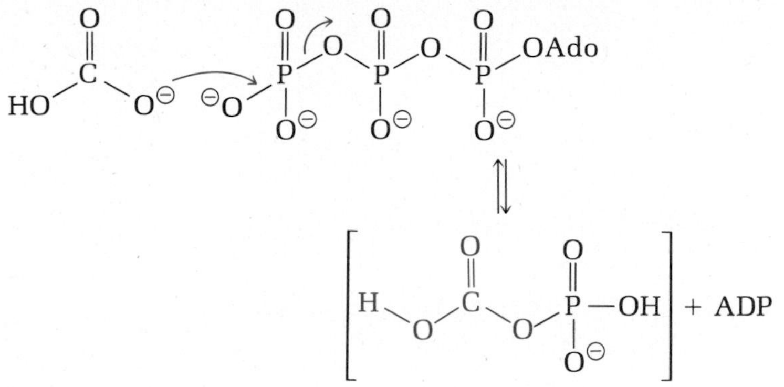

$$\left[\begin{array}{c} H-O-\overset{\overset{\displaystyle O}{\|}}{C}-O-\overset{\overset{\displaystyle O}{\|}}{\underset{\underset{\displaystyle O^{\ominus}}{|}}{P}}-OH \end{array} \right] + \text{ADP} \qquad (10.38)$$

Carbonyl phosphate

tive nitrogen because of severe steric hindrance at N-3 caused by the proximity of the R group.

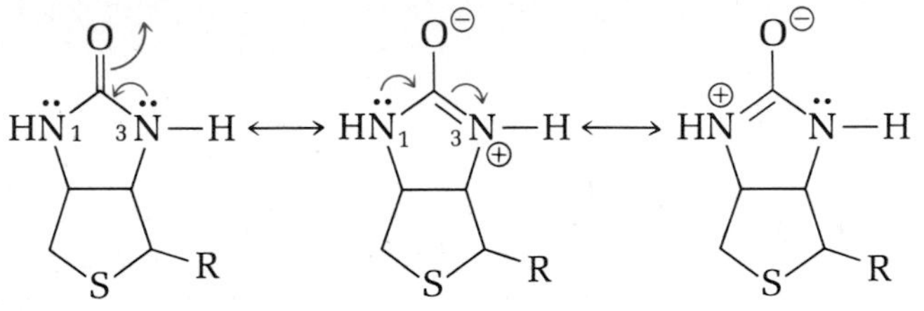

Resonance structures for biotin

The intermediate, carbonyl phosphate, is attacked by the electron pair on N-1 of biotin. Phosphate is the leaving group and N^1-carboxybiotin is formed (reaction 10.39).

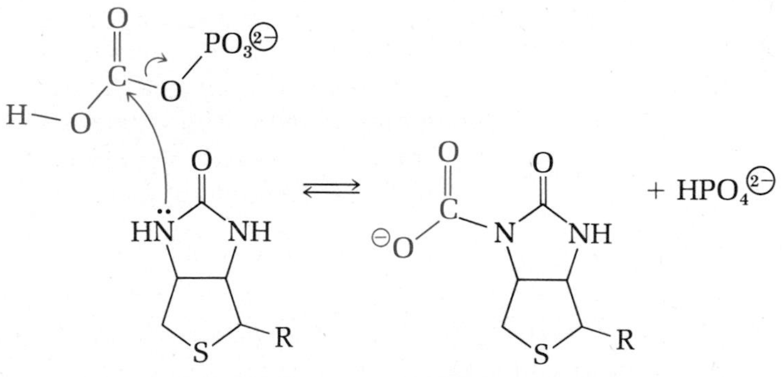

$$+ \text{HPO}_4^{\textcircled{2}\ominus} \qquad (10.39)$$

N^1-Carboxybiotin

Let us consider a typical reaction of a biotin carboxylase having an acetyl-CoA substrate. The enzyme abstracts the α-hydrogen (we recall that enolate anions of thioesters are more stable than enolate anions of carboxylate esters). The enolate anion then attacks the carboxyl group of N^1-carboxybiotin. Biotin itself is the leaving group and the product is a β-keto acid (reaction 10.40).

In our discussion of transcarboxylase we showed a proposal for a concerted transfer of CO_2 from N^1-carboxybiotin to methylmalonyl coenzyme A to produce propionyl coenzyme A. The question whether or not carboxylations involving biotin are concerted has not, however, been resolved. Both transcarboxylase and propionyl coenzyme A carboxylase catalyze elimination of HF from β-fluoropropionyl coenzyme A to form acrylyl coenzyme A without formation of carboxylated products. Elimination occurs at the same

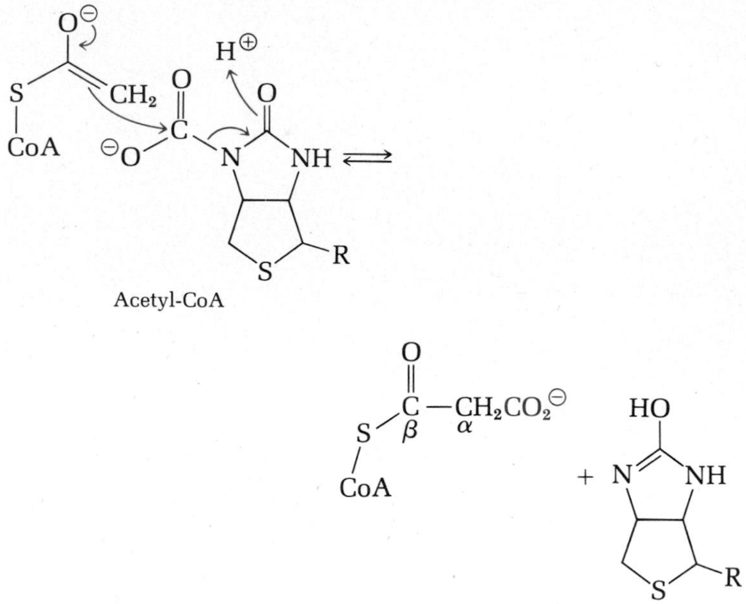

Acetyl-CoA

$$\text{(10.40)}$$

Malonyl-CoA
(β-keto acid)

rate as carboxylation. These observations have led to a proposal that biotin-dependent carboxylations normally occur with intermediate formation of a carbanion, rather than as concerted processes (Figure 10.25).

10–8 TETRAHYDROFOLIC ACID AND ONE-CARBON METABOLISM

The transfer of one-carbon fragments at the oxidation level of methanol, formaldehyde, and formic acid is catalyzed by tetrahydrofolic acid–dependent enzymes. Tetrahydrofolic acid, FH_4, is composed of a *pteridine ring system*, *p*-aminobenzoic acid, and L-glutamate. The active coenzyme, tetrahydrofolate, is produced by enzymatic reduction of the 5, 6, 7, and 8 positions of the vitamin *folic acid*. The structure shown has a single glutamate, but the active coenzyme can have anywhere from one to seven glutamates polymerized with the *p*-aminobenzoic acid residue. The reduction of folic acid to

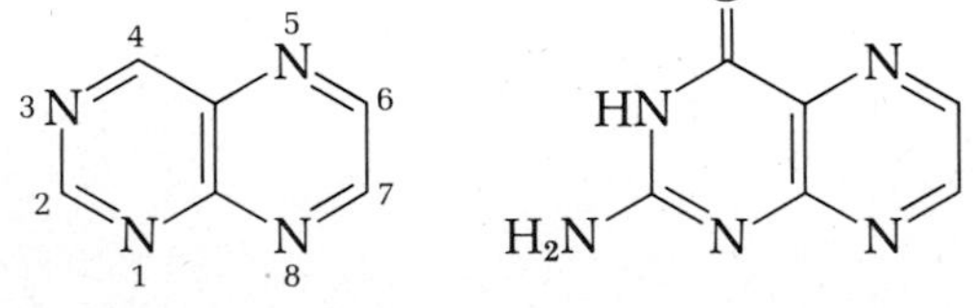

Pteridine ring system

Pterin
(2-amino-4-oxopteridine)

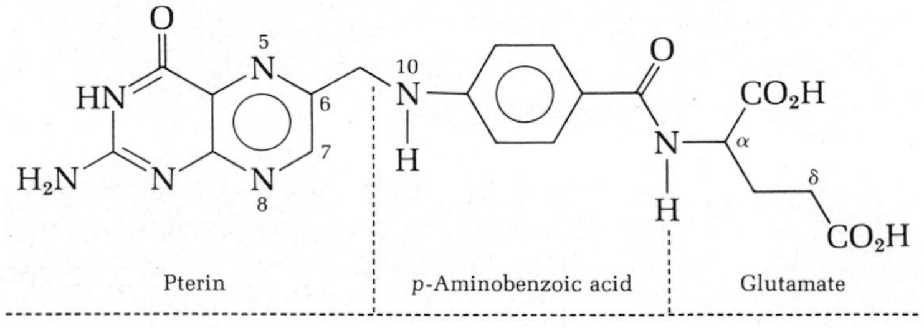

Pterin   *p*-Aminobenzoic acid   Glutamate

Folic acid

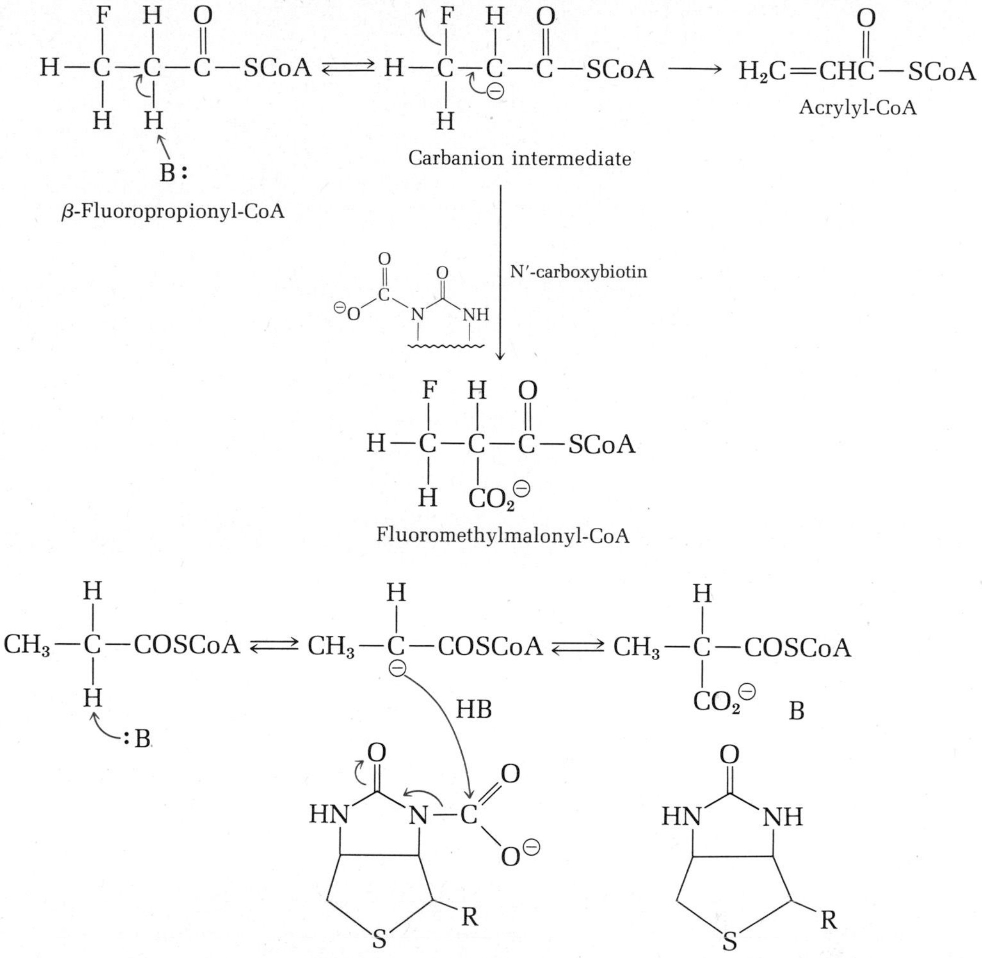

Figure 10.25 Transcarboxylase and propionyl-CoA carboxylase catalyze elimination of HF from β-fluoropropionyl-CoA to form acrylyl-CoA. No carboxylated products are detected, and elimination occurs at the same rate as the usual carboxylation reactions. This result implies that a carbanion intermediate forms not only in the elimination reactions but in the usual carboxylation reaction as well, and that the mechanism is not concerted. [From J. A. Stubbe, S. Figh, and R. H. Abeles, *J. Biol. Chem.*, 255, 236–242 (1980).]

the active coenzyme tetrahydrofolate occurs in two steps and is catalyzed by the NADPH-dependent enzyme dihydrofolate reductase.

The molecular complex of dihydrofolate reductase and the anticancer agent *methotrexate* is shown in Figure 10.26. Methotrexate differs from tetrahydrofolate by a methyl group at N-10 and by the replacement of the 4-oxo group with an amino group. Methotrexate binds the reductase about 10^3 times more tightly than tetrahydrofolate itself. The inhibition constant, K_i, for the *E. coli* enzyme is $2.4 \times 10^{-9} M$. Methotrexate therefore effectively inhibits biochemical reactions that are dependent upon tetrahydrofolate. One of these reactions is catalyzed by *thymidylate synthetase* (dTMP synthetase), discussed later in this chapter. The inhibition of thymidylate synthetase slows DNA biosynthesis or stops it altogether, thereby slowing the growth of cancer cells.

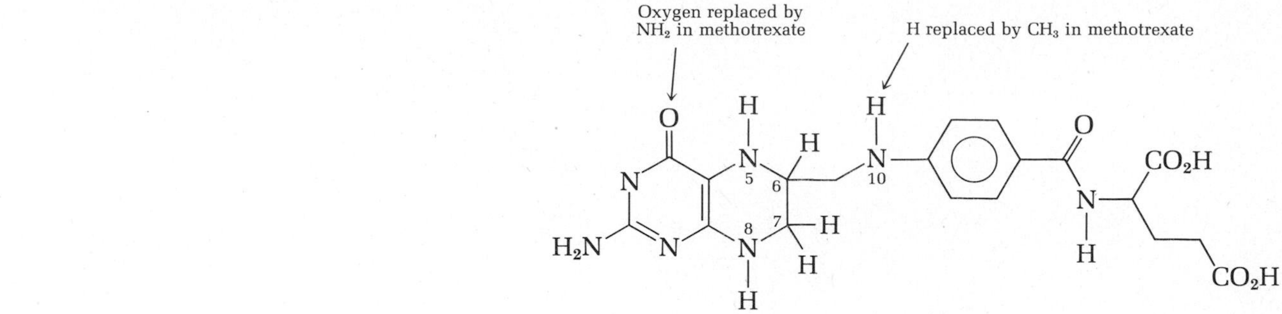

Oxygen replaced by NH_2 in methotrexate

H replaced by CH_3 in methotrexate

5,6,7,8-Tetrahydrofolic acid (FH_4)

The one-carbon transfer reactions of FH_4 involve a covalent bond between N-5 and N-10 or both and the one-carbon fragment. The one-carbon fragment is then transferred to another substrate. The structures of the one-carbon derivatives of FH_4 are shown in Table 10.8.

The chemistry of FH_4 is illustrated by its reaction with formaldehyde. First, N-5 is more basic and more nucleophilic than N-10, since the electron pair of N-10 can be delocalized into the aromatic ring, whereas the electron pair of N-5 cannot be delocalized. N-5 attacks formaldehyde to produce a carbinolamine intermediate. Loss of water gives an iminium ion which is then attacked by N-10 to give N^5,N^{10}-*methylenetetrahydrofolate* (Figure 10.27). This reaction occurs rapidly in the absence of enzyme. In aqueous solution at pH 7.0 the equilibrium constant for formation of the methylene adduct is greater than 10^4. Among the large number of reactions possible with FH_4 derivatives we will consider the synthesis of N^5,N^{10}-methylenetetra-

Table 10.8 Structures of the one-carbon derivative of tetrahydrofolate

N^5-Formyl-FH_4

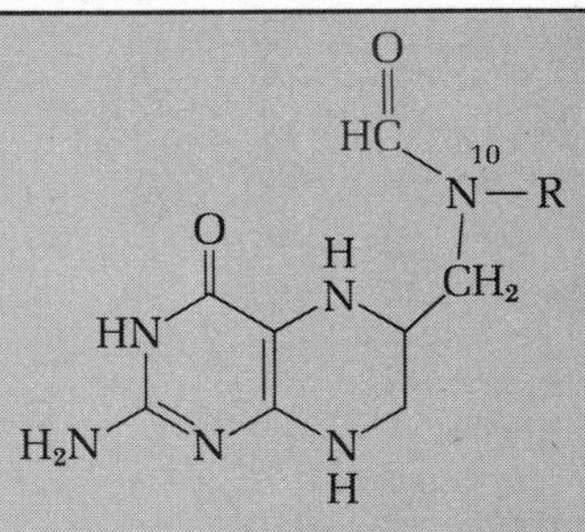

N^{10}-Formyl-FH_4

N^5-Formimino-FH_4

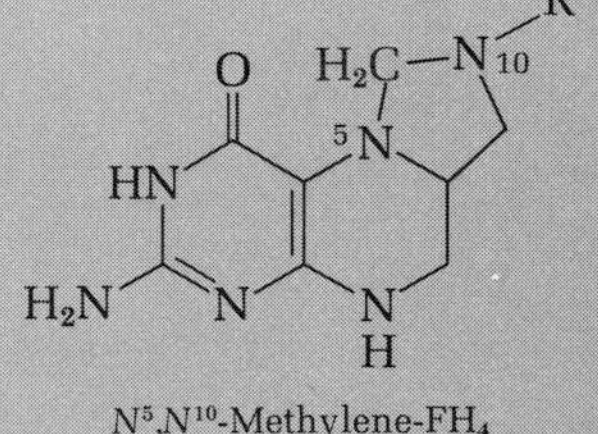

N^5,N^{10}-Methylene-FH_4

N^5,N^{10}-Methenyl-FH_4

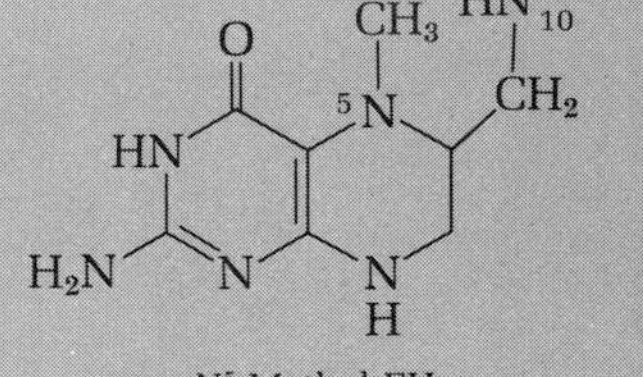

N^5-Methyl-FH_4

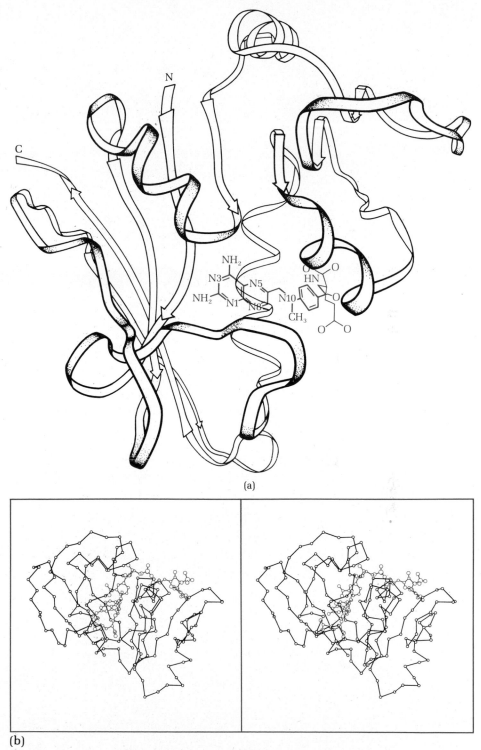

(a)

(b)

Figure 10.26 (a) Structure of the complex between the anticancer agent methotrexate and the NADPH-dependent enzyme dihydrofolate reductase. [From D. Matthews et al., *Science*, *197*, 452 (1977). Copyright 1977 by the American Association for the Advancement of Science.] (b) Stereo view of the complex.

hydrofolate catalyzed by *serine hydroxymethyltransferase*. Then we shall see how the derivative is used in the biosynthesis of thymidylate from dUMP.

Serine hydroxymethyltransferase catalyzes the interconversion of serine and glycine (reaction 10.41).

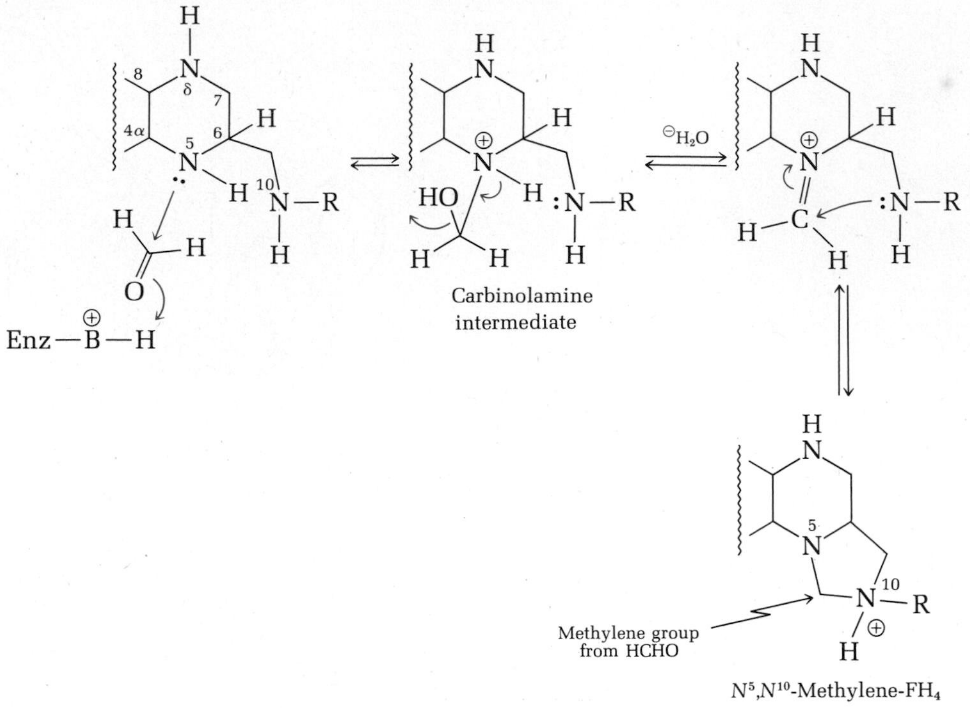

Figure 10.27   Reaction of tetrahydrofolate with formaldehyde.

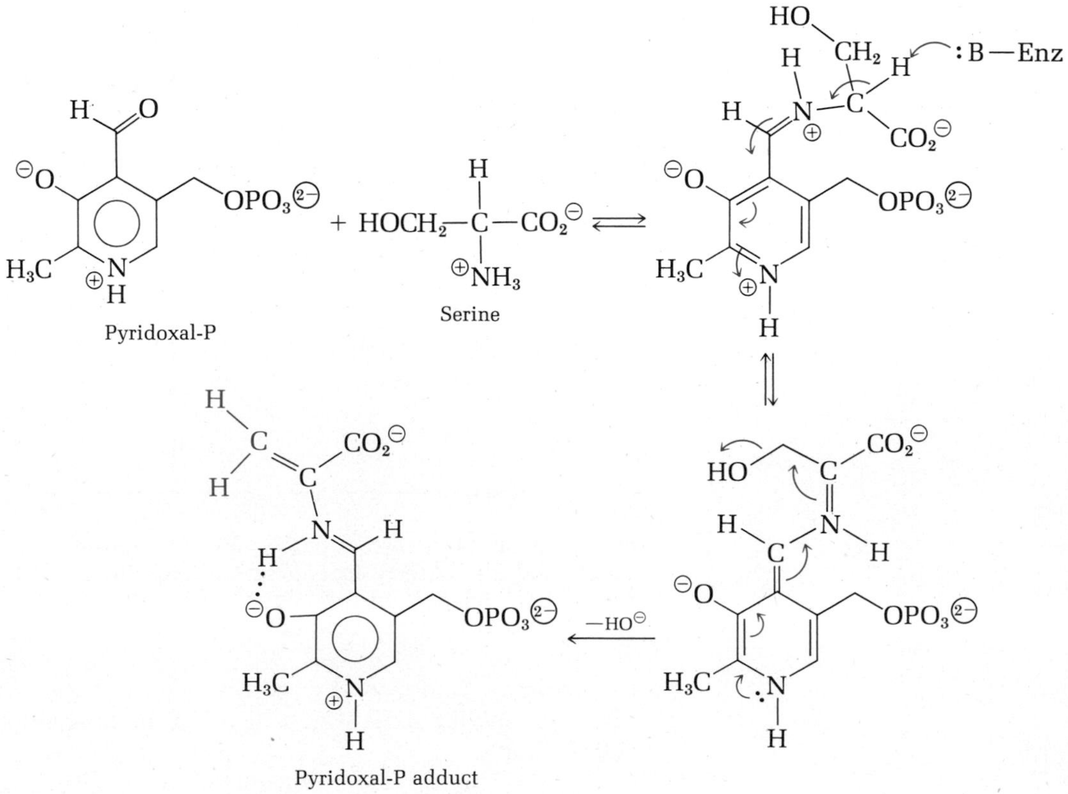

Figure 10.28   Reaction of serine with pyridoxal phosphate.

$$\text{L-serine} + FH_4 \rightleftharpoons \text{glycine} + N^5,N^{10}\text{-methylene-}FH_4 \quad (10.41)$$

Pyridoxal phosphate is also a coenzyme in the reaction. In the first step serine forms an aldimine with pyridoxal phosphate. Abstraction of the α-proton of serine and loss of the serine hydroxyl group generates the methylene group that will eventually find its way to FH_4 (Figure 10.28). Attack by N-5 of FH_4 on the methylene carbon of the pyridoxal-P derivative gives an adduct in which both pyridoxal-P and FH_4 are bound to the methylene group. Next N-10 attacks the methylene group, "PLP-glycine" is released, and the product is N^5,N^{10}-methylenetetrahydrofolate (Figure 10.29).

Thymidylate synthetase uses N^5,N^{10}-methylenetetrahydrofolate in the conversion of dUMP to dTMP as shown below (reaction 10.42).

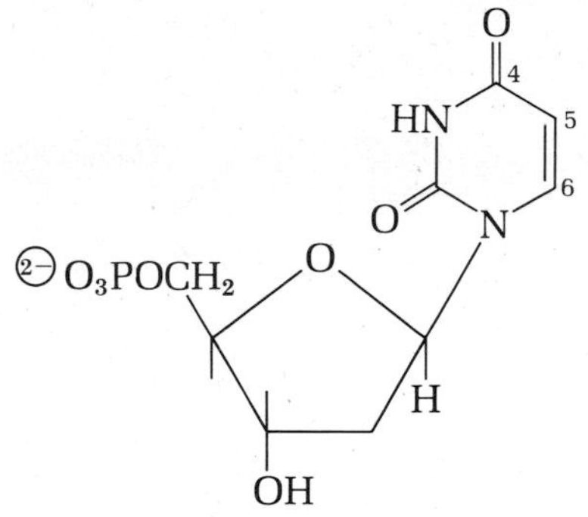

$+ N^5,N^{10}$-methylene-FH_4 $\longrightarrow$

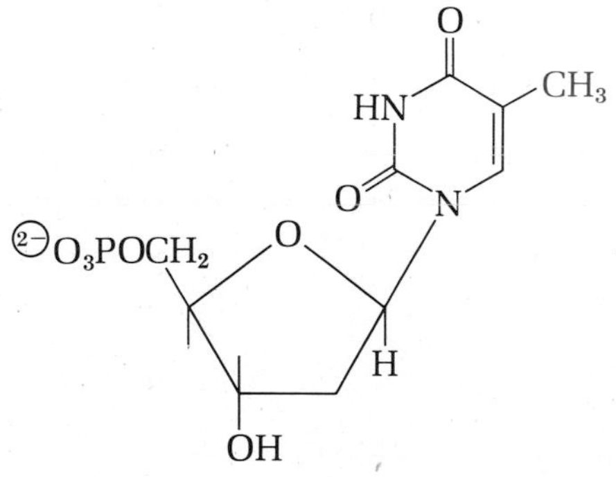

+ 7,8-Dihydrofolate (10.42)

The reaction begins by taking advantage of the α,β-unsaturated ketone concealed in C-4, C-5, and C-6 of dUMP. The β position, C-6, is susceptible to attack by nucleophiles. The enzymatic reaction is a *Michael addition* of a nucleophile to an α, β unsaturated ketone (reaction 10.43).

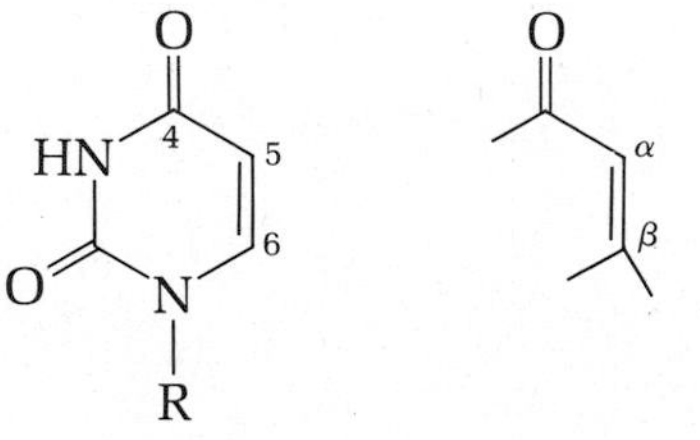

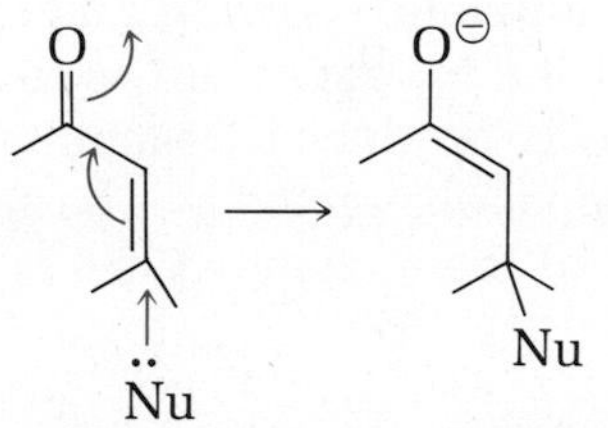

(10.43)

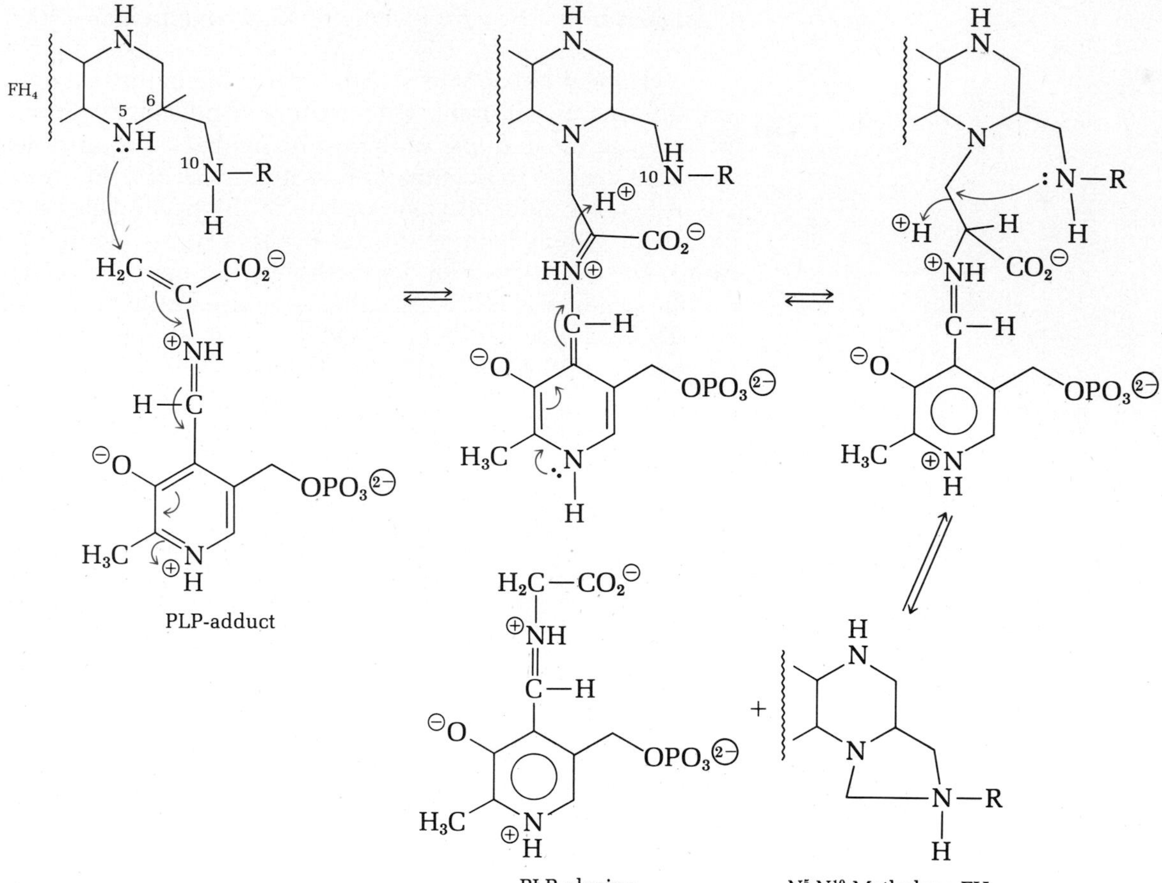

Figure 10.29 Reaction of the adduct between pyridoxal phosphate and serine with tetrahydrofolate.

The enolate anion which forms initially attacks the methylene group of N^5,N^{10}-methylenetetrahydrofolate (reaction 10.44).

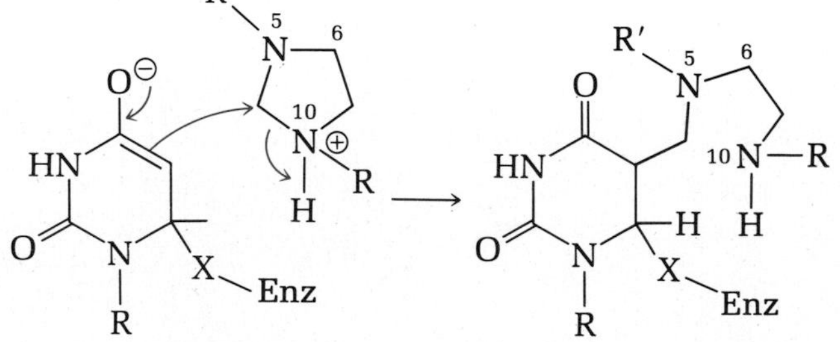

$$(10.44)$$

When tritium (^{3}H) is incorporated at C-6 of FH_4, it ends up in the methyl group of thymidylate, showing that the next step of the reaction is a 1,3 hydride shift of the hydrogen at C-6 to the methylene group. In this step 7,8-dihydrofolate is produced (reaction 10.45). The acidic hydrogen α to the carbonyl group is abstracted by a basic group on the enzyme, and the nucleophilic group of the enzyme at C-6 is displaced to produce dTMP (reaction 10.46).

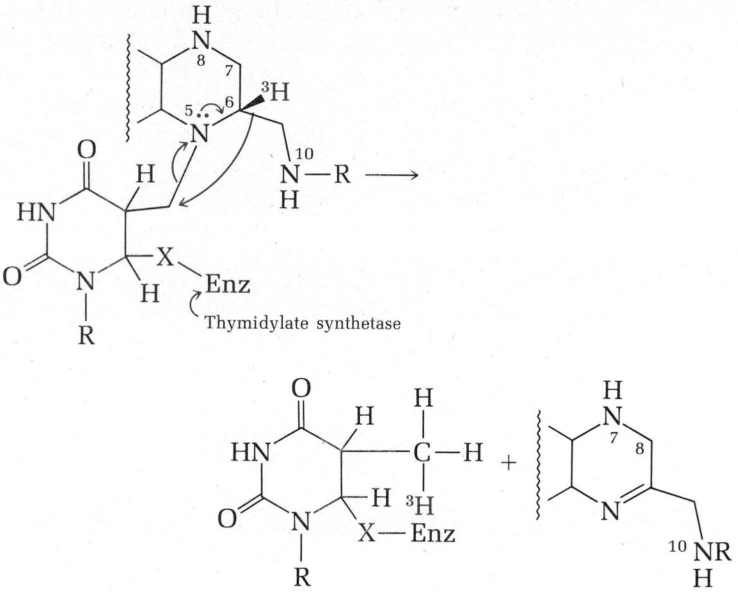

Thymidylate synthetase

7,8-Dihydrofolic acid

$$(10.45)$$

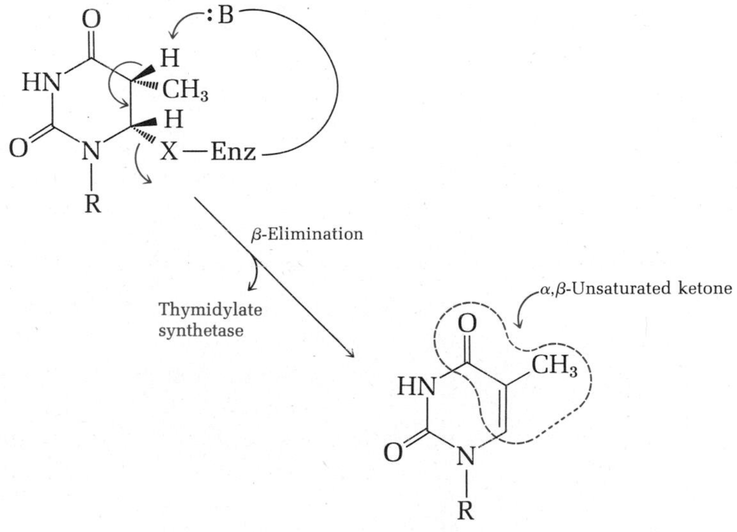

β-Elimination

Thymidylate
synthetase

α,β-Unsaturated ketone

$$(10.46)$$

dTMP

10–9 SULFANILAMIDE ANTIMETABOLITES

Mammals cannot synthesize the pteridine ring system and must obtain their folic acid from intestinal microorganisms. These microorganisms, in turn, require *p*-aminobenzoic acid for growth. Sulfanilamide, a structural analog of *p*-aminobenzoate, inhibits bacterial growth. The inhibition is strictly compet-

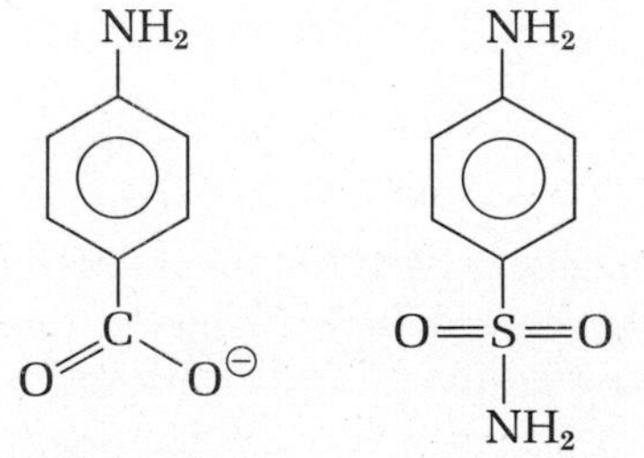

p-Aminobenzoate   Sulfanilamide

itive and can be overcome by addition of excess p-aminobenzoate. In 1940 D. D. Woods and P. Fildes proposed that sulfanilamide (and its derivatives) is an *antimetabolite* that blocks a specific enzyme reaction. This hypothesis is correct. The bacterial biosynthesis of *dihydropteroic acid*, the direct precursor of folic acid, requires p-aminobenzoate (reaction 10.47).

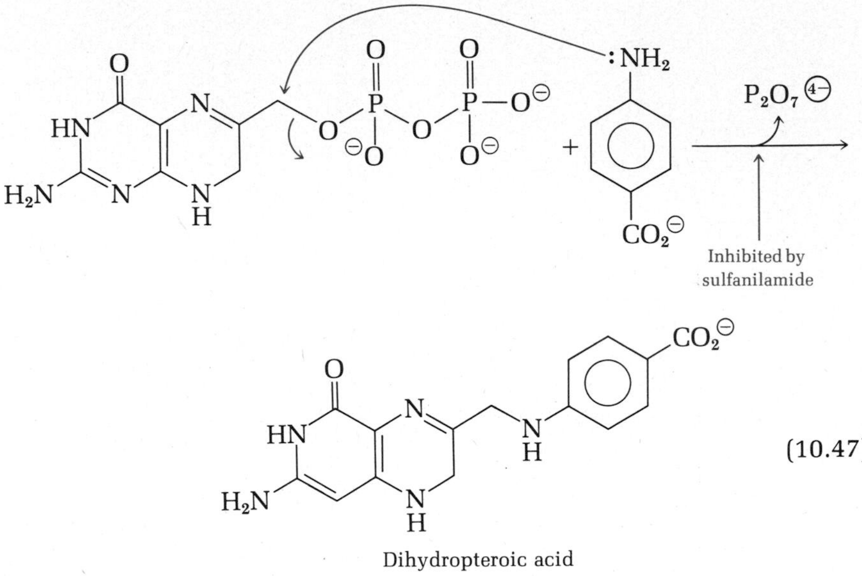

Inhibited by sulfanilamide

(10.47)

Dihydropteroic acid

10–10 COENZYME B₁₂

Cobalamin, vitamin B_{12}, was discovered in 1948, 22 years after it had been postulated as an "animal protein factor" missing in persons suffering from pernicious anemia. In 1956 its structure was determined, and in 1972 the total synthesis of vitamin B_{12} was accomplished by R. B. Woodward and his collaborators. Four pyrrole rings surround the central Co(II) ion and the metal complex is called a *corrin* ring system. The axial ligands of the 6-coordinate cobalt ion are cyanide ion and 5,6-dimethylbenzimidazole.[4]

Coenzyme B_{12} is produced in an unusual reaction in which FAD reduces Co(III) to Co(I) in a two-electron reduction. Co(I) is a powerful nucleophile which displaces a 5′-triphosphate group of ATP to give active, adenylated coenzyme B_{12} (Figure 10.30). The most unusual feature of coenzyme B_{12} is the carbon-cobalt bond. The Co–C bond length is 0.205 nm. This rather long bond is quite labile, and it is intimately connected to the biological activity of coenzyme B_{12}. The types of reactions catalyzed by coenzyme B_{12}–dependent enzymes are listed in Table 10.9. Most of these reactions conform to a single type: the migration of a group between adjacent carbons (reaction 10.48).

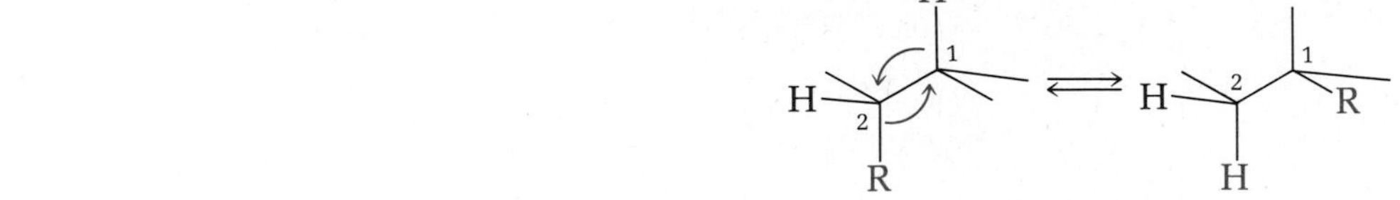

(10.48)

In our discussion of the mechanisms of coenzyme B_{12} reactions we shall assume that they are the same in their overall features, although this is undoubtedly an oversimplification.

[4] The form of vitamin B_{12} first isolated contained cyanide and is called cyanocobalamin. Natural B_{12} contains no cyanide. Although cyanocobalamin is an artifact of the isolation procedure, it has become accepted as the "standard structure."

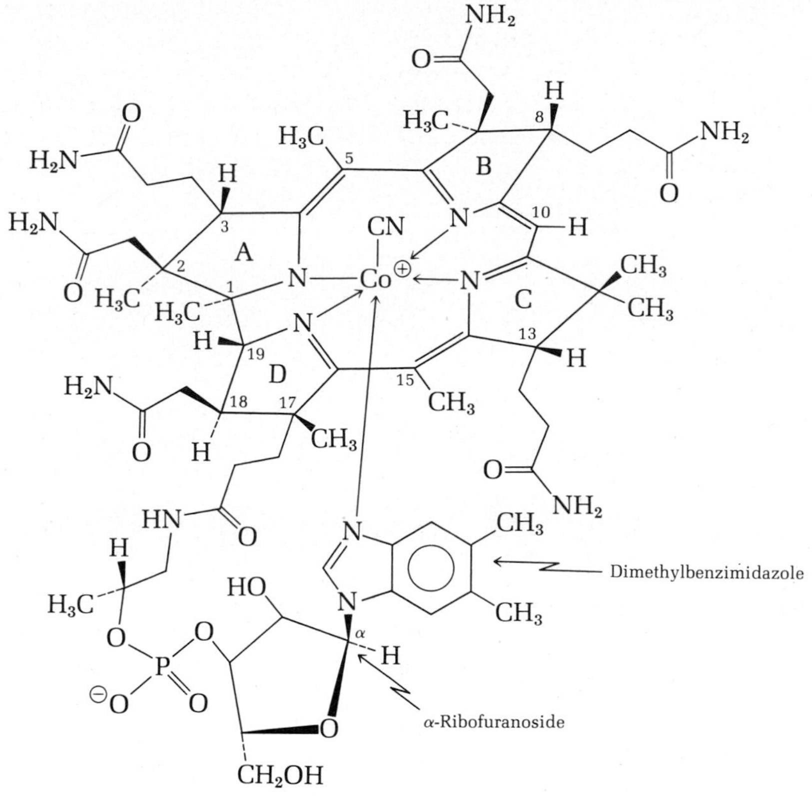

Vitamin B$_{12}$

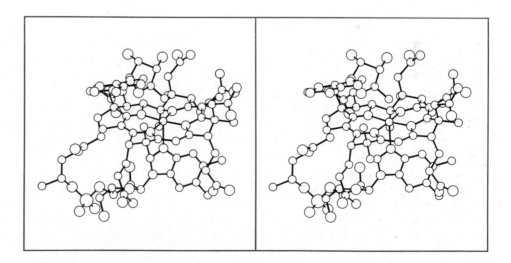

Propanediol dehydrase catalyzes the coenzyme B$_{12}$–dependent conversion of 1,2-propanediol to propanal (reaction 10.49).

$$CH_3-\overset{\overset{\displaystyle OH}{|}}{CH}-CH_2OH \xrightarrow[\text{dehydrase}]{\text{Propanediol}} CH_3CH_2\overset{\overset{\displaystyle O}{\|}}{C}\diagdown_H \ + \ H_2O \qquad (10.49)$$

When 1R,2R-1-[^{2}H]-propanediol is treated with the enzyme only (2S), 2-[^{2}H]-propanal is produced (reaction 10.50).[5]

---

[5] The *RS* configurational nomenclature is described in Section 10.3C. Recall that the symbols ^{2}H and D represent deuterium.

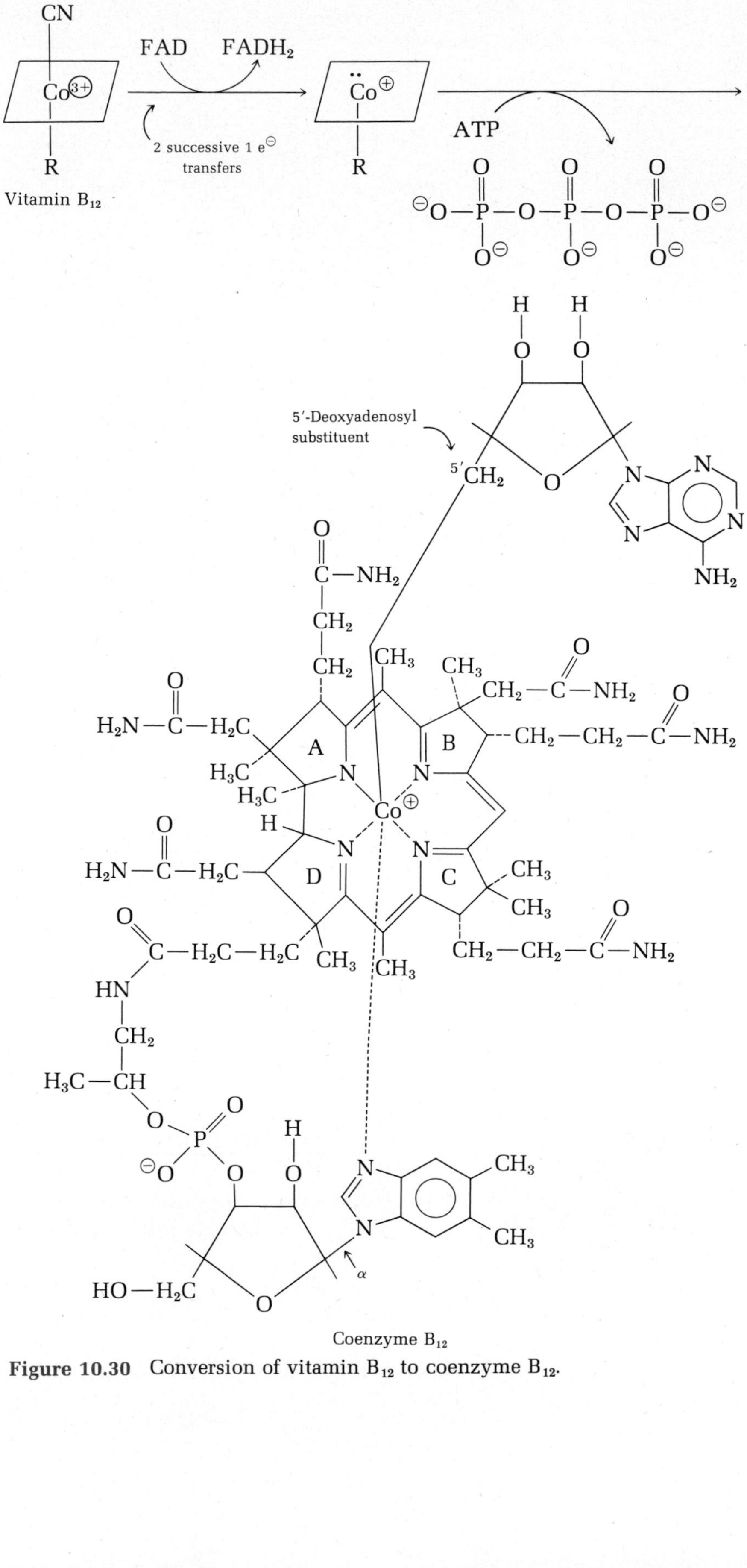

Figure 10.30 Conversion of vitamin B_{12} to coenzyme B_{12}.

Table 10.9 Examples of hydrogen-transfer reactions catalyzed by coenzyme B_{12}–dependent enzymes

Reaction	Enzyme

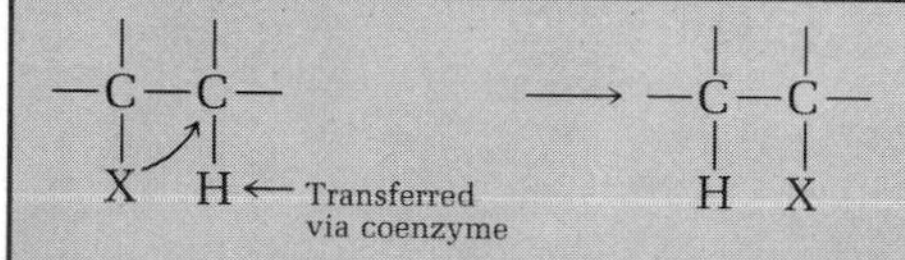

General reaction (migrating group is colored)

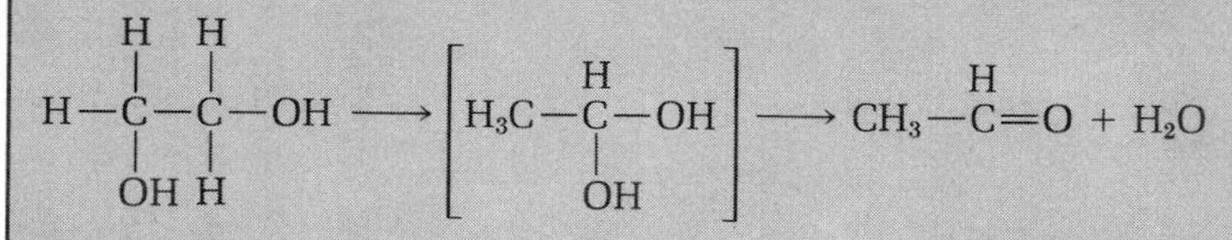

Diol dehydratase, glycerol dehydratase

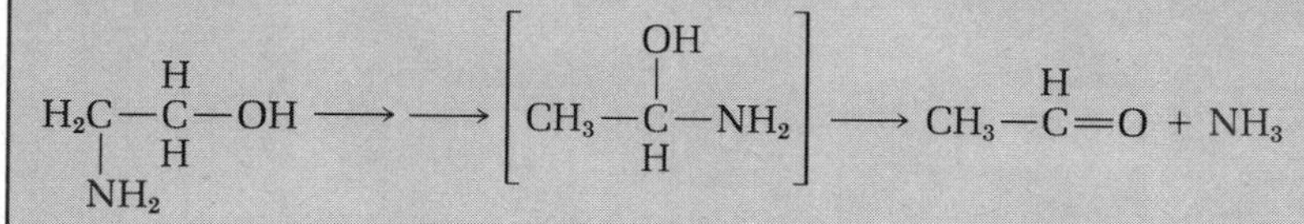

Ethanolamine ammonia-lyase

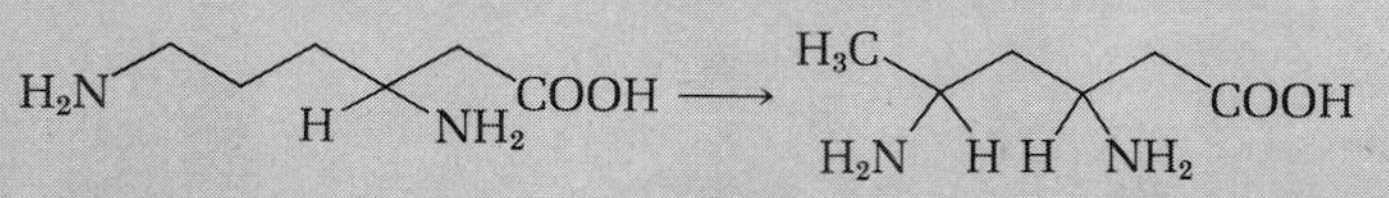

L-β-Lysine mutase, D-α-lysine mutase, ornithine mutase

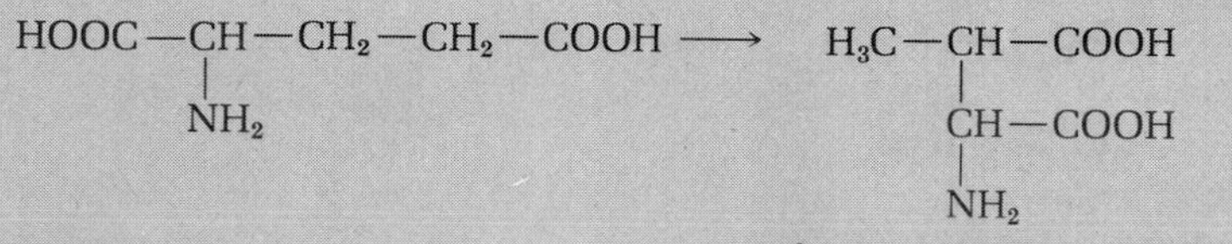

Glutamate mutase

threo-β-Methyl-L-aspartate

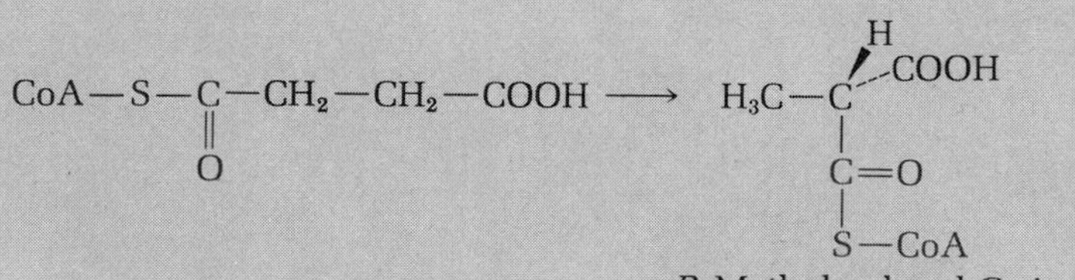

Methylmalonyl-CoA mutase

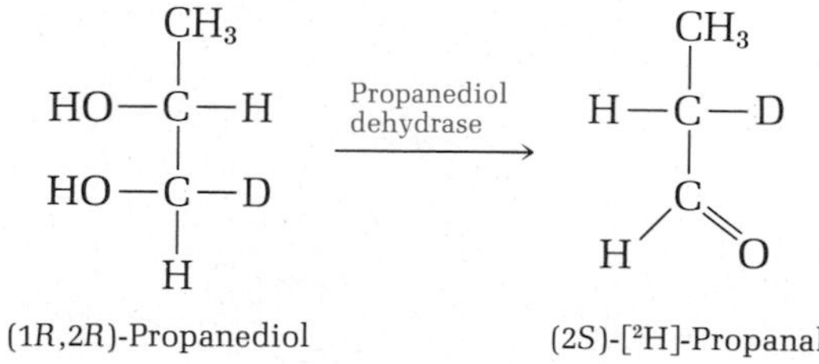

R-Methylonyl-CoA

(1R,2R)-Propanediol → (2S)-[²H]-Propanal

$$ (10.50) $$

The reaction thus proceeds with inversion of configuration at C-2. Since only deuterium migration is observed, the reaction is also stereospecific at C-1. On the other hand, 1R,2S-1-[²H]-propanediol reacts with the dehydrase to give only 1-[²H]-propanal (reaction 10.51). In this reaction only the hydrogen migrates. Thus, the migrating groups are reversed when the chirality at C-2 is reversed. Reaction of 2S-1-[¹⁸O]-1,2-propanediol with propanediol dehydrase proceeds via a *1,1-gem diol* and gives propanal in which 88% of the ¹⁸O in the reactant is retained in the product (reaction 10.52).

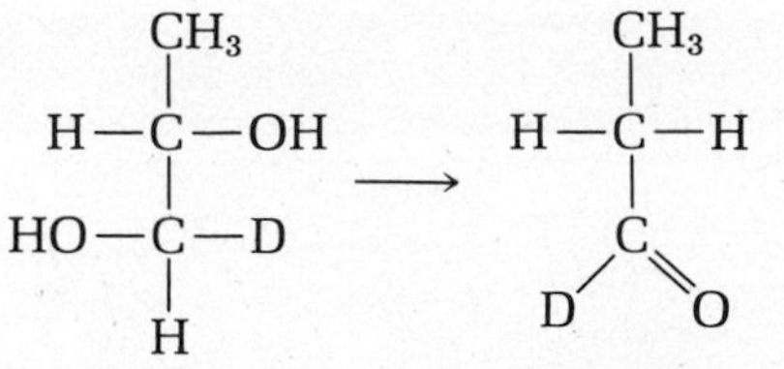

$$
\begin{array}{ccc}
& CH_3 & \\
& | & \\
H-&C&-OH \\
& | & \\
HO-&C&-D \\
& | & \\
& H &
\end{array}
\longrightarrow
\begin{array}{c}
CH_3 \\
| \\
H-C-H \\
| \\
C \\
\diagdown \\
D \quad O
\end{array}
\qquad (10.51)
$$

1R,2S | 1-[²H]-Propanal

$$
\begin{array}{c}
CH_3 \\
| \\
H-C-OH \\
| \\
H_S-C-H_R \\
| \\
^{18}OH
\end{array}
\longrightarrow
\begin{array}{c}
CH_3 \\
| \\
H-C-H \\
| \\
H-C-OH_R \\
| \\
^{18}OH_S
\end{array}
\xrightarrow{\text{Loss of pro-R } ^{16}OH}
\begin{array}{c}
CH_3 \\
| \\
H-C-H \\
| \\
C \\
^{18}O \diagup \diagdown H
\end{array}
\qquad (10.52)
$$

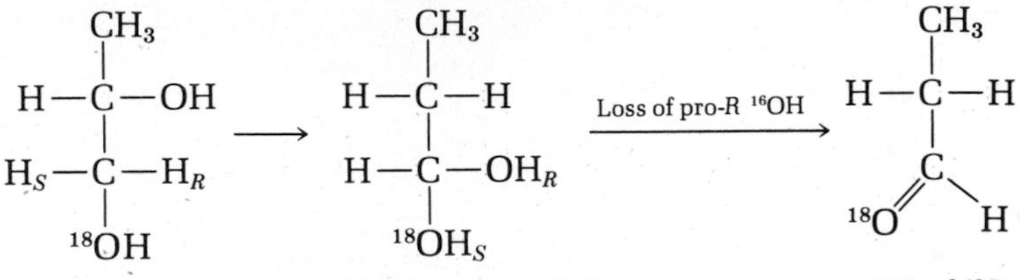

1-[¹⁸O]-1-Propanediol
("gem diol") | 88% of ¹⁸O retained

When 1-[³H]-propanediol and ethylene glycol, also a substrate for the enzyme, are both in the reaction mixture with propanediol dehydrase, the products *both* contain tritium, [³H] (reaction 10.53).

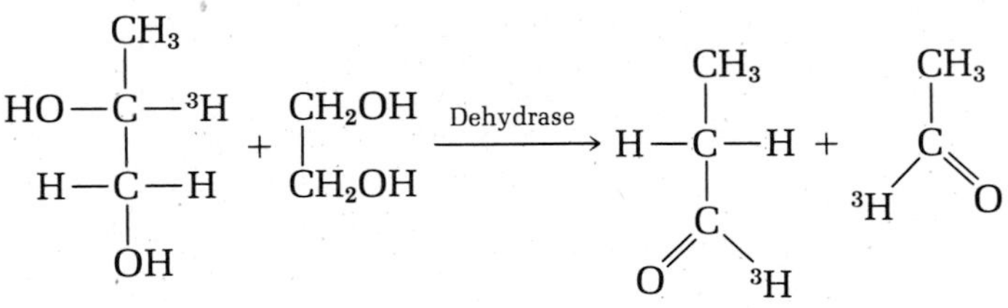

$$
\begin{array}{c}
CH_3 \\
| \\
HO-C-{}^3H \\
| \\
H-C-H \\
| \\
OH
\end{array}
+
\begin{array}{c}
CH_2OH \\
| \\
CH_2OH
\end{array}
\xrightarrow{\text{Dehydrase}}
\begin{array}{c}
CH_3 \\
| \\
H-C-H \\
| \\
C \\
O \diagup \diagdown {}^3H
\end{array}
+
\begin{array}{c}
CH_3 \\
| \\
C \\
{}^3H \diagup \diagdown O
\end{array}
\qquad (10.53)
$$

Tritium is apparently transferred to coenzyme B_{12} and then retransferred to the substrate. Coenzyme B_{12} isolated from the above reaction mixture contains tritium at C-5 of the adenosyl ligand bound to cobalt.

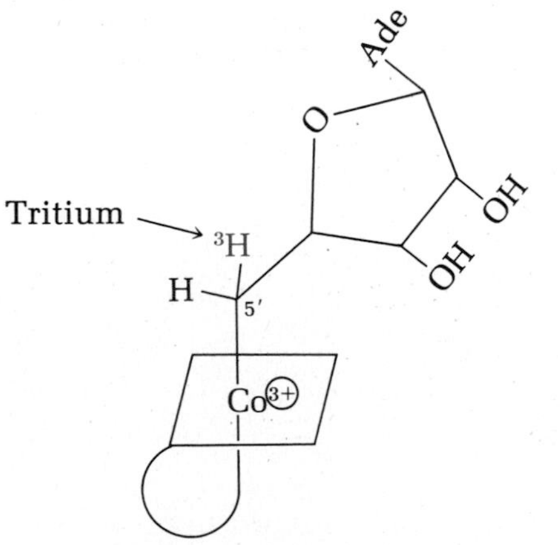

The enzyme does not distinguish between the two prochiral hydrogens at C-5 of the adenosyl ligand. Since these two hydrogens are equivalent, and since a free radical can be detected in the reaction, the following general mechanism has been proposed for the reaction. First, hydrogen transfer from the substrate to the methylene group of the adenosyl ligand generates a substrate-free radical and expels a methylriboside from coenzyme B_{12}. Then the free radical combines with the reactive coenzyme B_{12} intermediate (Fig-

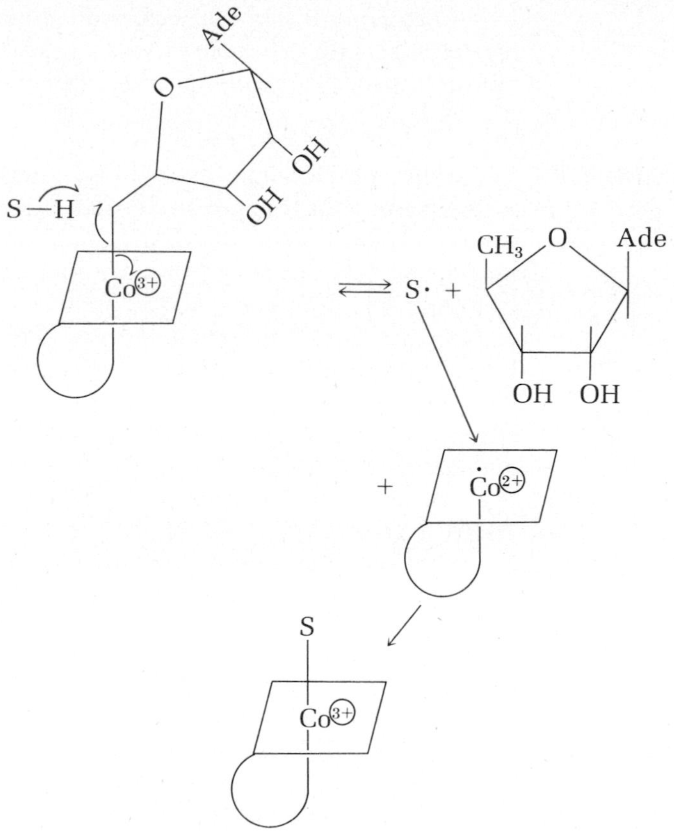

Figure 10.31 Proposed mechanism for the transfer of a hydrogen atom from a substrate to the 5′ position of the adenosyl ligand of coenzyme B$_{12}$, followed by radical recombination of the substrate and coenzyme B$_{12}$.

ure 10.31). The covalent substrate–coenzyme B$_{12}$ intermediate must rearrange, and the process by which this occurs is unknown ("reaction" 10.54).

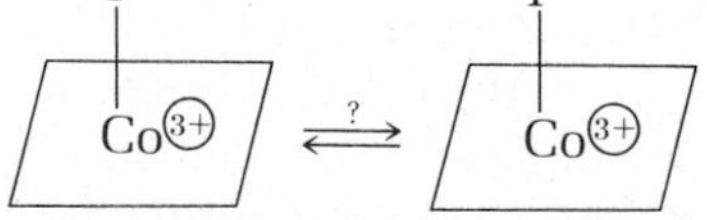

$$(10.54)$$

Next, the C-5 hydrogen of 5′-methyladenosine is transferred to the product in another poorly understood process (reaction 10.55).

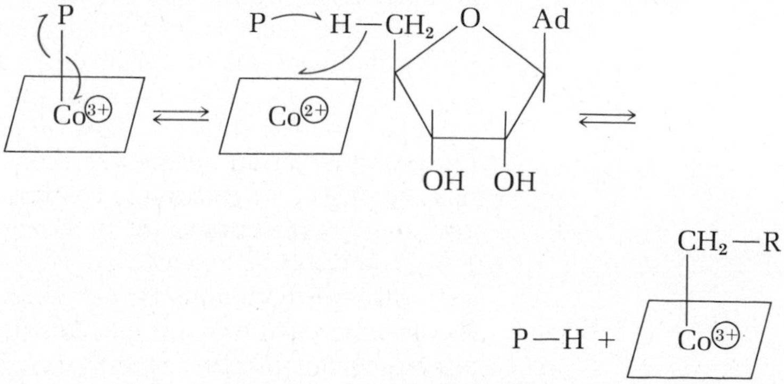

$$(10.55)$$

Coenzyme B_{12} is the least well understood of the coenzymes. Many mysteries remain to be solved, and the future of coenzyme B_{12} research holds out the prospect of many surprising discoveries.

10–11 SUMMARY   The water-soluble vitamins are biosynthetic precursors of many coenzymes. A coenzyme is a true substrate that is required for an enzyme reaction to occur. A coenzyme is chemically modified in one enzyme reaction and then recycled in another reaction. It can therefore be used repeatedly. A cofactor is a nonprotein molecule or ion, such as a metal ion, that is required for an enzyme reaction. A prosthetic group, such as heme, is required for catalysis, but is more tightly bound to the enzyme than a cofactor or coenzyme.

Several coenzymes contain adenine nucleotide moieties as part of their structures. Coenzyme A is comprised of β-mercaptoethanolamine, adenosine-3′,5′-diphosphate, and the vitamin pantothenic acid. Coenzyme A serves as an acyl carrier, and it also promotes the condensation of thioesters.

The nicotinamide coenzymes, NAD and NADP, contain vitamin B_3, nicotinic acid, and a nucleotide phosphate. These coenzymes catalyze oxidation-reduction reactions. In most cases the reactions of these coenzymes occur by direct hydride transfer to or from a given substrate. NAD is predominantly involved in catabolic reactions, and NADP is primarily involved in biosynthetic reactions. Enzymes that require NAD or NADP are able to distinguish between the prochiral hydrogens at C-4 of the reduced coenzymes and between the prochiral *re* and *si* faces of the nicotinamide rings of the oxidized coenzymes.

The flavin coenzymes, FAD and FMN, contain vitamin B_2, riboflavin, as part of their structures. The isoalloxazine rings of FAD and FMN are reversibly oxidized and reduced in flavin-dependent oxidation-reduction reactions. The mechanisms of action of flavin-dependent enzymes are variable and may involve either one-electron or two-electron transfers.

Thiamine pyrophosphate (TPP) is the coenzyme derived from vitamin B_1, thiamine, and a pyrophosphoryl group originating in ATP. The quaternary nitrogen of the thiazolium ring of TPP is an electron sink that enhances the acidity of the vinylic hydrogen at C-2 of the thiazolium ring. The mechanism of action of TPP-dependent enzymes involves formation of a dipolar ion called an ylid or a nucleophilic carbene at C-2. The reactions of TPP-dependent enzymes involve the synthesis or cleavage of bonds to a carbonyl carbon and may be either oxidative or nonoxidative.

Lipoic acid is found in two multienzyme complexes—pyruvate dehydrogenase and α-ketoglutarate dehydrogenase. In both cases it serves as an acyl carrier and as a two-electron carrier.

Pyridoxal phosphate (PLP) is the coenzyme derived from vitamin B_6, pyridoxine. At pH 7.0 pyridoxal-P is a dipolar ion in which the C-3 hydroxyl group is ionized, and the pyridine nitrogen is protonated. The pyridinium ring of pyridoxal-P acts as an electron sink, that is, as an electrophilic catalyst in the reactions of pyridoxal-P. The primary function of pyridoxal-P in its action as a coenzyme is stabilization of carbanions. Among the types of reactions catalyzed by pyridoxal-P are (1) transamination, (2) racemization, (3) decarboxylation of the α-carboxyl group and decarboxylation of the carboxyl group at the β-carbon of aspartate; (4) other reactions at the β-carbon of amino acids (discussed in the context of amino acid metabolism in Chapters 21 and 22).

Biotin, or vitamin H, catalyzes transfer of carboxyl groups from one substrate molecule to another. Most of these processes are ATP-dependent, although a few biotin-dependent enzymes do not require ATP. The reactions of the ATP-dependent biotin carboxylases occur with formation of N^1-carboxybiotin as an intermediate.

The vitamin folic acid is the precursor of the coenzyme tetrahydrofolic acid, a participant in one-carbon metabolism. The anticancer agent methotrexate acts by inhibiting the binding of tetrahydrofolate to thymidylate synthetase. As a result DNA synthesis is slowed or stopped completely, and the growth of cancer cells is also slowed accordingly. The one-carbon transfer reactions catalyzed by tetrahydrofolate are varied, but share the common feature of a covalent bond between the one-carbon fragment and N-5, N-10, or both.

The sulfanilamide antibiotics act by inhibiting the biosynthesis of dihydropteric acid, a precursor of folic acid. Deprived of this essential coenzyme, the bacteria cannot survive.

Coenzyme B_{12}, derived from vitamin B_{12}, catalyzes a variety of often unusual intramolecular rearrangements. One of the most unusual features of coenzyme B_{12} is a carbon-cobalt bond. Many aspects of the mechanisms of action of coenzyme B_{12}–dependent enzymes have remained unsolved, and it is the least well understood of the coenzymes.

REFERENCES

Babior, B. M., Ed., "Cobalamin," Wiley, New York, 1975.

Blakley, R. L., "The Biochemistry of Folic Acid and Related Pteridines," North-Holland, Amsterdam, 1969.

Gubler, C. J., Ed., "Thiamine," Wiley, New York, 1976.

Kamin, H., Ed., "Flavins and Flavoproteins," University Park Press, Baltimore, Md., 1971.

Krampitz, L. O., "Thiamin Diphosphate and Its Catalytic Functions," Dekker, New York, 1970.

McCormick, D. B., and Wright, L. D., Eds., "Methods in Enzymology," Vol. 18 (Parts A, B, and C), Academic Press, New York, 1970.

Pratt, J. M., "Inorganic Chemistry of Vitamin B_{12}," Academic Press, New York, 1972.

Smith, E. L., "Vitamin B_{12}," 3rd ed., Methuen, London, 1965.

Walsh, C., "Enzymatic Reaction Mechanisms," Freeman, San Francisco, 1979.

Walsh, C. "Flavin Coenzymes: At the Crossroads of Biological Redox Chemistry," *Acc. Chem. Res.*, 13, 148–155 (1980).

Problems

1. Write a mechanism for the reaction of urea with N^1-carboxybiotin. In what respect is this mechanism similar to the mechanism with an acyl coenzyme A?

2. Alanine racemase is irreversibly inhibited by L-β-chloroalanine. Write a mechanism that explains this inhibition.

$$Cl-CH_2-\overset{\overset{\displaystyle H}{|}}{\underset{\underset{\displaystyle \oplus NH_3}{|}}{C}}-CO_2^{\ominus}$$

L-β-Chloroalanine

3. A metal ion, which forms a complex with the carbonyl group of biotin, is required for acetyl coenzyme A carboxylase activity. How might the metal ion facilitate the reaction of the enolate anion of acetyl coenzyme A with N^1-carboxybiotin? (Hint: See section 6.2 D.)

4. "Schiff base catalysis" is common in biochemical systems. What is the advantage of this type of catalysis compared to the corresponding uncatalyzed nucleophilic addition to a carbonyl group?

5. Why do amines stimulate the decarboxylation of β-ketoacids? How does pyridoxal phosphate stimulate the decarboxylation of β-amino acids?

6. The thiazole derivative (compound 1) catalyzes hydrogen-deuterium exchange of aldehydes. (a) After what biocatalyst is the compound 1 Modeled? (b) Write a step-by-step mechanism for the exchange reaction.

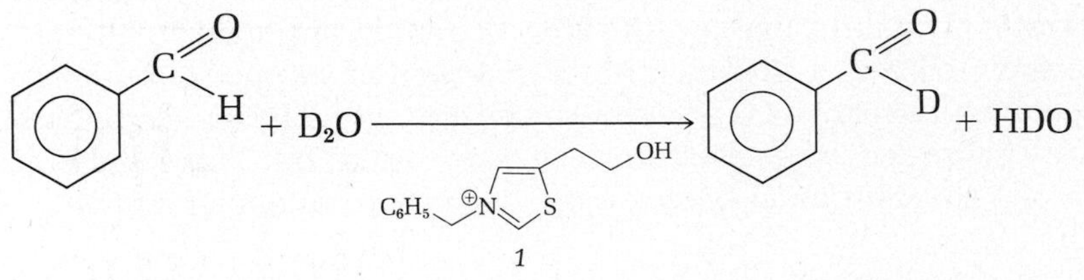

7. L-Alanine transaminase can catalyze the exchange of H and D in the α-position of alanine. Write a mechanism to show how this process occurs.

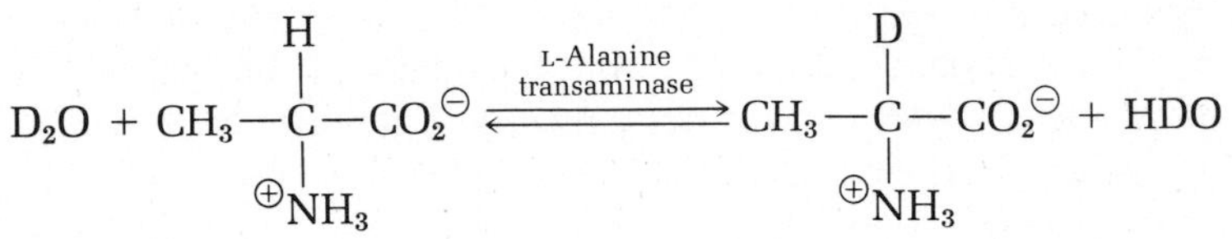

8. L-Alanine transaminase can also catalyze the exchange of H and D in the α-methyl group of alanine. Write a mechanism to show how this process occurs.

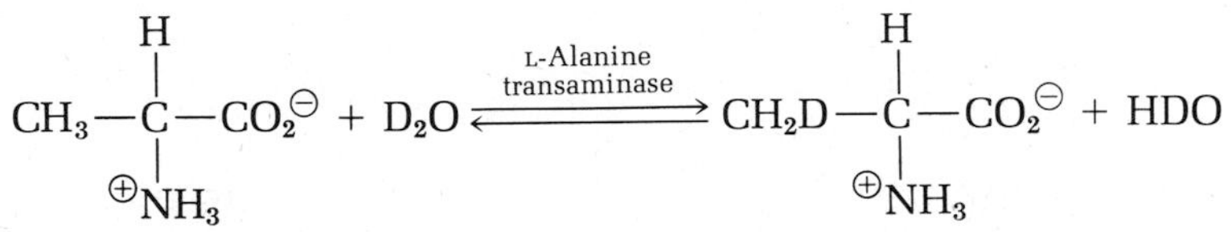

9. Formate dehydrogenase catalyzes the reaction

$$NAD + HCO_2^{\ominus} \rightleftharpoons CO_2 + NADH$$

The kinetic isotope effect for the reaction, k_H / k_D, is 2.8. Propose a mechanism for this reaction, and show how your hypothetical transition state accounts for the observed isotope effect. [Reference: J. S. Blanchard and W. W. Cleland, *Biochemistry*, *19*, 3542–3550 (1980).]

10. The pyridoxal phosphate–dependent enzyme methionine γ-lyase catalyzes the formation of methanethiol, α-ketobutyrate, and ammonia from L-methionine.

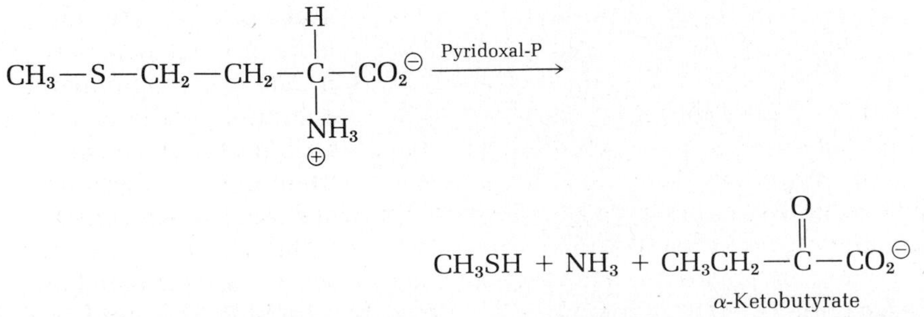

Propose a mechanism of action for this enzyme. [Reference: M. Johnston, R. Raines, M. Chang, N. Eksaki, K. Soda, and C. Walsh, *Biochemistry*, *20*, 4325–4333 (1981).]

Chapter 11

Lipids and biological membranes

11-1 INTRODUCTION

The lipids are an extremely heterogeneous group of molecules that participate in a wide variety of cellular functions. The simplest group of lipids consists of the long-chain saturated and unsaturated fatty acids. Certain of these fatty acids are components of the complex lipids of biological membranes. Others, such as arachidonic acid, are precursors of the prostaglandins, potent stimulants of smooth muscle contraction. The lipid components of biological membranes are phosphoglycerides, sphingolipids, and sterols. The steroid lipids have a wide range of functions. Cholesterol is a component of biological membranes, the bile salts are important in lipid digestion, and the steroid hormones have a multitude of functions that influence many aspects of the life of higher organisms by regulating cellular metabolism. Vitamin D, a steroid derivative, participates in calcium and phosphate metabolism. The terpenes, another large class of molecules, include the lipid-soluble vitamins A, E, and K.

Within the past few years biological membranes have become the focus of an enormous volume of biochemical research. Several convergent factors have led to the rebirth of membrane research. It became clear with the novel hypothesis of respiration proposed by P. Mitchell in the early 1960s that a detailed understanding of biological membranes would be required to understand respiration. The transport of metabolites across cell membranes, a process that affects many aspects of metabolism, also requires an understanding of membrane structure.

The currently accepted general model of membrane structure was proposed in the early 1970s by S. J. Singer and G. Nicolson. The key to the formulation of their "fluid-mosaic" model of membrane structure was the thermodynamic hypothesis that membrane structure is dominated by the hydrophobic effect. (The long dominant model proposed by Danielli in the 1930s required severe alterations to be thermodynamically acceptable.) The hydrophobic effect is a generalized principle that largely controls the structures of proteins, nucleic acids, and biological membranes. It is one of the major unifying forces in the biological world.

11−2 LIPIDS

Biological membranes are complex structures that are a composite of proteins and lipids. Membrane lipids can be divided into two broad classes, each having several subdivisions. The *amphipathic lipids* are the phospholipids which include *glycolipids, phosphoglycerides,* and *sphingolipids.* They have ionic or polar, water-soluble "heads" and nonpolar hydrocarbon "tails." These lipids are called amphipathic because they combine the "sympathy" of the polar heads for water with the "antipathy" of the hydrocarbon tails for water. The second broad class of lipids is the *sterol family,* whose principal representative in animal membranes is *cholesterol.* It, too, is amphipathic, since its hydroxyl group is polar and the steroid ring system and hydrocarbon "tail" are nonpolar. However, the affinity of the hydroxyl group for water is much less than the affinity of an ionic head for water, and the amphipathic properties of cholesterol do not dominate its physical properties.

Cholesterol

A. Triacylglycerols

The most abundant lipids are esters of *glycerol.*

$$CH_2OH$$
$$|$$
$$CHOH$$
$$|$$
$$CH_2OH$$

Glycerol

If the glycerol is esterified at all three positions, the product is a *triacylglycerol* or *triglyceride.* Triacylglycerols, also called neutral lipids, are the major components of adipose tissue and plant oils, but they are not found in membranes.

B. Fatty Acids

The most common "tails" of amphipathic lipids are esters of long chain fatty acids. These carboxylic acids usually contain an even number of carbon

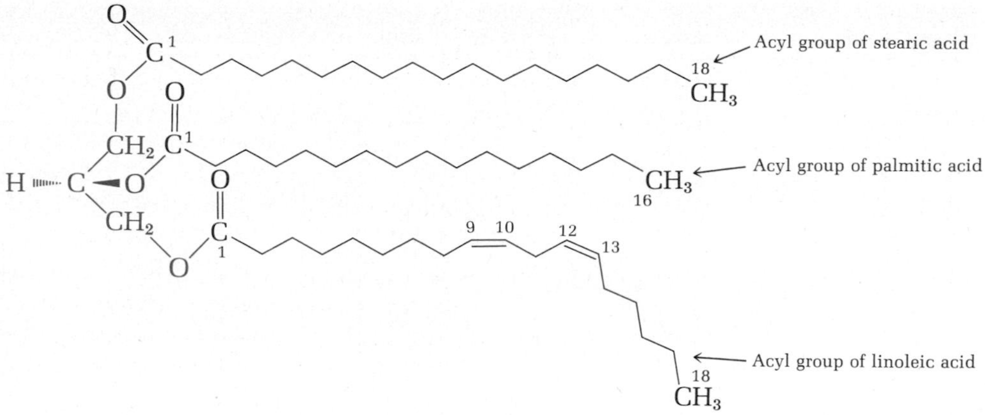

Structure of a triglyceride

atoms. If the hydrocarbon is an alkane, the fatty acid is *saturated*; if it is an

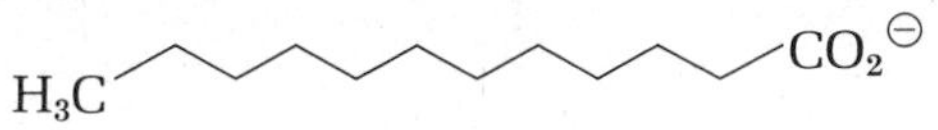

Dodecanoate, the conjugate base of a saturated fatty acid

alkene, the fatty acid is *unsaturated*. Table 11.1 lists the names and formulas of the saturated fatty acids. The saturated fatty acids can exist in extended conformations that can be tightly packed in well-ordered crystals. Therefore, they have relatively high melting points: the homologs above C-12 are solids at room temperature. Table 11.2 lists the names and formulas of the most common unsaturated fatty acids.

The double bonds of unsaturated fatty acids are indicated by the symbol $\triangle^n$, where the $\triangle$ stands for the double bond and the superscript gives its location with respect to the carboxyl group. Double bonds may be either

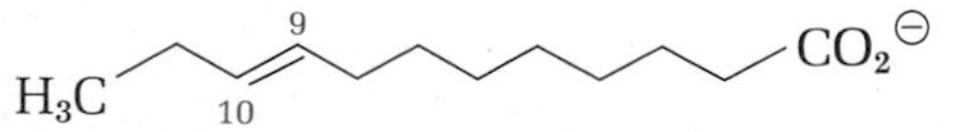

Trans-Δ^9-Dodecenoate, the conjugate base of an unsaturated fatty acid

cis or *trans* in unsaturated fatty acids, with *cis* isomers predominating. A *cis* double bond puts a "kink" in the hydrocarbon chain.

Table 11.1	Saturated fatty acids		
Acid	Number of C atoms	Formula	Melting point (°C)
Butyric	4	$CH_3(CH_2)_2CO_2H$	−4.7
Isovaleric	5	$(CH_3)_2CHCH_2CO_2H$	−51.0
Caproic	6	$CH_3(CH_2)_4CO_2H$	−1.5
Caprylic	8	$CH_3(CH_2)_6CO_2H$	16.5
Capric	10	$CH_3(CH_2)_8CO_2H$	31.3
Lauric	12	$CH_3(CH_2)_{10}CO_2H$	43.6
Myristic	14	$CH_3(CH_2)_{12}CO_2H$	58.0
Palmitic	16	$CH_3(CH_2)_{14}CO_2H$	62.9
Stearic	18	$CH_3(CH_2)_{16}CO_2H$	69.9
Arachidic	20	$CH_3(CH_2)_{18}CO_2H$	75.2
Behenic	22	$CH_3(CH_2)_{20}CO_2H$	80.2
Lignoceric	24	$CH_3(CH_2)_{22}CO_2H$	84.2
Cerotic	26	$CH_3(CH_2)_{24}CO_2H$	87.7

Table 11.2 Unsaturated fatty acids

Acid	Number of C atoms	Formula	Melting point (°C)
Δ^9-Decylenic	10	$CH_2{=}CH(CH_2)_7CO_2H$	
Stillingic	10	$CH_3(CH_2)_4CH{=}CHCH{=}CHCO_2H$ (cis, trans)	
Δ^9-Dodecylenic	12	$CH_3CH_2CH{=}CH(CH_2)_7CO_2H$	
Palmitoleic	16	$CH_3(CH_2)_6CH{=}CH(CH_2)_7CO_2H$ (cis)	
Oleic	18	$CH_3(CH_2)_7CH{=}CH(CH_2)_7CO_2H$ (cis)	13, 16
Ricinoleic	18	$CH_3(CH_2)_6CH(OH)CH_2CH{=}CH(CH_2)_7CO_2H$ (cis)	50
Petroselinic	18	$CH_3(CH_2)_{10}CH{=}CH(CH_2)_4CO_2H$ (cis)	30
Vaccenic	18	$CH_3(CH_2)_5CH{=}CH(CH_2)_9CO_2H$ (cis and trans)	
Linoleic	18	$CH_3(CH_2)_4CH{=}CHCH_2CH{=}CH(CH_2)_7CO_2H$	−5
Linolenic	18	$CH_3CH_2CH{=}CHCH_2CH{=}CHCH_2CH{=}CH(CH_2)_7CO_2H$	−11
Eleostearic	18	$CH_3(CH_2)_3(CH{=}CH)_3(CH_2)_7CO_2H$ (cis, trans, trans)	49
Punicic	18	$CH_3(CH_2)_3(CH{=}CH)_3(CH_2)_7CO_2H$ (cis, trans, cis)	44
Licanic	18	$CH_3(CH_2)_3(CH{=}CH)_3(CH_2)_4CO(CH_2)_2CO_2H$	75
Parinaric	18	$CH_3CH_2(CH{=}CH)_4(CH_2)_7CO_2H$	86
Gadoleic	20	$CH_3(CH_2)_9CH{=}CH(CH_2)_7CO_2H$	
Arachidonic	20	$CH_3(CH_2)_4(CH{=}CHCH_2)_4(CH_2)_2CO_2H$	
5-Eicosenic (65%)	20	$CH_3(CH_2)_{13}CH{=}CH(CH_2)_3CO_2H$ (cis)	
5-Docosenic (7%)	22	$CH_3(CH_2)_{15}CH{=}CH(CH_2)_3CO_2H$ (cis)	
Cetoleic	22	$CH_3(CH_2)_9CH{=}CH(CH_2)_9CO_2H$	
Erucic (13%)	22	$CH_3(CH_2)_7CH{=}CH(CH_2)_{11}CO_2H$ (cis)	33.5
5,13-Docosadienic (10%)	22	$CH_3(CH_2)_7CH{=}CH(CH_2)_6CH{=}CH(CH_2)_3CO_2H$ (cis, cis)	
Selacholeic or nervonic	24	$CH_3(CH_2)_7CH{=}CH(CH_2)_{13}CO_2H$ (cis)	**39**

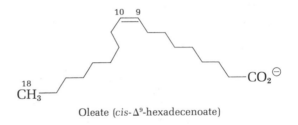

Oleate (cis-Δ^9-hexadecenoate)

The kinked tails cannot form closely packed, well-ordered crystals as well as saturated fatty acids, and the unsaturated fatty acids have lower melting points than their saturated counterparts.

The degree of unsaturation of the hydrocarbon tails of membrane lipids affects the properties of membranes because the fluidity of membranes is determined by the proportions of saturated and unsaturated fatty acids. About 50% of the fatty acids isolated from membrane lipids from all sources are unsaturated. Bacteria grown at different temperatures have different ratios of saturated and unsaturated fatty acids in their membranes, enabling them to maintain the same fluidity at different temperatures. The percentage of unsaturated fatty acids in membranes is inversely proportional to the temperature of the growth medium or environment. The body temperatures of mammals, on the other hand, are more uniform, and their fatty acid compositions are less sensitive to temperature changes. The reindeer provides an interesting exception. The proportion of unsaturated fatty acids in membranes in the reindeer leg increases closer to the hoof. The lower melting points and greater fluidity of the unsaturated fatty acids enable them to function in the low temperatures of ice and snow to which the hoof is exposed.

It has long been known that three of the fatty acids of Table 11.2— *linoleic acid, linolenic acid,* and *arachidonic acid*—are essential in the human diet. The function of these fatty acids was revealed rather recently, when it was found that they are required for the biosynthesis of the *prosta-glandins*, hormones that elicit many physiological responses (Section 11.7).

Straight-chain fatty acids predominate in plants and animals, but microorganisms often contain branched-chain fatty acids. *Tuberculostearic acid* is one example of a branched fatty acid. The methyl side chains lower the melting points of lipid bilayers. The methyl groups serve the same purpose as the double bonds in the straight-chain fatty acids of plants and animals. Polyunsaturated fatty acids, by contrast, have not been found in bacteria.

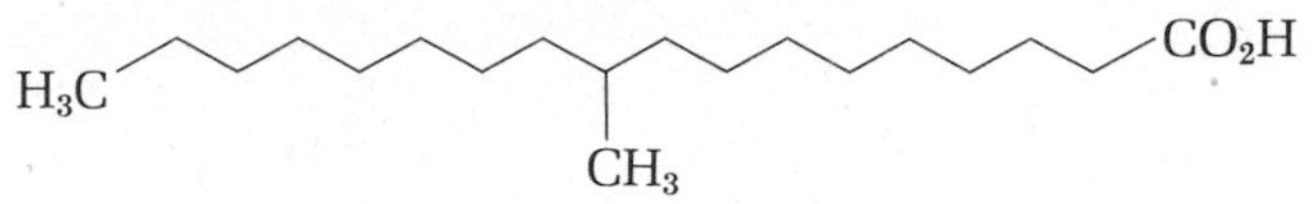

Tuberculostearic acid, a branched-chain fatty acid

Certain bacteria also produce fatty acids that contain cyclopropane rings, cyclopropene rings, and even cyclopentene rings. *Lactobacillic acid*, for example, is an important component of the membrane of lactobacilli. The cyclopropane ring puts a kink in the hydrocarbon chain, lowering the melting point of the cell membrane.

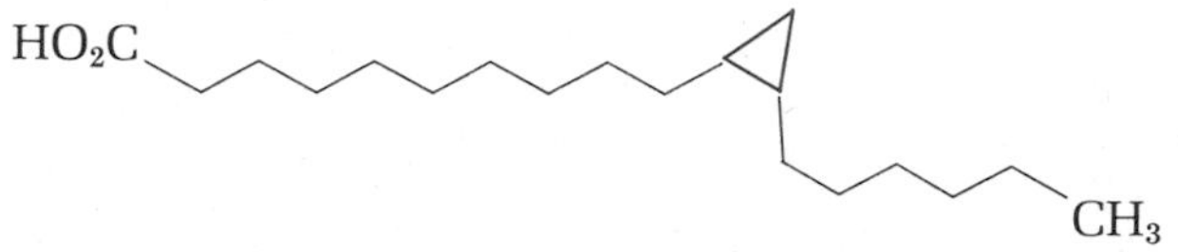

Lactobacillic acid

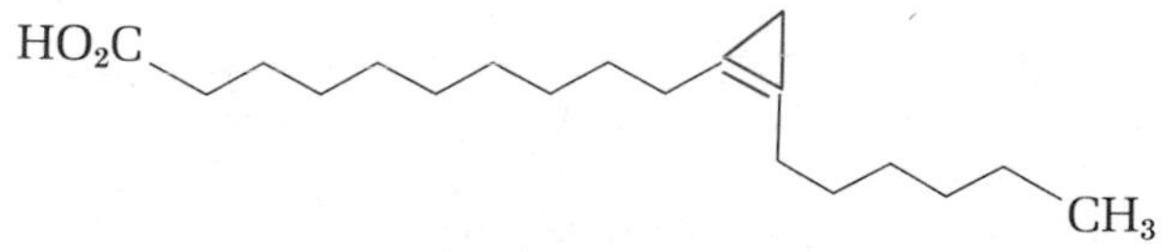

Sterculic acid

C. Phosphoglycerides

The most abundant membrane lipids are derived from a phosphate ester of glycerol. The two —CH_2OH groups of glycerol are prochiral.

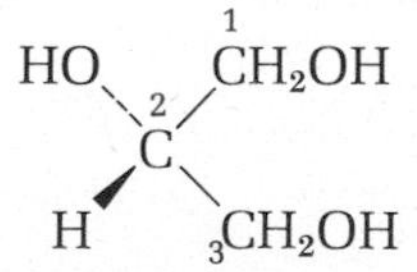

pro-S

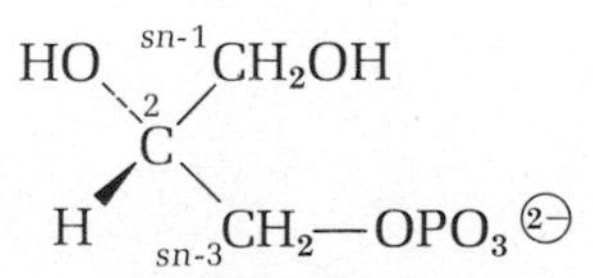

sn-Glycerol 3-phosphate
(L-α-glycerol-phosphate)

In the *stereochemical numbering system*, the pro-S carbon is assigned position 1 and is designated *sn*-1. The phosphate ester of phosphoglycerides in this nomenclature is *sn*-glycerol 3-phosphate. The same molecule is also known as L-α-glycerol phosphate.

The phosphoglycerides contain acyl groups at C-1 and C-2 of *sn*-glycerol 3-phosphate and have the general structure shown on page 455. The simplest phosphoglyceride contains a free phosphate group and is called a *phosphatidic acid* or *phosphatidate*. Further esterification of the phosphate esters produces entire new classes of phosphoglycerides. Table 11.3 lists the most important phosphoglycerides, and Figure 11.1 shows a few of their many possible structures.

Table 11.3 Substituents in glycerolipids

Structure	Name of substituent group	Type of intact molecule
R_1, R_2 $CH_3(CH_2)_nC-$ (with C=O)	Fatty ester, often saturated in R_1, unsaturated in R_2; n usually = 10–20	If $R_3 = H$, diglyceride
$CH_3(CH_2)_nCH_2-$	Fatty ether, n usually > 10; usually at R_1	Glycerol ether
$CH_3(CH_2)_nCH=CH-$	Vinyl ether; n usually > 9; usually at R_1	Plasmalogen
R_3 $CH_3(CH_2)_nC-$ (with C=O)	Fatty ester	If R_1 and R_2 are esters, triglyceride
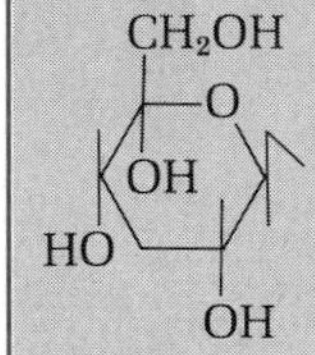 sugar (CH_2OH, OH, HO, OH)	Glucoside or other mono- or polysaccharide	Glycosyl glyceride (a glycolipid)
$^{\ominus}O-P-$ (phosphate with O and OH)	Phosphate ester	Phosphatidic acid
$\overset{\oplus}{H_3N}CH_2CH_2OP-$ (phosphate with O and $O^{\ominus}$)	Ethanolamine phosphate ester	Phosphatidylethanolamine
$^{\ominus}O_2CCHCH_2OP-$ with $\overset{\oplus}{NH_3}$ (phosphate with O and $O^{\ominus}$)	Serine phosphate ester	Phosphatidylserine
$(CH_3)_3\overset{\oplus}{N}CH_2CH_2OP-$ (phosphate with O and $O^{\ominus}$)	Choline phosphate ester	Phosphatidylcholine (lecithin)
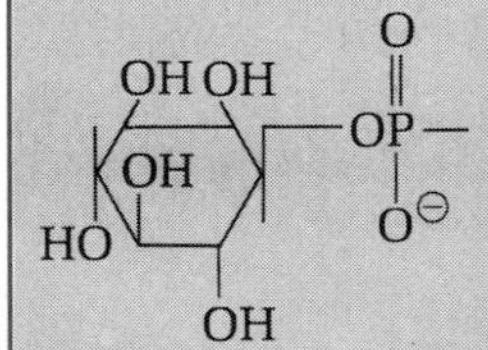 inositol ring (OH OH, OH, HO, OH) $-OP-$ (phosphate with O and $O^{\ominus}$)	Inositol phosphate ester	Phosphatidylinositol
$HOCH_2CHCH_2OP-$ with OH (phosphate with O and $O^{\ominus}$)	Glycerol phosphate ester	Phosphatidylglycerol

(continued)

Table 11.3 Substituents in glycerolipids (continued)

Structure	Name of substituent group	Type of intact molecule

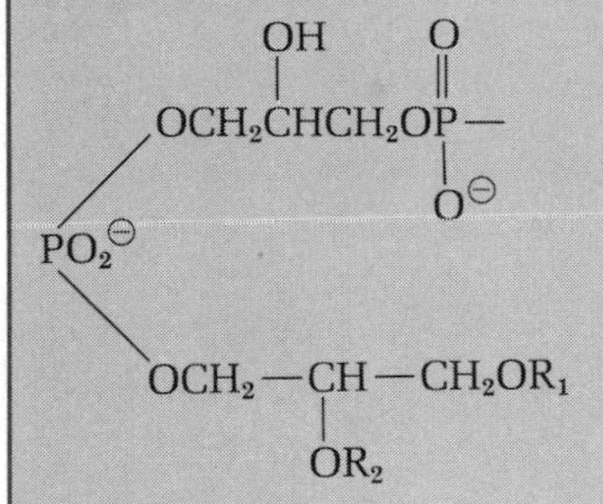 OH   O
OCH$_2$CHCH$_2$OP—
        O$^{\ominus}$
PO$_2$$^{\ominus}$
OCH$_2$—CH—CH$_2$OR$_1$
      OR$_2$
	Phosphatidylglycerol phosphate ester	Diphosphatidylglycerol (cardiolipin)

O H$_3$$\overset{\oplus}{\text{N}}CH_2CH_2$P— O$^{\ominus}$	Aminoethylphosphonate ester	Phosphonolipid

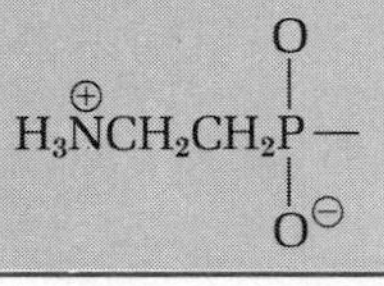

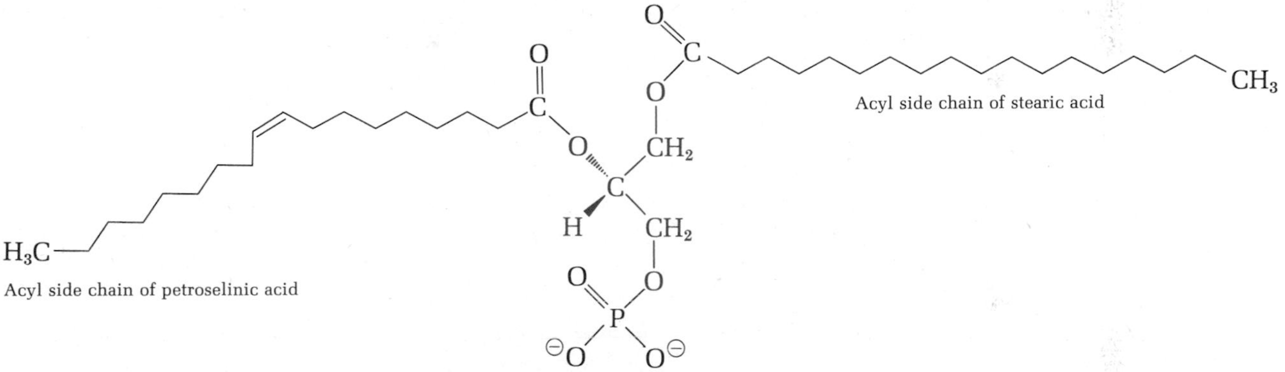

Acyl side chain of stearic acid

Acyl side chain of petroselinic acid

Phosphatidate

When C-1 of *sn*-3-glycerol phosphate contains a vinyl ether group rather than an acyl group, the resulting family of phosphoglycerides are called *plasmalogens*. Ethanolamine is commonly esterified to the phosphate group of plasmalogens. The plasmalogens are found in large amounts in muscle and nerve tissue, accounting for about 10% of the lipids in the human central nervous system.

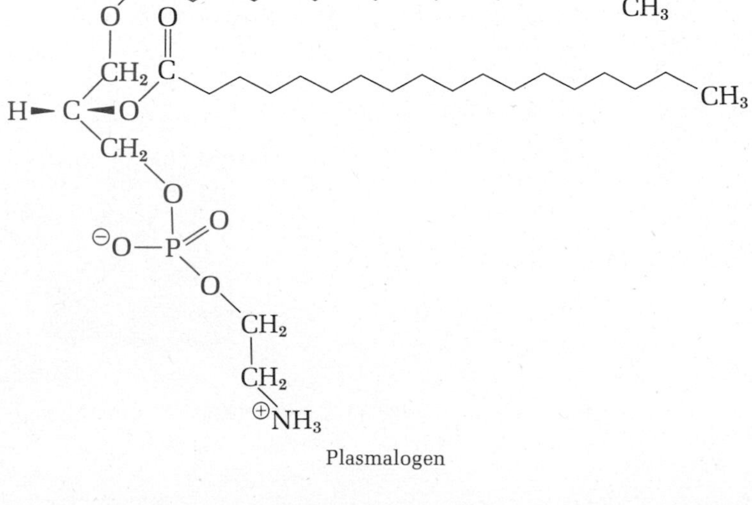

Plasmalogen

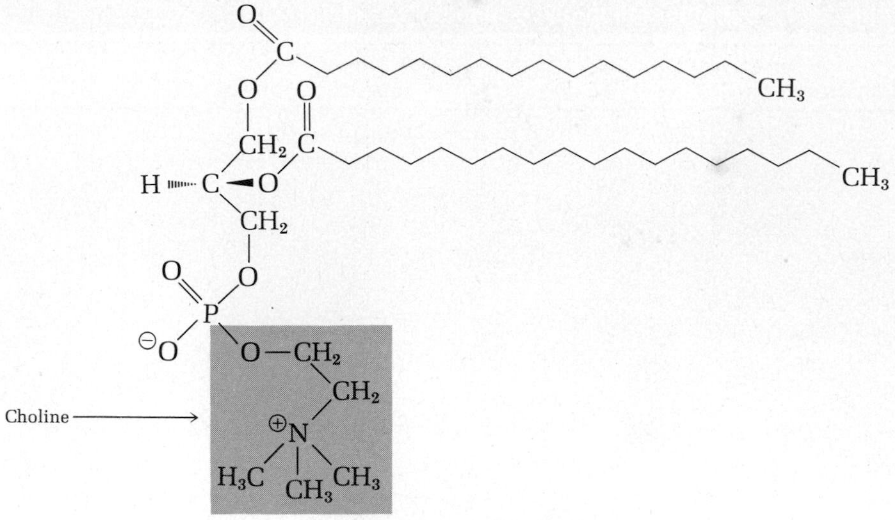

Phosphatidylcholine (lecithin)

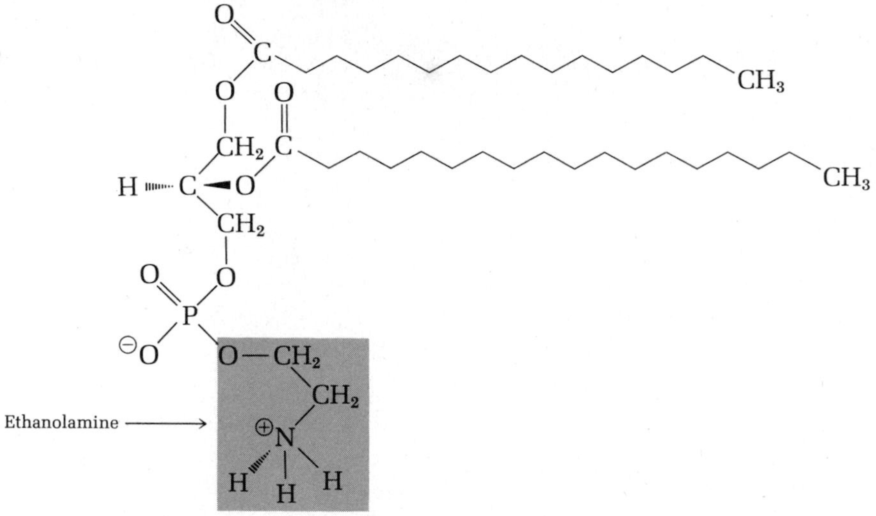

Phosphatidylethanolamine (cephalin)

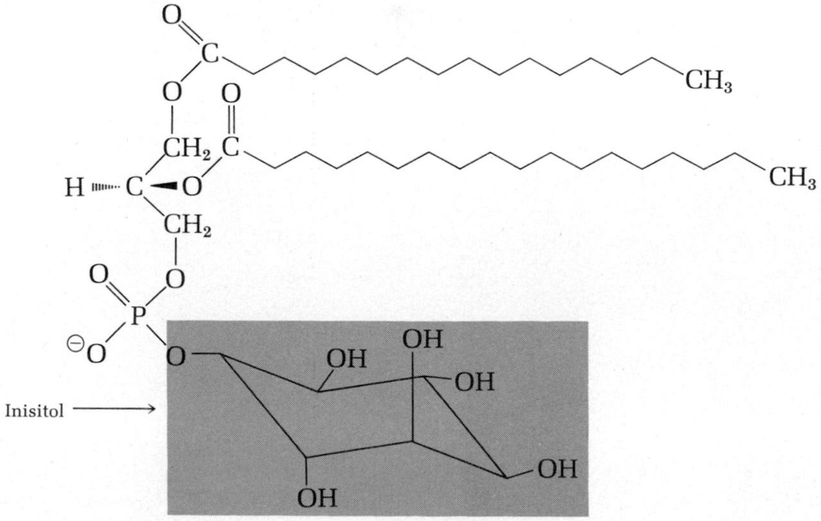

Phosphatidylinisitol

Figure 11.1 Structures of some phosphoglycerides.

Any class of phosphoglycerides or of the other lipids is a heterogeneous group of molecules because a variety of fatty acids can be esterified at C-1 and C-2 of the molecules. The relative proportion of the major classes of lipids in four membranes is given in Table 11.4.

D. Sphingolipids (Sphingomyelins, Glycolipids) and Sterols

A second group of phospholipids contain the long chain amino alcohol sphingosine as the central unit. These sphingolipids are the second major lipid component of animal membranes. They are constructed along the same lines as the phosphoglycerides.

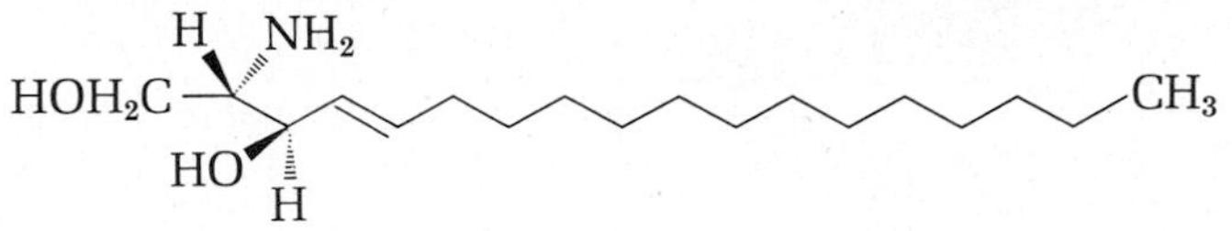

Sphingosine

In the simplest sphingosine derivates, known as *ceramides*, a single fatty acid is linked by an amide bond to the amino group of sphingosine. When a choline phosphate group is bound to the primary alcohol, the product is a *sphingomyelin*.

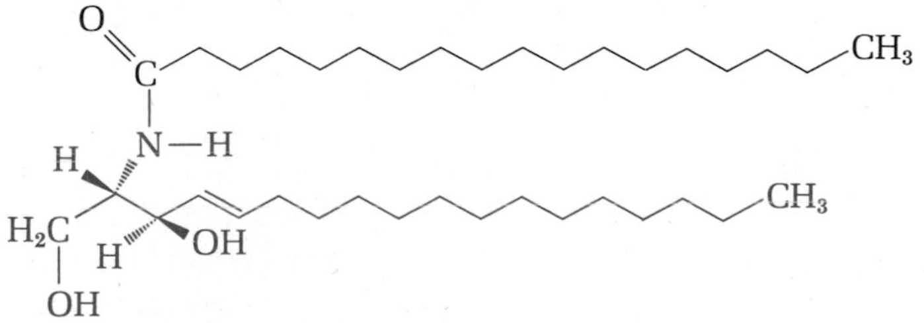

Ceramide

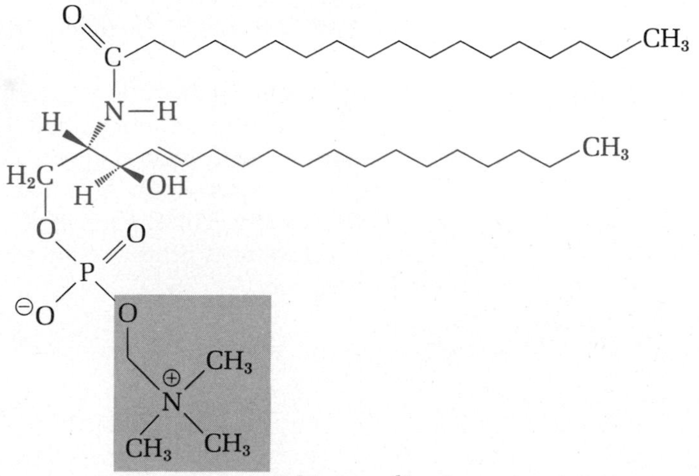

Sphingomyelin

When a glycoside is bound to the primary alcohol, the product is a glycolipid. Most *glycolipids* are relatives of sphingomyelin in which the choline phosphate group is replaced by a glycoside. The most common glycosides, called *cerebrosides*, contain glucose or galactose as their sole carbohydrates.

Table 11.4 Percentage of major classes of lipids in biological membranes

	Human erythrocyte	Human myelin	Beef heart mitochondria	*E. coli*
Phosphatidic acid	1.5	0.5	0	0
Phosphatidylcholine	19.0	10.0	39.0	0
Phosphatidylethanolamine	18.0	20.0	27.0	65.0
Phosphatidylglycerol	0	0	0	18.0
Phosphatidylinositol	1.0	1.0	7.0	0
Phosphatidylserine	8.5	8.5	0.5	0
Cardiolipin	0	0	22.5	12.0
Sphingomyelin	17.5	8.5	0	0
Glycolipids	10.0	26.0	0	0
Cholesterol	25.0	26.0	3.0	0

SOURCE: From C. Tanford, "The Hydrophobic Effect," Wiley, New York, 1973, p. 97.

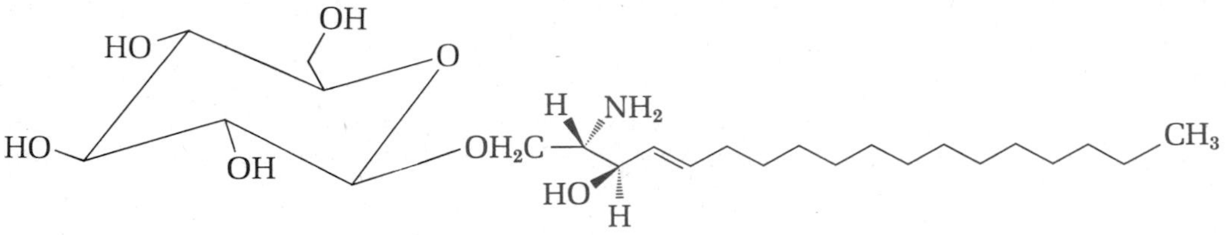

Cerebroside with glucose as the carbohydrate

The most common substituents of the sphingolipids are listed in Table 11.5. The heads of the phosphoglycerides and sphingomyelins are either neutral, dipolar ions, or anionic. Cationic head groups are extremely rare, and have been found only in a few bacteria. The surfaces of biological membranes therefore bear a net negative charge.

The third major lipid component of animal membranes is the sterol, *cholesterol*, a highly hydrophobic molecule whose maximum solubility in water is 10^{-8} M. Cholesterol helps maintain the structure of membranes by providing rigidity. Procaryotic cells (which have cell walls) do not have cholesterol in their cell membranes, an observation that supports the view that cholesterol is needed to provide rigidity.

E. Phospholipases

Phospholipases hydrolyze the ester and phosphodiester bonds of phospholipids. Such lipases are named by the bond of the phosphoglyceride that they cleave. The A_1 lipases cleave the acyl bond at the sn-1 position; the A_2 phospholipases cleave the acyl bond at C-2 and require the S (or L) configuration for activity. The C phosphodiesterases produced by bacteria and the D phosphodiesterases produced by plants cleave phosphodiester bonds as indicated below.

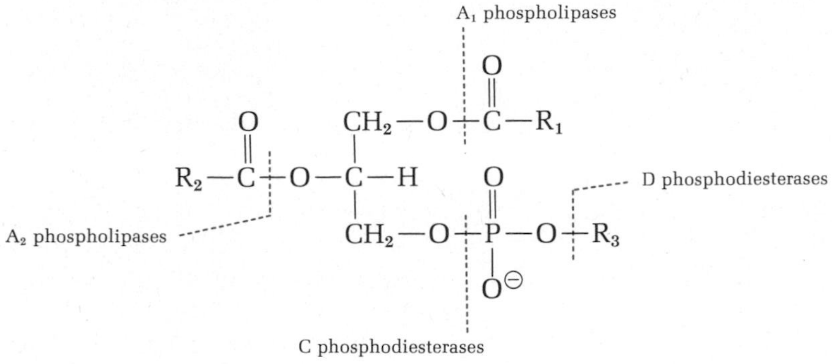

The properties of a few phospholipases are summarized in Table 11.6.

Table 11.5 Sphingolipids

Structure	Name of substituent group	Type of intact molecule
R_1 $CH_3(CH_2)_nCH-$ $\qquad\qquad\ \vert$ $\qquad\qquad OH$		If $n = 14$, $R_2 = R_3 = H$, dihydrosphingosine
$CH_3(CH_2)_nCH-CH=CH$ $\qquad\qquad\qquad\qquad\ \vert$ $\qquad\qquad\qquad\qquad OH$		If $n = 12$, $R_2 = R_3 = H$, sphingosine
$CH_3(CH_2)_nCH-CH-$ $\qquad\qquad\ \vert\qquad\ \vert$ $\qquad\qquad OH\quad OH$		If $n = 13$, $R_2 = R_3 = H$, phytosphingosine
$R_2\qquad\quad O$ $\qquad\qquad\ \Vert$ $CH_3(CH_2)_nC-$	Fatty amide, $n = 14-24$	
$\qquad\qquad\quad O$ $\qquad\qquad\quad \Vert$ $CH_3(CH_2)_nCHC-$ $\qquad\qquad\ \vert$ $\qquad\qquad OH$	α-Hydroxy fatty amide	
$R_3\qquad\qquad\qquad\ O$ $\quad\ \overset{\oplus}{\qquad}\qquad\qquad \Vert$ $(CH_3)_3NCH_2CH_2OP-$ $\qquad\qquad\qquad\quad\ \vert$ $\qquad\qquad\qquad\quad O^{\ominus}$	Choline phosphate ester	Sphingomyelin
$\qquad\qquad\qquad O$ $\qquad\qquad\qquad \Vert$ $\overset{\oplus}{H_3}NCH_2CH_2OP-$ $\qquad\qquad\qquad\ \vert$ $\qquad\qquad\qquad O^{\ominus}$	Ethanolamine phosphate ester	
$\qquad\qquad\qquad\ O$ $\qquad\qquad\qquad\ \Vert$ $Oligosaccharide-OP-$ $\qquad\qquad\qquad\quad\ \vert$ $\qquad\qquad\qquad\quad O^{\ominus}$	Oligosaccharide phosphate ester	Phytoglycolipids
Glucose or galactose	Glucoside or galactoside	Cerebrosides
Oligosaccharide	Oligosaccharide	Hematosides, globosides, blood group substances, gangliosides
H		If R_2 = fatty amide, ceramide

Table 11.6 Properties of some phospholipases

Enzyme	Substrates	Cofactors	Sources
Pancreatic lipase	Triglycerides, diglycerides	Ca^{2+}, bile salts	Mammalian pancreas
Fungal lipases	Glycerides		Yeast, molds
Phospholipase A_1	Phospholipids, glycerides	None	Widely distributed
Phospholipase A_2	Phospholipids	Ca^{2+}	Snake venom, mammalian pancreas, bee venom
Phospholipase C	Phospholipids	Ca^{2+}	Bacteria
Phospholipase D	Phospholipids	Ca^{2+}	Plants
Phosphalidate phosphohydrolase	Phosphatidic acid	None	Widely distributed, particulate

SOURCE: From J. H. Law and W. R. Snyder, in C. F. Fox and A. Keith, Eds. "Membrane Molecular Biology," Sinauer, Stamford, Conn., 1973, p. 19.

F. Plasma Lipoproteins

In humans acyl glycerols and phospholipids are transported through the blood as complexes with a heterogeneous class of proteins called *plasma lipoproteins*. The sizes of these complexes range from particles having diameters of 10 nm and masses of 400,000 daltons to particles having diameters of 1 μm and masses of 40×10^6 daltons. Although the lipids themselves are only sparingly soluble in water, lipoprotein complexes are water soluble.

Lipoprotein complexes are roughly spherical particles whose cores are lipids that have few or no hydrophilic groups, such as triacylglycerols. Lipids that have hydrophilic groups, such as phosphatidyl choline are on the surface of the complexes. The protein snakes its way through the complex with its hydrophobic groups in contact with the lipid phase and its hydrophilic groups in contact with the aqueous phase.

There are four major classes of human plasma lipoproteins (Table 11.7). The *chlyomicrons* carry triacylglycerols from the intestine to other tissues, except kidney. The remaining lipoproteins are classified by their densities. The *very low density lipoproteins* (VLDL) bind triglycerides synthesized in the liver. The *low density lipoproteins* (LDL) are produced by degradation of the lipid portion of VLDL. LDL's may regulate cholesterol synthesis in tissues other than the liver. The *high density lipoproteins* (HDL) are bound to most of the plasma cholesterol. HDL may stimulate transport of cholesterol from various tissues to the liver. A general scheme for transport of lipids in lipoprotein complexes between tissues is summarized in Figure 11.2.

Low density lipoprotein receptors (LDL receptors) outside of the liver govern the uptake of cholesterol by various tissues. The complex between the LDL receptor and lipoprotein, bound to cholesterol, is taken up by the cell by endocytosis. The lysosomes then digest the entire complex to release cholesterol. Cholesterol itself inhibits its own biosynthesis, activates an enzyme that stores cholesterol in cholesterol ester droplets, and inhibits the synthesis of LDL receptors to assure that the cell will not take up too much cholesterol. In persons who have a genetic defect in the gene coding for the receptor, LDL-cholesterol accumulates in the plasma. Persons who are heterozygous for the mutant gene produce about half of the normal number of

Table 11.7 Properties of plasma lipoproteins

Property	Chylomicra	Very low density (VLDL)	Low density (LDL)	High density (HDL)	Very high density (VHDL)
Density	<0.95	0.95–1.006	1.006–1.063	1.063–1.210	>1.21
Diameter (nm)	30–500	30–75	20–25	10–15	10
Amount (mg/100 ml plasma)	100–250	130–200	210–400	50–130	290–400
Approximate composition (%)					
Protein	2	9	21	33	57
Phosphoglyceride	7	18	22	29	21
Cholesterol					
Free	2	7	8	7	3
Ester	6	15	38	23	14
Triacylglycerol	83	50	10	8	5
Fatty acids	—	1	1		
Lipid characteristic	Mainly triacylglycerol	Mainly triacylglycerol; phosphatidylcholine and sphingomyelin main P components	High in cholesteryl linoleate	High in phosphatidylcholine and cholesteryl linoleate	

[Adapted from A. White, P. Handler, E. C. Smith, R. L. Hill, I. R. Lehman, "Principles of Biochemistry," 6th ed., McGraw-Hill, N.Y., 1978, p. 573.]

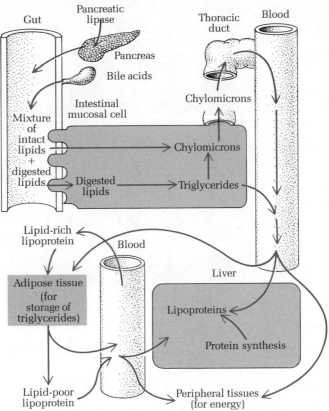

Figure 11.2 Lipid transport in humans. The liver is the site of formation of lipoprotein complexes that carry triglycerides to adipose tissue for storage. [From R. D. Ellefson and W. T. Caraway, in N. T. Tietz, Ed., "Fundamentals of Clinical Chemistry," Saunders, Philadelphia, 1976, p. 484.]

LDL receptors, and persons who are homozygous for the mutant gene produce almost no LDL receptors. This excess concentration of plasma cholesterol is then deposited in the artery wall, inducing *atherosclerosis*.

The liver LDL receptors are quite similar to nonhepatic LDL receptors, at least *in vitro*. The liver receptors enable large amounts of cholesterol to be removed from the blood, thus ensuring low concentrations of dietary cholesterol in plasma. Other factors being equal, the person with the most lipoprotein receptors will be the least vulnerable to a high cholesterol diet and will have the least likelihood of developing atherosclerosis.

11–3 LIPID BILAYERS Phospholipids and glycolipids that contain some unsaturated fatty acids spontaneously form *lipid bilayers* (given the right experimental conditions) in which the polar head groups are exposed to the aqueous phase and the hydrocarbon tails are buried in van der Waals contact with one another (Figure 11.3). The hydrophobic effect provides the driving force for the formation of lipid bilayers and biological membranes. These lipid bilayers form closed *vesicles* that are similar in many respects to biological membranes. The head groups of the lipids lie at roughly right angles to the plane of the bilayer surface, and provide it with an ionic coat (Figure 11.4). The absence of strong attractive forces in the interior of the bilayer makes it fluid and deformable and the hydrocarbon tails have considerable conformational mobility.

X-ray diffraction studies of lipid bilayers show that the terminal methyl groups in the center of the bilayer are quite far apart relative to the length of

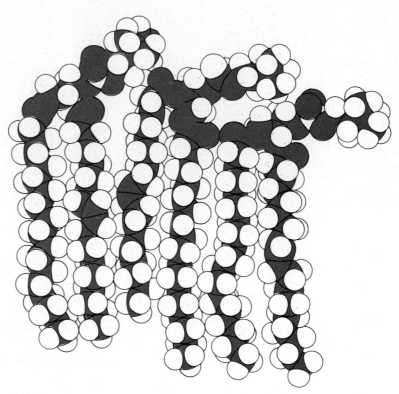

Figure 11.3 Lipid bilayer. (a) Schematic diagram. (b) Space-filling model of lecithin in a lipid bilayer.

a covalent bond. The hydrocarbon tails interact through van der Waals contacts. The addition of cholesterol to the bilayer raises its melting point, although the interior of the membrane remains fluid. Cholesterol probably alters the conformation of the hydrocarbon tails favoring a completely extended hydrocarbon chain, rather than the coiled conformations that predominate when cholesterol is absent. Since the extended chains pack better, a more crystalline array is formed in the presence of cholesterol, and the melting point of the bilayer increases (Figure 11.5).

In lipid vesicles consisting of mixtures of phospholipids and in biological membranes, the phospholipids are distributed asymmetrically between

Figure 11.4 Polar head groups of a lipid bilayer or membrane. The "heads" lie at approximately right angles to the longest axes of the hydrocarbon "tails." [From P. L. Yeagle, *Acc. Chem. Res.*, 9, 321–327 (1978).]

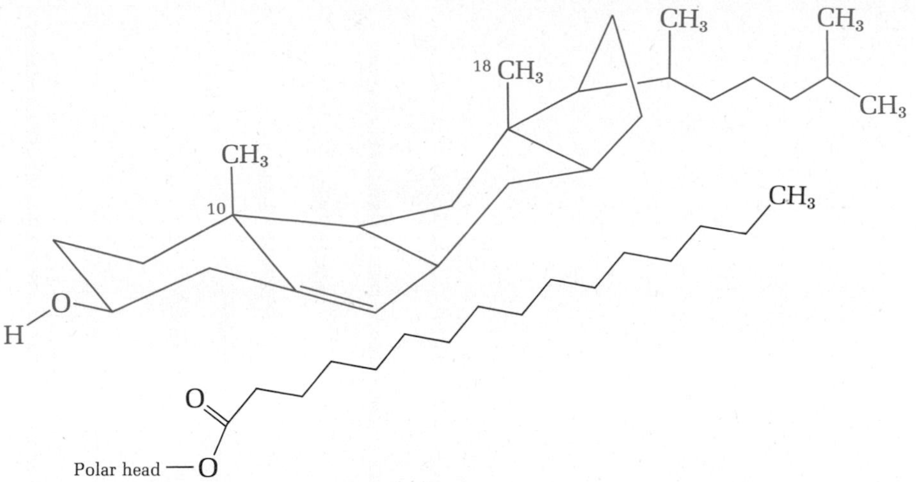

Figure 11.5 Effect of addition of cholesterol to bilayer. Cholesterol stabilizes extended chain conformations of adjacent hydrocarbons in lipid bilayers by van der Waals interactions.

the inner and outer halves of the bilayer. This asymmetry results partly from steric considerations and partly from electrostatic effects. Since the bilayer (or membrane) is highly curved, there is less room in the inner half than in the outer half. The ionic heads therefore repel one another more on the inner surface of the membrane than on the outer surface, and the tails are more tightly packed in the inner half of the bilayer. There is considerable lateral movement in the plane of the bilayer, but there is little "flip-flop" of lipids from one half to the other. Flip-flop is a highly endergonic event. The ionic heads have to burrow through the nonpolar interior of the membrane, where there is no water to stabilize their charge. Such transverse diffusion therefore seldom occurs. The distribution of phospholipids between the inner and outer halves of the membrane of human red blood cells and of the protoplasmic membrane of *Micrococcus lysodeikticus*[1] is given in Figure 11.6. The asymmetric distribution of lipids parallels, and may contribute to, the asymmetric functions of membranes. *Any asymmetric function or process requires an asymmetric structure.*[2]

11–4 MEMBRANE PROTEINS

Biological membranes contain proteins as well as lipids. The relative amounts of proteins and lipids vary from species to species or from tissue to tissue within an organism (Table 11.8). There is a rough correlation between the metabolic activity of a membrane and its protein content. The myelin sheath of nerves, for example, primarily serves as an insulator and contains relatively little protein, whereas the metabolically active inner mitochondrial membrane contains a large amount of protein.

The protein component of membranes also varies in heterogeneity. For example, membranes from rod outer segments are about 50% protein by weight, and this protein is nearly all *rhodopsin,* the visual receptor protein (Section 11.7). Red blood cell membranes, on the other hand, have about the same relative amount of protein, but contain many types of protein. Each

---

[1] By the most recent bacterial classification scheme, this species is now known as *Micrococcus luteus.* However, much biochemical work has been published using the older name.

[2] The general physical principle, known as the *Curie symmetry principle,* states that if a process—whether an enzyme-catalyzed reaction, membrane transport, or any other phenomenon—is asymmetric, the structure responsible for the process must be at least as asymmetric along the axis of the process.

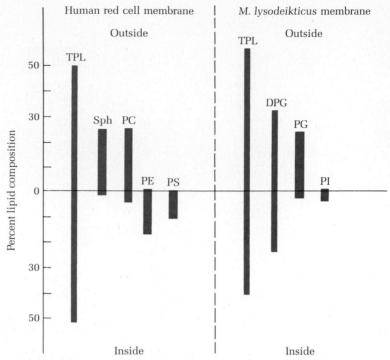

Figure 11.6 Distribution of lipids in human red cell membranes and in membranes from *M. lysodeikticus*. Abbreviations: TPL, total phospholipids; Sph, sphingomyelin; PC, phosphatidylcholine; PE, phosphatidylethanolamine; PS, phosphatidyl-L-serine; DPG, diphosphatidylglycerol; PI, phosphatidylinositol. [From L. D. Bergelson and L. I. Barsukov, *Science, 197,* 224 (1977). Copyright 1977 by the American Association for the Advancement of Science.]

type of membrane, then, has the particular protein complement suited to its function.

Membrane proteins can be divided into two types, *integral* and *peripheral*. The criteria used for distinguishing the two types are summarized in Table 11.9. Peripheral proteins constitute about 25% of membrane proteins. They are bound electrostatically to the ionic head groups on the membrane surface. Therefore, they can be isolated in high purity by a mild technique, such as increasing the ionic strength of the medium. For example, cytochrome *c* can be isolated from the inner mitochondrial membrane and *spectrin* can be isolated from red blood cells by this means.

The integral proteins, however, are much more difficult to isolate and purify. Draconian techniques that disrupt membranes, such as treatment

Table 11.8 Lipid and protein content of several membranes		
	% Dry weight	
Membrane	**Protein**	**Lipid**
Myelin	18	79
Human erythrocyte	49	43
Bovine retinal rod	51	49
Mitochondria (outer membrane)	52	48
Mycoplasma laidlaieii	58	37
Sarcoplasmic reticulum	67	33
Gram-positive bacteria	75	25
Mitochondria (inner membrane)	76	24

SOURCE: From G. Guidotti, *Ann. Rev. Biochem., 41,* 731 (1972). Reproduced with permission. © 1972 by Annual Reviews Inc.

Table 11.9 Criteria for distinguishing integral and peripheral proteins

Property	Peripheral	Integral
Requirement for dissociation from membrane	Mild: sonication or high ionic strength	Hydrophobic disruption: detergent or organic solvents
Association with lipids when free	Not bound to lipids	Usually bound to lipids
Solubility in water when free	Soluble	Insoluble

with detergent or exposure of the membranes to ultrasonic vibrations, are required. Even then, the proteins isolated often have lipid tightly bound to them. Worse yet, the integral proteins are often denatured or otherwise significantly altered when they are removed from the membrane.[3] To a large extent the function of integral membrane proteins depends upon the structural integrity of the membrane.

11–5 MEMBRANE STRUCTURE

A. Some Experimental Observations

We have seen that phospholipids spontaneously form bilayers, that cholesterol is incorporated in bilayers, and that there are proteins in membranes. We might suppose, then, that some type of lipid-protein complex would make a significant contribution to membrane structure and that the proteins would dominate the structure of membranes, at least in those membranes containing over 50% protein. This is not the case. Every study has shown that the basic structure of membranes is a lipid bilayer of the type that forms in the absence of proteins. Consider, for example, the x-ray diffraction pattern obtained from the membranes of retinal rod outer segments (Figure 11.7). (We recall that this membrane is about 50% protein.) Comparing Figure 11.7 with the results for x-ray diffraction of lipid bilayers, Figure 11.8, we see that there is almost no difference. Where, then, is the rhodopsin located? Mild treatment does not remove it. In fact, rhodopsin is an integral membrane protein by all of the criteria of Table 11.9. If rhodopsin is *mobile*, however, then the observed x-ray diffraction pattern makes sense, since the time average position of the protein is so broadly dispersed that the x-ray pattern misses it altogether. Treatment of the membrane with an antibody to rhodopsin "immobilizes" the rhodopsin and dramatically changes the x-ray pattern. This observation indicates both that the mobility of rhodopsin is decreased and that part of the rhodopsin is exposed to the external medium. Many other experiments support the general concept that the membrane is a fluid lipid bilayer. Proteins are able to float freely in the plane of the membrane.

The positions of proteins in the membrane can be detected by *freeze fracture electron microscopy*. In this technique a membrane in aqueous solution is rapidly frozen. It is placed in a vacuum chamber where it is fractured. In the vacuum water molecules evaporate from the exposed surface. The surface can be covered with a deposit of metal, such as platinum, by evaporating the metal off a hot filament (Figure 11.9). The "coat," or shadow, provides a replica of the surface which can be examined in the electron microscope.

The fracture occurs along the planes having the weakest intermolecular

---

[3] Recent advances in nuclear magnetic resonance instrumentation make it possible to study the reactions of integral membrane proteins in living cells. Entire metabolic pathways, such as oxidative phosphorylation (Chapter 16), that depend upon the structural integrity of the membrane, can be studied by these techniques.

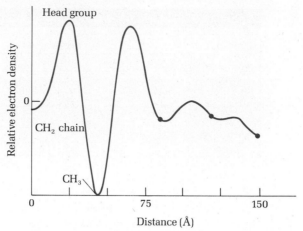

Figure 11.7   Electron density profile for a single membrane from rod outer segments. The maxima correspond to the ionic head groups. The internal methylene groups have a lower electron density, and a minimum is reached at the terminal methyl group inside the membrane. [From J. M. Corless, *Nature, 237,* 229 (1972). Reprinted by permission. Copyright © 1972 Macmillan Journals Limited.]

interactions, and therefore the membrane splits through the lipid portion, opening like a sandwich. The rough surface along the fracture plane has "bumps" and depressions corresponding to the integral membrane proteins (Figure 11.10). Replicas obtained from pure phospholipid bilayers and the myelin sheath are smooth, whereas replicas from membranes rich in proteins look like a cratered lunar landscape (Figure 11.11).

   Membrane proteins are distributed across the bilayer asymmetrically. In red blood cells only two proteins are accessible to chemical reagents, but when the membrane is disrupted, all of the proteins react with the same reagents. One of the erythrocyte proteins, *glycophorin,* is a glycoprotein that contains 200 amino acid residues in a single polypeptide chain whose molecular weight is 50,000. About one-half of its mass is carbohydrate. The primary structure of glycophorin is shown in Figure 11.12. All of the carbohydrates are located in the N-terminal region of the protein, and all are ex-

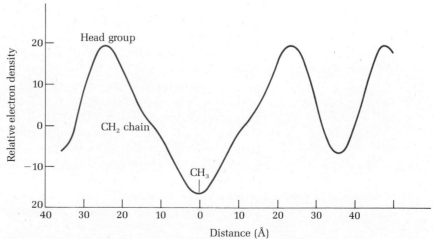

Figure 11.8   Electron density profile for a lipid bilayer of dipalmitoyl phosphatidylcholine. [From Y. K. Levine, A. I. Bailey, and M. H. F. Wilkens, *Nature, 220,* 578 (1968). Reprinted by permission. Copyright © 1968 Macmillan Journals Limited.]

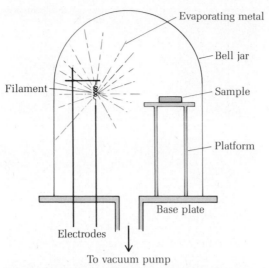

Figure 11.9 Apparatus for forming replicas of samples for electron microscopy. A shadow is cast on the sample by metal atoms that are boiled off a hot filament.

posed to the external medium. The C-terminal region of glycophorin, which protrudes into the cell, contains many charged residues and is highly hydrophilic. The interior of glycophorin contains a string of 20 hydrophobic residues that span the entire membrane. The hydrophilic regions of glycophorin are exposed to the aqueous phase (Figure 11.13). Some membrane proteins are exposed only to the external side of the membrane, and others are exposed only to the internal side of the membrane.

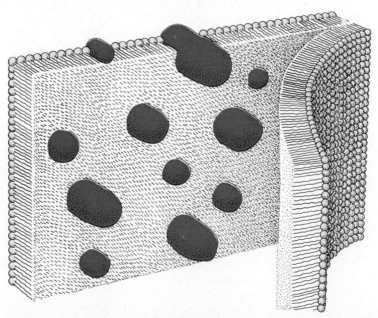

Figure 11.10 Replica of a freeze-fractured membrane. A lipid bilayer can be split between the hydrocarbon tails parallel to the surface of the membrane by freeze fracturing. Integral proteins are detected as protrusions or cavities in the replica. [Figure by B. Tagawa, from S. J. Singer, in G. Weismann and R. Claiborne, Eds., "Cell Membranes; Biochemistry, Cell Biology, and Pathology," HP Publishing Co., Inc., New York, 1975, p. 38. Reproduced with permission.]

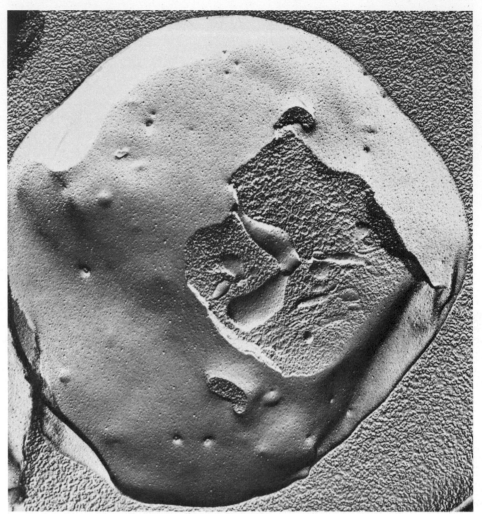

Figure 11.11 Electron micrograph of a freeze-fractured erythrocyte membrane. The bumps on the surface are thought to be integral membrane proteins. [Courtesy of Dr. Dorothea Zucker-Franklin, New York University School of Medicine.]

B. Fluid Mosaic Model of Membrane Structure

In the fluid mosaic model of membrane structure, proposed in 1972 by S. J. Singer and G. Nicolson, the lipids maintain the structural integrity of the membrane and the proteins are responsible for the specialized functions of the membrane. The fluid mosaic model is supported by an experiment carried out by L. D. Frye and M. Edidin who labeled specific proteins in human and mouse cell membranes with fluorescent dyes of different colors. The human and mouse cells were then fused—the new cell is called a *heterokaryon*—and observed through a light microscope. At first the red and green patches of the fluorescent dyes were localized within their original cells. An hour later the color patches were uniformly distributed between the two cells (Figure 11.14). This experiment dramatically illustrates that the lipid phase of the membrane is a fluid in which the proteins float freely (Figure 11.15). This movement is lateral. Proteins do not flip-flop across the membrane at appreciable rates.

There are many gaps in our knowledge of membrane structure. For example, the tertiary structures of proteins in membranes are unknown, as are the details of lipid-protein interactions. Inorganic ions seem to be required to maintain membrane structure, an observation not easily explained by the fluid mosaic model. (They might merely form ion pairs with the ionic heads to minimize electrostatic repulsion, thus stabilizing the bilayer.)

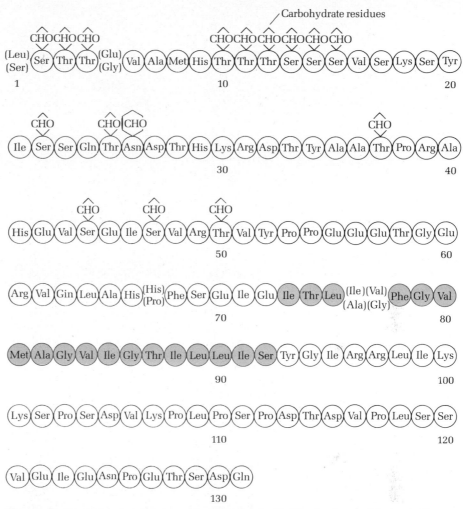

Figure 11.12 Primary structure of glycophorin A. The hydrophobic part of the molecule, which spans the hydrophobic interior of the membrane, is indicated by dark circles.

11–6 TRANSPORT ACROSS CELL MEMBRANES

Membranes control not only the rate of entry of metabolites into the cell, but also the partition of metabolites among various organelles, each of which has its own specialized membrane. The resulting compartmentation of metabolic activity is an important mechanism for metabolic regulation. Membrane transport often requires energy, so the energy economy of the cell is linked to transport. We shall see in future chapters that membrane transport is central to many cellular processes.

A. Passive Diffusion

Phospholipid bilayers and biological membranes are excellent permeability barriers to the diffusion of ions and polar molecules. Water is a conspicuous exception to this generalization (but, its concentration is enormous compared to the concentration of other metabolites), and it rapidly diffuses through phospholipids and bilayers. Table 11.10 gives the rates of passive diffusion of a few metabolites. Simple diffusion is biologically important only for water and for gases such as oxygen and carbon dioxide.

B. Facilitated Diffusion

The simplest transport mechanism involving a membrane protein is flow of substrate down a concentration gradient in a process called *facilitated diffusion*. No energy is consumed and the direction of transport depends upon the concentrations of the metabolite on each side of the membrane.

Glucose and glycerol are transported by facilitated diffusion. The rate

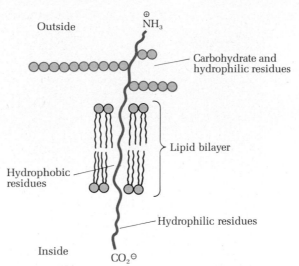

Outside

$\overset{\oplus}{NH_3}$

— Carbohydrate and hydrophilic residues

Hydrophobic residues —

— Lipid bilayer

— Hydrophilic residues

Inside $CO_2{}^{\ominus}$

Figure 11.13   Distribution of glycophorin A across the bilayer. Glycophorin A spans the red blood cell membrane. The carbohydrate moieties bound to the protein are all on the external side of the membrane. Those residues that are buried in the membrane are hydrophobic, and those exposed to the aqueous environment are hydrophilic.

of glucose transport as a function of glucose concentration is shown in Fig. 11.16. The hyperbolic shape of the curve reminds us of Michaelis-Menten kinetics, and, in fact, the rate law for facilitated diffusion has the same form as the Michaelis-Menten equation. The hyperbolic shape of the curve also implies that the transport system is saturable, that is, that the maximum rate of transport has an upper limit dictated by the number of transport proteins. The rate law for unidirectional flux of a substance by facilitated diffusion is

$$J_{12} = \frac{J_{max}\,C_1}{K_m + C_1} \tag{11.1}$$

where J_{12} is the flux from compartment 1 to compartment 2, J_{max} is the rate of maximum flux, K_m is the Michaelis constant (with the same meaning it has in enzyme kinetics), and C_1 is the concentration of substrate in compartment 1.

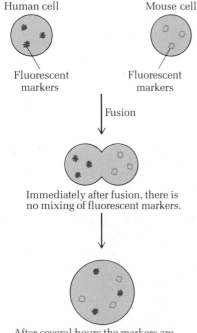

Human cell

Mouse cell

Fluorescent markers

Fluorescent markers

Fusion

Immediately after fusion, there is no mixing of fluorescent markers.

After several hours the markers are completely mixed indicating that proteins are mobile in cell membranes.

Figure 11.14   Demonstration that membranes are fluid and that proteins in membranes move freely in the plane of the bilayer.

Table 11.10	Diffusion coefficients for passage of ions and small molecules across lipid bilayers

Ion or molecule	Diffusion coefficient (cm² sec⁻¹)
H_2O	5×10^{-3}
Indole	3×10^{-4}
Urea	3×10^{-6}
Glycerol	3×10^{-6}
Tryptophan	1×10^{-7}
Glucose	5×10^{-8}
$Cl^{\ominus}$	7×10^{-10}
$K^{\oplus}$	5×10^{-12}
$Na^{\oplus}$	1×10^{-12}

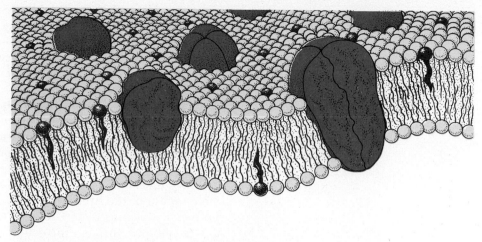

Figure 11.15 Fluid mosaic model of membrane structure. Membrane proteins are imbedded in the lipid bilayer. Some span the bilayer, some are exposed to the inside of the cell, and some are exposed to the outside of the cell. [Figure by B. Tagawa, from S. J. Singer, in G. Weismann and R. Claiborne, Eds., "Cell Membranes: Biochemistry, Cell Biology, and Pathology," HP Publishing Co., Inc., New York, 1975, p. 37. Reproduced with permission.]

The kinetic behavior predicted by equation 11.1 and shown in Figure 11.16 is one of the principal criteria for identifying facilitated diffusion.

The glucose transport system, and most others, is stereospecific. A hydroxyl group at C-2 in an equatorial position is required. The glucose transport system of red blood cells will transport both glucose and galactose, and each acts as a competitive inhibitor of the other. A facilitated transport system is normally subject to competitive inhibition by analogs of the substrate being transported.

In many cells the rate of transport affects the overall metabolic rate of

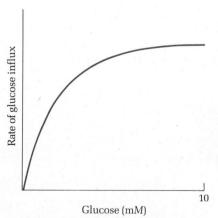

Figure 11.16 Rate of transport vs. concentration gradient in facilitated diffusion. The concentration-dependence of glucose influx into human erythrocytes is hyperbolic, indicating that a carrier is required. The rate of transport is maximal when all of the binding sites on the transport proteins are occupied.

the cell. For example, under some circumstances glycolysis is not only regulated by enzymes of the metabolic pathway, it is also limited by the rate of glucose transport. *Insulin*, a hormone synthesized by the islet β-cells of the pancreas, increases the maximum rate of glucose transport by a factor of three to four leading to a faster rate of glycolysis and a faster overall metabolic rate.

Visualizing transport within the context of the fluid mosaic model of membrane structure limits the types of mechanisms that are possible for facilitated diffusion. The model of the transport system as a "revolving door" which accepts substrate at one face of the membrane, turns around in the membrane, and releases the substrate on the other side is impossible because it requires that the hydrophilic ends of the protein move through the hydrophobic region of the lipid bilayer (Figure 11.17).

A more plausible mechanism is transport through pores (Figure 11.18). A helical pore can form through the center of a protein. (The helix in such a case cannot be an α-helix, which has no pore.) Such pores have been demonstrated for synthetic peptides having the sequence (Leu-Ser-Leu-Gly)$_n$ where $n = 6$, 9, and 12. The exteriors of these peptides are hydrophobic and their interiors are hydrophilic. The end view of the pore in the peptide with $n = 12$ is shown in Figure 11.19. The permeability of ions through phospholipid bilayers is greatly increased when the synthetic peptide is added to make an artificial membrane. Such pores permit passage of water, and biological membranes are freely permeable to water. The charge and geometry of the pores would determine which molecules would be transported. We recall that a pore in the center of hemoglobin is complementary to the charge and geometry of 2,3-diphosphoglycerate. By analogy we expect the pores of transport proteins to have their own unique properties.

C. Active Transport

Active transport is mediated by a specific transport protein, but, unlike facilitated diffusion, active transport operates against a concentration gradient and requires energy. The free energy required to transport 1 mole of substrate from compartment 1 to compartment 2 is

$$\Delta G = -2.303 \, RT \log \frac{C_1}{C_2} \tag{11.2}$$

where ΔG is the free-energy change, R is the gas constant, T is the absolute temperature, and C_1 and C_2 are the concentrations in compartments 1 and 2, respectively. If $C_1 > C_2$, then the log term is positive, ΔG is negative, and

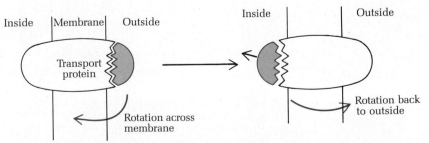

Figure 11.17 "Revolving door" model for transport by facilitated diffusion. This mechanism is forbidden because it requires that the transport protein flip-flop across the membrane, a process that is known to be quite slow.

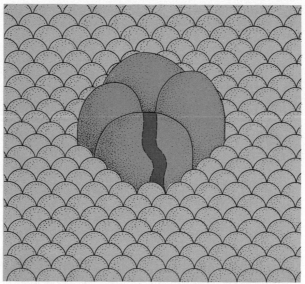

Figure 11.18 Pore model for transport by facilitated diffusion. The charge and conformation of the pore determines the specificity of the transport protein. [From S. J. Singer, in G. Weismann and R. Claiborne, Eds., "Cell Membranes: Biochemistry, Cell Biology, and Pathology," HP, New York, 1975, p. 42.]

Figure 11.19 Structure of a synthetic peptide that forms a pore through which ions pass by facilitated diffusion. The peptide chain is coiled in a helix of four turns. Each turn of the helix spans 12 residues. The exterior of the helix is hydrophobic, and the interior pore of the helix is hydrophilic. [Courtesy of Dr. H. R. Besch and Dr. Stephen J. Kennedy.]

transport is spontaneous from compartment 1 to compartment 2. If $C_1 < C_2$, then $\triangle G$ is positive, and energy must be supplied to "pump" the substrate "uphill" to the region of higher substrate concentration. Active transport makes severe demands upon the cellular energy pool. In a resting human between 30 and 40% of the total energy consumption is used for active transport.

All cells maintain ion gradients of many kinds. We will consider the sodium-potassium ion exchange pump of the erythrocyte membrane as a paradigm of active transport. The sodium-potassium ion pump operates asymmetrically, or *vectorially*. Sodium ions are pumped out of the cell and potassium ions are pumped in against concentration gradients. The intracellular concentrations of $Na^{\oplus}$ and $K^{\oplus}$ are about 10 mM and 100 mM, respectively, a $Na^{\oplus}$ to $K^{\oplus}$ ratio of 1:10. Extracellular fluids, on the other hand, are about 140 mM in $Na^{\oplus}$ and 5 mM in $K^{\oplus}$, a $Na^{\oplus}$ to $K^{\oplus}$ ratio of 28:1. (The $Na^{\oplus}$ to $K^{\oplus}$ ratio of 28:1 is about that found in seawater where life is thought to have emerged.) The $Na^{\oplus}$-$K^{\oplus}$ pump operates against this concentration gradient. Hydrolysis of ATP (Section 12.4C) supplies the energy required to drive the pump. The transport of $K^{\oplus}$ is coupled to the transport of $Na^{\oplus}$ and one will not occur without the other. In erythrocytes three $Na^{\oplus}$ ions are exported for every two $K^{\oplus}$ ions that enter the cell. The ADP and phosphate produced by ATP hydrolysis do not leave the cell, and adding ATP to the external medium does not affect the pump. The $Na^{\oplus}$-$K^{\oplus}$ transport ratio of 3:2 establishes an electrochemical gradient as well as a concentration gradient across the membrane.

By using ATP labeled with ^{32}P in the γ position a phosphorylated protein has been identified as an intermediate in the process. If no $K^{\oplus}$ is present in the medium, the ^{32}P-labeled protein can be isolated. Diisopropylphosphofluoridate (Section 6.3) inhibits the formation of the phosphoenzyme. The hydrolysis of ATP is coupled to phosphorylation of the transport protein. The hydrolysis of the phosphoprotein, perhaps coupled to a conformational change that opens a pore, then drives the transport of $Na^{\oplus}$ and $K^{\oplus}$ (reac-

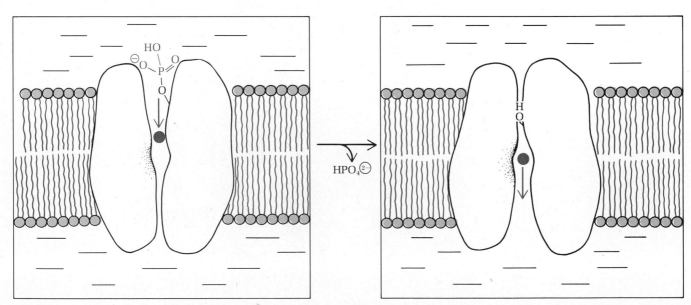

Figure 11.20 Model for the transport of $K^{\oplus}$ into a cell against a concentration gradient. The transport protein is phosphorylated, with ATP as the phosphoryl donor, and the phosphorylated protein then hydrolyzes in an exergonic step that drives transport by opening a pore. This model does not account for the exchange of 3 $Na^{\oplus}$ ions for 2 $K^{\oplus}$ ions, and the pump must be considerably more complex that our diagram suggests.

tions 11.3 and 11.4). The conformations of the $Na^{\oplus}$ and $K^{\oplus}$ binding sites might alter when the transport protein is phosphorylated (Figure 11.20).

$$ATP + transport\ protein \longrightarrow$$
$$ADP + transport\ protein—OPO_3^{\text{2}\ominus} \quad (11.3)$$

$$transport\ protein—OPO_3^{\text{2}\ominus} + H_2O \longrightarrow$$
$$transport\ protein + HPO_4^{\text{2}\ominus} \quad (11.4)$$

$$3\ Na^{\oplus} \underset{\text{In}}{\overset{\text{Out}}{\rightleftharpoons}} 2\ K^{\oplus}$$

Using equation 11.2 we calculate that a free-energy change of 9.3 kJ mol^{-1} will drive a concentration gradient of 50:1 uphill. This free-energy change is about what would be released in the hydrolysis of the phosphoprotein.

In a medium containing no $K^{\oplus}$ and a high $Na^{\oplus}$ concentration the erythrocyte pump reverses direction and moves $Na^{\oplus}$ inward and $K^{\oplus}$ outward. The hydrolysis of ATP reverses, and the "ATPase" synthesizes ATP. The combination of a concentration gradient and an electrochemical gradient drives the synthesis of ATP. The implications of this observation for oxidative phosphorylation will become apparent in Chapter 16.

11–7 LIPID-SOLUBLE VITAMINS, COENZYME Q, AND PROSTAGLANDINS

When we discussed the water-soluble vitamins in Chapter 10 we were able to consider the chemistry of the cellular reactions in which they participated. The lipid-soluble vitamins are another story (Table 11.11). Until quite recently the physiological function, to say nothing of the chemistry, of some of them was unknown. This situation is changing rapidly, and the outline of the biochemistry of the fat-soluble vitamins is beginning to emerge.

A. Vitamin A

Vitamin A, or *retinol*, an essential component of the visual cycle, is produced by oxidation of the plant pigment β-carotene (reaction 11.5). Retinol is oxidized to 11-*trans*-retinal, also called all-*trans* retinal, by retinol dehydrogenase. The 11-*trans* double bond is then converted to 11-*cis* by retinal isomerase (reactions 11.6 and 11.7). The outer segment of retinal rod cells contains *opsin*, a protein that binds to 11-*cis*-retinal through an ϵ-amino group of a lysyl side chain to give rhodopsin (reaction 11.8). Opsin does not absorb visible light, but rhodopsin absorbs strongly with an absorption spectrum that closely matches the ultraviolet and visible emission spectra of the sun (Figure 11.21).

There are two types of photoreceptors in humans: the *rod cells* and the *cone cells*. The cone cells are located near the center of the retina and are responsible for color vision. The rod cells can function in dim light, cap-

Table 11.11	Lipid-soluble vitamins
Vitamin	**Function**
Vitamin A	Photochemistry of vision
Vitamin D	Regulation of Ca(II) metabolism
Vitamin E	Antioxidant (free-radical scavenger), required for fertility in rats and perhaps in humans, regulates prostaglandin biosynthesis
Vitamin K	Required for carboxylation of prothrombin and other proteins in blood clotting

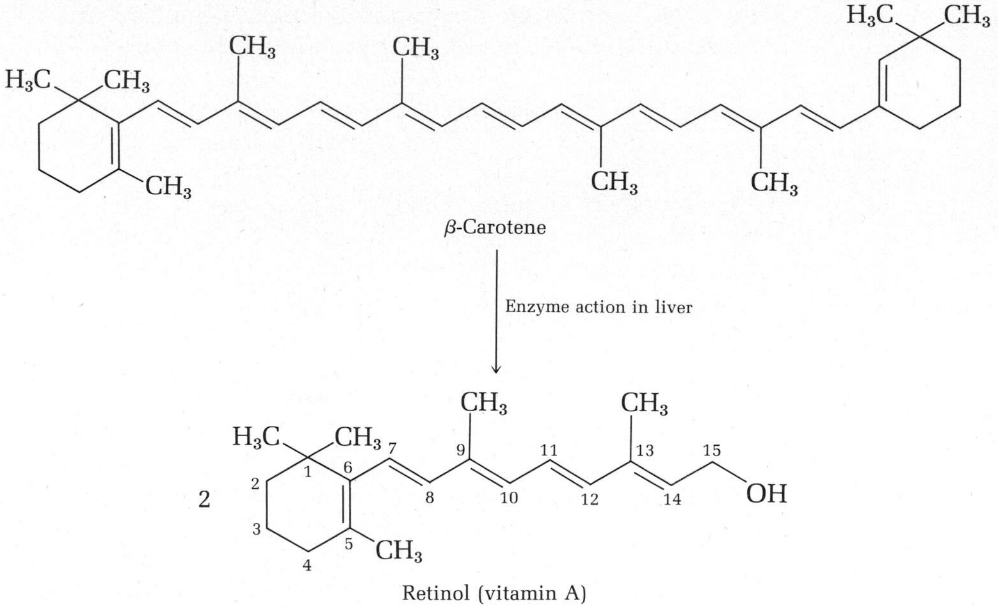

β-Carotene

Enzyme action in liver

(11.5)

Retinol (vitamin A)

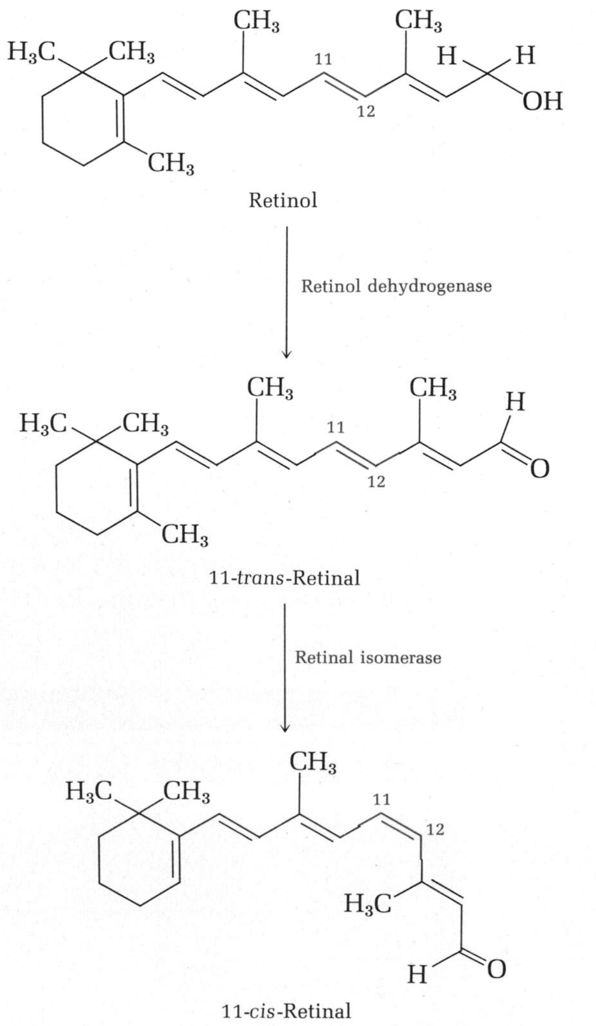

Retinol

Retinol dehydrogenase

(11.6)

11-*trans*-Retinal

Retinal isomerase

(11.7)

11-*cis*-Retinal

11-*cis*-Retinal

$\rightarrow H_2O$

(11.8)

Imine (Schiff base)

Rhodopsin

turing virtually every photon that falls upon them, and they are responsible for night vision. They derive their name from their shape (Figure 11.22). Each rod cell is divided into three sections: the photosensitive rod-like, outer segment contains about 500 flat discs enclosed in a double membrane that is about 60% rhodopsin and 40% lipid; the inner segment and the nucleus carry out the metabolism of the cell; and the synaptic body connects the rod cell to the central nervous system (Figure 11.23).

Irradiation of rhodopsin with visible light causes *photoisomerization* of the 11-*cis*-retinal to 11-*trans*-retinal and dissociation of 11-*trans*-retinal from rhodopsin (reaction 11.9). In its new configuration retinal cannot bind rhodopsin, the imine spontaneously hydrolyzes, and the chromophore dissociates leaving behind opsin. The entire process requires about one millisecond. During that interval a nerve impulse is generated. At least five intermediate species exist during the time interval between absorption of the photon and the release of 11-*trans*-retinal. Each of them is characterized by a different absorption maximum.

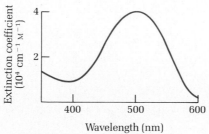

Figure 11.21 Visible absorption spectrum of rhodopsin.

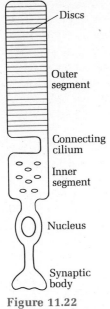

Figure 11.22 Schematic diagram of a rod cell from retinal outer segments.

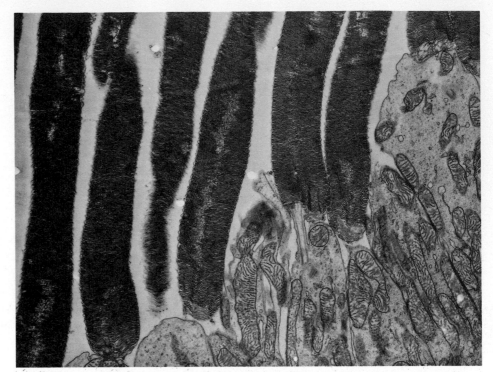

Figure 11.23 Electron micrograph of retinal rod cells. The densely packed dark lines are the stacked discs that contain rhodopsin. (Courtesy of Dr. K. Porter.)

Rhodopsin

Visible light
(photoisomerization)

$+ H_2N$

Opsin

11-*trans*-Retinal

(11.9)

In the dark there is a continual flow of $Na^\oplus$ ions into rod outer segments which is driven by an $Na^\oplus$-$K^\oplus$ transport system similar to the one we discussed in the preceding section. Absorption of a photon by rhodopsin stops this flow of $Na^\oplus$ ions into the outer segment. The membrane becomes more negative on the inside, creating an electrical imbalance across the *hyperpolarized* cell membrane. Photoexcitation releases $Ca^{2\oplus}$ ions that block the pores through which $Na^\oplus$ ions are transported. The synapse of the rod cell is stimulated by the hyperpolarized membrane and a nerve fires (Figure 11.24).

A deficiency of vitamin A can have terrible consequences. In children lack of vitamin A leads to *xerophthalmia,* an eye disease that results first in *night blindness* and eventually in total blindness. The disease can be prevented by an adequate dietary or supplementary supply of vitamin A. Unfortunately, in countries that have suffered from cruel famines, such as Bangladesh and Cambodia, the burdens of malnutrition and disease lead to total blindness in thousands of children.

B. Vitamin K and Blood Clotting

The chain of events by which a chemical or mechanical stimulus causes blood clotting to occur is an intricate process involving a dozen or more serum glycoproteins. A scheme for blood coagulation is shown in Figure 11.25. The activation of the various blood clotting "factors" (about half of

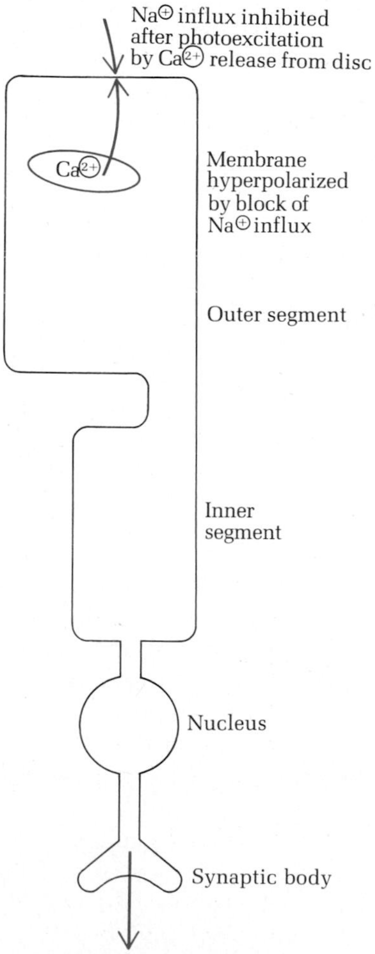

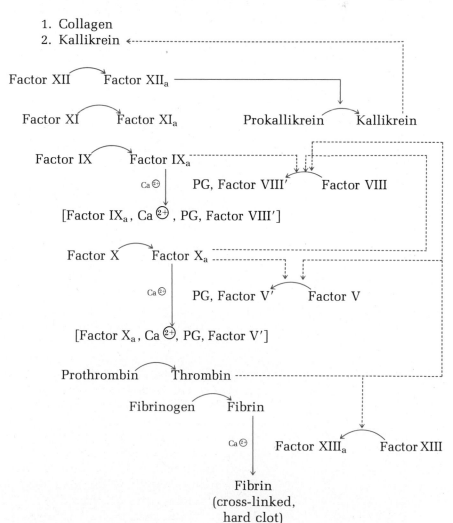

Figure 11.24 Rod cell synapse. Release of $Ca^{2\oplus}$ ions after photoexcitation blocks normal $Na^\oplus$ influx, the membrane becomes hyperpolarized, and the nerve is fired—all within a millisecond.

Figure 11.25 Intrinsic system for blood coagulation. Factors VII, IX, X, prothrombin, and at least one more as yet unidentified protein are affected by vitamin K.

which are serine proteases) is a serial activation process, or *cascade* phenomenon. In an enzymatic cascade, the rate of enzymatic catalysis provides a large multiplying effect at each stage of the process, and a small initial stimulus, or signal, is progressively amplified.

Vitamin K is an essential nutrient in animals because it is required for operation of the blood clotting cascade. Vitamin K has two prominent struc-

Vitamin K$_1$

tural features. The 1,4-diketone is called a *quinone*. Since the quinone ring is fused to benzene, the entire system is a naphthoquinone. In the structure of Vitamin K$_1$ as shown, the quinone is oxidized. A two-electron reduction of the quinone produces a reduced 1,4-*dihydroquinone* as shown in reaction 11.10.

$$+ 2H^{\oplus} + 2e^{\ominus} \rightleftharpoons \qquad\qquad (11.10)$$

Quinone (oxidized)   1,4-Dihydroquinone (reduced)

The second striking structural feature is the polyene side chain composed of repeating *isoprenyl units*.

Isoprenyl unit

Vitamin K occurs in two general forms. The K$_1$ vitamins are synthesized in plants and are known as the phylloquinones (from the Greek *phyllon*, "leaf"). The term *phytylmenaquinone* is now often used instead. These compounds have a double bond only in the isoprenyl unit closest to the ring and have four isoprenyl units in the chain. The K$_2$ vitamins, synthesized by intestinal bacteria in animals, typically have five isoprenyl groups with a double bond in each. They are known as *menaquinones* or *multiprenylmenaquinones*.

The only certain, but highly significant, function of vitamin K is associated with the blood clotting cascade (Figure 11.25). Five proteins of the cascade are known to be affected by vitamin K: factors VII, IX, X, prothrombin, and another unidentified serum protein. The effect of vitamin K on prothrombin is by far the best understood. In clot formation prothrombin molecules form aggregates with phospholipids, a process aided by calcium ions. Prothrombin binds to about 10 to 12 Ca(II) ions, an affinity that drops precipitously if vitamin K is absent. This intriguing observation began to be explained when it was discovered that within the first 40 N-terminal residues of prothrombin, which contains about 560 residues altogether, there are 9 γ-carboxyglutamate (Gla) residues (7, 8, 15, 20, 21, 26, 27, 30, and 33). This amino acid had not previously been identified in any other protein. It has since been found in several calcium-binding proteins (Section 13.7). Since it decarboxylates under acidic conditions, γ-carboxyglutamate will not be detected if acid hydrolysis is employed to determine the amino acid composition of a protein.

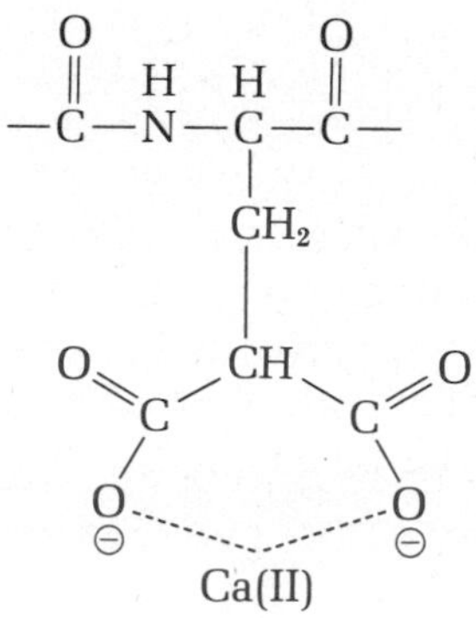

γ-Carboxyglutamate complexed with Ca(II)

The vicinal carboxyl groups of γ-carboxyglutamate comprise a *bidentate* ("having two teeth") ligand for Ca(II). If prothrombin lacks γ-carboxyglutamate residues, it loses most of its ability to bind calcium ions. The reduced, 1,4-dihydro form of vitamin K is required for carboxylation of prothrombin, a reaction that also requires molecular oxygen. One mechanistic possibility, as yet only a hypothesis, is that CO_2 reacts directly with the reduced form of vitamin K to give a carboxylated intermediate, analogous to N^1-carboxybiotin, as shown in reaction 11.11.

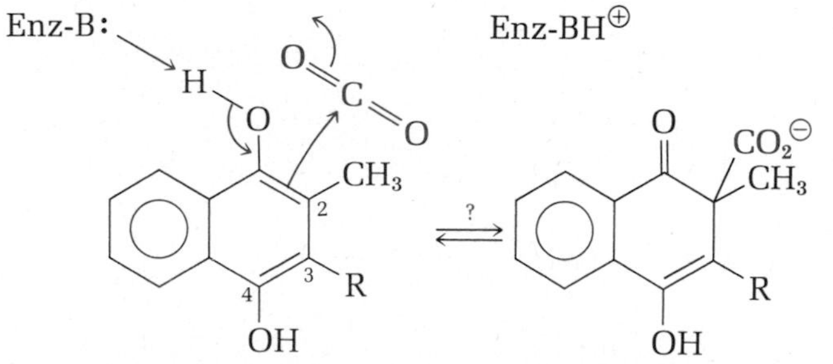

$$(11.11)$$

Vitamin K, reduced form   Carboxylated vitamin K intermediate

The carboxylation of prothrombin is still almost a complete mystery, but the reaction is so important physiologically that we shall probably not have long to wait for the outlines of a mechanism of carboxylation to appear.

C. Coenzyme Q   Coenzyme Q is synthesized by most organisms, including humans. It is not a vitamin, but its structure is so similar to the structure of vitamin K, that we shall discuss it here. Coenzyme Q is also a quinone, differing from vitamin K by the substitution of two methoxy groups for the fused benzene ring. It is also called *ubiquinone*, a pun on its apparently ubiquitous occurrence in nature. Mammalian ubiquinones typically have 10 isoprenyl groups in the side chain, and those from bacteria typically have 6. Plants also synthesize close relatives of coenzyme Q called *plastoquinones* because they are found in chloroplasts. The most abundant plastoquinone has 9 isoprene units.

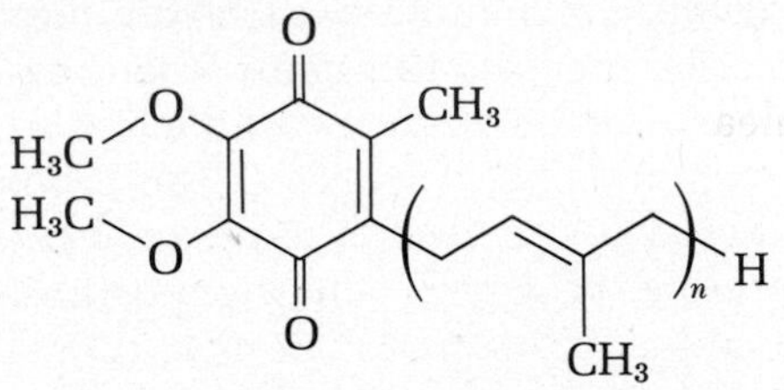

Coenzyme Q, n = 6 − 10 (ubiquinone)

Since the structures of coenzyme Q and vitamin K are so similar we might expect to find similar functions, yet so far as is known, coenzyme Q cannot substitute for vitamin K in the carboxylation of prothrombin and other blood-clotting factors. Coenzyme Q is a lipid-soluble carrier of protons and electrons that participates in the respiratory chain (Chapter 16). It undergoes facile one-electron reduction reactions with free radicals as intermediates (reaction 11.12).

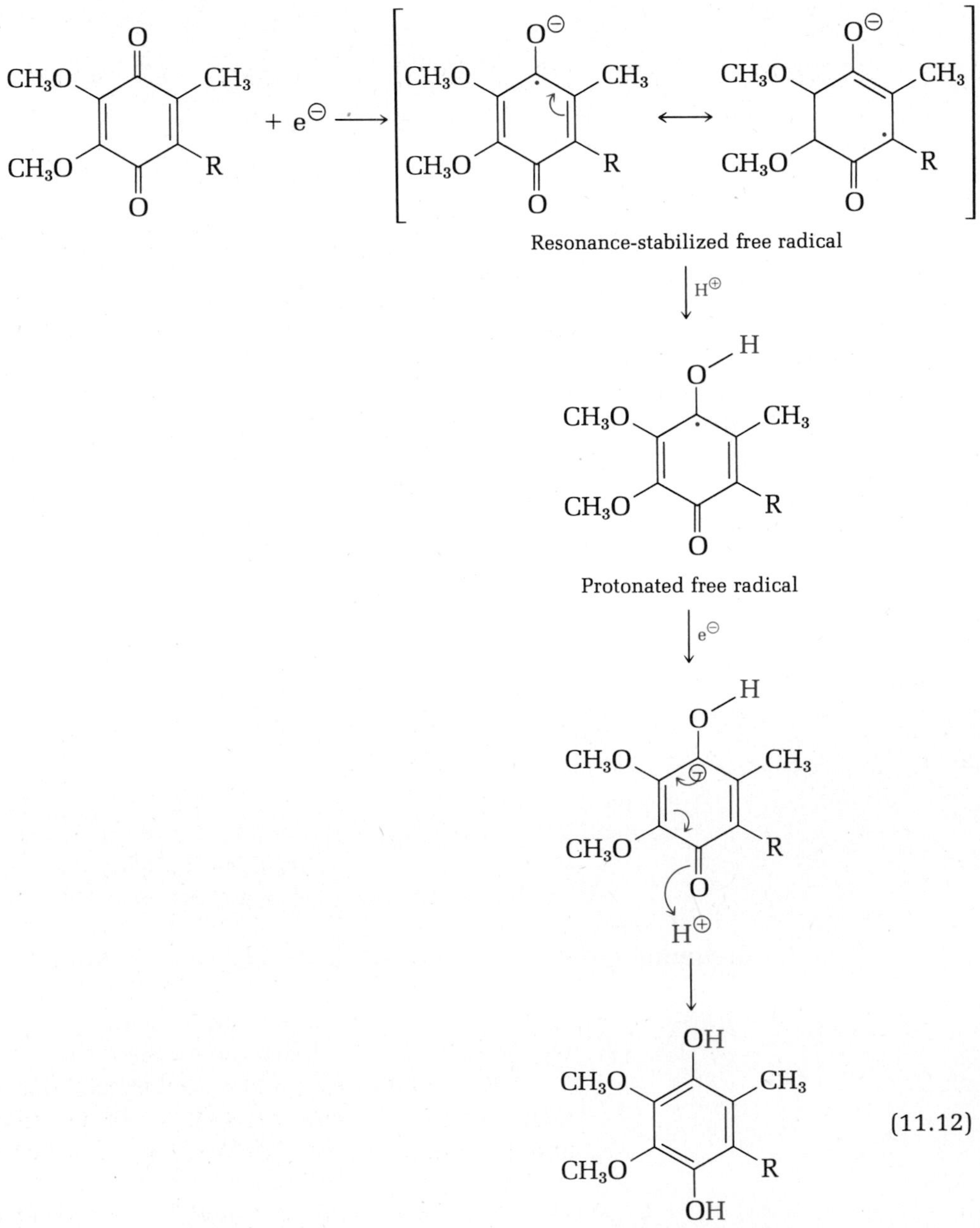

Resonance-stabilized free radical

Protonated free radical

(11.12)

1,4-Dihydroubiquinone (reduced)

Free radicals are often unstable, but the one generated by addition of an electron to ubiquinone is stabilized by resonance and by the electron-releasing substituents on the ring.

D. Vitamin D   Vitamin D_3, *cholecalciferol*, is required for the proper formation of bone and teeth. Liver and fish oils are rich in this vitamin, and the cod liver oil traditionally administered to New England children until recently had great nutritional value in the long New England winters. A deficiency causes *rickets*, a disease that cruelly afflicts children with soft, deformed, and poorly calcified bones. Vitamin D deficiency is almost totally confined to children, who require an additional 20 mg per day in their diet. The action of sunlight on skin produces adequate amounts of vitamin D in adults, for whom no dietary supplement is required.

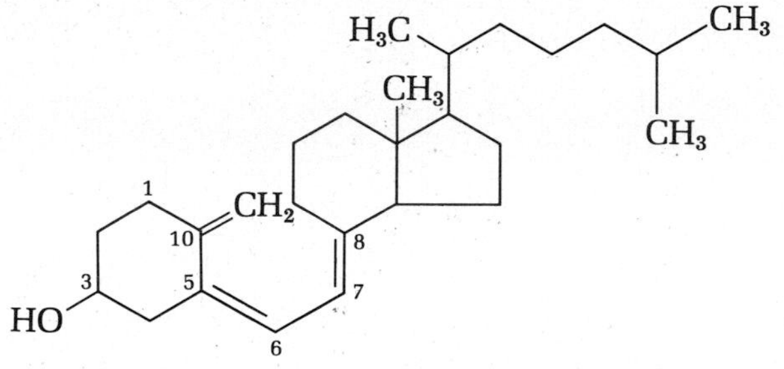

Vitamin D_3 (cholecalciferol)

Vitamin D_3 is produced by the action of sunlight upon $\triangle^{5,7}$-cholesterol in the skin. The first reaction in the conversion yields precalciferol (reaction 11.13). (This reaction is a photochemically allowed 4 + 2 conrotatory electrocyclic process reaction 11.13).[4] This process is shown schematically in reaction 11.14.

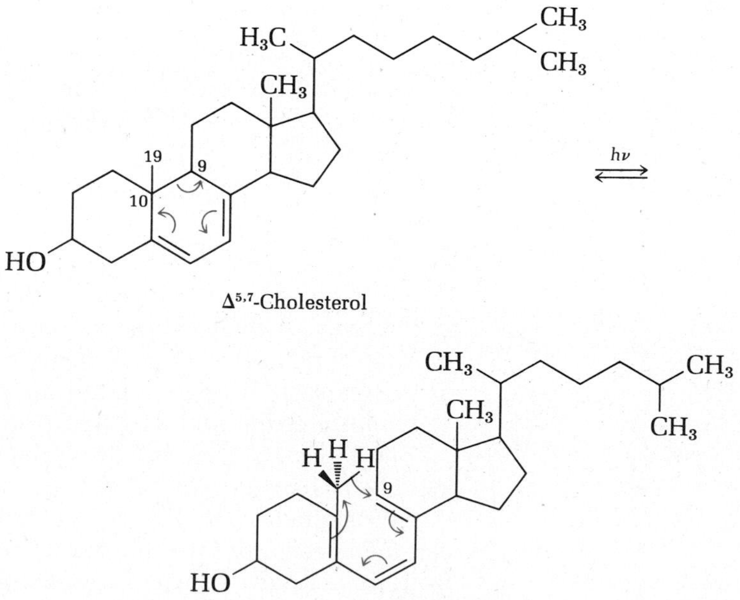

$\triangle^{5,7}$-Cholesterol

(11.13)

Precalciferol

The second reaction in the production of cholecalciferol is truly amazing. A hydrogen on the C-19 methyl group migrates to C-9 in a concerted motion. The migration occurs from the *top* of one end of the con-

---

[4] The theory of these reactions is discussed in most introductory organic chemistry texts.

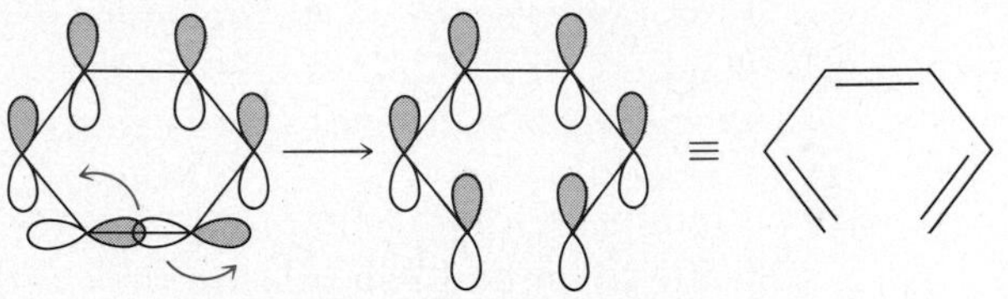

(11.14)

jugated π system to the *bottom* of the other end. This is called a *1-7 antarafacial sigmatropic shift*, and it is a photochemical reaction (reaction 11.15).

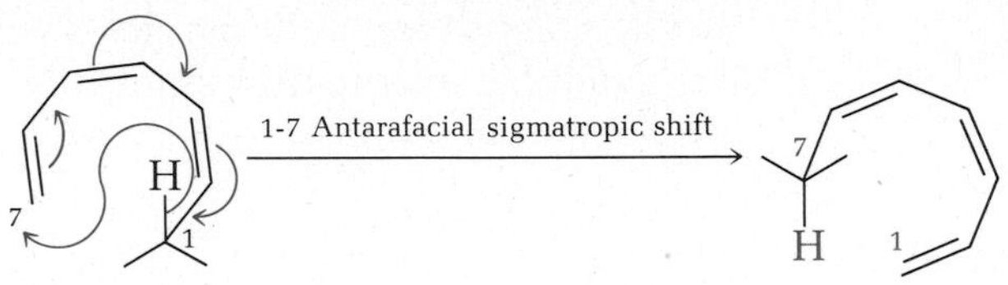

(11.15)

The conversion of precalciferol to cholecalciferol proceeds as in reaction 11.16.[5]

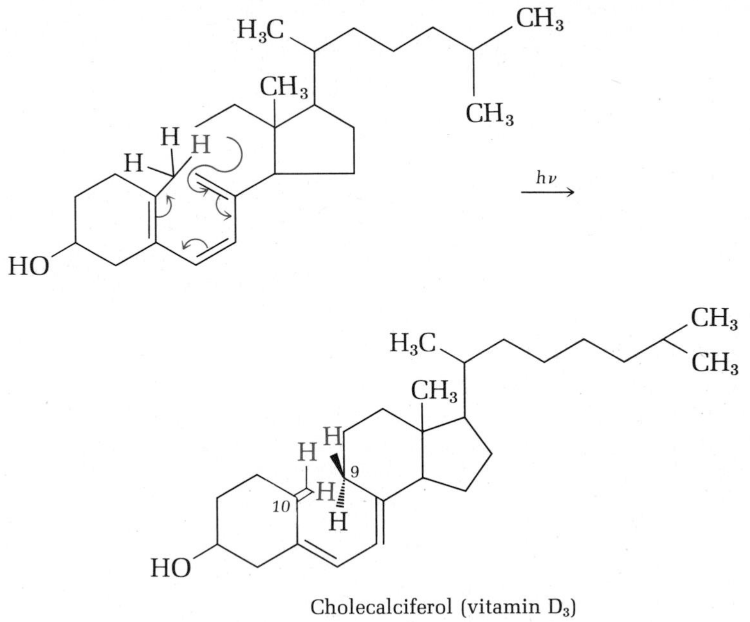

(11.16)

Cholecalciferol (vitamin D₃)

Vitamin D₃ is a *prohormone* that undergoes successive hydroxylations in the body. A liver hydroxylase produces 25-hydroxycholecalciferol. This molecule, not yet active, is then transported to the kidney where it undergoes a second hydroxylation to give 1,25-dihydroxycholecalciferol, the active form of the hormone (reactions 11.17 and 11.18).

We classify 1,25-dihydroxycholecalciferol as a hormone because it is synthesized in one tissue, but acts in other tissues. In the intestine 1,25-dihydroxycholecalciferol regulates transport of Ca(II) and phosphate across cell membranes. If for some reason the plasma calcium concentration drops from its normally constant concentration of 2.5 mM, *parathyroid hormone* is released. The parathyroid stimulates production of 1,25-dihydroxycholecalciferol. The newly synthesized hormone stimulates reabsorption of Ca(II) from bone and renal reabsorption of Ca(II) from plasma to raise the Ca(II) concentration. A low plasma phosphate concentration also stim-

---

[5] The deuterion isotope effect for reaction 11.15 (in which deuterium is the atom transferred) is a staggering 44.

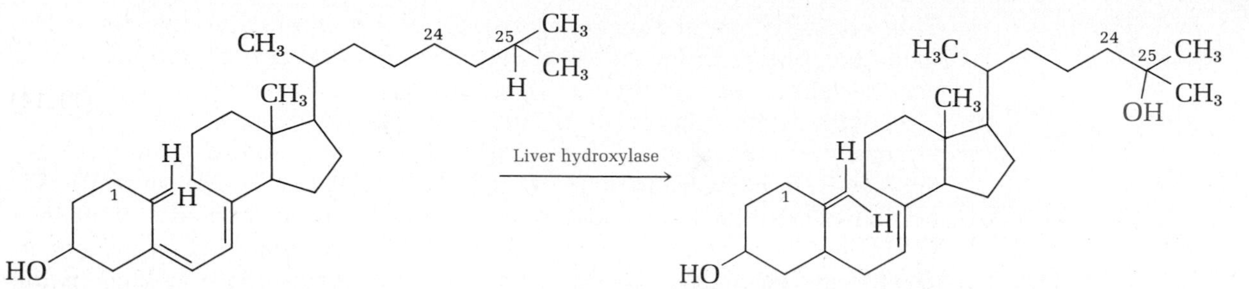

Cholecalciferol

Liver hydroxylase

25-Hydroxycholecalciferol

(11.17)

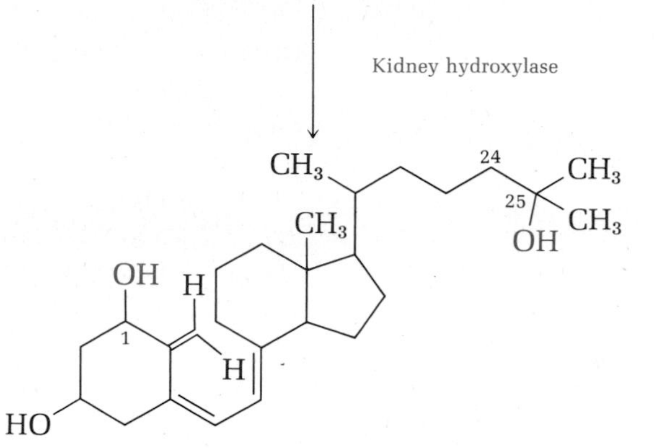

Kidney hydroxylase

(11.18)

1,25-Dihydroxycholecalciferol

ulates synthesis of the 1,25-dihydroxycholecalciferol without release of parathyroid hormone. The intestinal mucosa contains a receptor protein for 1,25-dihydroxycholecalciferol. The hormone-receptor complex is thought to stimulate copying of the genes that code for the Ca(II) and phosphate transport proteins. Only about 1% of the body's calcium exists outside of bone, but regulating Ca(II) concentration is critical because calcium ions are involved in many physiological processes from blood clotting to muscle contraction, including heart function.

Persons suffering from chronic kidney failure often lose the ability to synthesize 1,25-dihydroxycholecalciferol. This results in painful bone deterioration. Administration of the hormone alleviates this condition. Certain families transmit a genetic form of rickets characterized by a lack of the kidney enzyme that hydroxylates the 1 position of 25-hydroxycholecalciferol. This genetic disease can be controlled by doses of 1,25-dihydroxycholecalciferol as small as 1 μg per day. The clinical applications of synthetic analogs of 1,25-dihydroxycholecalciferol appear to have great potential in the treatment of diseases related to vitamin D deficiencies.

E. Vitamin E   Vitamin E (*α-tocopherol*) is a disguised version of a 1,4-dihydroquinone in which one of the hydroxy groups has been converted to an ether.

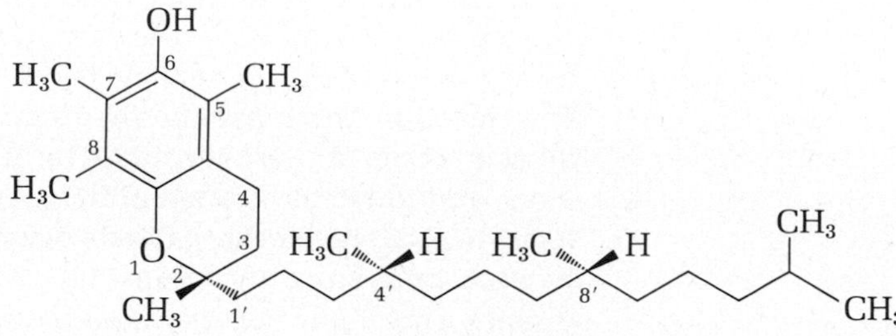

Vitamin E (α-tocopherol)

This constellation of functional groups is a *6-chromanol* structure. The name tocopherol (from the Greek words *tokos,* "offspring" and *phero,* "to bear") defines one of its few documented functions: rats deprived of α-tocopherol are infertile. Rats and other laboratory animals deprived of vitamin E also develop the degenerative disease *muscular dystrophy.* A deficiency in humans leads to fragile red blood cells that hemolyze easily. Vitamin E also inhibits the oxidation of Vitamin A; thus a deficiency of vitamin E can lead to deficiency in vitamin A also. Since vitamin E is present in green leafy vegetables, wheat germ, and rice, it is seldom lacking in the human diet.

The properties of vitamin E as an *antoxidant* are important biologically, as the vitamin A case above demonstrates. Free-radical reactions wreak havoc on cellular molecules. Unsaturated lipids and other unsaturated functional groups are especially vulnerable to destruction. Free radicals, such as hydroperoxide, can be generated by the metabolism of drugs, as intermediates in enzyme-catalyzed reactions, and by other pathways.

Consider the effect of a free radical on an unsaturated fatty acid or other alkene. The first step in the reaction of radicals with alkenes is abstraction of an allylic hydrogen to give a resonance-stabilized allylic radical (reaction 11.19).

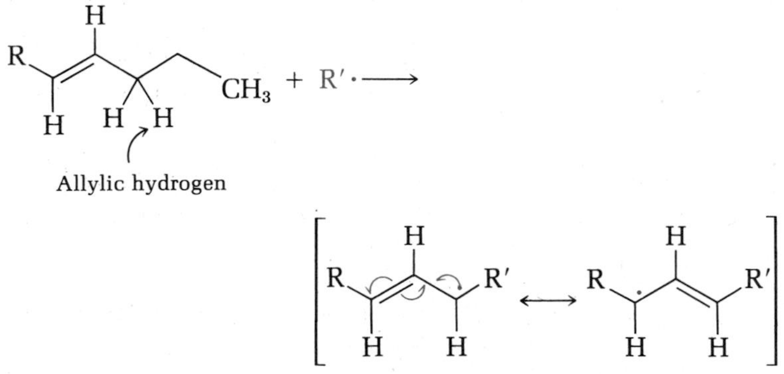

(11.19)

The allylic radical can react with another unsaturated fatty acid (reaction 11.20).

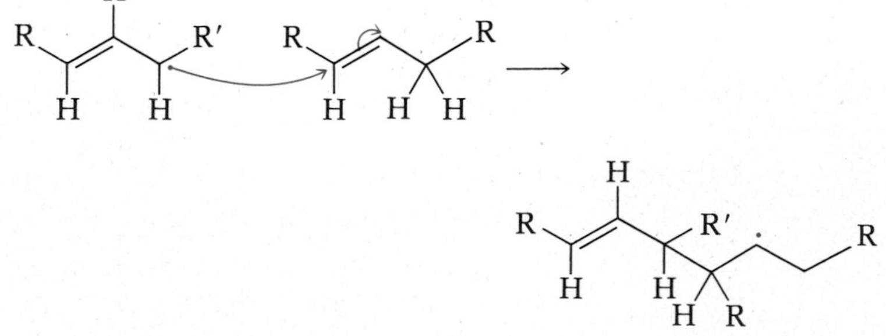

(11.20)

This reaction is simply the first step in a free-radical polymerization. The consequences of these reactions for unsaturated membrane lipids are disastrous, and the deterioration of the membranes of the endoplasmic reticulum and other cell membranes in rats deprived of vitamin E might well be caused by such free-radical reactions. Oxygen uptake increases when vitamin E is absent. This, too, can be explained by free-radical reactions. Oxygen is extremely reactive toward free radicals. The product is a hydroperoxy radical (reaction 11.21).

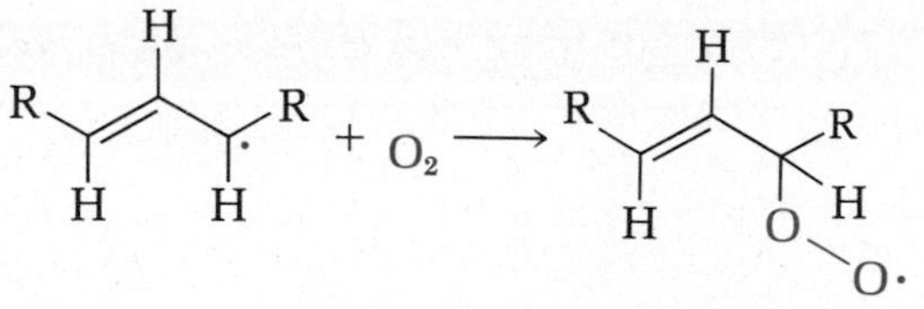

$$\text{(11.21)}$$

Hydroperoxy radical

The hydroperoxy radical is itself a reactive species. It can extract a proton from the allylic position of another alkene and then disproportionate, giving the very dangerous hydroxyl radical (reaction 11.22).

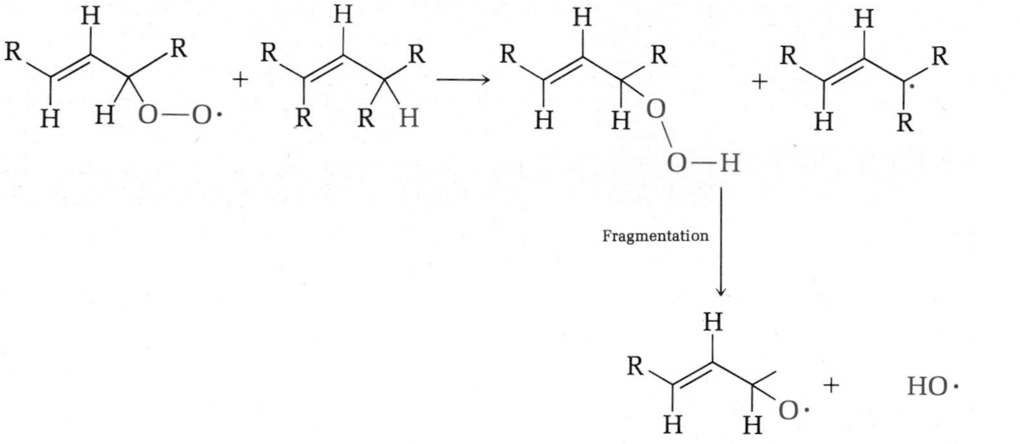

Fragmentation

$$\text{(11.22)}$$

Alkoxy radical   Hydroxyl radical

Free radicals such as these are thought to play a significant part in the aging process.

Vitamin E interrupts free-radical chain reactions by capturing free radicals. It can act as a free radical scavenger because it forms a relatively stable hydroquinone radical (reaction 11.23).

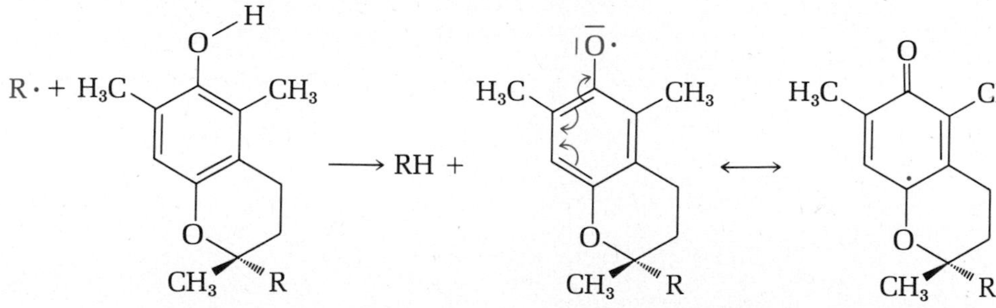

$$\text{(11.23)}$$

Resonance-stabilized radical

Since the free radical derived from vitamin E is relatively stable, it does not run amuck and disrupt cellular chemistry. Various enzyme systems that catalyze one-electron reductions, such as the flavin dehydrogenases, can regenerate vitamin E. The net result of vitamin E's action is inhibition of free-radical reactions. How this activity is related to infertility in rats lacking vitamin E is not clear. Of all the vitamins vitamin E is the one about which the least detailed biochemistry is known.

F. Prostaglandins   The prostaglandins were so named because they were first isolated from seminal fluid and were assumed to be synthesized in the prostate gland. They are an extraordinary family of compounds that can be regarded as derivatives of the hypothetical compound prostanoic acid, itself derived from ara-

Table 11.12 Structures of the major prostaglandins

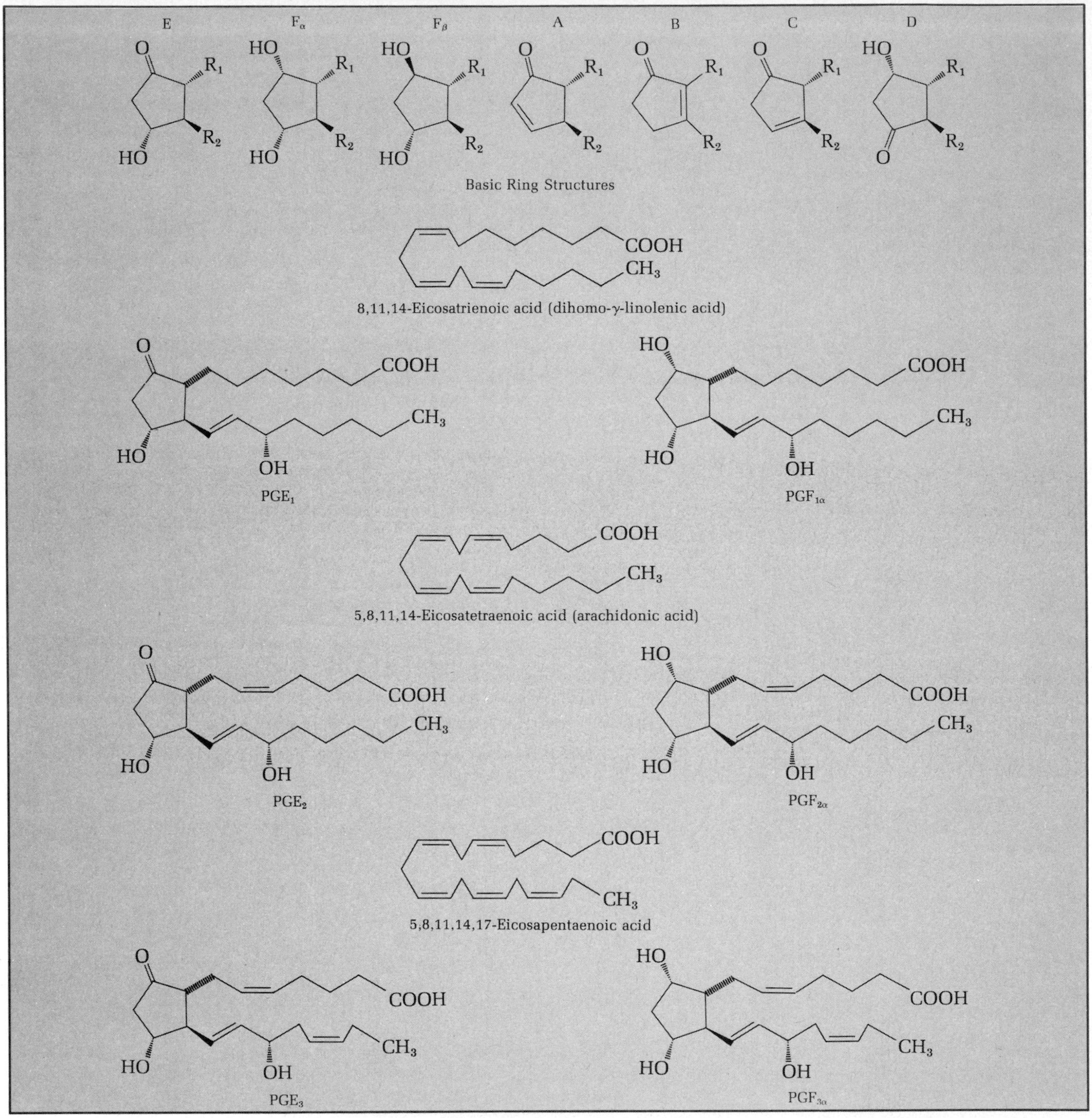
E
Fα
Fβ
A
B
C
D
Basic Ring Structures
COOH
CH3
8,11,14-Eicosatrienoic acid (dihomo-γ-linolenic acid)
PGE1
PGF1α
5,8,11,14-Eicosatetraenoic acid (arachidonic acid)
PGE2
PGF2α
5,8,11,14,17-Eicosapentaenoic acid
PGE3
PGF3α

chidonic acid, one of the essential fatty acids. The basic ring structure and the structures of the prostaglandins, which were elucidated in the 1960s, are given in Table 11.12.

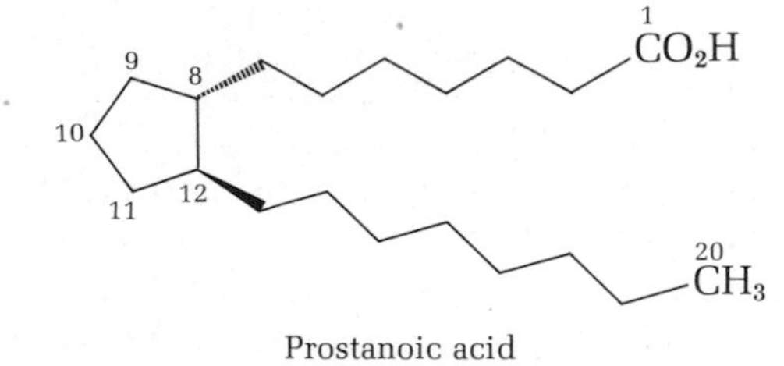
1
CO2H
9
8
10
11
12
20
CH3

Prostanoic acid

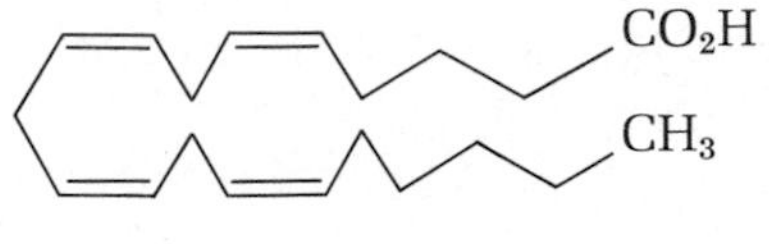
CO2H
CH3

Arachidonic acid

The range of biological activities of the prostaglandins is staggering. Nearly every tissue contains prostaglandins, and nearly every biological process seems to be affected by minute amounts of these compounds. A few examples hint at the range of their action. Prostaglandins act as *pyrogens* (producing fire, that is, inducing fever) in the brain. They are also inflammatory and pain-sensitizing. Aspirin and aspirinlike drugs have inhibitory actions on different enzymes involved in prostaglandin biosynthesis that parallel their effectiveness against the above three effects of prostaglandins. The antiinflammatory action of *cortisone* (Section 20.7B) and other steroids may also be related to their inhibition of prostaglandin synthesis. Another physiological effect of prostaglandins is the lowering of blood pressure. Still another general property is the stimulation of contraction of smooth muscle, such as the uterine muscle. The prostaglandins PGF_1, PGF_2, and PGF_3 are powerful agents for inducing labor and are sometimes used to induce therapeutic abortions. Some prostaglandins are licensed in Europe for stimulating uterine contractions, and two in the United States are at the experimental stage.

11–8 SUMMARY

The most abundant lipids are the esters of glycerol, known as triacylglycerols or triglycerides, which are the major components of fat. The fatty acids that are esterified in triacylglycerols and in the lipids found in biological membranes contain an even number of carbon atoms. The unsaturated fatty acids also contain an even number of carbons and one or more double bonds. The geometry about the double bond may be either *cis* or *trans*, although the *cis* isomers predominate. The percentage of unsaturated fatty acids in membrane lipids determines membrane fluidity. The polyunsaturated fatty acids linoleic acid, linolenic acid, and arachidonic acid are essential in the human diet. These fatty acids are required for the biosynthesis of prostaglandins, a group of hormones that elicits a wide range of physiological functions, among which are stimulation of smooth muscle contraction and lowering of blood pressure.

The most abundant membrane lipids are phosphoglycerides derived from *sn*-glycerol 3-phosphate. Phosphoglycerides that contain fatty acyl groups at C-1 and C-2 are called phosphatidates. The plasmogens are phosphoglycerides in which a vinyl ether replaces an acyl group at C-1 of *sn*-glycerol 3-phosphate. Some of the most important phosphoglycerides are phosphatidyl choline (lecithin), phosphatidyl elthanolamine (cephaline), and phosphatidyl inositol. Sphingosine is the core of the sphingolipids. The simplest sphingolipids, known as ceramides, contain a single fatty acyl group bound to the amino group of sphingosine. If a phosphocholine moiety is bound to the primary alcohol of a ceramide, the product is sphingomyelin. In the glycolipids called cerebrosides the phosphocholine group is replaced by a glycoside. Cholesterol is the third major component of animal membranes. The primary function of cholesterol is to provide structural rigidity to the membrane. Phospholipases are a class of enzymes that cleave the acyl bonds of phosphoglycerides. The plasma lipoproteins transport phospholipids and glycerol lipids through the bloodstream.

The hydrophobic effect provides the driving force for the formation of lipid bilayers in which the polar heads are exposed to water and the nonpolar tails are buried out of contact with the bulk solvent. Cholesterol stabilizes extended chain conformations of adjacent hydrocarbon tails by van der Waals interactions. Lateral motion of lipids in the bilayer is facile, but flip-flop is thermodynamically forbidden. Membrane proteins are divided into two major classes. The peripheral proteins are bound by electrostatic interactions to the polar head groups and are readily removed from the membrane by increasing the ionic strength of the medium. The integral proteins, on the other hand, are imbedded in the membrane and cannot be removed without destroying the lipid bilayer.

In the fluid-mosaic model of membrane structure the lipids maintain the structural integrity of the membrane, and the proteins float in the membrane as "icebergs" in a lipid sea. The proteins may move laterally, but may not flip-flop across the plane of the lipid bilayer.

Transport of metabolites across cell membranes can occur by passive diffusion, but the rate of passive diffusion for most metabolites are extremely low. Transport is therefore mediated by proteins. In facilitated diffusion a metabolite diffuses down its concentration gradient without input of energy. The rate of transport follows an equation whose form is the same as that of the Michaelis-Menten equation. A facilitated transport system can be competitively inhibited by substrate analogs. The pore through which the metabolite travels may be assumed to have a conformation and electric charge that are complementary to the transported metabolite. In active transport the transported metabolite moves against a concentration gradient in a process that is driven by a source of energy such as hydrolysis of a phosphoanhydride bond of ATP.

The lipid-soluble vitamins have diverse functions. Vitamin A, or retinol, is a vital component of the visual cycle. Vitamin K is associated with the intrinsic system of blood coagulation. Five proteins of the blood-clotting cascade are affected by vitamin K. The reduced form of vitamin K is required for carboxylation of prothrombin to convert certain glutamate residues near the N-terminus into γ-carboxyglutamate. Coenzyme Q, also a quinone, whose structure is quite similar to that of vitamin K, is a lipid-soluble component in the mitochondrial respiratory chain. It undergoes facile one-electron transfer reactions that are an important part of the respiratory chain. Vitamin D_3 is a prohormone that is successively hydroxylated in the liver and then transported to the kidney where it is further hydroxylated to give the active hormone. Vitamin D regulates the calcium ion and phosphate ion concentration of plasma. Vitamin E, α-tocopherol, is an antioxidant that is the least well understood of the lipid-soluble vitamins. Vitamin E may exert its effects as an antioxidant by scavenging such dangerously reactive free radicals as the hydroxyl radical and peroxy radicals.

The prostaglandins are derivatives of arachidonic acid. They are involved in a wide range of physiological processes. For example, they are pyrogenic, inflammatory, and pain-sensitizing. They lower blood pressure and stimulate smooth muscle contraction.

REFERENCES

Anderson, H. C., Probes of Membrane Structure, *Ann. Rev. Biochem.*, *47*, 359 (1978).

Bangham, A. D., Lipid Bilayers and Biomembranes, *Ann. Rev. Biochem.*, *41*, 753 (1972).

Dahl, J. L., and Hokin, L. E., The Sodium-Potassium Adenosintriphosphatase, *Ann. Rev. Biochem.*, *43*, 327 (1974).

Finean, J. B., Coleman, R., and Mitchell, R. H., "Membranes and Their Cellular Functions," Wiley, New York, 1974.

Fox, C. F., and Keith, A. D., Eds., "Membrane Molecular Biology," Sinauer, Stamford, Conn., 1972.

Guidotti, G., Membrane Proteins, *Ann. Rev. Biochem.*, *41*, 731 (1972).

Op den Kamp, J. A. F., Lipid Asymmetry in Membranes, *Ann. Rev. Biochem.*, *48*, 47–71 (1979).

Oxender, D. C., Membrane Transport, *Ann. Rev. Biochem.*, *41*, 777 (1972).

Rothfield, L. I., Ed., "Structure and Function of Biological Membranes," Academic Press, New York, 1975.

Singer, S. J., The Molecular Organization of Membranes, *Ann. Rev. Biochem.*, *43*, 805 (1974).

Singer, S. J., and Nicolson, G., The Fluid Mosaic Model of the Structure of Cell Membranes, *Science, 175*, 720 (1972).

Tanford, C., "The Hydrophobic Effect," Wiley, New York, 1973.

Tanford, C., The Hydrophobic Effect and the Organization of Living Matter, *Science, 200*, 1012 (1978).

Weissman, G., Ed., "Cell Membranes, Biochemistry, Cell Biology and Pathology," HP, New York, 1975.

Yeagle, P. L., Phospholipid Headgroup Behavior in Biological Assemblies, *Acc. Chem. Res., 11*, 321 (1978).

Problems

1. Imidoesters react under mild conditions (pH 8.0, 30°C) with primary amines to form amidines and alcohols.

$$\overset{\overset{\oplus}{NH_2}}{\underset{\|}{CH_3-C-OR}} + RN_2H \longrightarrow \overset{\overset{\oplus}{NH_2}}{\underset{\|}{CH_3-C-NHR}} + ROH$$

The imidoester in which R is C_2H_5 (ethyl acetimidate, EAI) readily penetrates the erythrocyte membrane, whereas that in which R is $CH_2CH_2SO_3^{\ominus}$ (isothionyl acetimidate, IAI) is nonpenetrating.

Assume that IAI labeled with carbon 14 in the methyl group and EAI labeled with tritium in the methyl group are available. (The tritium and carbon 14 contents of isotopically labeled protein can be measured simultaneously.) Design one or more experiments that make use of these labeled compounds *and a proteolytic enzyme* to decide whether an erythrocyte membrane protein is located only on the outer surface, is located only on the inner surface, is located on both surfaces (separate copies), or spans the membrane.

You are permitted to treat *only* the intact erythrocyte with these reagents. Show how your proposed experiment(s) lead to a different predicted result for each of four possibilities. (Reference: *J. Mol. Biol., 87,* 541 (1974) with special attention to the cases of gels A and B.)

2. The *transition temperatures* of biological membranes are the temperatures in which the ordered structure of the bilayer is disrupted. These transition temperatures are analogous to the melting points of DNA that we discussed in Section 8.10. An *E. coli* mutant, lacking an enzyme for synthesizing unsaturated fatty acids, is grown on the following fatty acid supplements.

Oleic acid, Δ^9-C_{18} (*cis*) (found in some bacteria instead of olefinic acids)

$$\overset{\overset{\displaystyle CH_2}{\diagup\diagdown}}{CH_3(CH_2)_7CH-CH(CH_2)_7CO_2H}$$
Lactobacillic acid

Linoleic acid, $\Delta^{9,12}$-C_{18} (*cis, cis*)
Cetoleic acid, Δ^{11}-C_{22} (*cis*)

(a) Rank the order of transition temperatures that will be caused by the addition of these fatty acids to the membrane. (b) Which two of the fatty acids are likely to have similar transition temperatures?

3. Red blood cells contain a glucose "pump" that couples glucose transport to the hydrolysis of ATP. Assume that $\triangle G'$ for ATP hydrolysis is -40 kJ mol^{-1} under prevailing physiological conditions. What is the maximum gradient against which glucose can be transported by hydrolysis of ATP? (In other words, what is the extracellular to intracellular ratio that can be driven by this pump? Assume 1 mole of glucose is transported per mole of ATP hydrolyzed.)

4. Some amphipathic compounds cause red blood cells to form small bubbles on their surface. This process is called *crenation* and these compounds are referred to as *crenators*. Other amphipathic compounds cause the formation of invaginations and pits on the red cell surface. These compounds are called *cup formers*. Give an explanation for the following ob-

servations (made at pH 7.4). (*Hint:* Charged compounds do not pass readily through the red cell membrane.)

a. Compounds of the type $R(CH_2)_3N^{\oplus}(CH_3)_3$ are crenators of intact red cells, but cup formers with unsealed red blood cell "ghosts" (ghosts are red blood cells which have been lysed to release hemoglobin). R is an aromatic moiety.

b. Compounds of the type $R(CH_2)_3NH(CH_3)_2$ are cup formers with *both* intact red blood cells and unsealed red blood cell ghosts. The pK for the system $R_3NH \rightleftarrows R_3N + H^{\oplus}$ is about 10.

Part II

Metabolism

Chapter 12

Bioenergetics II: The hydrolysis of ATP and other "energy-rich" metabolites

Energy is eternal delight.

William Blake

12–1 INTRODUCTION

In this chapter we shall apply the fundamental principles of thermodynamics to a study of a group of metabolites whose hydrolysis liberates substantial amounts of free energy. The equilibria for these reactions lie far on the side of products. Among such metabolites, adenosine triphosphate (ATP) is by far the most important. In fact, if a single molecule could be said to dominate metabolism, ATP would enjoy that distinction since nearly all biochemical processes involve ATP. Substances which upon hydrolysis liberate as much or more energy than ATP are called *energy-rich metabolites*. The hydrolysis of energy-rich metabolites, and especially of ATP, provides a focal point for an analysis of much of metabolism.

We will consider the biochemistry of ATP and other energy-rich metabolites from two perspectives. First, we will focus on thermodynamics and equilibrium constants. We will analyze the structural properties of these molecules so that we can relate their structures to their *relative* free energies of hydrolysis. Once we have analyzed the equilibrium constants, we will be able to analyze many of the coupled reactions of metabolism. Second, we will examine some of the enzymes that act upon energy-rich metabolites. The mechanisms of action of these enzymes may be considered independently of thermodynamics, but, of course, the two subjects are interrelated in the sense that the action of the enzymes provides the pathway through which the metabolites' thermodynamic potential is exploited.

12−2 BIOLOGICAL STANDARD STATE

In Section 2.9 we defined the standard state of a solute in solution as 1.0 M. In formulating solution standard states we also singled out the hydrogen ion concentration, defining the standard free energy of formation of 1.0 M hydronium ion as zero (equation 12.1).

$$\Delta G_f^\circ(1.0 \ M \ [\text{H}^\oplus]) \equiv 0 \tag{12.1}$$

A 1.0 M hydronium ion solution has a pH of zero. This standard state for hydronium ions is not very meaningful in biological systems. However, since the assignment of standard states is arbitrary, we define a "biological standard state" as one in which the standard free energy of formation of 10^{-7} M hydronium ion is zero. The symbol for the biological standard state is $\Delta G^{\circ\prime}$ (equation 12.2).

$$\Delta G_f^{\circ\prime}(10^{-7} \ M \ [\text{H}^\oplus]) \equiv 0 \tag{12.2}$$

The solvent for many biochemical reactions is water. The standard state of water is defined as 1.0 M.[1] Some highly important metabolic processes occur within biological membranes or at interfaces between the aqueous phase and the membrane. These reactions are extraordinarily difficult to analyze thermodynamically, as we will discover in later chapters.

Let us now consider the hydrolysis of ethyl acetate, a model for the important hydrolytic reactions of metabolism (reaction 12.3).

$$\underset{\text{CH}_3-\overset{\overset{\text{O}}{\|}}{\text{C}}-\text{O}-\text{CH}_2\text{CH}_3}{} + \text{H}_2\text{O} \underset{}{\overset{K}{\rightleftharpoons}} \text{CH}_3\text{CO}_2^\ominus + \text{CH}_3\text{CH}_2\text{OH} + \text{H}^\oplus \tag{12.3}$$

The equilibrium constant for this reaction, written with the term for $[\text{H}^\oplus]$ extracted, is given by equation 12.4.

$$K = \frac{[\text{CH}_3\text{CH}_2\text{OH}][\text{CH}_3\text{CO}_2^\ominus]}{[\text{CH}_3\text{CO}_2\text{CH}_2\text{CH}_3][\text{H}_2\text{O}]} \times [\text{H}^\oplus] \tag{12.4}$$

The standard free-energy change. $\Delta G^{\circ\prime}_{\text{hydrolysis}}$, for reaction 12.3, again written with the term for the hydrogen ion concentration set to one side, is

$$\Delta G^{\circ\prime}_{\text{hydrolysis}} = -2.303RT \log \frac{[\text{CH}_3\text{CH}_2\text{OH}][\text{CH}_3\text{CO}_2^\ominus]}{[\text{CH}_3\text{CO}_2\text{CH}_2\text{CH}_3]}$$
$$- 2.303RT \log [\text{H}^\oplus] \tag{12.5}$$

---

[1] Actually it is the *activity*, or "effective concentration," of pure water whose standard state is 1.0. For dilute solutions concentrations can be used instead of activities.

If the standard state for hydronium ion is 1 M, the term containing $[H^{\oplus}]$ in equation 12.5 equals zero (the log of 1 is zero). At pH 7.0 and $T = 298$ K, however, the term $-2.303RT \log 10^{-7}$ makes a contribution of -39.5 kJ mol^{-1} to the free energy of hydrolysis. Therefore,

$$\Delta G^{\circ\prime} = \Delta G^{\circ} - 39.5 \text{ kJ mol}^{-1} \tag{12.6}$$

or

$$\Delta G^{\circ} = \Delta G^{\circ\prime} + 39.5 \text{ kJ mol}^{-1} \tag{12.7}$$

From equation 12.6 we conclude that the hydrolysis of ethyl acetate, or any reaction that liberates hydrogen ions, is *more* spontaneous at pH 7.0 than at pH 0. We must be careful to use the balanced equation in calculating $\Delta G^{\circ\prime}$. If 2 moles of hydronium ions are produced in the reaction, then the term containing $[H^{\oplus}]$ in the equilibrium constant must be squared. We must also express all species in their correct ionic forms. In the hydrolysis of ethyl acetate, for example, the acetate ion concentration appears in equation 12.5 rather than the concentration of acetic acid, since acetic acid is almost totally ionized at pH 7.0. Keeping these caveats in mind, it is as easy to calculate $\Delta G^{\circ\prime}$ as to calculate ΔG°.

12–3 STANDARD FREE-ENERGY CHANGES IN COUPLED REACTIONS

When the free energy released in one reaction is used to "drive" a second, nonspontaneous reaction, the two reactions are *coupled*. Two reactions can be chemically coupled if they share a common intermediate, as illustrated by reactions 12.8 and 12.9.

$$A + B \underset{}{\overset{K_1}{\rightleftarrows}} C \qquad \Delta G_1^{\circ\prime} \tag{12.8}$$

$$C + D \underset{}{\overset{K_2}{\rightleftarrows}} E \qquad \Delta G_2^{\circ\prime} \tag{12.9}$$

The sum of these two reactions is

$$A + B + D \underset{}{\overset{K_1 \times K_2 = K_3}{\rightleftarrows}} E \qquad \Delta G_3^{\circ\prime} = \Delta G_1^{\circ\prime} + \Delta G_2^{\circ\prime} \tag{12.10}$$

The total standard free-energy change for the process, $\Delta G_3^{\circ\prime}$, is the sum of the standard free-energy changes for the individual steps. Suppose that reaction 12.8 is endergonic ($\Delta G_1^{\circ\prime} > 0$), and the equilibrium lies far to the left. The net reaction (12.10) can still be exergonic, resulting in net formation of product E, if the standard free-energy change for reaction 12.9 is sufficiently exergonic to overcome the unfavorable standard free-energy change of reaction 12.8. In such a case the two reactions are coupled, and the second reaction provides the "driving force" for the first.

The thermodynamic driving force for the coupled reaction 12.10 is reflected in the equilibrium constant, K_3, which is greater than 1.0 if the second step is more exergonic than the first one is endergonic. The general coupled reaction is a manifestation of the law of mass action. We need only to consider the product $K_1 \times K_2$. If this product is greater than 1.0, the second reaction drives the first one. The extent to which E forms is dependent upon the relative magnitudes of K_1 and K_2.

Let us consider the conversion of malate to aspartate (reactions 12.11 and 12.12).

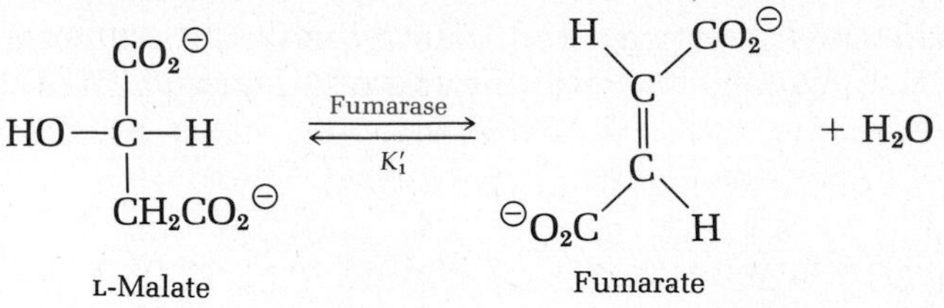

$$\Delta G_1^{\circ\prime} = +3.76 \text{ kJ mol}^{-1} \quad (12.11)$$
$$K_1' = 0.21$$

$$NH_4^{\oplus} + \text{fumarate} \underset{K_2'}{\overset{\text{Asparatase}}{\rightleftarrows}} {}^{\ominus}O_2C-CH_2\overset{\overset{H}{|}}{\underset{\underset{\oplus NH_3}{|}}{C}}-CO_2^{\ominus}$$

$$\Delta G_2^{\circ\prime} = -15.4 \text{ kJ mol}^{-1} \quad (12.12)$$
$$K_2' = 5.28 \times 10^2$$

L-Aspartate

The sum of these two reactions is

$$NH_4^{\oplus} + \text{L-malate} \overset{K_3'}{\rightleftarrows} \text{L-aspartate} + H_2O$$

$$\Delta G_3^{\circ\prime} = -11.7 \text{ kJ mol}^{-1} \quad (12.13)$$
$$K_3' = 1.15 \times 10^2$$

The conversion of L-malate to fumarate in reaction 12.11 has an unfavorable standard free-energy change and an equilibrium constant of less than 1.0, but the conversion of fumarate to aspartate in reaction 12.12 has a large equilibrium constant. Reaction 12.11 is therefore coupled to reaction 12.12, and there is a net conversion of malate to aspartate.

Let us now consider the conversion of glucose to glucose 6-phosphate, a reaction in glycolysis in which ATP is involved.

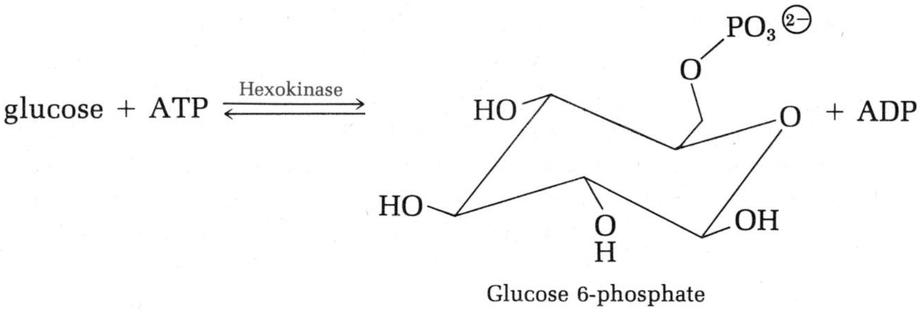

$$\text{glucose} + \text{ATP} \overset{\text{Hexokinase}}{\rightleftarrows} $$

$$+ \text{ADP} \quad (12.14)$$

Glucose 6-phosphate

In Sections 12.5 through 12.7 we will consider the mechanism of action of hexokinase, but since a thermodynamic analysis depends only upon the initial and final states of the system, a knowledge of the enzyme mechanism is not necessary for a thermodynamic analysis. Reaction 12.14 can be treated as the sum of two reactions: the phosphorylation of glucose and the hydrolysis of ATP (reactions 12.15 and 12.16).

$$\text{glucose} + HPO_4^{2-} \overset{K_1'}{\rightleftarrows} \text{glucose 6-phosphate} + H_2O$$

$$\Delta G_1^{\circ\prime} = +13.8 \text{ kJ mol}^{-1} \quad (12.15)$$
$$K_1' = 3.8 \times 10^{-3}$$

$$\text{ATP} + H_2O \overset{K_2'}{\rightleftarrows} \text{ADP} + HPO_4^{2-}$$

$$\Delta G^{\circ\prime} = -30.7 \text{ kJ mol}^{-1} \quad (12.16)$$
$$K_2' = 2.4 \times 10^5$$

When reaction 12.14 is viewed as phosphorylation coupled to hydrolysis, we see that ATP possesses enough energy to drive the phosphorylation reaction far to completion. The equilibrium constant for reaction 12.14 ($K_1' \times K_2'$) equals 9.1×10^2, corresponding to a standard free-energy change ($\Delta G^{\circ\prime} = \Delta G_1^{\circ\prime} + \Delta G_2^{\circ\prime}$) of -16.9 kJ mol^{-1}.

12–4 ANALYSIS OF SOME ENERGY-RICH METABOLITES

In this section we consider the free energies of hydrolysis of many of the most significant molecules in metabolism: those possessing sufficient free energies of hydrolysis to drive endergonic reactions to completion. For each class of molecules we will first examine the structural properties responsible for the relative amount of free energy released upon hydrolysis. Then we will examine the ability of the metabolite in question to drive coupled reactions in metabolism.

A. Hydrolysis of Carboxylic Esters and Anhydrides

Let us compare the free energies of hydrolysis of ethyl acetate and acetic anhydride, models for the hydrolysis of many other esters and anhydrides.

$$CH_3-\overset{\overset{\textstyle O}{\|}}{C}-OCH_2CH_3 + H_2O \longrightarrow CH_3CO_2^{\ominus} + H^{\oplus} + CH_3CH_2OH \qquad \Delta G^{\circ\prime}_{hydrolysis} = -19.7 \text{ kJ mol}^{-1} \qquad (12.17)$$

$$CH_3\overset{\overset{\textstyle O}{\|}}{C}-O-\overset{\overset{\textstyle O}{\|}}{C}-CH_3 + H_2O \longrightarrow 2 \text{ } CH_3CO_2^{\ominus} + 2 \text{ } H^{\oplus} \qquad \Delta G^{\circ\prime}_{hydrolysis} = -91.2 \text{ kJ mol}^{-1} \qquad (12.18)$$

The functional groups attached to the acetyl group are ethoxide and acetate, respectively. The two reactions are *group transfer reactions*, and they involve transfer of an acetyl group to water. Why is hydrolysis much more exergonic for the anhydride than for the ester? First, consider the relative stabilities of the products in the two reactions. For ethyl acetate the hydrolysis products are ethanol and acetate; for acetic anhydride the products are 2 moles of acetate. Acetate is resonance stabilized, but ethanol is not. Anhydride hydrolysis is more exergonic because its products are more resonance stabilized than the products of ester hydrolysis.

$$CH_3-\overset{\overset{\textstyle O}{\|}}{C}-OCH_2CH_3 + H_2O \longrightarrow CH_3-\overset{\overset{\textstyle O}{\|}}{C}-O^{\ominus} + CH_3CH_2OH + H^{\oplus} \qquad (12.19)$$

$$CH_3-\overset{\overset{\textstyle O^{\ominus}}{|}}{C}=O$$

Resonance-
stabilized acetate

Second, since both products of anhydride hydrolysis are charged, the products of anhydride hydrolysis are better solvated than the products of ester hydrolysis. The solvation of products is at least as important as the resonance stabilization of products as a driving force for the reaction.

Now consider the relative stabilities of the reactants ethyl acetate and acetic anhydride. Ethyl acetate is slightly resonance stabilized by overlap of the lone pair electrons of the acyl oxygen with the carbonyl carbon.

$$CH_3-\overset{\overset{\textstyle O}{\|}}{C}-OCH_2CH_3 \longleftrightarrow CH_3-\overset{\overset{\textstyle O^{\ominus}}{|}}{C}=\overset{\oplus}{O}CH_2CH_3$$

The acyl oxygen of acetic anhydride *cannot* simultaneously satisfy the demand of the two carbonyl carbons for electrons. Thus resonance is *inhibited* in anhydrides, a phenomenon known as *competing resonance*.

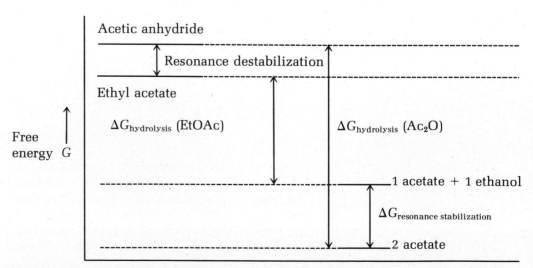

When comparing two different reactions, we must analyze "differences between differences." The anhydride is destabilized by resonance relative to the ester. The products of anhydride hydrolysis are stabilized relative to the products of ester hydrolysis. The standard free energy of hydrolysis of anhydrides is therefore more exergonic than that of esters (Figure 12.1).

These differences are applicable to coupled reactions. Given its free energy of hydrolysis, acetic anhydride can drive coupled reactions that ethyl acetate cannot drive. For example, consider the hydrolysis of acetic anhydride coupled to the synthesis of ethyl acetate.

$$CH_3CH_2OH + CH_3CO_2H \longrightarrow CH_3\overset{O}{\overset{\|}{C}}-OCH_2CH_3 + H_2O \qquad \Delta G^{\circ\prime} = +19.7 \text{ kJ mol}^{-1} \qquad (12.20)$$

$$CH_3\overset{O}{\overset{\|}{C}}-O-\overset{O}{\overset{\|}{C}}CH_3 + H_2O \longrightarrow 2CH_3CO_2H \qquad \Delta G^{\circ\prime} = -91.2 \text{ kJ mol}^{-1} \qquad (12.21)$$

$$CH_3CH_2OH + CH_3\overset{O}{\overset{\|}{C}}-O-\overset{O}{\overset{\|}{C}}CH_3 \longrightarrow CH_3\overset{O}{\overset{\|}{C}}-OCH_2CH_3 + CH_3CO_2H \qquad \Delta G^{\circ\prime} = -73.4 \text{ kJ mol}^{-1} \qquad (12.22)$$

The compound with the more negative $\Delta G^{\circ\prime}$ of hydrolysis transfers its group to the compound with the less negative $\Delta G^{\circ\prime}$ of hydrolysis.

Figure 12.1 Two factors primarily responsible for the greater free energy of hydrolysis of acetic anhydride than of ethyl acetate. (1) Acetic anhydride is less resonance stabilized than ethyl acetate because of competing resonance. (2) The products of hydrolysis of acetic anhydride, 2 moles of acetate, are more resonance stabilized than the 1 mole each of acetate and ethanol that are the products of hydrolysis of ethyl acetate.

B. Hydrolysis of Phosphate Esters The standard free-energy changes for hydrolysis of phosphate esters such as glucose 6-phosphate, sn-glycerol 3-phosphate, and adenosine monophosphate (AMP) are typically in the range of -8 to -16 kJ mol^{-1} (reactions 12.23, 12.24, and 12.25).

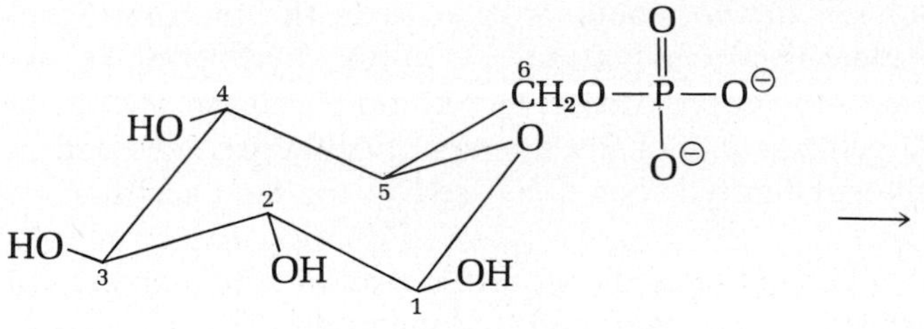

Glucose 6-phosphate

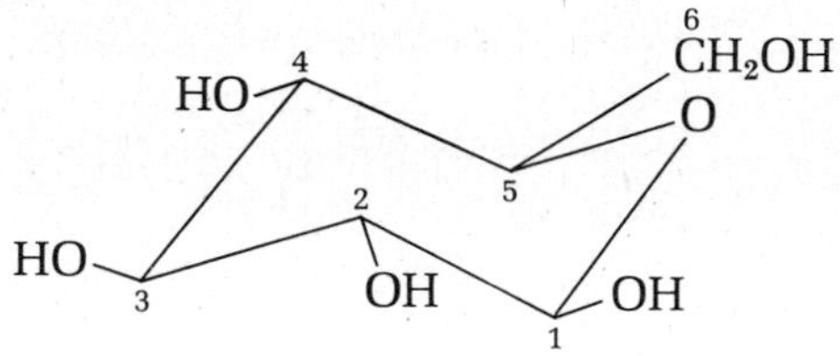

$+ HPO_4^{\,2\ominus}$ $\Delta G^{\circ\prime} = -13.8$ kJ mol^{-1} (12.23)

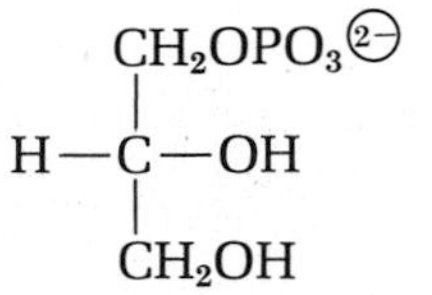

sn-Glycerol 3-phosphate

$+ H_2O \longrightarrow$ glycerol $+ HPO_4^{\,2\ominus}$ $\Delta G^{\circ\prime} = -9.6$ kJ mol^{-1} (12.24)

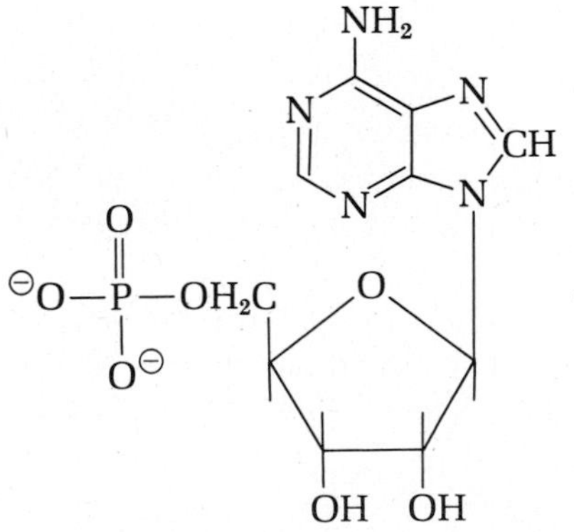

AMP

$+ H_2O \longrightarrow$

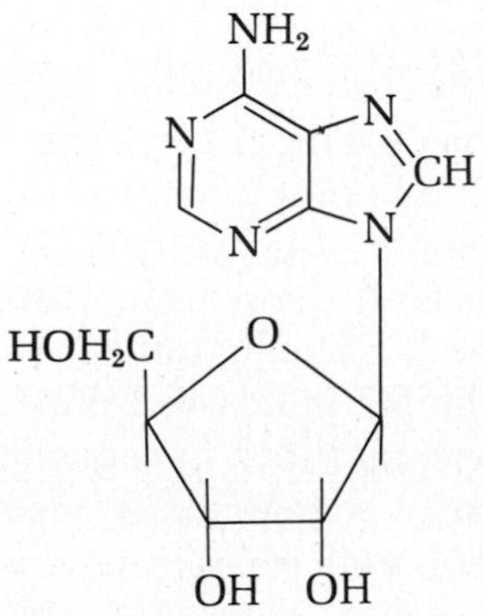

$+ HPO_4^{\,2\ominus}$ $\Delta G^{\circ\prime} = -8.4$ kJ mol^{-1} (12.25)

Adenosine

These standard free-energy changes are slightly less than the free energy of hydrolysis of carboxylate esters, and the hydrolysis of phosphate esters does not provide enough energy to drive most biochemical reactions.

The driving force for the hydrolysis of phosphate esters is small because there is little resonance stabilization of products relative to reactants, little electrostatic repulsion in the reactants relative to the products, and probably not much difference in solvation. However, hydrolysis is exergonic, and we can regard the approximately 12 kJ mol^{-1} release of free energy as a "fundamental component" of the free-energy changes observed in the hydrolysis of the phosphoanhydrides discussed in the next section.

C. Hydrolysis of ATP

The hydrolysis of ATP is one of the most important reactions of metabolism because the energy released drives many coupled reactions. The magnitude of the standard free-energy change for hydrolysis of ATP depends on which products are formed. The standard free-energy change for hydrolysis of ATP to AMP and pyrophosphate (PP$_i$),[2] -32.5 kJ mol^{-1} (reaction 12.26).

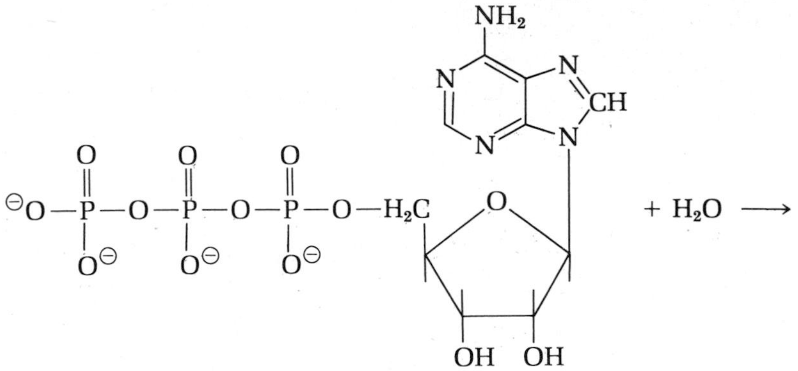

Adenosine 5'-triphosphate (ATP^{4-})

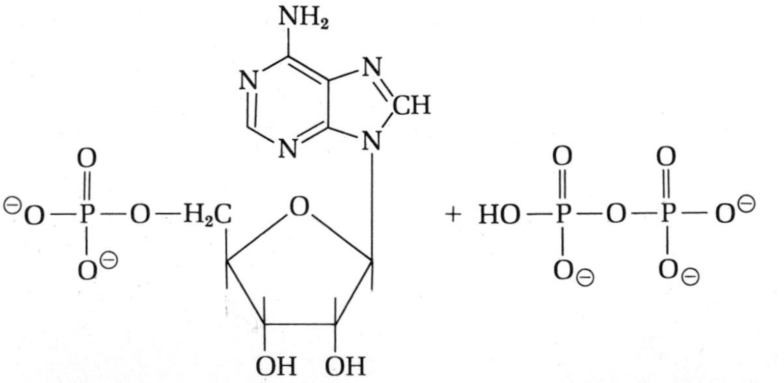

$$+ H_2O \longrightarrow$$

Adenosine 5'-monophosphate (AMP)   Pyrophosphate (PP$_i$)

(12.26)

ATP can also hydrolyze to give ADP and orthophosphate (reaction 12.27). This reaction has a standard free energy of hydrolysis of -30.5 kJ mol^{-1}. It is the more common one in metabolism.

The free-energy change for ATP hydrolysis depends upon the pH of the solution because all the reactants and products have ionizing groups with pK$_a$ values near 7.0. The hydrolysis of ATP becomes nearly 20 kJ mol^{-1} more exergonic as the pH increases from 4.0 to 10.0 (Figure 12.2). The values of $\Delta G^{\circ\prime}_{\text{hydrolysis}}$ reflect equilibrium concentrations of reactants and products. The actual free-energy change ($\Delta G'$) depends upon the concentrations of ATP, ADP, phosphate, and pH. These concentrations vary with metabolic con-

[2] The abbreviation PP$_i$ refers to the mixture of ionic forms of pyrophosphic acid: $H_2P_2O_7^{2-}$ (pK$_a'$ = 6.1), $HP_2O_7^{3-}$ (pK$_a'$ = 9.0), and $P_2O_7^{4-}$.

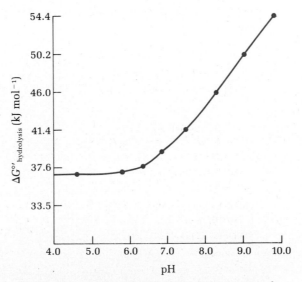

$$\text{Adenosine 5'-triphosphate (ATP}^{\textcircled{4-}}) + H_2O \longrightarrow$$

$$\text{Adenosine diphosphate (ADP}^{\textcircled{3-}}) + HPO_4^{\textcircled{2-}} + H^{\textcircled{+}} \qquad (12.27)$$

Orthophosphate

ditions. Considerable variation in the free-energy change is therefore possible.

Why does cleavage of the P—O—P bond in ATP liberate more energy than cleavage of the C—O—P bond of phosphate esters? The three major factors involved are electrostatic repulsion, solvation, and competing resonance. ATP has an ionizable hydrogen with a pK_a of 6.95. At pH 7.0 ATP exists as 53% ATP$^{\textcircled{4-}}$ and 47% ATP$^{\textcircled{3-}}$ (reaction 12.28).

Figure 12.2 Effect of pH on the free energy of hydrolysis of Mg-ATP. The Mg(II) concentration is 10^{-4} M. [Replotted from R. A. Alberty, *J. Biol. Chem.*, **243**, 1337–1343 (1968).]

$$\text{Ado-5}'-\text{O}-\overset{\overset{\displaystyle O}{\|}}{\underset{\underset{\displaystyle O^{\ominus}}{|}}{P}}-\text{O}-\overset{\overset{\displaystyle O}{\|}}{\underset{\underset{\displaystyle O^{\ominus}}{|}}{P}}-\text{O}-\overset{\overset{\displaystyle O}{\|}}{\underset{\underset{\displaystyle O^{\ominus}}{|}}{P}}-\text{O}-\text{H} \underset{\longleftarrow}{\overset{pK_a = 6.95}{\longrightarrow}}$$

ATP ③⁻ (47% at pH 7.0)

$$\text{Ado-5}'-\text{O}-\overset{\overset{\displaystyle O}{\|}}{\underset{\underset{\displaystyle O^{\ominus}}{|}}{P}}-\text{O}-\overset{\overset{\displaystyle O}{\|}}{\underset{\underset{\displaystyle O^{\ominus}}{|}}{P}}-\text{O}-\overset{\overset{\displaystyle O}{\|}}{\underset{\underset{\displaystyle O^{\ominus}}{|}}{P}}-\text{O}^{\ominus} + \text{H}^{\oplus} \qquad (12.28)$$

ATP ④⁻ (53% at pH 7.0)

The high charge density on the triphosphate group destabilizes ATP by electrostatic repulsion. This repulsion is less in ADP, which has an ionizable hydrogen with a pK_a of 6.7 (reaction 12.29).

$$\text{Ado-5}'-\text{O}-\overset{\overset{\displaystyle O}{\|}}{\underset{\underset{\displaystyle O^{\ominus}}{|}}{P}}-\text{O}-\overset{\overset{\displaystyle O}{\|}}{\underset{\underset{\displaystyle O^{\ominus}}{|}}{P}}-\text{O}-\text{H} \underset{\longleftarrow}{\overset{pK_a = 6.7}{\longrightarrow}}$$

ADP ②⁻ (32% at pH 7.0)

$$\text{Ado-5}'-\text{O}-\overset{\overset{\displaystyle O}{\|}}{\underset{\underset{\displaystyle O^{\ominus}}{|}}{P}}-\text{O}-\overset{\overset{\displaystyle O}{\|}}{\underset{\underset{\displaystyle O^{\ominus}}{|}}{P}}-\text{O}^{\ominus} + \text{H}^{\oplus} \qquad (12.29)$$

ADP ³⁻ (68% at pH 7.0)

ADP therefore exists as 32% ADP②⁻ and 68% ADP③⁻ at pH 7.0.

The second factor that favors hydrolysis of ATP is solvation of the products. ADP and orthophosphate together are better solvated than ATP. Solvation of the products helps to "pull" the reaction to completion.

The third major factor in ATP hydrolysis is competing resonance. Because of inhibition of resonance in the phosphoanhydride, the anhydride is less resonance stabilized than the phosphate ester. This means that structures 1a and 1b contribute less to 1 than 2a contributes to 2. The phosphoanhydride (1) is therefore less resonance stabilized than the phosphate ester (2).

(1) Phosphoanhydride   1a

1b

(2) Phosphate ester   2a

The free energy of hydrolysis of ATP is also considerably altered by divalent metal ions such as Mg(II). More than 90% of cellular ATP is bound to Mg(II), and Mg(II) also binds to orthophosphate and to ADP (reactions 12.30–12.32).

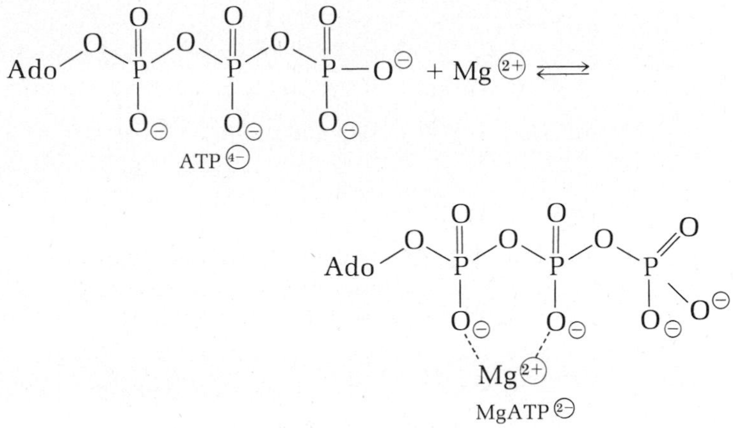

$$ATP^{4-} + Mg^{2+} \rightleftharpoons MgATP^{2-} \tag{12.30}$$

$$K' = \frac{[MgATP^{2-}]}{[ATP^{4-}][Mg(II)]} = 3.8 \times 10^2$$

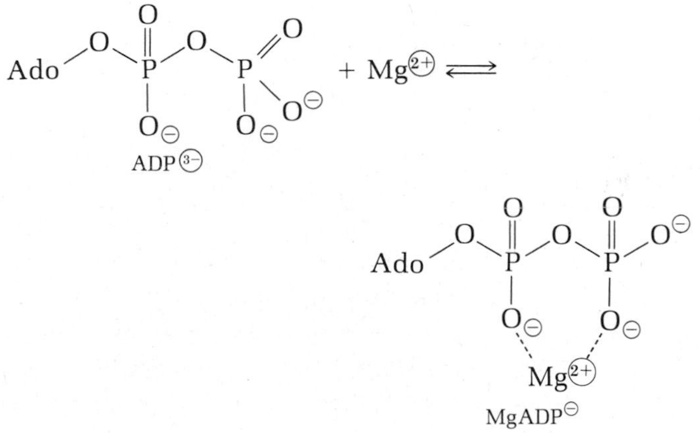

$$ADP^{3-} + Mg^{2+} \rightleftharpoons MgADP^{-} \tag{12.31}$$

$$K' = \frac{[MgADP^{-}]}{[ADP^{3-}][Mg(II)]} = 2.2 \times 10^3$$

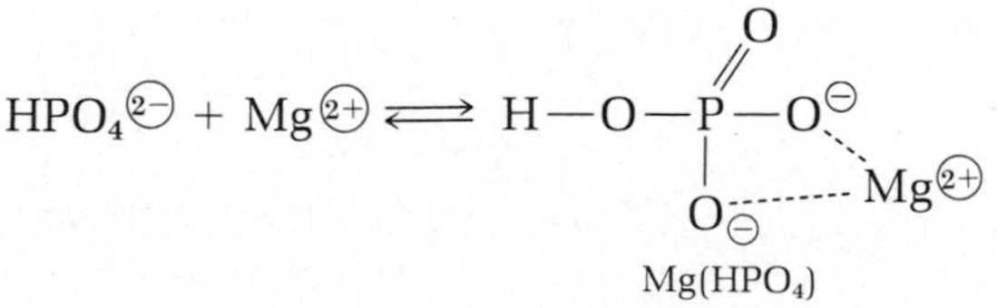

$$HPO_4^{2-} + Mg^{2+} \rightleftharpoons Mg(HPO_4) \tag{12.32}$$

$$K = \frac{[Mg(HPO_4)]}{[Mg(II)][HPO_4^{2-}]} = 1.13 \times 10^2$$

As we see from these equilibria and their equilibrium constants, magnesium ions tend to pull the hydrolysis of ATP toward ADP and orthophosphate. The affinity of ADP for magnesium ions is about six times that of ATP, and ATP and orthophosphate have about the same affinity for magnesium ions. This effect is complex, however, as may be seen from the dependence of the standard free energy of hydrolysis as a function of the concentration of magnesium ions in Figure 12.3.

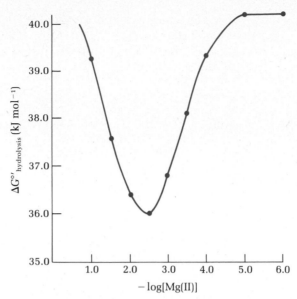

Figure 12.3 Effect of magnesium ions on the free energy of hydrolysis of ATP at pH 7.0. [Data from R. A. Alberty, *J. Biol. Chem., 243,* 1337–1343 (1968).]

D. Hydrolysis of Acyl Phosphates

Acyl phosphates are *mixed anhydrides* whose general structure and hydrolysis are shown in reaction 12.33.

$$R-\overset{O}{\underset{\parallel}{C}}\diagdown_{O}\overset{O}{\underset{\parallel}{P}}-O^{\ominus} + H_2O \longrightarrow RCO_2^{\ominus} + HPO_4^{2\ominus} + H^{\oplus} \qquad (12.33)$$

Acyl phosphate

The acyl phosphates, like the other anhydrides we have discussed, have large standard free energies of hydrolysis. Their hydrolysis liberates more free energy than is released by ATP hydrolysis, so hydrolysis of acyl phosphates can drive the phosphorylation of ADP.

a. Acetyl Phosphate

The standard free energy of hydrolysis of acetyl phosphate (acetyl-P) is −43.0 (reaction 12.34).

$$CH_3-\overset{O}{\underset{\parallel}{C}}\diagdown_{O}\overset{O}{\underset{\parallel}{P}}-O^{\ominus} + H_2O \longrightarrow CH_3CO_2^{\ominus} + HPO_4^{2\ominus} + H^{\oplus}$$

Acetyl phosphate

$$\Delta G^{\circ\prime}_{\text{hydrolysis}} = -43.0 \text{ kJ mol}^{-1} \qquad (12.34)$$

Reaction 12.34, and all the other phosphate hydrolyses discussed in this chapter, is a *phosphoryl* ($-PO_3^{2\ominus}$) *group transfer* to water. Acetyl-P is an important intermediate in certain lactic acid bacterial fermentations where its hydrolysis is coupled to ATP synthesis (reaction 12.35).

b. Carbamoyl Phosphate

Carbamoyl phosphate, the anhydride of carbamic acid and orthophosphoric acid, is another energy-rich acyl phosphate whose standard free energy of hydrolysis is −49.1 kJ mol^{-1} (reaction 12.36).

$$CH_3-\overset{\overset{\displaystyle O}{\|}}{C}-O-PO_3^{\circled{2-}} + ADP \rightleftharpoons CH_3CO_2^{\ominus} + ATP$$

Acetyl phosphate

$$\Delta G^{\circ\prime} = -12.5 \text{ kJ mol}^{-1} \qquad (12.35)$$

$$H_2N-\overset{\overset{\displaystyle O}{\|}}{C} \quad \overset{\overset{\displaystyle O}{\|}}{\underset{\underset{\displaystyle O_{\ominus}}{|}}{\overset{|}{P}}}-O^{\ominus} + H_2O \longrightarrow H_2N-\overset{\overset{\displaystyle O}{\|}}{C}-O^{\ominus} + HPO_4^{\circled{2-}} + H^{\oplus}$$

Carbamoyl phosphate

Carbamate

$$\Delta G^{\circ\prime}_{hydrolysis} = -49.1 \text{ kJ mol}^{-1} \qquad (12.36)$$

Carbamoyl phosphate is an intermediate in several metabolic pathways, including the urea cycle (Section 21.2) and the biosynthesis of pyrimidines (Section 23.6). Since its standard free energy of hydrolysis is highly exergonic, it can drive otherwise unfavorable biosynthetic reactions.

c. 1,3-Diphospho-D-glycerate

1,3-Diphospho-D-glycerate is an intermediate in glycolysis (Section 14.4) and the precursor of the allosteric inhibitor of hemoglobin, 2,3-diphosphoglycerate. Its standard free energy of hydrolysis is -49.4 kJ mol^{-1} (reaction 12.37).

$$\begin{array}{l} \overset{\overset{\displaystyle O}{\|}}{C}-OPO_3^{\circled{2-}} \\ | \\ H-C-OH \\ | \\ CH_2OPO_3^{\circled{2-}} \end{array} + H_2O \longrightarrow \begin{array}{l} CO_2^{\ominus} \\ | \\ H-C-OH \\ | \\ CH_2OPO_3^{\circled{2-}} \end{array} + HPO_4^{\circled{2-}} + H^{\oplus}$$

1,3-Diphosphoglycerate
(1,3-DPG)

3-Phosphoglycerate

$$\Delta G^{\circ\prime}_{hydrolysis} = -49.4 \text{ kJ mol}^{-1} \qquad (12.37)$$

This exergonic reaction drives the first energy-yielding reaction of glycolysis (reaction 12.38).

$$\text{1,3-DPG} + ADP \longrightarrow ATP + \text{3-phosphoglycerate}$$
$$\Delta G^{\circ\prime} = -18.9 \text{ kJ mol}^{-1} \qquad (12.38)$$

E. Hydrolysis of Guanidine Phosphates

The two N-phosphoryl guanidine derivatives, or *phosphagens*, creatine phosphate and arginine phosphate, are energy-storage molecules in muscle and nervous tissue. Creatine phosphate serves this function in vertebrates and arginine phosphate in invertebrates.

$$H_2\overset{\oplus}{N} \quad \overset{\overset{\displaystyle H}{|}}{N}-\overset{\overset{\displaystyle O^{\ominus}}{|}}{\underset{\underset{\displaystyle O_{\ominus}}{|}}{P}}=O$$
$$\overset{|}{C}$$
$$CH_3-N-CH_2CO_2^{\ominus}$$

Creatine phosphate

$$H_2\overset{\oplus}{N} \quad \overset{\overset{\displaystyle H}{|}}{N}-\overset{\overset{\displaystyle O}{\|}}{\underset{\underset{\displaystyle O_{\ominus}}{|}}{P}}-O^{\ominus}$$
$$\overset{|}{C}$$
$$NH$$
$$|$$
$$(CH_2)_3$$
$$|$$
$$H-C-NH_3^{\oplus}$$
$$|$$
$$CO_2^{\ominus}$$

Arginine phosphate

The standard free energy of hydrolysis of both metabolites is -43.9 kJ mol^{-1} (reaction 12.39).

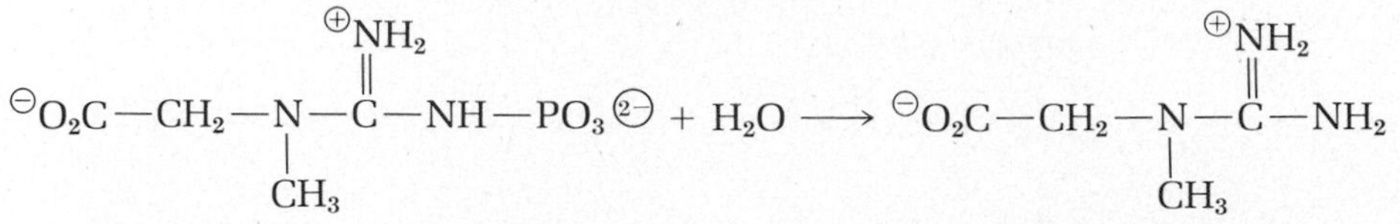

$$\overset{\ominus}{O_2C}-CH_2-\underset{\underset{CH_3}{|}}{N}-\overset{\overset{\overset{\oplus}{NH_2}}{\|}}{C}-NH-PO_3{}^{\small{2\ominus}} + H_2O \longrightarrow \overset{\ominus}{O_2C}-CH_2-\underset{\underset{CH_3}{|}}{N}-\overset{\overset{\overset{\oplus}{NH_2}}{\|}}{C}-NH_2$$

Creatine phosphate (or arginine phosphate)   Creatine (or arginine)

$$+ HPO_4{}^{\small{2\ominus}} + H^{\oplus} \qquad \Delta G^{\circ\prime}_{\text{hydrolysis}} = -43.9 \text{ kJ mol}^{-1} \qquad (12.39)$$

This large free-energy change is probably due almost entirely to resonance stabilization of the products. The following resonance structures are possible for the guanidinium group of the product with either creatine or

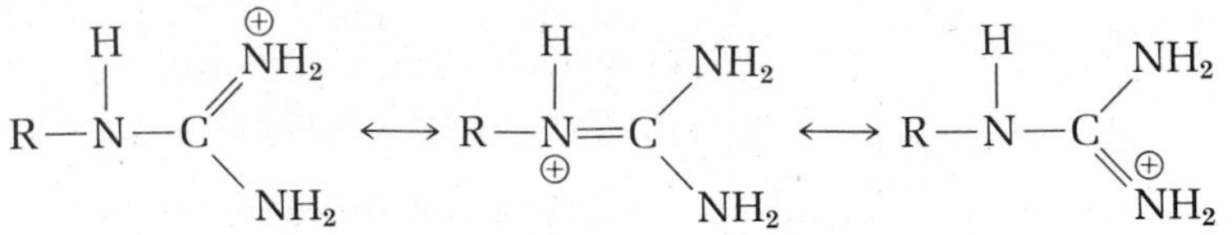

Resonance structures in the guanidinium ion product

arginine as the phosphate carrier. The stability of the guanidinium ion is due to its high symmetry. The positive charge on the ion is equally distributed among three nitrogen atoms, each of which bears a charge of $+\frac{1}{3}$.

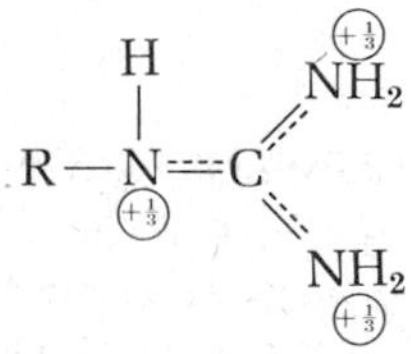

Resonance hybrid of
guanidinium ion

Such a resonance structure would be inhibited in the reactants, either creatine phosphate or arginine phosphate, because it would have two adjacent positive charges.

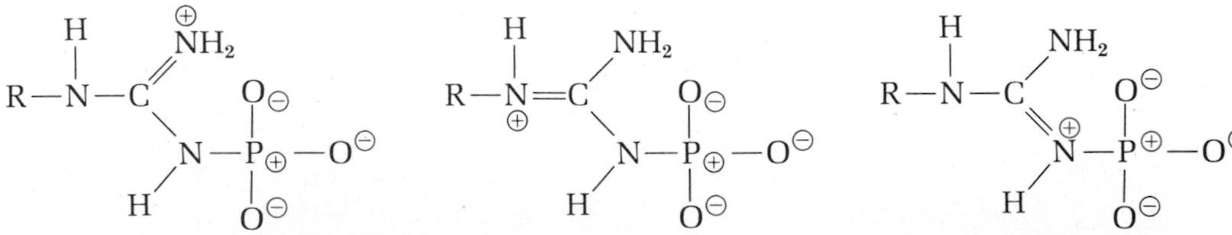

Unfavorable contributing structures to the resonance hybrid of the reactant

The hydrolysis of either of these phosphagens can be coupled to the phosphorylation of ADP as illustrated for creatine phosphate in reaction 12.40.

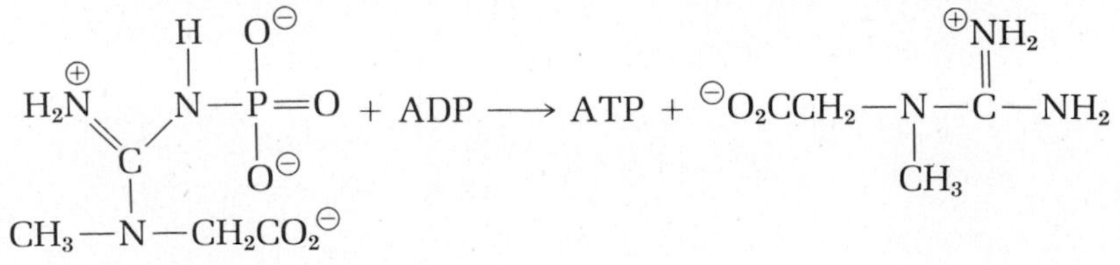

Creatine phosphate

$$\Delta G^{\circ\prime} = -13.4 \text{ kJ mol}^{-1} \qquad (12.40)$$

ATP is continually produced by metabolism, and it maintains creatine phosphate at an intracellular concentration of about 20 mM. Creatine phosphate provides an immediate and local source of ATP for muscle contraction. Arginine phosphate plays the same role in invertebrates.

F. Hydrolysis of Phosphoenolpyruvate

Phosphoenolpyruvate (PEP), an intermediate in carbohydrate metabolism, is an *enol ester* whose hydrolysis is highly exergonic. The hydrolysis of PEP can be divided into two steps. First, the enol ester hydrolyzes, produc-

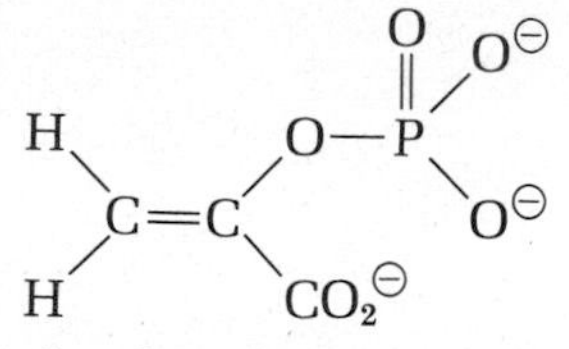

Phosphoenolpyruvate (PEP)

ing phosphate and a resonance-stabilized enolate anion. The standard free-energy change for this step is estimated to be -36.8 kJ mol^{-1} (reaction 12.41).

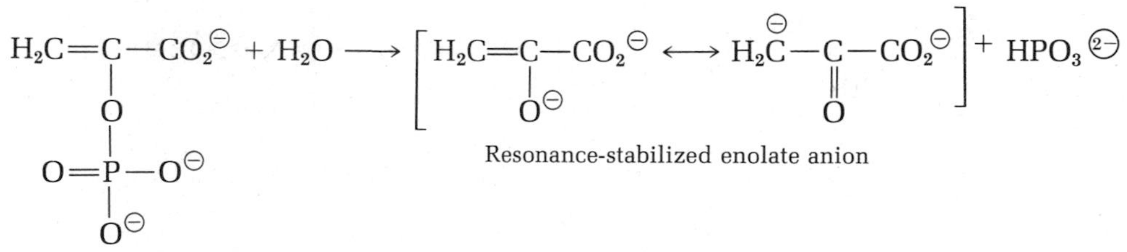

Resonance-stabilized enolate anion

$$\Delta G^{\circ\prime} = -36.8 \text{ kJ mol}^{-1} \qquad (12.41)$$

The standard free-energy change for the first step is about 21 kJ mol^{-1} more exergonic than the standard free energy for hydrolysis of simple phosphate esters, reflecting resonance stabilization of the enolate anion. The products of hydrolysis of simple phosphate esters do not have this resonance stabilization.

The enolate anion is unstable in aqueous solution, since the pK_a of the α-hydrogen of pyruvate is about 20, and it is rapidly protonated to give the enol (reaction 12.42).

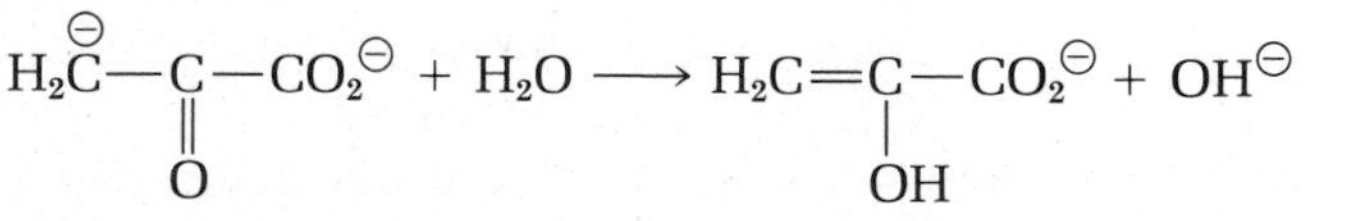

$$(12.42)$$

Enol pyruvate

In the second step of the reaction the enol rearranges to the keto acid, pyruvate. The tautomeric equilibrium favors the keto acid by about 25.1 kJ mol^{-1} (reaction 12.43).

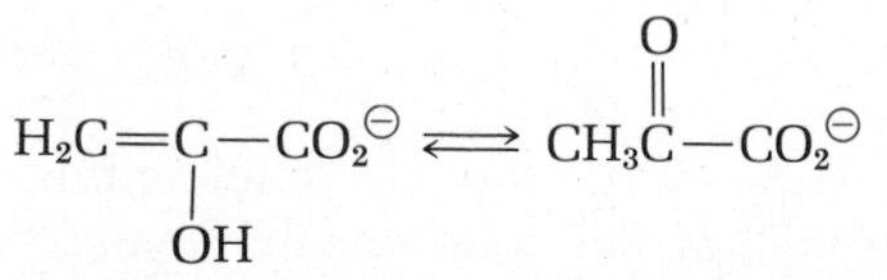

$$\Delta G^{\circ\prime} = -25.1 \text{ kJ mol}^{-1} \qquad (12.43)$$

Pyruvate

Thus, the overall free-energy change for hydrolysis of phosphoenolpyruvate is -62 kJ mol^{-1}, by far the most exergonic hydrolysis reaction we have considered (Figure 12.4).

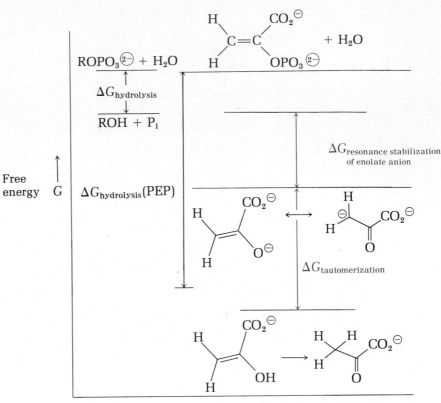

Figure 12.4 Comparison of the free energy of hydrolysis of phosphoenolpyruvate (PEP) and phosphate esters. ΔG^0 is much greater for PEP because the enolate anion is resonance stabilized and the tautomeric equilibrium strongly favors the keto form, pyruvate.

G. Group Transfer Potentials

The free-energy changes for hydrolysis of the metabolites we have discussed are summarized in Table 12.1. The conspicuous role of ATP in metabolism is reflected in this table. The free energy of hydrolysis of ATP lies about midway between the most exergonic reaction and the least exergonic reaction. ATP is therefore well placed for phosphoryl group transfer in coupled reactions. The energy released in the hydrolysis of metabolites lying above ATP in the table can drive the synthesis of ATP, and the energy released in the hydrolysis of ATP can drive the synthesis of metabolites lying below ATP in the table (assuming that all reactants and products are in their standard states).

The metabolites of Table 12.1 transfer a phosphoryl group ($-PO_3^{2-}$), *not* a phosphate group, to water when they hydrolyze. We define the *group transfer potential* of the phosphoryl group as the *negative* of the standard free energy of hydrolysis.

$$\text{group transfer potential} \equiv -\Delta G^{\circ\prime}_{\text{hydrolysis}}$$

Group transfer potentials for phosphoryl groups are listed in the right-hand column of Table 12.1.

Water is not the only nucleophile that can accept a phosphoryl group from ATP or the other metabolites of Table 12.1. Group transfer potentials are applicable to many reactions in metabolism. The hydrolysis of thioesters discussed in the next section and of certain other metabolites that we shall encounter in metabolism, such as 3'-5' cyclic AMP, aminoacyl tRNA molecules and S-adenosylmethionine can also be treated in terms of group transfer potentials. The general rule is that any metabolite can transfer a group,

Table 12.1 Free energy of hydrolysis and group transfer potential for selected phosphates

Compound	$\Delta G^{\circ\prime}_{\text{hydrolysis}}$ (kJ mol^{-1})	Phosphate group transfer potential (kJ mol^{-1})
Phosphoenolpyruvate		
pH 7.0	−61.9	61.9
pH 7.4	−53.5	53.5
Carbamoyl phosphate, pH 9.5	−51.5	51.5
1,3-Diphosphoglycerate, pH 6.9	−49.4	49.4
Acetyl phosphate, pH 7.0	−43.0	43.0
Creatine phosphate, pH 7.0, 37°	−43.0	43.0
Arginine phosphate, pH 8.0, excess Mg(II)	−32.2	32.2
ATP to AMP + PP, pH 7.0, excess Mg(II)	−32.2	32.2
ATP to ADP + P$_i$		
pH 7.0, 37°, excess Mg(II)	−30.5	30.5
pH 7.4, 25°, 10^{-3} M Mg(II)	−36.8	36.8
pH 7.4, 25°, no Mg(II)	−40.2	40.2
Pyrophosphate		
pH 7.0	−33.5	33.5
pH 7.0, 5 × 10^{-3} M Mg(II)	−18.8	18.8
Glucose 1-phosphate pH 7.0, 25°	−20.9	20.9
Glucose 6-phosphate, pH 7.0	−13.8	13.8
Glycerol 1-phosphate		
pH 8.5, 38°	−9.2	9.2
pH 5.810, 38°	−10.9	10.9

SOURCE: Data from W. P. Jencks, in G. D. Fasman, Ed., "Handbook of Biochemistry and Molecular Biology," 3rd ed., CRC Press, Cleveland, Ohio, 1976. Reprinted with permission. Copyright: The Chemical Rubber Co., CRC Press, Inc.

such as phosphoryl, to another nucleophile with a favorable equilibrium constant if the product has a lower group transfer potential than the reactant.

H. Hydrolysis of Thioesters

We have seen that coenzyme A functions as an acyl carrier (Section 10.3A). The hydrolysis of acetyl-CoA and other thioesters also is highly exergonic (reaction 13.44). The standard free energy of hydrolysis is about the same as that of ATP. The thioesters have large group transfer potentials for acyl groups. Thus, coenzyme A not only carries acyl groups, but is also able to transfer them to other acceptors. These twin properties make acyl-CoA derivatives highly important in coupled metabolic reactions.

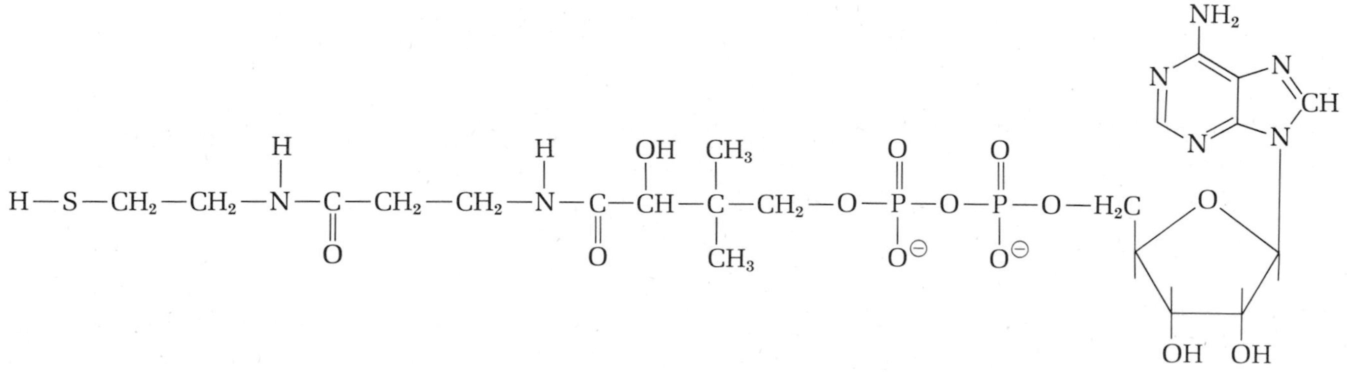

Coenzyme A (CoA—SH)

(12.44)

$$CH_3-\overset{\overset{\displaystyle O}{\|}}{C}-S-CoA + H_2O \longrightarrow CH_3CO_2^{\ominus} + CoA-SH$$

Acetyl coenzyme A

$$\Delta G^{\circ\prime} = -31.4 \text{ kJ mol}^{-1}$$

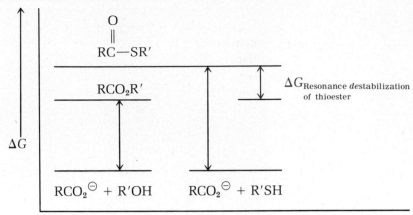

Figure 12.5 Free energy of hydrolysis of thioesters and carboxylate esters. Thioesters liberate more free energy upon hydrolysis primarily because the carboxylate esters are more resonance stabilized than the thioesters. The products have about the same energy in both cases.

As we discussed in Section 10.3A, thioesters are less resonance stabilized than carboxylate esters. There is a greater difference in resonance stabilization between thioesters and their hydrolysis products than between carboxylate esters and their hydrolysis products (Figure 12.5).

In fatty acid metabolism (Chapter 19) we shall encounter two other acyl carriers: carnitine and the acyl carrier protein. Both these molecules form thioesters, and their group transfer potentials are about the same as the group transfer potential of acetyl-CoA.

12–5 GROUP TRANSFER REACTIONS: THE KINASES AND RELATED ENZYMES

In our discussion of the hydrolysis of energy-rich metabolites we focused on the differences in energy between reactants and products. We now turn to the enzymes responsible for adenosyl transfer, adenylyl transfer, and phosphoryl transfer.

A. Adenosyl Transferases

Adenosyl transfer reactions involve attack by a nucleophilic group in one substrate molecule upon the 5′ adenosyl carbon of ATP with displacement of triphosphate (reaction 12.45).

These are rare processes: coenzyme B_{12}–catalyzed reactions (Section 10.9, Figure 10.28) and methionine biosynthesis are the only known examples.

B. Adenylyl (and other Nucleotidyl) Transferases

Nucleophilic attack by a group in one substrate molecule upon the α-phosphorus of ATP (or other nucleoside triphosphate) results in transfer of an adenylyl (or other nucleotidyl) group to the attacking substrate and displacement of pyrophosphate (reaction 12.46).

$$(12.46)$$

DNA polymerase and RNA polymerase catalyze nucleotidyl group transfer. In these cases the "nucleophile" of reaction 12.46 is a growing DNA or RNA chain. The conformations of the substrates in the enzyme-substrate complex and a possible reaction mechanism are shown in Figure 12.6. The metal ion is critical to catalysis. It not only functions as an electrophile, or electron sink, that increases the electrophilic character of the phosphorus, but also holds the reactants in the correct orientation for reaction, and thus plays two catalytic roles simultaneously.

In certain cases adenylyl transfer affects the kinetic properties of an enzyme and thus is a component in metabolic regulation. For example, an adenylyl transferase whose substrate is the E. coli enzyme *glutamine synthetase* (Section 22.4B) regulates the activity of this synthetase. Adenylylation virtually abolishes catalytic activity by increasing K_m values for the substrates of glutamine synthetase. A similar fate awaits the RNA polymerase of E. coli when that organism is invaded by bacteriophage T4. The bacteriophage directs the synthesis of an adenylyl transferase that immediately adenylylates the E. coli RNA polymerase, and inactivates protein synthesis in the bacterium. The phage then directs the bacterium to synthesize a bacteriophage RNA polymerase that participates in the synthesis of new proteins for future virus particles.

C. Pyrophosphoryl Transferases

Attack by a nucleophilic group of one substrate upon the β-phosphoryl group of ATP results in pyrophosphoryl group transfer with AMP as the "leaving group" (reaction 12.47). One important example of pyrophosphoryl group transfer occurs in the biosynthesis of 5-phosphoribosyl 1-pyrophosphate (PRPP), an intermediate in the biosynthesis of histidine, and in the biosynthesis of purine and pyrimidine nucleotides (reaction 12.48).

D. Kinases

Nucleophilic attack by a group in one substrate upon the γ-phosphorus of ATP leads to phosphoryl group transfer (reaction 12.49). The enzymes responsible for phosphoryl group transfer when the nucleophile is not water are called *kinases*. The substrates of kinase reactions are ATP and the meta-

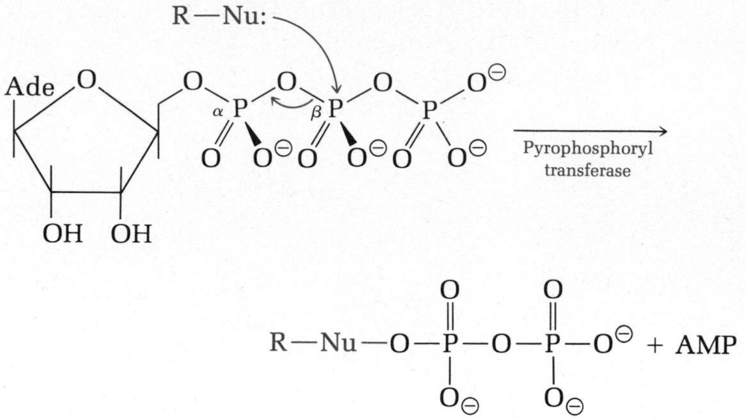

$$R-Nu-O-\overset{O}{\underset{O_\ominus}{P}}-O-\overset{O}{\underset{O_\ominus}{P}}-O^\ominus + AMP \qquad (12.47)$$

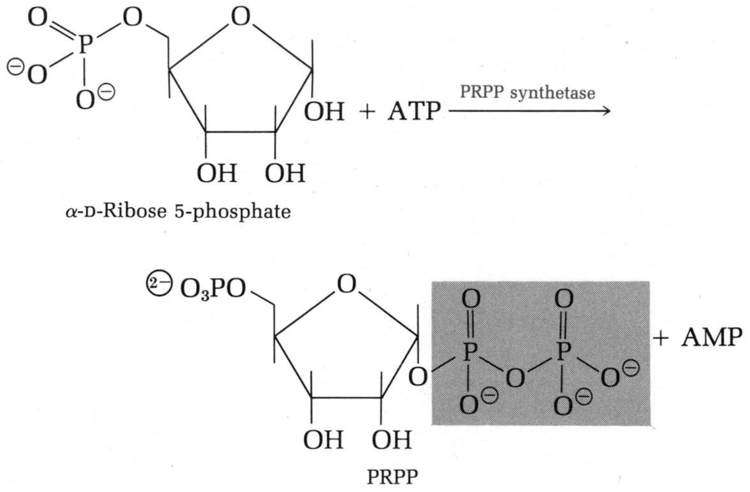

α-D-Ribose 5-phosphate

PRPP

$$\xrightarrow[\text{Kinase}]{M^{2+}}$$

$$R-Nu-\overset{O}{\underset{O_\ominus}{P}}-O^\ominus + ADP \qquad (12.49)$$

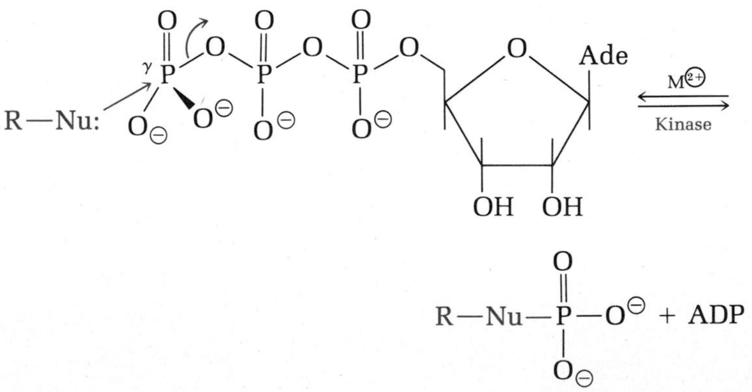

bolites (many of them also energy-rich) whose free energies of hydrolysis we have just discussed. ATP is thermodynamically unstable ($K'_{\text{hydrolysis}} \gg 1$), but it is kinetically stable. A specific kinase is required to lower the activation energy of a given reaction for the energy stored in ATP to be accessible.

Kinases require metal ions for activity. The metal ion requirement is complex. A divalent metal ion, usually Mg(II) or Mn(II), is often required to form a complex with ATP before ATP can bind to the enzyme active site. In such cases it is the ATP–metal ion complex that is the reactive substrate. In other cases the ATP reacts with a metalloenzyme complex. In pyruvate kinase two Mg(II) atoms are required per enzyme active site. The metal ions

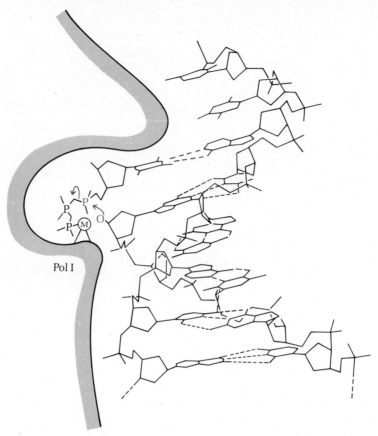

Pol I

Figure 12.6 Conformations of the substrates in the enzyme-substrate complex of DNA polymerase I (Pol I). The metal ion is required to maintain the correct conformation of the substrate at the active site of the enzyme. [From A. S. Mildvan, *Acc. Chem. Res.*, **10**, 251 (1977).]

bound to ATP decrease the negative charge at the γ-phosphorus making it more susceptible to nucleophilic attack. Metal ions also hold the substrates in place in the correct orientation for reaction, a function that we also noted in our discussion of adenylyl transferases. *Pyruvate kinase*, for example, catalyzes the transfer of a phosphoryl group from phosphoenolpyruvate (PEP) to ADP (reaction 12.50).

$$
\begin{array}{c}
\overset{\displaystyle OPO_3{}^{2\ominus}}{\underset{\displaystyle PEP}{CH_2{=}\overset{\textstyle |}{C}{-}CO_2{}^{\ominus}}} + ADP \xrightarrow[\text{M}^{2\oplus},\,\text{M}^{\oplus}]{\text{Pyruvate kinase}}
\end{array}
$$

$$
\underset{\text{Pyruvate}}{CH_3{-}\overset{\displaystyle O}{\overset{\|}{C}}{-}CO_2{}^{\ominus}} + ATP \qquad (12.50)
$$

A metal ion holds pyruvate as shown in Figure 12.7. Since the group transfer potential of PEP is so much higher than the group transfer potential of ATP, reaction 12.50 lies overwhelmingly on the side of ATP.

Many kinases require monovalent cations, such as $K^{\oplus}$ and $NH_4{}^{\oplus}$, as well as divalent cations for activity. The monovalent cation often binds the enzyme at its active site. Pyruvate kinase and acetate kinase are among the enzymes that require a monovalent cation for activity.

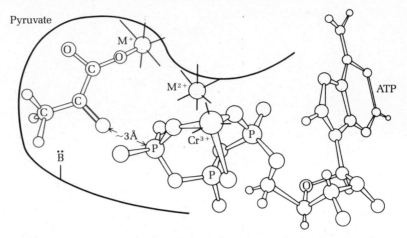

Figure 12.7 Conformations of the substrates at the active site of pyruvate kinase. The chromium (III)-ATP complex is called a substitution inert complex. It is not active in catalysis, but it does aid the enolization of pyruvate. By using the inert complex a stable enzyme-substrate analog complex forms that can be analyzed for interatomic distances. [From A. S. Mildvan, *Acc. Chem. Res.*, 10, 250 (1977).]

12-6 MECHANISMS OF PHOSPHORYL GROUP TRANSFER REACTIONS: HEXOKINASE AND GLYCEROL KINASE

We have discussed the bioenergetics of phosphoryl group transfer from ATP to glucose and from ATP to glycerol in the reactions catalyzed by hexokinase and glycerol kinase, respectively. Next let us consider some of the details of the mechanism of action of these enzymes. J. Knowles and his

$$\text{Ado}-\text{O}-\overset{\displaystyle \text{O}}{\underset{\displaystyle \text{O}^{\ominus}}{\overset{\|}{\text{P}}}}-\text{O}-\overset{\displaystyle \text{O}}{\underset{\displaystyle \text{O}^{\ominus}}{\overset{\|}{\text{P}}}}-\text{O}-\overset{^{16}\text{O}}{\underset{^{17}\text{O}}{\text{P}}}\overset{^{18}}{\text{O}}$$

γ-[(S)-^{16}O,^{17}O,^{18}O]-ATP

coworkers synthesized chiral γ-[(S)-^{16}O,^{17}O,^{18}O]-ATP and showed that the chiral γ-phosphate group undergoes *inversion of configuration* in reactions catalyzed by glycerol kinase (reaction 12.51) and hexokinase (reaction 12.52).

$$\text{Ado}-\text{O}-\overset{\displaystyle \text{O}}{\underset{\displaystyle \text{O}^{\ominus}}{\overset{\|}{\text{P}}}}-\text{O}-\overset{\displaystyle \text{O}}{\underset{\displaystyle \text{O}^{\ominus}}{\overset{\|}{\text{P}}}}-\text{O}-\overset{^{16}\text{O}}{\underset{^{17}\text{O}}{\text{P}}}\overset{^{18}}{\text{O}} \quad + \quad \overset{\displaystyle \text{CH}_2\text{OH}}{\underset{\displaystyle \text{CH}_2\text{OH}}{\overset{\displaystyle |}{\underset{\displaystyle |}{\text{H}-\text{C}-\text{OH}}}}}$$

γ-[(S)-^{16}O,^{17}O,^{18}O]-ATP

| Glycerol kinase

Inversion of configuration at chiral phosphorus →

$$\text{ADP} + \quad \overset{\displaystyle \text{CH}_2-\text{O}-\overset{^{17}\text{O}}{\underset{^{16}\text{O}}{\text{P}}}\overset{^{18}}{\text{O}}}{\underset{\displaystyle \text{CH}_2\text{OH}}{\overset{\displaystyle |}{\underset{\displaystyle |}{\text{H}-\text{C}-\text{OH}}}}} \tag{12.51}$$

sn-Glycerol 3-[(R)-^{16}O,^{17}O,^{18}O]-phosphate

γ-[(S)-^{16}O,^{17}O,^{18}O]-ATP

Hexokinase → ADP

← Inversion of configuration at transferred phosphoryl group

(12.52)

6-[(R)-^{16}O,^{17}O,^{18}O]phospho-α-D-glucose

(a)

Pentacovalent,
trigonal bipyramid
intermediate

Inversion of
configuration

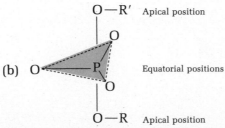

(b)

O—R′ Apical position

Equatorial positions

O—R Apical position

Pentacovalent intermediate

Figure 12.8 (a) In-line mechanism for addition-elimination in nucleophilic substitution at tetracovalent phosphorus. (b) Pentacovalent intermediate in the in-line mechanism with the entering and leaving groups occupying apical position.

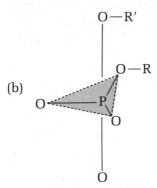

Pentacovalent,
trigonal bipyramid
intermediate

Trigonal bipyramid
intermediate

Figure 12.9 (a) Adjacent mechanism for nucleophilic substitution at tetracovalent phosphorus. The nucleophile enters on the same side as the leaving group. (b) Trigonal bipyramidal intermediate. The entering and leaving groups are not initially on opposite sides of the intermediate, and it must rearrange to place them in apical positions before RO^- can leave.

Substitution at tetracovalent phosphorus proceeds by addition-elimination mechanisms analogous to addition-elimination mechanisms for carbonyl compounds. Two types of mechanisms are possible: in-line and adjacent. In the *in-line mechanism* the nucleophile enters and the leaving group departs from opposite sides of a pentacovalent, trigonal bipyramid intermediate in which the nucleophile and the leaving group each occupy an apical position (Figure 12.8). In the *adjacent* mechanism the nucleophile enters on the same side as the leaving group (Figure 12.9). In both cases a trigonal bipyramid intermediate forms. In the adjacent mechanism it must undergo *pseudorotation*, a process in which the equatorial and apical positions interchange to

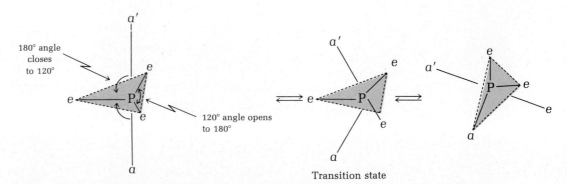

Transition state

Figure 12.10 Pseudorotation in the adjacent mechanism for nucleophilic substitution at tetracovalent phosphorus. Pseudorotation involves closing the 180° bond angle by which the original apical substitutents are separated and opening the 120° bond angles by which the equatorial substituents are separated. In the process one of the apical groups becomes equatorial and one of the equatorial groups becomes apical.

bring the entering nucleophile and the leaving group into apical positions (Figure 12.10). The pseudorotation rearrangement is required by the *principle of microscopic reversibility,* that is, the mechanism of a reaction in the forward direction must be the same as the mechanism of reaction in the reverse direction. If the nucleophile enters from an apical position, the leaving group must leave from an apical position. The in-line mechanism, like an S_N2 reaction, proceeds with inversion of configuration. The adjacent mechanism involves racemization because pseudorotation interchanges the equatorial and apical positions, making them equivalent. Neither hexokinase nor glycerol kinase follows an adjacent mechanism, and an in-line mechanism seems highly probable.

12–7 CONFORMATIONAL CHANGES IN HEXOKINASE INDUCED BY SUBSTRATE BINDING

In Section 6.2 we noted that enzymes are conformationally mobile. One hypothesis, introduced by D. E. Koshland, postulates that substrate binding induces a conformational change in the enzyme that brings the catalytic groups of the active site close to the substrate. Such a conformational change can also aid catalysis if the enzyme induces strain in the substrate to make it more reactive. A model of yeast hexokinase in the absence of substrate is shown in Figure 12.11. The most significant feature of the molecule is a large cleft, the binding site for substrate, that divides hexokinase into two lobes. A model of yeast hexokinase in the presence of glucose is shown in Figure

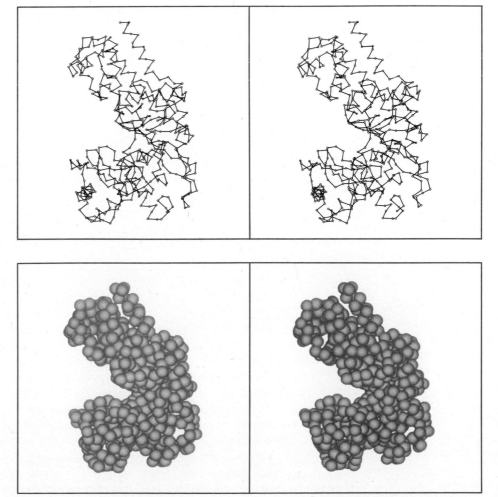

Figure 12.11 Space-filling model of yeast hexokinase. The binding site for substrate is located in the cleft that divides the molecule into two lobes.

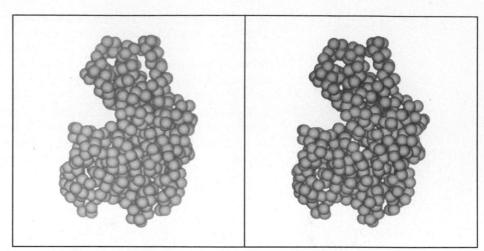

Figure 12.12 Space-filling model of heast hexokinase complexed with glucose. The cleft between the two lobes has closed on the substrate by a rotation of 12° of the upper lobe relative to the lower one.

12.12. The lobes have closed on the substrate by a rotation of the upper lobe of 12° relative to the lower lobe. This conformational change is necessary for catalysis. ATP also binds to hexokinase with its γ-phosphate situated directly over the cleft ready for an in-line transfer of phosphate from ATP to glucose.

All kinases whose structures are known have clefts similar to the one in hexokinase, and it seems likely that these kinases also undergo substrate-induced closing of the cleft. The kinases have quite different tertiary structures, and the clefts are formed in different ways. The kinases are not as similar as the closely related serine proteases (Section 6.3), suggesting that their structural similarities are related to kinase function rather than to divergent evolution.

12–8 SUMMARY

The reactions of metabolism usually occur at a pH of 7.0, and we therefore define a biological standard state ($\Delta G^{\circ\prime}$) in which the standard free energy of formation of 10^{-7} M hydronium ion is zero. The biological solvent is water, whose standard state is defined as 1.0 M. The shift of the standard state of hydronium ion from 1.0 M to 10^{-7} M corresponds to a shift of -39.5 kJ mol^{-1}. It is therefore imperative that the correct standard state be chosen for making biochemical calculations.

A sufficient (though not always necessary) condition for the coupling of two metabolic reactions is that they share a common intermediate. The standard free-energy changes, either ΔG° or $\Delta G^{\circ\prime}$ depending upon the choice of standard states, of coupled reactions are additive, and their equilibrium constants are multiplicative.

Metabolites whose hydrolysis releases as much or more energy than the hydrolysis of ATP ($\Delta G^{\circ\prime}_{\text{hydrolysis}} \cong -32$ kJ mol^{-1}) are said to be energy rich. The energy-rich metabolites are capable of driving endergonic biochemical reactions and physical processes. In comparing the standard free energies of hydrolysis of metabolites we analyze the differences in stability between the reactants and the differences in stabilities of the products. Three factors that are usually important in determining the magnitude of the free energy of hydrolysis are the resonance stabilities of the reactants and products in each of the pairs being compared, their relative electrostatic properties, and the extent to which the products and reactants are solvated. The phosphate group transfer potential is defined as $-\Delta G^{\circ\prime}_{\text{hydrolysis}}$. The singular importance of ATP in metabolism is partly derived from its group transfer potential, which lies about midway between the highest and lowest group transfer potentials. ATP is, therefore, ideally placed to mediate coupled biochemical reactions since ADP can accept a phosphoryl group from a metabolite with a higher phosphate group transfer potential, and ATP can transfer a phosphoryl group to a metabolite having a lower group transfer potential.

Adenosyl transferases, adenylyl transferases (and other nucleotidyl transferases), and phosphoryl transferases, better known as kinases, catalyze group transfer reactions. Adenosyl transferases are rare. They are found only in certain coenzyme B_{12}—dependent reactions and in methionine biosynthesis. Adenylyl and nucleotidyl transferases catalyze nucleophilic displacement of pyrophosphate from ATP or other nucleoside triphosphates. DNA and RNA polymerases are nucleotidyl transferases. Pyrophosphoryl transferases catalyze nucleophilic substitution at the β-phosphorus of ATP with AMP as a leaving group. The biosynthesis of 5-phosphoribosyl 1-pyrophosphate, an intermediate in the biosynthesis of histidine, and of purine and pyrimidine nucleotides involves the action of a pyrophosphoryl transferase. The kinases catalyze phosphoryl group transfer from ATP to another substrate. The kinases are metalloenzymes, and they require a divalent cation for activity. A monovalent cation is also required by some kinases.

Hexokinase and glycerol kinase are among the kinases that are known to catalyze phosphoryl group transfer by an in-line mechanism. A pentacovalent, trigonal bipyramidal intermediate is formed during the catalytic process.

Hexokinase undergoes a large conformational change when it binds glucose. This conformational change aids catalysis by bringing catalytic groups into the correct orientation for reaction.

REFERENCES

Boyer, P. D., Ed., "The Enzymes," 3rd ed., Vol. 8, Academic Press, New York, 1973.

Bray, H. G., and White, K., "Kinetics and Thermodynamics in Biochemistry," 2nd ed., Academic Press, New York, 1966.

Ingraham, L. L., and Pardee, A. B., Free Energy and Entropy in Metabolism, in D. M. Greenberg, Ed., "Metabolic Pathways," 3rd ed., Vol. 1, Academic Press, New York, 1967.

Jencks, W. P., Free Energies of Hydrolysis and Decarboxylation, in G. D. Fasman, Ed., "Handbook of Biochemistry and Molecular Biology," 3rd ed., Vol. 1, CRC Press, Cleveland, Ohio, 1976.

Klotz, I. M., "Energy Changes in Biochemical Reactions," Academic Press, New York, 1967.

Van Holde, K. E., "Physical Biochemistry," Prentice-Hall, Englewood Cliffs, N.J., 1971.

Problems

1. 3′-Phosphoadenosine-5′-phosphosulfate (PAPS) is an intermediate in the biosynthesis of sulfate esters. Explain why PAPS has a large negative free energy of hydrolysis in terms of its structure.

3′-Phosphoadenosine-5′-phosphosulfate

2. The hydrolysis of PEP in glycolysis is coupled to the synthesis of ATP from ADP and monohydrogen phosphate. Using the information given below, calculate the standard free-energy change and the equilibrium constant for the coupled reaction.

$$ATP + H_2O \rightleftharpoons ADP + HPO_4^{\ominus} \qquad \Delta G^{\circ\prime} = -30.5 \text{ kJ mol}^{-1}$$
$$PEP + H_2O \rightleftharpoons pyruvate + HPO_4^{\ominus} \qquad \Delta G^{\circ\prime} = -59.8 \text{ kJ mol}^{-1}$$

3. The standard free-energy change for the hydrolysis of ATP to ADP and orthophosphate is -32.2 kJ mol^{-1}. Under cellular conditions, however, the actual free-energy change, $\Delta G'$, can be as large as -58 kJ mol^{-1}. What cellular ratio of ATP/ADP is required to give this free-energy change for the hydrolysis of ATP?

4. Adenosine 5-O-R-[3-^{18}O,3-thio]-phosphate is a substrate for glycerol kinase as shown below. Write the mechanism for thiophosphoryl transfer, including the correct stereochemistry for the product.

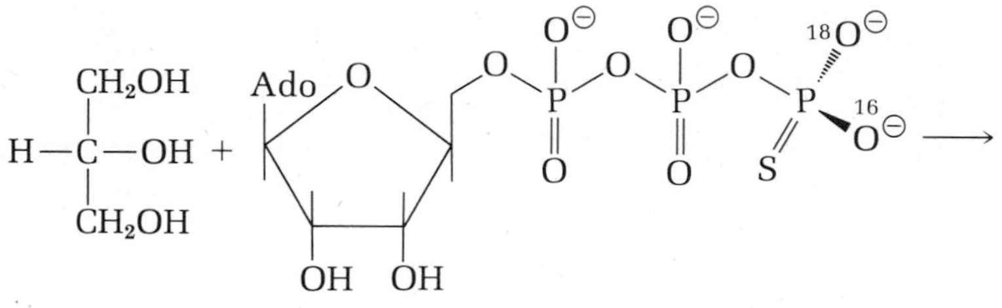

Adenosine 5'-O-R-[3-^{18}O,3-thio]-phosphate

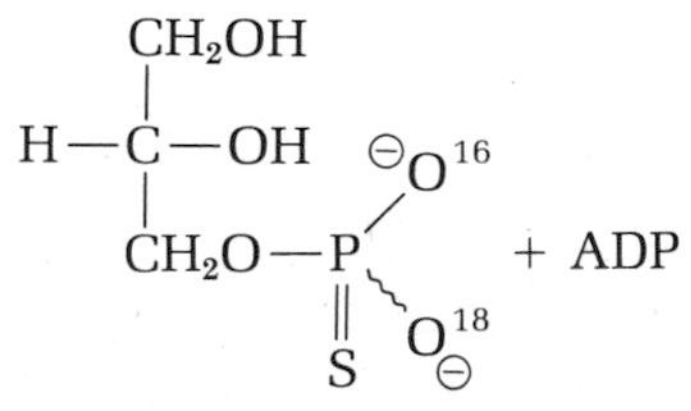

5. Pyruvate carboxylase, a biotin-dependent enzyme (Section 10.7), catalyzes the conversion of pyruvate to oxaloacetate. The standard free energy of decarboxylation is -25.9 kJ mol^{-1}. The standard free-energy change for decarboxylation of enzyme-biotin-CO$_2$ is -19.6 kJ mol^{-1}. (a) From these data calculate the standard free-energy change for the carboxylation of pyruvate catalyzed by pyruvate carboxylase. (b) What is the equilibrium constant for carboxylation of pyruvate?

$$enzyme\text{-}biotin\text{-}CO_2 \longrightarrow enzyme\text{-}biotin + HCO_3^{\ominus} \qquad \Delta G^{\circ\prime} = -19.6 \text{ kJ mol}^{-1}$$

6. The bacterium *Pseudomonas saccharophila* catalyzes the breakdown of sucrose according to the equation shown below. For the hydrolysis of sucrose to glucose and fructose, $\Delta G^{\circ\prime} = -29.3$ kJ mol$^-$; for the hydrolysis of glucose 1-phosphate, $\Delta G^{\circ\prime} = -20.9$ kJ mol^{-1}.

$$sucrose + P_i \rightleftharpoons glucose \text{ 1-phosphate} + fructose$$

a. Calculate the standard free-energy change ($\Delta G^{\circ\prime}$) and the equilibrium constant for the overall reaction.

b. Would you expect the bacterium to be able to accomplish this transformation in two successive steps (as shown below) or in one coupled reaction? Explain your answer.

$$\text{sucrose} \rightleftharpoons \text{glucose} + \text{fructose}$$
$$\text{glucose} + P_i \rightleftharpoons \text{glucose 1-phosphate}$$

7. The following reactions occur together in solution:

$$\text{ATP} + \text{3-phosphoglycerate} \underset{\text{pH 7.0, 38°C}}{\overset{K_{eq} = 3.1 \times 10^{-4}}{\rightleftharpoons}} \text{ADP}$$

$$+ \text{ 1,3-diphosphoglycerate}$$

$$\text{3-phosphoglycerate} \underset{\text{pH 7.0, 38°C}}{\overset{K_{eq} = 0.24}{\rightleftharpoons}} \text{2-phosphoglycerate}$$

a. Calculate the standard free-energy change for the formation of ADP and 1,3-diphosphoglycerate from ATP and 2-phosphoglycerate.
b. Assuming the concentrations of 2-phosphoglycerate and 1,3-diphosphoglycerate to be equal, calculate the minimum ratio of ATP to ADP required to maintain equilibrium. How would this ratio need to change in order to form 1,3-diphosphoglycerate from 2-phosphoglycerate under these conditions?

8. The stereochemical course of the ribosome-dependent GTPase reaction of elongation factor G from *E. coli* has been determined. When guanosine 5'-O-S-[3-^{17}O,^{18}O,3-thio]phosphate is hydrolyzed the stereochemistry of the released chiral [^{16}O,^{17}O,^{18}O]thiophosphate can be either *R* or *S*. Only the product in which the chiral phosphorus is *R* is observed. Provide a mechanism that accounts for this observation. [Reference: M. R. Webb and J. F. Eccleston, *J. Biol. Chem.*, *256*, 7734–7737 (1981).]

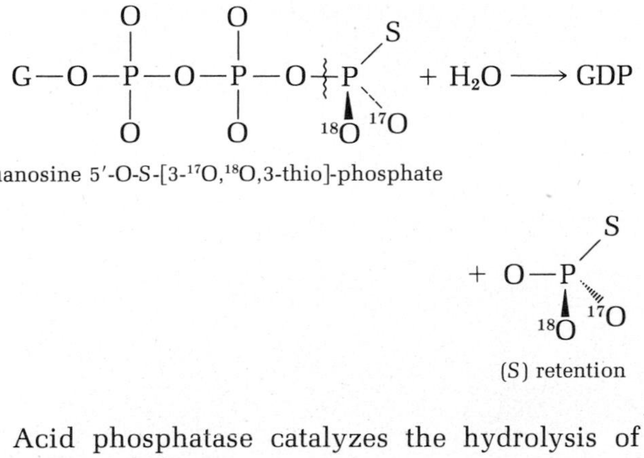

Guanosine 5'-O-S-[3-^{17}O,^{18}O,3-thio]-phosphate

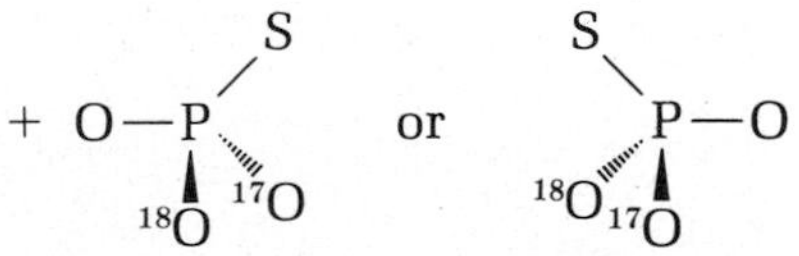

(S) retention   (R) inversion

9. Acid phosphatase catalyzes the hydrolysis of phosphate monoesters, as shown below.

$$\text{R—OPO}_3{}^{2\ominus} + H_2O \xrightarrow{\text{Acid phosphatase}} \text{ROH} + \text{HPO}_4{}^{2\ominus}$$

The question arises whether this reaction occurs with formation of a covalent phosphoenzyme intermediate: E-(X)-PO$_3{}^{2\ominus}$. An alcohol can substitute for water as the nucleophile in acid phosphatase-catalyzed reactions, and the transphosphorylation of a chiral phosphate ester has recently been studied by treating phenyl-R-(^{16}O,^{17}O,^{18}O) phosphate with acid phosphatase in the presence of S-propane 1,2-diol. The product that forms is 2Spropanol-3-R-(^{16}O,^{17}O,^{18}O) phosphate. The reaction has occurred with 90% *retention* of configuration in the final product. Explain whether this observation supports the hypothesis that a covalent phosphoenzyme intermediate forms during the reaction. [Reference: M. S. Saini, S. L. Buchwald, R. L. van Etten, and J. R. Knowles, *J. Biol. Chem.*, *256*, 10453–10455 (1981).]

10. Studies of enzymatic reaction mechanisms typically require pure preparations of enzyme. The important enzyme adenylate cyclase catalyzes the conversion of adenosine triphosphate to 3'-5' cyclic adenosine monophosphate. The enzyme has not yet been purified; it is tightly bound to cell membranes, and such proteins are difficult to purify. Nevertheless, its mechanism of action has been studied by following the stereochemical course of the conversion of adenosine to adenosine 3'-5' cyclic adenosine thiophosphate, as shown below. The only product that is isolated from the reaction mixture has the R configuration at phosphorous. Propose a mechanism for the reaction. [Reference: F. Eckstein and P. R. Romanink, *J. Biol. Chem.*, *256*, 9118–9120 (1981).]

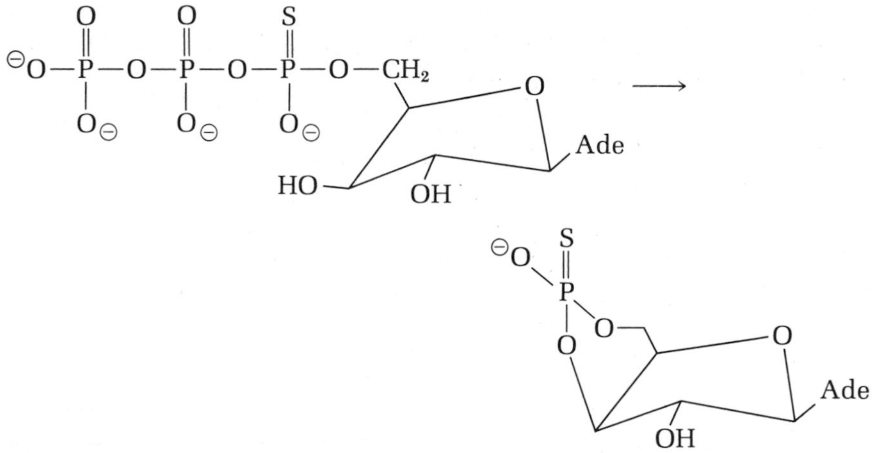

Chapter 13

The design and regulation of metabolic pathways

The kind of stability that is displayed by the living organism is of a nature somewhat different from the stability of atoms or crystals. It is a stability of process or function rather than the stability of form.

W. Heisenberg

13–1 INTRODUCTION The chemical reactions of the cell, its metabolism, constitute an extraordinary system. Within a volume of less than a microliter, the cell carries out thousands of enzymatic reactions. The thousands of separate, stereospecific reactions are each a necessary part of a larger pathway which is itself part of a larger process. The last fifty years have witnessed the gradual elucidation of the major metabolic pathways, and attention is more and more focused on the regulation of metabolism that provides the stable steady state known as homeostasis.

<table>
<tr><td>**13−2 PRODUCT-
PRECURSOR
RELATIONS IN
METABOLIC PATHWAYS**</td></tr>
</table>

A metabolic pathway consists of a series of reactions in which a precursor is converted to a product in one or more steps. A diagram showing all the pathways together resembles a city street map. It would be clear from such a map that a trip from point A to B might be made by more than one route, parts of some routes could be bypassed, and a given "intersection" might interconnect several pathways. Studying pathways separately is at least partly a matter of organizational convenience.

The flow of metabolites through a metabolic pathway can be followed by tracing either chemical markers or isotopic labels, such as ^{14}C, through the sequence of steps by which a molecule is synthesized or degraded. For example, as early as 1904 F. Knoop fed to dogs fatty acids, chemically labeled with a phenyl group in the ω position. Analysis of the urinary excretion products showed that when the phenyl group was bound to the ω end of fatty acids containing an even number of carbon atoms, phenylacetic acid was produced (reaction 13.1).

$$C_6H_5-\underset{\omega\ end}{CH_2}(CH_2)_{n=2,4,6...}-CO_2H \longrightarrow \underset{Phenylacetic\ acid}{C_6H_5-CH_2CO_2H} \tag{13.1}$$

When odd-chain fatty acids labeled with a phenyl group in the ω position were metabolized, benzoic acid was produced (reaction 13.2).

$$C_6H_5-CH_2(CH_2)_{n=1,3,5,...}CO_2H \longrightarrow \underset{Benzoic\ acid}{C_6H_5CO_2H} \tag{13.2}$$

These results established that fatty acids are degraded two carbons at a time (Section 19.3). Although the results do not reveal the nature of the intermediates, they define the relation of the precursor to the product. To delineate the entire pathway, each intermediate must be isolated, identified, and placed in relation to its immediate precursor and immediate product.

Isotopic labeling techniques are more subtle than chemical labeling techniques, since the isotopically labeled material possesses the same chemical properties (disregarding kinetic isotope effects) as unlabeled material. Radioisotopically labeled compounds can be followed in minute concentrations since radioisotope detection, principally by liquid scintillation counting, its extremely sensitive. For example, the metabolism of acetate, whose many transformations are summarized in Figure 13.1, was determined by tracing the flow of ^{14}C-labeled acetate.

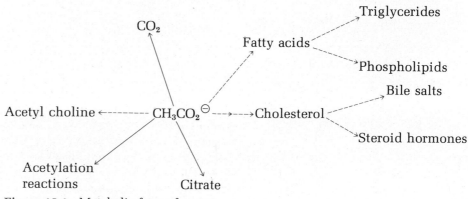

Figure 13.1 Metabolic fates of acetate.

Isotopic labeling has also served as a tool in the study of polymer biosynthesis. We would like to know, for example, whether proteins grow from their C- or their N-terminus, whether polysaccharides grow from their reducing or nonreducing ends, and whether nucleic acids grow from their 3′ or 5′ termini. The direction of chain growth can be classified as *tail growth* or *head growth*. In tail growth an unactivated end of the polymer attacks the activated end of the monomer, while in head growth an unactivated monomer attacks the growing, activated end of the polymer. Protein synthesis on a ribosome (Chapter 26), fatty acid synthesis in the fatty acid synthetase complex (Section 19.7), and O-antigen biosynthesis (Section 19.5) are examples of head growth. DNA biosynthesis and polysaccharide biosynthesis, on the other hand, are examples of tail growth (Figure 13.2).

In *pulse labeling* experiments an aliquot of a monomer is added in a short pulse. The isotopically labeled monomer is incorporated at the initiating end of the chain. Polymers formed during a pulse time shorter than the time required for complete synthesis will be labeled in the region synthesized last, but not in the region synthesized first. If the pulse time is longer than the time required for complete synthesis, all regions of the polymer will contain isotopic label, but the initiating end of the polymer will contain the least isotopic label. The direction of synthesis can therefore be determined by analyzing the radioactivity contained in fragments obtained by hydrolysis of the polymer. Fragments of manageable size are provided for nucleic acids by restriction endonucleases (Section 9.5D) and for proteins by proteolytic enzymes (Section 1.9D). Pulse labeling has shown that chain growth in DNA is strictly in the 5′ to 3′ direction, as we noted above.

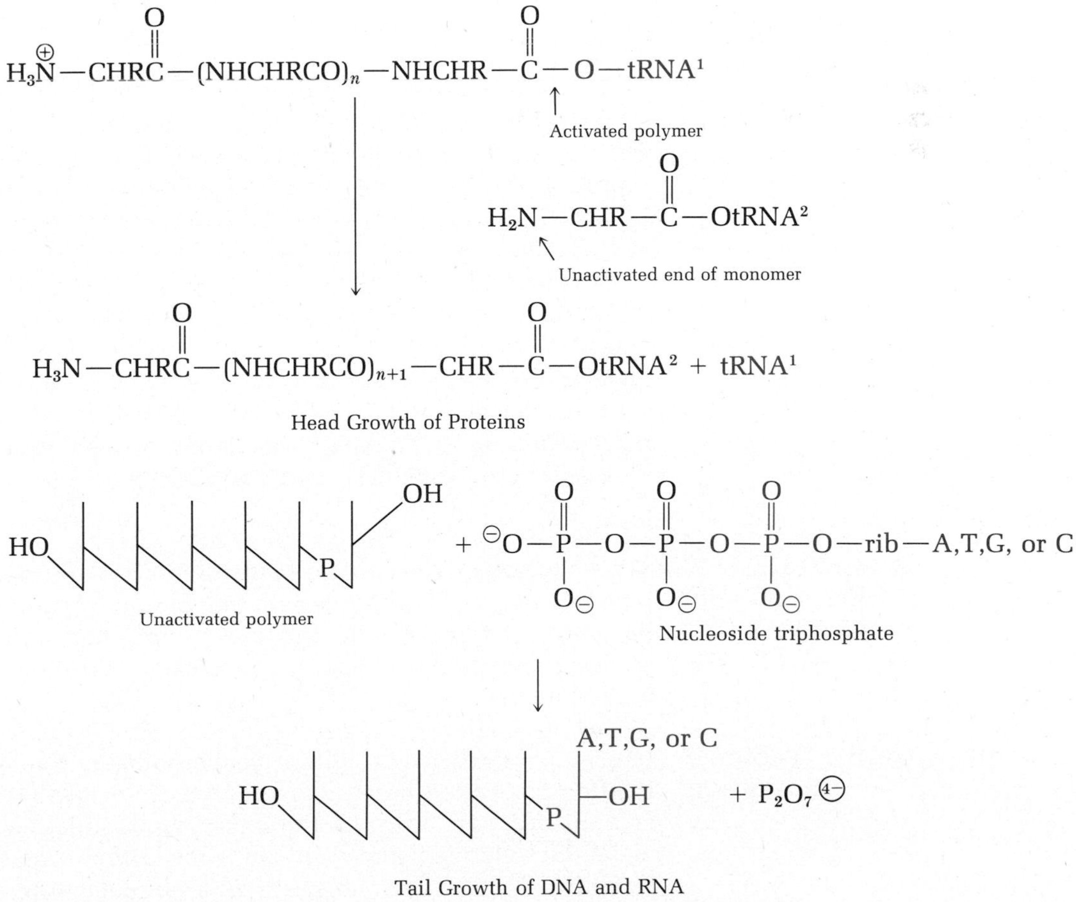

Figure 13.2 Examples of head growth and tail growth of polymers.

Most metabolic pathways are unidirectional; the net flux of a given metabolite in a given pathway is heavily toward the end product of the pathway. Although enzyme-catalyzed reactions are thermodynamically reversible, the overall equilibrium constant for the pathway lies overwhelmingly on the side of the final product. There are usually one or more reactions in which there is a large free-energy change, and these reactions "pull" a pathway to completion. The first unique step in the pathway which involves such a large free-energy change is called the *committing step*. This committing step can be controlled at three levels: (1) the synthesis of the enzyme responsible for the committing step can be controlled; (2) the steady state concentration of the metabolite involved in the committing step can be controlled by an allosteric enzyme; and (3) the availability of substrate can be controlled through regulation of membrane transport.

Catabolic (degradative) and biosynthetic processes usually occur by different pathways. Since the pathways are distinct, the two processes can be independently regulated. In eucaryotic cells the separation of pathways is often reinforced by *compartmentation*, a simple regulatory mechanism in which the enzymes for catabolism and biosynthesis are physically separated by membrane-bound cell organelles.

13–3 GENERAL FEATURES OF METABOLISM

If we reduce metabolism to its barest outlines, as in Figure 13.3, we see that catabolic pathways converge upon intermediates of the citric acid cycle, upon pyruvate, and upon acetyl-CoA. Before the convergence each pathway goes through a preliminary stage: polysaccharides are converted to monosaccharides, proteins to amino acids, and lipids to fatty acids and glycerol. The reactants and the products are dissimilar in the three processes, but each is an example of an *hierarchical pathway,* and each conforms to the same general pattern.

The catabolism and biosynthesis of fatty acids, monosaccharides, and amino acids occur by *linear* pathways. The electron transport chain of respiration coupled to phosphorylation of ADP (Chapter 000) is a linear pathway that is similar to several other less elaborate oxidative pathways. Oxidative phosphorylation is strikingly similar to photophosphorylation in photosynthesis. Linear pathways are regulated by a variety of negative feedback mechanisms (Section 13.4).

Many metabolic pathways are *cyclic.* The citric acid cycle (Chapter 15) and the urea cycle (Section 21.3) are examples. We might also describe some metabolic pathways as *spiral.* These pathways are neither linear nor cyclic, but share some features of both. The same sets of enzymes are used over and over for step-by-step lengthening or breakdown of a given molecule. Protein synthesis and fatty acid synthesis and degradation are examples of spiral pathways.

13–4 FEEDBACK REGULATION OF ENZYMIC ACTIVITY

The enzymes of a metabolic pathway do not all operate at their maximum rates. Rather, the rates of certain key *regulatory enzymes* are controlled by the concentrations of either the final end product of the pathway or by a critical intermediate. The regulatory enzymes control the rate of the entire metabolic pathway.

A. Negative Feedback Inhibition

In *negative feedback inhibition* the committing step of a metabolic pathway is often controlled by an end product of the pathway. Control of the first committing step prevents the excess accumulation of both product and metabolic intermediates. And, since the committing step often involves the hydrolysis of ATP or another energy-rich metabolite, control of the committing step prevents wasteful depletion of the cellular energy supply. The product of a metabolic pathway usually bears little resemblance to the metab-

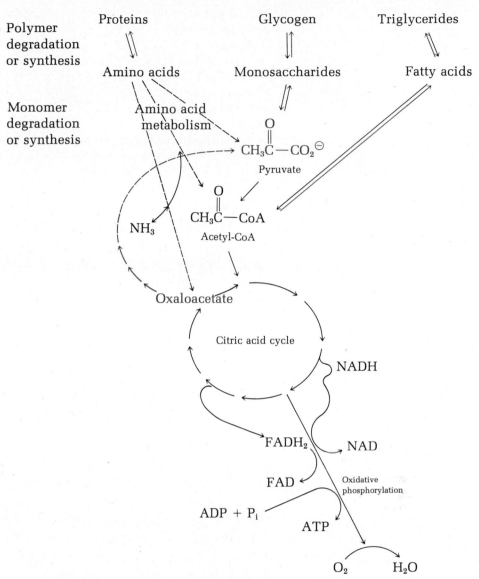

Figure 13.3 General features of metabolism. Degradative processes generate ATP from ADP and phosphate, whereas biosynthetic processes are driven by the hydrolysis of ATP. Catabolic processes are oxidative and generate NADH from NAD. Biosynthetic processes are reductive and use NADPH as their reducing agent.

olite involved in the committing step, and regulation by negative feedback inhibition in an *allosteric effect*. The kinetic effects of allosteric inhibition are widely varied: they may be competitive, noncompetitive, or uncompetitive. These terms imply a mechanism of inhibition, an implication we wish to avoid in the far more complex allosteric cases. Those inhibitors that change K_m, but not V are called K *systems*. Inhibitors that alter the apparent V, but do not alter K_m are called V *systems*. The terms K system and V system imply nothing whatever about the mechanism of the inhibition; these refer only to the observed kinetic effects.

B. Control of Branched Pathways

Many metabolic pathways are branched, that is, one of the intermediates, the branch-point metabolite, can be converted into two or more products (Figure 13.4). If one of the end products completely inhibits the first step, its presence could prevent, or severely restrict, the formation the other end product. Several mechanisms have evolved to control branched pathways.

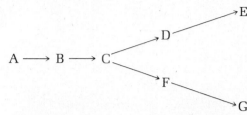

Figure 13.4 Branched metabolic pathway. C is the branch-point metabolite, E and G are end products, and the conversion of A to B is the first committing step of this pathway.

a. Parallel Enzymes

Consider a case in which the first reaction in a branched pathway is catalyzed by two different enzymes acting "in parallel" (Figure 13.5). The first enzyme is inhibited by one end product of the branched path, and the second enzyme is inhibited by the other end product. Each end product then serves as a negative feedback inhibitor of its own formation, but its synthesis is not curtailed by formation of the other products. The *aspartokinases* of *E. coli* provide one example of such parallel enzymes (Section 21.6C).

Branched pathways may also be regulated at the first step after a branch point. In this type of control, the first step that is *unique* to each pathway is inhibited, and in many cases the first committing step is also inhibited. The concentration of the branch-point metabolite is therefore regulated twice: its formation is regulated in the first committing step and its disappearance is regulated in the first step *after* the branch-point. The biosynthesis of aromatic amino acids is one example of this phenomenon (see Figure 22.8); purine biosynthesis is another (Section 23.2).

b. Cumulative Feedback Inhibition

In some cases a high concentration of any one of the products of a branched pathway partially inhibits the first committing step of the sequence. In Figure 13.6, E and G each partially inhibits the conversion of A to B. Each

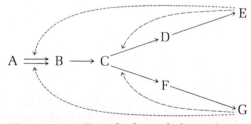

Figure 13.5 Branched metabolic pathway in which two enzymes catalyze the conversion of A to B in the first committing step of the pathway. Black arrows represent enzyme-catalyzed reactions in the pathway, and colored arrows represent inhibition. E inhibits one of the enzymes, and G inhibits the other. A high concentration of E *or* G therefore slows the rate of production of C, and a high concentration of E *and* G nearly eliminates production of C. The first reaction *after* the branch point is also inhibited. E inhibits its own formation a second time by slowing the rate of conversion of C to D, while G inhibits the rate of conversion of C to F.

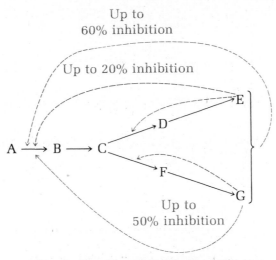

Figure 13.6 Cumulative feedback inhibition.
Black arrows represent enzyme-catalyzed reactions
in the pathway, and colored arrows represent
inhibition. Suppose that the enzyme that catalyzes
the conversion of A to B is saturated with A, that
saturating the allosteric binding sites of this
enzyme with E inhibits the rate of conversion of A
to B by 20%, and that saturating the allosteric
binding sites for G inhibits the rate of conversion of
A to B by 50%. The maximum velocity for the
conversion of A to B is then $0.8V \times 0.5V = 0.4V$.

inhibitor acts independently of the others, and the effects are *cumulative*.
The inhibition caused by a mixture of products can be calculated by the inhibition that each causes separately. Suppose that saturating concentrations of
end product E cause the maximum rate of the conversion of A to B to decrease from V to 0.8 V (a 20% decrease in V), and that saturating concentrations of G cause the rate of conversion of A to B to decrease from V to 0.5 V (a
50% decrease in V). In the presence of saturating, or "infinite," concentrations of end products E *and* G, the velocity for the conversion of A to B is
$0.8\ V \times 0.5\ V = 0.4\ V$. Under normal metabolic conditions neither E nor G
will be present in infinite concentrations, but since the effects of the inhibitors are independent of each other, the inhibition increases as the concentrations of E and G increase.

If there are many products in a branched pathway, as there are in biosynthetic pathways leading from glutamine (Section 22.4B), cumulative feedback inhibition progressively decreases production of the branch-point metabolite. Since each end product inhibits its own branch, the diminished
supply of branch-point metabolite is diverted to the branch that is still
operating.

c. Sequential Feedback Inhibition

In sequential feedback inhibition the branch-point metabolite inhibits
the first committing step by negative feedback inhibition. The end products
are negative feedback inhibitors of the first reaction after the branch point
along the pathway leading to their own formation (Figure 13.7). The inhibition is *sequential* because accumulation of E, for example, causes a decrease
in the rate at which C is converted to D. C therefore accumulates and inhibits
its own formation by negative feedback inhibition.

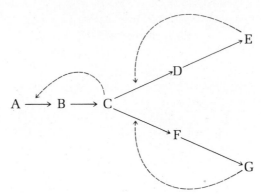

Figure 13.7 Sequential feedback inhibition. Black arrows represent enzyme-catalyzed reactions, and colored arrows represent inhibition.

d. Concerted Feedback Inhibition

When the branch-point intermediate of a branched pathway inhibits the first committing step and when a combination of end products act together to inhibit the first committing step, the inhibition is called *concerted feedback inhibition* (Figure 13.8). The products of each branch inhibit the first step after the branch. The biosyntheses of the products of each branch are coupled to each other. The presence of one product is not able to inhibit the first common, committing step, so the branch-point metabolite is continuously synthesized.

Only when both products accumulate will the first common step be inhibited.

e. Synergistic Feedback Inhibition

In synergistic feedback inhibition the end products of the branches inhibit the first committing step of the pathway, and the total inhibition caused by the mixture of the products is greater than the sum of the inhibitions caused by each product acting separately. The inhibition is therefore *synergistic* (Figure 13.9). The end products also inhibit their own formation by inhibiting the first reaction after the branch point.

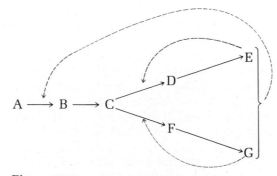

Figure 13.8 Concerted feedback inhibition. Black arrows represent enzyme-catalyzed reactions, and colored arrows represent feedback inhibition. Only the action of E and G in concert can inhibit the first committing step; either one alone has no effect on the enzyme that catalyzes the conversion of A to B.

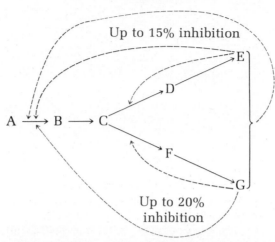

Up to 100% inhibition

Up to 15% inhibition

Up to 20% inhibition

Figure 13.9 Synergistic feedback inhibition. Black arrows represent enzyme-catalyzed reactions, and colored arrows represent inhibition. The total inhibition caused by E and G acting together is greater than the sum of the inhibition caused by each acting separately. The inhibition is synergistic because E inhibits both the formation of G and its own formation by acting on the enzyme which converts A to B, and G acts similarly on E and itself. The percents inhibition represents a hypothetical case that would be observed for saturating concentrations of allosteric effectors E and G and saturating concentrations of the substrate A.

13–5 ADENYLATE ENERGY CHARGE AND PHOSPHORYLATION STATE RATIO

ATP, ADP, and AMP are allosteric effectors for many enzymes. Their concentrations also reflect the energy state of the cell. In some respects the cellular energy economy resembles an electric storage battery in which the state of the *adenylate system* — ATP, ADP, AMP, phosphate, and Mg(II) — is analogous to the charge of the battery. A cell is "fully charged" when the adenylate system consists only of ATP and is completely discharged when it consists only of AMP and phosphate. The *adenylate energy charge* is defined as

$$\text{Adenylate energy charge} = \frac{[\text{ATP}] + \frac{1}{2}[\text{ADP}]}{[\text{AMP}] + [\text{ADP}] + [\text{ATP}]} \qquad (13.3)$$

ADP is considered half-charged because it has one-half as many phosphoanhydride bonds as ATP. The adenylate energy charge varies between 0, when only AMP is present, and 1.0, when only ATP is present. It has been suggested, and widely accepted, that the adenylate energy charge plays an important role in metabolic regulation. The rates of catabolic, ATP-regenerating pathways (R pathways) and biosynthetic, ATP-utilizing pathways (U pathways) as a function of the adenylate energy charge are plotted in Figure 13.10. The two curves cross when the energy charge equals about 0.9. In most cells at most times the energy charge is "buffered" between 0.7 and 0.9. This is also the region in which the slopes of the curves for R and U pathways are steepest. Therefore, a small change in the energy charge has a large effect on the rates of R and U pathways in the region of energy charge between 0.7 and 0.9.

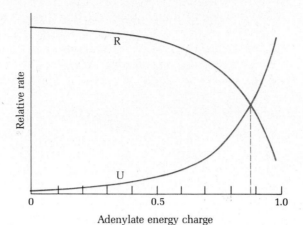

Figure 13.10 Effects of the adenylate energy charge on the relative rates of ATP-regenerating pathways (R) and ATP-utilizing pathways (U). [From D. E. Atkinson, *Biochemistry*, 7, 4030 (1980). Reprinted with permission. Copyright 1980 American Chemical Society.]

Let us consider the effects of ATP, ADP, and AMP on metabolic regulation in greater detail. We must consider not only the ratio that defines the adenylate energy charge, but also the concentrations of each of the adenosine nucleotides, the pH of the system, and the concentrations of the non-adenylated products. The state of the adenylate system also depends upon *adenylate kinase*[1] (reaction 13.4).

$$\text{ATP} + \text{AMP} \underset{}{\overset{\overset{\text{Adenylate}}{\text{kinase}}}{\rightleftharpoons}} 2\ \text{ADP} \tag{13.4}$$

The mass action ratio, $[\text{ADP}]^2/[\text{ATP}][\text{AMP}]$, can vary from 0.2 to 1.0 depending upon pH and the concentration of Mg(II). Over this range $[\text{ATP}]/[\text{AMP}]$ and $[\text{ATP}]/[\text{ADP}]$ ratios can vary by about a factor of 2 at constant energy charge. More experimental evidence is required to establish the role of the adenylate energy charge in metabolic regulation.

The adenylate energy charge is not a thermodynamic parameter. For a thermodynamic parameter that reflects the state of the adenylate system we turn to the *phosphorylation state ratio*, R_p (equation 13.5).

$$R_p = \frac{[\text{ATP}]}{[\text{ADP}][\text{P}_i]} \tag{13.5}$$

R_p is a thermodynamic parameter since it can be related to the free energy of hydrolysis of ATP (equation 13.6).

$$\text{ATP} + \text{H}_2\text{O} \longrightarrow \text{ADP} + \text{HPO}_4^{2-} \tag{13.6}$$

The free-energy change for reaction 13.6 is

$$\Delta G' = \Delta G^{\circ\prime} + 2.303RT \log \frac{[\text{ADP}][\text{P}_i]}{[\text{ATP}]} \tag{13.7}$$

Substituting in equation 13.7, we obtain

$$\Delta G' = \Delta G^{\circ\prime} - 2.303RT \log R_p \tag{13.8}$$

---

[1] Adenylate kinase is also sometimes known as myokinase.

The phosphorylation state ratio can therefore be used in thermodynamic calculations, whereas the adenylate energy charge cannot. Under certain conditions the phosphorylation state ratio can climb as high as 10^5, a contribution of -22.5 kJ mol^{-1} to the free energy of hydrolysis of ATP. A high phosphorylation state ratio corresponds to a high adenylate energy charge.

13–6 REGULATION OF ENZYME ACTIVITY BY PROTEIN MODIFICATION

A wide variety of cellular regulatory agents—neurotransmitters, steroid hormones, insulin, calcium ions, and others—exert their effects through a protein phosphorylation system (Figure 13.11). These chemical messengers act through a *second messenger*, adenosine 3',5'-monophosphate (cyclic AMP or cAMP).

A general diagram of the protein phosphorylation system, capable of diverse metabolic and physiological activities, is shown in Figure 13.12. The first stage in this cascade of enzyme activities is activation of a *cAMP-dependent protein kinase*. A hormone or neurotransmitter binds to a specific receptor protein located on the *outer* side of the plasma membrane. The first messenger never enters the target cell.

The hormone-receptor complex diffuses to *adenylate cyclase* in the plane of the bilayer. Adenylate cyclase contains at least two subunits. One possesses catalytic activity, and the other is a regulatory protein. The regulatory protein binds GTP. Only then does the catalytic protein become active, forming cAMP from ATP (reaction 13.9).

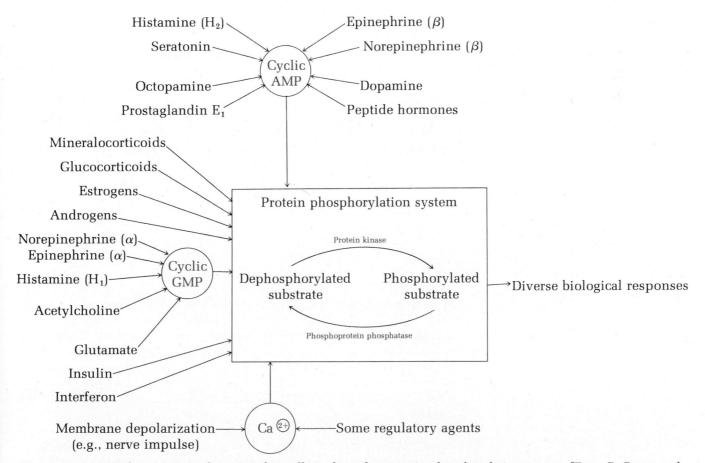

Figure 13.11 Regulatory agents that exert their effects through a protein phosphorylation system. [From P. Greengard, *Science*, **199**, 148 (1978). Copyright 1978 by the American Association for the Advancement of Science.]

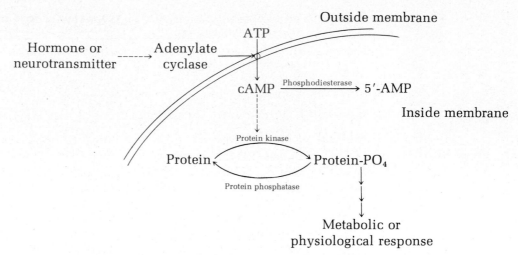

Figure 13.12 Schematic diagram of the role played by protein phosphorylation in mediating the biological effects of hormones and neurotransmitters. [From P. Greengard, *Science, 199,* 148 (1978). Copyright 1978 by the American Association for the Advancement of Science.]

The regulatory protein can hydrolyze GTP, and once GTP has been hydrolyzed, the catalytic subunit is no longer active. An enormous range of hormones also alter the activity of adenylate cyclase in higher organisms. These hormonal effects are summarized in Table 13.1.

$$\text{(13.9)}$$

Cyclic AMP (cAMP)

The second stage in the enzyme cascade is activation of a protein kinase by cAMP. Cyclic AMP–dependent protein kinases are tetramers composed of two catalytic and two regulatory subunits. Cyclic AMP binds to the regulatory subunits, and the inactive protein dissociates into active catalytic subunits (C) and regulatory subunits (R) bound to cAMP (reaction 13.10).

$$\underset{\text{Inactive}}{R_2C_2} + 2\ cAMP \longrightarrow \underset{\text{Active}}{2\ C} + (R\text{-}cAMP_2)_2 \qquad \text{(13.10)}$$

The newly phosphorylated protein produces a physiological or metabolic response. A *phosphatase* that is specific for the protein phosphorylated by protein kinase reverses the effects of the cAMP-dependent protein kinase. The metabolic details of protein phosphorylation have been elucidated for

Table 13.1 Adenylate cyclase systems influenced by hormones

Tissue or cells	Hormone	Effect on adenyl cyclase system
Erythrocytes	Catecholamines (β)	Stimulation
Blood platelets	Catecholamines (α)	Inhibition
Melanocytes	MSH	Stimulation
	Catecholamines (α)	Inhibition
Parotid gland	Catecholamines (β)	Stimulation
Pineal gland	Catecholamines (β)	Stimulation
Spleen	Epinephrine	Stimulation
Lung	Epinephrine	Stimulation
Liver	Catecholamines (β), glucagon	Stimulation
Heart muscle	Catecholamines (β), glucagon, T_4	Stimulation
	Acetylcholine	Inhibition
Skeletal muscle	Catecholamines (β)	Stimulation
Smooth muscle	Catecholamines (β)	Stimulation
	Catecholamines (α)	Inhibition
Brain	Catecholamines (β), serotonine	Stimulation
Pituitary	Hypothalamic releasing hormones	Stimulation
Pancreas	Glucagon, catecholamines (β)	Stimulation
	Catecholamines (α)	Inhibition
Adipose tissue (white)		
Rat	Catecholamines (β), ACTH, glucagon, LH, TSH, secretin	Stimulation
Rabbit	ACTH, MSH (α,β), LPH (β), glucagon, catecholamines (β)	Stimulation
Human	Catecholamines (β)	Stimulation
	Catecholamines (α)	Inhibition
Adipose tissue (brown)		
Rat	Catecholamines	Stimulation
Kidney	PTH, calcitonin, OT, VP	Stimulation
Bone	PTH, calcitonin	Stimulation
Ovary	LH	Stimulation
Testis	FSH, LH, hCG	Stimulation
Adrenal	ACTH	Stimulation
Thyroid	TSH	Stimulation

Abbreviations: MSH, melanocyte-stimulating hormone; T_4, thyroxine; ACTH, adrenocorticotripin; LH, luteinizing hormone; TSH, thyroid-stimulating hormone; LPH, lipotropic hormone; PTH, parathormone; OT, oxytocin; VP, vasopressin; FSH, follicle-stimulating hormone; LCG, human choriogonadotropin.

SOURCE: From T. Braun and L. Birnbaumer, in M. Florkin and E. H. Stotz, Eds., "Comprehensive Biochemistry," Vol. 25, Elsevier, New York, 1975, p. 70.

several systems, including glycogen metabolism (Section 17.4) and fatty acid degradation (Section 19.2).

Cyclic nucleotide phosphodiesterase converts cAMP to AMP, limiting the lifetime and activity of cAMP (reaction 13.11).

All major classes of steroid hormones—mineralocorticoids, glucocorticoids, androgens, and estrogens—stimulate protein phosphorylation systems mediated by cyclic AMP. The steroid hormones stimulate protein synthesis (Chapter 26) through the general scheme outlined in Figure 13.13. Certain animal viruses (RNA tumor viruses) elicit production of viral protein kinases that act on other viral proteins. Protein phosphorylation thus plays a fundamental role in biological regulation in normal cells, but viral protein phosphorylation might disrupt homeostasis in rapidly multiplying tumor cells.

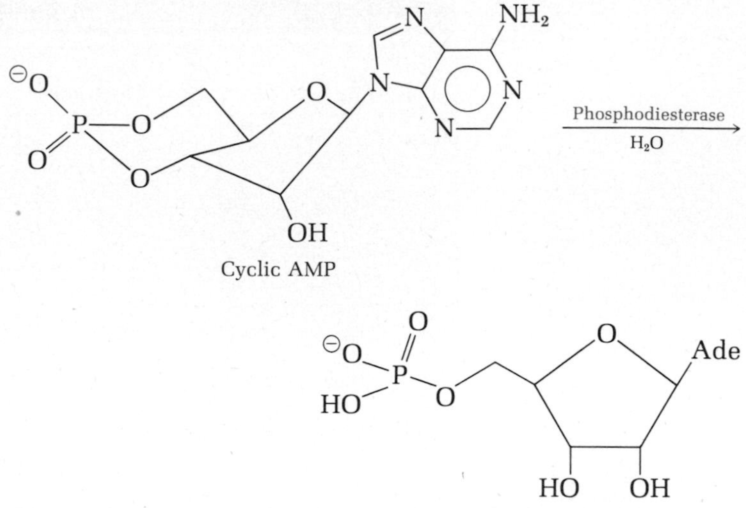

Cyclic AMP

$$\xrightarrow[\text{H}_2\text{O}]{\text{Phosphodiesterase}}$$

(13.11)

5′-Adenylic acid (AMP)

Protein phosphorylation is but one of more than a hundred known covalent modifications of proteins. Many of these covalent modifications alter metabolic activity. The protein phosphorylation scheme we have discussed is the best understood general system of metabolic regulation by covalent modification.

13–7 REGULATION OF ENZYME ACTIVITY BY Ca(II) IONS: CALMODULIN

Many biological phenomena are triggered by transient increases in the Ca(II) ion concentration, among them glycogen and lipid degradation, the release of chemical transmitters by nerves, muscle contraction, cell division, and the beating of the cilia and flagella that propel single-celled organisms and sperm.

All eucaryotic cells contain the calcium-binding protein *calmodulin*, so named because it modulates the effects of Ca(II) ions on the cell (Figure 13.14). Among the enzymes known to be directly affected by calmodulin are cyclic nucleotide phosphodiesterase, brain adenylate kinase, the Ca(II)-dependent ATPase of the plasma membrane, myosin light chain kinase, phosphorylase *b* kinase, phospholipase A$_2$, and plant NAD kinase, the en-

Steroid + receptor
↓
Steroid-receptor complex (cytoplasm)

Steroid-receptor complex (nucleus)
↓
Messenger RNA synthesis
↓
Protein synthesis
↓
Regulatory substance
↘
cAMP-dependent protein kinase activity
↙↓↘
Diverse physiological responses

Figure 13.13 Chain of events by which steroid hormones influence cAMP-dependent protein kinase activity. [Adapted from P. Greengard, *Science, 199,* 149 (1978). Copyright 1978 by the American Association for the Advancement of Science.]

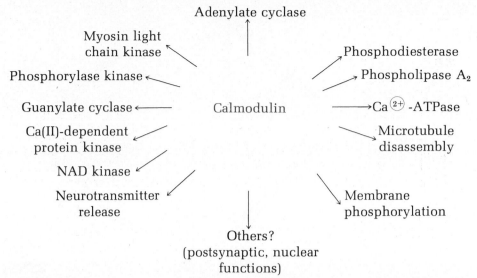

Figure 13.14 Enzymes and physiological processes regulated by calmodulin. [From W. Y. Cheung, *Science, 207,* 22 (1980). Copyright 1980 by the American Association for the Advancement of Science.]

zyme that converts NAD to NADP. The list of proteins affected by calmodulin is growing steadily as knowledge about this protein continues to expand.

Calmodulin is a small protein (mol wt 16,790). The amino acid sequences of calmodulin from all sources are virtually identical. Figure 13.15 shows the primary structure of bovine brain calmodulin, including regions of α-helix and the calcium ion–binding sites. Each calcium ion-binding domain, contains 12 amino acid residues. Domains I and III and domains II and IV are especially closely related. This similarity in primary structure is reflected in the tertiary structure. Each domain consists of two helices arranged in a *hand* arrangement as shown in Figure 13.16. The dissociation constants for the four calcium ion–binding sites are in the range of 10^{-6} M. Binding of Ca(II) induces large conformational changes in calmodulin. Although the dissociation constants for the 4 Ca(II) ions are in the same range, they seem to differ from one another. The α-helical content of the protein increases by 5 to 10% in going from a state in which no Ca(II) ions are bound to the fully occupied state. Each new Ca(II) ion that binds to calmodulin induces its own conformational change, and perhaps it is this property that enables calmodulin to affect so many diverse enzymes.

The interaction of calmodulin with the enzymes that it regulates occurs in at least two steps. First, binding of Ca(II) generates an activated intermediate (reaction 13.11).

$$\underset{\substack{\text{Calmodulin}\\\text{(inactive}\\\text{conformation)}}}{\text{CaM}} + n\,\text{Ca(II)} \rightleftharpoons \text{CaM-Ca(II)}_n \rightleftharpoons \underset{\substack{\text{Calmodulin}\\\text{(active}\\\text{conformation)}}}{\text{CaM*-Ca(II)}_2} \qquad (13.11)$$

Second, enzyme activation occurs when the active conformation of calmodulin binds to a specific enzyme, a process that results in a new conformation in the target enzyme (reaction 13.12).

$$\underset{\substack{\text{Inactive}\\\text{enzyme}}}{\text{enzyme}} + \text{CaM*-Ca(II)}_2 \rightleftharpoons \underset{\substack{\text{Active enzyme-}\\\text{calmodulin complex}}}{\text{enzyme*-CaM*-Ca(II)}_2} \qquad (13.12)$$

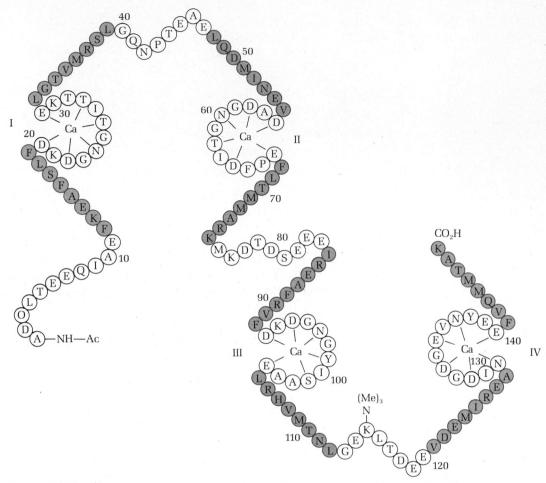

Figure 13.15 Primary structure of calmodulin. The sequence is shown with one-letter abbreviations for amino acid residues: A, Ala; D, Asp; E, Glu; F, Phe; G, Gly; H, His; I, Ile; K, Lys; L, Leu; M, Met; N, Asn; P, Pro; Q, Gln; R, Arg; S, Ser; T, Thr; V, Val; Y, Tyr. The four Ca(II)-binding domains are indicated. Colored circles are regions of α-helix. There is considerable sequence homology between domains I and III and between domains II and IV. [From C. B. Klee, T. H. Crouch, and P. G. Richman, *Ann. Rev. Biochem.*, 49, 496 (1980). Reproduced with permission. © 1980 by Annual Reviews Inc.]

Calmodulin binds Ca(II)-dependent cyclic nucleotide phosphodiesterase. The phosphodiesterase is a dimer each of whose subunits binds to one molecule of calmodulin. Cyclic nucleotide phosphodiesterase inactivates cAMP by hydrolysis to AMP, and in some tissues calmodulin acts in opposition to the protein phosphorylation scheme we described. In brain tissue, however, calmodulin stimulates *both* adenylate cyclase and cyclic nucleotide phosphodiesterase. This situation, in which the same protein stimulates enzymes that act counter to each other, is less paradoxical than it seems because calmodulin has different affinities for the two enzymes. By unequally stimulating cAMP production and cAMP hydrolysis, calmodulin can maintain a relatively constant cAMP concentration. Since the phosphodiesterase has different K_m values for cAMP ($K_m = 10^{-4}\ M$) and cGMP ($K_m = 2 \times 10^{-5}\ M$), the relative amounts of cAMP and cGMP are altered, and this too may be important in metabolic regulation.

Most of the effects of cAMP in mammals are mediated by protein phosphorylation. The function of the nervous system, however, is intimately related to Ca(II) metabolism, whose effects are modulated by calmodulin.

The release of Ca(II) from the membrane of nerve cells or from the sarcoplasmic reticulum of muscle cells activates calmodulin, which in turn ac-

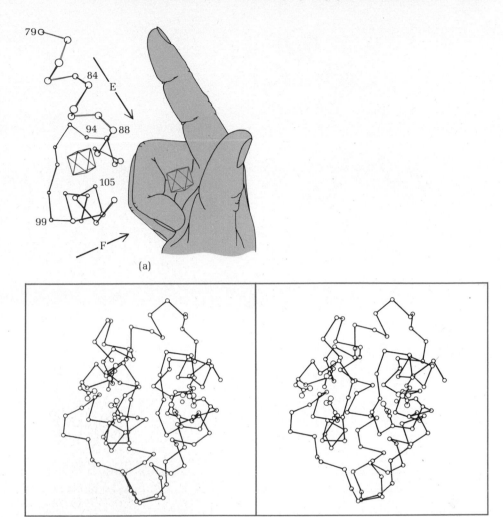

(b)

Figure 13.16 (a) Diagram of the Ca^{+2} binding site found in many calcium ion-binding proteins. One helix corresponds to the thumb, and the second helix to the index finger. The octahedron is the calcium-binding site. [From *Science, 208,* 275 (1980). Copyright 1980 by the American Association for the Advancement of Science.]

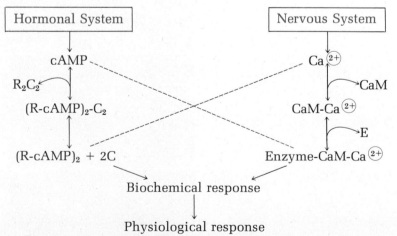

Figure 13.17 Interaction between hormonal systems and the nervous system. Hormonal systems are regulated by cAMP-mediated protein phosphorylation. The nervous system is regulated by calmodulin, whose effects are also mediated by cAMP. [From W. Y. Cheung, *Science, 207,* 25 (1980). Copyright 1980 by the American Association for the Advancement of Science.]

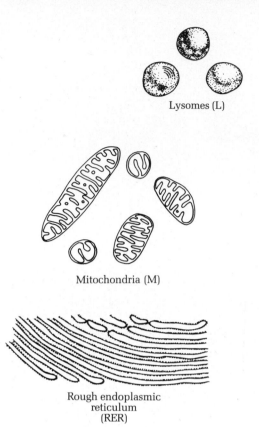

Lysomes (L)

Mitochondria (M)

Rough endoplasmic
reticulum
(RER)

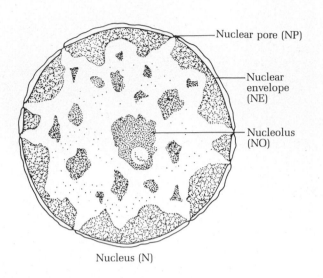

Nuclear pore (NP)

Nuclear
envelope
(NE)

Nucleolus
(NO)

Nucleus (N)

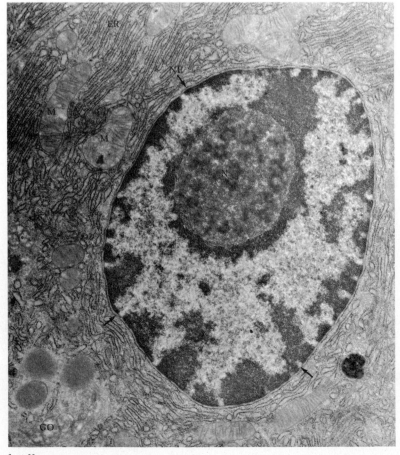

Figure 13.18 (a) Major structural features of animal cells.

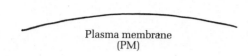

Plasma membrane
(PM)

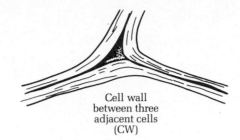

Cell wall
between three
adjacent cells
(CW)

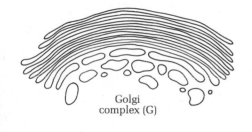

Smooth endoplasmic
reticulum
(SER)

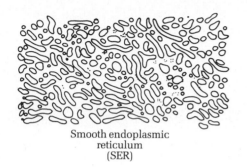

Golgi
complex (G)

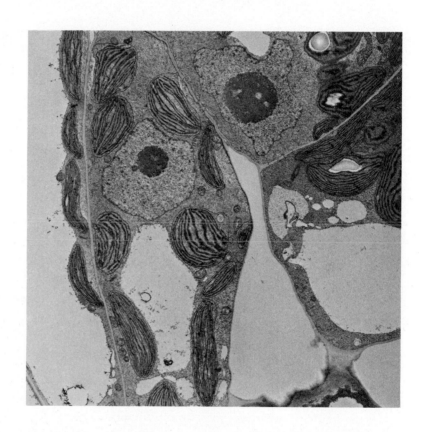

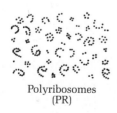

Polyribosomes
(PR)

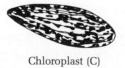

Chloroplast (C)

Figure 13.18   (b) Major structural features of plant cells.

tivates various enzymes. The nervous system can act independently, cooperatively, or antagonistically to the hormonal system stimulated by adenylate cyclase. The related effects of cAMP and Ca(II) are summarized in Figure 13.17.

13–8 COMPARTMENTATION OF METABOLIC ACTIVITY

The interior of eucaryotic cells contains many specialized, membrane-bound cell organelles that divide the labor of metabolism. Figure 13.18 shows the major features common to plant and animal cells. The compartmentation of metabolic activity provides a simple means of metabolic control since it isolates processes that might otherwise interfere with one another.

A. Nucleus

The nucleus is the "control center" of the eucaryotic cell. It contains over 95% of the cellular complement of DNA. A double-membrane structure, called the *nuclear envelope*, separates the nucleus from the cytosol. This membrane structure controls the balance of $Na^{\oplus}$ and $K^{\oplus}$ ions in the nucleus. It also controls the transport of proteins, RNA, and nucleic acid–protein complexes of enormous molecular weight. The transport of these large particles is controlled by octahedral arrays of proteins called *pore complexes* (Figure 13.19). The DNA is coiled in the nucleus in a dense mass called *chromatin*. In *interphase*, the period of the cell cycle during which the cell is not dividing, the nucleus carries out the synthesis of RNA transcripts of chromatin. This is the first step of protein synthesis and is the major metabolic activity of the nucleus. We shall discuss the structure of chromatin further in Section 24.5A.

Another major feature of the nucleus is a region called the *nucleolus*. The nucleolus is responsible for biosynthesis of ribosomal RNA.

B. Mitochondria and Chloroplasts

The mitochondrion (Section 16.3) is the organelle responsible for most cellular energy production. The interior or *matrix space* of the mitochondrion is enclosed by two membranes, separated by an intermembrane space (Figure 13.20). The enzymes responsible for oxidative phosphorylation (Chapter 16) are located in the inner membrane. The matrix space is the site of the enzymes responsible for the citric acid cycle (Chapter 15) and β-oxidation of fatty acids (Chapter 19) and of transaminases that couple amino acid catabolism to carbohydrate degradation (Chapter 21). The mitochondrion is partially autonomous. It has its own genetic system, and synthesizes some of its own proteins.

The chloroplast is the major energy-producing organelle of higher plants, certain algae, ferns, and mosses. The membrane systems of chloroplasts (Figure 13.21) contain the enzyme system responsible for light-driven phosphorylation of ADP (Chapter 18).

We shall discuss the structures of mitochondria and chloroplasts and their relation to metabolic function in detail in later chapters. At present we note only that both of these organelles require intact membrane structures for biological activity.

C. Endoplasmic Reticulum and Golgi Bodies

An extensive membrane network called the *endoplasmic reticulum* (ER) wends its way through most eucaryotic cells (Figure 13.22). The interior space of this membrane is called the *cisterna*. The cisternal space provides a channel through which newly synthesized material travels. Most cells contain regions of densely packed membranes, apparently continuous with the

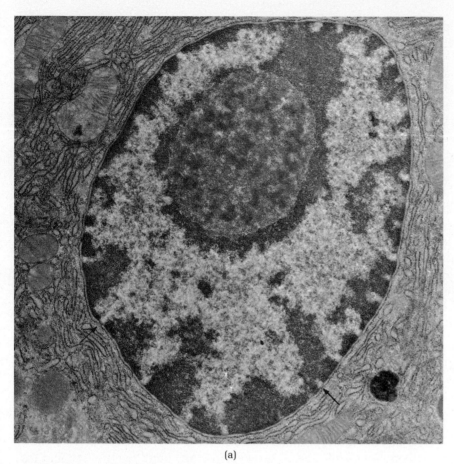

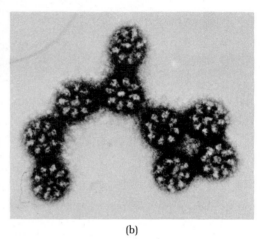

(a)

(b)

Figure 13.19   (a) Electron micrograph of the nucleus. The arrows point at pore complexes. (b) Electron micrograph of pore complexes.

endoplasmic reticulum, called *Golgi bodies*, that store newly synthesized molecules for eventual secretion.

Some regions of the endoplasmic reticulum (rough endoplasmic reticulum) are studded with ribosomes, the site of protein synthesis. Regions without ribosomes (smooth endoplasmic reticulum) are the site of lipid biosynthesis (Chapter 20). The smooth endoplasmic reticulum is also the site of covalent modifications of proteins that are synthesized prior to their secretion. A general flow sheet of protein modification and excretion is shown in Figure 13.23. Centrifugation of cell homogenates at forces of about 100,000 × g yields pellets that contain the rough and smooth endoplasmic reticulum plus Golgi bodies. The components of this fraction are collectively known as *microsomes*. The microsomal fraction is a rich source of many enzymes.

D. Lysosomes and Vacuoles   Most animal cells contain digestive organelles called *lysosomes* containing a potent array of hydrolytic enzymes that digest proteins, nucleic acids, glycogen, and mucopolysaccharides. Lysosomes digest bacteria or other particles ingested by the cell by *phagocytosis*. The membrane of the lysosome fuses with the membrane surrounding the captured particle to form a secondary lysosome which is responsible for the initial degradation of the particle. Once the harmful captive has been hydrolyzed, the monomers can be catabolized by other metabolic pathways outside of the lysosome. When a eucaryotic cell dies, its lysosomes rupture, and the enzymes that are released hydrolyze the dead cell.

The plant organelles that contain digestive enzymes and in that respect resemble lysosomes are vacuoles. Vacuoles, whose name unfortunately implies that they are empty, also contain the pigments that give plants their characteristic colors.

E. Microbodies: Peroxisomes and Glyoxysomes

The term microbody encompasses the similar cell organelles known as peroxisomes and glyoxysomes. About 40 enzymes have been isolated from microbodies from various sources. Most of the reactions that occur in microbodies are oxidation reactions that produce toxic hydrogen peroxide and superoxide, $O_2^{\ominus}$. The enzymes catalase and superoxide dismutase rapidly degrade these toxic metabolic by-products. The sequestering of oxidative pathways within specialized cell organelles enables the cell to product itself from highly toxic intermediates. Microbodies also contain many other oxidases and carry out such metabolic interconversions as fatty acid oxidation and the glycoxylate cycle.

F. Cytosol

The *cytosol,* or cell sap, is the fluid in which cell organelles are bathed. The cytosol contains the enzymes responsible for glycolysis (Chapter 14), gluconeogenesis, glycogen metabolism, and the pentose phosphate pathway (Chapter 17), and the enzyme complex responsible for fatty acid biosynthesis (Chapter 19). The enzymes responsible for many other processes, including heme biosynthesis, catabolism of some amino acids, and pyrimidine catabolism, are also found in the cytosol. Although we are tempted to view the cytosol as a structureless fluid, it may have some delicate structural features that are destroyed by even the mildest techniques of cell fractionation.

13-9 SUMMARY

A metabolic pathway consists of a set of reactions in which a precursor is converted to a product. The individual steps of a pathway may be delineated by following a chemical marker or a radioisotopic label, such as ^{3}H or ^{14}C, through the various intermediates of the pathway. Most metabolic pathways are unidirectional, and while each enzyme-catalyzed reaction is thermodynamically reversible, the overall equilibrium constant for the pathway, the product of the equilibrium constants of the individual steps of the pathway, lies overwhelmingly on the side of the final product of the pathway. This unidirectionality prevents a futile cycling of metabolites and the loss of free energy such a cycling would entail. A committing step is identifiable in most metabolic pathways. The first committing step of a pathway is usually the first step that involves a large free-energy change. Biosynthetic and catabolic pathways are distinct in most cases. In eucaryotic cells pathways are sometimes further separated by being compartmentalized in different cell organelles. The oxidation of fatty acids, for example, occurs in the mitochondrion

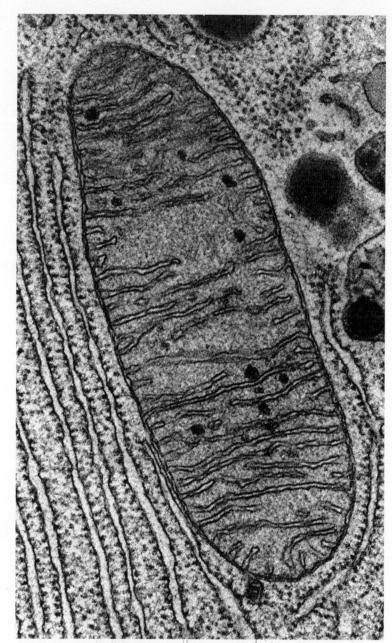

Figure 13.20 Electron micrograph of a mitochondrion. [Courtesy of Dr. Keith R. Porter.]

and the synthesis of fatty acids occurs in the cytosol. Catabolic pathways, in general, are convergent, and many precursors are channeled into a few central metabolites such as acetyl-CoA, pyruvate, and a few intermediates of the citric acid cycle. Biosynthetic pathways, by contrast, are highly branched. Acetyl-CoA, for example, is a precursor in the biosynthesis of a wide variety of metabolites.

The regulatory enzymes of many metabolic pathways are controlled by

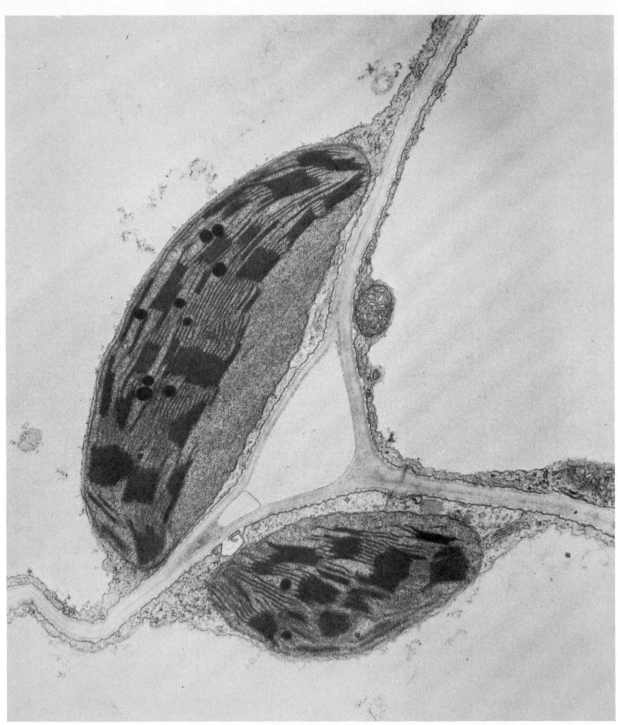

Figure 13.21 Electron micrograph of a chloroplast. The densely folded inner membrane is the site of the enzymes that carry out the photophosphorylation of ADP.

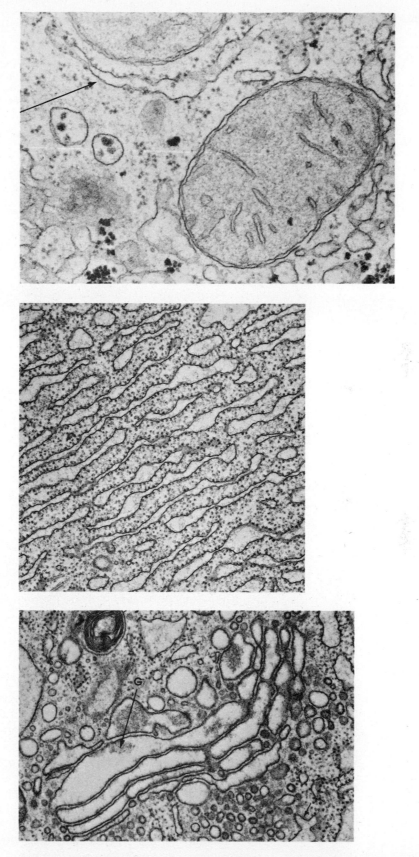

Figure 13.22 Electron micrographs of (a) the smooth endoplasmic reticulum (region indicated by arrow), the site of lipid biosynthesis; (b) the rough endoplasmic reticulum, where proteins are synthesized; and (c) Golgi bodies (G), which store proteins for export and in which certain covalent modifications of proteins are carried out.

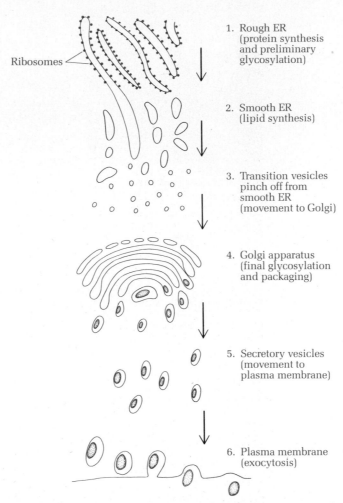

1. Rough ER
(protein synthesis
and preliminary
glycosylation)

2. Smooth ER
(lipid synthesis)

3. Transition vesicles
pinch off from
smooth ER
(movement to Golgi)

4. Golgi apparatus
(final glycosylation
and packaging)

5. Secretory vesicles
(movement to
plasma membrane)

6. Plasma membrane
(exocytosis)

Ribosomes

Figure 13.23 Sequence of steps by which proteins are synthesized, modified and stored for secretion by the action of enzymes located in the rough endoplasmic retidulum, the smooth endoplasmic reticulum, and Golgi bodies. [From D. L. Kirk, "Biology Today," 3rd ed., Random House, New York, 1980, p. 315.]

feedback inhibition by the final product of the pathway. Branched pathways may be regulated through catalysis by two parallel enzymes in the first committing step. The product of one branch of the pathway regulates one of the parallel enzymes and the product of the other branch regulates the second. In cumulative feedback inhibition each end product of a branched pathway partially inhibits the first committing step of the pathway. In sequential feedback inhibition each end product of a branched pathway inhibits the first unique step after the branch-point. The branch-point metabolite accumulates as a result, and it in turn inhibits the first committing step of the pathway. In concerted feedback inhibition each product of a branched pathway inhibits the enzyme catalyzing the first unique step leading to its own formation and each end product also inhibits the first committing step of the pathway. Synergistic feedback inhibition differs from concerted feedback inhibition since the total inhibition caused by the end products is greater than the sum of the inhibitors caused by the end products acting separately.

The adenylate system, consisting of ATP, ADP, AMP, phosphate, and Mg(II), is analogous to a storage battery that is fully charged when only ATP is present. The adenylate energy charge is defined as the sum of the ATP concentration and one-half the ADP concentration divided by the sum of the ATP, ADP, and AMP concentrations. The value of the adenylate charge can vary from 0 to 1.0. In most cells at most times the adenylate energy charge is "buffered" at about 0.9. The adenylate energy charge is not a thermodynamic parameter, but the phosphorylation state ratio R_p is. A high phosphorylation state ratio corresponds to a high adenylate energy charge.

Covalent modification of proteins by phosphorylation is an important regulatory mechanism for many eucaryotic cells. Protein phosphorylation is often mediated by hormonal activation of adenylate cyclase. The product of adenylate cyclase activity, cAMP, mediates the activity of a protein phosphorylation system consisting of a kinase and a phosphorylase.

All eucaryotic cells contain the calcium-binding protein calmodulin. The calcium ion concentration regulates an extraordinary range of physiological phenomena. Calmodulin binds to a calcium ion–dependent cyclic nucleotide phosphodiesterase which then hydrolyzes cAMP. In many cases calmodulin thus acts in opposition to the general scheme of protein phosphorylation that is triggered by synthesis of cAMP.

Metabolic activity in eucaryotic cells is often compartmentalized. The nucleus is the control center of the cell and contains most of the cell's complement of DNA. Most cellular mRNA is synthesized in the nucleus during interphase. Most cellular energy production occurs in mitochondria. Oxidative phosphorylation, fatty acid oxidation, the citric acid cycle, and several transaminases which couple amino acid catabolism to carboxylate metabolism are among the metabolic pathways located in mitochondria. Chloroplasts are plant cell organelles that carry out the photophosphorylation of ADP. The endoplasmic reticulum contains many membrane-bound enzymes. The interior channel of the endoplasmic reticulum is a conduit through which newly synthesized molecules travel. Golgi bodies are densely packed membranes that store newly synthesized molecules for eventual secretion. Lipid biosynthesis occurs in the smooth endoplasmic reticulum. The rough endoplasmic reticulum, studded with ribosomes, is the site of protein synthesis. Lysosomes contain many hydrolytic enzymes that are capable of degrading virtually all types of biological macromolecules. Vacuoles are plant cell organelles that contain both digestive enzymes and the pigments that give plants their colors. Microbodies (peroxisomes and glyoxysomes) catalyze many biological oxidations. The cytosol, the fluid in which the various cell organelles are immersed, contains enzymes involved in many metabolic pathways.

REFERENCES

Boyer, P. D., Ed., "The Enzymes," 3rd ed., Vol. 1, Academic Press, New York, 1970.

Builder, S. E., Beavo, J. A., and Krebs, E. G., The Mechanism of Activation of Bovine Skeletal Muscle Protein Kinase by Adenosine 3'-5'-monophosphate, *Proc. Nat. Acad. Sci. U.S.*, 255, 3514–3519 (1980).

Cheung Wai Yiu, Calmodulin Plays a Pivotal Role in Cellular Regulation, *Science*, 207, 19–27 (1980).

Chock, P. B., Rhee, S. G., Stadtman, E. R., Interconvertible Enzyme Cascades in Cellular Regulation, *Ann. Rev. Biochem.*, 49, 813–845 (1980).

Greengard, P. Phosphorylated Proteins as Physiological Effectors, *Science*, 199, 146–152 (1978).

Klee, C. B., Crouch, T. H., and Richman, P. G., Calmodulin, *Ann. Rev. Biochem.*, 49, 489–517 (1980).

Purich, D. L., and Fromm, D. L., Additional Factors Influencing Enzyme Responses to the Adenylate Energy Charge, *Proc. Nat. Acad. Sci. U.S.*, 248, 461–466 (1980).

Ross, E. M., and Gilman, A. G., Biochemical Properties of Hormone-Sensitive Adenylate Cyclase, *Ann. Rev. Biochem.*, 49, 533–565 (1980).

Problems

1. Adenylate kinase catalyzes phosphoryl transfer between MgATP and AMP.

$$\text{MgATP} + \text{AMP} \rightleftharpoons \text{MgADP} + \text{ADP}$$

The apparent equilibrium constant for this reaction is

$$K_{app} = \frac{[\text{ATP}_{total}][\text{AMP}_{total}]}{[\text{ADP}_{total}]^2}$$

a. The equilibrium constant for the adenylate kinase reaction has been variously reported between 0.236 and 1.0. Since enzymes do not alter the equilibrium position of the reactions they catalyze, and since equilibrium constants are true constants, how is such a difference to be explained?

b. How will increasing the magnesium ion concentration affect the adenylate kinase equilibrium? (Recall equations 12.30–12.32.)

2. Consider an enzyme that is subject to cumulative feedback inhibition. Referring to Figure 13.6, calculate the percent inhibition obtained if a saturating concentration of G causes 50% inhibition and a saturating concentration of E causes 40% inhibition.

3. The standard free energy of hydrolysis of ATP at pH 7.0 in the presence of 1 mM Mg(II) is -36.8 kJ mol^{-1}. What phosphorylation state ratio (R_p) is required to drive the hydrolysis of ATP in the direction of net synthesis of ATP?

Chapter 14

Glycolysis

14–1 INTRODUCTION

In the summer of 1896 M. Hahn was attempting to separate proteins from yeast by grinding the yeast in a mortar with fine sand and diatomaceous earth and then isolating the yeast extract by filtering the mixture through cheesecloth. The extract was not easy to preserve, and Hans Buchner, remembering that fruit preserves are made by adding sugar, suggested adding glucose to the yeast extract. When Hahn went on vacation, Eduard Buchner, Hans' brother, arrived in the laboratory for a busman's holiday of experimentation with yeast extracts. When he added glucose to his yeast extract, he noticed bubbles evolving from the solution. He concluded that the extracts were carrying out cell-free fermentation. Now such an observation may not strike us today as especially profound, but at that time it was nearly universally believed that "life processes" were the exclusive prerogative of living cells. Buchner found, however, that the most draconian procedures, thoroughly fatal to the yeast, did not prevent his cell-free extracts from performing fermentation. This series of observations represented a complete break with the past and ushered in the molecular science of life: biochemistry.[1] Our knowledge has advanced far beyond Buchner's hypotheses about the activity of "zymase," his name for the enzyme he believed was responsible for fermentation. Yet the secrets of glucose metabolism are still not completely revealed, and not a week passes without the publication of some new experimental data about some aspect of glycolysis.

14–2 AN OVERVIEW OF GLYCOLYSIS

D-Glucose, the most abundant organic molecule in the biosphere, is a source of energy for nearly all organisms. The pathway by which glucose is converted to pyruvate is called *glycolysis*, a word coined from the Greek words for sweet, *glycos*, and splitting, *lysis*. The 10 steps by which glucose is converted to pyruvate are summarized in Figure 14.1.

The net reaction of glycolysis is

$$\text{D-glucose} + 2\ HPO_4^{2\ominus} + 2\ ADP + 2\ NAD \longrightarrow$$

$$\underset{\text{Pyruvate}}{CH_3\overset{\overset{\displaystyle O}{\|}}{C}-CO_2^{\ominus}} + 2\ ATP + 2\ NADH + 2H^{\oplus} \qquad (14.1)$$

The enzymatic reactions of glycolysis, and their standard free-energy changes at 298 K are summarized in Table 14.1. Glycolysis results in the *net* production of 2 moles of ATP and 2 moles of NADH per mole of glucose catabolized. Looking at the individual reactions in Table 14.1, we see that two of them involve phosphoryl group transfer from ATP to another substrate.

$$\text{D-glucose} + ATP \xrightarrow{\text{Hexokinase}} \text{D-glucose 6-phosphate} + ADP \qquad (14.2)$$

$$\text{D-fructose 6-phosphate} + ATP$$

$$\xrightarrow{\text{Phosphofructokinase}} \text{D-fructose 1,6-diphosphate} + ADP \qquad (14.3)$$

The reactions leading from glucose to fructose 1,6-diphosphate (FDP) are driven by ATP hydrolysis. The purpose of these "pump-priming" reactions is the synthesis of a substrate for aldolase, the enzyme responsible for cleaving the hexose into two trioses (reaction 14.4).

$$\underset{\text{Fructose 1,6-diphosphate (FDP)}}{\begin{array}{c} CH_2OPO_3^{2\ominus} \\ | \\ C=O \\ | \\ HO-C-H \\ | \\ H-C-OH \\ | \\ H-C-OH \\ | \\ CH_2OPO_3^{2\ominus} \end{array}} \xrightarrow{\text{Aldolase}}$$

$$\underset{\text{D-glyceraldehyde 3-phosphate}}{\begin{array}{c} CHO \\ | \\ H-C-OH \\ | \\ CH_2OPO_3^{2\ominus} \end{array}} + \underset{\substack{\text{Dihydroxyacetone} \\ \text{phosphate (DHAP)}}}{\begin{array}{c} CH_2OH \\ | \\ C=O \\ | \\ CH_2OPO_3^{2\ominus} \end{array}} \qquad (14.4)$$

[1] Buchner received the Nobel Prize for his work. His happiness was, however, short-lived. He died on the Rumanian front in the First World War.

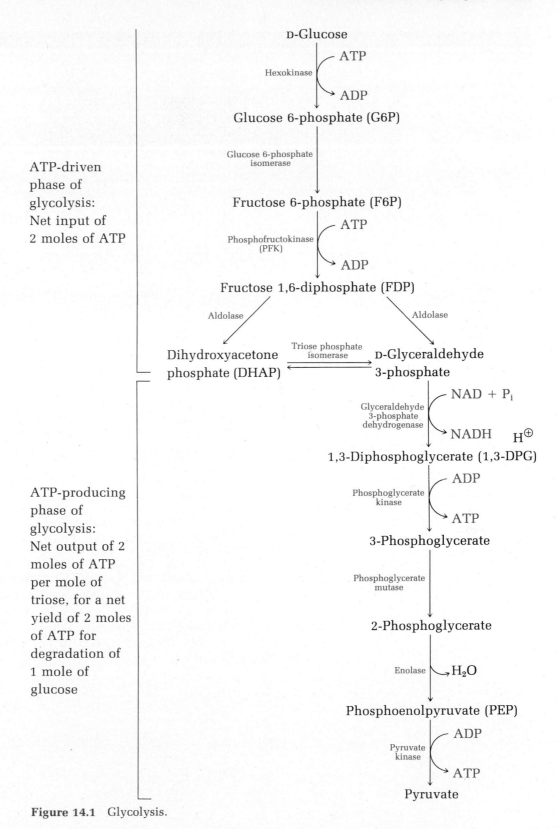

ATP-driven phase of glycolysis: Net input of 2 moles of ATP

ATP-producing phase of glycolysis: Net output of 2 moles of ATP per mole of triose, for a net yield of 2 moles of ATP for degradation of 1 mole of glucose

Figure 14.1 Glycolysis.

Table 14.1   Individual reactions of glycolysis and their standard free-energy changes

Enzyme	Reaction	$\Delta G^{\circ\prime}$ (kJ mol^{-1})
Hexokinase	D-glucose + ATP $\rightleftarrows$ glucose 6-phosphate + ADP	-16.74
Glucose 6-phosphate isomerase	glucose 6-phosphate $\rightleftarrows$ fructose 6-phosphate	$+1.67$
Phosphofructokinase	fructose 6-phosphate + ATP $\rightleftarrows$ fructose 1,6-diphosphate + ADP	-14.2
Aldolase	fructose 1,6-diphosphate $\rightleftarrows$ D-glyceraldehyde 3-phosphate + dihydroxy-acetone phosphate	$+23.8$
Triose phosphate isomerase	glyceraldehyde 3-phosphate $\rightleftarrows$ dihydroxyacetone phosphate	$+14.6$
Glyceraldehyde 3-phosphate dehydrogenase	glyceraldehyde 3-phosphate + P$_i$ + NAD $\rightleftarrows$ 1,3-diphosphoglycerate + NADH + H$^{\oplus}$	$+6.3$
Phosphoglycerate kinase	1,3-diphosphoglycerate + ADP $\rightleftarrows$ 3-phosphoglycerate + ATP	-18.8
Phosphoglyceratemutase	3-phosphoglycerate $\rightleftarrows$ 2-phosphoglycerate	$+4.6$
Enolase	2-phosphoglycerate $\rightleftarrows$ phosphoenolpyruvate + H$_2$O	$+1.7$
Pyruvate kinase	phosphoenolpyruvate + ADP $\rightleftarrows$ pyruvate + ATP	-31.8

$$\Delta G^{\circ\prime}_{glycolysis} = -28.5$$

The standard free-energy change for the conversion of D-glucose to the two trioses—dihydroxyacetone phosphate and D-glyceraldehyde 3-phosphate—is -5.4 kJ mol^{-1}. As Figure 14-1 indicates, these two trioses are equilibrated by an isomerase. D-Glyceraldehyde 3-phosphate is the minor product of this equilibrium. It is also the product that continues in glycolysis. Adding the standard free-energy change for this isomerization to the standard free-energy change for the first four steps gives a total standard free-energy change for the first five steps of 9.2 kJ mol^{-1}, corresponding to an equilibrium constant of 2.4×10^{-2} for the net reaction of the first five steps of glycolysis (reaction 14.5).

$$\text{D-glucose} + 2 \text{ ATP} \longrightarrow 2 \text{ D-glyceraldehyde 3-phosphate}$$
$$\Delta G^{\circ\prime}_{step\ 1 \rightarrow step\ 5} = +9.2 \text{ kJ mol}^{-1}; \quad K'_{1 \rightarrow 5} = 2.4 \times 10^{-2} \quad (14.5)$$

These five reactions constitute the "ATP-driven phase of glycolysis." We will consider the details of the individual steps in Section 14.3.

The ATP-producing phase of glycolysis follows by converting D-glyceraldehyde 3-phosphate into energy-rich metabolites that transfer phosphoryl groups to ADP. There are two such phosphoryl group transfers (reactions 14.6 and 14.7).

1,3-Diphospho-D-glycerate
(1,3-DPG)

3-Phospho-D-glycerate

$$(14.6)$$

$$\underset{\substack{\text{Phosphoenolpyruvate}\\\text{(PEP)}}}{\overset{\displaystyle ^{\ominus}_3\text{OPO}\diagup\overset{\displaystyle \text{C}}{\underset{\displaystyle \text{C}}{\parallel}}\diagdown\text{CO}_2^{\ominus}}{\underset{\diagup\;\diagdown}{}}} \;+\; \text{ADP} \;\xrightarrow{\text{Pyruvate kinase}}\; \underset{\text{Pyruvate}}{\overset{\displaystyle \text{O}\diagdown\diagup\text{CO}_2^{\ominus}}{\underset{\displaystyle \text{CH}_3}{\overset{\displaystyle \text{C}}{\mid}}}} \;+\; \text{ATP} \qquad (14.7)$$

The conversion of glucose to dihydroxyacetone phosphate and glyceraldehyde 3-phosphate consumes 2 moles of ATP. The steps from glyceraldehyde 3-phosphate to pyruvate yield 2 moles of ATP. Since 1 mole of glucose yields 2 moles of glyceraldehyde 3-phosphate, the degradation of 1 mole of glucose by glycolysis occurs with net production of 2 moles of ATP.

Examining Figure 14.1 and Table 14.1 we find that NADH is produced by a dehydrogenase in step six of glycolysis. The fate of this NADH depends upon the organism and upon the metabolic conditions. Under anaerobic conditions the NADH is recycled in a reductive step in fermentation (Section 14.5). Under aerobic conditions the NADH is reoxidized by respiration (Chapter 16).

The sum of the standard free-energy changes for the coupled reactions of glycolysis is -28.5 kJ mol^{-1}, corresponding to an overall equilibrium constant for glycolysis of 9.7×10^4.

14–3 ENZYMATIC REACTIONS OF THE ATP-DRIVEN PHASE OF GLYCOLYSIS

A. Hexokinase

The first reaction of glycolysis is phosphoryl group transfer from ATP to the 6-hydroxyl group of glucose catalyzed by hexokinase (reaction 14.8).

β-D-Glucose

$\xrightarrow[\text{MgATP}^{\circleddash}]{\text{Hexokinase}}$

β-D-Glucose 6-phosphate (G6P)

$+\; \text{MgADP}^{\ominus} \qquad (14.8)$

Hexokinase, as its name implies, will accept various hexoses as substrates, including mannose and fructose as well as glucose. The related enzyme *glucokinase* is specific for glucose. However, glucokinase does not participate in glycolysis, but is involved in the synthesis of glycogen (Section 17.4B).

Having discussed the mechanism of action of kinases in Section 12.5 and the conformational change that occurs when hexokinase binds to glucose in Section 12.6, let us now turn to the energy changes involved in the phosphoryl group transfer from ATP to glucose.

The phosphorylation of glucose can be written as

$$\text{D-glucose} + \text{HPO}_4^{2\ominus} \longrightarrow \text{D-glucose 6-phosphate} + \text{H}_2\text{O}$$
$$\Delta G^{\circ\prime} = +12.6 \text{ kJ mol}^{-1}$$
$$\text{ATP} + \text{H}_2\text{O} \longrightarrow \text{ADP} + \text{HPO}_4^{2\ominus}$$
$$\Delta G^{\circ\prime} = -30.5 \text{ kJ mol}^{-1}$$
$$\overline{\text{D-glucose} + \text{ATP} \longrightarrow \text{D-glucose 6-phosphate} + \text{ADP}}$$
$$\Delta G^{\circ\prime} = -17.9 \text{ kJ mol}^{-1} \qquad (14.9)$$

The phosphorylation of glucose by hexokinase is highly exergonic, but the free-energy change implies nothing about the mechanism of the reaction (which we considered in Section 12.5).

The rate of glycolysis in mammals is limited by the rate of the hexokinase reaction. Since the hexokinase reaction is highly exergonic, and since it is rate-limiting in mammalian glycolysis, we might be tempted to designate it as the first committing step of glycolysis. This would be too hasty a conclusion, however, since glucose 6-phosphate is a branch-point metabolite that need not be degraded by glycolysis. In fact, we shall see that it is the first *unique* step of glycolysis, catalyzed by phosphofructokinase, that is the committing step of glycolysis.

An inherited defect of glycolysis in red blood cells, called *hexokinase deficiency*, illustrates the complexity of metabolic interrelationships. Persons with low hexokinase activity in their blood produce little D-glucose 6-phosphate. The concentrations of other glycolytic intermediates also fall as a result. One of these intermediates is 1,3-diphosphoglycerate (1,3-DPG), a precursor of 2,3-diphosphoglycerate (2,3-DPG). We recall (Section 4.6) that 2,3-DPG lowers the affinity of hemoglobin for oxygen. If the concentration of 2,3-DPG is low, the affinity of hemoglobin for oxygen is too high and oxygen transport is impaired.

B. Glucose 6-Phosphate Isomerase

The second step of glycolysis is the conversion of D-glucose 6-phosphate (G6P) to D-fructose 6-phosphate (F6P) catalyzed by glucose 6-phosphate isomerase.

β-D-Glucose 6-phosphate (G6P)

Glucose 6-phosphate isomerase
K = 0.3

(14.10)

β-D-Fructose 6-phosphate (F6P)

The equilibrium constant for the reaction is 0.3. The ratio of fructose 6-phosphate to glucose 6-phosphate in red blood cells and in intact muscle cells is almost the same as the equilibrium value, indicating that this step is essentially at equilibrium in glycolysis.

Opening of the pyranose ring is the first step of the reaction catalyzed by glucose 6-phosphate isomerase. The α-anomer of the pyranose ring is the preferred substrate, perhaps because a stereochemically favored *anti* elimination can occur for the α-anomer (reaction 14.11).

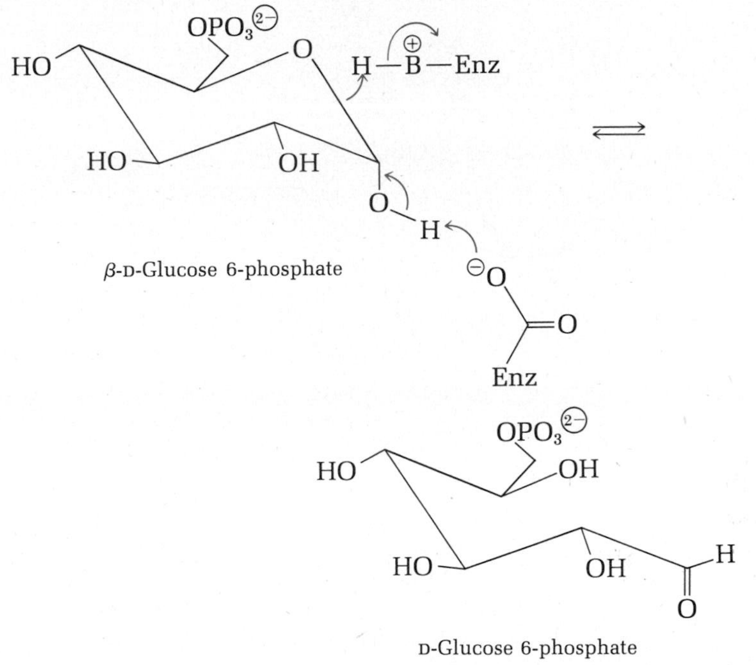

β-D-Glucose 6-phosphate

D-Glucose 6-phosphate

(14.11)

X-ray crystallographic evidence suggests that a carboxylate group of an active site glutamate participates in ring opening, although this is not firmly established.

In the second step of catalysis the actual isomerization begins by generation of a *cis*-1,2-enediol intermediate, a pattern repeated by other aldose-ketose isomerases (reaction 14.12).

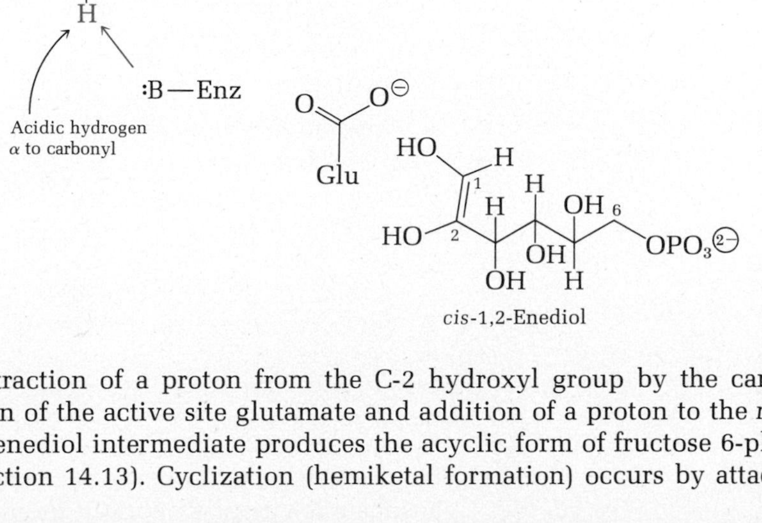

Acidic hydrogen
α to carbonyl

cis-1,2-Enediol

(14.12)

Abstraction of a proton from the C-2 hydroxyl group by the carboxylate anion of the active site glutamate and addition of a proton to the *re* face of the enediol intermediate produces the acyclic form of fructose 6-phosphate (reaction 14.13). Cyclization (hemiketal formation) occurs by attack of the

C-5 hydroxyl group on the *si* face of the carbonyl group to give α-D-fructo*furanose* 6-phosphate (reaction 14.14).

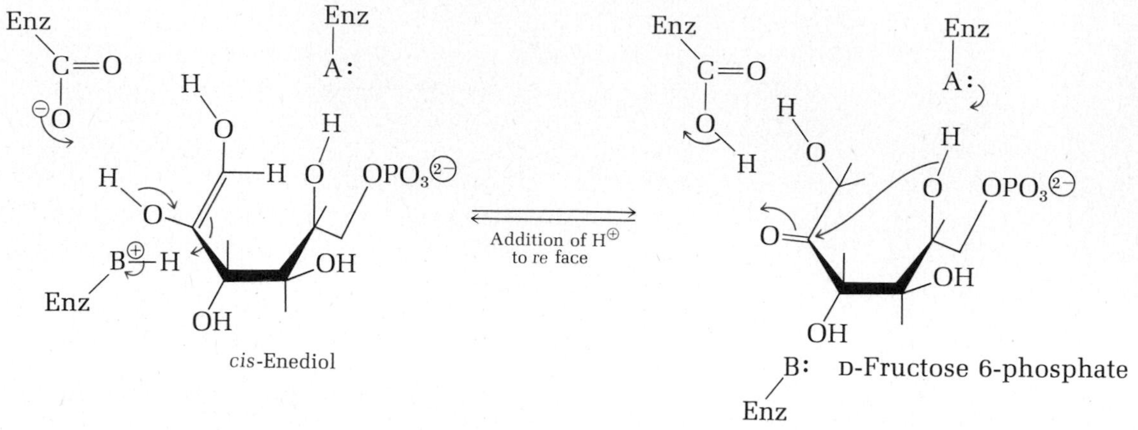

cis-Enediol

B: D-Fructose 6-phosphate

(14.13)

Addition of H⊕ to re face

Addition of OH to si face of carbonyl

Enz—B:

α-D-Fructofuranose 6-phosphate

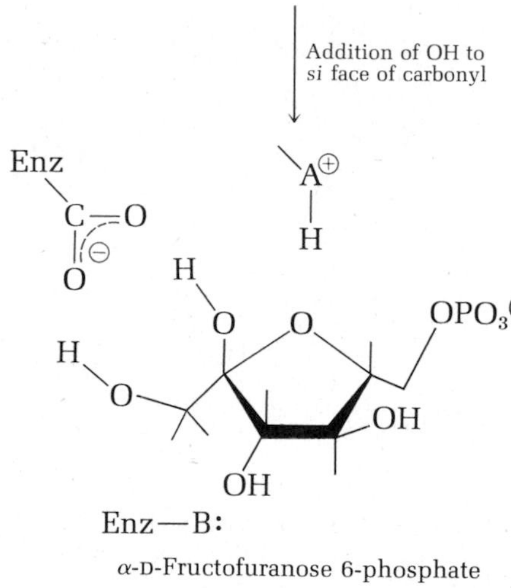

(14.14)

C. Phosphofructokinase

The committing step in glycolysis is phosphoryl group transfer from ATP to fructose 6-phosphate catalyzed by *phosphofructokinase* (reaction 14.15).

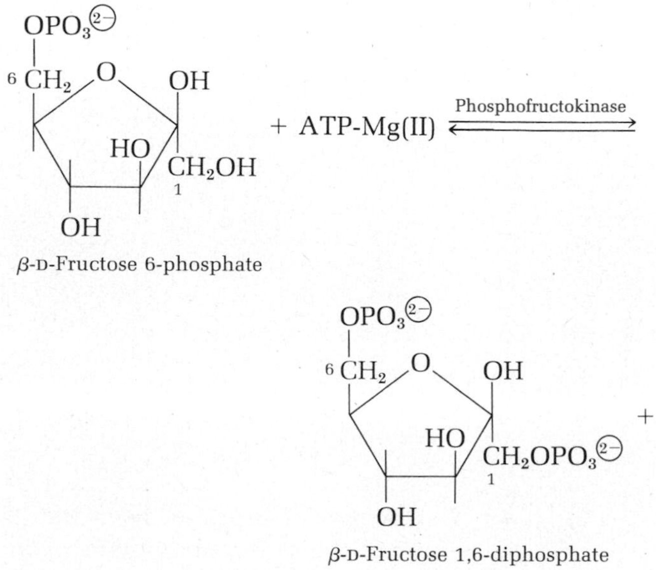

β-D-Fructose 6-phosphate

$+ \text{ATP-Mg(II)} \xrightleftharpoons{\text{Phosphofructokinase}}$

$+ \text{ADP}$ (14.15)

β-D-Fructose 1,6-diphosphate

The reaction is quite exergonic, and the cellular supply of fructose 6-phosphate is virtually depleted in this step. Phosphofructokinase catalyzes the first unique step of glycolysis, and is the critical enzyme in the regulation of glycolysis.

Phosphofructokinase is an allosteric enzyme that is subject to activation and inhibition by many metabolites (Table 14.2). The mass of data in Table 14.2 can be reduced to three major effects. (1) ATP is an allosteric inhibitor of phosphofructokinase that acts by increasing the K_m for fructose 6-phosphate, the second substrate. (2) Citrate, an intermediate in the citric acid cycle, is an allosteric inhibitor of phosphofructokinase. Since citrate is a participant in the citric acid cycle, it couples the two pathways. (3) Inhibition of phosphofructokinase by ATP and citrate is counteracted by the allosteric activators AMP, 3'-5' cyclic AMP, and the product of the reaction, fructose 1,6-diphosphate.

Phosphofructokinase is a tetramer. In skeletal muscle it is exclusively

Table 14.2 Action of effectors on phosphofructokinase (PFK) from various sources

Source	Shape of saturation curve for fructose 6-phosphate (F6P)	Effector[a] ATP	Citrate	AMP	Cyclic AMP	ADP	Fructose 1,6-diphosphate (FDP)	P_i
Mammalian								
Muscle	Sigmoidal	I	I	A	A	A	A	A
Brain	Sigmoidal	I	I	A	A	A	A	A
Heart	Sigmoidal	I	I	A	A	A	A	A
Liver	Sigmoidal	I	I	A	A	A	A	—
Human erythrocyte	—	I	I	—	—	—	—	A
Kidney cortex	Sigmoidal	I	I	A	—	A	A	—
Adipose tissue	—	I	I	A	A	A	—	A
Lens	Hyperbolic[b]	I	Weak I	Weak A	A	Weak A	—	Weak A
Sperm	Sigmoidal	I	I	A	A	A	A[c]	A
Jejunal mucosa	Sigmoidal	I	—	A[d]	—	A[d]	—	A[d]
Ascites tumor	—	I[e]	I[e]	A	—	—	None	A
Plant								
Pea seed	Double hyperbola[f]	I[g]	I	Weak I	None	I	—	A
Carrots	Sigmoidal	I	I	Weak I	None	I	—	A
Brussels sprout leaf[h]	Sigmoidal	I	I	I	—	I	—	A
Corn	—	I	I	I	—	I	—	A[i]
Microorganisms								
N. crassa	Sigmoidal	I	—	Weak I	—	I	—	—
A. crystallopoietes	Hyperbolic	None	None	None	None	None	—	—
D. discoideum	Hyperbolic	None	None	None	None	I[j]	I[j]	None
C. pasteurianum	Sigmoidal[k]	None	—	Weak A	None	A	None	None
F. thermophilum	Hyperbolic[f]	None	None	—	None	A	—	None
E. coli	Sigmoidal[k]	I	None	A	None	A	—	—
L. casei	Hyperbolic	None	I	None	None	None	—	A
L. plantarum	Hyperbolic	None	I	I	I	I	I	—
Yeast	Sigmoidal	I	I	A[l]	None	Weak A	None	None

[a] Abbreviations: I, inhibitor; A, activator.
[b] Kinetics of enzyme are normal Michaelis–Menten with respect to F6P. This experiment was performed at pH 8 and the inhibition by ATP was independent of pH or F6P concentration.
[c] FDP does not reverse citrate inhibition of sperm PFK. However, it does activate the enzyme in the presence of ATP.
[d] The maximum activation of this enzyme occurs when all three activators are present simultaneously.
[e] Inhibition of enzyme by ATP and citrate shows a lag in onset of inhibition of about 2 min.
[f] For this enzyme the curve becomes sigmoidal in the presence of inhibitory concentrations of PEP.
[g] The most effective inhibitor of the enzyme is PEP. A plot of inhibition vs. PEP concentration is sigmoidal and has a slope greater than one in the Hill plot, suggesting a cooperative interaction of PEP with the enzyme.
[h] Degree of regulation of activity appears to be greatest with PFK from young leaves.
[i] In the presence of low ATP levels, P_i is an inhibitor. Activation by P_i is observed at inhibitory ATP concentrations.
[j] Both of these products were competitive inhibitors with respect to their paired substrate, i.e., FDP with F6P and ADP with ATP.
[k] The sigmoidal kinetics of this enzyme are not pH dependent.
[l] Apparently only the enzyme from brewer's yeast is activated by AMP. Baker's yeast is not activated by AMP.

SOURCE: From D. P. Bloxman and H. A. Lardy, in P. D. Boyer, Ed., "The Enzymes," Vol. 8, Academic Press, New York, 1973, pp. 262–263.

composed of one type of subunit, called the M type. In liver it is composed of another type of subunit, called the L type. The symbols of these isomeric enzymes, known as *isozymes*, are M_4 and L_4 in muscle and liver, respectively. In red blood cells all possible combinations of isozymes are found: M_4, M_3L_4, M_2L_2, ML_3, and L_4. These enzymes all catalyze the same reaction, but differ in their K_m values for fructose 6-phosphate. For the M_4 enzyme, the K_m for fructose 6-phosphate is 0.24 mM, and for the L_4 isozyme the K_m for fructose 6-phosphate is 0.89 mM.

In certain rare genetic muscle diseases phosphofructokinase activity is entirely absent in muscle and half absent in red blood cells because the gene that codes for the M subunit, and hence the M_4 isozyme, is absent. The liver enzyme in afflicted persons is normal. Since phosphofructokinase is necessary for glycolysis, its absence in muscle severely impairs carbohydrate metabolism, and the muscle is quite weak.

D. Aldolase Aldolase catalyzes aldol cleavage of D-fructose 1,6-diphosphate between C-3 and C-4 to produce D-glyceraldehyde 3-phosphate and dihydroxyacetone phosphate (DHAP) (reaction 14.16).

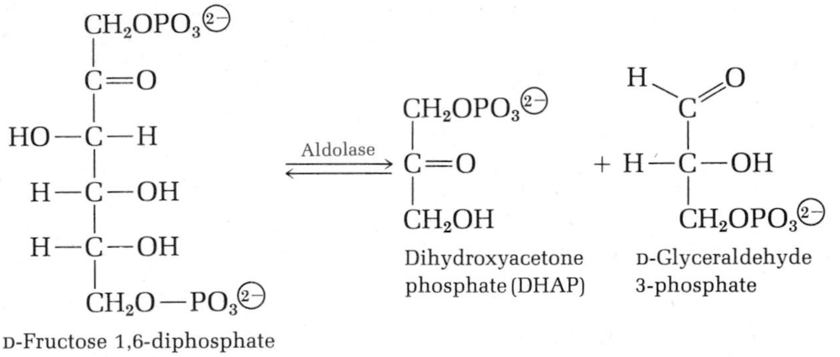

(14.16)

D-Fructose 1,6-diphosphate

Dihydroxyacetone phosphate (DHAP)

D-Glyceraldehyde 3-phosphate

Before discussing the enzyme-catalyzed reaction, let us briefly review aldol condensations and aldol cleavages.

a. Base-Catalyzed Aldol Condensations and Aldol Cleavages

The aldol condensation occurs by nucleophilic addition of the enolate anion of an aldehyde or ketone to the carbonyl group of an aldehyde (reaction 14.17).

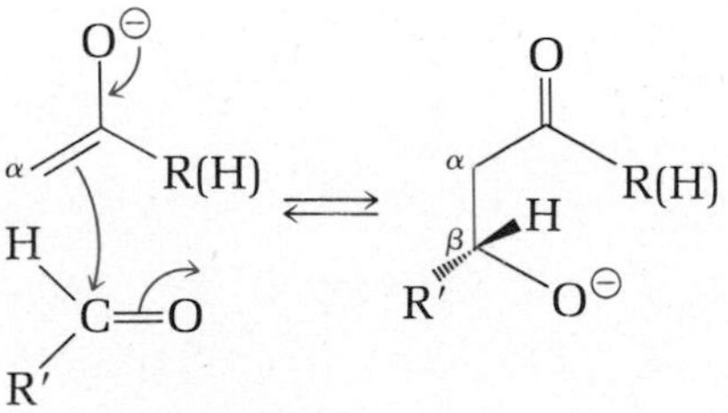

(14.17)

The enolate anion is generated by abstraction of the proton α to the carbonyl group by a strong base such as hydroxide ion (reaction 14.18).

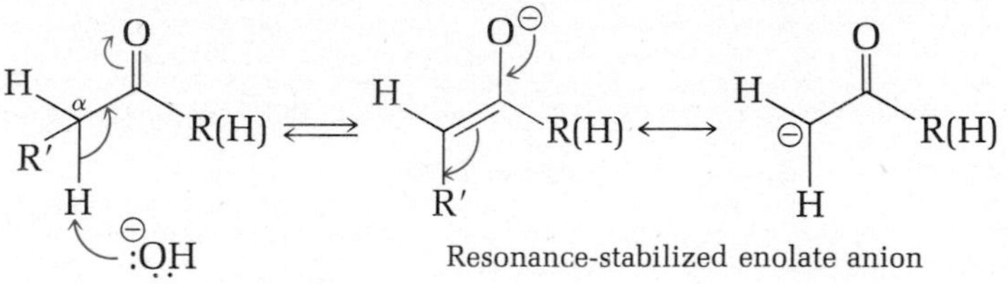

(14.18)

Resonance-stabilized enolate anion

The reaction between glyceraldehyde 3-phosphate and the enolate anion of DHAP is shown in reaction 14.19.

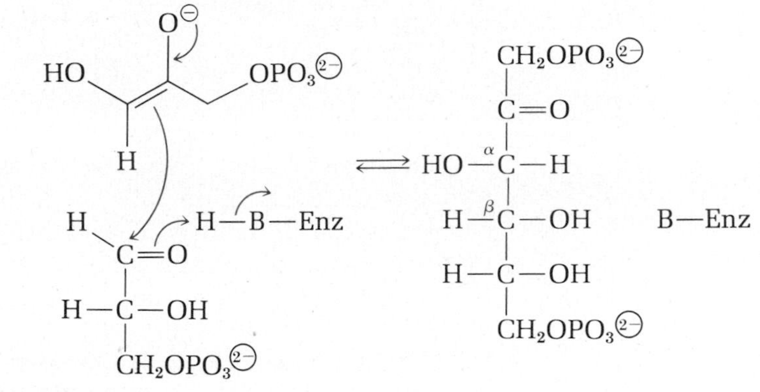

$$(14.19)$$

The first step in an aldol cleavage is abstraction of the hydrogen of the β-hydroxyl group followed by elimination of an enolate anion. This is shown for fructose 1,6-diphosphate in reaction 14.20.

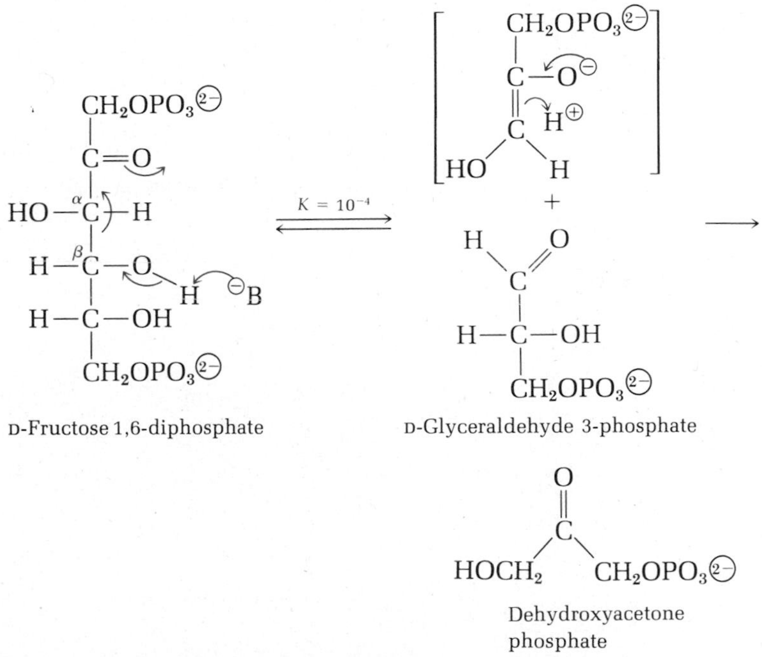

D-Fructose 1,6-diphosphate

D-Glyceraldehyde 3-phosphate

$K = 10^{-4}$

$$(14.20)$$

Dehydroxyacetone phosphate

b. Amine-Catalyzed Aldol Cleavage and the Mechanism of Action of Aldolase

In the aldol cleavage shown above electrons flow from the β-hydroxyl oxygen toward the carbonyl oxygen. Protonation of the carbonyl oxygen increases the rate of reaction. The pK_a of the conjugate acid of the carbonyl group is about -2, and at pH 7 the concentration of the conjugate acid is extremely low.

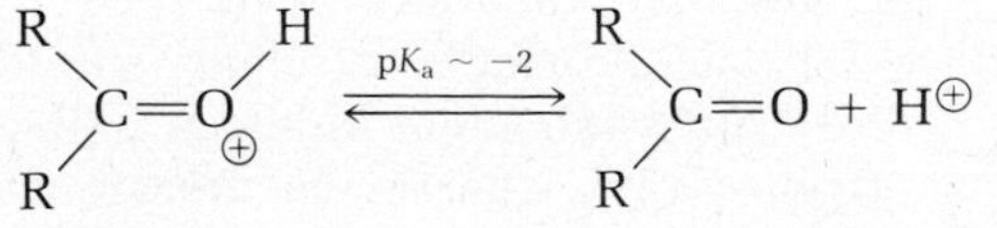

$$pK_a \sim -2 \qquad (14.21)$$

In the presence of an amine, however, the carbonyl group reacts to form an iminium ion.

$$\underset{R}{\overset{O}{\overset{\|}{\underset{}{C}}}}\underset{R}{} + RNH_2 \rightleftharpoons \underset{R}{\overset{H\overset{\oplus}{N}R}{\overset{\|}{\underset{}{C}}}}\underset{R}{} + HO^{\ominus} \qquad (14.22)$$

Iminium ion
($pK_a \sim 7$)

The pK_a of an iminium ion is about 7, and therefore at pH 7 imines are about 50% protonated. Iminium ions are excellent electron sinks, and aldol cleavage is catalyzed by amines. The acidity of the β-hydroxyl proton is increased by the presence of an iminium ion since it exerts a much greater inductive effect than a carbonyl group. The mechanism of an amine-catalyzed aldol cleavage is shown in Figure 14.2.

There are two classes of aldolases. Class I (animal) aldolases are inactivated by sodium borohydride ($NaBH_4$) in the presence of substrate, do not require a metal ion, and are not inhibited by EDTA. Class II aldolases (bacteria and fungi) are not inactivated by $NaBH_4$. They contain an active site metal ion (usually $Zn^{2\oplus}$), and are inactivated by EDTA. The class II aldolase of yeast does not become covalently bound to the substrate during catalysis. Class II aldolases rely upon an active site $Zn^{\oplus}$ ion to polarize the carbonyl bond and to act as an electrophilic catalyst. Some blue-green algae and flagellates contain both classes of aldolase.

The mechanism of action of class I aldolases follows the same principles as the reaction in Figure 14.2 and is shown in Figure 14.3. An ϵ-amino group of a lysyl residue of class I aldolases forms an imine with the keto group of fructose 1,6-diphosphate. The intermediate can be trapped by reduction with sodium borohydride which converts the imine to an amine (reaction 14.23).

$$(14.23)$$

Borohydride anion   Substrate irreversibly
bound to aldolase

Through the use of a ^{14}C-labeled substrate it has been determined that an active site lysyl residue, number 227 in the 361 amino acid residue rabbit enzyme, becomes covalently bound to the substrate.

Aldolase is completely stereospecific. In the presence of hydroxyacetone phosphate and D-glyceraldehyde 3-phosphate only fructose 1,6-diphosphate is produced, even though two chiral centers (at C-3 and C-4) are created and four diastereoisomers are theoretically possible. The enzyme can thus distinguish between the pro-R and pro-S hydrogens of C-3 in dihydroxyacetone phosphate and between the re and si faces of the carbonyl group of glyceraldehyde 3-phosphate.

The anomeric specificity of aldolases from different sources is markedly different. Liver aldolase will not bind the α-anomer of fructose 1,6-diphosphate, whereas the muscle enzyme will bind the α-anomer, but does not catalyze its aldol cleavage. The yeast enzyme, by contrast, both binds the α-anomer, and catalyzes its cleavage. The β-anomer and the keto form of fructose 1,6-diphosphate are substrates for both muscle and liver aldolases. The yeast enzyme can use either the α- or the β-anomer because it catalyzes

Figure 14.2 Mechanism of an amine-catalyzed aldol cleavage.

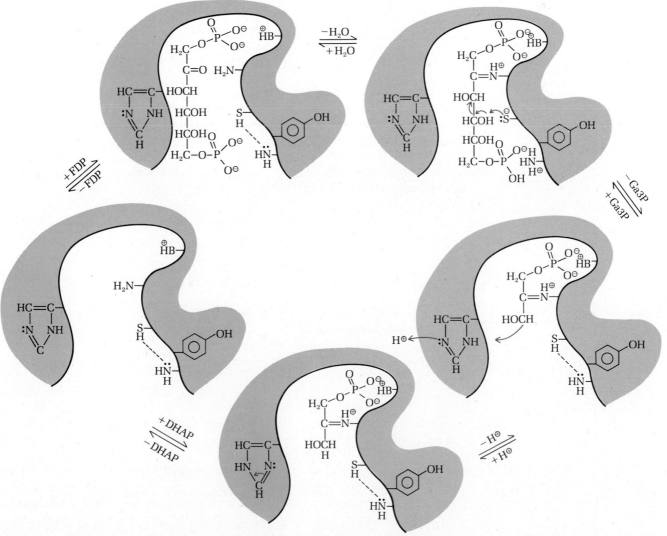

Figure 14.3 Mechanism of action of aldolase. [From B. L. Horecker, O. Tsolase, and C. Y. Lai, in P. D. Boyer, Ed., "The Enzymes," 3rd. ed., Vol. 6, Academic Press, New York, 1972, p. 233.]

their rapid anomerization. The α form is cleaved. The liver enzyme lacks this anomerization activity.

E. Triose Phosphate Isomerase

Triose phosphate isomerase catalyzes interconversion of dihydroxyacetone phosphate and D-glyceraldehyde 3-phosphate. The equilibrium constant for formation of D-glyceraldehyde 3-phosphate is 2.7×10^{-3} (reaction 14.24).

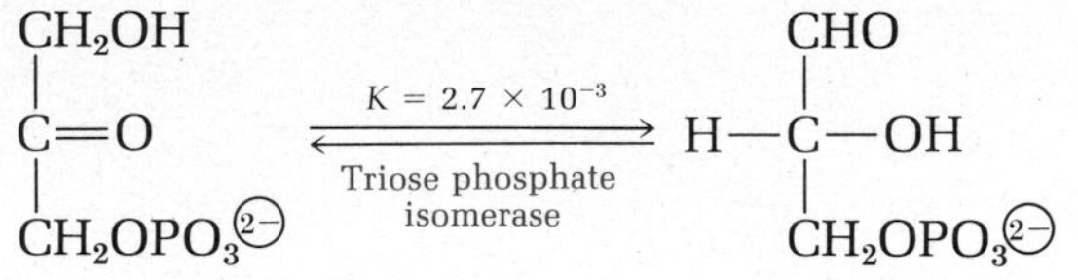

$$\text{CH}_2\text{OH} \atop \underset{\substack{| \\ \text{CH}_2\text{OPO}_3^{2-}}}{\text{C}=\text{O}} \underset{\substack{\text{Triose phosphate} \\ \text{isomerase}}}{\overset{K = 2.7 \times 10^{-3}}{\rightleftharpoons}} \text{CHO} \atop \underset{\substack{| \\ \text{CH}_2\text{OPO}_3^{2-}}}{\text{H}-\text{C}-\text{OH}}$$

(14.24)

Dihydroxyacetone phosphate (DHAP) — D-Glyceraldehyde 3-phosphate

We have written the structures of DHAP and glyceraldehyde 3-phosphate with "free" carbonyl groups. In aqueous solution, however, DHAP is 45% hydrated and glyceraldehyde 3-phosphate is 96.7% hydrated (reaction 14.25).

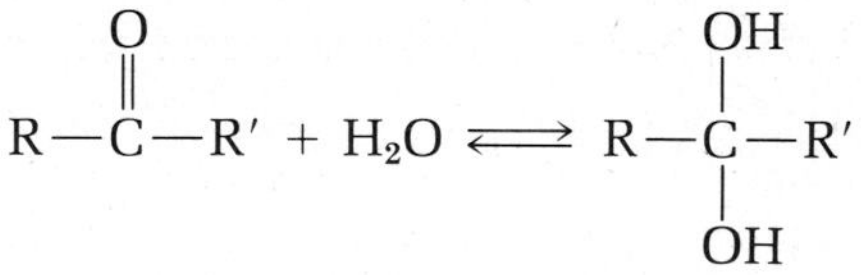

$$\underset{\substack{\text{Unhydrated} \\ \text{carbonyl group}}}{\text{R}-\overset{\overset{\text{O}}{\|}}{\text{C}}-\text{R}' + \text{H}_2\text{O}} \rightleftharpoons \underset{\substack{\text{Hydrated} \\ \text{carbonyl group}}}{\text{R}-\overset{\text{OH}}{\underset{\text{OH}}{\text{C}}}-\text{R}'}$$

(14.25)

The actual substrates for triose phosphate isomerase are unhydrated.

The mechanism of action of triose phosphate isomerase has been determined by experiments involving incorporation of tritium and deuterium in the reactants and products. When the reaction is carried out in D_2O, deuterium is stereospecifically incorporated in the product, implying formation of an enediol intermediate that stereospecifically abstracts a deuterium from the solvent (reaction 14.26).

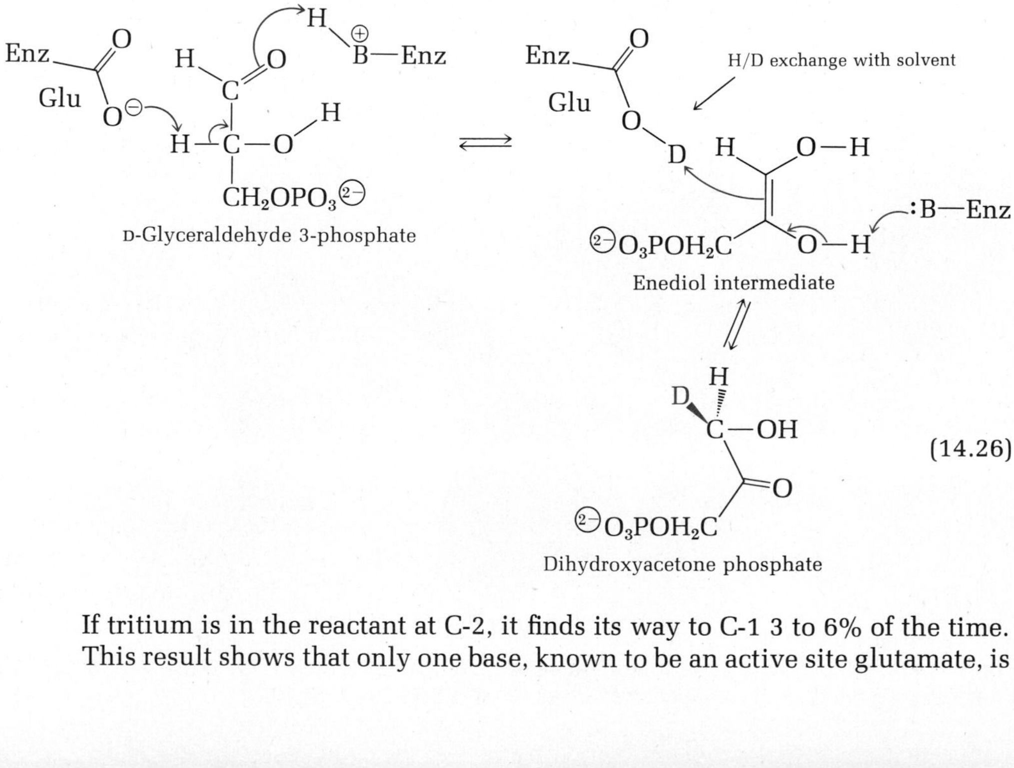

D-Glyceraldehyde 3-phosphate

Enediol intermediate

Dihydroxyacetone phosphate

(14.26)

If tritium is in the reactant at C-2, it finds its way to C-1 3 to 6% of the time. This result shows that only one base, known to be an active site glutamate, is

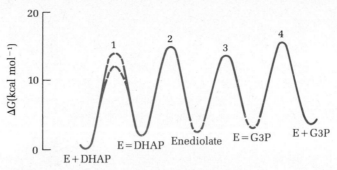

Figure 14.4 Free energy profile for the reaction catalyzed by triose phosphate isomerase. The energy barriers, all about equal, are about the magnitude expected for dissociation of the enzyme-substrate complex. The rate has therefore evolved to a maximum and k_{cat}/K_m is a maximum. The rate-limiting step of the reaction is not bond breaking, but the diffusion of the substrate to the active site of the enzyme. [From W. J. Albery and J. R. Knowles, *Biochemistry*, 15, 5558, 5627 (1976). Reprinted with permission. Copyright 1976 American Chemical Society.]

involved in the proton transfer reaction. If there were two bases, the tritium would get lost in the shuffle, and there would not be any intramolecular transfer. In other words, all of the protons transferred to C-1 would originate in the solvent. A single base also accounts for the stereospecificity of proton transfer. The base removes a proton from one side of the reactant and delivers it to the same side of the enediol.

J. Knowles and his coworkers have made a large number of isotope exchange studies of triose phosphate isomerase culminating in a complete free energy profile for the reaction (Figure 14.4). The rate of isomerization occurs at almost the diffusion controlled limit. The rate-determining step in the reaction is the rate of encounter of enzyme and substrate, not any subsequent chemical step. Albery and Knowles suggest that this enzyme "has reached the end of its evolutionary development." It is a "perfectly evolved" enzyme whose rate can increase no further.

At this point in glycolysis glucose has been acted on by five enzymes. Two moles of ATP have been consumed in the formation of fructose 1,6-diphosphate. Driven by ATP hydrolysis, the formation of FDP is overwhelmingly favored. The next two reactions, catalyzed by aldolase and triose phosphate isomerase, have equilibrium constants much less than 1.0. Nevertheless, through the first five steps the standard free-energy change is only slightly endogonic. As a result of the high group transfer potential of ATP, the equilibrium constant for the first five coupled steps is about 0.5.

14–4 ENZYMATIC REACTIONS OF THE ATP-PRODUCING PHASE OF GLYCOLYSIS

A. Glyceraldehyde 3-Phosphate Dehydrogenase

The oxidation of D-glyceraldehyde 3-phosphate to 1,3-diphospho-D-glycerate is catalyzed by glyceraldehyde 3-phosphate dehydrogenase, which requires NAD as a coenzyme (reaction 14.27). The equilibrium constant for reaction 14.27 is 0.65 at pH 7.1. The cellular ratio of NADH/NAD is about 10^{-3}, and the phosphate concentration is about 1.5 mM. Under these conditions the [1,3-diphospho-D-glycerate]/[D-glyceraldehyde 3-phosphate] ratio is about 1.0.

The first step of the reaction is nucleophilic attack of a cysteine sulfhydryl group on the carbonyl group of the substrate giving a hemithioacetal intermediate (reaction 14.29). The next step is direct hydride transfer from the hemithioacetal to NAD, a reaction in which histidine participates as a general base and whose net result is conversion of the hemithioacetal to a thioester (reaction 14.30).

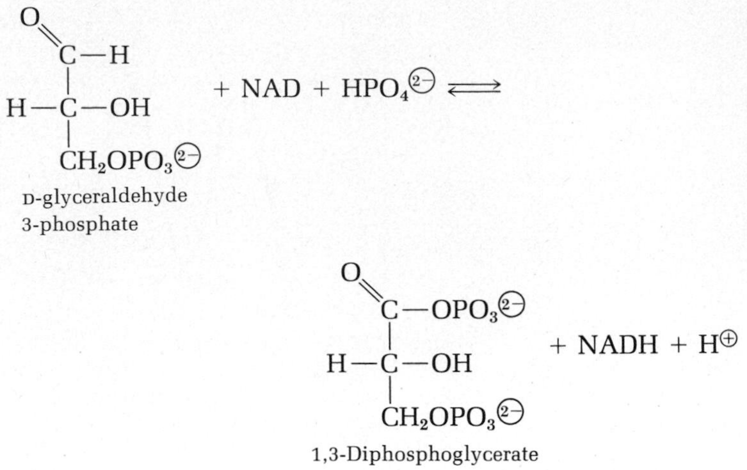

$$
\begin{array}{c}
\text{O} \\
\parallel \\
\text{C}-\text{H} \\
\mid \\
\text{H}-\text{C}-\text{OH} \\
\mid \\
\text{CH}_2\text{OPO}_3^{2\ominus}
\end{array}
\quad + \text{ NAD} + \text{HPO}_4^{2\ominus} \rightleftharpoons
$$

D-glyceraldehyde
3-phosphate

$$
\begin{array}{c}
\text{O} \\
\parallel \\
\text{C}-\text{OPO}_3^{2\ominus} \\
\mid \\
\text{H}-\text{C}-\text{OH} \\
\mid \\
\text{CH}_2\text{OPO}_3^{2\ominus}
\end{array}
\quad + \text{ NADH} + \text{H}^{\oplus} \qquad (14.27)
$$

1,3-Diphosphoglycerate

$$
K = \frac{[\text{NADH}][\text{1,3-diphosphoglycerate}][\text{H}^{\oplus}]}{[\text{NAD}][\text{glyceraldehyde 3-phosphate}][\text{HPO}_4^{2\ominus}]} = 0.65 \qquad (14.28)
$$

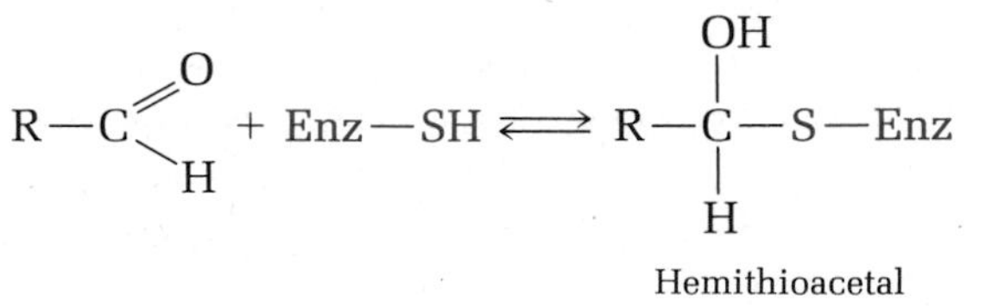

$$
\begin{array}{c}
\text{O} \\
\parallel \\
\text{R}-\text{C} \\
\mid \\
\text{H}
\end{array}
+ \text{ Enz}-\text{SH} \rightleftharpoons
\begin{array}{c}
\text{OH} \\
\mid \\
\text{R}-\text{C}-\text{S}-\text{Enz} \\
\mid \\
\text{H}
\end{array}
\qquad (14.29)
$$

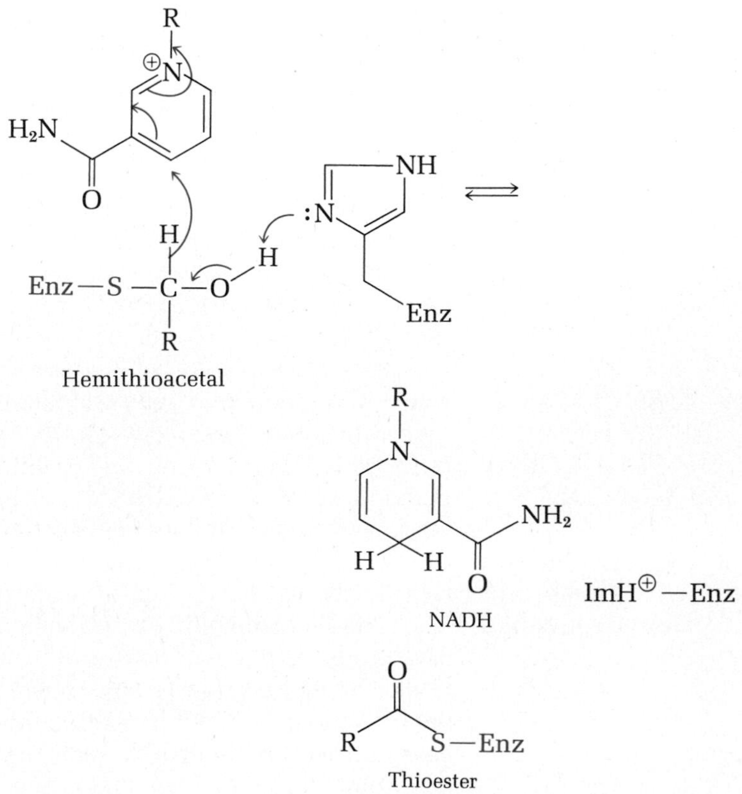

Hemithioacetal

Hemithioacetal

NADH

$\text{ImH}^{\oplus}-\text{Enz}$ \qquad (14.30)

$$
\begin{array}{c}
\text{O} \\
\parallel \\
\text{R} \quad \text{C} \quad \text{S}-\text{Enz}
\end{array}
$$

Thioester

The thioester is an energy-rich intermediate bound to the enzyme. Nucleophilic displacement by phosphate gives another energy-rich product, the mixed anhydride 1,3-diphosphoglycerate (reaction 14.31).

$$
R-C{\overset{O}{\underset{S-Enz}{}}} + HPO_4^{2\ominus} \rightleftharpoons \left[R-C{\overset{O^{\ominus}}{\underset{\underset{PO_3^{2\ominus}}{O}}{}}}{\overset{}{\underset{}{S-Enz}}} H^{\oplus} \right]
$$

$$
\Updownarrow
$$

$$
\begin{array}{l}
O{\overset{}{\diagdown}} \\
\quad C-OPO_3^{2\ominus} \\
H-C-OH \\
\quad CH_2OPO_3^{2\ominus}
\end{array} \qquad (14.31)
$$

1,3-Diphosphoglycerate

Glyceraldehyde 3-phosphate dehydrogenase is a tetramer. Glyceraldehyde 3-phosphate binds to each of the four subunits, as does NAD. In bacteria and rabbit muscle dehydrogenases the NAD binds with *negative cooperativity* (recall Section 5.10C). The dissociation constants for NAD decrease from 10^{-11} for the first mole of NAD bound to 10^{-5} for the fourth mole bound. Data from x-ray crystallographic studies strongly suggests that NAD binding induces a conformational change in the protein, and we therefore conclude that NAD adds sequentially to glyceraldehyde 3-phosphate dehydrogenase.

B. Phosphoglycerate Kinase

Phosphoglycerate kinase catalyzes phosphoryl group transfer from the energy-rich, mixed anhydride 1,3-diphosphoglycerate to ADP (reaction 14.32).

$$
\begin{array}{l}
O{\overset{}{\diagdown}} \\
\quad C-OPO_3^{2\ominus} \\
H-C-OH \\
\quad CH_2OPO_3^{2\ominus}
\end{array} + ADP \underset{}{\overset{\text{Phosphoglycerate kinase}}{\rightleftharpoons}}
\begin{array}{l}
CO_2^{\ominus} \\
H-C-OH \\
\quad CH_2OPO_3^{2\ominus}
\end{array} + ATP \qquad (14.32)
$$

1,3-Diphosphoglycerate   3-Phosphoglycerate

Since 1 mole of ATP is synthesized per triose, 2 per mole of glucose, in this step the cell recovers the energy it expended in producing fructose-1,6-diphosphate. The standard free-energy change; $\Delta G°$, for transfer of phosphate from 1,3-diphosphoglycerate to ADP is -17.5 kJ mol^{-1}.

Phosphoryl transfer depends upon the Mg(II) concentration since it is the MgADP complex, not ADP itself, that binds to phosphoglycerate kinase. We recall that the MgADP complex has a high formation constant. The cellular concentration of Mg(II) therefore has a strong influence upon the fraction of ADP bound to Mg(II).

$$
ADP^{3\ominus} + Mg(II) \longrightarrow MgADP^{\ominus} \qquad K_{assn} = 2.2 \times 10^3 \qquad (14.33)
$$

When the cellular Mg(II) concentration is low, the rate of phosphoglycerate kinase is also low, and the concentration of fructose 1,6-diphosphate increases, especially if the ATP/ADP ratio is also high.

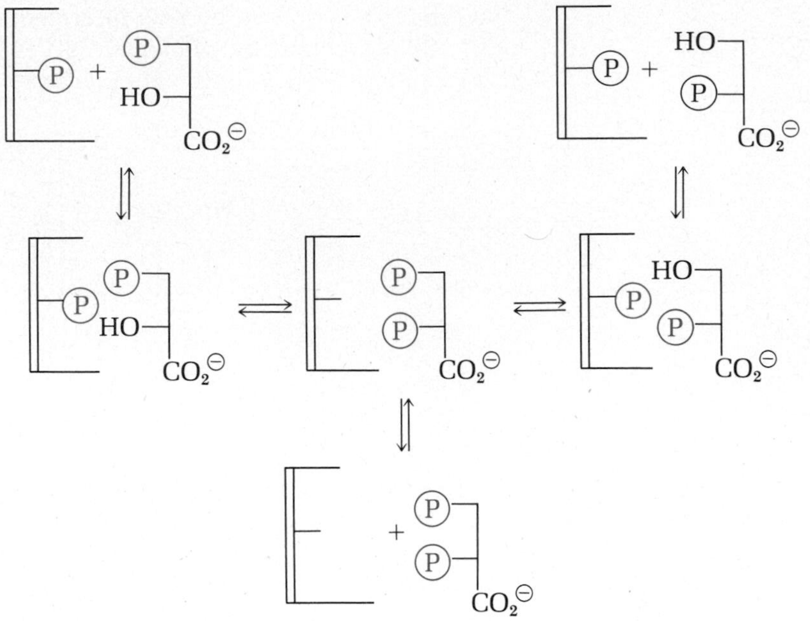

Figure 14.5 Pathway for phosphoryl transferin the reaction catalyzed by rabbit muscle and yeast phosphoglycerate mutase. [From W. A. Blättler and J. R. Knowles, *Biochemistry, 19,* 738–743 (1980). Reprinted with permission. Copyright 1980 American Chemical Society.]

C. Phosphoglycerate Mutase

Phosphoglycerate mutase interconverts 3-phosphoglycerate and 2-phosphoglycerate (reaction 14.34).

$$\underset{\text{3-Phosphoglycerate}}{\mathrm{H{-}C(CO_2^{\ominus})(CH_2OPO_3^{2\ominus}){-}OH}} \xrightleftharpoons[K\,=\,6.3]{\text{Phosphoglycerate mutase}} \underset{\text{2-Phosphoglycerate}}{\mathrm{H{-}C(CO_2^{\ominus})(CH_2OH){-}OPO_3^{2\ominus}}} \qquad (14.34)$$

In most cells, under most conditions, the phosphoglyceric acids are present in equilibrium concentrations.

Two types of phosphoglycerate mutases are known. The enzymes from yeast and rabbit muscle require 2,3-diphosphoglycerate (2,3-DPG) as a cofactor and catalyze *intermolecular* phosphoryl group transfer (Figure 14.5). The wheat germ enzyme, on the other hand, does not require 2,3-diphos-

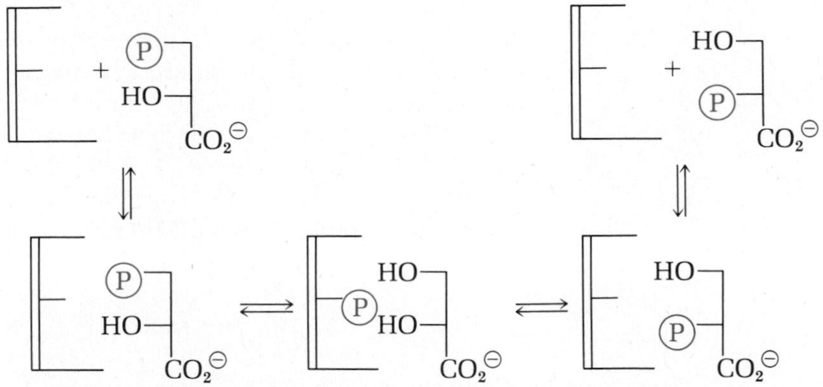

Figure 14.6 Pathway for phosphoryl transfer in the reaction catalyzed by wheat germ phosphoglycerate mutase. [From W. A. Blättler and J. R. Knowles, *Biochemistry, 19,* 738–743 (1980). Reprinted with permission. Copyright 1980 American Chemical Society.]

phoglycerate and catalyzes *intramolecular* phosphoryl group transfer (Figure 14.6). In both cases the reaction occurs with retention of configuration in the migrating phosphoryl group, implying a double displacement mechanism. Both enzymatic reactions proceed through a phosphoenzyme intermediate.

The details of phosphoryl group transfer in the yeast enzyme are summarized in Figure 14.7. Two nearly parallel imidazole side chains of histidyl residues flank the substrate. One of them bears a phosphate group, donated by 2,3-diphosphoglycerate. This phosphate group is transferred to the C-2 oxygen giving an enzyme-bound 2,3-diphosphoglycerate intermediate. A second phosphoryl group transfer from the intermediate to the second histidine regenerates the phosphoenzyme and produces 2-phosphoglycerate.

D. Synthesis of 2,3-Diphosphoglycerate, an Allosteric Inhibitor of Hemoglobin

Phosphoglycerate mutase is not a source of 2,3-diphosphoglycerate because in the mutase the diphospho intermediate never leaves the enzyme surface. Biosynthesis of 2,3-DPG occurs by another pathway. The energy-rich acyl-phosphate 1,3-diphosphoglycerate does not always donate its phosphate

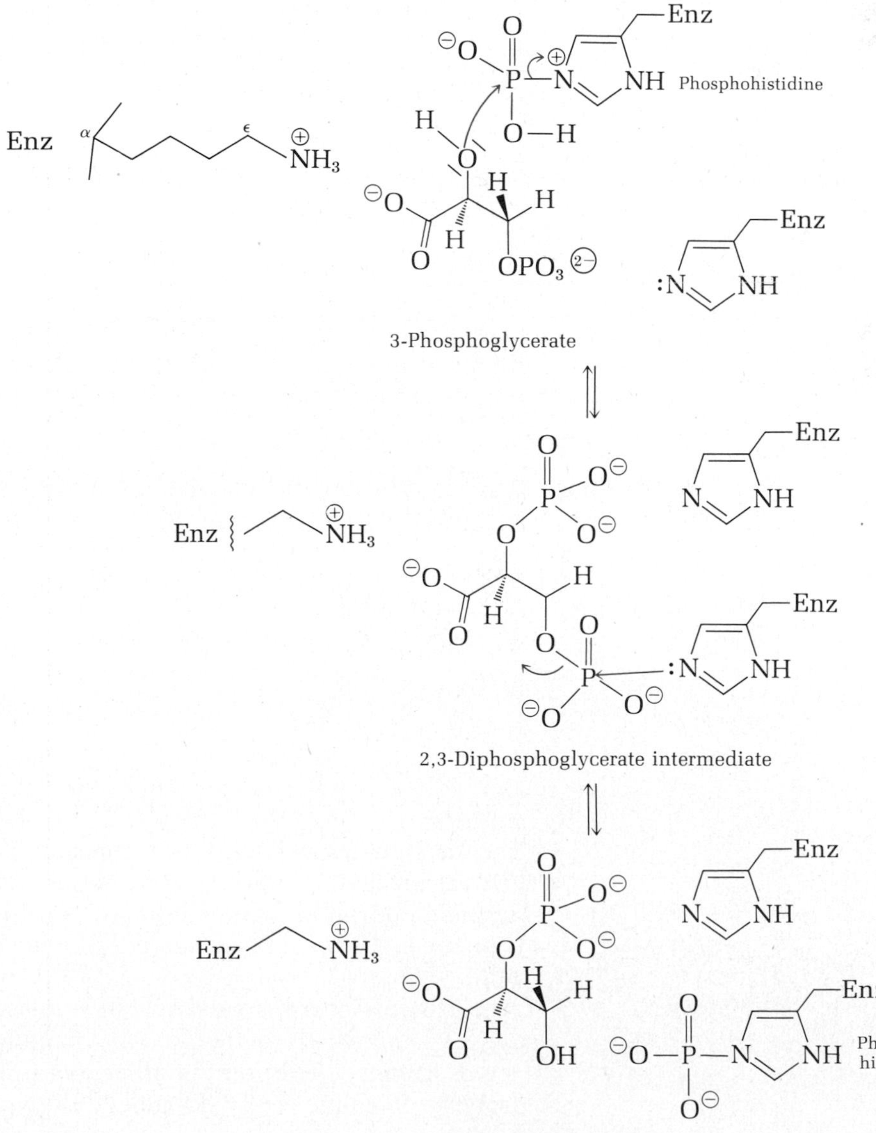

Figure 14.7 Mechanism of action of yeast phosphoglycerate mutase.

group to ADP. In red blood cells diphosphoglycerate mutase interconverts 1,3-diphosphoglycerate and 2,3-DPG (reaction 14.35).

$$
\begin{array}{c}
\underset{\text{1,3-Diphosphoglycerate}}{
\begin{array}{c}
O \\
\parallel \\
C-O-PO_3^{2\ominus} \\
\mid \\
H-C-OH \\
\mid \\
CH_2OPO_3^{2\ominus}
\end{array}}
\quad\xrightarrow[\text{mutase}]{\text{Diphosphoglycerate}}\quad
\underset{\text{2,3-Diphosphoglycerate}}{
\begin{array}{c}
CO_2^{\ominus} \\
\mid \\
H-C-OPO_3^{2\ominus} \\
\mid \\
CH_2OPO_3^{2\ominus}
\end{array}}
\end{array}
\tag{14.35}
$$

Although present in only trace amounts in most cells, in red blood cells the concentration of 2,3-DPG is about 4 mM. Hydrolysis of the 2-phosphate ester of 2,3-DPG, catalyzed by a phosphatase, produces 3-phosphoglycerate that reenters glycolysis (reaction 14.36).

$$
\underset{\text{2,3-DPG}}{
\begin{array}{c}
CO_2^{\ominus} \\
\mid \\
H-C-OPO_3^{2\ominus} \\
\mid \\
CH_2OPO_3^{2\ominus}
\end{array}}
+ H_2O
\xrightarrow[\text{Phosphatase}]{\text{2,3-DPG}}
$$

$$
\underset{\text{3-Phosphoglycerate}}{
\begin{array}{c}
CO_2^{\ominus} \\
\mid \\
H-C-OH \\
\mid \\
CH_2OPO_3^{\ominus}
\end{array}}
+ HPO_4^{2\ominus}
\tag{14.36}
$$

The synthesis of 2,3-DPG in red blood cells is therefore a shunt along the main path of glycolysis that is intimately related to oxygen transport in higher mammals (Figure 14.8). This shunt, however, dissipates the free energy that can otherwise be used to drive the synthesis of ATP.

E. Enolase   Enolase catalyzes the conversion of 2-phospho-D-glycerate to phosphoenol pyruvate (PEP) (reaction 14.37). The enzyme requires Mg(II) ions for activity.

$$
\underset{\text{2-Phospho-D-glycerate}}{
\begin{array}{c}
CO_2^{\ominus} \\
\mid \\
H-C-OPO_3^{2\ominus} \\
\mid \\
CH_2OH
\end{array}}
\quad\underset{\text{Mg(II)}}{\overset{\text{Enolase}}{\rightleftharpoons}}\quad
\underset{\substack{\text{Phosphoenolpyruvate} \\ \text{(PEP)}}}{
\begin{array}{c}
{}^{2\ominus}O_3PO \qquad CO_2^{\ominus} \\
\diagdown \quad\diagup \\
C \\
\parallel \\
C \\
\diagup \quad\diagdown \\
H \qquad H
\end{array}}
\tag{14.37}
$$

The reaction has an equilibrium constant of 6.7, and 2-phosphoglycerate and PEP are present in approximately equilibrium concentrations in cells. PEP is formed by *anti* elimination of water.[2] Rapid formation of a carbanion appears to be the first step of the reaction, followed by slow loss of hydroxide (Figure 14.9).

The loss of water catalyzed by enolase has an important result for glycolysis. The standard free energy of hydrolysis of 2-phosphoglycerate is -17.5 kJ mol^{-1}. PEP, on the other hand, has a standard free energy of hydrolysis of about -53.5 kJ mol^{-1}. Dehydration has therefore created an

---

[2] In an *anti* elimination the dihedral angle between the leaving groups is 180°.

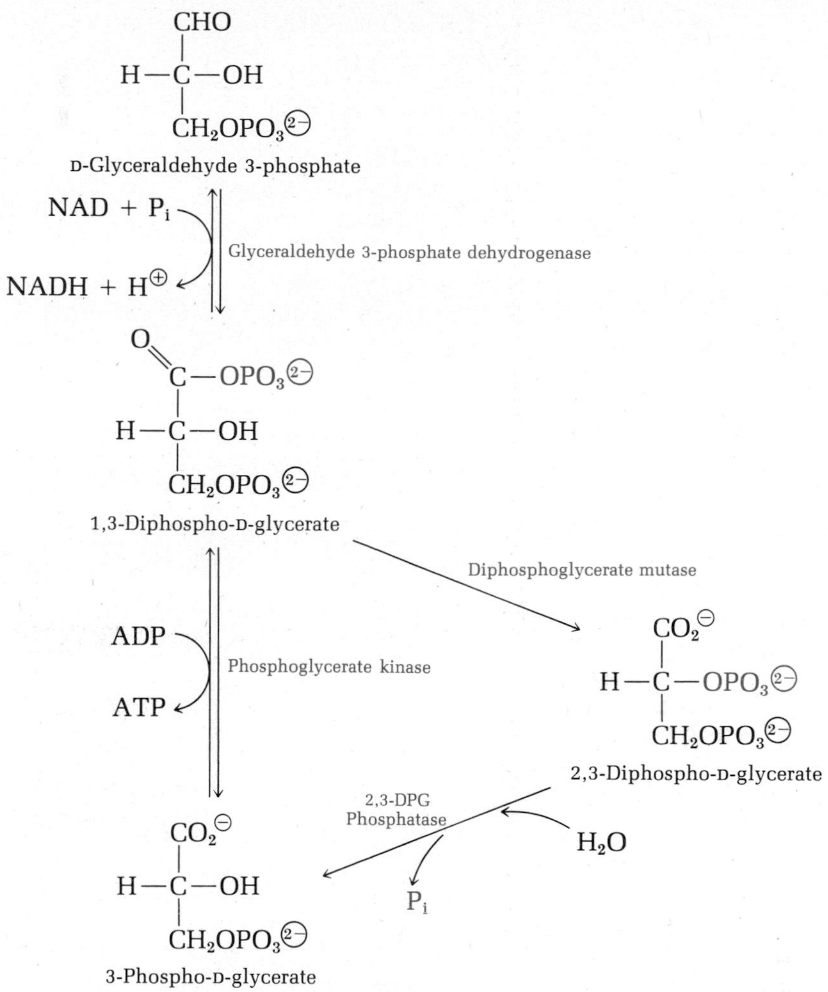

$$\underset{\text{D-Glyceraldehyde 3-phosphate}}{\begin{array}{c} \text{CHO} \\ | \\ \text{H---C---OH} \\ | \\ \text{CH}_2\text{OPO}_3^{\,2\ominus} \end{array}}$$

NAD + P_i

NADH + $H^{\oplus}$

Glyceraldehyde 3-phosphate dehydrogenase

$$\underset{\text{1,3-Diphospho-D-glycerate}}{\begin{array}{c} \text{O} \\ \| \\ \text{C---OPO}_3^{\,2\ominus} \\ | \\ \text{H---C---OH} \\ | \\ \text{CH}_2\text{OPO}_3^{\,2\ominus} \end{array}}$$

Diphosphoglycerate mutase

ADP

Phosphoglycerate kinase

ATP

$$\underset{\text{2,3-Diphospho-D-glycerate}}{\begin{array}{c} \text{CO}_2^{\ominus} \\ | \\ \text{H---C---OPO}_3^{\,2\ominus} \\ | \\ \text{CH}_2\text{OPO}_3^{\,2\ominus} \end{array}}$$

2,3-DPG Phosphatase

H_2O

P_i

$$\underset{\text{3-Phospho-D-glycerate}}{\begin{array}{c} \text{CO}_2^{\ominus} \\ | \\ \text{H---C---OH} \\ | \\ \text{CH}_2\text{OPO}_3^{\,2\ominus} \end{array}}$$

Figure 14.8 Shunt in glycolysis that leads to synthesis of 2,3-diphospho-D-glycerate (2,3-DPG) in red blood cells.

energy-rich metabolite that will donate a phosphoryl group to ADP in the next step of glycolysis.

F. Pyruvate Kinase Pyruvate kinase catalyzes phosphoryl group transfer from PEP to ADP (reaction 14.38). The reaction occurs by an in-line phosphoryl group transfer (recall Sections 12.5 and 12.6).

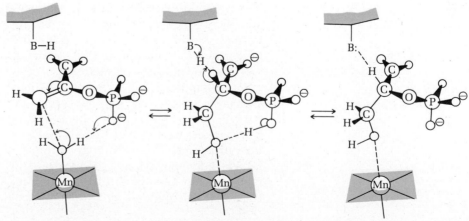

Figure 14.9 Mechanism of action of enolase. [From A. S. Mildvan, *Ann. Rev. Biochem., 43,* 382 (1974). Reproduced with permission. © 1974 by Annual Reviews Inc.]

$$
\begin{array}{c}
\underset{\text{Phosphoenolpyruvate}}{\underset{\text{(PEP)}}{\overset{\displaystyle ^{\scriptsize \textcircled{2}\ominus}O_3PO}{\underset{H}{\overset{C}{\underset{\|}{C}}}}\overset{CO_2^{\ominus}}{}}}
\end{array}
+ \text{ADP} \xrightarrow{\underset{\text{kinase}}{\text{Pyruvate}}} \underset{\text{Pyruvate}}{CH_3-\overset{\displaystyle O}{\overset{\|}{C}}-CO_2^{\ominus}} + \text{ATP} \qquad (14.38)
$$

The reaction catalyzed by pyruvate kinase strongly favors the formation of ATP since the group transfer potential of PEP is so much higher than the group transfer potential of ATP (reaction 14.39).

$$
\begin{array}{ll}
\text{PEP} + H_2O \longrightarrow \text{pyruvate} + HPO_4^{\textcircled{2}\ominus} & \Delta G^{\circ\prime} = -53.5 \text{ kJ mol}^{-1} \\
\underline{\text{ADP} + HPO_4^{\textcircled{2}\ominus} \longrightarrow \text{ATP} + H_2O} & \underline{\Delta G^{\circ\prime} = +36.8 \text{ kJ mol}^{-1}} \\
\text{PEP} + \text{ADP} \longrightarrow \text{pyruvate} + \text{ATP} & \Delta G^{\circ\prime} = -16.7 \text{ kJ mol}^{-1}
\end{array}
$$

$$(14.39)$$

Pyruvate kinase is found in three isozymic forms in higher animals. All are composed of four subunits. Type M is found in muscle and brain; type L is found in parenchymal liver cells,[3] to a minor extent in the kidney, and also in erythrocytes. Type K (also designated in a profusion of nomenclature as A or M_2) is found in the kidney and is also the major enzyme of nonparenchymal liver cells. The erythrocyte and liver type L isozymes appear to be under the same genetic control.

The type L isozyme is subject to complex regulation. A diet high in carbohydrates stimulates synthesis of the L_4 isozyme. The phenomenon of protein synthesis in the presence of a metabolite is known as *induction*. The concentration of the L_4 isozyme can fluctuate by a factor of 10. Liver is a major storage site for glycogen, and the dietary stimulation of the liver isozyme seems to be a means to meet extraordinary energy demands. For example, long distance runners, in need of energy, eat vast amounts of carbohydrates before racing.

The activity of pyruvate kinase is further regulated by a protein phosphorylation system that responds to the peptide hormone *glucagon* and to the steroid hormones known as *glucocorticoids* (Figure 14.10). A cAMP-dependent protein kinase is hormonally activated. The active kinase transfers a phosphoryl group to pyruvate kinase. Phosphorylated pyruvate kinase has a much higher K_m for PEP than the physiological concentration of PEP, and the phosphorylated pyruvate kinase is therefore relatively inactive. When pyruvate kinase is inactive, PEP is diverted from glycolysis to the reverse pathway of gluconeogenesis, or the resynthesis of glucose (Section 17.3). A specific phosphatase hydrolyzes phosphorylated pyruvate kinase to restore normal catalytic activity. The muscle and kidney isozymes are not affected by hormones. In muscle all PEP is converted to pyruvate, producing the ATP required for muscle contraction.

A genetic defect in some persons leads to a deficiency of pyruvate kinase. In affected persons glycolytic intermediates accumulate, including 2,3-DPG in red blood cells. Since DPG inhibits oxygen binding by hemoglobin, persons having inordinately high levels of 2,3-DPG have oxygen transport problems. This is the opposite problem from the one caused by hexokinase deficiency where too low a 2,3-DPG level causes inefficient transfer of oxygen from hemoglobin to myoglobin. A delicate balance must be maintained for adequate oxygen transport.

---

[3] The epithelial covering of an organ is known as its *parenchyma* (Greek, "something poured in").

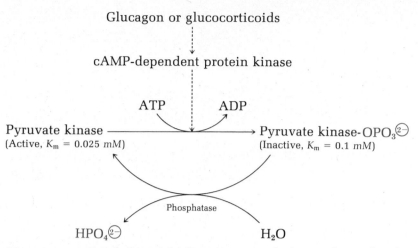

Figure 14.10 Regulation of the liver isozyme of pyruvate kinase by glucagon and glucocorticoids. These hormones stimulate a cAMP-dependent protein kinase, resulting in phosphorylation and inactivation of pyruvate kinase.

14–5 THE ANAEROBIC FATES OF PYRUVATE

Pyruvate is a branch-point metabolite that can undergo many transformations in various organisms under either aerobic or anaerobic conditions. In this section we consider only two fates of pyruvate in anaerobic metabolism: reduction to ethanol in yeast and reduction to lactate in animals. The many microbial transformations so important in microbiology are beyond the scope of this text. In Section 14.8 we will consider the aerobic metabolism of pyruvate.

A. Fermentation in Yeast

Under anaerobic conditions yeast cells convert pyruvate to ethanol and carbon dioxide in two steps. First, the thiamine pyrophosphate–dependent enzyme *pyruvate decarboxylase* (Section 10.4A) converts pyruvate to acetaldehyde and carbon dioxide. Second, the NAD-dependent enzyme *alcohol dehydrogenase* reduces acetaldehyde to ethanol, regenerating the NAD required by glyceraldehyde 3-phosphate dehydrogenase for the continued functioning of glycolysis (reaction 14.41 and Figure 14.11).

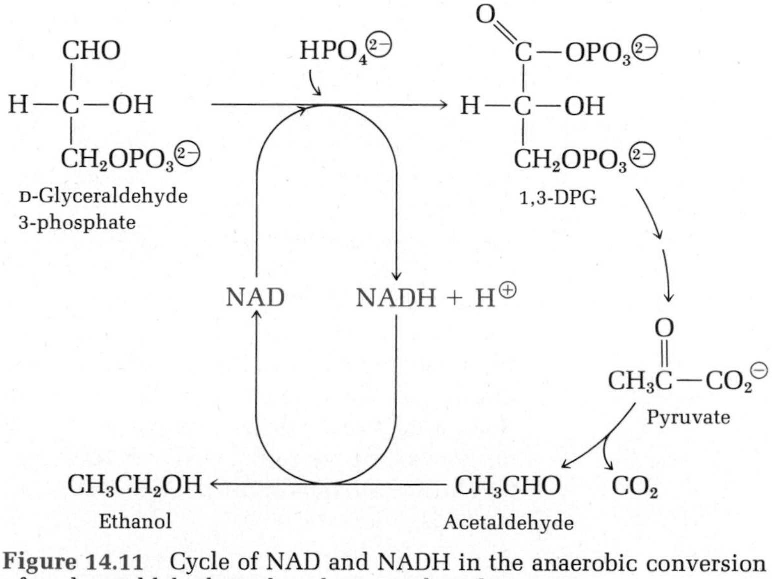

Figure 14.11 Cycle of NAD and NADH in the anaerobic conversion of D-glyceraldehyde 3-phosphate to ethanol in yeast.

$$\underset{\text{Pyruvate}}{CH_3\overset{\overset{\displaystyle O}{\|}}{C}-CO_2^{\ominus}} \xrightarrow{\text{Pyruvate decarboxylase}} \underset{\text{Acetaldehyde}}{CH_3CHO} + CO_2$$

$$\xrightarrow[\text{Alcohol dehydrogenase}]{\text{NADH}} \text{NAD}$$

$$\underset{\text{Ethanol}}{CH_3CH_2OH}$$

(14.40)

No net oxidation of glucose occurs under these anaerobic conditions, and the net reaction for the conversion of glucose to ethanol is

$$\text{glucose} + 2ADP + 2HPO_4^{\ominus} \longrightarrow 2\text{ethanol} + 2CO_2 + 2ATP$$
$$\Delta G^{\circ\prime} = -166 \text{ kJ mol}^{-1} \quad (14.41)$$

Although anaerobic, the degradation of glucose is nevertheless strongly exergonic. L. Pauling has calculated that the resonance stabilization of the carbon dioxide is -151 kJ mol^{-1}, and it seems reasonable to conclude that this resonance stabilization is the driving force for anaerobic production of ethanol and carbon dioxide from glucose in yeast.

B. Reduction to Lactate in Animals

Under anaerobic conditions pyruvate is reduced to L-lactate in animals. For example, an abundance of lactate is produced in diving seals once they have exhausted their oxygen storage capacity, in white fish muscle, and in human tissue that is poorly supplied with blood, such as the cornea of the eye. When rapidly contracting skeletal muscle outstrips the available oxygen supply, it, too, produces lactate. In fact, anyone who has felt muscle cramps during strenuous exercise has experienced the effects of anaerobic metabolism in muscle in which the pyruvate resulting from glycolysis has been converted to the "muscle poison" lactate. Glycolysis is the major source of energy in anaerobic tissues.

L-Lactate is formed from pyruvate by the action of the NAD-dependent enzyme *lactate dehydrogenase* (reaction 14.42).

$$H^{\oplus} + NADH + \underset{\text{Pyruvate}}{CH_3\overset{\overset{\displaystyle O}{\|}}{C}-CO_2^{\ominus}} \xrightarrow{\text{Lactate dehydrogenase}}$$

$$\underset{\text{L-Lactate}}{CH_3-\overset{\overset{\displaystyle OH}{|}}{\underset{\underset{\displaystyle H}{|}}{C}}-CO_2^{\ominus}} + NAD$$

(14.42)

Once again no net oxidation of glucose has occurred in the conversion of glucose to lactate since the NADH produced by the action of glyceraldehyde 3-phosphate dehydrogenase is consumed by the reaction of lactate dehydrogenase. By recycling NAD, muscle is able to obtain energy for contraction for short periods of time by glycolysis. The net reaction for the conversion of glucose to lactate is

$$\text{D-Glucose} + 2HPO_4^{\ominus} + 2ADP \longrightarrow 2 \text{ L-lactate} + 2H^{\oplus} + 2ATP$$
$$\Delta G^{\circ\prime} = -127 \text{ kJ mol}^{-1} \quad (14.43)$$

The conversion of glucose to lactate is less exergonic than the conversion of glucose to ethanol and carbon dioxide, presumably because lactate is less resonance stabilized than carbon dioxide.

14–6 GLYCOLYSIS IN LIVING CELLS

Throughout the preceding discussion we have considered experiments performed upon pure enzymes or upon cell-free extracts that contain all of the enzymes required for glycolysis. Recent advances in nuclear magnetic resonance (nmr) spectroscopy have made it possible to study metabolism in living cells. These studies, particularly of processes in *E. coli*, provide an important link between our reductionist approach to the study of individual enzymes and the vastly more complex interactions of the living organism.

When the bacterial cells are starved of glucose, the concentration of glycolytic intermediates is low. After the addition of glucose, the concentration of β-D-fructose 1,6-diphosphate increases dramatically as glycolysis "takes off." Fructose 1,6-diphosphate is the major metabolic intermediate during glycolysis; its concentration rises from nearly zero to about 13 mM. Once all of the glucose has been consumed, about two minutes are required to completely deplete the supply of fructose 1,6-diphosphate. Meanwhile the nucleotide pool, which is high in ADP and low in ATP before the addition of glucose, changes over to a high phosphorylation state ratio. Finally, there is a *net flux* of material through aldolase from D-fructose 1,6-diphosphate to the trioses D-glyceraldehyde 3-phosphate and dihydroxyacetone phosphate. During glycolysis in living cells, the aldolase-catalyzed reaction is therefore *not* at equilibrium. This is the major significant departure from our preceding discussion, in which we treated the entire enzyme system of glycolysis as if it were at equilibrium. Equilibrium does not exist in living cells, which are *open* to the environment and through which there is a steady flow of metabolic intermediates and products.

14–7 PHOSPHOFRUCTO-KINASE: THE MAJOR REGULATORY ENZYME OF GLYCOLYSIS

As we indicated in our discussion of the individual enzymes of glycolysis and of glycolysis in living *E. coli* cells, phosphofructokinase is the key regulatory point in glycolysis. Phosphofructokinase (PFK) is regulated by a half-dozen metabolites (see Table 14.2) that collectively reflect the metabolic state of the system. A high ATP concentration, for example, indicates an adequate energy supply, reflected in a high adenylate energy charge or phosphorylation state ratio. If the energy supply is adequate, there is no reason for a high rate of glycolysis, and so ATP inhibits PFK. On the other hand, a high ADP concentration indicates a need for energy, ADP stimulates PFK, and the rate of glycolysis increases.

PFK from muscle, brain, and heart is activated by low concentrations of fructose 1,6-diphosphate, especially in the transition from anaerobic to aerobic metabolism and the shift from rest to exercise in muscle. This is an especially elegant regulatory mechanism, since the onset of muscle contraction, for example, will be accompanied by a high demand for ATP, and glycolysis is preparing itself for this demand for energy through the activation of PFK by fructose 1,6-diphosphate. By this means glycolysis can maintain a high ATP/ADP ratio under conditions when the energy supply is high, but the demand for energy is also high. Phosphofructokinase therefore operates in bursts stimulated by fructose 1,6-diphosphate.[4] This is an example of *activation by product* rather than negative feedback inhibition. Suppose that glycolysis is proceeding at a steady state rate, a condition approached when phosphofructokinase activity is about one-fourth the rate of ATP consump-

---

[4] Notice how this oscillation of fructose 1,6-diphosphate is corroborated by *in vivo* studies of glycolysis (Section 14.6).

tion. A burst of phosphofructokinase activity, caused by a sudden demand for energy and accompanied by a burst in fructose 1,6-diphosphate, results in an oscillation in the rate of glycolysis. The result is that fructose 1,6-diphosphate causes a higher than normal ATP/ADP ratio for short periods of high energy demand.

Citrate is an allosteric inhibitor of phosphofructokinase. A high citrate concentration signals an active citric acid cycle. And, since the citric acid cycle leads to a much more efficient production of ATP than glycolysis, a

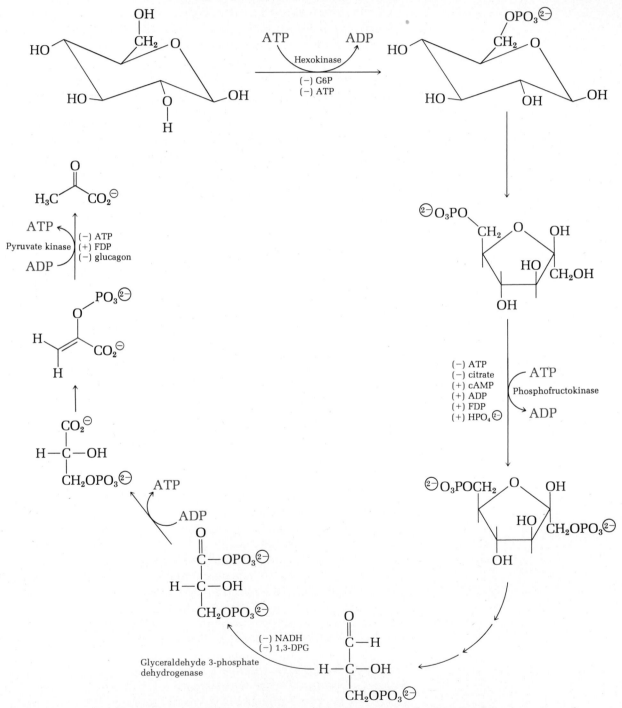

Figure 14.12 Regulatory features of glycolysis. Phosphofructokinase is the key regulatory enzyme of glycolysis. Hexokinase, glyceraldehyde 3-phosphate dehydrogenase, and pyruvate kinase are also subject to allosteric regulation. Metabolites that are inhibitors of glycolysis are indicated by (−); activators are indicated by (+).

high citrate concentration is therefore a signal that the cell has an adequate energy supply.

The major regulatory features of glycolysis are summarized in Figure 14.12.

14–8 AEROBIC FATE OF PYRUVATE: CONVERSION TO ACETYL COENZYME A

Under aerobic conditions pyruvate undergoes oxidative decarboxylation, and the two-carbon fragment that remains emerges as the acetyl group of acetyl coenzyme A. The net reaction is

$$\text{pyruvate} + \text{CoA} + \text{NAD} \longrightarrow \text{acetyl-CoA} + \text{NADH} + CO_2 \qquad (14.44)$$

Three enzymes, assembled in a multienzyme complex, are responsible for this reaction. First, *pyruvate decarboxylase* (Section 10.4) converts pyruvate to hydroxyethylthiamine pyrophosphate (reaction 14.45).

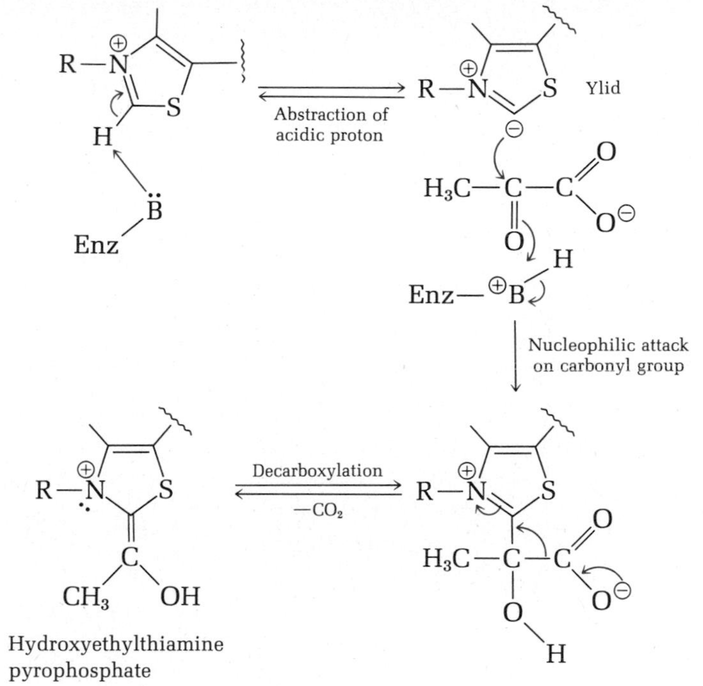

$$(14.45)$$

Second, *dihydrolipoyl transacetylase* shifts the two-carbon fragment from thiamine pyrophosphate to lipoic acid and thence to coenzyme A (reaction 14.46).

hydroxyethylthiamine pyrophosphate + CoA ⟶

acetyl-CoA + thiamine pyrophosphate   (14.46)

We discussed the mechanism of this two-step acyl transfer reaction in Section 10.5. Third, the flavoprotein *dihydrolipoyl dehydrogenase* (Section 10.5) reoxidizes the reduced form of lipoic acid (reaction 14.47). The $FADH_2$

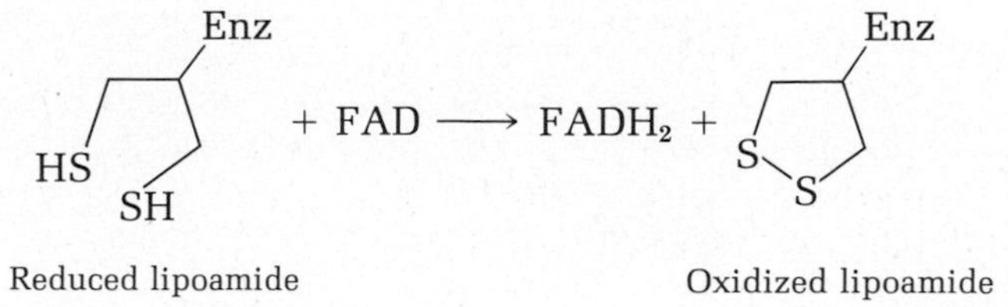

$$(14.47)$$

Reduced lipoamide   Oxidized lipoamide

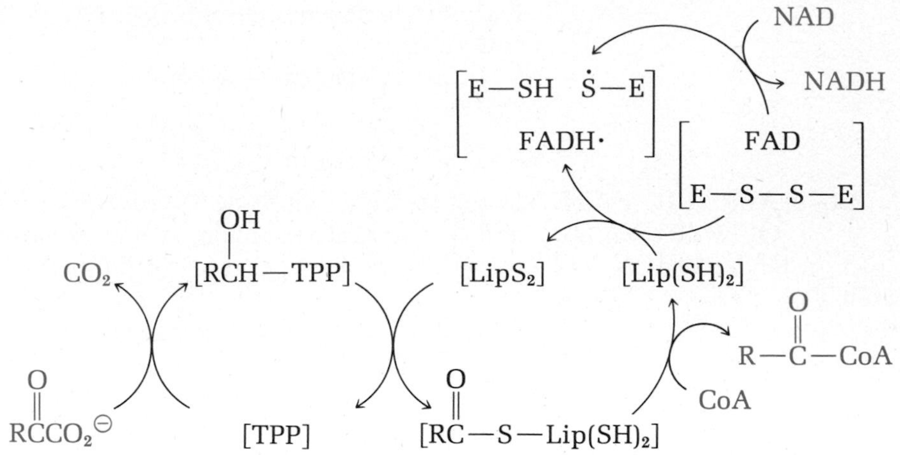

Figure 14.13 Sequence of reactions catalyzed by pyruvate dehydrogenase and αketoglutarate dehydrogenase multienzyme complexes. R is methyl in the pyruvate dehydrogenase complex; it is $^{\ominus}O_2C-CH_2-CH_2-$ in the α-ketoglutarate dehydrogenase complex. [From L. J. Reed, *Acc. Chem. Res., 7,* 40–46 (1974).]

produced in this step then reduces NAD (reaction 14.48), completing the action of pyruvate dehydrogenase (Figure 14.13).

$$FADH_2 + NAD \longrightarrow FAD + NADH + H^{\oplus} \qquad (14.48)$$

The pyruvate dehydrogenase complex from *E. coli* contains a total of 60 polypeptide chains whose total molecular weight is 4 million. Pyruvate decarboxylase is represented by 24 subunits, in 12 dimers, of molecular weight 183,000; dihydrolipoyl transacetylase contributes 24 subunits, each with a molecular weight of 70,000; and dihydrolipoyl dehydrogenase adds 6 dimers (12 subunits), each with a molecular weight of 112,000. The composition of the *E. coli* complex, including coenzymes, is summarized in Table 14.3. The complex can be separated into inactive subunits, and it spontaneously reassembles when the components—pyruvate decarboxylase, dihydrolipoyl transacetylase, and dihydrolipoyl dehydrogenase—are mixed at pH 7.0.

The transacetylase lies at the center of the complex with its subunits arrayed in a symmetrical cube. Surrounding this core enzyme are the 24 pyruvate decarboxylase subunits and the 24 dihydrolipoyl dehydrogenase subunits (Figure 14.14). The entire structure is a "microscopic crystal" with 60 faces.

We have drawn attention to the "long arm" by which lipoic acid is attached to dihydrolipoyl transacetylase. This enzyme lies at the center of the complex, and it has been proposed that the arm picks up the β-hydroxyethyl group bound to thiamine pyrophosphate in pyruvate decarboxylase and then

Table 14.3 Components of *E. coli* pyruvate dehydrogenase complex

Enzyme	Coenzyme	Number of subunits	Molecular weight	Number of subunits per complex
Pyruvate de-carboxylase	Thiamine pyrophosphate	2	192,000	24 (12 dimers)
Dihydrolipoyl transacetylase	Lipoic acid, coenzyme A	24	1,700,000	24
Dihydrolipoyl dehydrogenase	FAD, NAD	2	112,000	12 (6 dimers)

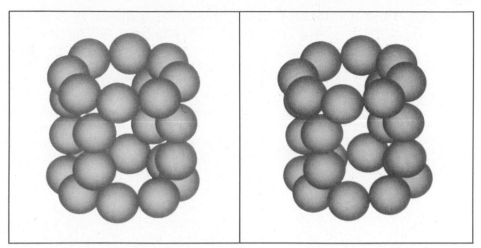

Figure 14.14 Stereo view of a schematic diagram of the structure of the pyruvate dehydrogenase complex. The 24 subunits of dihydrolipoyl transacetylase lie at the center of the complex. A dimer of the flavoenzyme dihydrolipoyl dehydrogenase binds on each face of the cube formed by the 24 transacetylase subunits, and a dimer of pyruvate decarboxylase binds at each of 6 identical sites on the edges of the central cube.

swings it round to the site where the acetyl group is transferred to coenzyme A. The arm then moves the reduced disulfide to dihydrolipoyl dehydrogenase where the disulfide is regenerated (Figure 14.15). The net charge on the lipoyl group during this process goes from 0 to -1 to -2. The change in charge might provide the driving force for movement of the arm.

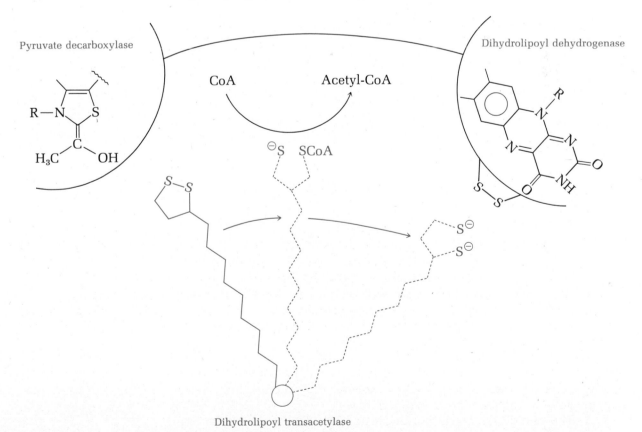

Figure 14.15 Lipoyllysyl group of the transacetylase. The group rotates from the decarboxylase to the dehydrogenase during the course of the reactions catalyzed by the pyruvate dehydrogenase complex. [Redrawn from L. J. Reed, *Acc. Chem. Res.*, **7**, 43 (1974).]

14–9 SUMMARY Glycolysis is the pathway for the catabolism of glucose that leads to pyruvate. Two moles of ATP are produced in glycolysis. The production of ATP is preceded by an ATP-driven series of reactions in which 2 moles of ATP are expended in the conversion of glucose to fructose 1,6-diphosphate. Aldolase converts fructose 1,6-diphosphate to the trioses dihydroxyacetone phosphate and D-glyceraldehyde 3-phosphate. The latter is converted to pyruvate. The acyl phosphate, 1,3-diphospho-D-glycerate, donates a phosphoryl group to ADP in the first energy-producing step of glycolysis. The formation of 1,3-diphospho-D-glycerate is also accompanied by reduction of NAD to NADH. The second energy-yielding reaction of glycolysis is phosphoryl group transfer from phosphoenol pyruvate to ADP, catalyzed by pyruvate kinase. Since 2 moles of ATP are produced per mole of triose, the net yield of glycolysis is 2 moles of ATP per mole of glucose.

Phosphofructokinase is the key regulatory enzyme of glycolysis. This allosteric enzyme is controlled by several metabolites that reflect the energy state of the cell. The ATP/ADP ratio is especially important in the regulation of phosphofructokinase. When the adenylate energy charge (or phosphorylation state ratio) is low, the rate of glycolysis is high, and vice versa.

Under anaerobic conditions the NAD required for the continued operation of glycolysis can be regenerated by various means. In yeast pyruvate is converted to acetaldehyde and then to ethanol, and NAD is regenerated from NADH by alcohol dehydrogenase. In muscle pyruvate is converted to lactate by lactate dehydrogenase, and NAD is regenerated from NADH.

In aerobic organisms or tissues pyruvate is oxidized by the multienzyme complex pyruvate dehydrogenase. Acetyl-CoA is produced by pyruvate dehydrogenase.

REFERENCES

Boyer, P. D., Ed., "The Enzymes," Vols. 5–9, Academic Press, New York, 1972. (The various volumes contain discussions of all glycolytic enzymes.)

Florkin, M., and Stotz, E. H., "Comprehensive Biochemistry," Vol. 17, Elsevier, New York, 1967.

Larner, J., "Intermediate Metabolism and Its Regulation," Prentice-Hall, Englewood Cliffs, N.J., 1971.

Newsholme, E. A., and Start, C., "Regulation in Metabolism," Wiley, New York, 1974.

Shulman, R. G., Brown, T. R., Ugurbil, K., Ogawa, S., Cohen, S. M., and den Hollander, J. A., *Science, 205,* 160–166 (1979).

Stanier, R. Y., Douderoff, M., and Adelberg, E. A., "The Microbial World," 3rd ed., Prentice-Hall, Englewood Cliffs, N.J., 1970.

Problems

1. Fluoride inhibits enolase by reaction with Mg(II) to form the precipitate $MgFPO_3$. What metabolic intermediate will accumulate when fluoride is added to cell-free extracts that contain all glycolytic enzymes?

2. What metabolic intermediate will accumulate in cell-free extracts capable of glycolysis if lactate dehydrogenase is inhibited?

3. In yeast extracts capable of fermentation, what intermediate will accumulate if alcohol dehydrogenase is inhibited?

4. Ehrlich ascites tumor cells have an abnormally high rate of glycolysis. An "ATPase" (an enzyme that hydrolyzes ATP) bound to the plasma membrane is unusually active in these tumors. Why should this abnormal enzyme activity increase the rate of glycolysis?

5. The rate of glycolysis is slow in red blood cells where the Mg(II) concentration is about $10^{-4} M$. How does the low Mg(II) concentration lower the rate of glycolysis?

6. What is the effect of each of the following metabolites on the overall

rate of glycolysis: (a) glucose 6-phosphate, (b) fructose 1,6-diphosphate, (c) Mg(II), (d) citrate?

7. Discuss two ways in which the ATP/ADP ratio can affect the rate of glycolysis.

8. Iodoacetate is a potent inhibitor of glyceraldehyde 3-phosphate dehydrogenase. (a) What chemical reaction occurs between iodoacetate and the enzyme? (b) What intermediate will accumulate if iodoacetate is added to cell-free extracts capable of glycolysis?

$$I-CH_2-CO_2^{\ominus}$$
Iodoacetate

9. Write chemical equations which show that there is no net oxidation or reduction in anaerobic glycolysis.

10. In the aldolase-catalyzed condensation of D-glyceraldehyde 3-phosphate and dihydroxyacetone phosphate either the pro-R or the pro-S hydrogen must be stereospecifically abstracted by the enzyme. Which one must be removed to give the observed product: fructose 1,6-diphosphate?

11. Draw the four stereoisomers which would result from forming two new chiral centers at C-3 and C-4 in the product of the aldol condensation catalyzed by aldolase if the enzyme were not stereospecific.

12. Suggest an experiment which would allow you to determine whether or not the pro-R or the pro-S hydrogen is being abstracted from dihydroxyacetone phosphate by aldolase.

13. If glucose labeled with ^{14}C at C-1 is incubated with the necessary glycolytic enzymes and cofactors, where would the label be found on the isolated pyruvate? Assume that the interconversion of glyceraldehyde 3-phosphate and dihydroxyacetone phosphate is very rapid.

14. Enolase isomerizes 2-phospho-3-butenoate to 2-phosphoenol-ketobutyrate. Write a mechanism for this process.

$$H_2C=CH-\underset{\underset{OPO_3^{\ominus}}{|}}{CH}-CO_2^{\ominus} \underset{\text{Enolase}}{\rightleftarrows} CH_3CH=C\underset{OPO_3^{\ominus\ominus}}{\overset{CO_2^{\ominus}}{<}}$$

2-Phospho-3-butenoate   2-Phosphoenolbutyrate

15. D-Glyceraldehyde 3-phosphate dehydrogenase catalyzes phosphate exchange with acetyl phosphate. ^{32}P is used as a marker. What does this observation imply about the mechanism of action of the enzyme?

$$^{\ominus}O-\overset{OH}{\underset{\underset{O}{\|}}{\overset{|}{{}^{32}P}}}-O^{\ominus} + CH_3\overset{O}{\overset{\|}{C}}-OPO_3^{\ominus\ominus} \rightleftarrows CH_3\overset{O}{\overset{\|}{C}}-O-{}^{32}PO_3^{\ominus\ominus} + HPO_4^{\ominus\ominus}$$

16. Class II aldolases catalyze aldol cleavage of fructose 1,6-diphosphate. Instead of forming an imine with the substrate, the class I aldolases employ a Zn(II) ion at their active sites as an electrophilic catalyst. Propose a mechanism that involves electrophilic catalysis.

17. Rat brain hexokinase has two binding sites for glucose 6-phosphate. One has a dissociation constant for glucose 6-phosphate of 2.5×10^{-6} M. The other has a far lower affinity. The low affinity site is the active site of the enzyme, and the high affinity site has a regulatory function. Suggest a role for the high affinity site in the regulation of glycolysis. [Refer-

ence: P. A. Lago, A. Solo, and J. E. Wilson, *J. Biol. Chem.*, *255*, 7548–7551 (1980).]

18. If [1-^{13}C]-glucose is the substrate for glycolysis in yeast cells, the labeled ^{13}C soon appears in C-1 and C-6 of fructose 1,6-diphosphate. Write enzymatic reactions that explain this result. [Reference: J. A. den Hollander, T. R. Brown, K. Ugurbil, and R. G. Shulman, *Proc. Nat. Acad. Sci. U.S.*, *76*, 6096–6100 (1979).]

19. A certain person was found to be incapable of prolonged intense exercise. Analysis of his enzymes revealed that all of the glycolytic muscle enzymes except phosphoglycerate mutase were present in normal concentrations. (a) What metabolic intermediate will accumulate in the relative absence of phosphoglycerate mutase? (b) How will lactate production be affected by the absence of this enzyme? [Reference: S. DiMauro et al., *Science, 212*, 1277–1279 (1981).]

Chapter 15

The citric acid cycle

15-1 INTRODUCTION

If scientific texts were chronicles that traced the tortuous progress of science from one theory to the next, up one blind alley and down another, they would be more faithful to the reality of scientific progress than the orderly narratives they present. The history of the discovery of the citric acid cycle is a case in point. Citric acid was discovered in lemon juice by K. Scheele in 1784, and early in this century it gradually became apparent that citric acid

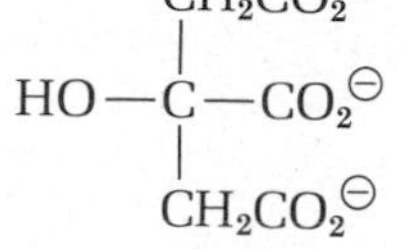

$$
\begin{array}{c}
\mathrm{CH_2CO_2^{\ominus}} \\
| \\
\mathrm{HO-C-CO_2^{\ominus}} \\
| \\
\mathrm{CH_2CO_2^{\ominus}}
\end{array}
$$

Citrate

is found not only in citrus fruits, but is widespread in nature. But it was not until the late 1940s that the universal role of the citric acid cycle was fully understood.

The unraveling of the citric acid cycle began in earnest in the 1930s with the discovery that addition of succinate, fumarate, and malate to

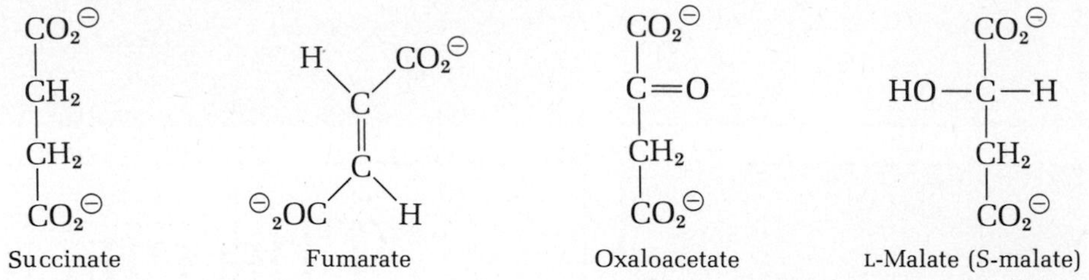

Succinate   Fumarate   Oxaloacetate   L-Malate (S-malate)

minced muscle increased the rate of oxygen uptake. Oxaloacetate was added to the list of these dicarboxylic acids, when it was found to be formed under aerobic conditions from pyruvate. In 1936 A. Szent-Gyorgi proposed that two pairs of the dicarboxylic acids, fumarate and succinate, and oxaloacetate and malate, were interconverted by dehydrogenases and that these dehydrogenases were coupled to respiration (Figure 15.1).

Citric acid was introduced into the picture when C. Martius showed that it was converted to succinate by way of *cis*-aconitate, isocitrate, and α-ketoglutarate. The path from citrate to isocitrate is nonoxidative, but the conversion of isocitrate to α-ketoglutarate requires a dehydrogenase. Citrate is thus converted to oxaloacetate (Figure 15.2).

The considerable complexities of respiration, when superimposed on the as yet unknown citric acid cycle, presented investigators with accumulating observations, but still no way to relate them to one another. One further step provided the solution which unified metabolism at a single stroke. In 1937 H. A. Krebs and W. A. Johnson showed that citrate was derived from pyruvate and oxaloacetate, and the first version of a citric acid cycle could be written (Figure 15.3). Fully a decade was required to work out the details of the cycle. For example, it is acetyl-CoA, not pyruvate, that combines with oxaloacetate to form citrate. Considerable effort was also required to show that the citric acid cycle is a feature of all aerobic cells.

The complete citric acid cycle with all of its cofactors is shown in Fig-

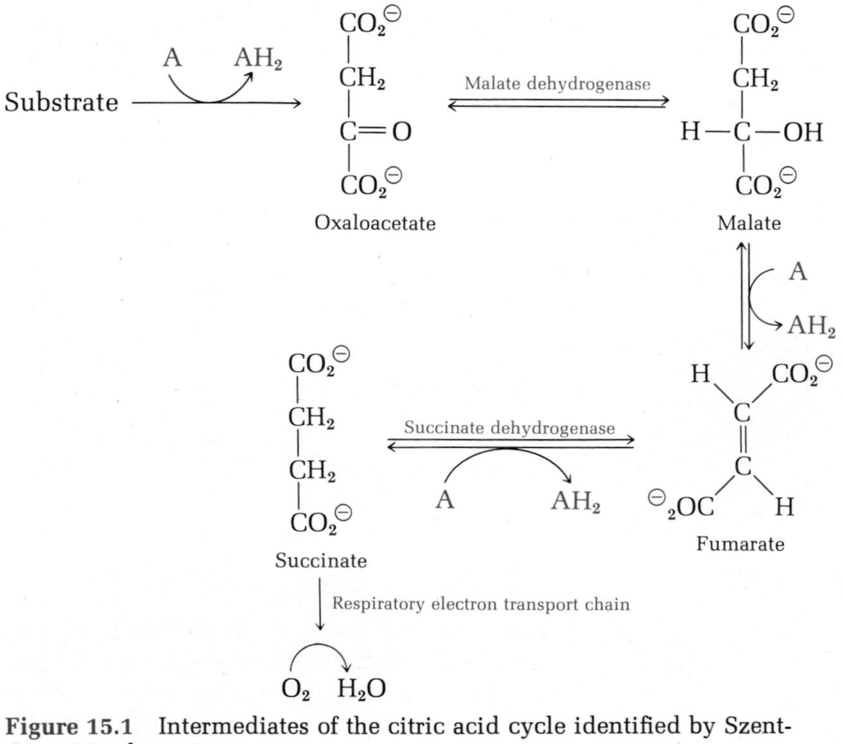

Figure 15.1 Intermediates of the citric acid cycle identified by Szent-Gyorgi in the 1930s.

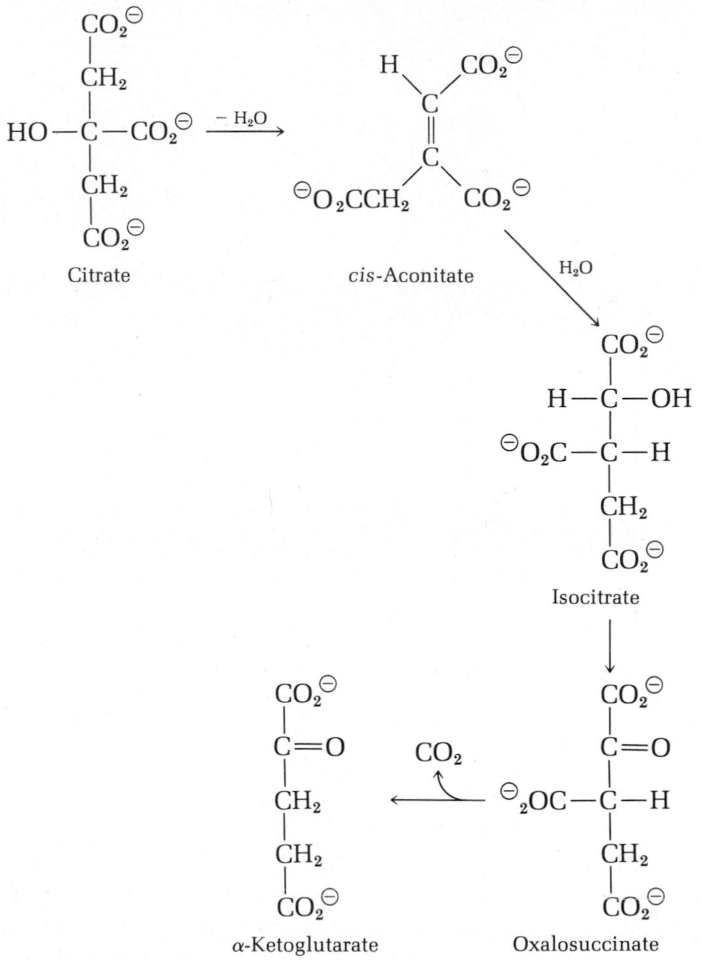

Figure 15.2 Conversion of citrate to α-ketoglutarate demonstrated by Martius in the 1930s.

ure 15.4. We can regard the citric acid cycle as the "hub" of the metabolic wheel. It is the final oxidative stage in the catabolism of carbohydrates, fatty acids, and amino acids, the central pathway of aerobic metabolism.

15–2 AN OVERVIEW OF THE CITRIC ACID CYCLE

A. Energy Production

Perhaps the most striking feature of aerobic metabolism in contrast with anaerobic metabolism is the enormous increase in energy production that is realized by the oxidations of the citric acid cycle and respiration. The conversion of 1 mole glucose to pyruvate by glycolysis yields 197 kJ mol^{-1} (reaction 15.1).

$$\text{glucose} \xrightarrow{\text{Glycolysis}} 2 \text{ pyruvate} \qquad \Delta G^{\circ\prime} = -197 \text{ kJ mol}^{-1} \qquad (15.1)$$

Complete oxidation, however, yields 2,870 kJ mol^{-1} (reaction 15.2)

$$\text{glucose} + 6O_2 \longrightarrow 6CO_2 + 6H_2O \qquad \Delta G^{\circ\prime} = -2,870 \text{ kJ mol}^{-1} \qquad (15.2)$$

In fact, oxygen does not enter the citric acid cycle at all, but is the final electron acceptor of the respiratory electron transport chain. The citric acid cycle produces 3 moles of NADH and 1 mole of FADH$_2$ for every mole of acetyl-CoA oxidized. The oxidation of these reduced coenzymes by the electron transport chain is coupled to the phosphorylation of ADP. Three moles of ATP are produced for each mole of NADH oxidized, and 2 moles of ATP are

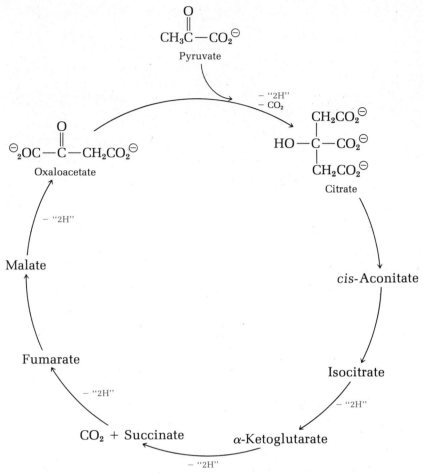

Figure 15.3 The "original" citric acid cycle proposed by Krebs and Johnson in 1937.

produced by oxidation of $FADH_2$. Thus the 3 moles of NADH produced by the citric acid cycle, the NADH produced by glyceraldehyde 3-phosphate dehydrogenase in glycolysis and the NADH produced by the action of pyruvate dehydrogenase yield a total of 34 moles of ATP per mole of glucose oxidized. When added to the 2 moles of ATP produced by glycolysis, and 2 moles of ATP produced by substrate-level phosphorylation in the citric acid cycle, this gives a total of 38 moles of ATP produced by the complete oxidation of glucose (Figure 15.5).

B. Major Chemical Features   The enzymatic reactions of the citric acid cycle are complex. In studying them we shall be concerned with several interrelated subjects: the stereochemistry of the enzymatic reactions, the energy yield of the individual steps, and the regulation of each enzyme individually and of the pathway as a whole.

Consider oxaloacetate (Figure 15.4). This intermediate is the key to the citric acid cycle, since it is a carrier of the two-carbon fragment donated by acetyl-CoA to form citrate. In the presence of an excess of acetyl-CoA the rate of the citric acid cycle is limited by the availability of oxaloacetate, and, conversely, in the presence of excess oxaloacetate the rate of the citric acid cycle is limited by the availability of acetyl-CoA.

Next consider citrate (again referring to Figure 15.4). Citrate is formed from oxaloacetate and acetyl-CoA by a Claisen condensation catalyzed by *citrate synthetase*. Perhaps the most puzzling aspect of the citric acid cycle,

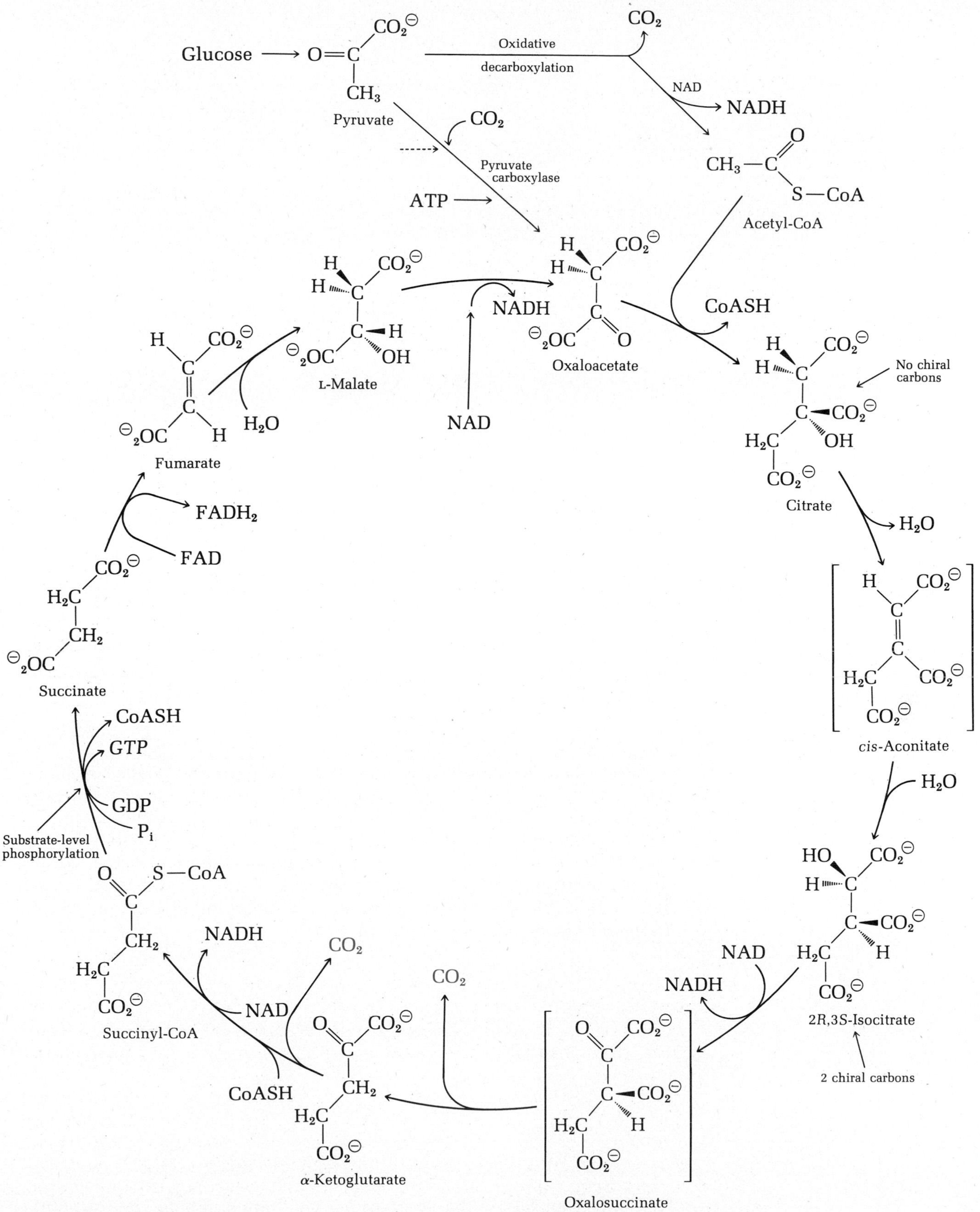

Figure 15.4 The citric acid cycle.

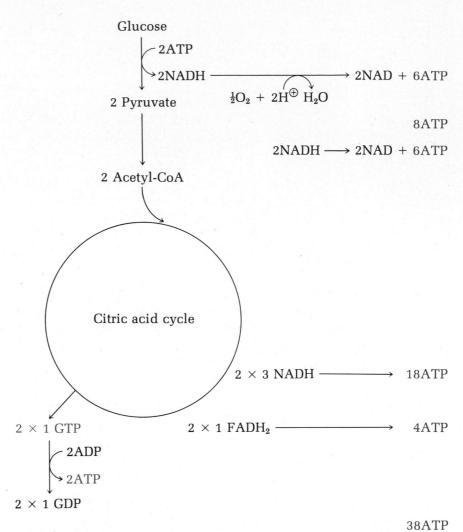

Figure 15.5   Energy production of the citric acid cycle and respiration. The complete oxidation of 1 mole of glucose leads to production of up to 38 moles of ATP.

and the major reason that a decade was required to advance from the original cycle of Krebs and Johnson to the modern cycle, is the fate of the two prochiral arms of citrate. Citrate, a symmetrical molecule, is converted to isocitrate, a molecule having two chiral centers. Only one of the four possible diastereoisomers is formed (reaction 15.3).

The stereochemistry of the reactions leading from citrate to 2R,3S-isocitrate is one of the most important chemical features of the entire pathway.

The conversion of isocitrate to succinyl-CoA occurs in successive oxidative decarboxylations, each one resulting in formation of NADH. The first oxidative decarboxylation is catalyzed by *isocitrate dehydrogenase*; the second is catalyzed by the multienzyme complex α-*ketoglutarate dehydrogenase*, which resembles the pyruvate dehydrogenase complex.

The conversion of succinyl-CoA to succinate is accompanied by the phosphorylation of guanosine diphosphate (GDP). The thioester of succinyl-CoA is energy rich, and the overall reaction resembles the phosphorylation of ADP in the reaction of phosphoglycerate kinase in glycolysis (Section 14.4B). Phosphoryl group transfer to GDP is not coupled to the respiratory chain and is known as a *substrate-level phosphorylation*.

The membrane-bound enzyme *succinate dehydrogenase* catalyzes the oxidation of succinate to fumarate. FAD is the coenzyme of succinate dehydrogenase and $FADH_2$ is its reduced counterpart. Succinate dehydrogenase is also a component of the electron transport chain (Section 16.5B).

The regeneration of oxaloacetate from fumarate occurs in two steps. First, the hydration of fumarate catalyzed by *fumarase* yields L-malate. Second, malate is oxidized by *malate dehydrogenase* to give oxaloacetate. In this fourth oxidative step of the citric acid cycle NADH is produced.

All of the reactions of the citric acid cycle occur within the mitochondria, in contrast to the cytostolic reactions of glycolysis. This compartmentation of metabolic pathways is one of the simplest means of metabolic regulation. Some citrate diffuses out of the mitochondrion and, by its allosteric interaction with phosphofructokinase, mediates the interactions of the citric acid cycle and glycolysis.

15–3 ENZYMES OF THE CITRIC ACID CYCLE

A. Citrate Synthetase

Citrate synthetase catalyzes a Claisen condensation of oxaloacetate (OAA) and acetyl-CoA. The α-carbanion of a thioester is unusually stable; its pK_a of the α-proton is about 8.5 (Section 10.3A). Citrate synthetase abstracts an α-proton from acetyl-CoA, and the carbanion attacks the keto group of oxaloacetate to give *citryl-CoA* (reaction 15.4).

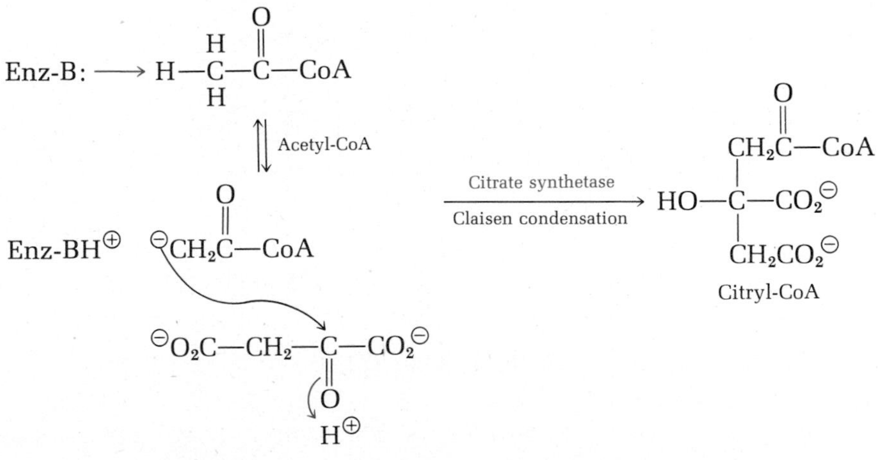

The driving force for the reaction is hydrolysis of the energy-rich thioester of citryl-CoA, $\Delta G^{\circ'}_{hydrolysis} = -30$ kJ mol^{-1} (reaction 15.5).

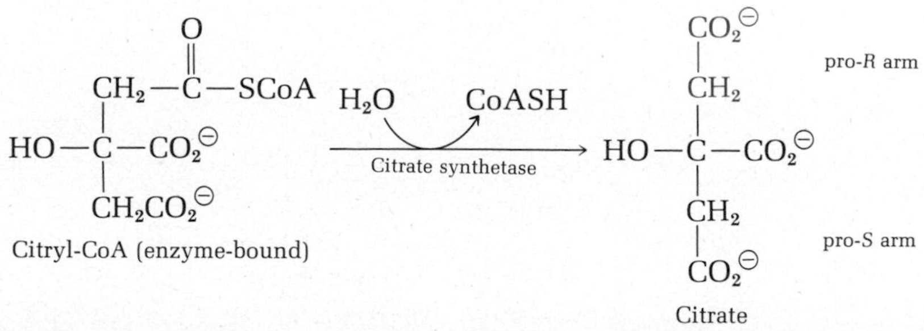

Citrate is a symmetrical molecule whose two —$CH_2CO_2^{\ominus}$ groups are prochiral. If acetyl-CoA labeled with ^{14}C in the methyl position is the substrate for citrate synthetase, only S-[2-^{14}C]citrate is produced in the reaction. This means that the *si* face of the carbonyl group of oxaloacetate is exclusively attacked by the methyl carbon of acetyl-CoA. The stereochemistry of this reaction is even more subtle. When R-α-[^1H,^2H,^3H]acetyl-CoA is the substrate for the reaction, the product has *two* chiral centers. One of them results from inversion of configuration at the chiral methyl group of acetyl-CoA. This inversion of configuration occurs when the enzyme stereospecifically removes ^1H from R-α-[^1H,^2H,^3H]acetyl-CoA, and the carbanion undergoes pyramidal inversion as it attacks the carbonyl carbon of oxaloacetate (Figure 15.6).

B. Aconitase and the "Ferrous Wheel"

The conversion of citrate to isocitrate catalyzed by *aconitase* involves an equilibrium among citrate, *cis*-aconitate, and isocitrate (reaction 16.6).

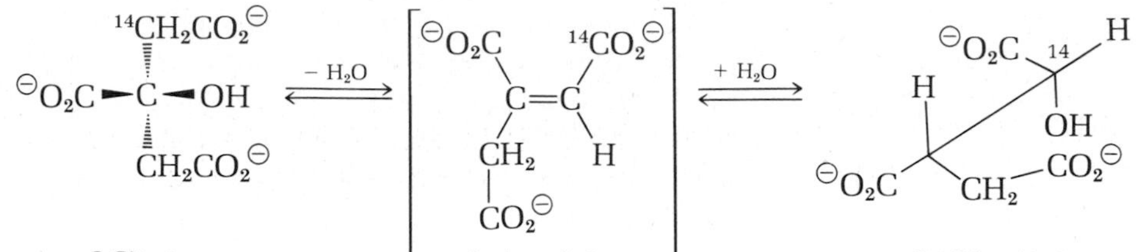

(15.6)

S-Citrate *cis*-Aconitate 2R,3S-Isocitrate

Loss of water from citrate gives *cis*-aconitate. Addition of water to *cis*-aconitate with the reverse orientation gives isocitrate. At equilibrium the ratio of citrate to *cis*-aconitate to isocitrate is about 90:4:6. In this process symmetrical citrate is converted to chiral isocitrate. Of the four possible diastereoisomers of isocitrate, only one is produced.

Citrate has two pairs of prochiral hydrogens, one pair in the pro-R arm and one pair in the pro-S arm. Aconitase removes the pro-R hydrogen from

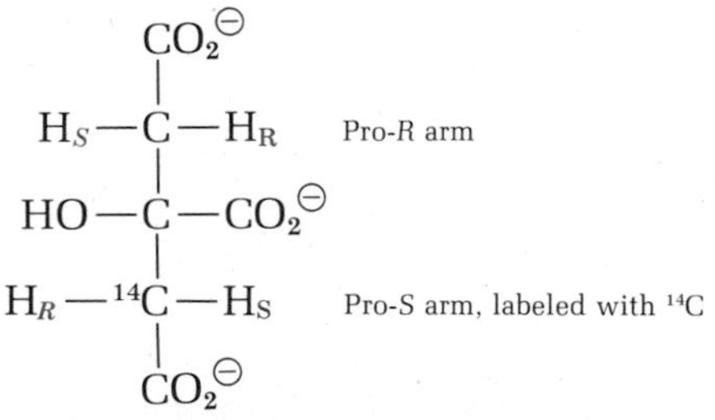

S-[2-^{14}C]citrate

the pro-R arm in an *anti* elimination of water to give *cis*-aconitate (reaction 15.7).

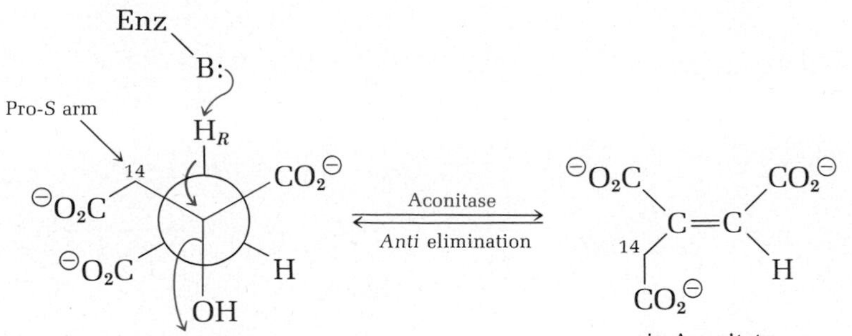

(15.7)

S-[2-^{14}C]citrate *cis*-Aconitate

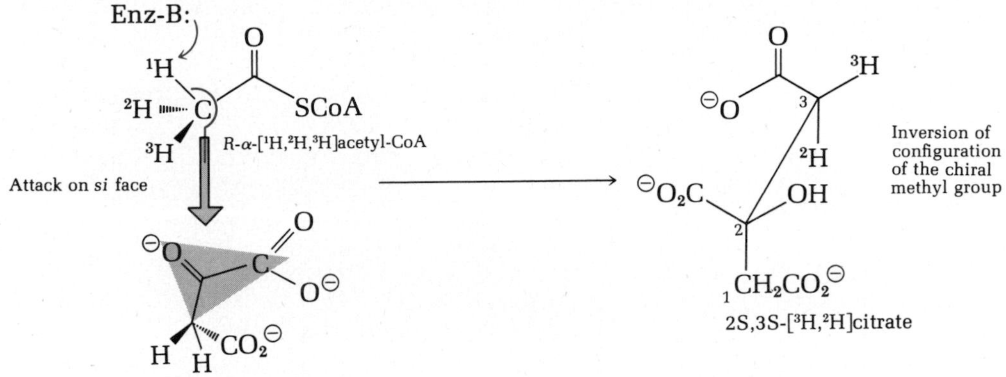

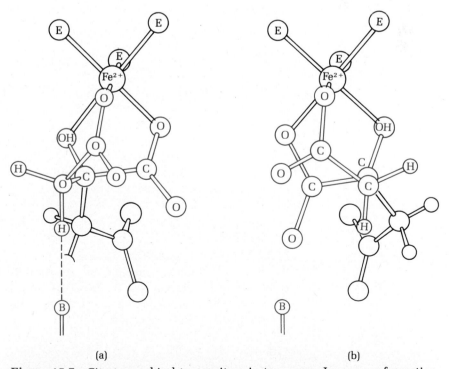

Figure 15.6 Stereochemistry of the reaction catalyzed by S-citrate synthase. The reaction occurs by stereospecific transfer of a hydride anion from the methyl group of acetylCoA to the *si* face of oxaloacetate. The configuration of the chiral methyl group is inverted in the course of the hydride transfer.

An Fe(II) atom at the active site of aconitase holds citrate in position for reaction by forming a complex that involves two carboxyl groups and a hydroxyl group. The two prochiral hydrogens of the pro-R arm are in different environments in this complex. Only one of the two complexes places the pro-R hydrogen at C-2 of the pro-R arm *anti* to the hydroxyl group in the conformation required for *anti* elimination. The other complex can form, but is not correctly positioned for the elimination reaction (Figure 15.7).

Cis-aconitate has two prochiral *sp²* carbons. The two faces of the planar system are named by indicating the chirality at each prochiral carbon (Fig-

Figure 15.7 Citrate can bind to aconitase in two ways. In one conformation (a), the prochiral methylene proton that must be removed to give *cis*-aconitate is correctly positioned for abstraction. In the other conformation (b), this proton is too far away to be abstracted. The enzyme is therefore able to distinguish between prochiral protons and between prochiral —$CH_2CO_2^{\ominus}$ groups. [Courtesy of Dr. Jenny P. Glusker.]

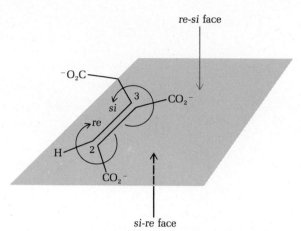

Figure 15.8 Stereochemical designations of the prochiral centers of *cis*-aconitate. C-2, viewed from above, has the *re* configuration, and C-3, also viewed from above, has the *si* configuration. The top face is therefore *re-si*.

ure 15.8). Aconitase adds water to *cis*-aconitate with $OH^{\ominus}$ attacking C-2 from the *si-re* face and $H^{\oplus}$ adding at C-3 on the *re-si* face producing 2R,3S-isocitrate (reaction 15.8).

$$\text{(15.8)}$$

cis-Aconitate   2R,3S-Isocitrate

In the mechanism we have described a proton is extracted from one side of citrate and returned from the opposite side. Several possibilities have been proposed for this reaction. We will describe one of them. The *cis*-aconitate produced by loss of water from citrate must rotate from its "citrate-like" conformation to an isocitrate conformation. This rotation is the result of a change of places by water and a carboxylate anion bound to the active site iron. *Anti* addition of water to the "isocitratelike" conformation gives isocitrate (Figure 15.9). This mechanism, with its rotation of isocitrate around Fe(II), has been called the "ferrous wheel" by an incorrigible punster.

C. Isocitrate Dehydrogenase

Isocitrate dehydrogenase catalyzes oxidative decarboxylation of isocitrate in the first oxidative step of the citric acid cycle. The reaction occurs in two steps. First, isocitrate is oxidized to *oxalosuccinate* with simultaneous reduction of NAD (reaction 15.9).

$$\text{+ NAD} \xrightarrow[\text{dehydrogenase}]{\text{Isocitrate}} \text{+ NADH} \qquad (15.9)$$

2R,3S-Isocitrate   3S-Oxalosuccinate

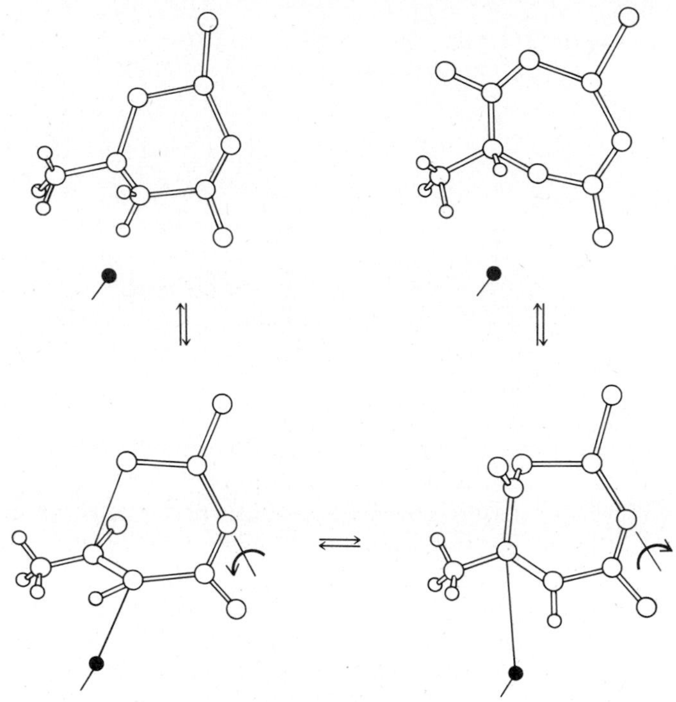

Figure 15.9 "Ferrous wheel" mechanism of action of aconitase. The elimination and addition of water both occur from an *anti* conformation of the substrate, which is held in place by bonds to an active site Fe(II) ion. [Courtesy of Dr. Jenny P. Glusker.]

Oxalosuccinate does not dissociate from the enzyme, and is decarboxylated in the second step of the reaction to give α-ketoglutarate (reaction 15.10).

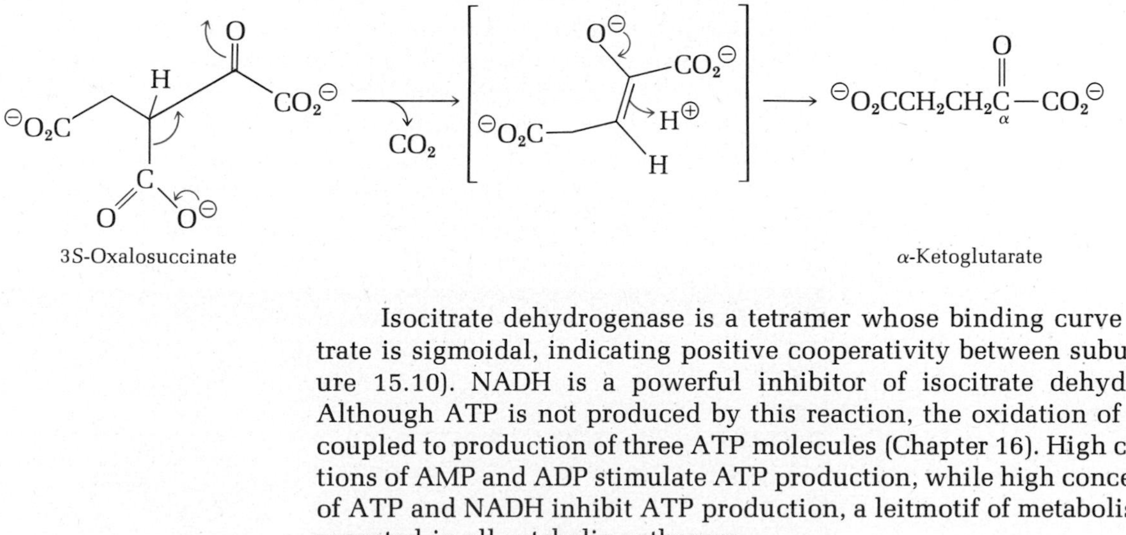

(15.10)

3S-Oxalosuccinate α-Ketoglutarate

Isocitrate dehydrogenase is a tetramer whose binding curve for isocitrate is sigmoidal, indicating positive cooperativity between subunits (Figure 15.10). NADH is a powerful inhibitor of isocitrate dehydrogenase. Although ATP is not produced by this reaction, the oxidation of NADH is coupled to production of three ATP molecules (Chapter 16). High concentrations of AMP and ADP stimulate ATP production, while high concentrations of ATP and NADH inhibit ATP production, a leitmotif of metabolism that is repeated in all catabolic pathways.

D. α-Ketoglutarate Dehydrogenase

The second oxidative step of the citric acid cycle is the conversion of α-ketoglutarate to succinyl-CoA catalyzed by α-ketoglutarate dehydrogenase (reaction 15.11). Reaction 15.11 is analogous to the reaction catalyzed by pyruvate dehydrogenase. The α-ketoglutarate dehydrogenase complex, like the pyruvate dehydrogenase complex, requires thiamine pyrophosphate, lipoic acid bound to a transacetylase, and FAD. The composition of the α-ketoglutarate dehydrogenase complex is summarized in Table 15.1.

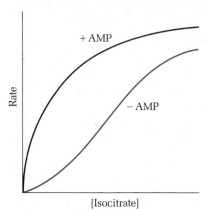

Figure 15.10 Activation of the allosteric enzyme isocitrate dehydrogenase by AMP. [From J. A. Hathaway and D. E. Atkinson, *J. Biol. Chem., 238*, 2875 (1963).]

$$^{\ominus}O_2CCH_2CH_2-\overset{\overset{\displaystyle O}{\|}}{C}-CO_2^{\ominus} + NAD + CoA-SH \xrightarrow{\overset{\alpha\text{-Ketoglutarate}}{\text{dehydrogenase}}}$$

$$^{\ominus}O_2CCH_2CH_2\overset{\overset{\displaystyle O}{\|}}{C}-\overset{\overset{\displaystyle O}{\|}}{C}-SCoA + NADH + CO_2 \qquad (15.11)$$

Succinyl-CoA

To this point in the citric acid cycle 2 moles of CO_2 have been lost, and 2 moles of NADH have been produced. In one turn of the cycle the carbon dioxide is derived from the original molecule of oxaloacetate. If ^{14}C-labeled acetyl-CoA is the substrate for citrate synthetase, one trip around the citric acid cycle would not release any $^{14}CO_2$. This observation demonstrates that aconitase can in fact distinguish between the two prochiral arms of citrate. Otherwise equal amounts of $^{12}CO_2$ and $^{14}CO_2$ would be produced by the successive decarboxylations of the citric acid cycle.

E. Succinate Thiokinase

The energy stored in the energy-rich thioester succinyl-CoA is conserved as the energy-rich phosphoanhydride of guanosine triphosphate (GTP) in a reaction catalyzed by succinate thiokinase (reaction 15.12).

$$^{\ominus}O_2CCH_2CH_2-\overset{\overset{\displaystyle O}{\|}}{C}-SCoA + GDP + HPO_4^{\textcircled{2\ominus}} \underset{\xrightarrow{\hspace{1cm}}}{\overset{\text{Succinate thiokinase}}{\xleftarrow{\hspace{1cm}}}}$$

$$^{\ominus}O_2CCH_2CH_2CO_2^{\ominus} + GTP + CoA\text{-}SH \qquad (15.12)$$

Succinate

The standard free-energy change is -3.3 kJ mol^{-1} for this *substrate-level phosphorylation* (so named because it is not coupled to an oxidation).

Succinate thiokinase contains a phosphohistidine at its active site. Phosphorylation of the active site histidine apparently occurs at the same time as cleavage of the thioester bond, since when ^{18}O-labeled phosphate is a substrate, ^{18}O is found in the succinate isolated from the reaction. The rate of formation of succinyl phosphate is, however, much slower than the overall reaction rate, and it has been suggested that the catalytic mechanism is concerted (Figure 15.11). The synthesis of GTP occurs by phosphoryl group transfer from the active site phosphohistidine to GTP. The GTP

Table 15.1 Components of *E. coli* α-ketoglutarate dehydrogenase complex

Enzyme	Coenzyme	Number of subunits	Molecular weight	Number of subunits per complex
α-Ketoglutarate dehydrogenase	Thiamine pyrophosphate	2	95,000 each 190,000 total	12
Dihydrolipoyl transsuccinylase	Lipoic acid, coenzyme A	24	42,000 each 1,000,000 total	24
Dihydrolipoyl-succinyl dehydrogenase (a flavoprotein)	FAD, NAD	2	56,000 each 112,000 total	12

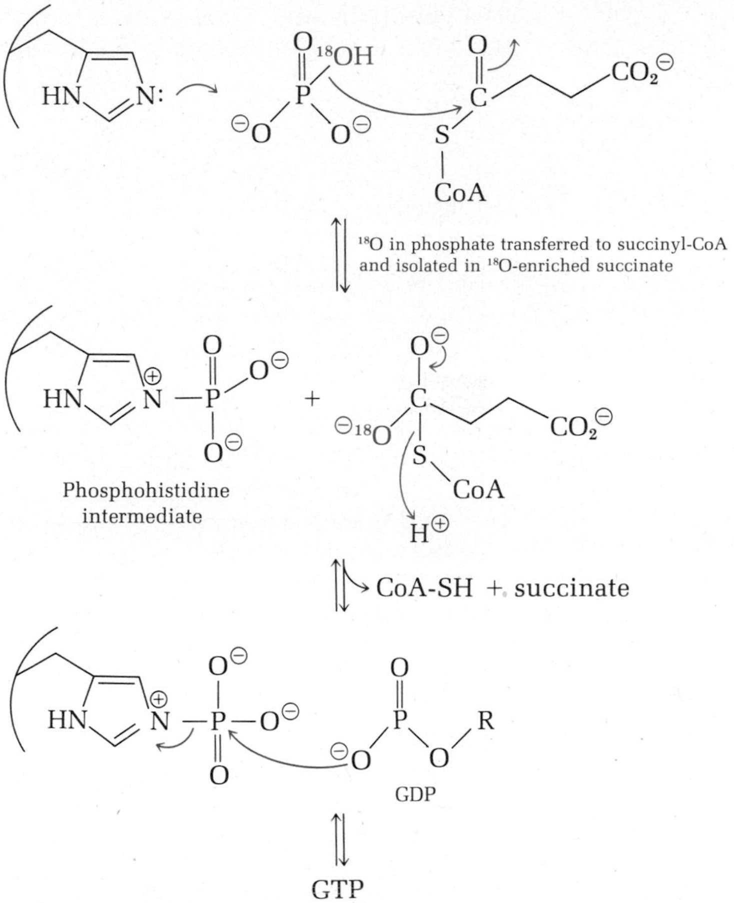

Figure 15.11 Possible concerted reaction mechanism for succinate thiokinase. [Adapted from J. L. Robinson, R. W. Benson, and P. D. Boyer, *Biochemistry*, 8, 2503 (1979). Reprinted with permission. Copyright 1979 American Chemical Society.]

formed by succinate thiokinase is coupled to synthesis of ATP by *nucleoside diphosphate kinase* (reaction 15.13).

$$\text{GTP} + \text{ADP} \underset{\text{Nucleoside diphosphate kinase}}{\rightleftharpoons} \text{ATP} + \text{GDP} \tag{15.13}$$

GTP itself is an important energy-rich metabolite that plays a part in protein synthesis. Succinate is also a versatile metabolite, and it serves as one of the precursors in the biosynthesis of heme.

F. Succinate Dehydrogenase

Succinate dehydrogenase, a component of both the citric acid cycle and the electron transport chain, catalyzes the oxidation of succinate to fumarate, reducing FAD to $FADH_2$ in the process (reaction 15.14).

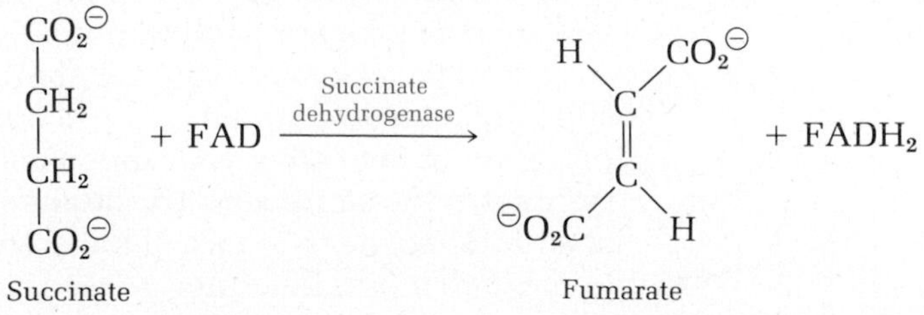

Succinate dehydrogenase is an integral protein of the inner mitochondrial membrane. It is one of about a dozen flavoproteins in which the flavin is covalently bound to the protein. In mammalian succinate dehydrogenase the flavin is covalently bound to a histidyl residue by its 8-α-methyl group.

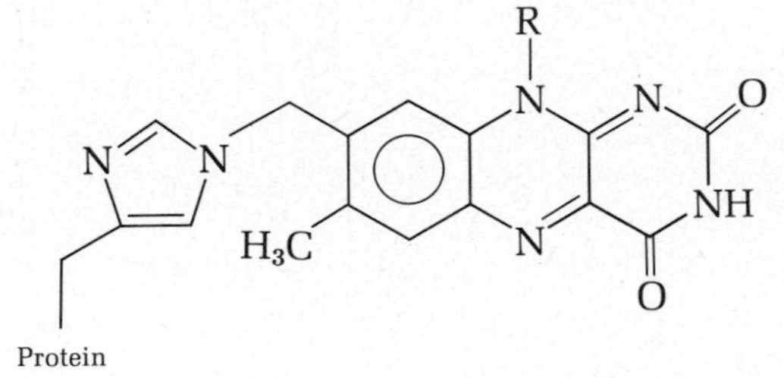

Histidyl-8-α-FAD

Succinate dehydrogenase is a dimer whose subunits have molecular weights of 70,000 and 30,000. The larger subunit is bound to FAD, and both subunits contain an "iron-sulfur center" (Section 16.3B).

Succinate is a symmetrical molecule having two pairs of prochiral hydrogens. Succinate dehydrogenase abstracts the pro-R proton at one carbon,

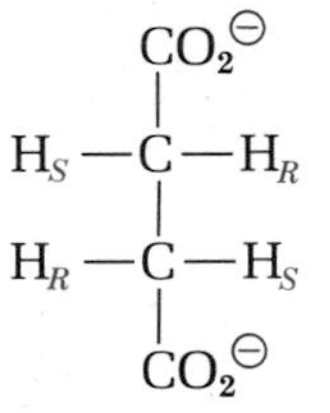

Succinate

and transfers the other hydrogen, with its electron pair, to FAD (Figure 15.12). We have written the mechanism as a concerted *anti* elimination, but a carbanion intermediate has not been experimentally excluded. It is not known whether the pro-R or pro-S hydrogen is transferred to FAD.

G. Fumarase

Fumarase catalyzes the *anti* addition of water to fumarate to give S-malate (reaction 15.16). The equilibrium constant for this readily reversible reaction is about 4.

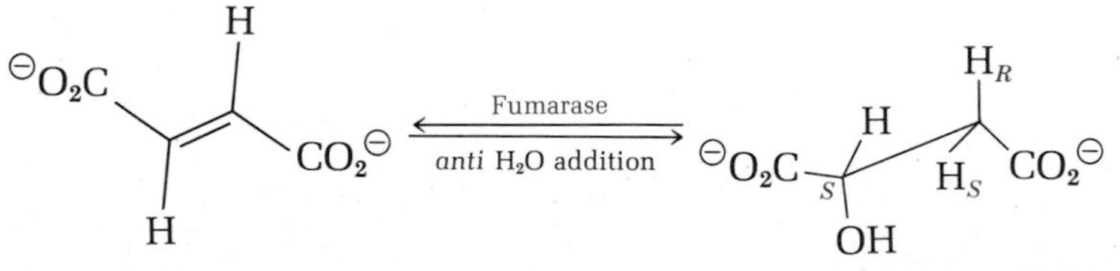

$$(15.15)$$

S-Malate

A carbonium ion has been proposed as an intermediate in the reaction of fumarase. Although carbonium ions are planar, sp^2 hybridized species, and their reactions in solution nearly always lead to racemization, on the surface of the enzyme water can approach the positive charge only from one side. Rotation around the C-2—C-3 bond is prevented by stereospecific complexation of the carboxylate anions of the substrate with positively charged residues at the active site. The observed stereochemistry arises from addition of a proton to the *re-re* face of the double bond followed by addition of water to the carbonium ion at its *si* face (Figure 15.13).

H. Malate Dehydrogenase

Malate dehydrogenase completes the citric acid cycle by oxidizing S-malate to oxaloacetate. NAD is reduced in the process (reaction 15.16).

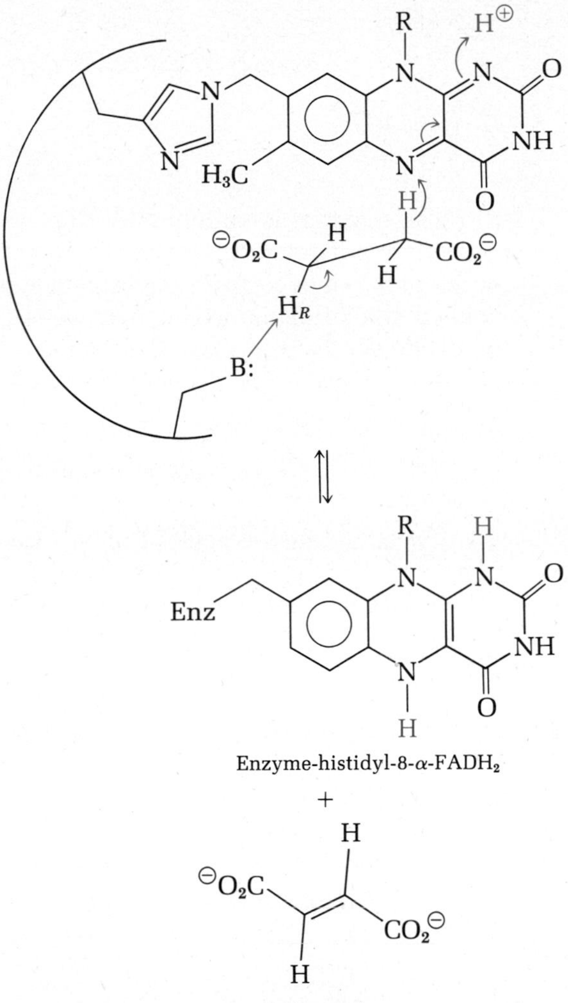

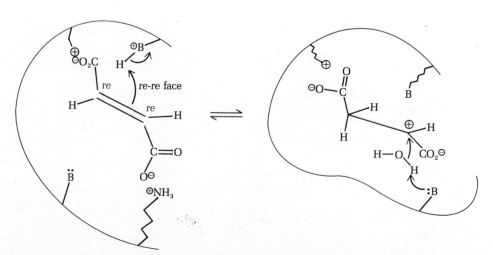

Figure 15.12   Possible concerted reaction mechanism for succinate dehydrogenase. It is known that the pro-*R* proton is abstracted, but it is not certain that the reaction is concerted.

Figure 15.13   Mechanism of action of fumarase. First, a proton adds to the *re-re* face of the double bond. Then, water adds to the *si* face of the trigonal (planar) carbocation.

$$HO-\underset{\underset{\overset{|}{CH_2CO_2^\ominus}}{\overset{|}{|}}}{\overset{\overset{CO_2^\ominus}{|}}{C}}-H \quad + NAD \xrightarrow{\text{Malate}} {}^\ominus O_2CCH_2\overset{\overset{O}{\|}}{C}CO_2^\ominus + NADH + H^\oplus \quad (15.16)$$

S-Malate   Oxaloacetate

The reaction is endergonic, $\Delta G^{\circ\prime} = 29$ kJ mol^{-1}, and the equilibrium constant is much less than one. The reaction proceeds *in vivo* because oxaloacetate is removed from the system when it condenses with acetyl-CoA. Even so, the mitochondrial concentration of oxaloacetate is less than its K_m for citrate synthetase. Viewed thermodynamically, the driving force for the oxidation of malate is cleavage of citryl-CoA in the next step of the cycle (reaction 15.17).

$$\text{malate} + \text{NAD} \longrightarrow \text{oxaloacetate} + \text{NADH}$$
$$\Delta G^{\circ\prime} = +29.3 \text{ kJ mol}^{-1}$$
$$\text{oxaloacetate} + \text{acetyl-CoA} \longrightarrow \text{citrate} + \text{CoA}$$
$$\Delta G^{\circ\prime} = -33.5 \text{ kJ mol}^{-1}$$
$$\overline{\text{malate} + \text{NAD} + \text{acetyl-CoA} \longrightarrow \text{NADH} + \text{CoA} + \text{citrate}}$$
$$\Delta G^{\circ\prime} = -4.2 \text{ kJ mol}^{-1} \quad (15.17)$$

15–4 CONTROL OF THE CITRIC ACID CYCLE

The major regulatory effectors for the citric acid cycle are the concentration of oxaloacetate, the concentration of acetyl-CoA, the NADH/NAD ratio, and the ATP/ADP ratio.

A. Pyruvate Carboxylase and Control of Oxaloacetate Concentration

In animals oxaloacetate is synthesized from pyruvate and carbon dioxide by the biotin-dependent enzyme *pyruvate carboxylase* (Section 11.7) (reaction 15.18).

$$CH_3\overset{\overset{O}{\|}}{C}CO_2^\ominus + \quad \text{N}^1\text{-Carboxybiotin} \xrightarrow{\text{Pyruvate carboxylase}} {}^\ominus O_2CCH_2\overset{\overset{O}{\|}}{C}CO_2^\ominus \quad (15.18)$$

Pyruvate   N^1-Carboxybiotin   Oxaloacetate

Pyruvate carboxylase requires the allosteric effector acetyl-CoA for activity. A high concentration of acetyl-CoA reflects a demand for energy and simultaneously ensures a supply of oxaloacetate.

B. Pyruvate Dehydrogenase Complex and Control of Acetyl-CoA Concentration

The pyruvate dehydrogenase complex plays an important role in regulating the supply of acetyl-CoA available for oxidation by the citric acid cycle. *E. coli* pyruvate dehydrogenase is subject to manifold metabolic regulation. Acetyl-CoA is an inhibitor of the transacetylase component of the complex, and coenzyme A is an activator. NADH is an inhibitor of the dihydrolipoyl dehydrogenase component, and NAD is an activator. ATP is an allosteric inhibitor of pyruvate dehydrogenase, and AMP is an allosteric activator. As a result, the activity of the pyruvate dehydrogenase complex reflects the adenylate energy charge (or phosphorylation state ratio), the NADH/NAD ratio, and the concentrations of acetyl-CoA and coenzyme A.

The mammalian pyruvate dehydrogenase of bovine kidney adds another level of regulation to the allosteric regulation observed for the *E. coli* system. The bovine multienzyme complex is also regulated by covalent modification. *Pyruvate decarboxylase kinase*, present in about five copies per multienzyme complex (Table 15.2), phosphorylates a seryl residue of one of the subunits of pyruvate decarboxylase. ATP is the phosphoryl donor (reaction 15.19).

$$\text{pyruvate decarboxylase-(Ser-OH)} + \text{MgATP}$$

$$\downarrow \text{\small Pyruvate decarboxylase kinase}$$

$$\text{pyruvate decarboxylase-(Ser-OPO}_3{}^{\ominus\ominus}) + \text{MgADP} \qquad (15.19)$$

The inactive, phosphorylated decarboxylase is reactivated by a specific phosphatase (reaction 15.20).

$$\text{pyruvate decarboxylase-(Ser-OPO}_3{}^{\ominus\ominus}) + \text{H}_2\text{O}$$

$$\downarrow \text{\small Phosphatase}$$

$$\text{pyruvate decarboxylase-(Ser-OH)} + \text{HPO}_4{}^{\ominus\ominus} \qquad (15.20)$$

The kinase and the phosphatase are themselves subject to metabolic regulation (Figure 15.14). Pyruvate and ADP both inhibit the inactivation of pyruvate decarboxylase by the kinase. The concentration of Mg(II) also regulates the activity of the phosphatase and the kinase. The concentration of free Mg(II) varies as the ATP/ADP ratio varies because Mg(II) has different affinities for these two nucleotides. At a high concentration of Mg(II) the kinase is inactive and the phosphatase is active. Mg(II) ions act by promoting the binding of the phosphatase to the multienzyme complex and by simultaneously inhibiting binding of the kinase to the multienzyme complex. A concentration of about 2 *mM* Mg(II) gives half-maximal phosphatase activity, and only 0.02 *mM* MgATP is required for half-maximal kinase activity.

Metabolic regulation similar to that observed for pyruvate dehydrogenase is also a feature of two other dehydrogenases of the citric acid cycle. Isocitrate dehydrogenase is inhibited by NADH, which competes with NAD for the coenzyme binding site. It is also inhibited by ATP and activated by ADP. Once more, then, a high ATP/ADP ratio inhibits the citric acid

Table 15.2	**Components of bovine kidney pyruvate dehydrogenase complex**					

Enzyme	Coenzyme	Number of subunits	Molecular weight	Number of subunits per complex
Pyruvate decarboxylase	Thiamine pyrophosphate	2	(2) 41,000 each	40
		2	(2) 36,000 each 154,000 total	40
Dihydrolipoyl transacetylase	Lipoic acid, coenzyme A	60	52,000	60
Dihydrolipoyl dehydrogenase	FAD, NAD	2	55,000 each 110,000 total	10
Pyruvate decarboxylase kinase		?	50,000	~5
Phosphatase		1	100,000	~5

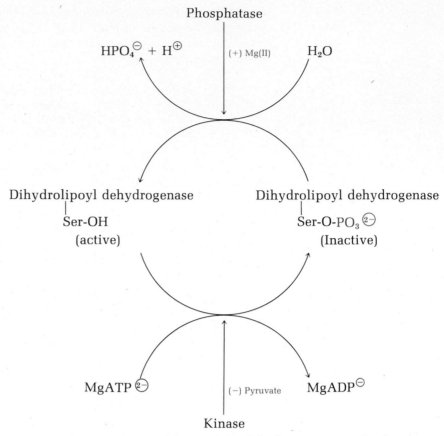

Figure 15.14 Regulation of the pyruvate dehydrogenase complex by phosphorylation and hydrolysis of the phosphate ester of dihydrolipoyl dehydrogenase. Allosteric activation is indicated by (+); allosteric inhibition is indicated by (−). [Redrawn from L. J. Reed, *Acc. Chem. Res., 7,* 45 (1974).]

cycle, as does a high NADH/NAD ratio. α-Ketoglutarate dehydrogenase is inhibited by succinyl-CoA and NADH and activated by ADP. These modes of metabolic regulation of the citric acid cycle are summarized in Figure 15.15.

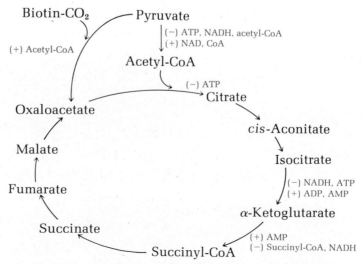

Figure 15.15 Regulation of the citric acid cycle. Allosteric activation is indicated by (+); allosteric inhibition is indicated by (−).

15–5 GLYOXALATE CYCLE   In plants and microorganisms the *glyoxalate cycle* converts fats to carbohy-

$$\begin{array}{c} CHO \\ | \\ CO_2^{\ominus} \end{array}$$

Glyoxalate

drates, a biosynthetic capacity absent in animals (Figure 15.16). The glyoxalate cycle is especially important in germinating seeds and in microorganisms subsisting with acetyl-CoA as their sole carbon source.

Three of the enzymes of the glyoxalate cycle—citrate synthetase, aconitase, and malate dehydrogenase—are shared by the citric acid cycle. The unique enzymes of the glyoxalate cycle are *isocitrate lyase* and *malate synthetase*. In plant cells these two enzymes are found in subcellular organelles called *glyoxosomes*. Both isocitrate lyase and malate synthetase are present only in low concentrations when plants are not synthesizing glucose from fatty acids. Enzymes such as these, that are synthesized upon metabolic demand, are said to be *inducible*.

Isocitrate lyase catalyzes aldol cleavage of isocitrate producing succinate and glyoxalate (reaction 15.21).

$$
\text{Enz-B:} \quad
\begin{array}{c}
CO_2^{\ominus} \\
| \\
CH_2 \\
| \\
H-C-C{\nwarrow}{}^{O}_{O^{\ominus}} \\
| \\
H-O-C-H \\
| \\
CO_2^{\ominus}
\end{array}
\rightleftharpoons
\begin{array}{c}
O \\
\diagdown \\
{}^{}C-CO_2^{\ominus} \\
/ \\
H
\end{array}
+
\begin{array}{c}
CO_2^{\ominus} \\
| \\
CH_2 \\
| \\
CH_2 \\
| \\
CO_2^{\ominus}
\end{array}
\qquad (15.21)
$$

Isocitrate   Glyoxalate   Succinate

The succinate produced by this reaction is eventually converted to oxaloacetate and then to glucose.

When isocitrate lyase is incubated with isocitrate in D_2O, deuterium is stereospecifically incorporated to yield $2S$-[^2H]succinate. The glyoxalate formed in the reaction contains no deuterium. The hydrogens at C-2 and C-3

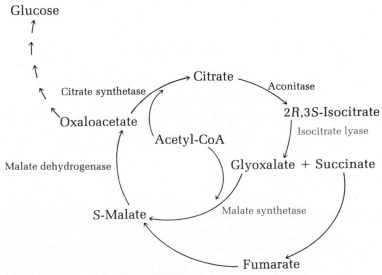

Figure 15.16   The glyoxalate cycle.

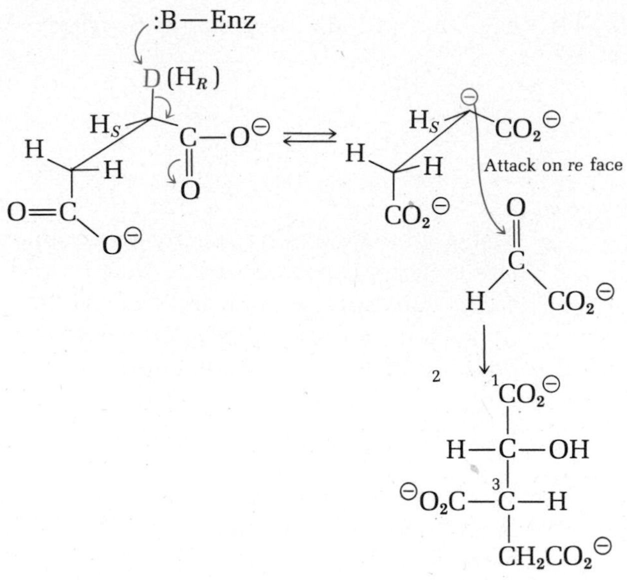

Figure 15.17 Mechanism of the aldol condensation of succinate and glyoxalate catalyzed by isocitrate lyase.

of succinate are prochiral. In the reverse reaction—the aldol condensation of succinate and glyoxalate—the pro-R hydrogen is abstracted (deuterium in the isotopically labeled product of the forward reaction). Only $2R,3S$-isocitrate is formed in this reaction, indicating that the enolate anion of succinate exclusively attacks the re face of glyoxalate (Figure 15.17).

Malate synthetase, the second unique enzyme of the glyoxalate cycle, catalyzes the Claisen condensation of acetyl-CoA and glyoxalate to produce S-malate (reaction 15.22).

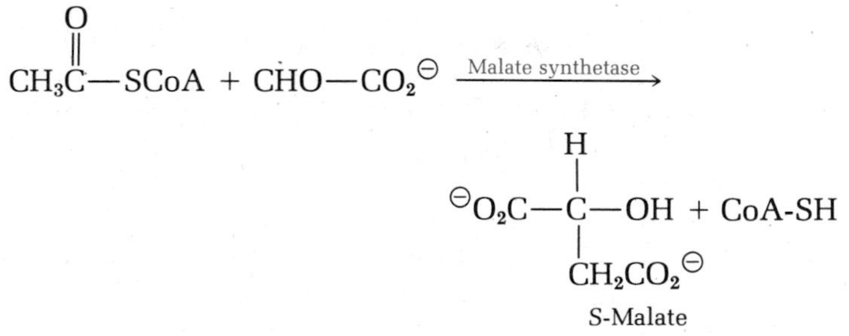

When acetyl-CoA with a chiral methyl group is a substrate, the reaction occurs with inversion of configuration at the chiral carbon, and S-acetyl-CoA is incorporated in $2S,3R$-$[3$-$^3H]$malate. We noted a similar result in our discussion of citrate synthetase. The mechanism of malate synthetase involves formation of a thioenolate anion that attacks the si face of the aldehyde carbon of glyoxalate. Figure 15.6 applies to malate synthetase as well as to S-citrate synthetase; it is only necessary to change oxaloacetate to glyoxalate. Hydrolysis of the thioester provides the driving force for the reaction.

15–6 SUMMARY   The citric acid cycle is the final stage of the aerobic catabolism of carbohydrates, amino acids, and fatty acids. The bridge from pyruvate to the citric acid cycle is provided by the pyruvate dehydrogenase multienzyme complex that converts pyruvate to acetyl-CoA. The enzymes of the citric acid cycle and pyruvate dehydrogenase complex are located in the mitochondria.

Complete oxidation of glucose by the citric acid cycle and the electron transport chain releases 2,870 kJ mol^{-1}, as opposed to the 197 kJ mol^{-1} released by anaerobic conversion of glucose to pyruvate in glycolysis. Two moles of ATP are produced by glycolysis, whereas up to 38 moles of ATP are produced by the complete oxidation of glucose.

Oxaloacetate condenses with acetyl-CoA to form citrate in the step that initiates the cycle. The driving force for this reaction is hydrolysis of the thioester of the intermediate citryl-CoA. The conversion of citrate to isocitrate catalyzed by aconitase converts symmetrical citrate to chiral 2R,3S-isocitrate. Successive oxidative decarboxylations catalyzed by isocitrate dehydrogenase and α-ketoglutarate dehydrogenase produce 1 mole of NADH each, leading to succinyl-CoA. Both moles of carbon dioxide produced in these steps originate in oxaloacetate. Succinyl-CoA is converted to succinate by succinate thiokinase in a reaction that results in phosphorylation of GDP. The resulting GTP can transfer a phosphoryl group to ADP in a reaction catalyzed by nucleoside diphosphate kinase. This reaction, which is not coupled to an oxidation, is known as a substrate-level phosphorylation. The membrane-bound enzyme succinate dehydrogenase, also a participant in the respiratory electron transport chain, oxidizes succinate to fumarate with concomitant formation of FADH$_2$. Hydration of fumarate catalyzed by fumarase produces S-malate (L-malate). Finally, in the fourth oxidative step of the citric acid cycle, S-malate is oxidized to oxaloacetate, with concomitant formation of NADH. Each mole of NADH produced in the mitochondria by the citric acid cycle leads to formation of 3 moles of ATP by oxidative phosphorylation. The FADH$_2$ produced is reoxidized with a yield of 2 moles of ATP.

The citric acid cycle is regulated in several ways. The concentration of oxaloacetate is controlled by pyruvate carboxylase, an enzyme that contains an allosteric binding site for acetyl-CoA. Unless acetyl-CoA is bound to pyruvate carboxylase, the enzyme is inactive. The concentration of acetyl-CoA is governed by the pyruvate dehydrogenase complex. The products acetyl-CoA and NADH inhibit the complex. ATP is also an allosteric inhibitor and ADP an allosteric activator. Bovine pyruvate dehydrogenase is also subject to regulation by covalent modification. Isocitrate dehydrogenase is inhibited by NADH, stimulated by ADP, and inhibited by ATP, as is α-ketoglutarate dehydrogenase, which is also inhibited by succinyl-CoA. The net result of this metabolic regulation is that the citric acid cycle operates at a high rate when the NADH/NAD and the ATP/ADP ratios are low and is inhibited when these ratios are high. Thus, the citric acid cycle is sensitively tuned to the energy state of the cell.

The glyoxalate cycle, characteristic of plants and microorganisms, provides a means for converting fatty acids to glucose. The unique enzymes of the glyoxalate cycle, not found in animals, are isocitrate lyase and malate synthetase. The other three enzymes of the glyoxalate cycle, citrate synthetase, aconitase, and malate dehydrogenase, are shared by the citric acid cycle and the glyoxalate cycle.

REFERENCES

Boyer, P. D., Ed., "The Enzymes," 3rd ed., Vol. 5 (reviews of aconitase and enolase); Vol. 11 (reviews of dehydrogenases), Academic Press, New York, 1975.

Goodwin, T. W., Ed., "The Metabolic Roles of Citrate," Academic Press, New York, 1968.

Lowenstein, J. M., Ed., "Citric Acid Cycle: Control and Compartmentation," Dekker, New York, 1969.

Popjak, G., Stereospecificity of Enzymic Reactions, in P. D. Boyer, Ed., "The Enzymes," 3rd ed., Vol. 2, Academic Press, New York, 1970.

Problems

1. Yeast can grow both aerobically and anaerobically on glucose. When yeast which has been maintained under anaerobic conditions is exposed to oxygen the rate of glucose consumption decreases. Why?

2. What is the effect of increasing the ADP/ATP ratio on the activity of isocitrate dehydrogenase?

3. How will decreasing the acetyl-CoA concentration affect the incorporation of ^{14}C-labeled HCO_3^{2-} into citric acid cycle intermediates?

4. Trace the fate of ^{14}C in $^{14}CH_3$-labeled acetyl-CoA through two turns of the citric acid cycle. (Assume label only enters in first turn.)

5. How will decreasing the concentrations of each of the following metabolites affect the rate of the citric acid cycle: (a) NADH, (b) acetyl-CoA, (c) citrate?

6. If ^3H-labeled oxaloacetate is reduced to malate with $NaBH_4$, the ^3H-malate is incubated with fumarase for an extended period, and all of the fumarate that is formed is trapped, 50% of the ^3H is found in fumarate. On the other hand, if ^{14}C-acetyl-CoA is added to respiring cells, the resulting ^{14}C-malate is isolated, and the isolated ^{14}C-malate is reincubated with fumarase (as above), all of the ^{14}C is converted to ^{14}C-fumarate. Explain these observations.

7. Certain poisonous plants produce fluorocitrate. The poisonous properties of this molecule are manifest after it has been converted to $2R,3R$-fluorocitrate, an inhibitor of aconitase. (a) Explain this inhibition.

$$\begin{array}{c} CO_2^{\ominus} \\ | \\ H-C-F \\ | \\ {}^{\ominus}O_2C-C-OH \\ | \\ CH_2CO_2{}^{\ominus} \end{array}$$

2R,3R-Fluorocitrate

[Use Figures 15.7 and 15.8 as a guide.] (b) Of the four possible diastereoisomers of fluorocitrate only the $2R,3R$ isomer is an inhibitor. Explain why the others do not inhibit aconitase.

8. ATP, ADP, and AMP are competitive inhibitors of NADH binding to malate dehydrogenase. What is the structural basis for this inhibition?

9. Pyruvate labeled with ^{18}O in the ketone oxygen is incubated with a cell-free system capable of undergoing the citric acid cycle. (a) Show what the fate of this label would be in one turn of the cycle. Show all intermediates and circle the labeled oxygen. (b) Describe the *specific step* where labeled CO_2 would be generated using as many turns of the cycle as necessary. (Assume that the label enters only in the first turn of the cycle.)

Chapter 16

Oxidative phosphorylation

Concepts without observations are empty; observations without concepts are blind.

—Kant

16–1 INTRODUCTION

The oxidative reactions of the citric acid cycle generate $FADH_2$ and NADH. The reduced coenzymes are oxidized in the electron transport chain of respiration. The free energy liberated in these oxidations is partially conserved in the phosphoanhydride bonds of ATP. For every mole of NADH that is oxidized by respiration, 3 moles of ATP are produced, and for every mole of $FADH_2$ that is oxidized by respiration, 2 moles of ATP are produced. These reactions occur within discrete multienzyme *respiratory chain complexes* located in the inner mitochondrial membrane. Oxidation and phosphorylation are coupled to each other, and in normal cells one does not usually occur without the other.

The theories that have been advanced to explain respiration date to the original proposals of a chain of electron carriers made by O. Warburg and D. Keilin in 1927. Their initial hypothesis, made well before all of the carriers had been discovered, has been transformed into the *chemiosmotic theory* of P. Mitchell as a result of over five decades of research. The chemiosmotic theory is not without competitors, however, and considerable controversy surrounds the theories that attempt to explain oxidative phosphorylation.

Let us consider the basic "flow sheet" of electrons in respiration, including the sites of ATP synthesis.

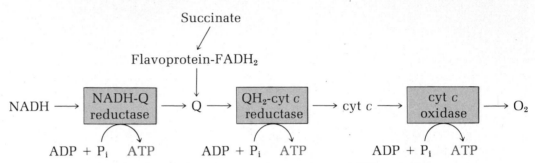

Flow of electrons in respiration and sites of ATP synthesis

It is readily apparent why the oxidation of NADH yields 3 moles of ATP, whereas the oxidation of $FADH_2$ yields but 2 moles. $FADH_2$ passes its electrons to coenzyme Q and further oxidation passes the electrons through two coupling sites, but NADH passes its electrons to NADH-CoQ reductase and then to coenzyme Q, and so the electrons originating in NADH pass through all three coupling sites.

| 16–2 ELECTROCHEMISTRY | The respiratory chain consists of a set of coupled oxidation-reduction or *redox* reactions, and we shall prepare the way for a discussion of these reactions by a brief summary of the fundamentals of electrochemistry. |

A. Redox Reactions and Electrochemical Cells

Oxidation is defined as loss of electrons, and reduction is defined as gain of electrons. In a *redox reaction* a reduced electron donor, A_{red}, transfers one or more electrons to an oxidized electron acceptor in a *reversible process* (reaction 16.1).

$$aA_{red} + bB_{ox} \rightleftharpoons aA_{ox} + bB_{red} \tag{16.1}$$

Consider, for example, the transfer of a pair of electrons from Cu(II) to Zn(s) (reaction 16.2).

$$\underset{A_{red}}{Zn(s)} + \underset{B_{ox}}{Cu(II)} \rightleftharpoons \underset{A_{ox}}{Zn(II)} + \underset{B_{red}}{Cu(s)} \tag{16.2}$$

The equilibrium constant and standard free-energy change for reaction 16.2 can be determined by means of an *electrochemical cell* (Figure 16.1). In one compartment a Zn(s) *electrode* is immersed in a solution of $ZnSO_4$. In the other compartment, separated from the first by a porous plug that permits the flow of ions while preventing bulk mixing, a Cu(s) electrode is immersed in a solution of $CuSO_4$. The two strips of metal are connected externally by a voltmeter. When the concentrations of the ions in the two solutions are 1 M, the system is called a *standard electrochemical cell*. The voltmeter indicates that electrons flow *spontaneously* from the zinc to the copper electrode. The

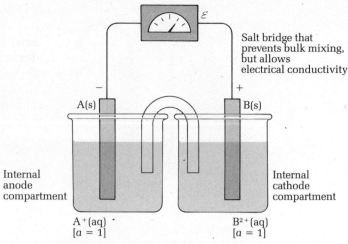

Spontaneous electrochemical
cell reaction: $2 A^+ (aq) + B (s) \rightarrow 2 A (s) + B^{+2} (aq)$

Figure 16.1 Diagram of an electrochemical cell. The spontaneous electrochemical cell reaction is: $2 A^{\oplus}(aq) + B(s) \rightarrow 2 A(s) + B^{2\oplus}(aq)$.

reading on the voltmeter is the *electromotive force*, or *emf*, of the cell. Two processes are occurring. (1) Zinc is being oxidized at the Zn(s) electrode (reaction 16.3). This reaction is called the *oxidation half-reaction*.

$$Zn(s) \rightleftharpoons Zn(II) + 2 e \qquad (16.3)$$

(2) Cu(II) ions are being reduced at the copper electrode in the *reduction half-reaction* (reaction 16.4).

$$Cu(II) + 2 e \rightleftharpoons Cu(s) \qquad (16.4)$$

B. Standard Electrode Potentials

We would like to be able to measure the potential at a single electrode in an electrochemical cell. This potential would measure the tendency of each half-reaction to occur. However, such a measurement is impossible, because an oxidation can occur if it is coupled to a reduction. To bypass this difficulty, we choose one electrode as a reference point whose emf is arbitrarily set at zero. The *hydrogen electrode* has been chosen as the standard electrode having an emf of 0.0 volts (Figure 16.2). In the hydrogen electrode a 1 M solution of hydrogen ions is in contact with an inert electrode (usually platinum). When a stream of H_2 at a pressure of 1.0 atm is bubbled across the surface of the platinum electrode, a reversible reaction occurs.

$$H_2 \rightleftharpoons 2H^{\oplus} + 2 e^{\ominus} \qquad (16.5)$$

If the reaction at the hydrogen electrode is an oxidation, as it is written in reaction 16.5, then the other half-reaction is a reduction. The electrons released by oxidation of H_2 flow to an electron acceptor through the external circuit of the electrochemical cell (Figure 16.3). When the solution in the other compartment of the electrochemical cell is 1 M, the measured emf is the *standard reduction potential* of the reduction half-reaction, E^0. A cell composed of a hydrogen electrode and copper–copper sulfate electrode (abbreviated Cu‖CuSO$_4$), therefore undergoes the following reactions.

$$
\begin{array}{ll}
Cu(II) + 2 e \rightleftharpoons Cu(s) & E^0 = 0.34 \text{ V} \\
\underline{2H^{\oplus} + 2 e \rightleftharpoons H_2} & \underline{E^0 \equiv 0.0 \text{ V}} \\
Cu(II) + H_2 \rightleftharpoons Cu(s) + 2H^{\oplus} & E^0_{cell} = 0.34 \text{ V}
\end{array} \qquad (16.6)
$$

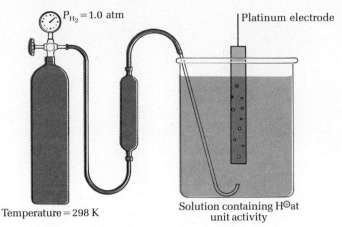

Figure 16.2 Hydrogen electrode. The thermodynamic "true" hydrogen electrode is an idealized device that can only be approximated experimentally.

The cell emf is identical to the emf for the reduction of Cu(II). In those cases in which the reduction at the second electrode is less spontaneous than the reduction of H_2, the standard reduction potential is negative. The *standard cell potential* must, however, be positive because electrons flow spontaneously. Consider, for example, a cell composed of a hydrogen electrode and an Sn(s)‖Sn(II) electrode.

$$
\begin{array}{lr}
\mathrm{Sn(II)} + 2\ e \rightleftharpoons \mathrm{Sn(s)} & E^0 = -0.14\ \mathrm{V} \\
\underline{2\mathrm{H}^{\oplus} + 2\ e \rightleftharpoons \mathrm{H_2}} & \underline{E^0_{\mathrm{red}} \equiv 0.0\ \mathrm{V}} \\
\mathrm{Sn(s)} + 2\mathrm{H}^{\oplus} \rightleftharpoons \mathrm{Sn(II)} + \mathrm{H_2} & E^0_{\mathrm{cell}} = 0.14\ \mathrm{V}
\end{array}
\tag{16.7}
$$

Were we to assume that Sn(II) is reduced in such a situation, then we would obtain a standard cell potential of -0.14 V. This result tells us that we erred in supposing Sn(II) would be reduced. The spontaneous reaction is, rather, the reduction of $H^{\oplus}$ and the oxidation of Sn(s), as we have written the net reaction 16.7.

The definition of the standard hydrogen electrode includes a 1 M solu-

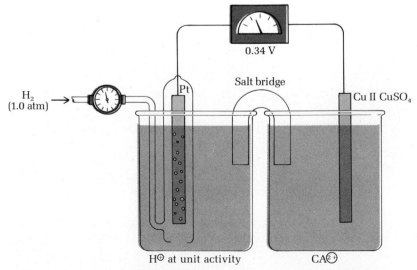

Figure 16.3 Standard electrochemical cell composed of a hydrogen electrode and a Cu‖CuSO₄ electrode. The single electrode potential for the reaction Cu(II) + 2 e ⇌ Cu(s) can be determined with such a cell.

tion of hydronium ions. Its pH is therefore zero. We prefer a standard system whose pH is 7.0, the most commonly encountered pH of biochemical systems. We therefore define *biological standard reduction potentials, $E^{0'}$*, as those observed at pH 7.0. This corresponds to a hydronium ion concentration at the hydrogen electrode of 10^{-7} M. The standard biological reduction potentials for some important substances are summarized in Table 16.1.

C. Thermodynamics of Electrochemical Cells

We have several times used the phrase "spontaneous change" in our description of the direction of electron flow in redox reactions. The driving force for reaction in an electrochemical cell is the cell emf, an electrical potential energy difference. The cell emf also corresponds to the *maximum work* available from the reversible operation of the cell. The amount of work that can be done by a reversible process at constant temperature and pressure is related to the Gibbs free energy by equation 16.8.

$$w_{max} = \Delta G \tag{16.8}$$

The free-energy change for the transfer of an electric charge from a given potential to another lower one is related to the Gibbs free energy by equation 16.9.

$$\Delta G = -nFE \tag{16.9}$$

Table 16.1 Standard biological reduction potentials

System	$E^{0'}$ (V)
$O_2 \rightarrow H_2O$	0.816
$Fe^{3+} \rightarrow Fe^{2+}$	0.771
Photosystem P700	0.43
$NO_3^- \rightarrow NO_2^-$	0.421
Cytochrome f, $Fe^{3+} \rightarrow Fe^{2+}$	0.365
Cytochrome a, $Fe^{3+} \rightarrow Fe^{2+}$	0.29
Cytochrome c, $Fe^{3+} \rightarrow Fe^{2+}$	0.254
Cytochrome c_1, $Fe^{3+} \rightarrow Fe^{2+}$	0.22
Hemoglobin, $Fe^{3+} \rightarrow Fe^{2+}$	0.17
$CoQ \rightarrow CoQH_2$	0.10
Cytochrome b (mitochondrial), $Fe^{3+} \rightarrow Fe^{2+}$	0.077
Myoglobin, $Fe^{3+} \rightarrow Fe^{2+}$	0.046
Fumarate $\rightarrow$ succinate	0.031
Cytochrome b_5 (microsomal), $Fe^{3+} \rightarrow Fe^{2+}$	0.02
Glyoxylate $\rightarrow$ glycolate	−0.09
Oxaloacetate $\rightarrow$ malate	−0.166
Pyruvate $\rightarrow$ lactate	−0.185
Acetaldehyde $\rightarrow$ ethanol	−0.197
$FMN \rightarrow FMNH_2$	−0.219
$FAD \rightarrow FADH_2$	−0.219
Glutathione $\rightarrow$ 2-reduced glutathione	−0.23
Lipoic acid $\rightarrow$ dihydropoic acid	−0.29
$NAD \rightarrow NADH$	−0.32
$NADP \rightarrow NADPH$	−0.324
Lipoyl dehydrogenase (FAD) $\rightarrow$ lipoyl dehydrogenase ($FADH_2$)	−0.34
Uric acid $\rightarrow$ xanthine	−0.36
Gluconolactone $\rightarrow$ glucose	−0.364
$H^+ \rightarrow H_2$	−0.421
Ferrodoxin (spinach), $Fe^{3+} \rightarrow Fe^{2+}$	−0.432
Gluconate $\rightarrow$ glucose	−0.44
Acetate $\rightarrow$ acetaldehyde	−0.581

Data from P. A. Loach, in G. D. Fasman, Ed., "Handbook of Biochemistry and Molecular Biology," 3rd ed., Vol. 1, CRC Press, Cleveland, Ohio, 1975, pp. 123–129. Reprinted with permission. Copyright The Chemical Rubber Co., CRC Press, Inc.

where n is the number of electrons transferred; F is the Faraday, the magnitude of the charge of 1 mole of electrons, or 96,500 C, or 96,500 kJ mol^{-1} v^{-1}; and E is the emf of the electrochemical cell.[1] Since a positive cell emf corresponds to a spontaneous process, a minus sign is required by the convention that spontaneous processes have negative ΔG values. For a standard cell the reactants have a concentration of 1 M, the emf is E^0, and

$$\Delta G^\circ = -nFE^0 \tag{16.10}$$

At pH 7.0 the standard emf is $E^{0'}$, and thus

$$\Delta G^{\circ'} = -nFE^{0'} \tag{16.11}$$

At equilibrium we recall that $\Delta G^{\circ'} = -RT \ln K'$. Therefore,

$$E^{0'} = \frac{RT}{nF} \ln K' \tag{16.12}$$

At 25°C (298 K)

$$E^0 = \frac{0.059}{n} \log_{10} K \tag{16.13}$$

Suppose that all reactants are not present in concentrations of 1 M in an electrochemical cell. In this case the observed emf, E', depends upon both the standard emf, $E^{0'}$, and the actual concentrations of reactants and products. For the general reaction

$$aA_{red} + bB_{ox} \rightleftharpoons aA_{ox} + bB_{red} \tag{16.1}$$

the standard free-energy change is related to the observed free-energy change by

$$\Delta G' = \Delta G^{\circ'} + RT \ln Q \tag{16.14}$$

where

$$Q = \frac{[A_{ox}]^a[B_{red}]^b}{[A_{red}]^a[B_{ox}]^b} \tag{16.15}$$

Substituting equations 16.10 and 16.11 in 16.14 we obtain the *Nernst equation.*

$$\Delta E' = \Delta E^{0'} - \frac{RT}{nF} \ln Q \tag{16.16}$$

At 298 K, in terms of base 10 logarithms, equation 16.16 reduces to

$$\Delta E' = \Delta E^{0'} - \frac{0.059}{n} \log Q \tag{16.17}$$

Having laid the groundwork for our future discussions of biological oxidations, let us apply the foregoing to the oxidation of malate to oxaloacetate

---

[1] The coulomb (C) is the mks unit of electrical charge, defined as the quantity of electricity transported in 1 second by a current of 1 ampere.

at pH 7.0.[2] Malate is oxidized, and the coenzyme NAD is reduced in this process (reaction 16.18).

$$\text{malate}^{\text{2}\ominus} + \text{NAD} \rightleftharpoons \text{oxaloacetate}^{\text{2}\ominus} + \text{NADH} + \text{H}^{\oplus} \tag{16.18}$$

The first step in analyzing such a reaction is conversion of the net reaction to two reduction half-reactions (reactions 16.19a and 16.19b).

$$\text{oxaloacetate}^{\text{2}\ominus} + 2\text{H}^{\oplus} + 2\ \text{e} \rightleftharpoons \text{malate}^{\text{2}\ominus} \qquad E^{0\prime} = -0.166\ \text{V} \tag{16.19a}$$

$$\text{NAD} + \text{H}^{\oplus} + 2\ \text{e} \rightleftharpoons \text{NADH} \qquad E^{0\prime} = -0.320\ \text{V} \tag{16.19b}$$

Since electrons flow from the half-reaction having the more negative standard reduction potential to the half-reaction having the more positive standard reduction potential, oxaloacetate will be reduced. The spontaneous reaction is therefore

$$\text{oxaloacetate}^{\text{2}\oplus} + \text{NADH} + \text{H}^{\oplus} \rightleftharpoons \text{malate}^{\text{2}\ominus} + \text{NAD} \tag{16.20}$$

The value of $E^{0\prime}_{\text{cell}}$ for reaction 16.20 is the algebraic difference between the $E^{0\prime}$ values for the reduction half-reactions, or 0.154 V. Two electrons are transferred. The standard free-energy change for the reaction is

$$\begin{aligned} \Delta G^{0\prime} &= -2(96{,}500\ \text{J V}^{-1})(0.154\ \text{V}) \\ &= -29.7\ \text{kJ} \end{aligned} \tag{16.21}$$

The equilibrium constant, K', is given by

$$\begin{aligned} \log K' &= \frac{n\Delta E^{0\prime}}{0.059} \\ &= 1.66 \times 10^5 \end{aligned} \tag{16.22}$$

16–3 THE MITOCHONDRION

In the 1940s E. Kennedy and A. Lehninger found that the enzymes responsible for respiration, the citric acid cycle, and fatty acid oxidation are located in the mitochondria, subcellular organelles that are readily visible in a light microscope. Many other metabolic pathways also occur in mitochondria (Table 16.2). These reactions provide most of the cell's energy, and the mitochondrion has been called the "power plant" of the cell. The unraveling of the complexities of respiration has been closely linked to studies of the structure of the mitochondrion, especially of its intricate membrane system.

A. Structure

The mitochondrion is a football-shaped organelle that ranges in size from 0.2 to 0.8 μm in diameter and from 0.5 to 1.0 μm in length. The number of mitochondria per cell spans the range from a single mitochondrion in certain algae to 100,000 in amphibian eggs. Vertebrates typically have 500 to 1,000 mitochondria in their cells.

The mitochondrion is surrounded by two membranes (Figure 16.4). The outer membrane serves primarily as a boundary between the mitochondrion and the cytosol. The inner membrane is a continuous, highly folded

[2] The reaction is catalyzed by malate dehydrogenase, but this is thermodynamically irrelevant since an enzyme does not change the equilibrium constant of a reaction.

Table 16.2 Major metabolic pathways in the mitochondrion

Pathway	Text reference
Citric acid cycle	Chapter 15
Oxidative phosphorylation	Chapter 16
Citrulline biosynthesis	Section 21.3B
Oxidation of fatty acids	Chapter 19
Pyruvate→ oxaloacetate→ phosphoenolpyruvate	Section 17.3
δ-Amino levulinate biosynthesis	Section 22.10
NADH-dependent electron transfer from succinate, proline, pyruvate, choline, sn-glycerol 3-phosphate, and choline to O_2	Section 16.4
Glucuronate pathway (in chicken kidney mitochondria, in smaller organelles in other species)	Section 17.2

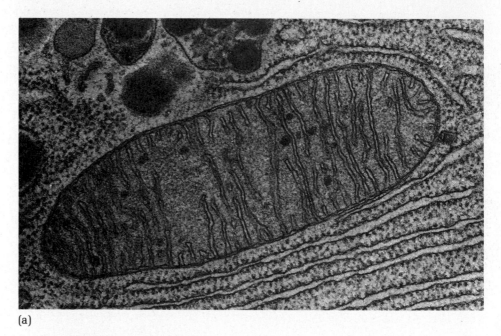

(a)

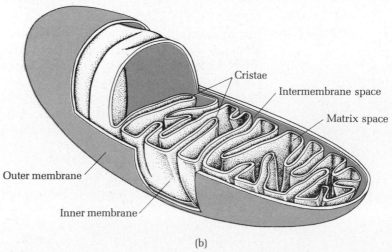

(b)

Figure 16.4 Structure of the mitochondrion. (a) Electron micrograph of part of a mitochondrion from an acinar cell of the pancreas. [From A. W. Ham and D. H. Cormack, "Histology," 8th ed., Lippincott, Philadelphia, 1979, p. 118.] (b) Schematic drawing of the mitochondrion. [From S. L. Wolfe, "Biology of the Cell," Wadsworth, Belmont, Calif., 1972, p. 104.]

structure that wends its ways through the interior of the mitochondrion. The infolds of the inner membrane are called *cristae*.

The proteins of the electron transport chain are bound to the inner mitochondrial membrane, as is the protein responsible for synthesis of ATP. Since respiration is a membrane-associated phenomenon, the net rate of respiration depends upon the surface area of the membrane, and the number of cristae is correlated to the respiratory activity of the cell. Heart muscle cells, for example, have a high respiratory rate and the mitochondria have densely packed cristae; the mitochondria of liver cells, which have a lower rate of respiration, have fewer cristae.

The region enclosed by the inner mitochondrial membrane, called the *matrix space*, contains pyruvate dehydrogenase and the enzymes responsible for the citric acid cycle (except the membrane-bound enzyme succinate dehydrogenase, which also participates in respiration) and those responsible for β-oxidation of fatty acids. The region between the inner and outer membranes, called the *intermembrane space*, contains few enzymes and is the site of little metabolic activity.

B. Proteins of Mitochondrial Membranes

The major proteins of the mitochondrial membranes are listed in Table 16.3. (We have already encountered one of the outer membrane proteins, phospholipase A, in Section 11.2E.) The proteins responsible for oxidative phosphorylation, except the peripheral protein cytochrome c, are imbedded in the inner membrane. The outer mitochondrial membrane is permeable to substances whose molecular weights are as high as 10,000. The inner membrane, by contrast, is freely permeable only to O_2, CO_2, and water. Transport proteins, called translocases, located in the inner membrane, are responsible for the control of materials entering and leaving the matrix space (Section 16.12).

The matrix side of the inner mitochondrial membrane is studded with particles that resemble lollipops (Figure 16.5). These structures are central figures in oxidative phosphorylation. Whether they exist in intact mitochondria as they appear in Figure 16.5 is debatable, since the knob-like projections are seen only in certain types of stained preparations. In functioning mitochondria their structure may be quite different. The enzyme activity associated with the knob of the projection (mol wt 85,000) is ATP hydrolysis. We shall later reintroduce this ATPase as coupling factor 1.

Table 16.3	Major proteins of the inner and outer membranes of the mitochondrion
Inner membrane	**Outer membrane**
Cytochrome b, c_1, c, a, a_3	Cytochrome b_5
NADH dehydrogenase	Cytochrome b_5 reductase
Succinate dehydrogenase	Monoamine oxidase
Ubiquinone	Kynurenine hydroxylase
Electron-transferring flavoprotein	Fatty acyl-CoA synthetase
ATPase	Fatty acid elongation system
β-Hydroxybutyrate dehydrogenase	Glycerophosphate acyl transferase
Carnitine-palmityl transferase	Choline phosphotransferase
Fatty acid elongation system	Phospholipase A
ADP-ATP translocase	Nucleoside diphosphokinase
P_i-OH-translocase	
Dicarboxylate translocase	
Tricarboxylate translocase	
α-Ketoglutarate translocase	
Pyruvate translocase	
Glutamate-aspartate translocase	

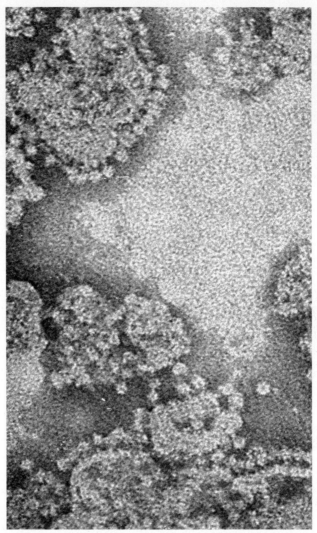

Figure 16.5 Electron micrograph of the inner mitochondrial membrane. The knob-like structures project into the matrix space. The heads are attached to the membrane by thin stalks. [From D. F. Parsons, *Science*, *140*, 985 (1963). Copyright 1963 by the American Association for the Advancement of Science.]

a. Iron-Sulfur Proteins

Certain proteins contain covalently bound clusters of iron and sulfur in a 1:1 ratio. The sulfur is "labile" and can be removed as H_2S by lowering the pH of the protein solution to 1. These *iron-sulfur proteins* are often called *nonheme iron proteins*. Many biological oxidations involve iron-sulfur proteins, and they are especially prominent in the respiratory electron transport chain and in the photosynthetic electron transport chain (Chapter 18). The respiratory electron transport chain contains iron-sulfur proteins having two or four iron atoms, Fe_2S_2 and Fe_4S_4.

The structure of the iron-sulfur center in the two-iron proteins is thought to be similar to that of the synthetic Fe_2S_2 complex shown in Figure 16.6. The two-iron ferredoxin from spinach chloroplasts is a one-electron acceptor in a chain of electron carriers in the photosynthetic electron transport chain. Its standard reduction potential ($E^{0\prime}$) of -0.420 V is quite negative compared with the standard reduction potential for aqueous Fe(III) of $+0.77$ V. The standard reduction potentials of iron-sulfur proteins vary widely, each reflecting the unique environment created by the polypeptide chain wrapped around the iron-sulfur center.

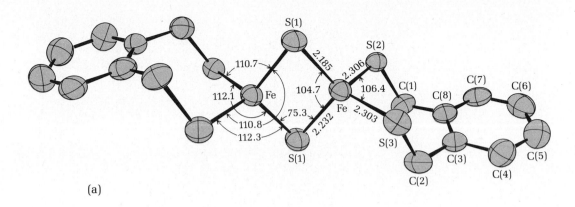

(a)

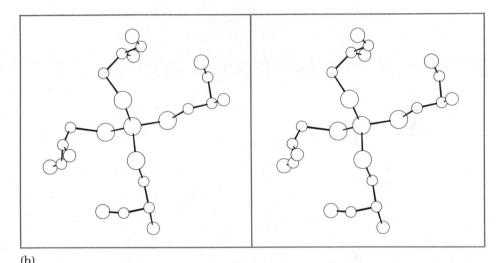

(b)

Figure 16.6 (a)Structure of an Fe_2S_2 center in this synthetic complex is thought to be similar to the Fe_2S_2 centers in iron-sulfur proteins. [From R. Holm, *Acc. Chem. Res.*, *10*, 427 (1977).] (b) Stereo view of an Fe-S center.

The four irons and four sulfurs of the Fe_4S_4 iron-sulfur proteins are arranged approximately cubically with irons and sulfurs sitting at alternate corners. Each of the four iron atoms is covalently bonded to a cysteinyl sulfhydryl group donated by the protein (Figure 16.7). Each of the four iron atoms can exist either in the Fe(III) or Fe(II) oxidation state, and a four-iron center thus can store four electrons when it is fully reduced. The standard reduction potentials for Fe_4S_4 iron-sulfur proteins range from -0.420 V for the so-called *low potential* proteins to $+0.3$ V for the *high potential* proteins. With such a wide range of reduction potentials, they can fit into most biological electron transport chains and can donate electrons to or accept electrons from a wide variety of substrates. This versatility certainly contributes to their widespread occurrence and importance.

b. Cytochrome *c*

Cytochrome *c*, as we noted in Section 1.10, is an ubiquitous component of aerobic organisms. It functions as a one-electron transfer agent in a wide variety of photosynthetic and respiratory electron transport chains (Table 16.4). The heme group of cytochrome *c* differs from the heme of hemoglobin and myoglobin by the conversion of two vinyl groups to thiomethyl groups. The iron atom of cytochrome *c* is reversibly oxidized and reduced (reaction 16.23).

$$\text{cytochrome } c \text{ (Fe}^{3+}） + \text{e} \rightleftharpoons \text{cytochrome } c \text{ (Fe}^{2+}） \qquad (16.23)$$

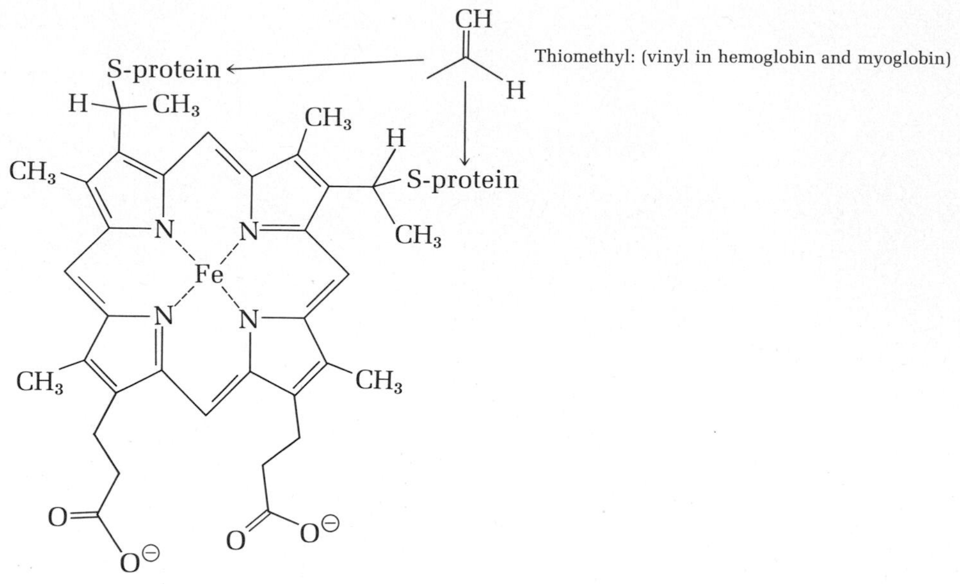

CH
Thiomethyl: (vinyl in hemoglobin and myoglobin)
H
S-protein
H
CH3
CH3
H
CH3
S-protein
CH3
N
N
Fe
N
N
CH3
CH3
O
O
O
O

Heme *c*

The standard reduction potentials for cytochrome *c* isolated from various sources range from -0.25 V to $+0.39$ V. It seems likely that the wide distribution of *c*-type cytochromes reflects their wide range of reduction potentials.

The historic classification of cytochromes was derived from their UV-visible spectra (Figure 16.8). Three characteristic bands—the γ or Soret band, the α band, and the β band—are present in all cytochromes, and the position of the bands in the spectrum indicates the type of cytochrome Table 16.5).

Table 16.4 Photosynthetic and respiratory electron transport chains in which cytochromes *c* are found

Metabolic pathway	Typical cytochromes
Sulfide-using photosynthesis: $H_2S \rightarrow S \rightarrow SO_4^{2-}$	
Green sulfur bacteria: *Chlorobium*	c_{555}
Purple sulfur bacteria: *Chromatium*	c_{553}, c_{550}
Sulfate respiration: $SO_4^{2-} \rightarrow H_2S$	
Desulfovibrio, Desulfotomaculum	c_3, c_{553}
Cyclic photosynthesis: no external reductant necessary	
Purple nonsulfur bacteria: *Rhodospirillum, Rhodopseudomonas*	c_2
Water-using photosynthesis: $H_2O \rightarrow O_2$	
Procaryotes: blue-green algae	c_{552} (*f*)
Eucaryotes: other algae, green plants	c_{552} (*f*)
Oxygen respiration: $O_2 \rightarrow H_2O$	
Procaryotes	
Purple nonsulfur bacteria (secondary importance)	c_2
Blue-green algae (secondary importance)	
Other nonphotosynthetic, respiring bacteria (nitrate respiration sometimes an alternative)	c_{550}, c_{551}, c_5, c_5, others
Eucaryotes	
Other algae (secondary importance)	*c*
Green plants (major importance)	*c*
Animals (essential)	*c*

From R. E. Dickerson, R. Timkovich, and R. J. Almassy, *J. Mol. Biol., 100,* 475 (1976). Reproduced with permission.

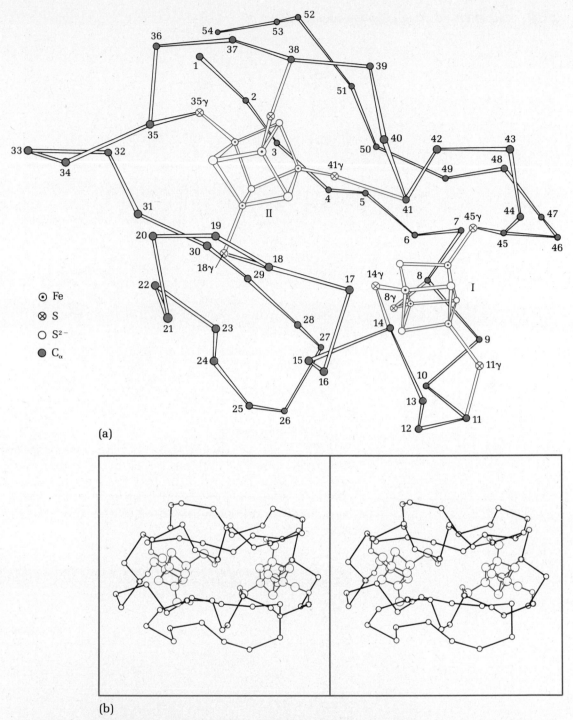

Figure 16.7 (a) Structure of the Fe_4S_4 ferredoxin from *Peptococcus aerogenes*: $(\ominus)$ = Fe, $(\otimes)$ = S, $(\bigcirc)$ = $S^{\ominus}$, $(\bullet)$ = C_α. [From E. T. Adam, L. C. Siekar, and L. H. Jensen, *J. Biol. Chem.*, 248, 3987–3996 (1973)]. (b) Stereo view of ferredoxin.

Table 16.5	Major absorption bands in the optical spectra of the cytochromes		
	Absorption band		
Heme protein	α **(nm)**	β **(nm)**	γ **(Soret) (nm)**
Cytochrome *c*	550–558	521–527	415–423
Cytochrome *b*	555–567	526–546	408–449
Cytochrome *a*	592–604	Absent	439–443
Oxyhemoglobin	577	542	415

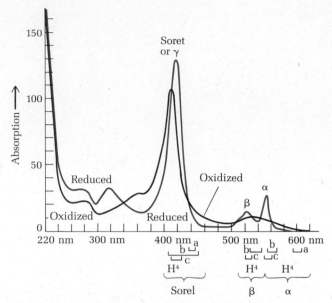

Figure 16.8 Optical spectra of oxidized and reduced horse cytochrome c. [From R. E. Dickerson and R. Timkovich, in P. D. Boyer, Ed., "The Enzymes," 3rd ed., Vol. 11, Academic Press, New York, 1975, p. 399.]

c. Cytochrome *b*

The family of cytochromes *b* has the same heme group as hemoglobin and myoglobin. Some of the *b*-type cytochromes and their functions are summarized in Table 16.6. Besides the classic cytochrome *b* of the mitochondrial respiratory chain, at least eight other *b*-type cytochromes have been found in mitochondria. Specific functions for all these cytochromes have yet to be found.

Table 16.6   *b*-Type cytochromes and their functions

Cyto-chrome	α band (nm)	Occurrence	Function and remarks
b	561–563	Mitochondria of animals, plants, and yeast; some bacteria	Electron carrier in respiratory system containing cytochrome aa_3
b_1	557–560	Bacteria (denitrifying bacteria in a limited sense)	Electron carrier in nitrate respiration system
b_2	557	Yeast mitochondria	Lactate dehydrogenase (FMN + protoheme)
b_3	559	Plant microsomes	Function uncertain
b_4			Type c cytochrome
b_5	555–556	Microsomes of animals and plants	Function uncertain; primary and ternary structures known
b_6	563	Chloroplasts	Electron carrier in photosynthetic system
b_7	560	Plant mitochondria, especially in spadix	Function uncertain; probably cyanide-insensitive terminal oxidase

From B. Hagihara, N. Sato, and T. Yamanaka, in P. D. Boyer, Ed., "The Enzymes," Vol. II, Academic Press, New York, 1975, p. 550.

16–4 AN OVERVIEW OF OXIDATIVE PHOSPHORYLATION

We have now introduced the elements necessary for a discussion of oxidative phosphorylation. First, the structure of biological membranes, and the inner mitochondrial membrane in particular, is central to oxidative phosphorylation. Sealed membrane systems, either intact mitochondria or sealed membrane vesicles, are required for oxidative phosphorylation. Second, we have considered the major electron carriers in oxidative phosphorylation: NAD, flavin coenzymes, coenzyme Q (abbreviated as Q or as QH_2 when reduced), the iron-sulfur proteins, and the cytochromes. Cu(II) ions also participate in one of the multienzyme complexes of the electron transport chain. The cytochromes, the iron-sulfur proteins, and Cu(II) ions can accept one electron at a time, whereas the various coenzymes can accept two electrons

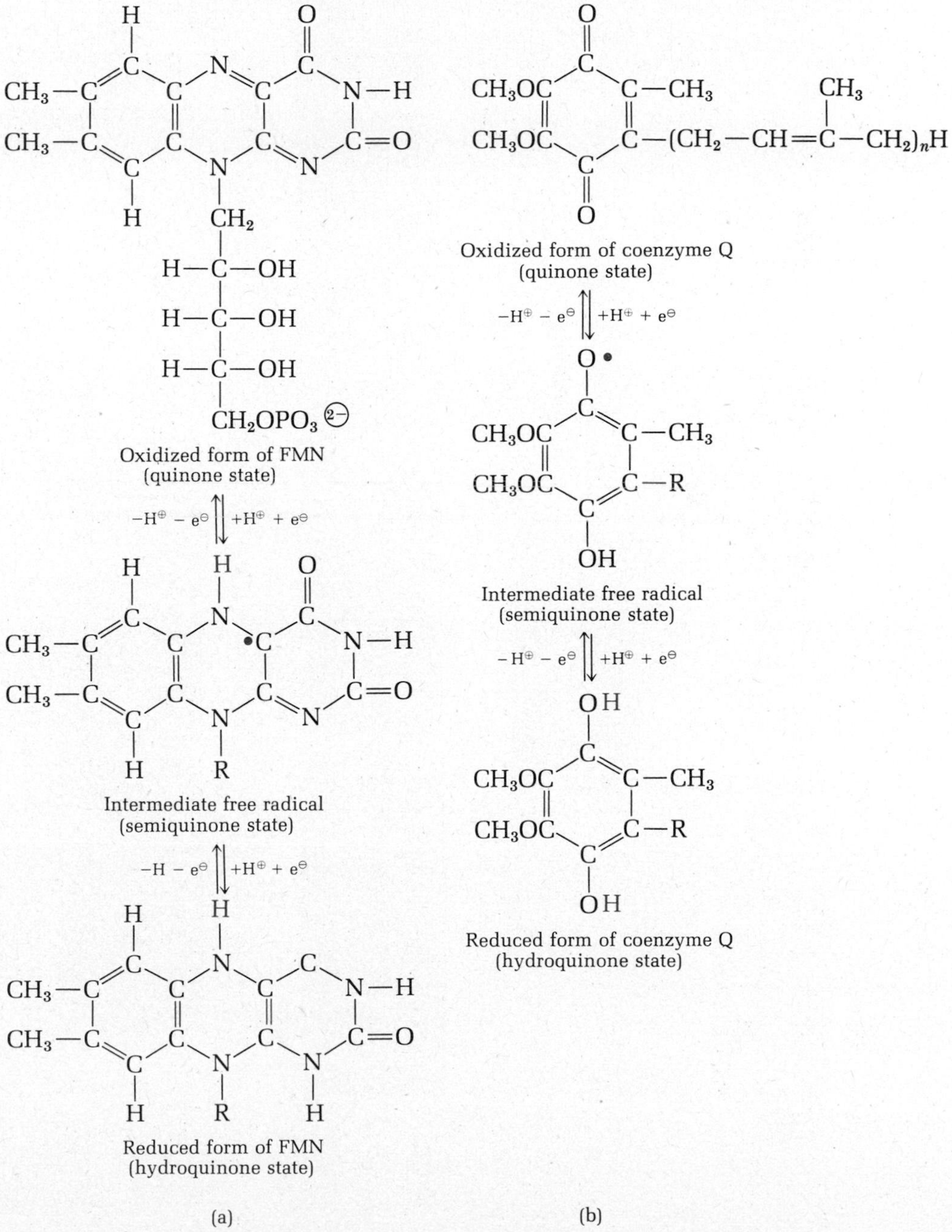

(a) (b)

Figure 16.9 Successive one-electron and one-proton transfer reactions to flavins (a) and coenzyme Q (b). Semiquinone free radical intermediates are formed in both cases.

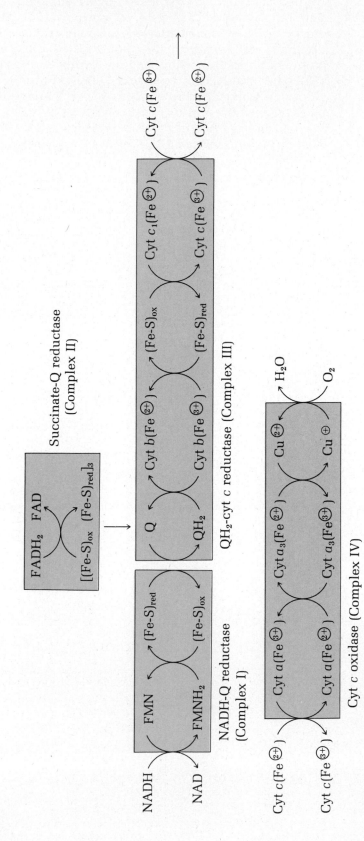

Figure 16.10 Respiratory electron transport chain.

at a time. The one-electron and two-electron transfers are mediated by free radicals. The flavin coenzymes and coenzyme Q can exist as relatively stable free radicals, and a single electron can thus be passed from, say, an iron-sulfur protein to coenzyme Q (Figure 16.9). The various electron carriers are part of multiprotein *respiratory chain complexes* imbedded in the inner mitochondrial membrane (Figure 16.10).

Consider the free-energy change that occurs for the flow of electrons from NADH to O_2. The standard reduction potential of NADH is -0.32 V (reaction 16.24).

$$NAD + H^{\oplus} + 2\ e \longrightarrow NADH \qquad E^{0\prime} = -0.32\ V \qquad (16.24)$$

The standard reduction potential for O_2 is $+0.82$ V (reaction 16.25).

$$\tfrac{1}{2}O_2 + 2H^{\oplus} + 2\ e \longrightarrow H_2O \qquad E^{0\prime} = +0.82\ V \qquad (16.25)$$

The spontaneous reaction is thus the oxidation of NADH and the reduction of O_2, a process whose standard cell potential is $+1.14$ V (reaction 16.26).

$$NADH + H^{\oplus} + \tfrac{1}{2}O_2 \longrightarrow NAD + H_2O \qquad E^{0\prime} = +1.14\ V \qquad (16.26)$$

Applying equation 16.11, we find that a potential energy change of 1.14 V corresponds to a standard free-energy change of -220 kJ mol^{-1} (equation 16.27).

$$\Delta G^{\circ\prime} = -nFE^{0\prime} = -(2)(96,000\ J\ V^{-1})(1.14\ V)$$
$$= -220\ kJ\ mol^{-1} \qquad (16.27)$$

The phosphorylation of 3 moles of ADP is coupled to the oxidation of NADH, so that 47% of the free energy released in the oxidation of NADH is conserved in the synthesis of the phosphoanhydride bonds of ATP. The conservation of energy occurs at those points where large drops in potential energy occur: the transfer of electrons from NADH through NADH-Q reductase to coenzyme Q, the transfer of electrons from reduced coenzyme Q through QH_2-cytochrome c reductase to cytochrome c, and the transfer of electrons from cytochrome c through cytochrome c oxidase to O_2 (Figure 16.11). The stoichiometry of oxidative phosphorylation when NADH is the reduced coenzyme is 3 moles of ATP production per *atom* of oxygen reduced, a quantity known as the *P/O ratio*. The transfer of electrons from $FADH_2$ through succinate-Q reductase to coenzyme Q does not liberate enough energy to drive the phosphorylation of ADP, and thus complete oxidation of $FADH_2$ yields 2 moles of ATP, a P/O ratio of 2.

16–5 RESPIRATORY CHAIN COMPLEXES

The electron transport chain has been resolved into four multienzyme complexes: NADH-Q reductase (complex I), succinate-Q reductase (complex II), QH_2-cytochrome c reductase (complex III), and cytochrome c oxidase (complex IV). Let us consider each of these in greater detail.

A. NADH-Q Reductase (Complex I): First Site of ATP Production

NADH-Q reductase transfers electrons from NADH to coenzyme Q (reaction 16.28). ATP synthesis is coupled to this step, and complex I also contains

$$NADH + H^{\oplus} + Q \xrightarrow[\text{(Complex I)}]{\text{NADH-Q reductase}} NAD + QH_2 \qquad (16.28)$$
$$ADP + P_i \qquad ATP$$

Coupling site I

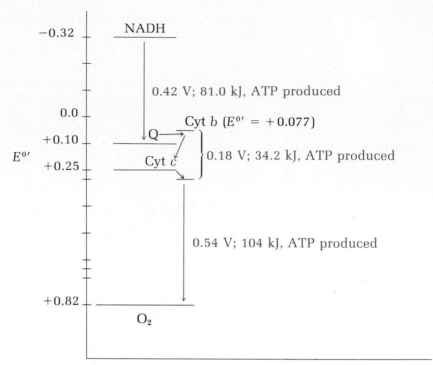

Figure 16.11 Free-energy changes for the flow of electrons through the respiratory chain.

coupling site I. NADH-Q reductase contains one FMN, four iron-sulfur centers, coenzyme Q, and about 20% lipid. The multiprotein complex can be separated into a subunit that contains FMN (mol wt 70,000) and two iron-sulfur proteins. The first step in the transfer of electrons from NADH to coenzyme Q is reduction of the FMN bound to the large subunit (reaction 16.29).

$$\text{NADH} + H^{\oplus} + \text{enzyme-FMN} \longrightarrow \text{NAD} + \text{enzyme-FMNH}_2 \quad (16.29)$$

Electrons then flow from the reduced flavin to the iron-sulfur center of the flavoprotein in successive one-electron transfers. Studies, carried out near the temperature of liquid helium (4 K), have shown that the standard reduction potentials of the iron-sulfur centers fall into three groups having $E^{0'}$ values of -0.33 V, -0.26 V, and -0.02 V (reaction 16.30).

$$\text{enzyme-FMNH}_2 \xrightarrow[2H^{\oplus}]{2e} \xrightarrow{-0.33\ V} (\text{Fe-S})_1 \longrightarrow (\text{Fe-S})_2 \xrightarrow{-0.26\ V}$$

$$(\text{Fe-S})_3 \longrightarrow \xrightarrow{-0.02\ V} (\text{Fe-S})_4 \qquad (16.30)$$
$$\Big|{-}2H^{\oplus},\ 2e$$
$$\downarrow$$
$$\underset{0.10\ V}{\text{QH}_2}$$

The standard reduction potential of the most positive center (-0.02 V) depends upon the concentration of ATP and upon the pH. This iron-sulfur center has been proposed as the site of ATP synthesis.

Complex I is inhibited by rotenone, amytal, and piericidin A (Figure 16.12). Specific inhibitors are used to study the sequence of electron carriers.

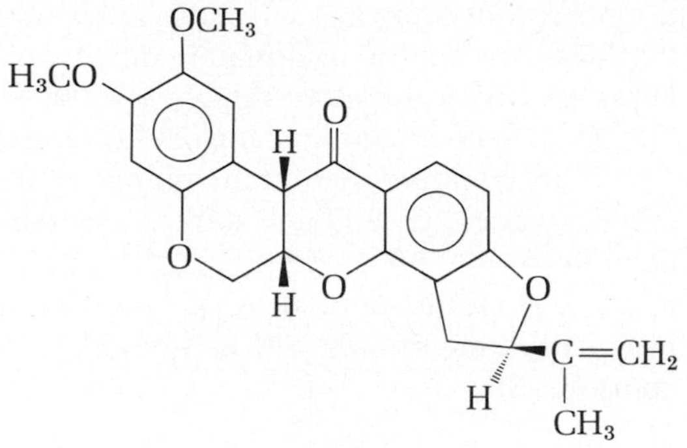

Rotenone

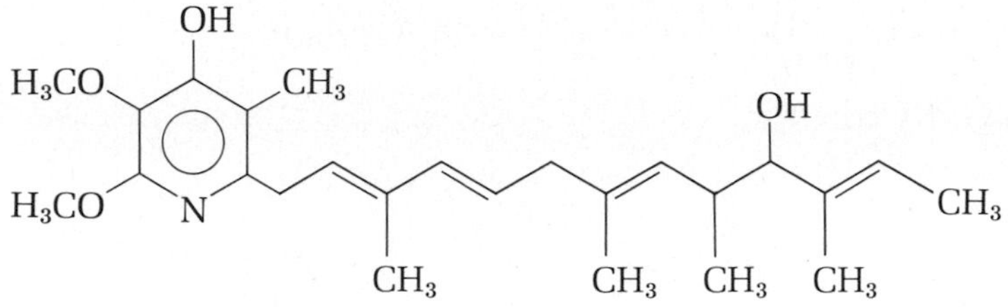

Piericidin A

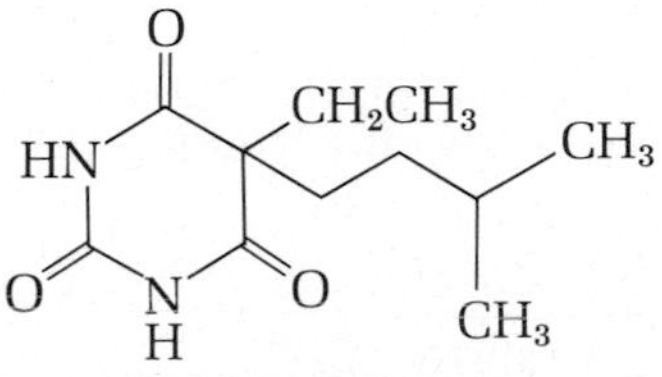

Amytal

Figure 16.12   Structures of rotenone, amytal, and piericidin A. All three are inhibitors of NADH-Q reductase.

When complex I is inhibited, electrons cannot be passed from NADH to coenzyme Q, but they can be passed from $FADH_2$ to coenzyme Q. Complex II is responsible for this reduction.

B. Succinate-Q Reductase (Complex II)

Complex II catalyzes transfer of electrons from succinate to enzyme-bound flavin and from the flavin to coenzyme Q (reaction 16.31).

$$\text{succinate + enzyme-FAD} \longrightarrow$$

$$\text{fumarate + enzyme-FADH}_2$$

$$\downarrow Q$$

$$\text{enzyme-FAD + QH}_2 \qquad (16.31)$$

Complex II is a multiprotein assembly containing four polypeptides of which at least three contain redox centers. The polypeptides have molecular weights of 70,000, 27,000, 15,000, and 13,000. The two largest polypeptides are the components of succinate dehydrogenase, whose activity we discussed in the context of the citric acid cycle (Section 15.4F). The smaller subunit of succinate dehydrogenase (mol wt 27,000) contains an Fe_4S_4 center. The two smallest polypeptides contain an Fe_2S_2 center. One of these two small polypeptides is also a b-type cytochrome whose α-band has an absorbance maximum of 560 nm. The standard reduction potential of complex II cytochrome b_{560} is -0.20 V, far more negative than the standard potential for reduction of fumarate to succinate of $+0.03$ V. Cytochrome b_{560} of complex II can therefore donate electrons to fumarate, but it cannot accept electrons from succinate, and its function in complex II is not known.

The standard reduction potential for transfer of electrons from succinate to coenzyme Q is $+0.07$ V, corresponding to a standard free-energy change of -13.5 kJ. Complex II does not, therefore, provide enough energy to drive the phosphorylation of ADP. Coenzyme Q, as we noted above, is soluble in the membrane and is thought to carry electrons from complex II to complex III.

C. QH₂-Cytochrome c Reductase (Complex III): Second Site of ATP Synthesis

QH_2-cytochrome c reductase catalyzes transfer of electrons from coenzyme Q to cytochrome c (reaction 16.32). ATP synthesis is coupled to this redox reaction at coupling site II.

$$QH_2 + 2 \text{ cyt } c(\text{Fe}^{3+}) \xrightarrow[\substack{P_i + ADP \quad ATP \\ \text{Coupling site II}}]{\substack{QH_2\text{-cyt } c \text{ reductase} \\ (\text{Complex III})}} Q + 2 \text{ cyt } c(\text{Fe}^{2+}) + 2H^{\oplus} \qquad (16.32)$$

Complex III contains b-type cytochromes, a c-type cytochrome called cytochrome c_1, an iron-sulfur protein, coenzyme Q, and lipid. The b-type cytochromes have different properties. The absorbance maxima of the α bands of the b-type cytochromes are 560 and 562 nm. A second nomenclature designates cytochrome b_{560} as cytochrome b_k and cytochrome b_{562} as cytochrome b_T. Cytochrome c itself is not part of complex III; it acts as an electron shuttle that carries electrons from complex III to complex IV. Electrons are passed through complex III as shown in reaction 16.33.

$$QH_2 \xrightarrow[2H^{\oplus}]{} \text{cyt } b_k \longrightarrow \text{cyt } b_T \longrightarrow \text{cyt } c_1 \longrightarrow (\text{Fe-S}) \longrightarrow \text{cyt } c \qquad (16.33)$$

The two-electron carrier, coenzyme Q, passes electrons to one-electron acceptor heme groups. The ability of coenzyme Q to exist as a free radical makes successive one-electron transfers possible. Cytochromes b_k and b_T have different reduction potentials. The standard reduction potential for cytochrome b_T increases from -0.03 V to $+0.24$ V as the concentration of ATP increases. Cytochrome b_T has therefore been proposed as the site of ATP synthesis (Figure 16.13).

The transfer of electrons from coenzyme Q to cytochrome c is specifically inhibited by *antimycin A*, an antibiotic produced by *Streptomyces*.

Antimycin A

D. Cytochrome *c* Oxidase (Complex IV): Third Site of ATP Synthesis

Cytochrome *c* oxidase is the terminal component in the respiratory chain. It catalyzes the *four-electron* reduction of water, a step that is also coupled to ATP synthesis (reaction 16.33).

$$O_2 + 4H^{\oplus} + 4e \xrightarrow[\substack{\text{Cyt } c \text{ oxidase} \\ \text{(Complex IV)}}]{} 2H_2O \qquad (16.33)$$

$$ADP + P_i \quad ATP$$
$$\text{Coupling site III}$$

Cytochrome *c* oxidase is an integral protein of the inner mitochondrial membrane. It completely spans the membrane. The enzyme from beef heart mitochondria contains six subunits. The enzyme complex contains two Cu(II) ions, cytochrome *a*, and cytochrome a_3.

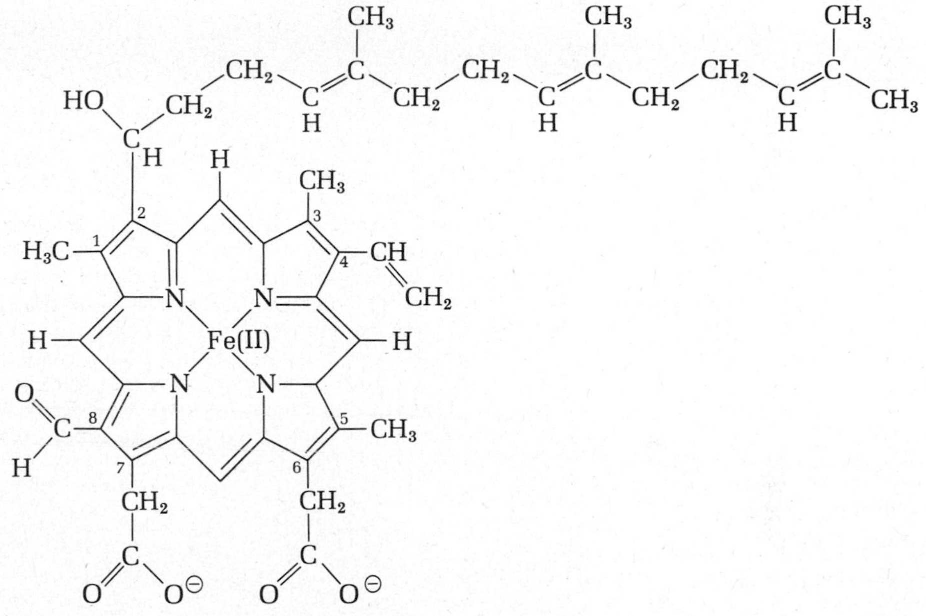

Heme *a*

The flow of electrons in complex IV is shown in reaction 16.34.

$$\text{cyt } c \ (Fe^{2+}) \longrightarrow \text{cyt } a \longrightarrow Cu^{2+} \longrightarrow \text{cyt } a_3 \longrightarrow O_2 \qquad (16.34)$$

The energy changes for this electron transport chain are illustrated in Fig. 16.14. As the concentration of ATP increases, the reduction potential of cytochrome a_3 decreases from 0.375 V to 0.115 V. Thus the concentration of ATP regulates the reduction potentials of proteins in complexes I, III, and IV.

Cytochrome *c* oxidase is specifically inhibited in cyanide anion, azide, and carbon monoxide.

E. Cytochrome *c*: The Mediator of Electron Transfer from Complex II to Complex III

Cytochrome *c* is a peripheral protein bound to the intermembrane side of the inner mitochondrial membrane. It accepts electrons from complex II and passes them to complex III. Cytochrome *c* is readily removed from the inner membrane by treatment with concentrated salt solutions, and once it has been removed, respiration and phosphorylation of ADP cease.

In Section 2.10 we noted that in the long evolution of aerobic organisms, the sequences of certain regions of cytochrome *c* are conserved. These sequences include the residues in contact with the heme (70 to 80) that are virtually identical in all species. This conservation of primary struc-

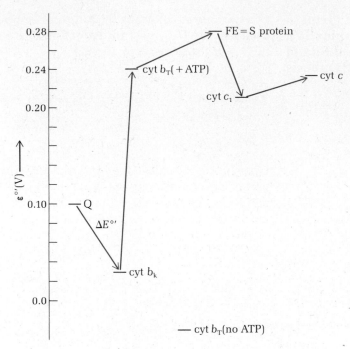

Figure 16.13 Energy changes for transfer of electrons from coenzyme Q to cytochrome c. The transfer is mediated by complex III. Coupling site II is associated with complex III. The reduction potential of cytochrome b_T is highly dependent upon the concentration of ATP. Changes in $\Delta E^{\circ\prime}$ which are algebraically negative indicate that the equilibrium constant is less than 1.0 under standard conditions. Under cellular conditions the reaction can occur with net formation of product because $\Delta E^\prime$ is the criterion of spontaneity.

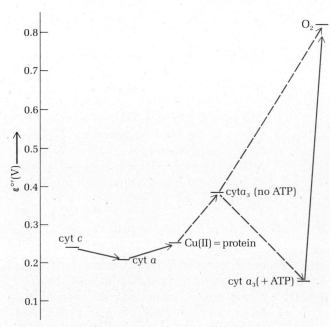

Figure 16.14 Energy changes for electron transfer reactions in complex IV. Note that the concentration of ATP influences the reduction potential of cytochrome a_3.

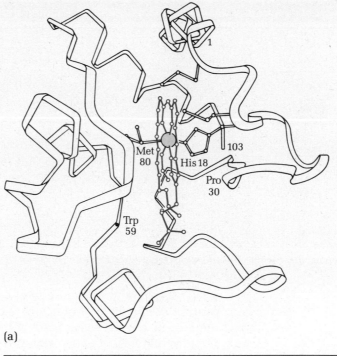

(a)

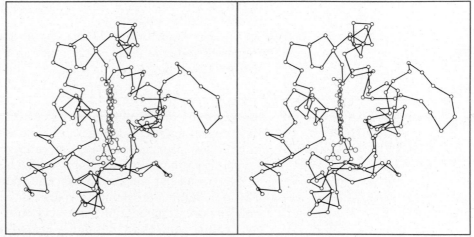

(b)

Figure 16.15   (a) Schematic representation of the structure of tuna heart mitochondrial cytochrome c. The heme is bound to the protein by histidine 18 and methionine 80. The most conserved residues are indicated in color. [From F. R. Salemme, *Ann. Rev. Biochem.*, 46, 303 (1977). Reproduced with permission. © 1977 by Annual Reviews Inc.] (b) Stereo view of tuna heart cytochrome c.

ture is reflected in the tertiary structure. The structures of mitochondrial cytochromes c are nearly identical, and the structure of cytochrome c has changed little since the emergence of eucaryotic organisms. The structure of the mitochondrial cytochrome c from tuna heart is shown in Figure 16.15.

All cytochromes c have a series of lysyl residues along one side of the heme. These residues undoubtedly influence the interaction of cytochrome c with both QH$_2$-cytochrome c reductase and with cytochrome c oxidase. Cytochrome c binds stoichiometrically to cytochrome c oxidase. The question how an electron at the c-heme buried in cytochrome c can "jump" to the heme of cytochrome a in cytochrome oxidase has received much study. A free radical mechanism in which tyrosyl residues participate in electron transfer initially seemed attractive, but a free radical mechanism is unlikely because of its large energy barrier. Direct transfer of an electron from one

Table 16.7 Composition of respiratory chain complexes and their location in the inner mitochondrial membrane

Enzyme complex	Molecular weight	Number of subunits	Prosthetic groups	Location of binding sites in membrane		
				Matrix side	Middle	Cytoplasmic side
NADH-Q reductase (complex I)	850,000	16	1 FMN 16−24 Fe-S (5−6 centers)	NADH	Q	
Succinate-Q reductase (complex II)	97,000	2	1 FAD 8 Fe-S	Succinate	Q	
QH$_2$-cytochrome c reductase (complex III)	280,000	6−8	2 b-type cyt cyt c_1 2 Fe-S		Q	cyt c
Cytochrome c	13,000	1	1 c-heme			cyt c_1 cyt a
Cytochrome c oxidase (complex IV)	200,000	6−7	2 a-hemes 2 Cu	O$_2$		cyt c

From J. W. DePierre and L. Ernster, *Ann. Rev. Biochem.*, 46, 215 (1977). Reproduced with permission. © Annual Reviews Inc.

heme to the other now seems more likely, but the question is not settled.[3] It is not known whether cytochrome c is simultaneously in contact with both cytochrome c_1 in the reductase and with cytochrome oxidase. Some evidence suggests that it is, but trace amounts of cytochrome c free on the membrane surface also catalyze electron transport, and the question awaits a more definitive answer.

The composition of the respiratory chain complexes, their locations in the inner mitochondrial membrane, and the membrane binding sites for various substrates are summarized in Table 16.7.

16−6 COUPLING OF OXIDATION TO PHOSPHORYLATION: COUPLING FACTOR 1

In our discussion of mitochondrial structure we noted the presence of stalk-like projections studding the inner surface of the inner membrane. The function of these particles has been determined by dissecting the mitochondrion. The outer mitochondrial membrane can be removed by radically increasing or decreasing the electrolyte concentration (ionic strength) of the medium. The remaining mitochondrial particle is still capable of oxidative phosphorylation. Treatment of the mitochondrion with ultrasonic waves, a process called *sonication*, produces sealed vesicles called submitochondrial particles. These, too, can carry out oxidative phosphorylation. Electron micrographs of sonicated submitochondrial particles shows the stalks on the outer surface of the vesicles, and thus the particles have been turned inside out. Treatment of the inside-out submitochondrial particles with urea removes the stalks from the vesicles (Figures 16.16 and 16.17). Vesicles bereft of stalks retain their electron transport chains intact and can catalyze the oxidation of NADH with O$_2$ as the ultimate electron acceptor. They cannot, however, phosphorylate ADP. It is the stalks that hydrolyze ATP; they are the ATPase to which we earlier referred. Reconstituting the vesicles regenerates a system capable of oxidative phosphorylation. The stalks are therefore the agents responsible for ATP synthesis, and they are known as ATP synthetase, as *coupling factor 1*, or simply F$_1$.

The mitochondrial ATPase is an extremely complex oligomer, com-

---

[3] Students who have studied quantum mechanics in physical chemistry will recall the phenomenon of "tunneling." A consequence of the uncertainty principle, tunneling is a mechanism by which an electron on one side of a potential barrier can reach the other side of the barrier without climbing over the top. If transfer of electrons to and from cytochrome c occurs directly, then it seems likely that it is by quantum mechanical tunneling that the process occurs.

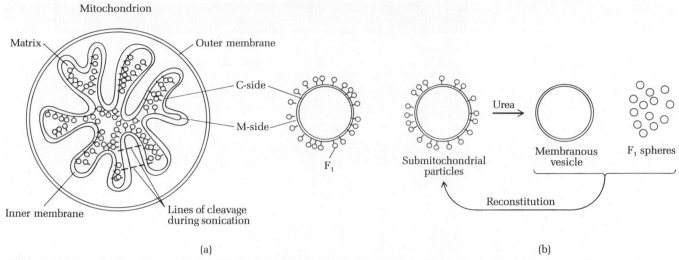

(a)   (b)

Figure 16.16   Dissection of the mitocondrion. (a) Removal of the outer mitochondrial membrane, followed by treatment with ultrasonic vibration, produces inside-out submitochondrial particles. (b) Treatment with urea yields sealed vesicles and F_1 particles. The sealed vesicles contain intact electron transport chains, but they cannot phosphorylate ADP. The isolated F_1 particles hydrolyze ATP. Reconstitution of the submitochondrial particles regenerates a system capable of oxidative phosphorylation. [From E. Racker, *Biochem. Soc. Trans.*, **3**, 789 (1975).]

posed of a dozen proteins, whose total molecular weight is 360,000 (Table 16.8). The ATPase is believed to be constructed as shown in Figure 16.18. The membrane sector contains a channel through which protons flow between the intermembrane space and the mitochondrial matrix. A protein "stalk" connects the membrane sector with the F_1 spherical particle that is responsible for ATP hydrolysis when separated from the membrane and for ATP synthesis when bound to the membrane. A variety of agents inhibit the ATPase activity of F_1 by specifically interacting with subunits of the as-

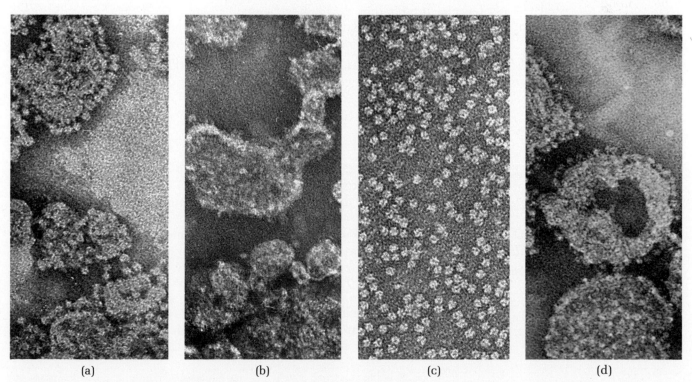

(a)   (b)   (c)   (d)

Figure 16.17   Electronmicrographs of (a) submitochondrial particles, (b) isolated vesicles, (c) F_1 particles, and (d) reconstituted submitochondrial particles. (Courtesy of Dr. E. Racker.)

Table 16.8	Composition, membrane location, and function of the mitochondrial ATPase				
Component subunits	**Molecular weight**	**Location**	**Function**	**Inhibitors**	
F_1	360,000	Peripheral, matrix side	ATP synthesis	Aurovertin	
α	53,000				
β	50,000				
γ	33,000				
δ	17,000				
ϵ	7,500				
F_1 inhibitor	10,000	Peripheral	Regulation of ATP synthesis		
OSCP	18,000	Peripheral, stalk	Binds F_1 to membrane		
Membrane sector proteolipids	29,000 22,000 12,000 7,800	Integral	Proton channel	Oligo-mycin Dicyclo-hexyl carbodii-mide (DCCD)	
Fc_2 (F6)	8,000	Stalk (?)			

From J. W. DePierre and L. Ernster, *Ann. Rev. Biochem., 46,* 216 (1977). Reproduced with permission. © 1977 by Annual Reviews Inc.

sembly. The antibiotic *oligomycin* uncouples oxidation and phosphorylation.

The *oligomycin-sensitivity-conferring protein* (OSCP) is a stalk protein that is required for the binding of F_1 to the membrane. Oligomycin does not bind to OSCP, but to a membrane sector protein. Oligomycin inhibits the translocation of protons through the membrane sector. The membrane sector is also inhibited by dicyclohexyl carbodiimide (DCCD). The antibiotic *auro-vertin* inhibits oxidative phosphorylation by binding to a subunit of the F_1 particle. Aurovertin inhibits binding of ADP to F_1. *Dicoumarol* and *2,4-dinitrophenol* (DNP) are two nonspecific uncoupling agents.[4]

2,4-Dinitrophenol (DNP)   Dicoumarol

16–7 CHEMIOSMOTIC HYPOTHESIS

In 1961 P. Mitchell proposed a *chemiosmotic hypothesis* in which an electrochemical gradient across the membrane provides the driving force for phosphorylation of ADP without the intervention of an energy-rich "covalent" intermediate. The electrochemical gradient is established by asymmetric ejection of protons into the intermembrane space. We have noted that the carriers of the electron transport chain are of two types: one-electron carriers, such as cytochrome *c* and iron-sulfur proteins, and carriers of both protons and electrons, such as coenzyme Q. In passing electrons from one of

[4] Dicoumarol has a second activity of some importance: it inhibits the action of vitamin K. Cattle that eat spoiled clover, which contains large amounts of dicoumarol, are fatally susceptible to hemorrhages.

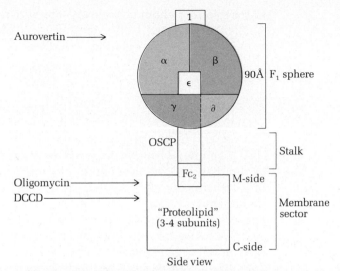

Figure 16.18 Schematic representation of the subunit composition of the mitochondrial ATPase. [Adapted from J. W. DePierre and L. Ernster, *Ann. Rev. Biochem.*, 46, 288 (1977). Reproduced with permission. © 1977 by Annual Reviews Inc.]

these carriers to, say, a cytochrome, the protons must be jettisoned. According to the chemiosmotic hypothesis, they move into the intermembrane space (Figure 16.19). The gradient of protons that arises by their ejection into the intermembrane space has two components: a pH gradient (ΔpH) and an electrical gradient, or membrane potential ($\Delta\psi$). The sum of these two terms is called the *proton motive force* (Δp). It is described by a relationship analogous to the Nernst equation (recall equation 16.16).

$$\Delta p = \Delta\psi - 2.303\frac{RT}{F}\Delta pH \qquad (16.35)$$

The term Δp, whose units are volts, corresponds to E'; $\Delta\psi$ is analogous to $E^{0'}$; and the term $-2.303\,(RT/F)\Delta pH$ is the Nernst potential, also in volts, that is generated by the ejection of protons into the intermembrane space. Expelling protons and creating gradients requires energy. The driving force for gradient formation is the energy released by the redox reactions catalyzed by

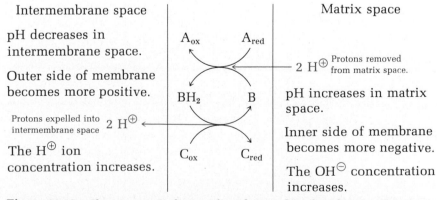

Figure 16.19 Chemiosmotic theory of oxidative phosphorylation. The ejection of protons from the mitochondrial matrix space during electron transport is equivalent to the formation of the elements of water on opposite sides of the membrane as the hydroxide ion concentration on the inner side of the membrane increases and the hydrogen ion concentration on the outside of the membrane increases.

the respiratory chain complexes. The function of oxidation is therefore only incidentally related to the synthesis of ATP; it provides the gradient. The reaction for phosphorylation of ADP is

$$ADP + HPO_4^{2\ominus} \rightleftharpoons ATP + H_2O \tag{16.36}$$

At a concentration of 1 mM Mg(II) ions the standard free energy of hydrolysis of ATP, $\Delta G^{\circ\prime}$, is -31.8 kJ mol^{-1}. Approximately this much energy must be provided by the gradient to drive ATP synthesis.

The chemiosmotic hypothesis is an extraordinary theory, and it has generated an enormous amount of work to test its predictions. The existence of a transmembrane potential, absolutely essential to the theory, has been demonstrated. The intermembrane space is more acidic than the matrix space in respiring mitochondria. A total transmembrane potential of 0.180 V to 0.220 V has been measured. The chemiosmotic theory proposes that two protons are ejected at each coupling site. Various experiments suggest that at least three, and perhaps as many as four, protons are ejected per site. This discrepancy may not be fatal to the chemiosmotic theory, and the hypothesis of a proton gradient is certainly supported, but the discrepancy may in the end lead to some revisions of the details of the theory.

It has been demonstrated that the three coupling sites of the respiratory chain—at complexes I, III, and IV—can each generate an electrochemical potential across the membrane that is negative on the inside. Each of the respiratory complexes must therefore span the inner mitochondrial membrane. The distribution of the components of the respiratory chain complexes is shown in Figure 16.20. Mitchell's original proposal is remarkably close to the results obtained from two decades of research.

In the chemiosmotic hypothesis the coupling between the proton gradient is *direct*. It has been proposed that the membrane sector of the mitochondrial ATPase is a conduit for protons, and that the flow of protons back into the mitochondria through the ATPase alters the active site of the ATPase leading to ATP synthesis (Figure 16.21). Oligomycin uncouples oxi-

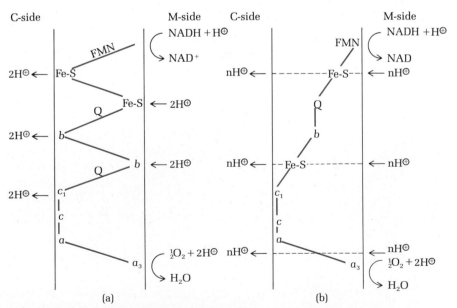

Figure 16.20 Arrangement of the electron carriers of the respiratory chain across the inner mitochondrial membrane. (a) The original proposal of P. Mitchell. (b) The distribution of carriers according to current evidence. [From J. W. DePierre and L. Ernster, *Ann. Rev. Biochem.*, **46**, 213 (1977). Reproduced with permission. © 1977 by Annual Reviews Inc.]

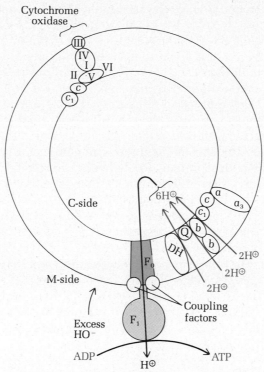

Figure 16.21 Flow of protons through coupling factor F_1. Proton motive force proves the energy to drive the synthesis of ATP in the chemiosmotic hypothesis. The process is shown for inside-out submitochondrial particles. [From E. Racker, *Ann. Rev. Biochem.*, 46, 1008 (1977). Reproduced with permission. © 1977 by Annual Reviews Inc.]

dation and phosphorylation by preventing protons from flowing through the membrane sector, a result that strongly suggests that the ATPase is involved in the transport of protons across the mitochondrial membrane. The chemical mechanism that might result in ATP synthesis from the direct interaction of ADP, phosphate, and protons is not at all clear. Here again, however, the theory has demonstrated its fertility. An artificially imposed proton gradient can drive the synthesis of ATP in the absence of electron transport. This result implies that the ATPase is, in fact, affected by the proton gradient. Reversing the gradient by making the external medium more basic than the matrix space leads to ATP hydrolysis.

The effects of uncouplers, such as 2,4-dinitrophenol, are readily explained by the chemiosmotic hypothesis. The 2,4-dinitrophenol is a weak acid that is soluble in the membrane when it is protonated, and it is soluble in water as the 2,4-dinitrophenolate anion. 2,4-Dinitrophenol is thus a transporter of protons across the membrane. It abolishes the proton gradient and uncouples oxidation from phosphorylation.

16–8 CONFORMATIONAL COUPLING HYPOTHESIS

In the *conformational coupling hypothesis* the proton gradient is *indirectly* coupled to the synthesis of ATP. According to the chemiosmotic hypothesis, the proton gradient is generated by three proton-translocating redox loops (Figure 16.20a). In the indirect mechanism these protons are translocated by amino acid side chains rather than by prosthetic groups of the respiratory chain complexes. The pK_a values of these side chains alter as redox reactions occur at the active sites of the respiratory chain complexes. The respiratory chain complexes must span the membrane for protons to be picked up on one side or released on the other to generate the gradient, but it is not necessary to have a vectorial arrangement of proteins in the membrane in the conformational hypothesis. The hypothesis of vectorial secretion of protons by vectorially arranged electron carriers is central to the chemiosmotic hypothesis. A second requirement of the vectorial *redox loops* proposed by the chemiosmotic hypothesis is that two protons be ejected per coupling site. The observation of more than two protons per site may argue against redox loops and in favor of an indirect coupling between electron transport and proton translocation. We have noted that the chemiosmotic hypothesis postulates that the intermembrane sector acts as a proton well in which there is a direct interaction between phosphate, ADP, and protons. This feature is not a part of the indirect conformational coupling hypothesis.

The conformational change occurs in the mitochondrial F_1 particle. The F_1 particle is believed to have two binding sites for ATP. In the conformational hypothesis one of the catalytic sites is on the matrix side of the membrane. The conformational interaction occurs between F_1 and the proton-translocating membrane sector. The synthesis of ATP is driven by a conformational change at one of the two ATP-synthesizing sites. This conformational change is driven by the proton motive force. As ATP is synthesized at one of these sites, a previously synthesized ATP is released at the other (Figure 16.22). The coupling site is therefore a two-cycle conformational engine.

The solutions to the still vexing problems of oxidative phosphorylation will require further knowledge of the structure of the inner mitochondrial membrane, of the stoichiometry of the entire electron transport chain, especially the number of protons ejected per site, and of the chemical events of ATP synthesis.

16–9 MITOCHONDRIAL ENERGY STATES AND RESPIRATORY CONTROL

It has long been known that mitochondria undergo dramatic structural changes when they switch from a resting state to a respiring state (Figure 16.23). In the respiring state the inner membrane is not folded into cristae; but seems to have shrunk upon itself leaving a much more voluminous intermembrane space. Five energy states of mitochondria have been defined (Table 16.9). In state 1 oxygen is present, but the supply of ADP is low, and limits the rate of oxidative phosphorylation. In state 2 oxygen is present and the ADP concentration is high, but the NADH/NAD ratio is again low, and the rate of respiration, limited by the available supply of NADH (or $FADH_2$), is low. In state 3, an experimentally and physiologically common one, NADH, oxygen, and ADP are all present in high concentrations, and respiration is rapid. If a system in state 3 continues until its ADP supply is limited,

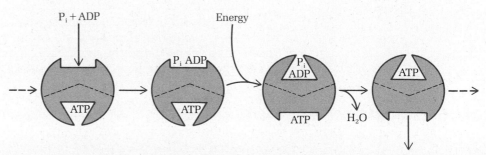

Figure 16.22 Conformational coupling hypothesis of oxidative phosphorylation. A two-cycle conformational engine is proposed as the agent of ATP synthesis. The rate-determining step is not the actual formation of ATP, but its release from the enzyme active site as an indirect consequence of the proton motive force. [From P. D. Boyer, *Ann. Rev. Biochem.*, 46, 964 (1977). Reproduced with permission. © 1977 by Annual Reviews Inc.]

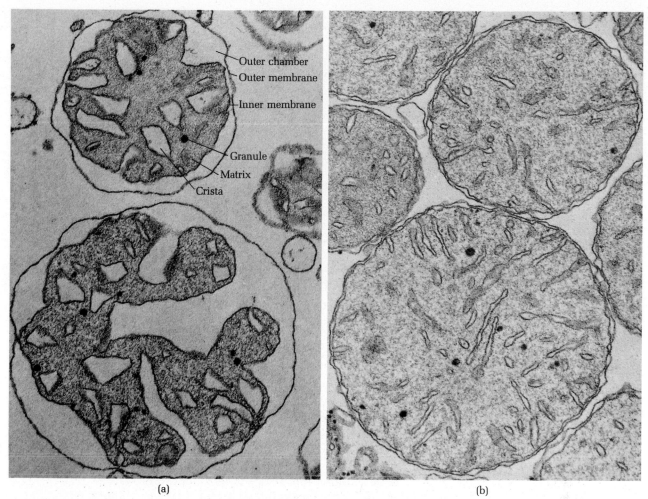

(a)   (b)

Figure 16.23   Electron micrographs of rat liver mitochondria in the active, respiring state (a) and the inactive, resting state (b). [Courtesy of Dr. C. R. Hackenbrock.]

the respiration rate slows, and the net rate is controlled by the concentration of ADP. If all of the oxygen is consumed, state 5 is reached, and respiration ceases.

These observations collectively define the *respiratory control* that is a property of *tightly coupled* mitochondria, that is, undamaged or uninhibited mitochondria in which oxidation cannot occur without phosphorylation of ADP. *Uncoupling* agents, such as DNP and dicoumarol, abolish respiratory control, and electron transport occurs without concomitant phosphorylation of ADP. In sum, respiration in tightly coupled mitochondria is con-

Table 16.9 Energy states of mitochondria					
	State 1	**State 2**	**State 3**	**State 4**	**State 5**
Characteristics	Aerobic	Aerobic	Aerobic	Aerobic	Anaerobic
ADP level	Low	High	High	Low	High
Substrate level	Low-endogenous	Approaching 0	High	High	High
Respiration rate	Slow	Slow	Fast	Slow	0
Rate limiting component	Phosphate acceptor (ADP)	Substrate	Respiratory chain	Phosphate acceptor (ADP)	Oxygen

From B. Chance and G. R. Williams, *J. Biol. Chem.*, **215**, 413 (1955).

trolled by (1) the NADH/NAD ratio, (2) the phosphorylation ratio, or phosphate potential, [ATP]/[ADP][P_i], and (3) the partial pressure of oxygen.

In state 4 the rate of respiration is limited by the phosphorylation ratio, and the entire respiratory chain is virtually at equilibrium (reaction 16.37).

$$NADH + 2cyt\ c(Fe^{3+}) + 2ADP + 2P_i \rightleftharpoons$$
$$NAD + 2cyt\ c(Fe^{2+}) + 2ATP \qquad (16.37)$$

The equilibrium constant for respiration in state 4 is 4.4×10^7 (equation 16.38).

$$K_{eq} = \frac{[NAD][cyt\ c^{2+}]^2[ATP]^2}{[NADH][cyt\ c^{3+}]^2[ADP]^2[P_i]^2} = 4.4 \times 10^7 \qquad (16.38)$$

The reduction potentials for the components of the electron transport chain are given by the Nernst equation.

$$E' = E^{0'} + \frac{RT}{nF} \ln \frac{[ox]}{[red]} \qquad (16.39)$$

The E' values for the respiratory chain components, also called midpoint potentials, and their relation to ATP synthesis are summarized in Figure 16.24. The E' values fall into four groups at -0.3 V, 0.0 V, $+0.22$ V, and $+0.40$ V. The $\Delta E'$ values between these potentials correspond to the energy required to synthesize ATP.

ATP dramatically alters the standard reduction potential of the iron-sulfur protein having the most positive standard reduction potential (-0.02 V). Complex I therefore responds directly to the ADP/ATP ratio, a manifestation of respiratory control in state 4 mitochondria. ATP also alters the standard reduction potential of cytochrome b_T in complex III, and the standard reduction potential of cytochrome a_3 in complex IV. Respiratory control can thus be explained in state 4 mitochondria in terms of allosteric interactions of ATP with each of the three coupling sites. By binding the respiratory chain complexes, ATP apparently alters the environments of the active sites of electron transfer and thus regulates the rate of its own production.

16–10 RELATIONSHIP BETWEEN GLYCOLYSIS, THE CITRIC ACID CYCLE, AND RESPIRATION

When oxygen is added to cells that are degrading glucose anaerobically by glycolysis, the rate of glycolysis drops precipitously, a phenomenon known as the *Pasteur effect*. Aerobic degradation of pyruvate by the citric acid cycle leads to abundant energy production by oxidative phosphorylation. Under aerobic conditions the oxidation of glucose yields up to 38 moles of ATP, versus 2 moles of ATP for the conversion of glucose to lactate by glycolysis. A slow rate of glycolysis is therefore sufficient to provide the cell with adequate energy under aerobic conditions. Viewed the other way around, under anaerobic conditions the rate of glycolysis increases to supply the ATP required when oxidative phosphorylation is not functional.

The Pasteur effect is mediated by the allosteric effects of various metabolites on the activities of glycolytic enzymes, chiefly phosphofructokinase. The phosphorylation of fructose 6-phosphate is a "metabolic crossover point." The onset of anaerobic conditions in cells carrying out glycolysis is shortly followed by a decrease in the concentrations of fructose 6-phosphate, glucose 6-phosphate, and glucose and an increase in the concentration of fructose 1,6-diphosphate. In nearly all cells, except brain and yeast cells,

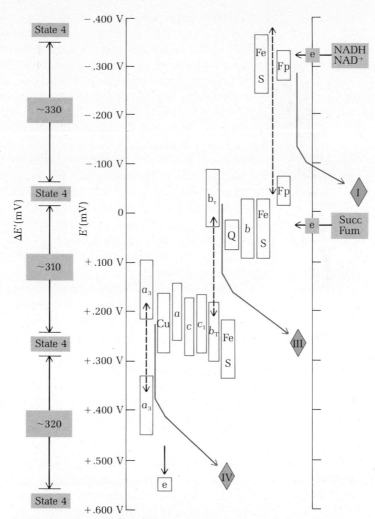

Figure 16.24 Relationship between the reduction potentials (E') of the components of the respiratory chain and the synthesis of ATP in mitochondrial energy state 4. The dashed lines indicate the effect of ATP on the reduction potentials of an iron-sulfur protein in complex I, cytochrome b_T in complex III, and cytochrome a_3 in complex IV. [From D. F. Wilson et al., *Acc. Chem. Res.*, **5**, 237 (1972).]

AMP and ADP increase the activity of phosphofructokinase, and ATP is an inhibitor.

Anaerobic conditions in brain cells are accompanied by increases in both phosphate and fructose 1,6-diphosphate, both activators of phosphofructokinase. The ATP level in brain cells can be maintained, at least for a while, by phosphorylation of ADP by phosphocreatine, itself a potent inhibitor of phosphofructokinase (reaction 16.40).

Phosphocreatine

creatine + ATP   (16.40)

In yeast the presence or absence of oxygen has little effect on the concentration of ATP, and the activity of phosphofructokinase is largely controlled by the allosteric inhibitor citrate. Under aerobic conditions citrate levels are relatively high and phosphofructokinase activity is low. The opposite conditions prevail under anaerobic conditions.

Pyruvate kinase and the pyruvate dehydrogenase complex are also regulated allosterically by the inhibitor ATP. The net effect of ATP is thus inhibition of glycolysis. The citric acid cycle is also inhibited by ATP through its allosteric effect upon isocitrate dehydrogenase.

NADH is also a significant regulatory metabolite. Because NADH inhibits isocitrate dehydrogenase and pyruvate dehydrogenase, it directly inhibits the citric acid cycle. The NAD/NADH ratio also affects the respiratory chain. Oscillations in the concentration of ADP and ATP are paralleled by changes in the concentrations of NADH and NAD, with ADP levels following NADH and ATP levels following NAD. These regulatory effects couple glycolysis, the citric acid cycle, and oxidative phosphorylation in a highly interactive system.

When glucose is added to aerobic cancer cells, oxygen consumption drops, a phenomenon known as the *Crabtree effect*. This high rate of *aerobic glycolysis* indicates that the normal regulatory mechanisms responsible for the Pasteur effect have gone awry. The lactate produced by anaerobic glycolysis is transported to the liver where it is converted to glucose by gluconeogenesis (Section 17.3). Six moles of ATP are required for the conversion of lactate to glucose. Since only 2 moles of ATP are produced by glycolysis, the cancer cells drain energy from the organism to support their own prolific consumption of glucose.

16–11 SHUTTLE MECHANISMS FOR CYTOSOLIC NADH "TRANSPORT"

The mitochondrion is permeable neither to NADH nor to NAD. The NADH generated in the cytosol can, however, undergo an oxidation-reduction cycle whose net result is the generation of NADH or $FADH_2$ inside the mitochondrion. The *glycerophosphate shuttle*, characteristic of insect flight muscle, consists of a cytosolic and a mitochondrial *sn*-3-glycerophosphate dehydrogenase. The cytosolic enzyme reduces dihydroxyacetone phosphate to *sn*-3-glycerophosphate, which is transported into the mitochondrion (reaction 16.41).

$$
\begin{array}{c}
CH_2OH \\
| \\
C{=}O \\
| \\
CH_2OPO_3^{2\ominus}
\end{array}
+ NADH
\xrightarrow[\text{dehydrogenase}]{\text{Glycerophosphate}}
\begin{array}{c}
CH_2OH \\
| \\
HO{-}C{-}H \\
| \\
CH_2OPO_3^{2\ominus}
\end{array}
+ NAD \qquad (16.41)
$$

Dihydroxyacetone phosphate   *sn*-3-Glycerophosphate

Reoxidation of *sn*-3-glycerophosphate by a membrane-bound flavoprotein dehydrogenase on the outer side of the inner mitochondrial membrane generates $FADH_2$ (reaction 16.42).

$$
\begin{array}{c}
CH_2OH \\
| \\
HO{-}C{-}H \\
| \\
CH_2OPO_3^{2\ominus}
\end{array}
+ FAD
\xrightarrow[\text{dehydrogenase}]{\text{Membrane flavoprotein}}
\begin{array}{c}
CH_2OH \\
| \\
C{=}O \\
| \\
CH_2OPO_3^{2\ominus}
\end{array}
+ FADH_2 \qquad (16.42)
$$

sn-3-Glycerophosphate

$FADH_2$ then passes its electron to coenzyme Q whose oxidation in the electron transport chain is coupled to the synthesis of 2 moles of ATP (Figure 16.25).

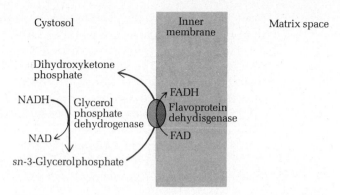

Figure 16.25 Glycerol phosphate shuttle in insect flight muscle.

Mammals have a more complicated shuttle mechanism that involves malate, oxaloacetate, a cytosolic enzyme pair (malate dehydrogenase and an aminotransferase), and a mitochondrial enzyme pair (malate dehydrogenase and an aminotransferase). The malate-aspartate shuttle is shown in Figure 16.26. The α-ketoglutarate carrier and the aspartate-glutamate carrier, discussed earlier, participate in this transport system. Since the oxidation of NADH yields 3 moles of ATP, the malate-aspartate shuttle is more efficient than the glycerophosphate shuttle of insect flight muscle. In fact, it was in anticipation of the malate-aspartate shuttle that we accounted for the production of 38 moles of ATP from the complete oxidation of 1 mole of glucose. In the absence of this shuttle, we would have to reduce the net ATP production by 6 moles, since NADH, barred from entering the mitochondrion, could not be oxidized by respiration.

16–12 MEMBRANE TRANSPORT IN MITOCHONDRIA

The inner mitochondrial membrane is impermeable to all but a few metabolites. Every substrate other than water, carbon dioxide, and oxygen that enters the mitochondria requires a specific transport protein. Many mitochondrial transport systems are linked to the proton motive force established by the electron transport chain, a striking instance of the power of the chemiosmotic hypothesis. All mitochondria contain transport proteins for phosphate, pyruvate, and fatty acids. Pyruvate, generated by glycolysis in the cy-

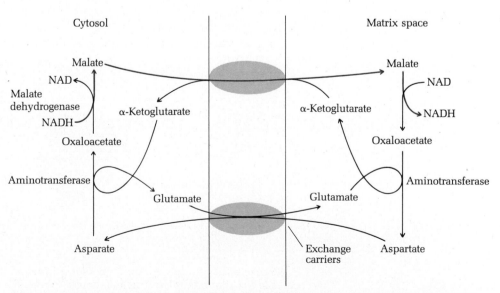

Figure 16.26 Malate-aspartate shuttle for transporting the hydride anion of NADH from the cytosol to the mitochondria of human skeletal muscle.

tosol, must be carried into the mitochondria before it can be acted upon by pyruvate dehydrogenase. β-oxidation of fatty acids occurs in the mitochondria, and phosphate must be carried into the mitochondria for use in ATP synthesis. Most of the enzymes that use ATP in biosynthetic reactions are in the cytosol, and the ATP generated in the mitochondrion is transported by a universal nucleotide carrier.

Carriers that exchange malate for α-ketoglutarate and L-glutamate for L-aspartate are also ubiquitous. These carriers constitute the malate-aspartate shuttle, an important system that has the effect of transporting hydride anion into the mitochondrion. Many other carriers are present in varying amounts in mitochondria from different sources.

A. Calcium Ion Transport

Calcium ions exert multiple effects in metabolism, as we noted in Section 13.7. Addition of small amounts of calcium ions to respiring mitochondria causes a burst in the rate of ATP synthesis that stops when all of the calcium ions have been absorbed. While Ca^{2+} is being absorbed, $H^{\oplus}$ ions are ejected, and the pH of the matrix space can increase by a unit or more. One Ca^{2+} is transported, via a specific transport protein, or *porter*, for every two electrons that pass each coupling site.

Calcium transport is inhibited by respiratory chain inhibitors and uncoupling agents. In terms of the chemiosmotic hypothesis, a membrane potential generated by the respiratory chain pulls the positive calcium ions toward the negative interior of the membrane. Thus, Ca^{2+} ions replace protons in neutralizing the proton motive force (Figure 16.27).

B. Ionophores

A large group of *peptide antibiotics* act as cation transporters. These peptides differ in their mechanisms of action and, of course, in their structures. We shall discuss three of them: valinomycin, nigericin, and gramicidin A. Although these antibiotics are bacteriocidal, they are too nonspecific for theraputic use. As probes of membrane structure and function, however, they are extremely valuable.

The peptide antibiotic *valinomycin* (Figure 16.28) specifically transports potassium ions across cell membranes. Valinomycin has a hydrophobic exterior, making it soluble in the membrane, and a hydrophilic core to which a single potassium ion binds. The ionic radius of sodium is 0.095 nm and that of potassium is 0.133 nm. However, valinomycin discriminates between the two not on the basis of size, but on the basis of hydration energy. Sodium ions have a higher charge-to-size ratio than potassium ions, and the hydration energy of sodium ions, -301 kJ mol^{-1}, is much larger than the hydration energy of potassium ions, -230 kJ mol^{-1}. Either of

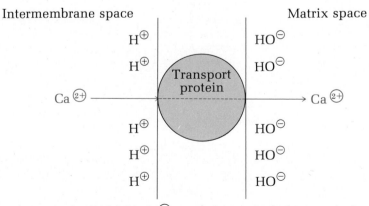

Figure 16.27   Model for Ca^{2+} transport in which the negatively charged interior side of the membrane "pulls" the cation to the inside.

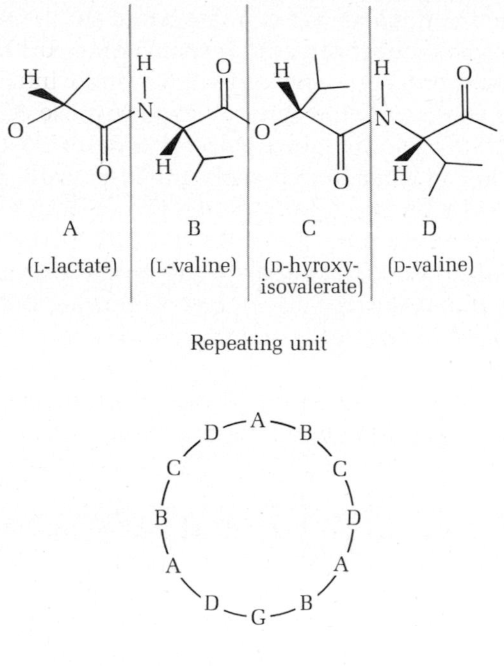

A	B	C	D
(L-lactate)	(L-valine)	(D-hyroxy-isovalerate)	(D-valine)

Repeating unit

Cyclic trimer

Figure 16.28 Valinomycin, a cyclic trimer with repeating unit A-B-C-D.

these ions must be desolvated to bind valinomycin, and the difference in solvation energy of -71 kJ mol^{-1} strongly favors the binding of potassium ions to valinomycin, tipping the balance overwhelmingly in favor of potassium ion transport. When a potassium ion binds to valinomycin, the carbonyl groups lining the center of valinomycin sequentially replace the water molecules to which the potassium ion is bound, and a large conformation change occurs (Figure 16.29).

The extent to which valinomycin equilibrates potassium ions across the cell membrane is dictated by the magnitude of the membrane potential. At low concentrations of potassium ions, the potassium ion gradient is a direct measure of the membrane potential. At high potassium ion concentrations, however, so much potassium enters the mitochondrial matrix that the electrochemical gradient across the membrane is abolished. The lipid soluble valinomycin-K$^{\oplus}$ complex has a net charge of $+1$, and the complex is pulled into the mitochondrial matrix by the negative potential on the inside of the membrane. The inward movement of potassium ions is therefore coupled to the ejection of hydrogen ions by the electron transport chain. When the potassium ion concentration is low, inward diffusion of potassium decreases $\Delta\Psi$, but actually increases ΔpH because the electron transport chain ejects more protons to restore the proton motive force.

The peptide antibiotic *nigericin* is also an ionophore that transports potassium ions. A carboxylate group of nigericin complexes with potassium. The nigericin-K$^{\oplus}$ complex is electrically neutral, and it therefore diffuses passively through the membrane. Since the carboxylate group of nigericin can also bind a proton, nigericin transports potassium ions into the mitochondria and hydrogen ions out of the mitochondria, a process known as *antiport*. Antiport decreases ΔpH across the membrane, but since one cation is exchanged for another, $\Delta\Psi$ is not altered. The combination of valinomycin and nigericin, however, abolished both ΔpH and $\Delta\Psi$, thereby uncoupling oxidative phosphorylation.

The antibiotic *gramicidin A* is a linear peptide that transports monovalent cations across membranes. Two molecules of gramicidin A interact to form a helical dimer that spans the membrane. In the dimer the N-termini of the monomers, each of which contains a formyl group, interact in the center of the membrane, and the C-termini are exposed to the aqueous phase (Figure 16.30). The helix of the gramicidin A channel has a diameter of 0.4 nm. This aqueous pore is surrounded by carbonyl groups of the peptide. The helical pore is a β-helix, formed by coiling two β-pleated sheets. The transport of ions through the gramicidin A channel is a pulsed phenomenon. The pulse rate corresponds to the formation and dissociation of the gramicidine A dimer. Each channel has a lifetime of about 1 s during which about 10^7 ions traverse the membrane.

C. Phosphate Transport

The transport of phosphate into mitochondria against a concentration gradient depends upon the membrane potential. The transport of phosphate is

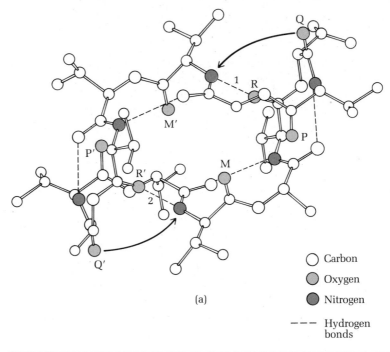

○ Carbon

● Oxygen (gray)

● Nitrogen (dark)

--- Hydrogen bonds

(a)

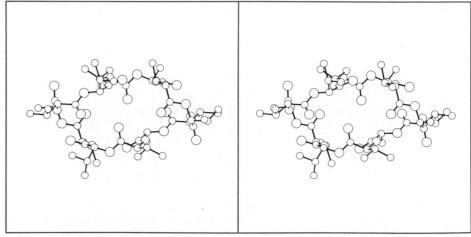

(b)

Figure 16.29 (a) Structure of valinomycin. (b) Stereo view. (c) Structure of the $K^{\oplus}$-valinomycin complex. The carbonyl groups P,P′, M,M′, Q,Q′, and R,R′ are involved in the complexation with $K^{\oplus}$. (d) Stereo view. [From D. Daux et al., *Science, 176*, 911 (1972). Copyright 1972 by the American Association for the Advancement of Science.]

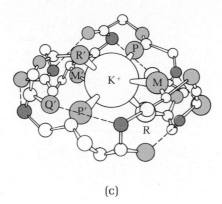

(c)

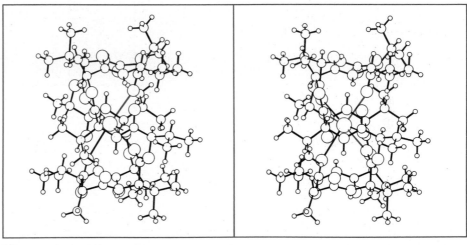

(d)

Figure 16.29 (*continued*)

an electrically neutral event that can occur in either of two ways. (1) Dihydrogen phosphate ($H_2PO_4^{\ominus}$) can exchange with a hydroxide ion by antiport. (2) Monohydrogen phosphate ($HPO_4^{\ominus}$) can be transported with a pair of protons, a process called *symport* or *cotransport*. Transport of hydrogen ions with another metabolite is common (Figure 16.31).

Phosphate transport is coupled to the transamination of glutamate; a proton carried in by glutamate is carried out by phosphate. Malate also exchanges with phosphate. These phosphate transport phenomena are part of a larger scheme that involves carboxylic acids, whose transport is the subject of the next section.

D. Transport of Carboxylic Acids

a. Pyruvate Transport

Pyruvate enters the mitochondrion by symport with protons. The ratio of pyruvate inside and outside the mitochondrion is directly proportional to the proton gradient. The process is symport because one proton disappears in the cytosol for each one that appears in the matrix space. The proton that enters with pyruvate is balanced by the exit of carbon dioxide (Figure 16.32). When the carbon dioxide reacts with water in the external medium, bicarbonate and a proton are produced.

b. Glutamate Transport

L-Glutamate is transported into mitochondria along with a proton by symport. The transamination of L-glutamate to α-ketoglutarate is followed by conversion of the latter to L-malate. The cycle is completed when L-malate exits in exchange for hydrogen phosphate (Figure 16.33). L-Gluta-

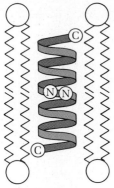

Figure 16.30
Schematic diagram of the dimeric, helical pore through a lipid bilayer in gramicidin A. Formyl groups at the N-termini of the monomers oppose one another, and the C-termini are exposed to the aqueous phase. [From S. Weinstein, B. A. Wallace, E. R. Blout, J. S. Morrow, and W. Veatch, *Proc. Nat. Acad. Sci. U.S.*, 96, 4230 (1979).]

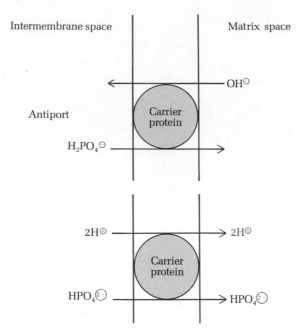

Figure 16.31 Phosphate transport system. Phosphate can be transported into the mitochondria by antiport, the exchange of an $H_2PO_4^{\ominus}$ ion for hydroxide, and by symport (cotransport) of two protons and $HPO_4^{\oplus}$.

mate transport is also directly related to the pH gradient across the inner mitochondrial membrane.

c. Dicarboxylate Carrier and Tricarboxylate Carrier

Two carriers have been found for dicarboxylic acids. The dicarboxylate transport system carries out antiport of substrates on opposite sides of the membrane. The substrates include phosphate, malonate, succinate, sulfate, sulfite, thiosulfate, and perhaps oxaloacetate. All are transported as

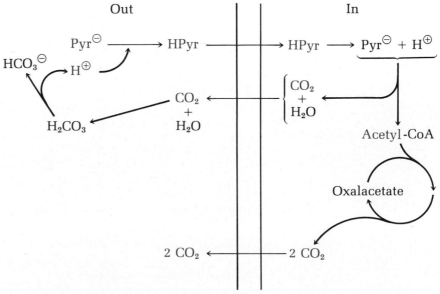

Figure 16.32 Pyruvate transport system. [From K. F. LaNoue and A. C. Schoolwerth, *Ann. Rev. Biochem.*, 48, 881 (1979). Reproduced with permission. © 1979 by Annual Reviews Inc.]

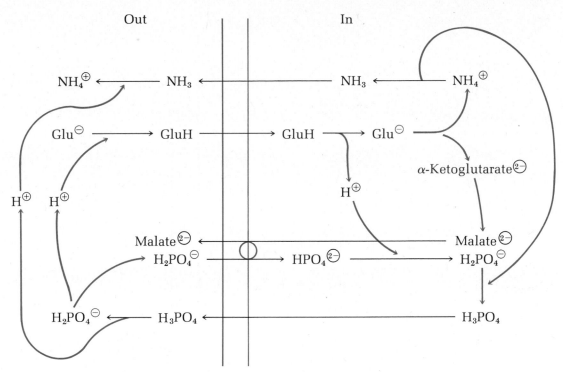

Figure 16.33 The glutamate transport system. [From K. F. LaNoue and A. C. Schoolwerth, *Ann. Rev. Biochem.*, *48*, 881 (1979). Reproduced with permission. © 1979 by Annual Reviews Inc.]

dianions. The process is therefore electrically neutral and does not depend upon a membrane potential. The K_m values for the carrier vary from 0.2 *mM* for L-malate to 1.5 *mM* for phosphate, but all ions are transported at the same maximum rate of 69 ± 6 nmol min^{-1} mg (carrier protein)$^{-1}$. The α-ketoglutarate transporter carries out antiport of α-ketoglutarate, L-malate, oxaloacetate, and succinate. This transport is electrically neutral and independent of the membrane potential. The K_m values for these substrates range from 0.05 *mM* for α-ketoglutarate up to 3.2 *mM* for succinate. The maximum rate of antiport is the same, 43 ± 2 nmol min^{-1} mg (carrier protein)$^{-1}$, for all substrates. The rapid rates of transport for both the dicarboxylate and the α-ketoglutarate carrier indicate that transport is not a rate-limiting factor in metabolism. The dicarboxylate carrier is abundant in liver and kidney mitochondria where biosynthesis of glucose from citric acid cycle intermediates is an important process. In heart mitochondria, where gluconeogenesis is not especially important, the carrier is present in only small amounts.

The tricarboxylate transport system carries out antiport of the citric acid cycle intermediates citrate, isocitrate, malate, and succinate. Phosphoenolpyruvate is also a substrate for this carrier. The K_m values for the carrier substrates vary considerably, but all are exchanged at the same rate. The tricarboxylate carrier is described as a proton-compensated electroneutral carrier because, in the exchange of citrate for malate, citrate enters as the dianion. Decarboxylation of citrate in the citric acid cycle results in the exit of carbon dioxide, whose reaction with water in the cytoplasm produces a proton (Figure 16.34).

The glutamate-aspartate carrier is coupled to the membrane potential and is not simply an electrically neutral case of antiport since both have the same electrical charge. For each molecule of aspartate that enters the matrix, one molecule of glutamate and one proton also enter. In rat heart mitochondria uncouplers of oxidative phosphorylation and valinomycin lead to the

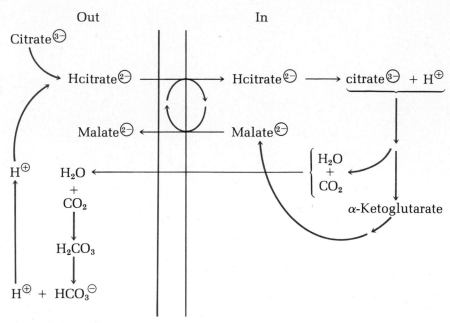

Figure 16.34 Ion balance in the tricarboxylate transport system. [From K. F. LaNoue and A. C. Schoolwerth, *Ann. Rev. Biochem. 48*, 882 (1979). Reproduced with permission. © 1979 by Annual Reviews Inc.]

accumulation of aspartate in the matrix space. The aspartate-glutamate shuttle is therefore driven by respiration.

E. Adenine Nucleotide Carrier

The adenine nucleotide carrier catalyzes exchange of ADP and ATP. The plant toxin *atractyloside* and the bacterial toxin *bongkrekic acid* (Figure 16.35) are potent inhibitors of the adenine nucleotide carrier. The carrier has a K_m value for ADP of 1 to 10 μM; its K_m value for ATP is around 1 μM in the absence of respiratory substrates, but may be as high as 150 μM in the presence of such substrates. During respiration the transport of ADP into the mitochondrion is much faster than inward transport of ATP, even at very high external ATP/ADP ratios. This is clearly a metabolic advantage since most of the biosynthetic reactions of the cell occur outside the mitochondria. The net result of the tendency for outward transport of ATP is that the ATP/ADP ratio is much higher in the cytosol than in the mitochondrion. Exchange of ATP for ADP is not electrically neutral since the major ionic species of ATP has a charge of -4 compared to a charge of -3 for the predominant species of ADP. Nevertheless, symport of ATP with a proton is not observed. We recall that there seem to be four protons expelled from the matrix space at each coupling site. Three of these protons reenter the matrix during ATP synthesis. One proton is left for ATP transport, leaving the membrane electrically balanced.

16–13 SUMMARY

The net energy yield for the complete combustion of glucose when it is degraded by glycolysis, the citric acid cycle, and respiration is 38 moles of ATP per mole of glucose catabolized. The points in the catabolic process in which either ATP (in one case GTP) or the reduced coenzymes NADH and FADH$_2$ are generated or consumed are summarized in Table 16.10. Of the total production of 38 ATP molecules, only 2 arise directly by substrate-level phosphorylation during glycolysis, and 2 more from substrate-level phosphorylation of GDP in the citric acid cycle. The other 34 ATP molecules produced by aerobic oxidation of glucose arise from oxidative phosphorylation.

The oxidation of NADH and FADH$_2$ by the mitochondrial electron

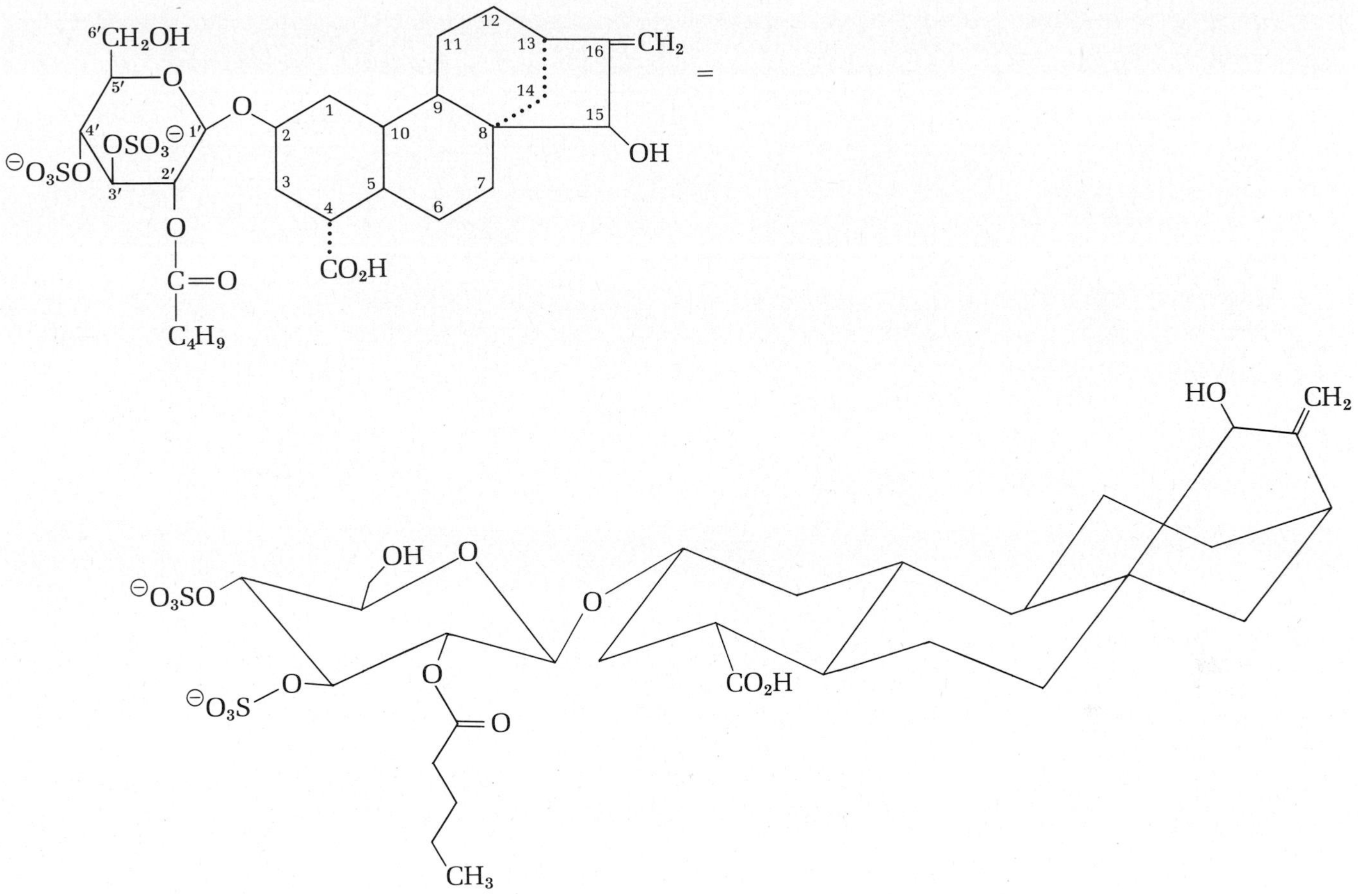

Atractyloside

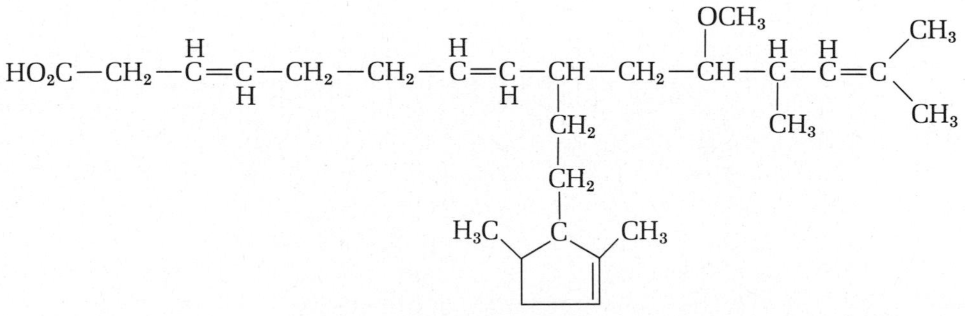

Bongkrekic acid

Figure 16.35 Structures of the plant toxin atractyloside and the bacterial toxin bongkrekic acid.

transport chain is coupled to the phosphorylation of ADP. For each mole of NADH that is oxidized, 3 moles of ATP are produced, and for each mole of $FADH_2$ that is oxidized, 2 moles of ADP are phosphorylated. Oxidation of NADH occurs within three multiprotein respiratory chain complexes: NADH-Q reductase passes electrons from NADH to coenzyme Q; reduced coenzyme Q is oxidized by QH_2-cytochrome *c* reductase as electrons are passed to the peripheral protein cytochrome *c* from QH_2; cytochrome *c* is oxidized by cytochrome *c* oxidase; and the final electron acceptor is molecular oxygen. Each of these three respiratory chain complexes is a site for ATP synthesis. Oxidation of $FADH_2$ yields 2 moles of ATP since electrons are passed from $FADH_2$ to coenzyme Q by succinate-Q reductase, bypassing NADH-Q reductase.

The mitochondrial ATPase is responsible for the synthesis of ATP. This

Table 16.10   Summary of ATP yield per mole of glucose for catabolism of glucose via glycolysis, the citric acid cycle, and respiration

Metabolic pathway	ATP yield	NADH yield	FADH₂ yield
Glycolysis (glucose → 2 pyruvate)	2	2	—
Glucose → glucose 6-phosphate	−1	—	—
Glucose 6-phosphate → fructose 1,6-diphosphate	−1	—	—
Glyceraldehyde 3-phosphate + NAD + P$_i$ → 1,3-diphosphoglycerate + NADH + H$^{\oplus}$	—	+1 (×2)	—
1,3-Diphosphoglycerate + ADP → 3-phosphoglycerate + ATP	+1 (×2)	—	—
Phosphoenolpyruvate + ADP → pyruvate + ATP	+1 (×2)	—	—
Pyruvate dehydrogenase (pyruvate → acetyl-CoA + CO₂)	—	+1 (×2)	—
Citric acid cycle	+2, via GTP		+2
Isocitrate dehydrogenase	—	+1 (×2)	—
α-Ketoglutarate dehydrogenase	—	+1 (×2)	—
Succinate dehydrogenase	—	—	+1 (×2)
Succinate thiokinase	+1, via GTP (×2)	—	—
Malate dehydrogenase	—	+1(×2)	—
Net ATP yield from substrate-level phosphorylation:	4 ATP		
Net ATP yield from oxidative phosphorylation:	34 via:	24(3 ATP per NADH)	4(2 ATP per FADH₂)

Total ATP yield for oxidation of glucose: 38 ATP
ATP from glycolysis: 5%
ATP from citric acid cycle: 5%
ATP from respiration: 90%

complex structure contains an intermembrane sector that can transport protons, a stalklike structure that contains a protein that confers sensitivity to oligomycin upon the ATPase, and a spherical particle, coupling factor 1 that is responsible for ATP synthesis when bound to the membrane. Protons are ejected from the matrix space to the intermembrane space during electron transport, generating an electrochemical gradient called the proton motive force. The proton motive force provides the driving force for the phosphorylation of ADP. According to the chemiosmotic hypothesis, the flow of protons back into the mitochondrion occurs through the membrane sector of the ATPase. The flow of protons down the concentration gradient provides the driving force for phosphorylation of ADP in a directly coupled process. In the conformational coupling hypothesis the proton gradient is indirectly related to phosphorylation of ADP. The rate-limiting step in the conformational coupling hypothesis is release of a previously synthesized ATP from one of two active sites on the ATPase. The proton gradient induces the ATPase to undergo a conformational change in which the previously synthesized ATP is released from the ATPase.

In normal, tightly coupled mitochondria, the phosphorylation of ADP is inseparable from electron transport. Respiratory control of mitochondria in energy state 4 is regulated by the ADP/ATP ratio. At low concentrations of ATP, respiration is stimulated. ATP regulates its own synthesis by binding to a specific protein in respiratory chain complexes I, III, and IV. Uncouplers such as DNP uncouple oxidative phosphorylation by eliminating the pH gradient that provides the driving force for ATP synthesis. In the glycerophosphate shuttle electrons are transferred from cytoplasmic NADH to mitochondrial FADH₂, resulting in formation of 2 moles of ATP when FADH₂ is oxidized in the electron transport chain. In the more complex and more efficient malate-aspartate shuttle, electrons in cytosolic NADH are indirectly transferred to mitochondrial NAD, yielding 3 moles of ATP by oxidative phosphorylation.

The inner mitochondrial membrane is virtually impermeable except to water, oxygen, and carbon dioxide molecules. All mitochondria contain transport systems for pyruvate, fatty acids, and phosphate. These systems provide the substrates for energy-producing reactions in the citric acid cycle, fatty acid oxidation, and respiration. The transport of calcium ions is driven by the membrane potential generated by the respiratory chain. Calcium ions exchange with protons. A group of peptide antibiotics act as cation transporters. The ionophore valinomycin transports potassium into the mitochondrion to an extent dictated by the magnitude of the membrane potential. Nigericin transports potassium ions into the mitochondrion and protons out in an electrically neutral process. The antibiotic gramicidin A transports monovalent cations across membranes in a pulsed process whose rate corresponds to the formation and dissociation of the gramicidin A dimer. The transport of phosphate and of carboxylic acids is also coupled to the proton motive force. The chemiosmotic hypothesis thus provides a unifying theme to both respiration and mitochondrial membrane transport.

REFERENCES

Baltscheffsky, H., and Baltscheffsky, M., Electron Transport Phosphorylation, *Ann. Rev. Biochem.*, 43, 871–899 (1974).

Boyer, P. B., Chance, B., Ernster, L., Mitchell, P., Racker, E., and Slater, E. C., Oxidative Phosphorylation and Photophosphorylation, *Ann. Rev. Bochem.*, 46, 955–1026 (1977).

Harold, F. M., Conservation and Transformation of Energy by Bacterial Membranes, *Bacteriological Reviews*, 36, 172–230 (1972).

Holian, A., and Wilson, D. F., Relationship of Transmembrane pH and Electrical Gradients with Respiration and Adenosine 5-triphosphate Synthesis in Mitochondria, *Biochemistry*, 19, 4213–4221 (1980).

Holm, R. H., Synthetic Approaches to the Active Sites of Iron-Sulfur Proteins, *Acc. Chem. Res.*, 10, 427–434 (1977).

LaNoue, K., and Schoolwerth, A. C., Metabolite Transport in Mitochondria, *Ann. Rev. Biochem.*, 48, 871–922 (1979).

Mitchell, P. Keilin's Respiratory Chain Concept and Its Chemiosmotic Consequences, *Science, 206*, 1148–1159 (1979).

Racker, E., Inner Mitochondrial Membranes: Basic and Applied Aspects, *Hosp. Prac.*, 9, 87 (1974).

Reinhart, G., and Lardy, H. A., Rat Liver Phosphofructokinase: Kinetic Activity Under Near Physiological Conditions, *Biochemistry*, 19, 1477–1484 (1980).

Shulman, R. G., et al., Cellular Applications of ^{31}P and ^{13}C Nuclear Magnetic Resonance, *Science, 205*, 160–166 (1979).

Wilson, D. B., Cellular Transport Mechanisms, *Ann. Rev. Biochem.*, 47, 933–965 (1978).

Wilson, D. F., et al., Mitochondrial Electron Transport and Energy Conservation, *Acc. Chem. Res.*, 5, 234–241 (1972).

Problems

1. The standard reduction potential for coenzyme Q is $+0.10$ V, and the standard reduction potential for FAD is -0.219 V. Using these values, show that the oxidation of $FADH_2$ by coenzyme Q theoretically liberates enough energy to drive the synthesis of ATP under standard conditions.

2. When the pH of the medium containing a suspension of mitochondria is greater than 7.5, the proton moving force depends only upon $\Delta\Psi$. Why?

3. Predict the effect of nigericin on the transport of pyruvate.

4. In a tightly coupled mitochondrion the following reactions take place:

$$\text{I. } \tfrac{1}{2}O_2 + H^{\oplus} + NADH + 3ADP + 3HPO_4^{\ominus} \longrightarrow$$
$$NAD + 3ATP + 4H_2O$$

$$\text{II. } \tfrac{1}{2}O_2 + \text{succinate} + 2ADP + 2HPO_4^{\ominus} \longrightarrow$$
$$\text{fumarate} + 2ATP + 3H_2O$$

The standard free energy of hydrolysis of ATP is -30 kJ mol^{-1}; the standard free-energy change for the reaction

$$NADH + \tfrac{1}{2}O_2 + H^\oplus \longrightarrow NAD + H_2O$$

is -219 kJ mol^{-1}; and the standard free-energy change for the reaction

$$succinate + \tfrac{1}{2}O_2 \longrightarrow fumarate + H_2O$$

is -151 kJ mol^{-1}.

a. Using the above information, calculate $\Delta G^{\circ\prime}$ for reactions I and II.

b. A mitochondrial preparation is incubated with an excess of O_2, succinate, and phosphate, but no additional substrates. What reaction occurs?

c. What change occurs if 2 μmoles of ADP are added to the preparation?

d. What causes the reaction to stop?

e. If 2,4-dinitrophenol is added to the reaction shortly after the addition of ADP, what happens (i) to the production of ATP and (ii) to the concentration of fumarate?

f. If 2,4-dinitrophenol is added to the medium after all of the ADP is consumed and the reaction mixture is analyzed shortly thereafter, no ATP is found in the medium and there are 2 μmoles of ADP. What enzymatic activity has 2,4-dinitrophenol induced in the mitochondria?

g. What is responsible for the mitochondrial activity you cited in (f)? What is this protein doing in tightly coupled mitochondria?

5. Hibernating mammals have a tissue known as brown fat. The mitochondria of brown fat tissue can oxidize succinate to fumarate in the absence of added ADP. In tightly coupled mitochondria the oxidation of substrate is coupled to the production of ATP. Why will a hibernating bear find brown fat useful?

6. In Chapter 11 we noted that the $Na^\oplus/K^\oplus$-ATPase can be reversed (with net synthesis of ATP) by reversing the concentrations of $Na^\oplus$ and $K^\oplus$ inside and outside the cell. How is this observation related to the chemiosmotic process and oxidative phosphorylation?

7. Nigericin has no effect on the efflux of aspartate in rat heart mitochondria. What does this imply about the driving force for transport by the aspartate-glutamate shuttle?

8. Hydrogen cyanide (HCN) blocks the transfer of electrons from cytochrome c to oxygen. If ADP levels increase as a result, what will be the immediate effect of HCN poisoning on blood levels of fructose 1,6-diphosphate?

9. Calculate the "efficiency of respiration" when FADH$_2$ is the substrate, that is, calculate the percent energy conservation for oxidation of FADH$_2$ with O_2 coupled to ATP production. ($\Delta G^{\circ\prime}_{hydrolysis(ATP)} = -30$ kJ mol^{-1}.)

10. When ascorbate is a substrate for the respiratory chain, it is oxidized by passing two electrons (one at a time) to cytochrome c. What is the P/O ratio for this process? The standard reduction potential of ascorbate is 0.08 V. What is the efficiency of oxidative phosphorylation when ascorbate is the original electron donor? ($\Delta G^{\circ\prime}_{hydrolysis(ATP)} = -30$ kJ mol^{-1}.)

11. $\Delta G'$ for the oxidation of NADH under the conditions actually prevailing in the cell is -213 kJ mol^{-1} when the oxygen pressure is 10^{-2} atm. What is the ratio of NAD/NADH under these conditions?

12. The half-reactions for the reduction of pyruvate and NAD are

$$\text{pyruvate} + 2\text{H}^{\oplus} + 2\text{ e} \longrightarrow \text{lactate} \qquad E^0 = -0.19 \text{ V}$$
$$\text{NAD} + \text{H}^{\oplus} + 2\text{ e} \longrightarrow \text{NADH} \qquad E^0 = -0.32 \text{ V}$$

Using these half-reactions, calculate the equilibrium constant, the standard free-energy change, and the standard electrochemical potential for the reduction of pyruvate by NADH. The actual ratio of NAD/NADH is 634:1 and the ratio of lactate to pyruvate is 14.2:1. Calculate the free-energy change, $\Delta G'$, that prevails under these cellular conditions.

13. What feature of the structure of NAD vis à vis NADH provides the driving force for the oxidation of NADH?

14. The transport of D-lactate in *E. coli* is coupled to the membrane potential, Δp. Addition of nigericin completely abolishes transport of D-lactate, but valinomycin increases the rate of transport by a factor of two. What do these results imply about the driving force for transport of D-lactate, that is, which component of the proton motive force is responsible for transport of D-lactate? [Reference: D. E. Robertson, G. J. Kacorowski, M. L. Garcia, and H. R. Kaback, *Biochemistry*, 19, 5692–5702 (1980).]

15. At pH 8.0 the pH gradient across the cell membrane of *E. coli* is zero, and the transmembrane potential is -0.142 V, interior negative. Lactose transport in *E. coli* is coupled to the proton motive force. Calculate the maximum concentration gradient of lactose that can be driven by a proton motive force of -0.142 V. [Hint: Convert both concentration gradients and membrane potentials to free energy differences. Assume that the distribution of lactose across the membrane is at equilibrium.]

Chapter 17

Carbohydrate metabolism

17–1 INTRODUCTION

The multiple branches of carbohydrate metabolism revolve around glucose and its derivatives. In this chapter we shall consider a pathway by which glucose, or one of its derivatives, is resynthesized from the lactate produced by anaerobic glycolysis, certain amino acids, and glycerol. This pathway, gluconeogenesis (the "new synthesis of glucose"), permits both the salvage of incompletely oxidized substrates such as lactate and glycerol, and the incorporation of the carbon skeletons of amino acids in carbohydrates. The pentose phosphate pathway, by contrast, is a catabolic pathway that can serve several purposes, such as production of required biosynthetic starting materials or the production of NADPH for biosynthetic reductions. Glycogen metabolism provides a pathway for the storage and retrieval of glucose 1-phosphate as an energy source. In the final section of the chapter we will consider some aspects of the biosynthesis of complex carbohydrates. The biosynthetic pathways for the synthesis of many complex sugars proceed through nucleotide sugars as biosynthetic intermediates, a common feature of otherwise diverse processes.

17–2 PENTOSE PHOSPHATE PATHWAY

The enzymes of the pentose phosphate pathway, also known as the *phosphogluconate pathway* and the *hexose monophosphate shunt,* are found in the cytosol of mammalian cells. The pentose phosphate pathway serves four purposes depending upon metabolic conditions. (1) It produces cytosolic *NADPH* for use in the reductive reactions in the biosynthesis of fatty acids (Chapter 19) and steroids (Chapter 20). In mammals these biosynthetic processes are particularly important in mammary glands, in adipose (fat) tissue, in the liver, and in the adrenal cortex. In these tissues the pentose phosphate pathway accounts for about 10% of the glucose metabolized. In contrast, it is essentially nonexistent in skeletal muscle. *NADPH and NADH have distinct functions: NADH is oxidized in the respiratory chain to produce ATP, and NADPH is a reducing agent in biosynthesis.* (2) The pentose phosphate pathway synthesizes ribose phosphate, a starting point in biosynthesis of nucleotides (Chapter 23). (3) If the pentose phosphate pathway is carried through to completion, the result is not only generation of NADPH, but the degradation of glucose to glyceraldehyde 3-phosphate, which can then be further oxidized to produce ATP. (4) In the dark reactions of photo-

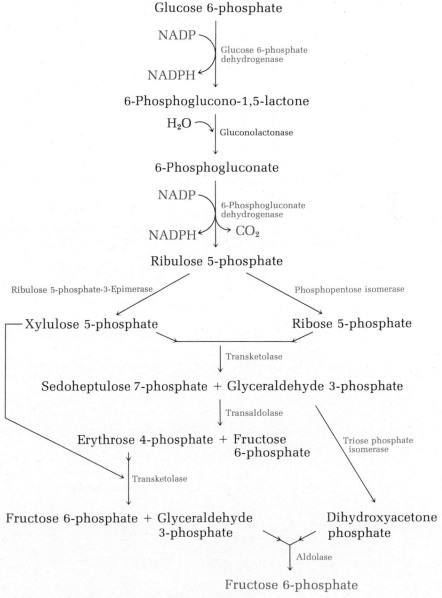

Figure 17.1 The pentose phosphate pathway.

Table 17.1 Reactions of the oxidative and nonoxidative portions of the pentose phosphate pathway

Reactions	Text reference
Oxidative reactions	Section 18.2A
glucose 6-phosphate + NADP → 6-phosphoglucono-1,5-lactone + NADPH	a
6-phosphoglucono-1,5-lactone + H_2O → 6-phosphogluconate	b
6-phosphogluconate + NADP → ribulose 5-phosphate + NADPH + CO_2	c
Nonoxidative reactions	Section 18.2B
ribulose 5-phosphate ⇌ ribose 5-phosphate	a
ribose 5-phosphate ⇌ xylulose 5-phosphate	b
xylulose 5-phosphate + ribose 5-phosphate ⇌	
sedoheptulose 7-phosphate + glyceraldehyde 3-phosphate	c
sedoheptulose 7-phosphate + glyceraldehyde 3-phosphate ⇌	
fructose 6-phosphate + erythrose 4-phosphate	c
xylulose 5-phosphate + erythrose 4-phosphate ⇌	
fructose 6-phosphate + glyceraldehyde 3-phosphate	c

synthesis (Chapter 18) a variation of the pentose phosphate pathway leads to synthesis of glucose.

The complex functioning of the pentose phosphate pathway can be divided into an oxidative part and a nonoxidative part (Figure 17.1). The oxidative part leads to generation of NADPH, and the nonoxidative part leads to either ribose 5-phosphate or to glyceraldehyde 3-phosphate, depending upon metabolic conditions (Table 17.1). Let us now turn to the individual steps of the pentose phosphate pathway, following which we shall consider regulation of the pathway and the metabolic factors that determine which function of the pathway is dominant.

A. Oxidative Reactions of the Pentose Phosphate Pathway

The oxidative steps of the pentose phosphate pathway occur within the span of three steps (Figure 17.2). *Glucose 6-phosphate dehydrogenase* catalyzes oxidation of the aldopyranose glucose 6-phosphate to a cyclic ester called a *lactone* with generation of NADPH; *gluconolactonase* opens the lactone ring; and a second dehydrogenase, *6-phosphogluconate dehydrogenase* cat-

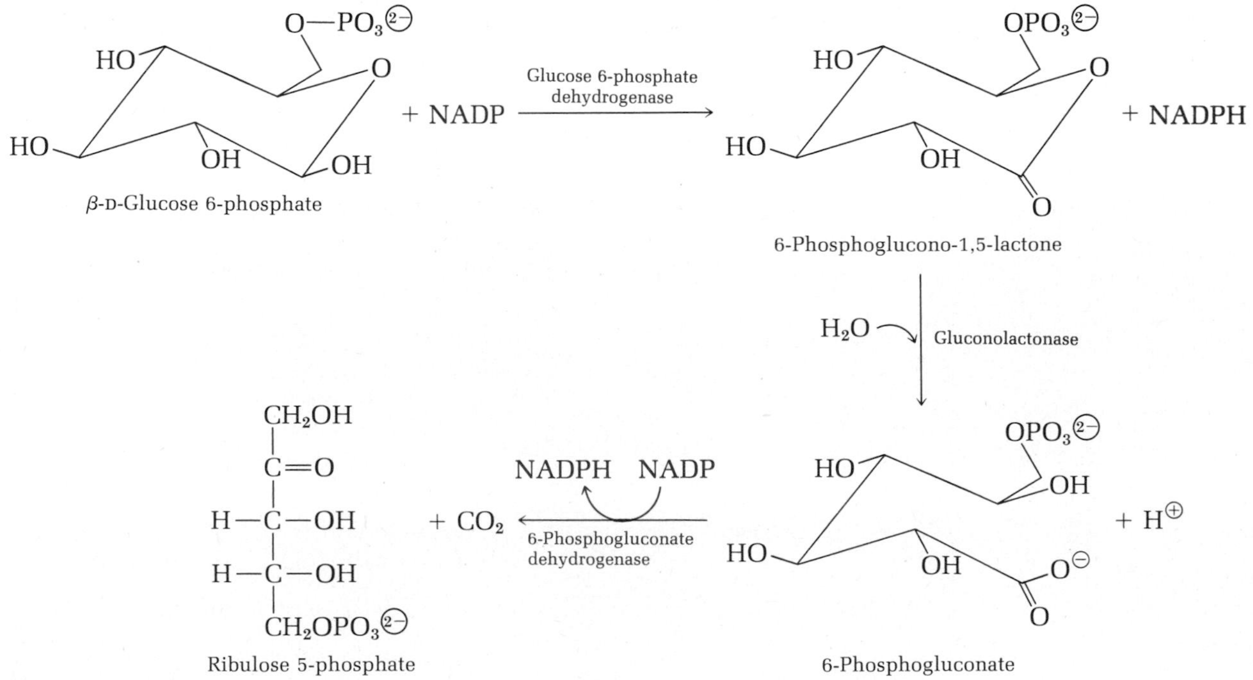

Figure 17.2 Reactions of the oxidative portion of the pentose phosphate pathway.

alyzes oxidative decarboxylation with generation of a second mole of NADPH and ribulose 5-phosphate. All subsequent reactions of the pathway are nonoxidative.

a. Glucose 6-Phosphate Dehydrogenase

Glucose 6-phosphate dehydrogenase catalyzes conversion of β-D-glucose 6-phosphate to 6-phosphoglucono-1,5-lactone. NADP is reduced in the process (reaction 17.1).

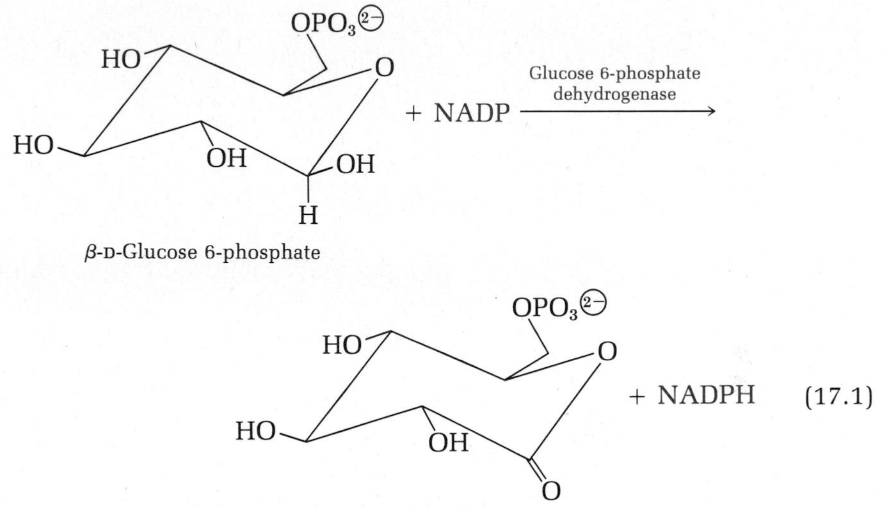

β-D-Glucose 6-phosphate

6-Phosphoglucono-1,5-lactone
(6-phosphoglucono-δ-lactone)

$$+ \text{NADP} \xrightarrow{\text{Glucose 6-phosphate dehydrogenase}} \qquad + \text{NADPH} \qquad (17.1)$$

Glucose 6-phosphate dehydrogenase is strongly inhibited by NADPH (product inhibition is a characteristic of most dehydrogenases, as we have now seen in several cases). NADPH is required for the reductive steps of fatty acid biosynthesis. Intermediates of fatty acid biosynthesis, such as stearyl-CoA, palmityl-CoA, and lauryl-CoA also inhibit glucose 6-phosphate dehy-

$$\underset{\text{Stearyl-CoA}}{CH_3(CH_2)_{16}\overset{\displaystyle O}{\overset{\|}{C}}\text{-CoA}} \qquad \underset{\text{Palmityl-CoA}}{CH_3(CH_2)_{14}\overset{\displaystyle O}{\overset{\|}{C}}\text{-CoA}} \qquad \underset{\text{Lauryl-CoA}}{CH_3(CH_2)_{10}\overset{\displaystyle O}{\overset{\|}{C}}\text{-CoA}}$$

drogenase. The active enzyme is a dimer, and the fatty acyl-CoA intermediates of fatty acid biosynthesis promote its dissociation into inactive monomers.

A deficiency of glucose 6-phosphate dehydrogenase in red blood cells leads to acute anemia, an extremely widespread genetic defect that affects

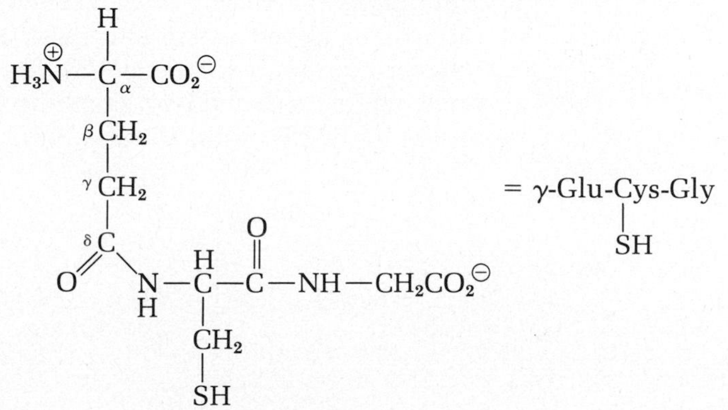

$= \gamma$-Glu-Cys-Gly

Glutathione (reduced)
(γ-glutamylcysteinylglycine)

about 11% of Black Americans, and over 100 million people worldwide. This anemic condition arises in a roundabout way. The tripeptide *glutathione* is required by red blood cells to maintain the cysteine sulfhydryl groups of hemoglobin in their reduced state. This reversible oxidation-reduction reaction is an example of disulfide exchange of the type we encountered in Chapter 2 (reaction 2.47) in our discussion of disulfide bonds in proteins (reaction 17.2).

$$2\gamma\text{-Glu-Cys-Gly} + \text{R-S-S-R} \longrightarrow \gamma\text{-Glu-Cys-Gly} + 2\text{RSH} \qquad (17.2)$$

with SH on the first tripeptide, and

$$
\begin{array}{c}
\text{S} \\
| \\
\text{S} \\
| \\
\gamma\text{-Glu-Cys-Gly}
\end{array}
$$

Glutathione must be reduced to catalyze disulfide exchange. Reduction of glutathione is catalyzed by *glutathione reductase*. NADPH is the reducing agent, and its only source in red blood cells is the pentose phosphate pathway (reaction 17.3).

$$
\begin{array}{c}
\gamma\text{-Glu-Cys-Gly} \\
| \\
\text{S} \\
| \\
\text{S} \\
| \\
\gamma\text{-Glu-Cys-Gly}
\end{array}
+ \text{NADPH} + \text{H}^{\oplus} \xrightarrow{\text{Glutathione reductase}} 2\gamma\text{-Glu-Cys-Gly} + \text{NADP} \qquad (17.3)
$$

with SH on the product.

Reduced glutathione is also required for maintaining the Fe(II) ions of hemoglobin in their reduced state and for maintaining the structure of red blood cells, though this latter function is mediated by an unknown mechanism. Glucose 6-phosphate dehydrogenase is therefore essential to red blood cells.

The gene for transmission of the dehydrogenase is inherited as a sex-linked character. Heterozygous females possess normal and glucose 6-phosphate dehydrogenase deficient red blood cells. Heterozygotes for glucose 6-phosphate deficiency are resistant to falciparum malaria, whose parasite requires NADPH, provided by the pentose phosphate pathway, for optimal growth. By this elegant adaptive mechanism, the activity of the pentose phosphate pathway is diminished by about 90% and the ravages of malaria are avoided at low cost to the oxygen transport system, a characteristic we also observed for sickle cell anemia.

Glutathione reductase contains domains for binding FAD, NADPH, and glutathione. The isoalloxazine ring of FAD and the NADP are stacked upon each other at the active site close to glutathione (Figure 17.3). Electron transfer thus can occur smoothly from NADPH to FAD to glutathione. The final step is mediated by a disulfide bridge in the protein which is first reduced by FADH_2, and then reduces the disulfide of glutathione (reaction 17.4).

$$
\begin{array}{ccccccc}
 & & & & \text{Cys 46-SH} & & \text{G-S} \\
\text{NADPH} & \searrow \nearrow & \text{FADH}_2 & \searrow \nearrow & \text{Cys 41-SH} & \searrow \nearrow & | \\
 & & & & & & \text{G-S} \\
 & \nearrow \searrow & & \nearrow \searrow & & \nearrow \searrow & \\
\text{NADP} & & \text{FAD} & & \text{Cys 46-S} & & \text{G-SH} \\
 & & & & \text{Cys 41-S} & & \text{G-SH}
\end{array} \qquad (17.4)
$$

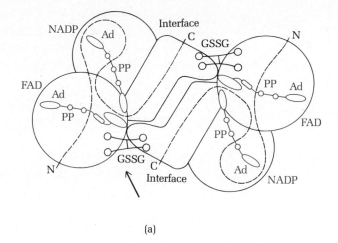

(a)

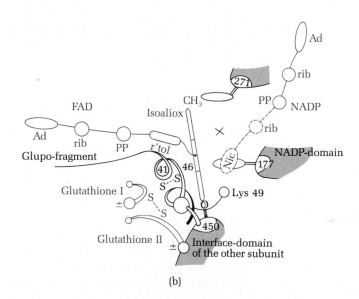

(b)

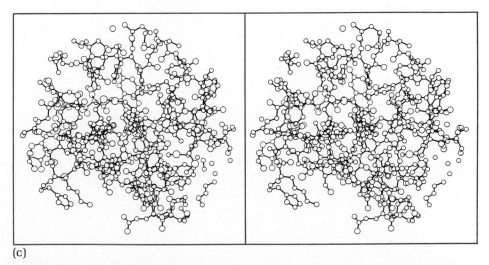

(c)

Figure 17.3 (a) Schematic diagram of the structure of glutathione reductase indicating the binding domains for NADP, FAD, and glutathione. (b) Expanded view of the catalytic site of glutathione reductase. [From G. E. Shultz, R. H. Schirmer, W. Sachsenheimer, and E. F. Pai, *Nature*, *273*, 123 (1978). Reprinted by permission. © 1978 Macmillan Journals Limited.] (c) Stereo view of glutathione reductase.

b. Gluconolactonase

D-Phosphoglucono-1,5-lactone hydrolyzes slowly in aqueous solution.[1] Gluconolactonase catalyzes rapid hydrolysis of 6-phosphoglucono-1,5-lactone producing 6-phosphogluconate (reaction 17.5).

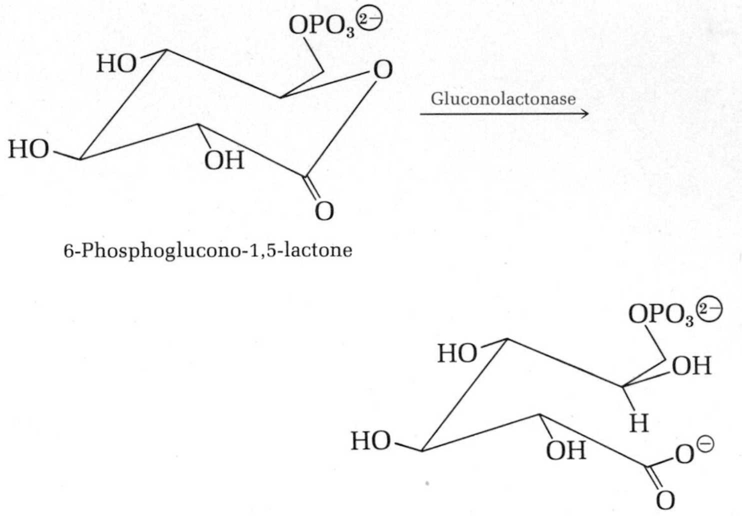

6-Phosphoglucono-1,5-lactone

(17.5)

6-Phosphogluconate

Gluconolactonase catalyzes the committing step in the diversion of glucose to the pentose phosphate pathway.

c. 6-Phosphogluconate Dehydrogenase

The second oxidative step of the pathway leads to production of a second mole of NADPH and to formation of ribulose 5-phosphate. The enzyme responsible for this reaction is 6-phosphogluconate dehydrogenase (reaction 17.6).

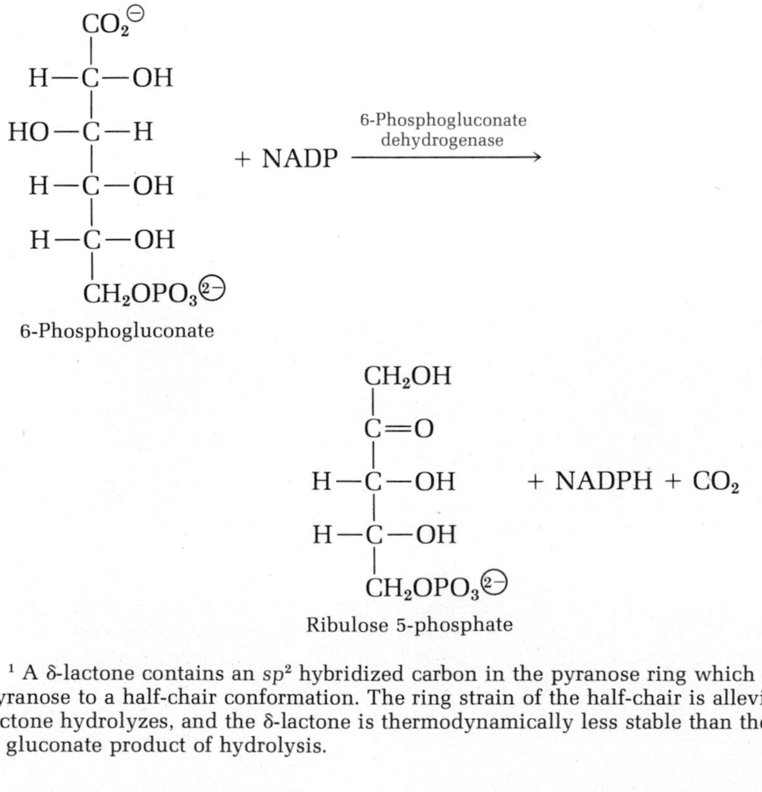

6-Phosphogluconate

(17.6)

Ribulose 5-phosphate

[1] A δ-lactone contains an sp^2 hybridized carbon in the pyranose ring which distorts the pyranose to a half-chair conformation. The ring strain of the half-chair is alleviated when the lactone hydrolyzes, and the δ-lactone is thermodynamically less stable than the open chain gluconate product of hydrolysis.

B. Nonoxidative Reactions of the Pentose Phosphate Pathway

Ribulose 5-phosphate undergoes a series of rearrangements without further oxidation in the pentose phosphate pathway (Figure 17.1). The ultimate product of the pentose phosphate pathway is fructose 6-phosphate, so the net result of this series of reactions is a detour around the first steps of glycolysis. Along the way erythrose 4-phosphate, a precursor of aromatic amino acids in *E. coli*, is produced. Ribose 5-phosphate, a precursor in nucleotide biosynthesis, is another important biosynthetic intermediate produced. Glucose can therefore be degraded by the pentose phosphate pathway with the production of metabolites required for biosynthesis.

a. Phosphopentose Isomerase

Phosphopentose isomerase catalyzes the conversion of ribulose 5-phosphate to ribose 5-phosphate (reaction 17.7).

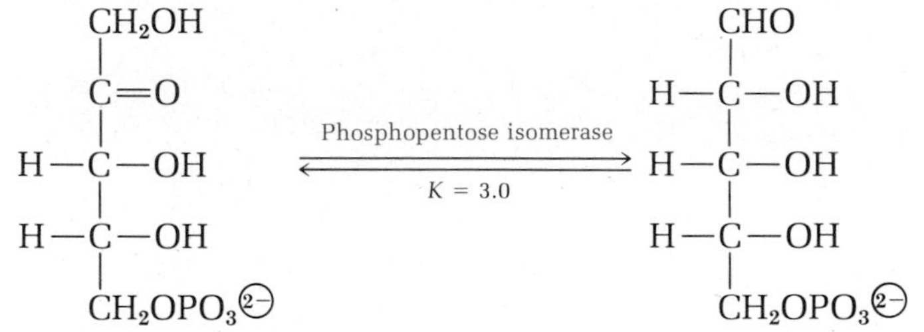

Ribulose 5-phosphate   Phosphopentose isomerase, $K = 3.0$ Ribose 5-phosphate   (17.7)

This is an aldose-ketose interconversion, and it has been proposed that the reaction occurs by formation of an enediol analogous to trisophosphate isomerase (Section 14.3E). The matter is not settled, however, and the mechanism of action of phosphopentose isomerase remains in doubt.

b. Ribulose 5-Phosphate-3-Epimerase

Ribulose 5-phosphate is epimerized at C-3 by ribulose 5-phosphate-3-epimerase to give xylulose 5-phosphate. The equilibrium constant for xylulose 5-phosphate formation is 0.67. The enzyme abstracts an acidic α-hydrogen at C-3 of ribulose 5-phosphate and returns it to the same carbon with inversion of configuration (reaction 17.8).

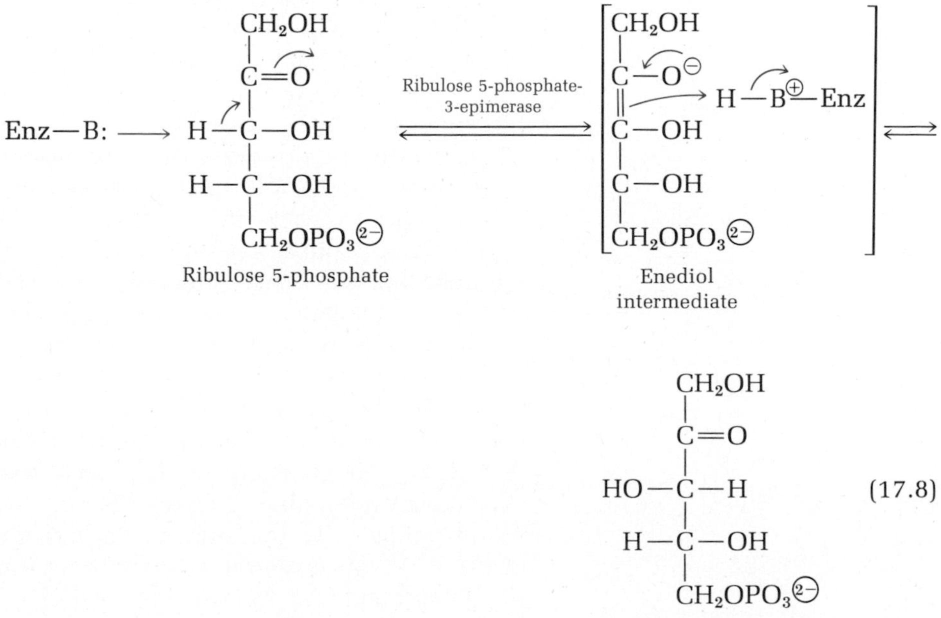

(17.8)

c. Conversion of Pentoses into Glycolytic Intermediates

The reactions of phosphopentose isomerase and ribulose 5-phosphate-3-epimerase create a pool of three pentoses: ribose 5-phosphate, ribulose 5-phosphate, and xylulose 5-phosphate. In the remaining steps of the pentose phosphate pathway these three pentoses are acted upon by *transketolase* and *transaldolase*. If the ribose 5-phosphate provided by the first several steps of the pathway far exceeds the biosynthetic requirements of the cell for NADPH, which is often the case, then transketolase and transaldolase provide a path to the glycolytic intermediates glyceraldehyde 3-phosphate and fructose 6-phosphate.

Transketolase transfers the two-carbon saccharose unit, and transaldolase transfers the three-carbon saccharose unit. Considering only the carbon

$$
\text{HO} - \text{CH}_2 - \underset{\underset{\text{O}}{\|}}{\text{C}} - \qquad\qquad \text{HO} - \text{CH}_2 - \underset{\underset{\text{O}}{\|}}{\text{C}} - \underset{\underset{\text{H}}{|}}{\overset{\overset{\text{OH}}{|}}{\text{C}}} -
$$

2-Carbon saccharose unit   3-Carbon saccharose unit

chains, we can summarize the reactions catalyzed by transaldolase and transketolase as

$$
\begin{aligned}
C_5 + C_5 &\xrightleftharpoons{\text{Transketolase}} C_7 + C_3 \\
C_7 + C_3 &\xrightleftharpoons{\text{Transaldolase}} C_4 + C_6 \\
\underline{C_5 + C_4} &\xrightleftharpoons{\text{Transketolase}} \underline{C_3 + C_6} \\
3\,C_5 &\rightleftharpoons 2\,C_6 + C_3
\end{aligned}
\tag{17.9}
$$

The net result of the action of the two enzymes is the conversion of three pentoses into two hexoses and a triose. The triose is glyceraldehyde 3-phosphate and the hexoses are 2 moles of fructose 6-phosphate. Both are intermediates of glycolysis whose catabolism can yield ATP. The pentose phosphate pathway thus provides a means for the cell to obtain both ATP and NADPH for biosynthesis.

i. Mechanism of Action of Transketolase Transketolase is a thiamine pyrophosphate-dependent enzyme that transfers a C_2 ketol group from xylulose 5-phosphate to ribose 5-phosphate (reaction 17.10). The three-carbon fragment left over from xylulose 5-phosphate becomes glyceraldehyde 3-phosphate, and the new seven-carbon sugar is sedoheptulose 7-phosphate.

Thiamine pyrophosphate has the same chemical function in transketolase as it has in pyruvate decarboxylase (Section 10.4A). Transketolase converts thiamine phosphate to a carbanion that attacks the carbonyl carbon of xylulose 5-phosphate, producing an adduct that cleaves to give a glycolylthiamine pyrophosphate carbanion and glyceraldehyde 3-phosphate. The glycolylthiamine pyrophosphate carbanion attacks the carbonyl carbon of ribose 5-phosphate to form sedoheptulose 7-phosphate (Figure 17.4).

ii. Transaldolase Transaldolase catalyzes the shift of a three-carbon fragment from sedoheptulose 7-phosphate to glyceraldehyde 3-phosphate (reaction 17.11). The mechanism of action of transaldolase closely resembles that of fructose diphosphate aldolase (Section 14.3D). Sedoheptulose 7-phosphate becomes bound to the enzyme through the ϵ-amino group of a lysyl side chain. Aldol cleavage releases erythrose 4-phosphate leaving a three-carbon fragment bound to the enzyme. Aldol condensation of the three-carbon

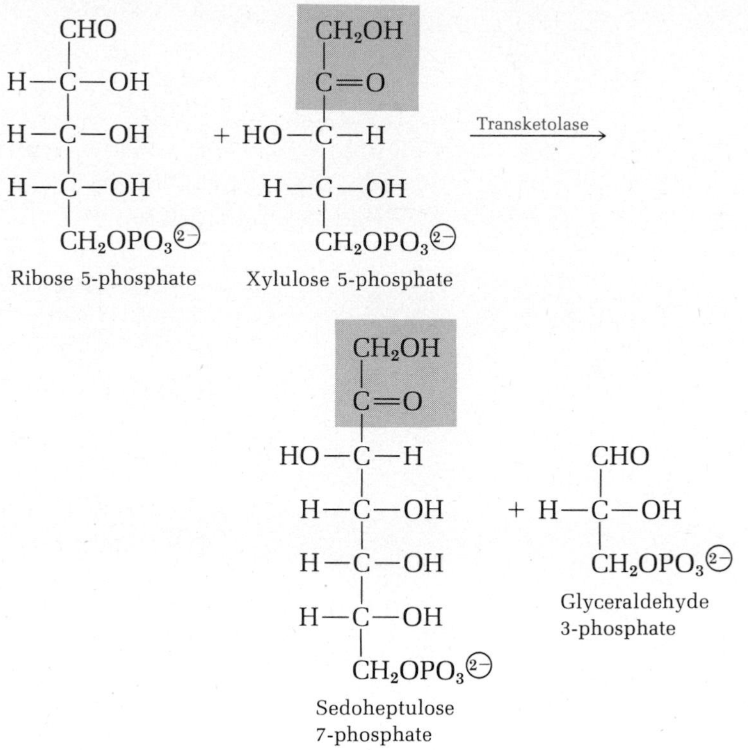

$$
\begin{array}{c}
\text{CHO} \\
\text{H—C—OH} \\
\text{H—C—OH} \\
\text{H—C—OH} \\
\text{CH}_2\text{OPO}_3^{2\ominus}
\end{array}
\quad + \quad
\boxed{\begin{array}{c}\text{CH}_2\text{OH} \\ \text{C=O}\end{array}}
\begin{array}{c}
\\ \\
\text{HO—C—H} \\
\text{H—C—OH} \\
\text{CH}_2\text{OPO}_3^{2\ominus}
\end{array}
\xrightarrow{\text{Transketolase}}
$$

Ribose 5-phosphate   Xylulose 5-phosphate

$$
\boxed{\begin{array}{c}\text{CH}_2\text{OH} \\ \text{C=O}\end{array}}
\begin{array}{c}
\\ \\
\text{HO—C—H} \\
\text{H—C—OH} \\
\text{H—C—OH} \\
\text{H—C—OH} \\
\text{CH}_2\text{OPO}_3^{2\ominus}
\end{array}
\quad + \quad
\begin{array}{c}
\text{CHO} \\
\text{H—C—OH} \\
\text{CH}_2\text{OPO}_3^{2\ominus}
\end{array}
\qquad (17.10)
$$

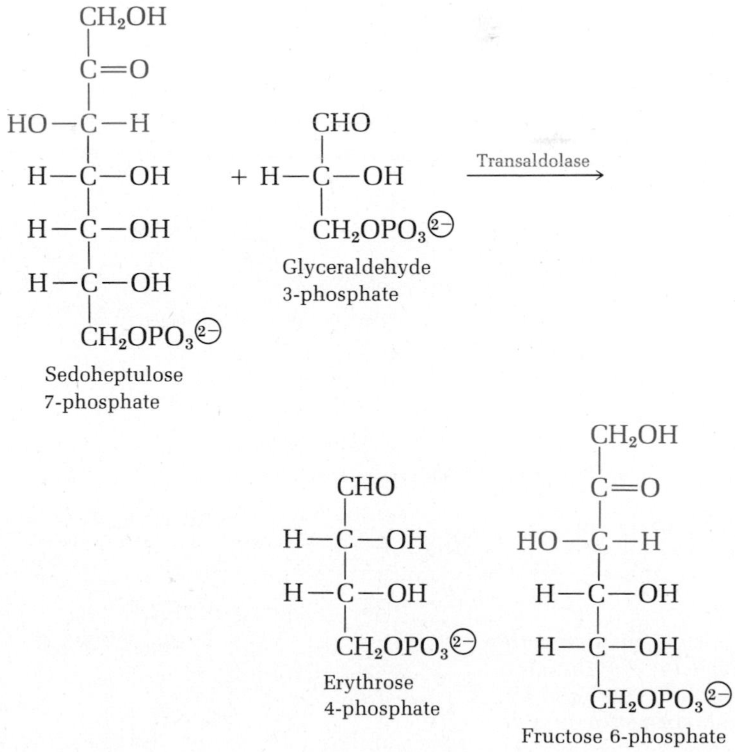

Sedoheptulose   Glyceraldehyde
7-phosphate   3-phosphate

$$
\begin{array}{c}
\text{CH}_2\text{OH} \\
\text{C=O} \\
\text{HO—C—H} \\
\text{H—C—OH} \\
\text{H—C—OH} \\
\text{H—C—OH} \\
\text{CH}_2\text{OPO}_3^{2\ominus}
\end{array}
\quad + \quad
\begin{array}{c}
\text{CHO} \\
\text{H—C—OH} \\
\text{CH}_2\text{OPO}_3^{2\ominus}
\end{array}
\xrightarrow{\text{Transaldolase}}
$$

Sedoheptulose   Glyceraldehyde
7-phosphate   3-phosphate

$$
\begin{array}{c}
\text{CHO} \\
\text{H—C—OH} \\
\text{H—C—OH} \\
\text{CH}_2\text{OPO}_3^{2\ominus}
\end{array}
\quad + \quad
\begin{array}{c}
\text{CH}_2\text{OH} \\
\text{C=O} \\
\text{HO—C—H} \\
\text{H—C—OH} \\
\text{H—C—OH} \\
\text{CH}_2\text{OPO}_3^{2\ominus}
\end{array}
\qquad (17.11)
$$

Erythrose   Fructose 6-phosphate
4-phosphate

fragment bound to the enzyme with glyceraldehyde 3-phosphate produces fructose 6-phosphate.

iii. Transketolase Transketolase reappears in the final step of the pentose phosphate pathway. A two-carbon fragment from xylulose 5-phosphate is transferred to erythrose 4-phosphate, producing another molecule of fructose 6-phosphate and glyceraldehyde 3-phosphate (reaction 17.12).

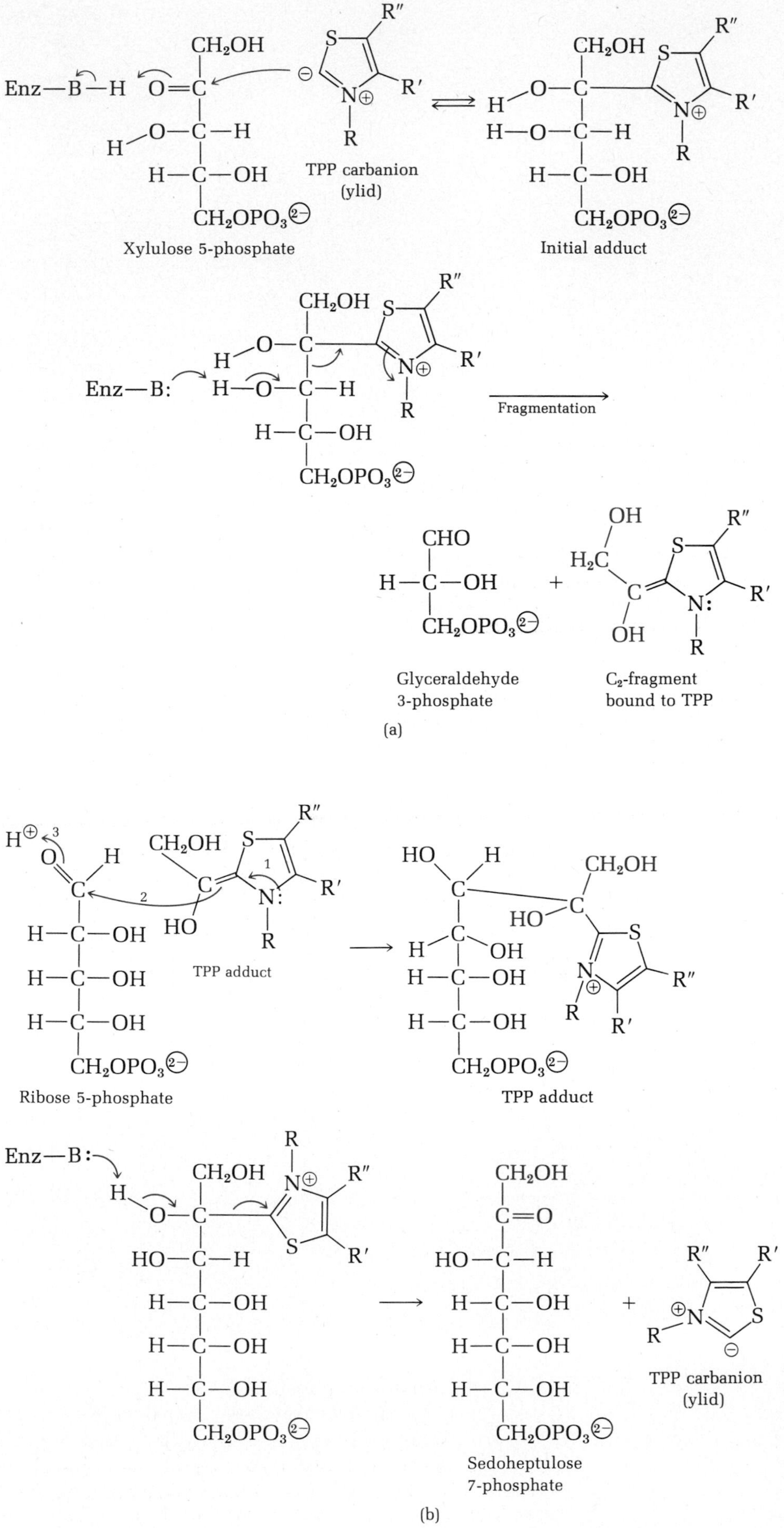

Figure 17.4 Mechanism of action of transketolase. (a) The initial step in the reaction catalyzed by transketolase is formation of a C_2 fragment that is bound to thiamine pyrophosphate. Glyceraldehyde 3-phosphate is produced in this step. (b) The second step in transketolase catalysis is nucleophilic attack of the C_2-adduct of TPP upon the carbonyl carbon of ribose 5-phosphate. The product is sedoheptulose 7-phosphate.

$$
\begin{array}{ccc}
\underset{\text{Xylulose 5-phosphate}}{\begin{array}{l}
\text{CH}_2\text{OH} \\
| \\
\text{C}=\text{O} \\
| \\
\text{HO}-\text{C}-\text{H} \\
| \\
\text{H}-\text{C}-\text{OH} \\
| \\
\text{CH}_2\text{OPO}_3^{2\ominus}
\end{array}}
&
\underset{\text{Erythrose 4-phosphate}}{\begin{array}{l}
\text{CHO} \\
| \\
\text{H}-\text{C}-\text{OH} \\
| \\
\text{H}-\text{C}-\text{OH} \\
| \\
\text{CH}_2\text{OPO}_3^{2\ominus}
\end{array}}
& \xrightarrow{\text{Transketolase}}
\end{array}
$$

$$
\underset{\text{Fructose 6-phosphate}}{\begin{array}{l}
\text{CH}_2\text{OH} \\
| \\
\text{C}=\text{O} \\
| \\
\text{HO}-\text{C}-\text{H} \\
| \\
\text{H}-\text{C}-\text{OH} \\
| \\
\text{H}-\text{C}-\text{OH} \\
| \\
\text{CH}_2\text{OPO}_3^{2\ominus}
\end{array}}
\;+\;
\underset{\begin{array}{c}\text{Glyceraldehyde}\\\text{3-phosphate}\end{array}}{\begin{array}{l}
\text{CHO} \\
| \\
\text{H}-\text{C}-\text{OH} \\
| \\
\text{CH}_2\text{OPO}_3^{2\ominus}
\end{array}}
\qquad (17.12)
$$

C. Net Reaction of the Pentose Phosphate Pathway Under Varying Metabolic Conditions

Consider the net result of oxidation of 6 moles of glucose 6-phosphate (reaction 17.13).

$$
\text{6 glucose 6-phosphate} + 12\text{NADP} + 6\text{H}_2\text{O} \longrightarrow
$$
$$
\text{6 ribose 5-phosphate} + 6\text{CO}_2 + 12\text{NADPH} + 12\text{H}^{\oplus} \qquad (17.13)
$$

The 6 moles of ribose 5-phosphate pass through a cycle involving phosphopentose isomerase, transaldolase, transketolase, the glycolytic enzymes triosephosphate isomerase and fructose diphosphate aldolase, fructose 1,6-diphosphate phosphatase, and glucose 6-phosphate isomerase (Figure 17.5). This set of reactions is dominant when the cellular demand for NADPH ex-

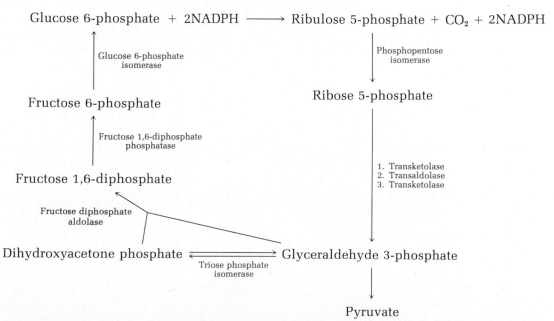

Figure 17.5 Cycling of the carbon skeleton of glucose 6-phosphate under conditions in which the demand for NADPH outstrips the requirement for ribose 5-phosphate. (See Table 17.2 for the reactions that lead to this cycle.) If glyceraldehyde 3-phosphate is not recycled to glucose 6-phosphate, it can be converted to pyruvate by glycolysis.

Table 17.2 Net reactions for oxidation of glucose 6-phosphate when cellular demand for NADPH exceeds need for ribose 5-phosphate

$$6 \text{ glucose 6-phosphate} + 6H_2O + 12NADP \longrightarrow 6 \text{ ribose 5-phosphate} + 6CO_2 + 12NADPH + 12H$$

$$6 \text{ ribose 5-phosphate} \longrightarrow 4 \text{ fructose 6-phosphate} + 2 \text{ glyceraldehyde 3-phosphate}$$

$$2 \text{ glyceraldehyde 3-phosphate} \longrightarrow \text{glyceraldehyde 3-phosphate} + \text{dihydroxyacetone phosphate}$$

$$\text{glyceraldehyde 3-phosphate} + \text{dihydroxyacetone phosphate} \longrightarrow \text{fructose 1,6-diphosphate}$$

$$\text{fructose 1,6-diphosphate} + H_2O \longrightarrow \text{fructose 6-phosphate} + P_i$$

$$5 \text{ fructose 6-phosphate} \longrightarrow 5 \text{ glucose 6-phosphate}$$

$$\text{glucose 6-phosphate} + 12NADP + 7H_2O \longrightarrow 6CO_2 + 12NADPH + 12H^{\oplus} + P_i$$

ceeds the demand for ribose 5-phosphate. The net result of this cycle is the complete oxidation of glucose 6-phosphate. The reactions leading to this result are summarized in Table 17.2.

Glyceraldehyde 3-phosphate can be converted to pyruvate instead of returning to glucose 6-phosphate. In this case the combination of the pentose phosphate pathway and glycolysis yields ATP and NADH as well as NADPH. The net reaction is

$$3 \text{ glucose 6-phosphate} + 6NADP + 5NAD + 5P_i + 8ADP$$

$$\downarrow$$

$$5 \text{ pyruvate} + 3CO_2 + 6NADPH + 5NADH + 8ATP + 2H_2O + 8H^{\oplus}$$

(17.14)

Suppose, however, that the opposite conditions prevail and that demand for ribose 5-phosphate is high and requirement for NADPH is low. In this case glycolysis is the predominant means for catabolism of glucose. Transketolase can convert glyceraldehyde 3-phosphate and fructose 6-phosphate to xylulose 5-phosphate and erythrose 4-phosphate. By the successive action of transaldolase and transketolase ribose 5-phosphate is produced without concomitant production of NADPH. This sequence of events (moving counterclockwise from glucose 6-phosphate to ribose 5-phosphate in Figure 17.5) thus completely bypasses the oxidative stage of the pentose phosphate pathway.

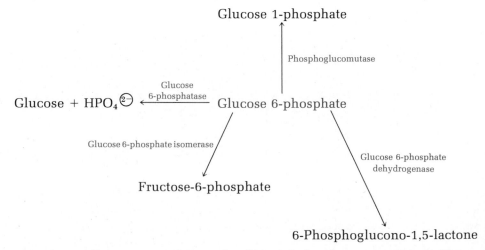

Figure 17.6 Four enzymes that require glucose 6-phosphate as a substrate. These enzymes are subject to strict metabolic control.

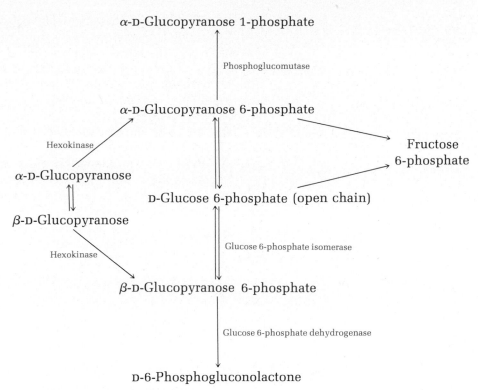

Figure 17.7 Anomeric specificity of the enzymes using glucose or glucose 6-phosphate as a substrate. [From M. Salas, E. Vinuela, and A. Sols, *J. Biol. Chem.*, *240*, 561 (1965).]

If the demand for ribose 5-phosphate is approximately equal to the demand for NADPH, only the oxidative portion of the pentose phosphate pathway is required (recall reaction 17.13).

D. Regulation of the Pentose Phosphate Pathway

The first intermediate of glycolysis, glucose 6-phosphate, is a substrate for glucose 6-phosphate isomerase, glucose 6-phosphate dehydrogenase, glucose 6-phosphatase, and phosphoglucomutase (Figure 17.6). The first of these enzymes, glucose 6-phosphate isomerase, produces a glycolytic intermediate. The second, glucose 6-phosphate dehydrogenase, catalyzes the first step in the pentose phosphate pathway. The third, glucose 6-phosphatase, gives free glucose. The fourth, phosphoglucomutase, couples both glycolysis and the phosphogluconate–pentose phosphate pathways to glycogen metabolism (Section 17.4).

The relative activities of these enzymes and the resulting fate of glucose 6-phosphate are the focus of strict metabolic control. The four enzymes that accept glucose 6-phosphate as a substrate have different affinities for the α and β anomers of the glucose 6-phosphate produced by hexokinase (Figure 17.7).[2] Phosphoglucomutase and glucose 6-phosphate isomerase prefer the α

α-D-Glucose 6-phosphate

β-D-Glucose 6-phosphate

[2] Hexokinase phosphorylates either the α or the β anomer of glucose.

anomer, and glucose 6-phosphate dehydrogenase prefers the β anomer. Glucose 6-phosphate isomerase has a mutarotase activity and catalyzes the conversion of α-D-glucose 6-phosphate to β-D-glucose 6-phosphate. This activity is strongly inhibited by erythrose 4-phosphate which is a feedback inhibitor of its own formation. A high erythrose 4-phosphate concentration deprives the first enzyme of the pathway, glucose 6-phosphate dehydrogenase, of a substrate to oxidize.

The rate of the pentose phosphate pathway is thus partially dictated by the anomeric specificity of the first reaction in which the β anomer is preferentially oxidized. Glucose 6-phosphate dehydrogenase is the key regulatory enzyme of the pentose phosphate pathway. It is strongly inhibited by NADPH. This inhibition is competitive, with NADP and NADPH competing for the substrate binding site. The acyl-CoA intermediates of fatty acid biosynthesis also inhibit the dehydrogenase. The dehydrogenases of the pentose phosphate pathway are inhibited by intermediates of all of the biosynthetic processes that use NADPH.

17–3 GLUCONEOGENESIS

The synthesis of glucose from the carbon skeletons of certain amino acids, from the lactate produced by anaerobic glycolysis, and from glycerol is known as *gluconeogenesis* (Figure 17.8). We recognize the intermediates of glycolysis in this diagram of gluconeogenesis, but gluconeogenesis is not simply a reversal of glycolysis. Three highly exergonic steps in glycolysis are bypassed in gluconeogenesis. The pathway provides four unique enzymes—pyruvate carboxylase, phosphoenolpyruvate (PEP) carboxykinase, fructose 1,6-diphosphatase, and glucose 6-phosphatase—to avoid steps in glycolysis that have equilibrium constants that are unfavorable to gluconeogenesis. The four enzymes provide detours around the reactions catalyzed by hexokinase, phosphofructokinase, and pyruvate kinase. The unique steps in gluconeogenesis and the glycolytic reactions they bypass are summarized in the following set of reactions.

1. In gluconeogenesis the path from pyruvate to phosphoenolpyruvate is catalyzed by pyruvate carboxylase (reaction 17.15) and by phosphoenolypyruvate carboxykinase (reaction 17.16). These steps bypass pyruvate kinase.

$$CH_3-\overset{O}{\underset{}{C}}-CO_2^{\ominus} + CO_2 + ATP + H_2O \xrightarrow{\text{Pyruvate carboxylase}}$$
Pyruvate

$$^{\ominus}O_2C-CH_2-\overset{O}{\underset{}{C}}-CO_2^{\ominus} + P_i + ADP + 2\,H^{\oplus} \qquad (17.15)$$
Oxaloacetate

$$^{\ominus}O_2C-CH_2-\overset{O}{\underset{}{C}}-CO_2^{\ominus} + GTP \xrightarrow{\text{PEP carboxykinase}} \qquad (17.16)$$
Oxaloacetate

$$CH_2=\overset{OPO_3^{2\ominus}}{\underset{}{C}}-CO_2^{\ominus} + GDP + CO_2$$
Phosphoenolpyruvate

The net standard free-energy change for reactions 17.15 and 17.16 is $+0.8$ kJ mol^{-1}, corresponding to an equilibrium constant for the conversion of pyruvate to phosphoenolpyruvate of 0.7. In contrast, the standard free-energy change for the reaction in glycolysis catalyzed by pyruvate kinase is -34 kJ mol^{-1} (reaction 17.17).

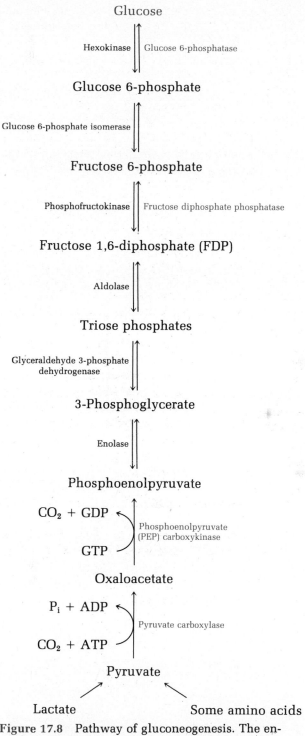

Figure 17.8 Pathway of gluconeogenesis. The enzymes indicated in color are unique to gluconeogenesis; the others are shared with glycolysis. The carbon sources leading into gluconeogenesis include certain amino acids, lactate, and glycerol.

The relatively favorable free-energy change for the conversion of pyruvate to phosphoenolypyruvate in gluconeogenesis requires the expenditure of two phosphoanhydride bonds, one from ATP and one from GTP.

From phosphoenolpyruvate to fructose 1,6-diphosphate the enzymes of glycolysis and gluconeogenesis are shared.

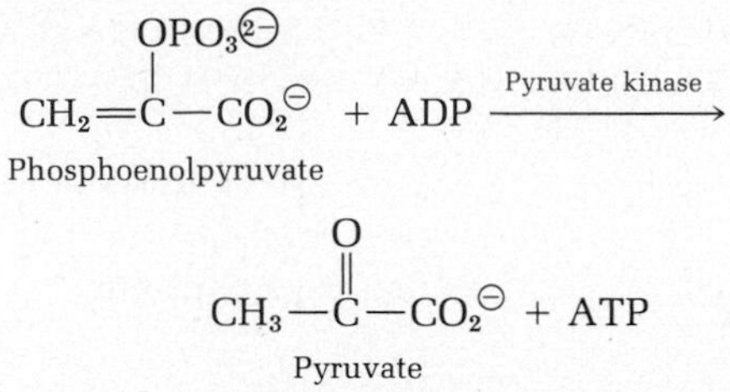

$$\underset{\text{Phosphoenolpyruvate}}{CH_2=\overset{\overset{\displaystyle OPO_3^{2\ominus}}{|}}{C}-CO_2^{\ominus}} + ADP \xrightarrow{\text{Pyruvate kinase}}$$

$$\underset{\text{Pyruvate}}{CH_3-\overset{\overset{\displaystyle O}{||}}{C}-CO_2^{\ominus}} + ATP \qquad \Delta G^{\circ\prime} = -34 \text{ kJ mol}^{-1} \qquad (17.17)$$

2. In gluconeogenesis the conversion of fructose 1,6-diphosphate to fructose 6-phosphate is catalyzed by fructose 1,6-diphosphatase (reaction 17.18). This step bypasses phosphofructokinase.

$$\text{fructose 1,6-diphosphate} + H_2O \xrightarrow[\text{1,6-diphosphatase}]{\text{Fructose}} \text{fructose 6-phosphate} + P_i \qquad (17.18)$$

3. The conversion of glucose 6-phosphate to glucose is catalyzed by glucose 6-phosphatase (reaction 17.19). This step bypasses hexokinase.

$$\text{glucose 6-phosphate} + H_2O \xrightarrow[\text{6-phosphatase}]{\text{Glucose}} \text{glucose} + P_i \qquad (17.19)$$

In mammals about 90% of gluconeogenesis occurs in the liver and 10% in the kidney.

The conversion of glucose 6-phosphate to glucose catalyzed by glucose 6-phosphatase is the last step in gluconeogenesis. In tissues that lack glucose 6-phosphatase, such as skeletal muscle, glucose 6-phosphate does not accumulate, and any lactate that forms must be exported to the liver to be converted to glucose or else metabolized by a pathway other than gluconeogenesis. In mammals about 90% of gluconeogenesis occurs in the liver and 10% in the kidney. The brain is utterly dependent upon glucose as a fuel, but it too lacks glucose 6-phosphatase. Glucose is transported to the brain through the blood, and low blood glucose levels stimulate gluconeogenesis.

A. Unique Enzymes of Gluconeogenesis and Their Regulation

a. Pyruvate Carboxylase

Pyruvate carboxylase is a biotin-dependent enzyme (recall Section 10.7). Biotin is covalently bound to the protein by an amide bond to an ϵ-amino group of a lysyl side chain. The net reaction catalyzed by pyruvate

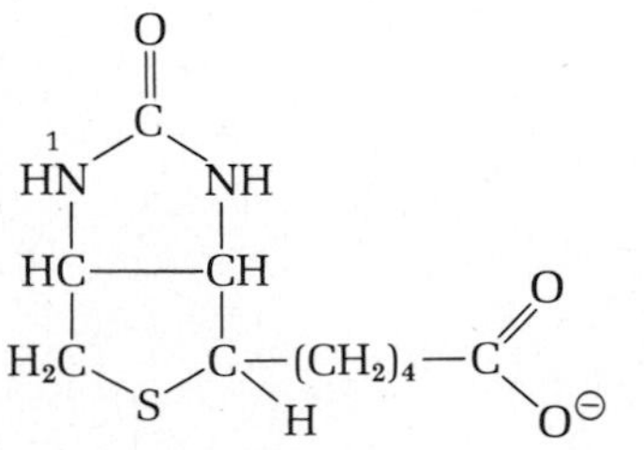

Biotin

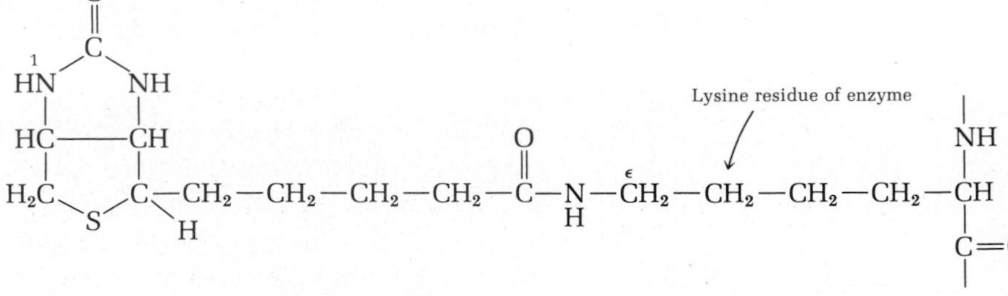

Biotin covalently bound to pyruvate carboxylase.

carboxylase occurs in two steps. First, biotin reacts with CO_2 to form N^1-carboxybiotin (reaction 17.20).

$$HCO_3^{\ominus} + \text{Enz-biotin} + \text{ATP} \longrightarrow$$

$$\text{(structure)} + \text{ADP} + P_i \qquad (17.20)$$

N'-Carboxybiotin-enzyme intermediate

Second, the carboxyl group bound to biotin is transferred to pyruvate at a second active site in the enzyme (reaction 17.21; Figure 17.9).

$$\underset{\text{Pyruvate}}{CH_3-\overset{\overset{\textstyle O}{\|}}{C}-CO_2^{\ominus}} + \text{Enz-biotin}-\overset{\overset{\textstyle O}{\|}}{C}-O^{\ominus} \rightleftharpoons$$

$$\overset{O}{\underset{\ominus O}{\diagdown}}C-CH_2-\overset{\overset{\textstyle O}{\|}}{C}-CO_2^{\ominus} + \text{Enz-biotin} \qquad (17.21)$$

Oxaloacetate

The carboxylation of biotin (reaction 17.20) occurs only if acetyl-CoA or an acyl-CoA is bound to the enzyme. We recall from our discussion of control of the citric acid cycle (Section 15.4A) that the rate of the citric acid cycle is limited by the concentration of the catalytic intermediate oxaloacetate, which is also a precursor of glucose in gluconeogenesis. When the citric acid cycle has produced enough ATP to meet present requirements, the excess

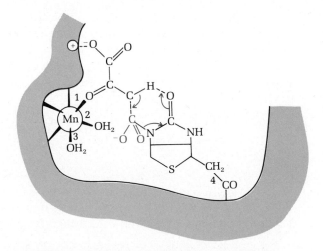

Figure 17.9 Possible concerted mechanism for transfer of a carboxyl group from N^1-carboxybiotin to pyruvate. An active site Mn(II) holds the substrate in the correct orientation for reaction, and also acts electrophilic catalyst. [From M. C. Scrutton and M. R. Young, in P. D. Boyer, Ed., "The Enzymes," 3rd. ed., Vol. 6, Academic Press, New York, 1972, p. 14.]

oxaloacetate is converted to glucose. Thus the function of pyruvate carboxylase varies with metabolic needs. It may act as an *anaplerotic* enzyme (from the Greek, "filling up") in the synthesis of a regenerating substrate for the catalytic citric acid cycle or as a biosynthetic enzyme in the anabolic pathway for the synthesis of other cell constituents.

Extensive studies of the structure and mechanism of pyruvate carboxylase have shown that the mammalian enzymes are tetramers composed of identical subunits, whose total molecular weight is between 440,000 and 530,000. Each subunit contains a covalently bound biotin at its carboxylation subsite and an allosteric site that binds acetyl-CoA. The carboxyl transfer site contains a tightly bound Mn^{2+} ion (four Mn^{2+} ions per tetramer). These distinct catalytic subsites are indicated by product inhibition studies, by isotopic exchange experiments, and by ping-pong kinetics (recall Sections 7.12D, E, F). The kinetic scheme for a ping-pong mechanism is shown below.

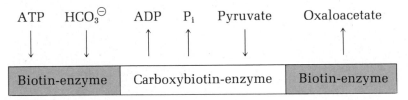

b. Phosphoenolpyruvate Carboxykinase

Pyruvate carboxylase is located in the mitochondrion, but phosphoenolpyruvate carboxykinase is a cytosolic enzyme. Once oxaloacetate has been synthesized in the mitochondrion in a slightly exergonic reaction, it is reduced to malate by mitochondrial malate dehydrogenase. We recall that the oxidation of malate in the citric acid cycle is strongly endergonic and proceeds only because oxaloacetate is removed from the system by citrate synthetase. The reverse reaction is therefore highly exergonic (reaction 17.22).

$$\text{oxaloacetate} + \text{NADH} + \text{H}^{\oplus} \xrightarrow[\text{Mitochondrion}]{\text{Malate dehydrogenase}} \text{malate} + \text{NAD}$$
$$\Delta G^{\circ\prime} = -28.0 \text{ kJ mol}^{-1} \qquad (17.22)$$

Malate is transported out of the mitochondria by the dicarboxylate carrier in exchange for another dianion such as phosphate. (We recall that the dicarboxylate carrier is suspected by transporting oxaloacetate, but that that is not certain.) In the cytosol another NAD-dependent dehydrogenase reconverts malate to oxaloacetate (reaction 17.23).

$$\text{malate} + \text{NAD} + \text{H}^{\oplus} \xrightarrow[\text{Cytosol}]{\text{Malate dehydrogenase}} \text{oxaloacetate} + \text{NADH}$$
$$\Delta G^{\circ\prime} = +28.0 \text{ kJ mol}^{-1} \qquad (17.23)$$

Once oxaloacetate has made its way to the cytosol, it is converted to phosphoenolpyruvate in a reaction that involves cleavage of a phosphoanhydride bond of GTP and decarboxylation (reaction 17.24).

$$\overset{\ominus}{}\text{O}_2\text{C}-\text{CH}_2-\overset{\overset{\text{O}}{\|}}{\text{C}}-\text{CO}_2^{\ominus} + \text{GTP} \longrightarrow$$

$$\text{CH}_2{=}\overset{\overset{\text{OPO}_3^{2-}}{|}}{\text{C}}-\text{CO}_2^{\ominus} + \text{GDP} + \text{CO}_2 \qquad (17.24)$$

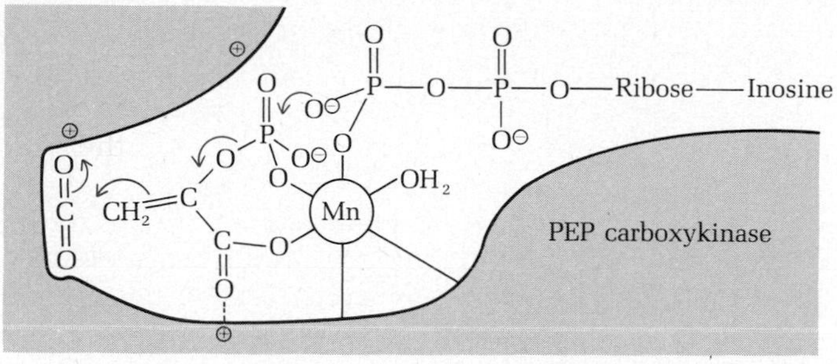

Figure 17.10 Possible concerted mechanism for the action of phosphoenolpyruvate carboxykinase. [From M. F. Utter and M. Kolenbrander, in P. D. Boyer, Ed., "The Enzymes," 3rd ed., Vol. 6, Academic Press, New York, 1972, p. 167.]

The reaction is slightly exergonic; $\Delta G^{\circ\prime} = -1.25$ kJ mol^{-1}. Although phosphoenolpyruvate is more energy rich than GTP, the reaction is greatly facilitated by the release of CO_2, which probably provides about half the driving force for the reaction.

The mechanism of action of phosphoenolpyruvate carboxykinase appears to involve concerted decarboxylation and phosphoryl group transfer (Figure 17.10). We might expect that a reaction involving a nucleoside triphosphate would be subject to metabolic regulation, but extensive studies of phosphoenolpyruvate carboxykinase have not been at all conclusive on this point. At present, it does not appear that this enzyme is subject to metabolic control.

c. Fructose 1,6-Diphosphatase and the Simultaneous Regulation of Glycolysis and Gluconeogenesis

The interconversion of fructose 6-phosphate and fructose 1,6-diphosphate (FDP) is the key regulatory reaction of both glycolysis and gluconeogenesis. In glycolysis phosphofructokinase catalyzes the phosphorylation of fructose 6-phosphate, and fructose 1,6-diphosphatase hydrolyzes the 1-phosphate group, reversing the effects of phosphofructokinase (reaction 17.25).

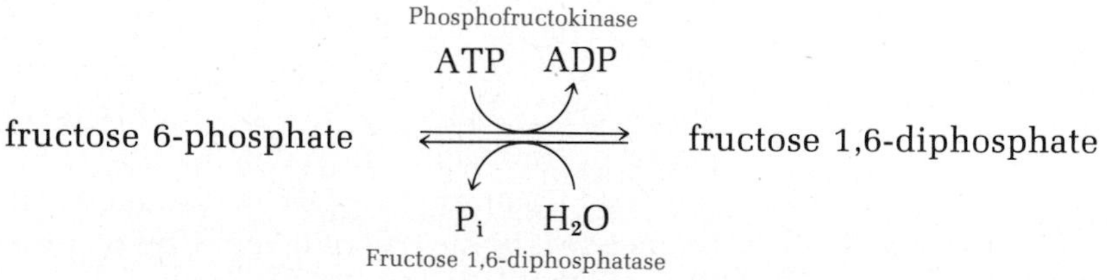

$$\text{fructose 6-phosphate} \underset{\text{Fructose 1,6-diphosphatase}}{\overset{\text{Phosphofructokinase}}{\rightleftharpoons}} \text{fructose 1,6-diphosphate} \qquad (17.25)$$

ATP ADP

P$_i$ H$_2$O

Since ATP is consumed in the reaction of phosphofructokinase, and since ATP has been expended in gluconeogenesis to convert oxaloacetate to fructose 1,6-diphosphate, both reactions are regulated simultaneously to avoid a futile cycle of ATP hydrolysis. AMP and citrate are allosteric activators of phosphofructokinase and allosteric inhibitors of fructose 1,6-diphosphatase. The energy state of the cell therefore regulates the two pathways. When ATP is required, glycolysis is favored, and when there is a surfeit of ATP, gluconeogenesis is favored.

Fructose 1,6-diphosphate, a product of phosphofructokinase and a substrate for fructose 1,6-diphosphatase, activates phosphofructokinase and inhibits fructose 1,6-diphosphatase, while at the same time greatly enhancing the inhibition caused by AMP (Figure 17.11).

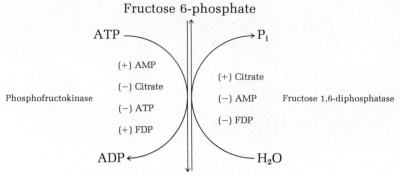

Figure 17.11 Simultaneous regulation of glycolysis and gluconeogenesis by phosphofructokinase and fructose 1,6-diphosphatase. Metabolites that act as activators are indicated by (+); inhibitors are indicated by (−).

d. Glucose 6-Phosphatase

Glucose 6-phosphatase catalyzes the final step of gluconeogenesis (reaction 17.26).

$$\text{glucose 6-phosphate} + H_2O \xrightarrow{\text{Glucose 6-phosphatase}} \text{glucose} + P_i \qquad (17.26)$$

Of all known enzymes of carbohydrate metabolism, glucose 6-phosphatase is the only one associated with the endoplasmic reticulum. The enzyme is found in mammalian kidney and liver tissues, which are sites of gluconeogenesis, but it is absent in skeletal muscle and brain tissues, in which gluconeogenesis cannot occur.

The net reaction of gluconeogenesis, including the cytoplasmic oxidation of malate which consumes NADH, is

$$2 \text{ pyruvate} + 4ATP + 2GTP + 2NADH + 2H^{\oplus} + 6H_2O$$
$$\downarrow$$
$$\text{glucose} + 6HPO_4^{2\ominus} + 4ADP + 2GDP + 2NAD \qquad (17.27)$$

The standard free-energy change is -37.7 kJ mol^{-1}. The standard free-energy change for the *reversal* of glycolysis would be $+44.1$ kJ mol^{-1}. The substitution of four new enzymes converts a highly unfavorable process—the reversal of glycolysis—to a highly favorable one—gluconeogenesis—at a cost of 6 phosphoanhydride bonds and 2 moles of NADH. Two moles of ATP and 2 moles of NADH are produced in the net reaction of glycolysis, so the net cost to the cell of converting oxaloacetate to glucose is 4 phosphoanhydride bonds.

B. Relation Between Gluconeogenesis and Glycolysis in Skeletal Muscle and Liver: The Role of Lactate Dehydrogenase

Glycolysis is the predominant mode of glucose catabolism in skeletal muscle, and gluconeogenesis is very active in the liver. One of the starting materials for gluconeogenesis is pyruvate and the end product of anaerobic glycolysis is lactate. Pyruvate and lactate are interconverted by lactate dehydrogenase (reaction 17.28). This reaction requires NADH and generates the NAD required to keep glycolysis going. The plasma membranes of muscle and liver cells are permeable to both pyruvate and lactate, and both are carried by the blood to the liver for gluconeogenesis. The newly synthesized glucose is then poured back into the bloodstream and recirculated to skeletal muscle for further glycolysis. The flow of glucose from liver to muscle and of lactate from muscle to liver is called the *Cori cycle* (Figure 17.12). This sequence of events is related to the activity of lactate dehydrogenase.

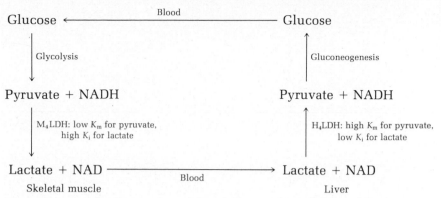

Figure 17.12 The Cori cycle and its relation to the activities of the isozymes of lactate dehydrogenase (LDH).

$$CH_3-\overset{\overset{O}{\|}}{C}-CO_2^{\ominus} + NADH \xrightarrow{\quad \text{Lactate dehydrogenase} \atop (LDH)\quad}$$

Pyruvate

$$CH_3-\underset{\underset{H}{|}}{\overset{\overset{OH}{|}}{C}}-CO_2^{\ominus} + NAD \qquad (17.28)$$

Lactate

Mammalian lactate dehydrogenases are tetramers composed of two types of subunits, designated M and H. Five combinations of subunits are possible: H_4, H_3M, H_2M_2, HM_3, M_4. These five forms of lactate dehydrogenase are isozymes that catalyze the same reaction, but they are found to different extents in different tissues. In liver and heart muscle the predominant isozyme has subunit composition H_4. In skeletal muscle the predominant isozyme is M_4.

The M_4 and H_4 isozymes have different K_m values for pyruvate. The skeletal muscle isozyme (M_4) binds pyruvate tightly, and the liver isozyme (H_4) has a higher K_m for pyruvate. Lactate inhibits both isozymes, but is a much more potent inhibitor of the liver isozyme, $K_i = 26$ mM, than of the skeletal muscle isozyme, $K_i = 130$ mM. These two sets of kinetic parameters are well accommodated to metabolic requirements. In the changeover from resting to anaerobic muscle, the concentration of lactate rises from 1.5 mM to 20 to 48 mM, but skeletal muscle requires NAD and at these concentrations of lactate the skeletal muscle isozyme is not significantly inhibited. In liver, on the other hand, increased concentrations of lactate have a much greater inhibitory effect, and the reduction of pyruvate to lactate is discouraged, to the benefit of gluconeogenesis.

17–4 GLYCOGEN METABOLISM

A steady supply of glucose is an absolute requirement for human life. The brain alone catabolizes 5 to 6 g of glucose per hour, to say nothing of the requirements of other organs and tissues. Since the total amount of glucose in the blood is only about 20 g, a steady supply must be produced to maintain the body. The storehouse of glucose is the highly branched polysaccharide glycogen (Figure 17.13). The total glycogen in a 70-kg adult amounts to about 200 g, enough to supply the metabolic needs for about a day. Most of the body's glycogen is stored in the liver and in skeletal muscle. In times of strenuous exercise gluconeogenesis can also provide glucose from noncarbohydrate sources.

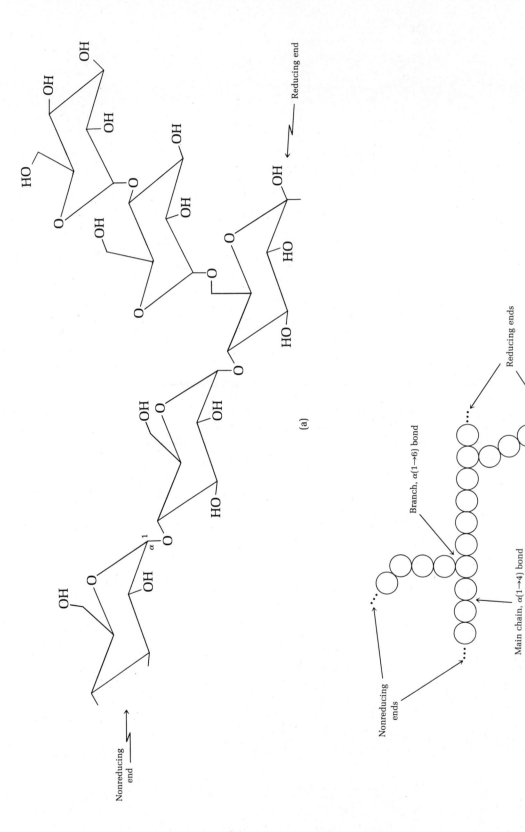

Figure 17.13 Covalent structure of glycogen. (a) Part of the covalent backbone. (b) Schematic structure. Main chain glucosyl residues are linked α(1 4), and branches are linked α(1 6).

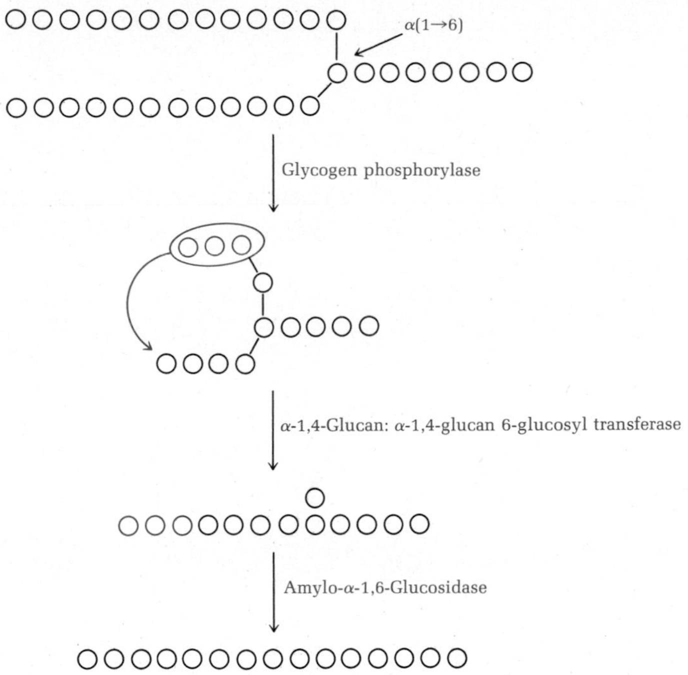

Figure 17.14 Enzymes responsible for glycogen degradation.

The pathway for glycogen degradation is almost deceptively simple (Figure 17.14). We shall see, however, that the enzymes responsible for glycogen degradation are subject to strict and complex metabolic regulation. Glycogen synthesis (Figure 17.15), is coordinated with glycogen degradation. Catabolism and biosynthesis occur by separate pathways and are reciprocally regulated.

A. Glycogen Degradation

a. Glycogen Phosphorylase

Glycogen phosphorylase catalyzes phosphorolysis of a glucosyl residue at the nonreducing end of a glycogen molecule (reaction 17.29). The smallest substrate for phosphorylase is the tetramer maltotetrose. Since α-D-glucose 1-phosphate is formed, the reaction proceeds with retention of configuration. A shielded, resonance-stabilized oxocarbonium ion that is

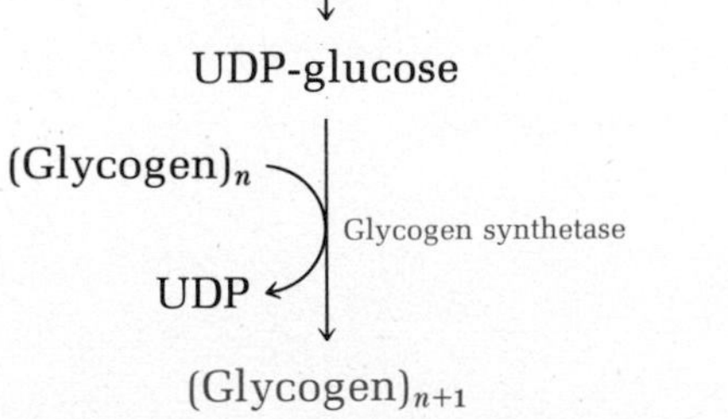

Glucose 6-phosphate

Phosphoglucomutase

Glucose 1-phosphate

UTP

Uridine diphosphate
glucose pyrophosphorylase

PP_i

UDP-glucose

$(Glycogen)_n$

Glycogen synthetase

UDP

$(Glycogen)_{n+1}$

Figure 17.15 Enzymes directly responsible for glycogen synthesis.

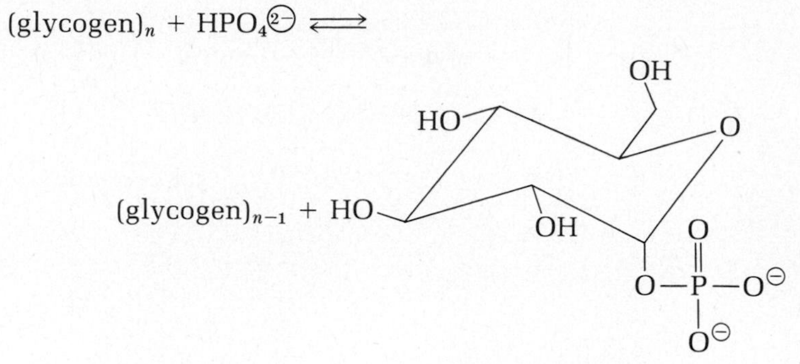

$$(\text{glycogen})_n + \text{HPO}_4^{\,2\ominus} \rightleftharpoons$$

$$(\text{glycogen})_{n-1} + \text{HO} \qquad\qquad (17.29)$$

α-D-Glucose 1-phosphate

attacked exclusively from the α side by phosphate seems to be an intermediate in the reaction (reaction 17.30).

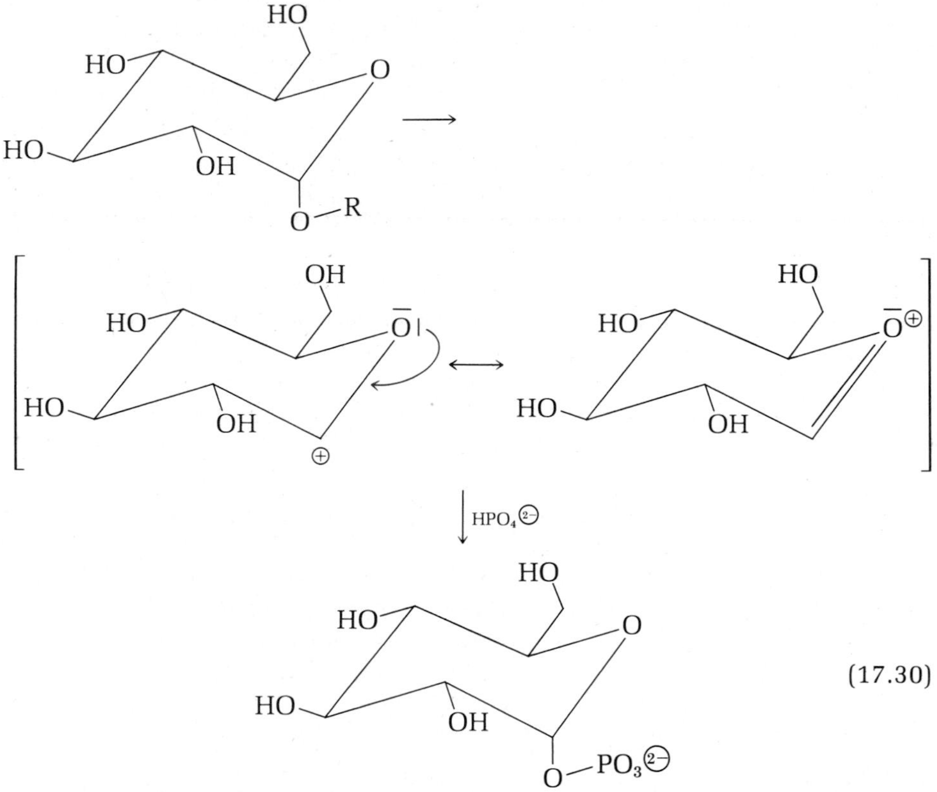

$$(17.30)$$

The mechanism of action of phosphorylase therefore is similar to the mechanism of action of lysozyme (Section 8.7A). D-Glucono-1,5-lactone (a δ-lactone) is a potent inhibitor of phosphorylase whose structure resembles the carbonium ion intermediate. We can regard the lactone as a transition state analog (Section 7.11B).

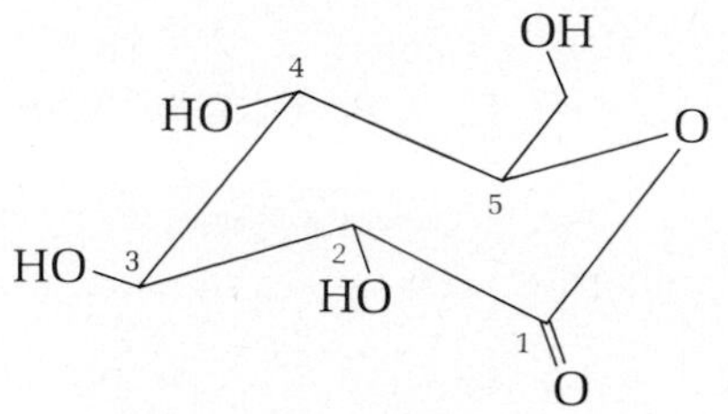

Glucono-1,5-lactone, a transition
state analog for phosphorylase

Phosphorylase *a* contains a pyridoxal phosphate residue bound to the enzyme by the ε-amino group of a lysyl side chain. Treatment of the enzyme with sodium borohydride (NaBH₄) reduces the imine to an amine. Reduction

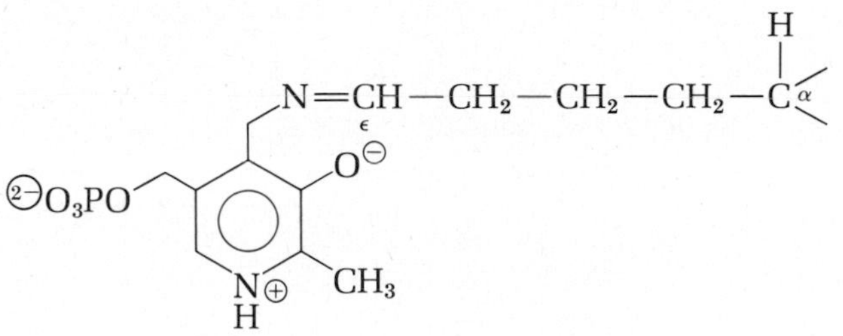

Pyridoxal phosphate bound to phosphorylase by
ε-amino group of a lysine residue of phosphorylase

abolishes activity in all known pyridoxal phosphate–dependent enzymes, *except* phosphorylase, which is still active following reduction. Considerable effort has been expended to determine the function of pyridoxal phosphate in phosphorylase. It seems likely that the pyridoxal phosphate is necessary to maintain the catalytically active conformation of the enzyme. The catalytic function of the pyridoxal phosphate, if any, appears to revolve around its phosphate group, which apparently acts as a base catalyst. It is buried in a hydrophobic region of the enzyme about 0.7 nm from the catalytic site. The hydrophobic nature of the catalytic site is probably an important factor is promoting capture of the carbonium ion intermediate by phosphate rather than by water.

The equilibrium constant[3] for glycogen degradation is

$$K = \frac{[\text{glucose 1-phosphate}]}{[\text{HPO}_4^{2-}]} \qquad (17.31)$$

At pH 6.8 the equilibrium constant of phosphorolysis is 0.28, corresponding to a standard free-energy change of $+3.2$ kJ mol^{-1}. The major bioenergetic advantage afforded by phosphorylase is the production of a phosphorylated glucose without expenditure of ATP.

Phosphorylase *a* degrades glycogen until a *limit dextrin* (Section 8.7B) is produced. Glycogen phosphorylase stops four residues from a branch point (reaction 17.32).

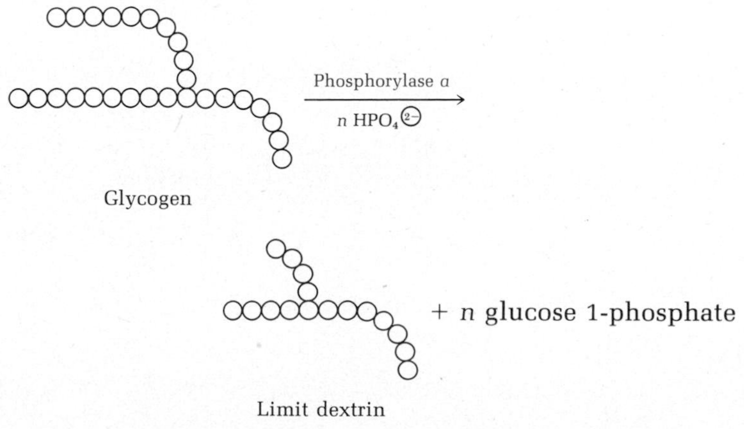

Glycogen

Limit dextrin

$+\ n$ glucose 1-phosphate   (17.32)

---

[3] In thermodynamic equilibrium constants the activities of solid phases are conventionally defined as unity.

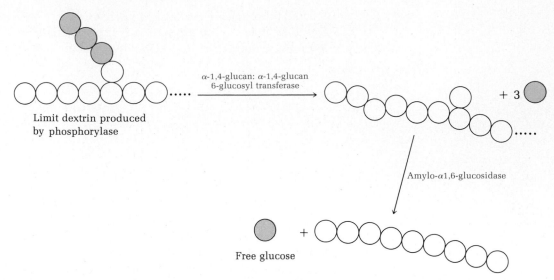

Figure 17.16   Degradation of glycogen. α-1,4 Glucan: α-1,4-glucan 6-glucosyl transferase and amylo-α1,6-glucosidase debranch the limit dextrin produced by the action of glycogen phosphorylase.

b. Debranching System

Glycogen is "debranched" by the combined action of two enzymes. The first enzyme, *α-1,4 glucan: α-1,4-glucan 6-glucosyltransferase*, shifts a trimer, maltotriose, from a branch to the end of the chain, leaving a branch with a single glucose residue bound to the chain by an $\alpha(1 \rightarrow 6)$ glycosidic bond. The second enzyme, *amylo-α-1,6-glucosidase*, hydrolyzes the lone remaining residue of the branch to give free glucose and glycogen (Figure 17.16). The net result of the action of these two enzymes is conversion of a four-residue branch to three molecules of glucose 1-phosphate and one molecule of free glucose. The glycogen degradation system allows glycogen to be completely degraded, but the total degradation of glycogen would only occur in extremes of starvation.

c. Phosphoglucomutase

Glucose 1-phosphate is converted to glucose 6-phosphate by *phosphoglucomutase* (reaction 17.33).

In the first step of the reaction, whose equilibrium constant is about 7, a phosphoryl group is transferred from a seryl residue of the enzyme to the C-6

(17.33)

hydroxyl group of glucose 1-phosphate, producing glucose 1,6-diphosphate as an intermediate that remains bound to the enzyme. If it dissociates, the enzyme becomes inactive. The dissociation constant is small, $K_{diss} = 10^{-8}$, and is much smaller than K_m for glucose 1,6-diphosphate, $K_m = 10^{-5}$. Addition of glucose 1,6-diphosphate to the inactive enzyme reactivates it.

In the second step the phosyl group at C-1 is transferred to the enzyme, producing glucose 6-phosphate and a phosphoryl enzyme. Since the same serine donates a phosphoryl group to the C-6 hydroxyl group and retrieves one from the C-1 position, a dramatic conformational alteration of either the enzyme or the substrate or both is required to bring the two groups close enough together for reaction (Figure 17.17).

d. Interconversion of Phosphorylase *a* and Phosphorylase *b*

Skeletal muscle phosphorylase is a dimer of identical monomers having 841 residues. The dimer can exist in two interconvertible forms, designated phosphorylase *a* and phosphorylase *b*. Each of the two forms can exist in two conformational states.

Phosphorylase *a* is formed by phosphoryl group transfer from ATP to seryl residue 14 in each monomer of the phosphorylase *b* dimer (reaction 17.34). This reaction is catalyzed by *phorphorylase kinase*. The equilibrium constant for interconversion of the inactive form of phosphorylase *a* and the active form is between 3 and 13 (reaction 17.35). Phosphorylase *a* also differs from phosphorylase *b* in its response to allosteric effectors. Phosphorylase *b* is inactive except in the presence of high concentrations of AMP. Phosphorylase *b* is also inhibited by ATP, ADP, and glucose 6-phosphate. Therefore, phosphorylase *b* will catalyze the degradation of glycogen when the

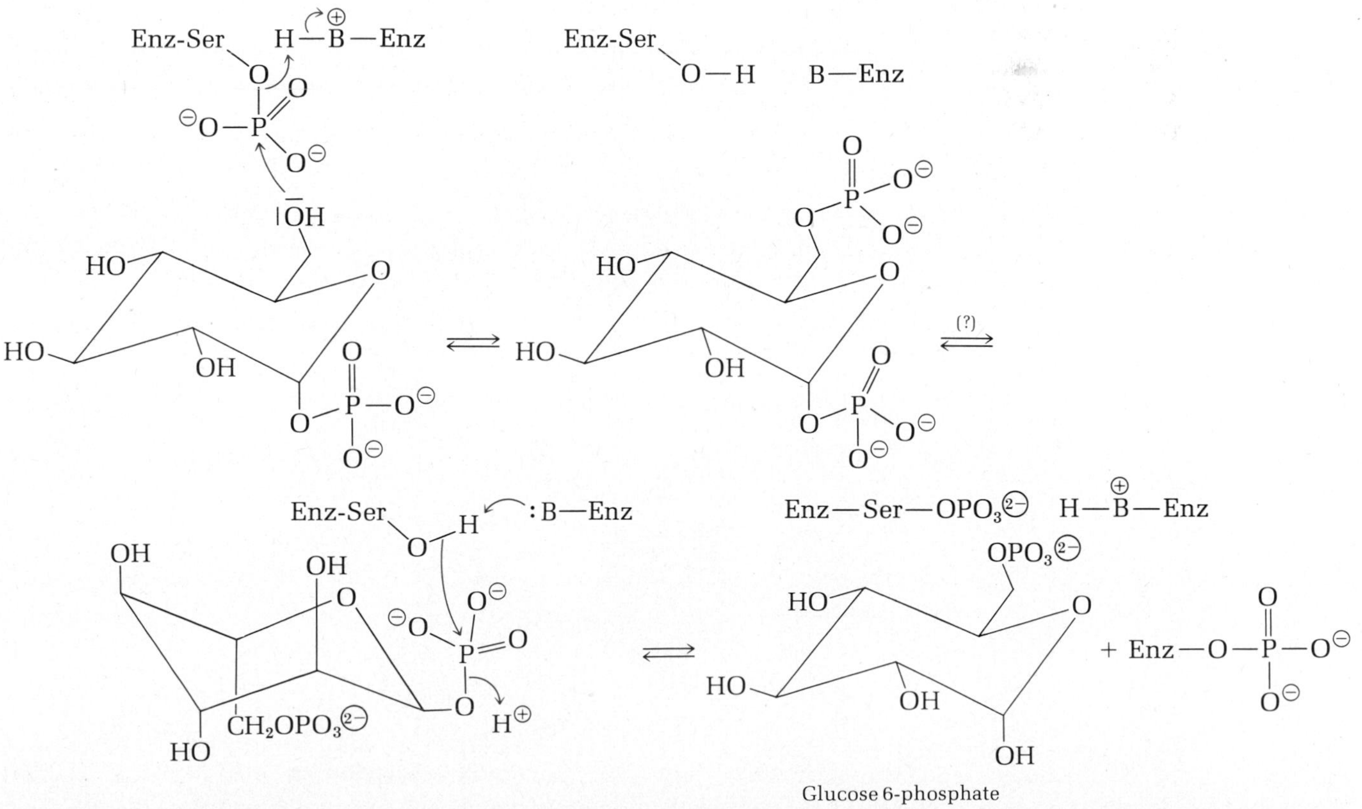

Figure 17.17 Mechanism of action of phosphoglucomutase. The enzyme transfers a phosphoryl group to glucose 1-phosphate and retrieves a phosphoryl from the intermediate glucose 1,6-diphosphate. The intermediate, the enzyme, or both must undergo a conformational change for the second phosphoryl transfer to occur. In the drawing the intermediate is shown in a new chair conformation, but this has not been established experimentally.

$$
\underset{\text{Phosphorylase } b}{\begin{array}{c} \text{Ser}_{14}\text{—OH} \\ \boxed{\diagdown} \\ \text{Ser}_{14}\text{—OH} \end{array}} + 2\,\text{ATP} \xrightarrow[\text{Mg(II)}]{\text{Phosphorylase kinase}}
$$

$$
\underset{\text{Phosphorylase } a}{\begin{array}{c} \text{Ser}_{14}\text{—OPO}_3^{\,2-} \\ \boxed{\diagdown} \\ \text{Ser}_{14}\text{—OPO}_3^{\,2-} \end{array}} + \text{ADP} \qquad (17.34)
$$

$$
\underset{\substack{\text{Phosphorylase } a \\ \text{(inactive state)}}}{\begin{array}{c} \text{Ser}_{14}\text{—OPO}_3^{\,2-} \\ \blacksquare \\ \text{Ser}_{14}\text{—OPO}_3^{\,2-} \end{array}} \underset{K = 3 - 13}{\rightleftharpoons} \underset{\substack{\text{Phosphorylase } a \\ \text{(active state)}}}{\begin{array}{c} \text{Ser}_{14}\text{—OPO}_3^{\,2-} \\ \bullet \\ \text{Ser}_{14}\text{—OPO}_3^{\,2-} \end{array}} \qquad (17.35)
$$

energy supply of the cell is low, as reflected in the adenylate energy charge or phosphorylation potential.

A schematic diagram of the phosphorylase b dimer is shown in Figure 17.18. The binding sites for substrate and pyridoxal phosphate and the catalytic site are symmetric with respect to 180° rotation about the center of symmetry of the dimer; that is, the dimer has a C_2 axis of symmetry. The allosteric activator AMP acts by inducing a conformational change in phosphorylase b.

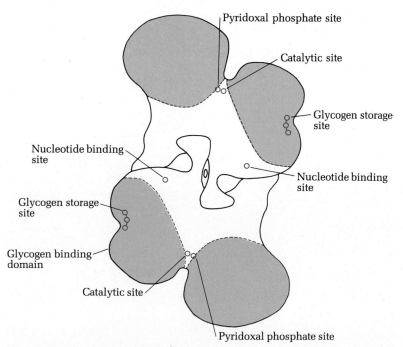

Figure 17.18 Schematic diagram of the phosphorylase b dimer. Note that the catalytic site and the binding sites for various ligands are symmetrical with respect to a 180° rotation around the center of symmetry of the dimer. [From T. Weber, L. N. Johnson, K. S. Wilson, D. G. R. Yeats, D. L. Wild, and J. A. Jenkins, *Nature*, 274, 435–436 (1978). Reprinted by permission. © 1978 Macmillan Journals Limited.]

Phosphorylase a, by contrast, is fully active at all concentrations of AMP, ADP, ATP, and glucose 6-phosphate. The structure of glycogen phosphorylase a has been determined by x-ray crystallography (Figure 17.19).

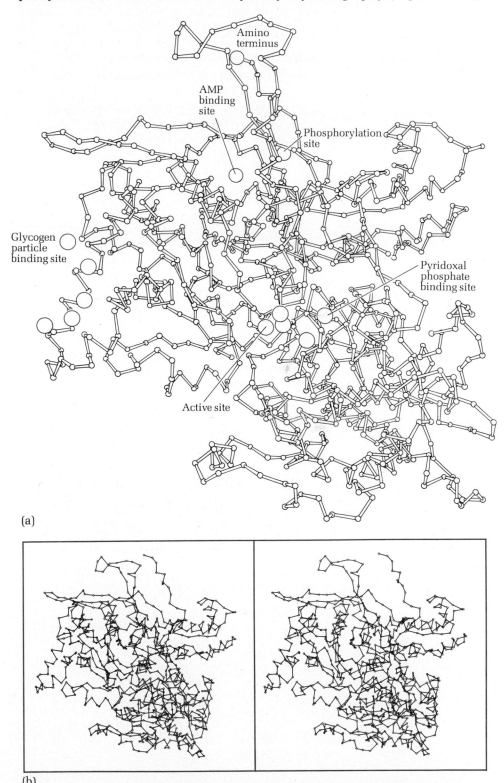

(a)

(b)

Figure 17.19 (a) α-Carbon positions of glycogen phosphorylase a. The glycogen binding site, the catalytic site, and the serine that is phosphorylated are indicated. [From S. Sprang and R. J. Fletterick, *J. Mol. Biol., 131*, 538 (1979). Reproduced with permission. Copyright: Academic Press Inc. (London) Ltd.] (b) Stereo view of phosphorylase a.

The enzyme contains a binding site for glycogen that is distinct from the catalytic site and separated from the catalytic site by some 3 nm. Since glycogen remains bound to phosphorylase, the enzyme always has a high local concentration of substrate, thus increasing its catalytic efficiency. The binding sites for pyridoxal phosphate, the position of the phosphorylated serine residue, and the location of pyridoxal phosphate are also indicated in Figure 17.19.

Glycogen degradation is controlled through the enzymes that interconvert phosphorylase a and b. The phosphorylation of phosphorylase b is regulated by an enzyme cascade that may be traced to the influence of the hormone *epinephrine* (Figure 17.20). We can distinguish four steps in this cascade. Let us consider the details of each stage of the amplified hormonal signal.

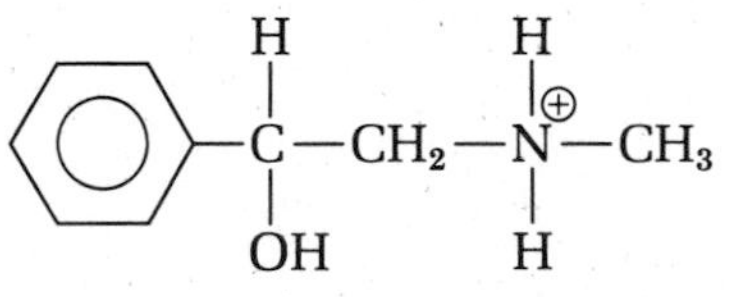

Epinephrine (adrenaline)

i. Step 1: Epinephrine stimulates adenylate cyclase which converts ATP to 3'-5' cAMP Epinephrine is released by the adrenal medulla and it binds to a receptor protein on the outer side of the plasma membrane. The protein is known as the *β-adrenergic receptor* because of its interaction with substances resembling adrenaline (epinephrine). The epinephrine–receptor protein complex activates adenylate cyclase, imbedded on the inner side of the plasma membrane, leading to synthesis of 3'-5' cAMP (reaction 17.36). In the cyclization reaction the 3'-hydroxyl group displaces pyrophosphate.

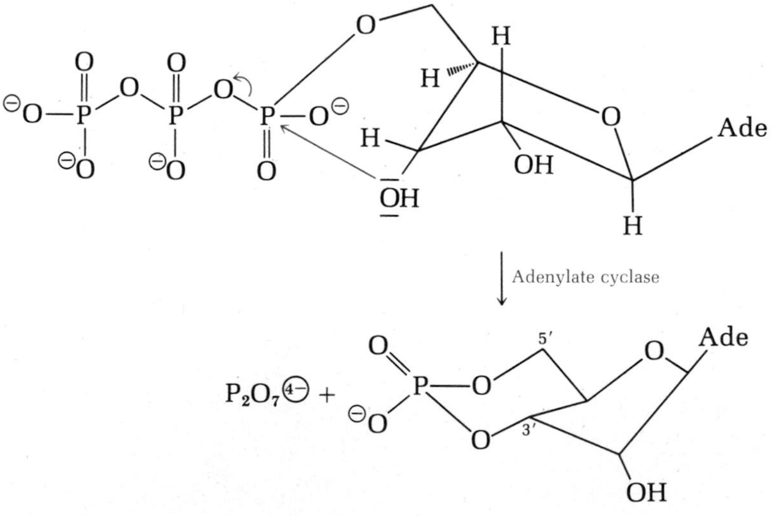

$$P_2O_7^{4\ominus} + \quad \text{3'-5' cAMP} \tag{17.36}$$

Pyrophosphate is hydrolyzed by *pyrophosphatase* (reaction 17.37). This hydrolysis provides the thermodynamic driving force for the synthesis of cAMP.

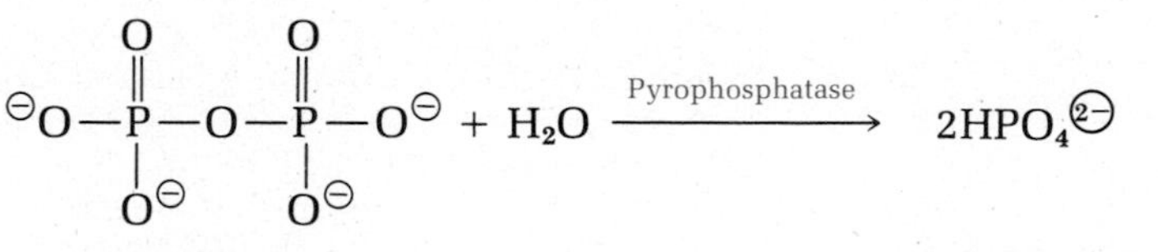

$$\ominus O - \overset{\displaystyle O}{\underset{\displaystyle O_\ominus}{\overset{\|}{P}}} - O - \overset{\displaystyle O}{\underset{\displaystyle O_\ominus}{\overset{\|}{P}}} - O^\ominus + H_2O \xrightarrow{\text{Pyrophosphatase}} 2HPO_4^{2\ominus} \tag{17.37}$$

Pyrophosphate   Orthophosphate

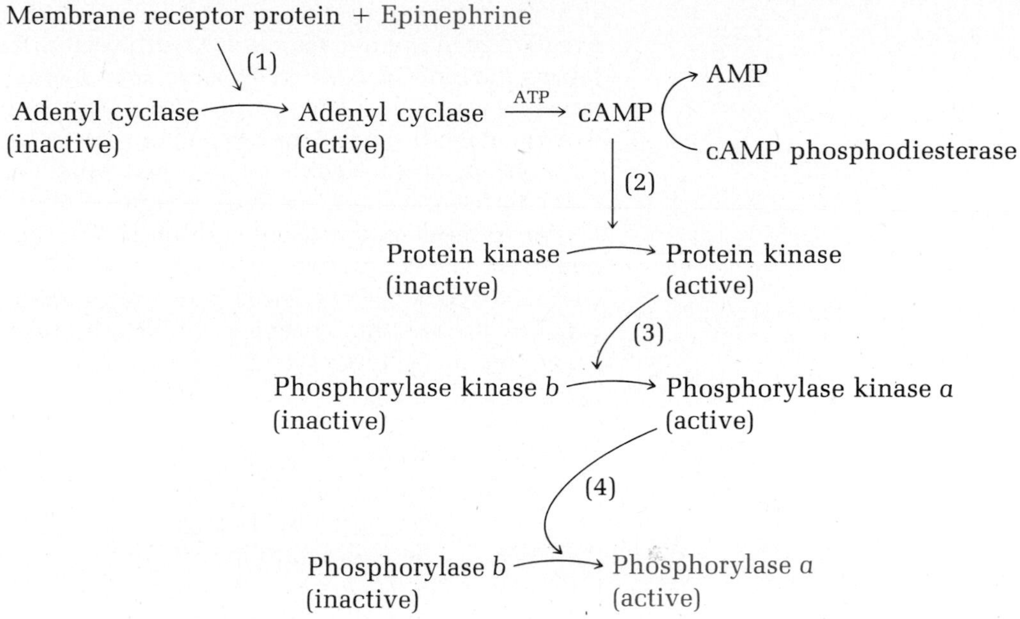

Membrane receptor protein + Epinephrine

Figure 17.20 Conversion of phosphorylase b to phosphorylase a. The conversion is stimulated by epinephrine through a four-step cascade of enzyme activation.

Cyclic AMP is rapidly hydrolyzed by *cAMP phosphodiesterase* (reaction 17.38). Hydrolysis of cAMP ensures that its potent effects will be felt only during the short time they are needed.

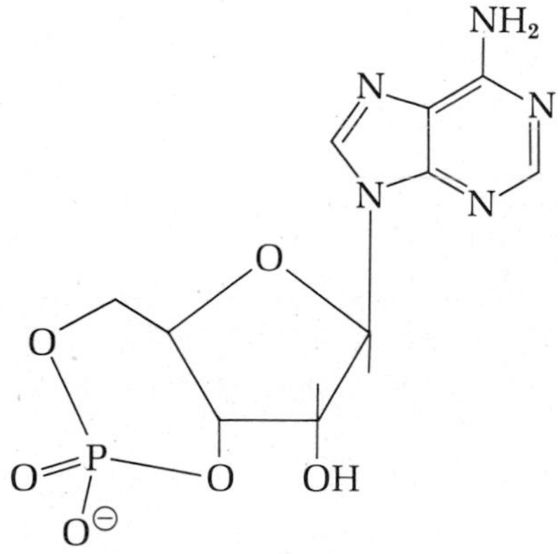

$$+ \; H_2O \xrightarrow{\text{cAMP phosphodiesterase}} AMP \qquad \Delta G^{\circ\prime} = -49.8 \text{ kJ mol}^{-1} \qquad (17.38)$$

cAMP

The stimulants *caffeine* and *theophylline* act in concert with cAMP. Both inhibit cAMP phosphodiesterase.[4]

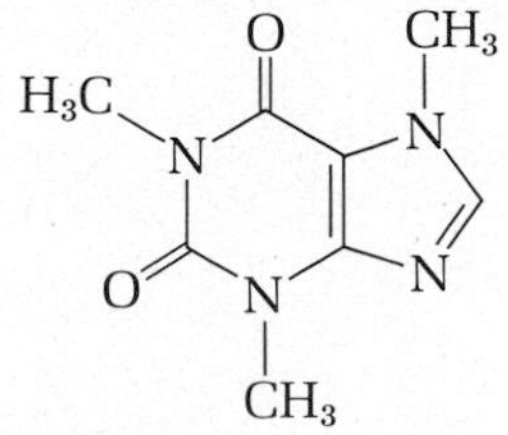

Caffeine
(1,3,7-trimethylxanthine)

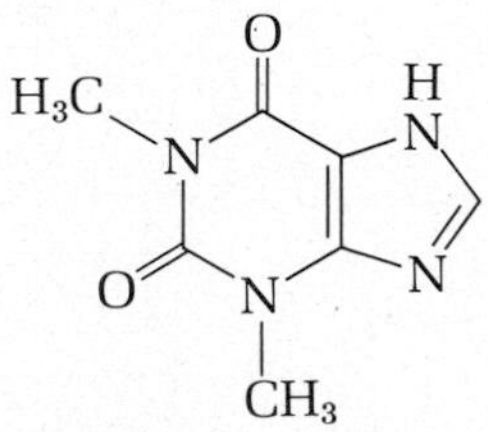

Theophylline
(1,3-dimethylxanthine)

[4] The inhibition of cAMP phosphodiesterase may not be the major effect of caffeine. In the brain adenosine depresses the firing of many nerves by binding to certain membrane receptors. Caffeine, a structural analog of adenosine, is now thought to block these adenosine receptors, and it therefore acts as a stimulant. Theophylline has the same effects as caffeine on the adenosine receptors of the brain. The levels of caffeine required to inhibit cAMP phosphodiesterase are also much greater than the caffeine concentration required to effectively inhibit the brain's adenosine receptors.

The free energy of hydrolysis of cAMP is -49.8 kJ mol^{-1}, significantly more negative than the free energy of hydrolysis of ATP, and the hydrolysis lies far on the side of AMP. The large free energy of hydrolysis of cAMP is largely a consequence of the unstable *trans* ring fusion between the six- and five-membered rings of the phosphodiester.

Let us turn again to the activation of adenylate cyclase. The epinephrine–receptor protein complex interacts with a third protein, known as the *G protein,* that binds either GTP or GDP. When the G protein is bound to GTP, adenylate cyclase is active. When the G protein is bound to GDP, adenylate cyclase is inactive. The exchange of GDP for GTP is catalyzed by the hormone-receptor complex. Just as the effects of cAMP are nullified by the action of cAMP phosphodiesterase, the activity of the G protein–GTP complex is curtailed to prevent continuous production of cAMP. The inactivation of the G protein–GTP complex is autocatalytic: the G protein has GTPase activity and hydrolyzes the GTP bound to it, simultaneously inactivating adenylate cyclase. The activity of adenylate cyclase therefore depends upon the relative rates of GDP-GTP exchange and of GTP hydrolysis. When epinephrine is bound to the receptor protein, the rate of GTP-GDP exchange is faster than the rate of GTP hydrolysis, and adenylate cyclase is active. In the absence of hormone, the rate of hydrolysis of GTP by the G protein is greater than the rate of exchange, most of the G protein is bound to GDP, and adenylate cyclase is inactive (Figure 17.21).

The importance of the GTPase activity of the GTP protein is illustrated by a deadly example—the effects on the body of Asiatic cholera. Cholera, caused by the bacterium *Vibrio cholera,* is characterized by such massive diarrhea that within a few hours affected persons can lose as much as half their body fluid. If not replaced, this fluid loss is shortly followed by shock and death. The protein toxin of the bacterium, known as *choleragen,* is an oligomer of molecular weight 84,000. The oligomer is composed of three types of subunits. The A$_1$ subunit (mol wt 23,000) is joined by a disulfide bond to the A$_2$ subunit (mol wt 5,500). The remainder of the oligomer consists of five B subunits (mol wt 16,000). The B subunits bind to a specific glycolipid on the cell surface known as ganglioside G$_{Ml}$ (Section 20.5). Once bound to the cell surface, choleragen activates adenylate cyclase. This activation takes the form of inhibition of the GTPase catalytic activity of the G

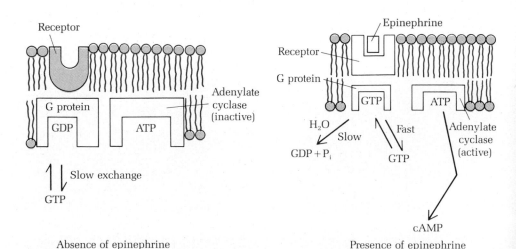

Figure 17.21 Effects of epinephrine upon adenylate cyclase mediated by the G protein. When bound to GTP, the G protein activates adenylate cyclase. The activity of adenylate cyclase depends upon the relative rates of GTP-GDP exchange, rapid when epinephrine is bound to receptor, and GTP hydrolysis catalyzed by the G protein itself.

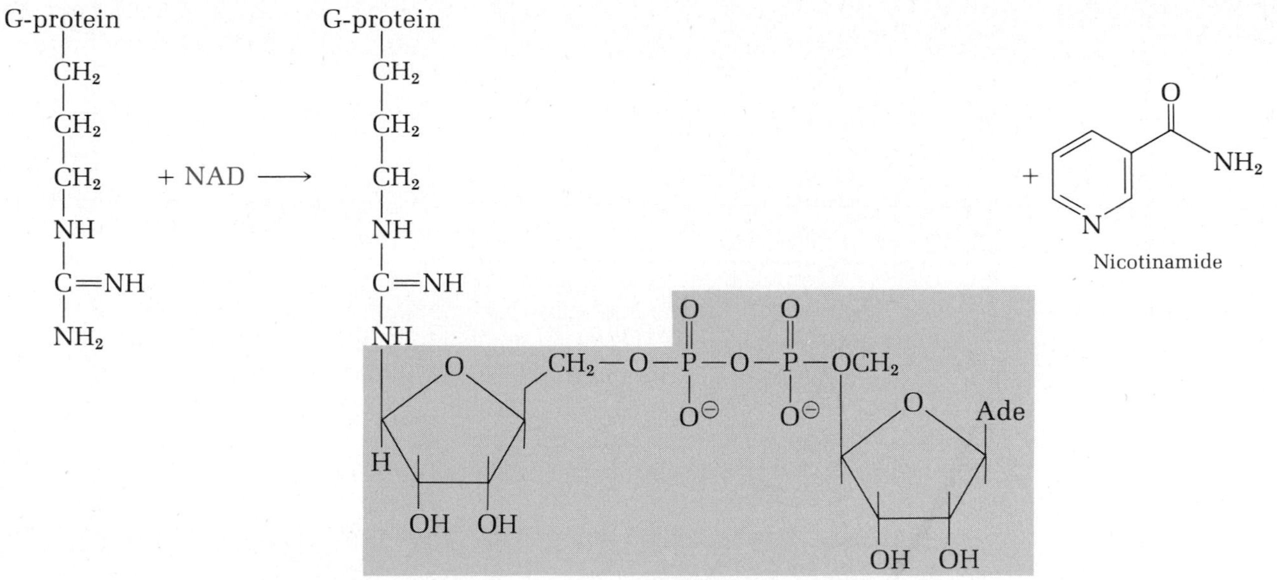

G-protein

CH$_2$

CH$_2$

CH$_2$ + NAD ⟶

NH

C=NH

NH$_2$

G-protein

CH$_2$

CH$_2$

CH$_2$

NH

C=NH

NH

ADP-ribosyl derivative of arginine residue in G protein

+ Nicotinamide

Figure 17.22 Effect of choleragen on the G protein. Choleragen inactivates the GTPase activity of the G protein by catalyzing an NAD-dependent ADP-ribosylation of an arginine residue in the G protein.

protein. The A$_1$ subunit of choleragen abolishes the GTPase activity of the G protein by catalyzing the NAD-dependent *ADP-ribosylation* of an arginine residue of the G protein (Figure 17.22). Once the G protein has been rendered permanently active by loss of its GTPase activity, cAMP is relentlessly produced by adenylate cyclase, even in the absence of epinephrine. The hydrolytic capacity of cAMP phosphodiesterase cannot keep up with cAMP production under these circumstances, and the many metabolic processes regulated by cAMP run amuck. Cyclic AMP stimulates active transport of Na$^{\oplus}$ ions across the epithelial cells of the intestinal mucosa. Once choleragen has done its work, the cAMP concentration rises dramatically, and the flow of Na$^{\oplus}$ ions and water into the gut increases apace. The result is the severe diarrhea that accompanies cholera.

ii. Step 2: Cyclic AMP activates *cAMP-dependent protein kinase*. Cyclic AMP binds to cAMP-dependent protein kinase with a K$_m$ of 10 *nM*. The kinase is a tetramer composed of two regulatory subunits and two catalytic subunits. The catalytic subunits (mol wt 49,000) are inactive when bound to the regulatory subunits (mol wt 82,000). When cAMP binds to the regulatory subunits, the tetramer dissociates into a regulatory protein–cAMP complex and two active, catalytic subunits (reaction 17.39).

$$R_2C_2 \; + \; cAMP \rightleftharpoons R_2\text{-}cAMP + \; 2C \qquad (17.39)$$

Inactive   Active

In every known case cAMP exerts its physiological effects by stimulating a specific protein kinase.

iii. Step 3: Active cAMP-dependent protein kinase activates phosphorylase *b*, converting it to phosphorylase *a*. The catalytic subunits of cAMP-dependent protein kinase phosphorylate inactive *phosphorylase kinase b*, converting it to active phosphorylase kinase *a*. Phosphorylase kinase *b* contains 16 subunits having the stoichiometry $(\alpha\beta\gamma\delta)_4$. The α and β subunits of phosphorylase kinase are phosphorylated by cAMP-dependent protein kinase (reaction 17.40).

$$\text{phosphorylase kinase } b + 8\,\text{ATP} \underset{}{\overset{\substack{\text{cAMP-dependent}\\\text{protein kinase}}}{\rightleftharpoons}}$$

$$\underset{\text{(Ser-OH)}_8}{|}$$

$$\underset{\text{(Ser-OPO}_3^{2\ominus})_8}{\overset{\displaystyle\text{phosphorylase kinase } a + 8\,\text{ADP}}{|}} \qquad (17.40)$$

Even in the absence of active cAMP-dependent protein kinase, a mechanism exists for partial activation of phosphorylase kinase b. The γ subunit of phosphorylase kinase is the calcium-binding protein *calmodulin* (recall Section 13.7). Calcium ion concentrations in the range of $0.1-0.3 \times 10^{-6}$ give half maximal activation of phosphorylase kinase b. This provides a rapid response mechanism for glycogen degradation. Calcium ions are released in muscle when a nerve impulse sweeps across the muscle cell membrane. This transient release of calcium ions signals that energy is needed for muscle contraction and the glycogen degradation system is stimulated even in the absence of a hormonal signal. Simultaneously, calcium ions inhibit glycogen synthesis.

The activation of phosphorylase kinase b is reversed by the action of *phosphorylase phosphatase* (reaction 17.41).

$$\underset{\text{(Ser-OPO}_3^{2\ominus})_8}{\overset{\displaystyle\text{phosphorylase kinase } a + \text{H}_2\text{O}}{|}} \xrightarrow[(-)\text{Ca}^{2\oplus},(-)\text{AMP}]{\substack{\text{Phosphorylase}\\\text{phosphatase}}}$$

$$\underset{\text{(Ser-OH)}_8}{\overset{\displaystyle\text{phosphorylase kinase } b + 8\text{HPO}_4^{2\ominus}}{|}} \qquad (17.41)$$

Phosphorylase phosphatase is inhibited by calcium ions and by AMP.

iv. Step 4: Phosphorylase kinase *a* phosphorylates phosphorylase *b*, leading to the degradation of glycogen. We have now traveled full-circle, returning to the interconversion of phosphorylase a and phosphorylase b. We need only add, to complete the picture, that phosphorylase phosphatase, the same enzyme we discussed above, hydrolyzes the seryl phosphates of phosphorylase a, stopping glycogen degradation (reaction 17.42).

$$\underset{\text{(Ser-OPO}_3^{2\ominus})_2}{\overset{\displaystyle\text{phosphorylase } a + 2\text{H}_2\text{O}}{|}} \xrightarrow{\substack{\text{Phosphorylase}\\\text{phosphatase}}}$$

$$\underset{\text{(Ser-OH)}_2}{\overset{\displaystyle\text{phosphorylase } b + 2\text{HPO}_4^{2\ominus}}{|}} \qquad (17.42)$$

Phosphorylase phosphatase has still another activity: it hydrolyzes a phosphate ester in glycogen synthetase, activating that enzyme. Thus phosphorylase phosphatase and cAMP-dependent protein kinase are the two enzymes that reciprocally regulate glycogen degradation and glycogen synthesis.

B. Glycogen Synthesis

The path from glucose 1-phosphate to glycogen involves the action of *uridine diphosphate glucose pyrophosphorylase* (UDP-glucose pyrophosphorylase) and *glycogen synthetase*. As with glycogen degradation, we again find that the apparently simple, two-step pathway of glycogen synthesis is actually quite complex because it is controlled by a cascade of enzyme activity initiated by epinephrine. The same enzymes responsible for regulation of the interconversion of phosphorylase a and phosphorylase b mediate the activity of glycogen synthetase.

a. UDP-Glucose Pyrophosphorylase

Glucose 1-phosphate, the product of glycogen degradation, is not the immediate substrate for glycogen synthesis. It must first be converted to UDP-glucose by UDP-glucose pyrophosphorylase (reaction 17.43).

Uridine diphosphate glucose (UDP-glucose)

$$\text{glucose 1-phosphate} + \text{UTP} \xrightarrow[\text{pyrophosphorylase}]{\text{UDP-glucose}} \text{UDP-glucose} + \text{PP}_i \qquad (17.43)$$

The equilibrium constant for the formation of UDP-glucose is 0.15. The reaction is pulled to completion by the hydrolysis of pyrophosphate (reaction 17.44).

$$\text{glucose 1-phosphate} + \text{UTP} \longrightarrow \text{UDP-glucose} + \text{P}_2\text{O}_7^{4-}$$
$$\Delta G^{\circ\prime} = 4.70 \text{ kJ mol}^{-1}$$
$$\text{P}_2\text{O}_7^{4-} + \text{H}_2\text{O} \longrightarrow 2\text{HPO}_4^{2-}$$
$$\Delta G^{\circ\prime} = -33.5 \text{ kJ mol}^{-1}$$

$$\text{glucose 1-phosphate} + \text{UTP} + \text{H}_2\text{O} \longrightarrow \text{UDP-glucose} + 2\text{HPO}_4^{2-}$$
$$\Delta G^{\circ\prime} = -28.8 \text{ kJ mol}^{-1} \qquad (17.44)$$

The synthesis of a biosynthetic intermediate driven by hydrolysis of pyrophosphate is often encountered, as we shall see in future chapters. Similarly, *nucleotide sugars*, such as UDP-glucose, are common intermediates in the biosynthetic pathways for complex lipopolysaccharides and carbohydrates.

b. Glycogen Synthetase

Glycogen synthetase, a tetramer whose subunits have a molecular weight of 90,000 each, catalyzes the transfer of a glucosyl residue from *uridine diphosphate glucose* to the nonreducing end of glycogen (reaction 17.45). A primer containing at least four glucosyl residues is required for glycogen synthetase activity. Glycogen itself is a much better substrate for the enzyme than the primer. Glycogen synthetase has a K_m value of 10^{-4} M for glycogen and a K_m value of 0.07 M for the primer.

Glycogen synthetase can exist in either of two interconvertible forms: glycogen synthetase a and glycogen synthetase b. Glycogen synthetase a is fully active. It is phosphorylated and inactivated by cAMP-dependent protein kinase (reaction 17.46). This enzyme also activates phosphorylase b. Thus, phosphorylation simultaneously inactivates glycogen synthetase and activates phosphorylase. The b form of glycogen synthetase is active only when the concentration of glucose 6-phosphate is high. At physiological

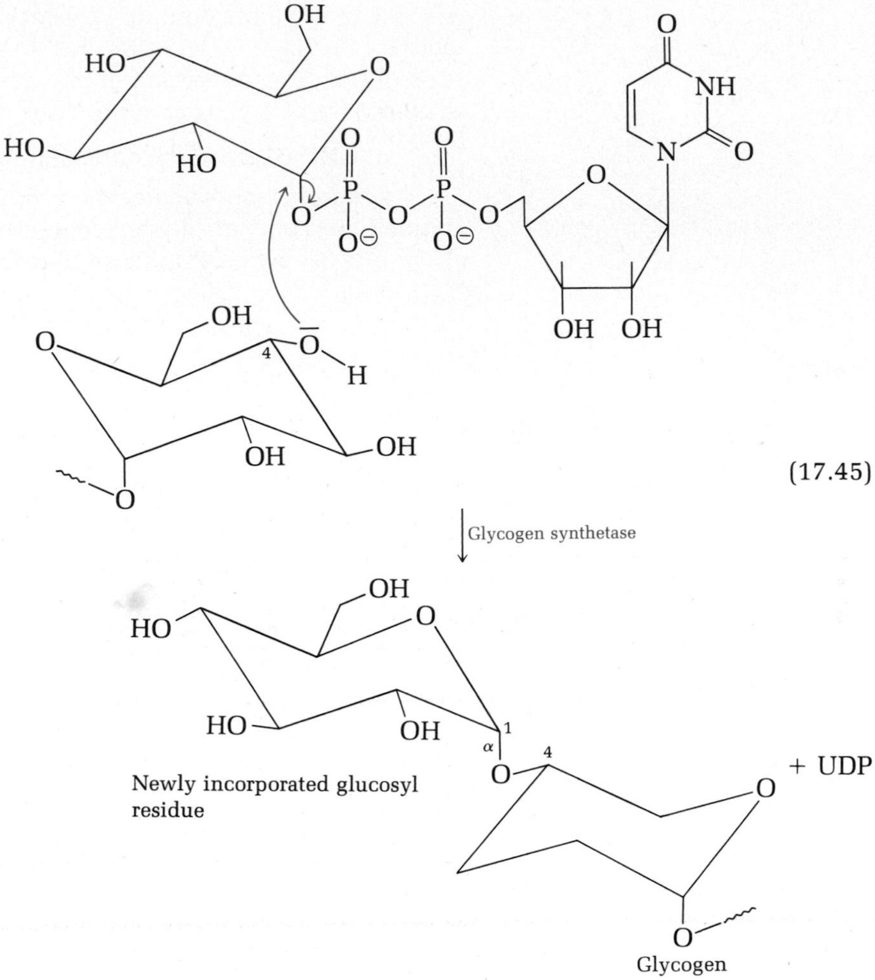

(17.45)

Glycogen synthetase

Newly incorporated glucosyl residue

+ UDP

Glycogen

$$\text{glycogen synthetase } a + 2\text{ADP} \xrightarrow[\text{protein kinase}]{\text{cAMP-dependent}}$$

|
(Ser-OH)$_2$

Active

$$\text{glycogen synthetase } b + 2\text{ATP} \qquad (17.46)$$

|
(Ser-OPO$_3$^{2⊖})$_2$

Inactive

concentrations of orthophosphate, however, this activity is completely abolished. Because it is dependent upon glucose 6-phosphate for activity, glycogen synthetase b is sometimes known as the D form of the enzyme; glycogen synthetase a is known as the I form, since its activity is independent of glucose 6-phosphate.

Glycogen synthetase b (the phosphoenzyme) is reactivated by hydrolysis of its phosphate groups. *Phosphorylase phosphatase* catalyzes this reaction.

$$\text{glycogen synthetase } b + \text{H}_2\text{O} \xrightarrow[\text{phosphatase}]{\text{Phosphorylase}}$$

|
(Ser-OPO$_3$^{2⊖})$_2$

$$\text{glycogen synthetase } a + \text{P}_\text{i} \qquad (17.47)$$

|
(Ser-OH)$_2$

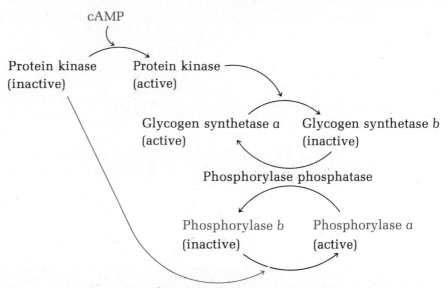

Figure 17.23 Control of glycogen synthesis and degradation. Protein kinase phosphorylates phosphorylase *b*, thereby activating it. Simultaneously, protein kinase phosphorylates glycogen synthetase *a*, switching glycogen synthesis "off". Phosphorylase phosphatase has exactly the opposite effects on phosphorylase and glycogen synthetase.

We therefore find reciprocal regulatory activities for phosphorylase phosphatase: it switches glycogen synthetase "on" and switches phosphorylase "off" (Figure 17.23).

C. Regulation of Glycogen Metabolism in the Liver by Regulation of Phosphorylase Phosphatase

The concentration of glucose in the blood is controlled by the liver. If the glucose concentration in the liver increases, the concentrations of UDP-glucose and glucose 6-phosphate both increase, as does the rate of glycogen synthesis. The activation of glycogen synthetase by glucose is preceded by a lag time of 2 to 3 minutes. During this time the activity of phosphorylase *a* decreases and that of glycogen synthetase *a* increases (Figure 17.24).

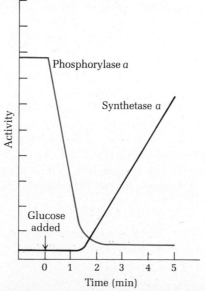

Figure 17.24 Sequential inactivation of glycogen phosphorylase and activation of glycogen synthetase following an increase in the concentration of liver glucose. [From H. G. Hers, *Ann. Rev. Biochem.*, 45, 176 (1976). Reproduced with permission. © 1976 by Annual Reviews Inc.]

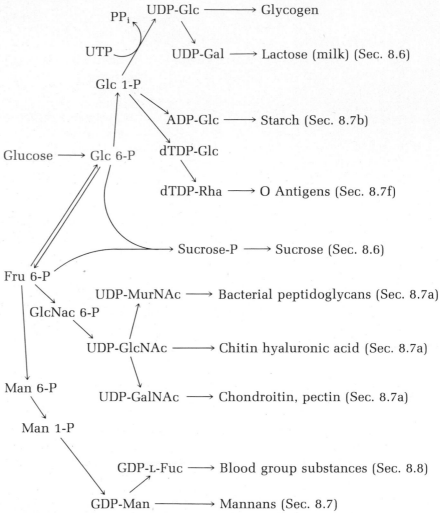

Figure 17.25 Some of the many fates of glucose 6-phosphate in the biosynthesis of complex carbohydrates.

The agent of these changes is phosphorylase phosphatase, which is activated by glucose. The active, *a* form of phosphorylase binds the phosphatase. Phosphorylase *b* does not bind the phosphatase. When the phosphatase hydrolyzes the seryl phosphates of phosphorylase *a*, it dissociates from the phosphorylase *b* it has just generated, and then it activates the inactive, phosphorylated glycogen synthetase *b*. Phosphorylase *a* is therefore the glucose receptor of the liver. Glucose induces an allosteric transition from the active state of phosphorylase *a* to the inactive state. This allosteric transition then uncovers the previously inaccessible phosphoryl group at serine 14, which is promptly cleaved by phosphorylase phosphatase.

17–5 GENERAL FEATURES OF THE BIOSYNTHESIS OF COMPLEX CARBOHYDRATES

Glucose 6-phosphate is the starting material for a large number of complex di-, oligo-, and polysaccharides (Figure 17.25). In surveying this figure we find most of the carbohydrates whose structures and biochemical importance we discussed in Chapter 8. Figure 17.25 also indicates that the biosynthesis of complex carbohydrates nearly always occurs with *nucleotide sugars*, such as the UDP-glucose we encountered in glycogen synthesis, as intermediates.

Two properties of nucleotide sugars make them excellent synthetic intermediates. (1) They are energy-rich metabolites. The standard free energy of hydrolysis of UDP-glucose is -30.5 kJ mol^{-1}, comparable to the free energy of hydrolysis of ATP. (2) The nucleotide is an excellent leaving group, and this is an important attribute in the synthesis of new glycosidic bonds. The large free energy of hydrolysis ensures favorable equilibrium constants for coupled reactions, and the excellent leaving group lowers the energy of activation for the coupled reaction. Both of these properties are features of the chemical synthesis of polypeptides and oligonucleotides, and both are general features of biosynthetic reactions. Rather than consider the vast number of biosynthetic reactions summarized by Figure 17.25, we shall focus upon general features of carbohydrate biosynthesis, picking out a couple of examples.

A. Synthesis of Nucleotide Sugars

We have already considered the biosynthesis of UDP-glucose from glucose 1-phosphate and UTP catalyzed by UDP-glucose pyrophosphorylase. Other pyrophosphorylases catalyze similar reactions to produce such nucleotide sugars as ADP-glucose, dTDP-glucose, and GDP-mannose. The general reaction is

α-D-Hexose 1-phosphate

Nucleotide sugar (an activated biosynthetic intermediate)

(17.48)

B. Nucleotide Sugars in the Biosynthesis of Glycosidic Bonds: Biosynthesis of Sucrose and Lactose

The standard free energy of hydrolysis of UDP-glucose is -30.5 kJ mol^{-1}. This may be compared with the standard free energy of hydrolysis of glycosidic bonds of -16.7 kJ mol^{-1}. The coupling of glycosidic bond formation to nucleotide sugar hydrolysis is therefore highly exergonic, with $\Delta G^{\circ\prime}$ of glycosidic bond formation being -13.8 kJ mol^{-1}. Glycosidic bond formation is also favored since the nucleotide is an excellent leaving group. The nucleotide sugars are *glycosyl donors* in reactions catalyzed by *glycosyl transferases*. We have seen one example in the action of glycogen synthetase, and the biosynthesis of sucrose and lactose provide two other examples.

a. Biosynthesis of Sucrose

Sucrose 6-phosphate synthetase catalyzes glycosidic bond formation between UDP-glucose and fructose 6-phosphate (reaction 17.49). A phosphatase hydrolyzes the 6-phosphate group of sucrose 6-phosphate (reaction 17.50).

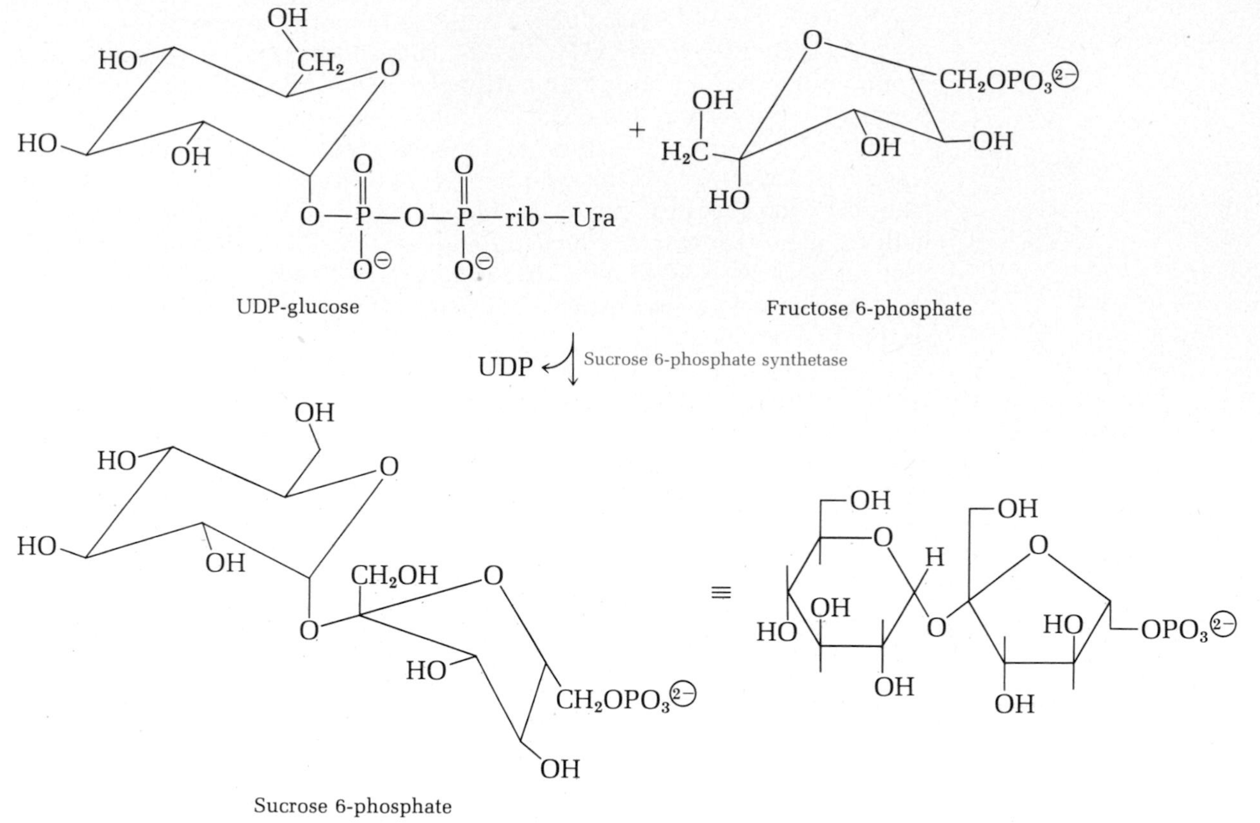

UDP-glucose

Fructose 6-phosphate

UDP ← Sucrose 6-phosphate synthetase

$$\equiv \qquad (17.49)$$

Sucrose 6-phosphate

$$\text{sucrose 6-phosphate} + H_2O \xrightarrow[\text{phosphatase}]{\substack{\text{Sucrose} \\ \text{6-phosphate}}} \text{sucrose} + HPO_4^{2-} \qquad (17.50)$$

b. Biosynthesis of Lactose

The biosynthesis of lactose is catalyzed by *lactose synthetase* in mammary gland cells during periods of lactation, but at no other time. The reaction proceeds by nucleophilic displacement of UDP from UDP-glucose by the C-4 hydroxyl group of glucose (reaction 17.51). Lactose synthetase has two components. The galactosyl transferase that carries out the reaction is present in most cells of plants and animals as a membrane-bound glycoprotein. In mammary cells the galactosyl transferase is bound to the endoplasmic reticulum and to the membranes of Golgi bodies. The second component of lactose synthetase is present only in mammary cells as a cytosolic protein during periods of lactation. This protein, *α-lactalbumin*, binds to the galactosyl transferase and promotes lactose synthesis. The α-lactalbumin changes the specificity of the galactosyl transferase by lowering its K_m for glucose from 2.5 M to the millimolar range. Sex hormones stimulate the synthesis of the α-lactalbumin.

The function of galactosyl transferase outside mammary gland cells is the synthesis of glycosidic bonds. The usual substrates for galactosyl transferase are UDP-galactose and either a free N-acetylglucosamine or an N-acetylglucosamine bound to a protein (reaction 17.52).

c. Metabolism of Galactose

Infants whose only food is milk receive virtually all their carbohydrates in the form of lactose. To enter the mainstream of carbohydrate metabolism, lactose must first be hydrolyzed to glucose and galactose. The enzyme that catalyzes this hydrolysis is *lactase* (reaction 17.53).

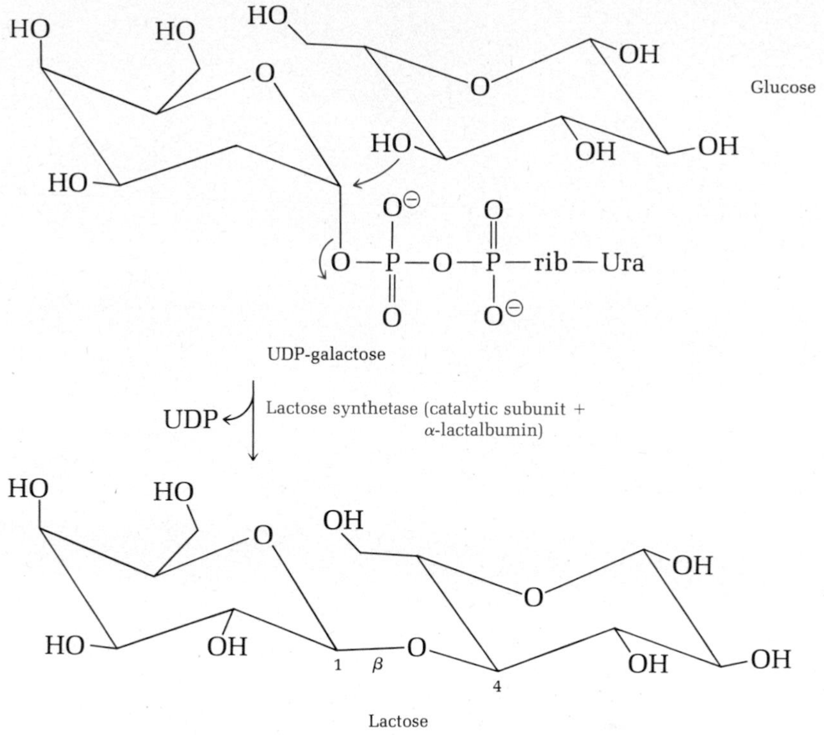
Glucose
UDP-galactose
UDP
Lactose synthetase (catalytic subunit + α-lactalbumin)
Lactose

(17.51)

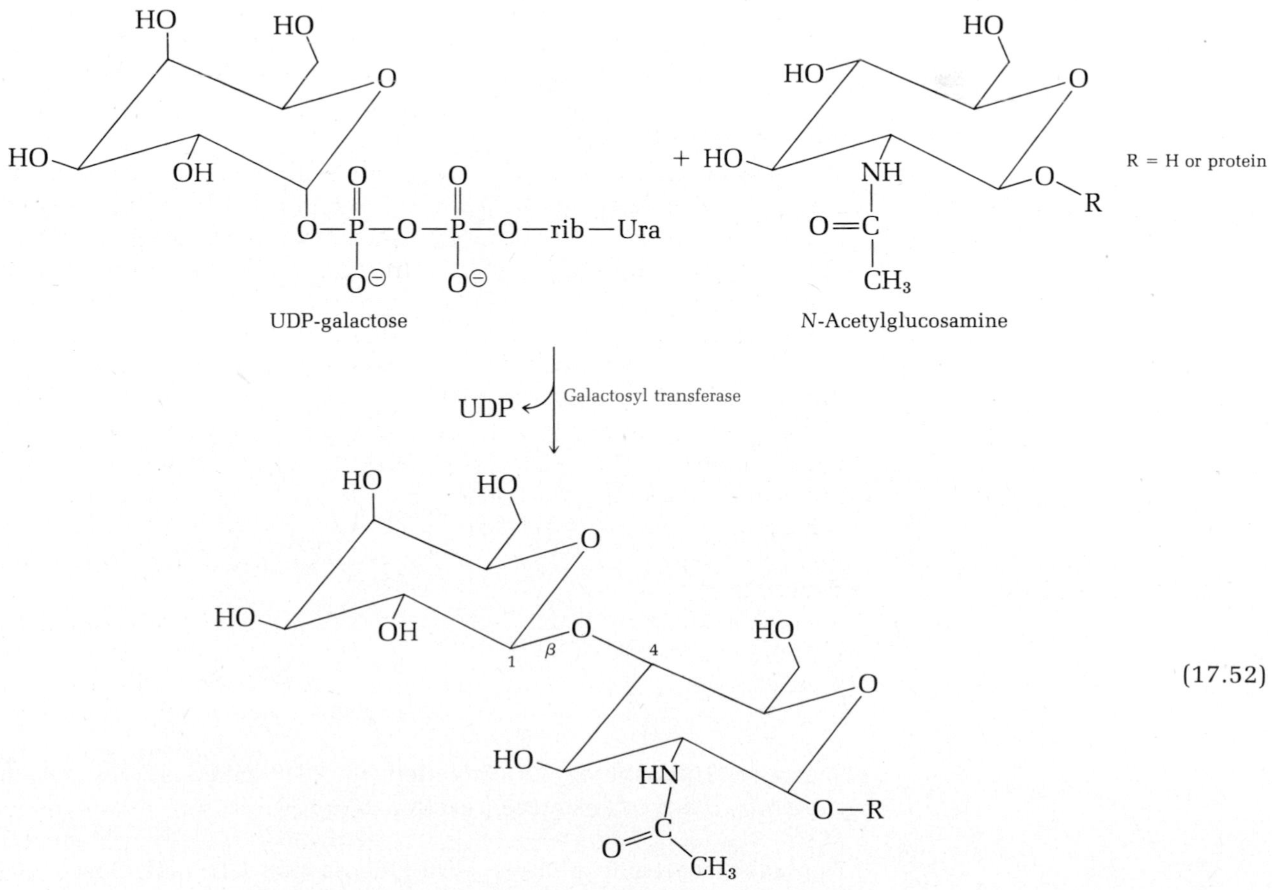
UDP-galactose
N-Acetylglucosamine
R = H or protein
UDP
Galactosyl transferase

(17.52)

Galactosyl-β-(1→4)*N*-acetylglucosamine [Gal-β-(1→4)NAcGlc]

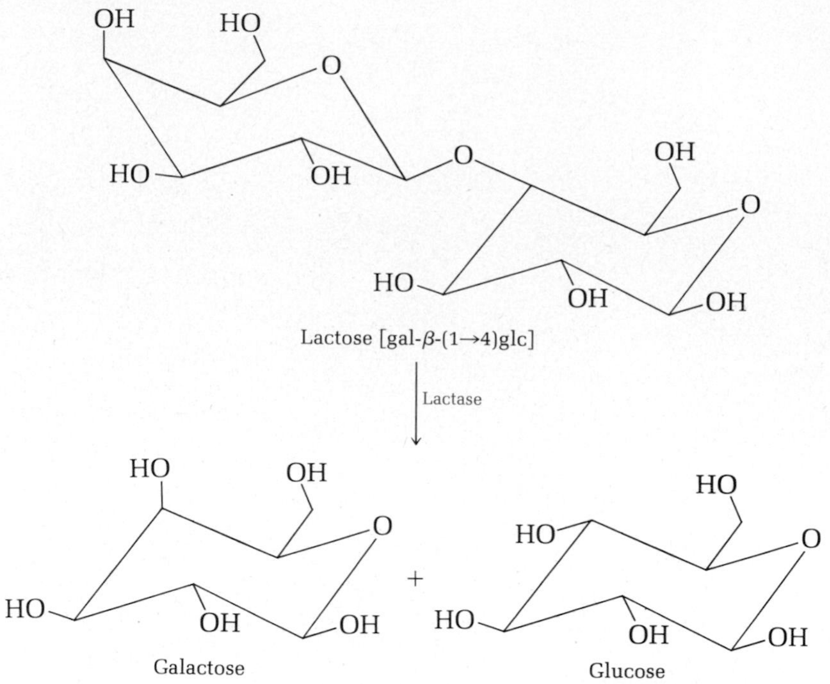

Lactose [gal-β-(1→4)glc]

| Lactase

Galactose + Glucose (17.53)

Most infants have a high level of lactase, but its biosynthesis slows considerably in most adults and may cease completely. Some adults, therefore, are unable to digest lactose, and consumption of milk and cheese leads to an accumulation of lactose in the intestine. The flow of fluid into the intestine leads to swelling of the intestine with all of its attendant discomforts.

Galactose is converted to glucose by the successive action of *galactokinase, galactose 1-phosphate uridyl transferase,* and *UDP-galactose 4'-epimerase* (Figure 17.26).

Galactokinase catalyzes the synthesis of galactose 1-phosphate from galactose and ATP (reaction 17.54).

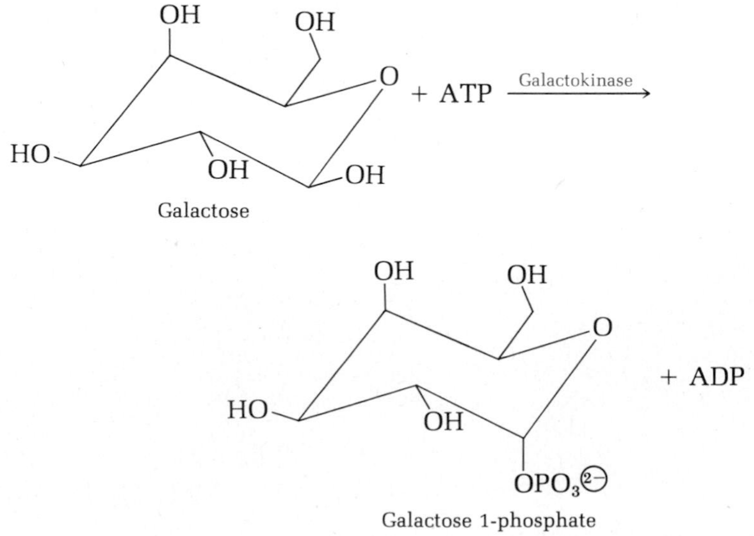

Galactose + ATP $\xrightarrow{\text{Galactokinase}}$

Galactose 1-phosphate + ADP (17.54)

Galactose 1-phosphate is converted to UDP-galactose by galactose 1-phosphate uridyl transferase (reaction 17.55).

galactose 1-phosphate + UDP-glucose $\rightleftharpoons$ UDP-galactose
+ glucose 1-phosphate (17.55)

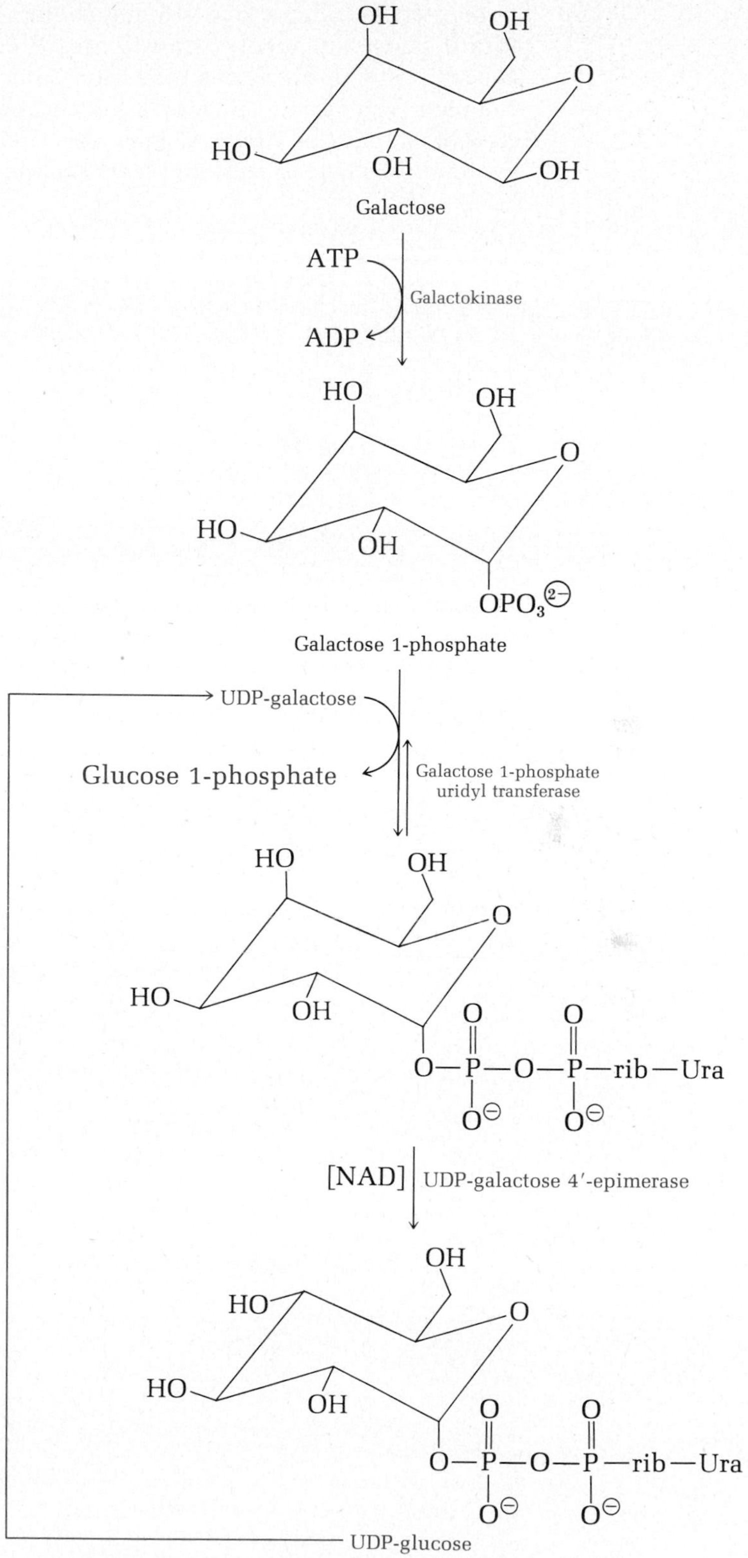

Figure 17.26 Entry of galactose into the mainstream of carbohydrate metabolism.

Some infants lack the ability to metabolize galactose because they lack galactose 1-phosphate uridyl transferase. This disorder, called *galactosemia*, causes liver damage which may lead to death. Severe mental retardation and cataracts occur in those who survive. Cataract formation is initiated by *dulcitol*, the reduced form of galactose. The effects of galactosemia can be avoided if lactose is excluded from the diet. This genetic defect, transmitted

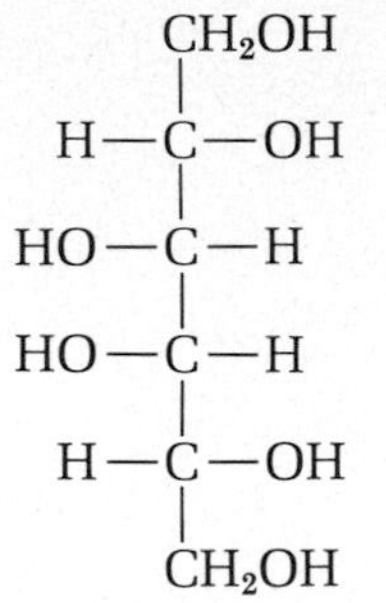

$$
\begin{array}{c}
CH_2OH \\
| \\
H-C-OH \\
| \\
HO-C-H \\
| \\
HO-C-H \\
| \\
H-C-OH \\
| \\
CH_2OH
\end{array}
$$

Galactitol (dulcitol)

as an autosomal recessive gene, can be detected at birth by assaying for the uridyl transferase in the red blood cells of the umbilical cord.

UDP-galactose 4'-epimerase interconverts UDP-glucose and UDP-galactose (reaction 17.56).

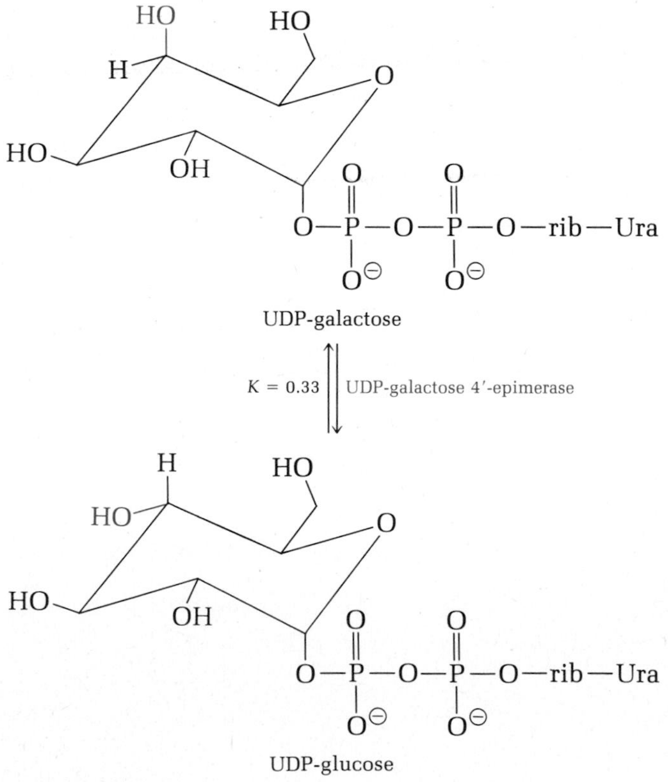

UDP-galactose

$K = 0.33$ UDP-galactose 4'-epimerase   (17.56)

UDP-glucose

The equilibrium constant for this reaction, [UDP-galactose]/[UDP-glucose], is 0.33.

NAD is required for epimerization, although no net oxidation occurs. A large body of evidence points to a mechanism in which a 4-keto intermediate is formed by transfer of hydride anion from C-4 to an enzyme-bound NAD. Reduction of the 4-keto group from the opposite side by the molecule of NAD that was originally reduced converts UDP-galactose to UDP-glucose.

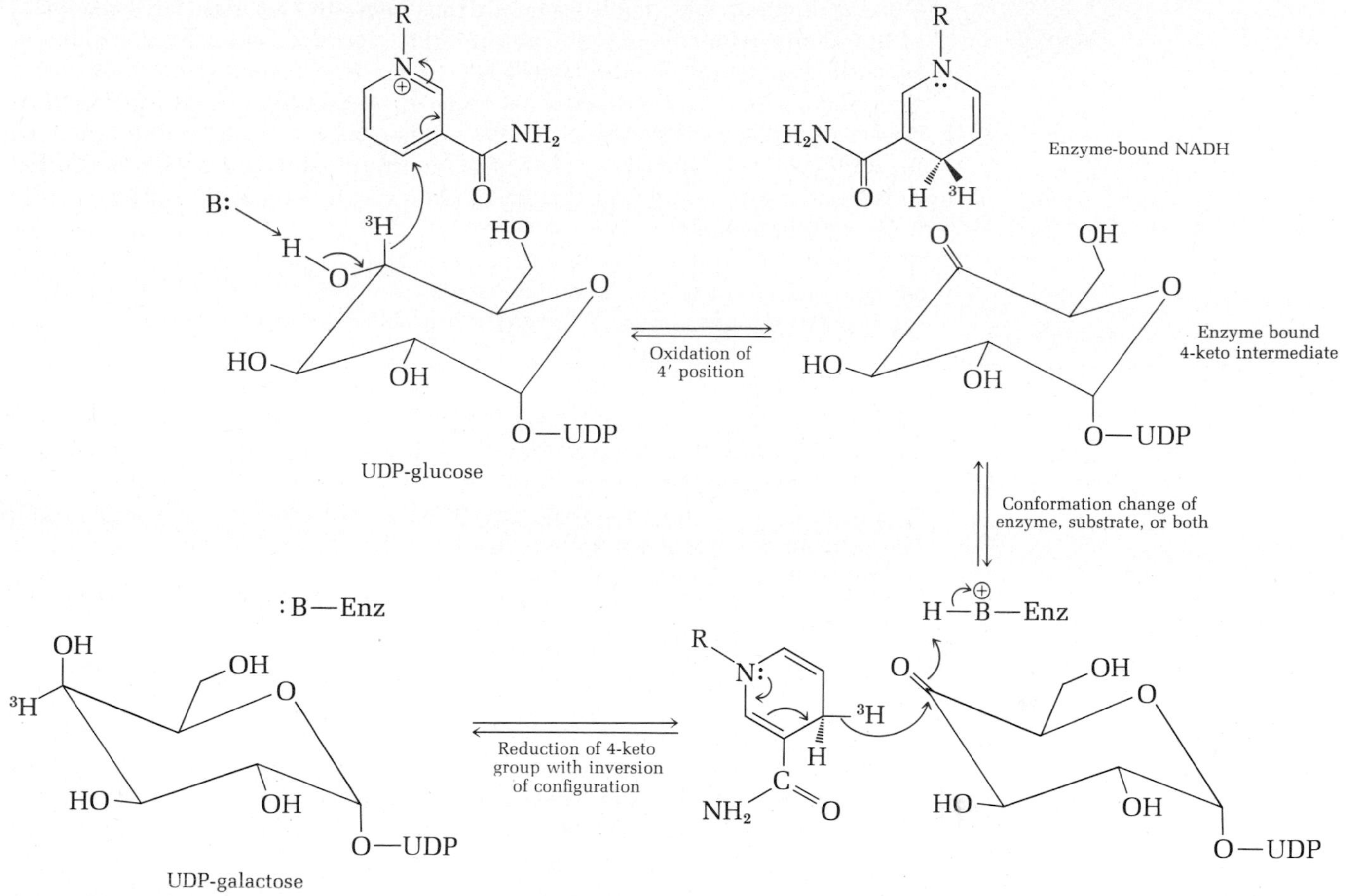

Figure 17.27 Mechanism of action of UDP-glucose 4′-epimerase.

The reduction step requires 180° rotation of the 4-keto intermediate at the active site or a large conformation change in the enzyme or both (Figure 17.27).

The net result of the action of galactokinase, galactose 1-phosphate uridyl transferase, and UDP-galactose 4′-epimerase is the conversion of galactose to glucose 1-phosphate. One mole of ATP is required for this reaction.

$$\text{galactose} + \text{ATP} \longrightarrow \text{glucose 1-phosphate} + \text{ADP} \qquad (17.57)$$

17–6 SUMMARY   The pentose phosphate pathway provides a detour around the reactions of glycolysis that lead to glyceraldehyde 3-phosphate. In the pentose phosphate pathway the biosynthetic reducing agent NADPH is produced as well as ribose 5-phosphate, a starting material in the biosynthesis of nucleotides and nucleic acids. The oxidative stage of the pentose phosphate pathway that produces NADPH consists of three enzymes: glucose 6-phosphate dehydrogenase, gluconolactonase, and 6-phosphogluconate dehydrogenase. If the cell requires about equal amounts of ribose 5-phosphate and NADPH, the net reaction of the pathway (including the action of phosphopentose isomerase) is

$$\text{glucose 6-phosphate} + 2\text{NADP} + \text{H}_2\text{O} \longrightarrow \text{ribose 5-phosphate}$$
$$+ 2\text{NADPH} + \text{CO}_2 + \text{H}^\oplus$$

Other combinations of cellular conditions bring the nonoxidative reactions of the pentose phosphate pathway into play. The successive action of transketolase, transaldolase, and transketolase results in conversion of ribose 5-phosphate into glucose 6-phosphate and glyceraldehyde 3-phosphate. If ribose 5-phosphate is not needed, but the demand for NADPH is high, the continuous cycling of the ribose 5-phosphate produced in the oxidative portion of the pathway through transketolase and transaldolase and back to fructose 6-phosphate leads to the complete oxidation of glucose 6-phosphate in a sequence of reactions whose stoichiometery is

$$\text{glucose 6-phosphate} + 12\text{NADP} + 7\text{H}_2\text{O} \longrightarrow 6\text{CO}_2 \\ + 12\text{NADPH} + 12\text{H}^{\oplus} + \text{P}_i$$

On the other hand, if ribose 5-phosphate is needed, but there is little demand for NADPH, the oxidative reactions of the pentose phosphate pathway are bypassed completely. Glucose 6-phosphate is converted to glyceraldehyde 3-phosphate and the reversible reactions catalyzed by transketolase and transaldose convert fructose 6-phosphate and glyceraldehyde 3-phosphate to ribose 5-phosphate. Since this set of reactions passes through fructose 1,6-diphosphate, ATP is required. The stoichiometry of the process is

$$\text{5 glucose 6-phosphate} + \text{ATP} \longrightarrow \text{6 ribose 5-phosphate} + \text{ADP} + \text{H}^{\oplus}$$

If more NADPH than ribose 5-phosphate is required, and if ATP is also required, then the oxidative portion of the pentose phosphate pathway generates NADPH, transketolase and transaldolase produce glyceraldehyde 3-phosphate, and glycolysis then converts glyceraldehyde 3-phosphate to pyruvate with net production of ATP. The net reaction for this process is

$$\text{3 glucose 6-phosphate} + 6\text{NADP} + 5\text{NAD} + 5\text{P}_i + 8\text{ADP} \\ \downarrow \\ \text{5 pyruvate} + 3\text{CO}_2 + 6\text{NADPH} + 8\text{ATP} + 2\text{H}_2\text{O} + 8\text{H}^{\oplus}$$

Gluconeogenesis provides a route for the synthesis of glucose from such metabolites as lactate, pyruvate, and glycerol. The net reaction of gluconeogenesis is

$$\text{2 pyruvate} + 4\text{ATP} + 6\text{GTP} + 2\text{NADH} + 2\text{H}_2\text{O} \\ \downarrow \\ \text{glucose} + 4\text{ADP} + 2\text{GDP} + 6P_i + 2\text{NAD}$$

Four unique enzymes are required for gluconeogenesis: the conversion of pyruvate to phosphoenolpyruvate requires the successive action of the biotin-dependent enzyme pyruvate carboxylase and of pyruvate carboxykinase. The pathway from phosphoenolpyruvate to fructose 1,6-diphosphate uses the enzymes of glycolysis. Fructose 1,6-diphosphatase, the third unique enzyme of gluconeogenesis, produces fructose 6-phosphate. In the final step of gluconeogenesis glucose 6-phosphatase cleaves the phosphate ester to give glucose. Muscle and brain do not contain glucose 6-phosphatase and are therefore incapable of gluconeogenesis, which occurs in both liver and kidney. Gluconeogenesis is regulated by metabolic control of fructose 1,6-diphosphatase. Fructose 1,6-diphosphatase is activated by citrate and inhibited by AMP; these metabolites exert exactly the opposite effects on phosphofructokinase. Therefore, glycolysis and gluconeogenesis are regulated at the same point: the interconversion of fructose 6-phosphate and fructose 1,6-diphosphate.

Glycogen degradation and glycogen synthesis occur by separate pathways. Three enzymes are directly responsible for glycogen degradation: phosphorylase a catalyzes pyrophosphorolysis of the $\alpha(1 \rightarrow 4)$ bonds of glycogen, stopping three glucosyl residues from a branch point. A three-unit fragment is transferred for a branch to the end of another chain by α-1,4 glucan: α-1,4 glucan 6-glucosyl transferase. The $\alpha(1 \rightarrow 6)$ bond is cleaved by amylo-α-1,6-glucosidase. Glycogen degradation is controlled by an enzyme cascade that interconverts active phosphorylase a and inactive phosphorylase b. The activation of phosphorylase b begins with the binding of epinephrine to a protein receptor on the outer surface of the plasma membrane. The hormone-receptor complex induces exchange of GDP for GTP on the G protein, which in turn leads to activation of adenylate cyclase. The cAMP produced by adenylate cyclase activates cAMP-dependent protein kinase. This protein phosphorylates glycogen synthetase, thereby stopping glycogen synthesis. The protein kinase also activates phosphorylate kinase, which in turn catalyzes the conversion of phosphorylase b to phosphorylase a. Phosphorylase a is the glucose receptor of liver cells. Phosphorylase phosphatase simultaneously inactivates phosphorylase a and activates glycogen synthetase b. The phosphatase binds to phosphorylase a, but once the binding of glucose has induced a conformation change in phosphorylase a, phosphorylase phosphatase hydrolyzes the phosphate ester, inactivating phosphorylase a, and then activates glycogen synthetase.

The biosynthesis of complex carbohydrates often occurs through activated nucleotide sugar intermediates. The biosynthesis of sucrose from UDP-glucose and fructose 6-phosphate is catalyzed by sucrose 6-phosphate synthetase and a phosphatase through the intermediate sucrose 6-phosphate. The biosynthesis of lactose is catalyzed by lactose synthetase in mammary gland cells during lactation when the galactosyl transferase subunit of the enzyme is bound to the regulatory subunit α-lactalbumin. UDP-galactose and UDP-glucose are interconverted by UDP-galactose 4'-epimerase, providing a pathway for galactose to enter the mainstream of glucose metabolism.

REFERENCES

Graves, D. J., and Wang, J. H., α-Glucan Phosphorylases—Chemical and Physical Basis of Catalysis and Regulation, in P. D. Boyer, Ed., "The Enzymes," Vol. 7, Academic Press, New York, 1972.

Greengard, P., Phosphorylated Proteins as Physiological Effectors, *Science*, 199, 146–152 (1978).

Hanson, R. W., and Mehlman, M. A., Eds., "Gluconeogenesis," Wiley-Interscience, New York, 1976.

Hers, H. G., The Control of Glycogen Metabolism in the Liver, *Ann. Rev. Biochem.*, 45, 167–189 (1976).

Horecker, B. L., in M. Florkin and E. H. Stotz, Eds., "Comprehensive Biochemistry," Vol. 15, Elsevier, New York, 1964.

Krebs, E. G., and Beavo, J. A., Phosphorylation and Dephosphorylation of Enzymes, *Ann. Rev. Biochem.*, 48, 923–959 (1979).

Moss, J., and Vaughan, M., "Activation of Adenylate Cyclase by Choleragen," *Ann. Rev. Biochem.*, 48, 581–600 (1979).

Pontremoli, S., and Grazi, E., in M. Florkin and E. H. Stotz, Eds., "Comprehensive Biochemistry," Vol. 17, Elsevier, New York, 1969.

Scrutton, M. C., and Young, M. R., Pyruvate Carboxylase, in P. D. Boyer, Ed., "The Enzymes," Vol. 6, Academic Press, New York, 1972.

Soling, H. D., and Williams, B., Eds., "Regulation of Gluconeogenesis," Academic Press, New York, 1971.

Stanbury, J. B., Wyngaarden, J. B., and Frederickson, D. S., Eds., "The Metabolic Basis of Inherited Disease," McGraw-Hill, New York, 1972.

Tsolas, O., and Horecker, B. L., Transaldolases, in P. D. Boyer, Ed., "The Enzymes," Vol. 7, Academic Press, New York, 1972.

Walsh, D. A., and Krebs, E. G., "Protein Kinases," in P. D. Boyer, Ed., "The Enzymes," Vol. 8, Academic Press, New York, 1973.

Problems

1. What is the effect of each of the following on the rates of glycogen synthesis and glycogen degradation: (a) Increasing the concentration of Ca (II) ions, (b) increasing the adenylate energy charge, (c) inhibiting ade-

nylate cyclase, (d) increasing the concentration of epinephrine, (e) increasing the concentration of AMP, (f) increasing the concentration of glucose 6-phosphate?

2. What are the enzymes responsible for the unique reactions of gluconeogenesis? How will increasing the adenylate energy charge affect each of these enzymes?

3. How are the allosteric effectors of phosphofructokinase releated to the allosteric effectors of fructose 1,6-diphosphatase?

4. What metabolic conversions would be inhibited if the formation of N^1-carboxybiotin is inhibited.

5. What are the mechanistic similarities of transaldolase and aldolase?

6. Which enzymes couple glycolysis to the pentose phosphate pathway?

7. Trace the fate of ^{18}O-*labeled in the 3 position* of ribulose-5-phosphate through the pentose phosphate pathway to glucose 6-phosphate.

8. What will be the fate of glucose 6-phosphate under each of the following metabolic conditions: (a) more NADPH than ribose 5-phosphate is required by the cell, (b) more ribose 5-phosphate than NADPH is required by the cell, (c) the need for ribose 5-phosphate and NADPH is balanced?

9. The antimalarial drug primaquine causes a decrease in glutathione in erythrocytes. How does this side effect influence each of the following: (a) the activity of glucose 6-phosphate dehydrogenase, (b) the NAD/NADPH ratio, (c) the rate of the pentose phosphate pathway, (d) the average lifetime of red blood cells?

10. Glucose 6-phosphate dehydrogenase catalyzes direct hydride transfer between glucose 6-phosphate and NADP. Propose a mechanism that is consistent with this observation and suggest an experiment that would support your mechanism.

11. The equilibrium constant for the oxidation of glucose 6-phosphate with NADP as hydride acceptor is 1.2. What structural property of 6-phosphoglucono-1,5-lactone makes it susceptible to facile reduction?

12. Guanosine $5'$-[β,γ-imido]triphosphate cannot be hydrolyzed to GTP by the G protein. What effect will the addition of this molecule have on the concentration of cAMP?

$$\text{GuO}-\overset{\overset{\displaystyle O}{\|}}{\underset{\underset{\displaystyle O_\ominus}{|}}{P}}-O-\overset{\overset{\displaystyle O}{\|}}{\underset{\underset{\displaystyle O_\ominus}{|}}{P}}-\overset{}{\underset{\underset{\displaystyle H}{|}}{N}}-\overset{\overset{\displaystyle O}{\|}}{\underset{\underset{\displaystyle O_\ominus}{|}}{P}}-O^\ominus$$

Guanosine $5'$-[β,γ-imido]triphosphate (Gpp[NH]p)

Chapter 18

Photosynthesis

18-1 INTRODUCTION

Though the ether is filled with vibrations, the world is dark. But one day man opens his seeing eye and there is light.

Ludwig Wittgenstein

The sun is the ultimate source of the free energy that drives all of the processes of living cells. The radiant energy of the sun is captured and converted to chemical energy by photosynthesis. The flow of carbon through the biosphere begins with photosynthesis. Organisms capable of photosynthesis produce carbohydrates, molecular oxygen from carbon dioxide and water, ATP and NADPH. The carbohydrates produced by photosynthesis serve as the energy source for other, nonphotosynthetic organisms. Carbohydrates are recycled to carbon dioxide and water by the combined action of glycolysis, the citric acid cycle, and respiration, which completes the cycle by reducing molecular oxygen to water as NADH is oxidized. The rising sun, whose energy powers this cycle, is therefore the symbol of continuously renewed life.

18–2 DIVISION OF LABOR IN PHOTOSYNTHESIS: LIGHT AND DARK REACTIONS AND THE TWO PHOTOSYSTEMS

The net reaction of photosynthesis is simplicity itself (reaction 18.1).

$$6CO_2 + 6H_2O \xrightarrow{\text{Photosynthesis}} (CH_2O)_6 + 6O_2 \tag{18.1}$$

The "fixation" of carbon dioxide in carbohydrates consists of two sets of reactions. The *light reactions of photosynthesis* are initiated by absorption of light by specialized pigments in the plant cell. These reactions generate NADPH and ATP and lead to the formation of molecular oxygen with water as the oxygen source. The general process, not written as a balanced reaction, is shown in reaction 18.2.

$$H_2O + NADP + HPO_4{}^{2\ominus} + ADP \xrightarrow{\text{Light}} O_2 + ATP + NADPH \tag{18.2}$$

NADPH and ATP provide "reducing power" and a source of free energy to drive a *reductive pentose phosphate pathway* that leads to synthesis of glucose from carbon dioxide. These reactions do not require light, though we shall see that they are regulated by light, and they are known as the *dark reactions of photosynthesis* (summarized in reaction 18.3).

$$CO_2 + NADPH + H^{\oplus} + ATP \xrightarrow{\text{Dark reactions}}$$
$$\text{glucose} + NADP + ADP + HPO_4{}^{2\ominus} \tag{18.3}$$

Both the light and the dark reactions of photosynthesis occur in *chloroplasts*, specialized, membrane-bound cell organelles.

The light reactions of photosynthesis are divided between *photosystem I* and *photosystem II*. Carbon dioxide is reduced when it is incorporated in carbohydrate. The reducing agent, NADPH, is generated by photosystem I. Since NADP has been reduced, something has to be oxidized, and one of the components of photosystem I is converted to a weak oxidizing agent when NADPH is produced. The weak oxidizing agent generated by photosystem I interacts with a weak reducing agent generated by photosystem II. The mild oxidizing agent produced by photosystem I accepts electrons from the mild reducing agent generated by photosystem II. This transfer of electrons occurs through an electron transport chain that is coupled to the synthesis of ATP. The *photoelectron transport chain* resembles the mitochondrial electron transport chain responsible for oxidative phosphorylation. Photosystem II also produces an extremely powerful oxidizing agent that is able to oxidize the oxygen atom of water to molecular oxygen (Figure 18.1). Notice the symmetry between the two photosystems: photosystem I generates a strong reducing agent, NADPH, and a weak oxidizing agent; photosystem II generates a powerful oxidizing agent and a weak reducing agent. The net result of the action of photosystem I is twofold: it passes electrons to the weak oxidizing agent generated by photosystem I, and it causes *photolysis* of water.

18–3 STRUCTURE OF CHLOROPLASTS

Photosynthesis in algae and green plants occurs in specialized cell organelles called *chloroplasts*. The chloroplasts are roughly ellipsoidal organelles having dimensions of about 5 μm $\times$ 10 μm. The alga *Micromonas* contains a single chloroplast, and higher plants contain from 100 to 150 chloroplasts per cell.

The chloroplast has an extensive and complex network of membranes (Figures 18.2 and 18.3). The chloroplast is bounded by a double membrane. The outer membrane of the chloroplast is freely permeable to many substances. However, the inner membrane is impermeable to most substances, and so a large number of transport proteins are required to carry metabo-

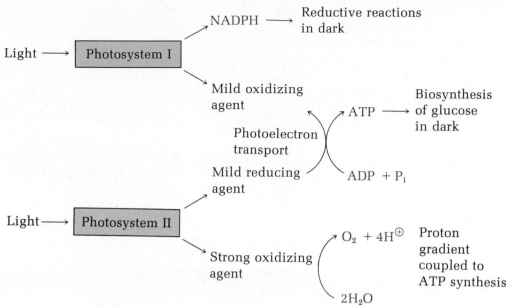

Figure 18.1 Photosystems I and II. The two photosystems of higher plants have a three-fold effect: (1) NADPH is generated by photosystem I for use in the dark reactions of photosynthesis. (2) Photoelectron transport between photosystems I and II generates ATP. (3) Photosystem II photolyzes water, and the proton gradient that is generated by this process is coupled to ATP synthesis.

lites across it. The space between the outer and inner membranes is called the *intermembrane space*, and the space enclosed by the inner membrane is called the *stroma*. Yet a third set of membranes is found within the interior of the chloroplast. These internal membranes are stacked in substructures known as *grana* that are connected by thin *fret membranes*. The stacked membranes of the grana are known as *thylakoid membranes*, and the space enclosed by the thylakoid membranes is called the *thylakoid space*. Each granum contains from 2 or 3 to more than 100 thylakoid membranes, and each chloroplast contains 40 to 60 grana.

The thylakoid membranes contain the proteins responsible for the light reactions of photosynthesis as well as the proteins responsible for photophosphorylation of ADP. The thylakoid membrane is composed of about 40% *galactosylglycerides* and *digalactosylglycerides*. The sulfolipid *sulfoquinovosyldiglyceride* accounts for about 4% of the total membrane lipid (Figure 18.4). Phospholipids, the major lipid component of most membranes, account for only 10% of the lipid of thylakoid membranes.

Chloroplasts contain some of their own DNA, ribosomes, and the associated proteins required for protein synthesis. They are therefore somewhat autonomous, but they cannot synthesize the proteins required for all of their functions.

18–4 PHOTOSYNTHETIC PIGMENTS AND THEIR FUNCTION IN PHOTOSYNTHETIC REACTION CENTERS

The light-gathering apparatus of photosynthesis is a complex of lipids and proteins imbedded in the thylakoid membranes of chloroplasts or in the plasma membrane of photosynthetic bacteria. A group of light-absorbing, colored molecules, or *chromophores*, is intimately associated with the membrane-bound proteins responsible for photosynthesis.

The photosynthetic pigments are responsible for efficient capture of the sun's radiation. Since the ozone layer of the upper atmosphere blocks most of the ultraviolet radiation, most of the radiation that reaches the earth's surface lies in the visible and infrared regions of the electromagnetic spectrum

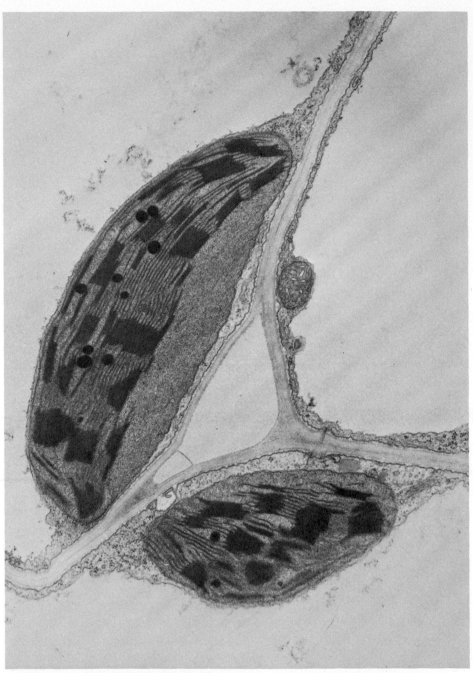

Figure 18.2 Electron micrograph of a sugar beet chloroplast. The grana are stacked membranes connected by fret membranes. The matrix space of the chloroplast is called the stroma. [Courtesy of Dr. W. M. Laetsch.]

(Figure 18.5). The photosynthetic pigments absorb electromagnetic energy over a range of frequencies that spans the visible region and extends into the infrared (Figure 18.6). *Chlorophyll a* is the major light-absorbing molecule in green plants, and *bacteriochlorophyll* is the major light-absorbing molecule of photosynthetic bacteria (Figure 18.7). These chlorophylls are closely associated with proteins. In photosystem I, for example, a protein having a molecular weight of 110,000 is bound to 14 molecules of chlorophyll a. The structure of the bacteriochlorophyll-protein complex of a green photosynthetic bacterium has been determined by x-ray crystallography. The protein complex is a trimer whose identical subunits have molecular weights of

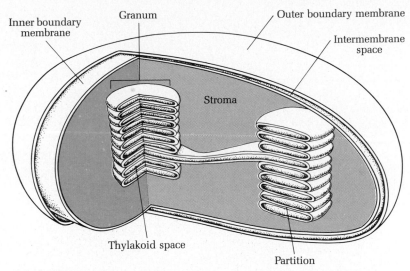

Figure 18.3 Diagram of the membrane structures of the chloroplast. [From S. L. Wolfe, "Biology of the Cell," 2nd ed., Wadsworth, Belmont, Calif., 1981, p. 130. © 1981 by Wadsworth, Inc. Reprinted by permission of Wadsworth Publishing Company.]

50,000. Each monomer is bound to seven molecules of bacteriochlorophyll (Figure 18.8). The orientation of the bacteriochlorophylls and the distances between them permit transfer of energy from one excited state chromophore to another, a process known as *exciton transfer*.[1]

Accessory pigments, including the *carotenoids* and the *phycobolins* (Figure 18.9), act as *antennae* that absorb light and transmit energy to chlorophyll *a* or to bacteriochlorophyll. The combination of primary acceptors and accessory pigments, all as part of multiprotein complexes, has been compared to the arrays of optical telescopes in which light is collected by each and the information sent to a central computer for processing and integration. We may envision a photosynthetic unit as such an integration center. The light gathered by the accessory pigments, which include antennae molecules of chlorophyll *a*, is transmitted by exciton transfer to a *photoreaction center*. The accessory chlorophylls far outnumber the few chlorophylls at a reaction center. In photosystems I and II the ratio of accessory chlorophylls to reaction center chlorophylls is about 200:1. The antenna chlorophylls are such efficient gatherers of light that they can capture a single photon of appropriate energy. Their excited states lie at higher energies than the excited state of the primary reaction center. Energy thus flows from the antenna chlorophylls to the reaction center.

The components of a reaction center for a photosynthetic bacterium have been isolated from the accessory pigments. The bacterial reaction center contains four bacteriochlorophyll molecules, two molecules of *bacteriopheophytin* in which the Mg(II) ion has been replaced by two protons, three proteins having molecular weights of 21,000, 23,000, and 28,000, two molecules of ubiquinone, and one atom of iron (Figure 18.10). This isolated reaction center is able to carry out the light reactions of bacterial photosynthesis. The more complex reaction centers have not yet been purified in a form that contains no accessory pigments, but the general principle of photoreaction centers also applies to photosystems I and II in higher plants.

---

[1] An expanded view of these chlorophyll molecules is shown on the cover of the text.

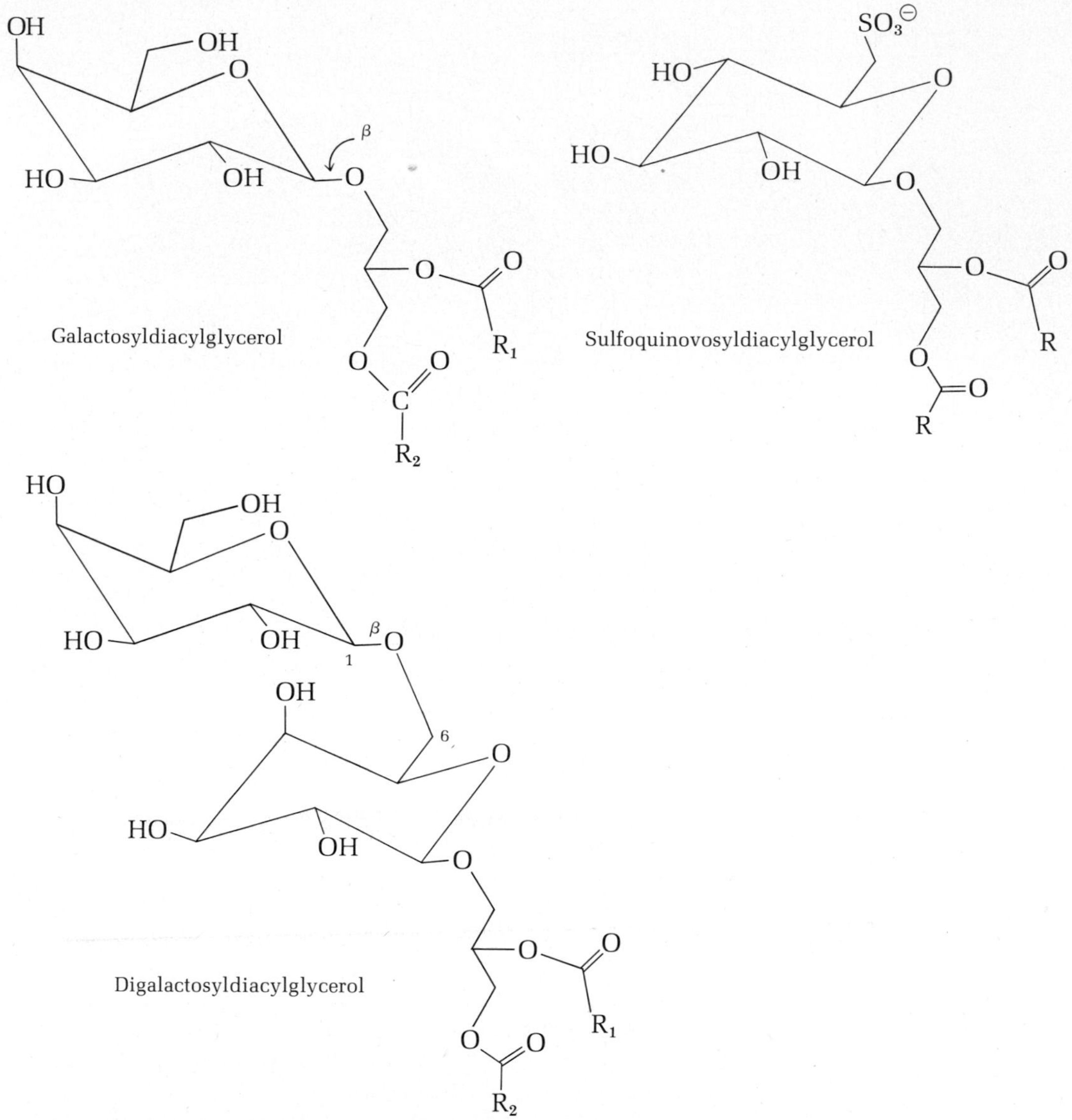

Galactosyldiacylglycerol

Sulfoquinovosyldiacylglycerol

Digalactosyldiacylglycerol

Figure 18.4   Lipid composition of the thylakoid membrane. The principle components of thylakoid membranes, other than phospholipids, are galactosylglycerides, digalactosyl-glycerides, digalactosylglycerides, and sulfoquinovosylglycerides.

18–5 PHOTOSYSTEMS I AND II

Photosynthesis in plants and algae requires the cooperation of two distinct molecular assemblies known as photosystem I and photosystem II. The combined action of these two systems accomplishes three tasks: (1) an oxygen atom in water in the -2 oxidation state is oxidized to molecular oxygen, (2) NADPH is produced, and (3) ATP is produced. Before considering the details of the two photosystems, let us examine some general features of each of these three processes.

One of the most fundamental features of photosynthesis is photolysis of water (reaction 18.4).

$$CO_2 + 2H_2O \xrightarrow{\text{Light}} (CH_2O) + O_2 + H_2O \tag{18.4}$$

We have chosen this stoichiometry to illustrate the parallel between bacterial photosynthesis and photosynthesis in higher plants. Photosynthetic bacteria are strictly anaerobic, and they use molecules such as H_2S as oxidizable substrates (reaction 18.5).

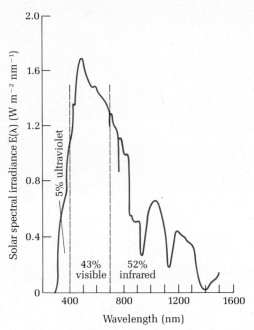

Figure 18.5 Distribution of energy from solar radiation at the earth's surface. [From J. R. Bolton, *Science*, *202*, 706 (1978). Copyright 1978 by the American Association for the Advancement of Science.]

$$CO_2 + 2H_2S \xrightarrow{\text{Light}} (CH_2) + 2S + H_2O \tag{18.5}$$

C. Van Neil noticed the similarity between bacterial and plant photosynthesis, and proposed a general scheme in which a hydrogen acceptor, CO_2, receives a hydride anion, or its equivalent, from a donor, H_2A. The donor is oxidized by the transfer of hydride anion to the hydrogen acceptor according to the general equation

$$\underset{\substack{\text{Hydrogen}\\\text{acceptor}}}{CO_2} + \underset{\substack{\text{Hydride}\\\text{donor}}}{2H_2A} \xrightarrow{\text{Light}} \underset{\substack{\text{Reduced}\\\text{acceptor}}}{(CH_2O)} + \underset{\substack{\text{Oxidized}\\\text{donor}}}{2A} + H_2O \tag{18.6}$$

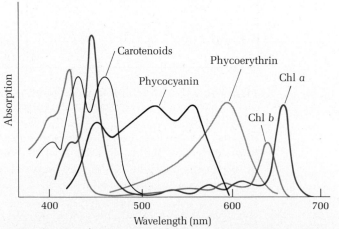

Figure 18.6 Absorption spectra of the major photosynthetic pigments. [From G. Govindjee and R. Govindjee, *Sci. Amer.*, December 1974, p. 132. Copyright © 1974 by Scientific American, Inc. All rights reserved.]

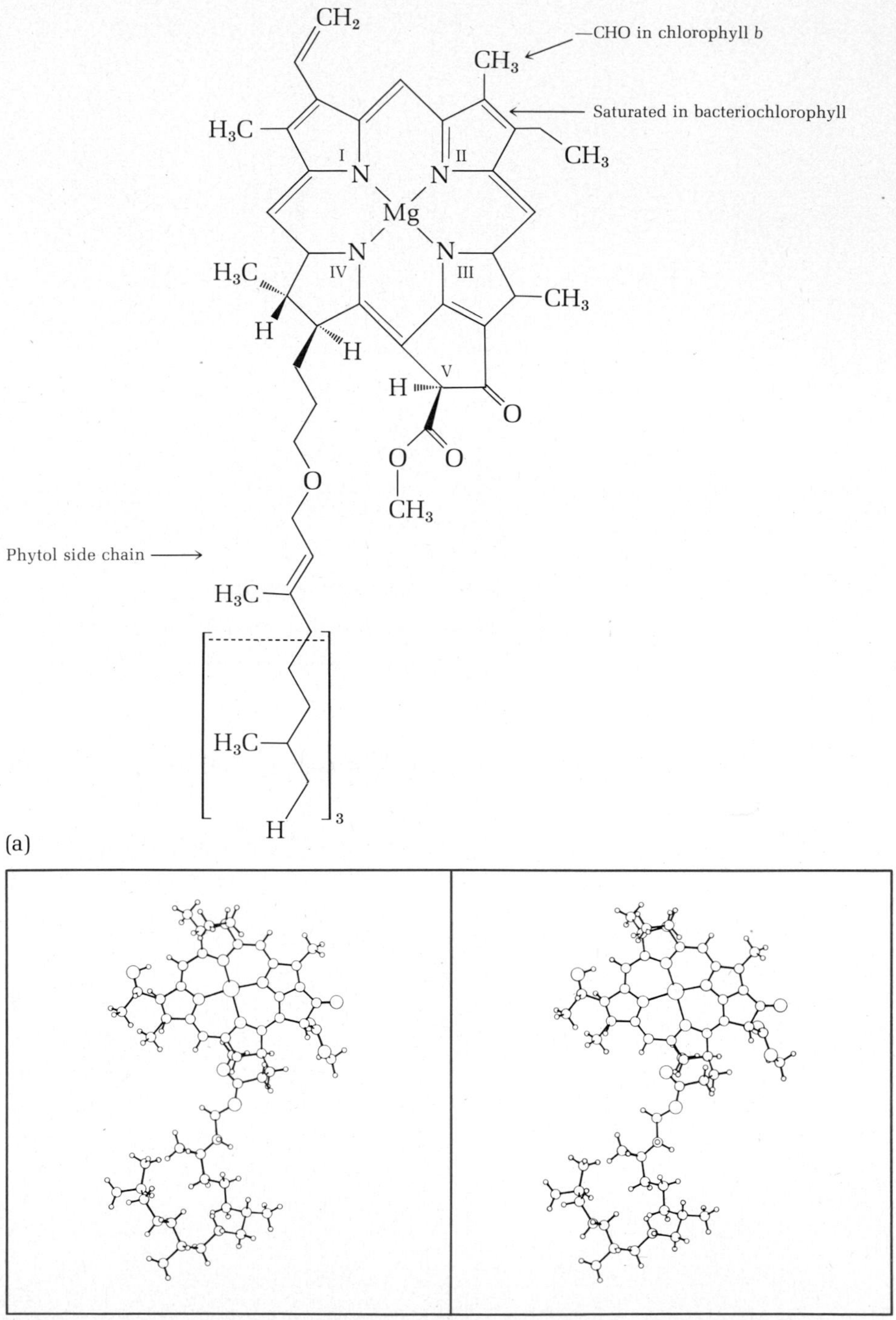

—CHO in chlorophyll *b*

Saturated in bacteriochlorophyll

Phytol side chain ⟶

(a)

(b)

Figure 18.7   (a) Structures of chlorophyll *a*, chlorophyll *b*, and bacteriochlorophyll. (b) Stereo view of bacteriochlorophyll.

Photosynthesis thus consists of coupled redox processes whose driving force is provided by the sun's radiation. If water is actually photolyzed during photosynthesis, then ^{18}O-labeled water should be oxidized to $^{18}O_2$. This is in fact observed (reaction 18.7).

$$CO_2 + H_2^{18}O \xrightarrow{\text{Light}} (CH_2O) + {}^{18}O_2 \qquad (18.7)$$

The isotopic labeling experiment excludes CO_2 as the source of molecular oxygen.

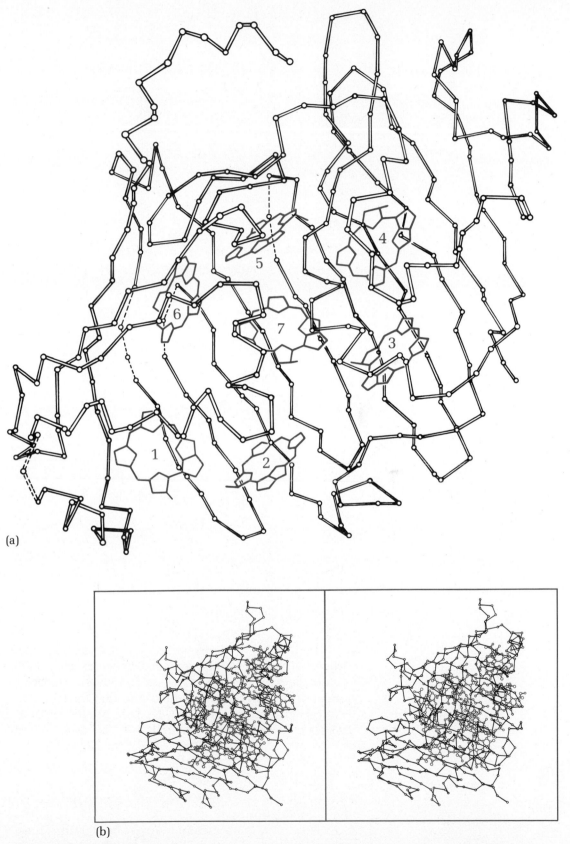

(a)

(b)

Figure 18.8 (a) Polypeptide backbone of bacteriochlorophyll protein and its seven bound molecules of bacteriochlorophyll. [From R. E. Fenna and B. W. Matthews, *Nature, 258,* 574 (1975).
Reprinted by permission. Copyright © 1975 Macmillan Journals Limited.] (b) Stereo view of
bacteriochlorophyll.

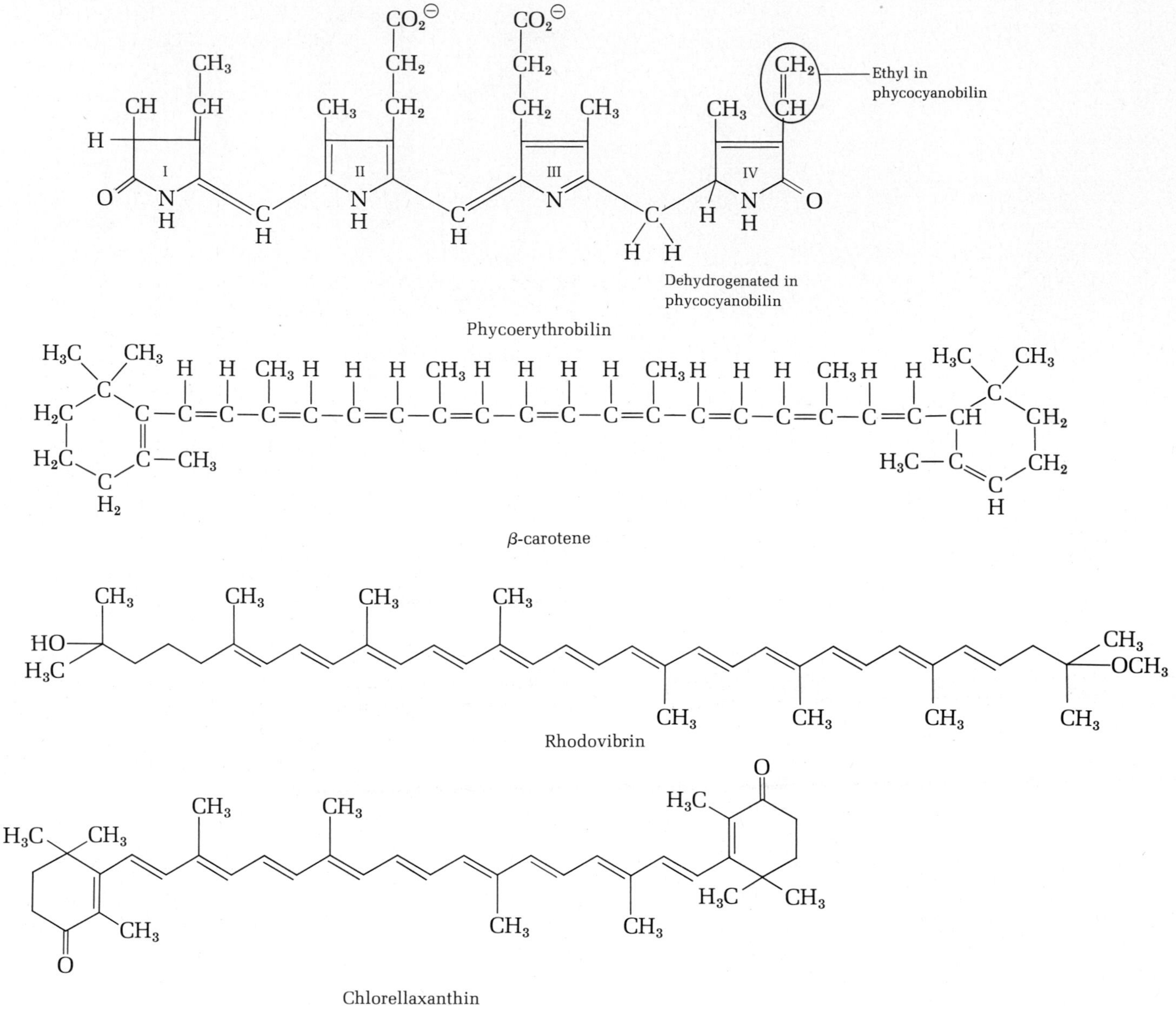

Ethyl in phycocyanobilin
Dehydrogenated in phycocyanobilin
Phycoerythrobilin
β-carotene
Rhodovibrin
Chlorellaxanthin

Figure 18.9 Structures of some accessory photosynthetic pigments.

In the general reaction of photosynthesis (equation 18.6) CO_2 is the hydrogen acceptor. If this general scheme is correct, then other molecules should be able to replace CO_2 as the electron acceptor. This was demonstrated in 1939 by R. Hill, who found that ferricyanide, $[Fe(CN)_6]^{3-}$, can be reduced to ferrocyanide, $[Fe(CN)_6]^{4-}$, by chloroplasts that are deprived of CO_2, a process known as the *Hill reaction* (reaction 18.8).

$$4[Fe(CN)_6]^{3-} + 2H_2O \xrightarrow{\text{Light}} [Fe(CN)_6]^{4-} + O_2 + 4H^{+} \qquad (18.8)$$

In the Hill reaction the iron atom of ferricyanide is reduced from the +3 to the +2 oxidation state.

Now, CO_2 is ultimately reduced by photosynthesis, but this reduction occurs through an intermediary. *NADPH is the primary reduced product of the light reactions of photosynthesis.* The photolysis of water is accomplished by photosystem II, and the reduction of NADPH is accomplished by photosystem I.

The third product of photosynthesis is ATP. ATP is not produced directly, but is the result of coupling electron flow through the carriers of the two systems to phosphorylation of ADP. As in the case of mitochondrial oxidative phosphorylation, the formation of ATP is mediated by a proton gradient. The chemiosmotic hypothesis therefore applies to photophosphorylation in chloroplasts.

With this introduction to the two photosystems, let us now turn to the molecular details of the light reactions of photosystems I and II.

A. Reduction of NADP by Photosystem I

The absorption of a photon by the reaction center of photosystem I initiates a series of electron transfer reactions that culminate in reduction of NADPH (Figure 18.11). Photosystem I is activated by the absorption of red light at a wavelength of 700 nm by a photoreaction center pigment, often abbreviated as P700. This pigment is a unique molecule of chlorophyll a surrounded by some 200 molecules of antennae chlorophyll a in the photosynthetic reaction center. Absorption of a photon by P700 promotes it to an excited state. The standard reduction potential ($E^{0\prime}$) of the ground state of P700 is $+0.4$ volts (V), but the standard reduction potential of the excited state of P700 is -0.6 V (reaction 18.9).

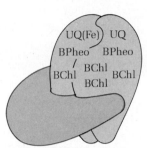

Figure 18.10 Hypothetical model of the photoreaction center of photosynthetic bacteria. The model shows the association of four molecules of bacteriochlorophyll (BChl), two molecules of bacteriopheophytin (BPheo), and ubiquinone (UQ), all bound in a complex with three peptides and an iron ion (Fe) [From K. Sauer, *Acc. Chem. Res.*, 7, 261 (1978).]

$$\underset{\substack{\text{Ground state}\\E^{0\prime} = +0.4\text{ V}}}{\text{Chl }a} + hV\ (700\text{ nm}) \longrightarrow \underset{\substack{\text{Excited state}\\E^{0\prime} = -0.6\text{ V}}}{(\text{Chl }a)^*} \tag{18.9}$$

We shall see that the photochemical pumping of a chlorophyll molecule to an excited state species that is a powerful reducing agent is the key to all that follows. (It is rather easy to become confused by the terms oxidizing agent and reducing agent within the context of the standard reduction potentials of the molecules in question. If a molecule can readily donate an electron to another species, it will have a negative standard reduction potential. When it loses an electron, it becoms oxidized.) A molecule of *bound ferredoxin* imbedded in the thylakoid membrane is reduced by the excited state P700. The bound ferredoxin has a molecular weight of 11,500 and contains an Fe_4S_4 center. Its absorbance maximum is 430 nm, and it is abbreviated P430. The reduction of P430 by the excited state chlorophyll is shown in reaction 18.10. We will consider the reduction of $(\text{Chl }a)^{\oplus}$ shortly.

$$\underset{\substack{\text{Strong}\\\text{reducing}\\\text{agent}}}{(\text{Chl }a)^*} + [\text{P430}]_{\text{ox}} \longrightarrow (\text{Chl }a)^{\oplus} + [\text{P430}]_{\text{red}} \tag{18.10}$$

Photosystem I also contains a molecule of *soluble ferredoxin* that is bound to the FAD-containing enzyme *ferredoxin-NADP-reductase*. The transfer of electrons in photosystem I is from bound ferredoxin to soluble ferredoxin to the FAD bound to the reductase to NADP (reaction 18.11).

$$\text{P700}^* \longrightarrow \text{P430} \longrightarrow \text{soluble ferredoxin} \longrightarrow$$
$$\text{FAD-enzyme} \longrightarrow \text{NADP} \tag{18.11}$$

The components of photosystem I are summarized in Table 18.1. This table lists three components of photosystem I that we have yet to discuss—cytochrome c_{552} (formerly known as cytochrome f), cytochrome b_{563} (formerly known as cytochrome b_6), and the protein *plastocyanin*. We shall soon discover that these proteins are involved both in cyclic electron flow in photosystem I (Section 18.5C) and that they couple photosystem I to photosystem II.

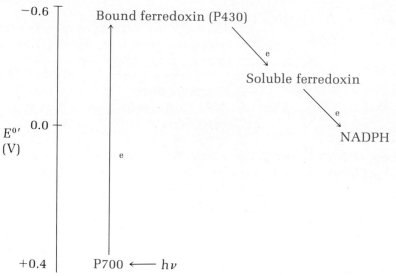

Figure 18.11 Electron flow from the photoreaction center chlorophyll a (P700) to NADPH.

B. Photolysis of Water and Reduction of Photosystem I by Photosystem II

Photosystem II is more complex than photosystem I, and it is less well understood. Photosystem II oxidizes water, producing O_2, and passes electrons to the oxidized chlorophyll a, $(Chl\ a)^{\oplus}$, that is created by the redox reactions of photosystem I (Figure 18.12).

The photoreaction center of photosystem II has an absorption maximum of 680 nm. This center contains about 200 chlorophyll a molecules, one of which uniquely participates in the actual photochemical reaction of photosystem II. The others are all antenna chlorophylls. The unique chlorophyll, abbreviated Chl a_{II}, absorbs an electron and is promoted to an excited state (reaction 18.12).

$$\underset{\text{Ground state}}{Chl\ a_{II}} + h\nu\ (680\ \text{nm}) \longrightarrow \underset{\text{Excited state}}{(Chl\ a_{II})^*} \qquad (18.12)$$

The excited state chlorophyll rapidly donates an electron to an acceptor, called Q because it *quenches* the fluorescence of $(Chl\ a_{II})^*$ (reaction 18.13).

$$(Chl\ a_{II})^* + Q \longrightarrow (Chl\ a_{II})^{\oplus} + Q^{\ominus} \qquad (18.13)$$

The identity of Q has not been unambiguously determined, but the weight of evidence strongly suggests that it is *plastoquinone*, a molecule whose struc-

Table 18.1 Components of photosystem I and their functions	
Component	**Function**
Antenna chlorophylls a and b	Light absorption
Carotenoids	Light absorption
Reaction center chlorophyll a (P700)	Primary photochemical event
Cytochrome c_{552} (f)	Cyclic electron flow; coupling of photosystem I to photosystem II
Cytochrome b_{563} (b_6)	Cyclic electron flow; coupling of photosystem I to photosystem II
Bound ferredoxin (P430)	Primary electron acceptor in photosystem I
Soluble ferredoxin	Electron acceptor from P430
Ferredoxin-NADP-reductase	Reduction of NADP

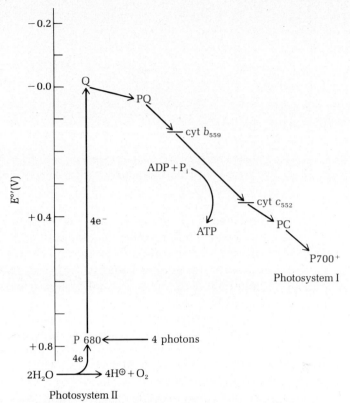

Figure 18.12 The oxidation-reduction process catalyzed by photosystem II involves photolysis of water and reduction of photosystem I. Four electrons are released by the oxidation of water, and all flow to photosystem I. (ATP production by photophosphorylation is discussed in Section 18.5D.)

ture closely resembles the structure of ubiquinone. It is well established that plastoquinone is a secondary acceptor of electrons in photosystem II. The

Plastoquinone

plastoquinone anion, $Q^{\ominus}$, is a weak reducing agent ($E^{0\prime}_{Q^{\ominus} \to Q} = +0.18$ V) that passes electrons through an electron transport chain consisting of plastoquinone, cytochrome b_{559}, cytochrome c_{552}, plastocyanin, and $P700^{\oplus}$.

Plastocyanin is a copper-containing protein whose molecular weight is 11,500. Its Cu(II) ion is bound to the protein by the imidazole rings of histidyl residues 57 and 37, by the sulfhydryl group of cysteine residue 84, and by the sulfur of methionine residue 92. The usual structure of tetracoordinate Cu(II) is a distorted tetrahedron. This is also its structure in plastocyanin. The reduction of tetracoordinate Cu(II) to Cu(I) gives a square planar species (Figure 18.13). The stereochemistry of reduction therefore involves only small atomic motions. According to the *principle of least motion*, those elementary reactions are favored that involve the least change in atomic position

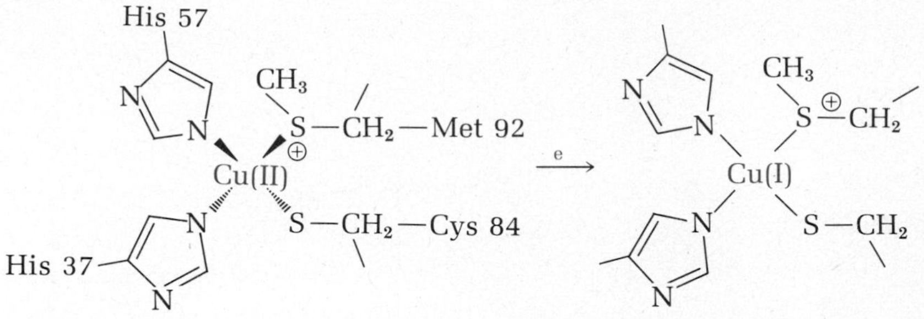

Figure 18.13 Tetrahedral Cu(II) ion in plastocyanin reduced to the square planar Cu(I) product in electron transport from photosystem II to photosystem I.

and electronic configuration. The evolutionary selection of copper as the ion in plastocyanin results in a system in which the principle of least motion applies. The principle of least motion confers a kinetic advantage upon the reduction that acts in combination with the thermodynamic advantage of a favorable reduction potential to give a molecule that undergoes facile redox reactions.

The oxidized photoreaction center of photosystem I is produced when P700* passes an electron to a ferredoxin (reaction 18.10). We may regard the cation P700$^\oplus$ as an electron "hole," analogous to the holes in a p-type semiconductor, that is "filled" by an electron ejected from the excited photoreaction center of photosystem II. The oxidized chlorophyll, (Chl a_{II})$^\oplus$, is also an electron hole. It is also an extremely powerful oxidizing agent. Its standard reduction potential has not been measured with great accuracy, but it is at least $+0.85$ V. Since the standard reduction potential of water is about $+0.82$ V, the oxidized photoreaction center can extract an electron from water. Oxidation of water appears to occur in four steps since four photons must be absorbed by photosystem II before water is oxidized (reaction 18.14).

$$2H_2O + 4 \text{ (Chl } a_{II})^\oplus \longrightarrow O_2 + 4H^\oplus + 4 \text{ Chl } a_{II} \tag{18.14}$$

Photosystem II contains 6 Mn(II) ions that are involved in the actual oxidation of water. The mechanism of oxidation of water is extremely complex and is not very well understood.

Since the four electrons released by photosystem II ultimately find their way to NADP, we can write the net reaction of the two photosystems as

$$2H_2O + 2NADP \longrightarrow O_2 + 2NADPH + 2H^\oplus \tag{18.15}$$

The protons released by the photosynthetic electron transport chain establish a gradient across the thylakoid membrane whose potential energy is coupled to the phosphorylation of ADP in a process called photophosphorylation, a topic we shall consider in Section 18.5D.

C. Cyclic Electron Flow in Photosystem I

We have seen that the cation P700$^\oplus$ generated in photosystem I by electron transfer to ferredoxin can be reduced by an electron transport chain that connects photosystem I to photosystem II. The electrons are produced by the oxidation of water. Photosystem I can also reduce itself by cyclic electron flow from bound ferredoxin back to the oxidized photoreaction center P700$^\oplus$ (Figure 18.14). A proton gradient is generated by cyclic electron flow that is coupled to ATP synthesis and the pathway of electron flow is therefore called *cyclic photophosphorylation.* Since this pathway is completely independent of photosystem II, no oxygen is evolved. NADPH inhibits

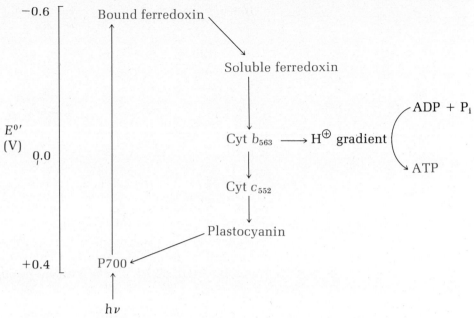

Figure 18.14 Electron flow in cyclic photophosphorylation. The proton gradient generated by electron transport is coupled to ATP synthesis.

ferredoxin-NADPH-reductase, and cyclic photophosphorylation occurs when the NADPH/NADP ratio is high. Under these conditions, ATP is generated in the absence of NADPH production. The combined systems of electron flow in photosystem I to NADPH, from photosystem II to photosystem I, and cyclic electron flow are summarized in Figure 18.15.

The sequence of electron carriers within and between photosystems II and I has been painstakingly established by the use of artificial electron donors and acceptors of electron transport. It is possible to use these agents

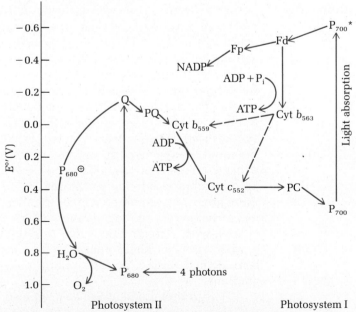

Figure 18.15 Electron transport in photosynthesis.
Pathways of electron flow that are not certain are indicated by a dotted line. Abbreviations: Fd, ferredoxin; Fp, flavoprotein; PC, plastocyanin; PQ, plastoquinone.

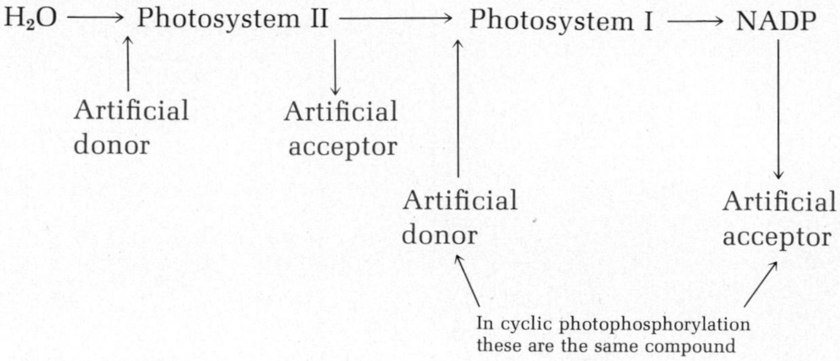

$$H_2O \longrightarrow \text{Photosystem II} \longrightarrow \text{Photosystem I} \longrightarrow \text{NADP}$$

Artificial donor

Artificial acceptor

Artificial donor

Artificial acceptor

In cyclic photophosphorylation these are the same compound

Figure 18.16 Use of artificial donors and acceptors to study portions of electron transport chain of photosystems I and II. [Adapted from A. Trebst, *Ann. Rev. Plant Physiol.*, 25, 424 (1974).]

to study any portion of the electron transport chain (Figure 18.16). One of the most widely used inhibitors is the herbicide *dichlorophenyldimethylurea* (DCMU), which blocks electron transport between photosystems II and

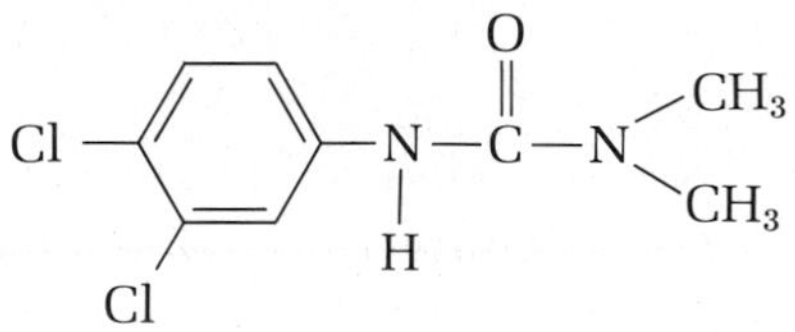

Dichlorophenyldimethylurea (DCMU)
[3-(3,4-dichlorophenyl-)-1,1-dimethylurea]

I by inhibiting transfer of electron from Q to plastoquinone. Ascorbate is one of the most widely used electron acceptors and ferricyanide one of the most commonly used electron donors (recall the Hill reaction).

Let us now turn to the details of ATP production in photosynthesis.

D. Mechanism of Photophosphorylation

Perhaps the strongest early evidence in favor of the chemiosmotic hypothesis (Section 16.7) was derived not from mitochondria, but from chloroplasts. In 1966 A. Jagendorf and his coworkers showed that chloroplasts could synthesize ATP in the dark if an artificial pH gradient was applied across the thylakoid membrane. The experiment itself was simple. First, chloroplasts were suspended in a pH 4.0 buffer and equilibrated in the dark for several hours. The acidified chloroplasts were then added to a buffer at pH 8.0 that contained ADP and phosphate. The fact that ATP was produced under these conditions indicates that (1) the photochemical events of photosynthesis are independent of phosphorylation of ADP and (2) the membrane potential is capable of driving the synthesis of ATP. This is what would be expected if the chemiosmotic hypothesis is correct (Figure 18.17). The thylakoid membrane is freely permeable to $Cl^{\ominus}$ and $Mg^{2\oplus}$ ions (unlike the inner mitochondrial membrane which is impermeable to these ions). Protons can be carried across the membrane with chloride ions or two protons can exchange with a single $Mg^{2\oplus}$ ion. In either case the formation of the proton gradient is electrically neutral. The electrochemical gradient is therefore zero, or nearly zero, and the proton motive force depends only on the pH gradient.

The proton gradient can be produced by any of three processes: (1) photolysis of water releases protons, (2) electron transport between photosystem II and photosystem I generates a proton gradient, or (3) cyclic electron flow in photosystem I generates a proton gradient. In chloroplasts the internal thylakoid space becomes more acidic than the external space. The free en-

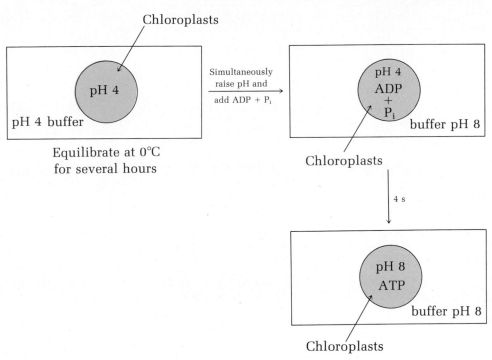

Figure 18.17 Synthesis of ATP in the dark. Chloroplasts can synthesize ATP in the dark if they are provided with a pH gradient and a supply of ADP and orthophosphate (P_i).

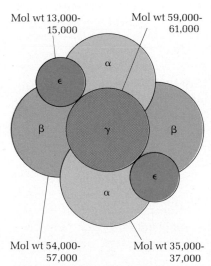

Figure 18.18 Proposed model for the structure of chloroplast coupling factor CF_1. The functions of the subunits have not been completely elucidated, but the α, β, and γ subunits have been implicated in ATP synthesis. The δ subunit is involved in binding CF_0 to CF_1, and the small ϵ subunit is an inhibitor of the ATPase activity of CF_1. [From R. E. McCarty, *Ann. Rev., Plant Physiol.*, *30*, 85 (1979). Reproduced with permission. © 1979 by Annual Reviews Inc.]

ergy required for ATP synthesis is derived by electron flow down the concentration gradient from the stroma to the thylakoid matrix space. A gradient of about 3.5 pH units is formed by electron transport. Between three and four protons are ejected from the stroma per molecule of ATP produced.

The coupling agent for phosphorylation of ADP in chloroplasts is remarkably similar to mitochondrial coupling factor F_1. Coupling factor CF_1 (the C differentiates it from its mitochondrial counterpart) is bound to the thylakoid membrane facing the stroma. CF_1 has ATPase activity when detached from the membrane, and it synthesizes ATP when bound to sealed vesicles. It is an oligomer of molecular weight 325,000. The oligomer is composed of five subunits whose stoichiometry is $\alpha_2\beta_2\gamma_1\delta_1\epsilon_2$ (Figure 18.18). A second coupling factor called CF_0 interacts with CF_1. CF_0 (the o indicates that it is sensitive to oligomycin) contains three types of subunits. One of them, a small polypeptide having a molecular weight of about 8000, apparently provides the channel by which protons traverse the thylakoid membrane.

It appears that the components of the electron transport chains of photosynthesis are asymmetrically distributed across the membrane. A detailed diagram of the relations of these proteins is not yet possible, but the general features of the flow of protons across the thylakoid membrane and the orientation of the ATPase in the thylakoid membrane are shown schematically in Figure 18.19. Although this scheme is provisional, it seems likely that a zig-zag arrangement of electron and proton carriers is required to generate a proton gradient. In short, a *vectorial* arrangement of the components of photoelectron transport is required since the thylakoid space becomes acidic. It is known that the electron acceptors of both photosystems are on the stroma side of the thylakoid membrane and that the electron donors of photosystem I are on the inner side of the thylakoid membrane. The ATPase, coupling factor CF_1, and ferredoxin-NADP-reductase are both on the stroma side of the thylakoid membrane. The entire picture is strikingly similar to the asym-

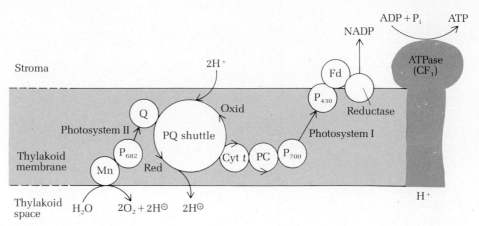

Figure 18.19 Hypothetical scheme of electron transport in which the thylakoid space becomes acidic as electrons flow from water to NADP.

metric distribution of electron and proton carriers in the inner mitochondrial membrane.

The ATP synthesized by coupling factor CF_1 is released into the stroma, as is the NADPH produced by photosystem I. The stroma is the site of the enzymatic reactions of the dark reactions of photosynthesis, which we will discuss in the next section.

18–6 FIXATION OF CARBON DIOXIDE BY THE DARK REACTIONS

The incorporation of carbon dioxide in carbohydrates, a process known as *fixation*, is essential for plant growth. The only known pathway for *net* incorporation of carbon dioxide is the *Calvin cycle* (Figure 18.20). As may be seen from the stoichiometry of carbon dioxide fixation, the incorporation of 15 moles of CO_2 in the pentose ribulose 1,5-diphosphate can be converted to 18 moles of ribulose 1,5-diphosphate. The number of moles of CO_2 acceptor thus increases. Cleavage of the hexose formed by incorporating CO_2 in ribulose 1,5-diphosphate provides triose phosphates that are eventually incorporated in starch and sucrose. The pool of ribulose 1,5-diphosphate steadily increases, like a snowball rolling downhill. Although small pieces continually split off, the snowball continues to grow by picking up more snow. "Snowball growth" is *autocatalytic*, the farther the ball rolls, the bigger it becomes. The *Calvin cycle* is a molecular phenomenon of the same sort; it, too, is autocatalytic. The Calvin cycle may be conveniently discussed in two phases. In the first phase ribulose 1,5-diphosphate becomes covalently bound to carbon dioxide and then undergoes a set of reactions that produce dihydroxyacetone phosphate. This triose may either become incorporated in carbohydrate or be converted to ribulose 1,5-diphosphate by the reactions of the second phase of the Calvin cycle.

A. First Phase of the Calvin Cycle: Conversion of Ribulose 1,5-Diphosphate to Dihydroxyacetone Phosphate

The advance of science has always been closely tied to advances in technology. Two unrelated discoveries provided the tools by which Calvin and his coworkers unraveled the pathway by which plants incorporate CO_2 in carbohydrates. The first was the discovery of the radioisotope of carbon ^{14}C. The half-life of ^{14}C is 5700 years. The previously used isotope of carbon, ^{11}C, has a half-life of only 21 minutes and was therefore much less conve- as a tracer. Another advantage of ^{14}C over ^{11}C is the smaller amount of energy released by its radioactive decay. ^{11}C can decay either by capturing an elec- electron from the innermost valence shell (K capture) or by emitting a posi- tron ($\beta^{\oplus}$) having an energy of 1.98 MeV (reaction 18.12).

$$^{11}C \xrightarrow{t_{1/2} = 20.3 \text{ min}} {}_5B^{11} + \beta^{\oplus}$$

(18.12)

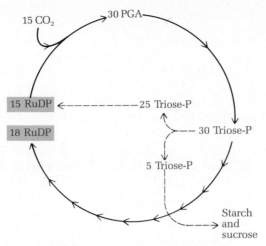

Figure 18.20 Incorporation of CO_2 in carbohydrates in the Calvin cycle. The incorporation of 15 moles of CO_2 in ribulose 1,5-diphosphate (RuDP) leads to 30 moles of triose phosphate (triose-P) in the first phase of the Calvin cycle. The trioses are glyceraldehyde 3-phosphate and dihydroxyacetone phosphate. These trioses can be partitioned with 5 moles incorporated in carbohydrate and 25 moles reconverted to ribulose 1,5-diphosphate. As an alternative, all 30 moles of triose can be converted to 18 moles of ribulose 1,5-diphosphate. [From G. J. Kelly, E. Latzko, and M. Gibbs, *Ann. Rev. Plant Physiol., 27*, 183 (1976). Reproduced with permission. © 1976 by Annual Reviews Inc.]

Radioactive decay of the long-lived ^{14}C occurs by emission of an electron (a $\beta^{\ominus}$ particle) with an energy change of 0.156 MeV (reaction 18.13).

$$^{14}C \xrightarrow{t_{1/2}\ =\ 5,730\ \text{yr}} {}_7N^{14} + \beta^{\ominus} \tag{18.13}$$

Since the energy released is so much less, ^{14}C is much safer to use.

The second technical advance seems almost prosaic, but its importance can scarcely by exaggerated. In 1944 R. Consden, A. H. Gordon, and A. J. P. Martin invented paper chromatography and applied it to separate mixtures of amino acids. M. Calvin, A. A. Benson, J. A. Basham, and many others then applied radioisotopic labeling and paper chromatography to identify the intermediates in the metabolic pathway of carbon dioxide fixation. In the first experiments suspensions of the unicellular green alga *Chlorella* were exposed to $^{14}CO_2$ for periods of 5 s to 90 s. The cells were then dropped in hot ethanol, and the ethanol extract was applied to a sheet of chromatography paper. After two-dimensional paper chromatography, the paper was exposed to a photographic plate. Development of the plate gave spots at all locations in which radioactive carbon was present. After exposure of a plant to $^{14}CO_2$ for 5 s, one compound was formed: 3-phosphoglycerate. If cells carried out photosynthesis for 90 s, a welter of spots was obtained (Figure 18.21). The identification of the compounds responsible for these spots and their arrangement in a cycle of metabolic reactions established the Calvin cycle.

The first phase of the Calvin cycle consists of five reactions in the chloroplast stroma (Figure 18.22). The first reaction is phosphoryl group transfer from ATP to ribulose 5-phosphate, catalyzed by *ribulose 5-phosphate kinase* (reaction 18.14).

$$
\begin{array}{c}
CH_2OH \\
| \\
C{=}O \\
| \\
H{-}C{-}OH \\
| \\
H{-}C{-}OH \\
| \\
CH_2OPO_3^{2-}
\end{array}
\;+\; ATP
\;\xrightarrow{\text{Ribulose 5-phosphate kinase}}\;
\begin{array}{c}
CH_2OPO_3^{2-} \\
| \\
C{=}O \\
| \\
H{-}C{-}OH \\
| \\
H{-}C{-}OH \\
| \\
CH_2OPO_3^{2-}
\end{array}
\;+\; ADP
\qquad (18.14)
$$

Ribulose 5-phosphate   Ribulose 1,5-diphosphate

The second reaction is incorporation of carbon dioxide in ribulose 1,5-diphosphate followed by cleavage of the 6-carbon product to give two moles of 3-phosphoglycerate. Both of these reactions are catalyzed by *ribulose 1,5-diphosphate carboxylase oxygenase* (reaction 18.15).

$$
\begin{array}{c}
CH_2OPO_3^{2-} \\
| \\
C{=}O \\
| \\
H{-}C{-}OH \\
| \\
H{-}C{-}OH \\
| \\
CH_2OPO_3^{2-}
\end{array}
\;+\; CO_2
\;\xrightarrow[\text{carboxylase oxygenase}]{\text{Ribulose 1,5-diphosphate}}\;
2\,\begin{array}{c}
CO_2^{-} \\
| \\
H{-}C{-}OH \\
| \\
CH_2OPO_3^{2-}
\end{array}
\qquad (18.15)
$$

Ribulose 1,5-diphosphate   3-Phosphoglycerate

Ribulose 1,5-diphosphate carboxylase oxygenase fixes atmospheric carbon dioxide into the metabolic intermediates leading to carbohydrate. The enzyme accounts for about 15% of the proteins in chloroplasts and is probably the most abundant protein on earth. It is composed of eight dimeric subunits. Each of the eight dimers consists of a catalytic subunit (mol wt 56,000) and a regulatory subunit (mol wt 14,000). A schematic diagram of

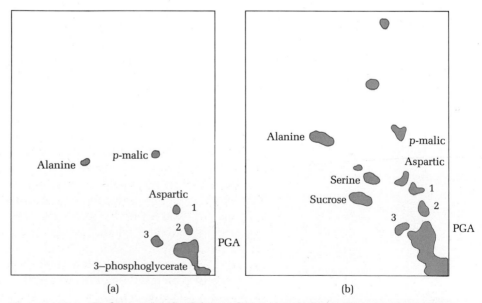

Figure 18.21   Radioautograph of a two-dimensional chromatogram of the photosynthetic intermediates that are produced by exposing the green alga *Chlorella* to $^{14}CO_2$ for (a) 5 seconds and (b) 90 seconds. The major product of the short incubation is 3-phosphoglycerate. The other intermediates of carbon fixation appear in the radiogram obtained after the 90-second incubation. [From M. Calvin and A. A. Benson, *Science*, *109*, 141 (1949). Copyright 1949 by the American Association for the Advancement of Science.]

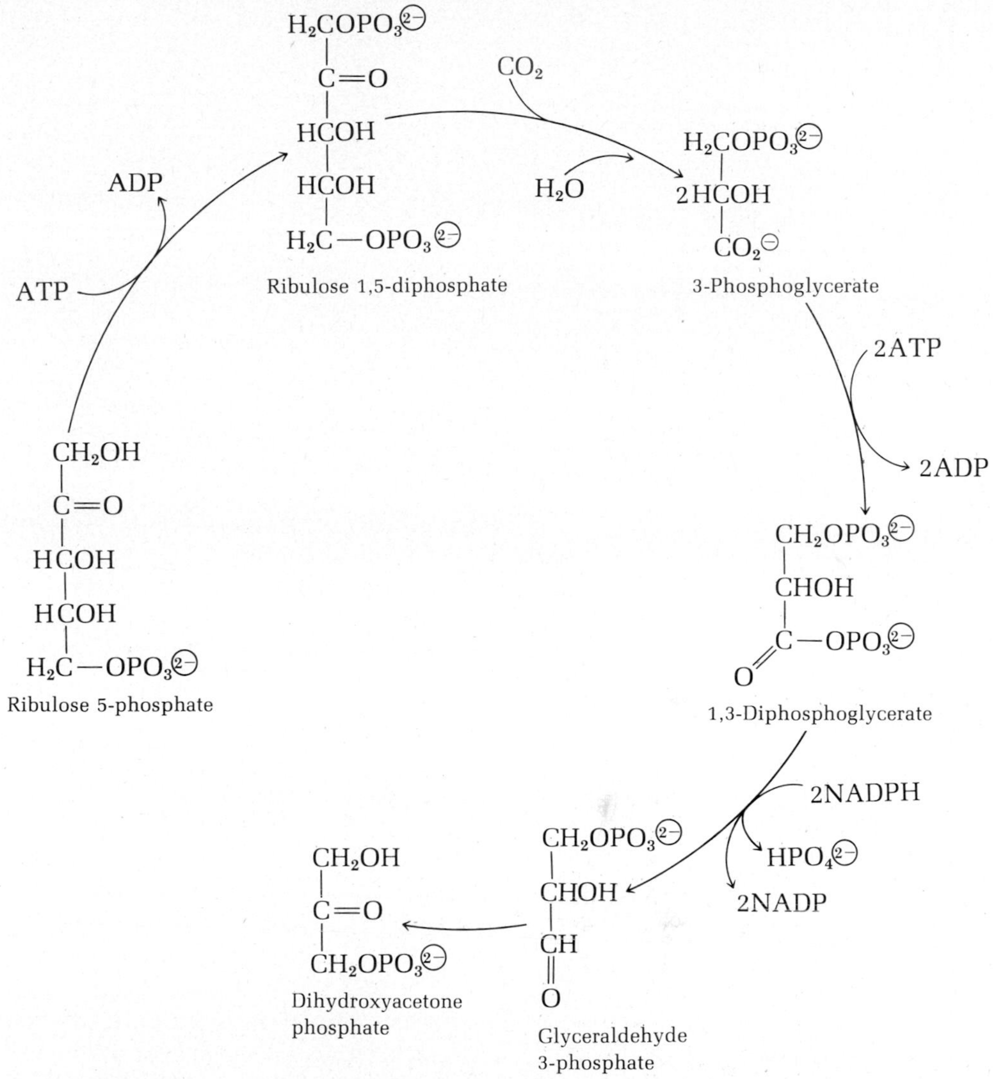

Figure 18.22 Five reactions of the first phase of the Calvin cycle. The first phase produces dihydroxyacetone phosphate. Part of it is converted to carbohydrate and part is used to regenerate ribose 5-phosphate in the second phase of the Calvin cycle.

the proposed arrangement of the subunits of ribulose 1,5-diphosphate carboxylase oxygenase is shown in Figure 18.23. Carbon dioxide itself is the substrate for the carboxylation reaction rather than carbonate or bicarbonate. The driving force for the net reaction is cleavage of the carbon-carbon bond of the initial carboxylated adduct which produces two resonance-stabilized carboxylate anions. The total free-energy change for synthesis of 3-phosphoglycerate from ribulose 1,5-diphosphate and carbon dioxide is -51.9 kJ mol^{-1}. The reverse reaction does not therefore occur to an appreciable extent. The rate-determining step in the formation of 3-phosphoglycerate is enolization of ribulose 1,5-diphosphate to an enediol. This intermediate attacks the electrophilic carbon of carbon dioxide giving a transient carboxylated intermediate. Attack by water on the carbonyl group of the β-keto acid followed by aldol cleavage produces two molecules of 3-phosphoglycerate (Figure 18.24).

In the third and fourth reactions phosphoglycerate kinase catalyzes the conversion of 3-phosphoglycerate to phosphate 1,3-diphosphoglycerate (reaction 18.16), and glyceraldehyde 3-phosphate dehydrogenase reduces the acyl phosphate to an aldehyde (reaction 18.17).

$$
\begin{array}{c}
\text{CO}_2^{\ominus} \\
| \\
\text{H}-\text{C}-\text{OH} \\
| \\
\text{CH}_2\text{OPO}_3^{\circled{2-}}
\end{array}
\quad + \text{ATP} \quad
\xrightarrow[\text{kinase}]{\text{Phosphoglycerate}}
\quad
\begin{array}{c}
\text{O} \\
\| \\
\text{C}-\text{OPO}_3^{\circled{2-}} \\
| \\
\text{H}-\text{C}-\text{OH} \\
| \\
\text{CH}_2\text{OPO}_3^{\circled{2-}}
\end{array}
\quad + \text{ADP} \qquad (18.16)
$$

3-Phosphoglycerate   1,3-Diphosphoglycerate

$$
\begin{array}{c}
\text{O} \\
\| \\
\text{C}-\text{OPO}_3^{\circled{2-}} \\
| \\
\text{H}-\text{C}-\text{OH} \\
| \\
\text{CH}_2\text{OPO}_3^{\circled{2-}}
\end{array}
\quad + \text{NADPH} \quad
\xrightarrow[\text{3-phosphate dehydrogenase}]{\text{Glyceraldehyde}}
\quad
\begin{array}{c}
\text{CHO} \\
| \\
\text{H}-\text{C}-\text{OH} \\
| \\
\text{CH}_2\text{OPO}_3^{\circled{2-}}
\end{array}
\quad + \text{NADP} + \text{HPO}_4^{\circled{2-}} \qquad (18.17)
$$

1,3-Diphosphoglycerate   Glyceraldehyde
3-phosphate

Net synthesis of glyceraldehyde 3-phosphate is favored in the chloroplast by the high phosphorylation potential, $[\text{ATP}]/[\text{ADP}][\text{P}_i]$, and by the high ratio of NADPH/NADP.

Glyceraldehyde 3-phosphate is converted to dihydroxyacetone phosphate in the fifth reaction of the first phase of the Calvin cycle (reaction 18.18).

$$
\begin{array}{c}
\text{CHO} \\
| \\
\text{H}-\text{C}-\text{OH} \\
| \\
\text{CH}_2\text{OPO}_3^{\circled{2-}}
\end{array}
\quad
\xrightarrow[\text{isomerase}]{\text{Triose phosphate}}
\quad
\begin{array}{c}
\text{CH}_2\text{OH} \\
| \\
\text{C}=\text{O} \\
| \\
\text{CH}_2\text{OPO}_3^{\circled{2-}}
\end{array}
\qquad (18.18)
$$

Glyceraldehyde   Dihydroxyacetone
3-phosphate   phosphate

We recall this reaction from glycolysis.

The fixation of carbon dioxide in the first phase of the Calvin cycle is

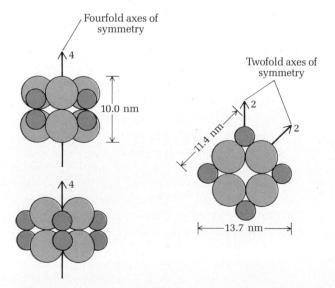

Figure 18.23 (a) Schematic diagram of the subunit composition of ribulose 1,5-diphosphate carboxylase. The arrows represent axes of symmetry. Rotation around the twofold axes generates the same structure. Rotation of 90° around the fourfold axes also generates the same structure. [From R. G. Jensen, *Ann. Rev. Plant Physiol.*, *28*, 383 (1977). Reprinted with permission. © 1977 by Annual Reviews Inc.] (b) Stereo view of the same schematic diagram (on page 725).

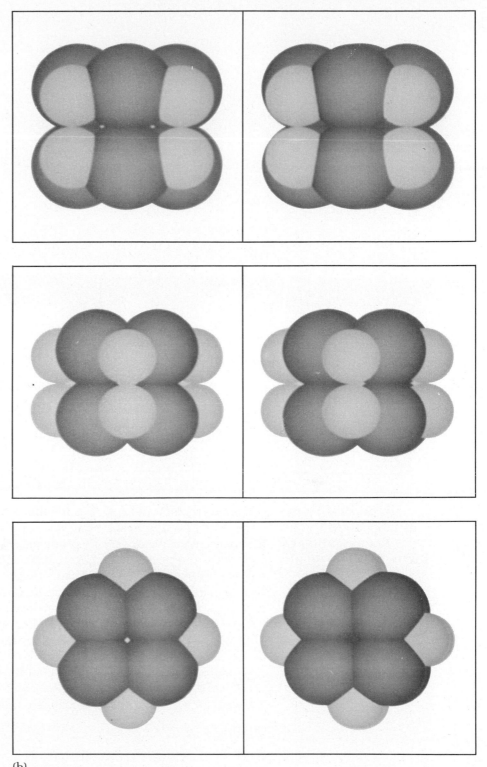

(b)

Figure 18.23 (continued)

energetically expensive, but the light reactions of photosynthesis provide a steady supply of NADPH and ATP. The net reaction for fixation of carbon dioxide is

$$CO_2 + \text{ribulose 5-phosphate} + 3ATP + 2NADPH \longrightarrow$$
$$2 \text{ dihydroxyacetone phosphate} \qquad (18.19)$$

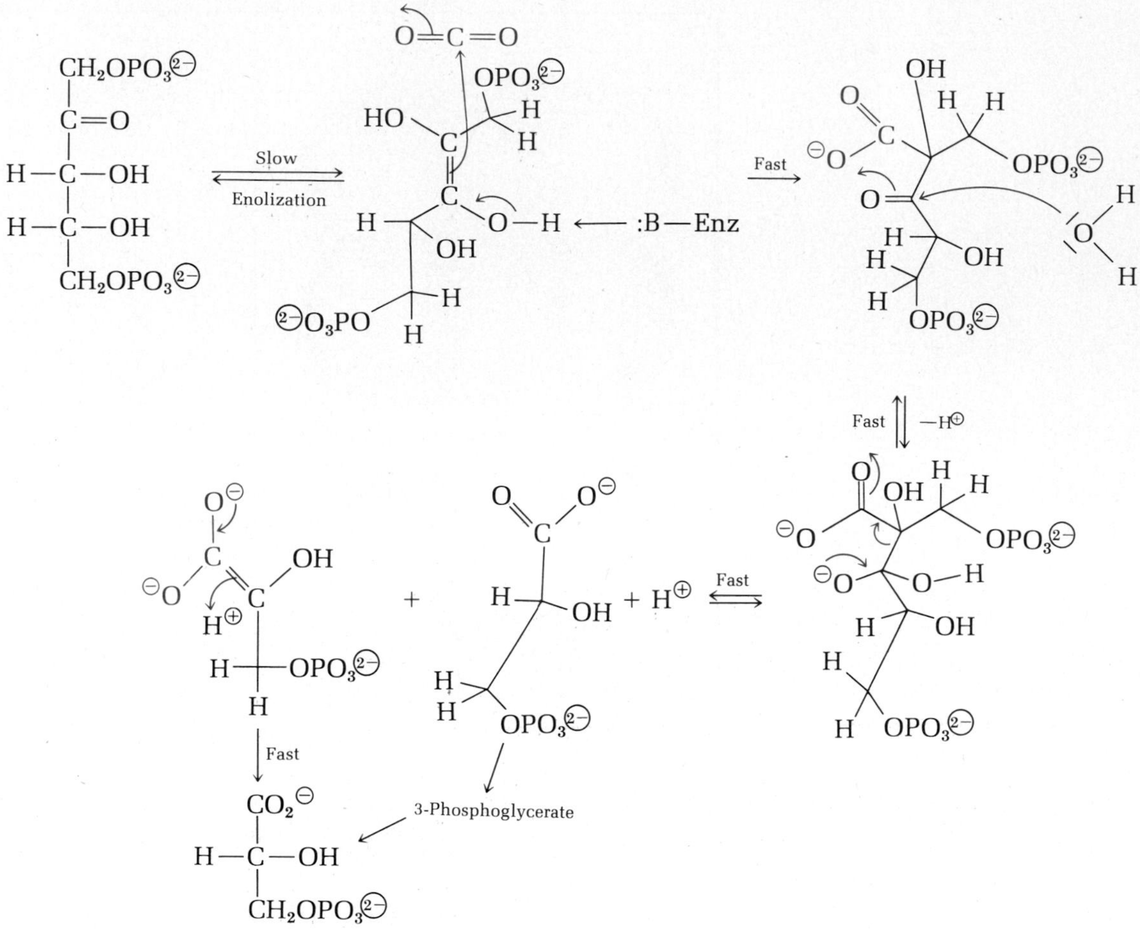

Figure 18.24   Possible mechanism for the action of ribulose 1,5-diphosphate carboxylase oxygenase.

B. Second Phase of the Calvin Cycle: Regeneration of Ribulose 5-Phosphate

Dihydroxyacetone phosphate can suffer two fates, as we noted earlier. For every 15 moles of ribose 5-phosphate that are carboxylated, 30 moles of dihydroxyacetone phosphate are produced. Five moles of triose can be directly converted to glucose by gluconeogenesis (Section 17.3). The remaining 25 moles of dihydroxyacetone phosphate are converted to 15 moles of ribose 5-phosphate in the second phase of the Calvin cycle. If no synthesis of glucose were to occur, the 30 moles of dihydroxyacetone phosphate could be converted to 18 moles of ribulose 5-phosphate. The reactions leading from dihydroxyacetone phosphate closely parallels the nonoxidative reactions of the pentose phosphate pathway (Section 17.2).

Transketolase (discussed in detail in Section 17.2B) catalyzes the conversion of glyceraldehyde 3-phosphate and fructose 6-phosphate to xylulose 5-phosphate and erythrose 4-phosphate (reaction 18.20).

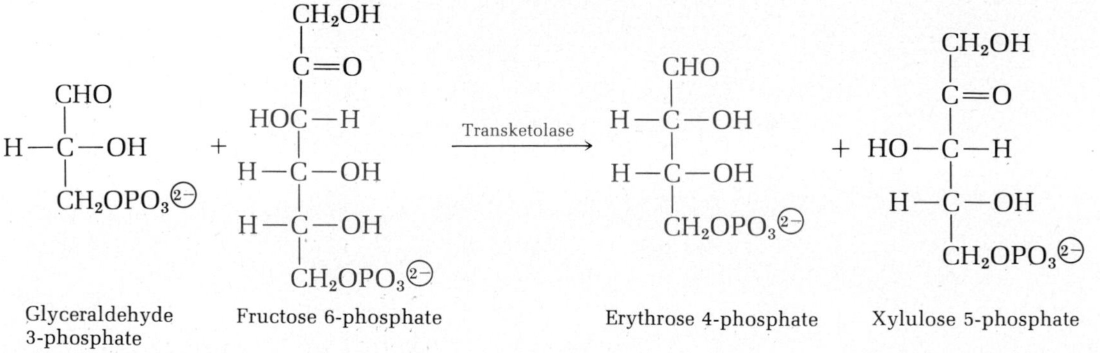

(18.20)

Glyceraldehyde 3-phosphate   Fructose 6-phosphate   Erythrose 4-phosphate   Xylulose 5-phosphate

The condensation of erythrose 4-phosphate and dihydroxyacetone phosphate, catalyzed by an aldolase, produces sedoheptulose 1,7-diphosphate (reaction 18.21). A phosphatase hydrolyzes the phosphate ester at C-1 producing sedoheptulose 7-phosphate (reaction 18.22).

$$
\begin{array}{c}
\text{CHO} \\
| \\
\text{H}-\text{C}-\text{OH} \\
| \\
\text{H}-\text{C}-\text{OH} \\
| \\
\text{CH}_2\text{OPO}_3^{2-}
\end{array}
\;+\;
\begin{array}{c}
\text{CH}_2\text{OH} \\
| \\
\text{C}=\text{O} \\
| \\
\text{CH}_2\text{OPO}_3^{2-}
\end{array}
\;\xrightarrow{\text{Aldolase}}
\qquad (18.21)
$$

Erythrose 4-phosphate   Dihydroxyacetone phosphate

$$
\begin{array}{c}
\text{CH}_2\text{OPO}_3^{2-} \\
| \\
\text{C}=\text{O} \\
| \\
\text{HO}-\text{C}-\text{H} \\
| \\
\text{H}-\text{C}-\text{OH} \\
| \\
\text{H}-\text{C}-\text{OH} \\
| \\
\text{H}-\text{C}-\text{OH} \\
| \\
\text{CH}_2\text{OPO}_3^{2-}
\end{array}
\;\xrightarrow{\text{Phosphatase}}
\begin{array}{c}
\text{CH}_2\text{OH} \\
| \\
\text{C}=\text{O} \\
| \\
\text{HO}-\text{C}-\text{H} \\
| \\
\text{H}-\text{C}-\text{OH} \\
| \\
\text{H}-\text{C}-\text{OH} \\
| \\
\text{H}-\text{C}-\text{OH} \\
| \\
\text{CH}_2\text{OPO}_3^{2-}
\end{array}
\qquad (18.22)
$$

Sedoheptulose 1,7-diphosphate   Sedoheptulose 7-phosphate

Transketolase also catalyzes the conversion of sedoheptulose 7-phosphate and glyceraldehyde 3-phosphate to ribose 5-phosphate and xylulose 5-phosphate (reaction 18.23).

$$
\begin{array}{c}
\text{CH}_2\text{OH} \\
| \\
\text{C}=\text{O} \\
| \\
\text{HO}-\text{C}-\text{H} \\
| \\
\text{H}-\text{C}-\text{OH} \\
| \\
\text{H}-\text{C}-\text{OH} \\
| \\
\text{H}-\text{C}-\text{OH} \\
| \\
\text{CH}_2\text{OPO}_3^{2-}
\end{array}
\;+\;
\begin{array}{c}
\text{CHO} \\
| \\
\text{H}-\text{C}-\text{OH} \\
| \\
\text{CH}_2\text{OPO}_3^{2-}
\end{array}
\;\xrightarrow{\text{Transketolase}}
$$

Sedoheptulose 7-phosphate   Glyceraldehyde 3-phosphate

$$
\begin{array}{c}
\text{CHO} \\
| \\
\text{H}-\text{C}-\text{OH} \\
| \\
\text{H}-\text{C}-\text{OH} \\
| \\
\text{H}-\text{C}-\text{OH} \\
| \\
\text{CH}_2\text{OPO}_3^{2-}
\end{array}
\;+\;
\begin{array}{c}
\text{CH}_2\text{OPO}_3^{2-} \\
| \\
\text{C}=\text{O} \\
| \\
\text{HO}-\text{C}-\text{H} \\
| \\
\text{H}-\text{C}-\text{OH} \\
| \\
\text{CH}_2\text{OPO}_3^{2-}
\end{array}
\qquad (18.23)
$$

Ribose 5-phosphate   Xylulose 5-phosphate

Ribose 5-phosphate is converted to ribulose 5-phosphate and xylulose 5-phosphate is converted to ribulose 5-phosphate by an isomerase and an epimerase, respectively, thus completing the reductive pentose phosphate pathway of photosynthesis (reaction 18.24). The entire process is summarized in Figure 18.25.

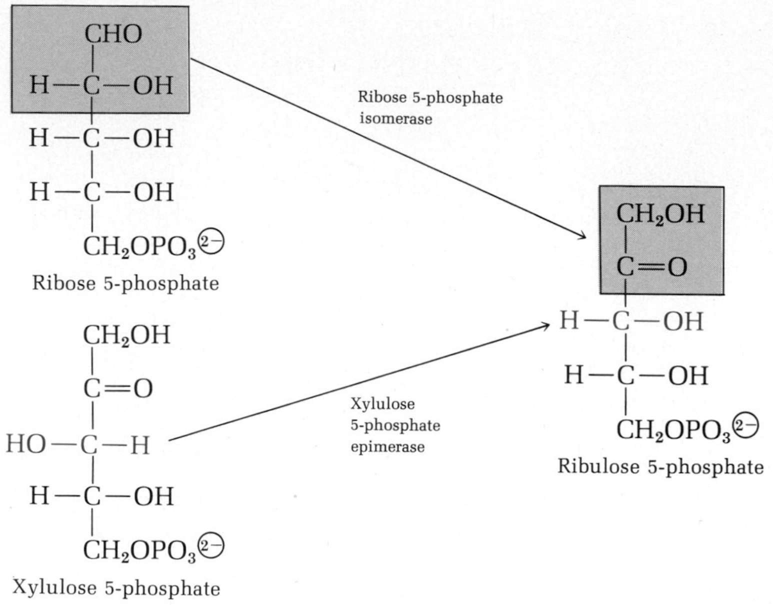

(18.24)

The net reaction of the Calvin cycle can be summarized in several ways. The formation of 2 moles of dihydroxyacetone phosphate from ribulose 5-phosphate and carbon dioxide requires 3 moles of ATP and 2 moles of NADPH (recall reaction 18.19). Since ribulose 5-phosphate is regenerated in each turn of the Calvin cycle, the net reaction for biosynthesis of glucose can be written without including ribulose 5-phosphate. That is glucose can be formed by incorporating 6 moles of CO_2 in ribulose 1,5-diphosphate in six turns of the Calvin cycle. After subtracting the recycled ribulose 1,5-diphosphate, we obtain reaction 18.25.

$$6CO_2 + 18ATP + 12NADPH + 12H_2O \longrightarrow$$
$$glucose + 18ADP + 18P_i + 12NADP \quad (18.25)$$

As we saw in reaction 18.19, 3 moles of ATP and 2 moles of NADPH are required to reduce 1 mole of CO_2 to its oxidation state in glucose. The standard free-energy change for reduction of CO_2 to (CH_2O), equivalent to reducing it to formaldehyde, is $+477$ kJ mol^{-1}.

C. Regulation of the Calvin Cycle

The activity of the Calvin cycle depends upon the fixation of CO_2 by ribulose 1,5-diphosphate carboxylase oxygenase, and this enzyme is the cycle's central regulatory enzyme. Its activity is closely related to the light reactions of photosynthesis. Upon illumination the stroma becomes more basic as protons flow into the thylakoid space. The activity of ribulose 1,5-diphosphate carboxylase oxygenase rises as the pH increases from 7 to 9. Thus, illumination increases its activity. NADPH, formed by photosystem I, is an allosteric activator of ribulose 1,5-diphosphate carboxylase oxygenase. The membrane potential of chloroplasts depends almost entirely upon the pH gradient. Two protons enter the thylakoid space and one Mg^{2+} ion is exported to the stroma. Mg^{2+} ions activate ribulose 1,5-diphosphate carboxylase oxygenase.

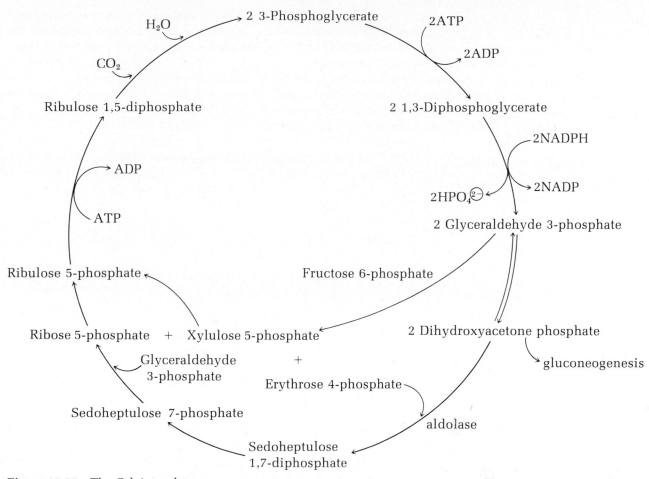

Figure 18.25 The Calvin cycle.

The regulation of ribulose 1,5-diphosphate carboxylase oxygenase by light is complemented by a second set of processes that control its activity less directly. Fructose 1,6-diphosphate inhibits the enzyme, whereas fructose 6-phosphate activates the enzyme. The concentration of fructose 1,6-diphosphate depends upon the activity of *fructose 1,6-diphosphatase* (reaction 18.26).

$$\text{fructose 1,6-diphosphate} \xrightarrow{\text{Fructose 1,6-diphosphatase}} \text{fructose 6-phosphate} + P_i \qquad (18.26)$$

The activity of this phosphatase in chloroplasts is controlled by light. This regulation is mediated by the soluble reduced ferredoxin generated by the light reactions of photosynthesis and by two regulatory proteins: *chloroplast thioredoxin* and *ferredoxin-thioredoxin reductase*. Thioredoxin is a low molecular weight protein that contains a cystine disulfide group that is reversibly interconverted with the sulfhydryl form by ferredoxin-thioredoxin reductase (reaction 18.27).

$$\underset{\underset{S—S}{|\quad|}}{\text{thioredoxin}} \xrightarrow{\text{Ferredoxin-thioredoxin reductase}} \underset{\underset{HS\quad SH}{|\quad|}}{\text{thioredoxin}} \qquad (18.27)$$

The electron source for the reduction of the disulfide bond is reduced soluble ferredoxin generated by electron transport in the light reactions of photosynthesis (Figure 18.26). Thioredoxin appears to activate three other enzymes of the Calvin cycle: sedoheptulose 1,7-diphosphatase, NADP-dependent glyceraldehyde 3-phosphate dehydrogenase, and ribulose 5-phosphate kinase.

During the day the plant receives its energy from the sun and uses the energy captured by the light reactions of photosynthesis to synthesize carbohydrates. At night the plant "becomes an animal" in a biochemical sense and obtains its energy by glycolysis and the oxidative pentose phosphate cycle. The day and night activities of the plant are not compartmentalized. Both types of reactions occur in the chloroplast. Fructose 1,6-diphosphatase is the critical regulatory enzyme of gluconeogenesis and phosphofructokinase is the key regulatory enzyme of glycolysis. Since light controls the activity of the phosphatase, it simultaneously prevents a "futile cycle" of phosphofructokinase and fructose 1,6-diphosphatase in which ATP is wasted. The effects of light on the daytime and nighttime reactions of the chloroplast are summarized in Figure 18.27.

18–7 CONCENTRATION OF CO₂ BY THE C₄ PATHWAY

In 1966 M. D. Hatch and C. R. Slack discovered a new set of dark reactions in photosynthesis. When certain plants are illuminated in the presence of $^{14}CO_2$, the first radioactive product that is formed is not 3-phosphoglycerate, but the "C_4 acid" oxaloacetate. This observation was at first taken as evidence for a new pathway for *net* fixation of carbon dioxide. Subsequent research has shown that the C_4 pathway does not result in net fixation of carbon dioxide, but that it serves as a mechanism for capturing carbon dioxide for delivery to the Calvin cycle. The C_4 pathway occurs in the cytoplasm of *mesophyll cells*. These cells are located on the leaf surface, in contact with atmospheric carbon dioxide. Phosphoenolpyruvate (PEP) is the CO_2 acceptor, and oxaloacetate is the carboxylated product. Oxaloacetate is reduced to malate or transaminated to aspartate by an aminotransferase (Section 21.2A). Transport of these metabolites into the *bundle sheath cells*, which are the major site of the Calvin cycle, is followed by decarboxylation. The carbon dioxide released enters the Calvin cycle (Figure 18.28). The C_3 product of decarboxylation is transported back to the mesophyll cells where phosphoenolpyruvate is regenerated. The net result of the C_4 pathway is concentration of carbon dioxide in the bundle sheath cells and a corresponding increase in the efficiency of the Calvin cycle. At first only certain tropical grasses were thought to possess the C_4 pathway, but now, as the list of plants that possess the C_4 pathway steadily grows, it is clear that it is a widespread phenomenon.

The carboxylation of phosphoenolpyruvate is catalyzed by *phosphoenolpyruvate carboxylase* (PEP carboxylase). Bicarbonate, formed from CO_2 by the action of carbonic anhydrase (Section 6.2D, reaction 6.11), is the substrate for PEP carboxylase (reaction 18.28). When ^{18}O-labeled bicarbonate is the substrate, two atoms of ^{18}O appear in oxaloacetate and one atom of ^{18}O is found in orthophosphate, suggesting the cyclic, concerted reaction mechanism shown in Figure 18.29.

Oxaloacetate is reduced to L-malate by an NADP-dependent malate dehydrogenase (reaction 18.29), and malate is transported to the bundle sheath cells.

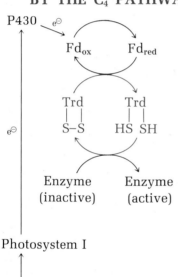

Figure 18.26 General scheme for the activation of fructose 1,6-diphosphatase, sedoheptulose 1,7-diphosphatase, ribulose 5-phosphate kinase, and glyceraldehyde 3-phosphate dehydrogenase by light. Electron transfer from photosystem I to soluble ferredoxin (Fd) leads to conversion of thioredoxin (Trd) to its disulfide form. In its reduced form it activates the enzymes.

$$CH_2{=}C{-}CO_2^{\ominus} + HCO_3^{2\ominus} \xrightarrow{\text{PEP carboxylase}} {}^{\ominus}O_2C{-}CH_2{-}\overset{\overset{\displaystyle O}{\|}}{C}{-}CO_2^{\ominus} + HPO_4^{2\ominus} \qquad (18.28)$$
$$\underset{\displaystyle OPO_3^{2\ominus}}{|}$$

Phosphoenolpyruvate   Oxaloacetate

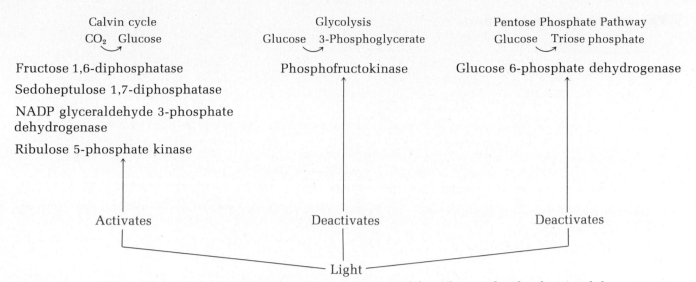

Figure 18.27 Effect of light on the activities of various components of the Calvin cycle, glycolysis, and the pentose phosphate pathway. [Adapted from B. B. Buchanan, *Ann. Rev. Plant Physiol. 31*, 365 (1980).]

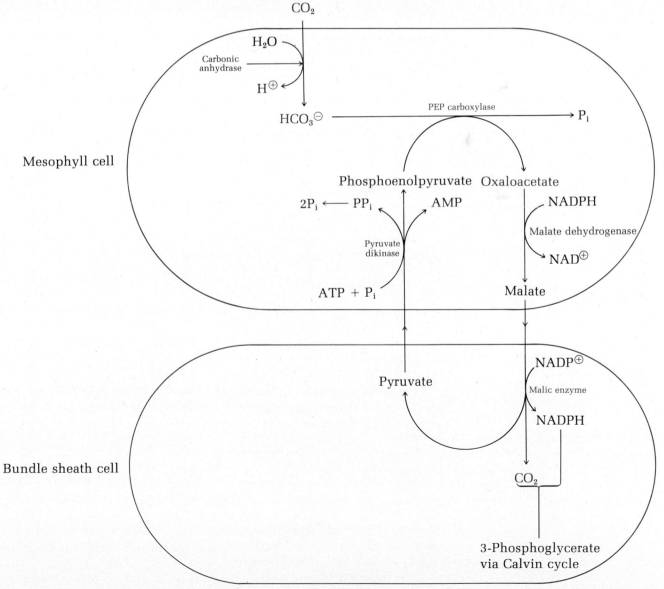

Figure 18.28 C$_4$ pathway. The pathway provides a mechanism for capturing CO$_2$ in mesophyll cells and carrying it to the bundle sheath cells for incorporation in carbohydrate by the Calvin cycle.

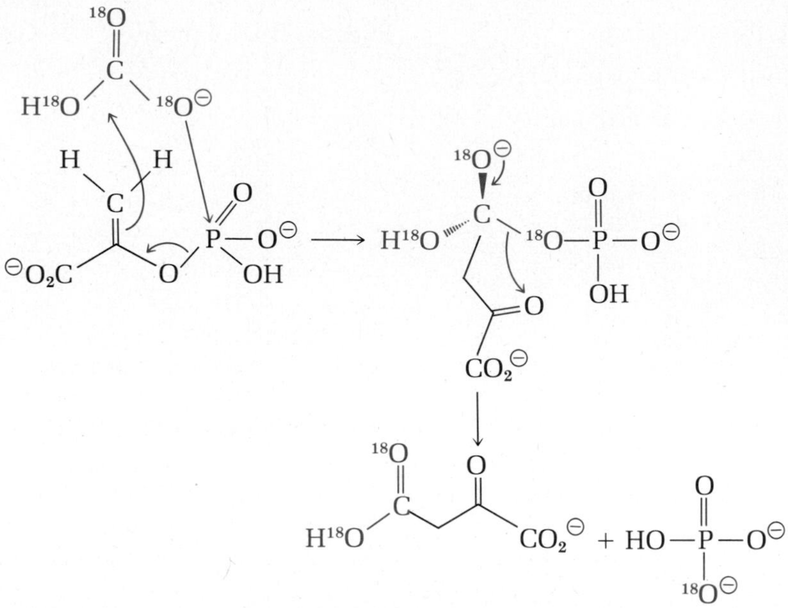

Figure 18.29 Possible mechanism for the action of PEP carboxylase in which two of the ^{18}O-labeled oxygens appear in oxaloacetate and one appears in phosphate. [From M. F. Utter and H. M. Kolenbrander, in P. D. Boyer, Ed., "The Enzymes," 3rd ed., Vol. 6, Academic Press, New York, 1972, p. 167.]

$$^{\ominus}O_2C-CH_2-\overset{\overset{\textstyle O}{\|}}{C}-CO_2^{\ominus} + NADPH \xrightarrow{\text{Malate dehydrogenase}} {}^{\ominus}O_2C-CH_2-\overset{\overset{\textstyle H}{|}}{\underset{\underset{\textstyle OH}{|}}{C}}-CO_2^{\ominus} + NADP \quad (18.29)$$

Oxaloacetate   L-Malate

Alternatively, oxaloacetate can be transaminated and aspartate is transported to the bundle sheath cells (reaction 18.30).

$$^{\ominus}O_2C-CH_2-\overset{\overset{\textstyle O}{\|}}{C}-CO_2^{\ominus} \xrightarrow[\underset{\text{Glu} \quad \alpha\text{-Ketoglutarate}}{}]{\text{Aminotransferase}} {}^{\ominus}O_2C-CH_2-\overset{\overset{\textstyle H}{|}}{\underset{\underset{\textstyle \oplus NH_3}{|}}{C}}-CO_2^{\ominus} \quad (18.30)$$

Oxaloacetate   Aspartate

Once these acids have reached the bundle sheath cells they are decarboxylated. In chloroplasts malate is decarboxylated by NADP-dependent *malic enzyme* (reaction 18.31).

$$^{\ominus}O_2C-CH_2-\overset{\overset{\textstyle O}{\|}}{C}-CO_2^{\ominus} + NADP \xrightarrow{\text{Malic enzyme}} CH_3-\overset{\overset{\textstyle O}{\|}}{C}-CO_2^{\ominus} + CO_2 + NADPH \quad (18.31)$$

Oxaloacetate   Pyruvate

Aspartate is transaminated to oxaloacetate in mitochondria, not in chloroplasts. The oxaloacetate is then reduced to malate, and malate is decarboxylated by mitochondrial malic enzyme.

The pyruvate formed by decarboxylation of oxaloacetate is transported back to mesophyll cells. Phosphoenolpyruvate in mesophyll cells is then regenerated by the action of *pyruvate phosphate dikinase* (reaction 18.32).

$$CH_3-\overset{O}{\underset{}{C}}-CO_2^\ominus + ATP + P_i \xrightarrow{\text{Pyruvate phosphate dikinase}} CH_2=\overset{OPO_3^{2\ominus}}{\underset{}{C}}-CO_2^\ominus + AMP + PP_i \quad (18.32)$$

Pyruvate   Phosphoenolpyruvate

Each cycle of the C_4 pathway thus requires two phosphoanhydride bonds, equivalent to requiring 2 moles of ATP. If all of the CO_2 incorporated in glucose by the Calvin cycle originated in the C_4 pathway, 30 moles of ATP would be required for the synthesis of glucose (reaction 18.33).

$$6CO_2 + 30ATP + 12NADPH + 12H_2O$$

$$\Big\downarrow \text{Combination of } C_4 \text{ pathway and Calvin cycle} \quad (18.33)$$

$$glucose + 30ADP + 12NADP + 30P_i + 18H^\oplus$$

18–8 PHOTORESPIRATION

Leaves of many plant species evolve carbon dioxide at a fairly rapid rate, a phenomenon called *photorespiration* to distinguish it from carbon dioxide evolution by mitochondrial respiration. Photorespiration occurs only in the light and is much slower in C_4 plants than in C_3 plants. Photorespiration severely lowers the efficiency of carbon dioxide fixation. In many species its rate is 50%, or more, as rapid as CO_2 uptake. When photosynthesis is studied in closed systems, both the light intensity and the CO_2 pressure are easily controlled. The level of illumination at which the rates of carbon dioxide fixation and photorespiration are equal is called the *light compensation point*. When the light intensity is greater than the light compensation point, a condition that would apply, say, at noon in a sugar cane field, the point at which carbon dioxide fixation and photorespiration are equal is called the CO_2 *compensation point*. C_3 plants have CO_2 compensation points in the range of 40 ppm CO_2 (40 parts per million CO_2 in the controlled atmosphere). For the efficient C_4 plants, the CO_2 compensation point hovers around 10 ppm CO_2. On a hot summer day, the C_4 plants grow at a faster rate than C_3 plants.

$$\overset{CO_2^\ominus}{\underset{CH_2OH}{|}}$$
Glycolate

The rate of photorespiration is closely related to the rate of *glycolate* synthesis. Glycolate metabolism is a complex phenomenon whose reactions are partitioned among chloroplasts, oxidative organelles known as *peroxisomes*, and mitochondria (Figure 18.30).

Ribulose 1,5-diphosphate carboxylase oxygenase is a bifunctional enzyme. We have considered the carboxylase function of the enzyme in CO_2 fixation. It also possesses oxygenase activity, and it converts ribulose 1,5-diphosphate to *phosphoglycolate* and 3-phosphoglycerate (reaction 18.34).

$$\overset{CH_2OPO_3^{2\ominus}}{\underset{CH_2OPO_3^{2\ominus}}{\overset{|}{\underset{|}{C=O}}}}\overset{}{\underset{}{\overset{H-C-OH}{\overset{|}{H-C-OH}}}} + {}^{18}O_2 + H_2O \xrightarrow{\text{Ribulose 1,5-diphosphate carboxylase oxygenase}} {}^\ominus O-\overset{{}^{18}O}{\underset{}{C}}-CH_2OPO_3^{2\ominus} + \overset{CO_2}{\underset{CH_2OPO_3^{2\ominus}}{\overset{|}{H-C-OH}}} \quad (18.34)$$

Ribulose 1,5-diphosphate   Phosphoglycolate   3-Phosphoglycerate

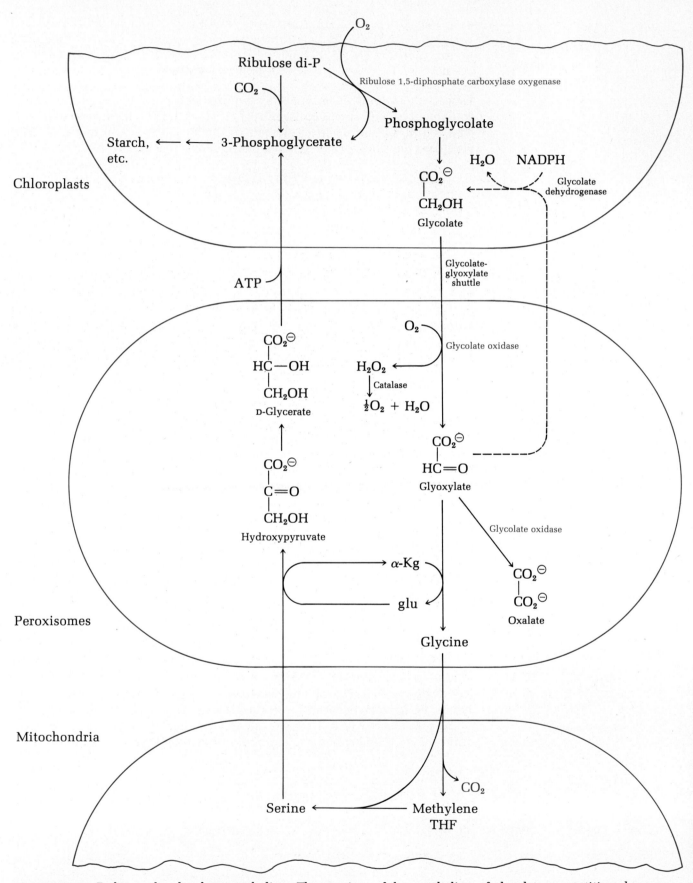

Figure 18.30 Pathways for glycolate metabolism. The reactions of the metabolism of glycolate are partitioned among chloroplasts, peroxisomes, and mitochondria.

Oxygen is a competitive inhibitor of carboxylation, and carbon dioxide is a competitive inhibitor of oxygenase activity. The enzyme has about the same K_m for each of these two substrates. The rates of the two competing reactions are not affected equally by increasing temperature, and as the temperature rises the ratio of phosphoglycolate production to carbon dioxide fixation increases.

Phosphoglycolate is converted to glycolate in chloroplasts by the action of glycolate phosphatase (reaction 18.35).

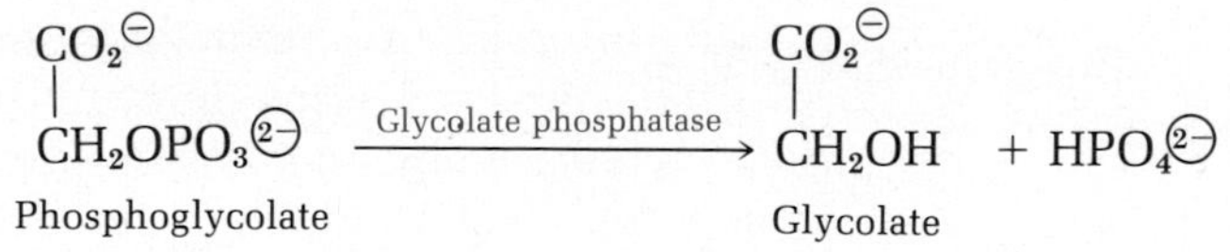

$$
\begin{array}{c}
CO_2^{\ominus} \\
| \\
CH_2OPO_3^{\small 2\ominus}
\end{array}
\xrightarrow{\text{Glycolate phosphatase}}
\begin{array}{c}
CO_2^{\ominus} \\
| \\
CH_2OH
\end{array}
+ HPO_4^{\small 2\ominus}
\qquad (18.35)
$$

Phosphoglycolate　Glycolate

Glycolate is then secreted from the chloroplasts, and glycolate metabolism continues in oxidative cell organelles called peroxisomes. The next reaction is catalyzed in the peroxisomes by the flavoprotein glycolate oxidase (reaction 18.36).

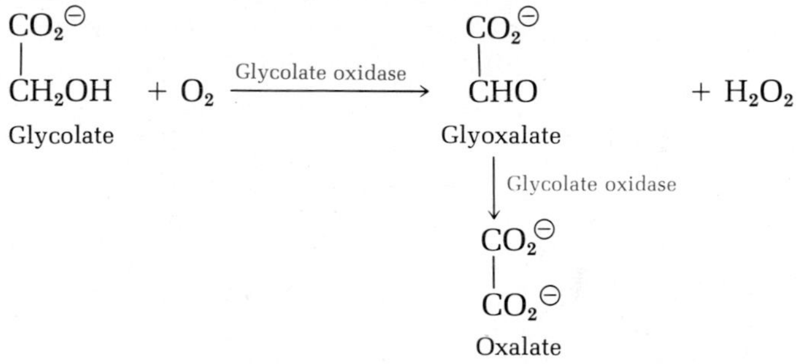

$$
\begin{array}{c}
CO_2^{\ominus} \\
| \\
CH_2OH
\end{array}
+ O_2
\xrightarrow{\text{Glycolate oxidase}}
\begin{array}{c}
CO_2^{\ominus} \\
| \\
CHO
\end{array}
+ H_2O_2
\qquad (18.36)
$$

Glycolate　　　　　　　　　Glyoxalate

$$
\downarrow \text{Glycolate oxidase}
$$

$$
\begin{array}{c}
CO_2^{\ominus} \\
| \\
CO_2^{\ominus}
\end{array}
$$

Oxalate

The hydrogen peroxide produced by glycolate oxidase is destroyed by catalase (reaction 18.37).

$$
H_2O_2 \longrightarrow \tfrac{1}{2}O_2 + H_2O \qquad (18.37)
$$

Hydrogen peroxide is a powerful oxidizing agent, and it is extremely toxic. To counter the deleterious effects that would result from an accumulation of hydrogen peroxide, catalase has evolved to an extremely efficient enzyme. Its k_{cat}/K_m value of 4×10^7 mol^{-1} s^{-1} is one of the highest known and ensures rapid destruction of hydrogen peroxide.

Not all the glyoxalate produced by glycolate oxidase is converted to oxalate, although in some leaves, such as spinach, oxalate concentrations are quite high. Some glyoxalate is transaminated to glycine by a glutamate-dependent aminotransferase (reaction 18.38).

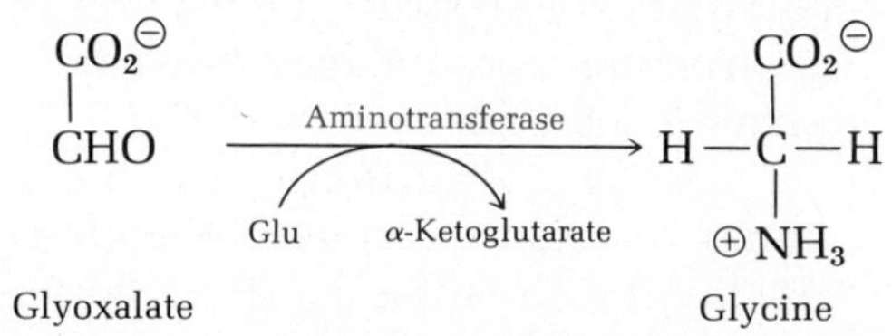

$$
\begin{array}{c}
CO_2^{\ominus} \\
| \\
CHO
\end{array}
\xrightarrow[\underset{\text{Glu}\quad\alpha\text{-Ketoglutarate}}{}]{\text{Aminotransferase}}
\begin{array}{c}
CO_2^{\ominus} \\
| \\
H-C-H \\
| \\
^{\oplus}NH_3
\end{array}
\qquad (18.38)
$$

Glyoxalate　　　　　　　　　Glycine

Some glyoxalate also is transported back to the chloroplast where it is reduced by an NADP-dependent dehydrogenase (reaction 18.39). This cycle is called the *glycolate-glyoxalate shuttle*.

$$\underset{\text{Glyoxalate}}{\overset{\overset{\displaystyle CO_2^{\ominus}}{|}}{\underset{|}{CHO}}} + NADPH \xrightarrow{\text{Glycolate dehydrogenase}} \underset{\text{Glycolate}}{\overset{\overset{\displaystyle CO_2^{\ominus}}{|}}{\underset{|}{CH_2OH}}} + NADP \qquad (18.39)$$

The glycine formed in peroxisomes is transported to mitochondria where it is converted to serine. Two moles of glycine are required for the biosynthesis of 1 mole of serine. One is converted to methylene tetrahydrofolate (Section 10.8) and CO_2. This decarboxylation is the source of the carbon dioxide released in photorespiration. The reactions leading from ribulose 1,5-diphosphate to serine thus undo the work of carbon fixation. In C_4 plants the CO_2 released by photorespiration can be captured in mesophyll cells and transported back to the mitochondria. Although this consumes energy (we recall that 2 moles of ATP are required to regenerate the CO_2 acceptor phosphoenolpyruvate), it is more efficient than losing carbon dioxide altogether.

Meanwhile, serine can be transported to peroxisomes where transamination converts it to hydroxypyruvate. Reduction of hydroxypyruvate yields glycerate that leaves the peroxisome and is phosphorylated in the cytoplasm to give 3-phosphoglycerate. Reentry of 3-phosphoglycerate into the chloroplast completes the cycle, since 3-phosphoglycerate is an intermediate of the Calvin cycle. This circuitous route from ribulose 1,5-diphosphate to 3-phosphoglycerate is wasteful because it is not accompanied by net incorporation of carbon dioxide and because NADPH is consumed in the glyoxalate-glycolate shuttle.

18–9 EVOLUTION OF ENERGY METABOLISM

So many similarities exist between oxidative phosphorylation and photophosphorylation that it is natural to ask whether this resemblance is merely fortuitous or whether it reflects common origins that have diverged over vast tracts of time. We saw the outlines of an approach to such questions when we considered the evolution of cytochrome c in Section 2.10. Evolutionary relationships emerge much more clearly from a comparison of the tertiary structures of proteins than from comparisons of their primary structures alone, since a shared type of polypeptide folding will be revealed in the absence of any clear sequence homologies. The presence of c-type cytochromes in both eucaryotic respiratory chains and procaryotic photosynthetic electron transport chains provides an opportunity to pursue the metabolic relationships between these widely separated types of organisms.

The tertiary structures of several c-type cytochromes have been determined. Tuna cytochrome c contains 103 amino acid residues, the cytochrome c_2 from *Rhodospirillum rubrum* contains 112 amino acid residues, and the cytochrome c_{550} of the nitrate-respiring bacterium *Paracoccus dentrificans* contains 134 amino acid residues. Nevertheless, the structures of the c-type cytochromes are similar (Figure 18.31). The hydrophobic environment of the heme, in particular, is nearly the same in all three proteins. In all, there are no structural features that distinguish these proteins. So strong a structural resemblance can hardly be a coincidence. The proteins must share a common ancestral origin.

In *Rhodospirillum capsulata* and in *R. spheroides* the same electron transport chain carries out both respiration and photosynthesis (Figure 18.32). If a mutation deprived these bacteria of photosynthesis, Figure 18.32 would look like a diagram of mitochondrial respiration. It therefore seems highly likely that respiration and photosynthesis share a common origin. The similarities between noncyclic electron flow in bacteria, cyclic electron flow in photosynthetic bacteria, photosystem I and photosystem II in algae and green plants, and mitochondrial electron transport are illustrated in Figure 18.33.

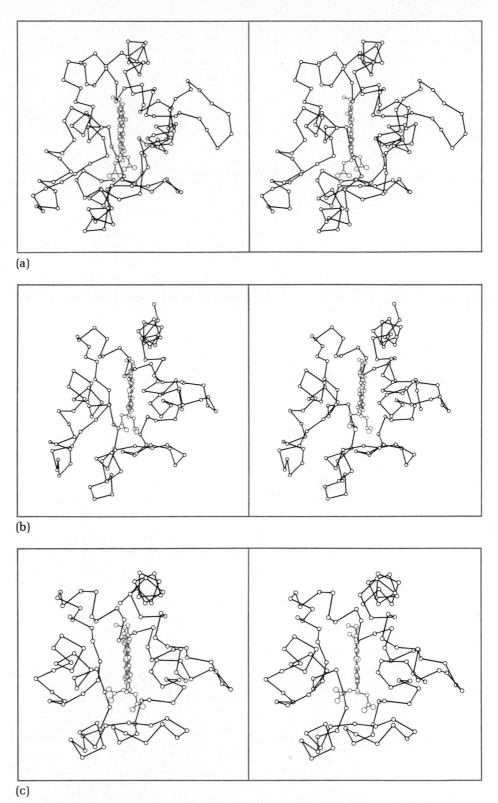

Figure 18.31 Structures of cytochromes *c* from (a) tuna, (b) *Rhodospirillum rubrum*, and (c) *Paracoccus dentrificans*.

An evolutionary tree of cytochrome *c*−containing electron transport chains is shown in Figure 18.34. The ancient primitive bacteria that existed over 3.5 billion years ago are thought to have resembled the present day *Clostridia*. These bacteria lack the cytochromes necessary for respiration and obtain their energy by anaerobic glycolysis and fermentation. The evolution

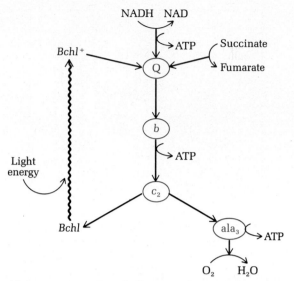

Figure 18.32 Shared photosynthetic and respiratory electron transport chains of *R. capsulata* and *R. spheroides.* [From R. E. Dickerson, R. Timkovich, and R. J. Almassey, *J. Mol. Biol., 100,* 485 (1976). Reproduced with permission. Copyright: Academic Press Inc. (London) Ltd.]

(a)

(b)

(c)

(d)

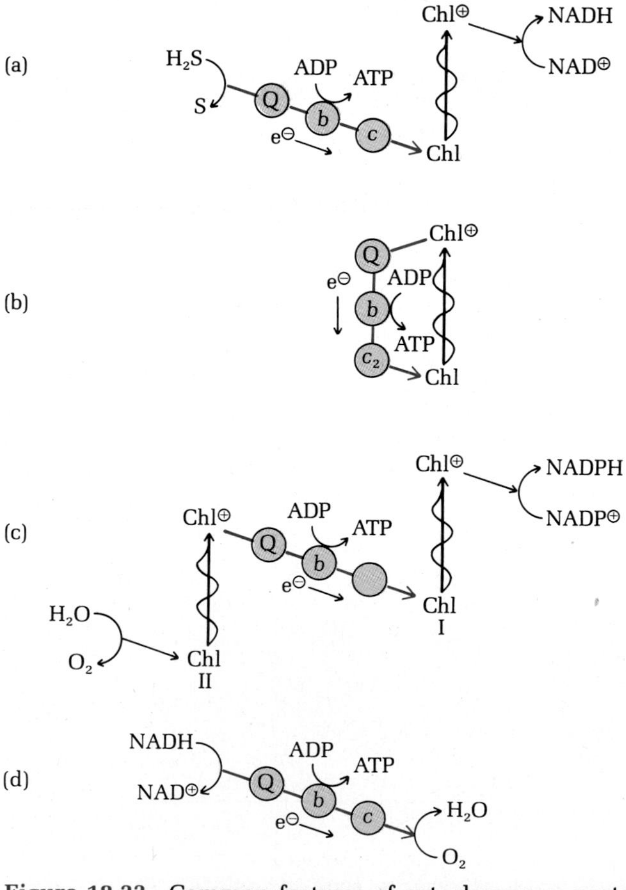

Figure 18.33 Common features of cytochrome *c*–containing electron transport chains. (a) Noncyclic bacterial photosynthesis. (b) Cyclic bacterial photosynthesis. (c) Two-center photosynthesis in algae and green plants. (d) Mitochondrial respiration. The dark arrow represents the shared evolutionary features of these pathways. [From R. E. Dickerson, R. Timkovich, and R. J. Almassey, *J. Mol. Biol., 100,* 484 (1976). Reproduced with permission. Copyright: Academic Press Inc. (London) Ltd.]

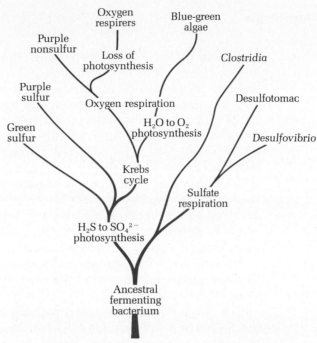

Figure 18.34 Evolution of energy metabolism. [From R. E. Dickerson, R. Timkovich, and R. J. Almassey, *J. Mol. Biol.*, *100*, 489 (1976). Reproduced with permission. Copyright: Academic Press Inc. (London) Limited.]

of the cytochromes led to photosynthesis by the green and purple bacteria such as *Chlorobium* and *Chromatium*. Hydrogen sulfide is oxidized in these organisms. The sulfate-reducing bacteria, on the other hand, reduce sulfate to sulfide leading to a sulfur cycle such as the one that exists today in *Clostridia* and *Desulfovibrio*. In this scheme of things, the citric acid cycle developed slowly in the sulfurbacteria, eventually producing a source of NADH independent of photosynthesis. The blue-green algae developed along lines leading from a single photosystem to one in which two photosystems combine to use water as a reducing agent. This development, the birth of green plant photosynthesis, had momentous consequences. These release of oxygen to the atmosphere changed the atmosphere from a reducing to an oxidizing environment. Oxygen respiration arose at this time to take advantage of the tremendous increase in energy available when substrates are oxidized rather than fermented. Photosynthesis was no longer necessary to organisms capable of aerobic respiration, and the respiring organisms evolved a new line, of which we are one of the products.

18–10 SUMMARY Photosynthesis occurs in chloroplasts. The chloroplast is bounded by two membranes. The outer membrane is freely permeable to most low molecular weight molecules. The inner membrane, however, permits only a few molecules to pass without the aid of protein transport systems. The space bounded by the inner membrane is called the stroma. Within the stroma is a set of stacked membranes known as grana. The stacked membrane discs of the grana are called thylakoid membranes, and the space they enclose is the thylakoid space.

Photosynthesis may be divided into light and dark reactions. The enzymes that carry out the light reactions of photosynthesis and the proteins responsible for ATP synthesis are bound to the thylakoid membrane. The dark reactions of photosynthesis occur in the stroma. The light reactions are partitioned between photosystem I and photosystem II. Light is absorbed by unique chlorophyll molecules that are tightly bound to protein at the photoreaction center. Besides the unique chlorophyll, many other chlorophylls and carotenes act as antennae that capture light. Energy is passed from the antennae molecules of the photoreaction center to the unique chlorophyll of the reaction center by radiationless transfer. Photosystem I generates NADPH by passing electrons through a bound form of the iron-sulfur protein ferredoxin, P430, through soluble ferredoxin, and then to ferredoxin-NADPH-reductase. The oxidized photoreaction center of photosystem I receives electrons either by cyclic electron flow, a process that is coupled to ATP synthesis, or by an electron transport chain that couples photosystem I to photosystem II. Photosystem II oxidizes water and passes electrons to the oxidized photoreaction center of photosystem I, $P700^{\oplus}$. Photosystem II generates a powerful oxidizing agent that oxidizes water. The electron transport chain coupling photosystem I to photosystem II is coupled to ATP synthesis.

The driving force for photophosphorylation is the proton motive force. Electron transport either by cyclic electron transport in photosystem I or by electron transport from photosystem II to photosystem I, as well as the photolysis of water generates a proton gradient across the thylakoid membrane, the thylakoid space being more acidic than the stroma. Since the thylakoid membrane is freely permeable to $Mg^{\oplus\oplus}$ and $Cl^{\ominus}$ ions, electron transport is electrically neutral. The thylakoid membranes contain coupling factors CF_1 and CF_0. The oligomeric proteins closely resemble their mitochondrial counterparts. They are responsible for synthesis of ATP. CF_1 can be dissociated from the thylakoid membrane. When not bound to the membrane, it is an ATPase; when bound, it is responsible for ATP synthesis.

Both ATP and NADPH are released into the stroma, the site of the dark reactions of carbon dioxide fixation by the reactions of the Calvin cycle. The key reaction of the Calvin cycle is fixation of carbon dioxide by ribulose 1,5-diphosphate carboxylase oxygenase. Incorporation of CO_2 in ribulose 1,5-diphosphate leads to production of 2 moles of 3-phosphoglycerate. The pathway from 3-phosphoglycerate to glucose is identical to gluconeogenesis, except that the chloroplast version of glyceraldehyde 3-phosphate dehydrogenase is specific for NADPH rather than NADH. Regeneration of ribose 5-phosphate consumes 25 of every 30 moles of glyceraldehyde 3-phosphate produced by the Calvin cycle. The steps leading from 3-phosphoglycerate to ribulose 5-phosphate involve transketolase- and aldolase-catalyzed reactions. These regenerative reactions parallel the reactions of the pentose phosphate pathway.

C_4 plants possess enzymes in mesophyll cells that capture carbon dioxide and deliver it to bundle sheath cells. Many plants evolve carbon dioxide by photorespiration in a series of reactions that are divided between chloroplasts, peroxisomes, and mitochondria. Photorespiration is linked to the production of glycolate. In C_4 plants photorespiration is relatively low.

REFERENCES

Alvarez, L. W., et al., Extraterrestrial Cause for the Cretaceous-Tertiary Extinction, *Science, 208,* 1095–1108 (1980).

Blankenship, R. E., and Parson, W. W., The Photosynthetic Electron Transport Reactions of Photosynthetic Bacteria and Plants, *Ann. Rev. Biochem., 47,* 635–654 (1978).

Buchanan, B. B., and Schurman, P., Ribulose 1,5-Diphosphate Carboxylase: A Regulatory Enzyme in the Photosynthetic Assimilation of Carbon Dioxide, *Curr. Topics Cell. Regul., 7,* 1–20 (1973).

Butler, W. L., Primary Photochemistry of Photosystem II in Photosynthesis, *Acc. Chem. Res., 6,* 177–184 (1973).

Clayton, R. K., and Sistrom, W. R., Eds., "The Photosynthetic Bacteria," Academic Press, New York, 1976.

Dickerson, R. E., Timkovich, R., and Almassy, R. J., The Cytochrome Fold and the Evolution of Bacterial Energy Metabolism, *J. Mol. Biol., 100,* 473–491 (1976).

Govindjee, R., Ed., "Bioenergetics of Photosynthesis," Academic Press, New York, 1975.

Gregory, R. P. F., "Biochemistry of Photosynthesis," 2nd ed., Wiley, New York, 1977.

Jensen, R. G., and Bahr, J. T., Ribulose-1,5-Bisphosphate Carboxylase-Oxygenase, *Ann. Rev. Plant Physiol., 28,* 379–400 (1977).

Junge, W., Membrane Potentials in Photosynthesis, *Ann. Rev. Plant Physiol., 28,* 503–536 (1977).

Kelly, G. J., and Latzko, E., Regulatory Aspects of Photosynthetic Carbon Metabolism, *Ann. Rev. Plant Physiol., 27,* 181–205 (1976).

McCarty, R. E., Roles of a Coupling Factor for Photophosphoryla-tion in Chloroplasts, *Ann. Rev. Plant Physiol., 30,* 79–104 (1979).

Sauer, K., Photosynthetic Membranes, *Acc. Chem. Res., 11,* 257–264 (1978).

Tolbert, N. E., Glycolate Biosynthesis, *Curr. Topics Cell. Regul., 7,* 21–50 (1973).

Trebst, A., Energy Conservation in Photosynthetic Electron Transport in Chloroplasts, *Ann. Rev. Plant Physiol., 25,* 423–458 (1974).

Zelitch, I., Pathways of Carbon Fixation in Green Plants, *Ann. Rev. Biochem., 44,* 123–145 (1975).

Problems

1. Indicate whether each of the following changes will increase, decrease, or leave unaltered the rate of the Calvin cycle: (a) increasing the stroma pH, (b) decreasing the stroma concentration of Mg^{2+}, (c) increasing the concentration of fructose 1,6-diphosphate, (d) increasing the concentration of fructose 6-phosphate, (e) increasing the oxygen pressure.

2. In the experiments of Jagendorf and his coworkers in which an artifically impressed pH gradient was shown to drive phosphorylation of ADP, the addition of DCMU removed the need to maintain the chloroplasts in the dark. Explain this observation.

3. Calculate the minimum pH gradient that is required to provide enough energy to drive the phosphorylation of ADP. Assume that the standard free energy of hydrolysis of ATP is -30.5 kJ mol^{-1}.

4. Agents that uncouple oxidative-phosphorylation in mitochondria also uncouple photoelectron transport and ATP synthesis. Explain this observation.

5. If [1-^{14}C]-ribulose 5-phosphate is the substrate for the dark reactions of photosynthesis, in which carbon of 3-phosphoglycerate will the label appear?

6. If ^{4}C-labeled carbon dioxide is the substrate for the Calvin cycle, the label first appears in the carboxyl group of 3-phosphoglycerate. After a short time, however, the label is found on C-1 of D-glyceraldehyde 3-phosphate and C-3 of 3-dihydroxyacteone phosphate. (a) Which enzyme(s) of the Calvin cycle are responsible for scrambling the label? (b) Where do the labels appear in fructose 1,6-diphosphate?

Chapter 19

Fatty acid metabolism

19–1 INTRODUCTION

Fatty acids are highly reduced sources of fuel for cells. Stored as triglycerides, the fatty acids provide an energy reserve during times of low food supplies. In mammals far more energy is stored in fat than in glycogen. Highly reduced fatty acids yield much more energy when they are oxidized than can be obtained from glycogen; about 36 kJ per gram is released by oxidation of fatty acids versus 16 kJ per gram for glycogen oxidation. Glycogen is about 70% water by weight, whereas fat contains virtually no bound water and is a compact source of energy.

Fatty acid metabolism repeats many of the themes we have already discussed. First, in eucaryotic cells fatty acid metabolism is compartmentalized; oxidation occurs in the mitochondrion and synthesis in the cytosol. Physical separation of the enzymes that catalyze these processes provides a simple means of metabolic control. Second, the pathways for oxidation and biosynthesis are different. Third, the final product of fatty acid oxidation and the starting material for fatty acid biosynthesis is acetyl-CoA. We might say with only slight hyperbole that "In metabolism all roads lead to, and from, acetyl-CoA". As Figure 19.1 indicates, acetyl-CoA is the final catabolic product of glucose degradation, of fatty acid degradation, and of fully half of the

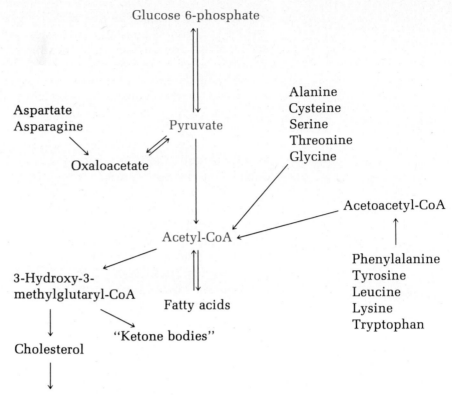

Figure 19.1 Metabolic processes in which acetyl-CoA is either an end product or a starting material.

amino acids. If we add to these catabolic processes the biosynthesis of fatty acids, cholesterol, steroid hormones, and a group of ketones known collectively as "ketone bodies," and if we further recall that pyruvate carboxylase is active only when bound to acetyl-CoA, we see that acetyl-CoA ranks with glucose 6-phosphate and pyruvate as one of the most important of all metabolic intermediates.

19–2 LIPASES Triacylglycerides (Section 11.2) are the principle storage depot for fatty acids. In mammals triacylglycerides are stored as droplets in the cytosol of

$$
\text{CH}_3(\text{CH}_2)_n-\overset{\overset{\displaystyle O}{\|}}{\text{C}}-\text{O}-\underset{\underset{\displaystyle \text{CH}_2\text{O}-\overset{O}{\underset{\|}{\text{C}}}-(\text{CH}_2)_n\text{CH}_3}{\overset{\displaystyle \text{CH}_2\text{O}-\overset{\overset{\displaystyle O}{\|}}{\text{C}}-(\text{CH}_2)_n\text{CH}_3}{\text{C}-\text{H}}}} \quad + \text{H}_2\text{O} \xrightarrow{\text{Lipase}}
$$

Triglyceride

$$
\text{CH}_3(\text{CH}_2)_n-\overset{\overset{\displaystyle O}{\|}}{\text{C}}-\text{O}-\underset{\underset{\displaystyle \text{CH}_2\text{O}-\overset{O}{\underset{\|}{\text{C}}}-(\text{CH}_2)_n\text{CH}_3}{\overset{\displaystyle \text{CH}_2\text{OH}}{\text{C}-\text{H}}}} \quad + \text{CH}_3(\text{CH}_2)_n\text{CO}_2^{\ominus} + \text{H}^{\oplus} \quad (19.1)
$$

Fatty acid

Diglyceride

adipose (or fat) cells. The first step in the release of fatty acids from triglycerides is hydrolysis of fatty acyl esters by *lipases* (reaction 19.1). Lipases are most efficient when their substrates are triglycerides, although they also hydrolyze di- and monoglycerides

Mammals contain three types of lipases: tissue lipases, milk lipases, and digestive tract lipases produced by the pancreas. The *pancreatic lipases* of animals hydrolyze a wide range of fatty acylglycerols. *Pancreatic esterases* hydrolyze the esters of fatty acids, especially those containing cholesterol. These lipases have an absolute requirement for bile salts for activity (Figure 19.2). Another pancreatic lipase, phospholipase A_2, is produced as an inactive zymogen, prophospholipase A_2 (recall Section 6.4). The zymogen is activated by trypsin. Phospholipase A_2 requires both bile salts and Ca(II) ions for activity. Ingested triglycerides are converted to 2-monoglycerides and fatty acids by pancreatic lipases. The fatty acids pass through the epithethelial membranes of mucosal cells as complexes with bile salts. The fatty acids are then reconverted to triglycerides that bind tightly to small particles with an average diameter of about 1 μm called *chylomicra*. The chylomicra exit the mucosa cells, enter lymphatic vessels, and enter the blood through the thoracic lymph duct (Figure 19.3).

The mechanism of action of phospholipase A_2 resembles the mechanism of action of chymotrypsin (Section 6.3). A histidine residue and two aspartate residues form a charge relay network, and an oxyanion pocket stabilizes the tetrahedral intermediate that forms during ester hydrolysis. A combination of chemical and x-ray crystallographic evidence suggests the mechanism shown in Figure 19.4.

Adipose tissue contains a *hormone-sensitive lipase* whose activity is regulated by epinephrine and by the peptide hormones glucagon, adrenocortotrophic hormone, and thyroid-stimulating hormone. This hormonal regulation is similar to the hormone-regulated degradation of glycogen (Section 17.3A). An inactive adenylate cyclase is activated by a hormone and produces cAMP. An inactive protein kinase, composed of catalytic and regula-

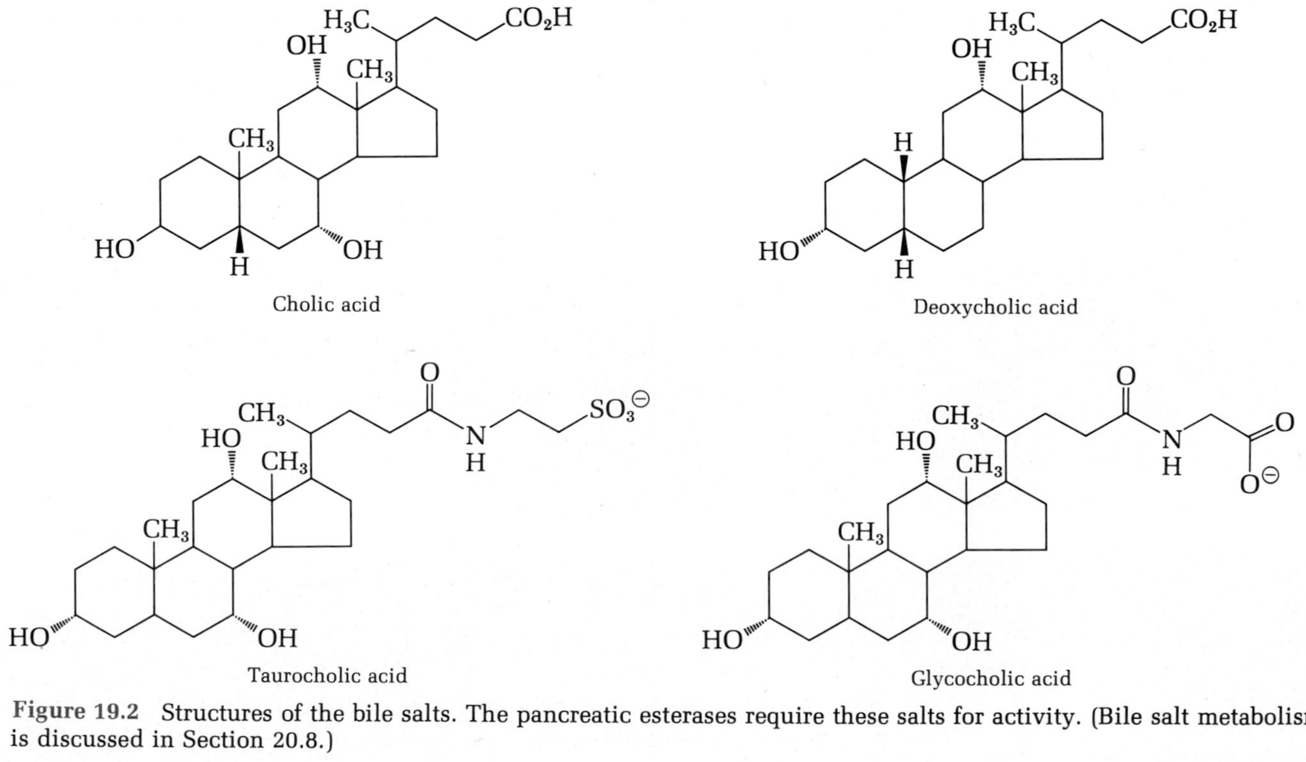

Figure 19.2 Structures of the bile salts. The pancreatic esterases require these salts for activity. (Bile salt metabolism is discussed in Section 20.8.)

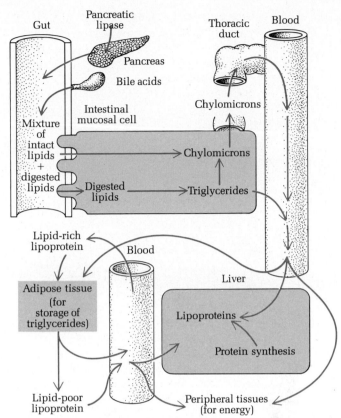

Figure 19.3 Lipid transport in the body. [Adapted from N. W. Tietz, Ed., "Fundamentals of Clinical Chemistry," Saunders, Philadelphia, 1976, p. 484.]

tory subunits, binds to cAMP and dissociates into free catalytic subunits and cAMP-bound regulatory subunits. Active protein kinase then activates hormone-sensitive lipase in a process which has the general features we noted for glycogen degradation (Figure 19.5). The details of the process have not yet been elucidated.

19–3 β-OXIDATION OF FATTY ACIDS

A. Overview of β-Oxidation

F. Knoop's experiments (recall Section 13.2) led him to propose that oxidation of fatty acids occurs by oxidation at the carbon β to the carbonyl group and that two-carbon fragments are produced as "acetate." He reached his hypothesis by examining the excretion products obtained by oxidation of ω-phenyl fatty acids. If ω-phenyl fatty acids containing an even number of carbon atoms are oxidized, the final product is phenylacetate (reaction 19.2).

$$\bigcirc\!\!\!\!-CH_2(CH_2)_{n=even}CO_2^{\ominus} \xrightarrow{\text{Oxidation}}$$

$$\bigcirc\!\!\!\!-CH_2CO_2^{\ominus} + "n\ CH_3CO_2^{\ominus}" \qquad (19.2)$$

Phenylacetate

On the other hand, oxidation of ω-phenyl fatty acids whose carbon chains contain an odd number of carbons yields benzoate (reaction 19.3).

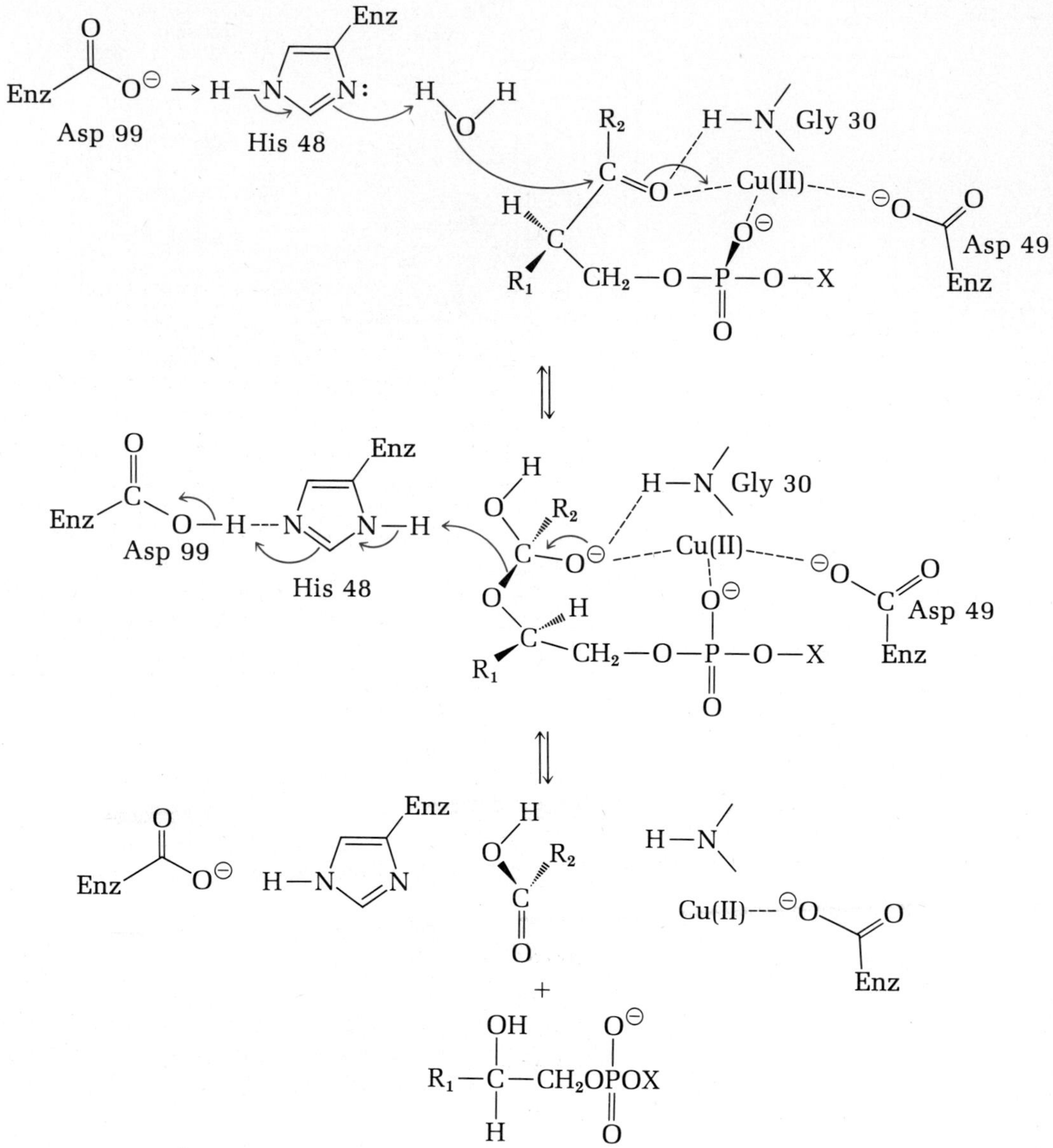

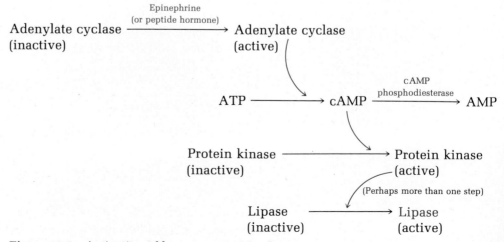

Figure 19.4 Mechanism of action of phospholipase A_2. [From H. M. Verheij et al., *Biochemistry, 19*, 743–750 (1980).]

Figure 19.5 Activation of hormone-sensitive lipase.

$$\langle \bigcirc \rangle - CH_2(CH_2)_{n=odd}CO_2^{\ominus} \xrightarrow{\text{Oxidation}}$$

$$\langle \bigcirc \rangle - CO_2^{\ominus} + \text{"n } CH_3CO_2^{\ominus}\text{"} \qquad (19.3)$$

Benzoate

The sequential removal of two-carbon fragments is called *β-oxidation*. Proof that the two-carbon fragments produced by *β*-oxidation were bound to coenzyme A, as acetyl-CoA, was not forthcoming until the early 1950s when F. Lynen and E. Reichert were able to isolate acetyl-CoA from yeast.

Six steps are required for the removal of each acetyl group from a fatty acid as acetyl-CoA. The six steps are listed here and summarized in Figure 19.6. The reactions are described in Section 19.3B.

1. The thioester of the fatty acid is synthesized by an *acyl-CoA synthetase* in an ATP-dependent step.

2. The formation of an acyl-CoA occurs in the cytosol. The other enzymes responsible for *β*-oxidation are located in the matrix space of the mitochondrion. Transport of the acyl-CoA across the inner mitochondrial membrkane is catalyzed by *carnitine-acyl transferases I and II (CAT I, II)*. All of the remaining reactions of the *β*-oxidation sequence occur in the matrix space of the mitochondrion.

3. Oxidation of the acyl-CoA to an unsaturated thioester (*trans-Δ^2-enoyl-CoA*) is catalyzed by a flavin-dependent *acyl-CoA dehydrogenase*.

4. Hydration of the *trans-Δ^2-enoyl-CoA* is catalyzed by enoyl-CoA hydratase (also called crotonase or 3-hydroxyacyl-CoA hydrolase).

5. S-3-hydroxyacyl-CoA is oxidized by NAD-dependent *S-3-hydroxyacyl-CoA dehydrogenase*.

6. 3-Ketoacyl-CoA is converted to acetyl-CoA and an acyl-CoA that has been shortened by two carbons by *thiolase*.

The three steps from the mitochondrial acyl-CoA to 3-ketoacyl-CoA parallel the three-step process in which succinyl-CoA is converted to oxaloacetate in the citric acid cycle (reaction 19.4).

$$\text{succinyl-CoA} \xrightarrow[\text{FAD} \quad \text{FADH}_2]{} \text{fumarate} \longrightarrow$$

$$\text{malate} \xrightarrow[\text{NAD} \quad \text{NADH}]{} \text{oxaloacetate} \qquad (19.4)$$

Citric acid cycle

$$\text{Acyl-CoA} \xrightarrow[\text{FAD} \quad \text{FADH}_2]{} \textit{trans-}\Delta^2\text{-Enoyl-CoA} \longrightarrow$$

$$\text{S-3-Hydroxyacyl-CoA} \xrightarrow[\text{NAD} \quad \text{NADH}]{} \text{3-Ketoacyl-CoA}$$

β-oxidation

Let us consider the net reaction for the complete conversion of palmitoyl-CoA, a C_{16}-fatty acyl-CoA, to acetyl-CoA. The net reaction for seven rounds of *β*-oxidation is

$$\text{palmitoyl-CoA} + 7\text{FAD} + 7\text{NAD} + 7\text{CoA} + 7\text{H}_2\text{O} \longrightarrow$$

$$8\text{acetyl-CoA} + 7\text{FADH}_2 + 7\text{NADH} + 7\text{H}^{\oplus} \qquad (19.5)$$

We recall from our discussion of oxidative phosphorylation that the oxidation of $FADH_2$ yields 2 molecules of ATP and that the oxidation of NADH yields 3 molecules of ATP. The 7 moles of $FADH_2$ and the 7 moles of NADH obtained by complete oxidation of palmitoyl-CoA therefore yield 35 moles of ATP. To this total we add the 96 moles of ATP obtained for complete

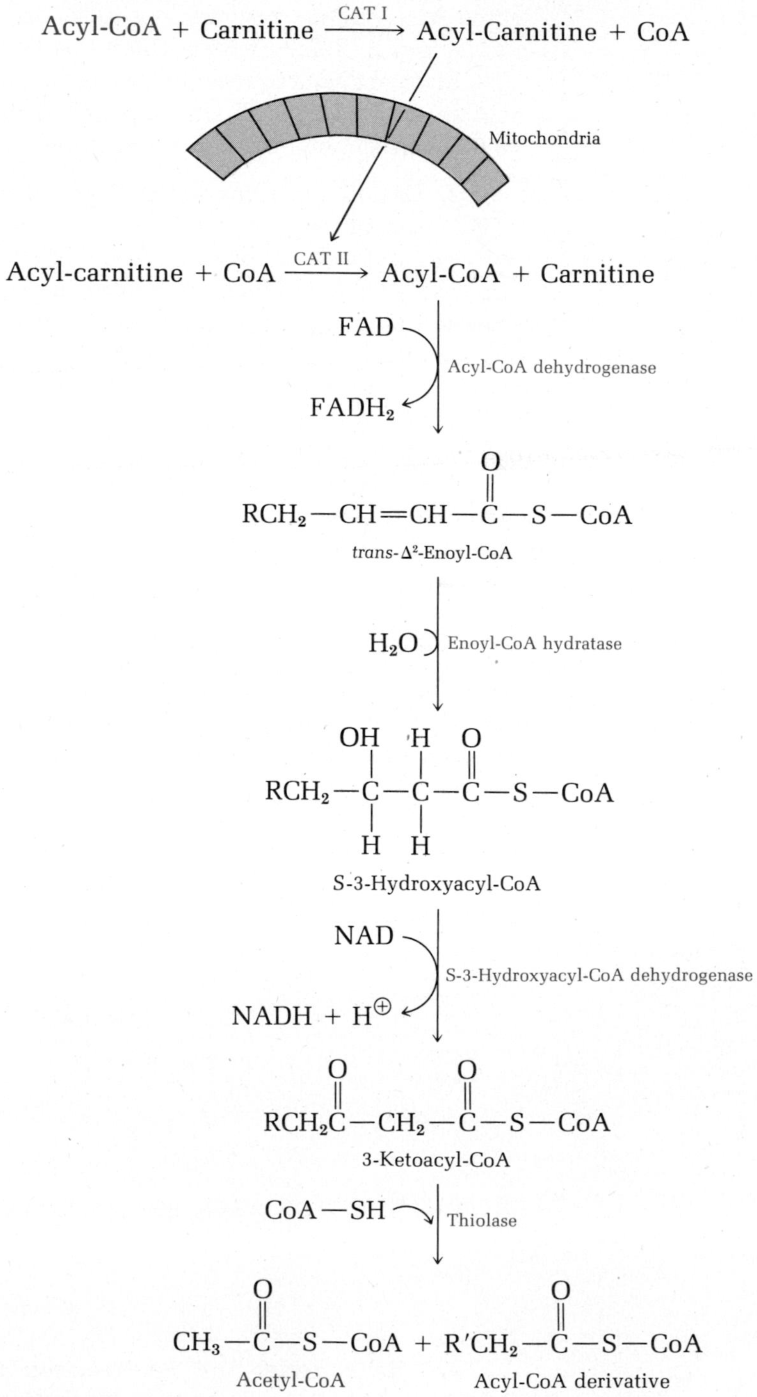

Figure 19.6 Pathway for β-oxidation of fatty acids through one sequence of oxidation.

oxidation of 8 moles of acetyl-CoA by the citric acid cycle and oxidative phosphorylation. This gives us a total of 131 moles of ATP produced for oxidation of palmitoyl-CoA. However, since two phosphoanhydride bonds are consumed in the formation of the acyl-CoA, the net yield is 129 moles of ATP. Since the standard free-energy change for complete combustion of palmitate is $-9{,}790 \text{ kJ mol}^{-1}$, and since the sum of the standard free-energy changes for complete hydrolysis of 129 moles of ATP is $-3{,}940 \text{ kJ mol}^{-1}$, the β-oxidation of fatty acids occurs with about 40% efficiency. The complete oxidation of glucose is also about 40% efficient.

Mammals are not able to convert the acetyl-CoA produced by β-oxidation into glucose since both carbon atoms of acetyl-CoA are lost as CO_2 as acetyl-CoA is carried through the citric acid cycle. Thus, while oxaloacetate is regenerated, there is no net increase in the concentration of oxaloacetate, and no net synthesis of glucose from oxaloacetate is possible. Plants, on the other hand, use the glyoxalate cycle (Section 15.5) to bypass the decarboxylations of the citric acid cycle and are therefore able to carry out net conversion of acetyl-CoA to oxaloacetate. The biosynthesis of glucose from the acetyl-CoA derived from β-oxidation of fatty acids is especially important in germinating seedlings.

B. Enzymes of β-Oxidation

a. Acyl-CoA Synthetases

The acyl-CoA synthetases catalyze the ATP- or GTP-dependent conversion of fatty acids to their acyl-CoA derivatives. The GTP-dependent enzymes are specific for chain lengths from C_4 to C_{12}, and the ATP-dependent enzymes act on longer chain fatty acids (reaction 19.6).

$$CH_3(CH_2)_n\!-\!CO_2^{\ominus} + ATP(GTP) + CoA\text{-}SH \xrightarrow{\text{Acyl-CoA synthetase}}$$

$$CH_3(CH_2)_n\!-\!\overset{\overset{\displaystyle O}{\|}}{C}\!-\!SCoA + AMP(GMP) + PP_i \qquad (19.6)$$

The first step in the reaction is formation of an activated intermediate, an *acyl adenylate*. The nucleophilic sulfur of CoA reacts with the activated intermediate to form the acyl-CoA (Figure 19.7).[1] The thermodynamic driving force for synthesis of the acyl-CoA, a thioester that is an activated intermediate, is provided by cleavage of both phosphoanhydride bonds of ATP. The first one is broken in the formation of the acyl adenylate intermediate. The second one is cleaved by the action of pyrophosphatase (reaction 19.7).

$$PP_i + H_2O \xrightarrow{\text{Pyrophosphatase}} 2P_i \qquad \Delta G^{0\prime} = -30 \text{ kJ mol}^{-1} \qquad (19.7)$$

This enzyme is so efficient that cellular concentrations of pyrophosphate are less than 10^{-6} M. The standard free energy of hydrolysis of an acyl-CoA is about -30 kJ mol^{-1}, about the same as the energy released for hydrolysis of a single phosphoanhydride bond. By cleaving two phosphoanhydride bonds, the cell ensures a large negative standard free-energy change and an equilibrium constant greater than 10^4 for the synthesis of an acyl-CoA. We saw in our overview of β-oxidation that the thioester is not cleaved throughout the entire process. It does, however, have an important chemical function. We recall that the thioester enhances the acidity of the protons bound at the

---

[1] We will see a striking parallel to this reaction when we discuss peptide bond formation in the biosynthesis of proteins.

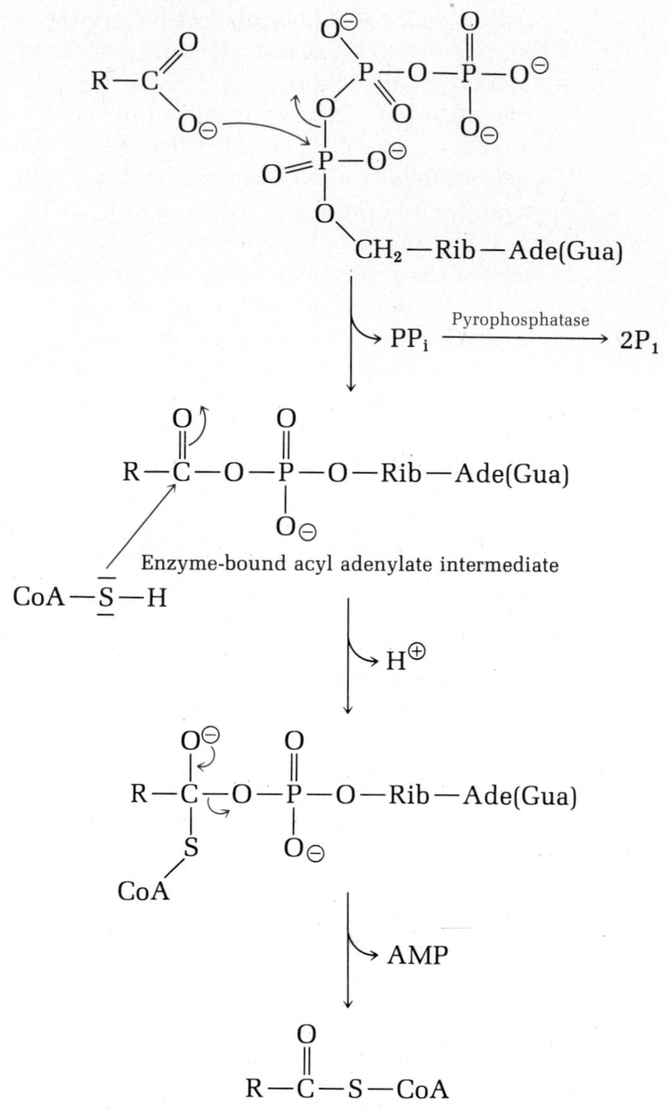

Figure 19.7 Mechanism of action of acyl-CoA synthetase. The enzyme-bound acyl adenylate intermediate is a highly reactive mixed anhydride. AMP is a good leaving group, and it is displaced by the sulfhydryl group of coenzyme A.

α-carbon by two orders of magnitude. This enhanced acidity will be exploited by the flavoprotein dehydrogenase in the next step and by thiolase.

b. Carnitine Acyl Transferases I and II

Acyl-CoA derivatives are not permeable to the inner mitochondrial membrane. Cytosolic acyl-CoA derivatives enter the mitochondrion by a shuttle system in which *carnitine* acts as an acyl carrier. Carnitine acyl transferase I, located on the outer surface of the inner mitochondrial membrane,

$$(CH_3)_3\overset{\oplus}{N}-CH_2-CH-CH_2CO_2^{\ominus}$$
$$|$$
$$OH$$

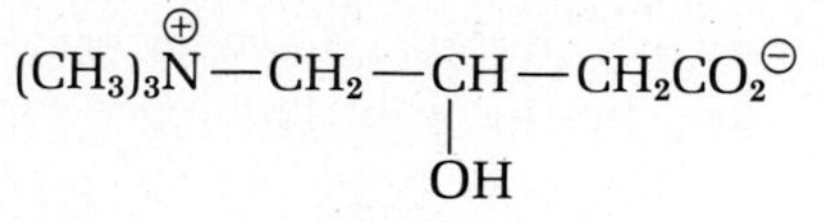

Carnitine

catalyzes acyl group transfer from coenzyme A to the hydroxyl group of carnitine (reaction 19.8).

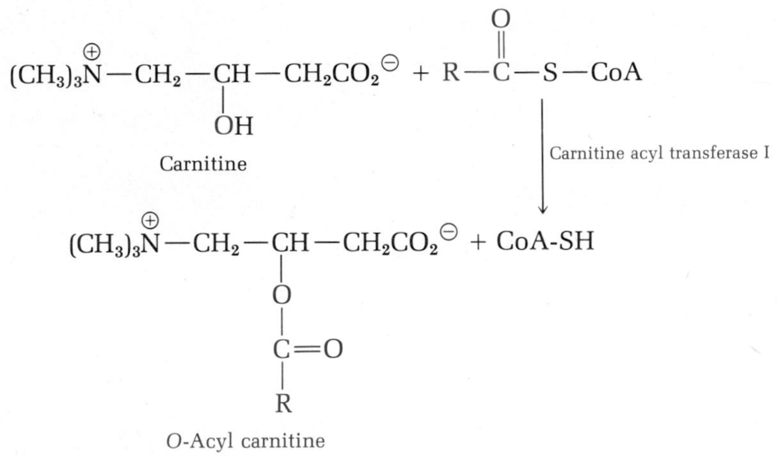

$$(CH_3)_3\overset{\oplus}{N}-CH_2-\underset{\underset{OH}{|}}{CH}-CH_2CO_2{}^{\ominus} + R-\overset{\overset{O}{\|}}{C}-S-CoA$$

Carnitine

Carnitine acyl transferase I

$$(CH_3)_3\overset{\oplus}{N}-CH_2-\underset{\underset{\underset{\underset{R}{|}}{C=O}}{\underset{|}{O}}}{CH}-CH_2CO_2{}^{\ominus} + CoA\text{-}SH \qquad (19.8)$$

O-Acyl carnitine

The *O*-acyl carnitine is transported into the mitochondrion where carnitine acyl transferase II catalyzes the reverse reaction (Figure 19.8). The combination of these two enzymes allows acyl-CoA derivatives to be transported across the inner mitochondrial membrane.

c. Acyl-CoA Dehydrogenase

The next step in β-oxidation of fatty acids is dehydrogenation of fatty acyl-CoA catalyzed by a flavoprotein dehydrogenase whose specificity depends upon the chain length of the fatty acid. The reaction produces the more stable *trans* double bond (reaction 19.9).

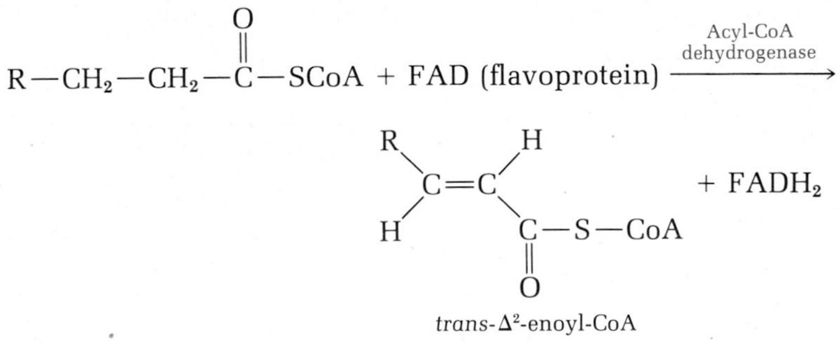

$$R-CH_2-CH_2-\overset{\overset{O}{\|}}{C}-SCoA + FAD \text{ (flavoprotein)} \xrightarrow{\overset{\text{Acyl-CoA}}{\text{dehydrogenase}}}$$

$$\underset{H}{\overset{R}{\diagdown}}C=C\underset{\underset{\underset{O}{\|}}{C-S-CoA}}{\overset{H}{\diagup}} + FADH_2 \qquad (19.9)$$

trans-Δ²-enoyl-CoA

The electrons transferred from the fatty acyl-CoA to $FADH_2$ flow through an electron transport chain to cytochrome *b* in the mitochondrial respiratory chain (Figure 19.9). This oxidation step therefore leads to the synthesis of 2 moles of ATP, making good the initial pump-priming expenditure of two phosphoanhydride bonds.

d. Hydration of *trans*-Δ²-enoyl-CoA

S-3-hydroxyacyl-CoA hydrolase stereospecifically adds water to the double bond produced by the preceding dehydrogenase (reaction 19.10). This hydration reaction is reminiscent of the stereospecific hydration of fumarate in the citric acid cycle (Section 15.3G, Figure 15.18).

Hydration of *cis*-Δ²-enoyl-CoA derivatives, formed in the degradation of unsaturated fatty acids, leads to R-3-hydroxyacyl-CoA derivatives. We will discuss the fate of the R enantiomer later in this section.

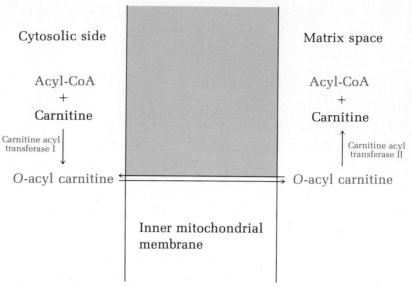

Figure 19.8 Transfer of acetyl-CoA across the inner mitochondrial membrane mediated by carnitine and two carnitine acyl transferases.

$$\underset{\text{trans-}\Delta^2\text{-enoyl-CoA}}{\begin{array}{c} R \\ \diagdown \\ C = C \\ \diagup \quad \diagdown \\ H \qquad C - S - CoA \\ \parallel \\ O \end{array}} + H_2O \xrightarrow{\begin{array}{c} \text{S-3-hydroxyacyl-CoA} \\ \text{hydrolase} \end{array}}$$

$$\underset{\text{S-3-Hydroxyacyl-CoA}}{R - \underset{\underset{H}{\mid}}{\overset{\overset{OH}{\mid}}{C}} - CH_2 - \overset{\overset{O}{\parallel}}{C} - S - CoA} \qquad (19.10)$$

e. S-Hydroxyacyl-CoA Dehydrogenase

S-3-hydroxyacyl-CoA is oxidized to 3-ketoacyl-CoA by an NAD-dependent dehydrogenase (reaction 19.11).

$$\underset{\text{S-3-Hydroxyacyl-CoA}}{R - \underset{\underset{H}{\mid}}{\overset{\overset{OH}{\mid}}{C}} - CH_2 - \overset{\overset{O}{\parallel}}{C} - S - CoA} + NAD \xrightarrow{\begin{array}{c} \text{S-3-Hydroxyacyl-CoA} \\ \text{dehydrogenase} \end{array}}$$

$$NADH + \underset{\text{3-Ketoacyl-CoA}}{R - \overset{\overset{O}{\parallel}}{C} - CH_2 - \overset{\overset{O}{\parallel}}{C} - S - CoA} + H^{\oplus} \qquad (19.11)$$

The NADH produced in this step is reoxidized in the electron transport chain with concomitant production of 3 moles of ATP.

f. 3-Ketothiolase

3-Ketothiolase (also called β-ketothiolase and thiolase) catalyzes Claisen cleavage of 3-ketoacyl-CoA, completing one cycle of β-oxidation (reaction 19.12).

$$R-\overset{\overset{\displaystyle O}{\|}}{C}-CH_2-\overset{\overset{\displaystyle O}{\|}}{C}-S-CoA + CoASH \xrightarrow{\text{Thiolase}}$$

3-ketoacyl-CoA

$$R'-\overset{\overset{\displaystyle O}{\|}}{C}-S-CoA + CH_3\overset{\overset{\displaystyle O}{\|}}{C}-SCoA \qquad (19.12)$$

Acyl-CoA   Acetyl-CoA

The equilibrium constant for this reaction is greater than 10^4, pulling β-oxidation to completion.

The 3-ketothiolase of β-oxidation is a mitochondrial enzyme. A second version of thiolase, found in the cytosol, is responsible for the synthesis of acetoacetyl-CoA. The mitochondrial reaction is a Claisen cleavage, while the cytosolic reaction is a Claisen condensation. Together, these enzymes represent two sides of the same mechanistic coin.

i. Claisen Condensation of Carboxyate Esters The base-catalyzed condensation of two carboxyate esters, called a Claisen condensation, is made possible by the acidic nature of the proton α to the carbonyl group. Abstraction of this proton generates a resonance-stabilized carbanion (reaction 19.13).

$$CH_3-\overset{\overset{\displaystyle O}{\|}}{C}-OR + B \rightleftharpoons {}^{\ominus}CH_2-\overset{\overset{\displaystyle O}{\|}}{C}-OR \longleftrightarrow$$

$$\left[CH_2{=}\overset{\overset{\displaystyle O^{\ominus}}{|}}{C}-OR\right] + BH^{\oplus} \qquad (19.13)$$

The carbanion attacks the electrophilic carbon of a second ester molecule giving a tetrahedral intermediate that expels alkoxide anion producing a 3-ketoester (reaction 19.14).

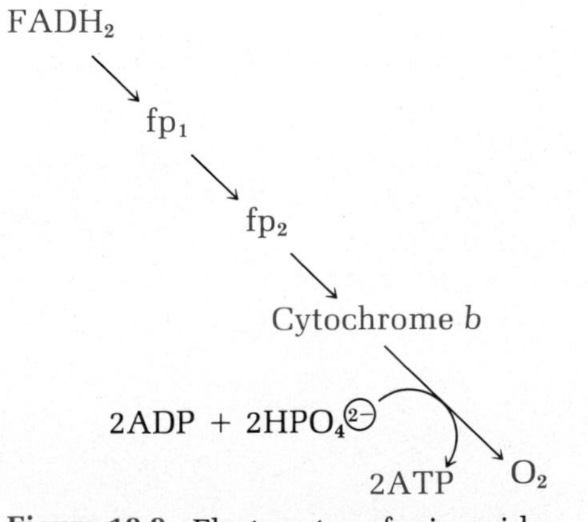

FADH$_2$

fp$_1$

fp$_2$

Cytochrome b

2ADP + 2HPO$_4^{2\ominus}$

2ATP   O$_2$

Figure 19.9 Electron transfer in oxidation of FADH$_2$ produced by the action of acyl-CoA dehydrogenase. The oxidation step in the respiratory electron transport chain yields 2 moles of ATP.

$$CH_3-\overset{\overset{\displaystyle O}{\|}}{C}-OR + {}^{\ominus}CH_2-\overset{\overset{\displaystyle O}{\|}}{C}-OR \longrightarrow$$

$$CH_3-\overset{\overset{\displaystyle O^{\ominus}}{|}}{\underset{\underset{\underset{\underset{OR}{|}}{C=O}}{\underset{|}{CH_2}}}{C}}-OR \longrightarrow CH_3-\overset{\overset{\displaystyle O}{\|}}{C}-CH_2-\overset{\overset{\displaystyle O}{\|}}{C}-OR + OR^{\ominus} \qquad (19.14)$$

3-Ketoester

The pK_a of the α-proton of carboxyate esters is about 10.5. This is not very acidic, and an enzyme would have trouble extracting the α-proton from carboxyate esters. In fact such esters are not substrates for enzyme-catalyzed Claisen condensations.

ii. Enzyme-Catalyzed Claisen Condensations *Acyl-CoA ligases,* such as thiolase, catalyze Claisen condensations of thioesters, reactions that are possible because thioesters have α-protons whose pK_a values are about 8.5. Abstraction of a proton gives a thioenolate anion (reaction 19.15).

$$R-\overset{\overset{\displaystyle H}{|}}{\underset{\underset{\underset{\ddot{B}\diagdown Enz}{}}{\underset{|}{H}}}{C}}-\overset{\overset{\displaystyle O}{\|}}{C}-S-CoA \rightleftharpoons$$

$pK_a = 8.5$

$$\left[R-\overset{\ominus}{C}H-\overset{\overset{\displaystyle O}{\|}}{C}-S-CoA \longleftrightarrow RCH=\overset{\overset{\displaystyle O^{\ominus}}{|}}{C}-S-CoA \right] + H-B^{\oplus} \qquad (19.15)$$

Thioenolate anion   Enz

In the reaction catalyzed by cytosolic thiolase the thioenolate anion of acetyl-CoA attacks the electrophilic carbonyl carbon of a second acetyl-CoA molecule (reaction 19.16).

$$CH_2=\overset{\overset{\displaystyle O^{\ominus}}{|}}{\underset{\underset{\underset{O}{\|}}{CH_3\overset{}{C}-S-CoA}}{C}}-S-CoA \longrightarrow CH_3-\overset{\overset{\displaystyle CH_2-\overset{\overset{\displaystyle O}{\|}}{C}-S-CoA}{}}{\underset{\underset{H^{\oplus}}{S-CoA}}{C\diagdown O^{\ominus}}} \xrightarrow[\text{(cytosol)}]{\text{Thiolase}}$$

$$CH_3-\overset{\overset{\displaystyle O}{\|}}{C}-CH_2-\overset{\overset{\displaystyle O}{\|}}{C}-S-CoA \qquad (19.16)$$

Acetoacetyl-CoA

The condensation of acetoacetyl-CoA with acetyl-CoA that produces S-3-hydroxymethylglutaryl-CoA (Figure 19.14) is also a Claisen condensation. In our discussion of ketone bodies we shall see that it can be converted to acetoacetate and that it is also the starting material in the biosynthesis of a variety of polyisoprenoid molecules, such as the terpenes and sterols.

The thiolase reaction we encounter in β-oxidation is a Claisen cleavage. A cysteine sulfhydryl group of thiolase adds to the 3-keto group of 3-ketoacyl-CoA. The tetrahedral intermediate fragments with elimination of the thioenolate anion of acetyl-CoA. Hydrolysis of the enzyme-thioester intermediate releases an acyl-CoA for another round of β-oxidation (reaction 19.17).

$$CH_3—\overset{O}{\overset{\|}{C}}—S\text{-CoA} + R\overset{O}{\diagup}S\text{-Enz} \xrightarrow{\text{CoA-SH}} R\overset{O}{\diagup}S\text{-CoA} \qquad (19.17)$$

Thioacyl enzyme   Acyl-CoA

19–4 OXIDATION OF UNSATURATED FATTY ACIDS

The enzymes of the β-oxidation pathway plus two novel reactions, involving an epimerase and an isomerase, convert unsaturated fatty acids completely to acetyl-CoA. Consider the β-oxidation of the doubly unsaturated linoleic acid. One round of oxidation produces acetyl-CoA and cis,cis-$\Delta^{7,10}$-hexadecadienoyl-CoA. Two more rounds of β-oxidation produce cis,cis-$\Delta^{3,6}$-dodecadienoyl-CoA and 2 moles of acetyl-CoA (Figure 19.10). Cis, cis-$\Delta^{3,6}$-dodecadienoyl-CoA is not a substrate for acyl-CoA dehydrogenase. However, in the first novel reaction cis-Δ^3-$trans$-Δ^2-$enoyl$ $isomerase$ produces a $trans$-Δ^2-cis-Δ^6-dodecadienoyl-CoA (reaction 19.18). The driving force for this reaction is the resonance stabilization that results when the double bond migrates into conjugation with the carbonyl group.

cis, cis-$\Delta^{3,6}$-hexadecadienoyl-CoA

$$(19.18)$$

$trans$-Δ^2-cis-Δ^6-Dodecadienoyl-CoA

The $trans$-Δ^2 double bond is now hydrated to give cis-Δ^6-S-3-hydroxyacyl-CoA. This intermediate is oxidized in the usual way to give cis-Δ^4-decenoyl-CoA and acetyl-CoA (Figure 19.11). Another round of β-oxidation produces cis-Δ^2-octenoyl-CoA and acetyl-CoA. The cis-Δ^2-acyl-CoA can be hydrated, but it gives R-3-hydroxyacyl-CoA which is not a substrate for S-3-hydroxyacyl-CoA dehydrogenase. In the second novel reaction of the oxi-

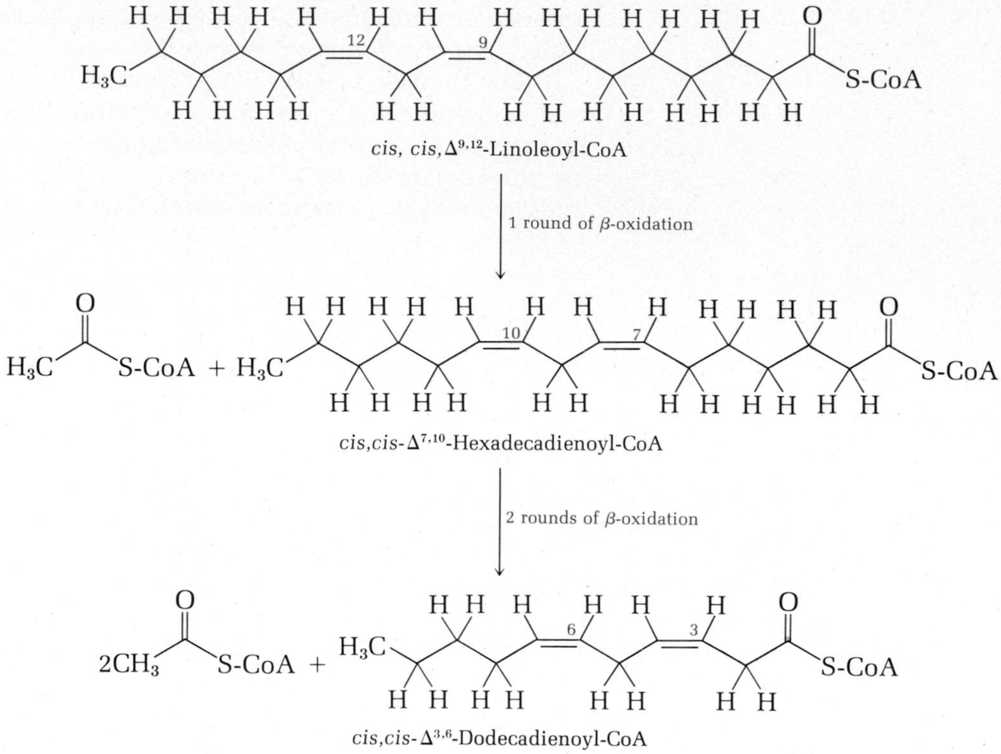

cis, cis,$\Delta^{9,12}$-Linoleoyl-CoA

1 round of β-oxidation

cis,cis-$\Delta^{7,10}$-Hexadecadienoyl-CoA

2 rounds of β-oxidation

cis,cis-$\Delta^{3,6}$-Dodecadienoyl-CoA

Figure 19.10 β-Oxidation of unsaturated fatty acids. One round of β-oxidation yields 1 mole of acetyl-CoA and a *cis,cis-*$\Delta^{7,10}$ acyl-CoA. Two more rounds produce 2 moles of acetyl-CoA and a *cis,cis-*$\Delta^{3,6}$ acyl-CoA.

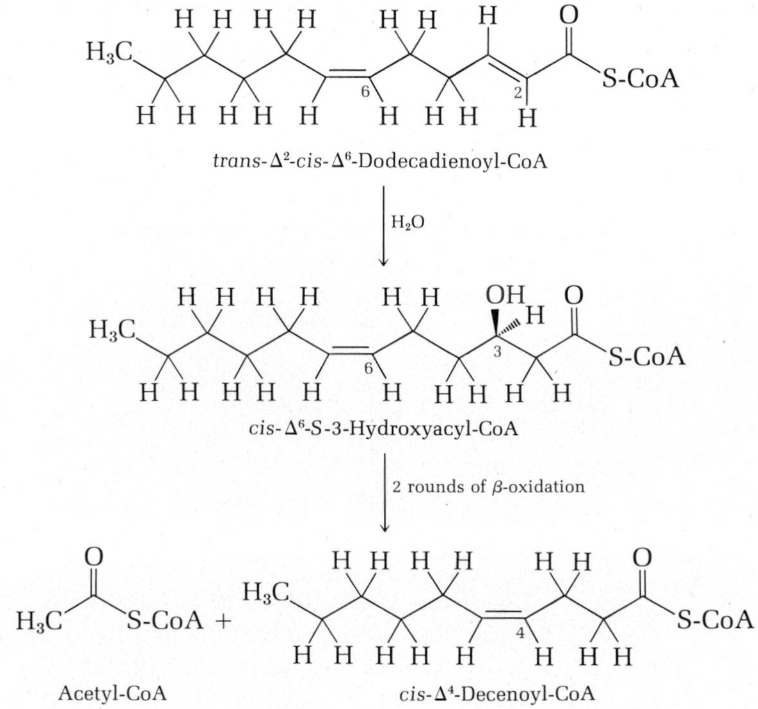

trans-Δ^2-*cis-*Δ^6-Dodecadienoyl-CoA

H_2O

cis-Δ^6-*S*-3-Hydroxyacyl-CoA

2 rounds of β-oxidation

Acetyl-CoA   *cis-*Δ^4-Decenoyl-CoA

Figure 19.11 Conversion of a *trans-*Δ^2-*cis-*Δ^6 acyl-CoA to a *cis-*Δ^4 acyl-CoA.

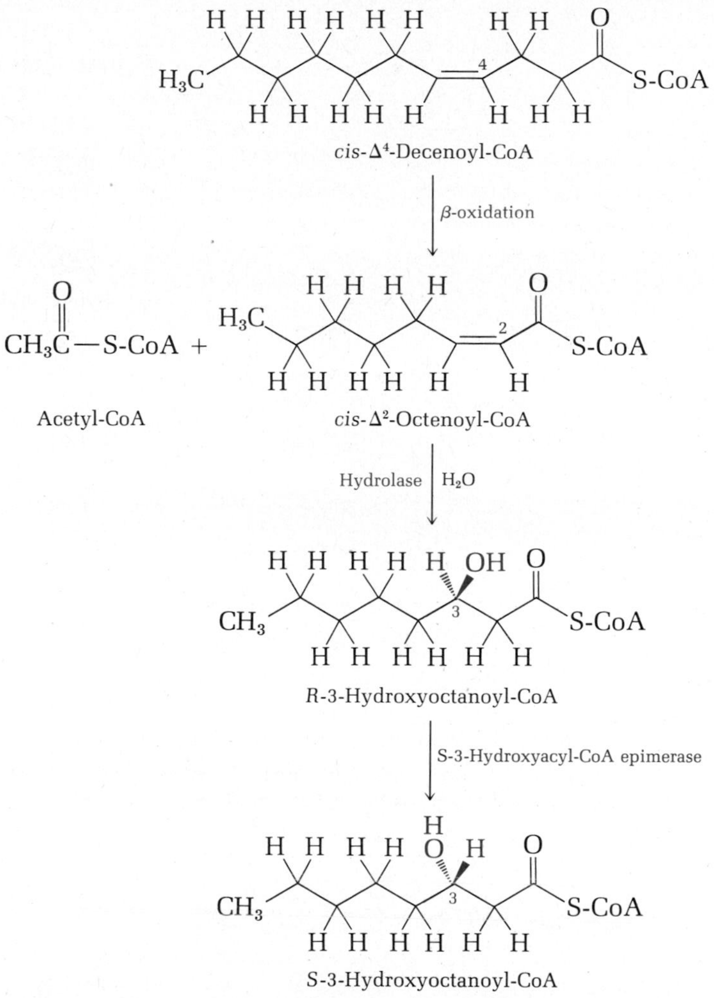

Figure 19.12 Conversion of a *cis*-Δ⁴ acyl-CoA to an S-3-hydroxy acyl-CoA.

dation of β-unsaturated fatty acids, *S-3-hydroxyacyl-CoA epimerase* inverts the configuration of the alcohol at C-3 to produce the S-enantiomer (Figure 19.12). Fatty acid oxidation now proceeds as usual.

Since unsaturated fatty acids have fewer hydrogens to transfer to FAD and NAD, their complete oxidation yields slightly less energy than the oxidation of saturated fatty acids.

19–5 OXIDATION OF ODD-CHAIN FATTY ACIDS

Fatty acids that contain an odd number of carbon atoms are oxidized by β-oxidation to acetyl-CoA and 1 mole of propionyl-CoA.[2] A three-step pathway involving *propionyl-CoA carboxylase*, *S-methylmalonyl-CoA racemase*, and *methylmalonyl-CoA mutase* converts propionyl-CoA to the citric acid cycle intermediate succinyl-CoA (Figure 19.13).

A. Propionyl-CoA Carboxylase

Propionyl-CoA carboxylase requires both ATP and biotin and incorporates bicarbonate in propionyl-CoA to produce S-methylmalonyl-CoA (reaction 19.19). We discussed the mechanism of this reaction in Section 10.7.

B. Epimerization of *S*-Methylmalonyl-CoA

The second step in the conversion of propionyl-CoA to succinyl-CoA is conversion of S-methylmalonyl-CoA to R-methylmalonyl-CoA. This reaction is

[2] The degradation of isoleucine and valine also produce propionyl-CoA (Section 21.4D).

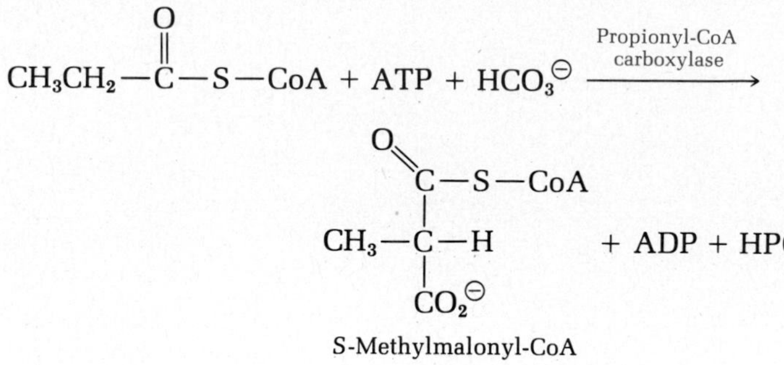

$$CH_3CH_2-\overset{\overset{\displaystyle O}{\|}}{C}-S-CoA + ATP + HCO_3^{\ominus} \xrightarrow{\text{Propionyl-CoA}\atop\text{carboxylase}}$$

$$\underset{\substack{\text{S-Methylmalonyl-CoA}}}{CH_3-\underset{\underset{\displaystyle CO_2^{\ominus}}{|}}{\overset{\overset{\displaystyle \overset{\displaystyle O}{\|}}{C}-S-CoA}{\underset{|}{C}}-H}} + ADP + HPO_4^{\textcircled{2}\ominus} \qquad (19.19)$$

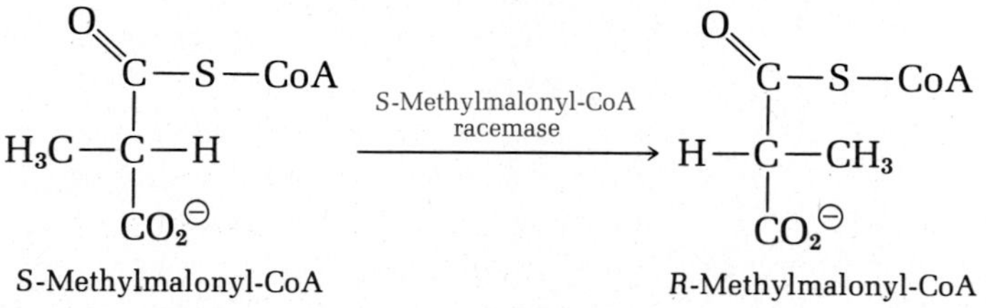

$$\underset{\text{S-Methylmalonyl-CoA}}{H_3C-\underset{\underset{\displaystyle CO_2^{\ominus}}{|}}{\overset{\overset{\overset{\displaystyle O}{\|}}{C}-S-CoA}{C}}-H} \xrightarrow[\text{racemase}]{\text{S-Methylmalonyl-CoA}} \underset{\text{R-Methylmalonyl-CoA}}{H-\underset{\underset{\displaystyle CO_2^{\ominus}}{|}}{\overset{\overset{\overset{\displaystyle O}{\|}}{C}-S-CoA}{C}}-CH_3} \qquad (19.20)$$

catalyzed by S-methylmalonyl-CoA racemase (reaction 19.20). The R-enantiomer connects the oxidation of odd-chain fatty acids to the citric acid cycle.

C. Methylmalonyl-CoA
Mutase

R-Methylmalonyl-CoA mutase converts R-methylmalonyl-CoA to succinyl-CoA in the final step of this short pathway. The enzyme requires coenzyme B_{12} to transfer the —COSCoA unit of R-methylmalonyl-CoA to the methyl

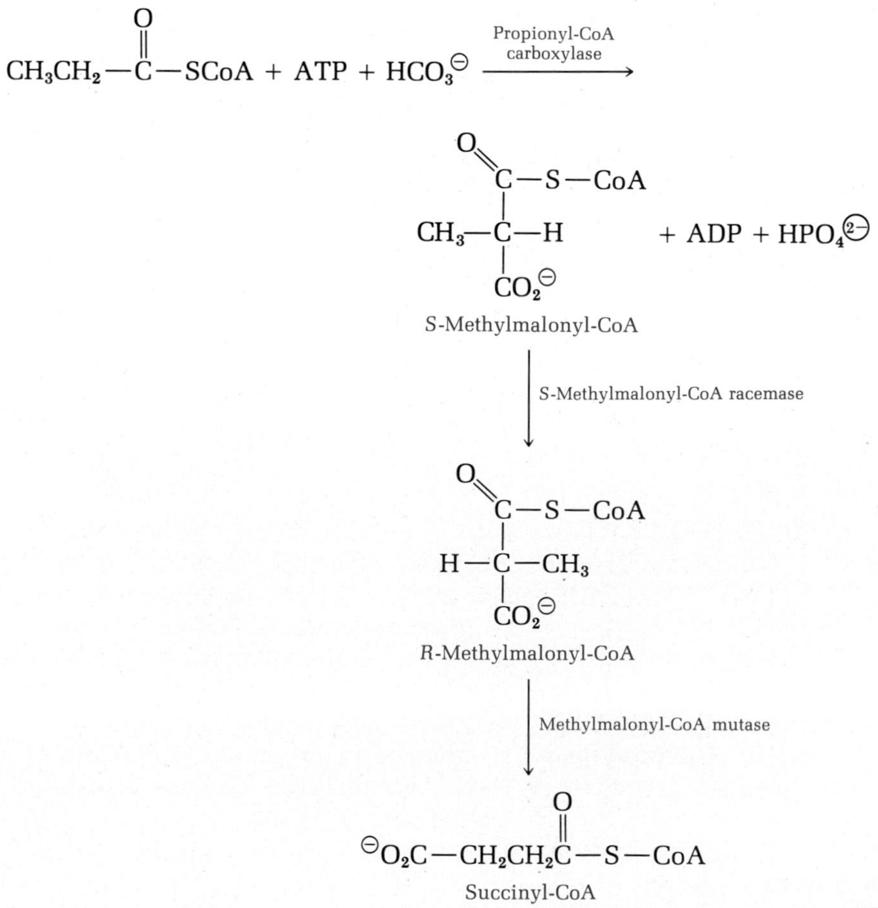

$$CH_3CH_2-\overset{\overset{\displaystyle O}{\|}}{C}-SCoA + ATP + HCO_3^{\ominus} \xrightarrow{\text{Propionyl-CoA}\atop\text{carboxylase}}$$

$$\underset{\text{S-Methylmalonyl-CoA}}{CH_3-\underset{\underset{\displaystyle CO_2^{\ominus}}{|}}{\overset{\overset{\overset{\displaystyle O}{\|}}{C}-S-CoA}{C}}-H} + ADP + HPO_4^{\textcircled{2}\ominus}$$

$$\Big\downarrow \text{S-Methylmalonyl-CoA racemase}$$

$$\underset{\text{R-Methylmalonyl-CoA}}{H-\underset{\underset{\displaystyle CO_2^{\ominus}}{|}}{\overset{\overset{\overset{\displaystyle O}{\|}}{C}-S-CoA}{C}}-CH_3}$$

$$\Big\downarrow \text{Methylmalonyl-CoA mutase}$$

$$\underset{\text{Succinyl-CoA}}{^{\ominus}O_2C-CH_2CH_2\overset{\overset{\displaystyle O}{\|}}{C}-S-CoA}$$

Figure 19.13 Pathway for the conversion of propionyl-CoA to succinyl-CoA.

group on the adjacent carbon, shifting a hydrogen to the methine carbon in the process (reaction 19.21).

$$
\begin{array}{c}
\underset{\text{R-Methylmalonyl-CoA}}{\text{H}_3\text{C}\overset{\displaystyle\text{H}\ \ \text{H}}{\underset{\displaystyle\text{H}}{-\text{C}-\text{C}}}-\text{CO}_2^{\ominus}} \xrightarrow[\text{Coenzyme B}_{12}]{\substack{\text{R-methylmalonyl-CoA}\\\text{Mutase}}}
\end{array}
$$

$$
\overset{\ominus}{\text{O}}_2\text{C}-\text{CH}_2-\text{CH}_2-\overset{\displaystyle\text{O}}{\overset{\|}{\text{C}}}-\text{S}-\text{CoA} \qquad (19.21)
$$

Succinyl-CoA

We discussed the mechanism of this reaction in Section 10.9.

19–6 FORMATION OF ACETOACETATE AND OTHER "KETONE BODIES"

The oxidation of the acetyl-CoA produced by the β-oxidation of fatty acids depends upon an adequate supply of oxaloacetate. If glycolysis is operating at a rate comparable to β-oxidation, there will be a steady supply of pyruvate which can be converted to oxaloacetate by pyruvate carboxylase. If, on the other hand, an adequate supply of oxaloacetate is not available, then acetyl-CoA is converted to R-3-hydroxybutyrate and to the so-called "ketone bodies," acetoacetate and acetone.

$$
\underset{\text{R-3-Hydroxybutyrate}}{\text{H}_3\text{C}\overset{\displaystyle\text{H}}{\underset{\displaystyle\text{OH}}{-\text{C}-}}\text{CH}_2-\text{CO}_2^{\ominus}} \qquad \underset{\text{Acetoacetate}}{\text{H}_3\text{C}-\overset{\displaystyle\text{O}}{\overset{\|}{\text{C}}}-\text{CH}_2-\text{CO}_2^{\ominus}} \qquad \underset{\text{Acetone}}{\text{H}_3\text{C}-\overset{\displaystyle\text{O}}{\overset{\|}{\text{C}}}-\text{CH}_3}
$$

Four enzymes are required for the formation of these products: (1) cytosolic 3-ketothiolase, discussed in Section 19.3B, (2) hydroxymethylglutaryl-CoA synthetase, (3) hydroxymethylglutaryl-CoA cleavage enzyme, and (4) R-3-hydroxybutyrate dehydrogenase (Figure 19.14). The first two of these enzymes catalyze Claisen condensations, and the third a Claisen cleavage. The fourth enzyme is a typical NAD-dependent dehydrogenase. Acetoacetate spontaneously (nonenzymatically) decarboxylates to give acetone. Diabetics typically have rather high blood concentrations of acetoacetate, and this decarboxylation is the origin of the "acetone breath" often noticed in diabetics.

Two conditions can lead to the formation of ketone bodies: extremely low carbohydrate intake or diabetes. Since diabetics have quite normal carbohydrate intake, diabetes amounts to starvation in the midst of plenty.

Acetoacetate and R-3-hydroxybutyrate are produced in the liver. Both molecules have carboxylate groups that make them water soluble. They diffuse into the bloodstream and are circulated to the heart, which oxidizes them. In fact, these two substances provide more fuel for heart muscle than does glucose. In the heart acetoacetate is converted to 2 moles of acetyl-CoA by the successive action of *CoA transferase*, an enzyme absent in liver, and thiolase. The source of CoA for CoA transferase is succinyl-CoA (reaction 19.22).

$$
\underset{\text{Acetoacetate}}{CH_3-\overset{\displaystyle O}{\overset{\|}{C}}-CH_2-CO_2^{\ominus}} \quad \xrightarrow[\substack{\text{succinyl-}\\ \text{CoA}}]{\text{CoA transferase}} \text{succinate}
$$

$$
\underset{\text{Acetoacetyl-CoA}}{CH_3-\overset{\displaystyle O}{\overset{\|}{C}}-CH_2-\overset{\displaystyle O}{\overset{\|}{C}}-S-CoA}
$$

$$
\Big\downarrow \text{Thiolase}
$$

$$
2CH_3-\overset{\displaystyle }{\underset{\displaystyle O}{\overset{\|}{C}}}-S-CoA \qquad (19.22)
$$

$$
\text{Acetyl-CoA}
$$

19–7 FATTY ACID BIOSYNTHESIS

Acetyl-CoA is the starting point of fatty acid biosynthesis, but the biosynthetic pathway is not a simple reversal of β-oxidation. First, fatty acid synthesis occurs in the cytosol, whereas β-oxidation occurs in the mitochondrion. Second, fatty acid catabolism produces NADH, whereas the reducing agent for fatty acid biosynthesis is NADPH. Third, fatty acid synthesis requires preliminary synthesis of malonyl-CoA from bicarbonate and acetyl-CoA by the ATP and biotin-dependent enzyme *acetyl-CoA carboxylase* (reaction 19.23).

$$
\underset{\text{Acetyl-CoA}}{CH_3\overset{\displaystyle O}{\overset{\|}{C}}-S-CoA} + ATP + HCO_3^{\ominus} \xrightarrow{\text{Acetyl-CoA carboxylase}}
$$

$$
\underset{\text{Malonyl-CoA}}{\overset{O}{\underset{\ominus O}{\diagdown}}C-CH_2-\overset{\displaystyle O}{\overset{\|}{C}}-S-CoA} + ADP + P_i + H^{\oplus} \qquad (19.23)
$$

Fourth, once malonyl-CoA has been synthesized, all further reactions of fatty acid biosynthesis are catalyzed by a six-enzyme complex called *fatty acid synthetase*. The intermediates of fatty acid biosynthesis are bound to an *acyl carrier protein* (ACP or ACP-SH), rather than to CoA. This small protein is part of the fatty acid synthetase complex. We recall that the phosphopantetheine moiety of CoA is bound to the 5′ position of adenosine 3′-phosphate. In the acyl carrier protein this moiety is bound to a hydroxyl group of a seryl side chain (Figure 19.15). The acetyl group of acetyl-CoA is transferred to the acyl carrier protein by *acetyl transacylase* (reaction 19.24).

$$
CH_3-\overset{\displaystyle O}{\overset{\|}{C}}-S-CoA + ACP-SH \xrightarrow{\substack{\text{Acetyl}\\ \text{transacylase}}}
$$

$$
\underset{\text{Acetyl-ACP}}{CH_3-\overset{\displaystyle O}{\overset{\|}{C}}-S-ACP} + CoA-SH \qquad (19.24)
$$

Similarly, the malonyl group of malonyl-CoA is transferred to the acyl carrier protein by *malonyl transacylase* (reaction 19.25).

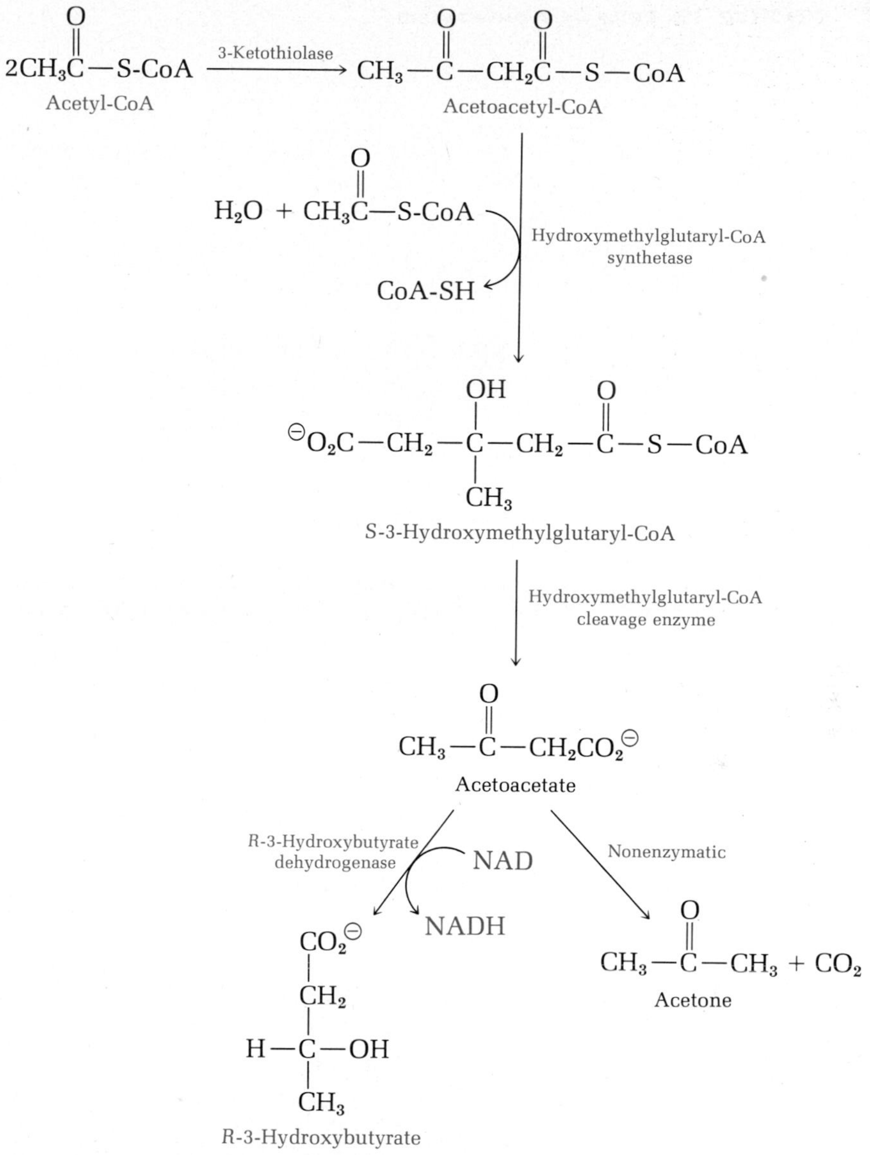

Figure 19.14 Pathway for formation of the ketone bodies acetoacetone, acetone and of R-3-hydroxybutanoate.

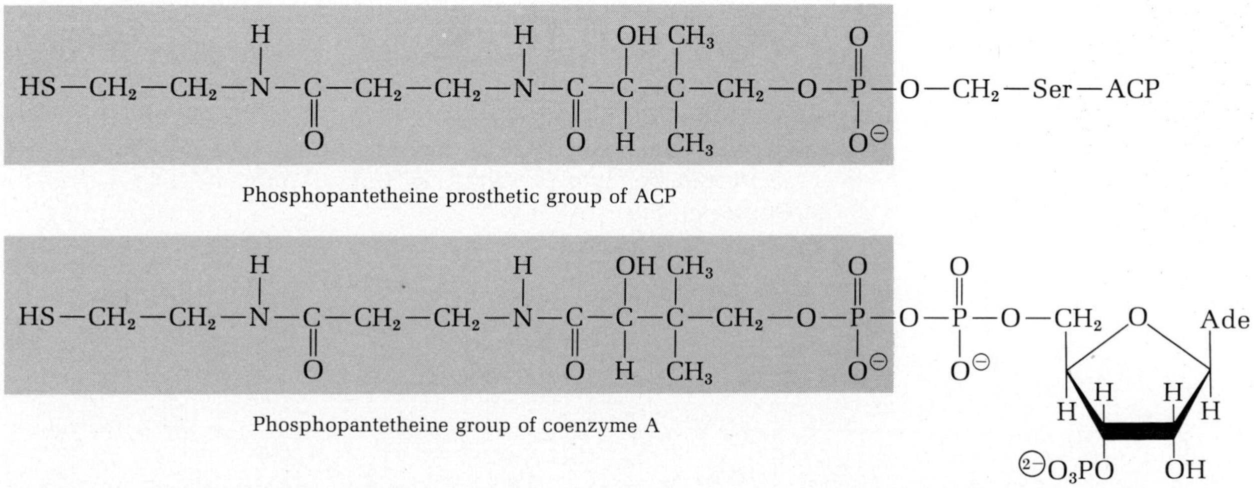

Phosphopantetheine prosthetic group of ACP

Phosphopantetheine group of coenzyme A

Figure 19.15 Phosphopantetheine group of acyl carrier protein and coenzyme A. The prosthetic group of ACP is bound to a seryl side chain. The same moiety is bound to the 5′ position of adenosine 3′-phosphate in coenzyme A.

$$\ominus O_2C-CH_2-\overset{\overset{\displaystyle O}{\|}}{C}-S-CoA \;+\; ACP-SH \;\xrightarrow[\text{transacylase}]{\text{Malonyl}}$$

$$\ominus O_2C-CH_2-\overset{\overset{\displaystyle O}{\|}}{C}-S-ACP \;+\; CoA-SH \qquad (19.25)$$

Malonyl-ACP

Fifth, once malonyl-ACP and acetyl-ACP have been synthesized, the remaining four components of fatty acid synthetase catalyze (1) condensation of malonyl-ACP and acetyl-ACP to give acetoacetyl-ACP, (2) reduction of acetoacetyl-ACP to R-3-hydroxybutyryl-ACP, (3) dehydration to crotonyl-ACP, and (4) reduction of crotonyl-ACP to butyryl-ACP (Figure 19.16). A single round of fatty acid synthesis therefore produces butyryl-ACP.

The next C_2 fragment is donated by malonyl-CoA, the entire set of reactions catalyzed by fatty acid synthetase is repeated, and hexanoyl-ACP is produced. This process repeats until palmitoyl-ACP is synthesized. Hydrolysis of the thioester of ACP then yields palmitate. Mammalian fatty acid synthetase is unable to synthesize chains longer than C_{16}. The net reaction for the synthesis of palmitate is

$$\text{acetyl-CoA} + 7\,\text{malonyl-CoA} + 14\text{NADPH} + 7\text{H}^{\oplus} \longrightarrow$$

$$\text{palmitate} + 7\text{CO}_2 + 14\text{NADP} + 8\text{CoA} + 6\text{H}_2\text{O} \qquad (19.26)$$

We can eliminate malonyl-CoA from the reaction by recalling that it is derived from acetyl-CoA and CO_2 (reaction 19.27).

$$7\,\text{acetyl-CoA} + 7\text{CO}_2 + 7\text{ATP} \longrightarrow$$

$$7\,\text{malonyl-CoA} + 7\text{ADP} + 7\text{P}_i + 7\text{H}^{\oplus} \qquad (19.27)$$

Adding equations 19.26 and 19.27 we obtain

$$8\,\text{acetyl-CoA} + 7\text{ATP} + 14\text{NADPH} \longrightarrow$$

$$\text{palmitate} + 14\text{NADP} + 8\text{CoA} + 6\text{H}_2\text{O} + 7\text{ADP} + 7\text{P}_i \qquad (19.28)$$

A. Sources of Acetyl-CoA and NADPH for Fatty Acid Synthesis

The enzymes that carry out fatty acid synthesis are located in the cytosol. However, the acetyl-CoA required for fatty acid synthesis is located in the mitochondrion, and since the inner mitochondrial membrane is impermeable to it, transport of acetyl-CoA to the cytosol does not occur directly. Acetyl-CoA and oxaloacetate are condensed by citrate synthetase to yield citrate. It is by transport of citrate out of the mitochondria by the tricarboxylate carrier that the elements of the acetyl group of acetyl-CoA enter the cytosol.

Once in the cytosol, citrate is converted to acetyl-CoA by the action of *ATP-citrate lyase* (reaction 19.29).

$$\begin{array}{l} \text{CH}_2-\text{CO}_2^{\ominus} \\ | \\ \text{HO}-\text{C}-\text{CO}_2^{\ominus} \\ | \\ \text{CH}_2\text{CO}_2^{\ominus} \end{array} \;+\; \text{CoA} + \text{ATP} \;\xrightarrow[\text{lyase}]{\text{ATP-citrate}}$$

Citrate

$$\text{CH}_3-\overset{\overset{\displaystyle O}{\|}}{C}-S-CoA \;+\; \ominus O_2C-CH_2\overset{\overset{\displaystyle O}{\|}}{C}-CO_2^{\ominus} \;+\; \text{ADP} + \text{HPO}_4^{\text{②}\ominus} \qquad (19.29)$$

Oxaloacetate

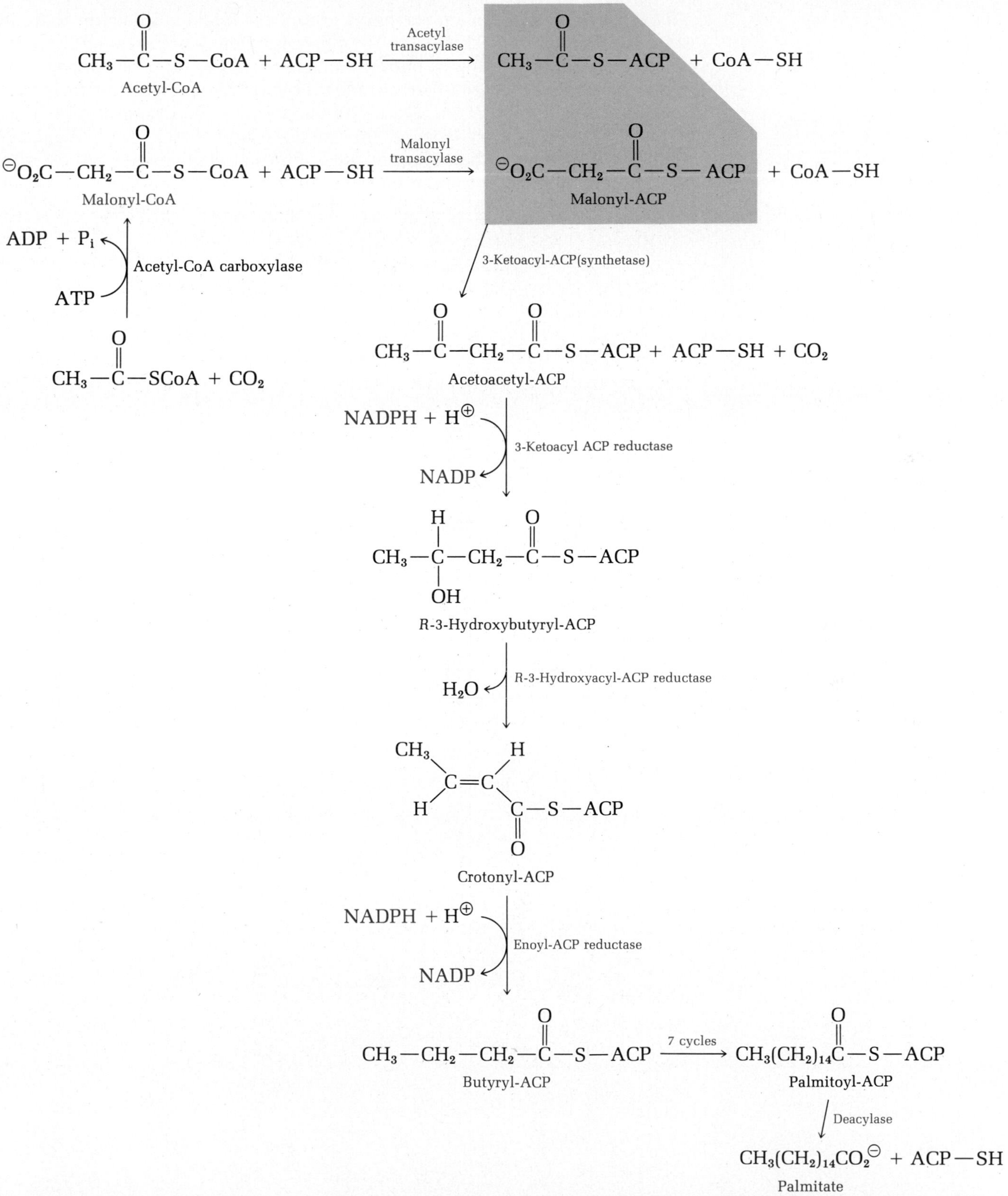

$$CH_3-\overset{\overset{\displaystyle O}{\|}}{C}-S-CoA + ACP-SH \xrightarrow{\text{Acetyl transacylase}} CH_3-\overset{\overset{\displaystyle O}{\|}}{C}-S-ACP + CoA-SH$$

Acetyl-CoA

$$^{\ominus}O_2C-CH_2-\overset{\overset{\displaystyle O}{\|}}{C}-S-CoA + ACP-SH \xrightarrow{\text{Malonyl transacylase}} {}^{\ominus}O_2C-CH_2-\overset{\overset{\displaystyle O}{\|}}{C}-S-ACP + CoA-SH$$

Malonyl-CoA Malonyl-ACP

Acetyl-CoA carboxylase
ADP + P$_i$
ATP

$$CH_3-\overset{\overset{\displaystyle O}{\|}}{C}-SCoA + CO_2$$

3-Ketoacyl-ACP(synthetase)

$$CH_3-\overset{\overset{\displaystyle O}{\|}}{C}-CH_2-\overset{\overset{\displaystyle O}{\|}}{C}-S-ACP + ACP-SH + CO_2$$

Acetoacetyl-ACP

NADPH + H$^{\oplus}$
3-Ketoacyl ACP reductase
NADP

$$CH_3-\overset{\overset{\displaystyle H}{|}}{\underset{\underset{\displaystyle OH}{|}}{C}}-CH_2-\overset{\overset{\displaystyle O}{\|}}{C}-S-ACP$$

R-3-Hydroxybutyryl-ACP

H$_2$O
R-3-Hydroxyacyl-ACP reductase

Crotonyl-ACP

NADPH + H$^{\oplus}$
Enoyl-ACP reductase
NADP

$$CH_3-CH_2-CH_2-\overset{\overset{\displaystyle O}{\|}}{C}-S-ACP \xrightarrow{\text{7 cycles}} CH_3(CH_2)_{14}\overset{\overset{\displaystyle O}{\|}}{C}-S-ACP$$

Butyryl-ACP Palmitoyl-ACP

Deacylase

$$CH_3(CH_2)_{14}CO_2^{\ominus} + ACP-SH$$

Palmitate

Figure 19.16 Reactions catalyzed by fatty acid synthetase. In *E. coli* seven separate enzymes are responsible for these reactions. In yeast and mammals two protein subunits of a single enzyme complex catalyze the same reactions.

Isotopic labeling studies have shown that the acetyl-CoA produced by ATP-citrate lyase is incorporated in fatty acids. ATP hydrolysis is required to drive the synthesis of the energy-rich thioester of acetyl-CoA. The mechanism of action of ATP-citrate lyase can be divided into five steps (Figure

19.17). (1) ATP donates a phosphoryl group to a nucleophilic side chain at the active site of the enzyme. (2) The phosphorylenzyme reacts with citrate to give an energy-rich acyl phosphate intermediate. (3) The enzyme displaces a phosphoryl group by an addition-elimination reaction to form a citryl-enzyme intermediate. (4) The nucleophilic sulfhydryl group of coenzyme A cleaves the citryl-enzyme intermediate to give citryl-CoA. We recall that this same intermediate is produced by the action of citrate synthetase in the mitochondrion. (5) Claisen cleavage of citryl-CoA produces acetyl-CoA

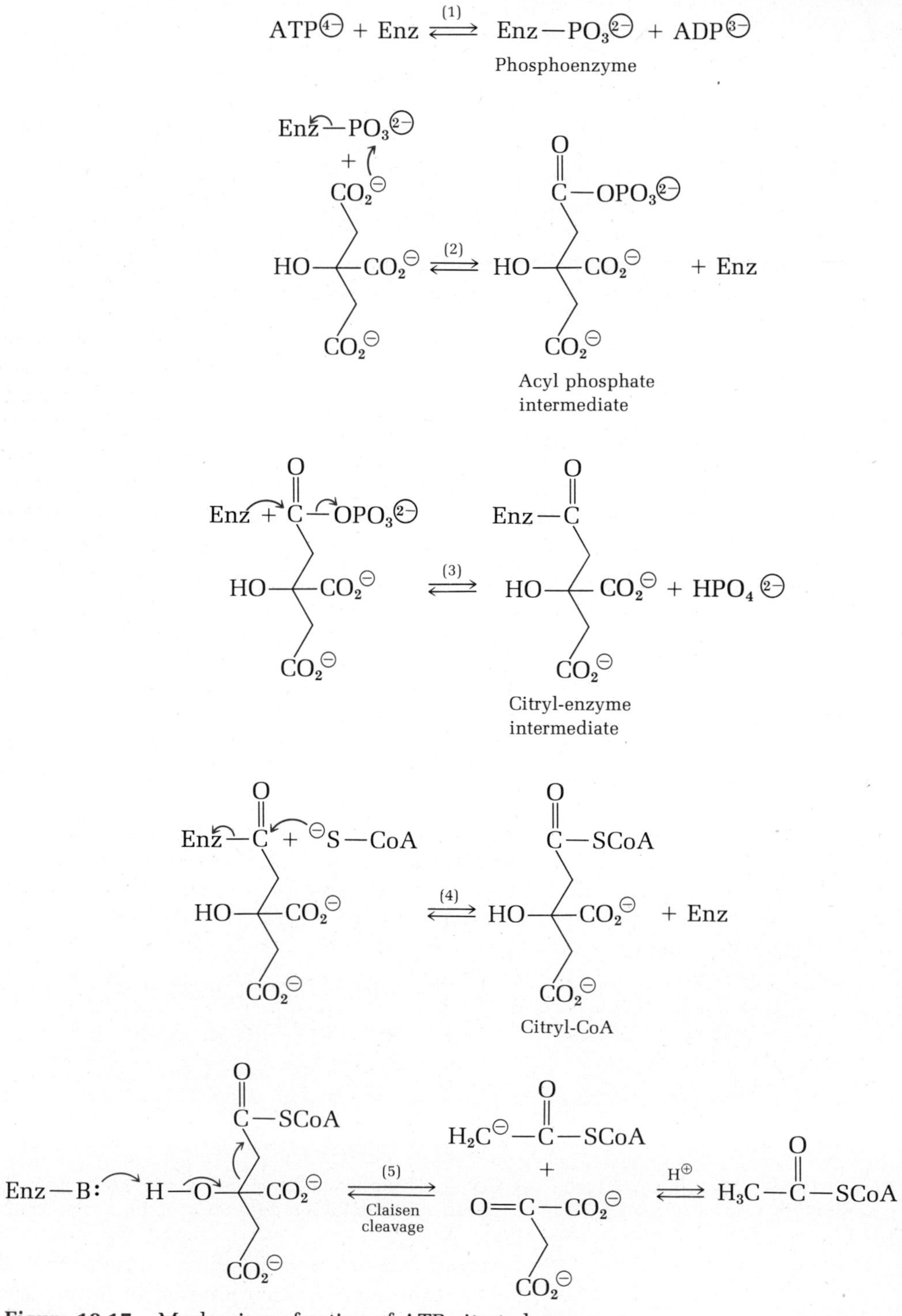

Figure 19.17 Mechanism of action of ATP-citrate lyase.

and oxaloacetate. The enzymic reaction is stereospecific, and the pro-S arm of citrate emerges in acetyl-CoA.

The NADPH required for fatty acid synthesis has two sources. The major one is the pentose phosphate pathway (Section 17.2). NADPH can also be supplied by the reduction of oxaloacetate by cytosolic *NAD-dependent malate dehydrogenase* (reaction 19.30).

$$^{\ominus}O_2C-CH_2-\overset{\overset{\displaystyle O}{\|}}{C}-CO_2^{\ominus} + NADH + H^{\oplus} \xrightarrow[\text{dehydrogenase}]{\text{Malate}}$$

Oxaloacetate

$$^{\ominus}O_2C-\overset{\overset{\displaystyle OH}{|}}{\underset{\underset{\displaystyle H}{|}}{C}}-CH_2-CO_2^{\ominus} + NAD \quad (19.30)$$

Malate

Oxidative decarboxylation by *NADP-dependent malic enzyme* then yields pyruvate and NADPH (reaction 19.31).

$$^{\ominus}O_2C-\overset{\overset{\displaystyle OH}{|}}{\underset{\underset{\displaystyle H}{|}}{C}}-CH_2-CO_2^{\ominus} + NADP \xrightarrow{\text{Malic enzyme}}$$

Malate

$$CH_3-\overset{\overset{\displaystyle O}{\|}}{C}-CO_2^{\ominus} + CO_2 + NADPH + H^{\oplus} \quad (19.31)$$

Pyruvate

The net result of reactions 19.30 and 19.31 is that an NADH is exchanged for an NADPH. No net oxidation occurs in the two reactions. The reactions by which acetyl-CoA and NADPH are provided for fatty acid synthesis are summarized in Figure 19.18.

B. Acetyl-CoA Carboxylase

For many years, it seemed that fatty acid synthesis and degradation occurred by the same pathways, one merely being the reverse of the other, a situation reminiscent of the state of knowledge of glycogen metabolism a few years earlier. The discovery of *acetyl-CoA carboxylase* in 1958 led to the discovery that fatty acid synthesis and degradation occur by different processes. Acetyl-CoA carboxylase converts acetyl-CoA to malonyl-CoA in a two-step reaction that requires ATP as an energy source (reaction 19.32).

$$ATP + HCO_3^{\,\ominus} + \text{biotin-enzyme} \longrightarrow \,^{\ominus}O_2C\text{-biotin-enzyme} + ADP$$

N^1-carboxybiotin

$$^{\ominus}O_2C\text{-biotin-enzyme} + CH_3\overset{\overset{\displaystyle O}{\|}}{C}-S\text{-CoA} \longrightarrow$$

$$^{\ominus}O_2C-CH_2-\overset{\overset{\displaystyle O}{\|}}{C}-S-CoA \quad (19.32)$$

Malonyl-CoA

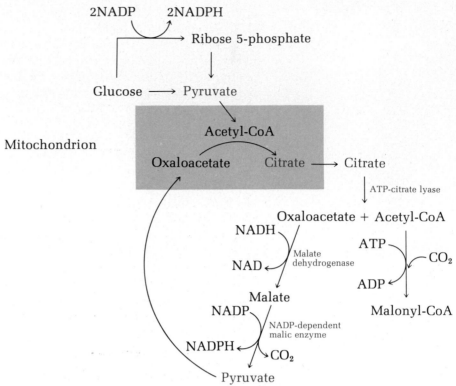

Figure 19.18 Sources of acetyl-CoA and NADPH for fatty acid synthesis. Acetyl-CoA is provided by transport of citrate across the inner mitochondrial membrane followed by the action of ATP-citrate lyase. NADPH is produced by the successive action of malate dehydrogenase and the NADP-dependent malic enzyme and by the pentose phosphate pathway.

The properties of acetyl-CoA carboxylase are quite similar to those of pyruvate carboxylase (Section 17.3A). Biotin is bound to the enzyme through its carboxylate group to a lysyl side chain on the enzyme. As we see in reaction 19.32, the synthesis of malonyl-CoA occurs in two half-reactions. N^1-carboxybiotin forms as a reaction intermediate in the first half-reaction, and CO_2 is transferred to acetyl-CoA in the second half-reaction. Ping-pong kinetics are also observed. In fact, if we recall the ping-pong mechanism for pyruvate carboxylase, we need only replace pyruvate by acetyl-CoA and oxaloacetate by malonyl-CoA to obtain the diagram for ping-pong kinetics with acetyl-CoA decarboxylase (see page 672).

We discussed the major points of the chemical details of acetyl-CoA carboxylase in Section 10.7. Perhaps the most striking property of acetyl-CoA carboxylase from adipose liver tissue and lactating mammary gland is its activation by citrate and isocitrate. These allosteric activators have a high affinity for acetyl-CoA carboxylase; K_m for citrate is 3×10^{-6} M. Both activators dramatically increase V for the enzyme. Acetyl-CoA carboxylase is an oligomer whose monomer (mol wt 410,000) is completely inactive. Citrate and isocitrate promote oligomer formation. The active enzyme is a gigantic polymer with a molecular weight of 8×10^6. In the presence of citrate 10 to 20 monomers assemble in a linear, filamentlike polymer 7 to 10 nm in diameter, whose length is up to 40 nm (Figure 19.19). The polymer is fully active. Acetyl-CoA is required for polymer stability, and in the absence of acetyl-CoA the polymer dissociates.

The effect of citrate on metabolism is astonishing. It has a regulatory function in glycolysis, the citric acid cycle, and fatty acid metabolism. These regulatory effects indicate the central role played by the citric acid cycle in metabolism. A high concentration of citrate simultaneously promotes both glycogen synthesis and fatty acid synthesis.

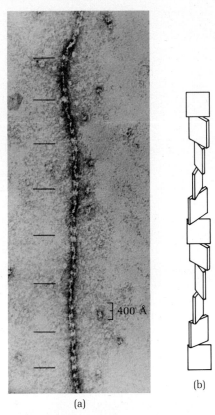

(a)

(b)

400 Å

Figure 19.19 (a) Electron micrograph of the enzymatically active linear polymer of acetyl-CoA carboxylase. (b) Schematic diagram. [Courtesy of Dr. M. Daniel Lane.]

C. Fatty Acid Synthetase

Once malonyl-CoA has been synthesized all remaining reactions in the biosynthesis of fatty acyl groups are catalyzed by *fatty acid synthetase* (recall Figure 19.16). The subunit compositions and molecular weights of fatty acid synthetases from a variety of sources are summarized in Table 19.1. Fatty acid synthetase of yeast and mammals consists of two subunits, each of which is a *multifunctional enzyme* (Table 19.2). Each subunit is encoded by one gene. In contrast, the fatty acid synthetase of *E. coli* is a tightly ag-

Source	Molecular weight of native fatty acid synthetase ($\times 10^{-3}$)	Molecular weight of subunit ($\times 10^{-3}$)
Table 19.1	**Molecular weights of fatty acid synthetases from various sources and their subunit compositions**	
Mammalian liver		
Human	410	
Rat	540	250
Rabbit	453	225
Mammalian mammary gland		
Rat	530	250
Rabbit	910	
Mammalian brain		
Mouse	500	
Avian liver		
Pigeon	450	220
Yeast	2200	180
		185

From K. Bloch, *Ann. Rev. Biochem., 46,* 272 (1977). Reproduced with permission. © 1977 by Annual Reviews Inc.

Table 19.2	**Multifunctional proteins of the *fas* 1 and *fas* 2 genes of yeast fatty acid synthetase**
***fas* 1 (subunit A)**	***fas* 2 (subunit B)**
Acetyl transacylase	4-Phosphopantetheine-binding region
Malonyl transacylase	3-Ketoacyl-ACP synthetase
Dehydratase	3-Ketoacyl reductase
Enoyl reductase	

gregated multienzyme complex that consists of seven separate enzymes. In both cases the growing fatty acyl chain is covalently bound to 4-phosphopantetheine. In the *E. coli* enzyme a separate protein, the acyl carrier protein to which we referred above, is the carrier of the fatty acyl group. In the mammalian polyfunctional enzymes the binding site for the prosthetic group is not a separate acyl carrier protein, but part of the B subunit. Covalent bond formation between the seryl side chain and 4-phosphopantetheine is catalyzed by holo-ACP synthetase. Coenzyme A is the source of 4-phosphopanthetheine for this reaction, and AMP is displaced (reaction 19.33).

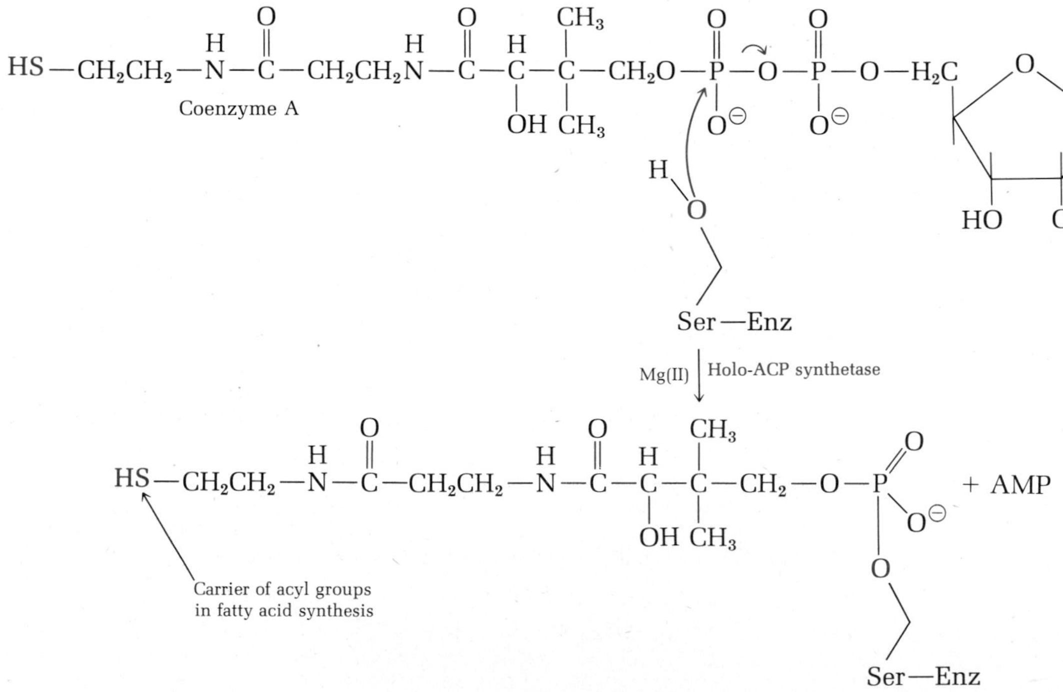

$$\tag{19.33}$$

In either case, the prosthetic group has a long arm that can swing the biosynthetic intermediates from one active site to the next. The synthesis of fatty acids in a multienzyme complex ensures both that intermediates will not accumulate and that they will not diffuse away from the site of their synthesis.

a. Transacylation

The first step of fatty acid synthesis is the formation of two acyl-ACP intermediates in reactions catalyzed by acetyl-CoA-ACP transacylase and malonyl-CoA-ACP transacylase (reactions 19.34 and 19.35).

b. Chain Elongation: 3-ketoacyl-ACP synthetase

The second step of fatty acid synthesis is formation of a 3-ketothioester by 3-ketoacyl-ACP synthetase (reaction 19.36).

$$CH_3-\overset{\overset{\displaystyle O}{\|}}{C}-S-CoA + ACP-SH \xrightarrow{\text{Acetyl-CoA-ACP transacylase}}$$

$$CH_3-\overset{\overset{\displaystyle O}{\|}}{C}-S-ACP + CoA-SH \qquad (19.34)$$
$$\text{Acetyl-ACP}$$

$$^{\ominus}O_2C-CH_2-\overset{\overset{\displaystyle O}{\|}}{C}-S-CoA + ACP-SH \xrightarrow{\text{Malonyl-CoA-ACP transacylase}}$$

$$^{\ominus}O_2C-CH_2-\overset{\overset{\displaystyle O}{\|}}{C}-S-ACP + CoA-SH \qquad (19.35)$$
$$\text{Malonyl-ACP}$$

$$CH_3-\overset{\overset{\displaystyle O}{\|}}{C}-S-ACP + {}^{\ominus}O_2C-CH_2-\overset{\overset{\displaystyle O}{\|}}{C}-S-ACP \xrightarrow{\text{3-Ketoacyl-ACP synthetase}}$$

$$CH_3-\overset{\overset{\displaystyle O}{\|}}{C}-CH_2-\overset{\overset{\displaystyle O}{\|}}{C}-S-ACP + ACP-SH + CO_2 \qquad (19.36)$$
$$\text{3-Ketobutyryl-ACP}$$

The rate of condensation increases with increasing chain length, and acyl-ACP intermediates do not accumulate. Acetyl-ACP acts as an electrophile, and malonyl-ACP acts as a nucleophile in a concerted condensation reaction that is driven to completion by decarboxylation (Figure 19.20).

This reaction reveals the logic of the ATP-driven synthesis of malonyl-CoA catalyzed by acetyl-CoA carboxylase. The large free energy of formation of carbon dioxide makes it an excellent leaving group, and the step catalyzed by 3-ketoacyl-malonyl synthetase is "anticipated" by the preliminary carboxylation of acetyl-CoA catalyzed by acetyl-CoA carboxylase.

c. Reduction of the 3-ketothioester to a 3-hydroxythioester

3-Ketoacyl-ACP reductase uses NADPH to convert the keto group to an alcohol. The equilibrium constant is 3.9×10^7 (reaction 19.37).

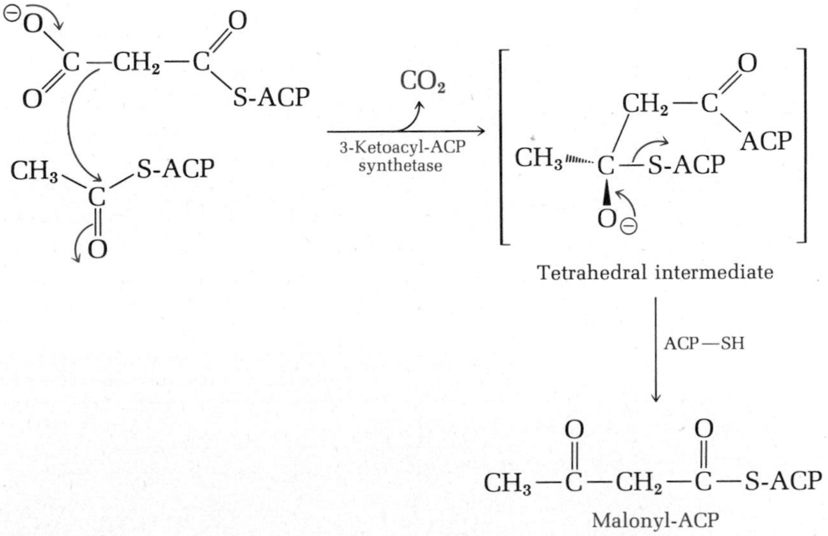

Figure 19.20 Condensation of malonyl-ACP and acetyl-ACP by a concerted reaction mechanism driven by loss of carbon dioxide.

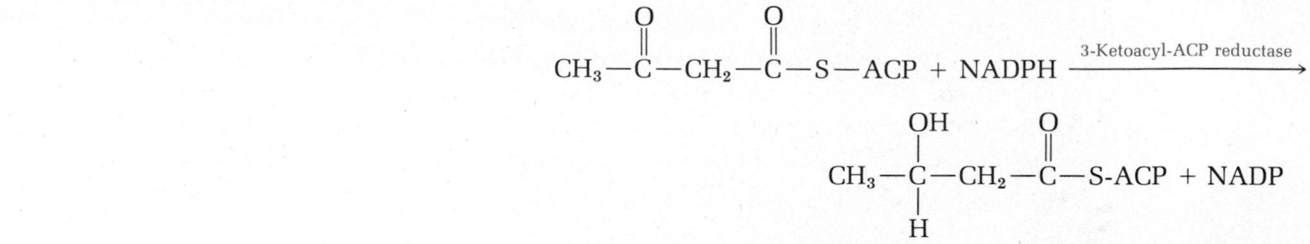

$$CH_3-\overset{\overset{O}{\|}}{C}-CH_2-\overset{\overset{O}{\|}}{C}-S-ACP + NADPH \xrightarrow{\text{3-Ketoacyl-ACP reductase}}$$

$$CH_3-\overset{OH}{\underset{H}{\overset{|}{C}}}-CH_2-\overset{\overset{O}{\|}}{C}-S\text{-}ACP + NADP \qquad (19.37)$$

R-3-Hydroxybutyryl-ACP

The reductase acts upon 3-ketoacyl-ACP substrates from C_4 to C_{16} in length.

d. Dehydration

A family of three R-3-hydroxyacyl-ACP dehydratases is found in *E. coli*. These enzymes all require the R-alcohol, and all produce *trans* products by stereospecific *anti* elimination of water (reaction 19.38).

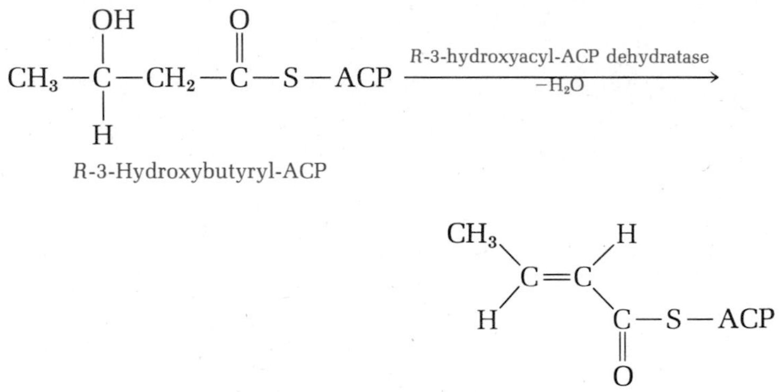

$$CH_3-\overset{OH}{\underset{H}{\overset{|}{C}}}-CH_2-\overset{\overset{O}{\|}}{C}-S-ACP \xrightarrow[-H_2O]{\text{R-3-hydroxyacyl-ACP dehydratase}}$$

R-3-Hydroxybutyryl-ACP

$$\underset{H}{\overset{CH_3}{>}}C=C\overset{H}{\underset{\underset{\overset{\|}{O}}{C}-S-ACP}{<}} \qquad (19.38)$$

The chain length specificities of the *E. coli* enzymes are given in Table 19.3.

e. Hydrogenation

The *trans* thioester is next reduced by *enoyl-ACP reductase*. Two *E. coli* enzymes can catalyze this reaction. One requires NADPH as its reducing agent and the other requires NADH. These enzymes are active for substrates whose chain length varies from C_4 to C_{16} (reaction 19.39).

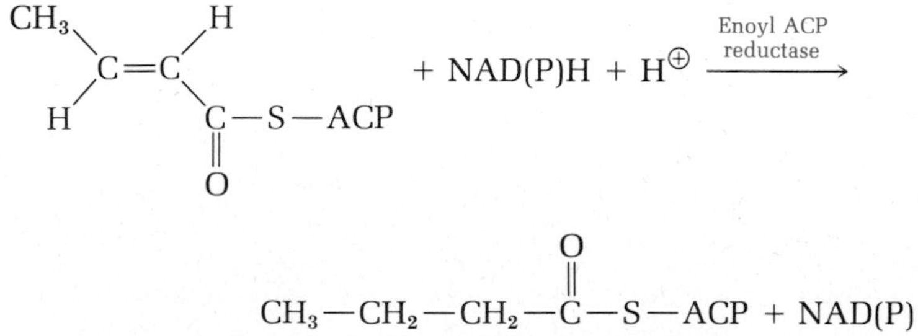

$$\underset{H}{\overset{CH_3}{>}}C=C\overset{H}{\underset{\underset{\overset{\|}{O}}{C}-S-ACP}{<}} + NAD(P)H + H^{\oplus} \xrightarrow{\text{Enoyl ACP reductase}}$$

$$CH_3-CH_2-CH_2-\overset{\overset{O}{\|}}{C}-S-ACP + NAD(P) \qquad (19.39)$$

Table 19.3 Chain length specificities of *E. coli* hydroxyacyl dehydratases

Enzyme	Optimal chain length
3R-Hydroxyacyl*butyryl*-ACP dehydratase	$C_4 > C_6 > C_8$ (no activity with C_{10})
3R-Hydroxyacyl*octanoyl*-ACP dehydratase	$C_6 < C_8 < C_{12} < C_{10}$
3R-Hydroxyacyl*palmitoyl*-ACP dehydratase	$C_{14} < C_{12} < C_{16}$

f. Termination: Palmitoyl Thioesterase

The repeated cycling of steps 1-5 leads to formation of palmitoyl-ACP. *E. coli* fatty acid synthetase will not catalyze further chain elongation, a process requiring additional enzymes. Palmitoyl thioesterase catalyzes the hydrolysis of palmitoyl-ACP (reaction 19.40).

$$H_2O + \text{palmitoyl-ACP} \xrightarrow{\underset{\text{thioesterase}}{\text{Palmitoyl}}} \text{palmitate} + H^\oplus + \text{ACP—SH} \qquad (19.40)$$

19-8 REGULATION OF FATTY ACID METABOLISM

Fatty acid synthesis is regulated by the control of the activity of acetyl-CoA carboxylase and fatty acid synthetase. The acetyl-CoA carboxylases from mammalian liver adipose tissue and lactating mammary gland are activated by citrate and isocitrate. The acetyl-CoA carboxylases from other sources are not affected by citrate and isocitrate.

Palmitoyl-CoA is also a potent regulator of fatty acid synthesis. It acts in at least 5 ways: (1) It inhibits acetyl-CoA carboxylases from many sources. The effects of citrate on liver and mammary gland acetyl-CoA carboxylases are reversed by palmitoyl-CoA. (2) It inhibits fatty acid synthetase. (3) It controls the concentration of citrate by inhibiting citrate synthetase. (4) By inhibiting the tricarboxylate anion carrier of mitochondria, it controls the exit of citrate to the cytosol. Therefore, it controls the supply of acetyl-CoA required for biosynthesis by limiting the availability of substrate for citrate lyase. (5) It inhibits glucose 6-phosphate dehydrogenase, thereby limiting the supply of NADPH required for fatty acid biosynthesis. This combination of effects integrates the citric acid cycle, fatty acid metabolism, and the pentose phosphate pathway.

Acetyl-CoA is the final degradative product of β-oxidation. It is also produced by the action of pyruvate dehydrogenase and can therefore be obtained from both carbohydrate catabolism and from fat. The bridge between glucose degradation and fatty acid biosynthesis is the conversion of acetyl-CoA to malonyl-CoA. If the acetyl-CoA derived from carbohydrates and the fatty acids derived from triglycerides are to be oxidized, they must enter the mitochondria. The enzyme responsible for fatty acyl group transport into the mitochondrion is carnitine acyl transferase I. Malonyl-CoA inhibits this enzyme and prevents the entry of acyl-CoA molecules, including acetyl-CoA, into the mitochondrion. If acetyl-CoA carboxylase is active, the supply of malonyl-CoA increases, the entry of fatty acyl groups into the mitochondrion is inhibited, β-oxidation stops for want of substrate, and fatty acid biosynthesis takes over. The relationship between fatty acid biosynthesis and β-oxidation is summarized in Figure 19.21.

19-9 CHAIN ELONGATION AND DESATURATION

Palmitate is the longest chain fatty acid that can be synthetized by fatty acid synthetase. Other pathways are required for synthesis of longer chains, up to 24 carbons, and for the synthesis of unsaturated fatty acids.

Chain elongation can occur either in the mitochondrion or in particles derived from the endoplasmic reticulum called microsomes. The small membrane-sealed vesicles are artifacts of experimental procedure, not cell organelles in their own right. The well-characterized mitochondrial system, the better understood of the two, adds C_2 fragments derived from acetyl-CoA in a series of reactions that parallels the reactions catalyzed by fatty acid synthetase.

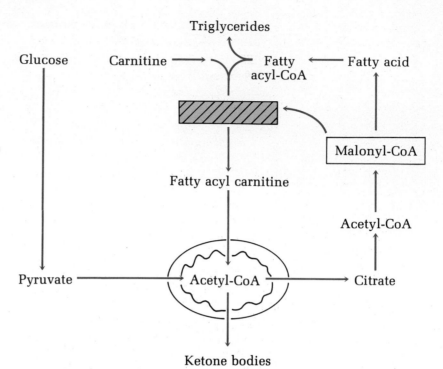

Figure 19.21 Regulation of fatty acid oxidation and fatty acid synthesis by the concentration of malonyl-CoA. Malonyl-CoA inhibits carnitine acyltransferase, thus inhibiting the entry of acyl groups into the mitochondrion. β-Oxidation therefore stops for want of substrate, and acetyl-CoA is incorporated in fatty acids. [From J. D. McGarry and D. W. Foster, *Ann. Rev. Biochem.*, 49, 411 (1980). Reproduced with permission. © 1980 by Annual Reviews Inc.]

A. Chain Elongation

Chain elongation occurs by a thiolase-catalyzed reaction between a long chain acyl-CoA and acetyl-CoA (reaction 19.41).

$$\text{R—CH}_2\text{=}\overset{\overset{\displaystyle O}{\|}}{C}\text{—S—CoA} + \text{CH}_3\text{—}\overset{\overset{\displaystyle O}{\|}}{C}\text{—S—CoA} \xrightarrow{\text{Thiolase}}$$

$$\text{R—CH}_2\text{—}\overset{\overset{\displaystyle O}{\|}}{C}\text{—CH}_2\text{—}\overset{\overset{\displaystyle O}{\|}}{C}\text{—S—CoA} + \text{CoA—SH} \qquad (19.41)$$

An acyl carrier protein is not required, and malonyl-CoA cannot substitute for acetyl-CoA.

B. Reduction

An NADH-dependent dehydrogenase, in contrast to the NADPH-dependent dehydrogenase of fatty acid synthetase, converts the 3-ketothioester to a R-3-hydroxyacyl-CoA (reaction 19.42).

$$\text{R—CH}_2\text{—}\overset{\overset{\displaystyle O}{\|}}{C}\text{—CH}_2\text{—}\overset{\overset{\displaystyle O}{\|}}{C}\text{—S—CoA} + \text{NADH} + \text{H}^{\oplus} \xrightarrow{\substack{\text{3-Ketoacyl-CoA} \\ \text{dehydrogenase}}}$$

3-Ketoacyl-CoA

$$\text{R—CH}_2\text{—}\underset{\underset{\displaystyle H}{|}}{\overset{\overset{\displaystyle OH}{|}}{C}}\text{—CH}_2\text{—}\overset{\overset{\displaystyle O}{\|}}{C}\text{—S—CoA} + \text{NAD} \qquad (19.42)$$

R-3-Hydroxyacyl-CoA

C. Dehydration Dehydration of the chain-lengthened acyl-CoA is catalyzed by R-3-hydroxyacyl dehydrase (reaction 19.43). This reaction is analogous to the dehydration reaction catalyzed by fatty acid synthetase, but a different enzyme is used.

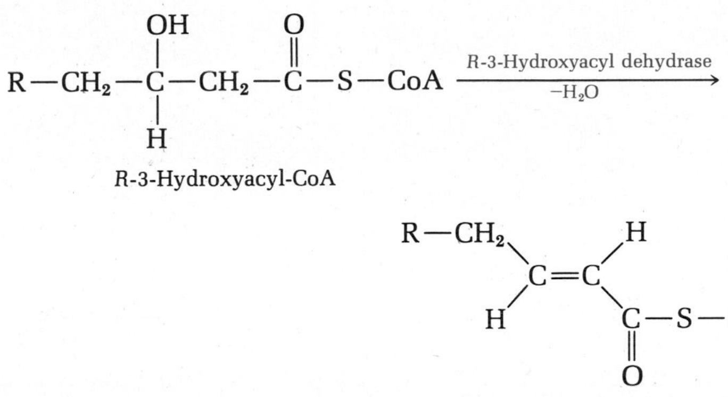

$$R-CH_2-\underset{\underset{H}{\overset{OH}{|}}}{C}-CH_2-\overset{\overset{O}{\|}}{C}-S-CoA \xrightarrow[-H_2O]{\text{R-3-Hydroxyacyl dehydrase}}$$

R-3-Hydroxyacyl-CoA

$$\underset{H}{\overset{R-CH_2}{\diagdown}}C=C\overset{H}{\underset{\underset{\overset{\|}{O}}{C-S-CoA}}{\diagup}} \qquad (19.43)$$

trans-Δ²-enoyl-CoA

D. Hydrogenation An NADPH-dependent enoyl reductase catalyzes the hydrogenation of the trans-Δ²-enoyl-CoA to complete one round of chain elongation (reaction 19.44).

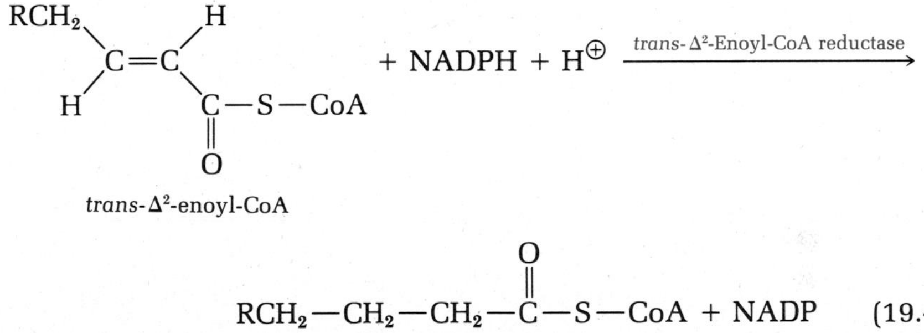

$$\underset{H}{\overset{RCH_2}{\diagdown}}C=C\overset{H}{\underset{\underset{\overset{\|}{O}}{C-S-CoA}}{\diagup}} + NADPH + H^{\oplus} \xrightarrow{\text{trans-Δ²-Enoyl-CoA reductase}}$$

trans-Δ²-enoyl-CoA

$$RCH_2-CH_2-CH_2-\overset{\overset{O}{\|}}{C}-S-CoA + NADP \qquad (19.44)$$
Chain-lengthened acyl-CoA

E. Desaturation of Fatty Acids In mammals desaturation of fatty acids is carried out by microsomal enzymes of the liver. An NADH-dependent stearyl-CoA desaturase system converts stearyl-CoA to oleyl-CoA. This system contains the flavoprotein dehydrogenase NADH-cytochrome b_5 reductase, cytochrome b_5, and an oxygen-dependent desaturase that contains a single nonheme iron atom. The electron transport chain by which stearyl-CoA is desaturated is summarized in Figure 19.22.

This desaturation system cannot provide all the unsaturated fatty acids required for human metabolism. The essential fatty acids linoleic acid and linolenic acid must be included in the diet.

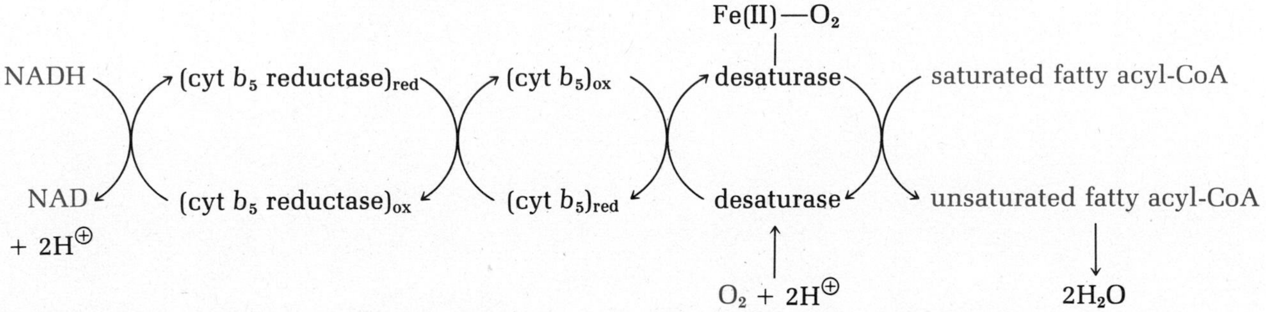

Figure 19.22 Microsomal electron transport chain by which stearyl-CoA is desaturated.

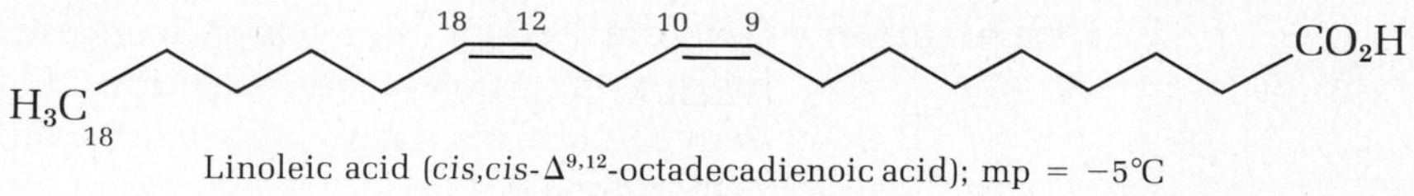

Linoleic acid (*cis,cis*-$\Delta^{9,12}$-octadecadienoic acid); mp = $-5°C$

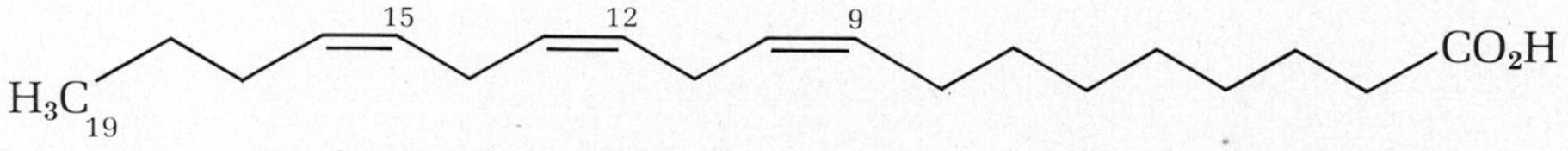

Linolenic acid (*cis*-$\Delta^{9,12,15}$ nonadecatrienoic acid); mp = $-11°C$

The pathway for the synthesis of arachidonoyl-CoA, derived from linoleoyl-CoA and a precursor of the prostaglandins (Section 11.7F) is

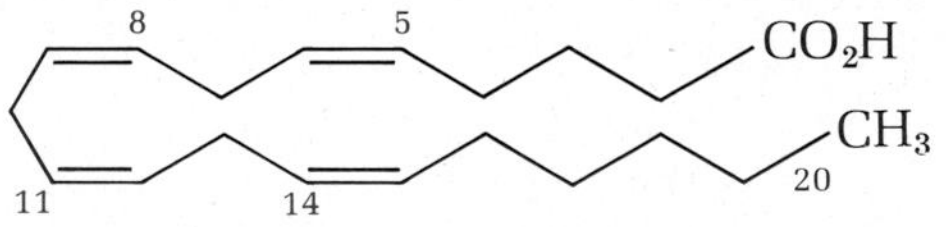

Arachidonic acid ($\Delta^{5,8,11,14}$-eicosatetraenoic acid)

shown in Figure 19.23. When linoleic acid is not present in the diet, an illness ensues called *essential fatty acid deficiency*. It seems likely that this

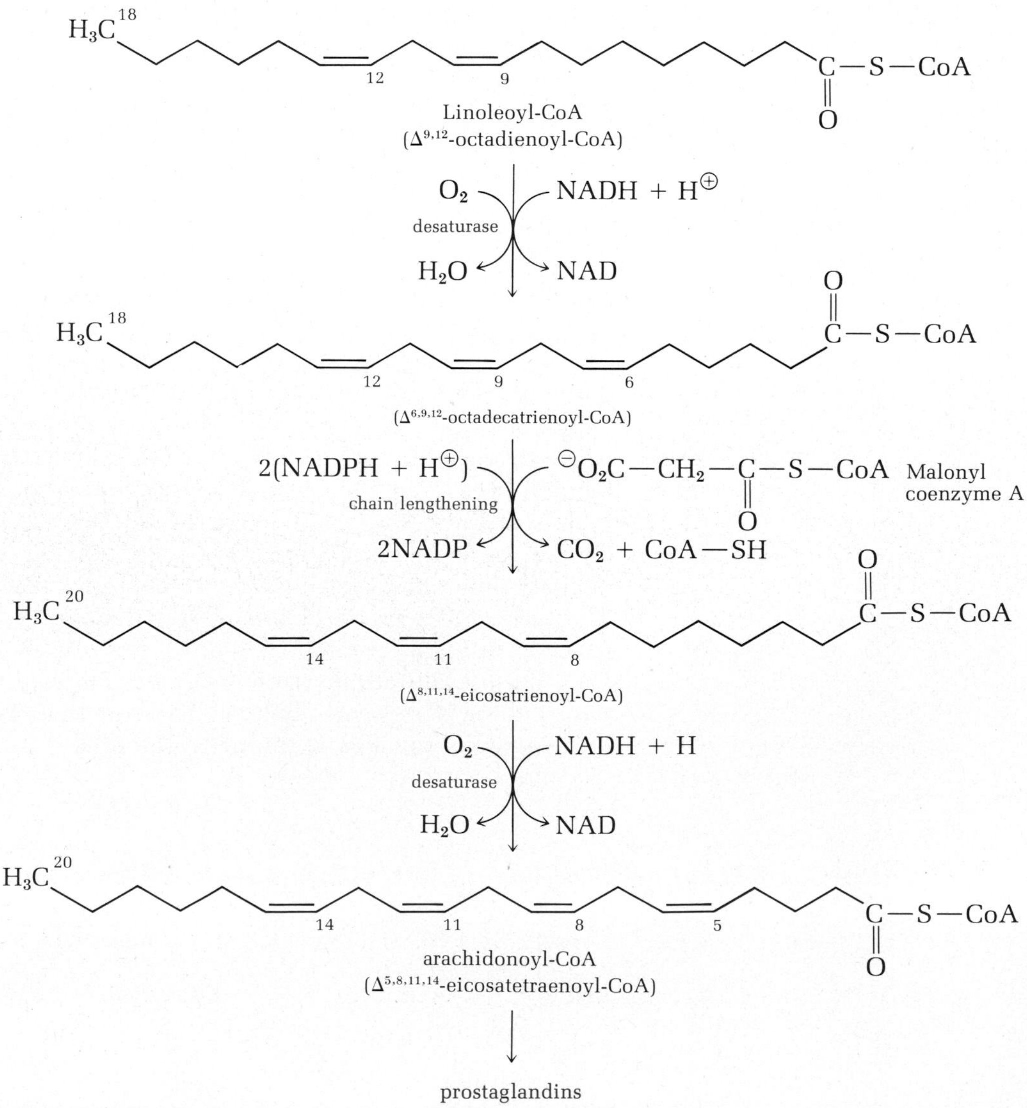

Figure 19.23 Pathway for the biosynthesis of arachidonoyl-CoA.

disease results from an inability to synthesize prostaglandins derived from arachidonic acid.

19–10 SUMMARY   Fatty acids are stored as triacylglycerides in fat droplets in the cytosol of adipose cells. Lipases are responsible for hydrolysis of triacylglycerides. Hormone-sensitive lipase is activated by an enzyme cascade that involves activation of adenylate cyclase and subsequent activation of a protein kinase. The protein kinase then activates the lipase. Fatty acids are degraded to acetyl-CoA by the β-oxidation pathway in the mitochondria of eucaryotic cells. Before entering mitochondria fatty acids are converted to derivatives of coenzyme A. Fatty acyl groups are transported into the mitochondrion as acyl carnitine derivatives, then reconverted to fatty acyl-CoA molecules. β-oxidation involves two oxidation steps, one generates $FADH_2$ and the second produces NADH. The complete oxidation of palmitoyl-CoA in the mitochondrion releases 7 moles of $FADH_2$, 7 moles of NADH, and 8 moles of acetyl-CoA. The citric acid cycle and oxidative phosphorylation of these products generates a net yield of 129 ATP molecules. Oxidation of unsaturated fatty acids requires the action of two additional enzymes; an isomerase that converts a cis-Δ^3-enoyl-CoA to the $trans$ geometrical isomer and an epimerase to convert an R-3-hydroxy-cis-Δ^5-enoyl-CoA to the S-enantiomer. All other steps in the β-oxidation pathway are the same as those for degradation of saturated fatty acids. The ketone bodies—acetoacetate, acetone, and 3-hydroxymethylglutaryl-CoA—are produced if fatty acid oxidation occurs at a greater rate than carbohydrate degradation since the entry of acetyl-CoA into the citric acid cycle depends upon an adequate supply of oxaloacetate, and the concentration of oxaloacetate in turn depends upon pyruvate as a substrate for pyruvate carboxylase.

Fatty acid synthesis occurs by a different pathway in the cytosol. The committing step in fatty acid biosynthesis is the conversion of acetyl-CoA to malonyl-CoA catalyzed by the biotin-dependent enzyme acetyl-CoA carboxylase. The acetyl-CoA carboxylase of mammalian adipose tissue and of lactating mammary gland is activated by citrate and inhibited by palmitoyl-CoA. The acetyl-CoA required for fatty acid synthesis is provided by the action of citrate lyase upon citrate transported out of the mitochondrion by the tricarboxylate carrier. The NADPH required for fatty acid biosynthesis is mostly provided by the pentose phosphate pathway. Mammalian fatty acid synthetases appear to be dimers of multifunctional enzymes each of which catalyzes several steps of fatty acid synthesis. The intermediates of fatty acid synthesis are bound to a 4-phosphopantetheine group and thus never become free during the entire synthesis. Once malonyl-CoA and acetyl-CoA have been bound to fatty acid synthetase through the 4-phosphopantetheine group, one round of synthetic reactions involves: (1) condensation of malonyl-CoA and acetyl-CoA, (2) reduction of the 3-keto group of the product, acetoacetyl-ACP, to an alcohol, (3) dehydration of the alcohol to a $trans$-Δ^2-enoyl-ACP, (4) reduction of the $trans$-Δ^2-alkene. After seven rounds of synthesis, the product, palmitoyl-CoA, is released from the enzyme complex by hydrolysis of the thioester. Longer chain fatty acids are synthesized in the mitochondrion and in the endoplasmic reticulum. Desaturation of fatty acids is carried out by microsomal enzymes in the liver. Fatty acid metabolism is regulated by malonyl-CoA which inhibits carnitine acyl transferase, preventing acyl groups from reaching the mitochondrion and permitting net synthesis of fatty acids.

REFERENCES

Bloch, K., and Vance, D., Control Mechanisms in the Synthesis of Saturated Fatty Acids, *Ann. Rev. Biochem.*, 46, 263–298 (1977).

Florkin, M., and Stotz, E. H., Eds., "Comprehensive Biochemistry," Vol. 18, Elsevier, New York, 1967.

Gehring, U., and Lynen, F., Thiolase, in P. D. Boyer, Ed., "The Enzymes," 3rd ed., Vol. 7, Academic Press, New York, 1972.

Greenberg, D. M., Ed., "Metabolic Pathways," Vol. 2, "Lipids, Steroids, and Cartenoids," 3rd ed., Academic Press, New York, 1968.

McGarry, J. D., and Foster, D. W., Regulation of Hepatic Fatty Acid Oxidation and Ketone Body Production, *Ann. Rev. Biochem.*, 49, 395–420 (1980).

Masoro, E. J., Lipids and Lipid Metabolism, *Ann. Rev. Physiol.*, 39, 301–321 (1977).

Wakil, S., Ed., "Lipid Metabolism," Academic Press, New York, 1970.

Problems

1. The acetyl group of acetyl-CoA derived from fatty acid oxidation cannot be used for net synthesis of glucose in mammals. Why not?

2. Explain how the acetyl group of acetyl-CoA derived from glucose can be converted to fatty acid with net formation of product.

3. Write a reaction mechanism which illustrates that thiolase (reaction 19.15) catalyzes a Claisen cleavage.

4. What are the parallels between the citric acid cycle, or any portion thereof, and β-oxidation of fatty acids?

5. The following reaction occurs in the biosynthesis of fatty acids.

$$CH_3-CH{=}CH-\overset{\overset{\displaystyle O}{\|}}{C}-S-ACP + NADPH + H^{\oplus} \longrightarrow$$

Crotonyl-ACP

$$CH_3-CH_2-CH_2-\overset{\overset{\displaystyle O}{\|}}{C}-S-ACP + NADP$$

Butyryl-ACP

The reduction half-reactions for crotonyl-ACP and NADPH are

$$\text{crotonyl-ACP} + 2H^{\oplus} + 2\ e^{\ominus} \longrightarrow \text{butyryl-ACP} \qquad E^{0\prime} = -0.015\ \text{V}$$
$$\text{NADP} + 2H^{+} + 2\ e^{\ominus} \longrightarrow \text{NADPH} + H^{\oplus} \qquad E^{0\prime} = -0.320\ \text{V}$$

What is $\Delta G^{\circ\prime}$ for this reaction? What is the equilibrium constant for the reaction?

6. Trace the fate of ^{14}C from the methylene group of the pro-S arm of citrate to malonyl-CoA.

7. How does the inhibition of citrate synthetase affect fatty acid synthesis?

8. To the extent that it occurs at all, fatty acid biosynthesis in heart occurs by reversal of β-oxidation. How will the NAD/NADH ratio in heart affect the direction of oxidation or synthesis of fatty acids? How will the NADP/NADPH ratio affect this process in heart?

Chapter 20

The biosynthesis of lipids, cholesterol, and steroid hormones

20–1 INTRODUCTION

Lipids derived from glycerol are the major components of eucaryotic cells. In terms of dry weight they account for anywhere between 10% and 90% of the total mass of the cell. Triacylglycerols are the major source of stored energy in eucaryotic organisms. We discussed the biosynthesis and the catabolism of the fatty acid components of lipids in Chapter 19, and in this chapter we will see how the fatty acids are incorporated in triacylglycerols. We will also consider the pathway for the biosynthesis of cholesterol, the precursor of the steroid hormones, and the paradigm of biosynthetic pathways leading from acetate to complex natural products.

20–2 BIOSYNTHESIS OF TRIACYLGLYCEROLS

Most of the fatty acids of mammalian cells are either esterified with glycerol in *triacylglycerols* or are components of membrane lipids such as the phosphoglycerides, plasmalogens, and sphingolipids. The cellular concentration of free fatty acids is very small.

$$
\begin{array}{c}
\qquad\qquad\qquad\qquad \underset{\textstyle O}{\overset{\textstyle O}{\parallel}} \\
\qquad\qquad CH_2O-\overset{O}{\overset{\parallel}{C}}-R_1 \\
R_2-\overset{O}{\overset{\parallel}{C}}-O-\overset{\mid}{\underset{\mid}{C}}-H \\
\qquad\qquad CH_2-O-\overset{}{\underset{\parallel}{C}}-R_3 \\
\qquad\qquad\qquad\qquad O
\end{array}
$$

Structure of a triacylglycerol

The biosynthesis of triacylglycerols occurs primarily on the cytoplasmic surface of the endoplasmic reticulum in adipose and liver cells. The overall pathway for biosynthesis of triacylglycerols is summarized in Figure 20.1. This pathway may be divided into three components: (1) the synthesis of sn-3-glycerol phosphate (L-α-glycerol phosphate) from either dihydroxyacetone phosphate or from glycerol, (2) the conversion of the triose phosphate to the diacylglyceride, and (3) the conversion of the diacylglyceride to the triacylglyceride. The C-2 hydroxyl group is esterified in the synthesis of the *monoacylglycerol phosphate*. The fatty acid is added to the hydroxyl group as its CoA derivative, not as the free carboxylate. In short, fatty acids must be activated before they can be incorporated in glycerol lipids. The diacylglycerol phosphates are called *phosphatidates*.

$$
\begin{array}{c}
\underset{\textstyle O}{\overset{\textstyle O}{\parallel}} \qquad CH_2-OH \\
R-\overset{O}{\overset{\parallel}{C}}-O-\overset{\mid}{\underset{\mid}{C}}-H \qquad \underset{\textstyle O}{\overset{\textstyle O}{\parallel}} \\
\qquad\qquad CH_2-O-\overset{O}{\overset{}{P}}-O^{\ominus} \\
\qquad\qquad\qquad\qquad O^{\ominus}
\end{array}
$$

Monoacylglycerol phosphate esterified at C-2

$$
\begin{array}{c}
\qquad\qquad\qquad\qquad \underset{\textstyle O}{\overset{\textstyle O}{\parallel}} \\
\qquad\qquad CH_2-O-\overset{O}{\overset{\parallel}{C}}-R_1 \\
R_2-\overset{O}{\overset{\parallel}{C}}-O-\overset{\mid}{\underset{\mid}{C}}-H \qquad \underset{\textstyle O}{\overset{\textstyle O}{\parallel}} \\
\qquad\qquad CH_2-O-\overset{}{\overset{}{P}}-O^{\ominus} \\
\qquad\qquad\qquad\qquad O^{\ominus}
\end{array}
$$

Diacylglycerol phosphate (phosphatidate)

Phosphatidates are the major phospholipid components of membranes.

Reduction of dihydroxyacetone phosphate to sn-glycerol 3-phosphate by NAD-dependent glycerol 3-phosphate dehydrogenase is the major source of the glycerol moiety of triacylglycerides (reaction 20.1).

$$
\begin{array}{ccc}
CH_2-OH & & CH_2OH \\
| & \xrightarrow{\substack{\text{Glycerol 3-phosphate} \\ \text{dehydrogenase}}} & | \\
C=O & & HO-C-H \\
| & \overset{}{\underset{NADH + H^{\oplus} \quad NAD}{}} & | \\
CH_2-OPO_3{}^{\ominus} & & CH_2-OPO_3{}^{\ominus}
\end{array}
\qquad (20.1)
$$

Dihydroxacetone phosphate — sn-Glycerol 3-phosphate (L-glycerol 3-phosphate)

Alternatively, glycerol itself may be phosphorylated by *glycerol kinase* (reaction 20.2).

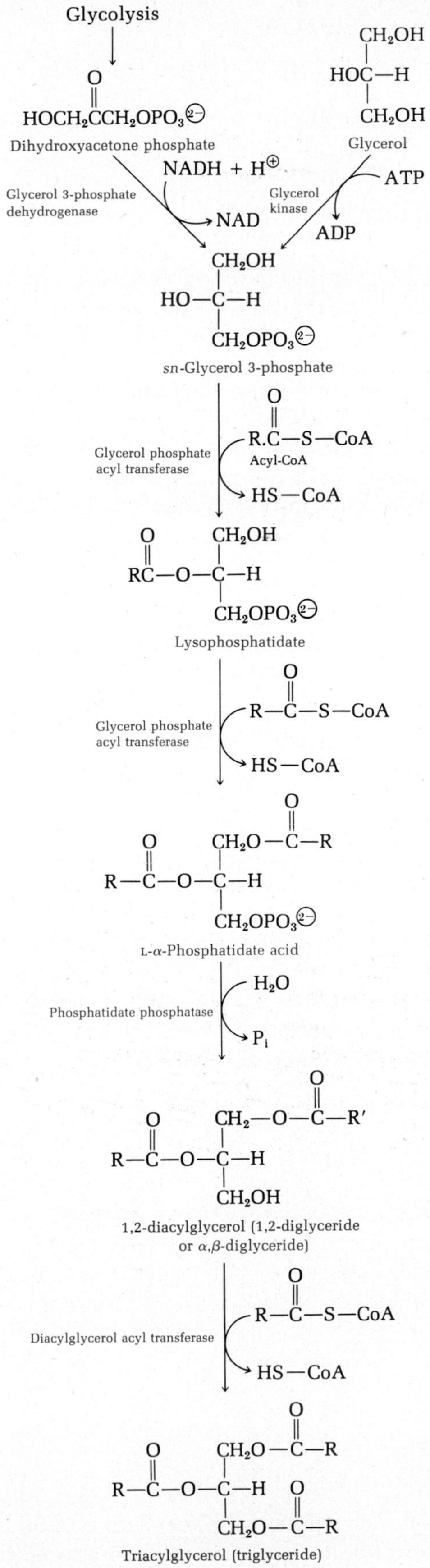

Glycolysis

$$\underset{\text{Dihydroxyacetone phosphate}}{HOCH_2\overset{\overset{\displaystyle O}{\|}}{C}CH_2OPO_3^{2\ominus}}$$

$$\underset{\text{Glycerol}}{\overset{\displaystyle CH_2OH}{\underset{\displaystyle CH_2OH}{|}}\overset{\displaystyle HOC-H}{}}$$

NADH + H$^{\oplus}$ ⟶ ATP

Glycerol 3-phosphate dehydrogenase ⟶ NAD   Glycerol kinase ⟶ ADP

$$\underset{\textit{sn}\text{-Glycerol 3-phosphate}}{\overset{\displaystyle CH_2OH}{\underset{\displaystyle CH_2OPO_3^{2\ominus}}{|}}HO-C-H}$$

$$\overset{\displaystyle O}{\overset{\|}{R.C}}-S-CoA$$
Acyl-CoA

Glycerol phosphate acyl transferase

HS—CoA

$$\underset{\text{Lysophosphatidate}}{RC-O-\overset{\displaystyle CH_2OH}{\underset{\displaystyle CH_2OPO_3^{2\ominus}}{\underset{|}{C}-H}}}$$

Glycerol phosphate acyl transferase

$$R-\overset{\overset{\displaystyle O}{\|}}{C}-S-CoA$$

HS—CoA

$$\underset{\text{L-}\alpha\text{-Phosphatidate acid}}{R-\overset{\overset{\displaystyle O}{\|}}{C}-O-\overset{\displaystyle CH_2O-\overset{\overset{O}{\|}}{C}-R}{\underset{\displaystyle CH_2OPO_3^{2\ominus}}{\underset{|}{C}-H}}}$$

Phosphatidate phosphatase

H$_2$O ⟶ P$_i$

$$\underset{\substack{\text{1,2-diacylglycerol (1,2-diglyceride}\\ \text{or }\alpha,\beta\text{-diglyceride)}}}{R-\overset{\overset{\displaystyle O}{\|}}{C}-O-\overset{\displaystyle CH_2-O-\overset{\overset{O}{\|}}{C}-R'}{\underset{\displaystyle CH_2OH}{\underset{|}{C}-H}}}$$

Diacylglycerol acyl transferase

$$R-\overset{\overset{\displaystyle O}{\|}}{C}-S-CoA$$

HS—CoA

$$\underset{\text{Triacylglycerol (triglyceride)}}{R-\overset{\overset{\displaystyle O}{\|}}{C}-O-\overset{\displaystyle CH_2O-\overset{\overset{O}{\|}}{C}-R}{\underset{\displaystyle CH_2O-\overset{\overset{O}{\|}}{C}-R}{\underset{|}{C}-H}}}$$

Figure 20.1 Pathways for the biosynthesis of triacylglycerols.

$$
\begin{array}{c}
\text{CH}_2\text{OH} \\
| \\
\text{HO}-\text{C}-\text{H} \\
| \\
\text{CH}_2\text{OH}
\end{array}
+ \text{ATP} \xrightarrow{\text{Glycerol kinase}}
$$

Glycerol

$$
\begin{array}{c}
\text{CH}_2\text{OH} \\
| \\
\text{HO}-\text{C}-\text{H} \\
| \\
\text{CH}_2\text{OPO}_3^{2\ominus}
\end{array}
+ \text{ADP} \qquad (20.2)
$$

sn-Glycerol 3-phosphate
(L-glycerol 3-phosphate)

Glycerol is a symmetrical molecule whose two —CH$_2$OH groups are prochiral. Glycerol kinase stereospecifically phosphorylates the pro-R —CH$_2$OH group. By convention, symmetrical molecules, like glycerol, are numbered with the pro-S carbon as C-1. In this case the pro-R hydroxyl group is phosphorylated, and the product is called sn-glycerol 3-phosphate (Figure 20.2). The designation sn-3 means "carbon number 3 numbering from pro-S carbon number 1."

Acylation of sn-glycerol 3-phosphate does not occur from a free fatty acid, but from an acyl-CoA. We recall that free fatty acids are converted to acyl-CoA derivatives by the action of acyl-CoA synthetase (Section 19.3B). An acyl-CoA synthetase bound to the endoplasmic reticulum catalyzes this reaction. This enzyme, known as acyl-CoA synthetase I, will accept substrates having from 14 to 18 carbon atoms and any degree of unsaturation (reaction 20.3).

$$
\text{R-CO}_2^{\ominus} + \text{CoASH} + \text{ATP} \xrightarrow[\text{synthetase I}]{\text{Acyl-CoA}}
\begin{array}{c}
\text{O} \\
\| \\
\text{R}-\text{C}-\text{S}-\text{CoA}
\end{array}
$$

Acyl-CoA

$$
+ \text{AMP} + \text{PP}_i \qquad (20.3)
$$

The acylation of sn-glycerol 3-phosphate is catalyzed by *glycerol phosphate acyl transferase* (reaction 20.4).

$$
\begin{array}{c}
\text{CH}_2\text{OH} \\
| \\
\text{HO}-\text{C}-\text{H} \\
| \\
\text{CH}_2\text{OPO}_3^{2\ominus}
\end{array}
+
\begin{array}{c}
\text{O} \\
\| \\
\text{R}-\text{C}-\text{SCoA}
\end{array}
\xrightarrow[\text{CoASH}]{\substack{\text{Glycerol phosphate} \\ \text{acyl transferase}}}
$$

sn-Glycerol 3-phosphate   Acyl-CoA

$$
\begin{array}{c}
\text{O} \quad\quad \text{CH}_2\text{OH} \\
\| \qquad\quad | \\
\text{R}-\text{C}-\text{O}-\text{C}-\text{H} \\
| \\
\text{CH}_2\text{OPO}_3^{2\ominus}
\end{array}
\qquad (20.4)
$$

Lysophosphatidate
(lysophosphatidic acid)

The monoacylglycerol phosphate formed by reaction 20.4 is known as a *ly-sophosphatidate*. A second acylation converts lysophosphatidate to phosphatidate (reaction 20.5).

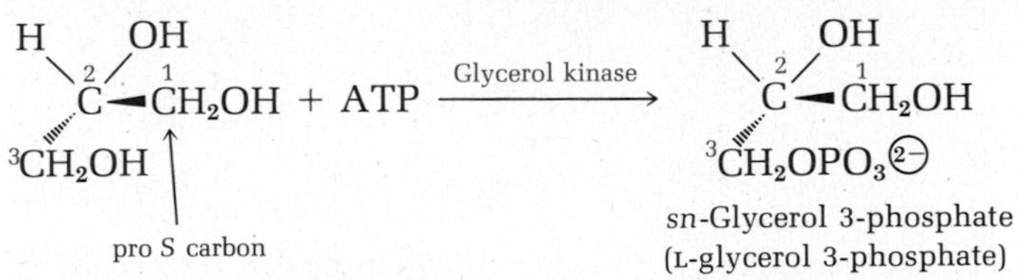

Figure 20.2 Phosphorylation by glycerol kinase of C-3 to produce *sn*-glycerol 3-phosphate (*L*-glycerol 3-phosphate). The pro-S carbon is designated as C-1 in the stereochemical numbering (sn) system.

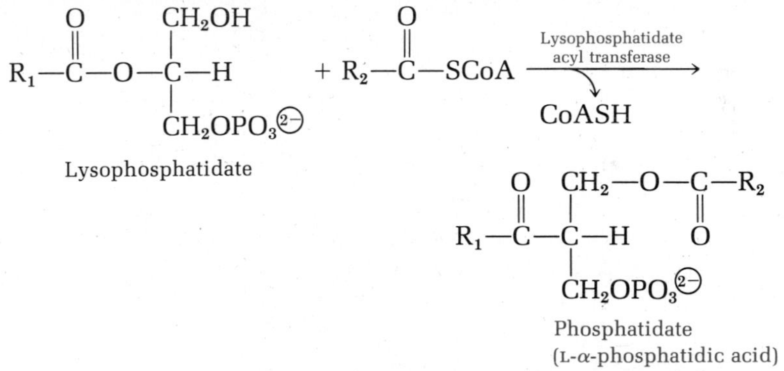

Whether C-1 or C-2 of glycerol 3-phosphate is acylated depends largely on the acyl-CoA group. Saturated fatty acyl groups are usually transferred to the C-1 hydroxyl group, and unsaturated fatty acyl groups are usually transferred to the C-2 hydroxyl group.

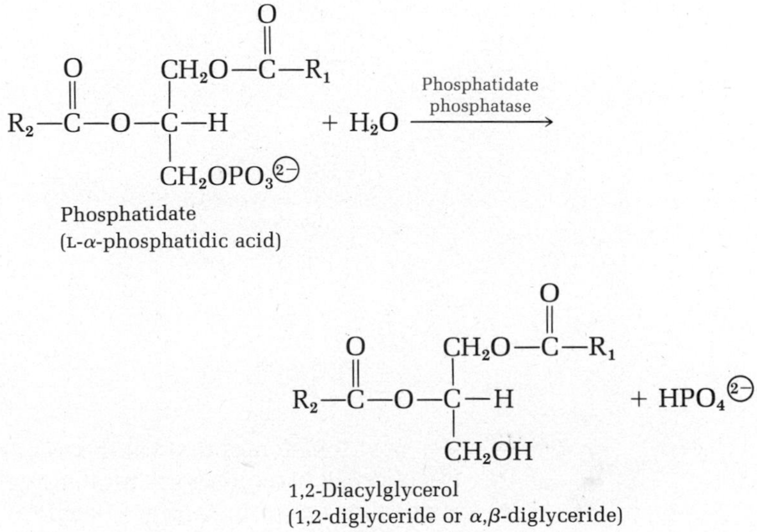

Phosphatidate phosphatase hydrolyzes the C-3 phosphate group (reaction 20.6), and a third acylation by *diacylglycerol acyl transferase* produces a triacylglycerol (reaction 20.7).

$$R_2-\overset{\overset{\displaystyle O}{\|}}{C}-O-\overset{\overset{\displaystyle CH_2-O-\overset{\overset{\displaystyle O}{\|}}{C}-R_1}{|}}{\underset{\underset{\displaystyle CH_2OH}{|}}{C}-H} \quad + \quad R_3\overset{\overset{\displaystyle O}{\|}}{C}-S-CoA \xrightarrow{\text{Diacylglycerol}\atop\text{acyl transferase}}$$

1,2-Diacylglycerol

$$R_2-\overset{\overset{\displaystyle O}{\|}}{C}-O-\overset{\overset{\displaystyle CH_2-O-\overset{\overset{\displaystyle O}{\|}}{C}-R_1}{|}}{\underset{\underset{\displaystyle CH_2-O-\overset{\displaystyle C}{\underset{\displaystyle O}{\|}}-R_3}{|}}{C}-H} \quad + \quad CoA\text{-}SH \qquad (20.7)$$

Triacylglycerol
(triglyceride)

The enzymes that carry out the acylations are part of a membrane-bound *triacylglycerol synthetase* complex. Lysophosphatidates are synthesized in liver by acylation of dihydroxyacetone phosphate followed by reduction of the keto group by an NADPH-dependent dehydrogenase, respectively (reactions 20.8 and 20.9).

$$\underset{\underset{\displaystyle CH_2OPO_3^{\,2\ominus}}{|}}{\overset{\overset{\displaystyle CH_2OH}{|}}{C}=O} \quad + \quad R_1-\overset{\overset{\displaystyle O}{\|}}{C}-S-CoA \xrightarrow{\text{Dihydroxyacetone phosphate}\atop\text{acyl transferase}}$$

Dihydroxyacetone
phosphate

$$\underset{\underset{\displaystyle CH_2OPO_3^{\,2\ominus}}{|}}{\overset{\overset{\displaystyle CH_2O-\overset{\displaystyle C}{\overset{\displaystyle \|}{}}-R_1}{|}}{C}=O} \quad + \quad CoA\text{-}SH \qquad (20.8)$$

Acyl dihydroxyacetone
phosphate

$$\underset{\underset{\displaystyle CH_2OPO_3^{\,\ominus}}{|}}{\overset{\overset{\displaystyle CH_2O-\overset{\displaystyle C}{\overset{\displaystyle \|}{}}-R_1}{|}}{C}=O} \quad + \quad NADPH + H^{\oplus} \xrightarrow{\text{Acyl dihydroxyacetone}\atop\text{phosphate dehydrogenase}}$$

Acyl dihydroxyacetone
phosphate

$$\underset{\underset{\displaystyle CH_2OPO_3^{\,2\ominus}}{|}}{\overset{\overset{\displaystyle CH_2O-\overset{\displaystyle C}{\overset{\displaystyle \|}{}}-R_1}{|}}{HO-C-H}} \quad + \quad NAD \qquad (20.9)$$

Lysophosphatidate

The acyl transferase that catalyzes acylation of dihydroxyacetone phosphate (reaction 20.8) requires a saturated acyl-CoA substrate. A second liver enzyme is specific for an unsaturated acyl-CoA substrate that reacts with lysophosphatidate to produce phosphatidate (reaction 20.10).

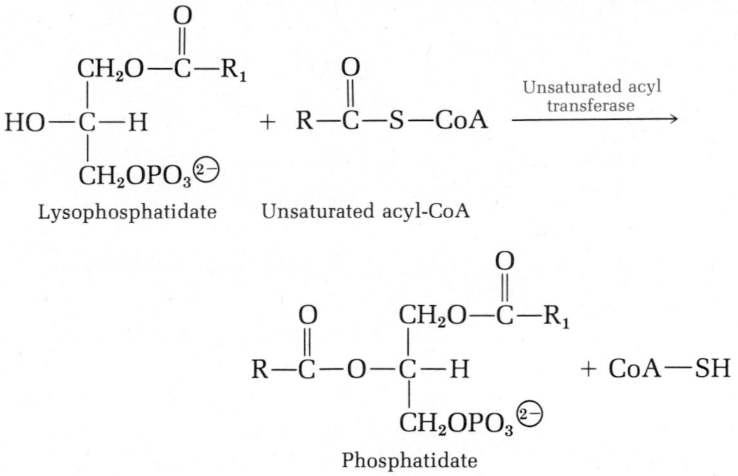

$$
\begin{array}{c}
\underset{\text{Lysophosphatidate}}{
\begin{array}{c}
\text{CH}_2\text{O}-\overset{\displaystyle\overset{\text{O}}{\|}}{\text{C}}-\text{R}_1 \\
| \\
\text{HO}-\text{C}-\text{H} \\
| \\
\text{CH}_2\text{OPO}_3{}^{\textcircled{2-}}
\end{array}}
\quad + \quad
\underset{\text{Unsaturated acyl-CoA}}{
\text{R}-\overset{\displaystyle\overset{\text{O}}{\|}}{\text{C}}-\text{S}-\text{CoA}}
\quad \xrightarrow[\text{transferase}]{\text{Unsaturated acyl}}
\end{array}
$$

$$
\underset{\text{Phosphatidate}}{
\begin{array}{c}
\text{R}-\overset{\displaystyle\overset{\text{O}}{\|}}{\text{C}}-\text{O}-\text{C}-\text{H} \qquad \text{CH}_2\text{O}-\overset{\displaystyle\overset{\text{O}}{\|}}{\text{C}}-\text{R}_1 \\
| \\
\text{CH}_2\text{OPO}_3{}^{\textcircled{2-}}
\end{array}}
\quad + \quad \text{CoA}-\text{SH} \qquad (20.10)
$$

The acyl group transfer reactions in the biosynthesis of glycerols, phosphatidate, and lysophosphatidate conform to a general type, outlined in Figure 20.3. A hydroxyl group of the glycerol substrate, acting as a nucleophile, attacks the carbonyl carbon of the acyl-CoA substrate to give a tetrahedral intermediate. Collapse of the tetrahedral intermediate and expulsion of CoA-SH, favored because sulfur is a good leaving group (recall Section 10.3A), gives the acyl glycerol.

20–3 BIOSYNTHESIS OF PHOSPHOGLYCERIDES

Phosphatidate is the precursor of many other phosphoglycerides, the major components of cell membranes (Section 11.2C). Nucleotide diphosphate derivatives, such as cytidine diphosphate glycerol (CDP-glycerol), are important intermediates in the biosynthesis of lipids, as they are in the biosynthesis of carbohydrates (recall Section 17.5).

A. Synthesis of Phosphatidylethanolamine and Phosphatidylcholine

In mammals *phosphatidylethanolamine* is synthesized in two steps from phosphatidate. Phosphatidate phosphatase hydrolyzes the 3-phosphate group (reaction 20.6), and the 1,2-diacylglycerol product then reacts with *cytidine diphosphate ethanolamine* producing phosphatidylethanolamine (Figure 20.4).

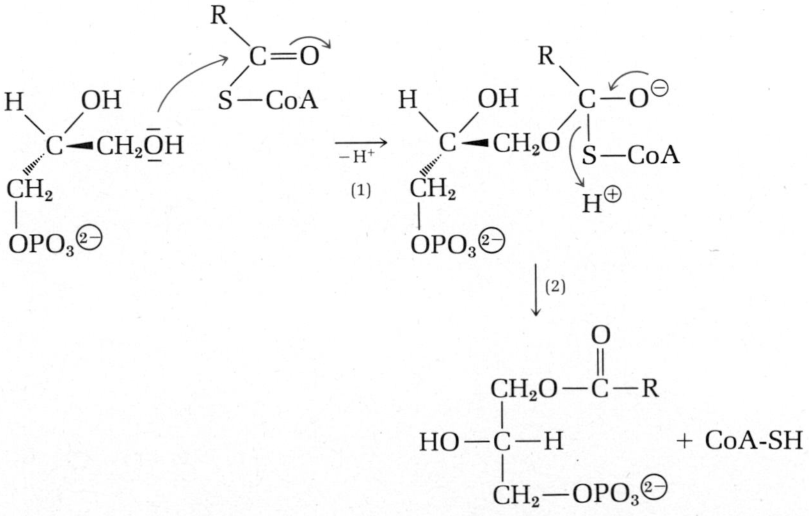

Figure 20.3 Mechanism of action of acyltransferases. The first step (1) is formation of a tetrahedral intermediate that collapses with expulsion of coenzyme A (step 2). The net result of the reaction is transesterification.

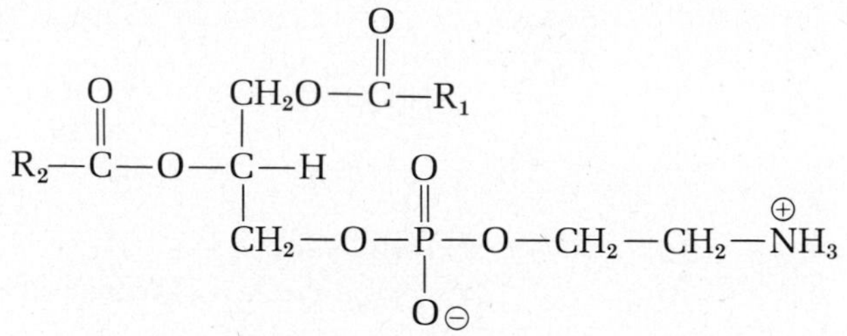

Phosphatidylethanolamine

Bacteria synthesize phosphatidylethanolamine by converting phosphatidate to cytidine diphosphodiacylglycerol in a reaction catalyzed by *phosphatidate cytidyl transferase* (Figure 20.5). The formation of the CDP-diglyceride in bacteria is analogous to the reaction of glucose 1-phosphate with UTP in the biosynthesis of glycogen (recall Section 17.4B). Cytidine diphosphodiacylglycerol reacts with ethanolamine phosphate to give phosphatidylethanoamine (Figure 20.6).

Mammals convert phosphatidylethanolamine to phosphatidylcholine by three consecutive methylations in which S-*adenosylmethionine* (SAM) is the methyl donor (Figures 20.7 and 20.8). Phosphatidylcholine can be synthesized by a second pathway in which choline is phosphorylated and con-

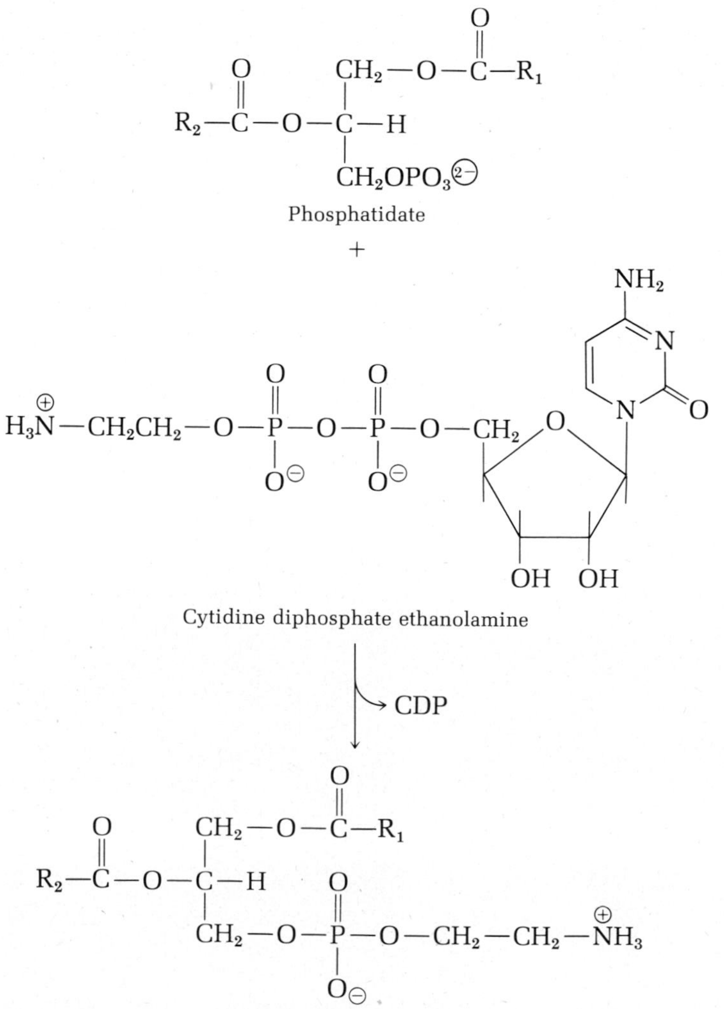

Phosphatidate

+

Cytidine diphosphate ethanolamine

CDP

Phosphatidylethanolamine

Figure 20.4 Pathway for the biosynthesis of phosphatidylethanolamine.

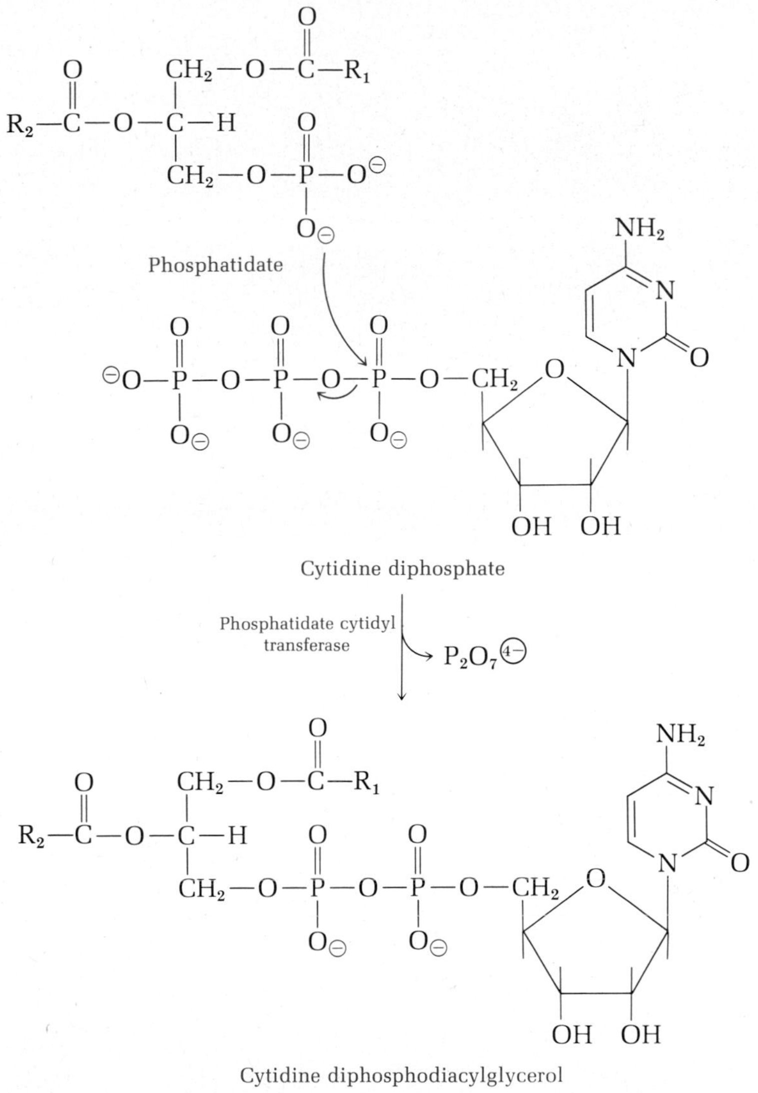

Figure 20.5 Pathway for the bacterial biosynthesis of cytidine diphosphodiacylglycerol, the immediate precursor of phosphatidylethanolamine.

verted to CDP-choline that then reacts with a 1,2-diacylglycerol to give phosphatidylcholine (Figure 20.9). This is an important pathway because the supply of phosphatidylethanolamine is too small to provide for the requirements of mammals for phosphatidylcholine.

B. *S*-Adenosylmethionine, the Major Biological Methylating Agent

S-Adenosylmethionine (see Figure 20.7), the major biological methylating agent, is synthesized by methionine adenosyl transferase in a nucleophilic substitution in which the electron pair of the sulfur of methionine attacks the 5′ carbon of ATP displacing triphosphate (Figure 20.10). This is one of the rare cases, coenzyme B_{12} providing the only other example, in which a nucleophile attacks the 5′ carbon of ribose rather than one of the phosphate groups of ATP. The positive charge on the sulfur atom of *S*-adenosylmethionine converts the unreactive methylthiol group of methionine into a powerful alkylating agent because the charge converts the *S*-adenosylhomocysteine moiety into an excellent leaving group. Alkylation, then, is simply bimolecular substitution (reaction 20.11).

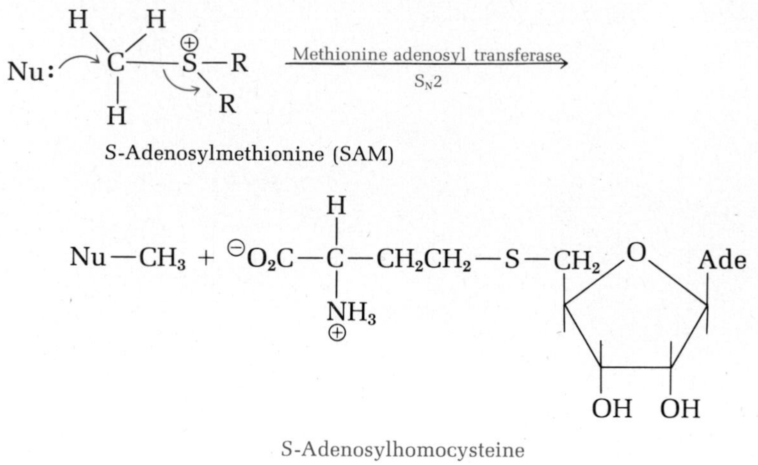

$$\text{Nu:} \quad \underset{\underset{H}{|}}{\overset{\overset{H}{|}}{C}} - \overset{\oplus}{S} - R \qquad \xrightarrow[\text{S}_N2]{\text{Methionine adenosyl transferase}}$$

S-Adenosylmethionine (SAM)

$$\text{Nu} - CH_3 + \ {}^{\ominus}O_2C - \underset{\underset{\oplus}{\overset{|}{NH_3}}}{\overset{\overset{H}{|}}{C}} - CH_2CH_2 - S - CH_2 \underset{\underset{OH \quad OH}{}}{\overset{O}{\diagup}} \text{Ade} \qquad (20.11)$$

S-Adenosylhomocysteine

S-Adenosylmethionine alkylates the bases of DNA and tRNA, the ε-amino group of lysine, the guanidino group of arginine, the nitrogens of the imidazole side chain of histidine, and many other substrates having nucleophilic substituents. In the synthesis of phosphatidylcholine the electron pair of the ethanolamine nitrogen is the nucleophile.

C. Synthesis of Phosphatidylserine

Phosphatidylserine is formed in mammals by a pyridoxal phosphate–catalyzed reaction in which serine displaces ethanolamine from phosphatidylethanolamine (reaction 20.12).

$$RO - CH_2CH_2 - \overset{\oplus}{NH_3} + HO - CH_2 - \underset{\underset{\oplus}{\overset{|}{NH_3}}}{\overset{\overset{H}{|}}{C}} - CO_2^{\ominus} \longrightarrow$$

Phosphatidylethanolamine   Serine

$$RO - CH_2 - \underset{\underset{\oplus}{\overset{|}{NH_3}}}{\overset{\overset{H}{|}}{C}} - CO_2^{\ominus} + H_3\overset{\oplus}{N} - CH_2CH_2 - OH \qquad (20.12)$$

Phosphatidylserine   Ethanolamine

Decarboxylation of the serine carboxyl groups of phosphatidylserine regenerates phosphatidylethanolamine (reaction 20.13).

$$RO - CH_2 - \underset{\underset{\oplus}{\overset{|}{NH_3}}}{\overset{\overset{H}{|}}{C}} - CO_2^{\ominus} + H^{\oplus} \longrightarrow$$

Phosphatidylserine

$$RO - CH_2CH_2 - \overset{\oplus}{NH_3} + CO_2 \qquad (20.13)$$

Phosphatidylethanolamine

This pathway provides a way for serine to be incorporated in lipids. It also stretches the limited supply of ethanolamine. Successive methylations, with S-adenosylmethionine as the methyl donor, produce phosphatidylcholine. The carbon framework and nitrogen of choline are derived from serine.

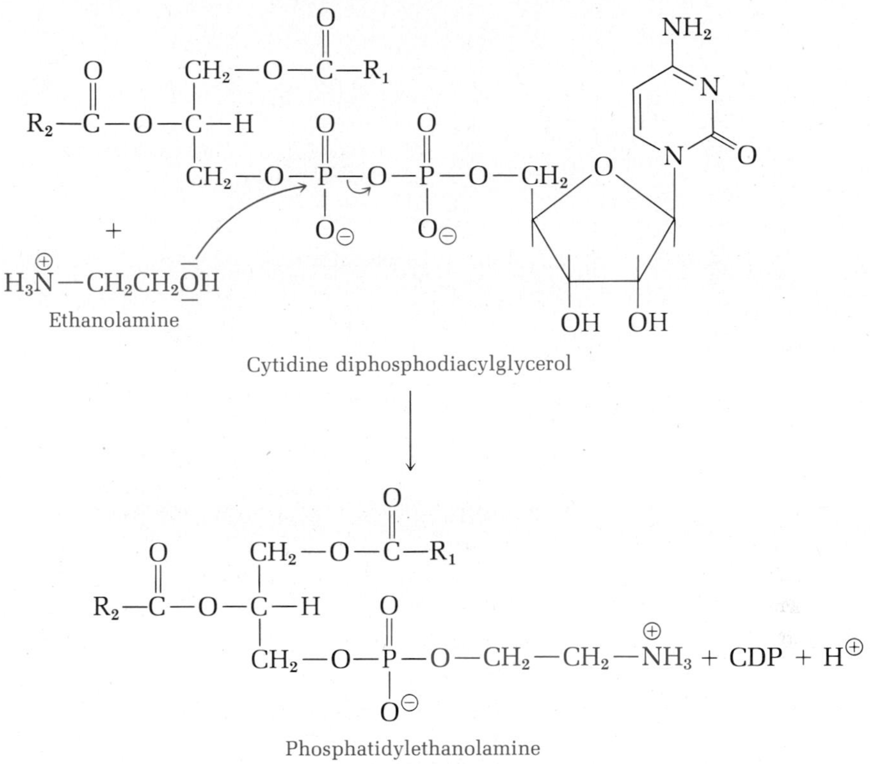

Figure 20.6 Final step in bacterial biosynthesis of phosphatidylethanolamine.

Bacteria, such as *E. coli*, form phosphatidylserine as they form most of their lipids, by means of cytidine diphosphodiacylglycerol (see Figures 20.5 and 20.6). The serine hydroxyl groups displace CMP producing phosphatidylserine, a reaction that parallels the synthesis of phosphatidylethanolamine.

D. Synthesis of Phosphatidylinositols and Phosphatidylglycerophosphates

In mammals cytidine diphosphodiacylglycerol can react with *myo*-inositol to give phosphatidyl-*myo*-inositol. Two kinases then successively phosphorylate the 4 and 5 positions of the *myo*-inositol portion of the lipid producing the triphospho-*myo*-inosotide (Figure 20.11).

Phosphatidylglycerols are formed by displacement of CMP from cytidine diphosphodiacylglycerol by the 1-hydroxyl group of *sn*-3-glycerol phosphate. Hydrolysis of the *sn*-3-phosphate group by a phosphatase gives

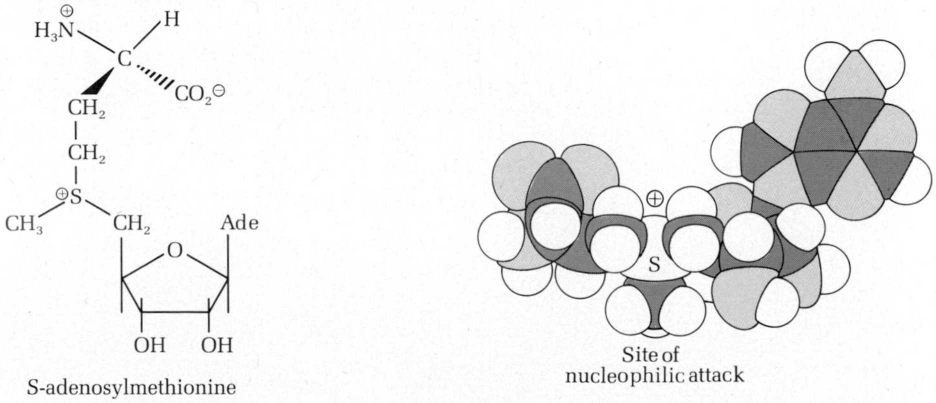

Figure 20.7 Structure of S-adenosylmethionine (SAM), a biological methylating agent. (a) Ball and stick structure. (b) Space-filling model. Nucleophiles attack the methyl carbon, displacing S-adenosylhomocysteine.

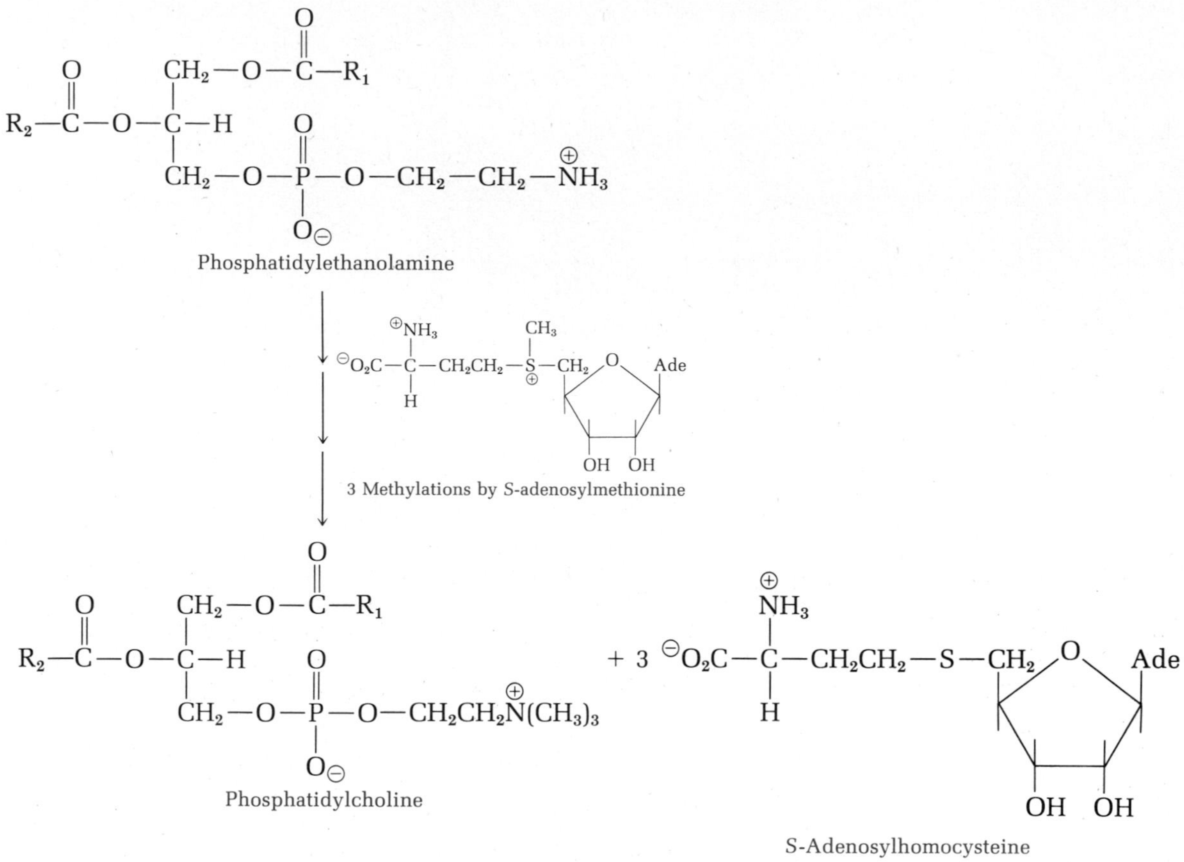

Phosphatidylethanolamine

3 Methylations by S-adenosylmethionine

Phosphatidylcholine

S-Adenosylhomocysteine

Figure 20.8 Pathway for the biosynthesis of phosphatidylcholine.

phosphatidylglycerophosphate. The 3-hydroxyl group of the glycerol moiety then displaces CMP from a second mole of cytidine diacyldiphosphoglycerol giving diphosphatidylglycerol, or *cardiolipin* (Figure 20.12). The cardiolipins account for about 10% of the lipids in the inner mitochondrial membrane.

20–4 BIOSYNTHESIS OF PLASMALOGENS

About 10% of the lipids in the human central nervous system are vinyl ether derivatives of phosphoglycerides called *plasmalogens* (Figure 20.13). The first step of plasmalogen biosynthesis is the formation of the fatty acyl derivative of dihydroxyacetone phosphate (reaction 20.8). An unusual reaction then follows in which the carboxylate group of acyl DHAP is displaced by the hydroxyl group of a long chain alcohol (Figure 20.14). The keto group of the DHAP portion of the lipid is next reduced by a dehydrogenase with either NADH or NADPH as coenzyme (reaction 20.14).

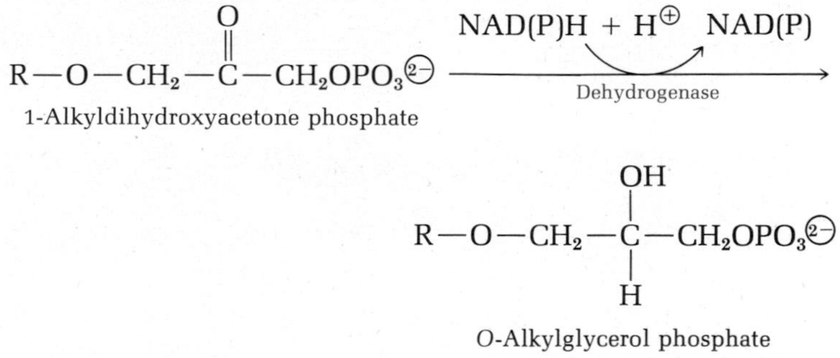

1-Alkyldihydroxyacetone phosphate

O-Alkylglycerol phosphate

(20.14)

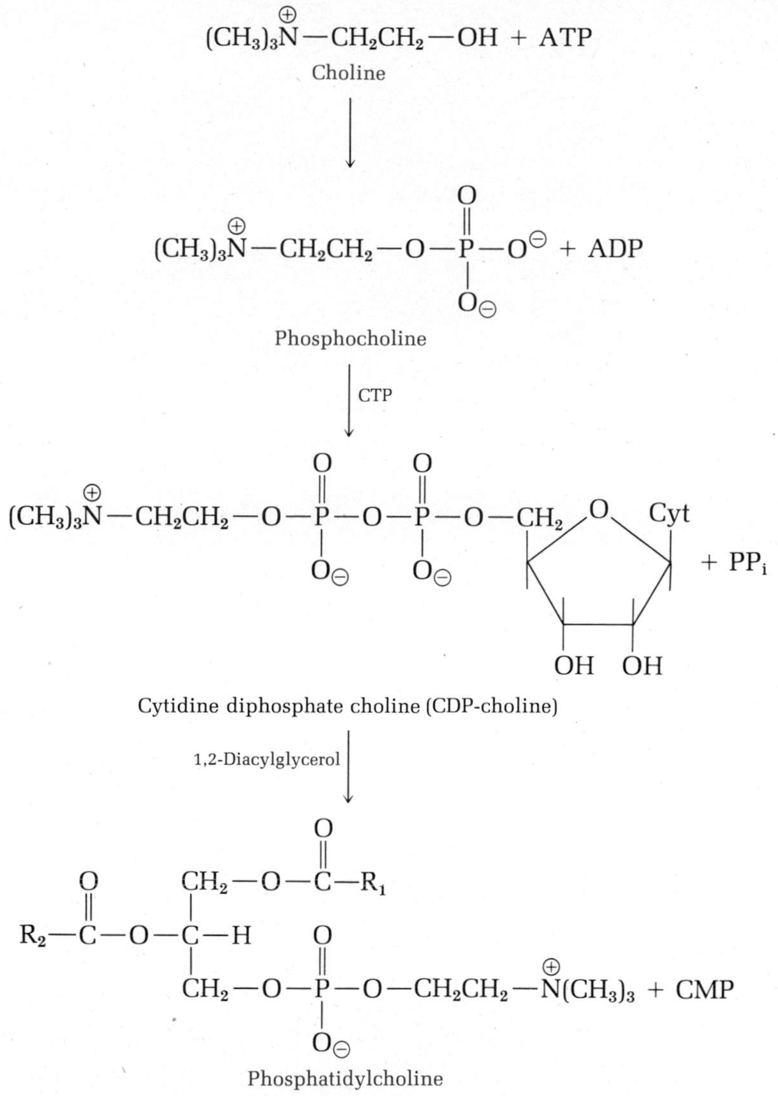

$$(CH_3)_3\overset{\oplus}{N}-CH_2CH_2-OH + ATP$$

Choline

$$(CH_3)_3\overset{\oplus}{N}-CH_2CH_2-O-\overset{\overset{O}{\|}}{\underset{\underset{O\ominus}{|}}{P}}-O\ominus + ADP$$

Phosphocholine

CTP

$$(CH_3)_3\overset{\oplus}{N}-CH_2CH_2-O-\overset{\overset{O}{\|}}{\underset{\underset{O\ominus}{|}}{P}}-O-\overset{\overset{O}{\|}}{\underset{\underset{O\ominus}{|}}{P}}-O-CH_2 \quad \text{Cyt} \quad + PP_i$$

OH OH

Cytidine diphosphate choline (CDP-choline)

1,2-Diacylglycerol

$$R_2-\overset{\overset{O}{\|}}{C}-O-\overset{\overset{\displaystyle CH_2-O-\overset{\overset{O}{\|}}{C}-R_1}{|}}{\underset{\underset{\displaystyle CH_2-O-\overset{\overset{O}{\|}}{\underset{\underset{O\ominus}{|}}{P}}-O-CH_2CH_2-\overset{\oplus}{N}(CH_3)_3 + CMP}{|}}{C}-H}$$

Phosphatidylcholine

Figure 20.9   Alternate pathway for the biosynthesis of phosphatidyl-choline.

An acyl-CoA transferase now esterifies the C-2 hydroxyl group (reaction 20.15).

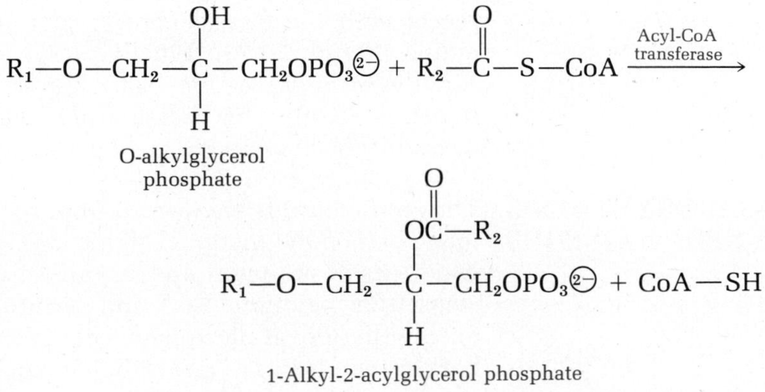

$$R_1-O-CH_2-\overset{\overset{OH}{|}}{\underset{\underset{H}{|}}{C}}-CH_2OPO_3^{2\ominus} + R_2-\overset{\overset{O}{\|}}{C}-S-CoA \xrightarrow{\text{Acyl-CoA transferase}}$$

O-alkylglycerol phosphate

$$R_1-O-CH_2-\overset{\overset{\displaystyle O\overset{\overset{O}{\|}}{C}-R_2}{|}}{\underset{\underset{H}{|}}{C}}-CH_2OPO_3^{2\ominus} + CoA-SH \qquad (20.15)$$

1-Alkyl-2-acylglycerol phosphate

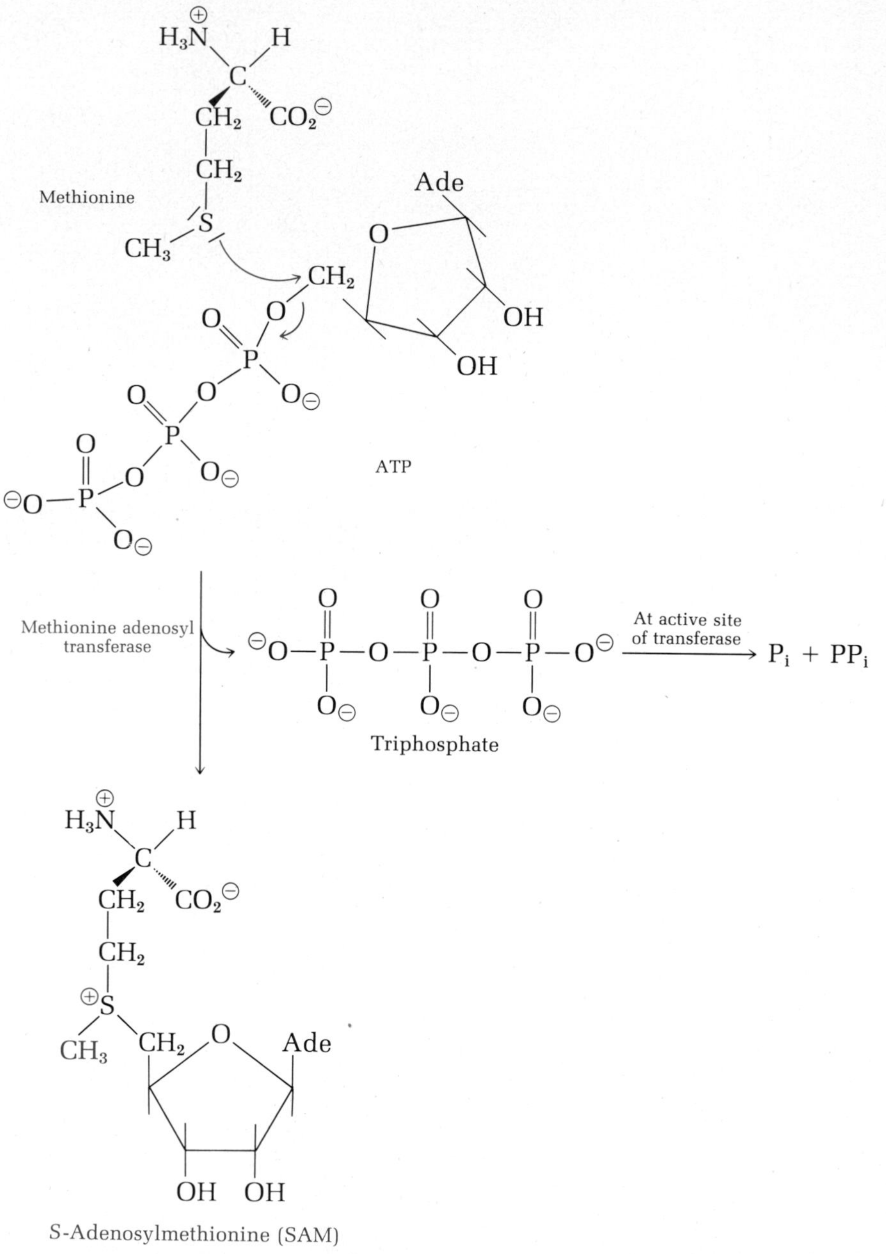

Figure 20.10 Pathway for the biosynthesis of S-adenosylmethionine.

A phosphatase hydrolyzes the phosphate group, and CDP ethanolamine reacts with the free alcohol to give a 1-alkyl-2-acylphosphatidylethanolamine (Figure 20.15). Finally, a specific oxidase converts the ether to a vinyl ether in a reaction that requires both molecular oxygen and NADPH (reaction 20.16). This reaction parallels the desaturation of fatty acids discussed in Section 19.9E.

20–5 BIOSYNTHESIS OF SPHINGOLIPIDS

The sphingolipids are formed from *sphingosine*, also called *sphingenine*, a long chain, polyfunctional amine, rather than from glycerol. The brain is the major site of sphingolipid biosynthesis. Sphingosine is synthesized in three steps from palmitoyl-CoA and serine (Figure 20.16). In the first step, requiring pyridoxal phosphate, coenzyme A is displaced, and the serine carboxyl group is lost to form 3-*dehydrosphinganine* (Figure 20.17). Reduction

$$R_2-\overset{\overset{\displaystyle O}{\|}}{C}-O-\overset{\overset{\displaystyle CH_2-O-CH_2-CH_2R_1'}{|}}{\underset{\underset{\displaystyle CH_2-O-\overset{\overset{\displaystyle O}{\|}}{\underset{\underset{\displaystyle O_\ominus}{|}}{P}}-O-CH_2CH_2NH_3^\oplus}{|}}{C}}-H \quad \xrightarrow[\text{NADPH, O}_2]{\text{Oxidase}}$$

1-alkyl-2-acylphosphatidyl ethanolamine

$$R_2-\overset{\overset{\displaystyle O}{\|}}{C}-O-\overset{\overset{\displaystyle CH_2-O-CH=CHR_1}{|}}{\underset{\underset{\displaystyle CH_2-O-\overset{\overset{\displaystyle O}{\|}}{\underset{\underset{\displaystyle O_\ominus}{|}}{P}}-O-CH_2CH_2NH_3^\oplus}{|}}{C}}-H \qquad (20.16)$$

Ethanolamine plasmalogen

of the 3-keto group by an NADPH-dependent dehydrogenase gives D-*sphinganine*, a common component of mammalian lipids. A flavoprotein then forms the $\Delta^{4,5}$ double bond of D-sphinganine.

Sphingosine (sphingenine)

The amino group of sphingosine is acylated by an acyl-CoA transferase to give an N-*acylsphingosine*, called a *ceramide* (reaction 20.17).

$$R_1-\overset{\overset{\displaystyle OH}{|}}{\underset{\underset{\displaystyle H}{|}}{C}}-\overset{\overset{\displaystyle}{}}{\underset{\underset{\displaystyle NH_2}{|}}{CH}}-CH_2OH + R_2-\overset{\overset{\displaystyle O}{\|}}{C}-SCoA \xrightarrow{\text{Acyl transferase}}$$

$$R_1-\overset{\overset{\displaystyle OH}{|}}{\underset{\underset{\displaystyle H}{|}}{C}}-\overset{\overset{\displaystyle}{}}{\underset{\underset{\underset{\underset{\underset{\displaystyle R_2}{|}}{C=O}}{|}}{NH}}{CH}}-CH_2OH + CoASH \qquad (20.17)$$

N-Acylsphingosine (ceramide)

Ceramides can be converted to *sphingomyelin* by reaction of their primary alcohol with CDP-choline. They are converted to *cerebrosides* by reaction of their primary alcohol with UDP-glucose.

The *gangliosides* are synthesized by addition of oligosaccharides that contain at least one mole of N-*acetylneuraminate*. The open chain form of N-acetylneuraminate is in equilibrium with the pyranose form. The equilibrium constant lies far on the side of the pyranose (reaction 20.18). The biosynthesis of ganglioside GM_2 is summarized in Figure 20.18. The carbohydrate moieties of gangliosides are added sequentially as their UDP derivatives, and N-acetylneuraminate is added as a CMP derivative via the C-1 hydroxyl group. In the intestine the ganglioside GM_1 is the receptor for the protein cholera toxin (Figure 20.19). Produced by *Vibrio cholerae*, this toxin causes

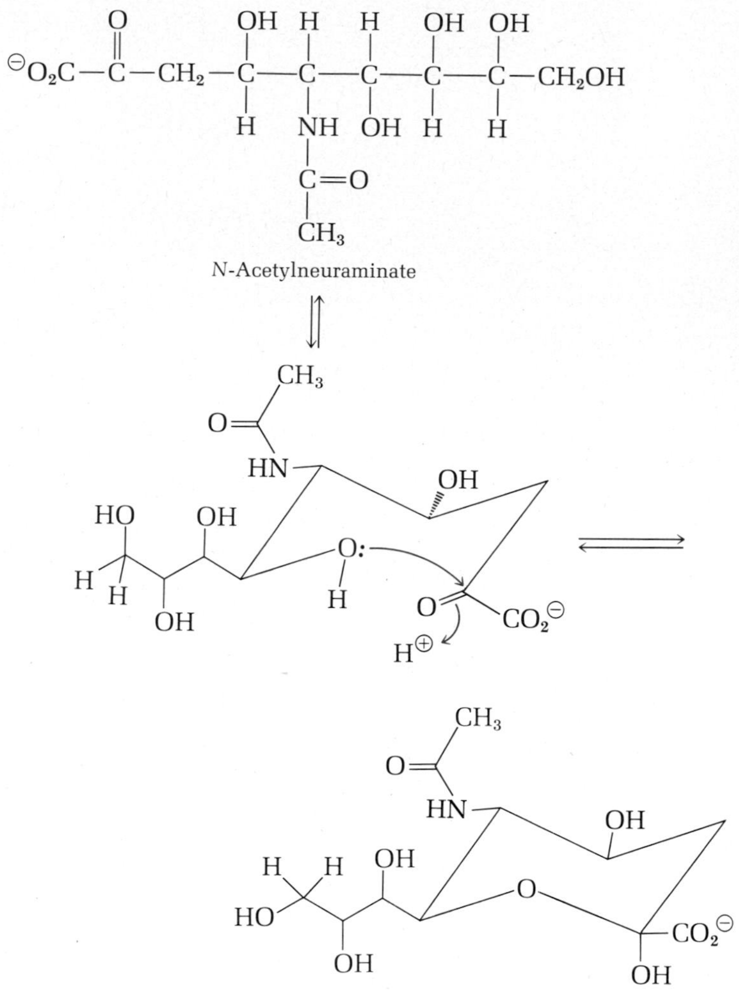

$$^{\ominus}O_2C-\overset{\overset{\textstyle O}{\|}}{C}-CH_2-\overset{\overset{\textstyle OH}{|}}{\underset{\underset{\textstyle H}{|}}{C}}-\overset{\overset{\textstyle H}{|}}{\underset{\underset{\textstyle NH}{|}}{C}}-\overset{\overset{\textstyle H}{|}}{\underset{\underset{\textstyle OH}{|}}{C}}-\overset{\overset{\textstyle OH}{|}}{\underset{\underset{\textstyle H}{|}}{C}}-\overset{\overset{\textstyle OH}{|}}{\underset{\underset{\textstyle H}{|}}{C}}-CH_2OH$$

$$\underset{\underset{\textstyle CH_3}{|}}{\overset{\overset{\textstyle C=O}{|}}{}}$$

N-Acetylneuraminate

(20.18)

severe diarrhea and loss of body salts by stimulating adenylate cyclase in the walls of the small intestine (recall page 686).

In normal cells gangliosides are continuously synthesized and degraded by the *lysosomes*, cell organelles containing digestive enzymes. If the pathways for degradation of these lipids are inhibited, gangliosides accumulate in the nervous system and the body "drowns in its own lipids." In diseases arising from imbalances of ganglioside metabolism (Table 20.1),

Table 20.1   Inherited disorders of ganglioside metabolism

Disease	Missing enzyme	Accumulated lipid
Sandhoff's disease	Hexosaminidases A and B	GM$_2$
Fabry's disease	α-Galactosidase	Gal-Gal-Glc-Cer
Lactosylderamidosis	β-Galactosyl hydrolase	Gal-Glc-Cer
Gaucher's disease	β-Glucosidase	Glc-Cer
Niemann-Pick disease	Sphingomyelinases	Sphingomyelin
Krabbe's leukodystrophy	Galactosylceramide-galactosyl hydrolase	Gal-Cer
Metachromatic leukodystrophy	Sulfatase	HSO$_3$-Gal-Cer
Generalized gangliosidosis	Galactosidase specific for GM$_1$	GM$_1$
Tay-Sachs disease	β-N-Acetylgalactosaminidase (Hexosaminidase A)	GM$_2$

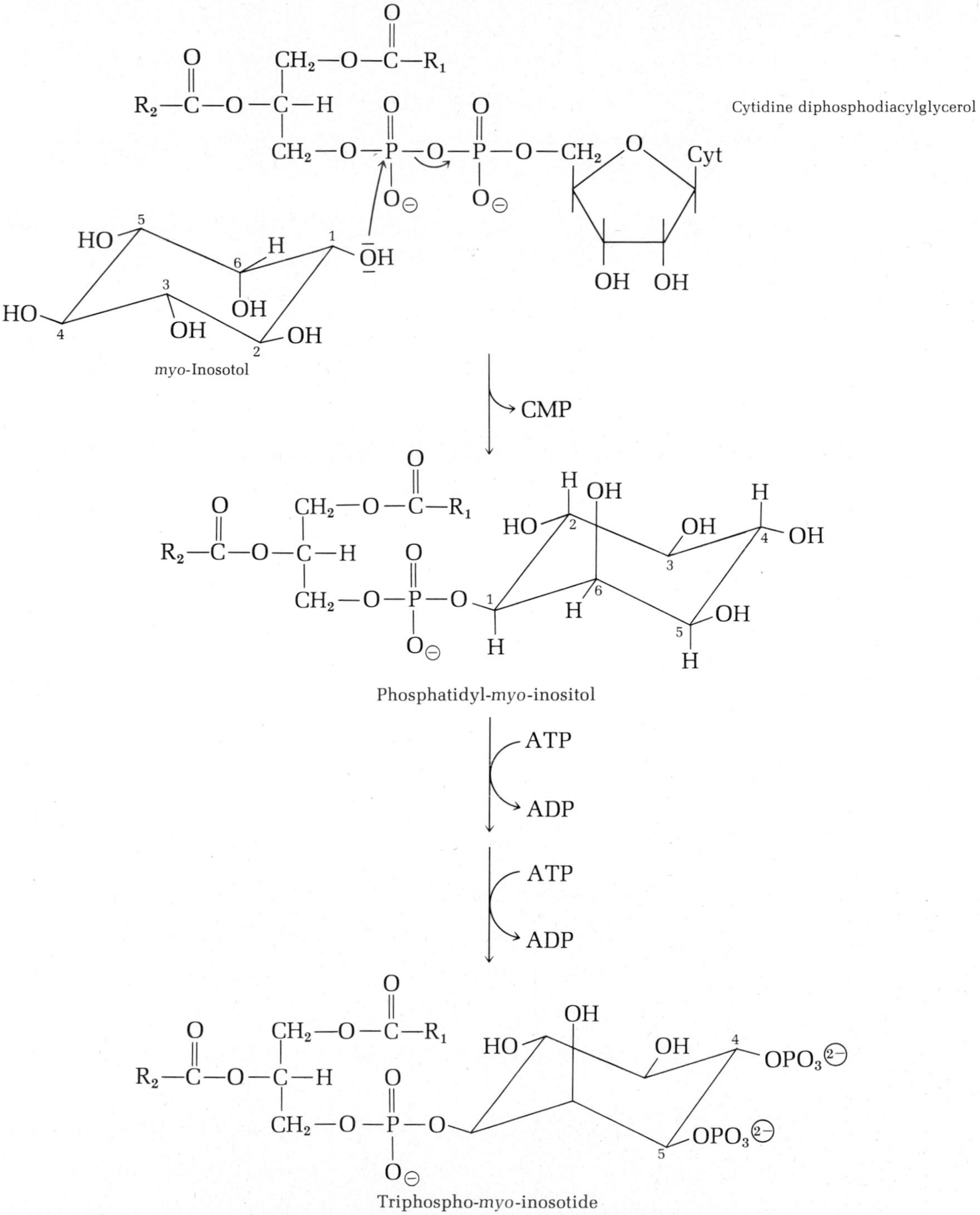

Figure 20.11 Pathway for the biosynthesis of triphospho-*myo*-inositides.

death usually occurs at an early age. For example, *Tay-Sachs disease* is usually fatal by age 3 and death is preceded by blindness and slow physical development. The concentration of ganglioside GM_2 is abnormally high because the enzyme normally responsible for removal of the terminal *N*-acetylgalactosamine of GM_2, *β-N-galactosaminidase* (*hexosaminidase A*) is absent. Tay-Sachs disease is a hereditary defect that is transmitted as an autosomal recessive gene. The absence of the critical enzyme can be detected during pregnancy by amniocentesis and assaying for *β*-N-acetylhexosaminidase activity.

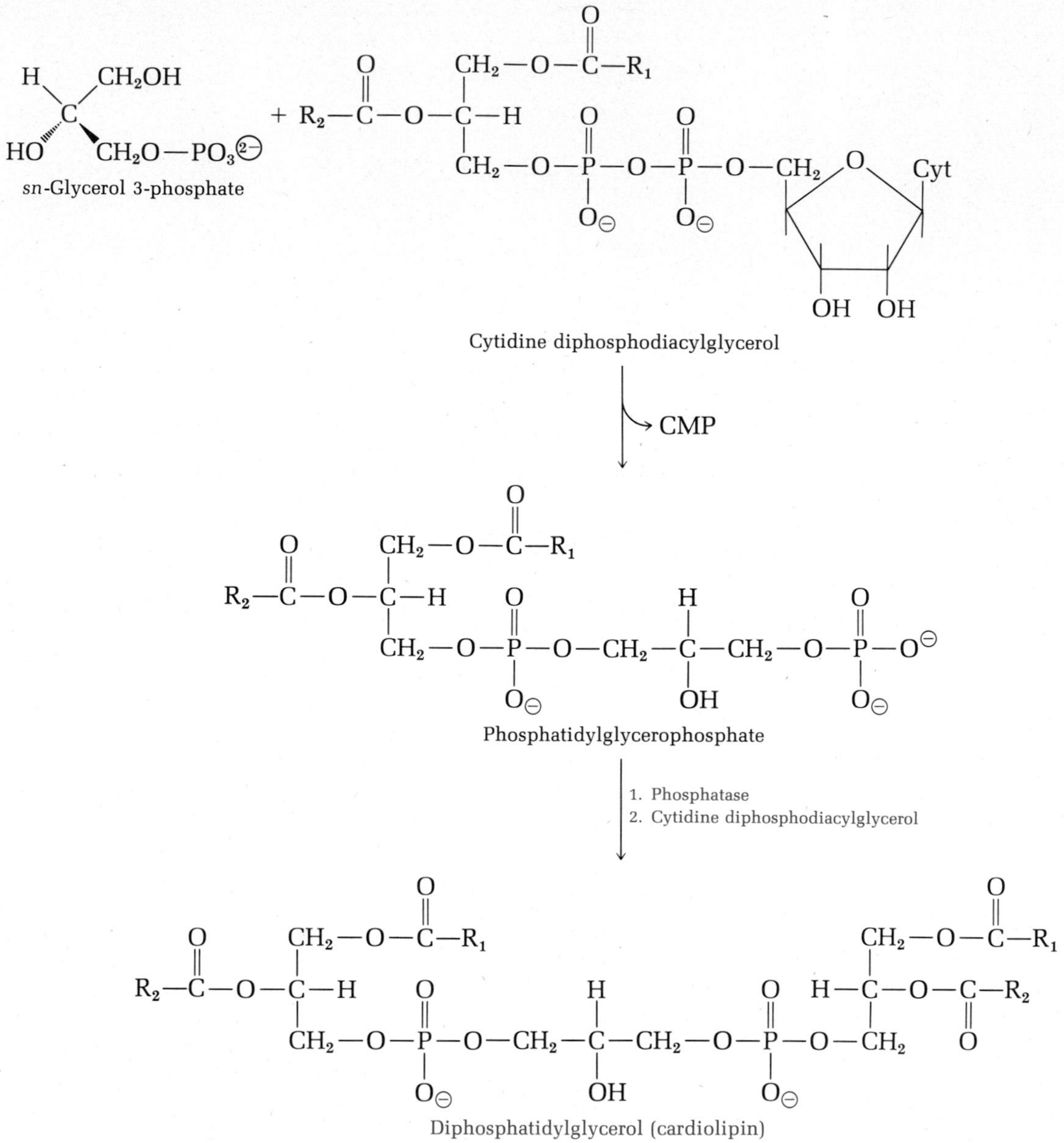

Figure 20.12 Pathway for the biosynthesis of cardiolipin.

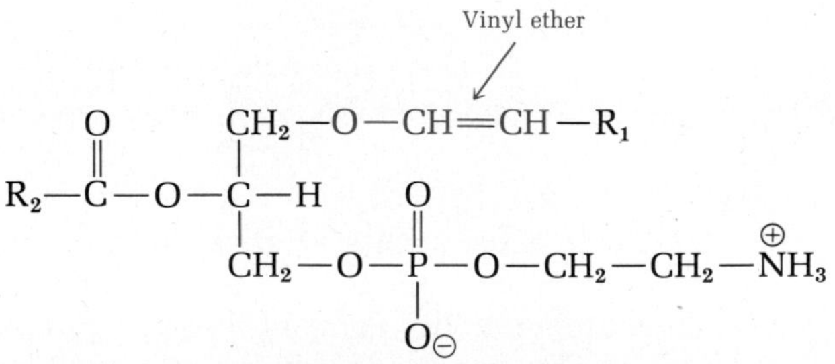

Figure 20.13 Structure of ethanolamine plasmalogen. The basic structural feature that distinguishes plasmalogens from phosphoglycerides is the vinyl ether group at C-1 of the glycerol portion of the lipid.

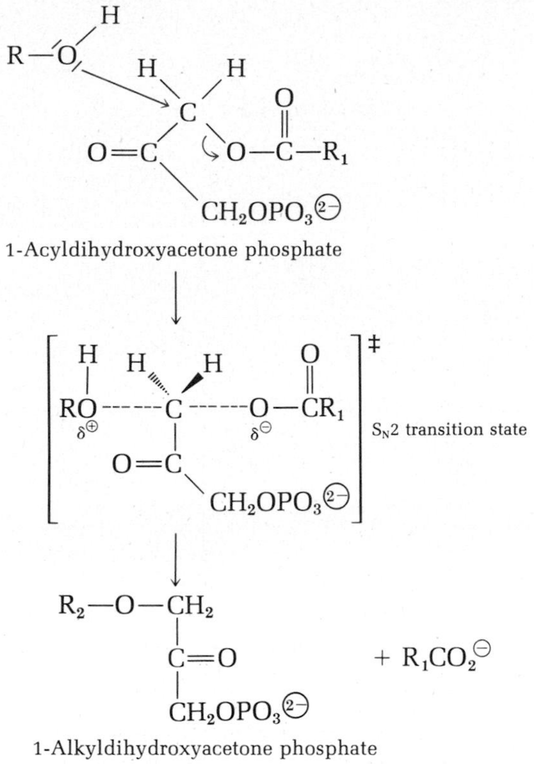

1-Acyldihydroxyacetone phosphate

S_N2 transition state

1-Alkyldihydroxyacetone phosphate

Figure 20.14 Conversion of 1-acyldihydroxy-acetone phosphate to 1-alkyldihydroxyacetone phosphate.

1-Alkyl-2-acylglycerol phosphate

1-Alkyl-2-acylglycerol

CDP-ethanolamine

1-Alkyl-2-acylphosphatidylethanolamine

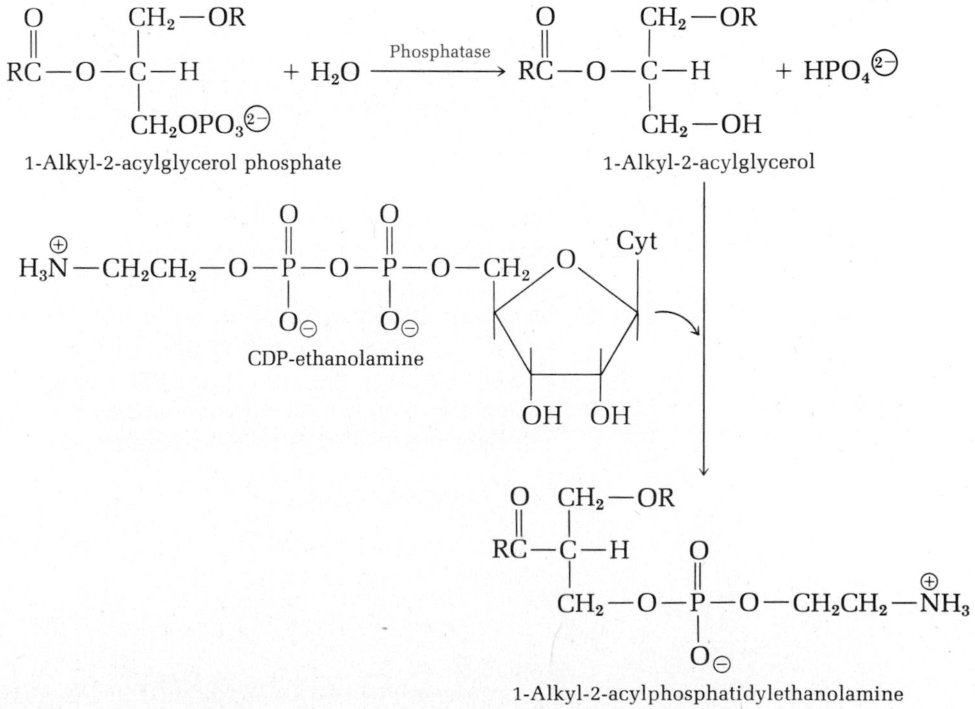

Figure 20.15 Formation of 1-alkyl-2-acylphosphatidylethanolamine. The net reaction is a nucleophilic substitution in which CDP, an excellent leaving group, is displaced by the 3-hydroxyl group of 1-alkyl-2-acylglycerol.

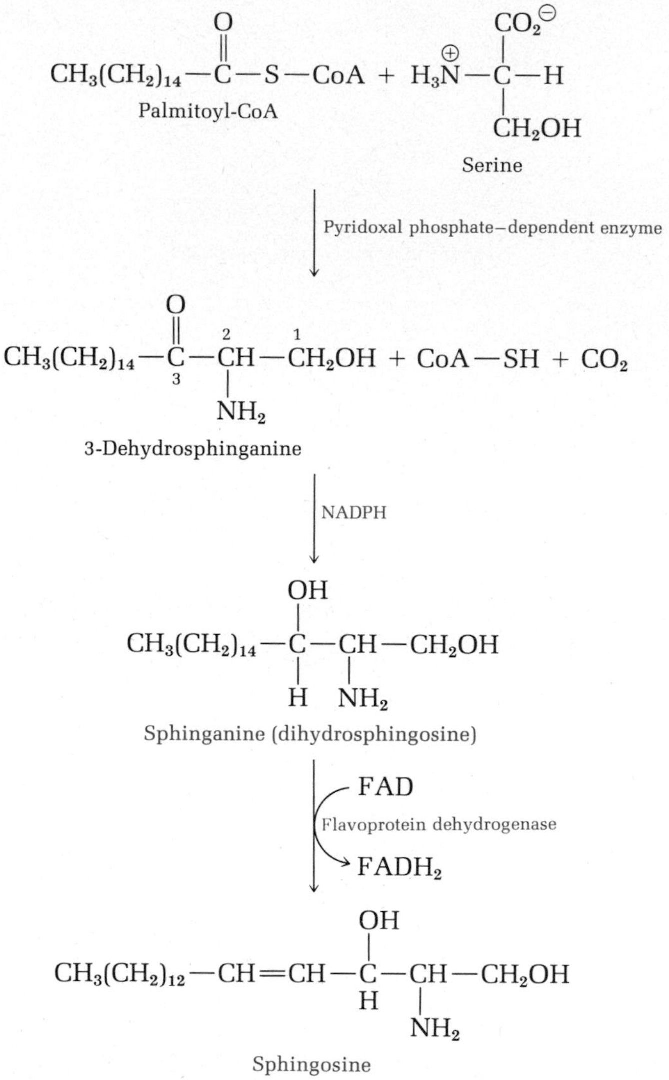

$$CH_3(CH_2)_{14}-\overset{\overset{\textstyle O}{\|}}{C}-S-CoA \;+\; \overset{\oplus}{H_3N}-\overset{\overset{\textstyle CO_2^{\ominus}}{|}}{\underset{\underset{\textstyle CH_2OH}{|}}{C}}-H$$

Palmitoyl-CoA   Serine

Pyridoxal phosphate–dependent enzyme

$$CH_3(CH_2)_{14}-\overset{\overset{\textstyle O}{\|}}{\underset{3}{C}}-\overset{2}{\underset{\underset{\textstyle NH_2}{|}}{CH}}-\overset{1}{CH_2OH} \;+\; CoA-SH \;+\; CO_2$$

3-Dehydrosphinganine

NADPH

$$CH_3(CH_2)_{14}-\overset{\overset{\textstyle OH}{|}}{\underset{\underset{\textstyle H}{|}}{C}}-\overset{}{\underset{\underset{\textstyle NH_2}{|}}{CH}}-CH_2OH$$

Sphinganine (dihydrosphingosine)

FAD

Flavoprotein dehydrogenase

FADH$_2$

$$CH_3(CH_2)_{12}-CH=CH-\overset{\overset{\textstyle OH}{|}}{\underset{\underset{\textstyle H}{|}}{C}}-\overset{}{\underset{\underset{\textstyle NH_2}{|}}{CH}}-CH_2OH$$

Sphingosine

Figure 20.16   Pathway for the biosynthesis of D-sphingosine.

20–6 BIOSYNTHESIS OF CHOLESTEROL

In humans the biosynthesis of cholesterol, a component of cell membranes (Section 11.3) and the precursor of the *steroid hormones,* occurs mainly in the liver. The basic ring system of cholesterol, its numbering system and its structure are shown in Figure 20.20. Viewed from above the planar representation, groups projecting toward the viewer are designated β, and those projecting away from the viewer are designated α. The points of fusion between the B and C and C and D rings are *trans*-diequatorial. The angular methyl groups, C-18 and C-19, are axial.

Isoprene   Isoprenyl

Cholesterol is synthesized from six *isoprene* units that polymerize in a series of *prenyl transfer reactions* involving the biosynthetic monomers *isopentenyl pyrophosphate* and its isomer *dimethylallyl pyrophosphate.*

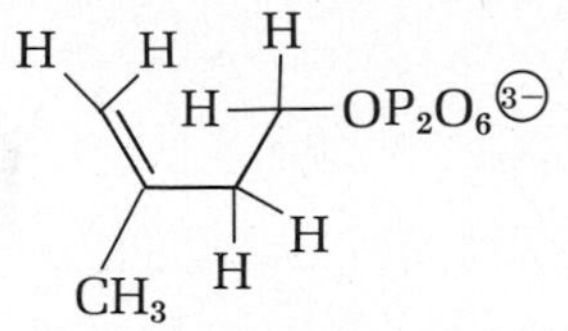

Isopentenyl pyrophosphate

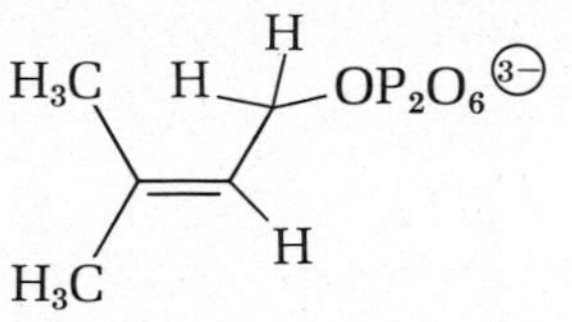

Dimethylallyl pyrophosphate

The synthesis of these monomers is a necessary prelude to cholesterol biosynthesis. In the 1940s it was shown that the carbon atoms of cholesterol are derived from acetate as shown in Figure 20.21, but it was many years before the discovery of prenyl transfer reactions opened the door to an understanding of cholesterol biosynthesis.

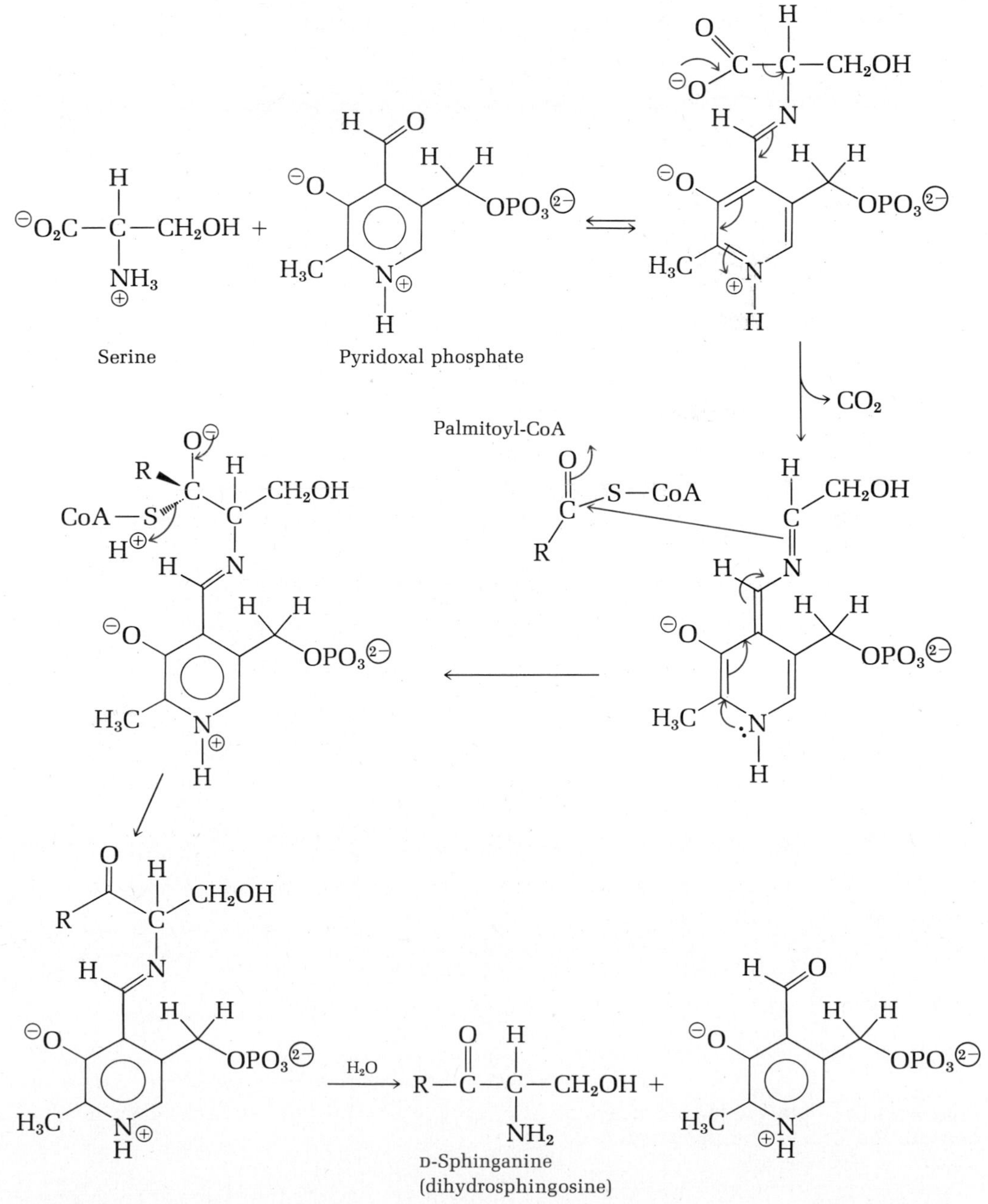

Figure 20.17 Mechanism of action of the pyridoxal phosphate–dependent biosynthesis of D-sphinganine.

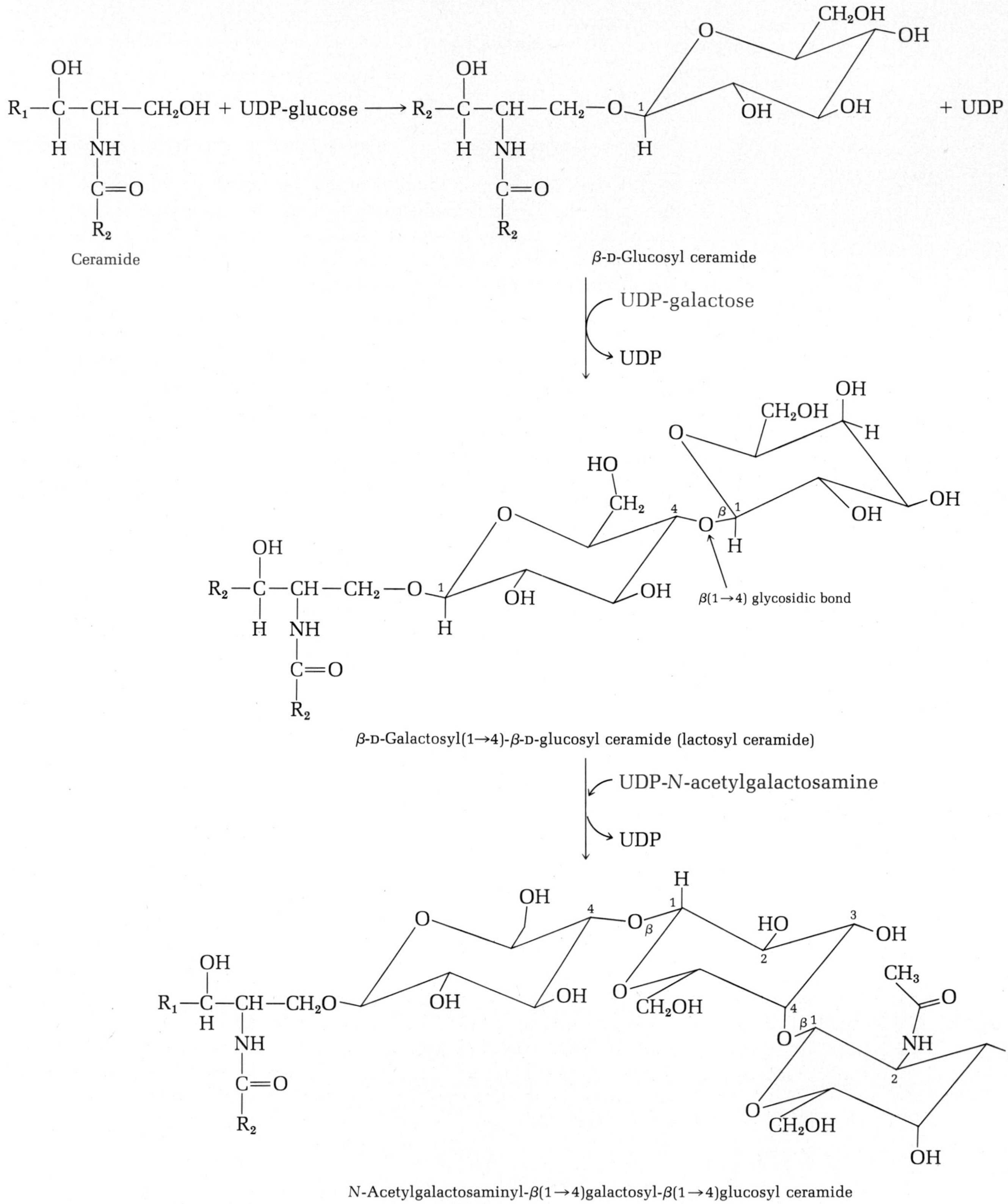
UDP-glucose
+ UDP
Ceramide
β-D-Glucosyl ceramide
UDP-galactose
UDP
β(1→4) glycosidic bond
β-D-Galactosyl(1→4)-β-D-glucosyl ceramide (lactosyl ceramide)
UDP-N-acetylgalactosamine
UDP
N-Acetylgalactosaminyl-β(1→4)galactosyl-β(1→4)glucosyl ceramide

Figure 20.18 Pathway for the biosynthesis of ganglioside GM_2, or Tay-Sachs ganglioside. The systematic name is given beneath the structure of the ganglioside.

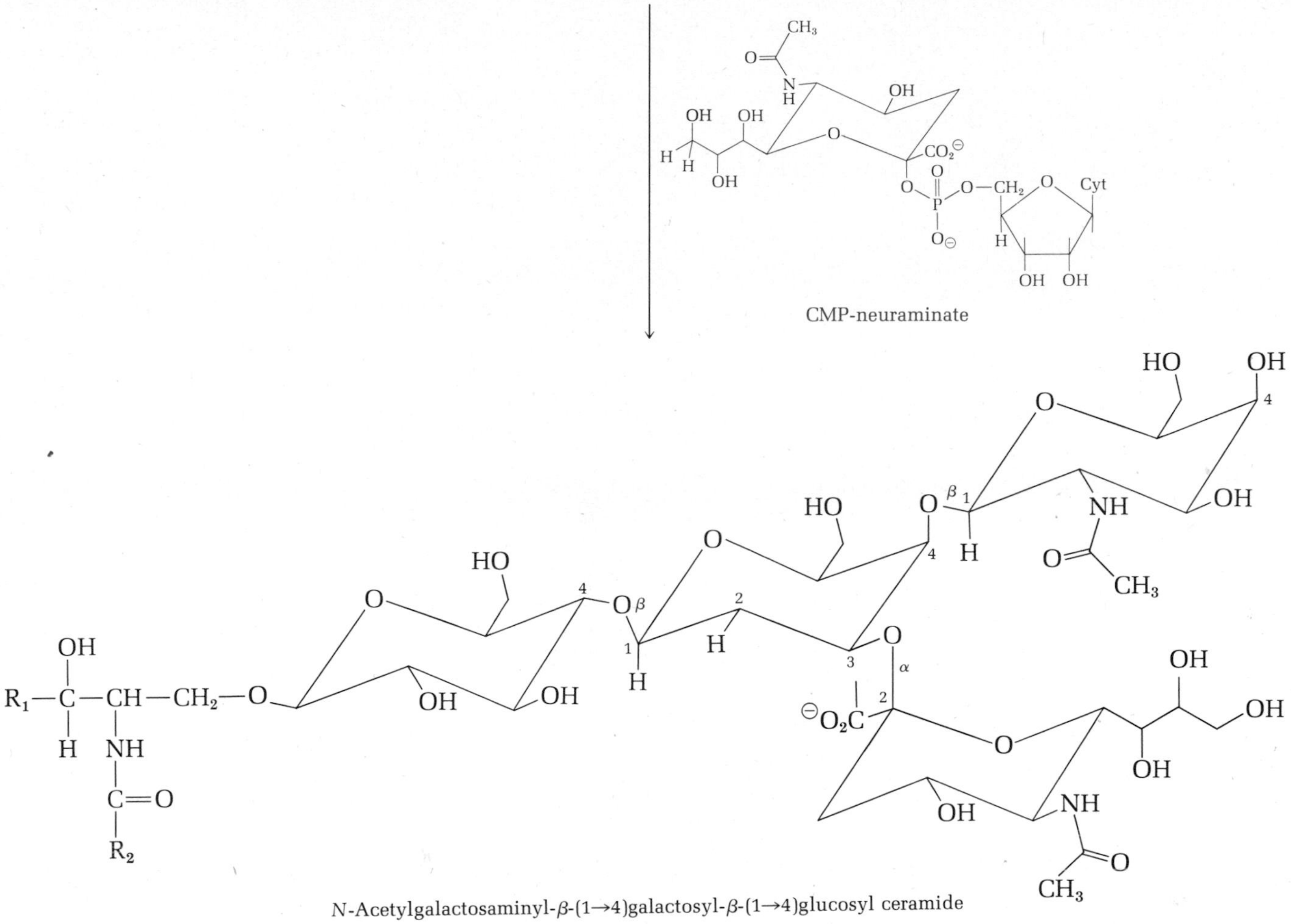

CMP-neuraminate

N-Acetylgalactosaminyl-β-(1→4)galactosyl-β-(1→4)glucosyl ceramide

$$\begin{array}{c} | \\ (2 \to 3) \\ | \\ \text{sialyl} \end{array}$$

Ganglioside GM$_2$

Figure 20.18 (*continued*)

A. Synthesis of Isopentenyl Pyrophosphate and Dimethylallyl Pyrophosphate

The carbon atoms of isopentenyl pyrophosphate and dimethylallyl pyrophosphate originate in acetyl-CoA and acetoacetyl-CoA. A Claisen condensation between these molecules produces 3-hydroxy-3-methylglutaryl-CoA. An NADH-dependent reductase then successively transfers 2 moles of hydride to reduce the thioester to a primary alcohol in the product 3-*R-mevalonate* (reaction 20.19).

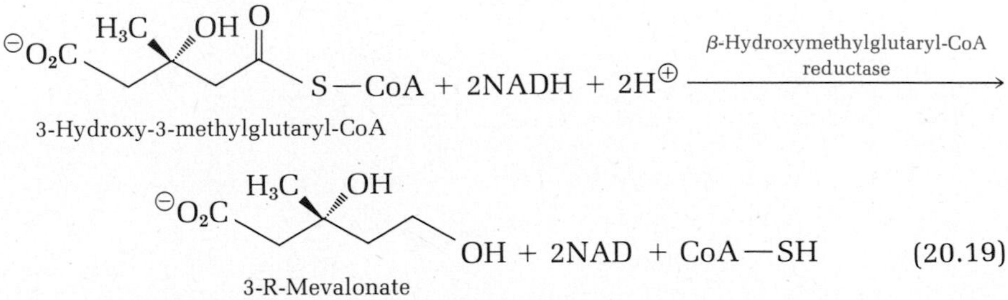

3-Hydroxy-3-methylglutaryl-CoA

$\xrightarrow[\text{reductase}]{\beta\text{-Hydroxymethylglutaryl-CoA}}$

3-R-Mevalonate

$\text{OH} + 2\text{NAD} + \text{CoA}—\text{SH}$ (20.19)

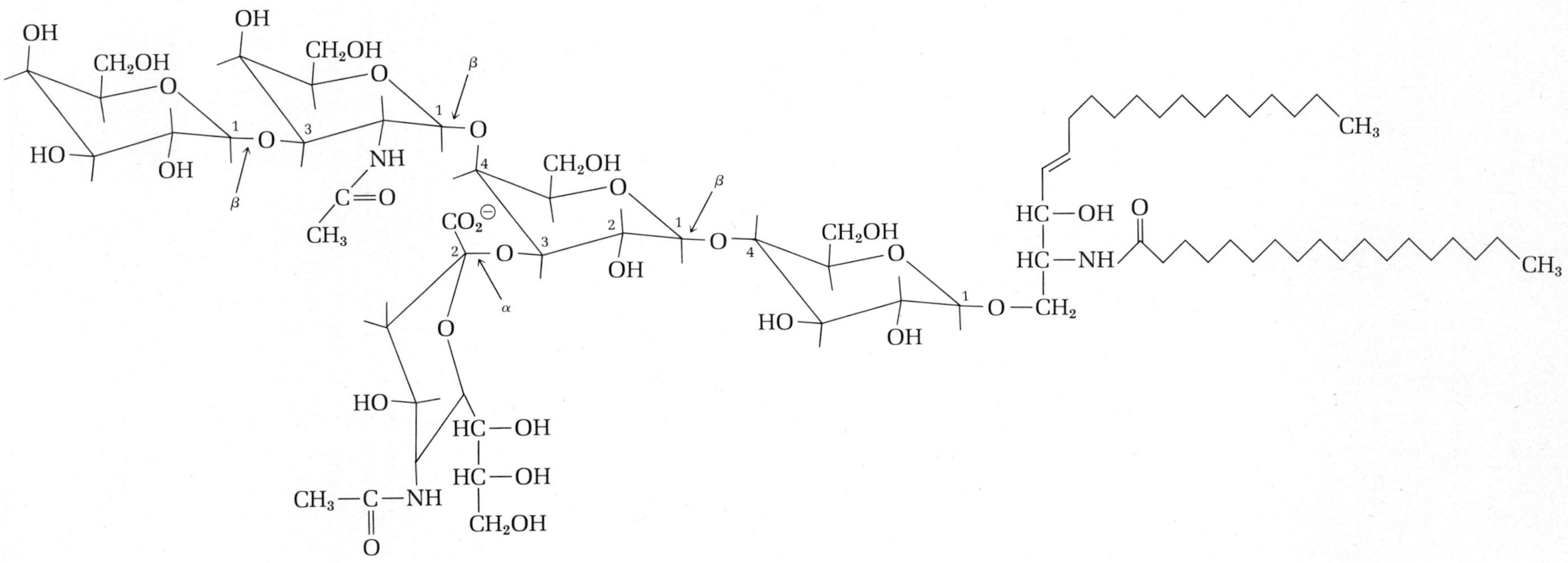

Ganglioside GM_1

(Galactosyl-β-(1→3)N-acetylgalactosaminyl-β-(1→4)galactosyl-β-(1→4)glucosyl ceramide)

—(2←3)—

sialyl

Figure 20.19 Structure of ganglioside GM1, the membrane receptor for cholera toxin.

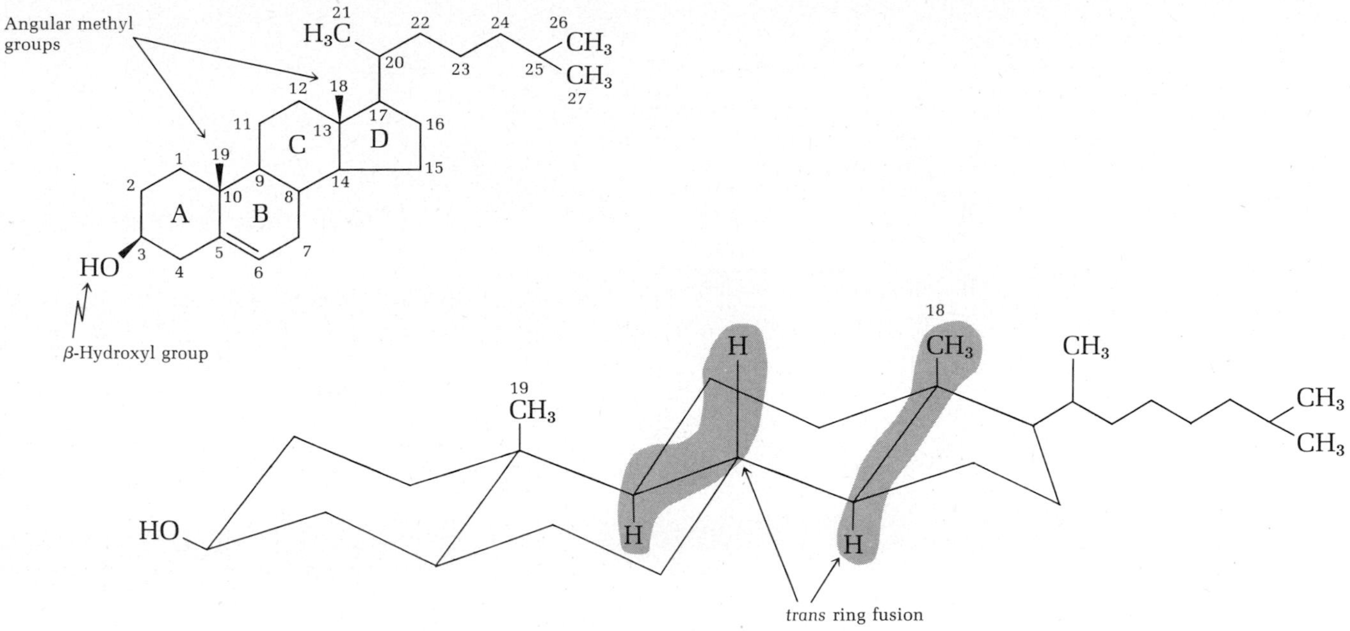

Figure 20.20 Structure and numbering system of cholesterol.

This reaction occurs in two steps with formation of a *hemithioacetal* intermediate that remains bound to the enzyme until it is reduced to the alcohol (Figure 20.22). The synthesis of the reductase is inhibited by cholesterol, and the biosynthesis of mevalonate is the committing step of cholesterol biosynthesis.

Mevalonate is phosphorylated by 3 kinases, and the final product is 3-phospho-5-pyrophosphomevalonate (reaction 20.20).

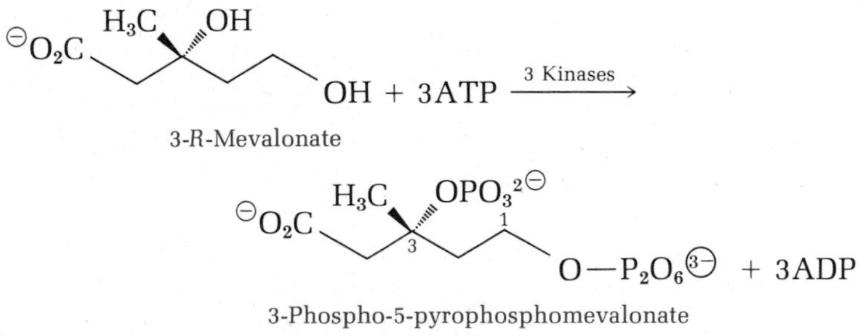

3-*R*-Mevalonate

3-Phospho-5-pyrophosphomevalonate

The 5-pyrophosphoryl group will eventually be the leaving group in prenyl transfer reactions. The C-3 phosphate group and carbon dioxide are lost in a 1,4 elimination that produces isopentenyl pyrophosphate (reaction 20.21).

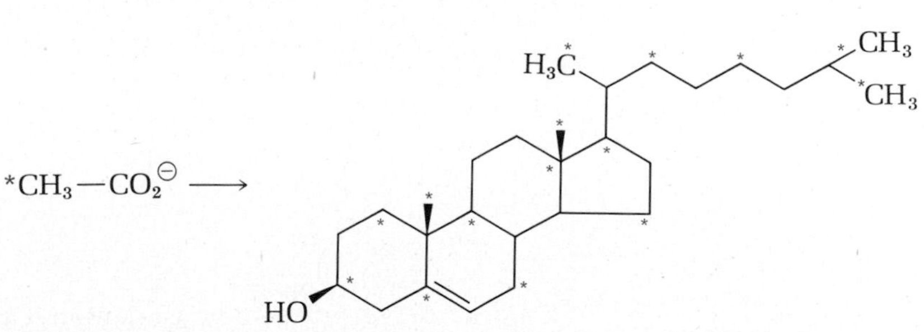

Figure 20.21 Incorporation of acetate in the carbon skeleton of cholesterol. An asterisk signifies a radioisotopic label.

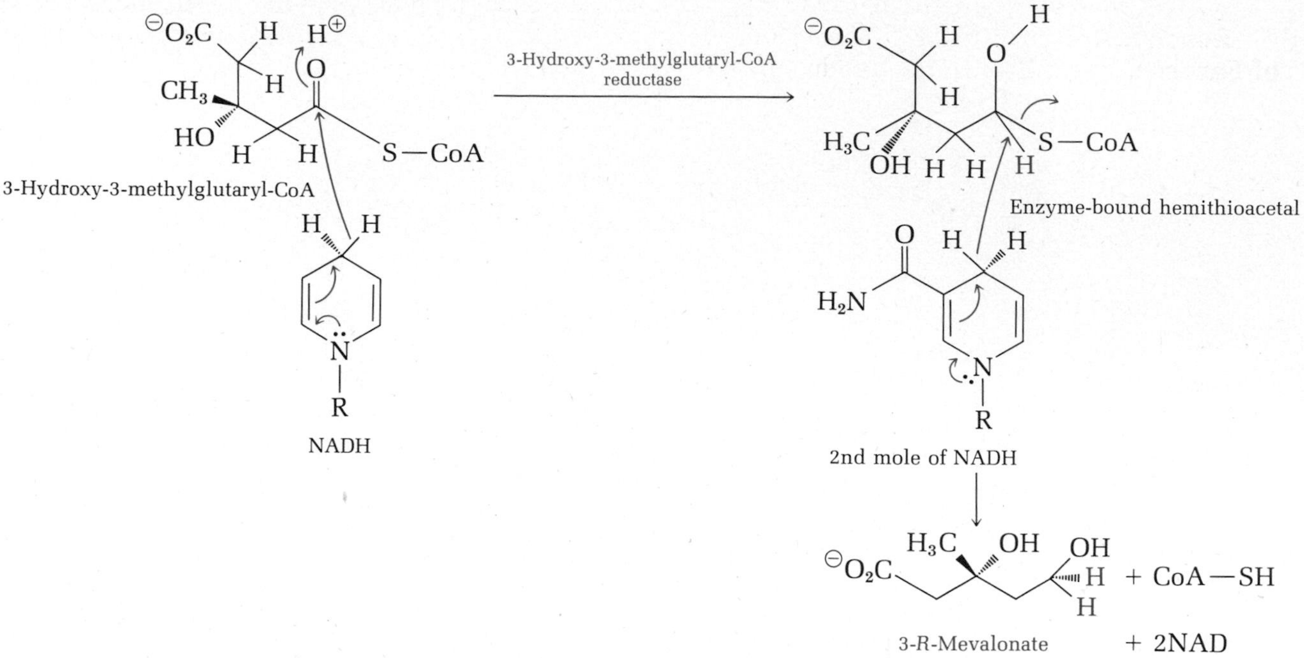

Figure 20.22 Biosynthesis of mevalonate by reduction of R-hydroxy-3-methylglutaryl-CoA. The reduction occurs by successive hydride transfers from NADH.

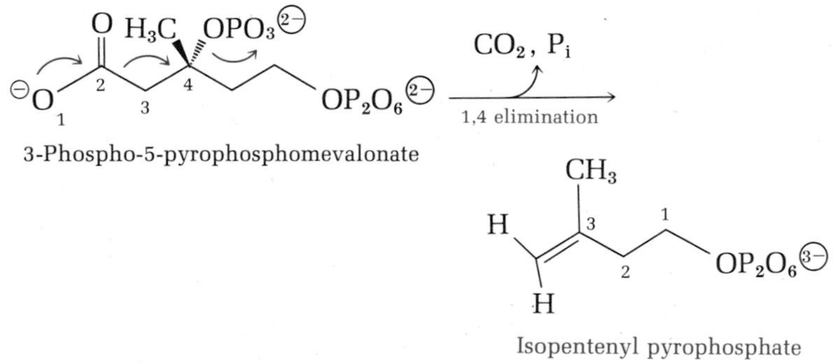

$$(20.21)$$

Isopentenyl pyrophosphate isomerase interconverts isopentenyl pyrophosphate and its allylic isomer, dimethylallyl pyrophosphate (reaction 20.22).

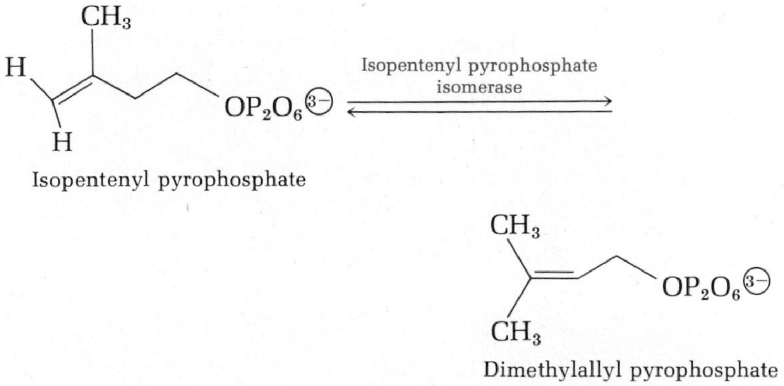

$$(20.22)$$

The equilibrium mixture of isopentenyl pyrophosphate and dimethylallyl pyrophosphate provides the substrates for prenyl transfer reactions.

B. Prenyl Transfer Reactions: Biosynthesis of Farnesyl Pyrophosphate

Prenyl transferase, isolated from pig liver, catalyzes the condensation of C-4 of isopentenyl pyrophosphate and C-1 of dimethylallyl pyrophosphate to produce *geranyl pyrophosphate* (reaction 20.23).

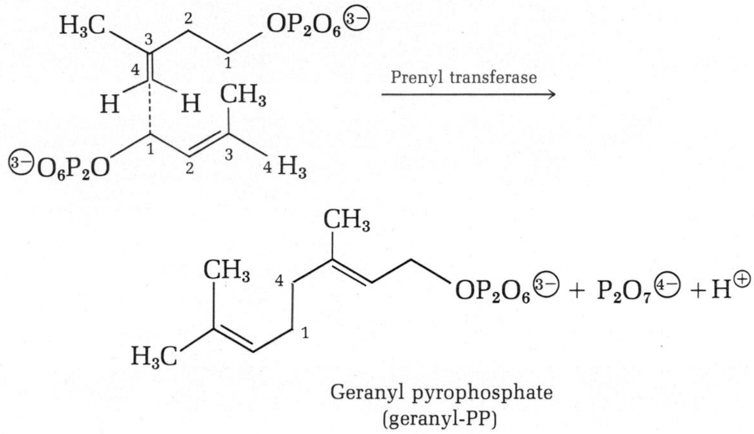

$$(20.23)$$

Geranyl pyrophosphate
(geranyl-PP)

The reaction is a *head-to-tail condensation*, with the pyrophosphate group of isopentenyl pyrophosphate being the tail and C-1 of dimethylallyl pyrophosphate being the head. In all prenyl transfer reactions the allylic substrate loses pyrophosphate from C-1, and the C-4 carbon of isopentenyl pyrophosphate forms a new carbon-carbon bond to elongate the chain of isoprenyl units. Thus, prenyl transferase catalyses the conversion of geranyl pyrophosphate to *farnesyl pyrophosphate* by addition of another prenyl group (reaction 20.24).

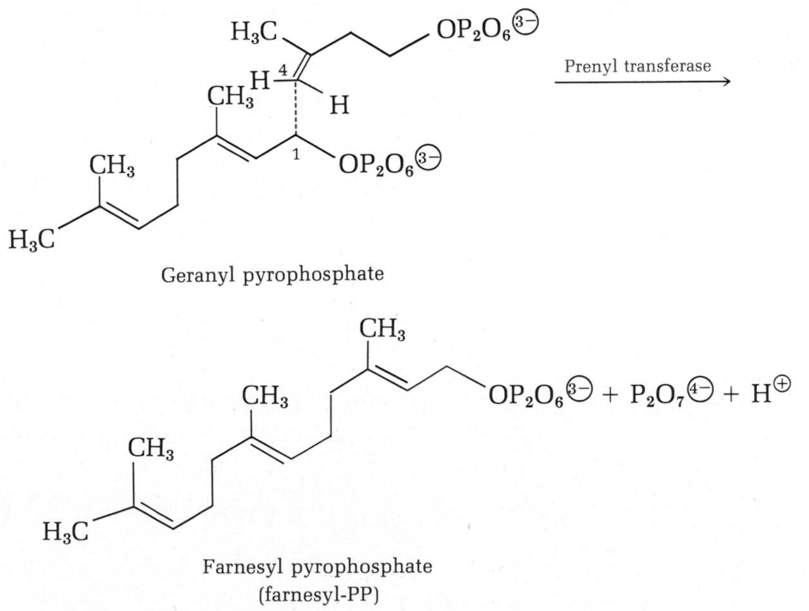

Geranyl pyrophosphate

$$(20.24)$$

Farnesyl pyrophosphate
(farnesyl-PP)

The question how such transformations occur is still unresolved. The bulk of evidence seems to support a carbonium ion mechanism in which the pyrophosphate group of the allylic substrate ionizes to give a resonance-stabilized allylic carbonium ion that is attacked by the π-electrons of isopentenyl pyrophosphate. Loss of a proton from the resulting tertiary carbocation gives the product (Figure 20.23).

C. Conversion of Farnesyl Pyrophosphate to Squalene

Two moles of farnesyl pyrophosphate are converted to squalene by *squalene synthetase* (reaction 20.25).

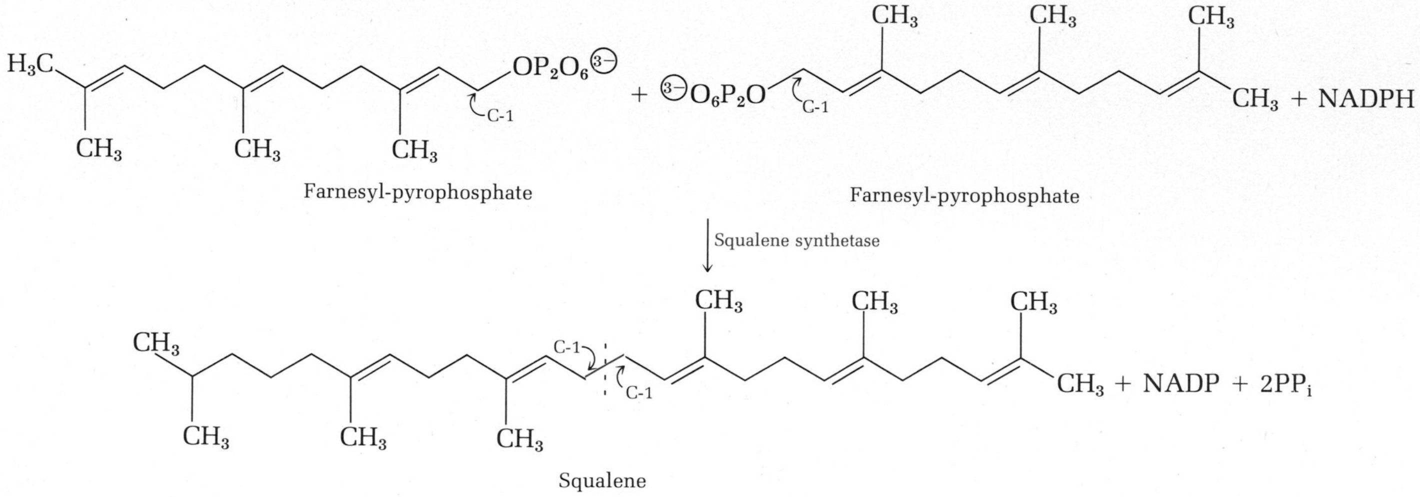

Farnesyl-pyrophosphate

Farnesyl-pyrophosphate

Squalene synthetase

Squalene

(20.25)

In this *tail-to-tail condensation* the two C-1 carbons of farnesyl pyrophosphate become bonded in the product. Two moles of pyrophosphate are released. Each of the C-1 carbons is reduced in the condensation, and NADPH is required for the reaction to occur. The mechanism of this extraordinary condensation is not known.

D. Conversion of Squalene to Cholesterol

Squalene is converted in two steps to *lanosterol* (reaction 20.26).

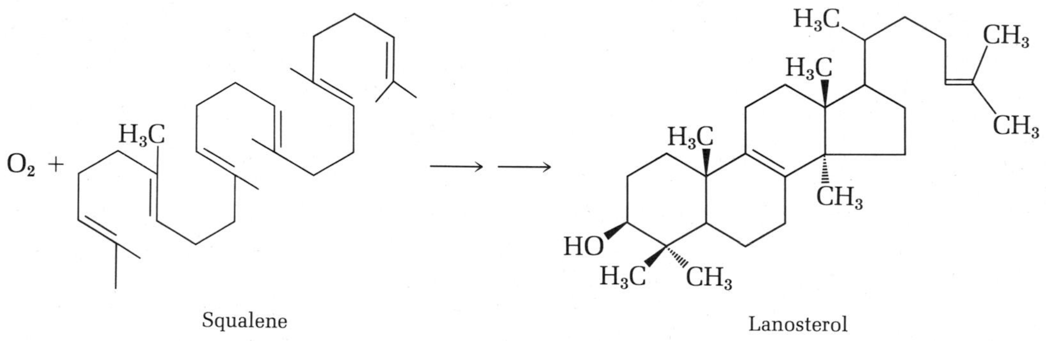

Squalene

Lanosterol

(20.26)

First, squalene is converted to *squalene-2,3-epoxide* by *squalene epoxidase*, an NADPH-dependent monooxygenase (reaction 20.27).

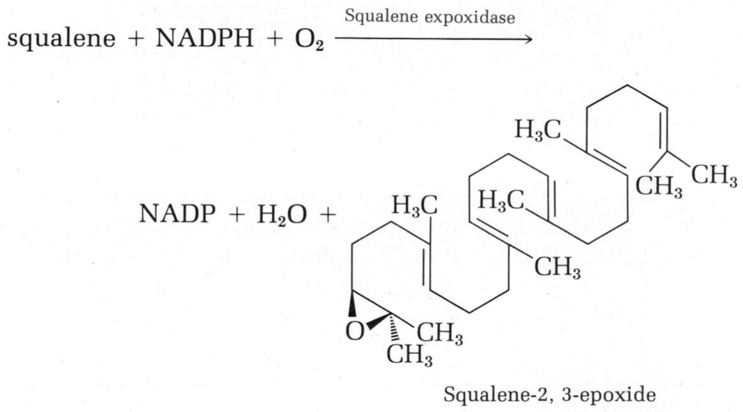

$$\text{squalene} + \text{NADPH} + O_2 \xrightarrow{\text{Squalene epoxidase}}$$

$$\text{NADP} + H_2O +$$

(20.27)

Squalene-2, 3-epoxide

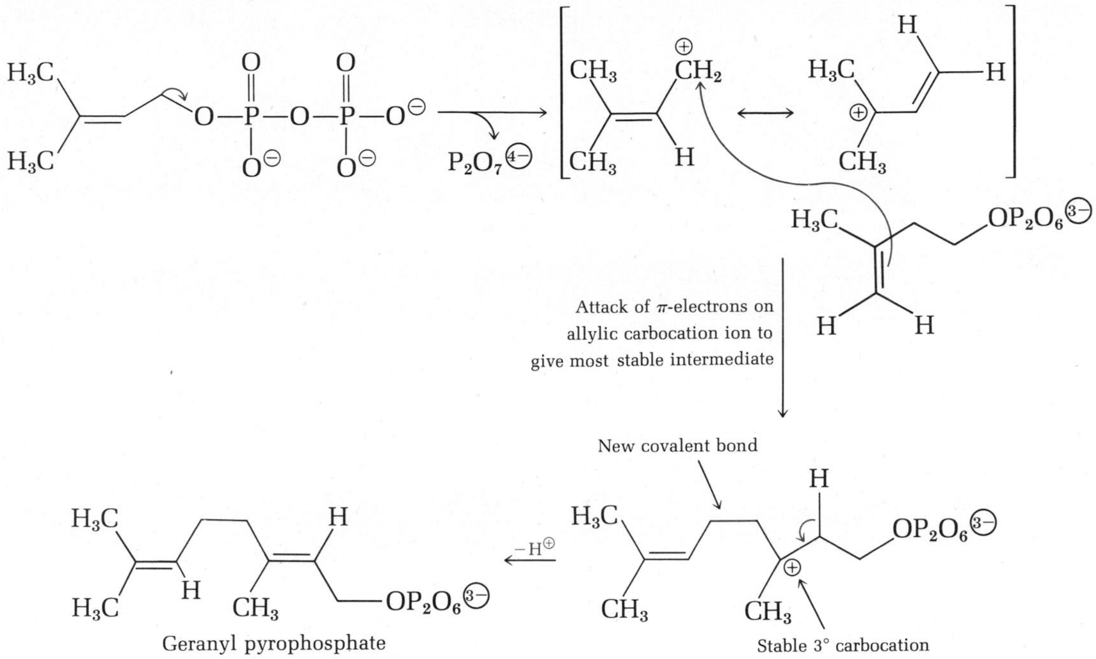

Figure 20.23 Biosynthesis of geranyl pyrophosphate by a prenyl transfer reaction.

Second, squalene-2,3-epoxide is converted to lanosterol by *squalene oxidocyclase* with protosterol as an intermediate product. The apparently converted process begins with nucleophilic attack by the enzyme at C-20 and terminates with the opening of the epoxide ring (reaction 20.28). All of the ring junctions of the first product, *protosterol,* are *trans*, and the hydroxyl group at C-3 is in the thermodynamically more stable equatorial position. The C-3 hydroxyl group projects toward the viewer when the ring system is viewed in a planar representation and thus is β (recall Figure 20.20).

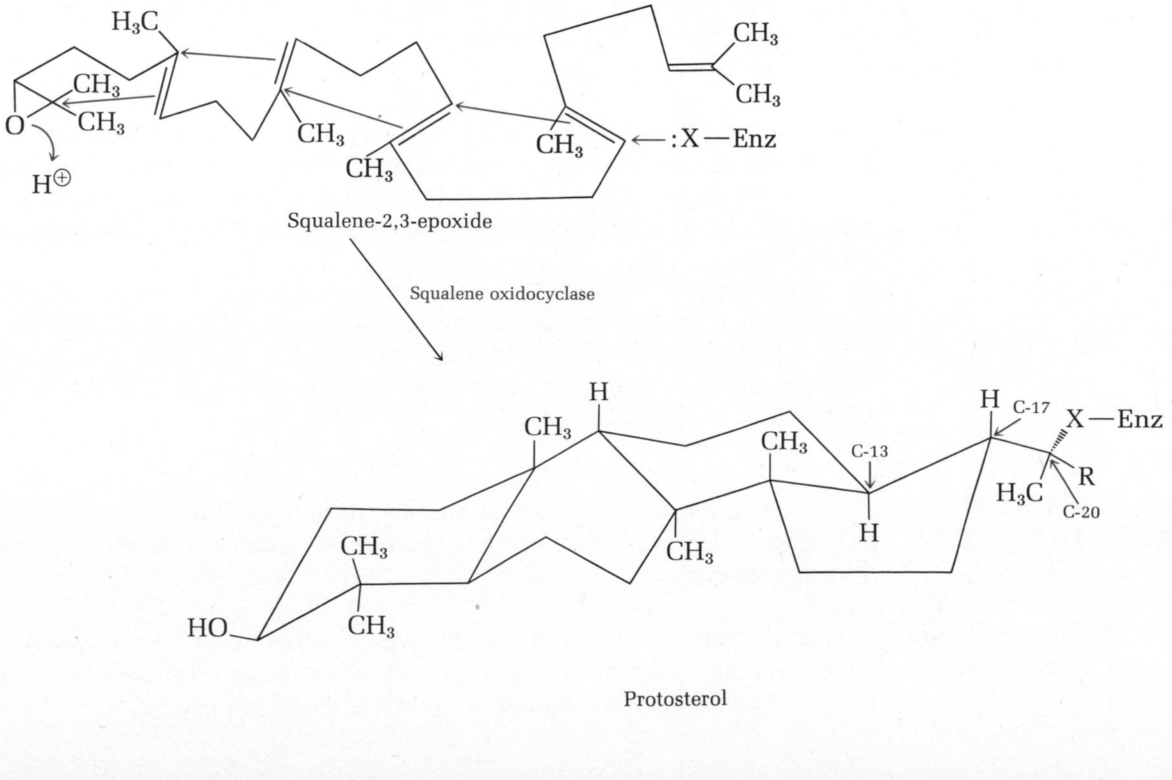

(20.28)

Protosterol

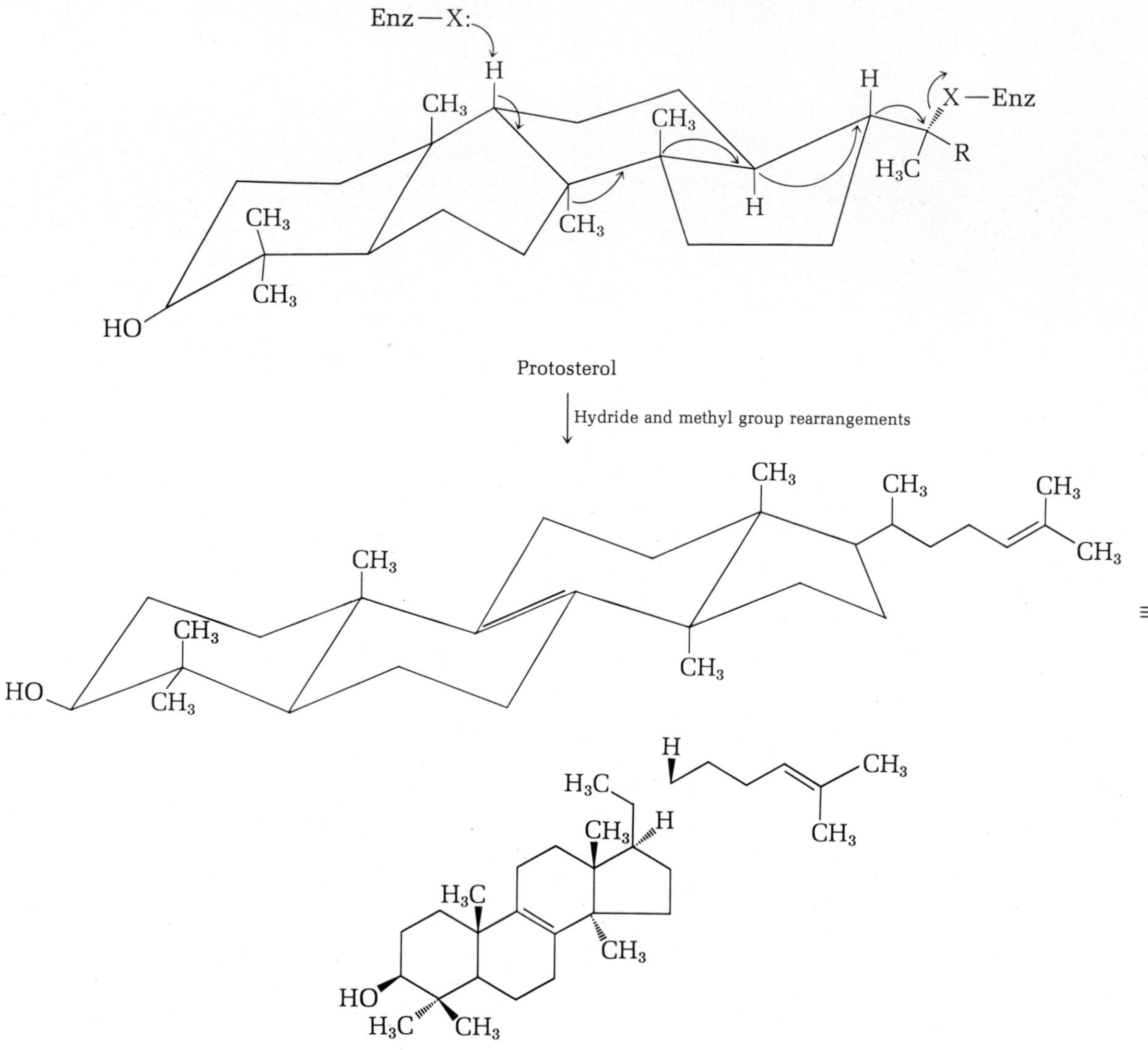

Figure 20.24 Conversion of protoserol to lanosterol.

The protosterol intermediate then rearranges in two 1,2 hydride shifts. The first is from C-13 to C-17, and the second is from C-17 to C-20. Two methyl group rearrangements occur simultaneously as the methyl group at C-8 moves to C-14, and the methyl group at C-14 moves to C-13. Lanosterol is the result (Figure 20.24).

Lanosterol is the precursor of cholesterol in all animals. At least 25 steps are required to convert lanosterol to cholesterol, many of them poorly, if at all, understood. Three methyl groups are lost giving *zymosterol*, rearrangement of a double bond gives *desmosterol*, and introduction of a new double bond gives cholesterol (Figure 20.25).

20–7 BIOSYNTHESIS OF STEROID HORMONES

There are three major classes of steroid hormones: the sex hormones, the progestins, and the adrenal cortical hormones, or cortocoids. All are derived from cholesterol.

A. Biosynthesis of Pregnenolone

The biosynthesis of all steroid hormones requires loss of a six-carbon fragment from the hydrocarbon "tail" bound to C-17 in the D ring of cholesterol. The C-21 product is *pregnenolone* (reaction 20.29).

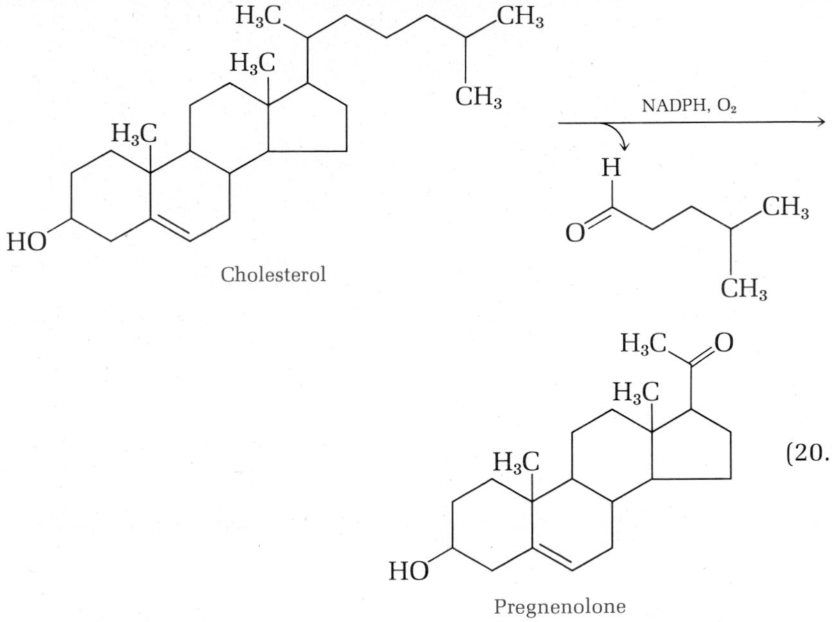

Cholesterol

NADPH, O₂

(20.29)

Pregnenolone

The cleavage of a saturated carbon-carbon bond with no functional groups nearby does not occur in one step. An electron transport chain with monooxygenase activity produces a vicinal diol at C-20 and C-22 by successive hydroxylations (reactions 20.30). The electron transport chain contains cytochrome P450, an iron-sulfur protein called *adrenodoxin*, and an NADPH-dependent flavoprotein dehydrogenase. Oxidation of the vicinal diol by the same electron transport system produces pregnenolone in an oxidative cleavage of the vicinol diol (reaction 20.31). The synthesis of pregnenolone is stimulated by the polypeptide *adrenocorticotrophic hormone* (ACTH), a 39-residue polypeptide synthesized by the anterior pituitary.

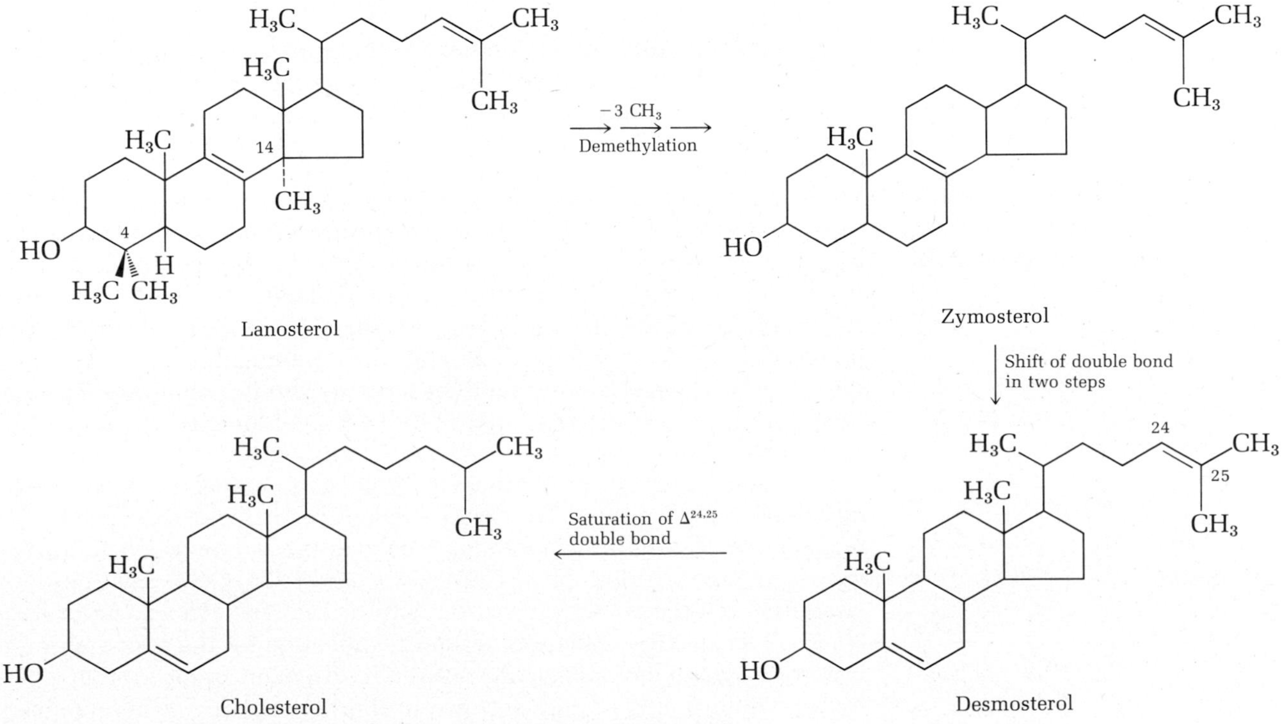

Lanosterol

$-3 CH_3$
Demethylation

Zymosterol

Shift of double bond
in two steps

Saturation of $\Delta^{24,25}$
double bond

Cholesterol

Desmosterol

Figure 20.25 Steps in the conversion of lanosterol to cholesterol.

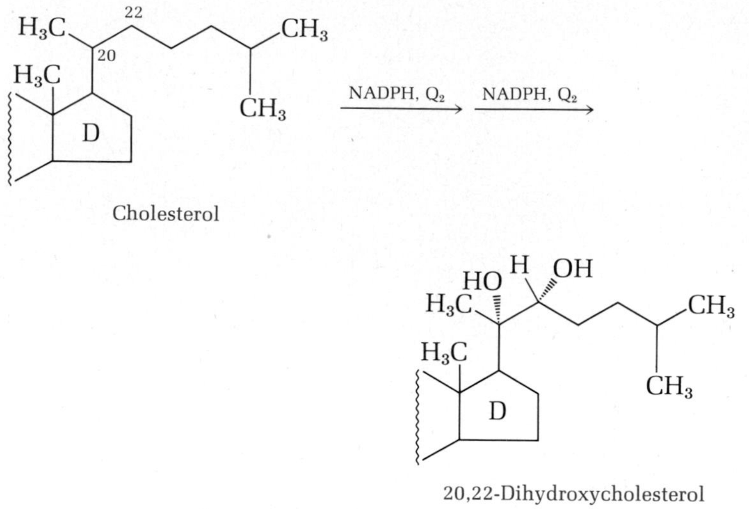

Cholesterol

$$\xrightarrow{\text{NADPH, Q}_2} \xrightarrow{\text{NADPH, Q}_2}$$

20,22-Dihydroxycholesterol

(20.30)

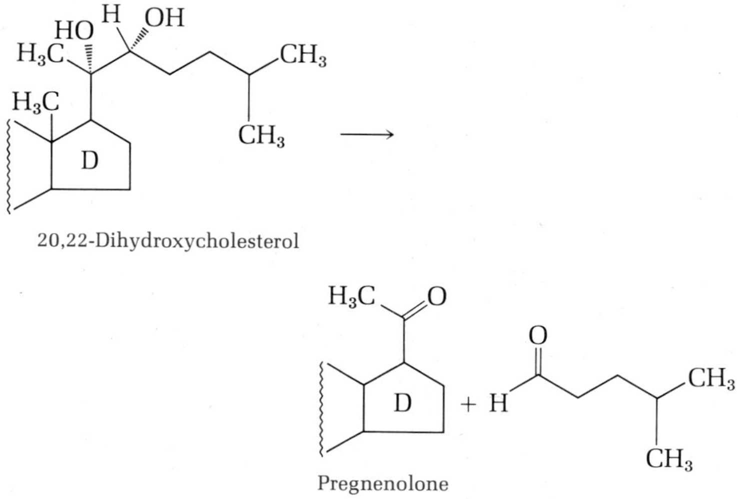

20,22-Dihydroxycholesterol

Pregnenolone

(20.31)

All mammalian ACTH sequences that have been determined are identical over the first 24 residues, the only ones required for biological activity. ACTH exerts its effects through a cAMP-dependent mechanism.

B. Biosynthesis of Progesterone and Corticosteroids

Progesterone prepares the lining of the uterus, called the *endometrium,* for the implantation of an ovum. It is also required for the maintenance of pregnancy. Two steps are required for the conversion of pregnenolone to progesterone. First, the C-3 hydroxyl group is oxidized to a ketone in a reaction catalyzed by an NADH-dependent dehydrogenase. Second, the β,γ double bond between C-5 and C-6 migrates into conjugation with the carbonyl group in an isomerization catalyzed by steroid-Δ-isomerase. The driving force for the reaction is resonance stabilization of the α,β-unsaturated ketone (reaction 20.32).

Cortisol and *cortisone* stimulate glycogen synthesis, gluconeogenesis, and protein degradation. They are called *glucocorticoids* because of their effect on carbohydrate metabolism. Progesterone is converted to cortisol by successive hydroxylations at C-17, C-11, and C-21. C-17 must be hydroxylated first, but the other two hydroxylations can occur in any order (reaction 20.33). The electron transport system responsible for the conversion of cholesterol to pregnenolone carries out the hydroxylation reaction at C-11. The hydroxylations at C-17 and C-21 are similar, but involve different sets of enzymes. All three hydroxylations require cytochrome P450.

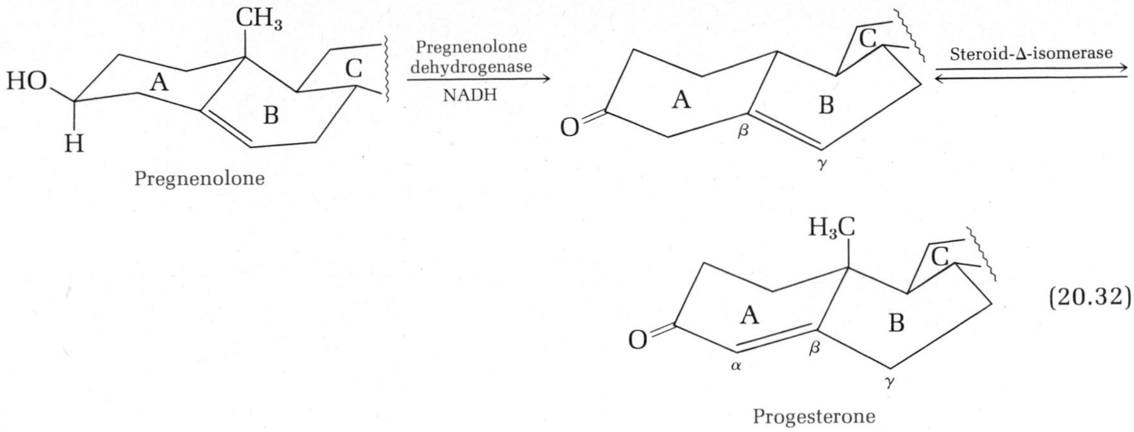

Pregnenolone

(20.32)

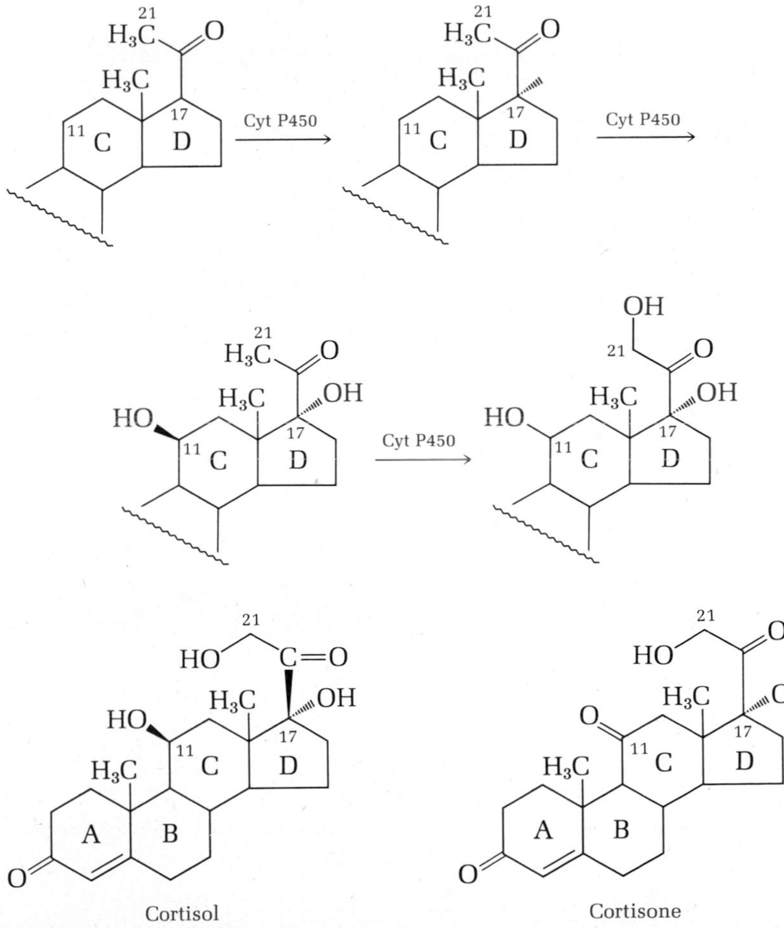

Progesterone

(20.33)

Cortisol

Cortisone

Progesterone can also be converted to aldosterone via corticosterone in successive hydroxylations at C-11 and C-21 (reaction 20.34). Oxidation of the C-18 methyl group of corticosterone to an aldehyde produces *aldosterone*, the major *mineralocorticoid* (reaction 20.35).

The mineralocorticoids alter blood volume and blood pressure by stimulating the kidney to absorb $Na^{\oplus}$, $Cl^{\ominus}$, and $HCO_3^{\ominus}$. Aldosterone is about twenty times more powerful than the next most potent mineralocorticoid, corticosterone. If the 21-hydroxylase is deficient, the glucocorticoids and mineralocorticoids are produced in inadequate quantities to meet physiological requirements. The adrenal gland, sensing an insufficient supply of corticoids, produces more ACTH, and more pregnenolone is synthesized.

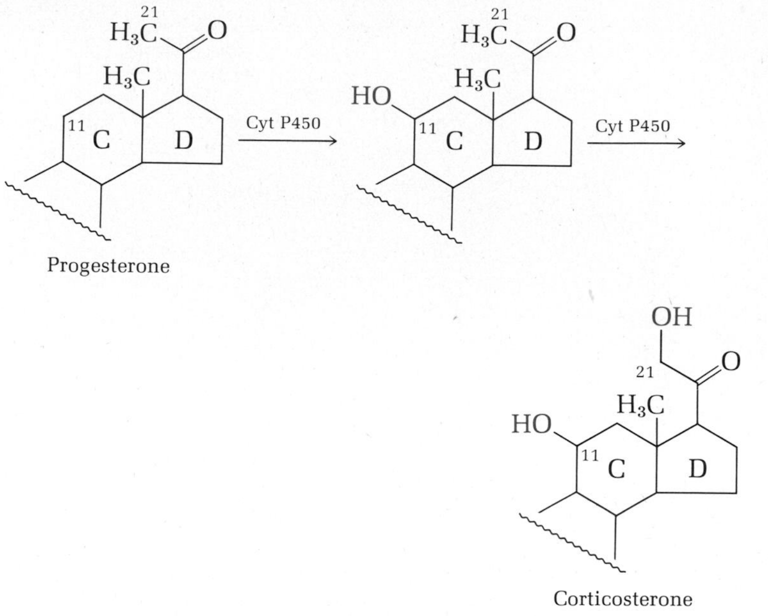

Progesterone → Cyt P450 → → Cyt P450 → (20.34)

Corticosterone

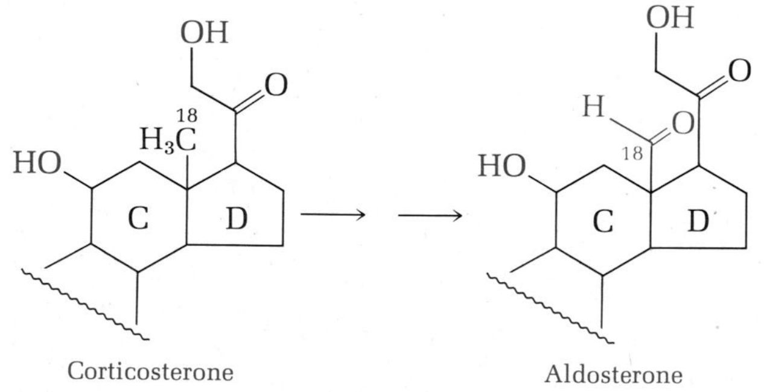

Corticosterone → → Aldosterone (20.35)

The male and female sex hormones are derived from pregnenolone. A high pregnenolone concentration results in a high progesterone concentration, and more male sex hormones are produced. Sexual precocity is the result in males, and virilization is the result in females. As a further complication, sodium ions are not absorbed by the kidney and are excreted in the urine since there is also an insufficient supply of mineralocorticoids. Hormone treatment, consisting of administration of glucocorticoids and mineralocorticoids, corrects these problems if treatment is begun before the child is 2 years old.

C. Biosynthesis of Androgens and Estrogens

Progesterone is converted into the male sex hormones, called *androgens*, by the pathway shown in Figure 20.26. The C-17 position of progesterone is hydroxylated in the first step producing 17-α-hydroxyprogesterone. Oxidative cleavage of the bond between C-17 and C-20 results in loss of acetaldehyde, and oxidation of the 17-hydroxyl group to a carbonyl group produces *androstenedione*. Reduction of the C-17 keto group to an alcohol gives *testosterone*, the androgen that is responsible for the development of secondary sex characteristics in males.

The female sex hormones, called *estrogens*, are synthesized mainly in the ovaries. The male sex hormone testosterone is the precursor of estrogens. The pathway for biosynthesis of *estradiol*, the principal female sex hormone, is shown in Figure 20.27. Successive hydroxylations by a mixed function oxidase using NADPH and O_2 convert the C-19 methyl group to a *gem*-diol. A

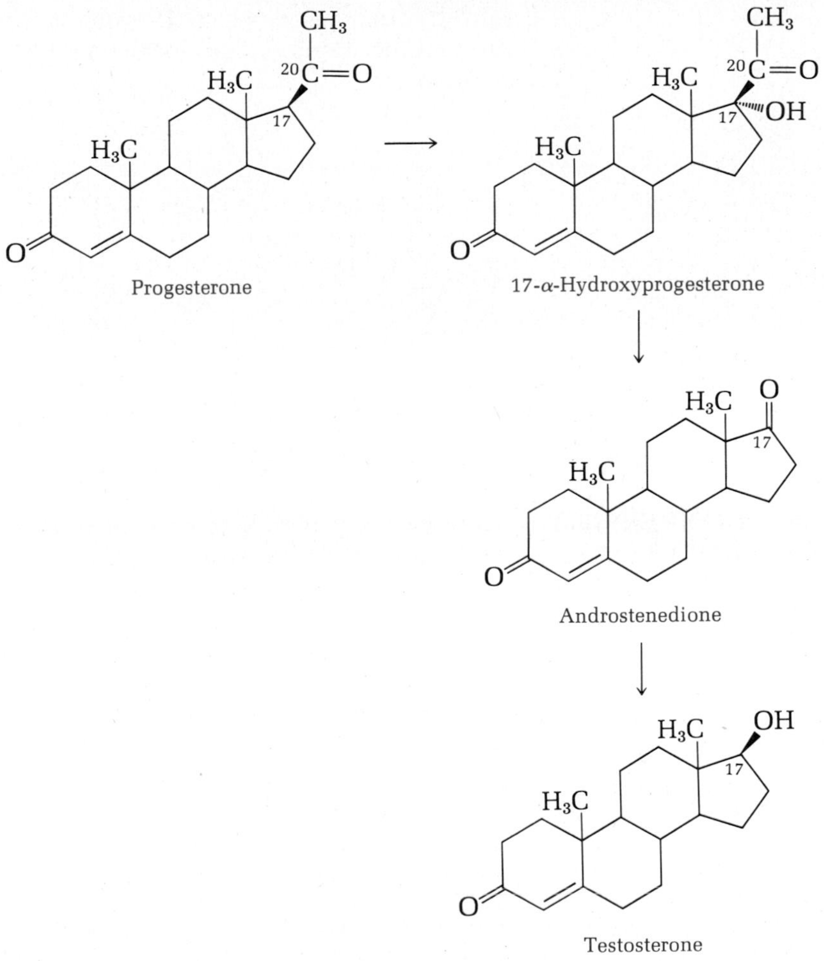

Figure 20.26 Pathway for the biosynthesis of testosterone.

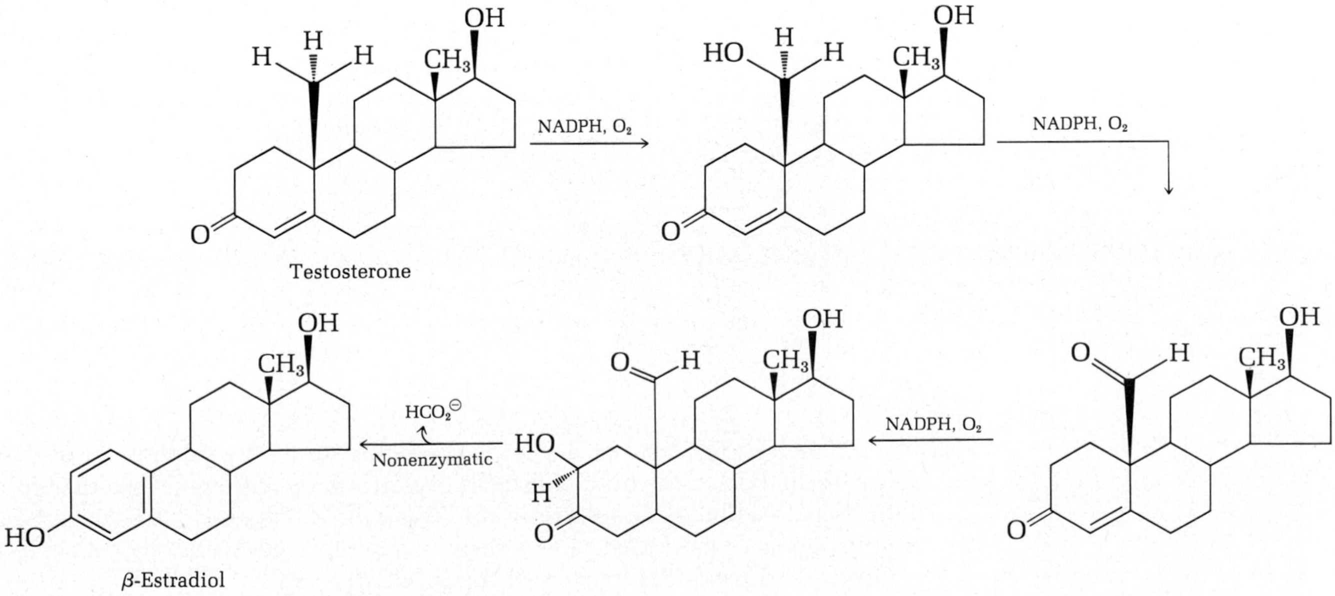

Figure 20.27 Pathway for the biosynthesis of estradiol.

third hydroxylation at C-2, the rate-determining step of the process, leaves a product that rapidly loses formic acid and aromatizes to estradiol (reaction 20.36).

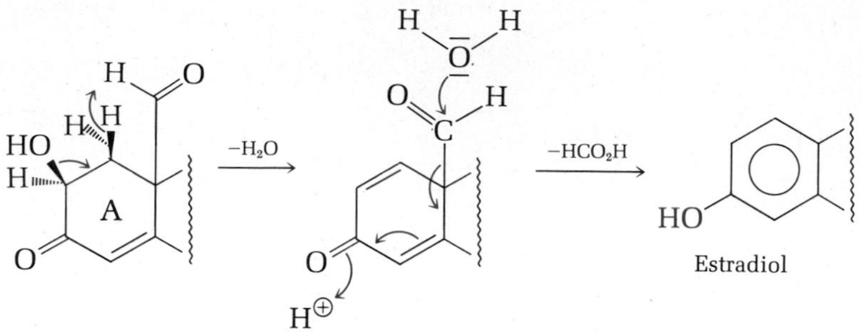

(20.36)

20–8 CONVERSION OF CHOLESTEROL TO BILE SALTS

The major metabolic fate awaiting cholesterol is not conversion to steroid hormones, but degradation to *cholic acid* and then to *glycholic acid* and other *bile salts*. The bile salts act as detergents to facilitate the digestion of lipids (recall Figure 19.3). Multiple pathways exist in humans for the conversion of cholesterol to cholic acid and other bile acids (Figure 20.28). The quantitative importance of each of these pathways has not been determined.

Cholyl-CoA can be converted either to glycocholate by reaction with glycine (reaction 20.37) or to taurocholic acid by reaction with taurine (reaction 20.38).

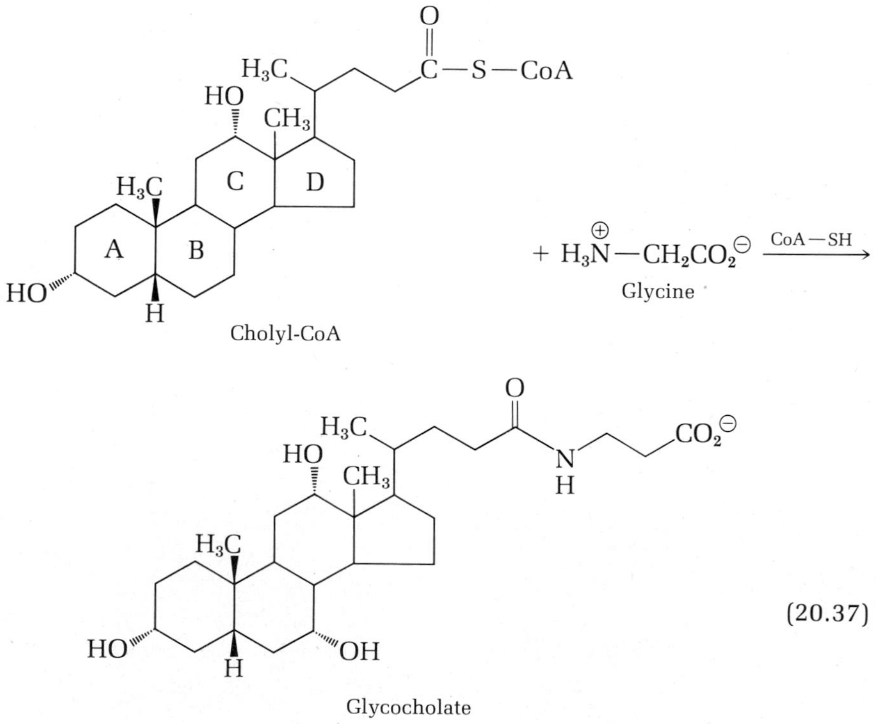

(20.37)

Taurine originates by an oxidative pathway in the catabolism of cysteine. Oxidation of the cysteine sulfhydryl group by a dioxygenase that contains Fe(II) yields *cysteine sulfinic acid* (reaction 20.39). Oxidative decarboxylation gives *hypotaurine,* and a further oxidation of the sulfinic acid to a sulfonic acid produces *taurine* (reaction 20.40).

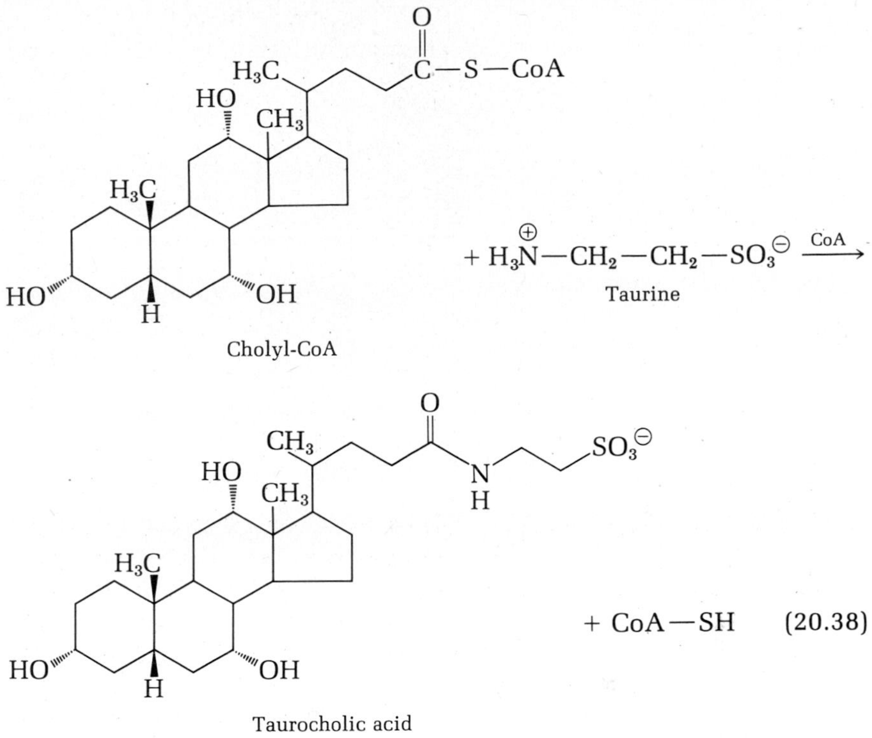

Cholyl-CoA

$+ H_3\overset{\oplus}{N}-CH_2-CH_2-SO_3^{\ominus} \xrightarrow{\text{CoA}}$

Taurine

Taurocholic acid

$+ CoA-SH \qquad (20.38)$

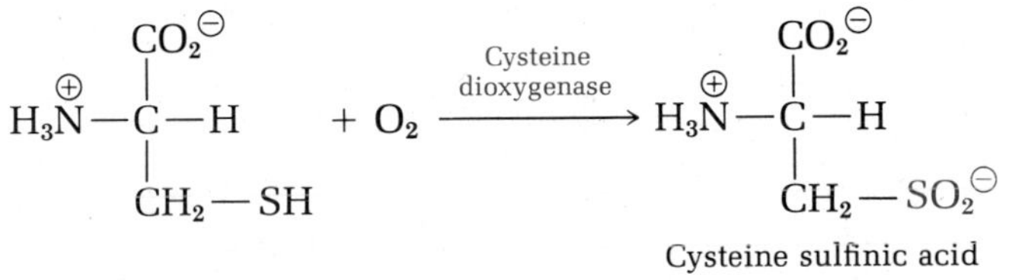

(20.39)

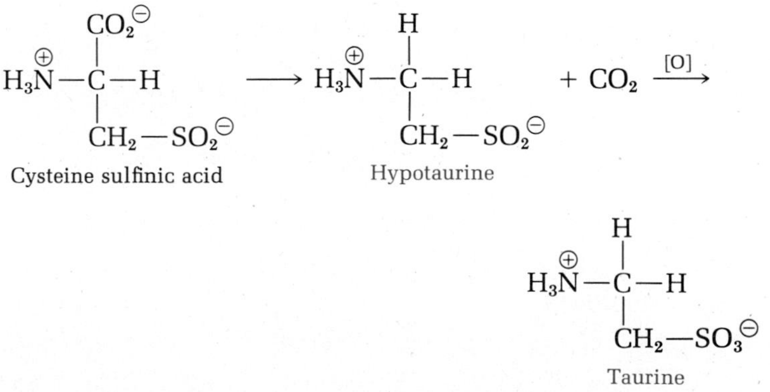

Cysteine sulfinic acid

Cysteine sulfinic acid   Hypotaurine

(20.40)

Taurine

20–9 ROLE OF ACETYL-CoA IN LIPID BIOSYNTHESIS

Acetyl-CoA is the focal point of lipid biosynthesis. It may be derived from catabolism of carbohydrates, from catabolism of amino acids, and from β-oxidation of fatty acids. We recall that acetyl-CoA is incorporated in fatty acids by the action of fatty acid synthetase. Since fatty acids are esterified in di- and triacylglycerides, a carbon atom that began in glucose can be incorporated in a membrane lipid. Fatty acids are the raw material of prostaglandins, and prostaglandins are therefore also derived from acetyl-CoA. The important biosynthetic intermediate R-3-hydroxy-3-methylglutaryl-CoA is the precursor of ketone bodies, cholesterol, steroid hormones, and bile acids. Since R-3-hydroxy-3-methylglutaryl-CoA is derived from acetyl-CoA. The relationship of acetyl-CoA to lipid biosynthesis is summarized in Figure 20.29.

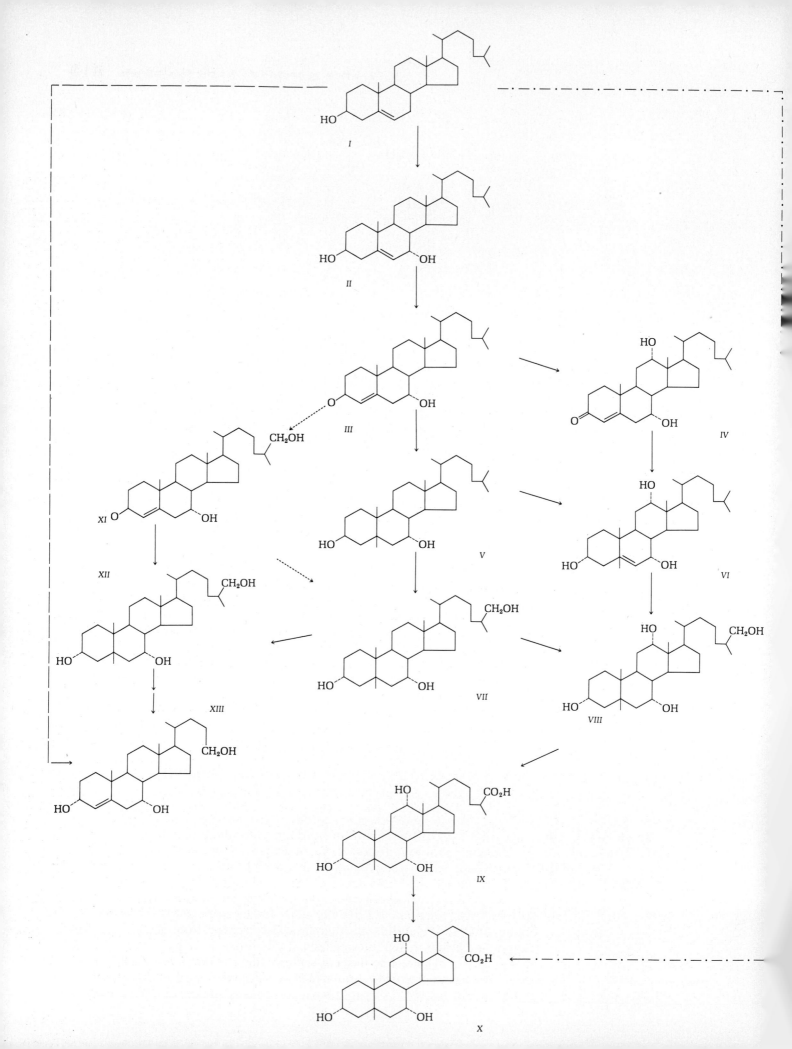

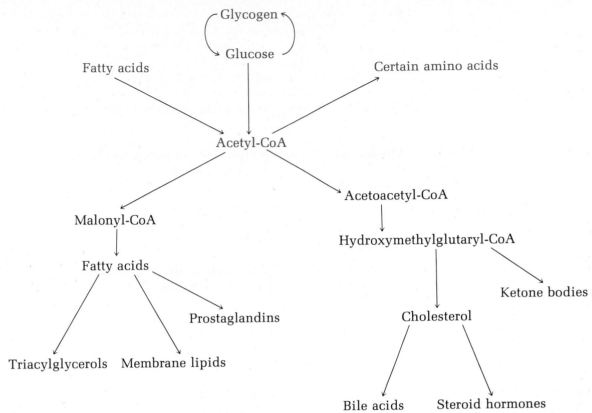

Figure 20.29 Biosynthetic pathways in which acetyl-CoA is the starting material.

20–10 SUMMARY   The biosynthesis of triacylglycerides and membrane lipids proceeds through the intermediate phosphatidate. The glycerol phosphate moiety of glycerides is derived either from dihydroxyacetone phosphate or by phosphorylation of the sn-3 position of glycerol itself. Acylation of sn-glycerol 3-phosphate occurs from the fatty acyl-CoA, not from the free acid. Phosphatidates are converted to triacylglycerides by the action of a phosphatase and by esterification of the sn-3 position of the resulting diglyceride. Phosphatidate is the precursor of the phospholipids phosphatidylethanolamine and phosphatidylcholine. Methylation reactions in lipid biosynthesis use S-adenosylmethionine as the methylating agent. Choline itself can be a precursor of phosphatidylcholine. In this pathway the choline moiety is phosphorylated and then converted to CDP-choline.

The plasmalogens are vinyl ether derivatives of phosphoglycerides. They are synthesized from the fatty acyl derivatives of dihydroxyacetone phosphate, and their formation requires an unusual nucleophilic substitution of an alkoxy group from a long chain alcohol for a long chain fatty acid. The formation of the vinyl ether is catalyzed by an oxidase that requires both molecular oxygen and NADPH.

The sphingolipids are formed from sphingosine, which is acylated to

◀ **Figure 20.28** Multiple pathways for the conversion of cholesterol to bile acids in humans. *I*, cholesterol; *II*, 7α-hydroxycholesterol; *III*, 7α-hydroxy-4-cholesten-3-one; *IV*, 7α,12α-dihydroxy-4-cholesten-3-one; *V*, 5β-cholesane-3α,7α-diol; *VI*, 5β-cholestane-3α,7α, 12α-triol; *VII*, 5β-cholestane-3α,7α, 26-triol; *VIII*, 5β-cholestane-3α,7α,12α,26-tetrol; *IX*, 3α,7α,12α-trihydroxy-5β-cholestanoic acid; *X*, 3α,7α,12α-trihydroxy-5β-cholanoic acid (cholic acid); *XI*, 7α,26-dihydroxy-4-cholesten-3-one; *XII*, 3α,7α-dihydroxy-5β-cholestanoic acid; *XIII*, 3α,7α-dihydroxy-5β-cholanoic acid (chenodeoxycholic acid). Dotted arrows represent the pathway to chenodeoxycholic acid and cholic acid by way of III; ----, bypass pathway to cholic acid; ---; bypass pathway to chenodeoxycholic acid. [From Z. R. Vlahcevic et al., *Proc. Nat. Acad. Sci. U.S.*, 77, 2925–2933 (1980).]

give a ceramide. Ceramides are converted to cerebrosides by glucosylation with UDP-glucose as the glucosyl donor. The gangliosides are glycolipids that are derived from cerebrosides by the addition of oligosaccharides that contain N-acetylneuraminate. Defects in ganglioside degradation are responsible for a variety of inherited metabolic diseases, many of them fatal.

Biosynthesis of cholesterol and the steroid hormones occurs by a succession of prenyl transfer reactions starting with isopentenyl pyrophosphate and dimethylallyl pyrophosphate. These intermediates are derived from R-3-hydroxy-3-methylglutaryl-CoA, itself derived from acetyl-CoA. Prenyl transferase catalyzes synthesis of geranyl pyrophosphate and then farnesyl pyrophosphate from successive condensations of dimethylallyl pyrophosphate and isopentenyl pyrophosphate. Two moles of farnesyl pyrophosphate then condense tail-to-tail to give squalene. The conversion of squalene to lanosterol requires two steps. At least 25 steps are required for the conversion of lanosterol to cholesterol.

The steroid hormones—androgens, estrogens, mineralocorticoids, and progestagens—are all derived from cholesterol. Pregnenolone is the first hormonal product derived from cholesterol, and it is the precursor of all the others. In the manifold reactions leading from pregnenolone the oxidative enzyme cytochrome P450 plays a critical part. Pregnenolone is converted to progesterone by oxidation and migration of a double bond into conjugation with the carbonyl group. A series of hydroxylations convert progesterone to cortisol, corticosterone, and aldosterone. Progesterone is the precursor of the male and female sex hormones testosterone and estradiol. Cholesterol is degraded by conversion to bile salts.

The fatty acyl components of all lipids are derived from acetyl-CoA. Ketone bodies, cholesterol, the steroid hormones, and the bile acids are also synthesized from the acetyl group of acetyl-CoA.

REFERENCES

Bell, R. M., and Coleman, R. A., The Enzymes of Glycerolipid Synthesis in Eucaryotes, *Ann. Rev. Biochem.*, 49, 459–487 (1980).

Bloch, K., The Biological Synthesis of Cholesterol, *Science*, 150, 19–23 (1965).

Brady, R. O., Inborn Errors of Lipid Metabolism, *Adv. Enzymol.*, 38, 298–316 (1973).

Brown, M. S., and Goldstein, J. L., "Familial Hypercholesterolemia: Defective Binding of Lipoproteins to Cultured Fibroblasts Associated with Impaired Regulation of 3-hydroxy-3-methylglutaryl Coenzyme A Activity," *Proc. Natl. Acad. Sci. 71*, 788–792 (1974).

Brady, R. O., Sphingolipidoses, *Ann. Rev. Biochem.*, 47, 687–714 (1978).

Danielsson, H., and Sjovall, J., Bile Acid Metabolism, *Ann. Rev. Biochem.*, 44, 233–253 (1975).

Gatt, S., and Barenholz, Y., Enzymes of Complex Lipid Metabolism, *Ann. Rev. Biochem.*, 42, 61–85 (1973).

Lennarz, W. J., Lipid Metabolism, *Ann. Rev. Biochem.*, 39, 359–388 (1970).

McMurray, W. C., and Magee, W. L., Phospholipid Metabolism, *Ann. Rev. Biochem.*, 41, 129–160 (1972).

Numa, S., and Yamashita, S., Regulation of Lipogenesis in Animal Tissues, *Curr. Topics Cell Regul.*, 8, 197–246 (1975).

Stanbury, J. B., Wyngaarden, J. B., and Frederickson, D. S., Eds., "The Metabolic Basis of Inherited Disease," 4th ed., McGraw-Hill, New York, 1980 pp. 615–638.

Stoffel, W., Sphingolipids, *Ann. Rev. Biochem.*, 40, 57–82 (1971).

Sweely, C. C., Ed., "Chemistry and Metabolism of Sphingolipids," North-Holland, Amsterdam, 1970.

Van Den Bosch, H., Phosphoglyceride Metabolism, *Ann. Rev. Biochem.*, 43, 243–277 (1974).

Wakil, S., Ed., "Lipid Metabolism," Academic Press, New York, 1970.

Weigandt, H., "Chemistry and Metabolism of Sphingolipids," Dekker, New York, 1969.

Problems

1. Trace the fate of the isotopic label from R-3-hydroxy-3[^{14}C]-methylglutaryl-CoA to geranyl pyrophosphate. Assume that isopentenyl pyrophosphate is labeled and that dimethylallyl pyrophosphate is not labeled.

2. Propose a biosynthetic pathway from *R*-3-hydroxy-3-methylglutaryl-CoA to the *terpene* myrcene.

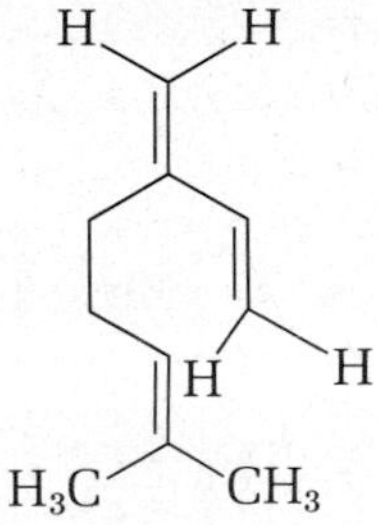

Myrcene

3. Geranyl pyrophosphate can undergo an enzyme-catalyzed isomerization to give *neryl pyrophosphate*. Assuming that neryl pyrophosphate ionizes to give the carbocation shown below, write reaction mechanisms leading to (a) borneol, (b) camphor, (c) α-terpineol (juniper oil), and (d) α-pinene (turpentine).

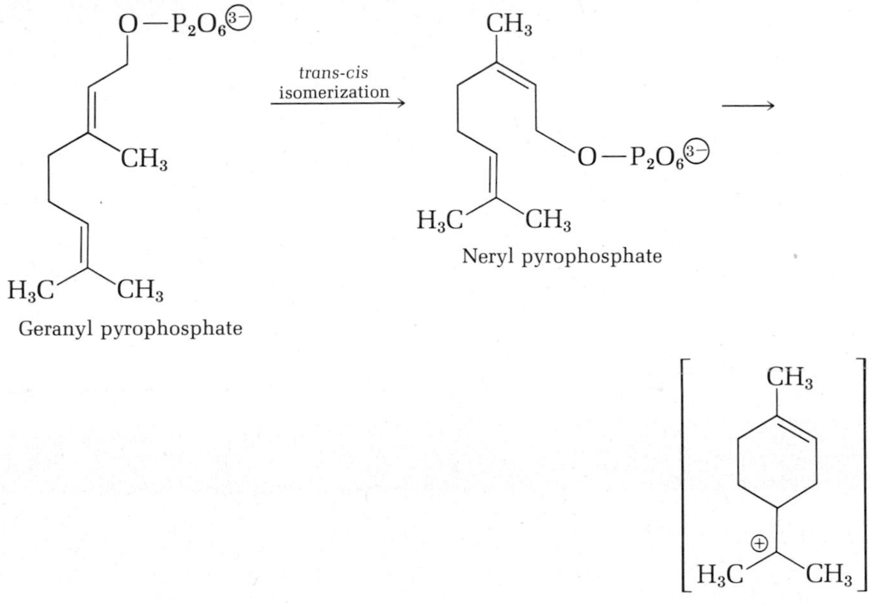

Geranyl pyrophosphate

trans-cis isomerization

Neryl pyrophosphate

Carbocation intermediate

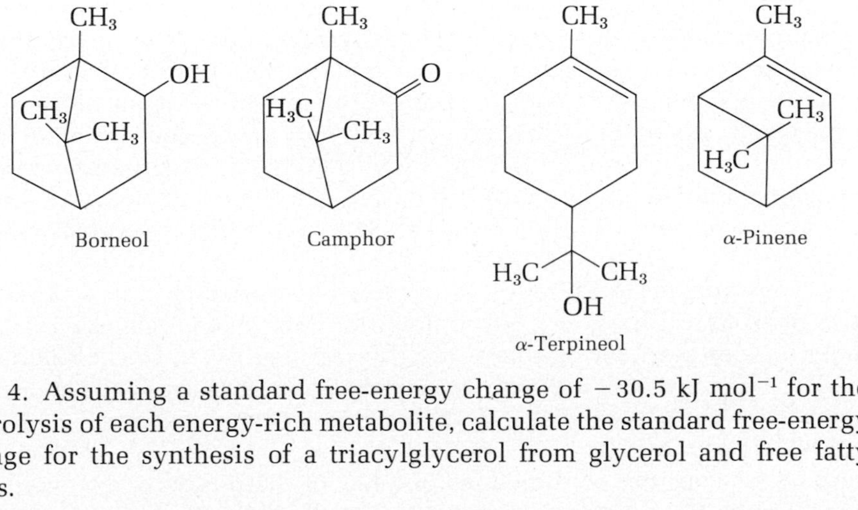

Borneol

Camphor

α-Terpineol

α-Pinene

4. Assuming a standard free-energy change of −30.5 kJ mol⁻¹ for the hydrolysis of each energy-rich metabolite, calculate the standard free-energy change for the synthesis of a triacylglycerol from glycerol and free fatty acids.

Chapter 21

Catabolism
of amino acids

21–1 INTRODUCTION

Virtually all amino acids are ingested as proteins. Following digestion by proteolytic enzymes (Sections 6.3–6.5), amino acids can be incorporated in new proteins or consumed by the oxidative pathways of the cell to provide ATP. If the demand for energy and the demand for a supply of amino acids for protein synthesis are both small, then the carbon skeletons of the amino acids can be stored either in glycogen or in fatty acids, depending upon the amino acid in question. To further complicate the picture, the protein complement of the cell is continuously recycled. The net result of all this activity is that in a person who weighs 70 kg, about 100 g of amino acids are oxidized per day.

The turnover of protein within cells is surprisingly rapid. The half-lives of proteins vary from a few minutes to a few weeks. Table 21.1 lists the proteins from rat liver that have the shortest half-lives. These rapidly degraded proteins carry out important metabolic functions, and the short half-life of these proteins enables a cell to respond quickly to changing metabolic conditions.

Let us consider the functions of a few of these rapidly degraded proteins. Ornithine decarboxylase catalyzes the rate-limiting step in the bio-

Table 21.1 Half-lives of most rapidly degraded rat liver proteins	
Enzyme	**Half-life (hr)**
Ornithine decarboxylase	0.2
δ-Aminolevulinate synthetase	
Soluble	0.33
Mitochondrial	1.1
RNA polymerase I	1.3
Tyrosine aminotransferase	2.0
Tryptophan oxygenase	2.5
Deoxythymidine kinase	2.6
β-Hydroxy-β-methylglutaryl coenzyme A reductase	3.0
Serine dehydratase	4.0
Amylase	4.3
PEP carboxykinase	5.0
Aniline hydroxylase	5.0
Glucokinase	12
RNA polymerase II	12
Dihydrooratase	12
Glucose 6-phosphate dehydrogenase	15
3-Phosphoglycerate dehydrogenase	15

SOURCE: From A. L. Goldberg and A. C. St. John, *Ann. Rev. Biochem. 45*, 751 (1976). Reproduced with permission. © 1976 Annual Reviews Inc.

synthesis of polyamines (Section 22.5E). Polyamine concentrations reflect the growth rate of the cell since they affect the rate of DNA replication, protein synthesis, and cell division. RNA polymerase I (Section 25.3) determines the rate of ribosomal RNA synthesis. The rate of degradation of serine, tryptophan, and tyrosine are controlled by serine dehydratase, tryptophan oxygenase, and tyrosine aminotransferase, respectively. These three enzymes are all subject to hormonal and nutritional control. Similarly, PEP carboxykinase catalyzes a key reaction in gluconeogenesis (Section 17.3A). In sum, many metabolically critical and highly regulated enzymes turn over rapidly.

The metabolism of amino acids commits them to a wide diversity of fates: their biosynthetic pathways lead to proteins and such biosynthetic products as the porphyrin ring system. The catabolism of amino acids can lead to their complete oxidation or to the incorporation of their carbon skeletons in either glucose or fatty acids.

The first phase in the degradation of all amino acids is deamination (Section 21.2). This process occurs in two steps: (1) the amino group is transferred to an α-keto acid by an amino transferase (transaminase) and (2) the amino group is lost as ammonium ion in an oxidative deamination. Since ammonium ions are toxic to all cells, they must be eliminated. This aspect of the degradation of amino acids is considered in Section 21.3.

The second phase of amino acid degradation revolves about the fate of the α-keto acids produced by transamination of the original amino acids. The degradation product into which the carbon skeleton of an amino acid is converted determines the "family" to which it belongs. The 20 α-keto acids produced by transamination of their parent amino acids are converted into pyruvate, acetyl-CoA, or to certain intermediates of the citric acid cycle: oxaloacetate, α-ketoglutarate, succinyl-CoA, and fumarate (Figure 21.1). Tyrosine and phenylalanine are placed in the combined fumarate and acetoacetate family. It is far simpler to learn the outlines of amino acid metabolism by grouping them into such families, than by attempting to memorize 20 pathways at random. We shall consider the pathways by which the various families of amino acids are degraded in Section 21.4.

Amino acids can also be classified by the metabolic pathways that their carbon skeletons can enter. Those converted to citric acid cycle intermediates or into pyruvate are said to be *glycogenic* because they can be converted to

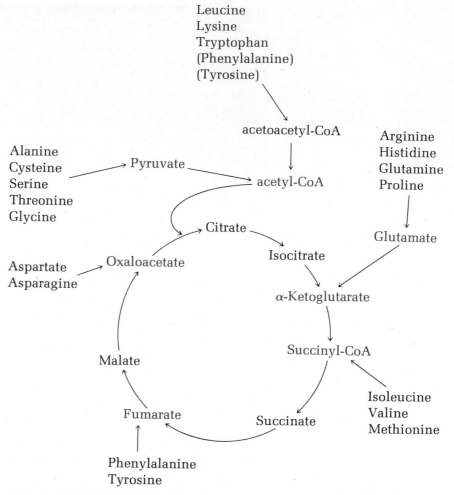

Figure 21.1 Families of amino acids based on a common degradation product. The carbon skeletons of each family can be completely oxidized in the citric acid cycle or converted to carbohydrates and fatty acids.

glucose by gluconeogenesis. Amino acids that are converted to acetyl-CoA or into acetoacetyl-CoA are *ketogenic*. They cannot be converted to carbohydrate, but their carbon skeletons can be incorporated in fatty acids and stored in triacylglycerides. A few amino acids are both glycogenic and ketogenic (Table 21.2).

Table 21.2 Glycogenic and ketogenic amino acids		
Glycogenic	**Ketogenic**	**Glycogenic and ketogenic**
Aspartate	Leucine	Isoleucine
Asparagine		Lysine
Alanine		Phenylalanine
Glycine		Tryptophan
Serine		Tyrosine
Threonine		
Cysteine		
Glutamate		
Glutamine		
Arginine		
Proline		
Histidine		
Valine		
Methionine		

Five amino acids are converted to α-ketoglutarate, and eleven are converted to acetyl-CoA. Phenylalanine and valine are degraded to acetyl-CoA and fumarate. The degradation of isoleucine yields succinyl-CoA and acetyl-CoA. Aspartate and asparagine are converted to oxaloacetate. Leucine is the only purely ketogenic amino acid. It can be converted to acetyl-CoA, but not to any citric acid cycle intermediates. In certain plants and microorganisms the glyoxalate cycle (Section 15.5) provides a mechanism for the conversion of acetyl-CoA to glucose, abolishing the distinction between glycogenic and ketogenic amino acids.

21–2 DEAMINATION OF AMINO ACIDS

Deamination, the first phase of amino acid catabolism, involves two processes. In the first step the amino group is removed by transamination to generate an α-keto acid (reaction 21.1).

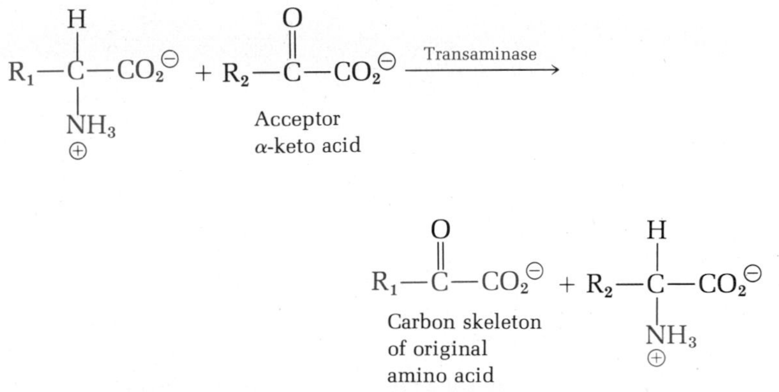

$$R_1 - \underset{\underset{\oplus}{NH_3}}{\overset{H}{\underset{|}{\overset{|}{C}}}} - CO_2^{\ominus} + R_2 - \overset{O}{\overset{\|}{C}} - CO_2^{\ominus} \xrightarrow{\text{Transaminase}}$$

Acceptor
α-keto acid

$$R_1 - \overset{O}{\overset{\|}{C}} - CO_2^{\ominus} + R_2 - \underset{\underset{\oplus}{NH_3}}{\overset{H}{\underset{|}{\overset{|}{C}}}} - CO_2^{\ominus} \qquad (21.1)$$

Carbon skeleton
of original
amino acid

The most common acceptor is the citric acid cycle intermediate α-ketoglutarate, whose transaminated product is glutamate. In the second step the transferred amino group, now part of glutamate, is released as ammonium ion by oxidative deamination. The α-ketoglutarate acceptor is regenerated in this reaction, ready to accept an amino group from another amino acid. This cycle of events is shown in Figure 21.2. The major sites of deamination in mammals are the kidney and the liver.

A. Amino Transferases

Over 50 pyridoxal phosphate–dependent amino transferases (transaminases) have been discovered. The general reaction catalyzed by these reactions is shown below, and the mechanism of action of pyridoxal

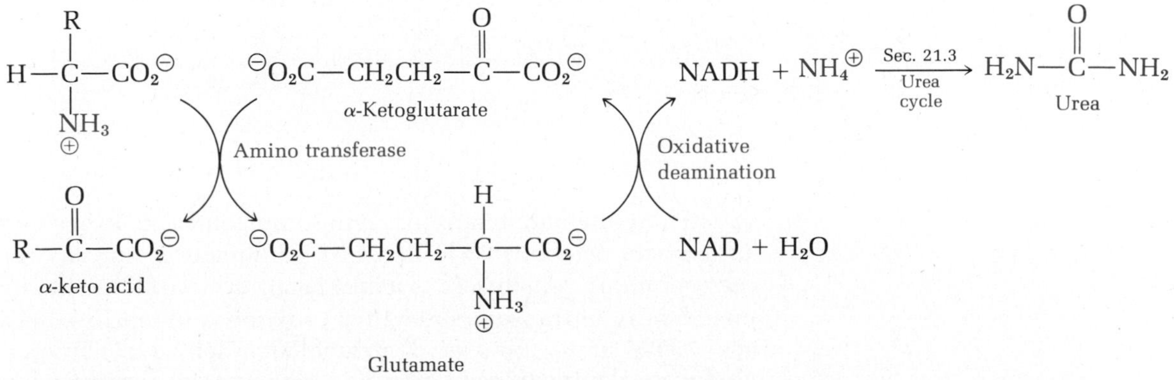

Figure 21.2 Deamination of amino acids. First, a transaminase converts the original amino acid to an α-keto acid. Second, the glutamate derived from α-ketoglutarate undergoes oxidative deamination. The ammonium ion that is produced in this step is incorporated in urea and excreted.

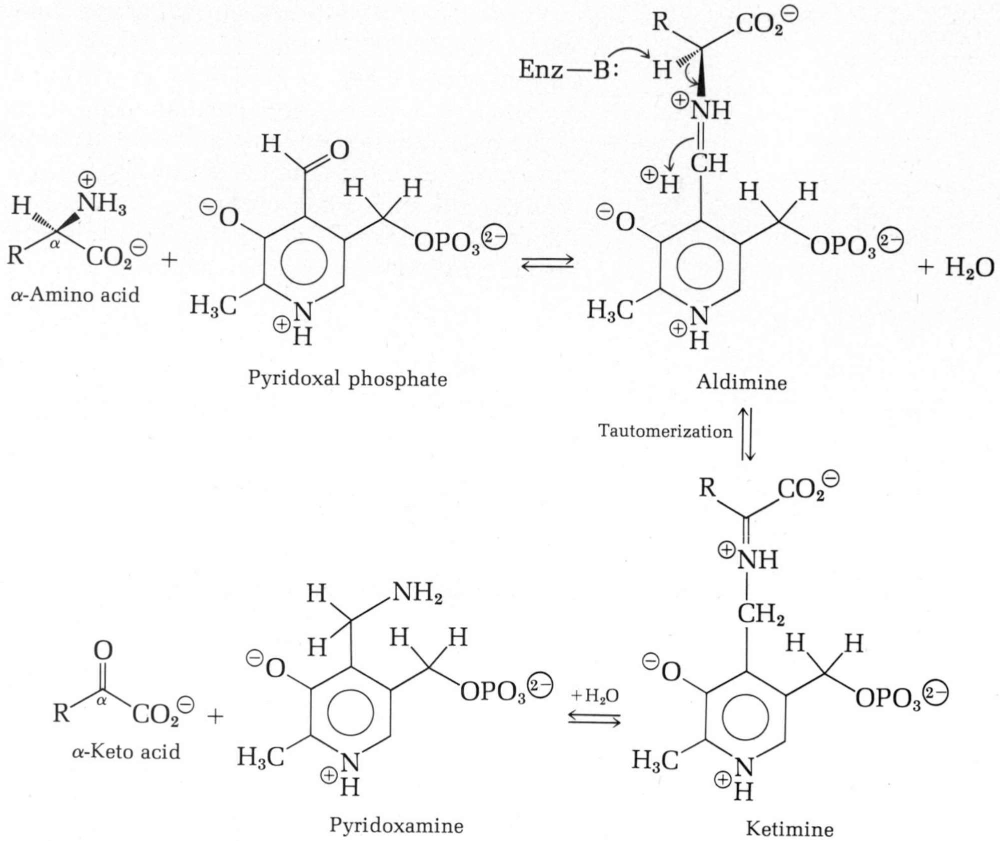

Figure 21.3 Mechanism of pyridoxal phosphate–dependent transamination. In the forward direction the amino acid to be degraded transfers its amino group to pyridoxal phosphate. In the reverse direction an α-keto acid, such as α-ketoglutarate, receives an amino group from pyridoxamine.

phosphate–dependent transamination, previously discussed in Section 10.6 A, is summarized in Figure 21.3.

$$\underset{\text{Amino acid to be degraded}}{R_1\text{—}\underset{\underset{\oplus}{NH_3}}{\overset{H}{C}}\text{—}CO_2^{\ominus}} + \underset{\text{Acceptor }\alpha\text{-keto acid}}{R_2\text{—}\overset{O}{\overset{\|}{C}}\text{—}CO_2^{\ominus}} \xrightleftharpoons[]{\text{Pyridoxal phosphate–dependent transaminase}} R_1\text{—}\overset{O}{\overset{\|}{C}}\text{—}CO_2^{\ominus} + R_2\text{—}\underset{\underset{\oplus}{NH_3}}{\overset{H}{C}}\text{—}CO_2^{\ominus} \qquad (21.2)$$

The glutamate family of amino transferases is especially important in amino acid catabolism (Table 21.3). Glutamate is so important because its corresponding α-keto acid is a citric acid cycle intermediate, and the glutamate family of transaminases thus provides a direct link between the citric acid cycle and amino acid metabolism. Let us consider two examples: glutamate-alanine transaminase and glutamate-aspartate transaminase. These two enzymes both convert amino acids to citric acid cycle intermediates,

Table 21.3 Family of glutamate transaminases

Transaminase	Source	Reaction catalyzed
Glutamate-alanine	Animals, plants	L-Glutamate + pyruvate $\rightleftarrows$ α-ketoglutarate + L-alanine
Glutamate-aspartate	Animals, plants, bacteria	L-Glutamate + oxalacetate $\rightleftarrows$ α-ketoglutarate + L-aspartate
Glutamate-cysteine	Animals (especially liver)	L-Glutamate + mercaptopyruvate $\rightleftarrows$ α-ketoglutarate + cysteine
Glutamate-glycine	Animals, plants, bacteria	L-Glutamate + glyoxylate $\rightleftarrows$ α-ketoglutarate + glycine
Glutamate-leucine	Animals, plants, bacteria	L-Glutamate + α-ketoisocaproate $\rightleftarrows$ α-ketoglutarate + L-leucine
Glutamate-phosphohistidinol	Molds	L-Glutamate + imidazole acetol phosphate $\rightleftarrows$ α-ketoglutarate + histidinol phosphate
Glutamate-tyrosine	Animals, bacteria, plants	L-Glutamate + p-hydroxyphenylpyruvate $\rightleftarrows$ α-ketoglutarate + L-tyrosine

and the carbon skeletons of aspartate and alanine can both be incorporated in glucose by gluconeogenesis.

Most animal tissues and plants possess glutamate-alanine transaminase, which catalyzes the reversible conversion of L-glutamate and pyruvate to L-alanine and α-ketoglutarate. Liver contains particularly large amounts of this enzyme.

$$^{\ominus}O_2C-CH_2-CH_2-\overset{\overset{\displaystyle H}{|}}{\underset{\underset{\displaystyle \oplus}{\underset{\displaystyle NH_3}{|}}}{C}}-CO_2^{\ominus} + CH_3-\overset{\overset{\displaystyle O}{\|}}{C}-CO_2^{\ominus} \xrightarrow{\text{Glutamate-alanine}\atop\text{transaminase}}$$

Glutamate   Pyruvate

$$^{\ominus}O_2C-CH_2-CH_2-\overset{\overset{\displaystyle O}{\|}}{C}-CO_2^{\ominus} + CH_3-\overset{\overset{\displaystyle NH_3^{\oplus}}{|}}{\underset{\underset{\displaystyle H}{|}}{C}}-CO_2^{\ominus} \qquad (21.3)$$

α-Ketoglutarate   Alanine

Gluconeogenesis in mammalian liver primarily recycles lactate, pyruvate, and glycerol to glucose. Glutamate-alanine transaminase provides a source of pyruvate for gluconeogenesis. Pyruvate in muscle is transaminated to give alanine, the alanine is transported to the liver and transaminated to give pyruvate, and the pyruvate is converted to glucose (Figure 21.4). The ammonia released in the liver is excreted as urea (Section 21.3). Alanine accounts for over half of the conversion of amino acids to glucose by gluconeogenesis in human liver.

Glutamate-aspartate transaminase is found in plants, bacteria, and most animal tissues. It catalyzes the reversible interconversion of glutamate and oxaloacetate to aspartate and α-ketoglutarate (reaction 21.4).

B. Oxidative Deamination of Amino Acids by Glutamate Dehydrogenase

The amino groups of all amino acids are collected in glutamate by the action of glutamate transaminases. Glutamate dehydrogenases have been found in every biological kingdom. The enzyme is located in the mitochondrion of animal cells. Glutamate dehydrogenase releases the amino group of glutamate as ammonium ion in an oxidative deamination in which either NAD or NADP can serve as the coenzyme (reaction 21.5).

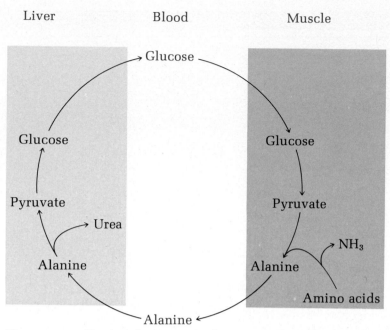

Figure 21.4 Alanine-glucose cycle. [From P. Felig, *Ann. Rev. Biochem.*, 44, 938 (1975). Reproduced with permission. © 1975 by Annual Reviews Inc.]

$$^{\ominus}O_2C-CH_2-CH_2-\underset{\underset{H}{|}}{\overset{\overset{NH_3^{\oplus}}{|}}{C}}-CO_2^{\ominus} + {}^{\ominus}O_2C-CH_2-\overset{\overset{O}{\|}}{C}-CO_2^{\ominus} \xrightarrow{\text{Glutamate-aspartate transaminase}}$$

Glutamate   Oxaloacetate

$$^{\ominus}O_2C-CH_2-\underset{\underset{H}{|}}{\overset{\overset{NH_3^{\oplus}}{|}}{C}}-CO_2^{\ominus} + {}^{\ominus}O_2C-CH_2-CH_2-\overset{\overset{O}{\|}}{C}-CO_2^{\ominus} \qquad (21.4)$$

Aspartate   α-Ketoglutarate

$$^{\ominus}O_2C-CH_2-CH_2-\underset{\underset{H}{|}}{\overset{\overset{NH_3^{\oplus}}{|}}{C}}-CO_2^{\ominus} + NAD(P) + H_2O \xrightarrow{\text{Glutamate dehydrogenase}}$$

L-Glutamate

$$^{\ominus}O_2C-CH_2-CH_2-\overset{\overset{O}{\|}}{C}-CO_2^{\ominus} + NH_4^{\oplus} + NAD(P)H \qquad (21.5)$$

α-Ketoglutarate

The first step in the conversion of glutamate to α-ketoglutarate is hydride anion transfer from the substrate to the coenzyme (reaction 21.6). A lysyl side chain of the enzyme then attacks the electron-deficient imino carbon displacing ammonia in a transimidation reaction. Hydrolysis then generates the α-keto acid and free enzyme (reaction 21.7).

The glutamate dehydrogenase of bovine liver is a hexamer of identical subunits. GDP and ADP are allosteric activators and GTP and ATP are allosteric inhibitors of glutamate dehydrogenase. Thus the enzyme is inhibited

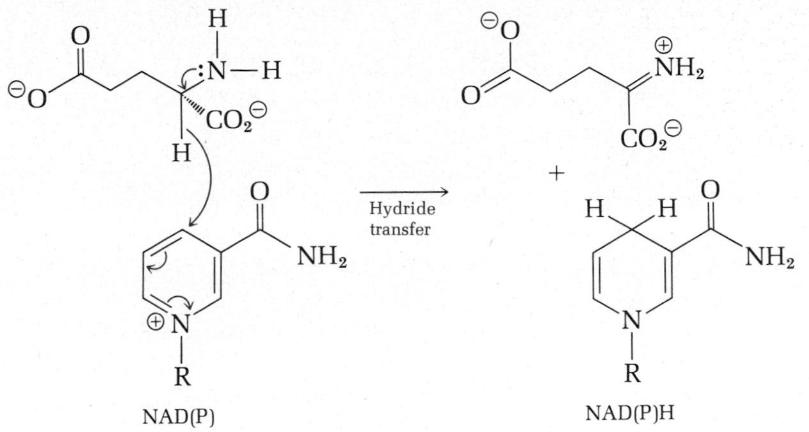

(21.6)

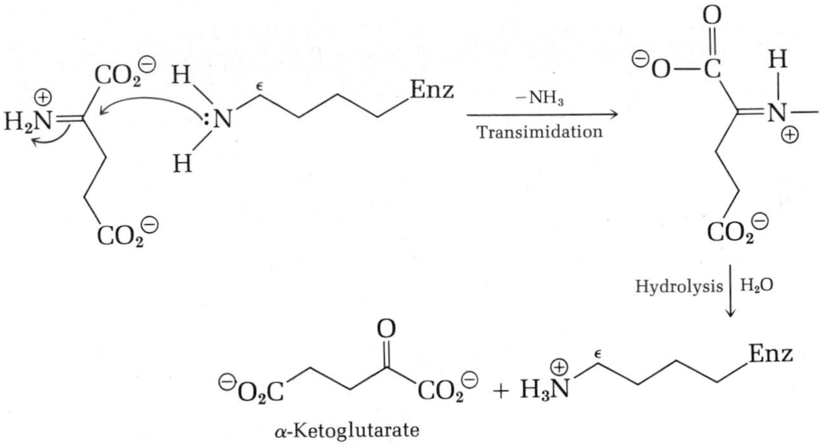

(21.7)

when the cell has a high energy supply and is activated when the energy supply is low.

The net reaction catalyzed by the successive action of a transaminase and glutamate dehydrogenase provides an α-keto acid derived from an α-amino acid, NADH (or NADPH), and ammonium ion (reaction 21.8).

$$\underset{\alpha\text{-Amino acid}}{R{-}\overset{\displaystyle H}{\underset{\displaystyle NH_3^{\oplus}}{C}}{-}CO_2^{\ominus}} + NAD(P) + H_2O \xrightarrow[\text{(2) Glutamate dehydrogenase}]{\text{(1) Transaminase}} \underset{\alpha\text{-Keto acid}}{R{-}\overset{\displaystyle O}{\overset{\|}{C}}{-}CO_2^{\ominus}} + NAD(P)H + NH_4^{\oplus} \qquad (21.8)$$

If deamination of amino acids generates an excess of ammonium ions, the excess is incorporated in urea by the urea cycle and the urea is excreted, as we shall find in the next section. The α-amino group can also be "stored" as the amido group of glutamine by the action of glutamine synthetase (Section 22.4B).

21-3 THE UREA CYCLE

A. Overview of the Urea Cycle

Ammonium ions are extremely toxic. The toxicity is caused in large part by the effect of ammonium ions on the equilibrium position of glutamate dehydrogenase. At high concentrations of ammonium ions, the position of equilibrium shifts toward glutamate, drastically lowering the concentration of α-ketoglutarate (reaction 21.9).

$$NH_4^{\oplus} + {}^{\ominus}O_2C-CH_2-CH_2-\overset{\overset{\displaystyle O}{\|}}{C}-CO_2^{\ominus} + NADH \xrightleftharpoons{\text{Glutamate dehydrogenase}}$$

$$NAD + H_2O + {}^{\ominus}O_2C-CH_2-CH_2-\overset{\overset{\displaystyle H}{|}}{\underset{\underset{\displaystyle \oplus}{NH_3}}{C}}-CO_2^{\ominus} \quad (21.9)$$

Without an adequate supply of α-ketoglutarate the citric acid cycle and respiration slow down. This is especially harmful for the brain, whose functioning is extremely sensitive to the concentration of ATP. The production of ammonium ions by glutamate dehydrogenase, their incorporation in glutamine by the action of glutamine synthetase, and the elimination of ammonium ions are all highly regulated processes. Animals which excrete $NH_4^{\oplus}$ directly are called *ammonotelic*. In *ureotelic* animals, ammonium ions are not excreted directly, but are first incorporated in the less toxic molecule urea by the urea cycle. *Uricotelic* animals excrete uric acid (see Section 23.4). The conversion to urea keeps toxic ammonium ions out of the bloodstream since the liver is the major site of both ammonium ion production by glutamate dehydrogenase and the urea cycle.

The incorporation of ammonium ions in urea requires a five-step, cyclic pathway (Figure 21.5). The enzymes of the urea cycle are distributed between the cytosol and the mitochondrion. The synthesis of carbamoyl phosphate and citrulline occur in the mitochondrion, and the remaining reactions occur in the cytosol. Three moles of ATP are consumed in the synthesis of urea. The net reaction for urea synthesis is

$$NH_4^{\oplus} + CO_2 + 3ATP + {}^{\ominus}O_2C-CH_2-\overset{\overset{\displaystyle \oplus NH_3}{|}}{\underset{\underset{\displaystyle H}{|}}{C}}-CO_2^{\ominus} + 2H_2O$$

Aspartate

$$\downarrow$$

$$H_2N-\overset{\overset{\displaystyle }{|}}{\underset{\underset{\displaystyle O}{\|}}{C}}-NH_2 + 2ADP + 2P_i + AMP + PP_i + \quad \overset{{}^{\ominus}O_2C}{\underset{H}{}}\overset{}{C}=\overset{}{C}\overset{H}{\underset{CO_2^{\ominus}}{}}$$

Urea

Fumarate

$$(21.10)$$

The amino groups of urea are both derived from amino acids: one of them is provided by aspartate and the ammonium ion is provided by the action of glutamate dehydrogenase. The carbonyl group of urea is provided by carbon dioxide. Fumarate couples the urea cycle to the citric acid cycle (Figure 21.6). In fact, the urea cycle was first proposed by H. A. Krebs and K. Henesleit some five years before Krebs and Johnson proposed the citric acid cycle.

B. Enzymatic Reactions of the Urea Cycle

a. Mitochondrial Reactions

i. Synthesis of Carbamoyl Phosphate Mitochondrial carbamoyl phosphate synthetase produces carbamoyl phosphate with ammonium ion as the source of its nitrogen and bicarbonate as the source of its carbonyl group (reaction 21.11).

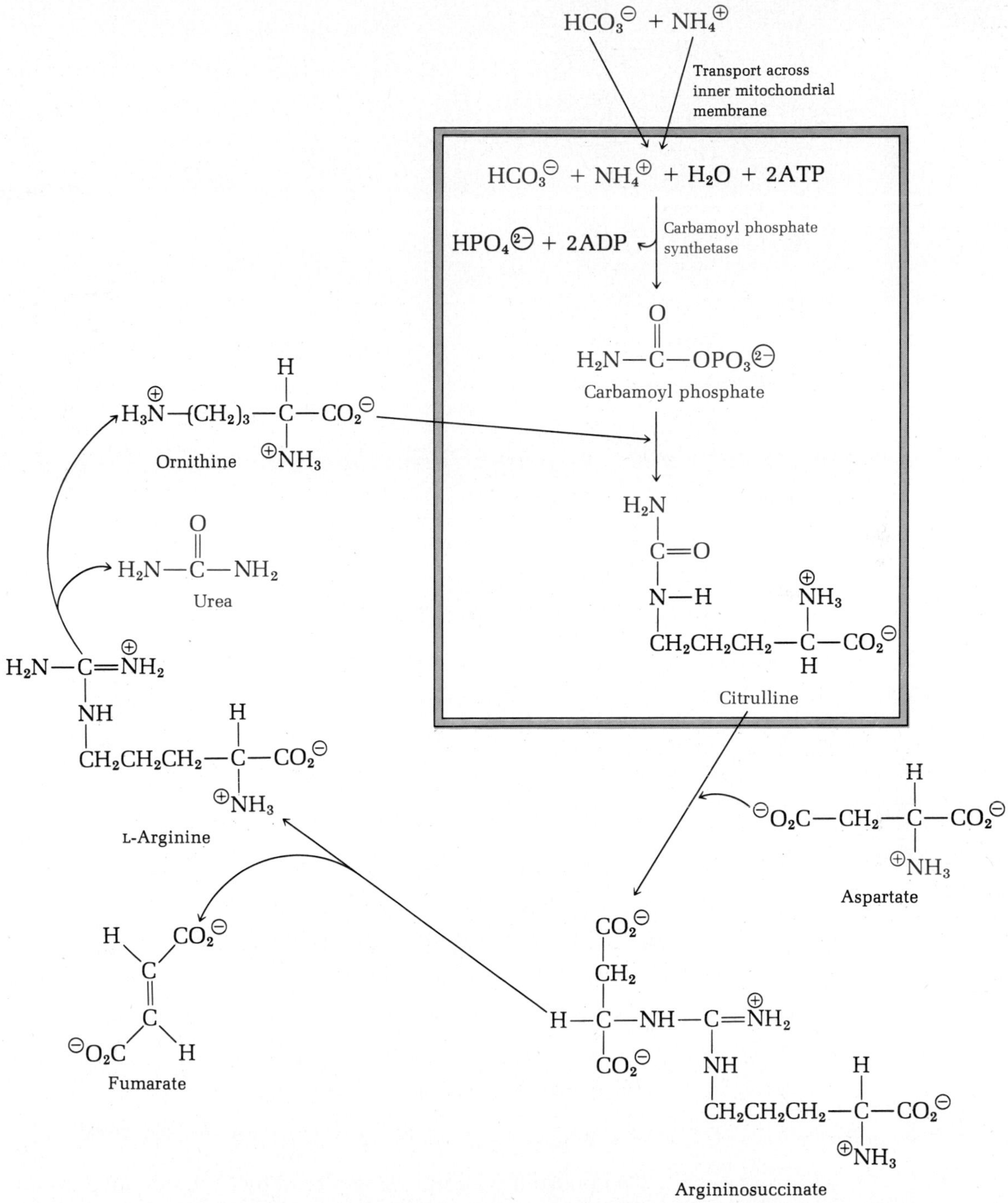

Figure 21.5 Compartmentalized reactions of the urea cycle. Specific carriers transport citrulline, ornithine, ammonium ion, and bicarbonate ion across the inner mitochondrial membrane.

$$NH_4^{\oplus} + HCO_3^{\ominus} + H_2O + 2ATP \xrightarrow{\text{Carbamoyl phosphate synthetase}}$$

$$H_2N\overset{\overset{\displaystyle O}{\|}}{-C}-OPO_3^{\text{②}\ominus} + HPO_4^{\text{②}\ominus} + 2ADP \qquad (21.11)$$

Carbamoyl phosphate

Two moles of ATP are required. The first is used to activate bicarbonate by formation of the enzyme-bound intermediate *carbonyl phosphate* (reaction 21.12).

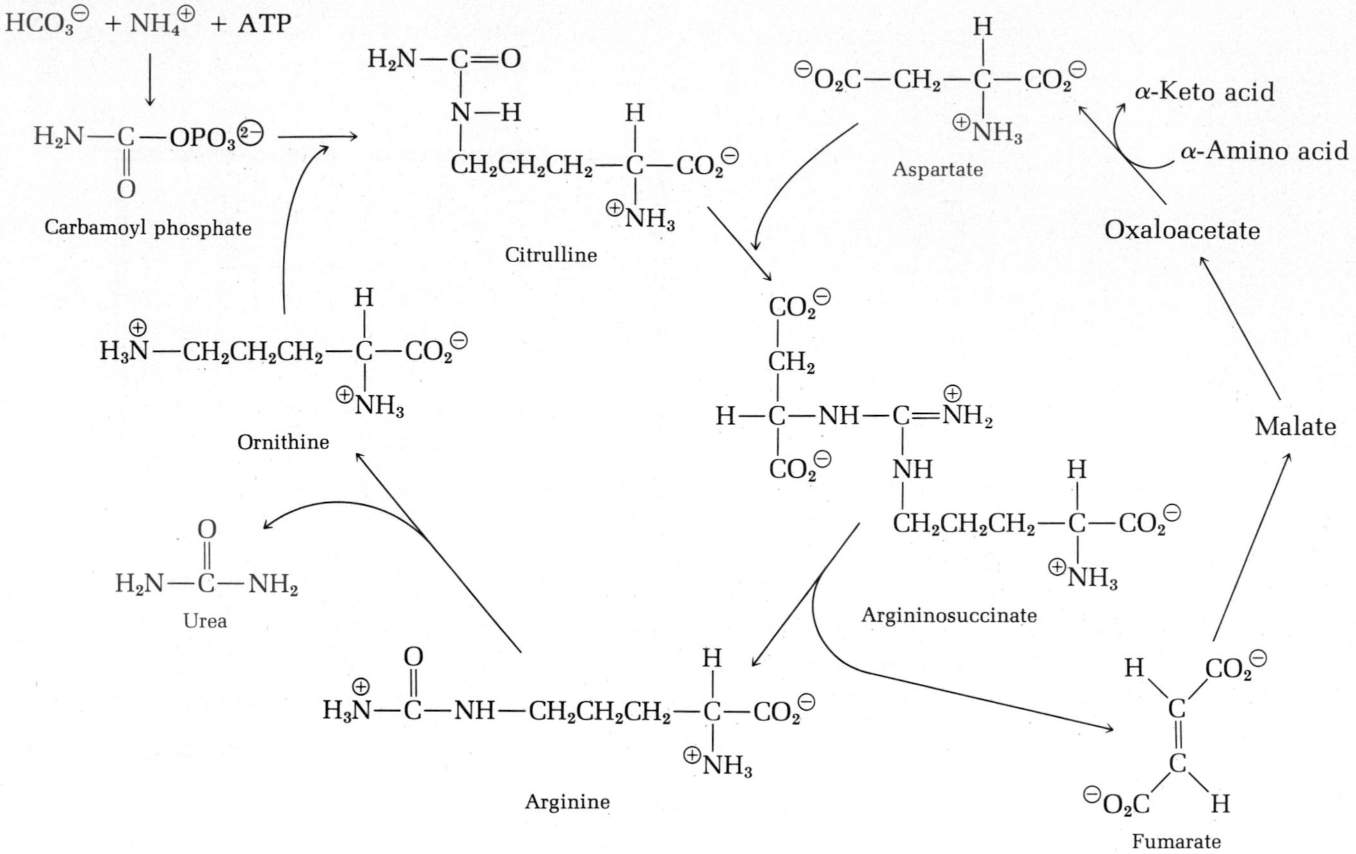

Figure 21.6 Interaction of the urea cycle with the citric acid cycle.

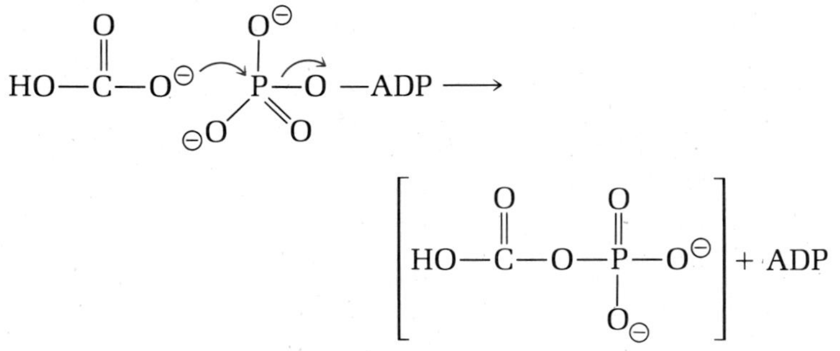

Carbonyl phosphate (21.12)

The phosphoryl group is displaced by ammonia to give carbamate (reaction 21.13).

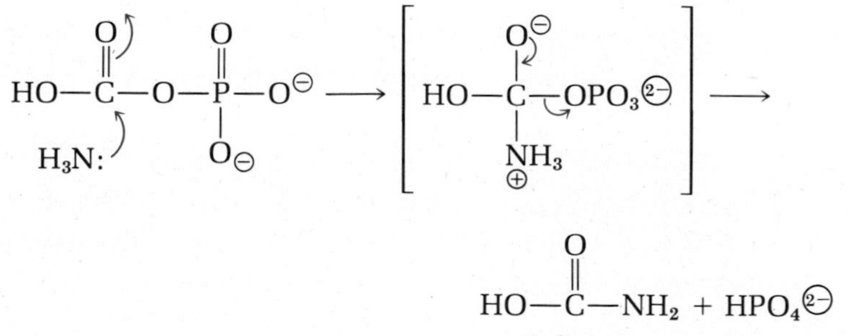

Carbamate (21.13)

The second mole of ATP is consumed by phosphoryl group transfer from ATP to carbamate, resulting in formation of carbamoyl phosphate (reaction 21.14).

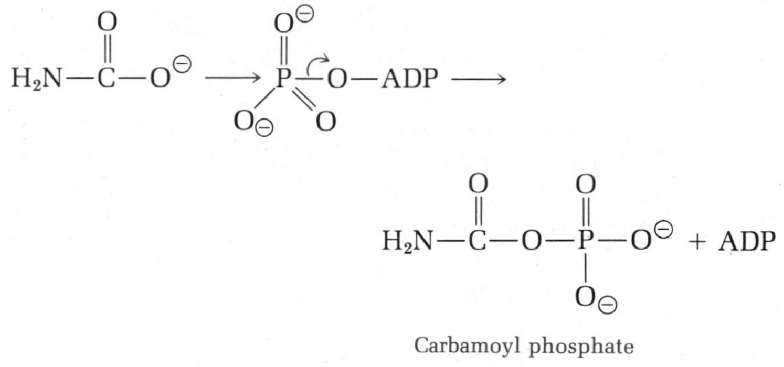

$$H_2N-\overset{\overset{\displaystyle O}{\|}}{C}-O^{\ominus} \longrightarrow \underset{\underset{\displaystyle O_{\ominus}}{}}{\overset{\overset{\displaystyle O^{\ominus}}{\|}}{P}}\overset{\frown}{}O-ADP \longrightarrow$$

$$H_2N-\overset{\overset{\displaystyle O}{\|}}{C}-O-\underset{\underset{\displaystyle O_{\ominus}}{|}}{\overset{\overset{\displaystyle O}{\|}}{P}}-O^{\ominus} + ADP \qquad (21.14)$$

Carbamoyl phosphate

ii. Synthesis of Citrulline Citrulline is formed from ornithine, an intermediate in the catabolism of lysine (Section 21.4F), and carbamoyl phosphate. The ornithine is produced in the cytosol, and a specific transport protein brings it across the inner mitochondrial membrane. *Ornithine carbamoyltransferase,* tightly bound to carbamoyl phosphate synthetase in a multienzyme complex, catalyzes a nucleophilic addition-elimination reaction in which the δ-amino group of ornithine adds to the carbonyl carbon of the energy-rich mixed anhydride, carbamoyl phosphate (reaction 21.15). Citrulline is then exported from the mitochondrion for the remaining reactions of the urea cycle.

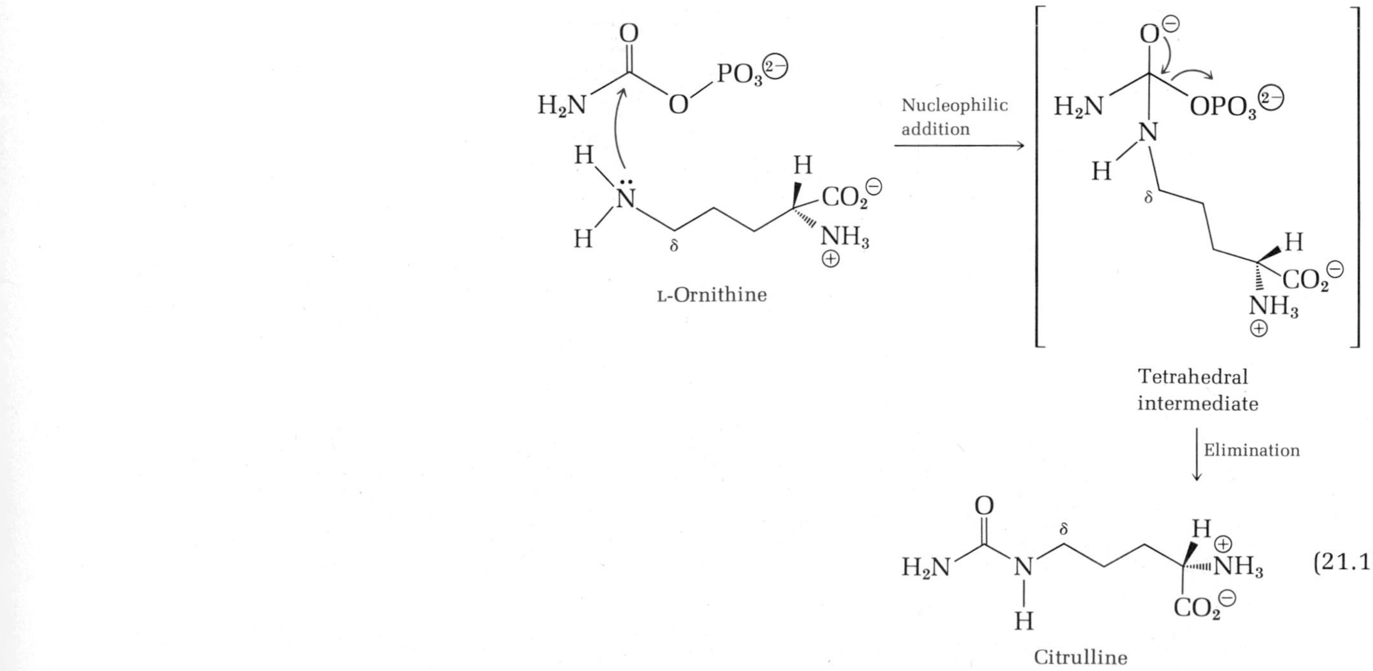

L-Ornithine

Nucleophilic addition

Tetrahedral intermediate

Elimination

Citrulline   (21.15)

b. Cytosolic Reactions

i. Synthesis of Argininosuccinate Once citrulline has been transported to the cytosol, it condenses with aspartate in an ATP-dependent reaction catalyzed by *argininosuccinate synthetase* (reaction 21.16).

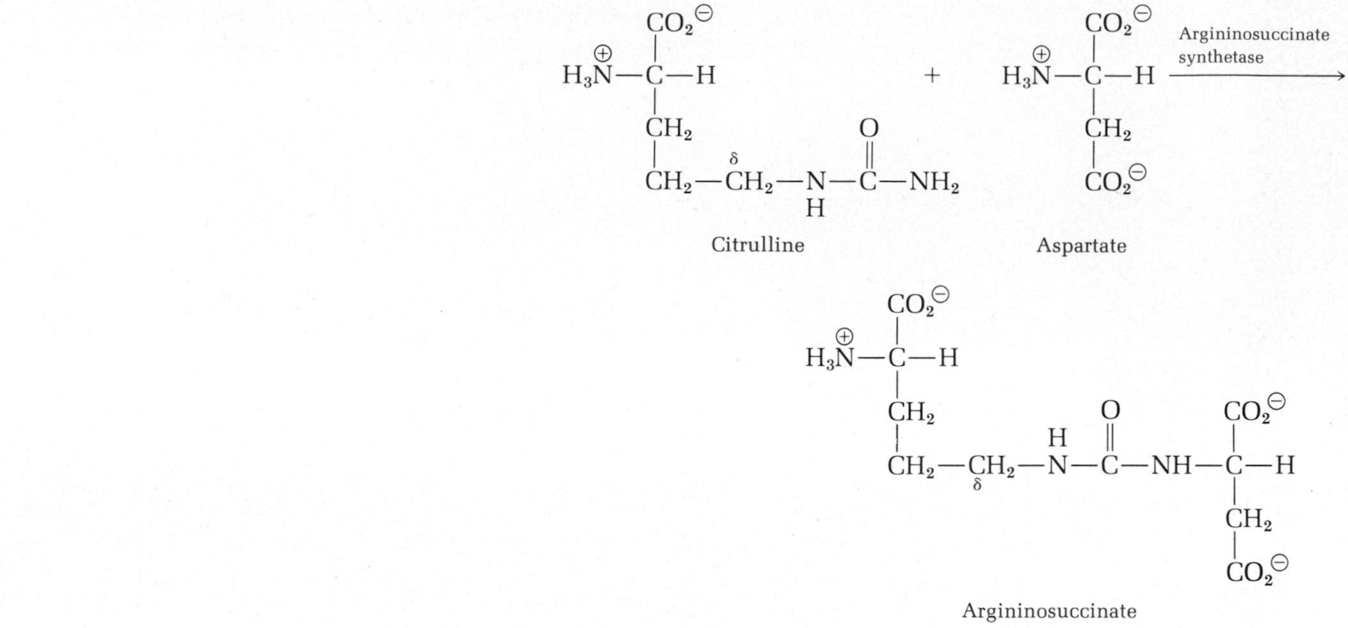

Citrulline   Aspartate

(21.16)

Argininosuccinate

The ATP requirement is concealed by the net reaction. Transfer of an adenylyl group from ATP to citrulline generates the activated intermediate adenylylcitrulline. Nucleophilic attack of the amino group of aspartate upon the enzyme-bound, activated intermediate displaces AMP and yields argininosuccinate. The elements of urea have now been incorporated in argininosuccinate: the carbonyl oxygen is derived from bicarbonate, one of the amino groups is provided by the action of glutamate dehydrogenase, and the other amino group is derived from aspartate (Figure 21.7). The carbon skeleton of aspartate will reemerge in the next step with elimination of fumarate, and urea will split out in the step after that.

ii. Elimination of Fumarate and the Concomitant Synthesis of Arginine *Argininosuccinase* catalyzes *anti* elimination of fumarate from argininosuccinate, generating arginine (reaction 21.17).

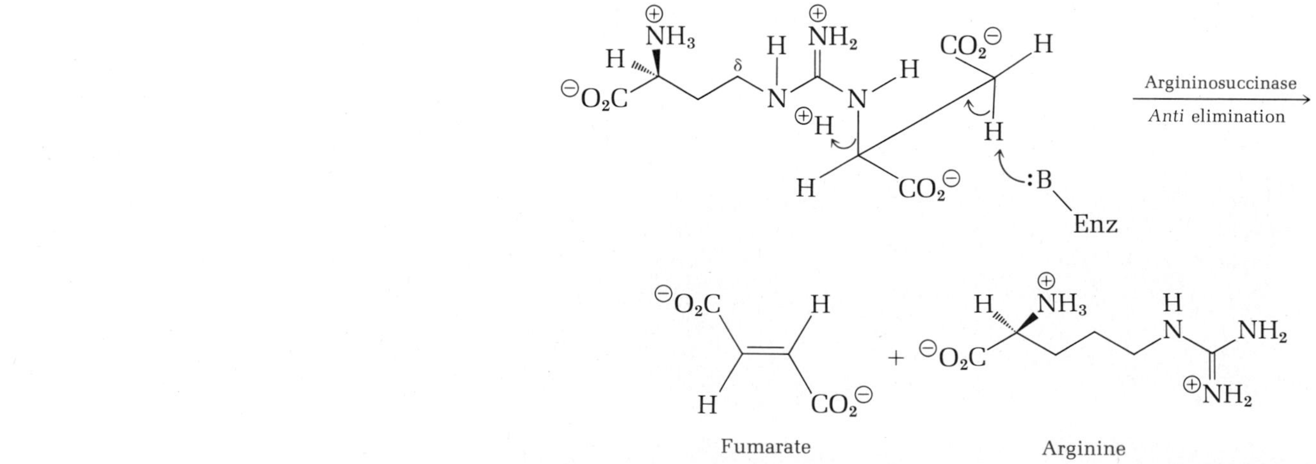

Fumarate   Arginine

(21.17)

This reaction provides the fumarate that couples the citric acid cycle and the urea cycle, and it is also part of the pathway for the biosynthesis of arginine (see Section 22.4D).

iii. Synthesis of Urea In the final step of the urea cycle arginase splits out urea from arginine and regenerates ornithine (reaction 21.18).

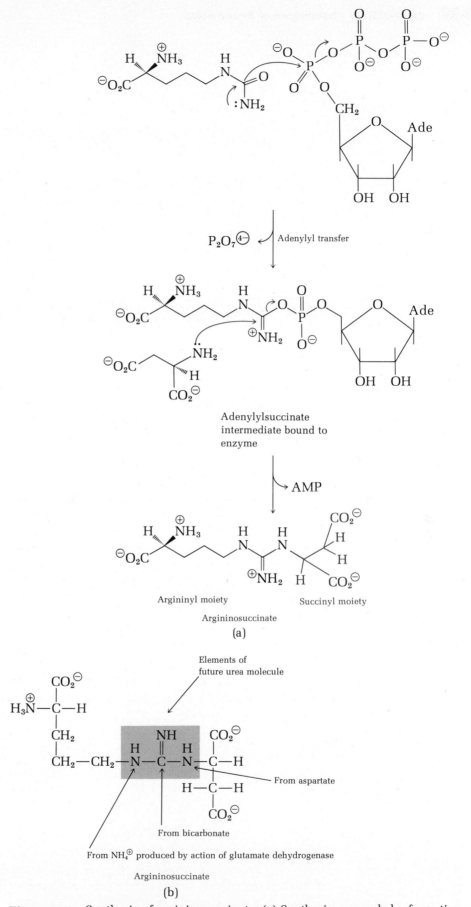

Adenylyl transfer

$P_2O_7^{4-}$

Adenylylsuccinate
intermediate bound to
enzyme

AMP

Argininyl moiety   Succinyl moiety

Argininosuccinate

(a)

Elements of
future urea molecule

From aspartate

From bicarbonate

From $NH_4^{\oplus}$ produced by action of glutamate dehydrogenase

Argininosuccinate

(b)

Figure 21.7 Synthesis of argininosuccinate. (a) Synthesis proceeds by formation of the energy-rich intermediate adenylylsuccinate. The intermediate enables the carbonyl oxygen of citrulline to depart as part of the leaving group, AMP. (b) The elements of the future urea molecule are incorporated in argininosuccinate.

$$\overset{H}{\underset{\ominus O_2C}{\Large\diagup}} \overset{\overset{\oplus}{NH_3}}{\Big|} \cdots \overset{\delta}{\underset{\oplus NH_2}{N}} \overset{H}{\underset{NH_2}{N}} + H_2O \xrightarrow{\text{Arginase}}$$

$$\underset{H_2N}{\overset{O}{\underset{\displaystyle \text{Urea}}{\diagdown C \diagup}}} \overset{}{NH_2} \;+\; \underset{\ominus O_2C}{\overset{\overset{\oplus}{NH_3}}{\underset{\displaystyle \text{Ornithine}}{\diagup}}} \cdots \overset{\delta}{\diagdown} NH_2 \qquad (21.18)$$

Enzymes catalyzing the synthesis of argininosuccinase are found in all mammalian tissues, but arginase, and hence the urea cycle, is found only in the liver.

21–4 DETAILED PATHWAYS FOR THE CATABOLISM OF AMINO ACIDS

The fates of the carbon skeletons of the α-keto acids that result from deamination will occupy us for the rest of this chapter. We shall consider the catabolism of the amino acid families based on final degradation products: oxaloacetate, pyruvate, α-ketoglutarate, succinyl-CoA, fumarate, and acetyl-CoA (recall Figure 21.2). The amino acids are funneled into the citric acid cycle or to metabolites that can easily enter the citric acid cycle. Depending upon metabolic conditions, the amino acids can thus either be oxidized with production of ATP or their carbon skeletons can be variously stored in carbohydrates or fatty acids.

In many respects the division between catabolism of amino acids and biosynthesis of amino acids is rather arbitrary. For example, is a glutamine-dependent transaminase that equilibrates aspartate and α-ketoglutarate with oxaloacetate and glutamate responsible for biosynthesis of glutamate, catabolism of aspartate, or both? To the extent that aspartate is converted to oxaloacetate and degraded in the citric acid cycle, the transaminase is catabolic with respect to aspartate. If the glutamate is later incorporated in a protein, the transaminase is biosynthetic with respect to glutamate, but if the glutamate is converted to α-ketoglutarate and degraded in the citric acid cycle, it is catabolic with respect to both amino acids. The blurred distinction between catabolism and biosynthesis will become even more apparent as we proceed.

A. Oxaloacetate Family: Aspartate and Asparagine

The pathways for catabolism of aspartate and asparagine are extremely simple. Aspartate is transaminated by a member of the glutamate family of transaminases yielding oxaloacetate (reaction 21.19).

$$\underset{\displaystyle \text{Aspartate}}{\ominus O_2C - CH_2 - \overset{\overset{\displaystyle H}{|}}{\underset{\underset{\oplus}{NH_3}}{C}} - CO_2^{\ominus}} \;+\; \ominus O_2C - CH_2CH_2 - \overset{\overset{\displaystyle O}{\|}}{C} - CO_2^{\ominus} \xrightarrow[\displaystyle \text{α-Ketoglutarate}]{\text{Transaminase}}$$

$$\underset{\displaystyle \text{Oxaloacetate}}{\ominus O_2CCH_2\overset{\overset{\displaystyle O}{\|}}{C} - CO_2^{\ominus}} \;+\; \underset{\displaystyle \text{Glutamate}}{\ominus O_2CCH_2CH_2 - \overset{\overset{\displaystyle H}{|}}{\underset{\underset{\oplus}{NH_3}}{C}} - CO_2^{\ominus}} \qquad (21.19)$$

Asparagine is catabolized in two steps. First, its amide is hydrolyzed by *asparaginase,* producing aspartate (reaction 21.20). Aspartate is then transaminated to oxaloacetate.

$$H_2N-\overset{\overset{O}{\|}}{C}-CH_2-\overset{\overset{H}{|}}{\underset{\underset{\oplus}{NH_3}}{C}}-CO_2^{\ominus} + H_2O \xrightarrow{\text{Asparaginase}}$$

Asparagine

$$^{\ominus}O_2C-CH_2-\overset{\overset{H}{|}}{\underset{\underset{\oplus}{NH_3}}{C}}-CO_2^{\ominus} + NH_3 \qquad (21.20)$$

Aspartate

B. Pyruvate Family: Alanine, Threonine, Glycine and Serine, and Cysteine

The pyruvate family contains five members: alanine, threonine, glycine, serine, and cysteine (Figure 21.8).

a. Alanine

Alanine is catabolized by glutamate-alanine transaminase, which converts it to pyruvate (reaction 21.21).

$$CH_3-\overset{\overset{H}{|}}{\underset{\underset{\oplus}{NH_3}}{C}}-CO_2^{\ominus} + {}^{\ominus}O_2CCH_2CH_2\overset{\overset{O}{\|}}{C}-CO_2^{\ominus} \underset{\xrightarrow{\hspace{1cm}}}{\overset{\text{Glutamate-alanine}}{\overset{\text{transaminase}}{\rightleftharpoons}}}$$

Alanine   α-Ketoglutarate

$$CH_3-\overset{\overset{O}{\|}}{C}-CO_2^{\ominus} + {}^{\ominus}O_2CCH_2CH_2-\overset{\overset{H}{|}}{\underset{\underset{\oplus}{NH_3}}{C}}-CO_2^{\ominus} \qquad (21.21)$$

Pyruvate   Glutamate

We discussed the importance of this reaction in the alanine-glucose cycle in Section 21.2A.

b. Threonine

Threonine is degraded by multiple pathways (Figure 21.9). In the first pathway *serine hydroxymethyl transferase,* a tetrahydrofolate-dependent enzyme (Section 10.8), converts threonine to glycine and acetaldehyde. As its name indicates, this enzyme also accepts serine as a substrate. *Aldehyde dehydrogenase* converts acetaldehyde to acetyl-CoA. The pathway from threonine to pyruvate flows through glycine, and we will pick up that thread shortly.

In the second pathway *threonine dehydratase* catalyzes the pyridoxal phosphate–dependent β-elimination of ammonium ion producing α-ketobutyrate. Oxidative decarboxylation of α-ketobutyrate yields propionyl-CoA. *E. coli* produce two versions of threonine dehydratase: one is responsible for catabolism of threonine, and the other participates in the biosynthesis of isoleucine. The catabolic enzyme is activated by AMP and

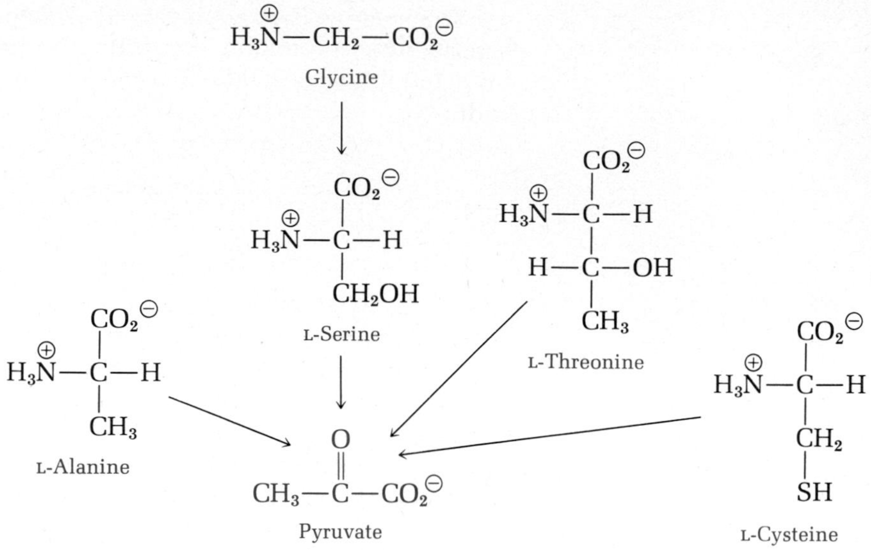

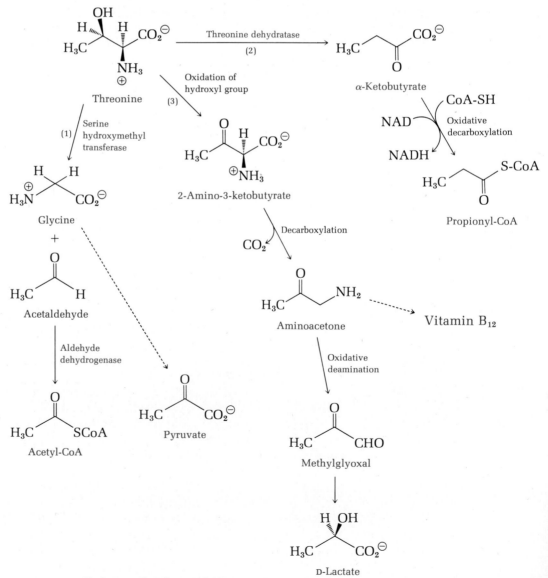

Figure 21.8 Pyruvate family of amino acids.

Figure 21.9 Pathways for the catabolism of threonine.

unaffected by isoleucine, while the biosynthetic enzyme is inhibited by isoleucine.

The third pathway for threonine catabolism involves oxidation of the hydroxyl group of threonine followed by decarboxylation. The product is aminoacetone, a biosynthetic precursor of vitamin B_{12}. Aminoacetone can also undergo oxidative deamination to methylglyoxal followed by conversion to D-lactate.

c. Glycine and Serine

Glycine is converted to serine in two steps (Figure 21.10). In the first step *serine hydroxymethyl transferase* converts glycine to serine. Since this reaction is readily reversible, it can also double as the pathway for biosynthesis of serine. *Serine dehydratase,* a pyridoxal phosphate–dependent enzyme, catalyzes β elimination of ammonia from serine producing pyruvate. We shall later see (Section 22.10) that glycine is involved in the biosynthesis of heme.

d. Cystine and Cysteine

Multiple pathways, all converging on pyruvate, are responsible for the catabolism of cysteine (Figure 21.11). Despite their diversity, they revolve about the fate of sulfur, and they all begin with the conversion of cystine to cysteine by *cystine reductase* (reaction 21.22).

In animals the major route of cysteine catabolism is the three-step pathway from cysteine to pyruvate (path 1 in Figure 21.11). Oxidation of cysteine by cysteine dioxygenase, an iron-sulfur protein, produces cysteinesulfinate; transamination of cysteinesulfinate yields β-sulfinylpyruvate; and loss of SO_2, a process analogous to decarboxylation of β-ketocarboxylic acids, leads to pyruvate.

Cysteinesulfinate is also a biosynthetic intermediate. Decarboxylation

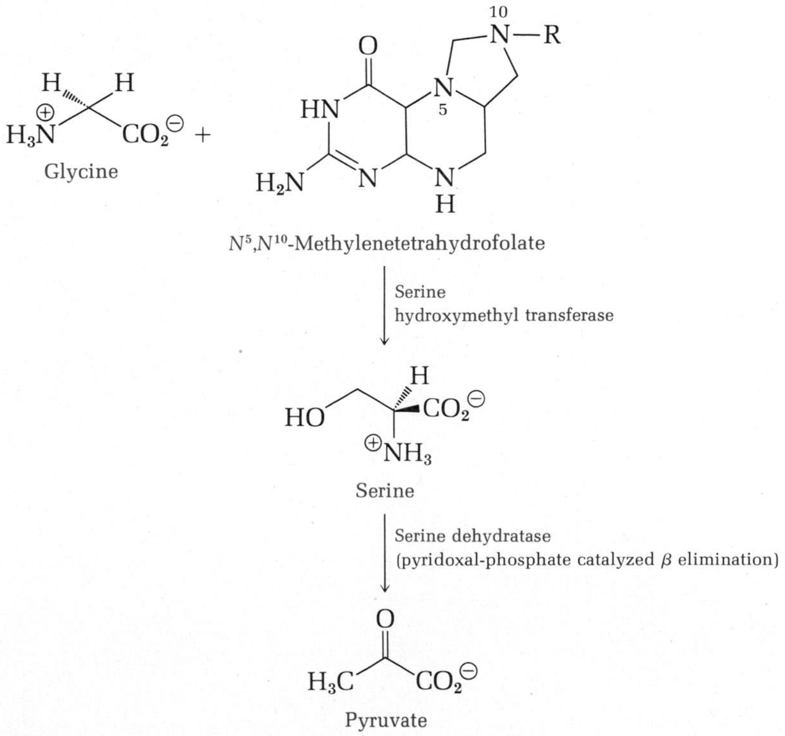

Figure 21.10 Catabolism of glycine and serine.

$$R-CH_2-S-S-CH_2-R + NADH + H^{\oplus} \xrightarrow{\text{Cystine reductase}}$$

Cystine

$$2\ HS-CH_2-\overset{\overset{\displaystyle H}{\displaystyle |}}{\underset{\underset{\displaystyle \oplus}{\displaystyle NH_3}}{C}}-CO_2^{\ominus} + NAD \qquad (21.22)$$

Cysteine

produces *taurine* (2-amino-ethansulfonate), a component of certain bile acids (Section 20.8).

In bacteria (path 2) cystathionine β-synthetase catalyzes the displacement of the hydroxyl group of homoserine (its side chain is CH_2CH_2OH) by the sulfhydryl group of cysteine to give cystathionine. Cystathioninase catalyzes 1,4 elimination producing pyruvate and homocysteine.

A third pathway in mammalian liver as well as in bacteria involves transamination of cysteine to give β-mercaptopyruvate. Desulfuration, catalyzed by *β-mercaptopyruvate transsulfurase*, yields pyruvate. In Figure 21.11 a sulfhydryl group is shown as the sulfhydryl acceptor. Cyanide, sul-

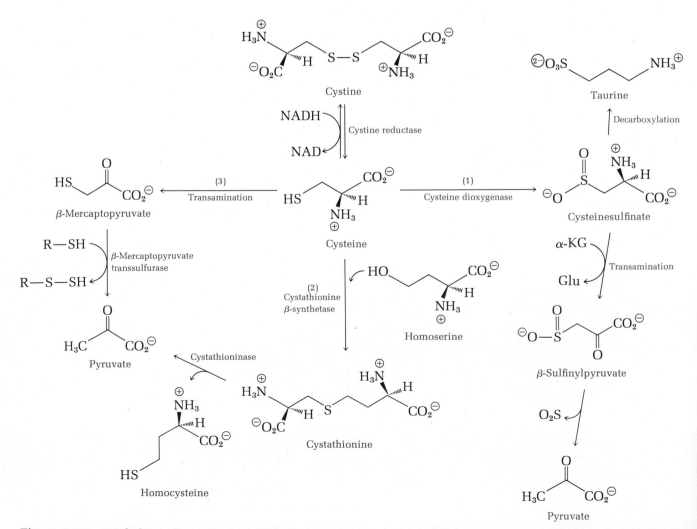

Figure 21.11 Catabolism of cysteine and cystine.

fite, thiosulfate, and a variety of other nucleophiles can also serve as sulfhydryl acceptors.

C. α-Ketoglutarate Family: Proline, Arginine, Histidine, Glutamine, and Glutamate

The α-ketoglutarate family contains five amino acids. Since four of them—proline, arginine, histidine, and glutamine—are converted to glutamate, this family might equally well be called the glutamate family (Figure 21.12).

a. Proline and Hydroxyproline

The pathways for degradation of proline and hydroxyproline are summarized in Figure 21.13. Proline is converted in three steps to glutamate. *Proline oxidase* produces an imine that spontaneously hydrolyzes to glutamate-γ-semialdehyde. An NAD-dependent dehydrogenase next converts the γ-semialdehyde to glutamate. An NAD-dependent dehydrogenase oxidizes proline to an imine that hydrolyzes nonenzymatically to give γ-hydroxyglutamate-γ-semialdehyde. Oxidation, transamination, and aldol cleavage occur in succession to produce pyruvate and glyoxalate.

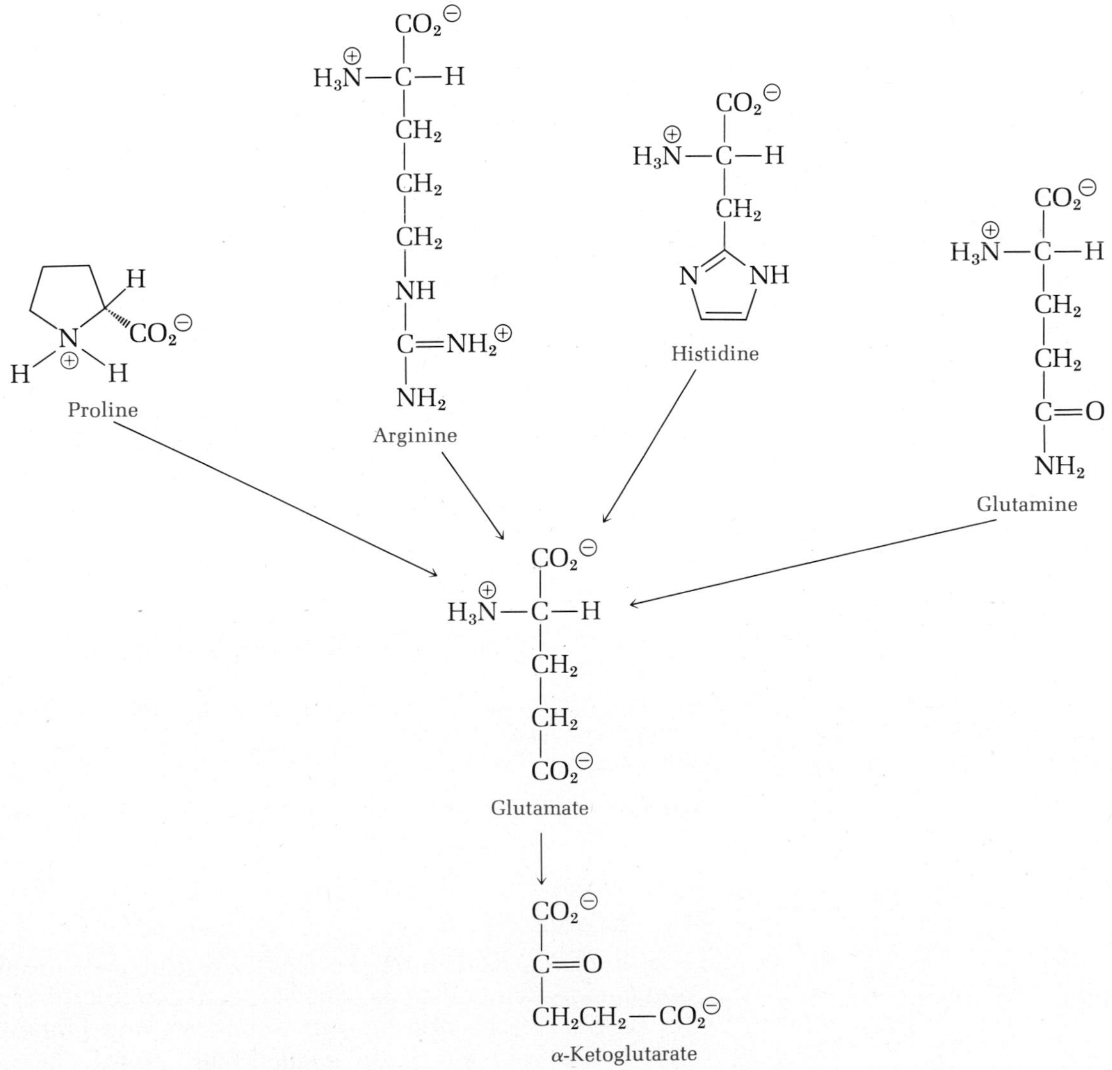

Figure 21.12 α-Ketoglutarate (or glutamate) family of amino acids.

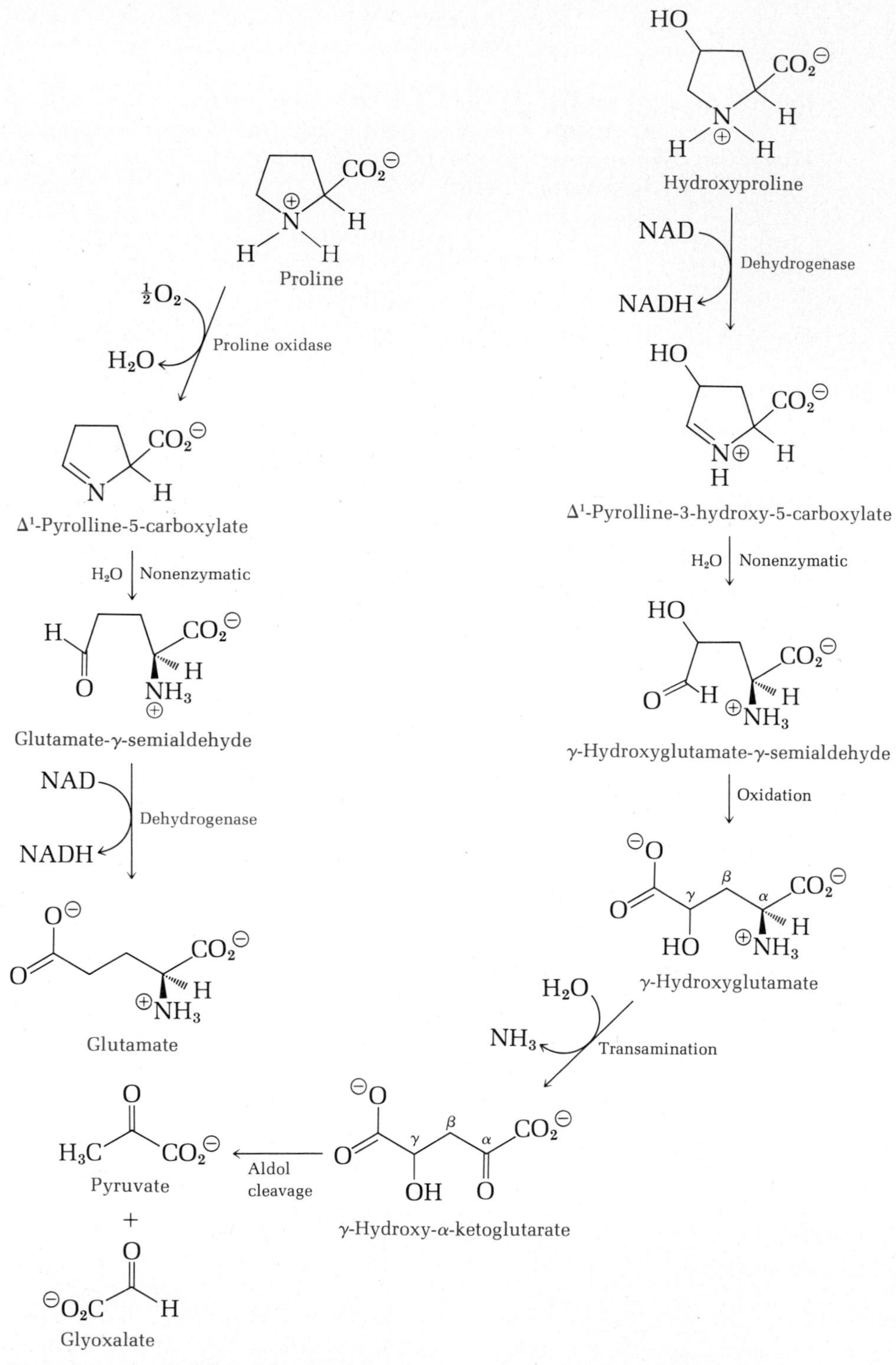

Figure 21.13 Pathways for the degradation of proline and hydroxyproline.

b. Arginine

In mammals arginine is converted to glutamate through the intermediate glutamate-γ-semialdehyde. The successive action of arginase, encountered in the urea cycle, ornithine transaminase, and glutamate-γ-semialdehyde dehydrogenase leads to glutamate (Figure 21.14).

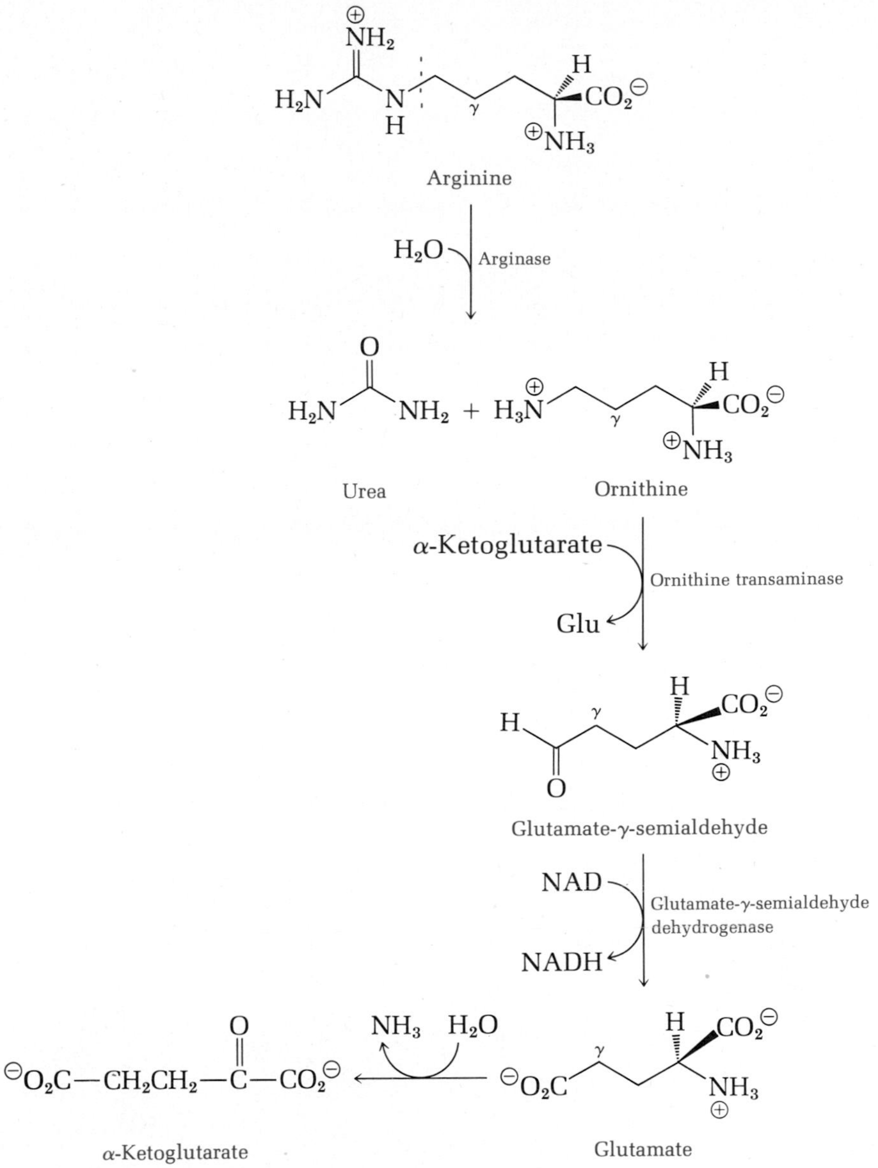

Figure 21.14   Pathway for the catabolism of arginine.

c. Histidine

The major mammalian pathway for the degradation of histidine is shown in Figure 21.15. The first step, catalyzed by *histidase* (*histidine-ammonia lyase*), is *anti* elimination of ammonia that produces *urocanate*. The 1,4 addition of water to urocanate, catalyzed by *urocanase*, produces 4-imidazolone-5-propionate. Hydrolysis gives *N*-formimino-L-glutamate in a reaction catalyzed by *imidazolone propionic acid hydrolase*. In the final step of the pathway the formimino group is transferred to tetrahydrofolate to give glutamate and N^5-formiminotetrahydrofolate.

Two of the enzymes of histidine catabolism catalyze novel reactions. The tetrameric enzyme *histidase* is a member of the class of enzymes called carbon-nitrogen lyases. The active site of histidase contains an N-terminal *dehydroalanine* residue, formed by elimination of water from serine (reaction 21.23).

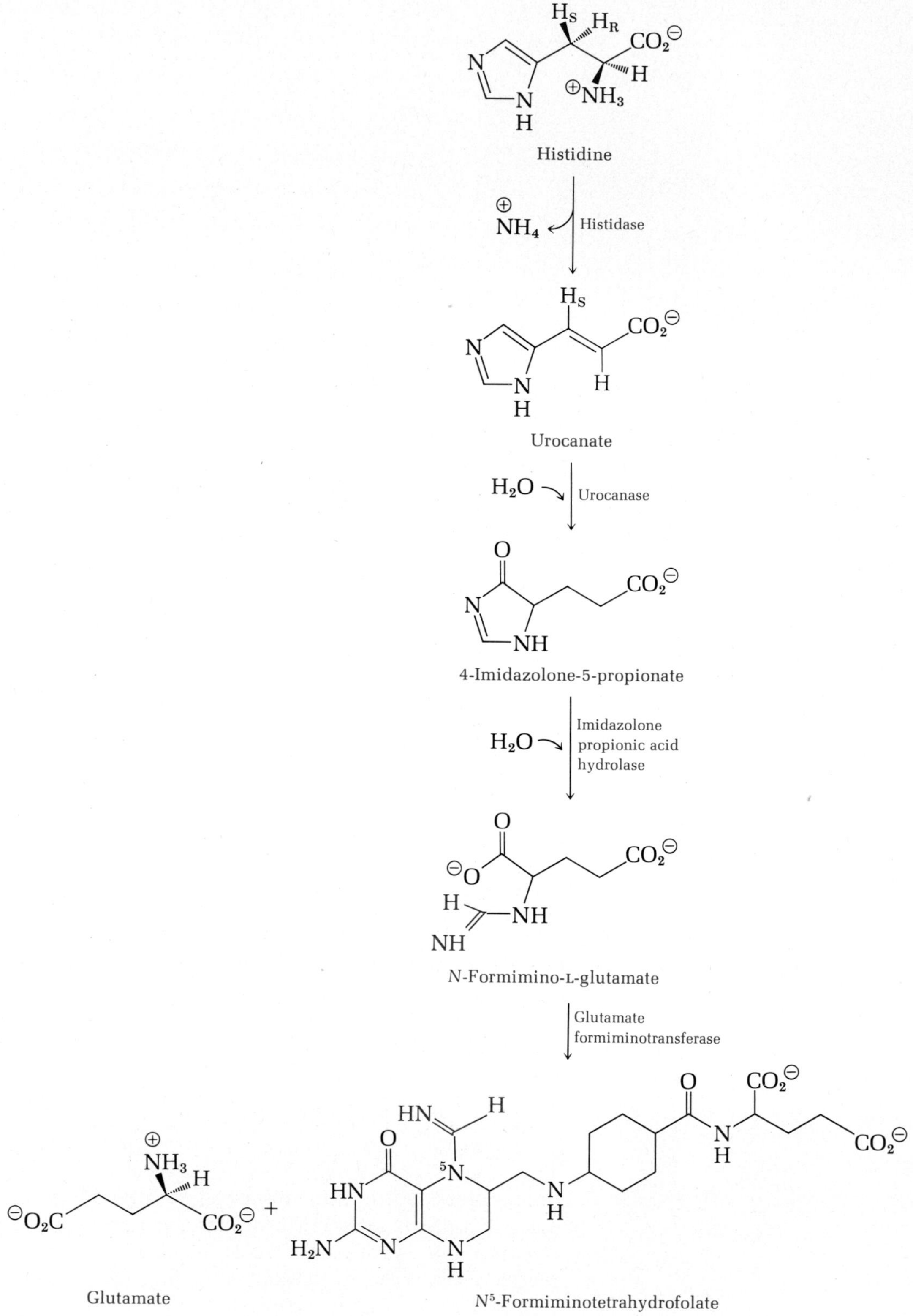

Figure 21.15 Major mammalian pathway for the catabolism of histidine.

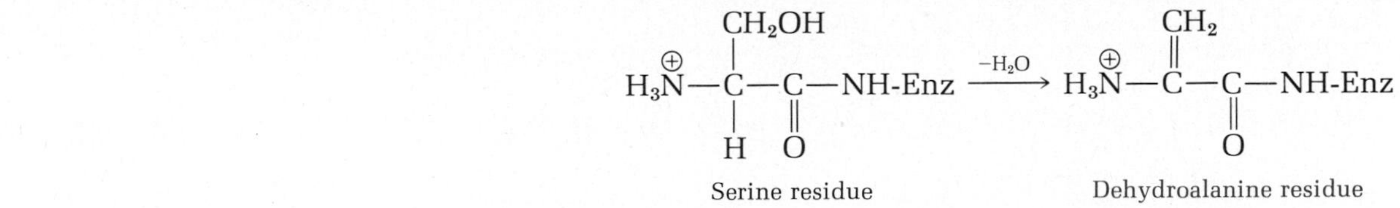

$$
\underset{\text{Serine residue}}{\overset{\displaystyle \overset{CH_2OH}{\underset{|}{\overset{|}{C}}}}{H_3\overset{\oplus}{N}-\underset{\underset{H}{|}}{C}-\underset{\underset{O}{\parallel}}{C}-NH\text{-}Enz}} \xrightarrow{-H_2O} \underset{\text{Dehydroalanine residue}}{H_3\overset{\oplus}{N}-\overset{\overset{CH_2}{\parallel}}{C}-\underset{\underset{O}{\parallel}}{C}-NH\text{-}Enz} \qquad (21.23)
$$

It has been proposed that the dehydroalanyl residue forms a covalent bond with a serine-aldehyde residue at the N-terminal position of a second subunit (reaction 21.24).

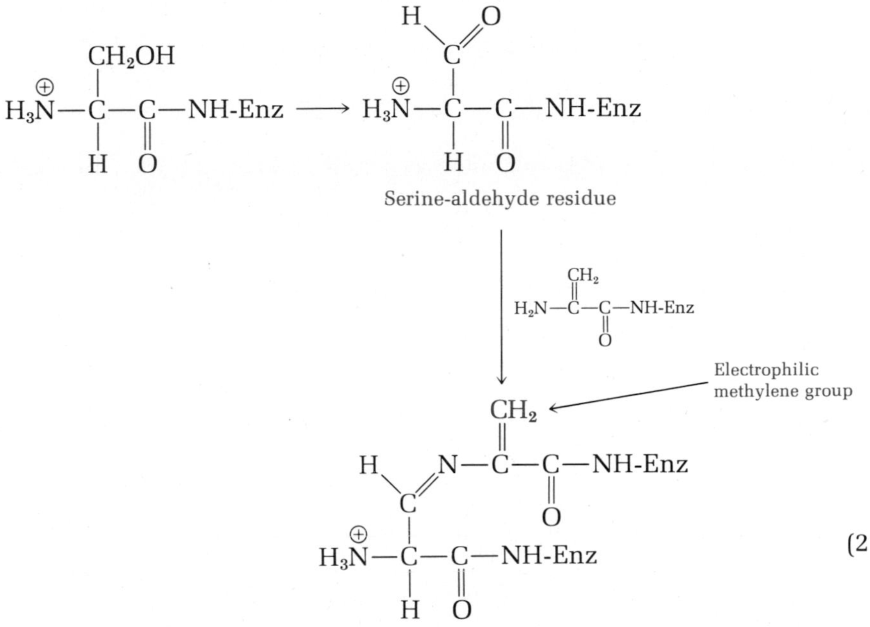

Serine-aldehyde residue

Electrophilic methylene group

Cross-linked subunits of histidase

$$(21.24)$$

This conjugated π system provides an electrophilic center that reacts with the α-amino group of histidine. β elimination then produces urocanate plus an enamine-enzyme that loses ammonia to regenerate the enzyme (Figure 21.16).

Urocanase catalyzes 1,4 addition of water to urocanate followed by a series of rearrangements that lead to 4-imidazolone propionate. The N-terminal residue of each subunit of the dimeric enzyme is bound to the unusual cofactor α-ketobutyrate. It has been suggested that the first step of catalysis is formation of an imine between the keto group of the α-keto group and N-1 of the imidazole ring. The imidazole ring is made more electrophilic when this imine forms, so that attack by water is facilitated. Once the hydroxyl group has been incorporated at C-4 of the imidazole ring, a series of rearrangements produce 4-imidazolone propionate (Figure 21.17).

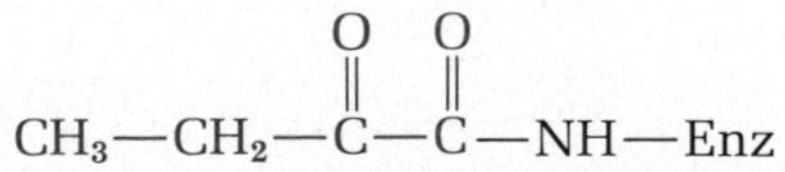

$$
CH_3-CH_2-\overset{\overset{O}{\parallel}}{C}-\overset{\overset{O}{\parallel}}{C}-NH-Enz
$$

α-Ketobutyryloyl adduct of N-terminal residue of urocanase

d. Glutamine and Glutamate

Glutamine is converted to glutamate in a single step by hydrolysis of the amide in a reaction catalyzed by *glutaminase* (reaction 21.25).

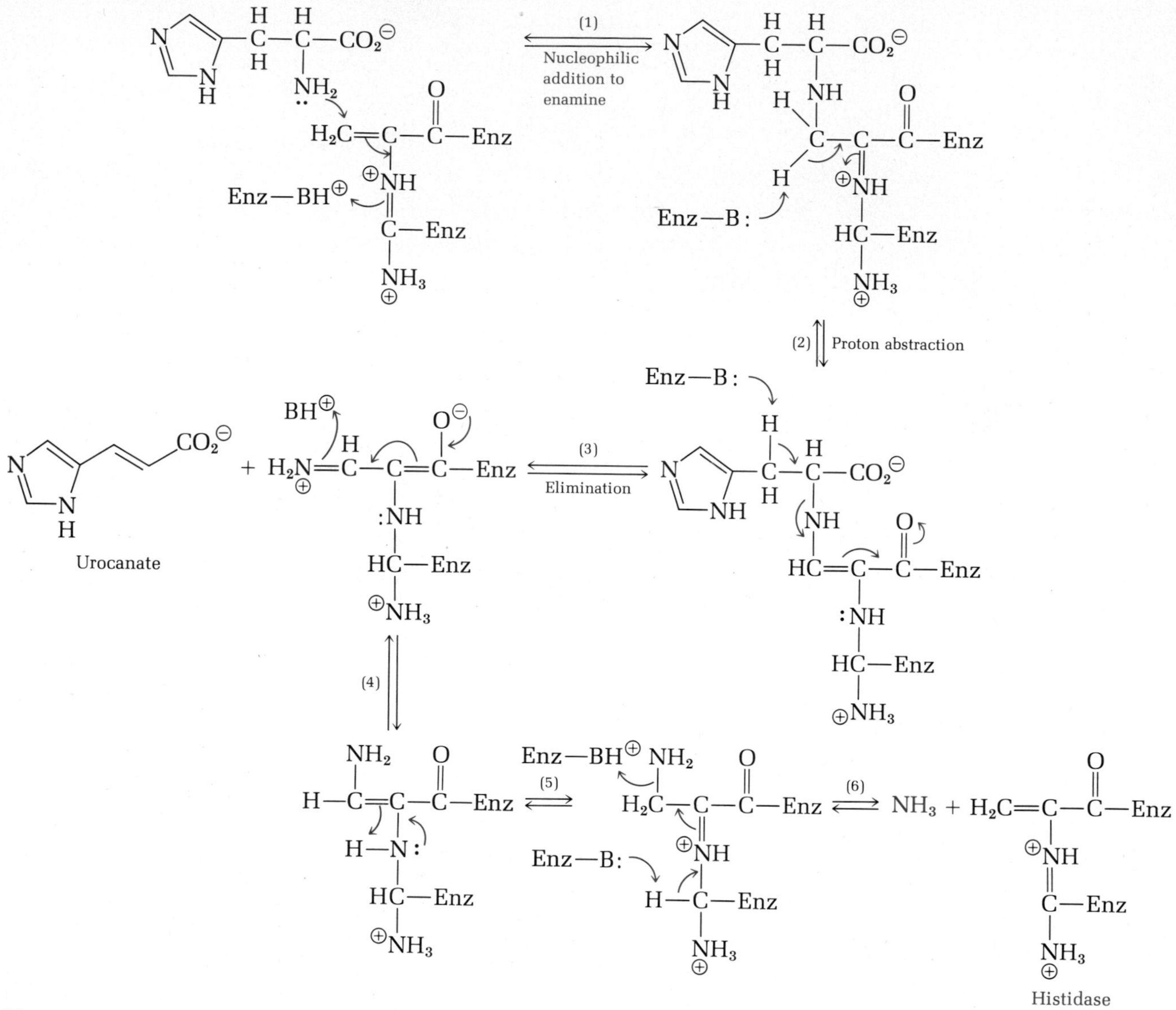

Figure 21.16 Mechanism of action of histidase. Steps 1–3 yield urocanate. Steps 4–6 regenerate histidase, releasing ammonia in the process.

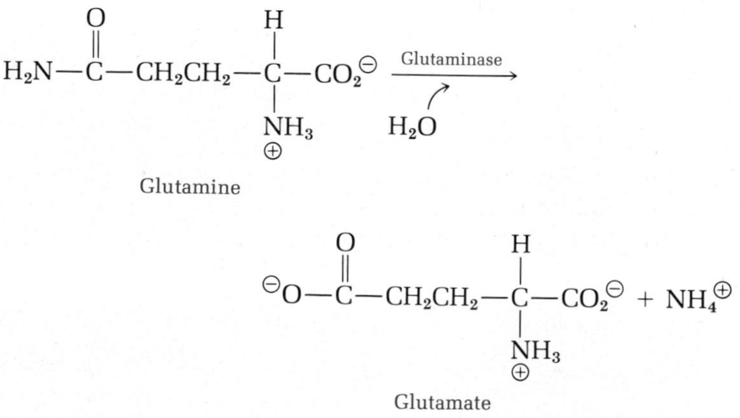

Glutamate is converted to α-ketoglutarate by transamination or by the action of glutamate dehydrogenase.

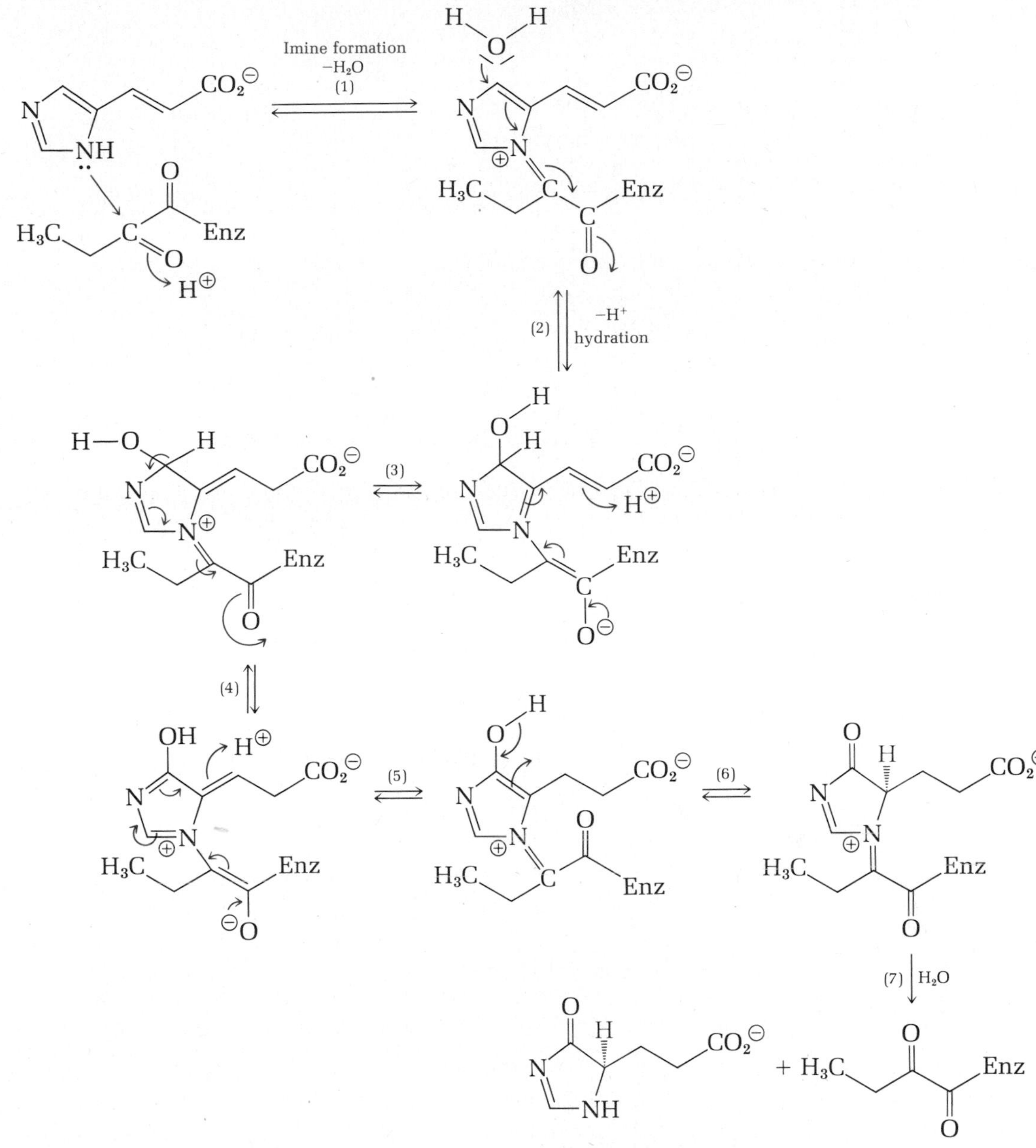

4-Imidazolone-5-propionate α-Ketobutyryoyl-enzyme

Figure 21.17 Mechanism of action of urocanase. Following imine formation and hydration (steps 1 and 2) a series of proton transfers and electronic rearrangements (steps 3–7) produce 4-imidazolone-5-propionate.

D. Succinyl-CoA Family: Valine, Isoleucine, and Methionine

Valine, isolecuine, and methionine are converted to the citric acid cycle intermediate succinyl-CoA (Figure 21.18). The part of the pathway leading from propionyl-CoA to succinyl-CoA is the same as the pathway for the final stages of degradation of odd-chain fatty acids (Section 19.5).

a. Valine

The pathway from valine to succinyl-CoA consists of eight steps (Figure 21.19). First, valine is transaminated to α-ketoisovalerate. The keto acid is converted to an acyl-CoA derivative, isobutyryl-CoA. Oxidation gives an α, β-unsaturated fatty acyl-CoA, methylacrylyl-CoA, that is hydrated to the β-hydroxy fatty acyl-CoA, β-hydroxyisobutyryl-CoA. Hydrolysis of the

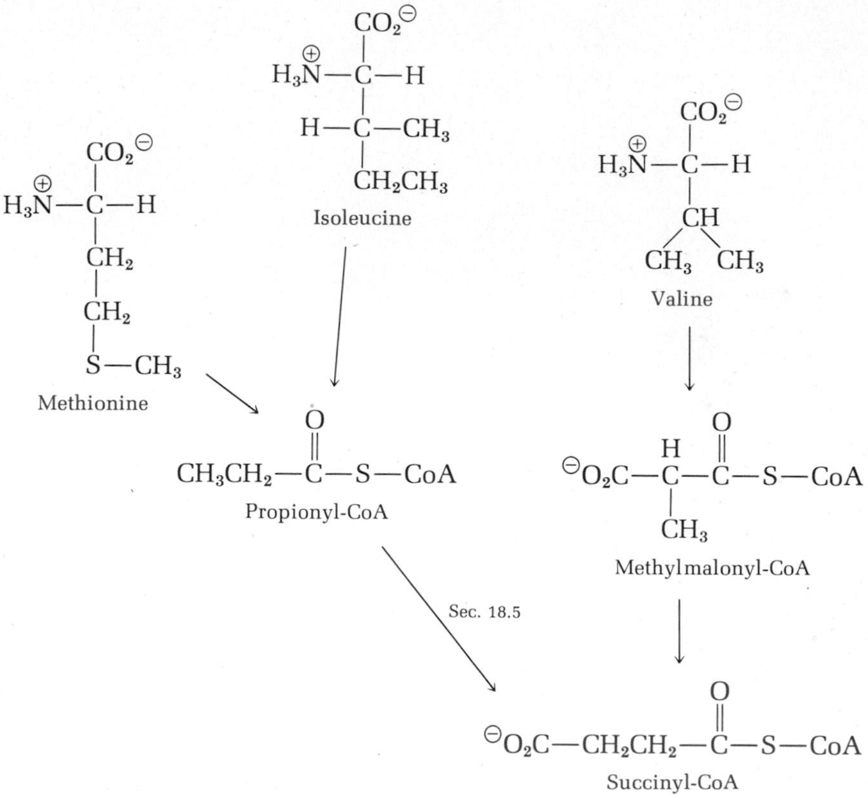

Figure 21.18 Succinyl-CoA family of amino acids. The pathway from propionyl-CoA to succinyl-CoA is identical to the last reactions in the degradation of odd-chain fatty acids.

thioester and oxidation produces methylmalonyl semialdehyde. Oxidation of the aldehyde, coupled to the formation of the thioester, yields methylmalonyl-CoA. The coenzyme B_{12}-dependent enzyme methylmalonyl-CoA mutase (Section 18.5C) catalyzes rearrangement to succinyl-CoA.

b. Isoleucine

Isoleucine is catabolized by the pathway shown in Figure 21.20. The first four steps are identical to the pathway for catabolism of leucine (Section 21.4F). In contrast to valine catabolism, the β-hydroxy-CoA product formed in step 4 is directly oxidized without an intermediate step for hydrolysis of the CoA thioester. Claisen cleavage of α-methylacetoacetyl-CoA, analogous to the reaction catalyzed by thiolase (Figure 21.21), produces acetyl-CoA and propionyl-CoA. Acetyl-CoA enters the citric acid cycle directly. Propionyl-CoA is converted to succinyl-CoA by the pathway we described in Section 18.5. The three steps required for this process are therefore summarized in Figure 21.22 without further comment. Since succinyl-CoA can be converted to glucose and acetyl-CoA can be converted to fatty acids, isoleucine is both glycogenic and ketogenic.

c. Methionine

The catabolism of methionine can be conveniently divided into two parts. The first consists of a sequence of six steps leading to propionyl-CoA (Figure 21.23), and the second consists of the three-step conversion of propionyl-CoA to succinyl-CoA. Since we have discussed the latter reactions in detail, only the conversion of methionine to propionyl-CoA need

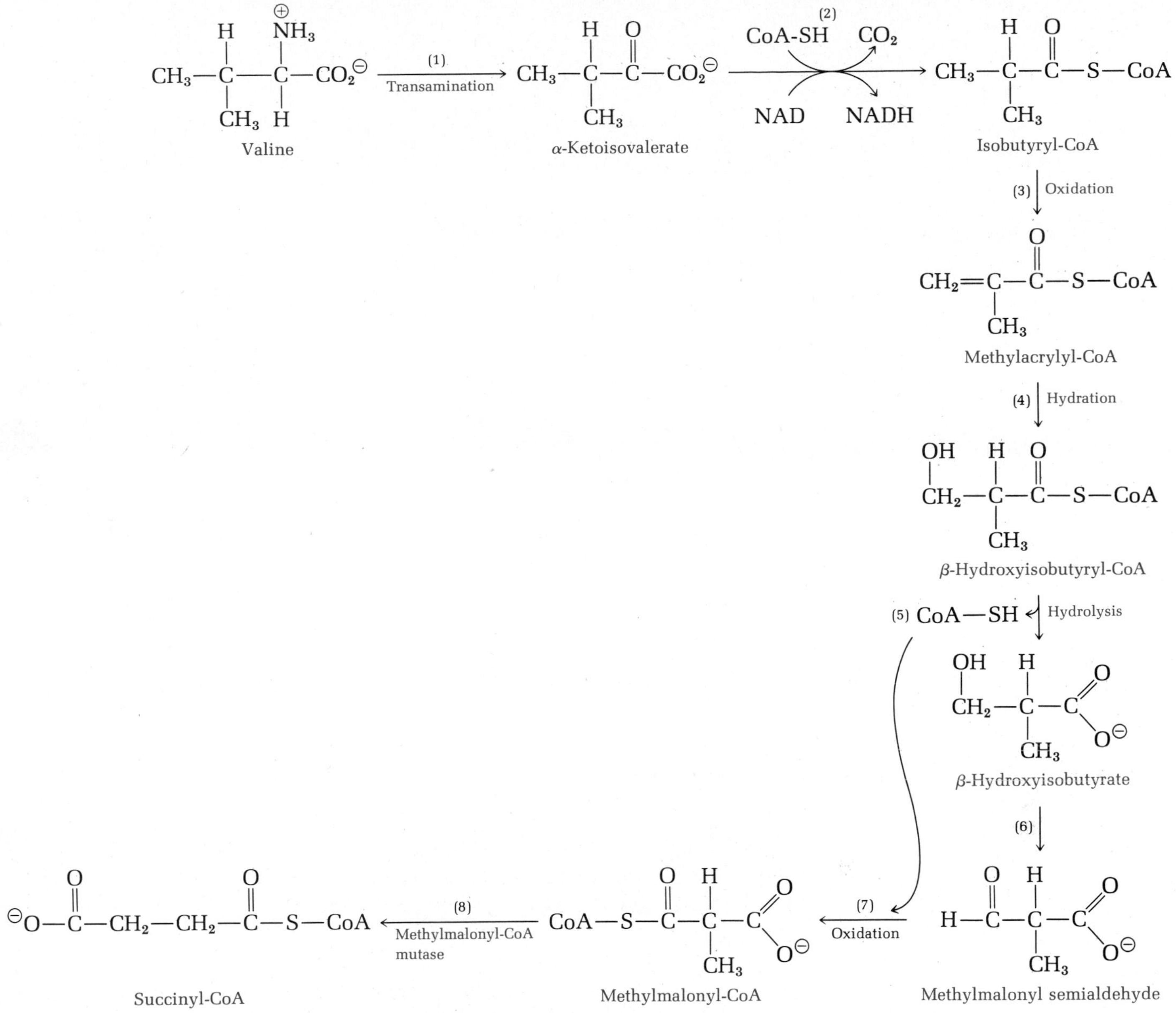

Figure 21.19 Pathway for the degradation of L-valine.

concern us here. In the first step of catabolism, methionine is adenylated by *methionine adenosyl transferase* (Section 20.3B). Transfer of a methyl group from S-adenosylmethionine to another acceptor gives S-adenosylhomocysteine in the second step. Nucleophilic attack of water on the 5′-carbon of S-adenosylhomocysteine gives adenosine and homocysteine. *Cystathionine β-synthetase* catalyzes nucleophilic displacement of the hydroxyl group of serine by the thiol group of homocysteine producing cystathionine. The pyridoxal phosphate–dependent enzyme *cystathioninase* catalyzes 1,4 elimination (γ elimination) of cystathionine to produce α-ketobutyrate and cysteine. The first catalytic step is formation of an aldimine between pyridoxal phosphate and the α-amino group of the homocysteine moiety of cystathionine. Tautomerization of the aldimine gives a quinoid, ketimine intermediate that undergoes 1,4 elimination releasing cysteine. Hydrolysis of the pyridoxal phosphate imine then produces α-ketobutyrate (Figure 21.24).

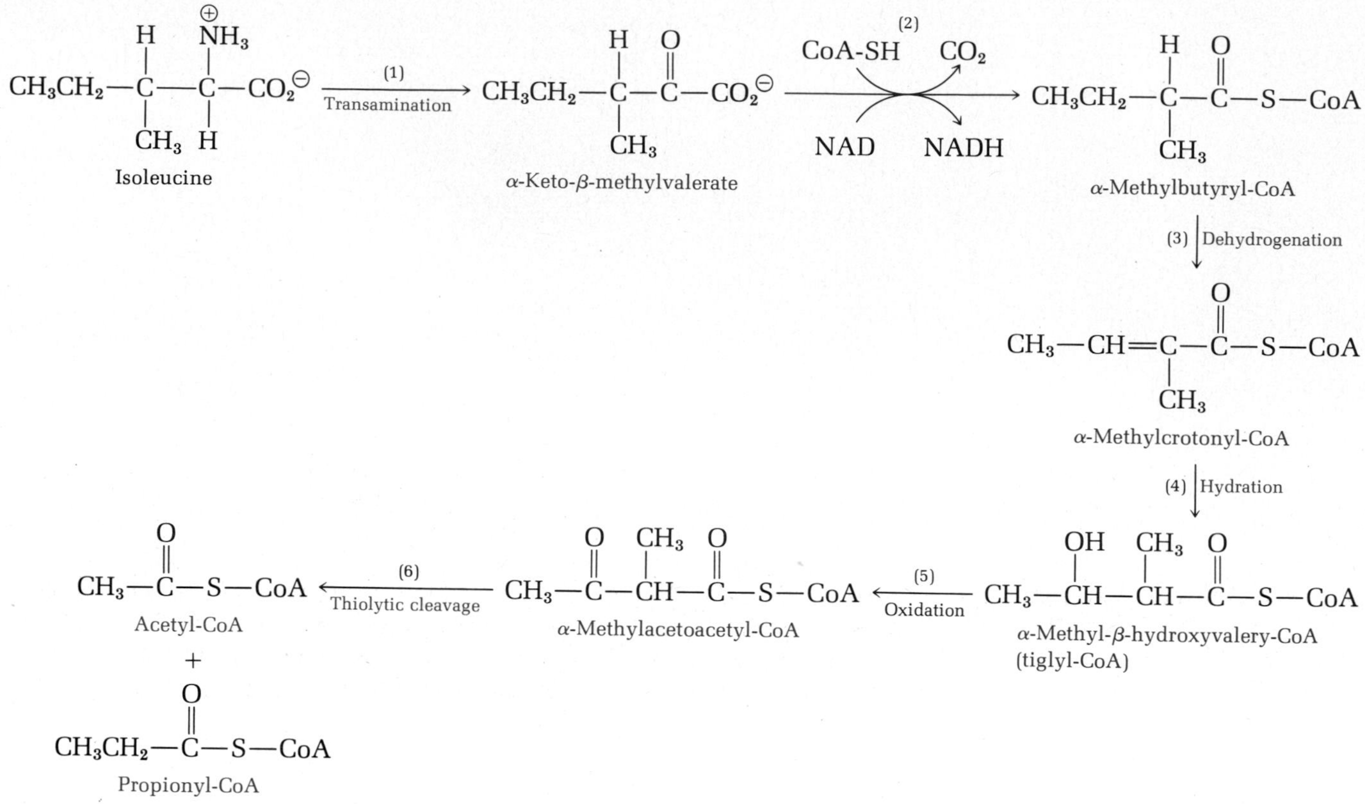

Figure 21.20 Pathway for the degradation of isoleucine.

E. Fumarate (and Acetoacetate) Family: Tyrosine and Phenylalanine

Tyrosine and phenylalanine are catabolized by the same pathway following the hydroxylation of phenylalanine to tyrosine (Figure 21.25). The *5,6,7,8-tetrahydrobiopterin*-dependent enzyme *phenylalanine-4-monooxygenase* catalyzes hydroxylation of phenylalanine. The active form of the cofactor is produced by reduction of 7,8-biopterin by an NADPH-dependent dehydrogenase (reaction 21.26).

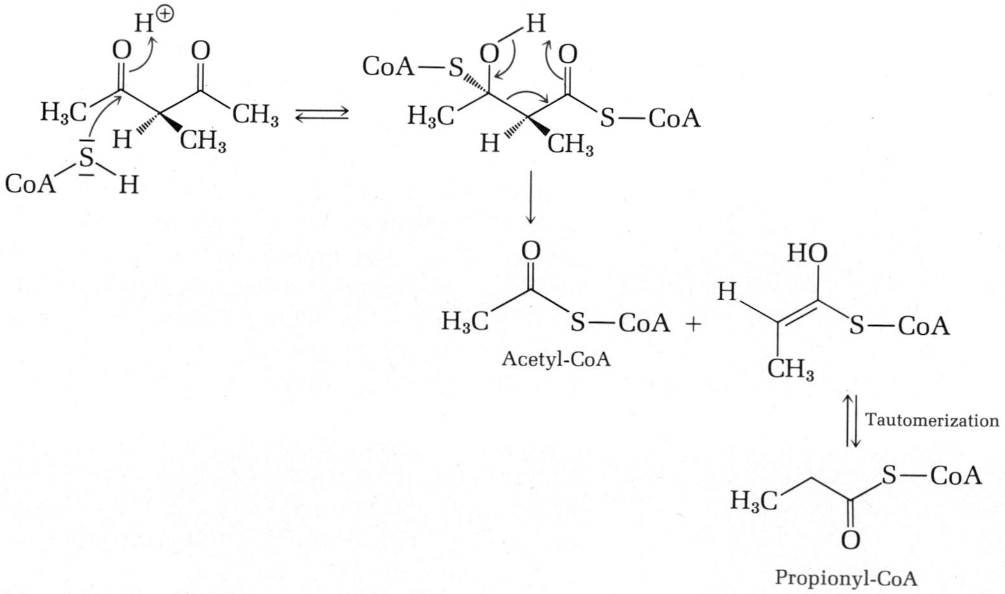

Figure 21.21 Thiolytic α cleavage of α-methylacetoacetyl coenzyme A produces acetyl coenzyme A and propionyl coenzyme A.

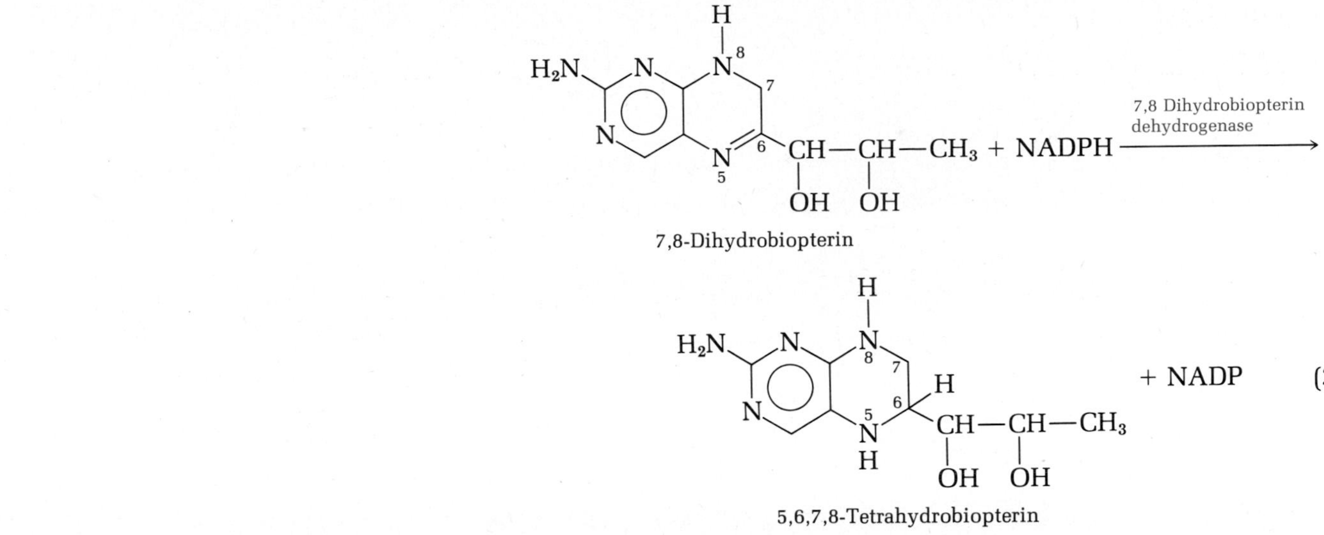

7,8-Dihydrobiopterin

$$+ \text{NADP} \qquad (21.26)$$

5,6,7,8-Tetrahydrobiopterin

Phenylalanine-4-monooxygenase is a dimer (mol wt 100,000) that contains an Fe(II) ion that binds to oxygen. The Fe(II)-O_2 complex is converted by the enzyme to an *hydroxy cation*, $OH^{\oplus}$. This powerful electrophile attacks the 4 position of phenylalanine. This step of catalysis is electrophilic aromatic substitution (reaction 21.27).

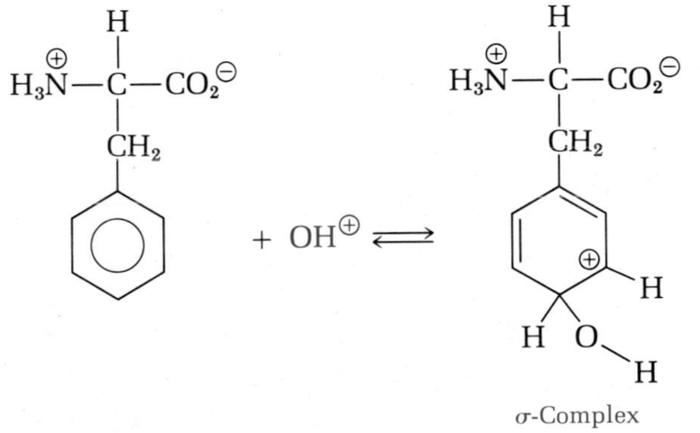

$$(21.27)$$

σ-Complex

The sigma complex collapses to an *arene oxide* that reopens to give a 4-keto intermediate (reaction 21.28).

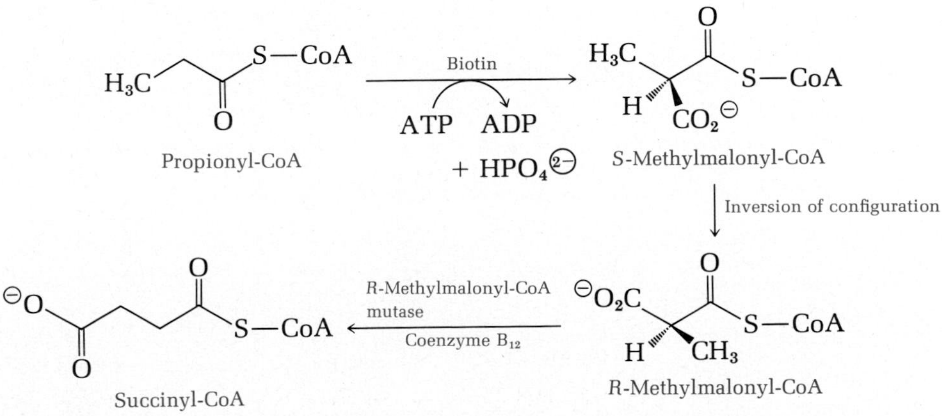

Figure 21.22 Conversion of propionyl coenzyme A to succinyl coenzyme A.

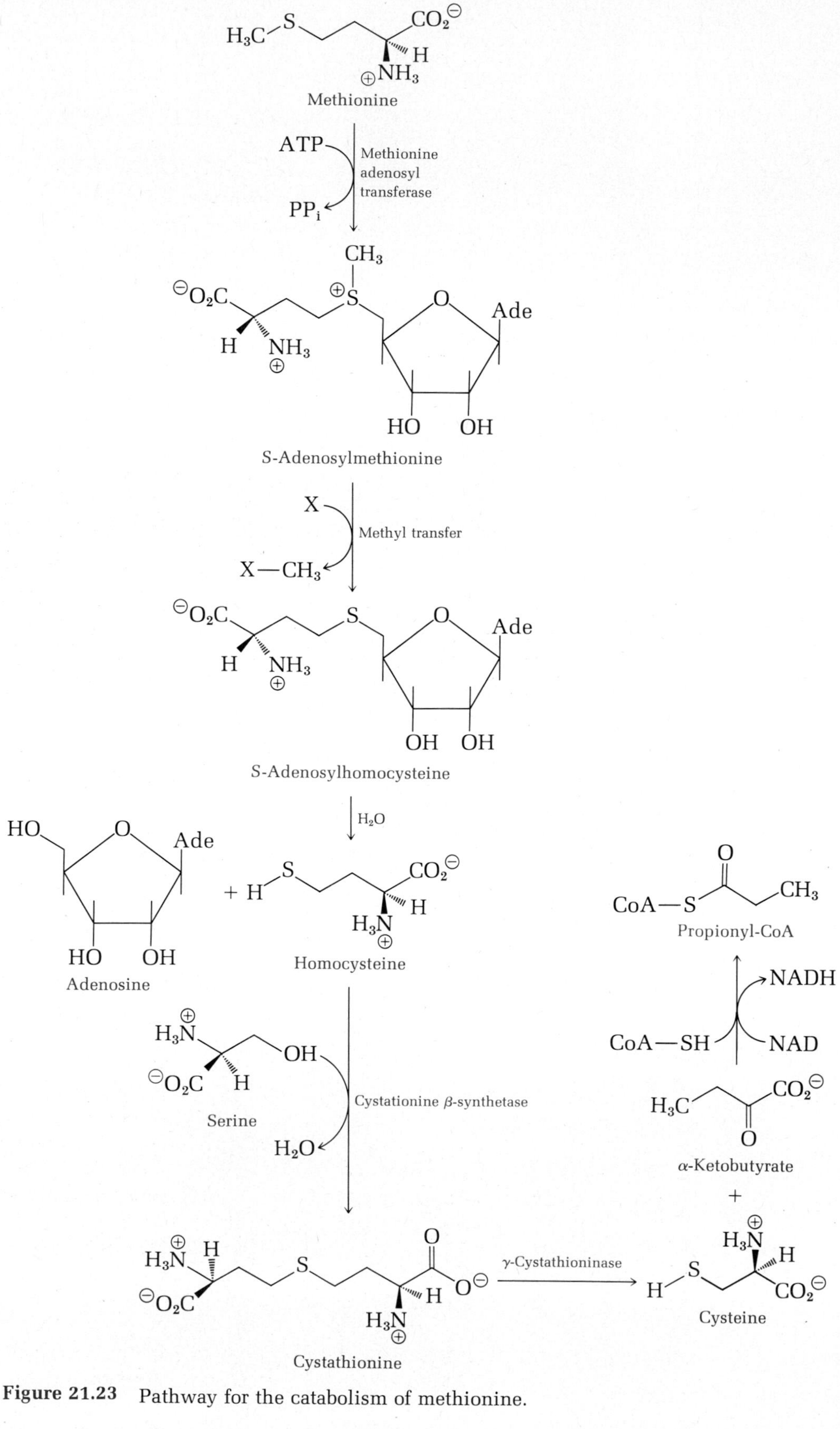

Figure 21.23 Pathway for the catabolism of methionine.

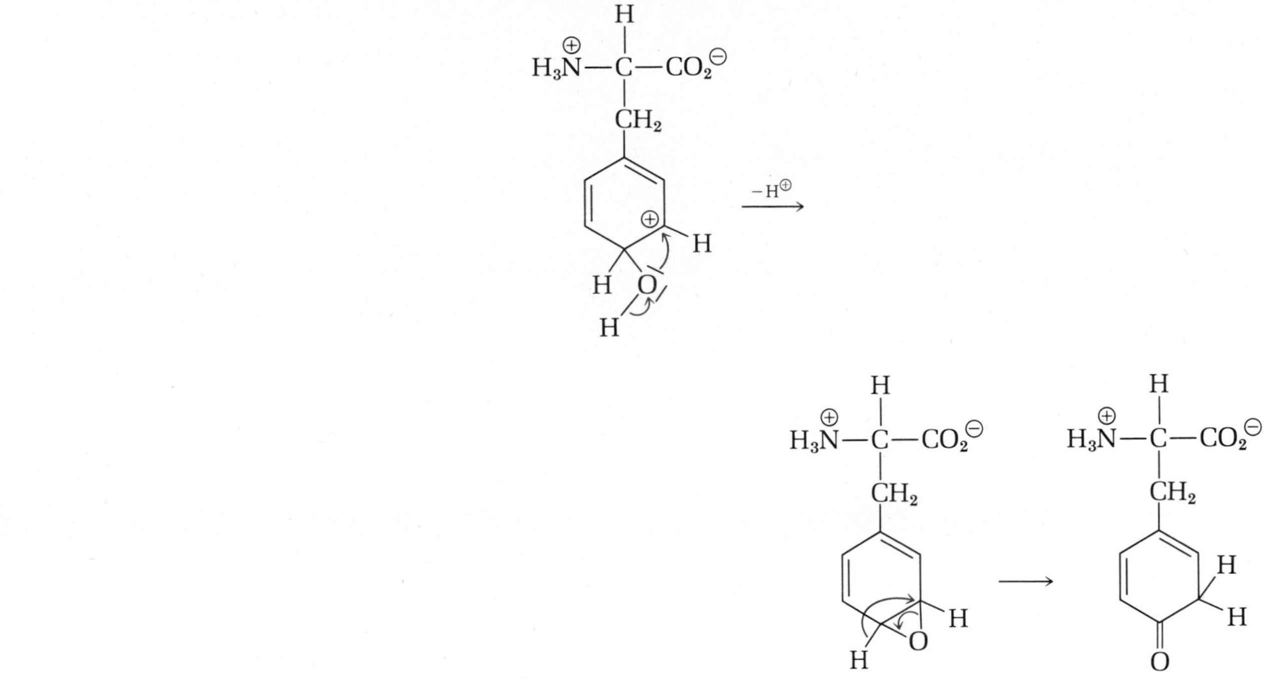

(21.28)

Arene oxide   4-Keto intermediate

The C-3 hydrogens of the 4-keto intermediate are prochiral. If phenylalanine labeled with tritium at C-4 is the substrate, only [3-^{3}H]-tryosine is produced. The final catalytic step in which the 4-keto intermediate rearranges to tyrosine is therefore completely stereospecific (reaction 21.29).

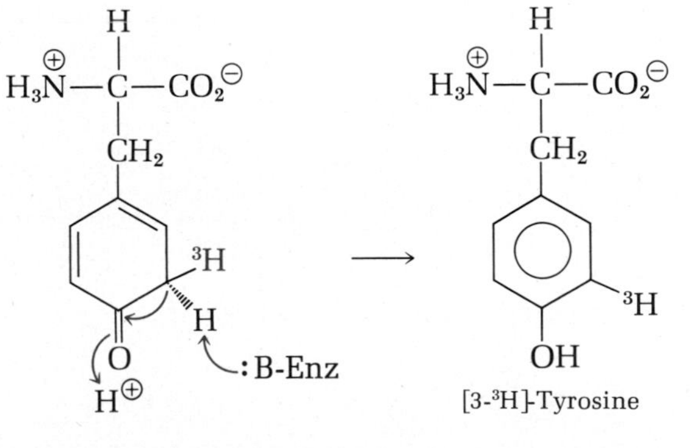

(21.29)

[3-^{3}H]-Tyrosine

The hereditary, metabolic disease phenylketonuria is caused by an absence of phenylalanine-4-monooxygenase. When this enzyme is absent phenylpyruvate, the α-keto acid derived by transamination of phenylalanine, accumulates and is present in high concentrations in the brain. For unknown reasons, phenylpyruvate, or one or more of its metabolic by-products, is toxic. The major effect of phenylketonuria is severe mental retardation. About 1 person in 20,000 lacks phenylalanine-4-monooxygenase, and the blood of newborns is routinely assayed for phenylpyruvate. The worst effects of phenylketonuria can be avoided if the disease is detected at birth and the diet of the afflicted infant is kept low in phenylalanine.

Tyrosine is transaminated to 4-hydroxyphenylpyruvate by tyrosine transaminase in a reaction requiring pyridoxal phosphate as a coenzyme (reaction 21.30). Homogentisate (homogentistic acid) is next produced by the action of 4-*hydroxyphenylpyruvate dioxygenase* (reaction 21.31). The oxygen of the new hydroxyl group is supplied by molecular oxygen, and an acetyl group migrates to the neighboring carbon on the aromatic ring.

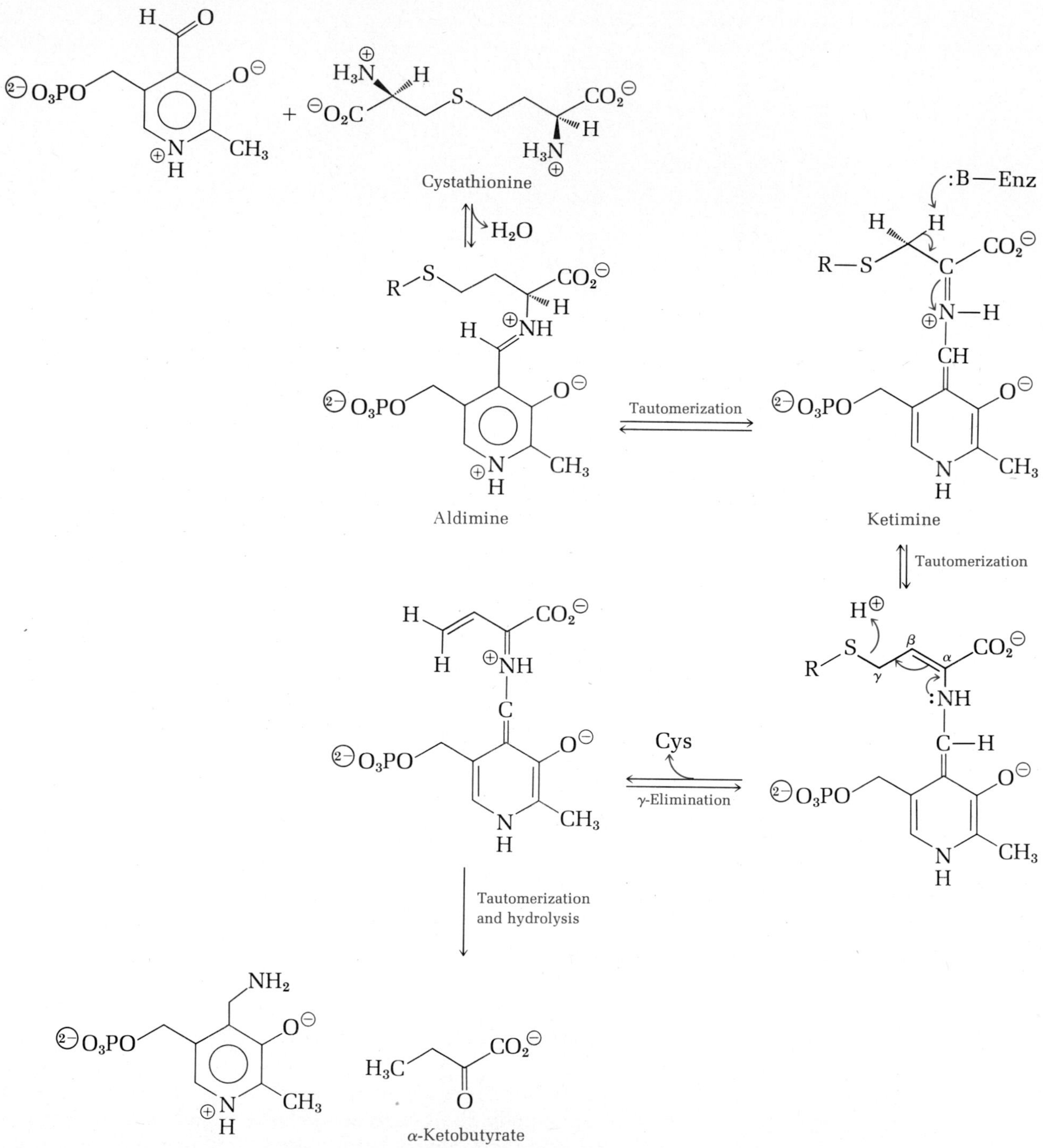

Figure 21.24 Mechanism of action of γ-cystathioninase.

The enzyme has an Fe(II) ion that forms a complex with molecular oxygen. Loss of carbon dioxide and formation of a peroxide seems to occur in a concerted step whose mechanism has not been elucidated. The decarboxylated peroxide then rearranges to homogentisate as shown in Figure 21.26.

Another dioxygenase, *homogentisate-1,2-dioxygenase*, catalyzes the antepenultimate step of the degradation of tyrosine: the conversion of homogentisate to maleylacetoacetate (reaction 21.32). Although several reaction mechanisms have been proposed, the mechanism of action of the enzyme is not known with any certainty.

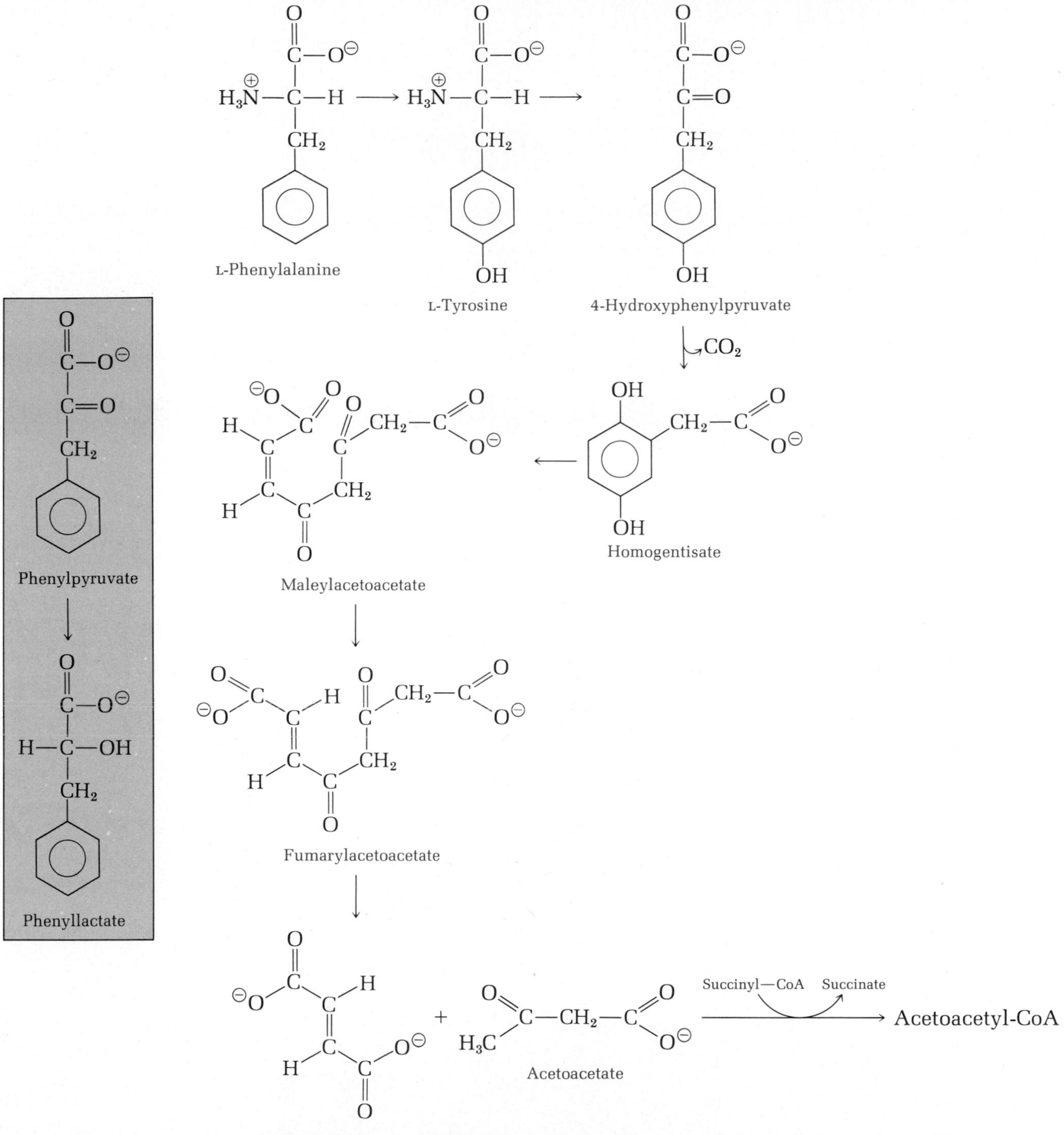

Figure 21.25 Pathways for the degradation of tyrosine and phenylalanine. Phenylpyruvate and phenyllactate, produced in the pathway on the left, accumulate in the metabolic disease phenylketonuria.

4-Maleylacetoacetate, isomerizes to fumarylacetoacetate in the penultimate step of the pathway in a reaction catalyzed by *maleylacetoacetate isomerase* (reaction 21.33). Cleavage of the β-diketone, 4-fumarylacetoacetate, by *fumarylacetoacetase* gives acetoacetate and fumarate (reaction 21.34). This process is similar to an aldol cleavage (Figure 21.27).

Since fumarate is a citric acid cycle intermediate that can be converted

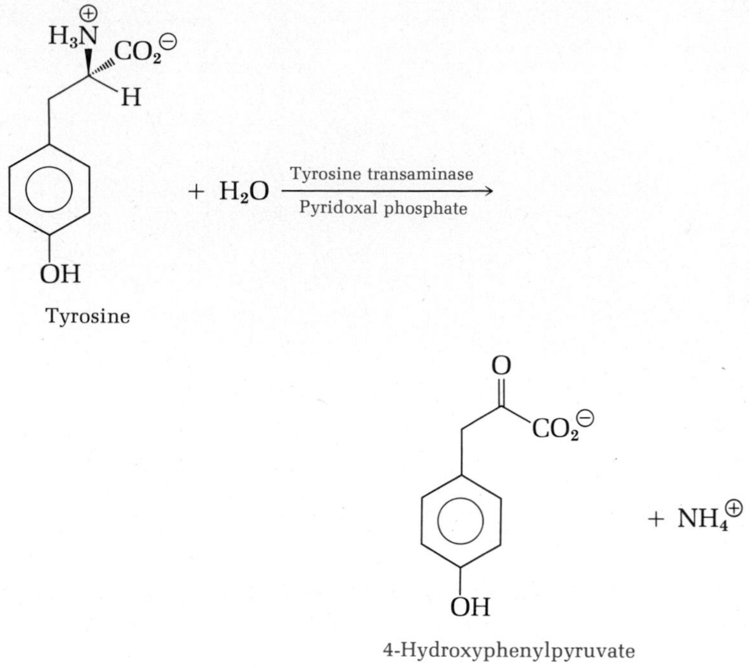

Tyrosine

$+ H_2O \xrightarrow[\text{Pyridoxal phosphate}]{\text{Tyrosine transaminase}}$

4-Hydroxyphenylpyruvate

$+ NH_4^{\oplus}$ (21.30)

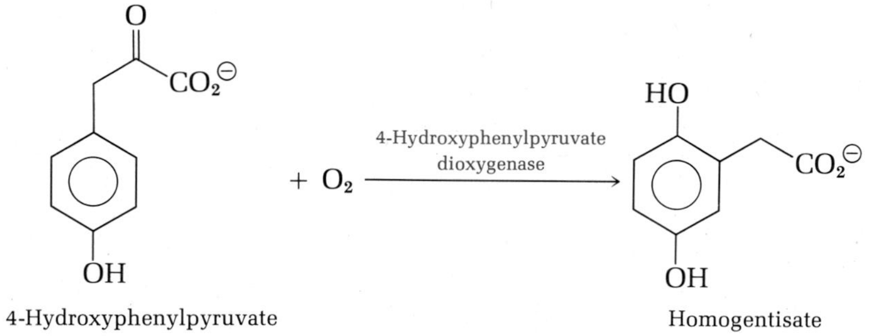

4-Hydroxyphenylpyruvate

$+ O_2 \xrightarrow[\text{dioxygenase}]{\text{4-Hydroxyphenylpyruvate}}$

Homogentisate

(21.31)

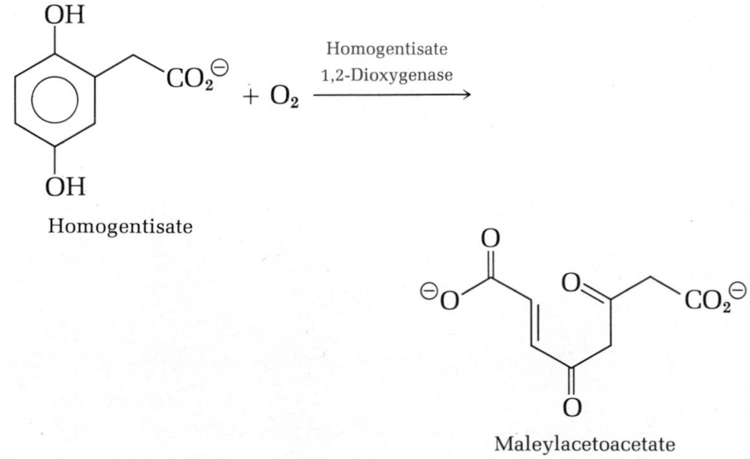

Homogentisate

$+ O_2 \xrightarrow[\text{1,2-Dioxygenase}]{\text{Homogentisate}}$

Maleylacetoacetate

(21.32)

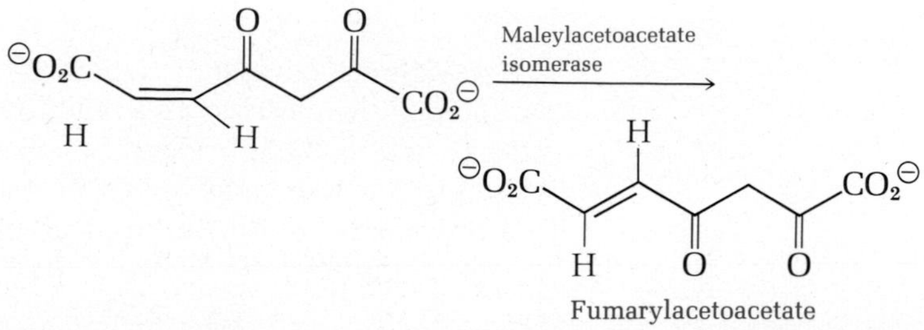

$\xrightarrow[\text{isomerase}]{\text{Maleylacetoacetate}}$

Fumarylacetoacetate

(21.33)

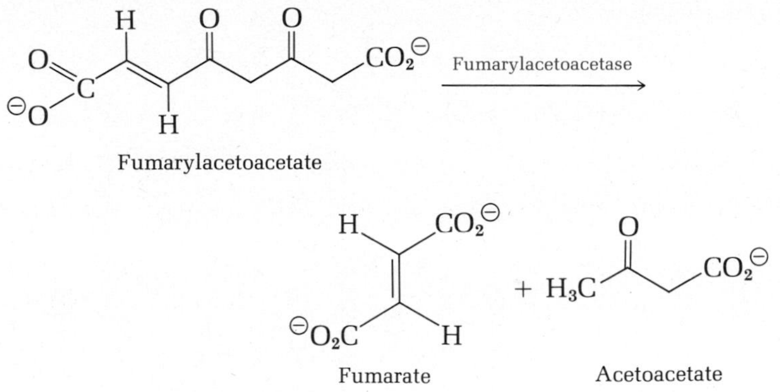

$$\text{Fumarylacetoacetate} \xrightarrow{\text{Fumarylacetoacetase}}$$

$$\text{Fumarate} \qquad \text{Acetoacetate} \qquad (21.34)$$

to carbohydrate and acetoacetate is a product of β-oxidation of fatty acids, tyrosine and phenylalanine both are glycogenic and ketogenic.

F. Acetyl-CoA Family: Leucine, Lysine, and Tryptophan

Leucine, lysine, and tryptophan belong to the large family of amino acids that are converted to acetyl-CoA. The final steps of the degradation of leucine, lysine, and tryptophan are similar, and so we shall consider these three as a related group. To a first approximation, the degradation of the carbon chains derived from all three resembles β-oxidation of fatty acids.

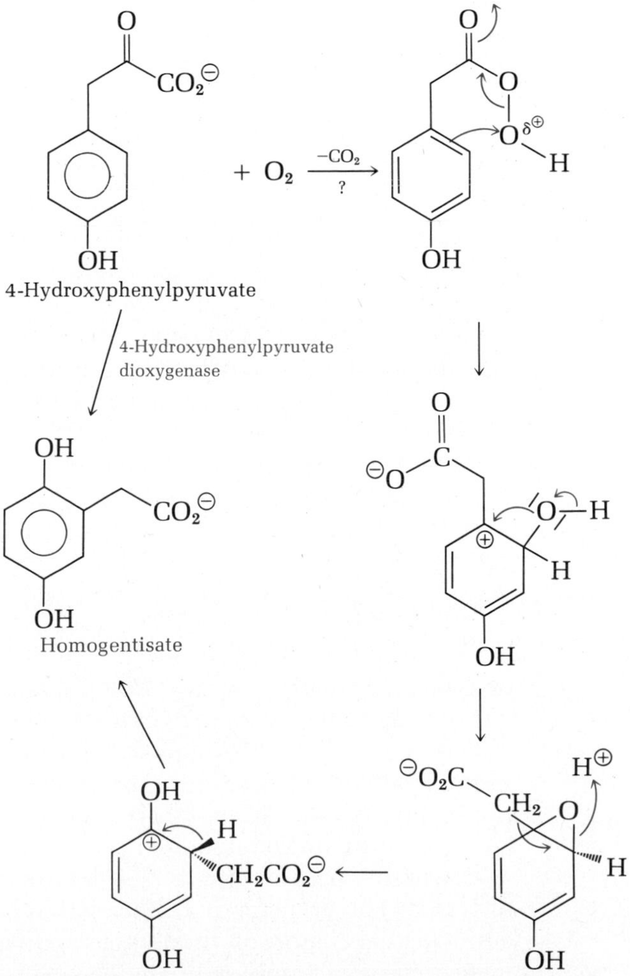

Figure 21.26 Conversion of 4-hydroxyphenylpyruvate to homogentisate. [Adapted from O. Hayaishi, M. Nozaki, and M. T. Abbot, in P. D. Boyer, Ed., "The Enzymes," 3rd ed., Vol. 12, Academic Press, New York, 1975, p. 185.]

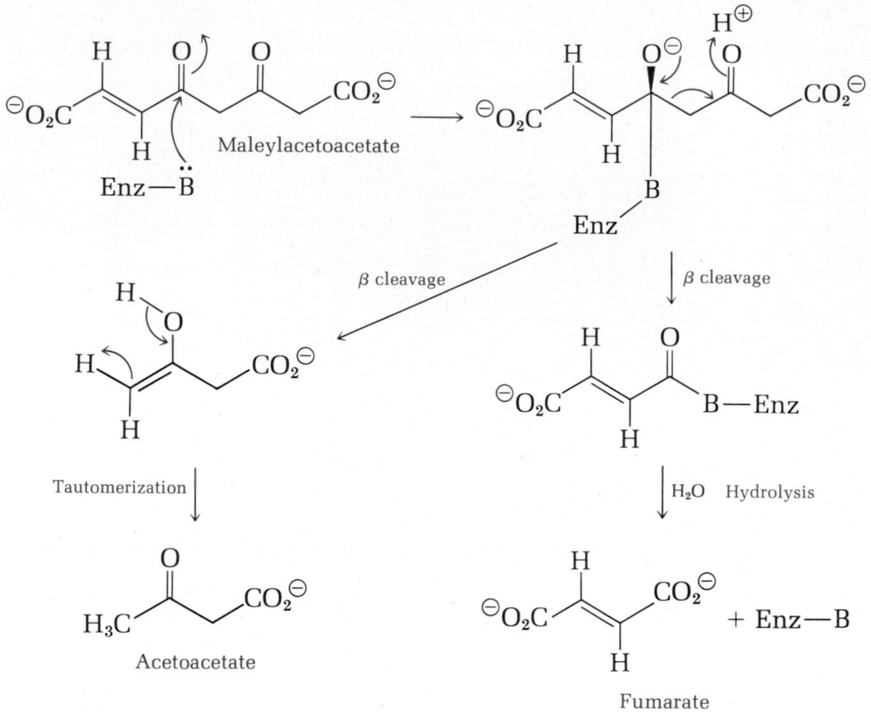

Figure 21.27 Conversion of maleylacetoacetate to acetoacetate and fumarate by fumarylacetoacetase.

a. The Catabolism of Leucine

The pathway for the conversion of leucine to acetyl-CoA and acetoacetate is shown in Figure 21.28. The first three steps of this pathway—transamination, oxidative decarboxylation coupled to synthesis of the fatty CoA derivative, and flavin-dependent dehydrogenation—are identical to the pathways for the catabolism of isoleucine and valine. In the fourth step β-methylcrotonyl-CoA carboxylase catalyzes a biotin-dependent carboxylation that is analogous to the reactions catalyzed by pyruvate carboxylase and acetyl-CoA carboxylase. The final two steps consist of hydration of the double bond of β-methylglutaconyl-CoA, an exact parallel to hydration in β-oxidation of fatty acids, and aldol cleavage of β-hydroxy-β-methylglutaryl-CoA to give acetoacetate and acetyl-CoA. The penultimate catabolic product, β-hydroxy-β-methylglutaryl-CoA is an important metabolite that is a starting material for cholesterol biosynthesis (Section 20.6).

b. Lysine

Multiple pathways exist for the catabolism of lysine in mammals. In one pathway (path 1, Figure 21.29) L-amino acid oxidase produces α-keto-ϵ-aminocaproate. Nonenzymatic cyclization produces Δ^1-piperidine-2-carboxylate. An NADPH-dependent reductase reduces the imine, and a flavoprotein dehydrogenase introduces it on the opposite side of the ring nitrogen producing Δ^1-piperidine-6-carboxylate. Nonenzymatic hydrolysis yields L-α-aminoadipate semialdehyde.

Another pathway (path 2) is the reverse of the pathway for biosynthesis of lysine. The first step is condensation of lysine with α-ketoglutarate. The adduct, L-saccharopine, undergoes oxidative cleavage, with loss of glutamate, to form L-α-aminoadipate. The net result of paths 1 and 2 is conversion of the ϵ-amino group of lysine to an aldehyde. Path 2 is prevalent in liver. Oxidation of α-aminoadipate semialdehyde produces α-aminoadipate, an intermediate also formed in the catabolism of tryptophan. The conversion of

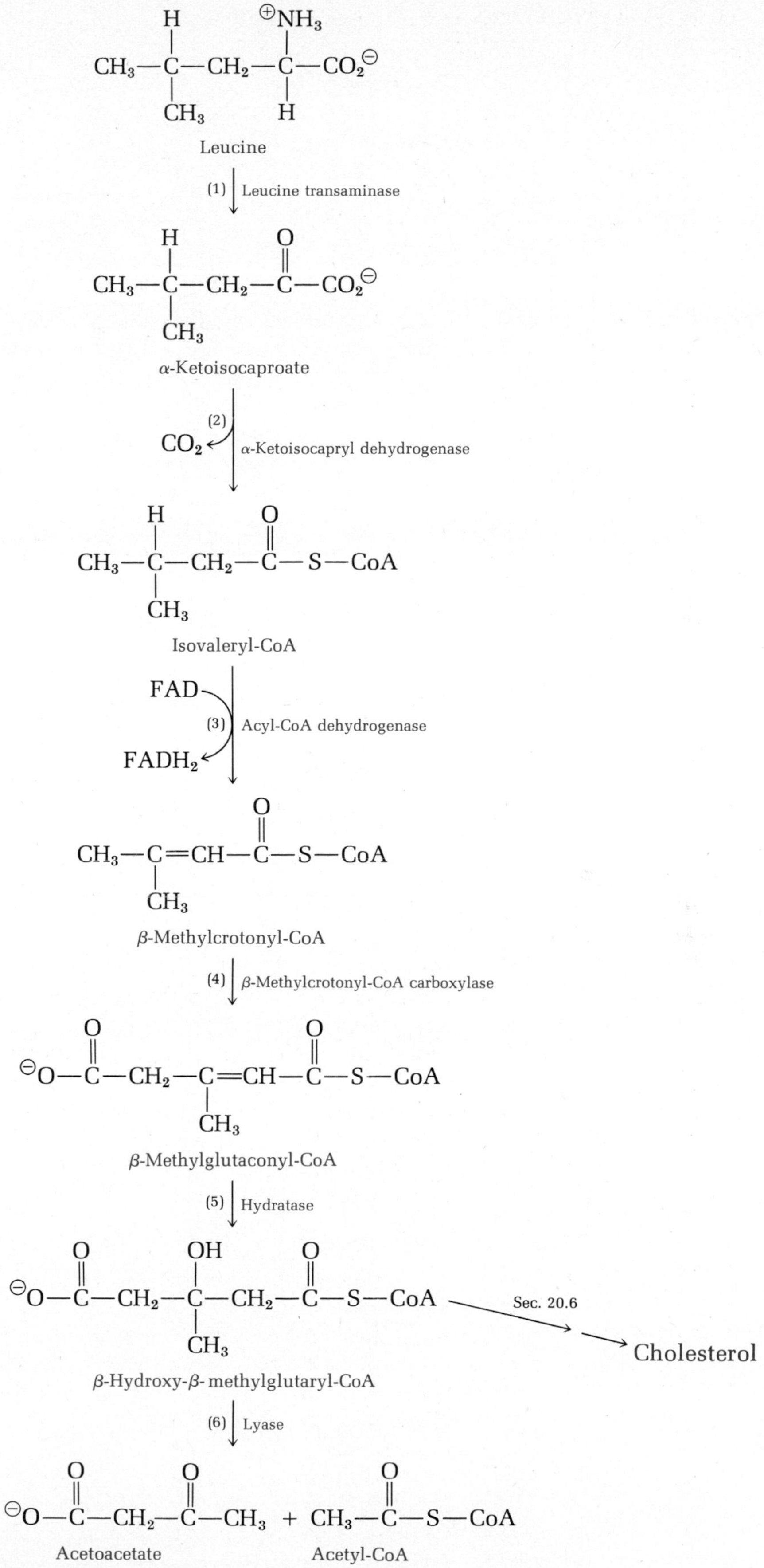

Figure 21.28 Pathway for the catabolism of L-leucine.

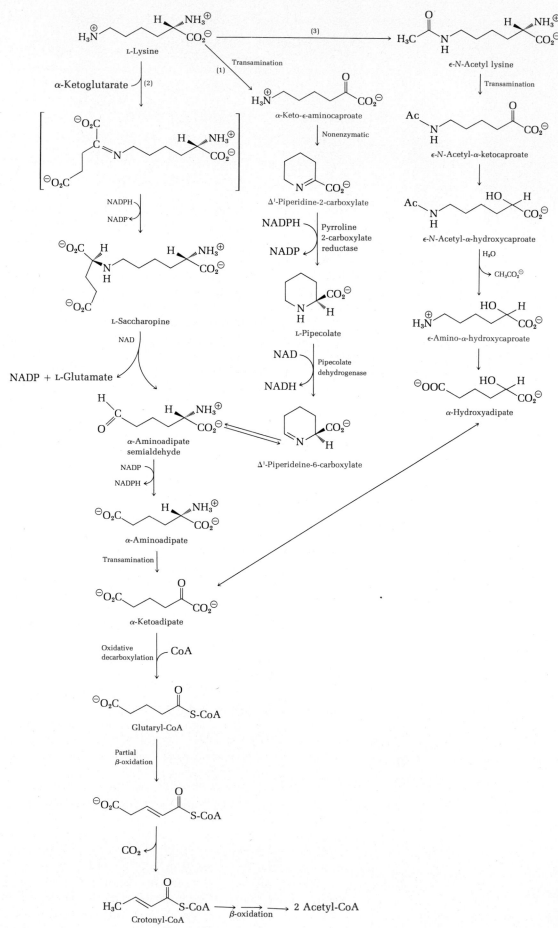

Figure 21.29 Pathways for the catabolism of lysine.

α-aminoadipate to acetyl-CoA follows a β-oxidation pathway parallel to β-oxidation of fatty acids.

Path 3 prefaces transamination by acylation of the ϵ-amino group. Transamination then produces α-keto-ϵ-acetamidocaproate. It cannot cyclize, but is converted to α-ketoadipate.

c. Tryptophan

The catabolism of tryptophan involves an extraordinary set of reactions (Figure 21.30). The first step of the pathway, catalyzed by *tryptophan 2,3-dioxygenase*, cleaves the heterocyclic ring to produce N-formylkynurenine. Steroid hormones, such as cortisol (Section 20.7) stimulate synthesis of tryptophan dioxygenase. In mammalian liver *kynurenine formylase* removes the formyl group as formate producing kynurenine. A monooxygenase hydroxylates the 3 position of the aromatic ring to give 3-hydroxykynurenine. The pyridoxal phosphate–dependent enzyme *kynureninase* catalyzes elimination of alanine with concomitant formation of 3-hydroxyanthranilate. The same enzyme also accepts kynurenine as a substrate producing alanine and anthranilate, the latter being excreted in the urine.

The fates of 3-hydroxyanthranilate are varied. In Section 22.8 we shall see that it is a precursor of NAD and NADP. In the catabolism of tryptophan 3-hydroxyanthranilate is converted to α-ketoadipate in three steps. First, *3-hydroxyanthranilate oxygenase* cleaves the aromatic ring incorporating both atoms of oxygen in the product, α-amino-β-carboxymuconic-δ-semialdehyde. Second, decarboxylation yields α-aminomuconate. Third, transamination produces α-ketoadipate, whose conversion to acetyl-CoA is the same as the pathway followed in lysine catabolism (Figure 21.29).

21–5 SUMMARY

The excess protein ingested by a cell can be metabolized by the oxidative pathways of the cell, and in periods of low demand for protein synthesis and low energy demand the carbon skeletons of the amino acids can be incorporated in glucose or in fatty acids. Those amino acids that can be converted to carbohydrate are called glycogenic, and those that can be converted to fatty acids are ketogenic. Most amino acids are purely ketogenic, a few are both glycogenic and ketogenic, and only one, leucine, is purely ketogenic.

The catabolism of amino acids can be divided into two stages. First, an amino acid is transaminated, usually by a member of the pyridoxal phosphate–dependent family of glutamate transaminases. The original amino acid becomes an α-keto acid, and α-ketoglutarate becomes glutamate in this step of the deamination process. Next, glutamate dehydrogenase catalyzes NAD(P)-dependent oxidative deamination, regenerating α-ketoglutarate and releasing ammonium ion.

Ammonium ions are toxic and must be eliminated. In terrestrial vertebrates ammonium ion is incorporated in urea in the liver by the action of the urea cycle. The carbon atom of urea is derived from bicarbonate, and the second amino group of urea is derived from the α-amino group of aspartate. Three moles of ATP are consumed in the production of urea by the urea cycle. One of the by-products of the urea cycle is fumarate, and the urea cycle is therefore coupled to the citric acid cycle. The urea cycle also provides the final steps in the pathway for the biosynthesis of arginine. Once urea has been synthesized in the liver, it is transported through the blood to the kidney and excreted in urine.

The α-keto acids formed by deamination of the amino acids are converted into pyruvate, acetyl-CoA, and the citric acid cycle intermediates α-ketoglutarate, succinyl-CoA, fumarate, and oxaloacetate. We have placed the amino acids in families whose shared property is a common degradation

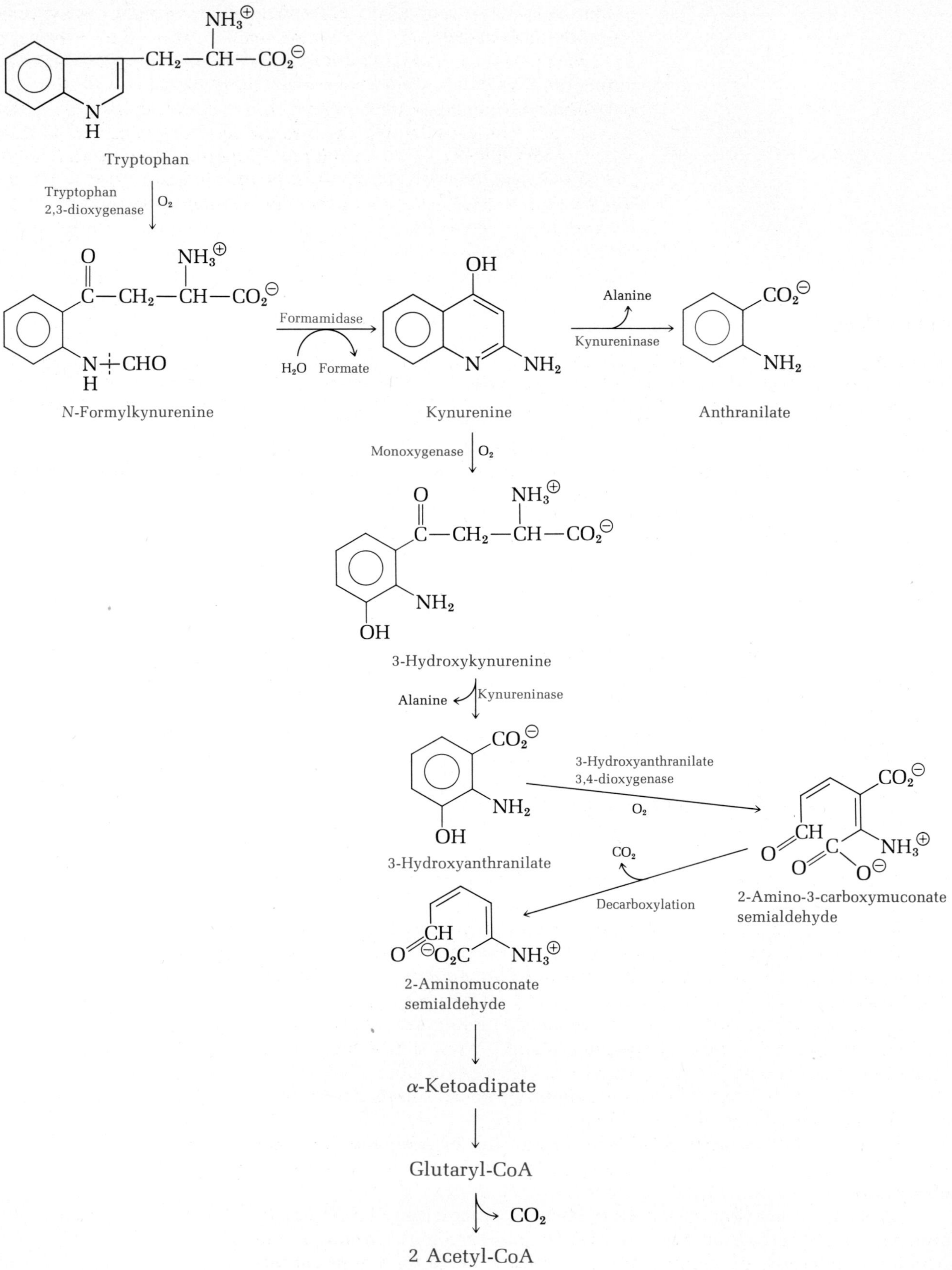

Figure 21.30 Pathway for the catabolism of tryptophan.

product. Thus the glycogenic amino acids aspartate and asparagine are converted to oxaloacetate; the glycogenic alanine, threonine, glycine, serine, and cysteine are converted to pyruvate; the glycogenic proline, arginine, histidine, and glutamine are converted to glutamate, and thence to α-ketoglutarate; the glycogenic methionine and valine as well as the glycogenic and ketogenic isoleucine, are converted to succinyl-CoA; the glycogenic and ketogenic tyrosine and phenylalanine are converted to fumarate and acetoacetate; the glycogenic and ketogenic lysine and tryptophan are converted to acetyl-CoA as is the purely ketogenic leucine.

REFERENCES

Bender, D. A., "Amino Acid Metabolism," Wiley, New York, 1975.

Braunstein, A. E., "Amino Group Transfer," in P. D. Boyer, Ed., "The Enzymes," 3rd ed., Vol. 9, Academic Press, New York, 1973.

Goldberg, A. L., and St. John, A. C., Intracellular Protein Degradation in Mammalian and Bacterial Cells: Part 2, *Ann. Rev. Biochem.*, 45, 747–803 (1980).

Grisola, S., Baguena, R., and Mayor, F., Eds. "The Urea Cycle," Wiley, New York, 1975.

Holzer, H., Control of Proteolysis, *Ann. Rev. Biochem.*, 49, 63–91, (1980).

Walsh, C., "Enzymatic Reaction Mechanisms," Freeman, San Francisco, 1979.

Further references to amino acid metabolism may be found at the end of Chapter 22.

Problems

1. The equilibrium constant for the reaction catalyzed by glutamate dehydrogenase is about 10^{-15}.

$$K = \frac{[\text{NAD(P)H}][\alpha\text{-ketoglutarate}][\text{NH}_4^{\oplus}][\text{H}^{\oplus}]}{[\text{L-glutamate}][\text{NAD(P)}]}$$

Since this is a very small equilibrium constant, how is *net* conversion of glutamate to α-ketoglutarate possible?

2. Write a plausible reaction mechanism involving NAD, thiamine pyrophosphate, coenzyme A, and lipoic acid for the conversion of α-keto-β-methylvaleric acid to α-methylbutyryl-CoA. [Hint: Remember pyruvate dehydrogenase.]

3. Write out a series of reactions for the β-oxidation of α-methylbutyryl-CoA to α-methylacetoacetyl-CoA. [Hint: Remember the β-oxidation of fatty acids.]

4. Write a mechanism for the pyridoxal phosphate–dependent transamination of phenylalanine.

5. When [3-^{14}C]-serine is added to mitochondrial preparations, the major isotopic product is [3-^{14}C]-alanine. Write a pathway that explains this observation.

6. Write a reaction mechanism for the hydrolysis of arginine by arginase to give ornithine and urea. [Hint: Identify the elements of urea which are concealed in arginine. Which carbon in arginine is most susceptible to nucleophilic attack by water?]

7. Trace the fate of the isotopic label from [5-^{14}C]-3-hydroxyanthranilate to [3-^{14}C]-glutaryl-CoA.

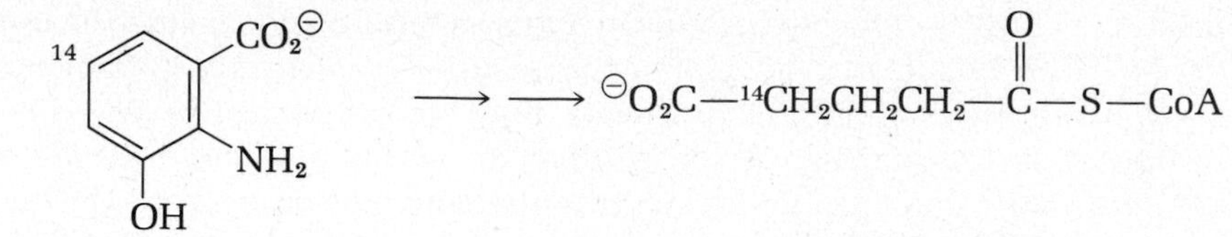

[1-^{14}C]-3-Hydroxyanthranilate

[3-^{14}C]-Glutaryl-CoA

$$\longrightarrow \longrightarrow \ ^{\ominus}O_2C-^{14}CH_2CH_2CH_2-\overset{\overset{\textstyle O}{\|}}{C}-S-CoA$$

8. Show how [2-^{14}C]-tryptophan can be converted to [2-^{14}C]-alanine.

9. Show where the isotopic label will appear if [2-^{14}C]-leucine is converted to palmitoyl-CoA derived solely from leucine.

10. Draw a path that traces the isotope from [3-^{14}C]-valine to [^{14}C]-glucose. In which carbon of glucose will the label appear?

11. Explain why leucine is purely ketogenic, whereas isoleucine is both glycogenic and ketogenic.

12. When [α-^{14}C]-α-methylbutyrate is metabolized, where will the ^{14}C-label appear in the degradation products?

Chapter 22

The biosynthesis of amino acids and related molecules

22–1 INTRODUCTION

The biosynthesis of amino acids and molecules, such as heme, derived from amino acids, is complex. Three separate aspects of amino acid biosyntheses must be considered.

Higher animals are unable to use N_2, nitrate, or nitrite as nitrogen sources for amino acids. They can, however, incorporate the nitrogen of ammonia. (Ruminants have bacteria in their stomachs that provide ammonium ions from nitrite and nitrate.) Higher plants are able to incorporate the nitrogen of nitrate, nitrite, or ammonium ions in amino acids. The first topic that we shall discuss, then, is *nitrogen fixation*.

We shall then turn to the pathways by which the carbon skeletons of amino acids are assembled. The pathways for biosynthesis of amino acids that share a common carbon source are related, but each of the 20 amino acids is synthesized by a separate pathway. The experimental determination of these pathways is a formidable task that relies heavily upon isotopic tracer studies to follow the course of a labeled carbon atom through a pathway. Microorganisms that lack the ability to synthesize a given amino acid, known as *auxotrophic mutants*, have also been extremely valuable in the study of amino acid biosynthesis. By studying microorganisms that lack a given en-

Table 22.1	Basic set of 20 amino acids required by humans (and albino rats)
Essential	**Nonessential**
Arginine	Alanine
Histidine	Asparagine
Isoleucine	Aspartate
Leucine	Cysteine
Lysine	Glutamate
Methionine	Glutamine
Phenylalanine	Glycine
Threonine	Proline
Tryptophan	Serine
Valine	Tyrosine

zyme in a biosynthetic pathway, each intermediate that accumulates at a "metabolic block" can be identified. The pathway for biosynthesis of an amino acid can therefore be elucidated by identifying the enzyme responsible for carrying out the last step in the pathway of the mutant microorganism.

Organisms vary greatly in their ability to synthesize amino acids. The bacterium *E. coli* can synthesize all of its amino acids, whereas *Leuconostoc mesenteroides* must be provided with 16 different amino acids to survive. Amino acids that cannot be synthesized and must therefore be obtained from an organism's diet are called *essential*; those that the organism can synthesize are called *nonessential*. The human shares with the albino rat the ability to carry out net synthesis 10 of the 20 amino acids required for normal growth (Table 22.1).

The regulation of amino acid biosynthesis is the third topic that we shall consider, although as an integral part of each pathway rather than as a separate phenomenon. Two types of regulation are important. The first is allosteric regulation by feedback inhibition of the committing step by the end product of the pathway. The second, to which we shall devote more attention in Chapter 27, is genetic regulation, which is exerted at the level of protein synthesis.

22–2 NITROGEN FIXATION

The earth's atmosphere is about 80% nitrogen, but N_2 is quite unreactive, and most organisms require a source of "chemically combined" nitrogen for their metabolism. The *fixation* of nitrogen—its conversion to some oxidized or reduced molecule—is therefore critical to the function of the entire biosphere.

Fixation of nitrogen occurs by both nonbiological and biological processes. Lightening catalyzes the formation of large amounts of nitrates and other oxides of nitrogen. Plants can use these oxidized products directly, but higher animals require reduced nitrogen in the form of ammonium ions for biosynthesis. (These ammonium ions are mostly derived from the amino group of aspartate and amide nitrogen of glutamine.) Other inorganic (although "biologically mediated") sources of large quantities of oxides of nitrogen are the combustion products of the ubiquitous automobile and the chemicals manufactured by the fertilizer industry.

However, the primary source of nitrogen fixation is biological. The reduction of nitrogen by bacterial *nitrogenase*, an enzyme system found in some soil bacteria, yields an estimated 1.7×10^{11} kg per yr, four times the amount produced by other means. Perhaps the most important of the nitrogen-fixing bacteria is *Rhizobium,* a symbiotic species that dwells in the roots of legumes, forming root nodules that reduce nitrogen for use by the plant.

A. Nitrogenase

Nitrogenase catalyzes the six-electron reduction of N_2 to ammonia (reaction 22.1). The reduction of nitrogen is highly endergonic and requires 12 moles of ATP.

$$N_2 + 6\,e + 12ATP + 12H_2O \xrightarrow{\text{nitrogenase}}$$
$$2NH_4^{\oplus} + 12ADP + 12P_i + 4H^{\oplus} \qquad (22.1)$$

The structure of nitrogenase is complicated, even by the standards that we are accustomed to applying to enzymes. It is composed of two oligomeric proteins. One component, called the *iron protein,* is a dimer that has identical peptide chains of molecular weight 30,000. The dimer contains 4 iron atoms in an iron-sulfur complex and 12 sulfhydryl groups. It has 2 binding sites for $MgATP^{\ominus}$ or MgADP. The second component, the *iron-molybdenum protein,* is a tetramer with the subunit composition $\alpha_2\beta_2$. The subunits have molecular weights of 51,000 and 60,000. The tetramer contains two molybdenum atoms whose oxidation state changes from Mo(IV) to Mo(VI) during the reduction of nitrogen. The iron-molybdenum protein also contains between 24 and 32 iron atoms, 24 disulfide groups, and 30 sulfhydryl groups. The structure of the immediate environment of the molybdenum ions is not known. Two possibilities are shown in Figure 22.1.

The reduction of N_2 requires an electron transport chain in which ferredoxin passes electrons to the iron protein, the reductase component of nitrogenase. Electron transfer then occurs from the reductase to the iron-molybdenum protein (Figure 22.2).

The mechanism of action of nitrogenase can be divided into four steps that indicate the direction of electron flow, but whose exact details are

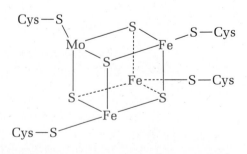

(a)

(b)

Figure 22.1 Two possible environments for the molybdenum sites of nitrogenase. In (a) the Mo atom is part of a bridged structure, and in (b) it substitutes for iron in a cubic cluster. The structure shown in (b) is similar to the complex found in iron-sulfur proteins. [Adapted from S. Cramer, K. Hodgson, W. Gillum, and L. Mortenson, *J. Amer. Chem. Soc., 100,* 3398 (1978).]

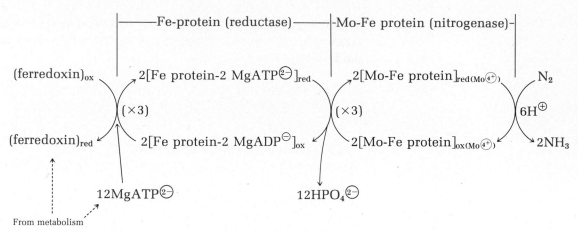

Figure 22.2 Pathway by which electrons are passed from an oxidized ferredoxin through nitrogenase to N_2. [Adapted from L. E. Mortenson and R. N. F. Thorneley, *Ann. Rev. Biochem.*, 48, 389 (1979). Reproduced with permission. © 1979 by Annual Reviews Inc.]

largely unknown. In the first step an electron donor, the reduced substrate, donates an electron to the iron protein (reaction 22.2).

$$\text{(Fe-protein)}_{ox} + e^{\ominus} \underset{}{\overset{(1)}{\rightleftharpoons}} \text{(Fe-protein)}_{red}$$

In the second step the reduced iron protein forms a complex with the iron-molybdenum protein and $MgATP^{2-}$ (reaction 22.3).

$$\text{(Mo-Fe protein)}_{ox} + \text{(Fe-protein)}_{red} + MgATP^{2-}$$

$$\updownarrow {\scriptstyle (2)}$$

$$[\text{(Mo-Fe protein)}_{ox} - \text{(Fe protein)}_{red} - MgATP^{2-}]_{complex} \qquad (22.3)$$

In the third step electrons are transferred from the iron protein to the iron-molybdenum protein, a reaction that is helped in part by a shift of the standard reduction potential of the iron protein from -0.25 to -0.40 V in the presence of $MgATP^{2-}$. ATP is hydrolyzed in this reduction step. A total of 12 moles of ATP are hydrolyzed for every mole of N_2 reduced. The mechanistic effect of ATP hydrolysis on the electron transfer reaction is not known. The electron transfer step is shown in reaction 22.4.

$$[\text{(Fe protein)}_{red} - MgATP^{2-} - \text{(Mo-Fe protein)}_{ox}]$$

$$(3) \downarrow \text{Internal electron transfer and ATP hydrolysis}$$

$$[\text{(Fe protein)}_{ox} - MgADP - \text{(Mo-Fe protein)}_{red}] + HPO_4^{2-} \qquad (22.4)$$

In the fourth and final step electrons are transferred from the reduced iron-molybdenum protein to N_2 or to other substrates of nitrogenase (reaction 22.5).

$$[\text{(Fe protein)}_{ox} - MgADP - \text{(Mo-Fe protein)}_{red}] + N_2$$

$$(4) \downarrow \text{Electron transfer to } N_2, H^{\oplus}$$

$$[\text{(Fe protein)}_{ox} - MgADP - \text{(Mo-Fe protein)}_{ox}] + NH_3 \qquad (22.5)$$

In reaction 22.5 MgADP is shown complexed with the iron protein. ADP is a potent inhibitor of the reduction; in a 1:1 ratio of ATP/ADP reduction is inhibited by 90%. Thus the replacement of ADP by ATP favors the reduction of nitrogen.

Reduction occurs in three discrete two-electron reduction steps. First, nitrogen is reduced to *diimide* (reaction 22.6). (If acetylene is the substrate, the final product is ethylene.)

$$N_2 + 2H^{\oplus} + 2\ e^{\ominus} \longrightarrow HN{=}NH \qquad (22.6)$$
$$\text{Diimide}$$

Diimide is then reduced to *hydrazine* (reaction 22.7), which is finally cleaved reductively to produce ammonia (reaction 22.8).

$$HN{=}NH + 2H^{\oplus} + 2\ e^{\ominus} \longrightarrow H_2N{-}NH_2 \qquad (22.7)$$
$$\text{Hydrazine}$$

$$H_2N{-}NH_2 + 2H^{\oplus} + 2\ e^{\ominus} \longrightarrow 2NH_3 \qquad (22.8)$$

Industrially, nitrogen is converted to ammonia by the Haber process. As we have noted, the reduction is quite endergonic. The commercial production of ammonia for use in fertilizers has therefore become increasingly expensive. Intense efforts are now being made to find ways of incorporating the gene for nitrogenase in plant cells. This research has received enormous impetus from the recent discoveries in the technology of recombinant DNA, in which genes can be transplanted from bacteria to eucaryotic cells and vice versa (see Section 24.9). If this research is successful, the extensive use of fertilizer may no longer be necessary.

B. Reduction of Nitrate

Nitrate enters the biosphere by the application of vast amounts of chemical fertilizer to the world's croplands and by lightening-catalyzed oxidation of atmospheric nitrogen. Various plants and microorganisms are able to reduce nitrate to ammonia. The eight-electron reduction of nitrogen from an oxidation state of $+5$ in nitrate to -3 in ammonia is catalyzed in two steps by *nitrate reductase* and *nitrite reductase*.

The standard reduction potential for conversion of nitrate to nitrite is large and positive, $E^{0\prime} = +0.421$ V. This corresponds to a standard free-energy change for reduction ($\Delta G^{\circ\prime}$) of -81.2 kJ mol^{-1} (reaction 22.9).

$$NO_3^{\ominus} + 2H^{\oplus} + 2\ e^{\ominus} \xrightarrow{\text{nitrate reductase}} NO_2^{\ominus} + H_2O \qquad E^{0\prime} = +0.421 \text{ V} \qquad (22.9)$$

The nitrate reductase of *Neurospora crassa* has been intensively studied. This protein (mol wt 230,000) contains Mo(VI), FAD, and cytochrome b_{557}, and it requires NADH for activity. Electrons flow from NADH to nitrate as shown below (reaction 22.10).

$$NADH \longrightarrow FAD \longrightarrow \text{cyt } b_{557} \longrightarrow Mo(VI) \longrightarrow NO_3^{\ominus} \qquad (22.10)$$

Nitrite reductase, also from *N. crassa*, catalyzes the reduction of nitrite to ammonia, a six-electron reduction in which 3 moles of NADPH are consumed.

$$3NADPH + NO_2^{\ominus} + 4H^{\oplus} \xrightarrow{\text{nitrite reductase}} NH_3 + 3NADP + 2H_2O \qquad (22.11)$$

The *N. crassa* enzyme contains FAD and an octacarboxylate-tetrahydroporphyrin known as *siroheme*. Nitrite binds to the siroheme iron and does not depart until it has been completely reduced to ammonia.

Siroheme

22–3 CARBON SOURCES OF THE AMINO ACIDS

The carbon sources for amino acid biosynthesis are derived from the citric acid cycle intermediates α-ketoglutarate and oxaloacetate, from the glycolytic intermediates pyruvate, phosphoenolpyruvate, and 3-phosphoglycerate, and from the pentose phosphate pathway intermediates ribose 5-phosphate and erythrose 4-phosphate. It is convenient to divide the amino acids into families whose members are descended from the same precursor. Thus, α-ketoglutarate is the carbon source for glutamate, glutamine, proline, and arginine. Oxaloacetate is the carbon source for aspartate, asparagine, methionine, threonine, isoleucine, and lysine. 3-Phosphoglycerate is the carbon source for serine, cysteine, and glycine. Pyruvate is the carbon source for alanine, valine, and leucine. Ribose 5-phosphate is the carbon source for histidine. And phosphoenolpyruvate and erythrose 4-phosphate are the carbon sources for the aromatic amino acids phenylalanine, tyrosine, and tryptophan (Figure 22.3).

22–4 BIOSYNTHESIS OF THE GLUTAMATE FAMILY

The pathways for the biosynthesis of amino acids derived from glutamate—glutamate, glutamine, proline, arginine—and the closely related biosynthesis of lysine in fungi are summarized in Figure 22.4. In most organisms lysine is a member of the aspartate family, but in fungi it is derived from glutamate.

A. Glutamate

Assimilation of nitrogen into proteins and nucleic acids is essential for the growth of all organisms. Almost all nitrogen assimilated by most bacteria is derived either from the amino group of glutamate or from the amido group of glutamine. *Glutamate dehydrogenase* catalyzes the reductive amination of α-ketoglutarate in a reversible reaction that uses either NADH or NADPH as the reducing agent and ammonia as a nitrogen source (reaction 22.12). Glutamate dehydrogenase can accept either NADH or NADPH as its reducing agent. In the catabolic reaction NADH is the reducing agent; in the biosynthetic reaction NADPH is the reducing agent.

B. Glutamine

The ATP-dependent conversion of glutamate to glutamine is catalyzed by *glutamine synthetase* (reaction 22.13). The amide nitrogen of glutamine is the source of nitrogen in many biosynthetic processes.

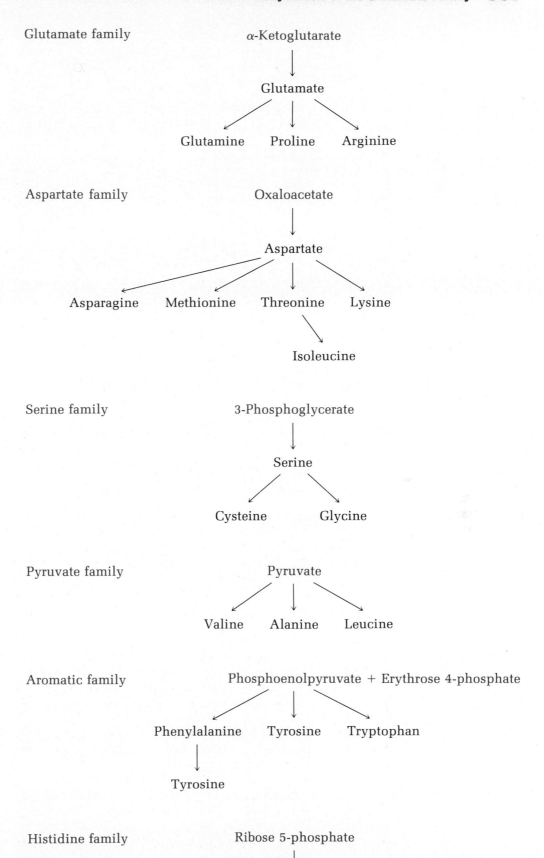

Figure 22.3 Families of amino acids based on a common biosynthetic precursor.

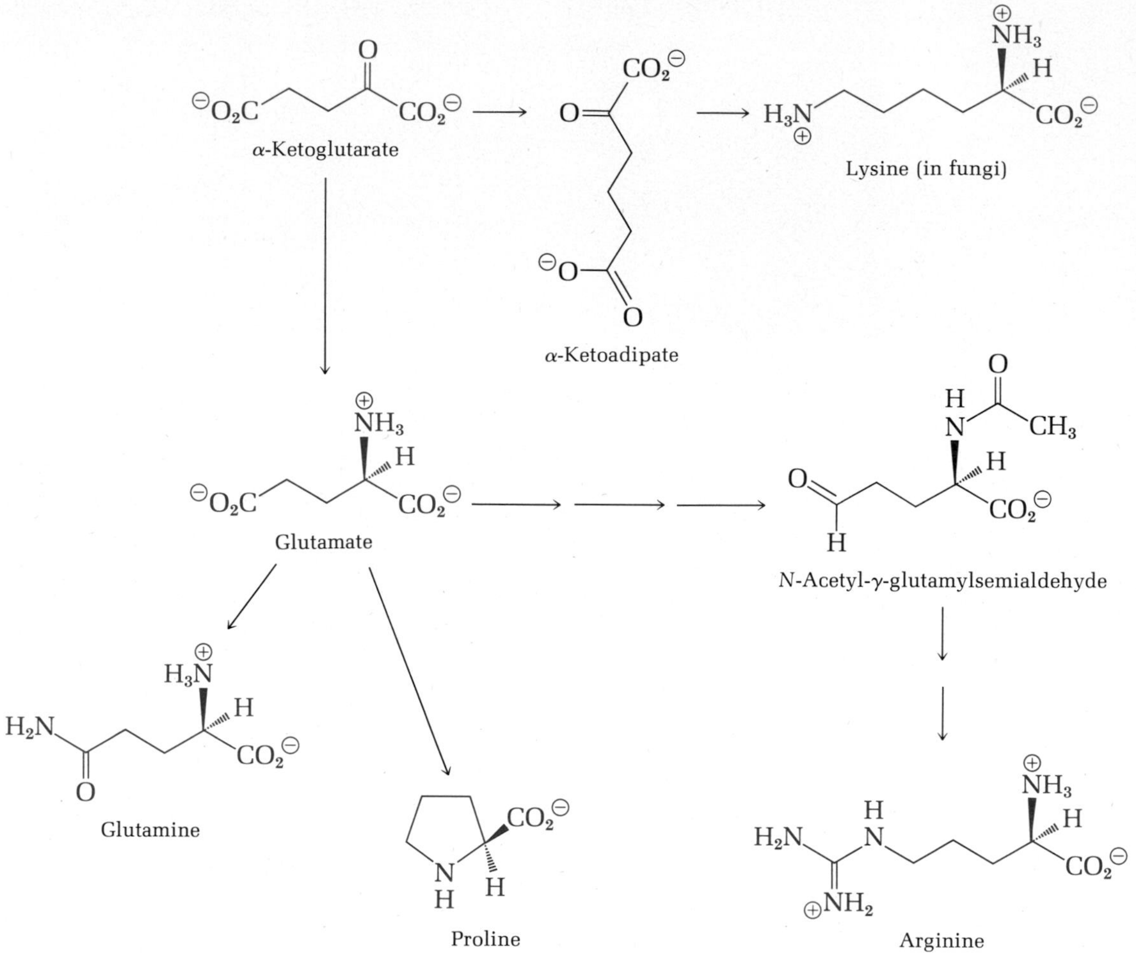

Figure 22.4 Pathways for the biosynthesis of the glutamate family of amino acids.

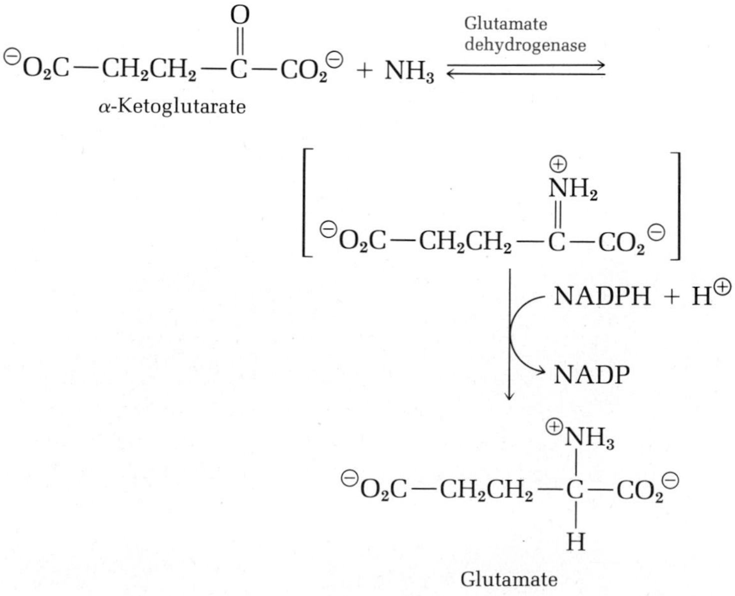

$$(22.12)$$

The conversion of the γ-carboxylate group of glutamate to an amide occurs via an acyl phosphate intermediate; the phosphate group is supplied by ATP. Displacement of phosphate by ammonia produces glutamine (reaction 22.14).

$$^{\ominus}O_2C-CH_2CH_2\overset{\overset{\displaystyle \oplus NH_3}{|}}{\underset{\underset{\displaystyle H}{|}}{C}}-CO_2^{\ominus} + NH_3 + ATP \xrightarrow{\text{Glutamine synthetase}}$$

Glutamate

$$\overset{\overset{\displaystyle O}{\|}}{H_2N-C}-CH_2CH_2\overset{\overset{\displaystyle \oplus NH_3}{|}}{\underset{\underset{\displaystyle H}{|}}{C}}-CO_2^{\ominus} + H_2O + ADP \qquad (22.13)$$

Glutamine

$$\text{Glutamate structure} + ATP \xrightarrow{\text{Glutamine syntheotase}}$$

$$\left[\overset{\overset{\displaystyle O \quad\quad O}{\| \quad\quad \|}}{^{\ominus}O-P-O-C} \cdots \overset{\oplus NH_3}{\underset{CO_2^{\ominus}}{}} \right] + ADP$$

Acyl phosphate intermediate

Nucleophilic substitution by addition-elimination $\overset{..}{N}H_3$

$$HPO_4^{\scriptsize\textcircled{2}\ominus} + H_2N\overset{\overset{\displaystyle O}{\|}}{-C} \cdots \overset{\oplus NH_3}{\underset{CO_2^{\ominus}}{}} \qquad (22.14)$$

Glutamine

We can discover the purpose of the phosphorylation step in reaction 22.14 by considering the second half-reaction, the displacement of phosphate by ammonia. The sole purpose of the phosphorylation of the acyl γ-carbon of glutamate is to convert the poor leaving group, $O^{\scriptsize\textcircled{2}\ominus}$, to an energy-rich, labile intermediate whose phosphate moiety is an excellent leaving group.

a. Regulation of *E. coli* Glutamine Synthetase by Covalent Modification

E. coli glutamine synthetase is composed of 12 identical subunits and has a total molecular weight of 600,000. Covalent modification radically alters its activity by raising its K_m for substrates. The covalently modified enzyme is also much more susceptible to negative feedback inhibition (discussed in the next section) than the unmodified enzyme. Fully active glutamine synthetase is deactivated by *glutamine synthetase adenylyl transferase* (ATase) by transfer of an adenylyl group from ATP to a tyrosyl residue on each of its 12 subunits (Figure 22.5). The subunits of glutamine synthetase are arranged in two stacked hexagonal arrays, and adenylylation is believed to occur on each of the subunits as indicated in Figure 22.6. The enzyme is progressively inactivated by adenylylation. The standard free-energy change for adenylyl transfer is -4.2 kJ mol^{-1}, so the adenylylation reaction is read-

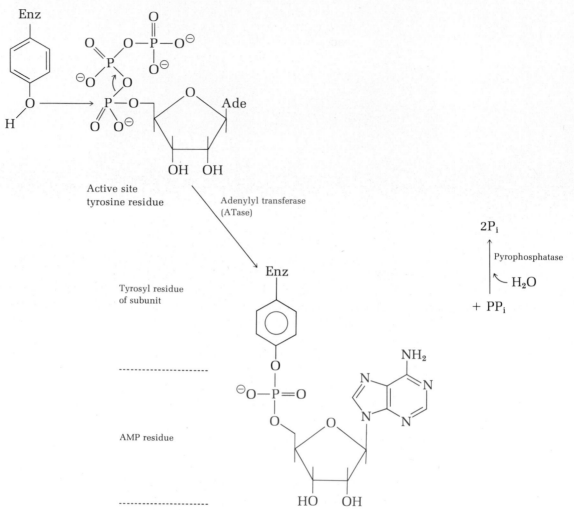

Figure 22.5 Inactivation of glutamine synthetase by ATase-catalyzed adenylylation of tyrosyl residues.

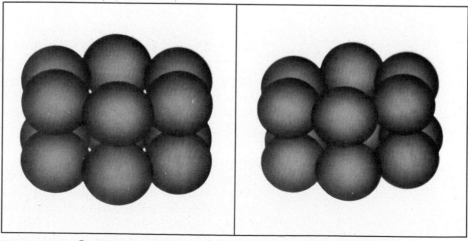

Figure 22.6 Stereo representation of the subunit structure of glutamine synthetase.

ily reversible. Adenylylation, however, is accompanied by release of pyrophosphate, and hydrolysis of pyrophosphate by pyrophosphatase pulls the reaction to completion.

Deadenylylation and reactivation of glutamine synthetase is also catalyzed by glutamine synthetase adenylyl transferase (reaction 22.15).

$$\text{glutamine synthetase-AMP} + \text{HPO}_4{}^{2\ominus} \underset{}{\overset{\text{ATase}}{\rightleftharpoons}} \text{glutamine synthetase}$$
(Inactive, adenylated enzyme)   (Active)
$$+ \text{ADP} \qquad (22.15)$$

The reaction is readily reversible, since $\Delta G^{\circ\prime} = -4.2 \text{ kJ mol}^{-1}$. The net result of adenylylation and deadenylylation is pyrophosphorolysis of ATP (reaction 22.16).

$$\text{ATP} + \text{HPO}_4{}^{2\ominus} \underset{}{\overset{\text{ATase}}{\rightleftharpoons}} \text{ADP} + \text{P}_2\text{O}_7{}^{4\ominus}$$
$$\Delta G^{\circ\prime} = -8.4 \text{ kJ mol}^{-1} \qquad (22.16)$$

The activity of ATase is regulated by glutamate, glutamine, and ATP. At high concentrations of α-ketoglutarate, the deadenylylation activity of ATase increases. Conversely, high concentrations of glutamine and ATP stimulate adenylylation.

The activity of glutamine synthetase adenylyl transferase depends upon another regulatory protein, designated P. This dimeric protein exists in two forms, designated P_A and P_D. The binding of P_A to glutamine synthetase adenylyl transferase yields a complex that catalyzes adenylylation. The complex of P_D and glutamine synthetase adenylyl transferase catalyzes deadenylylation. As we proceed into this labyrinth of metabolic regulation, we find that P_A is converted to P_D by *uridylylation* of a tyrosyl residue of both subunits of the regulatory protein. The attachment of UMP to P_D is catalyzed by a uridylyl transferase (UTase), which requires ATP, Mg(II) or Mn(II), and α-ketoglutarate for activity (reaction 22.17).

$$P_A + 2 \text{ UTP} \xrightarrow[\substack{\alpha\text{-Ketoglutarate,} \\ \text{ATP, Mg(II) or Mn(II)}}]{\text{UTase}} P_D(\text{UMP})_2 + 2 \text{ PP}_i \qquad (22.17)$$

The UMP moiety is removed from P_D by hydrolysis by the uridylyl removal enzyme (UR enzyme) (reaction 22.18).

$$P_D(\text{UMP})_2 + 2\text{H}_2\text{O} \xrightarrow[\substack{(+) \text{ Glutamine} \\ (-) \alpha\text{-Ketoglutarate} \\ (-) \text{ ATP}}]{\text{UR enzyme}} P_A + 2 \text{ UMP} \qquad (22.18)$$

The activities of the uridylyl transferase and the uridyl removal enzyme are regulated by ATP, α-ketoglutarate, and glutamine. The first two of these allosteric effectors activate uridylyl transferase, and glutamine activates uridyl removal enzyme. Allosteric regulation of glutamine synthetase adenylyl transferase and the uridylyl transferase–UR enzyme pair is summarized in Table 22.2. A schematic diagram of the regulation of glutamine synthetase is shown in Figure 22.7.

b. Allosteric Regulation of *E. coli* Glutamine Synthetase

E. coli glutamine synthetase responds to many feedback inhibitors: tryptophan, histidine, CTP, AMP, glucosamine-6-phosphate, carbamoyl phosphate, serine, glycine, and alanine. The inhibitors act by *cumulative feedback inhibition* of glutamine synthetase (Table 22.3). Each of the twelve monomers of glutamine synthetase has a separate binding site for each of the

Table 22.2 Effects of metabolites and the alternate forms of P on the activities of ATase and UT-URase

Effector	ATase catalyzed		UT-URase catalyzed		
			Uridylylation	Deuridylylation	
	Adenylylation	Deadenylylation		With Mg^{2+}	With Mn^{2+}
Mg^{2+} or Mn^{2+}	R	R	R		
P_A	+	N	R		
P_D	N	+		R	R
ATP	R	+	+	R	N
α-Ketoglutarate	−	+	+	R	N
Glutamine	+	−	−	N	N
UTP	−	+	R		
P_i	−	R^a	−	N	N
pH optimum	8.0	7.2	7.6	9.0	9.0

Abbreviations: R, required; N, no effect; −, inhibitor; +, activator.
^{a} P_i inhibits at high concentrations.
SOURCE: From E. R. Stadtman and A. Ginsburg, "Glutamine Synthetase of *E. coli*," in P. D. Boyer, Ed., "The Enzymes," 3rd ed., Vol. 10, Academic Press, New York, 1974, p. 790.

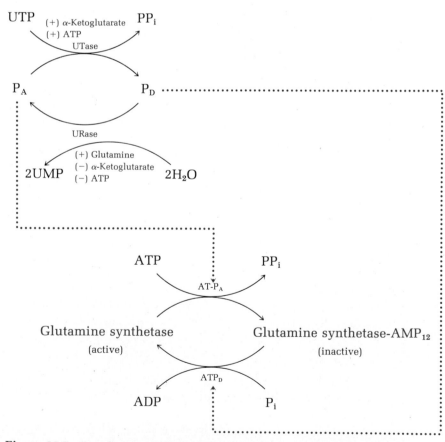

Figure 22.7 Regulation of glutamine synthetase by covalent modification through a cascade of enzyme activity. The complex of ATase and the regulatory protein P_A (AT–P_A) catalyzes adenylylation and inactivation of glutamine synthetase. The complex of ATase and P_D (AT–P_D) catalyzes the pyrophosphorolysis of the adenylyl group, reactivating glutamine synthetase. Allosteric activation is indicated by (+); inhibition is indicated by (−).

Table 22.3	Cumulative feedback inhibition of Glutamine synthetase from *E. coli*	

	Percent inhibition	
Inhibitors added	Observed	Calculated
Tryptophan (A)	84	
Carbamoyl phosphate (B)	86	
Glycine (C)	87	
AMP (D)	59	
(E) L-Alanine	52	
CTP (F)	37	
A + B	72	72
A + B + C	60	63
A + B + C + D	41	37
A + B + C + D + E	22	19
A + B + C + D + E + F	8	7

SOURCE: From E. R. Stadtman and A. Ginsburg, in P. D. Boyer, Ed., "The Enzymes," 3rd ed., Vol. 10, Academic Press, New York, 1974, p. 778.

inhibitors. The inhibitors are products of biosynthetic processes that require the amide nitrogen of glutamine (Figure 22.8). The biosynthesis of all the inhibitors therefore depends upon the activity of glutamine synthetase.

C. Proline The biosynthesis of proline occurs in three steps from glutamate (Figure 22.9). Proline's five carbon atoms are derived from glutamate, and the α-amino group of glutamate becomes its ring nitrogen. The first reaction of the pathway is conversion of glutamate to γ-glutamyl phosphate by *γ-glutamyl kinase*. This is the committing step of proline biosynthesis, and γ-glutamyl kinase is subject to feedback inhibition by proline. Reduction of γ-glutamyl phosphate by an NADPH-dependent dehydrogenase yields L-glutamate-γ-semialdehyde. The first reactions of proline biosynthesis are chemically analogous to the conversion of D-glyceraldehyde 3-phosphate to 1,3-diphosphoglycerate in glycolysis. The aldehyde group of L-glutamate-γ-semialdehyde is susceptible to nucleophilic attack by the α-amino group, and a rapid nonenzymatic reaction produces $L-\Delta^1$-pyrolline-5-carboxylate. Reduction of the imine by an NADPH-dependent dehydrogenase gives proline.

D. Arginine The first step in the biosynthesis of arginine from glutamate is the conversion of the γ-carboxyl group of glutamate to an aldehyde (see Figure 22.9). We recall that the product, L-glutamate-γ-semialdehyde, spontaneously cyclizes.

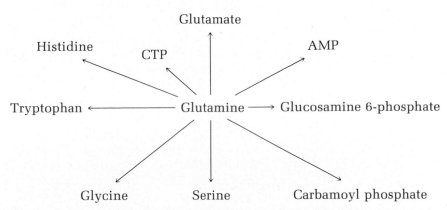

Figure 22.8 Products that derive their nitrogen from the amide nitrogen of glutamine and are also inhibitors of *E. coli* glutamine synthetase.

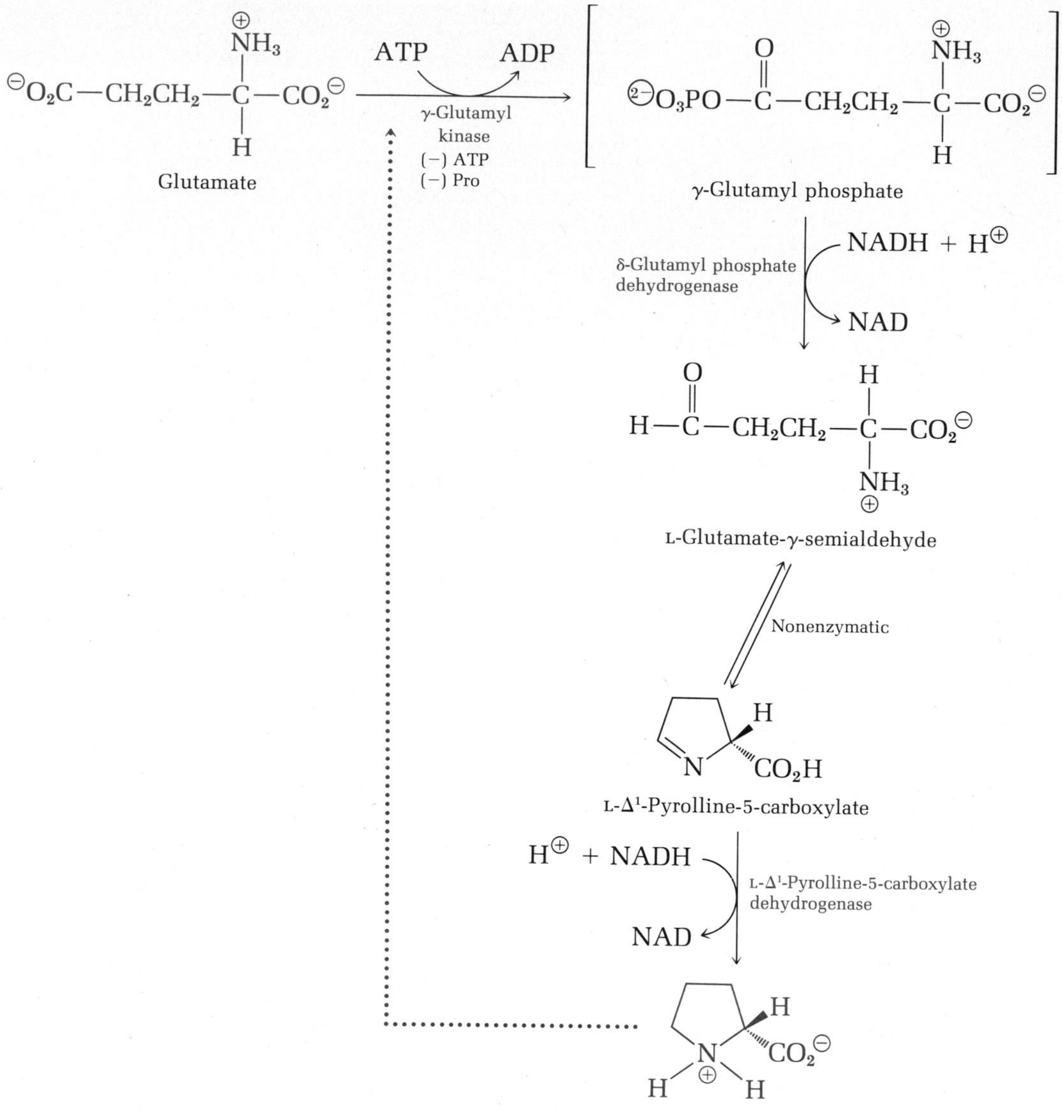

Figure 22.9 Pathway for the biosynthesis of proline. The first step of the pathway is inhibited allosterically by the end product of the pathway, proline itself.

To prevent this unwanted reaction, the cell blocks the α-amino group of the semialdehyde with an acetyl group derived from acetyl-CoA. Arginine is a feedback inhibitor of this step, the first *unique* step of its own biosynthesis. N-Acetylglutamate-δ-semialdehyde is transaminated to give α-N-acetylornithine, and hydrolysis of the acetyl groups gives ornithine (Figure 22.10). In Section 21.3 (Figure 21.5) we saw that ornithine is converted to arginine in the urea cycle.

E. Lysine in Fungi

Fungi derive their lysine from glutamate. Transamination of glutamate to α-ketoglutarate is followed by reactions that elongate the carbon chain by a single methylene group (Figure 22.11). These reactions parallel the conversion of pyruvate to α-ketoglutarate in the citric acid cycle. We shall encounter this sequence of reactions again when we consider the biosynthesis of leucine in the pyruvate family. The product of chain elongation is α-ketoadipate, an intermediate in lysine and tryptophan catabolism (Section 21.4F). Transamination of α-ketoadipate gives α-amino adipate. The ATP-dependent reduction of the ϵ-carboxylate group to α-aminoadipate-δ-semialdehyde parallels a similar reduction in the biosynthesis of proline.

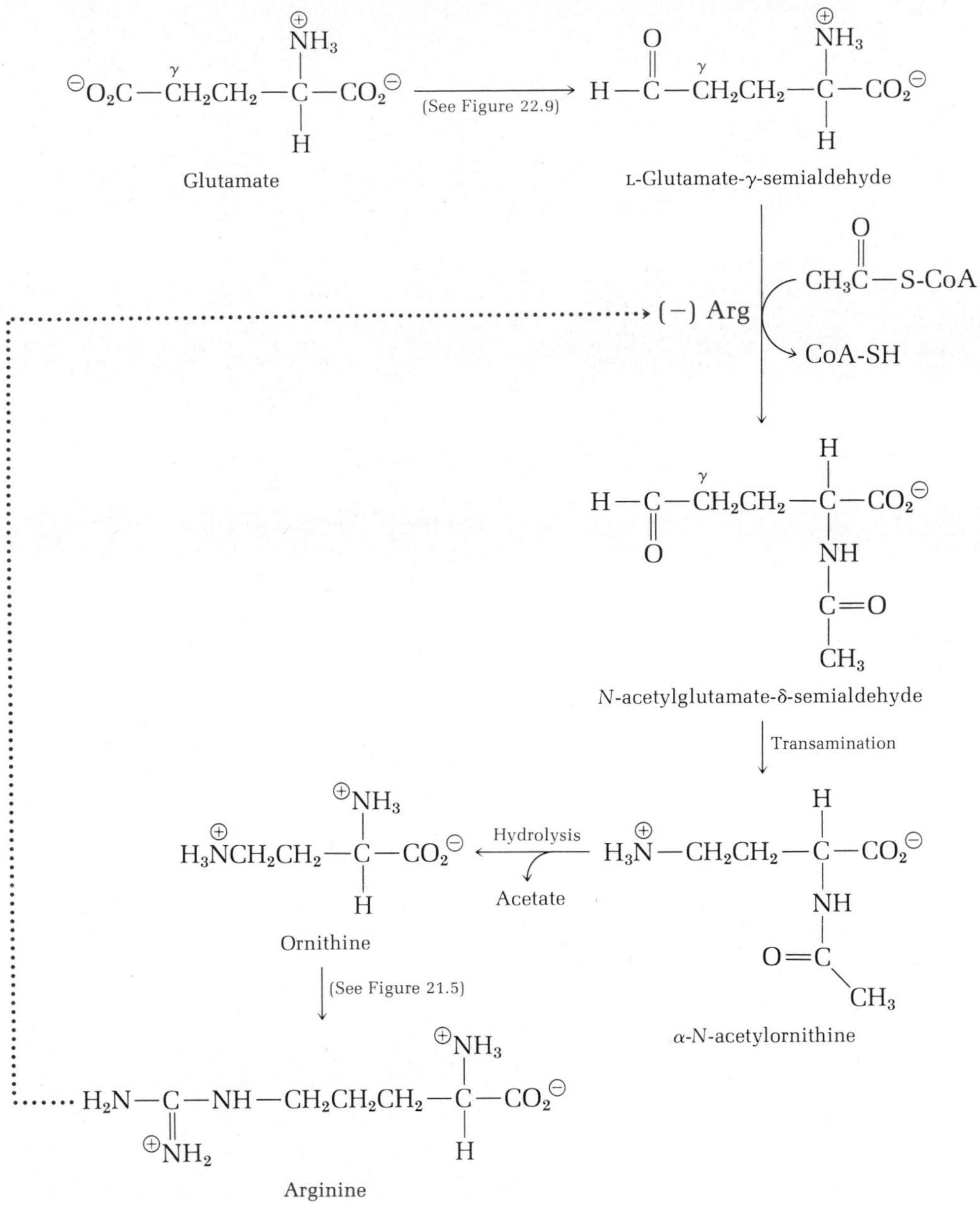

Figure 22.10 Pathway for the biosynthesis of arginine. Arginine inhibits the conversion of L-glutamate-γ-semialdehyde to N-acetylglutamate by negative feedback inhibition.

The semialdehyde is converted to lysine by way of saccharopine (Figure 22.12), a reversal of the pathway for catabolism of lysine.

22–5 BIOSYNTHESIS OF THE SERINE FAMILY

The serine family includes serine, glycine and cysteine.[1]

Many biosynthetic processes begin with either serine or glycine as their starting material (Figure 22.13). The tetrahydropyrrole ring system that forms the nucleus for porphyrins, chlorophylls, coenzyme B_{12}, the bile acids, and phospholipids are all derived from either glycine or serine. The tetrahydrofolate-dependent conversion of serine to glycine also provides a one-carbon pool that is used in the biosynthesis of purine nucleotides.

[1] The sulfur of methionine is derived from cysteine in plants and microorganisms, so methionine might be considered a member of this family. For our purposes, however, methionine is more conveniently discussed as a member of the aspartate family.

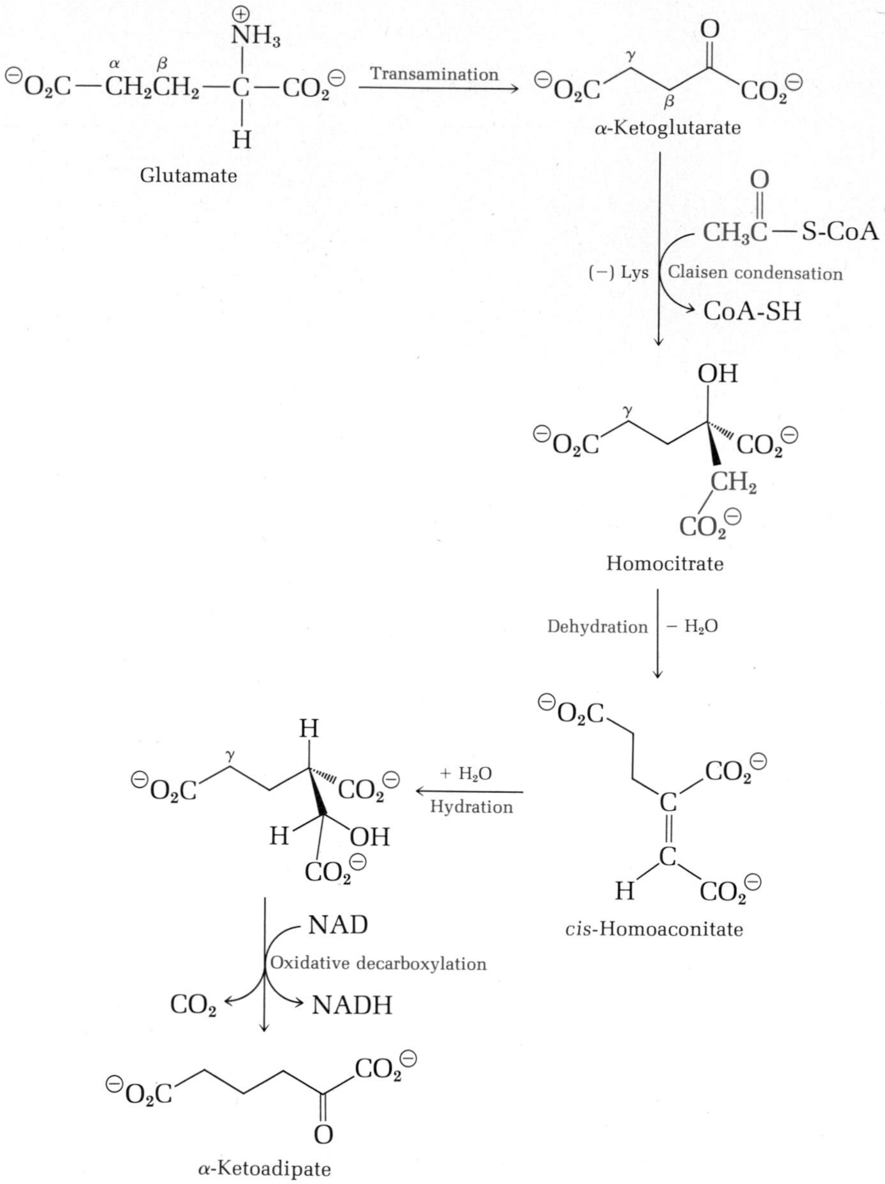

Figure 22.11 Conversion of glutamate to α-ketoadipate, part of the pathway for biosynthesis of lysine in fungi.

A. Sources of Sulfur for Cysteine Biosynthesis

The reductions of sulfate and sulfite to sulfide are two of the major sources of sulfur for cysteine biosynthesis. The reduction of sulfate to sulfite requires activation of the sulfate group by *ATP-sulfurylase*, an adenylyl transferase that catalyzes the formation of adenosine 5'-*phosphosulfate* (APS; also called adenylyl sulfate).

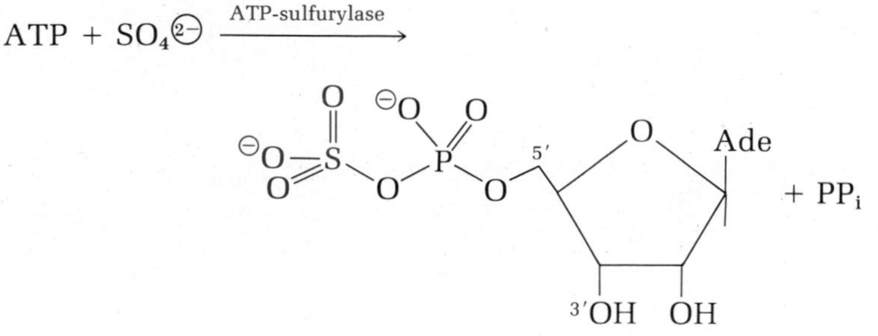

$$ATP + SO_4^{2\ominus} \xrightarrow{\text{ATP-sulfurylase}}$$

$+ PP_i$ (22.19)

Adenosine 5'-phosphosulfate (APS, or adenylyl sulfate)

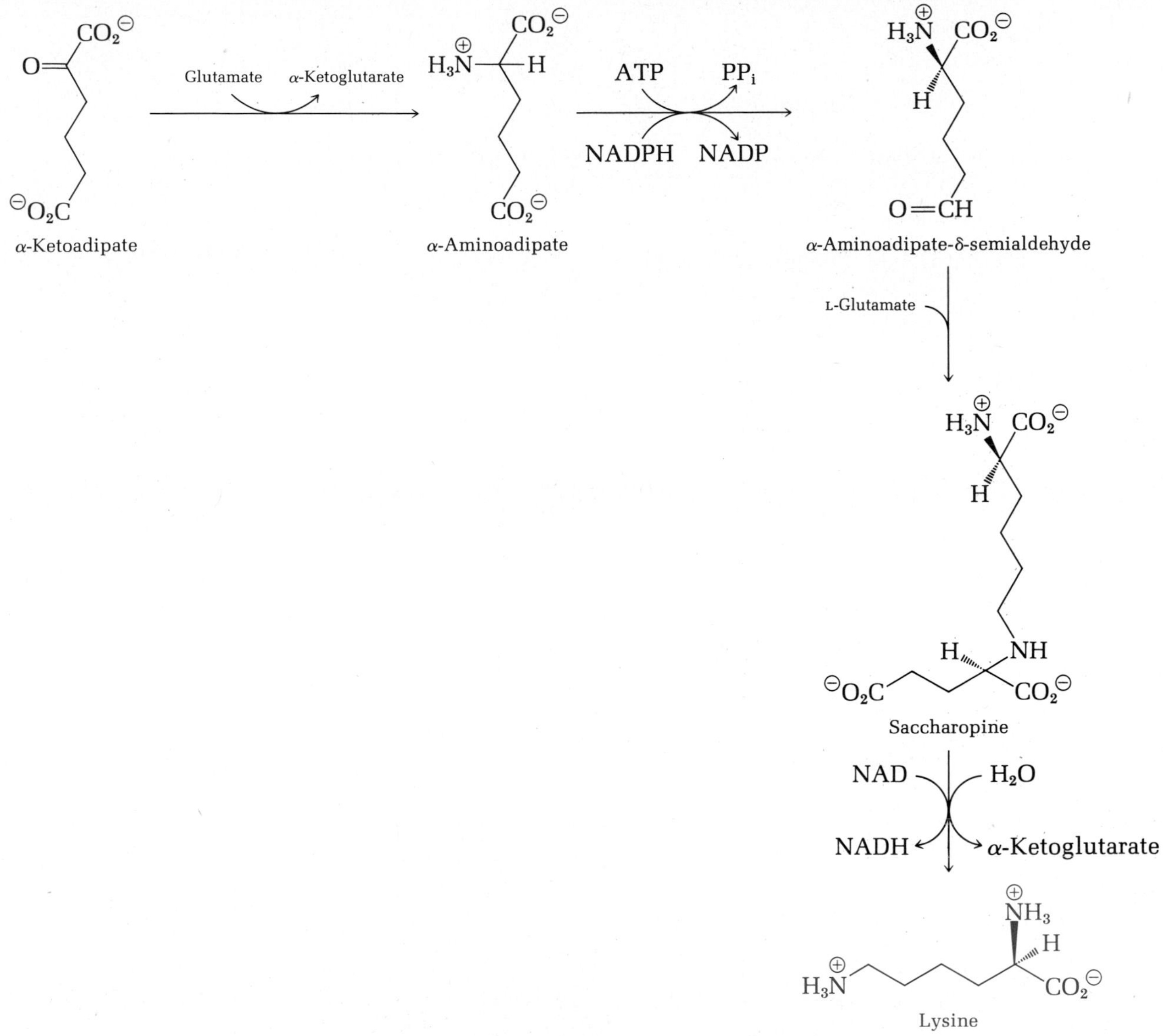

Figure 22.12 · Conversion of α-ketoadipate to lysine in fungi.

APS is reduced to sulfite by *APS-reductase*, a flavin-containing iron-sulfur protein. The electrons required for reduction are provided by a *c*-type cytochrome in *Desulfovibrio vulgaris* (reaction 22.20). An N^5-sulfite adduct of the flavin is formed during the reaction. This adduct is subsequently reduced.

The six-electron reduction of sulfite to sulfide is catalyzed by *sulfite reductase* (reaction 22.21).

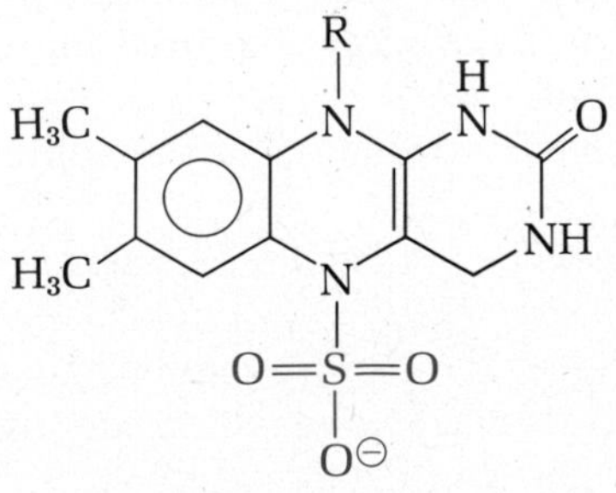

N^5-Sulfite adduct of flavin

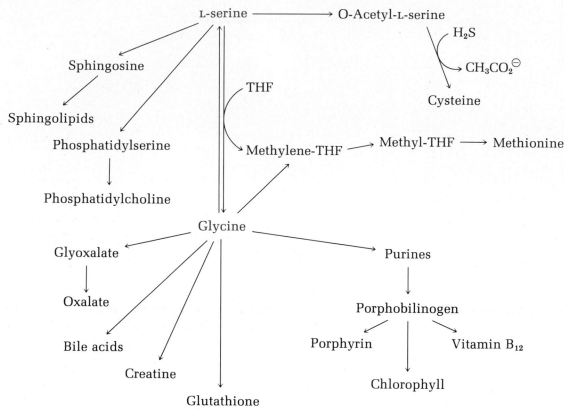

Figure 22.13 Biosynthetic pathways in which serine and glycine are important starting materials or intermediates.

$$\text{APS} + 2 \text{ cyt } c(\text{Fe}^{2+}) \xrightarrow{\text{APS-reductase}}$$

$$\text{AMP} + \text{cyt } c(\text{Fe}^{3+}) + \underset{\text{Sulfite}}{\overset{O^{-}}{\underset{^{-}O}{\overset{|}{\underset{\oplus}{:S}}}}O^{-}} \qquad (22.20)$$

$$\text{SO}_3^{2-} + 6 \text{ e}^{-} + 6\text{H}^{+} \xrightarrow{\text{Sulfite reductase}} \text{S}^{2-} + 3\text{H}_2\text{O} \qquad (22.21)$$

E. coli sulfite reductase has a molecular weight of 670,000 and contains 12 subunits, with subunit composition $\alpha_8\beta_4$. Each α subunit contains one FAD and one FMN, and each β subunit contains one iron-sulfur center (Fe_4S_4) and one siroheme prosthetic group. NADPH is the source of electrons for the reaction, and 3 moles are required for the complete reduction. Sulfite binds to the siroheme until it has been completely reduced. Electrons pass from NADPH to the siroheme-sulfite complex as shown in reaction 22.22.

$$\text{NADPH} \longrightarrow \text{FAD} \longrightarrow \text{FMN} \longrightarrow \text{Fe}_4\text{S}_4 \longrightarrow \text{heme} \longrightarrow \text{SO}_3^{2-} \quad (22.22)$$

Sulfate can also be reduced to sulfide by a pathway in which APS is first converted to *3'-phospho-5'-adenylyl sulfate* (PAPS) by phosphorylation of the 3' position catalyzed by APS kinase (reaction 22.23).

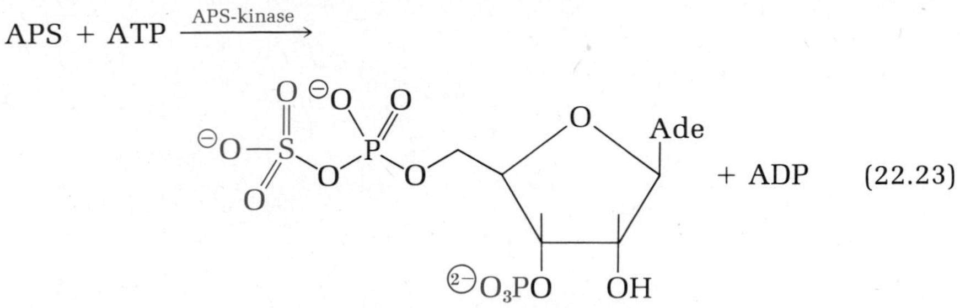

$$\text{APS} + \text{ATP} \xrightarrow{\text{APS-kinase}}$$

$$+ \text{ADP} \qquad (22.23)$$

3′-Phospho-5′-adenyl sulfate (PAPS)

The sulfite group of PAPS is then transferred to a sulfhydryl group of thioredoxin yielding *thioredoxin thiosulfonate*. *Sulfite reductase* then reduces the thiosulfonate to a *persulfide* which is the sulfur donor in cysteine biosynthesis (Figure 22.14).

B. Serine and Glycine The glycolytic intermediate 3-phosphoglycerate is the starting point of serine and glycine biosynthesis (Figure 22.15). The synthesis of serine requires three steps: (1) oxidation of the 2-hydroxyl group to a keto group, (2) transamination of the keto group of an amino group, and (3) hydrolysis of the 3-phosphate ester. Serine hydroxymethylase (serine hydroxymethyltransferase or serine transhydroxymethylase) (Section 10.8) converts serine to glycine in a single step.

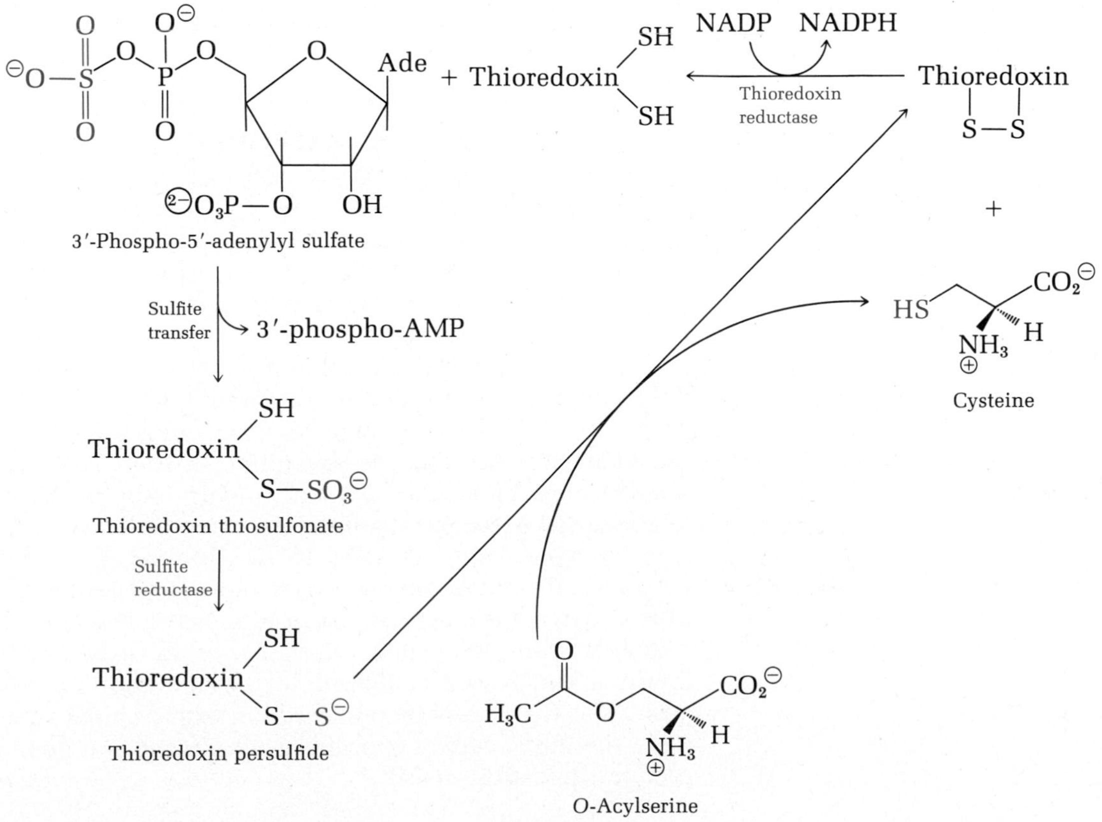

Figure 22.14 Role of thioredoxin in cysteine biosynthesis.

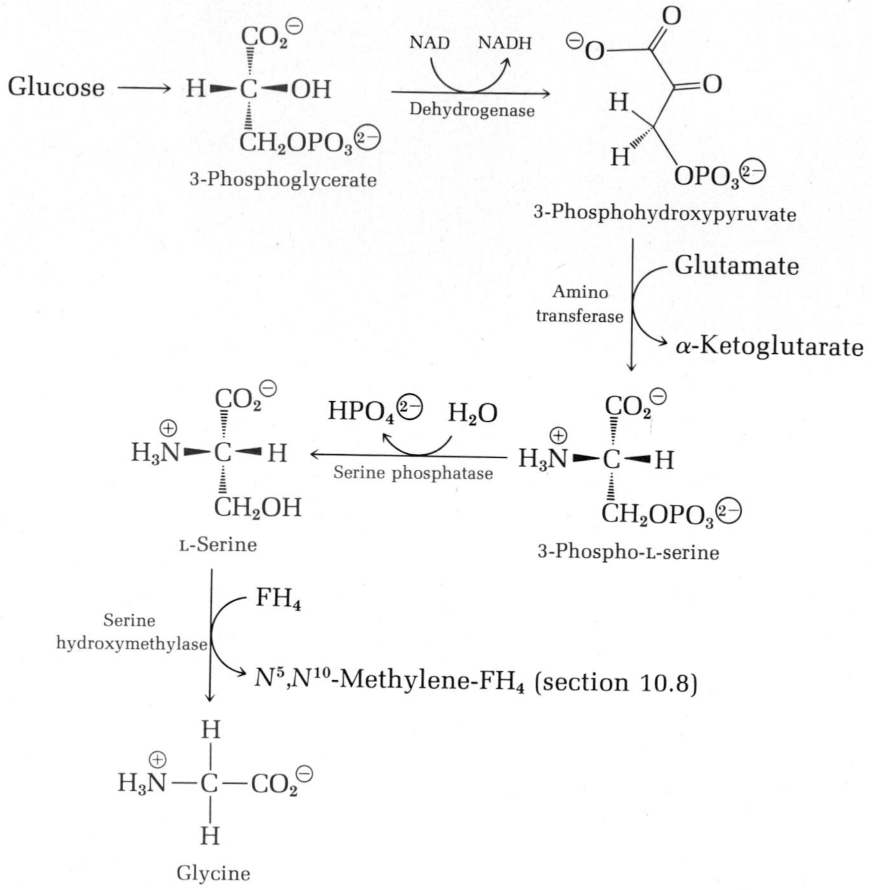

Figure 22.15 Pathway for the biosynthesis of glycine and serine.

C. Cysteine

The pathway for biosynthesis of cysteine consists of two steps. First, serine is converted to O-acylserine by *serine acyl transferase*. Acetyl-CoA is the acyl donor for this reaction. The acylation reaction converts the hydroxy group of serine to an excellent leaving group, acetate. Cysteine is a feedback inhibitor of serine acyl transferase. Cysteine biosynthesis is also controlled at the level of protein synthesis. Both cysteine and sulfide inhibit the synthesis of serine acyl transferase.

Second, sulfide displaces acetate to give cysteine (Figure 22.16). The conversion of O-acetylserine to cysteine can be regarded as thiolysis of an ester. In most cases cleavage of esters occurs by an addition-elimination mechanism. In this case, however, the cleavage of the ester occurs by an S_N2 displacement of acetate by attack of sulfide upon the methylene carbon of O-acylserine in an S_N2 displacement.

D. Creatine

Glycine and arginine are the carbon and nitrogen sources of *creatine*, a metabolite whose phosphate derivative is an energy storage molecule in muscle.[2] The first step in creatine biosynthesis is *amidino* group transfer from arginine to glycine catalyzed by *glycine-amidine transferase*. Amidino group transfer occurs by nucleophilic attack by the amino group of glycine upon the imine carbon of arginine. Ornithine and guanidinoacetate are the products (reaction 22.24).

---

[2] The "amino acids" creatine and ornithine are *not* incorporated in proteins.

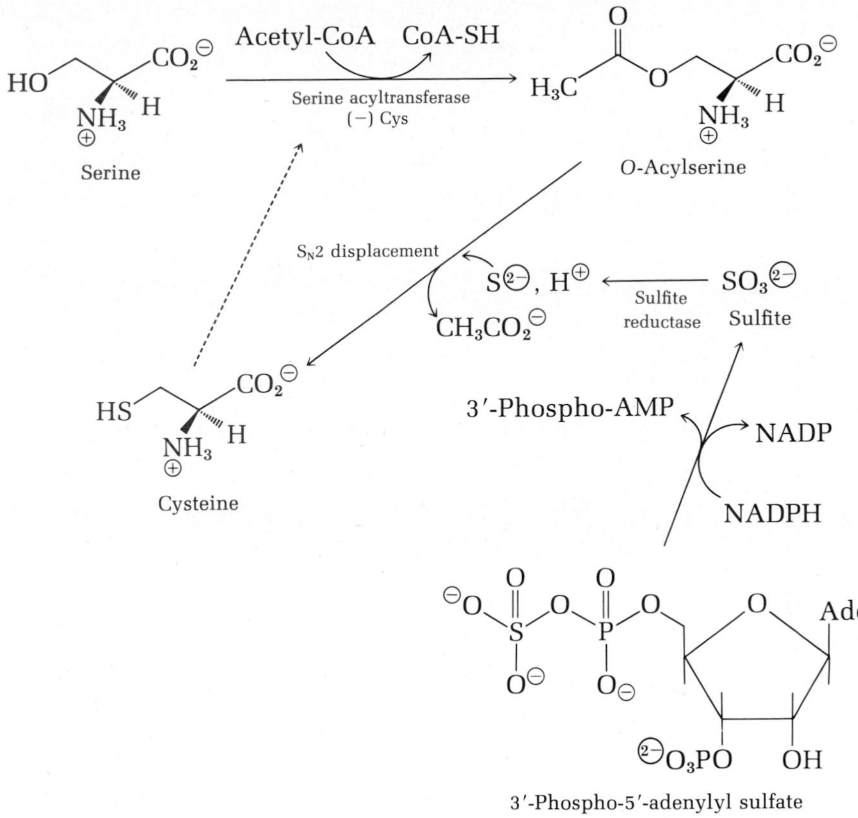

Figure 22.16 Pathway for the biosynthesis of cysteine. Cysteine is a negative feedback inhibitor of serine acyltransferase. Both cysteine and sulfide inhibit the biosynthesis of the acyltransferase at the genetic level of metabolic control.

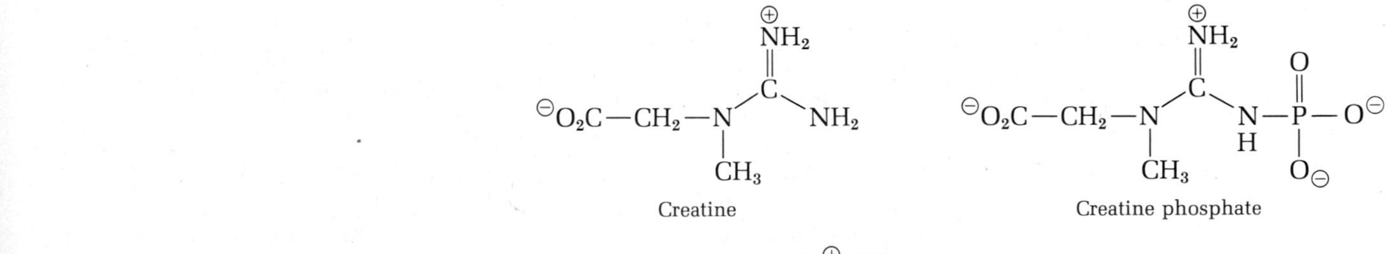

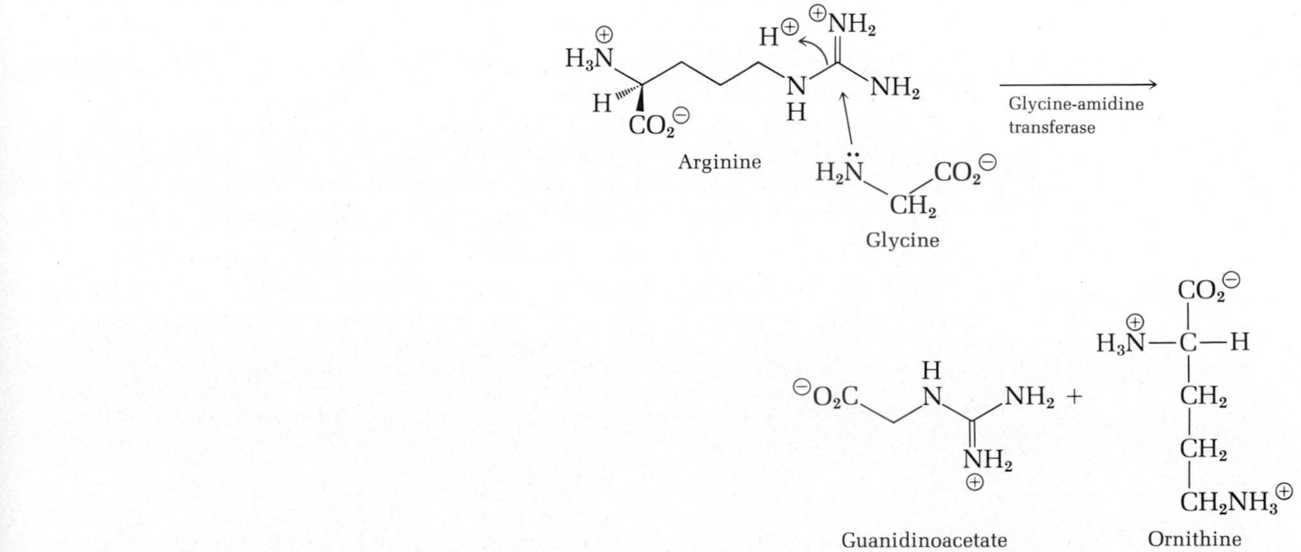

Methyl group transfer from S-adenosylmethionine to guanidinoacetate produces creatine (reaction 22.25).

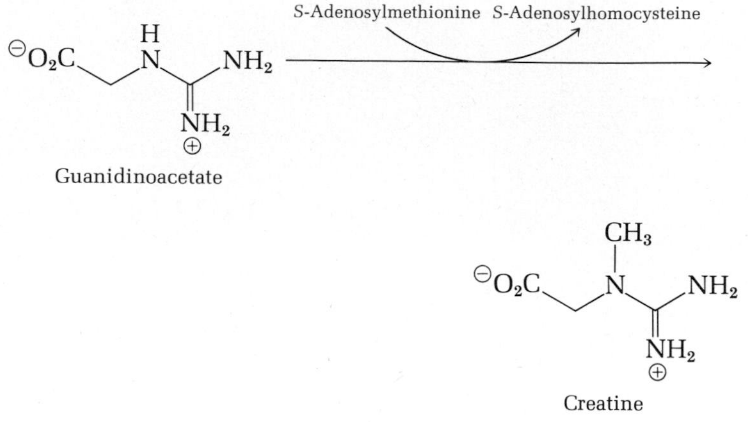

Guanidinoacetate

Creatine   (22.25)

Phosphoryl group transfer from ATP to creatine catalyzed by *creatine kinase* gives the energy-rich metabolite creatine phosphate (reaction 22.26).

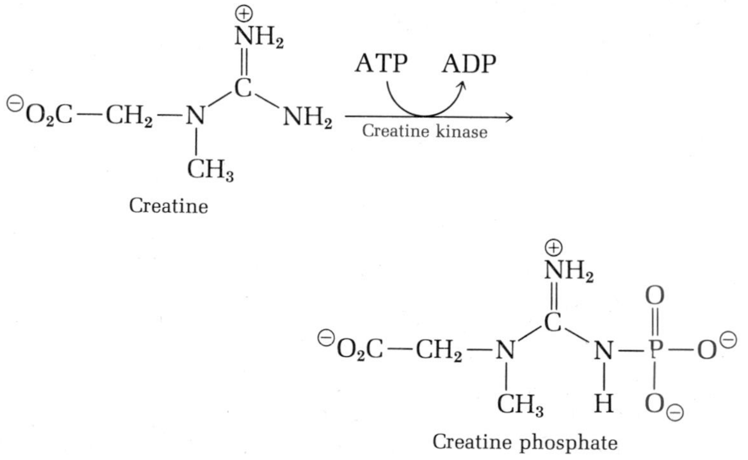

Creatine

Creatine phosphate   (22.26)

E. Polyamines   In animals and microorganisms the polyamines *spermine* and *spermidine* are derived from *putrescine* and S-adenosylhomocysteine, the latter being derived from decarboxylation of S-adenosylmethionine. An amino group of putrescine displaces 5′-thiomethylribose to give spermidine. Spermidine then reacts again with S-adenosylhomocysteine to give spermine (Figure 22.17).

The polyamines bind tightly to DNA, and they have regulatory roles in DNA replication, protein synthesis, and cell division. We have already noted that fully protonated spermine, whose charge is $+4$, binds to phosphate groups in tRNA, stabilizing the tertiary structure of the nucleic acid.

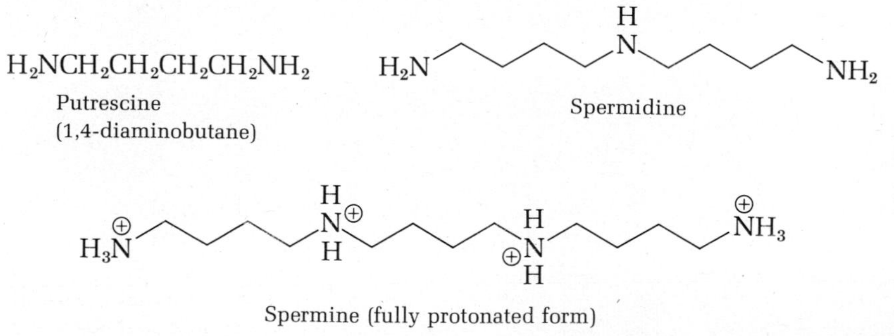

$H_2NCH_2CH_2CH_2CH_2NH_2$
Putrescine
(1,4-diaminobutane)

Spermidine

Spermine (fully protonated form)

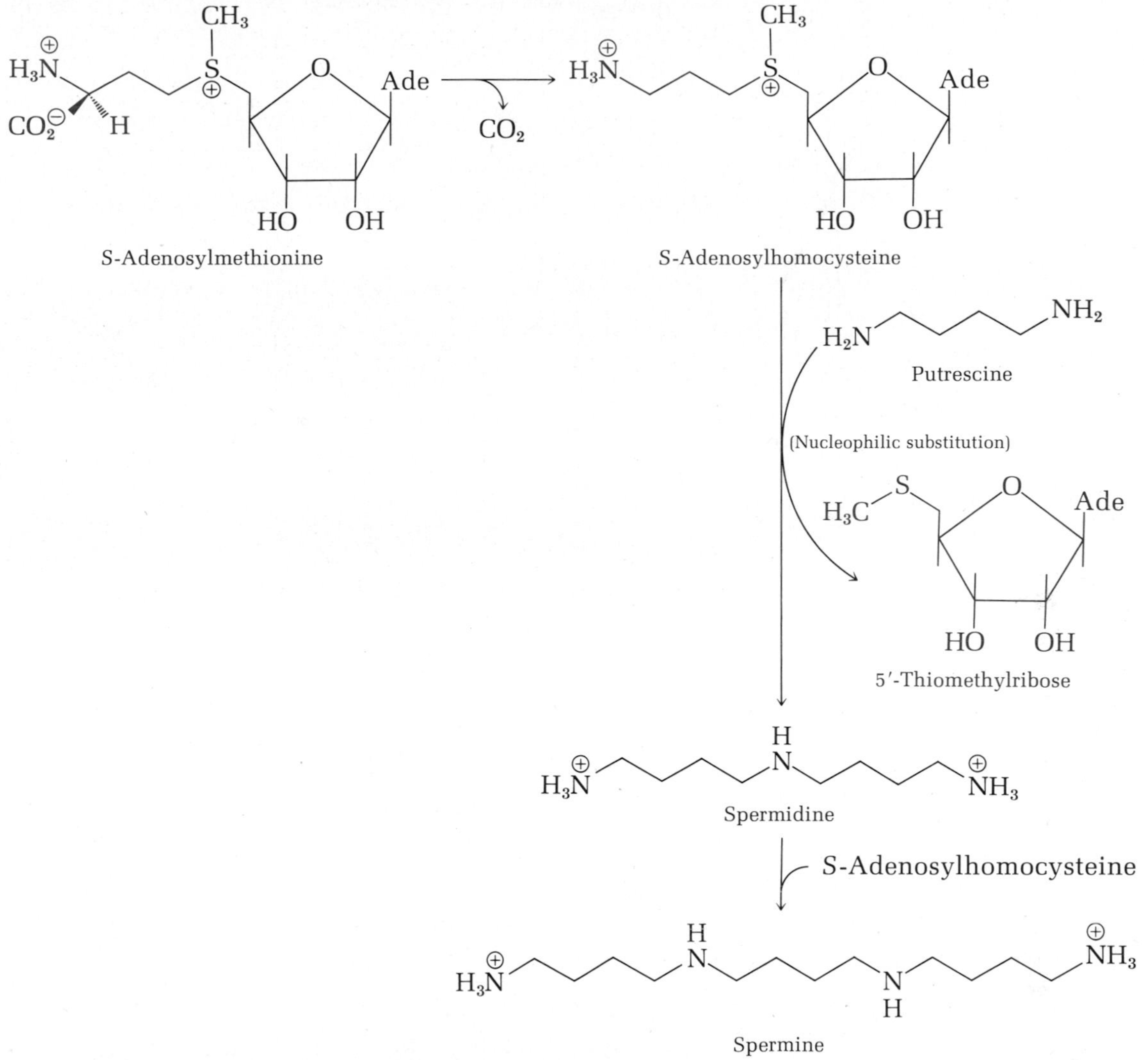

Figure 22.17 Pathway for the biosynthesis of the polyamines.

22–6 BIOSYNTHESIS OF THE ASPARTATE FAMILY

The aspartate family of amino acids consists of aspartate, asparagine, threonine, methionine, lysine (except in fungi), and isoleucine[3]. Aspartate derives its carbon framework from oxaloacetate by transamination (Section 21.4A). The rest of the family is then derived from aspartate. Aspartate is also the precursor of the pyrimidine nucleotides (Section 23.6).

Metabolic regulation of biosynthesis of amino acids derived from aspartate occurs by allosteric inhibition, inhibition of protein synthesis, or both. Inhibition may occur at the first step or at the production of the branch-point metabolite. All the amino acids derived from aspartate inhibit the first step of the pathway leading to their own formation. The conversion of aspartate to β-aspartyl phosphate catalyzed by aspartokinase is either inhibited allosterically or by inhibition of biosynthesis of aspartokinase. The branch-point metabolite homoserine can be converted to serine, threonine, isoleucine, or methionine. Each of these metabolites inhibits the synthesis of homoserine by allosteric inhibition, inhibition of protein synthesis, or both. The outline of the biosynthetic pathways of the aspartate family of amino acids and the points of metabolic regulation are summarized in Figure 22.18.

[3] Although isoleucine can be regarded as a member of the aspartate family, we shall find it more convenient to discuss isoleucine biosynthesis as part of the pyruvate family.

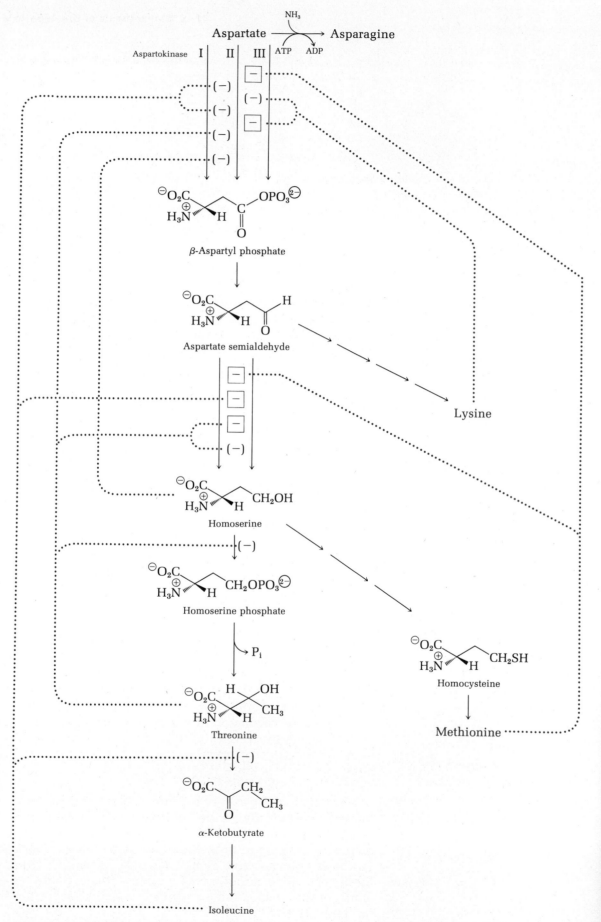

Figure 22.18 Pathways for the biosynthesis of amino acids derived from aspartate. Allosteric regulation by negative feedback inhibition is indicated by (−); inhibition of enzyme synthesis is indicated by ⊟.

A. Aspartate and Asparagine

In *E. coli* aspartate is formed from oxaloacetate by *aspartate-glutamate transaminase* (reaction 22.27).

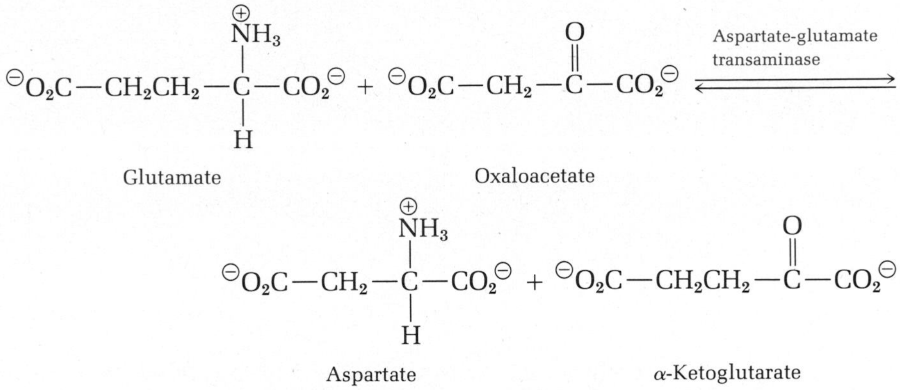

Glutamate   Oxaloacetate

Aspartate   α-Ketoglutarate

(22.27)

Aspartate is converted to asparagine by *asparagine synthetase*. In *E. coli* this ATP-dependent enzyme splits out pyrophosphate in the process of forming a β-aspartyladenylate intermediate. Nucleophilic displacement of adenylate by ammonia yields asparagine (reaction 22.28).

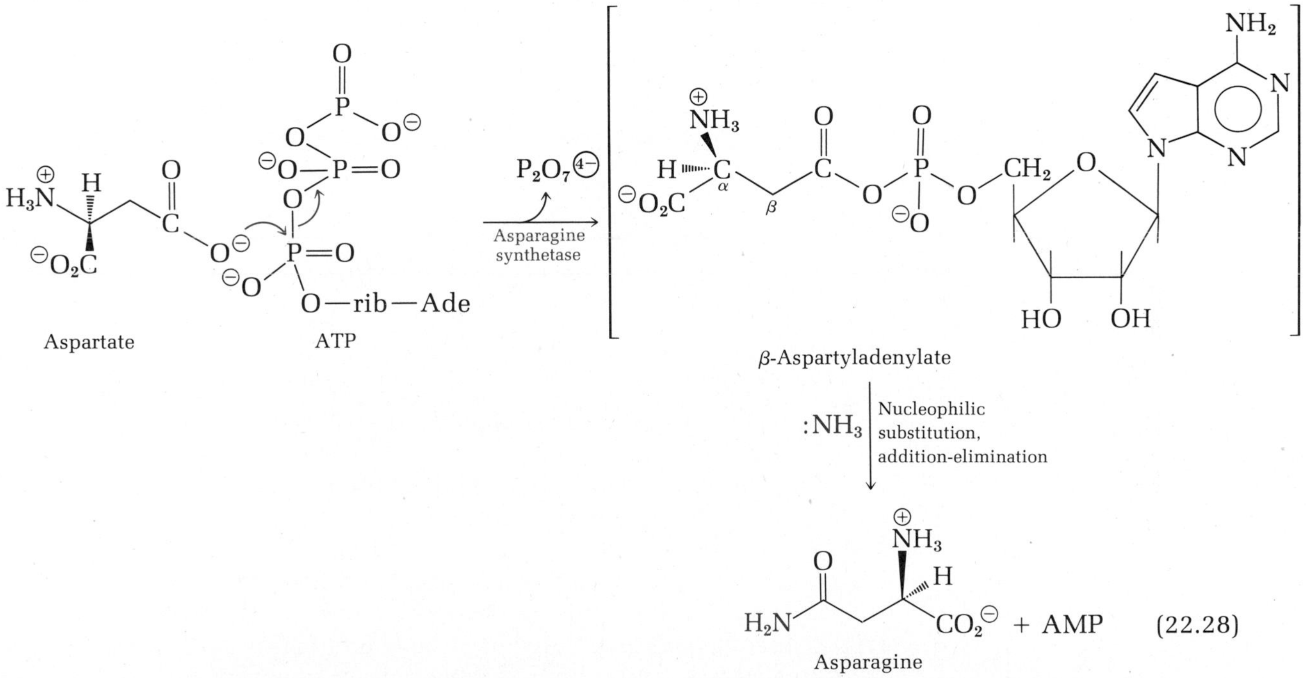

Aspartate   ATP

β-Aspartyladenylate

Asparagine   (22.28)

B. Conversion of Aspartate to the Branch-Point Intermediates β-Aspartate Semialdehyde and Homoserine

The branch-point intermediate β-aspartate semialdehyde is a precursor of lysine, threonine, methionine, and isoleucine. The pathways for biosynthesis of these four amino acids therefore share the two steps that convert aspartate to β-aspartyl semialdehyde.

This intermediate can lead to lysine in seven steps, or it can be converted to homoserine, a second branch-point metabolite that is a precursor of methionine, threonine, and isoleucine.

C. Aspartokinases of *E. coli*

The conversion of aspartate to β-aspartyl phosphate in *E. coli* is catalyzed by three *aspartokinases* (reaction 22.29).

Aspartokinase I is responsible for biosynthesis of threonine, and isoleucine. The enzyme is allosterically inhibited by threonine. Isoleucine and threonine both inhibit biosynthesis of aspartokinase I. This phenomenon is called *gene repression*, and threonine and isoleucine are *repressors*. Methionine represses the biosynthesis of aspartokinase II, but methionine is not

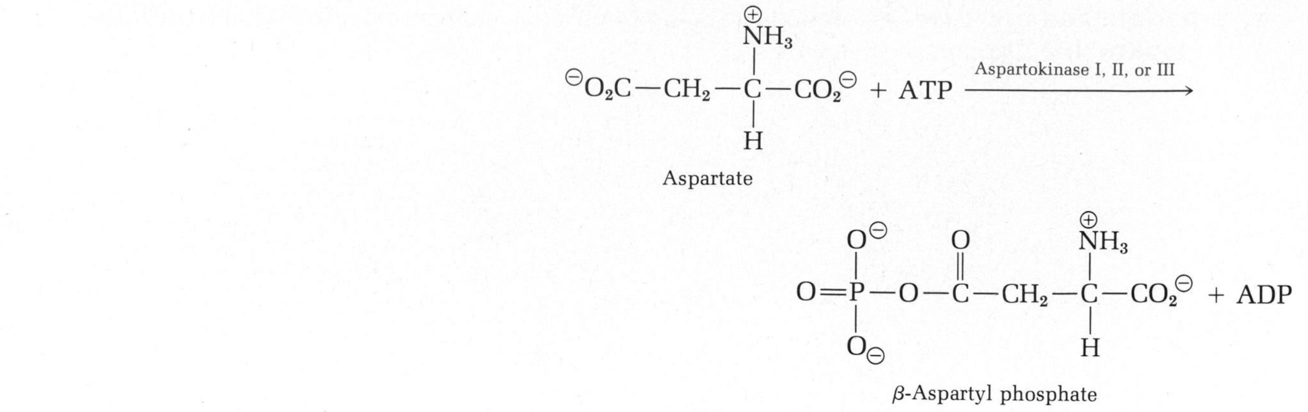

$$\overset{\oplus}{N}H_3$$
$$^{\ominus}O_2C-CH_2-\overset{|}{\underset{|}{C}}-CO_2^{\ominus} + ATP \xrightarrow{\text{Aspartokinase I, II, or III}}$$
$$H$$

Aspartate

$$O=\overset{O^{\ominus}}{\underset{O_{\ominus}}{\overset{|}{P}}}-O-\overset{O}{\overset{\|}{C}}-CH_2-\overset{\overset{\oplus}{N}H_3}{\underset{H}{\overset{|}{C}}}-CO_2^{\ominus} + ADP \qquad (22.29)$$

β-Aspartyl phosphate

an allosteric inhibitor of this enzyme. Lysine inhibits aspartokinase III allosterically by feedback inhibition. It also represses biosynthesis of aspartokinase III.

A single pool of β-aspartyl phosphate serves for the biosynthesis of threonine, isoleucine, methionine, and lysine. Reduction of β-aspartyl phosphate to aspartate semialdehyde is catalyzed by NADPH-dependent *β-aspartyl phosphate dehydrogenase* (reaction 22.30).

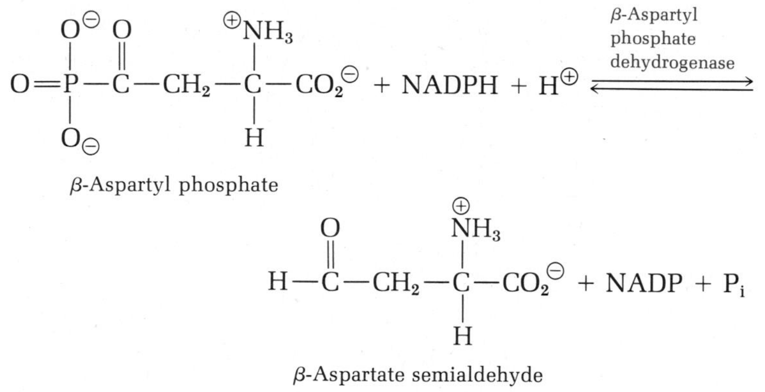

$$O=\overset{O^{\ominus}}{\underset{O_{\ominus}}{\overset{|}{P}}}-\overset{O}{\overset{\|}{C}}-CH_2-\overset{\overset{\oplus}{N}H_3}{\underset{H}{\overset{|}{C}}}-CO_2^{\ominus} + NADPH + H^{\oplus} \underset{\longleftarrow}{\overset{\text{β-Aspartyl phosphate dehydrogenase}}{\rightleftharpoons}}$$

β-Aspartyl phosphate

$$H-\overset{O}{\overset{\|}{C}}-CH_2-\overset{\overset{\oplus}{N}H_3}{\underset{H}{\overset{|}{C}}}-CO_2^{\ominus} + NADP + P_i \qquad (22.30)$$

β-Aspartate semialdehyde

Reduction of β-aspartate semialdehyde to homoserine is catalyzed by *homoserine dehydrogenase*, an NADPH-dependent enzyme that is tightly bound in a two-enzyme complex to aspartokinase. Aspartokinase I, a tetramer composed of identical subunits, is a *bifunctional enzyme*. Each subunit of the tetramer contains two catalytic sites. The N-terminal region is responsible for kinase activity, and the C-terminal region is responsible for dehydrogenase activity. Both of these activities are inhibited by threonine. Aspartokinase II is also a bifunctional enzyme, although it is not inhibited by methionine or any other amino acid that is a member of the aspartate family. Aspartokinase II–homoserine dehydrogenase II is a dimer of identical subunits. Its biosynthesis is repressed by methionine, and this is the major mechanism of regulation of this bifunctional enzyme.

D. Lysine   The conversion of β-aspartate semialdehyde to lysine requires seven steps (Figure 22.19). (1) Aldol condensation of pyruvate and β-aspartate semialdehyde yields an intermediate that immediately cyclizes with loss of water to

Figure 22.19   Pathway for the biosynthesis of lysine. Both the first unique step of lysine biosynthesis, the aldol condensation of aspartate semialdehyde with pyruvate, and the preceding conversion of aspartate to β-aspartyl phosphate are subject to negative feedback inhibition by lysine.

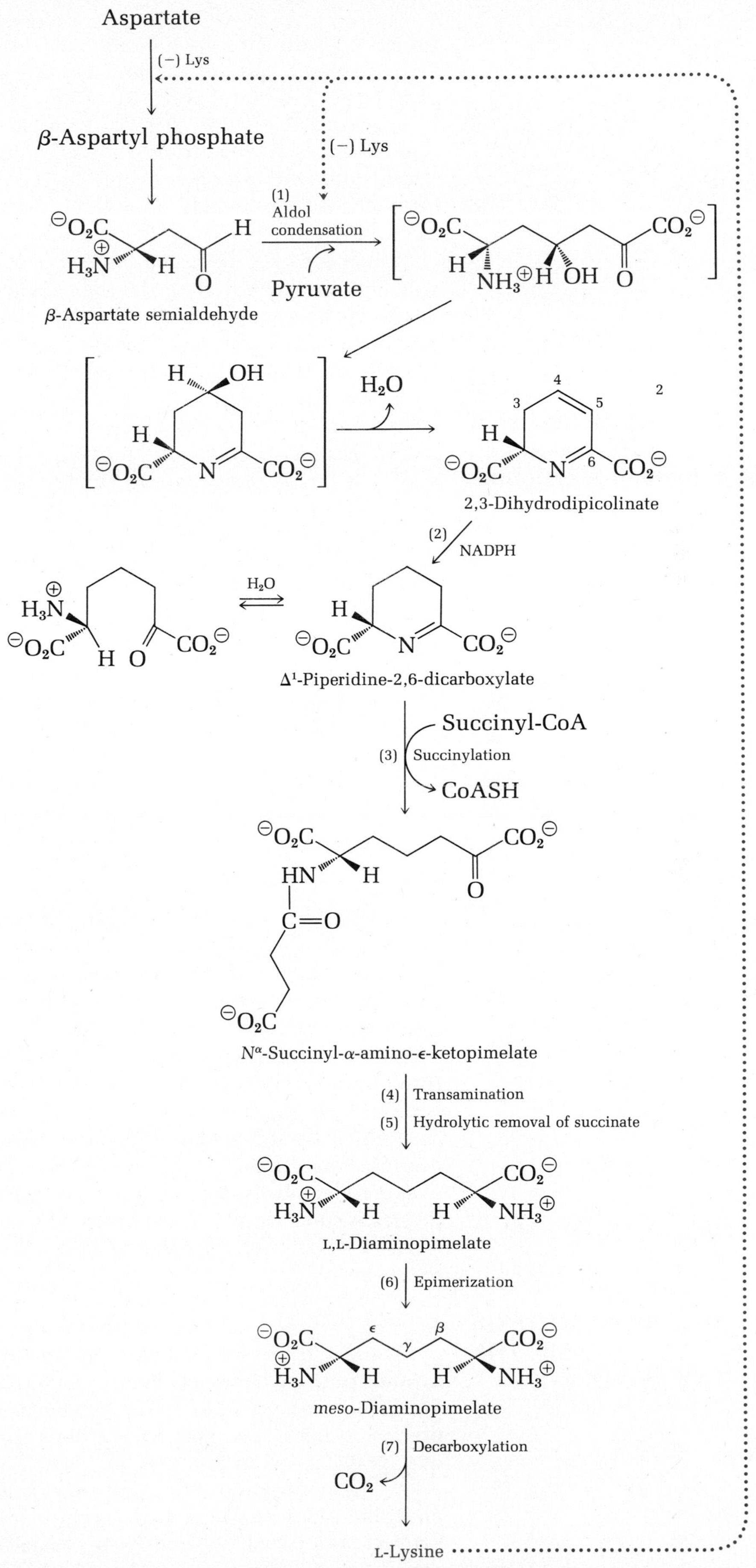

Aspartate

(−) Lys

β-Aspartyl phosphate

(−) Lys

(1) Aldol condensation

Pyruvate

β-Aspartate semialdehyde

H_2O

2,3-Dihydrodipicolinate

(2) NADPH

H_2O

Δ^1-Piperidine-2,6-dicarboxylate

Succinyl-CoA

(3) Succinylation

CoASH

N^α-Succinyl-α-amino-ϵ-ketopimelate

(4) Transamination

(5) Hydrolytic removal of succinate

L,L-Diaminopimelate

(6) Epimerization

meso-Diaminopimelate

(7) Decarboxylation

CO_2

L-Lysine

produce 2,3-dihydrodipicolinate. The condensing enzyme is inhibited by lysine. Thus, the first unique step of lysine biosynthesis and the conversion of aspartate to β-aspartyl phosphate catalyzed by aspartokinase III are inhibited by lysine. (2) Reduction of the $\Delta^{4,5}$ double bond by an NADPH-dependent reductase gives Δ^1-piperidine-2,6-dicarboxylate. This intermediate exists in equilibrium with an open chain dicarboxylic acid. (3) Succinylation of the open chain intermediate traps the open chain form, forming N^α-succinyl-α-amino-ϵ-ketopimelate. (4) Transamination and (5) hydrolysis of the succinamide give L,L-diaminopimelate, a component of certain bacterial cell walls. (6) Epimerization of the ϵ-amino group gives *meso*-diaminopimelate, which is also a component of the peptidoglycans of gram negative bacteria. Epimerization is required since the L,L-diastereoisomer is not a substrate for the decarboxylase that produces lysine in the final step (7) of the pathway.

E. Conversion of Homoserine to Threonine and Methionine

Homoserine is the common precursor of threonine and methionine (Figure 22.20). Acylation of the serine oxygen, by succinyl-CoA in enteric bacteria and by a variety of acyl groups including acetyl-CoA in higher plants, produces *O*-succinylhomoserine (or *O*-acetylhomoserine). Nucleophilic displacement of the succinyl group by the sulfhydryl group of cysteine gives cystathionine. Elimination of serine, catalyzed by γ-cystathioninase (Section 21.4B) gives homocysteine. In yeast direct displacement of succinate by H_2S provides an alternate route from *O*-succinylhomoserine to homocysteine. Methyl transfer from N^5-methyltetrahydrofolate gives methionine. In some strains of *E. coli* methyl transfer is mediated by coenzyme B_{12} (reaction 22.31). Some bacteria synthesize enzymes for both coenzyme B_{12}–dependent and coenzyme B_{12}–independent methylation of homocysteine.

$$\text{THF} \quad \overset{\overset{\textstyle CH_3}{\textstyle |}}{CO^{3+}} \quad \text{Homocysteine}$$
$$N^5\text{-methyl-THF} \qquad CO^{+1} \qquad \text{Methionine}$$

(22.31)

The conversion of homoserine to threonine requires isomerization from a primary to a secondary alcohol. This pathway occurs in two steps. (1) Phosphorylation of the primary alcohol is catalyzed by *homoserine kinase*. This step is competitively inhibited by threonine, and it controls the rate of threonine biosynthesis. (2) A pyridoxal phosphate–dependent enzyme catalyzes β,γ elimination of phosphate, followed by addition of water to the β,γ-dehydroamino acid adduct. Hydrolysis of the pyridoxal phosphate adduct yields threonine (Figure 22.21). Negative feedback inhibition by threonine controls the activities of aspartokinase I, homoserine dehydrogenase, and homoserine kinase in *E. coli*. Of the three enzymes, aspartokinase I is the most sensitive to feedback inhibition.

22–7 BIOSYNTHESIS OF THE PYRUVATE FAMILY

Alanine, valine, and leucine obtain their carbon skeletons directly from pyruvate. We saw in the last section that lysine derives three of its six carbons from pyruvate. Isoleucine, a member of the aspartate family, obtains four of its carbons from aspartate and has threonine as a more immediate precursor, but it also gets two carbons from pyruvate, and its biosynthesis will be discussed in this section.

The conversion of pyruvate to alanine occurs by a pyridoxal phosphate–dependent step that requires either glutamate or valine as the amino donor (recall Section 20.4C) (reactions 22.32 and 22.33).

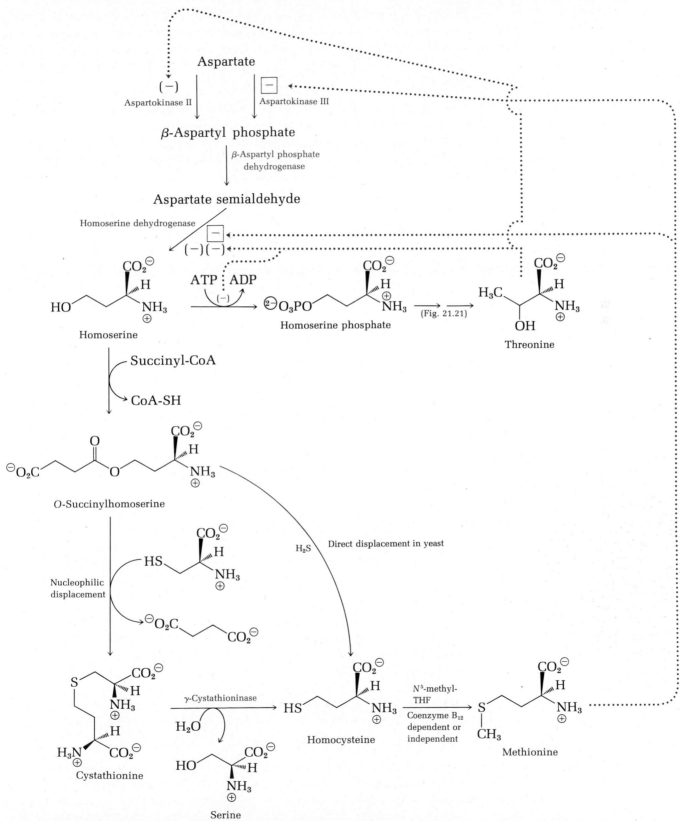

Figure 22.20   Pathways for the biosynthesis of threonine and methionine. Negative feedback inhibition is indicated by (−); inhibition by repression of enzyme synthesis is indicated by ⊟.

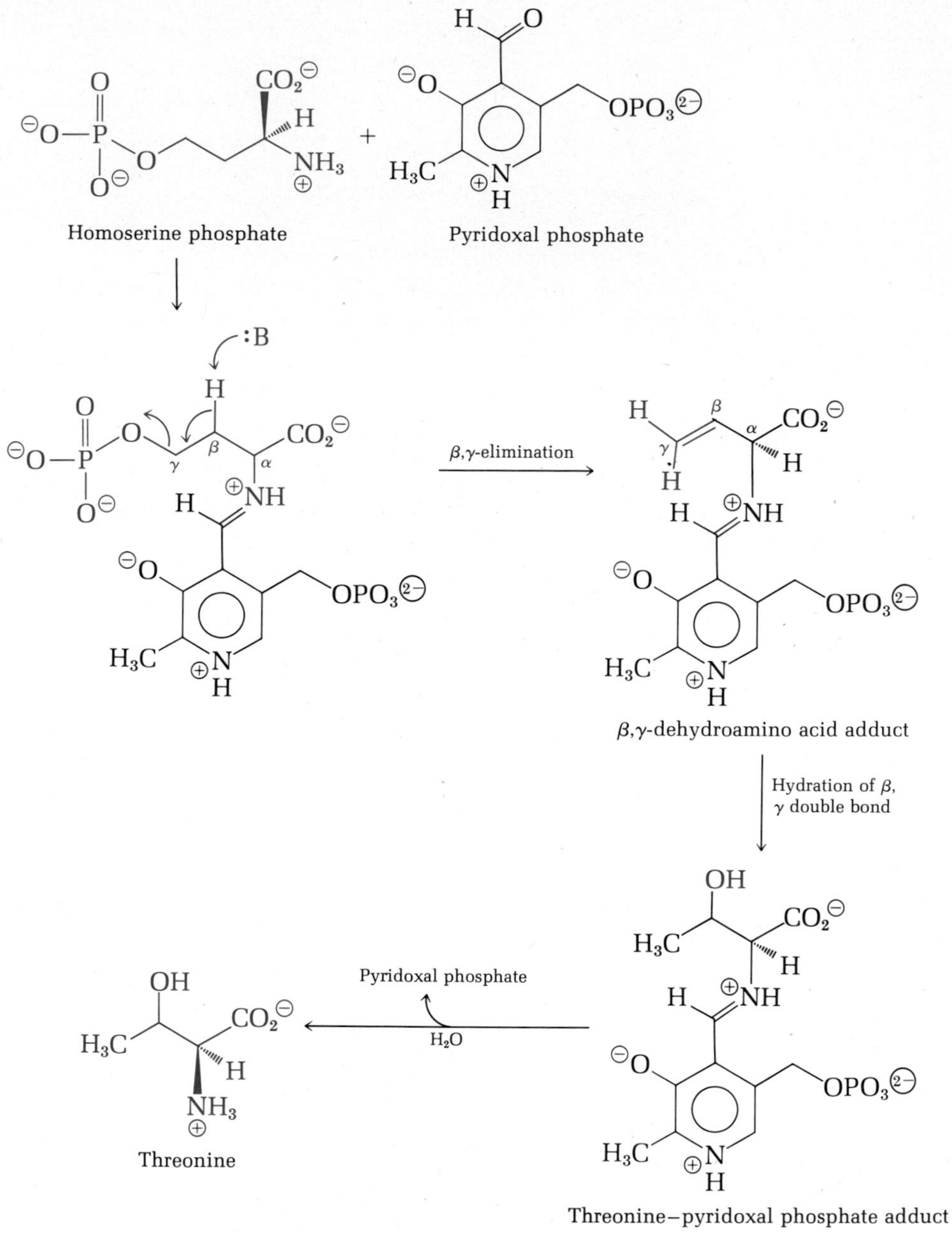

Figure 22.21 Conversion of homoserine phosphate to threonine.

$$\underset{\text{Pyruvate}}{CH_3-\overset{O}{\overset{\|}{C}}-CO_2^{\ominus}} + \underset{\text{Glutamate}}{{}^{\ominus}O_2C-CH_2CH_2-\overset{\overset{\oplus}{N}H_3}{\underset{H}{C}}-CO_2^{\ominus}} \rightleftharpoons$$

$$\underset{\text{Alanine}}{CH_3\overset{\overset{\oplus}{N}H_3}{\underset{H}{C}}-CO_2^{\ominus}} + \underset{\alpha\text{-Ketoglutarate}}{{}^{\ominus}O_2C-CH_2CH_2-\overset{O}{\overset{\|}{C}}-CO_2^{\ominus}} \qquad (22.32)$$

$$CH_3-\overset{\overset{\displaystyle O}{\|}}{C}-CO_2^{\ominus} + (CH_3)_2CH-\overset{\overset{\displaystyle \overset{\oplus}{N}H_3}{|}}{\underset{\underset{\displaystyle H}{|}}{C}}-CO_2^{\ominus} \rightleftharpoons$$

Pyruvate   Valine

$$CH_3-\overset{\overset{\displaystyle \overset{\oplus}{N}H_3}{|}}{\underset{\underset{\displaystyle H}{|}}{C}}-CO_2^{\ominus} + (CH_3)_2CH-\overset{\overset{\displaystyle O}{\|}}{C}-CO_2^{\ominus} \qquad (22.33)$$

Alanine   α-Ketoisovalerate

The biosynthesis of the branched chain amino acids isoleucine, leucine, and valine occur by closely related pathways (Figures 22.22 and 22.23). Valine and isoleucine are synthesized in parallel pathways. The same enzymes catalyze the biosynthesis of isoleucine from α-ketobutyrate and the biosynthesis of valine from pyruvate.

We will consider the two pathways in tandem.

In the first step of valine biosynthesis a two-carbon fragment, derived from pyruvate and bound to thiamine pyrophosphate as the "active aldehyde," adds to the carbonyl carbon of a second mole of pyruvate in a reaction catalyzed by *acetolactate synthetase* (Figure 22.24). In the pathway to isoleucine the same enzyme catalyzes conversion of α-ketobutyrate, arising from transamination of threonine, to α-aceto-α-hydroxybutyrate.

α-Aceto-α-hydroxybutyrate

The α-acetohydroxy acids rearrange and are reduced in a second step yielding α-dihydroxyisovalerate on the path to valine and α,β-dihydroxy-β-methylvalerate on the path to isoleucine. This pair of reactions is catalyzed by *acetohydroxy acid isomerase reductase*. The rearrangement occurs by 1,2 migration of a methyl group *with* its electron pair in a concerted process that is followed by an NADPH-dependent reduction of the α-keto group (reaction 22.34).

Loss of water from the vicinal diol gives an enol that rearranges to α-ketoisovalerate on the path to valine and to α-keto-β-methylvalerate on the path to isoleucine (reactions 22.35 and 22.36). Transamination of these two produces yields valine and isoleucine, respectively.

Leucine biosynthesis commences, as a unique pathway, from the branch-point metabolite α-ketoisovalerate. The leucine pathway follows the same set of chain lengthening reactions as the biosynthesis of lysine in

$$CH_3-\overset{\overset{\displaystyle CH_3}{|}}{\underset{\underset{\displaystyle H}{|}}{C}}-CH_2-\overset{\overset{\displaystyle O}{\|}}{C}-CO_2^{\ominus}$$

α-Ketoisocaproate

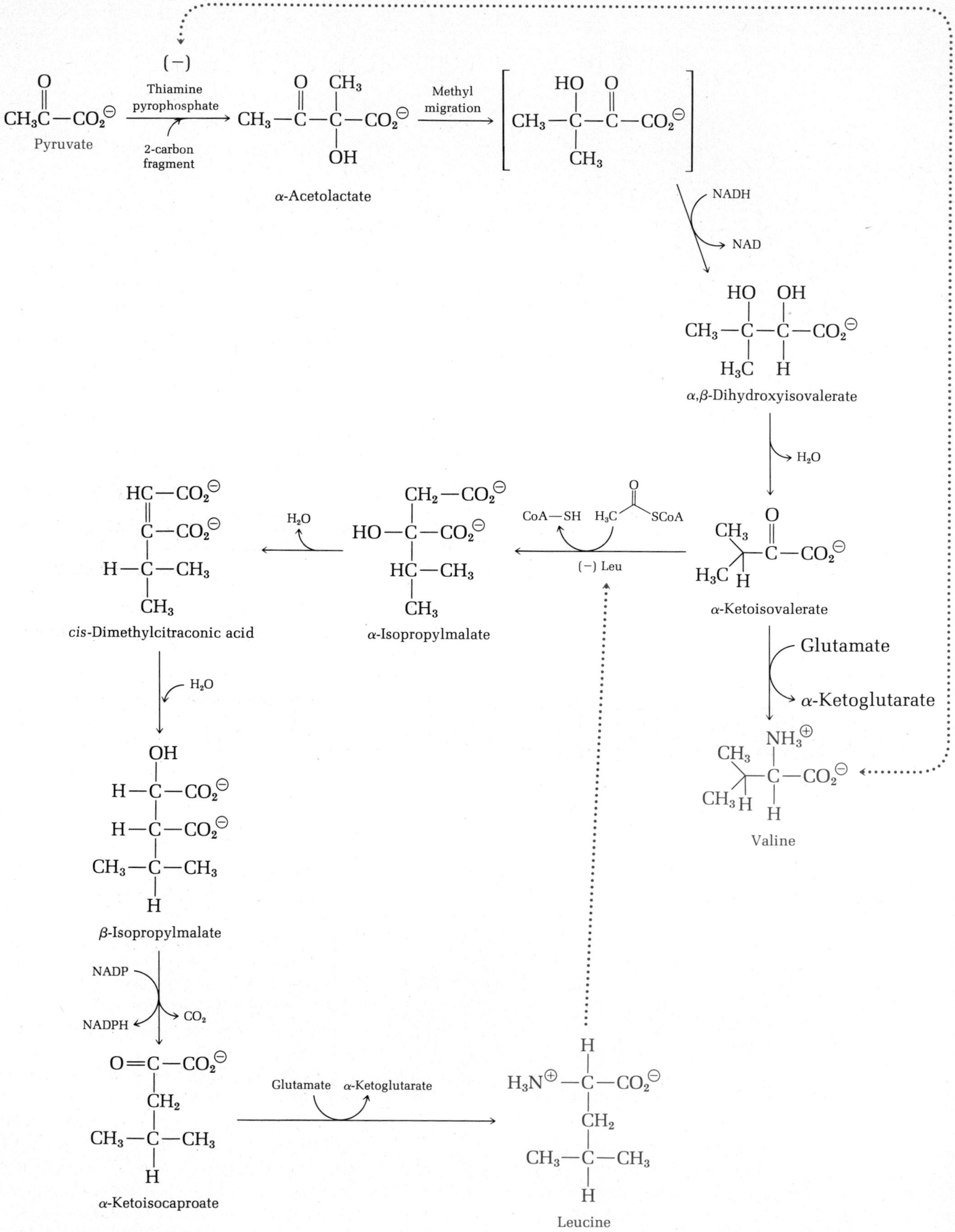

Figure 22.22 Pathways for the biosynthesis of leucine and valine. Negative feedback inhibition is indicated by (−).

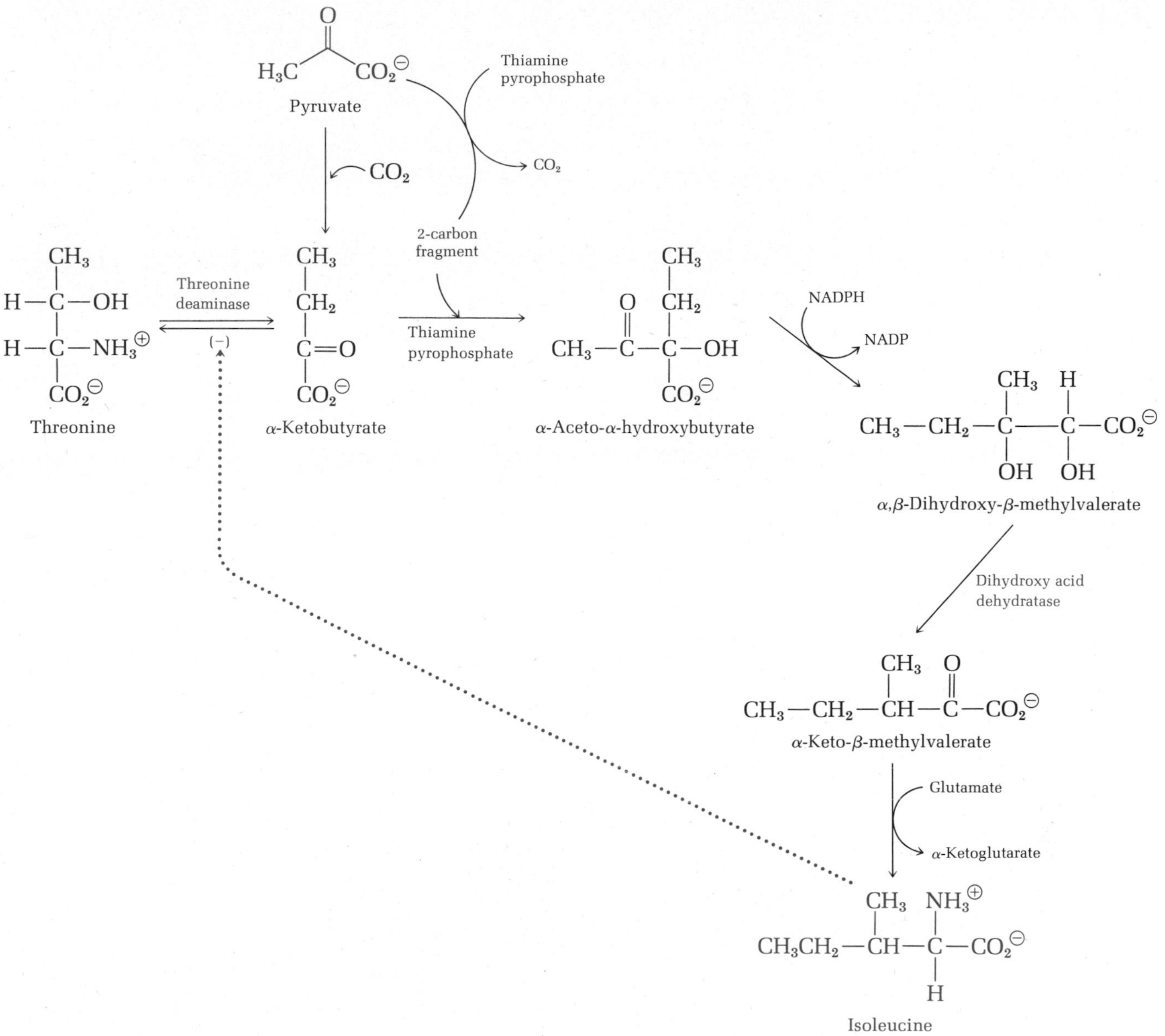

Figure 22.23   Pathway for the biosynthesis of isoleucine. The enzyme that catalyzes the conversion of α-aceto-α-hydroxy-butyrate to isoleucine also converts α-acetolactate to valine. Negative feedback inhibition is indicated by ($-$).

fungi. The chain-lengthened α-keto acid, α-ketoisocaproate, is transaminated with glutamate as amino donor to give leucine.

Control of biosynthesis of branched chain amino acids is elaborate, and varies from species to species. In *E. coli* biosynthesis is largely controlled by *threonine deaminase*, the first enzyme in the pathway. Isoleucine is a negative feedback inhibitor of the deaminase, while valine is a positive effector. In the presence of both threonine and isoleucine the subunits of threonine deaminase interact cooperatively. If valine and isoleucine are both present, however, the inhibition by isoleucine is abolished, and the threonine binding curve is hyperbolic. Valine biosynthesis is controlled by *α-aceto-α-hydroxybutyrate synthetase*. The enzyme from *Pseudomonas aeruginosa*, the only one that has been purified, is subject to cumulative feedback inhibition by valine, isoleucine, and leucine. The biosynthesis of leucine is regulated by end product inhibition of α-ketoisopropylmalate synthetase, the enzyme that catalyzes the first unique step in the biosynthesis of leucine.

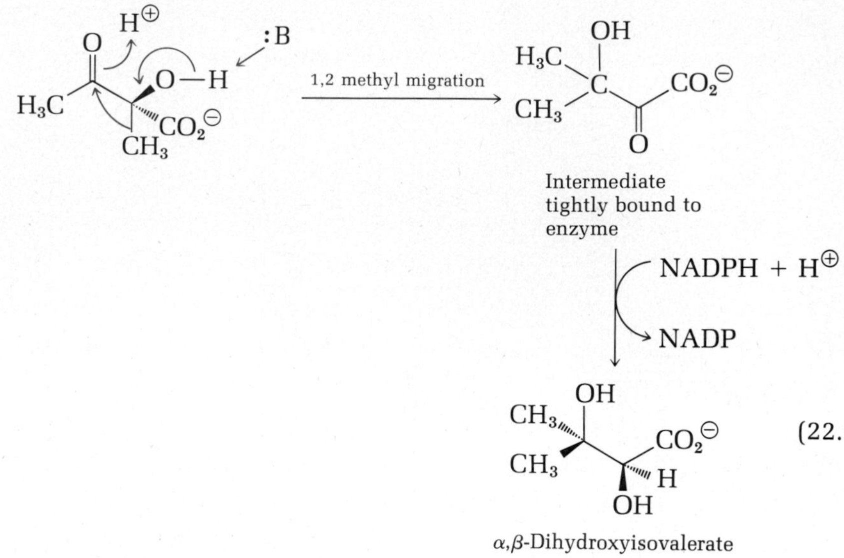
1,2 methyl migration
Intermediate tightly bound to enzyme
NADPH + H⊕
NADP
α,β-Dihydroxyisovalerate

(22.34)

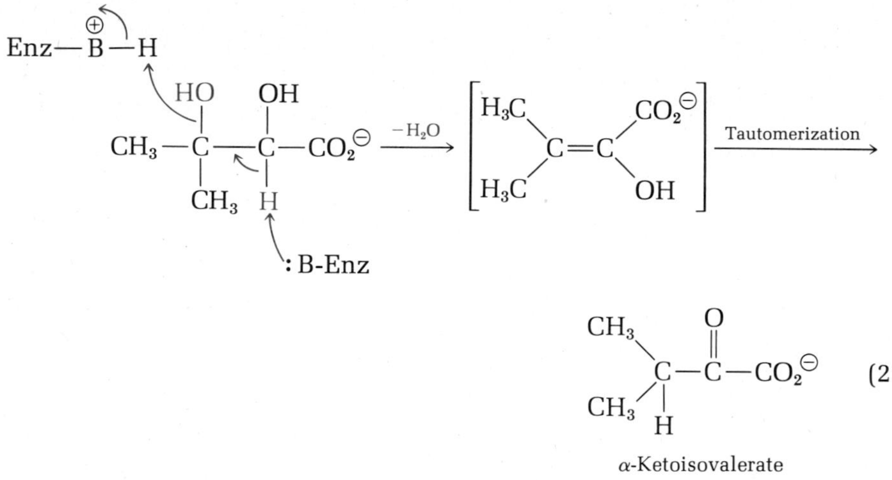
Enz—B—H
:B-Enz
−H₂O
Tautomerization
α-Ketoisovalerate

(22.35)

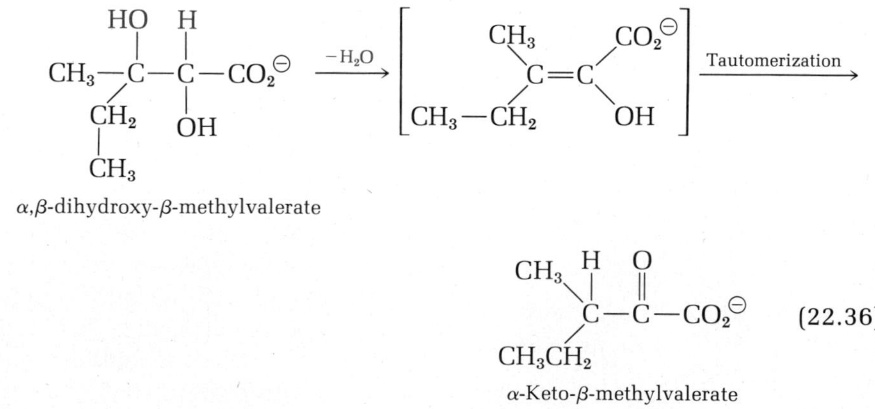
−H₂O
Tautomerization
α,β-dihydroxy-β-methylvalerate
α-Keto-β-methylvalerate

(22.36)

22–8 BIOSYNTHESIS OF THE AROMATIC FAMILY

The biosynthesis of the aromatic family of amino acids—phenylalanine tyrosine, and tryptophan—has been extensively studied in many microorganisms. Perhaps the most interesting feature of the enzymes of the aromatic pathways is their multifunctional nature. Many of the enzymes are capable of catalyzing two or more usually consecutive reactions. These multifunctional enzymes are in turn associated in multimeric enzyme complexes. Species differences can be quite large, however, and observations from one species cannot be guaranteed to apply to any others. Aromatic amino acids can be synthesized only by plants and microorganisms.

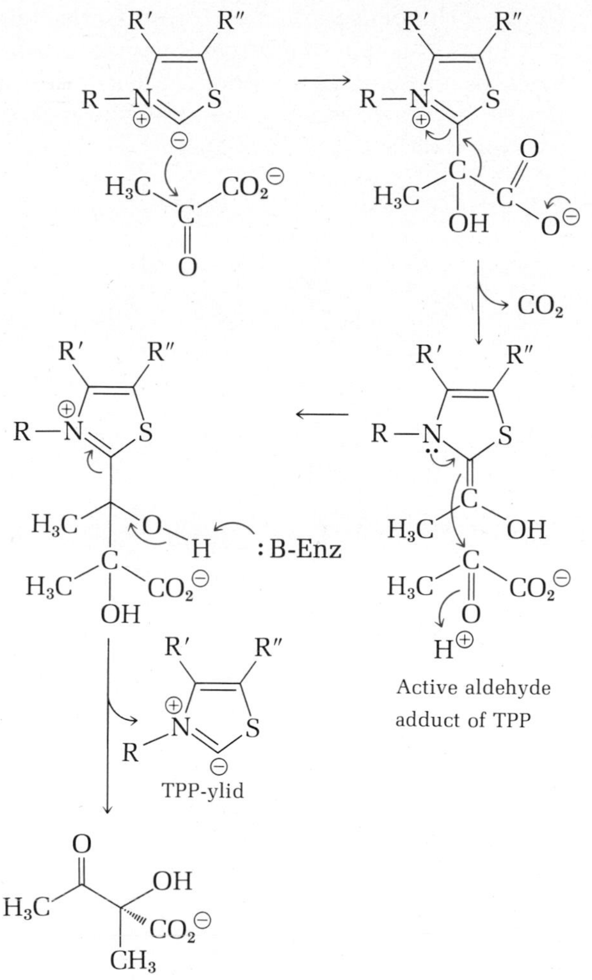

α-Acetolactate

Figure 22.24 Mechanism of action of acetolactate synthetase.

A. Chorismate Pathway

Phenylalanine, tyrosine, and tryptophan share a seven-step pathway originating with erythrose 4-phosphate, an intermediate in the pentose phosphate pathway, and phosphoenolpyruvate that leads to chorismate (Figure 22.25). Since shikimate is a precursor of aromatic amino acids, the common pathway is sometimes called the *shikimate pathway*.

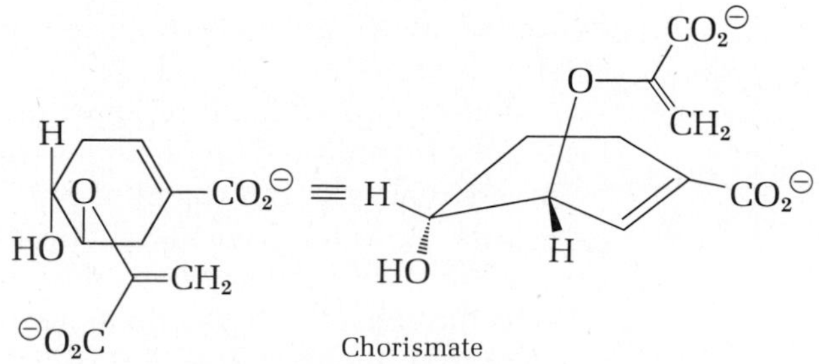

Chorismate

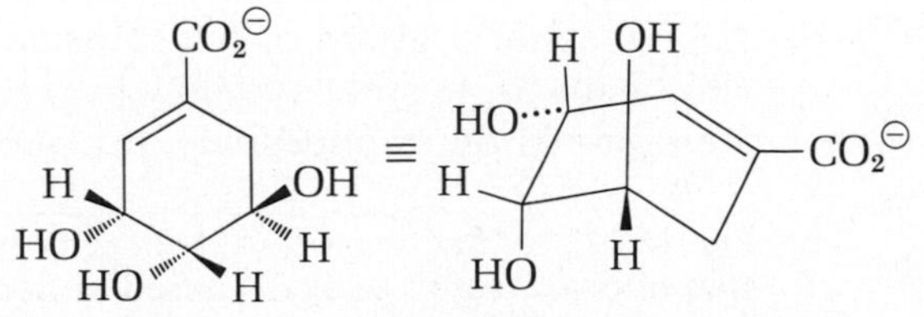

Shikimate

Condensation of erythrose 4-phosphate and phosphoenolpyruvate produces *3-deoxy-2-keto-D-arabinoheptulosonate 7*-phosphate (DAHP). The reaction resembles an aldol condensation since the electron pair of phosphoenolpyruvate attacks the aldehyde carbon of erythrose 4-phosphate (reaction 22.37).

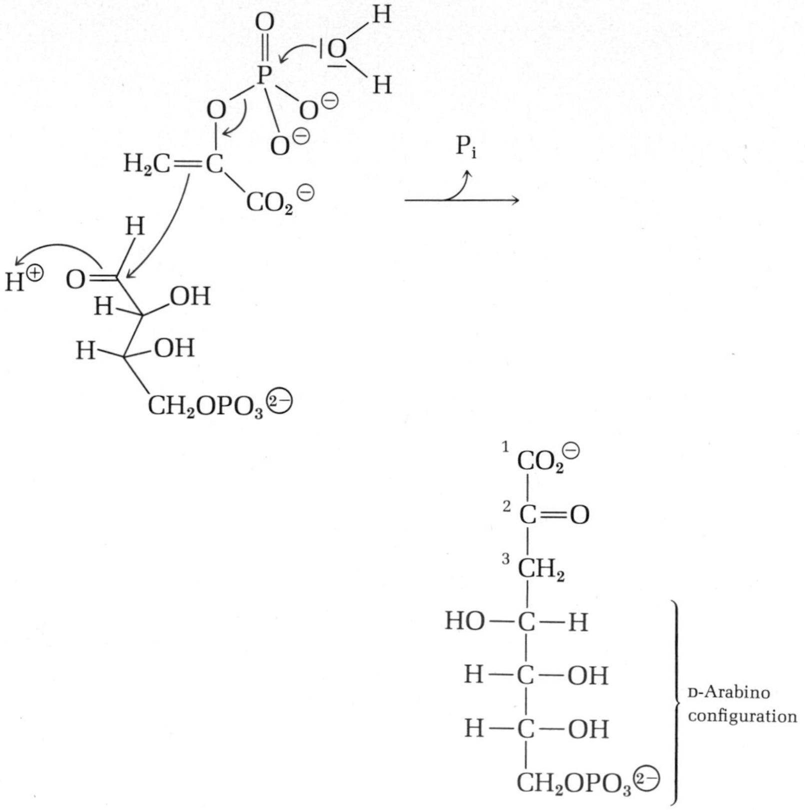

$$\begin{array}{l} {}^{1}\text{CO}_2{}^{\ominus} \\ {}^{2}\text{C}=\text{O} \\ {}^{3}\text{CH}_2 \\ \text{HO}-\text{C}-\text{H} \\ \text{H}-\text{C}-\text{OH} \\ \text{H}-\text{C}-\text{OH} \\ \text{CH}_2\text{OPO}_3{}^{\tiny 2\ominus} \end{array}$$ (22.37)

D-Arabino configuration

3-Deoxy-2-keto-D-arabinoheptulosonate 7-phosphate

Loss of phosphate and ring closure produces 5-dehydroquinic acid. No net oxidation or reduction occurs in this reaction, but NADH is required as a coenzyme. An internal oxidation-reduction occurs during the reaction in which the C-5 hydroxyl group is oxidized to a keto group. Enolization of the 5-keto intermediate and loss of phosphate produce a cyclic intermediate in which C-6 is nucleophilic and C-2 is electrophilic.

Cyclization produces an intermediate that is reduced at C-5 to give dehydroquinate (Figure 22.26). The enzyme-bound NAD therefore functions catalytically.

Both *E. coli* and *S. typhimurium* possess three dehydroquinate synthetases. One is inhibited by phenylalanine, one by tyrosine, and one by tryptophan, a situation reminiscent of the *E. coli* aspartokinases.

Loss of water from 5-dehydroquinate give 5-dehydroshikimate. The driving force for the reaction is the resonance stabilization of the conjugated α,β-unsaturated carbonyl system. Reduction of the ketone followed by phosphoryl group transfer from ATP to the newly formed hydroxyl group at C-5 gives shikimate 5-phosphate (Figure 22.27). Nucleophilic displacement of the phosphoryl group of phosphoenolpyruvate by the 3-hydroxyl group of shikimate 5-phosphate produces 3-enolpyruvylshikimate 5-phosphate. A 1,4 elimination of phosphate produces chorismate (reaction 22.38).

Figure 22.25   Pathways for the biosynthesis of aromatic amino acids. The regulation of these pathways is discussed in Section 22.8B. ▶

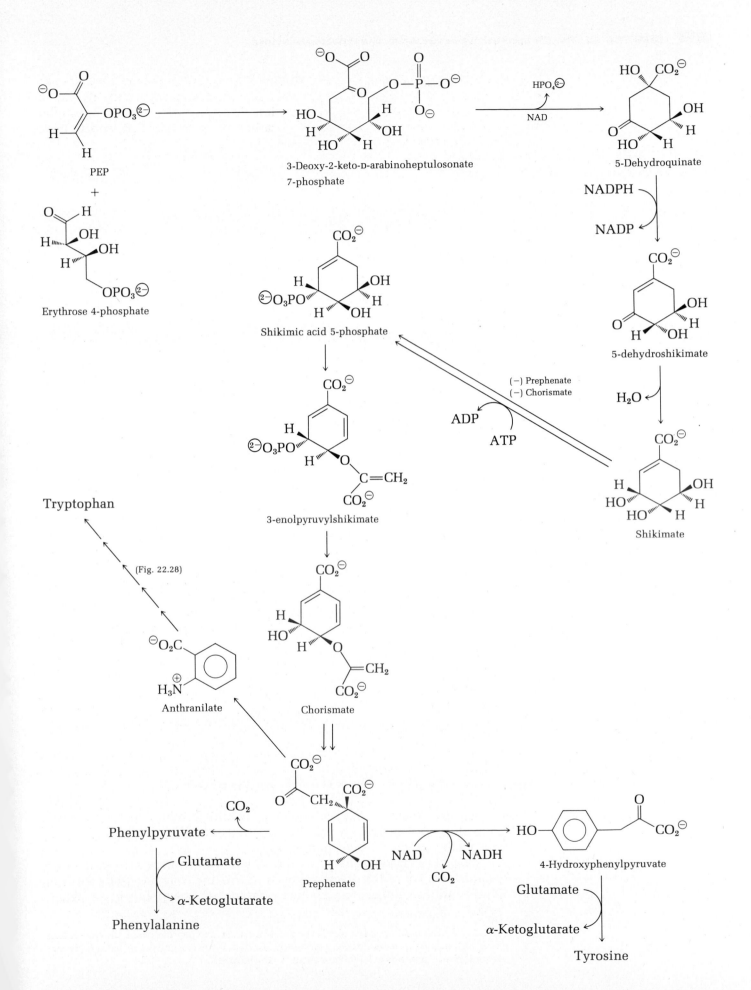

PEP

+

Erythrose 4-phosphate

3-Deoxy-2-keto-D-arabinoheptulosonate
7-phosphate

5-Dehydroquinate

NADPH

NADP

5-dehydroshikimate

H_2O

Shikimate

Shikimic acid 5-phosphate

3-enolpyruvylshikimate

(−) Prephenate
(−) Chorismate

ADP

ATP

Tryptophan

(Fig. 22.28)

Anthranilate

Chorismate

CO_2

Phenylpyruvate

Glutamate

α-Ketoglutarate

Phenylalanine

Prephenate

NAD   NADH

CO_2

4-Hydroxyphenylpyruvate

Glutamate

α-Ketoglutarate

Tyrosine

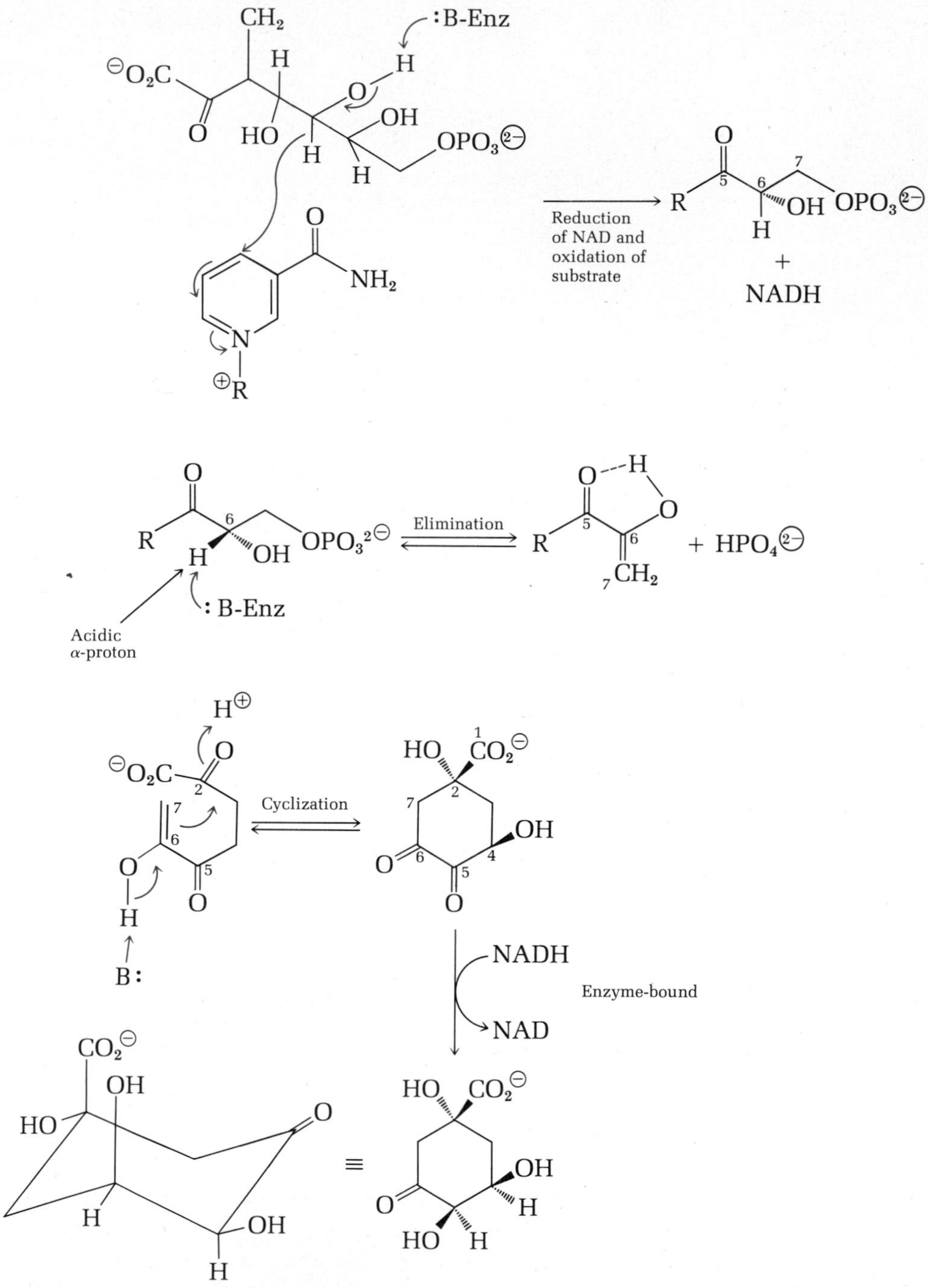

5-Dehydroquinate

Figure 22.26 Mechanism of action of dehydroquinate synthetase.

Chorismate is the central branch-point metabolite in the biosynthesis of aromatic amino acids. By one pathway it is converted to phenylalanine and tyrosine; by another it leads to tryptophan.

The chorismate pathway in the fungus *Neurospora crassa* is quite remarkable. Five of the seven reactions of this pathway are catalyzed by a single multifunctional enzyme that consists of one polypeptide chain[4]

[4] A multifunctional enzyme can consist of a single polypeptide chain which contains active sites for two or more different enzymatic reactions, or it can consist of a multisubunit complex whose subunits have different enzymatic activities.

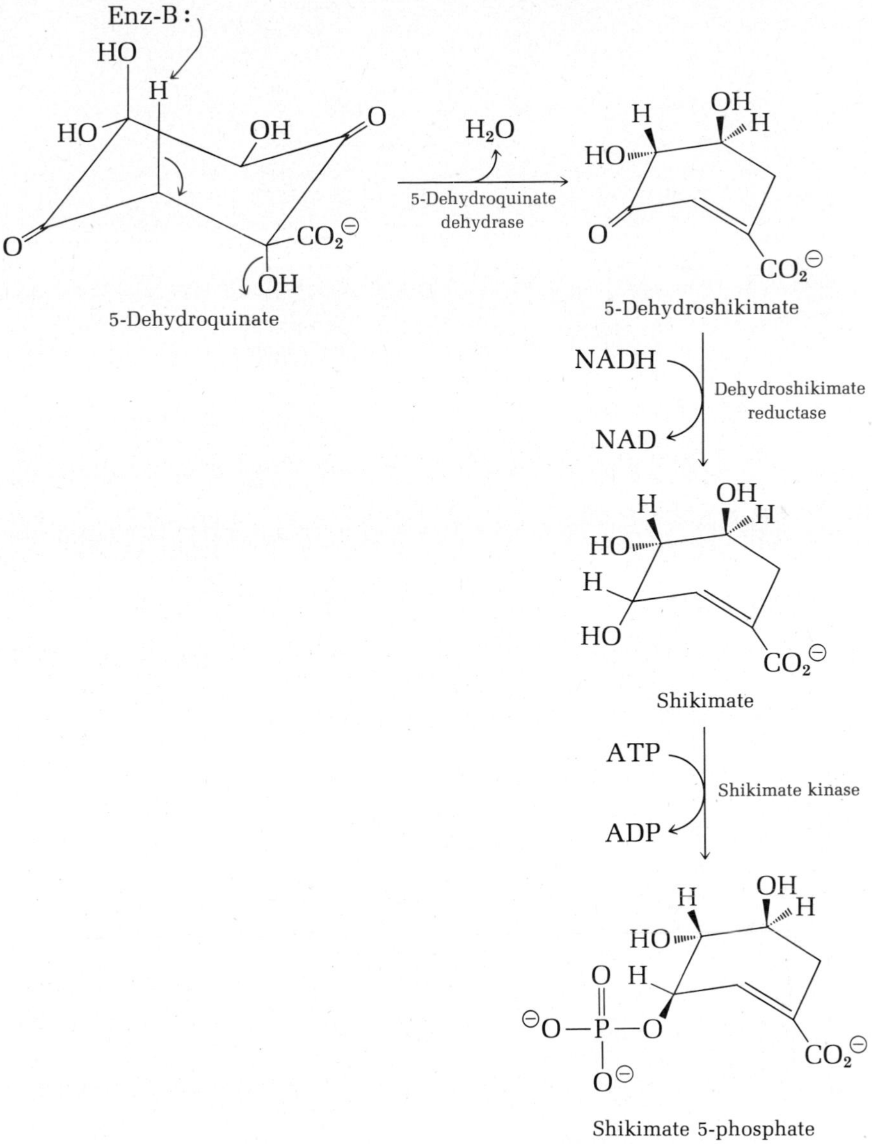

Figure 22.27 Conversion of dehydroquinate to shikimate 5-phosphate.

Chorismate synthetase, responsible for the seventh reaction, and 2-keto-3-deoxy-D-arabinoheptulosonate 7-phosphate synthetase, responsible for the first reaction, combine with the multifunctional enzyme to form a complex that can be isolated as a functional unit. Three enzymes of the biosynthetic pathway for tryptophan also bind to this synthetic unit. Mixing erythrose 4-phosphate, phosphoenolpyruvate, and one other enzyme to this complex leads to synthesis of anthranilate. The rate of formation of anthranilate is much faster when erythrose 4-phosphate and phosphoenolpyruvate are the substrates than when shikimate is the substrate.

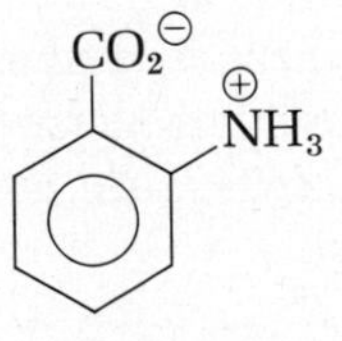

Anthranilate

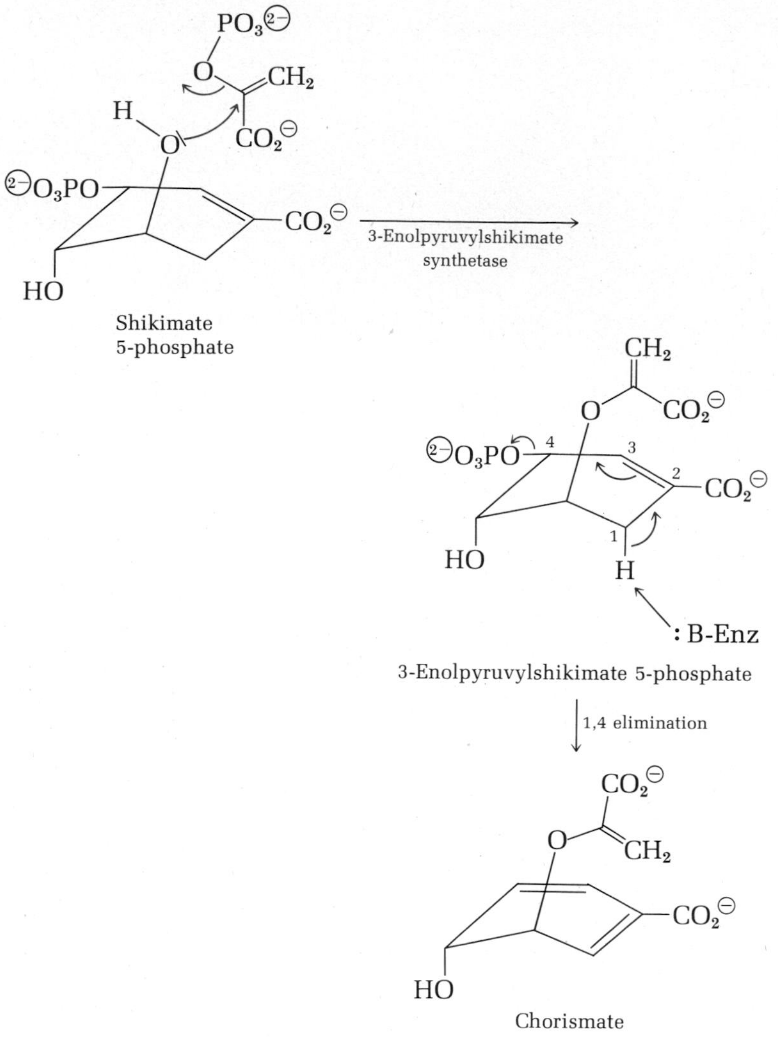

Shikimate
5-phosphate

3-Enolpyruvylshikimate
synthetase

3-Enolpyruvylshikimate 5-phosphate

1,4 elimination

(22.38)

Chorismate

B. Conversion of Chorismate to Phenylalanine and Tyrosine

The pathway from chorismate to phenylalanine and tyrosine in *B. subtilis* is shown in Figure 22.28. Prephenate and chorismate inhibit the first step of the pathway by negative feedback inhibition. The enzyme activities that produce 3-deoxy-2-keto-D-heptulosonate 7-phosphate (DAHP synthetase) and that convert chorismate to prephenate (chorismate mutase) are located on the same polypeptide chain. Prephenate inhibits both of these activities allosterically by feedback inhibition. Prephenate and chorismate also inhibit the conversion of shikimate to shikimate 5-phosphate. The branch-point metabolite prephenate can be converted to either phenylalanine or tyrosine (in the next section we shall see that it can also be converted to tryptophan). Each of these end products inhibits the first *unique* reaction of its own biosynthesis. This inhibition pattern is called *sequential feedback inhibition* (recall Section 13.4B). There are many variations of the inhibition pattern observed in *B. subtilis*. For example, the enteric bacteria *E. coli* and *Salmonella typhimurium* have three isozymes for DAHP synthetase,—one inhibited by tyrosine, one by phenylalanine, and one by tryptophan. This isozyme pattern is also found in *N. crassa* and is reminiscent of the isozyme pattern for the *E. coli* aspartokinases.

The conversion of chorismate to prephenate is a Claisen rearrangement catalyzed by *chorismate mutase* (reaction 22.39).

PEP + Erythrose 4-phosphate

3-Deoxy-2-keto-D-arabinoheptulosonate 9-phosphate

Shikimate

Shikimate 5-phosphate

Chorismate

Prephenate

Phenylpyruvate

p-Hydroxyphenylpyruvate

Glutamate

α-Ketoglutarate

Glutamate

α-Ketoglutarate

Phenylalanine

Tyrosine

Figure 22.28   Conversion of chorismate to phenylalanine and tyrosine. Prephenate inhibits the conversion of shikimate to shikimate 5-phosphate and the first step of the common aromatic pathway. The conversion of the branch-point metabolite prephenate to phenylpyruvate is inhibited by phenylalanine, and the conversion of prephenate to 4-hydroxyphenylpyruvate is inhibited by tyrosine. This is an example of sequential feedback inhibition.

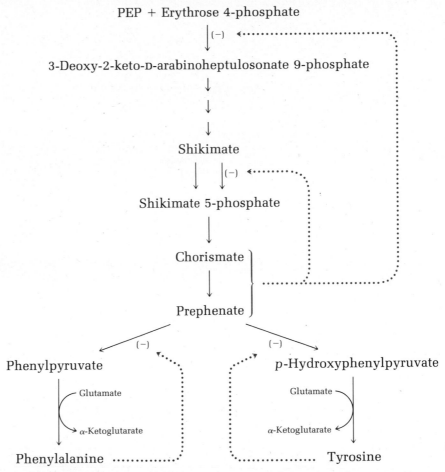

Chorismate mutase
Claisen rearrangement

Chorismate

Prephenate

(22.39)

This reaction, in which the distracting functional groups have been removed to show this process more clearly, is a Claisen rearrangement in which the new π bond is formed three atoms away from the broken σ bond—a process called a [3,3]-sigmatropic shift (Figure 22.29).

(a)

(b)

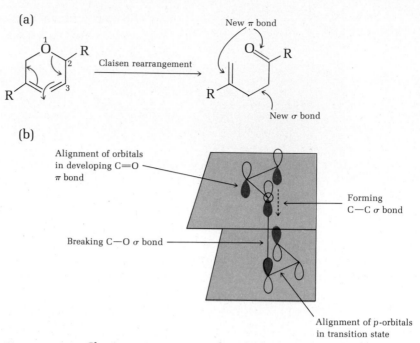

Figure 22.29   Chorismate mutase-catalyzed Claisen rearrangement, or [3,3]-sigmatropic shift, of chorismate to prephenate. (A) Flow of electrons. (b) Configuration of molecular orbitals in the transition state of the reaction.

The conversion of prephenate to phenylalanine occurs in two steps. The first is a 1,4 elimination driven by loss of carbon dioxide and the resonance stabilization of the aromatic ring. The final step is a glutamate-dependent transamination (reaction 22.40).

(22.40)

In some organisms, including humans, tyrosine can be formed from phenylalanine by an hydroxylase (reaction 22.41). The mechanism of this reaction is discussed in Section 22.8F.

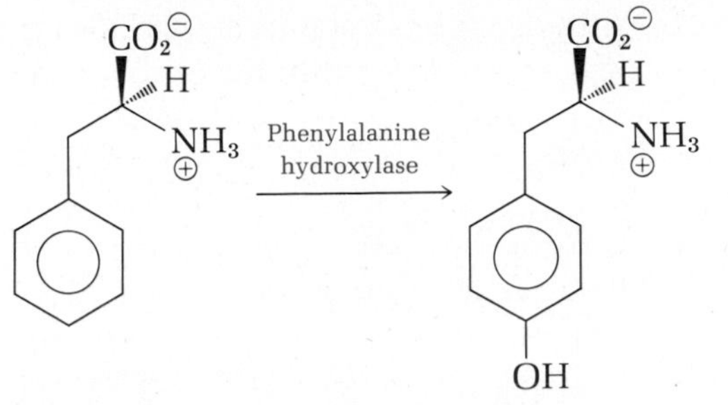

(22.41)

Phenylalanine Tyrosine

Phenylalanine is an essential amino acid in humans, but tyrosine is not, since phenylalanine can be converted to tyrosine by reaction 22.41.

C. Conversion of Tyrosine to Catecholamines

The catecholamines—*norepinephrine* (noradrenaline), *epinephrine* (adrenaline), and *dopamine*—are synthesized from tyrosine by the same pathway in the adrenal medulla and in nervous tissue (Figure 22.30). The catecholamines have many physiological functions. They transmit messages in the central nervous system and are therefore called *neurotransmitters*. They also stimulate glycogen and lipid degradation.

Tyrosine hydroxylase catalyzes the conversion of tyrosine to L-3,4-

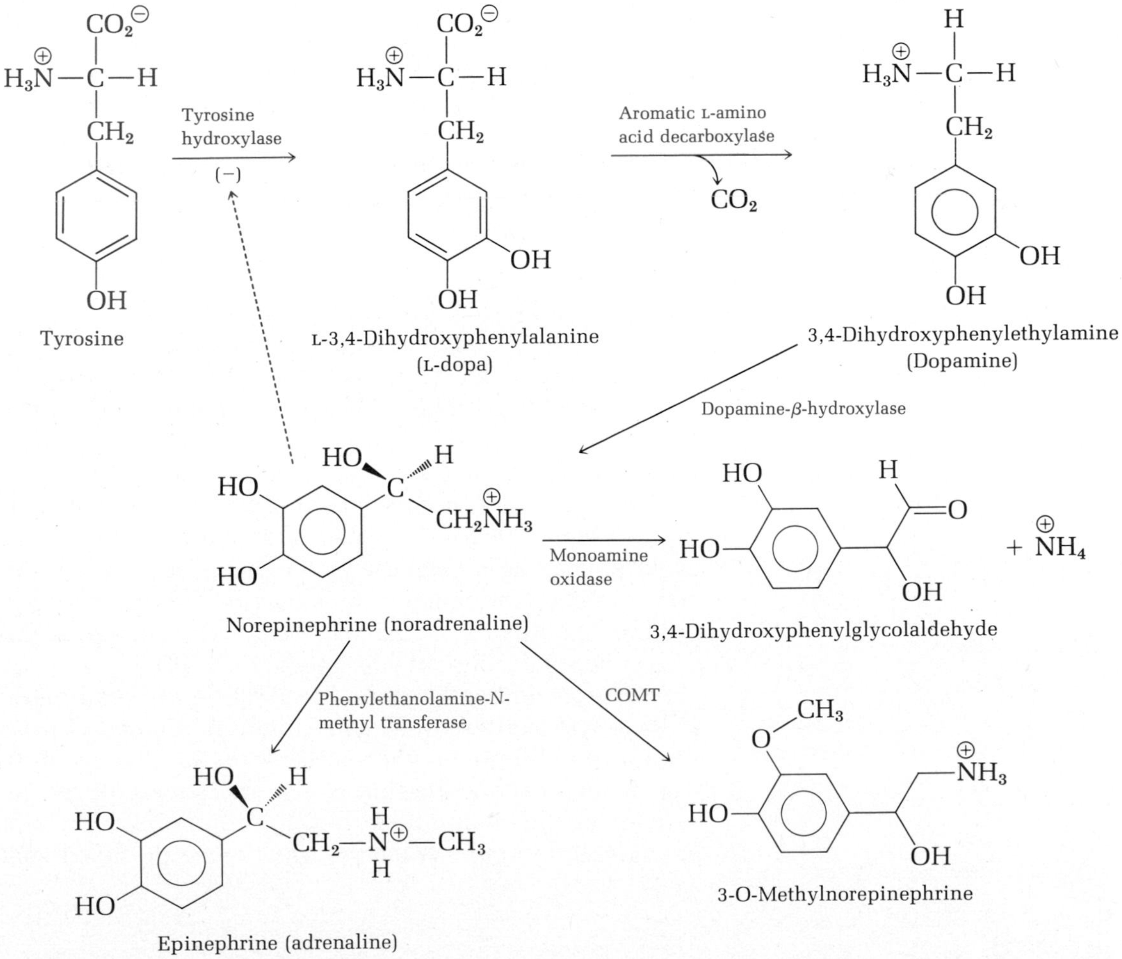

Figure 22.30 Conversion of tyrosine to the catecholamines and the inactivation of norepinephrine by conversion to 3,4-dihydroxyphenylglycolaldehyde or 3-*O*-methylnorepinephrine. Negative feedback inhibition is indicated by (−).

dihydroxyphenylalanine (*L-dopa*) (reaction 21.42). Norepinephrine is a negative feedback inhibitor of this reaction.

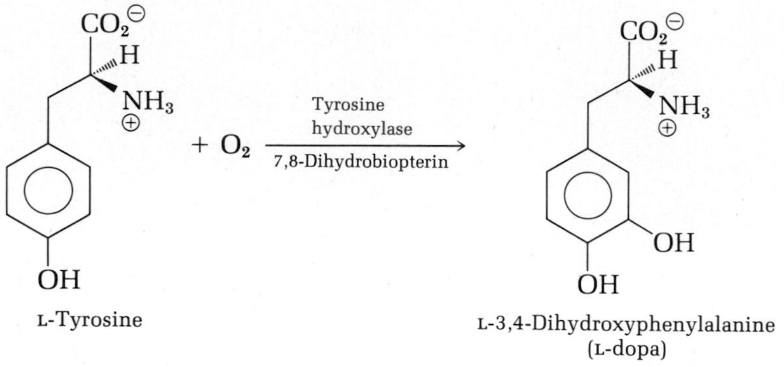

L-Tyrosine $+ O_2$ $\xrightarrow[\text{7,8-Dihydrobiopterin}]{\text{Tyrosine hydroxylase}}$ L-3,4-Dihydroxyphenylalanine (L-dopa)   (22.42)

L-Dopa is converted to *dopamine* by the pyridoxal phosphate–dependent enzyme *aromatic* L-*amino acid decarboxylase* (reaction 22.43).

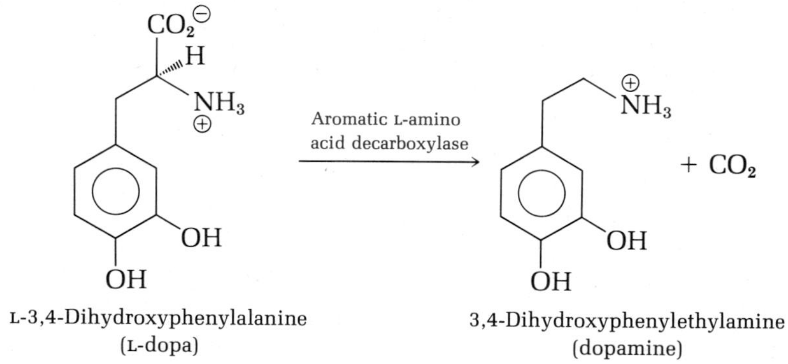

L-3,4-Dihydroxyphenylalanine (L-dopa) $\xrightarrow{\text{Aromatic L-amino acid decarboxylase}}$ 3,4-Dihydroxyphenylethylamine (dopamine) $+ CO_2$ (22.43)

Dopamine-β-hydroxylase catalyzes hydroxylation of the benzylic position of dopamine. Molecular oxygen is the source of the oxygen for this reaction, and the hydroxyl group of the product stereospecifically replaces the pro-S benzylic hydrogen. The enzyme contains a Cu(II) ion and requires ascorbate as a cofactor. The ascorbate is oxidized to dehydroascorbate during the reaction. When $^{18}O_2$ is the substrate, one atom of ^{18}O is found in norepinephrine, and one atom of ^{18}O appears in water (reaction 22.44).

Norepinephrine is a neurotransmitter that stimulates an adenylate cyclase, causing a decrease in the firing of nerves in the sympathetic nervous system. Norepinephrine is inactivated by two mechanisms. The flavoprotein *monoamine oxidase* converts norepinephrine to *3,4-dihydroxyphenylglycolaldehyde* (reaction 22.45).

The second pathway for inactivation of norepinephrine is methylation of its 3-hydroxyl group by *catecholamine-O-methyl transferase* (COMT). S-adenosylmethionine is the source of the methyl group (reaction 22.46).

S-Adenosylmethionine is also the source of the methyl group in the conversion of norepinephrine to epinephrine which is catalyzed by *phenylethanolamine-N-methyl transferase* in the adrenal medulla (reaction 22.47).

D. Conversion of Chorismate to Tryptophan

The branch-point metabolite chorismate is converted to tryptophan in five steps (Figure 22.31). The nitrogen atom of the indole ring of tryptophan is derived from ammonia and is introduced by nucleophilic substitution in an

S_N2' mechanism (Figure 22.32). Elimination of pyruvate produces anthranilate. Tryptophan inhibits anthranilate synthetase, thereby regulating its own rate of production.

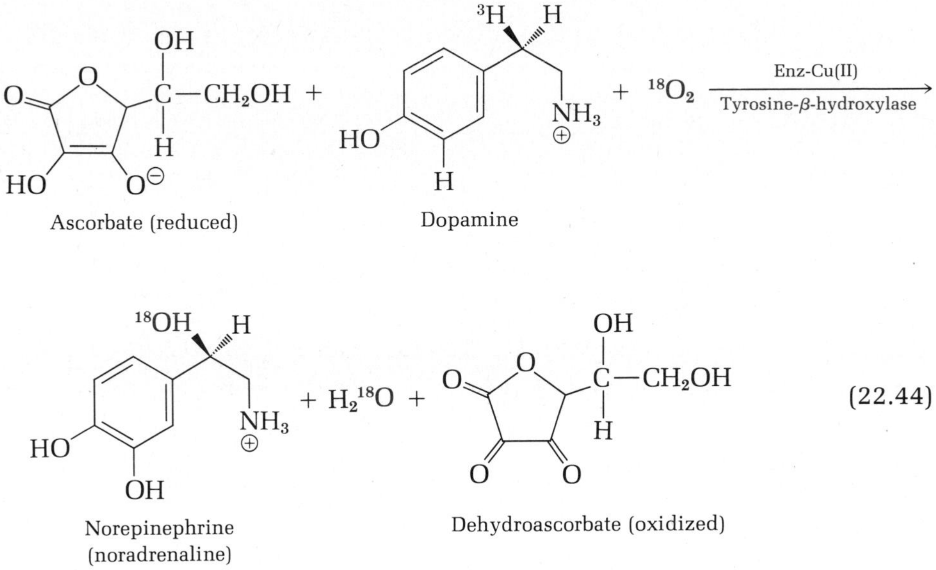
Enz-Cu(II)
Tyrosine-β-hydroxylase
Ascorbate (reduced)
Dopamine
Norepinephrine
(noradrenaline)
Dehydroascorbate (oxidized)

(22.44)

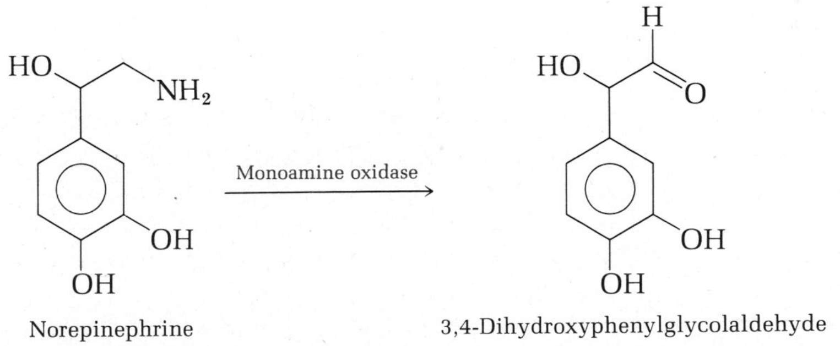
Monoamine oxidase
Norepinephrine
3,4-Dihydroxyphenylglycolaldehyde

(22.45)

The amino group of anthranilate displaces pyrophosphate from 5-phosphoribosylpyrophosphate (PRPP), producing N-(5-phosphoribosyl)-anthranilate. C-1 and C-2 of ribose provide the two additional carbons needed for formation of the indole ring. The furanose ring opens and then tautomerizes to *enol*-1-O-carboxyphenylamino-1-deoxyribulose phosphate.

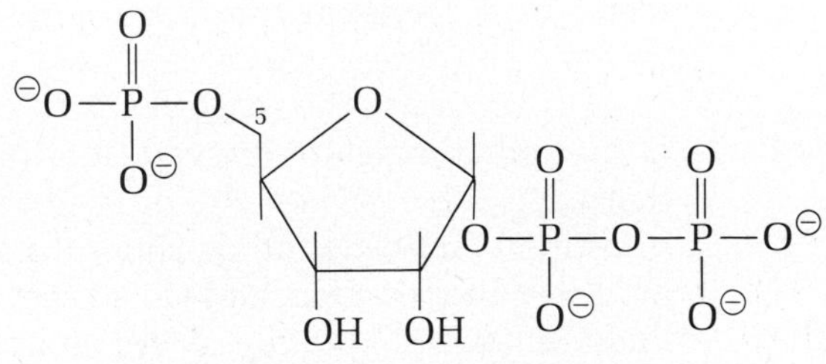

5-Phosphoribosylpyrophosphate (PRPP)

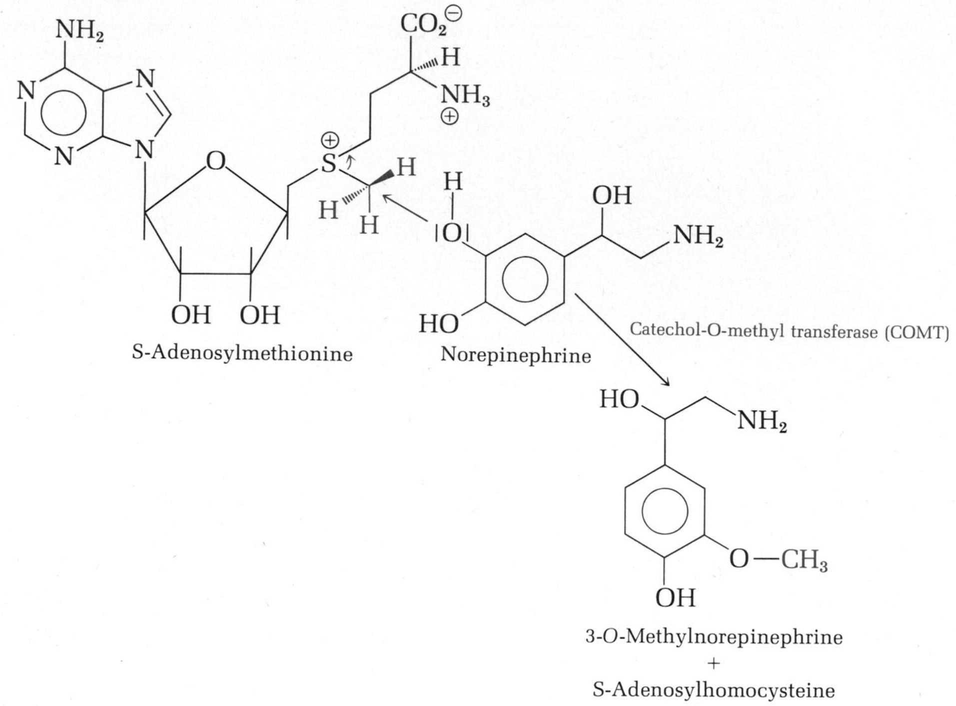

S-Adenosylmethionine

Norepinephrine

Catechol-O-methyl transferase (COMT)

3-O-Methylnorepinephrine
+
S-Adenosylhomocysteine

(22.46)

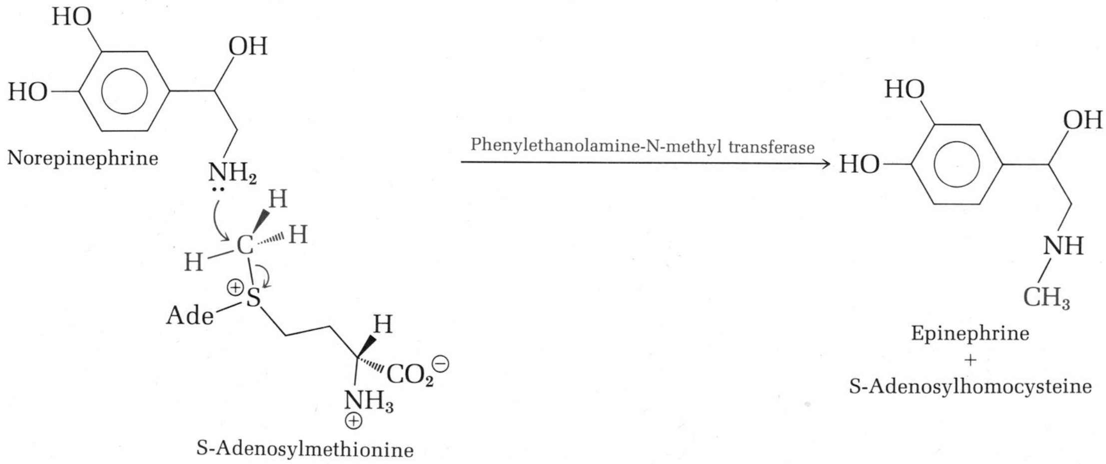

Norepinephrine

Phenylethanolamine-N-methyl transferase

S-Adenosylmethionine

Epinephrine
+
S-Adenosylhomocysteine

(22.47)

Cyclization yields indolglycerol-3-phosphate (Figure 22.33). The final step in tryptophan biosynthesis is reaction of indolglycerol-3-phosphate with the pyridoxal phosphate adduct of serine followed by elimination of glyceraldehyde 3-phosphate (Figure 22.34). The first step of this reaction can be visualized as attack by the enamine concealed in the indole ring upon the electrophilic methylene group of pyridoxal phosphate-bound dehydroserine. Aldol cleavage releases glyceraldehyde 3-phosphate. In *Neurospora crassa* the conversion of anthranilate to indolglycerol-3-phosphate is catalyzed by a single trifunctional enzyme. The final step, consisting of two enzyme activities, is also catalyzed by a single, bifunctional enzyme.

In retracing the reactions from the beginning we find that C-1 and C-6 of the indole ring are derived from phosphoenolpyruvate, by way of DAHP; that C-2, C-3, C-4, and C-5 of the indole ring are derived from erythrose 4-phosphate; that the indole ring nitrogen is derived from the amide nitrogen of glutamine; that C-1 and C-2 of the indole ring are derived from PRPP; and that the "alanyl" moiety of tryptophan is derived from serine (Figure 22.35).

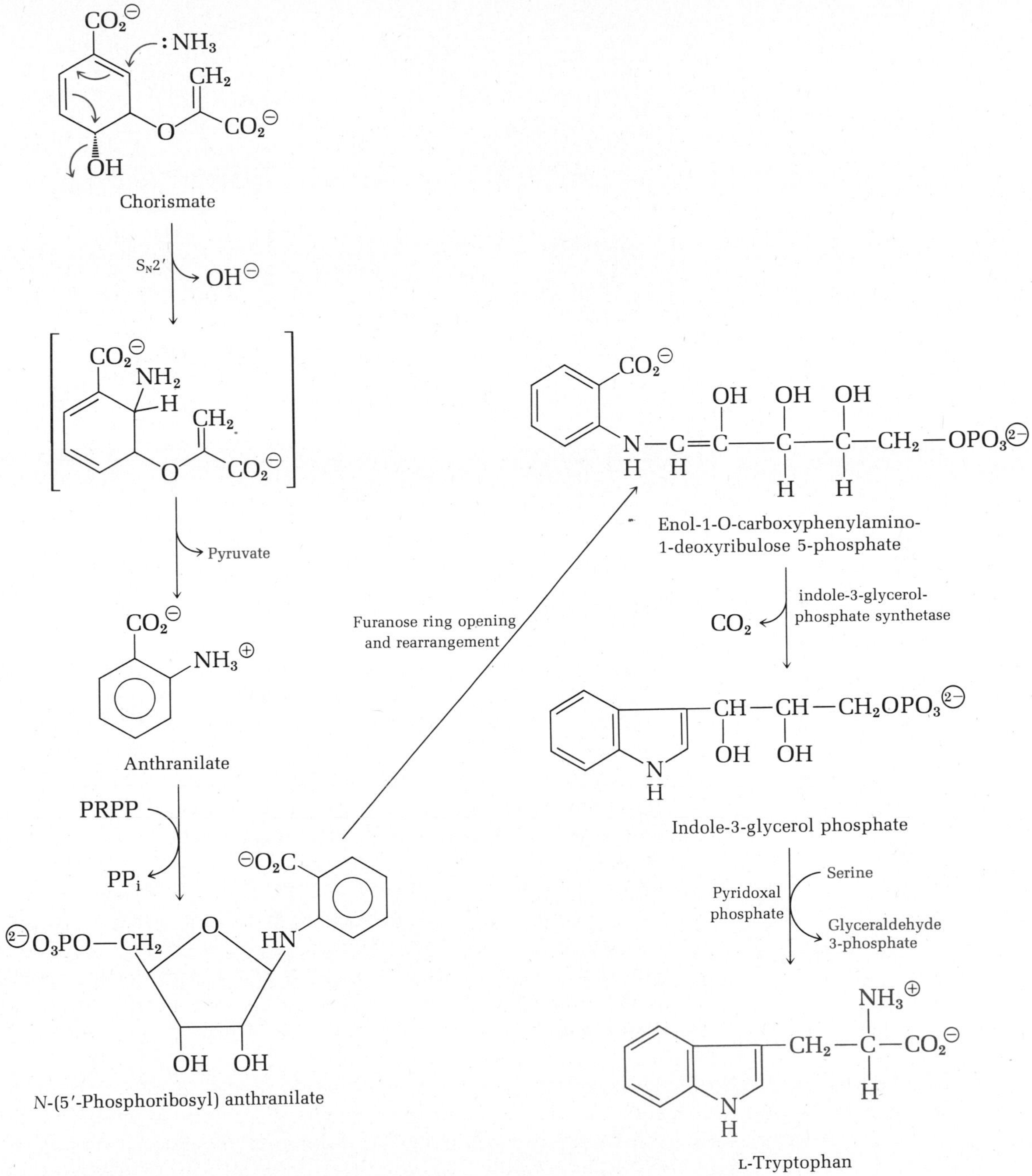

Figure 22.31 Conversion of chorismate to tryptophan.

E. Precursors and Products Related to Tryptophan

Tryptophan, its precursors, and its catabolic products, are a source of many important biosynthetic products. Among them are NAD and NADP, the plant hormone indole-3-acetate (*auxin*), and the neurotransmitter 5-hydroxytryptamine (*serotonin*). The plant alkaloids are also derived from tryptophan.

NAD and NADP are derived from tryptophan via the catabolic intermediate 2-amino-3-carboxymuconate semialdehyde. NAD is produced in four steps from this intermediate (Figure 22.36). (1) Ring closure of 2-amino-3-carboxymuconate semialdehyde gives quinolinate. (2) The nucleophilic nitrogen of the pyridine ring displaces pyrophosphate from PRPP, and the

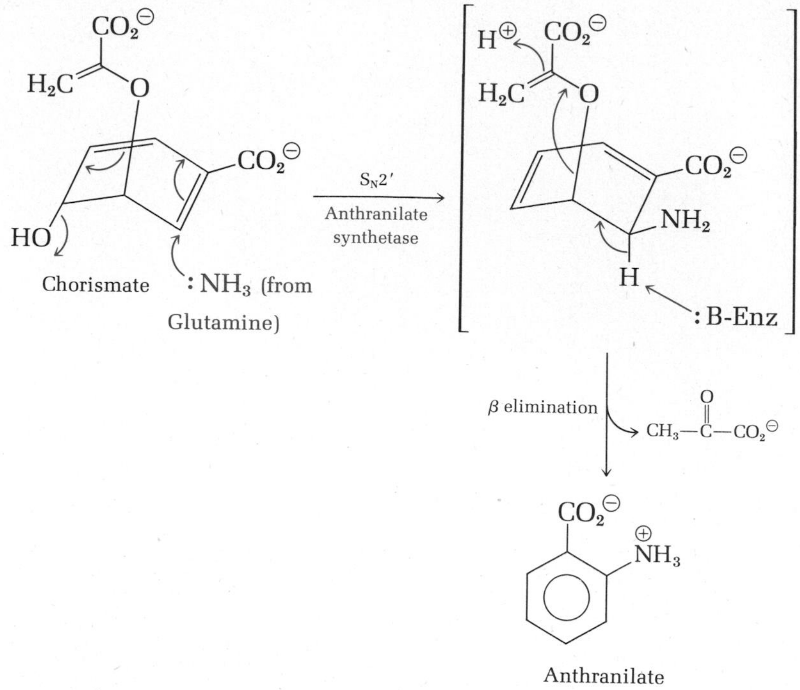

Figure 22.32   Conversion of chorismate to anthranilate.

2-carboxylate group is lost as CO_2 to form nicotinate mononucleotide. (3) The vitamin nicotinate also reacts with PRPP to yield the same product. In a similar reaction nicotinamide also reacts with PRPP yielding nicotinamide mononucleotide (NMN). (4) *NAD pyrophosphorylase* catalyzes conversion of nicotinate mononucleotide and nicotinamide mononucleotide to the dinucleotides, deamido-NAD and NAD, respectively. In the latter case the amino group is provided by the amido group of glutamine in an ATP-dependent transamidation. NAD may be converted to NADP by phosphoryl group transfer from ATP to the 3' position of the ribose.

Tryptophan leads directly to the plant hormone indole-3-acetate by transamination and decarboxylation. In another biosynthetic pathway a monooxygenase hydroxylates the 5 position of the indole ring, and decarboxylation gives the neurotransmitter serotonin (5-hydroxytryptamine). These pathways are also included in Figure 22.36.

F. Mechanism of Action of the Pterin-Dependent Amino Acid Hydroxylases

Three animal enzymes are known to require 5,6,7,8-tetrahydrobiopterin as a cofactor. These pterin-dependent enzymes—phenylalanine hydroxylase, tyrosine hydroxylase, and tryptophan hydroxylase—catalyze hydroxylation of aromatic amino acids. The pterin-dependent hydroxylases, or monoxygenases, all incorporate one atom of molecular oxygen in the hydroxylated product and one atom of molecular oxygen in water. Phenylalanine hydroxylase is the best understood of the three enzymes, and since it is likely that all have similar mechanistic characteristics, we shall limit our discussion to this enzyme.

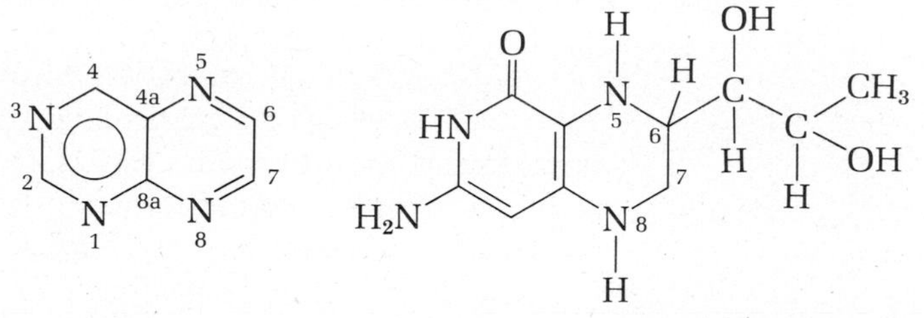

Pteridine

5,6,7,8-Tetrahydrobiopterin

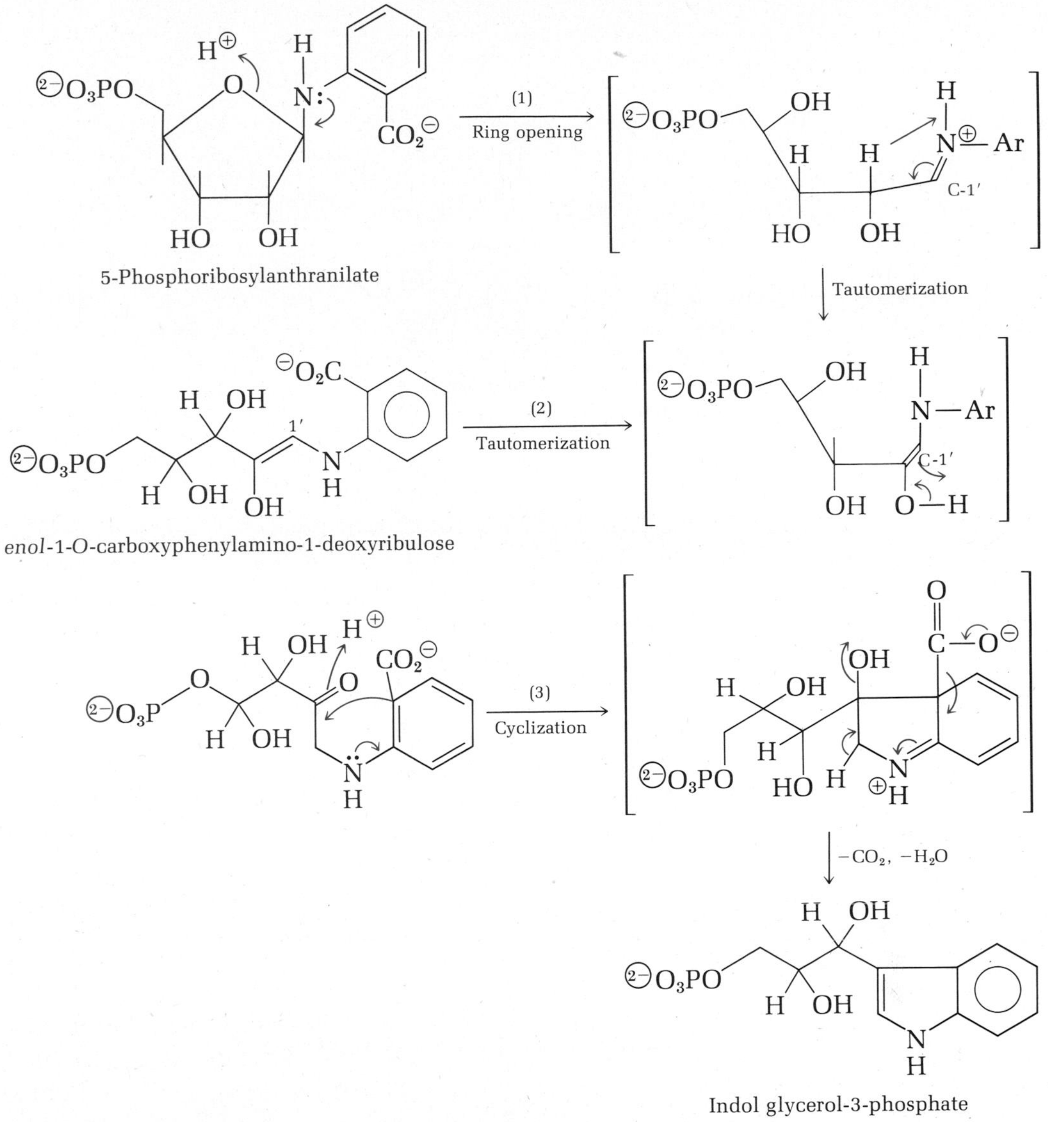

Figure 22.33 Conversion of anthranilate to indole glycerol phosphate. In N. *crassa* these reactions are catalyzed by a trifunctional enzyme.

Phenylalanine hydroxylase catalyzes the conversion of phenylalanine to tyrosine. Expanding upon our initial statement of this reaction in reaction 22.41, we find that 5,6,7,8-tetrahydrobiopterin is converted to *o-quinoid dihydrobiopterin* in the process (reaction 22.48).

Under physiological conditions *o*-quinoid dihydrobiopterin is in equilibrium with 7,8-dihydrobiopterin (reaction 22.49)

Neither of the dihydrobiopterin isomers is able to act as a cofactor in the enzymatic hydroxylation. The NADP-dependent dehydrogenase *dihydrobiopterin reductase* specifically reduces the *o*-quinoid isomer of dihydrobiopterin to 5, 6, 7, 8-tetrahydrobiopterin (reaction 22.50).

In the late 1960s workers at the National Institutes of Health who were studying phenylalanine hydroxylase found that phenylalanine labeled with tritium (^{3}H) at the *para* position migrates to the *meta* position when phenylalanine is hydroxylated (reaction 22.51). The rearrangement has entered the vernacular as the *NIH shift*. Since peroxides form epoxide intermediates that undergo rearrangement analogous to the NIH shift, perhaps phenylalanine hydroxylase also catalyzes electrophilic attack of oxygen as $OH^{\oplus}$ on the aromatic ring with formation of an epoxide or oxonium ion intermediate (reaction 22.52).

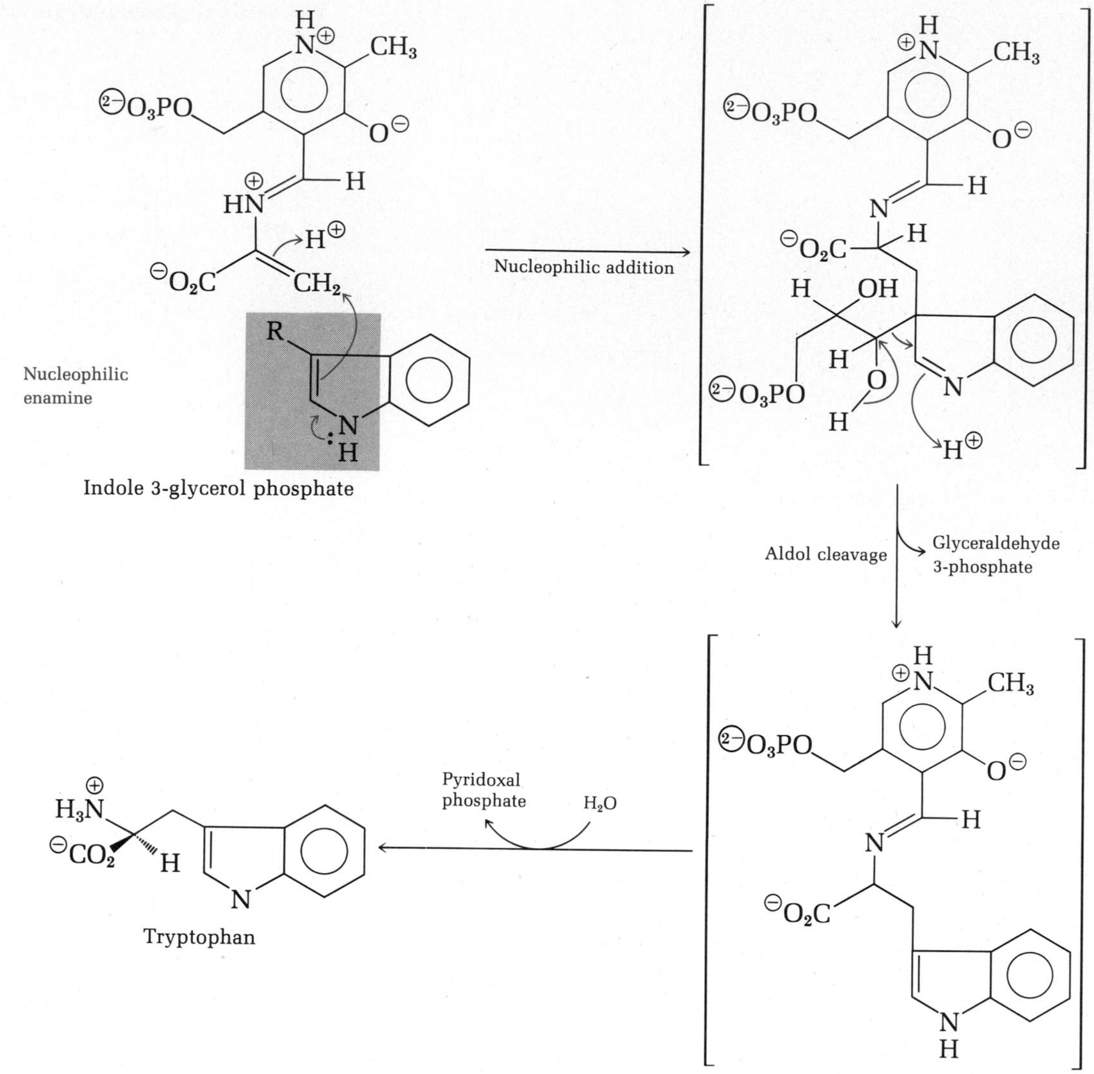

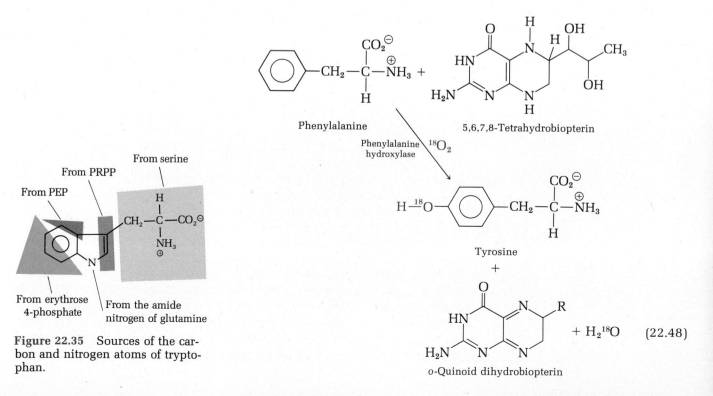

Figure 22.34 Conversion of indole 3-glycerol phosphate in tryptophan. In *N. crassa* these two reactions are catalyzed by a bifunctional enzyme.

Figure 22.35 Sources of the carbon and nitrogen atoms of tryptophan.

From serine

From PRPP

From PEP

From erythrose 4-phosphate

From the amide nitrogen of glutamine

Phenylalanine

5,6,7,8-Tetrahydrobiopterin

Phenylalanine hydroxylase $^{18}O_2$

Tyrosine

+

o-Quinoid dihydrobiopterin

$+ H_2{}^{18}O$ (22.48)

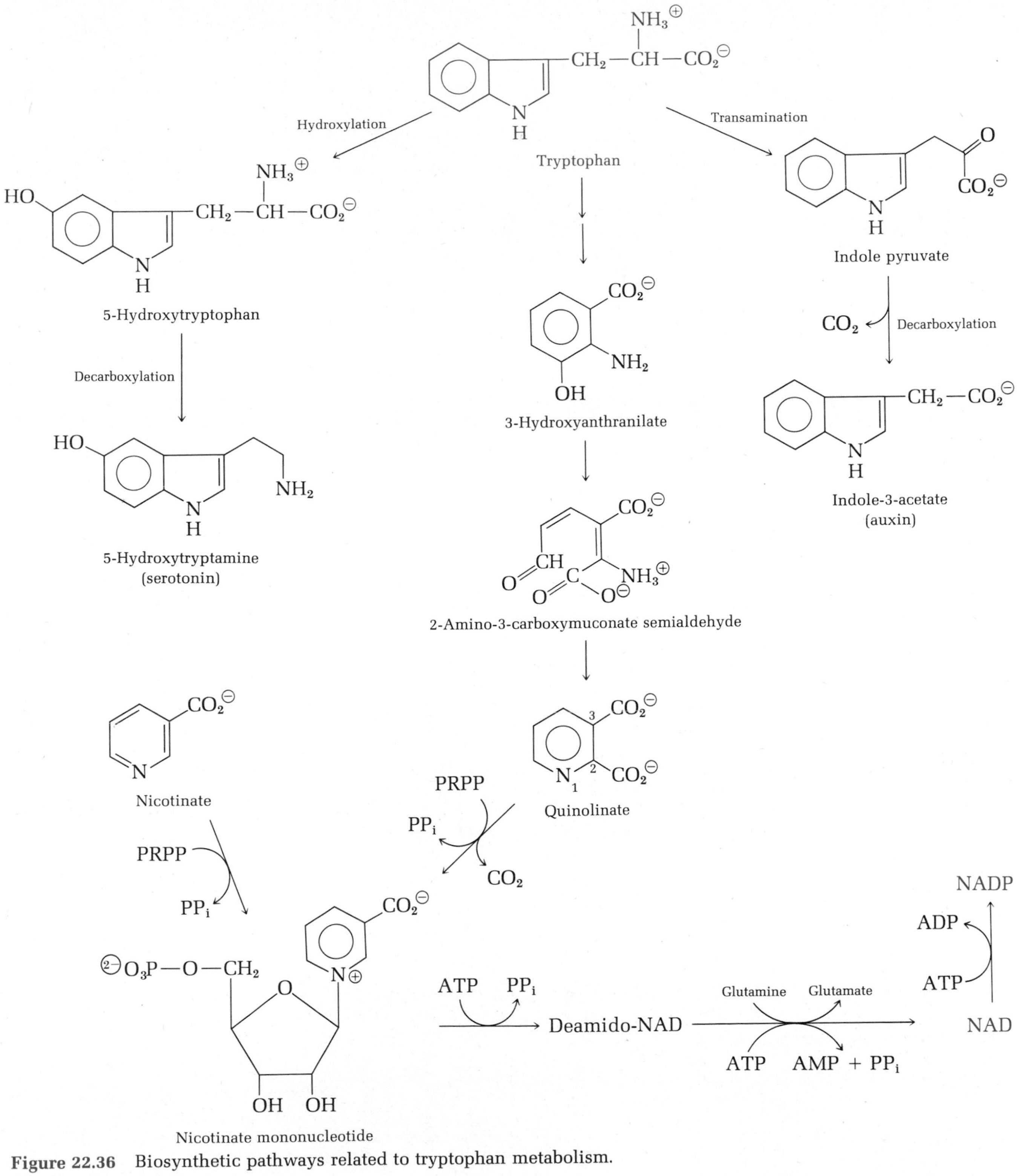

Figure 22.36 Biosynthetic pathways related to tryptophan metabolism.

22–9 BIOSYNTHESIS OF HISTIDINE

The biosynthesis of histidine requires 10 enzymatic steps (Figure 22.37). In the enteric bacteria S. typhimurium and E. coli these steps are catalyzed by nine enzymes. Most microorganisms and plants can synthesize histidine, but it is an essential amino acid for animals. Histidine is a negative feedback inhibitor of the first step of the pathway. This is the major mechanism of

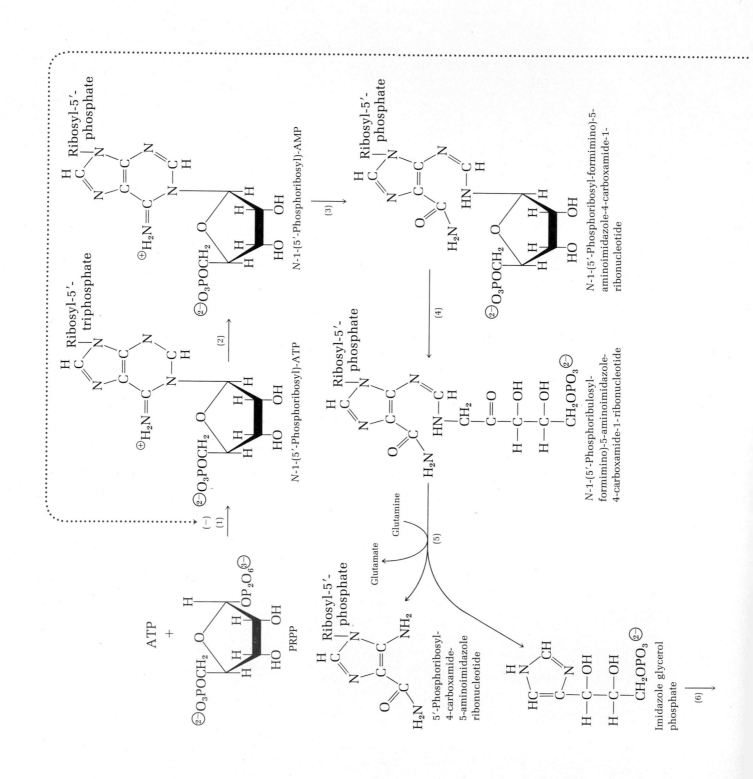

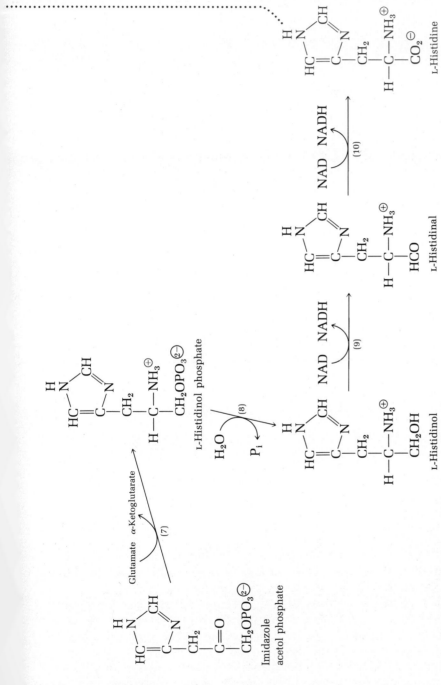

Figure 22.37 Pathway for the biosynthesis of histidine.

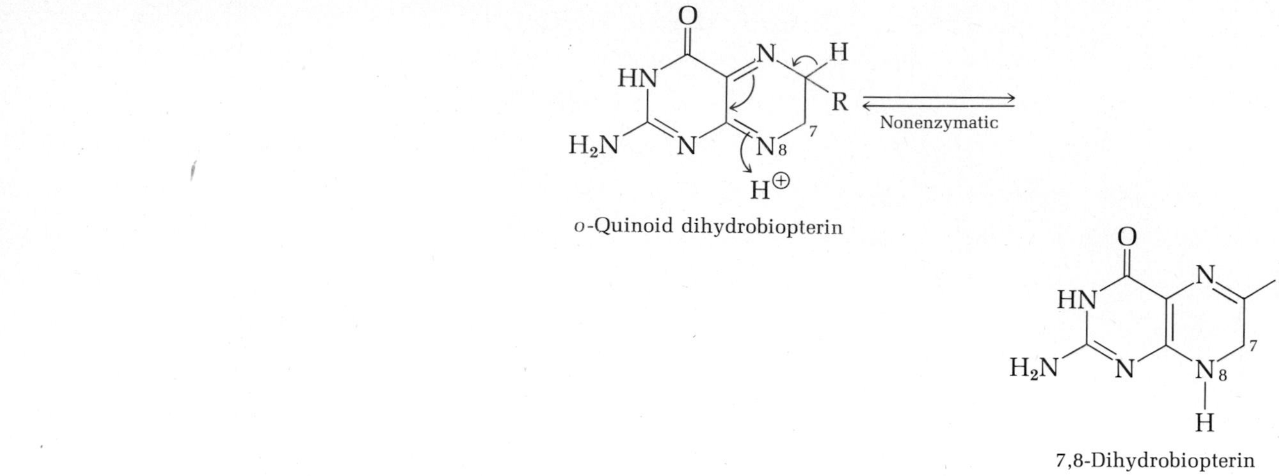
Nonenzymatic
o-Quinoid dihydrobiopterin
7,8-Dihydrobiopterin

(22.49)

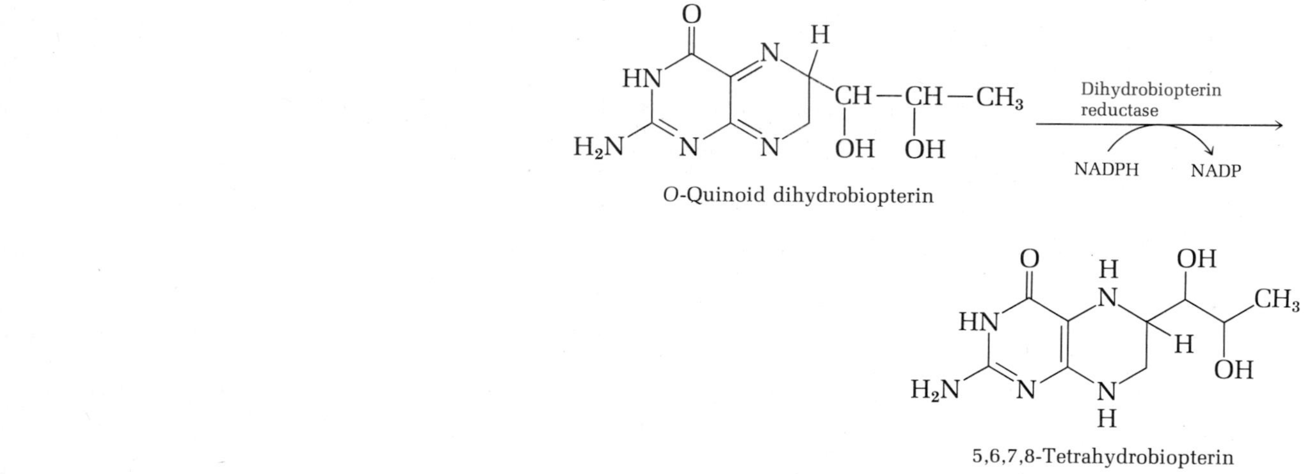
Dihydrobiopterin reductase
NADPH
NADP
O-Quinoid dihydrobiopterin
5,6,7,8-Tetrahydrobiopterin

(22.50)

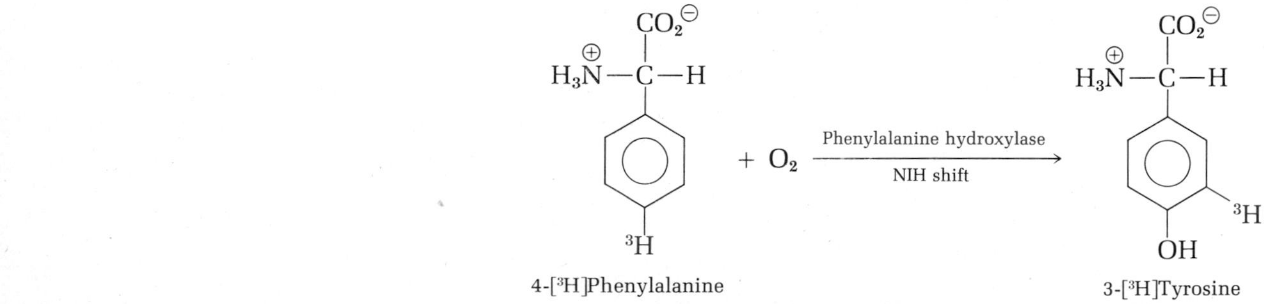
Phenylalanine hydroxylase
NIH shift
4-[³H]Phenylalanine
3-[³H]Tyrosine

(22.51)

metabolic control of histidine biosynthesis. We shall see in Chapter 27 that genetic regulation of synthesis of the enzymes responsible for histidine biosynthesis adds considerable complexity to this picture.

In step 1 of histidine biosynthesis, *ATP phosphoribosyl transferase* catalyzes the displacement of pyrophosphate from 5-phosphoribosyl 1-pyrophosphate (PRPP) by N-1 of ATP producing N-1-(5′-phosphoribosyl)-ATP. In step 2 hydrolysis of the pyrophosphate moiety of ATP gives N-1-(5′-phosphoribosyl)-AMP. C-6 of this product is bound to a strongly electron withdrawing iminium ion and is therefore quite susceptible to nucleophilic attack by water. We might expect hydrolysis to convert the iminium ion to a carbonyl group, but instead in step 3 *phosphoribosyl-AMP cyclohydrolase* catalyzes opening of the purine ring betwen N-1 and C-6, yielding N-1-(5′-phosphoribosyl-formimino)-5-aminoimidazole-4-carboxyamide-ribonucleotide. In step 4 an isomerase opens the ribofuranose ring that began as PRPP and isomerizes it to the ketose: N-1-(5′-phosphoribulosyl-

(22.52)

formimino)-5-aminoimidazole-4-carboxamide-ribonucleotide. In step 5 *glutamine amido transferase* catalyzes the synthesis of imidazole glycerol phosphate and 5-aminoimidazole-4-carboxamide ribonucleotide, an intermediate in the biosynthesis of purine nucleotides. The biosynthetic product that lies on the pathway to histidine is imidazole glycerol phosphate. In step 5 the amido group of glutamine, perhaps existing transiently as NH_3 at the enzyme active site, attacks the carbon atom of the formimino group. Cleavage of the formimino bond is followed by cyclization to the imidazole ring. The amido nitrogen of glutamine emerges as N-1 of the imidazole ring of histidine. N-2 and C-5 of the imidazole ring are derived from the original purine ring of ATP, and the remaining five carbon atoms of imidazole glycerol phosphate are derived from PRPP. Steps 4 and 5 are shown in Figure 22.38. In step 6 *imidazole glycerol phosphate dehydratase* catalyzes elimination of water. Tautomerization of the enol product yields imidazole acetol phosphate. In step 7 the glutamate-dependent enzyme *histidinol-phosphate transaminase* catalyzes transamination to give L-histidinol phosphate. In step 8 *histidinol phosphatase* hydrolyses the phosphate ester producing L-histidinol. Steps 9 and 10 are successive dehydrogenations catalyzed by the NAD-dependent enzyme *histidinol dehydrogenase*. The product of the first oxidation is L-histidinal, and the product of the second oxidation is L-histidine itself.

To recapitulate, we find that five of the carbon atoms are derived from the original substrate PRPP, the α-amino group is derived from glutamate, N-1 of the imidazole ring is derived from the amino group of glutamine, and N-3 and C-4 are derived from N-1 and C-2 of the purine ring of ATP (Figure 22.39).

22–10 HEME BIOSYNTHESIS

The heme prosthetic group of hemoglobin, myoglobin, cytochrome *c*, peroxidase, and catalase, hemes *a* and *b*, the chlorophylls, and the corrin ring system of vitamin B_{12} are all ultimately derived from glycine and succinyl-CoA. The first step in heme biosynthesis is the synthesis of δ-*aminolevulinate* from glycine and succinyl-CoA catalyzed by δ-*aminolevulinate synthetase* (reaction 22.53). This enzyme is located in the mitochondria of mammalian cells, and it requires pyridoxal phosphate for activity. The reaction begins by formation of the aldimine adduct of glycine and pyridoxal phosphate (step 1 in Figure 22.40). The enzyme then extracts the pro-R hydrogen from the glycine moiety of the adduct generating

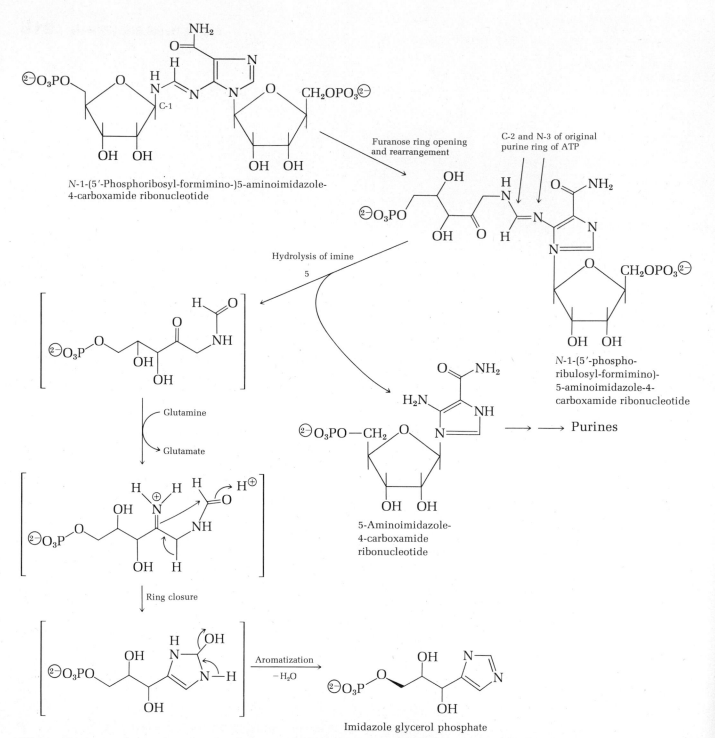

Figure 22.38 Synthesis of imidazole glycerol phosphate in steps 4 and 5 of histidine biosynthesis.

Figure 22.39 Sources of the carbon and nitrogen atoms of histidine.

$$H_3\overset{\oplus}{N}-CH_2-CO_2^{\ominus} + {}^{\ominus}O_2C-CH_2-CH_2-\overset{\overset{\displaystyle O}{\|}}{C}-S-CoA \xrightarrow{\text{δ-Aminolevulinate synthetase}}$$

$$\begin{array}{c} {}^{\ominus}O_2C \\ | \\ CH_2 \\ | \\ CH_2 \\ | \\ C=O \\ | \\ CH_2 \\ | \\ {}^{\oplus}NH_3 \end{array} \quad + \text{ CoA-SH} + CO_2 \qquad (22.53)$$

δ-Aminolevulinate

a resonance-stabilized carbanion (step 2). Nucleophilic attack by the carbanion displaces CoA-SH from succinyl-CoA (step 3). In the final steps of the reaction carbon dioxide is lost from the adduct (step 4), the resulting carbanion picks up a proton (step 5), and the aldimine hydrolyzes to give δ-aminolevulinate (step 6). The carbon dioxide lost in step 4 originated in

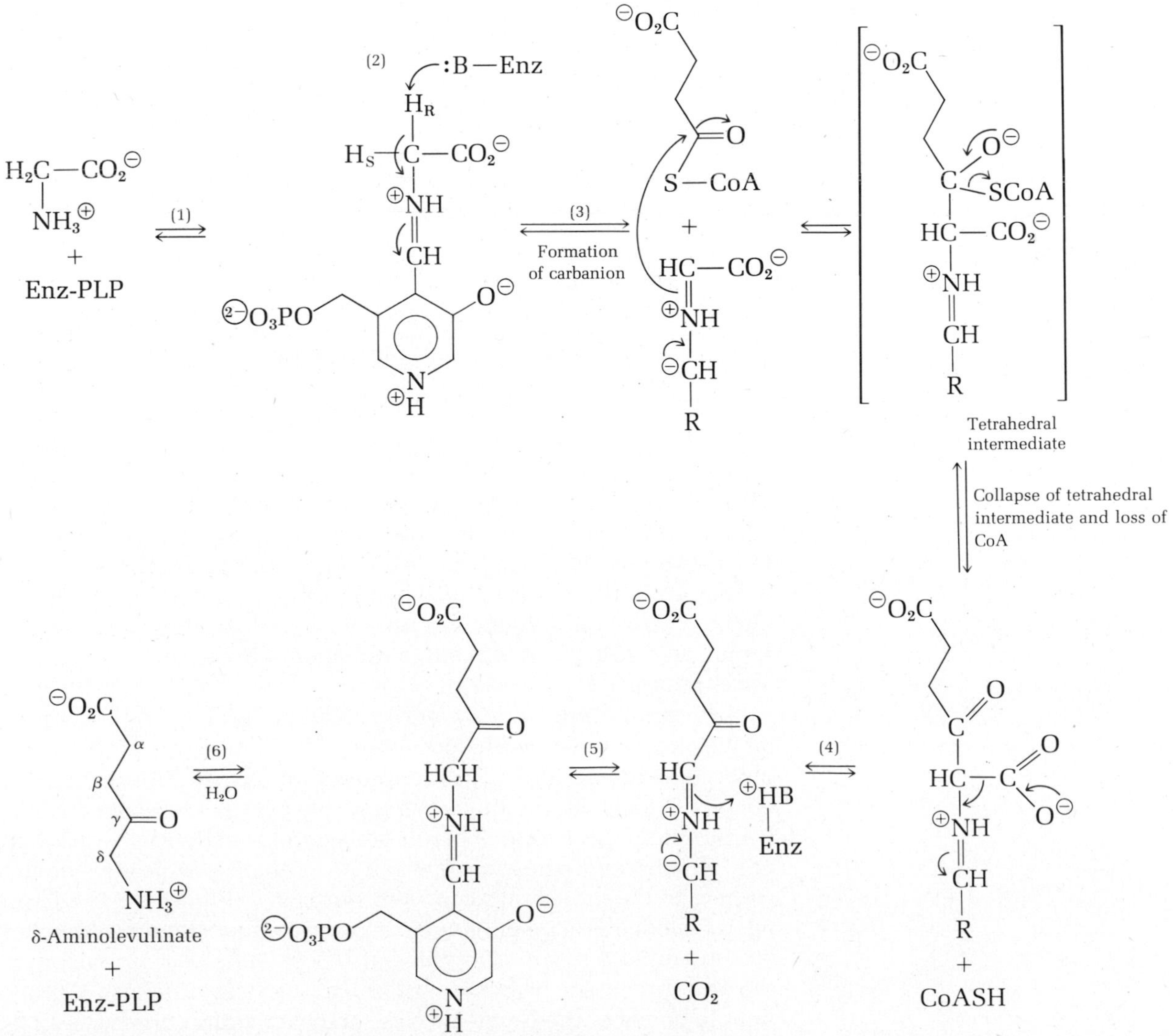

Figure 22.40 Mechanism of action of δ-aminolevulinate synthetase.

the carboxyl group of glycine. Thus δ-aminolevulinate contains four carbons derived from succinyl-CoA and the α-carbon and α-amino group of glycine.

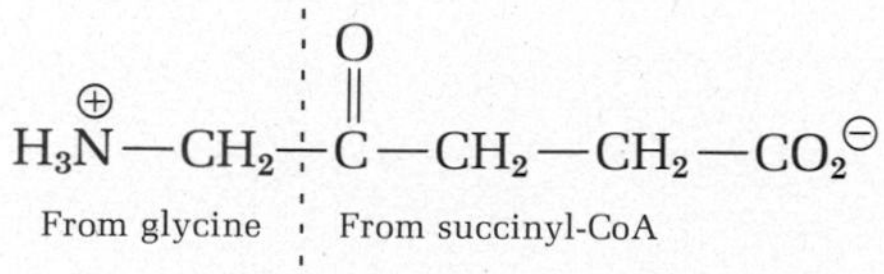

$$\overset{\oplus}{H_3N}-CH_2 \vdots \overset{\overset{\displaystyle O}{\|}}{C}-CH_2-CH_2-CO_2^{\ominus}$$

From glycine | From succinyl-CoA

Two moles of δ-aminolevulinate condense in the next step of the pathway to yield the *pyrrole* derivative *porphobilinogen* (reaction 22.54). (The asterisks denote atoms derived from glycine; all others are from the original succinyl-CoA.)

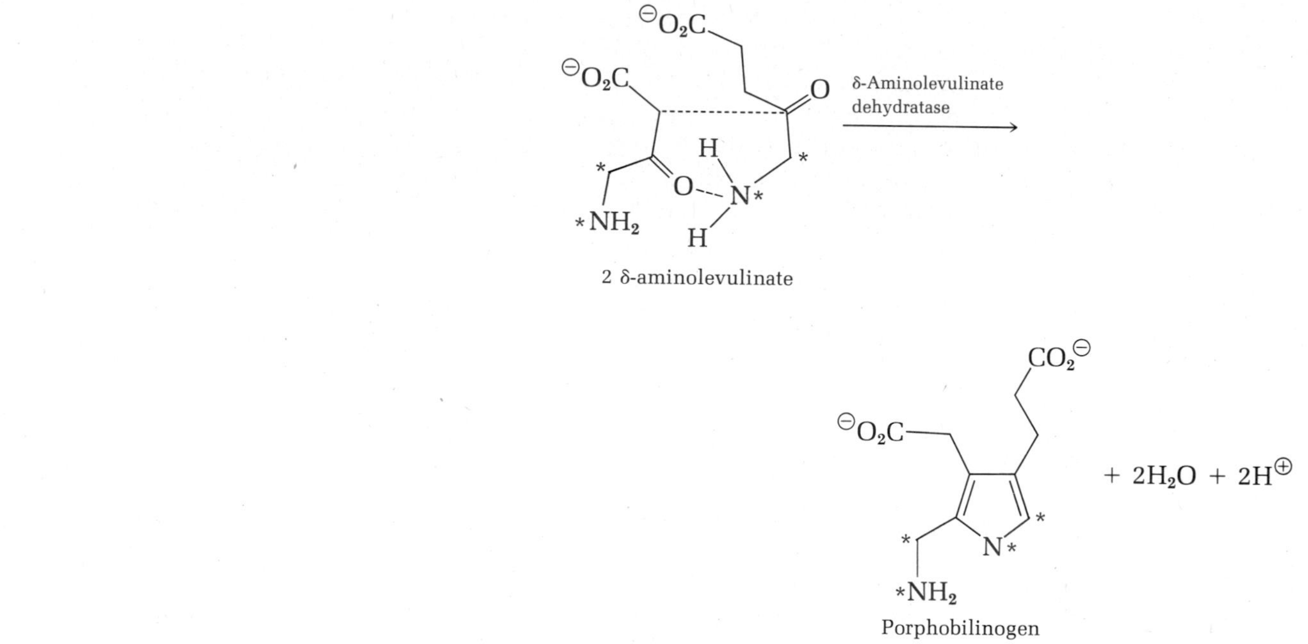

2 δ-aminolevulinate

δ-Aminolevulinate dehydratase

$$+ \ 2H_2O + 2H^{\oplus} \qquad (22.54)$$

Porphobilinogen

The reaction, catalyzed by *δ-aminolevulinate dehydratase*, proceeds in four steps. (1) An aldol condensation between 2 moles of δ-aminolevulinate occurs (Figure 22.41). (2) The initial aldol adduct loses water (Figure 22.42). (3) The conjugated acylic enzyme–bound intermediate formed in step 2 cyclizes with cleavage of the covalent bond to the enzyme to give a pyrrole imine product (Figure 22.43). (4) Abstraction of a proton α to the pyrrole nitrogen gives the product, porphobilinogen (Figure 22.44). The acetate side chain of porphobilinogen will become a methyl group in heme, and the propionic acid side chain will either remain unaltered, or be decarboxylated to a vinyl group.

Porphobilinogen deaminase catalyzes condensation of four molecules of porphobilinogen with formation of a *linear tetrapyrrole*. Each time a methylene bridge forms, an ammonium ion is lost (Figure 22.45). If no other enzyme is present, the linear tetrapyrrole cyclizes to give *uroporphyrinogen I*, a product whose double bonds are symmetrical with respect to two planes of symmetry and whose acetate and propionate groups are symmetrical with respect to the molecule's center of symmetry (Figure 22.46). Uroporphyrinogen I is not a precursor of heme. However, if porphobilinogen deaminase is accompanied by *uroporphyrinogen III cosynthetase*, uroporphyrinogen III, the true precursor of heme, is formed. The cosynthetase tautomerizes the double bonds of the D pyrrole ring, and cyclization produces an asymmetric

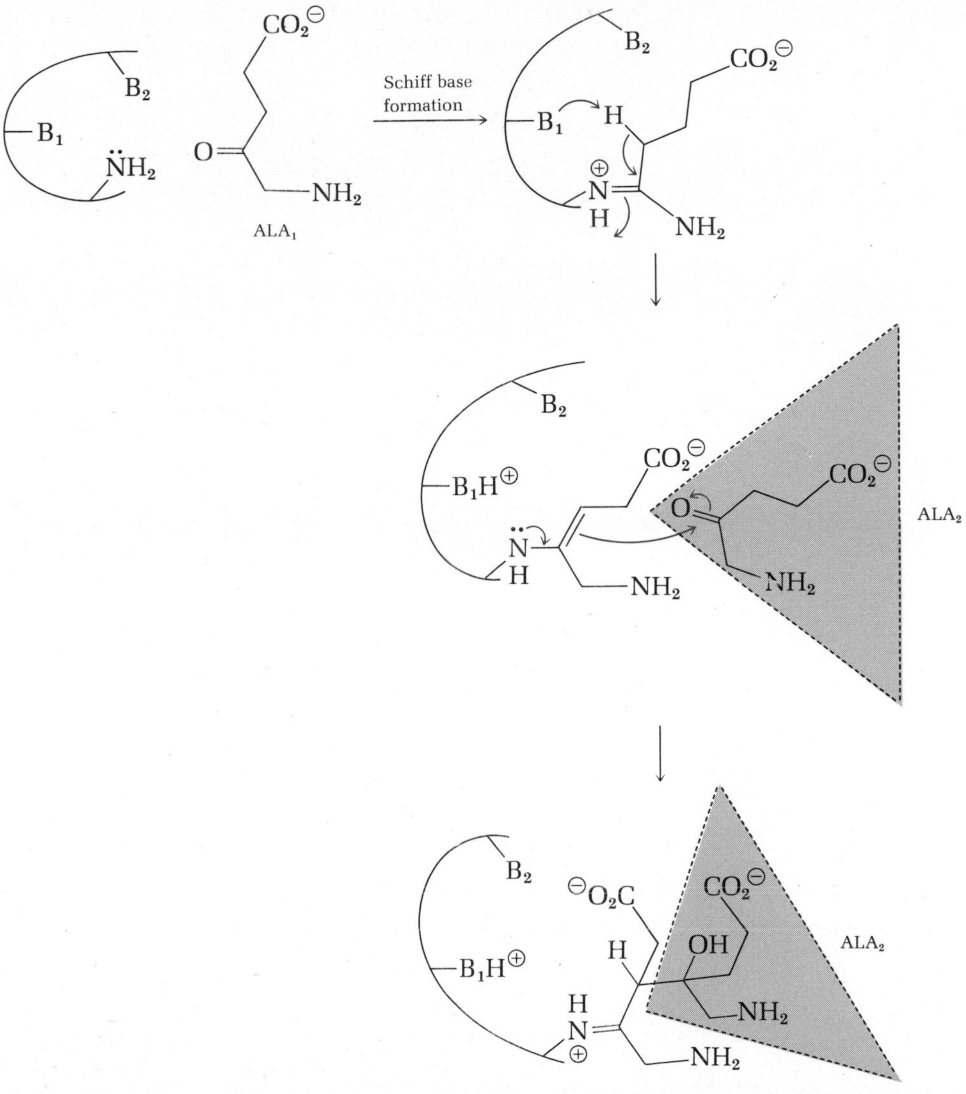

Figure 22.41 An aldol condensation is the first step in the synthesis of porphilobilinogen catalyzed by δ-aminolevulinate dehydratase. The 2 moles of δ-aminolevulinate are abbreviated as ALA$_1$ and ALA$_2$.

product whose acetate and propionate side chains are no longer symmetric with respect to the molecule's center of symmetry (Figure 22.47).

Protoporphyrin IX, the direct precursor of heme, results from a series of oxidation and decarboxylation reactions of uroporphyrinogen III. The iron(II) atom of heme is enzymatically inserted by *ferrochelatase* (reaction 22.55). Asterisks indicate the atoms originating in glycine.

Defects in heme metabolism are known collectively as *porphyrias.* There are 3 major classes of hereditary porphyria. In *erythropoietic uroporphyria* uroporphyrinogen III cosynthetase is present at only about one-third its normal concentration, and large amounts of uroporphyrin I are deposited in tissues and eventually excreted in urine and feces. This hereditary defect is transmitted as an autosomal recessive trait. The urine in affected persons is red, and their teeth and skin fluoresce strongly in UV light. In extreme cases the skin of affected persons is extremely photosensitive, and they may suffer from severe neurological attacks. *Erythropoietic protoporphyria,* the second class of hereditary porphyria, has quite similar symptoms. This disease is transmitted as an autosomal dominant trait.

Acute intermittent porphyria is an hereditary disease of the liver in

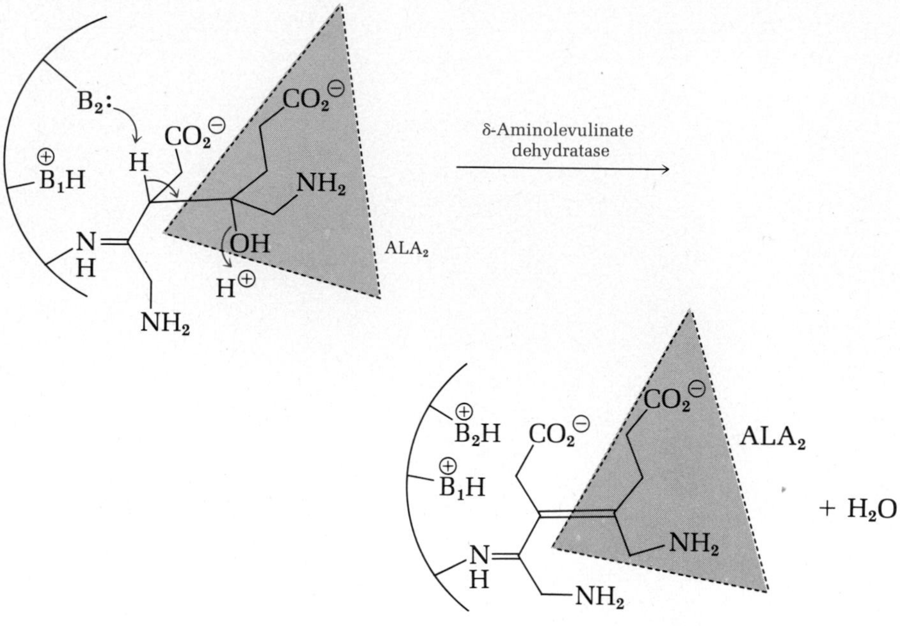

Figure 22.42 Loss of water from the aldol adduct formed between 2 moles of δ-aminolevulinate catalyzed by δ-aminolevulinate dehydratase.

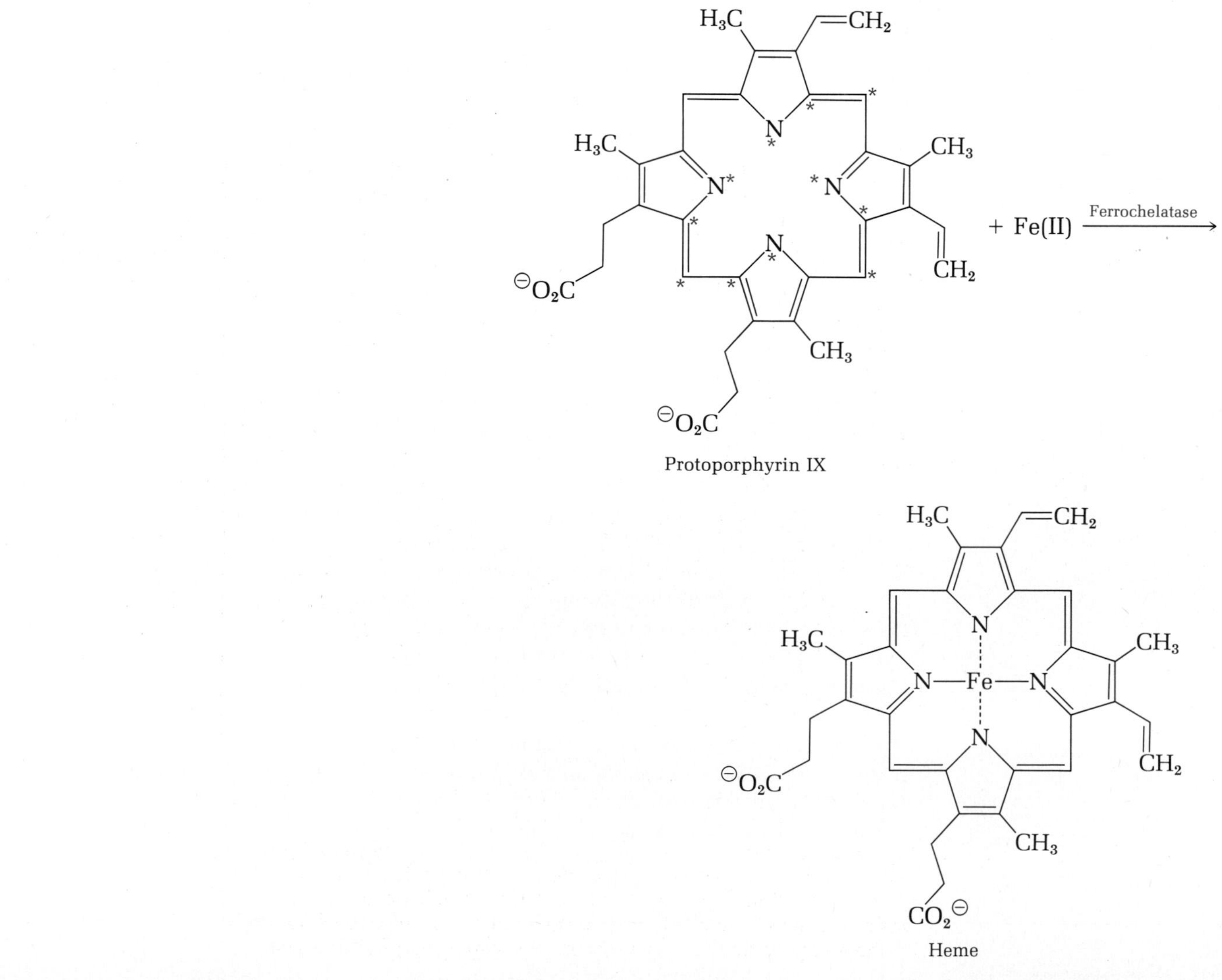

Protoporphyrin IX

(22.55)

Heme

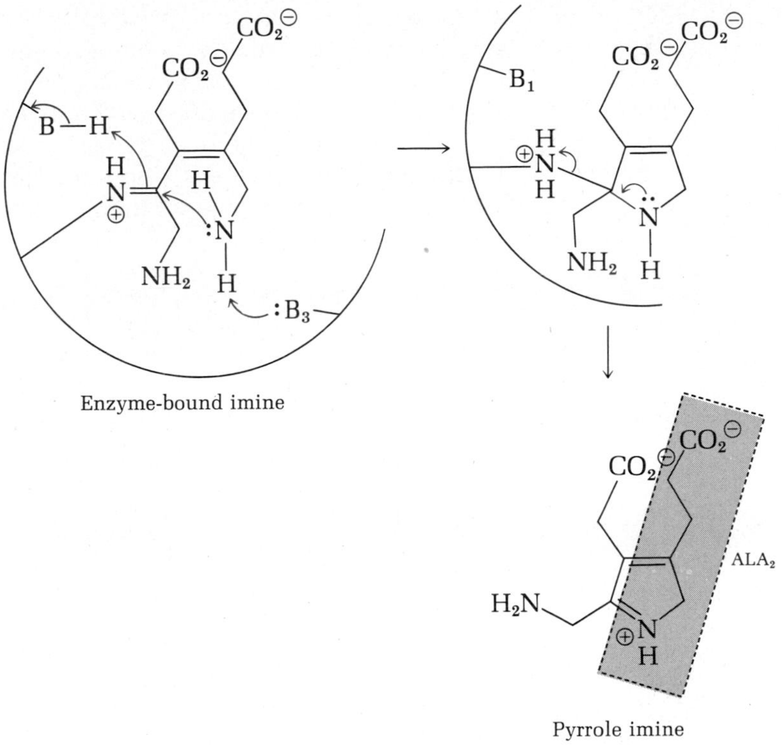

Enzyme-bound imine

Pyrrole imine

Figure 22.43 Cyclization of the acrylic, enzyme-bound intermediate to give a pyrrole imine product. The net result of the cyclization is *transaldiminization* since the enzyme-substrate adduct is an imine and the product is also an imine. The atoms in the box are derived from the first molecule of δ-aminolevulinate (ALA$_1$); those outside the box are derived from the second molecule (ALA$_2$).

which the activity of δ-aminolevulinate synthetase is abnormally active. Large amounts of δ-aminolevulinate and porphobilinogen are excreted in the urine. Wine red urine and neurological disorders are the result of this "inborn disease of metabolism." Barbiturates, estrogens, and other drugs can cause severe outbreaks of otherwise mild cases of acute intermittent porphyria by inducing a large increase in the rate of synthesis of δ-aminolevulinate synthetase. King George III of England almost certainly suffered from a variation of acute intermittent porphyria. He often experienced severe mania, and his irrational policies at critical points in the American struggle for independence might very well have been caused by biochemical defect over which he had no control.

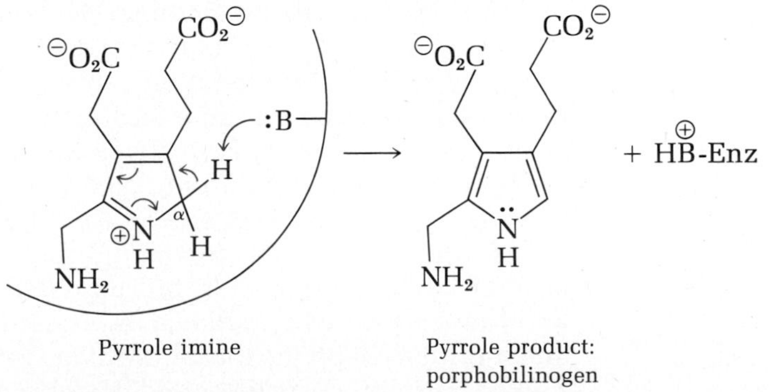

Pyrrole imine

Pyrrole product: porphobilinogen

Figure 22.44 Conversion of pyrrole imine to pyrrole to give porphobilinogen. The conversion is aided by the acidity of the protons α to the imine nitrogen, which are easily abstracted by a basic residue at the active site of the enzyme, and by the stability of the aromatic product, the pyrrole derivative porphobilinogen.

22–11 THE BILE PIGMENTS

Erthrocytes have a lifetime of about 120 days. After they die, the heme of hemoglobin is degraded by a *microsomal heme oxygenase* that requires O_2 and NADPH. This oxygenase is coupled to the cytochrome P450 electron transport system that we discussed in the biosynthesis of pregnenolone. The oxidase converts the methine bridge between the A and B rings of heme to a vinyl alcohol, and further oxidation gives the diketo derivative *biliverdin*, Fe(II), and carbon monoxide, derived from the methine bridge carbon (reaction 22.56).

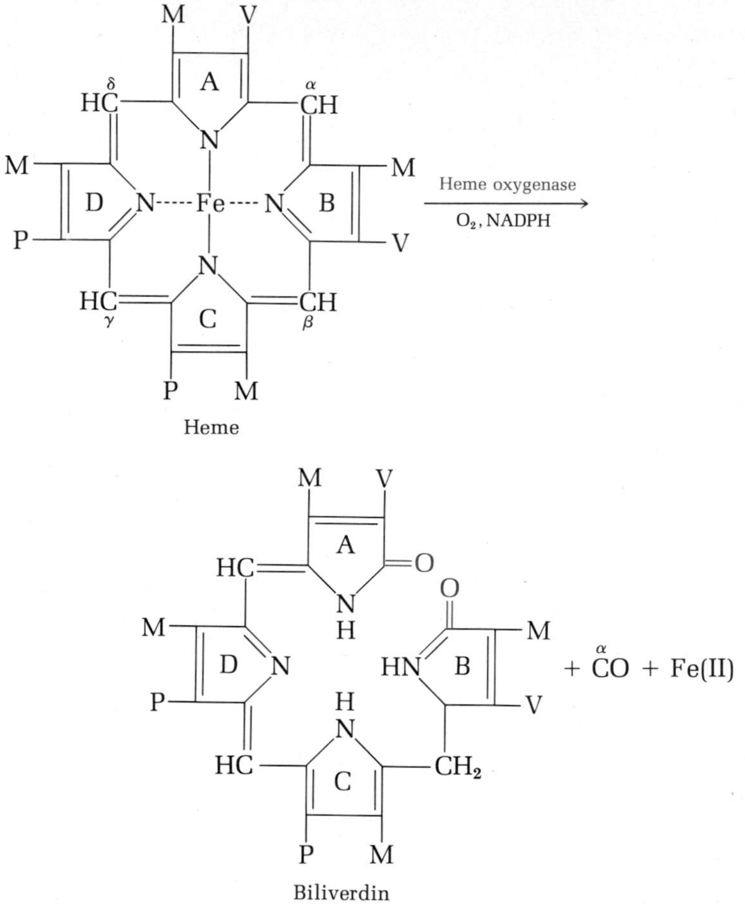

Heme

Heme oxygenase
O_2, NADPH

$$+ \overset{\alpha}{C}O + Fe(II) \qquad (22.56)$$

Biliverdin

The iron is transported by *plasma transferrin* to bone, where it is eventually incorporated in newly synthesized hemoglobin. Conversion of biliverdin to *bilirubin* is catalyzed by the NADPH-dependent enzyme *biliverdin reductase* (reaction 22.57). Bilirubin is transported to the liver in a complex with serum albumin. A propionic acid side chain of bilirubin is then conjugated to glucuronide, derived from UDP-glucuronate, in a reaction catalyzed by *bilirubin-UDP glucuronyl transferase* (reaction 22.58). A second glucuronide is bound to another propionic acid side chain of bilirubin yielding *bilirubin diglucuronide*, which is secreted into bile.

If the liver is unable to form bilirubin glucuronides or if there is excessive destruction of red cells, bilirubin may accumulate in the blood plasma and cause a yellowing of the skin, a condition called *jaundice*. In *cirrhosis* and *infectious hepatitis* the failure of the liver to keep up with excess production of bilirubin gives the jaundice characteristic of those diseases.

In red-green and in red algae open chain tetrapyrroles are part of the light-gathering apparatus of photosynthesis. The tetrapyrroles are derived from phycocerythrobilin by way of biliverdin (Figure 22.48).

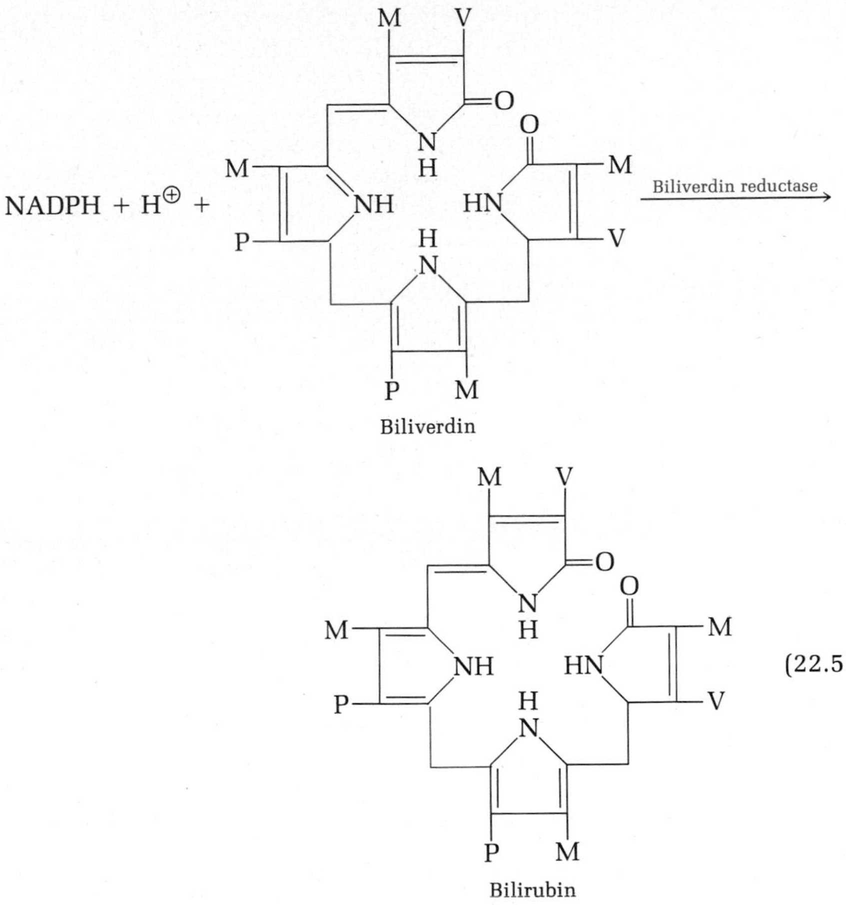

$$NADPH + H^{\oplus} + \quad \text{Biliverdin} \quad \xrightarrow{\text{Biliverdin reductase}}$$

Bilirubin

(22.57)

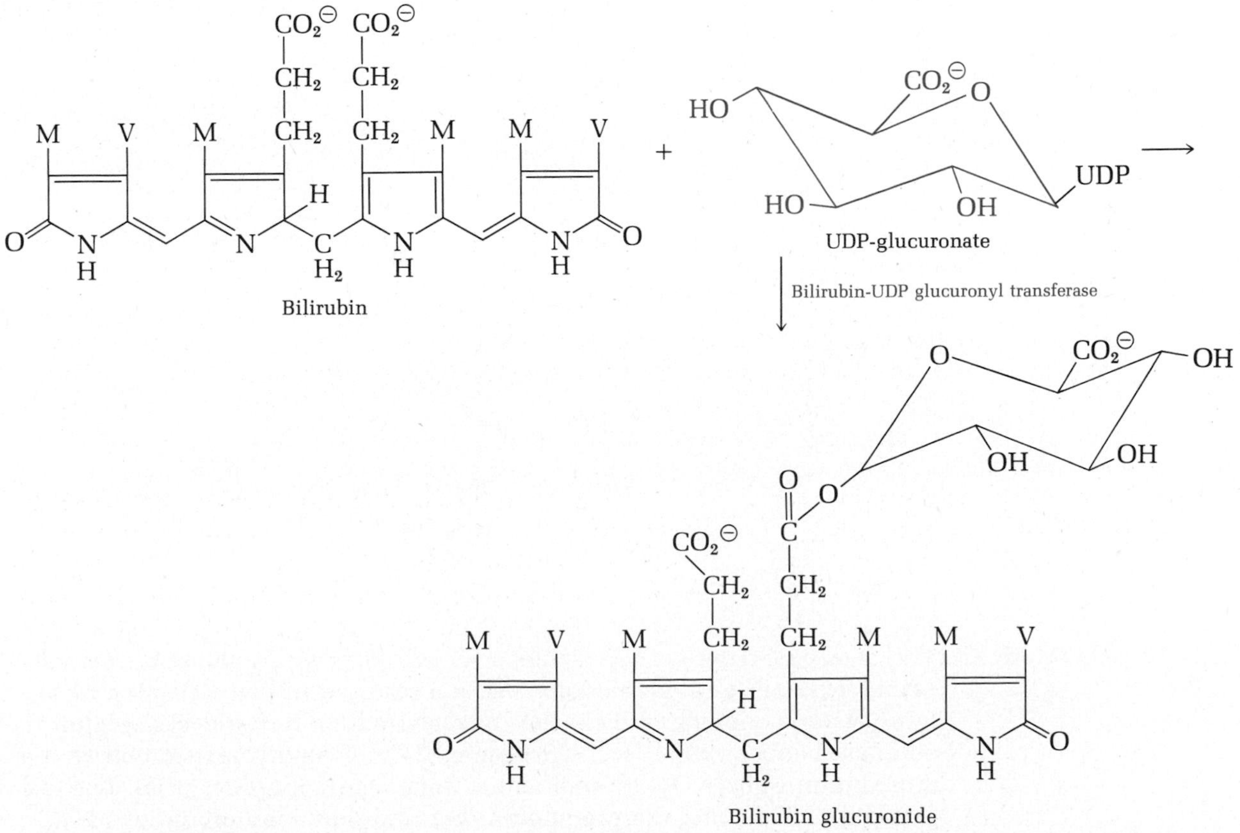

Bilirubin + UDP-glucuronate $\xrightarrow{\text{Bilirubin-UDP glucuronyl transferase}}$

Bilirubin glucuronide

(22.58)

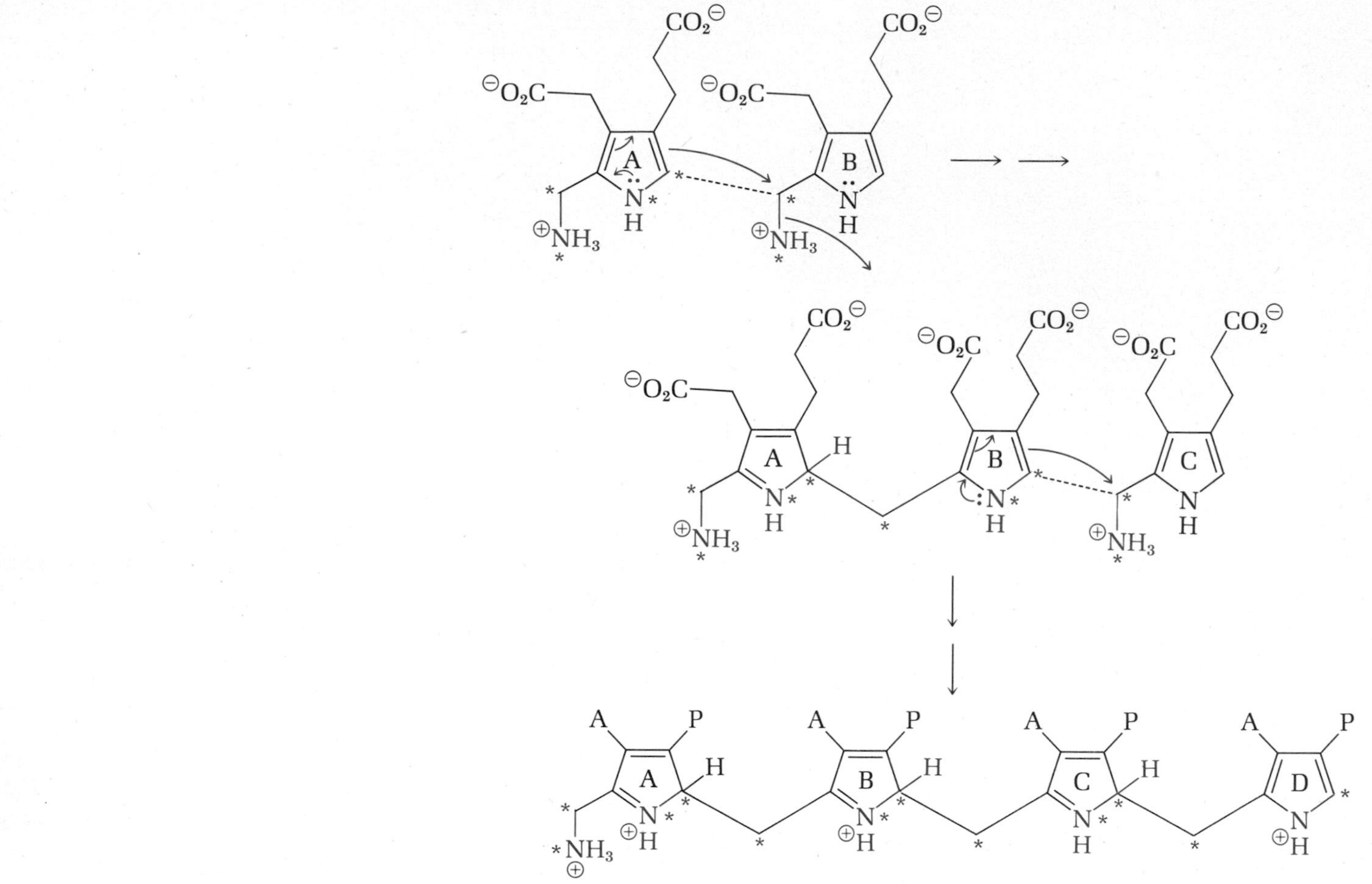

Figure 22.45 Polymerization of four molecules of porphobilinogen to yield a linear tetrapyrrole. Asterisks indicate atoms derived from glycine; all others are derived from succinyl-CoA. Abbreviations: A, acetate; P; propionate; A–D, rings in heme.

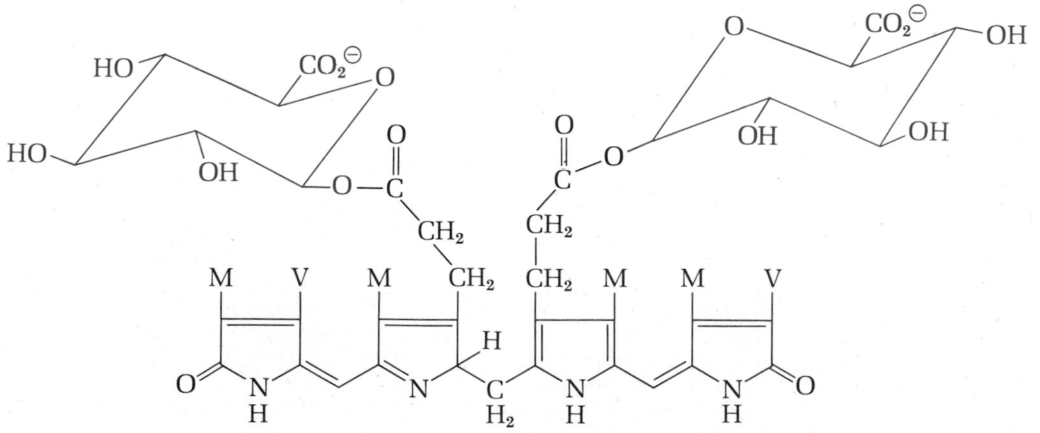

Bilirubin diglucuronide

22–12 SUMMARY

The nitrogen required for amino acid biosynthesis is obtained by most plants, animals, and microorganisms as a result of nitrogen fixation by certain bacteria, among which the nitrogen-fixing bacteria *Rhizobium* is perhaps the most important. Nitrogenase is the enzyme responsible for reducing atmospheric N_2 to ammonium ions. The reduction of N_2 is a six-electron process, and the production of 2 moles of ammonium ions from 1 mole of N_2 requires 12 moles of ATP. In higher organisms the major nitrogen

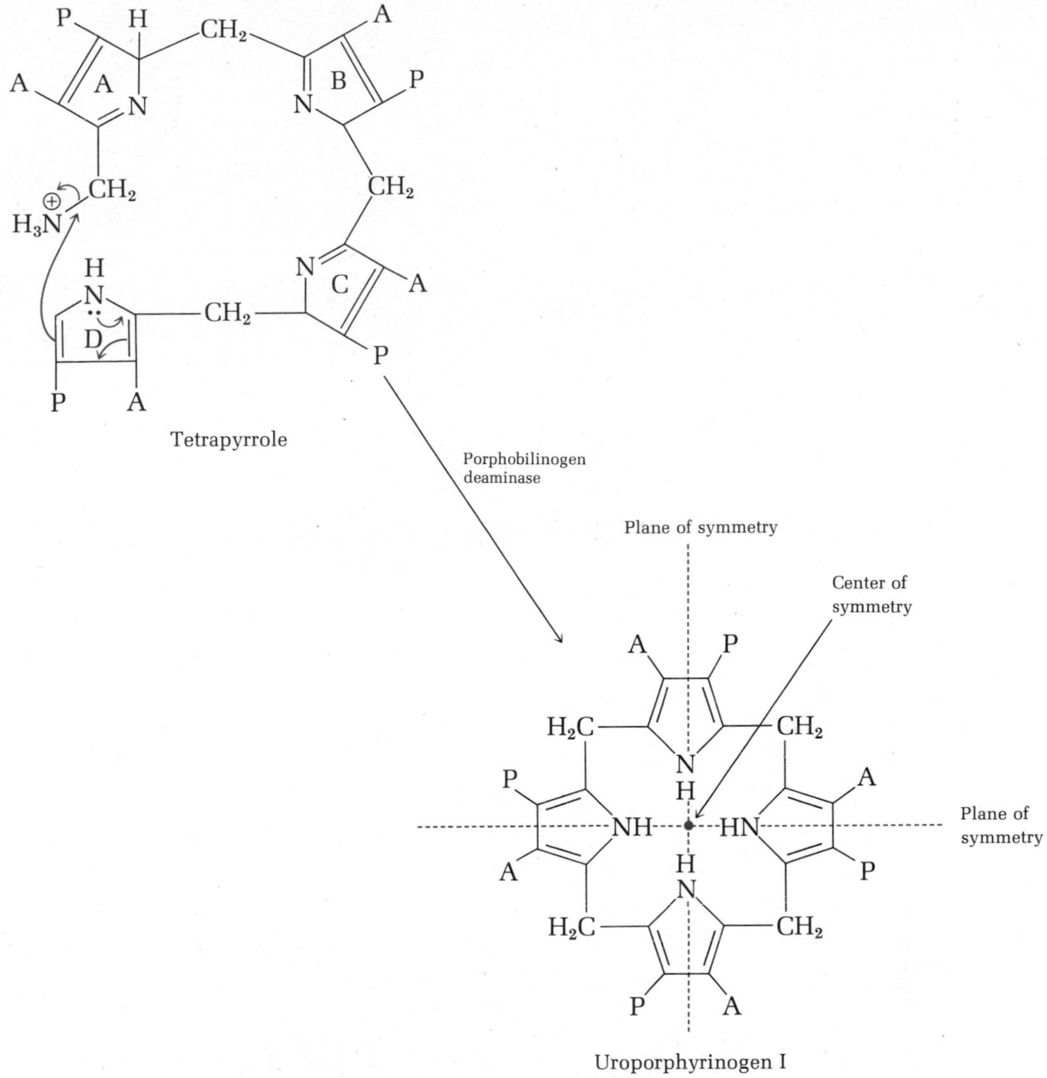

Figure 22.46 Cyclization of the linear tetrapyrrole in the presence of porphobilinogen deaminase to yield uroporphyrinogen I, a symmetrical molecule that is not a precursor of heme. Abbreviations: A, acetate; P, propionate, A–D, rings of heme.

sources for biosynthesis are the amido group of glutamine, the α-amino group of glutamate, and (as we shall see in the next chapter) carbamoyl phosphate rather than ammonium ions.

Organisms vary greatly in the number of amino acids that they can synthesize. *E. coli* can carry out net synthesis all 20 amino acids, whereas man and the albino rat can carry out net synthesis only 10. The remaining 10 must be supplied in the diet, and they are called essential amino acids. Those that can be synthesized are called nonessential.

The carbon sources for amino acid biosynthesis are the citric acid cycle intermediates oxaloacetate and α-ketoglutarate, the glycolytic intermediates phosphoenolpyruvate, 3-phosphoglycerate, and pyruvate, and the pentose phosphate pathway intermediates erythrose 4-phosphate and ribose 5-phosphate. These carbon sources provide a scheme for dividing pathways for amino acid biosynthesis into five families. The common character of the members of the same family is a common carbon source.

The pathways for the nonessential amino acids are quite simple, consisting for the most part of transamination of an α-keto acid or of a few simple steps. Thus the biosynthesis of L-alanine, L-glutamate, and L-aspartate occurs by transamination of either glycolytic or citric acid cycle

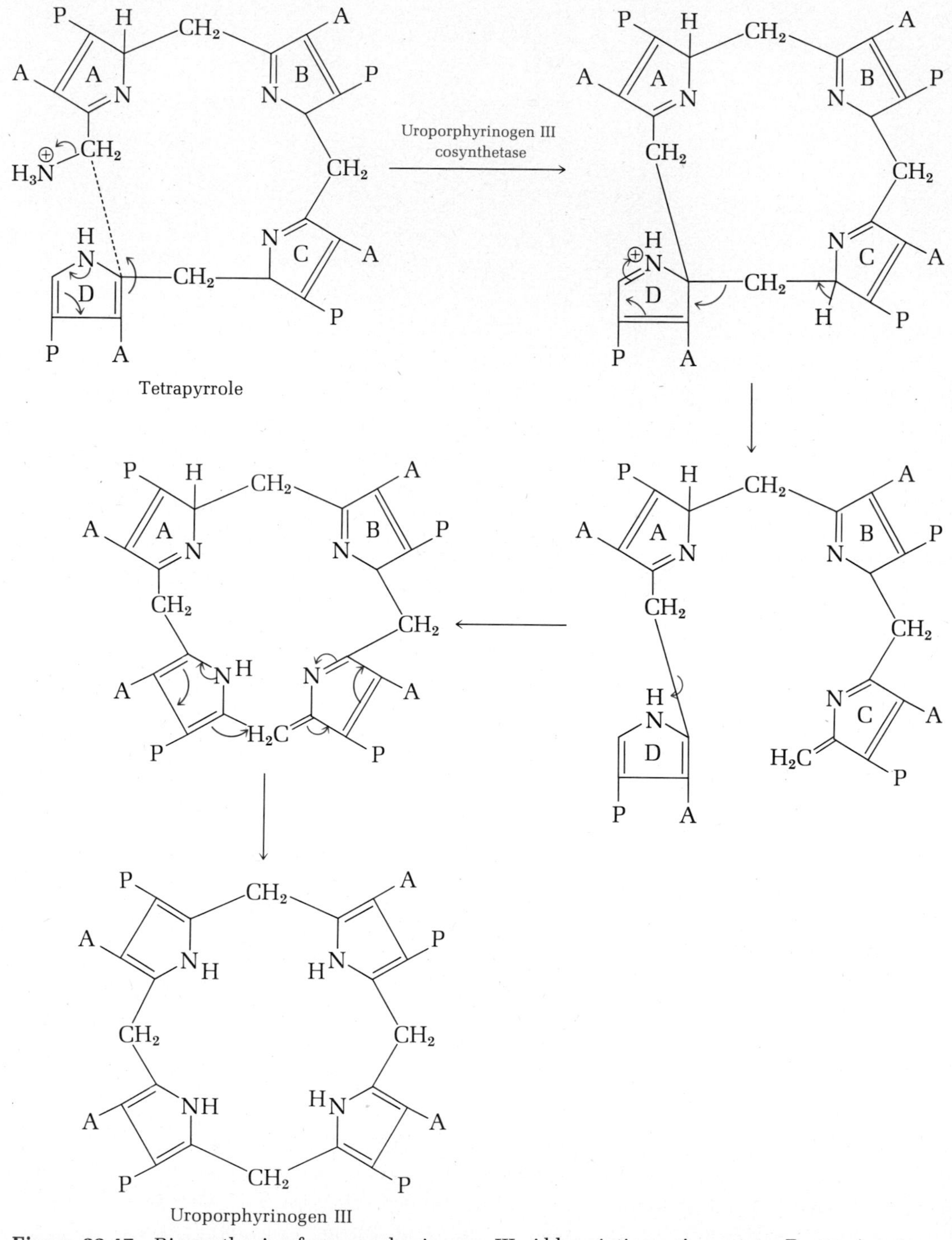

Figure 22.47 Biosynthesis of uroporphyrinogen III. Abbreviations: A, acetate; P, propionate; A–D, rings of heme.

intermediates. The biosynthesis of the amide derived from glutamate and asparagine is somewhat more complicated, but glutamine is synthesized from ammonium ions; the energy required to drive the reaction is supplied by ATP. Similarly, asparagine is synthesized from aspartate and ammonium ions in an ATP-driven process. The synthesis of glutamine is catalyzed by the oligomeric enzyme glutamine synthetase. Glutamine synthetase is subject to regulation by covalent modification and by feedback inhibition by the products of eight separate biosynthetic pathways. Proline is derived from glutamate and is formed by a four-step pathway. Cysteine is produced from

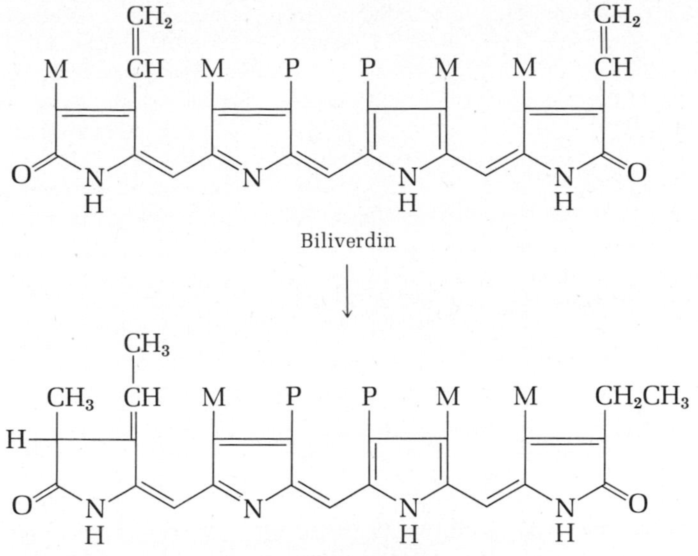

Figure 22.48 Conversion of biliverdin to phycoerythroblin, an antenna molecule in the light-gathering apparatus of photosynthesis.

serine in a two-step pathway in some organisms. Serine and glycine, both important precursors in the biosynthesis of molecules such as phospholipids and heme, are interconverted by serine hydroxymethyl transferase, an enzyme that requires tetrahydrofolate and pyridoxal phosphate as coenzymes. Tyrosine is produced by hydroxylation of phenylalanine.

The biosynthesis of the remaining ten essential amino acids involves a variety of complex metabolic steps. The common aromatic pathway leads from erythrose 4-phosphate and phosphoenolpyruvate to chorismate. Three separate pathways lead from chorismate to phenylalanine, tyrosine, and tryptophan. The biosynthesis of aromatic amino acids is regulated by sequential feedback inhibition. In general, the first step of pathways for amino acid biosynthesis is inhibited by the end product of the pathway, and more complex regulatory mechanisms are often superimposed on this basic pattern. Tyrosine is the precursor of the catecholamines norepinephrine and epinephrine, and tryptophan is the precursor of many biosynthetic products including auxin, serotonin, and the nicotinamide nucleotides NAD and NADP.

The biosynthesis of histidine involves incorporation of two atoms from the purine ring of ATP in the imidazole ring, five carbons from PRPP, the amido group of glutamine, and the amino group of glutamate in the final amino acid.

Glycine and succinyl-CoA are converted to δ-aminolevulinate in the first step of heme biosynthesis. Two molecules of δ-aminolevulinate are incorporated into porphobilinogen. The condensation of 4 moles of porphobilinogen gives a linear tetrapyrrole that cyclizes to protoporphyrin IX. Incorporation of Fe(II) gives heme. The bile pigments are obtained by metabolic reactions of heme.

REFERENCES

Adams, E., and Frank, L., Metabolism of Proline and the Hydroxyprolines, *Ann. Rev. Biochem.*, 49, 1005–1063 (1980).

Dagley, S., and Nicholson, D. E., "An Introduction to Metabolic Pathways," Wiley, New York, 1970.

Greenberg, D. M., Ed., "Metabolic Pathways," 3rd ed., Vols. 1–7, Academic Press, New York, 1967–1975.

Haslam, E., "The Shikimate Pathway," Butterworths, London, 1974.

Jordon, P. M., and Shemin, D., δ-Aminolevulinic Acid Synthetase, in P. D. Boyer, Ed., "The Enzymes," 3rd ed., Vol.7, Academic Press, New York, 1972.

Nyham, W. L., Ed., "Heritable Disorders of Amino Acid Metabolism," Wiley, New York, 1974.

Schepartz, B., "Regulation of Amino Acid Metabolism in Mammals," Saunders, Philadelphia, 1973.

Schmid, R., and McDonagh, A. F., The Enzymatic Formation of Bilirubin, *Ann. Rev. N. Y. Acad. Sci.*, 244, 533–552 (1975).

Shemin, D., δ-Aminolevulinic Acid Dehydratase, in P. D. Boyer, Ed., "The Enzymes," 3rd ed., Academic Press, New York, 1972.

Stadtman, E. R., Mechanisms of Enzyme Regulation in Metabolism, in P. D. Boyer, Ed., "The Enzymes," 3rd ed., Vol. 1, Academic Press, New York, 1970.

Stanbury, J. B., Wyngaarden, J. B., and Frederickson, D. B., Eds., "The Metabolic Basis of Inherited Disease," 3rd ed., McGraw-Hill, New York, 1972.

Umbarger, H. E., Amino Acid Biosynthesis and Its Regulation," *Ann. Rev. Biochem.*, 47, 533–606 (1978).

With, T. K., "Bile Pigments; Chemical, Biological, and Clinical Aspects," Academic Press, New York, 1968.

Problems

1. In the early 1930s H. A. Krebs showed that the rate of formation of glutamine in tissue slices depends upon rate of respiration. Write a sequence of metabolic reactions that explains this observation.

2. Certain mutants of *Neurospora crassa* lack the ability to synthesize methionine. This inability can be circumvented if cystathionine is provided, but not if homoserine or cysteine is provided. Suggest a pathway for methionine biosynthesis in these mutants.

3. If [1-^{14}C]-glucose is the sole source of carbon for yeast, the valine isolated from yeast proteins contains ^{14}C in its methyl carbons. Why does the label appear in this position?

4. If pyruvate is labeled with ^{14}C in its methyl carbon, where will the label appear in isoleucine?

5. Write a mechanism for the reaction in which the C-2 fragment derived from pyruvate is incorporated in α-ketobutyrate in the biosynthesis of isoleucine.

6. It has been shown that the carboxyl and methyl carbons of acetate are both incorporated in leucine, but that only the α- and β-carbons of lactate are found in leucine. (a) What does this observation imply about the carbon sources for leucine biosynthesis? (b) Write a metabolic pathway that accounts for the above labeling pattern.

7. When $CH_3{}^{14}CO_2{}^{\ominus}$ is incubated with yeast extracts capable of leucine biosynthesis, β-carboxy-α-hydroxyisocaproate is formed as an intermediate. Show where the ^{14}C-labeled carbon appears in this intermediate.

8. Pyruvate and aspartate-γ-semialdehyde are biosynthetic precursors of lysine in bacteria. If the methyl group of pyruvate is labeled with ^{14}C, where will the label appear in L,L-diaminopimelate?

9. What is the purpose of succinylation of Δ^1-piperidine-2,6-dicarboxylate in the biosynthesis of lysine?

10. If 3-deoxy-D-arabinoheptulosonate 7-phosphate is labeled with ^{14}C in positions 4, 5, 6, and 7, where will the labeled carbons be found in shikimic acid in the common pathway for biosynthesis of aromatic amino acids?

11. If $^{14}CH_2O$ and [1-^{13}C]-glycine are the substrates for serine biosynthesis, where will the labels appear in serine?

Chapter 23

Biosynthesis of nucleotides

23–1 INTRODUCTION

We have often seen that the adenine nucleotides—AMP, ADP, and ATP—are centrally involved in cellular metabolism. ATP is the biological "quantum" of energy, and the immediate source of energy for most cellular processes. The relative concentrations of AMP, ADP, and ATP, whether expressed as the adenylate energy charge or the phosphorylation state ratio, influence most metabolic pathways since these nucleotides are allosteric effectors of many key regulatory enzymes. The second messenger of many hormonally stimulated processes is 3′-5′ cyclic AMP, itself derived from ATP. The guanosine nucleotides are also important in many metabolic transformations. Nucleotide derivatives are activated intermediates in the biosynthesis of glycogen and complex polysaccharides as well as in the biosynthesis of lipids. Nucleoside triphosphates are also the activated biosynthetic precursors of nucleic acids. The nucleotide coenzymes NAD, NADP, and coenzyme A contain nucleotide moieties as part of their structures. The nucleotides are too important to burn up as fuel, and most organisms possess pathways to salvage their nucleotides as well as enzymes that are capable of *de novo* (from the beginning) nucleotide biosynthesis. In keeping with their extreme importance, the pathways for nucleotide biosynthesis are under strict regulatory control.

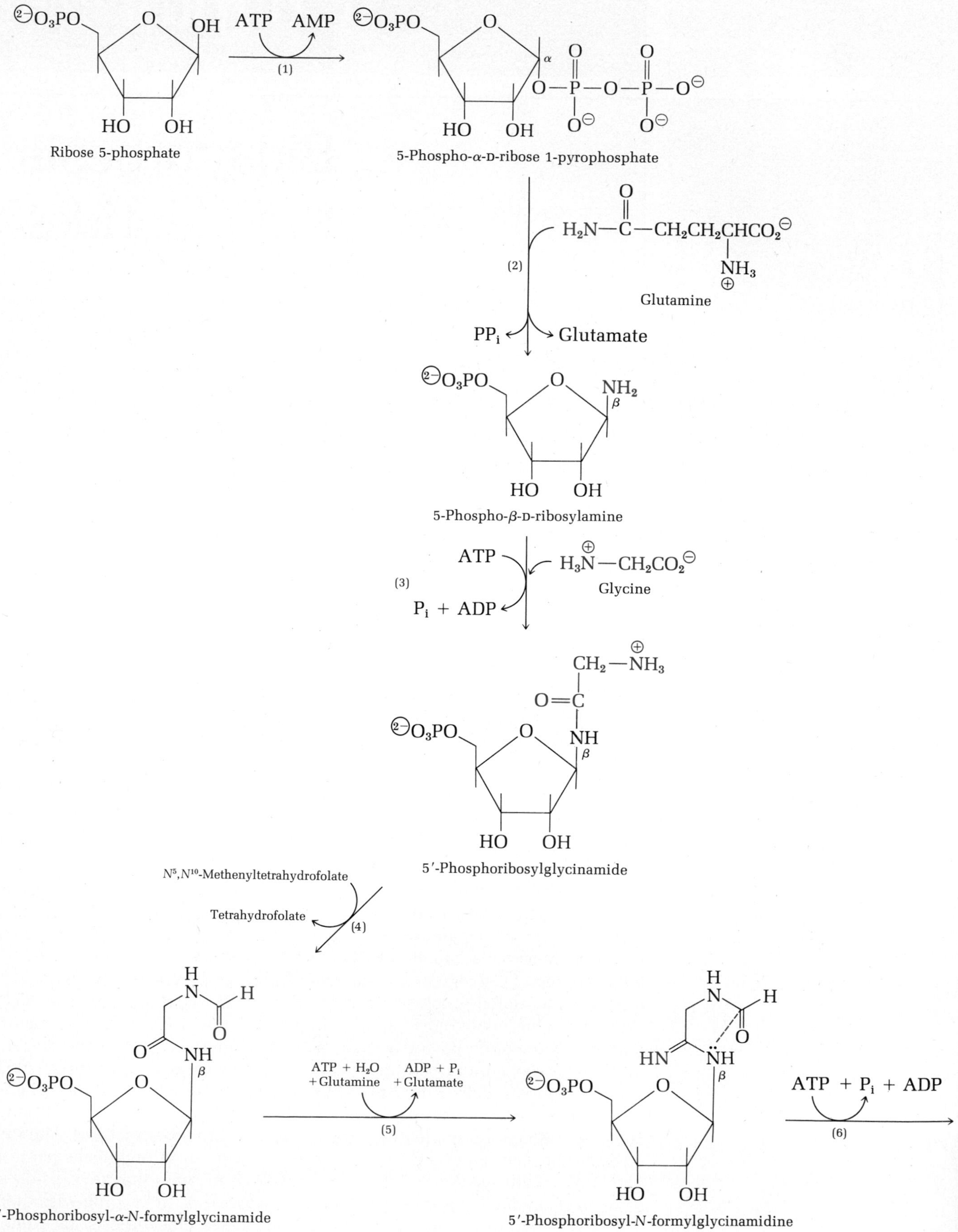
Ribose 5-phosphate
ATP AMP
(1)
5-Phospho-α-D-ribose 1-pyrophosphate
(2)
Glutamine
PPi
Glutamate
5-Phospho-β-D-ribosylamine
ATP
(3)
Pi + ADP
Glycine
5′-Phosphoribosylglycinamide
N5,N10-Methenyltetrahydrofolate
Tetrahydrofolate
(4)
5′-Phosphoribosyl-α-N-formylglycinamide
ATP + H2O + Glutamine
ADP + Pi + Glutamate
(5)
5′-Phosphoribosyl-N-formylglycinamidine
ATP + Pi + ADP
(6)

Figure 23.1 Pathway for the biosynthesis of inosine monophosphate (IMP). The individual enzyme-catalyzed reactions are discussed in Section 23.3B.

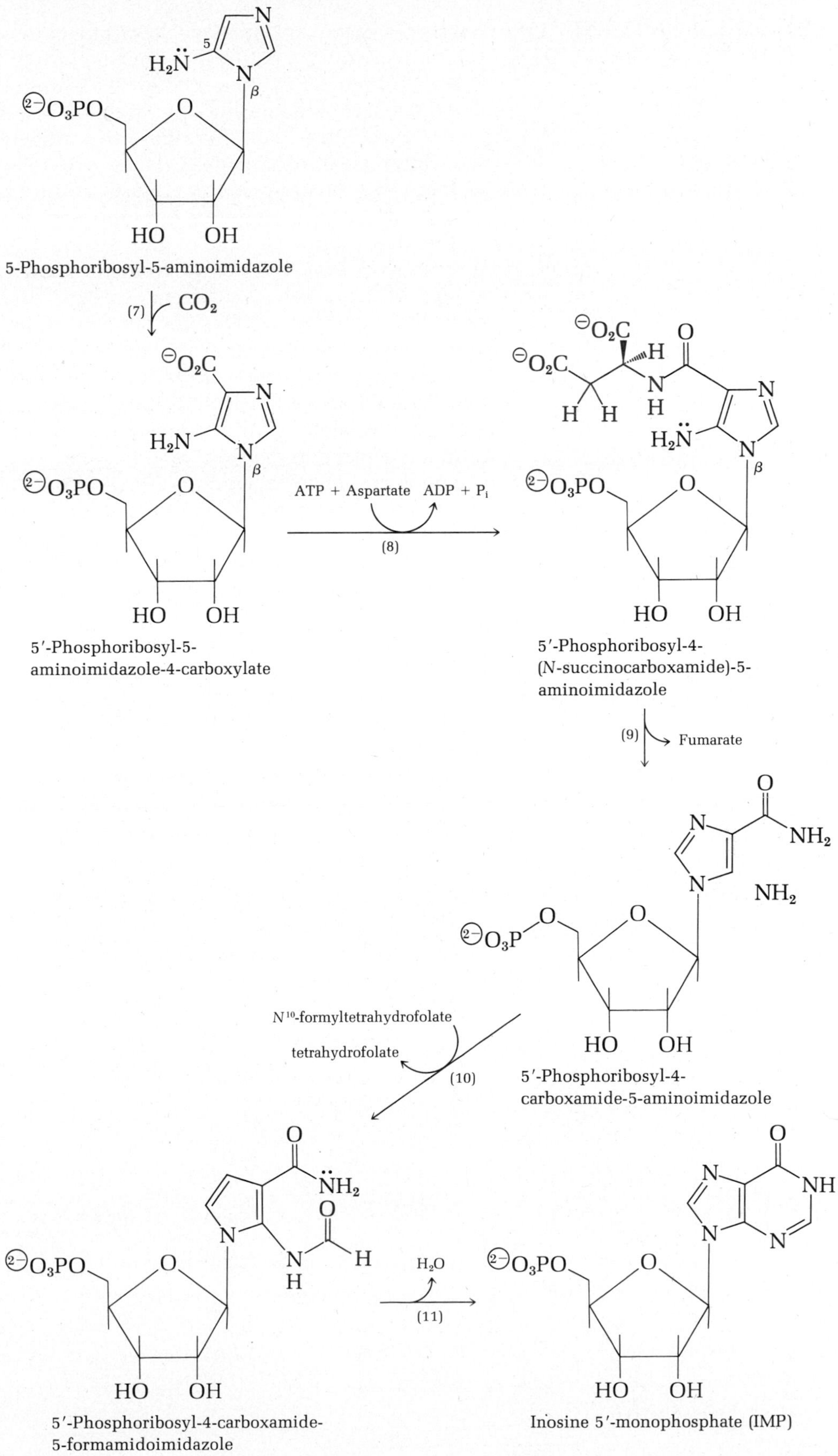

5-Phosphoribosyl-5-aminoimidazole

(7) CO_2

5′-Phosphoribosyl-5-
aminoimidazole-4-carboxylate

ATP + Aspartate   ADP + P_i

(8)

5′-Phosphoribosyl-4-
(N-succinocarboxamide)-5-
aminoimidazole

(9)   Fumarate

N^{10}-formyltetrahydrofolate

tetrahydrofolate

(10)

5′-Phosphoribosyl-4-
carboxamide-5-aminoimidazole

H_2O

(11)

5′-Phosphoribosyl-4-carboxamide-
5-formamidoimidazole

Inosine 5′-monophosphate (IMP)

Figure 23.1   (continued)

23–2 BIOSYNTHESIS OF PURINE NUCLEOTIDES

Nearly all organisms are capable of biosynthesis of purine nucleotides. In most organisms the pathway for purine biosynthesis consists of 11 steps as shown in Figure 23.1. In a preliminary examination of this pathway we see that the atoms of the purine ring originate in a wide variety of sources, summarized in Figure 23.2. The ribose ring moiety is provided by 5-phospho-α-D-ribose 1-pyrophosphate. Glycine provides N-7, C-1, and C-2; N^5,N^{10}-methenyl-tetrahydrofolate provides C-8, and the amido group of glutamine provides N-9, thus accounting for the atoms of the imidazole ring. The atoms of the pyrimidine portion of the purine ring are derived from several sources: N-3 from the amido group of glutamine, N-5 from the α-amino group of aspartate, C-6 from carbon dioxide, and C-4 from N^{10}-formyltetrahydrofolate. The biosynthesis of the purine ring is driven by the hydrolysis of seven phosphoanhydride bonds. The "strategy" of purine nucleotide biosynthesis is to incorporate the ring atoms by ATP-driven reactions in such a way that cyclization is favored between a nucleophilic amino group and an electrophilic carbon atom. The nucleophilic ring atoms are the nitrogen atoms derived from aspartate and glutamine, and the electrophilic carbon atoms are formyl groups derived from derivatives of tetrahydrofolate. Since the pathway for purine nucleotide biosynthesis is complicated, we shall reverse our usual procedure and examine regulation of purine nucleotide biosynthesis to obtain a better overview of the pathway first. Then we shall turn to the individual steps of the pathway.

A. Regulation of Purine Nucleotide Biosynthesis

A total of 11 steps are required for the synthesis of the mononucleotide inosine 5'-monophosphate (IMP). This product is then converted by separate pathways to AMP and GMP. The branched pathway is subject to negative feedback regulation by the nucleotide end products: AMP, ADP, ATP, GMP, GDP, and GTP (Figure 23.3). The first unique step in purine nucleotide biosynthesis is the displacement of the pyrophosphoryl group of 5-phospho-α-D-ribose 1-pyrophosphate (PRPP) by the amido group of glutamine in a reaction catalyzed by *amidophosphoribosyl transferase* (reaction 23.1).

5-Phospho-α-D-ribose
1-pyrophosphate

+

Glutamine

5-Phospho-β-D-ribosylamine

(23.1)

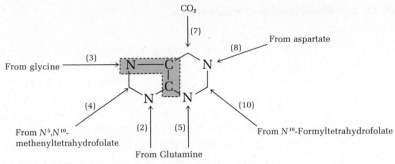

Figure 23.2 Sources of the atoms of the purine ring in purine nucleotide biosynthesis. Numbers refer to the steps of the pathway (Figure 23.1) in which the atoms are incorporated in the ring system.

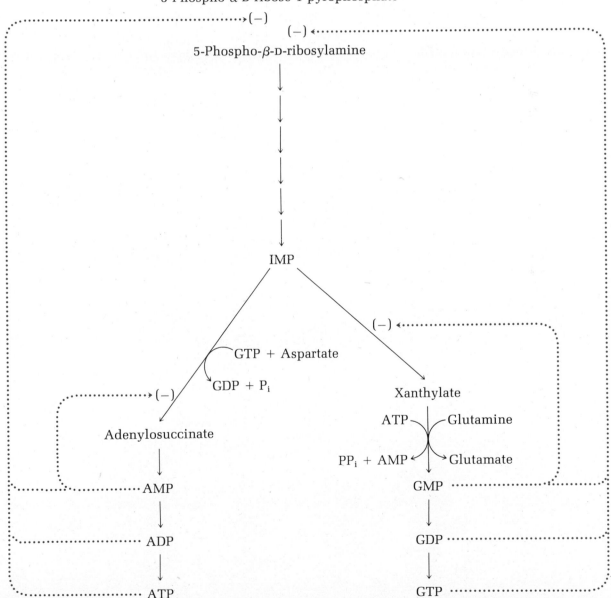

Figure 23.3 Regulation of purine nucleotide biosynthesis by negative feedback inhibition (−). The combination of an adenine and a guanine nucleotide exerts a greater inhibitory effect on the first step than the sum of the effects of each acting separately, an example of synergistic feedback inhibition. Since ATP is a substrate in a pathway leading to GMP, and GTP is a substrate is a pathway leading to AMP, an excess of one nucleotide leads to production of the other, and the rates of adenine and guanine nucleotide production are balanced.

This step is subject to *synergistic feedback inhibition* by all of the adenine and guanine nucleotides. Both types of nucleotides can independently inhibit the amido transferase, but the combination of AMP and GMP, for example, is far more effective than the combination of AMP and ADP or GMP and GDP. The unique pathways leading from IMP to GMP and AMP are subject to an interesting type of regulation. First, the end products—AMP, ADP, and ATP or GMP, GDP, and GTP—all inhibit the first step leading to their production. Second, GTP is a substrate in the first reaction of the pathway leading from IMP to AMP, and ATP is a substrate in the second step of the pathway leading from IMP to GMP. In each case a pyrophosphoryl group is hydrolyzed in the nucleoside triphosphate. Thus, an excess of GTP encourages the production of ATP, and an excess of ATP encourages the production of GTP. The rates of synthesis of the two nucleotides are therefore balanced.

B. Enzymatic Reactions of Purine Nucleotide Biosynthesis: Synthesis of IMP

a. Conversion of Ribose 5-Phosphate to PRPP

Ribose 5-phosphate, an intermediate in the pentose phosphate pathway, is the starting point of purine nucleotide biosynthesis. Pyrophosphoryl group transfer from ATP to ribose 5-phosphate is catalyzed by *PRPP synthetase*. The reaction occurs by nucleophilic attack of the C-1 hydroxyl group of ribose 5-phosphate upon the β-phosphorus of ATP. Since a pyrophosphoryl group is displaced by the amido group of glutamine in the next step, the activation of C-1 of the ribose sets the stage for the next reaction. Hydroxide, which would be produced if the amido group of glutamine were to directly attack C-1 of ribose 5-phosphate, is a very poor leaving group. The pyrophosphoryl group, however, is an excellent leaving group. The product, 5-phospho-α-D-ribose 1-pyrophosphate (PRPP), is an energy-rich intermediate, and displacement of a pyrophosphoryl group in the second step is thus favored thermodynamically as well as kinetically (reaction 23.2).

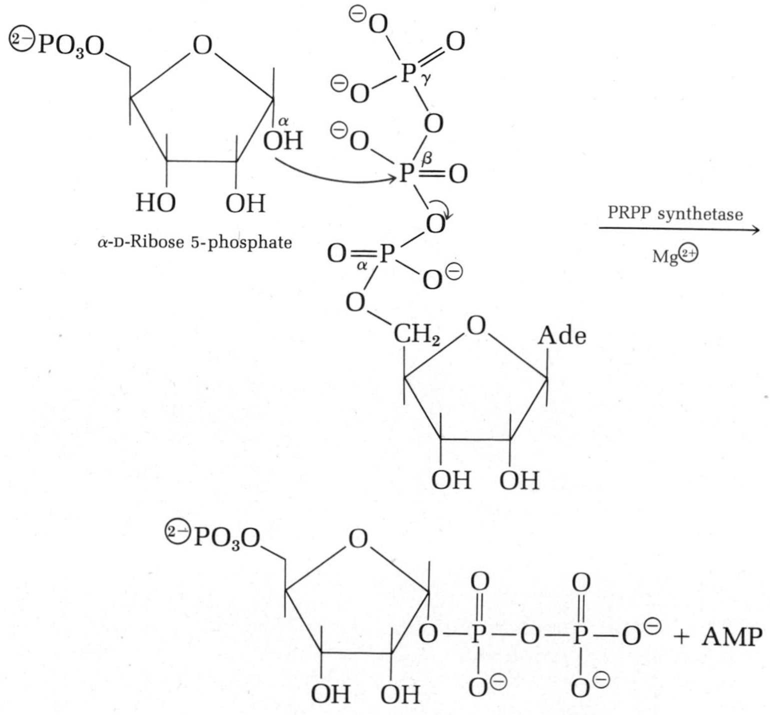

α-D-Ribose 5-phosphate

PRPP synthetase
Mg^{2+}

5-Phospho-α-D-ribose 1-pyrophosphate
(PRPP)

b. Conversion of PRPP to 5-Phospho-β-D-Ribosylamine

Amidophosphoribosyl transferase catalyzes the conversion of PRPP to 5-phospho-β-D-ribosylamine. The amide group of glutamine is a source of ammonia, generated in high local concentration at the active site of the enzyme. Ammonia itself is a substrate for the enzyme, but it is a poor one whose K_m of about 1.5 M is three orders of magnitude higher than K_m for glutamine. Ammonia is quite toxic in any case, and the enzyme provides a "safe" source of ammonia for the reaction. Displacement of the C-1 pyrophosphate group by ammonia in an S_N2 reaction occurs with inversion of configuration at C-1 and introduces the first atom of the purine ring (reaction 23.3).

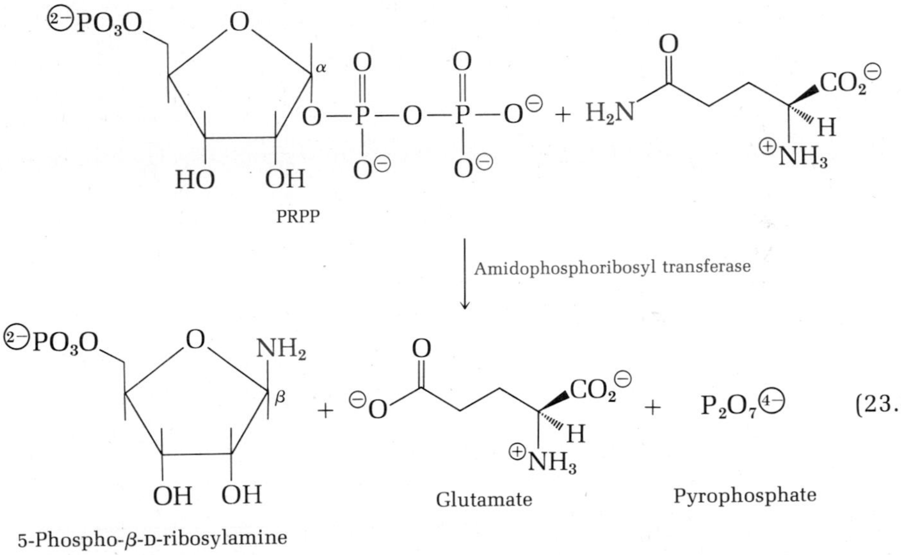

PRPP

Amidophosphoribosyl transferase

$$\text{(23.3)}$$

5-Phospho-β-D-ribosylamine Glutamate Pyrophosphate

This step of purine nucleotide biosynthesis is inhibited by the end products of the pathway.

c. Formation of 5′-Phosphoribosylglycinamide

Phosphoribosylglycinamide synthetase catalyzes amide formation between the C-1 amino group of 5-phospho-β-D-ribosylamine and the carboxyl group of glycine (reaction 23.4).

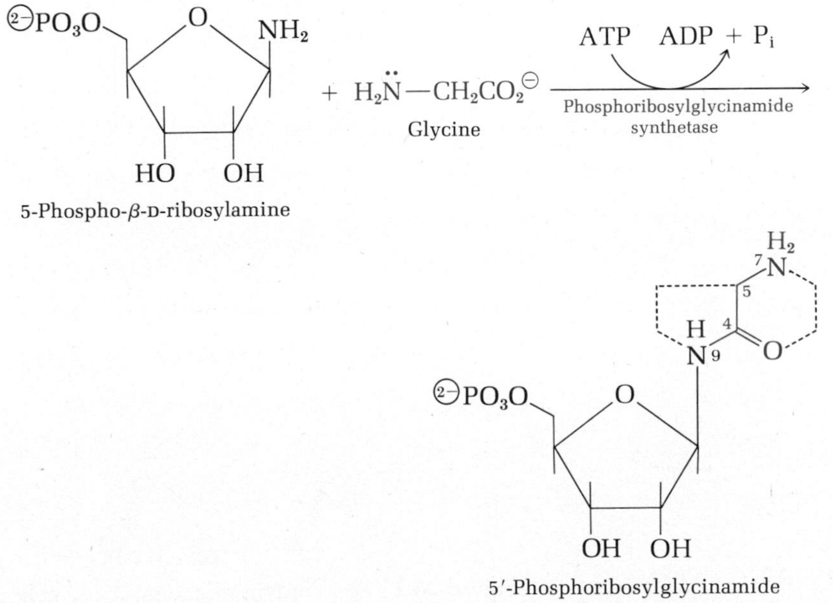

5-Phospho-β-D-ribosylamine

Glycine

ATP ADP + P_i

Phosphoribosylglycinamide synthetase

$$\text{(23.4)}$$

5′-Phosphoribosylglycinamide

The reaction probably occurs by formation of the reactive acyl phosphate of glycine at the active site, with ATP providing the phosphoryl group (Figure 23.4). After three steps of the synthesis four of the purine ring atoms are in place.

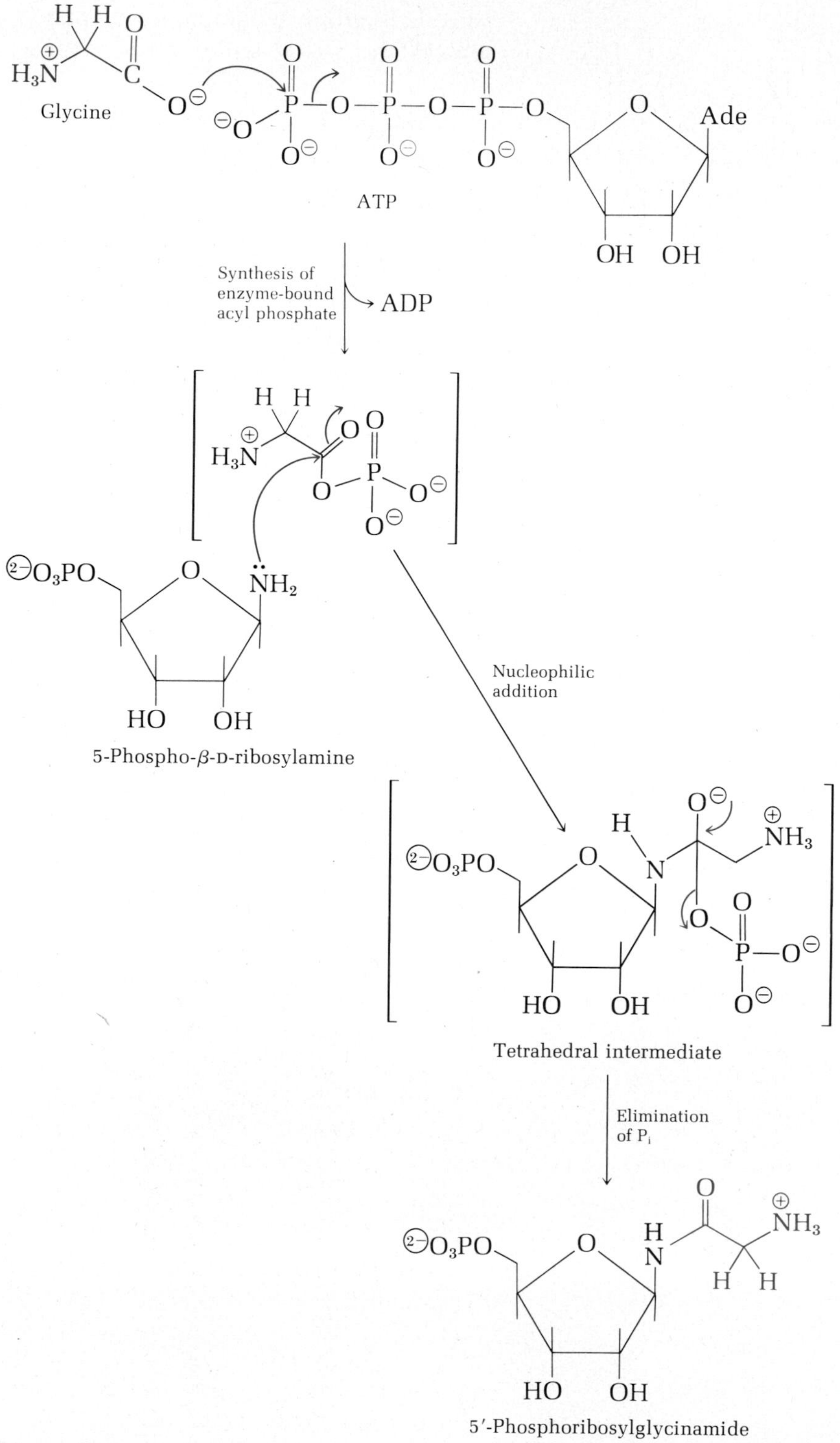

Figure 23.4 Mechanism of action of phosphoribosylglycinamide synthetase.

d. Synthesis of 5′-Phosphoribosyl-α-N-Formylglycinamide

The free amino group of 5′-phosphoribosylglycinamide reacts with the electrophilic methenyl carbon of N⁵,N¹⁰-methenyltetrahydrofolate (Section 10.8) in a reaction catalyzed by *phosphoribosylglycinamide formyl transferase* (reaction 23.5).

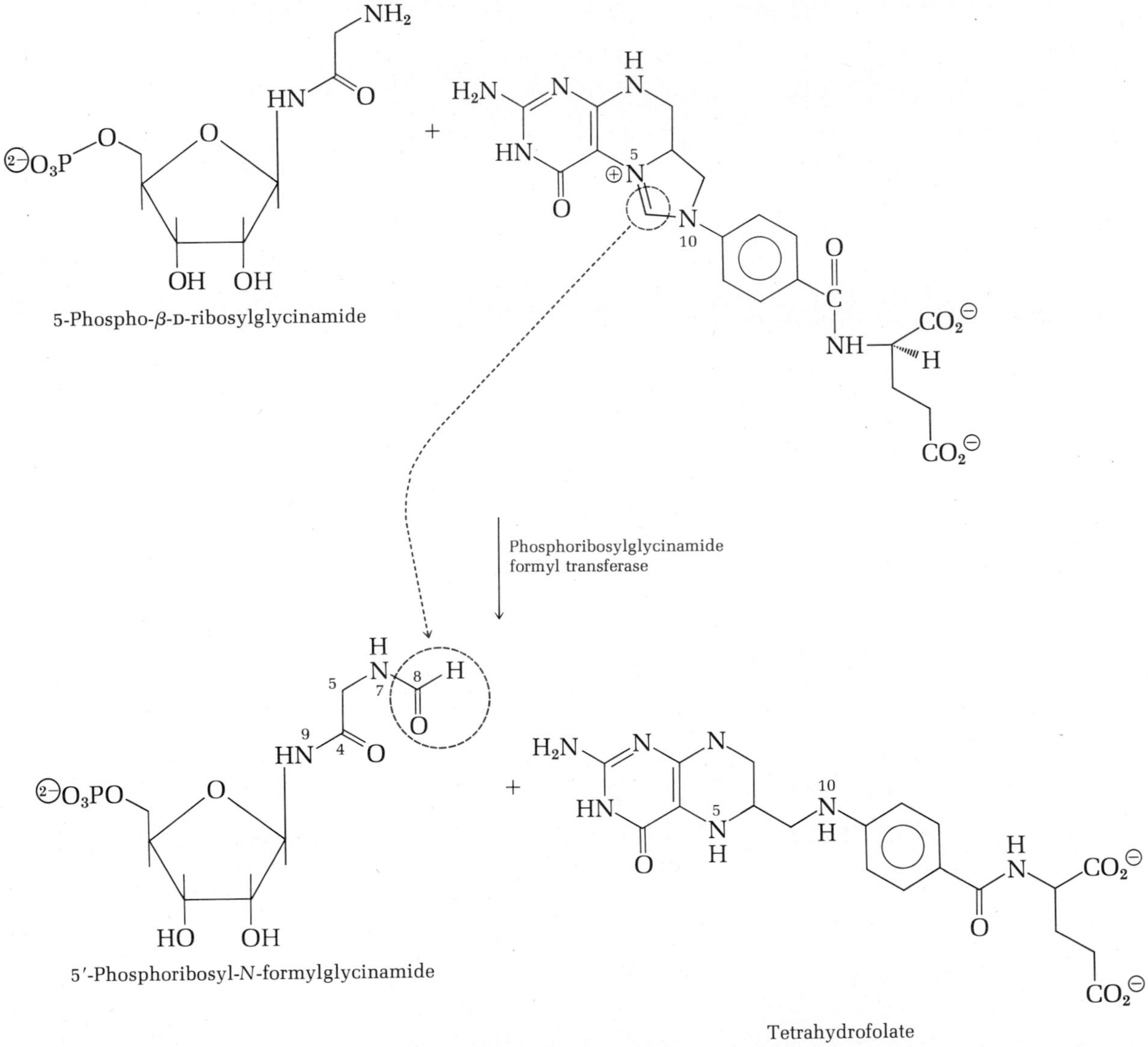

5-Phospho-β-D-ribosylglycinamide

Phosphoribosylglycinamide formyl transferase

5′-Phosphoribosyl-N-formylglycinamide

Tetrahydrofolate

(23.5)

The methenyl group of N⁵,N¹⁰-methenyltetrahydrofolate is susceptible to nucleophilic attack because of the electron sink provided by the positively charged N⁵-nitrogen of the coenzyme. The mechanistic details of the formyl transfer reaction are shown in Figure 23.5.

e. Formation of 5′-Phosphoribosyl-N-Formylglycinamidine

Conversion of the keto group of the amide to an imino group creates an amidine in the next step of the synthesis. Glutamine again is the amino donor, and ATP is the energy source (reaction 23.6). Phosphorylation of the amide oxygen by ATP generates a reactive intermediate. Ammonia, supplied by the glutamine amide group at the active site of the enzyme, displaces a phosphoryl group to produce the amidine (Figure 23.6). Since water is lost in this step, ATP acts as a "dehydrating agent."

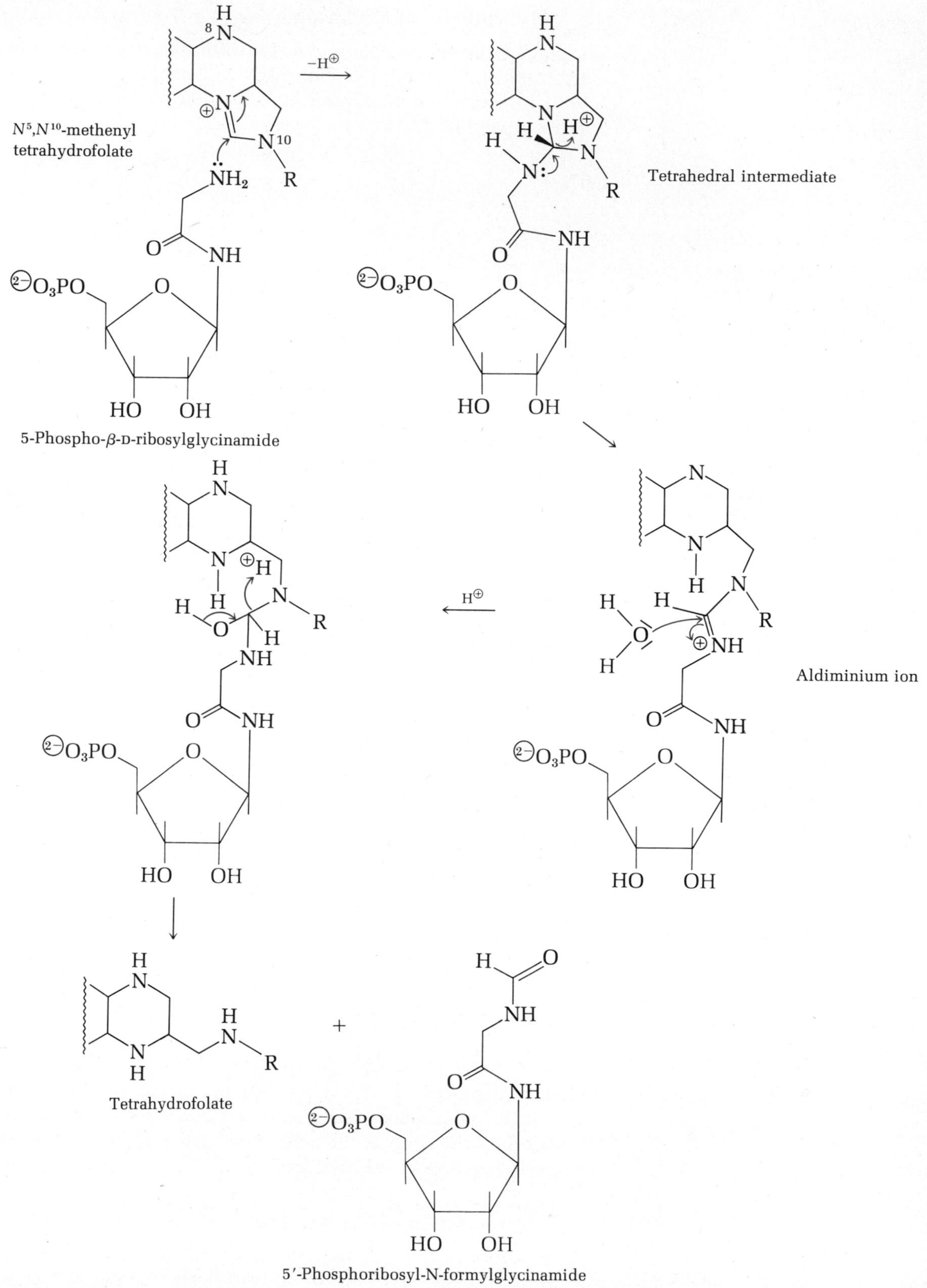

Figure 23.5   Mechanism of action of α-N-formylglycinamide ribonucleotide synthetase. [Adapted from S. Benkovic, *Ann. Rev. Biochem.*, 49, 241 (1980). Reproduced with permission. © 1980 by Annual Reviews Inc.]

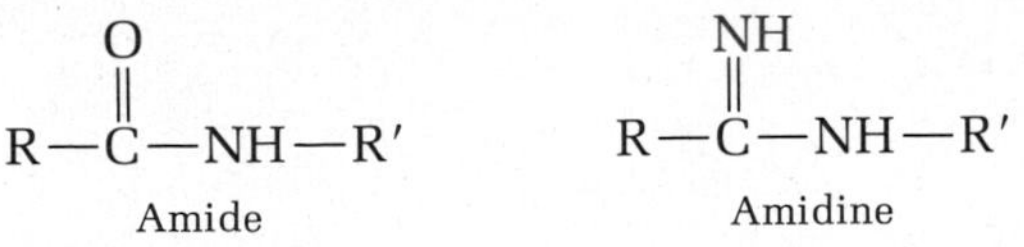

Amide

Amidine

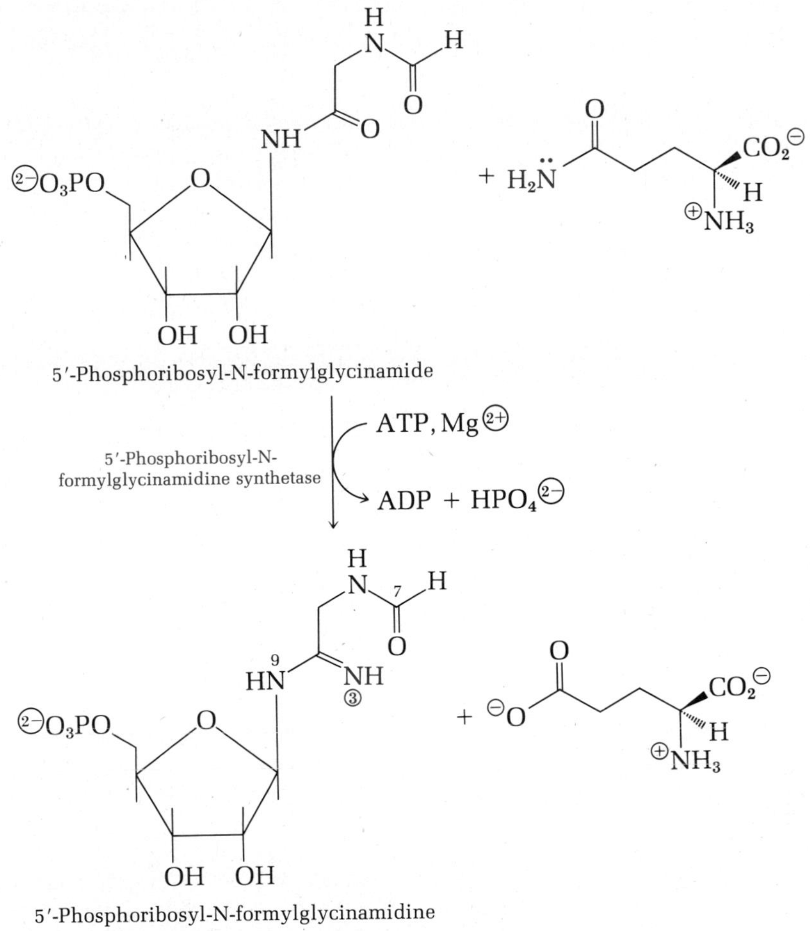

5′-Phosphoribosyl-N-formylglycinamide

5′-Phosphoribosyl-N-
formylglycinamidine synthetase

ATP, Mg^{2+}

ADP + HPO$_4^{2-}$

(23.6)

5′-Phosphoribosyl-N-formylglycinamidine

f. Formation of the Imidazole Ring

Phosphoribosylaminoimidazole synthetase catalyzes ring closure between the carbonyl carbon of the formyl group and N-9, the glycosidic nitrogen (reaction 23.7).

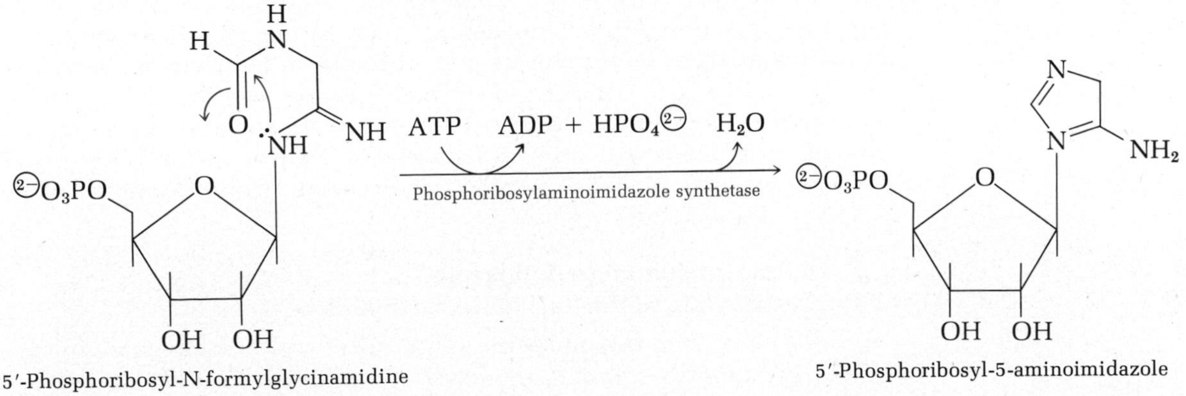

ATP   ADP + HPO$_4^{2-}$ H$_2$O

Phosphoribosylaminoimidazole synthetase

(23.7)

5′-Phosphoribosyl-N-formylglycinamidine

5′-Phosphoribosyl-5-aminoimidazole

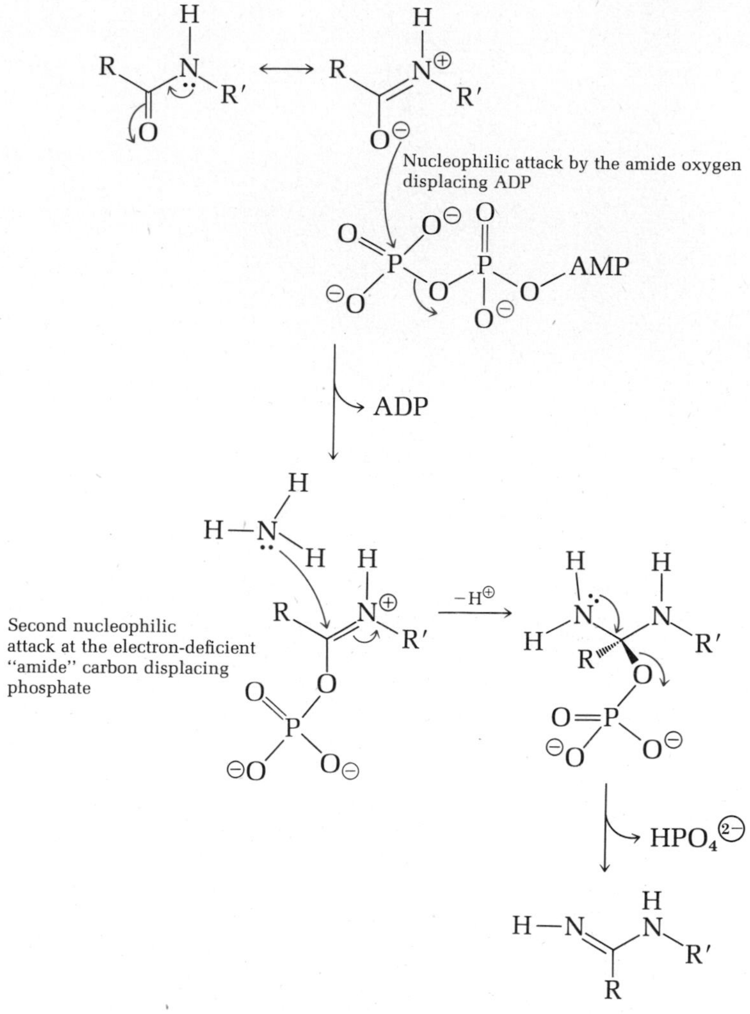

Figure 23.6 Mechanism of action of phosphoribosylglycinamidine synthetase. Nucleophilic attack by the amide oxygen builds a good leaving group, phosphate, into the substrate.

ATP once more phosphorylates the substrate, this time at the carbonyl oxygen of the formyl group. Phosphorylation aids ring closure because phosphate is a good leaving group. The reaction is further favored by resonance stabilization of the imidazole ring (Figure 23.7).

g. Carboxylation of the Imidazole Ring

Phosphoribosylaminoimidazole carboxylase catalyzes addition of carbon dioxide to C-2 of the imidazole ring, introducing C-6 of the purine nucleus (reaction 23.8). The carboxylation of the imidazole ring is somewhat unusual in that it involves neither biotin as a coenzyme nor ATP as an energy source. The structure of the reactant provides the clue to the mechanism. It is an enamine, and since enamines are already "activated nucleophiles," no further activation is necessary. The mechanism is simply attack of a nucleophilic enamine α-carbon upon the electrophilic carbon of carbon dioxide (Figure 23.8).

h. Conversion of the Imidazole Carboxylate to the Succinocarboxamide

The next step resembles the synthesis of argininosuccinate in the urea cycle. *Phosphoribosylaminoimidazole-succinocarboxamide synthetase* cat-

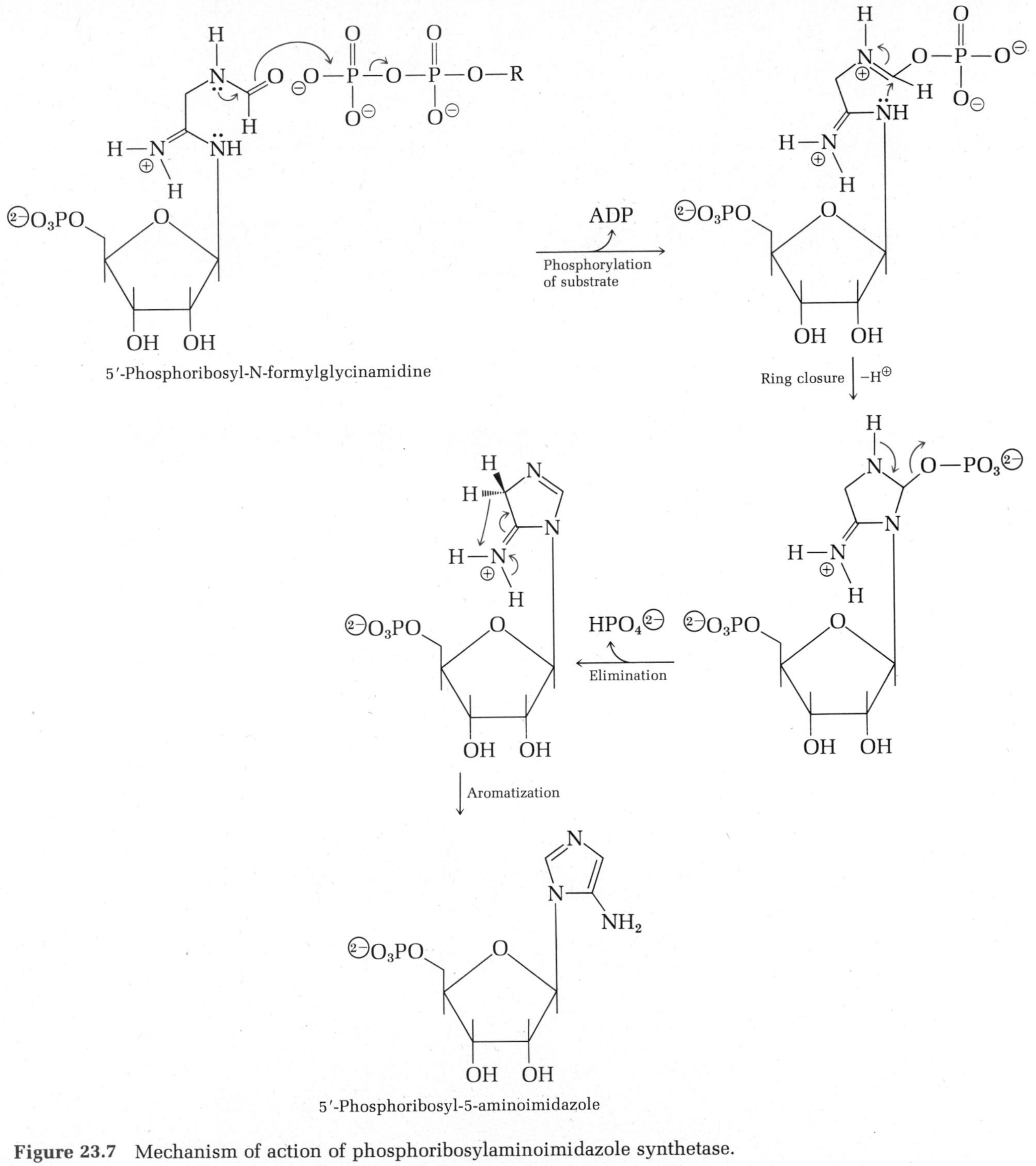

5'-Phosphoribosyl-N-formylglycinamidine

ADP

Phosphorylation of substrate

Ring closure | $-H^\oplus$

HPO$_4$

Elimination

Aromatization

5'-Phosphoribosyl-5-aminoimidazole

Figure 23.7 Mechanism of action of phosphoribosylaminoimidazole synthetase.

alyzes formation of an amide between the α-amino group of aspartate and the C-4 carboxylate formed in the preceding step in a reaction that requires ATP as an energy source (reaction 23.9). Once more a phosphoryl group of ATP is transferred to the carboxylate group, generating an electrophilic carbonyl carbon. Nucleophilic attack by the amino group of aspartate displaces phosphate to generate the product 5'-phosphoribosyl-4-(N-succino-carboxamide)-5-aminoimidazole (Figure 23.9).

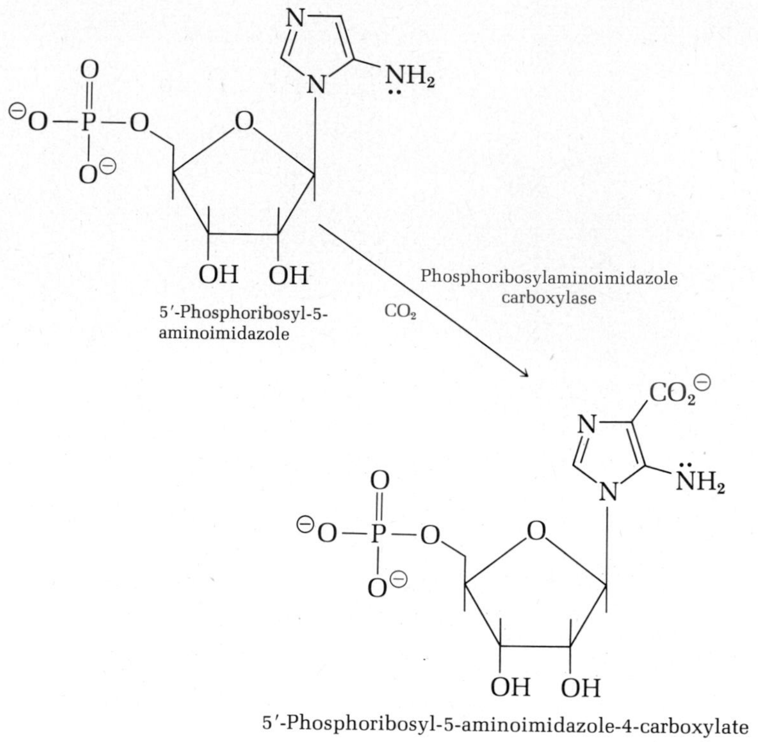

5′-Phosphoribosyl-5-
aminoimidazole

Phosphoribosylaminoimidazole
carboxylase

CO_2

(23.8)

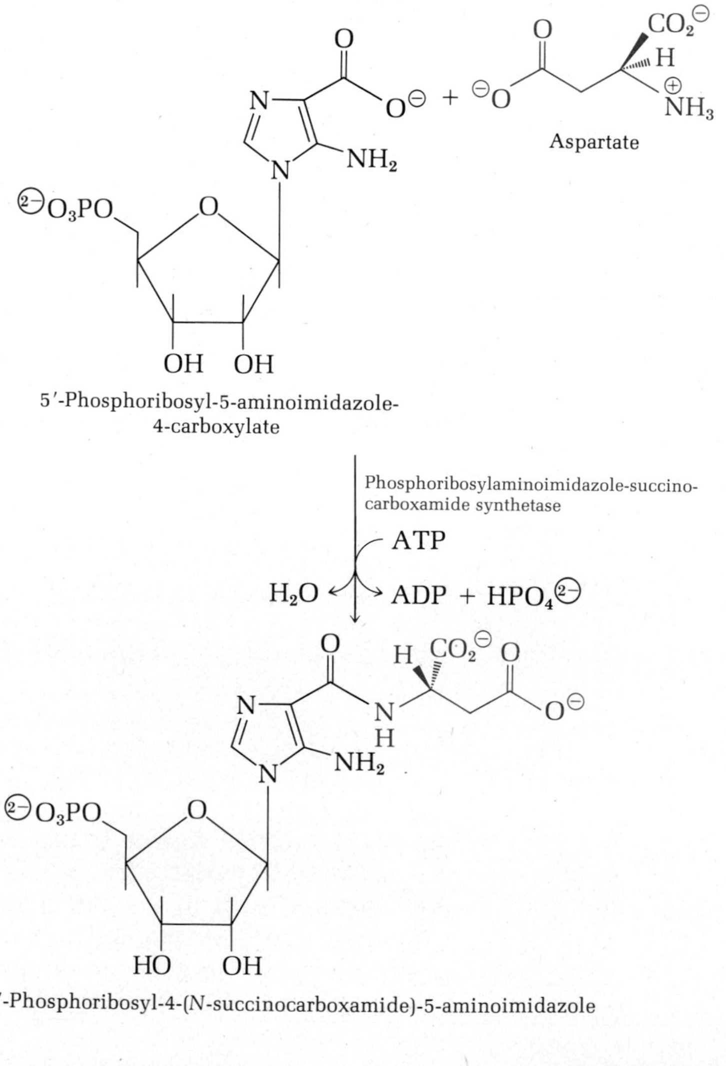

5′-Phosphoribosyl-5-aminoimidazole-4-carboxylate

Aspartate

5′-Phosphoribosyl-5-aminoimidazole-
4-carboxylate

Phosphoribosylaminoimidazole-succinocarboxamide synthetase

ATP

H_2O ⟵ ⟶ ADP + HPO_4^{2-}

(23.9)

5′-Phosphoribosyl-4-(N-succinocarboxamide)-5-aminoimidazole

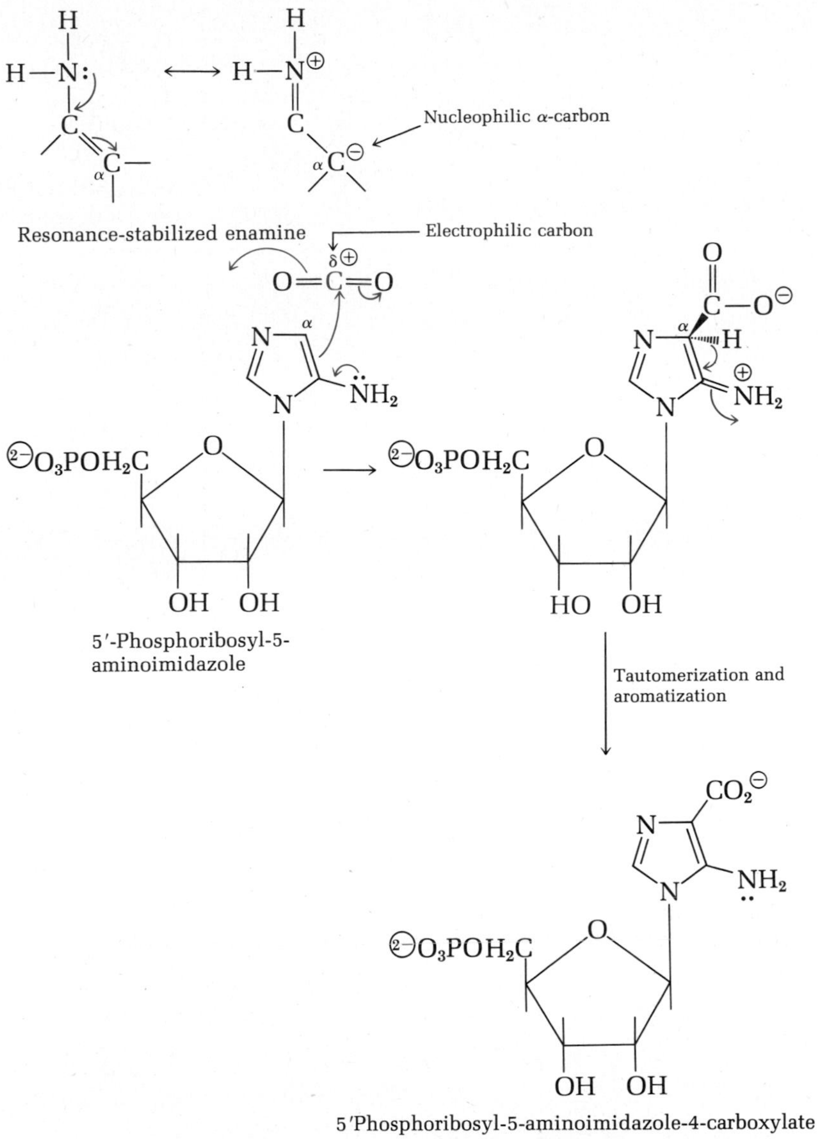

Figure 23.8 Nucleophilic addition of the electrophilic carbon of CO_2 to the nucleophilic enamine 5'-phosphoribosyl-5-aminoimidazole.

i. Elimination of Fumarate from the Succinocarboxamide

Adenylosuccinase catalyzes *anti* elimination of fumarate to give 5'-phosphoriboxyl-4-carboxamide-5-aminoimidazole (reaction 23.10). All but one of the purine ring atoms have now been set in place.

We encountered 5'-phosphoribosyl-4-carboxamide-5-aminoimidazole in our discussion of histidine biosynthesis (step 5, Figure 22.37). In organisms capable of histidine biosynthesis, intermediates can be salvaged for use in the biosynthesis of purine nucleotides, by-passing the first 9 steps of the pathway for *de novo* biosynthesis of purines.

j. Incorporation of the Final Ring Carbon

The final ring atom of the purine nucleus is donated by N^{10}-formyltetrahydrofolate in a reaction catalyzed by *phosphoribosylaminoimidazolecarboxamide formyltransferase*. This reaction is exactly parallel to the reaction in step 4 (reaction 23.11).

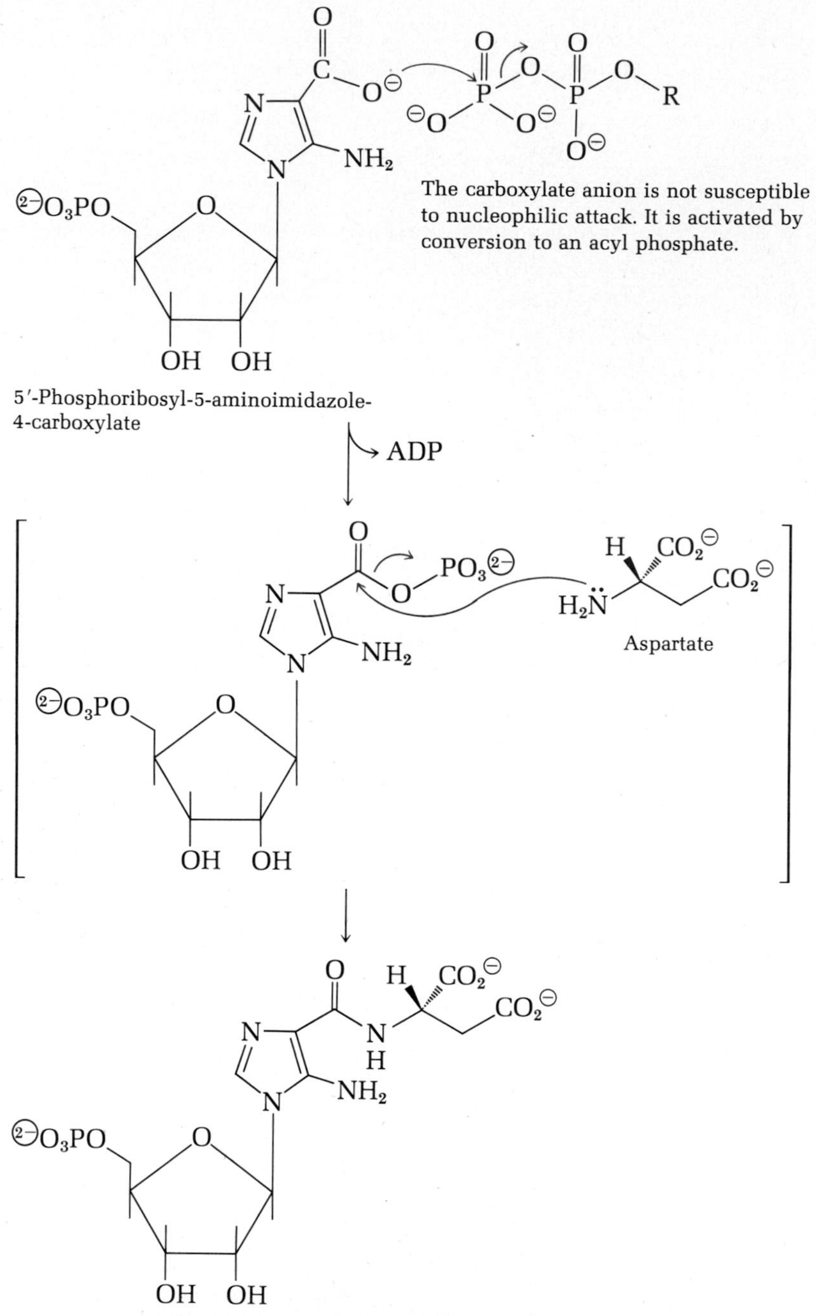

The carboxylate anion is not susceptible to nucleophilic attack. It is activated by conversion to an acyl phosphate.

5'-Phosphoribosyl-5-aminoimidazole-4-carboxylate

Aspartate

5'-Phosphoribosyl-4-(N-succinocarboxamide)-5-aminoimidazole

Figure 23.9 Synthesis of 5'-phosphoribosyl-4-(N-succinocarboxamide)-5-aminoimidazole.

k. Cyclization to IMP

With all of the purine ring atoms in place a cyclization catalyzed by *IMP cyclohydrolase* produces inosine 5'-monophosphate (reaction 23.12).

23–3
CONVERSION OF IMP TO ADENINE AND GUANINE NUCLEOTIDES

IMP is a branch-point metabolite that leads to adenine nucleotides along one pathway (Figure 23.10) and to guanine nucleotides along another pathway (Figure 23.11). AMP and GMP inhibit the first unique steps in their own biosynthesis and that the guanosine and adenosine nucleotides all inhibit the conversion of PRPP to 5-phospho-β-D-ribosylamine by synergistic feedback inhibition (recall Figure 23.3).

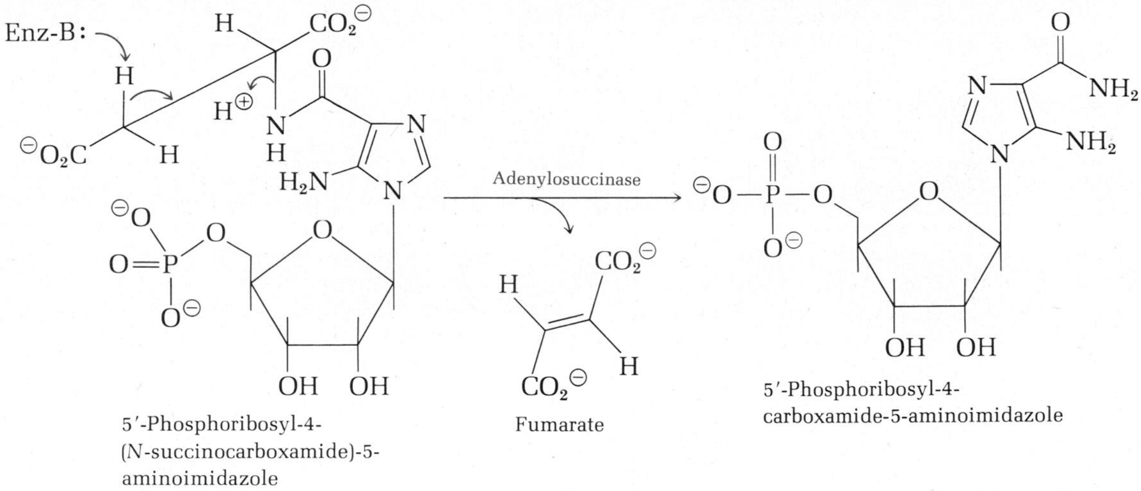

5′-Phosphoribosyl-4-
(*N*-succinocarboxamide)-5-
aminoimidazole

Adenylosuccinase

Fumarate

5′-Phosphoribosyl-4-
carboxamide-5-aminoimidazole

(23.10)

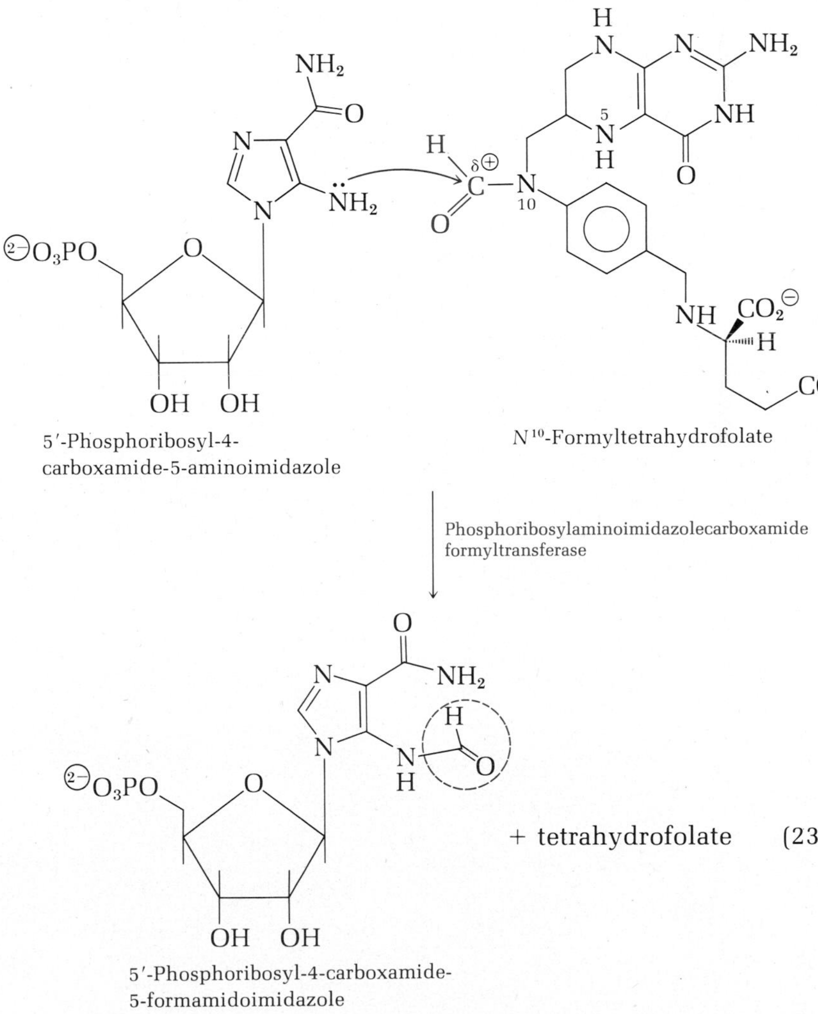

5′-Phosphoribosyl-4-
carboxamide-5-aminoimidazole

N^{10}-Formyltetrahydrofolate

Phosphoribosylaminoimidazolecarboxamide
formyltransferase

5′-Phosphoribosyl-4-carboxamide-
5-formamidoimidazole

+ tetrahydrofolate   (23.11)

A. Conversion of IMP to AMP   *Adenylosuccinate synthetase* catalyzes the conversion of IMP to adenylosuccinate (reaction 23.13). In yet another example of substrate activation by phosphorylation, a phosphoryl group is transferred from GTP to the C-6 oxygen of IMP, converting the carbonyl carbon to an electrophilic center that contains an excellent leaving group. The nucleophilic α-amino group of aspartate then attacks the electrophilic C-6 carbon of the purine ring,

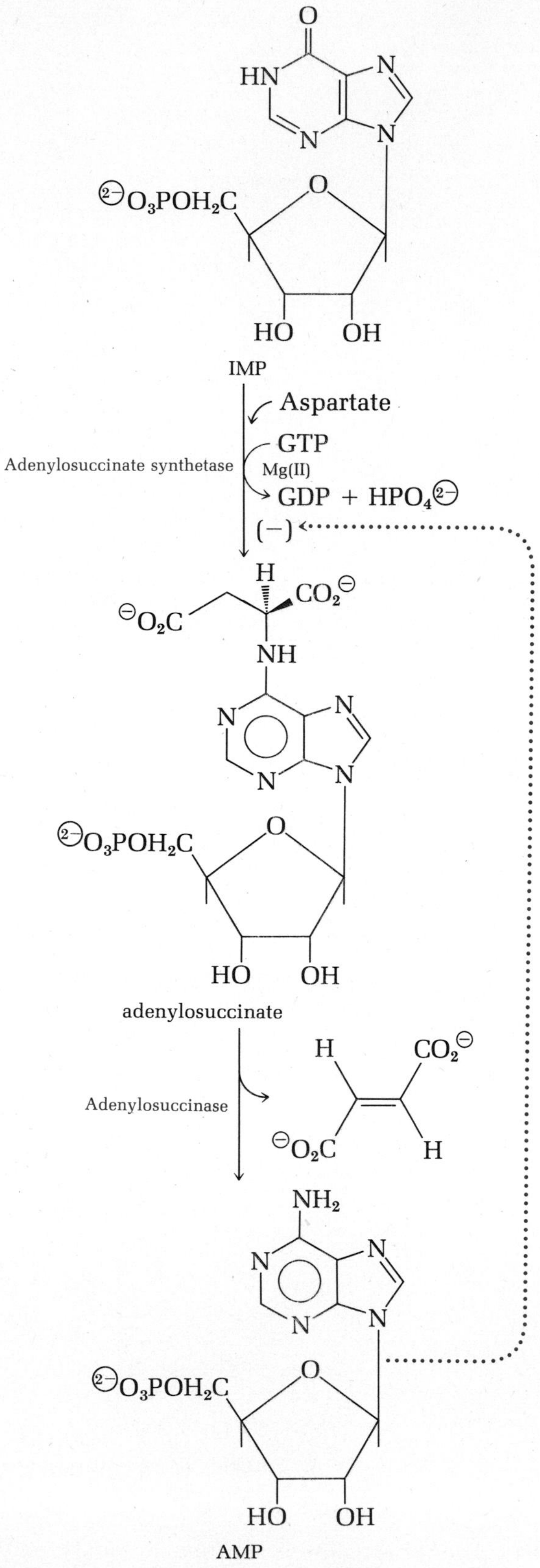

Figure 23.10 Conversion of IMP to AMP. Negative feedback inhibition is indicated by (−).

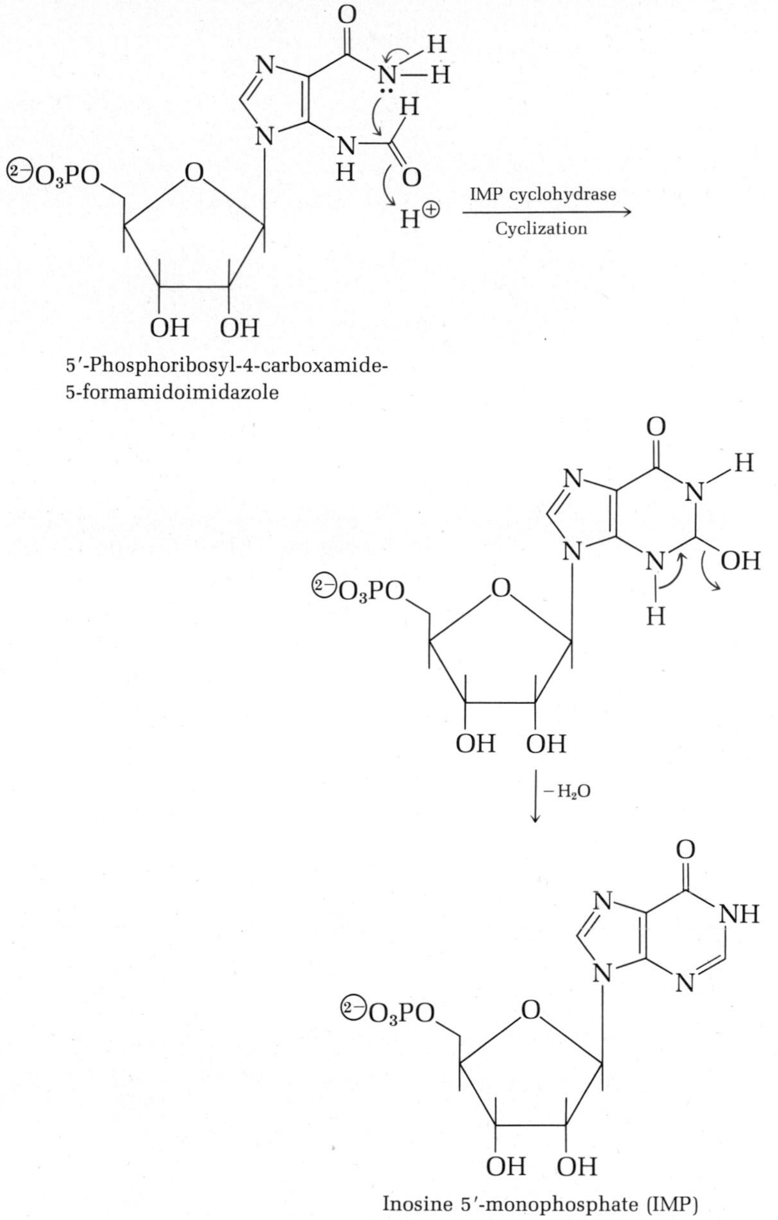
IMP cyclohydrase
Cyclization
5′-Phosphoribosyl-4-carboxamide-5-formamidoimidazole
$-H_2O$
Inosine 5′-monophosphate (IMP)

(23.12)

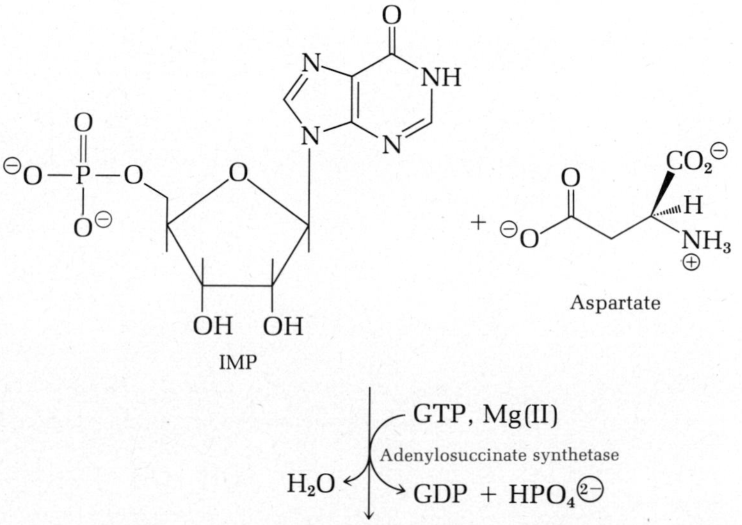
IMP
Aspartate
GTP, Mg(II)
Adenylosuccinate synthetase
H_2O
$GDP + HPO_4^{2-}$

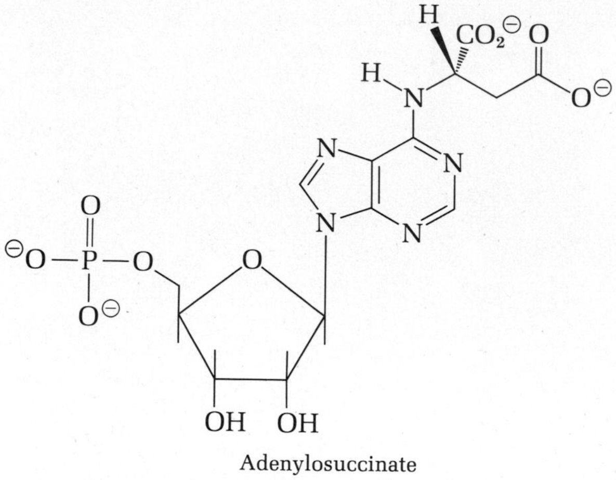

Adenylosuccinate

(23.13)

displacing phosphate (Figure 23.12). We have already noted that excess GTP encourages ATP production by providing an energy source for this reaction.

In the second step of this branch of IMP metabolism *adenylosuccinase,* the enzyme of step *c* of IMP synthesis and of the urea cycle, splits out fumarate to produce AMP (reaction 23.14).

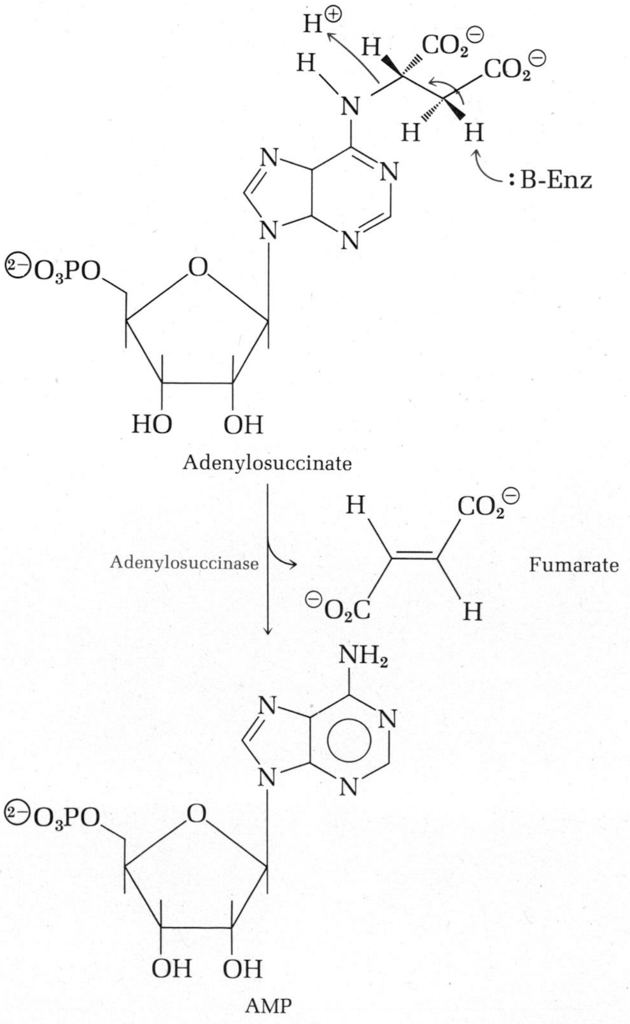

Adenylosuccinate

Fumarate

(23.14)

Adenylosuccinase

AMP

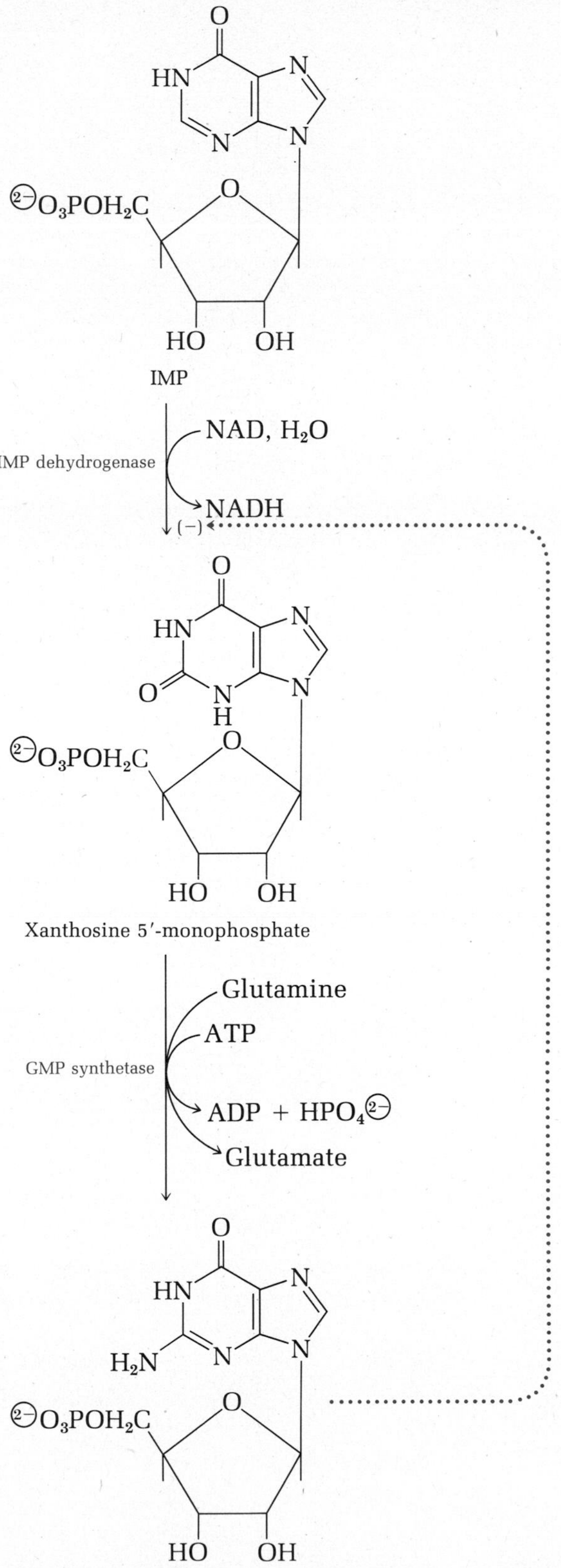

Figure 23.11 Conversion of IMP to GMP. Negative feedback inhibition is indicated by (−).

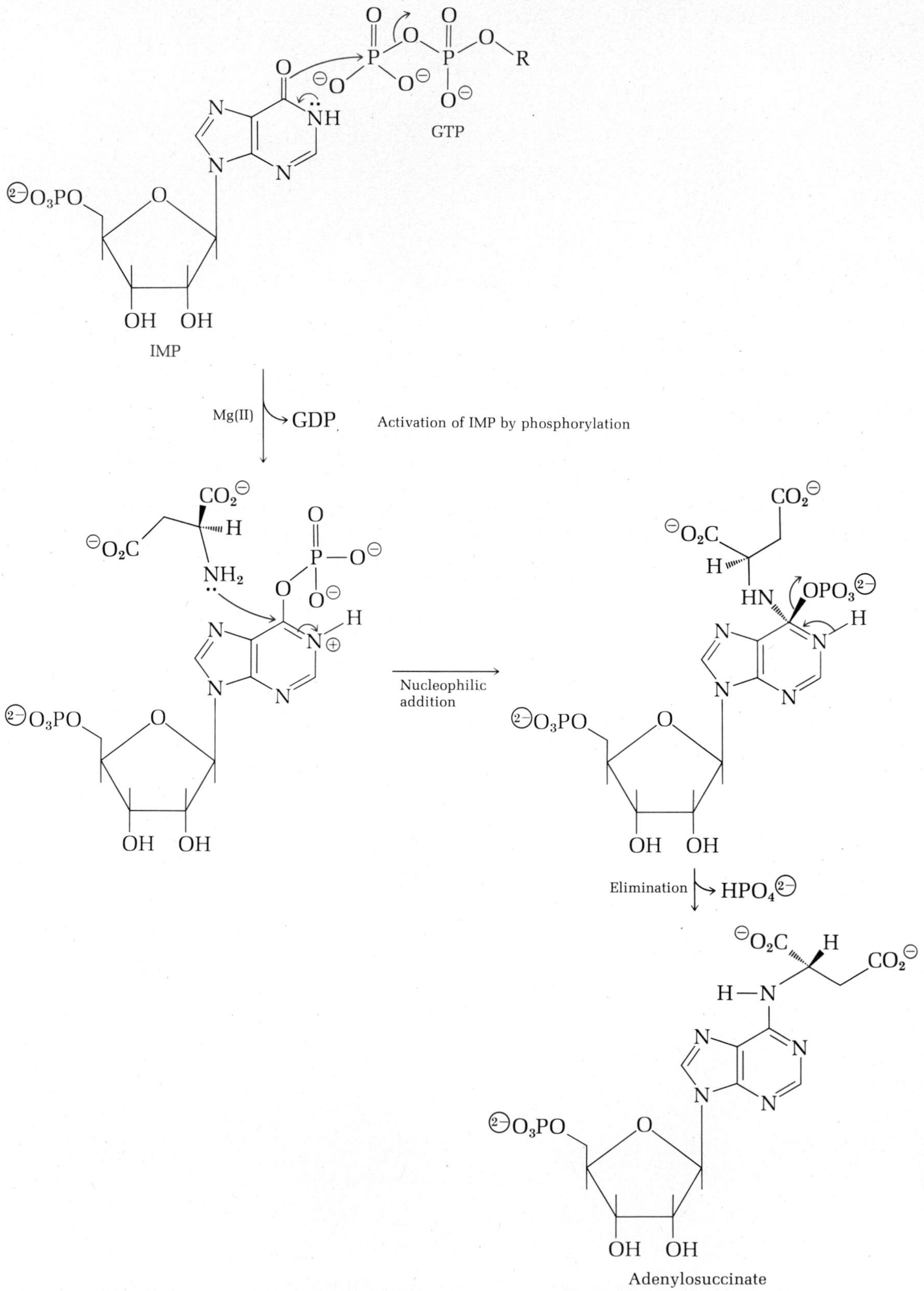

Figure 23.12 Conversion of IMP to adenylosuccinate catalyzed by adenylosuccinate synthetase.

B. Synthesis of GMP   IMP is oxidized to *xanthosine 5'-monophosphate*, or *xanthylic acid* (XMP), by NAD-dependent *inosine 5'-phosphate dehydrogenase* (reaction 23.15). The ring oxygen is provided by water.

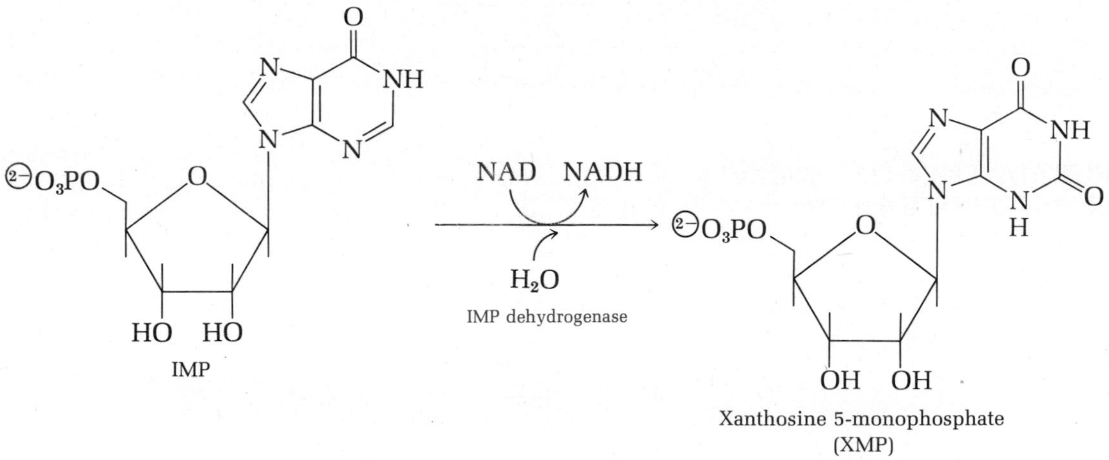

(23.15)

XMP is converted to GMP by *GMP synthetase* (reaction 23.16). Glutamine is the amino donor. The amido group of glutamine (ammonium ion in certain microorganisms) is transferred to C-2 of the purine ring. ATP phosphorylates the carbonyl oxygen at C-2 during the reaction. This reaction is analogous to step 5 in purine nucleotide biosynthesis (recall reaction 23.6 and Figure 23.6).

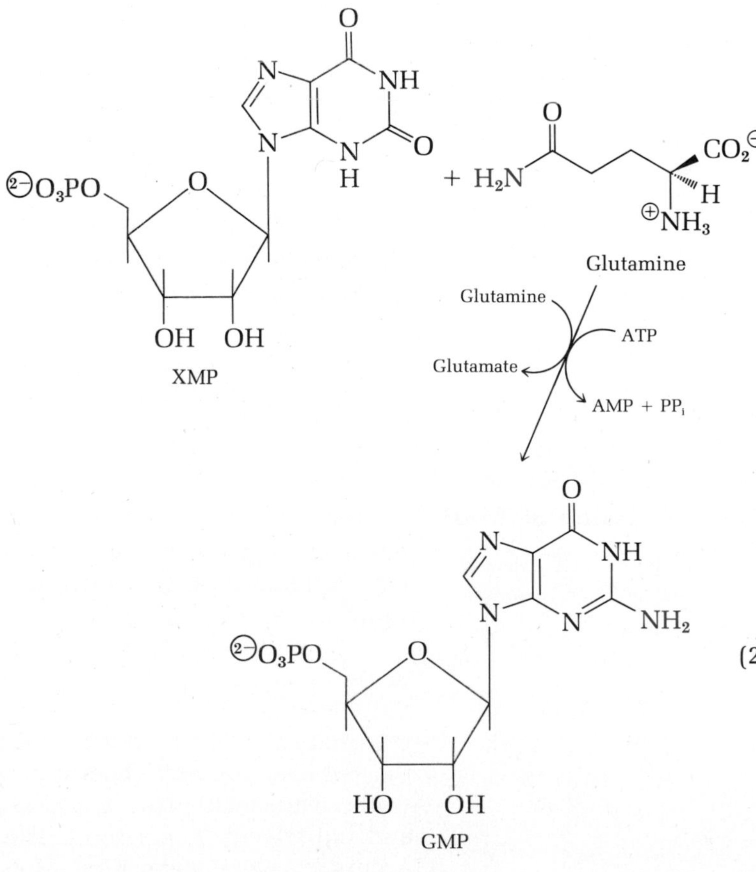

(23.16)

As we noted in Section 23.2A, ATP is a substrate for the synthesis of GMP, and excess ATP encourages biosynthesis of GMP.

C. Synthesis of Purine Nucleoside Di- and Triphosphates

Purine nucleotides can be interconverted by *nucleoside monophosphate kinase* and *nucleotide diphosphate kinase* (reactions 23.17 and 23.18).

$$\text{ATP} + \text{GMP} \underset{\text{Nucleoside monophosphate kinase}}{\rightleftarrows} \text{ADP} + \text{GDP} \tag{23.17}$$

$$\text{ATP} + \text{GDP} \underset{\text{Nucleoside diphosphate kinase}}{\rightleftarrows} \text{ADP} + \text{GTP} \tag{23.18}$$

23–4 DEGRADATION OF PURINE NUCLEOTIDES

The pathways for catabolism of purine nucleotides lead to *uric acid* (Figure 23.13). Uric acid can exist in either of two tautomeric forms (reaction 23.19).

Uric acid (lactam)   Uric acid (lactim)   (23.19)

The lactim form has two acidic hydrogens whose pK_a values are 5.4 and 10.3. At neutral pH the predominant species of the lactim tautomer is the urate anion (reaction 23.20).

$pK_a = 10.3$ $pK_a = 5.4$ Urate   (23.20)

Both the lactam and the protonated lactim are quite insoluble in water. Urea, in contrast, is highly water soluble. Birds and fishes excrete uric acid rather than urea as the end product of nitrogen catabolism. The urine of these *uricotelic* animals is acidic and contains a high proportion of crystalline uric acid. Uric acid is apparently the preferred excretion production because of its high insolubility in water at low pH. If these animals were to excrete urea, they would also have to excrete ruinous amounts of water. Conservation of water is therefore the main advantage of uric acid excretion.

A. Degradation of AMP

AMP is converted to hypoxanthine in successive hydrolyses of first the 5'-phosphoryl group, second the amino group at C-6, and third the glycosidic bond (Figure 23.13). The conversions of hypoxanthine to xanthine and of xanthine to uric acid are catalyzed by *xanthine oxidase* (reaction 23.21).

Xanthine oxidase is a dimer whose molecular weight is 260,000. Each of its identical subunits contains one atom of the heavy transition metal molybdenum, a flavin coenzyme, and an Fe_4S_4 center. Molecular oxygen is required as the electron acceptor for the reaction, and hydrogen peroxide is the reduced product. The oxygen incorporated in uric acid originates in water. The mechanism of action of xanthine oxidase is extremely complex and imperfectly understood. The oxidation of xanthine (or hypoxanthine) occurs in several steps. First, Mo(IV) is oxidized to Mo(VI). Xanthine itself is the source of the electrons for this oxidation step. It has been proposed that an unusual persulfide bond is formed between the enzyme and substrate in

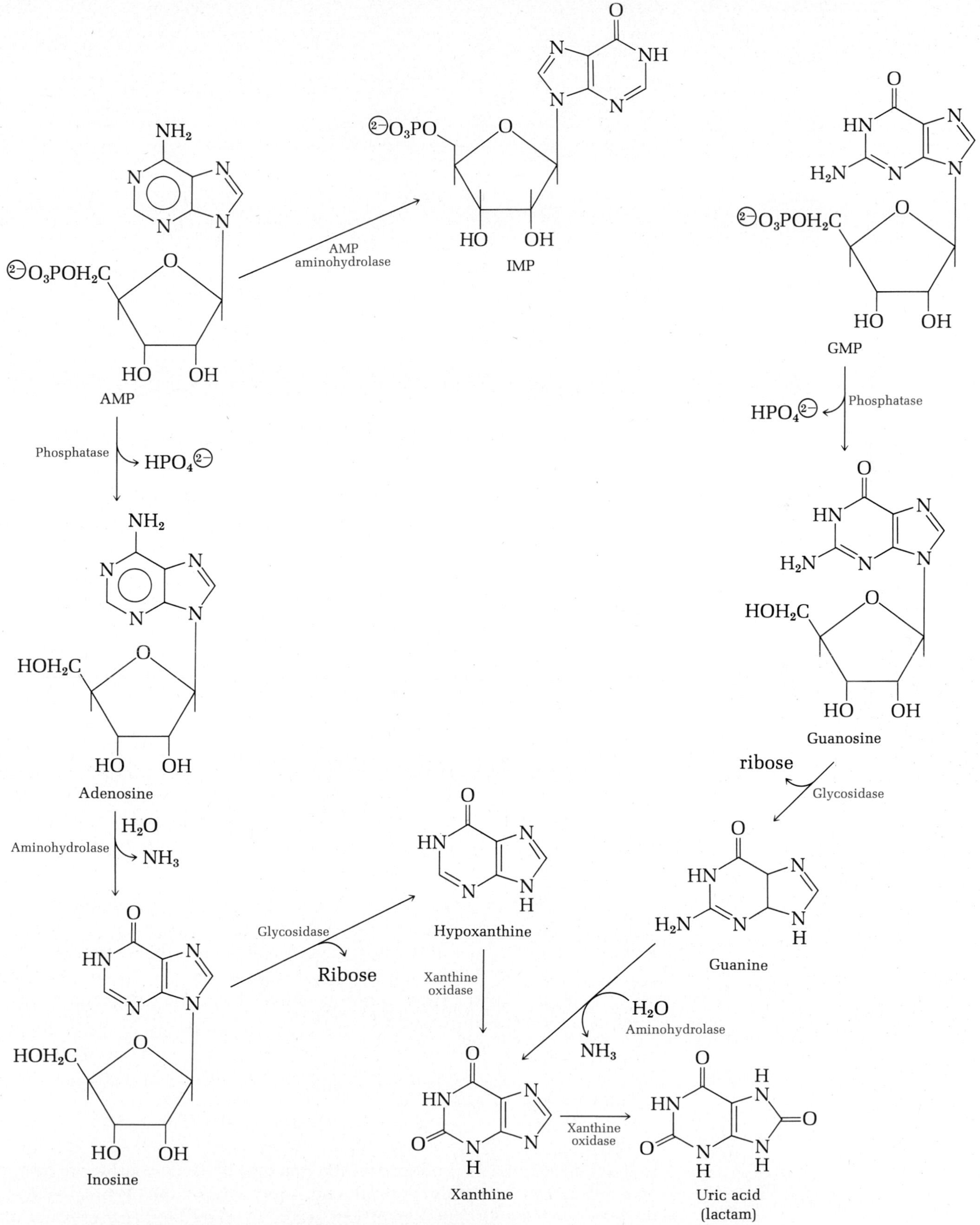

Figure 23.13 Pathway for the degradation of AMP and GMP.

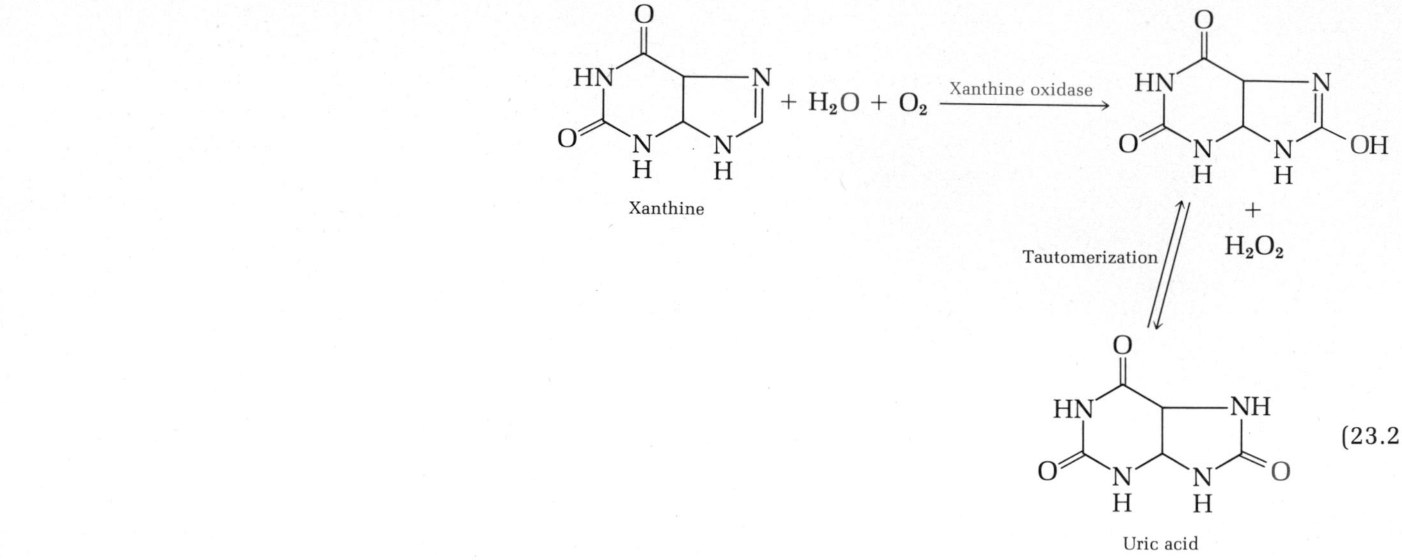

Xanthine

(23.21)

Uric acid

this step. Displacement of the persulfide by water then gives uric acid (reaction 23.22).

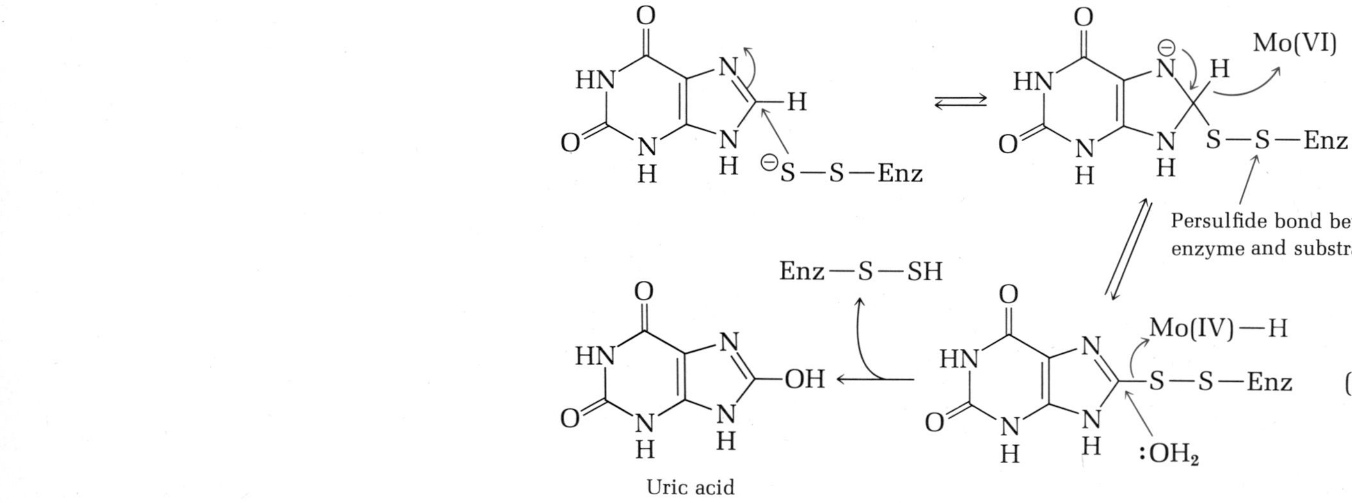

Uric acid

(23.22)

Persulfide bond between enzyme and substrate

To this point, molecular oxygen has played no role, but molybdenum has been reduced to Mo(IV), and in this oxidation state the enzyme is inactive. Molecular oxygen enters in the next step in a sequence of electron transfer steps involving the flavin and the iron-sulfur centers. In these electron transfer steps molecular oxygen is reduced to hydrogen peroxide and Mo(IV) is reoxidized to Mo(VI) (reaction 23.23).

$$\text{enzyme-Mo(IV)} + O_2 + 2H^{\oplus} \xrightarrow{\text{Flavin and Fe}_4\text{S}_4\text{ center}} \text{enzyme-Mo(VI)} + H_2O_2 \quad (23.23)$$

The hydrogen peroxide produced in the reaction is destroyed by catalase.

Gout is an extremely painful condition whose major biochemical symptom is excess production of uric acid. In extreme cases crystals of sodium urate precipitate in the synovial fluid of joints. The cause of excess production of uric acid has not been determined, but it is possible to treat gout by decreasing the production of uric acid by inhibiting xanthine oxidase. An analog of hypoxanthine called *allopurinol* is a potent inhibitor of xanthine oxidase. In allopurinol the positions of N-7 and C-8 are reversed. Allopurinol is a substrate for the first stage of the reaction in which the substrate is oxidized. But the product, alloxanthine, remains tightly bound to the active site of the enzyme in which Mo(IV) has been produced (reaction 23.24).

Allopurinol   Hypoxanthine

(23.24)

Allopurinol   Alloxanthine
                     (chelate with Mo^{4+})

In the complex of alloxanthine and xanthine oxidase the rate of reoxidation of Mo(IV) to Mo(VI) is extremely slow (the half-life for reoxidation is about 5 hours). This type of inhibition, in which a substrate analog is converted to an inhibitor at the enzyme active site, is called *suicide inhibition*.

B. Degradation of GMP

The degradation of GMP to uric acid closely parallels the degradation of AMP (see Figure 23.13). A phosphatase hydrolyzes the 5′-phosphate group of GMP producing guanosine; a glycosidase cleaves the β-glycosidic bond yielding guanine; and an aminohydrolase converts guanine to xanthine. Xanthine oxidase then converts xanthine to uric acid.

C. Degradation of Urate

In some species urate is not the final product of purine catabolism. Humans, primates, dogs, birds, snakes, and lizards excrete urate. Other mammals, turtles, and mollusks possess *urate oxidase* and they degrade urate and excrete allantoin. Certain teleost fishes possess *allantoinase* and take the degradation a step further to excrete allantoate. Other fishes and amphibia possess *allantoicase* and excrete urea. Some marine invertebrates carry purine degradation still further and convert urea to ammonium ions and carbon dioxide by the action of *urease*. These pathways for purine catabolism from urate are summarized in Figure 23.14.

23–5 PURINE SALVAGE PATHWAYS

The purine ring system is constructed at the cost of seven phosphoanhydride bonds, and it is therefore much more efficient for the cell to salvage the purines produced by hydrolysis of nucleic acids and nucleotides than to carry out *de novo* synthesis. Adenine, guanine, and hypoxanthine can all be converted to mononucleotides in a single step from the purine bases and PRPP. Thus, *adenine phosphoribosyl transferase* catalyzes synthesis of AMP from adenine and PRPP, and *hypoxanthine-guanine phosphoribosyl trans-*

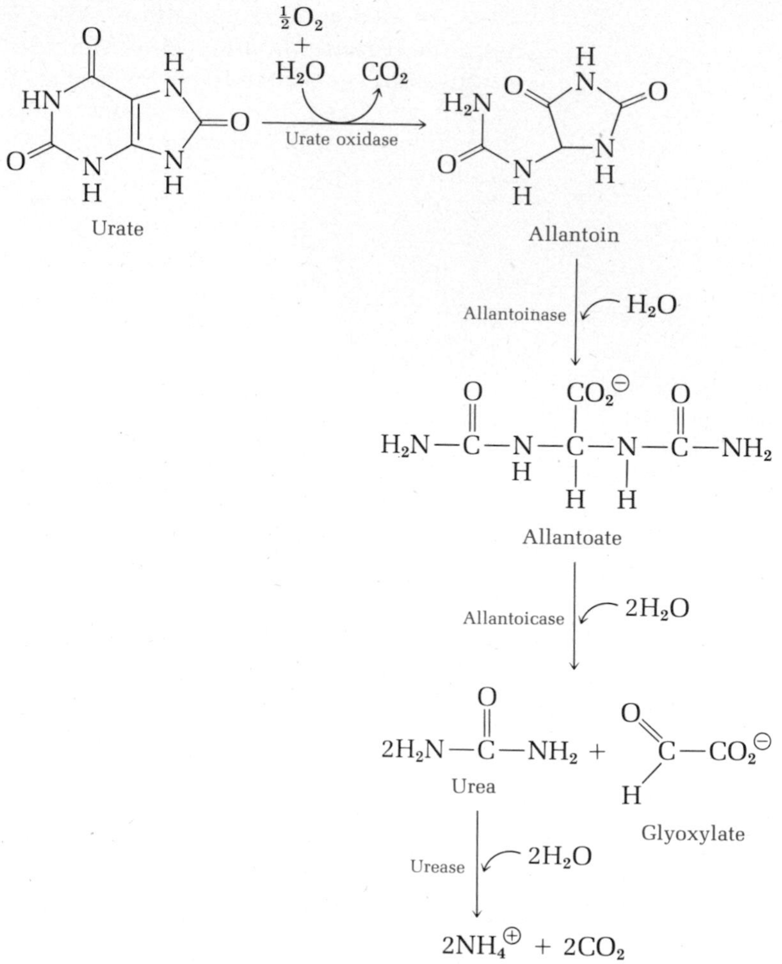

Figure 23.14 Pathway for the degradation of urate to ammonium ions and carbon dioxide.

ferase catalyzes the synthesis of IMP and GMP from the free bases and PRPP (reactions 23.25–23.27).

The *Lesch-Nyhan syndrome* is an inherited metabolic disease, linked to the X chromosome, in which hypoxanthine-guanine phosphoribosyl transferase is present at less than 1% of its normal concentration. This defect is transmitted as a sex-linked recessive trait. The failure to salvage guanine and hypoxanthine leads to an excess production of uric acid. The *de novo* and salvage pathways for purine biosynthesis are ordinarily balanced. Since the synthesis of 5-phospho-β-D-ribosylamine is inhibited by purine nucleotides, the absence of the salvage pathway leads to an increased rate of purine biosynthesis. The high rate of production of urate leads to kidney stones and gout at an early age. These symptoms can be alleviated by allopurinol, which inhibits xanthine oxidase. The more severe consequence of Lesch-Nyhan syndrome is a compulsive urge for self-mutilation, and this deranged condition is unaffected by allopurinol. Although we can see why increased production of urate leads to gout, it is not known why an absence of the salvage pathway leads to such destructive neurological conditions.

23–6 BIOSYNTHESIS OF PYRIMIDINE NUCLEOTIDES

Six enzymatic reactions are required for the biosynthesis of UMP (Figure 23.15). The pyrimidine ring system is constructed from aspartate, the amide nitrogen of glutamine, and carbon dioxide (Figure 23.16). The amide nitrogen and carbon dioxide are first incorporated in carbamoyl phosphate. The ribose moiety is derived from PRPP.

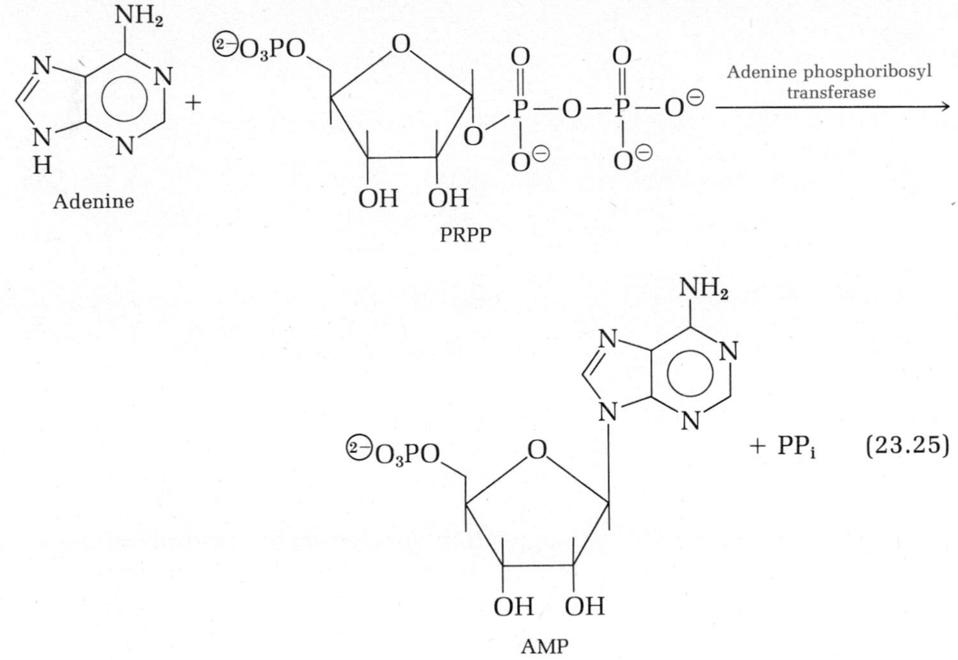

Adenine + PRPP $\xrightarrow{\text{Adenine phosphoribosyl transferase}}$

AMP + PP$_i$ (23.25)

Guanine + PRPP $\xrightarrow{\text{Hypoxanthine-guanine phosphoribosyl transferase}}$

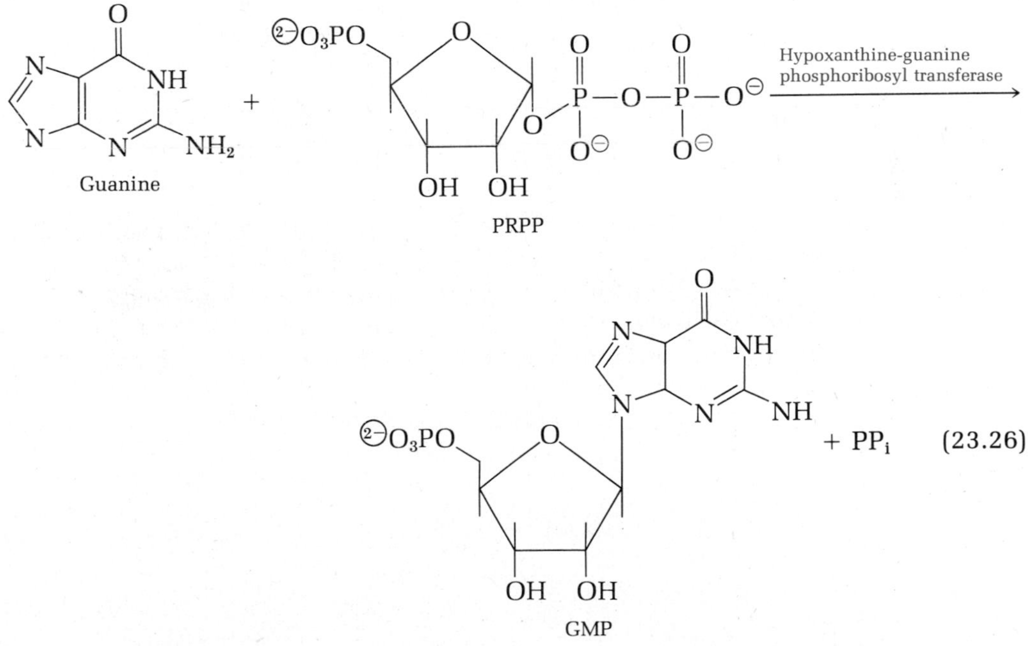

GMP + PP$_i$ (23.26)

Hypoxanthine + PRPP $\xrightarrow{\text{Hypoxanthine-guanine phosphoribosyl transferase}}$

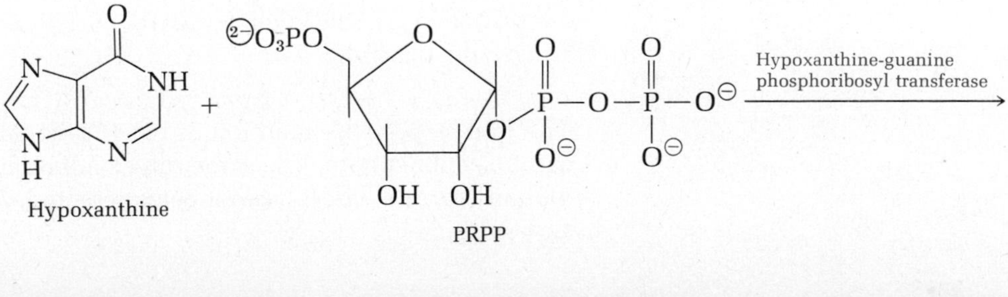

$$\text{[structure of IMP]} \qquad + \text{PP}_i \qquad (23.27)$$

IMP

Microorganisms require six separate enzymes for the biosynthesis of UMP. In animals, however, these enzymes are encoded by only three genes. The first three reactions of pyrimidine are catalyzed by a multienzyme polypeptide, located in the cytosol, whose molecular weight is about 200,000. The fourth step is catalyzed by an enzyme that is bound to the outer side of the inner mitochondrial membrane. The final two steps are catalyzed by a bifunctional cytosolic enzyme. In a normal mammalian cell the intermediates of pyrimidine biosynthesis do not accumulate; none are present at a concentration of greater than 1 μM. A biosynthetic product formed in one step in a multienzyme complex is immediately passed to the next step without dissociating, a phenomenon known as *channeling*. Since intermediates do not accumulate, they are not hydrolyzed by side reactions in the cytosol, with resulting loss of the energy required for their synthesis. We recall that the fatty acid synthetase complex is a multienzyme polypeptide, and that many steps in the biosynthesis of tryptophan also occur on a multienzyme polypeptide chain. It seems likely that the existence of multifunctional enzymes is a widespread phenomenon in eucaryotic cells.

A. Enzymatic Reactions of Pyrimidine Nucleotide Biosynthesis

a. Carbamoyl Phosphate Synthetase II

In our discussion of the urea cycle we encountered mitochondrial carbamoyl phosphate synthetase I. All mammalian cells capable of pyrimidine biosynthesis contain cytosolic *carbamoyl phosphate synthetase II* as part of a multienzyme polypeptide. This enzymatic activity catalyzes the biosynthesis of carbamoyl phosphate from the amido group of glutamine, bicarbonate, and ATP (reaction 23.28).

$$\underset{\text{Glutamine}}{\overset{\overset{\displaystyle O}{\|}}{H_2N-C-CH_2CH_2-\underset{\underset{H}{|}}{\overset{\overset{\oplus}{NH_3}}{C}}-CO_2^{\ominus}} \;+\; HCO_3^{\ominus} \;+\; 2ATP}$$

$$\Big\downarrow \text{Carbamoyl phosphate synthetase II}$$

$$\underset{\text{Carbamoyl phosphate}}{\overset{\overset{\displaystyle O}{\|}}{H_2N-C-O-PO_3^{\text{\tiny ②}\ominus}} \;+\; 2ADP \;+\; HPO_4^{\text{\tiny ②}\ominus} \;+\; \text{glutamate}} \qquad (23.28)$$

The first step in the reaction of the *E. coli* enzyme is probably formation of carbonyl phosphate. The same intermediate has been proposed for the mammalian enzyme, but it has not been detected. ATP is the phosphoryl donor in

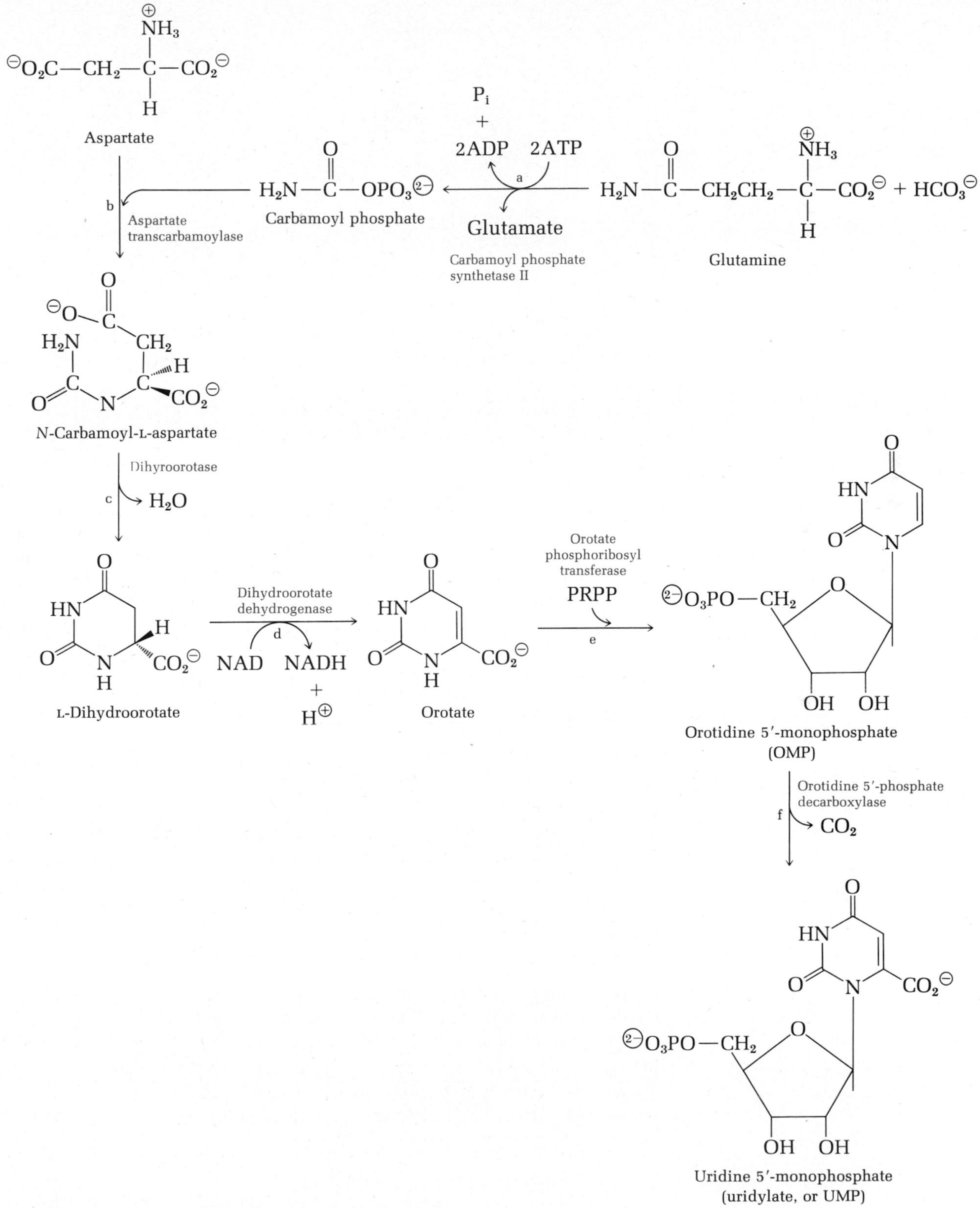

Figure 23.15 Pathway for the biosynthesis of UMP.

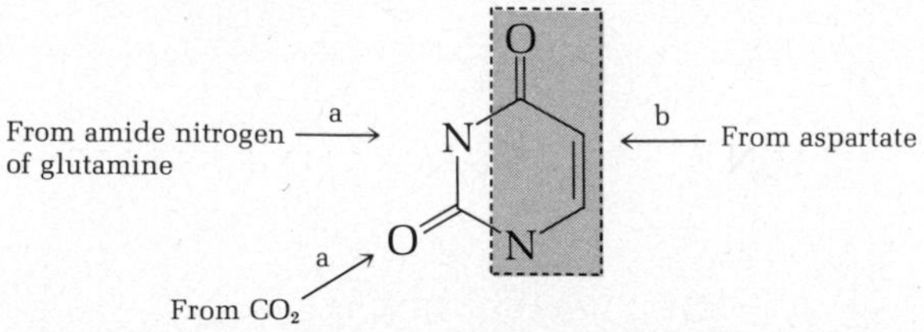

Figure 23.16 Sources of the ring atoms in pyrimidine nucleotide biosynthesis. Numbers refer to the steps of the pathway (Figure 23.15) in which the atoms are incorporated.

the synthesis of the energy-rich intermediate (an acyl phosphate) carbonyl phosphate, or carbonic phosphate anhydride. Bicarbonate is the source of the carbonyl group (reaction 23.29).

$$\text{HO}-\underset{\underset{O^\ominus}{|}}{\overset{\overset{O}{\|}}{C}} + \text{ATP} \longrightarrow \text{H}-\text{O}-\overset{\overset{O}{\|}}{C}-\text{O}-\text{PO}_3^{\,\text{2}\ominus} + \text{ADP} \qquad (23.29)$$

Carbonyl phosphate
(Carboxyphosphate)

Reaction of the acyl phosphate with ammonia, provided at the active site of the enzyme by glutamine, gives carbamate (reaction 23.30).

$$\underset{\text{Carbonyl phosphate}}{\text{HO}-\overset{\overset{O}{\|}}{C}-\text{OPO}_3^{\,\text{2}\ominus}} + \text{NH}_3 \longrightarrow \underset{\text{Carbamate}}{\text{HO}-\overset{\overset{O}{\|}}{C}-\text{NH}_2} \qquad (23.30)$$

The second mole of ATP is consumed in phosphoryl group transfer from ATP to the carboxylate oxygen producing carbamoyl phosphate (reaction 23.31).

$$\underset{\text{Carbamate}}{\text{H}_2\text{N}-\overset{\overset{O}{\|}}{C}-\text{O}^\ominus} + \text{ATP} \xrightarrow{\overset{\text{Carbamate}}{\text{kinase}}} \underset{\text{Carbamoyl phosphate}}{\text{H}_2\text{N}-\overset{\overset{O}{\|}}{C}-\text{OPO}_3^{\,\text{2}\ominus}} \qquad (23.31)$$

The mammalian enzyme is under strict metabolic control. UTP, UDP, UDP-glucose, CTP, dUDP, and dUTP are all feedback inhibitors. PRPP, on the other hand, is an activator. These allosteric effectors exert their effects by altering K_m for ATP. Glycine is a competitive inhibitor of glutamine binding. We have already noted that the biosynthesis of glutamine is itself under stringent control. The polyamines spermine and spermidine (Section 22.5E) are inhibitors of activation by PRPP.

b. Aspartate Transcarbamoylase

The phosphate group of carbamoyl phosphate is displaced by the α-amino group of L-aspartate to form N-*carbamoyl*-L-*aspartate* in a reaction catalyzed by *aspartate transcarbamoylase* (reaction 23.32).

In microorganisms the rate of pyrimidine biosynthesis is largely controlled by metabolic regulation of aspartate transcarbamoylase. However, the mammalian enzyme is not subject to allosteric regulation. *E. coli* aspartate transcarbamoylase is subject to allosteric inhibition by CTP, an end product

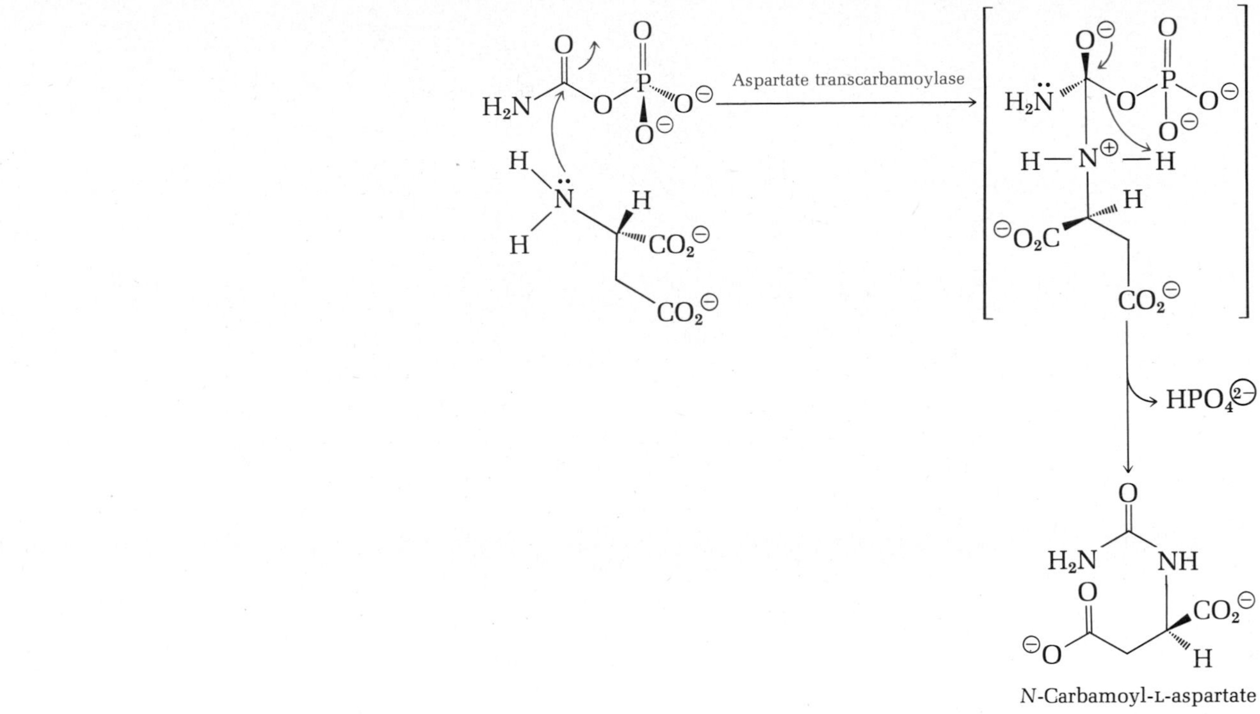

N-Carbamoyl-L-aspartate

(23.32)

of pyrimidine biosynthesis, and it is activated by ATP (Figure 23.17). CTP acts by increasing the K_m of the enzyme for aspartate, and ATP acts by decreasing the K_m for aspartate. ATP and CTP compete for the same regulatory binding site in aspartate transcarbamoylase. Thus, a high ATP concentration largely abolishes inhibition by CTP, and a high CTP concentration largely abolishes activation by ATP. The counterbalancing effects of a pyrimidine nucleotide and a purine nucleotide help to balance the rates of purine and pyrimidine biosynthesis.

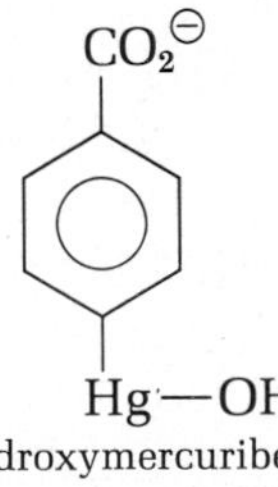

4-Hydroxymercuribenzoate

These allosteric effects are abolished when the enzyme is treated with mercurial compounds, such as 4-hydroxymercuribenzoate, a process known as *desensitization*. The desensitized enzyme retains full catalytic activity. Desensitization of aspartate transcarbamoylase is accompanied by dissociation of the enzyme into two types of subunits that can be separated by ion exchange chromatography. One of them exists as a trimer whose monomers have a molecular weight of 33,000. The trimer possesses the catalytic activity of aspartate transcarbamoylase. The catalytic subunits are designated C. The second type of subunit has a molecular weight of 17,000 and exists as a dimer. The dimer binds ATP and CTP and is responsible for the allosteric properties of aspartate transcarbamoylase. The regulatory subunits, designated R, each contain one Zn(II) ion and four sulfhydryl groups that are ligands of the Zn(II) ion. Mercurial reagents react with sulfhydryl groups, and disrupt the interaction of the sulfhydryl groups with the Zn(II) ion.

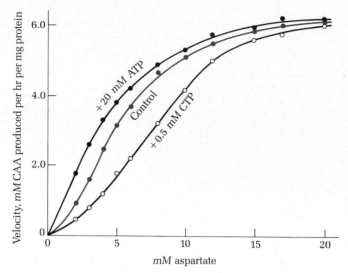

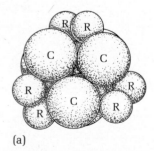

Figure 23.17 Response of aspartate transcarbamylase to the allosteric effectors CTP and ATP. CTP is a negative feedback inhibitor, and ATP is an allosteric activator. In the presence of ATP and CTP the inhibition of CTP is overcome by the activation of ATP. [From J. C. Gerhart, *Curr. Top. Cell Regul.*, 2, 275 (1970).]

When the isolated subunits are mixed, they spontaneously recombine to give native aspartate transcarbamoylase (reaction 23.33).

$$3R_2 + 2C_3 \xrightarrow{\text{Zn(II)}} R_6C_6 \tag{23.33}$$

Zn(II) ions must be present for the enzyme to be reconstituted. The reconstituted enzyme possesses full catalytic activity and has the same allosteric properties as the native enzyme.

The subunit structure of aspartate transcarbamoylase is shown in Figure 23.18. The two C_3 subunits are nearly eclipsed, and each catalytic monomer of one trimer is associated with a catalytic monomer of the second trimer through interactions with the R_2 dimer. A sigmoidal binding curve, such as the one for aspartate transcarbamoylase shown in Figure 23.17, implies a cooperative interaction of the subunits of the oligomer (recall Section 5.8), and in aspartate transcarbamoylase the cooperative interactions are mediated by the regulatory subunits. The allosteric binding site in the regulatory subunit is remote from the catalytic site, a phenomenon often observed in allosteric proteins. The active site of the catalytic subunit seems to be formed from residues contributed by two C chains, an unusual and interesting feature of the molecule.

(a)

Figure 23.18 (a) Subunit structure of aspartate transcarbamoylase. Regulatory subunits are indicated by R and catalytic subunits by C. [From H. L. Monsco, J. L. Crawford, and W. N. Lopscomb, *Proc. Nat. Acad. Sci. U.S.*, 75, 5277 (1978).] (b) Stereo views of aspartate transcarbamoylase (page 963).

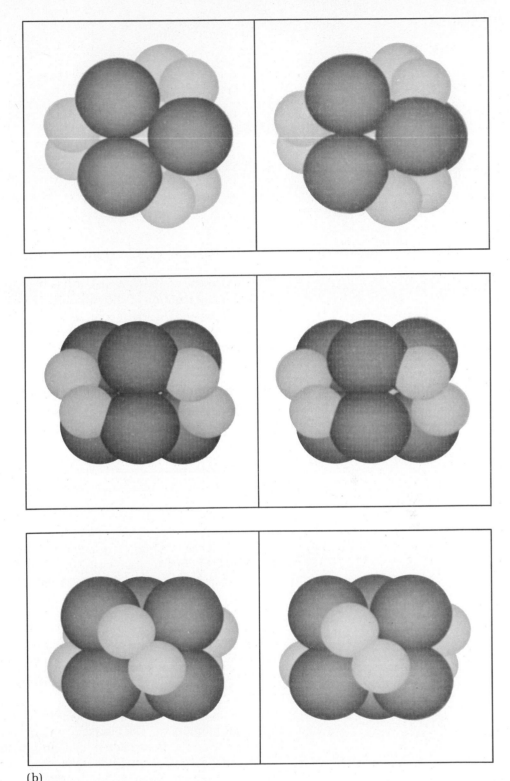

(b)

Figure 23.18 (continued)

c. Dihydroorotase

The first two steps of pyrimidine biosynthesis assemble the atoms of the pyrimidine ring. Dihydroorotase catalyzes ring closure to produce *L*-dihydroorotate (reaction 23.34).

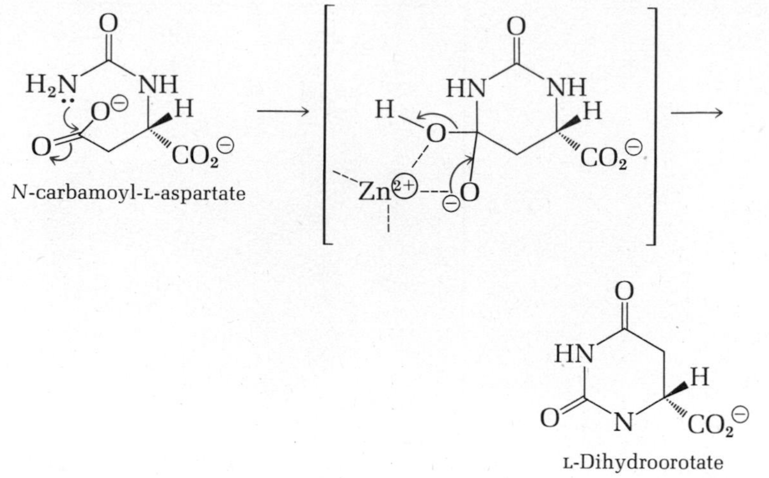

N-carbamoyl-L-aspartate

$$(23.34)$$

L-Dihydroorotate

Dihydroorotase contains Zn(II). It has been proposed that the mechanism of action for the reaction of dihydroorotase in the multienzyme polypeptide of mammals resembles the mechanism of action of carboxypeptidase (Section 6.2D). In such a mechanism Zn(II) would act as an electrophilic catalyst to convert the hydroxyl group of the tetrahedral intermediate to a good leaving group.

d. Dihydroorotate Dehydrogenase

Dihydryoorotate dehydrogenase catalyzes the NAD-dependent conversion of dihydroorotate to orotate (reaction 23.35).

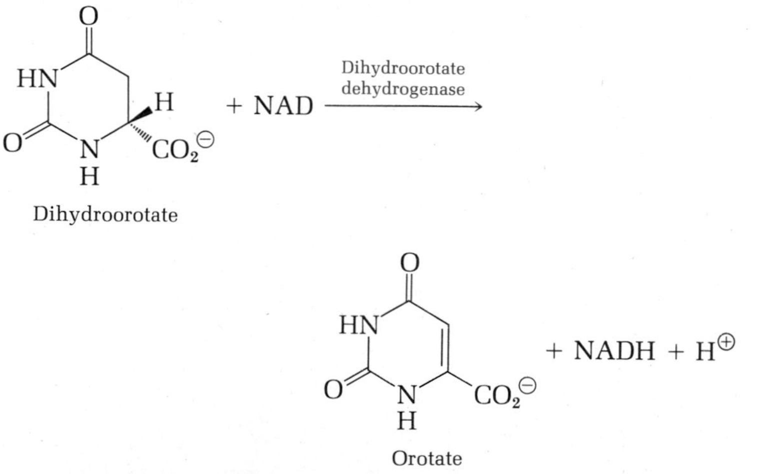

Dihydroorotate

$$+ \text{ NADH} + \text{H}^{\oplus} \quad (23.35)$$

Orotate

The mammalian enzyme is bound to the outer side of the inner mitochondrial membrane. The enzyme is a "dimer of dimers" (mol wt 115,000). Each dimer contains one active site, and each active site contains two iron atoms, an FMN, and an FAD. Like most of the other flavoprotein dehydrogenases we have discussed, the reaction probably occurs by an electron transport chain involving the iron atom, the FMN, and the FAD. Dihydroorotate dehydrogenase is the only mitochondrial enzyme of pyrimidine biosynthesis and the only one that is not a multifunctional enzyme.

e. Orotate Phosphoribosyl Transferase

In step 5 orotate phosphoribosyl transferase catalyzes displacement of the pyrophorphoryl group of PRPP by N-1 of orotate to give *orotidine 5'-monophosphate* (reaction 23.36).

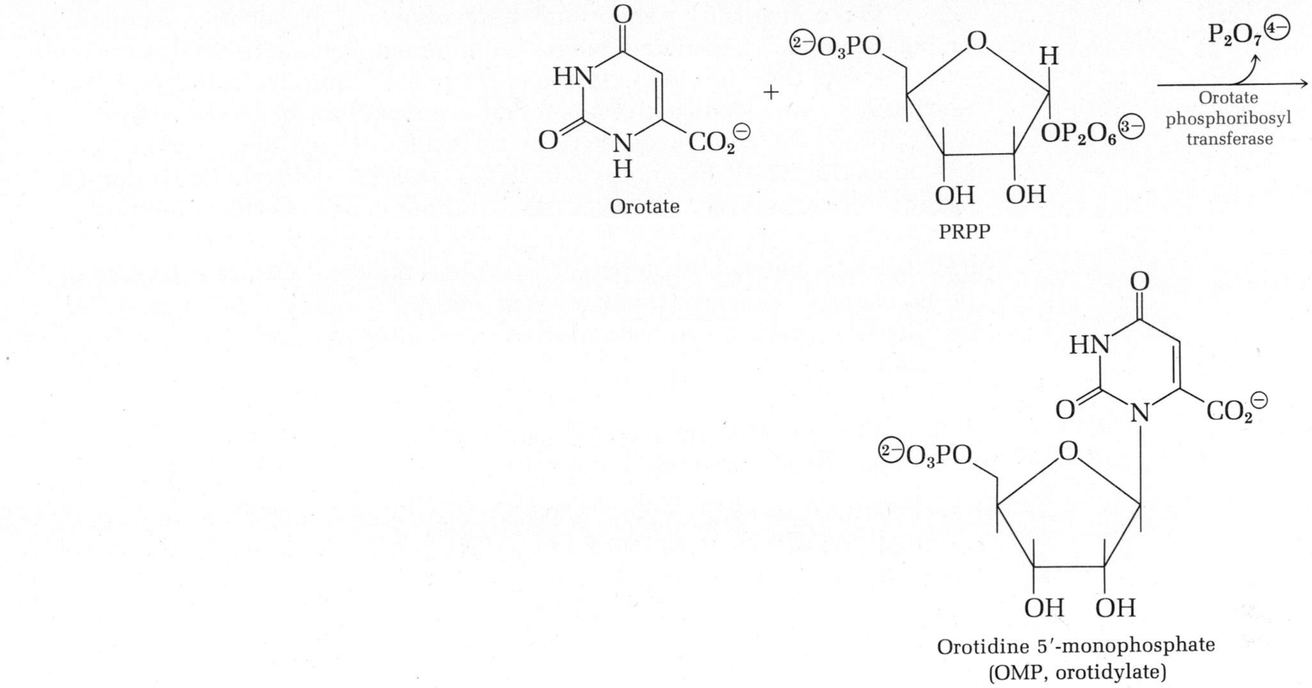

$$(23.36)$$

Orotidine 5′-monophosphate
(OMP, orotidylate)

f. Orotidine 5′-Phosphate Decarboxylase

The final step in the biosynthesis of UMP is decarboxylation of orotidine 5′-monophosphate catalyzed by orotidine 5′-monophosphate (or orotidylate) decarboxylase (reaction 23.37).

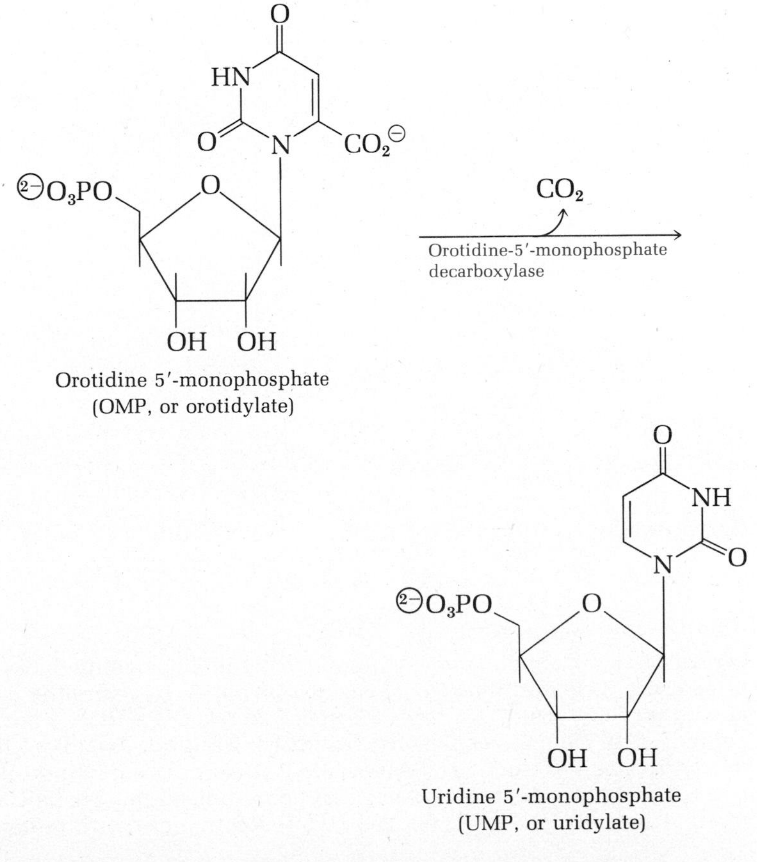

$$(23.37)$$

In microorganisms steps e and f are catalyzed by separate enzymes. Mammals, however, contain a single bifunctional enzyme to catalyze these reactions. The OMP formed in step e is not released into the solution, but is transferred to the second active site for decarboxylation, an example of channeling. Since the decarboxylase activity is about twice that of glycosidic bond formation, OMP does not accumulate. Since channeling is also observed for the multienzyme that catalyzes the first three steps of pyrimidine synthesis, none of the intermediates of pyrimidine biosynthesis accumulates. Figure 23.19 summarizes the apparent K_m values and the relative activities of the mammalian enzymes that carry out pyrimidine biosynthesis, and Figure 23.20 summarizes the regulation of pyrimidine nucleotide biosynthesis schematically.

B. Synthesis of UTP and CTP

UMP can be converted to UTP by the successive action of *uridylate kinase* and *nucleoside diphosphate kinase* (reactions 23.38 and 23.39).

$$\text{UMP} + \text{ATP} \underset{\text{Uridylate kinase}}{\rightleftharpoons} \text{ATP} + \text{UDP} \tag{23.38}$$

$$\text{UDP} + \text{ATP} \underset{\text{Nucleoside diphosphate kinase}}{\rightleftharpoons} \text{UTP} + \text{ADP} \tag{23.39}$$

No net formation or hydrolysis of phosphoanhydride bonds occurs in either of these freely reversible reactions.

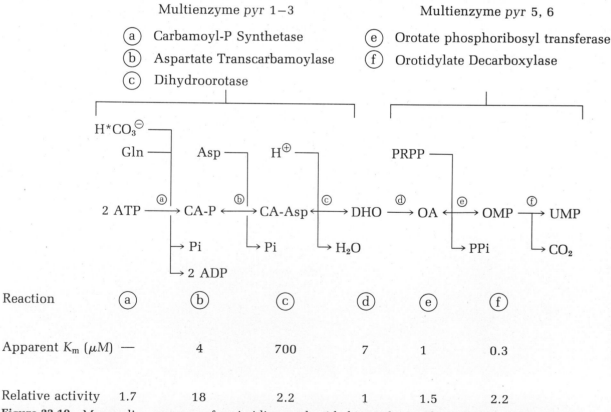

Figure 23.19 Mammalian enzymes of pyrimidine nucleotide biosynthesis. The enzymes are encoded by three genes. One multifunctional enzyme (*pyr 1–3*) catalyzes the first three steps of the synthesis, and a second (*pyr 5, 6*) catalyzes steps 5 and 6. The relative activities and apparent K_m values of the five enzymes are indicated. Double-headed arrows indicate freely reversible reactions; single-headed arrows indicate nearly irreversible reactions. Abbreviations: CA-P, carbamoyl phosphate; CA-Asp, carbamoyl aspartate; DHO, dihydroorotate; OA, orotate; OMP, orotidine 5'-phosphate; UMP, uridine 5'-phosphate. [From M. E. Jones, *Ann. Rev. Biochem.*, 49, 264 (1980). Reproduced with permission. © 1980 by Annual Reviews Inc.]

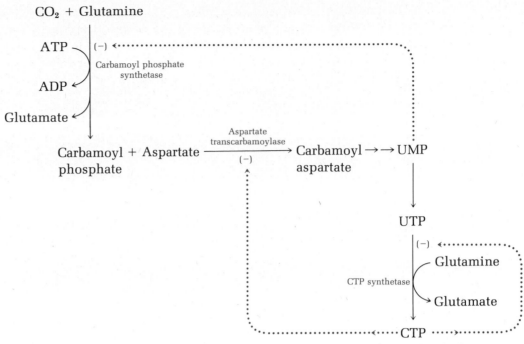

Figure 23.20 Regulation of pyrimidine biosynthesis in *E. coli*. CTP inhibits both aspartate transcarbamoylase and CTP synthetase, and UMP inhibits carbamoyl phosphate synthetase.

CTP is synthesized from UTP by *CTP synthetase*. The amide nitrogen of glutamine (required in yet another biosynthetic reaction) is transferred to C-4 of UTP in a reaction similar to the formation of 5-phospho-β-D-ribosylformylglycinamidine in purine nucleotide biosynthesis (reaction 23.40).

(23.40)

C. Pyrimidine Salvage Pathways

UMP can be synthesized in one step from uracil and PRPP by the action of uracil phosphoribosyl transferase (reaction 23.41). Cytidine is not a substrate for this enzyme.

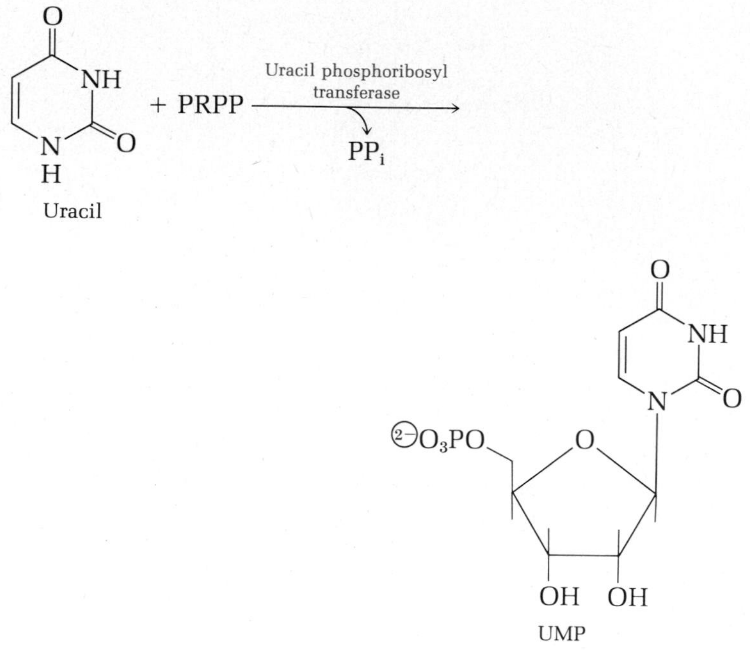

$$(23.41)$$

The nucleosides uridine and cytidine can be converted to UMP and CMP, respectively, by *uridine-cytidine kinase* (reactions 23.42 and 23.43).

$$\text{uridine} + \text{ATP} \xrightleftharpoons[]{\text{Uridine-cytidine kinase}} \text{UMP} + \text{ADP} \qquad (23.42)$$

$$\text{cytidine} + \text{ATP} \xrightleftharpoons[]{\text{Uridine-cytidine kinase}} \text{CMP} + \text{ADP} \qquad (23.43)$$

Uracil is converted to uridine by ribosyl transfer from ribose 1-phosphate in a reaction catalyzed by *uridine phosphorylase* (reaction 23.44).

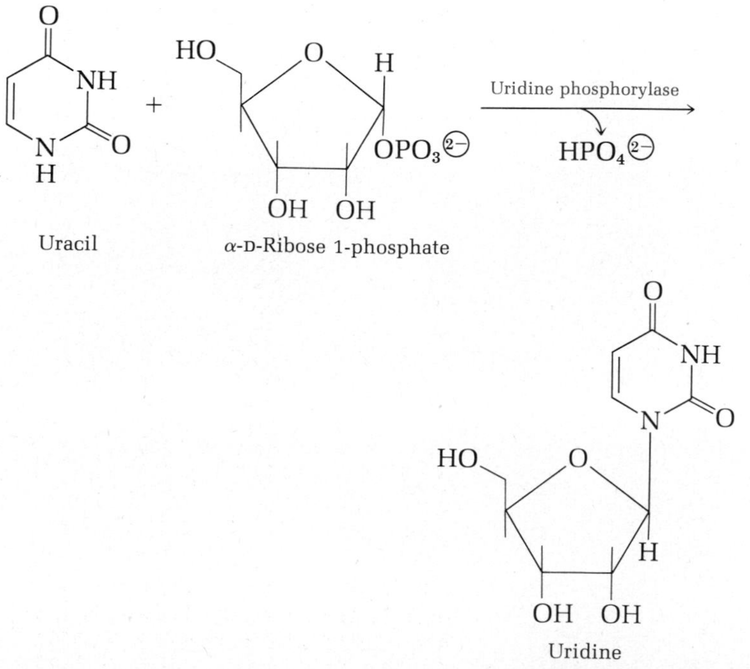

$$(23.44)$$

23-7 REDUCTION OF RIBONUCLEOTIDES

All organisms contain enzymes that are capable of reducing ribonucleoside diphosphates to 2'-deoxyribonucleoside diphosphates. In some microorganisms, such as *E. coli*, and in higher organisms the enzyme responsible for synthesis of 2'-deoxyribonucleotides is *ribonucleotide reductase*. (Certain microorganisms contain a version of this enzyme that requires coenzyme B_{12} as a cofactor. This enzyme is less widely distributed than the B_{12}-independent enzyme, and we shall not discuss it further.) NADPH is the reducing agent in the reaction. When the reaction is carried out in tritiated water (^3H_2O), the tritium replaces the 2'-hydroxyl group with retention of configuration (reaction 23.45).

Ribonucleotide diphosphate

$$+ \ ^3H_2O \xrightarrow[\text{NADPH} \quad \text{NADP}]{\text{Ribonucleotide reductase}}$$

(23.45)

2'-Deoxyribonucleotide diphosphate

The reduction of ribonucleotide diphosphates is the first unique step of DNA biosynthesis (Chapter 24), and its rate closely parallels the overall rate of DNA synthesis.

E. coli ribonucleotide reductase is composed of two nonidentical subunits known as proteins B1 and B2. A schematic diagram of ribonucleotide reductase is shown in Figure 23.21. Neither subunit by itself possesses catalytic activity. Protein B1 is a dimer whose molecular weight is 160,000. B1 contains identical binding sites for all four ribonucleoside diphosphate sub-

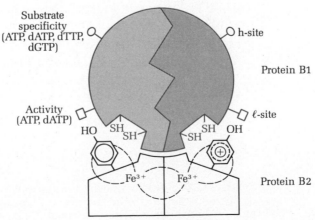

Figure 23.21 Schematic diagram of *E. coli* ribonucleotide reductase. [Adapted from L. Thelander and P. Reichard, *Ann. Rev. Biochem.*, 48, 139 (1979). Reproduced with permission. © 1979 by Annual Reviews Inc.]

strates and for the nucleoside triphosphates that act as allosteric effectors of the reductase. B1 also contains a pair of sulfhydryl groups that are part of the catalytic site. Protein B2 has a molecular weight of 78,000 and is composed of two identical subunits. B2 contains one Fe(III) ion in each of its polypeptide chains. Each subunit also contains a tyrosyl residue that is thought to participate in catalysis. A free radical mechanism involving the tyrosyl residues and the iron atoms of B2 plus the sulfhydryl groups of B1 has been proposed for the reduction, but the details of the mechanism of action of ribonucleotide reductase are not known.

Electrons are transferred from NADPH to ribonucleotide reductase by an electron transport chain that involves *thioredoxin*. This protein contains a disulfide bond that is reversibly oxidized by the flavoprotein *thioredoxin reductase* (reaction 23.46).

$$\text{Enzyme}-\text{FADH}_2 + \underset{\underset{\text{S}-\text{S}}{|\quad|}}{\text{thioredoxin}} \xrightarrow{\text{Thioredoxin reductase}}$$

$$\text{Enzyme}-\text{FAD} + \underset{\underset{\text{SH}\quad\text{SH}}{|\quad|}}{\text{thioredoxin}} \qquad (23.46)$$

The electron transport chain leading from NADPH involves the flavoprotein, thioredoxin reductase, thioredoxin, and ribonucleotide reductase. The terminal electron acceptor is the nucleoside diphosphate that is reduced (Figure 23.22). An *E. coli* cell contains from 10,000 to 20,000 molecules of thioredoxin to provide an electron source for the 1,500 to 3,000 molecules of ribonucleotide reductase that are also present in an *E. coli* cell.

We have noted that the same ribonucleotide reductase reduces all four ribonucleoside diphosphates. The substrate specificity of the enzyme is determined by allosteric regulation. The B1 subunit of ribonucleotide reductase has two types of binding sites for nucleotides. The low affinity sits, known as l-sites, bind weakly to nucleotides; the high affinity binding sites, known as h-sites, bind tightly to nucleotides. The l-sites bind either ATP or dATP, and the h-sites bind ATP, dATP, dTPP, and dGTP. The high affinity sites prefer deoxynucleotides, except ATP. The effects of allosteric effectors on the reduction of each of the four deoxyribonucleotides is summarized in Table 23.1. Four positive and one negative allosteric states exist. Let us first consider the positive states. (1) When ATP is bound at an l-site, GDP or UDP are reduced when ATP or dATP occupies an h-site. (2) When ATP is bound at an l-site and dTPP at the h-sites, GDP and ADP are reduced. (3) When ATP is bound at an l-site and dGTP at the h-sites, ADP and GDP are reduced. (4) When dTTP is bound at the h-sites and there is no ligand at the l-sites, all four ribonucleotides are reduced. The negative allosteric state exists when dATP is bound at an l-site, regardless of the allosteric effector bound at the

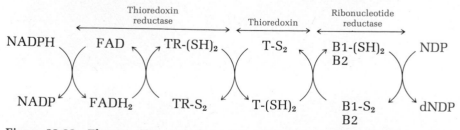

Figure 23.22   Electron transport chain from NADPH to nucleoside diphosphates in the reduction of ribonucleotides to 2′-deoxyribonucleotides catalyzed by ribonucleotide reductase. [Adapted from L. Thelander and P. Reichard, *Ann. Rev. Biochem.*, **48**, 141 (1979). Reproduced with permission. © 1979 by Annual Reviews Inc.]

Table 23.1 Regulation of _E. coli_ ribonucleotide reductase

Effector binding to		Reduction of			
l-sites	h-sites	CDP	UDP	GDP	ADP
0	ATP	+	+	0	0
0	dTTP	+	+	+	+
0	dGTP	0	0	+	+
ATP	ATP or dATP	+	+	0	0
ATP	dTTP	–	–	+	(+)
ATP	dGTP	nd	nd	(+)	+
dATP	Any effector	–	–	–	–

Abbreviations: 0, no effect; +, stimulation; –, inhibition; nd, not determined.

SOURCE: From L. Thelander and P. Reichard, _Ann. Rev. Biochem._, 48, 139 (1979). Reproduced with permission. © 1979 by Annual Reviews Inc.

h-site. Thus, dATP is a general inhibitor of ribonucleotide reductase. Note the subtle effects of the 2′-hydroxy group. With ATP bound at an l-site reduction is stimulated, but with dATP bound at an l-site reduction is inhibited. A general scheme for allosteric regulation of ribonucleotide reductase is shown in Figure 23.23. The net result of this complex allosteric regulation is that the reduction of ribonucleotides is balanced to meet the requirements of DNA synthesis, and one deoxyribonucleotide is not produced at the expense of the others.

23–8 BIOSYNTHESIS OF dTMP FROM dUMP

Thymidylate synthetase catalyzes the conversion of dUMP to dTMP by incorporating the methylene group of N^5,N^{10}-methylenetetrahydrofolate (Figure 23.24) in the 5 position of the pyrimidine ring of dUMP (reaction 23.47; see Section 10.8 for a discussion of the reaction mechanism).

The biosynthesis of dTMP is the slow step in the synthesis of DNA, and considerable effort has been expended to find specific inhibitors of thymidylate synthetase. One such inhibitor is the thymine analog 5-_fluorouracil_, an effective drug for some types of cancer. In the body 5-fluorouracil is converted to the deoxynucleotide 5-fluoro-2′-deoxyuridine. Inhibition of thymi-

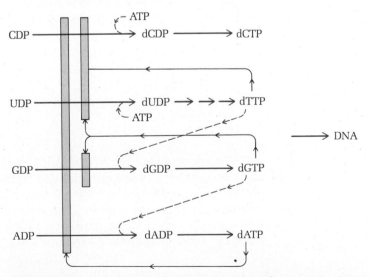

Figure 23.23 Regulation of ribonucleotide reductase in microorganisms and mammals. Open bars indicate metabolic inhibition; dash lines indicate metabolic activation of a given transformation. [From L. Thelander and P. Reichard, _Ann. Rev. Biochem._, 48, 153 (1979). Reproduced with permission. © 1979 by Annual Reviews Inc.]

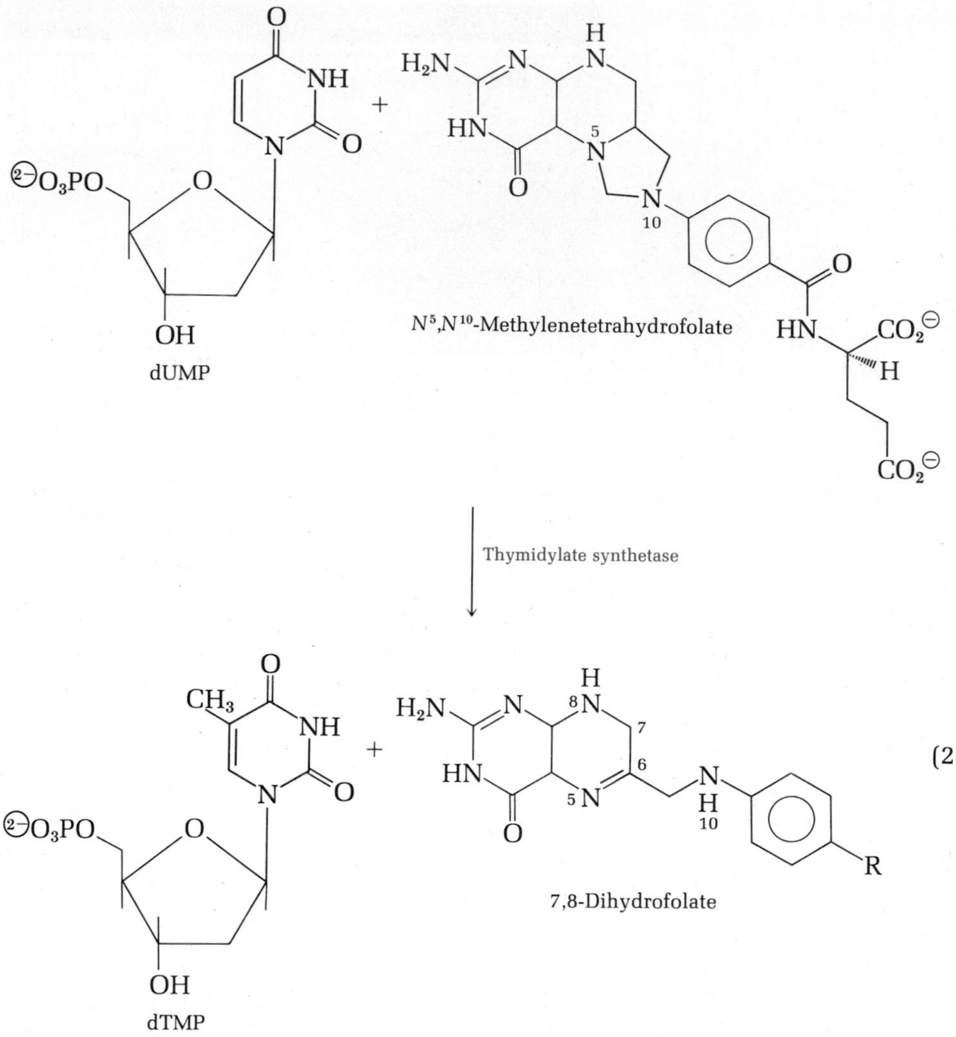

N^5,N^{10}-Methylenetetrahydrofolate

dUMP

Thymidylate synthetase

dTMP

7,8-Dihydrofolate

(23.47)

dylate synthetase by this nucleotide shuts off the supply of dTMP required for DNA replication in rapidly dividing cancer cells. Normal, slowly dividing cells are not affected by the inhibitor.

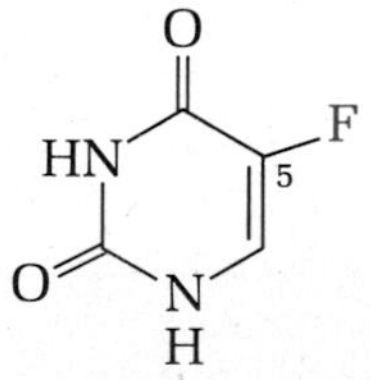

5-Fluorouracil

The methylene group of N^5,N^{10}-methylenetetrahydrofolate has the same oxidation state as formaldehyde. In the methylene transfer reaction catalyzed by thymidylate synthetase the methylene group gains an electron pair and is reduced to the oxidation state of methanol. Folate simultaneously undergoes a two-electron oxidation to 7,8-dihydrofolate, which is then reduced to tetrahydrofolate (reaction 23.48).

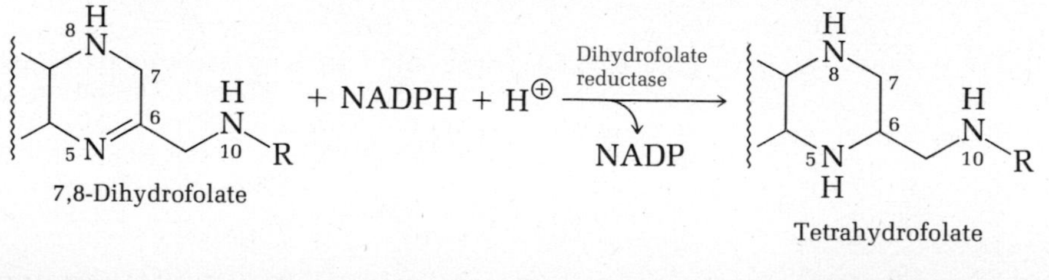

7,8-Dihydrofolate
$+ \text{NADPH} + \text{H}^\oplus \xrightarrow[\text{NADP}]{\text{Dihydrofolate reductase}}$
Tetrahydrofolate

(23.48)

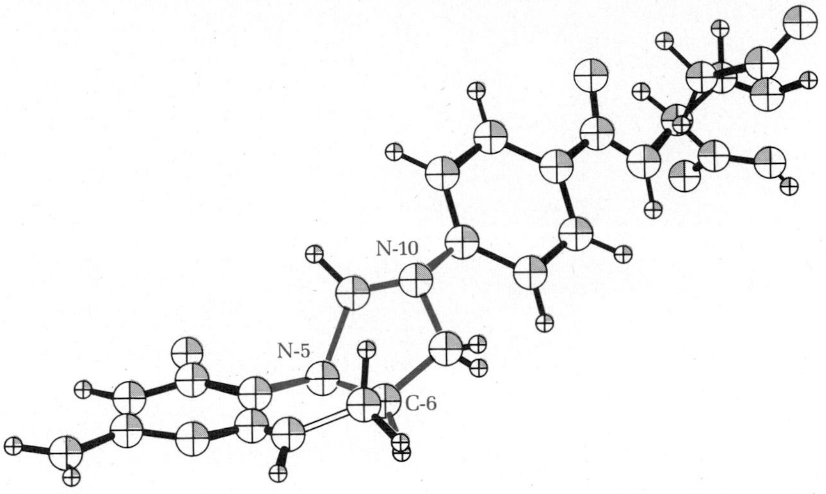

Figure 23.24 Structure of N^5,N^{10}-methylenetetrahydrofolate. [From S. Benkovic, *Ann. Rev. Biochem.*, **49**, 230 (1980). Reproduced with permission. © 1980 by Annual Reviews Inc.]

We recall (Section 10.8) that *methotrexate* inhibits this reaction by binding virtually irreversibly ($K_{\text{diss}} = 10^{-10}$) to dihydrofolate reductase. The closely related anticancer drug *aminopterin*, in which an amino group replaces the keto group at C-4 of 7,8-dihydrofolate, also inhibits dihydrofolate reductase.

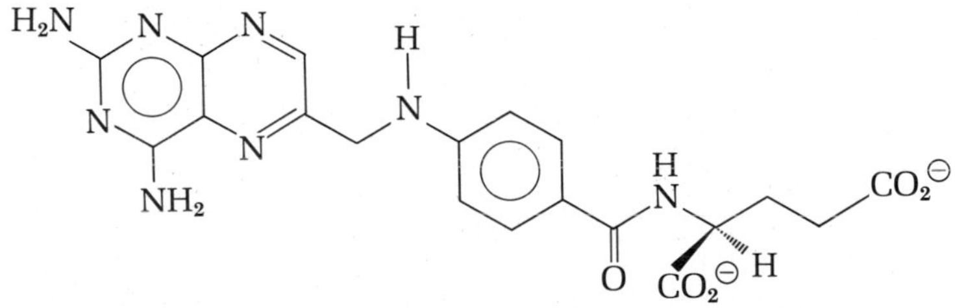

Aminopterin

23–9 BIOSYNTHESIS OF NUCLEOTIDE-CONTAINING COENZYMES

A. Coenzyme A

The biosynthesis of coenzyme A requires five steps beginning with the vitamin *pantothenate* (Figure 23.25). Plants and microorganisms are capable of synthesizing pantothenate, but animals require pantothenate in their diet. (1) In the first step ATP provides the phosphoryl group in the conversion of pantothenate to 4′-phosphopantothenate. (2) The thiol group of coenzyme A is derived from cysteine. Formation of an amide between the carboxyl group of 4′-phosphopantothenate and the α-amino group of cysteine yields 4′-phosphopantothenyl cysteine. ATP is the energy source for this reaction. (3) Decarboxylation yields 4′-phosphopantetheine. (4) The phosphoryl group of 4′-phosphopantetheine displaces a pyrophosphoryl group from ATP to give dephosphocoenzyme A. (5) Phosphoryl group transfer from ATP to the 3′-position of the ribosyl moiety of dephosphocoenzyme A produces coenzyme A.

B. NAD and NADP

In animals NAD and NADP may be derived from nicotinate, also known as vitamin B_3 or niacin. If niacin is not present in the diet, the nicotinamide

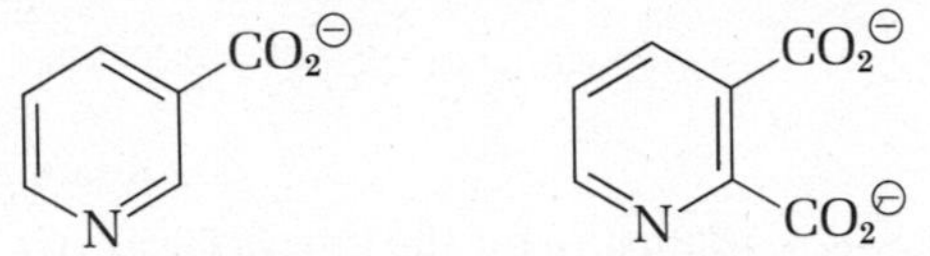

Nicotinate

Quinolinate

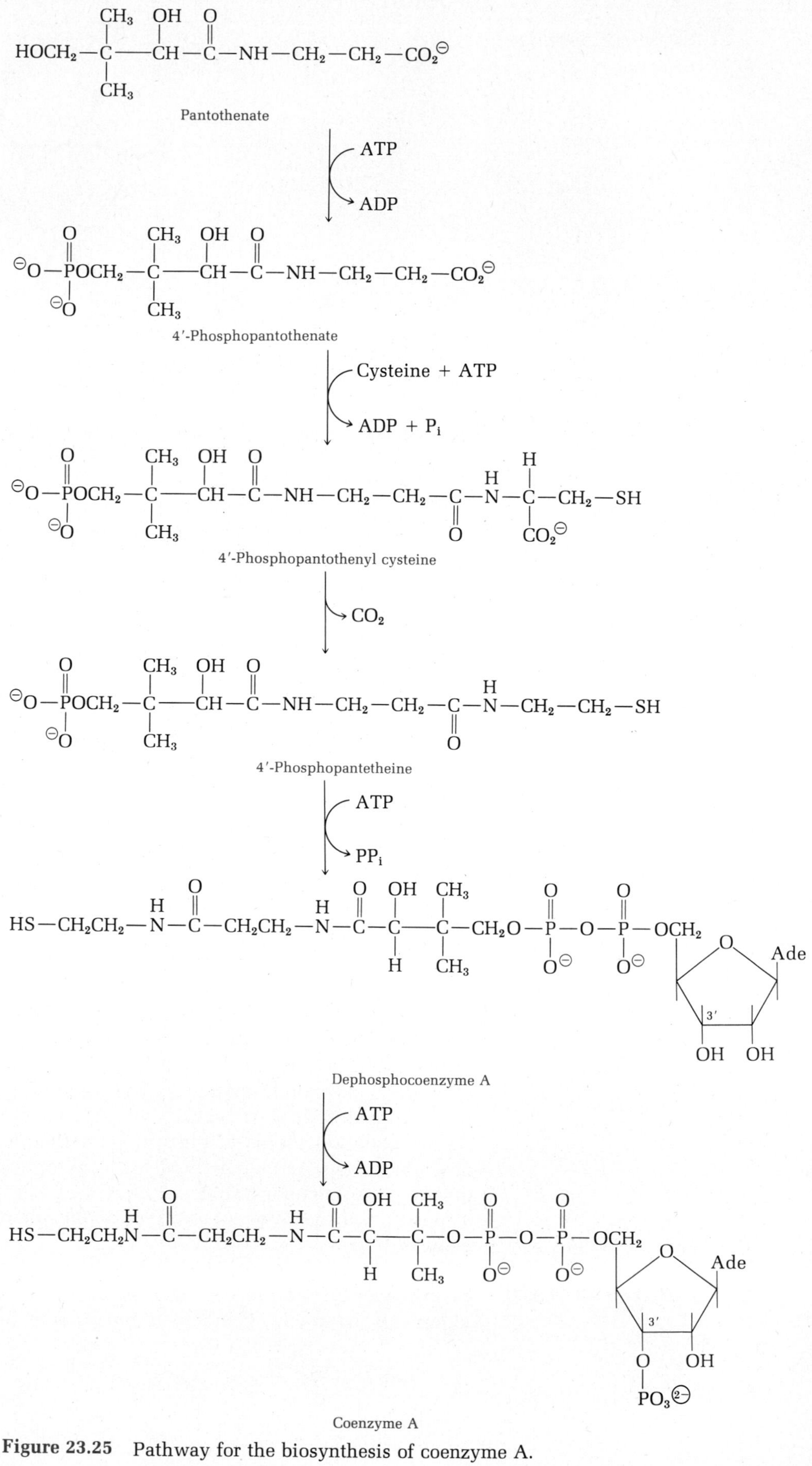

Figure 23.25 Pathway for the biosynthesis of coenzyme A.

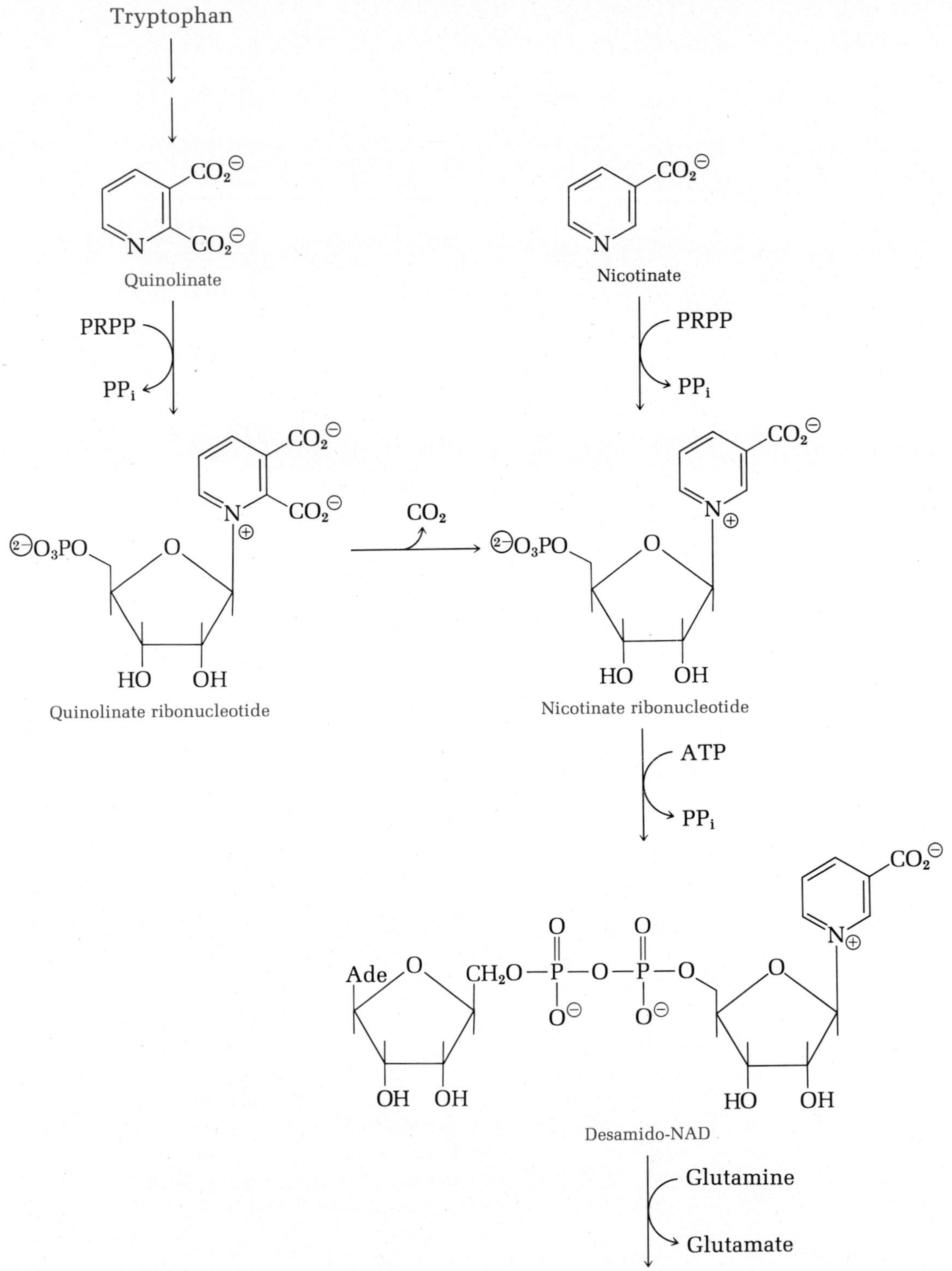

Figure 23.26 Pathway for the biosynthesis of NAD and NADP.

(continued)

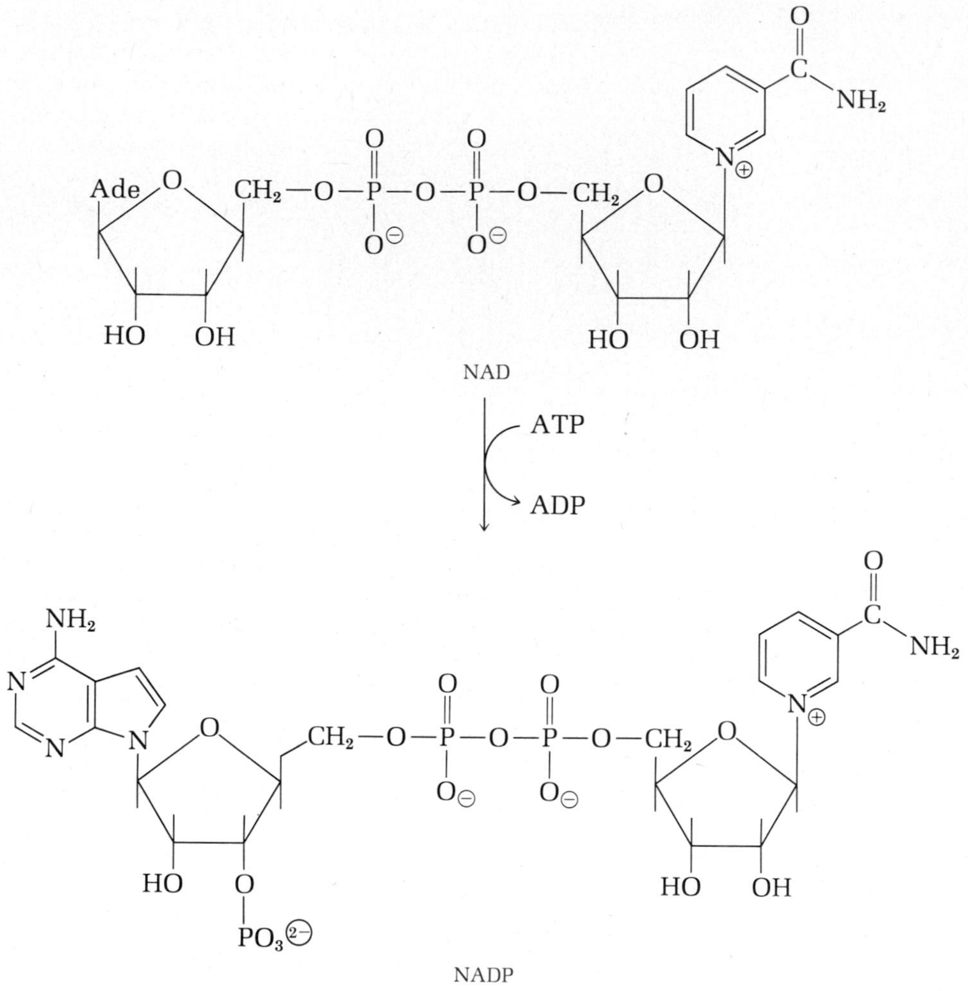

Figure 23.26 (continued)

ring system of NAD and NADP can also be obtained from quinolinate, a metabolic product of tryptophan metabolism, if the dietary supply of tryptophan is sufficient.

Three steps are required for the conversion of nicotinate to NAD, and four steps are required for the conversion of quinolinate to NAD (Figure 23.26). In either case, nicotinate ribonucleotide is an intermediate. PRPP is the source of the ribosyl moiety of the mononucleotide. The adenylate moiety of NAD is derived from ATP, and the amido group of NAD is derived from glutamine. Phosphoryl group transfer from ATP to the 2'-hydroxyl group of the adenylate moiety of NAD gives NADP.

23–10 SUMMARY

The purine ring system is assembled in 11 steps. The carbon and nitrogen atoms of purines are derived from several sources: glycine provides N-7, C-1, and C-2; N^5,N^{10}-methenyltetrahydrofolate provides C-8; and the amido group of glutamine provides N-9, thus accounting for the atoms of the imidazole ring. The pyrimidine ring portion of purines derives N-1 from glutamine, N-3 from aspartate, and C-7 from carbon dioxide. The ribosyl moiety of purine nucleotides is derived from PRPP. The biosynthesis of IMP is regulated by adenine and guanine nucleotides. All of them inhibit the first unique step of purine nucleotide biosynthesis catalyzed by amidophosphoribosyl transferase. The total inhibition of guanine and adenine nucleotides is greater than the sum of the inhibition of either of these groups of nucleotides acting separately, and purine nucleotide biosynthesis is therefore regulated by synergistic feedback inhibition. The first unique step of adenine nucleotide biosynthesis is inhibited by AMP, and the first unique step leading from IMP to guanine nucleotides is inhibited by GMP. Purine nucleotides can be synthesized by salvage pathways. Purine nucleotides are degraded to urate in humans, primates, dogs, birds, snakes, and lizards. Some organisms degrade urate further to allantoin, allantoate, urea, and ammonium ions.

The pyrimidine nucleotides are synthesized in six steps. Ring atoms N-1, C-6, C-5, and C-4 are provided by aspartate, N-3 is provided by the amido group of glutamine, C-2 is provided by carbon dioxide. PRPP again provies the ribosyl moiety. Steps a to c and steps e and f are each catalyzed by a multifunctional enzyme. In microorganisms all reactions are catalyzed by separate enzymes. The biosynthesis of pyrimidine nucleotides in microorganisms is regulated by allosteric control of aspartate transcarbamoylase. In mammals the first step of pyrimidine biosynthesis is subject to allosteric control. In microorganisms CTP inhibits both aspartate transcarbamoylase and CTP synthetase, and UMP inhibits carbamoyl phosphate synthetase.

Ribonucleotide reductase reduces all four nucleoside diphosphates to deoxyribonucleoside diphosphates. Thymidylate synthetase catalyzes the conversion of dUMP to dTMP by methylene transfer from N^5,N^{10}-methylenetetrahydrofolate. The substrate analog 5-fluorouracil inhibits thymidylate synthetase. Methotrexate and aminopterin inhibit dihydrofolate reductase. By preventing regeneration of tetrahydrofolate, they prevent synthesis of dTMP and thus inhibit DNA synthesis in rapidly dividing cancer cells.

REFERENCES

Benkovic, S., On the Mechanism of Action of Folate- and Biopterin-Requiring Enzymes, *Ann. Rev. Biochem.* **49**, 227–254 (1980).

Davidson, J. N. in R. P. L. Adams, R. H. Burdon, A. M. Campbell, and R. S. Smellie, Eds., "The Biochemistry of the Nucleic Acids," Academic Press, New York, 1977.

Greenberg, D. M., Ed., "Metabolic Pathways," vol. 4, Academic Press, New York, 1970, chaps. 19, 20.

Henderson, J. F., and Paterson, A. R. P., "Nucleotide Metabolism," Academic Press, New York, 1973.

Jones, M. E., Pyrimidine Nucleotide Biosynthesis in Animals: Genes, Enzymes, and the Regulation of UMP Biosynthesis, *Ann. Rev. Biochem.*, **49**, 253–281, (1980).

Stanbury, J. B., Wyngaarden, J. B., and Frederickson, D. S., Eds., "The Metabolic Basis of Inherited Disease," Part 6, Disorders of Purine and Pyrimidine Metabolism, McGraw-Hill, New York, 1972.

Thelander, L., and Reichard, P., Reduction of Ribonucleotides, *Ann. Rev. Biochem.*, **48**, 133–158 (1979).

Problems

1. Write a mechanism for steps 10 and 11 of purine nucleotide biosynthesis.

2. Write chemical equilibria which show that uric acid is really an acid.

3. The antibiotic L-azaserine inhibits formation of 5-phospho-β-D-ribosylamine. The enzyme glutamine pyrophosphoribosyl pyrophosphate amido transferase has a nucleophilic group at its active site. How does this group react with L-azaserine? Why is L-azaserine a substrate analog of glutamine?

L-Azaserine (O-(Diazoacetyl)-serine)

4. Treatment of aspartate transcarbamoylase (ATCase) with urea, heat, or mercurial reagents results in loss of inhibition by CTP. The plot of enzyme rate versus aspartate concentration is hyperbolic following these treatments, so enzyme activity is not lost. What is a plausible interpretation of these observations?

5. Sulfanilamides (Section 10.8) inhibit bacterial growth. What metabolic intermediate of purine biosynthesis will accumulate if a sulfanilamide is added to a colony of *E. coli*?

6. N-(Phosphonacetyl)-L-aspartate (PALA) is a powerful inhibitor of aspartate transcarbamoylase. It has been suggested that PALA is a transition state analog of aspartate transcarbamoylase. What structural features of PALA lead to ths conclusion?

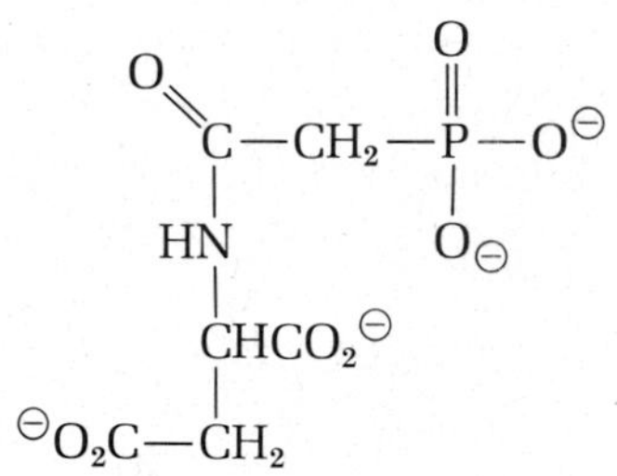

$$
\underset{\ominus O_2C-CH_2}{\underset{|}{\underset{CHCO_2^{\ominus}}{\underset{|}{\underset{HN}{\underset{|}{\overset{O}{\overset{\|}{C}}-CH_2-\overset{\overset{O}{\|}}{\underset{\underset{O^{\ominus}}{|}}{P}}-O^{\ominus}}}}}}
$$

N-(Phosphonacetyl)-L-aspartate
(PALA)

7. If rats are fed ^{15}N-labeled uracil, ^{15}N is not found in either the pyrimidines or purines of nucleic acids. What is the implication of this observation (made in 1944) for pyrimidine biosynthesis?

8. If ^{14}C-labeled carbamoyl phosphate is fed to rats, where will the ^{14}C appear in orotate?

Part III

Molecular biology

Chapter 24

DNA replication

24–1 INTRODUCTION

DNA is the genetic material for all organisms, except some viruses, and it represents, in a certain sense, the very idea of life. DNA contains the "program" that ultimately directs all cellular activity, the data that are manipulated as genes turn on and off, and the "language" that is transcribed into RNA and translated into protein.

The history of our understanding of the central function of DNA is long and tortuous. DNA was discovered in white blood cells in 1869 by the German biochemist F. Miescher. Nearly 80 years were to elapse before it became clear that all cells contain both DNA and RNA, DNA long having been believed to be present only in animal cells and RNA only in plants. Although hints about the function of DNA were scattered in the historical record, they flew in the face of the prevailing dogma that proteins were the only biopolymers complex enough to account for the transmission of the genetic message from one generation to the next. In 1944 O. Avery and his coworkers showed that purified DNA caused benign R-type pneumococci to become *transformed* to virulent S-type. This discovery clearly showed that DNA is the genetic material. The discovery was made in a theoretical vacuum, however, and its extreme importance was recognized only in the wake of other discoveries. In 1952 A. Hershey and M. Chase found that certain viruses infect bacteria by inserting their DNA into the bacterium, leaving their coats outside. Meanwhile E. Chargaff showed that the base composition of DNA varies widely from species to species, but that the adenine/thymine and guanine/cytosine ratios are nearly 1.00 for all species. By 1953 the stage was set for the discovery which, coming at exactly the right moment, revolutionized biology. In that year J. Watson and F. H. C. Crick, using x-ray data gathered by R. Franklin and M. Wilkins, deduced the double helical structure of DNA.

How DNA is synthesized remained unclear for many years after the structure was known, and its synthesis has turned out to be more complex than anyone would have guessed in the days that followed the discovery of its structure. At present, however, some of the mystery has been unraveled, and that is the topic of our chapter.

24–2
SEMICONSERVATIVE
REPLICATION OF DNA

It has not escaped our attention that the specific pairing [of bases in DNA] we have postulated immediately suggests a possible copying mechanism for the genetic material.

J. D. Watson and F. H. C. Crick [*Nature*, 171, 737 (1953)]

What is the fate of the parent double-helical DNA when DNA replication occurs? Does the double helix unwind with each single strand serving as a template upon which a complementary strand is synthesized? In this *semiconservative* mode of replication each "daughter" DNA duplex contains one strand of the parent. Or, does DNA replication occur by a *dispersive* mechanism in which the two parent strands break at random during replication, so that the daughter strands contain varying amounts of parent DNA? Or, is replication *conservative*, with the original parent remaining intact? In this mode of replication the daughter DNA duplex would contain two new strands, and the original duplex would be unaltered (Figure 24.1).

The mechanism of DNA replication to which Watson and Crick referred in the quotation above is semiconservative. Watson and Crick hypothesized that "each chain acts as a template for the formation onto itself of a new companion chain." Although *logically* semiconservative replication seemed the best of the three choices, experimental proof of the actual mode of DNA replication was not achieved until 1958. M. Meselson and F. W. Stahl grew *E. coli* cells in a medium with $^{15}NH_4^{\oplus}$ as the sole nitrogen source. The nonradioactive isotope of nitrogen was incorporated in all cellular DNA. When the cells containing ^{15}N-labeled DNA were added to a medium containing only the naturally abundant isomer of nitrogen as $^{14}NH_4^{\oplus}$ and were incubated for one cell division, all the isolated DNA contained a 50/50 mixture of ^{14}N and ^{15}N. After a second round of cell division one-half of the isolated DNA contained no ^{15}N, and one-half of the isolated DNA contained a 50/50 mixture of ^{15}N and ^{14}N (Figure 24.2). Thus both strands of the parent DNA serve as templates for replication, and *replication is semiconservative.*

Meselson and Stahl separated ^{15}N-labeled DNA from ^{14}N-labeled DNA by *equilibrium density centrifugation*, a technique they invented for the purpose with J. Vinograd (recall Section 9.7B). High-speed centrifugation (at 44,700 rpm in the Meselson and Stahl experiment) of a solution of cesium chloride generates a density gradient. Small amounts of ^{14}N-containing and ^{15}N-containing DNA were added to the CsCl and centrifuged for 24 hr. The DNA in the gradient moved into the region where the density of the gradient equaled the density of the CsCl (Figure 24.3). The more dense ^{15}N-labeled DNA formed a band in a region of higher density than the less dense ^{14}N-

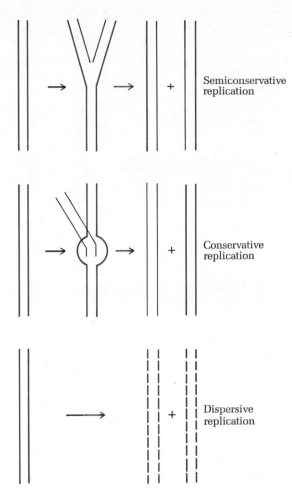

Figure 24.1 Possible mechanisms of replication. In semiconservative replication a daughter strand in synthesized on one strand of the parent, which serves as a template, so that each daughter duplex is a hybrid of a newly synthesized strand and a parental strand. In conservative replication the parent duplex remains intact and the daughter consists of two newly synthesized strands. In dispersive replication the template is destroyed, and varying amounts of both parental strands are found in the daughter strands. DNA replication is semiconservative.

labeled DNA, and identification of the labeled and unlabeled parent and daughter molecules was remarkably simple (Figure 24.4). This experiment is widely regarded as one of the most elegant in the history of biochemistry.

24–3 DNA POLYMERASE

DNA polymerase catalyzes polymerization of deoxyribonucleoside triphosphates when provided with a template strand of DNA and a primer (Figure 24.5). The template strand of DNA dictates the sequence of the newly synthesized chain since each adenine is always paired with thymine, and each guanine is always paired with cytosine. Since the template dictates the next nucleotidyl residue to be added to the nascent chain, DNA polymerase is *DNA-directed*. The primer serves as a clamp, like the clasp of a zipper, to provide a stable hydrogen-bonded complex between the nascent chain and the template. Despite extensive search no 3′-5′ polymerase has yet been discovered, so we assume that the chain grows only in the 5′-3′ direction.

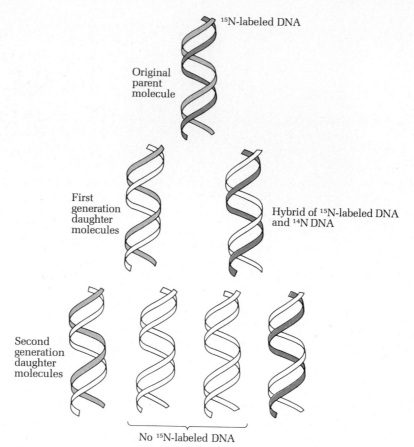

Figure 24.2   In semiconservative replication of DNA each daughter contains one of the parental chains after one round of replication. Continued replication produces two daughter duplexes that contain none of the original parent DNA and two "hybrid daughters." [Adapted from M. Meselson and F. W. Stahl, *Proc. Nat. Acad. Sci. U.S.*, 44, 671 (1958).]

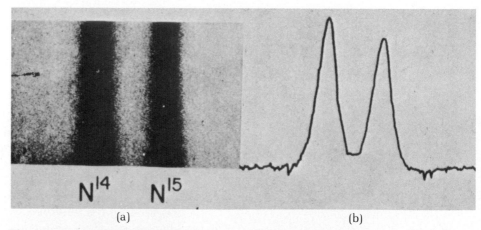

Figure 24.3   Resolution of DNA by density gradient centrifugation. Centrifugation at high speed in cesium chloride establishes a density gradient. If DNA is added to the cesium chloride and centrifuged, it forms a "band" where its density equals the density of the cesium chloride solution. (It "floats" on the CsCl.) (a) DNA that is enriched with ^{15}N has a higher density than DNA that contains only ^{14}N, therefore it moves to a different position in the cesium chloride gradient. The difference in density in the figure is 0.014 g cm^{-3}. (b) A microdensitometer trace shows the "density profile" of DNA distribution in (a). The area under each curve corresponds to the amount of DNA in the sample. [Adapted from M. Meselson and F. W. Stahl, *Proc. Nat. Acad. Sci. U.S.*, 44, 671 (1958).]

The phosphodiester bonds of the growing chain are formed by nucleophilic attack of the 3′-hydroxyl group of the terminal residue of the growing chain upon the α-phosphorus of the deoxynucleoside triphosphate, displacing pyrophosphate (Figure 24.6). The reaction, then, is an example of nucleotidyl transfer (recall Section 12.5B).

A. DNA Polymerases of *E. Coli*

Three DNA polymerases have been isolated from *E. coli*. Each *E. coli* cell has about 400 molecules of DNA polymerase I, 40 molecules of DNA polymerase II, and 10 of DNA polymerase III. DNA polymerases have more than one catalytic activity. Besides catalyzing phosphodiester bond formation in the 5′-3′ direction, DNA polymerases I and III catalyze 3′-5′ exonucleolytic degradation of DNA. They also catalyze 5′-3′ exonucleolytic and endonucleolytic degradation of DNA. DNA polymerase II possesses only 3′-5′ exonucleolytic activity. The major enzyme responsible for DNA replication *in vivo* is DNA polymerase III. DNA polymerase I contains 975 amino acid residues and has a molecular weight of 109,000. The enzyme contains a tightly bound Zn(II) ion and requires Mg(II) for activity. Proteolysis of DNA polymerase I by trypsin or subtilsin gives two polypeptides having molecular weights of 36,000 and 75,000. The smaller polypeptide possesses all of the exonuclease activity, and the large polypeptide contains all of the polymerase activity of DNA polymerase I. Thus, both activities are found within a single polypeptide chain in the native enzyme.

The 3′-5′ exonuclease activity of DNA polymerases I, II, or III has a proofreading function (Figure 24.7). If incorrect base pairing occurs in the growing DNA chain, that is, if the pairing does not follow the Watson-Crick rule of adenine with thymine and guanine with cytosine, the 3′-5′ exonuclease activity removes the mismatched base by hydrolysis of the phosphodiester bond. The error rate for DNA polymerase I is 1 base in 100,000. This extreme fidelity is a result of a two stage process. First, the correct base pairing is required for initial synthesis of the phosphodiester bond to occur at an appreciable rate. Second, if a mistake is made, the exonucleolytic activity of the polymerase corrects it. DNA polymerization is thus a self-correcting process.

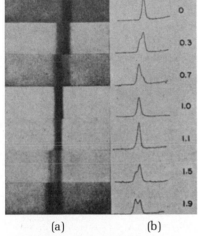

(a)　　　　(b)

Figure 24.4 Experiment of M. Meselson and F. W. Stahl showing that DNA is replicated semiconservatively. (a) Ultraviolet absorption photographs showing DNA bands resulting from density gradient centrifugation of lysates of bacteria sampled at various times after the addition of an excess of ^{14}N substrates to a growing ^{15}N-labeled culture. Each photograph was taken after 20 hours of centrifugation at 44,770 rpm. Regions of equal density occupy the same horizontal position on each photograph. (b) Microdensitometer tracings of the DNA bands in (a). The peak of intermediate density, due to the daughter strand, is centered at 50% of the distance between the ^{14}N and ^{15}N peaks, a result that demonstrates that replication is semiconservative. [Adapted from M. Meselson and F. W. Stahl, *Proc. Nat. Acad. Sci. U.S.*, 44, 671 (1958).]

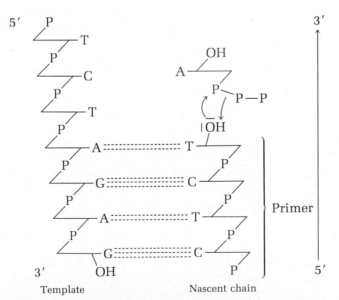

Figure 24.5 Template and primer required for activity by DNA polymerase. The template determines the sequence of the nascent chain, and the primer provides a stable, hydrogen-bonded complex between the template and the growing chain.

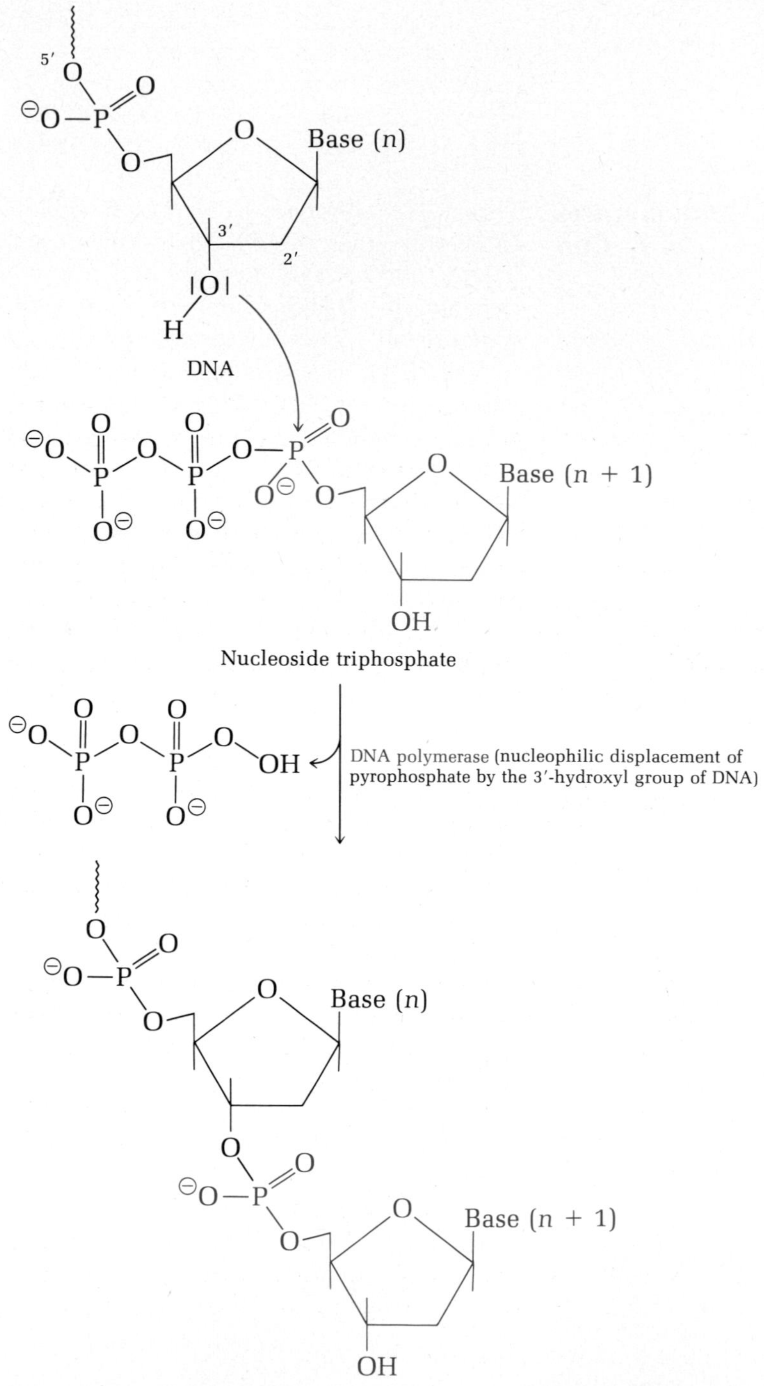

Figure 24.6 Nucleotidyl transfer reaction catalyzed by DNA polymerase. The transfer occurs by an in-line nucleophilic substitution.

The 5′-3′ nuclease activity also has a correcting function in DNA biosynthesis (Figure 24.8). The 5′-3′ exonuclease can cleave an oligonucleotide of up to 10 residues. (The 3′-5′ nuclease cleaves a single residue at a time.) The 5′-3′ endonuclease activity of the polymerase enables it to remove thymine dimers that result from ultraviolet radiation (Figure 24.9). The gap that remains after the 5′-3′ endonucleolytic activity removes the thymine dimer and surrounding residues is then filled in by DNA polymerase. The 5′-3′ nu-

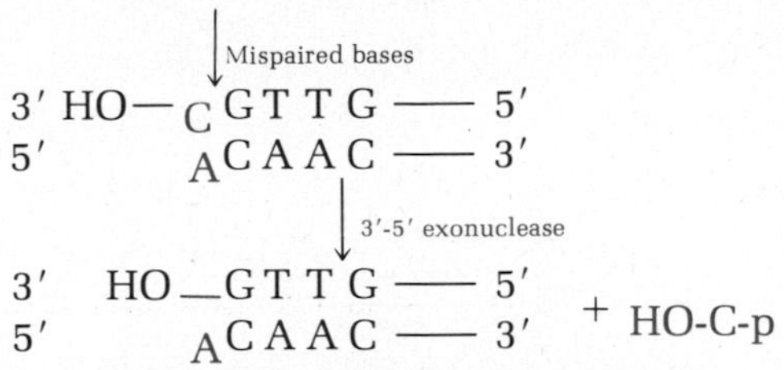

Mispaired bases

3′ HO—$_C$G T T G —— 5′
5′ $_A$C A A C —— 3′

3′-5′ exonuclease

3′ HO—G T T G —— 5′
5′ $_A$C A A C —— 3′ + HO-C-p

Figure 24.7 Proofreading function of 3′-5′ exonuclease activity of DNA polymerases I, II, and III. The activity removes bases that are incorrectly paired.

cleolytic activity of DNA polymerase carries out *excision repair*, in contrast to the proofreading repair carried out by the 3′-5′ exonuclease activity.

It has been found that DNA polymerase I is largely responsible for repair of DNA. The exonuclease and repair synthesis (polymerization) activities of DNA polymerase I are its predominant activities in vivo. DNA polymerase III is the major enzyme responsible for DNA replication *in vivo*. It is a very complex protein, often referred to as the *DNA polymerase III holoenzyme*. The holoenzyme contains nine subunits of seven types (Table 24.1 and Figure 24.10). The α subunit contains the polymerase activity. The functions of the other subunits are not known.

B. Eucaryotic DNA Polymerases

Animal cells contain three DNA polymerases, designated α, β, and γ. The α and β forms are found in the nucleus and the γ form in mitochondria. The mechanisms of action of the eucaryotic DNA polymerases are thought to be similar to the mechanisms of action of their procaryotic counterparts. DNA polymerase α is largely responsible for DNA biosynthesis in replicating chromosomes. This enzyme consists of two subunits whose molecular weights are 76,000 and 66,000. The template-primer for DNA polymerase α is a DNA dimer that contains gaps of about 20 nucleotidyl residues in one of the chains. Between 20,000 and 60,000 molecules of DNA polymerase α are found in the nucleus of a dividing cell, and *de novo* DNA replication can be attributed to this enzyme.

DNA polymerase β does not seem to be important in DNA replication, but it acts to repair defective DNA. It is only about a tenth as abundant as the α enzyme.

DNA polymerase γ, a monomer having a molecular weight of 150,000, is responsible for DNA replication in mitochondria. In contrast to DNA polymerases in procaryotic cells, none of the animal DNA polymerases possess exonucleolytic activity. Separate enzymes carry out exonucleolytic repair in animal cells.

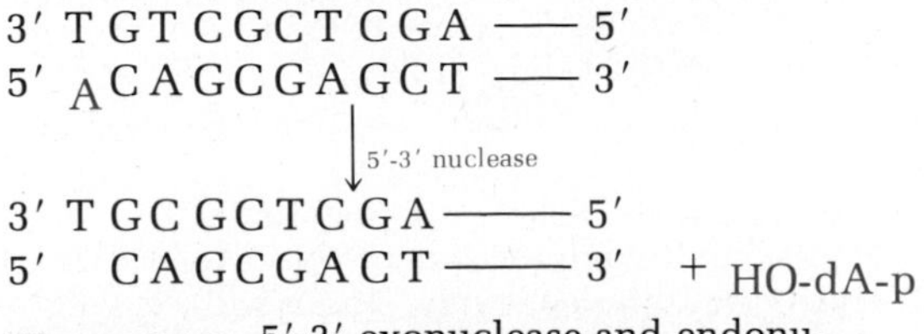

3′ T G T C G C T C G A —— 5′
5′ $_A$C A G C G A G C T —— 3′

5′-3′ nuclease

3′ T G C G C T C G A —— 5′
5′ C A G C G A C T —— 3′ + HO-dA-p

Figure 24.8 5′-3′ exonuclease and endonuclease activities of DNA polymerases I and III. Cleavage can occur up to five bases away from the 5′ terminus of one strand. A double-helical DNA is required for either exonuclease or endonuclease activity.

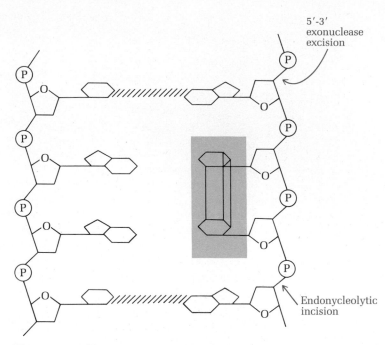

Figure 24.9 Excision repair of a thymine dimer by DNA by the 5'-3' exonuclease activity of DNA polymerase.

24–4 DNA REPLICATION IN *E. COLI*

Although the idea of semiconservative replication of DNA is simple, the biochemistry of this process is exceedingly complex. To understand it, we must consider not only DNA polymerase itself, but a host of other proteins. As a prelude to this discussion, let us briefly consider the chromosome structure of *E. coli*, the bacterial cell cycle, and some aspects of their relation to DNA replication.

A. Structure of the *E. coli* Chromosome and the Bacterial Cell Cycle

The main chromosome of *E. coli* consists of a single circular molecule of DNA having a molecular weight of about 2.6×10^9. This circular molecule is densely packed into the center of the bacterial cell in a region of space known as a *nucleoid*. The nucleoid is not bounded by a membrane (procaryotes contain no membrane-bounded cell organelles). The 1 mm long

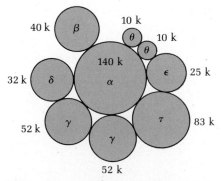

Figure 24.10 Schematic diagram of the subunit composition of DNA polymerase III holoenzyme. The α subunit contains the polymerase activity. The functions of the other subunits are not known. The molecular weights of the subunits are indicated in kdal (amu $\times 10^3$), abbreviated k.

Table 24.1	Subunit composition of DNA polymerase III holoenzyme	
Component (mw × 10³)	**Subunit and alternate designations**	
140	α	
25	ε } pol III (*dnaE, polC*)	
10	θ	
83	τ	} pol III
52	γ *dnaZ*	
32	δ Factor III } factor II	
40	β Factor I, copol III	

chromosome is tightly coiled in superhelical twists and compressed into a dense mass about 1 μm in diameter (Figure 24.11).

E. coli, and many other bacteria, also contain minor chromosomes called *plasmids*. They are also supercoiled, closed, circular duplexes. The plasmids contain genes that code for as few as three proteins or as much as 20% of the total cellular protein. They also contain the genes that confer resistance to antibiotics upon bacteria. We shall find that they are critical elements in the techniques used to create artificially recombinant DNA (Section 24.9). The plasmids can exist in up to 50 copies per cell, and plasmid DNA replicates independently of the main chromosome.

E. coli cells replicate every 20 to 60 minutes. The process of cell divi-

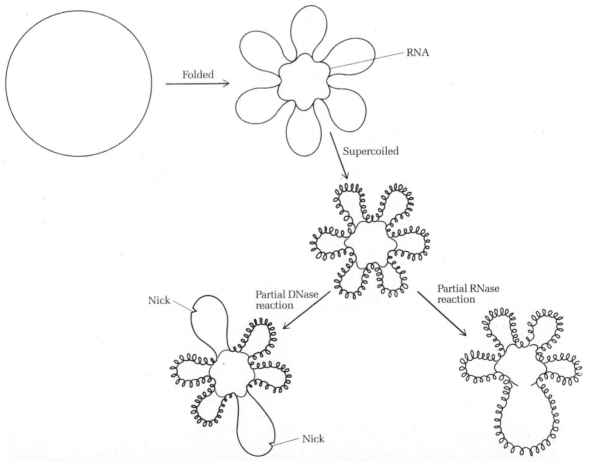

Figure 24.11 Schematic diagram of the highly folded *E. coli* chromosome. RNA is tightly bound to the chromosome. Treatment with RNase opens some of the folds, but does not affect the DNA. Treatment with DNase nicks the superhelical DNA, causing some unfolding. [Adapted from D. E. Peeijohn and R. Hecht, *Cold Springs Harbor Symp. Quant. Biol.*, *38*, 39 (1974).]

sion can be divided into several stages: (1) DNA replication, (2) separation of the two chromosomes, (3) physiological division, and (4) cell separation (Figure 24.12). Cell division occurs about 20 minutes after DNA replication at 37°. If for any reason DNA synthesis stops, cell division does not occur. DNA replication is therefore coordinated with the entire cycle of cell division.

B. Bidirectional Replication from a Unique Origin

We can imagine many possibilities for the semiconservative replication of the *E. coli* chromosome. For example, (1) the circular chromosome might open with DNA replication occuring linearly from one end of the opened circle to the other, (2) replication might begin at a point on the circular chromosome and continue in one direction until the entire chromosome has been copied, or (3) replication might originate at a single point in the chromosome and travel bidirectionally until the entire chromosome has been copied. These possibilities and many others were tested in the 1960s and 1970s in experiments employing autoradiography and in classical genetic studies.

When *E. coli* are grown for two generations in a medium containing [³H]-thymidine, all of the DNA in the first generation is radioactive in one strand since replication is semiconservative. In the second generation, however, some of the DNA is isotopically labeled in both strands. When the sample is exposed to a photographic emulsion, the autoradiograph reveals that the DNA formed in the second generation has twice the radioactivity, and hence twice the "grain density," as the first generation. The doubly labeled DNA is connected to the *replication fork*, and the bacterium's circular chromosome has a *theta* structure (Figure 24.13).

The autoradiograph does not, however, indicate whether replication occurs undirectionally or bidirectionally. This question was answered by a variation of the autoradiographic experiment in which cells were first grown in a weakly radioactive medium containing [³H]-thymidine. The cells were then transferred to a strongly radioactive medium containing [³H]-thymidine. The photographic plate showed that the regions nearest the replication forks contained the highest amount of radioactivity. This conclusion is to be expected if replication is bidirectional, since the last nucleotides incorporated in DNA had a higher radioactivity than those incorporated in DNA in the first part of the experiment (Figure 24.14).

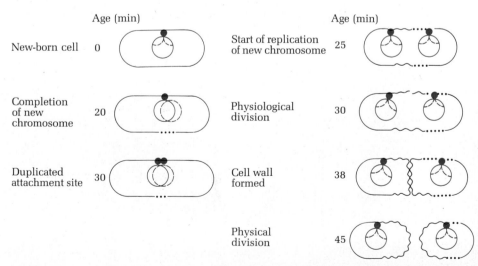

Figure 24.12 Schematic diagram of the *E. coli* cell division cycle. Dotted lines indicate regions where the cell membrane and cell wall are growing prior to duplication of the chromosome and its attachment point. Dashed lines indicate a replicated chromosome. Wavy lines indicate regions of cell membrane and cell wall growth that occur after DNA replication. [Adapted from D. J. Clark, *Cold Spring Harbor Symp. Quant. Biol., 33*, 825 (1968).]

DNA replication appears to originate at a unique point in the *E. coli* chromosome near the *ilv* gene. (This gene codes for the enzymes responsible for isoleucine and valine biosynthesis.) This point is known as the replication *origin*. Since DNA synthesis occurs at the same rate in the counterclockwise and clockwise directions, DNA synthesis terminates near the *trp* gene (which codes for the enzymes responsible for tryptophan biosynthesis) at a point diametrically opposed to the origin (Figure 24.15).

C. DNA Replication

DNA replication can be divided into five stages: (1) unwinding of the parental double helix, (2) synthesis of an oligonucleotide primer, (3) growth of the DNA chain in the 5′-3′ direction, (4) excision of the primer, (5) joining of the newly synthesized DNA chains by filling in the excision gap and sealing by phosphodiester formation.

a. Unwinding of Supercoiled, Duplex DNA

The 4 million base pairs of the *E. coli* chromosome replicate in about 40 min *in vivo* (at 37°). Since there are two replication forks, DNA biosynthesis occurs at a rate of 50,000 base pairs per minute. Each replication fork is thus unwinding at a rate of about 2,500 revolutions per minute. DNA is negatively supercoiled, and thus it is underwound. But supercoiling does not seriously alter the enormous rate of unwinding that is required for DNA replication.

Several proteins aid in this unwinding process. The *helix-destabilizing protein*, also known as *helicase*, promotes unwinding by binding extremely tightly to single-stranded DNA. The equilibrium between duplex DNA and the helix-destabilizing protein and single-stranded DNA and the helix-destabilizing protein lies on the side of single-stranded DNA-helix destabilizing protein (Figure 24.16). The *E. coli* helix-destabilizing protein is a tetramer whose identical monomers have molecular weights of 20,000. When bound to DNA, the tetramer spans a segment of 30 to 36 nucleotidyl residues. The binding of the helix-destabilizing protein to DNA is cooperative, and many helix-destabilizing protein molecules, bound to each strand of DNA, may help to open the template DNA at the replication fork.

DNA gyrase (Section 9.5D) puts negative supertwists in DNA. The negative supercoiling of DNA lowers the energy required to disrupt a base pair in the double helix by about 4.2 kJ mol^{-1}, and the action of DNA gyrase ahead of the replication fork thus makes it easier to open the double helix. After a new strand of DNA is synthesized on the template, DNA gyrase may also retwist the replicated DNA into its native, supercoiled conformation.

The *rep* protein, another helicase that was formerly known as *DNA-unwinding enzyme*, also participates in strand separation. The *rep* protein denatures DNA in an ATP-driven process. (The name *rep* protein is a genetic term whose abbreviation refers to a gene that is required in the *E. coli* chromosome for DNA replication to occur.) Disrupting an adenine-thymine base pair (A-T next to A-T) requires 15 kJ mol^{-1}; disrupting a guanine-cytosine base pair (G-C next to G-C) requires about 21 kJ mol^{-1}. The *rep* protein (mol wt 180,000) forms fibrous aggregates, and it has been suggested that it "invades" duplex DNA by a sliding motion of the monomers of the aggregate past each other.

The proteins involved in DNA uncoiling act in concert, and all are likely to be bound at or near the replication fork. A schematic diagram of their interaction with DNA is shown in Figure 24.17).

b. Synthesis of an Oligonucleotide Primer

DNA polymerase requires both a template and a primer for activity. Once duplex DNA has been unwound to expose the template, a specific *RNA*

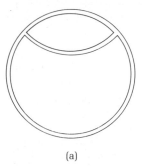

(a)

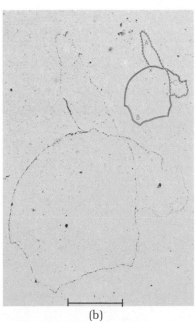

(b)

Figure 24.13 Schematic diagram of the theta structure of a replicating *E. coli* chromosome. (b) Autoradiograph of a chromosome *E. coli* labeled with titrated thymidine for two generations. The scale shows 100 μ. Inset, the same structure is shown diagrammatically and divided into three sections (A, B, and C) that rise at the two forks (X and Y). [Courtesy of Dr. John Cairns.]

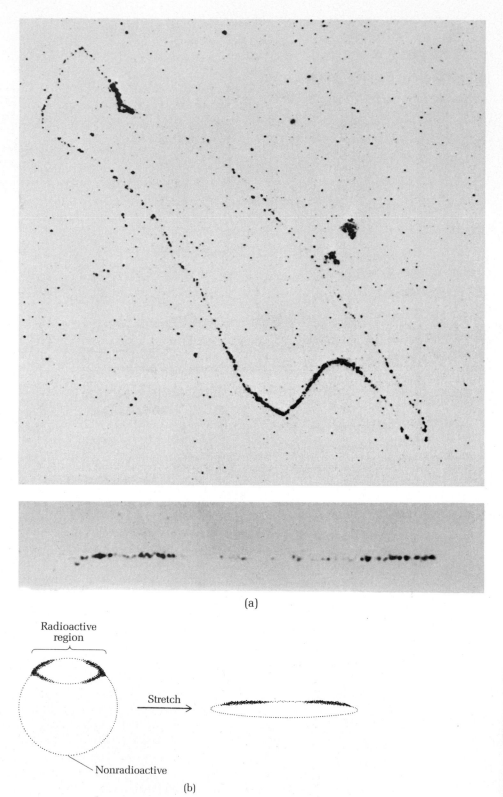

(a)

(b)

Figure 24.14 Autoradiograph of a replicating *E. coli* chromosome in which the highest radioactivity is found at the most recently replicated regions of the DNA. Since the two fork regions in the theta structure are the most densely labeled, we conclude that replication occurs bidirectionally, as indicated schematically in (b). [Courtesy of Dr. Raymond Rodriguez.]

polymerase, known as *primase* because of its unique function in DNA replication, synthesizes a short complementary strand to a specific region of chromosomal DNA that corresponds exactly to the origin of replication. The 3′-hydroxyl group of the terminal residue of the RNA primer acts as the nucleophile in the formation of the first phosphodiester bond of the nascent DNA chain (Figure 24.18). Primase is distinct from the RNA polymerase responsible for transcription of the genetic code (Section 26.10), and it has been identified as the *dnaG* protein of *E. coli.* Primase consists of a single polypeptide chain that has a molecular weight of 60,000. An *E. coli* cell contains between 50 and 100 molecules of primase.

The synthesis of an RNA primer at the origin of DNA replication is itself triggered by a protein designated the *dnaB* protein. This protein apparently has a mass of about 300,000 and is composed of six identical monomers. About twenty molecules of *dnaB* protein are present per cell. The *dnaB* protein binds to single-stranded DNA, and it is thought to bind specifically at the replication origin. Its presence at the replication origin might then signal primase to begin DNA replication (Figure 24.19). After the *dnaB* protein binds DNA and triggers the action of primase, it remains bound to the replication fork and travels down DNA propelled by the hydrolysis of ATP. At least five other proteins are involved in the prepriming phase of DNA replication, but the biochemical details of their functions have not yet been elucidated.

c. Discontinuous Synthesis at the Replication Fork

Once the duplex DNA has been unwound at each of the two replicating forks, and the prepriming and primase steps have been completed, DNA polymerization begins. Since the strands of parental DNA have opposite polarity, continuous DNA synthesis along both sides of the replication fork would require a polymerase with 3′-5′ activity on one strand and a polymerase with 5′-3′ activity along the other strand. Since all known DNA polymerases catalyze chain growth exclusively in the 5′-3′ direction, such continuous synthesis does not occur. DNA polymerase requires a free 3′-hydroxyl group in the primer, and continuous chain growth can occur in the 5′-3′ direction only on one strand, known as the *leading strand.* Chain growth in the other strand, known as the *lagging strand,* occurs *discontinuously* (Figure 24.20).

When bacteria are labeled with a *pulse* of radioactive thymidine, newly synthesized DNA molecules containing between 1,000 and 2,000 nucleotidyl residues may be isolated shortly thereafter, provided that the DNA has been denatured following its removal from the cells. These small polynucleotides, known as *Okazaki fragments* in honor of their discoverer, Reiji Okazaki, have been found in all cells in which DNA is replicating. (The Okazaki fragments found in eucaryotic cells contain from 100 to 200 nucleotidyl residues.) That Okazaki fragments are intermediates in DNA biosynthesis is demonstrated by a "pulse-chase" experiment. A pulse of radioactive thymidine is added to the cells. Shortly thereafter, the cells are transferred to a nonradioactive medium, which terminates the "pulse" of radioactivity. DNA replication is allowed to continue for a few minutes more. The radioactive DNA is isolated by sedimentation in a sucrose gradient. Radioactive DNA fragments are obtained after denaturation. The chase experiments revealed that all of the Okazaki fragments become incorporated into intact DNA. Okazaki fragments are therefore intermediates in DNA biosynthesis. It appears that DNA biosynthesis on the leading strand is also discontinuous, but the fragments in the leading strand are much larger than those in the lagging strand.

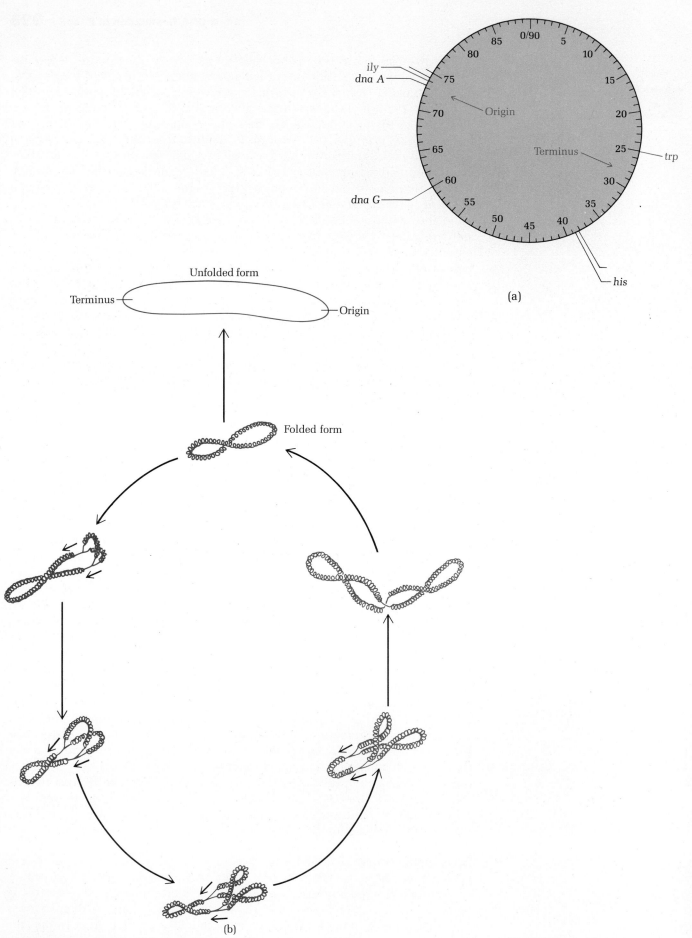

Figure 24.15 (a) Abbreviated genetic map of *E. coli* in which the origin and terminus of DNA replication are indicated. (b) Schematic diagram of replication of the *E. coli* chromosome.

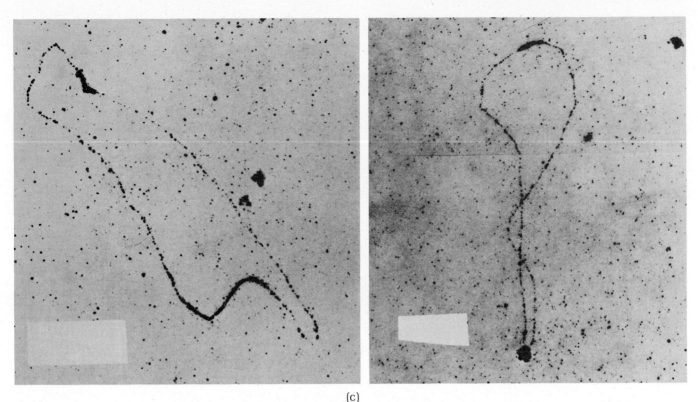

(c)

Figure 24.15 (*continued*) (c) Autoradiographs of replicating *E. coli* DNA. The heavy dark regions on opposite sides of the chromosome correspond to replication at the beginning and end of the replication cycle. [Courtesy of Dr. Raymond L. Rodriguez.]

d. Excision of Primer and Phosphodiester Formation

Once the Okazaki fragments have been synthesized three steps remain: (1) a ribonuclease removes the RNA primer, (2) DNA polymerase I fills in the small gap, and (3) *DNA ligase* "seals" the "nicks" that remain (Figure 24.21). *E. coli* DNA ligase consists of a single polypeptide chain having a molecular weight of 75,000. The substrate for DNA ligase is nicked, double-helical DNA; the ligase will not act upon single-stranded DNA. Synthesis of the phosphodiester bond between the 3′-hydroxyl group of one nucleotidyl residue and the 5′-phosphate ester of an adjacent residue by DNA ligase requires NAD as a cofactor in *E. coli* and ATP as a cofactor in eucaryotic organisms. Phosphodiester formation occurs in three steps. First, an adenylyl-enzyme

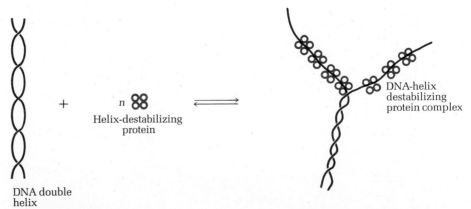

Figure 24.16 Formation of complex between the helix-destabilizing protein and single-stranded DNA. The free energy of binding the protein to a single strand of DNA offsets the lost stability resulting from breaking the hydrogen-bonded base pairs.

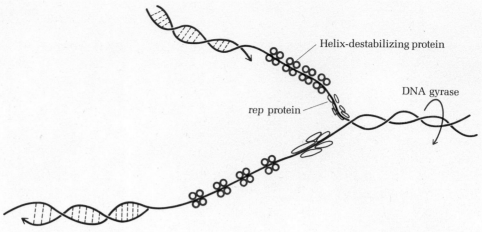

Figure 24.17 Possible roles for three proteins in unwinding DNA prior to replication. The helix-destabilizing protein stabilizes single-stranded DNA. The *rep* protein, or DNA-unwinding enzyme, and DNA gyrase, each of which hydrolyzes ATP, disrupt base pairing in duplex DNA.

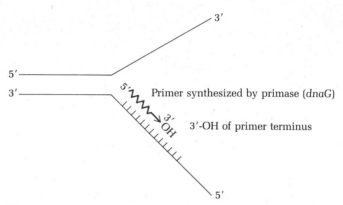

Figure 24.18 Primer synthesized by primase. The primer possesses a free 3′-hydroxyl group that is the nucleophile in the addition of the first residue added by DNA polymerase.

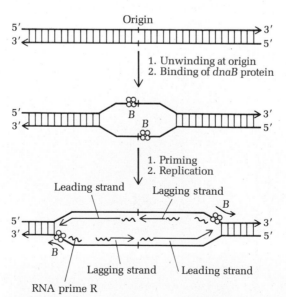

Figure 24.19 Model for the initiation of DNA replication in *E. coli*. The *dnaB* protein binds to the initiation origin as a prelude to the action of primase. [From A. Kornberg, "DNA Replication," W. H. Freeman and Company, San Francisco, Calif., 1980, p. 389. Copyright © 1980.]

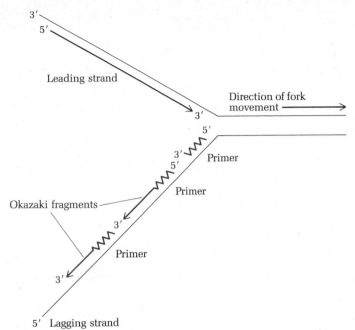

Figure 24.20 Discontinuous replication of DNA at a replication fork. Replication must occur in the 5′-3′ direction, and it occurs on both the leading strand and the lagging strand.

intermediate forms. Either ATP or NAD can be the source of the adenylyl group. A covalent bond is formed between an ε-amino group of a lysyl residue of the ligase and the phosphoryl group of AMP. Second, the 5′-phosphate group of DNA displaces ligase, giving an ADP-DNA adduct. Third, nucleophilic displacement by the 3′-hydroxyl group of a nucleotidyl group in DNA gives the final product (Figure 24.22).

The current scheme for enzyme functions in discontinuous DNA synthesis in *E. coli* is summarized in Figure 24.23, and the proteins required for DNA replication are summarized in Table 24.2.

24–5 REPLICATION OF THE EUCARYOTIC CHROMOSOME

The chromosomes of eucaryotic cells contain a thousand times as much DNA as the procaryotic chromosome. The eucaryotic chromosome, called *chromatin* because it becomes deeply colored in the presence of certain dyes, is not "naked" like the procaryotic chromosome, but is complexed with basic proteins called *histones*. The relation between the structure of chromatin and DNA replication is slowly being unraveled. We shall first consider the structure of chromatin and then turn to events at the replication fork in eucaryotic DNA.

A. Nucleosomes and the Structure of Chromatin

a. Determination of the Molecular Weight of Eucaryotic DNA

The DNA of eucaryotic chromosomes consists of a single linear polymer. In the common fruitfly, *Drosophilia melanogaster*, the molecular weight of the largest chromosome is a staggering 41×10^9. DNA molecules of this size, and even a great deal smaller, tend to fragment in the normal procedures of DNA isolation. B. Zimm and his coworkers have invented a special technique for measuring the molecular weight of eucaryotic DNA that exploits the *viscoelasticity* of the largest DNA molecules in solution.

The viscosity of a liquid is an index of its resistance to flow. (We might regard it as the equivalent of "friction" in the liquid phase.) If a fluid flows

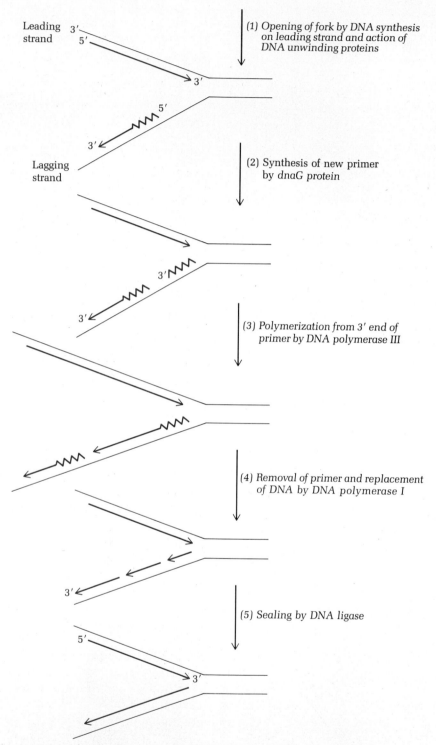

Figure 24.21 Schematic representation of discontinuous replication of DNA on the lagging strand of a replication fork.

through a capillary tube of radius R and length l, under a pressure P, the velocity of the liquid is not the same at all points in the capillary tube (Figure 24.24). The velocity of flow is greatest in the center of the tube and virtually zero at the walls of the tube. Sir Isaac Newton showed in the seventeenth century that the relation between the frictional drag, F, or the *shear stress*, and the difference in velocity between the center of the tube and the wall, or *velocity gradient*, dv/dr, is given by

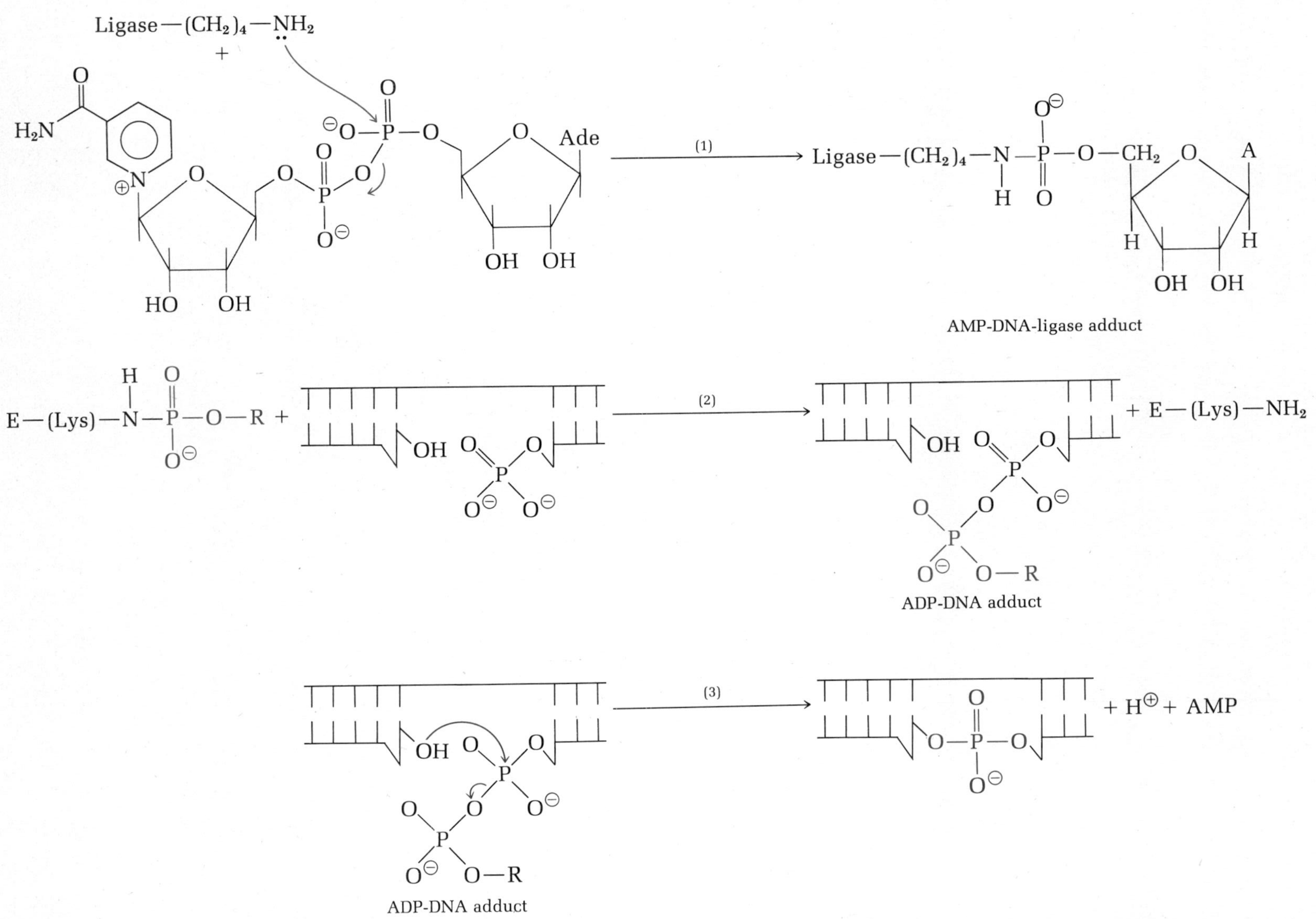

Figure 24.22 The mechanism of action of *E. coli* DNA ligase. (1) The ε-amino group of a lysyl residue displaces nicotinamide mononucleotide from NAD giving an AMP-ligase adduct. (2) The 5'-phosphate group of DNA displaces the enzyme giving an ADP-DNA adduct. (3) Nucleophilic displacement of AMP by the 3'-hydroxyl group of DNA gives a sealed DNA strand.

**Table 24.2 Summary of proteins involved in DNA replication in *E. coli*

Protein	Gene	Function
Protein i (X)		Prepriming
Protein n (Z)		Prepriming
Protein n' (Y)		DNA-dependent ATPase
		Prepriming
Protein n"		Prepriming
dnaC protein	*dnaC*	Formation of *dnaB-dnaC* complex
		Prepriming
dnaB protein	*dnaB*	DNA-dependent rNTPase
		Mobile promotor
		Prepriming
		Priming
Helix-destabilizing protein (DNA-binding protein; single-strand binding protein)	*ssb—1*	Binding to single-stranded DNA
Primase	*dnaG*	Primer synthesis
DNA polymerase III holoenzyme		DNA elongation
α (polymerase III)	*polC (dnaE)*	
θ		
ε		
τ		
γ (*dnaZ* protein)	*dnaZ*	
δ (EF III)		
β (EF I)		
DNA polymerase I	*polA*	Primer degradation, gap filling
DNA ligase	*lig*	Joining of short chains
DNA gyrase	*gyrA (nalA)*,	Supertwisting
		DNA-dependent ATPase
	gyrB (cou)	Relaxation of supercoils
rep protein	*rep*	DNA-dependent ATPase
		Strand separation

SOURCE: From T. Ogawa and T. Okazaki, *Ann. Rev. Biochem., 49,* 450 (1980). Reproduced with permission. © 1980 by Annual Reviews Inc.

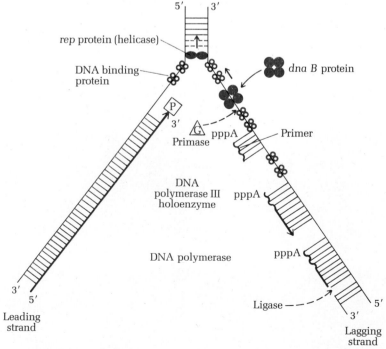

Figure 24.23 Schematic diagram of the functions of the various enzymes involved in DNA replication.

$$F = -\eta(2\pi lr)dv/dr \tag{24.1}$$

where η is the *coefficient of viscosity*, or simply the *viscosity*, $2\pi lr$ is the surface area of the inner cylinder (as in Figure 24.24), and dv/dr is the *shear gradient*. It is the shear stress that tears apart large DNA molecules, and the shear stress becomes larger as the shear gradient increases.

The viscosity of a solvent increases when DNA molecules (or other solutes) are added. Since DNA is much larger than the solute molecules, the friction between adjacent layers of moving liquid increases as the DNA concentration increases. By measuring the change in viscosity as a function of DNA concentration, the *average* molecular weight of the DNA molecules can be obtained. But it is the molecular weight of unbroken DNA that we require, not the average molecular weight of fragments. In measurements of viscoelasticity, the molecular weight of the largest molecules of DNA in solution are measured.

The viscoelastic determination of the molecular weight of DNA employs a *Zimm-Crothers* viscosimeter. This device consists of two concentric cylinders. The inner cylinder contains a steel pellet that is made to spin by an external magnet, which in turn causes the inner cylinder to spin. The spinning inner cylinder applies a shear stress to the DNA molecules in the annular gap between the two cylinders, causing them to become extended. When the shear stress is removed by turning off the external magnet, the extended DNA molecules begin to relax to their native conformation. Relaxation of the DNA molecules causes the direction of rotation of the inner cylinder to reverse (Figure 24.25). Eventually the inner cylinder stops rotating. The time required for rotation to stop is known as the relaxation time, $\tau°$. It can be shown that the relaxation time is related to the molecular weight, M, of the DNA by the equation

$$M = 1.56 \times 10^8(\tau°)^{0.60} \tag{24.2}$$

The molecular weights of DNA from several sources determined by viscoelasticity measurements are summarized in Table 24.3.

b. Histones

The structure of chromatin may be divided into two parts. The *nucleosomes* are particles of protein and DNA; the *linker* region is a span of DNA that connects nucleosomes. A small group of low molecular weight proteins called *histones* are bound to both the linker and nucleosome regions of chromatin and are essential components of chromatin. There are five types of his-

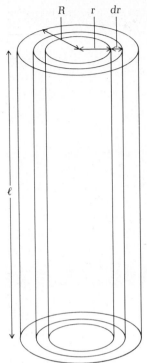

Figure 24.24 Schematic diagram of the flow of liquid through a capillary tube of length l and radius R. The rate of flow is greater in the interior of the capillary tube, radius r.

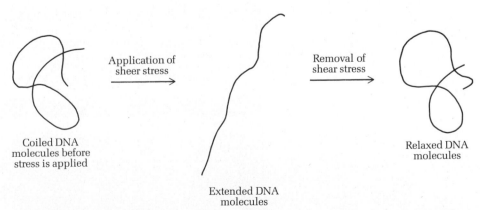

Coiled DNA molecules before stress is applied

Application of sheer stress →

Extended DNA molecules

Removal of shear stress →

Relaxed DNA molecules

Figure 24.25 Extension of DNA molecules by application of shear stress, followed by relaxation of the molecules when the stress is removed. The relaxation time is related to the molecular weight of the largest DNA molecules (equation 24.2).

Table 24.3 Molecular weights of DNA molecules from several sources as determined by viscoelasticity

DNA	Molecular weight
E. coli phage T7	$25 \pm 2 \times 10^6$
E. coli phage T2	$109 \pm 6 \times 10^6$
E. coli	2.7×10^9
Bacillus subtilis	2.0×10^9
Drosophila melanogaster	$41 \pm 3 \times 10^9$
Drosophila americana	$79 \pm 10 \times 10^9$

tones, designated H1, H2A, H2B, H3, and H4 (Table 24.4). H1 is associated with the linker region of chromatin, and the other four form part of the structure of nucleosomes. The nucleosomal histones are sometimes called "inner histones." The histones are all basic proteins with high arginine and lysine content. For example, histone H4 contains 26 basic residues and only 7 acidic residues in its 102-residue sequence. The protonated forms of these basic side chains bind electrostatically to the negative phosphate groups along the phosphodiester backbone of DNA, and by minimizing electrostatic repulsion, they make it possible for the DNA to fold into an extremely tight mass.

The histones display conservation of amino acid sequence. For example, the primary structure of calf histone H4 differs by only 2 out of 102 residues from that of pea seedling histone H4. Valine residue 60 in calf thymus histone H4 is replaced by isoleucine in the pea seedling, and arginine residue 77 is replaced by lysine in the pea seedling (Figure 24.26). The primary structure of calf histone H3 differs from that of pea seedling histone H3 by only 4 residues out of 104. It is estimated that the divergence of the evolutionary lines "leading to calves and peas" occurred some 1.2 billion years ago. The time required for an evolutionary divergence of 1%, known as the *unit evolutionary period*, is thus 600 million years for histone H4 and 300 million years for histone H3. This extreme conservation of primary structure implies that the detailed interactions of these histones with one another and with DNA are critical to the structure of chromatin. The other three histones also have relatively conserved primary structures, but they are far more variable than histones H_3 and H_4.

Histones are covalently modified after they have been synthesized, a phenomenon known as *posttranslational modification*. In histones H1, H2A, H3, and H4, the aminoterminal serine residue is acetylated. In the interior of the chain various ϵ-amino groups of lysyl residues are also acetylated. During the phase of mitosis when DNA replication occurs (S-phase), certain specific serine and threonine residues are phosphorylated. Histones can also

Table 24.4 The histones

Type	Number of Residues	Number of Basic Residues	Number of Acidic Residues
H1	215	65	9
H2A	129	30	9
H2B	125	39	8
H3	132	33	11
H4	102	27	8

10   20
Ac-Ser- Gly -Arg- Gly -Lys-Gly -Gly- Lys -Gly-Leu -Gly-Lys -Gly- Gly -Ala- Lys -Arg- His -Arg-Lys-

30   40
Val -Leu-Arg-Asp-Asn -Ile- Gln -Gly- Ile -Thr- Lys -Pro- Ala -Ile- Arg-Arg-Leu -Ala-Arg-Arg-

50   60
Gly -Gly- Val -Lys- Arg -Ile- Ser -Gly- Leu -Ile- Tyr -Glu-Glu-Thr-Arg-Gly- Val -Leu-Lys -Val-

70   77   80
Phe-Leu-Glu-Asn- Val -Ile- Arg-Asp-Ala -Val- Thr -Tyr -Thr-Glu- His -Ala- Lys -Arg-Lys -Thr-

90   100
Val -Thr- Ala -Met-Asp-Val- Val -Tyr- Ala -Leu- Lys -Arg-Gln -Gly- Arg-Thr-Leu -Tyr -Gly-Phe- Gly-Gly

Figure 24.26 Primary structure of calf histone H4. [From I. Isenberg, *Ann Rev. Biochem.*, 48, 162 (1979). Reproduced with permission. © 1979 by Annual Reviews Inc.]

be methylated at specific residues, and they may also undergo ADP-ribosylation (recall Section 16.4A). The detailed functions of these covalent modifications are not known, but preliminary evidence strongly suggests that the covalent modification is related to DNA replication.

c. Structure of Chromatin

Chromatin is a nucleoprotein complex of DNA and histones that consists of beadlike particles called *nucleosomes* and a connecting strand of DNA known as the *linker* (Figure 24.27). The linker region of chromatin consists of double-stranded DNA bound to histone H1. The amount of DNA in the linker region is variable, but typically contains 60 to 80 base pairs. Each nucleosome contains about 200 base pairs of double-helical DNA coiled around a complex of histones.

Treatment of nucleosomes with microbial nucleases removes about 60 base pairs, leaving behind a *nucleosome core particle* that contains 140 base pairs and 2 of each of the inner histones H2A, H2B, H3, and H4 (Figure 24.28). The size of the nucleosome core particle is about the same in all eucaryotes that have been studied. The core particle is roughly cylindrical with a diameter of about 11 nm and a height of about 5.5 nm. The histones lie at the center of the nucleosome, and the DNA is on the surface. The exact path

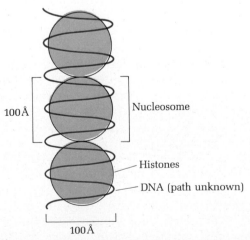

100Å — Nucleosome

Histones

DNA (path unknown)

100Å

Figure 24.27 Schematic diagram of the structure of a chromatin fiber. [Adapted from A. Kornberg, *Ann. Rev. Biochem.*, 46, 933 (1977). Reproduced with permission. © 1977 by Annual Reviews Inc.]

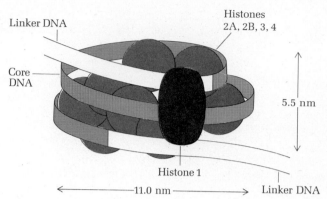

Figure 24.28   Structural organization of the nucleosome core particle. [Adapted from A. Kornberg, "DNA Replication," W. H. Freeman and Company, San Francisco, 1980, p. 294. Copyright © 1980.]

by which DNA traverses the nucleosome is not known, but it appears that that nucleosomal DNA is wound around the histones in two left-handed, negative superhelical twists. It has been estimated that about 100 kJ mol^{-1} are required to form this superhelix, whose pitch is about 2.8 nm.

A fully extended double helix containing 200 base pairs has a length of 68 nm. Coiling this DNA around a nucleosome reduces the length of the DNA to 10 nm. The degree of condensation, or *packing ratio*, is defined as the ratio of the length of a fully extended DNA strand to its actual length in chromatin. The packing ratio in a nucleosome is thus 68 nm/10 nm, or 6.8.

In its most densely packed state in metaphase, chromatin is condensed by a factor of about 1,000. This raises the question how nucleosomes are packed in the higher ordered structure of chromatin. It is a difficult question to answer because various methods of isolating chromatin may all involve some disruption of the higher ordered structure. Experimental evidence points to a *solenoid model* for chromatin structure in which a linear array of nucleosomes is supercoiled to form a cylindrical fiber some 30 nm in diameter and 11 nm long (Figure 24.29). Each turn of the helix of the cylindrical fiber contains 6 to 7 nucleosomes. The packing ratio of the solenoid is about 40. The linker region of DNA, complexed with histone H1, must be long enough and flexible enough to permit this packing. It has been proposed that histone H1 acts as a "lock" that helps to maintain the solenoid structure.

The cylindrical subunits of the solenoid structure of chromatin must themselves be packed in yet another higher ordered structure that is condensed by another two orders of magnitude.

B. Replication of Eucaryotic DNA

The period of active DNA replication in the life cycle of eucaryotic cells is called the S-phase (Figure 24.30). As in procaryotes, DNA replication in eucaryotic cells is semiconservative and occurs at replication forks traveling bidirectionally. DNA biosynthesis in eucaryotic cells is continuous in the leading strand and discontinuous on the lagging strand. In contrast to *E. coli* cells, DNA replication in eucaryotes has many replication origins. For example, the largest chromosome of *Drosophilia melanogaster* contains some 6,000 replication forks. The rate of fork movement in eucaryotic cells is much slower than in procaryotic cells, but since there are many replication forks, the net rate of synthesis is actually faster in eucaryotic cells. Unique DNA sequences apparently mark these replication origins, but little is known about the initiation sequences. The replication of chromatin also involves replication of nucleosomes. The nucleosome histones must therefore be synthesized and assembled to "package" the newly synthesized DNA. Nucleosome assembly thus adds enormous complexity to the already elaborate process of DNA biosynthesis.

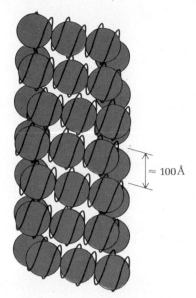

Figure 24.29 Solenoid model of chromatin structure. Each turn of the helix contains six or seven nucleosomes. DNA is shown as dark lines and nucleosomes as spheres. [Adapted from J. T. Finch and A. Klug, *Proc. Nat. Acad. Sci. U.S.*, 73, 1900 (1976).]

Many proteins are involved in eucaryotic DNA replication. Most of them have been less well characterized than the *E. coli* replication proteins. The general pattern of DNA replication in eucaryotic DNA biosynthesis is, however, similar to the sequence of events in *E. coli* DNA replication: specific proteins are required to unwind DNA at each replication fork, to synthesize RNA primers, to polymerize DNA, to excise the primers, to fill in the gaps, and to join the pieces of newly synthesized DNA. Perhaps the most intensively studied of the eucaryotic replication proteins are the DNA polymerases. The α enzyme is responsible for DNA replication in cell nuclei, the β polymerase is responsible for DNA repair, and the γ enzyme is responsible for DNA replication in mitochondria.

The general pattern of eucaryotic DNA replication is shown in Figure 24.31. DNA synthesis is continuous in the 5′-3′ direction on the leading strand. On the lagging strand DNA biosynthesis is discontinuous in the 5′-3′ direction. Discontinuous DNA synthesis consists of five steps: (1) a specific RNA polymerase synthesizes ribonucleotide primers whose average length is 9 ± 1 residues, (2) DNA polymerase α catalyzes the synthesis of Okazaki fragments whose average length is 135 base pairs, (3) the RNA primers are hydrolyzed, (4) DNA polymerase α fills in the gaps, (5) DNA ligase seals the "nicks" in the nascent DNA.

As DNA is replicating, nucleosomal DNA is unwinding ahead of the replication fork, and nucleosomes are being formed behind the replication fork (Figure 24.32). This process requires that biosynthesis of histones proceed at the same time as DNA biosynthesis. Histone synthesis and DNA synthesis occur not only concurrently, but at about the same rates. In each round of chromosome duplication the number of histones is doubled. It is possible that old nucleosomes distribute randomly between newly synthesized DNA and parental DNA, a process known as *random segregation*. It is also possible that histones bound to parental DNA remain bound to it, and that all new histones bind to the newly synthesized DNA. The second possibility is known as *conservative segregation*. It appears that newly synthesized histones associate with the lagging strand of the replication fork (as indicated in Figure 24.32). When DNA replication is studied in the presence of *cycloheximide*, an inhibitor of protein synthesis, a strand of "naked" DNA (DNA not bound to histones) is observed in electron micrographs (Figure 24.33). The naked DNA is the lagging strand of the replication fork, and the newly synthesized histones must therefore bind to the lagging strand. This in turn implies that the original histones remain with parental DNA, which is now segregated by semiconservative replication into two daughter molecules. Histone segregation is therefore conservative.

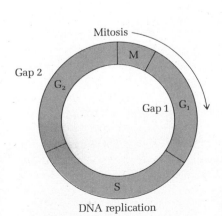

Figure 24.30 Life cycle of the eucaryotic cell.

Cycloheximide

DNA biosynthesis in eucaryotic cells appears to occur in stages involving the linking of progressively larger fragments. First, 130 to 200 residue Okazaki fragments are formed. Second, the Okazaki fragments are joined by merging of replication forks to give pieces containing 500 to 800 residues. Third, pieces containing 1,000 to 4,000 residues are formed (Figure 24.34). The formation of larger and larger pieces eventually leads to replication of the entire chromosome without the formation of discrete intermedi-

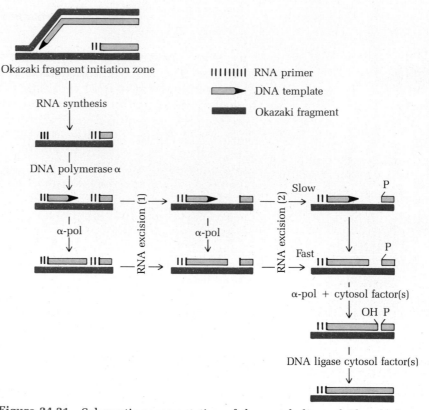

Okazaki fragment initiation zone

| | | | | | | | | RNA primer

▭➤ DNA template

████ Okazaki fragment

RNA synthesis

↓

DNA polymerase α

↓

α-pol → RNA excision (1) → α-pol → RNA excision (2) → Slow / Fast

α-pol + cytosol factor(s)

↓ OH P

DNA ligase cytosol factor(s)

Figure 24.31 Schematic representation of the metabolism of Okazaki fragments. Initiation of Okazaki fragments occurs within an "initiation zone" by synthesis of RNA primers (| | | | | | | |) complementary to the lagging strand of the DNA template (▭➤). Within this zone initiation events occur at preferred DNA sequences. The maturation of Okazaki fragments (████) is represented on the vertical coordinate in four distinct steps: synthesis of RNA primers, elongation of DNA by DNA polymerase α (α-pol), incorporation of the final deoxyribonucleotides (gap filling), and joining of the Okazaki fragments to growing DNA chains (ligation). Excision of RNA primers is shown on the horizontal coordinate in two distinct steps: (1) excision of the bulk of the RNA which is independent of Okazaki fragment synthesis and (2) removal of the p-rN-p-dN-(pdN)$_n$ junction which is facilitated by concomitant DNA synthesis. Excision of all RNA primers is assumed to occur concurrently. [Adapted from M. L. DePamphilis and P. M. Wasserman, *Ann. Rev. Biochem.*, *49*, 629 (1980). Reproduced with permission. © 1980 by Annual Reviews Inc.]

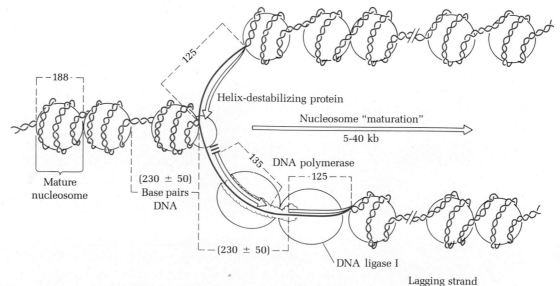

Figure 24.32 Replication fork of a eucaryotic chromosome. Abbreviations: RNA-primed Okazaki fragments | | | ████; helix-destabilizing protein, HD-prot.; DNA polymerase α, α-pol; DNA ligase I, lig-I; kilobase (1,000 nucleotide units), kb. [From M. L. DePamphilis and P. M. Wasserman, *Ann. Rev. Biochem.*, *49*, 631 (1980). Reproduced with permission. © 1980 by Annual Reviews Inc.]

ates of varying size. The discontinuous increase in the size of pieces of nascent DNA may result from the movement of replication forks toward one another. The rate of merging of replication forks is slower that the rate of DNA polymerization and seems to be the rate-limiting step in DNA replication.

24–6 MUTATIONS AND MUTAGENS

DNA replication normally occurs with astounding fidelity. On the average only about 1 error is made for every 10^9 to 10^{10} base pairs that are copied, but any error that alters the base sequence of DNA alters the genetic message. Besides suffering from errors of DNA replication, a cell is continuously exposed to agents such as ultraviolet light and a barrage of chemicals that can alter base structures in DNA. These changes in base sequence are called mutations.

Mutations are of two general types. A *base substitution mutation* (base pair switch or *point mutation*) is the result of the substitution of one base for another without changing the total number of bases. If a purine is replaced by a purine or a pyrimidine is replaced by a pyrimidine, the substitution is called a *transition*. If a purine is replaced by a pyrimidine or vice versa, the substitution is known as a *transversion*. The second general type of mutations is the *frame-shift mutation*, which is the insertion or deletion of one or more bases. These mutations are quite deleterious because they alter the sequence of bases that code for a protein.

Mutations that occur without any external environmental influence are called *spontaneous mutations*. Most spontaneous mutations are base substitutions. One source of a base substitution during DNA replication is lactam-lactim tautomerization of the bases (recall Section 9.2). For example, the rare lactim form of thymine pairs with guanine rather than with adenine (Figure 24.35). Although the equilibrium constant for lactam-lactim tautomerization is about 10^{-6}, the rate of spontaneous mutation caused by this tautomerization is several orders of magnitude lower. This low mutation rate probably reflects the proofreading abilities of DNA polymerase, which very effectively repairs mismatched base pairs during DNA replication.

An external influence that causes a mutation or increases the rate of appearance of mutations is called a *mutagen*. Ultraviolet light, x-rays, and

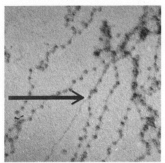

Figure 24.33 Electron micrograph of single-stranded DNA, not bound to histones, at the replication fork of a cell in which protein synthesis has been inhibited. The single-stranded DNA is the thin line to the right of the replication fork. DNA bound to histones extends upward to the right of the fork. [From D. Riley and H. Weintraub, *Proc. Nat'l. Acad. Sci.,* 76, 331 (1979).]

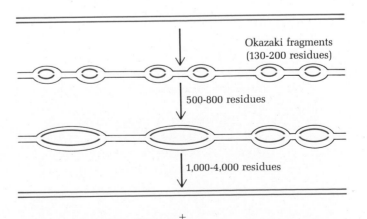

Figure 24.34 Stages of progressively larger fragments in DNA biosynthesis. First, Okazaki fragments that contain 130 to 200 residues are formed. Second, merging of replication forks gives fragments that contain 500 to 800 residues. Third, fragments containing up to 4,000 residues are formed.

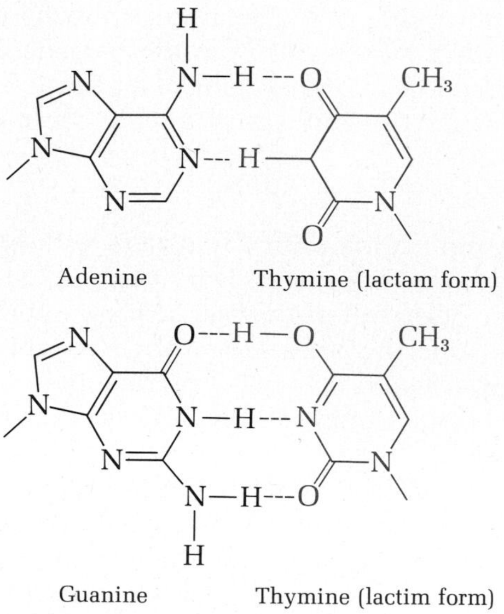

Figure 24.35 Base pairing of the tautomeric forms of thymine.

many chemicals are mutagenic. Consider, for example, the effects of compounds whose structures resemble those of the naturally occurring purines and pyrimidines, but that have altered base-pairing properties. These compounds, known as *base analogs,* can substitute for the naturally occurring bases in DNA. One such compound is *5-bromouracil,* whose bromine atom has about the same van der Waals radius as the methyl group of thymine. The mutagenicity of 5-bromouracil results from a shift of the lactam-lactim tautomeric equilibrium toward the side of the lactim (reaction 24.3).

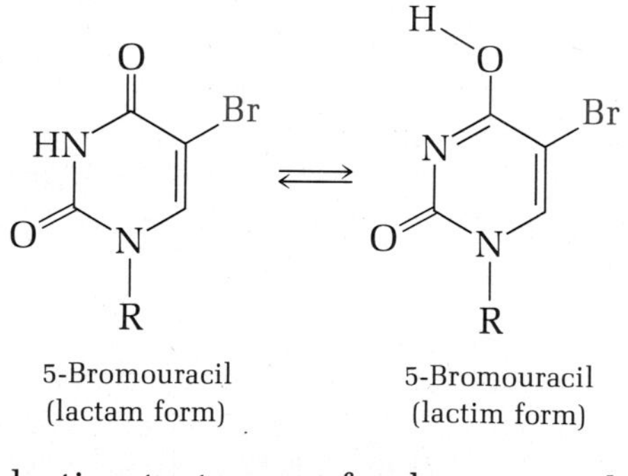

$$(24.3)$$

The lactim tautomer of 5-bromouracil pairs with guanine rather than adenine, and the substitution of 5-bromouracil in DNA leads to 5-bromouracil-guanine and 5-bromouracil-adenine base pairs when DNA is replicated.

Substances may act as mutagens also by chemically modifying the DNA bases in ways that disrupt their normal base pairing. For example, when DNA is exposed to sodium nitrite ($NaNO_2$) in acidic solutions, free nitrous acid forms (reaction 24.4).

$$NO_2^{\ominus}(aq) + H^{\oplus} \rightleftharpoons H-O-N=O \qquad (24.4)$$
$$\text{Nitrous acid (p}K_a = 3.3)$$

Nitrous acid causes deamination of the bases in DNA (Figure 24.36). Deamination of cytosine produces uracil. Since uracil pairs with adenine, the original cytosine-guanine base pair is disrupted.

Hydroxylamine (H_2N-OH) causes a base substitution mutation by reacting with cytosine to produce N⁴-hydroxycytosine (Figure 24.37). The

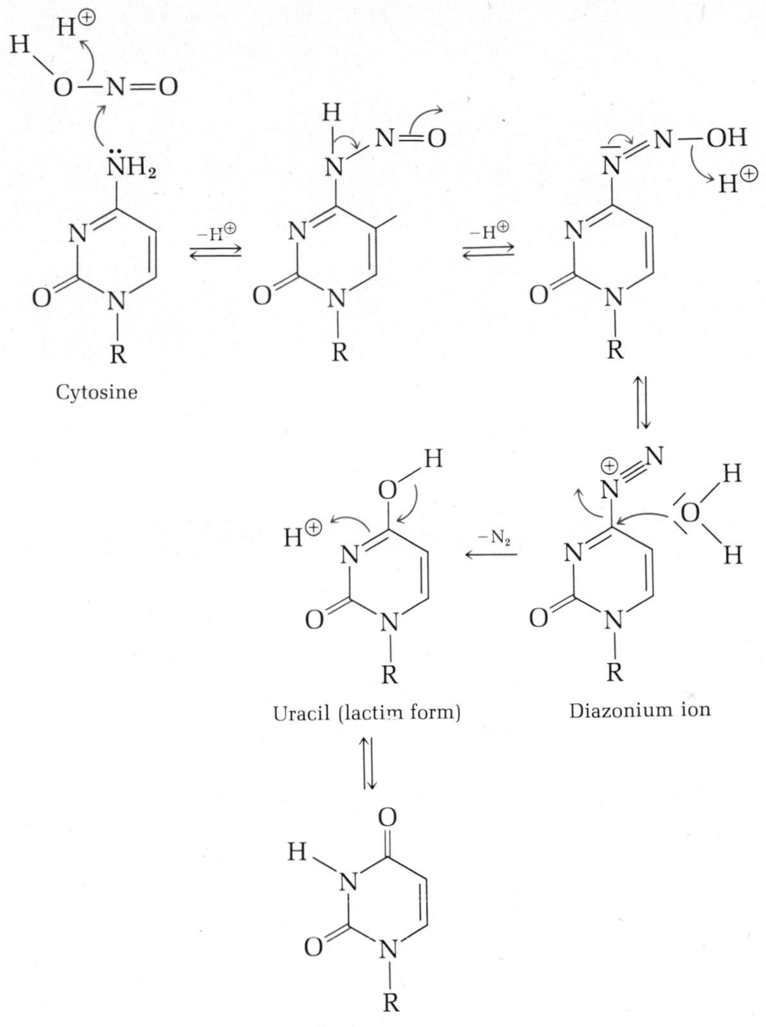

Figure 24.36 The mechanism of deamination of cytosine by nitrous acid.

N^4-hydroxylated derivative is most stable as the hydroxamic acid tautomer, which pairs with adenine instead of with guanine.

Ultraviolet radiation causes adjacent thymine bases to form a cyclobutyl dimer (Figure 24.38). Once such a dimer has formed, the bases can no longer form correct hydrogen bonds. In fact, the dimer does not fit into the double helix at all. DNA replication is effectively halted by thymine dimers, and a repair mechanism is necessary to restore the original double helix. We shall discuss the mechanism by which thymine dimers are repaired in the next section.

Alkylating agents are another class of mutagens. The N-7 position of guanine is especially susceptible to alkylation. N-7 is more nucleophilic than the nitrogens of the pyrimidine ring of purines, which are either involved in hydrogen bonding or sterically sheltered from alkylation. In the sulfur "mustard," *bis*-(2-chloroethyl)sulfide, the alkylating groups are cova-

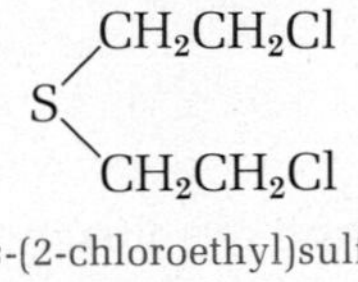

bis-(2-chloroethyl)sulfide

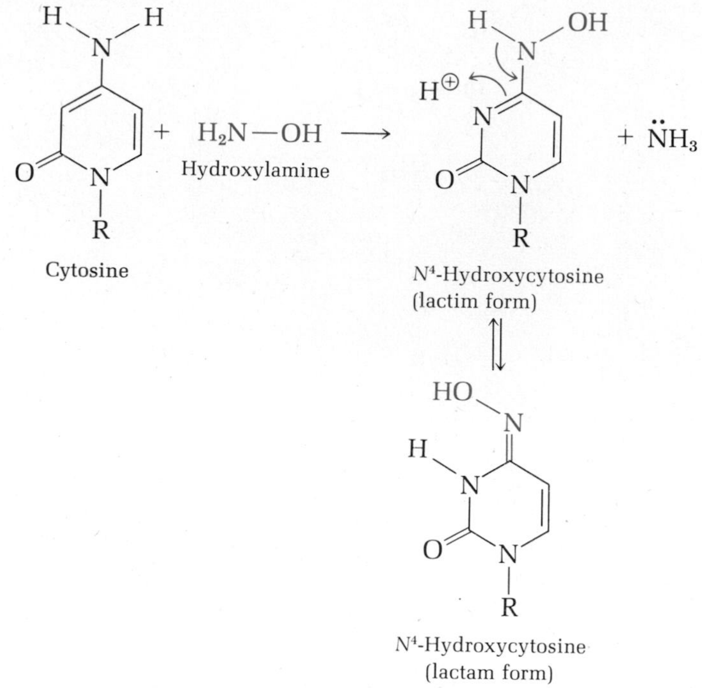

Figure 24.37 Formation of N^4-hydroxycytosine by the reaction of cytosine with hydroxylamine.

lently bonded, and this mustard can cross-link DNA strands. The first step in the reaction is an internal nucleophilic substitution reaction in which chloride is displaced and a cyclic sulfonium ion is formed (reaction 24.5).

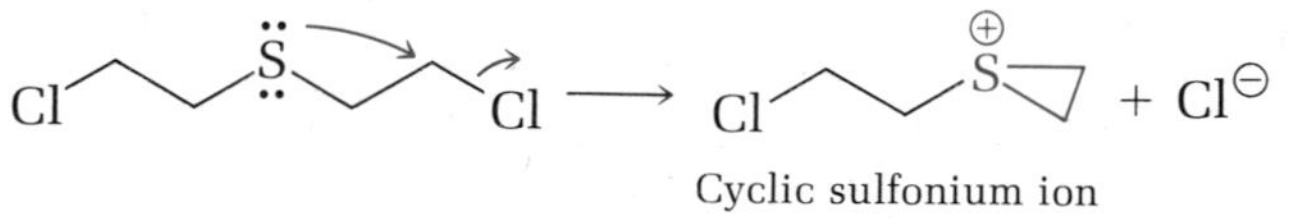

Cyclic sulfonium ion

(24.5)

The cyclic sulfonium ion is strained and is susceptible to nucleophilic attack by N-7 of a guanine moiety of a nucleotidyl residue, opening the ring (reaction 24.6). In this step one strand of DNA is alkylated. Repetition of reactions 24.5 and 24.6 produces a cross-linked DNA (Figure 24.39).

Many frame-shift mutagens are planar, fused ring systems (Figure 24.40) that intercalate between adjacent base pairs in DNA in a manner that is analogous to the intercalation of ethidium bromide into the double helix

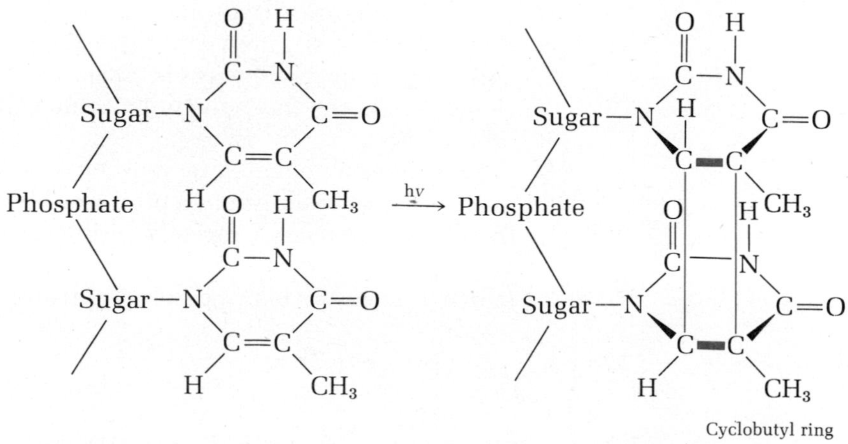

Cyclobutyl ring

Figure 24.38 Formation of a thymine dimer induced by ultraviolet light.

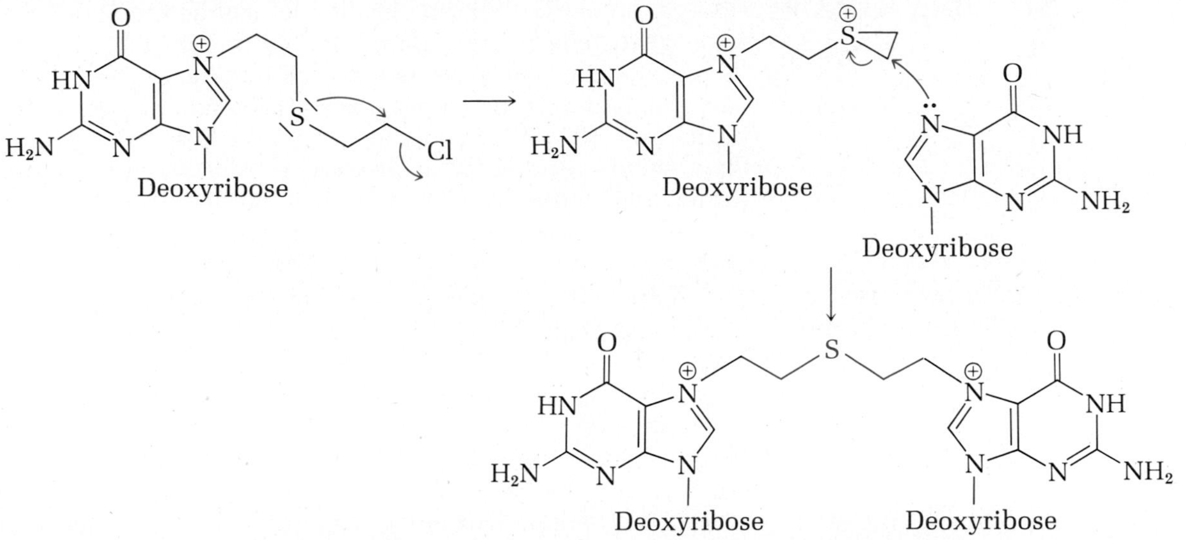

Figure 24.39 Cross-linking of the two strands of duplex DNA by *bis*-(2-chloroethyl)sulfide.

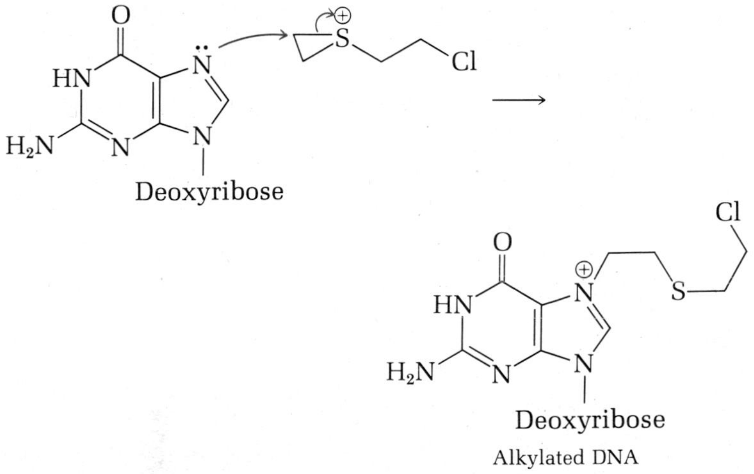

(24.6)

(recall Section 9.5D). An organism would soon be killed by accumulated lesions in its DNA were it not for the repair mechanisms to which we now turn.

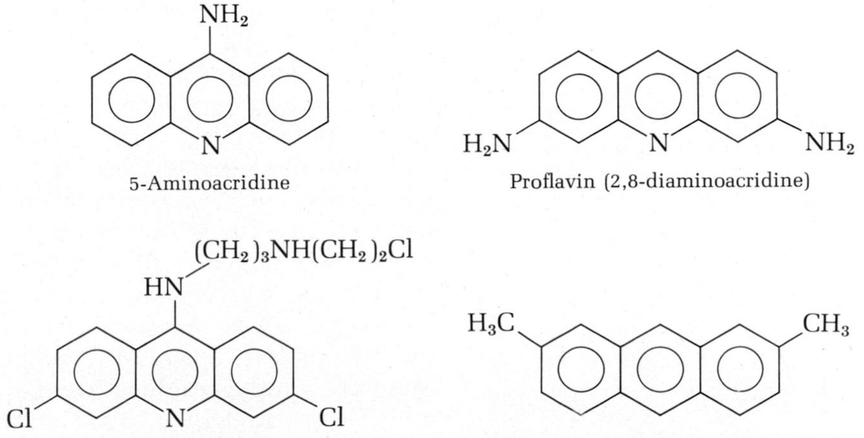

5-Aminoacridine

Proflavin (2,8-diaminoacridine)

ICR-191

7,12-Dimethylbenz(a)anthracene

Figure 24.40 Structures of some mutagens that cause additions and deletions of bases in DNA (frame-shift mutations).

24–7 DNA REPAIR

DNA is the only cellular molecule that can be repaired. The uniqueness of such repair reflects the extraordinary importance of DNA to cellular life. If the genetic message is garbled, everything that follows will also be garbled, and a mutation in the wrong place can be fatal. A variety of repair mechanisms are known. When combined with the high fidelity of DNA replication, they lead to a remarkably stable genome. A bacterial chromosome can be replicated 100 million times with a 50% probability that the last descendant will have the same genetic composition as the original parent.

A. Photoreactivation of Thymine Dimers

The thymine dimers that are occasionally produced by ultraviolet light can be repaired by an enzyme that reverses the dimerization reaction photochemically in a single step. The presence of a thymine dimer distorts the double helix from its normal conformation, and the photoreactivating enzyme binds to this distorted region. Light is not required for binding. The enzyme-DNA complex absorbs a photon in the wavelength range of 300 to 600 nm, and the dimerization reaction reverses. The enzyme isolated from *E. coli* (mol wt 35,000) does not possess any unique light-absorbing chromophores; it is the DNA-enzyme complex that absorbs light. Once the dimerization reaction has been reversed, the enzyme dissociates from DNA, and the adjacent thymines again form hydrogen bonds with their adenine partners in the double helix (Figure 24.41).

B. Excision Repair of DNA

Except for photoreactivation of thymine dimers, all known modes of DNA repair operate by *excision repair* (Figure 24.42). In the most general sense excision repair requires four steps: *incision*, that is, formation of a nick in the double-stranded DNA; *excision* by cleavage of a second phosphodiester bond to give a gap in double-stranded DNA; *synthesis* of new DNA to replace the deleted region; and *insertion* of the new DNA and *sealing* of the nick.

Incision is cleavage of a phosphodiester bond near a region of damaged DNA by specific endonucleases that recognize different types of lesions. In *E. coli* specific enzymes have been identified that cleave DNA adjacent to thymine dimers (Figure 24.43). In humans, the inability to nick defective DNA at thymine dimers leads to a variety of hereditary disorders of DNA repair. The best known of them is *xeroderma pigmentosa*. Persons lacking the specific endonuclease are hypersensitive to the sun and ultraviolet radiation and have a greatly increased incidence of skin cancer.

Once the defective region of DNA has been nicked, DNA polymerase continues the repair by synthesizing a complementary strand of new DNA in the nicked region and by catalyzing 5'-3' excision of the defective strand. Many *E. coli* enzymes are able to participate in excision of damaged DNA and the resynthesis of a new complementary strand. For example, three enzymes are capable of excising thymine dimers: the 5'-3' exonuclease activity of DNA polymerase I, the same exonuclease activity of DNA polymerase III, and an exonuclease that is specific for single-stranded DNA. All three DNA polymerases of *E. coli* are potentially capable of resynthesizing a complementary strand of DNA. DNA polymerase I seems the most likely repair enzyme since it binds to nicked regions of DNA better than the other two, and since it also possesses 5'-3' exonucleolytic activity. "Patches" of various lengths have been detected, and it seems likely that the three enzymes work on different size repairs. DNA polymerase I probably repairs short gaps of 10 to 30 nucleotidyl residues, and DNA polymerases II and III probably repair longer gaps, stretching to several hundred residues.

Excision and resynthesis in mammals is more complex since mammals do not possess any bifunctional enzymes that are capable of both excision and polymerization. A variety of endonucleases and polymerases probably operate in DNA repair, rather than a single "dedicated" pair of enzymes. The

Structural distortion
(thymine dimer)

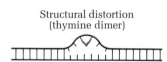

Enzyme-DNA complex

Absorption of light

Release of enzyme

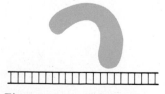

Figure 24.41 Schematic diagram of photoreactivation of a DNA molecule in which a thymine dimer has formed.

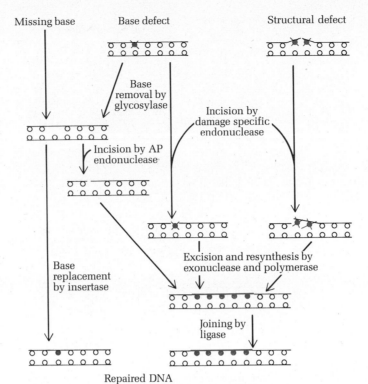

Figure 24.42 Pathways for excision repair of DNA. [Adapted from P. C. Hanawalt, P. K. Cooper, A. K. Ganesan, and C. A. Smith, *Ann. Rev. Biochem.*, 48, 785 (1980). Reproduced with permission. © 1980 by Annual Reviews Inc.]

problem in eucaryotic cells is further complicated by the structure of chromatin, and it is possible that different repair enzymes function in the nucleosome core and linker regions of chromatin.

The last step in excision repair is insertion of the new DNA to replace the excised region. It is joined to the existing strand and the nick is sealed by the action of a ligase.

As an example, consider the excision repair of the base-pairing defect caused by hydroxylamine in which uracil, formed by deamination of cytosine, pairs with adenine rather than with the guanine with which the original cytosine was paired (Figure 24.44). *Uracil-DNA glycosidase* hydrolyzes the glycosidic bond of the uridyl residue leaving an *apyrimidine* residue (reaction 24.7).

Uracil-substituted DNA $\xrightarrow{\text{Uracil-DNA glycosidase}}$ Apyrimidine residue + Uracil (24.7)

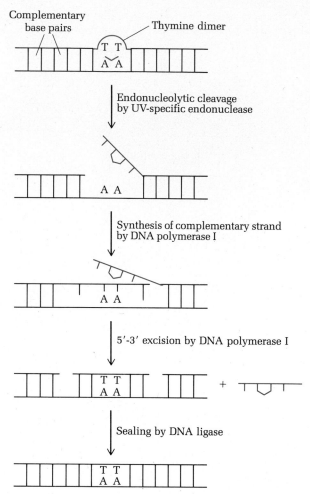

Figure 24.43 Repair of a thymine dimer by the combined action of UV-specific endonuclease, DNA polymerase I, and DNA ligase in *E. coli*. [From P. C. Hanawalt, *Endeavor, 31,* 83 (1972). Reproduced with permission. Copyright 1972, Pergamon Press, Ltd.]

The remaining DNA is defective, and an endonuclease creates a nick. The 5'-3' exonuclease activity of DNA polymerase I excises the apyrimidine residue and inserts a cytosine to complete the repair operation.

24–8 GENETIC RECOMBINATION

The term genetic recombination embraces a wide range of phenomena whose common denominator is exchange of DNA from one intact double-helical molecule of DNA to another. Genetic recombination can be divided into two types of processes. In *general recombination* an exchange of DNA occurs between chromosomes that pair with one another during meiosis. These paired chromosomes have the same general shapes, contain approximately the same genetic information, and are said to be *homologous*. General genetic recombination is particularly important in generating new combinations of genes during the production of eggs and sperm (gametes) in meiosis (Figure 24.45). In *site-specific recombination* the insertion of a piece of DNA occurs at a specific region in the genome. Examples of site-specific recombination are the integration of viral DNA into the chromosome of a bacterium and the insertion of a strand of eucaryotic DNA into a bacterial plasmid in the techniques used to generate synthetic recombinant DNA.

A. General Genetic Recombination

General recombination between homologous chromosomes can occur by a *single-strand transfer mechanism* as shown in Figure 24.46. The alignment

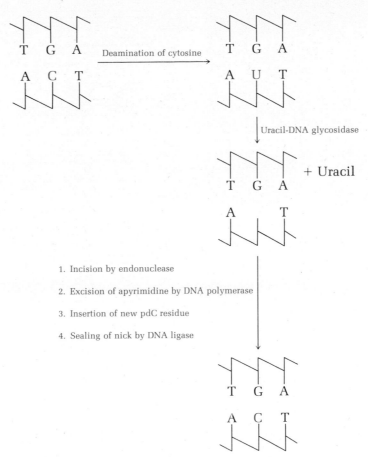

Deamination of cytosine

Uracil-DNA glycosidase

+ Uracil

1. Incision by endonuclease

2. Excision of apyrimidine by DNA polymerase

3. Insertion of new pdC residue

4. Sealing of nick by DNA ligase

Figure 24.44   Excision repair of a uracil-substituted DNA molecule.

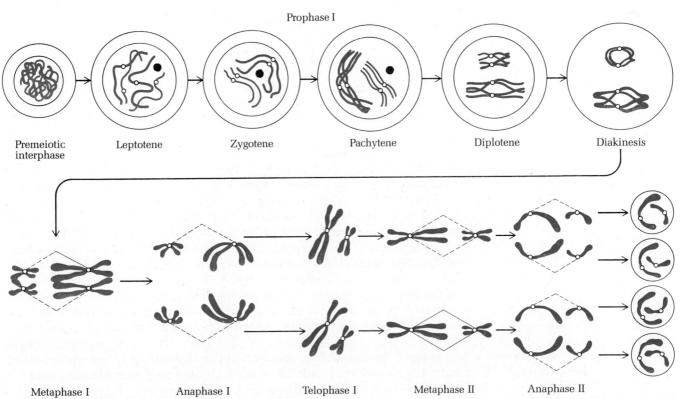

Prophase I

Premeiotic interphase   Leptotene   Zygotene   Pachytene   Diplotene   Diakinesis

Metaphase I   Anaphase I   Telophase I   Metaphase II   Anaphase II

Figure 24.45   Stages of meiosis.

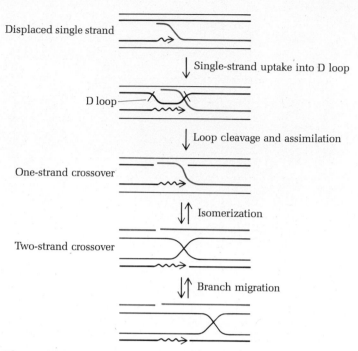

Figure 24.46 Mechanism of single-strand transfer in general genetic recombination. [From C. M. Radding, *Ann. Rev. Biochem.*, *47*, 854 (1978). Reproduced with permission. © 1978 by Annual Reviews Inc.]

of homologous sequences that is necessary for general recombination is extremely precise, since as many as 3×10^9 base pairs must be exactly aligned for strand transfer to occur without introducing mutations. In fact mutations seldom arise because of errors in general recombination.

In the first step of general recombination a nuclease puts a nick in one strand of one of the two double helices, and a second protein unwinds the nicked double helix. The driving force for unwinding is provided by the relief of strain in the superhelical, nicked DNA. Two *E. coli* proteins, designated *recB* (mol wt 140,000) and *recC* (mol wt 128,000) carry out these preliminary steps.

DNA polymerase I then synthesizes a complementary strand to the exposed region of the nicked, unwound DNA to give a three-stranded product. The *recA* protein (mol wt 40,000) then catalyzes the ATP-dependent assimilation of the free third strand into the second double helix to give a *D loop* structure, so named because of its appearance in electron micrographs (Figure 24.47). The single-stranded DNA that forms the D loop has a short life and is rapidly hydrolyzed by a dimeric enzyme, known as the *recAB* protein, consisting of one *recA* protein and one *recB* protein.

After the single strand of the D loop has been digested, the structure that remains has a one-strand crossover. Isomerization of this one-strand crossover gives a two-strand crossover. This process is essentially a switching of partners between the free 5' ends of the two nicked duplexes. In essence, it involves a rotation of 180° around an axis that runs parallel to the two double helices. Branch migration can then occur by the movement of the two-strand crossover. This process requires that the DNA of one strand exchange partners with the DNA of the other double helix.

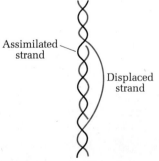

Figure 24.47 Schematic diagram of a D loop intermediate in the single-strand transfer mechanism of general genetic recombination.

B. Site-Specific Recombination

Site-specific recombination involves the incorporation of one molecule of DNA within another at a specific site, in contrast to the more or less random recombination sites observed in general recombination. This mode of recombination occurs by the *lap joint mechanism*. The basic principle is dia-

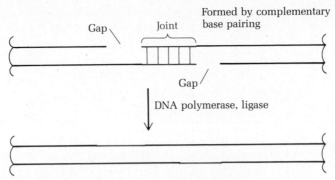

Figure 24.48 Site-specific recombination by a lap joint mechanism. Complementary base pairs are produced by the action of specific nucleases that hydrolyze DNA leaving single-stranded ends that possess complementary base pairs.

gramed in Figure 24.48. If two DNA molecules have exposed single-stranded ends, then these ends can anneal if their base sequences are complementary. The overlapping region is the so-called *joint* between the pieces. If gaps in the DNA exist to either side of the joint, they can be filled in by DNA polymerase, and the nicks can be sealed by DNA ligase.

This mechanism is the preferred route for recombination in the incorporation of the linear DNA of phage λ in the circular chromosome of *E. coli* (Figure 24.49). The basic procedure is simple. An exonuclease, called λ exonuclease, cleaves the host DNA in the 5'-3' direction. A 15-nucleotide se-

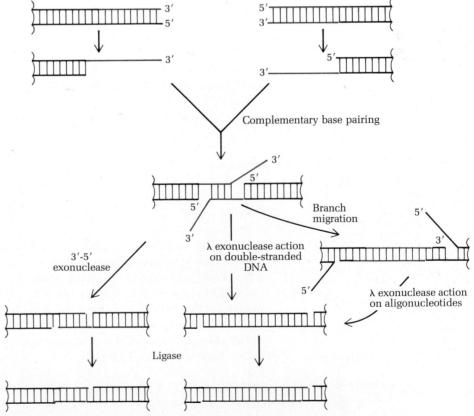

Figure 24.49 Site-specific recombination of the DNA of bacteriophage λ into the chromosome of *E. coli.*

quence that is exactly complementary to a 15-nucleotide region in the virus is excised by the exonuclease. This exonuclease also carries out hydrolysis, in the 5'-3' direction, of a sequence of 15 nucleotides in the λ DNA. The exposed single-stranded regions then anneal by forming complementary base pairs, λ exonuclease clips off the ends, and DNA ligase seals the remaining nicks.

24–9 LABORATORY SYNTHESIS OF RECOMBINANT DNA

The advent of techniques for the synthesis of DNA molecules in which a eucaryotic gene is "implanted" in a procaryotic chromosome has revolutionized biology for the second time in as many generations. The hybrid DNA molecules are somewhat poetically known as *chimeras*, mythological beasts of strange composition. The potential uses of recombinant DNA technology are virtually limitless, ranging from the synthesis of rare and expensive hormones to the invention of plants that contain the genes for nitrogenase. Pure science will also profit from recombinant DNA techniques. For example, such techniques provide a convenient way of studying eucaryotic genes, which are more complex and far more difficult to study than procaryotic genes. As with any new technology the potential benefits of synthetic recombinant DNA are in some sense "obvious," but the potential hazards are unknown. These hazards have been much discussed and appear, at present, to be minimal.

The synthesis of a DNA molecule in which a eucaryotic gene is inserted in a procaryotic DNA molecule is a multistep process (Figure 24.50). First, the DNA molecule to be inserted is isolated. Then, it is structurally modified and incorporated in a carrier molecule, such as a bacterial plasmid or a bacteriophage such as λ. The recombinant DNA molecule is then introduced into the host bacterium. Since not all bacteria will take up the plasmid, some experimental method, such as resistance to an antibiotic (conferred by a gene which is also conveniently located in a plasmid), is used to identify the bacteria that actually possess the recombinant DNA molecule. These selected bacteria are separated and allowed to reproduce. Progeny descended from the same ancestor are known as *clones*. The clones of the selected bacteria possess the recombinant DNA molecule, now *amplified* thousands of times. The clones are each little factories that produce copious quantities of the protein whose gene was inserted in the original bacterial plasmid or bacteriophage.

A. Synthesis of Complementary DNA (cDNA)

The genetic information of RNA tumor viruses is contained in RNA rather than in DNA. The viruses contain a gene that is translated by the host cell to produce *reverse transcriptase*. This enzyme uses RNA as a template to synthesize a DNA molecule, cDNA, which is complementary (in the Watson-Crick sense) to the RNA of the virus. In the translation of the genetic message (Chapter 26.1) DNA is first transcribed to an RNA copy known as messenger RNA (mRNA). If the mRNA copy of a gene is isolated, then a complementary DNA molecule can be synthesized *in vitro* by incubation of deoxyribonucleoside triphosphates with reverse transcriptase (Figure 24.51). The cDNA molecule that is synthesized in this way contains all of the information of the original gene in its single strand. Incubation of the cDNA with DNA polymerase and a deoxyribonucleoside triphosphate pool results in the synthesis of double-stranded DNA.

B. Chemical Synthesis of a Gene

If the sequence of the protein whose gene is to be cloned is known, then the genetic code (Section 26.10) can be used to deduce a primary structure for the gene. (The sequence is not unique because the genetic code is degenerate, that is, some amino acids are represented by more than one code word.) This

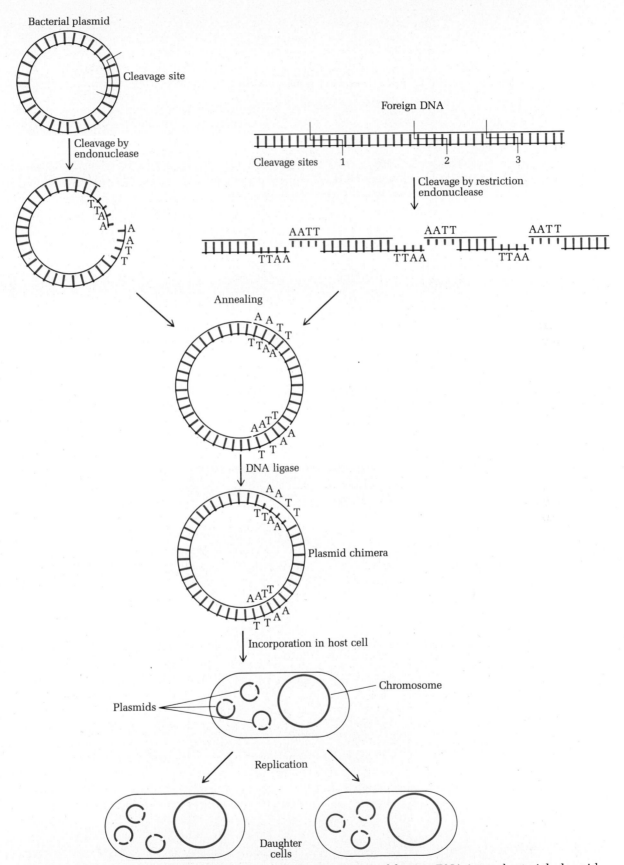

Figure 24.50 Sequence of steps required to incorporate a piece of foreign DNA into a bacterial plasmid. [Adapted from S. N. Cohen, *Scientific American*; July 1975, p. 30. Copyright © 1975 by Scientific American, Inc. All rights reserved.]

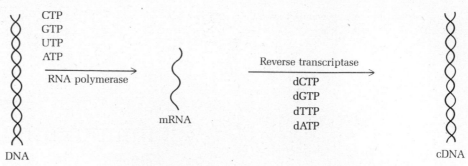

Figure 24.51 Synthesis of complementary DNA (cDNA) from an mRNA molecule catalyzed by reverse transcriptase.

gene can then be synthesized by chemical methods. This method has been used to synthesize the genes for somatostatin and insulin.

C. Cleavage of DNA by Restriction Endonucleases

If neither isolation of mRNA nor chemical synthesis of a given gene is practical or of particular interest, a eucaryotic gene can be cleaved into a few fragments by restriction endonucleases (recall Section 9.8D). The most important restriction endonucleases in recombinant DNA research are those that cleave DNA symmetrically, such as the enzyme *Eco*R1, which recognizes the sequence

3′ C T T A A G 5′

5′ G A A T T C 3′

Table 24.5	Specificities of some common restriction endonucleases
Enzyme	**Recognition sequence and cleavage points**
r. *Hind*II	3′ N C A R Y T G N 5′ 5′ N G T Y R A C N 3′
r. *Bsu*R	3′ N N C C G G N N 5′ 5′ N N G G C C N N 3′
r. *Hpa*1	3′ N C A A T T G N 5′ 5′ N G T T A A C N 3′
r. *Eco*RI	3′ N C T T A A G N 5′ 5′ N G A A T T C N 3′
r. *Eco*RII	3′ N N G G C C N N 5′ 5′ N N C C G G N N 3′
r. *Hin*dIII	3′ N T C G A A N 5′ 5′ N A A G C T T N 3′

SOURCE: K. Murray, *Endeavor*, **35**, 130 (1976). Reprinted with permission. Copyright 1976, Pergamon Press, Ltd.

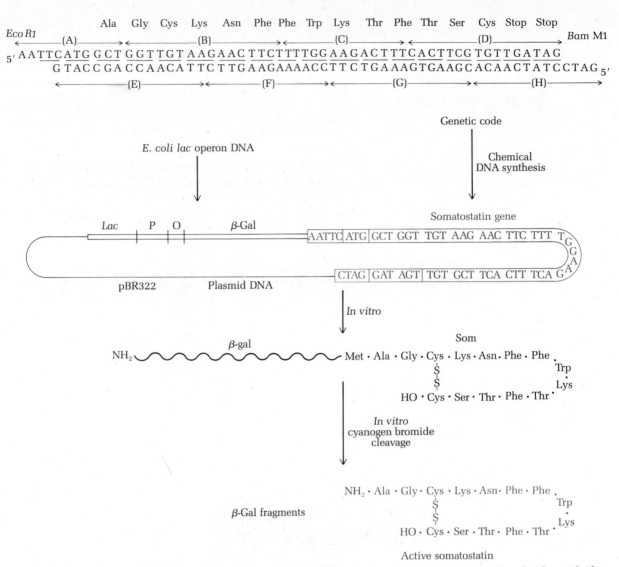

Figure 24.52 Schematic diagram of the incorporation of the somatostatin gene into an *E. coli* plasmid (designated pBR322). [Adapted from K. Itakura and A. D. Riggs, *Science, 209,* 1404 (1980). Copyright 1980 by the American Association for the Advancement of Science.]

and cleaves at the points indicated by arrows. These sequences are *palindromes.* (A palindrome is a word, series of words, or number that reads the same forward or backward, such as "MADAM I'M ADAM".) If the sequences of the two strands are read in the direction (5'-3'), they are the same. Table 24.5 indicates the specificities of a few widely used restriction endonucleases.

Suppose that an *E. coli* plasmid possesses a sequence that is recognized by the restriction endonuclease that was used to cleave the eucaryotic gene. Cleavage of the plasmid by this restriction endonuclease gives single-stranded ends whose base sequences are precisely complementary to the exposed ends of the eucaryotic DNA. Mixing the plasmid with the eucaryotic DNA leads to spontaneous annealing. The nicks are sealed, and now the bacterial plasmid contains the eucaryotic gene (see Figure 24.50). The same technique is employed to incorporate DNA produced by chemical synthesis or to incorporate cDNA in a plasmid. In these cases complementary strands are introduced enzymatically into both the cDNA and the plasmid. Annealing then follows. The technique by which the chemically synthesized gene for somatostatin was introduced in an *E. coli* plasmid is indicated in Figure 24.52. In this case the 14-residue polypeptide of somatostatin was covalently bound to β-galactosidase. Cleavage with cyanogen bromide (Section 2.9B) released free somatostatin. The product possessed full biological activity.

24–10 SUMMARY

DNA replication is semiconservative. Each strand of the parent molecule serves as a template for the synthesis of a complementary daughter strand. In the first round of replication, therefore, each DNA molecule contains one newly synthesized strand and one strand derived from the parent. DNA polymerase catalyzes the polymerization of deoxyribonucleoside triphosphates when provided with a primer. All known DNA polymerases catalyze chain growth exclusively in the 5'-3' direction. There are three major DNA polymerases in E. coli, designated I, II, and III. DNA polymerase III is the enzyme primarily responsible for DNA replication in vivo. All three DNA polymerases possess 3'-5' exonuclease activity that provides a proofreading function to remove any mispaired bases that have been incorporated during DNA replication. DNA polymerases I and III also possess 5'-3' endonuclease activity that is important in DNA repair. There are several eucaryotic DNA polymerases, designated α, β, and γ. The α form is responsible for DNA replication in the nucleus, the β form is primarily responsible for repair, and the γ form is responsible for replication of mitochondrial DNA.

The E. coli chromosome is a tightly coiled superhelical molecule of circular DNA that is associated with RNA. E. coli cells also contain small circular molecules of DNA called plasmids that contain a few genes, among them those responsible for conferring bacterial resistance to antibiotics. DNA replication occurs bidirectionally from a unique origin. At each replication fork the double helix is unwound by a complex containing several proteins, including the helix-destabilizing protein, DNA gyrase, and the rep protein. The dnaB protein binds at or near the replication origin and triggers a specific RNA polymerase, known as primase, which synthesizes an RNA primer. DNA polymerase catalyzes apparently continuous polymerization in the 5'-3' direction along the leading strand. Along the lagging strand polymerization occurs discontinuously. The small fragments of DNA produced in discontinuous DNA synthesis are called Okazaki fragments. Removal of the primer and filling of the gaps by DNA polymerase I is followed by sealing of the nicks by DNA ligase.

Eucaryotic DNA is tightly bound to basic proteins known as histones. The structure of chromatin consists of a string of nucleosomes joined by a strand of DNA known as the linker. Nucleosomes are octameric complexes of histones upon which DNA is coiled. The linker region consists of DNA bound to histone H1. In the solenoid model of chromatin structure nucleosomes are coiled in a helical structure. Each turn of the solenoid helix contains six nucleosomes. DNA replication of eucaryotic DNA follows the same general pattern as procaryotic DNA replication. There are, however, some differences. First, DNA replication begins at many loci along the chromosome. Second, histone synthesis occurs concurrently, and newly synthesized DNA is immediately packaged in histones.

Many chemicals and short wavelength radiation induce mutations in DNA. Ultraviolet light, for example, causes formation of thymine dimers. Repair of DNA is of two general types. The simplest is photoreactivation of thymine dimers in a single step. Other types of DNA repair fall collectively under the heading of excision repair. The general pathway involves an incision or nicking, excision to produce a gap, and filling in of the gap by DNA polymerase.

Genetic recombination during meiosis is a process in which exchange of DNA occurs between homologous chromosomes. In general genetic recombination new combinations of genes are formed by exchange of DNA at many, more or less random points along the chromosome. In spite-specific recombination a piece of DNA is inserted in a chromosome at a unique point.

The laboratory synthesis of artificially recombinant DNA represents one example of site-specific recombination. The messenger RNA of a foreign organism, which contains the information required for protein synthesis, can be copied by reverse transcriptase to give cDNA. This cDNA can then be converted to DNA for insertion into the plasmid. Genes can also be synthesized directly by chemical methods. Restriction endonucleases provide the molecular tools that prepare the DNA of the host and the foreign DNA for synthesis of a recombinant DNA molecule by exposing complementary strand segments in the host DNA and the foreign DNA. Mixing of the foreign DNA with the plasmid leads to spontaneous annealing by complementary base pairing. DNA ligase then seals the new piece of DNA in place. The plasmids that contain the recombinant DNA can then be incorporated in *E. coli*, and the host cells are cloned. The clones are microbial factories that produce large amounts of the protein whose gene was incorporated in the original plasmid. The cDNA does not contain the control regions for gene expression (Chapter 27), and this is a problem in some cases.

REFERENCES

Champoux, J., Proteins That Affect DNA Conformation, *Ann. Rev. Biochem.*, 47, 449–480 (1978).

Cozzarelli, N. R., The Mechanism of Action of Inhibitors of DNA Synthesis, *Ann. Rev. Biochem.*, 46, 641–668 (1977).

Drake, J. W., "The Molecular Basis of Mutation," Holden-Day, San Francisco, 1970.

Goulian, M., and Hanawalt, P. C., Eds., "DNA Synthesis and Its Regulation," W. A. Benjamin, Menlo Park, Calif., 1975.

Grobstein, C., "A Double Image of the Double Helix," Freeman, San Francisco, 1979.

Hanawalt, P. C., and Setlow, R. B., Eds., "Molecular Mechanisms for Repair of DNA," Plenum, New York, 1975.

Hollander, A., Ed., "Chemical Mutagens," Plenum, New York, 1971.

Kornberg, A., "DNA Replication,[22]" Freeman, San Francisco, 1980.

Radding, C. M., Genetic Recombination: Strand Transfer and Mismatch Repair, *Ann. Rev. Biochem.*, 47, 847–881 (1978).

Recombinant DNA, *Science*, 209, 1317–1438 (1980).

Schekman, R., Weiner, A., and Kornberg, A., Multienzyme Systems of DNA Replication, *Science*, 186, 987–993 (1974).

Sinsheimer, R., Recombinant DNA, *Ann. Rev. Biochem.*, 46, 415–438 (1977).

Tomizawa, Jun-ichi, and Selzer, G., Initiation of DNA Synthesis in *Escherichia Coli*, *Ann. Rev. Biochem.*, 48, 999–1034 (1979).

Weissbach, A., Eucaryotic DNA Polymerases, *Ann. Rev. Biochem.*, 46, 25–47 (1977).

Wickner, S. H., DNA Replication Proteins of *E. coli*, *Ann. Rev. Biochem.*, 47, 1163–1192 (1978).

Chapter 25

DNA transcription

Je cherche à comprendre.

Jacques Monod

25–1 INTRODUCTION With the discovery of the structure of DNA in 1953, it might have seemed that it would be a simple matter to show how the genetic message is expressed as the protein complement of the cell. But it was not so. Nearly a decade was required to show that the expression of the genetic information contained in DNA is a two-stage process. In the first stage, the topic of this chapter, the genetic message is *transcribed;* one strand of the DNA molecule serves as a template for the synthesis of an RNA molecule. The transcribed RNA was christened *messenger RNA* by F. Jacob and J. Monod. The information contained in this messenger is then *translated* into proteins on the complex scaffolds known as *ribosomes,* the topic of the next chapter.

Transcription also includes the synthesis of *transfer RNA*, the carrier of amino acid residues that acts as a molecular "adapter" in the translation of the genetic message that has been transcribed into mRNA and the synthesis of *ribosomal* RNA, an integral part of ribosomes involved in protein synthesis.

We shall also include under the heading of transcription the remarkable reactions by which newly synthesized RNA molecules are covalently modified, a phenomenon known as processing. We shall see that recent discoveries in RNA processing are among the most startling in the short, but eventful, history of molecular biology.

25-2 THE SEARCH FOR THE MESSENGER

The concept of messenger RNA received its first formal explication in a classic paper by F. Jacob and J. Monod in 1961. The authors postulated that the as yet undiscovered molecule should have the following properties.

1. The messenger should be a polynucleotide. This postulate proved correct.

2. The molecular weight of the messenger should be highly variable to reflect the variability of protein molecular weights. This was later found to be correct.

3. Messenger RNA should have a base composition that reflects the base composition of DNA. The crucial experiment that demonstrated this hypothesis is discussed below.

4. Since ribosomes are the sites of protein synthesis, messenger RNA should be found associated with ribosomes. It was soon found that newly synthesized mRNA is tightly bound to ribosomes on which protein synthesis is occurring.

5. Messenger RNA should have a short life. It has been found that the average lifetime of mRNA is about one-tenth the cell generation time, roughly 5 minutes in *E. coli*.

Rather than repeat the long experimental history of the search for messenger RNA, let us consider the culmination of this search, the formation of RNA-DNA duplexes, first isolated by S. Spiegelman. Since mRNA should have a base composition that reflects the base composition of DNA (postulate 3 above), any mRNA in a system carrying out protein synthesis should form complementary base pairs with single-stranded DNA to give a hybrid DNA-RNA double helix. Heating DNA above its "melting point" causes the double helix to unwind and slow cooling of the solution leads to renaturation of the DNA. If single-stranded RNA molecules that have complementary base sequences are present when the double helix unwinds, then renaturation will lead to the formation of DNA-RNA duplexes between RNA and single-stranded DNA (Figure 25.1).

When *E. coli* cells are infected with bacteriophage T4, the phage takes over the protein synthesis machinery of the cell. Since the phage DNA is translated into protein by the *E. coli*, any RNA present will be the transcription product of the phage DNA, that is, it will be messenger RNA. If ^{32}P-labeled phosphate is added a few minutes after the T4, newly synthesized RNA contains the isotopic label. This RNA is isolated and purified. Meanwhile, ^{3}H-labeled T4 DNA is prepared in a separate experiment. The tritiated phage DNA and the ^{32}P-labeled RNA are mixed, heated above the melting point of the DNA, and the solution is slowly cooled. The ^{32}P-labeled RNA is complementary to the phage DNA and forms a hybrid duplex (Figure 25.2). This mixture is then passed through a nitrocellulose filter, which retains the hydrid DNA-RNA duplex, but not single-stranded DNA or RNA (Figure 25.3). The ^{32}P-labeled RNA binds only to T4 DNA, not to the DNA of *E. coli* or other bacteria. This result can only mean that the base sequence of T4 RNA is complementary to the base sequence of T4 DNA, and that T4 DNA serves as a template for the synthesis of the RNA, which is none other than messenger RNA.

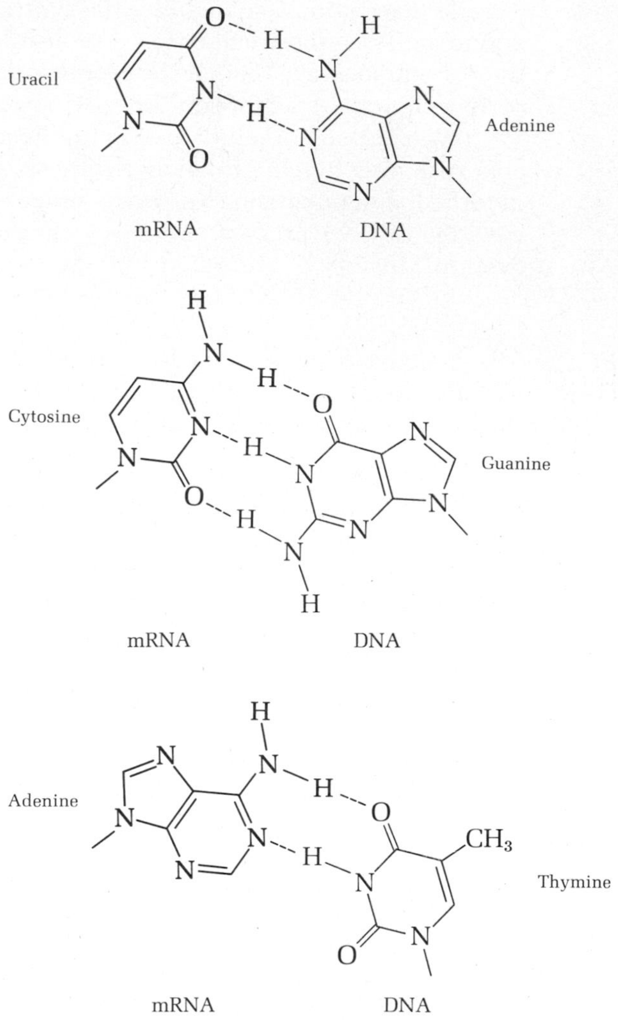

Figure 25.1 Base pairing of mRNA with DNA.

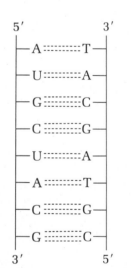

Figure 25.2 Schematic diagram of a DNA-mRNA duplex. In this duplex U of mRNA pairs with T of DNA, and G pairs with C. Such a duplex can form only between phage DNA and an mRNA molecule that is complementary to it. *E. coli* DNA does not form a hybrid duplex with phage mRNA.

DNA contains two complementary strands. Is only one of them is copied, or is an mRNA transcript is made of both? Since each gene produces only a single polypeptide chain, and since each strand of DNA would give two completely different mRNA molecules, each coding for a different protein, we expect that only one strand of DNA will be copied. The virus $\phi\chi174$ contains a single strand of DNA that invades its host cell. This strand, called the plus strand, is copied to give a double-stranded, circular DNA molecule. This duplex DNA is in turn transcribed to yield the phage mRNA molecules. If ^{32}P-labeled phosphate is added following infection, the newly synthesized mRNA contains ^{32}P. The radioactive mRNA is isolated and hybridized with previously prepared duplex DNA from $\phi\chi174$. The results of the hybridization experiment show that only the minus strand, the one copied from the plus strand that originally entered the cell, is the template for RNA synthesis. Similar results are observed for many other bacteriophages. In *E. coli* and phages λ and T4 the transcription process is more complex. Only one strand of the DNA is copied, but the template strand varies from one gene to the next.

25–3 RNA POLYMERASE

All RNA is synthesized by RNA polymerase, an enzyme that was discovered nearly simultaneously by S. Weiss and J. Hurwitz in 1960. The synthesis of mRNA is a DNA-directed process that can be divided into four steps (Figure 25.4). (1) The enzyme binds to a specific site on double-helical DNA. (2) Once bound to DNA, RNA polymerase catalyzes the coupling of a

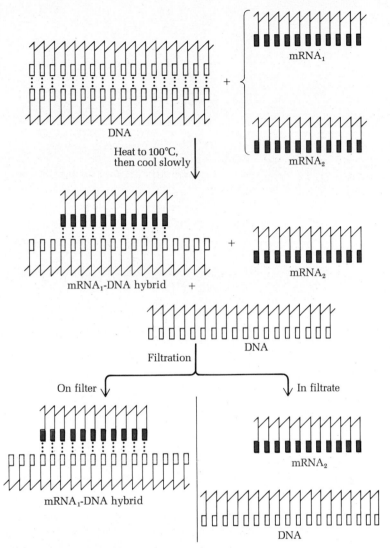

Figure 25.3 Formation of duplex hybrid between mRNA and DNA. If mRNA$_1$ from T4-infected *E. coli* and mRNA$_2$ produced by uninfected *E. coli* are added to a solution of T4 DNA, melting and annealing yields a hybrid duplex between the phage mRNA, and phage DNA, but no hybrid between mRNA$_2$ and phage DNA. The hybrid is readily isolated by filtration through nitrocellulose.

purine nucleoside triphosphate (either ATP or GTP) with a second ribonucleoside triphosphate. An example is shown in reaction 25.1 for ATP and UTP.

$$\text{ATP} + \text{UTP} \xrightarrow[\text{Mg(II)}]{\text{RNA polymerase}} \text{pppApU} + \text{PP}_i \qquad (25.1)$$

The dinucleotide remains bound to the RNA polymerase–DNA complex throughout the synthesis of RNA. (3) Chain elongation takes place. The nascent RNA chain grows exclusively in the 5′-3′ direction. The sequence of messenger RNA is dictated by base pairing with the DNA template: adenine in mRNA pairs with thymine in DNA, uracil in mRNA pairs with adenine in DNA, and cytosine in mRNA pairs with guanine in DNA. (4) Once the RNA has been synthesized, chain growth stops, and RNA polymerase is released. Let us consider each of these steps in detail, beginning with the structure of *E. coli* RNA polymerase.

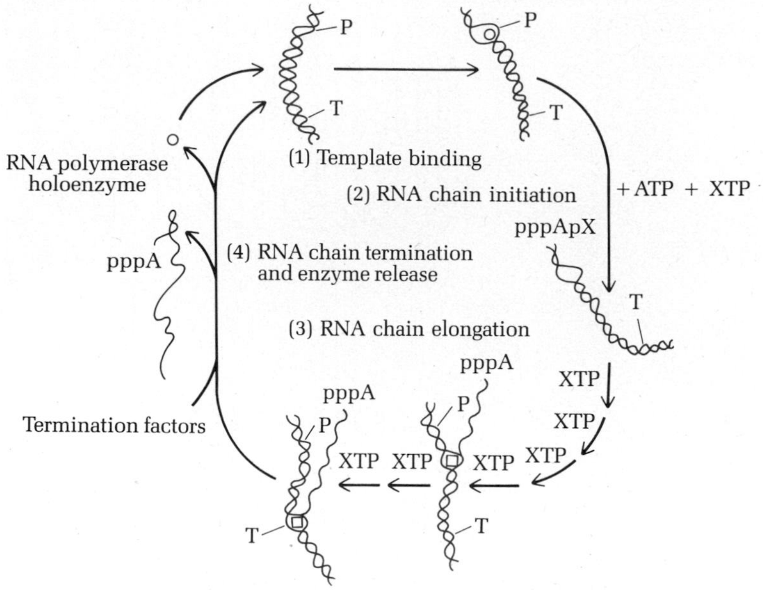

Figure 25.4 Schematic diagram of the steps involved in RNA transcription by a bacterial RNA polymerase. [Adapted from M. J. Chamberlin, in R. Losick and M. J. Chamberlin, Eds., "RNA Polymerase," Cold Spring Harbor Laboratory, Cold Spring Harbor, N.Y., 1976, p. 31.]

A. Structure of E. coli RNA Polymerase

The RNA polymerase of *E. coli* and other procaryotes is an oligomer composed (in almost all cases) of five types of subunits that are assembled in an $\alpha_2\beta\beta'\omega$ oligomer known as the *core enzyme*. A sixth subunit, designated σ, binds to the core enzyme to give the holoenzyme (Table 25.1). The functions of the subunits of RNA polymerase are for the most part unknown. The β' subunit is, however, known to possess the binding site for *rifamycin*, an inhibitor of protein synthesis, and to contain at least part of the catalytic site of the enzyme. The σ subunit, which readily dissociates from the core enzyme, is responsible for the binding of RNA polymerase holoenzyme to specific initiation sites on DNA. RNA polymerase has dimensions of about 6 nm by 15 nm, large enough to span three turns of the double helix.

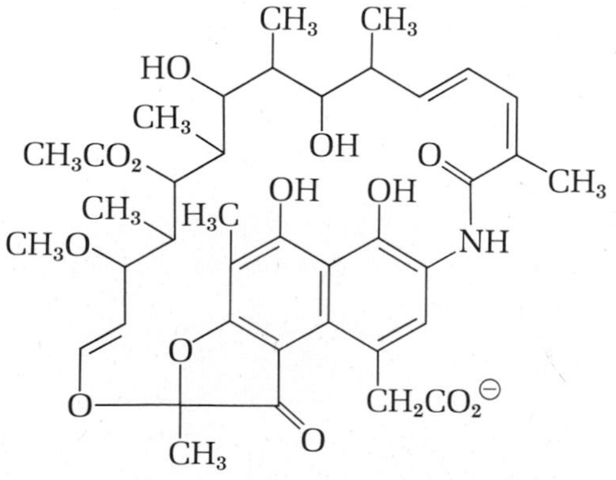

Rifamycin

B. Eucaryotic RNA Polymerases

Eucaryotic cells contain three RNA polymerases, designated I, II, and III, that synthesize different types of RNA. RNA polymerase I is located in the nucleolus and synthesizes 18S ribosomal RNA (rRNA), 5.8S rRNA, and 28S rRNA. RNA polymerase II catalyzes synthesis of mRNA and of the large precursors of eucaryotic mRNA known as *heterogeneous nuclear RNA* (HnRNA). RNA polymerase II is inhibited by the poisonous mushroom toxin α-amanitin, whose K_m for RNA polymerase is $10^{-8} M$. The elongation of RNA

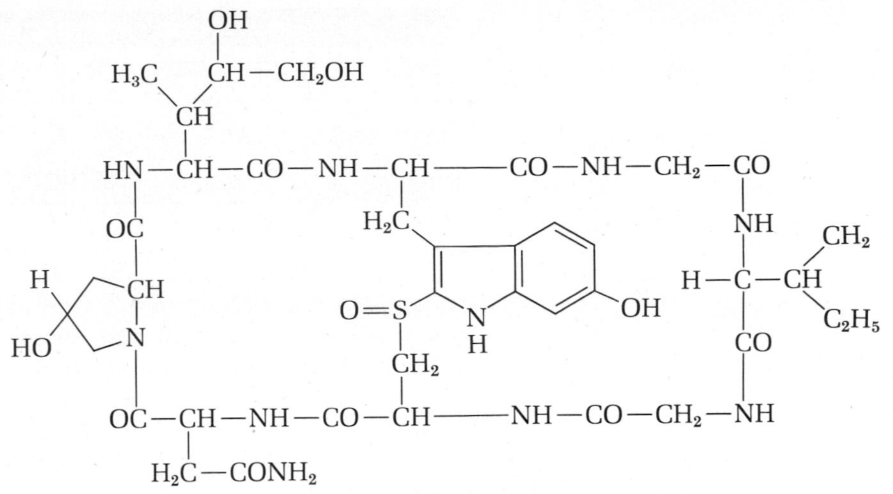

α-Amanitin

is blocked by α-amanitin. A hundredfold higher concentration of α-amanitin is required to inhibit RNA polymerase III, the enzyme responsible for synthesis of tRNA and 5S rRNA (Table 25.2). By using [³H]-α-amanitin to form a radioactive complex, it has been estimated that a given cell contains anywhere from 4,000 to 40,000 molecules of RNA polymerase II. Each of the eucaryotic RNA polymerases is a complex oligomeric protein, and little is known about the functions of the subunits.

C. Interaction of RNA Polymerase with the DNA Template and Synthesis of RNA

Duplex DNA contains many binding sites for RNA polymerase. The sites are apparently randomly distributed along the DNA, but some sites are stronger than others. The stronger binding sites at which RNA synthesis is initiated are known as *promoter sites*. The σ subunit of the holoenzyme is required for binding of RNA polymerase to the promoter sites. Table 25.3 shows the sequences of several promoter sites. Each of these sequences contains the seven-residue sequence

5′ T A T Pu A T G 3′
3′ A T A Py T A C 5′

RNA polymerase initially binds to duplex DNA, but shortly thereafter a local unwinding, or "melting," of the duplex occurs to provide access to the template DNA. This common seven-residue sequence may be the region that first comes apart during transcription since base pairing is weak in regions that are rich in A-T base pairs.

Once RNA polymerase binds to the promoter site, the dinucleotide pppPupNp forms and remains bound to the DNA–RNA polymerase complex throughout synthesis of RNA (recall reaction 25.1). The addition of all

Table 25.1	Subunit composition of *E. coli* RNA polymerase
Subunit	**Molecular weight**
α	40,000
β	150,000
β′	160,000
σ	90,000
ω	10,000

Table 25.2	Properties of eucaryotic RNA polymerases	
RNA polymerase	**Transcribed product**	**Location**
I	5.8S, 18S, 28S rRNA	Nucleolus
II	mRNA, HnRNA	Nucleoplasm
III	5S rRNA, tRNA	Nucleoplasm

other nucleotidyl residues occurs by displacement of pyrophosphate by the free 3'-hydroxyl group of the terminal nucleotidyl residue. These reactions, then, are examples of nucleotidyl group transfer (recall Section 12.6B). After RNA chain growth has been initiated, RNA polymerase remains bound until the entire RNA molecule has been transcribed.

D. Termination of Transcription

Transcription is terminated by specific DNA sequences. To further the parallel with initiation, a specific protein, known as *rho*, is usually a participant in termination, echoing the effect of σ in initiation.

Several termination sequences have been identified (Table 25.4). Two features of these sequences are striking. First, the terminal six to eight residues are very rich in U. In this respect the sequences for initiation and termination resemble each other. Second, immediately preceding the string of U's is a region that is rich in G-C base pairs. When RNA polymerase reaches one of these sequences it "pauses," and the newly synthesized RNA is released. RNA polymerase appears to unwind about one turn of the double helix prior to RNA synthesis, and about one turn of the double helix remains unwound as synthesis proceeds. The blockage of motion of RNA polymerase by the G-C rich region may be a consequence of the increased difficulty of unwinding DNA at this point. This region immediately precedes the string of U's, whose weak hydrogen bonding to the template may permit the newly synthesized RNA to dissociate from the DNA. The secondary structure of the transcript may also play a role in transcription termination. Each of the termination sequences listed in Table 25.3 can form a hairpin loop, and this loop may destabilize the interaction of RNA polymerase with the DNA template and with its transcript (Figure 25.5).

Many termination sites also require participation of the rho factor. *E. coli* cells contain about 20 molecules of this tetrameric protein (mol wt 200,000). Rho binds to RNA, DNA, and RNA polymerase. Rho acts catalytically by promoting the release of RNA from the ternary complex at a defined sequence on the DNA template. Rho also possesses an RNA-dependent ATPase activity. This mysterious property has given rise to a model of rho action in which the tetramer is propelled along the nascent RNA along with

Table 25.3	Promoter sequences for initiation of RNA transcription	
Source	**RNA polymerase binding site sequences**	
fd	T G C T T C T G A C T A T A A T A G A C A G G G T A A A G A C C T G A T T T T T G	
T7 A3	A A G T A A A C A C G G T A C G A T G T A C C A C A T G A A A C G A C A G T G A G T C	
T7 A2	A G T A A C A T G C A G T A A G A T A C A A A T C G C T A G G T A A C A C T A G C A G	
Lac-UV-5	G C T T C C G G C T C G T A T A A T G T G T G G A A T T G T G A G C G G A T A A C A A	
Lambda P_R	A C C T C T G G C G G T G A T A A T G G T T G C A T G T A C T A A G G A G G T T G	
SV40	T T T A T T G C A G C T T A T A A T G G T T A C A A A T A A A G C A A T A G C A T C	
Lambda P_L	A C C A C T G G C G G T G A T A C T G A G C A C A T C A G C A G G A C G C A C T G A C	
E. coli Tyr tRNA	C G T C A T T T G A T A T G A T G C G C C C C G C T T C C C G A T A A G G G A G C	
Lac wild-type	G C T T C C G G C T C G T A T G T T G T G T G G A A T T G T G A G C G G A T A A C A A	

The single colored base indicates the starting point of transcription. The colored seven residue sequences are rich in A-T base pairs where local unwinding of duplex DNA occurs.

Table 25.4 Nucleotide sequences in DNA and mRNA at sites of sequence termination

4S	mRNA	3'–OH G-G-C-G-U-U-U-U-U-A-(A) 5' 0–4 OH
	DNA	5' GGGCGTTTTTATTGGTGAGAATCGGCAGCAACTTGTCCGCGCCAATCG 3' 3' CCCGCAAAAATAACCACTCTTAGCGTCGTTGAACAGGCGCGGTTAGC 5'
6S	mRNA	3'OH C-G-G-G-A-U-U-U-U-U-A-(A) 5' 0–4 OH
	DNA	5' CGGGATTTTTATCTGCACAACAGGTAAG 3' 3' GCCCTAAAAATATAGACGTGTTGTCCATTC 5'
9S	mRNA	3'OH A-C-A-U-U-C-A-A-U-G-A-A OH 5'
	DNA	5' ACATTCAATCAATTGTTATCTAAGGAAAT 3' 3' TGTAAGTTAGTTAACAATAGATTCCTTTA 5'
trp leader	mRNA	3'OH C-A-G-C-C-C-G-C-C-U-A-A-U-G-A-C-G-G-G-C-U-U-U-U-U-U OH 5'
	DNA	5' CAGCCCGCCTAATGAGCGGGCTTTTTTTGAACAAAATTAGAGA 3' 3' GTCGGGCGGATTACTCGCCCGAAAAAACTTGTTTTAATCTCT 5'

Note that in each case a G-C rich region immediately precedes a sequence of U's.

SOURCE: From S. Adhya and M. Gottesman, *Ann. Rev. Biochem.*, *47*, 974 (1978). Reproduced with permission. © 1978 by Annual Reviews Inc.

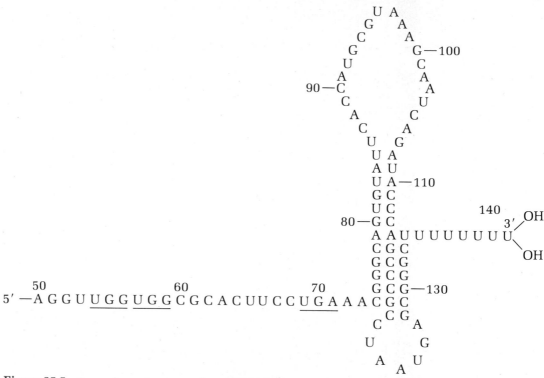

Figure 25.5   Secondary structure of an mRNA transcript that is part of the sequence of genes coding for biosynthesis of tryptophan. [Adapted from S. Adahya and M. Gottesman, *Ann. Rev. Biochem., 47*, 977 (1978). Reproduced with permission. © 1978 by Annual Reviews Inc.]

the polymerase. When RNA polymerase reaches a pause signal, rho binds RNA polymerase and RNA is released.

E. Inhibition of RNA Polymerase by Actinomycin D

RNA polymerase is inhibited by the antibiotic *actinomycin D*. This molecule consists of two identical cyclic peptides that are bound to a phenoxazine ring (Figure 25.6). The cyclic polypeptide contains a D-valine residue, an L-proline residue, an L-threonine residue, L-methylvaline, and sarcosine (N-methylglycine). Actinomycin D is one of the most potent antitumor agents known, but it is so toxic that it is clinically useless. A three-dimensional computer-generated model of actinomycin D is shown in Figure 25.7. Rotation of the molecule by 180° around the vertical axis (turn the page upside down) leads to an equivalent structure. Actinomycin D thus has a *twofold symmetry axis*. Actinomycin D intercalates between the stacked bases in DNA, just as ethidium bromide does. In contrast to ethidium, however, actinomycin binds to the base sequence dG-dC. If we read this sequence along one strand of DNA in the 5'-3' direction, then the sequence of the com-

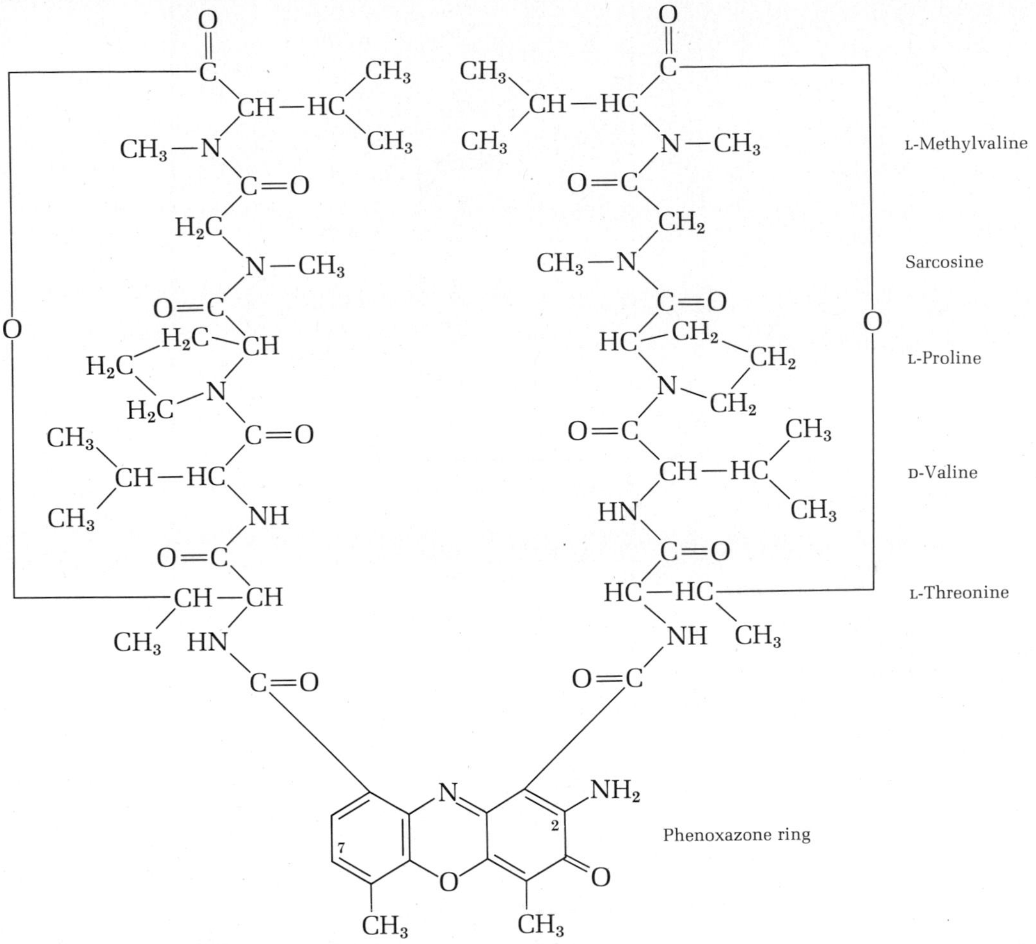

Figure 25.6 Structure of actinomycin D.

L-Methylvaline

Sarcosine

L-Proline

D-Valine

L-Threonine

Phenoxazone ring

plementary dinucleotide (in the 3′-5′ direction) is also dG-dC. Therefore, this base pair also possesses a twofold symmetry axis. Actinomycin D has a perfect stereochemical fit to dG-dC sequences in DNA. The binding of actinomycin D to the model compound deoxyguanosine and to DNA is shown in Figures 25.8 and 25.9. When actinomycin D is intercalated in

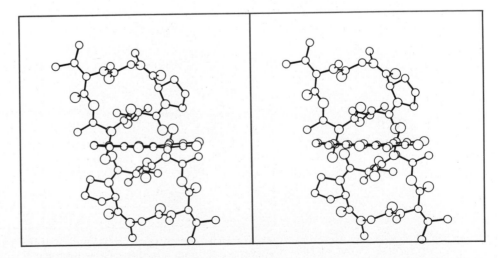

Figure 25.7 Computer-drawn projection of the three-dimensional structure of actinomycin D. If the page is turned upside down, the structure will appear to be the same, that is actinomycin D has a twofold symmetry axis.

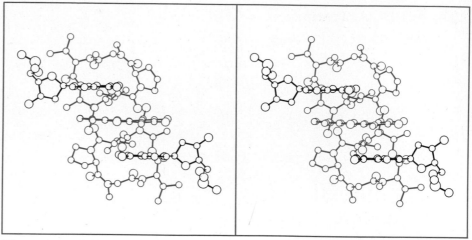

Figure 25.8 Structure of a 1:2 complex of actinomycin with deoxyguanosine.

DNA, RNA polymerase is unable to bind to the template DNA and is therefore inhibited.

25–4 RNA PROCESSING

Transcription involves copying of the genetic message contained in DNA, but the transcribed RNA molecules are often further modified after their initial synthesis, a phenomenon known as *processing*. RNA processing encompasses three general types of enzymatic reactions. (1) Nucleases trim the initial RNA transcripts to RNA molecules of shorter length. (2) Although trimming of newly synthesized RNA molecules is generally the rule, RNA processing can also involve the addition of new nucleotidyl residues. For example, eucaryotic mRNA molecules often contain polyadenylate at the 3′ terminus, and the 3′ terminus of tRNA bears the trinucleotide CCA that is sometimes added after transcription. (3) Covalent modification of the bases of RNA, especially tRNA, is common. More than 50 modified bases of tRNA have been isolated. In this section we shall discuss processing of rRNA and tRNA, leaving the fascinating discoveries of mRNA processing in eucaryotes until Section 25.5.

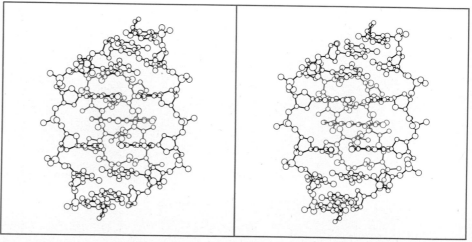

Figure 25.9 Structure of the actinomycin-DNA complex. The phenoxazine ring system of actinomycin intercalates between the base-paired dinucleotide sequence GC, while the peptide subunits of actinomycin lie in the narrow groove of the DNA helix and interact with the deoxyguanosine residues by hydrogen bonding.

**A. Processing
of Transfer RNA**

Let us briefly consider some of the properties of tRNA (discussed further in Section 9.6 and in Section 26.4). Transfer RNA serves as a carrier of activated amino acid residues. An *aminoacyl-tRNA* molecule transfers its aminoacyl residue to a growing peptide chain in protein synthesis. The "cloverleaf structure" and three-dimensional structure of phenylalanyl-tRNA, are shown in Figures 25.10 and 25.11. The triplet CCA is found at the 3′ termini of all functional transfer RNA molecules. The transfer RNA molecules of bacteria are synthesized as precursors that are trimmed to produce a product that is then covalently modified. The precursor RNA contains a "leader" at the 5′ end and a "trailer" at the 3′ end. At least two nucleases are known in *E. coli* which operate on the precursor to tRNA. First, *ribonuclease P* (RNase P) specifically cleaves the 5′ leader; then, *ribonuclease Q* (RNase Q), sometimes also known as ribonuclease PIII (RNase PIII), cleaves the 3′ trailer sequence (Figure 25.12).

H. Khorana and his coworkers have synthesized the 129-residue precursor of tyrosyl-tRNA (Figure 25.13). This precursor contains 44 more nucleotidyl residues than the final product. Ribonuclease P cleaves a 41-residue sequence from the 5′ end, and ribonuclease Q hydrolyzes three residues, AMP, CMP, and UMP, in that order, from the 3′ terminus. All tRNA molecules contain the 3′-terminal sequence CCA. In some tRNA molecules this triplet is added after transcription by a specific nucleotidyl transferase.

Once the leader and trailer have been excised, further processing consists of modifying the bases of the tRNA. Some modifications occur before the precursor RNA molecules have been trimmed, and others occur after the

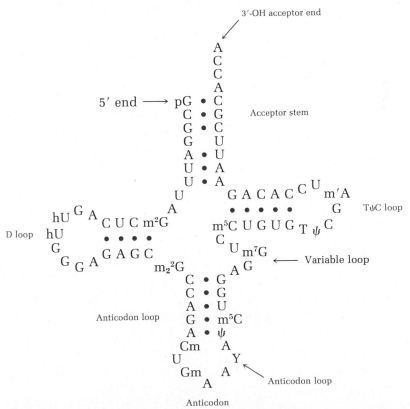

Figure 25.10 Cloverleaf structure of phenylalanyl tRNA. A similar cloverleaf structure can be written for all tRNA molecules. Abbreviations: •, hydrogen-bonded base pair; m$_2^2$G, ^{2}N-dimethylguanosine; hU, 5,6-dihydrouridine; m^{2}G, 2-methylguanosine; ψ, pseudouridine; m^{5}C, 5-methylcytidine; m^{1}A, 1-methyladenosine; m^{7}G, 7-methylguanosine. [From G. J. Quigley and A. Rich, *Science*, 194, 796-806 (1976). Copyright 1976 by the American Association for the Advancement of Science.]

initial transcript has undergone some processing. The most common modifications in both procaryotes and eucaryotes are methylations, producing, for example, ribothymidine, dimethyladenine, and many other methylated derivatives. The methyl donor in these reactions is S-adenosylmethionine (recall Section 20.3B). Replacement of the oxygen at C-4 in uridine with sulfur gives 4-*thiouridine*, s⁴U, and reduction of the 5,6 double bond gives dihydrouridine, H_2U. Perhaps the most remarkable of all these transformations is the conversion of uridine to *pseudouridine* (ψ) which requires cleavage of the glycosidic bond and formation of a carbon-carbon bond between C-1 of the ribosyl moiety and C-6 of uracil. At least two enzymes are required for the conversion of uridine to pseudouridine, but the details are not very well understood. The structures of some of the more than 50 modified bases that have been found in various tRNA molecules are shown in Figure 25.14.

tRNA processing also takes place in eucaryotic cells. One novel feature of eucaryotic tRNA molecules is the presence of *2'-O-methylribosyl* moieties in about 1 percent of the nucleotidyl residues. Many modified bases are also found in eucaryotic tRNA molecules. Most tRNA methylases have been found in the cytosol, whereas the trimming of large precursors of eucaryotic tRNA molecules occurs in the nucleus.

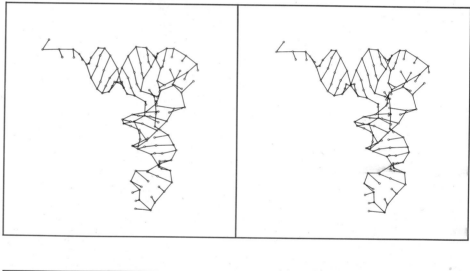

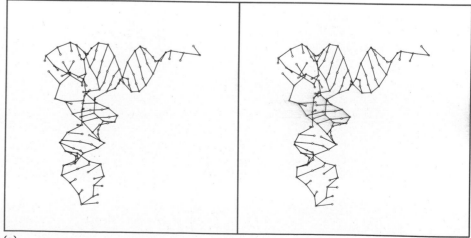

(a)

Figure 25.11 Two views of the folded backbone of yeast phenylalanyl tRNA. The sugar-phosphate backbone of the molecule is represented as a coiled tube, and the cross-rungs are the nucleotide base pairs. Short rungs indicate bases not involved in base pairing. (b) Space-filling model of the same molecule (page 1037).

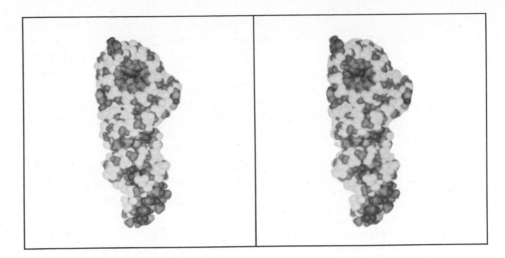

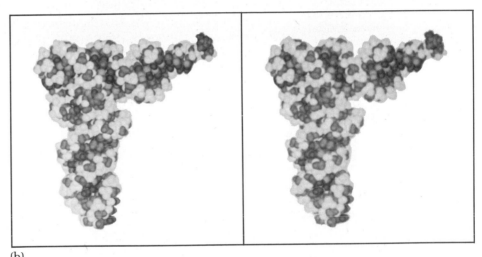

(b)

Figure 25.11 (*continued*)

HO — O — Base
HO — O — CH₃

2′-*O*-Methylribosyl moiety

B. Processing of Ribosomal RNA

Ribosomal RNA (rRNA) is an integral component of the subcellular particles called *ribosomes* which are the site of protein synthesis. Ribosomes are complexes of rRNA and proteins. The *E. coli* ribosome has a sedimentation coefficient of 70 S. The 70S ribosome, in turn, consists of two subunits whose sedimentation coefficients are 50S and 30S. The 50S subunit contains 32 different proteins and one 5S rRNA molecule (mol wt 4×10^4), and one 23S rRNA molecule (mol wt 1.2×10^6). The 30S subunit contains 21 different proteins and one 16S rRNA molecule (mol wt 6×10^5) (Figure 25.15).

The larger ribosomes of eucaryotic organisms have sedimentation coefficients of about 80S. The two subunits of eucaryotic ribosomes have sedimentation coefficients of 40S and 60S. The sedimentation coefficients of eucaryotic rRNA molecules are also larger than those of procaryotic RNA molecules. We shall discuss the structure and function of ribosomes in greater detail in Section 26.3.

The *E. coli* chromosome contains three regions that code for synthesis

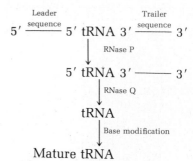

Figure 25.12 Steps in the processing of tRNA. RNase P cleaves the 5′ leader sequence of tRNA, and RNase Q cleaves the 3′ trailer sequence. Base modification is shown as the last step, but some base modification occurs before the leader and trailer have been excised.

of rRNA. Each of these genes is bracketed by a promoter sequence and a terminator sequence. The entire transcript contains the sequences of 16S rRNA, 23S rRNA, and 5S rRNA. 5S rRNA contains 120 nucleotidyl residues, 16S rRNA contains 1,520 nucleotidyl residues, and 23S rRNA contains in excess of 3,000 residues. This enormous transcript does not normally appear in *E. coli*. Once the sequence of 16S rRNA has been synthesized, ribonuclease III cleaves it to give a precursor of 16S rRNA that is denoted p16S rRNA. (In abbreviations of nucleic acid structure "p" represents a phosphoryl group. In the abbreviation p16S "p" stands for precursor.) Following cleavage of the p16S rRNA, RNA polymerase continues to transcribe the DNA template. At least 30 molecules of RNA polymerase are transcribing the DNA at any time. At peak efficiency about 60 molecules of 16S rRNA are synthesized per minute. If we were to follow the progress of a single RNA polymerase down the DNA template, we would find that it continues transcription by copying the DNA template for 23S rRNA. Once the sequence of 23S rRNA has been synthesized, ribonuclease III cleaves it to give p23S rRNA. Finally, the region coding for 5S rRNA is transcribed. It, too, undergoes preliminary trimming to give p5S rRNA. The precursor rRNA molecules bind almost immediately to ribosomal proteins, and a final trimming generates the "mature" rRNA molecules (Figure 25.16). Methylations of bases in 23S rRNA occur at an early stage of processing, and methylations of 16S rRNA occur later.

Eucaryotic organisms contain hundreds to thousands of copies of genes coding for ribosomal RNA, whose synthesis and processing occur in the nu-

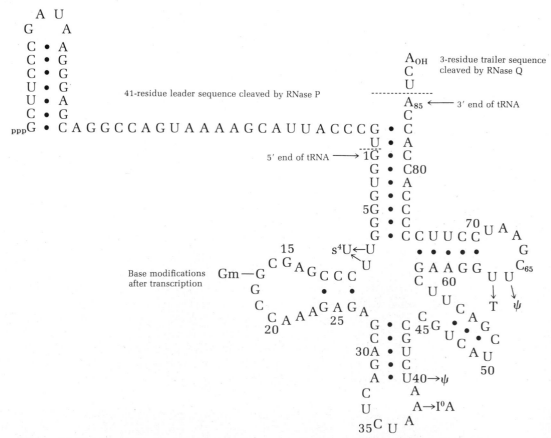

Figure 25.13 Steps in the processing of the 129-residue precursor of tyrosine tRNA.

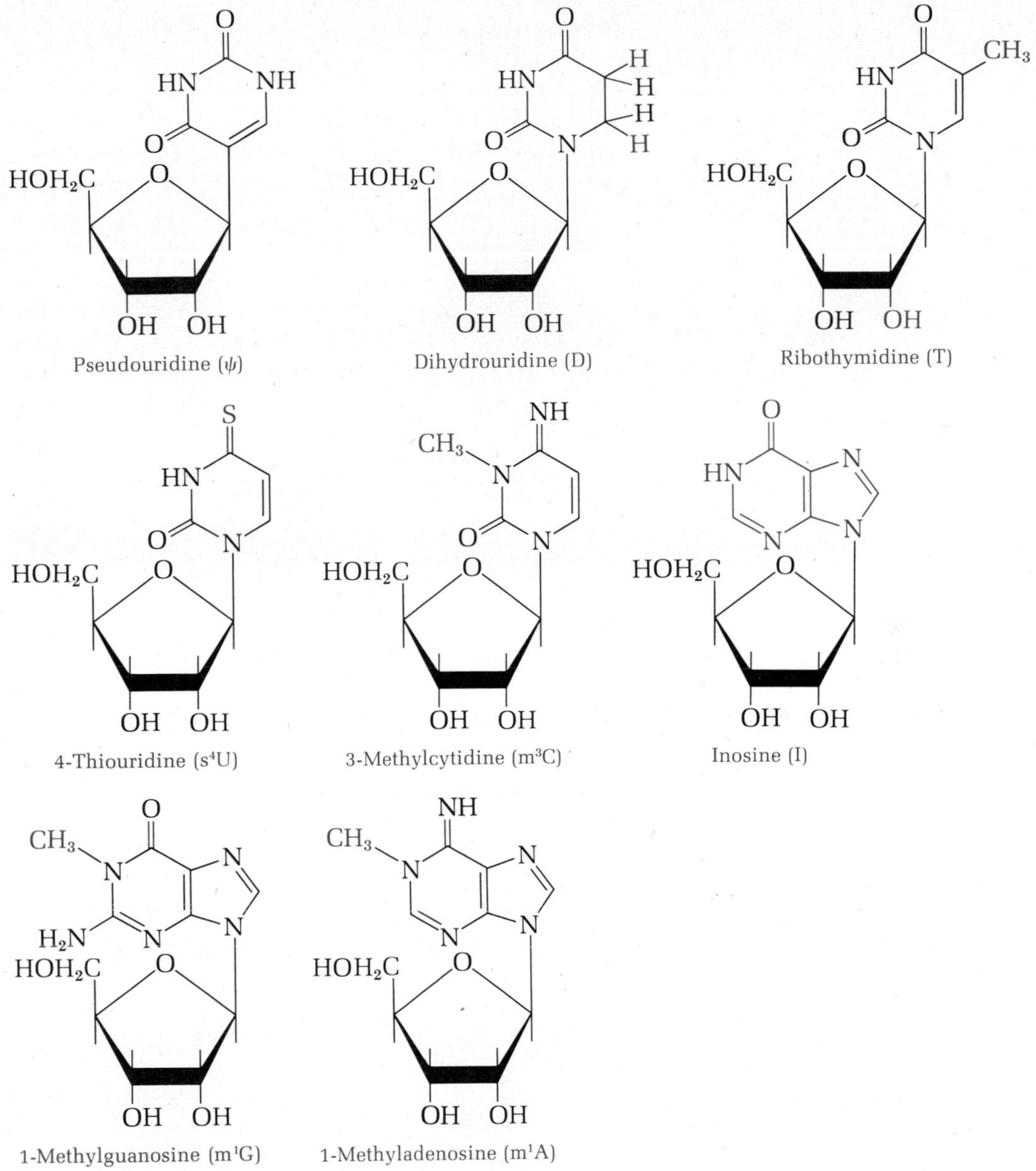

Pseudouridine (ψ)   Dihydrouridine (D)   Ribothymidine (T)

4-Thiouridine (s⁴U)   3-Methylcytidine (m³C)   Inosine (I)

1-Methylguanosine (m¹G)   1-Methyladenosine (m¹A)

Figure 25.14 Structures of some of the modified nucleosides that have been found in transfer RNA molecules.

cleolus. Each gene cluster contains three elements: a sequence of DNA that codes for the mature rRNA; a spacer sequence that is transcribed, but trimmed shortly thereafter; and a DNA sequence that is not transcribed at all. Eucaryotic ribosomal RNA molecules have sedimentation coefficients of 28S and 18S. The entire transcriptional unit has a sedimentation coefficient of 45S (mol wt 4.5×10^6) that consists of the sequence: spacer–18S rRNA–28S rRNA. This sequence spans some 14,000 residues, of which fully half are ex-

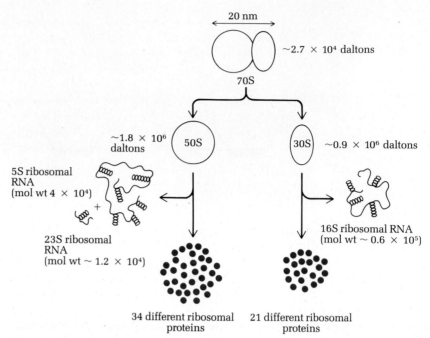

Figure 25.15 Schematic diagram of the structure of the *E. coli* ribosome.

cised from the mature rRNA molecules. The 60S subunit of eucaryotic ribosomes contains 28S rRNA, a molecule that contains about 5,000 nucleotides, and the 40S ribosomal subunit contains 18S rRNA, a molecule that contains 2,000 nucleotidyl residues.

Processing of the 45S rRNA precursor is a complex and imperfectly understood process. Trimming by a variety of endonucleases begins at the 5' end of the molecule and works its way from the spacer on the 5' side of the 18S rRNA to the 3' end of the transcript. Many intermediates are generated

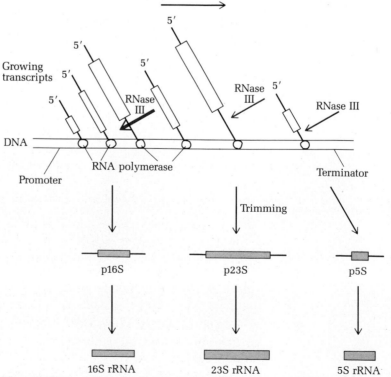

Figure 25.16 Steps in the processing of an *E. coli* rRNA molecule.

by processing, including RNA precursors whose sedimentation coefficients are 41S, 36S, 24S, and 20S. Eucaryotic RNA molecules contain 2′-O-methyl-ribosyl moieties. These methyl groups are added to the 45S RNA prior to trimming. Modification of bases, mostly by methylation, also occurs prior to trimming.

The transcriptional unit for procaryotic rRNA contains the sequence for 5S rRNA. In eucaryotic cells, however, the 5S sequence is located elsewhere in the chromosome and is transcribed separately. Since 5S rRNA is found in eucaryotic ribosomes, its synthesis ought to be coordinated to the synthesis of 18S and 28S rRNA. Little is known about either the transcription or processing of eucaryotic 5S rRNA.

25–5 PROCESSING OF MESSENGER RNA IN EUCARYOTIC ORGANISMS

A. Addition of "Caps" and "Tails" to mRNA

The initial transcript of mRNA in eucaryotic organisms undergoes so many processing reactions that the mRNA that leaves the nucleus bears little resemblance to the initial transcription product. All known eucaryotic mRNA molecules are "capped" at their 5′ termini by a *7-methylguanylate* residue that is bound to the terminal nucleotide of the original mRNA through a 5′-phosphoryl group rather than through the usual 3′ position. The 5′ termini of RNA molecules contain triphosphate residues since a nucleotide triphosphate is the primer for RNA polymerase (recall reaction 25.1). Hydrol-

7-Methylguanylate

ysis of the terminal phosphate gives a 5′-nucleotidyl diphosphate residue. The guanylate residue is added by nucleotidyl group transfer from GTP to the 5′-nucleotide diphosphate residue of the mRNA (Figure 25.17). The 7-methyl group is then added with S-adenosylmethionine as the methyl donor, and the product is designated cap O (Figure 25.18). The last two residues at the 3′ terminus of the capped mRNA are often methylated in their 2′-hydroxy positions. Addition of one more methyl group gives a product known as *cap 1*, and addition of another methyl group gives *cap 2* (Figure 25.19). A cap stabilizes mRNA by protecting it from exonucleases and phosphatases.

Many eucaryotic mRNA molecules contain *polyadenylate "tails"* at their 3′ termini as well as 5′ caps. These tails contain from 150 to 200 adenylate residues and are added to the mRNA before it leaves the nucleus. A 3′-terminal sequence of GC and an internal sequence AAUAA some 20 residues away apparently act as signals that lead to addition of a polyadenylate (poly A) tail. The poly A tail acts as a buffer that protects mRNA from nucleases and phosphatases. The tail is slowly hydrolyzed by the action of these enzymes in the cytosol. A few mRNA molecules do not have polyadenylate tails. Histone mRNA molecules are the most conspicuous, but not the only exception. Although several theories have been advanced, the function of the polyadenylate tail on mRNA remains obscure.

The major species of RNA in the nucleus consists of very large molecules of varying sizes (1,500 to 30,000 nucleotidyl residues) known as *heter-*

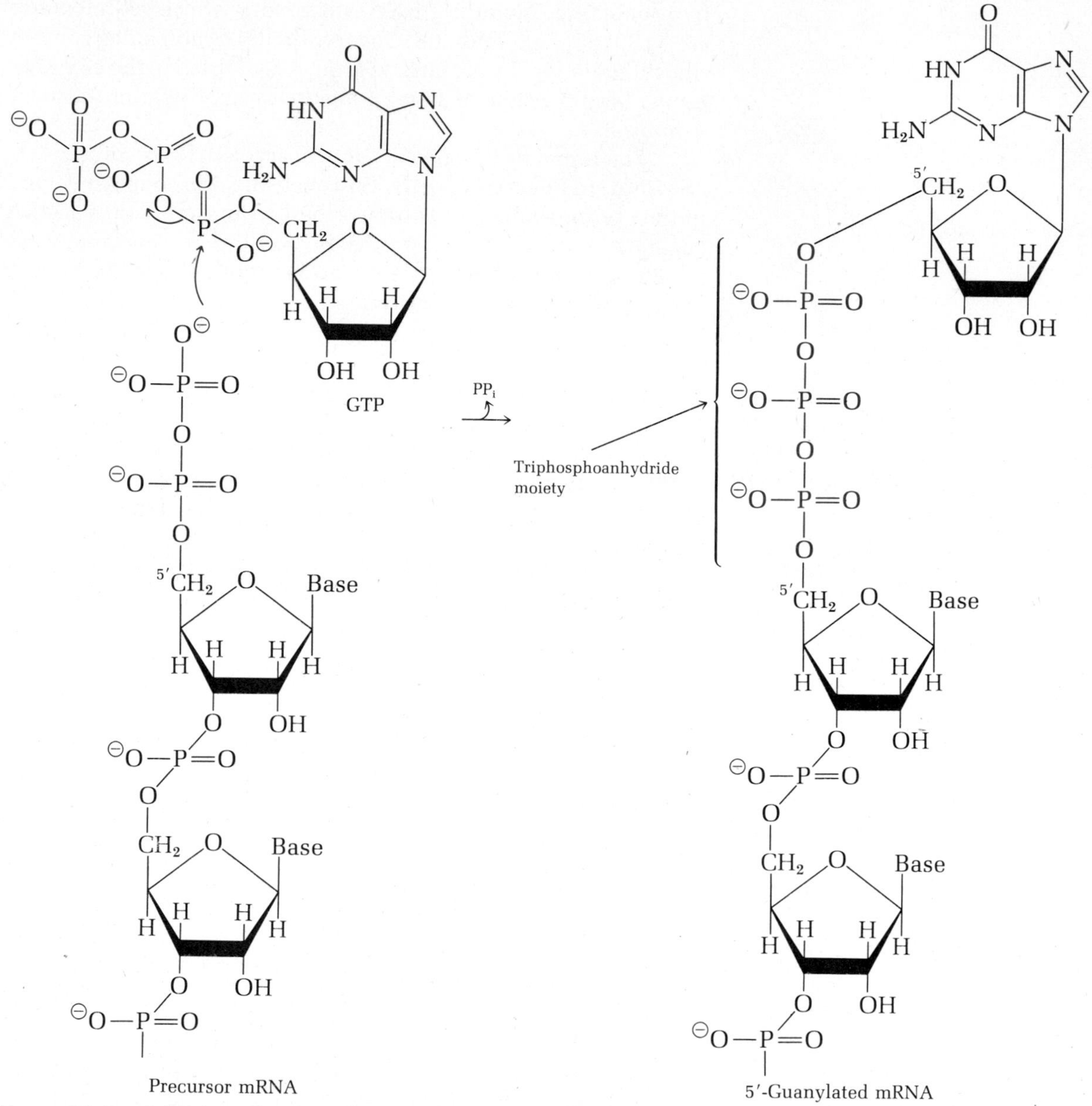

Figure 25.17 Addition of a guanylate residue to the 5′ terminus of precursor mRNA by nucleotidyl transfer. The guanylate residue is joined to the 5′ end of the mRNA by an unusual 5′-5′ triphosphoanhydride bridge.

ogeneous nuclear RNA (HnRNA). Most of HnRNA is short-lived, and virtually none of it leaves the nucleus. For a long time it has been thought that HnRNA is the precursor of mRNA, and in the next section we shall see that this hypothesis is probably correct.

B. Split Genes and RNA Splicing

Nil admirari.[1]

Horace

The genes of procaryotic organisms are continuous linear sequences of DNA. Until recently there was no reason to suppose that the situation is any different in eucaryotic organisms, despite the enormous complexity of chromatin, but in 1977 it was found that some eucaryotic genes are *discontin-*

---

[1] Be astonished at nothing.

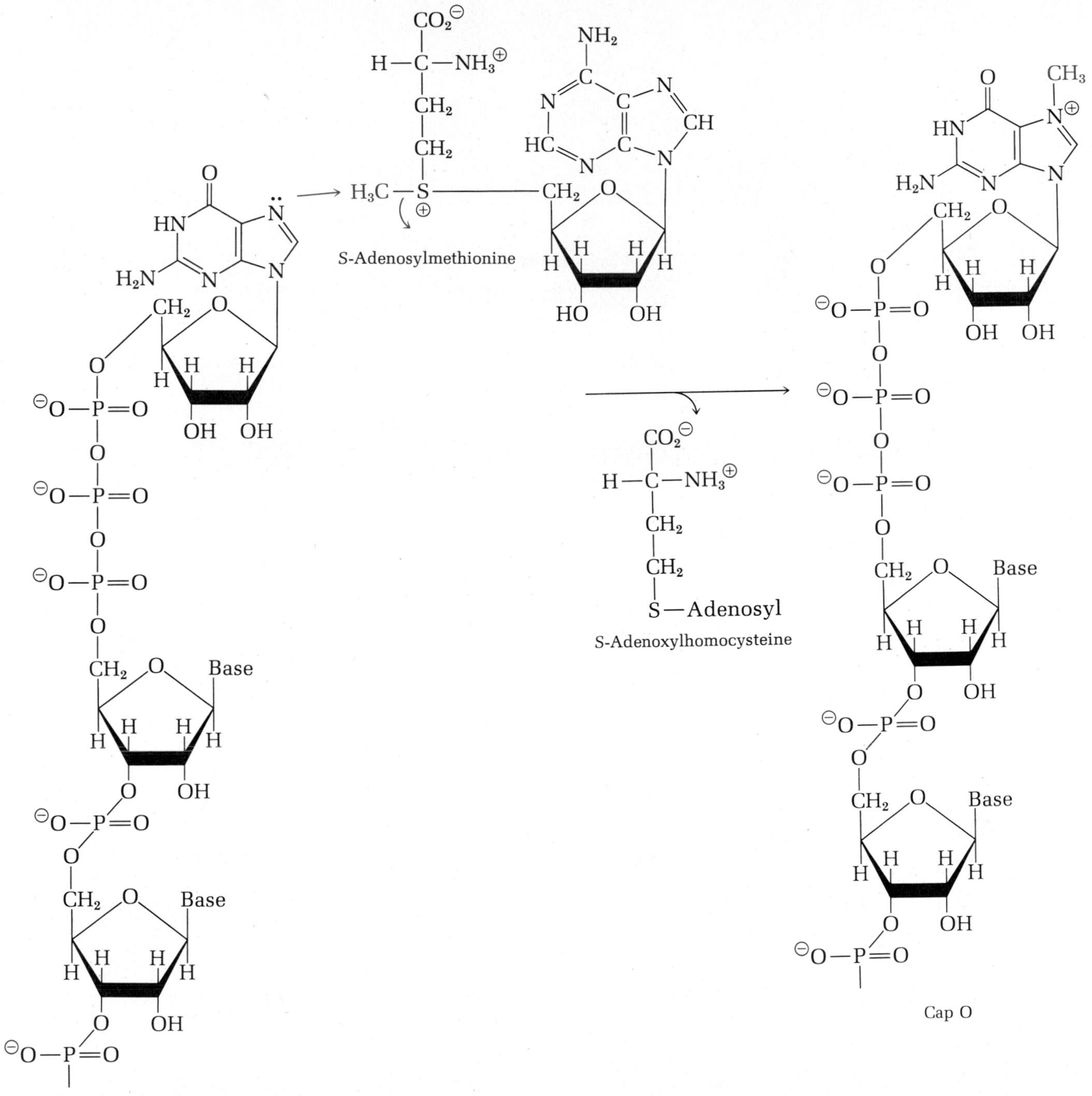

Figure 25.18 Addition of the 7-methyl group to the guanylate residue by methyl group transfer from S-adenosyl-methionine, giving the so-called cap 0, in which the next two residues are not methylated.

uous. The DNA sequence that is finally expressed as a protein is split into pieces that must be spliced to give the mRNA that is translated in the cytosol. It seems likely that the large precursors of mRNA in the first transcripts of the discontinuous genes constitute the pool of heterogeneous nuclear RNA. The discovery of split genes is one of the most important discoveries of recent years, and its ramifications are still far from understood.

The first split gene to be discovered codes for the β subunit of hemoglobin in mice. Techniques for incorporating eucaryotic DNA in microorganisms, by the methods for creating synthetic recombinant DNA that we dis-

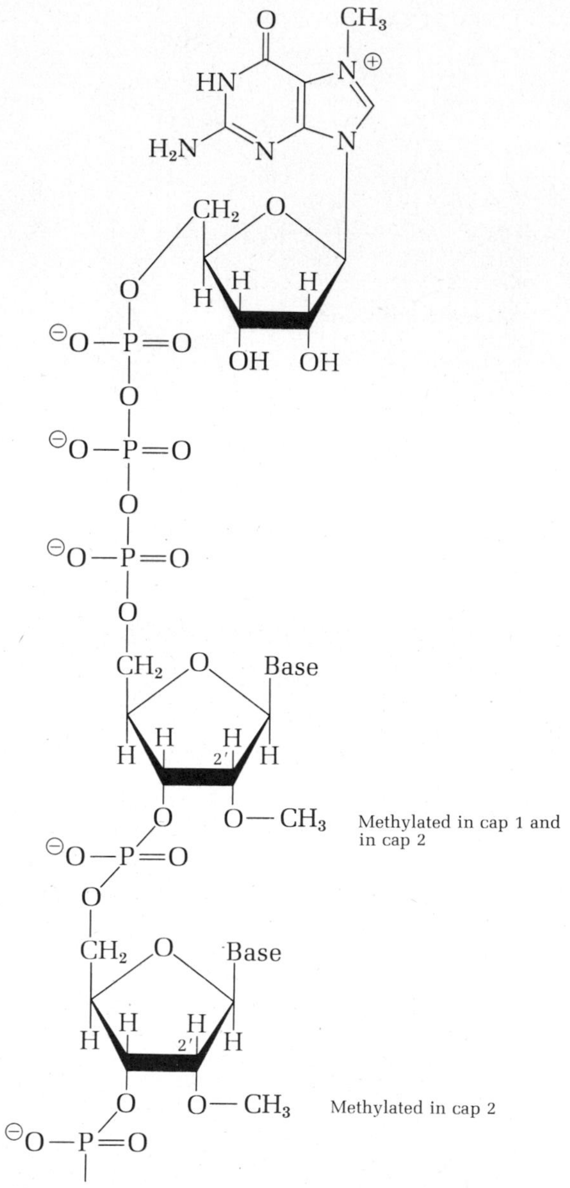

Figure 25.19 Structures of cap 1 and cap 2. In cap 1 the second residue is methylated in the 2'-O position. In cap 2, the second and third residues are methylated in the 2'-O position. The methyl groups are derived from S-adenosylmethionine.

cussed in Section 24.9, were used to obtain the cloned β hemoglobin DNA. This DNA was then partially denatured and hybridized with the globin mRNA, and the results were visualized by electron microscopy. If the sequences of the mRNA and the DNA are complementary and colinear, a loop would appear in the electron micrograph. This loop contains the DNA-RNA hybrid and a displaced piece of single-stranded DNA (Figure 25.20). The structure found in the electron micrograph, however, had *two* loops (Figure 25.21). Each loop corresponds to a hybrid between the DNA and mRNA. The two loops are separated by an *intervening sequence,* called an *intron,* that is not present in the mRNA. The intron contains 550 base pairs. A smaller intervening sequence containing 120 base pairs is found near the 5' end of the DNA encoding for the β globin. The gene coding for the β chain of mouse hemoglobin is thus split in three pieces (Figure 25.22).

The processing of the mRNA precursor passes through several stages.

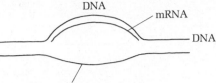

Figure 25.20 DNA-RNA hybrid and displaced single-stranded DNA. Under partially denaturing conditions, when the DNA double helix is partly un-wound, messenger RNA displaces one strand of the duplex to give a hybrid mRNA-DNA molecule and a displaced strand on DNA. If the mRNA and the DNA from which it is copied are co-linear, a single loop of displaced DNA is observed in electron micrographs.

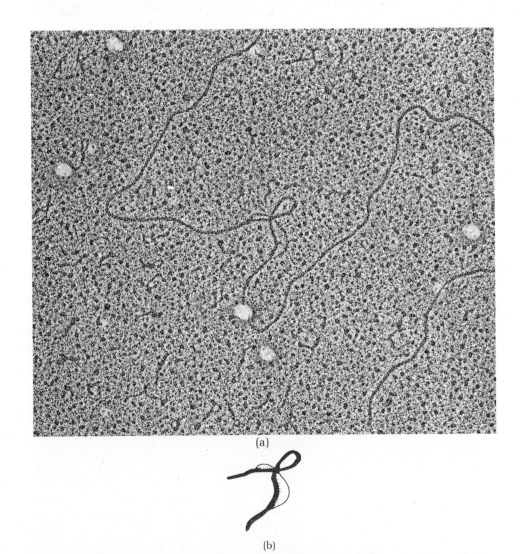

(a)

(b)

Figure 25.21 (a) Electron micrograph of the hybrid formed between mouse DNA and mature β globin mRNA. Arrows indicate the regions where the DNA-mRNA hybrid formed. (b) Schematic diagram of the electron micrograph shown in (a). The heavy dotted lines indicate the DNA-mRNA hybrid, the thin lines indicate single-stranded DNA, and the heavy loop indicates intervening sequences. [Courtesy of Dr. P. Leder.]

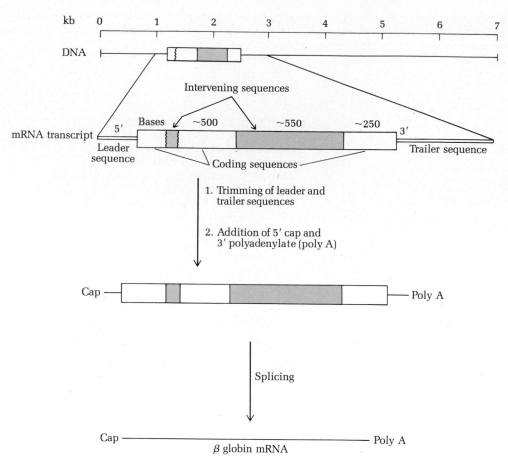

Figure 25.22 Schematic diagram of the structure and processing of the mRNA transcript of the split gene for mouse β globin.

After trimming of leader and trailer sequences, a 7-methylguanylate residue is added to the 5′ end of the transcript, and a polyadenylate tail is added to the 3′ end of the transcript. The covalently modified transcript is then spliced to give the final mRNA. The splicing reactions must bring together widely separated nucleotidyl residues. In some cases the intervening sequence is longer than the final mRNA product. This "gene splicing" must be extremely accurate, since a shift of a single base during the splicing process would alter the entire "reading frame" of the final mRNA and would result in production of a different, inactive protein. Such a result would be as deleterious as a mutation in the gene itself. The intervening sequences are split by the action of an endonuclease and tied together by a ligase. These two en-

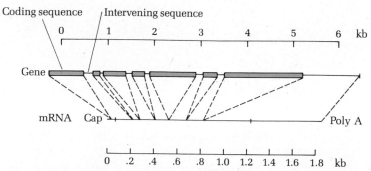

Figure 25.23 Split gene of chick ovalbumin containing seven intervening sequences that are spliced to produce the mature ovalbumin mRNA. [Adapted from J. Abelson, *Ann. Rev. Biochem.*, 48, 1049 (1979). Reproduced with permission. ©1979 by Annual Reviews Inc.]

Table 25.5 Sequences surrounding splicing points in split genes

	Intervening sequences	
	5′	3′
Ovalbumin	A U.A A A.U A A G|G U G A G C C U A	C A A U U A C A G|G U.U G U.U C G.C
	G A.A G C.U C A.G|G U A C A G A A A	U G U A U U C A G|U G.U G G.C A C.A
	A U.C C U.G C C.A|G U A A G U U G C	G C U U U A C A G|G A A.U U A U.C A
	A.G A C.A A A U G|G U A A G G U A G	U U C U U A A A G|G A A.U U A U.C A
	G U.G A C U G A G|G U A U A U G G G	G U U C U C C A G|C A A.G A A A.G C
	C U.U G A G C A G|G U A U G G C C U	U C C U U G C A G|C U U G A.G A G.U
Ovomucoid	U C C U C C C A G|G U G A G U A A C	U U C C C C A G|A U G C U G C C U
	G G G.G C U G A G|G U G A G A A A G	U U U G U C G A G|G U G G A.C U G C
	C.U A C A G C.A U|G U G U G U A C U	C C U C U U C A G|A G A A.U U U G G
	A C U.G U U C C U|G U A A G U G A A	C U U C A C A G|A U G A A C U G C
	C C A C A A A G U|G U U A U U G U A	U C C U U U C A G|A G A G.C A G G G
	G C U.G U G A G U|G U G A G U A G C	C U U U U G C A G|G U U G A.C U G C
	U.G C A G U C G U|G U A C G U A C A	C G C U U U C A G|G.G A A A G C A A
Human β globins	C.C U G G G C A G|G U U G G U A U C	C A C C C U U A G|G.C U G C U G G U
Rabbit β globins	C.C U G G G C A G|G U U G G U A U C	U U U U C U C A G|G.C U G C U G G U
Mouse β globins	C.C U G G G C A G|G U U G G U A U C	C U U U U U U A G|G.C U G C U G G U
Human β globins	A A C.U U C A G G|G U G A G U C U A	C C U C C A C A G|C U C.C U G G G C
Rabbit β globins	A A C.U U C A G G|G U G A G U U U G	U U C C U A C A G|C U C.C U G G G C
Mouse β globins	A A C.U U C A G G|G U G A G U C U G	U U C C C A C A G|C U C.C U G G G C
Rat insulin	A C.C C A C A A G|G U A A G C U C U	C C C U G G C A G|U G G C A.C A A C
Immunoglobulins		
λI L-VI	U.C A G C U C A G|G U C A G C A G C	U G U U U G C A G|G G G C C.A U U U
λI J1-C1	C U G.U C C U A G|G U G A G U C A C	C A U C C U G C G|G C C.A G C C C A
λII L-VIII	U.C U G C U C A G|G U C A G C A G C	U G U U U G C A G|G A G C C.A G U U
κ J1-1	A G C.U G A A A C|G U A A G U A C A	C U U C C U C A G|G G G C U G.A U G
κ J2-2	A A A U A A A A C|G U A A G U A G A	
κ J3-3	A A A U A A A A C|C U A A G U A C A	
κ J4-4	A A A U A A A A C|G U A A G U C U U	
κ J5-5	A A A U C A A A C|G U A A G U A G A	
μ J H1-CH1	U C.U C C U C A G|G U A A G C U G G	G U C C U C A G|A G.A G U C.A G
μ CH1-CH2	C C.A U U C C A G|G U A A G A A C C	U C A U U C C A G|C U G.U C.G C A
μ CH2-CH3	G U.G C U G C C.A|G U G A G U G G C	U G A C U G C A G|G U.C C C U C C
μ CH3-CH4	A A C.C C A A U G|G U A G G U A U C	C A U U U A C A G|A G.G U.G C A C
μ CH4-CM1	A G.U C C A C U G|G U A A A C C C A	C C U U C A U A G|A G G.G G.G A G G
μ CM1-CM2	C U G.U U C A A G|G U A G U A U G G	C A C C U G C A G|G U.G A A A U G A
γ1 -CH1		U U C U U G U A G|C C.A A A A C G.A
γ1 CH1-hinge	A G.A A A A U U.G|G U G A G A G G A	U C U C C A C A G|U G.C C C A G G.G
γ1 hinge-CH2	U A.U G U A C A.G|G U A A G U C A G	C A U C C U U A G|U C.C C A G A.A G
γ1 CH2-CH3	A A.A C C.A A A G|G U G A G A G C U	C A C C C A C A G|G C.A G A C.C G.A
γ2b -CH1		C U C U U G C A G|C C.A A A A C A.A
γ2b CH1-hinge	A A.A A A C U U.G|G U G A G A G G A	U C U C U G C A G|A G.C C C.A G C.G
γ2b hinge-CH2	A A.U G C.C C A G|G U A A G U C A C	C C U C A U C A G|C U C.C U A A C.C
γ2b CH2-CH3	A A.A U U A.A A G|G U G G G A C C U	A C C C C A C A G|G G.C U A G.U C.A
Consensus sequence	A G|G U R A G	Y - Y Y Y - C A G

Note that all intervening sequences begin with GU at their 5′ end and end with AG at the 3′ terminus.

SOURCE: From P. A. Sharp, *Cell*, *23*, 644 (1981).

zymatic activities are closely related, for once the intervening sequences have been excised the remaining pieces cannot be allowed to escape. However, little is known about the enzymes responsible for such splicing.

Every mammalian and avian gene that has been mapped thus far is split, with the exception of the histone genes. The intervening sequences of some genes are almost incredibly complex. The gene coding for chick *ovalbumin* contains 7 intervening sequences (Figure 25.23), and the gene

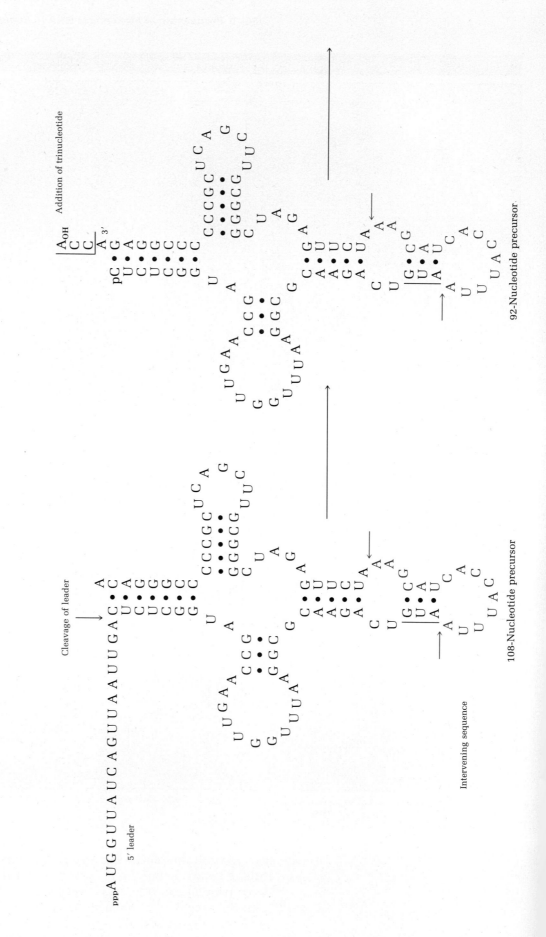

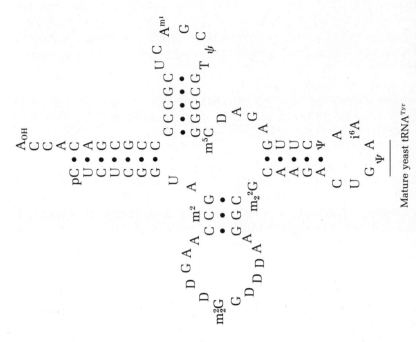

Figure 25.24 Stages in the processing of yeast tyrosyl-tRNA. A 19-residue *leader sequence at the 5′ terminus* is removed, the trinucleotide CCA is added to the 3′ terminus, and bases in the tRNA precursor are modified. Finally, the intervening sequence (in color) is excised, and the ends are joined to give the mature tRNA. The bases joined by a line are the anticodon; arrows indicate the sites of phosphodiester bond cleavage and ligation. [Adapted from E. M. DeRobertis and M. Olson, *Nature, 278,* 142 (1978). Reprinted by permission. Copyright © 1978 Macmillan Journals Limited.]

coding for α-collagen has about 50 intervening sequences. The intervening sequences within a gene have very little sequence homology, and homologies between the intervening sequences for the same gene in different species, say, mouse and rabbit β globin, are almost nonexistent. The only homology that has been found is that all of the intervening sequences that have been studied begin with GU and end with AG (Table 25.5). These sequences *might* be a common signal that determines where the endonucleases that cleave the intervening sequence will attack. At present, however, there is too little data for a confident generalization.

What is the purpose of gene splicing? It has been suggested that gene splicing allows different gene combinations to arise easily, and that this provides a mechanism for the evolution of new proteins arising from the new combinations of expressed regions of DNA. It has also been suggested that gene splicing is critical to gene expression in eucaryotic organisms. Let us consider some tantalizing observations. If the 3' end of one intervening sequence is tied to the 5' end of another intervening sequence and synthetically incorporated in a gene, the gene is expressed normally. The splicing sequences are apparently critical since all but about 20 of the bases of the intervening sequences can be removed and the gene is still processed normally. A virus that infects monkeys, known as SV40, contains a sequence of DNA that is processed in two ways. In one gene a certain base sequence is part of an intervening sequence. This same sequence is also part of the coding region of another gene. The mitochondrial gene of yeast cytochrome *b* is split into six parts. After the first intervening sequence of the gene has been spliced, the first coding region plus part of the next intervening sequence codes for a protein that carries out excision of the next intervening sequence. As these observations suggest, the phenomenon of gene splicing is as baroque as it is ubiquitous.

25–6 PROCESSING OF tRNA IN EUCARYOTIC ORGANISMS

Some of the tRNA molecules produced by eucaryotic organisms have also been found to have intervening sequences. When plasmids containing the gene for yeast tyrosyl-tRNA are injected into the living oocytes of the African clawed toad, *Xenopus laevis*, the yeast gene is both transcribed and processed by the oocytes. This in itself is quite remarkable since these species are so widely divergent. The transcribed tyrosyl-tRNA precursor contains 108 nucleotide residues. A 5'-leader sequence, analogous to the leader in procaryotic tRNA, that contains 19 nucleotide residues is removed first. Following this excision, the trinucleotide CCA is added to the 3' terminus of the precursor tRNA, and other bases within the precursor tRNA are modified. In the last step of processing oocyte enzymes cleave a 14-residue intervening sequence that begins on the 3' side of the anticodon. A ligase then joins the two pieces to give the mature DNA (Figure 25.24).

25–7 SUMMARY

Transcription is the process by which the genetic message in DNA is copied by RNA polymerase. In procaryotic organisms all types of RNA—messenger RNA (mRNA), transfer RNA (tRNA), and ribosomal RNA (rRNA)—appear to be transcribed by a single RNA polymerase. The sequences of these RNA molecules, all of which are single-stranded, are determined by the sequence of the DNA template. RNA polymerase is an oligomeric enzyme that catalyzes the synthesis of RNA in the 5'-3' direction. The *E. coli* enzyme contains six types of subunits. Five of them form the core enzyme, and the sixth subunit, designated σ, is responsible for binding of the holoenzyme to the specific sequence of bases in DNA known as the promoter site. The eucaryotic RNA polymerases are also oligomeric enzymes. Eucaryotic RNA polymerase I is responsible for the synthesis of 5.8S, 18S, and 28S rRNA; RNA polymerase II is responsible for the synthesis of mRNA; and RNA polymerase III is responsible for synthesis of tRNA and 5S rRNA. Transcription is terminated by specific sequences at which RNA polymerase pauses in its travel down the DNA template. A specific protein known as rho catalyzes termination of transcription by causing dissociation of the ternary complex between DNA, RNA polymerase, and the newly synthesized RNA. Actinomycin D inhibits RNA polymerase by intercalating in duplex DNA at the sequence dG-dC, and rifamycin inhibits RNA polymerase by binding the σ subunit of RNA polymerase.

RNA transcripts are subjected to a variety of covalent modification reactions known collectively as processing. Posttranscriptional processing consists of several types of enzymatic reactions. (1) Specific nucleases trim the initial transcript by removing nucleotide residues at the 3' terminus, the 5' terminus, or both. (2) Processing sometimes involves addition of new residues. (3) Many bases are methylated or otherwise modified after transcription. Over 50 modified bases have been found in tRNA molecules. In *E. coli* processing of tRNA involves removal of a 5'-leader sequence by RNAse P, followed by removal of a 3'-trailer sequence by RNAse Q. These reactions are then followed by base modification. Some eucaryotic tRNA molecules are transcribed with a 5' leader sequence that must be excised as well as an intervening sequence that is spliced to give the final, mature tRNA. The *E. coli* chromosome contains three regions that code for the synthesis of rRNA. Each of these genes is bracketed by an initiator and a terminator sequence. The transcript contains the sequences of 16S rRNA, 23S rRNA, and 5S rRNA, reading in the 5'-3' direction. These molecules are excised from the original transcript as it is synthesized.

All eucaryotic mRNA molecules contain a 7-methylguanylate residue at their 5' termini that is added after transcription. Many eucaryotic mRNA molecules, but not histone mRNA molecules, also contain polyadenylate tails at their 3' termini. Most eucaryotic genes are discontinuous and contain intervening sequences that are transcribed, but then excised following transcription. The pieces are then spliced to give the final mRNA. The precursors of mRNA, having large and variable molecular weights, form the pool of heterogeneous nuclear RNA. The biological function of split genes is unknown.

REFERENCES

Abelson, J., RNA Processing and the Intervening Sequence Problem, *Ann. Rev. Biochem., 48*, 1035–1069 (1979).

Adahya, S., and Gottesman, M., Control of Transcription Termination, *Ann. Rev. Biochem., 47*, 967–996 (1978).

Crick, F., Split Genes and RNA Splicing, *Science, 204*, 264–271 (1979).

DeRobertis, E. M., and Olson, M. V., Transcription and Processing of Cloned Yeast Tyrosine tRNA Genes Microinjected into Frog Oocytes, *Nature, 278*, 137–143 (1979).

Hall, B. D., and Spiegelman, S., Sequence Complementarity in T2-DNA and T2-specific RNA, *Proc. Nat. Acad. Sci. U.S., 47*, 137–146 (1961).

Jacob, F., and Monod, J., Genetic Regulatory Mechanisms in the Synthesis of Proteins, *J. Mol. Biol., 3*, 318–356 (1961).

Knapp, G., Beckman, J. S., Johnson, P. F., Fuhrman, S. A., and Abelson, J., Transcription and Processing of Intervening Sequences in Yeast tRNA Genes, *Cell, 14*, 221–236 (1978).

Losick, R., and Chamberlin, M., Eds., "RNA Polymerase," Cold Spring Harbor Laboratory, Cold Spring Harbor, N.Y., 1976.

Perry, R. P., Processing of RNA, *Ann. Rev. Biochem., 45*, 605–630 (1976).

Schimmel, P. R., Soll, D., and Abelson, J., Eds., "Transfer RNA," Cold Spring Harbor Laboratory, Cold Spring Harbor, N.Y., 1979.

Chapter 26

Translation of the genetic message: Protein synthesis and the genetic code

26–1 INTRODUCTION

Information flows in the cell from DNA to RNA to protein. We can imagine DNA as the "master tape" of the cell in which all of the information for the operation and construction of the cell is stored. Messenger RNA is a "second generation tape." It is played only a few times before enzymatic hydrolysis ends its short life. Transfer RNA reads the information contained in the mRNA and translates it, through the complex machinery of protein synthesis, into the protein complement of the cell.

As the word translation implies, protein synthesis is in one sense inseparable from the genetic code; we would otherwise have a medium without a message. However, we treat the two separately to preserve a modicum of order. We shall first examine the machinery required for protein synthesis. More than 150 macromolecules are involved, and many of the details are still unknown, but the overall process is well understood. Once we have examined protein synthesis, we shall turn to the code itself, an apparently universal language whose "vocabulary" of 64 words is sufficient to write the entire volume of cellular life.

26–2 AN OVERVIEW OF PROTEIN SYNTHESIS

The linear sequence in mRNA that is translated to protein contains four bases: adenine, uracil, guanine, and cytosine. These four "letters"—A, U, G, C—constitute the mRNA "alphabet." The translated genetic message is a protein, and the protein contains 20 letters, one for each amino acid. Translation thus consists of converting a base sequence message into an amino acid sequence: the protein product.

Amino acids do not bind to mRNA, and their sequence in the final product is mediated by transfer RNA molecules that act as adapters that bind to mRNA on the surface of a ribosome. The actual synthesis itself involves the formation of a peptide bond between two aminoacyl groups that are bound to adjacent tRNA molecules that are lined up on the mRNA template. A complex machinery is required to assure the proper alignment of the tRNA molecules on the mRNA template. The most complex aspect of protein synthesis is the web of noncovalent interactions between RNA molecules and proteins on the ribosome that are required for synthesis to occur.

We shall approach protein synthesis by considering the structure of ribosomes, the structure of tRNA, and the interaction of ribosomes and tRNA with mRNA. Once we have these molecules in place, we shall consider the activation of amino acid carboxyl groups by aminoacyl-tRNA synthetases, and the action of a variety of specific proteins that are required for initiation, chain elongation, and termination of protein synthesis.

26–3 STRUCTURE OF RIBOSOMES

The ribosomes are the particles upon which protein synthesis is carried out. Procaryotic ribosomes have a mass of 2.6×10^6 daltons and contain 5S, 16S, and 23S ribosomal RNA and about 52 different proteins. The intact ribosome has a sedimentation coefficient of 70S and is usually called the 70S ribosome. The 70S ribosome can be dissociated into two subunits having sedimentation coefficients of 30S and 50S.

The 30S subunit contains 21 different proteins (Table 26.1) and one molecule of 16 S rRNA, a molecule that contains 1,542 nucleotide residues. A model of the 30S ribosome is shown in Figure 26.1. The primary structure of 16S rRNA has been determined (Figure 26.2). The 16S ribosomal RNA molecule acts as a scaffolding around which the ribosomal proteins are arrayed. A tentative model of the interactions of ribosomal proteins with the 16S rRNA is shown in Figure 26.3.

The 50S ribosome contains a molecule of 5S rRNA (120 nucleotide residues) and a molecule of 23S rRNA (2,904 nucleotide residues). One molecule each of 34 different proteins is found in the 50S ribosome (Table 26.2). A model of the 50S ribosome is shown in Figure 26.4. Overall, the structures of these ribosomal subunits are highly asymmetric. A schematic diagram of the 70S ribosome is shown in Figure 26.5.

Both the 30S and the 50S ribosomes can be completely dissociated, and, in a spectacular example of spontaneous assembly, they can both be reconstituted when their components are mixed. The reassembly of the ribosome depends upon the order of addition of the ribosomal proteins. At least six proteins and perhaps a seventh—S4, S7, S8, S15, S17, S20, and perhaps S13—bind to 16S rRNA, and they must be present initially along with the rRNA for the ribosome to be reconstituted. Eventually nearly all 21 proteins must be present to obtain a functional ribosome. Most, if not all, of the ribosomal proteins interact directly with 16S rRNA, and the conformations of the ribosomal proteins are strongly influenced by their neighbors. In all, the ribosome is a highly interactive, cooperative association of RNA and proteins. The binding of ribosomal proteins must be highly specific since the 16S rRNA of yeast will not form a functional ribosome with the ribosomal proteins of E. coli.

Table 26.1 Proteins of the 30S ribosomes	
Protein	Molecular weight
S1	61,159
S2	26,613
S3	25,852
S4	23,138
S5	17,515
S6	15,704
S7K	19,732
S8	13,996
S9	14,569
S10	11,736
S11	13,728
S12	13,606
S13	12,969
S14	11,063
S15	10,001
S16	9,191
S17	9,573
S18	8,897
S19	10,299
S20	9,554
S21	8,369

SOURCE: From H. G. Wittmann, *Ann. Rev. Biochem., 51,* 157 (1982). Reproduced with permission. © 1982 by Annual Reviews Inc.

Table 26.2	Proteins of the 50S ribosomal subunit

Protein	Molecular weight
L1	24,599
L2	29,416
L3	22,258
L4	22,087
L5	20,171
L6	18,832
L7	12,207
L8	19,000
L9	15,531
L10	17,736
L11	14,874
L12	12,165
L13	16,019
L14	13,541
L15	14,981
L16	15,296
L17	14,364
L18	12,770
L19	13,003
L20	13,366
L21	11,565
L22	12,227
L23	11,013
L24	11,185
L25	10,694
L26 = S20	9,554
L27	8,994
L28	8,876
L29	7,273
L30	6,411
L31	6,971
L32	6,315
L33	6,255
L34	5,381
IF-1	8,119
IF-3	20,695
EF-Tu	43,225

SOURCE: From H. G. Wittmann, *Ann. Rev. Biochem., 51,* 157 (1982). Reproduced with permission. © 1982 by Annual Reviews Inc.

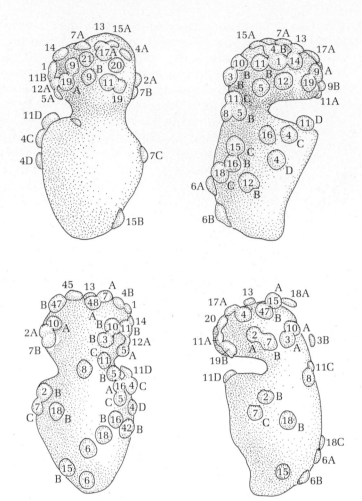

Figure 26.1 Three-dimensional model of the 30S ribosome. Numbers indicate the locations of the ribosomal proteins in the 30S subunit. [From R. Brimacombe, G. Stoffler, and H. G. Wittman, *Ann. Rev. Biochem., 47,* 217–249 (1978).]

In eucaryotic organisms ribosomes are located in the cytosol. Eucaryotic ribosomes have a sedimentation coefficient of 80S and can be dissociated into two subunits having masses of 40S and 60S. The 40S subunit contains about 30 proteins and a molecule of 18S rRNA; the 60S subunit contains about 40 proteins, a molecule of 28S rRNA, and one of 5S rRNA (Table 26.3). Many of the ribosomes in eucaryotic cells are bound to the rough endoplasmic reticulum. The membrane-bound ribosomes are primarily responsible for the synthesis of proteins that are exported to other cells.

26–4 ROLE OF TRANSFER RNA IN PROTEIN SYNTHESIS

Transfer RNA is at the hub of activity in protein synthesis. The L-shaped tRNA molecule acts as an adapter which interacts with mRNA and which is bound to an aminoacyl residue at its 3′ terminus. There is at least one transfer RNA molecule for each amino acid. This aminoacyl residue is transferred to the growing peptide chain in protein synthesis. First, let us consider the interaction of tRNA with mRNA. Each nucleotide sequence in mRNA, for example, CUU, either codes for incorporation of an amino acid in a protein or is a stop signal. The sequence of three bases is called a *codon*. A complementary sequence of nucleotides in tRNA, known as the *anticodon,*

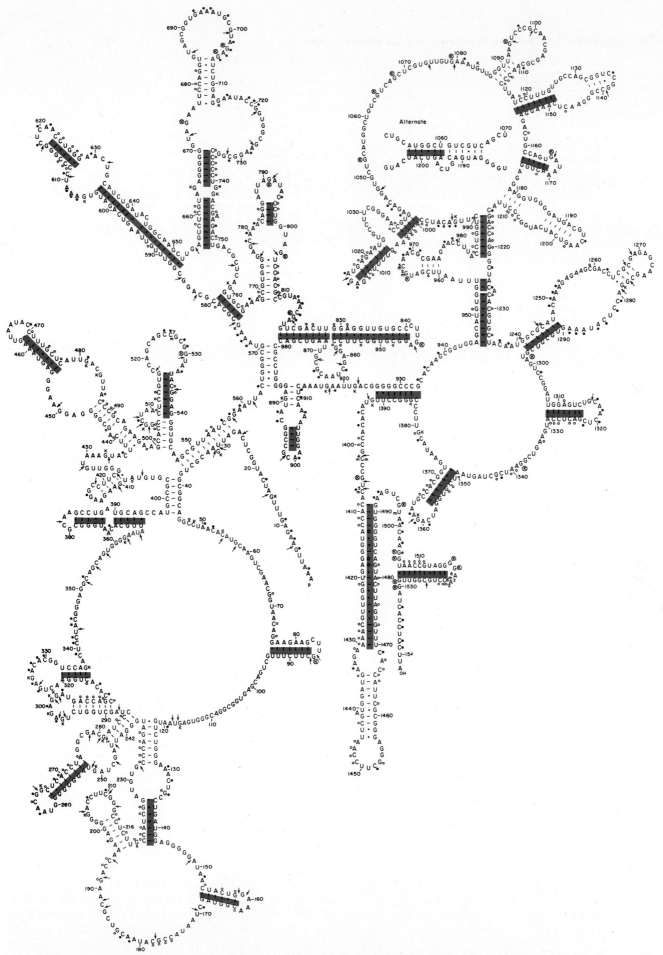

Figure 26.2 Primary and secondary structures of *E. coli* 16S rRNA. The molecule must be highly folded when it is bound to the ribosome, but its tertiary structure in the ribosome is not known. [From H. F. Noller and C. R. Woese, *Science, 212,* 405 (1981). Copyright 1981 by the American Association for the Advancement of Science.]

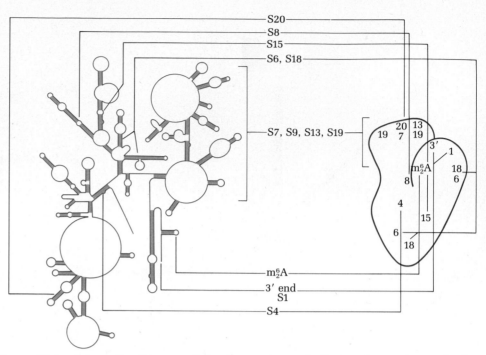

Figure 26.3 Schematic diagram of the interactions of ribosomal proteins with the 30S ribosomal subunit. [From H. F. Noller and C. R. Woese, *Science, 212,* 408 (1981). Copyright 1981 by the American Association for the Advancement of Science.]

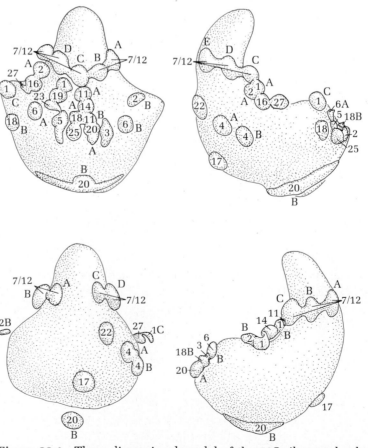

Figure 26.4 Three-dimensional model of the 50S ribosomal subunit. Numbers indicate the locations of 19 of the ribosomal proteins. [From R. Brimacombe, G. Stoffler, and H. G. Wittman, *Ann. Rev. Biochem., 47,* 221 (1978). Reproduced with permission. © 1978 by Annual Reviews Inc.]

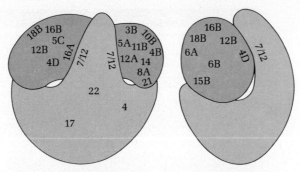

Figure 26.5 Two views of the 70S ribosome. The numbers indicate the locations of several ribosomal proteins. [From H. G. Wittman, *Eur. J. Biochem.*, *61*, 1–13 (1976).]

interacts with mRNA by complementary base pairing. The sequence CUU in mRNA, a "code word" for phenylalanine, binds to the anticodon GAA.

$$
\text{Anticodon}
$$

$$
\text{tRNA} \quad 5' \;-\text{G}-\text{A}-\text{A}-\; 3'
$$

$$
\text{mRNA} \quad 3' \;-\text{C}-\text{U}-\text{U}-\; 5'
$$

$$
\text{Codon}
$$

The anticodon is part of the seven-residue *anticodon loop*, whose sequence in all tRNA's is

$$
5' \text{ pyrimidine}-\text{U}-\text{X}-\text{Y}-\text{Z}-\text{modified purine } 3'
$$

$$
\text{Anticodon}
$$

The features of tRNA that are conserved in all tRNA molecules are indicated in a cloverleaf diagram in Figure 26.6. The presence of so many common features in the structures of all transfer RNA molecules is required since all tRNA molecules bind to ribosomes in approximately the same way. The subtle differences among tRNA molecules are equally important because they are responsible for the specific interactions of tRNA with mRNA and for the interactions of tRNA molecules with the enzymes that catalyze synthesis of aminoacyl-tRNA molecules (Section 26.5).

The major steps of protein synthesis, as we have outlined them to this point, are summarized in Figure 26.7.

26–5 AMINOACYL-tRNA SYNTHETASES

Aminoacyl-tRNA synthetases catalyze formation of an ester between the 3' acceptor end of a tRNA molecule and the carboxylate group of an amino acid (reaction 25.3).

amino acid[1] + tRNA[1] + ATP

$$
\xrightarrow{\text{Aminoacyl-tRNA synthetase[1]}} \text{aminoacyl[1]-tRNA[1]} + \text{AMP} + \text{PP}_1 \quad (26.3)
$$

This process is remarkably accurate. For example, valine differs from isoleucine by a single methylene group, but the incorporation of valine in human hemoglobin occurs with at least 99.97% accuracy.

We shall divide our consideration of aminoacyl-tRNA synthetases into four parts: (1) occurrence and structures of the enzymes, (2) recognition of the correct tRNA by a given aminoacyl-tRNA synthetase, a topic which contains part of the explanation for the specificity of the enzyme, (3) mechan-

Table 26.3 Proteins of the subunits of rat liver ribosomes

40S subunit		60S subunit	
Protein	Mol wt. $\times 10^{-3}$	Protein	Mol. wt. $\times 10^{-3}$
(Sa)	41.5	(La)	37.9
(Sc)	33.0	(Lb)	29.8
S2	33.1	(Lf)	14.6
S3	30.4	P1	16.1
S3a	32.0	P2	15.2
(S3b)	30.4	L3	37.8
S4	29.5	L4	41.8
S5	22.8	L5	32.5
(S5′)	21.5	L6	33.0
S6	31.0	L7	29.2
S7	22.2	L7′	28.7
S8	26.8	L8	28.4
S9	24.3	L9	24.7
S10	20.1	L10	24.2
S11	20.7	L11	21.3
S12	14.9	L12	18.7
S13	18.6	L13	26.3
S14	17.3	L13′	24.6
(S15)	19.6	L14	25.8
S15′	15.7	L15	24.5
(S16)	17.1	L16	18.7
S17	18.0	L17	22.1
S18	18.5	L18	24.5
S19	17.1	L18′	21.3
S20	16.5	L19	25.3
S21	12.3	L20	16.2
S23/S24	18.8	L21	20.3
S25	17.0	L22	16.1
S26	16.5	L23	15.6
S27	14.5	L23′	18.0
S27′	12.8	L25	17.5
S28	11.3	L26	18.6
S29	11.2	L27	17.8
		L27′	18.0
		L28	17.8
		L29	20.5
		L30	14.5
		L31	15.6
		L32	17.2
		L33	15.6
		L34	15.8
		L35	17.5
		L35′	13.7
		L36	14.3
		L36′	16.2
		L37	15.4
		L37′	12.8
		L38	11.5
		L39	11.6

SOURCE: From I. G. Wool, *Ann. Rev. Biochem., 48,* 725 (1979). Reproduced with permission. © 1979 by Annual Reviews Inc.

ism of action of the synthetase, and (4) proofreading or editing activity of the synthetase, which is also an important factor in the specificity of amino-acylation.

A. Occurrence and Structures of Aminoacyl-tRNA Synthetases

Each aminoacyl-tRNA synthetase transfers a single amino acid to a tRNA molecule. A single aminoacyl-tRNA synthetase for each amino acid has been found in bacteria. In eucaryotic cells, however, multiple synthetases for the same amino acid are found in the cytosol, and mitochondria and chloroplasts possess their own unique copies of these enzymes.

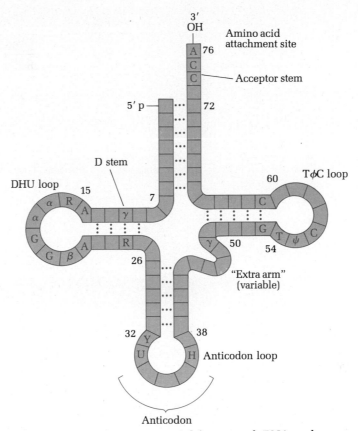

Figure 26.6   Common structural features of tRNA molecules. The number of nucleotidyl residues in each stem and loop is more or less constant, except in the variable loop and in the two parts of the DHU (dihydrouracil) loop marked α and β. Abbreviations: R, purine; Y, pyrimidine; H, highly modified purine. [From S. H. Kim, in P. R. Schimmel, D. Soll, and J. N. Abelson, Eds., "Transfer RNA," Cold Spring Harbor Laboratory, Cold Springs Harbor, N.Y., 1979, p. 84.]

We should like to be able to report that since aminoacyl-tRNA synthetases catalyze the same general esterification reaction, they all have the same general composition and they differ in their specificities because of subtle differences in their binding sites, resembling the serine proteases in this respect. Alas, the subunit compositions and molecular weights of these enzymes, as summarized in Table 26.4, suggest that they may be quite different. We find molecular weights ranging from 69,000 in a single polypeptide chain for *E. coli* glutaminyl-tRNA synthetase to a molecular weight of 267,000 in a tetramer composed of two types of subunits, designated $\alpha_2\beta_2$, in *E. coli* phenylalanyl-tRNA synthetase. It is possible, however, that sequence homologies and structural similarities will emerge for these enzymes as more structural data become available.

B. Recognition of tRNA by Aminoacyl-tRNA Synthetases

The translation of the genetic message depends upon the correct binding of the proper anticodon of a tRNA molecule to its cognate codon.[1] A second criterion for correct translation is the attachment of the proper amino acid to its cognate tRNA molecule, which is the subject of this section. The interaction between a given tRNA molecule and its cognate aminoacyl-tRNA synthetase can be divided into two parts: the binding of the enzyme to tRNA and the esterification of the amino acid to its tRNA.

---

[1] The word cognate is derived from the Latin verb meaning, roughly, born together.

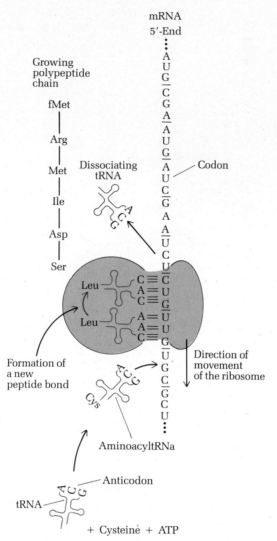

Figure 26.7 Major steps in protein synthesis. The anticodon of an aminoacyl-tRNA molecule binds to a codon on mRNA by complementary base pairing. Both the aminoacyl-tRNA and the mRNA molecule are bound to the ribosome. Formation of a peptide bond is catalyzed by peptidyl transferase, following which the tRNA molecule dissocates from the ribosome, a new aminoacyl-tRNA molecule binds to its specific codon on mRNA, and the process continues. [From H. Holler, *Angew. Chem. Int. Ed. Engl.*, *17*, 648 (1978).]

The binding of an aminoacyl-tRNA synthetase to its tRNA has an association constant of 10^8-10^9 M^{-1} (reaction 26.4).

aminoacyl-tRNA synthetase[1] + tRNA[1]

$$K_{assn} = 10^8 - 10^9 \ M^{-1}$$

(26.4)

[aminoacyl-tRNA synthetase[1]-tRNA[1]]

Aminoacyl-tRNA synthetases also bind the "wrong" tRNA, but the association constant is about a hundred times lower. Once the enzyme and tRNA are bound, discrimination also occurs in the rate of esterification. The incor-

Table 26.4 Molecular weight and quaternary structure of some aminoacyl-tRNA synthetases

Synthetase	Source	Number of subunits	Molecular weight	Subunit composition
Alanyl-	Yeast	1	128,000	α
	E. coli	1	63,000	α
Glutaminyl-	E. coli	1	69,000	α
Glutamyl-	E. coli	2	102,000	$\alpha_2 \beta_2$
			$\alpha = 56,000$	
			$\beta = 46,000$	
Isoleucyl-	E. coli	1	124,000	α
Lysyl-	E. coli	2	104,000	α_2
Methionyl-	E. coli	2	170,000	α_2
Phenylalanyl-	E. coli	4	267,000	$\alpha_2 \beta_2$
			$\alpha = 69,000$	
			$\beta = 75,000$	
Prolyl-	E. coli	2	94,000	α_2
Valyl-	E. coli	1	110,000	α

SOURCE: Adapted from P. R. Schimmel and D. Söll, *Ann. Rev. Biochem., 48*, 601 (1978). Reproduced with permission. © 1978 by Annual Reviews Inc.

poration of an incorrect amino acid when a noncognate tRNA binds to a tRNA synthetase is much slower than when a cognate tRNA-enzyme pair are bound. The recognition of a cognate pair is subtly dependent upon interactions between the enzyme and tRNA. The acceptor stem, the D stem, and the anticodon all lie on the same side of the tRNA. To varying extents these three regions are critical to correct binding. The variable loop is also implicated in correct binding. A given synthetase wraps around the tRNA, so that its distinct binding sites for each of the several regions of the tRNA can interact with the tRNA (Figure 26.8).

C. Mechanism of Action of Aminoacyl-tRNA Synthetases

The formation of an ester between either the 2'- or the 3'-hydroxyl group of the 3' terminal adenylate residue and an amino acid occurs in two steps. The amino acid reacts with MgATP$^{\ominus}$ to give an enzyme-bound *aminoacyl adenylate* plus pyrophosphate (Figure 26.9). ATP is a substrate only when it is bound to Mg(II). The aminoacyl adenylate is an energy-rich, mixed anhydride, with a group transfer potential of 58.5 kJ mol^{-1} (recall Section 12.5G). The group transfer potential of the aminoacyl-tRNA product, on the other hand, is 29 kJ mol^{-1} (roughly the same as the group transfer potential of ATP). The aminoacyl adenylate is bound extremely tightly to the enzyme, but the pyrophosphate diffuses into the solution, and its hydrolysis by pyrophosphatase shifts the equilibrium in the direction of aminoacyl adenylate formation.

As the group transfer potentials for the final product and the intermediate show, there is a strong driving force (29.5 kJ mol^{-1}) for conversion of the aminoacyl adenylate to an aminoacyl-tRNA. Nevertheless, transfer of the aminoacyl group from the energy-rich adenylate to either the 2'-hydroxyl group or the 3'-hydroxyl group of the 3'-terminal adenylate of tRNA is the rate-determining step of the reaction (Figure 26.10).

D. Proofreading Activity of Aminoacyl-tRNA Synthetases

The joining of the correct amino acid to a given tRNA occurs with great precision. Let us consider the case of mistaken incorporation of valine into isoleucine tRNA. Valine and isoleucine differ by a single methylene group. The increase in the binding energy of isoleucine to isoleucyl-tRNA synthetase that can be attributed to this methylene group is about 12 kJ mol^{-1}, equivalent to a factor of 200 to 300 in the equilibrium constant for binding isoleucine to the synthetase. Errors of translation are a thousandfold lower than

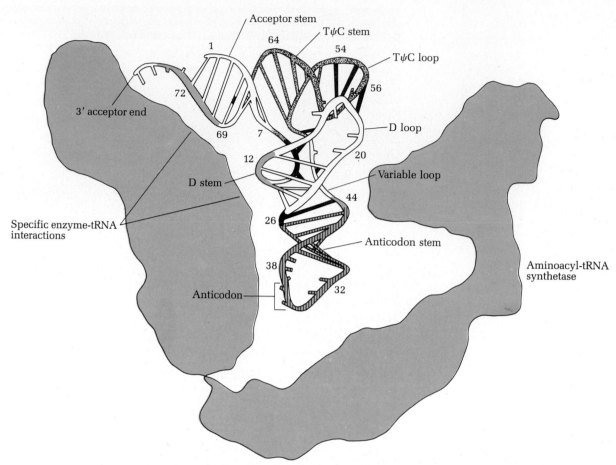

Figure 26.8 Schematic representation of specific interaction between an aminoacyl-tRNA synthetase and several regions of its tRNA substrate. The acceptor stem, the D stem, the anticodon, and the variable loop interact with the synthetase. Neither the conformation of the enzyme nor the conformation of the tRNA in the enzyme-substrate complex is known.

this, and the specificity of isoleucyl-tRNA synthetase cannot be solely a result of the difference in binding energy between isoleucine and valine. In some cases the specificity of an aminoacyl transferase might be attributed to steric constraints within the active site of the enzyme, but valine is smaller than isoleucine, and it could therefore bind to the enzyme without steric hindrance.

The specificity of isoleucyl-tRNA synthetase, which greatly exceeds the difference between the binding energies for valine and isoleucine, is the result of a proofreading activity. (We recall that DNA polymerases also possess proofreading activities.) If isoleucyl-tRNA synthetase (E^{ile}), which has already exhibited considerable specificity in binding its cognate tRNA, synthesizes valyl adenylate (E^{ile}-[val-AMP]), it quantitatively hydrolyzes the mistakenly formed energy-rich intermediate (reaction 26.5).

$$\text{valine} + \text{ATP} \xrightarrow{\text{Isoleucyl-tRNA synthetase}} E^{ile}\text{-[val-AMP]} + PP_i$$

$$\downarrow {\scriptstyle H_2O}$$

$$\text{valine} + \text{AMP} + E^{ile}$$

$$(26.5)$$

Similar editing functions have been found for valyl-tRNA synthetase, which discriminates between valine and threonine, and for methionyl-tRNA

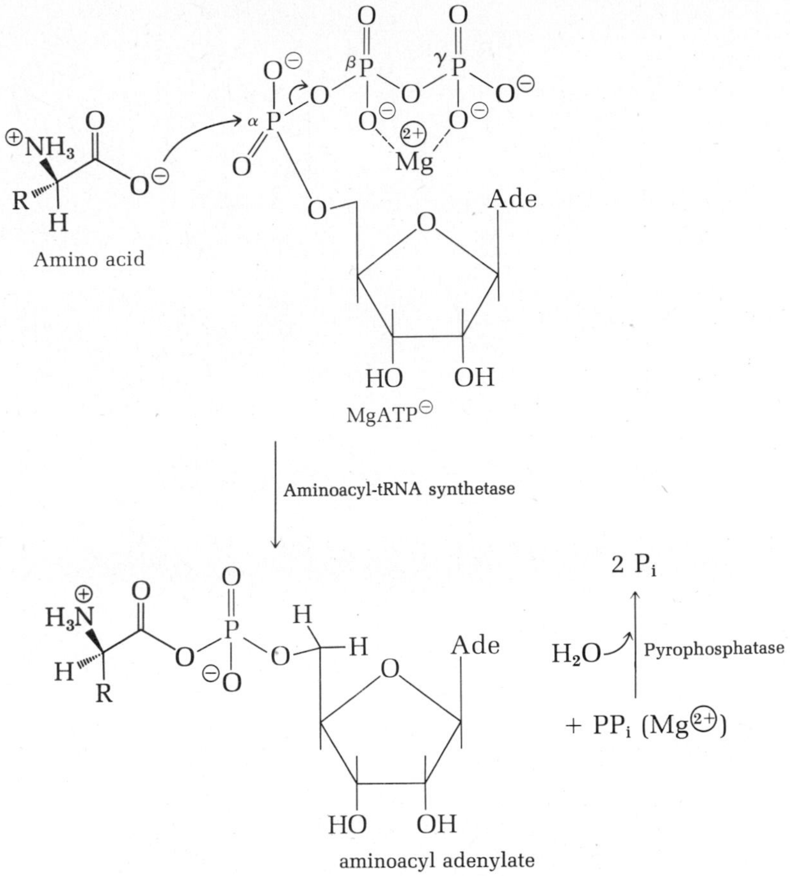

Figure 26.9 Activation of an amino acid by formation of an amino-acyladenylate, a mixed anhydride of adenylic acid and a carboxylic acid having a large, negative free energy of hydrolysis. The intermediate remains bound to the enzyme's active site, $K_{diss} = 10^{-9}M^{-1}$.

synthetase, which discriminates between methionine and homocysteine. In both of these cases, as well as in the case of isoleucyl-tRNA synthetase, an incorrect aminoacyl adenylate formed by incorrect binding of a smaller, competing substrate to the enzyme's active site is promptly hydrolyzed by the proofreading activity of the synthetase.

Editing has not been observed for all aminoacyl-tRNA synthetases, even when the substrates have about the same size. For example, cysteinyl-tRNA synthetase has so much more affinity for cysteine than for serine and alanine that no proofreading is required. The extreme specificity of aminoacyl-tRNA synthetases thus consists of two parts: first, recognition by the enzyme of its cognate tRNA and, second, discrimination by the enzyme's active site for substrates of the correct size and charge. If an error is made at the binding site, the synthetase corrects its own mistake by hydrolyzing the incorrectly formed aminoacyl adenylate.

26–6 DIRECTION OF POLYPEPTIDE-SYNTHESIS AND mRNA TRANSLATION

The direction in which mRNA is "read" in protein synthesis was determined by synthesizing a polyadenylate whose 3′ terminus is pCpC. It had been previously established that the triplet AAA is the codon for lysine and

5′ A A A A A A A A A———A C C 3′

Codon for lysine   Codon for asparagine

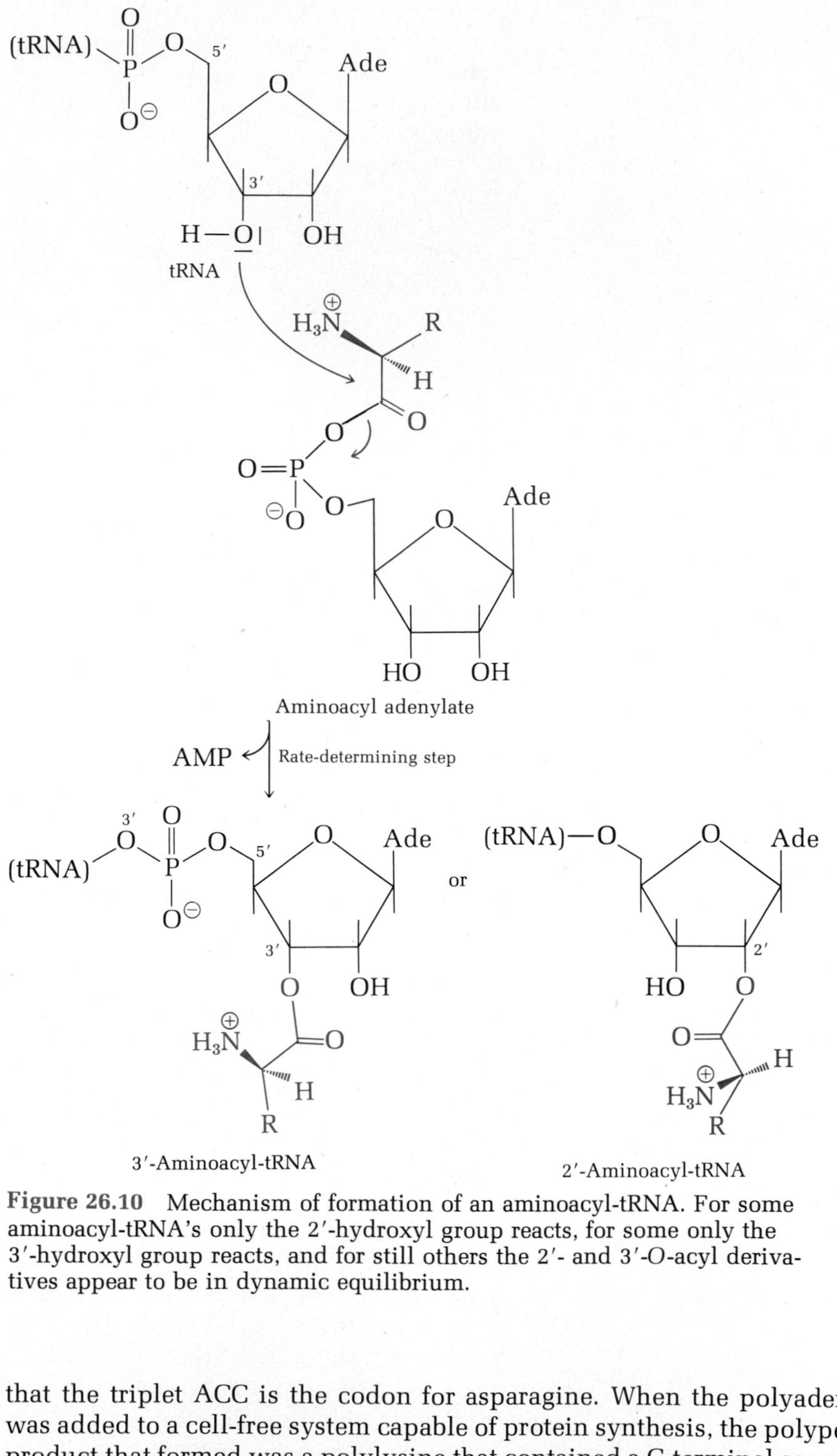

Figure 26.10   Mechanism of formation of an aminoacyl-tRNA. For some aminoacyl-tRNA's only the 2'-hydroxyl group reacts, for some only the 3'-hydroxyl group reacts, and for still others the 2'- and 3'-*O*-acyl derivatives appear to be in dynamic equilibrium.

that the triplet ACC is the codon for asparagine. When the polyadenylate was added to a cell-free system capable of protein synthesis, the polypeptide product that formed was a polylysine that contained a C-terminal asparagine residue. Messenger RNA is therefore translated in the 5'-3' direction.

$$\overset{\oplus}{\text{H}_3\text{N}}\text{-(Lys)-(Lys)}_n\text{-Asn-CO}_2^{\ominus}$$

The direction of polypeptide chain growth was determined by H. M. Dintzis in experiments that used [³H]-leucine to follow the incorporation of amino acids into hemoglobin. The tritiated leucine was added to immature red blood cells, known as *reticulocytes*, and protein synthesis was allowed to continue for periods of 4 to 60 min. Hemoglobin is the major protein synthesized by reticulocytes. The isotopically labeled hemoglobin was isolated from the reticulocytes, the α and β chains were separated, and the chains were digested with trypsin to give a set of peptides of known sequence. The ratio of labeled to unlabeled leucine (a quantity known as the *specific radioactivity*) in each peptide was then measured. Hemoglobin α and β chains, whose synthesis had already begun when the isotopic label was added, had a low specific radioactivity in their N-terminal tryptic peptides, but tryptic peptides near the C-termini had a relatively high specific radioactivity. The tryptic peptides thus contained a gradient of radioactivity that increased from their N-termini to their C-termini (Figure 26.11). This result shows that polypeptides grow sequentially, and that the N-terminal amino acid is the point at which chain growth is initiated.

26–7 INITIATION OF PROTEIN SYNTHESIS

N-Formylmethionine is the N-terminal amino acid residue of all procaryotic

N-Formylmethionine

proteins. (It is often removed by posttranslational processing after the nascent polypeptide chains have been synthesized.) This modified amino acid residue is synthesized in two steps. First, a specific N-formylmethionine-tRNA, $\text{tRNA}^{\text{f-met}}$, is converted to methionyl-$\text{tRNA}^{\text{f-met}}$, by methionyl-tRNA synthetase (reaction 26.6).

$$\text{methionine} \quad + \quad \text{tRNA}^{\text{f-met}}$$

$$\Big\downarrow \text{Methionyl-tRNA synthetase} \qquad\qquad (26.6)$$

$$\text{methionyl-tRNA}^{\text{f-met}}$$

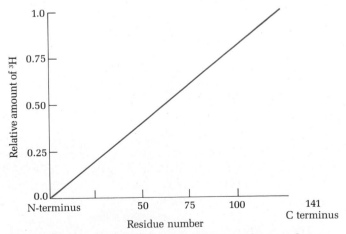

Figure 26.11 Gradient of radioactivity in the α and β chains of reticulocytes (immature red blood cells) when [³H]-leucine is added. More radioactivity is incorporated in the C-terminal peptides than in the N-terminal peptides. The greater radioactivity at the C-terminal ends of the chains shows that synthesis occurs sequentially from the N-terminal to the C-terminal amino acid residue.

Second, N^{10}-formyltetrahydrofolate donates its formyl group to methionyl-tRNA^{f-met} yielding N-formylmethionyl-tRNA^{f-met} (reaction 26.7).

Methionyl-tRNA^{f-met} N^{10}-Formyltetrahydrofolate

Methionine-tRNA^{f-met} formyl transferase

N-Formylmethionyl-tRNA^{f-met} Tetrahydrofolate

$$(26.7)$$

Procaryotic cells contain both tRNA^{f-met} and methionyl-tRNA, tRNA^{met}. The same synthetase adds a methionyl group to each of these tRNA molecules. The formyl transferase, however, does not formylate methionyl-tRNA^{met}. The tRNA^{met} and tRNA^{f-met} differ by only three residues: one in the anticodon loop, one in the TψC loop, and one in the acceptor stem (Figure 26.12). To further complicate the picture, both of these tRNA molecules have the same anticodon—CAU—yet only N-formylmethionyl-tRNA^{f-met} initiates protein synthesis. The detailed interactions of these tRNA molecules with the formyl transferase and with the ribosome have not yet been elucidated.

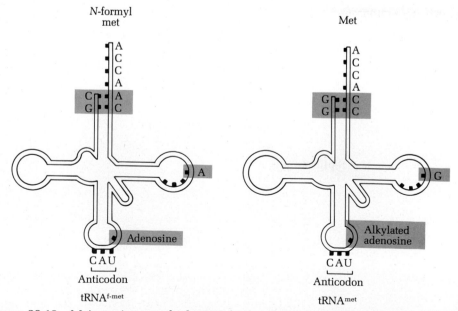

Figure 26.12 Major points at which tRNA^{f-met} and tRNA^{met} differ. [Adapted from J. D. Watson, "Molecular Biology of the Gene," 3rd ed., W. A. Benjamin, Menlo Park, Calif., 1976, p. 329. Used with permission from the Benjamin/Cummings Publishing Co.]

The mRNA codon which is complementary to the tRNA^{f-met} anticodon is AUG. Messenger RNA molecules, however, do not begin with AUG at their 5′ termini, and the first initiation codon is often found many residues toward the 3′ end of the mRNA. The sequence on the 5′ side of the initiation codon is responsible for specific binding to the ribosome. The initiation region of mRNA binds to the 16S rRNA of the 30S ribosomal subunit to correctly align the messenger RNA for translation (Table 26.5). The initiation sequence can be as long as 200 nucleotides, and there is little sequence homology between various mRNA leader sequences. One common feature, however, is the presence of the sequence 5′-AGGAGGU-3′ in the leader regions of all mRNA molecules. Complementary base pairing therefore plays a fundamental part in the recognition of the initiation site of mRNA by the ribosome.

In the presence of the cytosolic proteins known as *initiation factors*, IF-1, IF-2, and IF-3, the 30S ribosomal subunit binds to mRNA and N-formylmethionyl-tRNA^{f-met}. Formation of this initiation complex occurs in a fixed order. First, initiation factor IF-3 binds mRNA to a free 30S ribosomal unit which is present in the cytosol. Second, a complex of GTP and IF-2, perhaps also in association with IF-1 (the least understood of the three initiation factors), carries the N-formylmethionyl-tRNA^{f-met} and binds it to the IF-3–30S complex at the AUG initiation site of the mRNA (at the 5′ end of the message). IF-3 is released in this step. The 30S initiation complex then binds a free 50S ribosomal subunit to give the 70S initiation complex bound to the mRNA. Formation of this complex is driven by hydrolysis of the GTP bound to the 30S initiation complex. IF-2 catalyzes GTP hydrolysis in this step, whereupon IF-2 dissociates from the ribosome. The formation of the mRNA-bound 70S initiation complex is summarized in Figure 26.13. Initiation factors are also required for eucaryotic protein synthesis. The general pattern is the same, but less is known about eucaryotic initiation factors than about their procaryotic counterparts.

Table 26.5 Initiation regions of mRNA molecules and their complementary base pairing to *E. coli* 16S rRNA

Coliphage R17 maturation protein
5′ A U U C C U A G G A G G U U U G A C C U A U G 3′ (mRNA)
3′ HO A U U C C U C C A C U A 5′ (*E. coli* 16S RNA)

Coliphage R17 coat protein site
C C U C A A C C G G G G U U U G A A G C A U G
HO A U U C C C C A C U A
U

Coliphage R17 replicase site
A A A C A U G A G G A U U A C C C A U G
HO A U U C C U C C A C U A

E. coli lac operon
A C A C A G G A A A C A C G U A U G
HO A U U C C U C C A C U A

E. coli trpA protein
G A A A G C A C G A G G G G A A A U C U G A U G
HO A U U C C C C A C U A
U

SOURCE: From R. Mazumder and W. Szer, in M. Florkin, A. Neuberger, and L. L. M. van Deenen, Eds., "Comprehensive Biochemistry," Vol. 24, Elsevier, New York, 1977, p. 215.

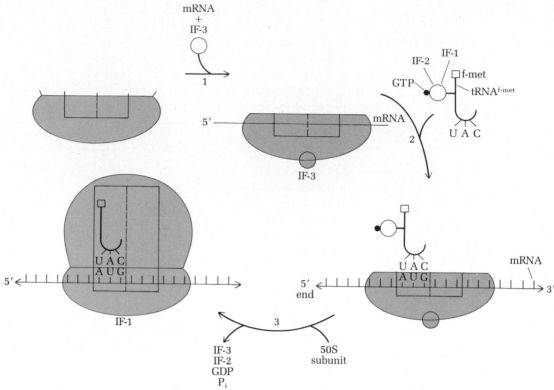

Figure 26.13 Steps in the formation of the 70S initiation complex in protein synthesis. [Adapted from G. Karp, "Cell Biology," McGraw-Hill, New York, 1979, p. 515. Used with permission of the McGraw-Hill Book Company.]

26–8 CHAIN ELONGATION

The 70S ribosome contains two binding sites for aminoacyl- (or peptidyl-) tRNA molecules. The N-terminus of the growing peptide chain, including the first residue, N-formylmethionine, is bound to the ribosome through the tRNA of the last amino acid. This binding site is known as the peptidyl site or *P site*. The tRNA that bears the aminoacyl group to be added to the growing peptide chain occupies the aminoacyl or *A site*. Peptide synthesis can be divided into three steps: (1) binding of an aminoacyl-tRNA to the A site of the mRNA−70S ribosome complex, (2) peptide bond synthesis, and (3) translocation of the newly formed peptide from the A site to the P site.

The first step in protein synthesis that follows formation of the 70S ribosome initiation complex is binding of an aminoacyl-tRNA to the still vacant A site. *Elongation factor* T is a dimeric protein that is designated Ts·Tu. The Tu subunit has a molecular weight of 47,000. It binds tightly to GTP and to the aminoacyl-tRNA at the A site. GTP is hydrolyzed to GDP and phosphate by Tu in this process. Tu will bind to all aminoacyl-tRNA molecules except N-formylmethionyl-tRNA^{f-met}. Once the A site is occupied, Tu dissociates from the ribosome.

Tu binds tightly to its hydrolysis product, GDP, and in this state it is unable to bind to another aminoacyl-tRNA. At this point subunit Ts, a membrane bound protein whose molecular weight is 44,000, comes into play by binding Tu and displacing GDP (reaction 26.8).

$$\text{Tu-GDP} + \text{Ts} \longrightarrow \text{Tu·Ts} + \text{GDP} \tag{26.8}$$

GTP then displaces Ts from the Tu·Ts complex, regenerating active Tu (reaction 26.9).

$$\text{Tu·Ts} + \text{GTP} \longrightarrow \text{Tu-GTP} + \text{Ts} \tag{26.9}$$

The 50S ribosomal subunit contains a peptidyl transferase, comprising two of its integral components (proteins L7 and L12). This enzyme catalyzes peptide bond formation between the aminoacyl moiety of the tRNA at the A site and the activated peptidyl residue bound to tRNA at the P site (Figure 26.14). Once the peptide bond has been formed, the P site is bound to a deacylated ("uncharged") tRNA molecule, and a peptidyl-tRNA (one residue longer) bound to the A site. In the next step the deacylated tRNA is ejected, and the peptidyl-tRNA is *translocated* to the P site as the ribosome moves along the mRNA by three nucleotides. Elongation factor G (EF-G) and GTP are required for this complex process, and GTP is hydrolyzed by EF-G during translocation. Each time a peptidyl group is translocated, a GTP molecule is hydrolyzed, but the details of this critical step in protein synthesis are not well understood. The steps involved in chain elongation are summarized in Figure 26.15.

The structure of the antibiotic *puromycin* closely resembles the structure of the 3′ terminus of an aminoacyl-tRNA (Figure 26.16), and it inhibits further translocation by randomly occupying part of the A site of the ribosome. Once bound to the ribosome, the free amino group of puromycin forms a peptide bound with the peptidyl-tRNA at the P site, and this product dissociates from the ribosome (Figure 26.17).

Until the advent of mass immunization in this century, diphtheria was a major cause of childhood death. In 1884 the bacterium responsible for diphtheria, *Corynebacterium diphtheria,* was identified, and four years later it was found that this bacterium releases a protein toxin that is the actual agent of the disease. Only those *C. diphtheria* that have incorporated the DNA of a certain lysogenic bacteriophage are able to synthesize the toxin. Many years elapsed before the molecular details of diphtheria toxin were discovered. The diphtheria toxin protein consists of a single polypeptide chain (mol wt 62,000). The toxin is extraordinarily potent; as little as 100 ng kg^{-1} is lethal. The toxin inhibits protein synthesis by inhibiting the action of eucaryotic elongation factor eEF-2 ("e" denotes eucaryotic).

The toxin can be split into two parts, known as the A fragment (mol wt

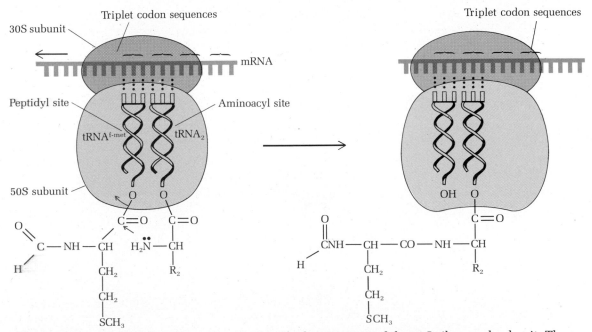

Figure 26.14 Peptidyl transferase that is an integral component of the 50S ribosomal subunit. The enzyme catalyzes peptide bond formation between residues at the A site and P site.

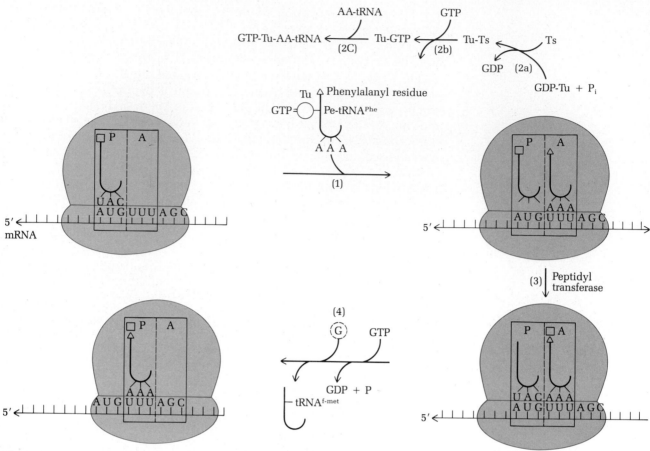

Figure 26.15 Steps in chain elongation. (1) Elongation factor Tu guides an aminoacyl-tRNA to the A site. (2) Tu is reactivated by Ts following the release of Tu from the ribosome. (3) Peptidyl transfer occurs. (4) EF-G hydrolyzes GTP as the peptidyl-tRNA is translocated from the A site to the P site. [Adapted from G. Karp, "Cell Biology," McGraw-Hill, New York, 1976, p. 516. Used with the permission of the McGraw-Hill Book Company.]

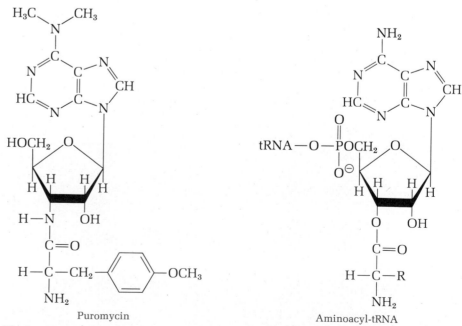

Figure 26.16 Structures of puromycin and the 3′ terminus of an aminoacyl-tRNA molecule.

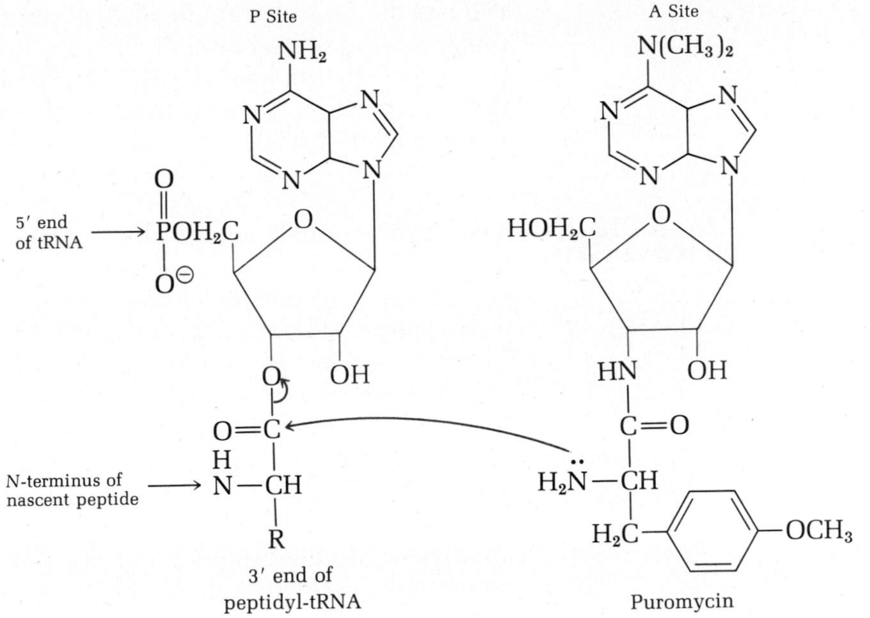

Figure 26.17 Formation of a peptide bond between puromycin bound to the A site of a ribosome and the peptidyl-tRNA bound to the P site. The product dissociates from the ribosome, terminating protein synthesis at random and yielding incomplete and inactive peptides.

21,000) and the B fragment (mol wt 40,000). The B fragment binds to ganglioside GM_1 (recall Section 21.5 in which we noted that this complex lipid is also the receptor for cholera toxin) on the cell surface. The B fragment of the toxin is required for transport of the toxin across the plasma cell membrane. The mechanism of this transport process has yet to be determined. The A fragment of the toxin catalyzes *ADP-ribosylation* of elongation factor eEF-2 (reaction 26.10).

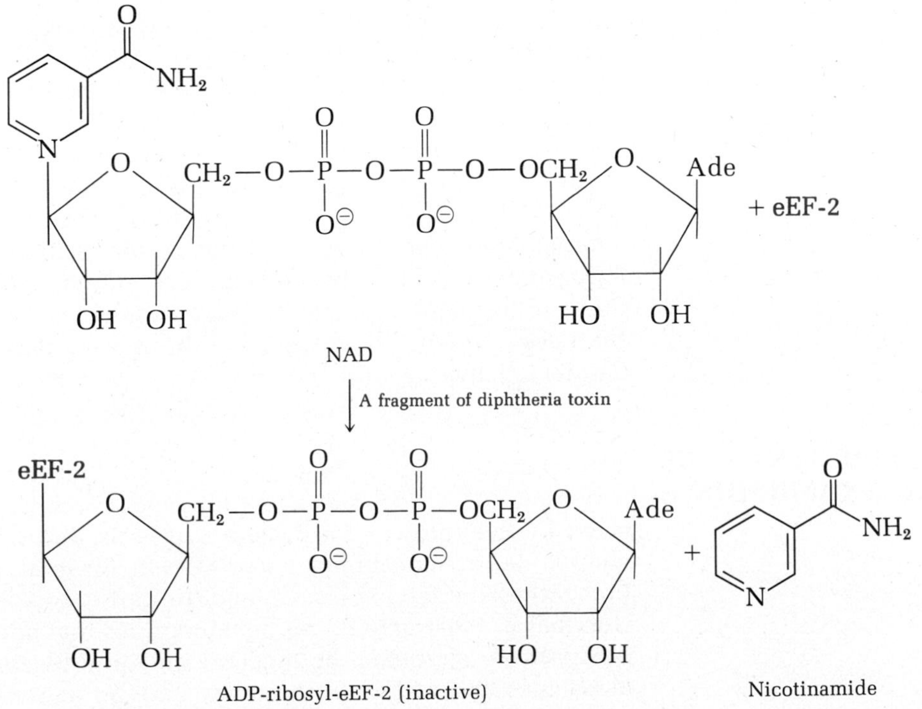

(26.10)

(This is the second example of ADP-ribosylation that we have encountered, the first being caused by cholera toxin, Section 17.4A.) The site of the initiation factor that undergoes ADP-ribosylation is not known, but the result is that translocation of the growing peptide chain from the P site to the A site is inhibited, and protein synthesis stops.

26–9 CHAIN TERMINATION

Termination of polypeptide chain growth is signaled by any of three *termination codons*: UAG, UAA, or UGA. Normally, no transfer RNA molecules contain an anticodon that is complementary to these codons. They are recognized by specific *release factors* that bind directly to the termination codons. Three release factors, designated RF-1, RF-2, and RF-3, are produced by *E. coli*. Release factor RF-1 (mol wt 44,000) binds to UAA and UAG. Release factor RF-2 (mol wt 47,000) binds to UAA and UGA. RF-3 is tightly bound to GTP. It does not bind to a codon, but enhances the activities of RF-1 and RF-2.

Termination of a polypeptide chain occurs when a termination codon appears within the reading frame at the A site of the ribosome, and the completed polypeptide chain, still bound to the tRNA of the C-terminal aminoacyl group in the chain, occupies the P site. The release factor alters the activity of the peptidyl transferase, which is an integral component of the 50S ribosomal subunit, causing it to hydrolyze the aminoacyl ester bond holding the C-terminal amino acid residue to the tRNA. Release factors are tightly bound to GTP, and release of the final polypeptide product is accompanied by GTP hydrolysis. Hydrolysis of GTP, catalyzed by RF-3 is followed by the dissociation of the release factors from the ribosome. At this point the ribosome dissociates into its 30S and 50S subunits, which enter the cytosol, released from the 5' end of the messenger RNA molecule. The 30S subunit can then be used to form a new 30S initiation complex at the 3' end of a messenger RNA molecule. The entire process then begins anew.

We have focused on a single ribosome as it travels down an mRNA molecule translating it into protein. In fact, protein synthesis involves the simultaneous action of many ribosomes on a single mRNA molecule (Figure 26.18). The complex between an mRNA and several ribosomes is known as a *polyribosome*, or *polysome*. An mRNA may begin to be translated almost as soon as it is transcribed, and by the time the mRNA has been completely transcribed it can be attached to many ribosomes. This situation is illustrated by Figure 26.19. Part of an *E. coli* chromosome is being transcribed, and the newly synthesized mRNA is bound to many ribosomes. Each ribosome synthesizes an entire polypeptide. About 1 ribosome is bound for every 80 nucleotide residues in mRNA. For example, the mRNA coding for hemoglobin contains about 500 nucleotide residues and is usually bound to 5 ribosomes during active hemoglobin synthesis. We have treated transcription and translation separately, but they are closely coupled processes, as we shall see more explicitly when we discuss regulation of gene expression in the next chapter.

26–10 BIOSYNTHESIS OF GRAMICIDIN S

The gramicidins are a group of peptide antibiotics that act as ionophores (recall Section 15.10B). These low molecular weight peptides are not synthesized by the pathway for protein synthesis that we have discussed in this chapter, but are assembled in a process that is in many respects similar to the biosynthesis of fatty acids. Gramicidin S (Figure 26.20) is a cyclic decapeptide that is constructed from pentapeptides that are linked head-to-tail. The biosynthesis of gramicidin S occurs on a multienzyme complex without the aid of messenger RNA or ribosomes, and the "activation" of amino acid resi-

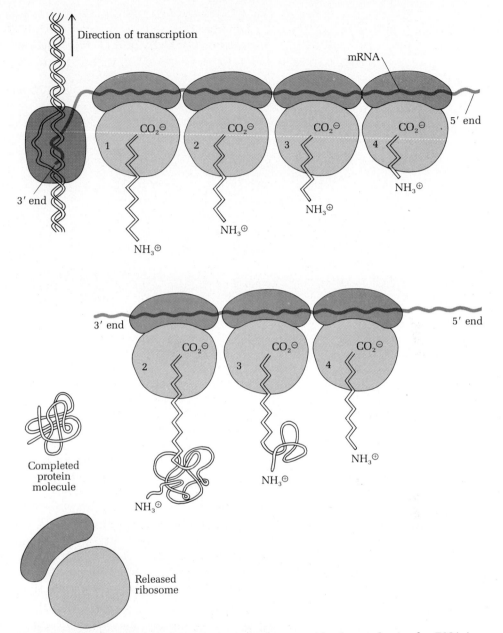

Figure 26.18   Schematic diagram of a polyribosome. Newly synthesized mRNA is translated by several ribosomes simultaneously. Each ribosome synthesizes an entire polypeptide. Following completion of the protein, the ribosome dissociates into its 30S and 50S subunits.

dues to be incorporated in gramicidin S does not involve an aminoacyl-tRNA.

The multienzyme complex that catalyzes the biosynthesis of gramicidin S consists of two components, designated E_I and E_{II}. E_I has a molecular weight of 180,000, and E_{II} has a molecular weight of 100,000.

E_I contains four sulfhydryl groups that form specific thioesters when treated with L-proline, L-valine, L-ornithine, and L-leucine. E_{II} racemizes L-phenylalanine and then forms a thioester with the D-amino acid. The thioesters formed with either E_I or E_{II} are aminoacyl adenylates. The activation of the amino acids that are incorporated in gramicidin S thus occurs in two steps (reactions 26.12 and 26.13).

$$H_3\overset{\oplus}{N}-\overset{\overset{\textstyle H}{|}}{\underset{\underset{\textstyle R_1}{|}}{C}}-CO_2^{\ominus} + ATP \longrightarrow H_3\overset{\oplus}{N}-\overset{\overset{\textstyle H}{|}}{\underset{\underset{\textstyle R}{|}}{C}}-\overset{\overset{\textstyle O}{\|}}{C}-O-AMP + PP_i \qquad (26.12)$$

Aminoacyl adenylate

$$H_3\overset{\oplus}{N}-\overset{\overset{\textstyle H}{|}}{\underset{\underset{\textstyle R}{|}}{C}}-\overset{\overset{\textstyle O}{\|}}{C}-O-AMP + E_{(I\ or\ II)}-SH \longrightarrow$$

$$H_3\overset{\oplus}{N}-\overset{\overset{\textstyle H}{|}}{\underset{\underset{\textstyle R}{|}}{C}}-\overset{\overset{\textstyle O}{\|}}{C}-S\text{-}E + AMP \qquad (26.13)$$

Thioester

The next step in gramicidin S biosynthesis is transfer of a phenylalanyl group from E_{II} to E_I (reaction 26.14).

$$E_{II}-S-\overset{\overset{\textstyle O}{\|}}{C}-Phe-\overset{\oplus}{N}H_3 + E_I-S-\overset{\overset{\textstyle O}{\|}}{C}-Pro-\overset{\oplus}{N}H_3 \longrightarrow$$

$$E_{II} + E_I-S-\overset{\overset{\textstyle O}{\|}}{C}-Pro-Phe-\overset{\oplus}{N}H_3 \qquad (26.14)$$

All subsequent steps in gramicidin S biosynthesis occur on E_I. We recall that E_I forms four thioesters to specific amino acids. Besides the dipeptide formed in reaction 26.14, E_I is also bound to valyl, ornithyl, and leucyl residues. The next steps of gramicidin S biosynthesis involve the sequential transfer of aminoacyl residues from one sulfhydryl site to another. Each step of this process leaves a "vacancy" at the given aminoacyl binding site (Figure 26.21). E_I contains a 4-phosphopantetheine moiety that is believed to act as a long arm that swings the growing peptide from one site on E_I to the next. This, we recall, is the function of the phosphopantetheine moiety in

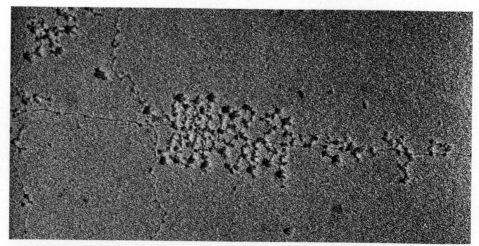

Figure 26.19 Electron micrograph of an *E. coli* chromosome that is being simultaneously transcribed and translated. The dark particles are polyribosomes. [Courtesy of Dr. J. B. Rattner.]

fatty acid synthetase (Section 19.7). In each step the free sulfhydryl group of 4'-phosphopantetheine picks up the peptidyl chain at a given site and swings it to the next position where the free N-terminal amino group displaces the sulfhydryl group of 4'-phosphopantetheine (Figure 26.22). At the end of these peptidyl transfer steps a pentapeptide is bound to E_I. A pentapeptidyl group bound to one molecule of E_I then combines head-to-tail with the pentapeptide bound to a second molecule of E_I to give the cyclic pentapeptide, gramicidin S (Figure 26.23). Similar mechanisms are used in the synthesis of many oligopeptides that contain up to 15 residues.

26–11 THE GENETIC CODE

A. Preliminary Considerations

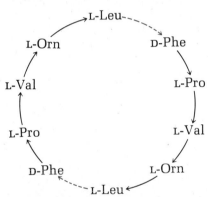

Figure 26.20 Schematic diagram of the structure of gramicidin S. Dotted lines indicate the peptide bonds formed in the final step of biosynthesis when two pentapeptides are joined head-to-tail. The arrows indicate the polarity of the peptide chain; the direction of the arrows indicates the bond between the N-terminus of one residue and the C-terminus of the next.

Suppose that you felt like sending a message to a friend in a secret code. A cryptographer, suspecting mischief, examines the code, detects certain clues, and is able to deduce the message from the cryptogram. This is the ordinary procedure; the decoding proceeds from the cryptogram to the message. Now imagine the situation is which the cryptographer possesses the message, but must guess the cryptogram. This is a problem of the first magnitude, since there are virtually unlimited possibilities for constructing a code from a large alphabet, such as English with its 26 letters, that is able to convey the known message. Cracking the genetic code is an example of this type of problem. The message was already known; it consists of the primary structures of proteins. But what was the code by which the messages are conveyed? The physicist G. Gamow proposed that since there are only 4 "letters" in the DNA alphabet—A, T, G, and C—and since there are 20 amino acids, the code must contain at least 3 letters. A 2-letter code constructed from any of 4 letters has a possible vocabulary of only 16 words (4^2), not enough for all 20 amino acids, while a 4-letter code gives 256 (4^4) words, far more than are needed. A 3-letter code, however, has a possible vocabulary of 64 (4^3) words, sufficient but not excessive.

If the code in fact contains 3-letter words, the problems of the "extra" words (44 in all) and how the code is read remain to be solved.

The code can, in principle, be either overlapping or nonoverlapping. If it is overlapping, then in the sequence shown here ATG can specify one

Overlapping

A T G | C A T | G C A | T G C |————

| amino acid$_1$ | amino acid$_2$ | amino acid$_3$ | amino acid$_4$ |

Nonoverlapping

amino acid, TGC another, and so forth. The ribosome would move along an mRNA molecule one or two nucleotide residues at a time in an overlapping code. If the code is nonoverlapping, ATC would specify one amino acid, CAT another, and so forth. The ribosome would move three residues each time a codon is translated, provided that the code is "comma-less." If "commas" were included in the code, the ribosome would be translocated by more than three nucleotide residues at a time during translation of a messenger RNA molecule. In fact, the genetic code is nonoverlapping and comma-less, as we know from our discussion of protein synthesis. Genetic experiments provided the initial evidence that the code is nonoverlapping. Experiments with tobacco mosaic virus (Section 4.8A) showed that changing a single base results in a change in one amino acid in the coat protein. This result is possible only with a nonoverlapping code. In an overlapping code a single base mutation would lead to a change in two or three amino acids. Extensive experiments by S. Bremner and F. Crick showed that the triplet

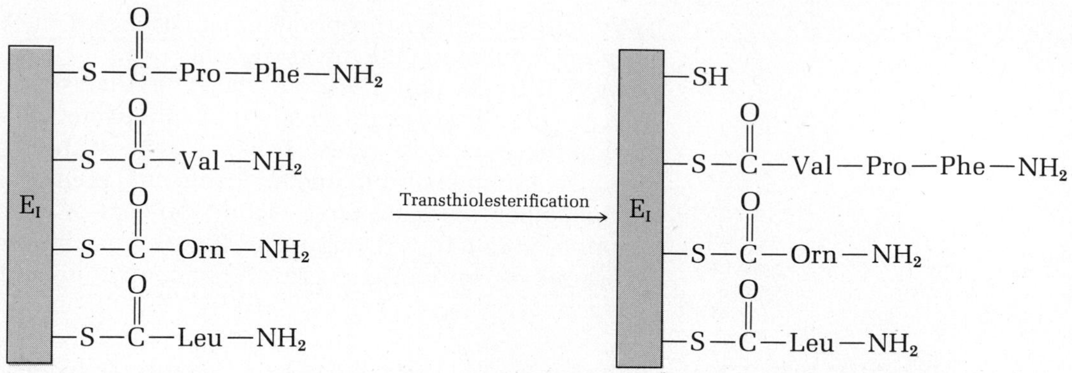

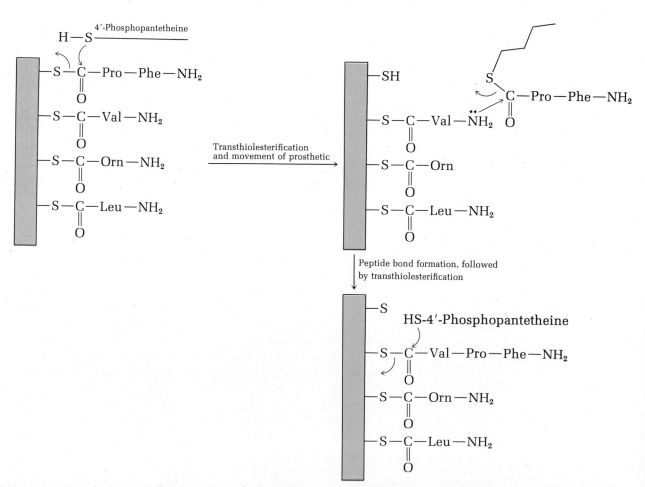

Figure 26.21 Schematic diagram of the polymerization steps of gramicidin S biosynthesis. E_I is bound to fire aminoacyl groups by thiolester bonds. The phenylalanyl residue bound to the prolyl residue at the first site was transferred from E_{II} in a transthiolesterification reaction in the preceding step. The transthiolesterification steps of the polymerization reactions are mediated by a 4′-phosphopantetheine moiety bound to E_f, as discussed below.

codons are read sequentially and that no intervening bases serve as "commas."

A nonoverlapping, unpunctuated code requires a specific initiation site to which the ribosome binds at the outset of protein synthesis. We have seen, in advance, the AUG, the codon for N-formylmethionyl-tRNA is the initia-

Figure 26.22 Schematic diagram of the role of 4′-phosphopantetheine in the polymerization steps of gramicidin S biosynthesis. In the first half-reaction the sulfhydryl group of 4′-phosphopantetheine displaces the original thioester. In the second half-reaction the amino group of the next residue displaces the thioester involving the prosthetic group.

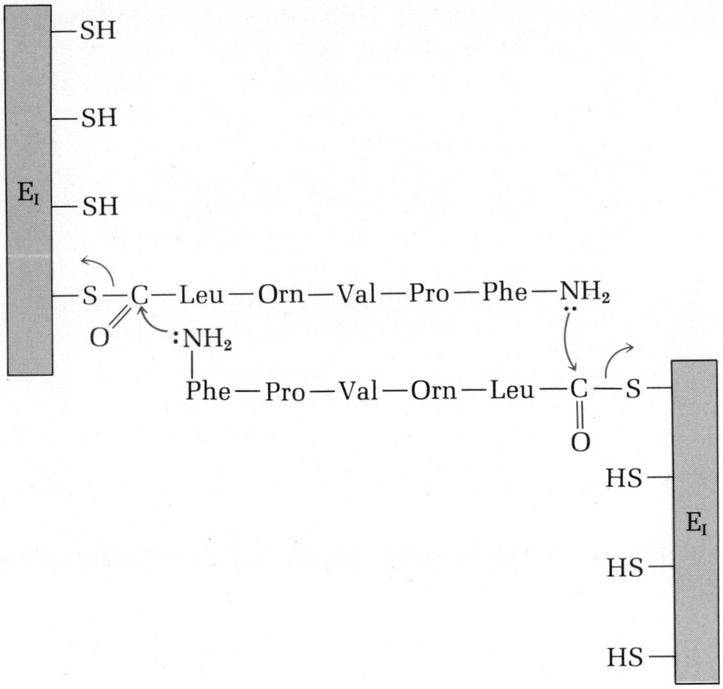

Figure 26.23 Head-to-tail joining of the two pentapeptides bound to E_I in the final step of gramicidin S biosynthesis.

tion codon (although in the 1950s, the time of these experiments, the identity of the codons was unknown). Genetic experiments showed that there is also at least one specific termination codon. So, some of the codons serve purposes other than specifying amino acids.

Another possible explanation of the "extra" words might be that some amino acids have more than one codon. Such a code is *degenerate*. F. Crick proposed that the code was degenerate since the base contents of DNA from microorganisms are widely variable (recall Section 9.5B), whereas the amino acid compositions of their proteins are quite similar. Different base sequences must therefore code for the same amino acids. Crick's hypothesis proved to be correct.

We can therefore summarize the state of knowledge of the genetic code in 1961 in a few brief statements. (1) The genetic code consists of triplet codons. (2) Nonoverlapping codons are read sequentially from a fixed starting point without interruption. (3) At least one codon leads to termination of protein synthesis. (4) The code is degenerate. These statements describe the type of code, but it was still not deciphered. The question what the three-letter codons spell had not yet been answered.

B. Deciphering the Genetic Code

The first major breakthrough in the deciphering of the genetic code came in 1961, partly as a result of an apparently unrelated discovery. In 1955 M. Grunberg-Manago and S. Ochoa discovered *polynucleotide phosphorylase*, an enzyme that catalyzes the synthesis of RNA from ribonucleoside diphosphates in the absence of a DNA template (reaction 26.11).

n ribonucleoside diphosphates

$\Updownarrow$ Polynucleotide phosphorylase

RNA + $n\,P_i$

(26.11)

M. W. Nirenberg and J. H. Matthaei used polynucleotide phosphorylase to synthesize polyuridylic acid (poly U) which they added to a cell-free system that contained ribosomes, a mixture of tRNA molecules, amino acids, and aminoacyl-tRNA synthetases. This system proceeded to synthesize polyphenylalanine. Since the genetic code was known to consist of triplet code words this experiment unambiguously showed that UUU is the codon for phenylalanine. Shortly thereafter it was found that polycytidylic acid (poly C) is translated as polyproline, and that polyadenylic acid (poly A) is translated as polylysine. CCC is therefore a codon for proline, and AAA is a codon for lysine. Polyguanylic acid (poly G) forms a triple helix that is not translated. Later work showed that GGG is a codon for glycine. The possibilities for homopolymers—poly U, poly A, poly C—were rather quickly exhausted, and efforts next turned to the study of proteins produced when random copolymers, poly AC, for example, were added to cell-free systems capable of polypeptide synthesis. The results for poly AC are summarized in Table 26.6. These experiments revealed the most probable compositions of various codons, but did not reveal the internal sequences of codons. H. G. Khorana and his coworkers surmounted this difficulty by synthesizing RNA polymers of known sequence as well as all 64 possible codons. Both synthetic polymers and synthetic trinucleotides proved extremely useful in deciphering the genetic code.

A happy discovery led to the rapid deciphering of many codons when it was found that trinucleotides stimulate the binding of specific aminoacyl-tRNA molecules to the ribosome. For example, when pUpUpU is added to ribosomes in the presence of a mixture of aminoacyl-tRNA molecules, only phenylalanyl-tRNA^{phe} binds to the ribosome. If pApApA is added to a system of ribosome and charged tRNA molecules, only lysyl-tRNA^{lys} binds to the ribosomes. This unexpected finding made it possible to identify many condons rather quickly. A given trinucleotide was added to a system containing ribosomes and aminoacyl-tRNA molecules that were labeled with a ^{14}C-aminoacyl group. The mixture was passed through cellulose nitrate filters. Unbound ^{14}C-aminoacyl-tRNA molecules passed through the filter, but ^{14}C-aminoacyl-tRNA molecules that were bound to ribosomes were re-

Table 26.6   Assignment of codon composition based on amino acid content of proteins synthesized from synthetic mRNA

Amino acid	Relative amino acid incorporation	Best assignment of triplets				Sum of triplet frequencies for each amino acid
		3A	2A1C	1A2C	3C	
A A:C = 5:1						
Asparagine	24.2		20			20
Glutamine	23.7		20			20
Histidine	6.5			4.0		4.0
Lysine	100	100				100
Proline	7.2			4.0	0.8	4.8
Threonine	26.5		20	4.0		24.0
B A:C = 1:5						
Asparagine	5.3		3.3			3.3
Glutamine	5.2		3.3			3.3
Histidine	23.4			16.7		16.7
Lysine	1.0	0.7				0.7
Proline	100			16.7	83.3	100
Threonine	20.8		3.3	16.7		20

SOURCE: From J. F. Spreyer et al, *Cold Spring Harbor Symp. Quant. Biol., 28,* 559 (1963).

tained. Identification of the isotopically labeled aminoacyl group attached to the tRNA revealed the identity of the codon. About 50 of the 64 codons were identified by this method (Table 26.7). The others could not be unambiguously determined because some codons bind two or more tRNA molecules about equally.

Synthetic polynucleotides of known sequence were also valuable tools in the cracking of the genetic code. Let us consider the polypeptides produced by translation of poly UC. The first codon is UCU, and the second codon is CUC. Translation of poly UC gave an alternating polymer of serine and leucine.

$$U\ C\ U\ |C\ U\ C\ |U\ C\ U\ |C\ U\ C\ |U\ C\ U\ |$$

amino | amino | amino | amino | amino
acid$_1$ | acid$_2$ | acid$_1$ | acid$_2$ | acid$_1$

With the trinucleotide-ribosome binding data, it could be concluded that UCU is a codon for serine, and CUC is a codon for leucine. (The identity of UCU and CUC is not unambiguous from the heteropolymer data alone since no reading frame is specified.)

The case for translation of polynucleotides containing three nucleotides is slightly more complex. In these cases three homopolymers are produced. Consider the case of poly AAG. If translation begins with the first

$$A\ A\ G\ A\ A\ G\ A\ A\ G\ A\ A\ G\ A\ A\ G\ A\ A\ G$$

letter, then AAG is translated repeatedly to give a homopolymer. If translation begins with the second letter, then AGA is translated repeatedly to give a different homopolymer. And if translation begins at the third letter, the codon is GAA, and yet another homopolymer will be synthesized. The three polymers formed from poly AAG are polylysine, polyglutamic acid, and polyarginine. Independent experiments established that AAG is a codon for lysine, AGA is a codon for arginine, and GAA is a codon for glutamic acid (Figure 26.24). The results obtained for a variety of polynucleotides of known sequence are summarized in Table 26.8.

C. Major Features of the Genetic Code

The genetic code is shown in Table 26.9. The genetic code has the same significance for biology that the periodic table has for chemistry: it reveals a "deep structure" in the world that transcends appearances.

Many marked patterns emerge from an analysis of the genetic code.

Table 26.7	Identification of codons by specific binding of trinucleotides to tRNA-ribosome complexes					
Trinucleotide						**tRNA bound**
5'UUU3'	UUC					Phenylalanine
UUA	UUG	CUU	CUC	CUA	CUG	Leucine
AAU	AUC	AUA				Isoleucine
AUG						Methionine
GUU	GUC	GUA	GUG	UCU		Valine
UCU	UCC	UCA	UCG			Serine
CCU	CCC	CCA	CCG			Proline
AAA	AAG					Lysine
UGU	UGC					Cysteine
GAA	GAG					Glutamic acid

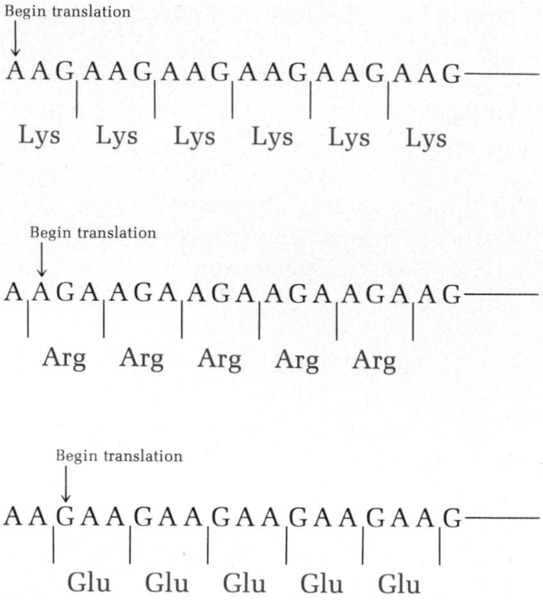

Begin translation
↓
A A G A A G A A G A A G A A G A A G ——

Lys Lys Lys Lys Lys Lys

Begin translation
↓
A A G A A G A A G A A G A A G A A G ——

Arg Arg Arg Arg Arg

Begin translation
↓
A A G A A G A A G A A G A A G A A G ——

Glu Glu Glu Glu Glu

Figure 26.24 Translation of poly AAG. Translation may begin in any of three reiterated reading frames, and three homopolypeptides are produced.

The code is highly degenerate. Methionine and tryptophan are the only amino acids that have only a single codon. All others have at least two codons, and serine and leucine each have six codons. There is a pattern to the degeneracy of the genetic code. Many amino acids have four codons. In these cases the first two bases define the amino acid, and the third base is variable. Mutations in the third position of the codon therefore often have no effect upon the amino acid eventually incorporated into a protein. Chemically similar amino acids often have similar codons. For example, a single base substitution converts any of the codons for serine into a codon for

Table 26.8	Incorporation of amino acids in proteins stimulated by synthetic polynucleotides having known repeating sequences	
Repeating sequence	**Codons in sequence**	**Amino acids incorporated**
UC	UCU-CUC	Ser-Leu
AG	AGA-GAG	Arg-Glu
UG	UGU-GUG	Cys-Val
AC	ACA-CAC	Thr-His
UUC	UUC; UCU; CUU	Phe; Ser; Leu
AAG	AAG; AGA; GAA	Lys; Arg; Glu
UUG	UUG; UGU; GUU	Leu; Cys; Val
CAA	CAA; ACA; AAC	Gln; Thr; Asp
GUA	GUA; UAG; AGU	Val; Ser
UAC	UAC; ACU; CUA	Tyr; Thr; Leu
AUC	AUC; UCA; CAU	Ile; Ser; His
GAU	GAU; AUG; UGA	Asp; Met
UAUC	UAU-CUA-UCU-AUC	Tyr-Leu-Ser-Ile
GAUA	GAU-AGA-UAG-AUA	None
UUAC	UUA-CUU-ACU-UAC	Leu-Thr-Tyr
GUAA	GUA-AGU-AAG-UAA	None

SOURCE: From H. G. Khorana et al., *Cold Spring Harbor Symp. Quant. Biol., 31,* 39 (1966).

Table 26.9 The genetic code

First position in codon (5' end)	Second position in codon				Third position in codon (3' end)
	U	C	A	G	
U	Phe	Ser	Tyr	Cys	U
	Phe	Ser	Tyr	Cys	C
	Leu	Ser	Stop	Stop	A
	Leu	Ser	Stop	Trp	G
C	Leu	Pro	His	Arg	U
	Leu	Pro	His	Arg	C
	Leu	Pro	Gln	Arg	A
	Leu	Pro	Gln	Arg	G
A	Ile	Thr	Asn	Ser	U
	Ile	Thr	Asn	Ser	C
	Ile	Thr	Lys	Arg	A
	Met	Thr	Lys	Arg	G
G	Val	Ala	Asp	Gly	U
	Val	Ala	Asp	Gly	C
	Val	Ala	Glu	Gly	A
	Val	Ala	Glu	Gly	G

threonine. Codons that have U in their second position invariably code for a hydrophobic amino acid. A mutation in the second position of these codons leads to incorporation of a similar amino acid at that position in the protein. The codons for phenylalanine and tyrosine also differ by a single base, and the codons for aspartic and glutamic acid begin with AG, differing only in the third position.

A point mutation is not, however, always harmless. Consider the closely related codons for valine and glutamic acid.

Valine: GUU, GUC, GUA, GUG
Glutamic acid: GAA or GAG

A point mutation A → U in the second nucleotide changes a codon for glutamic acid to a codon to valine. This is the mutation that leads to the substitution of valine for glutamic acid in position 6 of the β chains of human hemoglobin, resulting in sickle-cell hemoglobin (recall Section 4.5E). Taken as a whole, however, the genetic code is remarkably resistant to the synthesis of defective proteins as a result of point mutations.

The genetic code is almost completely universal. (Mitochondrial DNA molecules are translated by a slightly altered genetic code in which a couple of codons have been changed.) One theory that accounts for the existence of a universal genetic code holds that a "frozen accident" set the code in place. Any subsequent change in the code would alter the sequences of all of its proteins and would therefore be lethal to its possessor. There is thus a strong evolutionary tendency to maintain the genetic code in its same form across vast reaches of time.

D. Relation Between Codons and Anticodons

The codons of mRNA bind to the anticodons of various tRNA molecules by complementary base pairing. In the simplest scheme each codon would have its own specific tRNA anticodon. It was found, however, that many tRNA molecules recognize more than one of a set of synonymous codons. For example, yeast alanyl-tRNA binds to three of the four alanine codons: GCU,

Table 26.10 Rules for base pairing between the third position of the codon and the first position of the anticodon according to the wobble hypothesis

First position in anticodon (5′)	Third position in codon (3′)
U	A or G
C	G
A	U
G	U or C
I	U, C, or A

GCC, and GCA. Few tRNA molecules bind to one codon. F. Crick proposed that this variability in codon-anticodon binding was the result of *wobble* in the first position in the anticodon. The first two bases in the codon determine the binding of the anticodon, and there is considerable latitude in binding the third position. Table 26.10 gives the correspondence between the first anticodon base—the "wobble position"—and the third base in the codon.

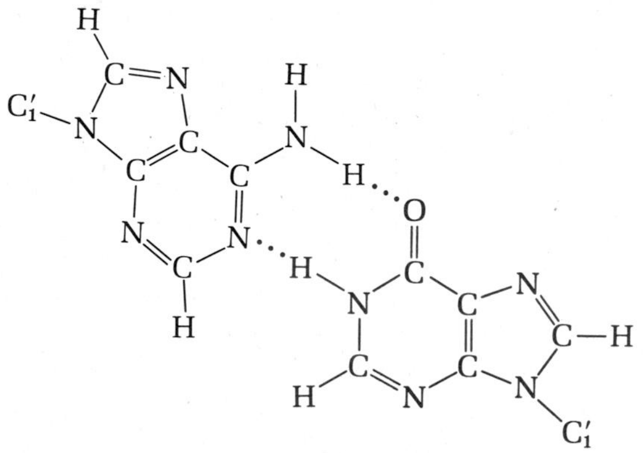

Inosine-adenosine
base pair

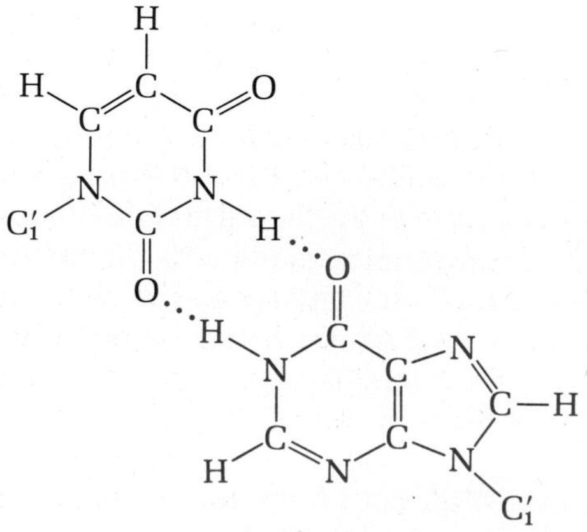

Inosine-uridine
base pair

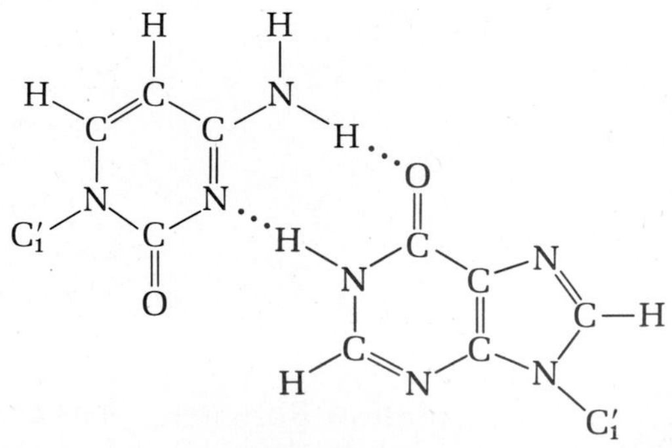

Inosine-cytidine
base pair

Figure 26.25 Base pairing between inosine in a tRNA anticodon and C, A, and U in the codon that occurs because of wobble in the third position of the anticodon.

In structural terms wobble is the result of conformational flexibility in the first position of the anticodon in tRNA.

The presence of inosine (which, we recall, results from deamination of adenosine) is particularly interesting. Inosine is often present in position 1 (the 5′ position) of the anticodon, and it pairs with position 3 (the 3′ position) of the codon. This is the case, for example, in yeast alanyl-tRNA.

Alanine tRNA anticodon

3′ C G I 5′   3′ C G I 5′   3′ C G I 5′
    | | | | | | | | |
5′ G C U 3′   5′ G C C 3′   5′ C G A 3′

Alanine codons

The substitution of a keto group for an amino group that results when adenosine is converted to inosine makes it possible for inosine to form hydrogen bonds with uridine, adenosine, and cytidine (Figure 26.25).

In the case of alanyl-tRNA binding to a codon, and in all other cases, an anticodon binds to a codon by Watson-Crick base-pairing rules in the first two codon positions. Part of the degeneracy of the genetic code in the third (3′) position is thus matched by wobble in the first (5′) position of the anticodon.

26–12 SUMMARY

Protein synthesis is carried out on ribosomes—macromolecular assemblies that contain ribosomal RNA and many proteins. Procaryotic ribosomes have a sedimentation coefficient of 70S, and they are composed of two subunits whose sedimentation coefficients are 30S and 50S. A molecule of 16S rRNA is an integral part of the 30S ribosome. This nucleic acid provides a scaffolding around which the ribosomal proteins are arranged. The 50S ribosome contains a molecule of 23S rRNA and a molecule of 5S rRNA, plus 34 different proteins. The slightly larger eucaryotic ribosome has a sedimentation coefficient of 80S; its two subunits have sedimentation coefficients of 40S and 60S.

Transfer RNA molecules act as adapters in protein synthesis. At their 3′ termini they are bound to an aminoacyl residue. At least 1 species of tRNA exists for each of the 20 amino acids. Transfer RNA also interacts with mRNA through a sequence of three bases known as the anticodon. All tRNA molecules have approximately the same L-shaped conformation. Aminoacyl-tRNA synthetases recognize the structural differences between tRNA molecules, and each nearly infallibly recognizes only its "own" tRNA. Aminoacyl-tRNA synthetases catalyze the synthesis of an aminoacyl-tRNA in two steps. First, an enzyme-bound aminoacyl adenylate intermediate is synthesized in a reaction that requires ATP. Second, the energy-rich aminoacyl adenylate is converted to the aminoacyl-tRNA.

Some aminoacyl-tRNA synthetases have a proofreading function. If an amino acid becomes bound to the wrong tRNA, the mistakenly formed aminoacyl-tRNA is hydrolyzed to a free amino acid and the tRNA. For example, this proofreading function prevents the tRNA for isoleucine from being converted to valyl-tRNA. This hydrolytic activity has not been found for all aminoacyl-tRNA synthetases, but is important for those amino acids that are chemically similar and which have similar sizes.

Messenger RNA is translated in the 5′ to 3′ direction, the same direction in which it is transcribed. Polypeptide synthesis occurs sequentially starting with the N-terminal amino acid residue.

N-Formylmethionine is the N-terminal amino acid in all procaryotes. Three initiation factors act in the presence of N-formylmethionyl-tRNA^{f-met} and the 30S ribosomal subunit to give an mRNA-bound 30S initiation complex which binds to the 50S ribosomal subunit, a process that is accompanied by GTP hydrolysis, to give an mRNA-bound 70S initiation complex. A base sequence in mRNA binds specifically to the 16S rRNA of the 30S particle to ensure that the initiation complex forms correctly.

Chain elongation requires elongation factors Tu and Ts to deliver an aminoacyl-tRNA to the A site of the ribosome, adjacent either to N-formylmethionyl-tRNA^{f-met} or to a growing peptidyl-tRNA. The peptidyl-tRNA occupies the P site of the ribosome. A peptidyl transferase that is a component of the 50S ribosomal subparticle catalyzes peptide bond synthesis. Elongation factor G hydrolyzes GTP as the peptide-tRNA is translocated from the A site to the P site. This sequence of events is repeated for each amino acid residue that is added to the peptide chain.

The codons UAG, UAA, or UGA signal termination of protein synthesis. Three release factors—RF-1, RF-2, and RF-3—are involved in termination of polypeptide chain synthesis. RF-1 and RF-2 bind to the termination codons. Release factors alter the activity of the peptidyl transferase, which hydrolyzes the final peptidyl-tRNA, liberating the completed protein. RF-3 hydrolyzes GTP, following which the release factors dissociate from the ribosome, and the 30S and 50S ribosomal particles dissociate from the mRNA. Each mRNA molecule is copied by many ribosomes simultaneously. One ribosome is bound for approximately every 80 nucleotide residues of mRNA.

The genetic code, which is the same for all organisms, consists of 64 nucleotide triplets, or codons. Three of these codons signal termination of protein synthesis, the other 61 code for amino acids. Since there are 20 amino acids, the genetic code is highly degenerate. Tryptophan and methionine each have a single codon, all of the others have two or more codons, and leucine and serine have six codons each.

The genetic code was deciphered by using synthetic RNA molecules and synthetic trinucleotides. The polynucleotide UUU was first used as a template for protein synthesis, leading to synthesis of polyphenylalanine. By using polynucleotides of known sequence the codons for many amino acids were determined. Trinucleotides stimulate the binding of their specific aminoacyl-tRNA molecules to ribosomes. Binding experiments using synthetic trinucleotides permitted identification of about 50 of the 64 codons.

Many tRNA molecules bind to more than one codon. The sequence of the first two bases of the codon determines the major specificity of the binding of a codon to an anticodon. The variability in the first position of the anticodon, which corresponds to the usual point of difference between synonymous codons, is a result of conformational flexibility in tRNA, or wobble.

REFERENCES

Altman, S., Ed., "Transfer RNA," MIT Press, Cambridge, Mass., 1978.

Brimacombe, R., Stoffler, G., and Wittmann, H. G., Ribosome Structure, *Ann. Rev. Biochem.*, **47**, 217 (1978).

Challberg, M. D., and Kelly, T. J., "Initiation Factors in Protein Biosynthesis," *Ann. Rev. Biochem.*, **51**, 869–907 (1982).

Cohn, W., Ed., "Progress in Nucleic Acid Research," Academic Press, New York.

Crick, F. H. C., Codon-Anticodon Pairing: the Wobble Hypothesis, *J. Mol. Biol.*, **19**, 548 (1966).

Crick, F. H. C., The Origin of the Genetic Code, *J. Mol. Biol.*, **38**, 367 (1968).

Crick, F. H. C., Barnett, L., and Brenner, S., The General Nature of the Genetic Code, *Nature*, **192**, 1227 (1961).

The Genetic Code, *Cold Spring Harbor Symp. Quant. Biol.*, **31**, (1966).

Lengyel, P., Speyer, J. F., and Ochoa, S., Synthetic Polynucleotides and the Amino Acid Code, *Proc. Nat. Acad. Sci. U.S.* **47**, 1936 (1961).

Lewin, B., "Gene Expression," Vols. 1, 2, Wiley, New York, 1974.

Losick, R., and Chamberlin, M., Eds., RNA Polymerase, *Cold Spring Harbor Symp. Quant. Biol.*, 41 (1976).

Mechanism of Protein Synthesis, *Cold Spring Harbor Symp. Quant. Biol.*, 34 (1969).

Momura, M., Tissieres, A., and Lengyel, P., Eds., "Ribosomes," Cold Spring Harbor Laboratory, Cold Spring Harbor, N.Y., 1974.

Nirenberg, M. W., and Matthaei, J. H., The Dependence of Cell-Free Protein Synthesis in *E. coli* upon Naturally Occurring or Synthetic Polynucleotides, *Proc. Nat. Acad. Sci. U.S.*, 47, 1588 (1961).

Salas, M., Smith, M. A., Stanley, W. M., Jr., Wahba, A. J., and Ochoa, S., Direction of Reading of the Genetic Message, *J. Biol. Chem.*, 240, 3988 (1965).

Transcription of Genetic Material, *Cold Spring Harbor Symp. Quant. Biol.*, 35 (1970).

Watson, J. D., The Involvement of RNA in the Synthesis of Proteins, *Science*, 140, 17 (1963).

Wittman, H. G., "Components of Bacterial Ribosomes," *Ann. Rev. Biochem.*, 51, 155–185 (1982).

Problems

1. Using the wobble rules, Table 26.9, predict the number of anticodons that correspond to the six serine codons.

2. Draw the base pairing in the wobble position for the interaction of the serine codons with the anticodon of tRNA.

3. Recalling the mechanism of thymidylate synthetase and other reactions that use N^{10}-formyltetrahydrofolate, write a reaction mechanism leading from methionyl-tRNA^{f-met} to N-formylmethionyl-tRNA^{f-met}.

4. Why is an aminoacyl adenylate an "activated" amino acid? What is the structural basis of the activation?

5. How might the flexibility of the genetic code and the redundancy of the code contribute to the effective transmittance of the information contained in the gene?

6. Using Table 26.8, predict the polypeptide that will be formed and assign a codon to each amino acid given the following polynucleotide sequences: (a) CUCUCU; (b) UGUGUG, (c) ACACAC, (d) AGAGAG.

7. How can the genetic code explain the observation that the guanosine-cytosine content of procaryotic DNA can be widely different, but the amino acid content of their proteins is quite similar?

Chapter 27

Regulation of gene expression

27–1 INTRODUCTION

The regulation of gene expression occurs at many levels. We have seen that certain proteins are synthesized as zymogens that must undergo proteolytic cleavage to become active, that the activity of some proteins is regulated by covalent modification, and that the activity of many proteins may be controlled allosterically by a variety of feedback mechanisms. The activity of certain proteins is also regulated by control of protein synthesis. Some aspects of regulation of gene expression in procaryotic organisms are understood in considerable detail, but the regulation of eucaryotic gene expression, a vastly more complex phenomenon, remains less well understood.

We shall first consider the regulation of two E. coli genes, one of them coding for enzymes of a catabolic pathway, the other coding for enzymes of a biosynthetic pathway. Then we shall briefly consider some aspects of regulation of eucaryotic gene expression.

27–2 REGULATION OF GENE EXPRESSION IN PROCARYOTIC ORGANISMS

It has been known since the turn of the century that some enzymes in microorganisms are synthesized only in the presence of their substrates, although in 1900 the phenomenon would not have been described in terms of "enzymes and substrates." Enzymes that are produced only in the presence of their substrates are said to be *inducible,* the substrates are *inducers,* and the process is known as *enzyme induction.* In many cases a given metabolite stimulates induction of several enzymes, all of which are involved in the metabolism of the inducer. Let us consider the induction of enzymes required for galactose metabolism in *E. coli:* β-galactosidase and galactose permease. (A third enzyme is produced along with these two, but its purpose in galactose metabolism is obscure.) In the absence of lactose, an *E. coli* cell contains about 10 molecules of β-galactosidase and lactose permease. β-Galactosidase catalyzes hydrolysis of the β glycosidic bond of lactose (reaction 27.1).

Galactose permease is responsible for transport of lactose across the cell membrane. A third enzyme, *thiogalactosyl acetyl transferase,* catalyzes the transfer of an acetyl group from acetyl-CoA to the C-6 hydroxyl group of a thiogalactoside (reaction 27.2). The transferase is not required for lactose metabolism, and its physiological function is unknown. The actual inducer of β-galactosidase is not lactose, but its structural isomer *allolactose.* The few β-galactosidase molecules already present in the cell catalyze the *trans-*

Allolactose
Gal[β(1→6)]glc

glycosylation that converts lactose to allolactose (reaction 27.3). In the presence of the inducer, allolactose, the number of β-galactosidase molecules increases from 10 per cell to about 10,000 per cell. The increase in the amount of β-galactosidase (and the permease and acetyl transferase) occurs within 3 min of addition of the inducer and stops almost immediately after the inducer is removed. Some β-galactosides, such as *isopropylthiogalactoside* (IPTG), induce synthesis of β-galactosidase, but are not substrates for the enzyme. These artificial inducers can be used to study the rate of synthesis of β-galactosidase at constant inducer concentration, and are valuable experimental tools. The slope of the line of Figure 27.1 is the ratio of the amount of β-galactosidase to the total bacterial protein. This slope is 0.066, showing that β-galactosidase synthesis accounts for 6.6% of the total protein synthesized by the cell in the presence of inducer. This is a remarkable figure when we recall that an *E. coli* cell contains about 4,000 different types of proteins.

Isopropylthiogalactoside (IPTG)

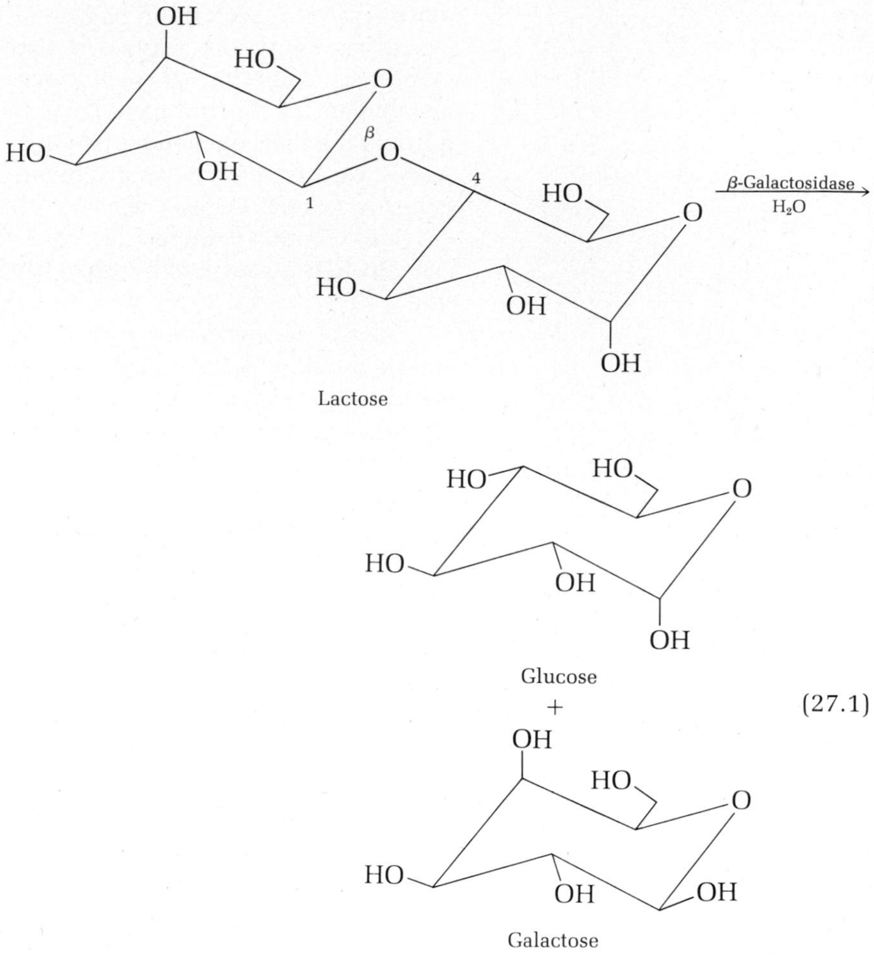

Lactose

Glucose

+

Galactose

(27.1)

A. The *lac* Operon Extensive genetic studies initiated by F. Jacob, J. Monod, and their coworkers in the 1940s and continued in many laboratories to the present day have shown that β-galactosidase, lactose permease, and the acetyl transferase genes are contiguous within the *E. coli* chromosome. They are designated the *lac* genes. The *lacZ* gene codes for β-galactosidase, the *lacY* gene codes for the permease, and the *lacA* gene codes for the acetyl transferase. In

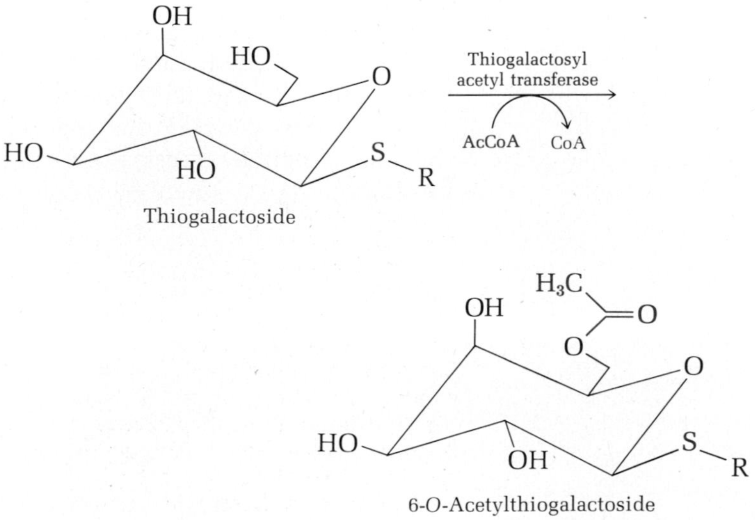

Thiogalactoside

6-O-Acetylthiogalactoside

(27.2)

Lactose

$$\beta(1\rightarrow4)$$

$$\xrightarrow{\beta\text{-Galactosidase}\ \text{(transglycosylation)}}$$

$$\beta(1\rightarrow6)$$

(27.3)

Allolactose

experiments with various *E. coli* mutants it was found that some *constitutive mutants* synthesize β-galactosidase, lactose permease, and the acetyl transferase in the absence of inducer. F. Jacob and J. Monod deduced that a single *regulatory gene*, designated *lacI*, controls the expression of the *lacZ*, *Y*, and *A* genes by coding for a specific regulatory protein called the *repressor*. The repressor protein binds to an *operator* region of the DNA, denoted *lacO*

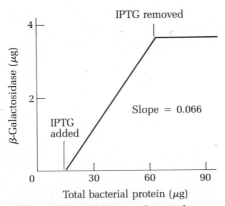

Figure 27.1 Addition of an inducer to *E. coli* cells resulting in a dramatic increase in the concentration of β-galactosidase. The increase parallels the rate of growth of a culture of *E. coli* cells. Enhanced protein synthesis (and cell growth) follows addition of the inducer by about 3 min and stops almost immediately when the inducer is removed. The slope of the line, 0.066, is the ratio of β-galactosidase synthesis to total protein synthesis.

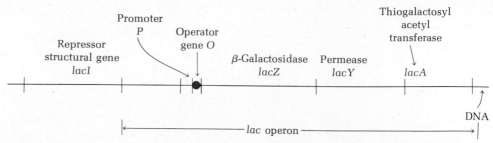

Figure 27.2 Sequence of genes in *E. coli* that code for β-galactosidase, lactose permease, and thiogalactosyl acetyl transferase. The *lacI* gene codes for a repressor protein that binds to the *lac* operator.

which is adjacent to the *lacZ* gene. The combination of the contiguous *lacO*, *Z*, *Y*, and *A* genes is known as the *lac* operon (Figure 27.2). Subsequent studies also showed that a promoter sequence, denoted *lacP*, lies between the regulatory gene, *lacI*, and the operator region, and is also part of the operon.

The *lac* repressor protein is a symmetrical tetramer, each of whose identical chains contains 360 amino acid residues. An *E. coli* cell contains about 10 molecules of the repressor protein. The repressor protein specifically binds to a sequence of 24 base pairs. Sixteen base pairs within this sequence form a partial palindrome, and therefore have a twofold axis of near symmetry (Figure 27.3). The N-terminal region of each chain of the repressor, about 50 residues, has the closest interaction with the operator, but the details of binding have not yet been elucidated. The dissociation constant for the operator–repressor protein complex is about 10^{-13} M (equation 27.4).

$$K_{\text{diss}} = \frac{[\text{operator}][\text{repressor protein}]}{[\text{operator–repressor protein}]} = 10^{-13} \text{ M} \tag{27.4}$$

The extremely tight binding of the repressor protein to the operator explains how 10 repressor molecules are able to keep the *lac* operon completely turned off, or repressed. The rate constant for formation of the operator–repressor protein complex, k_f, is 7×10^9 M^{-1} sec^{-1}. This rate constant approaches the diffusion controlled limit for a second order rate constant. It has been suggested that the repressor protein diffuses along the DNA until it recognizes the palindromic sequence of the operator gene. Since the palindrome possesses a twofold axis of symmetry, it will "look the same" if approached from either direction.

The inducer exerts its dramatic effects indirectly. It binds to the repressor, and the inducer–repressor protein complex is unable to bind to the operator gene. The *lac* operon also contains a promoter site, *lacP*, to which RNA polymerase binds. The promoter is separated from the *lacZ* gene by the operator. When the operator is bound to the repressor, DNA cannot be transcribed. Thus, *expression of the* lac *operon is controlled at the level of transcription.* This is a fairly general phenomenon in bacteria, whose mRNA

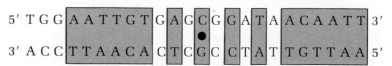

Figure 27.3 Partially palindromic sequence of the operator gene to which the *lac* repressor protein binds. The palindrome is enclosed in boxes; the twofold symmetry axis is indicated by a dot.

molecules have short lives, but are translated quickly once they have been produced.

Let us briefly summarize our discussion to this point. The *lac* operon consists of three structural genes encoding for β-galactosidase, lactose permease, and the acetyl transferase; a 24–base pair operator gene; and a promoter sequence that is recognized by RNA polymerase. A regulatory gene, *lacI*, encodes for a repressor protein that binds to the operator and inhibits transcription. An inducer (either allolactose or a thiogalactoside analog) binds to the repressor, causing it to dissociate from the operator. Transcription then occurs rapidly and is followed by translation. These events are summarized in Figure 27.4.

We have now described several cases in which palindromes in DNA are important: (1) restriction endonucleases recognize palindromes, (2) termination sequences for transcription are palindromes, (3) the peptide antibiotic actinomycin D binds to palindromes, and (4) the *lac* repressor protein binds to a partially palindromic sequence. The recognition of specific palindromes by proteins that bind to DNA is a phenomenon of widespread significance and almost complete generality.

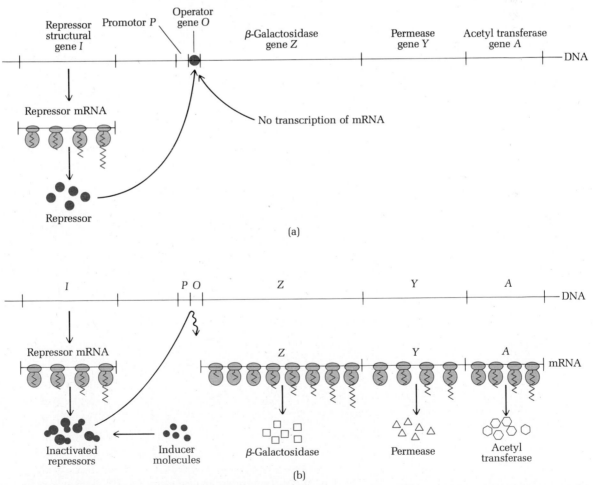

Figure 27.4 Regulation of the *lac* operon. (a) In the absence of inducer the repressor gene is expressed, and the repressor protein binds to the operator, blocking transcription of the genes that code for β-galactosidase, the premease, and the acetyl transferase. (b) In the presence of inducer the inducer-repressor complex dissociates from the operator, and the *lac* genes are expressed. [Adapted from F. J. Ayala and J. A. Kiger, Jr., "Modern Genetics," W. A. Benjamin, Menlo Park, Calif., 1980, p. 434. Used with permission from the Benjamin/Cummings Publishing Co.]

B. Catabolite Repression

If glucose and lactose are both present in the growth medium of *E. coli*, only glucose is catabolized, and the concentrations of β-galactosidase, lactose permease, and the acetyl transferase remain low. The lactose catabolite, D-glucose, therefore *represses* (though only indirectly as we shall see) the synthesis of the enzymes required for lactose catabolism. This phenomenon is known as *catabolite repression*. Glucose represses synthesis of β-galactosidase even in the presence of artificial inducers. Glucose also represses the synthesis of enzymes responsible for the catabolism of such sugars as maltose and arabinose. Catabolite repression provides the procaryotic cell with a mechanism for energy conservation. Glucose is readily catabolized, and there is no need for the cell to expend the energy necessary to synthesize the proteins that would be required to catabolize alternative substrates.

Genetic studies have shown that glucose itself is not the catabolite that is directly responsible for catabolite repression. The actual catabolite, whose identity is unknown, causes a decrease in the concentration of 3'-5' cAMP. It is possible that adenylate cyclase, cAMP phosphodiesterase, or both, are inhibited by this catabolite, but the mechanism by which glucose causes a decrease in the cellular concentration of cAMP is not known. The effect of cAMP is, however, well documented. A complex of cAMP and a *catabolite activator protein*, abbreviated CAP (a term not to be confused with the cap on messenger RNA), binds to a base sequence, yet another palindrome, in the promoter region of the *lac* operon. When the CAP-cAMP complex is bound to the promoter, transcription of the *lac* genes is stimulated. In sum, catabolite repression depends upon a cascade in which glucose is transformed to a catabolite that causes a decrease in the concentration of cAMP. With little or no cAMP to bind to the catabolite activator protein transcription of the *lac* genes does not occur. If the catabolite repressor is absent, however, cAMP binds to the catabolite activator protein, the cAMP-CAP complex binds to the promoter, and transcription is stimulated. The nucleotide sequences of part of the *lacI* gene, the promoter and operator, and part of the *lacZ* gene and the sites to which the *lac* repressor, the catabolite activator protein, and RNA polymerase bind are shown in Figure 27.5.

The catabolite activator protein binds to a 14-base pair sequence in the promoter. This sequence does not overlap the sequence that binds to the *lac* repressor protein, but it does overlap the RNA polymerase binding site. The CAP-cAMP complex therefore affects initiation of transcription by RNA polymerase. The structure of the dimeric CAP-cAMP complex has been determined (Figure 27.6). The N-terminal region of the catabolite activator protein binds cAMP, and the C-terminal region interacts with DNA. The x-ray structure of a CAP-cAMP-DNA complex has not been determined, but model building strongly suggests that the cAMP-CAP complex binds to *left-handed* double-helical DNA, rather than to the right-handed double helix in which DNA normally exists. The interaction of left-handed DNA with cAMP-CAP is shown in Figure 27.7. The CAP dimer interacts with 22 base pairs in this hypothetical complex. Activation of transcription is most effective when DNA is negatively supercoiled, a conformation in which the double helix is underwound and on its way to a left-handed double helix. Perhaps the cAMP-CAP complex "traps" the DNA in a left-handed double-helical conformation. The transition from right- to left-handed double helix might open the double helix transiently in the adjacent RNA polymerase site. Since RNA

Figure 27.5   Nucleotide sequence of the control region of the *lac* operon. Note ▶ that the operator region contains binding sites for both the catabolite activator protein (CAP) and RNA polymerase. The repressor binding site slightly overlaps the RNA polymerase binding site. [Adapted from R. C. Dickson et al., *Science,* 287, 27 (1975). Copyright 1975 by the American Association for the Advancement of Science.]

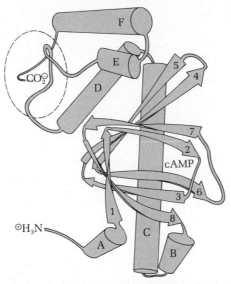

Figure 27.6 Schematic drawing of the catabolite activator protein–cAMP complex. Numbered arrows represent β-sheet; lettered cylinders represent α-helix. The cAMP binding site is labeled. [Adapted from D. B. McKay and T. A. Steitz, *Nature*, 290, 746 (1981). Reprinted by permission. Copyright © 1981 Macmillan Journals Limited.]

polymerase requires a single-stranded template, this unwinding stimulates transcription.

Transcription of the *lac* operon is thus subject to both positive and negative controls. In the presence of glucose the cAMP concentration is low, the catabolite activator protein does not bind to the promoter, and RNA polymerase, lacking a single-stranded template, does not initiate transcription. In the absence of glucose, the concentration of cAMP remains elevated, and it binds to the catabolite activator protein to stimulate initiation of transcription. By contrast, in the absence of inducer, the *lac* repressor protein remains bound to the operator gene, and initiation of transcription is inhibited. The concentrations of both glucose and lactose thus affect the initiation of transcription of the *lac* operon.

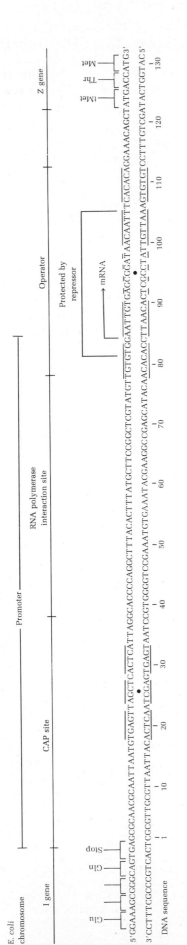

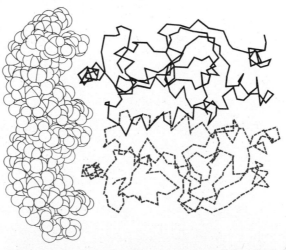

Figure 27.7 The α-carbon backbone of the CAP dimer interacts with 22 base pairs of left-handed, double-helical DNA. [From D. B. McKay and T. A. Steitz, *Nature*, 290, 747 (1981). Reprinted by permission. Copyright © 1981 Macmillan Journals Limited.]

C. Arabinose Operon

The arabinose (*ara*) operon illustrates a slight variation on the theme of regulatory genes. The arabinose operon consists of three contiguous genes, designated *araA*, *araB*, and *araD*, whose expression allows *E. coli* to use L-arabinose as a carbon source. These genes encode for L-arabinose isomerase, L-ribulokinase, and ribulose 4-epimerase. Acting sequentially, these enzymes lead from L-arabinose to D-xylulose 5-phosphate, an intermediate that is further catabolized in the pentose phosphate pathway (Figure 27.8). The arabinose operon also contains a promoter (*araI*), an operator (*araO*), and a regulatory gene (*araC*) (Figure 27.9).

The *ara* operon is activated by the same cAMP-CAP complex that activates the *lac* operon. The product of the arabinose regulatory gene, however, can act either as a repressor or as an activator. In the absence of arabinose the *araC* protein binds to the operator and represses transcription. Arabinose, however, binds to the *araC* protein, and the *araC* protein–arabinose complex binds to the promoter along with the cAMP-CAP complex. The combination of these two protein complexes bound to the promoter gives the maximal rate of transcription (Figure 27.10).

D. Tryptophan Operon

Transcription of many of the operons that contain genes encoding for enzymes involved in biosynthetic processes is under strict control. The tryptophan, histidine, and phenylalanine operons are perhaps the best understood of the biosynthetic operons. These operons are all controlled by similar mechanisms, and we shall begin by considering the tryptophan operon.

The tryptophan operon contains an operator gene, a promoter, a gene that encodes for a leader peptide, and the five genes encoding for the enzymes that convert chorismate to tryptophan (Figure 27.11). The leader peptide and the genes for the five biosynthetic enzymes are transcribed as a single mRNA transcript that contains 6,720 nucleotide residues. A regulatory gene, *trpR*, encodes for the *tryptophan repressor* protein. This gene is located far from the tryptophan operon and thus is not indicated on the map

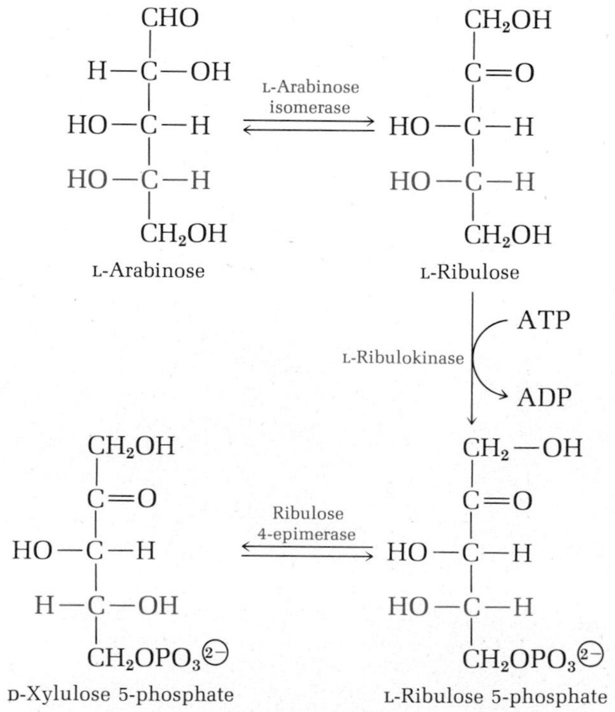

Figure 27.8 Enzymes of the arabinose operon that permit *E. coli* to metabolize L-arabinose.

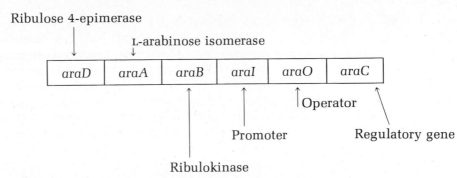

Figure 27.9 Arabinose operon and its regulatory gene.

of the tryptophan operon in Figure 27.11. The tryptophan repressor protein (mol wt 58,000) is a dimer of identical subunits that possess a binding site for tryptophan. The tryptophan–tryptophan repressor protein complex binds to the promoter and thereby inhibits transcription. The repressor protein alone does not bind to the promoter, and tryptophan is a *copressor* of transcription. Thus, tryptophan inhibits its own biosynthesis not only by negative feedback inhibition of the biosynthetic pathway (recall Section 22.8D), but also, in combination with the repressor protein, by inhibition of the synthesis of the mRNA that encodes for the enzymes responsible for the pathway. The promoter sequence to which the tryptophan repressor protein–tryptophan complex binds is a palindrome, echoing once more a general feature of protein binding to DNA (Figure 27.12).

Studies of *E. coli* mutants in which part of the promoter-operator sequence has been deleted transcribe the tryptophan operon at a far higher rate

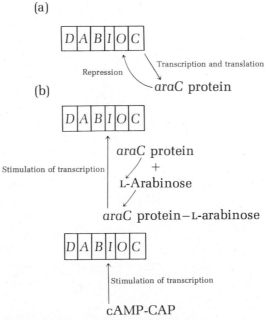

Figure 27.10 Regulation of the *ara* operon. (a) In the absence of arabinose the *araC* protein acts as a repressor. (b) In the presence of arabinose an *araC* protein–arabinose complex forms which binds to *araI* (the promoter) and, in combination with the cAMP-CAP complex, stimulates transcription.

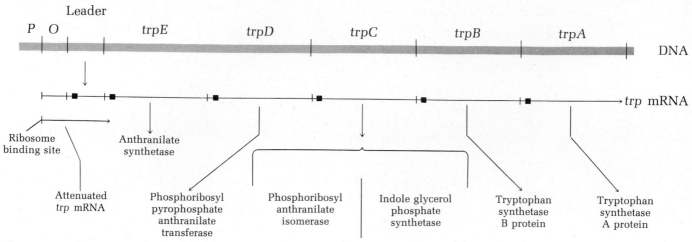

Figure 27.11 Map of the tryptophan operon of *E. coli*.

than bacteria possessing the entire promoter-operator sequence. The deletion that leads to this rather surprising result is located in a 162—base-pair sequence in the region encoding for the leader peptide. In "normal," or *wild-type, E. coli* transcription is prematurely terminated about 95% of the time in a base sequence known as the *attenuator site* (Figure 27.13). This nucleotide sequence contains a GC-rich segment followed by an AT-rich segment and a string of uridylate residues. We recall (Section 25.3D) that many terminator sequences possess similar constellations of residues. The attenuator site is therefore a sort of pre-echo (such as we often encounter in magnetic tape recordings) of the final terminator sequence (Figure 27.14). Transcription of the tryptophan operon is thus regulated by two independent mechanisms.

Termination of transcription by the attenuator site of the tryptophan operon is also closely coupled to translation, and the coupling agent is tryptophan itself. A few observations, that might at first appear unrelated, shall lead us to a model in which attenuation is reinforced at the level of translation. A 14-residue leader peptide (Figure 27.15) is translated. This leader peptide contains two adjacent tryptophan residues. At high concentrations of tryptophan translation is terminated, but at low concentrations of tryptophan translation of the leader peptide occurs. These effects of tryptophan are independent of interactions of the operator with the tryptophan repressor protein—tryptophan complex. These observations are in turn related to the conformation of the newly transcribed mRNA that encodes for the leader peptide. This mRNA is believed to fold into the secondary structure shown

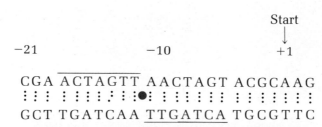

Figure 27.12 Nucleotide sequence of the tryptophan operator of *E. coli*. The dot indicates the twofold axis of symmetry; color indicates the palindromic sequences.

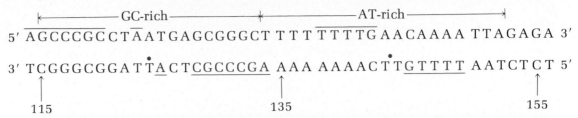

|←————GC-rich————→|←————AT-rich————→|

5′ AGCCCGC CTA̅A̅TGAGCGGGCT T T T T̅ T̅ T̅ T̅G A A C A A A A T T A G A G A 3′

3′ TCGGGCGGA T T̅A̅C T C̅G̅C̅C̅C̅G̅A A A A A A A A C T̅ T̅ G̅ T̅ T̅ T̅ T̅ A A T C T C T 5′

↑   ↑   ↑
115   135   155

Figure 27.13 Nucleotide sequence of the attenuator site of the *E. coli* tryptophan operon. Dots indicate the twofold symmetry axis of each palindrome; color indicates the palindromic sequences.

in Figure 27.16. This tightly packed structure assumes different conformations as a function of the tryptophan concentration. When the tryptophan concentration is high, the ribosome will cover up, or mask, the nascent mRNA in loops 1 and 2 (referring to Figure 27.16). Under these circumstances a hairpin loop forms between sections 3 and 4 of the mRNA. The ribosome cannot traverse the hairpin and dissociates from the mRNA, thereby terminating translation. In tryptophan-starved cells, however, the supply of tryptophanyl-tRNA becomes the limiting factor in translation. The ribosome "pauses" at codons 10 and 11 of the leader peptide to await delivery of the required tryptophanyl residue. During this hiatus a different secondary structure is assumed by the mRNA. *This* secondary structure does not contain a hairpin loop, and translation continues (Figure 27.17). This model for attenuation requires a tight coupling between translation and transcription and is still a hypothesis. The model is supported by the standard free-energy changes for hydrogen bond formation between the various loops. A hydrogen-bonded loop between segments 3 and 4 is favored by 87 kJ mol^{-1}. The hydrogen-bonded structure between loops 2 and 3, which must form for translation to continue, is stabilized by 50.2 kJ mol^{-1}. Thus, termination is thermodynamically favored most of the time.

Tryptophan biosynthesis is therefore controlled at three levels. (1) Tryptophan inhibits its own biosynthesis by negative feedback inhibition of the enzyme catalyzing the first unique step of its biosynthesis (recall Section 22.8D). (2) Tryptophan is a corepressor that acts in concert with the tryptophan regulatory protein to block transcription of the tryptophan operon. (3) The tryptophan attenuator site prematurely terminates transcription, and this effect is reinforced at the level of translation. The concentration of tryptophan, by affecting the concentration of tryptophanyl-tRNA, indirectly affects the passage of the ribosome across the leader sequence of the mRNA transcript.

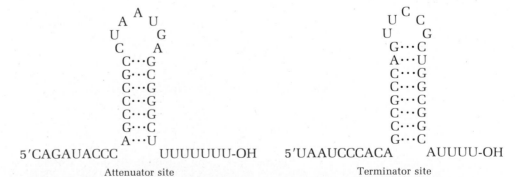

Figure 27.14 Nucleotide sequences of the attenuator and terminator sites of the *E. coli* tryptophan operon. Each sequence contains a segment that is rich in G-C base pairs, each ends with a string of U's, and each can form a hairpin secondary structure. [Adapted from A. Wu and T. Platt, *Proc. Nat. Acad. Sci. U.S.*, 75, 5442 (1978).]

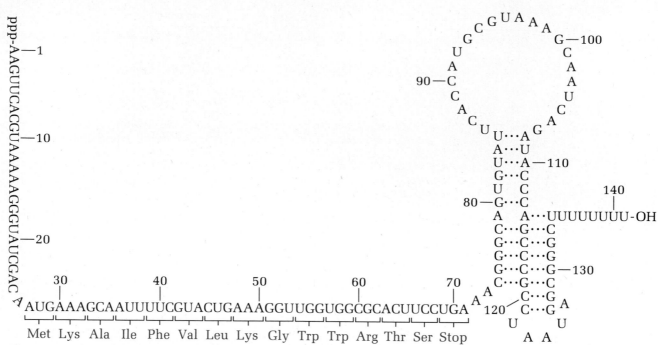

Figure 27.15 Nucleotide sequence and amino acid sequence of the 14-residue tryptophan leader peptide of *E. coli*. The leader peptide contains two adjacent tryptophan residues. The secondary structure of the nucleotide on the 3′ side of the 14-residue sequence determines whether the attenuator will be transcribed. [From F. Lee and C. Yanofsky, *Proc. Nat. Acad. Sci. U.S.*, 74, 4365 (1977).]

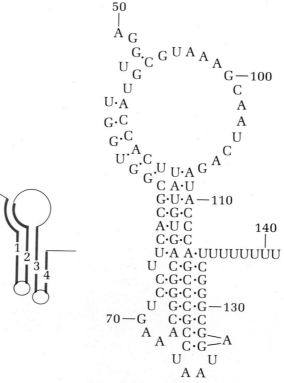

Figure 27.16 The proposed secondary structure of the 3′-OH portion of the terminated leader RNA transcript of the *E. coli* tryptophan operon. [Adapted from D. Oxender, G. Zurawski, and C. Yanofsky, *Proc. Nat. Acad. Sci., U.S.*, 76, 5526 (1979).]

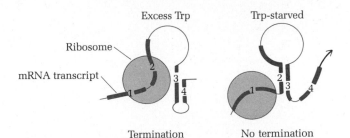

Figure 27.17 Hypothetical secondary structure of the mRNA transcript of the tryptophan operon. In the presence of excess tryptophan translation of the leader peptide is terminated by an unfavorable secondary structure which the ribosome cannot traverse. A different secondary structure in tryptophan-starved cells, however, permits translation of the leader peptide. [From D. Oxender, G. Zurawski, and C. Yanofsky, *Proc. Nat. Acad. Sci. U.S.*, 76, 5527 (1979).]

In the presence of excess tryptophan, when the genes encoding for tryptophan biosynthesis need not be expressed, the conformation of the mRNA in the leader peptide sequence prevents further translation.

Attenuator sites have also been discovered for the histidine and the phenylalanine operons. The concentration of the biosynthetic end product, histidine or phenylalanine, regulates translation by a mechanism that is similar to tryptophan control of attenuation. The sequence of the leader region of the phenylalanine operon is shown in Figure 27.18. A leader peptide of this operon is translated. This peptide contains 14 residues of which 7 are phenylalanines. Using the same hypothetical model for attenuation, when phenylalanine is scarce, the supply of phenylalanyl-tRNA is limited, the ribosome pauses as it traverses the leader sequence, and the "downstream" mRNA assumes a conformation that will eventually permit translation.

The histidine operon resembles the phenylalanine and tryptophan operons. The 14-residue leader peptide of the histidine operon contains 7 histidines in a row (Figure 27.19). The supply of histidine indirectly controls the rate of translation and contributes to the function of the attenuator site. In all three cases amino acid biosynthesis is controlled by negative feedback inhibition, by a regulatory gene, and by an attenuator.

27–3 REGULATION OF GENE EXPRESSION IN EUCARYOTIC ORGANISMS

The regulation of gene expression in eucaryotic organisms is not nearly so well understood as such systems as the *lac* operon in *E. coli*. The traditional methods of genetics are sorely taxed by the enormous complexity of the eucaryotic genome, but the biochemical methods for manipulating DNA that we have discussed are rapidly providing a wealth of new information about eucaryotic gene expression. These techniques have been described as "genetics by DNA analysis." The methods are new, enormously powerful, and relatively rapid. It seems likely that the biochemistry of eucaryotic gene expression will soon be an open book.

A. Gene Amplification

Most genes are present in but a single copy per cell, but several important classes of genes, including those encoding for rRNA, for histones, and for immunoglobulins, are present in multiple copies. Of these, the genes for rRNA in the oocytes of some organisms have the unique property of increasing dramatically at a certain stage of cell development. This phenomenon, known as *gene amplification,* is the topic of this section. Histone and immunoglobulin genes are discussed in succeeding sections.

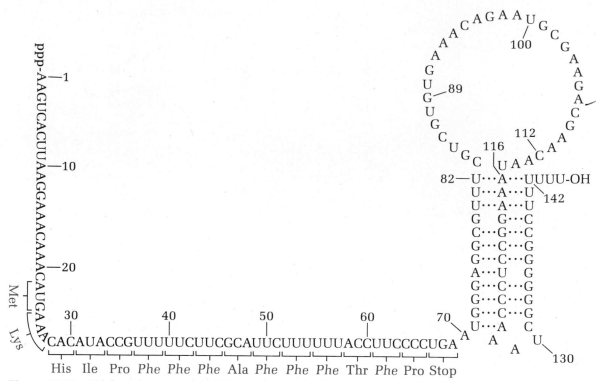

Figure 27.18 Nucleotide sequence of the attenuated leader of the phenylalanine operon of *E. coli.* The coding region for the leader peptide, which contains seven phenylalanine residues, is indicated. [From G. Zurakowski, K. Brown, D. Killingly, and C. Yanofsky, *Proc. Nat. Acad. Sci. U.S., 75,* 4273 (1978).]

The genes encoding for ribosomal RNA in the oocytes (egg cells) of the African clawed toad, *Xenopus laevis,* have been extensively studied. Three rRNA molecules—5.8S, 18S, and 28S—are encoded on a single transcript (recall Section 25.4B). The genes of this transcript are tandemly repeated many times (Figure 27.20). In *Xenopus laevis,* other amphibians, and fish the number of genes encoding for rRNA rises by a factor of 10^3, from about 500 copies to nearly 500,000 copies, during egg development, or *oogenesis.* The mechanism of this rapid *gene amplification* is not known, but it provides the cell with rRNA molecules for the enormous number of ribosomes (about 10^{12}) that are required for protein synthesis as the egg matures. The genes for 5S rRNA, located elsewhere on the chromosome, are *not* amplified during oogenesis. There are about 25,000 copies of 5S rRNA genes, and their number remains constant throughout development. To keep up with the thousandfold amplification of 18S and 28S rRNA genes, the synthesis of 5S rRNA commences considerably in advance of 18S and 28S rRNA synthesis. Certain 5S rRNA genes are expressed only in oocytes (the others are expressed both in oocytes and other cells). These genes are separated by a region of spacer DNA that is several times longer than the gene itself. A part of this spacer has a sequence that is quite similar to the transcribed region. This homologous piece of DNA is known as a *pseudogene,* but its significance is not known.

Met - Thr- Arg - Val - Gln - Phe - Lys - His - His - His - His - His - His - His - Pro - Asp-

5′ AUG ACA CGC GUU CAA UUU AAA CAC CAC CAU CAU CAC CAU CAU CCU GAC 3′

Figure 27.19 Nucleotide sequence and corresponding amino acid sequence of the leader peptide of the histidine operon of *E. coli.*

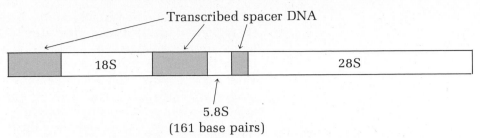

Figure 27.20 Map of the rRNA gene cluster of *Xenopus laevis*. [Adapted from E. O. Long and I. B. David, *Ann. Rev. Biochem.*, *49*, 739 (1980). Reproduced with permission. © 1980 by Annual Reviews Inc.]

B. Histone Genes

Having discussed the function of histones in maintaining the structure of chromatin (Section 24.5A), we now turn to the genes that encode for histones. The histone genes of the sea urchin, *Lytechinus pictus*, have been the most extensively studied, but it appears that the histone genes of other organisms are similar. Sea urchins contain from 300 to 1,000 copies of the genes encoding for the five histones. This large number of genes is required early in sea urchin development. Within 12 hours of fertilization the single fertilized egg, or *zygote*, undergoes 2^{10} (1,024) cell divisions. About 30% of the sea urchin's total protein synthesis is devoted to histone production during this period, and the capacity for histone synthesis is provided by multiple copies of histone genes. These genes are tandemly arranged in clusters that contain one gene for each of the five histones (Figure 27.21). Each gene is separated by a section of spacer DNA. The histone genes are among the few eucaryotic genes that are not interrupted by intervening sequences. We noted earlier (Section 24.5A) that the amino acid sequences of histones are remarkably conserved across evolutionary time spans. Histone genes also display very little heterogeneity; even the length of the spacer DNA, often highly variable in other gene clusters, is conserved in the histone genes. Detailed analysis of the sea urchin has revealed that some of the spacer DNA is highly conserved, while other regions of the spacer DNA are almost totally divergent.

C. Immunoglobulin Genes and the Origin of Antibody Diversity

In Section 4.9 we discussed the structure of immunoglobulins. Let us briefly summarize the major features of that discussion. Immunoglobulin G (IgG) consists of two light chains and two heavy chains that are joined by disulfide bonds in a Y-shaped structure (Figure 27.22). The heavy chains contain regions of constant (C_H) and regions of variable (V_H) sequences of amino acid residues. The constant region of the heavy chain can be divided into approximately equal lengths of three sequences, designated

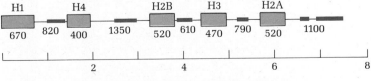

Thousands of base pairs (kb pairs)

Figure 27.21 Map of the histone gene cluster of the sea urchin *Lytechinus pictus*. Colored boxes indicate transcribed DNA. Lines between the boxes are spacer DNA. Homologous spacer DNA is indicated by thin lines, and nonhomologous spacer DNA is indicated by thick lines. The drawing is to scale, and numbers below the lines and boxes indicate the number of kilobase pairs in a given sequence. [From L. W. Kedes, *Ann. Rev. Biochem.*, *48*, 841 (1979). Reproduced with permission. © 1979 by Annual Reviews Inc.]

C_H1, C_H2, and C_H3. The variable region of the heavy chain may be subdivided into regions of variable and hypervariable sequences. Immunoglobulin light chains are of two types, designated kappa (κ) and lambda (λ). The light chains, too, contain regions of constant (C_L) and variable (V_L) sequences. As in the heavy chains, the variable regions of the light chains contain hypervariable sequences. The specificity of immunoglobulins depends upon the variable sequences. The antigen-binding sites are formed at the interfaces between the hypervariable regions of the light and heavy chains at the tips of the Y (Figure 27.23). With this reprise in mind, let us now turn to the question how the enormous range of antibodies, exceeding 10^6 different antibodies, is generated by the marvelously baroque structure of antibody genes. To tell this tale, we require yet another detour.

An organism develops from a single fertilized egg, the zygote. In the countless cell divisions that produce a heart, lungs, muscles, eggs, the whole anatomical catalog, the descendents of the ancestral cell become *differentiated*. The original zygote has the potential to become all of these 100 or so morphologically different cell types and is therefore said to be a *totipotential* cell. As the cells descended from the zygote become differentiated, they lose some of their potentiality. *Somatic* (body) cells lose totipotentiality, but *germ-line* cells that have not differentiated remain totipotential. Germ-line cells can multiply by mitosis and become differentiated cells, or they can undergo meiosis and develop into either eggs or sperm. The cells that produce antibodies start as germ-line cells and develop into somatic cells. Unique antibodies are synthesized at each stage of development of these cells.

Immunoglobulins are produced by a class of cells that are derived from *B lymphocytes*, found in lymphatic tissue. If B lymphocytes are exposed to an antigen, some of them develop the ability to produce antibodies. This initial response to the antigen is known as the *primary immune response*. Only those B lymphocytes that are "preprogrammed" to recognize a given antigen are able to synthesize antibodies against it. A few molecules of the antibody the B lymphocyte is programmed to synthesize are distributed over the cell

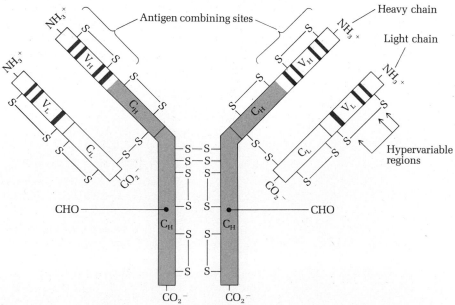

Figure 27.22 Schematic diagram of a human immunoglobulin G. The symbols C and V designate constant and variable regions within the light and heavy chains.

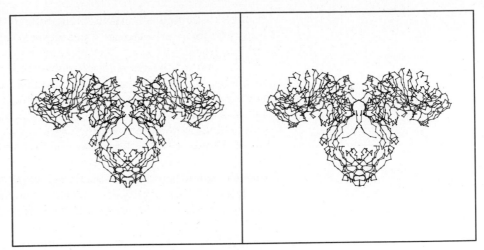

Figure 27.23   Structure of an immunoglobin G molecule.

surface. If the appropriate antigen is encountered, it binds to the antibody. This initial event triggers a series of events that culminates in vastly increased antibody production by the B lymphocyte, and the B lymphocyte divides many times. Its progeny, all having the same genetic makeup, are *clones* of the original B lymphocyte. These immature cells develop into *plasma cells,* which circulate in the blood and continue to be prolific producers of antibodies.

We can confidently predict that a rabbit as yet unborn will be able to produce an antibody against an organic compound not yet synthesized. Three different mechanisms have been proposed to account for this remarkable ability, which protects our future rabbit from some future environmental onslaught.

The *germ-line hypothesis* of antibody diversity proposes, in its most extreme form, that there are enough genes encoding for variable sequences of the light and heavy chains to account for *all* of the antibodies an organism can ever produce. A terrific amount of DNA would have to be devoted to immunoglobulin genes for this hypothesis to be correct. Methods for cloning eucaryotic genes (Section 24.9) and DNA sequence analysis (Section 9.9A) of immunoglobulin genes have shown that in the mouse there are two genes for λ chains and between 100 and 300 genes for κ chains. It is estimated that there are 500 to 1,000 genes coding for the variable regions of heavy chains. This is a prodigious number of genes, but less than would be required to generate 10^6 antibodies by random recombination between constant and variable regions of the light and heavy chains. The germ-line hypothesis of multiple copies of genes accounts for some antibody diversity, but it is not sufficient to account for it completely.

The *somatic-hypermutation* hypothesis proposes that immunoglobulin diversity results from point mutations in a variable portion of the gene. By itself, this theory does not account for the "concentration" of point mutations in the hypervariable regions of light and heavy chains. Point mutations—generated by site-specific recombination—do, however, account for some antibody diversity.

The *somatic recombination* hypothesis asserts that there are only a few genes encoding for antibodies, and that antibody diversity is generated by extensive general recombination. This theory cannot account for point mutations.

The actual mechanism for generating antibody diversity is a combination of all three mechanisms. Reduced to its essence, the generation of antibody diversity arises from a beautiful scheme of gene rearrangements and gene splicing.

A map of the gene encoding for mouse λ_1 chains in immature B lymphocytes and in differentiated plasma cells is shown in Figure 27.24. In embryonic DNA, the regions encoding for the variable (V_L) and constant (C_L) regions are widely separated. The coding sequence for the V_L region consists of a leader L and an intervening sequence (I_1) that contains 93 nucleotide residues. The V_L gene codes for 95 of the 108 residues of the variable portion of the λ_1 chain. A sequence of some 5,300 base pairs separates the first portion of the coding region from a second piece of the λ_1 gene known as the J gene (J for joining). This little gene codes for the last 13 amino acid residues of the variable region and joins the constant region to the variable region. An intervening sequence of 1,250 base pairs separates the J gene from the gene coding for the constant portion of the chain (C gene). When the embryonic cell develops into an antibody-producing plasma cell, the DNA rearranges by site-specific recombination. This gene rearrangement brings the V, J, and C genes together. Transcription and mRNA processing then yield the mRNA that is translated to give the λ_1 protein. Mouse κ chains display a similar pattern. All of the J_κ and J_λ DNA molecules whose sequences have been determined contain two blocks of sequences, of which one is a heptamer and the other a nonamer. These sequences are the sites of recombination. The J genes code for the last section of the hypervariable region of the κ or λ chains.

5′ C A T T G T G 3′   5′ G G T T T T T G T 3′

3′ G T A A C A C 5′   3′ C C A A A A A C A 5′

Heptamer sequence   Nonamer sequence

There are four J_κ and four J_λ genes. Each group of four is arranged in a tandem cluster. Since one of these J genes can combine with the V region, the J genes are a source of considerable antibody diversity. Further antibody diversity arises by imprecise genetic recombination between the J gene in the neighborhood of codon 95 of the V gene. Codon 95 is CCC (the codon for

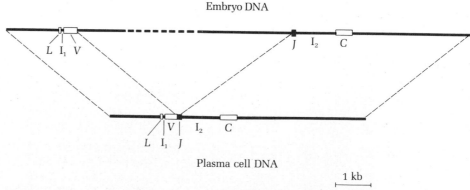

Embryo DNA

Plasma cell DNA

1 kb

Figure 27.24 Map of the mouse λ_1 gene in embryos and λ_1 chain-producing plasma cells. In the embryo the gene is in pieces. A leader is separated by 93 nucleotides (I_1) from most of the coding region of the variable chain (V). The gene for the rest of the variable region that joins the variable and constant regions, known as the J gene, is thousands of base pairs away. A second intervening sequence, I_2, separates the J gene from the coding region for the constant portion of the light chain (C). In differentiated plasma cells, the gene has rearranged, and the J gene is contiguous with the variable gene V. [From C. Brack et al. *Cell*, 15, 10 (1978).]

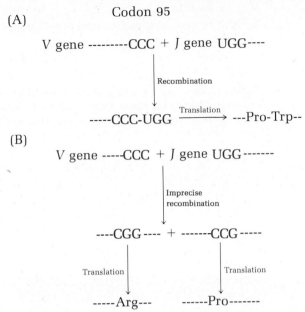

(A)   Codon 95

V gene ---------CCC + J gene UGG----

Recombination

-----CCC-UGG $\xrightarrow{\text{Translation}}$ ---Pro-Trp--

(B)

V gene -----CCC + J gene UGG-------

Imprecise
recombination

----CGG---- + -------CCG-----

Translation   Translation

-----Arg---   ------Pro-------

Figure 27.25   Variation in translated sequence in (a) by imprecise recombination in (b). Imprecise recombination can alter the translated sequence of the hypervariable region, thus adding to immunoglobulin diversity. Various possibilities can delete codons, insert codons, or form hybrid codons. (The codons in the figure are those for the mRNA transcript.)

proline).[1] Suppose that the first codon of the J gene, which is UGG (the codon for tryptophan), recombines next to CCC. If no overlapping occurs, the translated sequence will be proline-tryptophan. If recombination is imprecise, it can yield CGC, the codon for arginine. Imprecise recombination alters the reading frame of the J gene, and the result is that all 13 codons will be altered. This frame-shift mutation can change the entire translated sequence. (Changes in the third position of a codon, the wobble position, often give a codon for the same amino acid, so all 13 residues are not necessarily altered by imprecise recombination.) Imprecise recombination is thus a considerable source of antibody diversity (Figure 27.25).

Let us now turn to the genes that code for the heavy chains. We recall that the five classes of immunoglobulins—immunoglobulins G, A, M, D, and E—are formed by various combinations of light and heavy chains. The various classes of immunoglobulins are produced by a sequence that leads from immunoglobulin M, the first antibody produced, to immunoglobulin G and sometimes to either immunoglobulins A or E. The change from one class of immunoglobulin to another is a result of a change in the type of heavy chain present and is known as a heavy chain *class-switch*. The genes encoding for the heavy chains are in pieces. The variable region is produced by recombination of a V_H gene, a J_H gene, and another small gene known as the D gene (D for diversity DNA segment). The spliced sequence is V_H-D-J_H. In the mouse, the V_H gene codes for amino acid residues 1 to 101, the D gene codes for residues 102 to 106, a region of hypervariable sequence, and one of four J_H genes codes for residues 107 to 123. The N-terminal region of the V_H chain contains a leader peptide that contains 19 residues (Figure 27.26).

[1] Throughout this discussion we use the codons of the mRNA transcribed from DNA.

$$\overset{\oplus}{H_3N}\text{-Met-Lys-Leu-Trp-Leu-Asn-Trp-Val-Phe-Leu-Leu-}$$

$$\overset{}{10}$$

$$\text{Thr-Leu-Leu-His-Gly-Ile-Asn-Cys-}_{\uparrow}$$

15

Cleavage site

Figure 27.26 Amino acid sequence of the hydrophobic leader peptide of an immunoglobulin. After this peptide has helped the immunoglobulin enter the interior space of the endoplasmic reticulum, it is cleaved by a peptidase.

Most of the residues in this peptide are hydrophobic. The hydrophobic tail binds to the endoplasmic reticulum. It threads the immunoglobulin through a pore in the endoplasmic reticulum into the interior space of the endoplasmic reticulum. Finally, a peptidase hydrolyzes the leader peptide.

The genes for the constant portion of the heavy chain, are arranged tandemly to the 3′ side of the genes encoding for the variable region (Figure 27.27). In cells that produce both immunoglobulins D and M the entire region is transcribed. Splicing of the mRNA transcript then leads to deletion of the C_μ genes. Alternative splicing yields immunoglobulin M. The other classes of immunoglobulins are formed by gene rearrangements involving α, γ, δ, and ϵ chains (Figure 27.28). The class-switch from one type of immunoglobulin to another involves only changes in the constant region of the heavy chain. The variable regions of the heavy chain, and the κ and λ chains are unaltered by class-switching.

The mechanism of heavy chain class-switching is beginning to emerge. Class-switching appears to be mediated by so-called switching genes, designated S with a subscript to denote the particular heavy chain with which the switching gene is associated. These switching genes are the recognition sites for class-switching. A map of the switching region of mouse heavy chains, and a model for recombination mediated by the switching genes and the specific switching proteins that bind to the S genes is shown in Figure 27.28. Site-specific recombination between homologous regions of the S genes brings the C_α gene, originally far from the variable portion of the heavy chain, up to the *VDJ* gene of the heavy chain. Gene splicing then produces coding sequence that is transcribed and translated into immunoglobulin A. We have focused upon immunoglobulin A, but the same mechanism for class-switching is followed for the other heavy chains.

We noted above that three mechanisms for generating antibody diversity have been proposed. All three are partially correct. The germ-line hypothesis correctly predicts antibody diversity as a result of multiple copies

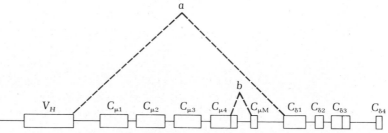

Figure 27.27 Map of the arrangement of the C_H genes that code for the various classes of immunoglobin heavy chains. Boxes represent transcribed regions of DNA, and lines represent intervening sequences. In cells that synthesize both immunoglobins D and M the entire region is transcribed. Gene splicing indicated by *a* yields immunoglobin D, and gene splicing indicated by *b* yields immunoglobulin M. [From M. Robertson and M. Hobart, *Nature*, 290, 543 (1981). Reprinted with permission. Copyright © 1981 Macmillan Journals Limited.]

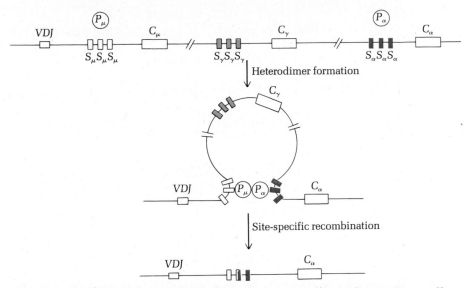

Figure 27.28 Model for class-specific heavy chain class-switching. The small boxes represent switching sequences (S) that bind to specific proteins that mediate class-switching. These coding proteins are denoted P_α and P_μ for the S_α and S_μ genes, respectively. Site-specific recombination between S_α and S_μ genes moves the C_α gene up to the VDJ coding sequence for the heavy chain. [From M. M. Davis, S. K. Kim, and L. E. Hood, *Science*, 290, 1364 (1980). Copyright 1980 by The American Association for the Advancement of Science.]

of genes encoding for the variable regions of light and heavy chains. The V, J, and D genes have evolved for eons and lead to antibodies that bind to common environmental antigens. Differentiation results in rearrangement of the V gene, and recombination with J genes gives considerable diversity to light chains. The heavy chains, with their various V-D-J combinations, generate even more antibody diversity. Somatic plasma cells, descended from germ-line B lymphocytes, can undergo further refinements of specificity. Mutations in the hypervariable region randomly lead to new antibodies of which some have greater specificity for a given antigen. The first antibody produced is immunoglobulin M. Further gene rearrangements in the somatic cell by heavy chain class-switching generate the other classes of immunoglobulins.

D. Regulation of Protein Synthesis in Eucaryotic Organisms

The messenger RNA molecules of procaryotic organisms are short lived, and control of protein synthesis is exerted mainly at the level of transcription. The messenger RNA molecules of eucaryotic organisms, however, are much longer lived—their half-lives range from 10 minutes to about 20 hours—and protein synthesis is also controlled at the level of translation. The initiation of eucaryotic protein synthesis is subject to strict regulation, and it is governed by a cascade of enzyme activation involving 3'-5' cyclic AMP and a protein kinase whose overall features resemble the regulation of glycogen metabolism (Section 17.4). In our discussion of procaryotic protein synthesis we noted the existence of three initiation factors. Initiation factors in eucaryotic protein synthesis number seven or eight. One of them, eucaryotic initiation factor eIF-2, is the target for the enzyme system that regulates protein synthesis. Initiation factor eIF-2 binds GTP and is responsible for carrying N-formylmethionyl-tRNA$^{\text{f–met}}$ to the 40S ribosomal subunit. If eIF-2 is prevented from carrying out this function, protein synthesis cannot begin.

Reticulocytes are immature red blood cells whose primary function is the synthesis of the globin moiety of hemoglobin. This process requires heme. If heme is not present, an inhibitor of protein synthesis is activated that prevents globin synthesis. The *heme-controlled inhibitor* of protein syn-

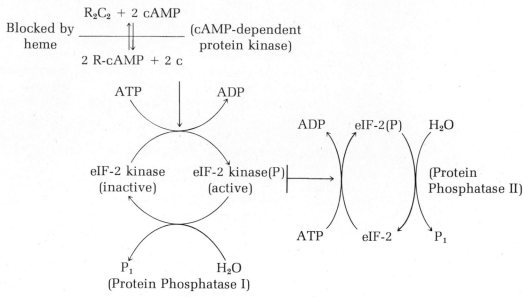

Figure 27.29 Model for the regulation of eIF-2. [From S. Ochoa and C. de Haro, *Ann. Rev. Biochem.*, **48**, 561 (1979). Reproduced with permission. © 1979 by Annual Reviews Inc.]

thesis is a cAMP-dependent protein kinase that catalyzes the phosphorylation of another enzyme, eIF-2 kinase. This kinase, in turn, phosphorylates eIF-2. The cAMP-dependent protein kinase is a tetramer of two catalytic subunits (C) and two regulatory subunits (R). The tetramer dissociates in the presence of cAMP to two catalytic monomers and two regulatory subunits. The regulatory subunits each contain a binding site for cAMP and a binding site for heme. The binding of heme to the regulatory subunit induces a conformational change that decreases the affinity of the regulatory subunit for cAMP (reaction 27.5).

$$R_2C_2 \xrightleftharpoons[\text{cAMP}]{\text{(-) Heme}} 2\ C\ +\ 2\ R\text{-cAMP} \tag{27.5}$$

The cascade of enzyme activity that follows release of the catalytic subunits from the cAMP-dependent protein kinase tetramer is summarized in Figure 27.29. The effects of the successive phosphorylation of eIF-2 kinase and eIF-2 are reversed by specific phosphatases. This cascade, then, resembles the regulation of phosphorylase kinase and reflects a general pattern of metabolic control by covalent modification.

27–4 SUMMARY

Gene expression in *E. coli* and other procaryotic organisms is primarily regulated at the level of transcription. Some of the genes of *E. coli* are organized as operons. An operon consists of an operator sequence, a promoter sequence to which RNA polymerase binds, and the structural genes that are translated into protein. A regulatory gene codes for a protein that binds to the operator to inhibit transcription. In most cases the nucleotide sequences to which repressor proteins bind are palindromes. In fact, most DNA-binding proteins recognize palindromes. The inducible enzymes of the *lac* operon are regulated by allolactose, a sugar derived from lactose. Certain synthetic galactosides also are able to induce transcription of the *lac* operon. The inducer binds to the repressor protein, which dissociates from the operator to permit initiation of transcription. Also, a catabolite activator protein binds cAMP, and the resulting cAMP-CAP complex binds to a sequence within the promoter and stimulates transcription. The tryptophan operon possesses an attenuator site that blocks transcription about 85% of the time by forming a hairpin loop that RNA polymerase cannot traverse. If some transcription does occur, the secondary structure of the mRNA determines whether the transcript will be translated. The phenylalanine and histidine operons are similarly regulated. The arabinose operon is regulated by a regulatory protein (the *araC* protein) that inhibits transcription by binding to the operator in the absence of arabinose, and stimulates transcription in the presence of arabinose by binding both arabinose and the promoter.

The genes for ribosomal RNA in eucaryotic organisms are tandemly arranged in clusters that are repeated many times. These genes are amplified a thousandfold in the oocytes of the African clawed toad, other amphibians, and fishes. The histone genes are also tandemly arranged in multiple copies.

Antibody molecules are composed of light (L) and heavy chains (H), each of which contains constant (C) and variable (V) regions. The antigen-binding site is formed at the interface between hypervariable regions of the light and heavy chains at the tips of a Y-shaped structure. Immunoglobulins are encoded by three families of genes—the λ and κ light chains and a heavy chain family. The light chains are encoded by two gene segments—V_L and J (joining). The V portion of the heavy chains are encoded by three gene segments—V_H, J_H, and D. The J and D genes of the heavy chains code most of the hypervariable region of the gene. There are at least 100 germ-line V_L and V_H gene segments, and there are 4 J genes for each variable gene class. J and V_L, and V_H, J_H, and D genes can be spliced at different points in their sequences by site-specific recombination to generate hybrid codons, codon insertions, and codon deletions. Various classes of immunoglobulins are interconverted by rearrangement of genes coding for the heavy chains. Site-specific recombination between switching genes (S) provides the mechanism for heavy chain class-switching.

Hemoglobin synthesis in reticulocytes is controlled at the translational level by a cascade of protein kinase activity that goes into effect in the absence of heme, which is required for globin synthesis. The protein kinase cascade culminates in phosphorylation of initiation factor eIF-2, which can then no longer initiate translation. The effects of phosphorylation are reversed by specific phosphatases. In the presence of heme the protein kinase cascade is inhibited at its origin, and hemoglobin synthesis proceeds.

REFERENCES

Brack, C., Hirama, M., Lenhard-Schuller, R., and Tonegawa, S., A Complete Immunoglobulin Gene Is Created by Somatic Recombination, *Cell*, *15*, 1–14 (1978).

Brown, D. W., Gene Expression in Eucaryotes, *Science*, *211*, 667–674 (1981).

Crawfod, I. P., and Stauffer, G. V., Regulation of Tryptophan Biosynthesis, *Ann. Rev. Biochem.*, *49*, 163–196 (1980).

Davis, M. S., Kim, K., and Hood, L. E., DNA Sequences Mediating Class Switching in α-Immunoglobulins, *Science*, *209*, 1360–1365 (1980).

Dickson, R. C., Abelson, J., Barnes, W. M., and Reznikoff, W. S., Genetic Regulation: The *Lac* Control Region, *Science*, *187*, 27–35 (1975).

Early, P., Huang, H., Davis, M., Calame, K., and Hood, L., An Immunoglobulin Heavy Chain Variable Region Gene Is Generated from Three Segments of DNA: V_H, D, and J_H, *Cell*, *19*, 981–992 (1980).

Jacob, F., and Monod, J., Genetic Regulatory Mechanisms in the Synthesis of Proteins, *J. Mol. Biol.*, *3*, 318–356 (1961).

Jelinek, W. R., and Schmid, C. W., "Repetitive Sequences in DNA and Their Expression," *Ann. Rev. Biochem.*, *51*, 813–845 (1982).

Kedes, L. H., Histone Genes and Histone Messengers, *Ann. Rev. Biochem.*, *48*, 837–870 (1979).

Lee, F., and Yanofsky, C., Transcription Termination at the trp Operon Attenuators of *Escherichia coli* and *Salmonella typhimurium*: RNA Secondary Structure and Regulation of Termination, *Proc. Nat. Acad. Sci. U.S.*, *74*, 4365–4369 (1977).

Long, E. O., and Dawid, I. B., Repeated Genes in Eucaryotes, *Ann. Rev. Biochem.*, *49*, 727–766 (1980).

McKay, D. B., and Steitz, T. A., Structure of Catabolic Activator Gene at 2.9Å Resolution Suggests Binding to Left-Handed B-DNA, *Nature*, *290*, 744–749 (1981).

Ochoa, S., and deHara, C., Regulation of Protein Synthesis in Eucaryotes, *Ann. Rev. Biochem.*, *48*, 549–580 (1979).

Oxender, D., Zurawski, G., and Yanofsky, C., Attenuation in the *Escherichia coli* Tryptophan Operon: Role of RNA Secondary Structure Involving the Tryptophan Coding Region", *Proc. Nat. Acad. Sci. U.S.*, *76*, 5524–5528 (1979).

Platt, T., Termination of Transcription and Its Regulation in the Tryptophan Operon of *E. coli*, *Cell*, *24*, 10–23 (1981).

Zurawski, G., Brown, K., Killingly, D., and Yanofsky, C., Nucleotide Sequence of the Leader Region of the Phenylalanine Operon of *Escherichia coli*, *Proc. Nat. Acad. Sci. U.S.*, *75*, 4271–4275 (1978).

Index